FIRE and MUD

Eruptions and Lahars
of Mount Pinatubo, Philippines

FIRE and MUD

Eruptions and Lahars
of Mount Pinatubo, Philippines

Edited by

CHRISTOPHER G. NEWHALL
Geologist, U.S. Geological Survey
Affiliate Professor, University of Washington Volcano Systems Center

RAYMUNDO S. PUNONGBAYAN
Director, Philippine Institute of Volcanology and Seismology

Philippine Institute of Volcanology and Seismology
Quezon City

University of Washington Press
Seattle and London

Printed in Hong Kong

A disk, prepared by the United States Geological Survey, accompanies the book. The disk contains two programs, PINAPLOT (for DOS) and VOLQUAKE (for Windows), both of which show the evolving seismicity before and after the 1991 eruptions of Pinatubo. VOLQUAKE also includes data sets for Mount St. Helens and Mount Spurr.

The book will be available on USGS/CVO and PHIVOLCS web sites.

Library of Congress Cataloging-in-Publication Data

Fire and mud : eruptions and lahars of Mount Pinatubo, Philippines / edited by
Christopher G. Newhall and Raymundo S. Punongbayan.
p.   cm.
ISBN 0-295-97585-7 (Seattle : alk. paper)
1. Pinatubo, Mount (Philippines) 2. Lahars—Philippines—Pinatubo, Mount, Region.
I. Newhall, Christopher G. II. Punongbayan, Raymundo.
QE523.P56F57   1996   96-33410
551.2'1'095991—dc20   CIP

The paper used in this publication meets the minimum requirements of American National Standard for Information Sciences—Permanence of Paper for Printed Library Materials, ANSI Z.48-1984.

CONTENTS

VOLCANIC DRAMA, HUMAN DRAMA

Overview of the eruptions

Edward W. Wolfe and Richard P. Hoblitt 3

Photographic record of rapid geomorphic change at Mount Pinatubo, 1991–94

Raymundo S. Punongbayan, Christopher G. Newhall, and Richard P. Hoblitt 21

Eruption hazard assessments and warnings

Raymundo S. Punongbayan, Christopher G. Newhall, Ma. Leonila P. Bautista, Delfin Garcia, David H. Harlow, Richard P. Hoblitt, Julio P. Sabit, and Renato U. Solidum 67

People's response to eruption warning: The Pinatubo experience, 1991–92

Jean Tayag, Sheila Insauriga, Anne Ringor, and Mel Belo 87

Assessment and response to lahar hazard around Mount Pinatubo, 1991 to 1993

Richard J. Janda, Arturo S. Daag, Perla J. Delos Reyes, Christopher G. Newhall, Thomas C. Pierson, Raymundo S. Punongbayan, Kelvin S. Rodolfo, Renato U. Solidum, and Jesse V. Umbal 107

Responses of Pampanga households to lahar warnings: Lessons from two villages in the Pasig-Potrero River watershed

Raoul M. Cola 141

The Mount Pinatubo disaster and the people of Central Luzon

Cynthia Banzon Bautista 151

ANCIENT AND MODERN HISTORY

Eruptive history of Mount Pinatubo

Christopher G. Newhall, Arturo S. Daag, F.G. Delfin, Jr., Richard P. Hoblitt, John McGeehin, John S. Pallister, Ma. Theresa M. Regalado, Meyer Rubin, Bella S. Tubianosa, Rodolfo A. Tamayo, Jr., and Jesse V. Umbal 165

Geothermal exploration of the pre-1991 Mount Pinatubo hydrothermal system

F.G. Delfin, Jr., H.G. Villarosa, D.B. Layugan, V.C. Clemente, M.R. Candelaria, and J.R. Ruaya 197

GEOPHYSICAL UNREST

Installation, operation, and technical specifications of the first Mount Pinatubo telemetered seismic network

Andrew B. Lockhart, Sergio Marcial, Gemme Ambubuyog, Eduardo P. Laguerta, and John A. Power 215

A PC-based real-time volcano-monitoring data-acquisition and analysis system

Thomas L. Murray, John A. Power, Gail Davidson, and Jeffrey N. Marso...................... 225

A comparison of preeruption real-time seismic amplitude measurements for eruptions at Mount St. Helens, Redoubt Volcano, Mount Spurr, and Mount Pinatubo

Elliot T. Endo, Thomas L. Murray, and John A. Power 233

Real-time seismic amplitude measurement (RSAM) and seismic spectral amplitude measurement (SSAM) analyses with the Materials Failure Forecast Method (FFM), June 1991 explosive eruption at Mount Pinatubo

Reinold R. Cornelius and Barry Voight.. 249

Preliminary observations of seismicity at Mount Pinatubo by use of the Seismic Spectral Amplitude Measurement (SSAM) system, May 13–June 18, 1991

John A. Power, Thomas L. Murray, Jeffrey N. Marso, and Eduardo P. Laguerta 269

Precursory seismicity and forecasting of the June 15, 1991, eruption of Mount Pinatubo

David H. Harlow, John A. Power, Eduardo P. Laguerta, Gemme Ambubuyog, Randall A. White, and Richard P. Hoblitt 285

Precursory deep long-period earthquakes at Mount Pinatubo: Spatio-temporal link to a basalt trigger

Randall A. White 307

Ground deformation prior to the 1991 eruptions of Mount Pinatubo

J.W. Ewert, Andrew B. Lockhart, Sergio Marcial, and Gemme Ambubuyog................. 329

Volcanic earthquakes following the 1991 climactic eruption of Mount Pinatubo: Strong seismicity during a waning eruption

Jim Mori, Randall A. White, David H. Harlow, P. Okubo, John A. Power, Richard P. Hoblitt, Eduardo P. Laguerta, Angelito Lanuza, and Bartolome C. Bautista 339

Relationship of regional and local structures to Mount Pinatubo activity

Bartolome C. Bautista, Ma. Leonila P. Bautista, Ross S. Stein, Edito S. Barcelona, Raymundo S. Punongbayan, Eduardo P. Laguerta, Ariel R. Rasdas, Gemme Ambubuyog, and Erlinda Q. Amin 351

Three-dimensional velocity structure at Mount Pinatubo: Resolving magma bodies and earthquake hypocenters

Jim Mori, Donna Eberhart-Phillips, and David H. Harlow 371

Computer visualization of earthquake hypocenters

Richard P. Hoblitt, Jim Mori, and John A. Power.. 383

Seismicity and magmatic resurgence at Mount Pinatubo in 1992

Emmanuel G. Ramos, Eduardo P. Laguerta, and Michael W. Hamburger...................... 387

A VOLATILE SYSTEM

Monitoring sulfur dioxide emission at Mount Pinatubo

Arturo S. Daag, Bella S. Tubianosa, Christopher G. Newhall, Norman M. Tuñgol,
    Dindo Javier, Michael T. Dolan, Perla J. Delos Reyes, Ronaldo A. Arboleda,
    Ma. Mylene L. Martinez, and Ma. Theresa M. Regalado 409

Preeruption vapor in magma of the climactic Mount Pinatubo eruption:
    Source of the giant stratospheric sulfur dioxide cloud

Terrence M. Gerlach, Henry R. Westrich, and Robert B. Symonds............................... 415

Evolution of a small crater lake at Mount Pinatubo

Nora R. Campita, Arturo S. Daag, Christopher G. Newhall, Gary L. Rowe, and
    Renato U. Solidum 435

OBSERVATIONS AND RECONSTRUCTIONS: THE 1991–92 ERUPTIONS

The west-side story: Observations of the 1991 Mount Pinatubo eruptions
    from the west

Julio P. Sabit, Ronald C. Pigtain, and Edwin G. de la Cruz....................................... 445

The preclimactic eruptions of Mount Pinatubo, June 1991

Richard P. Hoblitt, Edward W. Wolfe, William E. Scott, Marvin R. Couchman,
    John S. Pallister, and Dindo Javier 457

Tephra falls of the 1991 eruptions of Mount Pinatubo

Ma. Lynn O. Paladio-Melosantos, Renato U. Solidum, William E. Scott,
    Rowena B. Quiambao, Jesse V. Umbal, Kelvin S. Rodolfo, Bella S. Tubianosa,
    Perla J. Delos Reyes, Rosalito A. Alonso, and Hernulfo B. Ruelo 513

Dispersal of the 1991 Pinatubo tephra in the South China Sea

Martin G. Wiesner and Yubo Wang 537

Pyroclastic flows of the June 15, 1991, climactic eruption of Mount
    Pinatubo

William E. Scott, Richard P. Hoblitt, Ronnie C. Torres, Stephen Self,
    Ma. Mylene L. Martinez, and Timoteo Nillos, Jr. 545

Preeruption and posteruption digital-terrain models of Mount Pinatubo

John W. Jones and Christopher G. Newhall.. 571

Volume estimation of tephra-fall deposits from the June 15, 1991,
    eruption of Mount Pinatubo by theoretical and geological methods

Takehiro Koyaguchi 583

Infrasonic and acoustic-gravity waves generated by the Mount Pinatubo eruption of June 15, 1991

Makoto Tahira, Masahiro Nomura, Yosihiro Sawada, and Kosuke Kamo...................... 601

Worldwide observation of bichromatic long-period Rayleigh waves excited during the June 15, 1991, eruption of Mount Pinatubo

W. Zürn and R. Widmer.. 615

Meteorological observations of the 1991 Mount Pinatubo eruption

J. Scott Oswalt, William Nichols, and John F. O'Hara.................................... 625

Mount Pinatubo: A satellite perspective of the June 1991 eruptions

James S. Lynch and George Stephens 637

Growth of a postclimactic lava dome at Mount Pinatubo, July–October 1992

Arturo S. Daag, Michael T. Dolan, Eduardo P. Laguerta, Gregory P. Meeker, Christopher G. Newhall, John S. Pallister, and Renato U. Solidum 647

Secondary pyroclastic flows from the June 15, 1991, ignimbrite of Mount Pinatubo

Ronnie C. Torres, Stephen Self, and Ma. Mylene L. Martinez 665

A STORY IN THE ROCKS

Changing proportions of two pumice types from the June 15, 1991, eruption of Mount Pinatubo

Carlos Primo C. David, Rosella G. Dulce, Dymphna D. Nolasco-Javier, Lawrence R. Zamoras, Ferdinand T. Jumawan, and Christopher G. Newhall........ 681

Magma mixing at Mount Pinatubo: Petrographic and chemical evidence from the 1991 deposits

John S. Pallister, Richard P. Hoblitt, Gregory P. Meeker, Roy J. Knight, and David F. Siems 687

Mineral and glass compositions in June 15, 1991, pumices: Evidence for dynamic disequilibrium in the dacite of Mount Pinatubo

James F. Luhr and William G. Melson 733

Preeruption pressure-temperature conditions and volatiles in the 1991 dacitic magma of Mount Pinatubo

Malcolm J. Rutherford and Joseph D. Devine... 751

Petrology and geochemistry of the 1991 eruption products of Mount Pinatubo

Alain Bernard, Ulrich Knittel, Bernd Weber, Dominique Weis, Achim Albrecht, Keiko Hattori, Jeffrey Klein, and Dietmar Oles 767

Petrology and Sr, Nd, and Pb isotopic geochemistry of Mount Pinatubo
 volcanic rocks

Paterno R. Castillo and Raymundo S. Punongbayan 799

Occurrence and origin of sulfide and sulfate in the 1991 Mount Pinatubo
 eruption products

Keiko Hattori 807

Sulfur isotopic systematics of the June 1991 Mount Pinatubo
 eruptions: A SHRIMP ion microprobe study

Michael A. McKibben, C. Stewart Eldridge, and Agnes G. Reyes 825

Anhydrite-bearing pumices from the June 15, 1991, eruption of Mount
 Pinatubo: Geochemistry, mineralogy, and petrology

John Fournelle, Rebecca Carmody, and Arturo S. Daag 845

Highly oxidized and sulfur-rich dacitic magma of Mount Pinatubo:
 Implication for metallogenesis of porphyry copper mineralization
 in the western Luzon arc

Akira Imai, Eddie L. Listanco, and Toshitsugu Fujii 865

Relative timing of fluid and anhydrite saturation: Another consideration
 in the sulfur budget of the Mount Pinatubo eruption

Jill Dill Pasteris, Brigitte Wopenka, Alian Wang, and Teresa N. Harris 875

LAHARS, LAHARS,   Watershed disturbance and lahars on the east side of Mount Pinatubo
AND MORE LAHARS   during the mid-June 1991 eruptions

Jon J. Major, Richard J. Janda, and Arturo S. Daag 895

Flow and deposition of posteruption hot lahars on the east side of Mount
 Pinatubo, July–October 1991

Thomas C. Pierson, Arturo S. Daag, Perla J. Delos Reyes, Ma. Theresa M. Regalado,
 Renato U. Solidum, and Bella S. Tubianosa 921

The 1991 lahars of southwestern Mount Pinatubo and evolution of the
 lahar-dammed Mapanuepe Lake

Jesse V. Umbal and Kelvin S. Rodolfo 951

Channel and sedimentation responses to large volumes of 1991 volcanic
 deposits on the east flank of Mount Pinatubo

Kevin M. Scott, Richard J. Janda, Edwin G. de la Cruz, Elmer Gabinete, Ismael Eto,
 Manuel Isada, Manuel Sexon, and Kevin C. Hadley 971

Two years of lahars on the western flank of Mount Pinatubo: Initiation, flow processes, deposits, and attendant geomorphic and hydraulic changes

Kelvin S. Rodolfo, Jesse V. Umbal, Rosalito A. Alonso, Cristina T. Remotigue, Ma. Lynn Paladio-Melosantos, Jerry H.G. Salvador, Digna Evangelista, and Yvonne Miller 989

Instrumental lahar monitoring at Mount Pinatubo

Sergio Marcial, Arnaldo A. Melosantos, Kevin C. Hadley, Richard G. LaHusen, and Jeffrey N. Marso.. 1015

Rainfall, acoustic flow monitor records, and observed lahars of the Sacobia River in 1992

Norman M. Tuñgol and Ma. Theresa M. Regalado....................................... 1023

Observations of 1992 lahars along the Sacobia-Bamban River system

Ma. Mylene L. Martinez, Ronaldo A. Arboleda, Perla J. Delos Reyes, Elmer Gabinete, and Michael T. Dolan.. 1033

1992 lahars in the Pasig-Potrero River system

Ronaldo A. Arboleda and Ma. Mylene L. Martinez....................................... 1045

SELECTED IMPACTS   Building damage caused by the Mount Pinatubo eruption of June 15, 1991

Robin J.S. Spence, Antonios Pomonis, Peter J. Baxter, Andrew W. Coburn, Mark White, Manuel Dayrit, and Field Epidemiology Training Program Team.... 1055

Socioeconomic impacts of the Mount Pinatubo eruption

Remigio A. Mercado, Jay Bertram T. Lacsamana, and Greg L. Pineda........................... 1063

The 1991 Pinatubo eruptions and their effects on aircraft operations

Thomas J. Casadevall, Perla J. Delos Reyes, and David J. Schneider 1071

The atmospheric impact of the 1991 Mount Pinatubo eruption

Stephen Self, Jing-Xia Zhao, Rick E. Holasek, Ronnie C. Torres, and Alan J. King 1089

AUTHORS AND THEIR AFFILIATIONS... 1117

FOREWORD

Clear and present danger is a great unifier. When Mount Pinatubo devastated our land in 1991, it dissolved geographical, cultural, and economic barriers. Filipinos were one with each other and with the world as spontaneous aid poured into the Pinatubo region. Colleagues in the international scientific community came to help, not only to address the disaster of the moment, but also to resolve underlying scientific and technological problems of future hazards as well.

The significance of the Pinatubo experience lies not only in the successful forecasting of eruptions and lahar flows. It also highlighted many new challenges. As a circle of light increases, so does the circumference of darkness that surrounds it.

The Pinatubo event indeed shed plenty of light in volcanology. But it also highlighted many new questions: Will there be more devastating eruptions from Pinatubo? What is the volcano's eruption recurrence cycle? How do we deal with the Pinatubo lahars to minimize risk without stalling the region's development? What are the long-term impacts of volcanic deposits on the region's agricultural productivity? Which of our potentially active volcanoes will behave like Pinatubo and erupt after centuries of inactivity?

This monograph is a story of our tension, grief, insight, and relief. It also points to knowledge gaps that we must still bridge. It serves as a monument to scientific internationalism that has proven its worth. And through this monograph, we invite every reader to relive our experience at Pinatubo and to join us in taking up the new challenges ahead.

Fidel V. Ramos
President of the Philippines

MANILA
1995

PREFACE

The 1991 eruptions of Mount Pinatubo and subsequent widespread lahars are signal events both in volcanology and volcanic hazards mitigation. Accurate forecasting of the scale and date of an eruption is never easy, especially when the threatening volcano has not been monitored previously. Against considerable odds and with some critical logistical help, a small team of Philippine and U.S. scientists managed to avert a terrible disaster. Neither the eruption nor the lahars could be stopped, but, within the first 5 years of the crisis, hundreds of thousands of people have been warned and moved temporarily to safety.

The goals of this monograph are to capture a vast array of observations about the remarkable events before, during, and since the climactic eruption of Mount Pinatubo, to synthesize those observations into an understanding of Pinatubo as a dynamic, complex volcanic system (within an equally dynamic and complex social setting), and to make both observations and interpretations accessible to volcanologists around the world, especially those who must cope with similar crises in the future.

This last-mentioned goal bears special attention: Eruptions and lahars of Pinatubo's magnitude occur but a few times per century, and even less frequently in populated areas. But every geologist who studies volcanoes can identify scores of volcanoes at which such events are possible—some surrounded by even greater populations than Pinatubo. Because the options for precautionary actions are so limited around densely populated volcanoes, future forecasts must be as precise and accurate as possible. We might have only one chance to be right, for, after that, our credibility will be tenuous at best. This volume will be a vital reference for those who must make similarly difficult, high-stakes forecasts in the future. Even after such a large eruption has occurred, hazards from lahars and massive volumes of sediment will challenge the cleverest engineer, social scientist, and politician. Here, too, this volume will be a valuable reference.

We have not edited the volume specifically for student use, but instructors of volcanology will find within this monograph a selection of papers that illustrate great scholarly and practical value of viewing a volcano as a dynamic system, and, within such a dynamic system, speedy analogs of many otherwise hard-to-observe geologic processes. Advanced students will delight in seeing how their previous study in separate disciplines can be integrated and brought to bear on complex problems.

Another entire group of scientists—atmospheric scientists—also learned much from the disturbance that Pinatubo thrust into the atmosphere. Many exciting papers have resulted. These are beyond the scope of the present volume, but our last paper, by Self and others, bridges between the solid earth and atmospheric sciences and refers readers to literature of the latter.

International collaboration was notable at Mount Pinatubo—first, between Philippine and U.S. scientists, and later, spontaneously, involving scientists from at least 10 countries. An event of this magnitude would challenge any single country, and, because volcanological communities are typically small, it makes good sense for volcanologists to help each other in times of major crisis. Help given in one year is help received in the next. The Pinatubo effort epitomizes the spirit of the International Decade for Natural Disaster Reduction—international learning, from each other in a rare natural laboratory, with the ultimate goal of averting disasters wherever they might threaten.

Some of the papers in this monograph are by veterans of many volcanic crises; others are by younger scientists for whom Pinatubo was either a first eruption or for whom publication here is a first international publication. We take special pleasure in this mix, and thank both the veteran mentors and the younger scientists for their patience with each other and with the editors. Thirty-nine of the authors are or were students, including

11 first authors. Every veteran went back to school at Pinatubo, and many students are well along the way to becoming expert mentors.

Editors must make many choices on which papers and interpretations to accept. With an event like that of Pinatubo, some observations and interpretations simply cannot be confirmed or tested, either because of the extreme logistical or safety considerations or because a feature was ephemeral and is no longer the same or even present. Some data sets in this collection are regrettably sparse, but events cannot be restaged. Accordingly, as a rule of thumb, we have accepted papers that contain useful though sparse data. We have not accepted data or interpretations that are unlikely to be correct, but we have given the benefit of any remaining doubt to the author and have included some assertions that we cannot confirm.

We have worked with authors to reconcile discrepancies in observations and interpretations. Where that could be achieved, the results will be apparent as cross-references to mutually important points. Where it couldn't be achieved, we've allowed differences to stand and have noted them in the introduction to each section.

The schedule for preparation of a collection of this size, with authors around the world, poses special problems. Events like a giant eruption beg for early publication of results. However, there is also an argument for allowing the volcano enough time to show posteruption trends and allowing authors time to collect and analyze rich data sets and to communicate with each other. Some authors finished manuscripts within months; others were still in the field, struggling with ongoing monitoring responsibilities. Communication between authors will be evident in the papers; many, though perhaps not all, geographic and language barriers were overcome. Since asking for initial contributions in January 1993, we have pushed slower or busier authors hard, while begging for patience from those who were faster. Our goal was to balance timeliness with completeness.

Late in the preparation of this monograph, literally after type had been set, a budget crisis for the originally-intended publisher, the U.S. Geological Survey, forced us to choose between CD-ROM publication (only) and publication of this book by an alternate publisher. You can see that we chose the latter, and we will also make an electronic version available on the World-Wide Web. We welcome feedback from readers of both versions.

                                                        Chris Newhall, Seattle
                                                        Ray Punongbayan, Quezon City
                                                        May, 1996

ACKNOWLEDGMENTS

Many people have helped us to prepare this chronicle of Pinatubo. A simple thanks does injustice to their contributions; a more elaborate thanks, detailing each person's and group's contributions, would be a book unto itself. A Tagalog word, *bayanihan*, meaning an unreservedly cooperative effort to complete a big task, is an apt description for the outpouring of help that we have enjoyed.

Starting first with data collection, processing, and interpretation, we want to thank authors, with whom it has been our great pleasure and privilege to work. Your eyes and minds have captured a remarkable story. We also thank your many assistants—field assistants, lab assistants, secretaries, drafting assistants, and many more—who do not appear as authors but without whom the volume would be painfully thin. One co-author who deserves special thanks is Ma. Theresa Regalado, who served as the first editor's right hand, literally and figuratively, following an unscheduled injury.

Logistical assistance at Pinatubo required a special commitment, not only in hours, but in personal risk and sacrifice. Data sets in this collection, and successful mitigation measures that were based on those data, simply would not have been possible without sterling helicopter and fixed-wing support from the Philippine Air Force, U.S. Air Force, U.S. Navy, U.S. Marines, and the late Agustin Consunji. Drivers from USAID/Philippines, PHIVOLCS, the Philippine Mines and Geosciences Bureau, and the University of the Philippines not only kept us mobile on the ground but also doubled as field assistants, cooks, and other key support.

Critical reviews are essential for credibility. Reviewers are acknowledged in individual papers; we list them here, too, alphabetically, with our heartfelt thanks for a job well done. They were:

David Alexander, Fred Anderson, Onie Arboleda, Rich Bernknopf, Russell Blong, Marcus Bursik, Steve Carey, Mike Carroll, Tom Casadevall, Kathy Cashman, Pat Castillo, Bill Chadwick, Bernard Chouet, Toti Corpuz, John Costa, Doug Crowe, Art Daag, Mark Defant, PJ Delos Reyes, Rick Dinicola, Mike Dolan, Mary Donato, Mike Doukas, Dan Dzurisin, Jerry Eaton, Elliot Endo, John Ewert, Jon Fink, John Fournelle, Bruce Freeman, Elmer Gabinete, Terry Gerlach, Paula Gori, Marianne Guffanti, Dave Harlow, Brian Hausback, Wes Hildreth, Dave Hill, Todd Hinkley, Rick Hoblitt, Rick Holasek, Mark Holmes, Akira Imai, Chip Johnson, Juergen Kienle, Ulrich Knittel, Tak Koyaguchi, Dennis Krohn, Ed Laguerta, John Lahr, Rick LaHusen, Willie Lee, Jonathan Lees, Pete Lipman, Andy Lockhart, Jake Lowenstern, Jim Luhr, Jon Major, Steve Malone, Mylene Martinez, Mike McKibben, Steve McNutt, Bill Melson, Gerry Middleton, Dennis Mileti, C. Dan Miller, Seth Moran, Jim Mori, Vince Neall, Scott Oswalt, Bob Page, John Pallister, Jill Pasteris, Resty Pelayo, Tom Pierson, John Power, Manoling Ramos, Thess Regalado, Mark Reid, Jim Riehle, Alan Robock, Kelvin Rodolfo, Ed Roedder, Bill Rose, Gary Rowe, Mac Rutherford, Julio Sabit, Yoshihiro Sawada, Kevin Scott, Willie Scott, Steve Self, Pat Shanks, Haraldur Sigurdsson, Gary Smith, Kathleen Smith, Stu Smith, Rene Solidum, David Sussman, Don Swanson, Azadeh Tabazadeh, Makoto Tahira, Motoo Ukawa, Jess Umbal, Barry Voight, Richard Waitt, Lou Walter, Craig Weaver, Kelin Whipple, Randy White, Gerry Wieczorek, Stan Williams, and Ed Wolfe.

Papers in this volume reflect work done both for immediate hazard mitigation and for research. For the first purpose, we gratefully acknowledge team funding from Office of President Fidel V. Ramos, the Philippine Department of Science and Technology, U.S. Department of Interior, the Philippine National Disaster Coordinating Council, UNESCO, USAID/OFDA, USAID/ Philippines, and the U.S. Department of Defense. For the second purpose, and for preparation of this volume, we gratefully acknowledge funds from our respective departments and a wide variety of research funding sources, including the U.S. National Science Foundation, that are acknowledged in individual papers. Most USGS funds came from its Volcano Hazards Program; we also thank its Earthquake Hazards Reduction Program, Pacific Northwest region, for valuable support during final preparation of this monograph.

The publishers of this book—PHIVOLCS and the University of Washington Press—together with the originally-intended and now-assisting publisher, the U.S. Geological Survey—are new trailmates in new terrain. Budgetary circumstances forced the USGS to seek immediate help. The USGS' Branch of Eastern Technical Reports agreed to finish camera-ready copy and PHIVOLCS, which had already contributed many papers, arranged a subsidy from the Mount Pinatubo Commission (see special acknowledgments, next page). Naomi Pascal, Pat Soden, and Veronica Seyd of the University of Washington Press responded with much-appreciated trust and flexibility, accepting the book as written and agreeing to a concurrent World-Wide Web version. We look forward to hiking the rest of the trail together, despite the unusual circumstances under which we met!

We close this preface and acknowledgment with grateful thanks to the families of all who have worked at Pinatubo, for being supportive through initial risks and later exhaustion. We also recall, with enormous respect, wise guidance from the late Dick Janda that we use still today as Pinatubo lahars continue.

Chris Newhall and Ray Punongbayan, editors

SPECIAL ACKNOWLEDGMENTS

When a budget crisis in the U.S. Geological Survey forced us to change publishers, three organizations stepped forward with special help.

The USGS' Branch of Eastern Technical Reports (BETR), which had been preparing the monograph for USGS publication, was suddenly faced with the prospect that this report in which it had invested enormous amounts of time and skill might not get published. Two BETR employees in particular, Arlene Compher for illustrations and John Watson for text, had shaped many a gem in the rough into a polished product, with remarkable cheer and patience. Then, despite a painful reorganization and reduction-in-force, Arlene, John, Carolyn McQuaig for typesetting, Dave Murphy for final graphics work, and other BETR employees finished this huge job with great professional dedication.

The USGS also donated copies of its Mount St. Helens Professional Paper 1250, as in-kind publication support and to afford readers a chance to obtain this classic monograph together with the present Pinatubo volume.

The second group to whom we owe special thanks is the Mount Pinatubo Commission, chaired by the Hon. Salvador Enriquez (Antonio A. Fernando, Executive Director). The Mount Pinatubo Commission draws from government agencies and non-governmental organizations to manage relief and recovery from the Pinatubo eruption, including evacuation camps, resettlement areas, sediment control measures, and reconstruction of infrastructure. In recognition of the importance of Pinatubo events in volcanology and of a good understanding of volcanic events before those events can be mitigated, the Commission granted a generous subsidy that assured publication of the book.

The third source of critical support was PHIVOLCS' parent, the Department of Science and Technology (DOST) and its Secretary, Dr. William Padolina. A subsidy from DOST allowed us to print significantly more copies than would otherwise have been possible, and to offer this volume at an affordable price.

Thanks to these three organizations, many may learn the lessons of Pinatubo.

Children of the volcano: Aeta children in Pasbul, Kamias, Porac.

Prescient theater: advertisement for a play overtaken by real events, Clark Air Base.

Dark ash cloud over Clark's Base Operations building, June 22, 1991. Photograph by R.P. Hoblitt.

VOLCANIC DRAMA, HUMAN DRAMA

On the afternoon of June 15, 1991, Mount Pinatubo erupted between 3.7 and 5.3 km^3 of magma (8.4 to 10.4 km^3 of porous, pumiceous deposit), devastated more than 400 km^2, and blanketed most of Southeast Asia with ash (W.E. Scott and others; Paladio-Melosantos and others). For comparison, Mount St. Helens erupted only a tenth of this volume of magma but devastated and blanketed similar areas, more thinly. Within the 20th century, only the eruption of 13±3 km^3 of magma from Novarupta (Katmai) in 1912, in a remote part of Alaska, was larger.

Even more remarkable than the volume, area of impact, and eruption rate is that this was the first event of this size to be monitored in detail (Wolfe and Hoblitt). Many startling and fascinating phenomena were observed. Some, such as the occurrence of long-period earthquakes before explosive eruptions, were previously known but magnified at Pinatubo. Others, such as early separation of a volatile phase deep within dacite reservoirs, were noticed first at Pinatubo and are now being found elsewhere (Gerlach and others). In another example, a sudden, short-lived decrease in SO$_2$ emission shortly before an explosion was seen first at Pinatubo (Daag, Tubianosa, and others) and is now being discovered elsewhere. Still other phenomena remain unique for the moment. For example, deep long-period earthquakes marked the intrusion of basalt into a body of crystal-rich dacitic magma (White), and mixing of the magmas triggered the eruption (Pallister and others). Both processes were previously known on geologic grounds; neither had been constrained in time and space. Similarly, interaction between the eruption column and the ground generated tremor with periods of 200 to 300 s that was recorded around the world (Zürn and Widmer), and caldera collapse did *not* generate earthquakes larger than magnitude 5 (Bautista and others). Geomorphic changes that a geologist might normally expect to occur within millennia have occurred within days and years (Punongbayan, Newhall, and Hoblitt).

The most remarkable fact of all is that, although Pinatubo threatened 1,000,000 people, only a few hundred perished. The eruption was accurately predicted and tens of thousands of people were evacuated to safe distances. The 1991 eruption was the first to occur at Pinatubo in about 500 years, and the lack of baseline information for an eruption of Pinatubo and for other large eruptions made forecasts highly uncertain. Local disbelief that Pinatubo was even a volcano, much less one that could erupt, posed a horrific challenge for scientists and civil defense leaders. Indeed, some of this skepticism was never overcome (Tayag and others).

Long-term accumulation of volatiles within a capped, viscous dacitic magma, and sudden intrusion of that dacite by basalt, may have contributed to a surprisingly orderly and rapid progression from first precursors to climactic eruption. Only slightly more than 2 months elapsed from the first confirmed precursors to the massive eruptions, and the most diagnostic signs occurred within just 10 days of systematic escalation of unrest in early June 1991. A hazard map and a simple 5-level alert scheme, the primary warnings to public officials, proved to be remarkably accurate. In addition, a graphic video summary of volcanic hazards, made by the late Maurice Krafft for the International Association of Volcanology and Chemistry of the Earth's Interior, grabbed even the skeptics' attention (Punongbayan, Newhall, Bautista, and others). Fast action by civil defense and local leaders led to massive evacuations and saved many lives, and, especially on U.S. military bases, much property as well. Not to be overlooked, luck also played a role: the 1991 eruption was not as large as some prehistoric eruptions, and a series of moderate-scale eruptions that preceded the climax convinced skeptics and those clinging to their land and homes that evacuation was the only prudent option.

The muddy aftermath of the eruption, in which heavy rains continue to remobilize large volumes of 1991 deposit as lahars, still batters central Luzon. More than 2×10^9 m^3 of sediment have buried about 400 km^2 of lowland alluvial fans. Steaming hot debris flows with peak discharges of more than 1,000 m^3/s continue for hours after heavy rains and leave deposits of up to several tens of million cubic meters. Different watersheds yielded sediment at different rates, depending on factors such as slope, area versus thickness of deposit, and, later, vegetation recovery (Janda and others). Forecasts of sediment volume have been used for planning long-range mitigation projects; forecasts of imminent lahars have been used to trigger immediate evacuations. In principle, long- and short-range forecasts have been intended to help people to stay safely in their own homes and villages until they absolutely must move to higher or more distant ground; in practice, scientists, public officials, and citizens alike are still struggling to understand changing hazards and to neither underreact nor overreact. Hope runs eternal among those at risk from lahars, and people still delay their precautions until lahars are upon them (Cola). Technical accuracy in warnings is not enough to save lives; warnings must also convince those who are at risk to actually take the necessary precautions.

Dramatic changes to the face of the volcano are matched by dramatic cultural change. The minority Aeta population of about 20,000 was completely uprooted from their mountain life, wild fruits and animals, and supernatural spirits (C.B. Bautista). An even larger number of lowlanders were displaced, more physically than culturally, by the distal reaches of the eruption and by consequent lahars. Economic development of the region suffered severe short-term setbacks, and the large American military presence in the Philippines was brought to an early end.

In summary, the eruption of Mount Pinatubo in June 1991 and the subsequent lahars were the first of their magnitude to occur in a densely populated area among a people with the will and technological means to actually mitigate risks. With the benefit of good forecasts and generally constructive public responses, thousands of lives and untold amounts of property were saved. About 250 died during the eruption and a hundred more have died in subsequent lahars; without the forecasts and constructive public response, the toll would have been much greater.

Ash-covered newstand at Clark Air Base; headline tells the story. Photograph by Val Gempis, USAF.

Overview of the Eruptions

By Edward W. Wolfe[1] and Richard P. Hoblitt[1]

ABSTRACT

After 2½ weeks of locally felt earthquakes, steam explosions announced volcanic unrest at Mount Pinatubo on April 2, 1991. The unrest culminated 10 weeks later in the world's largest eruption in more than half a century. Volcanologists of the Philippine Institute of Volcanology and Seismology were joined in late April by colleagues from the U.S. Geological Survey. Together they successfully forecast the eruptive events and their effects, enabling Philippine civil leaders to organize massive evacuations that saved thousands of lives. The forecasts also led to the evacuation of Clark Air Base (U.S. Air Force), which is located just east of the volcano. Nevertheless, the climactic eruption, coincident with a typhoon on June 15, caused 200 to 300 deaths and extensive property damage, owing to an extraordinarily broad distribution of heavy, wet tephra-fall deposits.

The volcanic unrest and eruptions, which involved intrusion of basaltic magma into a reservoir of crystal-rich, vapor-saturated dacitic magma, evolved in stages: mid-March through May—felt earthquakes in March, phreatic explosions on April 2, and persistence of numerous volcano-tectonic earthquakes; June 1–7—localization of shallow earthquakes in a narrow pipelike zone near the volcano's summit; June 7–12—lava-dome growth, accompanied by increasing ash emission and seismic-energy release, including significant episodes of volcanic tremor; June 12–14—a series of four brief vertical eruptions accompanied by a profound buildup of long-period earthquakes; June 14–15—thirteen brief surge-producing eruptions that became progressively more closely spaced; June 15—the climactic eruption, which lasted approximately 9 hours and included collapse of the volcano's summit to produce a 2.5-kilometer-diameter caldera; June 15-middle or late July—decline and termination of continuous emission of a tephra plume from vents within the caldera and steady decline of volcano-tectonic earthquakes that began during the climactic eruption (intermittent small ash eruptions continued until early September); and July–October 1992—extrusion of a lava dome within the caldera.

---

[1]U.S. Geological Survey.

Runoff from monsoon and typhoon rains is eroding and redistributing the voluminous pyroclastic deposits emplaced during the eruption. Sedimentation from the resulting lahars continues to bury communities and valuable agricultural land over large areas in the lowlands surrounding the volcano.

INTRODUCTION

Mount Pinatubo is one of a chain of composite volcanoes (fig. 1) that constitute the Luzon volcanic arc. The arc parallels the west coast of Luzon and reflects eastward-dipping subduction along the Manila trench to the west (Defant and others, 1989). Mount Pinatubo is among the highest peaks in west-central Luzon. Its former summit (fig. 2), at 1,745 m elevation, may have been the crest of a lava dome that formed about 500 years ago during the most recent previous major eruptive episode (Newhall and others, this volume). The volcano's lower flanks, intricately dissected and densely sheathed in tropical vegetation prior to the 1991 eruptions (fig. 3), were composed largely of pyroclastic deposits from voluminous, explosive prehistoric eruptions.

Before the eruption, more than 30,000 people lived in small villages on the volcano's flanks. A much larger population—about 500,000—continues to live in cities and villages on broad, gently sloping alluvial fans surrounding the volcano. Clark Air Base lies to the east of the volcano, within 25 km of the summit, and Subic Bay Naval Station is about 40 km to the southwest (fig. 4).

This report provides a broad view of the eruptive activity as a context for the many topical papers that follow. It draws extensively on a summary report by the Pinatubo Volcano Observatory Team (1991) and on an account of the events by Wolfe (1992); no additional citations are given for those two reports. A detailed chronology of eruptive events during June 1991 is given in table 1 of Hoblitt, Wolfe, and others (this volume), and Harlow and others (this volume) review the seismic chronology.

A sequence of developmental stages of volcanic unrest and eruptions characterized the 1991–92 activity of Mount Pinatubo. A brief review of the major stages follows:

1. The initial stage, from mid-March through May, was characterized by felt earthquakes beginning on March 15, phreatic explosions on April 2, continuing emission of steam and minor ash, essentially constant rate of seismic-energy release dominated by persistent occurrence of volcano-tectonic earthquakes, and generally increasing emission of SO_2.

2 Increasing seismic-energy release in early June (fig. 4A of Harlow and others, this volume) and localization of shallow earthquake hypocenters near the volcano's summit culminated in a shallow intrusion that reached the surface on June 7 to initiate stage 3.

3. Extrusion of a lava dome, increasing emission of a dense but low ash plume, and increasing seismic-energy release, including significant episodes of volcanic

tremor, marked the interval between the first appearance of the dome on June 7 and the first large vertical eruption on June 12.

4. A series of 4 brief vertical eruptions and a profound buildup of long-period earthquakes occurred between the morning of June 12 and early afternoon of June 14. Growth of the dome may have continued through at least part of this interval.

5. A series of 13 pyroclastic-surge-producing eruptions occurred within the next 24 h. These became progressively more frequent, and seismic-energy release increased as the system evolved toward the climactic eruption.

6. An explosion at 1342 June 15 marked the beginning of the climactic eruption, which lasted approximately 9 hours and included collapse of the volcano's summit to form a 2.5-km-diameter caldera.

7. Gradually diminishing but continuous emission of tephra from vents in the summit caldera persisted for about another month, and small intermittent ash eruptions occurred through early September. An intense swarm of volcano-tectonic earthquakes that began during the climactic eruption steadily diminished as events decreased in both number and magnitude.

8. A lava dome was extruded within the caldera from July to October 1992.

PRE-1991 HINTS OF AN ACTIVE MAGMATIC SYSTEM

Many observations at Mount Pinatubo before March 1991 are, in retrospect, consistent with the presence of an active magmatic system beneath the volcano. Although no historic eruptive activity was known before the 1991 events,

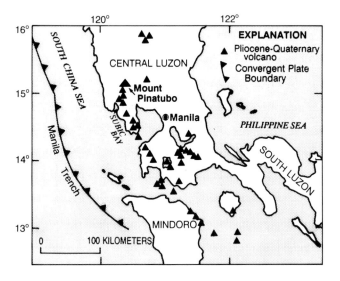

Figure 1. The location of Mount Pinatubo, within the Luzon volcanic arc.

Figure 2. Mount Pinatubo as seen from the building initially established as the Pinatubo Volcano Observatory (PVO) at Clark Air Base. Steam plume, to the right of the summit in this westward view, issued continuously from vents northwest of the summit after initial explosions on April 2, 1991. (Photograph by R.P. Hoblitt, June 6, 1991.)

Figure 3. Intricately dissected, densely vegetated upper north flank of Mount Pinatubo, which is underlain by prehistoric pyroclastic deposits. View north along the O'Donnell River from near Patal Pinto (fig. 4). (Photograph by R.P. Hoblitt, May 29, 1991.)

some Aeta—the indigenous people who lived on the mountain's flanks—had reported minor explosions (Newhall and others, this volume). Hot geothermal fluids were known to be present (Delfin, 1983; 1984), and following the 1991 eruptions, a magmatic-fluid component was identified in water samples obtained before 1991 from fumaroles and drill holes (Delfin and others, 1992).

Ground fracturing and steam emission related to a landslide high on the upper northwest flank of Mount Pinatubo occurred on August 3, 1990 (Isada and Ramos, 1990; Sabit and others, this volume), 2 to 3 weeks after the occurrence of a major earthquake (magnitude 7.8) about 100 km to the northeast. Hindsight suggests that these events may have been manifestations of a magmatic or hydrothermal disturbance that marked the reawakening of the Pinatubo magmatic system. Before April 2, 1991, however, none of these phenomena caused volcanologists to suspect that an eruption might occur in the near future.

PRECURSORY ACTIVITY AND HAZARD ASSESSMENT

Citizens of small communities on the lower northwest flank of Mount Pinatubo felt earthquakes beginning on March 15, 1991 (Sabit and others, this volume). Then, during the afternoon of April 2, 1991, a series of small explosions issued from a 1.5-km-long line of vents along a northeast-trending fissure on the upper north flank of the volcano (figs. 5 and 6). The explosions, which occurred over a period of several hours, stripped the vegetation from several square kilometers, deposited a meter or more of poorly sorted rock debris near the craters, and dusted villages 10 km to the west-southwest with ash. Vegetation was

not charred, blocks near the vents were heterolithologic, and some were hydrothermally altered; together these observations suggest that the April 2 explosions were hydrothermal rather than magmatic in origin. Felt earthquakes and sulfur odors continued after the explosions, and a line of active steam vents extended across the volcano's upper north flank west-southwest of the fissure and explosion pits (Sabit and others, this volume). Activity soon concentrated at three main fumaroles in narrow canyons at the heads of the Maraunot and O'Donnell River valleys (figs. 6 and 7). Light ash emission occurred intermittently through early June.

Concerned about the implications of the April 2 explosions and the ensuing felt earthquakes and sulfur odors, scientists from PHIVOLCS began to install portable seismographs near the northwest foot of Mount Pinatubo on April 5. The more than 200 small, high-frequency earthquakes recorded during the first 24 h of seismic monitoring led PHIVOLCS to recommend precautionary evacuation of areas within a 10-km-radius of the summit; approximately 5,000 residents evacuated. Through April and May, this network recorded between 30 and 180 high-frequency earthquakes per day (Sabit and others, this volume).

In late April, PHIVOLCS was joined by a group from the USGS, and the joint team installed a network of seven seismometers with radiotelemetry to an apartment at Clark Air Base. The apartment served as the initial home for the Pinatubo Volcano Observatory (PVO). Seismic data from the new network was gathered and processed at PVO on a DOS-based computer system (Lee, 1989). An electronic tiltmeter with radiotelemetry to PVO was also installed on the upper east flank of the volcano in late May.

Concurrently with installation of the telemetered seismic network, the PVO team undertook a rapid geologic

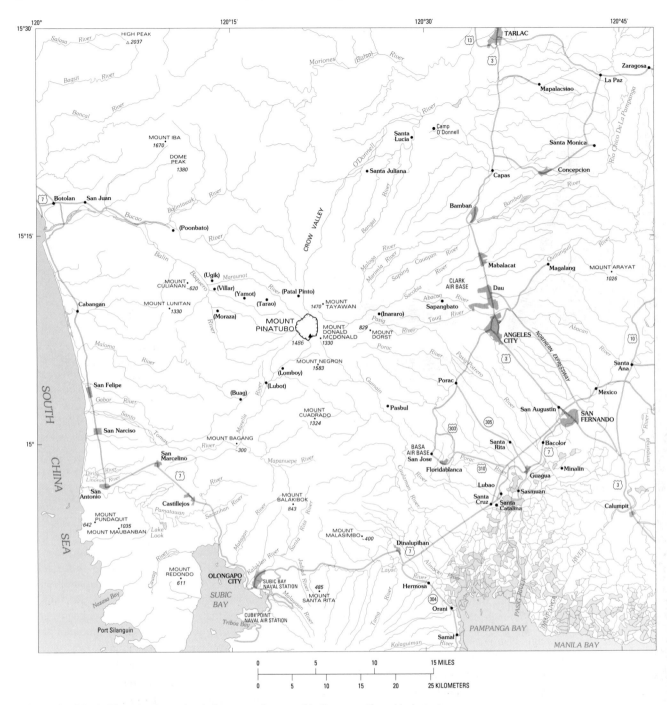

Figure 4. Mount Pinatubo, its major drainages, and geographic features referred to in text.

reconnaissance to determine the style and magnitude of past eruptions as the basis for a volcano-hazard map (fig. 6 of Punongbayan and others, this volume). This reconnaissance showed that terrain as far as about 20 km from the summit of Mount Pinatubo had been repeatedly engulfed by voluminous pyroclastic flows in the recent geologic past. Lahar deposits, probably composed of debris eroded from pyroclastic-flow deposits, extended down all major drainages well beyond the pyroclastic-flow deposits. Volcanic debris transported by lahars had probably inundated large parts of the low-gradient alluvial fans that originate at Mount Pinatubo.

Hasty ^{14}C analyses of charcoal collected from pyroclastic-flow deposits during the reconnaissance yielded preliminary new emplacement ages. The new ages, when combined with three previous ages (Ebasco Services, Inc., 1977), provided an important insight into Pinatubo's behavior through time: large explosive eruptions separated by

Figure 5. Fissure and line of new craters formed by explosions of April 2 on upper north flank of Mount Pinatubo. View is southwestward. Prominent fumarole in distance is in the head of the O'Donnell River drainage. (Photograph courtesy of C.G. Newhall, May 1991.)

repose periods lasting millennia. The three most recent major eruptions occurred approximately 500, 3,000, and 5,500 calendar years ago. Subsequent work has refined the record (Newhall and others, this volume), but the general pattern remains the same: episodes of voluminous explosive eruptions separated by centuries to millennia of repose.

The PVO team used the hazard map and analysis of the volcano's unrest to acquaint civil-defense officials and military commanders with the potential eruptive hazards: extensive voluminous pyroclastic flows, tephra falls, and lahars that could extend far beyond the reach of the pyroclastic flows. A preliminary version of a videotape illustrating volcano hazards, produced by filmmaker-volcanologist Maurice Krafft, enormously helped the team explain hazards foreign to people in an area lacking historic eruptions. Communication of information about Pinatubo's evolving state of unrest was simplified by a system of hazard-alert levels (fig. 4 of Punongbayan and others, this volume).

Numerous small, high-frequency earthquakes continued through May. They were generally too small to be felt, except locally. Magnitudes were less than 2.5 for all 1,800 located earthquakes between May 7 and June 1. These earthquakes were strongly clustered in a zone between 2 and 6 km deep, located about 5 km north-northwest of the volcano's summit (see fig. 5 in Harlow and others, this volume); relatively few scattered earthquakes occurred beneath the summit or the nearby area of vigorous fumaroles. The location of the earthquake cluster northwest of the area of active fumaroles was puzzling. Photogeologic interpretation indicated coincidence of the earthquake cluster with a zone of young faults. Initially it was unclear whether the cluster was recording localized adjustments of the crust to stresses generated by movement, growth, or pressurization of a magma body or whether the cluster was recording tectonic activity unrelated to volcanic processes. However, the increasing rate of SO_2 emission soon strongly supported the former hypothesis.

Measurements of the rate of emission of SO_2 from the fumaroles near the volcano summit provided compelling evidence for the presence of fresh magma beneath the volcano. Successive airborne measurements showed that the SO_2 flux increased from about 500 t/d when the first measurement was made on May 13 to more than 5,000 t/d on May 28 (Daag, Tubianosa, and others, this volume). In combination with the sustained seismicity, these data suggested at the time that magma beneath Pinatubo had risen to a level sufficiently shallow to permit substantial degassing of its dissolved volatile components. Subsequent volcanic-gas emission studies related to the 1992 eruption at Mount Spurr, Alaska (Doukas and Gerlach, in press) showed that hydrolysis of SO_2 in ground water masks emission of magmatic SO_2. This conclusion, combined with the observed vigor of the fumaroles formed in early April, suggests that the increasing rate of SO_2 emission during May at Mount Pinatubo reflected progressive evaporation of the volcano's hydrothermal system. Low permeability of Pinatubo's hydrothermal system (Delfin and others, 1992) may have inhibited its capacity to recharge boiled-off water fast enough to prevent or greatly diminish preeruption SO_2 emissions (Doukas and Gerlach, in press).

About June 1, a second cluster of earthquakes began to develop between the surface and a depth of 5 km, in the vicinity of the fuming vents, approximately 1 km northwest of the summit (see fig. 7 in Harlow and others, this volume). These earthquakes apparently recorded fracturing of rock as rising magma began to force open a conduit between the magma reservoir beneath the volcano and the surface. At the same time, the SO_2 emission rate was declining from the peak measured on May 28; the rate was about 1,800 t/d on May 30, 1,300 t/d on June 3, and 260 t/d on June 5 (Daag, Tubianosa, and others, this volume). The latter flux was about one-twentieth of the earlier (May 28) maximum. The reduction in SO_2 flux raised concern that the passages

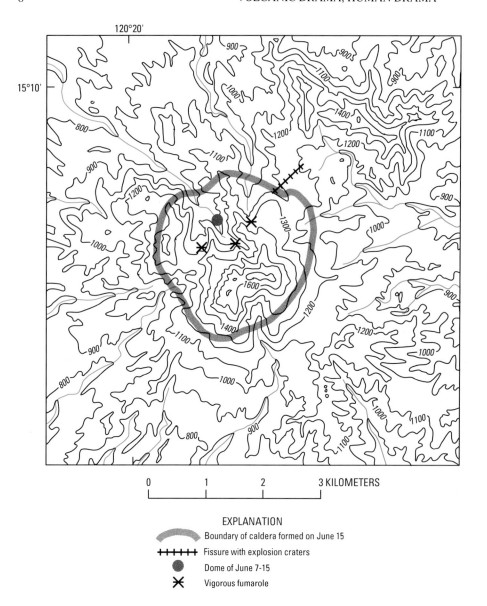

0   1   2   3 KILOMETERS

EXPLANATION

Boundary of caldera formed on June 15

++++++ Fissure with explosion craters

● Dome of June 7–15

✳ Vigorous fumarole

Figure 6. Major volcanic and hydrothermal features formed in the Mount Pinatubo summit region from April 2 through June 15, 1991. Preeruption topography modified from Mount Pinatubo 15-minute quadrangle (U.S. Defense Mapping Agency, Sheet 7073 II, Series S701, Edition 2–DMA). Contour interval is 100 m.

through which gas escaped to the surface might have become sealed, perhaps by the degassed tip of an ascending column of magma, and that rapidly increasing pressurization and an imminent explosive eruption might ensue.

A small explosion at 1939 on June 3 (Ewert and others, this volume; Harlow and others, this volume) initiated an episode of increasing volcanic unrest characterized by intermittent minor emission of ash, increasing seismicity beneath the vents, episodes of harmonic tremor, and gradually increasing outward tilt at a tiltmeter high on the volcano's east flank. In response to the growing restlessness of Mount Pinatubo, PHIVOLCS issued a level-3 alert on June 5, which indicated the possibility of a major pyroclastic eruption within 2 weeks.

DOME EXTRUSION AND RELATED UNREST

The electronic tiltmeter on Mount Pinatubo's upper east flank began to show accelerated outward tilt at about noon on June 6 (Ewert and others, this volume). Seismicity, as well as the outward tilt, continued to increase until late afternoon on June 7, when apparently increased emission generated a column of steam and ash 7 to 8 km high. Shortly thereafter, seismicity decreased, and the increase in outward tilt stopped. PHIVOLCS promptly announced a level-4 alert (eruption possible within 24 h) and recommended additional evacuations from the volcano's flanks.

Such outward tilt and increased shallow seismicity suggested that a shallow conduit was developing for

delivery of magma to the surface. This was confirmed the following morning when observers identified a small (50- to 100-m-diameter) lava dome northwest of the summit in the upper part of the Maraunot River canyon (fig. 6). The emission of a persistent, low, roiling ash cloud from the vicinity of the dome began at this time (fig. 8).

Initially the dome lay on the lower part of the northeast canyon wall and nearby part of the adjacent valley floor. Repeated aerial observations—generally under conditions of extremely poor visibility that were due to venting steam and tephra—indicated that the dome continued to grow at least until June 11. The southeastern sector of the dome was destroyed by the vertical eruptions of June 12–14. Despite the destruction, the dome margin continued to expand westward. When last seen, at about noon on June 14 (fig. 9), the dome was relatively flattopped and spanned the narrow upper Maraunot River valley. The expansion may have been due to continued extrusion or to postextrusion flowage as dome lava moved downhill under the influence of gravity.

A pyroclastic-flow deposit emplaced in the upper Maraunot valley on late on June 12 or on June 13 contains

← **Figure 7.** Three main fumaroles on north flank of Mount Pinatubo after explosions of April 2, 1991. View is approximately southward. Fumarole at left is at the head of the O'Donnell River valley; those at center and right are in headwater canyons of the Maraunot River valley (compare fig. 6). (Photograph courtesy of C.G. Newhall, May 1991.)

Figure 8. Parallel plumes of steam (left, white) and ash (right, gray) that issued from June 7 to June 12 from vents in the headwaters of the Maraunot River, about 1 km northwest of the summit of Mount Pinatubo. (Photograph by R.P. Hoblitt, June 10, 1991.)

Figure 9. Lava dome spanning the narrow upper Maraunot River canyon about 900 m northwest of the summit of Mount Pinatubo. View is southeastward. (Photograph by R.P. Hoblitt, approximately noon, June 14, 1991).

abundant blocks, some of them prismatically jointed. These blocks, which were sampled after the 1991 eruptions, are believed to represent the lava dome (Hoblitt, Wolfe, and others, this volume). They consist of hybrid andesite with undercooled and quenched basalt inclusions. Major- and trace-element data strongly support genesis of the andesite by mixing of basaltic magma, represented by the inclusions, with dacitic magma, represented by juvenile pumice of the climactic eruption (Pallister and others, 1992; Pallister and others, this volume).

The period from June 8 through early June 12 was marked by increasing ash emission, swarms of shallow earthquakes largely beneath the dome, and episodic harmonic tremor. The east wind carried the ash plume westward. At times, dilute ashy density currents flowed down the upper Maraunot valley. Seen from the distance, these ashy density currents resembled the ash clouds associated with pyroclastic flows, but they neither coincided with distinct seismic explosion signals nor left recognizable flowage deposits marking their tracks. Concern raised by both the escalating degree of volcanic unrest and observation of these ashy density currents led PHIVOLCS to raise the alert level to 5 (eruption in progress) on June 9. The radius of evacuation was extended to 20 km, and the number of evacuees increased to about 25,000.

On June 10, about 14,500 U.S. military personnel and their dependents evacuated Clark Air Base. They traveled by motorcade to Subic Bay Naval Station, from which most eventually returned to the United States. All remaining military aircraft except for three helicopters also left Clark that day. About 1,500 U.S. and a comparable of number of Philippine military personnel remained behind to provide

security and base maintenance. The PVO team of volcanologists also remained on the base. For an increased margin of safety, PVO moved its operations to a building near the east end of Clark Air Base, about 25 km from the volcano's summit, and continued round-the-clock monitoring there.

A burst of intense seismic tremor began at 0310 on June 12. High-amplitude tremor persisted for about 40 min, after which the amplitude gradually waned, returning to background level more than 2 h after it began. A small eruption signal with an onset time of 0341[2] was embedded in this tremor episode. Although no tephra fall was reported, the tremor burst probably coincided with an episode of increased ash emission. Daylight the next morning showed the plume of steam and ash rising about 3 km above the volcano, higher than during the previous few days, and aerial observations showed that small pyroclastic flows had coursed down the uppermost Maraunot and O'Donnell drainages and spawned small hot lahars.

LARGE VERTICAL ERUPTIONS AND BUILDUP OF LONG-PERIOD EARTHQUAKES

The first major explosive eruption began at 0851 on June 12, generating a column of ash and steam (see fig. 1A of Hoblitt, Wolfe, and others, this volume) that rose to at least 19 km, according to the weather-radar operators at Subic Bay Naval Station and Clark Air Base. At the onset, seismicity increased within a few seconds from low-amplitude tremor to a high-amplitude signal that saturated the records of all the seismic stations. The high-amplitude seismic signal and the rise of the eruptive column seemed to begin simultaneously. Interpretation of seismic records indicated that this event lasted about 38 min, although field observations suggest a duration of at least an hour (Hoblitt, Wolfe, and others, this volume). Beginning with the 0851 eruption, all seismic-station signals except from the station most distant from the volcano summit, at Clark Air Base, showed continuous seismic events and tremor until they were destroyed or their signal was otherwise lost.

Ash from the 0851 eruption was transported southwestward past communities north of Subic Bay, and a small pyroclastic flow traveled northwest from the vent in the headwaters of the Maraunot River. Six hundred of the remaining 1,500 U.S. military personnel at Clark Air Base were evacuated. The general evacuation radius was extended to 30 km, and the total number of evacuees increased to at least 58,000.

This was the first and longest of four vertical eruptions (fig. 10) from a vent on the southeast margin of the lava

---

[2]Times of explosive eruptions were determined from seismic-drum records, which recorded uncorrected local time.

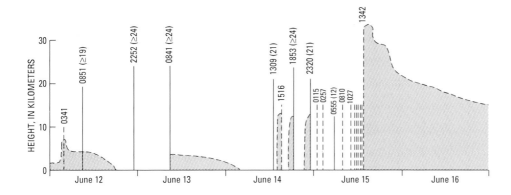

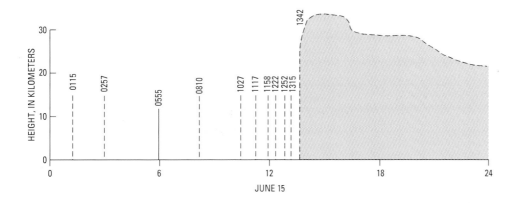

Figure 10. Chronology of explosive eruptions of June 12–15, 1991, determined from visual observations, weather-radar observations, seismic signatures, and, on June 15, data from a recording barograph in the Clark weather station. Events of June 15 are shown in the lower diagram with an expanded time axis. Vertical spikes correspond to individual brief explosions, each of which produced a tephra plume. Time given with each spike is the time of the explosion onset determined from seismic-drum records. Solid spikes record explosions for which weather-radar observers provided realtime tephra-column heights (given in parentheses). Upper limit of radar observations was 19 km at Cubi Point Naval Air Station and 24 km at Clark Air Base. Dashed spikes record explosions for which we received no realtime radar measurements of plume height. Shading portrays known continuous tephra emission.

dome. The other three—at 2252 on June 12, 0841 on June 13, and 1309 on June 14—were progressively shorter, with durations of 14, 5, and 2 min, respectively. Weather conditions were good during the period of vertical eruptions, and all three that occurred during daylight hours were visible from PVO (fig. 1 of Hoblitt, Wolfe, and others, this volume). Good visibility for these early eruptions was really fortunate because it provided an opportunity for the PVO volcanologists and the military weather-radar observers to correlate unmistakable eruptions with their seismic and radar signatures. We quickly developed a routine that proved indispensable during periods of poor visibility: PVO volcanologists recognized an incoming explosion signal on the seismographs and contacted the radar observers for confirmation and tracking of eruption-column development.

Ground-based observers to the west, at Poonbato, reported seeing pyroclastic flows during each of the three daytime eruptions (Sabit and others, this volume), and low tephra plumes were seen over the vent in the hours following the 0851 June 12 eruption and the June 13 eruption. Explosion plumes seen from the air along the upper 5 km of the Maraunot River valley after the 0851 June 12 eruption provided indirect evidence of new pyroclastic-flow deposits, but poor visibility prevented direct observation of the deposits themselves. Another observation flight shortly before the 1309 June 14 eruption documented new pyro-

clastic-flow deposits in the upper Maraunot valley. Later ground-based observations about 3.5 km downstream from the position of the 1991 dome showed deposits of only two pyroclastic flows beneath the June 15 deposits in the Maraunot valley (Hoblitt, Wolfe, and others, this volume).

Northeasterly winds prevailed during the vertical eruptions. Consequently, tephra falls extended southwestward across Subic Bay Naval Station and nearby communities to the north. No tephra fell on Clark Air Base during this period.

Unlike the initial eruption at 0851 on June 12, the eruptions at 2252 on June 12 and 0841 on June 13 were preceded by 2- to 4-h swarms of long-period earthquakes (Harlow and others, this volume), which enabled the PVO team to issue explicit advance warnings. The resumption of frequent long-period earthquakes during the evening of June 13 suggested that another eruption might be imminent, and another eruption warning was given. None occurred as expected, however, as the pattern was changing to one in which long-period earthquakes continued hour after hour.

In the early morning of June 14, clear weather gave a good view of Mount Pinatubo. In spite of the intensifying swarm (fig. 11; fig. 10B of Harlow and others, this volume) of long-period earthquakes, no visible ash and very little fume were being emitted. After 28 h with no major explosive activity, this long-period swarm culminated at 1309

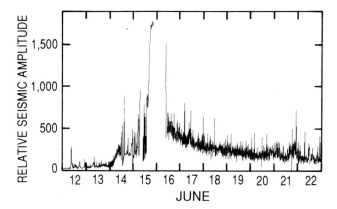

Figure 11. RSAM (realtime seismic-amplitude measurement; Endo and Murray, 1991) data for period of June 12-22, 1991. Vertical axis shows digital counts representing time-averaged voltage from the output of the seismic-data acquisition system. Plot shows individual explosions (compare with fig. 10) and, especially on June 14, increasing seismic-energy release related to increasing size and number of long-period earthquakes. There are no data for late June 15 and early June 16. Exponentially decreasing seismicity on and after June 16 reflects diminishing number and size of earthquakes recording structural adjustment of the volcano and the rock beneath it.

(see fig. 2A of Hoblitt, Wolfe, and others, this volume) in the fourth large vertical eruption.

Andesite scoria, similar in composition to the andesite of the dome (Pallister and others, this volume) is the dominant component in tephra-fall deposits from the vertical eruptions. Dacite, similar to that of the climactic eruption, was a lesser component; its abundance increased from just a few volume percent in the deposits of the 0851 June 12 eruption to approximately 35 volume percent in the tephra of the June 13 eruption (Hoblitt, Wolfe, and others, this volume).

More gas-rich andesitic magma ascending behind the degassed tip that formed the dome vesiculated to drive the initial vertical eruptions that began on June 12. Transient pressure relief in the conduit from each brief preclimactic eruption led to additional vesiculation of magma and progressive replacement of the limited volume of andesitic magma in the conduit by dacitic magma, as reflected in the progressive decrease of andesitic pyroclasts and complementary increase of dacitic pyroclasts in the deposits of the vertical and surge-producing eruptions.

SURGE-PRODUCING ERUPTIONS

A sequence of 13 explosive eruptions began shortly after the 1309 eruption on June 14 and continued for nearly 24 h. These became more closely spaced as time progressed (fig. 10), and, as discussed below, apparently reflected pyroclastic-surge production. Deteriorating weather conditions related to the approach of Typhoon Yunya obscured the

view of Mount Pinatubo from Clark Air Base so that only one of these events (at 0555 June 15) was seen directly from Clark. Several were seen, however, by observers to the northwest at Poonbato and to the north at Camp O'Donnell. In addition, two night eruptions (2320 on June 14 and 0115 on June 15) were recorded by an infrared-imaging device at Clark Air Base. All views were of large pyroclastic density currents sweeping the volcano's flanks.

The first indication of a change in style to surge-producing eruptions was an observation from the north at about 1410 on June 14 of small explosions or low fountaining that sent density currents down at least the Maraunot and O'Donnell drainages. A larger eruption at 1516, seen only from the northwest, produced a large pyroclastic density current that swept northwestward about 15 km. Observation of a hot lahar in the Sacobia River valley in the late afternoon suggests that a pyroclastic density current also entered this drainage.

Three main lines of evidence, largely evaluated in hindsight, suggest the changed nature of the eruptions between the last primarily vertical eruption at 1309 on June 14 and the onset of the climactic eruption at 1342 on June 15. First, limited visual observations and infrared images were of ground-hugging pyroclastic density currents sweeping down the volcano's flanks. Second, recording barographs at Clark Air Base (fig. 2B of Hoblitt, Wolfe, and others, this volume) and Cubi Point recorded an abrupt atmospheric compression coincident with each major explosion from approximately 1516 on June 14 through 1315 on June 15 and a subsequent more protracted rarefaction; the four vertical eruptions did not produce similar atmospheric responses. Hoblitt, Wolfe, and others (this volume) speculate that the rapid atmospheric compression recorded initial collapse of the eruptive column near the onset of each of the 13 large explosive eruptions. Third, later field observations showed a sequence of multiple pyroclastic-surge deposits emplaced after the initial vertical eruptions and before the climactic eruption (Hoblitt, Wolfe, and others, this volume).

None of the surge-producing eruptions was directly visible to the PVO team at Clark Air Base before the 0555 eruption on June 15, and much of the evidence for the change on June 14 to surge production was not yet in hand. Limited visibility at dawn on June 15 enabled PVO and Clark Air Base observers to see the pyroclastic density currents of the 0555 eruption (fig. 3 of Hoblitt, Wolfe, and others, this volume). The relatively low ash fountaining (radar operators reported a maximum height of 12 km)[3] and the voluminous density-current production were in marked contrast to the high vertical tephra columns seen earlier (last previous eruption cloud seen from Clark was the vertical column of 1309 on June 14). This apparent change in eruptive style suggested that a climactic eruption, perhaps accompanied by massive pyroclastic density currents, might be imminent. Consequently, the remaining U.S. Air Force personnel and the PVO volcanologists evacuated Clark Air Base. Second thoughts led the PVO staff and a small Air Force contingent to return later that morning, but the brief evacuation caused a 3-h gap in seismic recording.

Through June 14, winds blew virtually all of the ash west to southwest from Mount Pinatubo, causing ash fall in westernmost Luzon and over the South China Sea. However, wind patterns changed early on June 15, probably in response to the approaching typhoon. The view of the 0555 eruption cloud was quickly obscured by ash and rain that fell on Clark Air Base and over central Luzon throughout the day. Repeated explosions through midday were each followed 30 to 40 min later by total darkness at PVO, as winds blew ash eastward over Clark.

The abundance of dacite exceeds that of andesite in the pyroclastic-surge deposits, and the dominance of dacite increases upward; the stratigraphically highest surge deposits contain little or no andesite. As in the tephra-fall deposits from the vertical eruptions, fragments of these surge deposits have a broad range of density and a large proportion of dense clasts that represent relatively degassed magma (Hoblitt, Wolfe, and others, this volume).

Eruptions became progressively less vigorous and more closely spaced through the sequence of vertical and surge-producing eruptions, probably in response to increasing efficiency of the conduit system that delivered magma to the surface (Hoblitt, Wolfe, and others, this volume). By early afternoon of June 15 the supply rate had increased sufficiently to sustain the climactic eruption.

CLIMACTIC ERUPTION

An eruption at 1342 on June 15 initiated continuous high-amplitude tremor that saturated all operative seismic stations on Mount Pinatubo and, as we learned later, also initiated approximately 9 h of intense atmospheric-pressure variation (see barograph record, fig. 2B of Hoblitt, Wolfe, and others, this volume). By 1430, all seismometers but the one at Clark Air Base were inoperative (most were victims of pyroclastic density currents), and ash with pumice fragments as large as 4 cm in diameter was falling at PVO. The climactic eruption was clearly under way.

---

[3]Lynch and Stephens (this volume) report stratospheric eruptive plumes that persisted much longer than the eruptive events that generated them and in some cases reached much greater altitudes than indicated by the radar observations. For example, the satellite data show a plume at about 70,000 ft (21 km) for 1.5 h following the 0555 eruption. Seismic records show that this event was brief (3–5 min), and Clark Air Base weather-radar observers reported only a 40,000-ft (12-km) plume height at the time. We suspect that the satellite images record fine ash carried into the stratosphere by convection from denser eruptive columns "seen" by the radar or from hot ash clouds rising from pyroclastic density currents and that a tephra cloud maintains its identity in the stratosphere long after its parent eruption has ended.

Virtually blind as a result of destruction of the seismic net and near-zero visibility, the PVO staff and the few remaining Air Force personnel left Clark Air Base at about 1500 to join the remainder of the Clark Air Base evacuees at the Pampanga Agricultural College, about 38 km east of the summit of Mount Pinatubo on the western slopes of another volcanic cone, Mount Arayat. There the evacuees observed continuing ash fall and felt numerous earthquakes during the evening—small ones every minute or so, and large ones 10 to 15 min apart. The large earthquakes (M_b 4.3–5.7) were part of a series that began at 1539; the National Earthquake Information Center reported 29 events with $M_b \geq 4.5$ over the first 6 h (Mori, White, and others, this volume). At the time, the volcanologists speculated that the numerous felt earthquakes might record caldera formation. Indeed, subsequent views of the volcano showed that the original summit of the volcano was gone, and in its place was a new 2.5-km-diameter caldera (figs. 6, 12), the center of which is offset about 1 km northwestward from the preeruption summit. However, later analysis (B.C. Bautista and others, this volume) suggests that the largest earthquakes recorded crustal adjustments that occurred as strike-slip displacements on regional faults rather than caldera subsidence.

Satellite images showed that the climactic eruption cloud rose high into the stratosphere and expanded widely, umbrella-like. At 1540, it was 400 km in diameter, and shadow measurements against the surrounding white clouds indicate that its altitude was about 25 km at its eastern edge and 34 km at its center (Koyaguchi and Tokuno, 1993).

The Seismic Spectral Amplitude Measurement (SSAM) data (fig. 11 of Power and others, this volume) show a well-defined shift of the dominant seismic spectral peak from the 0.5- to 1.5-Hz band to the 1.5- to 3.5-Hz band approximately 3 h into the climactic eruption (at about 1630). Concurrently, the magnitude of the continuous, local, rapid atmospheric-pressure variation decreased (fig. 2B of Hoblitt, Wolfe, and others, this volume). Distant infrasonic stations also recorded approximately 3 h of intense activity correlative with the first 3 h of the climactic eruption, after which the infrasonic signal decreased in amplitude (see fig. 6 of Tahira and others, this volume).

The SSAM, barograph, and distant infrasonic data suggest collectively that the climactic phase was in full force for about 3 h and then began to wane. Structural adjustment of the volcano and the subjacent upper-crustal rocks to massive withdrawal of magma began after 2 h (1539) and accelerated an hour later (approximately 1630), as volcano-tectonic seismicity increased and long-period seismicity related to explosive volcanism declined. Unfortunately, our sole remaining seismometer failed for nearly 4 h beginning at about 2100 (fig 2A of Hoblitt, Wolfe, and others, this volume). However, the barograph record indicates that the rapid atmospheric-pressure fluctuations diminished to back-

Figure 13. Snowlike blanket of tephra deposit of June 15, 1991, about 9 cm thick, in the eastern part of Clark Air Base, about 25 km east of Mount Pinatubo. Numerals in lower right indicate date. (Photograph by R.P. Hoblitt.)

Figure 12. Caldera, 2.5 km in diameter, formed at crest of Mount Pinatubo on June 15, 1991. A continuous column of ash vented from the caldera floor during late June and July. Initially it filled the caldera, obscuring the caldera interior from view, but, by the time of this photograph, diminishing ash production permitted a view into the caldera. View is southwestward. (Photograph courtesy of J. Mori, June 26, 1991.)

ground level at about 2230. We interpret that time as the end of the climactic eruption.

We have no explicit evidence of the beginning and duration of caldera subsidence. It seems most likely that caldera formation occurred during the interval from 1630, when volcano-tectonic seismicity increased, to the end of the climactic eruption at 2230.

Typhoon Yunya made landfall at 0800 June 15 and decreased in intensity to Tropical Storm Yunya, which passed about 75 km northeast of Mount Pinatubo during the early vigorous hours of the climactic eruption (Oswalt and others, this volume). Owing at least in part to the atypical lower- and mid-level wind regime spawned by the passing storm, tephra was widely distributed in all directions around the erupting volcano. (Recall that prevailing winds had distributed essentially all earlier airborne tephra to the

southwest.) In addition, heavy rainfall accompanied the storm. As a result, a heavy, rain-saturated, snowlike blanket (fig. 13) of tephra a centimeter or more thick fell over about 7,500 km^2 of central and western Luzon, and almost all of the island received at least a trace (Paladio-Melosantos and others, this volume). The blanket has a remarkably consistent internal stratigraphy (fig. 14)—a thin, relatively fine, gray ash deposit overlain by a coarser, plinian pumice-fall deposit with conspicuous normal grading. The basal gray ash accumulated during the morning of June 15, and the overlying pumice-fall unit began accumulating during the afternoon, when pumice fragments were falling at PVO.

The greatest measured thickness of the June 15 tephra-fall deposits was 33 cm, at a locality 10.5 km southwest of the vent, and deposits 10 or more centimeters thick blanketed a densely settled area of about 2,000 km^2 (Paladio-Melosantos and others, this volume). The greatest bulk accumulated west to west-southwest of the volcano's summit (see fig. 7 of Paladio-Melosantos and others, this volume). Paladio-Melosantos and others (this volume) estimate that the on-land bulk volume of tephra from the climactic eruption within the 1-cm isopach is about 0.7 km^3. Even more fell into the South China Sea; indeed, tephra dusted parts of Indochina, more than 1,200 km away. The total bulk volume of tephra-fall deposits is estimated at 3.4 to 4.4 km^3 (Paladio-Melosantos and others, this volume).

The weight of the rain-saturated tephra, no doubt with assistance from repeated intense seismic shaking and buffeting by wind, caused numerous roofs to collapse in the Philippine communities around the volcano and on the two large U.S. military bases (figs. 15, 16). Between 200 and 300 people died during the eruption, most of them from collapsing roofs. Without Tropical Storm Yunya, the death toll would no doubt have been far less.

Voluminous pyroclastic flows were emplaced during the climactic eruption. They extended as far as 12 to 16 km from the vent down all sectors of the volcano and impacted an area of approximately 400 km² (fig. 1 of W.E. Scott and others, this volume), stripping vegetation and dramatically modifying the preexisting topography. Their passage is

Figure 14. Representative section of June 15, 1991, tephra deposit at a site about 36 km southeast of the summit of Mount Pinatubo near the village of Santa Catalina. The section is composed of a basal silt-size ash unit (overlying the pre-June 15 soil surface) that is 0.8 cm thick and an upper pumice-fall unit, 4.2 cm thick, which grades upward from coarse sand and granules to fine sand. (Photograph by E.W. Wolfe, June 18, 1991.)

recorded by a variety of pumiceous pyroclastic-flow deposits: massive deposits that filled preexisting valleys to depths locally as much as 200 m (fig. 2 of W.E. Scott and others, this volume); thinner, stratified deposits that veneer uplands and grade laterally into the massive valley-fill deposits; and ash-cloud deposits in distal areas that are interbedded with and adjacent to the massive deposits. Deposits of an additional lithic-rich facies cap the massive pumiceous deposits in the upper parts of several major drainages. Estimated bulk volume of the whole sequence is 5 to 6 km³ (W.E. Scott and others, this volume).

W.E. Scott and others (this volume) review stratigraphic relations that indicate that the plinian pumice fall and the production of pyroclastic flows during the climactic eruption were at least partly concurrent, as is shown by interlayering of distal, thin pyroclastic-flow and ash-cloud deposits with the upper graded part of the plinian pumice-fall deposit. Possibly the plinian fall began first, but the two processes occurred simultaneously for at least part of the climactic eruption, and the evidence suggests that the massive valley-filling flow deposits accumulated incrementally from numerous successive flow pulses.

A coarse, clast-supported, lithic-rich facies that overlies pumiceous pyroclastic-flow deposits in the upper parts of several major drainages (fig. 1 of W.E. Scott and others, this volume) contains angular to subangular clasts, tens of centimeters to 1 m long, fragments of Mount Pinatubo's old summit. W.E. Scott and others (this volume) conclude that the lithic-rich facies formed during caldera collapse and that once collapse began, pyroclastic-flow production waned and the intense phase of the eruption ended. A supporting observation is that there is very little pumice to be found within the caldera (C.G. Newhall, written commun., June 18, 1993), although ash erupted within the caldera after the climactic eruption may have concealed deposits of climactic-eruption pumice.

Figure 15. Schoolhouse in Balin Baquero River valley that collapsed from the weight of rain-saturated ash on June 15, 1991. Dark cloud in background is very fine (powdery) falling ash. (Photograph courtesy of J. Major, June 30, 1991.

Figure 16. Aircraft hangars at Clark Air Base that collapsed under the weight of rain-saturated ash. (Photograph by E.W. Wolfe, June 29, 1991.)

The dominant juvenile component of the climactic eruption is white, phenocryst-rich pumice derived from magma that was nearly 50 volume percent crystalline. Phenocryst-poor tan pumice, identical in bulk composition to the phenocryst-rich pumice, is a subordinate component (about 10–20 volume percent, David and others, this volume). The phenocryst-poor pumice apparently originated primarily by mechanical fragmentation of phenocrysts during ascent through the conduit system in the climactic eruption (Pallister and others, this volume).

Rain produced by Tropical Storm Yunya generated lahars on June 15 in all major drainages around the volcano (Major and others, this volume; Rodolfo and others, this volume; Pierson and others, 1992). Major and others point out that the rainfall alone was insufficient to generate lahars; the additional necessary condition was altered watershed hydrology from fresh mantling pyroclastic deposits. The preclimactic and climactic eruptions satisfied this second condition by depositing a thin mantle of new pyroclastic material over the various watersheds.

On the basis of stratigraphic evidence and eyewitness accounts, Major and others found that initiation of lahars on the east side of the volcano on June 15 migrated northward during the day, apparently tracking the northwestward progression of Tropical Storm Yunya. Accordingly, peak flows from the eastern and southeastern sector—along the mountain front from the Gumain River to the Sacobia River—preceded deposition of the plinian pumice-fall deposit. Farther north, from the Sacobia River to the O'Donnell River, peak flow followed the pumice fall. Multiple flows

occurred on June 15; thus, along the Sacobia-Bamban drainage as well as drainages farther south, lahars occurred both before and after the plinian pumice fall.

The June 15 lahars locally inundated arable land and homes. The most significant damage was caused by lahars along the Abacan River: dwellings upstream from Sapangbato were inundated (fig. 9 of Major and others, this volume); buildings and all bridges in Angeles City were undercut and destroyed by lateral bank erosion (figs. 10 and 12 of Major and others, this volume); and the Northern Expressway bridge downstream from Angeles was buried as the channel aggraded.

The large volumes of water responsible for the destruction along the Abacan may have been the consequence of stream capture that occurred before the climactic eruption. Before Pinatubo reawakened, a low divide separated the headwaters of the Abacan drainage from the upper reaches of the Sacobia drainage. Sometime during the night of June 14 or the morning of June 15, water from the Sacobia began to flow across the divide in the Abacan, probably because of aggradation along the Sacobia. Subsequent pyroclastic flows of the climactic eruption entered the head of the Abacan drainage by the same route (fig. 7 of Major and others, this volume).

POSTCLIMAX PHENOMENA

The PVO team returned to Clark Air Base and reestablished seismic recording on June 16. The one surviving seismometer initially recorded more than 150 tectonic earthquakes per hour larger than magnitude 2.0 (Mori, White, and others, this volume). The rate of seismic-energy release declined exponentially (fig. 11), and by the end of June, earthquake counts were 10 to 20 per hour. The decline continued for months, reflecting the adjustment of the volcano and the Earth's crust beneath it to the dramatic changes of mid-June. Reestablishment of a seismic network sufficient to locate earthquakes required about 2 weeks. Thereafter, earthquake locations largely defined outward-dipping trends enclosing a volume directly beneath the volcano that is relatively free of earthquakes (fig. 1 of Mori, White, and others, this volume) and may represent the reservoir that supplied the magma for the 1991 eruptions (Mori, White, and others, this volume).

The climactic eruption apparently left an open vent system, for ash billowed continually (fig. 12) from vents in the caldera for about another month. The resulting plume at times reached an altitude of 18 to 20 km. Winds caused significant fall of very fine powdery ash, especially southwest and northeast of the new summit caldera.

Episodic bursts of increased low-frequency seismic tremor, correlated with increased vigor of ash emission, occurred a few times per day during late June and most of July (Mori, White, and others, this volume). During these events, seismic amplitude increased for about 2 to 3 h and then declined to background level. From July 6 to July 11, these bursts of higher-amplitude tremor were periodic, with intervals of 7 to 10 h between events. As monsoon clouds and rain prevented direct observation of the volcano during late July, we know only that continuous ash emission had ceased by August 1; episodic ash emission continued until about September 1. By early September, the monsoon rains had formed a shallow caldera lake.

Partial filling of the valleys on Mount Pinatubo by the thick, loose, ash-rich pyroclastic-flow deposits thoroughly disrupted the drainage network that had become established during the 5 centuries since the previous eruption. Runoff from heavy rains began to reestablish drainageways by eroding the new deposits. Thus, beginning in July, the annual monsoon rains repeatedly generated lahars, many of them hot from incorporation of hot pyroclastic-flow material. The lahars progressively buried towns and agricultural land, destroyed homes and bridges, frequently cut roadways, and displaced tens of thousands of residents. By the end of October 1991 (after the end of the southwest monsoon season), about 0.9 km³ of new lahar deposits had buried 300 km² of lowland terrain, largely low-gradient alluvial fans formed of pyroclastic debris from previous eruptions of Mount Pinatubo (Pierson and others, this volume; Rodolfo and others, this volume). Rodolfo and others (this volume) estimate that lahars of the 1992 monsoon season delivered between 0.5 and 0.6 km³ of new sediment to the lowlands, which accords with estimates of Pierson and others (1992) that progressively diminishing volumes of

sediment are likely to be delivered to the lowland areas surrounding Mount Pinatubo for perhaps a decade. However, the 1991 and 1992 monsoon seasons were unusually mild. Sooner or later a typhoon will approach Mount Pinatubo, very likely triggering larger, more destructive lahars than have occurred so far (Rodolfo and others, this volume).

The Mount Pinatubo lahars have caused severe economic and social disruption. The dominant problem is continuing channel aggradation, which promotes lateral migration of the lahars across the low-gradient alluvial fans surrounding the volcano. Despite the resulting progressive burial of villages and valuable agricultural land, loss of life has been minimized by warning systems. Initially, warnings were provided by Philippine police and army personnel stationed at watchpoints along major river channels. Beginning in August 1991, that effort was supplemented by warnings from PVO based on realtime analysis of telemetered data from rain gauges and acoustic flow sensors (Janda and others, this volume; Marcial and others, this volume; Tuñgol and Regalado, this volume).

Numerous secondary explosions occurred during the 1991 and 1992 monsoon seasons from the interaction of water with hot pyroclastic-flow deposits. Such explosions generated tephra plumes, some as high as 18 km or more, that produced significant tephra falls and were at times mistaken by the press and some of the local populace as products of primary eruptions. Water gained access to hot pyroclastic-flow deposits by at least two different mechanisms: collapse of stream banks into streams, and invasion of ground water into hot deposits along buried stream channels. Explosions of the latter type commonly formed craters tens to hundreds of meters in diameter. Some generated secondary pyroclastic flows—hot, remobilized pyroclastic-flow deposits that traveled as far as 8 km downvalley and left prominent head scarps at their sources (Torres and others, this volume).

A lava dome formed within the new summit caldera from July through October 1992 (Daag, Dolan, and others, this volume). Its emplacement was heralded by increasing earthquakes and tremor in early July. By July 9, explosions near the center of the caldera lake had built a low pyroclastic cone that was about 70 to 100 m across and extended 5 m above the lake surface. Lava extrusion ensued; by July 14, a lava dome about 5 to 10 m high and 50 to 100 m in diameter had grown within the pyroclastic cone, which had widened to about 150 m. By July 23, the growing dome had completely buried the pyroclastic cone. Dome growth, which continued through October, was largely exogenous, marked by formation of a succession of extrusive lobes and accompanied by intermittent intense swarms of shallow earthquakes. At the end of October, the dome was 350 to 450 m in diameter and about 150 m high; its volume was about 4×10⁶ m³. A sharp decrease in seismicity on October 31 marked the apparent end of 1992 dome growth. Like the June 1991 dome, the 1992 dome consists of hybrid andesite indicative of magma mixing. Quenched basaltic andesite inclusions constitute 5 to 10 percent of the 1992 dome.

SIZE OF THE ERUPTION

A total bulk volume of 8.4 to 10.4 km^3 of eruption deposits is determined from a bulk volume of 5 to 6 km^3 of pyroclastic-flow deposits (W.E. Scott and others, this volume) and a bulk volume of 3.4 to 4.4 km^3 of tephra-fall deposits (Paladio-Melosantos and others, this volume). Allowing for uncertainty in the bulk density of the deposits, W.E. Scott and others estimate that the pyroclastic-flow and tephra-fall deposits together represent 3.7 to 5.3 km^3 of erupted magma (that is, dense-rock equivalent, or DRE).

The 1991 eruption of Mount Pinatubo was the world's largest in more than half a century and probably the second largest of the century. Its roughly 5 km^3 of erupted magma is an order of magnitude greater than the volume of magma erupted in 1980 from Mount St. Helens but is smaller than the 13±3 km^3 (DRE) of ignimbrite and fall deposits from the 1912 eruption of Novarupta, Alaska (Fierstein and Hildreth, 1992). Volumes of two other large plinian eruptions that occurred early in the 20th century approximate the lower limit of estimated volume for the Mount Pinatubo eruption: the 1902 eruption of Santa María volcano, Guatemala, produced approximately 8 km^3 (uncorrected volume, equivalent to 3 to 4 km^3 DRE) of fall deposits (Fierstein and Nathenson, 1992), and the 1932 eruption of Volcán Quizapu, Chile, produced approximately 4 km^3 (DRE) of fall deposits and minor ignimbrite (Hildreth and Drake, 1992).

The climactic eruption injected approximately 17 Mt of SO_2 into the atmosphere (Gerlach and others, this volume), generating atmospheric and climatic effects that are likely to persist for several years (Hansen and others, 1992). Gerlach and others (this volume) concluded that virtually all of this SO_2 as well as Cl, CO_2, and an appreciable volume of water (approximately 96 Mt) had accumulated prior to eruption in a vapor phase in volatile-saturated magma of a crustal reservoir. A large additional volume (about 6.25 wt%) of water was in solution in the melt phase of the magma reservoir. Gerlach and others estimate that, in addition to the measured 17 Mt of SO_2, the eruption of approximately 5 km^3 of magma was accompanied by release of at least 491 to 921 Mt of H_2O, 3 to 16 Mt of Cl, and 42 to 234 Mt of CO_2.

THE MAGMA RESERVOIR

Petrologic evidence suggests that the dacitic magma of the climactic eruption ascended from a reservoir that was highly crystalline (40 to 50 volume percent), relatively cool (approximately 780°C), volatile-saturated, and at an equilibrium pressure of approximately 200–220 MPa (Rutherford and Devine, this volume; Pallister and others, this volume). Using data from geothermal exploration and drilling (Delfin, 1983; Delfin and others, this volume), Pallister and others (this volume) equate 200 MPa to a depth of about 8 km beneath the pre-1991 summit of Mount Pinatubo (6 to 7 km below sea level).

Analysis of seismic results provides additional spatial evidence of the reservoir. Mori, Eberhart-Phillips, and Harlow (this volume) interpret a region of low P-wave velocity at a depth of 6 to 11 km below sea level as a magma reservoir. The low-velocity body is offset southward slightly from the summit of Mount Pinatubo. As appropriate for a body of magma, the inferred reservoir occurs in a nearly aseismic zone within the swarm of earthquake hypocenters that recorded crustal adjustments in the weeks after the climactic eruption (Mori, White, and others, this volume). The estimated volume of the low-velocity body is 40 to 90 km^3, and it is part of a still larger low-velocity body that extends as far south as Mount Negron. Mount Pinatubo's repeated eruption of virtually identical, voluminous, dacitic, pumiceous pyroclastic flows during the past 35,000 yr or more suggests that such a large, preexisting, crustal magma reservoir was repeatedly reactivated, most recently to produce the 1991–92 eruptions (Newhall and others, this volume).

The 1991 eruptions were apparently triggered by injection of basaltic magma into the dacitic reservoir (Pallister and others, 1992; Pallister and others, this volume), and the earliest erupted products, as well as the lava dome erupted in 1992, consist of hybrid andesite produced by mixing of intruded basalt with the dacitic magma of the reservoir. The occurrence of hybrid andesite, with mineralogic evidence of a basalt component in its parentage, in banded pumice fragments or as lithic clasts in the deposits of earlier eruptions of Mount Pinatubo, suggests that prehistoric eruptions were similarly triggered by intrusion of basalt into the dacitic magma reservoir (Newhall and others, this volume).

Basalt that intruded the dacitic reservoir mixed with dacite to produce a hybrid andesitic magma that ascended buoyantly (Pallister and others, this volume). The relatively degassed tip of the ascending andesite breached the surface on June 7, establishing a new eruptive conduit and supplying lava to the dome, which slowly grew during the following days.

Delicate disequilibrium textures preserved in samples of the June 1991 dome suggest that ascent and eruption of the hybrid andesite occurred within 4 days of the mixing event that formed it (Rutherford and others, 1993; Pallister and others, this volume). However, the preceding unrest—from mid-March through May—suggests that there was earlier magmatic activity. Possibly earlier intrusion of basalt into the dacitic reservoir produced deep-seated inflation that caused adjustments on faults, one of which intersected the hydrothermal system, or it produced an early hybrid magma that ascended sufficiently to penetrate the hydrothermal system but did not reach the surface.

Diminished magma volume in the reservoir caused structural adjustments including collapse of its roof. As indicated by the occurrence of lithic-rich breccia and pyroclastic-flow deposits at or near the top of the pumiceous pyroclastic-flow deposits, formation of the 2.5-km-diameter summit caldera was completed at the end of plinian eruption and production of pyroclastic flows.

Two essential conditions for the explosive 1991 eruptions were existence of vapor-saturated magma in the reservoir and intrusion of basalt into the reservoir. Eruption of hybrid andesite to form the 1992 dome implies that basalt intrusion has continued. Volatiles in the reservoir may have been sufficiently depleted by the 1991 eruptions that continued basalt intrusion has not led so far to renewal of explosive dacitic eruption (Daag, Dolan, and others, this volume).

ACKNOWLEDGMENTS

The 1991 volcanic unrest and eruptions of Mount Pinatubo caused volcanologists of the Philippine Institute of Volcanology and Seismology (PHIVOLCS), the U.S. Geological Survey (USGS), the University of the Philippines, and the University of Illinois, Chicago, to join in an effective response to extraordinary volcanic threat and devastation. Together, these volcanologists constituted the Pinatubo Volcano Observatory Team, which made the observations on which this overview is based.

REFERENCES CITED

Bautista, B.C., Bautista, M.L.P., Stein, R.S., Barcelona, E.S., Punongbayan, R.S., Laguerta, E.P., Rasdas, A.R., Ambubuyog, G., and Amin, E.Q., this volume, Relation of regional and local structures to Mount Pinatubo activity.

Daag, A.S., Dolan, M.T., Laguerta, E.P., Meeker, G.P., Newhall, C.G., Pallister, J.S., and Solidum, R., this volume, Growth of a postclimactic lava dome at Mount Pinatubo, July–October 1992.

Daag, A.S., Tubianosa, B.S., Newhall, C.G., Tuñgol, N.M, Javier, D., Dolan, M.T., Delos Reyes, P.J., Arboleda, R.A., Martinez, M.L., and Regalado, M.T.M., this volume, Monitoring sulfur dioxide emission at Mount Pinatubo.

David, C.P.C., Dulce, R.G., Nolasco-Javier, D.D., Zamoras, L.R., Jumawan, F.T., and Newhall, C.G., this volume, Changing proportions of two pumice types from the June 15, 1991, eruption of Mount Pinatubo.

Defant, M.J., Jacques, D., Maury, R.C., De Boer, J., and Joron, J-L., 1989, Geochemistry and tectonic setting of the Luzon arc, Philippines: Geological Society of America Bulletin, v. 101, p. 663–672.

Delfin, F.G., Jr., 1983, Geology of the Mt. Pinatubo geothermal prospect: unpublished report, Philippine National Oil Company, 35 p. plus figures.
———1984, Geology and geothermal potential of Mt. Pinatubo: unpublished report, Philippine National Oil Company, 36 p.

Delfin, F.G., Jr., Sussman, D., Ruaya, J.R., and Reyes, A.G., 1992, Hazard assessment of the Pinatubo volcanic-geothermal system: Clues prior to the June 15, 1991 eruption: Geothermal Resources Council, Transactions, v. 16, p. 519–527.

Delfin, F.G., Jr., Villarosa, H.G., Layugan, D.B., Clemente, V.C., Candelaria, M.R., Ruaya, J.R., this volume, Geothermal exploration of the pre-1991 Mount Pinatubo hydrothermal system.

Doukas, M.P., and Gerlach, T.M., in press, Airborne gas monitoring of SO2 and CO2 during the 1992 eruption at Crater Peak, Mount Spurr, Alaska, in Keith, T.E.C., ed., The 1992 eruptions of Crater Peak vent, Spurr Volcano, Alaska: U.S. Geological Survey Bulletin.

Ebasco Services, Inc., 1977, Philippine Nuclear Power Plant I, Preliminary Safety Analysis Report, v. 7, Section 2.5.H.1., Appendix, A report on geochronological investigations of materials relating to studies for the Philippine Nuclear Power Plant Unit 1: Washington, D.C., Nuclear Regulatory Agency.

Endo, E.T., and Murray, T., 1991, Real-time seismic amplitude measurement (RSAM): A volcano monitoring and prediction tool: Bulletin of Volcanology, v. 53, p. 533–545.

Ewert, J.W., Lockhart, A.B., Marcial, S., and Ambubuyog, G., this volume, Ground deformation prior to the 1991 eruptions of Mount Pinatubo.

Fierstein, J., and Hildreth, W., 1992, The plinian eruptions of 1912 at Novarupta, Katmai National Park, Alaska: Bulletin of Volcanology, v. 54, no. 8, p. 646–684.

Fierstein, J., and Nathenson, M., 1992, Another look at the calculation of fallout tephra volumes: Bulletin of Volcanology, v. 54, no. 2, p. 156–167.

Gerlach, T.M., Westrich, H.R., and Symonds, R.B., this volume, Preeruption vapor in magma of the climactic Mount Pinatubo eruption: Source of the giant stratospheric sulfur dioxide cloud.

Hansen, J., Lacis, A., Ruedy, R., and Sato, M., 1992, Potential climate impact of Mount Pinatubo eruptions: Geophysical Research Letters, v. 19, no. 2, p. 215–218.

Harlow, D.H., Power, J.A., Laguerta, E.P., Ambubuyog, G., White, R.A., and Hoblitt, R.P., this volume, Precursory seismicity and forecasting of the June 15, 1991, eruption of Mount Pinatubo.

Hildreth, W., and Drake, R.E., 1992, Volcán Quizapu, Chilean Andes: Bulletin of Volcanology, v. 54, no. 2, p. 93–125.

Hoblitt, R.P., Wolfe, E.W., Scott, W.E., Couchman, M.R., Pallister, J.S., and Javier, D., this volume, The preclimactic eruptions of Mount Pinatubo, June 1991.

Isada, M., and Ramos, A., 1990, Mt. Pinatubo not erupting: PHIVOLCS Observer, v. 6, no. 3, p. 6.

Janda, R.J., Daag, A.S., Delos Reyes, P.J., Newhall, C.G., Pierson, T.C., Punongbayan, R.S., Rodolfo, K.S., Solidum, R.U., and Umbal, J.V., this volume, Assessment and response to lahar hazard around Mount Pinatubo.

Koyaguchi, T., and Tokuno, M., 1993, Origin of the giant eruption cloud of Pinatubo, June 15, 1991: Journal of Volcanology and Geothermal Research, v. 55, nos. 1 and 2, p. 85–96.

Lee, W.H.K., ed., 1989, Toolbox for seismic data acquisition, processing, and analysis: International Association of Seismology and Physics of the Earth's Interior in collaboration with Seismological Society of America: El Cerrito, Calif., 283 p.

Lynch, J.S., and Stephens, G., this volume, Mount Pinatubo: A satellite perspective of the June 1991 eruptions.

Major, J.J., Janda, R.J., and Daag, A.S., this volume, Watershed disturbance and lahars on the east side of Mount Pinatubo during the mid-June 1991 eruptions.

Marcial, S.S., Melosantos, A.A., Hadley, K.C., LaHusen, R.G., and Marso, J.N., this volume, Instrumental lahar monitoring at Mount Pinatubo.

Mori, J., Eberhart-Phillips, D., and Harlow, D.H., this volume, Three-dimensional velocity structure at Mount Pinatubo, Philippines: Resolving magma bodies and earthquakes hypocenters.

Mori, J., White, R.A., Harlow, D.H., Okubo, P., Power, J.A., Hoblitt, R.P., Laguerta, E.P., Lanuza, L., and Bautista, B.C., this volume, Volcanic earthquakes following the 1991 climactic eruption of Mount Pinatubo, Philippines: Strong seismicity during a waning eruption.

Newhall, C.G., Daag, A.S., Delfin, F.G., Jr., Hoblitt, R.P., McGeehin, J., Pallister, J.S., Regalado, M.T.M., Rubin, M., Tamayo, R.A., Jr., Tubianosa, B., and Umbal, J.V., this volume, Eruptive history of Mount Pinatubo.

Oswalt, J.S., Nichols, W., and O'Hara, J.F., this volume, Meteorological observations of the 1991 Mount Pinatubo eruption.

Paladio-Melosantos, M.L., Solidum, R.U., Scott, W.E., Quiambao, R.B., Umbal, J.V., Rodolfo, K.S., Tubianosa, B.S., Delos Reyes, P.J., and Ruelo, H.R., this volume, Tephra falls of the 1991 eruptions of Mount Pinatubo.

Pallister, J.S., Hoblitt, R.P., Meeker, G.P., Knight, R.J., and Siems, D.F., this volume, Magma mixing at Mount Pinatubo: Petrographic and chemical evidence from the 1991 deposits.

Pallister, J.S., Hoblitt, R.P., and Reyes, A.G., 1992, A basalt trigger for the 1991 eruptions of Pinatubo Volcano?: Nature, v. 356, p. 426–428.

Pierson, T.C., Daag, A.S., Delos Reyes, P.J., Regalado, M.T.M., Solidum, R.U., and Tubianosa, B.S., this volume, Flow and deposition of posteruption hot lahars on the east side of Mount Pinatubo, July–October 1991.

Pierson, T.C., Janda, R.J., Umbal, J.V., and Daag, A.S., 1992, Immediate and long-term hazards from lahars and excess sedimentation in rivers draining Mount Pinatubo, Philippines: U.S. Geological Survey Water-Resources Investigations Report 92–4039, 35 p.

Pinatubo Volcano Observatory Team, 1991, Lessons from a major eruption: Mt. Pinatubo, Philippines: EOS, Transactions of the American Geophysical Union, v. 72, p. 545–555.

Power, J.A., Murray, T.L., Marso, J.N., and Laguerta, E.P., this volume, Preliminary observations of seismicity at Mount Pinatubo by use of the Seismic Spectral Amplitude Measurement (SSAM) system, May 13–June 18, 1991.

Punongbayan, R.S., Newhall, C.G., Bautista, M.L.P., Garcia, D., Harlow, D.H., Hoblitt, R.P., Sabit, J.P., and Solidum, R.U., this volume, Eruption hazard assessments and warnings.

Rodolfo, K.S., Umbal, J.V., Alonso, R.A., Remotigue, C.T., Paladio-Melosantos, M.L., Salvador, J.H.G., Evangelista, D., and Miller, Y., this volume, Two years of lahars on the western flank of Mount Pinatubo: Initiation, flow processes, deposits, and attendant geomorphic and hydraulic changes.

Rutherford, M.J., Baker, Leslie, and Pallister, J.S., 1993, Petrologic constraints on timing of magmatic processes in the 1991 Pinatubo volcanic system [abs.]: Eos, Transactions of the American Geophysical Union, v. 74, no. 43, p. 671.

Rutherford, M.J., and Devine, J.D., this volume, Preeruption pressure-temperature conditions and volatiles in the 1991 dacitic magma of Mount Pinatubo.

Sabit, J.P., Pigtain, R.C., and de la Cruz, E.G., this volume, The west-side story: Observations of the 1991 Mount Pinatubo eruptions from the west.

Scott, W.E., Hoblitt, R.P., Torres, R.C., Self, S, Martinez, M.L., and Nillos, T., Jr., this volume, Pyroclastic flows of the June 15, 1991, climactic eruption of Mount Pinatubo.

Tahira, M., Nomura, M., Sawada, Y., and Kamo, K., this volume, Infrasonic and acoustic-gravity waves generated by the Mount Pinatubo eruption of June 15, 1991.

Torres, R.C., Self, S., and Martinez, M.L., this volume, Secondary pyroclastic flows from the June 15, 1991, ignimbrite of Mount Pinatubo.

Tuñgol, N.M., and Regalado, M.T.M., this volume, Rainfall, acoustic flow monitor records, and observed lahars of the Sacobia River in 1992.

Wolfe, E.W., 1992, The 1991 eruptions of Mount Pinatubo, Philippines: Earthquakes and Volcanoes, v. 23, no. 1, p. 5–37.

Photographic Record of Rapid Geomorphic Change at Mount Pinatubo, 1991–94

By Raymundo S. Punongbayan,[1] Christopher G. Newhall,[2] and Richard P. Hoblitt[2]

ABSTRACT

Sequential photographs of various locations and features of Mount Pinatubo show dramatic geomorphic changes that resulted from the eruption of June 15, 1991. Owing to the large scale of events, only a few preeruption features of outcrop scale can be shown again in posteruption view. Instead, most photographs in this collection are aerial oblique, watershed-scale views. Rapid erosion and redeposition of 1991 deposits continues as of this time of writing; revegetation is also visible in many photographs.

INTRODUCTION

Geologic change is often thought to occur so slowly that even changes in civilizations and species seem rapid in comparison. Geologic change at volcanoes can be much faster—often occurring within minutes, days, months, and years. The most rapid changes are best captured on movie film or video; changes over days to years can be captured in sequential still images taken from the same site at various times through the course of that change. This paper presents 79 photographs in 28 sequences that span from preeruption to immediately after the eruptions and continue through the first four seasons of erosion and lahar deposition.

Sequential photographs of this paper emphasize geomorphic change and complement those of other papers in this volume. Other "before and after" views in this volume include:

- Preeruption and posteruption vertical SLAR images (courtesy of Intera Technologies, Ltd., reproduced in Newhall and others)
- Topographic and shaded relief maps (Jones and Newhall)
- Various stages of the eruption (Daag, Dolan and others; Hoblitt, Wolfe, and others; Wolfe and Hoblitt)
- Primary pyroclastic flows (W.E. Scott and others)
- Secondary pyroclastic flows (Torres and others)
- Evolution of the caldera lake (Campita and others)

- Lahar deposition (Rodolfo and others; Umbal and Rodolfo)
- Transient, pyroclastic-flow- or lahar-dammed lakes (Arboleda and Martinez; Umbal and Rodolfo)

Collectively, photographs in this and other papers of this volume tell a story of landscape change that ranks among the most rapid in the period of modern geology. We have been privileged witnesses.

METHODS

Ideally, photographs to document change are taken from a fixed tripod position by a camera with a fixed focal-length lens. Each image should be exactly the same as the previous except where change occurred. At Mount Pinatubo, harried working conditions, staff rotations, various cameras, inclement weather, the enormous scale of the event, and reliance on helicopters for overviews has made precise reoccupation of photo stations impossible. In most instances, change has been so great at outcrop scale that no sequential comparison is possible. Change has been so great that even watershed-scale features can be difficult to identify from one view to the next, but at least qualitative recognition of change is possible.

Photographs in this collection are from the three authors and several colleagues. Most are from helicopters that lacked global positioning system (GPS) navigational equipment, and time was generally too short to plot our positions on maps as we flew, so we know positions mainly from the photographs themselves. To reoccupy aerial vantage points, we sketched lines on maps for the pilots, but with frequent rotation of both scientists and pilots, and without intercoms or GPS equipment, our repeat flights were generally different from the original flights. Often, too, rain clouds or ash clouds prevented return to exactly the same place from which a previous photograph was taken.

Photographs were taken with a variety of cameras and lenses, including zoom lenses. All but two were originally shot as color slides.

[1] Philippine Institute of Volcanology and Seismology.
[2] U.S. Geological Survey.

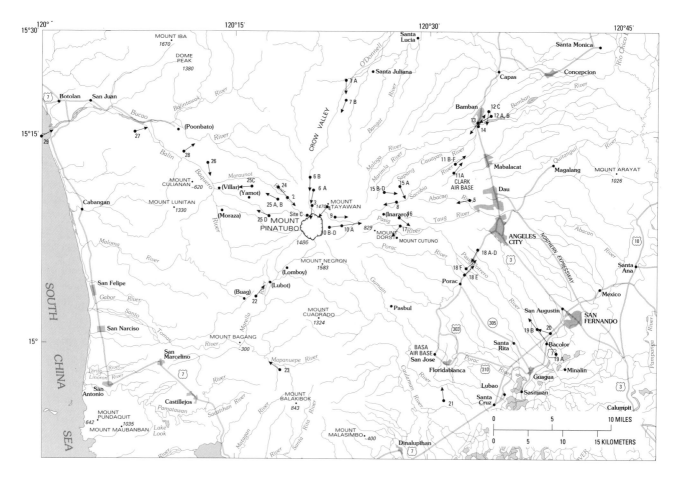

Figure 1. Approximate locations and directions of view of photographs in this paper. Each number represents a sequence of related photographs; letters within each sequence indicate time progression. Dots, approximate location of photographer; arrows, direction of view.

ORGANIZATION AND CONVENTIONS OF THIS COLLECTION

After we initially arranged photographs according to geologic feature and time period, we concluded that many of the sequences showed multiple geologic features and that sequences throughout the entire study period might be more interesting than pairs that bracketed only one geologic event. We therefore rearranged the photographs, here, according to watershed. The steep preeruption edifice and its later caldera are treated as one watershed; other watersheds are named by their principal river: O'Donnell, Sacobia-Bamban, Abacan, Pasig-Potrero, Gumain, Marella-Santo Tomas, Maraunot, Balin Baquero, and Bucao, in that order (clockwise, starting from north) (fig. 1). Within each watershed, the usual order, if good sequences were found, is from the headwaters of that watershed (Mount Pinatubo) to distal parts of each alluvial fan.

The approximate camera site and look angle of each photograph or set of photographs is shown on figure 1. Although some captions refer to the direction of view as "to" or "from," the arrows on figure 1 are all in the direction in which the camera was pointed. Banks of rivers are named as left or right according to standard hydrologic practice, as if one were looking downstream.

Most photographs in this collection were taken before or after an event, rather than during that event. Unless otherwise specified, dates in this paper are the dates of photography, not of events. Slightly different time-date formats on photographs reflect different cameras and photographers. Times that appear on some photos are not noted in the captions.

FUTURE PHOTOGRAPHY

Mount Pinatubo is photogenic, and more changes are yet to come. We encourage colleagues to try to reoccupy our photographic sites as those changes continue. Please carry a reprint of this paper with you as opportunities for flights and other visits to the area arise. Additional photographs that show change, taken previously or in the future, would be gratefully received by the authors.

ACKNOWLEDGMENTS

We thank Val Gempis of the U.S. Air Force, T.J. Casadevall, W.E. Scott, and E.W. Wolfe of the U.S. Geological Survey, and G.P. Yumul, Jr., of the University of the Philippines for permission to use their photographs.

We also thank the Philippine Air Force, U.S. Air Force, U.S. Navy, U.S. Marine Corps, and the late Agustin Consunji of Delta Aviation for helicopter support that was so vital to tracking events of such a large scale as at Pinatubo.

REFERENCES CITED

[all references are from this volume]

Arboleda, R.A., and Martinez, M.L., 1992 lahars in the Pasig-Potrero River system.

Campita, N.R., Daag, A.S., Newhall, C.G., Rowe, G.L., Solidum, R., Evolution of a small caldera lake at Mount Pinatubo.

Daag, A.S., Dolan, M.T., Laguerta, E.P., Meeker, G.P., Newhall, C.G., Pallister, J.S., and Solidum, R., Growth of a postclimactic lava dome at Mount Pinatubo, July–October 1992

Delfin, F.G., Jr., Villarosa, H.G., Layugan, D.B., Clemente, V.C., Candelaria, M.R., Ruaya, J.R., Geothermal exploration of the pre-1991 Mount Pinatubo hydrothermal system.

Ewert, J.W., Lockhart, A.B., Marcial, S., and Ambubuyog, G., Ground deformation prior to the 1991 eruptions of Mount Pinatubo.

Hoblitt, R.P., Wolfe, E.W., Scott, W.E., Couchman, M.R., Pallister, J.S., and Javier, D., The preclimactic eruptions of Mount Pinatubo, June 1991

Jones, J.W., and Newhall, C.G., Preeruption and posteruption digital-terrain models of Mount Pinatubo.

Newhall, C.G., Daag, A.S., Delfin, F.G., Jr., Hoblitt, R.P., McGeehin, J., Pallister, J.S., Regalado, M.T.M., Rubin, M., Tamayo, R.A., Jr., Tubianosa, B., and Umbal, J.V., Eruptive history of Mount Pinatubo.

Paladio-Melosantos, M.L., Solidum, R.U., Scott, W.E., Quiambao, R.B., Umbal, J.V., Rodolfo, K.S., Tubianosa, B.S., Delos Reyes, P.J., and Ruelo, H.R., this volume, Tephra falls of the 1991 eruptions of Mount Pinatubo.

Rodolfo, K.S., Umbal, J.V., Alonso, R.A., Remotigue, C.T., Paladio-Melosantos, M.L., Salvador, J.H.G., Evangelista, D., and Miller, Y., Two years of lahars on the western flank of Mount Pinatubo: Initiation, flow processes, deposits, and attendant geomorphic and hydraulic changes.

Scott, W.E., Hoblitt, R.P., Torres, R.C., Self, S, Martinez, M.L., and Nillos, T., Jr., Pyroclastic flows of the June 15, 1991, climactic eruption of Mount Pinatubo.

Torres, R.C., Self, S., and Martinez, M.L., Secondary pyroclastic flows from the June 15, 1991, ignimbrite of Mount Pinatubo.

Umbal, J.V., and Rodolfo, K.S., The 1991 lahars of southwestern Mount Pinatubo and the evolution of the lahar-dammed Mapanuepe Lake.

Wolfe, E.W., and Hoblitt, R.P., Overview of the eruptions.

MOUNT PINATUBO—STRATOVOLCANO, DOME(S), AND A NEW SUMMIT CALDERA

Figure 2A. Preeruption Mount Pinatubo, April 16, 1991. View from the northwest, up the Maraunot River valley. The river had become acidic and silty, owing to reactivation of the hydrothermal system and phreatic explosions of April 2, 1991 (the vents of which were just out of view at left edge of photograph). Steam was from 2-week-old fumaroles on the upper north slope of the volcano. The fumarole farthest to the right (behind a jagged ridge, right of the one visible on the valley floor) would later become the site of the preclimactic lava dome extrusion of June 7–12 (Hoblitt, Wolfe, and others, this volume). Mount Negron is behind and to the right of Pinatubo. (R.S. Punongbayan)

Figure 2B. Summit caldera and lake, with partly submerged relics (rocky islets) of a dome that grew between July and October, 1992. View is from the northwest, as in figure 2A, on October 5, 1994. The diameter of the caldera averages 2.5 km, rim to rim. Low point in the caldera rim (foreground) is the truncated valley of the Maraunot River. Mount Negron is in the background. (R.S. Punongbayan)

Figure 3A. Preeruption Mount Pinatubo, as viewed from the north in late April. Grayish-tan ash and several craters from the April 2 phreatic explosion craters are visible at the left. The road led to site C of the 1988–90 geothermal drilling (lower right; also indicated on fig. 1) discussed in Delfin and others (this volume); for reference, the same road is visible in figure 2A on the ridge northeast of the Maraunot River. Major drainage at lower left is the O'Donnell River; prominent fumarole at right center was in the Maraunot drainage. (V. Gempis)

Figure 3B. Summit caldera as seen on October 4, 1991, from the north-northeast. Much of the debris from the caldera walls had been washed onto the caldera floor, which was then submerged beneath a lake fed by ground water and rain water (Campita and others, this volume). (C.G. Newhall)

Figure 4A. Preeruption Mount Pinatubo, viewed from the northeast. The April 2, 1991, phreatic explosion craters (lower right, adjacent to '91 in date stamp) and the eventual location of June 7–12 dome extrusion (beneath the ash cloud) were aligned northeast-southwest across the north face of Mount Pinatubo (map of craters in Wolfe and Hoblitt, this volume). The ridge in the foreground may have been the wall of a prehistoric caldera that was only slightly larger than that which formed in 1991. (R.P. Hoblitt, June 9, 1991)

Figure 4B. Summit caldera, as seen August 1, 1991, from the northeast. The caldera formed by collapse during the June 15, 1991, climactic eruption. A small explosion had just occurred, forming the expanding ash cloud. Throughout the latter half of June and much of July, ash emission kept the caldera obscured; as continuous ash emission changed to intermittent explosions, the caldera became visible. See also, figure 1A of Campita and others, this volume. (T.J. Casadevall)

Figure 4C. Summit caldera viewed from the northeast, March 18, 1992. Row of fumaroles marks extension of the northeast-southwest-trending Sacobia lineament (figure 3 of Newhall and others, this volume) through the caldera, parallel to, and slightly south of the preeruption alignment of phreatic craters and fumaroles (fig. 4A). (R.P. Hoblitt)

Figure 5A. Mount Pinatubo, as seen from near the southwest end of the Clark Air Base runway. View is to the west, up the Sacobia valley. Mount Pinatubo is the light-colored (ash-covered) highest peak (center); prominent, darker ridges are relics of an ancestral Mount Pinatubo (Newhall and others, this volume). The highest dark ridge (right) is the relict northeastern rim of the 4×5-km-diameter Tayawan caldera, which formed in the summit of the ancestral Mount Pinatubo more than 35,000 years ago. (R.P. Hoblitt, June 14, 1991, 0722)

Figure 5B. Approximately the same view as in *A*, March 13, 1992. The highest peaks, light gray and stripped of vegetation, are the relict southeastern part of the preeruption Mount Pinatubo (left of center) and the previously dark ridge (Tayawan caldera rim) northeast of Pinatubo (skyline, immediately left of the siren tower). (R.P. Hoblitt)

O'DONNELL RIVER

Figure 6A. Upper O'Donnell River, late April 1991. View is from the north. Road on ridge in foreground led to geothermal drilling site C, located at right edge of photograph. (C.G. Newhall)

Figure 6B. Upper O'Donnell River, July 14, 1994. View is from the north, slightly farther from the volcano than the view in A. Low saddle on the skyline (left of center) is the ridge upon which site C was located; higher terrain that was obscured by or visible just beneath the cloud in A collapsed into the caldera. (C.G. Newhall)

Figure 7A. O'Donnell River (right) in Crow Valley, late July 1991. View is upstream, to the southwest. Circular targets and other features of a bombing range are visible on terraces of prehistoric lahar deposits. Lahars had already covered the lowest flood plain but had not yet covered higher terraces. Light-colored tephra-fall deposit partly covers all terraces, including those with the targets. Gently sloping distal pyroclastic-flow deposits of the June 15 eruption are dimly visible at the head of Crow Valley (background, right of center). (C.G. Newhall)

Figure 7B. O'Donnell River, as viewed on September 7, 1994, from above the small hill that is visible in the center foreground of *A*, between two circular bombing targets. All terraces in the foreground had been buried by fresh lahar deposits; several terraces in the background (unclear in this view) remained unburied. (C.G. Newhall)

SACOBIA-BAMBAN RIVER

Figure 8A. Fourteen-day-old undissected deposits of the Sacobia pyroclastic-flow field, east of Mount Pinatubo. View is west and upslope. Reestablishment of the drainage network had only just begun. (R.S. Punongbayan, June 29, 1991)

Figure 8B. Dissected Sacobia pyroclastic-flow deposits, June 1, 1994. Three years after emplacement, the Sacobia pyroclastic-flow deposits had a well-developed drainage system, and, locally, as much as 60% of the original deposit had been eroded. Resulting lahars had covered parts of the towns of Bamban and Concepcion in Tarlac Province, and Bacolor, Porac, Mabalacat, and Magalang in Pampanga Province. (R.S. Punongbayan)

Figure 9A. Undissected June 15, 1991, pyroclastic-flow deposits of the Sacobia pyroclastic fan, seen in this view to the east and downslope at 1043 on June 29, 1991. Mount Arayat is dimly visible in the background, left of center. (E.W. Wolfe)

Figure 9B. Extensive dissection of the June 1991 deposits of the Sacobia pyroclastic fan. View is similar to that of *A*, though slightly narrower; Mount Arayat is dimly visible in the center background. Clark Air Base lies between the pyroclastic fan and Mount Arayat in *A* and *B*. The lower Sacobia River, passing the north side of Clark Air Base, is at the upper left. (C.G. Newhall, August 30, 1994)

Figure 10A. Deep preeruption canyon of the Sacobia River, seen in this view to the east, downstream. Note sharp knob atop the right canyon wall, a feature that is common to B–D. Knob consists of lavas(?) of ancestral Mount Pinatubo (Newhall and others, this volume); badland topography (foreground and beyond the knob) consists of unconsolidated, prehistoric pyroclastic-flow deposits of modern Mount Pinatubo. The hamlet of Steding was at the foot of the knob; the road led upslope to geothermal sites and downslope to Angeles City. (Val Gempis, USAF, May or early June 1991)

Figure 10B. Early posteruption drainage from the upper part of the Sacobia pyroclastic fan, through the constriction in the midfield of this view, was into the Sacobia River (left). View is to the east, downstream. Only a small part of the upper watershed drained into the Pasig-Potrero River (right, past the constriction). (C.G. Newhall, September 1991)

Figure 10C. Drainage of the upper part of the Sacobia pyroclastic fan continued into the Sacobia River (left) through all of the 1991 and 1992 and most of the 1993 rainy seasons. As in *B*, only a small part of the upper watershed drained into the Pasig-Potrero River (far right). At the valley constriction, the channels of the two rivers were 600 m apart. (R.S. Punongbayan, February 18, 1993)

Figure 10D. Sacobia and Pasig-Potrero Rivers, approximately same view as in *B*, September 11, 1995. A secondary explosion in pyroclastic-flow deposits between the north and south forks of the Sacobia River, on October 5, 1993, generated a secondary pyroclastic flow that filled a 2-km-long segment of the Sacobia River channel and shunted flow from the upper Sacobia watershed into the channel of the Pasig-Potrero River. Subsequent deep erosion has entrenched flow into the Pasig-Potrero River (right side of valley) and greatly lessened lahar hazard in the Sacobia River valley (C.G. Newhall).

Figure 11A. An approximately 10-m-deep constriction in the lower Sacobia River at Barangay Maskup. View is downstream, June 23, 1991. Lahars during the 1991 rainy season buried the floor of the valley to a depth of about 5–6 m at Maskup; lahar deposition during 1992 and 1993 (C–F) aggraded the stream floor by an additional 6 m. (R.P. Hoblitt)

Figure 11B. Closeup view of the Maskup constriction, February 19, 1992. View is downstream; an antenna mast near the right bank transmitted data from a rain gage and tripwires to civil defense officials. About 5 to 6 m of posteruption aggradation had already occurred. (C.G. Newhall)

Figure 11C. The Department of Public Works and Highways constructed a house, near the antenna mast, for previously rain-soaked lahar observers of PHIVOLCS and nearby barangays. The house was completed in late August 1992. (C.G. Newhall)

Figure 11D. Lahars on August 29, 1992, threatened but did not destroy the observers' house; lahars on September 4–5 partly buried the house. During the worst of the September 4–5 lahars, observers moved to nearby, higher ground. Several other lahar watchpoints along banks of the Pasig-Potrero, Marella-Santo Tomas, and Bucao Rivers were also overrun or threatened and had to be relocated. (C.G. Newhall, September 5, 1992)

Figure 11E. Lahars in late 1992, and up to October 1993, completely buried the house and the lower one-third of the antenna mast. The bracket for the white rain gage is visible just above the low trees and bamboo in B. People standing atop the former location of the house are mostly PHIVOLCS observers from other Philippine volcanoes who are visiting Pinatubo to learn some of its lessons. The concrete pads in the foreground were footings for a pedestrian suspension bridge that was never completed. (C.G. Newhall)

Figure 11F. Broken wooden posts on concrete foundations of the former observers' house, exhumed shortly before September 5, 1994. The snapped stump of the antenna mast was behind the photographer. The last major lahars in this valley before this photograph was taken occurred during Typhoon Kadiang in October 1993. During that same typhoon, the Pasig-Potrero River captured most flow of the Sacobia River (fig. 10D); as a result, lahar deposition at Maskup in 1994 was exceeded by erosion and downcutting. (C.G. Newhall)

Figure 12A. Highway and railway bridges across the Bamban River, at Bamban, Tarlac, about July 17, 1991. View is upstream, to the southwest; immediately behind the bridges is the confluence of the Marimla River (directly to the rear) and the Sacobia River (joining from the left). Highway bridge spans fresh lahar deposits. Damage to the railway bridge predated 1991 and was not related to the eruption of Mount Pinatubo. (C.G. Newhall)

Figure 12B. Bamban River in mid-September, 1991, 1 month after lahars swept away the highway bridge (details in fig. 13A–C). The higher railway bridge (right of center) remains. Traffic in the lower right (beneath the date stamp) in this photograph is along the original path of a concrete road that crosses the right edge of A; debris that had covered the road was bulldozed into a makeshift levee. The same aggradation that set the stage for the demise of the Bamban highway bridge blocked the Marimla River. An impounded lake (middle background, right of center) broke out and partially drained on August 21. (C.G. Newhall)

Figure 12C. Lahars of August 21 and September 1991 breached protective levees and buried Barangay Lourdes, Bamban. View is to the south-southwest. Barangay Lourdes was off the lower right corners of *A* and *B*, downstream from the Bamban bridge. By the time of this photograph (January 1992), homeowners had salvaged their galvanized iron roofing and had moved into an evacuation camp on a nearby hill or to undamaged homes of relatives and friends. Mounds of sandy lahar debris (upper part of photograph) were being added to raise the left bank levee, for (unsuccessful) sediment control. (C.G. Newhall)

Figure 13*A.* Ground-level view of Bamban bridge, Bamban, Tarlac, July 5, 1991. This bridge carried most of the traffic between Manila and northern Luzon. The river bed was approximately 10–12 m below the bridge deck, and in preceding weeks the channel floor and piers had been scoured by water-rich lahars. For reference, note the cuesta (asymmetric, uptilted ridge) in the background. View is upstream, to the west-southwest. (R.S. Punongbayan)

Figure 13*B.* Bamban bridge at risk, August 16, 1991. The lahar event that buried the hut in figure 14A left a clearance of only 1 m under the Bamban bridge. Traffic remained undisrupted. (R.S. Punongbayan)

Figure 13C. The Bamban highway bridge was lifted and swept away by a lahar on August 21, 1991. Part of the railroad bridge (fig. 12A) is now visible at right center. The reference cuesta is in full view. (R.S. Punongbayan, August 25, 1991)

Figure 14A. A house by the Sacobia-Bamban River, Bamban, Tarlac, July 23, 1991. This quaint hut was located on the left bank of the river, downstream from the Bamban bridge. (R.S. Punongbayan)

Figure 14B. Only the roof of the hut remains unburied, August 16, 1991; nearly 9 m of sediment were deposited during a single lahar event on August 15, 1991. (R.S. Punongbayan)

ABACAN RIVER

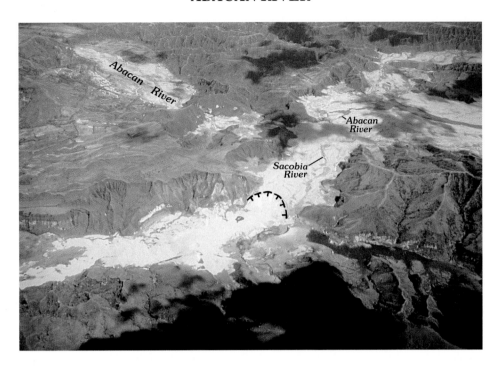

Figure 15A. Pyroclastic-flow deposits of June 15, 1991, filled the valley of the Sacobia River (foreground) to such a level that some of the earliest runoff and lahars (dark brown) passed through a pair of gaps in the hills (center), known informally among geologists as the "Abacan gap," and into the Abacan River (upper left). Prior to the eruption, the Abacan was fed only by its own small watershed below the Abacan gap; after the eruption, for most of the 1991 rainy season, a substantial part of the runoff from the Sacobia pyroclastic fan (upper right) flowed down the Abacan River. Downstream from the Abacan gap, pyroclastic-flow deposits in the Sacobia (lower left) blocked a tributary in which a new lake is seen here; opposite this lake is the headscarp of a secondary pyroclastic flow (Torres and others, this volume). Water from this lake might have contributed to early secondary explosions and generation of this secondary pyroclastic flow. View is to the south; all flow was from right to left. (R.P. Hoblitt, July 1, 1991)

Figure 15B. Abacan gap, October 1, 1991. View is to the east-southeast, downstream; flow in the Sacobia is from the foreground toward the left. The dark-gray top of the white pyroclastic-flow deposit is a several-meter-thick zone in which the deposit has been wetted and cooled enough to retain moisture. Erosion into pyroclastic-flow deposits led to secondary explosions (and further secondary pyroclastic flows?) and caused the headscarp shown in A to migrate upstream and outward toward the valley walls. Only a thin septum remained between the "upper Abacan" channel (above center, hugging preeruption hills) and the scalloped area within the Sacobia River watershed. The thinnest part of the septum is itself scalloped (center) and lacks the dark, moist surface layer. (C.G. Newhall)

Figure 15C. Abacan gap, March 18, 1992. View to the east shows that the "upper Abacan" (flowing from the upper right toward the center of the photograph) is on the verge of diversion into the Sacobia (lower left quadrant) and that very little remains of the original, flat surface of the June 15, 1991, pyroclastic-flow deposits. (W.E. Scott)

Figure 15D. Abacan gap, July 14, 1994. On April 4, 1992, a relatively large secondary pyroclastic flow that originated in the vicinity of the Abacan gap and flowed down the Sacobia removed the last of the septum that had kept water of the "upper Abacan" channeled out through the Abacan gap. Thereafter, runoff from the "upper Abacan" joined the Sacobia, as it had prior to the eruption. Rapid downcutting in 1992–94 reached and even incised the preeruption floor of the Sacobia in this area. Vegetation has noticeably recovered in areas where 1991 deposits were thin. (C.G. Newhall)

PASIG-POTRERO RIVER

Figure 16A. View to the southeast, down the north (Timbo) fork of the Pasig-Potrero River, to its confluence with the south (Papatak) fork. The lightest-colored material in the valley floor is a several-meter-thick pyroclastic-flow deposit from June 15, 1991. Slightly darker in color material is from early lahars across the 1991 pyroclastic-flow surface. Hills with columnar joints exposed along the valley walls are of the >35-ka, semi-welded Inararo pyroclastic-flow deposit (Newhall and others, this volume). Terraces in the background are underlain by post-Inararo lahar deposits. (V. Gempis, July 1991)

Figure 16B. Closeup view down the constricted section of the Timbo fork (right-center in *A*). By the time of this photograph (September 19, 1991), erosion by energetic, high-discharge lahars had cut through about 5(?) m of June pyroclastic-flow deposits and about 15 m into pre-1991 stream sediments. A concrete "sabo dam" that had been constructed a decade earlier, to trap sediments before they could reach and fill lowland channels, had been undercut and left hanging. The exact location of this sabo dam in *A* is at the small L-shaped extension of bedrock into the constricted channel; the right abutment of the sabo dam is against the long leg of the L, as can be seen in *B* and *C*. On figure 1, the sabo dam is near the tip of the arrow for sequence 16. (C.G. Newhall)

Figure 16C. Same view as in *B* on November 21, 1992. Erosion had cut about 5 m deeper than in *B*. (C.G. Newhall)

Figure 16D. Same view as in *A* on September 30, 1994. Rapid lahar deposition in August and September 1994, as well as deposition from a secondary pyroclastic flow, refilled the Timbo valley and reburied the sabo dam. Much of this rapid deposition was accomplished by continuous lahars with discharges of only 10 m³/s or less, related to reappearance of springs along a several-kilometer-long upstream reach, like springs that fed the river in preeruption time. The level of fill was about 15 m higher than in 1991; the L-shaped extension disappeared, as did the lowest two terraces of the background. (C.G. Newhall)

Figure 17A. Pyroclastic-flow deposits of June 15, 1991, in the valleys of the north (Timbo) and south (Papatak) forks of the Pasig-Potrero River. View is to the south, across the Timbo (foreground) to the larger Papatak (center); flow was from right to left. Deposits in the Papatak valley blocked a tributary (center), and a small lake had already begun to form behind this blockage by June 22. The lake grew larger through the ensuing rainy season and, on September 7, 1991, overtopped and breached its dam. The resulting large lahar killed several people downstream. A nearly identical series of events occurred in 1992. A secondary pyroclastic flow blocked the same tributary on July 13, and, on August 29, 1992, breaching caused 9 h of serious lahars downstream. Additional photographs of 1992 changes are in Arboleda and Martinez (this volume). (R.P. Hoblitt, June 22, 1991)

Figure 17B. Incipient formation of yet another lake in the same tributary as the lake in *A*, at the foot of Mount Cutuno, July 5, 1994. The principal impounding agent in 1994 was lahar deposition, though some secondary pyroclastic flows might also have contributed. (C.G. Newhall)

Figure 17C. Full development of the impounded lake, same location, August 30, 1994. (C.G. Newhall)

Figure 17D. Drained, eroded floor of the impounded lake, which broke out on the night of September 22, 1994, after moderate rainfall. Approximately 25 people were killed, mostly in Barangay Manibaug Pasig (fig. 18D). (C.G. Newhall)

Figure 18A. Pasig-Potrero River at barangay Mancatian, Porac, Pampanga, August 31, 1991. View is to the southwest; Mount Pinatubo's summit is about 22 km upslope (to the right). At this time, the channel of the river was relatively deep, and most lahar deposition was occurring downstream of Mancatian. The widespread light-gray areas are ash-fall deposits, mostly from the climactic eruption of June 15, 1991; brown strips are intrachannel and overbank lahar deposits. For reference, note red-roofed church near the top center of the picture (left side of road), the prominent bend of the road to the right after it crosses the river, and a factory complex (large dark buildings) on the left side of the road and the left (near) bank of the river. The last, known locally as "San Miguel," processed pumiceous sand from Mount Pinatubo for glass bottles. The distance from the glass factory to the church is 1.0 km. (R.S. Punongbayan)

Figure 18B. Fanhead of deposition had moved upstream to Mancatian by November 14, 1992. To prevent the spread of lahars, levees were built along both the right and left banks of the original channel (center), and deposition over the right levee, just downstream (left) of the village and four long chicken coops, had reached the right edge of a pre-1991 flood bypass (light green swath without houses, bounded on its far side by a small road, passing right to left across the upper part of photograph). The main road crosses the flood bypass on a low bridge about 100 m on the near side of the red-roofed church. (R.S. Punongbayan)

Figure 18C. The central part of Mancatian (near the four chicken coops) and the northeastern part of Mancatian (near the glass factory) were buried by lahars in late 1993. The right-bank levee upstream from Mancatian arrested the spread of lahar toward the church, but the left-bank levee was breached upstream from the barangay. The glass factory, at the bend in the road, is now surrounded by the lahar field. In the foreground is Barangay Manibaug Pasig of Porac town. (R.S. Punongbayan, February 26, 1994)

Figure 18D. Continued deposition, and continued raising of the levees, created a situation in which the Pasig-Potrero River was now building an elevated flood plain between levees, about 15 m above the surrounding countryside. Small-scale breaches of the right-bank levee in 1994 resulted in partial burial of the church and near total burial of nearby houses and a school (see also *E*). Larger breaches of the left-bank levee, on and before September 22, 1994, buried most of Barangay Manibaug Pasig (foreground). Breaches of the left-bank levee also directed the large, lake-breakout lahar of September 22 into Bacolor (fig. 20 *A,B*) (R.S. Punongbayan, October 1, 1994)

Figure 18*E*. The Mancatian-to-Manibaug Pasig crossing, as viewed from southwest to northeast (opposite to that of *A–D*). The church is buried to its front awning; a side building is buried to its roof. Just northeast of the church, at the location of the former right-bank levee, the road climbs up onto the elevated flood plain of the Pasig-Potrero, as described for *D*. Remains of the glass factory (circle) are visible just in front of the active channel. (C.G. Newhall, September 30, 1994)

Figure 18*F*. A second-generation, 10-m-tall buried telephone pole (the first generation was completely buried). Remains of the glass factory are in the background. The budding scientist in the photograph was spray painting marks at 2-m elevations on third-generation posts, such as that in the background. (C.G. Newhall, August 13, 1994)

Figure 19A. Town of Bacolor (foreground), the raised Gapan-Olongapo highway (right to left), and lahar-threatened barangays Santa Barbara, Parulog, and San Antonio, all of Bacolor. The levee-bounded Pasig-Potrero River formerly flowed off the left edge of the photograph. However, owing to several upstream breaches in its left-bank levee in 1991, lahars and stream flow covered a wide fan-shaped area east of the intended channel (virtually the entire field of view). By the time of this photograph, residents of Santa Barbara, San Antonio, and Parulog had begun to rebuild rice paddies and fishponds on top of 1991 lahar deposits. Breaches of the right bank in 1992 created the large gray area west (left) of the intended channel (covering Barangays Mitla, Balas, and several others), on which little recovery had been attempted. New lahar outbreaks in July 1994 had begun to encroach at the top right. (C.G. Newhall, July 27, 1994)

Figure 19B. Lahars from September 1994 through and after the date of this photograph (September 6, 1995) reburied Bacolor to depths of 5 m in the town proper and >10 m in some outlying villages. (C.G. Newhall)

Figure 20A. Lot for sale, Barangay Parulog, Bacolor, August 21, 1994. View is to the northwest, from the highway and across the prominent, flooded field, right center of figure 19A. (R.S. Punongbayan)

Figure 20B. Same lot as in A, but no longer for sale. Buried to depth of 0.5 to 1 m by lahar of September 22, 1994. Hundreds of square kilometers of land like this are being buried by distal overbank flows (here, 30 km from Pinatubo's summit). Houses on stilts can survive; fields are left temporarily unusable. (R.S. Punongbayan, January 15, 1995)

GUMAIN RIVER

Figure 21A. Abandoned meander of the Gumain River, May 28, 1991. (R.S. Punongbayan)

Figure 21B. Abandoned meander of the Gumain River, November 9, 1991. Now partly filled by lahars. Basa Air Base in the middle distance. (R.S. Punongbayan)

MARELLA-SANTO TOMAS RIVER

Figure 22A. Preeruption Marella River, June 3, 1991. View is to the northeast. Prominent hill in the center of the photograph rose above the fan of prehistoric pyroclastic debris. The flat-topped deposit that snaked its way down from the volcano (at least as far as the prominent hill) was a young lithic pyroclastic-flow deposit, probably from late in the 500-y-B.P. Buag eruptions. (R.P. Hoblitt)

Figure 22B. Posteruption Marella River, June 22, 1991. View is to the northeast. Same hill as in *A* appeared as an island (kipuka) in a sea of 1991 pyroclastic-flow deposit, the surface of which was undissected. (R.P. Hoblitt)

Figure 22C. Deeply dissected, medium-dark gray 1991 pyroclastic-flow deposits of the Marella River. The 1991 pyroclastic flows filled the narrow preeruption canyon (right of center in *A*) and deposited an additional 50-100 m on top of the preeruption, flat-surfaced, 500-year-old deposits (center in *A*). Erosion has removed well in excess of 100 m of 1991 deposit from the east edge of the prominent hill and might have reached to the preeruption canyon floor. Light-colored deposits in the main Marella channel (right) and in a lesser channel (left) are post-1991 lahar deposits. (C.G. Newhall, September 7, 1994)

MAPANUEPE RIVER

Figure 23A. The Mapanuepe River, which does not head on Mount Pinatubo, flowed west to its confluence with the Marella River, in the background, which flowed from right to left. The hill near the confluence is Mount Bagang. Heavy tephra fall in this area lent a gray color, but, except on June 14–15, no significant lahars had yet occurred. View is to the west. (C.G. Newhall, mid- to late-July, 1991).

Trench to drain the Mapanuepe Lake

Figure 23B. Mapanuepe Lake formed when lahars of the Marella River dammed the Mapanuepe River. The new lake submerged several villages, including one (Barangay Pili) in the foreground, left. Although sediment continues to aggrade (and prograde) the Marella-Mapanuepe delta, lake level was stabilized in late 1992 at approximately the same level seen here, by excavation of a trench through bedrock of the dark green, low peninsula that is directly in front of Mount Bagang and immediately south (left) of the confluence and impoundment. (C.G. Newhall, September 19, 1991)

MARAUNOT AND BALIN BAQUERO RIVERS

Figure 24A. Preeruption view southeast up the Maraunot River. The prominent steam cloud was issuing from a fumarole in the headwaters of the Maraunot River. (R.P. Hoblitt, June 6, 1991)

Figure 24B. Valley of the upper Maraunot River, June 14, 1991. Fresh pyroclastic-flow deposit of the June 12 or June 13 eruption (Hoblitt, Wolfe, and others, this volume) partially fills the valley. The remaining part of the June 7–12 dome is steaming in the background. (R.P. Hoblitt)

Figure 24C. Posteruption view southeast up the Maraunot River. Most of what remain of the preeruption edifice is now hidden behind hills (old domes?) northwest of Mount Pinatubo's summit. The flat surface is essentially that of 1991 pyroclastic-flow deposits; the stratified veneer on that surface consists of fall deposits from postclimactic ash emission (layer D of Paladio-Melosantos and others, this volume) and fall and surge deposits from nearby secondary explosions. Group is a graduate class in volcanology from the University of the Philippines, accompanied by a group of Aetas and nuns from LAKAS (Lubos na Alyansa ng mga Katutubong Ayta sa Sambales, or Negrito People's Alliance of Zambales). The LAKAS group formerly lived in this area. (G.P. Yumul, Jr., February 22, 1992)

Figure 25A. Maraunot River (foreground), flowing into the Balin Baquero River (middle background) and thence into the Bucao River (distant background) and the South China Sea, April 16, 1991. Most of the area in the foreground consisted of dissected pyroclastic-flow deposits from previous eruptions. (R.S. Punongbayan)

Figure 25B. Similar view as in *A*, on July 20, 1991, showing subdued topography in which prehistoric deposits were partly but not wholly buried beneath 1991 pyroclastic-flow deposits. Postclimactic tephra-fall deposits (layer D of Paladio-Melosantos and others, this volume) smoothly mantle the surface and are incised by shallow, incipient rills. (C.G. Newhall)

Figure 25C. Detail of an area near the prominent steaming in *B*. View is to the west-northwest and shows the confluence of the Maraunot River (prominent, unfilled channel at right) with the Balin Baquero River (top, flowing from left to right). Voluminous pyroclastic-flow deposits of June 15, 1991, had filled the main valley of the Maraunot River upslope from here, and also small channels of a broad area just south of the preeruption Maraunot. Barangay Villar was destroyed (right of center, on the interfluve between the voluminous deposits and the unfilled channel of the Maraunot). (R.P. Hoblitt, June 24, 1991)

Figure 25D. Similar view as in *A* and *B*, but from slightly south of *A*, 3.5 years after the eruption. New vegetation on what were thin 1991 tephra-fall and ash-cloud surge deposits contrasts with gray, largely unvegetated 1991 valley-filling pyroclastic-flow deposits. Runoff from the west side of Pinatubo, together with some that was diverted from upper Maraunot River (lower right, at the foot of the elongate green hills), has cut a new channel through the pyroclastic-flow field (straight down the field of view). Former site of Villar is between the old course of the lower Maraunot River (right center) and its temporary(?) new course (straight center). (R.S. Punongbayan, January 11, 1995)

Figure 26A. Balin Baquero River, Zambales, May 28, 1991. The Balin Baquero, a major tributary of the Bucao River, drains the western slopes of Mount Pinatubo. It is joined by the Maraunot River at the left edge of the photograph, about 15 km west-northwest of Mount Pinatubo's summit. View is to the south, upstream. (R.S. Punongbayan)

Figure 26B. Lahar deposits along the Balin Baquero River. The flood plains of the Balin Baquero and Maraunot Rivers were covered with lahar deposits up to 30 m thick; here, the deposit is about 10 m thick. A small lake at the right side of the picture formed when lahar deposits blocked drainage from the adjoining hillslope watershed. The road, its bend, and the village (Barangay Burgos) are the same in both A and B. (R.S. Punongbayan, November 22, 1992)

BUCAO RIVER

Figure 27A. Bridge to Poonbato, Botolan, Zambales, across the Bucao River, May 28, 1991. Access to the northwestern and western slopes of Mount Pinatubo was provided by this bridge. Barangay Poonbato was mostly right of the photograph, south of the bridge. View is to the east, upstream. Bucao River heads to the right, on Mount Pinatubo; Balintawak River heads on non-Pinatubo terrain in the background (R.S. Punong-bayan)

Figure 27B. Poonbato bridge was buried (but not swept away) by lahars of 1991 and 1992. Deposits are approximately 25 m thick. Barangay Poonbato (immediately to the right of the field of view) was buried. View is to the east, upstream. (R.S. Punongbayan, May 16, 1994)

Figure 28A. Bucao River, Zambales (view upstream), May 28, 1991. The Bucao River drains the northwestern slopes of Mount Pinatubo. This segment of the Bucao River is about 1 km downstream from sitio Magu-iguis, Botolan, Zambales (location shown by arrow 28 on figure 1), and about 12 km northwest of Mount Pinatubo's summit. Rice and corn were grown on the flood plain. (R.S. Punongbayan)

Figure 28B. Lahar deposits along the Bucao River, as of March 11, 1993. Lahar deposits completely covered the flood plain to depths as much as 25 m. (R.S. Punongbayan)

Figure 29A. Mouth of the Bucao River, April 16, 1991. White-sand beaches of Zambales Province, consisting of pumice and coralline debris, attracted many tourists. The ocean was clear blue; little sediment was carried by the Bucao River. (R.S. Punongbayan)

Figure 29B. Muddy water at the mouth of the Bucao River, October 1, 1991. Massive amounts of sediment were carried from Mount Pinatubo into the Bucao River valley; an unknown but relatively small percentage of that sediment is carried into the South China Sea. (R.S. Punongbayan)

Eruption Hazard Assessments and Warnings

By Raymundo S. Punongbayan,[1] Christopher G. Newhall,[2] Ma. Leonila P. Bautista,[1] Delfin Garcia,[1] David H. Harlow,[2] Richard P. Hoblitt,[2] Julio P. Sabit,[1] and Renato U. Solidum[1]

ABSTRACT

An urgent program to monitor unrest and to interpret the record of past eruptions at Mount Pinatubo led to remarkably accurate warnings of one of the largest eruptions of this century. Three cornerstones of the warnings were an interpretation of the origin of the unrest, a simple five-level warning scheme, and a hazards map based on a composite "worst-case" of prehistoric eruptions. Warnings were coupled with an intensive educational campaign to ensure that they were not only received but understood. Many lives were saved, as was much property.

Declining unrest in late 1991 and renewed unrest in 1992 required revised warnings. Partly because of the major eruptions and life-saving success in 1991, it was easier in 1992 to raise concern and more difficult to moderate it.

In hindsight, we should have been less concerned about overstating the hazard and more concerned about speeding preparations for evacuations. Pinatubo almost overtook us. We would also give stronger warnings about ash fall, especially in conjunction with typhoon rains, and we would modify our numerical alert scheme to be less specific about time windows and more specific about distinctions between large and small eruptions.

Credibility is an especially difficult issue at volcanoes that have not erupted in recent history. In general, volcanologists will lack credibility unless and until the volcano erupts and proves them right, and then it is too late. Several items might help, including in-person or videotaped accounts by officials who averted or failed to avert volcanic disasters elsewhere.

INTRODUCTION

Most of the **largest** eruptions of this century have occurred in sparsely populated areas, and death tolls have been correspondingly low. In contrast, most of the **deadliest** eruptions of this century have been relatively small but have occurred in populated areas and have been deadly more by virtue of surprise than by size (Blong, 1984; Tilling, 1989). The most dangerous situation of all is that of a large, unexpected explosive eruption from a long-dormant volcano in a densely populated area (Simkin, 1993).

As we struggled to understand the unrest and to warn of potential hazards at Mount Pinatubo in from April to early June 1991, we were acutely aware that premature or overstated warnings could lead to unnecessarily disruptive evacuations and serious loss of credibility for scientists—credibility that would be needed in case of true emergency. At the same time, we were acutely aware of how even small eruptions can lead to disaster (most recently, Nevado del Ruiz in Colombia in 1985; Hall, 1990; Voight, 1990). And, most worrisome of all, we could see from the geologic record of Pinatubo that its previous eruptions were so large that a recurrence would threaten several hundred thousand unsuspecting people. The ingredients for a colossal disaster were on hand.

Fortunately, Pinatubo gave us a brief but unmistakable warning and, before the worst of the eruptions, more than 60,000 people heeded warnings and fled to safety. Of these, more than 20,000 escaped certain death. The death toll from the eruption itself was between 200 and 300. We grieve over these deaths but also breathe a collective sigh of relief that this toll was small compared to what would have occurred without our warnings.

This paper describes warnings that were given before the eruptions of June 1991, the general philosophy and scientific basis for those warnings, and the complex socioeconomic and political context in which the warnings had to compete for the attention of those at risk. Our discussion includes questions that we were asking ourselves, or being asked by those at risk, in the hope that it might be a useful reference for those faced with similar situations in the future.

In addition, this paper describes advice and warnings, regarding the possibility of further eruptions, that were given to public officials in the months and first 2 years following the climactic eruptions. Although earthquake swarms and growth of a new lava dome since July 1992 are minor in comparison to the high drama of April–June 1991,

[1]Philippine Institute of Volcanology and Seismology.
[2]U.S. Geological Survey.

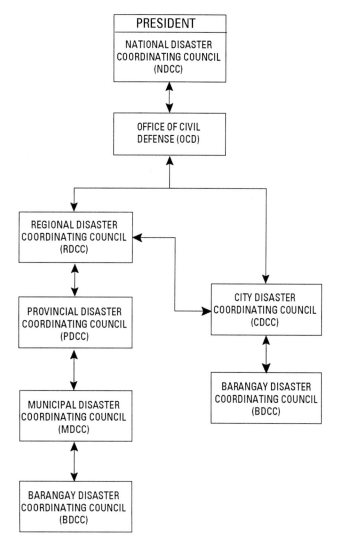

Figure 1. Organization of disaster management in the Philippines.

such swarms and dome growth remind us that the eruption might not be finished.

DISASTER MANAGEMENT IN A COMPLEX SOCIOPOLITICAL SETTING

Figure 1 illustrates the organization of disaster management in the Philippines. The National Disaster Coordinating Council (NDCC) is an umbrella body that is headed by the Department of National Defense (DND), and the Office of Civil Defense (OCD) is its administrative arm. Its members are agencies and organizations responsible for disaster warning, rescue, relief, and reconstruction. The Philippine Institute of Volcanology and Seismology (PHIVOLCS) became a member of NDCC in 1990 with responsibility for warning of earthquakes and volcanic eruptions. Parallel bodies exist at the regional (RDCC), provincial

(PDCC), municipal (MDCC), city (CDCC) and barangay or village (BDCC) levels, but, because there are no counterparts of PHIVOLCS on these lower level bodies, PHIVOLCS staff are called to provide technical advice at all levels.

Figure 2 is the preeruption flowchart for dissemination of Pinatubo warnings. PHIVOLCS was the source, and OCD, RDCC, the PDCC's, and the Philippine Air Force's Clark Air Base Command (CABCOM) were the first-level recipients of warnings. They, in turn, were charged with relaying messages to towns and various Philippine military groups, and the corresponding MDCC's were to pass warnings to barangays (villages) and to evacuation camps. Informally, the USGS team conveyed the same warnings to U.S. military commanders and the U.S. Embassy, and all of us conveyed warnings to local residents.

The formal structures for decisionmaking and communications hide a much more complicated reality. Unlike Mayon, Bulusan, or Taal Volcanoes, where we had worked with one governor and one provincial staff, Mount Pinatubo was at the apex of three provinces (Pampanga, Tarlac, and Zambales, fig. 3). We had to work with three governors and three provincial staffs. In addition, we had to warn three large cities (Angeles, San Fernando, and Olongapo), several tens of smaller towns, U.S. and Philippine military bases, and hundreds of barangay (village) captains, minor officials, and nongovernmental organizations. Each of the three provinces speaks its own dialect, and the indigenous mountain people, Aetas, speak yet a fourth dialect.

Two of the three governorships were in dispute, as was one important mayor's post, so we had to contend with local politics. Discussions between the Philippines and the United States over military bases were politically sensitive, and a guerrilla movement operated on Pinatubo itself. As a national election was approaching (May 1992), political decisions about Pinatubo precautions had to be weighed against electoral consequences. Other factors that complicated our efforts are noted in Punongbayan and others (1993).

In short, our warnings passed into theoretically simple, well-established channels for civil-defense communication and decisions. These channels did work, but slowly, because warnings had to filter through a myriad of political, social, and economic interests.

CHRONOLOGY OF NATURAL EVENTS, SCIENTIFIC RESPONSES, AND PUBLIC RESPONSES

JULY TO AUGUST 1990

On July 16, 1990, a M 7.8 earthquake occurred along the Philippine fault, about 100 km northeast of the volcano (Punongbayan and others, 1991). In the weeks thereafter,

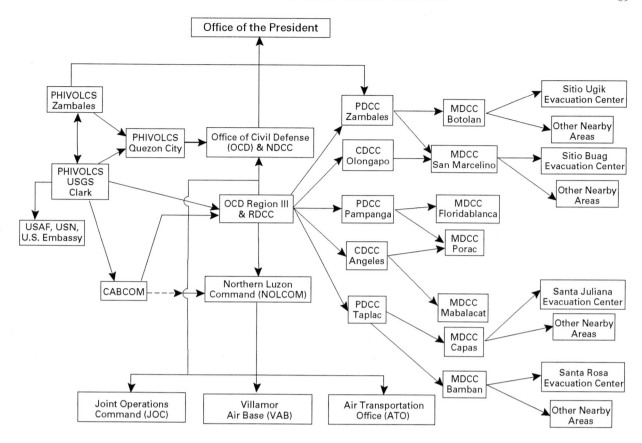

Figure 2. Flowchart for communicating warnings of Pinatubo hazard, May 1991. Dashed line indicates a secondary line of communication. Council acronyms are defined in figure 1.

numerous smaller earthquakes were felt by residents of the Pinatubo area. Some were undoubtedly distant aftershocks of the July 16 earthquake, but others were instrumentally located in the Pinatubo region. A few hours after the main shock, a M 4.8 earthquake occurred about 10 km southeast of Mount Pinatubo, and 5 other M>3 earthquakes occurred in the general vicinity of Pinatubo before the end of August (B.C. Bautista and others, 1991, this volume).

In early August, Aeta people from the northwest side of Pinatubo, led by Sister Emma Fondevilla, reported to PHIVOLCS rumbling sounds, ground cracks, and a landslide covering about 2 to 3 ha on the upper northwest face of the volcano. A PHIVOLCS Quick Response Team sent to Pinatubo on August 5 noted steaming and a landslide on the extreme northwestern part of the volcano, in a long-known solfataric area. The PHIVOLCS team reported four steaming vents with yellow sulfur deposits, hot spring discharge at 45°C, and ground temperature of 89°C. Because steaming and thermal activity were normal for this area, the team concluded, rightly or wrongly, that the landslide was caused by aftershocks of the strong tectonic earthquake and heavy rains in the area.

A memorandum sent on August 6 to the governor of Zambales Province, the mayor of Botolan town, and the provincial military commander stated that "Preliminary findings indicate that the phenomenon is not related with any volcanic activity…. The parameters necessary for deducing an approaching…volcanic eruption were not observed in the locality" (Ramos and Isada, 1990a). Local residents were reassured, despite the persistence of felt earthquakes weeks after the investigative team had left. A report to the Director of PHIVOLCS on August 10 followed an aerial survey on August 9 to map out the extent of the landslide; the report presented more details of the team's observations (Ramos and Isada, 1990b).

LATE MARCH TO APRIL 9, 1991

New rumbling sounds and felt earthquakes in late March, and explosions and the opening of new steaming vents on April 2, brought Sister Emma back to PHIVOLCS on April 3. Another Quick Response Team was immediately created, and, with the cooperation of the Office of Civil Defense and the Philippine Air Force, it conducted an aerial survey and found all that had been reported, plus a fissure and new craters at the northeast end of the line of steaming vents.

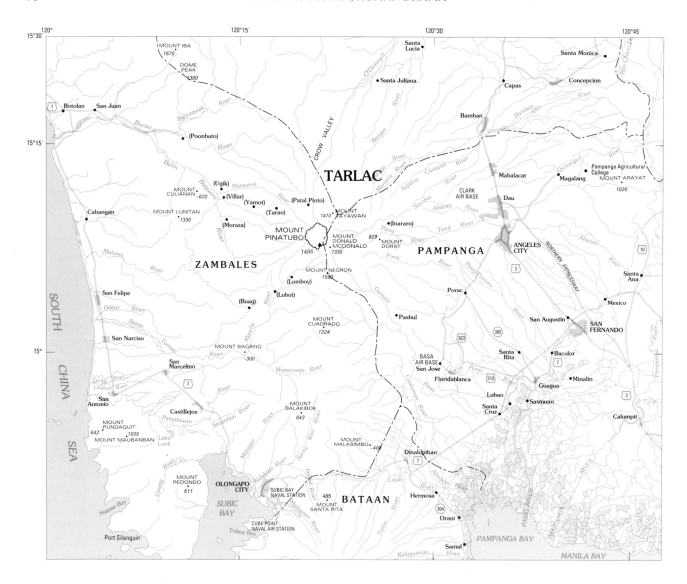

Figure 3. The three provinces, large cities, and numerous towns and villages to whom warnings of volcanic eruption had to be disseminated.

The team's initial thought was that the activity was purely hydrothermal. To determine the cause of these explosions, PHIVOLCS installed a portable seismograph the next day. About 500 earthquakes, some large enough to be felt at varying intensities at the temporary station, were recorded. Unable to say for sure whether the activity was purely hydrothermal or volcanic in origin, and whether it would culminate in a volcanic eruption, PHIVOLCS recommended a precautionary evacuation of villages within 10 km of the summit. The Department of Social Welfare and Development, together with several non-government organizations, established evacuation camps that quickly filled with about 5,000 Aetas.

PHIVOLCS then set up additional seismographs, in a simple, nontelemetered seismic network, and established a base camp where visual observations could be conducted

(Sabit and others, this volume). This initial seismic network, consisting of one analog seismograph and four digital seismographs, was installed on the northwest side of the volcano, where roads and unrestricted access were available. The digital instruments (which recorded data on disks that then had to be read and analyzed on playback units) were better suited for aftershock monitoring, where assessments and interpretation are done at a later date. Unfortunately, not much else could be done because these were the only instruments available at that time; other portable PHIVOLCS instruments were deployed at Taal Volcano, which was also restless at that time.

During these early days, volcano bulletins were prepared in the field and contained daily earthquake counts and visual observations for the preceding 24-hour period. Each bulletin also included a judgment about the overall condition of the volcano. Unsure of whether to use a warning scheme previously used for other Philippine volcanoes, we initially used the term "unstable" to describe the volcano's condition. Draft bulletins were radioed to the PHIVOLCS central office for review and release. From PHIVOLCS' central office, these bulletins were relayed to the NDCC via OCD and faxed to the Office of the President and the Departments of Science and Technology and National Defense. Bulletins were also radioed back to volcano monitoring field stations for local dissemination.

During the same early days of the Pinatubo crisis, PHIVOLCS and U.S. Geological Survey (USGS) geologists also began to compile what little was known about Pinatubo's geologic past. Hermes Ferrer and F.G. Delfin, Jr., of the Philippine National Oil Company and Mark Defant of the University of South Florida shared helpful knowledge of the geology and rock types of Pinatubo.

APRIL 10–APRIL 26, 1991

With Taal Volcano also restless at the same time, PHIVOLCS asked USAID/Philippines to obtain the assistance of the Volcano Crisis Assistance Team (VCAT) of the U.S. Geological Survey. Two weeks were needed to assess the potential seriousness of the unrest, ready equipment, delay previous commitments, and obtain official concurrence and token funding from USAID. Those hurdles overcome, a three-person USGS team arrived in the Philippines on April 23, bringing with it a large cache of equipment that had been developed or purchased specifically for emergencies such as this.

The new PHIVOLCS-USGS team immediately began to select sites for a network of radio-telemetered seismometers and a central recording base station. The team initially planned to work from a college or an apartment in Angeles City but changed its mind when the U.S. Air Force at Clark Air Base invited it to use Clark facilities and offered helicopter support. Given that the Philippine and U.S. governments were locked in negotiations over renewal of the bases agreement, and that Philippine government agencies were not generally allowed to operate on Clark Air Base, we were concerned that a move to Clark would compromise the political neutrality of our mission, be awkward for those of us from PHIVOLCS, and compromise the frequency or credibility of our interaction with other Philippine government authorities. However, the advantages of air support, other logistical support, communication facilities, ample housing and work space, and the fact that the USGS team could continue its work even though it was nearly out of money outnumbered the potential disadvantages, and we moved onto Clark on April 26. There, we created the Pinatubo Volcano Observatory (PVO) (Punongbayan and others, 1993), supplementing the temporary field station that had earlier been established on the northwest side of the volcano.

In retrospect, our communication with local authorities did not suffer much, as PHIVOLCS' main office in Quezon City remained the central focus for such communications and we made frequent trips to the offices of those officials. Although USGS scientists heard of some suspicion that their warnings were intended to favor the Americans in the bases negotiations, or that they would provide information to the Americans sooner than to the Philippine government, the entire PHIVOLCS-USGS team made it a firm policy to remain strictly apolitical and to provide all volcano information first to the Office of Civil Defense and then to all other interested parties including the U.S. Air Force. An added bonus—one that we did not anticipate at the time but which helped all parties in the end—was that the PHIVOLCS-USGS team on Clark Air Base was largely isolated from the news media and thus could work on technical issues and keep the media focus properly at PHIVOLCS' main office in Quezon City.

MAY 1991

In the shortest time possible, we needed to

- monitor and interpret seismicity, ground deformation, and gas emissions, and thus to understand the unrest;
- translate our understanding of the unrest into a simple warning scheme from which civil defense leaders could devise a simple "ready, set, go" set of evacuation and other plans;
- reconstruct from the geologic record a profile of Pinatubo's past eruptions;
- use the eruptive history to make a hazard map that would delineate areas of greater and lesser hazard during a "typical" and "worst-case" eruption; and
- communicate this information far and wide, to awaken both the public and its leaders to the fact that a large eruption **could** (at this point, no one could say "would") occur.

Each of these tasks is discussed below, prefaced by the main questions (*in italics*) that we were trying to answer.

UNDERSTANDING THE UNREST

Before we could offer much more advice to local and military authorities, we had to understand the origins of the unrest.

How frequently (if at all) did unrest like this occur at Pinatubo? How did this unrest compare to "baseline" activity of Pinatubo?

Was the unrest hydrothermal, tectonic, volcanic, or some combination of these?

If it was volcanic, what did the seismicity and other features of the unrest tell us about the likelihood and the likely date and magnitude of an eruption?

Without prior monitoring of Pinatubo, we had no real knowledge of how often swarms of small, unfelt earthquakes might occur beneath Pinatubo. Neither did we have any way to check whether baseline seismicity varied with steaming in the known thermal area and (or) with the monsoon season. Aeta residents told us that nothing like this had happened within their memories or oral traditions, so, at least, we knew that the unrest was substantially greater than anything of the past several decades.

If the unrest was of hydrothermal origin, we expected that all earthquake hypocenters would be shallow and that there would be minimal outgassing of SO_2, because the stable sulfur species would likely be H_2S. If the unrest was dominantly tectonic, it might be centered exclusively away from the volcano and release little or no SO_2. If the unrest was volcanic, it could produce small earthquakes at various depths beneath the volcano, with or without significant SO_2 emission at this stage.

The first priority for the PVO team at Clark was to establish the telemetered seismic network (Lockhart and others, this volume; Murray and others, this volume). The availability of seismometers, computers, telemetry radios, and related equipment, collected under the cooperative Volcano Disaster Assistance Program of USGS and USAID's Office of Foreign Disaster Assistance, was essential to this task, as was the availability of software developed by VCAT and by W.H.K. Lee and others of the USGS and the International Association of Seismology and Physics of the Earth's Interior (IASPEI). By May 7, the net was providing hypocenters and other seismic information in near-real time. We found two groups of hypocenters, one at about 4 to 6 km depth about 5 km northwest of Pinatubo (potentially tectonic, but too deep to be hydrothermal) and another at shallow depth beneath the north flank of Pinatubo (potentially hydrothermal, volcanic, or tectonic).

During this time, we worried that the large footprint of earthquake epicenters might indicate a large magma reservoir, but we also knew that similar widespread seismicity at Rabaul (Papua New Guinea), Long Valley (U.S.A.), and the Phlegraean Fields (Italy) had persisted for years without eruption. During May, we could not distinguish the seismicity recorded at those large systems from that which we were observing at Mount Pinatubo.

To constrain our interpretations of seismicity, we needed measurements of SO_2 emission, using a correlation spectrometer (COSPEC). Curiously, we had great difficulty in obtaining the use of a small, light plane for such measurements. Most military planes were unsuitable, and we lacked money to hire a private plane. Finally, on May 13, in desperation, we attempted measurements from a U.S. Air Force helicopter. Despite a terrible signal-to-noise ratio caused by

interference from the helicopter's main rotor blade, we determined that about 500 t/d of SO_2 was being emitted (Daag, Tubianosa, and others, this volume). From this we ruled out the tectonic explanation and tentatively ruled out the hydrothermal explanation as well. Magma appeared to be involved, and we interpreted a tenfold increase in SO_2 emission over the following 2 weeks to indicate that the magma was also rising (Daag, Tubianosa, and others, this volume). Subsequent analysis has suggested an alternative explanation for increasing SO_2 emission—boiling off of a hydrothermal system and thus progressively less absorption of SO_2 gas—but this was not our thinking at the time.

On May 19, we received a fax from an overseas colleague who had kindly analyzed data from the digital recorders used early in the crisis on the northwest side of Pinatubo. He judged that the earthquake swarm was similar to many tectonic swarms near arc volcanoes and geothermal areas, so he advised that "the swarm itself is not very dangerous, although close watching is necessary." Fortunately, we had several different data sets by that time, including that which showed high and increasing levels of SO_2 emission, so we set aside the suggested interpretation. In the Philippines, where the advice of a foreign expert is accepted politely and often with deference, this interpretation without our additional data could have been disastrous.

During the month of May, seismicity remained broadly constant, and SO_2 emissions showed a clearly defined and essentially linear increase. In the absence of any significant increase in seismicity, we guessed that magma was rising at a slow, constant rate, and we did not know whether or when it would erupt.

A SIMPLE WARNING SCHEME

Anticipating that it would be difficult to explain the unrest in terms simple enough for laymen to use for crisis decisions, we asked ourselves:

Would we be able to predict an eruption of Pinatubo with enough specificity and certainty for our prediction to be a reliable basis for evacuations?

If not, how many levels of unrest did we think we would be able to distinguish, and could those be used by public officials to trigger evacuations and other precautions even in the absence of specific predictions?

Because we had no baseline monitoring for Pinatubo, no information of any kind about the precursors of its previous eruptions, and practically no information about the precursors of eruptions as large as those apparent in the geologic record of Pinatubo, we concluded that we could not promise a specific prediction.

We did, though, think we could offer a simple, multi-level description of unrest (table 1), and a five-level scheme was introduced on May 13, 1991. Modified from schemes used at Rabaul caldera, Redoubt Volcano (Alaska), and

Table 1. Five-level alert scheme for Mount Pinatubo, May 13, 1991.

[Note that the criteria for each alert level are qualitative, not quantitative, and that the "meaning" is not strictly a forecast, but rather, a statement of what might occur. In practice, the latter distinction was largely lost.]

Alert level	Criteria	Interpretation
No alert	Background; quiet	No eruption in foreseeable future.
1	Low-level seismicity, other unrest.	Magmatic, tectonic, or hydrothermal disturbance; no eruption imminent.
2	Moderate level of seismicity, other unrest with positive evidence for involvement of magma.	Probable magmatic intrusion; could eventually lead to an eruption.
3	Relatively high and increasing unrest including numerous b-type earthquakes; accelerating ground deformation, increased vigor of fumaroles, gas emissions.	If trend of increasing unrest continues, eruption possible within 2 weeks.
4.	Intense unrest, including harmonic tremor and (or) many "long-period" (low-frequency) earthquakes.	Eruption possible within 24 hours.
5	Eruption in progress	Eruption in progress.

Stand-down Procedures:

In order to protect against "lull before the storm" phenomena, alert levels will be maintained for the following periods after activity decreases to the next lower level:

From Alert Level 4 to 3: wait 1 week.
From Alert Level 3 to 2: wait 72 hours.

Long Valley caldera (California), and generalized in UNDRO-UNESCO (1985), this scheme did not technically make predictions but simply noted increasing levels of unrest and correspondingly decreasing assurances that an eruption would **not** occur within a specified time period. Phrasing like "eruption possible within 2 weeks" was chosen carefully to mean that unrest had risen to such a level that an eruption **might** occur within that period. Perhaps predictably, the mass media and the general public misread the intent of the wording and concluded, first, that an eruption **would** occur 2 weeks from the date of the warning, and later, after our explanation, that an eruption **would** occur sometime within the 2 weeks following the warning. The intended distinction between descriptions of unrest and predictions vanished. In retrospect, use of the Pilipino "*ma'aaring mangyari*" ("**might** occur") would have been clearer than the English "possible."

Initially, the alert was placed at Alert Level 2 because of elevated seismicity, persistent felt earthquakes, rumbling sounds, and the possible recurrence of explosions like those of April 2, 1991. Even though many officials misunderstood

the subtleties of the alert levels, they understood clearly that Level 3 was more serious than Level 2 and required urgent preparations, Level 4 was more serious than Level 3 and, for people living near the volcano, required evacuation, and Level 5 was as serious as we could get. Indeed, Level 5 was originally intended for large explosive eruptions in progress, but in our simplifications we omitted the words "large explosive," a fact that we regretted later when the eruption began with dome extrusion and relatively small explosive eruptions.

One intent of the scheme was to provide a simple set of steps for which the Office of Civil Defense and military commanders could design corresponding response plans. Shortly after it was introduced, OCD, PDCC's, the U.S. Air Force at Clark and the U.S. Navy at Subic Bay, and CABCOM began to prepare contingency plans, each of which was loosely tied to alert levels. Most of the evacuations that were eventually ordered, including those of Aetas living on the lower flanks of Pinatubo and that of Clark on June 10, were based on this numerical scheme combined with hazard maps (described herein) and PHIVOLCS' recommendations of zones 10, 20, 30, and eventually even 40 km in radius from the volcano.

The scheme included automatic stand-down procedures to ensure that the alert would be lowered when unrest decreased but not before the danger had truly passed. Many volcanoes exhibit increasing unrest until magma reaches so close to the surface that no further rock fracturing is required and (or) until late-stage degassing and a corresponding increase in magma viscosity temporarily halt the final ascent of that magma (Newhall and Endo, 1987). At this point, seismicity decreases, but the hazard remains as high or higher than ever. Delays were built into the stand-down procedures to guard against such "lulls before the storm."

Details of how this scheme was applied are given later in the text for June 1991. One lesson bears mention here: schemes of this sort demand that one, and only one, person or consultative group has the authority for determining the level. This might have become a problem for us, because we were a distributed observatory with staff on both sides of the volcano and at PHIVOLCS' main office in Quezon City, and it was difficult to keep everyone's understanding of the changing state of the volcano exactly synchronous. However, the authority to issue alert levels lay solely with the Director of PHIVOLCS, in Quezon City, and eruptions overtook us before any differences of synchroneity could become a problem.

On May 17, we reviewed the situation and possible scenarios with selected officials (the Administrator of OCD and commanders of Clark Air Base) through use of a probability tree (fig. 4). By this time, we had nearly eliminated nonworrisome tectonic and hydrothermal explanations, and the primary questions were whether the inferred magma intrusion would lead to an eruption, how large that eruption

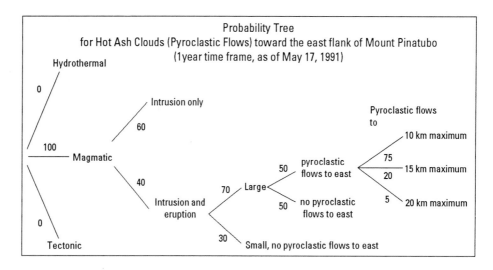

Figure 4. Probability tree for east flank Pinatubo hot ash clouds (pyroclastic flows), presented to Office of Civil Defense and Clark Air Base authorities to explain a range of possible origins and outcomes of the ongoing unrest (1-year timeframe). The probabilities indicated for individual events were subjective estimates based on our interpretation of unrest and our experience at other volcanoes. The probability of any of the outcomes at the right was the product of probabilities of each individual leg (expressed here as percent). It was explained that these estimates had high uncertainties and would change with new data and with changes at the volcano.

might be, whether pyroclastic flows would be restricted to the north and west or flow in all directions, and how far such pyroclastic flows might reach. The advantages of a probability tree were that it showed each of the possible outcomes, it showed the approximate relative likelihoods of each outcome, and it indicated fundamental points of uncertainty. The disadvantages were that it was too technical for most laymen to understand and that numbers (which were really subjective estimates) implied that we understood the situation better than we really did. Questions asked at the time suggested that it was readily understood by those to whom it was presented, but we do not know how (or whether) it was used in subsequent decisions.

Although most of our logistical support came from the U.S. command of Clark Air Base, we also sought out CAB-COM because that group could serve as a critical communications link between the PVO team at Clark and the RDCC in San Fernando, via Armed Forces of the Philippines phone lines and radios. After the initial briefing on May 21, CABCOM assigned liaison officers to stay in close touch with our monitoring operations and also prepared a base contingency plan for Filipino officials and dependents (never actually activated) and a communication plan that would link various agencies that would be involved during an actual eruption. As it turned out, CABCOM was a sup-

plemental rather than primary communications link between PHIVOLCS and the RDCC.

RECONSTRUCTING PINATUBO'S ERUPTIVE HISTORY, AND TRANSLATING THIS INTO A HAZARDS MAP

The past is often a key to the future, so we asked

What types and magnitudes of eruptions had occurred at Mount Pinatubo?

How often had eruptions occurred, and how long had it been since the latest eruption? How often did large eruptions occur, and how long had it been since the latest large eruption?

How far had prehistoric pyroclastic flows, lahars, and thick tephra falls reached into now-populated areas?

To what extent did the location of steam vents on the north side of Pinatubo limit hazard to the north side of the volcano?

In April 1991, only three radiocarbon ages were available for young Pinatubo volcanic deposits (Ebasco Services, Inc., 1977; see also Newhall and others, this volume). These ages—635±80 yr, 2,330±110 yr, and 8,050±130 yr—had been the basis for classifying Pinatubo as an active volcano (Punongbayan, 1987; PHIVOLCS, 1988). However, to estimate recurrence frequency and the length of the latest repose, we needed—and found—charcoal in several pyroclastic-flow deposits. These samples were sent by air courier to the lab of Dr. Meyer Rubin, USGS, Reston, Va., who analyzed them immediately. Within a week of delivery, we had preliminary ages—imprecise because there wasn't time to allow radon decay in the lab but invaluable because we got them while we were still trying to answer the question of recurrence frequency. Before the major eruptions of June, we had five new ages—enough to suggest that the recurrence frequency was on the order of 500 to 1,000 years and that 500 years had already passed since the latest eruption (ages are listed in Newhall and others, this volume). We didn't have enough data to say that Pinatubo was "overdue," or "not due yet," but we did have enough to say that an eruption was entirely plausible. Because these samples came from the east side of the volcano, where the pyroclastic fan was less prominent than on the north, west, and southwest, the ages also suggested that **all** eruptions sent pyroclastic flows down **all** sides of the volcano.

The next question was, how large and extensive might this eruption be? In our reconnaissance, we aimed only to define the major types of deposits and the outer limits of each. No attempt was made to distinguish any but the youngest deposits on maps; all other unconsolidated, relatively fresh-looking deposits were lumped into one unit and treated as products of a "worst-case" eruption. Although we worried that use of a composite "worst-case" might overstate the hazard, we opted to err on the side of safety. All of the prehistoric eruptions that we could recognize had been big, and we had no basis from the monitoring to argue that this eruption would be smaller than the norm for Pinatubo. In fact, during days of great uncertainty in May, we often told officials that we didn't know **whether** or **when** Pinatubo would erupt, but we did know that **if** it erupted, the eruption would almost certainly be big and serious.

Our mapping was a very hurried operation with two parts, aerial-photograph interpretation and field checking, and we did those in between the press of monitoring and public meetings. Typically, we would squeeze an hour or two per day of field geology into the schedule, taking hasty notes at only the most complete, best-exposed stratigraphic sections, and looking for charcoal by which we might get some idea of recurrence frequency. The paleomagnetic vectors of clasts in diamicts were checked in the field for consistency to help us distinguish between pyroclastic flows and lahars (Hoblitt and Kellogg, 1979).

One of us (R.S. Punongbayan) did the aerial-photograph interpretation; others checked questions from the aerial-photograph interpretation and did reconnaissance traverses down major river channels, especially near the outer limits of pyroclastic-flow deposits. Most work on the ground was on foot; a few outcrops were reached by helicopter. We also took advantage of helicopter trips to and

from monitoring sites to supplement our impressions from aerial photographs, topographic maps, and ground traverses. The aerial-photograph interpretation was generally consistent with ground observations, and by May 23 we released a preliminary hazard map (fig. 5).

In addition to the conventional aerial-photograph interpretation and field traverses, we also used a local industry—mining of pumice blocks—to help us find pyroclastic-flow outcrops. We could drive along roads, spot telltale sacks of pumice piled by the side of the road awaiting pickup, and find the outcrops nearby.

One group of deposits gave us special pause. Poorly sorted flowage deposits with clasts that showed partially coherent paleomagnetism were interpreted as deposits of either hot lahars or relatively cool pyroclastic flows. These deposits occurred just downslope from unequivocal pyroclastic-flow deposits, in Porac and along both the north and south sides of Clark Air Base, well into densely populated areas. Because these deposits flanked Clark Air Base, we informed base officials that there was a chance (small, but not zero) that the base would be swept by pyroclastic flows or hot lahars, either of which would be deadly.

In maps, we addressed this uncertainty, and the possibility of ash-cloud surges no longer represented in the geologic record, by adding a 1- to 2-km-wide buffer zone around the known extent of prehistoric pyroclastic flows. An internal suggestion at PVO to increase this buffer to 3 km was under consideration when the eruptions began. In retrospect, an even wider buffer zone (5–10 km) would have been needed to protect against surges associated with the largest eruptions of Pinatubo's history (see Newhall and others, this volume).

A preliminary, hand-sketched hazard map (fig. 5) showed the pyroclastic-flow hazard with the 1- to 2-km-wide buffer zone, the ash-fall hazard across the entire map area, and rivers down which mudflows (lahars) would be likely. Question marks along the Gumain River reflected the uncertainty we faced at that time about whether pyroclastic flows had truly jumped some high ridges to enter this drainage. (Field checking on June 1 showed that deposits in the Gumain canyon were indeed pyroclastic-flow deposits, but those on the nearby fan leading to Porac were probably lahar deposits.)

This hand-sketched map, complete with question marks, was photocopied and distributed to officials in Pampanga province during a briefing on May 23, 1991. A properly drafted copy of this map, in three sheets, was provided on the same day to President Corazon C. Aquino and her cabinet, including then-Defense Secretary and Chairman of the National Disaster Coordinating Council (NDCC), Fidel V. Ramos. A copy of this map was also printed in the Philippine Star after the Pampanga briefing but attracted little notice. The map was presented to Tarlac officials on May 27 and would have been presented to Angeles City officials on the same day had they attended a scheduled meeting; on

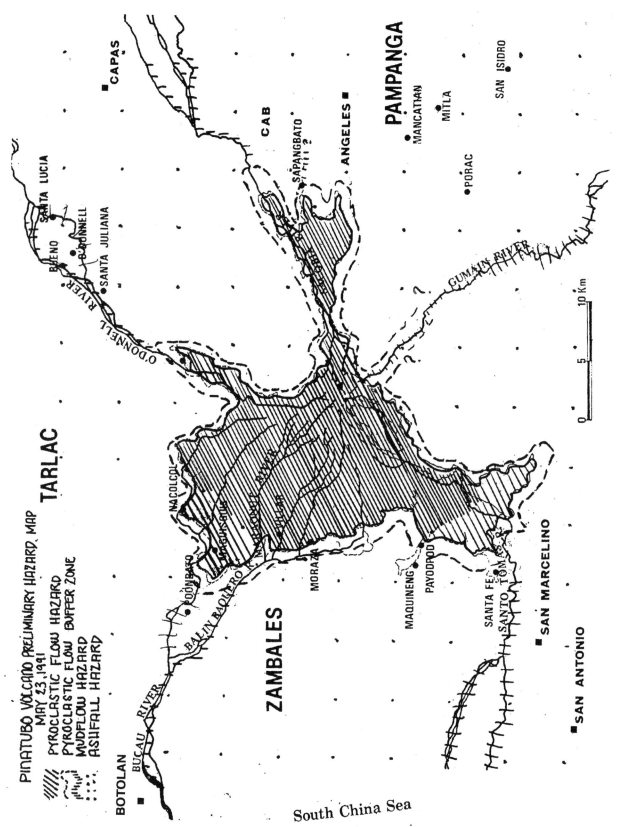

Figure 5. Hazard map distributed to provincial authorities beginning on May 23, 1991. All three major hazards—pyroclastic flows, ash fall, and lahars—are shown on this single map. Ash-fall hazard is not adjusted for prevailing winds. Question marks along the Gumain River reflect uncertainty about whether pyroclastic flows had jumped a high ridge near Mount Pinatubo to enter the drainage.

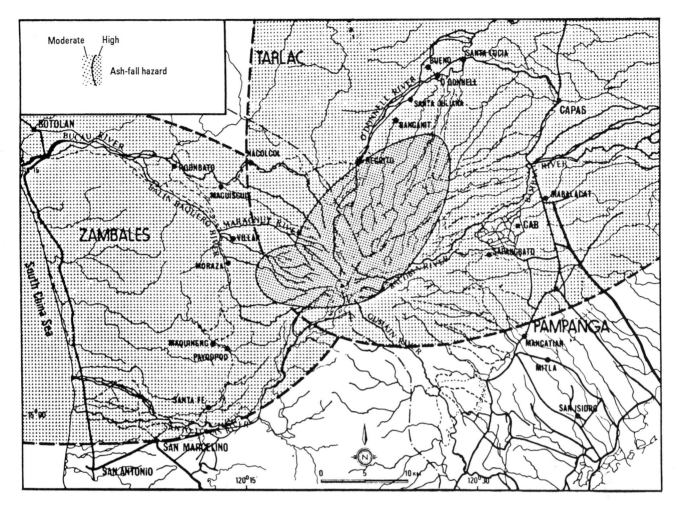

Figure 6. Ash-fall sheet of three-sheet hazard map issued by PHIVOLCS in late May 1991. Lobate areas are those most likely to be affected by ash fall, assuming normal prevailing winds.

May 28, PHIVOLCS and USGS scientists presented the map and a briefing to Zambales officials and concerned residents in Iba. On or about June 3, one of us (M.L.P. Bautista) gave a similar briefing to Angeles City officials and was in contact with those officials for the balance of the preeruptive period.

The cleanly drafted version differed from the sketched version in one significant aspect: the sketched version showed ash-fall hazard throughout the area, without regard to statistically prevailing winds; the cleanly drafted version limited ash-fall hazard to two lobes within which ash fall was most likely, based on prevailing winds (fig. 6). The sketch map was disseminated in Disaster Coordinating Council meetings from the regional down to the barangay levels, while the cleanly drafted map was disseminated at the national level including national media. We mention this discrepancy as a two-part caution to future workers. First, differences in hazard maps will confuse users. Second, representations of prevailing winds are valid as statistical probabilities, but winds on any single day of eruption can be different. Winds on the days of the eruptions made statisti-

cally unlikely sectors hazardous and many people who unexpectedly received ash fall concluded that the cleanly drafted map was wrong. We recommend against showing wind patterns by asymmetry of zones on a hazard map; a wind rose diagram in the text or margin of the map can convey the necessary information.

Both versions of the hazard map showed expected lahars along the main rivers draining Pinatubo, and residents of Sapangbato, Taconda, and other barangays along the Abacan River received direct briefings about the potential syneruptive lahar hazard. Neither map attempted to show the full area of the alluvial fans over which lahars might spread. Many populous towns and cities were built on these fans and, given the breadth and very gentle slopes of these fans, we had no simple way to know which of these towns, if any, were at high risk (see also, Janda and others, this volume). Until a typhoon approached Luzon just as Pinatubo was building toward its climactic eruption, lahars were of less concern to us than pyroclastic flows. In retrospect, we should have drawn more public attention to the hazard of syneruptive lahars, though we are not sure that these densely populated lahar-threatened areas could or would have been evacuated.

U.S. Air Force officials took a keen interest in a wall-size working copy of our hazard map, because they could see that Clark Air Base was right at the edge of the pyroclastic-flow hazard zone (fig. 5). There was, in effect, no margin of safety for the base. They also asked about tephra fall, as it might affect aircraft and other high-tech equipment, and about whether mudflows might affect the base or an evacuation to Subic Bay. Both ash fall and mudflows were possible and, indeed, likely.

INTENSIVE PUBLIC EDUCATION

Given our five-tiered warning scheme and the preliminary hazard map, how well did officials and the public understand the dangers that they faced?

If the answer was "little," as we feared, how could we explain these hazards clearly enough that officials would order, and residents would accept, evacuations and other necessary precautions?

How could we be forceful enough in our warnings that skeptics would pay attention yet not cause undue panic or overreaction and, sooner or later, destroy our still-limited credibility?

How could we convey the large uncertainties in our warnings, including the possibilities for both false alarms and unpredicted eruptions, without destroying that same, tenuous credibility?

The tragedy at Nevado del Ruiz, Colombia, occurred because a perfectly good scientific identification of hazard was poorly understood or not believed by key officials, who were accordingly slow to order precautions (Hall, 1990; Voight, 1990; Mileti and others, 1991; Peterson and Tilling, 1993). We were determined that such a tragedy would not be repeated, but we, too, faced a skeptical public (Tayag and others, this volume). Helpful tips to increase the credibility and effectiveness of warnings and thus prevent repetitions of the Nevado del Ruiz tragedy are available (for example, Sorensen and Mileti, 1987; Mileti and others, 1991), and we subconsciously attempted to follow lessons from that research. Frankly, though, it was difficult to ponder the most effective style of warnings when we were still struggling to anticipate the volcano's activity.

Our fieldwork attracted many curious residents. When told that we were studying deposits from the volcano, their usual reaction was disbelief that any volcano existed in their backyard, much less one that might erupt. Our briefings for government officials attracted many who wanted to do the right thing for their constituents but who found it difficult to even imagine the problem. Terms like pyroclastic flow (or even hot ash flow or hot blast), mudflow, and ash fall were completely new and meant little to them. Some government officials were not sure whether to believe heretofore unknown geologists.

We thought briefly about inviting officials and residents from more recently active volcanoes, like Mayon or Taal, to speak to Pinatubo groups, but no invitations were actually made. In retrospect, this was probably fortunate, because the problems at Pinatubo dwarfed those that visiting officials might have described.

Fortunately, we had an advance copy of a video entitled "Understanding Volcanic Hazards," produced by the late Maurice Krafft for the International Association on Volcanology and Chemistry of the Earth's Interior (IAVCEI). This video, made in response to the tragic misunderstanding and disaster at Nevado del Ruiz, shows graphic examples of hot ash flows, ash fall, volcanic mudflows, large volcanic landslides, volcanogenic tsunami, lava flows, and volcanic gases. Superb, sometimes shocking, footage and a simple text illustrate the nature of each phenomenon, how fast and far it travels, and its impact on people and houses.

Knowing the principal hazards of Pinatubo, we showed sections of the video on hot ash flows, ash fall, and mudflows to as many audiences as we could reach— ranging from then-President Aquino, then-Secretary of Defense Fidel Ramos and other Department Secretaries, to Governors, the chief of the RDCC, military base commanders, local officials, students, teachers, religious leaders, and barangay residents. We made perhaps 50 copies of the tape and left a copy with each group that we briefed; an untold number of second-generation copies was made. Initial response to the tape was typically shock and disbelief or denial. However, slowly but surely, the tape did convert some skeptics, and people did start to plan for a possible eruption.

One of our colleagues expressed concern that the video showed only dramatic, relatively large-scale examples and thus might have overstated hazards. We, too, were concerned about overstating hazards, but we judged then (and still judge) that strong images were needed to awaken people to the danger. In fact, our concern about overstating hazards led us to refrain from showing this video on broadcast television (except for a short segment that was shown too early in the crisis to catch anyone's attention); in retrospect, we wish we had sought broadcast time as alert levels were raised and the eruption neared.

We also made photocopies of technical reports about the potential impacts of ash on aircraft—both civilian and military—and on agriculture, public health, computers, electrical power systems, and other facets of daily life. Most of these reports were based on experience at Mount St. Helens and, in the case of aircraft, also experience from Indonesia and Alaska. Copies were given to all who asked questions, including OCD officials, a provincial health officer, and commanders at Clark Air Base and Subic Bay Naval Station. Distribution of these reports was not as wide as it might have been; ideally, one volcanologist should have been assigned to do nothing but disseminate technical information about likely impacts of an eruption.

We also had to ensure, somehow, that information and warnings we were disseminating to higher government and military officials were also being received by residents of villages on and around the volcano. A number of villages were outside the recommended evacuation zone but within or near the buffer zone for pyroclastic flows and near possible mudflow channels. In these places, we conducted an intensive information campaign that included showing the IAVCEI video and answering all questions posed by the groups. On the west side, PHIVOLCS staff were living in one of these villages and answered questions day and night. Although it would have been impractical for volcanologists to live in all of the villages at risk, there is no doubt that such arrangements greatly improved an educational campaign in those villages that were reached.

To reach further, the education campaign enlisted the national and local media. PHIVOLCS staff in the main office, and the PHIVOLCS field team in Zambales, were followed by persistent radio, print, and TV reporters who wanted to "scoop" each other. As might be expected, news reporting tended to the sensational and, at times, inaccurate, but it was certainly a key factor in rapid dissemination of volcano information to large numbers of people.

Two television interviews featuring PVO staff had more substance than most. One, on the Far East Network (FEN) on May 27, was aimed at U.S. military personnel and their dependents but was widely viewed by Filipinos throughout the area. It described the hazards briefly, the monitoring in more detail, and sought to reassure viewers that they would be warned if the situation became critical. At the time, we told viewers the situation was not yet critical. The second interview, on Manila TV the following evening, reached an even wider audience and explained the causes of the unrest, and potential hazards, as we knew them at the time. When unrest began to escalate only a few days later, Manila TV stations carried updated interviews with PHIVOLCS scientists, but FEN did not.

Two groups that might have helped with the public-education campaign were only barely reached: teachers and clerics. Although we spoke at a few schools, and to a special gathering of teachers on Clark Air Base, we ran out of time before talking to groups of teachers from the Department of Education and Culture (school was still in its summer recess) and groups of clerics. We recommend that in future crises, such groups be brought into the public-education campaign early.

The issue of credibility was a difficult one. Residents and their leaders were understandably skeptical about phenomena that they had never seen and that they either could not or did not wish to imagine. Furthermore, we as scientists were largely unknown to those at risk. We were suspected of various faults, chiefly ignorance and utter foolishness, but also of trying to make sensational headlines, pursuing an academic agenda cloaked in concern, conducting counterinsurgency reconnaissance, being dupes of the U.S. military, raising funds for our agencies, land-grabbing, and more. Some unknown persons or groups also used PHIVOLCS' name during this time to fraudulently seek relief goods. Such distractions might have been ignored were it not for the urgency of convincing skeptics, especially those in positions of leadership. We did not want to close minds through unduly aggressive messages; neither did we want to be so meek that we were ignored.

One leader who did take us seriously urged us to convince skeptics with data. That worked for a few technically conversant skeptics; it didn't work for others. We tried various approaches—convincing people with data and with scientific reasoning, building personal rapport and trust, relating our experiences from other volcanoes, rotating the job of briefings so that officials heard more or less the same story from different people, advertising our concerns through the media, and raising issues of political if not legal liability.

One approach that we did not try at Pinatubo but recommend for future crises would be to arrange in-person or videotaped accounts from leaders who had either averted or failed to avert a major volcanic disaster elsewhere. Examples could include Mont Pelée (1902; using readings), Mount St. Helens (1980), Nevado del Ruiz (1985), and Mount Pinatubo (1991). A mayor talking to another mayor is more effective than a scientist talking to a mayor.

The issue of uncertainty compounded the problem of credibility. We recognized large uncertainties in our assessment of the unrest but knew that too much mention of uncertainty would be mistaken for incompetence in the crisis situation. Even when we set aside strict scientific rigor, it was not until early June that we could make a categorical statement that Pinatubo would erupt (Harlow and others, this volume); until then, we weren't sure that it would erupt, and we didn't want to "cry wolf" lest we lose credibility that might be needed in the following months. Therefore, we stated uncertainties for those who could appreciate them, and we glossed over them for those who wanted only our "best guess." Most with whom we spoke wanted our best guess, partly because they believed a scientist would surely know the facts and not issue a false alarm and partly because they had no easy way to factor uncertainty into their decisionmaking process.

In the FEN interview of May 27, one of us (C.G. Newhall) noted uncertainties about future events, including the chance of a false alarm, and yet was chided by colleagues for being too optimistic about our ability to predict an eruption and was chided by officials for being too "wishy-washy." Perhaps equal chiding from both directions meant that a prudent balance had been achieved? Ultimately, the matter of credibility would be resolved by the volcano itself.

MAY 29–JUNE 4, 1991

By the end of May, an eruption still did not appear to be imminent, and we could say little or nothing about how much longer this unrest could continue before escalating or dying off. However, over the next several days, shallow seismicity and the amount of ash in the steam plume gradually increased.

On June 2, one of us (R.P. Hoblitt) expressed strong concern that the alert scheme did not consider the possibility that an eruption might occur suddenly, with little or no immediate warning beyond that which Pinatubo had already given us. (This was soon to happen, on June 12.)

On the evening of June 3, we discussed raising the alert from Level 2 to Level 3 because of the change in seismicity and ash emission and because SO_2 emission had reached a high level by the end of May. However, Pinatubo had been "teasing" us for weeks, and we decided to wait to see if these ash ejections would become more persistent and if earthquake hypocenters would become shallower.

Also by this time, communication networks for warnings were improving, thanks to government radio-telecommunications and especially to volunteer amateur radio operators. Officials planned to test a revised communication plan (fig. 2) on June 5, a date later moved to June 8. The message was:

"Exercise Pinatubo. Dry Run 1 (or 2 or 3) (meaning Alert Level 3 or 4 or 5). This is an exercise."

This was to test if the communication linkup between agencies would work if PHIVOLCS/PVO were to activate this message, but the test was never made, as events overtook our efforts.

JUNE 5–11, 1991

On June 5, increasing seismicity, an ominous, sudden decrease in SO_2 emission, gradually increasing ash emission, and a (false) report of a new lava spine prompted us to raise the alert to Level 3, indicating that an eruption was "possible within 2 weeks." However, the warning also acknowledged the possibility that we might have to back down to Level 2. About 10,000 Aetas, including some who had earlier evacuated but then returned home, moved from their homes high on Mount Pinatubo into the evacuation camps located in safer areas. On June 6, for their own safety and as an example, the PHIVOLCS team moved from Yamut, 7 km from the summit, to Poonbato, 22 km from the summit.

On June 7, further increases in seismicity, including seismic events that suggested dome growth or shallow intrusion, prompted us to declare Alert Level 4, meaning that an eruption was possible within 24 hours. A dome was confirmed the following morning. At this point, we began to question the wisdom of the five-level alert scheme, because dome growth was, technically, an eruption in progress. Twenty-four hours after declaration of Alert Level 4, an explosive eruption had still not occurred, and pressure on the team of scientists built as the news media and others complained that the "prediction" was wrong. One military commander misinterpreted Alert Level 4 as meaning that they had 24 hours in which to evacuate Clark Air Base; scientists and other commanders corrected the misunderstanding.

Raising of the alert level from 2 to 3 and then to 4 forced the RDCC, PDCC's, and the commanders of Clark Air Base and Subic Bay Naval Station to sharply accelerate their contingency planning. Within days, plans were to emerge that would help these officials to conduct relatively rapid evacuations and to anticipate and prepare for blocked transportation routes, interruption of electrical power and telephone service, and other impacts, even though there was not time to prevent them. From Clark Air Base, early evacuation was ordered for hospital patients who required more time and care than would be available in a general evacuation.

On June 9, when the monitoring team in Zambales reported the (false) sighting of a pyroclastic flow, the alert level was raised to 5, because to the observers the eruption certainly seemed to be in progress. Subsequent aerial reconnaissance indicated that the Zambales team had witnessed a ground-hugging pyroclastic density current rather than a true pyroclastic flow (Sabit and others, this volume; Hoblitt, Wolfe, and others, this volume), but the decision to raise the alert level to 5 had been made and released. In hindsight, even though the identification of the pyroclastic flow was in error, elevation of the alert level to 5 was an important factor in convincing more people to flee before the large eruptions actually began.

This increase to the highest alert level triggered even wider evacuations of Aeta villages than had already taken place, and one of the evacuation camps, Ugik, was moved even farther from the volcano (from Ugik to Poonbato) after we advised OCD that Ugik was situated within reach of possible pyroclastic flows. By this time, about 25,000 people, mostly Aetas, had been evacuated to safety.

On June 10, 14,500 U.S. personnel and their dependents were evacuated from Clark Air Base to Subic Bay Naval Station, and only a skeleton security force and the CABCOM personnel were left behind. The decision to evacuate Clark was courageous, because the economic and political price of leaving the base would surely be high. The mayor of nearby Angeles City said "they [the Americans] are overreacting" and "causing panic" (The Washington Post, June 11, 1991; The New York Times, June 12, 1991).

Also on June 10, three eruption scenarios were discussed within the scientific team and informally shared with concerned parties around us. A scenario in which a plinian eruption would occur and send pyroclastic flows only to the northwest was judged to have a "high" probability; the occurrence of an even larger eruption, in which pyroclastic flows would cross a topographic divide and also flow to the east, was judged to have a "moderate" probability. The possibility that dome growth and small explosions might continue for weeks to years before grading into a plinian eruption was also given a "moderate" probability.

This was a stressful time for the small team of PVO scientists. We were concerned about the volcano and the safety of those around it, including ourselves. We were also concerned about the serious consequences of a false alarm and whether we would have a second chance should the volcano not erupt as anticipated. Sleep was difficult, nerves were taut, and we were at our physical and emotional limits. We supported each other as best as we could with encouragement and humor. Lastly, we were concerned that the observatory should not have to move **during** an eruption, so on June 10 we moved our monitoring operations 5 km, from the center to the eastern edge of Clark Air Base. Our move to the east edge of the base was not planned as a strategy to overcome lingering skepticism, but it may have had that beneficial effect.

Looking back, we realized that the time windows we set in the alert level scheme were too limiting. Much later, after the 1992 eruption, we would change the terms "2 weeks" to "days to weeks," "24 hours" to "hours to days," and "eruption" to "large explosive eruption."

Details of the scientific reasoning behind each increase in alert level are given by Harlow and others (this volume) and Daag, Tubianosa, and others (this volume).

JUNE 12–15, 1991

As soon as large explosive eruptions began on June 12, the volcano spoke mainly for itself. Few additional warnings were required. However, as the number and magnitude of long-period earthquakes increased dramatically during this period, we continued to warn that "the big one is yet to come." This judgment was presented during emergency briefings of the NDCC and RDCC and was widely reported in the national news (for example, Manila Bulletin, June 13, 1991). Formal warnings during this period to the aviation community are documented in Casadevall and others (this volume).

On June 14, warnings were issued about the likelihood that an approaching typhoon (Yunya) would blow ash in many directions, not just those shown on the hazard map (see fig. 6). An aviation advisory issued at 1452 on June 14 warned that "...Yunya...will invite volcanic ash to move southeast to Manila from Mt. Pinatubo" (Casadevall and others, this volume). Similar warnings were given informally to various officials. Unfortunately, few of these warnings stressed the danger of added weight of ash when wetted by the anticipated typhoon rains and that roofs would likely collapse. Other, better publicized warnings

noted that lahars would be likely (Manila Bulletin, June 14, 1991).

By day's end on June 14, 45,000 to 50,000 people were in evacuation camps, in addition to the 14,500 who had moved from Clark Air Base to Subic Bay Naval Station. The zone of pyroclastic-flow hazard was entirely evacuated except for a handful of people who refused to leave; other, more distant villages were also evacuated, some on orders from the government and some on their own accord, out of fear about the volcano or desire to obtain relief food.

The climactic eruption occurred on June 15 and is described by several papers in this volume. The area swept by pyroclastic flows was very nearly that anticipated on the hazard maps. Lahar paths were also correctly anticipated by the hazard maps, except along the Abacan River. The most serious shortcoming of our warnings was that, even though the hazard maps anticipated ash fall, we treated it more as a nuisance than as a deadly hazard. We did not give adequate warning that the accumulated ash would be much heavier than expected because of rain from the typhoon and did not anticipate that roofs which were already burdened by ash would be strained even more by earthquakes during the latter stages of the eruption. About 180 people were killed by roofs that collapsed under the weight of wet ash; another 50 to 100 people died from various other causes, including several tens from pyroclastic flows and another several tens from lahars.

During the climactic eruption itself, all seismic stations were destroyed or disabled, we moved from Clark Air Base to the Pampanga Agricultural College (38 km east of Pinatubo), telephone and radio service was disrupted, and we knew little and could convey even less about the state of the volcano. Fearsome rumors spread, including one that a "3-mile-long fissure" had opened and that Olongapo would soon be hit by a giant lateral blast. Cellular telephones helped briefly, as long as their batteries lasted, but it was not until June 16 that we could tell the country that a caldera had already formed and that the climax of the eruption had probably passed. Gaps in communication between scientists in the field and those acting as spokespersons in Manila added to the confusion, as did lag times of a day or more between interviews and printing of newspapers. (Newspaper stories predicting a caldera-forming eruption were still hitting the street 2 days after the caldera had already formed, as in the Manila Bulletin, June 17, 1991.)

As eruptions from June 12 to June 15 had grown larger, PHIVOLCS had recommended progressively larger radii of evacuation. Until June 7 the recommendation stood at 10 km; at 1830 on June 7, the Zambales provincial government ordered evacuation on the west side of Pinatubo to a radius of 20 km. On June 9, together with its declaration of Alert Level 5, PHIVOLCS increased its recommendation for evacuation to 20 km radius on all sides of the volcano. On June 14, PHIVOLCS increased the radius to 30 km, and at 2000 on June 15, during the height of the climactic eruption, PHIVOLCS increased it yet again to 40 km. Even though the 40-km declaration was made primarily out of concern about pyroclastic flows and a lateral blast, it got many people out of their houses (where they would have been subject to roofs collapsing under ash fall) and onto the road and thereby saved some lives.

Because the 40-kilometer radius included Olongapo City, previously judged safe, Mayor Richard Gordon sought the advice of geologists. Unfortunately, the eruption and the typhoon had conspired to obscure the volcano as well as to destroy our monitoring system, so only guesses could be offered. Fortunately, geologists' advice and a perceptible lessening of earthquakes and ash fall by late evening of June 15 led the mayor to decide against an unnecessary and risky evacuation of the city. (Officially, the recommended evacuation was reduced to 20 km radius on June 18.)

Readers might ask why evacuations were of circular areas of specified radii rather than of areas known from the hazard map to be at risk from pyroclastic flows and lahars and perhaps the worst of the ash fall. Did circles not ignore our own hazard map? The answer is not simple. Traditionally, PHIVOLCS has recommended evacuations around volcanoes based upon a radius from the volcano's summit. In a crisis situation, it is easier to use a familiar procedure than to develop new ones. In addition, officials and the general public really didn't understand the nature of the hazards or hazard maps well enough to be confident that they were evacuating the right areas. Mr. Armando Duque, director of the Office of Civil Defense for Region III, told us that a radial hazard zone, scribed by a pencil on a string, would be the easiest zone to understand and implement. Furthermore, as the size of the eruption escalated and the typhoon and wet ash fall became a major concern, hazard zones on the maps that we had distributed in late May could not be readily expanded or otherwise changed. Circular evacuation zones could be easily expanded and seemed to be the only other option. The advantages of using the circles are noted above; the disadvantages were that some persons who would better have stayed at home to clean their roofs fled unnecessarily and that other persons who were truly in high-risk areas were unable to evacuate because transportation resources were spread too thinly.

JUNE 16, 1991, TO JUNE 1992

Even before the fine ash had settled from the climactic eruption of June 15, 1991, new questions were being asked:

How much longer would low-level eruptive activity persist?

Might there be another "big one"?

How will it be possible to know when the eruption is finished?

Psychological trauma from the June 15 experience, exacerbated by the fact that many earthquakes of magnitude 3 to 5.7 were occurring as "aftershocks" of the June eruptions, kindled fears of further eruptions. The PHIVOLCS-USGS team advised that eruptions could continue "for several years" but that it was unlikely that any new eruption in this episode would be as large as that of June 15. This advice was based on an analogy with Mount St. Helens, 1980–86.

Later, a review of the duration of other large eruptions (E. Santistevan, U.S. Geological Survey, unpub. data, 1991) led us to realize that very large eruptions like that of Pinatubo **tend** to be relatively short lived. However, there are enough instances of eruptions that continued for years that we opted not to change our advice. Also, secondary explosions and lahars were guaranteed to occur.

The last eruption of 1991, a small explosion, occurred on September 4, the same day the alert level was lowered from 5 to 3 and the 20-km-radius danger zone was reduced to 10 km in radius (Sabit, 1992). On December 4, 1991, the eruption alert was lowered to Alert Level 2, but the 10-km-radius danger zone was retained, on account of secondary explosions.

JULY TO NOVEMBER 1992

Seismic swarms and tremor beginning on July 3, 1992, prompted PHIVOLCS to raise the alert level to 3 on July 6. On July 8 or 9, small phreatic explosions began to build a small tuff cone in the caldera lake.

On July 14, rising magma finally reached the surface and a dome began to grow. Although not hazardous by itself, the dome growth was technically an eruption. In an internal debate, comparisons were drawn between long-period earthquakes, declining sulfur dioxide emission, and dome growth in early June 1991 (which were followed by the large explosive events) and similar, though not identical, earthquakes, similarly declining but smaller absolute values of SO_2 emission, and dome growth of the moment. None of us expected a repeat of the 1991 activity, but the apparently low-frequency character of earthquakes and the presence of a lake in the caldera required careful watching. The alert level was raised directly from 3 to 5 on July 14. Dome growth continued until October 30 (Daag, Dolan, and others, this volume; Ramos and others, this volume).

In retrospect, the decision to declare Alert Level 5 for dome growth proved to be awkward (Tayag and others, this volume). Seismicity remained high, and the dome grew rapidly for several months, so the alert level was kept at 5. However, residents of the area could see nothing: as far as they were concerned, Pinatubo was not erupting. Because this was also the rainy season, heavy rains and lahars would often trigger secondary explosions on the lower slopes, which look to the untrained eye like eruptions from the volcano. However, when asked whether these secondary explosions were the subject of Alert Level 5, PHIVOLCS had to answer "no." Worse than the awkwardness, however, keeping the alert level at 5 through the next several months probably eroded some of the high level of credibility that we had earned in 1991.

This experience of 1992 raised once again a shortcoming of the original 5-level alert scheme, namely, that it was designed for the early parts of a volcanic crisis, building up to a large explosive eruption, and that it is less appropriate for periods of prolonged low-level or declining activity. A revision of the original scheme was therefore undertaken (table 2), and instituted on December 9, 1992, at which time the alert level was lowered back to 2. In it, dome growth and other small-scale eruptive activity triggers only Alert Level 4, and Level 5 is reserved for large explosive eruptions—as originally intended but not clearly written!

DECEMBER 1992 TO THIS TIME OF WRITING (JULY 1994)

In February 1993, earthquakes and tremor were followed by small explosions in water-saturated sediments south and north of the 1992 dome. Renewed harmonic tremor on February 25, 1993, prompted us to raise the alert level to 3 on February 27, where it remained until lowered back to Alert Level 2 on May 12, 1993. Continued quiet allowed the alert level to be lowered to 1 on January 5, 1994, but renewed seismicity shortly thereafter forced us to raise it from 1 to 2 on January 19, 1994, and from 2 to 3 on February 11, 1994. On June 6, 1994, we lowered it again from 3 to 2, and on July 27, lowered it again to Alert Level 1. In general, Pinatubo is showing a gradual decline of activity, consistent with depletion of volatiles in the magma reservoir, but further dome growth is possible, indeed, likely, and even explosive eruptions remain possible. However, most of our warnings and concerns in the period since June 15, 1991, have been about lahars and are discussed in a separate paper (Janda and others, this volume).

CONCLUSION

Hazards assessments and warnings issued by PVO before the giant June 15, 1991, eruption succeeded in saving many lives and much property. Much credit goes to the PVO staff—on both sides of the volcano and in Quezon City—who worked long hours under difficult conditions. Cooperation from civil-defense and local officials was also critical; without their willingness to order emergency planning and precautionary evacuations, our warnings would not have been effective. Credit is also due to Pinatubo itself, which followed a remarkably straight and rapid course toward eruption, giving fair warning and avoiding false alarms.

Table 2. Five-level alert scheme for Mount Pinatubo, as revised on December 9, 1992.

[Principal changes were to be less specific about the time windows in which events might occur and more specific about types of eruptions, distinguishing here between dome growth or other relatively low-hazard eruptions and large explosive eruptions.]

Alert level	Criteria	Interpretation
No alert	Background; quiet	No eruption in foreseeable future.
1..........	Low-level seismic, fumarolic, or other unrest.	Magmatic, tectonic, or hydrothermal disturbance; no eruption imminent.
2..........	Moderate level of seismic or other unrest with positive evidence for involvement of magma.	Probable magmatic intrusion; could eventually lead to an eruption.
3..........	Relatively high unrest including numerous b-type earthquakes, accelerating ground deformation, increased vigor of fumaroles, gas emissions.	New or renewed eruption possible, probably within days to weeks.
4..........	Intense unrest, including harmonic tremor and (or) many "long-period" (low-frequency) earthquakes and (or) dome growth and (or) small explosions.	Magma close to or at the Earth's surface. Large explosive eruption likely, possible within hours to days.
5..........	Hazardous explosive eruption in progress, with pyroclastic flows and (or) eruption column rising at least 6 km above sea level.	Large explosive eruption in progress. Hazards in valleys and downwind.

Stand-down Procedures:

In order to protect against "lull before the storm" phenomena, alert levels will be maintained for the following periods after activity decreases to the next lower level:

From Alert Level 5 to 4: wait 12 hours after Alert Level 5 activity stops.
From Alert Level 4 to 3 or 2: wait 2 weeks after activity drops below Alert Level 4.
From Alert Level 3 to 2: wait 2 weeks after activity drops below Alert Level 3.

Note: Ash fall will occur from secondary explosions for several years after the major 1991 eruption, whenever rainfall and lahars come in contact with still-hot 1991 pyroclastic-flow deposits. These secondary explosions will occur regardless of alert level.

Lest readers get the mistaken impression that we had everything firmly in control, we reiterate that the assessment of unrest and hazards at a long-dormant, newly awakening volcano is fraught with difficulties. A large amount of equipment had to be taken to Pinatubo and deployed rapidly. Only the availability of the USGS/VCAT equipment cache, availability of IASPEI and other data-processing software, and availability of expertise in their use made this task possible. Even after the modern equipment was in use, many of our conclusions were tentative, based on only sketchy data and hasty scientific review. We did not know how much time we might have before an eruption, so we hurried to compile a minimum package of data and recommendations, with the plan that these could be refined if time permitted. There was no extra time. On June 3, only a few days after we completed this package, unrest began to escalate. On June 6 unrest escalated further, and only then were we confident that Pinatubo would erupt soon. The major explosive eruptions began only 6 days later, on June 12. Our warnings, and emergency preparations by civil-defense and other officials, were barely one step ahead of Pinatubo.

For dealing with future volcanic crises, we recommend most of the steps that we took. Each was critical and effective. If we had the chance to correct our mistakes, we would give stronger warnings about ash fall, especially in conjunction with typhoon rains and strong earthquakes during caldera formation, and we would modify the numerical alert scheme to be less specific about time windows and more specific about distinctions between large and small eruptions. We would present a documented worst-case scenario, as we did, but would add a margin of safety to cover the unseen and the unexpected. We would also ask that public officials and military commanders who were converted from skeptics to believers at Pinatubo help us address skeptics at the next such crisis.

ACKNOWLEDGMENTS

We gratefully acknowledge those who provided funding or in-kind support for our joint PHIVOLCS-USGS effort. Direct funding was from the Government of the Philippines' Departments of National Defense and Science and Technology, the U.S. Geological Survey, USAID/Office of Foreign Disaster Assistance (Volcano Disaster Assistance Program), USAID/Philippines, and later from the U.S. Department of Defense and UNESCO/ROSTSEA. In-kind support, including hundreds of hours of helicopter time, came from the Philippine Air Force, the U.S. Air Force, the U.S. Navy, and the U.S. Marine Corps; USAID/Philippines supplemented our vehicles and drivers with their own.

We also had the privilege to work with some exceptionally receptive and responsible public officials and military commanders. Four who merit special recognition are Engr. Fortunato Dejoras (Administrator of the Office of Civil Defense), Gen. Pantaleon Dumlao, Jr. (Chairman, Regional Disaster Coordinating Council, Region III), and Col. John Murphy and Col. Bruce Freeman (U.S. Air Force).

We also want to acknowledge, posthumously, the enormously important contribution of Maurice and Katia Krafft. As explained in this paper, they made a videotape of volcanic hazards that was invaluable in convincing officials and residents to take Pinatubo seriously.

REFERENCES CITED

Bautista, B.C., Bautista, M.L.P., Rasdas, A.R., Lanuza, A.G., Amin, E.Q., and Delos Reyes, P.J., 1991, The 16 July 1990 Luzon earthquake and its aftershock activity, in The July 16 1990 Luzon earthquake: A technical monograph: Manila, Department of Environment and Natural Resources, p. 81–118.

Bautista, B.C., Bautista, M.L.P., Stein, R.S., Barcelona, E.S., Punongbayan, R.S., Laguerta, E.P., Rasdas, A.R., Ambubuyog, G., and Amin, E.Q., this volume, Relation of regional and local structures to Mount Pinatubo activity.

Blong, R.J., 1984, Volcanic hazards: A sourcebook on the effects of eruptions: Sydney, Academic Press, 427 p.

Casadevall, T.J., Delos Reyes, P.J., and Schneider, D.J., this volume, The 1991 Pinatubo eruptions and their effects on aircraft operations.

Daag, A.S., Dolan, M.T., Laguerta, E.P., Meeker, G.P., Newhall, C.G., Pallister, J.S., and Solidum, R., this volume, Growth of a postclimactic lava dome at Mount Pinatubo, July–October 1992.

Daag, A.S., Tubianosa, B.S., Newhall, C.G., Tuñgol, N.M, Javier, D., Dolan, M.T., Delos Reyes, P.J., Arboleda, R.A., Martinez, M.L., and Regalado, M.T.M., this volume, Monitoring sulfur dioxide emission at Mount Pinatubo.

Ebasco Services, Inc., 1977, Philippine Nuclear Power Plant I, Preliminary safety analysis report, v. 7, Section 2.5.H.1, Appendix, A report on geochronological investigations of materials relating to studies for the Philippine Nuclear Power Plant Unit 1.

Hall, M.L., 1990, Chronology of the principal scientific and governmental actions leading up to the November 13, 1985 eruption of Nevado del Ruiz, Colombia: Journal of Volcanology and Geothermal Research, v. 42, p. 101–115.

Harlow, D.H., Power, J.A., Laguerta, E.P., Ambubuyog, G., White, R.A., and Hoblitt, R.P., this volume, Precursory seismicity and forecasting of the June 15, 1991, eruption of Mount Pinatubo.

Hoblitt, R.P., and Kellogg, K.S., 1979, Emplacement temperatures of unsorted and unstratified deposits of volcanic rock debris as determined by paleomagnetic techniques: Geological Society of America Bulletin, v. 90, p. 63–642.

Hoblitt, R.P., Wolfe, E.W., Scott, W.E., Couchman, M.R., Pallister, J.S., and Javier, D., this volume, The preclimactic eruptions of Mount Pinatubo, June 1991.

Lockhart, A.B., Marcial, S., Ambubuyog, G., Laguerta, E.P., and Power, J.A., this volume, Installation, operation, and technical specifications of the first Mount Pinatubo telemetered seismic network.

Janda, R.J., Daag, A.S., Delos Reyes, P.J., Newhall, C.G., Pierson, T.C., Punongbayan, R.S., Rodolfo, K.S., Solidum, R.U., and Umbal, J.V., this volume, Assessment and response to lahar hazard around Mount Pinatubo.

Mileti, D.S., Bolton, P.A., Fernandez, G., and Updike, R.G., 1991, The eruption of Nevado del Ruiz volcano, Colombia, South America, November 13, 1985: Washington, D.C., National Academy Press, Natural Disaster Series, v. 4, 109 p.

Murray, T.L., Power, J.A., March, G.D., and Marso, J.N., this volume, A PC-based realtime volcano-monitoring data-acquisition and analysis system.

Newhall, C.G., Daag, A.S., Delfin, F.G., Jr., Hoblitt, R.P., McGeehin, J., Pallister, J.S., Regalado, M.T.M., Rubin, M., Tamayo, R.A., Jr., Tubianosa, B., and Umbal, J.V., this volume, Eruptive history of Mount Pinatubo.

Newhall, C.G., and Endo, E.T., 1987, Sudden seismic calm before eruptions: illusory or real? [abs.]: Hawaii symposium on how volcanoes work [abstracts volume]: Hawaiian Volcano Observatory, U.S. Geological Survey, p. 190.

Peterson, D.W., and Tilling, R.I., 1993, Interactions between scientists, civil authorities, and the public at hazardous volcanoes, in Kilburn, C.R.J., and Luongo, G., eds., Monitoring active lavas: London, UCL Press, p. 339–365.

PHIVOLCS, 1988, Distribution of active and inactive volcanoes in the Philippines: Quezon City, PHIVOLCS, educational poster, 1 sheet.

Punongbayan, R.S., 1987, Disaster preparedness systems for natural hazards in the Philippines: An assessment, in Geologic hazards and disaster preparedness systems: Quezon City, PHIVOLCS, p. 77–101.

Punongbayan, R.S., Rimando, R.E., Daligdig, J.A., Besana, G.M., and Daag, A.S., 1991, The July 16 1990 Luzon earthquake ground rupture, in The July 16, 1990 Luzon earthquake: A technical monograph: Manila, Department of Environment and Natural Resources, p. 1–32.

Punongbayan, R.S., Sincioco, J.S., and Newhall, C.G., 1993, Pinatubo Volcano Observatory: Naples, WOVO News, Quarterly Newsletter, no. 1, p. 9–11.

Ramos, A.F., and Isada, M. G., 1990a, Emergency investigation of the alleged volcanic activity of Mount Pinatubo in Zambales Province: unpublished letter to Gov. Amor Deloso, Zambales Province, 1 p.

——1990b, Emergency investigation of the alleged volcanic activity of Mount Pinatubo in Zambales Province: Quezon City, PHIVOLCS, Memorandum Report, 2 p.

Ramos, E.G., Laguerta, E.P., and Hamburger, M.W., this volume, Seismicity and magmatic resurgence at Mount Pinatubo in 1992.

Sabit, J.P., 1992, Pinatubo Volcano's 1991 eruptions: Paper presented at the International Scientific Conference on Mt. Pinatubo, 27–30 May, 1992, Manila, 34 p.

Sabit, J.P., Pigtain, R.C., and de la Cruz, E.G., this volume, The west-side story: Observations of the 1991 Mount Pinatubo eruptions from the west.

Simkin, T., 1993, Terrestrial volcanism in space and time: Annual Review of Earth and Planetary Sciences, v. 21, p. 427–452.

Sorensen, J.H., and Mileti, D., 1987, Public warning needs, in Gori, P.L., and Hays, W.W., eds., The U.S. Geological Survey's role in hazards warnings: U.S. Geological Survey Open-File Report 87–269, p. 9–75.

Tayag, J., Insauriga, S., Ringor, A., and Belo, M., this volume, People's response to eruption warning: The Pinatubo experience, 1991–92.

Tilling, R.I., 1989, Volcanic hazards and their mitigation: Progress and problems: Reviews of Geophysics, v. 27, no. 2, p. 237–269.

UNDRO/UNESCO, 1985, Volcanic emergency management: Geneva, UNDRO/UNESCO, 86 p.

Voight, B., 1990, The 1985 Nevado del Ruiz volcano catastrophe: anatomy and retrospection: Journal of Volcanology and Geothermal Research, v. 44, p. 349–386 (initially published, with errors, v. 42, p. 151–188).

People's Response to Eruption Warning:
The Pinatubo Experience, 1991–92

By Jean Tayag,[1] Sheila Insauriga,[1] Anne Ringor,[1] and Mel Belo[1]

ABSTRACT

Two posteruption surveys, one in 1991 and another in 1992, assessed whether eruption warnings were received, understood, and used by citizens to take protective action. The 1991 survey showed that 71 percent of the total number of respondents (234) were forewarned; the remaining 29 percent learned of the hazard on June 12 by seeing the first large explosive events, a fact that indicates some weakness in the warning transmission.

Evacuation orders were issued by concerned Disaster Coordinating Councils or local government officials soon after danger zones were declared on April 7, June 7, and June 14–15, 1991. Eighty-six percent of the respondents received an evacuation order, but 30 percent of these people received it 2 or more days after it was issued.

Of those forewarned, 82 percent took protective action, including 46 percent who evacuated. However, within the group that evacuated, some waited two or more days after receipt of an evacuation order before moving, and some merely evacuated their women, children, and elderly or evacuated but returned. Some who did not evacuate as advised thought the eruption would not be strong enough to affect their places; others were reluctant to leave behind their houses and household effects, livestock, and crops, especially at harvest time; still others had no ready means of transport and could not walk long distances, or they believed that their God, Apo Namalyari, would not let them come to harm. Eventually, all but 5 of 234 respondents evacuated, either before or during the eruptions.

Communities in which LAKAS, an organization of the indigenous Aetas, was active showed the most exemplary operation of the system:transmission was total and response was consistently appropriate. These communities were reached by an information drive that featured the Maurice Krafft videotape on volcanic hazards, which he made for the International Association of Volcanology and Chemistry of the Earth's Interior (IAVCEI).

When Pinatubo threatened again to erupt in 1992, more than 90 percent of the respondents were forewarned and responded appropriately, indicating a marked improvement in the system. However, some overreaction was observed as an evacuation order was received by respondents who lived outside the danger zone. The errant evacuation order was traced to two sources: (1) some local government officials, who interpreted the Alert Level 5 released by the Philippine Institute of Volcanology and Seismology on July 14, 1992, to mean an eruption similar to that of June 12, 1991, and, hence, evacuation of the 20-kilometer-radius danger zone and (2) a popular radio announcer who broadcast that an eruption was imminent within 72 hours. The discrepancy between the warning message released by the source and that which was actually received appears to be a simple transmission problem. But other factors, including some features of the alert levels, may have inspired overexpectations and overreactions.

INTRODUCTION

If timely warning can be given of an impending disaster-causing event, the severity of the resulting disaster or adverse consequences can be reduced. The Mount Pinatubo 1991 eruption provides an excellent example of how accurate forecasting and timely warning saved lives from the destructive agents unleashed by a violent eruption. The number of casualties at the height of the June 1991 eruptions was small (only 200 to 300) despite the violence of the explosions and the vastness of the area affected. Early, perceptible signs from the volcano and prompt warning and mobilization of disaster-response officials minimized the human losses. It is precisely on account of its success that the Pinatubo warning system makes an interesting object of review. Its strengths, as well as its imperfections, provide insights on how other volcano-eruption warning systems could be developed or improved.

The degree to which the severity of the disaster can be reduced by warning depends on the interplay of the major components of a warning system, namely (1) the source and timing of the warning, (2) the warning message, (3) the

---

[1] Philippine Institute of Volcanology and Seismology.

warning transmission, and (4) the recipients' response (modified from UNDRO, 1986).

A team from the Philippine Institute of Volcanology and Seismology (PHIVOLCS) assessed all four aspects of the Pinatubo warning system to identify areas of success and those which needed improvement. The review involved two sampling surveys among the affected households: the first conducted within a month after the June 1991 major eruptions and the second during the month following the declaration of Alert Level 5 in July 1992.

METHODOLOGY

The 1991 survey was by stratified random sampling of respondents who had lived in the danger zones or zones recommended for evacuation. Respondents were selected from barangays that lay within 10 km, 10 to 20 km, and 20 to 40 km of the volcano's preeruption summit, radii that formally defined danger zones (fig. 1). 1990 census figures indicate that the barangays within the 10-km and 10- to 20-km danger zones had 7,653 households, or 41,100 residents; the 20- to 40-km danger zone, which included 106 barangays in 17 towns, had 58,696 households and more than 331,000 inhabitants (National Statistics Office, 1990).

The only recommendation for evacuation in 1992 was for the danger zone within <10 km of the summit. Very few people were affected, because most former residents of this zone had remained in evacuation camps or resettlement areas since 1991. Our survey was conducted among the next nearest population, from the 10- to 20-km danger zone. On the eastern side of the volcano, most barangays within the 10- to 20-km danger zone that were sampled in 1992 had only about half of their original pre-1991 eruption populations. The other residents had either relocated or were still in evacuation centers. On the western side, most of the former residents of the sample barangays in this zone were living (officially) in the relocation sites, but many were also spending days, weeks, or even months on their preeruption land planting and gathering food, whenever they felt it safe enough to do so. Some looked on the relocation site as a kind of "bakasyunan" or vacation home.

In both surveys, sampling size was determined by using a normal variable (z) value of 1.96 (see appendix 1 for the formula and computation). The survey covered only the survivors and is biased in favor of those who took precautions. However, those who died constituted a very small percentage of the population at risk, so the resulting bias is deemed insignificant. The respondents were of two types: *households* (with the household head or an adult household member as respondent) and *key informants* from among barangay and municipal officials. Household respondents were randomly selected from lists of household heads provided by barangay leaders, with substitutions when the original respondents were either unavailable or unwilling.

Interviews were conducted by PHIVOLCS staff and volunteers (local school teachers) with the aid of interview schedules (appendix 2) and, as needed, interpreters.

WARNINGS AND RESULTS

THE SOURCE AND TIMING OF WARNING

Normally, the source of eruption warning should be the entity tasked to study and monitor active volcanoes. In the case of the Philippines, this entity is PHIVOLCS. But when Mount Pinatubo started showing signs of restiveness in April 1991, PHIVOLCS had no monitoring at the volcano and, hence, no warning system for the area.

Consequently, it was not PHIVOLCS that recognized the first signs of volcanic unrest but, rather, indigenous Aetas who lived on the slopes of the volcano. Some of these Aetas, members of Lubos na Alyansa ng mga Katutubong Ayta sa Sambales (LAKAS) (Negrito People's Alliance of Zambales), reported their observations to PHIVOLCS through Sister Emma, a sister of the Franciscan Missionaries of Mary (FMM) who was doing missionary work among the Aetas.

Upon receipt of the LAKAS report, PHIVOLCS immediately began to monitor Pinatubo and, thenceforth, became the principal source of warnings. Details of the monitoring activities and chronology of preeruption events are given by Sabit and others (this volume) and Wolfe and Hoblitt (this volume); details of preeruption warnings are given by Punongbayan and others (this volume). Those warnings provided enough lead time for the beleaguered inhabitants to pack up and run away from the volcano.

THE WARNING MESSAGE

The warning message consisted of hazard zonation maps, alert levels, and "danger zones," which were zones of recommended evacuation, simplified from the hazard maps. Preliminary hazard zonation maps were disseminated by PHIVOLCS on and after May 23, 1991. These maps delineated the areas likely to be affected by the destructive agents, namely, pyroclastic flows, ash fall, and lahars. These maps illustrated the probable extent of the most probable hazards and served as guides for evacuation of endangered communities. Since the major eruption of June 15, 1991, the lahar hazard part of these maps has been updated several times.

Alert levels were designed to describe various levels of eruptive activity and danger. These provided information on the condition of the volcano, including whether its activities would likely culminate in an eruption. The alert levels were based on instrumentally derived data and daily visual observations. The original scheme of alert levels that was released on May 13, 1991, is shown in table 1.

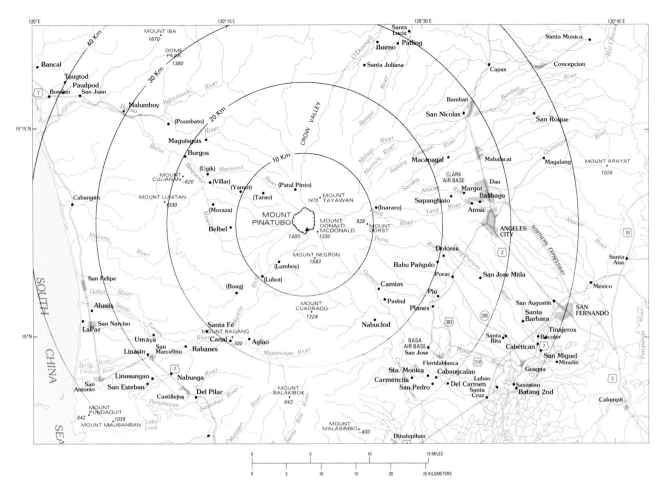

Figure 1.   Map of the Mount Pinatubo area showing recommended evacuation zones ("danger zones") of various radii, and barangays cited in the text.

Table 1. Alert levels for Mount Pinatubo, May 13, 1991.

Alert Level	Criteria	Interpretation
No alert 1	Background; quiet. Low-level seismicity other unrest.	No eruption in foreseeable future. Magmatic, tectonic, or hydrothermal disturbance; no eruption imminent.
2	Moderate level of seismicity, other unrest, with positive evidence for involvement of magma.	Probable magmatic intrusion; could eventually lead to an eruption.
3	Relatively high and increasing unrest including numerous b-type earthquakes, accelerating ground deformation, increased vigor of fumaroles, gas emissions.	If trend of increasing unrest continues, eruption possible within 2 weeks.
4	Intense unrest, including harmonic tremor and (or) many "long-period" (low-frequency) earthquakes.	Eruption possible within 24 hours.
5	Eruption in progress.	Eruption in progress.

Stand-down procedures:

In order to protect against "lull before the storm" phenomena, alert levels will be maintained for the
following periods after activity decreases to the next lower level:
    from Alert Level 4 to 3--Wait 1 week;
    from Alert Level 3 to 2--Wait 72 hours.

For explanations of the alert level scheme, see Punongbayan and others (this volume)

Daily volcano bulletins and special advisories on the volcano's condition always indicated the alert level and an associated danger zone that should be avoided and evacuated. PHIVOLCS' recommendations for evacuation were translated and transmitted by the concerned Disaster Coordinating Councils (DCCs) or local government officials as evacuation orders. A chronology of alert levels and danger zones declared in connection with Pinatubo's activity in 1991–92 is presented in table 2.

When the temporary seismic station installed near the volcano recorded high seismic activity on the first 3 days of operation, April 5–7, PHIVOLCS declared a danger zone of 10 km radius that was centered on the volcano's summit and advised evacuation of the residents from the area. Initially, volcanologists considered employing an alert level terminology used at other Philippine volcanoes but opted to design a new one for Pinatubo (table 1). The 10-km danger zone was reiterated when the alert level scheme shown in table 1 was officially adopted on May 13, 1991, and Alert Level 2 was raised. It was maintained even when Alert Level 3 was raised on June 5.

When Alert Level 4 was declared on June 7, the danger zone's radius was increased to 20 km. On June 14, this was further expanded to the 30-km radius. During the June 15

explosions, the danger zone was expanded to a 40-km radius to allow for the possibility of devastating, large-scale pyroclastic flows of a caldera-forming eruption. The danger zone was shrunk back to a 20-km radius on June 18, though Alert Level 5 remained. On September 4, the alert level was lowered to 3, and the danger zone was shrunk back to a 10-km radius. The alert level was further lowered to 2 on December 4. Alert Level 2 remained in effect until the volcano started manifesting a resurgence of activity in July 1992.

In 1992, renewed seismicity prompted PHIVOLCS to raise the Alert Level to 3 on July 9 and then to 5 on July 14, when viscous lava reached the surface and began to form a dome. The 10-km danger zone, in effect since September 1991, was maintained throughout the 1992 unrest.

The volcano's 1992 activities were entirely different from its 1991 eruptions. These were characterized by quiet effusion of lava and dome building punctuated by minor explosions and hence were not as explosive and hazardous as the 1991 events. Realizing the need to reflect these differences in the alert level scheme, PHIVOLCS revised the definitions of alert levels toward the end of the year. The first three alert levels were retained with only a slight revi-

Table 2. Alert levels and danger zones issued on Mount Pinatubo, 1991–92 (PHIVOLCS, variously dated).

Date	Alert Level	Danger Zone
1991		
7 April	volcano condition "unstable"	10-km radius centered on the volcano's summit
13 May	Alert Level 2	10-km radius
5 June	Alert Level 3	10-km radius
7 June	Alert Level 4	20-km radius
9 June	Alert Level 5	20-km radius
14 June	Alert Level 5	30-km radius
15 June	Alert Level 5	40-km radius (17 towns within this radius)
18 June	Alert Level 5	back to 20-km radius
4 September	Alert Level 3	back to 10-km radius
4 December	Alert Level 2	10-km radius
1992		
9 July	Alert Level 3	10-km radius
14 July	Alert Level 5	10-km radius
9 December	Alert Level 2	10-km radius

sion of Alert Level 3 interpretation, but Alert Levels 4 and 5 were substantially modified (table 3).

Alert Level 4 will be used only for impending hazardous explosive eruptions **or** for ongoing eruptive activity that involves only small explosions or lava dome extrusions. Alert Level 5 will be used only for large explosive eruptions in progress. Definitive time windows for the occurrence of an eruption, such as "eruption possible within 2 weeks" for Alert Level 3 and "eruption possible within 24 hours" for Alert Level 4, were modified to "within days to weeks" and "within hours to days," respectively.

TRANSMISSION OF WARNING

At the five most active volcanoes being monitored by PHIVOLCS—Mayon, Bulusan, Taal, Hibok-Hibok, and Canlaon—eruption warnings are usually passed through the appropriate DCC. When one of these volcanoes manifests abnormal behavior, PHIVOLCS interprets its changing behavior and decides whether or not to send warnings and, if so, when. As soon as PHIVOLCS decides to issue a warning, it notifies the Office of the President and the national and local DCCs, through Volcano Bulletins and advisories that explain the condition of the volcano and recommended

actions. The DCCs, in turn, transmit the warning to those at risk and respond in various other ways. Although transmission of warnings is officially the responsibility of the DCCs, PHIVOLCS observatory personnel help deliver warnings to nearby inhabitants. Later, PHIVOLCS' main office might release information to the media to clarify and explain the volcano's condition.

This warning procedure was modified in the case of Pinatubo. Warning messages were formulated at PHIVOLCS' main office and transmitted simultaneously through the DCC hierarchy, major national and local newspapers, radio and television stations, nongovernmental organizations (NGOs), and directly to the endangered inhabitants.

Multipath warning transmission has been found to create confusion, duplication, and administrative problems in some situations. This is why, at other monitored Philippine volcanoes, warnings and evacuation advice are passed, as much as possible, through the concerned DCCs. The effectiveness of the modified transmission procedure adopted at Pinatubo was assessed by use of two indicators: (1) consistency between the warning message released by the source (PHIVOLCS) and the message received by the recipients and (2) the time gap between issuance from the source and receipt by the target public.

Table 3. Revised alert levels for Mount Pinatubo (revised December 1992).

ALERT LEVEL	CRITERIA	INTERPRETATION
No alert	Background; quiet.	No eruption in foreseeable future.
1	Low-level seismic, fumarolic, or other unrest.	Magmatic, tectonic, or hydrothermal disturbance; no eruption imminent.
2	Moderate level of seismic or other unrest; positive evidence for involvement of magma.	Probable magmatic intrusion; could eventually lead to an eruption.
3	Relatively high unrest including numerous b-type earthquakes, accelerating ground deformation, increased vigor of fumaroles, gas emissions.	New or renewed eruption possible probably within days to weeks.
4	Intense unrest, including harmonic tremor and (or) many "long-period" (low-frequency) earthquakes and/or dome growth and (or) small explosions.	Magma close to or at the Earth's surface. Large explosive eruption likely, possible within hours to days.
5	Hazardous explosive eruption in progress, with pyroclastic flows and (or) eruption column rising at least 6 km above sea level.	Large explosive eruption in progress. Hazards in valleys and downwind.

Stand-down procedures:

In order to protect against "lull before the storm" phenomena, alert levels will be maintained for the following periods after activity decreases to the next lower level:

    From Alert Level 5 to 4--Wait 12 hours after Level 5 activity stops.
    From Alert Level 4 to 3 or 2--Wait 2 weeks after activity drops below Level 4.
    From Alert Level 3 to 2--Wait 2 weeks after activity drops below Level 3.

Note: Ash fall will occur from secondary explosions for several years after the major 1991 eruption, whenever rainfall and lahars come in contact with still-hot 1991 pyroclastic-flow deposits. These secondary explosions will occur regardless of alert level.

Respondents were asked if they received any eruption warning and (or) evacuation order, and, if so, when. The 1991 survey showed that 71 percent of the 234 respondents knew of the impending eruption before June 9, 1991, the date on which Alert Level 5 was issued, either through their own observation (9 percent) or through their own observation and forewarning from PHIVOLCS, media, local officials, or other people (62 percent). Before June 12, the date of the first large explosive events, 82 percent of the respondents knew of the danger. Subtracting the 9 percent (roughly) who knew only through their own observation, it appears that about one-fourth of the respondents were not reached by warnings before explosive eruptions began. By June 14, 99 percent of the respondents knew of impending danger, from continued warnings and, especially, from observing the preparoxysmal eruptions (Wolfe and Hoblitt, this volume).

Danger zones that were delineated by PHIVOLCS served as basis for the DCC's issuance of evacuation orders. Evacuation orders were transmitted soon after the danger zones were declared by PHIVOLCS, on April 7, June 7, and June 14–15. Those within the 10-km danger zone should have received their order as early as April 7; those within the 10- to 20-km danger zone, on June 7; and those within the 20- to 40-km danger zone, on June 14–15.

Two hundred and two (86 percent) of the 234 respondents received an evacuation order while 14 percent did not. Of the 32 respondents who did not receive an evacuation order, all except two (one from the <10 km danger zone and one from the 10- to 20-km danger zone) were from the more

Table 4.   Receipt of eruption warning and evacuation order by date.

[Household survey, 1991; number of respondents was 234; EW, eruption warning; EO, evacuation order; cum%, cumulative percentage of the 234 respondents]

Date	Number of respondents who received 10-km radius		10- to 20-km radius		20- to 40-km radius		Total EW	Cum %	Total EO	Cum %
	EW	EO	EW	EO	EW	EO				
				10 km radius danger zone declared						
April–May	35	23	60	22	36	9	131	56	54	23
June 1–6	2	2	4	9	8	3	14	62	14	29
				20 km radius danger zone declared						
June 7	0	0	2	8	4	2	6	65	10	33
June 8	0	2	11	23	5	3	16	72	28	45
June 9	2	0	10	3	10	6	22	81	9	49
June 10	0	1	0	4	2	2	2	82	7	52
June 11	0	0	1	0	0	0	1	82	0	52
June 12**	1	10	10	14	30	23	41	99	47	72
June 13	0	0	0	0	0	1	0	99	1	73
				30 km radius danger zone declared						
June 14	0	0	0	4	0	0	0	99	4	75
				40 km radius danger zone declared						
June 15	0	1	0	3	0	7	0	99	11	79
June 16	0	0	0	7	0	0	0	99	7	82
After June 16	0	0	0	0	1	10	1	100	10	86
Received EO	40	39	98	97	96	66	234		202	(86%)
Did not receive EO	0	1	0	1	0	30	0		32	(14%)
Total	40	40	98	98	96	96	234		234	

----
* Includes replies like "weeks to months before eruption"
**Includes replies like "when the first big eruption occurred"

distant, 20- to 40-km radius danger zone (table 4). Most received their evacuation orders on the day or 1 day after the danger zone was declared by PHIVOLCS; others received the evacuation order 2 or more days later, reflecting some delay in the transmission. Delays in transmission were reported in all the danger zones.

In 1992, 94 percent of the respondents learned of the impending eruption on or before July 14, the day PHIVOLCS issued Alert Level 5 (table 5). Almost all the respondents received the warning from multiple sources, with PHIVOLCS, the media, and military officials as the most common transmitters.

Throughout the 1992 activity, PHIVOLCS merely reiterated the continued enforcement of the 10-km danger zone. Therefore, evacuation was recommended only for those who had returned to the <10-km danger zone despite advice against reoccupation of the area. Because the 1992 survey was confined to the 10- to 20-km danger zone, no respondent was expected to have received an evacuation order. Surprisingly, 8 percent of the respondents (all from Pampanga) reported that they received an evacuation order

either from municipal or barangay officials or through the media.

One Municipal Disaster Coordinating Council (MDCC) official admitted that the council decided to order evacuation of barangays beyond the 10-km but within the 20-km radius (including one community that was already living in a relocation center) on the night of July 15. The council even provided vehicles to bring the evacuees to the evacuation centers. According to the town official interviewed, they wanted to play it safe. If the anticipated eruption would be similar to the June 12, 1991 eruption, then the 10- to 20-km danger zone would be affected, so why should they wait?

Respondents from the villages Sapangbato and Margot of Angeles City reported that sometime before July 14, 1992, a popular radio announcer, citing PHIVOLCS as his source, broadcast that Mount Pinatubo would erupt within 72 h. It is interesting to note that the PHIVOLCS alert levels do not include one that indicates that the volcano may erupt within 72 h. The signal with the closest time reference is Alert Level 4, which means that eruption is possible within

Table 5. Respondents who received preeruption warning and (or) false evacuation order.

[Household survey, 1992; number of respondents was 130; all respondents for the 1992 survey were from the 10- to 20-km danger zone]

	Number	Percent of total
Received warning before July 9, 1992	45	35
Received warning between July 9-14	77	59
Did not receive warning	8	6
Total	130	100
Received evacuation order*	11	8

*Based on replies to another question "What type of action advice did you receive?"

24 h. But Alert Level 4 was not used in 1992, as the Alert Level jumped from 3 to 5.

THE PEOPLE'S RESPONSE TO WARNINGS ISSUED

The final test of a warning system's effectiveness is the receipt of and appropriate response to the warning by the target recipients. The warning, no matter how timely, accurate, and precise, will not be of any value unless the recipient of the warning takes appropriate defensive action.

RESPONSE TO 1991 WARNINGS AND EVACUATION ORDERS

One hundred sixty-seven respondents, representing 71 percent of the total number of respondents (234), had forewarning of the eruption and were asked what they did when they learned that the volcano was going to erupt. About 81 percent of those who received forewarning took appropriate action, by evacuating immediately (the response that was called for in the case of those living within the 10-km radius as early as April and those living within the 10- to 20-km radius starting June 7) or by taking some other defensive action (such as preparing for evacuation, convening a meeting, disseminating the warning, seeking further information or confirmation, or observing for further signs, responses that were appropriate at radii of 10 to 20 km from April to June 7 and at radii of 20 to 40 km until June 15). Eight respondents from the <10-km danger zone who should have evacuated immediately merely took other precautionary actions like preparing for evacuation, seeking additional information, or watching out for further developments. Another 13 respondents (from the 20- to 40-km danger zone) overreacted by evacuating before they were ordered to do so.

In contrast, 13 percent of those who were forewarned either waited for the eruption or ignored the warning; and 5 percent ran without definite destination, prayed, or cried without taking any defensive action (table 6).

The responses to *evacuation orders* (a step beyond *warnings*) indicate that all the respondents except five (2 percent) eventually evacuated (table 7). Fifty-eight percent of all respondents evacuated when and as advised, and an additional 11 percent evacuated even without or before receiving evacuation order. But 23 percent delayed evacuation and 6 percent evacuated selectively. Households that evacuated selectively either (1) sent their women, children, sick and elderly to safety while the able-bodied adult males stayed behind or (2) evacuated all together but then allowed some member(s) of the household to return home (usually during daytime).

Table 8 lists some of the reasons given by those who dallied or evacuated selectively. Some thought that the eruption would not be strong enough to affect their place; some were reluctant to leave behind their property and livelihood, especially as it was harvest time; some had no ready means of transport and could not walk long distances; and some believed that their God, Apo Namalyari, would not let them come to harm. One respondent did not want to leave immediately because it was fiesta time.

RESPONSES TO 1992 WARNINGS

In 1992, PHIVOLCS advised the inhabitants of the 10- to 20-km danger zone to avoid the 10-km danger zone (where some residents would otherwise hunt or gather food or tend farm plots), be alert to possible deterioration in the volcano's condition, and prepare for this possibility. About 8 percent of the respondents received an evacuation order (table 5), some from their local officials, others through

Table 6. Response to preeruption warning in each of the danger zones.

[Household survey, 1991; number of respondents was 167]

Response	Number of respondents 10 km zone	10- to 20-km zone	20- to 40-km zone	Total	Percent of total
Evacuated immediately	24	39	13	76	46
Took defensive action prepared for evacuation convened a meeting disseminated information sought further information observed developments	8	27	25	60	36
Did nothing; ignored the warning; waited for the eruption; did not believe warning	4	7	11	22	13
Ran aimlessly, prayed/ cried	1	4	4	9	5
Total	37	77	53	167	100

Table 7. Response to evacuation order, 1991.

[Number of respondents was 234]

Action	Number of respondents 10 km zone	10- to 20-km zone	20- to 40-km zone	Total	Percent of total
Evacuated before any evacuation order	3	7	15	25	11
Evacuated immediately	25	59	51	135	58
Evacuated selectively (did not evacuate together as one household or at the same time, or evacuated but returned when still possible)	1	4	10	15	6
Delayed evacuation (evacuated one or more days after receipt of order)	11	28	15	54	23
Did not evacuate at all	0	0	5	5	2
Total	40	98	96	234	100

radio. Two percent of the respondents evacuated their entire households and another 3 percent evacuated selectively (table 9). The latter were from one community that was ordered by the municipal DCC to evacuate but, instead of complying fully, sent only its women, children, elderly, and sickly to the evacuation centers, where they stayed for about 3 months. Another community, about 15 km from the volcano, was ordered to evacuate and was even provided with

Table 8. Reasons for not evacuating immediately as when advised.

[Household survey, 1991; number of respondents was 69]

Reason	Number of persons giving this response *
Did not want to leave property unattended	25
Thought the eruption would not be strong enough to affect their place	25
It was harvest time; could not leave crops or harvest	15
Had no vehicle, no gas, no light; could not walk far	8
Did not know which evacuation center to go to or how to get there	3
Apo Namalyari would not let them come to harm	3
Stayed behind to look after barangay and things left behind by community members	2
Thought the eruption was over	1
It was fiesta time	1

* Some respondents gave multiple responses

vehicles for evacuation, but it refused to comply because residents believed that Apo Namalyari would protect them.

Had there been a real need for evacuation, the noncompliance of the recipients would have exposed them to danger. But, because the evacuation order was an overreaction on the part of the concerned officials and misinformation on the part of the radio announcer who broadcast "warning" of an imminent eruption within 72 h, their noncompliance led to no harm. Nevertheless, the more appropriate response on the part of the recipients of the evacuation order should have been to seek further information or verification of the order, instead of not to comply.

Among those who took other defensive action (table 9), some overreacted by suspending their normal activities, such as going to school or going to work, for 1 to several days. This overreaction occurred principally while waiting for the "imminent eruption within 72 hours" that was broadcast by the irresponsible radio announcer sometime before July 14, 1992.

Table 9. Response to preeruption warning and false evacuation order.

[Household survey, 1992; number of respondents was 130; all from the 10- to 20-km radius danger zone]

Action	Number	Percent of total
Evacuated		
immediately	3	2
selectively	4	3
Took other defensive action: prepared for evacuation disseminated information sought further information observed/waited for further developments	119	92
Did nothing/ignored warning (false evacuation order)	4	3
Total	130	100

DISCUSSION AND CONCLUSIONS

The fact that most respondents took appropriate defensive actions and evacuated as advised indicates that the warning system worked well enough in 1991. It performed even better in 1992. Nevertheless, some aspects could still be improved. For improvement, the following findings are particularly important:

1. The failure, in 1991, of 18 percent of those who were forewarned to take any defensive action and the delayed or selective evacuation of 34 percent of those who received evacuation orders indicate some failure to stimulate protective action.

Inhabitants who received warnings and evacuation orders but did not take defensive action obviously lacked appreciation of the magnitude of the dangers posed by the volcano. Even those who delayed evacuation or evacuated selectively showed lack of understanding of the gravity of the threat. Either the information drive launched by PHIVOLCS and other disaster response organizations before the eruption did not reach these respondents or the information campaign failed to drive home to them the magnitude of the threat and the urgency as well as the possibility of avoiding the volcano's fury.

It is worth pointing out that all of the respondents contacted by the LAKAS organization showed the exemplary appropriate response. All (except one old man who chose to die rather than leave his home) prepared and evacuated promptly. These respondents recounted that, before the eruptions, the eruption threat and the hazards posed by the volcano had been explained to them by

PHIVOLCS and other officials. They had also been shown the videotape on volcanic hazards produced by the late Maurice Krafft for IAVCEI (Punongbayan and others, this volume).

Some other Aetas did not fare as well. According to one informant (a Protestant pastor), a group of Aetas was about to evacuate (on the 15th of June) along with the others who were fetched by chartered buses. But these people changed their minds when they could not read sign boards on the buses that indicated which should be boarded by Villar residents, by Moraza residents, and so on. After boarding at random and being twice informed that they were in the wrong bus, they were so embarrassed that they decided to return to the mountain and seek refuge in the so-called caves, saying that Apo Namalyari would protect them.

Two other informants said half of the residents of sitio Lomboy were very reluctant to evacuate. Most of them did not want to leave their belongings, crops, and livestock and believed that Apo Namalyari would not let them come to harm. Many of them did not believe that the eruption would be strong enough to affect their places. Some feared lowlanders would burn their crops and homes. A Korean pastor was finally able to convince them to leave, but they put off their departure until the next morning and spent the night in some kind of natural shelter that they called caves. That night, pyroclastic flows buried the caves and killed those inside.

The findings of the survey corroborated news reports about the reluctance or refusal of some endangered inhabitants to leave the danger zones. In April, Aeta tribesmen who refused to move out reportedly said "they were afraid to leave their `precious belongings'" (Alcayde, 1991) or reasoned that they could not leave because their camote crops were due for harvesting. The latter group promised to leave as soon as the harvest was done (Gob, 1991). One Aeta leader who stayed behind was quoted to have said "Mahina lang siguro ang pagsabog dahil hindi naman ito narinig dito sa Belbel" ("The eruption was probably weak because it wasn't heard here at Belbel"), referring to the April 2 explosion (Empeno, 1991).

By June 9, Mayor Richard Gordon of Olongapo City was reported to have dispatched trucks to "clear" barangays within the 20-km danger zone where "there were still some Negritoes who chose to stay where they were, because of their livestock and other properties" (Villanueva and Dizon, 1991). One of the holdouts compromised by sending his family not to an evacuation center but to a place a bit farther away from the volcano, saying "Hindi naman daw kami aabutan ng pagputok ng bundok" ("We heard that the eruption will not reach us") (Cortes, 1991).

Hours after intensified ash emission on June 9, evacuees who earlier refused to leave were reported to have finally climbed into trucks brought by rescuers. Hundreds of Aetas with their belongings and work animals lined the roads, waiting for trucks to bring them down to evacuation centers. But there were still others who refused to be evacuated (Velarde and Bartolome, 1991). The mayor of San Marcelino reported that during rescue operations on June 9, 10 Aeta families opted to stay, believing that the eruption was nothing serious—"para lang daw malakas na bagyo'yan" ("it is just like a strong typhoon") (De Villa, 1991).

As late as June 11, Zambales Governor Deloso reported that some 200 tribesmen still refused to leave their settlements in Barangays Moraza, Nacolcol and Maguisguis. He added that the men may have wanted to stay to harvest their palay and camote crops so they could repay their loans to the Land Bank of the Philippines (anonymous, 1991a). An Aeta in Moraza who defied the evacuation order and stayed on to keep an eye on his home, farm, and carabao (water buffalo) was quoted to have said "We fear the volcano, but if we left our carabaos, we'll die" (Morella, 1991). In another barangay, Belbel, the barangay captain reported that some 252 tribesmen also refused to leave their homes (Anonymous, 1991b).

The anecdotes from the survey informants, and these news reports, highlight some of the communication and cultural problems with which the warning system had to contend. These problems indicate a need for hazard-awareness promotion that is more intensive and broader in outreach than was possible during the 2-month period from the time the volcano started showing signs of restiveness up to the time of the major explosions.

2. In 1991, the failure of 18 percent of the respondents to receive preeruption warning before June 12, the failure of 14 percent of the respondents to receive an evacuation order at any time (even after the June 15 eruption), and delay in the receipt of evacuation order by 26 percent of the respondents all indicate some deficiencies in the transmission system. In the <10- and 10- to 20-km danger zones, it is possible that the transmission network did not reach the most remote areas or the communities that were on the move.

Again, it is worth noting that in the case of the communities with a grassroots organization like LAKAS, warning transmission was total despite difficulties of transportation and terrain. The organization ensured that everyone received the warning and evacuation order.

The official warning system was unable to reach all residents of the large, 20- to 40-km danger zone during the short, hectic time that that zone was in effect (June 15–18). Communities in this zone are easily accessible but too numerous to reach in such a short time. Broadcast radio served as the principal channel for warning communities in this area. The warning communication system was improved in 1992 by the distribution of two-way radios to barangay leaders.

3. In 1992, the receipt of a false evacuation order by 8 percent of the respondents is a clear case of discrepancy between the warning message released by the source and the message transmitted to the concerned inhabitants. Issuance of this evacuation order for communities outside the official danger zone may have been a simple case of caution or overreaction on the part of the local officials. However, incentives for evacuation such as the availability of relief and emergency resources and the usual outpouring of sympathy might also have inspired the move.

The overreaction may also be traced, at least in part, to the warning messages released by PHIVOLCS. The United Nations Disaster Relief Office (UNDRO, 1986, 1987) advised, among other things, that warnings should be consistent in content and as specific as practicable in their information concerning the magnitude of the event, the place at which it is expected, and the time when it will occur. The 1991 alerts were indeed specific—in terms of expected magnitude, areas likely to be affected, and time of occurrence—but they were specific to the 1991 eruptive activities. The application of these same alerts to the less explosive and less hazardous 1992 events may have given rise to undue concern and inspired exaggerated media reporting.

After the 1992 experience, revision of the alert levels was in order. The revision removed the implication that eruptions could be predicted to the nearest hour or day, especially at volcanic systems in which the vent was already open. The original alert levels focused mainly on the imminence or occurrence of a large explosive eruption. The revised alert levels allow for differentiation of large and small eruptions. Missing still are the recommended actions for each of the alert levels and each of the danger zones. How to incorporate these without making the scheme of alert levels inflexible and too specific remains to be studied.

One specific aspect of the PHIVOLCS warning messages that went against UNDRO advice was the inconsistency in the danger zones associated with the various alert levels. In 1991, Alert Level 4 was associated with a 20-km danger zone, and Alert Level 5 was associated with both a 20-km and a 40-km danger zone. A 40-km danger zone was declared because there was concern about pyroclastic flows from a big eruption and the possibility that a caldera might form. The alert level-danger zone association, though not intentionally established, lingered, so that when Alert Levels 4 and 5 were released

in 1992, an understandable reaction was to react as in 1991 and evacuate the 10- to -20-km danger zone.

There are at least two options for rectifying this source of potential misunderstanding. One is to explore the possibility of striking some correspondence between alert levels and danger zones. The assumption is that it is possible to determine the areas likely to be endangered by each type and magnitude of activity referred to in each alert level. However, eruptions vary in style and intensity, so such a correspondence may not be feasible. The alternative is to consciously dissociate the alert levels from danger zones, define a permanent danger zone, and keep other danger zones open-ended and adjustable.

Another possible improvement in the alert level scheme would be to reword the "Interpretations" and specifically the phrase "eruption is possible within 2 weeks [or 24 hours]." That phrase was variously interpreted to mean "eruption will occur 2 weeks [or 24 h] hence" or that "an eruption *would* occur within 2 weeks [or 24 h]." Ironically, the wording was actually chosen to avoid making specific predictions. Rather, it was meant to define a window in which an eruption was *possible* and to indicate disappearing margins of safety. Thus, at Alert Level 2, an eruption within the next 2 weeks was judged unlikely, but at Alert Level 3, this was no longer true. At Alert Level 3, an eruption was unlikely within less than 24 h, but at Alert Level 4 all reassurances of safety was gone—an eruption could occur at anytime (C.G. Newhall, written commun., 1994). The Pilipino translation of the phrase "eruption is possible within…"— "maaaring mangyari ang pagputok sa loob ng…" would have conveyed the message that the authors meant to convey. A Pilipino version of the alert level scheme could be pilot tested the next time one of our volcanoes becomes restive.

The broadcast of a warning that an eruption was imminent within 72 h, falsely attributed to PHIVOLCS, triggered discussions on the wisdom of the modified warning transmission procedure adopted at Pinatubo. Critics of the multipath transmission procedure claimed that had PHIVOLCS stuck to the DCC channel instead of directly dealing with the media, reporters would not have been able to cite it as their source for their false or sensationalized reports. The traditional DCC channel would certainly minimize the warning source's need to deal with the media and make it easier to pinpoint responsibility for erroneous reporting. But it would also limit the area that could be reached, given a short lead time for warning dissemination. The institution of an emergency broadcast system might provide a mechanism for effectively involving media in warning transmission. The concept is not new, and there have been attempts to establish such a system. The idea is worth reviving. An emergency broadcast network could be identified, with media representatives officially identified and properly trained to handle warning and emergency response operations. This move might not eliminate exaggerated or fabricated reports but could minimize the effect of such reports if people listen to and believe only the official transmitters.

4. The fact that 94 percent of the respondents in 1992 knew of the impending eruption before Alert Level 5 was released and that 92 percent responded appropriately indicates improvement in warning transmission as well as in inducing optimal response. However, because no evacuation was required, the improvement in the percentage of appropriate response may be more apparent than real. The question remains, would this percentage be as large should there be a call for evacuation of areas beyond the 10-km radius?

Mount Pinatubo's continuing activity provides an excellent opportunity to continue the development of the eruption warning system. We hope the experience in evolving a suitable warning system for Pinatubo will yield some valuable lessons for issuing warnings at other active volcanoes.

ACKNOWLEDGMENTS

The authors are deeply indebted to Dr. Dennis Mileti and Dr. C. Dan Miller, whose comments and suggestions did not only enrich the final output but provided a rich source of learning for the authors as well, and to Dr. Chris Newhall for his patience, relentless prodding, and meticulous attention to detail.

REFERENCES CITED

Alcayde, Jerry, 1991, 1876 families move out of volcano area: The Philippine Star, April 22, 1991.

Anonymous, 1991a, 200 Aetas won't leave endangered settlements: Philippine Daily Inquirer, June 11, 1991.

Anonymous, 1991b, 'Big bang' looms, Yanks flee Clark: Philippine Daily Inquirer, June 11, 1991.

Cortes, Joseph, 1991, A roof or a wall they carried along, to remind them of home: The Manila Times, June 9, 1991.

De Villa, Arturo, 1991, Aetas may go hungry: Daily Globe [Manila], June 10, 1991.

Empeno, Henry, 1991, Mt Pinatubo already spewing lava, (say) Aetas: The Manila Times, April 23, 1991.

Gob, Fely, 1991, Rains bring death to Pinatubo evacuees: Daily Globe[Manila], April 22, 1991.

Morella, Cecil, 1991, Volano eruption displaces Aetas: Manila Bulletin, June 11, 1991.

National Statistics Office, 1990, 1990 census of population and housing, Report No. 2–74C (Pampanga); Report No. 2–95C (Tarlac); and Report No. 2–99C (Zambales), Population by City, Municipality and Barangay: NSO, Manila.

Philippine Institute of Volcanology and Seismology (PHIVOLCS), variously dated, Daily Volcano Bulletins, Pinatubo Volcano, April 7, 1991–December 1992: Quezon City, PHIVOLCS.

Punongbayan, R.S., Newhall, C.G., Bautista, M.L.P., Garcia, D., Harlow, D.H., Hoblitt, R.P., Sabit, J.P., and Solidum, R.U., this volume, Eruption hazard assessments and warnings.

Sabit, J.P., Pigtain, R.C., and de la Cruz, E.G., this volume, The west-side story: Observations of the 1991 Mount Pinatubo eruptions from the west.

United Nations Disaster Relief Organization (UNDRO), 1986, Social and sociological aspects, in Disaster prevention and mitigation, v. 12: New York, United Nations.

———1987, Public information aspects, in Disaster prevention and mitigation, v. 10: New York, United Nations.

Velarde, Cherry, and Bartolome, Noel, 1991, Pinatubo erupts: Malaya [Manila], June 10, 1991.

Villanueva, Marichu A., and Dizon, Romy, 1991, Mass evacuation starts in Pinatubo: Manila Standard, June 9, 1991.

Wolfe, E.W. and Hoblitt, R.P., this volume, Overview of the eruptions.

APPENDIXES

Appendix 1.   Computation of sampling size, household survey, 1991–92.

The following formula was used for determining the sample size:

$$n = \frac{NZ^2\text{p}(1\text{-p})}{Nd^2 + Z^2\text{p}(1\text{-p})}$$

where n = sample size;
  N = population,
  Z = 1.96 for a reliability of 95 percent,
  p = 0.50 (to produce the largest possible sample size),
  d = the maximum error deemed acceptable: 0.09-0.10.

N, based on 1990 census,

for 10- and 10- to 20-km radius = 7,187 households.
  20- to 40-km radius = 244,141 households.

$$s1 = \frac{7,187\,(1.96)^2 0.5(1\text{-}0.5)}{7,187(.09)^2 + (1.96)^2 0.5(1\text{-}0.5)} = 116.64$$

$$s2 = \frac{244,141(1.96)^2\,0.5(1\text{-}0.5)}{244,141(.1)^2 + (1.96)^2 0.5(1\text{-}0.5)} = 96$$

The sample compositions were as follows:

1991:

from the 10- and 10- to 20-km zones:
  116 household respondents and 22 key respondents;

from the 20- to 40-km zone:
  95 household respondents and one key respondent.

1992:

From the 20- to 40-km zone:
  116 household respondents and 14 key respondents.

Appendix 2.   Distribution of respondents, 1991 and 1992.

PROVINCE	Town	Barangay	Number of Households	Number of Respondents 1991		1992	
				Household	Key	Household	Key
Within 10-km radius of the summit							
ZAMBALES							
	Botolan						
		Moraza[1]	112	10	3		
		Villar[1]	230	20	7		
		Belbel	124	2	2		
Subtotal			466	32	12		
Within the 10- to 20-km radius (18 out of 22 barangays)							
ZAMBALES							
	Botolan	Burgos	145	9	1		
		Maguisguis	278	4	2	7	1
		Nacolcol	143	2	2	4	1
		Parel	145	2	1	4	1
	San Marcelino						
		Aglao	387	6	2	9	1
		Rabanes	396	3	-	5	1
PAMPANGA							
	Porac	Babo Pañgulo	278	4	-	6	1
		Camias	101	2	-	2	1
		Diaz	66	1	-	1	-
		Planas	313	5	-	7	1
		Cangatba	101	2	-	2	1
		Pio	399	6	-	10	1
	Angeles City						
		Margot	431	6	1	10	1
		Sapangbato	1,793	27	1	41	1
	Floridablanca						
		Nabuclod	154	2	-	3	1
Subtotal			5,351	84	10	116	14
Within the 20-40 km radius (45 out of 108 barangays)							
PAMPANGA							
	Mabalacat	Macapagal	227	1			
	Angeles City						
		Amsic	704	3			
		Balibago	6,500	25			
		Lourdes	1,245	5			
	Magalang	San Roque[2]	296	1			
	Porac	Poblacion[2]	364	2			
		San Jose Mitla	217	1			
	Floridablanca						
		Santa Monica	295	1			
		San Pedro	430	2			
		Carmencita	244	1			
		Cabangcalan	395	1			
		San Ramon	251	1			
	Lubao	Del Carmen	273	1			
	Sasmuan	Batang 2nd	222	1			
	Santa Rita	San Vicente	214	1			

Appendix 2.   Distribution of respondents, 1991 and 1992—Continued.

	Guagua	San Miguel	685	2			
		San Rafael	293	1			
	Bacolor	Cabetican	849	3			
		Santa Barbara	507	2			
		Tinajero	855	3			
	San Fernando						
		Poblacion[2]	230	1			
ZAMBALES							
	Botolan	Paudpod	223	1			
		Poonbato	483	2			
		Palis	79	1			
		Taugtod	130	1			
		Bancal	132	1			
		Paco	382	2			
		Tampo	297	1			
		Malomboy	208	1			
	Castillejos						
		Del Pilar	682	3			
		Nagbunga	262	1			
	San Marcelino						
		La Paz	226	1			
		Linasin	366	1			
		Linusungan	221	3	-	5	1
	San Antonio						
		San Esteban	282	1			
	San Narciso						
		Alusiis	261	1			
		Omaya	152	1			
	Subic	Asinan	124	1			
		Poblacion[2]					
TARLAC							
	Capas	Bueno	203	1			
		Santa Juliana	454	2			
		Patling	1,823	7			
		Maruglu	201	1			
	Bamban	San Nicolas	965	4			
		Poblacion[2]					
		Santo Nino	195	1			
			------	-----			
Total			23,426	95			

[1] The centers of Moraza and Villar are just outside the 10-km radius, but many of their sitios are inside that radius. Hence, they are included in this "close-in" category.

[2] Poblacion = central area of town

Appendix 3. Interview schedule.

Household survey on response to eruption warning, 1991.

Name:_____
Barangay before eruption:_____
                    now:_____
Sex:_____ Civil Status:_____ Age:_____
Educational Attainment:_____
Occupation:_____
Community group: Uplander_____   Lowlander:_____
Household size:_____

_____

For Barangay Officials only

Number of households in the barangay (before eruption):_____
Primary occupation of barangay residents:_____
Number of barangay residents who evacuated:_____
                    who did not evacuate:_____
Where (evacuation center/s) barangay residents took refuge:___
Reasons given by those who did not evacuate:_____
How many barangay residents were injured?_____
     How?_____
     How many died?_____How?_____

For all interviewees

How long have you been living in your (pre-eruption) barangay?
When did you first learn that Pinatubo was going to erupt?
     How?_____

For those who observed precursory signs,
     What signs did you observe?   When?

     _____   _____
     _____   _____

For those who received warning,
     Who warned you of the impending eruption?_____
     When you learned of the impending eruption, what did you do?
     Did you receive an evacuation order?_____

For those who received an evacuation order,
     When did you receive the order?_____
     Who gave you the order to evacuate?_____
     Did you evacuate?_____ When?
          Immediately when and as advised:_____
          One day after receiving the order:_____
          Two or more days after receiving the order:_____
     Did your whole household evacuate?_____
     Did you all stay at the evacuation center/refuge until the alert was
     lifted?_____
     If not, why?_____
     Where did you take refuge?/What evacuation center did you go
     to?_____
     What was your means of transportation to the refuge/evacuation
     center?_____
     Who/what organization(s) helped you evacuate?_____
     Did your household sustain any casualties?_____
     How many_____What happened to him/them?_____

For those who have returned to original residence,
     When did you return?_____
     Did you receive advice on whether or not and when it would be safe to
     return?_____
     From whom?_____

Name of Interviewer   _____ Date _____

Assessment and Response to Lahar Hazard around Mount Pinatubo, 1991 to 1993

By Richard J. Janda,[1] Arturo S. Daag,[2] Perla J. Delos Reyes,[2] Christopher G. Newhall,[1] Thomas C. Pierson,[1] Raymundo S. Punongbayan,[2] Kelvin S. Rodolfo,[3] Renato U. Solidum,[2] and Jesse V. Umbal[2][3]

ABSTRACT

Lahar hazard at Pinatubo is a function of prodigious sediment yield from Pinatubo's upper and middle slopes and the sediment storage capacity in the adjoining lowlands. Both are diminishing but at mismatched rates. Sediment yields set world records during the first three posteruption years, and yields in the Balin Baquero-Bucao and Marella watersheds may do so for several years more. In general, sediment yields peaked early and are decreasing rapidly in east-side watersheds, where the volume of 1991 pyroclastic-flow deposit is relatively low, deposits and streams are confined in a few steep-walled valleys, thin ash fall from secondary explosions is common, and vegetation recovery is fast. Sediment yields peaked later and are decreasing slowly in west-side watersheds, where pyroclastic-flow deposits are more voluminous, numerous small streams drain a broad, gently-sloping, unconfined pyroclastic apron, and vegetation recovery was initially slow.

We anticipate that slightly more than 3 cubic kilometers of sediment will move from the volcano's slopes into surrounding lowlands. By late 1993, almost two-thirds of this amount had already arrived in the lowlands; most of the remaining third will be from the Balin Baquero-Bucao and Marella watersheds. Sediment yield from the Gumain watershed virtually stopped in 1992, and that in the Abacan stopped because its 1991 headwaters were recaptured by the Sacobia in April 1992; 1991–93 yield in the Sacobia-Abacan-Pasig system was about three-fourths of the expected total.

Optimism that the worst is past, especially for the east side of Pinatubo, should be tempered by three factors. First, channels in the middle reaches of alluvial fans are filled, so the threat of lahars to several populated areas remains high. On unconfined alluvial fans of the Pasig-Potrero, Marella-Santo Tomas, and, perhaps, the Sacobia Rivers, sediment must continue to spread beyond present river channels, into catch basins, if the overall problem is to diminish. Towns on these fans that are beyond the reach of the most sediment-laden flows (debris flows) could be protected from more dilute hyperconcentrated flows and floods by relatively low ring dikes if flows are allowed to spread onto surrounding agricultural land, but that solution has not, to date, been politically acceptable. Second, unusually heavy rainfall, if sustained for several days or more, would temporarily reverse the declining sediment yield and cause serious over-bank lahars. Third, erosion and incorporation of sediment predating the 1991 eruptions into current lahars will add an as-yet-uncertain volume to deposits, and reworking of lahar deposits themselves will move sediment problems downstream and fill some distal reaches of channels that have not heretofore been filled.

Throughout the first 3 years of the Pinatubo crisis, assessments and warnings of lahar hazard, followed by mitigating actions, saved many lives and some property. Long-range warnings, including hazard maps and briefings, identified communities at high risk and led some residents to move transportable belongings—sometimes even houses—to safe, high ground. Immediate warnings from manned watchpoints, supplemented by rain gauges and flow sensors, alerted remaining people to flee villages when their lives were at risk. Information about the nature and magnitude of lahar hazard was also an important basis for planning projects, both sociopolitical and engineering, to help residents through this difficult time.

Despite some notable successes, not all warnings were perfect, nor were all warnings heeded. An early excess of false alarms made residents of some areas doubt all alerts; in other areas, dikes and other sediment-control structures offered a false sense of security that delayed evacuations. Some costly but futile efforts at lahar mitigation could have been avoided. In some of these instances, better scientific information and better presentation of that information could have reduced unnecessary losses; in other instances, competing political, economic, and social factors limited the acceptance of scientific information.

[1] U.S. Geological Survey.

[2] Philippine Institute of Volcanology and Seismology.

[3] Department of Geological Sciences M/C 186, University of Illinois at Chicago, 801 W. Taylor St., Chicago, IL 60607–7059.

Figure 1. *A*, Damage from lahars in Barangay Lourdes, Bamban, as seen looking upstream on January 29, 1992. Damage occurred in August 1991, when breakout of a newly impounded lake just upstream breached the dike that had earlier protected Lourdes. In the background, mounds of sediment have been bulldozed from the river channel in an attempt to fill the breach. *B*, Areas that were severely affected by lahars, 1991–93.

INTRODUCTION

Lahars from Mount Pinatubo have been flowing into densely populated areas of central Luzon since the major eruption of June 1991, taking a small toll of lives but causing enormous property losses and social disruption (fig. 1) (Mercado and others, this volume; C.B. Bautista, this volume). Bank erosion and thick lahar deposits have left more than 50,000 persons homeless, and flooding and isolation have affected more than 1,350,000 people in 39 towns and 4 large cities. More than 1,000 km^2 of prime agricultural land is affected by or at risk from lahars, flooding, and siltation. The lahar problem also delays economic recovery, as some potential investors wait until the lahar problem subsides.

Loss of life, property losses, and low investor confidence can be minimized by accurate and timely assessment of the lahar threat and corresponding prudent actions by those at risk. In this paper we examine approaches to the lahar hazard used in the 3 years following the 1991 eruption, with the goal to understand what was done well and what might yet be improved. We describe the evolution of our scientific understanding of the lahar hazard at Pinatubo, note specific warnings that were issued, and describe how those warnings were (and can still be) used to mitigate risks from Pinatubo lahars.

LAHAR HAZARD ASSESSMENT TEAMS

Preeruption and syneruption assessments of lahar hazard at Pinatubo involved scientists from the Philippine Institute of Volcanology and Seismology (PHIVOLCS), the University of Illinois at Chicago, and the U.S. Geological Survey (USGS). The Pinatubo Lahar Hazards Taskforce (PLHT) was formed under the leadership of K.S. Rodolfo to assess and warn of lahar hazards, principally on the west side of Pinatubo. Members of the team were drawn from the University of Illinois at Chicago, PHIVOLCS, the Philippine Mines and Geosciences Bureau (MGB), and the University of the Philippines' National Institute of Geological Sciences (UP-NIGS). In 1992, the PLHT became the Zambales Lahar Scientific Monitoring Group (ZLSMG). Lahar hazard assessment on the east side of Pinatubo, from 1991 to the present, was handled principally by PHIVOLCS with assistance from USGS scientists. This division of responsibility between the two groups and the two sides of the volcano began as a matter of logistical necessity and organizational autonomy; over time, the two teams may be merged into one.

Within days after the climactic eruption, PLHT organized systematic observations of lahars from a watchpoint at Dalanaoan, San Marcelino (15 km southwest of Pinatubo, along the Marella River; fig. 1*B*), and, while road conditions still permitted, from a watchpoint at Malumboy, Botolan (27 km northwest of Pinatubo, along the Bucao River). At the same time, topical but less systematic observations of lahars were begun on the east side of Pinatubo by PHIVOLCS and USGS scientists, who were often interrupted by the continuing eruptions. By 1992, both teams were making systematic observations and meeting to exchange insights as often as possible.

Both PHIVOLCS and PLHT have assessed lahar hazards and advised government officials on proposed mitigation measures. Their analyses have been similar, though PLHT and its successor, ZLSMG, have been more

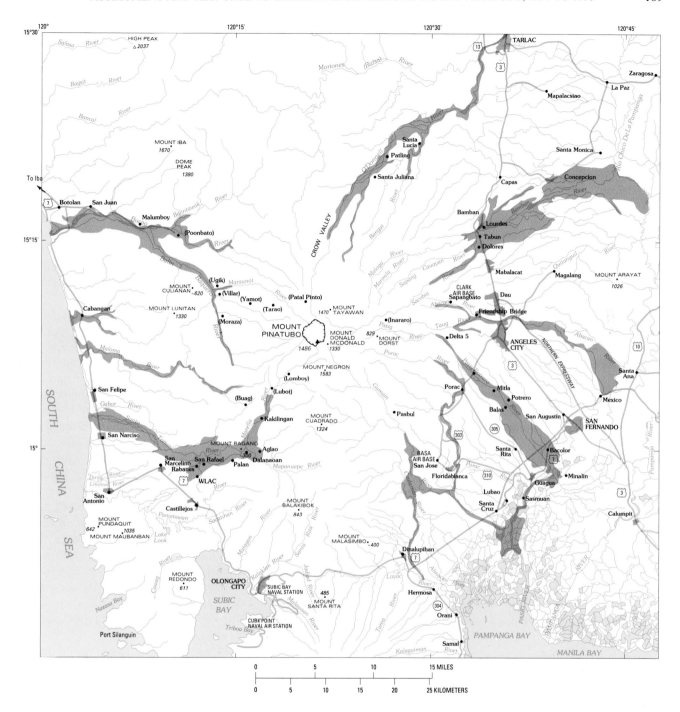

Figure 1.—Continued.

outspoken on proposed engineering countermeasures. At times, relatively minor scientific differences have arisen between the two groups, and media attention has focused more on the messengers than on constructive portrayal of the scientific differences.

Other organizations have made independent lahar hazard assessments, including the Philippine Bureau of Soils

and Water Management and the U.S. Army Corps of Engineers, as noted below.

Warnings by PHIVOLCS and others are provided to the Office of the President, the National Disaster Coordinating Council (NDCC), the Regional Disaster Coordinating Council for Region III (RDCC-III), the Department of Public Works and Highways (DPWH), the Department of

Social Welfare and Development (DSWD), and to Provincial, Municipal, and Barangay Disaster Coordinating Committees (PDCC's, MDCC's, and BDCC's) (fig. 2).

DEFINING THE THREAT

A lahar is a rapidly flowing mixture of volcanic rock debris and water, typically with 40–90 percent sediment by weight, and thus having a consistency ranging from muddy water to a dense slurry. We recognize two types of lahars at Pinatubo, defined in greater detail by Pierson and others (this volume). The first, **debris flow**, has high viscosity, notable yield strength, and sediment concentrations typically greater than 60 percent by volume. The second, **hyperconcentrated flow**, has moderate viscosity, low yield strength, and sediment concentrations of 20 to 60 percent by volume.

All lahars from Pinatubo are caused, directly or indirectly, by heavy, seasonal monsoon rainfall that is enhanced by rain from tropical cyclones. The rainy season can begin as early as May and continue through November; monsoonal rains normally coincide with the typhoon season and are heaviest during June through September, but early- or late-season tropical cyclones can extend the lahar season. The June 1991 eruption provided the missing ingredients for lahars: a severely disturbed landscape in which runoff would be high and an abundant supply of loose, easily erodible sediment. Most lahars of Pinatubo begin as surface runoff of rainfall; a few begin by the sudden release of standing water that has ponded on or against the margins of other deposits. The flowing water rapidly entrains loose sediment from channel beds and banks and is transformed into a sediment-rich flow with the sediment concentrations noted above.

These generalizations, based mainly on climatic data for the Pinatubo region and on experience at Mayon and at other volcanoes around the world, were understood at the time of the June 1991 eruption. In addition, we needed to learn:

What was the approximate range of velocities, discharge, sediment content, temperature, and flow behavior to be expected for the Pinatubo lahars?

How would lahars vary from one watershed to the next?

What critical amounts of rainfall would be needed to trigger lahars of various sizes?

Preliminary data about the character of Pinatubo lahars came quickly. The flows of June 15, though not observed by our scientific team, were soon reconstructed (Major and others, this volume), and additional flows were soon observed (Pierson and others, this volume; Rodolfo and others, this volume; K.M. Scott and others, this volume). A working hypothesis was sketched out for three types of

Pinatubo watersheds, based on their average slopes and intensity of eruption impact (see K.M. Scott and others, this volume). Other workers used oblique aerial photographs and preeruption topographic maps to estimate the volume of new pyroclastic deposits (then estimated to be 5–7 km³, now judged to be 5.5±0.5 km³, W.E. Scott and others, this volume), and Pierson and others (1992) estimated that between 40 and 50 percent of this debris would be transformed into lahars over the next decade. Clearly, the lahar problem would be massive and persistent.

Semiquantitative details about individual lahars were filled in during the first lahar season, comprising the period from June through November 1991. Table 1 summarizes observations of lahars in the Sacobia River; elsewhere around Pinatubo, lahars have roughly similar characteristics, scaled upward or downward according to the size of the watershed and the volume of sediment that can be entrained.

Information concerning the critical rainfall necessary to trigger lahars of various sizes came more slowly, as we had to install a network of rain gauges and observe enough flows to draw valid conclusions. During 1991 and 1992, about 6 mm of rainfall over 30 min (0.2 mm of rain per minute) was sufficient to trigger lahars in the Marella and Sacobia watersheds, and rain at double that rate triggered medium to large lahars, especially if that rainfall was sustained for several hours or if there had been other rain in the preceding days. Figure 5 of Tuñgol and Regalado (this volume) illustrates lahar-triggering thresholds for various intensities and durations of rainfall.

Total rainfall of about 2,000 mm (station MSAC) during 1992 resulted in production of about 1×10^8 m³ of lahar (water+sediment) (M.T.M. Regalado, unpub. data, 1994; Tuñgol and Regalado, this volume). For amounts of rainfall experienced in 1991 and 1992, the total volume of a lahar appeared to increase linearly with rainfall during the event (Regalado and Tan, 1992; Tuñgol and Regalado, this volume). In the Sacobia-Pasig watershed during 1992, lahar yield (*V*, sediment+water) was about 1×10^3 m³/mm of rainfall/km² of upland watershed. The same relation held for individual storms and for the lahar season as a whole. Lahar yields per millimeter of rainfall per square kilometer were generally lower in other watersheds; details of lahar yields will be discussed in the section "Long-Term Warnings: Hazards Assessment and Hazards Maps."

By 1992 and 1993, we began to detect several **changes** that could be projected as trends—in total sediment yield, sediment yield normalized for watershed area and rainfall, channel filling, upstream and downstream migration of the principal reaches of deposition, and other parameters. Most of these are noted and discussed later in this chapter; perhaps the most notable of these changes are the decreasing sediment yields normalized for rainfall and watershed area. Experience elsewhere suggests that this decrease will probably be exponential (Pierson and others, 1992), though

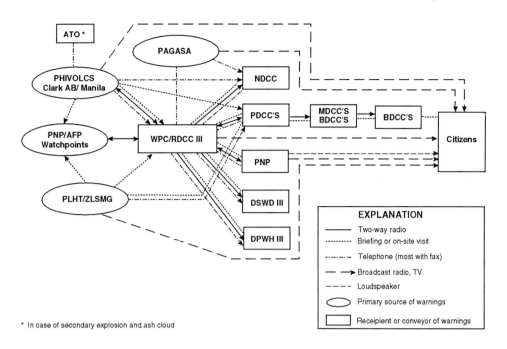

* In case of secondary explosion and ash cloud

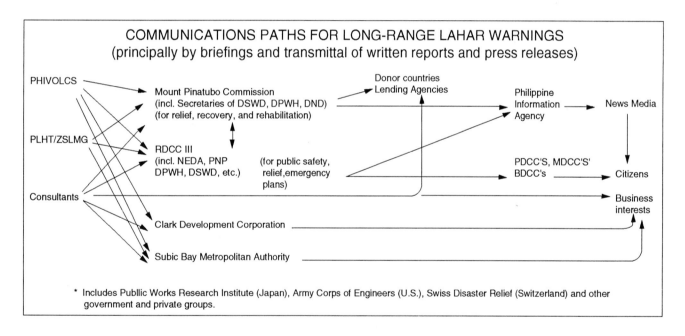

Figure 2. Communication paths for lahar warnings at Mount Pinatubo. *A*, Short-range warnings. *B*, Long-range warnings (principally by briefings, transmittal of written reports, maps, and press releases). Abbreviations: ATO, Air Transport Office, Manila; PHIVOLCS, Philippine Institute of Volcanology and Seismology; PNP-AFP, Philippine National Police-Armed Forces of the Philippines; PLHT/ZLSMG, Pinatubo Lahar Hazard Taskforce/Zambales Lahar Scientific Monitoring Group; PAGASA, Philippine Atmospheric, Geophysical, and Astronomical Administration; WPC/RDCC III, Watch Point Center/Regional Disaster Coordinating Council of Region III (operating agency=Office of Civil Defense, Region III); NDCC, National Disaster Coordinating Council (operating agency=Office of Civil Defense); PDCC, Provincial Disaster Coordinating Council; MDCC, Municipal Disaster Coordinating Council; BDCC, Barangay Disaster Coordinating Council; DPWH, Department of Public Works and Highways; DSWD, Department of Social Welfare and Development.

Table 1. Summary of lahar observations, Sacobia River, 1991–93.

Surface flow velocities at Mactan (upper end of Clark Air Base): 2–15 m/s.

Surface flow velocities at Maskup (lower end of Clark Air Base): 1–5 m/s.

Traveltime from Mactan to Maskup (8 km): 20–70 min (avg. velocity of flow front and peak=1.9 to 6.7 m/s, with large debris flows traveling fastest).

Peak discharge at Mactan: 10–2,000 m^3/s (possibly as high as 5,000 m^3/s in 1991).

Peak discharge at Maskup: 1–400 m^3/s measured, probably reaching as high as 1,000 m^3/s in large flows.

Attenuation of flows from Mactan to Maskup: small to moderate size are attenuated by 60–90 percent; large flows are attenuated by 30–50 percent.

Sediment contents: 40–70 percent by volume, highest in large debris flows.

Wet sample bulk density: 1.5–2.1, highest in large debris flows.

Temperatures of hot lahars: typically 50–85°C; highest recorded, 98°C.

Both erosion and deposition occur at Mactan and at Maskup, with net deposition at both locations. During single flows, there can be several meters of erosion or deposition or can even be alternation of several meters each of erosion and deposition. Strong lateral erosion by hyperconcentrated flows.

Debris flows typically had peak discharges of several hundred to 1,000 m^3/s and contained about 60–65 percent (rarely, 70 percent) sediment by volume; hyperconcentrated flows typically had peak discharges of several tens to several hundred m^3/s and contained about 50 percent sediment by volume. Hyperconcentrated flows are numerically more common, but the large debris flows carry a large part of the sediment that is deposited downstream.

perhaps not all the way back to preeruption levels (Pierson and Costa, 1994). Figure 3 contains computer-fit, manually adjusted curves for that decrease, projected back to preeruption yields of roughly 10^5 m^3/km^3/yr, still high in comparison to stable watersheds.

Lastly, we checked to see if our perception of the threat was consistent with what the geological record told us about the lahars from past eruptions of Mount Pinatubo, and it was. The general topography of gently sloping alluvial fans around Mount Pinatubo, soils maps of central Luzon, and inspection of exposures all suggest that much of the central valley of Luzon and all of the Santo Tomas plain owe their fertile soils and sediment to lahars and related floods of Pinatubo.

PUBLIC EDUCATION AND WARNINGS

PHIVOLCS and PLHT began to provide **general public education** about lahars, **long-range warnings** about what might be expected in coming months and years, and **short-range warnings** about lahars expected within minutes to days.

EDUCATION ABOUT THE THREAT

Before June 15, 1991, residents and local leaders of the Pinatubo area had only slight familiarity with lahars that had occurred at Mayon Volcano in 1984 and at Nevado del Ruiz, Colombia, in 1985. Fortunately, a videotape that was made by the late Maurice Krafft for the International Association of Volcanology and Chemistry of the Earth's Interior (IAVCEI), largely in response to the lahar tragedy in 1985 at Nevado del Ruiz, showed graphic, frightening, but realistic images of lahars and their effects on people. We showed this video to many decisionmakers and citizens, and the result was a noticeable increase in awareness and concern.

As described by Major and others (this volume), the first lahars formed from rainfall on preclimactic eruption deposits on the afternoon of June 14, almost 24 h before the climactic eruption. Much larger lahars occurred during that eruption, triggered by rain from passing Tropical Storm Yunya. On and shortly after June 15, people learned firsthand that lahars were raging torrents that destroyed bridges, eroded river banks, and flooded fields where riverbanks were overtopped. But most people still had little idea of the magnitude or persistence of the hazard that would be with them for years to come and had no concept of the insidious sediment buildup in channels that would soon lead to many more overbank lahars and floods.

Use of the term "lahar" was vigorously promoted on June 15, 1991, and in subsequent days by two of us (K.S. Rodolfo and J.V. Umbal), who were concerned that the previously introduced term "volcanic mudflow" misrepresented the material transported (mostly sand and coarse debris rather than mud), and, for that reason, gave people a dangerously understated sense of the threat (a similar concern was raised by Voight, 1988, 1990). Another purpose of introducing the term "lahar" was pedagogical: a catchy, unfamiliar term might (and did) get special attention. Indeed, the term lahar has now received so much attention that it has become a metaphor for practically any disaster in the Philippines.

Scientists targeted three groups for special education: news reporters, public officials (including civil defense officials and engineers), and police and army personnel assigned to lahar watchposts on the slopes of Mount Pinatubo. Instruction of the news media was provided in the course of day-to-day interviews and field trips, as reporters struggled to understand lahars. Educational posters were

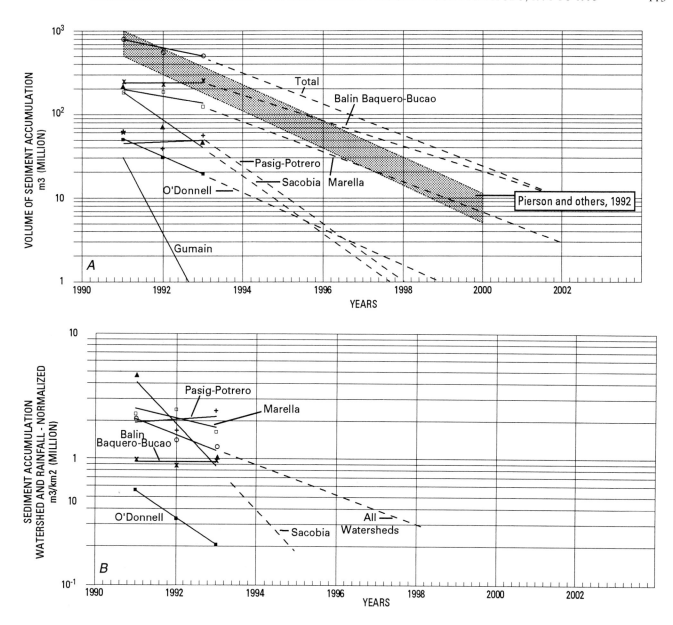

Figure 3. General decreases in sediment accumulation from 1991 to 1993. *A*, Volume of sediment accumulation (approximates sediment yield), per year. Symbols and lines for lahar years 1–3, measured values; dashed lines, projections, discussed in text. Shaded area, projection of Pierson and others (1992) for Pinatubo as a whole. *B*, Rates of sediment accumulation normalized for watershed area. *C*, Rates of sediment accumulation normalized for watershed area and rainfall.

prepared, as were, later, several videos and a primer booklet on Pinatubo lahars.

Formal press conferences on lahars were rare (perhaps, too rare); one-on-one and informal group interviews with scientists were common, as was press coverage of scientists giving briefings to emergency meetings of public officials. Most press contacts were at PHIVOLCS' Main Office in Quezon City; in Zambales with the Pinatubo Lahar Hazards Taskforce; and in San Fernando, the site of many emergency meetings for public officials. Most interviews were with the chiefs of PHIVOLCS and PLHT, though a number

of PHIVOLCS and PLHT staff served as spokespersons when needed.

Concern among both volcanologists and public officials was high. Throughout late June and July 1991, one of us (R.S. Punongbayan) briefed government officials, including then-President Corazon Aquino, about both the eruption and lahars. In mid-July, another of us (K.S. Rodolfo) met with officials of the Department of Public Works and Highways (DPWH) and stressed that past eruptions at Pinatubo and similar volcanoes had been followed by such large lahars and volumes of deposited sediment that

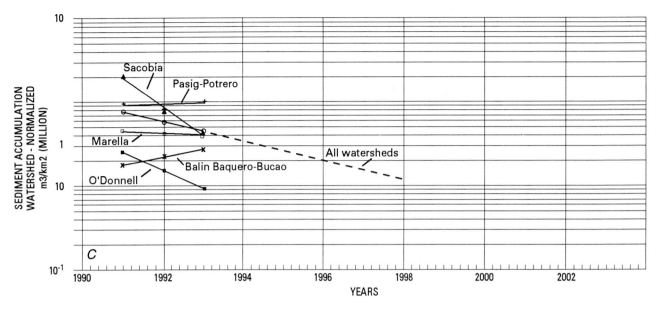

Figure 3.—Continued.

equally or more serious hazards lay ahead. A week later, another of us (R.J. Janda) met with the same group, outlined how sediment would move from the steep slopes of each watershed and fill lowland channels, and indicated how monitoring of erosion and channel filling in each watershed could provide information critical to the mitigation of the sediment problem. DPWH officials requested a ranking of barangays (villages) at risk and an estimation of the volumes of sediment that could be expected. On August 30, two of us (T.C. Pierson and K.S. Rodolfo) met with DPWH Secretary Jose P. de Jesus and his staff to further explain the lahar threat and to offer to work with design engineers on proposed engineering countermeasures.

For barangay leaders and the general public, illustrated flyers, posters, and leaflets were prepared and distributed by PHIVOLCS, the RDCC, the Philippine Information Agency, and others. One poster, "Mga Dapat Gawin Upang Maiwasan ang Pagiging Biktima ng Mudflow" ("How to Avoid Becoming a Victim of a Mudflow"), offered suggestions for what residents should do before, and when, they are warned of lahars (fig. 4). Paraphrased, the poster said:

1. Stay away from potential mudflow channels when it is raining at Mount Pinatubo and nearby hills;

2. If you live in a lowlying place, move immediately to high ground, keeping in mind that mudflows go to areas that are usually flooded during the rainy season. Keep in mind that some rivers and streams are already shallow from materials from Mount Pinatubo's eruption, so it is possible that these rivers will overflow;

3. Make your own "hill" at least 4 m high and widen the top to serve as an evacuation center;

4. Make barriers, if possible, but be aware that mudflows can be fast and strong;

5. Each group of houses should have a watchperson on a nearby hill to sound mudflow alarms;

6. Be ready with flashlights and a radio;

7. Listen for warnings and pay attention to the authorities;

8. Stay calm and don't be fooled by false news or rumors. WE WON'T HAVE TO WORRY IF WE STAY ALERT AND STAY TOGETHER.

In the absence of suitable coarse riprap to protect against erosion by lahars, the advice to "build your own high ground" was later concluded to be unwise, and the advice was withdrawn before any manmade hills were actually built.

During the second lahar season (May–November 1992), PHIVOLCS released four new educational tools. The first was a pamphlet, "A Technical Primer on Pinatubo Lahars," written to explain the general behavior of lahars to public officials, the news media, and other interested groups (Punongbayan and others, 1992a). Perhaps the most useful feature of this primer was a distinction between "malapot" (sediment-rich, viscous) and "malabnaw" (relatively dilute, less viscous) lahars, corresponding roughly to debris flows and hyperconcentrated flows, respectively. The distinction made it clear that lahars were fundamentally different from normal streamflow and much harder to control than normal floods. Another useful concept introduced in the primer was attenuation of flows, so that people would understand that warnings of "2-m-deep flow" past manned watchpoints upstream from populated areas might only be "0.5-m-deep flow" or less in distal areas.

Figure 4. Mga Dapat Gawin Upang Maiwasan ang Pagiging Biktima ng Mudflow ("What you should do to avoid becoming a victim of a mudflow"), a poster issued by the RDCC Region III, Philippine Information Agency, and Philippine Institute of Volcanology and Seismology in 1992. English translation is in the text.

A second educational tool, requested by participants in the May 1992 International Scientific Symposium on Mount Pinatubo, was a set of impact scenarios for three representative storms (Punongbayan and others, 1992b). In order of increasing severity, the storms were (1) normal afternoon rainshowers, (2) prolonged monsoonal rainfall ("siyam-siyam," literally, 9+9 days), and (3) intense typhoon rains. Scenarios of future lahars and their impacts were chosen from the most likely actual events, such as burial of Barangay Tabun of Mabalacat town. For example, starting on Day 5 of a "siyam-siyam," the scenario said:

In Mabalacat, sediment now fills to within ½m of the flat surface on which Tabun and Dolores are built . . . . [Day 6] Rain continues . . . more lahars, all rivers. Bgy. Tabun is now "Natabunan" (covered) with ½m of sediment. Before taking a shortcut through Tabun, lahars washed out the Bamban Bridge . . . . [Day 7] Unexpectedly great erosion of the banks of the Pasig-Potrero below Mancatian, coupled with rapid buildup of sediment in the channel just downstream, now threatens Santa Rita, Pampanga . . . . [Day 9] The Pasig-Potrero breaks out, but to the east, into Potrero, rather than into Santa Rita . . . .

The only event in this scenario that did not actually occur in 1992 was breakout on the east bank of the Pasig-Potrero; rather, flow broke over the southwest bank and buried Barangay Mitla of Porac and Barangay Balas of Bacolor (near Santa Rita). Many other events in the scenarios occurred in 1992 or 1993. Ironically, because these scenarios were so accurate, we later wondered whether we should have presented them as forecasts rather than as hypothetical scenarios. Our data were too sparse to have predicted these events with customary scientific certainty, but were we certain enough by laymen's standards? Would "forecasts" rather than "scenarios" have had a stronger, more constructive effect in preparing communities for lahars? The answer to all three questions is, in retrospect, "yes."

A third educational tool was a booklet in Tagalog titled "Ang Lahar" (The Lahar), designed principally for barangay leaders (Philippine Institute of Volcanology and Seismology, 1992b). The booklet contained nontechnical information about lahars, hazard zones, and the RDCC-PHIVOLCS lahar-warning system (the last is discussed under Short-Range Warnings).

The fourth and perhaps the most important tool was a video companion to the booklet, also titled "Ang Lahar," that provided basic information about lahars and showed graphic examples of 1991–92 damage to towns around Mount Pinatubo. Separate documentaries by Manila TV stations showed similar footage, with an emphasis on social impacts, and two new scientific videos ("Lahars of Pinatubo," by K.S. Rodolfo and H. Schaal, University of Illinois at Chicago; and "Pinatubo Volcano: Lahars and Other Volcanic Hazards," by M.T. Dolan, Michigan Technological University) were released in 1994. All of these videos have been wonderfully effective in conveying concepts that, for nonscientists, words and maps cannot convey.

LONG-RANGE WARNINGS: HAZARDS ASSESSMENTS AND HAZARDS MAPS

Information about topography, previous lahars, watershed and channel characteristics, volumes of erodible sediment, trends in current activity, and probable patterns of rainfall have been translated into **long-range** warnings about areas that will be at the greatest risk over coming months to years. This information has been presented in the form of hazard maps, tables of estimated sediment yield, or general statements of changing hazard, such as channel capture or upstream migration of avulsion points. Long-range warnings have influenced some decisions about relocation of towns, possible engineering countermeasures, and general emergency planning. They have also given residents a chance to move portable property—machinery, furniture, appliances, personal belongings, harvestable crops, farm animals, and even dismantled structures—out of the expected paths of future lahars. Some examples of this information are given in the following section.

LAHAR SEASON 1, JUNE THROUGH NOVEMBER 1991

On May 23, 1991, PHIVOLCS distributed a page-size map of volcanic hazards to civil defense and political leaders. Lahars (called "volcanic mudflows" on the map) were represented as hachures and later as bold lines along each of the major river valleys. Areas of potential overbank flow were not indicated because we lacked the fine-scale topographic maps needed to forecast likely points of overflow and paths of such flows across the alluvial fans. Clearly, the main preeruption channels were the most likely paths for lahars; we could not forecast whether and where lahars might fill channels and avulse.

Immediately after the June 1991 eruptions, questions about the long-term outlook for lahars at Pinatubo included:

How big is the lahar problem?

How long will the problem last?

What areas will be at greatest risk?

Over the next several years, will lahar deposits cover entire alluvial fans of Pinatubo, including all towns thereon, or just selected parts of those fans?

Between August and October 1991, PHIVOLCS and PLHT prepared the first of what was to become a series of mudflow (lahar) hazard maps. Two from PHIVOLCS were on a 1:100,000-scale base provided by the National Mapping and Resource Information Authority (NAMRIA) (Punongbayan and others, 1991a,b). The first, released in August, showed a zone judged to be "subject to mudflows" as of July 30. The second, released in October, showed lahars that had occurred as of September 15 and a significantly larger zone "subject to mudflows." The October hazard assessment was based on the lahars that had occurred to date, topography, and a subjective judgment about how widely lahars might spread during the next 1 to 5 years in the absence of engineering countermeasures. A simplified version of the October map is included as data layer A on figure 5. The October map was the first of the lahar hazard maps to be printed and widely distributed and, as such, was very influential even though the evolving crisis required later expansion of its hazard zones. Many of the evacuation sites, relocation sites, and other elements of recovery were based on that map and its subsequent revisions.

Also in August and September 1991, detailed lahar hazard maps (at a scale of 1:36,400) were issued by PLHT for the Santo Tomas and Bucao Rivers (Pinatubo Lahar Hazards Taskforce, 1991a,b). These maps were based on lahars to date, topography, a qualitative approximation of future lahar volumes, and projected channel avulsions. The map for the Santo Tomas River (dated August 23, 1991) showed lahars as of that date and four hazard zones, as follows: (a) areas subject to lahars and moderate to heavy flooding; (b) areas subject to overbank lahars and moderate to heavy flooding; (c) areas prone to minor or moderate flooding; and (d) areas subject to lahars escaping from filled irrigation canals. A September 5, 1991, map for the Bucao River used a similar zonation. In effect, these zonal categories ranked lahar hazard from highest to lowest. The relatively large map scale enabled officials and residents to judge whether their barangays, town streets, and secondary roads were in relatively high or relatively low danger.

Starting in September 1991 and continuing through the first half of 1992, the USGS, PLHT, and PHIVOLCS made semiquantitative estimates of probable volumes of lahars for the coming decade (Pierson and others, 1992). By then, it was apparent that sediment was being shed into surrounding lowlands at approximately 0.5 to 1.0×10^9 m^3/yr from a total upland source area of about 540 km^2, or at about 10^6 m^3/km^2 of upland watershed area per year, one order of magnitude faster than at Mount St. Helens and at the previous historical recordholder, Sakurajima Volcano, Japan (Janda and others, 1984). By analogy from Mounts Galunggung and Kelut in Indonesia, Pierson and others (1992) estimated that 40 percent of primary pyroclastic deposits on the east side of Pinatubo and 50 percent of the larger volume of primary deposits on the west side would be eroded and redeposited by lahars within roughly 10 years after the 1991 eruptions. Total sediment yield was projected to be approximately 2.5 to 3.6 km^3 of the estimated 4.8 to 7.0 km^3 of primary source material. Pierson and others (1992) also used aerial oblique photographs to estimate that 10 to 15 percent of most 1991 pyroclastic flow deposits had been eroded by September 10, 1991, and then used that volume of sediment as the first-year value on a sediment-delivery-rate decay curve, constructed from Mount Galunggung and Mount St. Helens data. Integration under that curve projected 1.2 to

2.5 km^3 of sediment yield during the first posteruption decade (fig. 3A).

Using the additional assumption that lahars would spread in fans and deposit to an average thickness of about 2 m, the projected 2.5 km^3 of sediment was translated into areas that **could** be covered by lahars during the coming decade (Pierson and others, 1992; also shown as data layer B on fig. 5). Potential hazards from future lahars (or from related backflooding and siltation) were mapped as being high or moderate, in relative rather than absolute terms. Whether lahars would actually impact this broad an area will depend in part on the actual advent and paths of typhoons, which deliver the most intense rains of lahar-generating duration. Preliminary copies of this 1:250,000-scale long-range hazards map were given to the Secretaries of National Defense and Public Works and Highways in mid-September 1991.

In October 1991, the Bureau of Soils and Water Management issued a GIS-based map of "Mudflow and Siltation Risk," showing a "high-risk area" subject to "moderate to severe mudflows and/or siltation" (data layer C on fig. 5), a "low risk area" subject to "low siltation," and a non-risk area (high ground) (Bureau of Soils and Water Management, 1991). The "high risk" zone of this map is much larger than that of any of the preceding maps, because it is a map of the distribution of sandy soils around Pinatubo. This map serves a useful purpose of reminding us that Pinatubo has, indeed, been the source of the fertile sandy loams throughout much of central Luzon, but it neglects to note that this sediment has been supplied by many eruptions over geologic time (>35,000 years, Newhall and others, this volume). Thus, risk appears to be exaggerated on this map. Given the relatively modest scale of the 1991 eruption in comparison to previous eruptions of Pinatubo, we do not expect the "high risk" zone of this map to be fully covered during the next decade.

One awkward role for PHIVOLCS geologists began during late 1991—responding to requests to "certify" that specific parcels of land were safe from lahars before new construction, loans, or other uses could proceed. Hazard maps were the primary guide for geologists, but, as has been indicated, several maps had been issued at various scales, and they were not understood by all applicants, not detailed enough for all purposes, and not legal documents. It was entirely appropriate for PHIVOLCS to examine sites of proposed evacuation camps and other major public projects, including one unsuitable resettlement site (Rabanes, San Marcelino) on which infrastructure was built, despite rejection of that site by PHIVOLCS, and which was destroyed by lahars in 1993. However, the task of checking hundreds of parcels has been burdensome, especially for parcels that are neither obviously safe nor obviously doomed. Risk is rarely "black or white," but present procedures in certification demand that it be considered so.

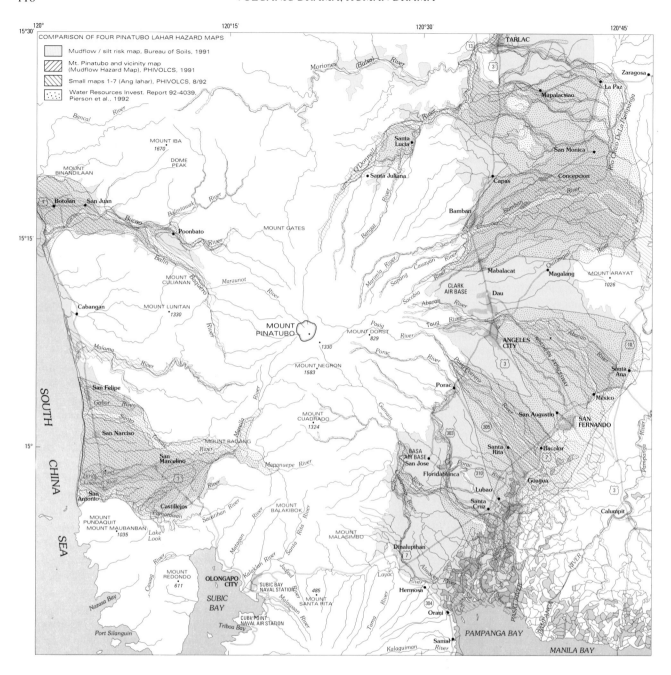

Figure 5. Comparison of four Pinatubo lahar hazard maps, 1991–92. Only the zone of highest lahar hazard from each map is shown. Differences in intended periods of applicability, in data sets, and in assumptions lead to different maps and, in some instances, confusion among users of these maps. Data layers are discussed in the text.

At the end of the first lahar season, PHIVOLCS and PLHT combined data from aerial photographs and field measurements and estimated 8×10^8 m^3 of deposited 1991 lahar sediment. This corroborated the estimate by Pierson and others (made in 1991, published in 1992) that between 5 and 10×10^8 m^3 of sediment would be moved into lowland areas in 1991. The estimated volumes of 1991 lahar deposits, categorized by river system, were provided to the Department of Public Works and Highways on March 24, 1992, for the purposes of design and decisions on engineering countermeasures.

On April 4, 1992, lahar hazards along the Abacan and Sacobia Rivers changed dramatically. A secondary explosion and pyroclastic flow from valley-filling 1991 deposits allowed the Sacobia River to capture drainage that had been entering the Abacan River (Martinez and others, this volume; Torres and others, this volume). Thus, the hazard along the Abacan decreased sharply, while that along the Sacobia increased. Some residents of villages along the Sacobia voiced suspicions that residents of Angeles City (along the Abacan) had caused the change, but the capture had been wholly natural, and it returned drainage to its preeruption pattern.

LAHAR SEASON 2, MAY THROUGH NOVEMBER 1992

During 1992, lahars did not reach as far as they had in 1991, and the loci of deposition migrated up most of the alluvial fans. The reasons for these two changes are still under study; factors may include faster and greater runoff in 1991 than in 1992, channel filling and thus an increasing number of channel avulsions, and distal decreases in stream gradient. The effect of this migration was that towns like Concepcion and Bacolor (figs. 1B, 5), hard hit during 1991, were spared in 1992, while upstream barangays of the towns of Mabalacat and Porac were hard hit in 1992 and 1993.

In July 1992, in response to ever-changing conditions and a request for more specific, barangay-by-barangay assessments of lahar hazard, PHIVOLCS released a revised lahar hazard map on a large-scale (1:50,000) ozalid base that showed many, though not all, barangays at risk (Philippine Institute of Volcanology and Seismology, 1992a) (data layer D in fig. 5). On this map, hazards zones for the next 1 to 2 years were estimated on the basis of lowland topography (in which contours were a relatively coarse 20 m, with some intermediate 10-m contours), proximity to nearly filled river channels, volume of sediments to be expected, and a subjective estimation of how far and how widely overbank flows might spread. Two hundred and sixty nine barangays that fell within these newly defined hazard zones were listed, in an effort to help those unaccustomed to reading maps. At least one breathless radio announcer told listeners that all 269 barangays were at dire risk during the

next rainstorm, and many people have found it difficult to understand that a hazard map with the words "subject to" is not an absolute prediction that all of the high hazard zone will get buried, but, rather, a geologist's graphic shorthand to indicate that actual flows over a period of years *could* but not necessarily *would* go anywhere within that zone.

Also on this map, hazard zones for the O'Donnell-Tarlac, Sacobia-Bamban, and Pasig-Potrero Rivers were larger than indicated in the October 1991 assessment, on the basis that 1992 lahars could have an equivalent volume to those of 1991 (8×10^8 m^3). Hazard zones for the Abacan and Gumain Rivers were reduced, reflecting the April 1992 capture of the upper Abacan by the Sacobia River and nearly complete erosion in 1991 of source materials in the Gumain drainage. Hazard zones for the Santo Tomas and the Bucao drainages were roughly the same as those in the August and September 1991 maps of PLHT (1991a,b). Page-size versions of these hazard maps were distributed with a corrected list of barangays at risk.

In late August 1992, the National Economic Development Authority (NEDA) and PHIVOLCS published a set of GIS-based maps of lahar hazard (Philippine Institute of Volcanology and Seismology and the National Economic Development Authority (PHIVOLCS-NEDA), 1992a). Hazard zones were modified only slightly from those of PHIVOLCS' July 1992 map (PHIVOLCS, 1992a), but the base map was substantially improved by including barangay and town boundaries. Important advantages of the GIS-based maps are that they can be revised quickly if conditions at and around the volcano change and that information about lahar hazard can be overlain by "data layers" for population, infrastructure, land use, evacuation routes, and myriad other parameters. An update of this PHIVOLCS-NEDA map, at a scale of 1:200,000, was issued on December 7, 1992 (PHIVOLCS-NEDA, 1992b).

The period after the 1992 lahar season was a time for scientists to analyze data they had gathered during 1991 and 1992 and to consider some new questions, including:

Was rainfall experienced during 1991 and 1992 below, near, or above the long-term average rainfall for these areas?

Had sediment yield at Pinatubo passed its peak and already begun to decline? (Sediment yield appeared to decline from 1991 to 1992, and we were asking whether we were past the period of peak sediment yields or just seeing an aberration due to unusually light rain.)

Unfortunately, discontinuation of rainfall monitoring at Clark Air Base in November 1991 and problems with telemetered rain gauges high on the west side of Pinatubo in 1992 prevent a direct comparison of rainfall for these 2 years. If we use rainfall at Dagupan as a proxy for that at Clark (table 2), we might conclude that rainfall on the east side during 1991 and 1992 was close to the long-term

Table 2A. Abbreviated comparison of rainfall (in millimeters) during 1991–93.

[–, not known; ( ), average derived from 1991–93 data only]

	East side		Southwest		West side
	Lowlands[1]	Highlands[2]	Lowlands[3]	Highlands[4]	highlands[4]
Annual rainfall					
1991	1,876	2,256	4,190	5,418	3,721
1992	2,442	2,215	2,667[3]	–	–
1993	–	2,492	7,165[3]	>3,929	3,465
Long-term mean	2,000	(2,321)	3,890	(>4,673)	(3,593)
Maximum 24-hour rainfall					
1991	>67	120	–	198	–
1992	–	199	–	189	–
1993	–	360	–	598[5]	428
Historical maximum	325	–	442	–	
Annual average 24-h maximum	154±78	(226±122)	265±86	(382±266)	–
Maximum monthly rainfall					
1991	>469	839	1,075	1,924	–
1992	794	925	1,163	1,083	–
1993	>355	689	3,102	1,122	1,027
Historical maximum	2,659	–	3,102	–	–
Annual average monthly maximum	670±380	(818±119)	1,346±738	(1,371±465)	–

[1] East-side lowlands are represented by Clark Air Base (1953–90) and Dagupan (1951–92) stations. Annual totals are averages of Clark Air Base and Dagupan; 24-h and monthly maxima are the maxima recorded at any single station.

[2] East-side highlands are represented by stations PI2, MSAC, FNG.

[3] Southwest lowlands are represented by Cubi Point Naval Air Station (1955–87). 1992–94 data courtesy of Jun Tonpong, PAGASA/SBMA weather station, Cubi Point. Rainfall for 1993 was a record high since data collection began at Cubi Point in 1955.

[4] Southwest highlands are represented by BUG, QAD; west-side highlands are represented by KAM (1993) and by average of BUG and PIE2 (1991–92). Annual values are averages of available data; monthly and 24-h values are maxima.

[5] During Typhoon Goring, 0406 June 26 to 0308 June 27, 1993, at QAD. An even greater 24-h amount, 744 mm, was recorded from 0600 July 25 to 0600 July 26, 1994 at the same station (QAD).

average, as was that on the west side (at Cubi Point). Rainfall high on the west slopes (fig. 8, stations BUG and QAD) was even higher than that of Cubi Point during 1991; we do not have comparable data for 1992, nor do we have any long-term average for BUG and QAD to tell whether the high rainfall in 1991 was above or below that average. Thus, we can say only that rainfall in 1991 and 1992 was not **demonstrably** different from long-term averages.

Field measurements during late 1992 suggest that, on the east side of Pinatubo, the volume of 1992 lahar deposits was about 40 percent that of 1991 (table 3). However, similar measurements on the west side suggest 1992 sediment volumes that were nearly equal to those of 1991. Thus, even though sediment yield on the east side appeared to have peaked and to have begun to decrease, we could not say the same for Pinatubo as a whole.

Was the difference between east and west related to rainfall? Again using Dagupan as a proxy for Clark and using Cubi Point to represent the west side, the ratio of *Rainfall*$_{1992}$/*Rainfall*$_{1991}$ was about 1.3 on the east and 1.0 on the west. This small, insignificant difference suggests that differences in rainfall **did not** account for the different sediment behavior from east to west during 1991 and 1992.

Rather, we think that differences in watersheds, including differences in volume of 1991 pyroclastic debris, slopes and stream channels, and vegetation recovery, were responsible for different patterns of sediment yield. Declining yields on the east side appeared to be the start of the anticipated exponential decline in sediment yield (Pierson and others, 1992).

LAHAR SEASON 3, MAY THROUGH NOVEMBER 1993

Among the questions we asked in 1993 was:

Did erosion of pre-1991 source materials, now apparent in some river valleys, add significantly to the volume of deposits to be anticipated?

Along some stream valleys, erosion had already cut through 1991 pyroclastic-flow deposits and into older deposits. Did sediment that was eroded from deposits predating 1991 add significantly to the volume of lahar deposits in 1992, above that anticipated in early studies? In the Sacobia-Abacan watershed, the volume of 1991–92 lahar deposits (270×10^6 m^3, table 3) exceeded the volume of 1991–92 erosion of the June 1991 pyroclastic fan (about 138×10^6 m^3, table 3). Could the difference represent the

Table 2B.   Rainfall (in mm) at selected PHIVOLCS/USGS rain gages on and near Mount Pinatubo, 1991–93.

[–, Months before or after recording, or with seriously incomplete data. Adjacent values in parentheses are the recorded, minimum values. Data compiled by Ma. Theresa M. Regalado, PHIVOLCS]

		O'Donnell (Station ODON)	Bangat (PIE2)	Middle Sacobia (MSAC)	Mount Cuadrado (QAD)	Sitio Buag (BUG)	Kamanggi (KAM)
1991	January	–	–	–	–	–	–
	February	–	–	–	–	–	–
	March	–	–	–	–	–	–
	April	–	–	–	–	–	–
	May	–	–	–	–	–	–
	June	–	–	–	–	–	–
	July	–	423	–	1,418	951	–
	August	–	839	–	1,924	1,908	–
	September	–	649	434	1,675	1,849	–
	October	–	193	260	414	408	–
	November	–	152	156	177	63	–
	December	–	0	0	50	0	–
	Total	–	>2,256	>850	>5,658	>5,179	–
1992	January	–	0	0	0	0	–
	February	–	0	0	0	0	–
	March	–	0	0	0	0	–
	April	–	2	74	0	1	–
	May	–	98	79	118	39	–
	June	–	321	95	422	–	–
	July	–	527	457	– (>11)	–	–
	August	–	925	1,005	–	1,083	–
	September	–	171	– (>48)	–	638	–
	October	–	135	198	336	–	–
	November	–	36	58	64	39	–
	December	–	0	0	6	9	–
	Total	–	2,215	– (>2,014)	– (>957)	– (>1,809)	–
1993	January	–	0	0	0	0	–
	February	–	0	0	0	0	–
	March	–	0	0	37	0	–
	April	–	0	0	12	0	–
	May	–	0	37	0	0	–
	June	179	463	441	691	–	506
	July	153	349	187	– (>47)	829	340
	August	1,010	689	743	391	1,122	1,027
	September	387	449	– (>80)	399	785	692
	October	464	319	615	– (>37)	879	740
	November	155	171	152	232	308	160
	December	45	52	28	43	6	0
	Total	2,393	2,492	>2,283	– (>1,889)	– (>3,929)	3,465

volume of debris predating 1991? An even greater contrast arose in the Pasig-Potrero watershed, where only 23×10^6 m^3 was estimated to have been eroded from the 1991 pyroclastic-flow fan, although about 90×10^6 m^3 of sediment was deposited downstream (table 3). If these values are extrapolated through 1993 and into the future, the ultimate volume of lahar deposits might be substantially greater than originally estimated. However, similar comparisons in the Balin Baquero-Bucao watershed show the opposite relation; that is, the volume of deposits is **less** than the volume eroded

from 1991 pyroclastic-flow deposits (table 3). Either the uncertainties in estimating volumes are so great that these differences are within normal estimation error or differences between watersheds on the east and the west sides of the volcano have led to earlier and faster erosion of pre-1991 materials on the east than on the west. Clearly, uncertainties in estimating volumes are large, but we think that advanced erosion in east-side watersheds has also cut significantly into underlying, pre-1991 deposits. Should deposits that predate 1991 on the west side of Pinatubo

Table 3. Volumes of source material, erosion, and lahar deposition, Mount Pinatubo watershed, 1991–93

Watershed	Area $(km^2)^1$		Volume of pyroclastic-flow deposits $(\times 10^6\ m^3)^2$	Volume of erosion, 1991–92 $(\times 10^6\ m^3)^3$	Volume of 1991 lahar deposit $(\times 10^6\ m^3)^4$	Volume of 1992 lahar deposit $(\times 10^6\ m^3)^4$	Volume of 1993 lahar deposit $(\times 10^6\ m^3)^4$	Revised total volume of 1991–93 lahar deposit $(\times 10^6\ m^3)^4$	Forecasts of sediment yield, 1994–2010 $(\times 10^6\ m^3)^5$
O'Donnell-Bangat	89	(10)	300	35	80	40	25	145	30
Sacobia+Pasig[6].........................	69	(22)	900	161	250	110	105	465	140–160
Sacobia-Bamban-Abacan	*46[7]*	*(15)*	*600[2]*	*138*	*200*	*70*	*50*	*320*	*40–110[7]*
Pasig-Potrero...................	*23[7]*	*(7)*	*300*	*23*	*50*	*40*	*55*	*145*	*40–110[7]*
Gumain	41	(2)	30	30	60	0	0	60	0
Marella-Santo Tomas, including Maloma.	79	(22)	1,260[2]	212	185	195	125	505	290–340
Balin Baquero- Bucao	262	(65)	3,000[8]	600	250	230	250	730	650–720
Total	540	(121)	5,490	1,038	825	575	505	1,915	1,110–1,250

[1]Upland area that contains, or drains through, valley-filling pyroclastic-flow deposits. In parentheses are areas of the valley-filling pyroclastic-flow deposits prior to erosion (after W.E. Scott and others, this volume).

[2]Volumes as estimated by W.E. Scott and others (this volume). In the O'Donnell and Sacobia-Abacan watersheds, Scott and others estimated slightly lower volumes than cited in Punongbayan and others (1994); in the Marella-Santo Tomas and O'Donnell watersheds, Scott and others inferred slightly larger volumes than estimated by Punongbayan and others. Estimates of Scott and others were made by sketching new levels of valley-filling pyroclastic-flow deposit onto preeruption topographic maps, estimating average cross-sectional thickness of 1991 deposit, and multiplying by the length of each reach.

[3]Volumes are of deep channels cut during 1991–92 into 1991 pyroclastic-flow fans and are estimated by the U.S. Army Corps of Engineers (1994).

[4]Most volumes are as estimated by PHIVOLCS and PLHT/ZLSMG lahar teams, from field measurements of accumulated sediment. Volumes for O'Donnell (1991–93) and Sacobia in 1991 are from USGS team members. All estimates have high uncertainties.

[5]Forecasts based on figures 3A (annual sediment yield) and 3C (annual sediment yield normalized for upland watershed area and rainfall). In figure 3A, we assume that sediment yield from the Balin Baquero-Bucao watershed will decay faster than at Mount St. Helens, at rates intermediate between those of the Balin Baquero-Bucao and Sacobia-Pasig watersheds from 1991–93. Annual sediment yield will decline to preeruption levels (in the order of $10^6\ m^3/yr$) in 20 yr. This assumption strongly influences the overall sediment forecast because of the large sediment contribution from the Balin Baquero-Bucao watershed. The decay rate for the Marella-Santo Tomas watershed was assumed to be intermediate between those of the Sacobia and the Balin

Baquero-Bucao watersheds. In figure 3C, we made a similar assumption, that the normalized rate of sediment yield from the Balin Baquero-Bucao, the slowest of any watershed, would reach 10 m³ $_{deposit}/mm_{rain}/km^2_{watershed}$ in 20 yr. Additional details are explained USGS-PHIVOLCS (1994).

[6]To understand watershed responses and the overall east-side hazard, we must treat the Sacobia, Abacan, and Pasig-Potrero as a single watershed. For engineering measures, they must be treated separately.

[7]Watershed areas are for conditions prior to secondary explosion in October 1993 that allowed the Pasig-Potrero to capture about 21 km² of the Sacobia watershed. After that capture, upland watershed for the Sacobia was about 25 km² and that for the Pasig-Potrero was about 44 km². We cannot say whether or when the Sacobia might recapture its own upper watershed. Therefore, our sediment forecasts for the Sacobia-Bamban and Pasig-Potrero are given as high and low estimates that apportion the expected total sediment forecast between the two rivers in 1/3:2/3 and 2/3:1/3 ratios, based on approximate proportions of upland watershed areas in the event of capture or recapture. The high estimate for the Sacobia and the low estimate for the Pasig assume recapture by the Sacobia early in 1994; the low estimate for the Sacobia and high estimate for the Pasig assume that recapture does NOT occur. Our forecast for the Sacobia-Pasig system as a whole is unaffected by whether recapture does or does not occur, though it might be an underestimate because new, larger runoff will tend to widen narrow channels of the Pasig-Potrero that were originally eroded by the much smaller runoff.

[8]Estimate is from U.S. Army Corps of Engineers (1994), based on photogrammetry. All other estimates of pyroclastic-flow volume by the Corps of Engineers were close to prior estimates of PHIVOLCS and those of W.E. Scott and others (this volume).

ultimately be eroded to the same extent, the lahar problem in west-side watersheds might be 50 percent larger than originally estimated.

Another question we asked in 1993 was:

Was the rate of sediment deposition in 1993, by watershed and total, declining from rates in 1991 and 1992?

The volume of lahar sediment deposited on the east side of Pinatubo during 1993 was approximately 130×10^6 m³, about 85 percent of that deposited in 1992 and one-third of that deposited in 1991 (table 3). The Pasig-Potrero sediment accumulation was anomalously high in comparison to other east-side yields, probably because the watershed's "clock" was reset by the large secondary pyroclastic flow of 1992 and because a large secondary explosion on October 6, 1993, captured a significant amount of drainage from the upper Sacobia. The volume of sediment deposited on the west side of Pinatubo during 1993 was approximately

375×10^6 m³, about 85 percent of that in both 1991 and 1992 (table 3). Thus, sediment yields declined noticeably on the east but slowly on the west. Similar trends appear when sediment yield is normalized for watershed area (fig. 3B).

Sediment yields per unit of rainfall (table 4) also show declining trends. Despite the fact that sediment yield per millimeter of rainfall (per square kilometer of watershed area) varied considerably from one drainage to the next (some of the apparent variability is surely the result of poorly constrained volume estimates), the **average** sediment yield of Pinatubo clearly decreased from 1991 to 1993 (from 1.5×10^6 m³$_{deposit}/km^2_{watershed}$ and ~460 m³$_{deposit}/mm_{rain}/km^2_{watershed}$ in 1991, to 0.9×10^6 m³$_{deposit}/km^2_{watershed}$ and ~300 m³$_{deposit}/mm_{rain}/km^2_{watershed}$ in 1993 (table 4, figs. 3B,C).

This general trend toward decreasing sediment yield at Pinatubo has two important corollaries: that both the numbers and sizes of individual lahars, and the annual

Table 4.   Sediment yields normalized for Pinatubo watershed area and rainfall, 1991–93.

[na, not applicable]

Watershed	Area (km^2)	Rainfall and gauge (mm)[1]	Measured volume of sediment ($\times 10^6$ m^3)[2]	Sediment yield per unit area ($\times 10^6$ m^3)[3]	Sediment yield, per unit of area and rainfall (m^3/mm/km^2)[3]
		1991			
O'Donnell-Bangat......................	89	2,250 (PIE2)	80	0.9	400
Sacobia-Pasig............................	69	2,250 (PIE2)	250	3.6	1,610
Gumain......................................	41	2,250 (PIE2)	60	1.5	650
Marella-Santo Tomas	79	5,200 (BUG)	185	2.3	450
Balin Baquero-Bucao.................	262	3,700 (BUG+PIE2)	250	.95	257
Total (or average)	540	(3,350)	825	(1.5)	(456)
		1992			
O'Donnell-Bangat......................	89	2,200 (PIE2)	40	.45	204
Sacobia-Pasig............................	69	2,000 (MSAC)	110	1.6	797
Porac-Gumain	41	2,000 (MSAC)	0	0	0
Marella- Santo Tomas	79	na	195	2.5	na
Balin Baquero-Bucao.................	262	na	230	.9	na
Total (or average)	540[4]	na	575	(1.1)	na
		1993			
O'Donnell-Bangat......................	89	2,400 (ODON)	25	.28	117
Sacobia-Pasig............................	69	2,500 (PIE2)	105	1.5	608
Porac-Gumain	41	2,500 (PIE2)	0	0	0
Marella-Santo Tomas	79	3,900 (BUG)	125	1.6	405
Balin Baquero-Bucao.................	262	3,500 (KAM)	250	.95	271
Total (or average)	540[4]	(3,150)	505	(.9)	(297)

[1] Rainfall data compiled by M.T.M. Regalado, PHIVOLCS. Although data on rainfall in the Balin Baquero-Bucao watershed are especially sparse, data from the Kamanggi station in 1993 suggest that rainfall in that watershed is intermediate between that of the southwest sector (BUG) and that of the northern sector (ODON and PIE2).

[2] Data sources as in table 3.

[3] Normalization for watershed area and rainfall is by simple division of sediment yield by those factors for the corresponding year.

[4] Totals and averages include the Gumain, even during 1992 and 1993. Rainfall average is weighted according to watershed area.

accumulations of sediment, will continue to decrease noticeably over the next several years. The precise rates at which they diminish will depend on storm events and annual rainfall, on channel captures such as that in 1993 which shifted some flow from the Sacobia watershed into the Pasig-Potrero, and on broad differences in watershed response. As regards differences in watershed response, three watersheds illustrate the range of possible behavior. The Gumain River, with a relatively small amount of 1991 pyroclastic flow-material in a generally steep watershed, was reamed out in 1991 and has produced very little sediment since 1991 (table 3, fig. 3A). Subsequent rapid recovery of vegetation in the Gumain watershed had little effect on sediment yield because source material was already exhausted. The Sacobia-Pasig watershed, with an intermediate amount of 1991 pyroclastic-flow material, large channels in a steep-walled valley, frequent secondary ashfall, and intermediate degrees of vegetation recovery, produced its peak sediment yield in 1991 and now shows a well-defined trend of decreasing sediment yield: from 1610

m$^3_{deposit}$/mm$_{rain}$/km$^2_{watershed}$ in 1991 to 608 m$^3_{deposit}$/mm$_{rain}$/km$^2_{watershed}$ in 1993. In contrast, sediment yield in the relatively gentle-terrain, pyroclastic-flow-rich, slowly revegetated Balin Baquero-Bucao watershed may or may not have peaked as of the end of 1993. Its future decline, though ultimately certain to occur, cannot yet be quantified. In the absence of any data to constrain its rate of decline, we assume that sediment yield will have declined to 1×10^6 m^3/yr (and to 10 m$^3_{deposit}$/mm$_{rain}$/km$^2_{watershed}$) after 20 years (dashed lines for Balin Baquero-Bucao, figs. 3A,C).

By using the computer-selected rates of declining sediment yield for the Sacobia-Pasig and O'Donnell Rivers, by choosing a rate of decline for the Marella that is intermediate between those of the Sacobia-Pasig and Balin Baquero-Bucao, and by assuming that further sediment yield from the Balin Baquero-Bucao watershed will decrease at the rates shown in figures 3A and 3C, we can make forecasts of eventual sediment yield for each watershed and for the volcano as a whole (last column of table 3). Though some assumptions are still required, these new forecasts are based increasingly on actual data, rather than on an assumed rate of decay that was adopted from volcanoes with much smaller eruptions than that of Pinatubo. Uncertainties are high in a forecast based upon only three annual measurements, but we think the forecasts are, nonetheless, an improvement over our original projections. The estimates are also consistent with observed differences in sediment yields from one watershed to the next. The new forecasts can serve for 1994 and will be revised as needed in subsequent years.

Interestingly, these new forecasts that are based on actual Pinatubo data are very similar to earlier forecasts that were based on an analogy with other volcanoes (Pierson and others, 1992). The only significant differences are in east-side watersheds where, apparently, incorporation of a significant volume of pre-1991 material is suggested by the abovementioned discrepancy between volume of erosion on the pyroclastic fan and volume of lahar deposits. Whether erosion of pre-1991 material on the west side will be as extensive as that on the east remains to be seen. If it is as extensive, the ultimate volume of sediment will be higher than that forecast here.

A third question we asked in 1993 was:

Was the upstream migration of sedimentation, noted in 1992, continuing?

Upstream migration of deposition continued in the Pasig-Potrero and Sacobia valleys through 1993, and avulsions above Mancatian on the Pasig-Potrero River brought lahars into that village for the first time. In contrast, some deposition in the Santo Tomas valley shifted downstream and toward the low southern margin of the Santo Tomas lahar field, largely as spillover from continuing deposition in the Marella valley. On July 1, 1993, ZLSMG released a revised, 1:50,000-scale lahar hazard map for the western Pinatubo area that identified points where lahars were likely to overtop and breach the dike that was being constructed along the south bank of the Santo Tomas River. This map accurately anticipated the site of an August 19 breach at the Western Luzon Agricultural College (WLAC). The map was modified after that breach and was modified again after lahars of October 4–7 breached the repaired dike at WLAC once again, as well as at other correctly predicted points at its eastern and western ends (Zambales Lahar Scientific Monitoring Group, 1993). Construction of dikes along the Santo Tomas River has resulted in faster, areally restricted deposition between Dalanaoan and San Marcelino proper than would have otherwise occurred, and, when breakouts have occurred, the artificially high origin points of the breakouts may have contributed to the distance traveled by breakout lahars.

All of the preceding long-range assessments of Pinatubo lahars were based on measured 1991–93 change at Pinatubo, supplemented by experience elsewhere. As a cross-check, we also asked:

Is the long-range assessment shown in the abovementioned hazard maps, based on events of just 3 posteruption years, consistent with what the geologic record tells us about entire periods of lahars following previous eruptions of Mount Pinatubo?

Pumiceous and sandy deposits in the lowlands of central Luzon are derived from lahars and related streamflow and overbank flooding. The sediment fans that surround Mount Pinatubo consist of layer upon layer of lahar and other stream deposits. Pinatubo sediments can be recognized beneath most cities and towns of Tarlac and Pangasinan Provinces, all the way northward to the Lingayen Gulf; beneath most cities and towns of Pampanga, all the way southward to Manila Bay; and beneath all towns of the Santo Tomas plain (Castillejos, San Marcelino, San Antonio, San Narciso, San Felipe) and Bucao plain (Botolan), in Zambales. Smaller parts of other provinces—Bataan, Bulacan, and Nueva Ecija—are also underlain by Pinatubo-derived sediment.

Prehistoric terraces throughout the Pinatubo area represent previous "high-stands" of sediment-rich flows. Examples include the terrace upon which Clark Air Base is built, one or more terraces immediately south of the town of Porac, several terraces along the Marella-Santo Tomas Rivers including those upon which the Aglao and the Palan evacuation camps are built, terraces along the Bucao River including that upon which San Juan and the main part of Botolan town are built, and terraces along the O'Donnell River upon which the old Crow Valley bombing targets were built (now, mostly buried) and on which Patling and Santa Lucia, Capas, are constructed.

Does this wide reach of Pinatubo sediments, areally and vertically, mean that all of these areas will be covered by lahar or flooding before the current crisis ends? We do not believe this is likely, and to draw finer distinctions, we asked:

What areas were covered after single previous eruptions of Pinatubo comparable in size to that of 1991 (specifically, after the penultimate "Buag" eruptions of 500 years ago)?

What are the ages of sediments in terraces that are still unburied, at various heights above present levels of fill, and were those sediments deposited only after much larger eruptions than that of 1991?

Which of these areas were covered by lahars, and which were covered by less lethal, but still troubling, floodwaters?

Radiocarbon ages for 25 lahar and related flood deposits (Newhall and others, this volume) help to distinguish between areas that were buried after the 500-yr B.P. Buag eruptions and areas that remained untouched after those eruptions. In general, in a clockwise direction from Tarlac

Province around to Zambales Province, low-lying areas of Bamban and Capas were flooded and (or) buried by lahar and fluvial sediment of the Buag eruptive period, while slightly higher land around Mapalacsiao (Hacienda Luisita) and Clark Air Base was not covered by sediment of that period. The youngest **known** lahars to cross Hacienda Luisita occurred during the Crow Valley age eruptions, 5,000 to 6,000 years ago, while the youngest **known** lahars to cover the **whole** of Clark Air Base occurred during the Maraunot eruptive period, about 2,700 years ago. Some lahars did cover the Friendship area, and perhaps part of Clark Air Base, during the Buag eruptive period. Information for lahar deposits of the Pasig-Potrero River, on either side of Porac, is less well defined, but we are sure that the entire Porac area was covered by Maraunot lahars (3,000-2,500 years ago), and we also suspect that some of the famed sand- and gravel-mining of Porac might have been of Buag deposits. Terraces of the Floridablanca area, including that upon which Basa Air Base is built, are even less well known, but it appears from pyroclastic-flow deposits high in the Gumain watershed that the latest voluminous fill of the Gumain valley (and hence the most likely age of the terraces, below) was during the Pasbul eruptive period, about 9,000 years ago. However, some of the terrace deposits of Floridablanca could have been derived by flow in the Porac River, which in the recent geologic past has captured flow from the Pasig-Potrero; therefore, we would not be surprised if the youngest terraces of the Floridablanca area are the same age as those near Porac, 3,000–2500 years or younger.

Along the Santo Tomas plain, only the highest prehistoric lahar terraces remain unburied as of this writing, including a narrow terrace near Sitio Palan, and one or more broad terraces above Aglao and below Kakilingan. Between the present channel of the Santo Tomas River and Castillejos, deposits of 3,000- to 2,900-year-old lahars are just a meter above the level of 1993 fill. Upslope from Kakilingan, in Sitio Buag, it appears that Buag-age lahars filled or nearly filled some canyons but did not bury relatively high terraces. On one of these high terraces, a pre-Hispanic settlement (Dr. E. Dizon, Philippine National Museum, written commun., 1992) was buried lightly by ashfall from Buag eruptions but was not inundated by Buag lahars.

In the valley of the Bucao River, Poonbato was built on two or three prominent terraces, and downstream barangays were built on terraces overlooking the active channel of the Bucao. The main part of Poonbato and its new church, built on deposits of the Crow Valley eruptive period (6,000–5,000 yr old), were buried in 1992. A Buag deposit that formed the south bank of the Bucao River just downstream from Malumboy was buried in 1993. There is no reassurance from the geologic past for the safety of Botolan town; without dredging or other intervention, Botolan will probably be buried.

Why are some parts of Pinatubo's lahar fans relatively high and apparently safer than others? The main factor is probably the scale of the eruptions that supplied sediment for the lahars. The volume of erupted products has been generally declining in Pinatubo eruptions, from 35,000 yr ago to the present (Newhall and others, this volume). The 1991 eruption and the Buag eruptions were about one-half or one-third the size of Maraunot and Crow Valley eruptions, and about one-fifth or less of the size of the >35,000-year-old Inararo eruptions. In general, the highest terraces around Pinatubo formed after the large Maraunot and Crow Valley eruptions, when the level of sediment fill rose to those relatively high levels. However, not every terrace of those older periods can be considered safe: Poonbato is now buried and several of the highest terraces of the Santo Tomas plain are dangerously close to being overtopped. Also, we found no lahar deposit or sediment terraces of Inararo age, associated with Pinatubo's largest eruptions; some might have been completely eroded, but most are probably buried beneath younger sediment.

In summary, lahar hazard maps that have been published in the 3 years since the 1991 eruption are broadly consistent with the geologic record. Lahars in the balance of the present crisis will not cover all areas around Pinatubo that have previously been covered by lahars, but they will threaten some areas that have not yet been buried, especially in the Santo Tomas, Bucao, and Pasig-Potrero watersheds. The areas threatened will depend partly on which areas are made more, or less, vulnerable by stream captures and by human intervention. Engineered dikes and other construction works have undoubtedly decreased risk to some areas but have probably increased risks elsewhere.

SHORT-RANGE WARNINGS

Information about approaching typhoons, rainfall on the volcano slopes, or lahars passing upper observations points was translated into **short-range warnings** of lahars expected within minutes to days. Some people chose to keep living in hazardous areas because they didn't believe the warnings, had no reasonable alternative, or simply chose to rely on short-range warnings for their safety (Cola, this volume). Those who chose to rely on short-range warnings took calculated risks that warnings would be issued early enough for them to escape. In return, they were able to remain in their own homes and on their own land for as long as possible and clung to the hope that their places would not be overrun by lahars.

LAHAR SEASON 1, JUNE THROUGH NOVEMBER 1991

The first short-range warnings were issued during the urgent, last-minute preparations for a possible eruption (Punongbayan and others, this volume). Lahars had

More mudflows, ashfalls expected

Pinatubo claims 2 more victims

By EDDEE RH. CASTRO

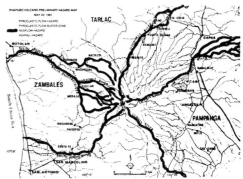

Mudflow paths

Heavylines in the map indicate the possible paths of mudflows to Zambales, Tarlac, and Pampanga from the crater of erupting Mt. Pinatubo. Volcanologists said mudflows may develop if heavy rains fall on the area and expressed concern over the approaching low-pressure area 'Diding.'

More mudflows, ashfalls, and volcanic quakes are expected from Mt. Pinatubo which has already had three major eruptions and spewed huge volumes of ashes and hot toxic gases, pumice stones, and boulders, the Philippine Institute of Volcanology and Seismology (Phivolcs) said yesterday.

Two more fatalities were reported.

The Philippine National Red Cross (PNRC) reported the first fatality. She was identified as Vinayag Butaytay, 70.

The second fatality was reported by Bulletin correspondent Jerry Uy. She was identified as Biniang Cosme, 55.

Both victims are residents of sitio Lomboy, barangay Sta. Fe, San Marcelino, Zambales. They were killed when they were hit by stones from the volcano.

The Red Cross also said that 19 persons, all residents of San Marcelino, Zambales, were injured. Their names, however, were not immediately available.

Latest reports from Zambales said that a total of 37 persons were injured as of last night. Three other persons were also reported missing, the same reports said.

Volcanologists monitoring Mt. Pinatubo reported that the latest major eruption, which occurred at 8:41 yesterday morning, had an intensity between 4 and 5.

Ashes and volcanic fragments fell on the western, northwestern, and southwestern sides of the volcano. Many pumice stones measuring three to five centimeters fell notably on sitio Ogik, the volcanologists assigned to the Zambales Observatory reported.

Disabled by Pinatubo

This Saudia Airline 717 jumbo jet was disabled as a result of the volcanic ashes that it sucked in while it was passing near Mt. Pinatubo Wednesday noon. It landed safely at the airport but failed to take off for its flight back to Saudi Arabia. (Louie Perez)

Figure 6. Headlines of the Manila Bulletin, June 14, 1991, warning of impending mudflows and showing the rivers along which they would likely pass.

previously been discussed as a possible adjunct to that eruption, on June 13, 1991, when weather forecasters recognized that Typhoon Diding (international name, Yunya) was heading toward central Luzon, and volcanologists realized that the storm would probably generate lahars from the fresh deposits of eruptions then in progress. New warnings about lahar hazard were issued to civil defense, local officials, and military commanders on June 13 and 14 and appeared as headlines, together with the "mudflow" part of the May 23 hazard map, in the Manila Bulletin (fig. 6), the Philippine Daily Enquirer, and other Manila papers on June 14.

The center of Typhoon Diding, downgraded to a tropical storm after it began crossing Luzon, passed within 100 km of Pinatubo's summit at about 1100 on June 15. Rain from Diding generated lahars from already-emplaced deposits and continued to generate lahars from fresh deposits of the climactic eruption that afternoon. Additional warnings of these lahars were issued on radio stations just before and throughout the climactic eruption, especially when reports arrived that bridges in Angeles City and in Barangay San Rafael, San Marcelino, had been destroyed by lahars. Some of these warnings originated from

PHIVOLCS and the Office of Civil Defense; others were based on reports from citizens or reporters.

By the end of July 1991, the Regional Disaster Coordinating Council (RDCC Region III) deployed police and army units to ten primary (upslope) lahar watchpoints, one each on the O'Donnell, Bangut, Sacobia, Abacan, Pasig-Potrero, Porac, Gumain, Marella-Santo Tomas, Maloma, and Bucao Rivers. Initially, lahar watchers were housed in tents and made relatively crude "eyeball" estimates of the height of lahars passing their watchpoints. Information was relayed to central points at Camp Olivas, San Fernando, Pampanga, and the Zambales Provincial Disaster Coordinating Council in Iba, Zambales. Those centers raised simple alerts: Lahar Alert 1 (rain is falling on Pinatubo; get ready); Lahar Alert 2 (rain is continuing and a lahar could form; get set); and Lahar Alert 3 (a lahar has been confirmed; go to high ground).

Lahar watchers were drawn from the large personnel pools of the Philippine National Police (PNP) and Armed Forces of the Philippines (AFP) and served for about 2 weeks before being relieved by new watchers. Each group received a small amount of training before deployment, but frequent rotation limited the expertise that could be developed. During the first part of the 1991 lahar season, RDCC

Figure 7. Mapanuepe Lake (left), impounded when lahars of the Marella River (lower right) blocked a tributary, the Mapanuepe River (left). Breakouts of this lake in 1991 prompted engineers to cut a channel through the ridge that lies south of (behind) the channel through which Mapanuepe Lake was still draining at the time of this photo. View is to the southwest, down the Santo Tomas River, which is the combined flow of the Marella and Mapanuepe Rivers. Much of the alluvial fan in the background (upper right) was buried by lahars in 1993.

III in Camp Olivas tended to declare Alert 3 for all lahar-threatened areas whenever a lahar hit a single channel in Tarlac or Pampanga. This was before it became clear to everyone that the lahar-triggering rains in July 1991 were being delivered to the east side of Pinatubo by the trade winds, whereas the western slopes remained dry and lahar free (Rodolfo, 1991). An unfortunate consequence was that people in Zambales, roused too often by an Alert 3 without a corresponding lahar, became distrustful of the warnings and were largely ignoring them when the southwest monsoon finally arrived and brought lahars to Zambales. Lahar warnings on a channel-by-channel basis were started by late September 1991 after several meetings with RDCC and PDCC officials. However, people remained skeptical of warnings aired over the radio. Another difficulty with the PNP-AFP watchpoints was that many were located so far above the lahar channels that, in inclement weather, they would be shrouded by clouds and mist. Furthermore, those on hilltops were subject to lightning strikes; on one occasion, two watchers were killed by lightning drawn to their radio antenna. (For the 1992 rainy season, primary watchpoints along the Porac, Gumain, and Maloma Rivers were abandoned, and all-weather buildings were constructed at the other watchpoints.)

Secondary watchpoints were established at each major bridge or river crossing, all in populated areas. These were manned by police, generally without special training about lahars but equipped with two-way radios to learn of lahars coming their way and to relay information about those lahars as they passed each secondary watchpoint. Information flowed to and from Camp Olivas (for all of Pampanga, Tarlac, and Zambales) and the Zambales Provincial Disaster Coordinating Council at the capital in Iba (for Zambales).

PHIVOLCS and PLHT staff coordinated closely with weather forecasters at Clark Air Base and Cubi Point Naval Air Station, particularly as typhoons approached. Although the primary responsibility for typhoon warnings remained with the Philippine Atmospheric, Geophysical, and Astronomical Service Administration (PAGASA), PLHT and PHIVOLCS staff were able to give extra verbal and graphical warnings to communities threatened by lahars, on the basis of detailed weather information.

USGS, PHIVOLCS, and PLHT scientists also provided warnings of lakes impounded by pyroclastic flows or lahars that might fail catastrophically. One such warning was issued by PHIVOLCS to the RDCC on July 25, 1991, about a lake impounded in the upper Pasig-Potrero watershed. Regrettably, the warning on July 25 could not be specific and was the last official mention of this lake before it broke out on September 7, 1991, and killed several people downstream. More specific warnings were issued by PLHT about the largest such impoundment, where the Mapanuepe River was impounded by lahar deposits from the Marella River (fig. 7). On August 25, 1991, at 1700, PLHT warned PNP watchers at Dalanaoan of an impending lake breakout from Mapanuepe within the next few hours. The warning, however, was not disseminated to town officials until 2200 because the watchers' radio ran out of power. When the warning was finally disseminated, the municipal officials of San Marcelino and San Narciso were very reluctant to act on it because of previous false Alert 3 signals and because most of the people were already asleep. Fortunately, the San Marcelino police did awaken people of Barangay San Rafael before the breakout actually occurred at 0400 the next day and destroyed 8 houses (Umbal and others, 1991; Rodolfo and Umbal, 1992; Umbal and Rodolfo, this volume). The only casualty was a person who suffered a heart attack.

In August 1991, after an emergency deployment of radio-telemetered rain gauges and acoustic lahar flow sensors (fig. 8) (LaHusen, 1994; Hadley and LaHusen, 1995; Marcial and others, this volume), a period of initial testing, and some initial difficulties with telephone and two-way radio links to civil defense officials, PHIVOLCS began to relay realtime instrument-based warnings to the Watch Point Center (WPC) of the RDCC at Camp Olivas, San Fernando, Pampanga. Instrumental records of flows were not yet calibrated by direct observations of those flows, nor did we know how much rain was needed to generate lahars, so warnings were limited to statements about rainfall and the occurrence of a strong signal on the instrumental flow monitors. Quiescence on the flow sensors was used to control frequent rumors of lahars. In general, WPC used PHIVOLCS' advice of lahar signals only to query the manned watchpoints; to our knowledge, WPC did not issue any lahar alert in 1991 solely on the basis of instrumental data from PHIVOLCS. One of the engineers who worked at WPC later told us that, because they didn't see the data themselves, they preferred to rely on reports from human observers.

An alternate lahar warning network, consisting of trip-wires and rain gauges (Iwakiri, 1992), was installed by DPWH and operated by staff of the WPC. Tripwires and rain gauges were installed at two sites, Barangay Dolores (Mabalacat) and Barangay Sapangbato (Angeles City), and rain gauges were installed at those sites and at Barangay Dalanaoan (San Marcelino); Sitio Ugik (Botolan); Sitio Pasbul (Camias, Porac); and at the PNP Delta 5 manned watchpoint (Porac) (fig. 8), between February 1992 and April 1992. Unfortunately, the large size of these units (principally, of the 10- to 15-m-high concrete and steel mast) limited their installation to sites that were accessible by road—usually at lower elevations and far from the lahar source area. The tripwires were located so close to the towns at risk, and were broken so frequently by lahars and also by people, and were later buried so deeply by lahars, that they have been of little use. The rain gauges have been more useful, though correlation between rainfall and lahars is always questionable for rain gauges that are far from the lahar source area. Data are telemetered in realtime to RDCC-III in Camp Olivas and to the PDCC in Iba, Zambales, where they are interpreted by engineers and relayed by radio operators to manned watchpoints for confirmation. During the height of the 1992 monsoon season, however, the rain gauges at Dalanaoan and Ugik stopped transmitting data, owing to power problems that were not remedied until after the lahar season was over.

LAHAR SEASON 2, MAY THROUGH NOVEMBER 1992

In 1992, to improve the technical accuracy of short-range warnings, we needed to answer the following questions:

What volumes of lahar discharge result from various amount of rainfall?

Can rainfall records and records from acoustic flow monitors (AFM's) serve as proxies, or even improvements, on direct observations from manned observation posts?

What levels of discharge are occurring, how and where is deposition occurring, and what channels are precariously full?

Is 1992 lahar activity changing relative to that of 1991, especially in any systematic way that can be of predictive value?

In June 1992, PHIVOLCS began systematic field monitoring of lahars in east-side channels, initially in the O'Donnell, Sacobia, Abacan, Pasig-Potrero, and Gumain channels and later reduced to the Sacobia and Pasig-Potrero. Staff reported peak discharges by radio, prepared simple hydrographs, recorded changes in the character of flows from start to finish, measured lahar temperatures, and took samples when practical (Arboleda and Martinez, this volume; Martinez and others, this volume). This activity was patterned after PLHT observations made on the Marella River in 1991; activity continued there and at the Bucao River in 1992 (Rodolfo and others, this volume; Umbal and Rodolfo, this volume).

Special effort was made to calibrate individual flow sensor records with observed discharge so that flow sensor data might be used as a substitute for manned observations. The first attempt at calibration, after lahar season no. 1, was for flows in the Marella watershed (Regalado and Tan, 1992). Similar work was done in the Sacobia watershed during lahar season no. 2 (Tuñgol and Regalado, this volume) and continued during lahar season no. 3 (Regalado and others, 1994). Though quantitative calibration of flow sensors in rivers other than the Sacobia is still ongoing, we can already use AFM records from all of these to distinguish the order of magnitude of discharge (for example, $10 m^3/s$, $100 m^3/s$, $1,000 m^3/s$, corresponding to small, medium, and large lahars). Qualitative information about the size of lahars is now included in PHIVOLCS' warnings to WPC/RDCC. In one early instrumental distinction of lahar size (August 1992), PHIVOLCS' advice of unusually strong lahar signals in the Sacobia was not used immediately by WPC/RDCC to order an evacuation, and 8 people downstream were killed by a major lahar. Since that time, somewhat greater credence has been given to instrumental indications of lahar size, but, even in the 1994 lahar season, people still find it much easier to visualize, and thus trust, a manned watchpoint's report of lahar depth (in feet) than

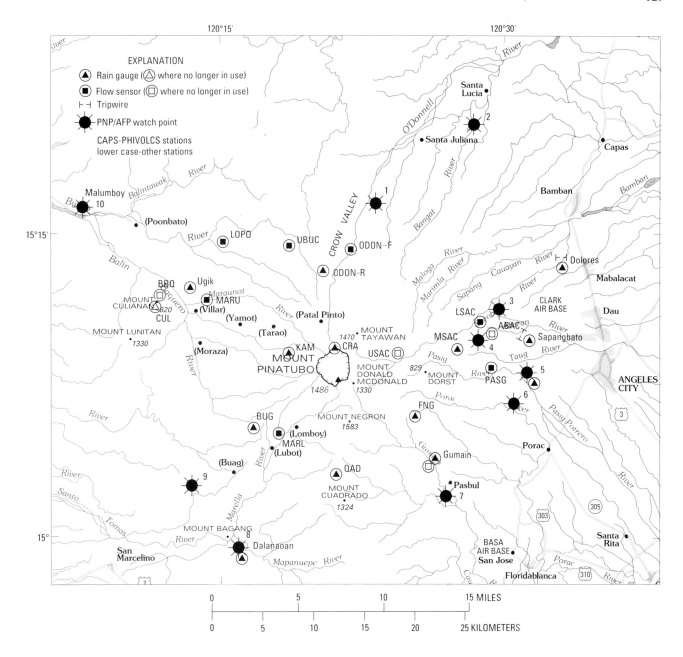

Figure 8. Sites of instruments and manned watchpoints for providing short-range warnings of lahar hazard around Pinatubo.

either a quantitative or qualitative report of discharge. Few among those receiving warnings seem to be bothered that flow depth depends strongly on the channel width.

Records of flow sensors and observed discharge were also correlated with rainfall in the Sacobia watershed during 1992 and 1993. The results have allowed us to give an even earlier alert to WPC/RDCC, even before flow is detected. The thresholds for lahar generation, measured during the 1992 lahar season and described under the section "Defining the Threat," are our present (1994) basis for issuing warnings to WPC/RDCC. Because these thresholds will rise as sediment yield (normalized to rainfall) decreases, we will need to revise thresholds for warning in future years.

Also during the 1992 lahar season, we briefly explored the possibility of computer modeling of lahars to improve our understanding of flow processes and to help forecast traveltimes and downstream discharges. However, the simplest model available to us (DAMBRK, PC version) had no rigorous way of adding or subtracting sediment, and we knew from field observations that these were important processes. We could force the model to replicate some aspects of observed flows, but we concluded that, at Pinatubo, the model had no predictive value (R. Dinicola, USGS, oral commun., 1992).

A special concern arose in July 1992 when Pinatubo threatened to erupt anew. Mindful of past disasters at Kelut Volcano in Indonesia, where a crater lake was repeatedly ejected by eruptions until engineers lowered the lake level through a series of tunnels, we made rough estimates of the volume of water in the caldera lake and how much of that water might be converted to lahars were the lake to be ejected suddenly. Estimates of lake volume ranged from 5×10^6 m^3 to 8×10^6 m^3 (November 1991 air photographs), depending mainly on assumptions about the lake's average depth. However, as a lava dome grew, our concern decreased, because the lake appeared shallow, perhaps as shallow as a few meters. The eastern half of the lake gradually disappeared during this period, from alluvial filling aided in small measure by uplift of the lake floor. (Note, at the time of this writing, July 1994, the lake is several tens of meters deeper and has a substantially greater volume than it did in November 1991.)

Late 1992 and early 1993 was also a time to reestablish a battered lahar warning system. PHIVOLCS-USGS rain gauges and flow sensors required cleaning, repair, and replacement, and several needed to be moved to higher elevations in the lahar source regions to obtain more direct information about events in the source regions, longer lead times in warnings, and reduced risk of vandalism. This work was completed just before the 1993 lahar season.

LAHAR SEASON 3, MAY THROUGH NOVEMBER 1993

The upgraded network of rain gauges and flow sensors operated successfully through the 1993 lahar season, with minimal periods of instrument failure. As expected, the threshold for lahar generation did rise slightly (Regalado and others, 1994), but not enough to change procedures for issuance of alerts. Alerts from PHIVOLCS to RDCC typically started with an alert that strong rain was falling on the upper slopes, followed by qualitative updates on the size of a resulting lahar ("small," "moderate," or "large," on the basis of the amplitude and duration of the AFM signal). Manned watchpoints provided similar information about the size of lahars, expressed as depth of the lahar (in feet). The latter reports were widely monitored by other watchpoints and local officials. Lahar watchers quickly realized that flows spread and peaks attenuate as they move downslope, so that by the time most flows reach downstream areas, reported "10-foot" lahars might be only a foot or two deep.

Some of the forward (higher elevation) manned watchpoints were discontinued in 1993, and more were discontinued in 1994, placing greater demands on PHIVOLCS instruments for initial warnings and on secondary (lower elevation) watchpoints for reports on arrival times and depths of lahars. As of this time of writing (July 1994), the formal lahar alert scheme from RDCC still retains the original 3 levels, with essentially the original meanings noted earlier. No formal distinction is made to indicate the size of an advancing lahar; rather, the depth of flow (in feet) and intensity of steaming are reported, with or without size information from PHIVOLCS, through a radio network of local observers at secondary watchpoints.

RESPONSE TO WARNINGS

The public's first and most effective response to lahar warnings was to move out of the way of the lahars. On June 14, 1991, PHIVOLCS' recommendation for general evacuation (principally for the eruption but also for lahars) was extended from 20 to 30 km in radius from Pinatubo's summit, and on June 15 it was extended again, to 40 km in radius (Punongbayan and others, this volume). During, and for weeks after, the climactic eruption, between 150,000 and 200,000 persons filled more than 300 evacuation camps (United Nations Disaster Relief Organization, UNDRO, unpub. data, 1991). Not everyone within 40 km of Pinatubo could or did evacuate; our impression, based only on anecdotal evidence, is that most of those who evacuated did so from fear of the eruption and associated earthquakes and lightning and not from the threat of lahars. But by so doing, some people also protected themselves from lahars.

After the eruption, evacuations were ordered every time Lahar Alert 3 (confirmed lahar—go to safe ground) was raised. Initially, evacuation recommendations were for all low-lying areas around the volcano; later, they were made on a river-by-river basis. Estimating the number of evacuees is complicated by having different barangays and families evacuating at different times, in some cases several times. The population of a reduced number of lahar-safe evacuation camps (~160) fluctuated seasonally, depending on rainfall. During periods of little rain and relatively low risk, the evacuee population dropped below 25,000 persons; during periods of heavy rain and strong lahars, the evacuee population swelled to more than 150,000 persons (UNDRO, unpub. data, 1992; Mercado and others, this volume). By the end of 1992, more than 50,000 persons had lost their homes entirely and another 150,000 had suffered some flooding or lahar damage (C.B. Bautista, this volume), and most of those people had been helped in evacuation centers.

As soon as the persistent character of the crisis became apparent, semipermanent and permanent resettlement sites were proposed as an alternative to evacuation camps. At about 10 previously undeveloped sites, the government built marketplaces, schools, health clinics, and buildings to attract and house manufacturing and other industry. Paved roads and utilities were, in many instances, of even higher standard than that in communities from which evacuees had come. Would-be settlers were offered low-interest loans to buy residential lots and to build houses. However, jobless, homeless families are reluctant to take on loans that they cannot repay, and for this and a variety of other reasons, displaced families have been slow to move into resettlement areas (C.B. Bautista, this volume). Virtually all of those who have moved from evacuation camps to more permanent resettlement areas have been replaced in the evacuation camps by families that are newly displaced by ongoing lahars.

Concurrently with operation of evacuation camps and development of resettlement areas, engineers from the Philippines, the United States, Switzerland, New Zealand, and Japan considered engineering measures to control the hazard itself. Politicians' reluctance to sacrifice any areas and engineers' training led them to try conventional channel maintenance measures—including construction of small "sabo dams" to trap sediment and stabilize base level, dredging of channels, and raising of dikes. The initial goal, more hopeful than realistic, was to contain the sediment within existing channels and other relatively small areas. Numerous small sabo dams were built but were promptly overrun (fig. 9). The dams were so hopelessly undersized that most scientists, and many engineers, recognized that they were little more than an employment program for local residents displaced by the eruption and visible evidence that the government was doing something rather than nothing.

Not surprisingly, the worst problems of lahar incursions into populated areas were on unconfined alluvial fans of the Pasig-Potrero, Sacobia-Bamban, and Marella-Santo

Tomas Rivers rather than in the deeper Bucao and O'Donnell valleys. Shallow channels on the alluvial fans filled quickly, and much overbank flow occurred. The dilemma for engineers quickly became apparent: the alluvial fans were not conducive to sediment storage, whereas the deeper valleys in which sediment could be trapped had relatively few people at risk.

When larger dikes were proposed to trap sediment on the fans of the Sacobia-Bamban, Pasig-Potrero, and Marella-Santo Tomas Rivers, we argued that the volumes of sediment would be so great that the only practical solution would be to let that sediment spread over relatively large "catch basins" outside current channels and in locations where sediment would naturally deposit. Most of the potential catch basins were in cultivation. In June 1992, PHIV-OLCS and DPWH proposed catch basins to residents of Bacolor and Santa Rita, Pampanga (fig. 10), but the proposal was angrily rejected. No one was prepared to allow his own land to be buried.

Neither an engineer nor nature waits for debate, and, from 1991 to this time of writing, dikes on all sides of Pinatubo have been built, and all have been breached at their weakest points or points with the lowest freeboard during actual lahars. Failure at any point along a dike renders downstream portions of that dike useless (Rodolfo and others, 1993). Most failures of dikes to date have stemmed from the simple geometric difficulty of containing very large volumes of sediment within narrow confines, a faulty assumption that sediment would be deposited evenly along the full length of a diked channel (it will not), or an underestimate of the erosive power of lahars on dikes. After the 1993 lahar seasons, new, larger and stronger dikes were constructed along the outer perimeter of many areas that had been buried by lahar. In effect, catch basins were created, though they were not called by this name. These larger, widely spaced dikes are surely more realistic measures than their predecessors, but they will still be tested by remaining lahars, especially in their upper reaches where sediment fill is the fastest.

One successful engineering measure was excavation of a spillway to stabilize the level of Mapanuepe Lake, newly impounded behind lahar deposits of the Marella River, and thus to prevent overflows and potentially catastrophic breakout floods. Meetings between PLHT, DPWH engineers and consultants, Dizon Mines management and engineers, and local officials led to an engineering solution in which a permanent spillway was excavated into bedrock at the southern margin of the lake, roughly 300 m away from the present edge of the lahar dam. Excavation started on October 23, 1992, and the spillway first opened on November 20, 1992, fixing lake level at or below 121 m in altitude.

In July 1992, PHIVOLCS had preliminary discussions with DPWH and Gov. M. Cojuangco of Tarlac Province to discuss a creative, promising combination of sociopolitical and engineering measures. In that combination, the

Figure 9. Overtopped and breached sediment dam ("sabo dam"), in the Sacobia River just northwest of Clark Air Base. *A*, Aerial view, looking downstream. Small hut and person can be seen for scale. *B*, View from the right bank of the downstream face of the dam.

government would temporarily lease land to be used as catch basins, until the new sediment fill could be returned to agricultural production. Towns around the margins of the catch basin would be protected by dikes. The scheme recognized that it would be much easier to protect the town of Bamban if sediment were allowed to spread around it; otherwise, efforts to confine sediment within dikes of the Bamban River were likely to assure burial of Bamban. Money to lease the land and to provide other incentives for relocation would come in part from money intended for engineering works. Regrettably, decisions were made to keep building

narrow, undersized sediment-containment dikes along the Bamban and other rivers. At the same time, with strong ties to family land, no strong incentives for relocation, and seemingly high and substantial new dikes in their backyards, most residents stayed in their villages, even in high-risk areas, and left only when or just before their communities were overrun by lahars. In 1992, lahars buried parts of the proposed leases in Tarlac Province before any formal leasing action was taken, and then the main path of lahars from the Sacobia shifted to the Pampanga side of the river, where government leasing of land had not been seriously

More 'catch basins' needed in lahar areas

By DING CERVANTES

SAN FERNANDO, Pampanga — The need to excavate more "catch basins" has cropped up anew to save populated areas and productive lands in Central Luzon from being buried by Mt. Pinatubo's lahar.

Public Works Secretary Jose de Jesus, in an interview here the other day, referred to it as "the political solution" to the lahar problem in the region, as he stressed that sabo dams and dikes are not sufficient safeguards to the recurring problem of lahar flowing from the slopes of the volcano.

·He urged the local officials to take the lead in enlightening their constituents on the urgent need for the basins to serve as receptacles for lahar sediments.

Sabo dams and dikes have been constructed by DPWH in the various lahar river channels in Zambales, Pampanga and Tarlac but catch basins are still needed to provide additional safeguards, De Jesus emphasized.

"Imagine a square whose sides each extends from Manila to Baguio and fill in this square with one-meter thick lahar," he said to illustrate his point. He said this space would accommodate only half of the seven billion cubic meters of volcanic materials resting on Mt. Pinatubo's slopes.

The proposed catch basins are areas already buried by lahar and abandoned by their residents last year. In Barangay Sta. Barbara in Bacolor, Pampanga, the proposal was strongly opposed by residents who have not returned to their buried homes up to now and are awaiting relocation to permanent and safer grounds.

Turn to Page 12

More 'catch basins'

From Page 1

"Our idea is for them to swap their already buried lands with the homelots they would be entitled to in the resettlement areas," De Jesus told newsmen.

De Jesus, however, lamented that the "political solution" to the lahar problem in Central Luzon has yet to be faced by local officials.

He said, however, that governor-elect Margarita "Tingting" Cojuangco of Tarlac and other local officials in Tarlac are now considering his proposal following the disclosure of government scientists that large volumes of volcanic deposits on the eastern slopes of Mt. Pinatubo would flow towards the Sacobia river affecting Bamban, Capas, and Concepcion in the province.

De Jesus described the response of the local officials as "an enlightening development."

Meanwhile, De Jesus also said that school buildings will still be used as evacuation centers in Central Luzon for "lack of choice."

He said that while the government is rushing the development of permanent resettlement areas in the region, they would not be enough to accommodate 89,000 more families expected to flee to existing evacuation centers this rainy season.

He said that by July 8, about 9,000 families still in temporary evacuation centers are expected to be transferred to permanent resettlement sites, the tent cities they would leave would not be enough for fresh evacuees.

De Jesus said that for an additional 50,000 families alone, the government would need P400 million for building temporary shelters.

The Mt. Pinatubo resettlement committee is now rushing the completion of "horizontal structures" such as roads and water facilities in 12 permanent resettlement sites in Zambales, Pampanga and Tarlac for lowland evacuees.

As soon as the structures are finished, the families would be transferred from their temporary shelters so that they could start building their homes at the sites from a P20,000 maximum loan from the government. The permanent sites have been subdivided into 150 square meter homelots

Figure 10. Newspaper headlines describing government efforts to convince local residents of need for sediment catch basins (Manila Star, June 21, 1992).

considered. Thus, on both sides of the river, residents lost their life savings and livelihood **and** lost any opportunity that the government would lease or buy their land.

Debate continued in late 1992 about the most effective long-term measures for mitigating lahar risk. Our impression is that debate became more polarized, between those who advocated engineering structures and those who advocated that available funds be spent for resettlement and other social measures. In an ongoing saga of trial and error, each side could now point to serious problems with the other—for example, failed dikes and unpopular, half-vacant resettlement areas. In defense, each side could point to its own successes. In some instances, media coverage misrepresented positions, setting up "pro-dike" and "anti-dike" sides when, in fact, principal points were that (1) dikes might, in some instances, be viable engineering options, and (2) dikes ought to be built only if they would be large enough and strong enough to actually work as intended and not inadvertently introduce a false sense of security or an unduly long-term maintenance problem (for example, Rodolfo, 1992). These are different statements of the same conclusion, but some media groups saw them as different conclusions.

Many of the practical, philosophical, and emotional issues that arose bore a marked similarity to issues raised by dike construction that keeps the lower Mississippi River from being captured by the Atchafalaya River (McPhee, 1989). In general, dike construction made more sense to those at risk than to disinterested observers.

In August and October of 1993, as in 1992, policymakers of the Mount Pinatubo Commission were forced by serious lahars to reexamine their long-range mitigation plans. Dredging and dike construction during the 1992–93 dry season was no match for the volume of 1993 lahars, so channels on the alluvial fans soon became choked with sediment, some of the dikes were overtopped, and several towns were badly damaged. Early lahars of the 1994 rainy season have already overtopped some dikes that were constructed or raised during the 1993–94 dry season, and greatly reduced freeboard at others. In a long-term view, undersized dikes have had little effect on where lahar sediment has been deposited, and most might just as well have not been built. They have postponed inevitable burial for some communities, and they might ultimately save a few communities right at the edge of where sediment would have naturally spread. But they have been built at great cost—roughly half of the monies available for coping with this disaster, plus loss of opportunities to use the same monies for nonstructural or even better structural measures, and losses attributable to a false sense of security until lahars were at hand.

Among the questions raised anew were:

What engineering solutions **can** *be technically viable? In order to be viable, what design specifications would they need?*

If an engineering solution saves a town for a year, or so, is it worth the associated costs? The direct cost of the dike is simple to calculate; indirect costs, and the value of saving a town for a year, are more subjective.

Are some engineering solutions actually aggravating lahar problems, for example, by initially helping lahars to reach far downstream, and now by causing avulsions to move upstream and thus damage towns on the edges of alluvial fans?

Politically, economically, and scientifically, what are the most practical options for catch basins, that is, land that can be used for sediment storage until the lahar crisis has passed?

We are neither engineers nor policymakers so we will not presume to answer these latter questions. Rather, we mention them to illustrate the continuing dilemma faced by all who seek to mitigate the lahar hazards of Mount Pinatubo: how to help people stay in their communities wherever possible and yet not waste resources where communities cannot be defended against lahar without spending an inordinate amount of money and (or) running a high risk that lahar defenses will ultimately fail. Scale and topography are major parts of the dilemma: solutions that might have worked where lahars are smaller, governments are wealthier, and valleys are deeper, are not readily transferable to Pinatubo. Denial of the hazard and difficulties in reaching political consensus are also important factors. Rather than asking that sediment be kept from all populated and cultivated areas, citizens and their leaders would do better to decide which areas must be protected and which can be temporarily sacrificed and to ensure that both hazard and compensation are fairly distributed.

SUGGESTIONS FOR FUTURE WARNINGS

Can lahar warnings at Pinatubo be improved? Yes, without question. Some improvements relate to incomplete data and uncertainties about process, and others relate to when and how warnings are conveyed. For technical improvements of long-range forecasts, we need to keep tracking:

- Sediment budgets, including how much sediment that predates 1991 is being incorporated into current lahars, how much sediment is flushing straight through the river systems, and, especially, how rapidly the annual sediment yields are declining.
- Details of topographic change in lowlands, with a contour interval of 5 m or less. Excellent maps have been made for limited areas, for a few flight dates, but to our knowledge there are no plans to make repeated, high-precision photogrammetric surveys of topographic change around the entire volcano.
- Reaches of lahar deposition, because those reaches will be the sites of the next channel avulsions and overbank lahar damage. This effort requires frequent remeasurements of cross sections of each major river draining Pinatubo.

Increasingly, GIS technology can be used to analyze, display, and disseminate current information about watershed response and risk. All long-range planning for lahar mitigation must anticipate and be adaptable to continued, rapid, major changes in the hydrologic system, such as stream captures. For example, captures of Abacan drainage by the Sacobia, and part of the Sacobia drainage by the Pasig-Potrero, are unlikely to be the last such events, and GIS-based hazard maps can be changed accordingly.

For technical improvements in short-range forecasts, we still need:

- Timely transmission of typhoon tracks and detailed evaluations of southwest monsoon air masses, from PAGASA to scientists and RDCC-III personnel at Pinatubo.
- Continued operation of rain gauges and flow sensors on the upper and middle slopes of Pinatubo, with commensurate technical support for those instruments and calibration of flow sensors to actual discharges.
- Better realtime communication with WPC/RDCC, perhaps aided by a duplicate data-receiving station at WPC to serve as a backup and to counteract skepticism that we encountered in 1991–1992. A more ambitious extension of this would be to link all lahar watchers—police, military, scientists, and others—by a robust radio communications network, supplemented where possible by voice telephone and facsimile machines.

Equally or even more important than technical refinements in forecasts, we need to strengthen the channels by which our scientific results are actually used in lahar mitigation measures. Four specific improvements in our presentation would be:

- Additional, highly visual presentations of hazard, such as videotapes and working models that are prepared in cooperation with those who are experienced in shaping public opinion and public policy.
- Warnings that speak clearly and firmly despite inescapable uncertainties in the outcome of such complex events as lahars. A graphic illustration of this need came during raging lahars in early September 1992, when several villages were being overrun by lahars and when the chief of RDCC Region III (Gen. Pantaleon Dumlao) said to one of us, "Yes, you warned us, but you didn't shout loudly enough." Scientists are trained to be conservative in their warnings, but there are times when we must make our best guesses loudly and clearly, even without a full set of data.
- Better use of scientific consensus in public statements. Differences of opinion among scientists are inevitable and have occurred. Some have concerned purely scientific matters; others have concerned recommendations for mitigation. Unfortunately, these differences have been played out in the newspapers rather than in private discussions among scientists, and we have therefore lost some credibility, as a group, in the eyes of decisionmakers. Development of *written* scientific consensus *before* public statements would help all parties.
- Equal attention to decreasing lahar hazard as to increasing hazard, partly to avoid undue alarm, partly to retain our credibility for those times when urgent warnings are needed, and partly to help people return to their land where such return is reasonably safe.

FUTURE LAHAR HAZARD

The nature of Pinatubo lahar threat was discussed in earlier sections. Here, we call special attention to three anticipated **changes** in hazard: extreme rainfall; diminution of annual sediment yield; and continued, even increasing, distal sedimentation.

EXTREME RAINFALL

Rainfall in the Pinatubo area is highly variable from one year and one month to the next, depending on typhoons and on sustained "siyam-siyam" southwest monsoon rainfall. Rainfall during 1991 and 1992 was close to the long-term mean (table 2). In contrast, that at Cubi Point in 1993 (7,165 mm, with 3,102 mm in August alone) was far above the mean and set new records for that station. Rainfall in excess of 2,500 mm/yr in Manila and Clark Air Base occurs, *on average*, about once every 10 yr, most recently in 1972 (>3,000 mm in the July–August period) and 1986 (>2,000 mm in August alone); none of our east-side stations approached these high levels yet, but, clearly, such levels can be reached.

Daily rainfall can also be heavier than has occurred since the eruption. We do not know the maximum daily rainfall during 1991 to 1993 at Clark Air Base or that at Cubi Point during 1993; during 1991 and 1992, the 24-h maximum at Cubi Point was 280 mm (J.S. Oswalt, oral commun., 1991). We have recorded higher amounts at many of our stations on Pinatubo itself (highest=744 mm in 24 h, from 0600 July 25 to 0600 July 26, 1994 at QAD), but we do not have a long-term average against which that could be compared.

Thus, in terms of the heavy rainfall that can generate severe lahars, rainfall in 1991 to 1992 was well below historical maxima, and within 1 standard deviation of average values. 1993 rainfall at Cubi Point was an historical maximum, while that at other west-side stations was probably normal. We expect that the next decade will include at least one additional year with heavy rainfall, >2,500 mm/yr on the east side and >5,000 to 6,000 mm/yr on the west side.

DIMINUTION OF ANNUAL SEDIMENT YIELD

Lahars and rapid sediment accumulation will diminish and eventually stop. In terms of sediment volume, we are at the time of this writing (July 1994) about two-thirds through the crisis (fig. 3; table 3); we are farther along in some east-side watersheds and not as far along on the west. Both the geographic scope and the timescale of further mitigation options, including resettlement and engineering measures, should recognize that sediment yields are diminishing rapidly and will soon become more manageable, perhaps without need to undertake massive measures.

CONTINUED DISTAL SEDIMENTATION

As vegetation recovers and as other aspects of the 1991 watershed disturbance heal, runoff, like sediment yield, will diminish. However, high rates of runoff may linger a few years longer than high rates of sediment yield. If so, two phenomena will result. First, bank erosion in downstream channels will be high, so maintenance of remaining dikes will remain a major task. Second, some of that erosion will be of lahar deposits themselves, and the remobilized sediment will be washed farther downstream. One can think of Pinatubo sediment as moving into distal lowlands in two steps—first as lahars that are deposited on alluvial fans around Pinatubo and second as gradual but persistent movement of that sediment into more distant channels, for years or even decades after lahars have ended. Low-gradient river channels, even far from Pinatubo, will experience gradual siltation and diminution of channel capacity. As that capacity diminishes, normal rains will produce overbank flooding. Dredging will be required as a long-term flood-control measure in those areas, and the need for distal dredging may soon outstrip the need for dikes and other structures nearer to Mount Pinatubo.

CONCLUSIONS

Erosion and lahars in posteruption Pinatubo watersheds have filled lowland channels at rates that are unprecedented in the history of volcano hazards mitigation or sediment control. Almost all of the sediment transport occurs as lahars, which are episodic events dependent on heavy rain. Sediment yields at Pinatubo, on the order of 10^6 m^3/km^2/yr during 1991–93, were an order of magnitude larger than at Sakurajima and at Mount St. Helens immediately after eruptions.

Rates of change in the lahar hazard are as remarkable as the initial magnitude of that hazard. Sediment yields are declining, on average by 20–30 percent per year and in some drainages by half or more each year. Sediment yields have decreased most sharply in east-side watersheds, where volumes of 1991 pyroclastic-flow deposits are small to moderate, slopes are relatively steep, and vegetation recovery is relatively rapid. In the Sacobia-Pasig watershed, three-fourths of the expected sediment has already arrived in the lowlands. Even on the west side, more than half of the expected sediment has arrived in the lowlands. Rates of sediment yield in 1993, normalized for watershed area and rainfall, ranged from one-third of 1991 values (in the Sacobia-Pasig system) to 100 percent (or more?) of 1991 values (in the Balin Baquero-Bucao) (table 4).

Despite long-range decline of sediment yield, days or weeks of extreme rainfall can still generate bank-full or overbank flow of several thousand cubic meters per second of dense lahar slurry. As channels are filled to capacity, each

new storm and rainy season, and especially extreme rainfall, will cause new avulsions and shift flows wildly from one side of an alluvial fan to the other. One day or one season a community might be relatively safe; the next, it might be buried in lahar deposits. Then, as quickly as it was buried, that area might be isolated from future hazard.

Scenarios of lahar hazard offered by PHIVOLCS and PLHT/ZLSMG—long-range and immediate—have been dire yet generally accurate. Undersecretary of Public Works and Highways T. Encarnacion said to us, in 1992, "You were right…unfortunately." Some of our warnings have been based on hard data, others on general experience and intuition. We have walked a fine line between issuing timely warnings (some were backed only by sparse data) and reliable warnings (backed by ample data). Most citizens and most public officials have responded constructively, being willing to make precautionary evacuations, even after a few false alarms. Mutual understanding, courage to risk being wrong, and tolerance in case of false alarms are essential.

Hazard maps, booklets, and videos have correctly identified the most hazardous areas; immediate warnings from rain gauges, flow sensors, and manned watchpoints alerted people who remained in these communities to move to high, safe ground at particularly critical times. Of the three sources of immediate warnings, rain gauges and manned watchpoints were immediately useful, while flow sensors are gradually gaining acceptance. Civil defense has tried with reasonable success to fine-tune evacuations so that they are neither too large nor too small. After serious problems with false alarms in 1991, warnings are now reliable enough that most people are choosing to evacuate when warned, albeit at the last minute.

Of the more than 50,000 people who lost their homes to lahars, an unknown but substantial percent would have been killed had there been no warnings. Actual deaths from lahars, in contrast, are approximately 100—many times fewer than might have been killed without warnings. Jeepneys (local public minivans) and dump trucks full of property—furniture, household goods, farm animals, and harvested crops—accompany the exodus of evacuees, so we judge that much portable property has also been saved, though poignant scenes of people digging through fresh mudflow deposits to salvage belongings indicate that much has also been lost.

Hazard maps and estimates of long-term sediment yield have been important input to some decisions on long-term lahar mitigation. Major relocation sites have been sited (or, in the case of Rabanes, canceled) on the basis of such maps. Year-to-year emergency plans have been made on the basis of such maps. But only some decisions about engineering control of lahars have utilized available scientific information. Many small dams and small and large dikes have been built in the face of scientific judgment that they are too small or in places where any structure will be

Table 5. Differences in perception and approach between various parties in Pinatubo lahar mitigation.

Geoscientists:
- approach natural hazards as forces to be understood and, when necessary, avoided, rather than as forces to be controlled.
- think over a wide range of spatial and time scales, from cubic centimeters to cubic kilometers and from minutes to millennia.
- think of natural hazards as complex, ongoing, ever-changing processes that need to be studied continuously. Frequent changes in facts and interpretations seem perfectly natural, indeed, good.
- must often be the bearers of bad news and must therefore be prepared for hostile receptions.

Engineers:
- are trained to control or otherwise design against natural hazards (floods, landslides, and others). To not do so is professionally awkward, just as it is for a doctor to "give up" on a patient.
- think principally on the scale of their structures and the design life of those structures.
- are trained to design with available or quickly obtainable data and to add a factor of safety for future change. Opportunities to change design after construction are limited

Residents:
- view natural hazards as God's will, largely beyond their control but possibly preventable by prayer and engineering works.
- must deal with harsh reality of finding food and shelter, which before a lahar is certainly easiest on their own land.
- have emotional as well as economic attachment to the places in which they were born and raised.

Politicians and other policymakers:
- view natural hazards as political and policy problems.
- need to make decisions on technical issues even when technical advice is conflicting or uncertain, and even when the natural situation is sure to change.
- are caught in a quandary between being sympathetic to public pressure for quick fixes, yet knowing that resources for coping with the lahar hazard are limited and must be used to everyone's advantage.
- are sensitive to issues of reelection and duration of their own responsibility.

overrun. Information about expectable volumes of sediment is only one and sometimes a relatively minor factor in mitigation decisions, in competition with differences in perception of both risk and solutions (table 5), peoples' reluctance to leave and "sacrifice" their land, a lack of attractive alternatives for those at risk, political competition, and cost of mitigation measures.

ACKNOWLEDGMENTS

We are pleased to acknowledge lively discussions with those to whom we provide warnings. They have constructively critiqued and used our warnings, sometimes even before we have decided exactly what the warning should be. We also thank those who have supported our work, financially and logistically, including His Excellency Fidel V. Ramos (as former Chairman of the National Disaster Coordinating Council and currently as President of the Republic of the Philippines), Secretary Renato de Villa (Chairman, National Disaster Coordinating Committee), Secretary Corazon Alma de Leon (Chairperson of the Mount Pinatubo Commission), Secretary Jose P. de Jesus and Secretary Gregorio Vigilar (Department of Public Works and Highways), Secretary Ricardo T. Gloria (Department of Science and Technology), Congressmen Rolando Andaya and Emigdio Lingad (Chairman and Vice-Chairman, respectively, of the House Committee on Appropriations), Gen. Pantaleon Dumlao (Chairman of the Regional Disaster Coordinating Committee, RDCC III), Mayor Richard J. Gordon (now Chairman of the Subic Bay Metropolitan Authority) and Congresswoman Katherine H. Gordon (the First District of Zambales), Mayor Sally Deloso of Botolan, Director Joel Muyco of the Mines and Geosciences Bureau, Dr. Eduardo Sacris of Dizon Mines (Benguet Corporation), faculty of the University of the Philippines' National Institute of Geological Sciences, other friends of the PLHT/ ZLSMG, USAID/Philippines, USAID/Office of Foreign Disaster Assistance, the Philippine Air Force, and U.S. military forces in the Philippines. Yvonne Miller of the University of Geneva and Mike Dolan of Michigan Technological University joined our efforts in 1992 and 1993 as student volunteers; Sheila Agosa of the University of Washington joined as a student volunteer in 1994. The work of PLHT in 1991 and ZLSMG in 1992–93 was supported by grants from the Earth Sciences and International Programs Divisions of the National Science Foundation. Numerous private individuals in the United States contributed to an emergency fund to support PLHT in 1991.

We acknowledge the seminal contributions of our late colleague, Dick Janda, and honor his memory with senior authorship. While most of us were still preoccupied with day to day events, Dick was already thinking in terms of watershed processes and how an understanding of those processes would lead to better forecasts.

REFERENCES CITED

Arboleda, R.A., and Martinez, M.L., this volume, 1992 lahars in the Pasig-Potrero River system.

Bautista, C.B., this volume, The Mount Pinatubo disaster and the people of Central Luzon.

Bureau of Soils and Water Management, 1991, Mudflow and siltation risk map: Bureau of Soils and Water Management, Quezon City, 1 sheet, scale 1:250,000.

Cola, R.M., this volume, Responses of Pampanga households to lahar warnings: Lessons from two villages in the Pasig-Potrero River watershed.

Hadey, K.C., and LaHusen, R.G., 1995. Technical manual for an experimental acoustic flow monitor: U.S. Geological Survey Open-File Report 95–114, 24 p.

Iwakiri, T., 1992, Photo album on lahar warning systems around Mount Pinatubo: unpublished report, Japan International Cooperation Agency, 27 p.

Janda, R.J., Meyer, D.F., and Childers, D., 1984, Sedimentation and geomorphic changes during and following the 1980–1983 eruptions of Mount St. Helens, Washington (parts 1 and 2): Shin Sabo, v. 37, no. 2, p. 10–21 and v. 37, no. 3, p. 5–19.

LaHusen, R.G., 1994, Real-time monitoring of lahar using ground vibrations [abs.]: Abstracts with Programs, Geological Society of America, Annual Meeting, Seattle, p. A–377.)

Major, J.J., Janda, R.J., and Daag, A.S., this volume, Watershed disturbance and lahars on the east side of Mount Pinatubo during the mid-June 1991 eruptions.

Marcial, S.S., Melosantos, A.A., Hadley, K.C., LaHusen, R.G., and Marso, J.N., this volume, Instrumental lahar monitoring at Mount Pinatubo.

Martinez, M.L., Arboleda, R.A., Delos Reyes, P.J., Gabinete, E., and Dolan, M.T., this volume, Observations of 1992 lahars along the Sacobia-Bamban River system.

McPhee, John, 1989, The control of nature: New York, Farrar Straus Giroux, 272 p.

Mercado, R.A., Lacsamana, J.B.T., and Pineda, G.L., this volume, Socioeconomic impacts of the Mount Pinatubo. eruption.

Newhall, C.G., Daag, A.S., Delfin, F.G., Jr., Hoblitt, R.P., McGeehin, J., Pallister, J.S., Regalado, M.T.M., Rubin, M., Tamayo, R.A., Jr., Tubianosa, B., and Umbal, J.V., this volume, Eruptive history of Mount Pinatubo.

Philippine Institute of Volcanology and Seismology, 1992a, Lahar hazards of Mount Pinatubo: unpublished map, 1 sheet, 1:50,000.

——1992b, Ang Lahar: PHIVOLCS, Quezon City, 19 p. + maps.

——1994, Lahar studies: Quezon City, PHIVOLCS Press, 80 p. [Report of studies sponsored by UNESCO.]

Philippine Institute of Volcanology and Seismology and the National Economic Development Authority, 1992a, Pinatubo lahar hazards map: National Economic Development Authority, Manila, 1 sheet at scale 1:200,000; 4 sheets, Quadrants 1–4, at scale 1:100,000.

——1992b, Pinatubo Volcano Lahar Hazards Map: National Economic Development Authority, Manila, 4 sheets at scale 1:200,000.

Pierson, T.C., and Costa, J.E., 1994, Trends in sediment yield from volcanoes following explosive eruptions [abs.]: Abstracts with Programs, Geological Society of America, Annual Meeting, Seattle, p. A–377.

Pierson, T.C., Daag, A.S., Delos Reyes, P.J., Regalado, M.T.M., Solidum, R.U., and Tubianosa, B.S., this volume, Flow and deposition of posteruption hot lahars on the east side of Mount Pinatubo, July–October 1991.

Pierson, T.C., Janda, R.J., Umbal, J.V., and Daag, A.S., 1992, Immediate and long-term hazards from lahars and excess sedimentation in rivers draining Mt. Pinatubo, Philippines: U.S. Geological Survey Water-Resources Investigations Report 92–4039, 35 p.

Pinatubo Lahar Hazards Taskforce (PLHT) (Umbal, J.V., and Rodolfo, K.S., compilers), 1991a, Lahar hazard map of the Santo Tomas river system, Pinatubo Volcano, Luzon, Philippines (as of August 23, 1991): unpublished map, 1 sheet, scale 1:36,400.

Pinatubo Lahar Hazards Taskforce (PLHT) (Paladio, M.L., Umbal, J.V., and Rodolfo, K.S., compilers), 1991b, Lahar hazard map of the Bucao river system, Pinatubo Volcano, Luzon, Philippines (as of September 5, 1991): unpublished map, 1 sheet, scale 1:36,400.

Punongbayan, R.S., Besana, G.M., Daligdig, J.A., Torres, R.C., Daag, A.S., and Rimando, R.E., 1991a, Mudflow hazard map: PHIVOLCS-NAMRIA, 1 sheet, scale 1:100,000 (released July 31, 1991).

——1991b, Mudflow hazard map: PHIVOLCS-NAMRIA, 1 sheet, scale 1:100,000 (released October 15, 1991).

Punongbayan, R.S., Newhall, C.G., Bautista, M.L.P., Garcia, D., Harlow, D.H., Hoblitt, R.P., Sabit, J.P., and Solidum, R.U., this volume, Eruption hazard assessments and warnings.

Punongbayan, R.S., Tungol, N.M., Arboleda, R.A., Delos Reyes, P.J., Isada, M., Martinez, M., Melosantos, M.L.P., Puertollano, J., Regalado, T.M., Solidum, R.U., Tubianosa, B.S., Umbal, J.V., Alonso, R.A., and Remotigue, C., 1994, Impacts of the 1993 lahars, long-term lahars hazards and risks around Pinatubo Volcano, in Philippine Institute of Volcanology and Seismology, Lahar Studies: Quezon City, PHIVOLCS Press, p. 1–40.

Punongbayan, R.S., Umbal, J., Torres, R., Daag, A.S., Solidum, R., Delos Reyes, P., Rodolfo, K.S., and Newhall, C.G., 1992a, A technical primer on Pinatubo lahars: PHIVOLCS, Quezon City, 21 p.

——1992b, Three scenarios for 1992 lahars of Pinatubo Volcano: unpublished report, PHIVOLCS, 15 p.

Regalado, M.T., Delos Reyes, P.J., Solidum, R.U., and Tuñgol, N.M., 1994, Triggering rainfall for lahars along the Sacobia-Bamban River in 1993, in Philippine Institute of Volcanology and Seismology, Lahar Studies: Quezon City, PHIVOLCS Press, p. 41–45.

Regalado, M.T., and Tan, R., 1992, Rain-lahar generation at Marella River, Mount Pinatubo: unpublished report, University of the Philippines, 6 p.

Rodolfo, K.S., 1991, Climatic, volcaniclastic, and geomorphic controls on the differential timing of lahars on the east and west sides of Mount Pinatubo during and after its June 1991 eruptions [abs.]: Eos, Transactions, American Geophysical Union, v. 72, no. 44, p. 62.

————1992, The continuing Pinatubo lahar threat: unpublished position paper, October 28, 1992, 7 p. Excerpts published in "At Large," Philippine Daily Inquirer, November 11–12, 1992.

Rodolfo, K.S., and Umbal, J.V., 1992, Catastrophic lahars on the western flanks of Mount Pinatubo, Philippines: Proceedings, Workshop on Effects of Global Climate Change on Hydrology and Water Resources at the Catchment Scale, February 3–6, 1992, Tsukuba, Japan, p. 493–510.

Rodolfo, K.S., Umbal, J.V., and Alonso, R.A., 1993, A rational analysis of the Santo Tomas-Pamatawan Plain and lahar system of southwestern Mount Pinatubo, Zambales Province, Philippines: unpublished report, Zambales Lahar Scientific Monitoring Group, 16 p.

Rodolfo, K.S., Umbal, J.V., Alonso, R.A., Remotigue, C.T., Paladio-Melosantos, M.L., Salvador, J.H.G., Evangelista, D., and Miller, Y., this volume, Two years of lahars on the western flank of Mount Pinatubo: Initiation, flow processes, deposits, and attendant geomorphic and hydraulic changes.

Scott, K.M., Janda, R.J., de la Cruz, E., Gabinete, E., Eto, I., Isada, M., Sexon, M., and Hadley, K.C., this volume, Channel and sedimentation responses to large volumes of 1991 volcanic deposits on the east flank of Mount Pinatubo.

Scott, W.E., Hoblitt, R.P., Torres, R.C., Self, S, Martinez, M.L., and Nillos, T., Jr., this volume, Pyroclastic flows of the June 15, 1991, climactic eruption of Mount Pinatubo.

Torres, R.C., Self, S., and Martinez, M.L., this volume, Secondary pyroclastic flows from the June 15, 1991, ignimbrite of Mount Pinatubo.

Tuñgol, N.M., and Regalado, M.T.M., this volume, Rainfall, acoustic flow monitor records, and observed lahars of the Sacobia River in 1992.

Umbal, J.V., and Rodolfo, K.S., this volume, The 1991 lahars of southwestern Mount Pinatubo and evolution of the lahar-dammed Mapanuepe Lake.

Umbal, J.V., Rodolfo, K.S., Alonso, R.A., Paladio, M.L., Tamayo, R., Angeles, M.B., Tan, R., and Jalique, V.,1991, Lahars remobilized by breaching of a lahar-dammed non-volcanic tributary, Mount Pinatubo, Philippines [abs.]: Eos, Transactions, American Geophysical Union, v. 72, no. 44, p. 63.

U.S. Army Corps of Engineers, 1994, Mount Pinatubo Recovery Action Plan: Long Term Report: Portland, Oregon, 162 p. + 5 appendices.

U.S. Geological Survey-Philippine Institute of Volcanology and Seismology (USGS-PHIVOLCS), 1994, Serious but rapidly declining hazards at Mount Pinatubo: unpublished progress report of PHIVOLCS-USGS collaboration, 1992-mid-1994, Quezon City, 59 p + 14 attachments.

Voight, B., 1988, Countdown to catastrophe: Earth and Mineral Sciences (Pennsylvania State University), v. 57, p. 17–30.

————1990, The 1985 Nevado del Ruiz volcano catastrophe: Anatomy and retrospection: Journal of Volcanology and Geothermal Research, v. 44, p. 349–386.

Zambales Lahar Scientific Monitoring Group (ZLSMG) (Alonso, R.A., Rodolfo, K.S., and Remotigue, C.T., compilers), 1993, Lahar hazard map of the Santo Tomas-Pamatawan plain, southern Zambales province, as of September 1, 1993: unpublished map, 1 sheet, scale 1:36,400.

Responses of Pampanga Households to Lahar Warnings: Lessons from Two Villages in the Pasig-Potrero River Watershed

By Raoul M. Cola[1]

ABSTRACT

Barangays Parulog and San Antonio, Bacolor, were hit by lahars in 1991 and by flooding in 1992. One hundred nineteen of 143 respondent families received lahar warnings, and, of those, all but 8 evacuated at least temporarily, at least once. However, false alarms and sometimes inadequate evacuation facilities raised skepticism in 1991, and a generally improving outlook in 1992 brought many evacuated families back to their homes.

Warnings could be improved by the addition of village-level, house-by-house hazard maps, vesting of sole warning responsibility in Barangay Disaster Brigades, identification and use of proven warning strategies, and recognition that warnings will only be heeded if adequate transportation, shelter, and livelihood can be provided to those who evacuate.

INTRODUCTION

Preparations made by households for the eventuality of a natural hazard are based on their knowledge of that hazard. Such knowledge is normally acquired from repeated encounters with a hazard, from which its observable precursors are identified, its recurrence frequency ("repeat time") is approximated, and practical steps toward its mitigation are learned. For example, people living at the foot of Mayon Volcano use folk precursors and a supposed 10-year repeat time to anticipate eruptions (Zarco, 1985), and people that live in typhoon-prone areas now build their houses to withstand wind (Cola, 1993).

However, households in central Luzon did not have such knowledge about lahars. No oral tradition survived from the previous eruptions and lahars of Pinatubo, and, initially, the populace could only respond in a haphazard and frantic manner. Subsequent warnings about the nature and occurrence of lahars have saved lives, but, even now, some weaknesses remain in the lahar warning process.

[1]College of Public Administration, University of the Philippines, Diliman, Quezon City, Philippines.

THE LAHAR WARNING SYSTEM: A BRIEF DESCRIPTION

Lahar warnings are issued for two timescales at Pinatubo: long range and short range (Janda and others, this volume). Long-range warnings consist of hazard maps, briefings, brochures, booklets, flyers, posters and leaflets, prepared and given (hopefully) long before actual lahars. The hazard maps indicate the degree of hazard to which each area is exposed. Some of these materials outline precautionary actions that can be taken to reduce one's risk.

In contrast, short-range warnings are issued minutes to days before the onslaught of lahars. Warnings are based on meteorological forecasts, realtime rainfall data, and instrumental and visual observations of lahars at forward (upslope) observation points. Short-term warnings are announced over the radio, through local authorities, or, in symbolic form, through the sound of gunshots and the pealing of church bells. They may also be attached to typhoon warnings. Three levels of short-range lahar warnings are issued at Pinatubo: level 1 (rainfall on Mount Pinatubo and vicinity), level 2 (rain has occurred for 30 min at critical intensities), and level 3 (lahars have been observed) (Philippine Institute of Volcanology and Seismology, 1993).

Long-range warning materials are issued mainly by the Philippine Institute of Volcanology and Seismology (PHIVOLCS), but the National Mapping and Resource Information Authority (NAMRIA), Bureau of Soils and Water Management (BSWM), and the National Economic and Development Authority (NEDA) have also produced materials. Short-range warnings are issued by the Office of Civil Defense (OCD), through local (regional, provincial, municipal, and barangay) disaster coordinating councils. Lahar watchpoints that are supervised by the OCD also directly disseminate warnings to those at risk.

METHODOLOGY

The primary data used in the study were gathered by students through interviews of household and key informants and by onsite observation (table 1). The survey

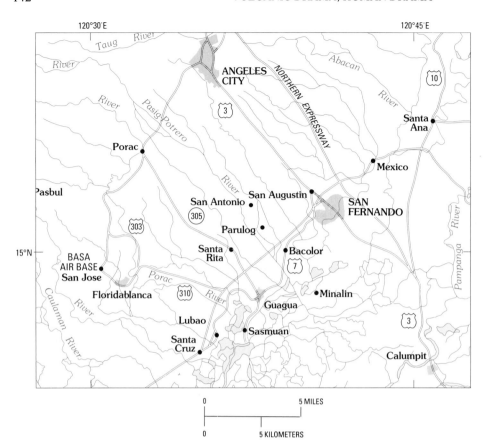

Figure 1. Location of barangays San Antonio and Parulog, Bacolor.

covered 143 households in Barangays Parulog and San Antonio, Bacolor, both of which are immediately east of the Pasig-Potrero River, and both of which were hit by lahars in 1991 and floods in 1992 (fig. 1).

According to the 1990 census, Parulog had a total of 321 households before the eruption of Mount Pinatubo, and San Antonio had 887 households (National Statistics Office, 1990). One hundred forty-three households (about one of every six) were asked about their receipt of and response to warnings about lahars. Any adult was allowed to serve as a respondent, but most of those who responded were either the father or the mother in a family. Interviews were in October and November 1992 in the affected barangays and in evacuation camps.

Seven key informants were also interviewed: two who never moved out; two who moved out and later returned; one who moved out and has not returned; and two who were familiar with the warning system in the site and willing to discuss it. Three of these key informants were recommended to the author as being articulate and knowledgeable on the topic; the others were interviewed because they were available during the author's field work and willing to tell their stories. Interviews with key informants were unstruc-

tured and aimed to draw out the experiences of the informants.

Although the questionnaire did not distinguish between lahars of 1991 and floods of 1992, a subsequent visit by the author to these barangays ascertained that responses referred to lahars and lahar warnings of 1991. The questionnaire did not distinguish between long-range and short-range warnings, either. Most respondents spoke of immediate, short-range warnings, though a few referred to warnings received weeks and months before lahars hit their barangays.

RECEIPT AND UNDERSTANDING OF LAHAR WARNINGS

Eighty-three percent of the households heard warnings of lahars before their barangays were hit. Long-range warnings that their barangays were at high risk had been issued as much as 3 months before lahars came, but only eight households reported receiving such warnings. Instead, most received warning within the week of the lahars. Apparently, long-range warnings lacked dissemination and (or)

Table 1. Questions asked of household heads, Parulog and San Antonio, Bacolor, and tubulated answers.

[Questions were asked in Tagalog]

Part I: Background

1. Ilang taon na po kayong naninirahan sa barangay na ito? [How many years have you lived in this barangay?]

Average number of years in the barangay (for heads of household): 35.5 years

2. Bakit dito po kayo naninirahan at hindi sa ibang lugar [Why do you live here, and not somewhere else?]

My parents/family have always lived in this place	16
I was born here	72
My spouse lives here	31
Our sources of livelihood	14
My relatives are here	1
The environment here is good	4
My children attend school here	1
I took refuge here from a massacre	1
No response	3
Total	143

3. Ano po ang inyong pangunahing hanap-buhay o pinagkakakitaan? [What is your primary source of income or livelihood?]

None	10
Assistant construction supervisor	1
Farming	29
Fishing	1
Business operation	19
Horse-drawn carriage driving	6
Engineer	1
Laundrywoman	6
Refrigerator technician/welder/mechanic/electrician	8
Sewing	1
Barber	1
Carpentry	19
Motor-vehicle driver	8
Teacher	5
Laborer	14
Livestock husbandry	4
Others (incl. security guard, domestic help)	10
Total	143

4. Ano po ang pinakamataas na antas ng par-aaral ang natopos ninyo [What is the highest level of education you completed?]

None	5
Grade 1	0
Grade 2	7
Grade 3	14
Grade 4	5
Grade 5	9
Grade 6	47
First-year high school	5
Second-year high school	12
Third-year high school	5
Fourth-year high school	12
One or two years of college	11
Four years of college	11
Total	143

Table 1. Questions asked of household heads, Parulog and San Antonio, Bacolor, and tabulated answers—Continued.

Part II: Reaction to the Disaster

1. Noong hindi pa natamaan ng lahar ang inyong barangay, kayo po ba ay nakatangap ng babala para lumipat ng tirahan or mag-evacuate? (Kung Oo, Ilang araw po ninyong natanggap ang babala bago kayo lumipat o nag-evacuate?) [Before lahars hit your barangay, did you receive any warning to move to another place or to evacuate? If yes, how many days after this warning did you move or evacuate?]

Respondents who heard a warning	119	
Moved out in less than 1 day	27	
1 day later	11	
2 days later	15	
3 days later	14	
4 days later	4	
5 days later	1	(?)
6 days later	7	
7 days later	14	
2 weeks later	15	
4 weeks later	7	
3 months later	1	
No response	3	
Respondents who did not hear a warning	24	

2. Saan po galing ang narinig ninyong babala? [From whom did you receive this warning?]

Barangay captain/officials	45
Church/parish priest	20
Police/soldiers	23
DSWD (Dept. of Social Welfare and Development) poster	1
Radio	8
Television	6
Neighbors	2
Municipal mayors/officials	22
Government	3
Bell	1
Siren	4

3. Noong nalaman ninyo na dadaan ang lahar sa inyong barangay, ano po kaagad ang ginawa ninyo? [When you learned that lahar would pass through your barangay, what did you do right away?]

Nothing	7
Immediately ran	50
Packed up things	33
Prepared to move out	8
Moved household items to higher location	12
Moved out children/family	3
Reinforced house against lahars	2
Moved to higher ground	11

4. (For households that received warning but did not evacuate): Bakit po kayo hindi lumipat ng tirahan o nag-evacuate? [Why did you not evacuate?]

Did not know where to go because government did not tell us	1
Thought the lahar would not really arrive	3
Did not want to leave house because things might get lost	2
Did not believe warning	1
Hoped that the lahar would stop soon	1

(Note: Some of these eight families that reportedly did not evacuate might have done so later, because responses to question 1, above, suggest that at least 116 of 119 who received warnings did evacuate. Alternatively, some of the responses to question 1 apply only to those members of a family that did evacuate.)

Table 1. Questions asked of household heads, Parulog and San Antonio, Bacolor, and tabulated answers.

5. (For households that evacuated immediately): Anu-ano po ang mga gamit na dinala ninyo doon sa lugar na inyong nilipatan? [What possessions did you carry with you in your evacuation?]

None	16
Clothes	90
Kitchen utensils	24
Food	6
Sleeping gear	6
Livestock and poultry	7
Rice	2
Appliances	29
Furniture	8
All household items	11
Livelihood implements	3
Child care items	3

6. May tumulong po ba sa inyo para lumipat ng tirahan o mag-evacuate? Kung Mayroon, sinu-sino po sila? [Did anyone help you to evacuate? If so, who?]

Respondents who were assisted in moving out	61
By relatives	43
By neighbors	16
By DSWD	7
By municipal officials	8
By barangay officials	12
Respondents who were not assisted in moving	50

7. Sa lahat po ng mga tumulong sa inyo, sino po ang nakapagbigay ng pinakamalaking tulong? [Of all those who helped you, who helped you the most?]

Relatives	41
Neighbors	10
Barangay captain	7
Municipal officials	3

8. Saan po kayo lumipat ng tirahan o nag-evacuate? [To what place did you evacuate?]

Relatives' house	55
Evacuation center	48
Another house owned by respondent	3
Friend's house	2
Rented house	3

9. (For people who went to an evacuation center): Mayroon po bang namahala sa evacuation center? Kung Meron, sinu-sino po sila? [Did someone take responsibility for you at the evacuation center? If so, who?]

DSWD	8
NGO (Non-governmental organization)	8
Barangay officials	8
School officials	2
Evacuees themselves	4
Municipal officials	1
Private groups	2
Government	3
Red Cross	3
Do not know	5

Table 1. Questions asked of household heads, Parulog and San Antonio, Bacolor, and tabulated answers.

10. Anu-ano po ang mga problema na naranasan ninyo sa evacuation center? [What problems did you experience at the evacuation center?]

None	23
No toilet	9
No lighting	2
No water	9
No cooking facilities	1
No medical facilities	1
Too hot during day; too cold during night	11
No sleeping space	1
Lack of food	17
Congestion	8
Occurrence of diseases	5
Unsanitary conditions	2
Others (flooding, mud, flies)	3

11. Mayroon po ba kayong kasamahang bahay na hindi tumuloy sa evacuation center? Kung Meron, bakit? [Did anyone from your household NOT go to an evacuation center? If yes, why?]

Household members who did not stay at evacuation center	14
To watch household belongings	11
To obtain relief goods	1
To watch over the farm	1
To avoid congestion at evacuation center	1
Every household member was in evacuation center	34

12. Mayroon po ba kayong matutuluyan pansamantala maliban sa evacuation center? Kung Meron, saan po? [Do you have any temporary alternative place to stay other than the evacuation center? If yes, where?]

With alternative place to stay	18
With relatives	5
Did not answer	13
Without alternative place to stay	30

attention-getting impact. One reason might be that the basis for long-range warnings is not readily observable, except by relatively mobile or media-exposed persons. This was suggested by the warning path to a professional in Barangay Parulog:

The first time I learned that we are in a very risky area was from a poster in the DSWD [Department of Social Welfare and Development] office in San Fernando, Pampanga, 3 months before the lahar came. But I thought that the lahar would not come, contrary to what the poster said. I later read about it in a newspaper. I wanted to move out from the barangay but I have my family, house, and land here. So I stayed.

The greater effectiveness of short-range warnings results from immediacy of the event and the manner in which the warnings are conveyed. Local officials themselves announced the short-range warnings. Forty-five households heard short-range warnings from barangay officials, while 22 additional households were warned by municipal officials. Twenty-three households received the warning from policemen and military personnel, and 20 households got the message from the parish priest of San Antonio, who rang the church bell as a warning. Others received warnings from television, radio, and other barangay members. Although a few were not reached by any warning, many received warning from two or three sources. The manner in which the warning was made in San Antonio is described by one resident:

The municipal government has a lahar monitoring team. The mayor has stationed persons along the river channel to observe if lahar is coming. These persons are equipped with hand-held radio and they call our parish priest and local officials [who were also provided with radios] when they spot the coming of lahar. The Barangay Council sounded the siren while the parish priest tolled the bell. My family moved out from the barangay. We brought some items for our immediate needs but not everything because sometimes we do not believe warnings.

The last-mentioned skepticism arose because a number of warnings had been issued before lahars actually reached their barangays. Thus, long-range warnings had limited coverage and impact, and short-range warnings had better coverage, but their effectiveness was partly compromised by prior false alarms.

RESPONSES TO THE WARNING

Among the households that heard a warning, 93 percent responded to it. The response depended on which warning they took seriously. Many of those who ignored previous warnings and responded only to the one issued shortly before the lahar came immediately ran to higher places bringing nothing with them. About 50 households responded in this manner. Among those who took heed of earlier warnings, 33 households managed to pack some of their things before moving out. Part of the preparation made by 12 families was to transfer their household items to the elevated portion of the house. Three families took their children to a relatively safe place. Two families reinforced their houses to withstand the force of lahars.

Whether a family heeded the warning depended on a number of factors. These include the degree to which the family's sources of livelihood are tied to its home and nearby farmland, their home's perceived vulnerability to lahars, the ease with which the family could move from the risk area, and the comfort available in a prospective shelter. Two different responses are illustrative. According to the head of one family in San Antonio that decided not to move out in spite of the warning,

The warning was too short. My family could not move out because the road to the highway was clogged with vehicles. Everybody tried to get out from the barangay. Besides, we have a number of goats and ducks that we could not bring with us. The evacuation site is too crowded. I was confident that, if the worst situation occurred, I could bring my family to my roof, which is made of galvanized iron. The lahar came about 7:15 in the evening, while we were watching a Filipino action movie on the television. We heard a rumbling sound and we all ran to the roof as the lights went out. But my neighbors also climbed onto my roof because it was higher than theirs. I estimated the lahar must have been about 15 feet high. There were about 50 of us on my roof. We stayed there until 11:00 in the evening because we waited for the rain to stop and the lahar to subside.

The decision to move as warned was swifter if a family felt vulnerable to the hazard, if the prospective shelter was comfortable, or if the family's livelihood would be unaffected by the transfer. These factors were apparent in the account of one family in Parulog:

We received the warning from the mayor. We used to plant vegetables here but we decided to move out early from the barangay because my mother is paralyzed. It would be very difficult for us to run from the lahars with her condition. Nor do we have a private vehicle to take us out at any time. We took public transport to Mexico, Pampanga, where my aunt has a house. My husband works in San Fernando, Pampanga, and he also urged us to move out. He knew that lahar would definitely come because he works with the Department of Public Works and Highways. We stayed with my aunt for 3 months and did not experience what our neighbors went through.

Households that moved earlier were able to bring along more than just their clothes. Twenty-four households brought with them their kitchen utensils, and 29 carried appliances to their temporary refuge. Some 43 households reported being helped by their relatives when they moved out. Some were able to take buses sent over by the mayor. Half of the families that moved out stayed temporarily with their relatives in other parts of central Luzon and in Metro Manila. A few opted to rent living quarters in San Fernando, Pampanga, or in Metro Manila.

Households that were without relatives willing to take them in or that lacked the means to rent living quarters moved to government- and NGO (nongovernmental organization)-run evacuation centers. Eleven respondents said the living quarters were too hot during days and too cold at night. Nine respondents were concerned about the lack of toilet facilities and eight found the evacuation centers too congested.

UPHEAVAL, COPING, AND DETERMINATION TO RETURN HOME

More than 60 percent of the residents of Parulog and San Antonio were born and had lived in those barangays until lahars came. Those who evacuated longed for the community that had nurtured them and had provided them with the sense of security produced by generations of patterned interactions. Leaving their community entailed changing one's economic base and also leaving a social world in which they were adept and comfortable. Lahars brought both physical displacement and social upheaval.

At the time of our interviews, Parulog and San Antonio were still partially depopulated. In Parulog, only 50 households remained out of the original 321, while in San Antonio, 528 of 887 remained. But some who moved out are trickling back to their homes. Those who remained are gradually rehabilitating their farmland and adapting to the physical reality of annual lahars and floods. Short-term crops such as peanuts and tomatoes are planted instead of annual crops so that these can be harvested before the annual onset of lahars. Thin lahar deposits at some distance away from the main channel are slowly being removed or mixed with underlying soil, and crops are planted and harvested during the dry season. Long-term investments like livestock and poultry raising are still being avoided.

The ability of those households to adapt to a new condition in the face of lahar threat is illustrated by one key informant's family that was living in a house half buried by lahar deposits. The condition, in 1992, was described by the housewife as follows:

The barangay captain urged us to move out, but we insisted on staying. The heat in the evacuation center is unbearable, and there is not enough space for my six children. Besides, we feel secure [mapalagay ang loob] in this village because everyone knows how to deal with us and we know how to deal with them. My children can play outside our yard, and my neighbors can keep an eye on them. The house can be left open when we are away and nothing is lost, because my neighbors can recognize any stranger who might come in. Most of our neighbors are related to us by blood or by affinity. We all grew up here, as did my parents and my neighbor's parents and grandparents.

My family put the sandbags around the house to minimize deposition. The jutebags were given to us by the municipal government. Floods last only for 20–30 minutes and we just stay on the roof when they occur. When the lahars are intense, we move to a safer area nearby. There is always a safe area because, unlike water, lahars have a new channel every time they flow down. After the lahars, we remove the water that seeped into the house. Our appliances are secured in the upper story which we constructed recently. When lahars cease, we resume our normal activities. Now I am growing sweet potato and cassava in the lahar deposits around the house. We can start to harvest these in 2 to 3 months.

The longing of other, former residents to return to the village is dramatized by a family that is now living in San Fernando, Pampanga. The mother provided this account:

On Sundays we still return to the village to hear Mass. I get to meet my old friends and neighbors on this occasion. I used to have a poultry farm in the village, but it was swept away by lahar flow. We now have a house in San Fernando, and my children are going to school there. But I still plan to go back to the village where I grew up. When I work there, I feel I am serving my own people.

RECOMMENDATIONS

The present warning system against lahars works, but it could be improved. First, village-level hazard maps should indicate those houses at greater and lesser risk so that families can make commensurate preparations. Such preparations can include arranging with relatives for temporary shelter, moving heavy and expensive items before lahars actually occur, formulating escape strategies with household members, and reinforcing houses against deposition and flooding.

Village-level hazard maps can also serve as a basis for a community evacuation plan. It can pinpoint households that need priority for evacuation, suggest the most practical escape routes, and identify relatively safe spots that can serve as pick-up points or temporary refuge for evacuees. Such maps would also show which households need to receive short-term warnings first. At the same time, village-level maps would identify any less-vulnerable households that no longer need to be evacuated. In this way, such households would not be inconvenienced, and the credibility of the warning system would be improved.

Another potential improvement would be to place the sole responsibility for warnings with the Barangay Disaster Brigade. Having one source of warning would prevent the spread of varying interpretations of the lahar situation, establish accountability for warnings, and facilitate audits to ensure that every concerned household receives warnings. The Barangay Disaster Brigade is best positioned to issue warnings because it has well-understood legitimacy and a major stake in the well-being of the village. Because its members are villagers themselves, the Barangay Disaster Brigade can readily assist households who evacuate, assign roving teams to safeguard deserted houses from looting, and eliminate unnecessary filters (government offices and mass media) that often cause miscommunications.

Third, continuous assessment of the lessons from each warning and response would help concerned agencies to sort out workable from nonworkable strategies. The assessment would be done by end users of the warnings, government personnel who manage the warning system, and scientists who provide the initial warnings.

Fourth, it is clear from the respondents of this study that the warning system does not work in isolation but as part of a continuum of measures intended to reduce destruction. A warning system cannot be effective if households are physically unable to move, if evacuation camps are unattractive alternatives to staying at home, if there are no alternate sources of livelihood, or if people do not feel that they are threatened. The warning system must be complemented by a well-conceived and well-implemented set of preparedness, rescue, and relief services.

Postscript: On the night of September 22–23, 1994, breakout of a lahar-impounded lake on the slopes of Mount Pinatubo generated a major lahar that once again buried San Antonio and Parulog. Many houses that had been only slightly buried in 1991–92 were buried up to the eaves of their roofs. Ample long-range warnings had been given; few or no immediate warnings were issued, and many residents were forced to take refuge on their rooftops.

REFERENCES CITED

Cola, R.M., 1993, Disaster warning in Metropolitan Manila: Content, communication, and consequence, in Disaster prevention and mitigation in the Manila Metropolitan area: Quezon City, PHIVOLCS Press, p. 134–153.

Janda, R.J., Daag, A.S., Delos Reyes, P.J., Newhall, C.G., Pierson, T.C., Punongbayan, R.S., Rodolfo, K.S., Solidum, R.U., and Umbal, J.V., this volume, Assessment and response to lahar hazard around Mount Pinatubo, 1991–93.

National Statistics Office, 1990, 1990 Census of population and housing: Population by city, municipality, and barangay (Pampanga): Manila, National Statistics Office, Report no.2–74C, December 1990.

Philippine Institute of Volcanology and Seismology (PHIVOLCS), 1993, Pinatubo Volcano lahars: Quezon City, PHIVOLCS Press, 31 p.

Zarco, R., 1985, Anticipation, reaction, and consequences: A case study of the 1984 Mayon Volcano eruption: Quezon City, Philippine Institute of Volcanology and Seismology, 21 p.

The Mount Pinatubo Disaster and the People of Central Luzon

By Cynthia Banzon Bautista[1]

ABSTRACT

The eruption of Mount Pinatubo and subsequent widespread and persistent lahars and flooding have taken a serious toll on the people of central Luzon. The most serious toll has been the displacement of more than 10,000 families (more than 50,000 persons) whose houses were destroyed and whose farmland or other source of livelihood was buried. Initially, the indigenous Ayta people were hardest hit, and many remain displaced from their livelihood and their cultural roots. Since the eruption, many lowlanders have also become evacuees, driven from their homes and land by lahars and floods.

Evacuations and damage from the volcano have undermined preeruption social standing and community leadership. Psychological stress is high. Decisions about how to mitigate lahar hazard have provoked suspicions between neighboring communities. Decisions about how to organize resettlement areas have engendered lively debate and, among some, concern that well-intended humanitarian relief is becoming a substitute for self-sufficiency. Viable resettlement options are badly needed, in which provisions for livelihood and social stability are given even more attention than matters of housing and visible public infrastructure.

INTRODUCTION

Since its major eruptions of June 12–15, 1991, Mount Pinatubo has changed the landscape of central Luzon, uprooted thousands of residents from their homes and means of livelihood, and affected the agenda not only of the local, regional, and national governments but also of nongovernment organizations. This paper aims to provide an overview of the social and psychological impact of the volcanic eruption and some of the issues and problems of resettlement.

METHODS

The discussion is based on several sources: documents of the now-defunct Mount Pinatubo Task Force and the Mount Pinatubo Resettlement and Development Commission, which was created by law in September 1992 to replace the task force; unpublished reports of the regional government agencies, specifically the National Economic Development Authority, the Department of Public Works and Highways, the Department of Agriculture, the Department of Agrarian Reform, and the Department of Social Welfare and Development; relevant clippings of all major newspapers from June 1, 1991, to December 31, 1992; papers read in scientific conferences and various forums; findings of ongoing studies; and other published materials on the Mount Pinatubo disaster.

Apart from written documents, the paper relies heavily on interviews with key informants in government and nongovernment institutions and discussions with victims in evacuation centers and resettlement sites. Some of the points in this overview first arose in a field-based multidisciplinary study of one municipality, Concepcion, Tarlac, which is located in the direct path of lahars (fig. 1) (C.B. Bautista, 1993). Our team of social scientists lived in the area for 3 months, from October to December 1991. The team's findings were later validated or qualified by key informants in other areas of central Luzon. In February 1992, a workshop involving researchers and key resource persons in the municipality was organized. A month later, some of the research findings were disseminated in a town assembly held on March 27, 1992.

Some of the team members went to Concepcion at regular intervals thereafter—twice a month from April to July 1992, about once a week from August to September, and biweekly in October and November to follow up developments there. While data were updated in Concepcion, another small research project was organized from October to December 1992 to gather information on the disaster in central Luzon. Devastated villages, resettlement sites, and evacuation centers were visited. Interviews with respondents in government agencies and discussions with victims

[1]Depatment of Sociology and Center for Integrative and Development Studies, University of the Philippines, Diliman, Quezon City, Philippines.

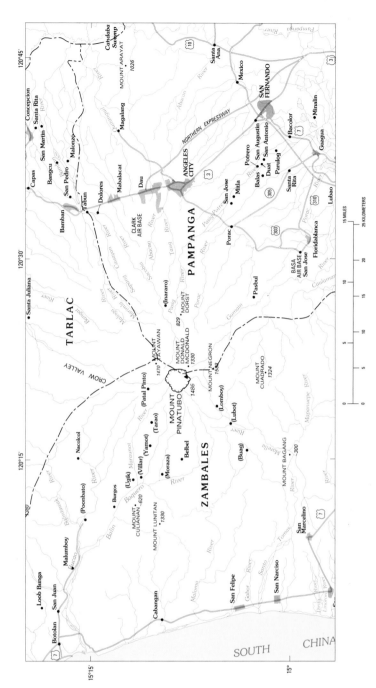

Figure 1. Locations of barangays, towns, cities, and provinces cited in this report. All of the barangays and smaller sitios that are shown in parentheses, except Poonbato, were destroyed by the eruption. Poonbato was destroyed by subsequent lahars.

Table 1. The distribution of families and persons affected by ash fall (Pardo de Tavera, 1992)

Province/City	Families	Percent of total	Persons	Percent of total
Bataan	7,551	3.5	31,322	3.1
Pampanga..............	113,640	52.6	529,578	51.9
Angeles City	15,688	7.3	62,770	6.2
Tarlac....................	11,371	5.2	61,633	6.0
Zambales...............	49,827	23.0	245,582	24.1
Olongapo City.......	17,815	8.2	88,935	8.7
Nueva Ecija...........	79	0.2	373	.0
Total	215,971	100.0	1,020,193	100.0

and their care-givers were conducted within this time period.

SOCIAL AND PSYCHOLOGICAL ASPECTS OF THE MOUNT PINATUBO DISASTER

The 1991 eruption of Mount Pinatubo and its muddy aftermath have affected hundreds of thousands of central Luzon's residents in varying degrees, depending upon the specific hazard and their physical vulnerability to it. For purposes of accuracy in assessing the social and psychological impact of Mount Pinatubo on the population, the effects of the volcanic eruption are discussed separately from the effects of lahars and floods.

VICTIMS OF THE 1991 ERUPTION

Ash fall from Mount Pinatubo's eruptions in June 1991 affected about a million people, half of whom were from the province of Pampanga (table 1) (Pardo de Tavera, 1992). A quarter of a million people remained displaced 1 week after the first major blasts, of whom about 3 percent were in formally organized evacuation camps (Department of Social Welfare and Development, unpub. disaster monitoring report, September 28, 1991). Tens of thousands of central Luzon residents fled to Metro Manila, and about 30,000 of these took refuge in the Amoranto Stadium in Quezon City. Though generally harmless in areas far from Mount Pinatubo, wet ash fall from 5 to 50 cm deep caused 189 deaths in areas near the volcano when roofs collapsed under its weight (Magboo and others, 1992; Spence and others, this volume). Ash fall also damaged public structures that housed social services. Ninety-eight hospitals and health centers, 18 public markets, 13 municipal buildings, and 70 other government buildings were destroyed (Department of Public Works and Highways, 1992a).

Of all persons affected, the hardest hit were the Aytas, an indigenous tribe (Shimizu, 1989). Around 7,800 Ayta

families, or 35,000 persons, were forced to flee their homes (Task Force Pinatubo, 1992). The Ayta's economic and cultural life before the eruption was rooted in Mount Pinatubo. They lived by the volcano's rhythm, timing the planting and harvesting of their crops by the volume of steam rising continuously from a natural vent on the upper slope. A relatively dense steam meant a good harvest; a thin one augured a sparse yield (Lubos na Alyansa ng mga Katutubong Ayta ng Sambales (Negrito People's Alliance of Zambales, LAKAS), 1991, p. 32). They hunted in the volcano's wooded slopes and fished in the rivers that drained it. The volcano was not only the source of the Ayta's livelihood but also the abode of Apo Namalyari, their God, and home to the spirits of their ancestors. For these reasons, the Ayta's evacuation from the volcano was especially disruptive and heart-rending. A vivid chronicle of their life and exodus is given in Eruption and Exodus (LAKAS, 1991).

From that book's account, earth tremors and heavy steam from the volcano summit at 1600 on April 2, 1991, caused panic-stricken Aytas from villages along the Zambales mountain slopes to flee their homes and converge in a village 12 km from the volcano's base. Those who ignored the first ominous signs on April 2 fled in waves within the next 2 weeks. By the third week of April, the number of evacuees in Poonbato, Botolan, Zambales, reached about 4,000.

Even Poonbato was not the last stop of the Aytas documented in the book. They changed sites with each extension of the danger zone from a 10- to 20-km radius of the volcano, from 20 to 30 km, and finally from 30 to 40 km. Some groups moved 9 times in 1991 before they found semipermanent relocation sites.

Fortunately for the Poonbato Aytas, whose experience was documented, the group was kept intact throughout the exodus. As organized constituents of the Lubos na Alyansa ng mga Katutubong Ayta ng Sambales (LAKAS), they were also in a better position to maintain their cultural and tribal bonds and to critically assess the options opened to them. The acronym of the federation, LAKAS, means power.

This was not the case, however, among the Aytas living on other parts of the volcano. Mount Pinatubo's eruption scattered them to various evacuation centers and disrupted their political and administrative structure. New leaders and factions began to emerge as the tribes separated (Tadem and Bautista, 1993).

Anticipating the need for Aytas to vacate the slopes of Mount Pinatubo, the government set evacuation procedures in place before the major eruptions in June 1991. Conditions in the evacuation sites, however, were extremely difficult for Aytas. The tents provided only minimal shelter from the elements. Evacuees suffered from extremely hot days and cold and damp nights in these tents. There was no basic sanitation. As a consequence, respiratory and gastrointestinal ailments were common. In early August, the Department of Social Welfare and Development reported that 156 Ayta children had died in evacuation centers in Tarlac and Zambales from various diseases such as measles, bronchopneumonia, and diarrhea.

Cultural differences accounted for the spread of some of these diseases. The measles outbreak is a case in point. A study conducted by the Department of Health reveals that its immunization campaign did not reach all vulnerable children in the camps. The cultural gap between the Aytas and health workers prevented the latter from reaching Ayta children (Magpantay and others, 1992; Surmieda and others, 1992). Some health workers believed that the Aytas eschewed Western medicine in favor of their own. However, the Sisters of the Franciscan Missionaries of Mary, who pioneered the literacy campaign among Aytas in Poonbato, claim that Aytas generally know when to ask for Western medical help. What was missing, at least initially, was rapport between health workers and Ayta evacuees that was based on an understanding of the Ayta culture. Malnutrition also contributed to high mortality among Ayta children (Surmieda and others, 1992).

Aside from getting more sick than their lowland counterparts in evacuation centers, Aytas were also disoriented by the surroundings far from their upland homes. They were unaccustomed to the sight of flatlands and plains, and they longed to roam the mountains and hills again in communion with nature and their God (LAKAS, 1991).

Thus, the Aytas were the prime victims of the volcanic eruption itself, before secondary effects like lahars and floods were taken into account. They were evacuated earlier than any other group, and throughout the ordeal of moving from one evacuation site to another, they were totally uprooted from their way of life. Even their resettlement sites, with high population density and town plaza complexes, were alien to their preeruption existence.

VICTIMS OF LAHARS AND FLOODS

AREAS MOST DAMAGED

The major eruptions and the plight of the Aytas in evacuation centers dominated newspaper headlines in June 1991. Since that time, lahars and floods have devastated many more villages and towns. Because preeruption river channels have been clogged by lahars, subsequent rainfall cannot drain away through those channels, so it floods adjoining lowlands.

By October 1992, most parts of 29 barangays (villages) had been buried by lahars to depths of several meters (table 2). Eleven of these were in only one municipality—Botolan, Zambales. Six were in Tarlac and 12 in the densely populated province of Pampanga. Half of Pampanga's abandoned barangays were in the municipality of Bacolor. The 29 severely affected barangays were home to almost 10,000 families, or about 53,000 people (1990 National Statistical Office population estimates for these 29 barangays, as

Table 2. Lahar-devastated barangays that were virtually abandoned in October 1992, listed by municipality and province.

[List of barangays is based on interviews with key informants in the municipalities, corroborated by newspaper accounts and field observations. The number of affected families and persons are taken from the 1990 National Statistics Office figures as stored in the data base of the National Economic Development Authority, Region III GIS Project]

Province	Municipality	Barangay	Families	Population
Pampanga			6,084	33,213
	Bacolor	Balas	308	1,771
		Duat	300	1,840
		Parulog	321	1,970
		Potrero	786	4,624
		San Antonio	887	5,718
		Santa Barbara	507	2,991
	Mabalacat	Cacutud	268	1,101
		Dolores	1,471	6,310
		Tabun	430	2,191
	Porac	Mitla	287	1,785
		San Jose Mitla	217	1,297
	Santa Rita	San Juan	302	1,615
Tarlac			1,832	10,634
	Bamban	Malonzo	128	811
		San Pedro	385	2,039
		Bangcu	35	216
	Concepcion	Malupa	230	1,463
		San Martin	178	1,220
		Santa Rita	876	4,885
Zambales			1,913	9,388
	Botolan	Villar	230	1,121
		Poonbato	483	2,553
		Moraza	112	569
		Belbel	124	534
		Burgos	145	669
		Cabatuan	77	330
		Malomboy	208	1,097
		Owaog-Nebloc	34	168
		Palis	79	328
		Nacolcol	143	693
		Maquisquis	278	1,326
Total			9,829	53,435

stored in the National Economic Development Authority Region III Geographic Information System data base). Even this high number is surely an underestimate, because it counts only those in barangays within which most sitios (hamlets) were buried. Families from devastated sitios of relatively less affected barangays are not counted.

Most of these families once lived in the densely populated barangays of Pampanga, the province that bore the brunt of lahars in the first 2 years. Porac, one of the most severely affected of central Luzon's once bustling towns, symbolizes Pampanga's travails. Many of its residents have left.

Table 3. Municipalities that were flooded, or isolated by flooding, as of late August 1992 (shown by asterisk) or October 1, 1992.

[All these municipalities were (and will continue to be) affected by clogging of river channels by lahars]

Province	Municipality
Pampanga...................................	Minalin
	Guagua
	City of San Fernando
	Santo Tomas
	Macabebe
	Floridablanca
	Mexico
	Bacolor
	San Simon (*)
	Sasmuan (*)
	Porac (*)
Zambales..................................	Cabangan
	San Marcelino
Bataan	Hermosa (*)
	Dinalupihan (*)
	Orani (*)
Tarlac.......................................	Bamban
	Concepcion

Floods caused by silted waterways have added to the misery of central Luzon's lowland victims. Compared to 1991, 1992 was the year of floods. Siltation of river channels (much by lahars) caused subsequent floods that submerged barangays in at least 18 municipalities (table 3). As of October 1, water from the September monsoon rains had not yet subsided in low-lying parts of these municipalities, and additional barangays (especially in Zambales) were also isolated by floods at this time. In the province of Bataan, adjoining the Pinatubo area, the municipalities of Dinalupihan, Hermosa, and Orani experienced persistent floods in 1992. Newspaper accounts on the week of August 20, 1992, cite the evacuation of residents of Hermosa from floodwaters of up to 5 to 7 feet (Philippine Daily Inquirer, August 20, 1992).

COMPARATIVE EFFECTS OF LAHARS AND FLOODS, 1991 AND 1992

Owing perhaps to the previous year's experience with lahars, there were very few deaths due to lahars and floods in 1992. Whereas 100 of the 932 disaster-related deaths in 1991 were due to lahars, fewer than 10 died from lahars during 1992 (1991 figures are from the Special Transition Report of Task Force Pinatubo, May 1992, and include deaths from illness in evacuation camps; 1992 data are from the Department of Social Welfare and Development Region III Disaster Monitoring Report, September 28, 1992).

Table 4. Distribution of affected families, listed by province and type of disaster.

[The 1991 figures were taken from DSWD (1992a). The 1992 data were taken from DSWD (1992b). Data are current up through the lahars and floods which occurred around August 15, 1992]

Province	Percent of Families Affected			
	Lahar, 1992	Floods, 1992	Lahar and flood combined	
			1992	1991
Bataan	–	17	15	5
Bulacan	–	2	2	
Nueva Ecija.....	–	<1	–	<1
Pampanga........	27	52	49	62
Tarlac..............	40	5	9	13
Zambales........	33	24	25	20
Total	100	100	100	100
Number	19,932	144,259	164,191	33,400

Sixteen deaths, however, were due to floods in 1992, as opposed to none in 1991.

Although the number of casualties decreased, significantly more people were affected by lahars and floods in 1992 than in 1991. A total of 33,400 families, or 159,939 persons, suffered the effects of lahars or floods in the first year (Department of Social Welfare and Development, 1992). In contrast, five times more families (164,191) and persons (802,742) were victimized by the end of August in the second year (Department of Social Welfare and Development, unpub. Disaster Monitoring data, September 28, 1992). The marked increase in the number of people affected was due primarily to massive lahar-induced floods in 1992. Viewing the data by type of hazard reveals that 144,259 families experienced floods in 1992, while 19,932 families were affected by lahars (table 4).

At least 3,140 houses were completely destroyed and 3,072 were seriously damaged in 1992. Of the former category, 54 percent were demolished by lahar and the rest by floods. These data are principally for Pampanga; additional houses were destroyed in Tarlac and Zambales.

A discussion of the effects of lahars and floods on people is not complete without mentioning damage to infrastructure and agricultural lands—damage that has profoundly affected livelihood and quality of life. Lahars inundated 15 km of roads, damaged 13 major bridges throughout the 1991–92 period, and threaten to affect another 10 bridges in the coming years (Department of Public Works and Highways, 1992b; Environment Management Bureau, 1992). Mudflows also breached 58 km of river dikes, half of which were along the Pasig-Potrero River.

Lahars and floods affected as much as 42 percent of the total cropland of Pampanga, Tarlac, and Zambales. By province, Tarlac had the highest percentage of its cropland damaged (52 percent of 841 km^2), followed by Pampanga

(41 percent of 618 km^2) and Zambales (13 percent of 236 km^2) (Department of Agriculture Region III, unpub. data, 1992).

In 1991, almost 9 out of 10 lahar-affected agricultural areas were planted to rice (Task Force Pinatubo, 1992). In 1992, that ratio remained practically the same; only a slight shift occurred toward sugarcane (Department of Agriculture Region III, unpub. data, September 28, 1992 Summary of Damage Report).

This observation also holds for agricultural lands inundated by floods. They were overwhelmingly rice based. Because floods induced by lahar deposition in river channels usually left a layer of mud in their wake, farmers in affected rice-producing areas had to rehabilitate their lands.

Beyond their direct effects on agricultural lands, lahars and floods affected 5 national and 176 communal irrigation systems that serviced 483 km^2 of farmlands and 25,476 farmers (Task Force Pinatubo, 1992). The latter figure includes many whose lands were spared from either lahars or floods.

In terms of the people associated with the land, a total of 11,540 farmers in Pampanga, Tarlac, and Zambales were affected by lahars in 1991. During this year, 85 percent of them were almost equally distributed between Pampanga and Tarlac. In 1992, however, the number of lahar-affected farmers declined by half, a finding consistent with the drop in the area of croplands newly affected by lahars.

Farmers in those areas covered with 7 to 15 cm of lahar were better off than their counterparts in places with more than 15 cm but less than 30 cm of mud. The land of farmers with 15 cm or less of lahar just needed to be plowed, while the land of those with thicker lahar had to be scraped. Still other land was either buried beyond rehabilitation by lahar or sunk in emergent lakes, and those who farmed such land must look for alternate land or leave farming.

Most of the affected farmers, especially in Pampanga and Tarlac, were land reform beneficiaries. Some are leaseholders, others amortizing owners, and still others are new owners who have fulfilled the state's requirement of land transfer. The loss of land owned through land reform, albeit temporary in a geologic sense, was a special blow to beneficiaries who have painstakingly paid in full and obtained legal ownership of their parcel of rice- or corn-producing lands. Among amortizing owners, those who have paid a significant amount of the cost of the land stood to feel the loss more than those who have defaulted on their payments.

SOME SOCIAL EFFECTS OF LAHARS AND FLOODS

For the former residents of barangays devastated by lahars, the disaster mitigated an internal process of social differentiation. Prior to the calamity, the population in these villages was divided into those with access to agricultural assets or opportunities for overseas employment and those without. Lahars buried agricultural lands, equipment, and houses in these communities regardless of the wealth of the owner. In some instances, residents who were relatively well-off lost so much that they found themselves in positions similar to those who were once their social inferiors. It is in this sense that ordinary folks have referred to Mount Pinatubo as the "Leveller" or the "Equalizer."

On the whole, however, class distinctions have not been eradicated, because many of those with financial resources managed to find new homes elsewhere. Residents who could not afford to go anywhere else were able to dig up housing materials, appliances, and furniture, thus preserving at least some of their position in the village's social hierarchy.

Although the basic social hierarchy has been maintained in those communities whose residents have remained intact in relocation sites, some new organizations of community life have emerged in relocation sites populated by residents hailing from different barangays. In addition, the social distance among groups in places that have not yet been severely affected by the disaster has been bridged by a significant increase in communal activities that have cut across classes. Religious rituals such as prayer sessions and processions around villages and along the river channels have been the most visible activities in different parts of central Luzon. The rituals have been conducted all year round, although they are practiced most frequently during the rainy season.

Apart from religious rituals, participation in activities to protect barangays from lahars (such as sandbagging) or to protest government decisions designating particular areas as catch basins have been documented. The explorations of the Disaster Coordinating Council to use the Candaba swamp as a catch basin for lahar was met by strong opposition from residents living in several municipalities in Pampanga.

In the summer of 1992, the Department of Public Works and Highways planned to make a sediment-trapping "sabo" dam in Maskup along the Sacobia River to contain lahars that were expected to flow into some parts of Mabalacat, Pampanga. Residents of this municipality resisted the plan, fearing that it would divert lahars to other parts of the town. Some of those interviewed claim that a few residents threatened the contractors who were beginning to survey the area. As a consequence, the plan was shelved.

The national government was not the only target of protests from angry residents who perceived efforts to protect some localities to threaten their own. Municipalities were pitted against each other as attempts to protect one political territory were deemed to be at the expense of a neighbor. The sandbagging operation along the boundary of San Fernando and Bacolor, Pampanga, is a case in point.

The latter's folks rejected the fortification efforts of the former because it will trap lahars in Bacolor (Daily Globe, September 16, 1992, verified by the author from key informants).

San Fernando's local officials were also caught in a dispute with those from Santa Rita, Pampanga. San Fernando officials were alleged to have asked the local court to stop residents of Santa Rita from putting up a sand-bag dike, lest the dike divert lahars and floods into San Fernando, Pampanga's capital (Philippine Star, September 25, 1992; Philippine Daily Inquirer, September 25, 1992). The provincial governor's mediation apparently led to the withdrawal of the suit on the same day.

The politics of lahar defenses also resulted in tension between neighboring communities. Sometimes, rumors were an outlet for the growing suspicion among neighboring barangays. Key informants in one municipality, for instance, cited ill will aroused by the differential impact of lahars on two sides of the river within that municipality. Those on the side that was spared were suspected of having breached the dike on the affected side. While such an operation would have been impossible to achieve without the affected area knowing about it, the impression held sway.

The fact that dike construction raised ill will reflects the extent to which lahar defense has become a highly politicized issue. A political culture in which the powerful are able to get away with practically anything, especially in rural areas, accounts for much of the cynicism and suspicion with which efforts to defend human settlements are viewed.

SOME PSYCHOLOGICAL EFFECTS OF LAHAR AND FLOODS

A cursory review of the graffiti on walls of abandoned homes reveals the angry and plaintive expressions of victims who, despite early warnings from the Philippine Institute of Volcanology and Seismology (PHIVOLCS), were caught off guard. Nothing in their individual and collective past prepared them for the disaster. As such, many of the victims suffered from psychological problems even long after their initial evacuation.

A study of victims and service providers in Tarlac (Jimenez, 1993) vividly described the evacuation process, which traumatized adults and children:

But whether it came by day or night, the sound and sight of the lahar was enough to frighten the people into immediate escape. The lahar was terrifyingly high and steaming hot, they reported. It swept along with it tree trunks and rocks so huge and heavy that it took five men to move them later on. Many believed it to be the end of the world and all thought they would die then. All thought of immediate escape. There were those who only had time to scoop in their infant children and run off, all the while shouting to their older children to run ahead. (Few) had enough time and presence of mind to scoop up ... belongings.

There was pandemonium as they ran, they recalled. People were screaming and crying as they ran, calling on their God for help and deliverance. Everyone was terrified and shouting for help. In their haste, they tripped or ran into each other, fell, picked themselves up and begun to run blindly again...there was a mad rush to get on the trucks. The women and children came off badly in this scramble, as they were pushed aside or thrown unceremoniously on.

Those who were caught in their houses only had enough time to rush up to their roofs. There, families huddled together in fear and for comfort— awaiting their certain death. All spent the night terrified, crying, and praying to God for help and mercy...."

Jimenez (1993) also reported several symptoms of stress among the victims who evacuated to the centers. Upon arrival in the evacuation sites, they trembled from cold and fear continuously. Some went into hysterical laughter. Even days later, victims found it difficult to sleep and did not have much appetite for food. Some of Jimenez's respondents judged that the Mount Pinatubo disaster affected males and females differently. Males tended to become more quiet and withdrawn than did the women, and spent time in all-male drinking sessions.

More than a year after their lives were uprooted, service providers in a relatively well-established resettlement site for farmers in Zambales cited sudden bouts of crying, irritability, and constant headaches among the resettlers, which could only be traced to the trauma of Mount Pinatubo. Symptoms of stress were not confined to those who left their homes. For those living along the potential corridors of lahars and floods, the monsoon season heralded sleepless nights, with families anxiously awaiting the warning to flee their homes. So intense was the stress that when the warning signals—church bells or successive gunshots— were raised, key informants reported incidents of residents who suffered from heart attacks.

Many of the psychological problems confronted by those who took flight from the perils of lahars and those who continue to live in natural catch basins could be attenuated by mass resettlement to areas that are not vulnerable to the disaster. Unfortunately, snags in the resettlement process and attachment to their original lands and homes have discouraged many would-be settlers from moving to resettlement areas. To the dismay of scientists who warn against remaining in danger zones, many of the potential victims have chosen to remain in high-risk areas because they have no viable alternatives.

ISSUES OF RESETTLEMENT

As noted earlier, the Mount Pinatubo disaster has displaced tens of thousands of people in Tarlac, Pampanga, and Zambales. A conservative estimate based only on the population of 29 most thoroughly buried barangays in 1992 is around 53,000 people, a figure that is bound to increase in 1993.

The U.S. Army Corps of Engineers Recovery Action Plan team projected in a briefing for the Mount Pinatubo Commission that about 1,900 km^2 of land in the three provinces may be buried beneath 2 m of lahar debris. The study, which basically supports the PHIVOLCS projections, prompted the commission to estimate that about 74,000 residents in the high risk areas could no longer be defended against lahar and might have to be evacuated by force (Mount Pinatubo Commission, 1993). These people, who were not victims during the 1991 and 1992 rainy seasons will add to the 53,000 dislocated victims.

Because of the scale of human displacement, the state poured massive financial resources into the development of various resettlement projects. Total expenditures during 1991 and 1992 were at least P2.5 billion (US $93 million) for evacuation and resettlement sites (Mercado and others, this volume). In addition, various civic groups, private relief agencies, and development-oriented nongovernmental organizations in some of the state's resettlement sites also infused private resources into these efforts. The Loob Bunga Resettlement Site in Zambales, for instance, stands out in terms of its private resources. At least 11 organizations extended food assistance, provided health and nutrition services, and promoted livelihood projects, as well as literacy and spring-water development projects.

TECHNOCRATIC TOP-DOWN PLANNING OR BOTTOM-UP PARTICIPATORY APPROACHES

An ongoing controversy over the state's resettlement efforts boils down to differences in the basic approach to the problem of resettlement. From the perspective of a technocrat, resettlement requires technical planning based on the principles of scale and efficiency. The logic underlying the technocratic perspective of the now-defunct Mount Pinatubo Task Force can be described as follows. The infrastructure and settlement patterns of some modern cities in the world were planned at critical junctures in their history. The present Tokyo, for example, developed from the ruins of the Great Kanto earthquake of 1923 and from the Second World War. While these calamities resulted in untold human misery, they also provided the occasion to plan the modern Tokyo.

Because the Pinatubo disaster dislocated tens of thousands of victims, the technocrats hope to address victims' needs while maximizing the rare opportunity to use planning principles. Their idea is to put up settlements bigger than the usual barangays in order to economize on basic services like schools, public markets, and hospitals. The concept for the physical layout of the envisioned towns drew from the sprouting subdivisions in Metro Manila and its suburbs, the plaza complex found in most towns, and ideas about the practicality of grid road networks versus the current linear pattern in rural areas.

In addition, because government will be building new towns anyway, the Mount Pinatubo Task Force's logic dictates that they might as well fit into the Regional Spatial Development Strategy of central Luzon. This strategy conceives Region III to be the transit lane between the resource-rich provinces of northern Luzon and the densely populated industrialized areas of Metro Manila. As such, central Luzon will "serve as a catchment area for population and industry spill-over from the metropolis, while maintaining its comparative advantage in agriculture in some places." The plans also project the region in the role of "providing the requirements of the Northern Luzon provinces in terms of processing and manufacturing of goods and their eventual shipment to areas of destination" (Mount Pinatubo Task Force, 1991).

The overall technocratic vision explains why new resettlement areas have gridded street systems, modern public buildings clustered around a plaza, productivity centers (large buildings intended for use as factories), and uniform houses made of either hollow concrete blocks or nipa (a palm).

However, these complexes have attracted much negative attention. Some critics are silent on the basic approach but object to aspects of the content or implementation of the program; other critics question the basic philosophy and the implementation of the plans. The first group of critics accepts the technocratic planning process but assails the state for its insensitivity to the plight of central Luzon's dislocated residents. Site development has been deemed too slow in the face of victims who have languished in evacuation centers for more than a year. In the case of O'Donnell, Capas, Tarlac, the most developed of the sites, millions of pesos had to be advanced by a cooperative headed by a private citizen to hasten the pace of its development.

Apart from the speed, the phasing of the project has also been questioned. Cemented roads and public buildings were put up before houses. To make matters worse, livelihood development efforts were relegated to the background, so some who moved into the resettlement areas decided to leave and others hesitated to move in. On the whole, these criticisms do not question the plaza complex or even the construction of public buildings as long as these are done after housing and livelihood needs are met.

The second group of critics questions not only the content of the plans but also the spirit and process of planning imbedded in them. Drawing from the principles of participatory development, these critics stress the importance of planning **with** and not **for** the affected people. In this approach, victims should participate actively in all stages of planning.

The participation of would-be resettlers is crucial for practical and psychological reasons. Victims will make sure that projects will meet their needs, and they will feel a pride of ownership. The resulting houses and emergent communities may not conform aesthetically to the technocratic vision, but they will be houses and a community in (and for) which the displaced families will work hard and succeed. Psychologically, the process of participation is as important as the visible outcomes of collective decisionmaking, because it enhances the self confidence of individual victims and the community's collective confidence in being able to rebuild its life in the new site.

The criticisms emanating from proponents of participatory development have two implications for resettlement. There can be no uniform design or blueprint for resettlement sites. This means some of the basic parameters in terms of sites can be set at the national or regional level, but the conceptualization of plans will have to be decentralized to the level of the communities involved. The second implication is that community organizational efforts must be an integral part of the resettlement process. It is easy enough to give lip service to community organizing, but it is hard to find capable people who have internalized the spirit of participatory development. Case studies of resettlement sites by the nongovernmental organization (NGO) Philippine Business for Social Progress, for instance, reveal that some NGO's that are committed in principle to participatory development encounter problems of finding enough good organizers.

COORDINATION

Meaningful and effective decentralized planning requires coordination with local and regional government agencies and NGO's. For all the criticisms hurled against the state in the last 2 years, it is the only institution that can mobilize all of the basic resources needed in resettlement work. Only the state is in a position to officially allocate land from the public domain and negotiate with private owners. Infrastructural works such as roads, power, sanitation, and schools are also better left with government.

Coordination with and among government agencies is necessary to speed up the process of resettlement. This is easier said than done. Even under a centralized scheme, conferences on relief and rehabilitation have been haunted by a recurring complaint—that government agencies continue to function within their own turfs. As such, the disaster-related programs have been far from integrated (Bonifacio, 1992). NGO's also have problems of duplicated effort and lack of communication. Key informants claim that very little is done to coordinate services rendered. Furthermore, when collaborative decisions are reached, there are no mechanisms for carrying them out.

In their book based on the lessons learned from disasters in different parts of the world, Anderson and Woodrow (1989) cautioned against too much stress on NGO coordination. Although they argue for coordination of services, they

raise basic questions. Who is in charge of coordination? Whose purpose does it serve? Is it intended to ease the work of NGO's and make logistical requirements run smoothly or is it to ensure the highest possible involvement of the victims in decisionmaking and planning?

LONG-TERM DEVELOPMENT VERSUS DEPENDENCY

Underlying the abovementioned questions is an argument for integrating disaster-related work with long-term development goals. If enhancing the capacities of victims and reducing their vulnerabilities is not kept in mind by development workers at every point of the rehabilitation and resettlement process, then well-intentioned attempts to improve coordination will merely add to a host of other emergency efforts that defer long-term development to the future.

In debates over models for development, all agree that the creation of economic and social structures, while necessary, is not a sufficient gauge of development. Over the long run, external agents cannot ensure a people's well being; only the people themselves can do that, by increasing their capacities and reducing their vulnerabilities.

Given the demands of ministering to the daily requirements of rehabilitation or resettlement, even the more development-oriented NGO's and committed government agents may fail to see how their humanitarian work can stifle the capacity of the victims to rise from the ashes. As their short-term emergency assignment becomes institutionalized in the field, they may be insensitive to the incipient dependence developed in the victims who are unwittingly made to rely on external agents for their needs.

Many of those involved with redevelopment around Pinatubo realize the dependence they have inadvertently created in the course of their work. NGO's operating in the Loob Bunga Resettlement site are themselves alarmed by the perpetuation of a culture of dependence and mendicancy among resettlers (Mondragon, 1992). While they all agree that food for work programs will have to end, they are prevented from focusing on rehabilitation by a lack of opportunities to sustain livelihood projects in the site.

The original concept of the new resettlement towns assumed industrial development in the central Luzon region. The planners hoped that multinational and domestic investors would see the prospect of employing resettlers in the newly built productivity centers. However, the uncertainty over the landscape of central Luzon in the next few years and the sluggish nature of overall Philippine economy has, to date, prevented investors from risking their fortunes in these centers. Naturally, concerns about livelihood have slowed acceptance of resettlement. Respondents to a survey by the Philippine Business for Social Progress (PBSP, 1993) revealed that most respondents gave livelihood a higher priority than housing. The relative absence of income sources in the resettlement sites accounts for the refusal of would-be settlers to move to the new sites. It also explains why some of those who moved in earlier have already left the sites.

The fact that some displaced victims have returned to their old homes has led the NGO's in Zambales to seriously consider internal repatriation. Supporters of repatriation point out that some of the victims who decided to stay put in their barangays at the height of the evacuations seem to have rebuilt their lives faster than anyone in the resettlement sites. It may be possible for some to return to their old homes during the dry season and to go to the resettlement sites when the rains begin. It may also be possible for some victims to evacuate to sites near their original barangays rather than to the resettlement areas.

CONCLUSIONS

The eruptions of 1991 and their muddy aftermath have taken an enormous toll on the people of central Luzon. Fewer than 1,000 lives have been lost, but more than 200,000 families and more than a million people have suffered some loss or dislocation as a result of ash fall, lahars, or flooding. Of these, barangays that were home to 9,800 families (53,000 people) were so severely buried or otherwise damaged that they have been virtually abandoned. Large areas of agricultural land have been covered, some beyond immediate rehabilitation, and additional areas have lost their supply of irrigation water.

Current victims of the Pinatubo disaster have not yet seen the end, and many others are still potential victims. The next several years will continue to bring untold misery to central Luzon. Those who are presently dislocated, and those who will be dislocated in the next several years, need resettlement options that provide livelihood and that facilitate psychosocial adjustments to the trauma of being uprooted. The task is urgent, because people in high-risk areas will agree to move away only if there are viable alternatives.

ACKNOWLEDGMENTS

The paper condenses the findings of a multidisciplinary team under the sponsorship of the College of Social Sciences and Philosophy, University of the Philippines—Centre for Asian Studies, University of Amsterdam.

REFERENCES CITED

Anderson, Mary, and Woodrow, Peter, 1989, Rising from the ashes: Boulder, Westview Press, and Paris, UNESCO Press, 300 p.

Bautista, Cynthia Banzon, ed., 1993, In the shadow of the lingering Mt. Pinatubo disaster: Quezon City, College of Social Sciences and Philosophy, University of the Philippines Faculty Book Series No. 2, 291 p.

Bonifacio, Manuel, 1992, Disaster in agriculture: A framework for an innovative approach to agricultural resources management for Region III: Paper presented to the Farming Systems and Soils Research Institute, College of Agriculture, University of the Philippines at Los Banos.

Department of Public Works and Highways, 1992a, Mt. Pinatubo Infrastructure, Rehabilitation, and Reconstruction Program': Paper presented at the International Scientific Conference on Mt. Pinatubo, May 27–31, 1992, Department of Foreign Affairs, Building, Manila, 35 p.

———1992b, Mt. Pinatubo Infrastructure, Rehabilitation and Reconstruction Program: Manila, unpublished report of September 15, 1992, 29 p.

Department of Social Welfare and Development (DSWD), 1992, DSWD and relief operations: Paper presented by the DSWD Secretary at the International Scientific Conference on Mt. Pinatubo, May 27–31, 1992, Department of Foreign Affairs, Manila, 12 p.

———1992b, Consolidated report on flashflood/lahar: Manila, Department of Social Welfare and Development, Region III, September 28, 1992.

Environment Management Bureau, Department of Environment and Natural Resources, 1992, A report of the Philippine environment and development: Issues and strategies: U.N. Conference on Environment and Development, Rio de Janeiro, Brazil, 260 p.

Jimenez, Maria Carmen, 1993, Stress and coping in difficult times: a study of evacuees and service providers in Concepcion, in Bautista, C.B., ed., In the shadow of the lingering Mt. Pinatubo disaster: Quezon City, College of Social Sciences and Philosophy Publications, Faculty Book Series No. 2, p. 131–162.

Lubos na Alyansa ng mga Katutubong Ayta ng Sambales (LAKAS), 1991, Eruption and exodus: Quezon City, Claretian Publications, 122 p.

Magboo, F.P., Abellanosa, I.P., Tayag, E.A., Pascual, M.L., Magpantay, R.L., Viola, G.A., Surmieda, R.S., Roces, M.C.R., Lopez, J.M., Carino, D., Carino, W., White, M.E., and Dayrit, M.M., 1992, Destructive effects of Mt. Pinatubo's ashfall on houses and buildings [abs.]: Abstracts, International Scientific Conference on Mt. Pinatubo, May 27–31, 1992, Department of Foreign Affairs, Manila, p. 22.

Magpantay, R.L., Abellanosa, I.P., White, M.E., and Dayrit, M.M., 1992, Measles among Aetas in evacuation centers after volcanic eruption [abs.]: Abstracts, International Scientific Conference on Mt. Pinatubo, May 27–31, 1992, Department of Foreign Affairs, Manila, p. 33.

Mercado, R.A., Lacsamana, J.B.T., and Pineda, G.L., this volume. Socioeconomic impacts of the Mount Pinatubo eruption

Mondragon, Gabriel, 1992, Resettlement and rehabilitation of Mt. Pinatubo victims: Problems and prospects for sustainable agricultural development. The Loob Bunga experience: Paper presented at the Symposium for Sustainable Agricultural Development of Mt. Pinatubo Affected Areas, National Institute of Biotechnology and Applied Meteorology, University of the Philippines at Los Banos.

Mount Pinatubo Commission, 1993, Pinatubo News Highlights: Manila, v. 1, no. 3, April, 6 p.

Mount Pinatubo Task Force, 1991, Rehabilitation and reconstruction program for Mt. Pinatubo affected areas: Manila, Government of the Philippines, October 1991, 62 p.

———1992, Rehabilitation of Mt. Pinatubo affected areas, special transition report: Manila, Government of the Philippines, May 1992, 43 p.

Pardo de Tavera, M., 1992, Department of Social Welfare and Development and Relief Operations: Paper presented by the DSWD Secretary at the International Scientific Conference on Mount Pinatubo, Department of Foreign Affairs, May 27–31, 1992.

Philippine Business for Social Progress, 1993, In search of alternatives for the Mt. Pinatubo victims, preliminary report: Manila, May 1993, 100 p.

Shimizu, Hiromu, 1989, Pinatubo Aytas: Continuity and Change: Manila, Ateneo de Manila Press, p. 6–14.

Spence, R.J.S., Pomonis, A., Baxter, P.J., Coburn, A.W., White, M., and Dayrit, M., this volume, Building damage caused by the Mount Pinatubo eruption of June 14–15, 1991.

Surmieda, M.R.S., Abellanosa, I.P., Magboo, F.P., Magpantay, R.L., Pascual, M.L., Tayag, E.A., Viola, Q.A., Diza, F.C., Lopez, J.M., Miranda, M.E.G., Roces, M.C., Sadang, R.A., Zacarias, N.S., Dayrit, M.M., and White, M.E., 1992, Mt. Pinatubo eruption: Disease surveillance in evacuation camps [abs.]: Abstracts, International Scientific Conference on Mt. Pinatubo, May 27–31, 1992, Department of Foreign Affairs, Manila, p. 34.

Tadem, E., and Bautista, C.B., 1993, Brimstone and ash: The Mt. Pinatubo eruption, in Bautista, C.B., ed., In the shadow of the lingering Mt. Pinatubo disaster: Quezon City, College of Social Sciences and Philosophy, University of the Philippines Faculty Book Series No. 2, p. 3–16.

ANCIENT AND MODERN HISTORY

An ancestral Mount Pinatubo began to grow at least 1 million years ago and produced a large but unremarkable stratovolcano. Eruptions of the modern Pinatubo began with an unusually large explosive event slightly more than 35,000 years ago and have occurred episodically (Newhall and others). That of >35,000 years ago was roughly ten times as large as that of 1991, and early reposes of thousands of years appear to be much longer than the most recent repose of 500 years. Burial and erosion of deposits from older, frequent, small-volume eruptions might create an artifact of change from big eruptions and long reposes to smaller eruptions and shorter reposes, but there is also a clear decrease in the size of Pinatubo's largest eruptions per eruptive episode. Except for slight variability in SiO_2 contents of the main dacitic products from one eruptive period to the next (SiO_2 is highest in the products of the oldest,

largest eruption of the modern Pinatubo), eruptive products have been remarkably uniform through time. This uniformity in eruptive products suggests that they are drawn from a single, large volume reservoir.

Exploratory geothermal drilling at Mount Pinatubo in 1988–90, fortuitous for science though not for the drillers, revealed high-temperature (261–336°C), highly acidic fluids and low permeability (Delfin and others). The prospect was abandoned for economic reasons; no one foresaw that most of those fluids would soon be erupted. Were it not for the requisite cementing those wells might have provided an unparalleled window into the temperature, pressure, and chemistry of a hydrothermal system as the volcano drew closer to its eruption! Not all was lost: highly acidic, high-Cl fluids at high elevations on a volcano are now understood to be a clear magmatic signature in geothermal fluids.

Stacked pyroclastic-flow and lahar deposits outside Clark Air Base, testimony to a long history of explosive eruptions.

Eruptive History of Mount Pinatubo

Christopher G. Newhall,[1] Arturo S. Daag,[2] F.G. Delfin, Jr.,[3] Richard P. Hoblitt,[1]
John McGeehin,[1] John S. Pallister,[1] Ma. Theresa M. Regalado,[2] Meyer Rubin,[1]
Bella S. Tubianosa,[1] Rodolfo A. Tamayo, Jr.,[4] and Jesse V. Umbal[2]

ABSTRACT

The eruptive history of Mount Pinatubo is divided into two parts—eruptions of an ancestral Pinatubo (~1 Ma to an unknown time before 35 ka) and eruptions of a modern Pinatubo (>35 ka to the present). Ancestral Mount Pinatubo was an andesite-dacite stratovolcano for which we have no evidence of large explosive eruptions. Modern Mount Pinatubo is a dacite-andesite dome complex and stratovolcano that is surrounded by an extensive apron of pyroclastic-flow and lahar deposits from large explosive dacitic eruptions.

Eruptions of the modern Pinatubo occur episodically, in eruptive periods that are short in comparison to intervening reposes of several centuries or millennia. Eruptions may be getting smaller with time, and repose periods shorter, though some of the apparent trend toward smaller eruptions and shorter reposes may be an artifact of erosion and (or) burial of older deposits. The explosive eruption of June 15, 1991, is one of the smallest we can identify in the geologic record, and the 500-year repose that preceded that eruption is relatively short for Pinatubo.

Rocks of prehistoric eruptions and the 1991 eruption of Mount Pinatubo are very similar and record evidence of repeated basaltic intrusion into a large, gas-charged dacitic body. Explosive eruptions followed, as did, in at least some instances, growth of late-stage lava domes.

INTRODUCTION

Before April 2, 1991, volcanologists knew Mount Pinatubo as an inconspicuous volcano, active within the past millennium, and the site of an aborted geothermal development. Indigenous people of the area, Aetas, knew Mount Pinatubo as their home, their hunting ground, and their haven from an ever-encroaching lowland population. For the Aeta, Mount Pinatubo was (and is) the home of Apo Namalyari, the Great Protector and Provider. A much larger population in the surrounding lowlands, including those in nearby military bases, barely knew of Mount Pinatubo at all, save for a few who climbed for a respite from the heat and hassles of lowland life, and for military personnel who received survival training from the Aetas.

Until June 1991, Mount Pinatubo (15°08.2' N., 120°21.1' E.) rose 1,745 m above sea level but only about 600 m above a gently-sloping apron and only 200 m above nearby mountains that largely obscured Mount Pinatubo from view (figs. 1, 2). The surrounding mountains are older volcanic centers and relics of an ancestral Mount Pinatubo, which is discussed below. The gently-sloping apron consists of thick pyroclastic-flow and lahar (volcanic mudflow) deposits. This paper summarizes what we learned of Pinatubo's eruptive history before and since the major eruption of June 1991.

GEOLOGIC SETTING

The geology of the Pinatubo region is described by Corby and others (1951), Roque and others (1972), de Boer and others (1980), Philippine Bureau of Mines and

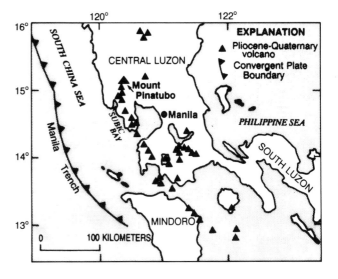

Figure 1. Location map for Mount Pinatubo and central Luzon.

[1] U.S. Geological Survey.
[2] Philippine Institute of Volcanology and Seismology.
[3] Philippine National Oil Corporation.
[4] University of the Philippines.

Figure 2.   Preeruption Mount Pinatubo, as seen from the north.

Geosciences (1982), Hawkins and Evans (1983), Schweller and others (1983), Delfin (1983, 1984), and Delfin and others (1992; this volume). Briefly, Pinatubo is flanked on the west (and partially underlain) by the Zambales Ophiolite Complex, an easterly-dipping slab of Eocene ocean crust uplifted during the late Oligocene (Villones, 1980). North, east, and southeast of Pinatubo, late Miocene to Pliocene marine, nonmarine, and volcaniclastic sediments of the Tarlac Formation (Roque and others, 1972) are believed to unconformably overlie the Zambales Ophiolite Complex (fig. 3).

The Zambales Ophiolite Complex west and northwest of Pinatubo is dominantly peridotite and gabbro; that north of Pinatubo is dominantly basalt (Hawkins and Evans, 1983). The Zambales ophiolite complex encountered southeast of Mount Pinatubo, from 1,100 m to 2,733 m below the surface in geothermal exploration hole PIN–1 (see Delfin and others, this volume), is dominated by microdiorite and diabase dikes, with lesser amounts of hornfels, basalt, gabbro, and monzodiorite (Aniceto-Villarosa and others, 1989). Interestingly, rocks from the ophiolite complex rarely occur

as xenoliths in Pinatubo volcanic rocks, despite their probable continuity beneath the volcano. The ophiolite, topographically high west and northwest of Mount Pinatubo, is apparently downdropped to the east across an unnamed north-south fault, about 10 km west of Mount Pinatubo (fig. 3; Delfin and others, this volume).

The lower part of the >1,000-m-thick Tarlac Formation is dominated by fossiliferous sandstone and siltstone and unconformably overlies the Zambales Ophiolite Complex. The upper part of the Tarlac Formation is dominated by conglomerate with sandstone lenses cut by andesite dikes. Near the southern end of Crow Valley (see fig. 7), ash-flow tuffs unconformably overlie sedimentary layers and, along the Malago and Marimla Rivers, sediments of the Tarlac Formation are overlain by lavas and intruded by the dome of Mount Mataba (fig. 3).

Volcanics of the Tarlac Formation originated from the same east-dipping subduction along the Manila trench that continues to the present. de Boer and others (1980) and Datuin (1982) report a general trend from tholeiitic rocks on the west to slightly alkaline rocks on the east, with calc-alkaline rocks dominating the main volcanic belt. Some Miocene and Pliocene volcanic centers have now been exhumed to the level of subvolcanic intrusions, such as the diorite of the Dizon Mine (fig. 3).

de Boer and others (1980) described the Iba fracture zone, a major northwest-trending fracture zone of which Delfin's (1983, 1984) Maraunot fault is perhaps a part (fig. 3). The same authors noted that Mount Pinatubo lies at the intersection of the Iba fracture zone and the volcanic arc. Defant and others (1991) described rocks of Mount Pinatubo in relation to others of the Luzon Arc.

GEOLOGY OF MOUNT PINATUBO

Our oldest references to volcanic rocks from Mount Pinatubo are accounts by von Drasche (1878), Oebbeke (1881), and Jordana y Morera (1885), describing the mineralogy and texture of pumice in tuffs in the vicinity of Porac, Magalang, and along the O'Donnell River. Von Drasche described a sample from Porac that was "spongy, brilliant white trachyte pumice, with large fractured sanidine and short column-shaped hornblende crystals in the vesicles." Oebbeke described the Magalang sample as pumice with green hornblende, plagioclase (An_{55}), and mica, the first two with glass and fluid inclusions. Jordana y Morera described "dolerite" (andesite) and gabbro, referring either to lavas and intrusive equivalents of Pinatubo area vents or to cobbles eroded from the Zambales Ophiolite Complex. None of these authors mention Mount Pinatubo, but Jordana y Morera clearly recognized a nearby volcanic source for these rocks, and von Drasche reported that his "trachyte nodules" (pumice clasts) increased in size toward the Sierra of Zambales (toward Pinatubo).

The first geologic commentary about Mount Pinatubo itself was by Smith (1909), who, after describing the "Aglao Valley" (Marella River valley) as once filled to a depth of 120 to 150 m with loose sand and boulders, then described Mount Pinatubo as devoid of "volcanic ash (and) any of the usual indications of volcanic activity." Smith concluded that "Mount Pinatubo is not a volcano and we saw no signs of it ever having been one, although the rock constituting it is porphyritic.... The region is quite unique and I have seen nothing in the Philippines quite like it."

Decades later, economic geologists noted Pinatubo's sulfur deposits (Jagolino, 1973) and Pinatubo's lahar deposits as a source for lightweight pumice aggregate (Cruz, 1981). Three radiocarbon ages and a number of chemical analyses of Pinatubo rocks were published in site-safety documents for the Philippine Nuclear Power Plant #1 (Ebasco Services, Inc., 1977); these same ages were reported, together with an interpretation of aerial photography, by Wolfe and Self (1982, 1983). More recently, the overall geology of Pinatubo was studied as part of a geothermal exploration program (Delfin, 1983, 1984).

Delfin (1983, 1984) recognized two suites of Pinatubo units, "old" and "young." To emphasize Pinatubo's life history more than absolute ages, we will call these same suites "ancestral" and "modern" (fig. 3). Discussion of each suite follows.

ANCESTRAL PINATUBO

Much of the rugged terrain surrounding and partially obscuring the modern Pinatubo consists of relics of the ancestral Pinatubo (fig. 4). Ancestral Pinatubo was an andesite and dacite stratovolcano whose center was in roughly the same location as the modern Pinatubo. Projection of dipslopes upward from existing terrain suggests that the ancestral Pinatubo, if it was an isolated, central cone, could have been as high as 2,300 m. If the summit was complex, or rounded, the peak would have been lower. A high-silica andesite intrusive on the east flank of Pinatubo (UTM 1675.7N, 220.5E) yielded a K-Ar age of 1.10 ± 0.09 Ma (Ebasco Services, Inc., 1977). A similar age of 1.09 ± 0.10 Ma (Bruinsma, 1983) was obtained from a hornblende andesite lava on the western slope of the volcano.

Today, ancestral Pinatubo is exposed in relict walls of an old 3.5×4.5-km caldera, recognized by Delfin (1983, 1984) and named here the Tayawan caldera (fig. 3). Prominent points on the relict rim of that caldera include Mount Donald Macdonald (southeast caldera rim) and Mount Tayawan (north-northeast caldera rim). Several patches of high, erosion-resistant terrain outside that caldera (for example, Mount Dorst) were part of the dip slopes of that ancestral Pinatubo (figs. 3, 4).

Lava flows of the caldera walls and rim are mostly two-pyroxene-hornblende andesite (augite+hypersthene± hornblende±biotite; plagioclase=An$_{60}$) and are interbedded with pyroclastic-flow and near-vent fall breccias. Lower on the volcano's flanks, deposits of ancestral Pinatubo are indurated, yellow-brown andesitic and dacitic breccias of mostly lahar origin.

Deep erosion in the Sacobia, Porac, Marimla, and Porac River valleys, coupled with deep weathering and induration of its products, suggests that activity of the ancestral volcano ended several tens of thousands of years (or more) before the caldera-forming eruption and initial growth of the modern Pinatubo.

ANCESTRAL SATELLITIC VENTS

Several nearby vents were active contemporaneously with ancestral Pinatubo. These include Mount Negron dome, Mount Cuadrado dome, the Mataba dome and adjoining Bituin plug, and the Tapungho plug (fig. 3) (Delfin, 1984). Mount Negron, situated 5 km south-southeast of Mount Pinatubo, is a 2×2.5 km, 800-m-high steep-sided dome built of glassy hornblende andesite lava with a K-Ar age of 1.27 ± 0.8 Ma (Bruinsma, 1983). Mount Cuadrado, 11 km south of Mount Pinatubo, is a 1.5-km-diameter, 600-m-high steep-sided dome of hornblende-pyroxene dacite, with an age of 1.59 Ma. The Mataba dome, 11 km northeast of Mount Pinatubo, is a 0.7×2 km, 300-m-high dome of biotite-hornblende andesite that intruded and upturned layers of the Tarlac Formation. The Bituin ("star") plug, 8.5 km north-northeast of Pinatubo and 3.5 km west of the Mataba dome, is about 400 m high, consists of hornblende andesite, and is either an eroded remnant of a volcanic neck or an intrusive plug. The plug at Tapungho, 14 km south-southwest of Pinatubo and 3.5 km southwest of Mount Cuadrado, is a small, 500-m-diameter, 300-m-high pyroxene-biotite-hornblende andesite feature of uncertain age.

MODERN PINATUBO

Our reconnaissance fieldwork on modern Pinatubo distinguished deposits of repeated, large explosive eruptions. Although a few deposits are lithologically, paleomagnetically, or morphologically distinctive, most are similar and are distinguished from each other in the field only by erosional breaks, paleosols, or widely different ^{14}C ages. Widespread tephra marker horizons are absent, probably lost to erosion.

Radiocarbon ages suggest that eruptions from the modern Pinatubo have been clustered in at least six and possibly as many as a dozen eruptive periods. Each eruptive period is short in relation to intervening repose periods (table 1, fig. 5). Each of these six eruptive periods of modern Mount Pinatubo is described below. More detailed, future mapping will probably add more eruptive periods to this scheme, especially in the period before 5,000 yr B.P.

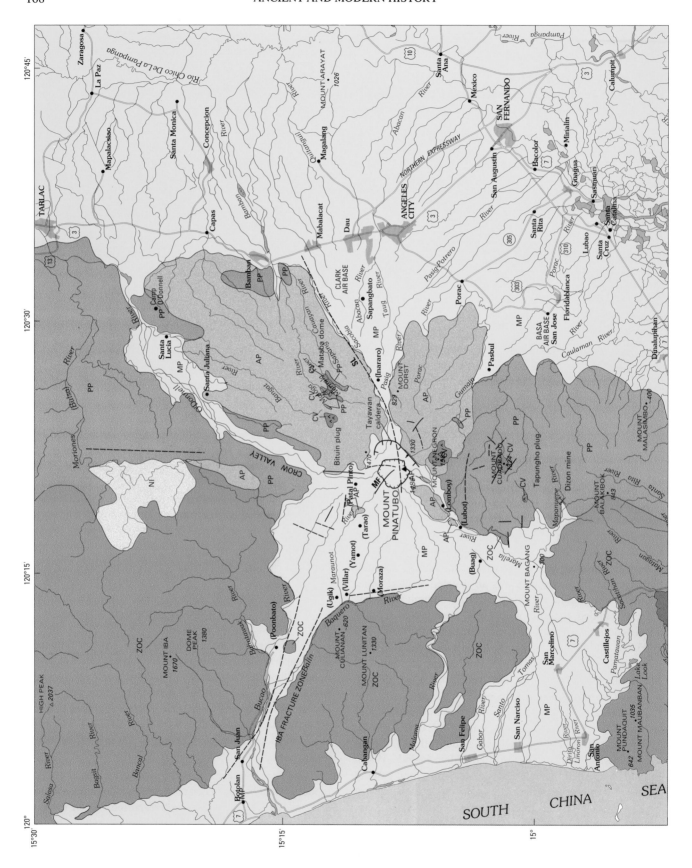

Figure 3. Generalized geologic map of Mount Pinatubo and environs (after Philippine Bureau of Mines (1963) and Delfin (1983, 1984)). Altitude in meters.

EXPLANATION OF MAP UNITS

Figure 3.—Continued.

MP **Volcanic rocks of modern Pinatubo (Holocene and late Pleistocene, 40 ka to present)**—Mainly dacitic pyroclastic-flow and lahar deposits

AP **Deposits of ancestral Pinatubo (Pleistocene)**—Andesitic and dacitic pyroclastic-flow and lahar deposits, and lava flows

CV **Ancestral satellite vent deposits (Pleistocene)**—Andesite and dacite domes and plugs, contemporaneous with deposits of ancestral Pinatubo

NI **Neogene intrusives (Pliocene? and Miocene)**—Granodiorite and diorite porphyry

PP **Pre-Pinatubo sedimentary and volcanic rocks, mostly Tarlac Formation (Early Pliocene and late Miocene)**

ZOC **Zambales Ophiolite complex (Eocene)**—Mainly peridotite, gabbro, and basalt

EXPLANATION OF MAP SYMBOLS

——— **Contact**—Dashed where inferred

——— **Fault**—Dashed where inferred; MF, Maraunot Fault; SL, Sacobia Lineament

╤╤╤ Rim of Tayawan Caldera

╥╥╥╥ Boundary of caldera formed on June 15, 1991

In addition, we show composite stratigraphic columns (fig. 6) for each of the 7 major watersheds of Pinatubo. River reaches to which these apply are shown by shading in figure 7, together with the locations of individual ^{14}C samples. Figure 7 also serves as a location map for localities mentioned in the text that follows. In the following discussion of eruptive periods, radiocarbon ages are referenced by their sample number in table 1 (1 through 51) and the stratigraphic column in which they appear (A–Q). For example, 50–G refers to sample number 50 of table 1 and stratigraphic column G of figure 6.

A word is necessary about the preeminent role that ^{14}C ages play in our distinction of eruptive periods of the modern Mount Pinatubo. Ideally, one would decipher a volcanic stratigraphy by the use of superposition and other stratigraphic relations, aided by correlations of distinctive lithologies. Then, ages would be determined for key units and the age span of an eruptive period would be that between the ages of its oldest and youngest deposits. However, lacking adequate field time and exposures, before and since the 1991 eruption, and also lacking obvious lithologic distinctions between products of the various eruptive periods, we

have defined eruptive periods on the basis of clusters of ^{14}C ages.

Given such strong reliance on ^{14}C ages, the most reliable age for an eruptive event is one obtained from the bark or outermost wood of in-situ trees that were rooted in soil underlying the deposit to be dated. However, only a few of our samples are of bark or the outermost, youngest parts of rooted trees. Most samples are from unknown parts of trees

Figure 4. (Following pages.) Airborne side-looking radar (SLAR) images of Mount Pinatubo in (A) March 1991 and (B) November 1991. In the preeruption view (A), the steep cone of the modern Pinatubo lies within the Tayawan caldera (dashed line) of ancestral Pinatubo. In the posteruption view (B), the 1991 caldera replaces the northern two-thirds of the preeruption edifice; the center of the caldera is about 750 m north-northwest of the broad preeruption summit; the vent of the June 7–12, 1991, lava dome was about 1,000 m north-northwest of the preeruption summit. Data acquired by Intera Information Technologies for the Government of the Philippines, Department of Public Works and Highways (DPWH), and the National Mapping and Resource Information Authority (NAMRIA). Used with permission.

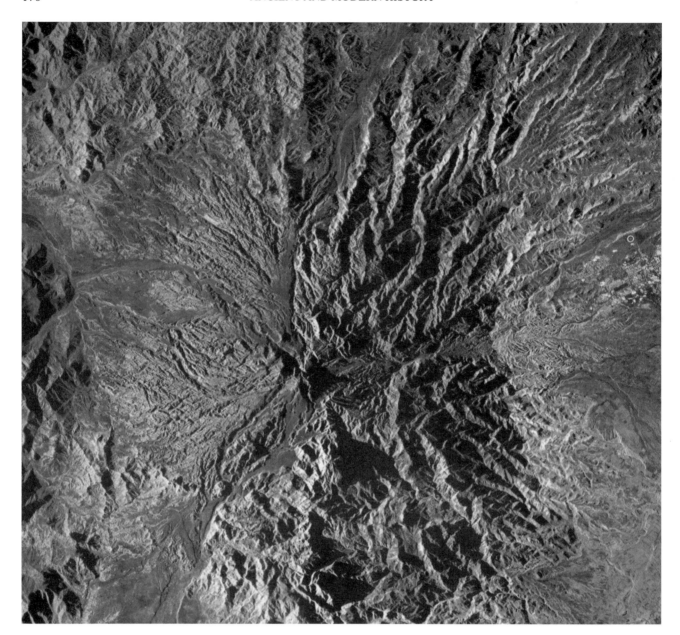

Table 1. Pinatubo radiocarbon data.

["Age" and "Standard dev." are the radiocarbon determination, in radiocarbon years. The ages of all but the oldest two eruptive periods were calibrated to calendar rather than radiocarbon years before 1950 by use of calibration curves developed by Stuiver and Reimer (1993). "Calib age" is the statistically-most-likely equivalent in calendar years before 1950, and "calib min" and "calib max" mark 1-sigma confidence limits for the calendar equivalence probability distribution. Sample locations are shown in fig. 7; the approximate stratigraphic position of each sample is shown in fig. 6]

Sample	Field number	Lab number	Latitude (°N.)	Longitude (°E.)	Material	Occurrence	Drainage	Section in figure 6	Age	Standard dev.	Calibration age	Calibration minimum	Calibration maximum	Significance
Deposits of Buag age														
1	NM-N1W4-1	WW-111	15°04'	120°16'	Charcoal	Cultural layer	Marella	N	397	70	476	318	511	Dates latest occupation; settlement on old lahar deposit.
2	CN-P-33	WW-26	15°16.8'	120°22.7'	Wood	Fluvial deposit.	O'Donnell	B	400	80	477	315	515	Latest major filling of Crow Valley.
3	CN-P-18	W-6478	15°09.2'	120°34.4'	Wood (in growth position).	Soil overlying stream gravel.	Abacan	I	410	55	485	331	511	After high-energy stream; before latest pre-1991 lahars.
4	Alonso et al.	W-6314	15°09'	ca. 120°24'	Charcoal	Pyroclastic-flow deposit	upper Sacobia.	G	460	30	509	501	518	Youngest pre-1991 pyroclastic flow in Sacobia.
5	TP-Sabo4	Beta	15°10.5'	120°29.5'	Wood (in growth position).	Post pyroclastic-flow soil.	Abacan	G	470	50	512	494	530	Start of latest pre-1991 fluvial deposition in Abacan.
6	JU-BL-2	W-6509	14°59.9'	120°15.5'	Charcoal	Lahar deposit	Marella	O	560	60	545	519	635	Covered by 1992 lahars.
7	CN-P-16	WW-25	15°8.2'	120°28.8'	Wood	Sediment of small lake.	Pasig-Potrero.	J	560	70	545	517	640	Latest fill in the north fork, Pasig-Potrero.
8	CN-P-17d	WW-21	15°09.7'	120°33.4'	Charcoal	Silt over lahar deposit.	Abacan	I	570	70	547	519	643	Age of terrace by Friendship Bridge, south edge, Clark Air Base.
9	JU-BL-1	W-6505	14°59.9'	120°15.6'	Charcoal	Lahar deposit.	Marella	O	600	60	617	539	648	Covered by 1992 lahars.
10	CN-P-46	WW-29	15°08.2'	120°28.7'	Uncharred root (in situ).	Fluvial deposits.	Pasig-Potrero.	J	630	70	597	545	658	After latest pre-1991 downcutting; before major siltation.
11	1275-17	Ebasco	15°03.3'	120°17.0'	Wood	Lahar deposits	Marella	N	635	80	593	543	661	Lahar just below toe of morphologically youngest pre-1991 pyroclastic-flow deposit.
12	CN-P-30	WW-28	15°18.5'	120°35.7'	Wood	Flood-plain deposits.	Bamban	F	660	80	648	550	668	Sedimentation across broad Bamban-Capas flood plain.
13	KSR-P-61	WW-107	15°16.03'	120°05.3'	Charcoal	Hyperconcentrated flow deposit.	Bucao	Q	730	80	665	577	710	Latest major sedimentation in Bucao; overtopped in 1993.
14	JU-AL-1	W-6504	14°59.9'	120°16.6'	Charcoal	Fluvial deposits.	Marella	O	760	60	673	657	721	Covered by 1993 lahars.
15	RPH-22592-2/1.	WW-31	15°08'	120°29'	Wood	Lahar deposit	Pasig-Potrero.	J	950	70	800	787	925	
Deposits of Maraunot age														
16	376-1G	Ebasco	15°03.3'	120°17.0'	Charcoal	Pyroclastic-flow deposit.	Marella	N	2,330	110	2,342	2,157	2,462	
17	RPH-91-2	W-6316	15°16.1'	120°22.2'	Charcoal	Pyroclastic-flow deposit.	O'Donnell	B	2,640	50	2,754	2,743	2,772	Same unit as sample 22; underlies sample 31.
18	RPH-3792-3/1	WW-32	15°11.2'	120°30.2'	Charcoal	Fluvial deposit.	Clark Air Base.	D	2,660	70	2,759	2,745	2,793	Maximum age for surface of Clark Air Base.

Table 1. Pinatubo radiocarbon data—Continued.

Sample	Field number	Lab number	Latitude (°N.)	Longitude (°E.)	Material	Occurrence	Drainage	Section in figure 6	Age	Standard dev.	Calibration age	Calibration minimum	Calibration maximum	Significance
						Deposits of Maraunot Age—Continued								
19	CN-P-45b	WW-13	15°10.7'	120°18.5'	Charcoal	Pyroclastic-flow deposit.	Maraunot	P	2,820	70	2,914	2,617	2,985	
20	KS-TP	Beta47404	15°09.6'	120°36.2'	Wood (in growth position).	Lahar deposit	Abacan	I	2,860	80	2,953	2,660	3,076	
21	CN-P-36	WW-23	15°06.8'	120°31.9'	Slightly charred wood.	Lahar deposit	Pasig-Potrero.	K	2,860	70	2,953	2,866	3,071	Maximum age for sediments forming Porac fan.
22	RPH-3592-1/4	WW-34	15°16.1'	120°22.2'	Charcoal (growth position.)	Pyroclastic-flow deposit	O'Donnell	B	2,870	100	2,958	2,856	3,147	Outer part of charred stump.
23	JU-CL-1	W-6508	15°58.3'	120°12.9'	Charcoal	Lahar deposit	Marella	O	2,880	70	2,971	2,877	3,103	Surface leading to Castillejos.
24	CN-P-22	WW-20	15°12.6'	120°31.0'	Charcoal	Fluvial deposits.	Sacobia	D	2,930	70	3,070	2,954	3,207	
25	JU-DL-1	W-6506	14°59.8'	120°16.0'	Charcoal	Sandy, fluvial silt.	Marella	O	2,930	80	3,070	2,949	3,209	Probably covered by 1993 lahars.
26	JU-PL-1	W-6507	14°59.5'	120°14.9'	Charcoal	Fluvial deposits.	Marella	O	2,950	50	3,084	2,993	3,206	1.7 m above 1992 lahars; 3 m below highest pre-1991 lahars.
27	CN-P-43	WW-19	15°10.1'	120°28.1'	Charcoal	Pyroclastic-flow deposit.	Sacobia	G	2,950	70	3,084	2,971	3,212	Latest big valley-filling pyroclastic flow in the Sacobia.
28	JU-Dal-1a	WW-112	15°00.0'	120°16.0'	Charcoal	Pyroclastic-flow deposit.	Marella	O	2,990	70	3,192	3,066	3,320	Largest pyroclastic flow known in the Marella drainage.
29	CN-P-2b	W-6311	15°10.2'	120°28.2'	Charcoal	Pyroclastic-flow deposit.	Sacobia	G	2,990	80	3,192	3,012	3,326	
30	CN-P-45a	WW-27	15°10.7'	120°18.5'	Charcoal	Pyroclastic-flow deposit.	Maraunot	P	3,030	30	3,214	3,167	3,320	Lowest exposed pyroclastic-flow deposit in Maraunot.
31	RPH-3592-1/2	WW-33	15°16.1'	120°22.2'	Charcoal	Pyroclastic-flow deposit.	O'Donnell	B	3,260	70	3,467	3,385	3,563	Questionable age; overlies sample 22.
32	CN-P-11	W-6375	15°11.0'	120°29.8'	Charcoal	Pyroclastic-flow deposit.	Sacobia	C	3,390	90	3,629	3,477	3,709	
33	CN-P-39	WW-12	15°02.7'	120°17.3'	Charcoal	Pyroclastic-flow deposit.	Marella	N	3,590	110	3,875	3,707	4,070	Toe of Marella pyroclastic-flow fan; apparently low in section.
						Deposits of Crow Valley Age								
34	CN-P-3a	W-6312	15°11.2'	120°28.8'	Charcoal	Pyroclastic-flow deposit.	Sacobia	C	4,450	70	5,004	4,875	5,259	
35	CN-P-12	WW-11	15°11.5'	120°29.3'	Charcoal	Pyroclastic-flow deposit.	Sacobia	C	4,800	60	5,503	5,468	5,597	Terrace outside Mactan Gate, Clark Air Base.
36	RHP-91-1	W-6315	15°13.6'	120°21.2'	Charcoal	Pyroclastic-flow deposit.	O'Donnell	A	4,890	80	5,612	5,583	5,716	
37	RHP-22392-4	WW-30	15°10.8'	120°28.5'	Charcoal	Pyroclastic-flow deposit.	Sacobia	C	4,890	70	5,612	5,586	5,705	
38	RHP-3592-1/5	WW-35	15°16.1'	120°22.2'	Charcoal	Pyroclastic-flow deposit.	O'Donnell	B	4,930	80	5,651	5,594	5,738	
39	CN-P-31b	WW-17	15°14.9'	120°22.0'	Charcoal	Pyroclastic-flow deposit.	O'Donnell	A	4,960	70	5,693	5,612	5,841	Overlies sample 42.

Table 1. Pinatubo radiocarbon data—Continued.

Sample	Field number	Lab number	Latitude (°N.)	Longitude (°E.)	Material	Occurrence	Drainage	Section in figure 6	Age	Standard dev.	Calibration age	Calibration minimum	Calibration maximum	Significance
							Deposits of Crow Valley Age—Continued							
40	CN-P-32b	WW-16	15°14.8'	120°21.8'	Charcoal	Pyroclastic-flow deposit.	O'Donnell	A	5,010	80	5,736	5,652	5,892	Big dissected pyroclastic-flow deposit, north side of Pinatubo.
41	CN-P-50	WW-104	15°25.0'	120°39.0'	Charcoal	Fluvial deposit	O'Donnell	B*	5,080	70	5,807	5,737	5,916	Major inundation of Tarlac Province.
42	CN-P-31a	WW-14	15°14.9'	120°22.0'	Charcoal	Pyroclastic-flow deposit.	O'Donnell	A	5,100	70	5,782	5,745	5,924	May overlie sample 40.
43	CN-P-41c	WW-22	15°15.7'	120°10.1'	Charcoal	Silt interbedded with lahar.	Bucao	Q	5,130	80	5,907	5,753	5,940	Major filling of the Bucao River valley.
44	CN-P-3b	W-6313	15°11.2'	120°28.8'	Charcoal	Pyroclastic-flow deposit.	Sacobia	C	5,140	100	5,911	5,749	5,985	
45	CN-P-51	WW-105	15°20'	120°40'	Charcoal	Fluvial deposit.	O'Donnell	B*	5,160	60	5,918	5,778	5,980	Major inundation of Tarlac Province.
46	CN-P-53	WW-106	15°25'	120°40'	Charcoal	Fluvial deposit.	O'Donnell	B*	5,280	70	6,079	5,939	6,176	Major inundation of Tarlac Province.
							Deposits of Pasbul Age							
47	975-9W-2	Ebasco	15°11.8'	120°14.8'	Charcoal	Pyroclastic-flow deposit.	Maraunot	P	8,050	130	8,979	8,655	9,187	
48	CN-P-42	WW-18	15°02.3'	120°27.2'	Charcoal	Pyroclastic-flow deposit.	Gumain	L	8,380	80	9,414	9,259	9,447	Deep pyroclastic-flow fill in Gumain; much deeper than in 1991.
							Deposits of Sacobia Age							
49	CN-P-21	WW-24	15°13.4'	120°32.2'	Charcoal	Silt overlying lahar deposit.	Sacobia	D	14,480	130	17,347	17,179	17,516	A debris-flow/fluvial sediment pair underlying Clark Air Base.
							Deposits of Inararo Age							
50	CN-P-27a	WW-15	15°10.5'	120°29.9'	Charcoal	Pyroclastic-flow deposit.	Abacan	G	30,390	890				Biggest pyroclastic flow of Pinatubo history.
51	WS-31392-1/1	W-6487	15°10.5'	120°29.25'	"Charcoal	Pyroclastic-flow deposit.	Abacan	G	>35,000					Biggest pyroclastic flow of Pinatubo history.

* Hacienda Luisita.

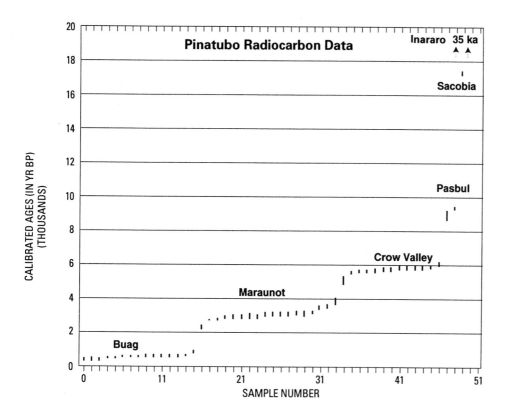

Figure 5. Radiocarbon ages of modern Pinatubo deposits, arranged in chronologic order (not necessarily in stratigraphic order). All ages except those of the three oldest samples are shown as their 1-sigma range in calibrated, calendar equivalence (calibrated minimum to calibrated maximum, from table 1). For sample 49, an arbitrary range of 1,000 yr was plotted; samples 50 and 51 are shown with open-ended arrows because the ages were beyond the limits of reliability for conventional radiocarbon determinations.

that may have fixed atmospheric ^{14}C several hundred years before the tree was actually killed by an eruption. For example, charcoal from one pyroclastic-flow deposit (22–B, table 1) is apparently 400 years younger than charcoal from an overlying pyroclastic-flow deposit (31–B), but the older sample is a loose fragment of charcoal while the younger sample is from the outer 1 cm of a standing, charred snag. In addition, pyroclastic flows and lahars can incorporate charcoal from older deposits; as a result, wide discrepancies can arise in ages within the same deposit. A possible example is the apparent, >900-yr difference between the ages of samples 34–C and 44–C, which are from 1 m apart in the same pyroclastic-flow unit. Some of the older ages for each eruptive period are possibly (indeed, probably) artifacts of inadvertently sampled pieces from the older, biologically inactive interior of trees, or of incorporation of charcoal from otherwise unsampled, older eruptive periods. A few ages might also be too young. For example, a younger age for sample 50–G than for 51–G, both from the same outcrop, may reflect air drying for sample 50–G versus oven drying for sample 51–G. Oven drying prevented the growth of mold that fixes modern carbon and might be hard to remove completely in preanalysis sample treatment. In the text of this report, a radiocarbon age is expressed in ^{14}C years. All other ages are calendar years.

INARARO ERUPTIVE PERIO —>35,000 ^{14}C YR B.P.

The largest eruption in the history of modern Pinatubo occurred before 35,000 ^{14}C yr B.P. and deposited up to

100 m or more of pumiceous pyroclastic-flow material on all sides of Mount Pinatubo. These deposits, once a thick, widespread apron around Pinatubo, are now found as distinctively eroded "pinnacles," erosional remnants within a deep dendritic drainage network between the Abacan and Pasig-Potrero Rivers, between 100 and 500 m above sea level (figs. 6 and 7, section G). In a few locations, notably between the north and south forks of the upper Pasig-Potrero River, and in a southern channel of the Sacobia River just upslope from its divergence with the Abacan, the deposit is incipiently welded and exhibits columnar jointing. The main deposit of this eruptive period is the basal and volumetrically dominant part of the "Inararo pyroclastic-flow deposit" of Delfin (1984, p. 16–17), named after a village, Inararo ("plowed"; location 29, fig. 7), that was destroyed by the 1991 eruption. Charcoal samples 50–G and 51–G, from a single outcrop and horizon (figs. 6, 7), have respective ages of 30,390±890 and >35,000 ^{14}C yr B.P.; we take the latter to be more reliable.

The Inararo pyroclastic-flow deposits contain pumice blocks of biotite-hornblende-quartz dacite with about 64–67% SiO_2 (table 2). Relatively abundant biotite and quartz phenocrysts, and the only occurrence of euhedral quartz in the Pinatubo sequence, distinguish these deposits from those of younger eruptive periods. Cummingtonite is present as rims on hornblende phenocrysts in some of the pumice blocks, a relation that indicates that preeruption magmatic temperature and pressure were similar to those in the 1991 magma body. The absence of cummingtonite in pumice blocks from some outcrops, slight differences in

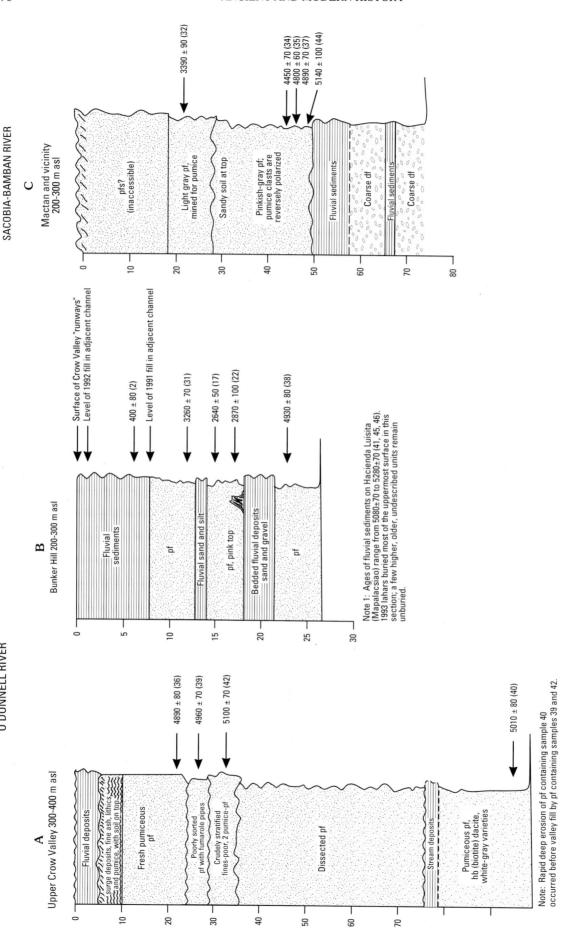

Figure 6. Composite stratigraphic sections for major rivers of Mount Pinatubo. The sections were constructed from stratigraphic relations and ^{14}C ages of multiple outcrops within the lettered boxes shown in figure 7. Where outcrops were discontinuous and stratigraphic correlations impossible, ^{14}C ages are the principal tie between individual outcrops and the composite section. Sections are mostly along rivers, as follow: A–B, O'Donnell; C–F, Sacobia-Bamban; G–I, Upper Sacobia-Abacan; J–K, Pasig-Potrero; L, Gumain; M–O, Marella-Santo Tomas; and P–Q, Maraunot-Balin Baquero-Bucao. Radiocarbon sample numbers (in parentheses) refer to table 1; locations of samples and stratigraphic sections are given in figure 7. Vertical scales in meters. Abbreviations: pf, pyroclastic-flow deposit; df, debris-flow deposit; hcf, hyperconcentrated-flow deposit; asl, above sea level; bio, biotite; qz, quartz; hb, hornblende; mt, magnetite.

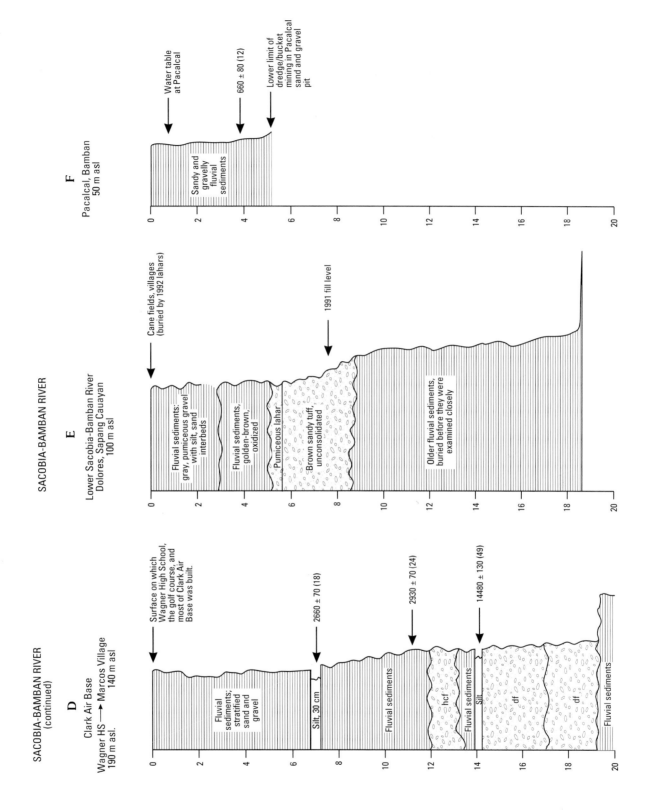

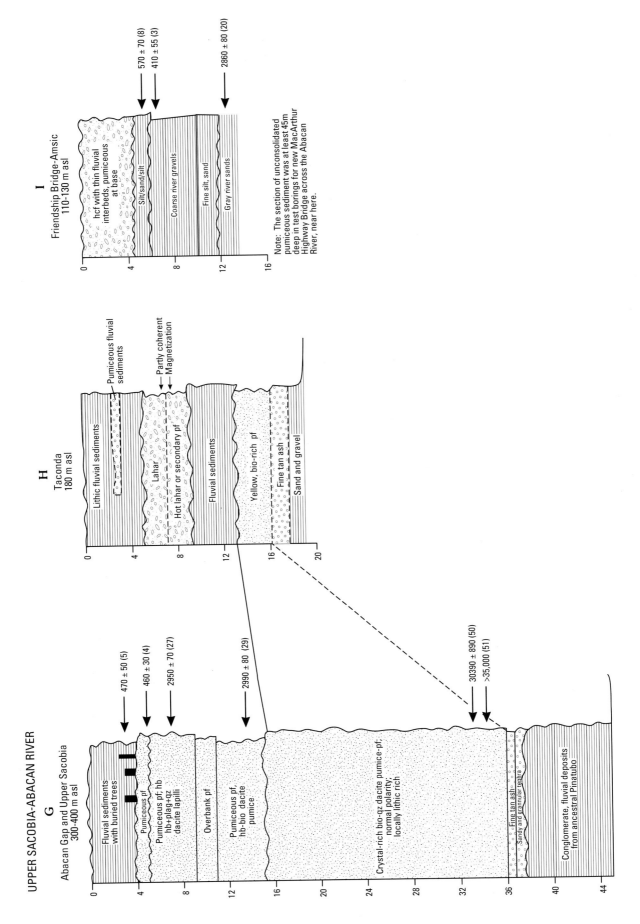

Figure 6.—Continued.

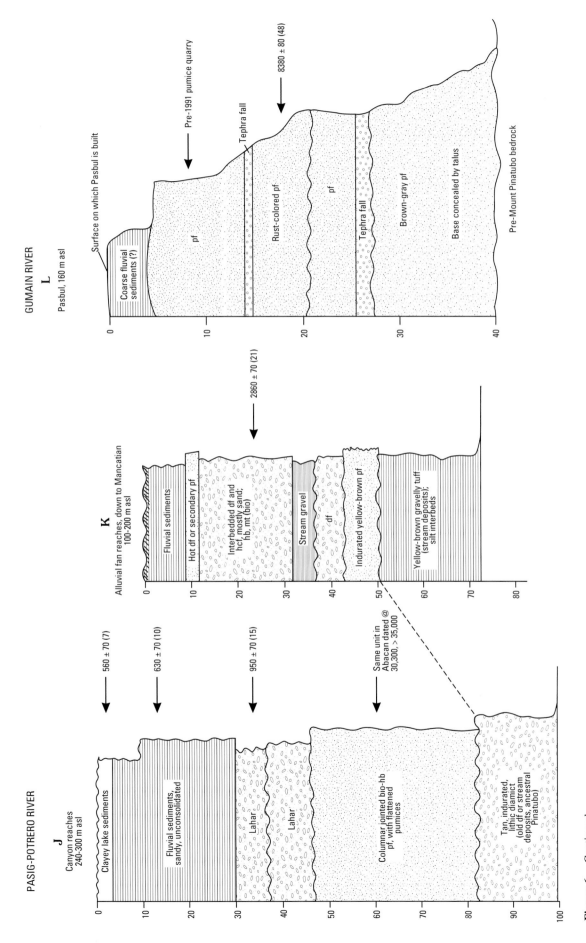

Figure 6.—Continued.

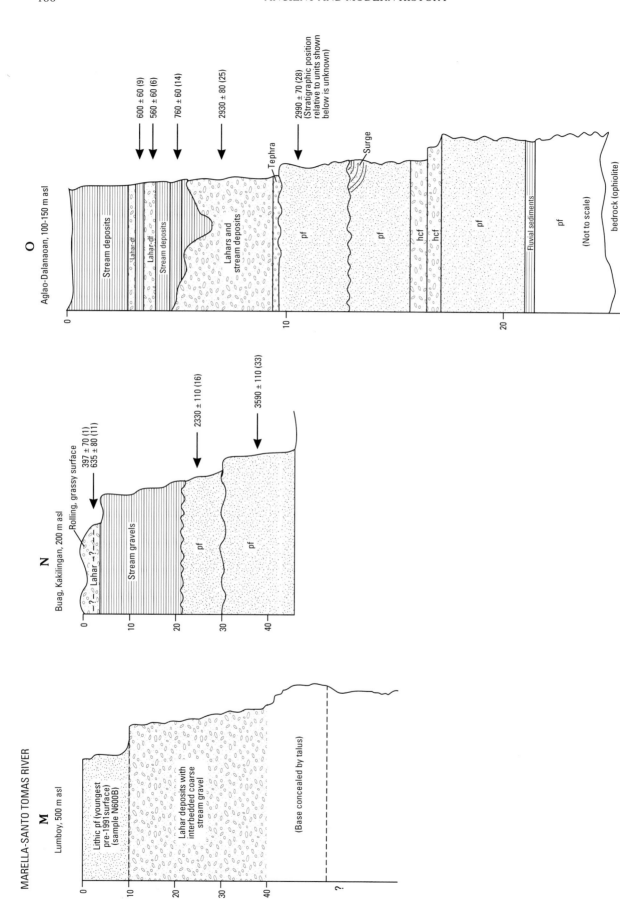

Figure 6.—Continued.

MARAUNOT-BUCAO RIVERS

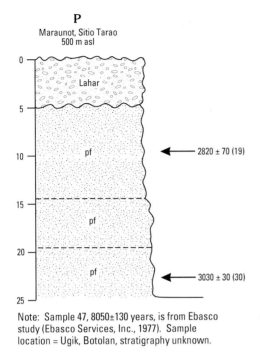

P

Maraunot, Sitio Tarao
500 m asl

← 2820 ± 70 (19)

← 3030 ± 30 (30)

Note: Sample 47, 8050±130 years, is from Ebasco
study (Ebasco Services, Inc., 1977). Sample
location = Ugik, Botolan, stratigraphy unknown.

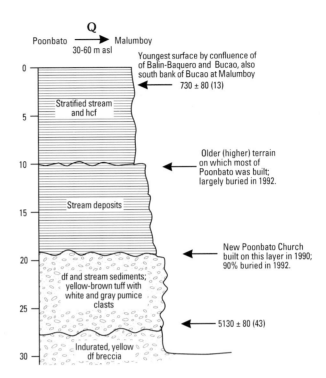

Q

Poonbato → Malumboy
30-60 m asl

Youngest surface by confluence of
of Balin-Baquero and Bucao, also
south bank of Bucao at Malumboy
← 730 ± 80 (13)

Stratified stream
and hcf

Older (higher) terrain
on which most of
Poonbato was built;
largely buried in 1992.

Stream deposits

New Poonbato Church
built on this layer in 1990;
90% buried in 1992.

df and stream sediments;
yellow-brown tuff with
white and gray pumice
clasts

← 5130 ± 80 (43)

Indurated, yellow
df breccia

Figure 6.—Continued.

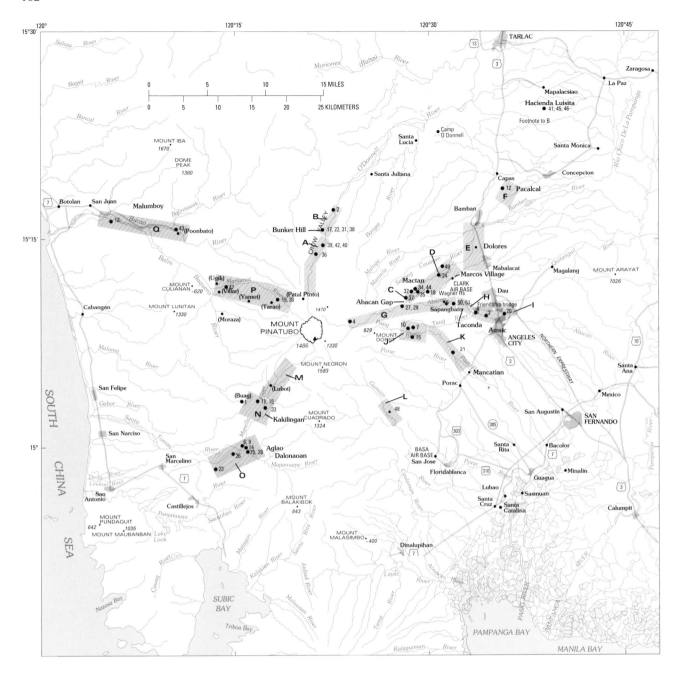

Figure 7. Radiocarbon sample sites. Lettered boxes show the locations of stratigraphic sections from figure 6. These letters are also used, in conjunction with radiocarbon sample numbers, to identify ¹⁴C samples in the text. For example, "36–A" in the text refers to sample 36 of table 1 and of composite section A (fig. 6 and this figure).

Table 2. Phenocryst minerals, whole-rock SiO$_2$ content, and notes for samples from pre-1991 eruptive periods.

[Abbreviations: pl, plagioclase; qz, quartz; hb, hornblende; cm, cummingtonite; bt, biotite; ol, olivine; ag, augite; hy, hypersthene. X, essential (>5%); x, varietal (<5%); t, trace; R, thick rims (>30 um); r, thin rims; c, cores of hornblende crystals; S, resorbed essential; s, resorbed varietal; B or b, broken crystal fragments; h, with hornblende rims. SiO$_2$ in wt% (volatile free). Samples are of the dominant juvenile material in dated deposits except for those samples that have a dash in the "Associated ^{14}C sample" column. Ages for dated deposits given in ^{14}C years; other ages are estimates based on stratigraphic relations or weathering]

Period	Age (years)	Field no.	Location	Associated ^{14}C sample (figs. 6, 7, table 1)	Notes	N. latitude	E. longitude	pl	qz	hb	cm	bt	ol	ag	hy	%SiO$_2$
Buag	~400	N600B	Marella River	—	Essential clast in lithic pyroclastic flow from summit dome	15°08'00"	120°22'00"	X	x	X		x				62.0
Buag	570±70	P17D	Clark A.B., Friendship Gate.	8	Phenocryst-rich dacite pumice from terrace deposit	15°08'42"	120°33'24"	X	x	X	r	x				65.7
Buag	950±70	P22592–2A	Pasig River	15	Dacite pumice in dated "gray lahar"	15°07'56"	120°28'59"	X	S	X	r	X				65.5
Buag	950±70	P22592–2B	Pasig River	15	Dacite pumice in dated "gray lahar"	15°07'56"	120°28'59"	X	S	X	r	x				65.2
Buag	950±70	P22592–1	Pasig River	15	Banded pumice from dated "gray lahar"	15°07'56"	120°28'59"	X	x	X	r	x	x			63.6
Buag?	950±70	P22592–3–bas.	Pasig River	—	Mafic inclusion in pumice. Not in place, possibly from "gray lahar."	15°07'56"	120°28'59"	X	S	s	r	x	s	s		52.0
Buag?		P5C	Marella River	—	Pumice from deposit beneath summit dome pyroclastic flow?	15°05'00"	120°20'00"	X	s	X	r	X				65.8
Buag?		P40–ii	Marella River	—	Mafic inclusion in dacite block. Deposit forms fluvial surface downslope from N600B.	15°08'00"	120°30'00"	no thin section								53.8
Buag?		P40–i	Marella River	—	Dense dacite, host for inclusion	15°15'00"	120°22'00"	X	S	X		x	h	t		63.6
Maraunot	2,660±70	P3792–3	Clark A.B., near TV tower.	18	Pumice lapilli in dated terrace deposit	15°11'18"	120°30'20"	X	s	X	R	c				65.8
Maraunot	2,870±100	P3592–1A	O'Donnell River	22	Phenocryst-rich dacite pumice from upper of 3 pyroclastic-flow deposits.	15°16'00"	120°22'00"	X	s	X	R	x				62.5
Maraunot	2,870±100	P3592–1B	O'Donnell River	22	Phenocryst-poor dacite pumice from upper of 3 pyroclastic-flow deposits.	15°16'00"	120°22'00"	B	b	b	r					62.5
Maraunot	2,870±100	P3592–1C	O'Donnell River	22	Darker, fine-grained pumice from upper of 3 pyroclastic-flow deposits.	15°16'00"	120°22'00"	X	s	X	R	x				63.4
Maraunot	2,990±80	P2Bi	Sacobia River	29	Phenocryst-rich dacite pumice from dated pyroclastic-flow deposit.	15°11'00"	120°30'00"	X	s	X	Rx	X				65.6
Maraunot	2,990±80	P2Bii	Sacobia River	29	Phenocryst-poor dacite pumice from dated pyroclastic-flow deposit.	15°11'00"	120°30'00"	B	b	b	t	t				64.2
Maraunot	3,260±70	P3592–2A	O'Donnell River	31	Phenocryst-rich dacite pumice from middle of 3 pyroclastic-flow deposits.	15°16'00"	120°22'00"	X	s	X	t	t				65.5
Maraunot	3,260±70	P3592–2B	O'Donnell River	31	Phenocryst-poor dacite pumice from middle of 3 pyroclastic-flow deposits.	15°16'00"	120°22'00"	B	b	B	t	t				64.9
Maraunot	3,260±70	P3592–2C	O'Donnell River	31	Darker, fine-grained pumice from middle of 3 pyroclastic-flow deposits.	15°16'00"	120°22'00"	X	s	X		t	t			62.1
Maraunot	3,260±70	P3592–2Fi	O'Donnell River	31	Mafic inclusion in phenocryst-rich dacite pumice from middle of 3 pyroclastic-flow deposits.	15°16'00"	120°22'00"	X		X		h	h			57.2
Maraunot	3,390±90	P11	Sacobia River	32	Phenocryst-rich dacite pumice from pyroclastic-flow deposit	15°11'00"	120°29'36"	X	s	X	R	t				66.1
Maraunot	3,590±90	P39A	Marella River	33	Phenocryst-rich dacite pumice from pyroclastic-flow deposit	15°05'00"	120°20'00"	X	s	X		t	h			62.8
Maraunot	3,590±90	P39B	Marella River	33	Dense phenocryst-rich dacite lithic from pyroclastic-flow deposit.	15°05'00"	120°20'00"	X	s	X	R	c				66.3
Maraunot	>1,000	P3792–2a	Clark A.B., Mabalacat River.	—	Pumice lapilli from tephra fall deposit, weathering suggests >1,000 yr B.P.	15°11'18"	120°30'20"	X	s	X	R					

Table 2. Phenocryst minerals, whole-rock SiO$_2$ content, and notes for samples from pre-1991 eruptive periods—Continued.

Period	Age (years)	Field no.	Location	Associated ^{14}C sample (figs. 6, 7, table 1)	Notes	N. latitude	E. longitude	pl	qz	hb	cm	bt	ol	ag	hy	%SiO$_2$
Crow Valley	4,800±60	P12A	Sacobia River	35	Dacite pumice from terrace deposit	15°11'00"	120°30'00"	X	S	X	R	t				64.9
Crow Valley	4,800±60	P12B	Sacobia River	35	Phenocryst-rich dacite pumice from terrace deposit	15°11'00"	120°30'00"	X	t	X	r	c				65.6
Crow Valley	4,890±70	P22392-4	Sacobia River	37	Dacite pumice from base of dated pyroclastic-flow deposit	15°10'55"	120°28'55"		s	X	R					65.6
Crow Valley	4,960±70	P31B	O'Donnell River	39	Dacite pumice from pyroclastic-flow deposit	15°14'54"	120°22'00"	no thin section								65.6
Crow Valley	4,930±80	P3592-3A	O'Donnell River	38	Phenocryst-rich dacite pumice from lower of 3 pyroclastic-flow deposits.	15°16'00"	120°22'00"	X		X	R					65.1
Crow Valley	4,930±80	P3592-3B	O'Donnell River	38	Phenocryst-poor dacite pumice from lower of 3 pyroclastic-flow deposits.	15°16'00"	120°22'00"	B	B	b	b	t				65.6
Crow Valley	5,010±80	P32B	O'Donnell River	40	Dacite pumice from thick pyroclastic flow on north side of drainage.	15°14'48"	120°21'48"	X		X	R					65.9
Crow Valley	5,100±70	P31A	O'Donnell River	42	Dacite pumice from pyroclastic-flow deposit	15°14'54"	120°22'00"	X	s	X	r	t				65.6
Crow Valley	5,130±80	P41B	Bucao River	43	Dacite pumice from lahar, ^{14}C age on interbedded silt	15°15'42"	120°10'06"	X	s	X	X	t,c				63.4
Pasbul	8,380±80	P42	Gumain River	48	Dacite pumice	15°02'03"	120°27'02"	X		X	t	c				63.9
Inararo	>35,000	P31392-1	Abacan River	51	Phenocryst-rich pumice from eroded pyroclastic-flow deposit.	15°10'25"	120°29'50"	X	S	X		X				66.7
Inararo	>35,000	P-1-16i	Abacan River	51	Dacite pumice lapilli from matrix of columnar-jointed pyroclastic-flow deposit.	15°11'00"	120°30'00"	X	Xs	X	R	X				64.4
Inararo	>35,000	P-1-16ii	Abacan River	51	Dacite pumice block from columnar-jointed pyroclastic-flow deposit.	15°11'00"	120°30'00"	X	Sx	X	R	X				65.8
Inararo	>35,000	P24D	Abacan River	—	Dacite pumice block from basal lahar or pyroclastic-flow deposit in Abacan River.	15°11'00"	120°30'00"	X	Xs	X		X				67.1
Inararo	>35,000	P44	Abacan River	—	Indurated tuff at base of sequence in Abacan River	15°11'00"	120°30'00"	X	s	X		x	t			64.4
Ancestral		P34	Pasig River	—	Dominant lithic type from indurated lahar	15°08'00"	120°30'00"	X	X			x			x	66.0

weathering characteristics, and apparent onlap relations suggest that at least two separate pyroclastic-flow units are present within deposits of the Inararo eruptive period.

The Inararo pyroclastic-flow deposits are more than 100 m thick in the proximal parts of the Sacobia–Pasig-Potrero pyroclastic fan, north to northeast of Mount Dorst, and are buried beneath alluvium at their distal end, at least 21 km from Mount Pinatubo's summit. Assuming an average thickness of 30 m over an area of 160 km^2 of inferred deposit, we estimate an original volume in the Sacobia–Pasig-Potrero fan of about 5 km^3 (bulk). The bulk volume of 1991 pyroclastic-flow deposit in the same fan is roughly 1 km^3 (W.E. Scott and others, this volume). 1991 deposits in the Sacobia–Pasig-Potrero system are about one-fifth of the total volume of 1991 pyroclastic-flow deposits so, assuming a similar azimuthal distribution then and now, the Inararo pyroclastic flows might have had a volume of 25 km^3. Nothing is known about the volume of additional tephra-fall deposit from this oldest eruption.

We speculate that the Tayawan caldera (described above) formed in response to the large Inararo eruption of dacite magma. The caldera's size (3.5×4.5 km) is that which would be expected from an eruption of 10 to 15 km^3 dense-rock equivalent (DRE) of magma (see Smith, 1979, fig. 2). Had the Tayawan caldera formed after one of the post-Inararo eruptions of Pinatubo, we would expect to find a larger, Inararo-age caldera outside the Tayawan caldera. We know of no evidence for a larger caldera.

Although Pinatubo had numerous eruptions after the Inararo eruptions, the Tayawan caldera was only partially filled—apparently because there were deep valleys in the preeruption edifice or local collapses along the caldera rim that allowed rivers and the pyroclastic flows of subsequent eruptions to flow through, and past, those segments of the caldera walls that remain high to this day. In the overall evolution of Pinatubo, the >35 ka Inararo dacitic eruption followed andesitic stratocone formation (ancestral Pinatubo) and an apparently long repose; the eruption also marked the birth of the modern Pinatubo.

SACOBIA ERUPTIVE PERIOD—~17,000 YR B.P.

An outcrop of two pumiceous, hot(?) debris-flow deposits capped by a silt layer, in the north bank of the Sacobia River opposite Clark Air Base (49–D, figs. 6, 7) is the only known remnant of what we speculate was an eruptive period about 17,000 years ago. An age of 14,480 ^{14}C yr B.P. (17,350 cal yr B.P.) was obtained from bits of charcoal in the silt.

The debris-flow deposits, like many other deposits of the modern Pinatubo, contain two distinct pumice types—a dominant, white, coarsely porphyritic hornblende pumice and a tan-gray, finer grained pumice with the same mineralogy. A check of paleomagnetic vectors of clasts in one of

these debris flows showed partial consistency from one clast to the next (8 out of 10 vectors were broadly north-pointing, with near horizontal inclination)—a pattern seen in perhaps half of the debris-flow units at this distance from Pinatubo. We interpreted this particular deposit as a debris-flow deposit on the basis of its interbedding with clearly fluvial units rather than on paleomagnetic grounds; alternatively, it and other units with similar, partial consistency of clast paleomagnetism, at this distance or less from Pinatubo, could have been emplaced by distal or secondary pyroclastic flows (see Torres and others, this volume).

Without additional outcrops, we can say little about eruption(s) of this period other than to guess that, because these debris-flow deposits and their included pumice clasts closely resemble those of the 1991 eruption, other aspects of the eruptions may also have been similar. The type section for this eruptive period was exposed before the 1992 rainy season in the scarp of a prominent terrace that extends several kilometers along the north bank of the Sacobia River, opposite and clearly visible from Marcos Village. As of this writing, more than one-half (>10 m) of the early 1992 exposure had been buried by 1992 and 1993 infilling.

PASBUL ERUPTIVE PERIOD—~9,000 YR B.P.

Pyroclastic-flow and tephra-fall layers exposed along the road between Sitio Pasbul, Camias, Porac, and the Gumain River, including one pyroclastic-flow deposit with an age of 8380±80 ^{14}C yr B.P. (sample 48–L; figs. 6, 7), represent a large eruption that overtopped the southeastern rim of the Tayawan caldera and nearly or completely filled the valley of the Gumain River. These deposits unconformably overlie pyroclastic-flow and debris-flow deposits of unknown modern Pinatubo age.

The dated pyroclastic-flow deposit near Pasbul contains two pumice types, a dominant hornblende plagioclase porphyritic dacite and a subordinate, gray, finer grained pumice. Biotite occurs only as cores within hornblende phenocrysts, and relict olivine xenocrysts occur as rare cores within hornblende-magnetite reaction clots. Relatively low SiO$_2$ (63–64%, table 2) and scarcity of quartz and biotite are distinguishing features of deposits of the Pasbul eruptive period.

1991 pyroclastic flows also overtopped the old caldera rim; at least one reached to within 2 km of Pasbul, but filled the Gumain canyon at its toe to a depth of only 2–4 m, much less than the 35–40 m of fill during the Pasbul eruptive period.

A pumiceous pyroclastic-flow deposit in the Maraunot River watershed near Sitio Ugik, Villar, Botolan (47–P, figs. 6, 7), sampled and dated by Ebasco Services, Inc. (1977), was emplaced about 8050±130 ^{14}C yr B.P. (calibrated age is about 8,980 yr B.P.). Chemical analyses of two clasts (samples 975–9w–4 and –5, Ebasco Services, Inc., 1977) indicate andesite and dacite; no details of mineralogy were given.

CROW VALLEY ERUPTIVE PERIOD— 6,000–5,000 YR B.P.

Deeply dissected pumiceous pyroclastic-flow deposits along both sides of upper Crow Valley, at and upstream from Bunker Hill (36,39,40,42–A; 38–B), form the most prominent exposures of major eruptions about 5,500 years ago (figs. 6, 7). These deposits form the basal and volumetrically largest part of Delfin's (1984) "Patal-Pinto pyroclastic-flow unit" but probably do not include younger deposits on which the village of Patal Pinto itself (destroyed in 1991) was built.

Along Crow Valley, and in small valleys just to its west, we found evidence for at least four thick pyroclastic flows separated by erosional unconformities. The two oldest units occur in dissected, topographically high deposits along the walls of the valley. Of these, the lower unit is a crudely stratified, 20-m-thick, fines-poor pyroclastic-flow deposit with abundant charcoal (sample 40, section A, figs. 6, 7). The upper unit, as much as 40 m thick, contains at least three pyroclastic-flow layers. We could not reach the upper parts of this unit but we could see moist partings that could be thin soils or beds of relatively fine-grained ash. Two additional pyroclastic-flow deposits are found topographically below but stratigraphically above (?) the aforementioned 40-m-thick deposits, in recently eroded cuts into the valley floor. Of these, the lower is notably fines poor and yielded a ^{14}C age of 5,100±70 years (42–A, figs. 6, 7, table 1); the upper is a typically massive, poorly sorted pyroclastic-flow deposit that yielded a ^{14}C age of 4,960±70 (39–A, figs. 6, 7, table 1).

Pyroclastic-flow deposits of the Crow Valley eruptive period also form prominent layers and terraces in the Sacobia River valley just outside the Mactan gate of Clark Air Base (34,35,37,44–C; figs. 6, 7). There, 20-m-thick deposits that underlie the terrace immediately below the gate on the south bank of the river correlate with a layer of similar thickness about two-thirds of the way up the northern valley wall, well above comparable 1991 deposits. Clasts in this layer show reversed polarity, presumably from self-reversal. Self-reversal occurred in both pumice types, coarsely and finely porphyritic, just as it did in 1991 pumice (R. Torres, oral commun., 1993). One and one-half kilometers upstream, the same unit is pinkish-gray, 20 to 24 m thick, with several flow units defined by concentrations of coarse pumice clasts.

Fluvial and debris-flow deposits of the Crow Valley eruptive period at Poonbato (sample 43–Q, figs. 6, 7) underlay a terrace upon which a new church for Poonbato was under construction at the time of the 1991 eruption. The church and these deposits were later buried by 1992 lahars. The deposits were yellow-brown colored, slightly indurated, well stratified, and rich in pumice. Clasts of white, quartz-hornblende-plagioclase pumice dominated, but gray, less porphyritic, less vesicular pumice fragments were also present. The one pumice lump from these fluvial units that we examined in thin section, and by chemical analysis, is a low-SiO$_2$ hornblende dacite, unlike other known pumice of the Crow Valley period and similar to that from the previous, Pasbul eruptive period.

Fluvial deposits that underlie cane fields of Tarlac Province, including those of Hacienda Luisita, Mapalacsiao, were emplaced during the early part of this eruptive period (samples 41, 45, 46, fig. 7). Voluminous pyroclastic-flow deposits in the headwaters of the O'Donnell River were the source of these fluvial sediments that, today, form rich sandy loams well suited to sugar cane production. Although our dated samples 41, 45, and 46 came from 6 m below today's land surface on Hacienda Luisita, we have no evidence suggesting this area has been reburied by fluvial sediments of any younger eruptive period.

To estimate the original volume of deposits of the Crow Valley eruptive period, we use the same method as used for the Inararo deposits. Pyroclastic-flow deposits of the Crow Valley eruptive period within the O'Donnell River watershed covered an area of about 20 km^2 to an average depth of about 30 m, for a volume of about 0.6 km^3. The 1991 pyroclastic-flow deposits in the same watershed were about 0.3 km^3, or 0.05 of the total 1991 pyroclastic-flow deposit. If we assume the same azimuthal distribution of deposits then as in 1991, the volume of Crow Valley pyroclastic-flow deposits was about 12, or, rounding, between 10 and 15 km^3. This estimate is approximate at best, but it is consistent with the observation that deposits of this eruptive period filled the lower Sacobia River valley to much greater depths than did those of 1991 and yet did not cover the more voluminous deposits of the Inararo eruptive period, between the Sacobia and Pasig-Potrero Rivers.

Almost all pumice fragments sampled from deposits of the Crow Valley eruptive period are cummingtonite-bearing hornblende dacite; several also contain resorbed quartz and trace amounts of biotite. Cummingtonite occurs as relatively thick (>30 µm) rims on hornblende phenocrysts. Two subtypes occur: phenocryst-rich white pumice and phenocryst-poor gray pumice, just as occurs in deposits of the Sacobia, Pasbul, and at least three younger eruptive periods. Both pumice types contain the same phenocryst minerals and are chemically similar, with 65–66% SiO$_2$ (table 2). The fine-grained gray type contains dominantly broken crystal fragments rather than the mostly intact phenocrysts seen in the coarser grained white samples.

Mingled pumices are a minor component in the deposits in Crow Valley proper; they are composed of white, cummingtonite-bearing dacite with large phenocrysts of plagioclase and hornblende interbanded with dark gray fine-grained hornblende andesite. Quench-textured andesite inclusions with sparse olivine xenocrysts are found in some pumice blocks.

MARAUNOT ERUPTIVE PERIOD— 3,900(?)–2,300 YR B.P.

The Maraunot eruptive period is defined by widespread, moderately well-preserved deposits of pyroclastic flows, lahars, and streamflow that have yielded ages from 3,590±110 to 2,330±110 ^{14}C yr B.P., or about 3,900–2,300 calendar yr B.P. (samples 16–33, sections B–D,G,I,K,O, and P of figs. 6, 7). We name this eruptive period for exposures of multiple pyroclastic flows along the Maraunot River, between 400 and 600 m above sea level (19,30–P). Pyroclastic flows also traveled far down the O'Donnell (17,22,31–B), Sacobia (27,29–G; 32–C), Abacan (fluvial sediments of this age occur at 20–I), Pasig-Potrero (lahar deposits of this age occur at 21–K), and Marella Rivers (28–O; related lahars at 23,25,26–O). Details on ages of this period are given in table 1.

Except for samples 16–N and 33–N, all other samples that we use to define the Maraunot period (samples 17–32) are statistically indistinguishable from the nearest other age. Thus, some of the apparent spread of ages in the Maraunot period may be an artifact of sampling or dating and **some, or even many,** eruptions of this period may have occurred over shorter, more discrete periods than the centuries-long activity suggested by figure 5.

Samples 16 and 33 make the Maraunot eruptive period seem about 600 yr longer than it would seem without those samples, and the apparent reposes between samples 16 and 17, and between samples 32 and 33, are of the same order as the repose from the Buag period to the present. Should these outlier ages be considered part of the Maraunot eruptive period? There is no clear petrologic affinity of pumice of unit 33–N to rocks of either the Maraunot or Crow Valley eruptive period; we do not have a rock sample from the 16–N unit. However, because sample 16 is much older than samples 1–15 (Buag period), and sample 33 is much younger than samples 34–46 (Crow Valley period), we leave them with the Maraunot period but alert readers to the possibility that the Maraunot period as we define it might actually include two or more shorter periods comparable to the Buag period.

About 3,000 yr B.P., a series of eruptions produced the largest known pyroclastic flows of the Marella drainage (28–O). During the same period, other pyroclastic flows filled the Sacobia valley to such a depth that lahars and fluvial sediments spilled onto and covered most of today's Clark Air Base (18,24–D).

The dominant pumice type in deposits of this eruptive period is cummingtonite-bearing hornblende dacite. Most of the analyzed samples have about 65% SiO_2, although three samples from ~3,000-yr-B.P. pyroclastic-flow deposits near Bunker Hill (samples P3592–A, P3592–1B, and P3592–2C, table 2) are distinctly less evolved, with 62.1–62.5% SiO_2. As in pumice of the preceding Crow Valley period, cummingtonite forms relatively thick (>30 μm) rims on hornblende crystals. Biotite is present as a trace phase in most samples, and resorbed quartz is a minor component of all the dacite samples. Dacite pumice fragments with mingled andesite bands are present in several pyroclastic-flow deposits of the Maraunot eruptive period; the andesite bands are finer grained and contain resorbed olivine xenocrysts rimmed by hornblende. As in the products of other eruptive periods, the dacite pumice occurs in two forms—a dominant, white, coarsely vesicular, coarsely porphyritic pumice, and a subordinate, gray, finely vesicular, finer grained pumice (fig. 8A). Also, as in deposits of the other eruptive periods, the two pumice types in any single deposit contain identical phenocryst minerals and are chemically similar, but the finer grained type contains abundant broken crystal fragments.

How voluminous were deposits of the Maraunot eruptive period? Their deposits cover and may also be more extensive than those of the Crow Valley period on the northwest side of Mount Pinatubo, yet they did not bury Crow Valley-age deposits in Crow Valley itself. Although we have not yet confirmed deposits of the Maraunot eruptive period above those of the Crow Valley eruptive period in the north wall of the Sacobia valley (34,35–C), we suspect that Maraunot-age fill of that valley was at least as deep as fill of the Crow Valley eruptive period, because water-reworked deposits of the Maraunot eruptive period overflowed onto and covered the area of Clark Air Base (18–D), and no comparable evidence of Crow Valley-age overflow has been found. Thus, the bulk volume of deposits of the Maraunot eruptive period was roughly equal to that of the Crow Valley eruptive period, an estimated 10 to 15 km^3.

BUAG ERUPTIVE PERIOD—~500 YR B.P.

The latest pre-1991 eruptive period emplaced pyroclastic-flow deposits in all major watersheds of Mount Pinatubo except those of the Gumain and Porac Rivers. During our hasty preeruption reconnaissance, we found only a few primary deposits of this period: a pumiceous pyroclastic-flow deposit high in the Sacobia drainage, a little-dissected lithic-rich pyroclastic-flow deposit in the Marella drainage, and little-dissected pyroclastic-surge or blast deposits in the O'Donnell and Bucao watersheds, near and beneath the village of Patal Pinto. The 1991 eruption covered all of these deposits and, as of our posteruption reconnaissance in late 1991 and early 1992, erosion had not yet reexposed them. From these three primary deposits, only the pumiceous unit yielded charcoal (sample 4–G), with an age of about 500 yr B.P.

The rest of our information about the Buag eruptive period comes from dated lahar, fluvial, and lake deposits

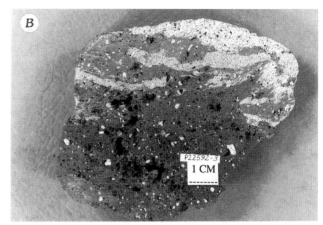

Figure 8. *A*, Phenocryst-rich (left) and phenocryst-poor pumice (right) from the Maraunot eruptive period. Eruptions of Pinatubo have repeatedly produced these same two types of pumice. *B* **and** *C*, Mingled andesite-dacite pumices from a pyroclastic flow deposit of the Buag eruptive period in the Pasig River channel. Dark bands are hybrid andesite with olivine and augite xenocrysts; light bands are cummingtonite- and biotite-bearing quartz hornblende dacite. Mingling of basaltic and dacitic magma occurred before many of the prehistoric eruptions of Pinatubo, and intrusion of basaltic magma into dacitic magma appears to be an important triggering mechanism of Pinatubo eruptions.

that occur in the O'Donnell, Sacobia, Abacan, Pasig-Potrero, Marella, and Bucao River valleys. At face value, radiocarbon ages of pumiceous lahar deposits imply that explosive eruptions of this period began at least by 600 yr B.P. and perhaps as early as 800 yr B.P. However, our only reliable, in-situ sample that is older than about 500 yr is sample 10–J, an uncharred root of a tree that was apparently growing on the bank or floor of a deep, narrow channel and was killed by a flood or lahar. If that flood was related to Buag eruptions, then those eruptions began at least by 600 yr B.P.; if the flood was of nonvolcanic origin, we cannot prove that the Buag eruptions lasted any longer than a few years or so.

During the Buag period, lahars filled channels of the abovementioned rivers to levels equal to or slightly higher than 1991–93 fill. Because lahars of the Buag eruptive period filled the Abacan valley, and yet no primary deposits of this period have been recognized beyond the limits of 1991 deposits, we judge that eruption(s) of the Buag period were approximately the same size as that of June 15, 1991.

In April and May 1990, archaeologists excavating a 13th–15th century site at Sitio Buag, Kakilingan, San Marcelino (sample 1–N, fig. 7), found pottery and other cultural relics, including charcoal from fires, mixed with ashy, pumice-bearing soil. This cultural layer was overlain by plowed soil and underlain by a second cultural layer and then several meters or more of "barren" ash and pumice, probably a lahar or pyroclastic-flow deposit (Zambales Archaeology Project, 1990; E. Dizon, National Museum, Manila, oral commun., 1992). A radiocarbon age of the charcoal (400±70 ^{14}C years, or about 500±50 years; sample 1–N, fig. 6, table 1) is consistent with the 13th to 15th century age determined independently by pottery typing. Our reading of Zambales Archaeology Project (1990) suggests that the site was built on pre-Buag deposits and abandoned but only lightly buried during Buag time, perhaps by a few tens of centimeters of ash fall. The site was covered by about 30 cm of tephra in 1991 but, as of this time of writing, can still be reexamined to learn the relation between Buag eruptions and the cultural occupation.

If we assume for now that the Buag eruptions were the same size as those of 1991, we might have expected to see a large crater or small caldera in the pre-1991 summit, similar to that which formed during the current episode. No such crater existed in April 1991; indeed, the most notable feature of pre-1991 Pinatubo topography was a domelike summit area, from which there had been one or more moderate-size rock avalanches in recent decades or centuries. Furthermore, Pinatubo's youngest appearing, least dissected pyroclastic-flow deposit (uppermost unit of section M in the Marella valley) was rich in porphyritic, prismatically jointed lithic andesite (field number N600B, table 2), suggesting an origin by collapse of a lava dome, not a plinian eruption. We explain both the absence of a crater and the lithic character of this youngest pyroclastic-flow deposit by postulating that a large, late-stage dome filled the crater left by the penultimate plinian eruption, 500 yr B.P. Avalanches from this dome, while it was growing, generated the final pre-1991 pyroclastic flow.

Regrettably, our only Buag rock samples are from dated lahars and the undated lithic pyroclastic-flow deposit. A lahar deposit of the Buag eruptive period in the Pasig River drainage contains blocks of biotite-, cummingtonite-, and quartz-bearing 65% SiO_2 dacite pumice (samples P22592–2A, 2B, table 2; unit 15–J). Cummingtonite is present as relatively thin (<30 μm) rims on hornblende crystals. Quartz and biotite are relatively abundant in several of dacite samples of the Buag eruptive period. However, in contrast to the biotite- and quartz- rich samples of the Inararo period, quartz phenocrysts in the Buag samples are resorbed. Mafic inclusions and mingled dacite pumice blocks (fig. 8 B,C) with dark olivine-augite bearing andesite bands and mafic inclusions are also present within this deposit. In contrast to the hybrid andesite of 1991, olivine crystals in samples of the Buag eruptive period lack hornblende reaction rims, their absence being suggestive of entrainment immediately prior to eruption.

The late-Buag summit dome, represented by sample N600B from the undated but morphologically young lithic pyroclastic-flow deposit in the Marella valley, is a biotite- and quartz-bearing high-silica andesite (SiO_2 62%) with highly resorbed xenocrysts of olivine and augite (both rimmed by hornblende). This disequilibrium mineral assemblage suggests that the pre-1991 summit dome represents a mixed magma, possibly analogous to the mixed andesite lava dome that, in 1992, began to refill the 1991 caldera (Daag, Dolan, and others, this volume).

If the entire steep-sided edifice of pre-1991 Mount Pinatubo grew in the final stages of the Buag eruptive period, most of the lava flows, domes, and moat(?) sediments exposed in the 1991 caldera would be products of late-Buag eruptions. We doubt, but cannot disprove, this. It seems more likely to us that a Buag caldera was a relatively small summit caldera, formed in roughly the same area as that of 1991, and that what we see in the walls of the 1991 caldera are proximal products of many eruptive periods, perhaps spanning the entire history of the modern Pinatubo.

PHOTOGEOLOGIC INTERPRETATION OF THE 1991 CALDERA WALLS

To date, we have been unable to sample and map units exposed in the walls of the new 1991 caldera. As an interim product, we offer a panoramic sketch (fig. 9), which was made from photos taken from various points inside the caldera from October 1991 to November 1992. From a distance, the 1991 caldera seems to lie in the summit of a stratocone (figs. 4A,B). From vantage points inside the caldera, many of the units appear stratified, with little apparent dip and considerable lateral continuity, some stretching almost halfway around the caldera (fig. 9). Stratified units are locally cut by domes and many small faults.

Tentatively, we recognize 3 main pre-1991 units in the caldera walls:

- a 100- to 600-m-thick sequence of poorly stratified to massive layers (unit L, fig. 9), inferred to be mostly breccia and lava flows, stretching from the north-northeast clockwise all the way around to the west. These are the myriad units of a stratovolcano, with their proximal parts now exposed in cross section. One 50- to 100-m-thick dark-gray band (subunit L_D) seems to extend over the eastern half of this arc;

- a set of well-stratified, subhorizontal layers on the northeast wall, several hundred meters thick (unit CF in fig. 9), inferred to be talus and sedimentary fill of an older, slightly larger caldera;

- at least two pre-1991 domes (unit D, fig. 9), one in the northwest wall, immediately west of the Maraunot River, and the other in the north wall.

Some of the 1991 explosion and collapse debris that coated the caldera walls in August 1991 (units T_1 and T_2 of fig. 9; see also, fig. 1A of Campita and others, this volume) is preserved as intricately rilled talus cones, and the balance has now been reworked into relatively undissected talus cones and broad, lake-filling deltas (T_3).

Numerous small faults, tens of meters long, occur but cannot be shown in figure 9. A major fault may cross the southeast wall, but its interpretation is complicated by a ridge and stepback in the wall that, in our photos, prevents distinction between a fault and an unfaulted, outward-dipping layer.

Until there is complete mapping and lithologic description of the lavas, domes, and other deposits of the caldera walls, we cannot attempt any detailed correlation of those units with distal deposits of the eruptive periods described earlier. Only a few general remarks can be made. The steep-sided edifice of the pre-1991 Mount Pinatubo, within which the 1991 caldera formed, postdates the giant Inararo eruptions and formation of the Tayawan caldera. We suspect that the apparently sedimentary layers on the northeast wall represent infilling of a subsequent, Maraunot to Sacobia eruptive period caldera, whose northeastern wall

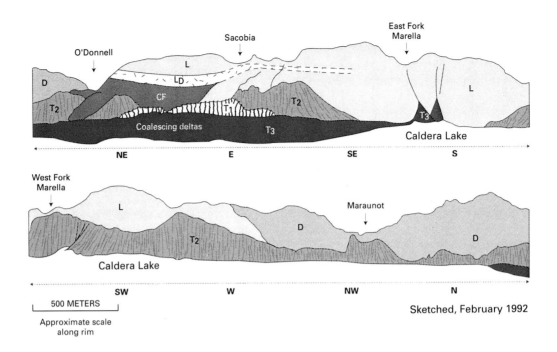

Figure 9. Panoramic sketch of the walls of Pinatubo's 1991 caldera as seen from the center of the caldera. View (left to right, top, then bottom) is 360°, clockwise starting in the northeast sector. Broad units are as follows: D, light-colored, massive dome(s); L, weakly stratified, gray lavas and volcanic breccias; L_D, a prominent, relatively dark breccia(?) within the larger L unit; CF, sub-horizontally layered pre-1991 caldera fill; T_1, 1991 explosion debris and talus, with subparallel rills being cut by rainfall and runoff on this unit only (that is, without influence by runoff from the larger, caldera walls); T_2, 1991 explosion debris or talus showing complex, dendritic drainage patterns influenced by runoff from the caldera walls; T_3, talus cones and deltas formed in late 1991, 1992, and 1993.

was slightly outboard of the present caldera wall. We also suspect, but cannot prove, that the dome that is exposed high on the northwest side of the caldera is relatively young, because it is not overlain by any other unit.

HISTORICAL RECORDS

The Aeta people have lived around Mount Pinatubo for at least 400 years (Perez, 1680), probably longer, and on the mountain itself for at least 300 years, having been forced into the highlands by Spanish attempts to control them (Reed, 1904; Fox, 1952; Panizo, 1967; Brosius, 1983; Corpuz, 1989; Shimizu, 1989). To our knowledge, they do not have an oral tradition of a previous eruption of Pinatubo. In Tagalog and in Sambal, the word "pinatubo" means "to make grow," so the name "Mount Pinatubo" might refer to a time when the mountain grew. Some of our Aeta field companions have mentioned this anecdotally, but it is difficult to know whether they would have suggested this origin for Mount Pinatubo's name *before* the dramatic events of 1991. More likely, "Pinatubo" refers to a fertile place where one can make crops grow.

Several Aeta residents have reported to us that their "grandparents" (elders) recall minor explosions in the past. Small phreatic or hydrothermal explosions could well have occurred in past decades and centuries. Thermal areas of Pinatubo were first reported by Andal and others (1965), and small explosions in areas like these are common.

The Philippine Commission on Volcanology (COMVOL) classified Mount Pinatubo as having solfataric activity in its 1981 Catalogue of Philippine Volcanoes and Solfataric Areas (COMVOL, 1981), and COMVOL's successor, the Philippine Institute of Volcanology and Seismology (PHIVOLCS), re-classified Pinatubo as "potentially active" (PHIVOLCS, 1988).

The only documented incident of unrest before 1991 occurred on August 3, 1990, about 2 weeks after a major, $M_s = 7.8$ earthquake occurred about 100 km to the northeast of Pinatubo. According to Isada and Ramos (1990), a large ground fracture and steam emission noticed by residents on August 3 resulted from a landslide on the upper northwest side of the volcano, along the Maraunot fault in an area of preexisting solfataric activity (see also, Sabit and others, this volume).

DISCUSSION

CONSISTENCY OF ERUPTIVE STYLE

Virtually all of the prehistoric deposits we have examined represent eruptive phenomena seen anew in 1991–92. Pumiceous pyroclastic-flow deposits, with minor volumes of lithic pyroclastic-flow deposits and pumiceous tephra-fall deposits, dominate deposits on the middle slopes of the

volcano. In the alluvial apron of Pinatubo, pumiceous lahar and streamflow deposits formed during every prehistoric eruptive period, as they do today. In addition, andesite to dacite lavas formed a number of prehistoric domes, especially late in eruptive periods, akin to that which began to form in 1992 (Daag, Dolan, and others, this volume). In short, Pinatubo stews, explodes, and heals itself in what, over millennia, may be a steady-state process. The volcano never grows very large, because it produces mostly unwelded, easily erodible deposits, and periodically destroys the viscous domes that fill its vents.

Will basaltic magma that intruded and mixed with dacitic magma before June 1991 be erupted? Probably not. Similar mixing events took place repeatedly throughout the history of modern Pinatubo, but did not lead to eruption of basaltic lavas that have been preserved in the geologic record. A large (~50–90 km^3) magma body is inferred from the distribution of hypocenters and tomographic modelling of 1991 earthquakes beneath Pinatubo (Mori, Eberhardt-Phillips, and Harlow, this volume). This large body could have existed throughout the modern history of the volcano. If so, this body has surely been sustained by the periodic influx of mafic magma, and has apparently blocked ascent of any significant volume of basaltic magma to the surface.

DURATION, AND NUMBER OF EXPLOSIVE ERUPTIONS, OF A TYPICAL ERUPTIVE PERIOD

For issues of public safety and planning, we need to ask how much longer the present eruption will continue and whether Mount Pinatubo can produce two or more large explosive eruptions within a single eruptive period, especially within the next several years to decades. In other words, is there anything in the geologic record that would corroborate, or argue against, the hope of residents that the worst of the present eruptive period has passed?

Taken at face value, radiocarbon ages in table 1 and figure 5 suggest that eruptive periods have lasted decades and perhaps as much as several centuries. Because many of the ages from the Crow Valley and Maraunot periods are of charcoal in pyroclastic-flow deposits and vary (from the beginning to the end of each period) by more than the 1-sigma standard deviation of their calibrated ages (fig. 5), one might infer that at least the youngest and oldest ages of each period represent separate eruptions.

However, due to the potential for interior wood or wood from an older deposit to be incorporated into a new deposit, discussed earlier, we cannot be sure that the older ages of each eruptive period are truly ages of eruptions. Thus, from the range of ^{14}C ages alone, we cannot infer that eruptive periods lasted any more than a few years or decades, and we cannot tell whether they included multiple large eruptions.

Several high, near-vertical gully walls show multiple pyroclastic-flow units in Crow Valley and in the Sacobia and Marella valleys. Thin partings between flow units are highlighted by grasses (and soil?). In the Sacobia, we obtained ages for the topmost units but not for the base; in Crow Valley we obtained ages for the basal units but couldn't reach the upper units. In both cases, a similarity of distribution and general appearance (and, in the case of Crow Valley units, similarity also in lithology; table 2) suggests that these might represent multiple eruptions of the same eruptive period, but we cannot prove this suggestion.

Only two field scenarios would be compelling evidence for multiple large explosive eruptions within a single eruptive period. The first would be an outcrop with multiple pyroclastic-flow or tephra-fall deposits and in-situ outer wood ages for the top and bottom units of that outcrop that are statistically different yet within the same eruptive period. The second would be an outcrop with multiple pyroclastic-flow or tephra-fall deposits that are separated by soils, yet having in-situ ages for the top and bottom units from the same eruptive period. We looked in vain for such evidence. Instead, we found numerous examples of multiple pyroclastic-flow deposits, some with intervening soils and some without, but none with datable wood or charcoal, much less in-situ datable material, top **and** bottom in the same outcrop or section, to tell us whether the top and bottom units were from separate eruptions of a single eruptive period.

In Crow Valley at least two and possibly three or more voluminous pyroclastic-flow deposits of the Crow Valley eruptive period, distinguished in the field, yield ages that are indistinguishable (38–B, 39,40A). Similarly, in the Maraunot River valley, ages of at least two pyroclastic-flow deposits (19,30–P) are statistically indistinguishable. The lack of confirmed paleosols between these units means that they **could** have been from single eruptions; at the same time, the uncertainty in ^{14}C ages also allows that they **could** have been from multiple eruptions within those two eruptive periods.

Thus, our reconnaissance geologic data do not **prove** that previous eruptive periods lasted any longer than the present one, nor do they tell us whether additional explosive eruptions are likely. The data do tell us of much more post-caldera dome growth than has occurred since 1991, and they hint of multiple explosive eruptions within single eruptive periods. Therefore, continued close volcano surveillance is still required as of this time of writing.

DECREASING ERUPTION SIZE THROUGH TIME?

Valley walls show evidence for cut and fill during all recent eruptive periods, and uniformitarianism argues that similar cut and fill has probably occurred throughout the history of Mount Pinatubo. However, rather than progressive aggradation with local cut and fill, and thus a concentration of younger units at the top of the volcanic pile, the deposits of the highest-standing fan surfaces and terraces tend also to be the oldest.

In the lower Sacobia, Pasig-Potrero, and Gumain Rivers, deposits of the Maraunot, Crow Valley, and Pasbul eruptive periods are exposed high on valley walls, beneath high terraces, well above those of 1991–93. Evidence of pyroclastic-flow runout suggests that pyroclastic flows of the Maraunot, Crow Valley, and Pasbul eruptive periods traveled farther than those of the Buag and 1991 eruptions. Together, depths of valley filling and distances of runout suggest that older **recognized** eruptions of Pinatubo really were larger than those of the Buag period and 1991.

Thus, we think the oldest eruption of modern Pinatubo was also its largest (Inararo, >35,000 yr B.P.; about 5 times as large as that of 1991). Those of the Pasbul period were as energetic, if not as voluminous, as the Inararo eruptions. Although smaller than the Inararo eruptions, those of the Crow Valley and Maraunot periods were still about 2 to 3 times as big as that of 1991. Eruptions of the Buag period were roughly the same size as those of 1991 and significantly smaller than those of the Inararo, Crow Valley, and Maraunot periods.

The word "recognized" is highlighted as a caution. On any volcano, erosion of some deposits and burial of others will bias the outcrops to those of younger eruptions. The older eruptions that are most likely to be recognized will be the largest ones. It is certainly possible that many eruptions of the size of that of 1991 or the penultimate, Buag eruptions also occurred during, or even between, the older eruptive periods and that their deposits are not recognized today. Uncertainties about erosion and burial prevent us from judging whether every eruption, or even the average size of eruptions, is getting smaller. What we can say with reasonable confidence is that the size of the **largest** eruption(s) of each eruptive period is diminishing.

CYCLIC MAGMATIC EVOLUTION AND POTENTIAL IMPACT OF PREVIOUS ERUPTIONS ON GLOBAL CLIMATE

The similarity in mineral assemblages, textures, and chemical compositions of dacites erupted at Pinatubo over the past >35,000 yr suggests that the 1991 magmas are similar to those of the past (Pallister and others, 1993). With few exceptions, the magma body cooled to the point that cummingtonite was stable (<800°C and P_{H_2O}<4 kbar; Rutherford and others, this volume) prior to each eruption. Shortly before and following each eruption, the body was apparently replenished and reheated by influx of basaltic magma (Pallister and others, 1993; this volume).

Given this similarity, might not the previous eruptions, many of which were likely larger than those of 1991, have also vented large amounts of SO_2 to the atmosphere? We think the answer may be "yes"; however, direct evidence of high sulfur contents is lacking. As has been noted by many workers, the June 15, 1991, eruption injected about 20 Mt of SO_2 into the stratosphere (Bluth and others, 1992), which has had a measurable impact on global climate (Hansen and others, 1992; Self and others, this volume). Yet, the evidence of a high sulfur content in 1991 eruption products is ephemeral. The 1991 magma was water saturated, and much of the sulfur was present in a separate gas phase, inferred from but no longer seen in erupted products (Westrich and Gerlach, 1992; Gerlach and others, this volume). In addition, any anhydrite that is present is quickly dissolved by ground water (Luhr and others, 1984; Bernard and others, 1991, this volume). Although the similarity in composition and mineral paragenesis suggests that many previous magmas at Pinatubo evolved under similar conditions, none of the pre-1991 samples we have examined is enriched in whole-rock sulfur content, and we have not identified anhydrite in thin sections of pumices. Additional work and new methods are needed to find a "fingerprint" of ancient sulfur-rich eruptions, possibly in the sulfur content of apatite crystals, and of glasses within apatite crystals (J. Luhr, Smithsonian Institution, written commun., 1993; Pallister and others, 1993; see also Imai and others, this volume).

CONCLUSIONS AND IMPLICATIONS FOR THE FUTURE

The eruptive history of Pinatubo is one of several guides to its future. Seven of the preceding observations are especially pertinent to judging future volcanic hazards.

- Pinatubo eruptions occur episodically and last for periods much shorter than the repose intervals between them. Some eruptive periods appear to include multiple large explosive eruptions, but we cannot resolve the time period over which they occurred.
- Pinatubo erupts large volumes of remarkably consistent, relatively cool and crystal-rich cummingtonite-bearing dacite magma, suggesting continual presence of a large and relatively shallow (<12km) magma body that is capable of eruption only when sufficient volatile- and crystal-rich magma evolves.
- Evidence of mixing of mafic magma with dacite is found in deposits of many of the previous eruptive periods and may have triggered those eruptions, just as inferred by Pallister and others (1991; this volume) for the 1991 eruption.
- The 1991 eruption of Pinatubo was among the smallest we can document in its geologic record.
- Eruptions appear to have been decreasing in size through the >35,000-yr history of modern Pinatubo, but this might be an artifact of erosion and (or) burial of older deposits. More certainly, the maximum size of eruptions in each eruptive period has been getting smaller.
- The 500-year repose between the Buag and present eruptive periods is among the shorter repose periods that we can recognize in the geologic record, but, again, this may be an artifact of erosion and (or) burial of older deposits.
- At least the penultimate Pinatubo eruption ended with growth of a large lava dome.

The inference of a large magma body beneath Mount Pinatubo and the observation that the 1991 eruption was relatively small by Pinatubo standards might be taken as evidence that additional, possibly larger, eruptions are possible. We cannot rule out that possibility. However, the observations of generally decreasing eruption size over time, and apparently shorter and shorter repose intervals, if not artifacts of preservation and exposure, suggest instead that the 1991 eruption was the biggest that could be mustered from the 1991 and present magma body, perhaps because it had been recharged with volatiles for only 500 years. Nonexplosive growth of a lava dome in 1992 corroborates that inference. The 1991 eruption might have been triggered "prematurely" by the July 16, 1990, Luzon earthquake (B.C. Bautista and others, this volume), before the usual large volume of magma could become volatile saturated. If so, no additional large explosive eruptions would be expected. An alternative interpretation of the general decrease in eruption size, namely that there has been prolonged overall cooling and a corresponding decrease in the volume of the magma body, would also lead to the same conclusion, which is that an explosive eruption on the scale of the June 15, 1991, eruption is unlikely to recur at Mount Pinatubo within our lifetimes.

Will the present eruptive period be short and, from now until its end, wholly nonexplosive? The late-Buag dome was large and, by comparison with other domes of the world, probably took several years if not decades or centuries to grow. To the extent that geologic history is a reliable guide to the future, we can expect more dome growth. Would such growth be wholly nonexplosive? The geologic record we have seen does not exclude the possibility of alternating dome growth and explosive eruptions during a single eruptive period, ending with dome growth. Indeed, alternating dome growth and explosive eruptions have been documented at Mount St. Helens (Pallister and others, 1992), where eruptions are smaller but otherwise similar to those of Pinatubo. We do not expect another explosive eruption as large as that of June 15, 1991, but we would not be surprised to see some moderate-size explosive events.

In summary, the geologic record suggests that the present eruption might not be finished, but it also suggests that the most likely style of eruption for the next several decades is continued dome growth, with or without moderate-size explosive interludes. As always, the key to shorter term forecasts will be careful seismic and related monitoring.

ACKNOWLEDGMENTS

We thank USAID/Philippines, USAID/OFDA, Mayor Richard Gordon of Olongapo City, the Philippine Air Force, and U.S. military forces in the Philippines for financial and logistical support. We also thank Ed Wolfe and William Scott for constructive reviews of this report, and Intera Information Technologies for preeruption and posteruption SLAR images (figures 4A and B).

REFERENCES CITED

Andal, A., Datuin, R., and Arciga, V., 1965, Investigation of solfataric field near Mt. Pinatubo: Quezon City, COMVOL [Commission on Volcanology] Annual Report for 1965, p. 111–112.

Aniceto-Villarosa, H.G., Lapuz, R.G., and Pagado, E.S., 1989, Geology and petrology of well PIN–1: Philippine National Oil Company, unpublished report, 23 p.

Bautista, B.C., Bautista, M.L.P., Stein, R.S., Barcelona, E.S., Punongbayan, R.S., Laguerta, E.P., Rasdas, A.R., Ambubuyog, G., and Amin, E.Q., this volume, Relation of regional and local structures to Mount Pinatubo.

Bernard, A., Demaiffe, D., Mattielli, N., and Punongbayan, R.S., 1991, Anhydrite-bearing pumices from Mount Pinatubo: Further evidence for the existence of sulfur-rich silicic magmas: Nature, v. 354, p. 139–140.

Bernard, A., Knittel, U., Weber, B., Weis, D., Albrecht, A., Hattori, K., Klein, J., and Oles, D., this volume, Petrology and geochemistry of the 1991 eruption products of Mount Pinatubo.

Bluth, G. J. S., Doiron, S. D., Schnetzler, C. C., Krueger, A. J., and Walter, L. S., 1992, Global tracking of the SO_2 clouds from the June, 1991 Mount Pinatubo eruptions: Geophysical Research Letters, v. 19, p. 151–154.

Brosius, J.P., 1983, The Zambales Negritos: Swidden agriculture and environmental change: Philippine Quarterly of Culture and Society, v. 11, p. 123–148.

Bruinsma, J.W., 1983, Results of potassium-argon age dating on twenty rock samples from the Pinatubo, southeast Tongonan, Bacon-Manito, Southern Negros, and Tongonan, Leyte areas, Philippines: Proprietary report by Robertson Research Ltd. to the Philippine National Oil Company-Energy Development Corporation, 24 p.

Campita, N.R., Daag, A.S., Newhall, C.G., Rowe, G.L., Solidum, R., this volume, Evolution of a small caldera lake at Mount Pinatubo.

Commission on Volcanology (COMVOL), 1981, Catalogue of Philippine volcanoes and solfataric areas: Quezon City, COMVOL, unpaginated.

Corby, G.W., and others, 1951, Geology and oil possibilities of the Philippines. Republic of the Philippines: Department of Agriculture and Natural Resources, Technical Bulletin 21, 363 p.

Corpuz, O.D., 1989, The roots of the Filipino Nation: Quezon City, AKLAHI Foundation, Inc., 2 v., 615 and 744 p.

Cruz, A., 1981, Pumice and other pumiceous materials in the Philippines: Philippine Bureau of Mines and Geosciences, Technical Information Series, Publication 43–81, 23 p.

Daag, A.S., Dolan, M.T., Laguerta, E.P., Meeker, G.P., Newhall, C.G., Pallister, J.S., and Solidum, R., this volume, Growth of a postclimactic lava dome at Mount Pinatubo, July–October 1992.

Datuin, R., 1982, An insight on Quaternary volcanoes and volcanic rocks of the Philippines: Journal of the Geological Society of the Philippines, v. 36, p. 1–11.

de Boer, J., Odom, L.A., Ragland, R.C., Snider, F.G., and Tilford, N.R., 1980, The Bataan orogene: Eastward subduction, tectonic rotations, and volcanism in the western Pacific (Philippines): Tectonophysics, v. 67, no. 3–4, p. 251–282.

Defant, M.J., Maury, R.C., Ripley, E.M., Feigenson, M.D., and Jacques, D., 1991, An example of island-arc petrogenesis: Geochemistry and petrology of the southern Luzon Arc, Philippines: Journal of Petrology, v. 32, p. 455–500.

Delfin, F.G., Jr., 1983, Geology of the Mt. Pinatubo geothermal project: Philippine National Oil Company, unpublished report, 35 p. plus figures.

————1984, Geology and geothermal potential of Mt. Pinatubo: Philippine National Oil Company, unpublished report, 36 p.

Delfin, F.G., Jr., Sussman, D., Ruaya, J.R., and Reyes, A.G., 1992, Hazard assessment of the Pinatubo volcanic-geothermal system: Clues prior to the June 15, 1991 eruption: Geothermal Resources Council, Transactions, v. 16, p. 519–527.

Delfin, F.G., Jr., Villarosa, H.G., Layugan, D.B., Clemente, V.C., Candelaria, M.R., Ruaya, J.R., this volume, Geothermal exploration of the pre-1991 Mount Pinatubo hydrothermal system.

Ebasco Services, Inc., 1977, Philippine Nuclear Power Plant I: Preliminary Safety Analysis Report, v. 7, section 2.5.H.1, Appendix, A report on geochronological investigations of materials relating to studies for the Philippine Nuclear Power Plant Unit 1.

Fox, R.B., 1952, The Pinatubo Negritos: their useful plants and material culture: Philippine Journal of Science, v. 81, no. 3–4, p. 173–414.

Gerlach, T.M., Westrich, H.R., and Symonds, R.B., this volume, Preeruption vapor in magma of the climactic Mount Pinatubo eruption: Source of the giant stratospheric sulfur dioxide cloud.

Hansen, J., Lacis, A., Ruedy, R., and Sato, M., 1992, Potential climate impact of Mount Pinatubo eruption: Geophysical Research Letters, v. 19, p.215–218.

Hawkins, J.W., and Evans, C.A., 1983, Geology of the Zambales Range, Luzon, Philippine Islands: Ophiolite derived from an island arc-back arc basin pair, in Hayes, D.E., ed., The tectonic and geologic evolution of Southeast Asian seas and islands, part 2: American Geophysical Union Monograph 27, p. 95–123.

Imai, A., Listanco, E.L, and Fujii, T., this volume, Highly oxidized and sulfur-rich dacitic magma of Mount Pinatubo: Implication for metallogenesis of porphyry copper mineralization in the Western Luzon arc.

Isada, M., and Ramos, A., 1990, Mt. Pinatubo not erupting: PHIVOLCS Observer, v. 6, no. 3, p. 6.

Jagolino, R., 1973, Geologic investigation of sulfur and other minerals, Mt. Pinatubo area, Zambales and Tarlac: Philippine Bureau of Mines, unpublished report.

Jordana y Morera, R., 1885, Bosquejo Geografico e Historico-Natural del Archipielago Filipino: Madrid, Imprenta de Moreno y Roxas, 461 p.

Luhr, J. F., Carmichael, I. S. E., and Varekamp, J. C., 1984, The 1982 eruptions of El Chichón volcano, Chiapas, Mexico: Mineralogy and petrology of the anhydrite-bearing pumices: Journal of Volcanology and Geothermal Research, v. 23, p. 69–108.

Mori, J., Eberhart-Phillips, D., and Harlow, D.H., this volume, Three-dimensional velocity structure at Mount Pinatubo, Philippines: Resolving magma bodies and earthquakes hypocenters.

Oebbeke, K., 1881, Beiträge zur Petrographie der Philippinen und der Palau-Inseln: Neues Jahrbuch für Mineralogie, Beilage-Band I, p. 451–501.

Pallister, J.S., Hoblitt, R.P., Crandell, D.R., and Mullineaux, D.R., 1992, Mount St. Helens a decade after the 1980 eruptions: Magmatic models, chemical cycles, and a revised hazards assessment: Bulletin of Volcanology, v. 54, no. 2, p. 126–146.

Pallister, J.S., Hoblitt, R.P., Meeker, G.P., Knight, R.J., and Siems, D.F., this volume, Magma mixing at Mount Pinatubo: Petrographic and chemical evidence from the 1991 deposits.

Pallister, J.S., Hoblitt, R.P., and Reyes, A.G., 1991, A basalt trigger for the 1991 eruptions of Pinatubo volcano?: Nature, v. 356, p. 426–428.

Pallister, J.S., Meeker, G.P., Newhall, C.G., Hoblitt, R.P., and Martinez, M, 1993, 30,000 years of the "same old stuff" at Pinatubo [abs.]: Eos, Transactions, American Geophysical Union, v. 74, no. 43, p. 667–668.

Panizo, Fr. A., O.P., 1967, The Negritos or Aetas: Unitas, v. 40, no. 1, p. 66–101.

Perez, Fr. Domingo, O.P., 1680, Relation of the Zambals, in Blair, E.H., and Robertson, J.A., 1903–1909, The Philippine Islands 1493–1898, v. 47: Cleveland, Arthur H. Clarke Co., p. 292.

Philippine Bureau of Mines, 1963, Geological map of the Philippines, Sheet ND–51, City of Manila: Manila, Philippine Bureau of Mines, scale 1:1,000,000.

Philippine Bureau of Mines and Geosciences, 1982, Geological map of the Sta. Juliana quadrangle, Tarlac province: 1:50,000, 1 sheet.

Philippine Institute of Volcanology and Seismology (PHIVOLCS), 1988, Distribution of active and inactive volcanoes in the Philippines: Quezon City, PHIVOLCS, educational poster, 1 sheet.

Reed, W.A., 1904, Negritos of Zambales: Manila, Ethnological Survey Publications, U.S. Department of Interior, v. 2, pt. 1, 90 p.

Roque, V.P., Jr., Reyes, B.P., and Gonzales, B.A., 1972, Report on the comparative stratigraphy of the east and west of the mid-Luzon Central Valley, Philippines: Mineral Engineering Magazine, Philippine Society of Mining, Metallurgical, and Geological Engineers, v. 24, p. 11–62.

Rutherford, M.J., and Devine, J.D., this volume, Preeruption pressure-temperature conditions and volatiles in the 1991 dacitic magma of Mount Pinatubo.

Sabit, J.P., Pigtain, R.C., and de la Cruz, E.G., this volume, The west-side story: Observations of the 1991 Mount Pinatubo eruptions from the west.

Schweller, W.J., Karig, D.E., and Bachman, S.B., 1983, Original setting and emplacement history of the Zambales ophiolite, Luzon, Philippines, from stratigraphic evidence, in Hayes, D.E., ed., The tectonic and geologic evolution of Southeast Asian seas and islands, part 2: American Geophysical Union Monograph 27, p. 124–138.

Scott, W.E., Hoblitt, R.P., Torres, R.C., Self, S, Martinez, M.L., and Nillos, T., Jr., this volume, Pyroclastic flows of the June 15, 1991, climactic eruption of Mount Pinatubo.

Self, S., Zhao, J-X., Holasek, R.E., Torres, R.C., and King, A.J., this volume, The atmospheric impact of the 1991 Mount Pinatubo eruption.

Shimizu, H., 1989, Pinatubo Aytas: Continuity and change: Quezon City, Ateneo de Manila Press, 185 p.

Smith, R.L., 1979, Ash-flow magmatism, in Chapin, C.E., and Elston, W.E., eds., Ash-flow tuffs: Geological Society of America Special Paper 180, p. 5–27.

Smith, W.D., 1909, Contributions to the physiography of the Philippine Islands: IV. The country between Subig and Mount Pinatubo: Philippine Journal of Science, v. 4(A), p. 19–25.

Stuiver, M., and Reimer, P.J., 1993, Extended ^{14}C database and revised CALIB radiocarbon calibration program: Radiocarbon, v. 35, p. 215–230.

Torres, R.C., Self, S., and Martinez, M.L., this volume, Secondary pyroclastic flows from the June 15, 1991, ignimbrite of Mount Pinatubo.

Villones, R., 1980, The Aksitero Formation: Its implications and relationship with respect to the Zambales ophiolite: Philippine Bureau of Mines and Geosciences, Technical Information Series, no. 16–80, 21 p.

von Drasche, R., 1878, Datos para un estudio geologico de la isla de Luzon (Filipinas): Boletin de la Comision de Mapa de España, v. 8, p. 269–342.

Westrich, H. R., and Gerlach, T. M., 1992, Magmatic gas source for the stratospheric SO_2 cloud from the June 15, 1991 eruption of Mount Pinatubo: Geology, v. 20, p. 867–870.

Wolfe, J.A., and Self, S., 1983, Structural lineaments and Neogene volcanism in southwestern Luzon, in Hayes, D.E., ed., The tectonic and geologic evolution of Southeast Asian seas and islands, pt. 2: American Geophysical Union Monograph 27, p. 157–172.

———1983, Subduction, arc volcanism and hydrothermal mineralization: The Manila Trench sector, Philippines: Philippine Journal of Volcanology, v. 1, no. 1, p. 11–40.

Zambales Archaeology Project, 1990, Archaeological report on Manggahan 2, Kakilingan, Buag, San Marcelino, Zambales: Manila, National Museum, unpublished report, 28 p.

Geothermal Exploration of the pre-1991 Mount Pinatubo Hydrothermal System

By F.G. Delfin,[1] Jr., H.G. Villarosa,[1] D.B. Layugan,[1] V.C. Clemente,[1]
M.R. Candelaria,[1] and J.R. Ruaya[1]

ABSTRACT

Geothermal exploration in Mount Pinatubo was conducted in two stages, surface investigations from 1982 to 1986 and deep exploratory drilling and well testing from 1988 to 1990. Discouraging results from the wells forced the abandonment of the prospect 13 months before the April 2, 1991 explosions.

Surface geological, hydrogeochemical, and geoelectrical studies indicated that Mount Pinatubo hosted a geothermal system with temperature of at least 200°C. That this brine system was most likely centered on the volcano's immediate northwestern flank was suggested by the better correlation of thermal features, faults, and resistivity anomaly in that sector. A possible magmatic input to this system was inferred from the high chloride contents of the high-altitude springs and solfatara pool. The hydrothermal features on the volcano's eastern slope were interpreted to be either leakages from the system on Pinatubo's northwestern flank or constituted a separate brine system.

Three exploratory wells drilled to depths ranging from 2,100 to 2,700 m encountered temperatures of 261° to 336°C. These boreholes were drilled through poorly permeable andesitic and dacitic volcanic rocks that had been altered to neutral-pH hydrothermal assemblages. Acid alteration minerals were not widespread, despite the bore fluids being acidic (pH = 2.3–4.3) Na-Cl waters. Low injectivity indices, low wellhead pressures, and the generally conductive temperature profiles of the wells confirmed the poor permeability of the hydrothermal system. This, together with the acid nature of the brine, made the wells noncommercial.

On the basis largely of the chemical and isotopic compositions of the deep reservoir fluids, significant magmatic input to the hydrothermal system is indicated. The deeper central part of the system was modeled to be a hot (>300°C) and generally impermeable two-phase zone produced by the condensation of magmatic volatiles. Two apparently separate brine systems with temperature of ~260°C convected above and on the flanks of the deep acid zone.

If our model of the pre-1991 Mount Pinatubo hydrothermal system applies to other systems associated with active or recently active stratovolcanoes, high-temperature acidic fluids are to be expected in the central parts of such reservoirs. Exploration efforts should be expended on the flanks of these systems, where neutral-pH brines are expected to form. Such efforts, however, must always include an assessment of the suitability for long-term development and exploitation of these "magmatic-hydrothermal systems."

INTRODUCTION

As part of an effort to locate and develop new electric power sources in Luzon island, the Philippine National Oil Company-Energy Development Corporation (PNOC-EDC) implemented in 1982 a geothermal exploration program at Mount Pinatubo, approximately 80 km northwest of Manila (fig. 1). The presence of numerous hydrothermal features in the volcanic complex and its proximity to industrial centers made it then a worthy geothermal prospect.

Prior to PNOC-EDC's studies, little was known about Mount Pinatubo. Reconnaissance geologic mapping, hydrogeochemical surveys, and geoelectrical investigations were carried out from 1982 to 1986. Different interpretations and assessments of the surface exploration data were made by the staff of PNOC-EDC, Kingston Reynolds Thom and Allardice (KRTA) consultants, and the New Zealand Department of Scientific and Industrial Research. Subsequently, three deep exploratory wells were drilled in 1988–89. Discouraging results from these wells led to the abandonment of the prospect by early 1990.

The principal objective of this paper is to present a summary of the results of the various stages of the exploration program. On the basis of these and other studies during and after the 1991 eruption, we propose a model of the pre-1991 Mount Pinatubo hydrothermal system. This model has important practical implications for geothermal exploration and development of recently active stratovolcanoes.

---

[1]Geothermal Division, PNOC-Energy Development Corporation Ft. Bonifacio, Metro Manila, Philippines 1201.

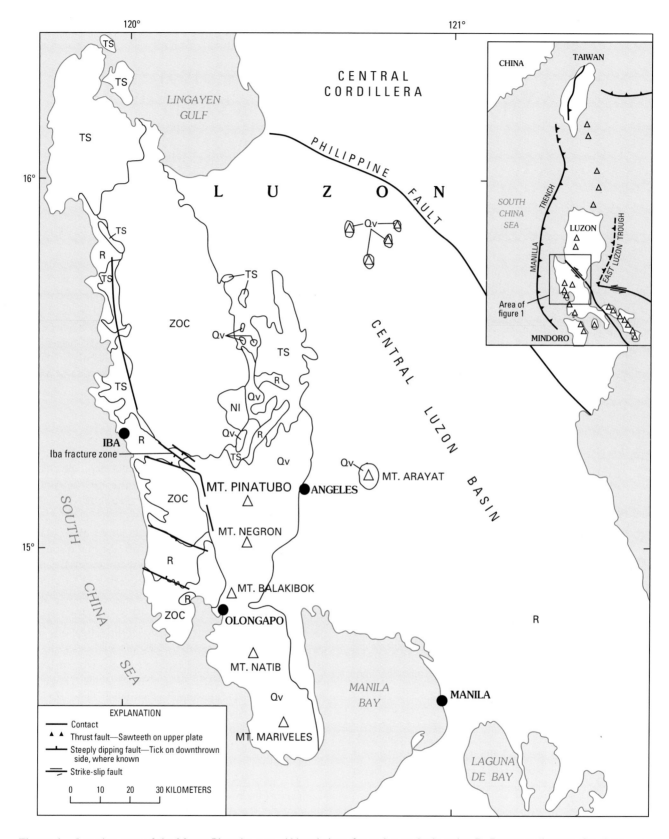

Figure 1. Location map of the Mount Pinatubo area. Abbreviations for major geologic units: R, Recent sediments; Qv, Quaternary volcanics; TS, Tertiary sediments; NI, Neogene intrusive; ZOC, Zambales Ophiolite Complex. Inset shows major tectonic features of Luzon. Triangles, volcanoes.

SURFACE EXPLORATION

GEOLOGY

Mount Pinatubo lies along the Luzon arc, a north-south-trending belt of late Tertiary to Quaternary volcanoes that extends northward from southern Luzon to Taiwan (Defant and others, 1989). The Luzon arc is associated with eastward subduction of the South China Sea oceanic crust along the Manila trench (fig. 1). In the Mount Pinatubo to Mount Mariveles segment of the arc, the volcanoes overlie the contact of the Zambales Ophiolite Complex (ZOC) and Tertiary sediments of the central Luzon basin.

The ZOC in the prospect area comprises variably serpentinized peridotites and pyroxenites, gabbros and related felsic intrusives, diabase dikes, and basaltic lavas. These rocks are exposed north, west, and southwest of Mount Pinatubo. North of the volcano (fig. 2), the ZOC unconformably underlies upper Miocene to lower Pliocene sandstones, siltstones, and conglomerate lenses (Delfin, 1984) and is intruded by a large body of Neogene diorite and quartz diorite (NI, fig. 1) (Philippine Bureau of Mines, 1963).

Mount Pinatubo is composed largely of hornblende andesite and dacite plus lesser amounts of pyroxene andesite and basalt emplaced as lava flows, pyroclastics, dikes, sills, and domes. The eruptive history of the volcano has been divided into two major periods (Delfin, 1984) termed here as ancestral and modern Mount Pinatubo (Newhall and others, this volume). Ancestral Mount Pinatubo was a Pleistocene andesite-dacite stratovolcano whose remnants are preserved in topographically elevated terrain surrounding the pre-1991 Pinatubo dome (fig. 2). Contemporaneous with this ancestral Mount Pinatubo stratocone are the domes of Mounts Negron, Cuadrado, and Mataba and the Bituin plug (Delfin, 1984). Modern Mount Pinatubo is a dacite-andesite dome complex surrounded by sheets of dacitic pyroclastic flows and related lahars. New ^{14}C dating and stratigraphic analyses (Newhall and others, this volume) reveal at least six distinct eruptive episodes of modern Pinatubo beginning with a caldera-forming event (Inararo episode) >35,000 ^{14}C yr B.P. The pre-1991 summit of Mount Pinatubo was a steep-sided, 2-km-wide hornblende andesite dome (Delfin, 1984) believed to have been emplaced during the Buag eruptive event about 800–500 cal yr B.P. (Newhall and others, this volume).

The Pinatubo volcanics are cut by predominantly northwest- and north-trending dip-slip faults. de Boer and others (1980) proposed the existence of a major northwest-striking structure, the Iba fracture zone, passing beneath Mount Pinatubo. Manifestations of this tectonic feature include the Marunot (or Maraunot) fault, the Dagsa fault, and the northwest-southeast alignment of surface hydrothermal discharges on Mount Pinatubo (Delfin, 1984). Shorter and less dominant northeast- and east-trending faults cut the major structures without any apparent offset. These shorter and easterly trending faults are particularly well defined near the summit region of the volcano (fig. 2).

DISTRIBUTION AND CHEMISTRY
OF THERMAL SPRINGS

Prior to the 1991 eruptions, twelve groups of hydrothermal discharges were found in Mount Pinatubo. Most of the springs were concentrated along a narrow, ~25-km-long belt extending from Nacolcol in the northwest to Cuyucut in the southeast (fig. 2).

Representative chemical analyses of the spring waters are listed in table 1. The thermal springs have been classified on the basis of their relative abundance of Cl, HCO_3, and SO_4 (fig. 3). The chemistry of the springs varies systematically with altitude and with distance from the volcano. Farthest from Mount Pinatubo is the Nacolcol chloride spring, 15 km northwest at an altitude of 220 m (fig. 2). The Cl-HCO_3 springs of Dagsa, Cuyucut, and Asin lie between ~300 and 400 m in altitude at distances of 6 to 10 km from Mount Pinatubo. The Kalawangan Cl-HCO_3 spring, however, is situated 4 km east of Mount Pinatubo at a higher altitude of 675 m. At still higher altitudes (>500–1,100 m) and closer to the vent are the HCO_3-SO_4 springs of Upper Maraunot, Dangey, Mamot, Pula, and Pajo. The highest feature is the Pinatubo solfatara and the Cl-SO_4 solfatara pool located 1.5 km northwest of Mount Pinatubo at an altitude of 1,180 m. The Lower Maraunot springs, immediately south of the solfatara, are likewise Cl-SO_4 in composition. The Pinatubo solfatara and the Upper Maraunot springs reached 95°C; the rest of the springs have discharge temperatures of <60°C. Except for the acid solfatara pool, all the Pinatubo thermal springs discharge neutral-pH waters. Cation and silica mixing-model geothermometers for the high-chlorine springs yielded subsurface temperatures of at least 200°C (Clemente, 1984; Villaseñor, 1984).

The composition of gases collected from the Pinatubo solfatara is shown in table 2. In terms of mole percent, the solfatara gases are composed of 86–90 percent CO_2, 8–12 percent H_2S, 0.92–1.3 percent N_2, and 0.005 percent H_2. A source temperature of 260°C for the Pinatubo solfatara was obtained by using the D'Amore and Panichi geothermometer (Villaseñor, 1984). The possibility of magmatic input to the solfatara gases (Bogie, 1984), however, makes this temperature estimate dubious.

RESISTIVITY SURVEYS

Schlumberger resistivity traversing (SRT) and vertical electrical soundings (VES) were employed at Mount Pinatubo to delineate the most likely location and depth of the geothermal resource (Esperidion, 1984; Apuada, 1988). SRT measurements were conducted at 522 stations by use of half-current electrode separation (AB/2) of 250 and

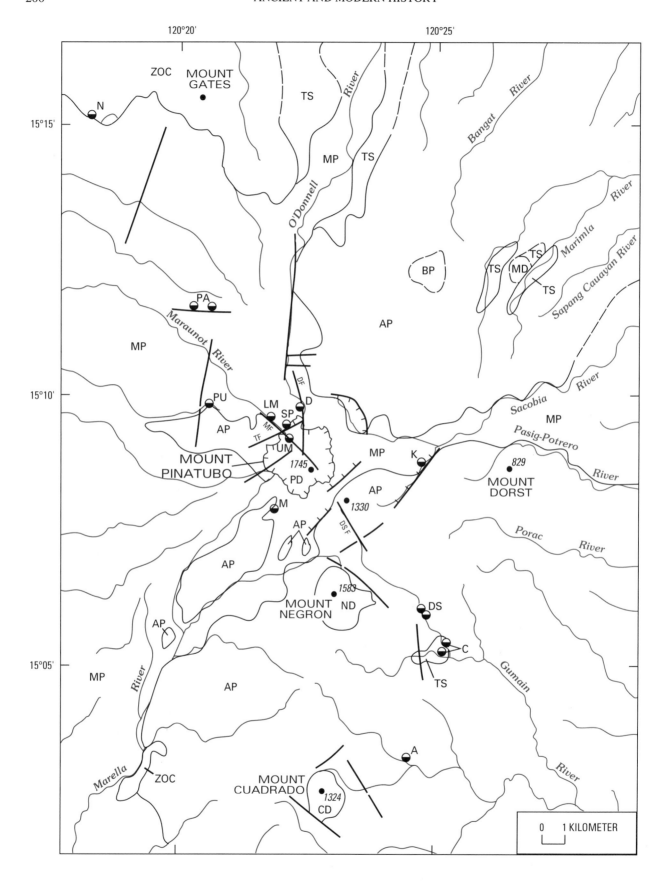

Table 1. Selected chemical analyses of Mount Pinatubo thermal waters.

[Abbreviations for sources appear on fig. 2. Concentrations are given in milligrams per kilogram.]

Source	Key	Elevation (m)	Date (year, month, day)	Temperature (°C)	pH	Li	Na	K	Ca	Mg	Cl	SO$_4$	HCO$_3$	B	SiO$_2$	δ^{18}O	δD
Solfatara pool	SP	1,180	830515	58	1.01	10.5	10,134	2,276	69.4	1,056	34,053	40,600	0	423	605		
Upper Maraunot	UM	1,075	821012	76	7.75	.12	57.7	13.9	186	55.1	37	613	259	5.4	200		
Upper Maraunot	UM	1,075	821021	69	7.88	.17	45.8	8.6	154	47.7	25	500	216	2.8	186		
Dangey	D	965	900326	39	7.99	.27	68.2	6.1	133	64.3	44	198	562	3.5	111	−9.38	−58.7
Dangey	D	965	830514	37	7.94	.43	149	12.9	146	96.3	80	421	595	2.0	127		
Mamot	M	900	900324	43	7.99	.44	158	11.4	85.6	68.5	127	373	382	6.4	120	−9.38	−62.1
Mamot	M	900	821013	48	7.77	.22	74.3	6.8	163	45.2	72	282	445	2.0	132		
Lower Maraunot	LM	900	900403	50	7.94	.38	157	14.8	62	62.7	265	260	165	16.2	157	−8.20	−51.3
Lower Maraunot	LM	900	821012	47	7.60	.46	369	27.6	82.4	155	787	391	64	23.3	158		
Kalawangan	K	675	900323	40	7.64	1.09	570	49.2	114	47.2	950	247	456	11.0	112	−8.36	−53.5
Pula	PU	600	821014	22	7.18	<.04	7.9	3.6	16.5	11.8	11	26	104	.7	103		
Pajo	PA	518	821028	27	7.66	<.04	21.4	4.9	44.7	27.7	5.0	37	288	.2	146		
Cuyucut	C	380	900329	47	8.30	3.19	1,720	184	26	60.8	2,387	413	481	33.5	142	−6.85	−50.2
Dagsa	DS	300	900327	44	7.83	.09	19.2	3.0	241	33.1	10	681	66	3.1	58	−8.78	−56.9
Dagsa	DS	300	830314	50	7.90	3.60	2,240	154	182	33.5	3,159	468	832	37.5	145		
Asin	A	295	900328	54	8.03	3.96	2,180	229	287	48.3	3,167	483	1,549	8.7	127	−5.79	−47.8
Asin	A	295	830513	47	7.96	3.03	2,152	159	163	47.7	3,138	285	1,490	17.1	125		
Asin	A	295	830513	45	8.21	2.81	2,061	152	181	46.1	2,960	275	1,365	15.6	105		
Nacolcol	N	220	900404	49	7.20	<.04	232	3.3	138	1.6	576	31	28	3.2	60	−8.30	−55.4

← **Figure 2.** Simplified geologic map of the Mount Pinatubo area. ZOC, Zambales Ophiolite Complex; TS, Tertiary sediments; AP, ancestral Mount Pinatubo; MP, modern Mount Pinatubo; PD, Pinatubo dome; ND, Negron dome; CD, Cuadrado dome; MD, Mataba dome; BP, Bituin plug. Heavy lines are faults. MF, Maraunot fault, TF, Tayawan fault, DS F, Dagsa fault, DF, Dangey fault. Outline of 1991 caldera from Newhall and others (this volume). Abbreviations for hot springs (◔) are keyed to table 1. Altitude in meters.

500 m. The isoresistivity of the area at AB/2=500 m is shown in figure 4. Three low-resistivity anomalies were delineated, the Maraunot, Mount McDonald, and Dagsa anomalies. The Maraunot anomaly is a northwest-trending resistivity low defined by the 50 Ω•m contour and enclosing the Maraunot, Pula, and Pajo thermal springs. The Dagsa anomaly is a broad low-resistivity feature traversing the entire eastern flank of Mount Pinatubo. Defined by the 50 Ω•m value, it includes pockets of 10–20 Ω•m lows that do not show any well-defined pattern. Between these two large anomalies is the Mount McDonald anomaly, a small northeast-trending 50 Ω•m low that is located southeast of the Pinatubo dome. It parallels the southeast caldera rim but is not associated with any surface hydrothermal feature.

Thirty-six VES measurements were undertaken to test the vertical persistence of shallow anomalies delineated by SRT. Maximum half-current electrode spread for the VES was 1,200 m. The contours of the interpreted resistivities of the VES bottom layers, derived from one-dimensional modeling, are broadly similar to what is shown in figure 4.

All three SRT anomalies persist at greater sounding depths. Both the Dagsa and Mount McDonald anomalies are characterized by almost uniform bottom resistivities of ≤50 Ω•m. On the other hand, the Maraunot anomaly shows further decrease in resistivity with depth attaining values of ≤30 Ω•m. The immediate vicinity of the Pinatubo dome, however, remained characterized by high resistivities of >100 to >1,000 Ω•m. These values suggest poor subsurface fluid circulation or minimal rock alteration around the Pinatubo dome, at least to depths penetrated by the VES.

PREDRILLING MODEL

At the conclusion of the surface exploratory studies in 1986, two conceptual models of the Mount Pinatubo hydrothermal systems were proposed. In the first model, Delfin (1984) and Villaseñor (1984) invoked two hot-water circulation systems. One of these systems, associated with Mount Pinatubo, was manifested by the Maraunot anomaly and by springs located on the western flank of the volcano. The Pinatubo solfatara and the numerous SO$_4$-HCO$_3$ springs at higher altitude were interpreted to indicate extensive boiling and steam condensation of upwelling geothermal fluids. The Maraunot fault and its subsidiary fractures were believed to be the main conduit of the hot water to the surface. A separate convective system, associated with Mount Negron, was invoked for the springs, altered rocks, and resistivity lows located on the eastern flank of the volcano. Of these two systems, the one associated with Pinatubo was deemed more promising but appeared small

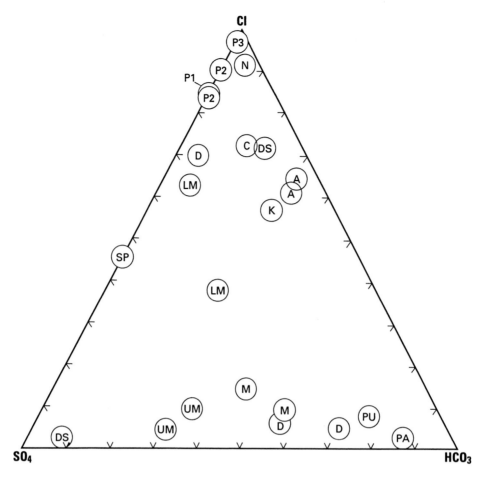

Figure 3. Cl-HCO₃-SO₄ ternary diagram of spring compositions, Mount Pinatubo. P1, P2, and P3 refer to water samples from PIN–1, PIN–2D, and PIN–3D, respectively. For spring names, see table 1.

Table 2. Representative gas compositions from the Pinatubo solfatara and wells.

[WHP = wellhead pressure; H = enthalpy; SP = sampling pressure; all concentrations in mmol/100 mol steam.]

Source	Date (year, month, day)	WHP (MPa)	H (kJ/kg)	SP (MPa)	CO₂	H₂S	NH₃	N₂	H₂	CH₄	Ar
Solfatara	840923	–	–	–	1,129	1,156	–	17	0.065	–	–
Solfatara	840929	–	–	–	2,030	227	–	21.1	.127	–	–
Solfatara	840929	–	–	–	493	44	–	5	.026	–	–
Solfatara	900103	–	–	–	879	221	–	16.9	.065	–	0.262
Solfatara	900103	–	–	–	1,230	213	–	19.3	.107	–	–
Solfatara	900103	–	–	–	2,677	236	–	30.2	.170	–	–
PIN–1	890407	0.173	1,659	0.158	2,653	93.7	0.01	12.14	2.78	0.13	.04
PIN–1	890506	.159	1,670	.151	3,116	75.4	–	3.17	.56	.08	.03
PIN–2D	890212	.723	1,907	.620	2,758	165	.22	14.37	6.12	.29	.01
PIN–3D	900103	.213	2,411	.137	10,051	124	.04	95.4	15.6	–	.05
PIN–3D	900116	.348	2,406	.264	12,224	174	.01	82.8	18.2	–	.03
PIN–3D	891227	.120	2,439	.113	43,597	651	–	446	64.8	1.87	.51

compared to other Philippine geothermal fields then explored.

The second model (Bogie, 1984) postulated a single large hydrothermal system in the complex. In this model, fluids from this huge system well up beneath Mount Pinatubo and flow out laterally in numerous directions to emerge as springs on both flanks of Mount Pinatubo. On the basis of the high chloride concentrations of the Maraunot

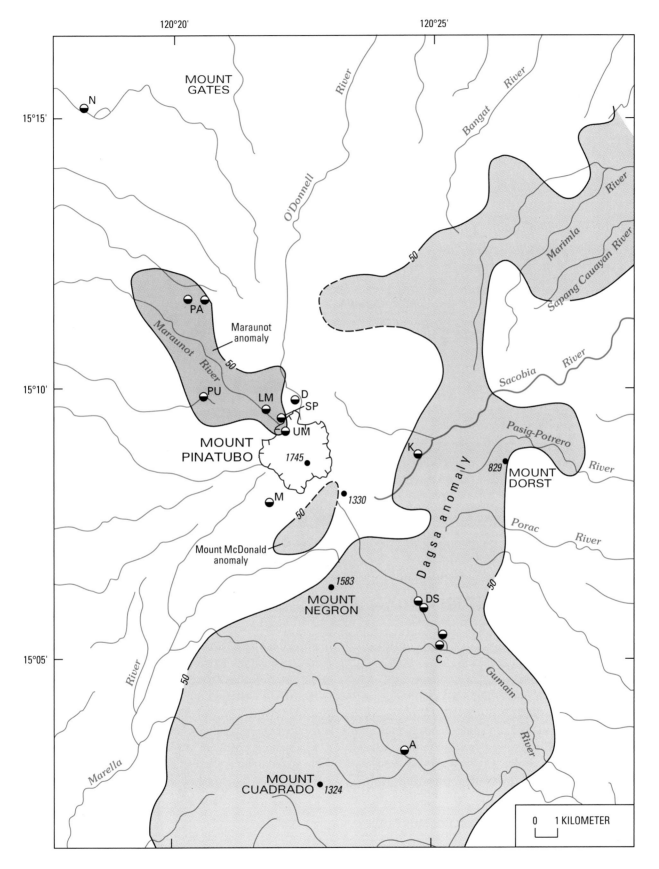

Figure 4. Isoresistivity map of the Pinatubo area obtained at half-current electrode separation (AB/2) of 500 m. The anomalies are delineated by 50-Ω•m contours. Abbreviations for hot springs (◕) are keyed to table 1. Altitude in meters.

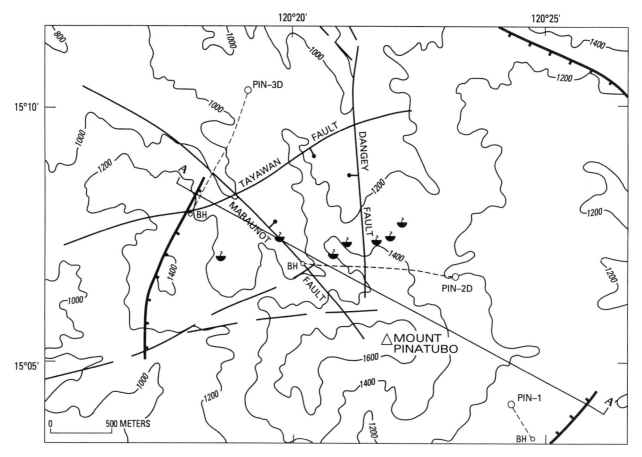

Figure 5. Well location map of Mount Pinatubo exploratory wells showing pre-1991 caldera (hachures), major faults (heavy lines, ball on downthrown side), and fumaroles formed after April, 1991 explosions. Dashed lines indicate well tracks; BH refers to well bottom. Fault names are the same as in figure 2. A–A' is section line shown in figures 6 and 7. Contour interval is 200 m. Well PIN–1 had a total vertical depth of 2,733 m; well PIN–2D, 2,216 m; well PIN–3D, 2,190 m.

springs and the solfatara pool, Bogie (1984) proposed a magmatic input to the hydrothermal system.

EXPLORATION DRILLING

In order to test the two exploration models and confirm the existence of a geothermal resource viable for electric power generation, three deep wells (fig. 5) were drilled between August 1988 and November 1989. Well PIN–1, southeast of Mount Pinatubo, reached a total vertical depth (TVD) of 2,733 m. PIN–2D (TVD of 2216 m) was located northeast of the dome and drilled directionally to the west. PIN–3D (TVD of 2190 m) was spudded from a pad northwest of the dome and deviated to the southwest. A summary of selected well data and bore outputs is shown in table 3.

DISCHARGE TESTS AND BORE OUTPUTS

PIN–1 was discharged by boiler stimulation after two attempts to discharge it by air compression failed (PNOC-EDC, 1990). This well had the longest flow test

(138 days), and sampled brines were largely free of drilling fluids. The maximum measured temperature of the well was 261°C at the bottom. Discharge enthalpy ranged from 1,050 to 1,700 kJ/kg. Low wellhead pressure (<0.40 MPag) did not allow the well to sustain discharge beyond 138 days.

After 15 days of heat-up, PIN–2D discharged by natural buildup of its wellhead pressure. This initial discharge had to be cut off after 3 days because of the low pH (2.3) of the brine. Stable output thus could not be obtained, but the initial enthalpy of 1,442 kJ/kg later increased to 2,145 kJ/kg. A second 3-day flow test yielded the same trend in fluid pH and discharge enthalpy. PIN–2D reached a maximum temperature of 336°C at the bottom, the highest measured in all the wells. Its mass flow of 24–69 kg/s was equivalent to a power output of 7.7 MWe.

PIN–3D discharged for 40 days after an 8-day heat-up period. The well reached 330°C at the bottom; its temperature profile was dominantly conductive from 500 to 1,400 m in depth, followed by a section of low gradient to the well bottom. Low wellhead pressure and low pH of the discharge brine made this well commercially nonproductive.

Table 3.   Selected Pinatubo well data and bore outputs.

[MD/VD = measured and vertical depths, in meters; BHT = maximum bottomhole temperature; WHP = wellhead pressure; NCG = non-condensible gases]

Well	Date completed (month, year)	Elevation (m)	Depth (MD/VD)	BHT (°C)	Injectivity (L/s-MPa)	Mass flow (kg/s)	WHP (MPa at gage)	Enthalpy (kJ/kg)	Remarks
PIN–1	10–88	1,150	2,771/2,733	261	12.7–24.4	13–23, cycling.	<0.4	1,050–1,700	Discharged 138 days; NCG = 8–16 wt%; pH = 4.2.
PIN–2D	01–89	1,230	2,622/2,216	336	9.7	24–69	.6	1,442–2,145	6-day test only; pH = 2.3; NCG = 6–12 wt%; 7MWe steam.
PIN–3D	11–89	1,098	2,553/2,190	330	7.3–16.3	2.5–10	<.2	1,950	40-day test; output unstable; pH = 3.4–4.3; NCG = 20 wt%.

All the wells had poor permeability, as indicated by their generally conductive temperature profile, low injectivity indices, and low wellhead pressures (table 3). The resulting low mass outputs (except for PIN–2D) combined with the acid nature of the brine made the wells noncommercial.

SUBSURFACE GEOLOGY

The subsurface stratigraphy of the volcano, based on deep drilling, is shown in figure 6. A pile of andesitic to dacitic volcanic rocks, 1,300 to more than 2,000 m thick, was intersected by the wells. The upper few hundred meters of this unit are largely dacite and andesite breccias and tuffs believed to be erupted from the modern Pinatubo. This sequence is underlain by dacitic lava flows and by a thicker sequence of waterlaid(?) pyroclastics presumed to be deposits of the ancestral Pinatubo. Dioritic dikes (D in fig. 6) of uncertain affinity crosscut the volcanics generally below 200 m in altitude. Only PIN–1 drilled through these volcanics to intersect the ZOC at mean sea level. In this borehole, the ZOC is a chaotic mixture of hornfels and moderately to completely altered microdiorite, micro-quartz monzodiorite, diabase, and gabbroic dikes.

Subsurface alteration is dominated by neutral-pH hydrothermal minerals. The distribution of hydrothermal clays at Mount Pinatubo (fig. 7) does not follow the sequential zonation from smectite, illite-smectite, illite, to biotite typically found in neutral-pH Philippine geothermal systems (Reyes, 1990). First, overlapping of some clay zones suggests that several hydrothermal regimes have existed within the volcanic field. Second, the smectite and illite-smectite zones persist to depths where temperatures are over 300°C. This occurrence is in marked contrast with other Philippine hydrothermal fields, where both smectite and illite-smectite zones are stable only up to temperatures of 180° and 230°C, respectively (Reyes, 1990). Such persistence of these two low-temperature alteration zones to high temperature at Pinatubo may be attributed to low permeability of the host rocks. Third, a well-developed biotite

subzone is present in the last 380 m of PIN–2D but is not found in PIN–1 and PIN–3D.

Acid alteration assemblages are restricted to major permeable zones associated with fault intersections (fig. 7). Acid minerals in the drillholes include alunite, diaspore, pyrophyllite, dickite, anhydrite, and pyrite. These minerals form discrete acid horizons 2 to 65 m thick. They occur at altitudes as shallow as 165 m to as deep as −514 m, where temperatures range from 164° to 302°C. The limited distribution of acid minerals, despite the widespread occurrence of acid fluids in the boreholes, implies restricted permeability within the hydrothermal system. The limited structural flowpaths of acid fluids in the reservoir is further supported by the persistence of calcite to more than 1,040 m below sea level (PNOC-EDC, 1990).

Fluid-inclusion heating and freezing measurements were conducted in vein minerals, principally anhydrite, collected from most of the permeable horizons. Homogenization temperatures (*Th*) are plotted on well temperature-depth curves in figure 8. There is a good correlation between modal *Th* values and measured stable temperatures in PIN–1 and, to some extent, in PIN–2D. At a depth of 2,000 m in PIN–3D, there is a bimodal distribution of *Th* values at 290° and 320°C. The latter value is consistent with the present-day measured temperature, implying that in the past, fluids cooler by at least 30°C have circulated in this part of the reservoir.

Measured salinities of fluid inclusions from freezing measurements range from 2,000 to 109,000 ppm Cl equivalent. About 90 percent of the inclusions yielded values of 10,000 to 40,000 ppm, although values of 40,000 to 109,000 ppm were obtained at discrete intervals in PIN–2D and PIN–3D. These latter values are generally higher than the actual reservoir chloride found in the wells. High gas (CO_2) concentrations in the parent fluid of the inclusions, which lower the freezing point depression (Hedenquist and Henley, 1985), may be the cause of their higher salinities compared to actual reservoir chloride. Another possible cause is direct contribution from magmatic HCl gas. In both cases, high gas content of the inclusions's parent fluid is implied.

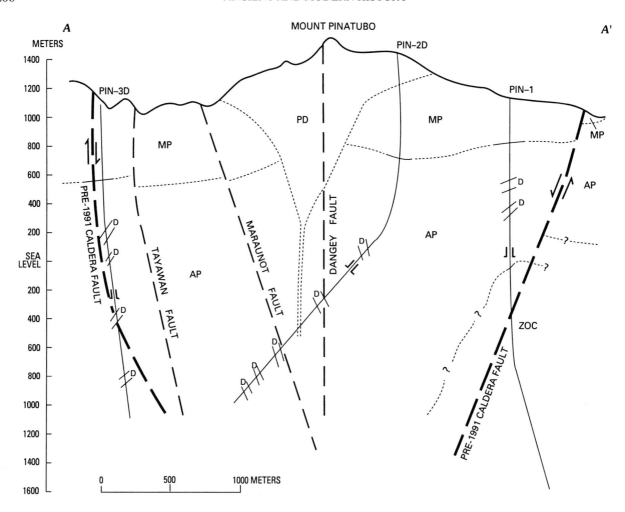

Figure 6.   Generalized subsurface stratigraphy of Mount Pinatubo volcano. Symbols are the same as in figure 2. D=diorite dike.

In general, the salinities increase with depth except in PIN–3D, where salinities have a bimodal peak at 2,000 and 23,000–26,000 ppm at –912 m. These salinities are consistent with the bimodal *Th* values measured at the same depth in the well, confirming that at least two distinct hydrothermal events have occurred in this part of the reservoir.

CHEMISTRY OF RESERVOIR FLUIDS

Selected water chemistry of the wells is shown in table 4. Discharge and downhole fluids are moderately to highly acidic Na-Cl waters. Calculated reservoir chloride concentrations range from 6,000 to 10,000 ppm in PIN–1, from 2,000 to 3,000 ppm in PIN–2D and from 5,000 to 20,000 ppm in PIN–3D. The lower values for PIN–2D, despite its high temperature, may be due to the contamination of the brine by low-chloride drilling fluids brought by the limited discharge period. Sulfate concentrations are high particularly in PIN-2D, where they reach 2,000 to 3,000 ppm. Mg and Fe are present in anomalously high concentrations.

Silica geothermometry for the well waters yields temperatures that are close to, but generally lower than, measured. Calculated temperatures are 224–281°C for PIN–1, 314°C for PIN–2D, and 310°C for PIN–3D.

On a Na-K-Mg ternary diagram (fig. 9), PIN–3D and PIN–1 waters fall on the partially equilibrated field, whereas those of PIN–2D, and nearly all the thermal springs, can be classified as immature waters. Two distinct trends are apparent in figure 9. PIN–1 falls on a tie line with the Dagsa, Asin, and Cuyucut springs and yields a Na-K temperature of 240°C, 20°C lower than measured. A second trend is defined by the waters of PIN–3D, PIN–2D, and the Pinatubo solfatara pool. Thus, the fluid supplying the Pinatubo solfatara is probably the same fluid tapped by PIN–2D and PIN–3D. The Na-K temperature for PIN–3D and PIN–2D ranges from 300 to 320°C, close to actual measured temperatures in both wells.

The geochemical trends above suggest two possibilities: one, that two separate convective systems exist in Mount Pinatubo; or two, that one large hydrothermal system is present but fluids are differentially cooled and diluted

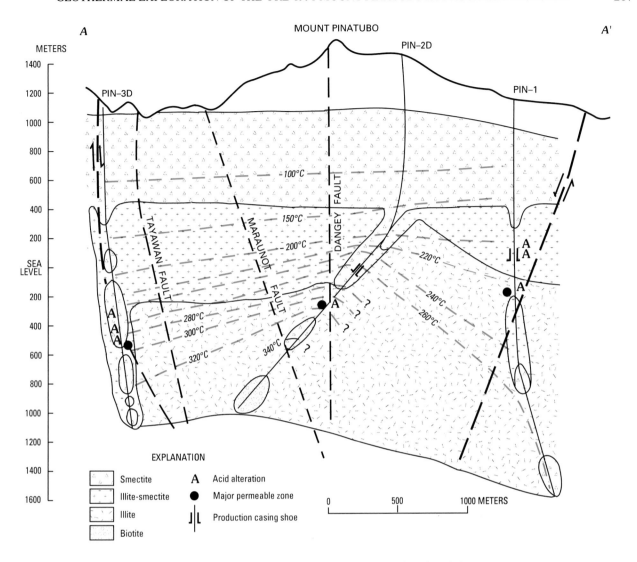

Figure 7. Hydrothermal clay zones in Mount Pinatubo wells. Isotherms are based on downhole surveys.

Table 4. Selected chemical analyses of discharge and downhole waters from Pinatubo wells.

[WHP = wellhead pressure; H = enthalpy. Concentrations in milligrams per kilogram]

Well no.	Date (year, month, day)	WHP (MPa)	H (kJ/kg)	pH	Li	Na	K	Ca	Mg	Fe	Cl	F	SO₄	HCO₃	B	SiO₂	δ¹⁸O	δD
PIN–1	890407	0.173	1,659	5.04	8.83	5,417	531	92.2	206	114	8,778	3.01	1,532	4.88	139	418		
PIN–1	890506	.169	1,670	4.46	9.95	6,378	617	99.5	189	106	10,650	2.09	1,567	1.99	162	573		
PIN–2D	890212	.723	1,907	2.32	5.69	2,717	492	39.2	233	407	4,768	2.74	2,627	–	147	1,129		
PIN–2D, 1,300 m.	890217	–	–	2.24	–	1,680	317	17.3	93	219	3,552	–	694	–	77.8	834	4.07	−27.9
PIN–2D, 1,450 m.	890216	–	–	2.39	–	1,514	300	10	95.3	426	4,417	–	975	–	120	860	4.27	−27.4
PIN–2D, 1,650 m.	890206	–	–	3.08	–	1,274	253	51.5	36.4	77.8	2,574	–	295	–	85.6	795	4.69	−27.3
PIN–2D, 1,650 m.	890217	–	–	3.18	–	937	185	42.5	61	1,468	3,574	–	216	–	96.9	1,141	4.33	−26.7
PIN–3D	900103	.213	2,411	4.19	26.30	17,400	3,540	2,207	133	525	37,760	2.78	500	–	399	1,660		
PIN–3D	900116	.348	2,486	3.71	32.20	21,330	4,770	2,730	222	679	45,860	2.42	240	–	729	1,829		
PIN–3D, steam.																	8.66	−41.40

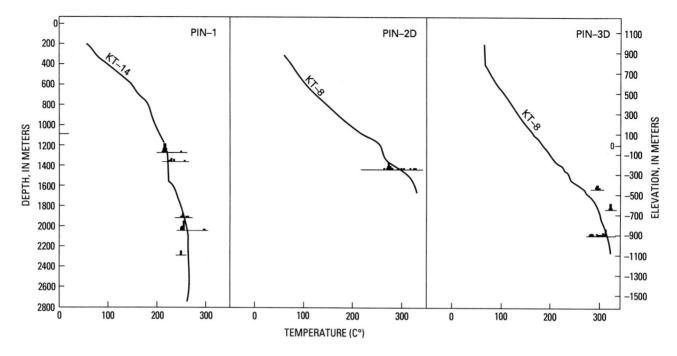

Figure 8. Temperature-depth plot of Mount Pinatubo wells. The horizontal range on histograms is the range of temperatures (scale at base of figure) at various depths, inferred from fluid inclusion homogenization. The height of each histogram is the frequency of each temperature reading, in increments of 1 to 10 readings (the latter is the highest value at 1,250 m depth in well PIN–1). KT–14 (PIN–1) and KT–8 (PIN–3D) temperature profiles were measured temperatures 84 and 5 days after completion, respectively. KT–8 (PIN–2D) profile was measured 15 days after flow test.

prior to their discharge (Ruaya and others, 1992). The immature character of the well waters may be due to the limited discharge period. While this may be true for PIN–2D, and to some extent to PIN–3D, it probably cannot apply to PIN–1, which was discharged for 138 days. Thus, nonattainment of equilibrium with the reservoir rocks appears to be an inherent characteristic of the Mount Pinatubo hydrothermal fluids. This characteristic most likely reflects the poor permeability of the hydrothermal system.

Selected gas chemistry data are listed in table 2. CO_2 is the principal gas specie in the steam, followed by H_2S. H_2, N_2, and Ar are present in varying amounts. On a CO_2-N_2-Ar ternary diagram (fig. 10), gases from the solfatara and the wells, PIN–2D in particular, have compositions similar to volcanic gases from White Island volcano (Giggenbach, 1987) and selected Japanese volcanic areas (Kiyosu, 1985; Kiyosu and Yoshida, 1988). The CO_2/N_2 ratio of PIN–2D (192) is comparable to those of White Island (180), while the N/Ar ratios of PIN–2D (1,437) and PIN–3D (874–2,760) fall between those of White Island (800) and selected Japanese volcanoes (4,250). Magmatic contribution to the Mount Pinatubo gases is thus implied.

STABLE ISOTOPE DATA

Figure 11 shows a plot of $\delta^{18}O$ versus δD for the thermal springs and wells PIN–2D and PIN–3D. Most of the thermal springs fall along or near the meteoric water line, affirming their meteoric or diluted steam-condensate origin. Although the 13 ‰$\delta^{18}O$ shift in PIN–2D may be attributed to rock-water interaction, the large shift in δD suggests otherwise. In fact, the isotopic composition of PIN–2D is close to the average values for high-temperature fumarolic steam from volcanic areas in Japan (Matsuo and others, 1975). PIN–3D fluids show a much smaller shift in δD compared to PIN–2D but still fall within the "primary magmatic water" area (Taylor, 1979).

The chloride-bearing springs of Asin and Cuyucut show small positive $\delta^{18}O$ and δD shifts and fall either on a mixing or an evaporation line connecting PIN–2D to the thermal springs. The relationship, if any, between PIN–2D and PIN–3D is less clear. Vaporization of PIN–2D waters at 310°C can shift the isotopic values of the residual liquid in the direction of PIN–3D in figure 11. The magnitude of such shift, however, is much too small to approximate PIN–3D's actual composition (fig. 11). A possible explanation for the isotopic composition of the PIN–3D steam is evaporation of fluids previously modified by rock-water interaction. This is consistent with the partially equilibrated character of PIN–3D waters (fig. 9).

In summary, the $\delta^{18}O$ and δD signatures of the well waters are comparable to those of high-temperature volcanic gases. This supports the contention that the hydrothermal system has a significant magmatic input.

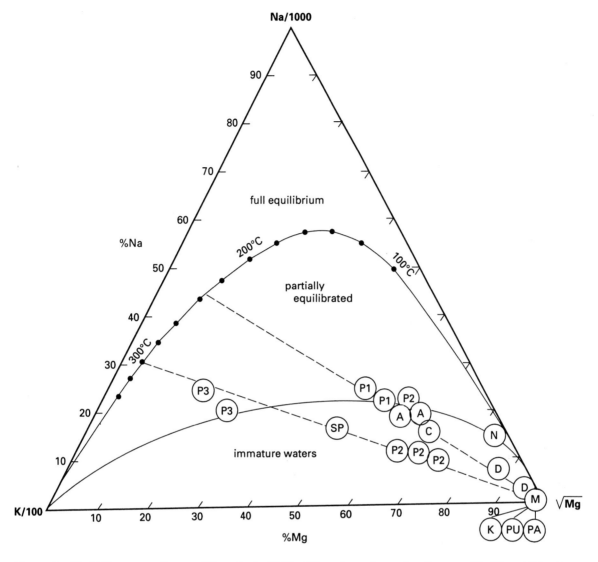

Figure 9. Na-K-Mg ternary diagram (Giggenbach, 1988) of Pinatubo spring and well waters. P1, P2, and P3 refer to water samples from PIN–1, PIN–2D, and PIN–3D, respectively. Spring abbreviations are given in table 1. Dashed lines represent trends discussed in text.

A MODEL OF THE PRE-1991 MOUNT PINATUBO HYDROTHERMAL SYSTEM

The geological, chemical, and isotopic data presented above show that the Mount Pinatubo hydrothermal fluids are not in chemical equilibrium with reservoir rocks dominated by neutral-pH alteration assemblages. Temperatures in excess of 300°C, moreover, suggest that a more efficient heat-transfer mechanism is operative than those associated with typical hot-water convective systems (Ruaya and others, 1992). A conceptual model of the pre-1991 Mount Pinatubo hydrothermal system is shown in figure 12. The temperature contours and the approximate location of the two-phase region are based on downhole observations.

We propose that the hydrothermal system was heated by a partially cooled or cooling magma chamber related to

the emplacement of the Pinatubo dome ~500 years ago (Delfin and others, 1992). Hot vapors released from this cooling magma rise through fractures, heating the volcanic rocks, and condense in deep recharge waters. This results in the production of a hot two-phase zone beneath the dome where PIN–2D and PIN–3D bottomed. Further ascent of these waters, and interaction with the rocks and shallow ground waters, results in neutral SO_4-HCO_3-Cl springs such as Mamot, Dangey, Pajo, Pula, and Kalawangan. The chemical and isotopic signatures of the original magmatic vapors, however, are retained in the gas ratios and isotopic composition of discharges from PIN–2D and PIN–3D.

The thermal energy of the convecting magmatic vapor induces the formation of convective cells of largely meteoric water on the flanks of Mount Pinatubo. At least two separate brine systems have evolved. One is located

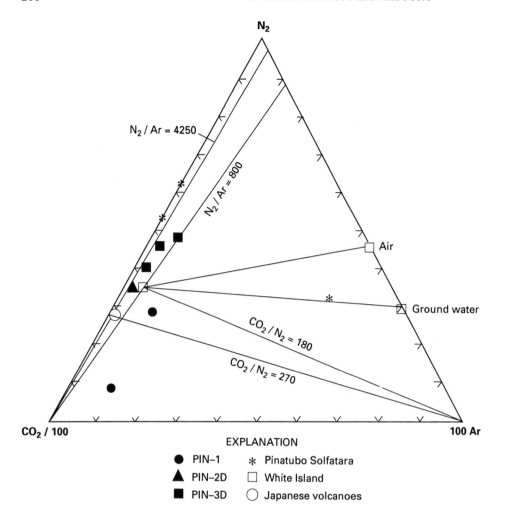

Figure 10. CO$_2$-N$_2$-Ar ternary diagram (Giggenbach, 1987) of Pinatubo solfatara and well gases.

EXPLANATION

● PIN–1   ✳ Pinatubo Solfatara

▲ PIN–2D   □ White Island

■ PIN–3D   ○ Japanese volcanoes

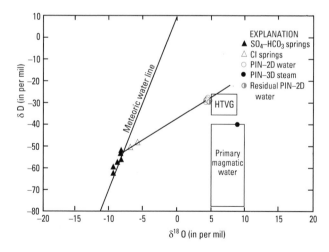

Figure 11. δ^{18}O versus 9.0 δD plot of Mount Pinatubo surface and well waters. Isotopic composition of high-temperature volcanic gases (HTVG) taken from Matsuo and others (1975). Field of primary magmatic water taken from Taylor (1979). Arrow indicates isotopic shift expected for evaporation of PIN-2D waters at 310°C. See text for discussion.

northwest of the pre-1991 dome, and part of this—probably near the intersection of the brine and two-phase region—was intersected by PIN–3D. The other brine system lies to the southeast. The northern and deeper portion of this brine system is represented by PIN–1 fluids, while its major lateral plume is manifested by the chloride-bearing springs at Dagsa, Asin, and Cuyucut.

The poor permeability in the wells is consistent with Giggenbach and others' (1990) model of "magmatic-hydrothermal systems," where an extensive, hydrothermally sealed carapace is likely to form. The formation of this carapace is attributed to such processes as evaporation of locally formed brines, deposition of solids from rising waters, and densification of volcanic rocks by alteration. The carapace of the Pinatubo system most likely lies along the boundary of the two-phase zone and the brine systems, or approximately near the 300°C contour in figure 12. At PIN–2D and PIN–3D, the depth corresponding to the 300°C isotherm is characterized by intense argillization and silicification (PNOC-EDC, 1990).

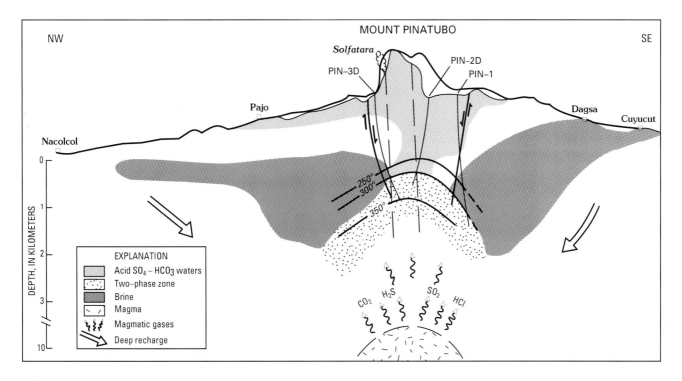

Figure 12. Conceptual model of the pre-1991 Mount Pinatubo hydrothermal system (modified from Ruaya and others, 1992).

THE HYDROTHERMAL SYSTEM AND THE 1991 ERUPTIONS

Mount Pinatubo came abruptly to life on April 2, 1991, with a series of phreatic to phreatomagmatic(?) explosions that formed three closely spaced craters on the northeast flank of the volcano (Punongbayan and others, 1991). At least six high-discharge fumaroles formed during or soon after these eruptions along an east-northeast-trending belt parallel to the Tayawan fault (fig. 5). On June 12-15, the volcano erupted violently, discharging 3.7–5.3 km^3 of dacitic magma (W.E. Scott and others, this volume).

The geothermal exploration conducted from 1982 to 1990 revealed that the Mount Pinatubo hydrothermal system had a significant magmatic input. This magmatic signature, unclear during the surface exploration stage, became distinct from the geochemical and isotopic signature of the deep reservoir fluids. Prior to April 2, 1991, however, there was little or no evidence of instability in the Mount Pinatubo hydrothermal system (Delfin and others, 1992). There were no significant changes in the hot spring and gas chemistry in the ~10 preceding years, nor were any historic phreatic explosion craters known in the area. The April 2 explosions and the ensuing intense seismic swarm were the first indications of a sudden magma intrusion event below the high-temperature hydrothermal system. Thus, prior to April 2, the accumulated information from geothermal studies did not indicate an imminent eruption.

EXPLORATION LESSONS

There is evidence from deep drilling at Mount Pinatubo that hot neutral-pH chloride waters, suitable for electric power generation, existed in the area but at some distance from those tested by deep drilling. If the model of the Mount Pinatubo hydrothermal system (fig. 12) is considered representative of most "magmatic-hydrothermal systems" in andesitic-dacitic volcanoes, high temperatures (≥300°C) may be expected in the central parts of such reservoirs. The model, however, clearly implies that acidic magmatic fluids, instead of neutral-pH chloride waters, are to be expected in the central region of such systems. Such acidic magmatic fluids, while attractive for their heat content, are not economically exploitable at present, owing to their high acidity and high gas content. Moreover, the limited permeability of the central part of such systems, as demonstrated in Mount Pinatubo, makes for poor fluid production.

A far more important lesson from Mount Pinatubo, as well as from recent eruptions of Nevado del Ruiz and El Chichón, concerns the need to assess the suitability for long-term development of geothermal systems associated with young volcanoes (Casadevall, 1992). Although volcanic hazard studies should be an important part of this assessment, other studies focusing on the degree of magmatic involvement in the hydrothermal system must be included. In this regard, the chemical and isotopic composition of waters and gases from wells and thermal features probably provide the best clues in determining magmatic input. Unfortunately, in most cases, such magmatic input becomes firmly established only after the expensive steps of drilling and testing deep wells have been completed.

ACKNOWLEDGMENTS

We would like to thank the management of PNOC-EDC for permission to publish this paper. Willie Carmen and Manny Teocson are thanked for drafting the figures. Reviews of this manuscript by Tom Casadevall, Marianne Guffanti, David Sussman, and Chris Newhall are gratefully acknowledged.

REFERENCES CITED

Apuada, N.A., 1988, Interpretation of the geo-electrical survey in Mt. Pinatubo geothermal prospect, Philippines: Pisa, International Institute for Geothermal Research, Diploma report, 29 p.

Bogie, I., 1984, Critique of Pinatubo geoscientific reports: Kingston Reynolds Thom and Allardice (KRTA) internal memorandum, 16 p.

Casadevall, T.J., 1992, Pre-eruption hydrothermal systems at Pinatubo, Philippines and El Chichón, Mexico: Evidence for degassing magmas beneath dormant volcanoes: Reports of the Geological Survey of Japan, v. 279, p. 35–38.

Clemente, V.C., 1984, A re-evaluation of the Mt. Pinatubo geochemistry: Philippine National Oil Company-Energy Development Corporation (PNOC-EDC) internal report, 16 p.

de Boer, J.Z., Odom, L.A., Ragland, P.C., Snider, F.G., and Tilford, N.R., 1980, The Bataan orogene: Eastward subduction, tectonic rotations, and volcanism in the western Pacific (Philippines): Tectonophysics, v. 67, p. 305–317.

Defant, M.J., Jacques, D., Maury, R.C., and de Boer, J.Z., 1989, Geochemistry and tectonic setting of the Luzon arc, Philippines: Geological Society of America Bulletin, v. 101, p.-663–672.

Delfin, F.G., 1984, Geology and geothermal potential of Mt. Pinatubo: Philippine National Oil Company-Energy Development Corporation (PNOC-EDC) internal report, 36 p.

Delfin, F.G., Sussman, D., Ruaya, J.R., and Reyes, A.G., 1992, Hazard assessment of the Pinatubo volcanic-geothermal system: clues prior to the June 15, 1991 eruption: Geothermal Research Council Transactions, v. 16, p. 519–527.

Esperidion, J.A., 1984, Resistivity traverse survey of Mt. Pinatubo: Philippine National Oil Company-Energy Development Corporation (PNOC-EDC) internal report, 20 p.

Giggenbach, W.F., 1987, Redox processes governing the chemistry of fumarolic gas discharges from White Island, New Zealand: Applied Geochemistry, 1987, p. 143–161.

———1988, Geothermal solute equilibria: derivation of Na-K-Mg-Ca geoindicators: Geochimica et Cosmochimica Acta, v. 52, p. 2749–2765.

Giggenbach, W.F., Garcia, P., N., Londoño C., A., Rodriguez V., L., Rojas G., N., and Calvache V., M.L., 1990, The chemistry

of fumarolic vapor and thermal-spring discharges from the Nevado del Ruiz volcanic-magmatic-hydrothermal system, Colombia: Journal of Volcanology and Geothermal Research, v. 42, p. 13–39.

Hedenquist, J.W., and Henley, R.W., 1985, Importance of CO_2 on freezing point measurements of fluid inclusions: Evidence from active geothermal systems and implications for epithermal more deposition: Economic Geology, v. 80, p. 1379–1406.

Kiyosu, Y., 1985, Variations in N_2/Ar and He/Ar ratios of gases from some volcanic areas in northeastern Japan: Geochemical Journal, v. 19, p. 275–281.

Kiyosu, Y. and Yoshida, Y., 1988, Origin of some gases from the Takinboue geothermal area in Japan: Geochemical Journal, v. 22, p. 183–193.

Matsuo, S., Suzuki, T., Kusakabe, M. Wada, H. and Suzuki, M., 1975, Isotopic and chemical compositions of volcanic gases from Satsuma-Iwojima, Japan: Geochemical Journal, v. 8, p. 165–173.

Newhall, C.G., Daag, A.S., Delfin, F.G., Jr., Hoblitt, R.P., McGeehin, J., Pallister, J.S., Regalado, M.T.M., Rubin, M., Tamayo, R.A., Jr., Tubianosa, B., and Umbal, J.V., this volume, Eruptive history of Mount Pinatubo.

Philippine Bureau of Mines, 1963, Geologic map of the Philippines: Manila, Philippine Bureau of Mines, scale 1:1,000,000, 9 sheets.

PNOC-EDC, 1990, Mt. Pinatubo resource assessment report: Philippine National Oil Company-Energy Development Corporation (PNOC-EDC) internal report, 46 p.

Punongbayan, R.S., Newhall, C.G., Ewert, J., Sussman, D., and Arevalo, E., 1991, Pinatubo-April 2, 1991 phreatic event and fumarolic activity summarized: Global Volcanism Network, v. 16, p. 12–13.

Reyes, A.G., 1990, Petrology of Philippine geothermal systems and the application of alteration mineralogy to their assessment: Journal of Volcanology and Geothermal Research, v. 43, p. 279–309.

Ruaya, J.R., Ramos, M.N., and Gonfiantini, R., 1992, Assessment of magmatic components of the fluids at Mt. Pinatubo volcanic-geothermal system, Philippines from chemical and isotopic data: Reports of the Geological Survey of Japan, v. 279, p. 141–151.

Scott, W.E., Hoblitt, R.P., Torres, R.C., Self, S, Martinez, M.L., and Nillos, T., Jr., this volume, Pyroclastic flows of the June 15, 1991, climactic eruption of Mount Pinatubo.

Taylor, H.P. Jr., 1979, Oxygen and hydrogen isotope relationships in hydrothermal mineral deposits, in Barnes, H.L., ed., Geochemistry of hydrothermal ore deposits (2d ed.): New York, Wiley-Interscience, p. 236–277.

Villaseñor, L.B., 1984, Summary of geochemical results—Pinatubo geothermal prospect: Philippine National Oil Company-Energy Development Corporation (PNOC-EDC) internal report, 9 p.

GEOPHYSICAL UNREST

Pinatubo was not seismically monitored before April 1991. Distant stations detected a few earthquakes in the vicinity of Mount Pinatubo following the magnitude 7.8 Luzon earthquake of July 1990. During April 1991, following steam explosions, several analog and digital seismometers were installed in a small network on the northwest side of Mount Pinatubo (Sabit and others). In late April and early May, the U.S. Geological Survey's Volcano Crisis Assistance Team (VCAT) helped the Philippine Institute of Volcanology and Seismology (PHIVOLCS) to install a network of seven radio-telemetered stations (Lockhart and others). Murray and others describe how data from these stations were processed on IBM-compatible PC's using software from the International Association of Seismology and Physics of the Earth's Interior (IASPEI). Earthquakes were located and examined for spectral characteristics, and the overall seismicity was examined in three dimensions and in plots of depth and magnitude versus time. High-frequency volcano-tectonic earthquakes 5 km northwest of Pinatubo and 5 km deep were succeeded in the first days of June by progressively shallower high-frequency and then low-frequency earthquakes beneath the north flank of Pinatubo. Escalating, shoaling high- and low-frequency events beneath the north flank became the main basis for successful forecasts of when magma might reach the surface; distinctive "long-period earthquakes," interpreted as qualitative pressure gauges for the volcanic system, provided continued warning after initial explosive events, until the climactic eruption on June 15 (Harlow and others).

Several refinements to the IASPEI software provided additional insights. Real-time Seismic Amplitude Measurement (RSAM, described by Endo and others), allowed tracking of the overall level of seismic energy release by one simple parameter. Seismic Spectral Amplitude Measurement (SSAM, described by Power and others) allowed tracking of changes in the spectral character of seismicity. SSAM was not used for forecasting during the crisis, owing principally to some cumbersome aspects of the software that have since been resolved, but SSAM did capture a distinct shift toward low-frequency events after June 1, 1991. Plotting of RSAM and SSAM data as inverse rates could have led to forecasts that were more precise at an earlier date than those based on non-inverted data (Cornelius and Voight).

In hindsight, R.A. White discovered deep long-period earthquakes (35 km deep) that are thought to record initial ascent of basaltic magma that eventually triggered the eruption. Although correlation of specific deep swarms with specific events near the surface might still be debated, a general correlation seems clear and represents the first seismic documentation of basalt intrusion and subsequent triggering of eruption of silicic magma.

Limited funds and logistical support in the preeruption period precluded detailed monitoring of ground deformation. The overall picture is unknown, but tiltmeters showed deformation precursors immediately before early dome growth (Ewert and others). In contrast, a quadrilateral array across the April 2 fissure showed no change above measurement error (about 1 cm in 20 m) between May 1 and May 28, even though this quadrilateral was just meters outside what would become the new caldera rim.

Posteruption seismic monitoring, by a reestablished network, showed exponentially decaying seismicity spread over wide area (Mori, White, and others). Episodic seismicity in late June and July 1991 probably correlated with small phreatomagmatic ash eruptions. Composite focal plane determinations by B. Bautista and others suggest that most preeruption, syneruption, and posteruption earthquakes were driven by regional east-west compression around the expanding and, later, partly vacated magma reservoir beneath Pinatubo.

An interesting outgrowth of the study of posteruption focal mechanisms was reexamination of changes in the regional stress field caused by the magnitude 7.8 1990 Luzon earthquake. Bautista and others, with coauthor R. Stein, conclude that compressive stress on the magma reservoir was about 1 bar and possibly enough to have triggered ascent of magma; alternatively, strong ground shaking associated with the earthquake might have triggered a chain of fault slippage that in turn allowed magma ascent.

Posteruption earthquakes also formed the basis for seismic tomography and a reevaluation of the three-dimensional velocity structure beneath Mount Pinatubo (Mori, Eberhart-Phillips, and Harlow). A low-velocity zone between 6 and 11 km beneath Pinatubo has an estimated volume of 40 to 90 km^3; an apparent extension beneath neighboring Mount Negron raises the total volume of low-velocity material to between 60 and 125 km^3. The improved three-dimensional velocity structure was also used to improve hypocenter locations. Two new programs by Hoblitt, Mori, and Power (PINAPLOT and VOLQUAKE) help visualize the seismicity of Pinatubo as it evolved, as well as seismicity during future volcanic crises.

Renewed seismicity and magma intrusion occurred in July through October 1992 (Ramos and others). As many as 1,400 low-frequency earthquakes occurred in a single day and included many multiplets (identical earthquakes), which are suggestive of repeated stress and failure at the same location. Were these earthquakes precursors to additional explosions, or were they the seismicity of a sluggish intrusion? They were from a sluggish intrusion, but they looked remarkably like events that in 1991 had preceded explosive eruptions. An interesting variety of tremor, possibly related to long-period seismic events, is also described by Ramos.

Ed Laguerta (PHIVOLCS), Andy Lockhart (USGS), Joey Marcial (PHIVOLCS), and John Power (USGS) installing a seismometer before the eruptions. Photograph by Val Gempis, USAF.

Installation, Operation, and Technical Specifications
of the First Mount Pinatubo Telemetered Seismic Network

By Andrew B. Lockhart,[1] Sergio Marcial,[2] Gemme Ambubuyog,[2]
Eduardo P. Laguerta,[2] and John A. Power[1]

ABSTRACT

In late April and early May 1991, staff from the Philippine Institute of Volcanology and Seismology and the U.S. Geological Survey / U.S. Agency for International Development Volcano Disaster Assistance Program installed a ruggedized nine-component telemetered seismic network to monitor Mount Pinatubo. They also installed a low-data-rate telemetry network of two tiltmeters and a rain gauge.

The Mount Pinatubo network provided real-time geophysical data to the on-site monitoring team, who could evaluate it immediately in the context of geologic observations. This permitted the team to give the civil and military authorities the best possible eruption forecasts quickly and contributed to the timely decisions to evacuate large areas around Mount Pinatubo, including Clark Air Base.

Seismic stations were located from 1 to 17 kilometers from the summit and in all quadrants surrounding the volcano. Our biggest installation problems were the difficulty of site access and a lack of viable repeater sites in the rugged, heavily vegetated terrain around Mount Pinatubo. We also had problems with poor batteries and cable, observatory wiring, and timing. The problems that we encountered with the Mount Pinatubo network did not greatly affect our ability to monitor the volcano for public safety but have placed some constraints on the usefulness of the data for posteruption studies. The network was destroyed by the June 15 eruption, except for one station we had installed in a safe, easily accessible place at a distance of 17 km from the volcano. Had there been thermal protection for the antenna cables, one or two of the lost stations might have been saved.

The Mount Pinatubo network demonstrated some important lessons for future volcano-monitoring networks. Networks must be reliable, because they are a key monitoring tool on which public safety may depend. They must be designed to capture a wide range of seismic events, from small long-period events and tremor to large rock-breaking events. It is important to keep at least one station in a safe, easily accessible place to be able to maintain rudimentary monitoring capabilities if the rest of the network is destroyed. The narrow bandwidth of the analog telemetry system can limit the usefulness of the seismic data. This limitation can be largely overcome with dual-gain voltage-controlled oscillators or by digital systems now under development.

INTRODUCTION

On April 2, 1991, Mount Pinatubo awoke from 6 centuries of tranquility with steam explosions and vigorous fumarolic activity. Staff from the Philippine Institute of Volcanology and Seismology (PHIVOLCS) immediately established an observatory west of Mount Pinatubo and began seismic monitoring with several portable seismic recorders (Sabit and others, this volume). On April 5, staff from the Volcano Disaster Assistance Program (VDAP), a cooperative program between the U.S. Geological Survey (USGS) and U.S. Agency for International Development (USAID) (Miller and others, 1993) based at Cascades Volcano Observatory (CVO) were alerted, and on April 23 a three-person USGS team flew to the Philippines to assist PHIVOLCS in monitoring Mount Pinatubo and to advise the U.S. Air Force at Clark Air Base, located at the eastern foot of the volcano. The USGS team carried seismic equipment designed around the standard USGS analog telemetry system and modified for use in volcanic monitoring. The USGS monitoring cache also included telemetered electronic inclinometers, which were incorporated into the network after the seismometers had been emplaced (Ewert and others, this volume).

The telemetered seismic network provided the joint Pinatubo Volcano Observatory (PVO) monitoring team with critical information that was used to forecast volcanic activity and which led to the successful evacuation of the area around Mount Pinatubo. This included evacuation of the majority of personnel from Clark Air Base 5 days before the

---

[1]U.S. Geological Survey.
[2]Philippine Institute of Volcanology and Seismology.

climactic eruption on June 15,1991. The portable seismic network and base-station equipment were installed in slightly less than 3 weeks. With the telemetered network in place, seismic energy levels, hypocentral locations, spectral information, and types of seismic events could be quickly analyzed by the monitoring team and the team's analysis could be acted upon by public officials during the crisis. After the climactic eruption we were able to continue monitoring seismic activity despite destruction of most of the network because one station (CAB) had been deployed at a greater distance from Mount Pinatubo and remained operational.

Along with those aspects of the seismic network that contributed to the successful prediction and evacuation, we had some problems with the network that limited its effectiveness: bad batteries and antenna radio-frequency (RF) connections that caused numerous helicopter trips for station maintenance; dependence on a military communications repeater site that was vandalized before the main eruption, leaving us unable to locate events in the last days of the buildup; the uncalibrated seismometers; and the narrow dynamic range of the telemetry system, insufficient to effectively transmit data across the full signal-amplitude range; the tendency of the event-recording software to ignore emergent events like long-period (LP) events and teleseisms (Murray and others, this volume); and the lack of thermal protection from hot ash clouds for exposed components at the field sites.

Both the successes and failures we encountered provide valuable lessons for future volcano-monitoring networks. In this paper we describe the seismic equipment installed around Mount Pinatubo and discuss the factors that influenced installation and configuration of the network. This is followed by a short description of the telemetry, repeater sites, observatory wiring, and the site problems we encountered in operating the Mount Pinatubo network. We conclude with a review of the Mount Pinatubo network and recommendations for improvements and alterations to seismic equipment for future rapid deployments at restless volcanoes.

SEISMIC EQUIPMENT AND NETWORK CONFIGURATION

In the 20 days between April 25 and May 14, 1991, we installed a seven-station, three-repeater, nine-component seismic network around Mount Pinatubo (fig. 1 and appendix 1) and set up a recording site at the observatory on Clark Air Base. Seismic stations were placed in each quadrant around Mount Pinatubo and in a line extending east from the volcano toward the observatory on Clark (fig. 1). The network was configured to yield high-quality hypocenter solutions in the expected zone of seismic activity beneath the volcano and to rely as little as possible on repeaters

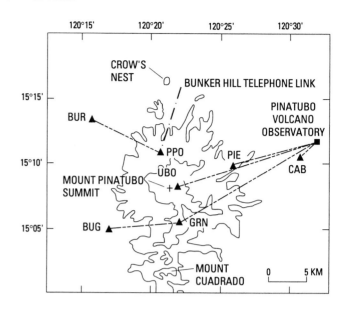

Figure 1. Station locations and radiotelemetry links in the Mount Pinatubo area.

located in rugged jungle terrain. The existing PHIVOLCS stations were not telemetered and were not routinely incorporated into this network. However, before the telemetered network became fully operational, S-minus-P phase arrival time differences from all stations around the volcano were used to determine rough locations.

HARDWARE AND INSTALLATION

The telemetered seismic equipment we installed at Mount Pinatubo came from the VDAP cache at CVO (Lockhart and others, 1992a) and had been especially designed to be both durable and portable for responding to volcanic crises worldwide. Because the telemetered seismic equipment had been prepared in advance for deployment in harsh environments (Lockhart and others, 1992b), site preparation and installation rarely took more than 2 or 3 hours for a three-person field crew. Derived from the standard USGS analog telemetry system (Eaton, 1976), the stations used Mark Products L–4 1-Hz and L–22D 2-Hz geophones and USGS seismic telemetry employing UHF radio links.

The voltage-controlled-oscillator (VCO) board and radio were housed together in a waterproof plastic instrument case. We used environmental connectors sealed to the instrument case for the power, geophone, and antenna connections. The VCO board itself was bolted inside a metal box inside the instrument case, for protection from stray RF signals. We did not encounter any problems with radio interference of the VCO circuitry. We placed desiccant in each of the instrument cases and placed them with the batteries in a larger, nonwatertight equipment enclosure, usually a large plastic cooler with a lid. The coolers at all remote sites were buried in soil, in part to avoid the extreme daytime heat but also for protection against theft and vandalism. We cut large holes in the bottoms of the coolers to drain condensed moisture. The soil was initially permeable, loose and well drained at all of the sites, and there were no problems with seepage before the eruption. However, the accumulated tephra caused poor drainage at some sites after the eruption and some equipment got wet, but not that in waterproof cases (T. L. Murray, USGS, oral commun., 1991).

In order to reduce effects of wind noise on the geophones, we buried them at least 0.5 m deep. At each site we encapsulated the geophone in a 10-in section of 3-in plastic tubing capped on both ends and used a gland connector to pass the cable through the top endcap. The 3-in diameter L–4 geophone fit tightly into the tubing, which was set into a small fast-setting cement pad for a solid connection to the ground; the geophone could still be removed from the tube once the cement had set. The plastic sheath provided electrical isolation of the geophone from the earth, eliminating a potential source of electrical ground loops.

The VCO boards were designed at CVO by T.L. Murray (T.L. Murray, USGS, unpub. data, 1991); he used the basic circuit design in the USGS J502 VCO (F. Fischer, USGS, unpub. data, 1986) but with a much simpler calibration circuit than the one used by the J502. Individual station gain settings (see appendix 1) were set so that the ambient seismic noise was 100 mV peak-to-peak (p-p) measured at the input of the VCO chip. The VCO boards drew 15 mA at 12 V. All sites used 100 mW UHF radios and transmitted through 9-dB-gain yagi antennas. The radios drew around 85 mA, so the total current drain for a single-component station was 100 mA, or 2.4 ampere-hours/day. To power remote sites, we used pairs of locally purchased 80-ampere-hour lead-acid automobile batteries and a regulated 18 W solar panel. The batteries had a high failure rate, the result of which was lost data and frequent maintenance trips to sites accessible only by helicopter.

Each site was grounded from the negative battery terminal to a ground rod that was driven into the soil next to the equipment enclosure. The radio outputs were protected from lightning by plasma-discharge tubes grounded to the negative battery terminal. The VCO signal input and output lines and power lines were all protected from transient voltage surges by avalanche diodes. Despite the many fierce lightning storms that passed over parts of the network, only one station was lost: BUG was damaged by what appeared to have been a nearly direct strike, which destroyed the radio, damaged the VCO, and left arc burns on the VCO board and the metal box that contained it.

STATION LOCATIONS

In this report we refer to the station locations by a three-letter code (such as CAB) and we refer to the recorded signal from that station by a four-letter code that indicates the orientation of the sensor; for example, CABZ is the signal from the vertical geophone at CAB.

We installed the first telemetered seismic station, LIL, immediately upon our arrival at Clark Air Base in order to quickly establish baseline seismic monitoring in a secure, accessible location. LIL was on Lily Hill at Clark Air Base near a large water tank. The site was selected because of ease of access and line-of-sight telemetry to our observatory. After a few days, it was clear that operation of the water pump in the tank generated excessive seismic noise, and the equipment was moved to a quieter site (CAB), which was still on the air base and readily accessible, 17 km from Mount Pinatubo. CAB was on a hill 100 m away from another water tank. The CAB site was not as noisy as LIL, but vibration noise of the pump for the nearby water tank limited CAB's usefulness until the evacuation on June 10 reduced water use on the air base, and, consequently, water-pump noise at CAB. We searched for alternative sites for the CAB station but found none better.

Nevertheless, CAB turned out to be one of the most critical stations in the network. During the most intense phase of volcanic activity, CABZ served as the most reliable indicator of Real-time Seismic Amplitude Measurement (RSAM) seismic energy (Endo and Murray, 1991; Endo and others, this volume) because CAB's great distance from the volcano resulted in smaller recorded amplitudes and less signal clipping than was observed at closer stations. It was the only station to survive the June 15 eruption, so it was the only seismic monitoring available to observatory staff until new stations could be installed some time after June 15.

We installed PIE and UBO as line-of-sight stations that, with CAB, formed an east-west line from Mount Pinatubo to Clark (fig. 1). PIE was installed on an elephant-grass-covered ridge 6 km east of Mount Pinatubo, at a site where we could land a helicopter, but which was otherwise inaccessible except by a long, steep hike through the jungle. UBO was located on an abandoned geothermal drill pad just east of the volcano's summit, 1.5 km from the vigorous fumaroles on the northeast side of the summit. Although there had been road access to this pad a few years before, landslides had made the road impassable, and the site was only accessible by helicopter. Burgos (BUR) and Patal Pinto (PPO) were sites north of Mount Pinatubo with road access. PPO was chosen because of road access, azimuthal coverage of the sector north of Mount Pinatubo, proximity to the active fumaroles, and the presence of a secure building we could use as a site. BUR site was selected because of its road access, azimuthal coverage of the sector to the northwest of Mount Pinatubo, and because it was line-of-sight to PPO. Mount Negron (GRN) was located on a sharp, brushy peak that provided azimuthal coverage on the south side of the volcano and that could serve as a repeater for stations to the southwest. At GRN, the site noise was very high until we re-set the geophone in a 1.5-m-deep hole. Sitio Buag (BUG) was chosen because it closed an azimuthal gap on the southwest side of the volcano, enjoyed helicopter access, and was line-of-sight to both GRN and Mount Cuadrado.

REPEATER SITES

Our observatory was east of Mount Pinatubo at Clark Air Base. Before the June 15, 1991, eruption, Mount Pinatubo appeared from Clark as a craggy summit visible through a broad notch in a north-south line of jungle-covered peaks and ridges forming a wall just east of the summit. To locate events beneath the volcano, we had to ring the volcano with seismometers and transmit signals from the west over the wall of peaks and ridges down to the observatory at Clark. There were no unforested, helicopter-landable repeater sites for the seismic stations north and northwest of Mount Pinatubo that were line-of-sight to Clark. The sector to the south and southwest of Mount Pinatubo was line-of-site to the summits of both Cuadrado (10 km from Mount Pinatubo) and Negron (5 km from Mount Pinatubo). We installed a seismic station and repeater on Negron (GRN site) because we thought it's proximity to Mount Pinatubo would make it a good site to colocate a seismic station and because of a fortuitous landing pad that had been constructed a few years before on the otherwise inaccessible peak.

It was very difficult to relay the signals from BUR and PPO, the stations north of Mount Pinatubo, because of the expanse of jungle-covered hills between them and Clark. Before the eruption, a telemetry link to Clark would have required several long and tenuous repeater shots. Initially we considered Crow's Nest as a possible repeater site, a peak 16 km north of Mount Pinatubo. Crow's Nest had been used as a temporary military site a few years before and was considered a high-risk site for theft and vandalism. This plus poor telemetry test results from Crow's Nest to PVO convinced us to use a microwave telephone link from a nearby military communications site, Bunker Hill, which the U.S. Air Force had provided for our use. The Bunker Hill site was line-of-sight to PPO, equipped with generator power and guarded around the clock. This seemed to be the best choice for repeating telemetry signals from the north side of Mount Pinatubo back to Clark. In fact, it served well until the guards were evacuated on June 10 and the generators were stolen. After this, attempts were made as late as June 14 to install an alternative repeater on Crow's Nest with a 5-W UHF radio transmitting yagi-to-yagi to a receiver some 60 ft up an antenna mast on Clark, but the signal could not be received; the radio signal path was not line-of-site.

After the eruption, there were many good repeater sites because the eruption had removed all the jungle from the hills (T.L. Murray, oral commun., 1993). After the destruction of the first telemetered seismic network at Mount Pinatubo on June 15, 1991, USGS and PHIVOLCS staff installed a second seismic network around the volcano. Some of the sites of the first network were re-occupied, and seismic repeaters were installed at sites that had been inaccessible to us before the eruption.

PPO TELEMETRY

Installed first as a single vertical-component seismic site, PPO was soon augmented to serve as a repeater for BUR. Later, we replaced the single vertical-component 1-Hz geophone with a three-component 2-Hz geophone. Then we added a two-component tiltmeter controlled by a low-data-rate (LDR) digital-telemetry platform (Murray, 1988, 1992, Ewert and others, this volume), which we patched into the seismic telemetry system. Next we added a rain gauge to the LDR platform (Marcial and others, this volume). With this addition there were seven data components transmitting over one radio link from PPO. As we had no multiple-input summing amplifier, signals from the three VCO's, the LDR platform, and the BURZ receiver were summed by wiring together the secondary windings of all the output transformers. The signal out from each platform was adjusted so that the amplitude of the summed signals was 1 V peak-to-peak and all signals were strong enough to break squelch on their respective discriminators. At PPO, the technique of hard-wiring the transformer secondaries worked as well as a multiple-input summing amplifier would have. In order to keep the LDR platform from spiking the seismic lines every time it transmitted, we kept the LDR board's modem chip turned on and transmitting a carrier tone at all times.

The PPO tiltmeter (Ewert and others, this volume) had initially been installed on the concrete floor of the hut at PPO on June 2. Wiring problems at PVO kept us from being able to analyze the PPO tilt data for a number of days; it became clear that our simple installation had been inadequate. Several more days passed while we worked on voice-communications problems between PVO and the west side of Mount Pinatubo. We returned to PPO to make a standard installation of the tiltmeter on June 10, but increasing activity at Mount Pinatubo cut our efforts short, so tilt data from the PPO tilt site were never reliable.

OBSERVATORY WIRING

The PVO observatory computerized data-acquisition system is described by Murray and others (this volume). We received the telemetered seismic signals with 9-dB-gain yagi antennas and UHF radios. Signals went from the radios to USGS J101 and J110 discriminators and from there to drum recorders and the computer system. As the telemetry system developed, the observatory wiring became more complicated. Soon there were three telemetered seismic signals received directly (PIEZ, UBOZ, CABZ) and one radio link with two signals (Negron line, GRNZ, BUGZ). A telephone link from Bunker Hill carried four seismic signals (PPON, PPOE, PPOZ, BURZ) and two signals from an LDR digital transmitter (PPO tilt and rain gauge). The two LDR signals from PPO and the UBO tilt LDR line went to an LDR-receiver computer modem. The Bunker Hill telephone link was patched into both the discriminator rack and through an isolation amplifier to the LDR-receiver computer modem.

OBSERVATORY TIMING

The network's location in the Philippines resulted in timing problems that we were not able to overcome completely. Although we had a GOES satellite clock for timing the seismic arrivals, the GOES satellite signals were not received over that part of the globe, so the satellite clock was not used. Instead, we timed the drums by manually synchronizing the drum's clocks to WWV broadcasts. Similarly, the PC clocks on both the seismic data-acquisition computer and the RSAM computer were manually synchronized to WWV. This was supposed to have been done daily, but often was not, especially during the first hectic weeks of the response. Manual timing to WWV gave a relative timing error no better than ±0.1 s. However, on several occasions we found the PC clock or RSAM clock to have drifted by several minutes, and this drift complicated posteruption studies. Either a GPS or Omega clock, which works anywhere in the world would have been a better solution. Even without a GPS or Omega clock, our timing problems would have been minimized had we had a stable time-code generator like that built by Jim Ellis for the USGS 5-day seismic recorder (Criley and Eaton, 1978).

INSTALLATION AND MAINTENANCE PROBLEMS

Battery failures and loose RF connections caused the majority of our site visits for maintenance. The batteries and some RF cable had been purchased locally. Because of airfreight restrictions against shipping batteries, we have always found it much easier to purchase batteries locally than to try to ship them by air to distant locations. We have had problems with battery quality in other places too, but at Mount Pinatubo poor-quality batteries cost us dearly in lost data and maintenance visits, which required a helicopter.

The battery problems that plagued the rest of the network were also a problem at CAB, but we were able to maintain CAB easily because of it's accessibility and proximity to the observatory. The water-pump noise reduced the quality of the seismic data from CAB, and as the rest of the network was sufficient to locate events and to monitor volcanic seismicity like fumarolic noise, CAB battery maintenance was a low priority until just before the most intense phase of volcanic activity. Then we replaced the batteries and brought CAB back online.

We used RG-8 coaxial cable for antenna cable and environmental 'N' type RF connectors. The locally purchased RG-8 turned out to be thinner than the cable we had been using, so the barrels of our crimp connectors were loose on the cable. On several occasions we had problems with the cable pulling out of the crimp connector barrels and ruining the connection; this problem required site visits. Later, we replaced suspect connectors with clamp connectors. Although the cable was still not held securely and could be pulled loose, heavy taping of the connector and first 6 inches of the cable with cold-shrink tape usually provided enough strain relief to make the connection serviceable.

SPECIAL PROBLEMS CAUSED BY THE ERUPTION

In the last days before June 15, the utility of the data was reduced because the recordings from increasing numbers of stations clipped on high signal amplitudes. With station gains set to ambient noise levels early in the precursory sequence, the overall seismic energy saturated our instruments just at the time when it was too dangerous to visit most sites and reset station gains. CAB was easily accessible, and it would have taken only minutes to drive to the site and reduce the gain. However, doing so would have changed the RSAM values and the appearance of the seismogram to which we had become visually "calibrated" in making quick estimates of levels of background seismicity. With the rapidly developing eruptive sequence, we did not want to have to recalibrate our eyes to a different, much quieter seismogram, so we did not reduce gain at any station.

Except for CAB, the Mount Pinatubo network was destroyed by the June 15 eruption. The Bunker Hill repeater, which tied into a U.S. Air Force microwave link to Clark (with the signals BURZ, PPOZ, PPON, PPOE, and the PPO tiltmeter and rain gauge) was lost to vandalism and theft on June 11. Station UBO was knocked out by a small volcanic ash cloud apparently related to an explosion signal at about 0341 local time June 12. Coincidentally, UBO seismic telemetry had failed several hours before the blast signal was registered at PIE. We collected data from PIE, BUG, and GRN until the climactic eruption on the afternoon of June 15, when pyroclastic flows or related surges damaged or annihilated these stations. We lost UBO, GRN, PIE, and PPO completely. UBO was at or near the edge of the caldera that formed on June 15. The PPO site was destroyed by the eruption. No remnants of the concrete hut were found during a visit to the site in early 1992 (R.P. Hoblitt, USGS, oral commun., 1993). GRN was either destroyed or buried. Nothing could have been done to save these sites. Two stations (BUG and BUR) were not destroyed but merely damaged, their RF cables melted through and shorted out. In addition, solar panels melted and the plastic insulators used on the yagi antennas melted. Neither the loss of the solar panels nor (probably) the antenna insulators would have been immediately fatal to the sites, but the melted antenna cables shorted out the radio signals to the antennas. After the climactic eruption, BUG required only a new antenna, antenna cable, and fresh battery (T.L. Murray, oral commun., 1993).

CONCLUSIONS: LESSONS FROM THE MOUNT PINATUBO TELEMETERED SEISMIC NETWORK

NETWORK DESIGN

The principal reason for monitoring volcanoes is the safety of the local populace, so seismic volcano monitoring requires seismic monitoring networks of great reliability. The network must also be flexible enough to capture a wide range of volcanic seismicity amplitudes, from the small but important rock-breaking and LP events and tremor to some of the large signals that accompanied the Mount Pinatubo eruption.

The destruction of all but one station in the Mount Pinatubo network during the major eruption raises the question of network design. The lesson from Mount Pinatubo is to maintain at least one station in a safe place far enough from the volcano to survive any likely eruption and record large-amplitude seismic activity on-scale. The station must be readily accessible. In anticipation of large eruptions like at Mount Pinatubo, it would be desirable to have an outer network of three or four stations in safe places up to 30 km from the volcano. This would allow location of large events during periods of high seismic activity, when the closer stations might be saturated or destroyed. At Mount Pinatubo the ability to locate events during the height of the eruption and afterward might not have had any immediate public-safety benefits, but the posteruption analysis of such data might have been of great scientific value.

We were very lucky to have had good helicopter support. The telemetered network may not have been possible at all without helicopter access to otherwise inaccessible sites. Because of the difficulties in telemetering seismic data from a widely spread network in mountainous terrain like that near the Mount Pinatubo area, more attention should be given to developing flexible and reliable alternatives to low-power ground-based line-of-sight telemetry networks. The major problem with low-power ground-based line-of-sight telemetry nets is the dependence on line-of-sight repeater sites, which are frequently difficult to access. Failure of an important telemetry link can eliminate a significant part of the network, as exemplified by the loss

of the Bunker Hill military microwave link on June 11 when a major eruption was clearly imminent. There were no line-of-sight repeater sites that could have replaced the Bunker Hill repeater.

Satellite telemetry is occasionally suggested for use in mountainous terrain like this, but systems like Argos (Collecte Localisation Satellites, 1992) and the Geostationary Operational Environmental Satellites (National Oceanic and Atmospheric Administration, 1984) transmit only a small amount of data as few as eight times daily. For a volcano observatory to receive the signals from data collection platforms that are not line-of-sight, one must either have a telephone link to an earth station, or purchase one at some expense. Because of these limitations, Argos and GOES platforms are better suited to monitor potentially active volcanoes rather than active volcanoes that threaten populations. We feel that low-power ground-based line-of-sight telemetry networks are currently a more effective monitoring tool than satellites. Another argument for the use of low-power ground-based line-of-sight telemetry networks is that the standard seismic telemetry techniques are widely known to technicians in geological institutions all over the world. Productive collaborations like that between the USGS and PHIVOLCS during the Mount Pinatubo crisis are much easier to coordinate and carry out when both groups use and know the same equipment.

A telemetered seismic network located on only one side of a volcano might be very useful as an alternative to the Pinatubo network configuration when monitoring a volcano where one side is inaccessible. Such a network trades hypocentral location accuracy for simplicity and ease of installation and might involve a tripartite array (see, for example, Ward and Gregersen, 1973), be a kilometer or more on a side, and use three-component geophones. It might be located off the volcano and be used in conjunction with other stations on the edifice in order to make better depth estimates. The tripartite array or some other design that maximizes the monitoring power of a network whose station sites are limited to one side of a volcano deserves further inquiry.

The most fundamental lesson from the telemetered Mount Pinatubo network was the importance of receiving the data locally so scientists could act on it immediately and advise the local officials in charge of the public's welfare. The presence of the monitoring team in the community near the volcano gave credibility and urgency to their statements and was a key factor in the pre-eruption evacuations. In turn, their minute-by-minute decisions would have been less well informed and less certain without the constant stream of telemetered tilt and seismic data. Realtime telemetered data were analyzed in the context of concurrent visual observations for rapid, informed predictions, which were then passed on to the officials in charge of the public's welfare. A monitoring scheme that transmitted seismic data by satellite to experts sitting in distant offices would not have worked nearly so well.

EQUIPMENT

Heat caused the failure of those stations that survived or could be found. The stations had been buried, so the geophones and electronics were unaffected by the heat, but at both UBO and BUG the antenna cables had been melted and shorted out by hot pyroclastic currents. Future installations should continue to protect the electronics from heat, and the antenna cables might be protected from hot pyroclastic currents by flexible metal conduit.

The Mount Pinatubo experience showed a need for more dynamic range at the seismic stations to cover the wide range of seismic amplitudes encountered during the evolution of volcanic activity. Early in the activity, station gains were set to detect the dominant small, shallow earthquakes and the LP's, both of which were important in assessments and predictions. As activity increased, the high-gain VCO transmissions went off-scale during the larger events. At Mount Pinatubo, it was impractical to change the station gains because of safety concerns and for consistency of the RSAM data and visual monitoring of the drums. Since returning from Mount Pinatubo, the USGS staff at CVO have addressed this problem by constructing seismic stations that use dual VCO's for volcano-monitoring, similar to a VCO design used at some of the California network stations (J.P. Eaton, USGS, written commun, 1993). The VCO boards constructed at CVO are piggybacked so that the preamplifier output of a VCO board using one carrier frequency is fed into a second board that uses another carrier frequency. The two boards are set at different gains: one is set for maximum attenuation (44 dB gain) and the other is set to a more standard gain of 62–74 dB. The output carrier frequencies of the two VCO boards are summed together at the secondaries of the output transformers and are fed into a single radio. The dual-VCO system doubles the telemetry load but in theory can increase the bandwidth of on-scale signals from around 40 dB to as much as 70 dB or so, which is nearly the 72 dB bandwidth of the 12-bit-plus-sign analog-to-digital converter used in the data acquisition system. The dual-VCO system retains the flexibility of analog telemetry to mix signals from different stations in the field and adds considerable bandwidth.

Digital systems are in use in some areas and are under development in others. In theory, they will be an improvement over analog telemetry, but analog telemetry will continue to be important in volcano monitoring for some time to come, especially where low cost, simplicity of design and ruggedness are concerns. Gain-ranging VCO's, such as are used in parts of Alaska (Rogers and others, 1980), are an alternative to dual-gain VCO's. Because of our reliance on RSAM values and rapid visual observation of drum records

in predicting eruptions, we would not have wanted an automatic gain-ranging VCO, which changes the telemetered seismic energy values and the appearance of the seismograms. Perhaps the next step in volcano-monitoring networks should be hybrid networks, with easily deployed and reliable analog stations to ensure good coverage of the volcano combined with a few digital stations for recording large events.

At Mount Pinatubo, we used four drum recorders for a quick analysis of the seismicity. Seismograms from a representative signal (PIEZ) were posted on a wall of the observatory where they made a dramatic display of the evolving seismic activity. Local officials who toured PVO could see this display of seismic records and understand the clear increase in seismic activity it showed. This simple display proved to be a powerful tool in educating local civil and military officials. Besides helping the monitoring team in public education, the seismograms were of great value in making rapid analyses of the state of the volcano. Critical volcano-seismic signals such as tremor and small, emergent LP's not captured by event recorders showed up on the seismograms, and as the system gains were kept constant, drum recorders permitted the overall level of seismic activity to be visually checked minute by minute. Event epicenters could be estimated rapidly by checking the order of first arrivals on the four drums.

Observatory wiring at PVO became a more complicated system as the number of telemetered instruments increased. Problems with the wiring at PVO combined with other factors to cost us the effective use of the tiltmeter installed at PPO. Wiring modifications consumed a lot of time as stations were added to the network. Ideally we would have liked the wiring to have been simple to minimize "down time" by making it easy to modify or troubleshoot. Since returning from Mount Pinatubo, CVO staff has simplified observatory wiring with printed-circuit-board switch panels (A.B. Lockhart, unpub. data, 1992). The switch panels route seismic signals between receivers and discriminators, and from there to output devices: drum recorders, RSAM, digital data-acquisition systems and auxiliary devices.

Although it did not affect our monitoring, the absence of a good timing system at Mount Pinatubo complicated the posteruption reconstruction of events. Because we were not tied into a global time standard, eruption events recorded elsewhere in the Philippines are not easily correlated with data from the Mount Pinatubo network. Another problem is that the timing common to the SSAM data and the events recorded on the data-acquisition system was independent from the RSAM timing.

ACKNOWLEDGMENTS

We would like to thank Tom Murray and John Ewert for their help installing and maintaining the telemetered network. Thanks are also due to the helicopter pilots and flight crews of the U.S. Thirteenth Air Force, Third Tactical Fighter Wing, especially Capt. Brett Nyander, for excellent logistical support in hazardous and uncertain conditions. This paper benefited mightily from discussions with Randy White and Tom Murray and from reviews by Jerry Eaton, John Lahr, and Emmanuel Ramos.

REFERENCES CITED

Collecte Localisation Satellites, 1992, Monitoring active volcanoes with ARGOS satellites; description of equipment: North American Collection and Location by Satellite, Inc., 9200 Basil Court, Suite 306, Landover, MD 20785, 16 p.

Criley, Ed, and Eaton, J.P., 1978, Five-day recorder system (*with a section on* the Time Code Generator, by Jim Ellis): U.S. Geological Survey Open-File Report 78–266, 86 p.

Eaton, J.P., 1976, Tests of the standard (30 Hz) NCER FM multiplex telemetry system, augmented by two timing channels and a compensation reference signal, used to record multiplexed seismic network data on magnetic tape: U.S. Geological Survey Open-File Report 77–884, 61 p.

Endo, E.T., and Murray, T.L., 1991, Realtime Seismic Amplitude Measurement (RSAM): A volcano monitoring and prediction tool: Bulletin of Volcanology, v. 53, p. 533–545.

Endo, E.T., Murray, T.L., and Power, J.A., this volume, A comparison of preeruption real-time seismic amplitude measurements for eruptions at Mount St. Helens, Redoubt Volcano, Mount Spurr, and Mount Pinatubo.

Ewert, J.W., Lockhart, A.B., Marcial, S., and Ambubuyog, G., this volume, Ground deformation prior to the 1991 eruptions of Mount Pinatubo.

Lockhart, A.B., Murray, T.L., Ewert, J.E., LaHusen, R.G., and Hadley, K., 1992a, A USGS equipment cache for responding to volcanic crises: Eos, Transactions, American Geophysical Union, v. 73, no. 43, p. 68.

Lockhart, A.B., Murray, T.L., and Furukawa, B., 1992b, Operating low-power telemetry networks in severe environments, *in* Ewert, J.W., and Swanson, D.A., eds., Monitoring volcanoes: Techniques and strategies used by the staff of the Cascades Volcano Observatory, 1980–1990: U.S. Geological Survey Bulletin 1966, p. 25–36.

Marcial, S.S., Melosantos, A.A., Hadley, K.C., LaHusen, R.G., and Marso, J.N., this volume, Instrumental lahar monitoring at Mount Pinatubo.

Miller, C.D., Ewert, J.E., and Lockhart, A.B., 1993, The USGS/USAID Volcano Disaster Assistance Program: The next five years, U.S. Geological Survey Open-File Report 93–379, 7 p.

Murray, T.L., 1988, A system for telemetering low-frequency data from active volcanoes: U.S. Geological Survey Open-File Report 88–0201, 28 p.

———1992, A low-data-rate digital telemetry system, *in* Ewert, J.W., and Swanson, D.A.,eds., Monitoring volcanoes: Techniques and strategies used by the staff of the Cascades Volcano Observatory, 1980-1990: U.S. Geological Survey Bulletin 1966, p. 11–23.

Murray, T.L., Power, J.A., March, G.D., and Marso, J.N., this volume, A PC-based realtime volcano-monitoring data-acquisition and analysis system.

National Oceanic and Atmospheric Administration, 1984, GOES data collection system and data processing system: Washington, D.C., National Environmental Satellite Data and Information System User Interface Manual, 132 p.

Rogers, J.A., Maslak, Sam, and Lahr, J.C., 1980, A seismic electronic system with automatic calibration and crystal reference: U.S. Geological Survey Open-File Report 80–324, 130 p.

Sabit, J.P., Pigtain, R.C., and de la Cruz, E.G., this volume, The west-side story: Observations of the 1991 Mount Pinatubo eruptions from the west.

Ward, P.L., and Gregersen, S., 1973, Comparison of earthquake locations determined with data from a network of stations and small tripartite arrays on Kilauea volcano, Hawaii: Bulletin of the Seismological Society of America, v. 63, no. 3, p. 679–711.

APPENDIX 1. NETWORK SYNOPSIS

A synopsis of the first telemetered Mount Pinatubo seismic network, April 25 to June 15, from A.B. Lockhart's field notes. Bracketed remarks are not from the notes.

LIL (**LIL**y Hill; LILZ) 15° 11.62 N., 120° 31.81 E.

4/25. Installed at Clark Air Base on Lily Hill adjacent to water tank. L–c4 buried 1 m deep in soil, set into Kwik-set pad. **Attenuation 30 dB.**

5/1. LIL removed.

PIE (Mount **PI**natubo **E**ast; PIEZ) 15° 10.00 N., 120° 25.73 E.

4/30. Installed in elephant grass on ridge 6 km east of summit. L–4 buried 1 m deep in soil, set into Kwik-set pad. **Attenuation 18 dB.**

6/15. PIE destroyed.

UBO (Mount Pinat**UBO**; UBOZ) 15° 08.33 N., 120° 21.76 E.

[So-named because map station location is in the center of the last letter ("o") of "Mount Pinatubo" label on the 1:50,000-scale DMA map of the Mount Pinatubo quadrangle].

5/1. Installed on geothermal drillpad 1 km east of summit. L–4 buried in rocky fill, set into Kwik-set pad. **Attenuation 36 dB.**

5/31. Site visit to install tiltmeter.

6/11. Telemetry signal lost, 2200 local time.

6/12. At 0341, explosion signal recorded on PIEZ record; pyroclastic current apparently dusted UBO, noted during 0600 flyby, signal still weakly transmitting. Antenna cable damaged?

BUR (**BUR**gos; BURZ) 15° 13.41 N., 120° 15.56 E. (May be as far west as 120° 15.41 E.).

5/5. Installed near Barangay Burgos on ridge overlooking bridge. L–4 buried 1 m deep in soil, set into Kwikset pad. Road access. **Attenuation 24 dB.**

5/7. Receiving BURZ at Clark.

6/11. Bunker Hill vandalized, telephone link to Clark lost.

PPO (**P**atal **P**int**O**; PPOZ, PPOE, PPON) 15° 10.95 N. 120° 20.50 E.

5/3. Installed in cinderblock guardhouse 2 km N of summit. in Aeta village (30 inhabitants). Road access. L–4 sits on floor inside structure, **Attenuation 36 dB.**

5/6. Receiving data at Clark.

5/12. Installed 3-component L–22D.

PPOZ, Attenuation 24 dB.

PPON, Attenuation 30 dB.

PPOE, Attn 30 dB.

5/30. Site visit, gains reset: PPOZ, **Attenuation 30 dB**.

6/11. Bunker Hill vandalized, telephone link to Clark lost.

BUG (Sitio **BU**a**G**; BUGZ) 15° 05.05 N. 120° 16.90 E.

5/11. Installed on grassy hilltop overlooking Sitio Buag, 10 km southwest of summit, L–4 buried 1 m deep in soil, set into Kwikset pad. **Attenuation 18 dB.**

5/14. Signal first repeated to base (through GRN).

5/24. Site visit; VCO, radio struck by lightning. Radio destroyed, VCO burned but reparable.

5/28. Site reinstalled: **Attenuation 18 dB.**

6/15. All exposed BUG components damaged by heat. Lose transmissions.

GRN (mt. ne**GR**o**N**; GRNZ) 15° 05.67 N., 120° 21.98E.

[Note: Installation notes on GRN were very sketchy, and the attenuation setting was never written down during any of the trips made to the site. After the eruption one of us, Gemme Ambubuyog, used the GRNZ VCO calibration signals recorded on seismograms to suggest an attenuation setting of 24 dB. The same analysis corroborated the relative attenuation settings for UBOZ, PIEZ and LILZ.]

5/14. Install repeater and L-4-based seismic station on peak of Negron, several kilometers southeast of Mount Pinatubo. Site is nearly always clouded in and is inaccessible on foot but has a helicopter pad on the summit. Off the pad, the peak of Negron is brushy and knife-edged.

5/16. VO-COM repeater installed at Negron.

[Early June: Geophone disinterred and reburied 1.5 m deep.]

6/15. GRN annihilated.

CAB (**C**lark **A**ir **B**ase; CABZ) 15° 10.76 N., 120° 30.61 E.

5/1. Installed on Clark Air Base on hill 200 m from water tank. Site is in concrete hut with L–4 buried adjacent to WW II concrete emplacement about 10 m away. Location is compromise between security (afforded by concrete hut), radio shot to apartment (afforded by hill), and noise of water tank/pump. It is hoped that the water tank noise will not be overwhelming.

5/5. Attenuation reduced from 30 dB to 24 dB 1245 local time.

[During the latter part of May and early June, CAB was offline due to battery problems. Because of noise from water tank, data are of poor quality anyway; consequently repair of the problem is assigned a low priority.]

6/11. Station brought back online with battery change. Now that the base has been evacuated, there is no water pump activity and station is quieter.

A PC-Based Real-Time Volcano-Monitoring Data-Acquisition and Analysis System

By Thomas L. Murray,[1] John A. Power,[1] Gail Davidson,[2] and Jeffrey N. Marso[1]

ABSTRACT

A system of networked personal computers provided real-time data acquisition and analysis during the 1991 eruption of Mount Pinatubo. The computers collected data telemetered from seismometers, tiltmeters, and lahar detectors. The seismic network provided earthquake location and magnitude information from the digitized event data, Real-time Seismic Amplitude Measurements and Seismic Spectral Amplitude Measurements. The seismic amplitude, seismic spectral, tiltmeter and lahar-detector data were processed automatically. Processing the seismic-event data required transferring the data from the acquisition computer to an analysis computer and then timing the events interactively. Data were sent via modem to other parties for further analysis. The modem connection also enabled scientists in the United States to troubleshoot the system. As a result of the Pinatubo experience, numerous improvements have been made to the system.

INTRODUCTION

The availability of low-cost, increasingly powerful microcomputers provides volcanologists with cost-effective tools for acquiring and analyzing geophysical data in real time. Lee (1989) described a system that used a personal computer to digitize signals from up to 16 seismic stations and store the records of the events. The digitized waveforms for the events were transferred via a Local-Area Network (LAN) to a second computer where they were timed and located. March and Power (1990) used such a system networked to a UNIX workstation to acquire and process seismic data during the 1989–90 Redoubt, Alaska, eruptions. Murray (1992a) described a low-data-rate telemetry system whose data can be received by a laptop computer with a 300-baud modem. This system enabled real-time acquisition and display of data from tiltmeters, temperature

[1] U.S. Geological Survey.

[2] Alaska Volcano Observatory, Alaska Division of Geological and Geophysical Surveys, 794 University Ave., Suite 2001, Fairbanks, AK 99709.

sensors and similar instruments. Using the data management and plotting program BOB (Murray, 1990a,b, 1992b), these data could be compared in real time with each other and seismicity as recorded by a Real-time Seismic Amplitude Measurement system (RSAM) (Endo and Murray, 1991). A mudflow-detection system developed at the Cascades Volcano Observatory (R.G. Lahusen, oral commun., 1990) during the 1989–90 Redoubt, Alaska, eruption also used a personal computer to acquire and plot the data in real time.

Together, the above items can constitute the core of a volcano observatory's data-acquisition and analysis system. However, as stand-alone systems, they do not foster interaction between different investigators or correlation of different data sets. The importance of being able to compare different data sets in near real time cannot be overstated. In reference to predictive methods used at Mount St. Helens, Swanson and others (1983) stated "the accuracy of our predictions depends on interactive use of all data by cooperating geophysicists, geologists, and geochemists." Therefore, it was necessary to integrate the stand-alone systems listed above into a system that would allow any of the computers being used for analysis to access any of the data. This was accomplished by providing paths to move data automatically from the different acquisition computers to the networked computers.

These paths consisted of both hardware (parallel-to-serial converters, the LAN, serial-port buffers) and software (data-conversion programs, modifications to existing programs). With the data stored on a network computer, each data-analysis computer, in addition to performing its specialty, can analyze the data via the network. Though all analysis could have been done on a single computer, we found that it was better to use three computers because more people could work simultaneously. One computer was dedicated to timing, locating, and archiving earthquakes. Between June 5 and June 12, 1991, this computer was in constant use at Mount Pinatubo because of the large number of earthquakes recorded. The other two computers shared the tasks of viewing the other data, report writing, outside communication, and system maintenance. Together, the computers formed the real-time volcano-monitoring data-

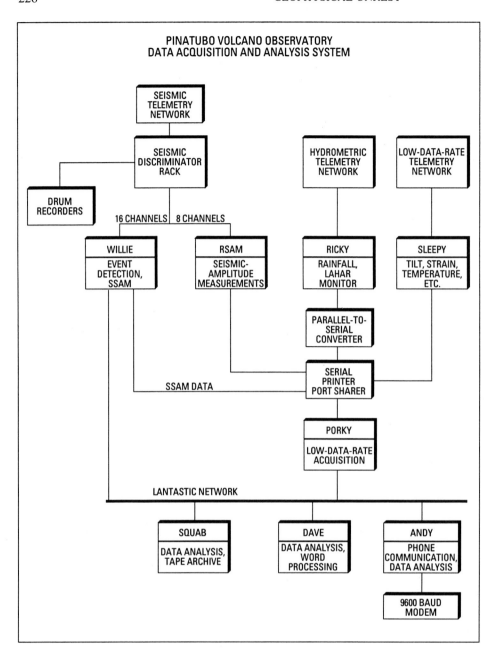

Figure 1. The Pinatubo Volcano Observatory's computer-based data-acquisition and analysis system. This system is described in detail in the text.

acquisition and analysis system used by the Pinatubo Volcano Observatory (PVO).

This report describes the system that was used during the 1991 Mount Pinatubo eruptions. As a result of our experience at Mount Pinatubo, new programs have been written to further speed the processing and analysis of data. These improvements will be summarized. We wish to emphasize that any use of trade, product, or firm names is for descriptive purposes only and does not imply endorsement by the U.S. Government.

HARDWARE CONFIGURATION

Figure 1 shows the hardware configuration for the data acquisition and analysis system used at PVO. Four

computers acquired the data. WILLIE, an IBM-AT compatible acquired the seismic-event and Seismic Spectral Amplitude Measurement (SSAM) data. RICKY and SLEEPY, both IBM-XT compatible computers, acquired data from the lahar-detector and tiltmeter networks, respectively. RSAM data were collected by a Tandy Model #100 computer (Endo and Murray, 1991).

At periodic intervals, the acquisition computers transmitted their data (except seismic-event data) via a print sharer to the serial port of PORKY, an IBM-AT compatible computer, where it was available for analysis. The SSAM data were transmitted every minute, tiltmeter and RSAM transmitted data every 10 min, and lahar data were transmitted whenever they were received from the field. The print sharer replaced a special circuit that allowed multiple devices to share a single serial port (Murray, 1992b). Though designed to allow multiple computers to share a single printer, in this configuration the print sharer buffered data sent by each of the acquisition computers and allowed each computer's data to be sent in turn to PORKY. Using the print sharer eliminated the need to assign each device its own transmission window during which it could transmit data without interfering with the other devices.

Three computers were used for data analysis. They accessed data on WILLIE and PORKY through a 2 mega-bit-per-second peer-to-peer network (Lee and others, 1989). All computers on the network were configured as servers, enabling any computer to access data or programs on any other computer. All data and programs moved between machines through the network, almost eliminating the need for the floppy disk drives, which might have quickly failed as ash from the eruption infiltrated every piece of equipment. Though each computer had a specialized task, each also served as a backup in the event that another failed. SQUAB, an 80386-based IBM-compatible computer, was used primarily for earthquake processing, analysis, and archiving. DAVE and ANDY, both IBM-AT compatible computers, were used for analysis and plotting of RSAM, SSAM, tiltmeter and lahar-detector data. They also were used for writing observatory communications and volcano advisories. A 9600-baud modem attached to ANDY provided access to the data-analysis computers from distant locations. The software package Remote (DCA Crosstalk Communications, Rosewell, Ga.) simplified connecting to the system. With Remote, the keyboard and monitor at the remote site behaved as though they were directly connected to ANDY. Commands entered on the remote keyboard cause ANDY to respond as if they were typed on its keyboard. Staff at the Cascades and Alaska Volcano Observatories were able to acquire data as well as troubleshoot the system from their offices in the United States.

USING LOCAL TIME INSTEAD OF GREENWICH MEAN TIME

At the time of the system's installation, all clocks were set to local time, not Greenwich Mean Time (G.m.t.). It was quickly discovered that the seismic acquisition program MDETECT (Tottingham and others, 1989) was time-tagging the SSAM data to the PC's clock while the event data were time-tagged to the PC's clock plus 8 hours (the time correction needed for G.m.t. in California, where the program had been written). It was several weeks before this discrepancy was resolved.

Unfortunately, at the same time that the time discrepancy was resolved, it was decided that all seismic data would be recorded in G.m.t. This was a mistake. Having to make the mental adjustments between G.m.t. and local time when correlating seismic events with other observations

and talking with local officials only adds to the confusion inherent to volcanic crises. The usual argument for using G.m.t. is that the data can be easily compared with data sets collected by other groups worldwide. These comparisons are done so seldom during a crisis that this argument cannot be justified. After the crisis, a simple program can adjust the time to G.m.t. if necessary. Also, the Philippines does not seasonally adjust its local time as the United States does with daylight savings. This factor eliminates the argument to use G.m.t. in order to have a consistent time base.

SEISMIC-DATA ACQUISITION AND ANALYSIS

Six short-period vertical seismic stations and one three-component seismic station radio-telemetered their data to PVO via a standard analog seismic telemetry system. The received data were recorded in four different ways: (1) analog drum recorders, (2) digitized waveforms of discrete events, (3) RSAM, and (4) SSAM. Each method has its limitations and strengths. The character of the seismicity, which may change through time, determined which technique was most useful. By using all four techniques, we were able to monitor seismic activity throughout the eruption. Monitoring began in early May, when there were only a few events per day, and continued through June 15, when only one station continued to transmit, and its record was largely saturated.

ANALOG DRUM RECORDERS

Signals from at least two seismometers were always being recorded on analog drum recorders. Despite advances in computer-based data acquisition, drum recorders were still the single most important technique for recording the seismic data. A glance at the drums gave a quick overall view of the current level of seismicity and its character. By displaying the records on a wall, one could qualitatively assess the change in seismicity through time. The records were also used to interpret data from other techniques, which in one way or another measure only a portion of the signal. Digitized events do not measure extended periods of tremor or constant activity. RSAM and SSAM can be contaminated with telemetry or cultural noise. The analog records, even when saturated, provided a qualitative context in which to interpret the other, more quantitative methods.

DIGITIZED EVENTS

Seismic-event data were acquired and processed on the computer WILLIE using the method described by Lee and others (1989). The program MDETECT (Tottingham and others, 1989) digitized seismic signals at 100 samples per second per channel using a 12-bit analog-to-digital converter. When MDETECT detected an event, the digitized

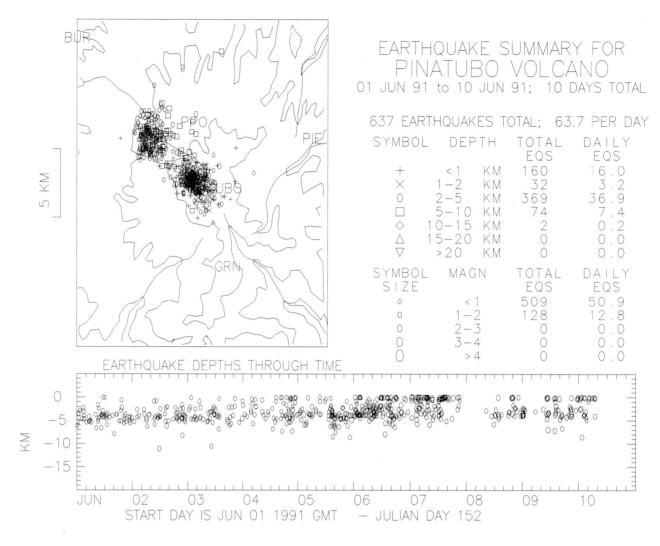

Figure 2. Plot produced by the program SEISPLOT (March and Murray, 1992). In the upper left-hand corner is an epicentral map with the seismic stations noted. A time-depth plot is shown along the bottom, and in the upper right-hand corner is a table summarizing the earthquake activity with respect to depth and magnitude.

data for the event were written to disk. Periodically the event data were transferred to SQUAB, where the events where timed with the program PCEQ (Valdes, 1989) and locations and magnitudes calculated by using HYPO71PC (Lee and Valdes, 1985; 1989). After being timed and located, the data were archived to tape.

Two programs were used to plot the data. The program AcroSpin (Parker, 1990) plotted the hypocenters in a three-dimensional format that could be rotated about any axis. An early version of the program SEISPLOT (March and Murray, 1992) provided a one-page summary of earthquake activity. The summary combined an earthquake epicenter plot, a time-depth plot, and a table summarizing the number, magnitude, and depth of the earthquakes recorded (fig. 2).

Processing the digitized events was the most time-consuming element in the entire system. Because of the large number of events recorded, SQUAB was in constant use

from about June 5 to June 12, either transferring and archiving events or timing and locating them. Also, lack of integration of the timing program and the location program made it difficult to check hypocenter solutions for accuracy or to re-time them. Despite these problems, the system worked remarkably well, seldom falling behind by more than 12 hours.

RSAM

The RSAM data were transferred to PORKY via the print sharer (fig. 1). A batch file was written so that a single command would cause the program BOB to quickly plot the latest RSAM data. During the period June 12 through early July (when there were insufficient seismic stations to locate events), RSAM was the most useful quantitative measure of seismicity.

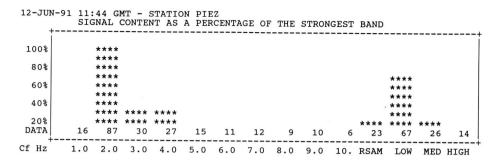

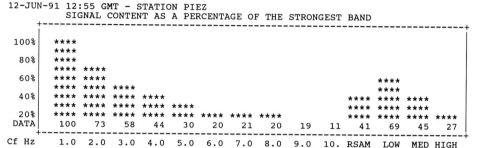

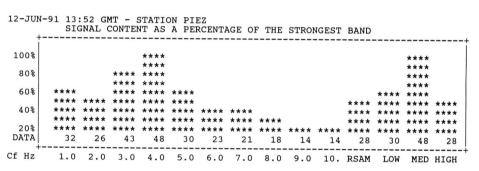

Figure 3. Example plot of SSAM data for station PIEZ at selected times produced by the program VU_FFT. The first 10 columns display 1-Hz-bandwidth data centered about the frequencies (cf) shown. The columns labeled RSAM, LOW, MED, and HIGH display data from wider frequency bands: 0.7–10.0 Hz, 1.5–3.0 Hz, 3.0–4.5 Hz, and 4.5–6.0 Hz, respectively. The data are arbitrary units of the signal amplitude in each frequency band.

SSAM

The version of MDETECT used at Pinatubo had been modified to calculate 1-minute averages of the spectral content of the seismic signals in 16 selected frequency bands (Rogers and Stephens, 1991). One-minute SSAM data were both stored on WILLIE and sent to PORKY via the print sharer (fig. 1). WILLIE's printer port, instead of its serial port, was used to transmit the data because the transmission time was faster and less likely to cause timing problems with MDETECT.

SSAM data were viewed with two programs (Marso and Murray, 1991). The first, VU_FFT, was a simple GWBASIC program that produced a bar-graph representation of the data for a specified time and station (fig. 3). VU_FFT was useful to see the spectral content of a specific event or to track shifts in spectral content during periods of extended tremor or during mudflows.

The program Surfer (Golden Software, Boulder, Colo.) was used to produce daily contour plots for each station (fig. 4). The plots were placed on a wall to produce a long-term view of SSAM. Each contour plot required about 15 minutes to produce. The process was delicate; a single mis-take would doom the plot with little or no indication why. However, the dramatic changes in the spectral content of the seismic signals could be clearly seen on the analog drum recorders (Power and others, this volume). This, in addition to the occasional difficulties in producing the plots, was the reason that SSAM was not a major factor in forecasting the Mount Pinatubo eruption.

TILTMETER DATA-ACQUISITION AND ANALYSIS

Data transmitted from two tiltmeters were logged on SLEEPY (Ewert and others, this volume). At 10-min intervals, the data were transmitted to PORKY via the print sharer (fig 1). Once PORKY received the data, the program BOB was used to view and compare the data with RSAM (fig. 5).

Though the telemetry system was designed to transmit data from such instruments as strainmeters, gas detectors, and ground-temperature sensors in addition to tiltmeters, lack of time prevented their installation. If installed, their

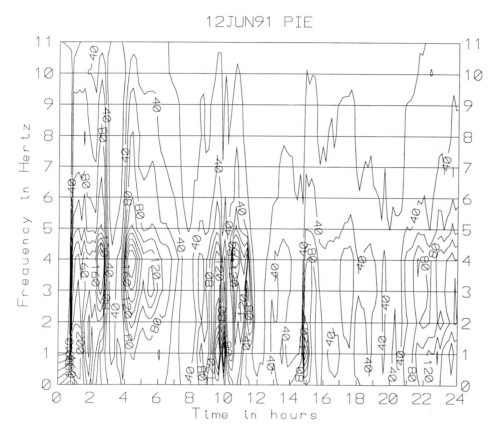

Figure 4. SSAM contour plot of station PIE for June 12, 1991, produced by the program Surfer. The x-axis represents time, and the y-axis represents the different frequency bands. The contours represent the amplitude of the seismic signal through time in the various bands, in the same manner that contours represent elevation on a standard topographic map. The numerical values of the contours are arbitrary units of the signal amplitude.

data would have been processed and available for analysis in the same manner as the tiltmeters'.

LAHAR DETECTION AND ANALYSIS

The lahar-detection system was first installed at PVO as a stand-alone system with no connection to other computers (Hadley and Lahusen, 1991). RICKY received, stored, and plotted the data. Later, RICKY was programmed to send the received data out its printer port and through a parallel-to-serial converter to PORKY via the print sharer (fig. 1). With the data on PORKY, analysis was done with BOB and plotted in conjunction with the RSAM and SSAM data.

IMPROVEMENTS TO THE SYSTEM

From the experience gained during the Mount Pinatubo crisis, several improvements have been made to the system. Among them are the following:

1. MDETECT has been replaced with a new version of XDETECT, a program originally described by Tottingham and Lee (1989). The most recent XDETECT, version 3.18 (Rogers, 1993) provides for subnetwork triggering, a continuous recording capability, recording the data in SUDS format (Ward, 1989), and sampling up

to 128 channels with a custom multiplexer circuit (Ellis, 1989).

2. New programs simplify the entire process of transferring the seismic-event data to the data-analysis computer, determining phase arrival times and coda lengths, and storing the phase and hypocenter data. The software also integrates the programs that pick the arrivals times and locate the events, so that hypocenters can be checked interactively. Many of the problems resulting from the volume of earthquakes recorded during the crisis would have been avoided if these programs had been available.

3. The general seismic and geophysical time-series data-plotting program, QPLOT, has been modified to run on an IBM-PC compatible (Murray and others, 1993). Whereas SEISPLOT provided only a single template with which to display the hypocentral information, PC-QPLOT is a flexible plotting package that allows a variety of different plotting templates to be constructed.

4. A program, VOLQUAKE, has been written to display simultaneously both an epicentral map and an associated cross-sectional plot of earthquakes through time (Hoblitt, Mori, and Power, this volume). This program combines many of the features of SEISPLOT and AcroSpin in an interactive, mouse-driven program.

5. Two programs have been written to plot SSAM data more quickly and easily than was possible with Surfer. One, SSAM_VU, can plot between 1 hour and 7 days of data. This program is described in detail by Power and

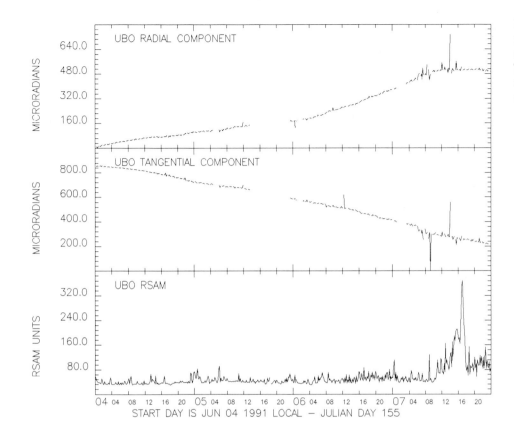

Figure 5. Plot produced by the program BOB showing RSAM and tilt data from station UBO. This plot shows the slowing in radial tilt that coincided with the seismic swarm on June 7.

others (this volume). Instead of using contours, it indicates amplitude through different colors on the computer monitor and a varying-density mesh on a laser printer to approximate gray scales. The other program, SCANS-SAM (John Rogers, USGS, written commun., 1992), can quickly scan 6 hours of data using SSAM files stored on WILLIE. It also uses color to indicate signal amplitude.

CONCLUSIONS

The system described here performed well during the 1991 Mount Pinatubo crisis, as it provided information that was essential in issuing the eruption forecasts. The only hardware component that failed was a hard-disk drive that began to exhibit erratic behavior in early July. Before it failed completely, it was replaced by a new disk that had been purchased in Manila.

The lack of hardware failures is even more remarkable, considering that the entire system was moved four times; (1) it accompanied the initial USGS response team from the United States to the Maryland Street apartment on Clark Air Base, (2) it was moved from the apartment to the Dau Complex on Clark Air Base as unrest intensified, (3) from Dau it was moved to Clark BaseOps just after the climactic eruption, and (4) in November 1991, the system was moved off Clark Air Base to Quezon City. All moves on Clark Air Base were accomplished in a matter of hours.

Many of the limitations of the system deployed in 1991 have been addressed; the current system has been streamlined. Also, the availability of ever more powerful computers at ever decreasing prices will continue to add to the speed and capability of the system such that more sophisticated analytical programs are feasible. The system provides a model for future systems, both at established, permanent observatories or, as in the case of Mount Pinatubo, in response to an unmonitored reawakening volcano.

ACKNOWLEDGMENTS

The system described in this report is the result of integrating the work of several different teams. The authors especially thank Willie Lee, U.S. Geological Survey, for his perseverance in the development and documentation of a PC-based seismic-event detection system; John Rogers, U.S. Geological Survey, for his assistance in modifying the program MDETECT to transmit SSAM data out the printer port; and John Lahr, U.S Geological Survey, for his suggestion of using a print-sharing device to buffer the low-frequency data transmitted to PORKY. We also acknowledge the assistance and patience of the Escuela Politecnica Nacional, Quito, Ecuador, and INGEOMINAS, Manizales and Pasto, Colombia. These groups patiently worked with early versions of the system and provided us with feedback regarding what worked well and what did not.

We also thank the Pinatubo Volcano Observatory staff; it was they who actually made the system work.

REFERENCES CITED

Ellis, J.O., 1989, Expanding the input multiplexer for the Data Translation, Inc. Model DT2821 analog-to-digital converter: U.S. Geological Survey Open-File Report 89–201, 5 p.

Endo, E.T., and Murray, T.L., 1991, Real-time seismic amplitude measurement (RSAM): A volcano monitoring and prediction tool: Bulletin of Volcanology, v. 53, p. 533–545.

Ewert, J.W., Lockhart, A.B., Marcial, S., and Ambubuyog, G., this volume, Ground deformation prior to the 1991 eruptions of Mount Pinatubo.

Hadley, K.C., and Lahusen, R.G., 1991, Deployment of an acoustic flow-monitor system and examples of its application at Mount Pinatubo, Philippines: Eos, Transactions, American Geophysical Union, v. 72, no. 44, supplement, p. 67.

Hoblitt, R.P., Mori, J., and Power, J.A., this volume, Computer visualization of earthquake hypocenters.

Lee, W.H.K., ed., 1989, Toolkit for seismic data acquisition, processing and analysis: El Cerrito, Calif., Seismological Society of America, International Association of Seismology and Physics of the Earth's Interior Software Library, v. 1, 284 p.

Lee, W.H.K., Tottingham, D.M., and Ellis, J.O., 1989, Design and implementation of a PC-based seismic data acquisition, processing, and analysis system, in Lee, W.H.K., ed., Toolkit for seismic data acquisition, processing and analysis: El Cerrito, Calif., Seismological Society of America, International Association of Seismology and Physics of the Earth's Interior Software Library, v. 1, p. 21–46.

Lee, W.H.K., and Valdes, C.M., 1985, HYPO71PC: A personal computer version of the HYPO71 earthquake location program: U.S. Geological Survey Open-File Report 85–749, 30 p.

_____1989, User manual for HYPO71PC, in Lee, W.H.K., ed., Toolkit for seismic data acquisition, processing and analysis: El Cerrito, Calif., Seismological Society of America, International Association of Seismology and Physics of the Earth's Interior Software Library, v. 1, p. 203–236.

March, G.D., and Murray, T.L., 1992, VOLPLOT; A PC-based program for viewing Cook Inlet Volcano-seismic data: U.S. Geological Survey Open-File Report 92–560–A, 6 p.

March, G.D., and Power, J.A., 1990, A networked computer configuration for seismic monitoring of volcanic eruptions: U.S. Geological Survey Open-File Report 90–422, 19 p.

Marso, J.N., and Murray, T.L., 1991, Real-time display of seismic spectral amplitude measurements: Examples from the 1991 eruption of Pinatubo volcano, Central Luzon, Philippines: Eos, Transactions, American Geophysical Union, v. 72, no. 44, supplement, p. 67.

Murray, T.L., 1990a, A user's guide to the PC-based time-series data-management and plotting program BOB: U.S. Geological Survey Open-File Report 90–56, 53 p.

_____1990b, An installation guide to the PC-based time-series data management and plotting program BOB: U.S. Geological survey Open-File Report 90–634, 25 p.

_____1992a, A low-data-rate digital telemetry system, in Ewert, J.W., and Swanson, D.A., eds., Monitoring volcanoes: Techniques and strategies used by the staff of the Cascades Volcano Observatory, 1980–90: U.S. Geological Survey Bulletin 1966, p. 11–24.

_____1992b, A system for acquiring, storing, and analyzing low-frequency time-series data in near-real time, in Ewert, J.W., and Swanson, D.A., eds., Monitoring volcanoes: Techniques and strategies used by the staff of the Cascades Volcano Observatory, 1980-90: U.S. Geological Survey Bulletin 1966, p. 37-43.

Murray, T.L., Power, J.A., and Klein, F.W., 1993, PC_QPLOT, an IBM-PC compatible version of the earthquake plotting program QPLOT: U.S. Geological Survey Open-File Report 93–22, 17 p.

Parker, D.B., 1990, User manual for AcroSpin, in Lee, W.H.K., ed., Toolkit for plotting and displaying seismic and other data: El Cerrito, Calif., Seismological Society of America, International Association of Seismology and Physics of the Earth's Interior Software Library, v. 2, p. 119–164.

Power, J.A., Murray, T.L., Marso, J.N., and Laguerta, E.P., this volume, Preliminary observations of seismicity at Mount Pinatubo by use of the Seismic Spectral Amplitude Measurement (SSAM) system, May 13-June 18, 1991.

Rogers, J.A., 1993, XDETECT version 3.18 user's reference guide: U.S. Geological Survey Open-File Report 93–261, 26 p.

Rogers, J.A., and Stephens, C.D., 1991, SSAM: a PC-based seismic spectral amplitude measurement system for volcano monitoring: Seismological Research Letters, v. 62, p. 22.

Swanson, D.A., Casadevall, T.J., Dzurisin, D., Malone, S.D., Newhall, C.G., and Weaver, C.S., 1983, Predicting eruptions at Mount St. Helens, June 1980 through December 1982: Science, v. 221, no. 4618, p. 1369–1376.

Tottingham, D.M., and Lee, W.H.K., 1989, User manual for XDETECT, in Lee, W.H.K., ed., Toolkit for seismic data acquisition, processing and analysis: El Cerrito, Calif., Seismological Society of America, International Association of Seismology and Physics of the Earth's Interior Software Library, v. 1, p. 89–118.

Tottingham, D.M., Lee, W.H.K., and Rogers, J.A., 1989, User manual for MDETECT, in Lee, W.H.K., ed., Toolkit for seismic data acquisition, processing and analysis: El Cerrito, Calif., Seismological Society of America, International Association of Seismology and Physics of the Earth's Interior Software Library, v. 1, p. 49–88.

Valdes, C.M., 1989, User manual for PCEQ, in Lee, W.H.K., ed., Toolkit for seismic data acquisition, processing and analysis: El Cerrito, Calif., Seismological Society of America, International Association of Seismology and Physics of the Earth's Interior Software Library, v. 1, p. 175–201.

Ward, P.L., 1989, SUDS: Seismic Unified Data System: U.S. Geological Survey Open-File Report 89–188, 123 p.

A Comparison of Preeruption Real-Time Seismic Amplitude Measurements for Eruptions at Mount St. Helens, Redoubt Volcano, Mount Spurr, and Mount Pinatubo

By Elliot T. Endo,[1] Thomas L. Murray,[1] and John A. Power[1]

ABSTRACT

Since 1985 we have had the opportunity to collect real-time seismic amplitude measurement (RSAM) data for preeruption periods at four different volcanoes. In this paper we introduce a technique to compare RSAM data corresponding to different magmatic eruptions. We normalized RSAM data and then used commercially available curve-fitting software for a personal computer to characterize RSAM data and simplify comparison. We found that the preeruption normalized RSAM data for three Mount St. Helens eruptions and one Mount Pinatubo dome-building eruption were best fit with an exponential equation of the type:

$$NORMALIZED\_RSAM\_COUNT\,(t) = a + b \exp\,(-\,t\,/\,c)$$

where $NORMALIZED\_RSAM\_COUNT\,(t)$ corresponds to a normalized RSAM value at time t in decimal hours, and a, b, and c are parameters determined by the curve fitting program. While the exponential function is not the only type of equation to provide a satisfactory fit to some RSAM data, it is a convenient function to describe the similar patterns of increasing RSAM value and suggests that similar preeruption processes occurred in most of the dome-building eruptions we studied. Where good observational information was available, we know that dome-building eruptions followed periods of exponential increase in volcano-tectonic or "B" type earthquakes and a decline in seismic activity. We speculate that these effects are related to the migration of magma from depth to a level just below the surface, followed by less seismically active migration to the surface.

Results of curve fitting for normalized RSAM data for explosive eruptions at Redoubt Volcano, Mount Spurr, and Mount Pinatubo were mixed. Of five eruptions studied that resulted in explosive activity, three showed premonitory exponential-like increases in long-period earthquakes and brief periods of reduced or constant seismic activity shortly before the eruptions. The June 1992 eruption of Mount Spurr differed by having an explosive eruption during an exponential increase in volcano-tectonic seismicity. A high background seismic-noise level, numerous small, explosive eruptions prior to the June 15 paroxysmal eruption at Mount Pinatubo, and loss of some RSAM data precluded meaningful analysis and interpretation of RSAM data leading up to the paroxysmal eruption.

The andesitic dome-building eruption of Mount Pinatubo on June 7, 1991, and the explosive eruptions of volatile-rich andesite magma at Redoubt Volcano in December of 1989 and at Mount Spurr in 1992 have shorter durations for exponential-like increases in seismicity than do high-SiO_2 dome-building eruptions. Estimated durations for the rapid increases in seismicity at Redoubt Volcano, Mount Pinatubo, and Mount Spurr were typically less than 10 hours, or about one order of magnitude shorter in duration than dome-building eruptions at Mount St. Helens.

INTRODUCTION

Real-time seismic amplitude measurement (RSAM) is a volcano-monitoring technique developed at the Cascades Volcano Observatory in Vancouver, Wash., that has been in use since 1985 (Endo and Murray, 1991). Since that time, preeruption RSAM data have been collected at Mount St. Helens, Redoubt Volcano, and Mount Spurr in the United States and at Mount Pinatubo in the Philippines. The RSAM technique has also been employed at the Mammoth Lakes area in California, at volcanoes on Hawaii, and in a number of countries in Central America and South America. The RSAM technique is a systematic electronic and computer method that provides a continuous measurement of average absolute seismic amplitudes for any number of seismic stations desired. Limitations of the technique are the number of seismic stations available for recording, electronics, and the computer hardware available. A potentially more serious limitation is that this simple technique does not discriminate

---

[1] U.S. Geological Survey.

between types of volcanic earthquakes, teleseismic events, regional earthquakes, wind, and other noise. Unlike the seismic spectral amplitude measurement (SSAM) technique, where user-defined spectral bands are monitored (Power and others, this volume; Stephens and others, 1994), all seismic signals are averaged and recorded. Briefly, the RSAM technique uses an analog-to-digital converter to convert analog seismic signals to a digital form suitable for computer storage and analysis. Sampled at a 50-Hz sampling rate, RSAM data are first averaged over 1 min and then averaged for a 10-min window for storage in a computer file. Each digital count represents 20 mV of analog seismic signal; thus, an average RSAM value is directly proportional to absolute average voltage of a seismic signal. The RSAM value is also proportional to the average ground velocity at the seismometer site. The digital form of the data and computer graphics provides a convenient method for near-real-time review of relative seismic activity. Details of the RSAM technique are described by Endo and Murray (1991) and Murray and Endo (1989).

Endo and others (1990) suggested that the migration of dacitic magma at Mount St. Helens was closely associated with the increase in RSAM counts, or the increase in the average amplitude of preeruption seismic signals (primarily volcano-tectonic events or "B" type earthquakes). Average ascent velocities were estimated for time periods defined by RSAM curves for two dome-building eruptions in 1986; however, there was no rigorous analysis of the RSAM data at that time. Assuming a linear relation and the approximate 24- to 48-h ascent times suggested by RSAM curves, Endo and others (1990) calculated a range of possible linear ascent velocities. Part of the problem in identifying the ascent time of magma was timing the onset of an eruption. Timed photography in October 1986 provided the best evidence for the onset of a dome-building eruption at Mount St. Helens. For that eruption there was a 12-hour interval between the first peak in RSAM data and the onset of the eruption. (Endo and others, 1990, reported a 6-h interval between a second peak in RSAM data and the onset of the eruption.) For other eruptions, the eruption onset was assumed to coincide with a substantial decrease in the rate of tilt close to the lava dome. Similar intervals, 12 to 18 h, between the peak in RSAM data and a rapid decrease in the rate of tilt adjacent to the lava dome (Endo and Murray, 1991) were observed for earlier eruptions in May 1986 and May 1985, respectively. The May 1986 dome-building eruption was preceded by almost 1 month of ash emissions and explosive activity.

The eruptions of Redoubt Volcano in 1989–90 (Power and others, 1994) provided the first opportunity after the October 1986 eruption of Mount St. Helens to evaluate RSAM data for another volcano. The swarm of long-period earthquakes (Chouet and others, 1994; Lahr and others, 1994; Stephens and others, 1994) that preceded the explosive eruption on December 14, 1989, produced an RSAM record that was at first difficult to analyze and interpret. A brief 6- to 8-h exponential-like increase starting December 13, 1991, at about 1030 local time was followed by about a 12-h linear increase in average RSAM amplitude. The linear increase was followed by a 5-h period of nearly constant amplitude, a 5-h period of decreasing average amplitude, and finally eruption of tephra (Power and others, 1994). This initial eruption was followed by many others during what Powers and other referred to as a "vent-clearing phase." This vent-clearing phase was followed by a dome-building phase that began about December 21, 1989. During the early part of this dome-building episode, volcano-tectonic events dominated. Starting around December 26, long-period earthquakes (Chouet and others, 1994) and tremor were dominant contributors to a rapid increase in RSAM amplitude counts. On January 2, 1990, this rapid increase leveled off a few hours before two large tephra eruptions at 1749 and 1927 local time (R. Page, written commun., 1993). The Redoubt RSAM data set from December 25, 1989, to January 2, 1990, differs from all other data sets of this report because it coincides with a period of known dome growth. However, owing to the rapid increase in cumulative RSAM counts, the data were used as a basis for issuing warnings before the eruptions on January 2.

RSAM data for volcano-tectonic earthquakes associated with a small dome-building eruption at Mount Pinatubo on June 7, 1991 (fig. 1), and RSAM data for a long-period earthquake swarm prior to an explosive eruption on June 14, 1991, had similarities to RSAM curves for eruptions at Mount St. Helens and Redoubt Volcano. Ewert and others (this volume), using tilt and seismic data, concluded that magma had been emplaced at shallow levels and possibly extruded on June 7, 1991. Visual observation on June 8 confirmed the presence of a new lava dome. The highest values in RSAM data for June 7 were probably related to a strong steam and ash emission (Ewert and others, this volume). RSAM data preceding the climactic eruption of Mount Pinatubo on June 15 are very complex, as a result of numerous smaller explosive eruptions starting on June 12 (Pinatubo Volcano Observatory Team, 1991; Hoblitt, Wolfe, and others, this volume) and did not lend themselves to detailed analysis and comparison to Mount St. Helens except for an increase in long-period seismicity (Power and others, this volume) prior to an explosive eruption at 1309 local time on June 14. The explosive events were part of a series that began on June 12 (Wolfe and Hoblitt, this volume).

The June 27, 1992, explosive eruption of Mount Spurr provided an example of RSAM data with a short but clear buildup in volcano-tectonic seismic activity. While not conspicuous, there was a discernible period of decrease in average seismic amplitudes about an hour prior to what was reported as the onset of the eruption (Power and others, in press). The August 18 eruption at Mount Spurr showed no

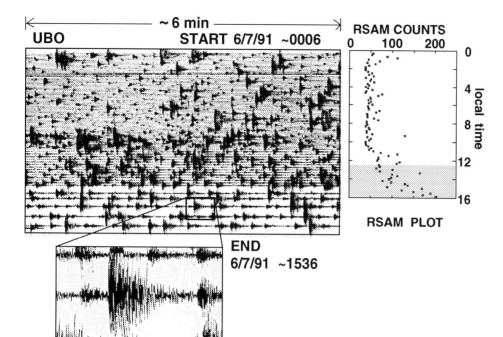

Figure 1. A 6-min-wide section of the UBO seismogram (15.5 h long) showing the increase in volcano-tectonic earthquakes associated with the emplacement of magma at shallow depth beneath Mount Pinatubo on June 7, 1991. Corresponding RSAM counts are shown on the plot to the right. The time scale for the shaded area is not the same for the seismogram because of two changes in the translation rate for the seismogram.

significant precursory increase in RSAM counts of earthquake activity, and the September 17 eruption had 3 h of low-level tremor preceding the onset of the eruption activity and a few discrete events.

The purpose of this paper is to compare preeruption RSAM data from four stratovolcanoes and to determine whether there are any characteristics common to all preeruption RSAM data. This is a preliminary examination and is not intended as a comprehensive examination of all RSAM data for every eruptive episode for volcanoes such as Redoubt (Powers and others, 1994) or Pinatubo, which have undergone numerous eruptive episodes (Hoblitt, Wolfe, and others, this volume). We attempt to account for the apparent fine differences between peaks of seismic activity and onsets of dome-building eruptions at Mount St. Helens or initial vent-clearing eruption of Redoubt Volcano in December of 1989. Our definition of an eruption for this paper requires magma breaking through the surface of the volcano. That breakthrough could be in the form of a quiet dome-building eruption or a phreatomagmatic or magmatic explosive eruption.

We do not attempt to draw any conclusions regarding the eruption prediction value of RSAM for the paroxysmal eruption on June 15 because that is done by Cornelius and Voight (this volume).

RSAM DATA AND ANALYSIS TECHNIQUE

DATA

The first preeruption RSAM data available for study were for the May 1985 dome-building eruption at Mount St.

Helens. Two subsequent dome-building eruptions at Mount St. Helens in May 1986 and October 1986 provided additional data. For the May 1985 dome-building eruption we used 1-h-average RSAM data from the GDN (Garden) seismic station. For the 1986 dome-building eruptions we used 1-h-average RSAM data from the YEL (Yellow Rock) seismic station for analysis. RSAM data for October 1986 were corrected for a minus 6-dB gain change in the field at 1300 local time) on October 21. GDN was located about 900 m north of the geometric center of the lava dome at Mount St. Helens (fig. 2) and YEL 1,200 m north.

In 1989–90, the eruption at Redoubt Volcano in Alaska provided the first pre-eruption RSAM data for a second stratovolcano (Power and others, 1994). For this study, we used 10-min-average RSAM data leading up to the initial vent-clearing phases on December 14, 1989, and the dramatic January 2, 1990, eruption. Owing to the preliminary nature of this paper, data for about 20 other episodes from January to April 1990 (Stephens and others, 1994) were not examined. RSAM data from the RED seismic station (fig. 3) were used because the automatic gain-ranging seismic amplifier at seismic station RDN presented uncertainty in the relative amplitude of seismic signals, particularly when the amplifier reverted back to normal gain (Chouet and others, 1994). Seismic station RED was located approximately 7 km south of the active crater at Redoubt Volcano.

In 1991, the activity at Mount Pinatubo, Philippines, provided preeruption RSAM data for a third volcano. For this paper, preeruption 10-min-average RSAM data from seismic stations UBO and CAB (fig. 4) were compared with RSAM data from other volcanoes. Stations UBO and CAB were located at approximately 1 km east and 17 km east,

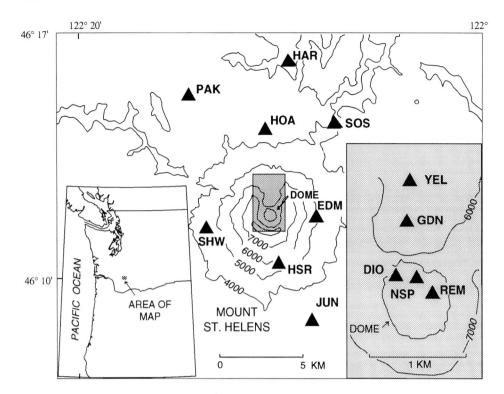

Figure 2. The Mount St. Helens area and the locations of seismic stations (black triangles). Contour interval 1,000 ft.

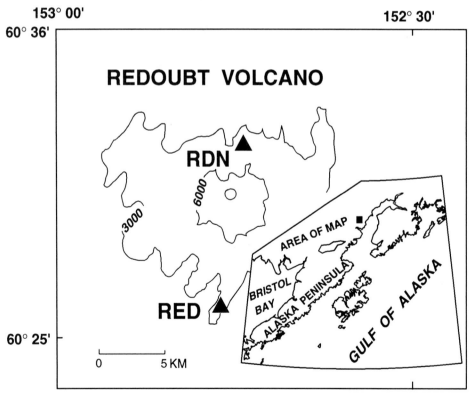

Figure 3. Redoubt Volcano and locations of the RED and RDN seismic stations (black triangles). (From Power and others, 1994). Contour interval 3,000 ft.

respectively, from the old summit (Lockhart and others, this volume). While RSAM data were available for other seismic stations (PIE, BUR, GRN, PPO, and BUG), RSAM data for the small dome-building event accompanied by a strong ash emission on June 7, 1991, were not usable at most of the other seismic stations because of the poor signal-to-noise ratio. RSAM data from CAB on June 14 provided the best example for an increase in seismicity that preceded well timed explosive eruptions.

The most recent preeruption RSAM data available for comparison were from Mount Spurr, Alaska. During 1992, there were three eruptions at Mount Spurr (Power and

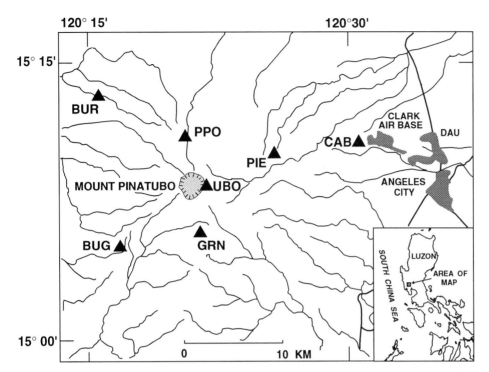

Figure 4. The Mount Pinatubo area showing locations of the UBO, CAB, and other seismic stations in operation before June 15, 1991.

others, in press; McNutt and others, in press). The first eruption, on June 27, 1992, produced a brief but gradual increase in RSAM counts. For the purpose of curve fitting, we used RSAM data that included the peak RSAM count, which was recorded about 3 h after the onset of the eruption (Power and others, in press). For Mount Spurr RSAM analysis, 10-min-average data were used for station BGL, located 7.5 km from the Crater Peak vent (fig. 5). RSAM data from CPK, located 400 m from the active vent, were not usable because of a gain-ranging problem with the amplifier.

The preeruption time window of RSAM data differed for each of the 10 eruptions studied. In each case, the starting time of a window was arbitrarily selected at some base level that corresponded to a period of relative seismic quiescence. For each of the Mount St. Helens eruptions and for two Mount Pinatubo eruptions, we looked at the preeruption data up to the time corresponding to the first peak in RSAM values. For other eruptions, we selected windows of data that appeared to have an exponential increase. For the June 27, 1992, eruption of Mount Spurr, we had to use RSAM data from the onset of the premonitory seismic swarm up to the peak in seismicity to have sufficient data for a fit. With the curve-fitting program it was important not to select points going into the period in which RSAM values were gradually decreasing. Table 1 provides the start and end times for each RSAM data set studied.

ANALYSIS

To facilitate comparison of RSAM data, we chose to normalize RSAM counts relative to their peak values.

Table 1. List of RSAM data start and end times for plots presented in this paper and minimum and maximum RSAM counts in each data set.

[Time is in Universal Time except for CAB and UBO, where local time was used. Minimum values are an estimated average in some cases. GDN and YEL data are for Mount St. Helens eruptions, RED for Redoubt Volcano, CAB and UBO for Mount Pinatubo, and BGL for Mount Spurr]

Station	Start Time		End Time		Minimum	Maximum
GDN	5/19/85	0000	5/27/85	1900	22–45	953
YEL	5/06/86	0600	5/08/86	2000	32	1,123
YEL	10/19/86	0900	10/22/86	0600	72–73	3,801
RED	12/13/89	1530	12/13/89	2130	38	111
RED	12/13/89	0000	12/15/89	0400	44	794
RED	12/25/89	0000	1/3/990	0230	42–64	361
CAB	6/13/91	1900	6/14/91	0900	45	252
CAB	6/1/91	0000	6/16/91	1900	19–20	1,759
UBO	6/5/91	1220	6/7/91	1700	39–46	369
BGL	6/27/92	0000	6/27/92	1820	59–61	1,331[1]
BGL	8/18/92	0000	8/19/92	2350	44–45	1,357
BGL	9/17/92	0000	9/17/92	2350	31–32	1,619

[1] BGL RSAM counts above 1,200 may correspond to a clipped seismic signal.

December 13–14, 1989, data for Redoubt Volcano were an exception to this procedure. For the December 13, 1989, RED RSAM data, we selected a time window that included only the initial 6 h of the swarm (Power, 1994) that preceded the eruption of December 14. After identifying a peak or maximum value, all points were divided by this value. Thus, the normalized peak or maximum value always had a value of 1. Absolute times for each of the RSAM average counts were converted to a simple decimal-hour scale starting at 0 for the first RSAM value selected for analysis.

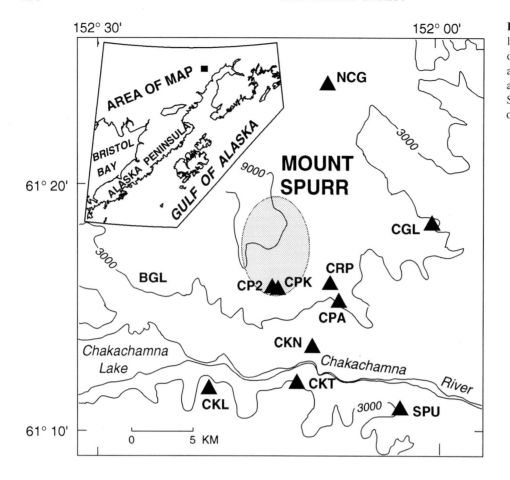

Figure 5. Mount Spurr and locations of the BGL, CPK, and other seismic stations (black triangles). The shaded region is the approximate position of the Spurr caldera (from Power and others, in press).

To characterize RSAM data objectively we used the commercially available curve-fitting software *Table Curve* from Jandel Scientific. This IBM PC-compatible program uses automated statistical techniques to process data for the best curve-fitting equations. Version 3.0 of *Table Curve* uses 3,320 linear and nonlinear equations for its curve-fitting equation selection. *Table Curve* orders the valid equations, or equations that produce some measure of fit to the data, by the user selected goodness-of-fit criterion. For this paper the r^2 coefficient of determination is shown for equation ordering. Normalized RSAM data were analyzed for 9 of the 12 preeruption and eruption periods for which RSAM data were available.

CURVE-FITTING RESULTS

MOUNT ST. HELENS

Using its automatic search procedure, the program *Table Curve* produced equation-curve fits for three-parameter exponential equations (figs. 6, 7, and 8) as the highest ranking equations that provided fits for all three pre-eruption RSAM data sets available for Mount St. Helens' dome-building eruptions. Polynomial functions were the next most common type of equation that produced high r^2 coeffi-

cients of determination or high F-statistic values. While the fit of a polynomial function to the May 1985 RSAM data showed a comparable r^2 coefficient of determination in comparison to the exponential function, examination of residuals showed a weaker fit to RSAM data during the first 100 h of preeruption data. The parameters for the exponential equation fits and the corresponding r^2 coefficients of determination are given in table 2. The May 1986 and October 1986 dome-building eruptions yielded similar parameter values.

REDOUBT VOLCANO

At first inspection, the December 13–15, 1989, RSAM data for Redoubt Volcano (RED seismic station) did not appear to have the potential for curve fitting like the Mount St. Helens data. A closer look at the data suggested a short exponential increase early in the long-period earthquake swarm that began on December 13. *Table Curve* selected a polynomial function for the highest ranked fit. For purposes of comparison to other RSAM data, we selected a fit for an exponential function. Results are shown in table 2 and figure 9. We refer to these author-selected fits as exponential-like increases. The RSAM counts for the period from December 25, 1989, to January 2, 1990, leading up to a

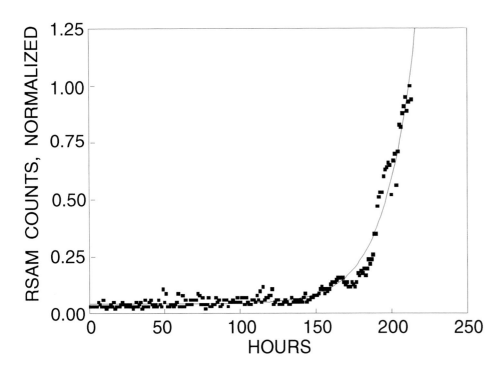

Figure 6. Preeruption RSAM data from the GDN seismic station for the May 1985 dome-building eruption at Mount St. Helens. The fitted exponential curve is shown as the solid line.

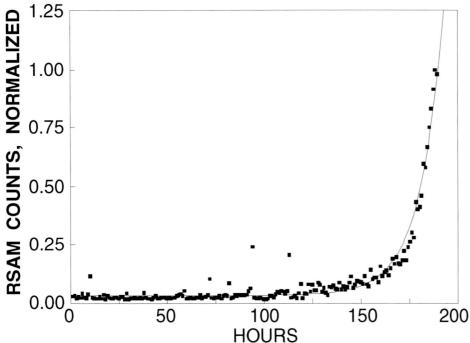

Figure 7. Preeruption RSAM data from the YEL seismic station for the May 1986 dome-building eruption at Mount St. Helens. The fitted exponential curve is shown as a solid line.

series of explosive eruptions (Power and others, 1994) produced an exponential fit (fig. 10). This particular result differs from all others presented here because the RSAM data represent long-period seismicity associated with a period of dome growth. The result is presented without further comment.

MOUNT PINATUBO

For the small dome-building eruption on June 7, preeruption RSAM data for the UBO seismic station pro-

duced a three-parameter exponential-equation fit by use of *Table Curve* (fig. 11). The RSAM data associated with the June 7 dome-building eruption had similar maximum counts (or amplitudes) to the RSAM data examined for Redoubt Volcano. However, owing to seismic activity 5 km north-northwest of Mount Pinatubo, RSAM data were noisy and produced a relatively poor fit ($r^2 < 0.9$). Initially viewed as complex, CAB RSAM data showed the possibility of an exponential increase for a 15-h window from June 13 to June 14 that did not appear to be contaminated by explosion

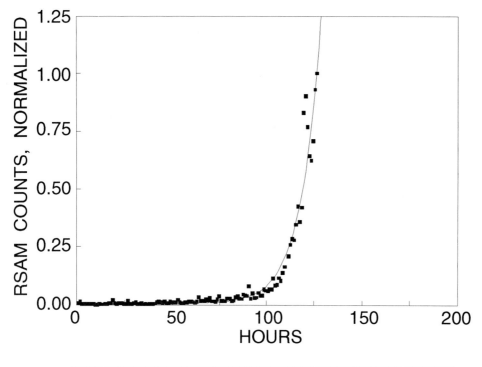

Figure 8. Preeruption RSAM data from the YEL seismic station for the October 1986 dome-building eruption at Mount St. Helens. The fitted exponential curve is shown as the solid line.

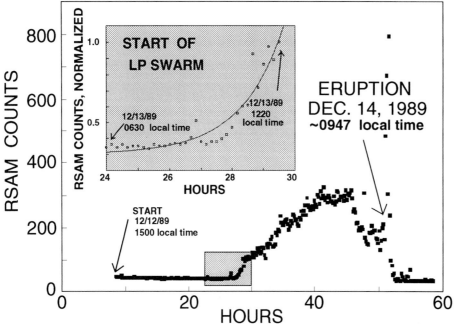

Figure 9. Preeruption RSAM data from the RED seismic station at Redoubt Volcano for the early period of the long-period (LP) earthquake swarm that preceded the explosive eruption on December 14, 1989. The shaded inset shows the fitted exponential curve for the shaded time window.

earthquakes or large volcano-tectonic events. *Table Curve* selected a polynomial function for the highest rank fit. As for the early Redoubt Volcano RSAM data, we selected the exponential function for a subsequent fit. Those results are shown in figure 12 and table 2. We suspect that the high coefficient of fit (table 2) is an artifact of the small RSAM data set and that the parameters for the exponential fit do not reflect the true character of the increase in seismicity preceding the explosive eruptions. The peak in RSAM

count on June 14 was followed by an explosive eruption about 4 h later (Power and others, this volume).

Preeruption fit to the June 11–15 CAB RSAM data leading up to the paroxysmal eruption on June 14 is complex and does not lend itself to simple analysis like the Mount Pinatubo data for June 7 and June 13–14. The r^2 coefficient of determination of about 0.7 (table 2) as well as residuals for the fit clearly indicate a poor fit. CAB RSAM data are shown in figure 12. However, the fitted curve is omitted. The *Table Curve* result is shown in table 2.

Table 2. Summary of parameters for the exponential fit of RSAM data for preeruption periods at Mount St. Helens (GDN, YEL) in 1985 and 1986, Redoubt Volcano (RED) in 1989–90, Mount Pinatubo (UBO, CAB) in 1991, and Mount Spurr (BGL) in 1992.

[As the plotted RSAM counts suggested by visual inspection, CAB RSAM data produced an unacceptable fit. a, b, and c are equation parameters determined by the program *Table Curve*. r^2 is the coefficient of determination. A value of 1.0 implies a perfect fit]

Station	Date	a	b	c	r^2
GDN	May 1985	0.043317	$1.58848{\times}10^{-5}$	−19.13048	0.96699
YEL	May 1986	0.044228	$3.39412{\times}10^{-6}$	−9.13283	0.97930
YEL	October 1986	0.007676	$4.52189{\times}10^{-6}$	−10.30416	0.94228
RED	December 13, 1989	0.241459	$3.01356{\times}10^{-22}$	−1.401031	0.89278
RED	December–January, 1989–90	0.135994	0.00512	−43.39740	0.96574
UBO	June 7, 1992	0.115703	$4.62922{\times}10^{-17}$	−4.119580	0.81192
CAB	June 13–14, 1992	0.00534	$5.16562{\times}10^{-19}$	−7.172232	0.95205
CAB	June 1–16, 1992	0.00709	$1.70051{\times}10^{-7}$	−21.89636	0.71476
BGL	June 27, 1992	0.05066	$2.33001{\times}10^{-6}$	−1.427477	0.98882

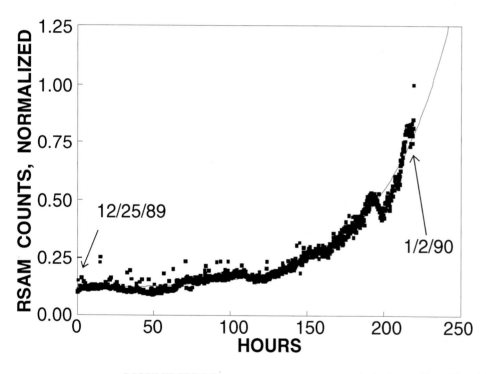

Figure 10. Preeruption (eruption was on January 2, 1990) RSAM data from the RED seismic station at Redoubt Volcano. The fitted exponential curve is shown as a solid line.

MOUNT SPURR

Mount Spurr provided some of the least noisy RSAM data available to date. RSAM data from the BGL station for the period before and during the June 1992 eruption were best fit by an exponential curve and gave the best r^2 coefficient of determination for all of the RSAM data analyzed (fig. 13, table 2). *Table Curve* also found acceptable fits to dozens of other equations that have curves similar to the exponential equation. Subsequent eruptions in August and September showed no similar gradual increase in seismicity at BGL. However, the September eruption showed a brief increase on the CPK and the CRP records (McNutt, written commun., 1993). Onsets of the latter two eruptions were

relatively sudden (fig. 14) on the BGL RSAM seismic record.

THE THREE-PARAMETER EXPONENTIAL EQUATION

The equation ranked number one by the r^2 coefficient of determination or F-statistic for all cases preceding dome-building eruptions was the exponential equation:

$$NORMALIZED\_RSAM\_COUNT(t) = a + b \exp\left(-t/c\right)$$

where *NORMALIZED_RSAM_COUNT(t)* corresponds to an RSAM value at time t in decimal hours for the

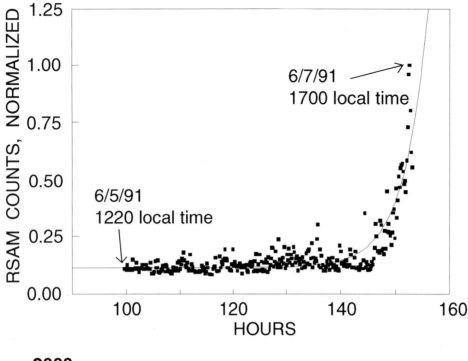

Figure 11. Preeruption (eruption was on June 7, 1991) RSAM data from the UBO seismic station at Mount Pinatubo. The highest RSAM values are probably associated with a strong gas and ash emission. The fitted exponential curve is shown as a solid line.

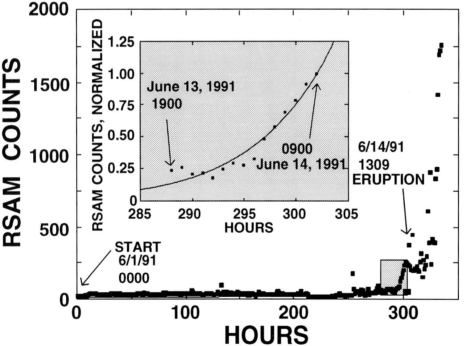

Figure 12. A 20-h window of preeruption normalized RSAM data from the CAB seismic station and fitted curve for the period preceding the first explosive eruption on June 14, 1991. The corresponding time window is shown as a shaded rectangle on the 0- to 350-h plot of CAB RSAM data. The fitted curve for the 0- to 350-h CAB RSAM data has been deleted. All times are local time.

RSAM value. a, b, and c are unknown parameters determined by *Table Curve*. For the purpose of comparison, the exponential equation was selected for fitting RSAM data or seismicity associated with explosive eruptions where the exponential function was not selected automatically by *Table Curve*.

The a parameter corresponds to the initial base level, or vertical offset, at 0 time. This parameter is related to the ratio of RSAM counts to the maximum RSAM value. A low

background noise level or low counts and a large number of counts for the peak RSAM value result in a relatively small a parameter. RED seismic station for Redoubt and UBO for Mount Pinatubo had the lowest ratio for peak signal to background noise level and, hence relatively higher a values (tables 1 and 2). Seismic stations at Mount St. Helens commonly had the highest ratio of maximum RSAM counts to background noise level, along with BGL at Spurr volcano, hence, low a values. High a values appear to have a

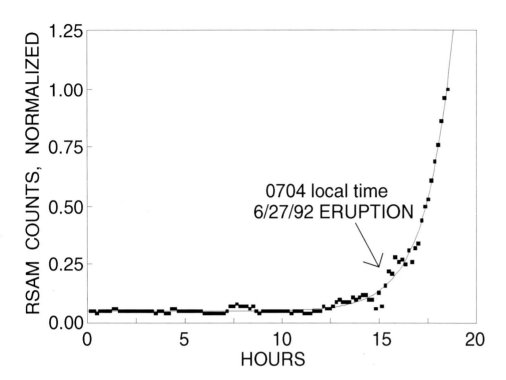

Figure 13. Preeruption and eruption-related RSAM (June 1992) data from the BGL seismic station for Mount Spurr. The fitted exponential curve is shown as a solid line.

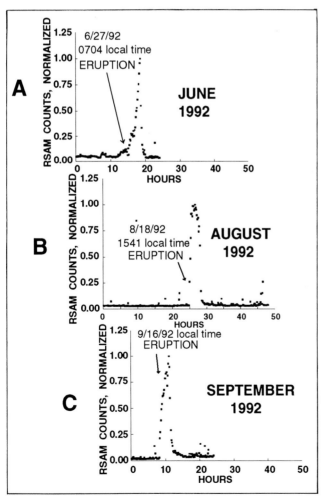

Figure 14. *A*, Graph of RSAM data from BGL seismic station, Mount Spurr, spanning the period of the June 1992 eruption. The rapid increase in RSAM counts, or average seismic amplitude, took place over a period of approximately 7 to 8 h. *B*, Graph of RSAM data from BGL seismic station spanning the period of the August 1992 eruption. No preeruption increase was observed for this eruption. *C*, Graph of RSAM data from BGL seismic station spanning the period of the September 1992 eruption. Low-level tremor prior to the August and September eruptions had almost no impact on RSAM counts.

negative impact on how well RSAM data define a particular type of function. An extreme case would be an *a* value almost equal to 1.0. That would indicate a signal-to-noise ratio of 1.0.

The *b* parameter is a scale factor related to the normalized RSAM data. If RSAM data were not normalized, this would be related to seismic station gain, proximity to the seismic source, background seismic noise level, seismic station site response, and seismic station instrumentation response. We have normalized RSAM data to allow meaningful comparison of RSAM data from different volcanoes.

Parameter *c* is related to the slope, or curvature, of the exponential curve and the time window for the RSAM data. Thus, a large absolute value indicates a relatively long time window, or run-up time, for RSAM data and relatively slow changes in RSAM counts from an early stage. Station RED for the period leading up to the eruption on January 2, 1990, is such an example. The time interval from a reference level to a peak value was about 220 h. A small *c* parameter indicates a short time window and a rapid increase in RSAM counts. The BGL RSAM data for Mount Spurr produced the smallest *c* parameter and one of the shortest run-up times. BGL RSAM data went from a base reference level to a peak value in less than 7 h. Mount St. Helens preeruption RSAM data for the two dome-building eruptions in 1986 produced similar *c* parameter values and run-up times. The *c* value determined for UBO RSAM data for the June 7, 1991, dome-building eruption at Mount Pinatubo differs by about a factor of two less. Results of all curve fits completed are shown in table 2.

The three-parameter exponential equation was not the only equation type that produced acceptable fits to the RSAM data selected for this study. The fact that it was a common equation type that fit a number of preeruption RSAM data does suggest an exponential component to the increase in average seismic amplitude and provides a basis for comparison of RSAM data. The three-parameter exponential equation provides a simple method for comparing RSAM data from different volcanoes and seems appropriate for the purpose of this paper because the exponential function is frequently used to describe natural processes such as growth of a forest, growth of the population of the Earth, and radioactive decay (Gellert and others, 1975).

DISCUSSION

While the increase in RSAM counts appears to have some value in forecasting or prediction of eruptions (Voight and Cornelius, 1991; Cornelius and Voight, 1994; Cornelius and Voight, this volume), the application of curve-fitting techniques to predicting precise times of eruptions needs to be viewed with caution. The process of magma migration to shallow levels in the crust and eruptions is complex. From detailed studies of drill core from the Inyo Domes, Califor-

nia, Westrich and others (1988) concluded that, at least for rhyolitic magma, there was a significant net increase in magma viscosity during the final 1 km of magma ascent owing to water exsolution and bubble growth that result from decompression. This change in bulk viscosity apparently was the reason that much of the Inyo dike failed to reach the surface. Besides having an impact on the ascent rate of magma, water exsolution and bubble growth have an impact on whether an intrusion results in explosive activity, relatively quiet dome building, or extrusion as a lava flow (Eichelberger and others, 1986). Westrich and others (1988) state, "The rate and extent to which exsolved water can escape the system control whether or not the magma fragments to form tephra or extrudes intact as lava."

The first explosive eruption on December 14, 1989, at Redoubt Volcano took place about 10 h after the peak in long-period earthquake RSAM counts and about 23 h after the onset of the swarm. Lahr and others (1994) reported hypocenter depths of 1.4 km below the crater floor for this group of long-period earthquakes. We suspect that the long-period swarm was a direct result of water exsolution and bubble growth. Chouet and others (1994) suggest interaction of magma with ground water. The explosive eruption at 1309 local time on June 14 at Mount Pinatubo was similarly preceded by an earthquake swarm dominated by long-period earthquakes. The explosive eruption took place about 4 h after a peak in RSAM counts. While weak, there is a suggestion of a decrease in volcano-tectonic seismic activity about 1 h prior to what has been called the onset of the eruption at Mount Spurr (Power and others, in press; McNutt and other, in press). For volatile-rich andesitic or dacitic magma there is a catastrophic disruption of the magma during magma ascent or following shallow emplacement because of substantial overpressures from volatiles and the onset of fragmentation at the upper levels in the conduit.

Dome-building eruptions like those at Mount St. Helens in 1985 and 1986 involve significantly degassed magma and apparently represent a low ratio of volatile exsolution to volatile escape from the system. In contrast, explosive eruptions like those of Redoubt Volcano in 1990 and Mount Pinatubo in 1991 represent a volatile-rich magma and a high ratio of volatile exsolution to volatile escape from the system between explosive episodes. The work of Westrich and others (1988) suggests that, for either case, magma ascent is slowed because of volatile exsolution and bubble growth resulting from decompression, particularly at depths above 1 km. That slowing down should not be confused with the approximately 12-h interval between the first peak in RSAM counts for the October 1986 dome-building eruption at Mount St. Helens and the actual extrusion that was confirmed by timed photography. For Mount St. Helens, that time interval of less seismic activity represents the time required for magma to ascend through the

Table 3. Volcanoes, eruption types, magma or lava types, dominant earthquake type, and approximate durations (run-up time) of rapid increase in precursory seismicity.

[MSH, Mount St. Helens; RV, Redoubt Volcano; MP, Mount Pinatubo; MS, Mount Spurr. Tokachi-dake Volcano precursory times were estimates from an amplitude-variation figure in Ikeda and others (1990). The Unzen Volcano precursory time is from Shimizu and others (1992). Low-frequency earthquakes at Tokachi-dake Volcano and Unzen Volcano may be "B" type rather than true long-period earthquakes described by Chouet and others (1994). VT, volcano-tectonic]

Volcano	Date	Eruption type	Magma type	Earthquake type	Run-up time (in hours)
MSH	May 1985	Dome-building	Dacite[1]	VT or "B"	~130
MSH	May 1986	Dome-building	Dacite	VT or "B"	~80
MSH	October 1986	Dome-building	Dacite	VT or "B"	~75
RV	December 1989	Explosive	Andesite[2]	Long-period	5–6
RV	January 1990	Explosive	Andesite+dacite[3]	Long-period	>200
MP	June 1991	Dome-building	Andesite[4]	VT or "B"	6–10
MP	June 1991	Explosive	Andesite+dacite	Long-period	~5
MP	June 1991	Explosive	Dacite	Long-period	
MS	June 1992	Explosive	Andesite[5,6]	VT	~7
Tokachi	1989	Explosive	Andesite[7]	Low-frequency	<0.2
Unzen	May 1991	Dome-building	Dacite[8]	Low-frequency	>200

[1] Swanson and Holcomb, 1990.
[2] Nye and others, 1994.
[3] The January 2, 1990, explosive eruption was preceded by dome building.
[4] Pallister and others, this volume.
[5] Harbin and others, in press.
[6] Nye and others, in press.
[7] Ikeda and others, 1990.
[8] Yanagi and others, 1992.

relatively weak pyroclastic material and the lava flows of the volcanic edifices.

Our results do not suggest that all eruptions have to be preceded by exponential or exponential-like increases in volcano-tectonic or long-period seismicity. Numerous explosive gas and ash emissions at Mount St. Helens in April and May of 1986 were not preceded by precursory seismic activity. Following the last dome-building eruptions in 1986, numerous ash emissions and explosions (Mastin, 1994) at Mount St. Helens typically did not show premonitory increases in seismic activity and RSAM counts. These eruptions did not involve the ascent of magma. Mastin (1994) hypothesized that "the explosion-like seismic events represent the transport of gas, probably of magmatic origin, in the shallow subsurface." Not all explosive eruptions at Redoubt Volcano were preceded by precursory long-period activity (Stephens and others, 1994). At Mount Pinatubo, several explosive eruptions before June 14 were preceded by significantly less seismicity (Hoblitt, Wolfe, and others, this volume). The August 1992 explosive eruption of Mount Spurr showed no signs of detectable earthquakes prior to the eruption, and the September eruption was preceded by some discrete events a few hours before the eruption. Episodes of tremor preceded the June, August, and September 1992 eruptions (AVO, 1993). During the 1988-89 eruptive activity at Mount Tokachi, Japan, many explosive eruptions did not show any detectable precursory earthquake activity (Nishimura and others, 1990). Only one-third of 23

observed explosive eruptions were preceded by small "low frequency" earthquakes (Okada and others, 1990). Eruption types, associated magma or lava types, dominant earthquake types, and approximate period of increased seismicity or an exponential-like buildup for each eruption are shown in table 3.

The primary intention of this paper was to investigate the possibility of the existence of common characteristics in the increase of preeruption RSAM data and precursory seismicity. We found that for most dome-building preeruption RSAM data, it was possible to find a function or equation type that fit the gradual increase in RSAM counts prior to an eruption. While it may appear remarkable that the exponential equation was the one ranked highest in all cases, the exponential equation was not the only type that produced acceptable fits to the RSAM data. However, it was the equation type that most commonly produced an acceptable fit (high r^2 coefficient) to most of the RSAM data and was the most convenient for comparing increases in RSAM counts.

The similarity of fits suggests that a common physical process is in operation at two volcanoes (Mount St. Helens and possibly Mount Pinatubo). At Mount St. Helens, we speculated (Endo and others, 1990) that the relative average seismic amplitude at a seismic station was directly related to the ascent of magma from a depth of about 1.4 km to just below the surface. If that is the case, then the exponential equation that fits the RSAM data analyzed for this paper is related to a magma velocity function. If the depth of origin for the magma is known, then a full expression for magma ascent velocity can be determined, and the normalized RSAM plot could be similar to a plot of time versus distance traveled. If the exponential increase is related to ascent of magma, the implication is that the initial ascent velocities are small and increase substantially over time. Unfortunately, the Mount St. Helens RSAM data suggest that the relation is far more complex. While the exponential equation parameters are nearly identical for the fit of normalized RSAM data for eruptions of similar volume in 1986 (Swanson and Holcomb, 1989), they differ for an eruption of smaller volume in 1985. Assuming no large differences in the conduit system, a possible explanation is that the smaller volume of eruptible magma ascended at a relatively slower velocity.

Redoubt Volcano, Mount Pinatubo (June 13–14), and Mount Spurr RSAM data were more difficult to analyze and interpret. In some cases the exponential equation was not selected automatically by the curve-fitting program. It is likely that the increases in RSAM data for the different eruptions studied are more complex. Furthermore, exponential increases in RSAM data for long-period earthquakes are probably not directly related to magma ascent. While the study is far from conclusive, there is a suggestion that the increases in seismicity (long-period or volcano-tectonic earthquakes) prior to explosive andesitic and dacitic eruptions are an order of magnitude shorter in duration (run-up time) than increases in seismicity associated with dacitic dome-building eruptions (table 3).

CONCLUSION

The main results of our comparison of preeruption seismicity by use of RSAM data follow.

1. The rapid increase in seismicity prior to dacitic and andesitic dome-building eruptions can be characterized by an exponential function.
2. For dacitic (Mount St. Helens) and andesitic (Mount Pinatubo, June 7, 1991) dome- building eruptions, we speculate that the exponential increase corresponds to the migration of magma from shallow depths (approximately 1.4 km at Mount St. Helens) to just below the surface. The exponential function is proportional to ascent velocity for magma. For Mount St. Helens, the time interval between the peak in RSAM counts and extrusion is believed to be the time required for dacitic magma to migrate through the relatively weak pyroclastic deposits of the edifice.
3. Not all andesitic eruptions are preceded by an increase in seismicity. For this case we speculate that magma migrates in well-established conduits with relatively few detectable earthquakes. Rapid exponential-like increases may precede the initial eruption or may occur in cases where the conduit is blocked.

4. Detectable long-period, volcano-tectonic, or "B" type earthquakes or tremor are not essential precursors to explosive gas and ash emissions.
5. Eruptions, dome-building or explosive, are sometimes preceded by a decrease in seismicity or a leveling off in seismicity an hour or longer before extrusions or explosive eruptions. For dome-building eruptions, that decrease, as at Mount St. Helens, may be related to the migration of magma into the volcano edifice. For explosive eruptions, the decrease may represent a transient equilibrium between volatile pressure and lithostatic pressure.

We believe accurate characterization of precursory seismic activity by using the RSAM monitoring technique is just one step forward in the understanding of the volcanic eruption process. Petrologic studies, gas-chemistry studies, experimental studies of bubble growth, tilt and other earth-deformation studies, studies of seismic source parameters, and so on, are all required to relate the increase in seismic activity better, or, in some cases, the absence of increased seismicity to volcanic eruptions so we can advance our understanding of the magma migration and eruption process.

REFERENCES CITED

Alaska Volcano Observatory, 1993, Mount Spurr's 1992 eruptions: Eos, Transactions, American Geophysical Union, v. 74, p. 217-222.

Brantley, S., ed., 1990, The eruption of Redoubt Volcano, Alaska, December 14, 1989–August 31, 1990: U.S. Geological Survey Circular 1061, 33 p.

Chouet, B.A., Page, R.A., Stephens, C.D., Lahr, J.C., and Power, J.A., 1994, Precursory swarms of long-period events at Redoubt Volcano (1989–1990), Alaska: Their origin and use as a forecasting tool, in Miller, T.P., and Chouet, B.A., eds., The 1989–1990 eruptions of Redoubt Volcano, Alaska: Journal of Volcanology and Geothermal Research, v. 62, p. 95–135.

Cornelius, R.R., and Voight, B., 1994, Seismological aspects of the 1989–1990 eruption at Redoubt Volcano, Alaska: The materials failure forecasting method (FFM) with RSAM seismic data: Journal of Volcanology and Geothermal Research, v. 62, p. 469–498.

Cornelius, R.R., and Voight, B., this volume, Real-time seismic amplitude measurement (RSAM) and Seismic spectral amplitude measurement (SSAM) analyses with the Materials Failure Forecast Method (FFM), June 1991 explosive eruption at Mount Pinatubo.

Eichelberger, J.C., Carrigan, C.R., Westrich, H.R., and Price, R.H., 1986, Non-explosive silicic volcanism: Nature, v. 323, p. 598–602.

Endo, E.T., Dzurisin, D., and Swanson, D.A., 1990, Geophysical and observational constraints for ascent rates of dacitic magma at Mount St Helens, in Ryan, M.P., ed., Magma transport and storage: John Wiley & Sons, p. 318–334.

Endo, E.T., and Murray, T., 1991, Real-time seismic amplitude measurement (RSAM): A volcano monitoring tool: Bulletin of Volcanology, v. 53, p. 533–545.

Ewert, J.W., Lockhart, A.B., Marcial, S., and Ambubuyog, G., this volume, Ground deformation prior to the 1991 eruptions of Mount Pinatubo.

Gellert, W., Gottwald, S., Hellwich, M., Kastner, H., Kustner, H., ed., 1975, The VNR concise encyclopedia of mathematics: New York, Von Nostrand Reinhold, 776 p.

Harbin, M.L., Swanson, S.E., Nye, C.J., and Miller, T.P., in press, Preliminary petrology and chemistry of proximal eruptive products: 1992 eruptions of Crater Peak, Mount Spurr, Alaska: U.S. Geological Survey Bulletin.

Hoblitt, R.P., Wolfe, E.W., Scott, W.E., Couchman, M.R., Pallister, J.S., and Javier, D., this volume, The preclimactic eruptions of Mount Pinatubo, June 1991.

Ikeda, Y., Katsui, Y., Nakagawa, M., Kawachi, S., Watanabe, T., Fujibayashi, N., Shibata, T., and Kagami, H., 1990, Petrology of the 1988–1989 essential ejecta and associated glassy rocks of Tokachi-dake volcano in central Hokkaido, Japan: Bulletin of the Volcanological Society of Japan, v. 35, no. 2, p. 47–163.

Lahr, J.C., Chouet, B.A., Stephens, C.D., Power, J.A., and Page, R.A., 1994, Earthquake classification, location, and error analysis in a volcanic environment: Implications for the magmatic system of the 1989–1990 eruptions at Redoubt Volcano, Alaska: Journal of Volcanology and Geothermal Research, v. 62, p. 137–151.

Lockhart, A.B., Marcial, S., Ambubuyog, G., Laguerta, E.P., and Power, J.A., this volume, Installation, operation, and technical specifications of the first Mount Pinatubo telemetered seismic network.

Mastin, L.G., 1994, Explosive tephra emissions at Mount St. Helens, 1989–1991: The violent escape of magmatic gas following storms: Geological Society of America Bulletin, v. 106, p. 175–185.

McNutt, S., Tytgat, G., and Power, J.A., in press, Preliminary analysis of volcanic tremor associated with the 1992 eruptions of Crater Peak, Mount Spurr, Alaska: U.S. Geological Survey Bulletin.

Murray, T.L., and Endo, E.T., 1989, A real-time seismic amplitude measurement system (RSAM): U.S. Geological Survey Open-File Report 89–684, 21 p.

Nishimura, Y., Miyamachi, H., Ueki, S., Nishimura, T., Shimizu, H., Ohmi, S., and Okada, H., 1990, Joint seismometrical observations by the National University Team during the 1988–1989 eruptive activity of Mt. Tokachi, Hokkaido: Bulletin of the Volcanological Society of Japan, v. 35, no. 2, p. 163–172.

Nye, C.J., Harbin, M.L., Miller, T.P., Swanson, S.E., Neal, C.A., in press, Whole-rock major- and trace-element chemistry of 1992 ejecta from Crater Peak, Mount Spurr, Alaska: A preliminary overview: U.S. Geological Survey Bulletin.

Nye, C.J., Swanson, S.E., Avery, V.F., and Miller, T.P., 1994, Geochemistry of the 1989–1990 eruption of Redoubt Volcano, pt. I: Whole-rock major and trace element chemistry:

Journal of Volcanology and Geothermal Research, v. 62, p. 429–452.

Okada, H., Nishimura, Y., Miyamachi, H., Mori, H., and Ishihara, K., 1990, Geophysical significance of the 1988–1989 explosive eruptions of Mt. Tokachi, Hokkaido, Japan: Bulletin of the Volcanological Society of Japan, v. 35, no. 2, p. 176–203.

Pallister, J.S., Hoblitt, R.P., Meeker, G.P., Knight, R.J., and Siems, D.F., this volume, Magma mixing at Mount Pinatubo: Petrographic and chemical evidence from the 1991 deposits.

Pinatubo Volcano Observatory Team, 1991, Lessons from a major eruption: Mount. Pinatubo, Philippines: Eos, Transactions, American Geophysical Union, v. 72, 4 p.

Power, J.A., Jolly, A.D., Page, R.A., and McNutt, S.R., in press, Seismicity and forecasting of the 1992 eruptions of Crater Peak vent, Mount Spurr, Alaska: An overview: U.S. Geological Survey Bulletin.

Power, J.A., Lahr, J.C., Page, R.A., Chouet, B.A., Stephens, C.D., Harlow, D.H., Murray, T.L., and Davies, J.N., 1994, Seismic evolution of the 1989–1990 eruption sequence of Redoubt volcano, Alaska: Journal of Volcanology and Geothermal Research, v. 62, p. 153–182.

Power, J.A., Murray, T.L., Marso, J.N., and Laguerta, E.P., this volume, Preliminary observations of seismicity at Mount Pinatubo by use of the Seismic Spectral Amplitude Measurement (SSAM) system, May 13–June 18, 1991.

Shimizu, H., Umakoshi, K., Matsuwo, N., and Ohta, K., 1992, Seismological observations of Unzen Volcano before and during the 1990–1992 eruption, in Yanagi, T., Okada, H., and Ohta, K., eds., Unzen Volcano the 1990–1992 eruption: Fukuoka, Japan, The Nishinippon and Kyushu University Press, p. 38–43.

Stephens, C.D., Chouet, B.A., Page, R.A., and Lahr, J.C., in press, Seismological aspects of the 1989–1990 eruptions at Redoubt Volcano, Alaska: The SSAM perspective: Journal of Volcanology and Geothermal Research.

Swanson, D.A., and Holcomb, R.T., 1989, Regularities in the growth of the Mount St. Helens dacite dome, 1980–1986, in Fink, J.H., ed., Lava flows and domes, emplacement mechanisms and hazards: Berlin, Springer, 34 p.

Voight, B., and Cornelius, R.R., 1991, Prospects for eruption prediction in near real-time: Nature, v. 350, p. 695–698.

Westrich, H.R., Stockman, H.W., and Eichelberger, J.C., 1988, Degassing of rhyolitic magma during ascent and emplacement: Journal of Geophysical Research, v. 93, p. 6503–6511.

Wolfe, E.W. and Hoblitt, R.P., this volume, Overview of the eruptions.

Yanagi, T., Nakada, S., and Maeda, S., 1992, Temporal variation in chemical composition of the lava extruded from the Jigokuato Crater, Unzen Volcano, in Yanagi, T., Okada, H., and Ohta, K., eds., Unzen Volcano the 1990–1992 eruption: Fukuoka, Japan, The Nishinippon and Kyushu University Press, p. 82–91.

Real-time Seismic Amplitude Measurement (RSAM) and Seismic Spectral Amplitude Measurement (SSAM) Analyses with the Materials Failure Forecast Method (FFM), June 1991 Explosive Eruption at Mount Pinatubo

By Reinold R. Cornelius[1] and Barry Voight[1,2]

ABSTRACT

Seismic and other unrest at Mount Pinatubo culminated in a major eruption that began on June 12, 1991, and peaked on June 15, when it generated ash plumes over 35 kilometers high. Precursory seismic activity was tracked by the scientists of the Philippine Institute of Volcanology and Seismology and the U.S. Geological Survey using the Real-time Seismic Amplitude Measurement (RSAM) and Seismic Spectral Amplitude Measurement (SSAM) systems, which provide consecutive averages of the absolute amplitudes of the seismic signal from several seismometers. RSAM and SSAM records are compared here for a variety of event types, involving combinations of volcano-tectonic seismicity, long-period swarms, tremor, and hybrid activity. Due to the strong signal and broad-band nature of the seismicity, RSAM patterns were generally similar to individual SSAM bands; unlike at Redoubt Volcano, Alaska, in 1990, at Pinatubo RSAM provided an authentic measure of overall seismic energy release throughout the precursory period. These data were used by the authors to test applicability of the Materials Failure Forecast Method (FFM), in which the time of failure (eruption) can be estimated by extrapolation of the inverse rate versus time curve toward the abscissa.

Graphical RSAM data for the period June 1 to 1600 on June 10 were sent to us via fax. Inverse RSAM, representing inverse seismic energy rates, was plotted against time and extrapolated according to conventional FFM procedures. The fit indicated impending failure about June 12 and allowed us to anticipate an explosive eruption in foresight for the first time with the FFM technique. In hindsight, inverse-rate analyses on June 7 would have suggested an eruption window from June 12 to 20, and analyses between June 12 and June 15 would have anticipated an eruption between June 14 and June 16.

[1]The Pennsylvania State University, Department of Geosciences, University Park, PA 16802.

[2]Also at U.S. Geological Survey.

INTRODUCTION

The recognition of unrest at Mount Pinatubo began on April 2, 1991, with steam explosions from a 1.5-km-long line of vents high on the north flank of the volcano, following a 500-year-long period of quiescence (Newhall and others, this volume). The unrest culminated in a major explosive eruption, beginning on June 12 and peaking on June 15, that generated extensive pyroclastic flows and ash plumes over 35 km high. Estimates of ash and gas emission rank this eruption as probably the world's largest since the eruption of Novarupta (Katmai) in 1912. A joint effort by the Philippine Institute of Volcanology and Seismology (PHIVOLCS) and the U.S. Geological Survey Volcano Crisis Assistance Team (USGS VCAT) produced effective hazard evaluation and forecast advisories, enabling decisionmakers to manage massive evacuations that possibly saved tens of thousands of lives (Punongbayan and others, this volume; Wolfe and Hoblitt, this volume).

Here, we show how the Materials Failure Forecast Method (FFM) was used, and could have been used, to track precursory volcanic activity at Pinatubo and to forecast forthcoming activity by fitting data according to an empirical rate-acceleration relation (Voight, 1988, 1989). This method uses numerical or graphical rate extrapolation toward an expected failure rate to forecast time windows within which eruptions are expected (Cornelius and Voight, 1989, 1994, 1995; Voight and Cornelius, 1991). Some of the series of FFM analyses on Pinatubo data reported here were performed in foresight, and others in hindsight.

We use Real-time Seismic Amplitude Measurements (RSAM) (Murray and Endo, 1989) as a measure of seismic-energy release rate and compare these data with the spectrally filtered energy release rate from the Seismic Spectral Amplitude Measurement system (SSAM) (Power and others, this volume).

The onsite eruption forecasts by scientists of PHIVOLCS and the USGS VCAT were also based on RSAM, as well as earthquake hypocenter migration, and waveform

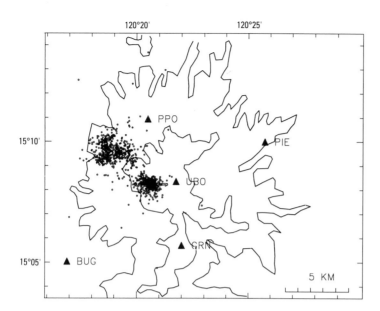

Figure 1. Locations of telemetered seismic stations in the Pinatubo network between May 13 and June 16, 1991. Epicentral zone northwest of Mount Pinatubo particularly active in May; epicentral zone near summit gained in activity after June 1. Figure courtesy of Mori, Eberhart-Phillips, and Harlow, this volume.

characteristics (Harlow and others, 1991; this volume), changes in gas flux (Daag, Tubianosa, and others, 1991; this volume), and other observations.

Following a brief presentation of event chronology, FFM methodology, and RSAM and SSAM monitoring tools, we present our foresight analysis based on RSAM through June 10, prior to the eruptive cycle of June 12 through 15. Hindsight analyses based on the original data are then presented and compared with foresight analyses. We compare RSAM records with comparable records for selected SSAM bands. Finally, we demonstrate with analyses of data through June 13 and 14 how the seismic development could have been tracked and forecasts made during a lull of explosive activity prior to climactic ejection on June 15. Our goal is to use the Pinatubo experience to evaluate FFM as a forecasting tool, to recognize limitations and possible pitfalls in applying FFM with RSAM and SSAM, and to provide perspectives for future near-real-time foresight applications.

CHRONOLOGIC SUMMARY

The following brief chronology is included to provide context to the discussions on forecasting. More detailed information is available elsewhere (Hoblitt, Wolfe, and others, this volume; Wolfe and Hoblitt, this volume). All times and dates in this paper are in Philippine local time, unless otherwise specified.

The climatic eruption on June 15 was preceded by at least 10 weeks of unrest. In response, PHIVOLCS installed several portable seismometers in early April; subsequently, a seven-station radiotelemetered seismic network was established during late April and early May that used

equipment from the USGS VCAT cache at the Cascades Volcano Observatory (CVO) (Lockhart and others, this volume). Two electronic tiltmeters were installed in late May (Ewert and others, this volume). The data were received and evaluated with the aid of a PC-based acquisition and analysis system established at Clark Air Base (Murray and others, this volume).

Numerous small, high-frequency volcano-tectonic (VT) earthquakes occurred during April and early May. Prior to late May, seismic activity was mainly located 4 to 8 km northwest of the phreatic vents of Pinatubo, at 3 to 6 km depth, with lesser activity under the summit (Harlow and others, this volume). The cause of this activity, whether due to hydrothermal, tectonic, or magmatic processes, was puzzling and ambiguous (Punongbayan and others, 1991; this volume). Epicentral clusters and seismic monitoring stations are shown in figure 1. The climb of SO_2 flux to a peak value of 5,000 t/d on May 28 suggested the rise of magma to a level that promoted gas exsolution and (or) drying and gas transmission through the hydrothermal system (Daag and others, 1991; Daag, Tubianosa, and others, this volume). After late May, the rate of VT seismicity increased markedly under the near-summit vents on Pinatubo. A pronounced decrease in SO_2 flux on June 5 was interpreted as near-surface sealing of the magma and led to speculation and concern about increasing gas pressurization (Daag, Tubianosa, and others, 1991; this volume).

A small steam explosion on June 3 marked increasing unrest, together with minor ash emission, heightening near-summit seismicity, tremor episodes, and outward tilt (Harlow and others, this volume; Ewert and others, this volume). In response, PHIVOLCS issued an Alert Level 3 notice on June 5, indicating "if trend of increasing unrest continues, eruption possible within 2 weeks" (Pinatubo Volcano Observatory Team, 1991; Wolfe, 1992; Punongbayan and others, this volume). The concern was for a "major pyroclastic eruption" (Wolfe and Hoblitt, this volume).

Outward tilt noticeably increased on June 6, and near-summit seismicity continued to grow until the afternoon on June 7, when enhanced steam and ash emission generated a column about 8 km high. Thereafter, seismicity and tilt decreased, and PHIVOLCS announced an Alert Level 4—"eruption possible within 24 h." The outward tilt and shallow seismicity suggested near-surface magma, and this was confirmed the following morning by observation of a 100-m-diameter dome on the northwest flank of the volcano. Shallow VT earthquake swarms under the dome, episodes of harmonic tremor, and weak ash emissions marked the period from June 8 through early June 12. At times "dilute ashy density currents…seen from (a) distance…resembled the ash clouds associated with pyroclastic flows, but they neither coincided with distinct seismic explosion signals nor left recognizable flowage deposits" (Wolfe and Hoblitt, this volume). Nevertheless, observation of these currents added to existing apprehensions and perceptions of unrest and led PHIVOLCS at 1700 on June 9 to raise the Alert Level to 5—"eruption in progress" (Wolfe and Hoblitt, this volume). The radius of evacuation was extended to 20 km, with evacuations increased to 25,000 (Punongbayan and others, this volume; Wolfe and Hoblitt, this volume). On June 10 about 14,500 military personnel and dependents were evacuated from Clark Air Base, and only volcanological staff and 1,500 security personnel remained.

A major eruptive vent-clearing phase started at 0851 on June 12, and this produced a sequence of three plinian eruption clouds that reached altitudes at least 24 km (table 1), accompanied by pyroclastic flows (Hoblitt, Wolfe, and others, this volume). The eruptions at 2252 on June 12 and 0841 on June 13 were preceded by 2- to 4-h episodes of long-period (LP) seismicity, which enabled volcanologists to announce advance warning (Harlow and others, this volume). The evacuation radius was extended to 30 km, and total evacuees increased to more than 58,000 (Punongbayan and others, this volume; Wolfe and Hoblitt, this volume).

After a hiatus of about 28 h, explosions recommenced on June 14, producing mostly surge deposits (Hoblitt and others, this volume). These explosions were preceded by complex seismicity, including VT and LP seismicity, and continued for about 24 h. Sweeping ash clouds were observed at dawn on June 15 from Clark Air Base, and increased concern led to evacuation of remaining base personnel and volcanologists and seismic acquisition hardware. Although volcanologists returned later, a 3-h gap was left in seismic recording (Harlow and others, this volume). At 1342 on June 15, the episodic activity culminated in 9 h of continuous cataclysmic ejection that produced a tephra plume exceeding 35 km high and heavy ash and pumice falls.

THE MATERIALS FAILURE FORECAST METHOD

FFM for eruption forecasting is based on an empirical law describing failure of materials (Voight, 1988, 1989). The concept of "failure" is broadly interpreted and may involve diverse processes such as rupture of solid or fluid-saturated porous rock, or critical points of fluid pressurization affecting fluid transport and (or) crack opening. The governing equation (eq. 1) relates rates, $\dot{\Omega}$, during accelerating activity, to their change in rate, $\ddot{\Omega}$, where the state variable Ω stands for any of several possible precursor observations such as ground displacement or tilt or seismic parameters. Constants A and α are empirical and α is dimensionless.

$$\ddot{\Omega} = A\,\dot{\Omega}^{\alpha} \qquad (1)$$

The solution of equation 1, in terms of rate (equation 2, α not equal to 1) has, for $\alpha > 1$, a singularity at time t_s, at which rates and accelerations (changes in rate) are infinite (fig. 2A shows rates versus time for various α values). The terms t_* and $\dot{\Omega}_*$ denote reference time and rate.

$$\dot{\Omega} = [A(1-\alpha)(t-t_*) + \dot{\Omega}_*^{(1-\alpha)}]^{1/(1-\alpha)} \qquad (2)$$

The time t_s is an upper limit for the time of failure t_f, with the assumptions that the described dynamic complex is terminated by some type of rock mass-fluid system failure or instability and that the onset of an eruption follows a relatively high, yet finite, rate. The critical point in time (t_s) can be estimated by extrapolating an inverse-rate data set to the time axis. The point t_f may be estimated with a predetermined intercept that represents the expected finite failure rate, $\dot{\Omega}_f$. As selection of this failure rate may be difficult in foresight, the default strategy is to solve for t_s. Both t_s and t_f may be used to estimate time of eruption.

The solution of equation 1 in terms of rate (eq. 2) leads to the convenient graphical procedure of "inverse-rate" plots. Inverse-rate trends are linear for the special case of $\alpha = 2$, which simplifies the extrapolation of data. For $\alpha < 2$, the inverse-rate curves are concave upward, and for $\alpha > 2$ convex upward (fig. 2B). The case $\alpha = 1$ implies exponential growth, and, in general, for $\alpha \le 1$ the inverse-rate curves are asymptotic to the time axis.

We distinguish between graphical and numerical applications of FFM. The graphical procedure relies on visual extrapolation of the inverse-rate data; extrapolation is not necessarily linear. In this paper the graphical approach is followed, commonly in conjunction with linear extrapolation that assumes $\alpha = 2$. Other regression procedures are also suitable with FFM (Cornelius and Voight, 1989; Cornelius, 1992). An overview of the method is given by Cornelius and Voight (1995), and software suitable for use with conventional PC hardware is available from the authors.

Table 1.   Eruptive episodes at Mount Pinatubo, June 1991.

[Data from various sources. Alerts issued by the Philippine Institute of Volcanology and Seismology (PHIVOLCS). Alert Level 3: If trend of increasing unrest continues, eruption possible within 2 weeks. Alert Level 4: eruption possible within 24 h. Alert Level 5: eruption in progress]

Date (1991)	Eruption (local time)	Maximum plume altitude (km)	Remarks
April 2.........			Ash/gas explosion, predominantly phreatic.
June 3...........	evening		Gas explosion, followed by harmonic tremor.
June 4..........	1200		Gas explosion, followed by harmonic tremor. Alert Level 3.
June 7..........	1700	8	Steam/ash explosion followed by lava extrusion. Alert Level 4.
June 9..........	0931	2	Continuing ash emission, continuous harmonic tremor.
	1700		Alert Level 5.
June 11........	1631	3.5	Voluminous ash-laden steam clouds.
	0341		Explosion, pyroclastic flows.
June 12........	0851	19	*First plinian eruption*, until 0926.
	2252	24	Explosion marked by long-period tremor.
June 13........	0841	24	Intense tremor and eruption column; pyroclastic flows.
June 14........	1309	21	Tephra column after break in eruptive activity.
	1410	15	Eruption in progress from multiple vents forming broad plume.
	1516	?	Explosion with atmospheric-pressure pulse.
	1853	24	Explosion with tremor and atmospheric-pressure pulse.
	2018	5	Low-level eruption.
	2320	21	Tephra plume gradually ascending; pyroclastic flow.
June 15........	0115	?	Onset of tremor and atmospheric-pressure pulse; pyroclastic flow.
	0257	?	Tremor and atmospheric-pressure pulse.
	0555	12	Tremor and atmospheric-pressure pulse. Tephra plume, pyroclastic flows.
	0629	?	Tremor; eruptive activity uncertain.
	0810	12?	Seismic-data gap. Atmospheric-pressure pulse and tephra plume.
	1027	15?	Tremor, atmospheric-pressure pulse, tephra plume.
	1117	?	Tremor, atmospheric-pressure pulse, eruption plume.
	1158	8?	Tremor, atmospheric-pressure pulses, eruption plume.
	1222	8	Tremor, atmospheric-pressure pulse, tephra plume.
	1252	?	Tremor and atmospheric-pressure pulse.
	1315	?	Tremor and atmospheric-pressure pulse.
	1342	>35	*Paroxysmal eruption and caldera collapse.*
	2230	28?	Waning phase of climactic eruption, until 0231; sustained tephra emission.
June 16........	0331	25?	Further waning of climactic eruption, until 0731.
	1031	5–6	
	1231	5–6	
	1431	5–6	
	2031	4–5	

Rate-time data are commonly evaluated by using values averaged at selected uniform time increments. An alternative is to consider time-integrated data over approximately uniform "rate-magnitude" increments instead of uniform time increments. This procedure yields an increasingly higher frequency of rate data toward the end of an accelerating time series and results in an end-weighted rate calculation that emphasizes the latest precursory developments.

A continuous (but not necessarily linear) inverse-rate downward trend may be interrupted by a sudden decrease in rate (positive slope in the plot of inverse rate versus time; solid dots in fig. 3) which is then followed by a sudden increase in rates (steepened negative slope in the plot of inverse rate versus time; open circles in fig. 3). The combined effect of such a pair of jogs in the trend may or may not offset the overall curve. We refer to a short-time acceleration of the rate (open circles) as an accelerating jog and to a short-time deceleration of the rate (solid dots) as a decelerating jog. Accelerating jogs may be analyzed individually by FFM (dashed arrows in fig. 3), or the overall trend may be analyzed (heavy solid line in fig. 3).

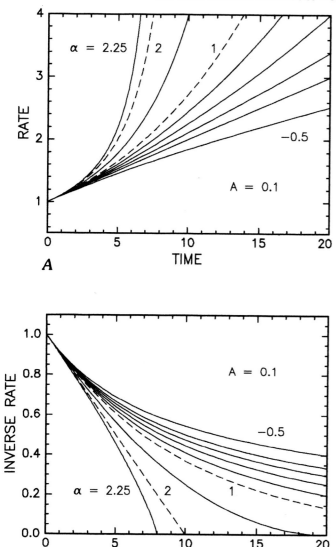

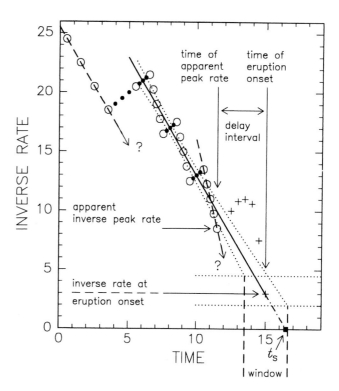

Figure 3. Schematic inverse-rate curve. Open circles define downward (accelerating) inverse-rate trends; solid dots define decelerating jogs. The overall trend may be time shifted by a decelerating jog. Decelerating jogs may cyclically alternate with accelerating jogs (dashed arrows) steeper than the overall trend (heavy solid line). Inverse-rate trends can be fit numerically, and a data envelope (angled dotted lines) can be constructed at a specific confidence level. The data envelope can be intersected with a range of expected rates near time of eruption (horizontal dotted lines). The resulting eruption window is shown by vertically dashed lines. Time of eruption may be separated from the time of apparent peak rate by a delay interval; the rate at time of eruption may or may not be larger than the apparent peak rate. Time of singularity (t_s) for curves with $\alpha > 1$ is the time at which the curve fit indicates an expected infinite rate.

Figure 2. Generic plots of curves governed by the Materials Failure Forecast Method equation for (A) rates and (B) inverse rates. Constants A and α are empirical, as used in eq. 1 in text. Curves calculated for A = 0.1 and $-0.5 < \alpha < 2.25$. Initial displacement (Ω_o) = 0.0 and initial rate ($\dot{\Omega}_*$) = 1.0; both at time $t_o = t_* = 0$. Values of α in increasing order are -0.5, 0.0, 0.25, 0.5, 0.75, 1.0, 1.5, 2.0, and 2.25.

As used here, the *eruption window* spans the time between the intercepts for a range of anticipated (guessed) rates at time of failure (fig. 3), usually with the upper and lower branches of a data envelope reflecting the 97.5% confidence level. It is important to recognize that this confidence level applies only to the mathematical extrapolation of a particular data set, and its use does not imply a *forecast* with the same confidence level. The envelope is derived from the linear least squares analysis, either by assuming α = 2 or by iterative linearization of equation 2 (Cornelius and Voight, 1995). The envelope is a measure of the error due to data scatter, and it increases in width with extrapolation over greater time.

Expected values of failure rate, $\dot{\Omega}_f$, may be estimated from previously recorded peak rates of seismic activity associated with the onset of earlier eruption events (for example, in RSAM units, as discussed below). A relatively conservative eruption window (dashed, heavy vertical lines in fig. 3) is established with the lower bound, defined as the intersection of the earlier (left) branch of the extrapolated data envelope with the lower expected finite failure rate, and with the upper bound, defined as the intersection of the later (right) branch of the data envelope with the higher expected finite failure rate.

A delay interval may separate the apparent peak rate from the time of eruption (Voight, 1988; fig. 3). A higher or lower rate than the previous apparent peak rate may occur at the onset of an eruption, but the intervening delay interval is

Figure 4. Cumulative RSAM and SSAM data for station →
PIEZ (10-min-interval data) from June 10 until the station was
destroyed: *A*, RSAM; *B*, SSAM band 4 (3.5–4.5 Hz); *C*, SSAM
band 3 (2.5–3.5 Hz); *D*, SSAM band 2 (1.5–2.5 Hz); *E*, SSAM
band 1 (0.5–1.5 Hz). RSAM data represent seismic-energy
release rates integrated over a broad frequency spectrum; low-
frequency SSAM (0.5–2.5 Hz) is characteristic of long-period
events and tremor buildup; higher-frequency SSAM (2.5–4.5
Hz) is characteristic of swarms of volcano-tectonic earthquakes or quasicontinuous volcano-tectonic earthquake activity. Explosive pyroclastic eruptions began on June 12 and
culminated in the cataclysmic eruption on June 15; sequence is
marked in *A* by arrows with lengths indicating plume altitude:
short, 10–19 km; medium, 20–24 km; and long, 25–40 km
(table 1).

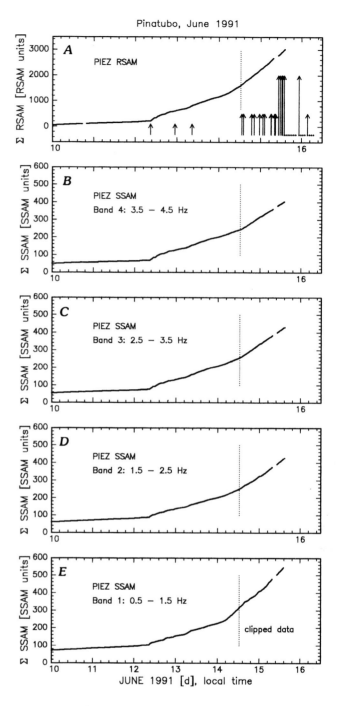

defined by drastically reduced rates: it is a period of apparent relative calm just prior to the eruption. It may be possible to establish a critical "failure rate" based on previously
observed apparent peak rates in a sequence of eruptions
rather than on the reduced rates sometimes observed at the
actual onset of previous eruption.

RSAM AND SSAM

Seismic activity was recorded at Pinatubo with the
recently developed Real-time Seismic Amplitude Measurement (RSAM) system (Murray and Endo, 1989; Endo and
Murray, 1991). RSAM provides consecutive 1- or 10-min
averages proportional to the absolute voltage output (amplitude) of a seismometer, representing ground velocity. The
advantage of RSAM is that it gives a measure of seismic
activity in near-real-time, even if separate seismic events
overlap in time or if volcanic tremor is present (figs. 4*A*,
5*A*).

The RSAM can be adjusted for geometric spreading
and instrument response to obtain "reduced displacement"
(Aki and others, 1977), and RSAM can be directly proportional to such commonly measured earthquake parameters
such as seismic moment rate or energy rate (Fehler, 1983).
Even if the proportionality constants are not precisely
known, RSAM counts can be used as $\dot{\Omega}$ with FFM without
further data manipulation; that is, the "inverse rate"
becomes equivalent to "inverse RSAM" (Voight and Cornelius, 1991; Cornelius and Voight, 1995). The precision of
proportionality is highest if the RSAM is dominated by a
single type of seismic event, because different event types
may have different constants.

In addition to RSAM, scientists used the relatively new
Seismic Spectral Amplitude Measurement system (SSAM)
(Rogers, 1989; Power and others, this volume; Stephens
and others, 1994). SSAM is based on the same principle as
RSAM in that it monitors mean amplitudes of ground
velocity, but it selectively filters the signal into 16 spectral
bands. During the 1989–90 eruption of Redoubt Volcano,

Alaska, SSAM was valuable for tracking swarms of LP
events and associated tremor that preceded many of the
tephra-producing eruptions (Stephens and others, 1994;
Cornelius and Voight, 1994). At Pinatubo (figs. 4*B*–*E*, 5*B*–
E), LP events and related tremor were observed to predominate the bands between 0.5 and 2.5 Hz, while VT events and
related broad-band tremor exhibited peaks between 2.5 and
4.5 Hz, and explosion eruptive-generated peaks occurred
between 0.5 and 1.5 Hz (Power and others, this volume). By
focusing on particular frequency bands, SSAM is able to
eliminate both signals and noise associated with other

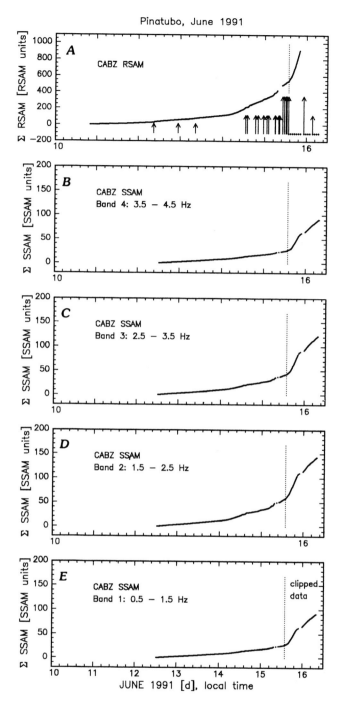

Pinatubo, June 1991

Figure 5. Cumulative RSAM and SSAM for station CABZ (10-min-interval data) from June 12 until station failure: *A*, RSAM; *B*, SSAM band 4 (3.5–4.5 Hz); *C*, SSAM band 3 (2.5–3.5 Hz); *D*, SSAM band 2 (1.5–2.5 Hz); *E*, SSAM band 1 (0.5–1.5 Hz). Explosive, pyroclastic eruptions began on June 12 and culminated in the cataclysmic eruption on June 15; sequence is marked in *A* by arrows with lengths indicating plume altitude: short, 10–19 km; medium, 20–24 km; and long, 25–40 km (table 1). Eruptions with plume altitudes less than 10 km, or with unconfirmed plume heights, are not shown.

frequencies. The result is to improve the signal-to-noise ratio for selected data sets and to enhance the suitability of the data for FFM analyses (Cornelius and Voight, 1994). SSAM was not used for foresight analyses at Pinatubo largely due to relatively "cumbersome processing" required at that time (Harlow and others, this volume; Power and others, this volume). More efficient processing routines are now available.

FORECASTING THE EXPLOSIVE ERUPTIONS OF JUNE 12–15, 1991

We present four groups of FFM analyses. The first group is the foresight analyses, in which we use the graphical RSAM data for the period June 1–10, which were received at Penn State University by fax from the USGS VCAT. The second group is the hindsight analyses of data over the same interval, but we use the digital RSAM data as recorded on a computer disk, with additional consideration of SSAM data. This period was the plinian buildup, from June 8 to 12. Third, we consider hindsight analyses during an earlier eruption phase, precursory explosions and dome emplacement, which was from June 3 to 7. All analyses of the first through third groups use basically the same analytical technique, which is inverse-rate plots with FFM data extrapolation in which a linear inverse-rate trend is assumed. A fourth group of analyses, representing in hindsight the cataclysmic buildup phase of June 12 to 15, considers nonlinear extrapolation techniques.

FORESIGHT ANALYSES: RSAM DATA BASE TO JUNE 10

Graphical RSAM output from two Pinatubo seismic stations with data from June 1 to 1600 on June 10 (fig. 6) was received by us at Penn State University by fax at about noon June 11, Eastern Daylight Time. The graphs were digitized and plotted as inverse RSAM the same day.

USGS VCAT also provided an earthquake summary for June 1 to 10, including an epicenter map and an epicenter depth-time plot, and data from June 5 through June 10 for the upper-flank UBO tiltmeter. VCAT noted two clusters of epicenters, one under the vent and the other under an intersection of youthful faults, and a trend toward fewer higher frequency earthquakes and more LP's.

During the previous year at Redoubt Volcano, Alaska, a major increase in shallow LP seismicity preceded the explosive January 2 eruptions. The LP character of the seismic swarms was interpreted to reflect pressurization of fluid-filled cracks (Chouet and others, 1994), and associated seismic rate changes were of sufficient consistency and duration to enable effective FFM analyses (in hindsight) by use of either RSAM or SSAM data (Voight and Cornelius, 1991; Cornelius and Voight, 1994).

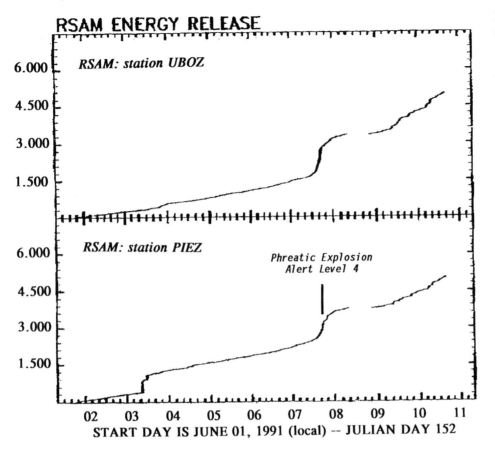

Figure 6. Cumulative RSAM energy release from two stations at Mount Pinatubo, UBOZ and PIEZ. Data shown through 1600 on June 10. Graphs were received at Penn State via fax from the U.S. Geological Survey Volcano Crisis Assistance Team.

The inverse-RSAM plots for two Pinatubo stations are shown in figure 7, referenced to Philippine local time. The seismic energy correlated with the gas/ash explosion of June 7 occurs as a spike on both plots. We knew about the cause of this spike and we knew about Alert Levels 3 and 4, issued for Pinatubo on June 5 and 7, respectively. About a day after this event, the data display a downward trend for station PIEZ (solid dots, fig. 7), which suggested the possibility of a forthcoming eruptive event. A linear least squares fit indicated a possible failure about June 12. We recorded the analogy between these results and our results from Redoubt, which had been published about 6 weeks before (fig. 4 of Voight and Cornelius, 1991). Although exhibiting more noise than station PIEZ, the inverse-RSAM downward trend at UBOZ (fig. 7B) crudely corroborated the analysis for station PIEZ.

The results were sent to USGS VCAT when our fax facility opened the next morning, but by then the eruption in fact had already begun. Therefore, our analysis, though made in foresight, did not contribute to hazard management at Pinatubo.

HINDSIGHT ANALYSES: RSAM AND SSAM DATA BASE TO JUNE 10

Next we recreate the foresight analysis of June 11 with the original digital data as stored on a computer disk. The "raw data," representing 10-min averages of the absolute voltage output of the seismometer, are averaged over 3-h intervals (fig. 8). Spikes in the RSAM output can be correlated to the phreatic explosions of June 3, 4, and 7 (compare table 1). The short data window from 2310 on June 8 to 1600 on June 10 (solid dots) allows a linear fit at both stations (figs. 8A,B). The time (t_s) when extrapolated rates become infinite—which represents a probable upper bound to the forecast eruption time—is 1830 on June 12 for station PIEZ and 2100 on June 12 for station UBOZ (fig. 8). The eruption windows (given by data extrapolation at the 97.5% confidence level) are large in both cases, but a lower bound is established for the morning of June 11—about 16 to 19 h after the last datum in the analysis. (As indicated previously, this confidence level reflects extrapolation statistics, not forecast probabilities.) For comparison, figures 8C and D show conventionally plotted RSAM data, together with fits

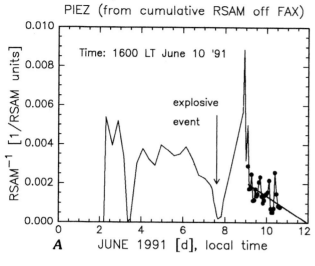

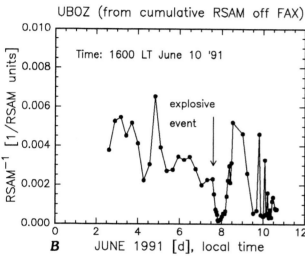

Figure 7. Inverse RSAM derived manually from graphs shown in figure 6, in foresight to the June 12 eruption. Data end at 1600 on June 10. RSAM data represent seismic-energy release rates, and inverse RSAM is expected to follow a downward trend prior to an eruption. Linear fit in *A* (PIEZ) to data between June 9 and 10 (solid dots) extrapolates the trend toward June 12 as the expected eruption time. In *B* (UBOZ), the trend is less distinct, but the overall trend between June 2 and 10 corroborates the analysis in *A*. The inverse-RSAM peak (low-point) on June 7 correlates with the high seismic energy release during a gas/ash explosion.

based on the extension of the inverse-RSAM data for figures 8*A* and *B*.

Additional insight into physical processes during this time period is gained by comparing RSAM and SSAM records over a 1-day period from June 9 to June 10 (fig. 9). The data comprise part of the precursory trend investigated in figures 6 and 8. The sequence is characterized by repetitive swarms of broad-band VT earthquakes over intervals of several hours; during these cycles, energy also occurred in the low-frequency bands. A strong 3-h signal starting at about 1100 on June 9 (June 9.46) is predominant

in low-frequency energy and was caused by LP seismicity including volcanic tremor (Power and others, this volume). A similar but less pronounced period of tremor occurred early on June 10 (fig. 9). Thus, the narrow frequency bands of SSAM allow the interpreter to distinguish between the buildup of energy release from different source mechanisms. These bands can be treated individually with FFM.

Figure 10 shows inverse SSAM calculated as 3-h averages from original 1-min PIEZ data files from June 1 to 1600 on June 10. The graphs can be compared to similar inverse-RSAM analyses for the same station (fig. 8*A*). The data involving higher frequency in band 4 (fig. 10*A*) reflect energy from VT activity that dominated the seismicity during this period. The inverse trend appears better defined for these data than for those of figure 10*B*, which includes energy from broad-banded VT events and for tremor. Comparison with the RSAM analysis of figure 8*A* suggests that if there is any advantage to evaluating the separate frequency band from 3.5 to 4.5 Hz, it is slight. In this particular case, RSAM was able to provide a realistic real-time "summary" of seismic development. This outcome was fortunate at Pinatubo, but such an outcome is not assured. At Redoubt Volcano after January 2, 1990 (Cornelius and Voight, 1994), where signal-to-noise ratio was low, the FFM results were sensitive to the source process evaluated. During this period at Redoubt, SSAM data in low-frequency bands were useful, but RSAM data were generally misleading.

HINDSIGHT ANALYSES: RSAM DATA BASE TO JUNE 7

The previous analyses represent the situation during the plinian buildup phase, when the possibility of an eruption was already clearly recognized (Harlow and others, this volume). The following analyses explore how FFM analyses might have supported forecasts at an earlier time.

Following the metastable phase of May, during which no trend in seismicity developed, a clear inverse trend could have been detected in early June, during the precursory explosions and dome emplacement phase, as shown in figures 11*A* and *B* for a data window chosen between 0010 on May 31 and 0900 on June 7 (solid dots). The trend is apparently suitable for a linear least squares fit. The resulting data envelope is better constrained (narrower) than in the analyses previously illustrated for June 10, chiefly because of the longer data window. Infinite rates are forecast at 0745 on June 16 for PIEZ and at 0030 on June 14 for UBOZ. The eruption window combining results from the two stations falls between 0845 on June 12 and 1945 on June 20. For comparison, figures 11*C* and *D* show conventionally-plotted RSAM data, together with conventional curve fits based on parameters derived from inverse-RSAM analyses. The gas/ash emission and dome extrusion of June 7 occurred just after the close of this data window; the graphs and fits

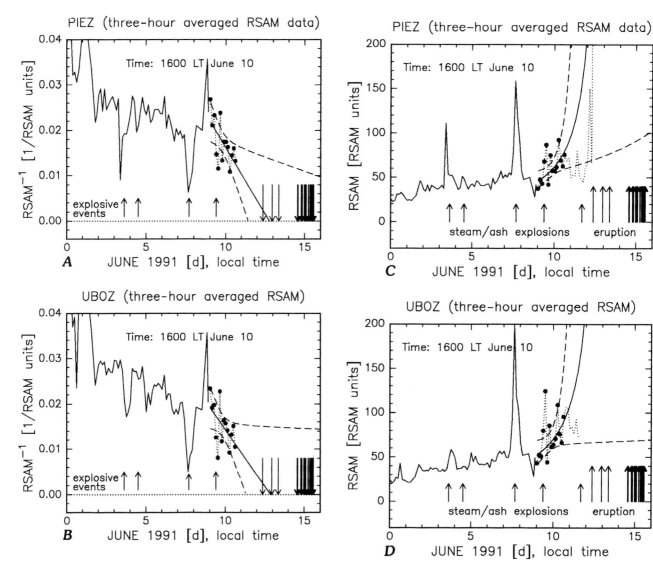

Figure 8. Inverse RSAM calculated as 3-h averages from original 10-min-interval digital data. Data end at 1600 on June 10, and the graphs are equivalent to the relatively crude digitized plots constructed in foresight and shown in figure 7. Linear fits to inverse-RSAM trends between 2310 on June 8 and 1600 on June 10 (solid dots) for *A* (PIEZ) and *B* (UBOZ) are shown as solid lines with error envelopes at the 97.5% confidence level (dashed lines). Explosive eruptions beginning on June 12 and culminating in the cataclysmic eruption on June 15 are marked by downward-pointing arrows. Short upward-pointing arrows mark ash/gas explosions. *C* and *D* show conventional, non-inverted RSAM together with the solid-line curve fits and dashed envelopes developed in *A* and *B*. Continuation of data is shown as dotted lines.

of figure 11 suggest that it would not have been anticipated by FFM for RSAM-data averaged over 3-h intervals.

However, more detailed representation with 10-min data on June 7 (fig. 12) suggests that recognizable precursory seismicity occurred prior to dome extrusion and gas emission. The inverse trend suggests event occurrence about 1648 on June 7 (June 7.7), near the time of the observed explosive event, and possibly reflecting the time of dome emergence. Due to weather conditions, the dome was not actually observed until the following day. The seismicity during this period was dominated by bursts of VT swarm activity, as reflected by a fairly even distribution of

energy throughout several frequency bands, with a slight maximum between 2.5 and 3.5 Hz. In such cases, with noise relatively low, RSAM provides an accurate overall measure of seismicity.

Cumulative RSAM based on 10-min interval data, from May 2 to mid-June, is shown in figure 13. The dashed curve is based on the inverse-RSAM linear fit for the data base to 1600 on June 10 (fig. 8); the dotted curve is based on the data base to 0900 on June 7 (fig. 11). The dotted and dashed vertical lines bracket the data bases used for June 7 and June 10 analyses, respectively. Observed values may be

PIEZ (10-minute data)

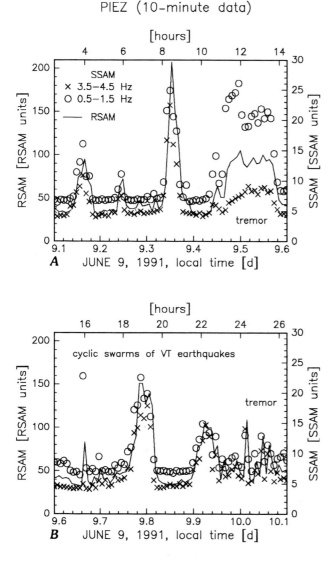

Figure 9. Comparison of RSAM record (solid line) with SSAM bands 1 (open circles) and 4 (×'s), both from station PIEZ, within the plinian buildup phase of seismicity as defined by Power and others (this volume), from 0224 on June 9 to 0224 on June 10. Note difference in RSAM and SSAM scales. The sequence is characterized by cyclic swarms of volcano-tectonic (VT) earthquakes at intervals of several hours. A strong, 3-h signal starting about 1100 on June 9 has predominantly lower frequencies and was caused by volcanic tremor (Power and others, this volume).

compared with extrapolated values beyond the data base limits.

These figures illustrate the advantage of the inverse-rate technique of FFM over conventional plots in amplifying changes in rates for data at subdued rate levels and in facilitating the extrapolation of accelerating trends, because the inverse trends are commonly nearly linear. Graphical curve fitting to conventionally displayed data is usually more ambiguous than curve fitting to inverse-rate data.

PIEZ (three-hour averaged SSAM data)

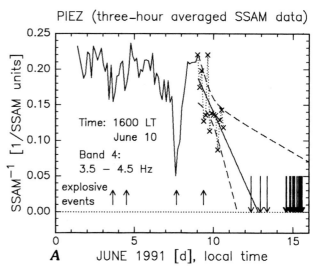

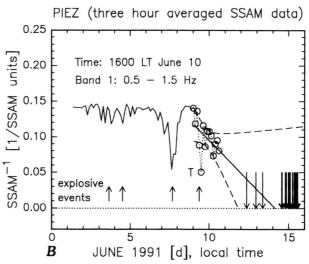

Figure 10. Inverse SSAM calculated as 3-h averages from original 1-min-interval data files: *A*, Band 4 (3.5–4.5 Hz); *B*, Band 1 (0.5–1.5 Hz). Data end at 1600 on June 10. Compare RSAM analysis, figure 8*A*. Linear fits are solid lines, and error envelopes at the 97.5% confidence level are dashed lines. Data of higher frequency, band 4 are dominated by VT energy. Tremor at midday on June 9 is marked as T (compare fig. 9). Explosive eruptions beginning on June 12 and culminating in the cataclysmic eruption on June 15 are marked by downward-pointing arrows. Upward-pointing arrows mark ash/gas explosions.

HINDSIGHT ANALYSES: RSAM AND SSAM DATA BASE BEYOND JUNE 10

Next we consider analyses for the later stages of the mid-June eruption. The major eruption had begun with a vent-clearing phase on June 12, which produced plinian columns exceeding 20 km in altitude. A 28-h break then occurred in explosive events, until 1309 on June 14. During this period the continuously updated RSAM and SSAM data might have been evaluated by FFM prior to the

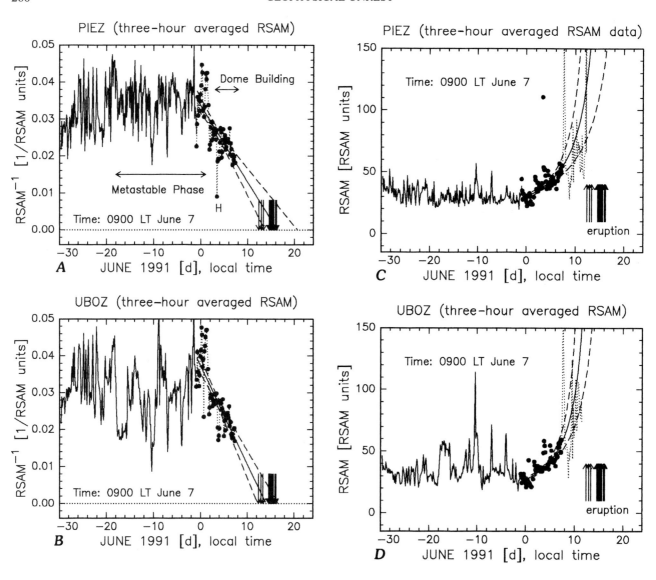

Figure 11. Inverse RSAM calculated as 3-h averages from original 10-min-interval data files between May 2 and June 7. Linear fits to inverse-RSAM trends between 0010 on May 31 and 0900 on June 7 (solid dots) for (*A*) PIEZ and (*B*) UBOZ are shown as solid lines and with error envelopes at the 97.5% confidence level (dashed lines). Inverse peak marked H was caused by helicopter noise (compare Power and others, this volume, fig. 3). Explosive eruptions beginning on June 12 and culminating in the cataclysmic eruption on June 15 are marked by arrows. *C* and *D* show conventional, non-inverted RSAM together with the solid-line curve fits and dashed envelopes developed in *A* and *B*. Continuation of data shown as dotted lines.

cataclysmic eruption phase of June 15, and we consider what might have been accomplished by this evaluation.

Figures 14*A* and 15*A* show 3-h averages of inverse RSAM from two stations from June 10 to 15 (compare figs. 4, 5). Data from distal station CABZ delineate an approximately continuous inverse trend beyond June 11. Data at station PIEZ exhibit a step function, with a prominent energy rate increase beyond the onset of the June 12 explosion. SSAM data for four bands are shown with 3-h-averaged data during this same period (figs. 14*B*–*E*, 15*B*–*E*). A significant precursory inverse-rate trend developed at PIEZ over the half-day preceding resumption of explosive activity on June 14 (fig. 14); this trend is evident both on

RSAM and all bands of SSAM and indicates the continued escalation of both VT swarms and LP seismicity that occurred during this period. SSAM spectrograms show enhanced peaks at 2.5 to 4.5 Hz, correlated with VT swarms and continuous VT seismicity, and also at 0.5 to 1.5 Hz, correlated with LP swarms and tremor (Power and others, this volume). Some of the tremor (noise) during this period (at 2–3 Hz) may be correlated with gas and ash emissions (low-level eruptive activity) rather than shallow hydrothermal events (Harlow and others, this volume). Relative band strength is also influenced by station location as well as by source characteristics; thus band 4 at CABZ is relatively weaker in comparison to other bands than band 4 at PIEZ

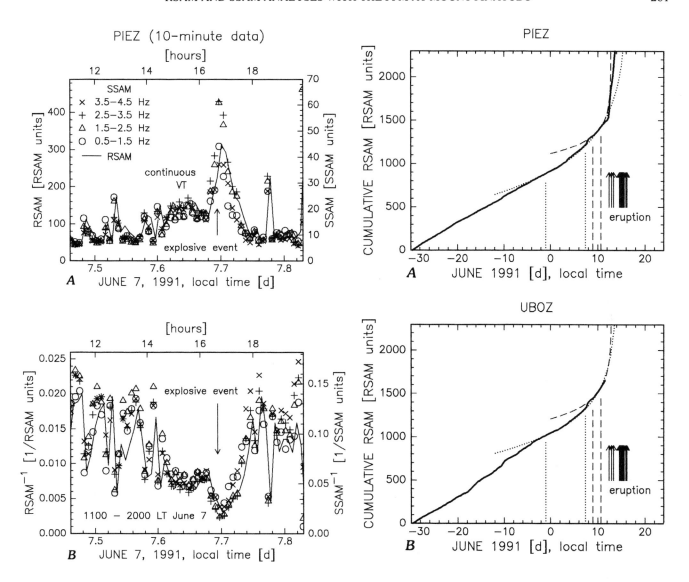

Figure 12. Comparison of station PIEZ RSAM record (solid line) of June 7 with SSAM bands 1, 2, 3, and 4 (symbols). *A*, conventional RSAM and SSAM; *B*, inverse RSAM and inverse SSAM. Note difference in RSAM and SSAM scales. A swarm of volcano-tectonic (VT) earthquakes occurs during the dome building phase (compare Power and others, this volume, fig. 2). Separate events overlap on helicorder records from 1600 onward, and the seismicity was classified as continuous VT (Power and others, this volume).

Figure 13. Cumulative RSAM (10-min-interval data) from May 2 to mid-June for (*A*) PIEZ and (*B*) UBOZ. Sequence of explosive eruptions beginning on June 12 and culminating in the cataclysmic eruption on June 15 is marked by arrows. Dashed curve is based on linear fits to inverse-RSAM data at 1600 on June 10 (fig. 8), and dotted curve is derived from linear fits at 0900 on June 7 (fig. 11). Dotted and dashed vertical lines mark the data window of the June 7 and 10 analyses, respectively.

(figs. 4, 5), and this difference probably reflects the selective attenuation of high-frequency signals.

The excellent match of patterns for SSAM bands 1 to 4 suggests that VT swarm seismicity may have dominated the patterns during this period, at least on the scale of 3-h-averaged data of figures 14 and 15. For example, there are no peaks or troughs in low-frequency band 2 not matched by those in higher frequency bands 3 and 4, an observation that suggests broad band activity. Yet from the perspective of helicorder records, a dominating aspect of this period was the buildup of LP activity from minor swarms to a series of

increasingly large events, with the largest of these approximately equivalent to magnitude 4 earthquakes (Harlow and others, this volume). It is possible that a significant proportion of the seismicity of this period may have been hybrid, reflecting source processes involving both VT and LP endmember types; this hybrid nature could help to explain the combination of broad band activity and multiple spectral peaks indicated by SSAM.

Analyses for RSAM from station CABZ were made in hindsight at intervals of 7 h for data windows beginning on June 11 and ending at 1930 on June 13 and at 0230 and

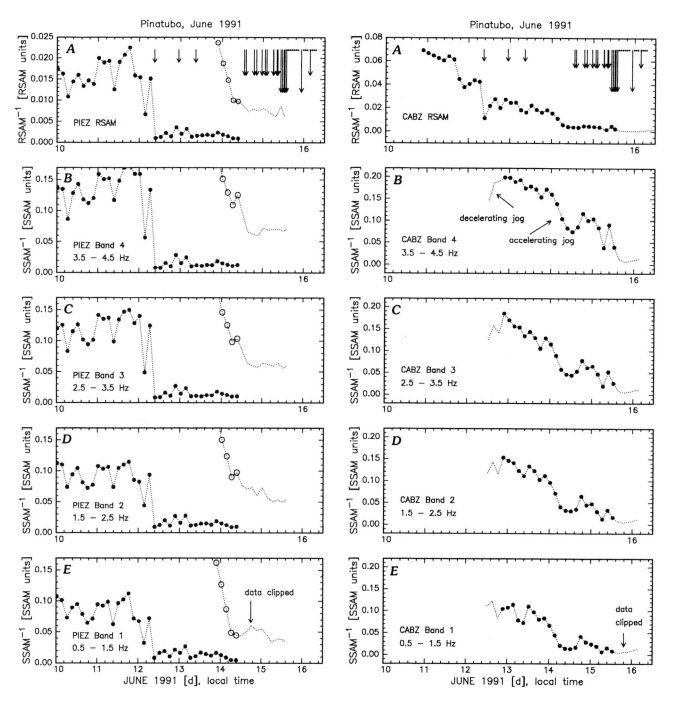

Figure 14. Inverse RSAM and inverse SSAM at station PIEZ calculated as 3-h averages from June 10 until the station was destroyed: *A*, RSAM; *B*, SSAM band 4 (3.5–4.5 Hz); *C*, SSAM band 3 (2.5–3.5 Hz); *D*, SSAM band 2 (1.5–2.5 Hz); *E*, SSAM band 1 (0.5–1.5 Hz). Open circles are inverse RSAM and inverse SSAM multiplied by a factor of 10. Many of the data beginning midday on June 14 were electronically clipped (dotted lines) and should not be used with FFM because of unknown distortions. Explosive eruptions beginning on June 12 are marked in *A* by arrows whose lengths indicate plume altitude: short, 10–19 km; medium, 20–24 km; and long, 25–40 km (table 1). Conventionally plotted cumulative data are shown in figure 4, at the same time scale.

← **Figure 15.** Inverse RSAM and inverse SSAM at station CABZ calculated as 3-h averages from June 10 until station failure. *A*, RSAM; *B*, SSAM band 4 (3.5–4.5 Hz); *C*, SSAM band 3 (2.5–3.5 Hz); *D*, SSAM band 2 (1.5–2.5 Hz); *E*, SSAM band 1 (0.5–1.5 Hz). Many of the data beginning midday on June 15 were electronically clipped (dotted lines at tail end of the time series) and should not be used with FFM because of unknown distortions. Inverse-SSAM trends of all bands (and to a lesser extent also inverse RSAM) define a decelerating jog in the aftermath of the June 12 eruption and an accelerating jog prior to renewed eruptions on June 14. Explosive eruptions beginning on June 12 are marked in *A* by arrows whose lengths indicate plume altitude: short, 10–19 km; medium, 20–25 km; and long, 25–40 km (table 1). Conventionally plotted cumulative data are shown in figure 5, at the same time scale.

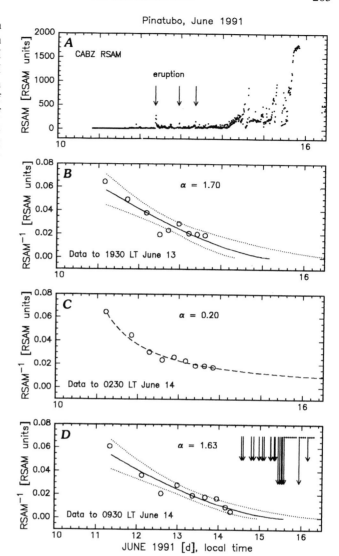

0930 on June 14 (fig. 16). Using the end-weighted FFM procedure as previously described, we calculated rates from integrated RSAM time series by linear interpolation over approximately constant RSAM increments. These constant RSAM increments were arbitrarily chosen to be 10% of the total encountered RSAM value, so each resulting rate series has nine data points, which occur with an increasing frequency as RSAM climbs toward the end of the data window. Spikes at times of the three eruptions within the data set (arrows, fig. 16) were removed prior to rate calculation so as not to contaminate precursory signal with syneruptive activity.

The first fit (fig. 16*B*) yields α=1.70 (A=0.0486) and gives 0400 on June 15 for the time of singularity with an eruption window from 0930 on June 14 to 2115 on June 16. The point t_f may also be estimated with a predetermined intercept that represents the expected finite failure rate. This failure value is unknown in this case, but to test the sensitivity of the solution, we arbitrarily have taken a rate one order of magnitude larger than the last data point in the data set. With this assumption, the time of failure is indicated as 1915 on June 14, and the eruption window is from 0315 on June 14 to 0700 on June 16. These extrapolations are consistent and in good agreement with the onset of the next explosive pulse at 1309 on June 14 and the cataclysmic phase at 1027 on June 15 (table 1).

Some short-term deceleration followed the accelerating jog correlated with the first eruption on June 12 (fig. 15). Also, the last two rate points of figure 16*B* suggest some deceleration toward the end of the data window, with respect to the curve fit. This deceleration trend continued through 0230 on June 14, the time of the second analysis (fig. 16*C*). The fit yields a trend with α=0.20 (A=7.988). This is a low α value, because α values recorded elsewhere typically fall between 1.5 and 2.2 (Cornelius, 1992; Voight, 1988). Interpretive concerns might arise with low α in a foresight situation, and an accurate failure rate would be required for event forecasting. Periods of deceleration near the end of a data window can cause such a shift toward

Figure 16. Nonlinear FFM analyses of RSAM data from station CABZ. 10-min-interval data of RSAM are shown in *A*. Spikes on June 12 and 13 (vertical arrows in *A*) reflect pyroclastic eruptions, and RSAM data during the eruptions are omitted to emphasize precursory changes. Data bases start at 2000 on June 10 and end at (*B*) 1930 on June 13, (*C*) 0230 on June 14, and (*D*) 0930 on June 14. Rates were calculated over constant RSAM increments (10% of total encountered RSAM value). *B*–*D* show best-fit curves (solid and dashed lines); *B* and *D* show error envelopes at the 97.5% confidence interval (dotted lines). α is empirical curve-fitting constant, as in equation 1 in text. Sequence of eruptions following the analyses is marked with arrows in *D*; lengths of arrows indicate altitude reached by eruption cloud, as in figure 4.

extreme α values. If an accelerating jog occurred soon thereafter, the overall trend might continue as before; but the interpreter needs to be vigilant about an alternative possibility, namely, that a deceleration in seismicity commonly precedes an eruption.

However, such an acceleration was observed by 0930 the same day (fig. 16*D*); the overall trend at this time was α=1.63 (A=0.0598), and forecast times returned to values similar to the first analysis: singularity at 1330 on June 15 with a window from 2315 on June 14 to 1600 on June 16. Assuming a finite failure rate as before yields 0800 on June 15, with the eruption window from 1845 on June 14 to 0815 on June 16, consistent with final phases of the eruption.

For perspective, we reemphasize that we view the FFM approach as one tool among several, a convenient method to track systematic, progressively accelerating energy release or deformation. Statements based on application of the method have to be interpreted within the context of volcano observations and other acquired information. At the time indicated for the above analysis, the morning of June 14, three plinian eruptions had already occurred (between 0851 on June 12 and 0840 on June 13) within the past 48 h. The continuing overall downward inverse trend within a quiet interval suggested that a very high rate (perhaps an eruption) was to be expected on June 15, and this turned out to be a correct inference. However, this did not exclude the possibility of eruptive activity at an earlier time, especially as the interpretation should consider the accelerating jog, which can be seen at the end of the last data window (fig. 16D, last three data points). In fact, the next eruptive pulse at 1309 on June 14 was heralded by a renewed spurt of seismic acceleration.

Details of trends on the morning of June 14 are shown in figure 17; error envelopes are shown at the 97.5% confidence level. The data are from an accelerating jog on the morning of June 14 (until 0830, dashed vertical line). The nonlinear analyses give somewhat varied results, implying imminent failure with RSAM or band 4 SSAM but failure later that day with band 1 SSAM. Overall, the trend extrapolations provided a fair indication for the renewed eruptive activity during the early afternoon (vertical arrow). A nonlinear curve fit to data from band 3 results in $\alpha < 0$ but is not accepted, because a roughly linear trend seems visually supportable for this time series; a linear fit is shown.

DISCUSSION

Pinatubo provided the first opportunity to compare RSAM and SSAM records in detail for a variety of event types. Previously, only data for the 1989–90 eruption of Redoubt Volcano, Alaska, had been thoroughly analyzed, and in that eruption precursory LP seismicity was dominant (Stephens and others, 1994). For the January 2 eruption at Redoubt, SSAM and RSAM provided comparable data; after January 2, when signal strength diminished relative to noise, RSAM data were typically misleading, although SSAM in low-frequency bands could detect eruption precursors (Cornelius and Voight, 1994; Stephens and others, 1994).

In contrast, at Pinatubo, RSAM was relied on throughout as a monitoring tool, and, unlike at Redoubt, the signal-to-noise ratio was large throughout the monitoring period.

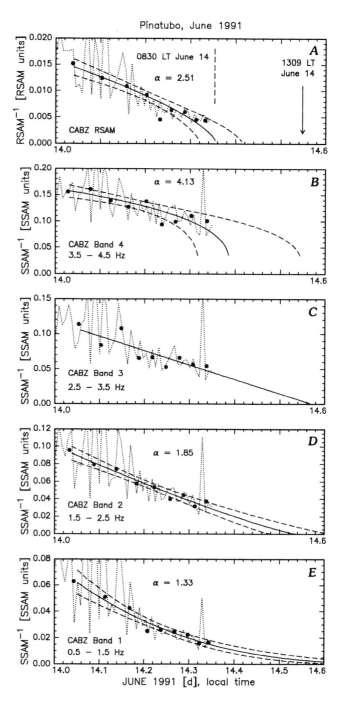

Figure 17. Nonlinear analyses of (A) RSAM and (B–E) SSAM data on June 14 from station CABZ with FFM. SSAM data shown for (B) band 4, (C) band 3, (D) band 2, and (E) band 1. Data window is from 0000 to 0830 (vertical dashed line), a period of accelerating seismicity. Inverse rates (solid dots) were calculated over constant RSAM or SSAM increments (10% of total encountered values). Inverse RSAM and inverse SSAM averaged over 10 min shown as dotted lines for comparison. Best-fit curves are solid lines; error envelopes are shown at the 97.5% confidence interval (dashed lines). α is empirical curve-fitting constant, as in equation 1 in text. Arrow in A marks eruption at 1309 on June 14.

Up to the time of dome emplacement, VT seismicity was dominant, and the individual SSAM bands virtually replicated the RSAM pattern (fig. 13). From June 8 to June 12, seismicity was characterized by cyclic swarms of VT seismicity, with a gradual increase in tremor. Most SSAM bands simulated the RSAM pattern, with some differences in low-frequency bands that reflect LP energy in addition to VT broad-band components (figs. 9, 10). From June 12 to 15 the seismicity increased in complexity, involving combinations of VT seismicity, LP swarms, and tremor. That many swarms may have been hybrid would help to explain the similar patterns displayed by RSAM and SSAM bands 1–4, despite two distinctive spectral peaks correlative with VT and LP seismicity. However, resolution of the question of hybrid seismicity will require detailed analyses of digital data and helicorder records.

On the whole, due to the broad-band nature of dominant seismicity, RSAM patterns were broadly simulated by each of the SSAM bands analyzed. Thus, unlike at Redoubt in 1990, RSAM at Pinatubo was able to provide an authentic measure of overall seismic energy release. This was fortunate, as the decisions based on seismic monitoring at Pinatubo, particularly after June 12, indeed relied on RSAM for this purpose (Harlow and others, this volume). SSAM played an important supporting role in pointing to source processes (LP and hybrid seismicity) suggestive of high eruptive potential.

In this paper we have recreated our foresight analyses on the basis of graphical RSAM data, and we presented several more FFM analyses to simulate what might have been done in foresight with digital data. By June 7 an eruption might have been anticipated around June 14–16, with an eruption window from June 12 to 20. At the time (June 5–7), the possibility of a major eruption was recognized by monitoring scientists from a multitude of pertinent volcanological observations; analyses using FFM could have been integrated with these data. Updated FFM analyses, such as those presented for June 10–14, provided information consistent with actual event occurrence and could have been integrated with the cumulative information base utilized for onsite eruption forecasts.

However, we recognize that the hazard management outcome of the Pinatubo eruption could not have been much improved by FFM, or, for that matter, by any additional technological device or methodology. The precursory phenomena were strong, and despite inevitable interpretive uncertainties, scientific advisories were effective in influencing prudent decisionmaking by public officials, who, to their credit, were willing to act decisively on the uncertain information available to them. In hindsight, hazard managers made virtually all the right moves.

As with all tools, experience is needed for a skillful application of the FFM approach to eruption prediction. The potential advantages of viewing accelerating processes with inverse-rate curves are significant. Small changes at low rates are amplified at the onset of any new development, and complex curves of conventionally plotted graphs may be nearly linearized by data inversion. However, irregularities involving precursory phenomena, such as decelerating and accelerating jogs (reflecting cyclic swarms), may complicate the interpretation. Perhaps not infrequently, profound departures from systematic progressive rate growth may render the FFM procedure inapplicable for certain cases. Under favorable circumstances a skilled observer may be able to use FFM to anticipate future developments and will make use of constantly renewed updates. Reliance of FFM on a feedback system between trend recognition and forecasting makes the procedure a potentially useful tracking method for physical processes.

Certain "conventions" of data treatment may help to define experience. Three such conventions have been demonstrated with the aforementioned analyses. First, eruption windows are based on data envelopes constructed at a specified confidence level to the linearized inverse-rate fit. The windows thus reflect data scatter, length of data window, and the time interval between the time of analysis and time of calculated failure. Second, in some sensitivity analyses used here, failure rates are arbitrarily assumed to be an order of magnitude larger than the last previously calculated rate. This alternative treatment is generally conservative, and its main effect is to counterbalance too-late forecasts founded on the simplifying assumption of infinite failure rates, especially for concave-upward inverse-rate curves ($\alpha<2.0$). The sensitivity of the approach can be judged by comparing values obtained using t_f and t_s. Third, rates may be calculated over constant data increments instead of constant time increments. Data increments of about 10% of the total encountered value produce a smoothed time series of rates that is favorably biased toward the latest developments (end-weighted).

Calculated eruption windows should be qualified by statements describing the observed situation. Ideally, eruption windows should combine analyses from several stations, and, where feasible, several data types. Forecast sensitivity should be tested in relation to choice of data-averaging intervals. In systems for which systematic behavior is dominant, FFM may provide the opportunity to quantify observed accelerating processes. However, it would be unwise to interpret inverse-rate curves without simultaneous and full consideration of other volcanologic information.

ACKNOWLEDGMENTS

We thank colleagues and U.S. Geological Survey staff for their assistance. We especially thank Dave Harlow, Dick Janda, Tom Murray, John Power, and Ed Wolfe for access to and assistance with data. Advance drafts of Pinatubo manuscripts were kindly provided by Dave Harlow, Chris

Newhall, John Power, and Randy White. This work was conducted under a cooperative project involving Penn State University and the Cascades Volcano Observatory (U.S. Geological Survey), with support by the National Science Foundation. Colleague review was provided by William Chadwick, Chris Newhall, and Don Swanson. Voight's involvement with Pinatubo was kindled during April 1991 in deliberations with Dick Janda on the cobblestoned streets of rococo Popayán, Colombia, and at a paramo tent camp on the shoulder of Cotopaxi, Ecuador, as an early VCAT overseas mobilization was being choreographed by Janda. With voluminous pumiceous pyroclastic flows hundreds of years old at an otherwise little-known and long-dormant volcanic complex, our concern for this densely populated region was that an event resembling but exceeding El Chichón might be in developmental stages. Janda would play a vital role in the Pinatubo crisis response. He is gone now.

The sea and the earth are unfaithful to their children: a truth, a faith, and a generation of men goes—and is forgotten, and it does not matter! Except, perhaps, to the few of those who believed the truth, confessed the faith—or loved the men.

REFERENCES CITED

Aki, K., Fehler, M., and Das, S., 1977, Source mechanism of volcanic tremor: Fluid-driven crack models and their application to the 1963 Kilauea eruption: Journal of Volcanology and Geothermal Research, v. 2, no. 3, p. 259–287.

Chouet, B.A., Page, R.A., Stephens, C.D., Lahr, J.C., and Power, J.A., 1994, Precursory swarms of long-period events at Redoubt Volcano (1989–1990), Alaska: Their origin and use as a forecasting tool, in Miller, T.P., and Chouet, B.A., eds., The 1989-1990 eruptions of Redoubt Volcano, Alaska: Journal of Volcanology and Geothermal Research, v. 62, p. 95–135.

Cornelius, R.R., 1992, Feasibility study for the materials science approach to volcano eruption prediction: University Park, Pa., The Pennsylvania State University, Ph.D. thesis, 470 p.

Cornelius, R.R., and Voight, B., 1989, Determination of eruption-prediction constants for accelerated creep or seismicity [abs.]: Geological Society of America Abstracts with Programs, v. 21, p. 13.

———1994, Seismological aspects of the 1989–1990 eruption at Redoubt Volcano, Alaska: The Materials Failure Forecast Method (FFM) with RSAM and SSAM seismic data, in Miller, T.P., and Chouet, B.A., eds., The 1989–1990 eruptions of Redoubt Volcano, Alaska: Journal of Volcanology and Geothermal Research, v. 62, nos. 1–4, p. 469–498.

———1995, Graphical and PC-software analysis of volcano eruption precursors according to the Materials Failure Forecast Method (FFM): Journal of Volcanology and Geothermal Research, v. 64, p. 295–320.

Daag, A.S., Tubianosa, B.S., and Newhall, C.G., 1991, Monitoring sulfur dioxide emissions at Pinatubo Volcano, central Luzon, Philippines [abs.]: Eos, Transactions, American Geophysical Union, v. 72, no. 44, Fall Meeting Supplement, p. 61.

Daag, A.S., Tubianosa, B.S., Newhall, C.G., Tuñgol, N.M, Javier, D., Dolan, M.T., Delos Reyes, P.J., Arboleda, R.A., Martinez, M.L., and Regalado, M.T.M., this volume, Monitoring sulfur dioxide emission at Mount Pinatubo.

Endo, E.T., and Murray, T.L., 1991, Real-time Seismic Amplitude Measurement (RSAM): A volcano monitoring and prediction tool: Bulletin of Volcanology, v. 53, p. 533–545.

Ewert, J.W., Lockhart, A.B., Marcial, S., and Ambubuyog, G., this volume, Ground deformation prior to the 1991 eruptions of Mount Pinatubo.

Fehler, M., 1983, Observations of volcanic tremor at Mount St. Helens volcano: Journal of Geophysical Research, v. 88, p. 3476–3484.

Harlow, D.H., Power, J.A., Laguerta, E.P., Ambubuyog, G., White, R.A., and Hoblitt, R.P., this volume, Precursory seismicity and forecasting of the June 15, 1991, eruption of Mount Pinatubo.

Harlow, D.H., Punongbayan, R.S., Ambubuyog, G., Laguerta, E., Power, J.A., Newhall, C.G., Hoblitt, R.P., Lockhart, A.B., Murray, T.L., Wolfe, E.W., and Ewert, J.E., 1991, Seismic activity and forecasting of the climactic eruption of Pinatubo Volcano, Luzon, Philippines on June 15, 1991 [abs.]: Eos, Transactions, American Geophysical Union, v. 72, no. 44, p. 61.

Hoblitt, R.P., Wolfe, E.W., Scott, W.E., Couchman, M.R., Pallister, J.S., and Javier, D., this volume, The preclimactic eruptions of Mount Pinatubo, June 1991.

Lockhart, A.B., Marcial, S., Ambubuyog, G., Laguerta, E.P., and Power, J.A., this volume, Installation, operation, and technical specifications of the first Mount Pinatubo telemetered seismic network.

Murray, T.L., and Endo, E.T., 1989, A Real-time Seismic Amplitude Measurement system (RSAM): U.S. Geological Survey Open-File Report 89–684, 21 p.

Murray, T.L., Power, J.A., March, G.D., and Marso, J.N., this volume, A PC-based realtime volcano-monitoring data-acquisition and analysis system.

Newhall, C.G., Daag, A.S., Delfin, F.G., Jr., Hoblitt, R.P., McGeehin, J., Pallister, J.S., Regalado, M.T.M., Rubin, M., Tamayo, R.A., Jr., Tubianosa, B., and Umbal, J.V., this volume, Eruptive history of Mount Pinatubo.

Pinatubo Volcano Observatory Team, 1991, Lessons from a major eruption: Mt. Pinatubo, Philippines: Eos, Transactions, American Geophysical Union, v. 72, p. 545–555.

Power, J.A., Murray, T.L., Marso, J.N., and Laguerta, E.P., this volume, Preliminary observations of seismicity at Mount Pinatubo by use of the Seismic Spectral Amplitude Measurement (SSAM) system, May 13–June 18, 1991.

Punongbayan, R.S., Ambubuyog, G., Bautista, L., Daag, A.S., Delos Reyes, P.J., Laguerta, E., Lanuza, L., Marcial, S., Tubianosa, B., Newhall, C.G., Power, J., Harlow, D.H., Ewert, J.W., Gerlach, T., Hoblitt, R.P., Lockhart, A.B., Murray, T.L., and Wolfe, E.W., 1991, Eruption precursors or a false alarm? A process of elimination [abs.]: Eos Transactions, American Geophysical Union, v. 72, no. 44, p. 61.

Punongbayan, R.S., Newhall, C.G., Bautista, M.L.P., Garcia, D., Harlow, D.H., Hoblitt, R.P., Sabit, J.P., and Solidum, R.U., this volume, Eruption hazard assessments and warnings.

Rogers, J.A., 1989, Frequency-domain detection of seismic signals using a DSP coprocessor board, in Lee, W.H.K., ed., Toolbox for seismic data acquisition, processing, and analysis: International Association of Seismology and Physics of the Earth's Interior in collaboration with Seismological Society of America, El Cerrito, Calif., p. 151–172.

Stephens, C.D., Chouet, B.A., Page, R.A., Lahr, J.C., and Power, J.A., 1994, Seismological aspects of the 1989–1990 eruptions at Redoubt Volcano, Alaska: The SSAM perspective, in Miller, T.P., and Chouet, B.A., eds., The 1989-1990 eruptions of Redoubt Volcano, Alaska: Journal of Volcanology and Geothermal Research, v. 62, p. 153–182.

Voight, B., 1988, A method for prediction of volcanic eruptions: Nature, v. 332, p. 125–130.

———1989, A relation to describe rate-dependent material failure: Science, v. 243, p. 200–203.

Voight, B., and Cornelius, R.R., 1991, Prospects for eruption prediction in near real-time: Nature, v. 350, p. 695–698.

Wolfe, E.W., 1992, The 1991 eruptions of Mount Pinatubo, Philippines: Earthquakes & Volcanoes, v. 23, no. 1, p. 5–35.

Wolfe, E.W. and Hoblitt, R.P., this volume, Overview of the eruptions.

Preliminary Observations of Seismicity at Mount Pinatubo by use of the Seismic Spectral Amplitude Measurement (SSAM) System, May 13–June 18, 1991

By John A. Power,[1] Thomas L. Murray,[1] Jeffery N. Marso,[1] and Eduardo P. Laguerta[2]

ABSTRACT

During the 1991 Pinatubo volcanic crisis, seismicity was monitored in part by a Seismic Spectral Amplitude Measurement (SSAM) system. This personal-computer-based system continuously monitors seismic amplitudes within 16 user-defined frequency bands. The SSAM system is particularly useful for monitoring the spectral character of seismic signals over time. In this paper, SSAM data are displayed in time-frequency plots where the spectral amplitude is represented by an 11-step gray scale that is produced by a computer program called SSAM_VU. The program allows for a number of user-defined parameters for producing the spectrograms. We review the various effects that each parameter can have on the appearance of the spectrogram and its interpretation.

The 1991 eruption of Mount Pinatubo produced a wide variety of seismicity including swarms of volcano-tectonic earthquakes, long-period events, volcanic tremor, and explosive eruptions. Swarms of volcano-tectonic earthquakes and periods of continuous volcano-tectonic seismicity exhibited strong spectral peaks between 2.5 and 4.5 hertz and occurred during the emplacement of a lava dome on June 5–7, prior to the onset of explosive activity (June 8–12), and following the cataclysmic eruption on June 15. Long-period events and volcanic tremor produced a strong signal between 0.5 and 2.5 hertz. Energy in this frequency range increased dramatically between June 1 and the cataclysmic eruption on June 15. Explosive eruptions generated signals with spectral peaks between 0.5 and 1.5 hertz. SSAM data from the cataclysmic eruption on June 15 shows a shift in the spectral peak from 0.5 to 1.5 hertz to between 1.5 and 3.5 hertz that is coincident with the onset of magnitude 4+ earthquakes. Although SSAM did not play a major role in forecasting eruptive activity at Pinatubo, analysis presented here, as well as experience at other volcanoes, indicates it is a valuable tool for quickly analyzing seismicity at restless volcanoes.

INTRODUCTION

The cataclysmic eruption of Mount Pinatubo on June 15, 1991, was preceded by at least 10 weeks of unrest characterized by increasing seismic activity, high SO_2 emissions, emplacement of a lava dome, and numerous smaller explosive eruptions (Punongbayan and others, this volume). In response to the initial unrest, a seven-station radiotelemetered seismic network was deployed around the volcano during late April and early May (Lockhart and others, this volume). A data acquisition and analysis system at Clark Air Base (Murray and others, this volume) received the data and provided rapid calculation of hypocenters, Real-time Seismic Amplitude Measurements (RSAM) (Endo and Murray, 1991), and Seismic Spectral Amplitude Measurements (SSAM) (Rogers, 1989).

Unlike earthquake location techniques and RSAM, SSAM is a relatively new technique in real-time volcano monitoring. At the time of this writing, eruptive cycles at only three volcanoes have been monitored by use of the SSAM system, and only data from the 1989–90 eruption of Redoubt Volcano, Alaska, have been analyzed in detail (Stephens and others, 1994). SSAM continuously monitors seismic signals and calculates 1-min average signal strengths in each of 16 specified frequency bands for each seismometer. It is ideal for monitoring changes in spectral character through time, particularly when a certain type of event with a given characteristic frequency dominates. During the Redoubt eruptions, SSAM was particularly valuable for detecting and tracking small swarms of long-period events and low-level volcanic tremor that preceded many of the tephra-producing eruptions (Stephens and others, 1994).

Our goal in this paper is to examine the precursory and eruption seismicity at Mount Pinatubo between May 13 and June 18, 1991, using SSAM measurements, and to evaluate the SSAM system as a monitoring tool. First we review the

---

[1]U.S. Geological Survey.

[2]Philippine Institute of Volcanology and Seismology.

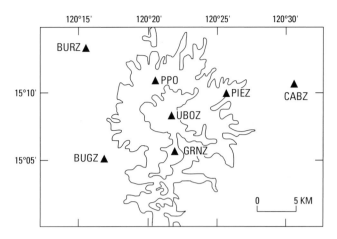

Figure 1. Locations of telemetered seismic stations (black triangles) in the Pinatubo network between May 13 and June 18, 1991.

seismic instrumentation at Pinatubo and the technical aspects of the SSAM system. This is followed by a discussion of methods and techniques for analyzing SSAM data. We then develop a chronology based on SSAM, helicorder records, and pertinent observations reported by others, which is followed by a brief discussion of the implications of SSAM observations for the source processes of seismicity during various time periods prior to and during the June 15 eruption. We conclude with an evaluation of the SSAM system as it was used at Pinatubo and recommendations for future SSAM deployments. The chronology presented here is only intended to provide context to the SSAM data and is intentionally kept brief. All times and dates are referenced to Philippine time (to convert to G.m.t., add 8 h). Detailed chronologies of the 1991 eruption are presented by Punongbayan and others, (this volume), Hoblitt, Wolfe, and others, (this volume), Harlow and others, (this volume), Wolfe, (1992), Wolfe and Hoblitt (this volume).

SEISMIC INSTRUMENTATION AND ACQUISITION

Following the initial precursory explosions in April 1991, a network of seven telemetered high-gain, short-period seismometers was installed around Mount Pinatubo. The installation was completed on May 13. The stations were from 1 to 19 km from the summit of the volcano, and except for one three-component 2-Hz station (PPO), all were single-component vertical 1 Hz seismometers (fig 1). Lockhart and others (this volume) describe the installation, operation, and technical specifications of the network. Data from this network were telemetered to Clark Air Base and recorded by a computer system described by Murray and others (this volume). The network remained intact until June 11, when eruptive activity destroyed station UBOZ and vandalism at a telemetry repeater caused the loss of sig-

nals from PPO and BURZ. PIEZ, GRNZ, and BUGZ were damaged by pyroclastic flows during eruptions on June 15. Only station CABZ survived the eruption, although recording of its signal was disrupted on June 15.

THE SSAM SYSTEM

The SSAM system acquires data on a PC/AT-compatible computer in conjunction with the program MDETECT (Tottingham and others, 1989) and a digital signal processor (DSP) board (Rogers, 1989). More recent versions of the SSAM system run in conjunction with the program XDETECT (Rogers, oral commun. 1992). Sixteen analog seismic signals are digitized at 100 samples per second. After 512 samples for each channel have been collected, the samples are transferred to the DSP board and a Fast-Fourier-Transform (FFT) is computed for each channel. As a result of limitations in the speed of the computer system used at Pinatubo, the amplitudes of individual spectral lines were approximated by use of a method described by Gledhill (1985). The average spectral amplitude is determined for a band by averaging the values for all of the spectral lines that fall within the band. Every minute the average FFT value for all 16 bands is calculated and stored on computer disk for each station. A comprehensive description of the SSAM system is given by Stephens and others (1994).

At Pinatubo, the first 10 SSAM bands were defined from 0.5 to 10.5 Hz in increments of 1 Hz. The remaining six bands each covered a broader spectrum in order to determine if SSAM could be used for specific applications such as replacing RSAM or detecting lahars. Data from these wider bands were not analyzed in this study.

As the SSAM system is a fairly recent development, its response characteristics are not completely understood. At the time of this writing we know little about the type and amount of distortion that is introduced if the seismic signal is electronically clipped. In theory, the square waves produced by electronic clipping should only introduce frequencies higher than those ordinarily of interest in volcanic environments, though the telemetry system may introduce some distortion (Chris Stephens, oral commun., 1993). Much of the data from stations close to the volcano (PIEZ, BUGZ, GRNZ) is continuously electronically clipped starting about 1230 on June 14. Data from CABZ began to clip continuously about 1345 on June 15.

SSAM DISPLAY AND ANALYSIS

A variety of techniques have been developed to display and analyze SSAM data. These include bar graphs of amplitude plotted against frequency for individual 1-min samples (see fig. 3 of Murray and others, this volume), time series of individual frequency bands (see Hoblitt, Wolfe,

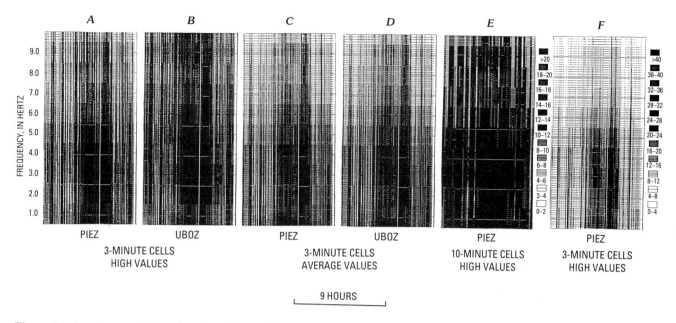

Figure 2. Spectrograms illustrating the effects of the various input parameters for SSAM_VU. Spectrograms are from stations PIEZ and UBOZ for the 9-h interval from 1100 to 2000 on June 7 during a swarm of volcano-tectonic earthquakes associated with dome emplacement. A and B show data from stations PIEZ and UBOZ with 3-min cells and high values. C and D show averaged values in 3-min cells. E shows highest values in 10-min cells at PIEZ, and F shows highest values in 3-min cells at a doubled amplitude interval. See text for further discussion.

and others, this volume), contoured two-dimensional or pseudo-three-dimensional time-frequency-amplitude plots (Marso and Murray, 1991), time-frequency plots that use a gray scale in which the density of dots is proportional to the spectral amplitude (Stephens and others, 1994), and time-frequency-amplitude plots where the spectral amplitude is displayed as a contoured surface (see fig. 4 of Murray and others, this volume).

The primary method of display in this paper is a time-frequency-amplitude plot in which the spectral amplitude corresponds to an 11-step gray scale. The different steps are constructed by varying the spacing between black lines. The lowest amplitude is represented by white (infinite space between lines), the highest amplitude by solid black (no space between lines). These spectrograms are produced by the computer program SSAM_VU, which runs on a PC/AT-compatible computer. Figures 2–12 are examples of spectrograms produced by SSAM_VU. When displayed on a computer monitor, a color scale is used instead of a gray scale. The gray scale interval is user defined and can range between 1 and 99 amplitude counts, which allows even very strong seismic signals to be analyzed. The program provides an option to display the spectrum from 0 to 5 Hz or 0 to 10 Hz and also allows a user-defined time scale that can range from 1 to 144 h (see fig. 9 for an example of a spectrogram with an expanded time scale). To overcome limitations in the resolution of printers and video displays, data are plotted for individual cells, the time represented by each cell can vary from 1 to 60 min. The program allows the user to specify whether the value represented by the cell is the

highest data value or an average of the values within the time-period of the cell and allows the user to select a greater cell duration if desired. Generally, in this paper, the highest data values are shown.

A number of physical and computational factors can greatly affect the appearance of spectrograms produced from SSAM data when SSAM_VU is used. The physical factors include the source process of the seismicity, the effects of attenuation between the source and receiver, and the response characteristics of the instrumentation. The computational factors include whether the high or average values are plotted for a given cell, the duration of the cell, and the amplitude interval used for defining the gray scale.

To illustrate the effect of the computational factors, figure 2 shows a number of spectrograms from the swarm of volcano-tectonic earthquakes on June 7 demonstrating the effect of different display parameters in SSAM_VU. Figures 2A and 2B show spectrograms for stations PIEZ and UBOZ with high values for each 3-min cell and an amplitude interval of 2. When high values are selected the spectrum is more heavily influenced by the frequency characteristics of the largest individual event in a given cell. Note that there is less high-frequency energy at PIEZ than at UBOZ. This is likely the result of attenuation of the higher frequencies caused by the greater distance from the volcano (10 km for PIEZ and 1 km for UBOZ). At Redoubt Volcano, significant spatial variations also were observed in swarms of long-period events at different stations in the network (Stephens and others, 1994). Choosing average values instead of highest values favors the overall character of the seismicity because the effects of distinct events are averaged over the duration of the cell. Figures 2C and 2D show the same swarm at PIEZ and UBOZ for averaged values in 3-min cells. Figure 2E shows highest values at station PIEZ plotted in 10-min cells. The longer cell puts greater emphasis on the spectral character of larger individual events because the cell represents the highest value in a longer time. Figure 2F shows high values at PIEZ in 3-min cells at an amplitude interval of 4, twice that of figures 2A–2E. Increasing the amplitude interval shows that the seismicity between 1500 and 1600 has only about half the energy as that between 1630 and 1730. An important consideration in viewing SSAM records when the amplitude scale is increased is that only the strongest portion of the spectrum is preserved. Consequently, if a very strong swarm of volcano-tectonic earthquakes with a strong spectral peak between 3 and 4 Hz and weak volcanic tremor peaked at 1.5 Hz are occurring simultaneously, the peak at 1.5 Hz may not rise above the lowest graduation.

THE PRECURSORY SEISMIC SEQUENCE

The seismic activity that preceded the cataclysmic eruption on June 15 is a complex sequence comprising volcano-tectonic earthquakes, long-period events, hybrid events, and dome-building and explosive eruptions (Harlow and others, this volume; White, this volume). In characterizing this seismicity, we have relied primarily on the SSAM and helicorder records. Earthquake hypocenters and waveforms are described by Harlow and others (this volume), White (this volume), and Mori, White, and others (this volume).

In describing the various waveforms and event types at Pinatubo, we use terminology and event classifications similar to that developed by Chouet and others, (1994), Lahr and others (1994), Power and others (1994), and Harlow and others (this volume). This classification is based on the present understanding of the physical processes associated with the seismic source. It identifies *volcano-tectonic* (VT) earthquakes as representing a purely elastic source and *long-period* (LP) events, which represent a more complex process involving gas and liquid phases (Chouet, 1992). We do not identify events as *hybrids* in this paper, as SSAM and helicorder records do not provide adequate spectral information for individual events. Volcanic tremor is generally defined as a continuous signal generated within the volcano (Koyanagi and others, 1987). Volcanic tremor frequently occurs in close association with long-period events (Chouet and others, 1994). In this paper we use volcanic tremor (or just tremor) to refer to continuous signals with frequencies between 0.5 and 3.5 Hz in which we cannot identify individual events on helicorder records. The term "*sustained long-period seismicity*" as developed by Stephens and others (1994) describes periods when both volcanic tremor

and long-period events were occurring simultaneously. We use the term "*continuous volcano-tectonic seismicity*" to describe periods when VT earthquakes are occurring so rapidly the individual events became indistinguishable. Those events associated with the forcible ejection of steam and ash from the volcano are referred to as explosive eruptions.

MAY 13–JUNE 3, 1991

Between April 4 and June 3, the activity at the volcano was characterized by vigorous fumarolic activity, high gas output, and VT earthquakes (Punongbayan and others, this volume). Seismicity during this interval consisted primarily of VT earthquakes and indistinct periods of volcanic tremor that occurred in association with a few LP events. Most of the earthquakes during this period were located on the northwest flank of the volcano between 2 and 7 km depth. Harlow and others (this volume), have described this as the metastable phase, because the seismic activity was elevated but relatively stable. SSAM records from this period are fairly quiet because the individual events were too infrequent to generate a sustained signal.

JUNE 3–JUNE 7, 1991

Between roughly June 1 and June 3, locatable VT earthquakes began to occur beneath the summit of Mount Pinatubo (Harlow and others, this volume). At 1939 on June 3 a small explosion occurred which produced minor amounts of ash (Punongbayan and others, this volume). This event was followed by 30 to 40 mins. of volcanic tremor (fig. 3). Figure 3 also shows a variety of microseismic and cultural noise. The occurrence of volcanic tremor and the rate of VT earthquakes increased over the next several days and by early on June 6 had reached the point where their spectral character was recorded well on the SSAM system. Data from station PIEZ for June 6 and 7 show that most of the energy associated with these earthquakes had a fairly broad spectrum (between 0.5 to 9.5 Hz), and peaked between 2.5 and 4.5 Hz (fig. 4). Figure 2 shows this period of seismicity in greater detail. These spectral characteristics are generally attributed to VT earthquakes (Lahr and others, 1994). Near the end of June 6 and for the first 3 h on June 7, the rate of VT earthquakes fluctuated several times, reaching a rate of about 1 per min by 1100. By 1600 they were occurring so frequently they formed a continuous signal on the helicorder records that we call continuous volcano-tectonic seismicity. Throughout this period of continuous volcano-tectonic seismicity the spectrogram shows a fairly strong signal that is present at all frequencies and peaks between 2.5 and 4.5 Hz. This seismicity culminated in a small explosion at 1700, which was followed by roughly 8 h of increased activity that slowly returned to background levels (figs. 4 and 5).

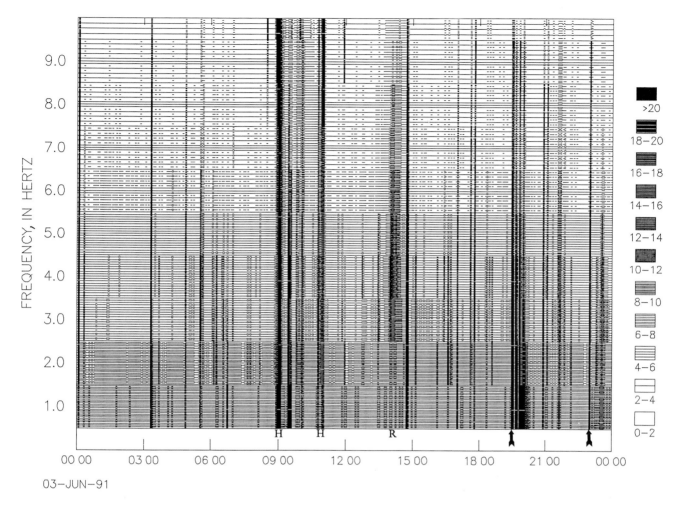

Figure 3. SSAM record from station PIEZ on June 3 plotted with high values in 3-min cells at an amplitude interval of 2. Strong pulses noted by the letter H at roughly 0900 and 1100 reflect helicopter landings near the station (note increased signal in higher bands). Activity between pulses reflects geophysicists' movements while servicing the station (battery change for radio transmitter). Diffuse signal between roughly 1350 and 1430 with a slight peak in the 3.5- to 4.5-Hz band (noted by letter R) results from a strong rainstorm. Arrow at 1928 notes onset of explosion which was followed by approximately 45 min of volcanic tremor. Additional spikes during the day result from large volcano-tectonic earthquakes, short bursts of tremor and entrained long-period events, and a second small explosion at 2258 (noted by second arrow).

Early on the morning of June 8 (0800) a lava dome was seen on the northwest flank of the volcano (Hoblitt, Wolfe, and others, this volume). Tilt measurements suggest that extrusion of lava began about 1700 on June 7 (Ewert and others, this volume), coinciding with the height of the continuous volcano-tectonic seismicity.

JUNE 8–JUNE 12, 1991

From June 8 to June 12, the seismicity was characterized by discrete swarms of VT earthquakes and a gradual increase in tremor throughout the phase. Small amounts of ash were continuously emitted from the margins of the lava dome, and visual observations suggested that the lava dome continued to grow until it was destroyed on June 12 (Hoblitt, Wolfe, and others, this volume).

Following the emplacement of the lava dome on June 7, seismic activity returned to near pre-June levels for much of June 8. Several short swarms of VT earthquakes occurred on June 9, the strongest of which was an hour-long swarm at roughly 0800, and this rivaled the strength of any previously recorded based on SSAM records. Starting at approximately 1100 a very strong 3-h period of volcanic tremor began with a dominant low-frequency signal in the 0.5- to 1.5-Hz band (fig. 5). Following this energetic low-frequency tremor the volcano entered a period of cyclic swarms of VT earthquakes that continued until the first plinian eruption at 0851 on June 12. On June 9 these swarms occurred at intervals of approximately 2 to 3 h; each episode lasting about an hour. These swarms form strong vertical stripes on SSAM spectrograms (figs. 5 and 6), but lack some of the higher frequency energy (>4.5 Hz) associated with the swarms observed prior to the dome

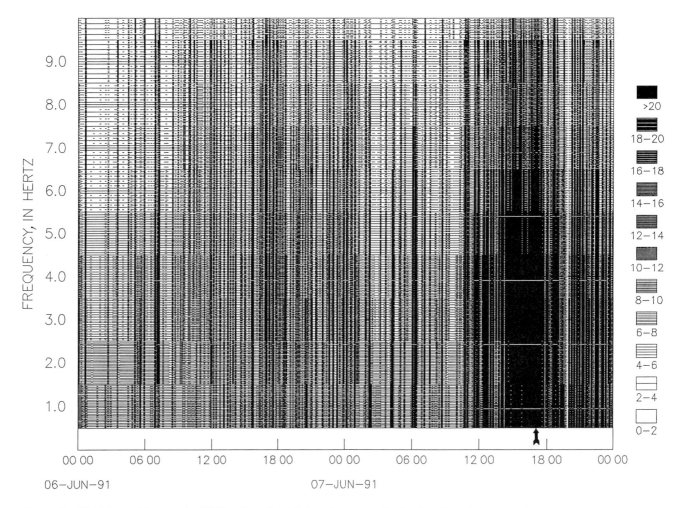

Figure 4. SSAM record from station PIEZ on June 6 and 7 plotted with high values in 7-min cells at an amplitude interval of 2. Spectrogram shows signature of short swarms of volcano-tectonic earthquakes on June 6. These intensified to form a continuous volcano-

tectonic signal on June 7 that preceded the extrusion of magma. A small explosion at 1700 (noted by arrow) coincided with the onset of magma extrusion.

emplacement on June 7. A less intense but more protracted swarm occurred between approximately 2145 on June 9 and 0300 on June 10. SSAM measurements indicate that this swarm contained a higher percentage of energy below 2.5 Hz, which is typical of volcanic tremor. The intensity and frequency of these swarms increased on June 10, while the duration of individual swarms declined (fig. 6).

The protracted earthquake swarm starting at 2145 on June 9 initiated a gradual increase in LP seismicity that continued until the plinian eruption on June 12 (fig. 6). A more distinct increase in the level of long-period seismicity occurred at about 2205 on June 11, coincident with the destruction of seismic station UBOZ by a small pyroclastic flow (Lockhart and others, this volume). A small explosive eruption at 0330 was followed by roughly 2 h of low-level volcanic tremor. Then a 4-h period of relative seismic quiescence immediately preceded the first plinian eruption at 0851 on June 12 (fig. 6).

JUNE 12–JUNE 15, 1991

Following the initial plinian eruption on June 12, the general level of seismicity increased dramatically. The June 12 eruption initiated a series of explosive eruptions that increased in rate of occurrence until the cataclysmic eruption on June 15. The onset times of many of these explosions correlate closely with atmospheric pressure waves observed on a microbarograph at Clark Air Base. A comparison of pressure waves and the seismic record indicates that large explosive eruptions occurred at 2252 on June 12, 0841 on June 13, 1309, 1410, 1516, 1853, 2218, 2320, on June 14, 0115, 0257, 0555, 0810, 1027, 1117, 1158, 1222, 1252, 1315 on June 15 (Hoblitt, Wolfe, and others, this volume). Between each of these eruptions, large VT earthquakes, swarms of LP events, and volcanic tremor occurred. Seismicity during this interval is very complex, and the record is often difficult to interpret, as a variety of event types were occurring simultaneously.

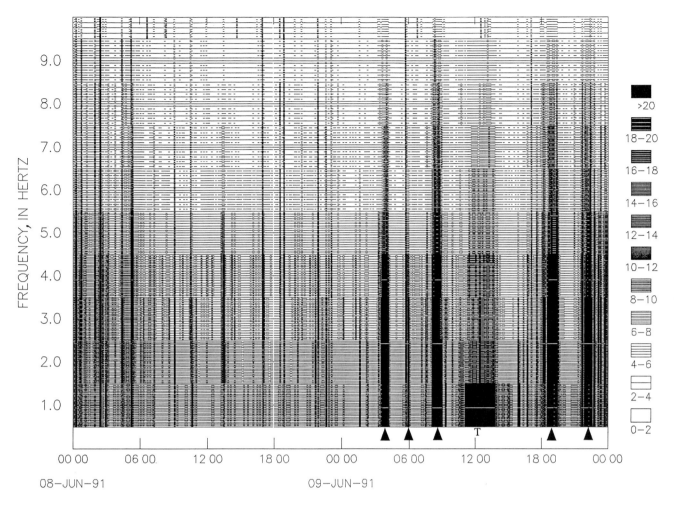

Figure 5. SSAM record from station PIEZ on June 8 and 9 plotted with high values in 7-min cells. Strong swarms of volcano-tectonic earthquakes form the broad spectral signals noted by the solid triangles. A strong period of tremor between approximately 1100 and 1400 on June 9 forms the prominent signal in the 0.5- to 1.5-Hz band noted by the letter T.

Because of the increase in seismic intensity, the amplitude scale on spectrograms produced by SSAM_VU needed to be increased by a factor of 6 to stay on scale (figs. 7 and 8). In viewing spectrograms from this period, recall that only the strongest portion of the spectrum is preserved when the amplitude scale is increased. Spectrograms from PIEZ for June 12 through 15 (figs. 7 and 8) show that the spectrum was dominated by two prominent peaks. The first is between 2.5 and 4.5 Hz and correlates with swarms of VT earthquakes on the helicorder records and intervals of continuous volcanic-tectonic seismicity. The second peak is in the 0.5- and 1.5-Hz band and correlates with episodes of volcanic tremor, swarms of LP events, and intervals of sustained long-period seismicity.

Two strong swarms of VT earthquakes followed the 0851 eruption on June 12 (fig. 7). Following the 2252 eruption on June 12, the volcano-tectonic bands began a gradual increase that persisted at station PIEZ until it was destroyed by a pyroclastic flow at 1409 on June 15. The gradual increase was punctuated by several more intense swarms on June 13 and 14 (figs. 7 and 8).

Swarms of LP events and tremor followed a similar pattern from June 12 to 15 (figs. 7 and 8). A strong burst of tremor coincided with a strong swarm of VT earthquakes between roughly 1530 and 1700 on June 12 (fig. 7). LP seismicity also began a gradual increase following the 2252 eruption, which continued to intensify until PIEZ was destroyed on June 15.

A very unusual seismic signal began at about 1830 on June 12 as a continuous string of VT earthquakes with a strongly peaked signal in the 3.5- to 4.5-Hz band at station PIEZ. The signal lasted a little over 1.25 h, and three times the peak in seismic energy shifted to progressively lower frequency bands for short periods. During approximately the last 20 min of the signal, the spectral peak shifted from the 3.5- to 4.5-Hz band to the 2.5- to 3.5-Hz band, and then to the 1.5- to 2.5-Hz band (fig. 9). The signal was localized at the volcano, as it did not record well at station CABZ, and there was no known associated eruptive activity.

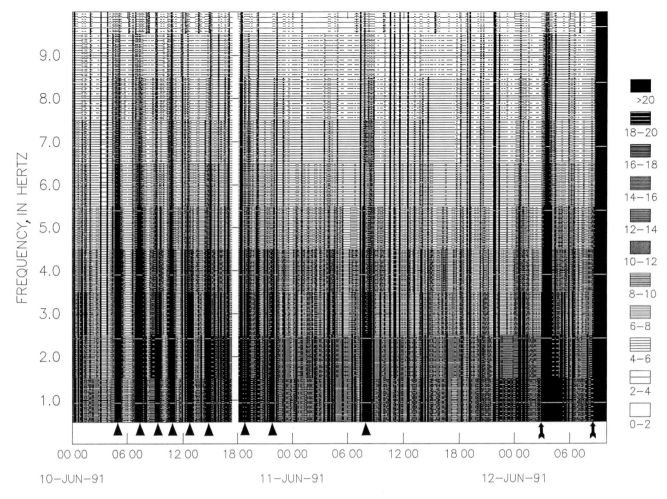

Figure 6. SSAM record from station PIEZ on June 10, 11, and first 10 h of June 12. Plot reflects high values in 8-min cells at an amplitude interval of 2. Swarms of volcano-tectonic earthquakes are noted by solid triangles. A small explosion at 0330 on June 12 and the first plinian eruption at 0851 on June 12 are noted by arrows.

Throughout the analysis of the Pinatubo seismicity we have referred to the this event as "the groan."

To gain further insight into the seismicity between June 12 and 18 we turn to station CABZ, which, at a distance of 19 km was the only station to survive the eruption on June 15 (figs. 10, 11, and 12). Fortunately, CABZ was also the station with the lowest gain (Lockhart and others, this volume); therefore, it was more suited to record the rapidly escalating seismicity. The signal from station CABZ did not experience significant electronic clipping until June 15.

The individual explosive eruptions generally form signals having energy from 0.5 to 10.5 Hz on SSAM spectrograms (figs. 6, 7, 10, and 12), with the strongest spectral peak in the 0.5- to 1.5-Hz band (fig. 10). The SSAM record suggests that relative seismic quiescence preceded the explosive eruptions at 0851 and 2252 on June 12 and at 0115 on June 15. In contrast, increases in LP seismicity preceded explosive eruptions at 0841 on June 13, 1309, 1516, 1853, 2218, on June 14, and 0257 on June 15.

Following the 0257 eruption on June 15, the level of background seismicity remained elevated, and it is difficult to establish a clear relation between seismicity and later individual explosive eruptions.

JUNE 15–JUNE 18, 1991

At approximately 1342 on June 15, the volcano began to erupt continuously, eventually forming a 2.5-km-wide caldera where the previous summit had been. Volume estimates of erupted material range between 3.7 and 5.3 km^3 of dense magma. (W.E. Scott and others, this volume). The intense seismicity associated with the eruption lasted over 9 h, and recording of data from CABZ was interrupted when various components of the acquisition system were temporarily removed during an evacuation (0730 to 1012 June 15) and when power to the acquisition system was lost (2014 to 2047 and 2102 to 2343 June 15) (figs. 8, 10, 11, 12). A sequence of large (M_b 4.3 to 5.7) earthquakes began at 0739 on June 15. Most of these events occurred between 1616 on

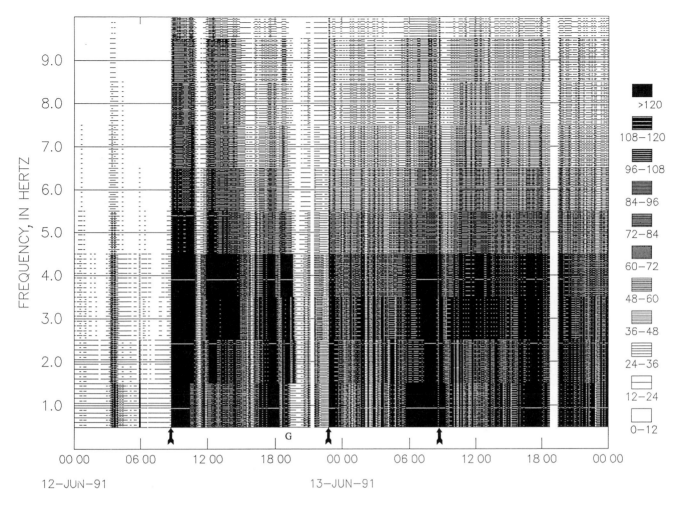

Figure 7. SSAM record for June 12 and 13 plotted with high values in 7-min cells and an amplitude interval of 12. The amplitude interval is increased by a factor of 6 from spectrograms shown in the previous figures. Strong swarms of volcano-tectonic earthquakes form the signals in the 2.5- to 3.5- and 3.5- to 4.5-Hz bands (noted by the solid triangles). Increased occurrences of volcanic tremor generate the strong signals in the 0.5- to 1.5-Hz band. Note the gradual escalation in both volcano-tectonic and long-period seismicity throughout June 12 and 13. Explosive eruptions at 0851 and 2252 on June 12 and at 0841 on June 13 are marked with arrows. "The groan" is noted with the letter G.

June 15 and 0032 on June 16, when their approximate rate was 1 event every 15 min. Forty-eight events were located by the worldwide seismographic network on June 15 and 16 (U.S. Geological Survey, 1991). Additional aspects of the seismicity as observed at distant stations are discussed by Kanamori and Mori (1992), and Zürn and Widmer (this volume).

The SSAM record for June 15 shows that the energy at the onset of the cataclysmic eruption was concentrated in the lowest frequency band and later abruptly shifted to higher bands. The eruption shows as the dominant wide vertical black stripe in figure 10. By increasing the amplitude scale by a factor of five (fig. 11) we see that the eruption began with a spectral peak in the 0.5- to 1.5-Hz band. At approximately 1630 the energy in the lowest band declined and a new dominant peak developed between 1.5 and 3.5

Hz. This change roughly coincides with the onset of large VT earthquakes, as detected on the worldwide network.

The spectral peak between 1.5 and 3.5 Hz at station CABZ dominated throughout much of the remainder of June (fig. 12). helicorder records from June 16 to 18 show large VT earthquakes which occur at a rate of roughly 1 per minute, as well as strong volcanic tremor.

DISCUSSION

The SSAM records from Pinatubo provide numerous examples of a wide variety of signals common at active volcanoes. In this section we review the character of the SSAM record and discuss the physical implications for the various processes active at the volcano.

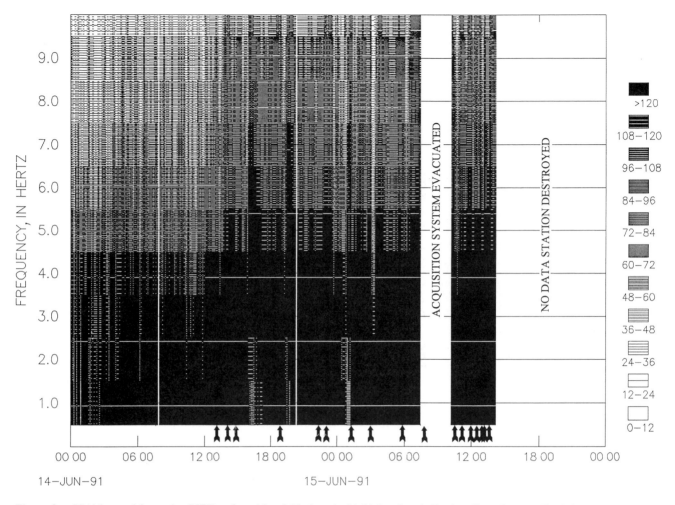

Figure 8. SSAM record for station PIEZ on June 14 and 15 plotted with high values in 7-min cells and an amplitude interval of 12. This plot shows the continued escalation of both volcano-tectonic earthquakes and long-period seismicity on June 14 and 15. Station PIEZ was destroyed by a pyroclastic flow at 1409 on June 15. Arrows correspond to times of known explosions. See text for further discussion.

The Pinatubo eruption provided the first opportunity for an SSAM system to record strong swarms of VT earthquakes and what we have called continuous volcano-tectonic seismicity. At stations PIEZ, UBOZ, and CABZ, VT seismicity had a broad spectrum generally ranging from 1.5 to 9.5 Hz with a well-defined peak generally between 1.5 and 4.5 Hz. Using SSAM makes it easy to distinguish between episodes of continuous VT seismicity and episodes of LP seismicity. The SSAM record is dominated by VT seismicity during the June 5–7 episode of dome emplacement, during the cataclysmic buildup, and after about 1630 on June 15, during the cataclysmic eruption. All of these episodes take place when we would expect changes in magmatic pressure to exert excessive stress on the brittle rock surrounding the Pinatubo magmatic system. The VT earthquakes that preceded the extrusion of the lava dome represent the forceful passage of magma to the surface. This interpretation is supported by the observed shoaling of hypocenters (Harlow and others, this volume) and ground deformation (Ewert and others, this volume). The VT

earthquakes during June 12 to 15 reflect the destruction, induced by the magma's destabilization, of the brittle rock above the Pinatubo magma body. Most located earthquakes during this period occur at shallow depth (Harlow and others, this volume). The numerous VT earthquakes initiated by the June 15 eruption are related to the adjustment of stresses resulting from the evacuation of material from the Pinatubo magma chamber (Mori, White, and others, this volume).

LP seismicity has been attributed to the dynamics of pressurized fluids associated with the magma (Chouet and Shaw, 1991; Chouet, 1992; and Chouet and others, 1994). Long-period seismicity became increasingly prevalent on the SSAM record between June 3 and 15. Tremor and LP events generally formed a strong signal in the 0.5- to 1.5-Hz band (figs. 5, 7, and 11). Episodes of LP seismicity may reflect time periods when pressurized fluids could accumulate. The increased occurrence of LP seismicity is somewhat expected following the emplacement of the lava dome. The magma associated with the lava dome is thought to

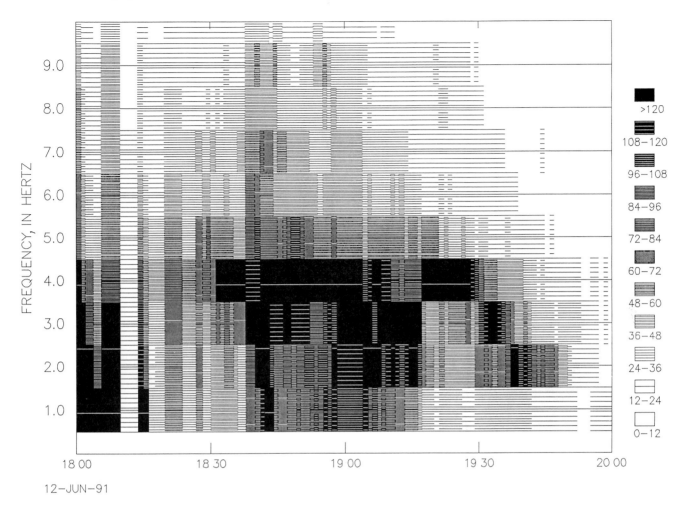

Figure 9. SSAM record from station PIEZ between 1800 and 2000 on June 12 plotted with data in 1-min cells at an amplitude interval of 12. This spectrogram shows the strong period of continuous volcano-tectonic seismicity that began at roughly 1830, which we have called "the groan." The peak in energy shifted to lower frequency bands several times during the signal. During the last 20 min the signal shifted in a stepwise manner to lower frequency bands. See text for further discussion.

have formed a seal on the magmatic system behind which volatiles could accumulate (White and others, this volume; Hoblitt, Wolfe, and others, this volume). This interpretation is supported by observations of decreased gas flux from the volcano beginning several days prior to the extrusion of magma (Daag, Tubianosa, and others, this volume). Between June 12 and 15, increased LP seismicity generally preceded explosive eruptions (fig. 10). This seismicity likely represents the pressurization of fluids that were trapped behind temporary barriers that were destroyed in the ensuing eruption.

The SSAM signals from individual explosive eruptions peak strongly in the 0.5- to 1.5-Hz band (figs. 7, 10 and 11), which is much like the LP seismicity preceding the events. This peak dominates throughout the duration of the eruption signals. SSAM measurements of explosive eruptions at Redoubt Volcano, Alaska, also show that LP seismicity and eruptions share similar spectral peaks. Stephens

and others (1994) suggest that the similarity in spectra is an indication that the source mechanism of the precursory LP seismicity continued to be active throughout each eruption. They suggest that the stronger signals observed during eruptions may be produced by the increased flow of magmatic fluids through cracks as obstructions at the vent are removed. This physical interpretation agrees with the observations of explosive eruptions at Pinatubo between June 12 and 15.

The signal between 1830 and 1940 on June 12, which we have called "the groan", is dominated by energy concentrated in the 3.5- to 4.5-Hz band. We feel this signal represents the movement of magma and associated volatiles into a new system of cracks and passageways. The continuous VT seismicity represents the brittle failure of competent rock as new cracks were forced open. The occasional shifts to lower frequencies may reflect the movement of fluids into the new cracks and passages.

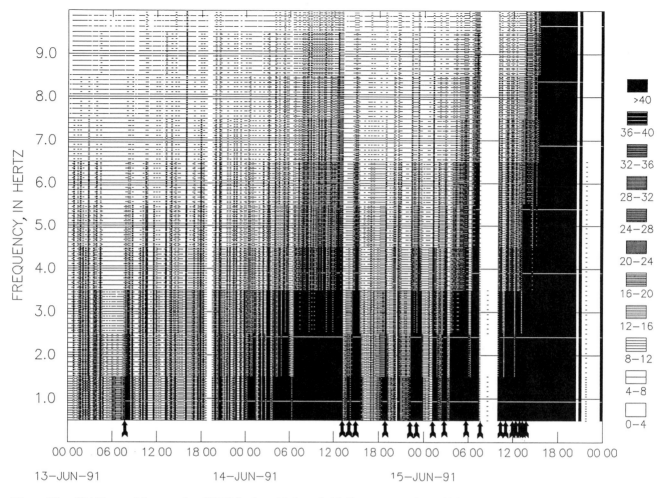

Figure 10. SSAM record from station CABZ for June 13 through 15. Spectrogram shows highest values in 10-min cells at an amplitude interval of 4. Arrows note the times of known explosive eruptions.

The cataclysmic eruption on June 15 began with 3 h of seismicity with a strong spectral peak in the 0.5- to 1.5-Hz band (fig. 11). The SSAM data on June 15 must be viewed with some caution, as the data were electronically clipped. This LP seismicity likely represents increased flow of fluids through cracks resulting from the removal of obstructions at the vent. The onset of strong energy in the VT bands (1.5- to 3.5-Hz) at roughly 1630 likely represents the onset of large stress adjustments in response to the removal of magma from beneath Mount Pinatubo. The onset of large earthquakes ($M_b4.3+$) suggests this is approximately the time that the summit of Mount Pinatubo was destroyed and the caldera began to form. The coincident decline in LP events would result from the destruction of the cracks and passageways through which the fluids were transported.

The strong VT peak between 1.5 and 3.5 Hz continued beyond June 16 (fig. 12). The continued occurrence of VT earthquakes represents stress adjustments in response to the removal of magma from beneath Mount Pinatubo. That this signal continued in the bands from 1.5 to 3.5 Hz suggests

that the clipped signals on June 15 did not drastically affect the data in these bands.

CONCLUSIONS AND RECOMMENDATIONS

During the 1989–90 eruption of Redoubt Volcano, SSAM measurements proved valuable both for detecting small swarms of LP events that preceded many of the tephra-producing eruptions and in reconstructing the details of the seismic record during intense periods of seismic activity. At Pinatubo, SSAM played only a minor role in formulating eruption forecasts. The strength of the various seismic signals was much greater than at Redoubt; consequently, it was much easier to recognize these signals on standard helicorder records. Additionally, at the time of the Pinatubo crisis, analysis software for SSAM data on PC/AT-compatible computers was not as flexible as that which now exists. These advances in software design should make the incorporation of SSAM data in real-time interpretation

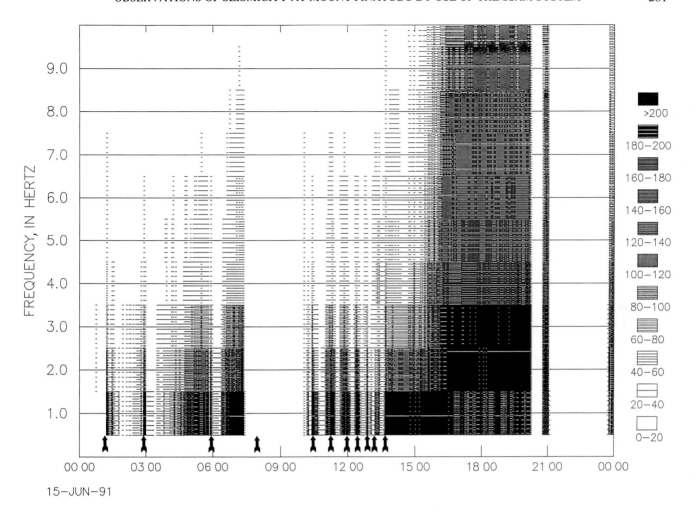

Figure 11. SSAM record for station CABZ for June 15 plotted with high values in 3-min cells at an amplitude interval of 20. This spectrogram shows the eruptions early on June 15 and the onset of the cataclysmic eruption at roughly 1342 (as noted by last arrow). The first 3 h of the cataclysmic eruption are characterized by a strong signal in the 0.5- to 1.5-Hz band. The onset of strong energy at roughly 1630 approximately corresponds to the onset of large volcano-tectonic earthquakes recorded on distant stations as the caldera formed. See text for further discussion.

much easier in the future. In studying the Pinatubo seismicity, SSAM has again proven to be a valuable tool for characterizing changes in the seismic spectrum through time.

In deploying SSAM systems at future active volcanoes, bands should be chosen that provide the greatest resolution in those areas of the spectra where critical activity is most likely to occur. Sustained seismic activity during the Pinatubo eruption covered a broad range of frequencies and resulted from periods of sustained LP seismicity as well as the continuous occurrence of VT earthquakes. At Redoubt Volcano, critical activity was associated with swarms of LP events and volcanic tremor concentrated between 0.9 and 1.9 Hz. To provide greater resolution in this range, the SSAM bands were adjusted during the Redoubt eruptions. At Pinatubo, the first nine bands provided adequate coverage from 0.5 to 10.5 Hz but lacked the resolution that the Redoubt bands provided. In future deployments of SSAM systems, bands should initially provide coverage between 0.1 and 12.0 Hz and have as much resolution as possible

between 0.1 and 8 Hz, because most events commonly associated with active volcanism fall within this range of frequencies (Lahr and others, 1994). It is important to dedicate a few bands above this range, as many signals such as storm noise frequently have significant energy at higher frequencies, and SSAM data help with their identification. Suggested band definitions based on experience with SSAM systems from both Pinatubo and Redoubt are contained in table 1. Established bands can be modified to monitor more closely a given type of seismicity once the seismic style and eruptive character of a given volcano are established. Band redefinition should be accomplished by widening some bands and narrowing others instead of deleting bands; it is important to keep the entire spectrum monitored, as seismicity may change unexpectedly. As is the case in any monitoring situation, care must be taken in redefining SSAM bands so that the continuity of baseline measurements is not disrupted at a critical time.

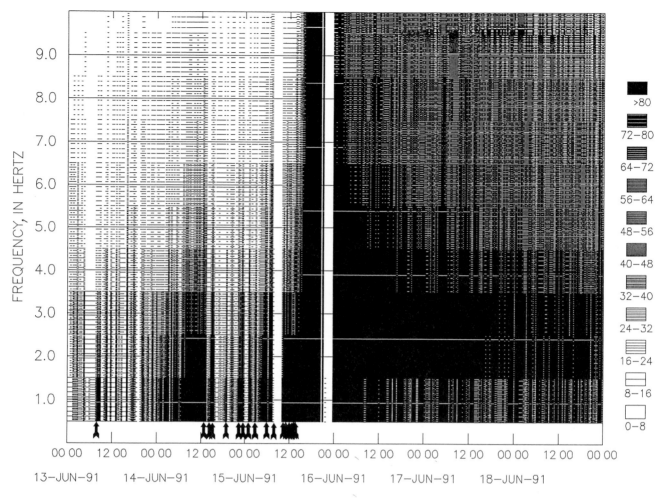

Figure 12. SSAM record for station CABZ for June 13 through 18 plotted with high values in 21-min cells at an amplitude interval of 8. This spectrogram provides a synoptic view of the temporal changes in seismicity before and after the cataclysmic eruption on June 15. The precursory explosions on June 13–15 are noted by arrows. The strong signal crossing all bands corresponds to the cataclysmic eruption. The strong peak between 1.5 and 3.5 Hz on June 16–18 reflects the vigorous volcano-tectonic earthquake activity resulting from the stress adjustments in response to the removal of magma.

Table 1. Suggested spectral band definitions as based on experience with SSAM systems from Mount Pinatubo and Redoubt Volcano.

Band	Frequency limits (Hz)
1	0.1–0.5
2	0.5–1.0
3	1.0–1.5
4	1.5–2.0
5	2.0–2.5
6	2.5–3.0
7	3.0–3.5
8	3.5–4.0
9	4.0–4.5
10	4.5–5.0
11	5.0–5.5
12	5.5–6.0
13	6.0–7.0
14	7.0–8.0
15	8.0–10.0
16	10.0–12.0

ACKNOWLEDGMENTS

We thank the many individuals involved in the Pinatubo response who assisted with data acquisition and reduction. This paper benefited greatly from discussions with Andy Lockhart, Randy White, Dave Harlow, Rick Hoblitt, Chris Stephens, Bernard Chouet, and John Ewert. Randy White, E.G. Ramos, and Robert Page provided formal reviews of the manuscript.

REFERENCES CITED

Chouet, B.A., 1992, A seismic model for the source of long-period events and harmonic tremor, *in* Gasparini, P., Scarpa, R., and Aki, K., eds., Volcanic sesimology: IAVECI Proceedings in Volcanology, Springer-Verlag, Berlin.

Chouet, B.A., and Shaw, H.R., 1991, Fractal properties of tremor and gas piston events observed at Kilauea Volcano, Hawaii Journal of Geophysical Research, v. 96, p. 10177–10189.

Chouet, B.A., Page, R.A., Stephens, C.D., Lahr, J.C., and Power, J.A., 1994, Precursory swarms of long-period events at Redoubt Volcano (1989–1990), Alaska: their origin and use as a forecasting tool: Journal of Volcanology and Geothermal Research, v. 62, p. 95–135.

Daag, A.S., Tubianosa, B.S., Newhall, C.G., Tuñgol, N.M, Javier, D., Dolan, M.T., Delos Reyes, P.J., Arboleda, R.A., Martinez, M.L., and Regalado, M.T.M., this volume, Monitoring sulfur dioxide emission at Mount Pinatubo.

Endo, E.T., and Murray, T.L., 1991, Real-time seismic amplitude measurement (RSAM), a volcano monitoring and prediction tool. Bulletin of Volcanology, v. 53 p. 533–545.

Ewert, J.W., Lockhart, A.B., Marcial, S., and Ambubuyog, G., this volume, Ground deformation prior to the 1991 eruptions of Mount Pinatubo.

Gledhill, K.R., 1985, An earthquake detector employing frequency domain techniques, Bulletin of the Seismology Society of America, v. 75, p. 1827–1835.

Harlow, D.H., Power, J.A., Laguerta, E.P., Ambubuyog, G., White, R.A., and Hoblitt, R.P., this volume, Precursory seismicity and forecasting of the June 15, 1991, eruption of Mount Pinatubo.

Hoblitt, R.P., Wolfe, E.W., Scott, W.E., Couchman, M.R., Pallister, J.S., and Javier, D., this volume, The preclimactic eruptions of Mount Pinatubo, June 1991.

Kanamori, H., and Mori, J., 1992, Harmonic excitation of mantle rayleigh waves by the1991 eruption of Mount Pinatubo, Philippines: Geophysical Research Letters, v. 19, no. 7, p. 721–724.

Koyanagi, R.Y., Chouet, B.A., and Aki, K., 1987, Origin of volcanic tremor in Hawaii, Part 1, data from Hawaiian Volcano Observatory 1969–1985, *in* Decker, R.W., Wright, T.L., and Stauffer, P.H., eds., Volcanism in Hawaii: U.S. Geological Survey Professional Paper 1350, p. 1121–1259.

Lahr, J.C., Chouet, B.A., Stephens, C.D., Power, J.A., and Page, R.A., 1994, Earthquake location and error analysis procedures for a volcanic sequence: Application to 1989–1990 eruptions: Journal of Volcanology and Geothermal Research, v. 62, p. 137–152.

Lockhart, A.B., Marcial, S., Ambubuyog, G., Laguerta, E.P., and Power, J.A., this volume, Installation, operation, and technical specifications of the first Mount Pinatubo telemetered seismic network.

Marso, J.N., and Murray, T.L., 1991, Real-time display of seismic spectral amplitude measurements: examples from the 1991 eruption of Pinatubo volcano, central Luzon, Philippines: Eos, Transactions, American Geophysical Union, 72, no. 44, p. 67.

Mori, J., White, R.A., Harlow, D.H., Okubo, P., Power, J.A., Hoblitt, R.P., Laguerta, E.P., Lanuza, L., and Bautista, B.C., this volume, Volcanic earthquakes following the 1991 climactic eruption of Mount Pinatubo, Philippines: Strong seismicity during a waning eruption.

Murray, T.L., Power, J.A., March, G.D., and Marso, J.N., this volume, A PC-based realtime volcano-monitoring data-acquisition and analysis system.

Power, J.A., Lahr, J.C., Page, R.A., Chouet, B.A., Stephens, C.D., Harlow, D.H., Murray, T.L., and Davies J.N., 1994, Seismic evolution of the 1989–90 eruption sequence of Redoubt Volcano, Alaska: Journal of Volcanology and Geothermal Research, v. 62, p. 69–94.

Punongbayan, R.S., Newhall, C.G., Bautista, M.L.P., Garcia, D., Harlow, D.H., Hoblitt, R.P., Sabit, J.P., and Solidum, R.U., this volume, Eruption hazard assessments and warnings.

Rogers, J.A., 1989. Frequency-domain detection of seismic signals using a DSP co-processor board, in Lee, W.H.K., ed., IASPEI software volume 1: Toolbox for seismic data acquisition, processing, and analysis. International Association of Seismology and Physics of the Earth's Interior; p. 151–172.

Scott, W.E., Hoblitt, R.P., Torres, R.C., Self, S, Martinez, M.L., and Nillos, T., Jr., this volume, Pyroclastic flows of the June 15, 1991, climactic eruption of Mount Pinatubo.

Stephens, C.D., Chouet, B.A., Page, R.A., Lahr, J.C., and Power, J.A., 1994, Seismological aspects of the 1989–1990 eruptions at Redoubt Volcano, Alaska: The SSAM perspective: Journal of Volcanology and Geothemal Research, v. 62, p. 153–182.

Tottingham, D.M., Lee, W.H.K., and Rogers, J.A., 1989. User manual for MDETECT, *in* Lee, W.H.K., ed., IASPEI software volume 1: Toolbox for seismic data acquisition, processing, and analysis. International Association of Seismology and Physics of the Earth's Interior; p. 49–88.

United States Geological Survey, 1991, Preliminary determination of earthquake epicenters, June 1991: U.S. Geological Survey, National Earthquake Information Center, p. 28.

White, R.A., this volume, Precursory deep long-period earthquakes at Mount Pinatubo, Philippines: Spatio-temporal link to a basalt trigger.

Wolfe, E.W., 1992, The 1991 eruption of Mount Pinatubo, Philippines: Earthquakes and Volcanoes, v. 23, no. 1, p. 5–38.

Wolfe, E.W. and Hoblitt, R.P., this volume, Overview of the eruptions.

Zürn, W., and Widmer, R., this volume, Worldwide observation of bichromatic long-period Rayleigh waves excited during the June 15, 1991, eruption of Mount Pinatubo.

Precursory Seismicity and Forecasting of the June 15, 1991, Eruption of Mount Pinatubo

By David H. Harlow,[1] John A. Power,[1] Eduardo P. Laguerta,[2] Gemme Ambubuyog,[2] Randall A. White,[1] and Richard P. Hoblitt[1]

ABSTRACT

Seismic monitoring was the primary tool used to assess the evolving eruptive potential of Mount Pinatubo prior to a climactic eruption on June 15, 1991. We used a seven-station seismic network and a portable, PC-based data acquisition and analysis system to track seismic activity in real time and near-real time. We divide seismic activity prior to the eruption into five distinct phases: (1) Metastable (through May 31)—characterized by 50 to 150 volcano-tectonic events per day; (2) Predome (June 1 to 7)—escalating seismic activity below the summit of Mount Pinatubo in early June which evolved into a strong swarm of volcano-tectonic earthquakes that culminated in the extrusion of a lava dome at the surface; (3) Preexplosive Buildup (June 8 to 12)—characterized by a variety of seismic activity leading up to the first explosive eruption on June 12, including volcano-tectonic earthquakes accompanying the extrusion of a dome, tremor episodes, and the occurrence of hybrid and long-period events; (4) Long-Period Buildup (June 12 to 14)—characterized by explosive eruptions with small pyroclastic flows and a remarkable progression of increasingly large long-period events; and (5) Preclimactic (June 14 to 15)—a 24 hour period of high-amplitude, widely diverse seismic activity that coincided with a series of increasingly frequent explosive eruptions; these produced large pyroclastic flows that evolved into the continuous climactic eruption that began about 1342 on June 15.

We were able to forecast the escalating eruptive potential on the basis of shifts in the locus of earthquake hypocenters, increases in total seismic energy release, changes in the character of earthquake waveforms, escalating nonseismic precursory activity, and an interpretation of magmatic processes during the precursory phenomena. The key changes in seismic activity on which we based our forecasting evaluations included (1) a shift in the locus of the dominant seismic source region during late May and early June

from a cluster 5 kilometers northwest of the summit to a cluster just beneath the summit, (2) an intense swarm of volcano-tectonic earthquakes culminating in the extrusion of a dome, (3) the increase in amplitude of continuous background tremor prior to the first explosive eruption on June 12, and (4) the dramatic increase in seismic energy release and in the number and magnitude of long-period events prior to the climactic eruption on June 15.

The successful forecast of the Mount Pinatubo eruption is a confirmation of our current capability to quickly install a network of monitoring instruments at remote volcanoes, analyze the data, and then provide an adequate interpretation of ongoing magmatic processes for eruption forecasting. The evolution of seismic activity leading up to the Mount Pinatubo eruption has critical implications for eruption forecasting at large volcanic systems where crisis-level seismicity persists for as long as years but has not led to eruptions. At Mount Pinatubo, only 1 week elapsed from the time we recognized that seismic activity differed significantly from crisis-level seismicity at other large volcanic systems to the time of the first explosive eruption on June 12.

INTRODUCTION

The value of seismic data for forecasting volcanic eruptions was recognized as early as 1910 for eruptions of Usu Volcano, Japan (Omori, 1911) and Mauna Loa Volcano, Hawaii (Wood, 1915). Long-term seismic monitoring of volcanoes led to the discovery of a diverse range of seismic signatures and patterns associated with eruptive activity (Sassa, 1935, 1936). A widely used classification scheme developed by Minakami (1960) was based on the characteristics of seismic event signatures and their hypocentral locations. This classification system was used for eruption forecasting by correlating the progression of various parameters of each event type and then matching emerging data patterns with seismicity patterns that preceded eruptions at the same and other volcanoes. The drawback to this technique, however, is that the wide variety of seismic activity

[1]U.S. Geological Survey.
[2]Philippine Institute of Volcanology and Seismology.

and different eruption styles can make the application of pattern recognition for eruption forecasting difficult and unreliable.

Over the last decade, significant improvements have been developed in using the different types and characteristics of seismic activity to identify the causative processes. This approach has evolved from theoretical modeling of source mechanism of different event types (Chouet, 1992) and the study of data from eruptions at "instrumented" volcanoes including Kilauea (Klein and others, 1987), Mount St. Helens (Endo and others 1981; Malone and others, 1981), Augustine (Power, 1988), Redoubt (Power and others, 1994), and Spurr (Power and others, in press) in the United States and Nevado del Ruiz (Nieto and others, 1990) in Colombia. While precursory seismicity at those volcanoes varied in duration, intensity, and character, a synthesis of those data sets provides us with a set of practical guidelines for identifying seismic activity associated with critical preeruptive processes.

At Mount Pinatubo (fig. 1), we were able to forecast eruptive activity by closely observing earthquake hypocenters, the waveform character of individual events, seismic energy release, and seismicity rate. The overall lessons learned from our Pinatubo crisis response have been summarized by the Pinatubo Volcano Observatory Team (1991). We focus here on the interpretation of seismic data from Mount Pinatubo and discuss how this experience can contribute to the overall forecasting effort. Specifically, we describe the temporal and spatial evolution of seismic activity leading to the climactic eruption of Mount Pinatubo on June 15, 1991, and discuss how the seismic data were used for eruption forecasting and volcano hazard management. We begin with a description of the seismic network and the data acquisition and analysis system used to monitor the volcano and then discuss the various types of seismic events observed at Mount Pinatubo. We follow with a chronology of seismic events and related phenomena observed before the climactic eruption and conclude with a discussion of the implications of preeruptive seismicity for the evolution of the Mount Pinatubo eruption. In this narrative we include insight into the forecasting deliberations as they occurred, without the comforting clarity of hindsight.

INSTRUMENTATION, DATA ACQUISITION, AND ANALYSIS

Shortly after the phreatic explosions on April 2, several portable seismographs were deployed on the west side of Mount Pinatubo (Sabit and others, this volume). In late April and early May, a network of seven radio-telemetered seismic stations was installed as part of a joint Philippine Institute of Volcanology and Seismology (PHIVOLCS)/ U.S. Geological Survey Volcano Crisis Assistance Team effort. The stations were located at distances of 1 to 19 km

from the volcano's summit (fig. 2). Single-component vertical seismometers with a natural frequency of 1-Hz were used at all stations except PPO, which used a 2-Hz three-component geophone. The first station was installed on April 28, followed by three more during the next 9 days, allowing the reliable calculation of earthquake hypocenters by May 7. Installation of the network was completed on May 13. The network operated satisfactorily until shortly after 2200 (all dates and times are Philippine local time; to convert to G.m.t. subtract 8 h) on June 11, when station UBO failed. Vandals subsequently disabled the radio-relay site for stations PPO and BUR on June 12. Stations PIE, GRN, and BUG were lost during the climactic eruption on June 15. Thus, CAB was the only station to operate continuously throughout the eruptions of Mount Pinatubo, although its recording was interrupted for several hours on June 15 when power to the acquisition system was lost. The installation, operation, and technical specifications of this network are described in Lockhart and others (this volume).

The seismic signals from these stations were transmitted to Clark Air Base and recorded on a PC-based data acquisition system as well as a number of drum recorders. The networked system of PC's represents a significant improvement in our ability to monitor remote volcanoes effectively, analyze the data quickly, and provide accurate eruption forecasts. The system provides for the digital recording of seismic waveforms, hypocenter and magnitude calculation, Real-time Seismic Amplitude Measurements (RSAM) (Endo and Murray, 1991), and Seismic Spectral Amplitude Measurements (SSAM) (Stephens and others, 1994). The acquisition system is described in detail by Murray and others (this volume).

Seismic data from the Pinatubo network were processed by a variety of computer programs. For each seismic event detected and saved by the acquisition system, phase arrival and signal duration data from individual stations were determined by use of the program PCEQ (Valdez, 1989). These data were then used to calculate earthquake hypocenters and coda magnitudes through the use of the program HYPO71 (Lee and Valdez, 1989) with a horizontally layered velocity model. The large volume of data and severe time constraints allowed only a rapid first pass through the data analysis scheme during the crisis response in 1991. Low-quality hypocenter solutions, due to poor or incorrect phase data, were rejected and not reanalyzed. The earthquake hypocenters and magnitudes presented in this paper were recalculated by use of more sophisticated analysis techniques than were available at Clark Air Base at the time of the eruption. Phase-arrival times were repicked by use of the program XPICK (Robinson, 1992), and hypocenters were again calculated using a layered structure. This posteruption processing reduced the dispersion of earthquake hypocenters and thereby make the earthquake clusters appear more compact than they appeared during our crisis response. Hypocenters were then recalculated by use

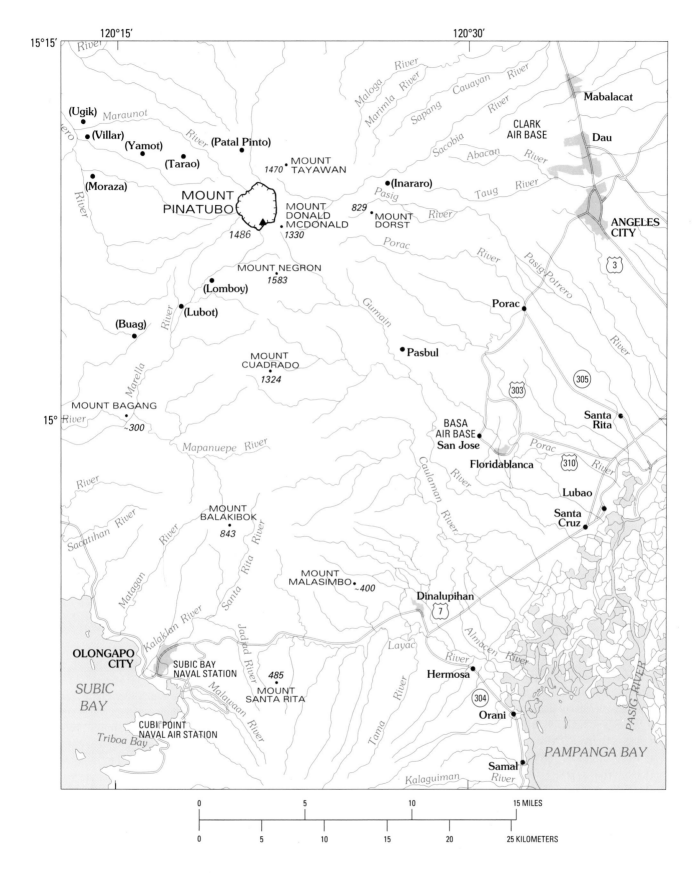

Figure 1.   Location map of the vicinity of Mount Pinatubo.

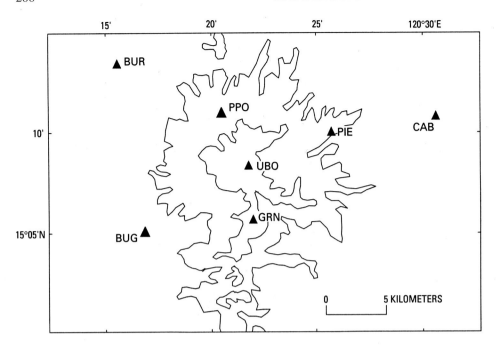

Figure 2. Map of radio-telemetered seismic stations installed around Mount Pinatubo, April–June 1991.

of a three dimensional-velocity model derived by Mori, Eberhart-Phillips, and Harlow (this volume).

The computer program ACROSPIN (Parker, 1990) permitted a two-axis rotation of plots on a computer screen either in a continuous or stepwise mode. During the crisis response, we gained an important impression of the three-dimensional development of seismic activity by using ACROSPIN to rotate earthquake hypocenter plots. Following the eruption Hoblitt, Mori, and Power (this volume) developed a computer program called VOLQUAKE, which displays the evolution of earthquake hypocenters throughout the Pinatubo eruption on a computer screen. The program uses colored symbols to represent earthquakes at various depths and allows the user to view the hypocenters in three dimensions from any perspective. This program offers what is perhaps the best representation of the development of earthquake activity at Pinatubo through the 1991 eruption sequence.

The RSAM system provides 10-min averages of the absolute seismic amplitude for each seismic station. RSAM data are a particularly useful tool for monitoring the overall level of seismic activity during periods of high and rapidly changing activity. RSAM data have been successfully used as an evaluation and forecasting tool at Mount St. Helens (Endo and Murray, 1991) and during the 1989–90 eruptions of Redoubt Volcano, Alaska (Power and others, 1994). RSAM data proved to be particularly valuable during the Pinatubo crisis, especially during the final days of the precursory seismic sequence, when activity escalated and a number of stations were destroyed or disabled. RSAM data were analyzed and displayed in realtime by use of the program BOB (Murray, 1990).

A Seismic Spectral Amplitude Measurement (SSAM) system was also used at Mount Pinatubo. This relatively new tool (Stephens and others, 1994) was developed during the 1989–90 eruptions of Redoubt Volcano, Alaska, and was successfully used to forecast some eruptions of Redoubt Volcano in the spring of 1990. SSAM is a refinement of the RSAM system and provides 1-min averages of the spectral amplitudes within several narrow frequency bands for each seismic station. SSAM results were not incorporated in formulating eruption forecasts at Mount Pinatubo because only cumbersome software was available to display the data in 1991. The SSAM data, however, have proved valuable in reconstructing the spectral character of seismicity preceding the Pinatubo eruption (Power and others, this volume) and promise to be a valuable tool for future eruption forecasting.

EVENT CLASSIFICATION

Chouet and others (1994) argue convincingly for classifying the diverse seismic signatures produced by volcanic activity on the basis of the physics of their source processes. The advantage of a process-oriented scheme is that the observed seismic activity is directly linked to ongoing volcanic processes. Chouet and others (1994) separate volcanic seismicity into two basic families of processes. The first family includes events caused by the brittle failure of rock as a result of stresses induced by magmatic processes. The second family consists of sources in which fluid plays an active role in the generation of seismic waves. Included in this family are long-period events, tremor, and signatures produced by degassing activity. We follow a process-based

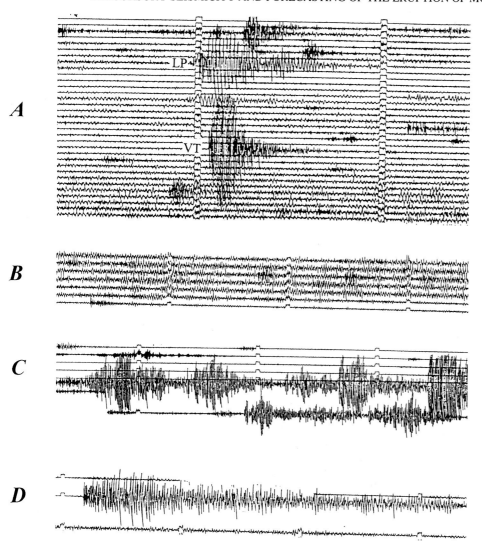

Figure 3. Examples of seismic event types observed at Mount Pinatubo (station PIE). *A*, Volcano-tectonic earthquake (VT) and long-period event (LP). *B*, Tremorlike episode of closely-spaced long-period events. *C*, Harmonic tremor. *D*, Explosive eruption at station CAB. Time marks represent 1-min interval.

classification system for the Pinatubo seismic data and describe below the various types of events observed.

Volcano Tectonic (VT) Earthquakes.—The signature of a typical VT earthquake is illustrated in figure 3*A*. Such earthquakes are characterized by sharp, mostly impulsive onsets, and their spectra are typically broad, between 1 and 15 Hz (Lahr and others, 1994). They are called VT earthquakes because their signatures are indistinguishable from those of tectonic earthquakes, although the stresses that trigger them are derived from magmatic processes rather than large-scale tectonic movements (Chouet and others, 1994). VT earthquakes occur as single events as well as sequences of rapidly occurring events with overlapping codas, called flurries by Hill and others (1985).

Long-period (LP) Events.—The signatures of LP events are characterized by a high-frequency onset followed by a lower frequency quasimonochromatic coda (fig. 3*A*). The spectra of LP events typically show a strongly peaked response in contrast to the broad spectra of VT earthquakes (Lahr and others, 1994). This spectral criterion is used to distinguish an LP event from the classic B-type earthquake

(Minakami, 1960), which is a VT earthquake occurring at a depth of 1 km or less. Because path effects can significantly modify their signatures, such VT earthquakes are often confused with true LP events. However, the broad spectra of VT earthquakes serve to distinguish them from LP events. Theoretical modeling by Chouet (1985, 1988, 1992), Ferrazzini and others (1990), and Chouet and others (1994) suggests that the signal characteristics of LP events are comparable to the synthetic signatures generated by the excitation of a fluid-filled crack or conduit in response to pressure transients. Thus, LP events are considered to be indicators of pressurization in volcanic conduits and, as such, represent critical precursors to volcanic eruptions. The source of LP events and tremor are in many cases thought to be closely related (Chouet, 1992; McNutt, 1992), and for this reason we refer to periods when both individual and LP events are occurring as LP seismicity.

An example of an LP event is shown in figure 3*A*, and the distinct character of its signature is evident compared with that of a VT earthquake (also shown in fig. 3*A*). Another illustration of LP activity from Mount Pinatubo is the sequence of events shown in figure 3C. Posteruption analysis revealed that this sequence of events originated at depths of more than 30 km (White, this volume).

Hybrid Events.—Hybrid events exhibit signatures that combine characteristics of both LP and VT earthquakes (Chouet and others, 1994; Lahr and others, 1994, Power and others, 1994). These events generally have broad spectra with a distinct spectral peak similar to an LP event. Hybrid events are thought to be generated by brittle fracturing either through or near fluid-filled cracks, which thereby generates seismic waves from source processes associated with both end member types (Lahr and others, 1994).

Volcanic Tremor.—Tremor denotes a wide range of irregular amplitude seismic signals with durations that can range from minutes to days, weeks, and even longer. The source mechanisms for tremor vary widely (Chouet, 1988, 1992; McNutt, 1992), and several types of tremor tied to specific source processes are found in the literature (Aki and others, 1977; Aki and Koyanagi, 1981; Chouet, 1985, 1988, 1992; Chouet and Shaw, 1991; Chouet and others, 1994; Koyanagi and others, 1987; Ferrazzini and others, 1990; Shaw and Chouet, 1991). Harmonic tremor refers to continuous signals of quasi-monochromatic appearance with dominant frequencies of a few hertz (figs. 3B and 3C). Volcanic tremor and LP events are intimately linked (Koyanagi and others, 1987), and the latter have been interpreted as the impulse response of the tremor-generating source (Chouet, 1985). Figure 3C shows a tremor episode composed of a closely spaced series of LP events that occurred at an unusual depth of 35 km (White, this volume).

Eruption Signals.—The signatures of eruptions generally have emergent onsets and extended codas, although the onset can be quite sudden in some cases (Power and others, 1994). At Pinatubo, the durations of eruption signals varied greatly; minor hydrothermal or ash eruptions were barely perceptible on the seismic records of the nearest high-gain stations and lasted only a few seconds, whereas the climactic eruption on June 15 produced a continuous signal for more than 12 h at regional stations at distances of more than 300 km. The signature of an explosive eruption at Mount Pinatubo is shown in figure 3D.

SEISMIC CHRONOLOGY

We divide the period April 5 to June 15, prior to the climactic eruption, into five phases based on the level, type, character, and intensity of the seismic activity. These seismic phases roughly coincide with the first five of eight eruptive phases outlined by Hoblitt, Wolfe, and others (this volume). The precursory phases are (1) Metastable—through May 31, (2) Predome—June 1 to 7, (3) Pre-explosive Buildup—June 8 to 12, (4) Long-Period Buildup—June 12 to 14, and (5) Preclimactic—June 14 to 15. Thus, the seismic phases have respective durations of a

minimum of 2 months (beginning with seismic monitoring on April 5), 7 days, 5 days, 3 days, and 1 day. In the following section we describe the seismicity that typified each of these phases.

METASTABLE PHASE: TO MAY 31

The Metastable Phase is characterized seismically by a high but roughly constant rate of VT earthquake activity. Epicenters of most located events cluster together roughly 5 km northwest of the volcano's summit (fig. 4). During this phase the fumarolic activity on the volcano's flanks, which began with the small explosions on April 2, remained roughly constant.

We do not know when seismic activity first reached a minimum level indicative of volcanic unrest. When the first portable seismographs were installed on April 5, seismic activity was already at an elevated level that would persist through the Metastable Phase (40 to 200 VT earthquakes per day). Sabit and others (this volume) note reports of felt earthquakes in the vicinity of Mount Pinatubo beginning on March 15, 1991.

Almost all of the recorded seismic activity during the Metastable Phase consisted of VT earthquakes with magnitudes ranging from less than 1 to 3. The majority of VT earthquakes at Mount Pinatubo occurred as separate events or in small flurries of events with overlapping codas lasting from 1 to a few minutes. The most prolonged flurry occurred on May 29 and lasted for approximately 20 min. The most active seismic source region during the Metastable Phase was a cluster of earthquakes located approximately 5 km northwest of the summit (fig. 4). At least 75 percent of the earthquakes occurred in that northwest cluster, over a range in depth from 1 to 10 km, with the majority of events occurring at depths of 3 to 7 km. Only a few events were detected under the energetic fumaroles that persisted after the April 2 phreatic eruptions or beneath what would eventually become the crater.

Sporadic episodes of low-level tremor with frequencies between 1 and 8 Hz were also recorded during this phase. The amplitude of these tremor episodes decreased rapidly with distance from the summit of Mount Pinatubo, suggesting a shallow source. Such tremor appears to have been associated with shallow hydrothermal activity coincident with the vigorous fumaroles observed on the north flank of Pinatubo.

A more vigorous episode of shallow tremor was associated with a small explosion that occurred between approximately 1800 on May 26 and 0200 on May 27. Several shallow LP events also occurred during this episode. Posteruption analysis showed that the shallow tremor and LP events roughly coincided with the onset of deep LP events and that tremor from a deep source also occurred during this time (see figs. 3A and 12 of White, this volume). Deep LP

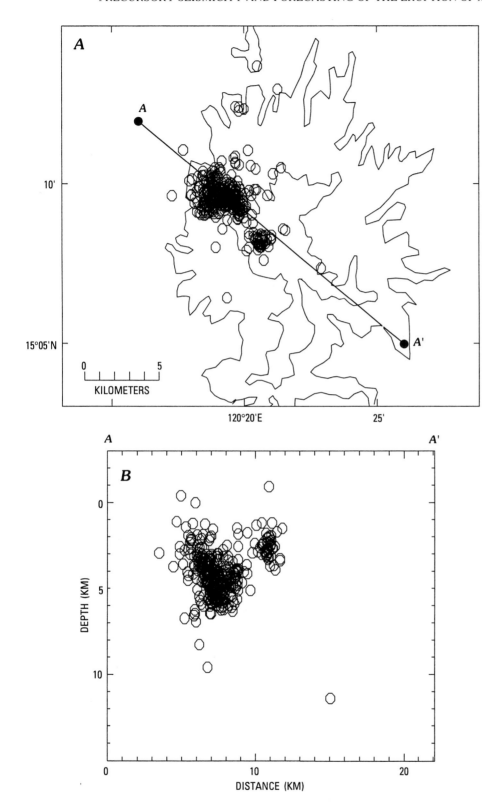

Figure 4. *A*, Epicenter map of events located during the interval from May 6 to May 31, 1991. *B*, Vertical cross section through *A–A'* shown in *A*.

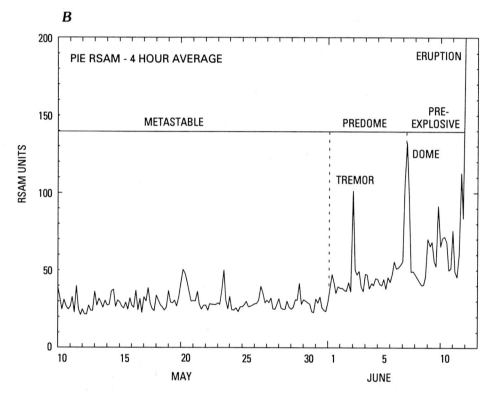

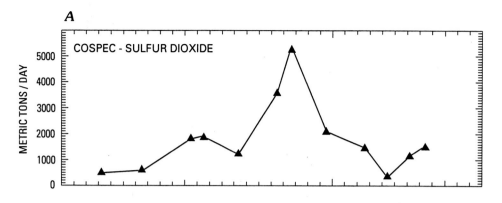

Figure 5. Plot of (*A*) the SO_2 volumes from May 10 to June 12, 1991, estimated from COSPEC measurements and (*B*) 4-h average RSAM values during the same time interval showing the marked division between the Metastable and Preexplosive Buildup seismic phases.

events continued until 1210 on May 28. A second occurrence of deep LP events has also been identified (White, this volume). We did not recognize the deep LP events as they occurred, due to limitations in our acquisition and analysis software and in the time available to analyze the data.

Although the number of earthquakes varied somewhat about a stable average during the Metastable Phase, SO_2 emissions changed drastically (fig. 5*A*). The emission rate of SO_2 determined from COSPEC measurements increased tenfold, from 500 t/d to 5,000 t/d, between May 13 and May 28. The next measurement on May 30 indicated that SO_2 had dropped abruptly to 1,400 t/d. SO_2 emissions continued to drop, reaching a low of only 260 t/d on June 5 (Daag and others, this volume). We note that the shallow LP event illustrated in figure 3*A* occurred on May 26 during the rapid buildup in SO_2 emission. This event and several other small

LP events occurred during an episode of phreatic activity that lasted several hours. This activity was significantly more energetic than had been observed previously. No other shallow LP events were recorded until June 9.

PREDOME PHASE: JUNE 1–7

Distinct changes in seismic activity began in early June that included an increase in the number of locatable VT earthquakes beneath the active fumaroles, an increase in small explosions, and an increase in the intensity and durations of episodes of tremor. Seismic activity during this phase eventually evolved into an intense swarm of shallow VT earthquakes that heralded the beginning of dome growth on the northwest flank of the volcano (Hoblitt, Wolfe, and others, this volume). RSAM values varied

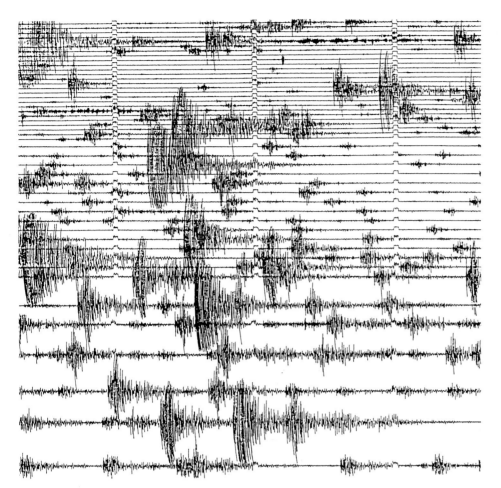

Figure 6. Drum record from station PIE showing a sample of the swarm of volcano-tectonic earthquakes associated with the dome extrusion on June 7. Time marks represent 1-min interval.

slightly around a low constant level until June 1, when they began to increase steadily and manifest larger fluctuations (fig. 5B). We use this change in RSAM values from station PIE to identify the transition from the Metastable to the Predome Phase. The Predome Phase begins at the change in the trend of RSAM averages on June 1 and ends at about 2300 on June 7, with a rapid drop in the high level of seismicity that we assume coincided with the inception of dome growth at the surface.

Small explosions and tremor episodes began to increase in number during the first few days of June. The largest explosions occurred at 1939 and 2258 on June 3. The relative amplitudes of this tremor on various stations in the network suggested it was associated with surficial

hydrothermal activity. On June 4, a 22-min-long episode of tremor (fig. 3C) occurred with a dominant frequency of about 2 Hz and roughly equal amplitudes at each seismic station. At the time, we speculated that this tremor episode occurred at between 2 and 5 km in depth. Posteruption analysis, however, reveals that this tremor consists of a series of overlapping LP events and occurs at a depth of about 35 km (White, this volume).

At about 0700 on June 6, VT activity beneath the active fumaroles began to increase rapidly (fig. 6). This earthquake swarm continued to intensify until 1630 on June 7, when it evolved into an hour-long episode of sustained activity. Earthquakes in that swarm ranged in depth from −1 to 3 km. This swarm continued until 2300 on June 7, when

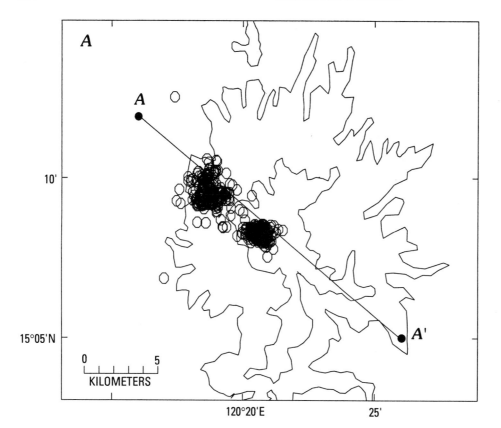

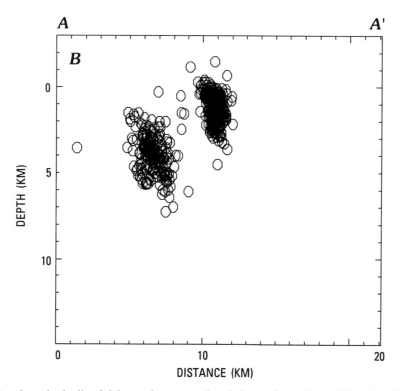

Figure 7. *A*, Epicenter map of events located between June 1 and June 7, 1991. *B*, Vertical cross section through *A–A'* shown in *A*.

activity abruptly declined. Map and cross-sectional views of earthquake hypocenters from June 1 through June 7 are shown in figure 7 and illustrate the concentration of seismic activity associated with the swarm beneath the summit of Mount Pinatubo. Figure 8 shows the number of earthquakes located each day in the northwest cluster and beneath the summit dome. Visual observations made early on June 8 established that a small dome had been extruded near a

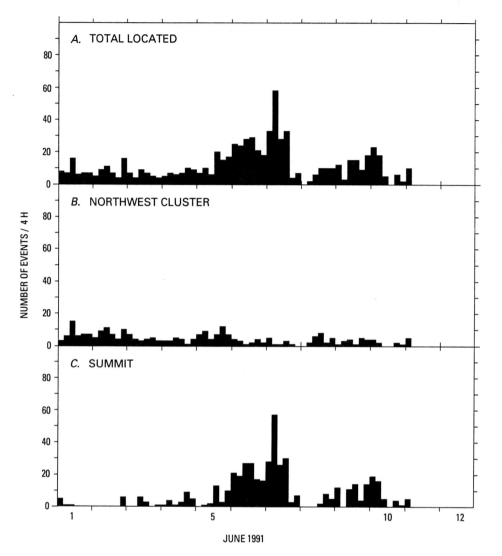

Figure 8. Number of located seismic events per 4-h intervals between June 1 and June 12 from (*A*) the entire Pinatubo network, (*B*) the cluster of seismic activity 5 km northwest of the summit of Mount Pinatubo, and from (*C*) the region beneath the summit. Events in the northwest cluster were of M≥1.2; events beneath the summit were of M≥0.4.

weak fumarole on the north-northwest flank of the volcano (Hoblitt, Wolfe, and others, this volume).

PREEXPLOSIVE PHASE: JUNE 8 TO 0851 JUNE 12

The Preexplosive Phase is characterized by swarms of VT earthquakes, a gradual increase in the incidence and intensity of volcanic tremor, and the reappearance of a few shallow LP events. This phase begins with a 33-hour period of relative seismic quiescence and ends with the first major explosive eruption on June 12.

The number of earthquakes detected and located in the northwest cluster continued more or less unchanged from the predome phase, at 20 to 25 per day (fig. 8). In contrast, seismicity beneath the summit decreased dramatically on June 8 and consisted only of occasional VT earthquakes of less than magnitude 1.5, 1- to 4-min episodes of tremor, and a few small explosion events that were likely produced by activity on the new dome. Visual observations suggest the

dome continued to grow throughout this phase, and small amounts of ash were continuously emitted from the dome's margins (Hoblitt, Wolfe, and others, this volume). The level of seismicity beneath the new dome began to escalate again at about 0800 on June 9. Earthquakes occurred in episodic swarms lasting from 0.5 to 2.0 h separated by 2- to 4-h intervals of quiescence. Most events during this period are concentrated at shallow depth beneath the dome (fig. 9). Preliminary spectral analysis of the earthquakes recorded on June 9 suggests that some events exhibit characteristics of hybrid earthquakes, while a few are LP events. This style of earthquake activity continued until the first explosive eruption on June 12. More detailed analysis will be required to determine the relative rate of occurrence of hybrid and LP events during this period.

Tremor episodes also began to appear on June 9 with a narrower frequency content than previously observed. The most vigorous episode began at 1030 on June 9 and lasted about 4 h. This tremor exhibited predominant frequencies in the 2- to 3-Hz range, and recorded amplitudes appeared to

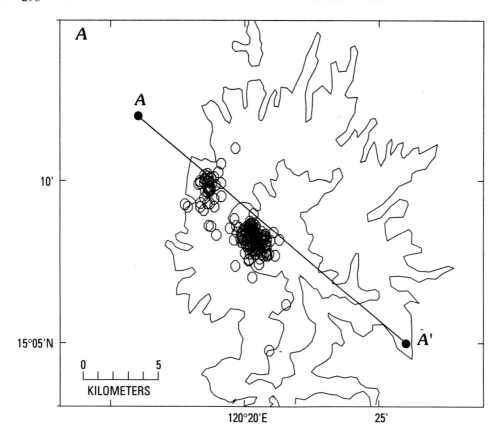

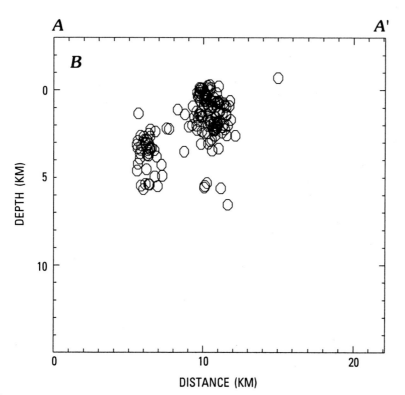

Figure 9. *A*, Epicenter map of events located between June 8 and June 12, 1991. *B*, Vertical cross section through *A–A'* shown in *A*.

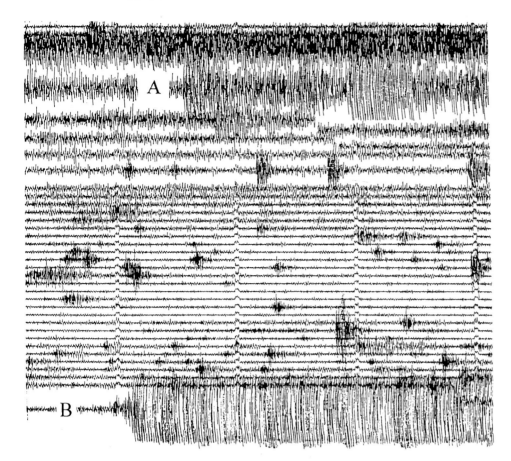

Figure 10. Station PIE drum record showing (*A*) the eruption signals for the small eruption at 0341 and (*B*) the first explosive eruption at 0851 on June 12, 1991. Time marks represent 1-min intervals; the pen was manually offset following each explosion.

correlate with the vigor of the continuous, low-volume ash and steam emissions from the dome. We interpret tremor during this period as reflecting low-level ash emission rather than hydrothermal boiling that preceded extrusion of the dome. On June 9, the seismic energy release began to show bigger fluctuations than had been observed previously and its average level began to increase steadily, as reflected in the RSAM values shown in figure 5. The steady increase in RSAM values was caused by the onset of sustained low-level tremor that began at about 1300 on June 10 and continuously increased in amplitude until early on June 12.

An episode of high-amplitude tremor with a broad spectrum began at about 0310 on 12 June and lasted for about 2 h (fig. 10). An eruption event was recorded at 0341 during that tremor episode, but because of darkness no visual observations of the eruption were made. An aerial observation at approximately 0700 on June 12 confirmed that a small eruption had occurred during the night (Hoblitt, Wolfe, and others, this volume). Low amplitude tremor began again at 0841 on June 12, followed 10 min later by a rapid increase in amplitude that signaled the onset of the first explosive eruption at 0851.

LONG-PERIOD BUILDUP PHASE: 0851 JUNE 12 TO 1309 JUNE 14

The Long-Period Buildup Phase is characterized seismically by a strong increase in the number and size of LP events, episodes of strong tremor, swarms of VT earthquakes, as well as signals from explosive eruptions. The overall level of seismic activity increased dramatically following the first large explosive eruption at 0851 on June 12.

By this time, seismic stations UBO, BUR, and PPO had been lost. Of the four remaining stations, the closest three, BUG, GRN, and PIE, recorded continuous seismic activity, and their seismic signals were often electronically off scale. The high level of seismic activity saturated the detection algorithm in the computer system, and in the absence of continuous seismic recording, most events were not recorded digitally. For the few events that were recorded, the poor geometry of the remaining four stations relative to the earthquake source region resulted in severe degradation in location quality. Seismic signals from the most distant station, CAB, were still on scale after the 0851 eruption. Thus, after that eruption, the identification of

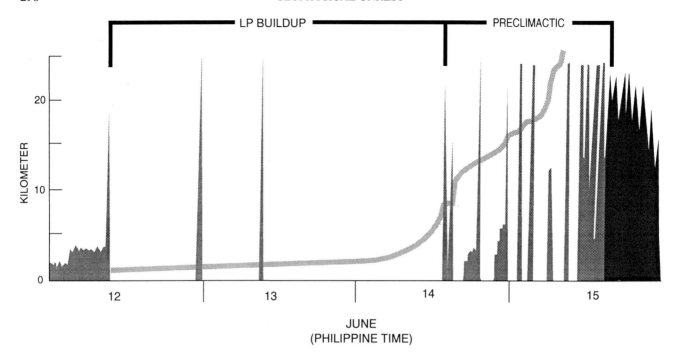

Figure 11. Estimated column heights (spikes) versus time for explosive eruptions that occurred between June 12 and the onset of the climactic eruption on June 15, 1991. The cumulative RSAM values (solid line) are shown for comparison.

seismic events for eruption forecasting depended almost exclusively on drum records of the CAB station.

Additional explosive eruptions occurred at 2252 on June 12, 0841 on June 13, and 1309 on June 14 (fig. 11). The durations of the signals for the four eruptions during this phase are 45, 20, 8 and 4 min, respectively, as measured on the CAB drum record. All of these eruptions produced vertical ash columns; at least two of them produced minor pyroclastic flows. From the thickness of their ash-fall deposits, the successive eruptions became progressively less vigorous (Hoblitt, Wolfe, and others, this volume). The decreasing pyroclast production is consistent with the decreasing durations of their eruption signals.

Following the striking increase in seismicity after the 0851 eruption, and in contrast to the decreasing vigor of the first four explosive eruptions, the rate of seismic energy release continued to increase throughout the Long-Period Buildup Phase. This increase is most evident in the RSAM data shown in figure 11. A remarkable buildup of LP events, from which this phase derives its name, evolved from 1- and 2-h swarms, respectively, of LP events preceding the explosive eruptions at 2252 on June 12 and at 0841 on June 13, to a sustained series of increasingly larger LP events that dominated seismic activity. The rapid increase in the RSAM values early on June 14 (fig. 11) shows the intense buildup of LP seismicity. The increasing size of LP events between June 12 and 14 is illustrated in figure 12 for events at station CAB. Note that the signature of the explosive eruption at 1309 in figure 12 is not significantly different from the signatures of the preceding LP events. The

largest of these events were recorded on regional seismic stations at distances of up to 350 km, and preliminary estimates suggest that the largest events were roughly equivalent to earthquakes of magnitude 3.5 to 4.

Although the Long-Period Buildup Phase was characterized by the LP seismicity, an uncountable number of VT earthquakes were also recorded. Few VT earthquakes were large enough to be recorded at station CAB, and most have magnitudes between 1 and 2.

PRECLIMACTIC PHASE: 1309 JUNE 14 TO 1342 JUNE 15

Following the eruption at 1309 on June 14, the character of seismicity changed to a diverse mix of large VT earthquakes, and LP and eruption events. This phase coincided with a significant change in eruptive style from convectively driven vertical eruption columns at intervals of 10 to 24 h to increasingly frequent eruptions that generated pyroclastic surges. The seismic and pressure record (Hoblitt, Wolfe, and others, this volume) indicates that explosive eruptions occurred at 1410, 1516, 1853, 2218, and 2320 on June 14 and 0115, 0257, 0555, 0810, 1027, 1117, 1158, 1222, 1252, and 1315 on June 15 before the climactic eruption began at 1342.

RSAM average amplitudes show the escalation in overall seismic activity throughout the phase and indicate that the rate of activity increased at about 0300 on June 15. A portion of the CAB drum record is shown in figure 13,

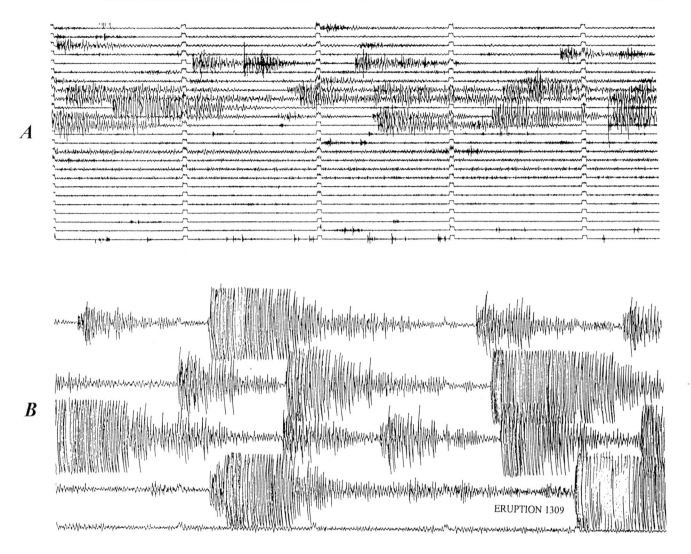

Figure 12. Sections of station CAB drum records showing the increase in size of long-period events between (*A*) June 12 and (*B*) June 14, 1991. Time marks represent 1-min intervals; pen was manually offset in *B*.

which illustrates the diverse seismic character during the phase. SSAM measurements indicate that seismic activity with both VT and LP spectral characteristics increased during this phase (Power and others, this volume).

DISCUSSION

In this section we review the role that seismological observations played in formulating eruption forecasts and evaluating volcanic hazards and review the factors that influenced our interpretations at the time. The overall forecasting strategy used at Pinatubo relied on synthesizing data from a number of monitoring techniques, which included a variety of seismic measurements, visual observations, gas emissions, and geologic observations. The successful forecasts and evacuations prior to the onset of explosive activity validate this approach. Our seismic interpretations were based largely on monitoring hypocentral locations, tracking seismicity rate, and energy release, and associating the various event types with possible source processes. Recent advances in computer technology allowed us to apply a number of relatively sophisticated analysis techniques, even at the relatively remote field locations.

During the Pinatubo crisis we used a numbered alert system to communicate our interpretations and forecasts of volcanic activity to civilian and military leaders, as well as the general public. A detailed description of this system and its application throughout the 1991 eruption crisis is given by Punongbayan and others (this volume) and Wolfe (1992).

In the following subsections, we discuss our interpretations of the seismic data and their implications for evaluating magmatic processes and anticipating future eruptions prior to the climactic eruption of Mount Pinatubo at 1342

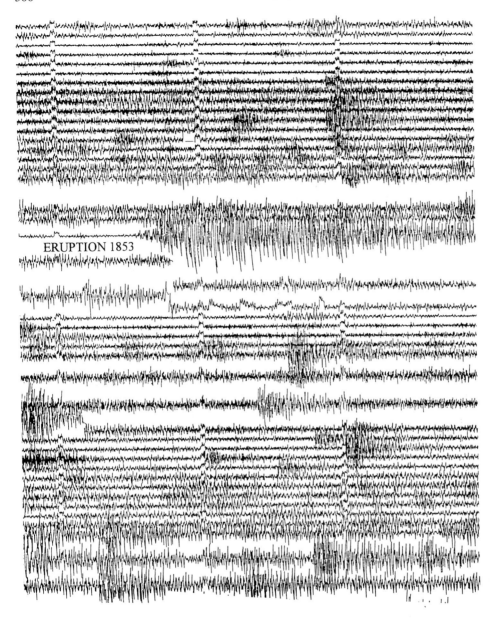

ERUPTION 1853

Figure 13. Station CAB drum record showing diverse character of seismic activity during the Preclimactic Phase on June 14, 1991. Time marks represent 1-min intervals.

on June 15. Our discussion follows the order of successive seismic phases presented earlier.

METASTABLE PHASE: TO MAY 31

Our interpretation of preeruption seismicity necessarily begins when seismic stations were installed on April 5, 3 days after the initial phreatic eruptions. By that time as many as 200 VT earthquakes a day were being recorded, and the Metastable Phase was already in progress. The significance of the elevated level and widespread distribution of VT earthquake activity was the focus of our initial evaluations of eruptive potential. Experiences at other volcanoes indicate that, although a high level of VT earthquake activity is a clear indication of volcanic unrest, VT earthquakes are not always a reliable indicator of an impending volcanic

eruption (Newhall and Dzurisin, 1988). For example, a marked increase in the rate of VT earthquakes preceded the 1986 eruption sequence at Augustine Volcano, Alaska (Power, 1988) and the initial eruption of Mt. Spurr, Alaska (Power and others, in press), by 9 and 10 months, respectively. However, in large volcanic systems such as Rabaul, New Britain, Papua-New Guinea (McKee and others, 1984), Campi Flegrei, Italy (Aster and others, 1992), and Long Valley, Calif. (Hill and others, 1985; Rundle and Hill, 1988), VT earthquakes have been observed to continue at elevated levels for months to years without (or, in the case of Rabaul, before) leading to eruptive activity. At each of these large complexes, evidence suggests that magma intrusions have taken place, but no eruptions have occurred.

Additionally, we were concerned that most of the seismic energy released during the Metastable Phase was associated with the cluster of hypocenters located about 5 km northwest of the summit of Mount Pinatubo (fig. 4). We considered the possibility that the earthquake activity resulted from tectonic processes that had merely perturbed the Pinatubo hydrothermal system. A rapid examination of aerial photographs indicated that the northwest cluster lay beneath an area of geologically young fault traces, and this fact raised the possibility that these earthquakes were tectonic in origin. The rationale for such a conclusion comes from a study of 35 destructive near-surface earthquakes along the volcanic chain of Central America made by White and Harlow (1993). They found that eruptive activity was associated with only two of the earthquakes, whereas the rest occurred on active faults in the vicinity of the volcanic chain and were associated with tectonic activity rather than magmatic or volcanic processes.

The COSPEC measurements that indicated that a large volume of SO_2 was being emitted from the volcano left little doubt that we were dealing with an active magmatic system. The rapid increase and later decrease in SO_2 emission were interpreted at the time as evidence that magma was rising and that gas escape was temporarily choked off (Daag, Tubianosa, and others, this volume).

Through the end of May, therefore, we estimated that there was very roughly a 50 percent probability that this particular episode of unrest would end in eruptive activity. This assessment was conditioned by several points: (1) the previous observations of prolonged seismic unrest exhibited by other large volcanic systems that did not evolve into eruptions (Newhall and Dzurisin, 1988), (2) the confusing situation seen at Mount Pinatubo where most of the seismic energy was being released 5 km northwest of the April 2 phreatic vent area, and (3) the large SO_2 flux observed during May.

PREDOME PHASE: (JUNE 1 TO 7)

Seismic activity evolved beyond the relatively steady occurrence of VT earthquakes in the early days of this interval to episodes of tremor and an intense swarm of VT earthquakes associated with a dome extrusion on June 7. The activity leading to June 7 suggested to us the increased probability of an eruption. Both the pace and the gravity of our decisions increased dramatically.

During the first days of June, the changes in seismic activity were subtle and we were unsure of their significance. The RSAM level, which began to increase on June 1 (fig. 5), is not visually obvious on the analog drum recorders and affirms the value of this straightforward tool to detect small changes in seismicity level. During our necessarily rapid assessment of seismicity, we failed to identify a tremor episode (fig. 3B) on June 4 as a sequence of LP events at depths of more than 30 km (White, this volume). We did, however, realize from a visual inspection of the drum records that this tremor episode was deeper than

previous tremor episodes. That conclusion was based on the roughly equal signal amplitudes at all stations of the sequence of deep LP events compared to the rapid decrease of signal amplitude with distance for shallow tremor. Therefore, we considered this a significant change in the seismic character.

The shift in the locus of VT earthquake activity from the northwest cluster to beneath the active fumaroles also began gradually in early June. By June 6, however, that shift had become obvious with the onset of the swarm of VT earthquakes that preceded the emplacement of the dome (fig. 7). In evaluating this activity on June 6 and 7, we were cognizant of the VT nature of the swarm. Strong swarms of VT earthquakes have frequently been observed in association with shallow intrusions of magma and dome formation (Fremont and Malone, 1987, Klein and others, 1987). Measured ground deformation from tiltmeters recently installed near the volcano (Ewert and others, this volume) strongly supported this interpretation.

The fivefold jump in SO_2 emission between May 24 and May 28, followed by the dramatic decrease by May 30 (fig. 5) (Daag, Tubianosa, and others, this volume), was possibly the initial step in a process that led to the extrusion of magma at the surface. Following the decrease in SO_2 emissions at the end of May, VT earthquake activity began to increase beneath the fumarole vents. Hypocenters from the VT earthquake swarm on June 6 and 7 suggested that the final seismic step in the dome formation process was an episode of intense brittle fracturing in a zone extending from a depth of about 3 km to the surface. The appearance of the dome on June 7 proved that activity at Mount Pinatubo was being driven by magmatic processes.

The decision to move to a higher alert level, which declared that the volcano could erupt at any time (Punongbayan, and others, this volume), was made during the most intense phase of the swarm associated with dome formation and before the dome had been visually observed. The decision was made in spite of the absence of LP seismicity. We speculated that magma had begun ascending a conduit system formed by the phreatic activity on April 2 and that this action was responsible for the pulse of SO_2 observed at the end of May. We speculated further that, as the upper tip of the magma column began to degas and cool, the system progressively sealed itself and caused the drop in SO_2 emissions observed after May 28.

PREEXPLOSIVE PHASE: JUNE 8 TO 0851 JUNE 12

Significant seismic changes during the Preexplosive Phase included continued swarms of VT earthquakes at shallow depths beneath the still growing lava dome, episodes of strong tremor, a gradual increase in low-level tremor that occurred throughout the phase, and the reappearance of a few shallow LP events as well as hybrid events. Interestingly the phase began with roughly 30 hours of relative seismic quiescence.

In our reflections during this period, some of us felt we had overreacted in raising the alert level on June 7, while others were confident that seismic activity would reappear and increase to higher levels than had yet been observed. This interval of seismic quiescence strongly suggested to us, however, that prior to the dome extrusion, magma movement was inducing high strain rates over a broad area. Correspondingly, the extrusion of the dome had acted as a temporary strain relief valve.

When seismic activity resumed on June 9, we observed that more seismic activity was occurring beneath the dome rather than in the now seismically subdued northwest cluster (fig. 9). This shift in seismicity also suggested that a fundamental change from the predome pattern of strain release had taken place. Continuous background tremor began on June 10, and we became increasingly concerned that an explosive eruption might be imminent, particularly when tremor increased and when we received the results of the June 10 measurement of SO_2 flux, which had risen sharply to 10,000 t/d (Daag, Tubianosa, and others, this volume).

We failed to recognize that many of the events recorded after June 9 were hybrid events, nor did we fully appreciate the appearance of a few LP events, because we were fatigued and increasing demands had been made on our time by civil defense officials and base commanders as evacuation plans were made and carried out.

Between June 9 and June 12 we grew more convinced that a large explosive eruption was imminent. This confidence was based on (1) the shift in the locus of seismic energy release from the northwest cluster to depths of 1 to 3 km beneath the dome (this suggested that the magma was inducing strain over a wide area, in which case we inferred that the volume of magma involved was likely to be large), (2) the steadily increasing level of background tremor after June 9, (3) the volume of SO_2 emanating from the volcano, (4) the emergence of a dome, which proved conclusively that the system was magmatic, and (5) geologic information on recent eruptive style (Newhall and others, this volume). We communicated our observations and interpretations to responsible officials frequently throughout this period. Additional communities close to the volcano, as well as nonessential personnel from Clark Air Base were evacuated on June 10, 2 days before the first explosive eruption.

LONG PERIOD BUILDUP PHASE:
0851 JUNE 12 TO 1309 JUNE 14

Seismic activity during this phase is characterized by increasing LP seismicity, as well as continued swarms of VT earthquakes and eruption events. Following the first large explosive eruption at 0851 on June 12, our interpretations of the Mount Pinatubo seismicity relied on the RSAM system to indicate the relative level of seismic energy release and on the waveform character of the seismic events as identified on station CAB drum records. As mentioned earlier, the abrupt increase in seismic activity observed at this time all but overwhelmed the earthquake detection algorithm, and the severely degraded accuracy and precision of calculated earthquake hypocenters resulting from the loss of three seismic stations prevented us from following the spatial development of earthquake hypocenters.

The large number of VT earthquakes following the first explosive eruption indicated that intense brittle rock fracture was occurring. These earthquakes were interpreted to be the result of readjustment of the system following the sudden evacuation of magma. At Mount St. Helens (Weaver and others, 1981), Nevado del Ruiz (Munoz and others, 1990), and at Redoubt Volcano (Power and others, 1994), significant increases in the number and distribution of VT earthquakes followed rather than preceded eruptive activity. In those cases, VT earthquakes were attributed to the readjustment of brittle rock following magma withdrawal. Similar VT earthquakes followed the June 15 eruptions of Mount Pinatubo (Mori, White, and others, this volume). That the level of VT activity remained high during this phase, however, suggests that at some point rock fracturing also may have been affected by the influx of new magma.

The dramatic buildup of LP events that dominated this phase suggested to us that the system was moving toward more energetic eruptive activity. Swarms of LP events occurred before the explosive eruptions of 2252 on June 12 and 0841 on June 13. Those swarms were minor in terms of duration and average event amplitude in comparison to the large buildup of LP events that began on June 13 and that became increasingly energetic by early June 14.

As a result of this buildup in LP seismicity, we were expecting an eruption larger than had yet occurred. We were surprised, therefore, that the eruption of 1309 on June 14 was a relatively small event compared to the previous explosive eruptions. Although the 1309 eruption was not as large as we expected, it was the last preclimactic eruption characterized mainly by vertical ash emission.

PRECLIMACTIC PHASE:
1309 JUNE 14 TO 1342 JUNE 15

After the 1309 eruption on June 14, the character of seismic activity changed from one dominated by LP events to one characterized by a diverse mix of VT earthquakes, LP events, tremor, and eruption signals (fig. 13). RSAM values show that seismic energy release continued to increase during this phase (fig. 11).

With this change in seismicity came a change in eruption style from explosive eruptions with vertical ash columns to eruptions generating large pyroclastic surges (Hoblitt, Wolfe, and others, this volume). The first large pyroclastic surge was observed from the west of the volcano at about 1510 on June 14. An infrared video system recorded two night eruptions that generated pyroclastic density currents at 2320, on June 14 and 0114, on June 15. The first observation of a large pyroclastic density current from Clark Air Base occurred later that morning during the 0555 eruption. After the 0555 eruption, clouds from the leading edge of Typhoon Yunya obscured the volcano and precluded further visual observations from Clark Air Base.

During the period between 1500, on June 14 and the onset of the climatic eruption at 1342, June 15, our interpretations were based solely on visual inspections of station CAB drum records and on the level of seismic energy release as depicted by RSAM values. At this point, as civil defense decisions had become the most critical, RSAM again proved to be an invaluable tool. Exhausted from the intense efforts needed to react to the escalating volcanic activity of the previous few days, we relied upon RSAM as an indicator of overall seismic energy release. This was all the information we had time to use and, indeed, all we could absorb. RSAM was an easy concept to grasp, and the data were automatically updated and displayed on a computer screen in an easy-to-understand graph.

The early dawn view of large pyroclastic density currents moving in all directions from Mount Pinatubo at 0555 on June 15 made us apprehensive about our safety. Once the view of the volcano was concealed, with RSAM values increasing, we decided to move to a more distant site. At 0730, our monitoring team and the remaining military personnel abandoned Clark Air Base. We reevaluated this decision a few kilometers from the base and elected to return. The team and a small military detachment returned to Clark Air Base at about at 1000. Although we could not see the volcano, eruption signals observed on the station CAB drum record were becoming more frequent, and RSAM values increased further. We retreated again at about 1500 after Mount Pinatubo had been in continuous eruption for more than an hour.

SUMMARY AND CONCLUSIONS

Our success in identifying the escalating eruptive potential between late April and mid-June can be credited to four factors: (1) our ability to deploy a seismic network around the volcano rapidly, (2) the capability to analyze seismic data from the network in near-real-time and to combine these results into geologic observations and measurements of other volcanic phenomena (3) our experience in working with seismic data from other recently active volcanoes, and (4) real-time synthesis of seismic and a number of other data sets and observations. On the basis of these forecasts, over 60,000 people were evacuated from high-risk areas surrounding the volcano, and an enormous loss of life was averted. The forecasts also prevented the loss of

hundreds of millions of dollars of equipment, such as the military aircraft that were evacuated.

The pivotal changes in seismic activity that led us to believe that a large eruption was increasingly imminent included (1) a shift in seismic energy release from the northwest cluster to beneath the summit during early June, (2) a swarm of VT earthquakes associated with the extrusion of a dome, (3) increasing seismic activity beneath the summit and the appearance of continuous background tremor between June 9 and the first explosive eruption on June 12, (4) the remarkable buildup of LP events between June 12 and 14, and (5) the dramatic increase in seismic energy release as indicated by RSAM values between June 14 and the onset of the climactic eruption on June 15.

A critical context for interpreting seismic data was formed by results from geologic reconnaissance work, and measurements of SO_2 flux, as well as ground-deformation measurements. Geologic information indicated that volcanism at Mount Pinatubo was distinguished by infrequent, large eruptions that produced extensive pyroclastic-flow deposits. Thus, if the seismic data indicated that an eruption was imminent, then there was a high probability that the eruption would be large and generate widespread pyroclastic flows.

The evolution of seismic activity during the Predome and Preexplosive Buildup Phases has critical implications for the early preparation of hazards maps, risk assessments, and civil defense plans at other large volcanic systems. Until June 1, seismic activity at Mount Pinatubo had been indistinguishable from that recently observed at other large volcanic complexes such as Long Valley, Calif., Rabaul, Papua New Guinea, and Campi Flegri, Italy, only one of which (Rabaul) has erupted. Two to 3 days passed before the subtle changes in the character of seismic activity that began on June 1 caught our attention, and those changes became significant enough for us to declare on June 5 that an eruption was possible within weeks. On June 7, in the midst of the swarm of earthquakes associated with the emergence of a dome, we declared that an eruption was possible within days. Thus, there was only about 1 week between the time we could state with confidence that an eruption was imminent, a confidence that wavered somewhat during the seismic quiescence on June 8, and the first explosive eruption on June 12. This short time span forcefully demonstrates that volcanic hazard studies and civil defense plans need to be completed well in advance.

During the 24 h interval preceding the onset of the climatic eruption, the value of being able to immediately display the level of seismic activity as depicted by RSAM values cannot be overstated. At a certain point, responsible officials needed to make evacuation decisions quickly, and the easily understood RSAM format proved invaluable. We strongly recommend, therefore, that computer software be developed to display clearly all data being collected in an easily understood and comparative format.

ACKNOWLEDGMENTS

We thank the many individuals who helped to make the response to the 1991 eruptions at Pinatubo a success. In particular, Ray Punongbayan, Chris Newhall, Richard Janda, C. Dan Miller, and members of the 13th Air Force. This paper benefited greatly from discussions and interactions with A. Lockhart, T. Murray, J. Ewert, P.J. Delos Reyes, B. Tubianosa, Jim Mori, Bernard Chouet, W.H.K. Lee, J. Lahr, S. Marcial, and M.L.P. Bautista. Jim Mori and Donna Eberhart-Phillips provided the hypocenter data used in figures 4, 7, 8, and 9. Bernard Chouet, E.G. Ramos, and Barry Voight provided helpful reviews of the text and figures.

REFERENCES CITED

Aki, K., Fehler, M., and Das, S., 1977, Source mechanism of volcanic tremor: Fluid driven crack models and their application to the 1963 Kilauea eruption: Journal of Volcanology and Geothermal Research, v. 2, p. 259–287.

Aki, K., and Koyanagi, R., 1981, Deep volcanic tremor and magma ascent mechanism under Kilauea, Hawaii: Journal of Geophysical Research, v. 86, p. 7095–7109.

Aster, R.C., Meyer, R.P., Denatale, G., Zollo, A., Martini, M., Del Pezzo, E., Scarpa, R., and Iannaccone, G., 1992, Seismic investigation of Campi Flegrei: A summary and synthesis of results, in Gasparini, P., Scarpa, R., and Aki, K., eds., Volcanic seismology: International Association of Volcanology and Chemistry of the Earth's Interior (IAVCEI) Proceedings in Volcanology, Berlin, Springer-Verlag, p. 462–483.

Chouet, B.A., 1985, Excitation of a buried magmatic pipe: A seismic source model for volcanic tremor: Journal of Geophysical Research, v. 90, p. 1881–1893.

———1988, A seismic source model for the source of long-period events and harmonic tremor: Journal of Geophysical Research, v. 93, p 4373–4400.

———1992, A seismic model for the source of long-period events and harmonic tremor, in Gasparini, P., Scarpa, R., and Aki, K., eds., Volcanic seismology: International Association of Volcanology and Chemistry of the Earth's Interior (IAVCEI) Proceedings in Volcanology, Berlin, Springer-Verlag, p. 133–156.

Chouet, B.A., and Shaw, H.R., 1991, Fractal properties of tremor and gas piston events observed at Kilauea Volcano, Hawaii: Journal of Geophysical Research, v. 96, p. 10177–10189.

Chouet, B.A., Page, R.A., Stephens, C.D., Lahr, J.C., and Power, J.A., 1994, Precursory swarms of long-period events at Redoubt Volcano (1989–1990), Alaska: Their origin and use as a forecasting tool: Journal of Volcanology and Geothermal Research. v. 62, p. 95–135.

Daag, A.S., Tubianosa, B.S., Newhall, C.G., Tuñgol, N.M, Javier, D., Dolan, M.T., Delos Reyes, P.J., Arboleda, R.A., Martinez, M.L., and Regalado, M.T.M., this volume, Monitoring sulfur dioxide emission at Mount Pinatubo.

Endo, E.T., Malone, S.D., Nosen, S.D., and Weaver, C.S., 1981, Locations, magnitudes, and statistics of the March 20–May 18 earthquake sequence, in Lipman, P.W., and Mullineaux, D.R., eds., The 1980 eruptions of Mount St. Helens, Washington: U.S. Geological Survey Professional Paper 1250, p. 93–107.

Endo, E.T., and Murray, T.L., 1991, Real-time seismic amplitude measurement (RSAM), a volcano monitoring and prediction tool: Bulletin of Volcanology, v. 53, p. 533–545.

Ewert, J.W., Lockhart, A.B., Marcial, S., and Ambubuyog, G., this volume, Ground deformation prior to the 1991 eruptions of Mount Pinatubo.

Ferrazzini, V., Chouet, B.A., Fehler, M., and Aki, K., 1990, Quantitative analysis of long-period events recorded during hydrofracture experiments at Fenton Hill, New Mexico: Journal of Geophysical Research, v. 95, p. 21871–21884.

Fremont, M., and Malone, S.D., 1987, High precision relative locations of earthquakes at Mount St. Helens, Washington: Journal of Geophysical Research, v. 92, p. 10233–10236.

Hill, D.P., Bailey, R.A., and Ryall, A.S., 1985, Active tectonic and magmatic processes beneath Long Valley Caldera, Eastern California: an overview: Journal of Geophysical Research, v. 20, p. 111–120.

Hoblitt, R.P., Mori, J., and Power, J.A., this volume, Computer visualization of earthquake hypocenters.

Hoblitt, R.P., Wolfe, E.W., Scott, W.E., Couchman, M.R., Pallister, J.S., and Javier, D., this volume, The preclimactic eruptions of Mount Pinatubo, June 1991.

Klein, F.W., Koyanagi, R.Y., Nakata, J.S., and Tanagawa, W.R., 1987, The seismicity of Kilauea's magma system, in Decker, R.W., Wright, T.L., and Stauffer, P.H., eds., Volcanism in Hawaii: U.S. Geological Survey Professional Paper 1350, p. 1019–1186.

Koyanagi, R.Y., Chouet, B.A., and Aki, K., 1987, Origin of volcanic tremor in Hawaii, Part 1, Data from Hawaiian Volcano Observatory 1969–1985, in Decker, R.W., Wright, T.L., and Stauffer, P.H., eds., Volcanism in Hawaii: U.S. Geological Survey Professional Paper 1350, p. 11121–1259.

Lee, W.H.K., and Valdes, C.M., 1989, User manual for HYPO71PC, in Lee, W.H.K., ed., Toolbox for seismic data acquisition, processing, and analysis, IASPEI software volume 1:. Seismological Society of America, International Association of Seismology and Physics of the Earth's Interior (IASPEI), El Cerrito, Calif., p. 203–236.

Lahr, J.C., Chouet, B.A., Stephens, C.D., Power, J.A., and Page, R.A., 1994, Earthquake location and error analysis procedures for a volcanic sequence: Application to 1989–1990 eruptions: Journal of Volcanology and Geothermal Research. v. 62, p. 137–152.

Lockhart, A.B., Marcial, S., Ambubuyog, G., Laguerta, E.P., and Power, J.A., this volume, Installation, operation, and technical specifications of the first Mount Pinatubo telemetered seismic network.

Malone, S.D., Endo, E.T., Weaver, C.S., and Ramey, J.W., 1981, Seismic monitoring for eruption prediction, in Lipman, P.W., and Mullineaux, D.R., eds., The 1980 eruptions of Mount St. Helens, Washington: U.S. Geological Survey Professional Paper 1250, p. 803–814.

McKee, C.O., Lowenstein, P.L., de Saint Ours, P., and Talai, B., 1984, Seismic and deformation crisies at Rabaul Caldera—prelude to an eruption: Bulletin of Volcanology, v. 47, no. 2, p. 397–410.

McNutt, S.R., 1992, Volcanic tremor, Encyclopedia of earth science system, Volume 4: San Diego, Calif., Academic Press, p. 417–424.

Minakami, T., 1960, Fundamental research for predicting volcanic eruptions. I—Earthquakes and crustal deformation originating from volcanic activities: Bulletin of the Earthquake Research Institute, University of Tokyo, v. 38, p. 497–544.

Mori, J., Eberhart-Phillips, D., and Harlow, D.H., this volume, Three-dimensional velocity structure at Mount Pinatubo, Philippines: Resolving magma bodies and earthquakes hypocenters.

Mori, J., White, R.A., Harlow, D.H., Okubo, P., Power, J.A., Hoblitt, R.P., Laguerta, E.P.,Lanuza, L., and Bautista, B.C., this volume, Volcanic earthquakes following the 1991 climactic eruption of Mount Pinatubo, Philippines: Strong seismicity during a waning eruption.

Munoz, F.C., Nieto, A.E., Hansjurgen, M., 1990, Analysis of swarms of high frequency seismic events at Nevado del Ruiz volcano, Colombia (January 1986–August 1987): Development of a procedure: Journal of Volcanology and Geothermal Research, v. 41, p. 327–354.

Murray, T.L., 1990, A user's guide to the PC-based time-series data management and plotting program BOB: U.S. Geological Survey Open-File Report 90–56, 53 p.

Murray, T.L., Power, J.A., March, G.D., and Marso, J.N., this volume, A PC-based realtime volcano-monitoring data-acquisition and analysis system.

Newhall, C.G., Daag, A.S., Delfin, F.G., Jr., Hoblitt, R.P., McGeehin, J., Pallister, J.S., Regalado, M.T.M., Rubin, M., Tamayo, R.A., Jr., Tubianosa, B., and Umbal, J.V., this volume, Eruptive history of Mount Pinatubo.

Nieto, A.H., Brandsdottir, B., and Munoz, F.C., 1990, Seismicity associated with the reactivation of Nevado del Ruiz, Colombia, July 1985–December 1986: Journal of Volcanology and Geothermal Research, v. 41, p. 315–326.

Newhall, C.G., and Dzurisin, D., 1988, Historical unrest at large calderas of the world: U.S. Geological Survey Bulletin 1855, 1108 p.

Omori, F., 1911, The Usu-san eruption and earthquake and elevation phenomena: Bulletin of the Imperial Earthquake Investigation Committee, v. 5, no. 1, p. 1–38.

Parker, D.B., 1990, User manual for Acrospin, in Lee, W.H.K., ed., Toolbox for seismic data acquisition, processing, and analysis, IASPEI software volume 1: Seismological Society of America, International Association of Seismology and Physics of the Earths Interior (IASPEI), El Cerrito, Calif., p. 119–164.

Pinatubo Volcano Observatory Team, 1991, Lessons from a major eruption: Mount Pinatubo, Philippines: Eos, Transactions, American Geophysical Union, v. 72, p. 545–555.

Power, J.A., 1988, Seismicity associated with the 1986 eruption of Augustine Volcano, Alaska: M.Sc. thesis, University of Alaska, Fairbanks, 142 p.

Power, J.A., Jolly, A.D., Page, R.A., McNutt, S.R., in press, Seismicity and forecasting of the 1992 eruptions of Crater Peak vent, Mount Spurr volcano, Alaska: An overview, in Keith, T.E.C., ed., U.S. Geological Survey Bulletin.

Power, J.A., Lahr, J.C., Page, R.A., Chouet, B.A., Stephens, C.D., Harlow, D.H., Murray, T.L., and Davies J.N., 1994, Seismic evolution of the 1989–90 eruption sequence of Redoubt Volcano, Alaska: Journal of Volcanology and Geothermal Research. v. 62, p. 69–94.

Power, J.A., Murray, T.L., Marso, J.N., and Laguerta, E.P., this volume, Preliminary observations of seismicity at Mount Pinatubo by use of the Seismic Spectral Amplitude Measurement (SSAM) system, May 13–June 18, 1991.

Punongbayan, R.S., Newhall, C.G., Bautista, M.L.P., Garcia, D., Harlow, D.H., Hoblitt, R.P., Sabit, J.P., and Solidum, R.U., this volume, Eruption hazard assessments and warnings.

Robinson, M.R., 1992, XPICK user's manual V4.2: Seismology lab of the Geophysical Institute, University of Alaska, Fairbanks, 119 p.

Rundle, J.B., and Hill, D.P., 1988, The geophysics of a restless caldera; Long Valley, California: Annual Review of Earth and Planetary Sciences, v. 16, p. 251–271.

Sabit, J.P., Pigtain, R.C., and de la Cruz, E.G., this volume, The west-side story: Observations of the 1991 Mount Pinatubo eruptions from the west.

Sassa, K., 1935, Volcanic micro-tremors and eruption earthquakes, (Part 1 of the Geophysical Studies on the volcano Aso): Memoirs of the College of Science, Kyoto Imperial University, Kyoto, p. 255–293.

———1936, Micro-seismic study on eruptions of the volcano Aso (Part 2 of the Geophysical studies on the volcano Aso): Memoirs of the College of Science, Kyoto Imperial University, Kyoto, p. 11–54.

Shaw, H.R., and Chouet, B.A., 1991, Fractal hierarchies of magma transport in Hawaii and critical self organization of tremor: Journal of Geophysical Research, v. 96, p. 10191–10207.

Stephens, C.D., Chouet, B.A., Page, R.A., Lahr, J.C., and Power, J.A., 1994, Seismological aspects of the 1989–1990 eruptions at Redoubt Volcano, Alaska: The SSAM perspective: Journal Volcanology and Geothermal Research. v. 62, p. 153–182.

Valdez, C.M., 1989, User manual for PCEQ, in Lee, W.H.K., ed., Toolbox for seismic data acquisition, processing, and analysis, IASPEI software volume 1: Seismological Society of America, International Association of Seismology and Physics of the Earths Interior (IASPEI), El Cerrito, Calif., p. 175–202.

Weaver, C.S., Grant, W. C., Malone, S.D., and Endo, E.T., 1981, Locations, magnitudes, and statistics of the March 20–May 18 earthquake sequence, in Lipman, P.W., and Mullineaux, D.R., eds., The 1980 eruptions of Mount St. Helens, Washington: U.S. Geological Survey Professional Paper 1250, p. 109–121.

White, R.A., this volume, Precursory deep long-period earthquakes at Mount Pinatubo, Philippines: Spatio-temporal link to a basalt trigger.

White, R.A., and Harlow, D.H., 1993, Destructive upper-crustal earthquakes of Central America since 1900: Bulletin of the Seismological Society of America, v. 83, p. 1115–1142.

Wolfe, E.W., 1992, The 1991 eruption of Mount Pinatubo, Philippines: Earthquakes and Volcanoes, v. 23, no. 1, p. 5–38.

Wood, H.O., 1915, The seismic prelude to the 1914 eruption of Mauna Loa: Bulletin of the Seismological Society of America, v. 5, p. 39–50.

Precursory Deep Long-Period Earthquakes at Mount Pinatubo: Spatio-Temporal Link to a Basalt Trigger

By Randall A. White[1]

ABSTRACT

About 600 deep long-period (DLP) earthquakes occurred beneath Mount Pinatubo in late May and early June 1991. This number is higher than the combined total number of such earthquakes previously reported at all convergent-margin volcanoes worldwide. The DLP earthquakes occurred in two episodes of roughly similar total energy release, from 1700 May 26 to 1210 May 28 and from 2114 May 31 to 1510 June 8. During these same periods, at least 25 hours of very low amplitude DLP tremor also occurred, in 1- to 10-hour-long episodes. The DLP earthquakes exhibit clear P- and S-phases on three-component records. A P-S converted phase, which apparently converts at the base of the dacite pluton at about 14 kilometers in depth, is also observed. DLP earthquake spectra are dominated by very narrow-bandwidth signals with a quality factor of 20 to 50. The dominant frequency for the vast majority of the events, including the 10 largest, is at 2.0 hertz. A few events with dominant frequencies of 3.2–3.3 or 3.6–3.9 hertz were also observed, usually at the beginning of swarms which contain large-amplitude 2.0 hertz events. The Reduced Displacement for the largest event is about 3,100 square centimeters (magnitude ~3.7), making it the largest DLP earthquake ever reported. The events locate from 28 to 35 (possibly 40) kilometers below, and about 6 kilometers northwest of, the summit and apparently shallow with time.

There is a striking temporal correlation between the two episodes of DLP activity and geological and seismological changes at the surface. For example, the onset of DLP seismicity on May 26 was accompanied just 1 hour later by the onset of shallow long-period earthquakes, and within 4 hours by continuous, shallow long-period tremor and a large steam emission. The highest DLP moment release occurred on June 4 and was followed within 3 hours by shallow long-period earthquakes and increased steam emission. Significant DLP activity continued through June 7, accompanied by inflation of the summit and a rapid acceleration in shallow seismicity beneath the summit which gradually shoaled with time. DLP seismicity waned with the emergence of a dome containing inclusions of a very primitive, freshly quenched olivine basalt that had recently arrived (several days to a few weeks prior) from the deep crust. The spatio-temporal development of the DLP seismicity and its temporal correlation with subsequent surficial activity, especially the emergence of the primitive basalt, is taken as evidence that the DLP seismicity is the elastic manifestation of the injection of deep-seated basaltic fluids into the base of the magma chamber. As such, the DLP seismicity provides the first direct evidence for the location and timing of such injections prior to a major eruption and strongly supports the notion that these basalt injections triggered the eruptive sequence at Mount Pinatubo. If basaltic injections trigger all eruptions of such size, early recognition and quantification of large-scale DLP seismicity may provide one of the best tools for predicting the timing and size of such destructive eruptions.

INTRODUCTION

Long-Period (LP) earthquakes and LP tremor have been observed at many volcanoes (for example, Koyanagi and others, 1987). The vast majority of LP events reported in the literature originated at depths of less than 3 km. Such shallow LP events have preceded eruptions at many volcanoes (for example, Chouet and others, 1994). LP events deeper than 10 km are less commonly observed. Such earthquakes have been noted in California beneath Long Valley caldera (Hill and Pitt, 1992) and the nearby Mono cones (Mitch Pitt, USGS, oral commun., 1994), Mt. Lassen and Medicine Lake volcanoes (Steve Walter, 1988, 1991), and the Clear Lake volcanic complex (Steve Walter, USGS, oral commun., 1994). Other deep LP earthquakes occur regularly under Kilauea volcano, Hawaii (Koyanagi and others, 1987), many have been observed under Mount Spurr, Alaska (Power and Jolly, 1994), one at Izu-Ooshima, Japan (Ukawa and Ohtake, 1987), and at several other active volcanoes and low-velocity in Japan (Hasegawa and others, 1991).

[1]U.S. Geological Survey.

Table 1. Reported deep long-period earthquakes worldwide.

[Under "Location," the data for "Northern Honshu, Japan" are for the combined data for nine volcanoes and low-velocity zones. Under "Quantity," "reported" indicates the total number of DLP earthquakes counted, including those too small to locate. For "Kilauea," these numbers are taken from the number and duration of tremor episodes given in Shaw and Chouet (1989), where they assumed that tremor there was produced by the sustained occurrence of LP earthquakes at the rate of three per min. "Time Span" is for the whole period of time for which records have been scanned, rather than the duration of a particular episode or active period, for example. M_{MAX} is the estimated amplitude magnitude, except where duration magnitude is noted by M_D, of the largest known DLP earthquake at that location. Note that LP events within the 5–15 km depth range are called "intermediate depth" by Aki and Koyanagi (1981)]

Location	Quantity located/ reported	Time span	M_{max}	Depth (km)	F_0 (Hz)	Singly/ swarms	Reference
			Convergent margin				
Mount Pinatubo..................	11/400	1991–1991	3.8	28–35(40?)	2–3.8	swarms	This paper.
Long Valley, Calif.	50/?	1989–1992	2.2	10–20	1.5–3	both	Mitch Pitt (USGS, oral commun., 1994).
Mono Cones, Calif.	2/?	1989–1992	1.5	25–35	1–3	singly	Mitch Pitt (USGS, oral commun., 1994).
Mt. Lassen, Calif................	25/50	1984–1992	2.4	13–22	1–3	singly	Walter (1991).
Medicine Lake, Calif.	2/3	1984–1992	$2.9M_D$	16	1–3	singly	Walter (1991).
Clear Lake, Calif................	15/?	1984–1992	$2.7M_D$	16–25	2	both	S.R. Walter (USGS, oral commun., 1994).
Mt. Shasta, Calif.	0/0	1984–1992	none	none	none	none	S.R. Walter (USGS, oral commun., 1994).
Mt. Spurr, Alaska..............	100/250	1991–1992	1.8	15–40	1–3	both	A. Jolly (USGS, oral commun., 1994).
Mt. Redoubt, Alaska.........	0/0	1989–1992	none	none	none	none	Power and others (1993).
Mt. St. Helens, Wash.........	4/4	1980–1992	1.6	11–33	?	singly	A. Qamar (USGS, oral commun., 1994).
Izu-Ooshima, Japan...........	1/1	1985–1986	2.7	29	1	singly	Ukawa and Ohtake (1987).
Mt. Moriyoshi, Japan........	20/20	? – ?	?	27–37	?	?	Hasegawa and others (1991).
Northern Honshu, Japan....	?/?	? – ?	2.5	25–40	1.5–3.5	?	Hasegawa and other (1991).
			Intraplate				
Kilauea, Hawaii.................	?/>48,000	1962–1983	3.2?	30–50	2–10	swarms	Shaw and Chouet (1988).
Kilauea, Hawaii.................	?/>18,000	1962–1981	?	5–15	?	swarms	Shaw and Chouet (1988).
Yellowstone, Wyo.	0/0	1973–1981	none	none	none	none	Mitch Pitt (USGS, oral commun., 1994).

Table 1 lists some basic parameters of deep long-period (DLP) events. Note that, apart from Mount Pinatubo, fewer than 400 DLP earthquakes have ever been observed at all convergent-margin volcanoes combined. At Kilauea, however, about 540 DLP "tremor episodes" occurred during 1962-83 alone (Shaw and Chouet, 1991), with each episode estimated to contain 3 to 300 DLP earthquakes (Shaw and Chouet, 1989). Aki and Koyanagi (1981) reported statistics on 200 of those episodes and showed that the LP nature of the signal is governed by the source process, not by path or receiver effects. Koyanagi and others (1987) showed that both shallow LP earthquakes and shallow LP tremor at Kilauea have identical spectral properties and that both phenomena must, therefore, have a similar source mechanism. Aki and Koyanagi (1981) showed that the process responsible for the DLP events beneath Kilauea has been "a generally steady process which does not seem to be significantly affected by major eruptions and large earthquakes" in the vicinity over at least 17 years.

Nowhere have DLP earthquakes been shown to correlate with an impending eruption. At Izu-Ooshima, the only reported DLP earthquake (Ukawa and Ohtake, 1987) was followed about 1 year later by an eruption, but any connection there is tenuous. At Kilauea, DLP activity and eruptions both have been frequent since the early 1960's (Aki and Koyanagi, 1981), but any correlation is unclear. I report the observation of about 400 DLP earthquakes beneath Mount Pinatubo, including the largest DLP earthquake ever observed, which immediately precede, by 1 to 3 weeks, the cataclysmic eruption of the volcano. Peaks in the DLP energy release rates are observed to precede, by 1 h to a few days, major geological and seismological changes near the surface, including the extrusion of an andesite dome containing inclusions of freshly quenched olivine basalt (Pallister and Hoblitt, 1992) that had only very recently (several days to a few weeks prior) arrived from the deep crust.

The evidence is compelling that the DLP earthquakes were produced by the flow of a deep-seated basaltic magma moving upward from near the base of the crust into a magma chamber containing a dacitic residuum and that this magma-mixing led directly to the cataclysmic eruption. I interpret DLP seismicity in terms of a model proposed by Chouet and others (1994), which invokes the choked flow of a mixed-phase magmatic fluid through a rectangular conduit. This model adequately describes the important features observed in DLP waveforms at Mount Pinatubo and can explain other features such as the unusual pattern of

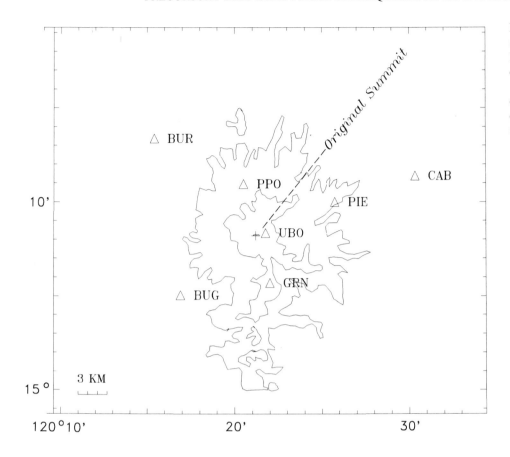

Figure 1. The seven high-gain seismometer station locations at Mount Pinatubo prior to the eruption. Contours are 500 and 1,000 m. The original summit elevation was 1,745 m. Map produced using PC-QPLOT (Murray and others, 1993).

DLP energy release over time and the various depth ranges of DLP events at other volcanoes.

THE 22-MIN DLP SEQUENCE OF JUNE 4, 1991

Following the occurrence of a series of steam blasts on the north flank of Mount Pinatubo on April 2, a network of seven telemetered high-gain seismometers was installed on and around the volcano (fig. 1). For a discussion of the network configuration and history, see Harlow and others (this volume) and Lockhart and others (this volume). For the purposes of this study, the network began producing marginally useful records on May 1 and relatively good records on May 10.

From the inception of the network through May 31, data show that a few volcano-tectonic events per hour were originating from a region about 5 km northwest of the summit, while a few volcano-tectonic events per day were originating from another region beneath the summit. On June 4, a 22-min sequence of relatively large-amplitude seismic signals, later determined to be DLP earthquakes, was recorded on the analog record from station PIE. Part of this record is shown in Figure 2. The largest swarm of shallow LP earthquakes occurred at this time, ash and steam emissions occurred 1 day later, and 2 days later the rate of shal-

low volcano-tectonic seismicity suddenly increased beneath the summit. The following day a hybrid andesite dome emerged (for a detailed description of shallow preeruption seismic and geologic activity, see Harlow and others, this volume; Hoblitt, Wolfe, and others, this volume).

Two features of the 22-min sequence were initially recognized as unusual, as compared with other seismicity prior to that time: (1) the events appeared monochromatic with a frequency of 2 Hz within the spindle-shaped envelope of individual events and (2) ground displacements were of similar amplitudes at stations near and far from the summit. Regarding the first feature, shallow LP earthquakes and tremor with monochromatic appearance had been noted beginning May 26, but the dominant frequency of this shallow LP activity was 0.8 Hz. A few periods of very low amplitude tremor with dominant frequencies higher than 2 Hz had also been noted, but the signals are broader band and generally correlated with the occurrence of strong rain showers or steam emissions. Ground displacements were of similar amplitudes at station PIE, located 9 km from the peak, and station UBO, located only 1 km from the peak. To the Pinatubo Volcano Observatory Team, which was focussing its attention on shallow seismicity (depths of less than about 10 km), this indicated a deep source for those events. These events were otherwise ignored until well after the cataclysmic eruption.

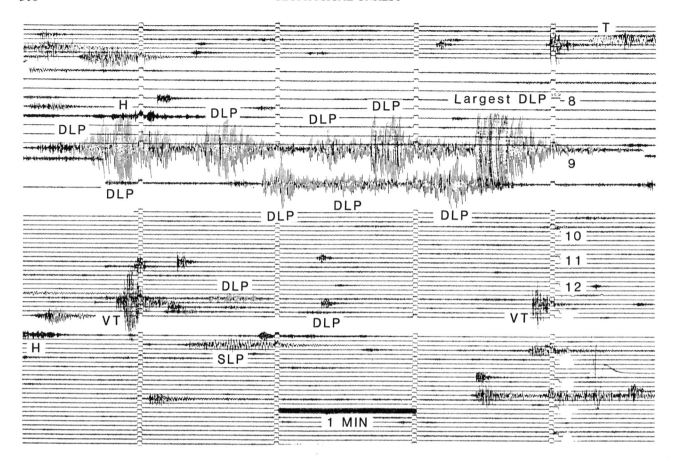

Figure 2. Analog Helicorder record from station PIE showing part of the 22-min deep long-period (DLP) earthquake sequence on June 4, 1991. This is the most energetic DLP earthquake sequence recorded at Mount Pinatubo. Note the monochromatic nature of the waveforms and the swarm-like pattern of energy release within the sequence. Other signals shown include a teleseism (T), shallow volcano-tectonic earthquakes (VT), shallow long-period earthquake (SLP), and passing helicopters (H). Tick marks are 1 min apart on each line. The local time (hour) is indicated by "8," "9," "10," etc. Each line is nearly 5 min long.

In late 1991, a search through the collection of about 1,000 digitally recorded waveforms from May 10 through June 12 initially produced two approximately 1-min-long sections of record beginning at 08:59 and 09:08 on June 4, each containing one well-recorded DLP earthquake. The event at 08:59 was later discovered to be the largest DLP event ever reported anywhere in the world. Waveforms are shown for this event at all stations in figure 3A. The other digitally recorded DLP event, shown in figure 3B, contains a third, much smaller DLP earthquake within its coda. Both of the well-recorded events themselves begin, in fact, within the coda of previous DLP events.

Note the nearly simultaneous arrival of the initial wave front at all stations in figure 3, suggesting an approximately planar wave front with nearly vertical incidence. The P-wave first motion is emergent and down at all six stations. Maximum horizontal S-phase amplitudes are observed to be about 5 times larger than maximum P-phase amplitudes. The vertical components of ground motion at the three outermost stations, PIE, BUG, and BUR, which have an angular separation of 25° to 29°, as viewed from the hypocenter,

were compared two at a time and found to have maximum correlation coefficients less than 0.4 and thus are uncorrelated. The S-wave arrival is very difficult to pick from the vertical component records but is most clear at GRN and on the horizontal components at PPO. This difficulty owes partly to the near-vertical incidence of the S-phase and partly to the arrival of a converted phase about 1.6 s prior to the S-phase arrival on some records. Figure 4 shows the three-component record and particle motions at PPO for the P- and S-phase arrivals and the converted phase. The converted phase is apparently a vertically arriving shear wave (P to S). The traveltime for the converted phase indicates that the conversion occurred at 13 to 14 km in depth, a location compatible with the base of the dacite pluton, according to Pallister and others (this volume).

By using P-arrival picks from each of the six vertical component stations and an S pick from the horizontal components at station PPO, I obtained preliminary hypocenter locations for both events at depths of 34 to 35 km beneath the edifice. For final solutions, additional S-arrival picks were incorporated and checked by running the location

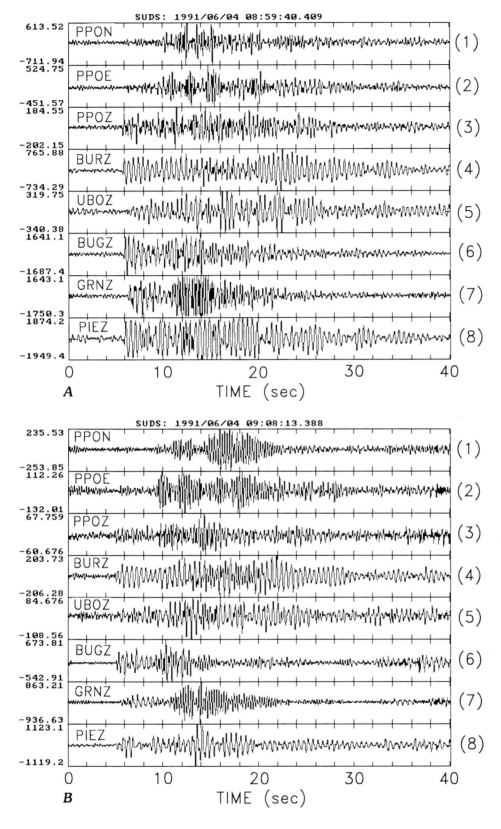

Figure 3. Digital seismograms for two events during the 22 min-long sequence on June 4, 1991. Maximum amplitudes in counts are shown for each trace at left. *A*, This event occurred at 0859 local time. It is the largest deep long-period earthquake ever reported anywhere in the world. Note that the record is offscale (clipped) at stations BUG, GRN, and PIE. Also note the nearly simultaneous arrivals and lack of surface waves at all stations, indicating a deep event from directly beneath the network. *B*, This event occurred at 0908 and is the largest onscale deep long-period earthquake recorded at Mount Pinatubo. Figure produced using PITSA (Scherbaum and Johnson, 1992).

program with different combinations of S picks and then running the final set of picks used with a 3-dimensional velocity model with stations delays determined from travel-time inversion (for details, see Mori, Eberhart-Phillips, and Harlow, this volume). These additional measures yielded hypocenters with depths of 33.5 and 33.6 km below, and

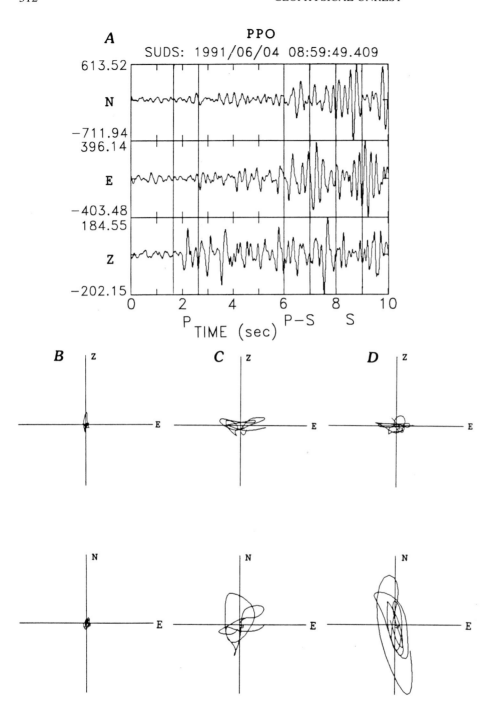

Figure 4. Particle motions at station PPO. *A*, 10-s-long record from station PPO showing the locations of the 1-s-long windows, containing the P-, P-S, and S-wave arrivals, used for the particle motions. Maximum amplitudes in counts are shown for each trace at left. *B*, Upper plot shows radial (Z) versus tangential (E) particle motion, while lower plot shows the particle motion in the horizontal plane, for the P-wave arrival. *C*, Same as *B*, for the P-S converted-phase arrival. *D*, Same as *B*, for the S-wave arrival. Figure produced using PITSA (Scherbaum and Johnson, 1992).

about 5 km northwest of, the pre-June 1991 summit. The relative precision at the 95% confidence level is ±0.7 km vertically and ±2.5 km horizontally. Owing to uncertainties in the seismic velocity model below 20 km in depth, the absolute depth of these hypocenters may actually vary from the calculated values by as much as 3 km. Figure 5 shows the location of these two events.

The 22-min sequence contains 7 of the 9 largest DLP earthquakes (with amplitudes ≥12 mm at station PIE) recorded at Mount Pinatubo. The coda duration of the largest DLP earthquake is about 150 s long, as measured from

the analog record from station PIE. This duration is similar to the duration of tectonic earthquakes of such amplitude, distance, and depth. The magnitude of this event is estimated to be 3.7, making it the largest DLP earthquake ever reported anywhere in the world. Other unusual aspects of the DLP sequence are: (1) the distinctly nontectonic temporal pattern of energy release and (2) the very narrow-band spectral content, with a dominant frequency of either 2.0, 3.25, or 3.7–3.8 Hz, depending to a large degree on the temporal location of the event within the sequence and (or) magnitude of the event.

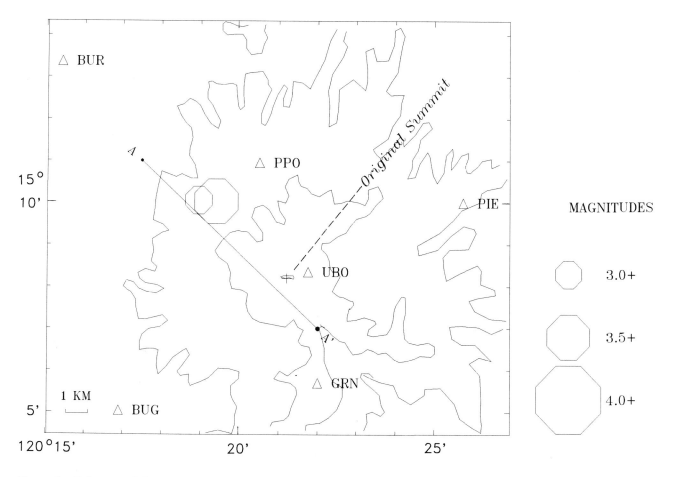

Figure 5. Epicenters of the two best-recorded and best-located deep long-period earthquakes. These events occurred at 0859 and 0908 June 4 during the 22-min sequence of deep long-period activity. The first event is the larger of the two. The horizontal location error, at the 95% confidence level, is ±2.5 km. The events locate at 33.5 and 33.6 km, respectively, below the elevation of the summit. The vertical location error is ±0.7 km. Contours are 500 and 1,000 m. Figure produced using PC-QPLOT (Murray and others, 1993).

The 22-min sequence is composed of 3 temporal clusters of 5, 7, and 10 events, respectively, although there may be a few additional small events hidden within the codas of identified events. The clusters have durations of 4, 5, and 7 min, respectively, with a 1.5-min interval of quiet between the first and second cluster and about 5 min of quiet between the second and third cluster. Figure 6 shows the amplitude of each event as a function of time. Of particular interest is the fact that the largest event is never the first event in a cluster. In fact, the events more or less increase in amplitude within each cluster, with the largest event occurring last within the first two clusters. Note also that the longest interval between events follows the largest event of the entire 22-min sequence. To our knowledge, the only other instance of such a pattern of seismic energy release is for DLP earthquakes at Long Valley caldera, Calif. (Mitch Pitt, USGS, oral commun., 1993). This pattern is certainly contrary to typical mainshock-aftershock patterns commonly observed for tectonic sources, in which the largest event occurs at or very near the onset of the sequence and

the number of events decreases logarithmically with time thereafter.

Each of the three clusters within the 22-min sequence begins with small amplitude events characterized by a dominant frequency of 3 to 4 Hz. These events are followed by larger amplitude events characterized by a dominant frequency (F_0) of 2.0 ±0.03 Hz. Amplitude spectra were computed for a 40-s window taken near the beginning of the first cluster, for each station. The spectra were then normalized and stacked. The result is shown in figure 7A. One can see a very narrow-band spectral peak located at 3.27 ±0.03 Hz. Figure 7B shows the spectrum, similarly computed for the largest event of the second cluster (the largest of all 400+ events). The spectral peak for this event is located at 2.03 ±0.05. The quality factor, $Q = F_0/\Delta F$, of the spectral peaks for these two events ranged from 20 to 50, at each station. Figure 7C shows the spectrum for an event near the middle of the third cluster. For this event there are two spectral peaks of similar amplitude located at 2.03 ±0.04 and 3.74 ±0.04, and Q for both peaks ranges from 20 to 70 at each station.

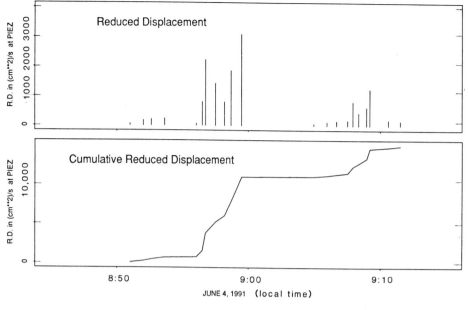

Figure 6. Reduced Displacement (*RD*) and cumulative *RD* versus time for the 22-min sequence of June 4, shown in part in figure 2. Note the three subgroups (top), with small events occurring first and the largest event occurring at or near the end of each subgroup.

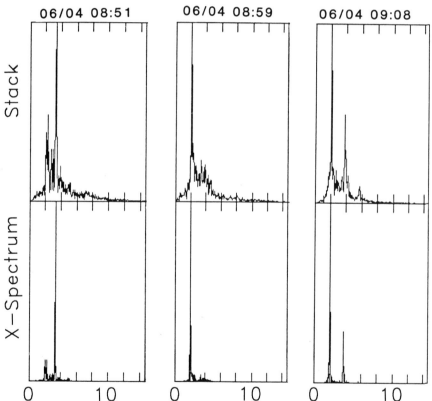

Figure 7. Stack of normalized (vertical) amplitude spectra from all stations and cross-spectra for two of those stations, BUR and PIE, for a 20-s window beginning with the first arrival for three well-recorded deep long-period earthquakes during the 22-min sequence. X-spectrum, cross-spectrum. Spectral units (horizontal axes) are in hertz. All three events have significant energy at 2.03 ±0.04 Hz. *A*, Small first event at onset of first subgroup; dominant frequency is 3.27 ±0.03 Hz. *B*, The largest deep long-period earthquake ever recorded, which occurred at 0859, at end of second subgroup (waveforms for this event are shown in fig. 3*A*). *C*, Event near middle of third subgroup, at 0908 (waveforms for this event are shown in fig. 3*B*); this event has a strong secondary peak at 3.74 ±0.04 Hz. Figure produced using PITSA (Scherbaum and Johnson, 1992).

Table 2 lists the individual events identified within the 22-min sequence along with the maximum peak-to-peak amplitude and dominant frequency for each event. Note the great similarity between this sequence at Mount Pinatubo and observations of deep tremor made in Hawaii by Aki and Koyanagi (1981): "A typical major episode starts with high-frequency...with moderate amplitude, grows into violent motion with low frequency (1.5–3 Hz) dominating, and becomes gradually weaker...near the end of the episode."

COMPREHENSIVE SEARCH FOR OTHER DLP SEISMICITY

Once the existence and basic characteristics of DLP earthquakes under Mount Pinatubo, such as a monochromatic 2- to 4-Hz signal with an S-P arrival-time interval (when the S-phase arrival was clear) of 4 to 6 s on the station PIE records and similar ground displacements at all stations, were recognized, an exhaustive search was made for

Table 2. Chronology of the 22-min deep long-period earthquake sequence of June 4, 1991.

[Sequence is grouped into three temporal clusters. "Dominant frequency at station PIE" was estimated from analog records. Where two dominant frequencies were found, the more dominant frequency is listed first. For those events for which digital records exist, "Dominant Frequency of stacked spectra" was calculated for the stack of normalized spectra for the vertical components at all six stations. This same pattern of initially small-amplitude, higher frequency events followed by larger amplitude, lower frequency events is also observed during 10 other DLP earthquake sequences during June 1–6.]

Time (local)	Amplitude (peak-to-peak) at station PIE (mm)	Dominant frequency at station PIE (Hz)	Dominant frequency of stacked spectra (Hz)
08:51:00	1.4	3.7 ?	–
08:52:13	1.4	3.25, 2.0	3.27, 2.03 ±.03
08:52:40	3.8	2.00	–
08:53:30	2.8	2.0	–
08:54:30	3.8	2.0	–
08:56:12	4	3.4–3.8	–
08:56:40	36	2.0	–
08:57:30	23	2.0	–
08:58:12	13	2.0	–
08:58:40	26	2.0	1.98 ±.04
08:59:50	50	2.00	2.00 ±.03
09:06:40	2	3+	–
09:07:40	2	3.7	–
09:07:55	13	3.83, 1.98	3.78 ±.05
09:08:18	6	2.00, 3.75	2.03, 3.74 ±.04
09:08:50	6	1.98	2.00 ±.04
09:09:12	15	2.0	–
09:10:10	3.3	3.7	–
09:11:30	4	2.0	–
09:12:20	1.2	2.0	–

other preeruption DLP events. We began with the simple expedient of overlaying the analog record from station UBO upon the record from station PIE over a light table. Because station UBO, located 1 km from the summit, was operated at 18 dB lower gain (8 times lower magnification at 2 Hz) than station PIE, located 9 km from the peak, signals of similar or larger amplitude on station UBO must have originated at shallow depth near the summit. All other signals were marked as potential DLP events. Many of these signals were obviously related to regional earthquakes, teleseisms, cultural noise, especially helicopters, or transient meteorological conditions near station PIE (wind and rainfall often vary greatly at different locations around the volcano). Such events were discarded from consideration. A type of signal found difficult to distinguish from the DLP earthquake signal was from occasional magnitude ~3 tectonic earthquakes originating at distances of a few hundred kilometers. At this distance, the spectral content of these events is somewhat peaked within the 1- to 3-Hz band, owing to the attenuation of higher frequencies with

distance, and the angle of incidence is steep enough to make the S-phase difficult to recognize on vertical component records. Larger regional earthquakes from this distance are readily distinguished by the S-P interval. An additional feature useful for recognizing DLP earthquakes was their unusual tendency to occur in groups in which the largest events occurred between the middle and end of the sequence, whereas tectonic events at regional distances tend to appear either singly or in groups of several with the largest event at or near the beginning of the sequence.

A list of 440 potential DLP earthquakes was eventually compiled for the interval from May 1 through the end of June 11. The earthquakes fell into two temporal groups, the first from 1700 May 26 to 1210 May 28 and the second from 2114 May 31 to 1510 June 8. Most of these events occurred either in large clusters lasting a few hours, with the largest event in a cluster occurring near the middle or end of the cluster, or in small clusters of events with similar amplitudes. For both types of clusters, there usually was at least one event in each cluster with large-enough amplitude on the PIE record that an S-P arrival-time interval of 5–6 s could be observed clearly. I feel confident that at least 400, or 90 percent, of these potential DLP signals are true DLP earthquakes. Because the identification problems are the greatest among events with small amplitudes, the erroneous inclusion of a few small-amplitude regional events is likely in our list but should have very little effect on our conclusions. For example, the inclusion of 40 such events would inflate the cumulative Reduced Displacement, discussed below, by less than 1 percent. Of the 350 potential DLP earthquakes with amplitudes ≥1 mm at station PIE, I believe that at least 330, or 94 percent, of these are in fact DLP earthquakes, and I am confident that virtually all 110 events with amplitudes ≥3 mm are DLP earthquakes.

To quantify LP earthquakes, it is convenient to use Reduced Displacement (*RD*) as defined by Aki and Koyanagi (1981). For fixed distance to the source, *RD* is proportional to the displacement, so that the logarithm of *RD* is similar to the magnitude of tectonic earthquakes. The *RD* for the DLP earthquakes and tremor at depths of 34 to 35 km in depth under Mount Pinatubo is given by (see appendix A):

$$RD = \text{Amplitude at PIE} \times 600 \ [\text{cm}^2]$$

where the amplitude is measured in centimeters. RD for the observed DLP earthquakes ranges from 30 to 3,100 cm^2 (M ~1.7–3.7), and the cumulative RD totals about 68,000 cm^2, roughly equal to a single M 5 earthquake.

In addition to the 400-plus DLP's, there are several long periods of time during which nearly monochromatic signals of about 2 Hz were apparently sustained with amplitude near background-noise levels. Because these long signals are of 2 Hz and because DLP earthquakes are embedded within most of these periods of tremor, these

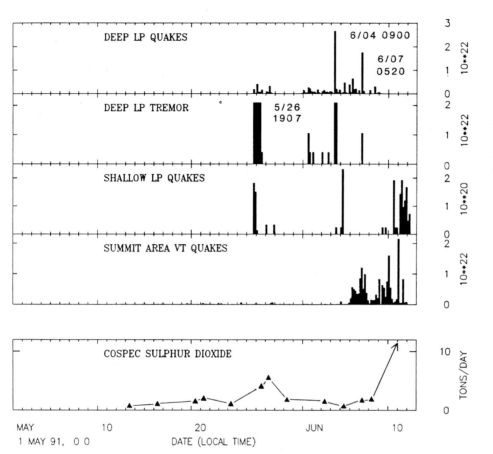

Figure 8. Moment release, as calculated for double-couple earthquakes (proportional to energy), per 4 h for four types of seismicity and SO₂ emission. *A*, Deep long-period earthquakes; *B*, Deep long-period tremor; *C*, Shallow long-period earthquakes; *D*, shallow volcano-tectonic earthquakes beneath the dome; *E*, SO₂ emission in 1,000 metric tons per day. Figure produced using PC-QPLOT (Murray and others, 1993).

signals were determined to be DLP tremor. These periods of apparently continuous, deep tremor occurred at the following times: 1800 May 26–0400 May 27, 1200–1300 May 27, 1600–1900 June 1, 0200–0300 June 2, 0100–0200 and 1700–1800 June 3, 1000–1600 June 4, and 0600 to 0800 June 7. If we assume, as do Shaw and Chouet (1989) for DLP tremor at Kilauea, that the tremor is composed of three DLP earthquakes per minute, and that each of these DLP earthquakes has an average *RD* of 20 cm² (M ~1.5), the additional cumulative *RD* due to observed DLP tremor is about 90,000 cm², roughly equivalent to a single earthquake of M 5.2.

The estimated moment release during these DLP earthquakes and during periods of DLP tremor over the interval May 1–June 12, is shown in figure 8. Also shown for the same time interval is the estimated moment release during SLP earthquakes and during volcano-tectonic earthquakes beneath the dome, as well as SO₂ emission.

Of the 400-plus DLP events identified, all or part of only 18 were recorded digitally. Only 10 of these were recorded well enough to be located. Our inability to record a greater percentage of the DLP earthquakes is due to two main factors: (1) stations located on the flanks of the volcano were operated at low gain settings because of the high levels of seismic noise originating from within or very near the volcanic edifice and (2) the trigger algorithm was set to recognize only shallow high-frequency events. An improved trigger algorithm has since been developed that triggers much more reliably on the low-frequency signals often present at volcanoes (Evans, 1992).

Both pulses of DLP seismicity apparently become shallower with time, as estimated from S-P arrival-time intervals from the PIE analog record. Both pulses seem to begin with events near 40 km in depth and are followed by a majority of events located at about 33 to 34 km in depth. Near the end of the second pulse, several events originated at about 28 km in depth. Figure 9 shows the P- and S-phase arrivals for 7 well-located, onscale DLP earthquakes at station GRN, a vertical-component-only station, located relatively far from the summit, for which S-phase arrivals are particularly clear. Note that the S-P arrival-time interval for the last event is noticeably shorter than for those above, indicating a distinctly shallower origin. Figure 10 shows the locations of the digitally well-recorded DLP earthquakes, all of which occurred during the second pulse of DLP seismicity, during June 3–6. At the 95% confidence limits, the horizontal errors are within 3 km, and the vertical errors are within 1.2 km for all of the events shown. Of the

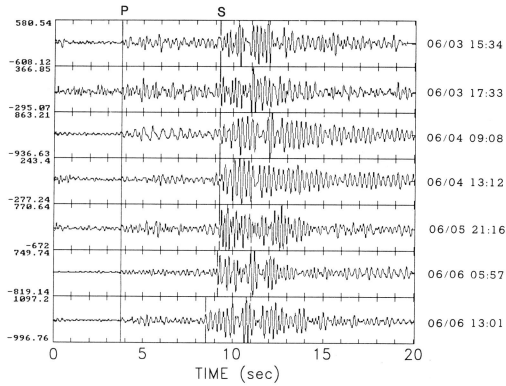

GRN(Z)

Figure 9. S-phase arrivals for seven well-recorded deep long-period earthquakes at the vertical component station GRN. Maximum amplitudes in counts are shown for each trace at left. Date and time of each event is shown at right. Records have been low-pass filtered with a 4-pole Butterworth filter with a high-frequency cutoff at 6 Hz. Vertical line at 3.7 s indicates the P-phase arrival. Note that the S-phases arrive at similar times except for the bottom trace, where it clearly arrives earlier. Figure produced using PITSA (Scherbaum and Johnson, 1992).

10 hypocenters shown, 8 are statistically indistinguishable from, and include, the 2 well-located events discussed earlier (largest octagons) that occurred within the 22-min sequence. Figure 10D shows the depths of these well-located events versus time.

Amplitude spectra were computed for 11 DLP earthquakes all together, including the three shown in figure 7, for the three outermost digitally recorded stations, BUG, BUR, and PIE. The spectra were normalized and then stacked for each event. Of these events, five are nearly monochromatic at 2.0 Hz, and two are nearly monochromatic at 3.25 Hz. The remaining four events have dominant frequencies of both 2.0 and 3.75 Hz, and have a more broad-band spectral content, similar to the spectra of "hybrid" earthquakes (Lahr and others, 1994), with considerable additional energy over the band from 3 to 6 Hz. The two events located at 28 km in depth are both of this type. Stacked spectra for one each of the first two types of events and two of the hybrid type are shown in figure 11. At least one event of each of these three types locates at 33 to 34 km in depth and about 5 km northwest of the summit. Because the travel path between the source and the receiver is virtually identical for these events, the differing spectral contents must be attributable to the source process rather than path or site effects.

TEMPORAL LINK WITH NEAR-SURFACE GEOLOGICAL AND SEISMOLOGICAL ACTIVITY

In this section, I briefly describe some of the major geological and seismological changes at or near the surface of the volcano that followed shortly after high rates of DLP energy release. The temporal correlation between the two is so striking that it seems inescapable that the process which produced the DLP activity deep beneath the volcano also directly influenced processes within the uppermost portions of the magma chamber.

The first of the two principal pulses of DLP seismicity began at 1700 May 26 and lasted about 2 days. Figure 12 shows part of the analog record from station PIE for the beginning of this period. The activity began with a 30-min swarm of DLP earthquakes that included 7 events with RD of 90 to 270 cm². Another swarm of at least 16 DLP earthquakes began about 8 h later. These 2 swarms appear to initiate and terminate a 10-h period of continuous deep tremor. The first shallow LP earthquake recorded during the month of May (others are seen on records from April 8–10) occurred 1 h after the beginning of the DLP activity and was followed by at least four smaller shallow LP earthquakes within the next hour. About 2 h later, there began a 4-h period of continuous shallow LP tremor, with a

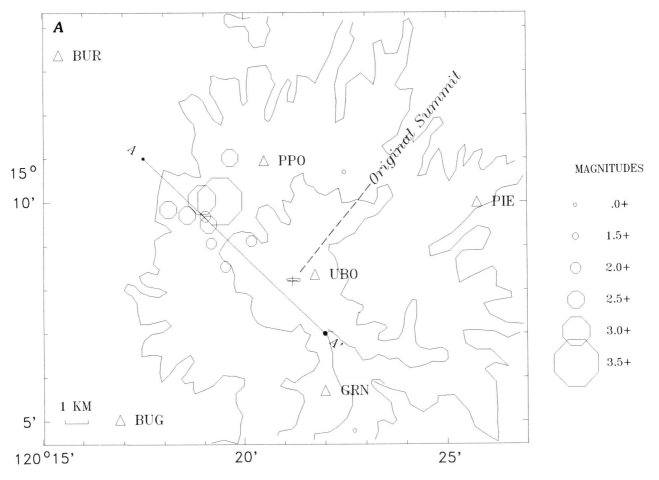

Figure 10. Locations of all 10 locatable deep long-period earthquakes and shallow volcanic-tectonic seismicity for the period from May 26 through June 7, 1991. Triangles are seismic stations. Contours are 500 and 1,000 m. *A*, deep long-period earthquake epicenters. *B*, Shallow volcano-tectonic earthquake epicenters; the large cluster of shallow volcano-tectonic earthquakes coincides with the location of the new dome first observed on June 7.

C, Cross section along *A–A'* showing both deep long-period earthquakes and shallow volcano-tectonic earthquakes. Magnitudes as in *A*. *D*, Depth versus origin time for the deep long-period earthquakes. "?" indicates events for which the depth was estimated from station PIE analog records. Magnitudes as in *A*. Figure produced using PC-QPLOT (Murray and others, 1993).

dominant frequency of 0.8 Hz, that was superimposed upon the 2-Hz DLP tremor. This period of shallow tremor was accompanied by a major increase in steam emission from vents high on the north flank of the volcano, at a location where the dome later emerged. Measurements of SO_2 gas emissions taken at about 0800 on May 27 showed a threefold increase over the previous measurement taken on May 24 (fig 8E). This was the largest increase in SO_2 emission observed prior to June 10 (see Daag, Tubianosa, and others, this volume).

After nearly 3 days of subsequent DLP quiescence, the second pulse of intense DLP activity began in earnest on June 1 and lasted into the morning of June 7, followed by a few additional events through June 8. The highest rate of DLP energy release occurred very early on June 4 and included the 22-min swarm with the largest single DLP (*RD* =3,100 cm²). The second major episode of SLP earthquakes began about 3 h after the 22-min swarm (see fig. 2), and the

steam emissions became much more intense about this time. During a 36-h period from 2100 June 5 through 0900 June 7, high-amplitude DLP's were frequent; this period includes 16 events with *RD* ≥350 cm². The largest event during that period, with *RD* of 2,300 cm², occurred near the end of the period. Early on June 6, shallow high-frequency volcano-tectonic seismicity increased dramatically beneath the summit and gradually shoaled from 2 km depth toward the surface (Harlow and others, this volume). This shallow seismicity culminated with the emergence of a dome on June 7. The dome was composed primarily of hybrid andesite but contained inclusions of freshly quenched olivine basalt (Pallister and Hoblitt, 1992) which, according to Pallister (USGS, oral commun., 1992), had a probable residence time within the upper portion of the magma chamber of no more than a few days.

By 2030 June 8, DLP earthquakes and tremor apparently ceased, though two small events may have occurred

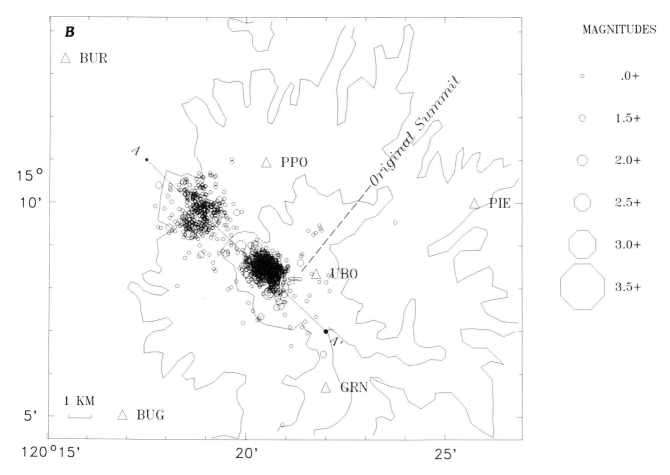

Figure 10.—Continued.

on June 9. One week later, Mount Pinatubo erupted in the largest eruption the world had seen in almost 80 years. Note that recognition of possible DLP seismicity after June 11 is essentially impossible, owing to (1) very high levels of shallow activity from June 12–15, (2) destruction of the network by the cataclysmic eruption on June 15, and (3) the generally 12-dB lower gain of the replacement network installed at the end of June. The 2-week-long period from May 26 through June 8, 1991, contained both the most energetic, and largest concentration, by far of DLP earthquakes ever recorded beneath any plate margin volcano worldwide.

SUMMARY OF OBSERVATIONS OF DLP EVENTS

APPEARANCE OF DLP EARTHQUAKE WAVEFORMS

On three-component records, DLP earthquake waveforms exhibit clear P-, P-S-, and S-phases. Spectra are dominated by very narrow bandwidth low-frequency signals with $Q=20$–50. The wave field is uncorrelated over the diameter of our network, 16 to 19 km, which represents an angular separation of 25 to 29°. These observations are similar to those made at Kilauea volcano by Aki and Koyanagi (1981) and Redoubt volcano by Chouet and others (1994). DLP coda lengths at Mount Pinatubo are not significantly longer than coda lengths of tectonic events of similar amplitude and distance.

VARYING DOMINANT FREQUENCIES BETWEEN DLP EARTHQUAKES

Most DLP events were found to have a dominant frequency of 2.0 Hz. Some sequences that contained large amplitude events, however, were observed to begin with small amplitude events with dominant frequencies of 3.2–3.3 or 3.7–3.8 Hz. At least 11 sequences from June 1 through June 6 began in this manner and were followed by larger 2-Hz events within 1 min. This observation is very similar to that made by Aki and Koyanagi (1981). A few isolated events were also observed with a 2-Hz dominant frequency, but these also contained strong additional energy in the 3- to 6-Hz range and appear similar to the "hybrid" events of Lahr and others (1994).

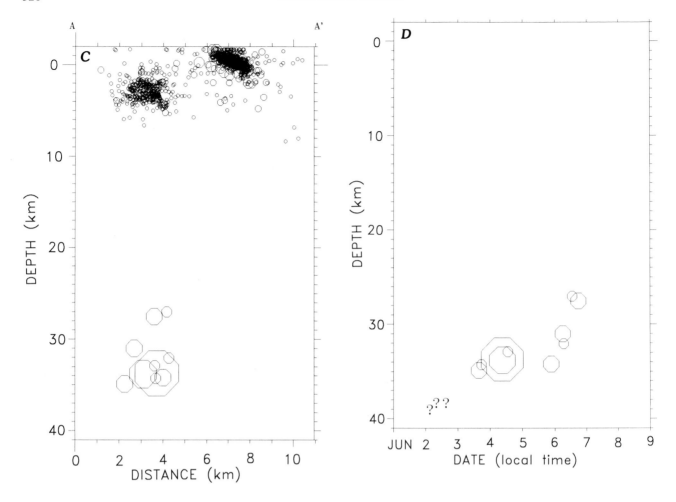

Figure 10.—Continued.

DEPTH AND RESTRICTED
VOLUME OF HYPOCENTERS

Of 11 locatable events, which span the interval June 3–6 and which include events with the four types of spectral signatures described previously, 9 originated from within a small source volume located about 33 to 34 km beneath, and about 5 km northwest of, the summit near the base of the crust. These events originated within a maximum volume defined by the associated 95% error ellipses: an oblate spheroid about 5 km across horizontally and about 2 km in vertical extent. The actual source volume for the events may be much smaller. The two remaining locatable events (the two that occurred last) locate at about 5.5 km shallower than the previous nine events. S-P arrival-time intervals from the PIE analog records indicate that, within each of the two pulses of DLP seismicity, the earliest events originated at depths as great as 40 km and the events generally became shallower with time.

ENERGY RELEASE PATTERN

Reduced displacements of the 400-plus observed DLP earthquakes ranged from 30 to 3,100 cm^2 (M ~1.7–3.7). The 25 h of DLP tremor had an average reduced displacement of 20 cm^2. Within well-recorded DLP earthquake swarms, amplitudes are observed to generally increase with time, with the largest event occurring at or near the end of the swarm. When the largest event occurs before the end, it is generally followed by the longest inter-event interval within that swarm. To our knowledge, the only other instance of such a pattern of seismic energy release is for DLP earthquakes at Long Valley caldera, California (Mitch Pitt, USGS, oral. commun., 1993).

CORRELATION OF DLP ACTIVITY
WITH NEAR SURFACE PHENOMENA

The initial burst of DLP activity on May 26 was followed 1 h later by the onset of shallow (depth <2 km) long-

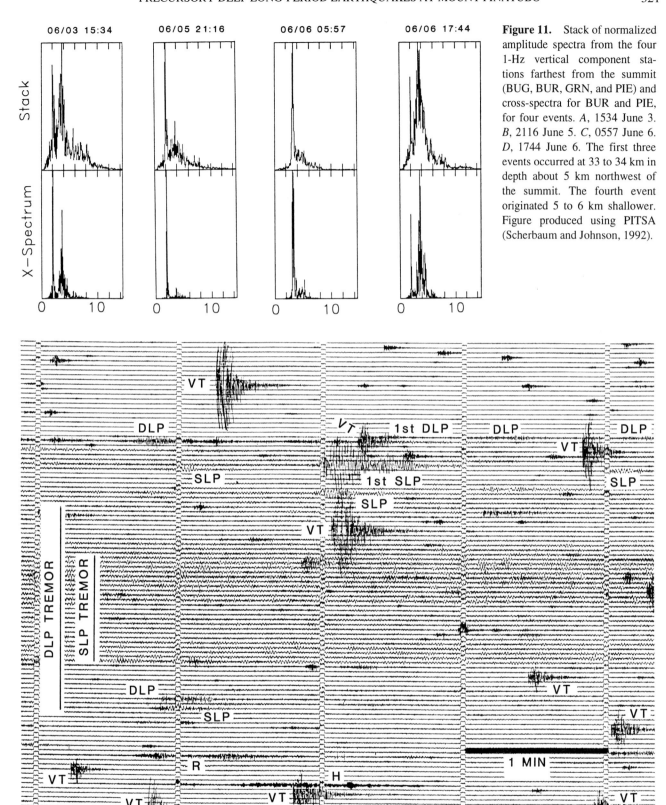

Figure 11. Stack of normalized amplitude spectra from the four 1-Hz vertical component stations farthest from the summit (BUG, BUR, GRN, and PIE) and cross-spectra for BUR and PIE, for four events. *A*, 1534 June 3. *B*, 2116 June 5. *C*, 0557 June 6. *D*, 1744 June 6. The first three events occurred at 33 to 34 km in depth about 5 km northwest of the summit. The fourth event originated 5 to 6 km shallower. Figure produced using PITSA (Scherbaum and Johnson, 1992).

Figure 12. Helicorder record from May 26, 1991, showing the first deep long-period earthquakes (DLP) and tremor (DLP TREMOR), the first shallow long-period earthquakes (SLP), and tremor (SLP TREMOR). The first DLP earthquake occurred at 1807. The first SLP earthquake occurred at 1907. Other signals on this record are shallow volcano-tectonic earthquakes (VT) from an area 5 km northwest of the summit, regional earthquakes (R), and a passing helicopter (H).

period earthquakes. Shallow long-period tremor and large steam emissions followed within 2 h, and the subsequent measurement of SO_2 showed a threefold increase over previous levels for the month to that date. The most intense DLP activity, on June 4, was followed within 3 h by shallow long-period earthquakes and increased steam emissions. DLP seismicity continued at a relatively high rate for 3 more days and was accompanied by inflation of the summit and a dramatic increase in shallow volcano-tectonic seismicity which shoaled with time. The DLP seismicity apparently tapered off after the emergence of the andesite dome on June 7.

A MODEL FOR DLP EARTHQUAKES

In this section the characteristics of DLP earthquakes and tremor observed above are interpreted in terms of a model proposed recently by Chouet and others (1994). The model is shown to be compatible with both the range of DLP event depths observed at Mount Pinatubo and the variety of generally wider depth ranges observed at other volcanoes. In the section following this, the model is used to interpret the seismological and geological events leading up to the cataclysmic eruption of Mount Pinatubo.

The observations summarized in the section above place strong constraints upon possible models:

1. The narrow-bandwidth, long-period characteristic cannot be ascribed to either a path or receiver effect but must be an effect of the source process.
2. Over a 1-week interval, from June 1–7, this source repeatedly produced DLP sequences that began with events having dominant frequencies of 3.2–3.8 Hz followed by larger events having dominant frequencies of 2.0 Hz; this pattern implies a repetitive, nondestructive source process of fixed-scale length.
3. This source process seems to be confined to a relatively small volume, less than a few kilometers in diameter, located near the base of the crust and almost directly beneath the volcano's summit; this small source volume appears to rise with time.
4. The energy release pattern within swarms, wherein DLP earthquakes generally increase in amplitude and the largest is followed by the longest period of quiescence, implies a process in which the timing is governed by a feature akin to a pressure relief valve.
5. Most importantly, this process near the base of the crust immediately preceded major geological and seismological effects at the surface, more than 30 km above, within 1 h to a few days, and therefore implies a causal relationship.

The combination of these observations provides compelling evidence that the DLP earthquakes and tremor were produced by flow, near the base of the crust, of basaltic fluids upward into the magma chamber. I believe that remnants of these fluids were preserved as inclusions of basalt within the andesite dome that emerged on June 7. These basalt inclusions have characteristics compatible with a deep-crustal origin and a probable residence time within the upper magma chamber of days to a few weeks at most (J.S. Pallister, USGS, oral commun., 1992).

In Chouet and others (1994), a mixed-phase magmatic fluid is forcefully injected through a crack and, upon reaching choked-flow conditions, excites the crack to resonance. In this model, the choked-flow condition is met when the flow speed exceeds Mach 1 locally. Such condition can be achieved at flow speeds near a constriction in the conduit if the magmatic fluid contains a free gas phase. Constriction can greatly increase flow speeds locally, while the compressibility of a mixed-phase fluid can greatly reduce the sound speed of an otherwise gas-free fluid (for example, Kieffer, 1977). Constrictions would be plentiful in a magma-transport structure consisting of magma-filled tensile cracks as proposed by Hill (1977) and Shaw and Chouet (1991).

Note that the choked-flow condition may be met only when the void fraction lies within a narrow range: (1) enough bubbles must have formed to lower the sound speed of the fluid to flow velocity, yet (2) the volume percent of gas bubbles must remain small enough that the magma remains somewhat incompressible, else choked flow ceases and (or) the capacity of the fluid to sustain resonance is destroyed through acoustic absorption. Though the required range of the void fraction for sound speed minima, and therefore choked flow, in basaltic magma is unknown, it almost certainly depends on pressure and possibly bubble radius (Kieffer, 1977). The void fraction for basaltic fluid at most times and places along the path of ascent undoubtedly lies outside this range and implies that basaltic fluid mostly rises aseismically.

The model requires that a free gas phase exist, at least temporarily, at our DLP source depth of 28 to 35 (possibly 40) km below sea level. To be a reasonable candidate, this gas must have a very low solubility and exist in relatively large quantity within basaltic magmas. CO_2 is the leading candidate for such a gas because it has by far the lowest solubility and exists in the greatest quantity among volatiles commonly present in basaltic magmas (Bottinga and Javoy, 1990). Gerlach and Graber (1985) showed that CO_2 is present in large quantities (0.3 to 0.6 wt%) in basaltic magmas from Kilauea and "is transported predominantly as a fluid saturating the parental melt from depths of ~40 km." Stolper and Holloway (1988) showed that basaltic CO_2 probably begins to exsolve in the 30- to 50-km depth range. Evidence for the existence of CO_2 as a separate gas phase at this depth range is provided by Willshire and Kirby (1989) and Kirby (1990), who found that xenoliths, which originated at upper mantle depths, showed clear evidence of healed fractures which once contained CO_2 gases.

DLP earthquakes have been reported to occur over certain depth ranges at some volcanoes but very different depth ranges at other volcanoes (see table 1). In particular, at Long Valley caldera, Mount Lassen, and Clear Lake, DLP earthquakes are reported from 10 to at most 25 km in depth while at Kilauea Volcano, most DLP events occur between 30 and 50 km in depth. Although, as noted above, depths of DLP earthquakes can be very difficult to estimate without particle motion studies using three-component data, the variety of depth ranges is most likely real. Because the production of DLP seismicity depends critically on the existence of a small volume percent of a free-gas phase, which depends on the solubility of that gas, which in turn depends on magma composition and on pressure, the variety of depth ranges of DLP events should reflect differences in magma composition between volcanoes and different pressure-temperature profiles along the deep magma conduits beneath different types of volcanoes. Significantly different depth ranges of DLP events should be expected from contrasting tectonic regimes such at Hawaii, where oceanic crust overlies a mantle plume, versus northern California, where thick continental crust overlies a subducting slab, owing to corresponding differences in magma composition. For example, the magma at Kilauea is likely more basaltic, with a higher CO_2 content, than that of plate margin volcanoes, so CO_2 saturation should occur at a greater average depth there.

INTERPRETATION AND IMPLICATIONS

Several effects are predicted as a result of our model: (1) a pressure transient produced by the choked flow itself, (2) a hydraulic pressure wave produced by the addition of a fluid volume at the base of the magma column, and (3) advective overpressure produced by gas bubbles rising in the column.

The pressure transient, ΔP, is related to the behavior of the shock wave produced by the choking, that is, the flow velocity exceeding Mach 1. ΔP is given, according to this model, by the formula derived in appendix 2. The largest DLP earthquake during the initial 1 hr of DLP activity occurred about 6 min after the onset of the activity. RD for this event is 240 ±50 cm^2, and, therefore, ΔP is about 80 ±17 bar. This pressure transient travels at acoustic velocity and should have a duration on the order of the coda length of the DLP earthquake.

Hydraulic pressure originating from the addition of a fluid volume at the base of the magma column should propel a corresponding volume of magma upward into the base of the magma chamber and thereby increase the pressure in the chamber proportionately. This effect is long term, but because the volume of the magma column is tiny compared with the volume of the magma chamber and the average flow velocity is fairly slow, the effect is negligible over days to weeks and may be impossible to detect.

Bubbles within the rising magma column will also increase the pressure by means of advective overpressure. Steinberg and others (1989) have pointed out that, within a closed system with constant volume and temperature, a single perfect gas bubble rising by distance Δh through an incompressible fluid of density ρ will increase the pressure everywhere within the system by $\Delta P = \rho g \Delta h$, where g is the force of gravity. Under these assumptions, and using $\rho = 3.2$ g/cm^3, and g=980 cm/s^2, a bubble rising 1 km would increase the pressure everywhere along the interconnected conduit system, including the upper portion of the magma chamber, by 330 bar. Allowing for compressibility of the fluid owing to multiple bubbles, this value is reduced by one-third to 11 bar per kilometer of rise. If the bubbles are entrained within the basaltic magma, then they will rise at, and therefore the pressure will increase with, the flow velocity. At the average flow rate of about 120 m/h (see below), the pressure within the system should increase at an average rate of only 11 bar/h. If the magma column is not strictly continuous, but, rather, comprises a plexus of magma-filled tensile cracks offset by faults, as proposed by Hill (1977) and Shaw and Chouet (1991), the pressure increase may take some time to propagate upward through the system.

One or more of the above effects may account for the observation that, on May 26, the first DLP seismicity recorded since May 1 was followed 1 h later by the first shallow (depth <4 km) LP earthquakes recorded since May 1. This same effect was also observed about 3 h after the 22-min sequence of DLP earthquakes on June 4. In the case of the May 26 events, communication between the DLP source region and the near surface was remarkably rapid, travelling at an average velocity of 8–12 m/s over 30–37 km vertically.

Finally I speculate that the batch of basaltic magma that produced the first DLP earthquakes on May 26 may have been the same batch of basaltic magma that reached the surface as inclusions in the dome that emerged on June 7. If so, the average ascent rate of that batch, assuming a starting depth of 35–40 km to the surface over 11.5 days, is 111–126 m/h, or nearly 3 km/day. A similar ascent rate, over a greater depth range, has been estimated for Hawaii (Decker, 1987). At this rate, the batch of basaltic fluid that produced the May 26 DLP seismicity should have reached the base of the dacite within the magma chamber at about 14 km in depth (Pallister and others, this volume) by about June 2. After a few days of magma-mixing there, the mixed-magma product extruded as a dome on June 7. By the above reasoning, the second batch would have begun reaching the base of the dacite by about June 8 and would have continued to arrive there for the following week. It may have been this second batch of basalt which pushed the system to the plinian stage. Because little hard-rock evidence of basalt was found in the products of the June 12–15 eruptions, other than from remnants of the dome itself, however, this second batch may have led to the final eruptive stages principally by contributing pressure, heat, and volatiles.

To summarize the events leading up to the cataclysmic eruption, as interpreted in terms of this model, I propose that (1) the system was leaky and relatively unpressurized prior to May 26; (2) on May 26, the first batch of basaltic magma began migrating upward from the base of the crust, producing the first DLP activity; this briefly pressurized the system enough to produce minor SLP activity, which, in turn, opened the system further and released enough volatiles to reduce the pressure gradient and stop further SLP activity temporarily; (3) during June 1–4, this first batch of basaltic magma reached the base of the dacite pluton as the second batch of basaltic magma began migrating upward from the base of the crust; the system was still somewhat leaky at this time, as evidenced by the fumarolic activity and lack of shallow LP activity prior to this time; (4) by June 4, pressure within the magma chamber had built sufficiently for the renewal of minor SLP seismicity and, by June 7, had built sufficiently to produce the emergence of a dome; (5) by June 11, the second batch of basaltic magma reached the base of the dacite pluton, dome growth had sealed the system shut, the volatiles became trapped, and pressure began to build at shallow depths below the dome, as manifested by a reemergence of SLP activity at that time, and led to the first plinian eruption, on June 12; (6) after the dome was partially destroyed, during the first plinian eruption, we begin to observe large SLP earthquakes that marked the onset of rapid acceleration of the pressure buildup toward the cataclysmic eruption on June 15 (Harlow and others, this volume).

The DLP seismicity that occurred during the 3 weeks leading to the eruption of Mount Pinatubo is the greatest ever reported, both in terms of number of events and energy released, for any plate margin volcano and was immediately followed by the largest eruption in the world in almost 80 years. If such large eruptions are triggered primarily by basalt injections such as occurred at Mount Pinatubo, marked by DLP seismicity, early recognition and quantification of large-scale DLP seismicity may provide one of the best tools for predicting the timing and size of such destructive eruptions.

The reader is cautioned that DLP events are usually difficult to record, owing to masking from surficial noise at the volcano. DLP events at Mount Pinatubo were, in general, very poorly recorded by stations nearest the summit. DLP events may have occurred at other volcanoes during the last few decades and gone undetected. In order to monitor DLP seismicity effectively, care should be taken to install several high-gain stations at quiet sites at least 10 to 15 km from the volcano summit.

CONCLUSIONS

1. About 400 deep long-period (DLP) earthquakes and 25 h of DLP tremor occurred beneath Mount Pinatubo, in two pulses, during May 26–28 and May 31–June 8. This DLP seismicity is greater than previously reported beneath all convergent margin volcanoes worldwide. Though two orders of magnitude more DLP earthquakes have been recorded at Kilauea, Hawaii, over the last 30 years, DLP seismicity apparently released more seismic energy at Mount Pinatubo during the 2-week-long period from May 26 to June 8, 1991, than at Kilauea during the entire 30 years.

2. Early events in each pulse may have originated as deep as 40 km. Eight of the locatable DLP earthquakes originated at depths of 33 to 34 km beneath a point about 6 km northwest of the summit. Two later locatable events originated at about 28 km in depth beneath the same area. Reduced Displacement for the largest event is about 3,100 cm^2, an amount that makes it the largest deep long-period earthquake yet reported.

3. DLP earthquake waveforms exhibit clear P- and S-phases. Spectra are dominated by very narrow bandwidth signals with $Q=20$–50. The dominant frequency for most of the events, including the 10 largest, is 2.0 Hz. A few of the smaller events have dominant frequencies of 3.25 and 3.75 Hz.

4. The first DLP earthquake recorded by the seismograph network, on May 26, was followed 1 h later by the first shallow (less than 3 km deep) long-period earthquakes and 2 h later by a shallow tremor and large steam emission. The high rate of DLP seismicity during June 4–7 was accompanied by inflation of the summit area, a rapid increase in shallow seismicity beneath the summit, and the eventual extrusion of a dome containing inclusions of a very primitive, freshly quenched, olivine basalt only recently arrived (days to a few weeks at most) from the deep crust.

5. The DLP earthquakes and tremor were likely produced by the forceful injection of mixed-phase basaltic fluids upward through cracks from near the base of the crust into the upper magma chamber, which contained a dacitic residuum. The pressure, heat, and volatiles injected into the dacite pluton by the first pulse of basaltic magma led to magma mixing and the dome extrusion on June 7, which sealed the system. The almost continual arrival of additional basaltic magma at the base of the pluton pressurized the system and led to the destruction of the dome on June 12. Depressurization, degassing, and vesiculation of magma at the top of the reservoir, following the vent-clearing eruptions, together with pressure and volatiles from the arrival of yet more basaltic magma, led to the paroxysmal dacitic eruptions of June 15.

ACKNOWLEDGMENTS

I wish to acknowledge the many members of the Philippine Institute of Volcanology and Seismology (PHIVOLCS) and the Volcano Crisis Assistance Team (VCAT) of the U.S. Geological Survey who labored day and night under very stressful and dangerous conditions to install and maintain the seismograph network and the data collection apparatus. From PHIVOLCS, I especially wish to thank Ed Laguerta, Gemme Ambubuyog, S. Marcial, and A. Melosantos. From VCAT, I especially wish to thank Dave Harlow, John Power, John Ewert, Andy Lockhart, and Tom Murray. Discussions with Bernard Chouet concerning the choked-flow model and its implications were critical. This manuscript was greatly improved by reviews from Motoo Ukawa, Bernard Chouet, Chris Newhall, Dave Hill, and Manoling Ramos.

REFERENCES CITED

Aki, K., and Koyanagi, R., 1981, Deep volcanic tremor and magma ascent mechanism under Kilauea, Hawaii: Journal of Geophysical Research, v. 86, p. 7095–7109.

Bottinga, Y., and Javoy, M., 1990, Mid-ocean ridge basalt degassing; bubble nucleation: Journal of Geophysical Research, B, Solid Earth and Planets, v. 95, no. 4, p. 5125–5131.

Chouet, B., Koyanagi, R.Y., and Aki, K., 1987, Origin of volcanic tremor in Hawaii, part 2, in Decker, R.W., Wright, T.L., and Stauffer, P.H., eds., Volcanism in Hawaii: U.S. Geological Survey Professional Paper 1350, v. 2, p. 1259–1280.

Chouet, B.A., Page, R.A., Stephens, C.D., Lahr, J.C., and Power, J.A., 1994, Precursory swarms of long-period events at Redoubt Volcano (1989–1990), Alaska: their origin and use as a forecasting tool: Journal of Volcanology and Geothermal Research, v. 62, p. 95–135.

Daag, A.S., Tubianosa, B.S., Newhall, C.G., Tuñgol, N.M, Javier, D., Dolan, M.T., Delos Reyes, P.J., Arboleda, R.A., Martinez, M.L., and Regalado, M.T.M., this volume, Monitoring sulfur dioxide emission at Mount Pinatubo.

Decker, R.W., 1987, Dynamics of Hawaiian volcanoes: an overview, in Decker, R.W., Wright, T.L., and Stauffer, P.H., eds., Volcanism in Hawaii: U.S. Geological Survey Professional Paper 1350, v. 2, p. 997–1018.

Evans, J.R., 1992, The TDETECT program, in Lee, W.H., and Dodge, D.A., eds., A course on PC-based seismic networks: U.S. Geological Survey Open-File Report 92–441, p. 152–164.

Gerlach, T. and Graber, E., 1985, Volatile budget of Kilauea volcano: Nature, v. 313, p. 273–277.

Harlow, D.H., Power, J.A., Laguerta, E.P., Ambubuyog, G., White, R.A., and Hoblitt, R.P., this volume, Precursory seismicity and forecasting of the June 15, 1991, eruption of Mount Pinatubo.

Hasegawa, A., Zhao, D., Hori, S., Yamamoto, A., and Horiuchi, S., 1991, Deep structure of the northeastern Japan arc and its relationship to seismic and volcanic activity: Nature, v. 352, p. 683–689.

Hill, D.P., 1977, A model for earthquake swarms: Journal of Geophysical Research, v. 82, p. 1347–1352.

Hill, D.P., and A.M. Pitt, 1992, Long period earthquakes at mid-crustal depths beneath the western margin of Long Valley caldera, California: Eos, Transactions, American Geophysical Union, v. 73, p. 343.

Hoblitt, R.P., Wolfe, E.W., Scott, W.E., Couchman, M.R., Pallister, J.S., and Javier, D., this volume, The preclimactic eruptions of Mount Pinatubo, June 1991.

Kieffer, S.W., 1977, Sound speed in liquid-gas mixtures: water-air and water-steam: Journal of Geophysical Research, v. 82, p. 2895–2904.

Kirby, S.H., 1990, CO_2-H_2O Fluids evolved from partial melting and ascent of magic magmas: Possible roles in mantle earthquakes beneath Hawaii and other deep volcanic centers: EOS, Transactions, American Geophysical Union, v. 71, p. 1587.

Koyanagi, R.Y., Chouet, B. and Aki, K., 1987, Origin of volcanic tremor in Hawaii, part 1, in Decker, R.W., Wright, T.L., and Stauffer, P.H., eds., Volcanism in Hawaii: U.S. Geological Survey Professional Paper 1350, v. 2, p. 1221–1258.

Lahr, J.C., Chouet, B.A., Stephens, C.D., Power, J.A., and Page, R.A., 1994, Earthquake classification, location, and error analysis in a volcanic environment: Implications for the magmatic system of the 1989–1990 eruptions at Redoubt Volcano, Alaska: Journal of Volcanology and Geothermal Research, v. 62, p. 137–151.

Lockhart, A.B., Marcial, S., Ambubuyog, G., Laguerta, E.P., and Power, J.A., this volume, Installation, operation, and technical specifications of the first Mount Pinatubo telemetered seismic network.

Mori, J., Eberhart-Phillips, D., and Harlow, D.H., this volume, Three-dimensional velocity structure at Mount Pinatubo, Philippines: Resolving magma bodies and earthquakes hypocenters.

Murray, T.L., Power, J.A., and Klein, F.W., 1993, PC-QPLOT, an IBM-PC compatible version of the earthquake plotting program QPLOT: U.S. Geological Survey Open-File Report 93–22–A, 16 p.

Pallister, J.S., and Hoblitt, R.P., 1992, A basalt trigger for the 1991 eruptions of Pinatubo volcano?: Nature, v. 356, p. 426–428.

Pallister, J.S., Hoblitt, R.P., Meeker, G.P., Knight, R.J., and Siems, D.F., this volume, Magma mixing at Mount Pinatubo: Petrographic and chemical evidence from the 1991 deposits.

Power, J.A., and Jolly, A.D., 1994, Seismicity at 10- to 45-km depth associated with the 1992 eruptions of Crater Peak vent, Mount Spurr, Alaska [abs.]: Eos, Transactions, American Geophysical Union, v. 75, no. 44, p. 715.

Power, J.A., Lahr, J.C., Page, R.A., Chouet, B.A., Stephens, C.D., Harlow, D.H., Murray, T.L., and Davies, J.N., 1993, Seismic evolution of the 1989-90 eruption sequence of Redoubt Volcano, Alaska: Journal of Volcanology and Geothermal Research, v. 62, p. 69–94.

Scherbaum, F., and Johnson, J., 1992, Programmable Interactive Toolbox for Seismological Analysis (PITSA), in Lee, W.H.K., ed., IASPEI Software Library, v. 5, International Association of Seismology and Physics of the Earth's Interior in collaboration with Seismological Society of America, El Cerrito, Calif., 269 p.

Shaw, H.R., and Chouet, B., 1989, Singularity spectrum of intermittent seismic tremor at Kilauea Volcano, Hawaii: Geophysical Research Letters, v. 16, 195–198.

———1991, Fractal hierarchies of magma transport in Hawaii and critical self-organization of tremor: Journal of Geophysical Research, v. 96, 10191–10207.

Steinberg, G.S., Steinberg, A.S., and Merzhanov, A.G., 1989, Fluid mechanism of pressure growth in volcanic (magmatic) systems: Modern Geology, v. 13, 257–265.

Stolper, E., and Holloway, J., 1988, Experimental determination of the solubility of carbon dioxide in molten basalt at low pressure: Earth and Planetary Letters, v. 87, 397–408.

Ukawa, M., and Ohtake, M., 1987, A monochromatic earthquake suggesting deep-seated magmatic activity beneath the Izu-Ooshima volcano, Japan: Journal of Geophysical Research 92, p. 12649–12663.

Walter, S.R., 1988, Long period earthquakes at southern Cascade volcanic centers [abs]: Seismological Research Letters, V. 59, p. 30.

———1991, Ten years of earthquakes at Lassen Peak, Mount Shasta, and Medicine Lake volcanoes, northern California: 1981–1990 [abs.]: Seismological Research Letters, v. 62, p. 25.

Wilshire, H.G. and Kirby, S.H., 1989, Dikes, joints, and faults in the mantle: Tectonophysics, v. 161, p. 23–31.

APPENDIX 1. CALCULATION OF REDUCED DISPLACEMENT

The formula for Reduced Displacement, which is the RMS displacement corrected for geometrical spreading and instrument response, is given by formula 2 of Aki and Koyanagi (1981):

$$RD = Ar/(2M\sqrt{2})$$

where A is the peak-to-peak amplitude in centimeters of the largest body wave phase, usually the shear wave, averaged over several cycles, r is the hypocentral distance (in centimeters), and M is the instrument magnification to ground displacement at the dominant frequency.

Shear waves from the DLP events beneath Mount Pinatubo arrived at our digitally recording stations within $15°$ of vertical. At the stations, the particle motion of these shear waves would have been confined essentially to the horizontal plane so that true shear-wave amplitudes would be best measured off horizontal-component recordings. Unfortunately, all but one of the digitally recording seismograph stations, including PIE, the helicorder record upon which the comprehensive catalog was based and the amplitudes taken, contained only vertical-component sensors. Worse yet, the three-component station was operated at the lowest magnification of any station in the network and recorded only a few of the largest DLP events. To find out by what factor vertical-component recordings underestimated true shear-wave amplitude, I compared vertical- and horizontal-component recordings of several DLP earthquakes at our three-component station. I found that shear-wave amplitudes recorded by the horizontal components (A_H) averaged 3 times the amplitudes recorded by the vertical component (A_V). Therefore, for estimates of A, I use A_V measured from the station PIE and multiply that value by this factor of 3.

To calculate the RD of DLP earthquakes or tremor from A_V for body waves at station PIE, I take (1) the depth to be 35 ± 1 km; (2) the epicenter to be the average of the best located events, at $15°10'$ N. latitude, $120°19.2'$ E. longitude, yielding $r = 36.7$ km $= 36.7 \times 10^5$ cm; and (3) the magnification at 2 Hz of $M = 6,250$. Note that Lockhart and others (this volume) estimate $M = 5,000 \pm 20\%$, assuming a sensor sensitivity of 1V/cm/s and neglecting site amplification. Probable site amplification at station PIE, however, causes calculations of RD to average about 25% too high compared with estimates of RD based on data from other stations, so the higher value for M will be used. This will lead to more conservative (lower) estimates of RD.

For DLP events: $RD_{PIE} = 3A_V r/(2M\sqrt{2}) = A_V \times 600 \pm 20\%$ [cm^2].

For the largest DLP earthquake in the catalog, recorded at 0859.5 June 4, A_V was estimated to be 5 ± 1 cm at station PIE, so $RD = 3,100 \pm 700$ cm^2.

For the largest DLP on May 26, $RD = 240 \pm 50$ cm^2.

APPENDIX 2. CALCULATION OF EXCESS PRESSURE

The pressure fluctuation, ΔP_F, produced by DLP earthquakes over the volume of the source is related to RD by formula 12 of Chouet and others (1987):

$$RD = \Delta P_F (1/2\pi)(1/Q_s)^{1/2}(V/(2f\alpha\rho b))^{1/2}$$

where Q_s is the quality factor corresponding to radiation loss, V is the volume of the conduit, f is the dominant frequency, α is the P-wave velocity, ρ is the density, and b is the bulk modulus of the basaltic fluid in the crack. This formula relates RD, calculated from the maximum P-wave amplitude (hereafter referred to as RD_P), to ΔP_F. However, because I have calculated RD from the maximum horizontal component of the S-wave (hereafter referred to as RD_S), ΔP_F based upon RDS will be too large by some constant factor. By calculating both RD_P and RD_S for the five most well-recorded events, I find that factor to be 4, that is, $RD_P = RD_S/4$. Thus, in order to use the above formula to calculate ΔP_F, the values for RD, as calculated according to appendix 1, must be divided by 4.

I assume, as Chouet and others (1987) did for Hawaiian DLP tremor, that the quantity $(1/Q_s)^{1/2}$ is on the order of 0.1, $\alpha = 8$ km/s, and $\rho = 3$ g/cm^3. I assume a small (~1 vol%) bubble content, although the bubbles will be nearly incompressible at 35 km depth. I assume, for the magmatic fluid, that b is ~10^{10} dyne/cm^2 and we shall use V, as estimated by Shaw and Chouet (1991) for deep fractures beneath Kilauea, to be $V = 2 \times 10^{10}$ cm^3. The dominant frequency for the vast majority of the DLP events, including all of the largest events, is $f = 2$ Hz. Inserting these values and rearranging gives $\Delta P_F = 138,000\ RD_P = 34,500\ RD_S$. Following this reasoning, the smallest DLP events in our catalog produced pressure fluctuations of about 10 bar, while the largest event, (with an RD_S~ 3,100 cm^2) produced a pressure fluctuation of about 1 kbar.

Note that the ambient pressure, due to the overburden, at the top of the magma chamber at about 6 km below the surface (Mori, Eberhart-Phillips, and Harlow, this volume) is about 2 kbar and at the DLP earthquake source depth of 34 km is probably about 11 Kbar.

Ground Deformation Prior to the 1991 Eruptions of Mount Pinatubo

By J.W. Ewert,[1] Andrew B. Lockhart,[1] Sergio Marcial,[2] and Gemme Ambubuyog[2]

ABSTRACT

Measurements of ground deformation during several weeks prior to the climactic eruption of Mount Pinatubo came from two telemetered tiltmeters located on and near the volcano and from a small quadrilateral array of points across the chain of craters that opened on April 2, 1991. Overall, the deformation data are sparse, owing to logistical constraints and hazards considerations. Data from one of the tiltmeters provided information on the timing of edifice inflation and on the development of a magma conduit to the surface. On June 7, 1991, tilt and seismic data were used to reach the conclusion that magma had reached very shallow levels or even the surface. On the basis of this conclusion, the Pinatubo monitoring team raised the alert level.

No measurable changes occurred at the quadrilateral array throughout the month of May, and measurements ceased at the end of May, owing to the high hazard of working at the site. Beginning June 4, a change in the locus of volcano-tectonic seismicity from an area 5 kilometers northwest of Pinatubo to the area beneath the summit dome and active fumarolic vents was accompanied by deformation registered by the closest tiltmeter. The tiltmeter detected about 50 microradians of cumulative tilt between June 4 and June 7 that ended when magma presumably reached the surface and a lava dome began to form. No further cumulative deformation was recorded by the tiltmeter before it was destroyed by the third large explosive eruption on June 13. The tiltmeter recorded only deformation caused by the formation of the conduit from the top of the magma body (5 to 7 kilometers deep) to the surface, not the emplacement of the magma body.

INTRODUCTION

Deformation measurements made prior to the climactic 1991 eruption sequence at Mount Pinatubo, although sparse, proved useful in the analysis of preeruptive unrest. These measurements came from two telemetered tiltmeters

---

[1]U.S. Geological Survey.
[2]Philippine Institute of Volcanology and Seismology.

on or near the volcano and from a small quadrilateral array of points across the chain of craters that opened on April 2.

Following the onset of unrest at Mount Pinatubo on April 2, 1991, the Philippine Institute of Volcanology and Seismology (PHIVOLCS) began monitoring seismic activity in the area northwest of the volcano. The U.S. Geological Survey (USGS) Volcano Crisis Assistance Team (VCAT) responded to a request from PHIVOLCS for monitoring assistance in the third week of April. Uncertain funding arrangements surrounding the VCAT response led to a "staged response" wherein a three-man team, a portable seismic monitoring system, and a correlation spectrometer (COSPEC) were dispatched to the Philippines to assist PHIVOLCS in assessing the volcanic hazards at Pinatubo and to determine further monitoring needs. Measurements of the quadrilateral array began on May 1, with three subsequent measurements taken that month. By mid-May a telemetered nine-component seismic net had been installed with the central recording and analysis site on Clark Air Base. Constantly increasing seismic activity, a high SO_2 emission rate, and a better understanding of the magnitude of previous eruptions convinced us that a more extensive monitoring effort was warranted. By the third week of May additional equipment and personnel arrived at Pinatubo to undertake work that included deformation monitoring.

Two telemetered tiltmeters were installed during the last week of May, and one proved useful in interpreting activity prior to the climactic eruptive phase. Data from the installation nearest the summit tracked the formation of the magma conduit from which the lava dome was extruded on June 7 and showed interesting responses at the beginning of the climactic eruption phase. Various instrumental and logistical problems, along with the increasing volcanic hazard, prevented us from getting many useful data out of the second tiltmeter.

VCAT TILTMETER SYSTEMATICS

Telemetered tiltmeters are the first deformation monitors VCAT installs in a crisis response because they provide near-real time, microradian-resolution data, they operate in all weather conditions, and can be installed in high-hazard areas that may quickly become too dangerous to revisit.

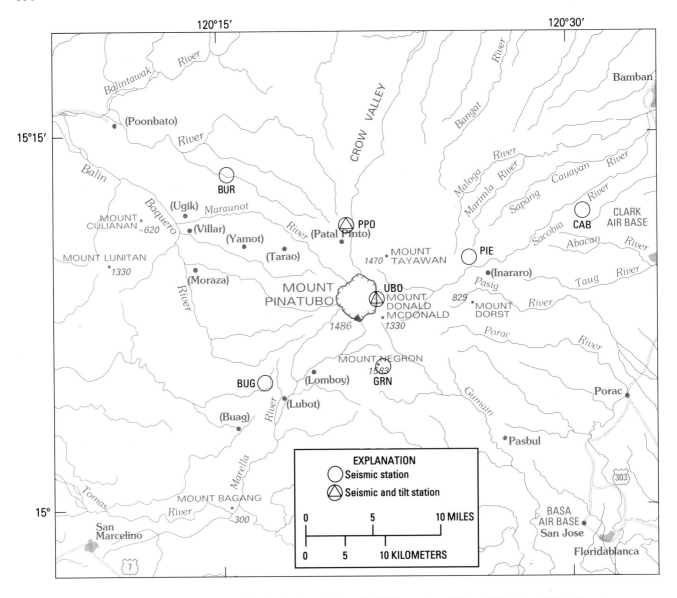

Figure 1. Telemetered tilt and seismic stations in place before the June 15, 1991 eruption of Mount Pinatubo. Altitude in meters.

Telemetered deformation data provide a nearly continuous record that allows volcanologists to track changes at a volcano that may occur in a shorter timeframe (minutes to hours) than can be reliably monitored by use of more standard geodetic monitoring techniques such as electronic distance meters or leveling measurements.

VCAT uses bubble-type, biaxial platform tiltmeters with a resolution of 0.1 μrad, and a digital data telemetry system designed at the Cascades Volcano Observatory (CVO) (Murray, 1988, 1992a). Telemetered data are received and processed every 10 min at a central monitoring site, which in this case is at Clark Air Base, by use of software developed at CVO that permits the tilt data to be compared easily and rapidly to other monitoring data (Murray, 1990a,b, 1992b).

We transmit four quantities from the tilt site: two perpendicular tilt axes (radial tilt and tangential tilt), temperature at the tiltmeter, and battery voltage. The temperature measurement allows us to look for temperature effects, especially diurnal variations in the tilt values. Tiltmeters are sensitive to even slight changes in temperature, so burial 1 to 2 m deep is necessary in order to isolate the instrument from diurnal temperature changes (Dzurisin, 1992). The battery-voltage measurement allows us to look for supply-voltage effects on the data quality.

To simplify logistics, tiltmeters were colocated with seismometers at the PPO and UBO sites (fig. 1).

Figure 2. The UBO site, an abandoned geothermal well pad on the east side of Mount Pinatubo summit dome.

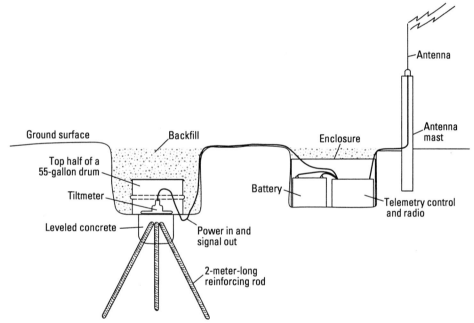

Figure 3. Schematic of the UBO tiltmeter installation on the east side of the Mount Pinatubo summit dome.

UBO SITE

The UBO site was at approximately 1,200 m altitude on the east side of the summit dome, approximately 1,000 m southeast of the active fumaroles, on an abandoned geothermal well pad (fig. 2). The tiltmeter installation at UBO is a typical VCAT tilt site. This style of tilt installation requires about two days to construct (fig. 3).

To prepare the site we dug a hole approximately 1.5 m deep, which passed through the well-pad fill into the undisturbed soil below. After driving half-inch reinforcing rod (rebar) about 2 m long through the bottom of the hole, we poured concrete around the exposed rebar ends and leveled

it. The next day, after the concrete had set, the instrument was placed in the center of the pad and electronically leveled, the top half of a 55-gallon drum was placed over it, and the hole was backfilled. Line-of-sight telemetry was established to Clark Air Base and the first data were received on May 31.

PPO SITE

The PPO seismic station was installed at the Aeta village of Patal Pinto, 4 km north of Pinatubo, on May 27 (fig. 1). The instrument was installed in an unused guard house,

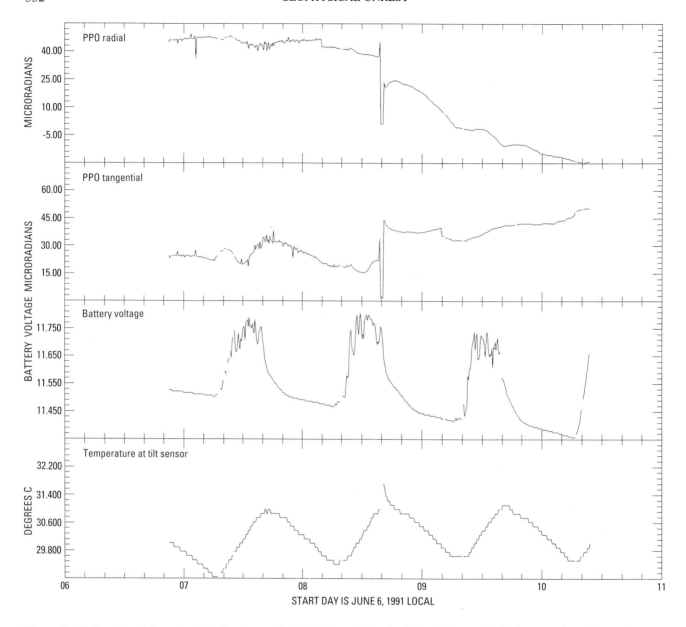

Figure 4. Entire data set from the PPO tiltmeter, north side of Mount Pinatubo. Diurnal changes in the battery voltage (due to the solar panel) and temperature can be seen. Radial and tangential are two perpendicular tilt axes.

placed there by the U.S. Air Force for the nearby Crow Valley bombing range. Time constraints and the likelihood of vandalism prevented us from making a standard tiltmeter installation, as at UBO. The tiltmeter was placed on the concrete floor of the guardhouse in hopes of avoiding the need for the laborious standard installation process. The tilt signal was summed onto the PPO seismic telemetry and transmitted to Clark Air Base via a microwave telephone link from the Crow Valley bombing range (Lockhart and others, this volume). The tilt station was installed on May 27, but problems encountered in the field and at Clark Air Base prohibited us from receiving data from the tiltmeter until June 6. Owing to the short time span and active state of the volcano during its operation, we cannot be sure whether

reliable data might have been obtained from this type of installation. Figure 4 shows diurnal temperature fluctuations of about 2°C, and diurnal battery-voltage fluctuations of about 0.2 V (owing to the use of a solar panel to charge the station batteries). As these fluctuations are not expressed in any regular manner on the tilt signals, the apparent changes might be real. However, we cannot interpret the PPO tiltmeter data, because the record is too short. As interesting as these tiltmeter data may be, they are probably scientifically useless without an adequate baseline to compare changes against. On the morning of June 10, shortly after the Clark Air Base personnel were evacuated from the Crow Valley installation, the microwave telephone link crucial to reception of the north side of the telemetry

Figure 5. The rift created on the north-northeast side of Mount Pinatubo in the April 2, 1991, explosion. Letters indicate rift-monitoring station measurement points. View is from the north. The preeruption summit is out of view, to the upper right.

net (including the PPO tiltmeter) was vandalized, and no further data were received.

RIFT-MONITORING STATION

On May 1, the USGS-PHIVOLCS team established a quadrilateral array of points across the rift that included the chain of craters created in the April 2 explosive event (fig. 5). Access to this area was by helicopter only. The measurement points consisted of meter-long pieces of rebar driven into the April 2 tephra; only the last centimeter of the rebar remained above the surface. A fifth rebar was driven in about 35 m away from the rift to provide a check on the stability of the marks near the rift (fig. 5). The distance between points across the rift was approximately 20 m. Measurements were made with a steel tape measure held by hand on the tops of the rebars. Typical precision for taping measurements on level ground is ±1/5,000 (Davis and others, 1981). On the basis of this figure, measurement precision of ±4 mm might have been possible. However, owing to time constraints on the use of the helicopter, there was insufficient ground time to make repeated measurements, and there was no means available to apply uniform tension each time the array was measured. Measurement precision was estimated at ±10 mm (±1/2,000) by the field crew (C.G. Newhall, written commun., 1993).

Four sets of measurements were made (table 1), the last on May 28.

If movement on the rift occurred between May 1 and May 28, it was too small to be measured accurately with this method. Accelerating unrest and increasing demands placed on the monitoring group's time prohibited further measurements of this array.

Table 1. Data from rift-monitoring station, north-northeast side of Mount Pinatubo.

[Line designations refer to figure 5; measurements in meters]

Line	May 1	May 11	May 16	May 28	Cumulative Difference
A–B	23.915	23.910	23.905	23.903	−0.012
A–C	25.643	25.635	25.638	25.640	−0.003
A–D	12.960	12.950	12.945	12.949	−0.011
B–C	22.875	22.886	22.873	22.880	+0.005
B–D	29.840	29.820	29.835	29.830	−0.010
B–E	26.140	26.135	26.130	26.140	0.000
C–D	19.870	19.860	19.865	19.863	−0.007
E–F			12.913	12.910	−0.003

TILT DATA

MAY 31–JUNE 3

The radial axis of the UBO tiltmeter was oriented nearly east-west (N.86°W.) such that increasing values would indicate west up, or edifice inflation. The tangential axis was oriented nearly north-south (N.4°E.), such that increasing values would indicate south up. Starting almost immediately after installation, the UBO tiltmeter record was reasonably stable. No diurnal changes were seen in the temperature or tilt data. The first 3 days of operation showed a stable radial component and a slight north-up trend on the tangential component (fig. 6).

These steady trends were interrupted on June 3 when a heavy rainstorm came through the area at about 1400 local time, and apparently caused significant tilt and temperature excursions (fig. 6). Two explanations of these excursions are reasonable: a local site response to the rainstorm or an

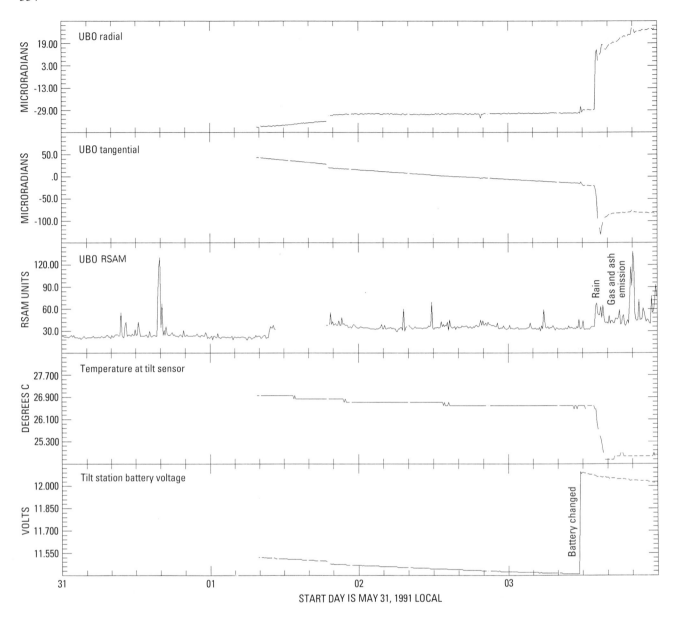

Figure 6.   Data from the UBO tiltmeter and seismometer on the east side of the Mount Pinatubo summit dome. Rainstorm on June 3 is indicated, as is the ash emission that followed. RSAM, Real-time Seismic Amplitude Measurement.

instrumental or electronic effect in which moisture from the rain affected the equipment. We feel that the tilt and temperature excursions are probably due to local site effects and are not instrumental or electronic. Rapid changes in tilt after heavy rainstorms have also been observed at Mount St. Helens (Dzurisin, 1992).

Earlier in the day (1130 local time) we had changed the batteries at the UBO tilt installation. The telemetered battery reading shows the jump in voltage. The radial and tangential axes show minor stepwise changes at this time, possibly resulting from site settling caused by our activity around on the installation. If settling occurred, then the

heavy rain a few hours later could readily have caused additional settling of the site.

Later on June 3, a small ash emission occurred that was accompanied by a high-frequency (4–6 Hz) tremor signal on the UBO seismic station. A small transient inflationary spike on the radial axis and a smaller north-down spike on the tangential axis coincide with this event. Soon after the small explosion, the locus of volcano-tectonic earthquakes shifted from about 5 km northwest of Pinatubo to the area immediately beneath the summit dome (Harlow and others, this volume). At this time the tangential tilt component returned to its previous slow, north-up trend,

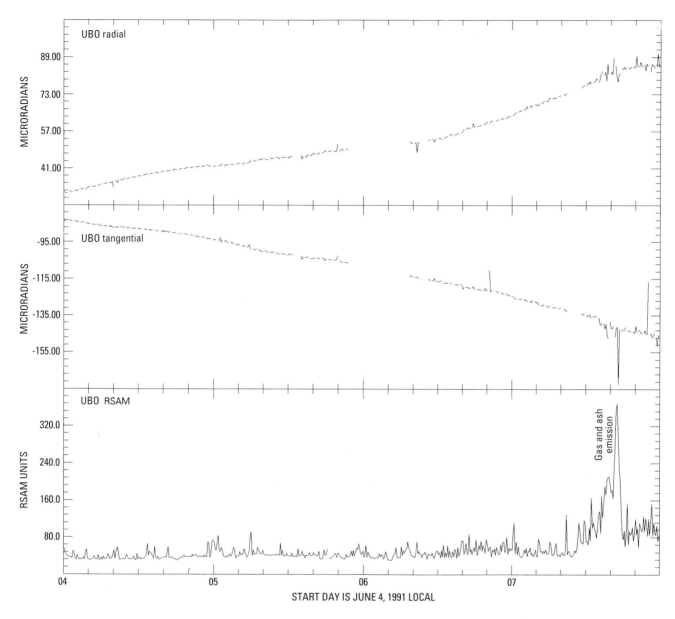

Figure 7. Radial and tangential components of the UBO tiltmeter compared to UBO seismometer Real-time Seismic Amplitude Measurement (RSAM) during days preceding dome extrusion, east side of the Mount Pinatubo summit dome.

and the radial component displayed a more decided inflationary trend.

JUNE 4–JUNE 7

The gradual inflationary tilt trend on the radial axis continued until about 1200 local time on June 6. Then it began to accelerate, coincident with intensifying shallow seismicity beneath the summit dome (fig. 7). Over the next 30 h more than 2,000 volcano-tectonic earthquakes occurred 0 to 3 km beneath the summit dome. In the last 4 h of the seismic swarm, the tilt signal on both axes became increasingly noisy, but the tilt trend remained clear; this

indicates that the tilt signal was noisy owing to seismic ground shaking, not to electronic problems. At about 1700 local time, a sudden gas and ash emission sent a plume to about 4,000 m. Immediately thereafter, the radial tilt signal leveled off but remained noisy for about another 8 h.

The tangential tilt axis at UBO continued tilting toward the south at a constant rate until about 2000 local time on June 6, when the tilt rate accelerated slightly, lagging the response of the radial axis by some 8 h. During this period, the tangential axis showed a response similar to that of the radial axis but did not mirror it, as might have been the case if electrical crosstalk had occurred between the electronics of the two axes. In addition, neither the battery-voltage nor temperature channels showed changes mirroring those seen on the tilt axes. Thus, we feel confident that the June 6–7 tilt excursion of the radial tilt axis reflects a real tilt away from the summit area and, possibly, a lesser tilt toward the south.

We interpret the increasing tilt from late on June 3 through the occurrence of the sudden steam and ash emission on June 7 as marking the formation of the magma conduit to the surface, and we interpret the abrupt flattening of the tilt signal as marking the time when magma broke through to the surface (see figures 5–8 in Harlow and others, this volume). At the time these events occurred we interpreted the combined seismic and tilt data to mean that, at the very least, magma had reached very shallow levels or even the surface. On the basis this conclusion an Alert Level 4 was declared (Ewert and others, 1991) (see Punongbayan and others, this volume, for a description of the alert level system). Visual observations on the morning of June 8 confirmed the presence of a new lava dome.

JUNE 8–JUNE 13

After the dome extrusion, ash emission became more voluminous and episodes of tremor more frequent and of longer duration, but no more cumulative tilt events occurred (fig. 8). Transient tilt events continued until the first large eruption on June 12. These events were about 10 μrad in amplitude and were associated with short-lived swarms of high- frequency earthquakes but not with tremor episodes. Tilt transients are also associated with explosions at 0341 and 0851 local time on June 12 (fig. 8). The tilt signal returned to baseline at the end of each swarm.

In the June 8–12 time period, the character of seismicity changed from being dominated by volcano-tectonic events and periods of high-frequency (3–6 Hz) tremor to dominantly long-period seismicity with periods of low-frequency tremor (<1 Hz). We see no correlation of tilt events with long-period swarms or tremor during this time period.

DISCUSSION AND SUMMARY

Beginning on June 4, a change in the locus of volcano-tectonic seismicity from 5 km northwest of Pinatubo to beneath the summit dome and active vents accompanied deformation registered by the UBO tiltmeter. This shift in seismicity probably marked the formation of the magma conduit to the surface (Harlow and others, this volume). The tiltmeter detected about 50 μrad of cumulative tilt between June 4 and June 7, ending when magma presumably reached the surface and a lava dome began to form. From June 7 to June 13, the tilt signal from UBO was noisy during periods of increased seismicity, probably due to ground shaking, but the signal always returned to its background level.

Measurements across the April 2 rift were confined to the month of May, and no deformation signal was apparent. We had planned to install a tiltmeter at this site during the first week of June, but the rapidly evolving unrest and other priorities associated with the response ultimately prevented the installation. Both the UBO tilt/seismic site and the rift-monitoring station were very close to what would become the caldera margin on June 15.

No signs of gross deformation (ground cracking or faulting) were observed in the rift area or the summit region during helicopter inspections between June 7 and 14. These observations, coupled with the small tilt signal observed from the UBO tiltmeter, may seem surprising given the large volume of erupted material (3.7–5.3 km^3 of dense rock equivalent) (W.E. Scott and others, this volume) and the even larger inferred volume (40–90 km^3) of the magma chamber (Mori, Eberhart-Phillips, and Harlow, this volume). However, Mount Pinatubo has been the site of recurring large silicic eruptions throughout the Holocene, the most recent only 600 years ago. (Newhall and others, this volume). Smith (1979) showed that silicic magma chambers erupt only about 10 percent of their volume during any one eruptive period, and Pallister and others (this volume) show that the trigger for the 1991 eruption of Mount Pinatubo was an injection of basaltic magma into a preexisting dacitic magma chamber. Therefore, most of the magma volume was probably already in place long before monitoring began.

From the small tilt changes and the lack of any obvious ground deformation prior to the Plinian phase of the eruption, we infer that no cryptodome was emplaced near the surface and that the eruptions proceeded from the overpressurized magma chamber, the top of which was about 5 to 7 km deep (Mori, Eberhart-Phillips, and Harlow, this volume; Pallister and others this volume). Thus, the tilt changes recorded only the formation of the conduit from the magma chamber to the surface and did not record the accumulation of magma beneath the summit.

ACKNOWLEDGMENTS

We would like to thank Art Daag (PHIVOLCS), Bella Tubianosa (PHIVOLCS), Chris Newhall (USGS), and Rick Hoblitt (USGS) for making the measurements at the rift station. Thanks are also due to the helicopter pilots and flight crews of the U.S. Thirteenth Air Force, Third Tactical Fighter Wing, for excellent logistical support in trying circumstances, and to Don Swanson and Dan Dzurisin for helpful comments during the preparation of this manuscript.

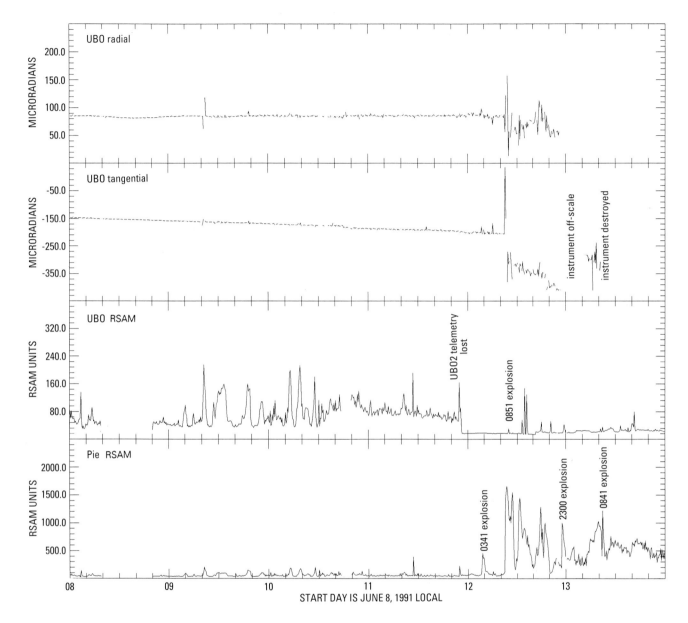

Figure 8. Last 5 days of operation of the UBO tiltmeter with Real-time Seismic Amplitude Measurement (RSAM) data from UBO and PIE (fig.1) seismometers, east side of Mount Pinatubo. Note generally flat tilt from June 8 through June 11 compared to RSAM data on same days. Peaks in RSAM data correspond to periods of tremor.

REFERENCES CITED

Davis, R.E., Foote, F.S., Anderson, J.M., and Mikhail, E.M., 1981, Surveying theory and practice: New York, McGraw-Hill, 992 p.

Dzurisin, D., 1992, Electronic tiltmeters for volcano monitoring: Lessons from Mount St. Helens, *in* Ewert, J.W., and Swanson, D.A., eds., Monitoring volcanoes: Techniques and strategies used by the staff of the Cascades Volcano Observatory, 1980–90: U.S. Geological Survey Bulletin 1966, p. 69–83.

Ewert, J.W., Lockhart, A.B., Hoblitt, R.P., and Harlow, D.H., 1991, Ground tilt events and the critical June 5–7, 1991, precursory volcanic activity at Mt. Pinatubo, Philippines [abs.],

Eos Transactions, American Geophysical Union, v. 72, no. 44, p. 61.

Harlow, D.H., Power, J.A., Laguerta, E.P., Ambubuyog, G., White, R.A., and Hoblitt, R.P., this volume, Precursory seismicity and forecasting of the June 15, 1991, eruption of Mount Pinatubo.

Lockhart, A.B., Marcial, S., Ambubuyog, G., Laguerta, E.P., and Power, J.A., this volume, Installation, operation, and technical specifications of the first Mount Pinatubo telemetered seismic network.

Mori, J., Eberhart-Phillips, D., and Harlow, D.H., this volume, Three-dimensional velocity structure at Mount Pinatubo, Philippines: Resolving magma bodies and earthquakes hypocenters.

Murray, T.L., 1988, A system for telemetering low-frequency data from active volcanoes: U.S. Geological Survey Open-File Report 88–201, 28 p.

———1990a, A user's guide to the PC-based time-series data-management and plotting program BOB: U.S. Geological Survey Open-File Report 90–56, 53 p.

———1990b, An installation guide to the PC-based time-series data-management and plotting program BOB: U.S. Geological Survey Open-File Report 90–634–A, 25 p.

———1992a, A low-data-rate digital telemetry system, *in* Ewert, J.W., and Swanson, D.A., eds., Monitoring volcanoes: Techniques and strategies used by the staff of the Cascades Volcano Observatory, 1980–90: U.S. Geological Survey Bulletin 1966, p. 11–23.

———1992b, A system for acquiring, storing, and analyzing low-frequency time-series data in near-real time, *in* Ewert, J.W., and Swanson, D.A., eds., Monitoring volcanoes: Techniques and strategies used by the staff of the Cascades Volcano Observatory, 1980-90: U.S. Geological Survey Bulletin 1966, p. 37–43.

Newhall, C.G., Daag, A.S., Delfin, F.G., Jr., Hoblitt, R.P., McGeehin, J., Pallister, J.S., Regalado, M.T.M., Rubin, M., Tamayo, R.A., Jr., Tubianosa, B., and Umbal, J.V., this volume, Eruptive history of Mount Pinatubo.

Pallister, J.S., Hoblitt, R.P., Meeker, G.P., Knight, R.J., and Siems, D.F., this volume, Magma mixing at Mount Pinatubo: Petrographic and chemical evidence from the 1991 deposits.

Punongbayan, R.S., Newhall, C.G., Bautista, M.L.P., Garcia, D., Harlow, D.H., Hoblitt, R.P., Sabit, J.P., and Solidum, R.U., this volume, Eruption hazard assessments and warnings.

Scott, W.E., Hoblitt, R.P., Torres, R.C., Self, S, Martinez, M.L., and Nillos, T., Jr., this volume, Pyroclastic flows of the June 15, 1991, climactic eruption of Mount Pinatubo.

Smith, R.L., 1979, Ash-flow magmatism, *in* Chapin, C.E. and Elston, W.E., eds., Ash-flow tuffs: Geological Society of America Special Paper 180, p. 5–25.

Volcanic Earthquakes following the 1991 Climactic Eruption of Mount Pinatubo: Strong Seismicity during a Waning Eruption

By Jim Mori,[1] Randall A. White,[1] David H. Harlow,[1] P. Okubo,[1] John A. Power,[1] Richard P. Hoblitt,[1] Eduardo P. Laguerta,[2] Angelito Lanuza,[2] and Bartolome C. Bautista[2]

ABSTRACT

One of the stronger sequences of volcanic earthquakes this century (cumulative seismic energy of 6.3×10^{13} joules) started during the June 15 eruption of Mount Pinatubo. The locations of these events were spread over a larger area (out to 20 kilometers from the summit) and greater depth extent (25 kilometers) than the preeruption seismicity. The intense rate of high-frequency volcano-tectonic earthquakes decreased rapidly over the first 2 weeks, following a smooth exponential decay. The locations of many of the events fall on outward-dipping trends that surround a region with relatively few earthquakes, which may be the magma reservoir that provided much of the material for the eruption. As the volcano-tectonic events diminished in late June, the low-frequency seismicity increased to levels similar to those during the preeruption period. The increase of low-frequency seismicity accompanied the transition in eruptive style from a steady outpouring of ash to more intermittent explosive activity. Episodic seismic activity developed into regular 7- to 10-hour intervals of increased low-frequency events often accompanied by large ash columns. Both the smooth decay of the high-frequency events and the periodicity of the low-frequency seismicity were interpreted as signs that the volcano was not building toward another large eruption.

INTRODUCTION

During the 1991 paroxsymal eruption of Mount Pinatubo on June 15, the seismic activity significantly increased and changed in character compared to the preeruption period. For the next several weeks the seismicity was dominated by high-frequency earthquakes that were distributed over a large volume extending out to 20 km from the volcano and to depths of 25 km. This was a marked change from the seismicity that was localized closer to the volcano summit at shallower depths during the preeruption period (Harlow and others, this volume).

The seismicity was the main source of information used to evaluate the state of the volcano, as was the case during the preeruption period (Harlow and others, this volume). Following the destruction of most of the seismic network during the eruption, the station on Clark Air Base (CAB) was the only local source of seismic data for the first 2 weeks following the climactic eruption. Although the seismic monitoring was severely limited, this single station located 17 km northeast of Mount Pinatubo (fig. 1) still proved useful in assessments of the possibility for future strong eruptions following the climactic eruption. The data from this instrument were particularly important during the first week, when very few visual observations could be made of the volcano because of the dense ash cloud that was constantly present and precluded visual observations. By the end of June, a seismic network of seven stations had been reinstalled, and the capability for recording and locating earthquakes was comparable to the preeruption period.

This paper summarizes the seismicity for the 2 months following the eruption of June 15. We note observations made during this period of declining seismicity and eruptive activity that we interpret as a waning eruption. During this time, the eruptive activity was small compared to June 15; however, the volcano still produced substantial ash columns to heights of over 18,000 m and small pyroclastic flows.

CHRONOLOGICAL SUMMARY

We present a summary of the seismic and eruptive activity from the time the strong sequence of earthquakes began on June 15 through the end of the eruptive activity in September 1991. We use terminology similar to Harlow and others (this volume) in describing the high-frequency volcano-tectonic earthquakes and the low-frequency activity, which includes long-period and the hybrid events as well as continuous volcanic tremor. Figure 2 shows a consistent

[1]U.S. Geological Survey.
[2]Philippine Institute of Volcanology and Seismology.

POSTERUPTION Jun 29 – Aug 16

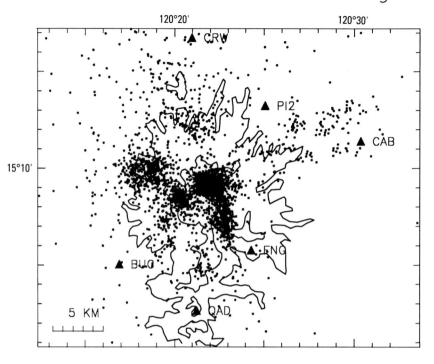

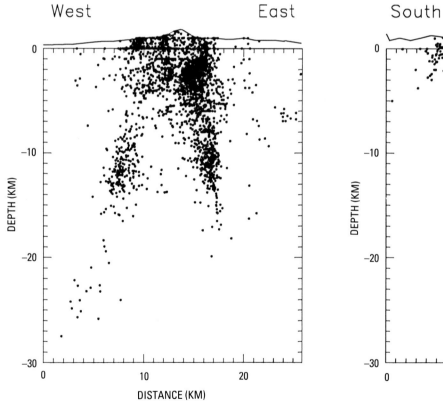

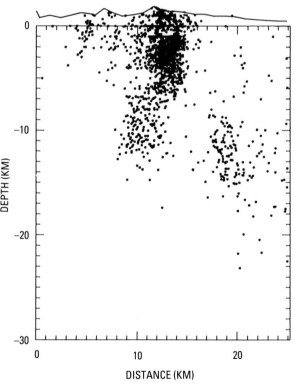

Figure 1. Locations of over 3,500 events recorded from June 29 through August 16, 1991. Hypocenters were calculated using a three-dimensional velocity model (Mori, Eberhart-Phillips, and Harlow, this volume). The west-east cross section shows earthquakes that were located within 5-km-wide slices centered on the summit. The south-north cross section shows earthquakes in a 3-km slice centered on the summit. Triangles mark seismic stations.

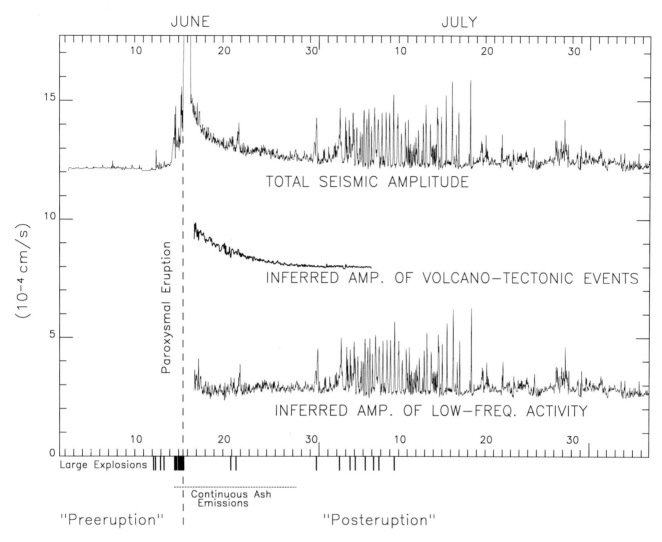

Figure 2. Hourly averages of seismic amplitudes at station CAB. Top trace is the total seismic amplitude recorded on the Real-time Seismic Amplitude Measurement (RSAM) system. Middle trace is the amplitudes of volcano-tectonic earthquakes inferred from the hourly event counts. Bottom trace is the amplitudes of the low-frequency activity obtained by subtracting the high-frequency amplitudes from the total amplitudes. The record of large explosions is incomplete for the posteruption period.

measurement of seismic amplitudes at station CAB and summarizes the seismic activity for June and July. Also included is information about the eruptive activity, although the record of large explosions is largely incomplete for the posteruption period.

The strong sequence of earthquakes started with a magnitude 5 event at 1539 (all times are Philippine local time), several hours into the paroxysmal eruption that began at 1342 on June 15 (Hoblitt, Wolfe, and others, this volume). There are no local seismograms of this activity because all seismic instruments, except CAB, had been destroyed by the eruption, and the recording of this last station at Clark Air Base stopped when the scientific team abandoned the Pinatubo Volcano Observatory (PVO) at around 1500. Some of the automated seismic amplitude measurements continued to operate after the team left; these measurements provided some information during the time

PVO was unmanned (Power and others, this volume). U.S. Geological Survey (USGS) National Earthquake Information Center (NEIC) reported 29 events with body-wave magnitudes (m_b) ≥ 4.5 and six events with m_b ≥ 5.0 within the next 6 hours. Earthquakes were being felt at the rate of nearly one per minute at the Arayat evacuation site, 35 km east of Mount Pinatubo. Probably sometime during this intense earthquake activity, the summit area collapsed to form the caldera, which is 2.5 km in diameter (W.E. Scott and others, this volume). The teleseismic locations of the large events have horizontal uncertainties of 10–15 km (Waverly Person, USGS, oral commun., 1994) and are not accurate enough to resolve if all of the earthquakes occurred in the immediate vicinity of the volcano or if some were associated with nearby regional faults (B. Bautista and other, this volume). The largest event reported by NEIC was an m_b 5.7 earthquake at 1911 on June 15 that was located close to the volcano summit.

After we returned to PVO on June 16, continuous recording on a heliocorder resumed on the remaining seismic station (CAB) starting from 1100. Earthquakes of magnitudes 2.0 or greater were occurring at the rate of more than 150 per hour, but the rate was declining rapidly following a smooth decay curve (fig. 2). During this first week following the climactic eruption, there were only a few observable low-frequency events among the thousands of high-frequency volcano-tectonic events. At this time the volcano was continuously venting large amounts of ash without producing large explosions. On June 21 from 1500 to 2300, the low-frequency events significantly increased for the first time since June 15 (fig. 2). The seismogram recorded at station CAB during this period shows approximately equal numbers of high-frequency and low-frequency events, 20 to 50 events per hour over about magnitude 2.0. Associated with the low-frequency seismic activity were large explosions with ash columns up to 12,000 m.

Following the June 21 episode, volcanic tremor began to appear more frequently on the CAB record. Smaller amplitude tremor probably occurred continuously from June 15, but it could not be distinguished with the one station at CAB during the intense sequence of high-frequency events. Over the next 10 days the tremor increased steadily in duration and amplitude. By the end of June, the tremor was a continuous feature on the seismic record. During this time, individual low-frequency events also became more numerous. As the tremor and the low frequency events increased, the rate of volcano-tectonic earthquakes continued to decline smoothly (fig. 2). This change in the character of the seismicity correlated with a change in the character of the eruption. For the first week after June 15, the volcano was in a continuous mode of spewing out ash. There were only a few large explosions observed, and the ash column was fairly constant. In contrast, toward the end of June the continuous ash emissions became more intermittent, and more discrete, stronger explosions were observed (fig. 2).

Seismic stations were redeployed around Mount Pinatubo in late June (fig. 1), and by June 29 there were 6 operational stations to locate earthquakes. The hypocenters showed that the earthquakes were occurring throughout a much larger volume than before June 15. During the preeruption period, the earthquakes were restricted to two clusters of events within 7 km of the summit at depths from the surface to 8 km (figs. 4, 7, 9 in Harlow and others, this volume). In contrast, the posteruption volcano-tectonic earthquakes were distributed out to distances of more than 20 km from the summit and to depths greater than 25 km (fig. 1).

On June 30, low-frequency seismicity again increased to levels similar to those of June 21 and started a sequence of episodic activity. From July 7, the occurrence of the low-

Figure 3. Ash column of July 7, 1991, at 1700, during one of the periodic peaks in low-frequency seismicity.

frequency activity became strikingly regular at 7- to 10-hour intervals. Low-frequency events that could be located all occurred at shallow depth near the summit area. During times that the volcano was visible, the peaks in the seismic activity correlated with increased eruptive activity. The near-predictability of these episodes enabled observers to photograph some of the larger ash columns that rose to over 20,000 m (fig. 3). The regular occurrences of the seismic and eruptive activity continued through July 11, although at increasing time intervals.

From the second week of July, the continuous tremor and episodes of low-frequency events began to occur at more irregular intervals. There were still episodes of large-amplitude tremor and low-frequency events, and throughout July the overall level of amplitudes for the tremor and low-frequency events remained comparable to the most intense preeruption activity on June 14 and 15 (fig. 2). The eruptive activity during this time was declining as explosions from the crater became fewer. As the rains of the monsoon season

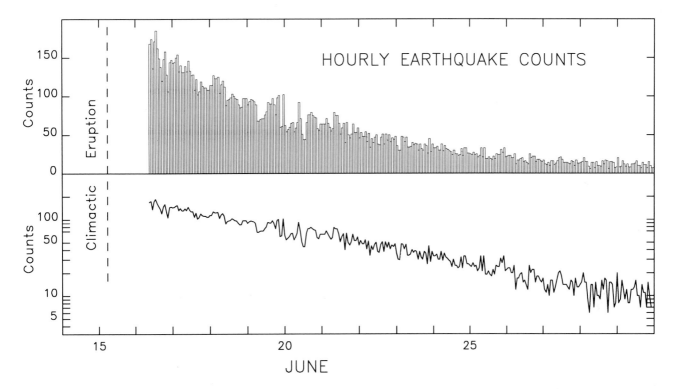

Figure 4. Hourly event counts of high-frequency volcano-tectonic earthquakes tabulated from CAB. Top panel plots the data on a linear scale. Bottom panel plots the logarithm of the event counts.

increased, the emphasis of the hazard evaluations concentrated less on the actual eruption and more on the numerous lahars (Pierson and others, this volume). From late July the seismicity began a slow decline that lasted through the next several months, reflecting the waning of the 1991 eruption. The eruptive activity had essentially stopped by early August, and the last reported explosion was on September 4, although that event could have been a secondary explosion. Occasional periods of continuous tremor and low-frequency activity were observed throughout 1991 and 1992 (Ramos and others, this volume).

VOLCANO-TECTONIC EARTHQUAKES

In order to estimate the rate of seismic activity, we tabulated events from the paper recordings of station CAB. Events with maximum trace amplitudes of greater than .2 cm were counted (fig. 4). This would correspond to magnitudes of 2.0 to 2.5 near the summit area and smaller magnitude for events closer to Clark Air Base. The high rate of earthquakes during June 16 to 19 made counting the events difficult, so the numbers in figure 4 are probably underestimates during this period. The smooth decline of the volcano-tectonic earthquakes qualitatively resembles an aftershock sequence, but the shape of the decay is different. Aftershock rates following large earthquakes follow a power law that decays proportional to time (t^{-p}), where p is close to 1.0 (Utsu, 1961). The rate at which the earthquakes diminished at Pinatubo follows an exponential decay, not a

power-law decay. This is illustrated by the linear trend in the bottom panel of figure 4 which shows the logarithm of the event counts versus time. A power-law decay would produce a concave upward trend to the data in this plot. We can fit the data to an exponential curve,

$$N(t) = Ke^{-qt}$$

where $N(t)$ is the event count, K is a constant, t is time in days. The decay constant (q) was calculated to be 0.0095. If we force the data of the first 9 days to fit the power-law decay of aftershocks, using the method of Ogata (1983),

$$N(t) = K(t+C)^{-p}$$

we obtain a p value of 2.42±0.79 and a c value of 7.79±3.76. Both of these of values are much higher than observed for earthquake aftershocks. Using longer time windows will yield even higher p values. An exponential decay or a power-law decay with large p value, points out that the rates of the volcano-tectonic earthquakes at Pinatubo diminished much faster than rates for aftershock sequences. This suggests differences in the geometry or mechanism for earthquakes following large volcanic events compared to aftershocks following large tectonic earthquakes.

Magnitudes for volcano-tectonic events in the range of 3.2 to 5.0 were estimated by using coda durations, and over 2,500 events were recorded in this range from June 16 to August 12 (table 1). We calculated a b-value for the

Table 1. Magnitude (*M*) distribution of volcano-tectonic earthquakes.

	M≥3.0	M≥4.9	M≥5.5	M≥5.7
June 16–20	1,750	19	4	2
June 21–30	580	4	0	0
July 1–31	120	9	0	0
August 1–12	26	0	0	0

sequence using these data along with magnitudes for the larger events (M> 5.0) which were taken from body-wave magnitudes determined by NEIC. The b-value calculated by use of the maximum likelihood method was 1.1±0.3, which is typical of crustal tectonic earthquake sequences (Utsu, 1961). There were no significant differences in the rate of occurrence of larger events compared to smaller events; that is, no large changes in b-value were observed during the intense activity of high-frequency events in June. We also used the magnitude data (M_L, local-wave magnitude) to estimate the radiated energy (E) by using a revised energy-magnitude relation from Kanamori and others (1993),

$$\log E = 1.96 \, M_L + 9.05$$

The cumulative radiated energy for the sequence from June 16 to August 12 was 6.3×10^{13} J. For comparison, this is larger than the radiated seismic energy for eruptions at Mount St. Helens, Washington but smaller than for Mount Katmai, Alaska (table 2).

Earthquake hypocenters were determined using a three-dimensional P-wave velocity structure determined from an inversion of arrival times from 298 selected events. Details about the velocity inversion and earthquake relocations are given in Mori, Eberhart-Phillips, and Harlow (this volume). The locations using this more complicated velocity structure, rather than a simple one-dimensional model, should be more accurate, considering the complex pattern of velocities often associated with volcanoes. The strengths of this method are especially useful around regions where there may be large low-velocity (magma) bodies. Mori, Eberhart-Phillips, and Harlow (this volume) show comparisons of the locations using the three-dimensional structure and a best-fitting one-dimensional structure. Locations plotted in figure 1 are for over 3,500 events recorded at more than five stations from June 29 through August 16. The horizontal uncertainties of the locations are less than 1 km, with vertical uncertainties of 1 to 3 km. The resultant earthquake locations are spread diffusely over a large area around the volcano with concentrations in the summit area (fig. 1). From the resumption of recording on June 16, the high-frequency events showed substantial variations in the relative arrival time between the S- and P-waves, indicating that the earthquakes were located at varying distances from the station and scattered throughout the volume. In cross section, these concentrations form trends that dip steeply away from Mount Pinatubo. The strongest clusters of

Table 2. Comparison of seismic activity for several significant eruptions of the century.

[Values are from the following references: Mount Katmai—Abe, 1992, Hildreth, 1983; Mount Pinatubo—W.E. Scott and others, this volume; Fernandina—Filson and others, 1973, Simkin and Howard, 1970; Mount St. Helens—Weaver and others, 1981, Moore and Albee, 1981. Seismic energies were recalculated by using the magnitude-energy relation from Kanamori and others (1993)]

	Mount Katmai, Alaska, 1912	Mount Pinatubo, 1991	Fernandina, Galapagos Islands, 1968	Mount St. Helens, Washington State, 1980
Largest earthquake (M_s)	7.0	5.7	5.1	5.2
Cumulative energy ($\times 10^{13}$ J)	1,570	6.3	2.0	0.43
Cumulative moment ($\times 10^{17}$ Nm)	1,400	23.4	9.8	1.1
b-value	0.90	0.95	0.68–1.91	0.67
Erupted volume Magma eq. (km³)	15	3.7–5.3	< 0.2	0.2
Collapse volume (km³)	5	2.5	2	2.7*

* includes both collapse and landslide volume at Mount St. Helens.

earthquakes to the east and west of the summit clearly show trends that dip outward away from the volcano (east-west cross section in fig. 1). A group of earthquakes to the north of the summit also appears to dip steeply to the north, although the pattern is not as clearly defined for these events (north-south cross section of fig. 1). There are many earthquakes in the region directly below the summit from the surface to depths of about 4 km and then a striking lack of the earthquakes in the deeper central area.

The spatial distribution of the volcano-tectonic earthquakes can be described as three legs of a tripod around a volume relatively free of earthquakes. The tripod of earthquake locations may be partially outlining a magma reservoir that supplied much of the material for the eruption. This interpretation is supported by the observation of a very large low-velocity zone that corresponds to the central region between the earthquakes (Mori, Eberhart-Phillips, and Harlow, this volume).

LOW-FREQUENCY EARTHQUAKES AND CONTINUOUS TREMOR

Following the decline of the volcano-tectonic events in late June, episodes of low-frequency seismic activity dominated the seismic records from about July 1 (fig. 2). The low-frequency activity consisted of discrete events that resembled those recorded during the preeruption period and often grade into continuous tremor, which remained at a high level during episodes of increased activity. There was always some level of continuous tremor observable on the

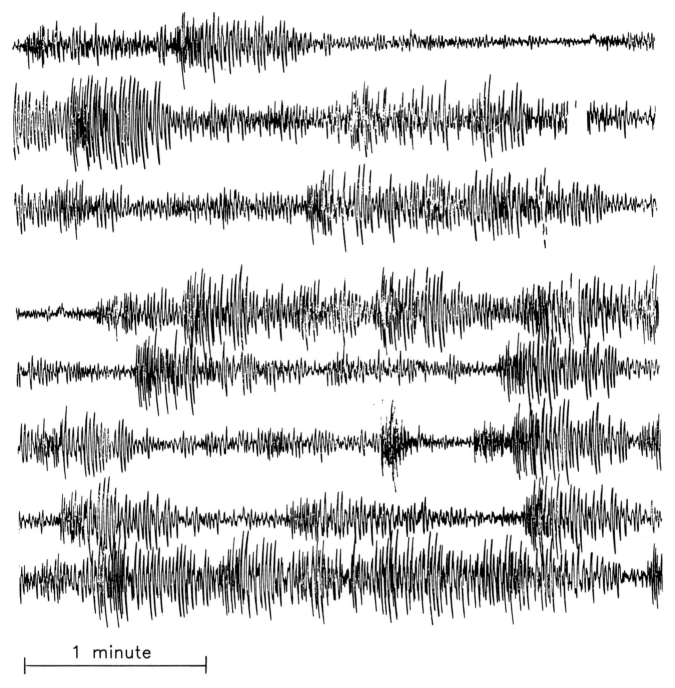

1 minute

Figure 5. Examples of low-frequency events taken from a section of a heliocorder record from seismic station CAB during one of the periodic peaks in activity on July 1, 1991, at 0900. Each line shows about a 3-minute portion of the 15-minute revolution of the recording drum.

CAB (17 km from the summit) record from June 20 through about July 7. A few low-frequency events that had discernible P-waves were located at shallow (less than 3 km) depths near the summit. The shallow depth is also consistent with the relative amplitudes recorded on the network, which decrease rapidly with distance from the summit.

When the summit area was visible, the increases in the low-frequency seismic activity were correlated with periods of large ash explosions from the volcano. Some of ash columns produced during the periods of stronger seismicity were quite large, reaching heights of 18,000 m (fig. 3). There is a similarity in the amplitudes (fig. 2) and character (fig. 5) of the low-frequency waveforms recorded at CAB during the posteruption with those recorded during the large ash explosions on June 14. This is further evidence that much of the low-frequency activity is associated with the large ash eruptions.

The episodes of increased activity can be seen most easily on the Real-time Seismic Amplitude Measurement (RSAM) data shown in the top trace of figure 2. The RSAM data is a cumulative measure of the seismic amplitude for a specified time window (Murray and Endo, 1989). The data in figure 2 are from station CAB averaged over 1-hour time windows. From June 16 through late June the amplitude is dominated by the high-frequency earthquakes. With the gradual increase of the continuous tremor starting around June 24 and the decline of the high-frequency earthquakes, the RSAM is measuring predominantly the low-frequency seismic activity by July 1. Large-amplitude peaks on June 21, June 30, and the many subsequent ones are caused by the episodes of low-frequency earthquakes that were associated with increased ash emissions when the volcano was visible. Since some of the largest amplitudes are offscale, the heights of the peaks during the strongest activity are underestimated. However, the duration of the clipped signals is relatively short compared to the hour or more duration of the low-frequency activity, so the underestimate may not be significant.

We tried to separate the amplitudes of the volcano-tectonic earthquakes from the low-frequency earthquakes in the RSAM data by using the event counts shown in figure 4. We assumed that during the first 10 days following the June 15 eruption, the RSAM record is dominated by the volcano-tectonic events and that the shape of the time series matches the shape of the event-count curve. Theoretically, this should be the case for a b-value of 1.0 and amplitudes that add constructively. The event-count curve was fit to the RSAM curve to find an amplitude constant between the two curves for that time period. The amplitude-corrected event-count curve was subtracted from the RSAM curve to give a resultant time series that represents amplitudes of the low-frequency events and continuous tremor (bottom trace of fig. 2). The inferred low-frequency amplitudes (fig. 2) illustrate the progression of low-frequency activity described above. There were single peaks of activity on June 21 and June 30 and an increase in activity to the sequence of regularly spaced episodes starting on July 1.

The amplitudes of the low-frequency events and continuous tremor throughout most of July were relatively large and were comparable to the amplitudes measured during the intense low-frequency activity that occurred on June 14, 1 day before the climactic eruption (fig. 2). The large amplitudes for the low-frequency activity during the first 2 weeks of July (fig. 2) correspond to reduced displacements (Fehler, 1983) on the order of 102 cm, assuming the predominant frequencies are 2 Hz with wavelengths of 1 km. This estimate of reduced displacement and the corresponding explosive activity (fig. 3) are consistent with results of McNutt (1994) which show that the amplitude of tremor is indicative of the size of the corresponding eruptions.

This low-frequency activity during the posteruption period is similar in amplitude and character to the preeruption period after 1309 on June 14 (compare fig. 5 to fig. 13 of Harlow and others, this volume). The seismic activity on June 14 is associated with a time when the eruptive activity became more continuous, with pyroclastic flows emanating from multiple vents (Hoblitt, Wolfe, and others, this volume). This type of activity may be characteristic of an open system in which much of the low-frequency seismicity is directly associated with eruptive activity. This is in contrast to the sequence of large, discrete, long-period events observed earlier on June 14 (see fig. 12B in Harlow and others, this volume), which may be more indicative of pressurization inside the volcano (Chouet and other, 1994).

PERIODIC ACTIVITY

One striking feature of the low-frequency events and continuous tremor during the posteruption period was the regular peaking of activity observed in early July. During this period, the seismic activity increased to relatively high levels for about 2 to 3 hours and then declined to background levels. From July 7 through 11, the pattern became nearly periodic with increased activity occurring at 7- to 10-hour intervals. These regular intervals can be seen clearly in the RSAM data recorded for 10-min time windows from five stations (fig. 6). The time interval between the peaks was not strictly periodic but increased with time. On July 7, the time interval was about 7 hours and by July 11 had increased to about 10 hours. All the peaks in figure 6 can be correlated from station to station, except for the small peak of July 9 seen on BUG, which was produced by a burst of high-frequency amplitudes. This peak followed a period of heavy rain and was inferred to be a lahar signal.

Since the regular increases in seismicity were also associated with increases in ash production, the recognition of the periodic pattern made it possible to anticipate several of the larger ash columns. Weather conditions allowed good visibility from PVO for three of the eruptive periods (July 6 at 0600 and July 7 at 0500 and 1700). Concurrent visual and seismic observation showed that the ash column could attain a significant height (9,000–12,000 m) before the increases in seismic activity were observed. On August 1, there was a similar observation that the appearance of an ash column preceded the increases of low-frequency seismic activity by a few minutes.

Since the occurrence of low-frequency seismic events sometimes follows the beginning of ash emissions, we suggest that they may be triggered by the release of the overburden pressure. This is consistent with interpretations that the low-frequency events are associated with the process of gas vesiculation and may provide an explanation for the low level of low-frequency activity during the first week following the climactic eruption. During that time, the newly formed conduit system was wide open, as a constant stream of outpouring ash that did not allow any significant buildup

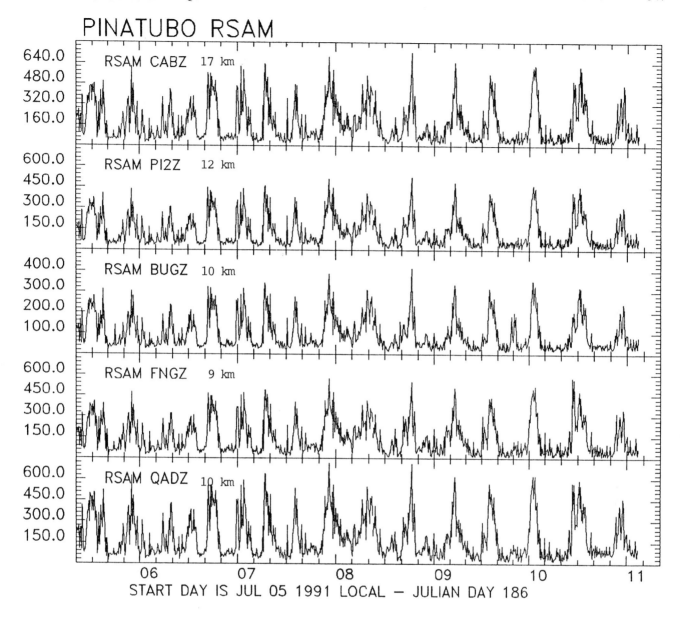

Figure 6. Real-time Seismic Amplitude Measurement (RSAM) data from five stations showing the regularity of the low-frequency seismicity during the first week in July. Amplitudes are in counts which are proportional to the cumulative amplitudes recorded on the velocity sensors. Distances to the summit are shown after the station name.

of gas pressures. As the vent area cooled, small volumes within the volcano became sealed and provided regions where the pressure could increase to sufficient levels to produce small explosions and associated low-frequency events. In this model, the appearance of low-frequency activity is an indication of the transition from continuous ash emissions to periods of discrete eruptive episodes.

DISCUSSION

INDICATIONS OF A WANING ERUPTION

The occurrence of many felt earthquakes during the few days after the June 15 eruption raised concern of further strong eruptions; however, the smooth decay in the number of earthquakes (fig. 4) suggested the opposite. A systematic decay in the rate of seismicity that follows a smooth decay indicates that the earthquakes were a time-dependent response to a large perturbation, analogous to the occurrence of many aftershocks following a large mainshock. The removal of several cubic kilometers of material from under the volcano drastically changes the local stress conditions, and it is not surprising that a high level of seismicity was induced around that volume of material. Even though people near Mount Pinatubo were being shaken by many more earthquakes following the eruption compared to before, this seismicity was interpreted to be a response to the previous eruption and not precursory to further activity.

The decreasing rate of earthquakes following the June 15 eruption is superficially similar to an aftershock sequence; however, the decay law is different. Typical aftershock sequences decay proportional to t^{-1} (Utsu, 1961; Kisslinger and Jones, 1991). In contrast, the Pinatubo data show a more rapid exponential fall-off. This disagreement might be attributed to a difference in the source region or physical process. Hirata (1987) suggested that aftershocks associated with a fractured surface follow a power-law decay, while intact volumes of material subjected to stress produce seismic activity with exponential decays. The seismicity that occurred in the large volume of material around Pinatubo is more similar to the latter case, rather than aftershocks associated with a fault surface. An exponential decay of seismic activity was also observed following eruptions at Usu Volcano, Japan (Yokoyama and others, 1981) and may be a general characteristic of seismicity following volcanic activity.

The periodicity of the low-frequency seismicity and accompanying ash eruptions was interpreted as another sign that the volcano was not building toward more strong eruptions. Periodic activity usually implies a system that is near equilibrium with recurrent events that are controlled by repeating processes. Examples of these systems would be eruption cycles of geysers (Kieffer, 1984) or the banded tremor observed at Karkar Volcano, Papua New Guinea (McKee and others, 1981). The banded tremor at Karkar was eventually followed by a moderate eruption, but the periodicity was probably due to a local hydrothermal system. At Pinatubo, the regularity in eruptions was interpreted to mean that the pressures were at a generally stable level and were not building toward another large eruption. Furthermore, the time interval between eruptive episodes, which was lengthening slightly, suggested that the rates at which the pressures accumulated to produce the eruptions were decreasing.

One observation that raised concern was that the levels of low-frequency activity during the beginning of July were as large or larger than during much of the preeruption period. The RSAM levels (fig. 2) show that the amplitudes were comparable to the intense low-frequency activity on June 14 preceding the climactic eruption. Also, the character of the low-frequency activity in the posteruption period is similar to the seismicity observed after 1309 on June 14. However, we suggested above that this type of low-frequency seismicity be may characteristic of an open system and associated with ongoing eruptive activity, as distinguished from discrete long-period events, which may be more indicative of pressurization inside the volcano. Therefore, despite the large amplitude of the low-frequency seismicity, this activity was not interpreted to be an imminent sign of a large eruption.

POSSIBLE RING FAULT

The spatial distribution of the volcano-tectonic earthquakes was described as a tripod shape around a volume relatively free of earthquakes. The interpretations from the three-dimensional velocity inversion (Mori, Eberhart-Phillips, and Harlow, this volume) indicate that there may be a large magma body located within this boundary of earthquakes. The observed pattern of seismicity that partly surrounds a large magma body suggests the beginning of a ring-fault structure. The clusters of earthquakes seen in figure 1 would bound a block that has east-west dimensions of 5 to 8 km and north-south dimensions of 10 to 12 km. This raises the possibility that subsequent eruptions could continue to contribute to the formation of a ring structure, and eventually the pieces would coalesce into a complete ring fault and lead to formation of a large caldera during a future eruption of Pinatubo.

Outward-dipping trends of earthquakes have been observed at several other volcanoes, such as Rabaul caldera, Papua New Guinea (Mori and McKee, 1987), Mount St. Helens, Washington State (Scandone and Malone, 1985), and Mount Spurr, Alaska (Jolly and others, 1994), although only Rabaul caldera is associated with a well-identified ring fault. If these trends reflect the geometry of faults, they would be consistent with orientations that would form in response to a dilatational source (Anderson, 1936). Following an eruption, the source region of the magma could be viewed as a dilatational source because of the volume change from the loss of the erupted magma. The depths of the earthquakes at Mount Pinatubo, Mount St. Helens, and Mount Spurr, which extend to depths greater than 20 km, may indicate the extent of the dilatational source, suggesting relatively deep source regions for the erupted magma. For the smaller eruptions with smaller volume changes, such as at Mount St. Helens and Mount Spurr, the regional stress field can also be an important factor (Barker and Malone, 1991).

COMPARISON TO OTHER ERUPTIONS

The occurrence of large earthquakes (M>5) during volcanic eruptions is often associated with large-scale deformation such as caldera formation or sector collapse (Okada, 1983). The activity at Pinatubo was one of the stronger seismic sequences related to a volcanic eruption this century and is probably directly or indirectly associated with the collapse of the summit area to form the small caldera (W.E. Scott and others, this volume), although it is unknown if the earthquakes were triggers or results of the collapse. The caldera with a diameter of 2.5 km has a collapse volume of about 2.5 km^3.

Table 2 shows comparisons of the maximum size earthquake, total seismic energy, collapse volume, and other estimates of source parameters for several other large eruptions plus the smaller 1980 Mount St. Helens eruption. The cumulative energies in table 2 were recalculated by using the magnitude-energy relationship from Kanamori and others (1993) to make them consistent with the energy estimate for Pinatubo. The seismic energy releases at Katmai (Alaska) Pinatubo, and Fernandina (Galapagos Islands) occurred primarily during the few days of or following the climactic eruptions. The Pinatubo activity was stronger than the seismic sequence associated with a 2-km^3 caldera collapse at Fernandina in 1968 (Filson and others, 1973) but much smaller than the series of large earthquakes that occurred at the 5-km^3 collapse of Mount Katmai in 1912 (Abe, 1992). Abe (1992) suggests that there is a correspondence between the seismic energy and the collapse volume, which appears to be roughly supported by table 2; however, there have been some notable exceptions, such as the 1914 eruption of Sakurajima, Japan, which had an associated earthquake with a surface wave magnitude of 7.0 but no significant collapse (Abe, 1979). Furthermore, it seems unlikely that there can be a direct relation between surficial collapse volume and seismic energy of earthquakes at depth under the volcano. Large earthquakes following volcanic eruptions may be controlled more by local stress changes induced by volume changes of the magma chamber rather than by collapse volumes of the volcano. The seismic energy associated with the Mount St. Helens eruption is much smaller and may have a different mechanisms, since almost all of the energy release occurred during the 2 months prior to the cataclysmic eruption.

The eruptive activity at Pinatubo declined relatively quickly after June 15. The duration of strong activity following climactic eruptions at other large explosive volcanoes has also generally been short (Simkin and others, 1981), Krakatau (6 months), Santa María, Guatemala (1 month), Katmai (2 months). In many instances, the relatively short length of these larger eruptions makes them easier to deal with in terms of issuing hazard evaluations, compared to smaller eruptions that can continue at a sustained or irregular level of activity for a year or more, such as recent activity at Unzen, Japan, or Galeras, Colombia.

CONCLUSIONS

The strong sequence of volcano-tectonic earthquakes that began during the climactic eruption of June 15 is thought to reflect the local response to a large volume change in the magma chamber underlying the volcano. The large areal and extended depth distribution of the earthquakes suggests that there were significant volume changes down to depths of 20 km. The smooth exponential decay in the rate of these events supports the idea that the high-frequency earthquakes were a response to the eruption and not

indications of increasing volcanic activity. The smooth decay of the seismicity was similar to the decreasing rates of aftershocks following a large tectonic earthquake, except that the decay curve at Pinatubo was exponential, in contrast to power-law decays observed in aftershocks. Much of the seismicity is located on outward-dipping trends that surround a volume directly under the volcano that is relatively free of earthquakes. This region at 7 to 20 km in depth may be the magma reservoir that supplied much of the material for the 1991 eruption (Mori, Eberhart-Phillips, and Harlow, this volume).

Strong increases in the low-frequency activity observed in late June were comparable to levels observed during the preeruption period. These episodes of low-frequency events were often accompanied by large ash columns. The low-frequency seismicity may reflect the transition from continuous ash emissions during the first week following June 15 to periods of discrete eruptive episodes. During the first week of July, the episodic behavior became very regular, having increases in seismicity and eruptive activity at 7- to 10-hour intervals. The near-periodicity of the activity and the gradual lengthening of the time interval were also thought to be indications that the volcano was relatively stable and not building toward another large eruption.

The decrease of high- and low-frequency seismicity along with the decline of the eruptive activity throughout July indicates an end to the 1991 eruption in early August, excepting one possible small explosion in September.

ACKNOWLEDGMENTS

The data and interpretations presented in this paper were made possible because of the considerable efforts of T.L. Murray, F. Fischer, L. Bautista, E.T. Endo, J.W. Ewert, A.B. Lockhart, J. Lockwood, J.N. Marso, A. Miklius, C.G. Newhall, E. Ramos, E.W. Wolfe, and others. We thank U.S. Air Force personnel at Clark Air Base for their support. Helpful comments on the manuscript were provided by S. McNutt, C.G. Newhall, and 2 anonymous reviewers.

REFERENCES CITED

Abe, K., 1979, Magnitudes of major volcanic earthquakes of Japan 1901 to 1925: Journal of the Faculty of Science, Hokkaido University, Series VII (Geophysics) v. 6, p. 201–212.

———1992, Seismicity of the caldera-making eruption of Mount Katmai, Alaska 1912: Bulletin of the Seismological Society of America, v. 82, p. 175–191.

Anderson, E.M., 1936, The dynamics of the formation of cone-sheets, ring-dykes, and caldron-subsidences: Proceedings of the Royal Society of Edinburgh, v. 56, p. 128–156.

Barker, S.E., and Malone, S.D., 1991, Magmatic system geometry at Mount St. Helens modeled from the stress field associated with posteruptive earthquakes: Journal of Geophysical Research, v. 96, p. 11883–11894.

Bautista, B.C., Bautista, M.L.P., Stein, R.S., Barcelona, E.S., Punongbayan, R.S., Laguerta, E.P., Rasdas, A.R., Ambubuyog, G., and Amin, E.Q., this volume, Relation of regional and local structures to Mount Pinatubo activity.

Chouet, B.A., Page, R.A., Stephens, C.D. Lahr, J. C., and Power, J.A., 1994, Precursory swarms of long-period events at Redoubt Volcano (1989–1990), Alaska: Their origin and use as a forecasting tool: Journal of Volcanology and Geothermal Research, v. 62, p. 95–135.

Fehler, M, 1983, Observations of volcanic tremor at Mount St. Helens Volcano: Journal of Geophysical Research, v. 88, p. 3476–3484.

Filson, J., Simkin, T., and Leu, L.-K., 1973, Seismicity of a caldera collapse: Galapagos Islands: Journal of Geophysical Research, v. 79, p. 8591–8622.

Harlow, D.H., Power, J.A., Laguerta, E.P., Ambubuyog, G., White, R.A., and Hoblitt, R.P., this volume, Precursory seismicity and forecasting of the June 15, 1991, eruption of Mount Pinatubo.

Hildreth, W., 1983, The compositionally zoned eruption of 1912 in the Valley of Ten Thousand Smokes, Katmai National Park, Alaska: Journal of Volcanology and Geothermal Research, v. 18, p. 1487–1494.

Hirata, T., 1987, Omori's power law aftershock sequences of microfracturing in rock fracture experiments: Journal of Geophysical Research, v. 92, p. 6215–6221.

Hoblitt, R.P., Wolfe, E.W., Scott, W.E., Couchman, M.R., Pallister, J.S., and Javier, D., this volume, The preclimactic eruptions of Mount Pinatubo, June 1991.

Jolly, A.D., Page, R.A., and Power, J.A., 1994, Seismicity and stress in the vicinity of Mt. Spurr volcano, south central Alaska, Journal of Geophysical Research, v. 99, p. 15305–15318.

Kanamori, H., Mori, J., Hauksson, E., Heaton, T.H., Hutton, L.K., and Jones, L.M., 1993, Determination of earthquake energy release and M_L using TERRAscope: Bulletin of the Seismological Society of America v. 83, p. 330–346.

Kieffer, S.W., 1984, Seismicity at Old Faithful Geyser: An isolated source of geothermal noise and possible analogue of volcanic seismicity: Journal of Volcanology and Geothermal Research, v. 22, p. 59–95.

Kisslinger, C., and Jones, L.M., 1991, Properties of aftershock sequences in southern California: Journal of Geophysical Research, v. 96, p. 11947–11958.

McKee, C.O., Wallace, D.A, Almond, R.A., and Talai, B., 1981, Fatal hydro-eruption of Karkar volcano in 1979: Development of a maar-like crater, in Johnson, R.W., ed., Cooke-Ravian volume of volcanological papers: Geological Survey of Papua New Guinea Memoir 10, p. 63–84.

McNutt, S. R., 1994, Volcanic tremor amplitude correlated with eruption explosivity and its potential use in determining ash hazard to aviation, in Casadevall, T.J., ed., Proceedings at the First International Symposium on Volcanic Ash and Aviation Safety: U.S. Geological Survey Bulletin 2047, p. 377–385.

Moore, J.G., and Albee, W.C., 1981, Topographic and structural changes at Mount St. Helens: Large-scale photogrammetric data, in Lipman, P.W., and Mullineaux, D.R., eds., The 1980 eruptions of Mount St. Helens, Washington: U.S. Geological Survey Professional Paper 1250, p. 123–134.

Mori, J., and McKee, C.O., 1987, Outward-dipping ring-fault structure at Rabaul Caldera as shown by earthquake locations: Science, v. 235, p. 193–195.

Mori, J., Eberhart-Phillips, D., and Harlow, D.H., this volume, Three-dimensional velocity structure at Mount Pinatubo, Philippines: Resolving magma bodies and earthquake hypocenters.

Murray, T.L., and Endo, E.T., 1989, A real-time seismic amplitude measurement system (RSAM): U.S. Geological Survey Open-File Report 89–684, 21 p.

Ogata, Y., 1983, Estimation of the parameters in the modified Omori formula for aftershock sequences by the maximum likelihood procedure: Journal of Physics of the Earth, v. 31, p. 115–124.

Okada, H., 1983, Comparative study of earthquake swarms associated with major volcanic activities, in Shimozuru, D., and Yokoyama, I., eds., Arc volcanism: Physics and tectonics: Tokyo, Terra Scientific Publishing Company, p. 43–61.

Pierson, T.C., Daag, A.S., Delos Reyes, P.J., Regalado, M.T.M., Solidum, R.U., and Tubianosa, B.S., this volume, Flow and deposition of posteruption hot lahars on the east side of Mount Pinatubo, July–October 1991.

Power, J.A., Murray, T.L., Marso, J.N., and Laguerta, E.P., this volume, Preliminary observations of seismicity at Mount Pinatubo by use of the Seismic Spectral Amplitude Measurement (SSAM) system, May 13–June 18, 1991.

Ramos, E.G., Laguerta, E.P., and Hamburger, M.W., this volume, Seismicity and magmatic resurgence at Mount Pinatubo in 1992.

Scandone, R. and Malone, S.D., 1985, Magma supply, magma discharge, and readjustment of the feeding system of Mount St. Helens during 1980: Journal of Volcanology and Geothermal Research, v. 23, p. 239–262.

Scott, W.E., Hoblitt, R.P., Torres, R.C., Self, S, Martinez, M.L., and Nillos, T., Jr., this volume, Pyroclastic flows of the June 15, 1991, climactic eruption of Mount Pinatubo.

Simkin, T. and Howard, K.A., 1970, Caldera collapse in the Galapagos Islands, 1968: Science, v. 169, p. 429–437.

Simkin, T., Siebert, L., McClelland, L., Bridge, D., Newhall, C., and Latter, J.H., 1981, Volcanoes of the World, a regional directory, gazetteer, and chronology of volcanism during the last 10,000 years: Stroudsburg, Pa., Hutchinson Ross Publishing Company, 232 p.

Utsu, T., 1961, A statistical study on the occurrence of aftershocks: Geophysical Magazine, v. 30, p. 521–605.

Weaver, C.S., Grant, W.C., Malone, S.D., and Endo, E.T., 1981, Post-May 18 seismicity: Volcanic and tectonic implications, in Lipman, P.W., and Mullineaux, D.R., eds., The 1980 eruptions of Mount St. Helens, Washington: U.S. Geological Survey Professional Paper 1250, p. 102–122.

Yokoyama, I., Yamashita, H., Watanabe, H., and Okada, Hm., 1981, Geophysical characteristics of dacite volcanism—the 1977–1978 eruption of Usu volcano: Journal of Volcanology and Geothermal Research, v. 9, p. 335–358.

Relationship of Regional and Local Structures to Mount Pinatubo Activity

By Bartolome C. Bautista,[1] Ma. Leonila P. Bautista,[1] Ross S. Stein,[2] Edito S. Barcelona,[1] Raymundo S. Punongbayan,[1] Eduardo P. Laguerta,[1] Ariel R. Rasdas,[1] Gemme Ambubuyog,[1] and Erlinda Q. Amin[1]

ABSTRACT

The spatial and temporal proximity of the 1990 M_s 7.8 Luzon earthquake and reawakening of Mount Pinatubo in 1991 hints at the possibility of a relation between the two events. Composite focal-plane solutions for preeruption and posteruption microearthquakes, fault plane solutions for M>5 syneruption events, and information about the local and regional faults suggest that movement along these regional faults preceded, accompanied, and followed the major eruption of Mount Pinatubo in June 1991.

Changes in Coulomb failure stress along faults of the Pinatubo area, as a result of the Luzon earthquake, were on the order of 0.1 bar and probably were not a cause of Pinatubo's reawakening. However, compressive stress on the magma reservoir and its roots was about 1 bar, possibly enough to squeeze a small volume of basalt into the overlying dacitic reservoir. Alternatively, strong ground shaking associated with the Luzon earthquake might have done the same or triggered movement along previously stressed faults that in turn allowed magma ascent.

INTRODUCTION

When Mount Pinatubo reawoke in 1991 from a 500-yr slumber, many scientists asked, "Was the 1991 eruption triggered by, or otherwise related to, the M_s 7.8 Luzon earthquake that had occurred on July 16, 1990?" One way to address this question is through the study of earthquakes that occurred after the July 16, 1990, Luzon earthquake and before, during, and after the big eruption. To test for a relation, we asked:

1. Did aftershocks of the 1990 Luzon earthquake occur at and near Mount Pinatubo?

2. How did seismicity that led to, accompanied, and followed the eruption relate to local and regional faults and to pre-1990 seismicity?

3. Was predicted stress change at Pinatubo from the July 16 earthquake of a sense and magnitude that might have triggered magma ascent?

REGIONAL AND LOCAL STRUCTURES

Mount Pinatubo is a part of the Luzon arc, whose volcanism is related to the activity of the Manila trench, located about 120 km west of the volcano (fig. 1). The trench is a product of the active subduction of the South China Sea Plate beneath Luzon island. Convergence is roughly east-west.

The northwest-trending, left-lateral Philippine fault passes northeast of Mount Pinatubo, and on July, 16, 1990, a 125-km-long segment of this fault ruptured and produced a M_s 7.8 earthquake. The epicenter of this earthquake was about 100 km northeast of Mount Pinatubo.

Closer to Mount Pinatubo, de Boer and others (1980) suggested a northwest-trending Iba fracture zone, possibly related to differential movement of segments in subduction beneath the Manila trench and Zambales Province. Delfin (1984) mapped a local structure, the Maraunot fault (fig. 1), that is probably an extension of the Iba fracture zone beneath Mount Pinatubo. The Maraunot fault trends N.30°W. and generally follows the trend of the Maraunot River and the thermal springs (now covered by the 1991 deposits) on the northwest flank of the volcano.

A northeast-trending lineament of comparable geomorphic expression, here called the Sacobia lineament, was discovered from SLAR imagery in 1991 (Newhall and others, this volume). Subparallel to the Sacobia lineament and <1 km to its north, having a slightly more northerly trend, was a 3-km-long, northeast-trending fracture that opened across the north face of Mount Pinatubo during the initial explosions in April 1991 (Ewert and others, this volume). Extension across the latter fracture occurred on April 2 but apparently not thereafter; tape measurements of a

[1] Philippine Institute of Volcanology and Seismology.
[2] U.S. Geological Survey.

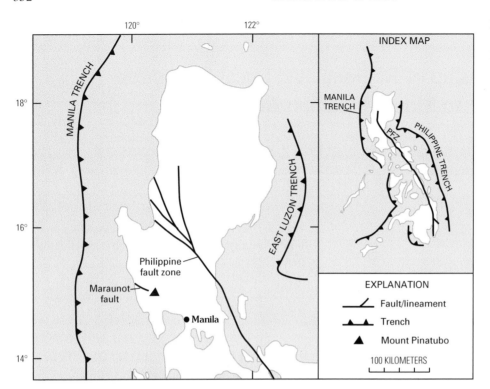

Figure 1. Regional tectonic setting of Mount Pinatubo. PFZ, Philippine fault zone.

quadrilateral across the fracture from May 1 to May 30 did not detect any additional movement along this fracture (Ewert and others, this volume). The Maraunot fault and the Sacobia lineament intersect at Mount Pinatubo.

Delfin (1984) described a subcircular caldera that fully enclosed the pre-1991 Pinatubo dome. The caldera, now termed the Tayawan caldera, measures 4.5 km by 3.5 km and is believed to have been formed >35,000 yr B.P. (Newhall and others, this volume). Delfin (1984) postulated that formation of the Tayawan caldera by "piecemeal and chaotic" collapse resulted in an ill-defined caldera wall.

A new, 2.5-km-wide caldera formed during the June 15, 1991, eruption. Clearly, this structure was not involved in preeruption seismicity; it might have been involved in syneruption and posteruption seismicity. Blocks may have collapsed along ring faults, or, alternatively, collapse may have been in such small pieces that no major faults formed.

METHODOLOGY

Earthquakes that occurred in the Pinatubo area during 1991 and 1992 were recorded digitally from the local PHI-VOLCS-USGS telemetry network (Murray and others, this volume; Lockhart and others, this volume). The PCEQ picking program of Valdes (1989) was used to obtain the phase arrival times of the events, initial motions of the primary waves (P-waves), and duration and quality of the arrival times. A PC version of the Hypo71 location program (Lee and Lahr, 1975; Lee, 1989) was also utilized to obtain hypocentral locations of events, magnitudes and other

parameters such as quality and fitness of the determination, and focal mechanism parameters including take-off angle, azimuth of station, and polarity of the P-wave initial motion.

We plotted hypocentral data on maps and sections, together with digitized topographic and geologic data, in order to determine the spatial distribution of seismicity and possible correlation with local structures in the area. Separate maps were drawn for preeruption seismicity (fig. 2) and various clusters of posteruption seismicity (see fig. 6).

We tried to plot double-couple focal mechanism solutions of individual events but found it difficult to obtain well-constrained focal mechanism solutions, owing to the very small number of seismic stations. Instead, we wrote software (Bautista and Narag, 1992) that will read the print file (PRT) output of Hypo71pc and display composite first-motion plots (upper hemisphere stereographic projection) for every cluster of seismicity. We then used the known fault geometry of Newhall and others (this volume), supplemented by any linear patterns of epicenters, to select the most likely nodal planes.

Uncertainties in composite focal mechanisms are high, because they require an assumption that earthquakes used in the composite originated from the same source region and have similar mechanisms (Rivera and Cisternas, 1990). With small-magnitude earthquakes (M<2.5), there is a possibility that events may represent deformation due to complex interaction of small faults and not from deformation caused by the regional stress field (Zoback, 1992). There is also the danger that events with two differing mechanisms

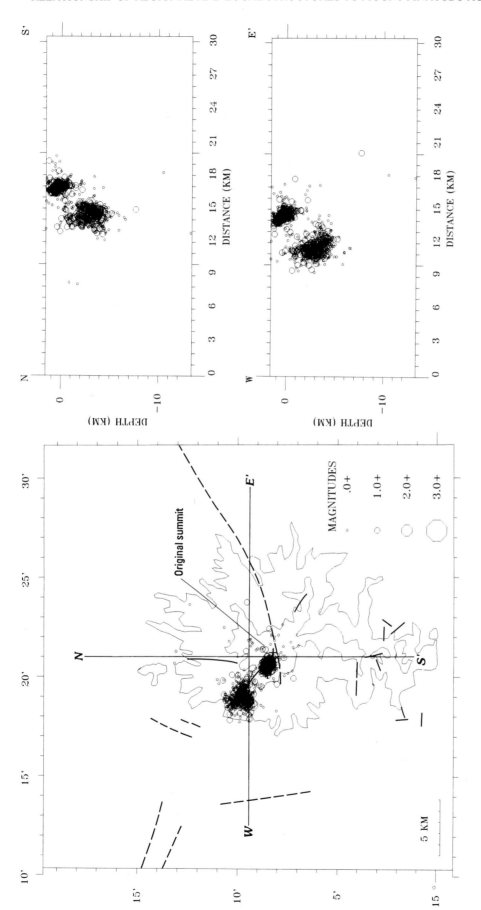

Figure 2. Preeruption seismicity at and near Mount Pinatubo, May 18–June 8, 1991. Events from 0 to 30 km in depth are shown in map view (scale 1:250,000); events from 0 to 15 km in depth are shown in cross sections. Sections show events to 10 km on each side of N–S and E–W. Zero depth indicates sea level. All events shown have RMS≤0.15. Unbroken and broken heavy lines are faults or lineaments from Newhall and others (this volume). Compare with figures 6 and 13.

may be combined and may result in an erroneous third mechanism. We would have preferred, of course, to have had simultaneous recording of single events at many stations and thus to have avoided use of composite mechanisms. However, small events and the limited network were all that we had.

To minimize the previously mentioned uncertainty, we designed plotting software in such a way that the spatial (latitude, longitude, and depth range) and time sampling range can be iteratively varied. This allowed us to observe how the mechanism of events was varying spatially as the sampling window was moved within the cluster being analyzed and also how the mechanism was changing through time. By plotting only events with impulsive first motions, we minimized the possibility of plotting incorrectly read first-motion polarity. Then, to prevent events with erroneous azimuth and take-off angles from obscuring the composite plots, we filtered out events with >1-km horizontal and >2-km vertical errors and >0.5 root mean square (rms) in hypocentral location. Any differences in mechanism between large and small earthquakes were also examined by varying the magnitude range.

The best-fit solutions were then manually correlated with mapped structures and linear features in the distribution of hypocenters, and the most likely nodal planes were selected. The program was then rerun, and events that did not fit the solution were discarded. The inconsistency ratio (Xu and others, 1992), an expression of the fitness of the solution, was then computed. Small inconsistency ratios (<0.3) indicated a reasonably reliable solution. Theoretical analysis of stress change at Mount Pinatubo resulting from the 1990 Luzon earthquake was by the method of Stein and others (1992).

LUZON EARTHQUAKE AFTERSHOCKS NEAR PINATUBO

Most aftershocks of the Luzon earthquake were concentrated north and northwest of the epicenter, and only six apparent aftershocks are known to have occurred in the Pinatubo area (table 1) (Bautista and others, 1992). A few hours after the M_s 7.8 mainshock of the July 16, 1990, Luzon earthquake, an M 4.8 "aftershock" occurred about 10 km southeast of Pinatubo's summit dome. The precise location is not known; uncertainty in National Earthquake Information Center (NEIC) locations for this area is about ±15 km (W. Person, USGS, oral commun., 1991). During the following days, several other earthquakes were recorded and felt around the volcano (table 1). Three weeks later, a nun who was working with the Aeta minority group of Zambales reported rumbling sounds and a landslide measuring about 2–3 ha on the upper northwest face of the volcano (Sabit and others, this volume). The rumbling sounds and landslide were attributed to aftershocks of the tectonic

Table 1. List of earthquakes around Mount Pinatubo area immediately after the July 16, 1990, Luzon earthquake.

[Local time = G.m.t. + 8 h; NEIC, National Earthquake Information Center, USGS, Denver, Colo.; PHIVOLCS, Philippine Institute of Volcanology and Seismology, Quezon City, Philippines; 33 km is a NEIC default for crustal depths]

Date	Time (G.m.t.)	Coordinates (lat/long)	Magnitude	Depth (km)	Data source
7/16/90	0825	15.46N/120.76E	4.6	33	NEIC
7/16/90	1015	15.11N/120.41E	4.8	33	NEIC
7/22/90	0221	15.38N/120.61E	4.3	33	NEIC
7/23/90	2025	15.39N/120.63E	3.0	6	PHIVOLCS
8/8/90	1938	15.16N/120.52E	4.4	10	NEIC
8/26/90	1058	15.29N/120.40E	3.6	6	PHIVOLCS

earthquake (Ramos and Isada, 1990). However, the geographic distance of events in table 1 from the mainshock and from most other aftershock activity suggests, alternatively, that they might have been earthquakes akin to those triggered in distant volcanic areas by the M 7.4 Landers earthquake, apparently by transient interaction of seismic waves and previously stressed faults, especially in the presence of pore water or magma (Hill and others, 1993).

COMPOSITE FOCAL MECHANISMS

PREERUPTION SEISMICITY

In early April 1991, in response to reports of unusual activity, PHIVOLCS installed a small network consisting of four portable digital seismographs and found an epicentral cluster located approximately 6 km northwest of the summit dome (Sabit and others, this volume). A network of 7 telemetered seismometers was installed during late April and the first week of May (Lockhart and others, this volume), and results from the first month of telemetry data confirmed the same northwest cluster (Power and others, this volume; Harlow and others, this volume) (fig. 2). Then, from May 27 to June 5, 1991, another cluster of hypocenters developed 1 to 4 km below the volcano summit (fig. 2). Seismic activity in the northwest cluster slowed but did not stop entirely. Meanwhile, seismicity below the summit intensified (Power and others, this volume; Harlow and others, this volume).

A comparison of focal mechanisms from the northwest and summit clusters could help us understand their origins and explain why only the summit cluster culminated in an eruption. Figure 3 is a composite first-motion plot for events of the northwest cluster, excluding those with emergent onsets and those whose solutions did not fit a preliminary match of solutions and known faults. The sampling window and other parameters used in the plots are displayed at the left of the figure. Because this cluster is bisected by the northwest-trending Maraunot fault

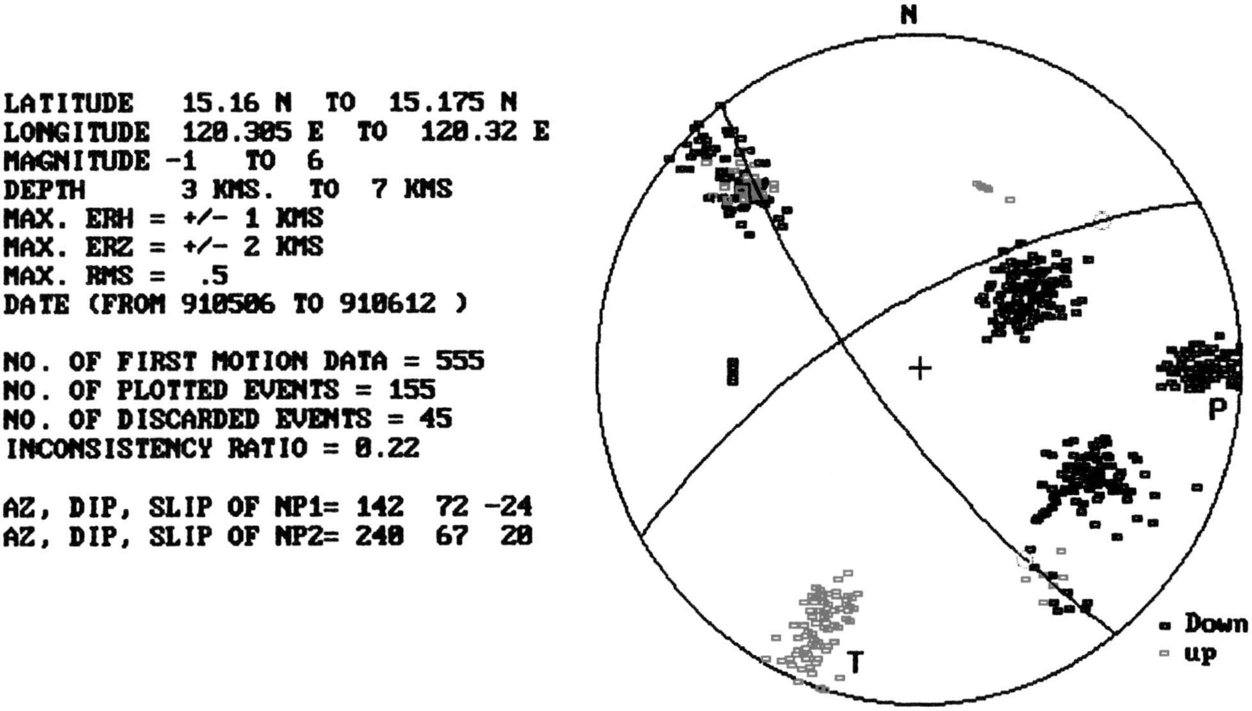

```
LATITUDE    15.16 N  TO   15.175 N
LONGITUDE  120.305 E  TO  120.32 E
MAGNITUDE  -1   TO   6
DEPTH       3 KMS.  TO   7 KMS
MAX. ERH = +/- 1 KMS
MAX. ERZ = +/- 2 KMS
MAX. RMS = .5
DATE (FROM 910506 TO 910612 )

NO. OF FIRST MOTION DATA = 555
NO. OF PLOTTED EVENTS = 155
NO. OF DISCARDED EVENTS = 45
INCONSISTENCY RATIO = 0.22

AZ, DIP, SLIP OF NP1= 142  72  -24
AZ, DIP, SLIP OF NP2= 240  67   20
```

Figure 3. Focal mechanism solutions for 155 preeruption earthquakes in the northwest cluster (see fig. 2). Plots are upper hemisphere stereographic projections. P, axis of maximum compressional stress; T, axis of minimum compressional (maximum tensional) stress. N, geographic north. NP1, first nodal plane; NP2, second nodal plane; ERH, horizontal error; ERZ, vertical error; RMS, root mean square.

(Delfin, 1984; Delfin and others, this volume), we constrained the first nodal plane (NP1) in figure 3 to be parallel to the fault trend (azimuth about 142°), with a steep trial dip of 72° northeast. Interestingly, the plane cuts through the two clusters of station first-motion plots with highly mixed polarities. Mixed polarities are commonly observed when the location of the station is aligned with the trend of the causative fault. However, points outside these clusters have either negative or positive first motions, in an interpretable pattern. Tentatively, we infer left-lateral strike slip along the northwest-trending Maraunot fault, with the maximum and minimum stress axes oriented east-west and north-south, respectively.

Figure 4 is a plot for the summit cluster. In figure 4, all six station clusters showed highly mixed first-motion polarities. Thus, events of the summit cluster have diverse mechanisms, which are characteristic of earthquakes resulting from magmatic intrusion.

SYNERUPTION SEISMICITY

The major explosive eruptions of Mount Pinatubo started on June 12 and peaked on June 15. Starting June 12, the remote stations of the seismic network began to fail from thick ash fall and pyroclastic flows. Only two seismographs on Pinatubo's east slope (PIEZ and CABZ) remained in operation until the start of the climactic

eruption on June 15, and, by 1415 on June 15, only CABZ remained.

Starting at about 1530 on June 15, quakes began to be felt one after another at Magalang, Pampanga, located about 30 km east of the volcano. Signals recorded by most of PHIVOLCS national seismic stations were saturated with eruption signals and thus failed to give good P-wave arrival data for epicentral or first-motion determinations. Fortunately, the NEIC was able to determine the epicenters of 32 events during the June 15–16 period by using data from participating seismic stations in the world, including those from the Philippines. All events could have been within 40 km of Mount Pinatubo. The biggest of the June 15 events occurred at 1915 (M_b 5.5).

NEIC determined focal mechanisms of the seven largest events (1539, 1841, 1911, 1915, and 2025 on June 15; 0348 and 0358 on June 16). Six of these events (all but that at 0348 on June 16) had predominantly strike-slip movement (fig. 5) with a minor vertical (normal or reverse) component. The event at 0348 on June 16, however, was normal faulting with a large strike-slip component. All events were poorly constrained because of the sparsity of the data and because first-motion reports were mostly emergent in nature. In addition, for NEIC to make reliable teleseismic fault-plane solutions, earthquakes generally need to have magnitudes of at least 5.8.

These results, although poorly constrained, suggest that the largest quakes on June 15 were strike-slip events

DATE.. 910610

LATITUDE   15.14 N   TO   15.16 N
LONGITUDE   120.325 E   TO   120.35 E
MAGNITUDE -1   TO   6
DEPTH   1 KMS.   TO   6 KMS
MAX. ERH = +/- 1 KMS
MAX. ERZ = +/- 2 KMS
MAX. RMS = .5
DATE (FROM 910506 TO 910612)

NO. OF FIRST MOTION DATA = 290
NO. OF PLOTTED EVENTS = 100
NO. OF DISCARDED EVENTS = 0
INCONSISTENCY RATIO = 0.00

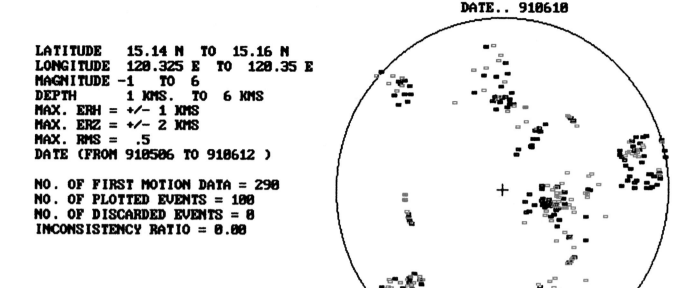

Figure 4. First-motion plots for 100 preeruption earthquakes in the summit cluster (see fig. 3). Upper hemisphere projections; conventions as in figure 3. Emergent first-motions removed. No consistent focal mechanism determined.

consistent with the east-west maximum regional stress. This also implies that regional tectonic adjustments accompanied and immediately followed the caldera-forming eruption. Apparently, tectonic adjustments occurred in response to the abrupt drop in pressure inside the magma chamber.

Contrary to our interpretation at the time and some interpretations elsewhere (for example, Filson and others, 1973; Hirn and others, 1991), the largest earthquakes were not due to caldera collapse. Among the largest earthquakes, only the 0348 event on June 16 might have been a collapse event, but independent stratigraphic evidence (W.E. Scott and others, this volume) suggests that collapse began in the late afternoon of June 15. However, there were also many smaller earthquakes that were difficult to locate and whose first motions are difficult to decipher because of an emergent character of the waveform, and some of these quakes could have been associated with caldera collapse.

POSTERUPTION SEISMICITY

After restoration of the seismic telemetry network (by June 29), we noticed a shift in locations of seismicity relative to preeruption sources. Earthquakes were distributed over a wide area outside and away from the active volcanic center (Mori, White, and others, this volume). Seven prominent posteruption earthquake clusters, labeled A–F, are shown in figure 6.

Cluster A—Northwest Cluster.—A cluster of earthquakes (A, fig. 6) occurred northwest of Mount Pinatubo but slightly south and deeper than the preeruption northwest cluster. We cannot say whether this shift is real or is due to the change in the configuration of the newly restored network. However, the large change in the hypocentral range and the presence of a deeper clusters of hypocenters suggest to us that the shift was real and that earthquakes were occurring in a new source zone.

The absence of a station northwest of Pinatubo in the new network increased uncertainty in first-motion solutions, especially that for cluster A. Of several solutions that we tried, that which gave the lowest inconsistency ratio (fig. 7) has one nodal plane bisecting the two first-motion clusters with mixed polarities and implies strike-slip motion with a moderate amount of normal slip. The maximum stress axis is oriented east-northeast. Mechanisms of earthquakes with M<1 gave the same solution as those with M>1.

Cluster B—Summit Cluster.—The preeruption summit cluster was still present after the eruption but was tightly concentrated in a small area close to the southwest rim of the new caldera (B, fig. 6). Figure 8 is the first-motion plot for the posteruption summit cluster. As during the preeruption period, the earthquakes in this cluster show very diverse mechanisms, and no predominant mechanism can be recognized from the first-motion plot.

Cluster C—East-northeast Cluster.—The most pronounced cluster of all (C, fig. 6), developed 2 to 3 km east-northeast of the new caldera between 0 and 5 km depth. This cluster completely masked the location of the northern trace of the Tayawan caldera, the northeast-trending fracture zone that formed on April 2, 1991, and that part of the Sacobia lineament (Newhall and others, this volume) that is just northeast of the 1991 caldera.

Starting August 25 (and especially after September 9), 1991, there was a noticeable shift in the polarity of first motions at several stations that suggests a change of mechanism. In figure 9A, we eliminated all events after this shift and found a strike-slip solution with east-west compression. After August 25, the solution shifted slightly (fig. 9B).

The above results indicate that right-lateral movement along one or more northeast-trending faults was responsible for this cluster. Some of that movement might have been along the fracture that opened on April 2, 1991, but it is unclear why there was no seismicity along this feature between April 5 and the June eruptions. Alternatively, movement might have been along the Sacobia lineament (fig. 6).

Cluster D—Southeast Cluster.—About 5 km southeast of the 1991 caldera rim, another dense cluster extends from 7 to 15 km in depth (D, fig. 6). If we view this cluster in east-west cross section, as is also shown in figure 6, the outwardly dipping structure resembles a dike that, if projected toward the surface, intersects the southeast trace of the Tayawan caldera. If this activity is related to the movement along the old caldera ring fracture, it apparently was limited to a short segment of that ring fracture, unlike movement along the whole ring structure of Rabaul caldera, Papua New Guinea (McKee and others, 1984; Mori and McKee, 1986). The focal mechanism for this cluster is not well constrained but is probably normal faulting with a small strike-slip component (fig. 10A) or pure normal faulting (fig. 10B). Choosing nodal plane 1 in figure 10B as the fault plane (strike 195°, dip 60°W.) will make it consistent with the spatial trend of the earthquake cluster.

Cluster E—South-southeastern Cluster.—Another prominent cluster is located approximately 7 km south-southeast of the 1991 caldera, in the shallow depth range of 0 to 4 km (E, fig. 6). There are at least three possible solutions for this cluster. One possible solution is a strike-slip mechanism (fig. 11A) with P-axis oriented east-west. The second solution (fig. 11B) is movement along an east-northeast striking normal fault. The third solution (fig. 11C) is also along a normal fault, but this time the strike is north-northeast. We have no basis on which to choose any one of these solutions over another, except to say that the first is the most consistent with mechanisms observed in other clusters.

Cluster F—Northeast, Clark Air Base Cluster.—The Clark Air Base cluster (F, fig. 6) consists of events that were scattered along a northeast-trending elongate area about 15 km northeast of the summit. Most of the earthquakes in this cluster were felt at the Pinatubo Volcano Observatory on Clark Air Base. Figure 12 shows a strike-slip solution with maximum stress axis oriented close to east-west. The above result suggests right-lateral movement along a northeast-trending fault, probably that seen as a strong lineament in SLAR imagery of the Sacobia River (fig. 4 of Newhall and others, this volume).

Starting in January 1992, the character of seismicity at Pinatubo changed (fig. 13). The activity in some of the clusters that were very active in 1991 diminished. In contrast, the seismicity at the summit cluster further intensified before and during growth of a dome in July–October 1992 (Ramos and others, this volume), and again at the time of this writing (February–May 1994).

THEORETICAL ANALYSIS OF STRESS CHANGE RESULTING FROM 1990 LUZON EARTHQUAKE

The magnitude of stress changes resulting from the Luzon earthquake is shown in figure 14 (Coulomb stress change). For an assumed coefficient of friction of 0.4, the rise in Coulomb stress would have been between 0.1 and 0.2 bar along optimally oriented faults of the Pinatubo area (thrust faults striking north-south and dipping 25–35°E. or 30–40°W.; vertical strike-slip faults striking N.25–35°E. and N.25–35°W. This is one order of magnitude higher than tidal stress and somewhat lower than that thought to have triggered aftershocks of the Landers earthquake at a comparable distance (Stein and others, 1992).

In contrast, volumetric stress change (fig. 15) would have subjected Pinatubo to a 1-bar increase in compressive stress. This is about 50 times the daily tidal pressure change.

Were such changes in Coulomb or volumetric stress sufficient to trigger magma ascent beneath Pinatubo? We know of no well-defined threshold of stress change that can be said to trigger magma ascent. Empirically, there are a number of instances of eruptions shortly after large regional earthquakes (some are listed in Newhall and Dzurisin, 1988), so we think that a threshold can, apparently, be exceeded. We can imagine that an increase in compressive stress might have squeezed magma upward, either from the main reservoir into the overlying volcanic edifice or from a deeper, possibly disseminated reservoir of basaltic magma into the base of the main dacitic body. The first of these cases was suggested by Nakamura (1971), though questioned by Rikitake and Sato (1989). The small diameter of Pinatubo's conduit (approximately 100 m; Hoblitt, Wolfe, and others, this volume) relative to the diameter of the magma reservoir (about 5 km, Mori, Eberhardt-Phillips, and

6/15/91 LUZON, PHILIPPINE IS.
07:39:09.5   5.1 mb

	AZIMUTH	PLUNGE
P axis	75.2	18.8
T axis	171.3	17.3
B axis	301.0	64.0
X axis	123.5	26.0
Y axis	33.0	1.0

123.00	89.00	-26.00
213.49	64.00	-178.89

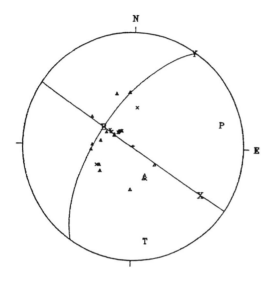

6/15/91 LUZON. PHILIPPINE IS.
10:41:14.2   5.5 mb

	AZIMUTH	PLUNGE
P axis	263.9	4.2
T axis	171.9	26.0
B axis	2.4	63.8
X axis	221.0	21.2
Y axis	125.0	15.0

215.00	75.00	158.00
310.97	68.76	16.12

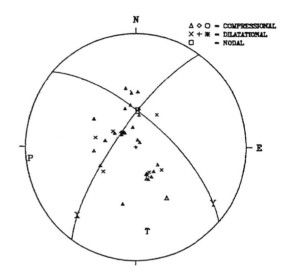

△ ◇ □ = COMPRESSIONAL
✕ + ✳ = DILATATIONAL
□ = NODAL

6/15/91 LUZON,PHILIPPINE IS.
11:11:18.4   5.0 mb

	AZIMUTH	PLUNGE
P axis	262.7	1.6
T axis	172.0	23.2
B axis	356.5	66.7
X axis	219.8	17.4
Y axis	125.0	15.0

215.00	75.00	162.00
309.81	72.63	15.73

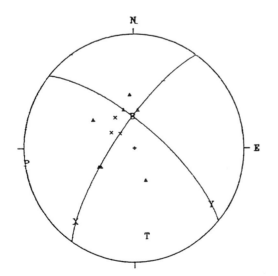

6/15/91 LUZON, PHILIPPINE IS..
11:15:28.0   5.7 mb

	AZIMUTH	PLUNGE	AZIMUTH	PLUNGE
P axis	263.9	4.2	99.0	0.0
T axis	171.9	26.0	9.0	0.0
B axis	2.4	63.6	0.0	90.0
X axis	221.0	21.2	54.0	0.0
Y axis	125.0	15.0	144.0	0.0

215.00	75.00	158.00	234.00	90.00	-180.00
310.97	68.79	16.12	324.00	90.00	0.00

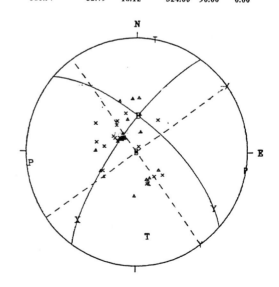

6/15/91 LUZON, PHILIPPINE IS.
12:25:30.4   5.0 mb

	AZIMUTH	PLUNGE
P axis	85.1	1.6
T axis	176.2	34.7
B axis	352.8	55.2
X axis	226.1	22.5
Y axis	125.0	25.0

215.00	65.00	155.00
316.15	67.48	27.23

6/15/91 LUZON. PHILIPPINE IS.
19:48:53.5   5.0 mb

	AZIMUTH	PLUNGE
P axis	94.0	50.8
T axis	330.8	24.1
B axis	226.5	28.9
X axis	13.9	56.8
Y axis	128.0	15.0

218.00	75.00	-120.00
103.85	33.23	-28.19

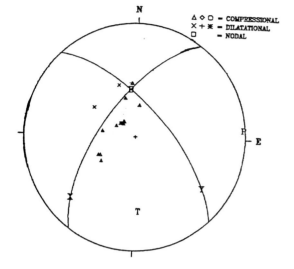

6/15/91 LUZON, PHILIPPINE IS.
19:58:35.1   4.9 mb

	AZIMUTH	PLUNGE
P axis	76.7	16.1
T axis	170.2	12.0
B axis	295.3	69.8
X axis	33.0	2.8
Y axis	124.0	20.0

214.00	70.00	183.00
122.97	87.18	339.97

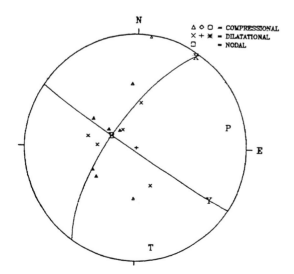

Figure 5. Focal mechanism solutions for seven syneruption earthquakes. Data and solutions from National Earthquake Information Center, USGS. Times are G.m.t.; local time=G.m.t. plus 8 h. Except for the event at 1948 G.m.t., all other events were strike-slip events along northwest- or northeast-trending faults.

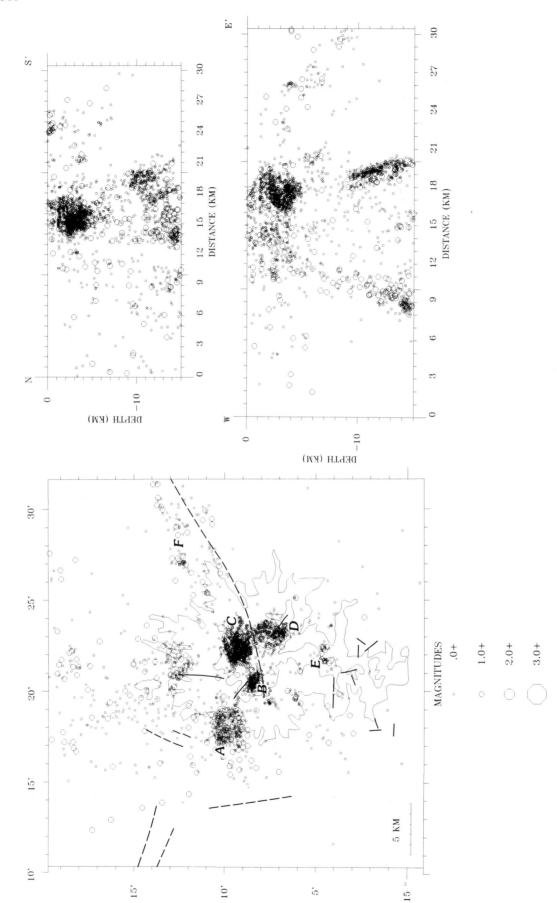

Figure 6. Posteruption seismicity at and near Mount Pinatubo, June 26 to August 19, 1991. Events from 0 to 30 km in depth are shown in map view; events from 0 to 15 km in depth are shown in cross sections. Zero depth is at the vent. All events shown have RMS≤0.15. Heavy lines are faults or lineaments from Newhall and others (this volume). Six clusters, labeled A–F, are described individually in the text and in figures 7–12. Cross sections are shown in figure 2.

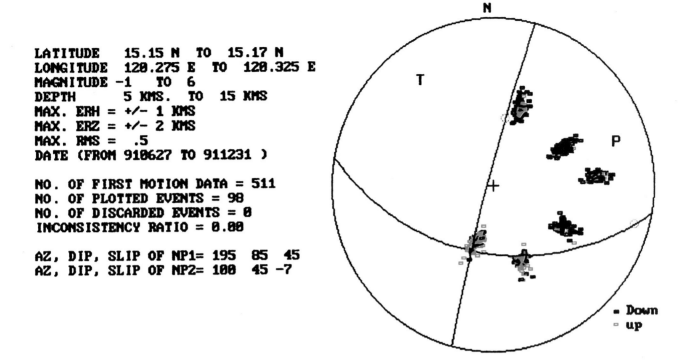

```
LATITUDE    15.15 N  TO  15.17 N
LONGITUDE  120.275 E  TO  120.325 E
MAGNITUDE -1   TO  6
DEPTH       5 KMS.  TO  15 KMS
MAX. ERH = +/- 1 KMS
MAX. ERZ = +/- 2 KMS
MAX. RMS =  .5
DATE (FROM 910627 TO 911231 )

NO. OF FIRST MOTION DATA = 511
NO. OF PLOTTED EVENTS = 98
NO. OF DISCARDED EVENTS = 0
INCONSISTENCY RATIO = 0.00

AZ, DIP, SLIP OF NP1= 195  85  45
AZ, DIP, SLIP OF NP2= 100  45 -7
```

Figure 7. Focal mechanism solutions for posteruption cluster A. Conventions as in figure 3.

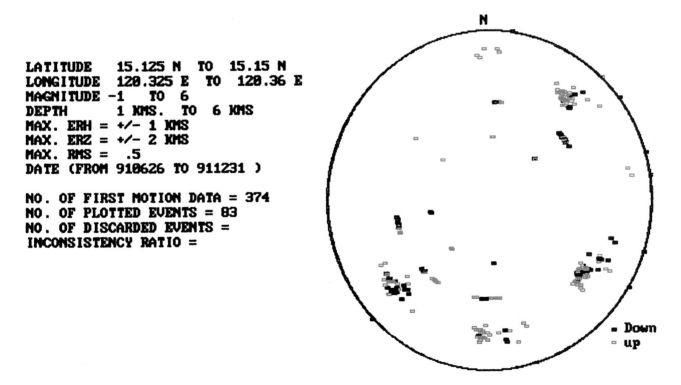

```
LATITUDE    15.125 N  TO  15.15 N
LONGITUDE  120.325 E  TO  120.36 E
MAGNITUDE -1   TO  6
DEPTH       1 KMS.  TO  6 KMS
MAX. ERH = +/- 1 KMS
MAX. ERZ = +/- 2 KMS
MAX. RMS =  .5
DATE (FROM 910626 TO 911231 )

NO. OF FIRST MOTION DATA = 374
NO. OF PLOTTED EVENTS = 83
NO. OF DISCARDED EVENTS =
INCONSISTENCY RATIO =
```

Figure 8. First-motion plot for posteruption cluster B. Conventions as in figure 3. No consistent mechanism emerged.

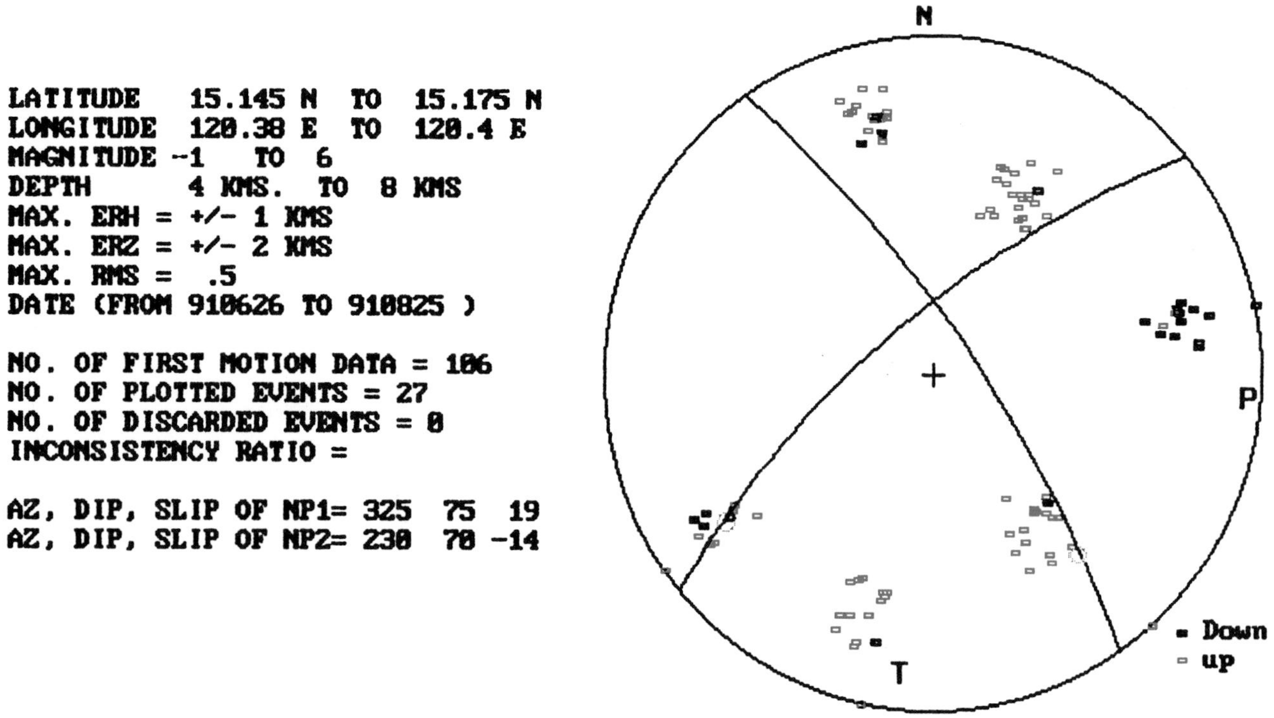

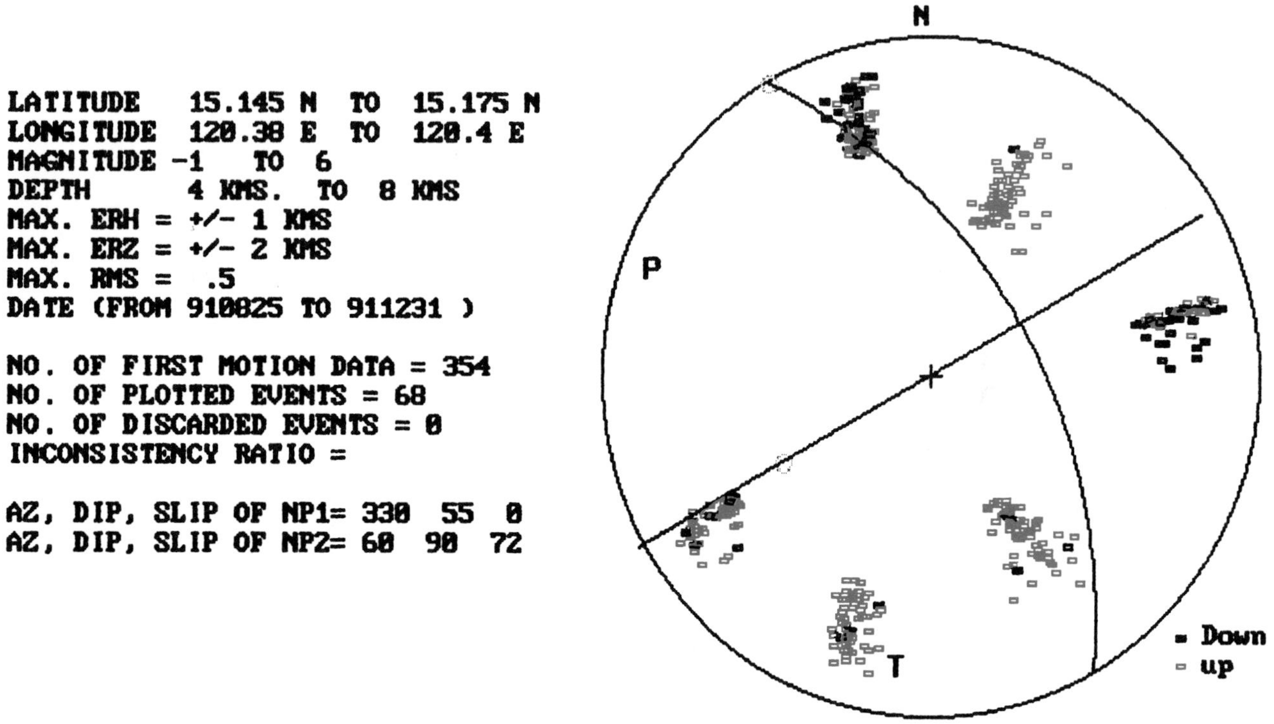

```
LATITUDE    15.145 N   TO   15.175 N
LONGITUDE  120.38 E   TO   120.4 E
MAGNITUDE -1    TO  6
DEPTH      4 KMS.  TO  8 KMS
MAX. ERH = +/- 1 KMS
MAX. ERZ = +/- 2 KMS
MAX. RMS =  .5
DATE (FROM 910626 TO 910825 )

NO. OF FIRST MOTION DATA = 106
NO. OF PLOTTED EVENTS = 27
NO. OF DISCARDED EVENTS = 0
INCONSISTENCY RATIO =

AZ, DIP, SLIP OF NP1= 325   75   19
AZ, DIP, SLIP OF NP2= 230   70  -14
```

```
LATITUDE    15.145 N   TO   15.175 N
LONGITUDE  120.38 E   TO   120.4 E
MAGNITUDE -1    TO  6
DEPTH      4 KMS.  TO  8 KMS
MAX. ERH = +/- 1 KMS
MAX. ERZ = +/- 2 KMS
MAX. RMS =  .5
DATE (FROM 910825 TO 911231 )

NO. OF FIRST MOTION DATA = 354
NO. OF PLOTTED EVENTS = 68
NO. OF DISCARDED EVENTS = 0
INCONSISTENCY RATIO =

AZ, DIP, SLIP OF NP1= 330   55   0
AZ, DIP, SLIP OF NP2= 60   90   72
```

Figure 9.   First-motion and focal mechanism plots for posteruption cluster C. Conventions as in figure 3. *A*, Events from June 26 to August 25, before an apparent shift in focal mechanisms for this cluster; *B*, Events from August 25 through December 31.

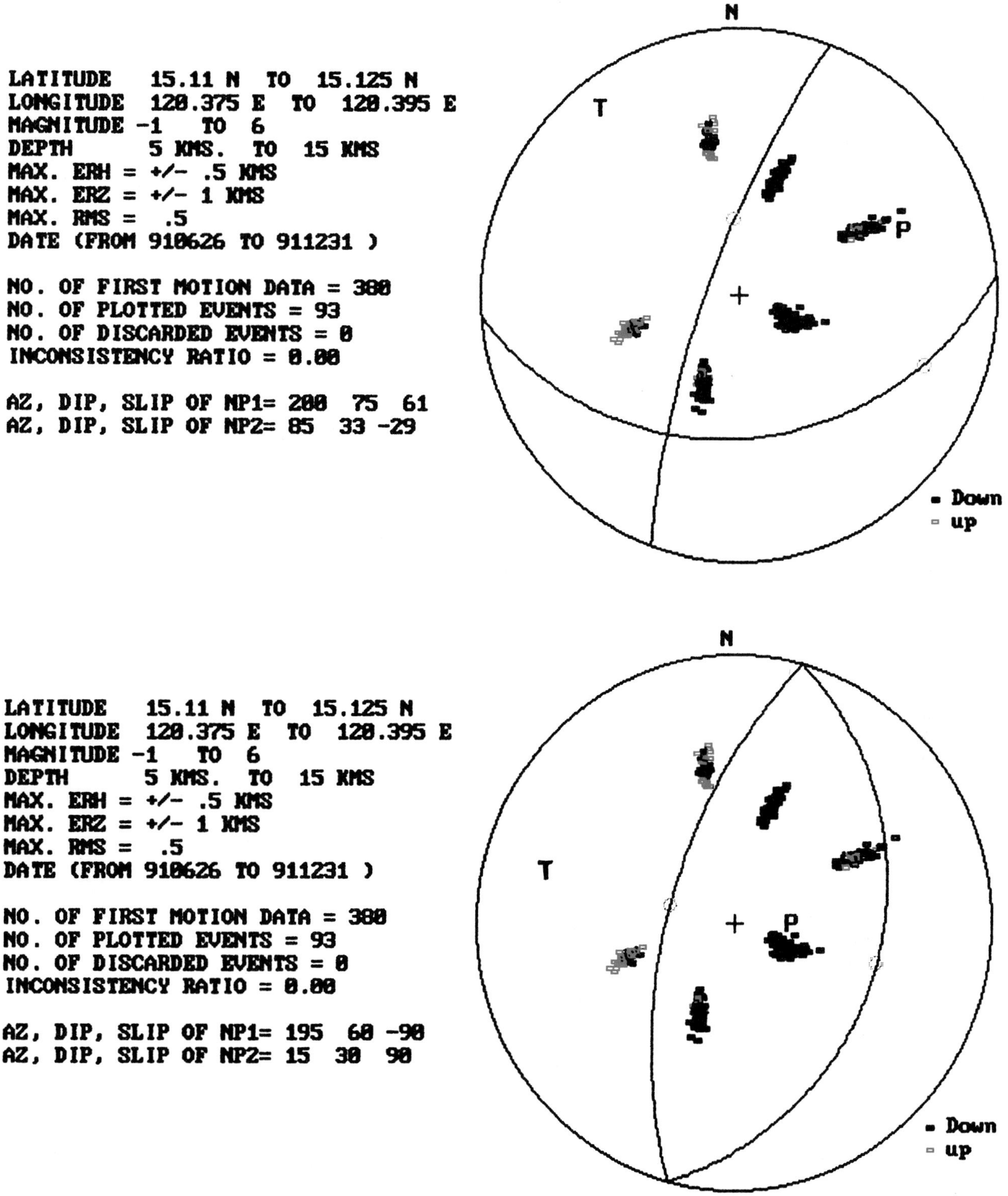

```
LATITUDE    15.11 N  TO  15.125 N
LONGITUDE  120.375 E  TO  120.395 E
MAGNITUDE -1   TO  6
DEPTH       5 KMS.  TO  15 KMS
MAX. ERH = +/- .5 KMS
MAX. ERZ = +/- 1 KMS
MAX. RMS =  .5
DATE (FROM 910626 TO 911231 )

NO. OF FIRST MOTION DATA = 380
NO. OF PLOTTED EVENTS = 93
NO. OF DISCARDED EVENTS = 0
INCONSISTENCY RATIO = 0.00

AZ, DIP, SLIP OF NP1= 200  75  61
AZ, DIP, SLIP OF NP2= 85  33 -29
```

```
LATITUDE    15.11 N  TO  15.125 N
LONGITUDE  120.375 E  TO  120.395 E
MAGNITUDE -1   TO  6
DEPTH       5 KMS.  TO  15 KMS
MAX. ERH = +/- .5 KMS
MAX. ERZ = +/- 1 KMS
MAX. RMS =  .5
DATE (FROM 910626 TO 911231 )

NO. OF FIRST MOTION DATA = 380
NO. OF PLOTTED EVENTS = 93
NO. OF DISCARDED EVENTS = 0
INCONSISTENCY RATIO = 0.00

AZ, DIP, SLIP OF NP1= 195  60 -90
AZ, DIP, SLIP OF NP2= 15  30  90
```

Figure 10. Focal mechanism solutions for posteruption cluster D. Conventions as in figure 3. *A,* All events; *B,* Same as *A,* with alternative focal plane solution.

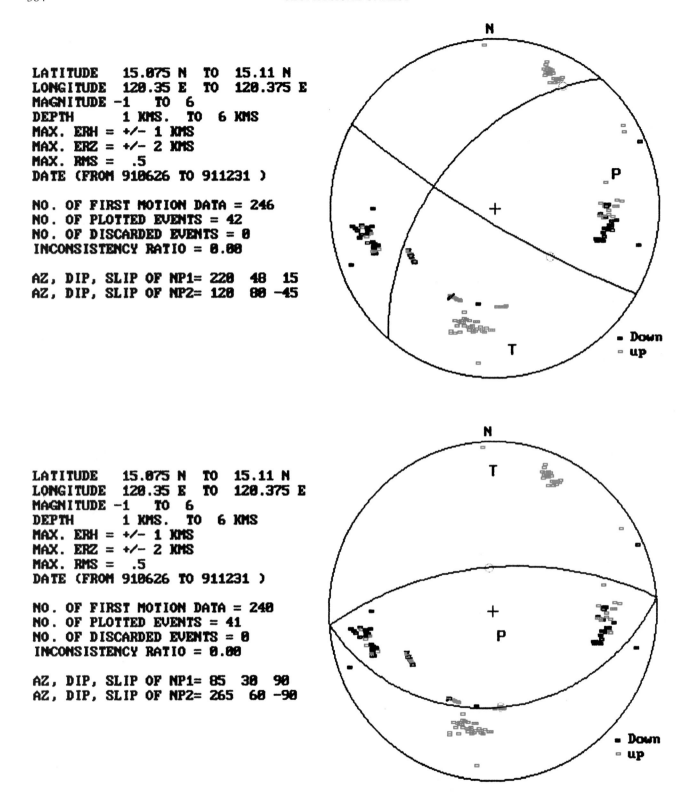

```
LATITUDE    15.075 N  TO   15.11 N
LONGITUDE  120.35 E   TO  120.375 E
MAGNITUDE -1    TO  6
DEPTH       1 KMS.  TO  6 KMS
MAX. ERH = +/- 1 KMS
MAX. ERZ = +/- 2 KMS
MAX. RMS =  .5
DATE (FROM 910626 TO 911231 )

NO. OF FIRST MOTION DATA = 246
NO. OF PLOTTED EVENTS = 42
NO. OF DISCARDED EVENTS = 0
INCONSISTENCY RATIO = 0.00

AZ, DIP, SLIP OF NP1= 220   48   15
AZ, DIP, SLIP OF NP2= 120   88  -45
```

```
LATITUDE    15.075 N  TO   15.11 N
LONGITUDE  120.35 E   TO  120.375 E
MAGNITUDE -1    TO  6
DEPTH       1 KMS.  TO  6 KMS
MAX. ERH = +/- 1 KMS
MAX. ERZ = +/- 2 KMS
MAX. RMS =  .5
DATE (FROM 910626 TO 911231 )

NO. OF FIRST MOTION DATA = 240
NO. OF PLOTTED EVENTS = 41
NO. OF DISCARDED EVENTS = 0
INCONSISTENCY RATIO = 0.00

AZ, DIP, SLIP OF NP1= 85   30   90
AZ, DIP, SLIP OF NP2= 265  60  -90
```

Figure 11. Three possible focal mechanism solutions (*A*, *B*, and *C*) for posteruption cluster E. Conventions as in figure 3. None are well constrained; the strike-slip solution in figure 12*A* is the most consistent with mechanisms observed in other clusters.

LATITUDE   15.075 N   TO   15.11 N
LONGITUDE   120.35 E   TO   120.375 E
MAGNITUDE -1   TO   6
DEPTH   1 KMS.   TO   6 KMS
MAX. ERH = +/- 1 KMS
MAX. ERZ = +/- 2 KMS
MAX. RMS =   .5
DATE (FROM 910626 TO 911231)

NO. OF FIRST MOTION DATA = 246
NO. OF PLOTTED EVENTS = 42
NO. OF DISCARDED EVENTS = 0
INCONSISTENCY RATIO = 0.00

AZ, DIP, SLIP OF NP1= 15   25   90
AZ, DIP, SLIP OF NP2= 195 65 -90

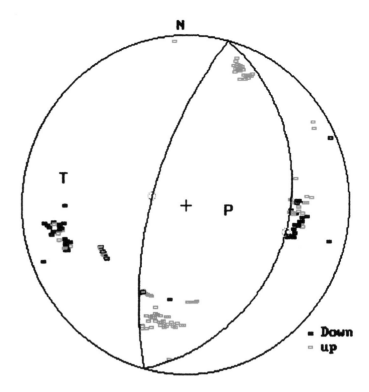

Figure 11.—Continued.

LATITUDE   15.18 N   TO   15.22 N
LONGITUDE   120.44 E   TO   120.46 E
MAGNITUDE -1   TO   6
DEPTH   4 KMS.   TO   10 KMS
MAX. ERH = +/- 2 KMS
MAX. ERZ = +/- 3 KMS
MAX. RMS =   .5
DATE (FROM 900101 TO 931230)

NO. OF FIRST MOTION DATA = 83
NO. OF PLOTTED EVENTS = 23
NO. OF DISCARDED EVENTS = 18
INCONSISTENCY RATIO = 0.44

AZ, DIP, SLIP OF NP1= 60   90   72
AZ, DIP, SLIP OF NP2= 330 70   0

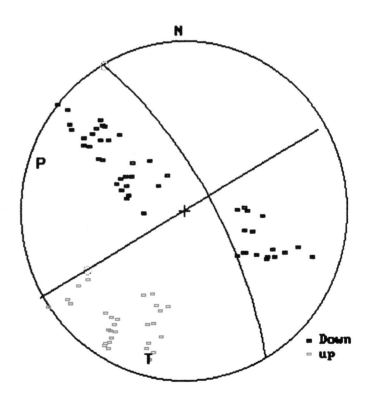

Figure 12.   Focal mechanism solutions for posteruption cluster F. Conventions as in figure 3.

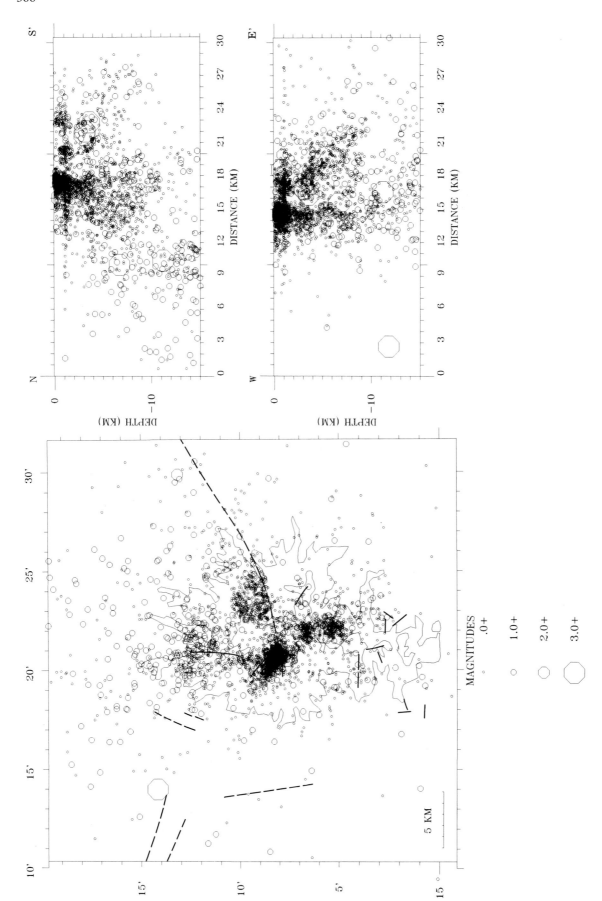

Figure 13. Seismicity at and near Mount Pinatubo, January–December 1992. Events from 0 to 30 km in depth are shown in map view; events from 0 to 15 km in depth are shown in cross sections. Zero depth is at the vent. All events shown have RMS≤0.15. Note that most 1992 seismicity was beneath or near the caldera (associated with renewed dome growth) and in posteruption cluster E (fig. 6). Cross sections are shown in figure 2.

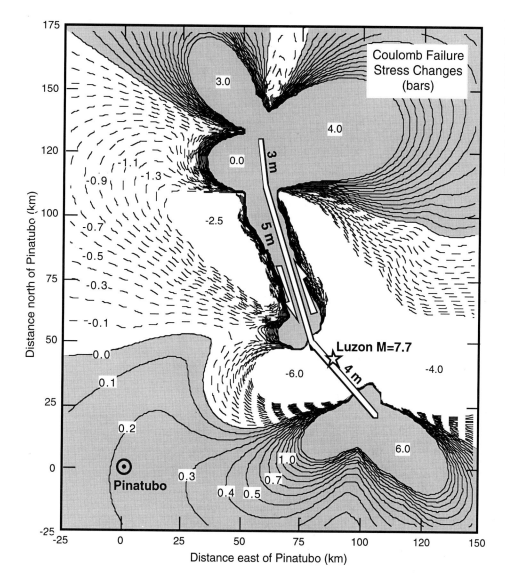

Figure 14. Coulomb stress change on optimally oriented strike-slip or thrust faults to 10 km in depth. Assumptions: Luzon earthquake left-lateral strike-slip on a 20-km-deep vertical fault; east-west compressive regional stress of 100 bar; coefficient of friction=0.4. Small, 0.1- to 0.2-bar increase in Coulomb stress at Pinatubo.

Harlow, this volume) would have been conducive to such a mechanism. The second case would be more consistent with inferences that the Pinatubo eruption was triggered by intrusion of basaltic magma from depth into the base of the dacitic reservoir (Pallister and others, this volume; White and others, this volume).

Alternatively, strong ground shaking during the July 16 earthquake itself might have squeezed a small amount of basaltic magma into the dacitic reservoir or caused pore-pressure changes and thus fault movement that allowed magma ascent. In either of these mechanisms, relatively small and transient effects from the Luzon earthquake itself would have triggered a cascading series of other changes that ultimately led to more basaltic intrusion into the dacitic reservoir, and the full eruption.

Because stress changes caused by the Luzon earthquake are small in relation to a typical, 100-bar compressive stress associated with subduction, that earthquake caused little change in the orientation of stress in the Pinatubo area. If we assume 100-bar east-west compression before the Luzon earthquake, stress after that earthquake is changed only very slightly, to 100.41 bar with a 2.4° counterclockwise rotation. Most focal mechanisms are so controlled by preexisting faults that we would not expect mechanisms to change even with stronger and azimuthally distinct stress changes (J. Mori, USGS, written commun., 1994), and certainly we have no evidence from focal mechanisms determined in 1991 and 1992 of any change from the inferred pre-1990 stress field.

CONCLUSIONS

Occurrence of several earthquakes in the Pinatubo region within days of the main July 16, 1990, Luzon earthquake suggests a **possible** causal relation between Pinatubo

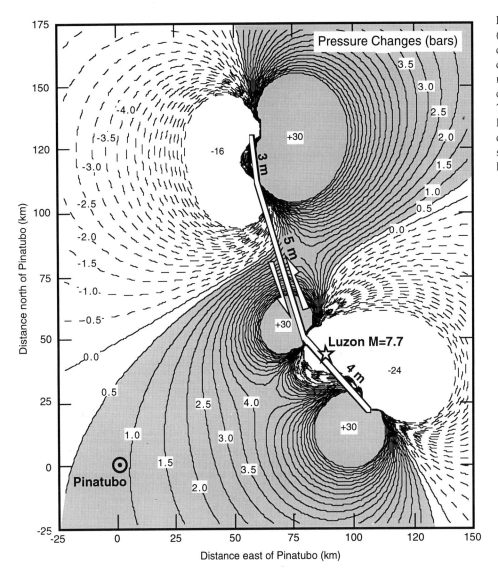

Figure 15. Volumetric stress (pressure) changes at 10 km depth caused by the Luzon earthquake. Assumptions: Luzon seismic moment (M_0) = 3.5 × 10²⁷ dyne-cm (moment magnitude = 7.7); event approximated by left-lateral strike-slip on a 20-km-deep vertical fault. Volumetric stress increase of about 1 bar at Pinatubo.

events and the mainshock. However, Pinatubo is outside the zone of most aftershocks.

The first earthquakes of the 1991 volcanic unrest, defining a northwest cluster, were caused by left-lateral movement along the northwest-trending Maraunot fault, consistent with preexisting, regional east-west compression. The preeruption steaming vents and fissure may have been along a conjugate, northeast-trending fault, also consistent with east-west compression. In contrast, the composite plot of the preeruption summit earthquakes showed highly mixed polarities for each station, suggesting diverse mechanisms that are characteristic of earthquakes caused by magmatic intrusion. Crustal readjustments continued during the posteruption stage, as reflected by several clusters of hypocenters with a wide range of depths and distances from the volcanic center. Such readjustment occurred on structures consistent with regional east-west compressional stress.

Several moderately large syneruption strike-slip earthquakes apparently occurred along preexisting regional faults, in response to volume changes in the magma chamber during the eruption. None of these showed a pure gravity mechanism. Thus, from seismic evidence, the 1991 Pinatubo caldera formed (1) too slowly to generate earthquakes or (2) by the collapse of many small pieces with associated small-magnitude earthquakes.

Changes in Coulomb failure stress as a result of the July 16, 1990, earthquake were small (at most, 10 times larger than tidal stress); compressive stress on the magma reservoir itself was about 1 bar, possibly enough to squeeze basaltic magma into the dacitic reservoir or dacitic magma upward from that reservoir. Alternatively, strong ground shaking during the July 16 earthquake triggered all that followed.

While this study was able to answer some of the Pinatubo puzzles by use of seismic methods, it also opened up a number of difficult questions that we cannot yet address. Some of these questions are:

- What triggered shallow (5-km-deep) seismicity along the Maraunot fault during April and May 1991? Did regional stress trigger magma ascent, or did magma ascent or inflation of a deep magma reservoir trigger movement along a previously stressed fault?

- Why was the northeast-trending fracture that formed on April 2, 1991, seismically inactive during the period of preeruption seismic monitoring (April 5 through June 12)?

- If 1990 movement along the Philippine fault did not induce a strong stress field in the Pinatubo area, what caused aftershocks in that area immediately after the earthquakes (table 1)? Were the structures along which aftershocks occurred already highly stressed prior to the July 16 Luzon earthquake, allowing strong ground motion from the July 16 earthquake to trigger secondary movements?

- Lastly, do past movements along nearby segments of the Philippine fault (as yet, mostly undated) correlate with dates of past eruptions? Slip along the Philippine fault in A.D. 1645 (Hirano and others, 1986) does not seem to correlate with the penultimate Buag eruptions of Pinatubo (600 to 500 yr B.P.; Newhall and others, this volume). Alternatively, do earthquakes associated with the Manila trench correlate with past eruptions (Javelosa, 1994)? Regrettably, data on prehistoric earthquakes of this region, from the Philippine fault, Manila trench, and other sources are, as yet, too sparse for us to check this correlation.

ACKNOWLEDGMENTS

We would like to acknowledge the help of R.A. White, J. Mori, C.G. Newhall, and an anonymous reviewer for encouragement and for painful but constructive suggestions. We give special thanks to Russ Needham, USGS/NEIC, for determining fault plane mechanisms of syneruption earthquakes, B.J.T. Punongbayan for the initial drafts of the figures and to T. Ohkura, E.G. Ramos, and R.A. White for helping us to produce the revised figures.

REFERENCES CITED

Bautista, B.C., Bautista, M.L.P., Rasdas, A.R., Lanuza, A.G., Amin, E.Q., and Delos Reyes, P.J., 1992, The 16 July 1990 earthquake and its aftershock activity, in The July 16, 1990 Luzon Earthquake: A Technical Monograph: Manila, Department of Environment and Natural Resources and the Interagency Committee for Documenting and Establishing Database of the July 1990 Earthquake, p. 81–118.

Bautista, B.C., and Narag, I.C., 1992, FOCPLOT—Computer Program for Plotting Composite Focal Mechanisms: unpublished computer program, Quezon City, PHIVOLCS.

de Boer, J.Z., Odom, L.A., Ragland, P.C., Snider, F.G., and Tilford, N.R., 1980, The Bataan orogene: Eastward subduction, tectonic rotations, and volcanism in the western Pacific (Philippines): Tectonophysics, v. 67, p. 305–317.

Delfin, F.G., Jr., 1984, Geology of the Mt. Pinatubo Area: unpublished report, Manila, Philippine National Oil Company, 36 p.

Delfin, F.G., Jr., Villarosa, H.G., Layugan, D.B., Clemente, V.C., Candelaria, M.R., Ruaya, J.R., this volume, Geothermal exploration of the pre-1991 Mount Pinatubo hydrothermal system.

Ewert, J.W., Lockhart, A.B., Marcial, S., and Ambubuyog, G., this volume, Ground deformation prior to the 1991 eruptions of Mount Pinatubo.

Filson, J., Simkin, T., and Leu, L-K., 1973, Seismicity of a caldera collapse: Galapagos Islands 1968: Journal of Geophysical Research, v. 78, p. 8591–8622.

Harlow, D.H., Power, J.A., Laguerta, E.P., Ambubuyog, G., White, R.A., and Hoblitt, R.P., this volume, Precursory seismicity and forecasting of the June 15, 1991, eruption of Mount Pinatubo.

Hill, D.P., Reasenberg, P.A., Michael, A., Arabaz, W.J., Beroza, G., Brumbauch, D., Brune, J.N., Castro, R., Davis, S., dePolo, D., Ellsworth, W.L., Gomberg, J., Harmsen, S., House, L., Jackson, S.M., Johnston, M.J.S., Jones, L., Keller, R., Malone, S., Munguia, L., Nava, S., Pechmann, J.C., Sanford, A., Simpson, R.W., Smith, R.B., Start, M., Stickney, M., Vidal, A., Walter, S., Wong, V., and Zollweg, J., 1993, Seismicity remotely triggered by the M 7.3 Landers, California earthquake: Science, v. 260, p. 1617–1623.

Hirano, S., Nakata, T., and Sangawa, A., 1986, Fault topography and Quaternary faulting along the Philippine fault zone, central Luzon, Philippines: Journal of Geography [Japan], v. 95, p. 71–93.

Hirn, A., Lepine, J-C., Sapin, M., and Delorme, H., 1991, Episodes of pit-crater collapse documented by seismology at Piton de la Fournaise: Journal of Volcanology and Geothermal Research, v. 47, p. 89–104.

Hoblitt, R.P., Wolfe, E.W., Scott, W.E., Couchman, M.R., Pallister, J.S., and Javier, D., this volume, The preclimactic eruptions of Mount Pinatubo, June 1991.

Javelosa, R.S., 1994, Active Quaternary environments in the Philippine mobile belt: unpublished Ph.D. thesis, Enschede, The Netherlands, International Institute for Aerospace Survey and Earth Sciences (ITC), 179 p.

Lee, W.H.K., ed., 1989, Toolbox for seismic data acquisition, processing, and analysis: IASPEI Software Library, v. 1, Seismological Society of America.

Lee, W.H.K., and Lahr, J.C., 1975, HYPO71 (revised): A computer program for determining hypocenter, magnitude, and first motion pattern of local earthquakes: U.S. Geological Survey Open-File Report 75–311, p. 1–116.

Lockhart, A.B., Marcial, S., Ambubuyog, G., Laguerta, E.P., and Power, J.A., this volume, Installation, operation, and technical specifications of the first Mount Pinatubo telemetered seismic network.

McKee, C.O., Lowenstein, P.L., De Saint Ours, P., Talai, B., Itikarai, I., and Mori, J., 1984, Seismic and ground deformation crises at Rabaul Caldera: Prelude to an eruption?: Bulletin of Volcanology, v. 47, no. 2, p. 397–411.

Mori, J., Eberhart-Phillips, D., and Harlow, D.H., this volume, Three-dimensional velocity structure at Mount Pinatubo, Philippines: Resolving magma bodies and earthquakes hypocenters.

Mori, J., and McKee, C., 1986, Outward-dipping ring-fault structure at Rabaul caldera as shown by earthquake locations: Science, v. 235, p. 193–195.

Mori, J., White, R.A., Harlow, D.H., Okubo, P., Power, J.A., Hoblitt, R.P., Laguerta, E.P., Lanuza, L., and Bautista, B.C., this volume, Volcanic earthquakes following the 1991 climactic eruption of Mount Pinatubo, Philippines: Strong seismicity during a waning eruption.

Murray, T.L., Power, J.A., March, G.D., and Marso, J.N., this volume, A PC-based realtime volcano-monitoring data-acquisition and analysis system.

Nakamura, K., 1971, Volcano as a possible indicator of crustal strain: Bulletin of the Volcanological Society of Japan, v. 16, p. 63–71. (In Japanese with English abstract.)

Newhall, C.G., Daag, A.S., Delfin, F.G., Jr., Hoblitt, R.P., McGeehin, J., Pallister, J.S., Regalado, M.T.M., Rubin, M., Tamayo, R.A., Jr., Tubianosa, B., and Umbal, J.V., this volume, Eruptive history of Mount Pinatubo.

Newhall, C.G., and Dzurisin, D., 1988, Historical unrest at large calderas of the world: U.S. Geological Survey Bulletin 1855, 1,108 p.

Pallister, J.S., Hoblitt, R.P., Meeker, G.P., Knight, R.J., and Siems, D.F., this volume, Magma mixing at Mount Pinatubo: Petrographic and chemical evidence from the 1991 deposits.

Power, J.A., Murray, T.L., Marso, J.N., and Laguerta, E.P., this volume, Preliminary observations of seismicity at Mount Pinatubo by use of the Seismic Spectral Amplitude Measurement (SSAM) system, May 13–June 18, 1991.

Ramos, A.F., and Isada, M. G., 1990, Emergency investigation of the alleged volcanic activity of Mount Pinatubo in Zambales Province: PHIVOLCS, Memorandum Report, 2 p.

Ramos, E.G., Laguerta, E.P., and Hamburger, M.W., this volume, Seismicity and magmatic resurgence at Mount Pinatubo in 1992.

Rikitake, T., and Sato, R., 1989, Up-squeezing of magma under tectonic stress: Journal of Physics of the Earth, v. 37, p. 303–311.

Rivera, L., and Cisternas, A., 1990, Stress tensor and fault plane solutions for a population of earthquakes: Bulletin of the Seismological Society of America, v. 80, no. 3, p. 600–614.

Sabit, J.P., Pigtain, R.C., and de la Cruz, E.G., this volume, The west-side story: Observations of the 1991 Mount Pinatubo eruptions from the west.

Scott, W.E., Hoblitt, R.P., Torres, R.C., Self, S, Martinez, M.L., and Nillos, T., Jr., this volume, Pyroclastic flows of the June 15, 1991, climactic eruption of Mount Pinatubo.

Stein, R.S., King, G.C.P., and Lin, J., 1992, Change in failure stress on the southern San Andreas fault system caused by the 1992 magnitude=7.4 Landers earthquake: Science, v. 258, p. 1328–1332.

Valdes, C., 1989, PCEQ, A computer program for picking earthquake phases, in IASPEI software library: Seismological software for IBM-compatible computers: International Association of Seismology and Physics of the Earth's Interior, and Seismological Society of America.

White, R.A., this volume, Precursory deep long-period earthquakes at Mount Pinatubo, Philippines: Spatio-temporal link to a basalt trigger.

Xu, Z., Wang, S., Huang, Y., and Gao, A., 1992, Tectonic stress field of China inferred from a large number of small earthquakes: Journal of Geophysical Research, v. 97, no. B8, p. 11867–11877.

Zoback, M.D., and Zoback, M.L., 1991, Tectonic stress field of North America and relative plate motions, in Slemmons, D.B., Engdahl, E.R., Zoback, M.D., and Blackwell, D.D., eds., Neotectonics of North America: Boulder, Geological Society of America, Decade Map Volume 1, p. 339–366.

Zoback, M.L., 1992, First- and second-order patterns of stress in the lithosphere; the world stress map project: Journal of Geophysical Research, v. 97, no. B8, p. 11703–11728.

Three-Dimensional Velocity Structure at Mount Pinatubo: Resolving Magma Bodies and Earthquake Hypocenters

By Jim Mori,[1] Donna Eberhart-Phillips,[1] and David H. Harlow[1]

ABSTRACT

P-wave arrivals from 298 events were used to determine the three-dimensional velocity structure in the region of Mount Pinatubo. The inversion results show several well-resolved regions of low-velocity (5 to 10 percent) which are inferred to be locations of magma bodies. There is a large low-velocity region between depths of 6 and 11 kilometers that is estimated to have a volume of 40 to 90 cubic kilometers. An extension of this low-velocity region reaches toward the surface under Mount Negron, which could represent magma that supplied eruptions at that vent. Another shallow low-velocity zone is located below the northwest flank of Mount Pinatubo, under the small caldera that formed during the June 15 eruption. If all of these low-velocity regions are interpreted as magma bodies, we estimated a total volume of 60 to 125 cubic kilometers for the magma system under Mount Pinatubo and Mount Negron, which is one of the largest active magma chambers. The three-dimensional velocity structure was also used to relocate all of the earthquakes recorded from May 5 to August 16, 1991. Compared to the locations done with a one-dimensional model, the hypocenters tend to be more tightly clustered and show more clearly the linear trends in both the preeruption and posteruption seismicity.

INTRODUCTION

The seismic activity associated with the 1991 eruptive activity of Mount Pinatubo provided a data set of many closely spaced earthquakes that can be used, through seismic tomography, to determine the three-dimensional P-wave velocity structure of the volcano. These results provide some of the few observations about the structure at depth under this volcano and are of particular interest for inferring the location and size of magma bodies that may have been the source of the large eruption of June 15, 1991. Three-dimensional velocity studies at other volcanoes (Iyer,

1992) have identified low-velocity zones that have been interpreted as magma bodies, such as Kilauea, Hawaii (Thurber, 1983), Mount St. Helens, Washington (Lees, 1992), Phlegrean Fields, Italy (Aster and Myer, 1988), Medicine Lake, California (Evans and Zucca, 1988), Newberry Volcano, Oregon (Achauer and others, 1988). Most of the magma bodies identified in these papers have been relatively small, with dimensions of only a few kilometers. The size of the 1991 Mount Pinatubo eruption of about 3.7–5.3 km^3 of magma (W.E. Scott and others, this volume) suggests there is a large source that may be imaged with the tomographic methods. Eruptions are thought to expel only a small percentage of the magma chamber volume, so the feature for which we are searching may have a volume on the order of 100 km^3. Some other studies have searched for large magma chambers under calderas at Long Valley, California (Sanders, 1984, Romero and others, 1993) and Yellowstone Caldera, Wyoming (Clawson and others, 1989), but there is still relatively little direct evidence for active magma reservoirs of 100 km^3 in size.

After determining the three-dimensional velocity structure, this velocity model can be used to relocate the earthquakes more accurately compared to the hypocenters calculated by using one-dimensional structures. The goal of this paper is to search for low-velocity regions that might be the magma source for the 1991 eruption and to investigate locations of the low-velocity regions in relation with the accurately relocated seismicity. This information provides valuable insights into the "plumbing" structure of the volcano and the processes that were occurring before and during the eruption.

DATA

This study used a set of 298 small earthquakes (M 2.0–4.0) recorded on the short-period network at Mount Pinatubo (Lockhart and others, this volume, Murray and others, this volume). There were 70 events from the preeruption period of May 16 to June 12 and 228 events from the posteruption period of June 29 to August 16 that were chosen to be relatively well-distributed in space

---

[1]U.S. Geological Survey.

371

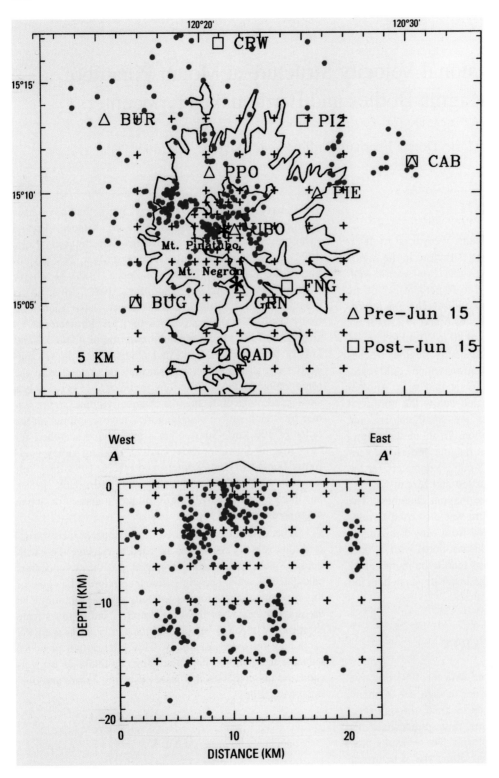

Figure 1. Map view of the Mount Pinatubo area showing the grid of node points (crosses) and earthquakes (small circles) used for the inversion. Stations of the preeruption network are shown by triangles, and stations of the post-eruption network are shown by squares. One station of the posteruption network is off the figure to the north. The topographic contour interval is 500 m. The west-to-east cross section shows all the hypocenters on the map and the depth distribution of the node points.

around the volcano (fig. 1). The earthquakes were located in an irregular volume that extended from the summit to distances of 10 to 20 km and ranged in depth from the surface to 20 km. The data for the tomographic inversion were the P-wave arrival times that were carefully picked from the digital seismograms with estimated uncertainties of 0.02 to 0.1 s. S-wave arrivals, especially those picked from

horizontal components at PPO (Patal Pinto) and PIE (Pinatubo east), were used to help the location of the earthquakes, but there were an insufficient number to be used for an S-wave velocity inversion. All but two of the seven sites that recorded the arrivals of the preeruption events were completely destroyed by the eruption so, there is a different configuration of seven stations for the posteruption events (fig. 1). The two stations that were common to both time periods are CAB (Clark Air Base) and BUG (Sitio Buag).

Having the two different station configurations for the preeruption and posteruption periods is an advantage to the tomographic study because it provides a more varied distribution of source-station ray paths. The preeruption seismicity located within several kilometers of Mount Pinatubo and recorded with nearby seismic stations can resolve the velocity in the shallow region close to the summit. The posteruption seismicity was spread over a much larger area and depth and therefore enabled us to estimate the velocity structure in a much larger volume around the volcano. However there were no stations close to the summit for the posteruption period to resolve the shallow summit structure. The two data sets are complementary and, when combined, allow a velocity inversion of both the shallow region near the summit and also the larger surrounding volume. One potential problem is that the velocity structure may have changed significantly during the June 15 eruption. If this is the case, combining these two data sets may be mixing data from two different velocity structures. This problem will cause errors in the shallow portion of the velocity structure, but the deeper portions of the model should still accurately reflect the velocities of the posteruption period. The preeruption earthquakes were all shallow and close to the summit, so ray paths to the stations do not travel at very large depths through the model and therefore should not affect the results in that region.

METHOD

A computer program originally written by Thurber (1981, 1993) and modified by Eberhart-Phillips (1993) was used to invert the arrival times for the three-dimensional velocity structure. This program can simultaneously solve for the velocity structure and the hypocentral locations by using an iterative damped least-squares inversion. The velocity structure is parameterized by a three-dimensional grid of node points. The velocity structure is derived by minimizing the traveltime residuals r_{ij} for event i from station j with respect the earthquake hypocentral parameters (t_i, x_i, y_i, z_i) and the n node point velocities (v_n).

$$r_{ij} = \Delta t_i + \frac{\partial t_{ij}}{\partial x_i}\Delta x_i + \frac{\partial t_{ij}}{\partial y_i}\Delta y_i + \frac{\partial t_{ij}}{\partial z_i}\Delta z_i + \sum_{n=1}^{N}\frac{\partial t_{ij}}{\partial v_n}\Delta v_n \quad (1)$$

For the calculation of traveltimes through the structure, linear velocity gradients are used between the node points, and rays are traced by using an approximate three-dimensional algorithm that produces curved nonplanar ray paths (Um and Thurber, 1987). This method results in a final velocity model with smooth velocity variations within the volume rather than sharp velocity contrasts.

Table 1. Starting velocity model (left) and best one-dimensional model (right) from inversion.

Depth (km)	P-wave velocity (km/sec)	P-wave velocity (km/sec)
0.0	3.30	3.89
1.0	4.00	4.59
4.0	5.00	5.72
7.0	5.70	5.44
10.0	6.00	5.48
15.0	6.70	5.70

In this study we started with a one-dimensional velocity structure (table 1) and first inverted the data to find the best one-dimensional model that fits the data (table 1). This model already reflects some of the anomalous velocity structure near the volcano, with a low-velocity zone at 7 km depth. The one-dimensional model was used as our starting velocity structure for the three-dimensional inversion. The node points for the three-dimensional inversion were spaced at 1- and 3-km intervals horizontally and at depths of 0, 1, 4, 7, 10, and 15 km, as shown in figure 1. There are relatively more ray paths from the distribution of earthquakes and stations that cross in the central region of the grid near the volcano summit, so the closer spacing of the node points allows finer resolution of the velocity structure in the summit area.

For the velocity inversion, P-wave data were used with weighted arrivals, depending on the uncertainty of the phase pick. One weakness of this data set is that there are relatively few arrivals for each earthquake. There can be a maximum of only seven, and sometimes only five or six stations recorded an earthquake. If we solve simultaneously for the velocity structure and the earthquake locations, there are relatively few data points compared to the number of unknowns, and the inversion can be unstable. For this reason, we used an iterative procedure of fixing the earthquake locations and solving for the velocity structure; then we fixed the derived velocity structure and relocated the earthquakes. The procedure minimized residuals (equation 1), alternately holding fixed the partial derivatives with respect to the hypocentral parameters and then holding fixed the partials with respect to the velocities. In the location step we included the S-wave arrivals, which helped to constrain the depths of the earthquakes. We used a constant ratio of the P- to S-wave velocity ($v_p/v_s = 1.8$). Three iterations of this process caused the inversion to converge to a solution.

RESULTS

THREE-DIMENSIONAL VELOCITY STRUCTURE

The results of the velocity inversion are shown in figure 2 as horizontal slices through the model at depths of

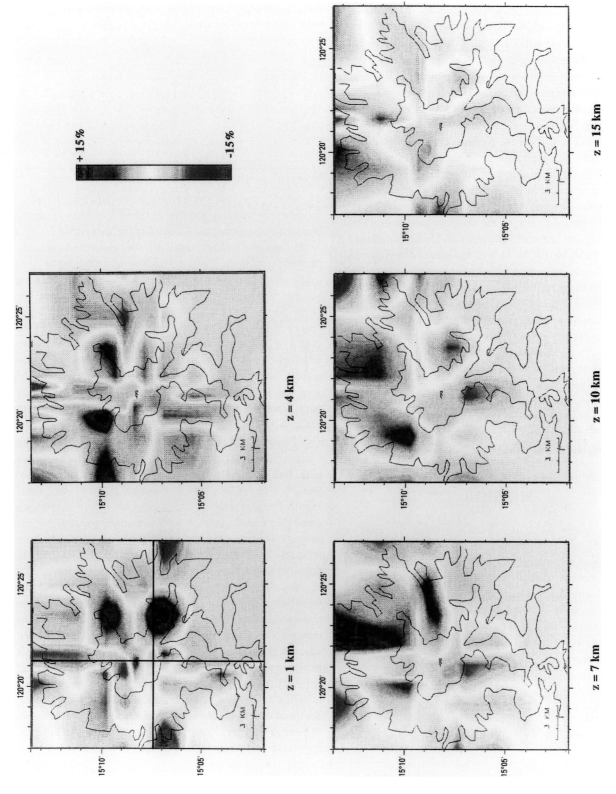

Figure 2. Depth slices of the three-dimensional velocity structure derived in this study. The colors show velocity differences from the average velocity at a given depth. Hotter/colder colors correspond to slower/faster velocities. Lines on the 1-km depth slice show the positions of the cross sections in figure 3.

1, 4, 7, 10, and 15 km. For the final velocity model and hypocenters, the root-mean-square fit of the predicted traveltimes to the observed traveltimes for the 296 events was 0.069 s, compared to 0.178 s for the best one-dimensional model. The colors reflect velocity differences from the average velocity at each depth, with slower/faster velocities shown by hotter/colder colors. The velocities generally range ±10% from the average value for each depth. The colors have been smoothed to show graduated velocity changes throughout the model, although very strong contrasts between adjacent nodes are reflected in the blockiness of some parts of the images. The inversion solves for discrete velocities at the nodes and assumes a linear gradient between the nodes for calculation of the raytracing and the velocity partial derivatives for the inversion. Therefore, the smoothed color images are representative of the smooth velocity variations that were determined in the model.

The pattern of velocity variations tends to be more complicated for the shallower depth slices at 1 and 4 km, which have closely spaced regions of fast and slow velocities. This might be expected in the vicinity of a volcano where the shallow crustal structure could have a complexity caused by magma intrusions, lava flows, ash deposits, and variations in degree of fracturing. Also, parts of the shallow velocity structure away from the summit are not well resolved, and the strong variations could be reflecting instabilities in the inversion. This is particularly true for the strong perturbations along the edges of the depth slices. At greater depth the model has a simpler image, which may be partly due to more homogeneous structure and partly to better stability of the model in a region where there are more crossing ray paths. One conspicuous feature in the 7- and 10-km depth slices is the strong low-velocity region south of the summit area. This region is 2 to 3 km across in the east-west direction and 4 to 5 km in the north-south direction, with velocities that are 5 to 10% slower than the surrounding material. At 15 km and greater depth, there is little resolution of the model because of the small number of crossing ray paths.

In the top portion of figure 3, the velocity model is plotted in a south-to-north cross section through the summit area. The black dots show the locations of the preeruption seismicity from May 7 to June 11. This image clearly shows the large low-velocity region south of the summit that was seen in the 7- and 10-km depth slices of figure 2. The low-velocity body is 4 to 5 km across at depths between 6 and 11 km, with an extension from the southern edge that reaches toward the surface. Another low-velocity region, possibly connected to the deeper one, is seen at shallow depth (1 to 3 km) below the northern flank of the old Pinatubo summit. The 1-km depth slice in figure 2 shows that this area is actually on the northwest flank of Mount Pinatubo, under the small caldera that formed during the June 15 eruption.

The bottom portion of figure 3 is a west-to-east cross section through the strong low-velocity zone 2 km south of the summit. The black dots are the posteruption earthquake locations. The low-velocity region falls within the outward trending limbs of the seismicity starting from about a depth of about 7 km and extending to the bottom of the model. There are relatively few earthquakes located within the low-velocity region between the two outward-trending limbs of seismicity.

We estimated the volume of the large low-velocity region between 6 and 11 km in depth to be 40 to 90 km^3 and the extension toward Mount Negron as 15 to 25 km^3. The smaller low-velocity region at shallow depth under the northwest flank of Mount Pinatubo was estimated to be 6 to 9 km^3. The sum of these three volumes gives a total of 60 to 125 km^3.

RELOCATED EARTHQUAKES

Using the three-dimensional velocity structure derived in this study, we recalculated hypocenters for all of the available data from May through August 16. To show the improvement in the hypocenters, we also relocated the earthquakes using the best one-dimensional structure (table 1). Figure 4 compares the posteruption locations determined using the one- and three-dimensional velocity structures for the preeruption period. Figure 5 shows the same for the posteruption period. For the three-dimensional structure, the hypocenters are more tightly clustered and appear to form more distinct spatial patterns, particularly in the cross sections. Dipping trends of earthquakes in both the preeruption and posteruption locations are more apparent for the locations when using the three-dimensional structure. The sharper images of earthquake distributions produced by the three-dimensional velocity model suggest that the locations are more accurate. The most significant difference was for the cluster of the earthquakes northwest of the summit area in the posteruption data. When the three-dimensional structure is used, locations of these earthquakes form a clear trend that dips outward from the summit area, matching a similar trend of earthquakes that dips outward on the east side of the summit (west-east cross sections in fig. 4). When the one-dimensional structure is used, the westward dipping trend is much more diffuse and harder to recognize. This area, where the largest difference between the locations calculated in the one- and three-dimensional structures exist, is also the region where the station coverage is the worst. Since most of the stations are to the east, the majority of ray paths from the earthquakes pass through the region of extreme velocity heterogeneity. This demonstrates that in an area where significant lateral variations in velocities can affect the traveltimes, the problems of locating earthquakes using a one-dimensional velocity structure are magnified when there is poor azimuthal station coverage.

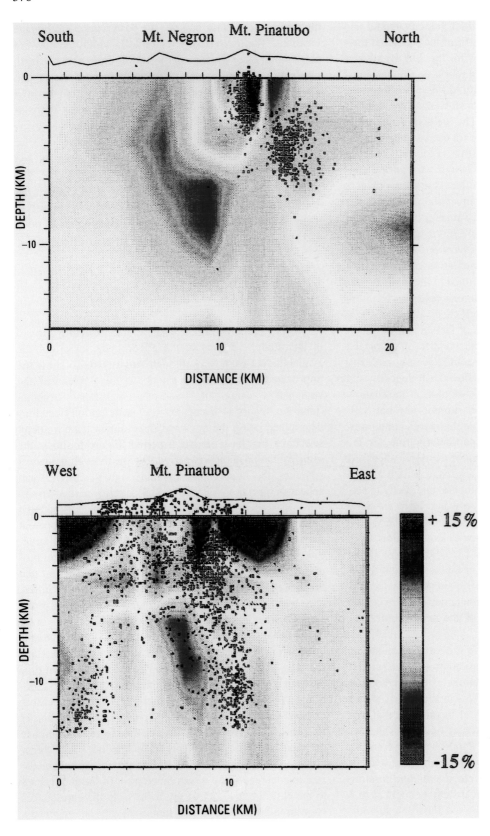

Figure 3. Cross sections of the velocity perturbations from the average velocity at a given depth. Top, South-to-north cross section of the velocity structure through the summit area. The black dots are all the preeruption seismicity recorded from May 7 to June 11. Bottom, West-to-east cross section of the velocity structure through an area 2 km south of the summit. The black dots are the posteruption seismicity from June 29 to August 16 in the region 3 km north and south of the summit.

1–D Locations for Preeruption

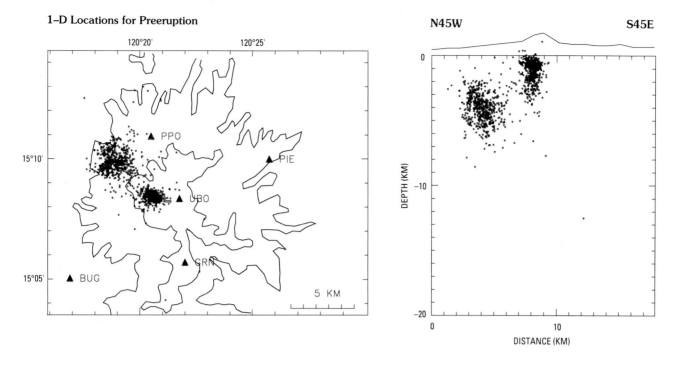

3–D Locations for Preeruption

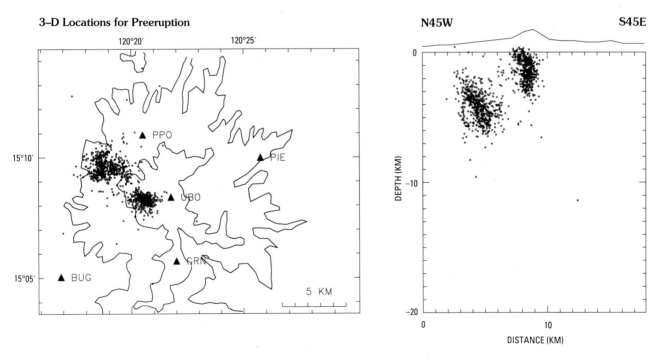

Figure 4. Comparison of the all preeruption seismicity recorded from May 5 to June 11 located with a one-dimensional velocity structure (top) and a three-dimensional velocity structure (bottom). The cross-section shows all the events projected on a plane oriented N.45°W. Triangles are stations of the preeruption network.

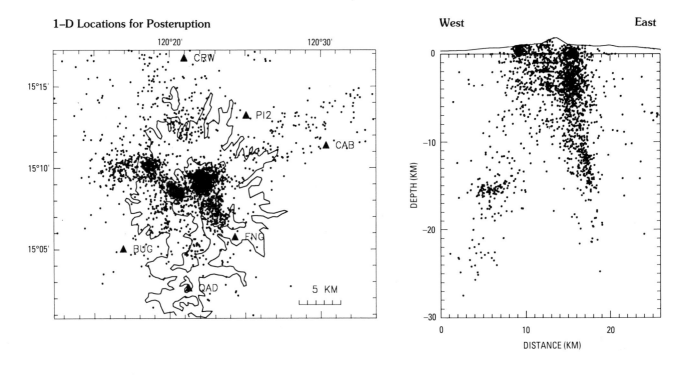

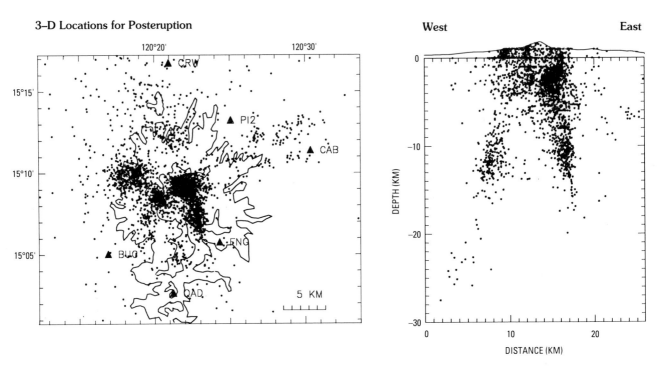

Figure 5. Comparison of all the posteruption seismicity recorded from June 29 through August 16 located with a one-dimensional velocity structure (top) and a three-dimensional velocity structure (bottom). The cross-section is oriented west to east and shows events within a 10-km-wide strip through the summit region. Triangles are stations of the posteruption network.

The plots of earthquake locations shown and discussed in this volume by Harlow and others (this volume) and Mori, White, and others (this volume) use hypocenters located using the three-dimensional velocity structure.

DISCUSSION

UNCERTAINTIES IN VELOCITY MODEL

We have attributed importance to the low-velocity regions seen in the cross sections of figure 3, and in order to ensure that these areas are well resolved by the velocity inversion, we examined the model resolution matrix from the inversion. The spread of values around the jth diagonal elements is an indication of the degree of resolution for the velocity value of the corresponding node. This spread (S_j) can be estimated by using the expression of Michelini and McEvilly (1991),

$$S_j = \log\left[|s_{jj}|^{-1} \sum_k \left(\frac{s_{kj}}{|s_{jj}|} \right)^2 D_{jk} \right], \qquad (2)$$

where s_{kj} are elements of the resolution matrix weighted by their distance (D_{ij}) from the node point. Small values of the width indicate that the values surrounding the eigenvalue are relatively small and the velocity is well resolved. Higher values represent poorer resolution and indicate that there may be smearing of the velocity values among adjacent nodes. These values are shown in figure 6 for the velocity node points used to construct the cross sections of figure 3. The smaller spreads are shown by darker colors, representing better resolved velocities. The white areas are regions where few or no ray paths crossed, so there is little information about the velocity.

Figure 6 shows that the low-velocity regions are located in regions that are well resolved by the model, so the large size of the low-velocity region is not a result of smearing a small anomaly over a large volume. The model is weakest in the deeper portions, especially east and south of Mount Pinatubo, where there were few earthquakes to provide information. The lack of resolution in the deeper regions directly under the low-velocities areas leaves open the possibility that any inferred magma body could extend to greater depth and have larger volumes than we estimated.

Another uncertainty in the velocity model calculation is due to the coarseness of the grid (1 to 3 km). Small-scale features such as pipes or dikes that are a few hundred meters across cannot be resolved with the present data, because there are relatively few stations. Ideally, seismic monitoring on volcanoes would include sufficient station density not only for locating earthquakes but also for making three-dimensional velocity studies that can resolve the location of the magma. More S-wave data from three-component stations would be useful for incorporation into the inversion procedure and for identifying S-wave shadows (i.e. Matumoto, 1971) to confirm the existence of the large magma bodies.

MAGMA BODIES

The images in figures 3 and 4 show some strong low-velocity regions, particularly near and beneath the summit region. It is common to interpret low-velocity zones in volcanic areas as magma bodies; however, they could also be porous units of incompetent rock or unconsolidated sediments (Achauer and others, 1988; Evans and Zucca, 1988; Romero and others, 1993) or areas dominated by fluid saturation (Romero and others, 1993; O'Connell and Johnson, 1991). For the deeper low-velocity regions, we prefer the magmatic interpretation because it seems unlikely that unconsolidated material would be present at the lithostatic pressures associated with depths of 6 to 11 km. Vera and others (1990) observed a large low-velocity zone at 5 to 9 km in depth under the East Pacific Rise that they infer is not a magma chamber because the seismic amplitudes indicate that it behaves as a solid. Similar arguments might be made for Pinatubo, where S waves are sometimes observed for ray paths that cross the large low-velocity regions. Alternatively, these observations may indicate that the large magma bodies are not single reservoirs of magma but instead a region of dikes and sills intruded into competent rock. If this is the case, the volumes of the low-velocity bodies estimated above would be upper limits on the volume of magma.

Identifying the smaller low-velocity zone under the northwest flank of Mount Pinatubo as a magma body is more uncertain, because there could easily be unconsolidated sediments with low seismic velocities at this shallow depth. However, the formation of the small caldera (2.5 km diameter) during the June 15 eruption is consistent with the inference that there was a significant magma body here prior to the eruption. The caldera (2.5 km^3 collapse volume, W.E. Scott and others, this volume) is centered about 1 km northwest of the old summit, almost directly over the shallow low-velocity zone on the northwest flank of Mount Pinatubo (see 1-km depth slice of fig. 2). This feature could have formed as a collapse in response to volume loss in the shallow magma body.

The pattern of seismicity is also suggestive that the low-velocity regions are magma bodies. Because the located seismicity consists mostly of high-frequency volcano-tectonic events, these events should have hypocenters in the areas of competent material. As the large magma bodies inflate or deflate, there would be significant volumetric strains surrounding the magma bodies, and one might expect most of the seismicity to occur in the regions of hard rock close to the magma bodies. The top portion of figure 3 shows that the preeruption seismicity occurred in two areas

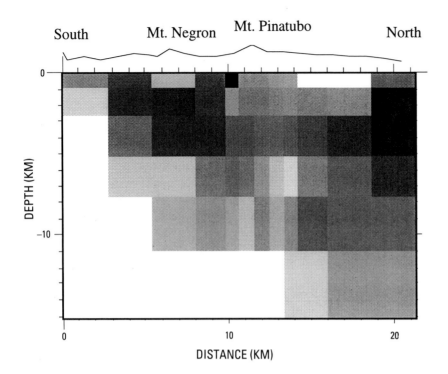

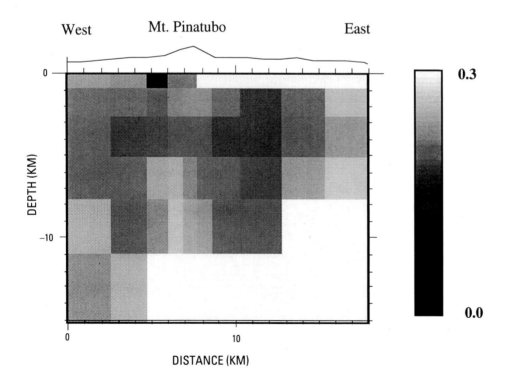

Figure 6. Estimates of the resolution for velocity values shown in figure 3, as measured by the spread in values about the diagonal elements of the model resolution matrix.

adjacent to the shallow low-velocity region under the northwest flank. The posteruption seismicity, shown in the bottom portion of figure 3, consists of outward-dipping limbs surrounding the strong low-velocity zone that is inferred to be the large magma chamber at depth. In both cross sections of figure 3, the earthquakes tend to occur in the areas of higher (blue) velocity, possibly indicating regions of competent rock.

Petrologic data indicate that most of the erupted material came from depths of 7 to 11 km below the surface

(Rutherford and Devine, this volume; Pallister and others, this volume; Fournelle and others, this volume). This depth coincides with the large low-velocity region and suggests that the main source of the magma is this volume beneath the ridge that connects Mount Negron and Mount Pinatubo. This could be the main magma reservoir, with a volume of 40 to 90 km^3, that supplies eruptions at Mount Pinatubo. The red Y-shaped pattern in the south-north cross section of figure 3 could be interpreted as a large magma reservoir at depth connected by two branches to volcanic vents at Mount Pinatubo and Mount Negron. If this system supplied the magma for past eruptions at both Mount Pinatubo and Mount Negron, it would have to be a long-enduring feature since the only available date for an eruption at Negron is 1.3 to 1.2 Ma (Delfin and others, this volume). The magma would have to travel a relatively long distance to reach the vents, but there is evidence from other volcanoes of large "plumbing" systems that can connect reservoirs and vents over long distances. One example is the 1912 eruption of Mount Katmai, Alaska, where the main eruption vent was 10 km from the summit collapse (Hildreth, 1983). Another example is the simultaneous eruptions at two vents 6 km apart at Rabaul caldera, Papua New Guinea, in 1937 (Fisher, 1939) and 1994.

The large volume of the inferred magma system (total of 60 to 125 km^3) under Mount Pinatubo and Mount Negron suggests there is a potential of future eruptions being larger than the 1991 activity. Large magma reservoirs of this size may be responsible for caldera-forming eruptions, and we speculate that an eruption at Pinatubo could be large enough to reduce the volume of the magma body significantly. The ridge between Pinatubo and Negron could lose support from beneath and result in collapse around a ring fault to form a large caldera. There is a hint in the posteruption seismicity of a developing ring fault system surrounding the large magma reservoir (Mori, White, and others, this volume). Geological data, however, indicate that there is a recent trend of diminishing sizes in the Pinatubo eruptions (Newhall and others, this volume). Whether this system evolves into a large caldera event or continues with decreasingly smaller eruptions, the large size of the inferred magma reservoir indicates that eruptive activity will probably extend significantly into the future.

CONCLUSIONS

We inverted the P-wave arrival times for earthquakes around Mount Pinatubo to produce a three-dimensional velocity structure in the region around the volcano. The velocity structure enabled better locations for the earthquakes recorded during the preeruption and posteruption periods. Hypocenters calculated with the improved velocity model show more clearly dipping trends of earthquakes for both the preeruption and posteruption seismicity.

The velocity model was also used to infer the physical structure under the volcano, particularly in identifying low-velocity regions that are interpreted to be locations of magma bodies. A large region between depths of 6 and 11 km under the ridge that connects Mount Pinatubo and Mount Negron may be the magma reservoir (40 to 90 km^3) that has supplied eruptions at these two vents. Another low-velocity region extends from the reservoir upward toward Mount Negron with a volume of 15 to 25 km^3. A small (6 to 9 km^3), shallow low-velocity region on the northwest flank of Mount Pinatubo is located under the crater that formed during the June 15 eruption. Volume loss in this shallow magma body may have initiated the collapse that formed the small summit caldera. We have inferred a large system of magma bodies under the Mount Pinatubo-Mount Negron complex with total volume estimated to be 60 to 125 km^3. The deeper portions of the velocity model are poorly resolved, and the volume could be larger if the magma body extends below 11 km. This large volume of magma may be typical of volcanoes that produce eruptions of the size that occurred at Mount Pinatubo in 1991 and indicates there is the potential for future eruptions to be comparable or larger.

ACKNOWLEDGMENTS

Data used in this study were the result of hard work by J.A. Power, T.L. Murray, F. Fischer, P. Okubo, R.A. White, L. Bautista, E.T. Endo, J.W. Ewert, A.B. Lockhart, J. Lockwood, J.N. Marso, A. Miklius, E.P. Laguerta, B. Bautista, C.G. Newhall, E. Ramos, E.W. Wolfe, and others. R.P. Hoblitt and J.S. Pallister provided many useful discussions. We thank U.S. Air Force personnel at Clark Air Base for their support. Helpful comments on the manuscript were provided by A. Pelayo, J. Lees, and C.G. Newhall.

REFERENCES CITED

Achauer, U., Evans, J.R., and Stauber, D.A., 1988, High-resolution seismic tomography of compressional wave velocity structure at Newberry Volcano, Oregon Cascade Range: Journal of Geophysical Research, v. 93, p. 10135–10147.

Aster, R.C., and Myer, R.P., 1988, Three-dimensional velocity structure and hypocenter distribution in the Campi Flegrei caldera, Italy: Tectonophysics, v. 149, p. 195–218.

Clawson, S.R., Smith, R.B., and Benz, H.M., 1989, P wave attenuation of the Yellowstone caldera from three-dimensional inversion of spectral decay using explosion source seismic data: Journal of Geophysical Research, v. 94, p. 7205–7222.

Delfin, F.G., Jr., Villarosa, H.G., Layugan, D.B., Clemente, V.C., Candelaria, M.R., Ruaya, J.R., this volume, Geothermal exploration of the pre-1991 Mount Pinatubo hydrothermal system.

Eberhart-Phillips, D., 1993, Local earthquake tomography: Earthquake source regions, in Iyer, H.M., and Hirahara, K., eds.,

Seismic tomography: Theory and practice: London, Chapman and Hall, p. 613–643.

Evans, J.R. and Zucca, J.J., 1988, Active high-resolution seismic tomography of compressional wave velocity and attenuation structure at Medicine Lake Volcano, Northern California Cascade Range: Journal of Geophysical Research, v. 93, p. 15016–15036.

Fisher, N.H., 1939, Geology and vulcanology of Blanche Bay, and the surrounding area, New Britain: Territory of New Guinea Geological Bulletin, v. 1, 68 p.

Fournelle, J, Carmody, R., and Daag, A.S., this volume, Anhydrite-bearing pumices from the June 15, 1991, eruption of Mount Pinatubo: Geochemistry, mineralogy, and petrology.

Harlow, D.H., Power, J.A., Laguerta, E.P., Ambubuyog, G., White, R.A., and Hoblitt, R.P., this volume, Precursory seismicity and forecasting of the June 15, 1991, eruption of Mount Pinatubo.

Hildreth, W., 1983, The compositionally zoned eruption of 1912 in the Valley of Ten Thousand Smokes, Katmai National Park, Alaska: Journal of Volcanology and Geothermal Research, v. 18, p. 1487–1494.

Iyer, H.M., 1992, Seismological detection and delineation of magma chambers: Present status with emphasis on the western USA, in Gasparini, P, Scarpa, R., and Aki, K., eds., Volcanic seismology: Berlin, Springer-Verlag, p. 299–338.

Lees, J.M., 1992, The magma system of Mount St. Helens: Nonlinear high-resolution P-wave tomography: Journal of Volcanology and Geothermal Research, v. 53, p. 103–116.

Lockhart, A.B., Marcial, S., Ambubuyog, G., Laguerta, E.P., and Power, J.A., this volume, Installation, operation, and technical specifications of the first Mount Pinatubo telemetered seismic network.

Matsumoto, T., 1971, Seismic body waves observed in the vicinity of Mt. Katmai, Alaska and evidence for the existence of molten chambers: Geological Society of America Bulletin, v. 82, p. 2905–2920.

Michelini, A., and McEvilly, T.V., 1991, Seismological studies at Parkfield: I. Simultaneous inversion for velocity structure and hypocenters using cubic B-splines parameterization: Bulletin of the Seismological Society of America, v. 81, p. 524–552.

Mori, J., White, R.A., Harlow, D.H., Okubo, P., Power, J.A., Hoblitt, R.P., Laguerta, E.P., Lanuza, L., and Bautista, B.C., this volume, Volcanic earthquakes following the 1991 climactic eruption of Mount Pinatubo: Strong seismicity during a waning eruption.

Murray, T.L., Power, J.A., March, G.D., and Marso, J.N., this volume, A PC-based realtime volcano-monitoring data-acquisition and analysis system.

Newhall, C.G., Daag, A.S., Delfin, F.G., Jr., Hoblitt, R.P., McGeehin, J., Pallister, J.S., Regalado, M.T.M., Rubin, M., Tamayo, R.A., Jr., Tubianosa, B., and Umbal, J.V., this volume, Eruptive history of Mount Pinatubo.

O'Connell, D.R.H. and Johnson, L.R., 1991, Progressive inversion for hypocenters and P wave and S wave velocity structure: Application to the Geysers, California, geothermal field: Journal of Geophysical Research, v. 96, p. 6223–6236.

Romero., A.E., Jr., McEvilly, T.V., Majer, E.L., and Michelini, A., 1993, Velocity structure of the Long Valley caldera from the inversion of local earthquake P and S travel times: Journal of Geophysical Research, v. 98, p. 19869–19879.

Rutherford, M.J., and Devine, J.D., this volume, Preeruption pressure-temperature conditions and volatiles in the 1991 dacitic magma of Mount Pinatubo.

Pallister, J.S., Hoblitt, R.P., Meeker, G.P., Knight, R.J., and Siems, D.F., this volume, Magma mixing at Mount Pinatubo: Petrographic and chemical evidence from the 1991 deposits.

Sanders, C., 1984, Location and configuration of magma bodies beneath Long Valley, California, determined from anomalous earthquake signals: Journal of Geophysical Research, v. 89, p. 8287–8302.

Scott, W.E., Hoblitt, R.P., Torres, R.C., Self, S, Martinez, M.L., and Nillos, T., Jr., this volume, Pyroclastic flows of the June 15, 1991, climactic eruption of Mount Pinatubo.

Thurber, C.H., 1981, Earth structure and earthquake locations in the Coyote Lake area, central California: Cambridge, Mass., Massachusetts Institute of Technology, Ph.D. thesis.

——1983, Seismic detection of the summit magma complex of Kilauea volcano, Hawaii: Science, v. 233, p. 165–167.

——1993, Local earthquake tomography: Velocities and Vp/Vs—theory, in Iyer, H.M., and Hirahara, K., eds., Seismic tomography: Theory and practice: London, Chapman and Hall, p. 563–583.

Um, J., and Thurber, C.H., 1987, A fast algorithm for two-point seismic ray tracing: Bulletin of the Seismological Society of America, v. 77, p. 972–986.

Vera, E.E., Mutter, J.C., Buhl, P., Orcutt, J.A., Harding, A.J., Kappus, M.E., Detrick, R.S., and Brocher, T.M., 1990, The structure of 0- to 0.2-m.y.-old oceanic crust at 9°N on the east Pacific Rise from expanded spread profiles: Journal of Geophysical Research, v. 95, 15529–15556.

Computer Visualization of Earthquake Hypocenters

By Richard P. Hoblitt,[1] Jim Mori,[1] and John A. Power[1]

ABSTRACT

We present two computer programs—PINAPLOT and VolQuake—to simplify the task of analyzing the locations of earthquakes associated with volcanic activity. Both programs require an IBM-PC compatible computer. PINA-PLOT runs under DOS; VolQuake requires Microsoft Windows, version 3.1. Earthquake locations are plotted in chronologic order, in both map and cross-sectional view, much like a movie. Consequently, the user can experience seismic data as they were received.

INTRODUCTION

In many, perhaps most, scientific pursuits, data interpretation can be a leisurely affair. In contrast, data acquired for the purpose of hazard mitigation must be processed and interpreted quickly. When Mount Pinatubo reawakened, earthquake hypocenters were displayed by preparing a series of monochrome epicenter maps (plan view) and hypocenter plots (cross-sectional views). The size of plot symbols was keyed to earthquake magnitude, and the symbol type was keyed to hypocentral depth. Epicenter maps were examined in chronologic order to gain an appreciation of how activity was evolving through time. A program called ACROSPIN (Parker, 1990) allowed hypocenters to be rotated about three orthogonal axes and so provided a three-dimensional perspective. Although this approach was workable, it was rather cumbersome. The analyst had to prepare and examine a collection of graphs—essentially "snapshots" of activity through time. Earthquake data had to be reformatted before ACROSPIN would accept them. While none of these deficiencies is serious under normal circumstances, each is magnified in a crisis situation in which personnel are fatigued and time is at a premium.

Following the climactic eruption of Mount Pinatubo on June 15, 1991, Mori wrote an unadorned computer program (PINAPLOT) to display the earthquake hypocenters in a somewhat different manner. Epicenter maps and hypocenter plots were displayed as before, but the time between the plotting of two hypocenters was roughly proportional to the actual elapsed time between the two earthquakes. This simple change markedly improved an observer's sense of time. In addition, different colors were used to indicate depth intervals, and each new symbol was heralded by a brief flash of light and a beep. Rather than a series of static "snapshots," we essentially had a motion picture of earthquake locations. The value of this program soon became apparent, as it allowed us to quickly review and track continuing seismic activity. Furthermore, it was extremely easy to use: a brief explanation would prepare virtually anyone to run the program.

The reawakening of Mount Spurr in 1992 again presented the need to expeditiously interpret seismic data. PINAPLOT—hastily written specifically for Mount Pinatubo—was modified by Hoblitt to accept data from Mount Spurr. The modified program performed the same useful functions that it had during the Mount Pinatubo eruptions. Despite its usefulness, PINAPLOT had limitations. Opportunities for user input were limited, and code modifications were necessary before the program could be used for different volcanoes. Further, data still had to be reformatted before they could be examined in ACROSPIN. We needed a user-friendly version of the original program that could (1) be employed quickly at restless volcanoes without modification, (2) provide the functionality of ACROSPIN, and (3) permit data sets from previous eruptions to be compared easily. Following the Mount Pinatubo and Mount Spurr experiences, a new program, VolQuake, was written by Hoblitt, with the assistance of Mori and Power, to overcome the limitations of PINAPLOT while retaining its useful features.

PINAPLOT reads data in the HYPOELLIPSE format (Lahr, 1989); VolQuake reads data in the HYPOELLIPSE and the HYPO71 formats (Lee and Valdés, 1989). In addition to the original references, the formats are described within VolQuake's built-in Help system.

Both PINAPLOT and VolQuake are provided on the accompanying floppy disk, along with data files for Mount Pinatubo (1991), Mount St. Helens (1980–92) and Mount Spurr (1992). PINAPLOT will accept only the Pinatubo data; VOLQUAKE will accept all three data sets.

---

[1]U.S. Geological Survey.

PINAPLOT

HARDWARE REQUIREMENTS

- IBM or IBM-compatible personal computer
- EGA color (640 × 350 pixel) or VGA color (640 × 480 pixel) graphics adapter and display monitor

INSTALLATION

PINAPLOT may be run from a floppy disk or a hard disk. The required files are provided in compressed form within PPSETUP.EXE, a self-extracting file in directory PINAPLOT. There is insufficient room on the distribution disk to contain the PINAPLOT files in an uncompressed form. You must therefore copy PPSETUP.EXE to a hard disk or floppy disk. From the DOS prompt, in the drive and directory to which to you have copied PPSETUP.EXE, type "PPSETUP", then press the Enter key; this will decompress the following files:

    PINAPLOT.EXE
    PINA.MAP
    PINAB4.DAT
    PINAAFTR.DAT
    README.DOC

These decompressed files, plus the self-extracting file, occupy about 0.5 megabytes.

OPERATION

From the DOS prompt, in the drive and directory containing the PINAPLOT files, type "PINAPLOT", then press the Enter key. The program will request some information. Type an answer to each question; press the Enter key after each answer. Press only the Enter key to input the default value for a given question.

VOLQUAKE

HARDWARE REQUIREMENTS

- IBM PC/AT or completely compatible equivalent certified for Microsoft Windows 3.1 (386, 486, or later CPU), at least 4 megabytes of RAM, and a hard disk
- VGA or better color (640 × 480 pixels) graphics adapter and display monitor

OPTIONAL HARDWARE

- Printer
- fax modem
- Sound board

SOFTWARE REQUIREMENTS

- Microsoft Windows 3.1

INSTALLATION

Insert the disk in drive A (or B). From Program Manager in Windows, choose Run from the File menu. Type A:\VQ\SETUP (or B:\VQ\SETUP) and press the Enter key. If you wish to install VolQuake in the default drive and directory (C:\VQ), then click the Continue button. Otherwise, type the drive and directory of your choice, then press the Enter key. Once they are decompressed, VolQuake files occupy about 2.1 megabytes.

OPERATION

Position the mouse pointer over the volcano icon and double-click the left mouse key; that is, depress the mouse key twice in rapid succession. Shortly after a brief introductory screen, a Menu will appear. In the present configuration, three topics are listed: Mount St. Helens (1976–92), Mount Pinatubo (1991), and Mount Spurr (1991–92) are listed. Each topic refers to an earthquake data set. To explore a given data set, select that set with the mouse pointer, then click the OK button. Alternatively, the desired topic may be selected by double-clicking. As soon as a topic is selected, two windows will appear. The window on the left shows earthquake locations around a volcano as seen in map view; this is called an epicenter map. The window on the right side of the screen shows earthquake locations as seen in cross section; this is called a hypocenter plot. Brief flashes of light and persistent symbols show earthquake locations. As an earthquake is plotted, its time of occurrence appears within a small box in the upper central part of the screen. A cartoonlike animation showing the volcano's outward behavior appears in this same box. Small boxes in the upper right and left corners of the screen explain the plot symbology. When the entire data set has been plotted, VolQuake will beep and display the number of earthquakes in the data set at the bottom of the screen. This is the standard presentation.

Numerous other presentations are available if you click the buttons along the bottom of the screen; you may click these buttons at any time. Probably the easiest way to learn how these buttons operate is to experiment with them.

Note that seismic data are unavailable for Mount Pinatubo between June 11 and June 29, 1991, due to eruption-related loss of seismometers. Within this time interval, VolQuake will display the eruption animation but will not plot any earthquake hypocenters.

Documentation of all aspects of VolQuake is contained within the built-in Help system. One way to enter the Help system is to click the Help buttons located on many of

VolQuake's windows. This will produce a summary Help screen that pertains to the window that contains the Help button. An alternative way to enter the Help system is to press the F1 key on your keyboard. The Help system is context sensitive, which means Help topics are linked to VolQuake's various elements. When you press the F1 key, the topic that appears in the Help window depends on which VolQuake element is active when you press F1. The active button can be recognized by the small dots that surround the text on the button. You make a button active by repeatedly pressing the Tab key. Not all VolQuake elements can become active. If you want help on some part of VolQuake that you can't activate with the Tab key, press the F1 key, then use the browse keys (<<, >>) on the Help window until the desired topic appears. If this fails, click the Search key on the Help window, then follow the instructions to conduct a keyword search.

There are four ways to close a Help window.

1. On your keyboard, press the Alt key, followed by the F4 key.
2. Click the negative sign in the upper left-hand corner of the window, then click "Close."
3. Double-click the negative sign in the upper left-hand corner of the window.
4 Click "File" on the upper left of the window, then click "Exit."

SUMMARY

PINAPLOT and VolQuake allow the user to experience seismic data as they were received. By using the Pause button in VolQuake, you can stop plotting at any time and ask yourself how you might have interpreted the data at that point or what advice you might have given to those at risk. Suggestions for future revisions are welcome.

ACKNOWLEDGMENTS

We wish to thank Steven Malone for kindly providing the Mount St. Helens data set for inclusion with VolQuake.

REFERENCES CITED

Lahr, J.C., 1989, HYPOELLIPSE/Version 2.0: A computer program for determining local earthquake hypocentral parameters, magnitude, and first motion pattern: U.S. Geological Survey Open-File Report 89–116.

Lee, W.H.K., and Valdés, C.M., 1989, User manual for HYPO71PC, Chapter 9, in Lee, W.H.K., ed., IASPEI Software Library Volume 1, Toolbox for seismic data acquisition, processing, and analysis: El Cerrito, Calif., International Association of Seismology and Physics of the Earth's Interior in collaboration with Seismological Society of America, p. 203–236.

Parker, D.B., 1990, User manual for ACROSPIN, Chapter 6, in Lee, W.H.K., ed., IASPEI Software Library Volume 2, Toolbox for plotting and displaying seismic and other data: El Cerrito, Calif., International Association of Seismology and Physics of the Earth's Interior in collaboration with Seismological Society of America, p. 119–184.

Seismicity and Magmatic Resurgence at Mount Pinatubo in 1992

By Emmanuel G. Ramos,[1][2] Eduardo P. Laguerta,[1] and Michael W. Hamburger[2]

ABSTRACT

Seismicity at Mount Pinatubo in 1992, related to a resurgence of magma after the 1991 eruptions, can be classified into two distinct types of earthquakes and several types of volcanic tremor. **High-frequency earthquakes** have characteristic waveforms dominated by distinct P- and S-phases and rapidly decaying coda, similar to tectonic earthquakes. The number of high-frequency earthquakes steadily increased from the start of seismic observation at Pinatubo in April 1991 until the cataclysmic eruptions of June 1991. During the second half of 1991 and the first half of 1992, the number of high-frequency earthquakes decreased steadily from the June 1991 peak. In contrast, **low-frequency earthquakes** are characterized by emergent long-period waveforms and extended codas. Numerous low-frequency earthquakes preceded the cataclysmic eruption of June 1991; after the 1991 eruption, low-frequency earthquakes became infrequent and constituted only 1 in 10 recorded events.

A resurgence of low-frequency earthquakes on July 7, 1992, was followed the next day by phreatic explosions through the caldera lake and then a 4-month period of dome growth. In the later stages of dome growth, swarms of low-frequency earthquakes were common, peaking in mid-October with more than 1,400 events per day. The low-frequency events in these swarms occurred mostly as multiplets—groups of earthquakes with strikingly similar amplitude and waveform characteristics and with pronounced spatial and temporal clustering.

Volcanic tremor at Pinatubo is of three principal types: (1) low-amplitude, low-frequency harmonic tremor, observed occasionally throughout the posteruptive period; (2) large-amplitude, bimodal-frequency tremor, observed only once prior to the eruption in 1991 but frequently during the early months of 1992; and (3) high-frequency tremor that lacks distinct dominant frequencies and exhibits irregular, "spiky" envelopes. This high-frequency tremor was caused by several internal and external processes at the volcano, including lahars, secondary explosions, and eruptions.

Bimodal-frequency tremor that occurred almost daily from April to July 1992, low-frequency earthquakes that briefly reappeared in early July 1992 before the phreatic explosions, 1992 dome formation inside the 1991 caldera, and swarms of low-frequency multiplet earthquakes during late stages of dome growth are all inferred to be the result of renewed magma intrusion, 16 months after the paroxysmal eruption.

INTRODUCTION

The work of Minakami and others (1950) at Usu Volcano, Japan, was among the first attempts to systematically classify volcanic earthquakes and to relate earthquake signal character to volcanic processes. Since then, the unusual and diverse characteristics of volcanic earthquakes and tremor have been widely documented and used to infer conditions of many active volcanoes. The 1991 eruption of Mount Pinatubo was preceded by at least 2 months of seismicity, during which earthquakes intensified in both number and size as the time of the eruption neared (Harlow and others, this volume). After the climactic eruption, the seismicity became increasingly diverse, not just because the period of observation grew longer but because of extraordinarily complex processes in the interior of the volcano. In this paper, we present an overview of the different types of seismic activity at Mount Pinatubo that focuses primarily on the events that occurred in 1992. Our preliminary classification of volcanic earthquakes is based mainly on their features on analog seismograms, supplemented by digital records. Two types of earthquakes and several classes of tremor are described, as are the possible source mechanisms of each type of seismic event.

SUMMARY OF RECENT ACTIVITY AT MOUNT PINATUBO

1990–91 ACTIVITY

To understand the seismicity at Pinatubo in 1992, it is useful to place this activity in a broader context (see papers by Harlow and others, this volume; Mori, White, and others,

---

[1] Philippine Institute of Volcanology and Seismology.

[2] Department of Geological Sciences, Indiana University, Bloomington, IN 47405.

this volume; White and others, this volume; Punongbayan and others, this volume). Table 1 presents a brief chronology of the events at Pinatubo from the first signs of activity in July 1990 until the seismic swarms in July through October 1992. The first indication of abnormality at Mount Pinatubo was in August 1990, 2 weeks after the July 1990 Luzon earthquake (M_s 7.8), when local residents reported frequent earthquakes and "smoke" coming from the summit of the volcano (Ramos and Isada, 1990). The site was investigated but, owing to the requirements of aftershock monitoring closer to the large earthquake's epicenter, the Philippine Institute of Volcanology and Seismology (PHIVOLCS) was unable to deploy seismic instruments at the volcano.

Seismic monitoring at Mount Pinatubo started in April 1991, when phreatic explosions and landslides were observed near the summit. Initial observations found that the rate of seismicity was abnormally high and that earth-

Table 1. Chronology of events at Mount Pinatubo, July 1990 to November 1992.

1990	
July 16	M_s 7.8 Luzon earthquake, epicenter ~100 km northeast of Pinatubo. Building collapses, faulting, liquefaction, landslides extensive over most of central and northern Luzon.
August 3	"Smoke" from cracks and tremors reported by residents on northwest flank of Mount Pinatubo. Five families evacuated. PHIVOLCS sends team on August 5, reports landslides near the summit, allayed fears of eruption.
1991	
April 2	Small phreatic explosions along 1.5-km-long northeast-southwest aligned vents. PHIVOLCS recommends precautionary evacuation of 4,000 people.
April 5–6+	PHIVOLCS records hundreds of high-frequency earthquakes at Yamut, 7 km northwest of the summit. Low-level steam and ash emissions observed until June 6. Seismographs borrowed from Taal Volcano, which was also having swarms of earthquakes. Assistance from USGS sought. Earthquakes continuously recorded at high levels (>100 per day), most from beneath the northwestern flank at depths of 6 to 35 km.
April 23	USGS team arrives. PC-based telemetered seismic network installed over following 2 weeks.
May 13	Five-level alert system for Mount Pinatubo adopted. Level 2 raised, indicating probable magmatic intrusion.
Late May 1991	Preliminary hazards maps for pyroclastic flows, airfall ash, and lahars prepared. Highest airborne SO_2 value measured. Earthquakes are shallower and migrated from northwest toward summit.
June 4	Short burst of low-frequency "deep" tremor observed.
June 5	Increase in size and number earthquakes. More phreatic explosions. SO_2 decrease detected; possible magmatic sealing of vents inferred. Alert Level 3 raised: Eruption possible within 2 weeks.
June 6	Small swarm of high-frequency earthquakes.
June 7	Shallow long-period tremor. Long-period earthquakes recorded. Inflationary tilt detected. Ash explosion reaching 8,000 m accompanied by tremor. Alert Level 4 issued: Eruption possible within 24 h.
June 8	Lava dome was observed on the northern flank.

1991—Continued	
June 9	Ash emission intensifies. Alert Level 5 issued: Eruption in progress.
June 10	14,500 personnel and dependents of Clark Air Base were evacuated.
June 11	Explosions with possible pyroclastic surge. Seismic station UBO, high on the east flank of Pinatubo, destroyed.
June 12–14	Major eruption phase began with explosions producing 25-km-high columns and, on June 14–early June 15, pyroclastic surges or blasts (Hoblitt, Wolfe, and others, this volume). Swarms of long-period earthquakes.
June 15	Peak of eruption with pyroclastic flow in all directions of the volcano. Ash columns reached as much as 35–40 km high. 2.5-km-diameter caldera was formed. More than 200,000 people evacuated to safer grounds. Rains caused lahars to form from the freshly deposited volcanic materials. About 280 people died, mostly from collapse of buildings under ash fall.
June 16+	Intensity of eruption slowly weakened, and seismicity slowly decreased in intensity.
June 29–July 18	Pulses of eruptions lasting about 2 h and separated by about 7–9 h of repose. Seismicity continued to decrease. Lahars caused by rains.
August–September	Lahars caused major damage, even far from the volcano. Last explosion reported from the caldera was on September 4, after which a lake formed on the caldera floor. Seismicity declined further.
1992	
May 3–July 19	Series of large amplitude, low-frequency volcanic tremor events resembling the "deep" tremor recorded on June 4, 1991.
July 6–8	Small swarm of long-period volcanic earthquakes. A small tuff cone from a series of small phreatic explosions formed in the crater lake. Alert Level 3 issued.
July 14	Lava dome 100 m in diameter replaced the small tuff cone. Alert Level 5 issued.
August–October	A series of swarms of long-period volcanic earthquakes, lasting five to 18 days, each including 50 to about 1,300 events per swarm. Lava dome grew until end of October.

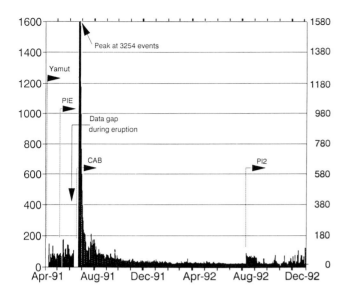

Figure 1. Histogram of daily counts of high-frequency earthquakes at Mount Pinatubo before and after the 1991 eruption, taken from various discontinuous but overlapping analog records. Records are from Yamut (April 14 to 30, 1991), PIE (May 1 to June 15, 1991), CAB (June 16, 1991, to August 4, 1992), and PI2 (August 5 to December 31, 1992). Left-hand scale, for Yamut, PIE, and PI2; right-hand scale, for CAB.

quake epicenters were located several kilometers northwest of the summit. The earthquakes were similar to tectonic earthquakes, with distinctly observable high frequency P- and S-phases. In the next few weeks, evidence of volcanic unrest mounted, and the cataclysmic eruptions of June 12–15, 1991, were successfully anticipated (PVO Team, 1991; Punongbayan and others, 1992; Wolfe, 1992). In addition to the high-frequency earthquakes, the other seismic events that preceded the volcanic eruption were low-frequency earthquakes and various types of tremor.

Figure 1 shows the history of seismicity associated with the 1991 eruption. The high rate of seismicity is evident in the periods immediately preceding and after the cataclysmic eruption. The seismicity rate declined rapidly after the eruption, even as explosions continued to recur at decreasing intensity. The last explosions from the crater took place in early September 1991. Most of the seismicity at that time was characterized by high-frequency earthquakes, which, by then, were gradually becoming less frequent. A shallow lake formed inside the new caldera. Some residents of villages near the volcano returned to their homes, although many still remained in evacuation centers as lahars continued to cause destruction around the slopes of the volcano.

1992 ACTIVITY

Four important events occurred at Mount Pinatubo in 1992 that interrupted the general decline of posteruptive

seismicity. First, large-amplitude tremor occurred almost daily from May to July. These tremor events resembled tremor of June 4, 1991, immediately prior to the explosive eruptions. Second, a brief swarm of small-amplitude, low-frequency earthquakes abruptly appeared from July 6 to 8, 1992. Third, a new island formed within Pinatubo's caldera lake by small phreatic explosions that were too small to be recorded by the seismographs. Alert Level 3 was issued (PHIVOLCS, unpub. daily updates, 1992), and, on July 14, a lava dome began to grow (Daag, Dolan, and others, this volume), even though seismicity still remained low. The fourth important event lasted from August to October 1992, when four intense swarms of low-frequency volcanic earthquakes were recorded by the Pinatubo seismic network. Lasting from 7 to 25 days, each of these swarms contained low-frequency earthquakes of nearly uniform magnitudes and strikingly similar waveforms. The swarms peaked on October 16, 1992, when 1,422 events were recorded in 1 day, dwarfing the levels of seismicity that preceded dome formation and even that recorded before the 1991 eruption. At the time of these swarms, no significant growth of the dome was observed, and the SO_2 flux measured with the correlation spectrometer (COSPEC) showed dips in values similar to those observed before the 1991 eruption (Daag, Tubianosa, and others, this volume). Dome growth and the swarms of low-frequency earthquakes prompted PHIVOLCS to raise Alert Level 5 until December 9, 1992 (PHIVOLCS, unpub. daily updates, 1992). Concern about renewed explosive activity slowly waned as both the swarms of low-frequency earthquakes and the dome-related activity inside the crater appeared to subside.

DATA AND METHODS

In response to the developing emergency at Mount Pinatubo in early 1991, PHIVOLCS, with the subsequent assistance of the U.S. Geological Survey (USGS), deployed seismic instruments to a total of 23 different sites around the volcano. Analog portable seismographs (Kinemetrics Ranger and Sprengnether MEQ 800 Seismographs with 1-s Ranger seismometers) recording on smoked paper were the first instruments deployed. When the high rate of seismicity was confirmed, digital event data recorders (Kenkei Systems EDR 1000 with 1-s L2 seismometer) were also deployed. All of these instruments were operating independently until late April 1991, when a telemetered, PC-based seismic monitoring system was installed by the USGS to centralize the system and facilitate data analysis. The seismometers used in the telemetered system were 1-s natural period Mark Products L4 vertical and horizontal sensors (Lockhart and others, this volume; Murray and others, this volume). Murray and others (this volume) provide a summary of the PC-based data processing. Routine data analysis included manual phase picking for hypocentral locations

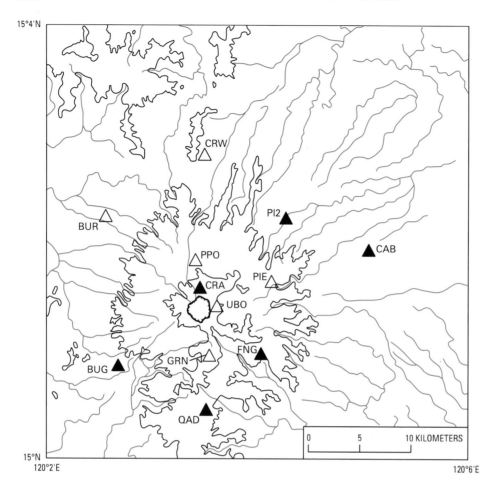

Figure 2. Location of telemetered seismic stations at Mount Pinatubo. Most preeruption sites of analog recorders were located farther from the summit, outside of the boundaries of this map. Open triangles are stations destroyed by the eruption, and closed triangles are stations operating in 1992. Signals were initially transmitted to CAB until early 1992, after which time they were transmitted to the PHIVOLCS office in Quezon City. Contour lines of 500- and 1,000-m elevations show preeruption topography. Hatched circle marks 1991 caldera.

and measuring coda amplitudes and durations for magnitude determinations. The analog data were collected from the same network by tapping the input lines to the digital data acquisition system and recording on pen-ink drum recorders. Only one of the original seismic stations (CAB) survived the eruption; after the eruption, CAB and one other original station (BUG) were repaired, and three new stations (PI2, FNG, and QAD) were added. In August 1991, recording on portable seismographs was discontinued, leaving only the telemetered, PC-based system in operation. In August 1992, a new seismic station (CRA) was installed near the rim of the 1991 caldera (fig. 2).

In our analysis of the earthquakes, we have made extensive use of the analog records obtained from the telemetered network. The completeness and continuity of these records are especially useful for studying small-magnitude events and those with emergent onset that do not ordinarily trigger the digital network. The analog records are also useful for studying temporal relations between events, because they show a continuous data stream. There are also significant disadvantages in using these analog records, however. These disadvantages include (1) inaccuracy in the timing of the analog seismograms, since the analog recorders relied on independent timing systems that were not synchronized

with the central digital system; (2) difficulty in making spectral measurements, particularly for high-frequency parts of the records; (3) limitation on the dynamic range of amplitude recorded on the analog seismograms, constrained by the physical limits of the pen-ink recording system which thus made it impossible to measure actual amplitudes of large events; (4) limitations on the number of analog recorders connected to the digital system (only two stations were connected consistently to analog recorders before July 1992, when more analog recorders became available, so station coverage of the analog records is inferior to that of digital records); and (5) the unavailability of calibration data for the analog recorders, which prevented us from using these records for magnitude calculations. Analog seismograms from stations PIE and UBO for the preeruption period (May–June 1991), and seismograms from station PI2 for the period from January to August 1992, were used to classify events and to measure phase arrival time, amplitude, and coda duration. All events exceeding 4 mm in amplitude on recording charts were noted; spectral frequencies of tremor were measured with an optical loupe.

Earthquake hypocentral locations obtained from routine digital data analysis were used to define the spatial distribution of the events. Coda-magnitude determinations from the digital seismograms were available for events detected by the digital system. Spectral analysis of the digital records of the two types of earthquakes was accomplished by using the International Association of Seismology and Physics of the Earth's Interior's (IASPEI's) program PCEQ (Valdez, 1989).

Analysis of the volcanic tremor presented an unusual difficulty, since none of these events were recorded in the digital system. Our analysis was entirely dependent on 1992 analog records that were available only for two stations, PI2 and FNG (fig. 2), for most of the study period. For tremor events preceding the 1991 eruption, we have also referred to the seismograms gathered by the portable analog seismographs deployed around the volcano.

CLASSIFICATION OF SEISMIC EVENTS AT MOUNT PINATUBO

The seismic activity at Mount Pinatubo is of two distinct varieties: discrete short-duration earthquakes and long, usually emergent, slowly decaying volcanic tremor. The earthquakes can be classified further into high-frequency and low-frequency events, according to the dominant frequency content of the signal. We also have classified tremor according to frequency, namely: low frequency (monochromatic or harmonic), bimodal frequency, and high frequency. High-frequency tremor was classified further into genetic subdivisions.

HIGH-FREQUENCY EARTHQUAKES

Waveform description.—The most common type of seismicity at Mount Pinatubo is the high-frequency (short-period) earthquake (fig. 3). The waveform of these events resembles that of typical tectonic earthquakes recorded near the source. The signature is dominated by distinct body wave P- and S-phases. The high-frequency type corresponds to Minakami's "A-type" (Minakami and others, 1950; Minakami, 1974), to the "volcano-tectonic earthquakes" of Latter (1979), and to the "type h" of Malone (1983). The onset of the high-frequency earthquakes is often very impulsive, although emergent arrivals that are occasionally observed at individual stations may be due to attenuation of the P-phases along raypaths traversing or skirting magma chambers (Latter, 1984) or possibly to a particular orientation of the energy radiation pattern at the earthquake source. Figure 4A shows a seismogram of a high-frequency earthquake and the corresponding displacement spectrum of the signal. The peak of the spectra for these high-frequency events tends be nearly flat between 1 and about 6 Hz.

Like tectonic earthquakes, the waveform of high-frequency earthquakes is also dependent on magnitude,

depth of hypocenter, and source-to-station distance. High-frequency earthquakes of larger magnitude, shallow depth, or distant sources tend to be dominated by lower frequencies because of high-frequency attenuation and (or) the generation of surface waves. Events classified by Minakami and others (1950) as "B-type," and the "m-type" events at Mount St. Helens (Malone, 1983), may reflect these variations in the high-frequency earthquakes (Hamada and others, 1976; Chouet, 1988).

Space-time distribution.—The high-frequency earthquakes have a wide range of spatial, temporal, and magnitude distribution. High-frequency earthquakes include the cluster of events prior to the 1991 eruption that were located to the northwest of the volcano at depths of about 4 to 10 km and include the more extensive posteruptive seismicity that ranged in depth from the surface to about 18 km (Mori, White, and others, this volume). The lateral distribution of high-frequency earthquakes is also wide, extending well beyond the edifice of Mount Pinatubo proper. The northwest cluster of preeruptive seismicity and the more extensive seismicity after the eruption consist mostly (>80%) of high-frequency earthquakes (Mori and others, 1991).

The number of high-frequency earthquakes increased sharply as the 1991 eruption neared and then decayed rapidly afterwards. In the first half of 1992, the high-frequency earthquakes continued their slow decrease (fig. 1). Prior to dome formation in 1992, the number and magnitude of high-frequency events changed. A brief and subtle swarm occurred from June 20 to 22 (fig. 5, top). A noticeable decrease in both number of events and magnitude followed, during the period of about 8 days starting around June 23, 1992. The level of high-frequency seismicity again increased before a short swarm of low-frequency seismicity and the dome formation in July 1992.

Magnitude distribution.—The magnitude distribution (based on coda measurements) of high-frequency earthquakes for the 16 months from April 1991 to August 1992 shows that the events ranged in size from magnitude $M_L=0$ to greater than $M_L=4$ ($>M_L$ 5.5 on June 15, 1991). This broad magnitude spectrum is observed for both long and short time duration. Figure 5 illustrates the wide magnitude distribution of high-frequency events over the entire study period; figure 3 shows a similarly wide range of earthquake sizes over the one-day period of a single seismogram. Mori, White, and others (this volume) calculated a b-value (Richter, 1958) of 1.1 ± 0.3 for this type of Pinatubo earthquakes.

LOW-FREQUENCY EARTHQUAKES

Waveform description.—The second type of discrete event observed at Pinatubo is the low-frequency earthquake (fig. 6). Harlow and others (this volume) and Murray and others (this volume) use the term long-period earthquake for this type of event at Pinatubo in 1991. The low

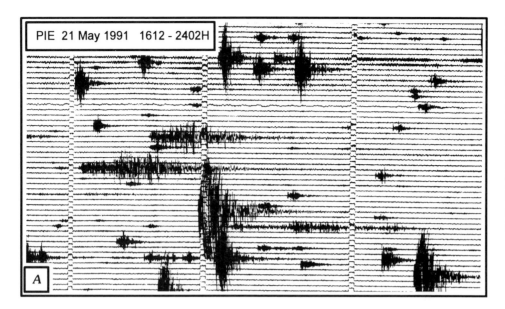

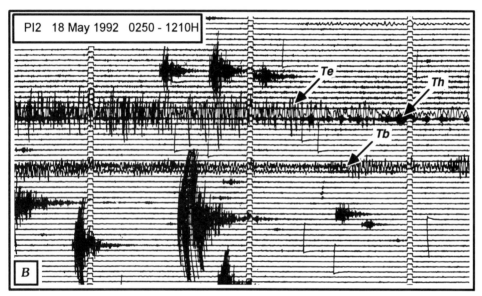

Figure 3. Seismograms from PIE (from PI2 after the 1991 eruption) showing high-frequency earthquakes of various size and from different hypocentral locations, as recorded in May 1991 (*A*) and May 1992 (*B*). *B* shows three other types of events, a tectonic earthquake (Te), harmonic tremor (Th), and bimodal-frequency tremor (Tb). Time runs from top to bottom, left to right. Time marks are 60 s apart. Date and time shown are local.

frequency signal peaks at about 1 Hz (fig. 4*B*), and the onset of these earthquakes is usually emergent and lacks distinct P- and S-phases. These events tend to have long coda duration with respect to their amplitudes, adding to their distinction from high-frequency earthquakes. The typical waveform of low-frequency events shows some pulses of energy (or pseudophases) that seem to persist from one event to another. These pseudophases are not easily correlatable between stations, and their overall characteristics seem to differ from those of typical P- and S-phases, being of low frequency and lacking the pulse of energy associated with the arrival of direct body phases. On some large events, particularly on seismograms recorded near the crater (station CRA), the early part shows as an impulsive phase, although even this is usually preceded by an emergent, low-amplitude, high-frequency signal resembling a refracted phase arrival. A high-frequency component is often observed early in the low-frequency earthquake signal, similar to that described by Fehler and Chouet (1982) and Chouet (1992) at other volcanoes. Clearly, some high-frequency content is present in the earthquake source and is released in the early part of the event.

Space-time distribution.—Unusual, deep, low-frequency events occurred on May 26 and June 4, 1991 (White, this volume). In the succeeding week, low-frequency events were overshadowed by shallow high-frequency events; then, during the period of explosive eruptions between June 12 and 15, shallower low-frequency events became prominent (Harlow and others, this volume). During the remainder of 1991, high-frequency earthquakes were about 10 times as numerous as low-frequency earthquakes.

In 1992, low-frequency earthquakes had an unusual temporal distribution. Low-frequency events began in early

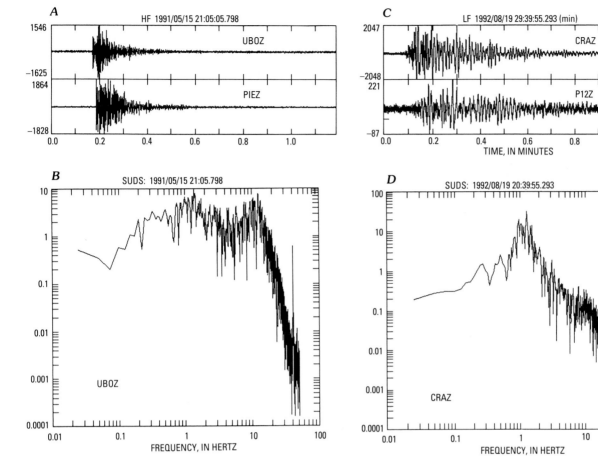

Figure 4. *A,* Seismograms of a typical high-frequency earthquake at Mount Pinatubo, May 15, 1991, as recorded at stations UBO and PIE. Amplitudes are normalized for comparison of waveforms. Time marks in seconds are shown below each seismogram. *B,* Instrument-corrected amplitude spectrum from the UBO signal for the same event as in A. Note the broad, relatively flat spectral signature. *C,* Seismograms of a typical low-frequency earthquake at Mount Pinatubo, August 19, 1992, as recorded at stations CRA and PI2. *D,* Instrument-corrected amplitude spectrum from the CRA signal for the event shown in *C.* All records are from vertical seismometers, hence the Z suffix on station names.

May and then became markedly absent for a period of about 8 days, from June 24 to about July 2, coincident with reduced high-frequency seismicity (fig. 5). On July 7 and 8, immediately preceding the phreatic explosions and the start of dome formation, a small swarm of low-frequency earthquakes occurred. The rate of occurrence of low-frequency earthquakes was slightly elevated, averaging about 15 per day, when, on July 14, the dome began to grow. This number of events gradually increased by the end of July, developing into a swarm with about 120 events per day. This was the first of four intense swarms recorded in 1992, each swarm lasting from 7 to 25 days (fig. 7). The peak number of events in each swarm also increased. In the last swarm, on October 24, the highest number of earthquakes in the history of Mount Pinatubo was recorded when 1,422 events were recorded in a day.

Hypocenters of the low-frequency earthquakes were within a relatively small region, mostly near or beneath the caldera, usually within a radius of 2 km from the growing dome. The events were always shallow, usually between 0 and 3 km depth, and rarely deeper than 5 km (fig. 8).

Magnitude distribution.—The range of magnitudes of low-frequency earthquakes is much narrower than that of high-frequency earthquakes. Although some low-frequency events in 1991 were recorded as far as 300 km away (PVO Team, 1991), the majority of 1992 events did not exceed a magnitude of 1. These earthquakes tend to occur as groups of earthquakes with uniform sizes, as illustrated by the seismograms in figure 6. In the August–October 1992 swarms, the low-frequency earthquakes changed size in unison, from one cluster of earthquakes to the next, with each size being retained only for a few hours to a few days. The b-value we have calculated for low-frequency earthquakes is 1.46 ± 0.04.

Earthquake multiplets.—One very unusual feature of the recent swarms of low-frequency earthquakes is the uniformity of their waveform. These earthquakes occurred as multiplets—earthquakes with very similar signatures.

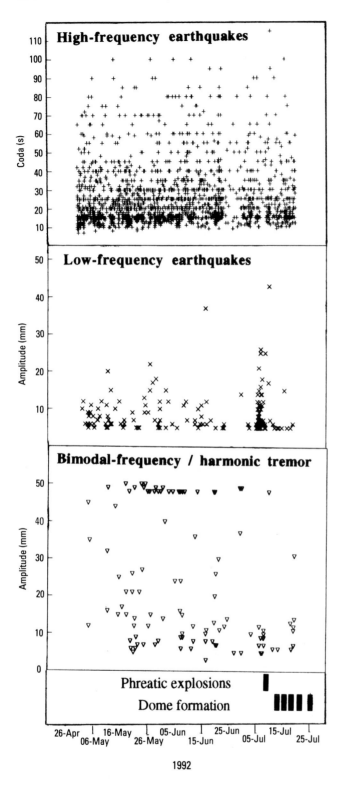

← **Figure 5.** Duration and maximum amplitude of earthquake codas, and maximum amplitude of bimodal-frequency and harmonic tremor, as recorded at station PI2 before the dome-forming episode in 1992. All counted earthquake codas were ≥4 mm in amplitude on PI2's analog record. All tremor events with amplitudes ≥ 20 mm are bimodal-frequency tremor. Clipping amplitude was 48 mm. Note the reduced number and size of both high- and low-frequency earthquakes prior to the July 7–8 swarm of low-frequency events and the dome formation.

signature of the earthquakes is almost identical, with peaks and troughs matching from one event to the other. Multiplets have been noted in many other sites, including Usu Volcano, Japan (Minakami and others, 1950; Okada and others, 1981), Mount St. Helens, United States (Fremont and Malone, 1987; Neri and Malone, 1989), Ruapehu Volcano, New Zealand (Dibble, 1974), and sites away from volcanoes (for example, Cole and Ellsworth, 1992; Sherburn, 1992). Okada and others (1981) termed events with such behavior as belonging to "earthquakes families." Minakami and others (1950) described events at Usu Volcano with similar repetition of waveforms and classified the events as "type C." The similarity in waveform of multiplets indicates that the events occur within a very limited region of space (Dibble, 1974; Fremont and Malone, 1987; Neri and Malone, 1989; Cole and Ellsworth, 1992) and that the earthquakes share the same source mechanism (Fremont and Malone, 1987). A significant feature of low-frequency earthquake multiplets in volcanoes is that they frequently precede or accompany dome formation (Fremont and Malone, 1987; S.D. Malone, written commun., 1993).

VOLCANIC TREMOR

Volcanic tremor, in contrast to the earthquakes, lasts for extended periods of time (Malone and Qamar, 1984); at Mount Pinatubo, tremor lasted from a few minutes to a few days. Tremor at Mount Pinatubo is classified into low-frequency (harmonic), bimodal-frequency, or a broad category of high-frequency tremor.

HARMONIC TREMOR

The type of tremor with the narrowest frequency spectra is the harmonic tremor. Every event of this tremor is monochromatic, occurring with a very characteristic narrow frequency that remains constant throughout the duration of the events (fig. 10A,B). The dominant frequency may vary from event to event, but the range of frequency is still narrow, from 0.7 to 2.4 Hz. One feature of this tremor is its pulsed amplitude that results in regularly spaced lobate envelopes of uneven sizes. This tremor is emergent, initially appearing from the background as small pulses and gradually emerging as symmetrical, smoothly formed trains of

Figure 9 shows the first 30 s of the low-frequency earthquakes recorded at stations PI2 and FNG in the first days of the August 1992 swarm. These events were randomly picked from the hundreds of earthquakes that occurred that day, and virtually all of them showed strikingly similar waveforms. From the start to the end of their codas, the

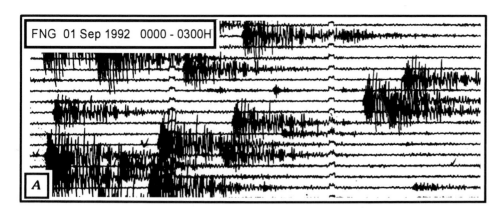

FNG 01 Sep 1992 0000 - 0300H

A

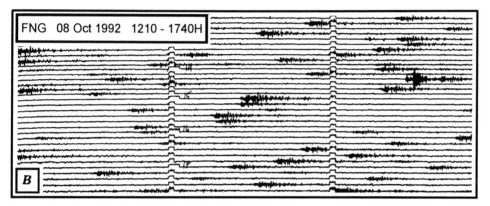

FNG 08 Oct 1992 1210 - 1740H

B

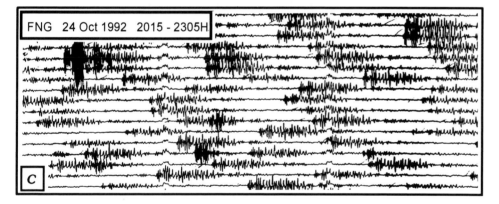

FNG 24 Oct 1992 2015 - 2305H

C

Figure 6. Seismograms recorded at station FNG during the swarms of low-frequency earthquakes in (*A*) September 1, 1992, (*B*) October 8, 1992, and (*C*) October 24, 1992. Instrument setting was the same for all records. Note uniform size of the earthquakes in each sequence. The similarity of the waveforms, most noticeable in *C*, indicates that many of these events are multiplets. Time marks as in figure 3.

lobes, then gradually fading into the background. The duration of harmonic tremor is highly variable and is inversely proportional to its amplitude. The large-amplitude harmonic tremor may be as short as a few minutes, while the small-amplitude events may last for a few hours to a few days, remaining as background microseism while other events occur. The longest period of harmonic tremor recorded at Pinatubo began on June 5, 1992, when a continuous background tremor was visible on the record for 72 h. In some events, the tremor repeatedly occurs after brief periods of repose, creating contrasting dark and light bands on the seismogram that are similar to those described by McKee and others (1981). In 1992, the harmonic tremor seemed to be most frequent in the 2 months preceding the dome formation (fig. 5). A crude estimate of tremor source depths can be made by comparing the relative amplitudes of tremor

at various stations to the relative amplitudes of well-located earthquake signals at the same stations. Using this procedure, we judge that some of the harmonic tremor originates between 5 and 10 km below the caldera.

BIMODAL-FREQUENCY TREMOR

Another type of tremor recorded at Pinatubo has waveforms characterized by the superimposition of two signals with two distinctly different frequencies, one high and the other low (fig. 11). The high-frequency portions of the tremor is always of low amplitude and usually starts earlier and ends later than the low-frequency segment, forming a "head" and a "tail" of the event (fig. 12). The peak frequency in the head and tail portions is relatively high, at

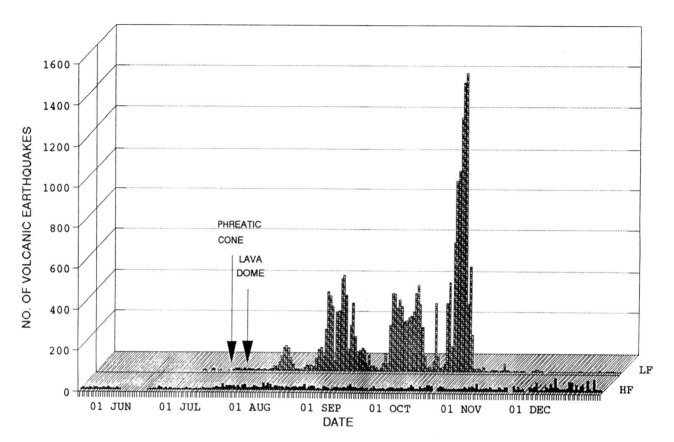

Figure 7. Histogram of high-frequency (foreground) and low-frequency earthquakes recorded at station FNG from June to December 1992.

about 2.4 Hz. This high-frequency signal appears to persist throughout the duration but is overridden by the low-frequency component in the central part of the tremor; it is only visible when the amplitude of the low-frequency portion subsides. Near the end of the signal, the low-frequency portion ends, and again, the high-frequency signal dominates until finally fading into the background. The low-amplitude high frequency head starts an average of 1.6 min (sometimes, 25 min) earlier and the high frequency tail finishes 1 to 11 min later than the low-frequency body, in proportion to the total duration of the whole tremor. Both the head and the tail portions gradually emerge from and fade to the background. Maximum amplitude of the head and tail average about one-tenth of the large-amplitude low-frequency signals.

The low-frequency segment (or body) of this bimodal-frequency tremor is usually monochromatic and emerges from the high-frequency head (figs. 12A,B). The dominant frequency is constant in all events at about 1.0 Hz. The amplitude is usually large, often clipping at PI2, and, as such, is much more vigorous than the earlier mentioned harmonic tremor. The body is composed of sinusoids of

smoothly but constantly varying peaks that form into pulsed amplitude-symmetric envelopes. Compared to the harmonic tremor, the bimodal-frequency is less monochromatic, exhibiting less symmetry along the time axis (that is, the trailing part does not mirror the leading part), and may be caused by the presence of high-frequency components in the tremor. The duration of the low-frequency body may be as short as 30 s and as long as 22 min, generally proportional to the total length of the tremor. On some bimodal events with low amplitude, the low-frequency portions appear to resemble harmonic tremor, but the presence of the high-frequency component in bimodal-frequency tremor distinguishes these from the monochromatic tremor (compare the top and bottom parts of fig. 10B, for example). The overall duration of the bimodal-frequency tremor, including the high-frequency head and tail, ranges from 90 s to 112 min.

Pinatubo's bimodal-frequency tremor usually precedes low-frequency earthquakes. This tremor may have appeared before the eruption, on June 4, 1991 (White, this volume), but was not observed during the eruption or in the next 11 months. On May 3, 1992, it began to occur almost daily

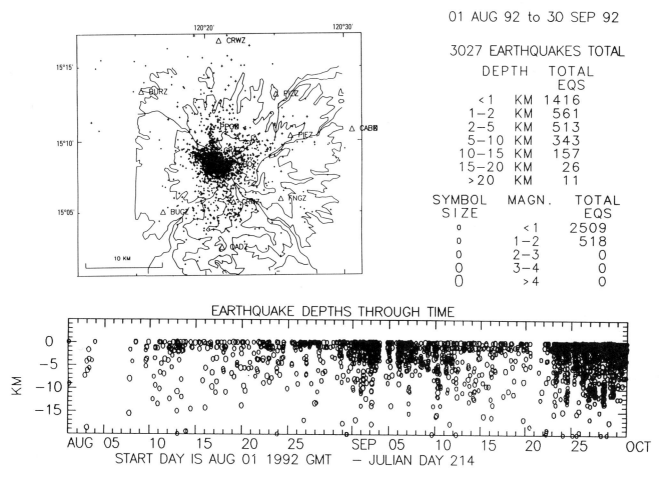

Figure 8. Map of low-frequency earthquakes in August and September 1992. Most of the better located events cluster around the crater, have shallow depths, and have magnitudes (MAGN.) less than 1.

until July 19, 1992. Despite their large amplitudes, the emergent nature of these events prevented them from being detected by the digital system and thus are only recorded on the analog records.

Figure 11*B* and *C* show seismograms of a bimodal-frequency tremor recorded at stations PI2 and FNG. Characteristic of the tremor shown by these seismograms is the similar frequency content of the signal recorded at the two stations and the slight resemblance of the tremor envelope recorded at the two stations. This consistency in the frequency and the resemblance of waveform between stations suggest that these features are caused at the source and not by the wavepath (Aki and Koyanagi, 1981). We infer the probable source region for the bimodal-frequency tremor to be deep, probably greater than the half-aperture of the seismic network, in order for these events to preserve not only their frequency but also their waveform as the waves travel to the various seismic stations. This interpretation is consistent with the observation of White and others (this volume) of deep, long-period earthquakes or tremor on June 4, 1991. Tremor amplitudes on two analog records indicate at least two probable sources for these tremor, one source generat-

ing signals that are consistently larger at FNG than PI2 (average ratio 1:2) and another generating larger signals at PI2 than FNG (average ratio 1:4). Initial estimates of source depths, based on comparisons with located high-frequency earthquakes, suggest that tremor with amplitudes largest at PI2 originates between 4 and 8 km in depth. The absence of any known surface activity at Pinatubo's new crater that might correlate with the occurrence of this type of tremor (fig. 11) further supports this inference of a nonsurface source for the bimodal-frequency tremor.

HIGH-FREQUENCY TREMOR

Other tremor at Mount Pinatubo does not exhibit distinct dominant frequencies and has higher and wider frequency content than the harmonic and bimodal-frequency types. Most is spasmodic, composed of randomly distributed peaks without any well-formed envelopes. Amplitude can increase or decrease abruptly, unlike smooth changes in amplitude in the two previous types of tremor. Much of this high-frequency tremor is genetically associated with lahars, steam emission, and explosive venting of ash. In detail, we

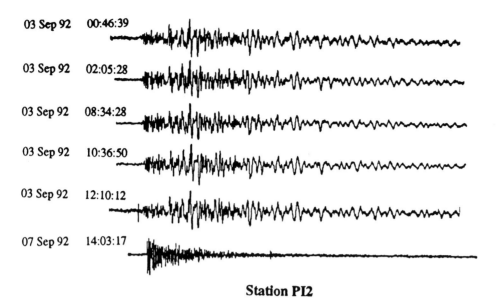

03 Sep 92   00:46:39

03 Sep 92   02:05:28

03 Sep 92   08:34:28

03 Sep 92   10:36:50

03 Sep 92   12:10:12

07 Sep 92   14:03:17

Station PI2

Figure 9. Seismograms of the first 30 s of multiplet low-frequency earthquakes on September 3, 1992, randomly chosen from the hundreds of events in ongoing swarm. Only FNG and PI2 traces are shown, although all stations recorded the events with identical signature from onset to the end. A high-frequency earthquake of September 7 is shown for comparison.

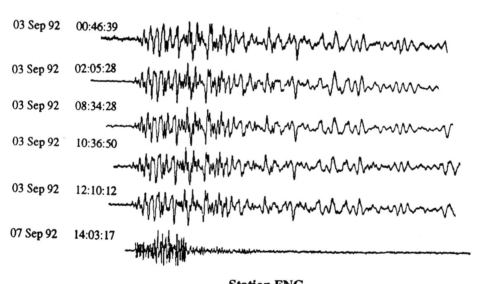

03 Sep 92   00:46:39

03 Sep 92   02:05:28

03 Sep 92   08:34:28

03 Sep 92   10:36:50

03 Sep 92   12:10:12

07 Sep 92   14:03:17

Station FNG

recognize the following three subtypes of high-frequency tremor.

Lahar-induced tremor.—Like the head of bimodal-frequency tremor, the onset of lahar-induced tremor is an emergent, high-frequency signal (fig. 10C). The amplitude of the lahar signal grows more slowly and more steadily than does the amplitude of bimodal-frequency tremor, and high-amplitude portions tend to contain especially high-frequency content. Some short portions of lahar signals appear to be low-frequency surface waves generated by lahar flow or signals from secondary phreatic explosions (described below). Episodes of lahar-induced tremor usually last from about 20 min to a few hours.

Explosion events.—Explosions from Pinatubo's crater usually contain substantial energy and produce high-frequency signals that reach most stations. Figure 13A shows the seismic trace of an early, small, phreatic explosion in May 1991 from a vigorously steaming fumarole on the north flank of Mount Pinatubo, and figure 13B shows seismicity before and at the start of large explosions of June 12, 1991.

Explosions in hot, 1991 pyroclastic-flow deposits are also sources of tremorlike signals. Secondary explosions are caused by near-surface interaction of water and hot pyroclastic materials. These explosions can generate surface waves that appear as low-frequency pulses on seismograms

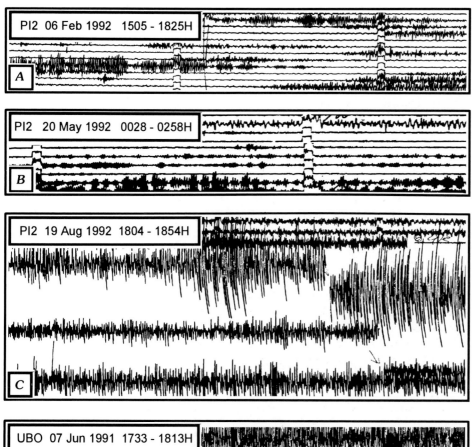

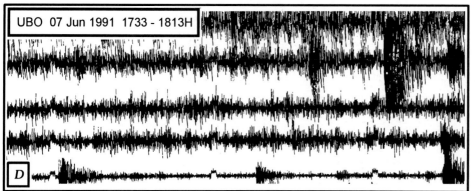

Figure 10. Seismograms of (*A*, *B*), harmonic tremor, (*C*), a lahar signal, and (*D*), high-frequency tremor. The signal on upper part of *B* is the tail of some bimodal-frequency tremor. In *C*, the pen was manually offset twice, as shown by arrows, to avoid overprinting of one rotation's signal on the next. All of the large-amplitude signal is of lahar. In *D*, the high-frequency tremor at station UBO, recorded prior to dome growth in 1991, faded to the background 2 h after the segment shown. Time marks as in figure 3.

(figs. 10*C*, 13*C*). These low-frequency pulses are brief and are recorded only at stations close to the source. Unlike bimodal-frequency tremor, signals from secondary explosions are neither monochromatic nor contained within a smooth amplitude envelope. Tremor from these explosions often blends with longer, lahar-generated tremor, but, because water percolation into the pyroclastic deposits may be delayed or be independent of lahars, secondary explosions can also occur on their own (fig. 13*C*).

Shallow high-frequency tremor.—On June 6, 1991, shortly before the preclimactic dome began to grow, very high-frequency, low-amplitude tremor started to occur in the seismic records of the near-summit station UBO. This tremor was emergent and gradually increased (fig. 10*D*), coincident with increased steam and ash emission from the most vigorous vents. The fact that this tremor was not recorded at stations farther away from the crater indicates that the source of this tremor was probably shallow.

SEISMIC SOURCE MECHANISMS OF THE 1992 EVENTS

A summary of the frequency and coda duration characteristics of the various types of seismic events at Pinatubo is presented in figure 14. Significant overlap in the frequency and coda duration of the various events suggests that a classification scheme based entirely on these two characteristics would be insufficient to completely separate different classes of events. Table 2, a summary of the types of seismic events observed at Mount Pinatubo, includes additional parameters for improved discrimination between event

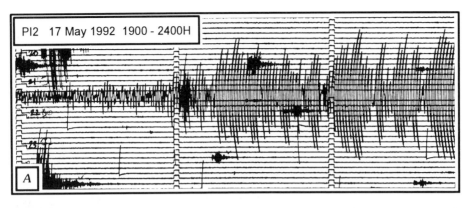

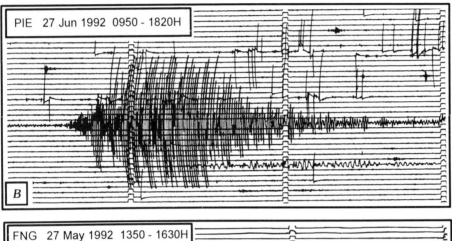

Figure 11. Seismograms of portions of bimodal-frequency tremor. *A*, Sustained bimodal-frequency tremor about 2120 on May 17, 1992, as recorded at PI2. *B, C*, Records of a short-lived bimodal-frequency tremor as recorded at two stations, PI2 and FNG. Record in *C* covers a subset of the time represented in *B*. The high-frequency portion of the tremor is more visible in the early or low-amplitude portions of the tremor. Time marks as in figure 3.

types. In table 2, we indicate the degree to which observable seismologic features of various events are common to seismograms recorded at different seismic stations, and we indicate the degree to which seismograms of a single event resemble earlier or later events of the same type. In general, events with high interstation similarity may be assumed to occur at greater depth, because these signals arrive at a steep angle and are less strongly affected by the raypath and local station conditions. High-frequency earthquakes and the low-frequency (harmonic) tremor are examples of these events. Events with high interevent similarity may originate from limited regions of space and may have easily repeatable source processes. Lahars and multiplet low-frequency earthquakes exhibit these features.

High-frequency earthquakes, being similar to tectonic earthquakes in many ways, probably represent shear failure on a fault or fracture under stress (Weaver and others, 1981; Malone, 1983). Frequent earthquakes after the 1991 eruption were vertically and laterally more extensive than the preeruptive seismicity (Mori, White, and others, this volume) and were probably along local and regional faults, in

response to deflation of the magma chamber and loading by new pyroclastic deposits on the surface.

Low-frequency (or "long-period") earthquakes can be generated by magma and related fluids ascending through existing channels (Shaw and Chouet, 1989; Chouet and others, 1994) or intruding into competent rocks (Latter, 1984). The shallow depths of most low-frequency events at Pinatubo, the existence of high-frequency signals at their onset, and their occurrence prior to surface magmatic activity corroborate the involvement of magmatic fluids in their generation. Multiplet low-frequency earthquakes at Mount Pinatubo during the last months of 1992 suggest that the earthquake source was in a very limited volume and that the earthquake-generating process was easily repeated. Multiplet low-frequency earthquakes can, therefore, be interpreted as repeated, shallow-level intrusions of magma.

We have classified volcanic tremor at Pinatubo according to its characteristics on seismograms recorded at various stations. On the basis of different waveform characteristics at different stations, we infer that the source mechanisms and possibly the source regions for the bimodal-frequency

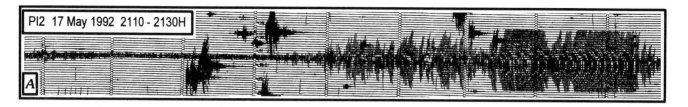

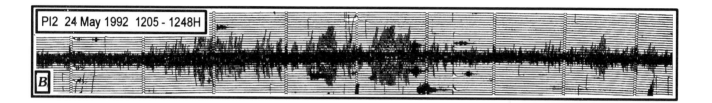

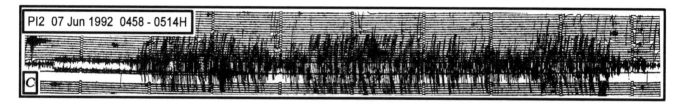

Figure 12. Seismograms of bimodal frequency tremor events in May and June 1992, all recorded at station PI2. Note that the short-duration, large-amplitude low frequency sections are preceded and trailed by long-duration, low-amplitude, high frequency signals.

Time marks are 1 min apart and the seismograms cover almost the whole length of the 10-min-wide helicorder chart. The recording pen was manually shifted while writing seismogram 12C to reduce overlapping of signals.

Table 2. Summary of waveform, amplitude, spatial, and temporal characteristics of Mount Pinatubo's earthquakes and tremor.

[Near-surface events have low interstation but high interevent correlation]

Event type	Spectral character[1]	Amplitude range	Spatial distribution	Temporal distribution	Interstation signal relation[2]					Interevent signal relation[2]				
					P	F	W	A	D	P	F	W	A	D
High-frequency earthquake.	1.0 to 6.0 Hz; broad, flat.	Broad	Extensive	Extensive; minor swarms.	P	S	S	P	P	V	S	V	V	V
Low-frequency earthquake.	0.7 to 4.0 Hz; peaked	Narrow, changing.	Limited	Rare or swarms and pulses.	S	S	S	P	P	P	S	P	S	S
Harmonic tremor	0.7 to 2.4 Hz; sharply peaked	Very narrow	Unknown	Broad swarms	S	P	P	P	P	P	P	S	S	V
Bimodal-frequency tremor.	0.7–1.5; >2.2 Hz; two peaks.	Narrow	Unknown	Rare; broad swarms	S	S	S	P	P	S	P	S	S	V
Lahar signals	1.2 to > 6 Hz; very broad, flat.	Wide, variable	Surface	Rain dependent	V	S	V	V	P	S	S	S	V	V
Explosion signals	0.7 to > 6 Hz; broad	Wide, variable	Shallow	Explosion related	S	S	S	S	S	S	S	S	V	V
Shallow tremor	> 3 Hz; broad	Narrow	Extensive	Eruption related	V	P	S	P	P	V	P	S	P	V

[1] Spectral character of tremor was determined from analog seismograms.
[2] Abbreviations used to describe coherence or preservation of signal in space (interstation) and time (interevent):

Phases/pulses:	Present / constant	Semblance	Variable/absent
Frequency:	Preserved	Slightly constant	Variable
Waveform:	Preserved	Slight resemblance	Variable
Amplitude:	Proportional	Slightly proportional	Variable
Duration:	Proportional	Slightly proportional	Variable

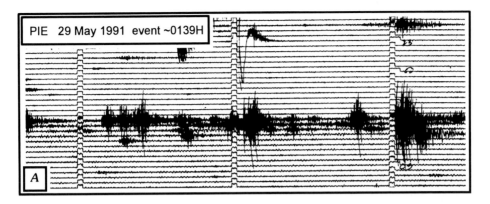

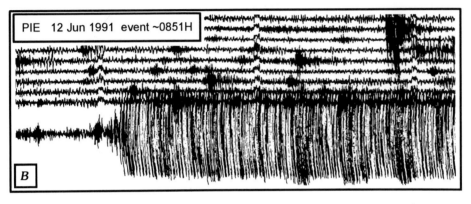

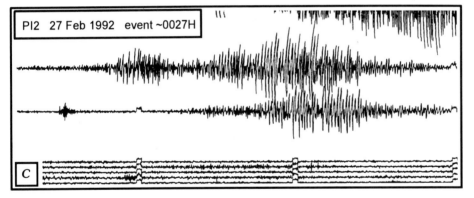

Figure 13. Seismograms of explosions from (*A, B*) Mount Pinatubo's crater and (*C*) 1991 pyroclastic deposits. *A* is part of a series of phreatic explosions from the 1991 pre-eruption summit vents. *B* shows the microseisms, earthquakes, and the start of preliminary explosions in June 1991, prior to the cataclysmic eruption. *C* is from a large phreatic explosion on a pyroclastic deposit accompanying a small lahar in the Sacobia River.

tremor are different from those of low-frequency (harmonic) tremor. A nonsurface origin of the bimodal-frequency events is inferred from (1) uniformity of the frequency and the similarity of its waveform envelopes at the various recording stations; (2) similarity of its waveform features with the low-frequency earthquakes, which often had source depths between 0 and 7 km; (3) their occurrence before the dome extrusion and its associated seismicity; and (4) the absence of correlatable activity at the crater at the time of these tremor. We tentatively infer that the source for bimodal-frequency tremor is deeper than that of the low-frequency earthquakes because the former usually precedes the latter. Thus, we interpret the bimodal-frequency tremor to be related to magmatic movements at depth, possibly during its ascent to shallower levels.

Low-frequency (harmonic) tremor at Mount Pinatubo may be caused by a different mechanism. The low-frequency content of the harmonic tremor partly overlaps that of the low-frequency earthquakes and of bimodal-frequency tremor (fig. 14). Many of the seismograms we have reviewed contained all three events adjacent to one another. The events, although sometimes occurring at similar amplitudes, nonetheless remained distinguishable from one another and preserved their own distinct waveforms. The consistent absence of high-frequency signals in Pinatubo's harmonic tremor, and consistently low amplitude of harmonic tremor (<20 mm at PI2; fig. 5), suggests that it is generated by a process different from that which causes the low-frequency earthquakes and the bimodal-frequency tremor. From the monochromatic waveform, we infer that

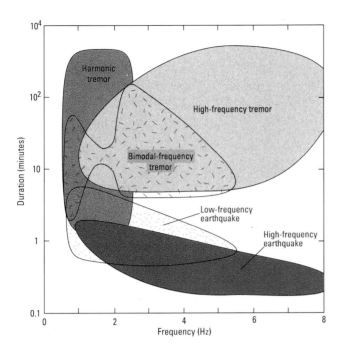

Figure 14. Duration-frequency distribution of various types of Pinatubo's volcanic seismicity. Data on tremors are from frequency counts using optical loupe on analog records, and those for earthquakes are from the Fast Fourier Transform of digital earthquake seismograms.

the source of the harmonic tremor can be some form of resonance of a magmatic conduit of some geometry. The requirements for producing such resonance are the presence of a triggering transient that initiates the vibration (Ferrick and St. Lawrence, 1984; Chouet, 1988) and some means of sustaining a vibration for extended duration (Steinberg and Steinberg, 1975; Ukawa and Ohtake, 1987; Chouet, 1988; Shaw and Chouet, 1989). The transients to trigger the event can be provided by Pinatubo's high-frequency earthquakes or some energetic pulses in magmatic movement. The geometrical and material elasticity requirements needed to sustain the resonance can be provided by the volcano's plumbing system. The similarity of the frequency and waveform of these earthquakes with volcanic tremor allows volcanic tremor to be considered as repeated occurrences of low-frequency earthquakes and the earthquakes to be considered as the impulse response of the tremor-generating process (Shaw and Chouet, 1989; Chouet and others, in press).

A schematic illustration of the spatial interrelation of the various seismic and volcanic processes at Mount Pinatubo is shown in figure 15.

IMPLICATIONS OF THE 1992 EVENTS

We expected that the 1991 eruption of Mount Pinatubo would be followed by dome formation because previous

eruptions of Pinatubo had culminated in dome growth (Newhall and others, this volume), as had explosive eruptions at Mount St. Helens (Fremont and Malone, 1987) and Usu (Okada and others, 1981). No dome formed immediately after Pinatubo's 1991 explosive events, but a dome did form 1 year later, in mid-1992, after repeated occurrence of bimodal-frequency tremor. We associate these bimodal-frequency tremor events with renewed magmatic intrusions because of their waveform, timing, and inferred depth. A fundamental question, then, is whether the dome that formed in 1992 represents the culminating phase of Mount Pinatubo's 1991 eruption or, alternatively, an early phase of further eruptions.

Two models can be constructed from the recent chronology of events. The first assumes that the swarms of low-frequency multiplet earthquakes in 1992 represent shallow-level intrusion of magma, after a deeper level influx of new magmatic material that was marked by bimodal-frequency tremor. In this scenario, the 1992 dome formed by fresh intrusion of basaltic magma into residual 1991 dacitic magma, mixing of basaltic and dacitic magma, and ascent of the mixture to the surface, similar to events that occurred in April through June 1991. In 1991, ascent of the dome-forming mixture triggered explosive eruption of volatile-rich dacitic magma (Pallister and others, 1992; Pallister and others, this volume). Volatiles in the remaining dacitic magma may have been so depleted after the 1991 eruption that renewed intrusion in 1992 produced another dome but could not trigger renewed explosive eruptions. In this model, activity in 1992 represents a dying phase of the cataclysmic eruption of 1991, and renewed explosive eruptions are considered unlikely in the near future.

An alternative model, featuring similar intrusions, is that following the 1991 eruption, the presence of a low-confining-pressure, relatively open-vent environment may have eased the ascent of magma to shallow levels beneath the volcano. The formation of the dome in 1992 may have capped this low-pressure system, and continuing low-frequency earthquakes and volcanic tremor, which have continued intermittently through early 1994, may represent ongoing intrusion and pressurization of the shallow-level magmatic plumbing system. If this model proves correct, further explosive activity remains possible. More detailed study, and further events themselves, will reveal which is the better model.

CONCLUSIONS

We have classified 1991 and 1992 seismic events at Mount Pinatubo into two types of earthquakes and several types of tremor, summarized as follows:

1. High-frequency (volcano-tectonic) earthquakes: tetonic-like events that are related to structural adjustments of

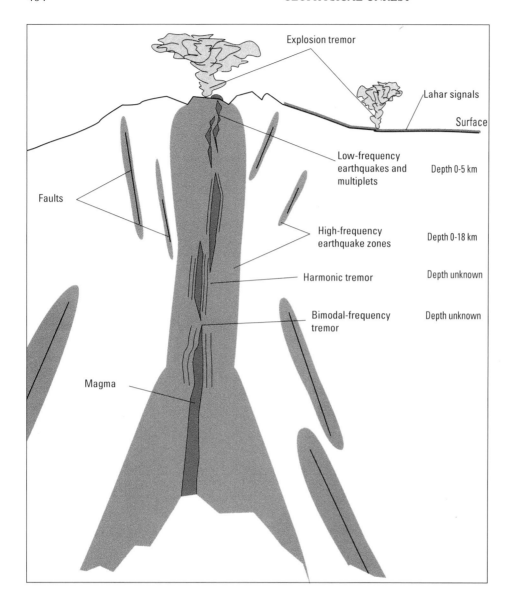

Figure 15. Schematic illustration of the spatial interrelation of Pinatubo's seismicity and volcanic processes.

the crust and volcanic edifice, especially during heightened volcanic activity.

2. Low-frequency (long-period) earthquakes: events with long-period signals that likely are associated with magmatic intrusion. Swarms of multiplet low-frequency earthquakes indicate that the source process of the low-frequency earthquakes is easily repeatable.

3. Low-frequency (harmonic) tremor: monochromatic signals that probably are generated by the resonance within the magmatic plumbing system of Pinatubo.

4. Bimodal-frequency tremor: tremor of mixed high- and low-frequency signals that are probably associated with magmatic transport below the source region of low-frequency earthquakes.

5. High-frequency tremor: high-frequency events with a variety of possible sources that may or may not be directly related to magmatic activity. The various sources

include lahars, explosions from vents and from fresh pyroclastic deposits, and degassing from active fumaroles.

Our classification of seismic events at Mount Pinatubo is preliminary and will require additional refinements as the immense volume of data available from Pinatubo is scrutinized in greater detail. In the process of trying to relate features of volcanic tremor from Pinatubo to those of other volcanoes, we encountered considerable uncertainty in comparing waveform descriptions, either because the available literature does not provide enough detail about the waveforms or because very few seismograms of tremor have been published. In general, our classification of volcanic earthquakes and tremor into various genetic types requires that the events associated with the seismicity be confirmed by other forms of observation, a requirement that can only be met under ideal observational conditions.

ACKNOWLEDGMENTS

The data we have used were gathered by the collective efforts of the PHIVOLCS and USGS personnel involved in the operations of the Pinatubo Volcano Observatory. The USGS provided the PC-based seismic instruments during the 1991 crisis, and the support of the USAID's Office of Foreign Disaster Assistance made it possible for PHIVOLCS to retain the system. The continued involvement and interest of USGS personnel, particularly Chris Newhall, Dave Harlow, Jim Mori, Paul Okubo, and John Power, were helpful in providing insight into the volcanic and seismic activity. Steve Malone of the University of Washington guided us through his work at Mount St. Helens, particularly his studies of multiplets and long-period earthquakes and tremor. Ray Punongbayan provided encouraging support, and the National Science Foundation provided funds, under NSF grant No. EAR–9121566, which made this research possible. Constructive reviews by Steve Malone, Motoo Ukawa, and Chris Newhall greatly improved our appreciation of events described in this paper.

REFERENCES CITED

Aki, K., and Koyanagi, R., 1981, Deep volcanic tremor and magma ascent mechanism under Kilauea, Hawaii: Journal of Geophysical Research, v. 86, no. B8, p. 7095–7109.

Chouet, B.A., 1988, Resonance of a fluid driven crack: Radiation properties and implications for the source of long period events and harmonic tremor: Journal of Geophysical Research, v. 93, p. 4373–4400.

———1992, A seismic model for the source of long-period events and harmonic tremor, in Gasparini, P., Scarpa, R., and Aki, K., eds., Volcanic seismology: Berlin, Springer-Verlag, IAVCEI Proceedings in Volcanology no. 3, p. 133–156.

Chouet, B.A., Page, R.A., Stephens, C.D., Lahr, J.C., and Power, J.A., 1994, Precursory swarms of long period events at Redoubt volcano (1989–1990), Alaska: Their origin and use as a forecasting tool: Journal of Volcanology and Geothermal Research, v. 62, p. 95–135.

Cole, A.T., and Ellsworth, W.T., 1992, Determining earthquake kinship from earthquake catalogs [abs.]: Eos, Transactions, American Geophysical Union, v. 73, no. 43, p. 344.

Daag, A.S., Dolan, M.T., Laguerta, E.P., Meeker, G.P., Newhall, C.G., Pallister, J.S., and Solidum, R., this volume, Growth of a postclimactic lava dome at Mount Pinatubo, July–October 1992.

Daag, A.S., Tubianosa, B.S., Newhall, C.G., Tuñgol, N.M, Javier, D., Dolan, M.T., Delos Reyes, P.J., Arboleda, R.A., Martinez, M.L., and Regalado, M.T.M., this volume, Monitoring sulfur dioxide emission at Mount Pinatubo.

Dibble, R.R., 1974, Volcanic seismology and accompanying activity of Ruapehu Volcano, New Zealand, in Civetta, L., Gasparini, P., Luongo, G., and Rapolla, A., eds., Physical volcanology: Amsterdam, Elsevier Scientific Publishing Co., p. 49–85.

Fehler, M., and Chouet, B.A., 1982, Operation of a digital seismic network on Mount St. Helens volcano and observations of long-period seismic events that originate under the volcano: Geophysical Research Letters, v. 9, no. 9, p. 1017–1020.

Ferrick, M.G., and St. Lawrence, W.F., 1984, Comments on "Observation of volcanic tremor at Mount St. Helens Volcano" by Michael Fehler: Journal of Geophysical Research, v. 89, no. B7, p. 6349–6350.

Fremont, M-J., and Malone, S.D., 1987, High-precision relative location of earthquakes at Mount St. Helens, Washington: Journal of Geophysical Research, v. 92, no. B10, p. 10223–10236.

Hamada, N., Jingu, H., and Ikumoto, K., 1976, On the volcanic earthquakes with slowly decaying coda waves: Bulletin of the Volcanological Society of Japan, v. 21, no. 3, p. 167–183.

Harlow, D.H., Power, J.A., Laguerta, E.P., Ambubuyog, G., White, R.A., and Hoblitt, R.P., this volume, Precursory seismicity and forecasting of the June 15, 1991, eruption of Mount Pinatubo.

Hoblitt, R.P., Wolfe, E.W., Scott, W.E., Couchman, M.R., Pallister, J.S., and Javier, D., this volume, The preclimactic eruptions of Mount Pinatubo, June 1991.

Latter, J.H., 1979, Volcanological observations at Tongariro National Park: 2. Types and classification of volcanic earthquakes, 1976–1978: Department of Scientific and Industrial Research Geophysics Division Report no. 150, New Zealand, 60 p.

———1984, Short notes on volcano seismology: Unpublished lecture notes, UNESCO training for volcano observers, Legaspi, Philippines, 96 p.

Lockhart, A.B., Marcial, S., Ambubuyog, G., Laguerta, E.P., and Power, J.A., this volume, Installation, operation, and technical specifications of the first Mount Pinatubo telemetered seismic network.

Malone, S.D., 1983, Volcanic earthquakes: Examples from Mount St. Helens, in Kanamori, H., and Boschi, E., eds., Earthquakes: Observation, theory and interpretation: Bologna, Società Italiana di Fisica, p. 436–455.

Malone, S.D., and Qamar, A., 1984, Repetitive microearthquakes as the source of volcanic tremor at Mount St. Helens [abs.]: Eos, Transactions, American Geophysical Union, v. 65, no. 45, p. 1001.

McKee, C.O., Wallace, D.A., Almond, R.A., and Talai, B., 1981, Fatal, hydro-eruption of Karkar Volcano in 1979: Development of maar-like crater, in Johnson, R.W., ed., Cooke-Ravian volume of volcanological papers: Memoir 10, Geological Survey of Papua New Guinea, p. 63–85.

Minakami, T., 1974, Seismology of volcanoes in Japan, in Civetta, L., Gasparini, P., Luongo, G., and Rapolla, A., eds., Physical volcanology: Amsterdam, Elsevier Scientific Publishing Co., p. 1–27.

Minakami, T., Ishikawa, T., and Yagi, K., 1950, The 1944 eruption of volcano Usu in Hokkaido, Japan: History and mechanism of formation of the new dome Showa Sinzan: Bulletin Volcanologique, v. 11, p. 45–157.

Mori, J., Okubo, P.G., Lockwood, J.P., Power, J.A., White, R.A., Hoblitt, R.P., Murray, T.L., Wolfe, E.W., Endo, E.T., Laguerta, E., Lanuza, A., and Ramos, E., 1991, Seismicity following the climactic eruption of Pinatubo Volcano, Luzon,

Philippines [abs.]: Eos, Transactions, American Geophysical Union, v. 72, no. 44, p. 62.

Mori, J., White, R.A., Harlow, D.H., Okubo, P., Power, J.A., Hoblitt, R.P., Laguerta, E.P., Lanuza, L., and Bautista, B.C., this volume, Volcanic earthquakes following the 1991 climactic eruption of Mount Pinatubo: Strong seismicity during a waning eruption.

Murray, T.L., Power, J.A., March, G.D., and Marso, J.N., this volume, A PC-based realtime volcano-monitoring data-acquisition and analysis system.

Neri, G., and Malone, S.D., 1989, An analysis of the earthquake time series at Mount St. Helens, Washington, in the framework of the recent volcanic activity: Journal of Volcanology and Geothermal Research, v. 38, p. 257–267.

Newhall, C.G., Daag, A.S., Delfin, F.G., Jr., Hoblitt, R.P., McGeehin, J., Pallister, J.S., Regalado, M.T.M., Rubin, M., Tamayo, R.A., Jr., Tubianosa, B., and Umbal, J.V., this volume, Eruptive history of Mount Pinatubo.

Okada, Hm., Watanabe, H., Yamashita, H., and Yokoyama, I., 1981, Seismological significance of the 1977–1978 eruptions and magma intrusion process of Usu Volcano, Hokkaido: Journal of Volcanology and Geothermal Research, v. 9, p. 311–334.

Pallister, J.S., Hoblitt, R.P., Meeker, G.P., Knight, R.J., and Siems, D.F., this volume, Magma mixing at Mount Pinatubo: Petrographic and chemical evidence from the 1991 deposits.

Pallister, J.S., Hoblitt, R.P., and Reyes, A.G., 1992, A basalt trigger for the 1991 eruptions of Pinatubo volcano?: Nature, v. 356, p. 426–428.

Pinatubo Volcano Observatory Team, 1991, Lessons from a major eruption: Mt. Pinatubo, Philippines: Eos, Transactions, American Geophysical Union, v. 72, p. 545, 552–553, and 555.

Punongbayan, R.S., Newhall, C.G., Bautista, M.L.P., Garcia, D., Harlow, D.H., Hoblitt, R.P., Sabit, J.P., and Solidum, R.U., this volume, Eruption hazard assessments and warnings.

Punongbayan, R.S., Newhall, C.G., and Listanco, E.L., 1992, Brief notes on the 1990–1991 Pinatubo Volcano events and corresponding scientific responses: Bulletin of the Volcanological Society of Japan, v. 37, no. 1, p. 55–59.

Ramos, A.F., and Isada, M.G., 1990, Emergency investigation of the alleged volcanic activity of Mount Pinatubo in Zambales Province: unpublished PHIVOLCS Memorandum Report, 2 p.

Richter, C.F., 1958, Elementary seismology: San Francisco, W.H. Freeman and Co., 768 p.

Shaw, H.R., and Chouet, B.A., 1989, Singularity spectrum of intermittent seismic tremor at Kilauea Volcano, Hawaii: Geophysical Research Letters, v. 16, no. 2, p. 195–198.

Sherburn, S., 1992, Characteristics of earthquake sequences in the Central Volcanic Region, New Zealand: New Zealand Journal of Geology and Geophysics, v. 35, p. 57–68.

Steinberg, G.S., and Steinberg, A.S., 1975, On possible causes of volcanic tremor: Journal of Geophysical Research, v. 80, no. 11, p. 1600–1604.

Ukawa, M., and Ohtake, M., 1987, A monochromatic earthquake suggesting deep-seated magmatic activity beneath the Izu-Ooshima Volcano, Japan: Journal of Geophysical Research, v. 92, v. B12, p. 12649–12663.

Valdez C., 1989, PCEQ, A computer program for picking earthquake phases, in Seismological software library: Seismological software for IBM-compatible computers: San Francisco, Calif., Seismological Society of America.

Weaver, C.S., Grant, W.C., Malone, S.D., and Endo, E.T., 1981, Post-May 18 seismicity: Volcanic and tectonic implications, in Lipman, P.W., and Mullineaux, D.R., eds., The 1980 Eruptions of Mount St. Helens, Washington: U.S. Geological Survey Professional Paper 1250, p. 109–121.

White, R.A., this volume, Precursory deep long-period earthquakes at Mount Pinatubo, Philippines: Spatio-temporal link to a basalt trigger.

Wolfe, E.W., 1992, The 1991 eruptions of Mount Pinatubo, Philippines: Earthquakes and Volcanoes, v. 23, no. 1, p. 5–37.

A VOLATILE SYSTEM

Though seismicity was ambiguous throughout April and May 1991 and might have been just a tectonic swarm, SO_2 emissions demonstrated that magma was involved, and, probably, rising (Daag, Tubianosa, and others). In retrospect, the increases in SO_2 emission may have been controlled more by depletion (by boiling) of the hydrothermal system than by decreasing gas solubility in rising magma, but the hazards implications were the same. A sudden drop in SO_2 emission, perhaps the result of quench sealing of the cap of the rising column of magma, was an immediate precursor to eruption at Pinatubo and has since been noted at Galeras (Colombia) and Colima (Mexico).

Traditionally, volcanologists have assumed that magmatic volatiles are dissolved in the silicate melt until near-surface exsolution and vesiculation occur. Gerlach and others make a strong argument for volatile saturation, development of a separate, H_2O-dominated vapor phase, and strong partitioning of SO_2 into that vapor phase, all while the dacitic magma was still in its reservoir, >6 km deep. This separation facilitated the release of voluminous SO_2 to the atmosphere and attendant effects on global climate. Because of early volatile separation, use of a common petrologic method in which the volatile content of glass inclusions is compared to that in matrix glass would have greatly underestimated the sulfur abundance in the magma and its climatic impact.

Some debate continues as to whether this separate volatile phase was the sole source of SO_2 to the atmosphere, or whether breakdown of anhydrite during magma ascent was also significant (as suggested by L. Baker, cited in Rutherford and Devine). Isotopic analysis by M.A. McKibben and others (in progress) of atmospheric sulfate from Pinatubo will help to resolve this issue.

A small caldera lake began to form during September 1991 and is now rising faster than sediment from the caldera walls can fill it. The lake's pH dropped from near-neutral in October 1991 to strongly acidic by December 1992 as the lake absorbed acidic fluids from the volcano (Campita and others).

Phreatic explosion craters (lower left) and ash-laden steam plumes (upper right) on the north flank of Mount Pinatubo, June 9, 1991. Photograph by R.P. Hoblitt.

Monitoring Sulfur Dioxide Emission at Mount Pinatubo

By Arturo S. Daag,[1] Bella S. Tubianosa,[1] Christopher G. Newhall,[2]
Norman M. Tuñgol,[1] Dindo Javier,[1] Michael T. Dolan,[3] Perla J. Delos Reyes,[1]
Ronaldo A. Arboleda,[1] Ma. Mylene L. Martinez,[1] and Ma. Theresa M. Regalado[1]

ABSTRACT

Correlation spectrometer measurements of sulfur dioxide (SO_2) emission were an important contributor to successful prediction of the June 1991 eruption of Mount Pinatubo. Our first measurement in mid-May (500 tonnes per day) indicated that unrest involved intrusion of magma; a tenfold increase in SO_2 output by late May implied that (1) magma was rising and (or) that (2) a hydrothermal system that absorbed volcanic gases was being boiled and thus removed, allowing more SO_2 to reach the surface. A sudden, short-lived drop of SO_2 output to 260 tonnes per day on June 5, even as seismicity was increasing, may have been caused by plugging or sealing of magma and fractures through which gas was escaping. On June 7, emissions rose again as a new dome was extruded, and our last preparoxysmal measurement on June 10 was more than 13,000 tonnes per day.

Postparoxysmal emissions from July to August 1991, before formation of a caldera lake, ranged from 1,200 to 5,000 tonnes per day, as magma remaining in the conduit degassed. A second dome began to grow in July 1992. Measurements from July to November 1992 ranged from 200 to 600 tonnes per day with some isolated peaks of 1,000 tonnes per day. These measurements suggested that the magma was largely degassed or gas poor compared to that which was the source for preparoxysmal emissions (of 1991).

INTRODUCTION

During volcanic eruptions, the dominant magmatic gases are water, carbon dioxide, sulfur gases, and hydrogen chloride. The major sulfur species in volcanic gases are sulfur dioxide (SO_2) and hydrogen sulfide (H_2S). Although we did not know in advance that magma of Pinatubo is unusu-

ally sulfur rich, this fact and nearly a month of preeruption SO_2 measurements aided greatly in successful prediction of the eruption.

METHODS

We measured SO_2 emissions before and after the 1991 eruption of Mount Pinatubo by use of a correlation spectrometer (COSPEC IV, herein referred to as COSPEC, Barringer Instruments, Toronto). In the past two decades, COSPEC measurements have been part of routine monitoring at a number of volcanoes worldwide, including Kilauea (Casadevall and others, 1987), Mount St. Helens (Casadevall and others, 1983), and Redoubt (Casadevall and others, 1994) in the United States, Nevado del Ruiz (Williams and others, 1990) and Galeras in Colombia, and Merapi (Volcanological Survey of Indonesia, unpub. data, 1993) in Indonesia.

The COSPEC uses solar ultraviolet light transmitted and scattered by the Earth's atmosphere as an illumination source. Pointed skyward during a traverse beneath a volcanic plume (fig. 1), the COSPEC compares the amount of ultraviolet light absorbed by SO_2 in the plume to that absorbed by internal, SO_2-filled calibration cells. Concentration-pathlength is expressed in units of parts per million-meters (ppm-m). A profile of SO_2 concentration-pathlength across the plume is integrated over the width of the plume and multiplied by the plume speed (meters per second) to yield the emission rate of SO_2 (metric tonnes per day, or t/d).

Measured differences in SO_2 emission rates may reflect actual differences in emission rate but can also reflect fluctuations in the wind speed and direction, change in conditions of cloud cover, change of sun angle and amount of solar ultraviolet radiation, and variations in plume opacity due to atmospheric cloud or suspended ash particles. Additional uncertainty in COSPEC measurements arises from several operational factors, including instrument calibration (± 2 percent) and chart reading error (± 4 percent), variation in aircraft speed during the measurement traverse (± 5 percent), operator variance affecting instrument

---

[1] Philippine Institute of Volcanology and Seismology.
[2] U.S. Geological Survey.
[3] Michigan Technological University.

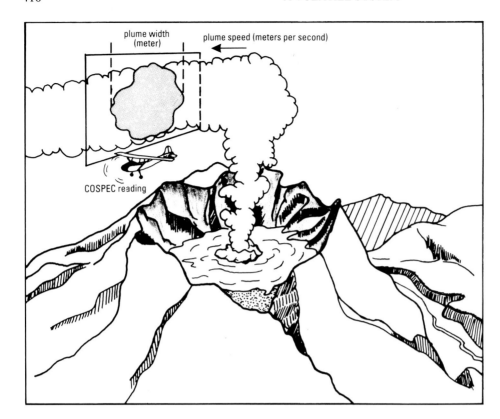

Figure 1. To estimate SO_2 flux, an aircraft carrying the COSPEC flies beneath and perpendicular to the plume to measure a cross-sectional profile of concentration-pathlength (ppm-m), which is integrated over the width of the plume and multiplied by the wind speed to get the rate of SO_2 emission.

operation and reduction of the data (±5 percent), and measurement of wind speed and direction (±10–40 percent, often a significant factor) (Casadevall and others, 1981; Stoiber and others, 1983).

After a frustrating and unsuccessful attempt to borrow a light plane, surveys were begun at Mount Pinatubo on May 13, 1991, from a UH–1 helicopter provided by the U.S. Air Force at Clark Air Base. Beginning in early June, surveys were made from a T–41 Cessna provided by the Philippine Air Force. Readings made from the helicopter were noisy due to interference between a rotating disk within the COSPEC and the overhead rotor of the helicopter. The interference was minimized by aiming the COSPEC at a forward angle, shooting upward but avoiding the helicopter's main rotor blade. Data gathered from the fixed wing aircraft show a considerably better signal-to-noise ratio.

Flying conditions were a frequent constraint, especially during the rainy season (June to November) and during a period of heavy ash fall for several months after the main eruption (June to September 1991). After the main eruption, the plane had to come from an air base more than 100 km from Mount Pinatubo, and, because fuel was unavailable at Clark Air Base, flying was limited to one to two passes before the plane had to return to its base.

The largest uncertainty in our SO_2 measurements is the wind-speed determination (the rate the plume is moving, in meters per second), which greatly affects the computation of SO_2 output. Winds over Mount Pinatubo are quite variable, even over the short period that measurements are taken, and especially with altitude. We used forecasts of winds aloft and our own differencing of the aircraft's ground speed as we flew upwind and downwind (wind speed is the difference between upwind and downwind ground speed, divided by two). For winds of 10 knots or more, our uncertainty is probably less than 30 percent, but for light winds of 2 or 3 knots, our estimate of wind speed and thus of SO_2 emission may be off by as much as 50 to 100 percent.

PREERUPTION SULFUR DIOXIDE OUTPUT (APRIL 2, 1991–JUNE 7, 1991)

On April 2, 1991, phreatic explosions north-northwest of the Pinatubo summit ejected steam and ash. People in the vicinity smelled hydrogen sulfide (H_2S). On April 4, 1991, a Philippine Institute of Volcanology and Seismology (PHIVOLCS) team conducted an aerial survey and found a line of vigorously steaming vents across the north face of the volcano. The team judged that the unrest was of hydrothermal origin. Vents at the northeastern side (where the phreatic explosions had occurred) were short lived; by the time COSPEC surveys were started, only five vents nearer to the summit were actively steaming.

Seismic monitoring was started on April 5, and seismicity from that time until mid-May was characterized by high-frequency earthquakes (impulsive events) in a dense

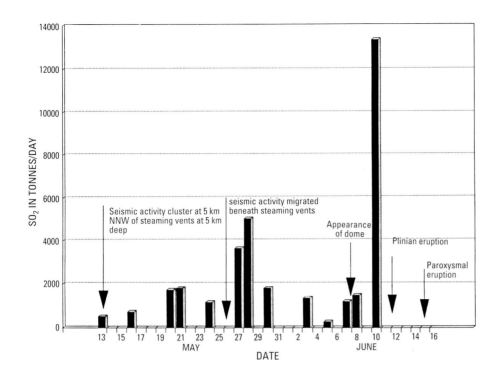

Figure 2. SO$_2$ emission from May 13 to June 10, 1991, precursory to the paroxysmal eruption. Key developments in seismicity and dome growth are noted.

cluster between 3 to 5 km in depth, 5 km northwest of Pinatubo's summit. We considered the possibility that this clustering might be controlled by the presence of local structures in the northwest and that the unrest was of hydrothermal, magmatic, or tectonic origin.

The initial COSPEC measurement was nearly 500 t/d (May 13), which told us that the activity involved magmatic intrusion 2 weeks before shallow earthquakes began to look magmatic. (Much later, after the eruption, deep earthquakes suggesting magma ascent from depth were discovered in seismic records of late May; White and others, this volume.) Without the COSPEC data, we might have concluded that the 3–5-km-deep earthquakes were of tectonic origin. By May 28, SO$_2$ output had increased tenfold, to 5,000 t/d, and we inferred that magma was not only involved but rising. In retrospect, an alternative explanation of this increase in SO$_2$ emission, compatible with magma rising, is that the hydrothermal system which would have been absorbing magmatic SO$_2$ was being boiled away, allowing more and more SO$_2$ from the magma to reach the surface.

Seismicity during the last week of May 1991 increased and began to shift to shallower levels beneath the summit. Earthquakes took on an unmistakably volcanic character, and our earlier inference from COSPEC data, that magma was involved and rising, was confirmed.

In the first week of June 1991, increasingly frequent earthquakes defined a narrow cylindrical volume that extended from a depth of 5 km to the surface (Ambubuyog, 1992; Harlow and others, this volume). In sharp contrast, SO$_2$ emission **decreased**, reaching a measured minimum of 260 t/d on June 5, 1991. Our interpretation was that a rising plug of viscous magma had either pinched shut cracks

through which gas had been escaping or had cooled and (or) lost volatiles and developed an impermeable carapace, shutting off the escape of gas in either case (fig. 2).

SYNERUPTION SULFUR DIOXIDE OUTPUT (JUNE 7, 1991– SEPTEMBER 7, 1991)

On June 7, 1991, a lava dome began to grow at the site of one vigorous steam vent. Concurrent light to moderately heavy ash ejections drifted westward. SO$_2$ output on June 7 and 8 was between 1,200 and 1,500 t/d. On June 10, the value jumped off scale (>13,000 t/d). Clearly, gases could once again escape. Owing to ash in the plume and associated hazards to aircraft, no further COSPEC measurements were made until after July 1991.

The Total Ozone Mapping Spectrometer (TOMS), carried in the National Aeronautics and Space Administration's (NASA's) Nimbus-7 satellite, measured 110,000 t of SO$_2$ in the June 12, 1991, eruption plume, and 20,000,000 t on June 15–16 (Bluth and others, 1992). Clearly, the amounts of SO$_2$ released during the main explosive events were several orders of magnitude greater than emission before or after the main explosive events.

POSTPAROXYSMAL SULFUR DIOXIDE OUTPUT (JULY 1991–JUNE 1992)

Continuous ash emission from the new caldera and from secondary explosions, and pilots' new-found reluctance to fly near Mount Pinatubo, limited us to three COSPEC surveys between July 5 and August 7. SO_2 fluxes were on the order of 1,200 to 3,200 t/d, still moderately high as magma continued to degas. Ash emission from the crater continued until September 4, 1991.

After the last ash emission from the crater on September 4, 1991, rainy weather persisted and a lake was formed on the new caldera floor. The next COSPEC measurement was on October 3, 1991, when only 20 t/d was being emitted. The only visible fumaroles were on the southern caldera wall, and, when sampled in late September, these fumaroles had a maximum temperature of 180°C and were emitting mostly water vapor with minor amounts of volcanic gases (J. Durieux; oral commun., 1991). A different set of fumaroles sampled by the authors in February 1992 (maximum temperature of 105°C) showed the same: vaporized ground water with very little magmatic gas (W. Giggenbach, oral commun., 1992). Gas discharging from vents on the caldera floor during September 1991 to June 1992 was apparently dissolving in the new lake and thus increasing its acidity (Campita, Daag, and others, this volume).

SO_2 OUTPUT DURING RENEWED DOME GROWTH (JULY–NOVEMBER 1992)

In July 1992, after a brief period of precursory seismicity, a new dome began to grow in the center of the caldera lake. Initially, small phreatic or phreatomagmatic explosions built a tuff cone; the dome then grew over, and completely covered, that tuff cone (Daag, Dolan, and others, this volume). SO_2 gas emission during this period of dome growth (table 1) was variable but generally between 200 and 1,000 t/d, virtually all from fumaroles on the dome itself. Owing to logistical difficulties described earlier, our data are sparse, and uncertainties in wind speeds are high.

Despite uncertainty about individual measurements, we think that SO_2 output during the period of dome growth was generally lower than in 1991 and showed a roughly inverse relation to seismicity (fig. 3). In August 1992, SO_2 output began to decline, and, at the same time, the number of shallow, "low-frequency" earthquakes increased. Many of these low frequency earthquakes resulted from shallow intrusion of magma (Ramos and others, this volume); some may have even occurred within the dome.

Although we were concerned at various times during July and August 1992 that renewed dome growth might be followed by renewed explosive eruptions, as in June 1991, the overall level of SO_2 flux and, ultimately, the lack of explosive eruptions indicated a relatively degassed or gas-poor magma body (fig. 4). The inverse relation between seismicity and SO_2 emission in 1992 may reflect relatively easy (and relatively aseismic) extrusion of slightly more volatile-rich lava during the early stage of dome growth,

Table 1. COSPEC measurements of SO_2 at Mount Pinatubo, May 1991 to November 1992.

Date	No. of Passes	SO_2 Flux (t/d)	Standard Deviation (if 3 or more passes)
1991			
May 13......................	2	490	–
May 16......................	8	720	86
May 20......................	7	1,720	363
May 21......................	6	1,800	355
May 24......................	7	1,160	410
May 27......................	7	3,650	211
May 28......................	6	5,020	785
May 30......................	3	1,760	326
June 3......................	6	1,340	401
June 5......................	6	260	30
June 7......................	6	1,180	97
June 8......................	6	1,480	117
June 10......................	1	>13,000	–
July 5	2	5,170	384
July 9	4	1,240	49
August 7......................	1	3,240	–
October 3	4	20	–
1992			
July 9	2	250	94
July 10	3	180	30
July 14	1	130	-
July 15	5	560	40
July 16	3	280	8
July 17	7	1,094	214
July 23	4	910	274
July 31	2	260	25
August 3....................	2	830	218
August 6	3	240	6
August 7....................	1	250	–
August 8....................	1	310	–
August 13..................	4	840	92
August 14..................	4	1,150	143
August 16..................	4	640	152
August 23..................	3	220	45
September 25	4	290	57
November 10	1	300	–

followed by more sluggish, more seismic, extrusion of volatile-poor lava after August.

CONCLUSIONS

COSPEC measurements of SO_2 were invaluable in anticipating the 1991 eruptions of Mount Pinatubo, giving evidence of rising magma and (or) progressive boiling and thus depletion of hydrothermal water. The initial SO_2 measurement of 500 t/d indicated that the unrest was of magmatic origin 2 to 3 weeks before shallow seismicity began

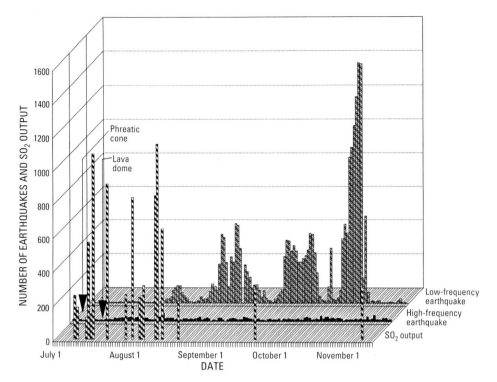

Figure 3. SO$_2$ emission and seismicity, July–November 1992, a period of renewed dome growth. Note the generally inverse relation between seismicity and SO$_2$ emission. SO$_2$ emission in t/d.

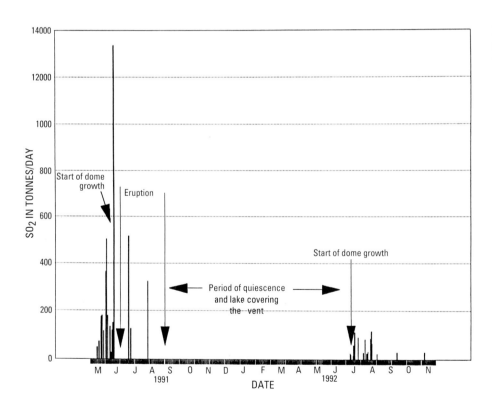

Figure 4. Comparison of SO$_2$ flux in 1991 and 1992. Fluxes in 1992 were notably lower than those in 1991.

to look volcanic. The tenfold increase from mid- to late-May 1991 told us, a week or more before earthquake hypocenters began to migrate upward, that magma was already rising and (or) boiling off the hydrothermal system.

The enormous SO_2 output of the paroxysmal eruption and retrospective evidence for high sulfur concentration in Pinatubo magma make it easy to understand the pronounced increase of SO_2 emission in May 1991. However, similar levels of SO_2 emission (5,000 t/d) have been observed at other volcanoes and were not followed by large explosive eruptions (for example, Nevado del Ruiz, Galeras, San Cristobal, and Etna). The daily flux of 5,000 t/d, by itself, did not tell us the magnitude of the impending eruption of Pinatubo. Rather, the geologic record of past eruptions suggested the most likely magnitude, and the pattern of increasing SO_2 emission suggested that an eruption was imminent, especially after the sudden decrease and even greater increase in early June 1991.

Postparoxysmal SO_2 emissions are consistent with the idea that magma remaining at shallow depth beneath Mount Pinatubo is gas poor in comparison to that present before the June 15, 1991, eruption. Measurements during July and August 1991 hinted of declining emission, until development of the caldera lake effectively captured all of the emission. Even during renewed dome growth in 1992, when fumaroles were once again venting to the atmosphere, emission rates barely exceeded 1,000 t/d. Therefore, despite the possibility of further intrusions of basalt or hybrid andesite into the main body of dacitic magma (Pallister and others, this volume), we now suspect that further large explosive eruptions are unlikely.

ACKNOWLEDGMENTS

We thank Ken McGee of the Cascades Volcano Observatory for lending the COSPEC instrument, which was used in measuring the SO_2 emissions at Pinatubo volcano, and Michael Doukas of the Alaska Volcano Observatory for providing a worksheet used to calculate SO_2 fluxes. We also thank the U.S. 13th Air Force and the Philippine Air Force (Lipa Air Base and CABCOM) for aircraft support during the survey.

REFERENCES CITED

Ambubuyog, G., 1992, Pre-eruption seismic interpretations of Mount Pinatubo: Unpublished report, PHIVOLCS, 36 p.

Bluth, G.J.S., Doiron, S.D., Krueger, A.J., Walter, L.S., and Schnetzler, C.C., 1992, Global tracking of the SO_2 clouds from the June 1991 Mount Pinatubo eruptions, Geophysical Research Letters, v. 19, no. 2, p. 151–154.

Campita, N.R., Daag, A.S., Newhall, C.G., Rowe, G.L., Solidum, R., this volume, Evolution of a small caldera lake at Mount Pinatubo.

Casadevall, T.J., Stokes, J.B., Greenland, L.P., Malinconico, L.L., Casadevall, J.R., and Furukawa, B.T., 1987, SO_2 and CO_2 emission rates at Kilauea Volcano, 1979–1984, in Decker, R.W., Wright, T.L., and Stauffer, P.H., eds., Volcanism in Hawaii: U.S. Geological Survey Professional Paper 1350, v. 1, p. 771–780.

Casadevall, T.J., Johnston, D.A., Harris, D.A., Rose, W.I., Jr., Malinconico, L.L., Jr., Stoiber, R.E., Bornhorst, T.J., Williams, S.N., Woodruff, L., and Thompson, J.M., 1981, SO_2 emission rates at Mount St. Helens from March 29 through December, 1980, in Lipman, P.W., and Mullineaux, D.R., eds., The 1980 eruptions of Mount St. Helens: U.S. Geological Survey Professional Paper 1250, p. 193–200.

Casadevall, T.J., Doukas, M.P., Neal, C.A., McGimsey, R.G. and Gardner, C.A., 1994, Emission rates of sulfur dioxide and carbon dioxide from Redoubt volcano, Alaska during the 1989–1990 eruptions, in Miller, T.P., and Chouet, B., eds., The 1989–1990 eruptions of Redoubt Volcano, Alaska: Journal of Volcanology and Geothermal Research, v. 62, p. 519–530.

Casadevall, T.J., Rose, W.I., Jr., Gerlach, T., Greenland, L.P., Ewert, J., Wunderman, R., and Symonds, R., 1983, Gas emissions and the eruptions of Mount St. Helens through 1982: Science, v. 221, p. 1383–1385.

Daag, A.S., Dolan, M.T., Laguerta, E.P., Meeker, G.P., Newhall, C.G., Pallister, J.S., and Solidum, R., this volume, Growth of a postclimactic lava dome at Mount Pinatubo, July–October 1992.

Harlow, D.H., Power, J.A., Laguerta, E.P., Ambubuyog, G., White, R.A., and Hoblitt, R.P., this volume, Precursory seismicity and forecasting of the June 15, 1991, eruption of Mount Pinatubo.

Pallister, J.S., Hoblitt, R.P., Meeker, G.P., Knight, R.J., and Siems, D.F., this volume, Magma mixing at Mount Pinatubo: Petrographic and chemical evidence from the 1991 deposits.

Ramos, E.G., Laguerta, E.P., and Hamburger, M.W., this volume, Seismicity and magmatic resurgence at Mount Pinatubo in 1992.

Stoiber, R.E., Malinconico, L.L., Jr., and Williams, S.N., 1983, Use of the correlation spectrometer at volcanoes, in Tazieff, H., and Sabroux, J-C, eds., Forecasting volcanic eruptions: Amsterdam, Elsevier, p. 425–444.

White, R.A., Precursory deep long-period earthquakes at Mount Pinatubo, Philippines: this volume, Spatio-temporal link to a basalt trigger.

Williams, S.N., Sturchio, N.C., Calvache V., M.L., Mendez F., R., Londoño C., A., and Garcia P., N., 1990, Sulfur dioxide from Nevado del Ruiz volcano, Colombia: Total flux and isotopic constraints on its origin: Journal of Volcanology and Geothermal Research, v. 42, p. 53–68.

Preeruption Vapor in Magma of the Climactic Mount Pinatubo Eruption: Source of the Giant Stratospheric Sulfur Dioxide Cloud

By Terrence M. Gerlach,[1] Henry R. Westrich,[2] and Robert B. Symonds[1]

ABSTRACT

The climactic June 15, 1991, eruption of Mount Pinatubo injected a *minimum* of 17 Mt (megatons) of SO_2 into the stratosphere—the largest stratospheric SO_2 cloud ever observed. This study is an investigation of the *immediate* source of the sulfur for the giant SO_2 cloud. Approximately 100 electron microprobe analyses show no significant differences, at the 95 percent confidence level, in S or Cl contents between glass inclusions and matrix glasses of the erupted dacite. These results indicate that there was no significant degassing of S or Cl from melt during ascent and eruption. Furthermore, the 17-Mt SO_2 cloud contained over an order of magnitude more sulfur than could have been dissolved in the quantity of erupted silicate melt at the preeruption conditions. A major source of "excess sulfur" is therefore required to account for the SO_2 cloud. Degassing of melt in non-erupted dacite as a source of the excess sulfur implies volumes of non-erupted dacite larger than the estimated volume of the magma reservoir beneath the Mount Pinatubo region. Direct degassing of excess sulfur from a basalt source seems unlikely, since the June 15 eruption products lack evidence of mixed or commingled contemporaneous basalt. Anhydrite decomposition rates at atmospheric pressure and expected eruption temperatures are extremely slow and grossly incapable of generating 17 Mt of SO_2 by anhydrite breakdown in the eruption cloud. Anhydrite breakdown during ascent decompression is too slow to keep pace with conduit travel times, which were considerably less than 8 minutes. Flash vaporization of sulfate-rich Pinatubo hydrothermal fluids during the eruption could have caused sulfate mineral deposition but virtually no SO_2 production.

It is proposed that the dacite erupted on June 15 was vapor-saturated at depth prior to eruption, and that an accumulated vapor phase in the dacite provided the *immediate* source of excess sulfur for the 17-Mt SO_2 cloud.

[1]U.S. Geological Survey, Cascades Volcano Observatory, 5400 MacArthur Blvd., Vancouver WA 98661.

[2]Sandia National Laboratory, Geochemistry Dept. 6118, Albuquerque NM 87185.

Investigations based on exploration drilling for geothermal energy suggest that magmatic volatiles were discharged into the Pinatubo hydrothermal system from the vapor-saturated dacite prior to the 1991 eruption. Experimental studies, geobarometer results, and the H_2O and CO_2 contents of glass inclusions indicate that the Pinatubo dacite was saturated with water-rich vapor before ascent and eruption. Models for the composition of the preeruption vapor suggest that it contained a *minimum* of approximately 96 Mt H_2O, 42 Mt CO_2, and 3 Mt Cl, in addition to 17 Mt of SO_2. The mole fraction composition of the vapor was $X_{H_2O}=$ 0.80–0.83, $X_{SO_2}=$0.01–0.04, $X_{CO_2}=$0.15, and $X_{Cl}=$0.01, indicating that the vapor was not excessively SO_2-rich. The volume and density of the vapor at depth prior to eruption were $\geq$0.25 km^3 and about 0.6 g/cm^3, respectively. Vapor comprised at least 5 volume percent of the preeruption dacite at depth; the bulk density of the preeruption dacite was less than 2.3×10^{12} kg/km^3. Solubility modeling indicates that the total amount of volatiles contained in the preeruption vapor and melt of the erupted dacite could not have been dissolved initially in completely molten dacite at the magma reservoir pressures, suggesting some process of preeruption vapor accumulation at depth.

The climactic eruption released the accumulated vapor in the estimated 5 km^3 of erupted dacite. About 6.25 wt percent of dissolved water was also degassed from melt during ascent and eruption; scaling to the volume of erupted dacite implies an additional release of 395 Mt of H_2O. Additional yields of SO_2, CO_2, and Cl from degassing of melt were minor to insignificant during ascent and eruption. Thus, the *minimum* volatile emissions for the climactic eruption— from preeruption vapor phase and degassing of melt—were 17 Mt SO_2, 42 Mt CO_2, 3 Mt Cl, and 491 Mt H_2O.

This study underscores the need for *both* petrologic measurements and emission measurements to constrain the quantity of dissolved volatiles *and* preeruption vapor in magma at depth. If explosive volcanism commonly involves magmas with substantial accumulated vapor, the volatile contents of glass inclusions alone are not a sufficient basis for inferring the total preeruptive volatile contents of magma and for predicting volatile emissions. Consequently, conventional petrologic estimates of SO_2 emissions during explosive eruptions of the past may be far too low and significantly underestimate their impacts on climate and the chemistry of the atmosphere.

INTRODUCTION

Scientists have used remote sensing to measure volcanic SO_2 emissions by correlation spectrometry (COSPEC) since 1972 (Stoiber and Jepsen, 1973) and by Total Ozone Mapping Spectrometry (TOMS) from the Nimbus-7 satellite since 1978 (Bluth and others, 1993). Petrologic analysis of eruption products is used to estimate SO_2 emissions and to assess atmospheric and climatic impacts of eruptions occurring before COSPEC and TOMS data were available. The conventional petrologic method assumes that the volatile contents of glass inclusions trapped in crystals are representative of the preeruption melt, and that the volatile contents of coexisting degassed matrix glass represent the melt after eruption. The difference in the volatile contents of the glass inclusions and matrix glasses is taken as a measure of volatile degassing. Scaling up volatile degassing for the mass of erupted melt yields estimates of volatile emissions during eruption (Johnston, 1980; Devine and others, 1984; Sigurdsson and others, 1985; Palais and Sigurdsson, 1989; Sigurdsson, 1990).

The petrologic method may tend to underestimate volatile emissions, since assumptions inherent in the method, if not fully satisfied, lead to low results (Devine and others, 1984; Palais and Sigurdsson, 1989). Comparisons of COSPEC and TOMS results with petrologic estimates for explosive eruptions from subduction zone volcanoes suggest that petrologic estimates of sulfur emissions are low by more than an order of magnitude (Luhr and others, 1984; Sigurdsson, 1990; Sigurdsson and others, 1990; Williams and others, 1990; Andres and others, 1991; Gerlach and others, 1994), although comparisons have been limited to explosive eruptions with a Volcano Explosivity Index (VEI) ≤5. An important issue is whether or not the level of agreement between petrologic and remote sensing results improves for larger eruptions (Sigurdsson, 1990). The June 15, 1991, climactic eruption of Mount Pinatubo permits a comparison of petrologic emission estimates for SO_2 with remote sensing determinations for a larger eruption (VEI=6) that had a clear impact on climate and stratospheric ozone (Dutton and Christy, 1992; Hofman and others, 1992; Prather, 1992; Gleason and others, 1993; Halpert and others, 1993; Solomon and others, 1993; McCormick and others, 1995).

This study is an investigation of the sources of sulfur for the giant SO_2 cloud injected into the stratosphere on June 15, 1991, during the climactic eruption of Mount Pinatubo. It is confined to *immediate* sources—that is, sources of sulfur potentially available at the moment the eruption began. We show that conventional petrologic estimates fail unconditionally in predicting the quantity of SO_2 released during the climactic eruption. Contrary to expectation, the preeruption melt was not a significant source of sulfur for the giant cloud. We analyze several alternative sources of sulfur and conclude that the SO_2 of the stratospheric cloud was the product of vapor saturation and accumulation in the Pinatubo magma at depth prior to eruption.

SO_2 EMISSION ESTIMATES BY REMOTE SENSING METHODS

Table 1 summarizes several attributes of the climactic June 15 eruption. The climactic event got underway at 1342 and lasted ~9 hours until 2230 (Wolfe and others, this volume; Hoblitt and others, this volume). It erupted 4–5 km^3 dense-rock equivalent (DRE) (W.E. Scott and others, this volume) of dacite magma (Pallister and others, this volume) and produced the largest SO_2 cloud ever observed in the stratosphere. Three determinations of the SO_2 content of the stratospheric cloud are now available from remote sensing data. The TOMS determination is 20 ±6 Mt (20 megatons= 20×10^9 kg) (Bluth and others, 1992). This result may include ~0.5 Mt of SO_2 from explosions preceding the climactic event (Bluth and others, 1992), an amount well within the estimated error. The Microwave Limb Sounder (MLS) experiment on the Upper Atmosphere Research Satellite (UARS), launched into orbit by the Space Shuttle on September 12, 1991, measured ~0.9 Mt of residual SO_2 100 days after the climactic eruption (Read and others, 1993). A best-fit decay line to this and subsequent MLS/UARS measurements recording the conversion of SO_2 to sulfuric acid gives an extrapolated initial SO_2 injection of 17 Mt (Read and others, 1993). The total error in this extrapolated value is not known but is assumed to be greater than ±1 Mt (W. G. Read, oral commun., January 28, 1994). The third remote sensing determination comes from spectral scan data of the SBUV/2 instrument on the NOAA-11 satellite. SBUV/2 data give an initial SO_2 injection of 13.5 ±1.5 Mt (McPeters, 1993). The unweighted mean of these results is 17 Mt with an estimated error in the mean ($s/n^{1/2}$) of ±2 Mt.

In this report, we take the 17 Mt mean remote sensing value for the SO_2 content of the cloud injected into the stratosphere as a *minimum* estimate of the *total* SO_2 emission during the climactic eruption. We assume that early removal of SO_2 may have occurred by scavenging on ash. It is estimated, for example, that nearly 30% of the vaporous sulfur in the 1982 eruptions of El Chichón Volcano (Varekamp and others, 1984) and over 30% of the sulfur in the 1974 eruptions of Fuego Volcano (Rose, 1977) were scavenged by ash within the plume.

Table 1. Attributes of the climactic June 15, 1991, eruption, Mount Pinatubo.

[MPa, megapascal; Mt, megaton, 10^9 kg; DRE, dense rock equivalent; NNO, O_2 fugacity of the nickel-nickel oxide buffer]

Parameter	Value	Basis
Eruption duration (t_e)	9 hours	Seismic and barograph records[1]
DRE volume erupted magma (V)	5 km^3	Studies of eruption deposits and topographic changes[2]
Total SO_2 emission (E_{SO_2})	17±2 Mt (minimum)	TOMS, MLS, and SBUV/2 remote sensing[3]
Magma O_2 fugacity (f_{O_2})	10^{-12} MPa (NNO+3.2)	Fe-Ti oxides and experimental calibration[4]
Magma temperature (T)	780±10°C	Fe-Ti oxides and experimental calibration[4]
Preeruption pressure (P)	220±50 MPa	Al-content hornblende and experimental phase equilibria[4]
DRE melt volume fraction (ϕ_m)	0.55	Groundmass volume fraction of white pumice by point-count, vesicle-free basis[5]
Melt density (ρ_m)	2.3×10^{12} kg/km^3	Rhyolite glass with 6 weight percent water[6]
Dacite density (ρ)	2.4×10^{12} kg/km^3	Estimated for crystal-glass of dacite, vesicle-free basis[2]
Magma reservoir depth (d)	6–11 km	Inversion of seismicity data[7]

[1] Wolfe and others (this volume).

[2] W.E. Scott and others (this volume).

[3] Mean of Bluth and others (1992); Read and others, 1993; and McPeters, 1993; the uncertainty is the estimated error in the mean ($s/n^{1/2}$); 17 Mt is a minimum because of potential scavenging by ash, as discussed in text.

[4] Rutherford and Devine (this volume).

[5] This study. Pallister and others (1992; this volume).

[6] Silver and others (1990).

[7] Mori, Eberhart-Phillips, and Harlow (this volume).

SO_2, CL, AND H_2O EMISSION ESTIMATES BY THE PETROLOGIC METHOD

SAMPLES AND PROCEDURES

The correlation of the giant stratospheric SO_2 cloud with the dacite magma erupted on June 15 implicates the dacite as the likely source of the SO_2. We therefore focus on the June 15 dacite in our determination of petrologic emission estimates for SO_2. The climactic June 15 eruption produced two dacite lithologies (Pallister and others, 1992; this volume). The dominant lithology is a white, phenocryst-rich (45 volume percent crystals) hornblende dacite pumice with 64–65 wt percent SiO_2 (anhydrous whole-rock basis). The subordinate lithology is a grayish-tan, phenocryst-poor dacite pumice with a similar whole-rock composition but a texture characterized by abundant micrometer-sized broken crystals suggesting mechanical fragmentation and shattering of former phenocrysts. The phenocryst-rich dacite comprises ~85 percent of the June 15 pyroclastic flow deposits. Inclusions of the phenocryst-rich pumice are common in deposits of the phenocryst-poor pumice, and some deposits contain the two types intermingled as banded pumice blocks. In addition to similar whole-rock compositions, the two dacite pumice types have similar phenocryst assemblages, although the phenocryst content is much lower in the fragmented phenocryst-poor pumice. Hornblende, with cummingtonite rims on some crystal faces, and plagioclase are the main phenocrysts in both dacites, and both types

also contain anhydrite, magnetite, quartz, and apatite phenocrysts; sulfides are rare (Imai and others, 1993; Fournelle and others, this volume; Hattori, this volume). See Pallister and others (1992; this volume) for modal petrographic data.

Our samples were collected from rapidly cooled tops of pyroclastic flow deposits within six days of the climactic June 15 eruption. They include both the phenocryst-rich and phenocryst-poor dacite pumice of pyroclastic flow deposits located northeast of Mount Pinatubo in the Sacobia River valley and to the northwest in the Bucao River valley. The glass inclusion analyses reported here come from both the phenocryst-rich and phenocryst-poor dacite pumice. The matrix glass analyses are restricted to the phenocryst-rich dacite, since it contains clear groundmass glass suitable for electron microprobe analysis. Microscopic crystal fragments make electron microprobe analysis of matrix glass difficult in the phenocryst-poor dacite (Pallister and others, this volume).

Light crushing and sieving followed by suspension in heavy liquids allowed separation and concentration of phenocrysts from the glassy matrix of the pumice samples. Phenocrysts were mounted in epoxy and their glass inclusions were exposed by grinding and polishing. To identify the glass inclusions, we examined the polished phenocryst surfaces with reflected light and backscattered electron microscopy. Glass inclusion-bearing phenocrysts include hornblende, magnetite, plagioclase, and quartz. Most glass inclusions are clear to brown in color and vary in size from

Table 2. Analyses of glass inclusions and matrix glasses from the June 15, 1991, dacite pumice, Mount Pinatubo.

[Values in weight percent unless otherwise indicated. Values are means; parentheses contain standard deviations. GI, glass inclusion from hosts in phenocryst-rich and phenocryst-poor pumice; MG, matrix glass from phenocryst-rich pumice; Hnb, hornblende; Mgt, magnetite; Plag, plagioclase; Qtz, quartz. FeO analyzed as ferrous iron. Na_2O data are not adjusted for sodium-loss during electron beam excitation. Sulfur analyzed as SO_3; ppm S calculated by multiplying percent SO_3 by 4,000. N is number of electron microprobe analyses. N_{H_2O} is number of ion probe analyses for H_2O. H_2O (max) is the maximum water content determined in the N_{H_2O} ion probe analyses.]

Oxide	Hnb GI	Mgt GI	Plag GI	Qtz GI	MG
SiO_2	77.06 (0.56)	76.59 (1.42)	75.15 (1.11)	75.61 (1.28)	78.53 (0.94)
TiO_2	0.14 (0.03)	0.29 (0.03)	0.11 (0.03)	0.10 (0.03)	0.13 (0.03)
Al_2O_3	12.88 (0.27)	12.44 (0.21)	12.44 (0.56)	11.83 (0.37)	12.86 (0.12)
FeO	1.11 (0.11)	1.27 (0.12)	0.71 (0.13)	0.61 (0.09)	1.01 (0.15)
MnO	0.05 (0.02)	0.03 (0.02)	0.03 (0.02)	0.04 (0.02)	0.04 (0.02)
MgO	0.19 (0.03)	0.19 (0.01)	0.17 (0.05)	0.15 (0.03)	0.21 (0.01)
CaO	1.27 (0.09)	1.17 (0.03)	1.04 (0.11)	1.08 (0.06)	1.21 (0.04)
K_2O	2.94 (0.10)	2.79 (0.12)	2.96 (0.09)	2.82 (0.08)	2.99 (0.07)
Na_2O	3.17 (0.24)	2.84 (0.22)	2.45 (0.24)	2.70 (0.41)	3.18 (0.21)
P_2O_5	0.04 (0.02)	0.03 (0.02)	0.03 (0.01)	0.02 (0.01)	0.03 (0.02)
SO_2	0.017 (0.010)	0.018 (0.008)	0.022 (0.017)	0.022 (0.014)	0.015 (0.008)
Cl	0.109 (0.010)	0.093 (0.009)	0.106 (0.025)	0.088 (0.014)	0.112 (0.012)
H_2O	1.34 (0.88)	3.51 (1.54)	3.09 (0.87)	4.37 (1.03)	0.31 (0.15)
Total	100.32	101.26	98.31	99.44	100.63
N	25	12	10	18	21
N_{H_2O}	8	5	4	10	3
S (ppm)	68 (40)	72 (32)	88 (68)	88 (56)	60 (32)
H_2O (max)	3.30	4.55	3.52	6.56	0.43

<80 μm in hornblende, magnetite, and plagioclase to 300 μm in quartz.

Major element analyses of glass inclusions and matrix glasses were obtained on a JEOL JXA-8600 electron microprobe at an accelerating voltage of 15 kV, a beam current of 10 nA, a beam diameter of 10 μm, and standard data reduction techniques (Bence and Albee, 1968). A higher beam current (20 nA) and longer counting times (80–120 s) improved detection limits for chlorine (Cl, 40 ppm) in matrix glass and glass inclusions and for sulfur (S, 55 ppm) in glass inclusions. Analytical precision for Cl and S was better than ±10 percent. Wavelength measurements of S-Kα X-rays gave the proportions of sulfur present as sulfate and sulfide (Carroll and Rutherford, 1988). A Cameca IMS 3f ion microprobe with a 1 nA mass-analyzed primary beam of $^{16}O^-$ ions provided analyses of selected glasses for H_2O and several trace elements. Secondary ions from a 20 μm spot were collected using a voltage offset (Hervig and Williams, 1988). The ion probe detection limit for water in glasses is 0.15 wt percent.

GLASS ANALYSES

Table 2 presents mean compositions based on analyses of 86 glasses, including matrix glasses and glass inclusions in hornblende, magnetite, plagioclase, and quartz phenocrysts. The glass compositions are similar regardless of the sample site, host phenocryst, or dacite lithology they represent. When normalized to anhydrous compositions, all glasses are high silica rhyolite compositions. Their similarity, even at trace element levels (Westrich and Gerlach, 1992), indicates a comagmatic origin for the glass inclusions and the matrix glasses, and suggests that crystallization was not significant after glass inclusion entrapment. The variability of SiO_2 in the glass inclusion data (table 2) reflects differences in water contents, as discussed below.

Most glasses have sulfur contents too low for precise measurement of the wavelength of S-Kα X-rays. A few glasses with higher than average sulfur, however, indicate sulfur is dissolved as sulfate (Carroll and Rutherford, 1988). This result is consistent with oxygen fugacity estimates (Rutherford and Devine, this volume) for the preeruption June 15 dacites of 3.2 log units above the nickel-nickel oxide buffer (NNO+3.2, table 1) (Huebner and Sato, 1970). We therefore report the sulfur data (table 2) as wt percent SO_3, but also show the data converted to ppm (wt) S, which are the units commonly used in this report. Figure 1 summarizes the sulfur data from table 2 and includes 15 additional sulfur analyses. Figure 2 summarizes the Cl data from table 2 and includes 12 additional Cl analyses. Table 2 presents the mean and maximum H_2O concentrations based on ion probe analysis.

PETROLOGIC EMISSION ESTIMATE FOR SO_2

The glass inclusion and matrix glass data for sulfur together with melt volume fractions, melt density, and volumes of erupted magma constrain petrologic SO_2 emission

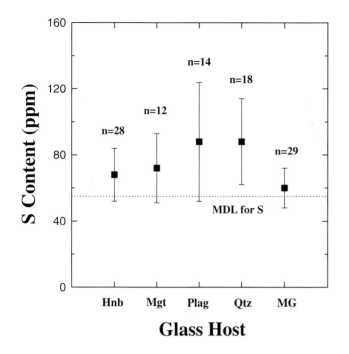

Figure 1. Sulfur concentrations for the June 15 pumice matrix glasses (MG) and glass inclusions in host phenocrysts of hornblende (Hnb), magnetite (Mgt), plagioclase (Plag), and quartz (Qtz). Dotted line at 55 ppm S is the microprobe detection limit (MDL). Solid squares are mean values; error bars are ±2 standard errors ($s/n^{1/2}$); and n is the number of samples. The mean values of glass inclusions and matrix glasses are not significantly different at the 95-percent confidence level, thus implying zero petrologic emission estimates for SO_2.

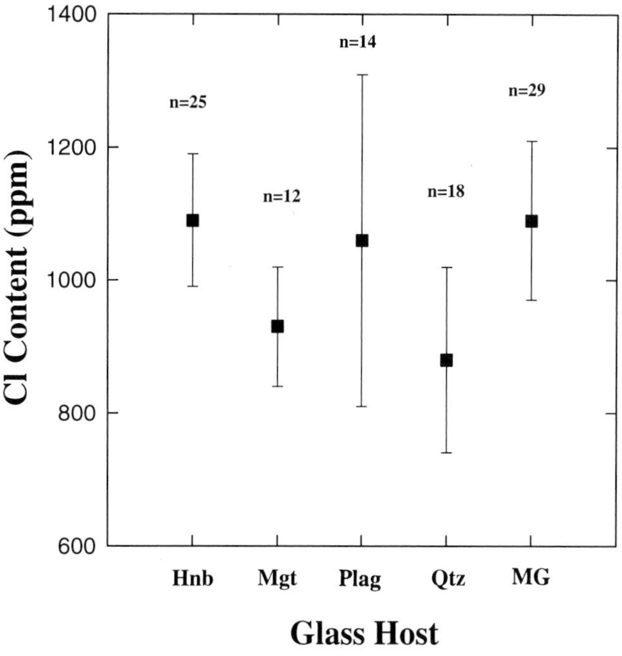

Figure 2. Chlorine concentrations for the June 15 pumice matrix glasses (MG) and glass inclusions in host phenocrysts of hornblende (Hnb), magnetite (Mgt), plagioclase (Plag), and quartz (Qtz). Solid squares are mean values; error bars are ±1 standard deviation; and n is the number of samples. The mean values of glass inclusions and matrix glasses are not significantly different at the 95 percent confidence level, thus implying zero petrologic emission estimates for chlorine.

estimates for comparison with the 17-Mt remote sensing result. This comparison assumes that sulfur was degassed almost entirely as SO_2 during the eruption, or if not, that the degassed sulfur species converted rapidly to SO_2, the sulfur gas species detected by TOMS and MLS. The predominance of SO_2 in degassed sulfur seems likely, since the June 15 dacite had an exceptionally high oxygen fugacity (NNO+3.2, table 1), and SKα X-ray data for the glasses indicate a predominance of sulfate over sulfide in the melt. Furthermore, thermodynamic calculations for the vapor phase reaction

$$H_2S(g) + 3/2O_2(g) = H_2O(g) + SO_2(g) \qquad (1)$$

indicate an equilibrium SO_2/H_2S fugacity ratio >55 during degassing from preeruption conditions (table 1) to atmospheric pressure (fig. 3). Similar calculations for the reaction

$$SO_2(g) + \frac{1}{2}O_2(g) = SO_3(g) \qquad (2)$$

indicate an equilibrium SO_2/SO_3 fugacity ratio of 3.5×10^5 during degassing. These calculations support the assump-

tion that sulfur was degassed predominantly as SO_2 during the June 15 eruption.

The equation for calculating petrologic emission estimates for SO_2 (E_{SO_2}) is as follows:

$$E_{SO_2} = 2(10^{-15}) \Delta S_m \rho_m \phi_m V \qquad (3)$$

where ΔS_m is the melt S-loss during eruption taken as the difference in the S contents of glass inclusions and matrix glasses in ppm, ρ_m is the bubble-free melt density in kg/km^3, ϕ_m is the bubble-free melt volume fraction, and V is the DRE volume of erupted magma in km^3. The constant 10^{-15} gives E_{SO_2} in Mt; the constant 2 takes into account the difference in the gram-formula-weights of SO_2 and S. Tables 1 and 2 provide data for the parameters, all of which are well-constrained, except V. W.E. Scott and others (this volume) report a range of 4–5 km^3 DRE for V. Since the distal volume of tephra fall remains uncertain and may be too low, we will use the 5 km^3-end of the range in this report. (All eruption product volumes in this report are DRE volumes.)

The usual practice for calculating ΔS_m is to subtract the mean S concentration of the matrix glass from the mean

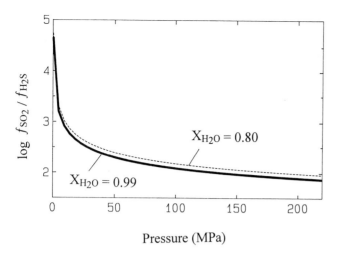

Figure 3. f_{SO_2}/f_{H_2S} versus total pressure during degassing of the June 15 dacite from a preeruption pressure of 220 MPa to atmospheric pressure at the dacite preeruption temperature (780°C) (table 1). The results indicate that f_{SO_2}/f_{H_2S} was always >55 during degassing. The fugacity ratio at equilibrium during degassing is given by

$$\log (f_{SO_2}/f_{H_2S}) = \log K_1 + 1.5 \log (f_{O_2}) - \log (\gamma_{H_2O}) - \log (X_{H_2O}) - \log P$$

where $K_1 = 10^{21.63}$ is the equilibrium constant (Chase and others, 1985) for reaction (1) at 780°C, f_{O_2} is the oxygen fugacity of the preeruption dacite ($10^{-12.1}$ MPa, table 1), γ_{H_2O} is the fugacity coefficient for H_2O in the gas, X_{H_2O} is the mole fraction of H_2O in the gas, and P is the total pressure during degassing. Note that f_{SO_2}/f_{H_2S} increases as X_{H_2O} decreases. γ_{H_2O} estimated from fugacity coefficients for pure H_2O (Burnham and others, 1969).

S concentration of glass inclusions in a host phase (ΔS_m = mean ppm $S_{inclusion, phase\ x}$ − mean ppm S_{matrix}). For the June 15 Pinatubo dacites, however, there is no significant difference between the mean sulfur concentration of the matrix glasses and that of glass inclusions in any of the host phases (table 2, fig. 1). Statistical analysis of the data shows that for all glass inclusion hosts, ΔS_m is not significantly different from zero at the 95 percent confidence level; indeed, it is not significantly different from zero at the 90 percent confidence level, except for glass inclusions in quartz. Thus, the petrologic method predicts an SO_2 emission (E_{SO_2}) for the climactic June 15 eruption that is not significantly different from zero but is significantly different indeed from the 17-Mt remote sensing result.

PETROLOGIC EMISSION ESTIMATE FOR Cl

The petrologic emission estimate for Cl (E_{Cl}) is obtained from the Cl analog of equation (3). Since the melt-Cl loss, ΔCl_m (= mean ppm $Cl_{inclusion}$ − mean ppm Cl_{matrix}), is not significantly different from zero either (table 2, fig. 2), a zero petrologic emission estimate is also implied for Cl.

DISCUSSION OF ZERO PETROLOGIC EMISSION RESULTS FOR SO₂ AND Cl

In making petrologic emission estimates for SO_2 and Cl, we assume that the analyzed glass inclusions represent the sulfur and chlorine contents of the melt from which their host crystals precipitated, and that they retain this sulfur and chlorine contents after entrapment. The inclusions may have lost sulfur and chlorine, however, resulting in the zero values for ΔS_m, ΔCl_m, E_{SO_2}, and E_{Cl}. Water-loss from glass (melt) inclusions by diffusive re-equilibration with external melt through quartz crystals is feasible (Qin and others, 1992); a similar process may cause sulfur- and chlorine-loss. Sulfur and chlorine could be lost by degassing if inclusions vesiculated due to crystallization, decompression, or thermal contraction after entrapment (Anderson, 1991; Skirius and others, 1990; Tait, 1992). Decompression during ascent and eruption may crack host crystals, also leading to loss of sulfur and chlorine (Tait, 1992).

We did not see textural evidence of glass crystallizing inside the inclusions nor at the glass-host crystal boundaries, although the latter is difficult to identify. Some glass inclusions are cracked, but they are a minority and restricted to quartz hosts, while rounded inclusions of clear glass with no visible cracks containing relatively large bubbles are common. Nearly all glass inclusions contain bubbles (Westrich and Gerlach, 1992), commonly with volumes larger than expected from melt shrinkage during cooling (~1–2 percent). Bubbles in hornblende-hosted glass inclusions have voids of up to 60 volume percent. Hornblende phenocrysts made poor glass inclusion containers, and they apparently underwent sufficient dilation stress upon decompression and subsequent vapor exsolution to deform along cleavage planes (Westrich and Gerlach, 1992). This explains the larger void volumes and significantly lower H_2O contents of hornblende-hosted glass inclusions compared to inclusions in stronger containers like quartz phenocrysts (table 2). The S and Cl contents of glass inclusions in hornblende are, nevertheless, similar to the S and Cl contents of inclusions in other hosts, despite the clear differences in H_2O contents (table 2). Perhaps the glass inclusions leaked sulfur and chlorine down to matrix glass-like concentrations the moment a bubble or crack was available. Because of their size and ubiquity, we suggested that the bubbles may represent a deep magmatic vapor phase present during entrapment (Westrich and Gerlach, 1992). Although this origin for the bubbles is difficult to confirm (Lowenstern and others, 1991; Lowenstern, 1993), bubbles of primary magmatic vapor within entrapped melt would presumably acquire sulfur and chlorine from degassing of the coexisting entrapped melt during syn- and post-eruptive decompression and cooling, thus facilitating sulfur- and chlorine-loss from the glass inclusion.

Loss of sulfur and chlorine from the glass inclusions may be the cause of the zero ΔS_m and ΔCl_m and, thus, the zero petrologic emission estimates for SO_2 and Cl. However, preliminary results from analyses of glass inclusions in plagioclase and quartz phenocrysts of air-fall deposits, which are expected to be less affected by post-eruption vesiculation because of more rapid quenching upon eruption (Skirius and others, 1990), show the same range of sulfur and chlorine contents as the present data set from pyroclastic flow deposits (H.R. Westrich, unpub. data, 1993). We stress, moreover, that the sulfur contents of the glass inclusions and matrix glasses (60–88 ppm, table 2) are virtually identical to experimentally determined sulfur solubility values (Carroll and Rutherford, 1987) for evolved, anhydrite-saturated, hydrous silicate melts at temperatures (800°C), pressures (200 MPa), and oxygen fugacities (>NNO+1) similar to those of the June 15 dacite before eruption (table 1). Indeed, recent experiments on anhydrite-saturated, hydrous June 15 dacite melts at T, P, and f_{O_2} similar to the preeruption values (table 1) yield a sulfur solubility limit of ~60 ±20 ppm (Rutherford and Devine, this volume). Therefore, an alternative explanation of the zero ΔS_m and E_{SO_2} values is that sulfur did not degas significantly from the melt of the June 15 dacite during ascent and eruption. Consequently, both matrix glasses and glass inclusions have similar S contents, and there is no need to suggest significant loss of S from the glass inclusions. Similarly, the zero ΔCl_m may also reflect the lack of Cl degassing from melt during ascent and eruption. By this alternative view, matrix glasses and glass inclusions of the June 15 dacite behaved consistently: both show evidence of water loss during eruption, yet both appear to have lost insignificant S and Cl. It is unclear why this should be so, but perhaps the exsolution of S and Cl from rhyolite melt is much slower than water exsolution. The rapid decompression and dewatering of the melt (accompanied by a steep rise in viscosity) during ascent and eruption may have been too fast to permit reduction of melt sulfate and diffusion of SO_2 to sites of growing water-rich bubbles (or, as the case may be, diffusion of sulfate to bubble-growth sites and reduction there to SO_2). The lack of eruptive Cl degassing may also be related to exsolution kinetics; decrease in the diffusion rate of chlorine in rhyolitic melt as water exsolved may be a factor (Bai and Koster van Groos, 1994). Perhaps the decrease in the value of the vapor/melt distribution coefficient (D_{Cl}), as pressure drops, is involved (Shinohara and others, 1989; Webster and Holloway, 1990; Webster, 1992a).

It is unlikely that the glass inclusions ever contained enough sulfur to account for the 17-Mt SO_2 cloud, however much sulfur they may or may not have lost. That would require a ΔS_m of 1,344 ppm S for a V of 5 km³ (from equation (3) with $E_{SO_2}=17$ Mt). Adding the average matrix glass sulfur content of 60 ppm (table 2) implies ~1,400 ppm S for the glass inclusions before the conjectured sulfur loss. This concentration is not credible for the high-silica rhyolite melt phase of the dacite at preeruption conditions (table 1). It

exceeds by over 23 times the ~60-ppm measured sulfur solubility of the June 15 dacite melts at conditions similar to those of the preeruption dacite (Rutherford and Devine, this volume).

PETROLOGIC EMISSION ESTIMATE FOR H_2O

The difference between the water contents of the glass inclusions and matrix glasses is significant (table 2). Water clearly did degas from the melt during ascent and eruption. The water contents of glass inclusions in the various host phenocrysts are variable. The difference is greatest between glass inclusions in hornblende and quartz. As discussed above, the hornblende phenocrysts made relatively weak glass inclusion containers compared to quartz. Thus, glass inclusions in hornblende vesiculated and lost more water than the quartz-hosted inclusions when decompressed during ascent and eruption. We therefore take the highest water content measured for glass inclusions in quartz as the best estimate of the water contents prior to any volatile loss. The difference between this value (6.56 wt percent) and the water content of the matrix glasses (0.31 wt percent) gives an eruptive H_2O-loss for the melt, $\Delta(H_2O)_m$, of 6.25 wt percent. This result together with the data in table 1 gives petrologic emission estimates for water of 395 Mt from the H_2O analog to equation 3:

$$E_{H_2O} = 10^{-11} \Delta(H_2O)_m \rho_m \phi_m V. \qquad (4)$$

PROBLEMATIC SOURCES OF EXCESS SULFUR

Conventional petrologic emission estimates for SO_2 assume that the sulfur dissolved prior to eruption in the silicate melt of the erupted magma, as represented by the S concentrations in glass inclusions, is the sole source of emitted SO_2. The June 15 climactic event, however, produced a quantity of SO_2 requiring an amount of sulfur ~23 times in excess of what could be dissolved in the volume of erupted melt at the preeruption T, P, and f_{O_2}. Thus, the 17-Mt SO_2 cloud requires a major *immediate* source(s) of "excess sulfur"—that is, sulfur in addition to that dissolved in the preeruption melt of the erupted dacite and available at the moment the eruption began. Several sources of excess sulfur possible at the time of the eruption can be envisioned; however, most are problematic.

MELT IN NON-ERUPTED DACITE

The degassing of melt in non-erupted dacite is a possible immediate source of excess sulfur. The volume, V, of non-erupted dacite required can be calculated from equation (3) constrained by the amount of S needed to make up the

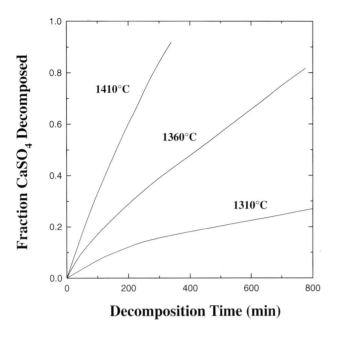

Figure 4. Rates for the anhydrite decomposition reaction

$$CaSO_4(s) \rightarrow CaO(s) + SO_2(g) + \tfrac{1}{2}O_2(g)$$

at various temperatures in air at atmospheric pressure. Data are from Hanic and others (1985). Over 2 hours are required for 10 percent decomposition at 1310°C. Decomposition times increase sharply at lower temperatures and thus suggest that anhydrite breakdown to SO_2 in the Pinatubo eruption cloud is unlikely.

difference between the remote sensing and petrologic emission estimate for SO_2 (17 Mt) and by data for ρ_m, ϕ_m, and ΔS_m (tables 1 and 2). Since ΔS_m is not significantly different from zero for the erupted dacite, it is unlikely to be greater than zero for the non-erupted dacite. Thus, the required volume of non-erupted dacite is infinitely large. If we ignore the statistical significance of dispersion in the glass S data and use the nominal mean ΔS_m (19 ppm, calculated from data in table 2) the implied volume of degassed non-erupted dacite is 350 km^3. However, the magma reservoir under the summit area of Mount Pinatubo, as defined by low velocity regions, is only 40–90 km^3 (Mori, Eberhart-Phillips, and Harlow, this volume); if low velocity regions extending out 5 km south-southeast of the summit are included, the reservoir is just 60–125 km^3. Even assuming, for purposes of illustration, that all the glass inclusions leaked during/after eruption and that before leaking they contained 120 ppm S (twice the S solubility limit at preeruption conditions), the calculated volume of non-erupted dacite (110 km^3) is still unrealistically large. We conclude that the degassing of melt in non-erupted dacite is not a viable source of excess sulfur for repairing the imbalance between remote sensing and petrologic emission estimates for SO_2.

BASALTIC MAGMA

Basalt magma was injected and mixed with the dacite a short time before the eruption to form the June 7–12 hybrid andesite lava dome (with basaltic inclusions) and the June 12 hybrid andesite scoria (Pallister and others, 1992; this volume). It is thus a candidate for directly supplying excess sulfur for the 17-Mt SO_2 cloud. We doubt, however, that direct degassing of excess sulfur from basalt magma was significant. Fragments of basalt mixed or commingled with dacite are absent in the June 15 eruption products. The occasional xenocrystic olivines scattered throughout the dacite apparently are from mixing with a small amount of basalt at an earlier time (Bernard and others, this volume; Pallister and others, this volume). It seems unlikely that basalt would be absent in the eruption deposits if it contributed most of the 17 Mt of SO_2, which would require ~2 km^3 of basalt even for a generous S release of 2,000 ppm. It is difficult to understand how basalt underlying the dacite or intruding and commingling with it could degas SO_2 explosively in large quantities but supply none of the erupted magma. Furthermore, the ubiquitous presence of anhydrite throughout the large volume of erupted dacite indicates an enhanced f_{SO_2} since long before the latest basalt mixing event (Rutherford and Devine, this volume).

Although we reject basalt degassing as the direct and immediate source of excess sulfur for the SO_2 cloud, we do not reject suggestions that basalt was the source of sulfur in the dacite on a longer time scale (Hattori, 1993; this volume; Matthews and others, 1992; Pallister and others, 1992; this volume). The recent injection and mixing of basalt with dacite, suggests the possibility of a long-term process involving migration of SO_2 (and other gases) upward into the dacite from intruding and underplating basalt. The high oxygen fugacity of the dacite (table 1) is problematic in this regard (Westrich and Gerlach, 1992), although recent work shows that gases containing SO_2 and released from basalt at 1200°C can become increasingly oxidizing with cooling (Gerlach, 1993a).

ANHYDRITE

The decomposition of anhydrite during eruption is a hypothesis often invoked as a source of excess sulfur (Devine and others, 1984; Sigurdsson, 1990). It seems unlikely, however, that anhydrite decomposition during degassing accompanying the ascent and eruption of the Pinatubo dacite provided a significant fraction of the 17-Mt SO_2 cloud, despite suggestions to the contrary (Rutherford and Devine, 1991; this volume). This much SO_2 would require the decomposition of an amount of anhydrite about equal to half that present in the erupted dacite (Westrich and Gerlach, 1992), as represented by 0.13–0.48 wt percent SO_3 in whole-rock analyses (Bernard and others, 1991; this volume; Pallister and others, this volume). The implied decomposition of a substantial fraction of preeruption anhydrite is inconsistent with the appearance of anhydrite in the deposits.

Bernard and others (1991) report, "Most [anhydrite] crystals [in the dacite] are euhedral to subhedral with sharp contacts against the groundmass; no reaction coronae have been observed." And later: "The absence of reaction coronae, as well as the observation of frequent inclusions within anhydrite phenocrysts, strongly suggests that anhydrite was in equilibrium with the silicate melt at the time the magma erupted." Fournelle and others (this volume) found some anhydrites with rounded corners, possibly indicating decomposition. They comment that several anhydrite inclusions in plagioclase phenocrysts are fully rounded, although these surely formed before ascent and eruption. Pallister and others (this volume) note that anhydrite forms euhedral crystals in the June 15 dacite and the June 12 hybrid andesite scoria. But they describe anhydrite in the hybrid andesite of the June 7–12 dome, as highly resorbed anhedral masses with dusty reaction rims containing oxide and sulfide minerals, which they attribute to breakdown during slow cooling in degassed dome rocks—not to breakdown during ascent and eruption. Nothing like this is seen in the more rapidly cooled and degassed dacite. Finally, a critical step in decomposing anhydrite to produce SO_2 during ascent and eruption is the reduction of S^{6+} in anhydrite to S^{4+} in SO_2. We know of no evidence for syn-eruptive oxidation of phases in the June 15 dacite.

The anhydrite breakdown hypothesis involves other problems as well. Over 2 hours are required for just 10 percent decomposition of anhydrite to CaO, SO_2 and O_2 in air at atmospheric pressure and 1,310°C (fig. 4) (Hanic and others, 1985). Furthermore, the decomposition rate decreases sharply with falling temperatures. The decomposition rate data indicate that anhydrite breakdown at atmospheric pressure and temperatures at or below the 780°C preeruption temperature of the Pinatubo dacite would be insignificant, perhaps even undetectable. This severely weakens the case for anhydrite decomposition to SO_2 in the Pinatubo eruption cloud as a source of excess sulfur.

Experimental data suggest that coupled anhydrite breakdown and sulfate reduction could have occurred in a 3- to 6-hour decompression ascent of the Pinatubo dacite from depth (Baker and Rutherford, 1992; Rutherford and Devine, this volume). The dacite ascent times, however, were probably greatly less than an hour during the climactic eruption. Consider an erupting magma rising within a cylindrical conduit of radius, r, from a reservoir depth, d. Let ϕ_v denote the average volume fraction of vapor, ρ the crystal-melt density, V the DRE volume of erupted crystal-melt products, and t_e the duration of the eruption. The average mass flux, M, of crystal-melt products per unit cross-sectional conduit area is

$$M = \rho V / t_e \pi r^2 \qquad (5)$$

and the average rise velocity of the magma in the conduit is

$$v = M / \rho (1 - \phi_v) \qquad (6)$$

(Jaupart and Allegre, 1991). The average conduit travel time, t_c, is thus

$$t_c = d/v = \pi r^2 d t_e (1 - \phi_v)/V \qquad (7)$$

To maximize the calculated values of t_c, ϕ_v was assumed to be zero, and a larger than expected conduit diameter of 100 m was assumed. Observed conduit diameters are more like ~10–30 m (Chadwick and others, 1988; Swanson and Holcomb, 1989; Jaupart and Allegre, 1991). The initial sighting of the June 7–10 lava dome revealed a plug ~100 m across, suggesting a conduit diameter less than this amount (R. Hoblitt and C. Newhall, oral commun., September 22, 1993). The initial post-eruption dome had a diameter of only several tens of meters (C. Newhall, oral commun., September 22, 1993). The average depth to tapped magma, d, was taken to be 9 km (table 1). Estimates for the values of parameters t_e, V and ρ are listed in table 1. The calculated conduit travel time is 7.6 minutes. The estimated travel times, moreover, are undoubtedly too long, because of the values assumed for r and ϕ_v. Furthermore, since plinian activity and pyroclastic flow production seem to have occurred largely during the first 3 hours of the eruption (Wolfe and others, this volume), the 9-hour t_e value (table 1) may be too large by a factor of ~3. If so, conduit travel times could have been <150 s and in good agreement with travel times predicted from recent models for two-phase flow of gas-particle mixtures through eruption conduits (Sparks and others, 1994). If anhydrite decomposed this rapidly to produce the 17-Mt SO_2 cloud, it is unclear why the anhydrite remaining in the pumice is predominantly as non-reacted euhedral and subhedral crystals. Perhaps the time needed for decompression breakdown of anhydrite could be satisfied if the erupted dacite decompressed and degassed in the magma chamber for 3 to 6 hours before ascent (Rutherford and Devine, this volume), but it is unclear how a decompression of the required magnitude (100–200 MPa) and duration could have occurred at 6–11 km; presumably such an event would also favor breakdown of cummingtonite and hornblende. Finally, we note that adiabatic cooling during rapid ascent (Wilson and others, 1980) would tend to enhance anhydrite stability.

HYDROTHERMAL FLUID

Another possible source of the 17-Mt SO_2 discharge is the sulfate-rich fluid of the Mount Pinatubo hydrothermal system (Delfin and others, 1992; this volume). SO_2 gas is not produced in significant amounts from dissolved sulfate

Table 3. Composition of a hydrothermal fluid sample from Pinatubo geothermal exploration well PIN–2D

[Data from Delfin and others (1992). Units are in ppm (wt). Sample collected 13 February 1989.]

Species	Concentration
pH	2.3
Na	1153
K	208
Ca	18
Mg	96
Li	2.4
Cl	1914
SO$_4$	1129
B	67
SiO$_2$	454

when hydrothermal systems boil; it is nearly undetectable in the steam released from geothermal well discharges (Giggenbach, 1980). SO$_2$ gas is unstable in the presence of hydrothermal liquid, relative to dissolved sulfate, which it tends to form by disproportionation reaction with water (Giggenbach, 1987; Williams and others, 1990). It is conceivable, however, that significant amounts of SO$_2$ could be released if large amounts of sulfate-rich fluid came in contact with magma during the eruption and flashed to superheated vapor.

If we assume that the sulfate-rich samples from geothermal exploration wells (Delfin and others, 1992; this volume) are characteristic of the Pinatubo hydrothermal system, then hydrothermal fluids such as those observed in exploration well PIN-2D containing 1,129 ppm sulfate (table 3) give a bulk SO$_2$ content of 753 ppm potentially available as an excess sulfur source. Even at this concentration level, at least 22,580 Mt of hydrothermal fluid, or ~23 km^3 liquid-volume-equivalent (at STP), are required to produce 17 Mt of SO$_2$, assuming all the sulfate is converted to SO$_2$. We doubt that the heat required to flash this volume of hydrothermal fluid could have been transferred from 5 km^3 of dacite within the 9 hours of June 15 eruption. Thermodynamic modeling of the flash-vaporization products indicates, moreover, that the potentially available SO$_2$ would not be realized, even if 23 km^3 of fluid did flash.

We used GASWORKS, a thermochemical equilibrium computer code (Symonds and Reed, 1993), to compute heterogeneous equilibrium distributions of gas, liquid, and mineral species as a function of temperature and pressure for complete flash-vaporizing of 22,580 Mt of Pinatubo hydrothermal fluid, with the composition of the PIN-2D sample (table 3), to superheated vapor. Since PIN-2D has a slight negative net charge, we subtracted small amounts of chloride, the most abundant anion, from the analysis to obtain charge balance in the starting fluid composition for the calculations. We neglected the small amounts of Li and B reported in the analysis (table 3). For the 9-element

Table 4. Calculated equilibrium distributions of gas species and minerals for flash vaporization of a 22,580 Mt quantity of Pinatubo hydrothermal fluid containing 17 Mt bulk SO$_2$.

[Based on hydrothermal fluid sample from geothermal well PIN–2D, table 3. Units are in Mt, megatons. Only SO$_2$ and gas species more abundant than SO$_2$ are shown. –, mineral not present at this temperature.]

Species	800°C	400°C
Gases		
H$_2$O	22,470	22,470
HCl	14.0	13.3
(NaCl)$_2$	23.6	7×10^{-6}
NaCl	11.5	5×10^{-6}
KCl	7.9	4×10^{-6}
O$_2$	.9	0.8
(KCl)$_2$	1.0	8×10^{-6}
SO$_2$	.004	6×10^{-9}
Minerals		
Na$_2$SO$_4$[1]	37.7	36.2
MgSiO$_3$[2]	7.9	8.9
SiO$_2$[3]	4.2	4.9
(Ca,Mg)SiO$_3$[4]	2.2	–
NaCl (halite)	–	36.4
KCl (sylvite)	–	8.9
CaSO$_4$ (anhydrite)	–	1.4

[1] Hexagonal form of anhydrous sodium sulfate.
[2] Orthoenstatite at 800°C and clinoenstatite at 400°C.
[3] β-quartz at 800°C and α-quartz at 400°C.
[4] Diopside.

system, calculations included 117 gas species and 65 minerals. We began the flash-vaporization model by calculating the equilibrium distribution of gas and mineral species that form with complete instantaneous vaporization of the hydrothermal fluid to superheated vapor at 800°C. We then decreased temperature and recalculated the equilibrium distributions at 10°C decrements down to 400°C. In these calculations, the pressure was atmospheric (0.1 MPa), and the system was treated as a closed system—no mineral fractionation or external buffering of components was allowed during cooling.

The quantities of gas and mineral species calculated in the equilibrium distributions at 800°C and 400°C are summarized in table 4. The results show that >99.9 percent of the hydrothermal sulfur precipitates as Na$_2$SO$_4$(s) and that very little sulfur remains in the gas phase as SO$_2$ (<0.004 Mt). SO$_2$ is the main sulfur gas species in the gas phase over the temperature range 800°C to 400°C, but its concentration is always very low (≤10^{-8} mole fraction), as shown in figure 5. Repeating the calculations at higher pressures up to the hydrostatic pressure of ~16 MPa expected at the 1,600 m bottom depth of the exploration well again showed that >99.9 percent of the hydrothermal sulfur precipitates as Na$_2$SO$_4$(s). Repeating the calculations for open system conditions allowing minerals to fractionate with cooling produced nearly identical abundances of SO$_2$ to those of the closed system model.

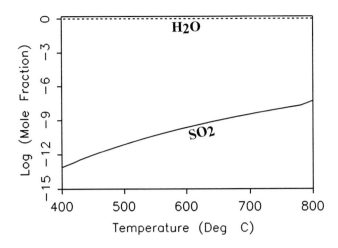

Figure 5. Mole fraction of SO_2 for closed system flash-vaporization of PIN-2D well fluid (table 3) from 800°C to 400°C at atmospheric pressure. Concentration of H_2O is shown for reference. SO_2 is the main sulfur gas species, but its concentration is always $<10^{-8}$.

These calculations indicate it is not possible to produce 17 Mt of SO_2 by flash-vaporization of Pinatubo hydrothermal fluid to superheated vapor. To get the 17 Mt of SO_2 at 800°C, absurd quantities of $\sim10^5$ km^3 of fluid need to be flashed; at 400°C, the volume of fluid required balloons to $>10^{10}$ km^3! These results suggest further that a magmatic source—not flash-vaporization of the hydrothermal system—supplied the high SO_2 emissions (up to 5,000–13,000 tons/day) observed in the weeks before the June 15 eruption (Daag, Tubianosa, and others, this volume).

PREERUPTION VAPOR SOURCE OF EXCESS SULFUR

We show below that the dacite erupted on June 15 was vapor-saturated at depth prior to eruption, and propose that an accumulated vapor phase in the dacite provided the *immediate* source of excess sulfur for the 17-Mt SO_2 cloud (Westrich and Gerlach, 1992; Gerlach, 1993b). As used here, "vapor" denotes a single volatile-rich phase having a distinctly lower density than that of the silicate liquid (melt) phase of the June 15 dacite. It is possible the dacite contained multiple (immiscible) volatile-rich phases, but lacking evidence to the contrary, we assume that a single volatile-rich phase predominated at depth. Fluid unmixing may have developed, however, during decompression ascent (Pasteris and others, this volume). It is also assumed that the vapor phase was dispersed as bubbles throughout the body of erupted dacite, although we recognize that the distribution may not have been uniform and that a vapor cap or foam layer (Jaupart and Vergniolle, 1989) may have existed near the top of the body.

VAPOR SATURATION

Low velocity regions defined by seismicity data (Mori, Eberhart-Phillips, and Harlow, this volume) permit a depth range of 6–11 km for the magma reservoir. Geothermal exploration and drilling data suggest that the volcano is underlain by about 2 km of non-lithified dacite (density=$\sim2.0\times10^{12}$ kg/km^3) and 2 km of lithified dacite and intrusions (density=$\sim2.6\times10^{12}$ kg/km^3), which overlie ophiolitic rocks (density=$\sim2.9\times10^{12}$ kg/km^3) (Delfin, 1983; Pallister and others, this volume). We therefore assume an average crustal density of 2.7×10^{12} kg/km^3, which gives a pressure range of ~160–300 MPa for the magma at 6–11 km in depth. This range is consistent with the preeruption pressure of 220 ± 50 MPa obtained by Al-in-hornblende geobarometry for the rims of hornblende phenocrysts (Rutherford and Devine, 1991; this volume); the uncertainty of ±50 MPa is the estimated error (M. Rutherford, oral commun., December, 8, 1993). These pressures also agree with experimentally determined stability limits for the phenocryst assemblage of the Pinatubo dacite at the preeruption temperature (780°C) (Rutherford and Devine, 1991; this volume). Thus, vapor saturation prior to eruption is plausible if vapor pressures significantly >160–170 MPa can be demonstrated.

The volatile contents of glass inclusions and the results of phase equilibria experiments are consistent with preeruption vapor saturation for the June 15 Pinatubo dacite (fig. 6). As discussed above, we take the highest water contents (6.1–6.6 wt percent) determined in this study for glass inclusions in quartz hosts as the best estimate for the water content of inclusions prior to any volatile loss. These water contents correspond to P_{H_2O} of ~210–250 MPa according to experimental solubility data for water in hydrous rhyolitic melt at the preeruption temperature (Silver and others, 1990). Rutherford and Devine (this volume) conclude from similar results that the water contents of glass inclusions correspond to P_{H_2O} of ~220 MPa for hydrous rhyolitic melt. They also report that water pressures of ~200–220 MPa are required to reproduce the composition of the melt in equilibrium with the natural phenocrysts of the June 15 dacite (Rutherford and Devine, this volume), and these pressures are presumably indicative of P_{H_2O} a short time before the eruption. The above values for P_{H_2O} are sufficient to make vapor saturation of the preeruption dacite plausible, even if water is assumed to have been the only volatile present. Thermodynamic constraints for C-O-H-S-volatile bearing magma require additional vapor pressure contributions, principally from CO_2 and SO_2. Rutherford and Devine (this volume) report that infrared spectroscopic analysis of glass inclusion-bearing quartz and plagioclase shows CO_2 below detection (20 ppm). Wallace and Gerlach (1994), on the other hand, report infrared spectroscopic analyses that show 280–420 ppm CO_2 for intact glass inclusions in quartz phenocrysts from splits of the samples used in this study. Their

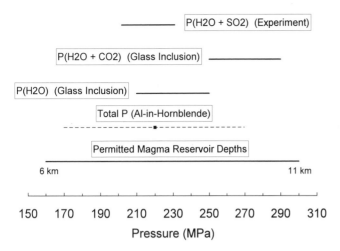

Figure 6. Total pressure and vapor pressure estimates for the June 15 dacite. Permitted magma reservoir depth range is based on seismic velocity data (Mori, Eberhart-Phillips, and Harlow, this volume) and an estimated average crustal density of 2.7×10^{12} kg/km^3, as discussed in text. Total pressure determined from the Al-in-hornblende geobarometer is 220 MPa with an estimated error of ±50 MPa (Rutherford and Devine, 1991; this volume). Several observations are consistent with vapor saturation and suggest vapor pressures in excess of 210 MPa. Water contents of 6.1–6.6 wt percent for glass inclusions in quartz hosts (Rutherford and Devine, this volume; this study) correspond to P_{H_2O} of ~210–250 MPa for water-rhyolitic melt systems at the preeruption temperature (Silver and others, 1990). Wallace and Gerlach (1994) report 280–420 ppm CO_2 and 6.1 to 6.6 wt percent water for intact glass inclusions in quartz phenocrysts from splits of samples used in the present study, implying $P_{H_2O} + P_{CO_2}$ of 250–290 MPa (mean=270 MPa). Experimental studies (Rutherford and Devine, this volume) indicate that water pressures of ~200–220 MPa are required to reproduce the composition of the melt in equilibrium with the natural phenocrysts of the June 15 dacite (Rutherford and Devine, this volume), presumably indicative of P_{H_2O} a short time before the eruption. Other experimental data (Baker and Rutherford, 1992; Rutherford and Devine, this volume) indicate that anhydrite stability at the preeruption conditions would have required an additional vapor pressure contribution from P_{SO_2} of 2–11 MPa; thus, experimental data suggest that the total vapor pressure (excluding P_{CO_2}) just prior to eruption was at least 202–231 MPa.

inclusions contain 6.1 to 6.6 wt percent water and ~70 ppm S. The water and carbon dioxide contents of these glass inclusions imply $P_{H_2O} + P_{CO_2}$ of 250–290 MPa, based on the method of Newman and others (1988) (revised to include recent experimental data for H_2O and CO_2 solubility in rhyolitic melt (Silver and others, 1990; Blank and Stolper, 1993) and a modified Redlich-Kwong equation of state (Holloway, 1977) for modeling the properties of the vapor). The mean $P_{H_2O} + P_{CO_2}$ is 270 MPa. Experimental data indicate that anhydrite stability at the preeruption conditions would have required yet an additional vapor pressure contribution from P_{SO_2}, which, although uncertain at this time, is thought to be in the range 2–11 MPa (Baker and

Rutherford, 1992; Rutherford and Devine, this volume). Finally, the 60–88 ppm S dissolved as sulfate in the June 15 glass inclusions and matrix glasses is compatible with a water-rich vapor phase. As noted above, experimental solubility data (Rutherford and Devine, this volume) show that the melt phase of June 15 dacite dissolves sulfur at this concentration level and oxidation state when saturated with anhydrite and water-rich vapor at the preeruption T, P, and f_{O_2}. Considering all the above results (fig. 6), we are persuaded that the total vapor pressure of the volatile components (mainly $P_{H_2O} + P_{CO_2} + P_{SO_2}$) of the June 15 Pinatubo dacite was sufficient to cause saturation with H_2O-rich vapor at depth in the magma reservoir prior to ascent and eruption.

Exploration studies for geothermal energy at Mount Pinatubo strongly support the suggestion of vapor saturation and buildup in the dacite of the Pinatubo magma reservoir prior to the 1991 eruption. Three deep exploration wells were drilled into the Pinatubo hydrothermal system between August 1988 and November 1989. Samples and observations from these wells indicate "a significant magmatic input" into the hydrothermal system, and provide the basis for a model invoking magma as a source of volatiles observed in the hydrothermal fluids (Delfin and others, 1992; this volume; Ruaya and others, 1992). Chemical and stable isotope data show that the acidic hydrothermal fluids and gases encountered in the wells have strong similarities to high-temperature volcanic gases; these similarities suggest volatile discharge from the underlying magma (Casadevall, 1992; Delfin and others, 1992; this volume; Ruaya and others, 1992). Absorption of the magmatic volatile discharges in the hydrothermal system produced the acidic waters and generated the high sulfate fluids (table 3) (Delfin and others, 1992; this volume), presumably by disproportionation of magmatic SO_2.

Finally, we speculate that the ubiquitous presence of bubbles in the Pinatubo glass inclusions (Westrich and Gerlach, 1992) supports a vapor saturation model. The pressure difference between entrapped melt and matrix melt is greater during decompression if the entrapped melt was vapor-saturated (Tait, 1992). Thus, entrapped vapor-saturated melt would have a greater tendency to deform host crystals and vesiculate during eruption. This would promote deformation of phenocrysts and the development of glass inclusions with relatively large bubble volumes, as commonly observed in the June 15 dacite (Westrich and Gerlach, 1992).

VAPOR COMPOSITION

We attempt to model the approximate composition of the preeruption vapor present a short time before the climactic eruption. The model is constrained by the mass abundance of SO_2 in the vapor phase and constraints on the vapor pressures. The mass abundance of SO_2 in the vapor phase is assumed to be 17 Mt, based on the remote sensing data, which is regarded as a minimum estimate, as discussed previously. P_{CO_2} is taken to be 40 MPa, based on experimental solubility measurements for rhyolitic liquids and the dissolved CO_2 content of glass inclusions (Wallace and Gerlach, 1994). We have used a water pressure of 220 MPa to be consistent with experimental P_{H_2O} values that reproduce the composition of the melt in equilibrium with the natural phenocrysts of the June 15 dacite (Rutherford and Devine, this volume), and that presumably reflect the vapor composition a short time before the eruption. The value of P_{SO_2} a short time before eruption is not as well constrained. P_{SO_2} values used here are based on experimental data for anhydrite stability in the June 15 dacite (Baker and Rutherford, 1992). The initial analyses of experimental run products indicated a P_{SO_2} of ~11 MPa, which leads to a total (model) pressure of ~275 MPa (including P_{Cl}, as discussed below). Recent analyses suggest a P_{SO_2} of ~2 Mpa and give a total (model) pressure of ~265 MPa. Both values for P_{SO_2} are used in the model calculations that follow, however we have reservations about the results constrained by the 2-MPa P_{SO_2} value, as discussed below.

The molar abundances of H_2O and CO_2 in the vapor phase are calculated from simple equations of the form

$$N_i = P_i (N_{SO_2}/P_{SO_2}) \qquad (8)$$

where N_i is the molar abundance of species i in the vapor phase. This equation links intensive parameters (P_{SO_2}) evaluated from petrologic experiments and extensive parameters (N_{SO_2}) obtained from remote sensing measurements. The resulting molar and mass abundances are given in table 5 and indicate a H_2O-rich vapor. Experimental studies of natural and synthetic rhyolites at Pinatubo-like conditions of 800°C and 200 MPa show that Cl partitions preferentially into aqueous vapor relative to melt (Webster and Holloway, 1990; Webster, 1992a, b; Metrich and Rutherford, 1992). These studies suggest that for a preeruption melt Cl content of 0.102 wt percent in the Pinatubo dacite, based on the average Cl content of the glasses (table 2), the vapor-melt distribution coefficient for Cl, D_{Cl}, given by

$$D_{Cl} = (\text{wt percent Cl}_{vapor})/(\text{wt percent Cl}_{melt}) \qquad (9)$$

is roughly 20, indicating ~2 wt percent Cl in the vapor phase. Estimated total Cl abundances in the vapor (mainly as NaCl, KCl, and HCl) are included in table 5.

The composition of the inferred preeruption vapor is summarized in table 5. The mole percent results for the upper- and lower-P_{SO_2} values are similar. The vapor phase was not excessively SO_2-rich. It contained only ~4 mole percent or less SO_2 despite the impressive 17-Mt SO_2 emission. The vapor also contained ~80–83 mole percent H_2O, ~15 mole percent CO_2, and ~1 mole percent Cl (table 5).

Table 5. Composition of the preeruption vapor present in the dacite of the climactic June 15, 1991, eruption, Mount Pinatubo.

[Mt, megaton; Tmol, teramole (10^{12}); MPa, megapascal]

Component	Mass[1] (Mt)	Moles[1] (Tmol)	Mole Percent	Pressure (MPa)
P_{SO_2} = 11 MPa				
H_2O	95.6	5.3073	80.1	220
SO_2	17.	0.2654	4.0	11
CO_2	42.5	0.9650	14.5	40
Cl	3.2	0.0911	1.4	3.8
Totals ....:	158.3	6.6288	100.0	274.8
P_{SO_2} = 2 MPa				
H_2O	525.9	29.1901	82.9	220
SO_2	17.	0.2654	0.8	2
CO_2	233.6	5.3073	15.1	40
Cl	16.1	0.4560	1.2	3.4
Totals ....:	792.6	35.2188	100.0	265.4

[1] Minimum estimates, as discussed in the text; extra significant figures for Moles to avoid round-off errors.

We stress that the molar and mass abundances of H_2O, CO_2, and Cl given in table 5 are minimum values, since they are linked to the minimum emission estimate of 17 Mt for SO_2.

The concentration of sulfur obtained for the preeruption vapor compared to its concentration in the melt indicates a high volatility for sulfur in the June 15 dacite. The distribution coefficient, D_S, for sulfur between the preeruption vapor and melt of the dacite can be calculated from an analog to equation (9), and provides a measure of sulfur volatility. The wt percent concentration of S in the preeruption vapor is equal to half that of SO_2 (table 5). The preeruption S content of the melt is taken to be 0.0075 wt percent, based on the average S content of the glasses (table 2). Thus, D_S is ~720 (P_{SO_2} = 11 MPa) or ~140 (P_{SO_2} = 2 MPa), underscoring a strong preference for sulfur to reside in the vapor as SO_2 compared to dissolution in the melt as sulfate, despite the highly oxidizing preeruption conditions.

VAPOR VOLUME AND DENSITY

The ideal gas law gives an estimate of the approximate total volume of vapor at depth, V_{vapor}, directly from the 17-Mt SO_2 emission (N_{SO_2} = 0.2654×10^{12} moles, table 5), the preeruption P_{SO_2} (2 and 11 MPa), and T (780°C), as follows:

$$V_{vapor} = (N/P)RT = (N_{SO_2}/P_{SO_2})RT \qquad (10)$$

The results are 0.21 km^3 (P_{SO_2} = 11 Mpa) and 1.2 km^3 (P_{SO_2} = 2 MPa). The corresponding ideal gas density estimates for the vapor masses given in table 5 are 0.75 g/cm^3 (P_{SO_2} = 11 MPa) and 0.68 g/cm^3 (P_{SO_2} = 2 MPa).

An ideal mixing model for the vapor volume, based on endmember volumes obtained from *PVT* data for H$_2$O (Burnham and others, 1969) and from a modified Redlich-Kwong equation of state (Holloway, 1977, 1981) for CO$_2$ and SO$_2$, gives V_{vapor} of 0.25 km^3 (P_{SO_2} = 11 MPa) and 1.35 km^3 (P_{SO_2} = 2 MPa). (The negligible Cl vapor was treated as ideal gas.) The respective density estimates are 0.63 g/cm^3 and 0.59 g/cm^3.

The vapor phase volume results are relatively insensitive to total pressure and vapor abundance. This is because ideal and nonideal values for V_{vapor} tend to be a strong function of the quantities N/P and RT, which are fixed by geothermometry data for T (780°C, table 1) and by remote sensing data and anhydrite stability data for N_{SO_2}/P_{SO_2} (for example, 0.2413×10^{11} mole/MPa for P_{SO_2} = 11MPa). As a result, changes in pressure (P) and vapor abundance (N) are coupled such that N/P remains fixed, which tends to stabilize V_{vapor}.

DACITE VAPOR FRACTION AND DENSITY BEFORE ERUPTION

We estimate the volume fraction of vapor at depth in the preeruption dacite from the ideal mixing vapor volume (above) and 5 km^3 of erupted dacite. Figure 7 shows the results expressed as volume percent vapor in the preeruption dacite at 780 °C and pressures of 265–275 MPa; the pressure range corresponds to the P_{SO_2} values of 2 and 11 MPa. The vapor fraction ranges from 4.8 volume percent (P_{SO_2} = 11 MPa) to 21 volume percent (P_{SO_2} = 2 MPa). The vapor fraction increases sharply for P_{SO_2} less than 5 MPa.

The volume fractions shown in figure 7 may be minimum estimates because of their dependence on the minimum SO$_2$ emission estimate of 17 Mt. Nevertheless, it is possible they may be too high, since it is assumed that all vapor was derived only from the *erupted* dacite. Although derivation of vapor by syn-eruptive degassing of *melt* in non-erupted dacite was apparently insignificant, as discussed above, our analysis does not rule out syn-eruptive escape of the preeruption vapor phase from non-erupted (vapor-saturated) dacite. It seems unlikely to us, however, that large amounts of such vapor could segregate and escape from non-erupted dacite fast enough to supply the explosive eruption without simultaneous ejection of the source material.

The preeruption bulk density of the dacite can be estimated from the sum of the vapor mass (table 5) and the erupted mass of crystals and glass (from data in table 1) divided by the sum of the ideal mixing vapor volume (V_{vapor}) and the DRE volume of erupted dacite. The results (fig. 8) range from 2.31×10^{12} kg/km^3 (P_{SO_2} = 11 MPa) to 2.01×10^{12} kg/km^3 (P_{SO_2} = 2 MPa).

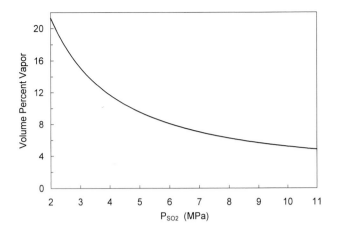

Figure 7. Volume percent vapor at depth in the preeruption dacite. The curve is for vapor compositions at 780°C from P_{SO_2} = 2 MPa (P = 265 MPa) to P_{SO_2} = 11 MPa (P = 275 MPa). The results assume ideal mixing in the vapor of end member volumes that are based on pressure-volume temperature data for H$_2$O (Burnham and others, 1969) and a modified Redlich-Kwong equation of state for CO$_2$ and SO$_2$ (Holloway, 1977, 1981).

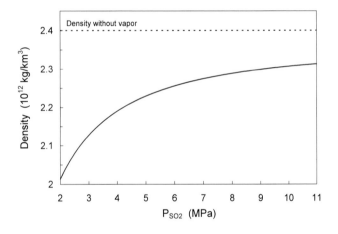

Figure 8. Bulk density of the preeruption dacite. (Note: 10^{12} kg/km^3 = 1 g/cm^3.) The solid curve is for vapor compositions at 780°C from P_{SO_2} = 2 MPa (P = 265 MPa) to P_{SO_2} = 11 MPa (P = 275 MPa). Densities are calculated from the sum of the vapor mass (table 5) and the erupted mass of crystals and glass (from data in table 1) divided by the sum of the ideal mixing vapor volume (V_{vapor}) and the DRE volume of erupted dacite (table 1). Dotted line is the estimated density of the dacite without vapor, based on density estimates for vesicle-free dacite (table 1).

VAPOR ACCUMULATION

It is impossible to dissolve the total amount of volatiles contained in the preeruption vapor and melt in a completely molten, 5 km^3 quantity of dacite at the magma reservoir pressures (≤300 MPa, fig. 6). Consider the mass of water, M_{H_2O}, dissolved in the rhyolite melt of the dacite and given by

$$M_{H_2O} = 10^{-11} (H_2O)_m \, \rho_m \, \phi_m \, V \qquad (11)$$

where M_{H_2O} is in Mt, $(H_2O)_m$ is the wt percent water dissolved in the rhyolite melt; the other terms are defined as for equation 4 above and have values given in table 1. On the basis of a preeruption melt concentration of 6.56 wt percent water (table 2), M_{H_2O} is 415 Mt. Including the 95.6 Mt of water in the preeruption vapor based on P_{SO_2} of 11 MPa (table 5) implies bulk water contents for completely molten dacite of 4.2 wt percent. Similarly, the estimated 42.5 Mt CO_2 in the preeruption vapor (table 5) together with ~0.03 wt percent dissolved CO_2 (Wallace and Gerlach, 1994) imply bulk carbon dioxide contents for completely molten dacite of 0.36 wt percent. Calculations based on the revised method of Newman and others (1988), indicate that to dissolve these amounts of water and carbon dioxide in *rhyolite* would require pressures >400 MPa. A vapor phase composition based on P_{SO_2} of 2 MPa (table 5) requires the dissolution of 7.4 wt percent H_2O and 1.8 wt percent CO_2, which would necessitate still higher pressures. Thus, it is unlikely that the total amount of water, carbon dioxide, and sulfur dioxide could have been dissolved in the erupted quantity of *dacite* at reservoir pressures, even if completely molten, since the solubilities of water and carbon dioxide used here are for rhyolite and would decrease in higher temperature, less evolved dacitic melts (Holloway, 1981; Fogel and Rutherford, 1990; Pan and others, 1991). The effect of SO_2 on the vapor saturation pressure, which has been neglected in this analysis, would presumably further increase the pressure required for complete volatile dissolution. Finally, the pressures required to achieve dissolution are very sensitive to the amounts of CO_2, which are minimal in the above calculations because of linkage to the minimum emission estimate of 17 Mt for SO_2.

The total volatile load contained in the vapor and dissolved in the melt of the erupted June 15 dacite apparently requires some process(es) of vapor accumulation either by open system volatile migration or by closed system crystallization/decompression (Wallace and Gerlach, 1994). We suggest that long term, open system migration of vapor upward from subjacent dacite and (or) mafic magmas intruding and underplating the magma chamber may have provided the main supply of excess volatiles for the accumulated vapor in the magma that erupted on June 15. Alternatively, the volatile load may have been dissolved in higher temperature, less crystallized dacite at depths greater than 11 km, with subsequent exsolution and accumulation as vapor during crystallization and decompression associated with cooling and emplacement at 6–11 km. If this were the case, however, at least some glass inclusions would be expected with more primitive and sulfur-rich compositions than are observed.

VOLATILE EMISSION ESTIMATES FOR THE CLIMACTIC ERUPTION

The climactic eruption released the accumulated vapor of the erupted dacite and resulted in emissions of 17 Mt SO_2, 96–526 Mt H_2O, 42–234 Mt CO_2, and 3–16 Mt of Cl (table 5). Additional yields of SO_2 and Cl from degassing of melt were minor to insignificant during ascent and eruption. Degassing of melt during ascent and eruption may have released an additional 2 Mt CO_2, depending on the kinetics of CO_2 exsolution. Melt degassing of water was significant during ascent and eruption, and yielded an additional 395 Mt of H_2O. Thus, the total H_2O emission was approximately 491 Mt (P_{SO_2} = 11 MPa) to 921 Mt (P_{SO_2} = 2 MPa). The total volatile emission for the climactic eruption was 555–1190 Mt, of which 90 to 95 mole percent was water.

DISCUSSION AND CONCLUSIONS

We have more confidence in the vapor composition model constrained by a P_{SO_2} of 11 MPa than in the model based on a P_{SO_2} of 2 MPa (table 5). The latter model implies a C/S—that is, the molar ratio of total C ($CO_2 + CO + ...$) to total S ($SO_2 + H_2S + 2S_2 + ...$)—for the vapor of ~20, compared to a C/S of ~3.6 in the former model. C/S values of high-temperature volcanic gases (>500°C) collected from active convergent plate volcanoes, however, are <20, except for a few samples collected from degassed domes (e.g., Showa-Shinzan) several years after dome emplacement (Marty and Le Cloarec, 1992; Williams and others, 1992; Symonds and others, 1994). A recent compilation of volcanic gas data (Symonds and others (1994) shows C/S values <20 for volcanic gases from volcanoes of all tectonic environments; the mean of the average C/S values for active convergent plate volcanoes is 6 with a standard error ($s/n^{1/2}$) of 2. Thus, the C/S of ~20 implied by the 2-MPa P_{SO_2} vapor composition model is unusually S-depleted, which seems unlikely for a magma that yielded 17 Mt of SO_2, primary anhydrite, and 1,300–4,800 ppm SO_3 in whole-rock analyses.

Several other implications of the 2-MPa P_{SO_2} vapor composition model also give us reservations. These include (1) vapor fractions for the preeruption dacite of up to 21 volume percent, (2) preeruption dacite bulk densities as low as 2×10^{12} kg/km^3, (3) bulk water contents up to 7.4 wt percent, and (4) total volatile emissions for the climactic eruption of up to 1,200 Mt. It is unclear how the dacite could acquire such high vapor fractions and low bulk densities without having erupted long ago, although recent work suggests similar vapor fractions and densities for the Bishop Tuff prior to eruption (Wallace and Anderson, 1994). In any event, this study underscores the need to examine the potential importance of a significant preeruption vapor fraction to the dynamics of explosive eruptions. Current models of the dynamics of magma degassing and explosive eruptions typically assume an initial condition of volatile undersaturation or of volatile saturation with only a few bubbles present (Sparks and others, 1994).

We conclude that the dacite magma erupted in the climactic June 15, 1991, eruption of Mount Pinatubo was vapor-saturated and contained an accumulation of at least 5 volume percent, water-rich vapor (~80 mole percent H_2O) at depth prior to eruption. The accumulated vapor made the dacite buoyant and able to respond rapidly to decompression, thus increasing its "eruptability" and sensitivity to triggering mechanisms. The accumulated vapor was the chief *immediate* source of excess sulfur for the giant 17-Mt SO_2 cloud injected into the stratosphere, although it was not excessively SO_2-rich, containing only ~4 mole percent or less SO_2. Other potentially available immediate sources of excess sulfur (non-erupted dacite, mafic magma, anhydrite decomposition, sulfate-rich hydrothermal fluids, and degassing of ascending melt) made relatively unimportant contributions to the 17-Mt SO_2 cloud. Accumulated preeruption vapor also contributed emissions of at least 96 Mt of H_2O, 42 Mt CO_2 and 3 Mt Cl during the climactic eruption. Melt degassing of water was significant during ascent and eruption and yielded an additional 395 Mt of H_2O, which undoubtedly played an important role in the dynamics of the eruption.

Conventional petrologic estimates based on glass inclusions predict statistically insignificant emissions for both SO_2 and Cl during the climactic eruption. The low petrologic estimates for these volatiles result from their relatively high volatility and enrichment in the vapor phase of the dacite prior to eruption, coupled with apparently negligible degassing of SO_2 and Cl from melt during ascent and eruption of the June 15 dacite. This study reinforces growing evidence, based on smaller eruptions (VEI ≤5), that petrologic emission estimates for SO_2 are many times lower than estimates based on remote sensing, and indicates that petrologic estimates may also seriously underestimate the SO_2 emissions of larger explosive eruptions. If explosive volcanism commonly involves vapor-saturated magma containing accumulated vapor, petrologic estimates for SO_2 emissions during explosive eruptions of the past may be far too low and significantly underestimate their impacts on climate and the chemistry of the atmosphere. Thus, an improved technique is sorely needed to infer the SO_2 yields and potential global impacts of pre-TOMS eruptions.

Conventional petrologic emission estimates for Cl and CO_2 probably are significantly low also, since these volatiles partition strongly into a vapor phase, while those for H_2O may be only marginally low. Because it cannot be assumed that volatiles will invariably accumulate in preeruption melt like most other incompatible elements, the volatile contents of glass inclusions alone do not, in general, provide a sufficient basis for predicting either the total preeruption volatile contents of magma or the volatile emissions of explosive eruptions. When volatile determinations on glass inclusions are combined with volatile measurements on emission clouds, however, the results are complementary and provide improved constraints for the preeruptive volatile contents of magmas, including magmas containing large quantities of accumulated vapor. This study illustrates the potential utility of obtaining values for *both* intensive parameters (for example, T, P, P_{H_2O}, P_{CO_2}, P_{SO_2}) from petrologic measurements and extensive parameters (for example, N_{SO_2}) from emission measurements to improve estimates of the preeruption volatile contents of explosive magma.

We stress the need to make greater use of remote sensing techniques to measure volcanic emissions and to infer subsurface magmatic conditions. Applications of techniques for measuring CO_2 emissions during eruptions, ideally in conjunction with glass inclusion studies, are especially encouraged. Because of its low solubility in silicate melts and the usual absence of carbonate minerals in magma, CO_2 should be strongly enriched in the vapor phase of vapor-saturated magmas (Holloway, 1976). Thus, measurements of CO_2 emissions in future eruptions, together with solubility data, can provide critical tests of the vapor saturation and accumulation model (see Gerlach and others, 1994).

Finally, we note that the apparent large distribution coefficient for sulfur between preeruption vapor and melt of the Pinatubo dacite (D_S >100) suggests the importance of measuring D_S experimentally for vapor-saturated dacitic and rhyolitic melts at a range of temperatures, pressures, and oxygen fugacities. Measurement of D_S at appropriate experimental conditions would directly test the validity of our proposal that excess sulfur for the 17-Mt stratospheric SO_2 cloud resided in a vapor phase at depth in the Pinatubo dacite prior to the climactic eruption.

ACKNOWLEDGMENTS

We are indebted to many people who helped us in the course of this study. Rick Hoblitt, John Pallister, and Ed Wolfe provided samples, and Rick Hervig at the Arizona State University's Center for Solid State Sciences assisted with ion probe analyses. Sally Newman provided calculations for estimating vapor pressures of rhyolitic melts containing carbon dioxide and water. Several authors of papers in this volume generously supplied their draft and revised versions of manuscripts containing critical data for our use. In this regard, we especially thank Alain Bernard, Joseph Devine, John Pallister, Malcolm Rutherford, Willie Scott, and Ed Wolfe. We are grateful to numerous colleagues who shared ideas with us, and we benefited greatly from our conversations with Gregg Bluth, Arlin Krueger, Jim Luhr, John Pallister, Bill Rose, Malcolm Rutherford, Steve Self, and Paul Wallace. We thank our reviewers Fred Anderson,

Michael Carroll, Chris Newhall, Bill Melson, Haraldur Sigurdsson, and Ed Wolfe for prompt and careful reviews. Fred Anderson's comments were especially helpful in improving the conceptual framework of the manuscript. Finally, we wish to acknowledge John Watson of the USGS Branch of Eastern Technical Reports for his skill, enthusiasm, and patience in carrying out several manuscript revisions. The USGS Global Change and Climate History Program and the U.S. Department of Energy BES Geosciences Research Program funded the investigation.

REFERENCES CITED

Anderson, A.T., 1991, Hourglass inclusions: Theory and application to the Bishop Rhyolitic Tuff: American Mineralogist, v. 76, p. 530-547.

Andres, R.J., Rose, W.I., Kyle, P.R., deSilva, S., Francis, P., Gardeweg, M., and Moreno Roa, H., 1991, Excessive sulfur dioxide emissions from Chilean volcanoes: Journal of Volcanology and Geothermal Research, v. 46, p. 323–329.

Bai, T. B., and Koster van Groos, A. F., 1994, Diffusion of chlorine in granitic melts: Geochimica et Cosmochimica Acta, v. 58, p. 113–123.

Baker, L., and Rutherford, M. J., 1992, Anhydrite breakdown as a possible source of excess sulfur in the 1991 Mount Pinatubo eruption [abs.]: Eos, Transactions, American Geophysical Union, v. 73, p. 625.

Bernard, A., Demaiffe, D., Mattelli, N., and Punongbayan, R. S., 1991, Anhydrite-bearing pumices from Mount Pinatubo: Further evidence for sulfur-rich silicic magmas: Nature, v. 354, p. 139–140.

Bernard, A., Knittel, U., Weber, B., Weis, D., Albrecht, A., Hattori, K., and Oles, D., this volume, Petrology and geochemistry of the 1991 eruption products of Mount Pinatubo (Luzon, Philippines).

Bence, A.E., and Albee, A.L., 1968, Empirical correction factors for the electron microanalysis of silicates and oxides: Journal of Geology, v. 76, p. 382–403.

Blank, J. G., Stolper, E. M., 1993, Solubilities of carbon dioxide and water in rhyolitic melt at 850°C and 750 bars: Earth and Planetary Science Letters, v. 119, p. 27–36.

Bluth, G. J. S., Doiron, S. D., Schnetzler, C. C., Krueger, A. J., and Walter, L. S., 1992, Global tracking of the SO_2 clouds from the June, 1991 Mount Pinatubo eruptions: Geophysical Research Letters, v. 19, p. 151–154.

Bluth, G. J. S., Schnetzler, C. C., Krueger, A. J., and Walter, L. S., 1993, The contribution of explosive volcanism to global atmospheric sulfur dioxide concentrations: Nature, v. 366, p. 327–329.

Burnham, W. C., Holloway, J., R., and Davis, N. F., 1969, Thermodynamic properties of water to 1,000°C and 10,000 bars: The Geological Society of America, Special Paper Number 132, p. 1–96.

Carroll, M.R., and Rutherford, M. J., 1987, The stability of igneous anhydrite: Experimental results and implications for sulfur behavior in the 1982 El Chichón trachyandesite and other evolved magmas: Journal of Petrology, v. 28, p. 781–801.

————1988, Sulfur speciation in hydrous experimental glasses of varying oxidation state: Results from measured wavelength shifts of sulfur X-rays: American Mineralogist, v. 73, p. 845–849.

Casadevall, T. J., 1992, Preeruption hydrothermal systems at Pinatubo, Philippines and El Chichón, Mexico: Evidence for degassing magmas beneath dormant volcanoes: Geological Survey of Japan Report, no. 279, p. 35–38.

Chadwick, W., Jr., Archuleta, R., and Swanson, D. A., 1988, The mechanics of ground deformation precursory to dome-building extrusions at Mount St. Helens 1981–1982: Journal of Geophysical Research, v. 93, no. B5, p. 4351–4366.

Chase, M. W., Davies, C. A., Downey, J. R., Frurip, D. J., McDonald, R. A., and Syverud, A. N., 1985, JANAF thermochemical tables: Journal of Physical and Chemical Reference Data, v. 14, Supplement no. 1, p. 1–1856.

Daag, A., Tubianosa, B., Newhall, C., Tungol, N., Javier, D., Dolan, M., Delos Reyes, P. J., Arboleda, R., Martinez, M., and Regalado, M. T. M., this volume, Monitoring sulfur dioxide emissions at Mount Pinatubo volcano.

Delfin, F. G., 1983, Geology of the Mount Pinatubo geothermal prospect: unpublished Philippine National Oil Company report, 35 p.

Delfin, F. G., Sussman, D., Ruaya, J. R., and Reyes, A. G., 1992, Hazard assessment of the Pinatubo volcanic-geothermal system: Clues prior to the June 15, 1991 eruption: Transactions Geothermal Resources Council, v. 16, p. 519–528.

Delfin, F. G., Villarosa, H. G., Layugan, D. B., Clemente, V. C., Candelaria, M. R., and Ruaya, J. R., this volume, Geothermal exploration of the pre-1991 Pinatubo hydrothermal system.

Devine, J. D., Sigurdsson, H., Davis, A. N., and Self, S., 1984, Estimates of sulfur and chlorine yield to the atmosphere from volcanic eruptions and potential climatic effects: Journal of Geophysical Research, v. 89, p. 6309–6325.

Dutton, E. G., and Christy, J. R., 1992, Solar radiative forcing at selected locations and evidence for global lower tropospheric cooling following the eruptions of El Chichón and Pinatubo: Geophysical Research Letters, v. 19, p. 2313–2316.

Fogel, R. A., and Rutherford, M. J., 1990, The solubility of carbon dioxide in rhyolitic melts: A quantitative FTIR study: American Mineralogist, v. 75, p. 1311–1326.

Fournelle, J., Carmody, R., and Daag, A. G., this volume, Mineralolgy and geochemistry of Mount Pinatubo anhydrite- and sulfide-bearing pumices from the SO_2-rich eruption of June 1991.

Gerlach, T. M., 1993a, Oxygen buffering of Kilauea volcanic gases and the oxygen fugacity of Kilauea basalt: Geochimica et Cosmochimica Acta, v. 57, p. 795–814.

————1993b, A magmatic vapor saturation and accumulation model for the 20-Megaton SO_2 cloud from the June 15, 1991, climactic eruption of Mount Pinatubo [abs.]: Eos, Transactions, American Geophysical Union, v. 74, no. 43, p. 104.

Gerlach, T. M., Westrich, H. R., Casadevall, T. J., and Finnegan, D. L., 1994, Vapor saturation and accumulation in magmas of the 1989–1990 eruption of Redoubt Volcano, Alaska: Journal of Volcanology and Geothermal Research, v. 62, p. 317–337.

Giggenbach, W. F., 1980, Geothermal gas equilibria: Geochimica et Cosmochimica Acta, v.44, p. 2021–2032.

————1987, Redox processes governing the chemistry of fumarolic gas discharges from White Island, New Zealand: Applied Geochemistry, v. 2, p. 143–161.

Gleason, J. F., Bhartia, P. K., Herman, J. R., McPeters, R., Newman, P., Stolarski, R. S., Flynn, L., Labow, G., Larko, D., Seftor, C., Wellemeyer, C., Komhyr, W. D., Miller, A. J., and Planet, W., 1993, Record low global ozone: Science, v. 260, p. 523–526.

Halpert, M. S., Ropelewski, C. F., Karl, T. R., Angell, J. K., Stowe, L. L., Heim, R. R., Jr., Miller, A. J., and Rodenhuis, D. R., 1993, 1992 brings return to moderate global temperatures: Eos, Transactions, American Geophysical Union, v. 74, no. 38, p. 433, 437–439.

Hanic, F., Galikova, L., Havlica, J., Kapralik, I., and Ambruz, V., 1985, Kinetics of the thermal decomposition of $CaSO_4$ in air: British Ceramics Transactions Journal, v. 84, p. 22–25.

Hattori, K., 1993, High-sulfur magma, a product of fluid discharge from underlying mafic magma: Evidence from Mount Pinatubo, Philippines: Geology, v. 21, p. 1083–1086.

Hattori, K., this volume, Occurrence and origin of sulfide and sulfate in the 1991 Pinatubo eruption products.

Hervig, R. L., and Williams, P., 1988, SIMS microanalysis of minerals and glasses for H and D, in Huber, A. M., and Werner, A. W., eds., Secondary ion mass spectrometry, SIMS IV: New York, John Wiley & Sons, p. 961–964.

Hoblitt, R. P., Wolfe, E. W., Scott, W. E., Couchman, M.R., Pallister, J., and Javier, D., this volume, The preparoxysmal eruptions, Mount Pinatubo, Philippines.

Hofman, D. J., Oltmans, S. J., Harris, J. M., Deshler, T., and Johnson, B. J., 1992, Observation and possible causes of new ozone depletion in Antarctica in 1991: Nature, v. 359, p. 283–287.

Holloway, J. R., 1976, Fluids in the evolution of granitic magma: Consequences of finite CO_2 solubility: Geological Society of America Bulletin, v. 87, p. 1513–1518.

Holloway, J. R., 1977, Fugacity and activity of molecular species in supercritical fluids, in Fraser, D., ed., Thermodynamics in geology: Boston, D. Reidel, p. 161–181.

Holloway, J. R., 1981, Volatile interactions in magmas, in Newton, R. C., Navrotsky, A., and Wood B. J., eds., Thermodynamic of Minerals and Melts: Springer-Verlag, p. 273–293.

Huebner, J. S., and Sato, M., 1970, The oxygen fugacity-temperature relationships of manganese oxide and nickel oxide buffers: American Mineralogist, v. 55, p. 934–952.

Imai, A., Listanco, E. L., and Fujii, T., 1993, Petrologic and sulfur isotopic significance of highly oxidized and sulfur-rich magma at Mt. Pinatubo, Philippines: Geology, v. 21, p. 699–702.

Jaupart, C., and Allegre, C. J., 1991, Gas content, eruption rate and instabilities of eruption regime in silicic volcanoes: Earth and Planetary Science Letters, v. 102, p. 413–429.

Jaupart, C., and Vergniolle, S., 1989, The generation and collapse of a foam layer at the roof of a basaltic magma chamber: Journal of Fluid Mechanics, v. 203, p. 347–380.

Johnston, D. A., 1980, Volcanic contribution of chlorine to the stratosphere: More significant to ozone than previously estimated?: Science, v. 209, p. 491–493.

Lowenstern, J. B., 1993, Evidence for a copper-bearing fluid in magma erupted at the Valley of Ten Thousand Smokes, Alaska: Contributions to Mineralogy and Petrology, v. 114, p. 409–421.

Lowenstern, J. B., Mahood, G. A., Rivers, M. L., and Sutton, S. R., 1991, Evidence for extreme partitioning of copper into magmatic vapor: Science, v. 252, p. 1405–1409.

Luhr, J. F., Carmichael, I. S., and Varekamp, J. C., 1984, The 1982 eruption of El Chichón volcano, Chiapas, Mexico: Mineralogy and petrology of the anhydrite-bearing pumices: Journal of Volcanology and Geothermal Research, v. 23, p. 69–108.

Marty, B., and Le Cloarec, M-F., 1992, Helium-3 and CO_2 fluxes from subaerial volcanoes estimated from polonium-210 emissions: Journal of Volcanology and Geothermal Research, v. 53, p. 67–72.

Matthews, S. J., Jones, A. P., and Bristow, C. S., 1992, A simple magma-mixing model for sulphur behaviour in calc-alkaline volcanic rocks: mineralogical evidence from Mount Pinatubo 1991 eruption: Journal of the Geological Society, London, v. 149, p. 863–866.

McCormick, P.M., Thomason, L.W., and Trepte, C.R., 1995, Atmospheric effects of the Mt. Pinatubo eruption: Nature, v. 373, p. 399–404.

McPeters, R. D., 1993, The atmospheric SO_2 budget for Pinatubo derived from NOAA-11 SBUV/2 spectral data: Geophysical Research Letters, v. 20, p. 1971–1974.

Metrich, N., and Rutherford, M. J., 1992, Experimental study of chlorine behavior in hydrous silicic melts: Geochimica et Cosmochimica Acta, v. 56, p. 607–616.

Mori, J., Eberhart-Phillips, D., and Harlow, D., this volume, Three-dimensional velocity structure at Mount Pinatubo, Philippines: Resolving magma bodies and earthquake hypocenters.

Newman, S., Epstein, S., and Stolper, E., 1988, Water, carbon dioxide, and hydrogen isotopes in glasses from the ca. 1340 A.D. eruption of the Mono Craters, California: Constraints on degassing phenomena and initial volatile content: Journal of Volcanology and Geothermal Research, v. 35, p. 75–96.

Palais, J. M., and Sigurdsson, H., 1989, Petrologic evidence of volatile emissions from major historic and pre-historic volcanic eruptions, in Kidson, J. W. ed., Understanding climate change: American Geophysical Union Monograph 52, p. 31–53.

Pallister, J. S., Hoblitt, R. P., and Reyes, A. G., 1992, A basalt trigger for the 1981 eruptions of Pinatubo volcano?: Nature, v. 356, p. 426–428.

Pallister, J. S., Hoblitt, J. S., Meeker, G. P., Newhall, C. G., Knight, R. J., and Siems, D. F., this volume, Magma mixing at Mount Pinatubo volcano: Petrographic and chemical evidence from the 1991 deposits.

Pan, V., Holloway, J. R., and Hervig, R. L., 1991, The temperature and pressure dependence of carbon dioxide solubility in tholeiitic basalt: Geochimica et Cosmochimica Acta, v. 55, p. 1587–1595.

Pasteris, J. D., Wopenka, B., Wang, A., and Harris, T. N., this volume, Relative timing of fluid and anhydrite saturation: Another consideration in the sulfur budget of the Mount Pinatubo eruption.

Prather, M., 1992, Catastrophic loss of stratospheric ozone in dense volcanic clouds: Journal of Geophysical Research, v. 97, p. 10,187–10,191.

Qin, Z., Lu, F., and Anderson, A. T., 1992, Diffusive reequilibration of melt and fluid inclusions: American Mineralogist, v. 77, p. 565–576.

Read, W. G., Froidevaux, L., and Waters, J. W., 1993, Microwave limb sounder measurement of stratospheric SO_2 from the Mt. Pinatubo volcano: Geophysical Research Letters, v. 20, p. 1299–1302.

Rose, W. I., 1977, Scavenging of volcanic aerosol by ash: Atmospheric and volcanic implications: Geology, v. 5, p. 621–624.

Ruaya, J. R., Ramos, M. N., and Gonfiantini, R., 1992, Assessment of magmatic components of the fluids at Mt. Pinatubo volcanic-geothermal system, Philippines from chemical and isotopic data: Geological Survey of Japan Report, no. 279, p. 141–151.

Rutherford, M. J., and Devine, J. D., 1991, Preeruption conditions and volatiles in the 1991 Pinatubo magma [abs.]: Eos, Transactions, American Geophysical Union, v. 72, p. 62.

Rutherford, M. J., and Devine, J. D., this volume, Preeruption P-T conditions and volatiles in the 1991 Mount Pinatubo magma.

Scott, W. E., Hoblitt, R. P., Torres, R., Martinez, M., Nillos, T., and Self, S., this volume, Pyroclastic flows of the June 15, 1991, paroxysmal eruption of Mount Pinatubo volcano, Philippines.

Shinohara, H., Iiyama, J. T., and Matsuo, S., 1989, Partition of chlorine compounds between silicate melt and hydrothermal solutions: I. Partition of NaCl-KCl: Geochimica et Cosmochimica Acta, v. 53, p. 2617–2630.

Sigurdsson, H., 1990, Assessment of the atmospheric impact of volcanic eruptions, in Sharpton, V. L., and Ward, P. D., eds., Global catastrophies in Earth history: Geological Society of America Special Paper 247, p. 99–110.

Sigurdsson, H., Carey, S., Palais J. M., and Devine, J. D., 1990, Preeruption composition gradients and mixing of andesite and dacite magma erupted from Nevado del Ruiz volcano, Colombia in 1985: Journal of Volcanology and Geothermal Research, v. 41, p. 127–151.

Sigurdsson, H., Devine, J. D., and Davis, A. N., 1985, The petrologic estimation of volcanic degassing: Jokull, v. 35, p. 1–8.

Silver, L. A., Ihinger, P. D., and Stolper, E., 1990, The influence of bulk composition on the speciation of water in silicate glasses: Contributions to Mineralogy and Petrology, v. 104, p. 142–162.

Skirius, C. M., Peterson, J. W, and Anderson, A. T., 1990, Homogenizing rhyolitic glass inclusions from the Bishop Tuff: American Mineralogist, v. 75, p. 1381–1398.

Solomon, S., Sanders, R. W., Garcia, R. R., and Keys, J. G., 1993, Increased chlorine dioxide over Antarctica caused by volcanic aerosols from Mount Pinatubo: Nature, v. 363, p. 245–248.

Sparks, R. S. J., Barclay, J., Jaupart, C., Mader, L., and Phillips, J.C., 1994, Physical aspects of magmatic degassing I. Experimental and theoretical constraints on vesiculation, in Carroll, M. R., and Holloway, J. R., eds., Volatiles in magmas: Mineralogical Society of America Reviews in Mineralogy, v. 30, p. 413–445.

Stoiber, R. E., and Jepsen, A., 1973, Sulfur dioxide contributions to the atmosphere by volcanoes: Science, v. 182, p. 577–578.

Swanson, D. A., and Holcomb, R. T., 1989, Regularities in growth of the Mount St. Helens dacite dome, 1980–1986: IAVCEI Proceedings in Volcanology, v. 2, p. 1–24.

Symonds, R. B., and Reed, M. H., 1993, Calculation of multicomponent chemical equilibria in gas-solid-liquid systems: Calculation methods, thermochemical data, and applications to studies of high-temperature volcanic gases with examples from Mount St. Helens: American Journal of Science, v. 293, p. 758–864.

Symonds, R. B., Rose, W. I., Bluth, G., J. S., Gerlach, T. M., 1994, Volcanic-Gas studies: methods, results, and applications, in Carroll, M. R., and Holloway, J. R., eds., Volatiles in Magmas: Mineralogical Society of America Reviews in Mineralogy, v. 30, p. 1–66.

Tait, S., 1992, Selective preservation of melt inclusions in igneous phenocrysts: American Mineralogist, v. 77, p. 146–155.

Varekamp, J. C., Luhr, J. F., and Prestegaard, K. L., 1984, The 1982 eruptions of El Chichón volcano (Chiapas, Mexico): Character of the eruptions, ash-fall deposits, and gas phase: Journal of Volcanology and Geothermal Research, v. 23, p. 39–68.

Wallace, P., and Anderson, A. T., 1994, Preeruptive gradients in H_2O, CO_2, and exsolved gas in the magma body of the Bishop Tuff [abs.]: Eos, Transactions, American Geophysical Union, v. 75, no. 44, p. 719.

Wallace, P., and Gerlach, T., M., 1994, Magmatic vapor source for sulfur dioxide released during volcanic eruptions: Evidence from Mount Pinatubo: Science, v. 265, p. 497–499.

Webster, J. D., 1992a, Fluid-melt interactions involving Cl-rich granites: Experimental study from 2 to 8 kbar: Geochimica et Cosmochimica Acta, v. 56, p. 659–678.

Webster, J. D., 1992b, Water solubility and chlorine partitioning in Cl-rich granitic systems: Effects of melt composition at 2 kbar and 800°C: Geochimica et Cosmochimica Acta, v. 56, p. 679–687.

Webster, J. D., and Holloway, J. R., 1990, Partitioning of F and Cl between hydrothermal fluids and highly evolved granitic magmas, in Stein, H. J., and Hannah, J. L., eds., Ore-bearing granite systems: Petrogenesis and mineralizing processes: Geological Society of America Special Paper 246, p. 21–34.

Westrich, H. R., and Gerlach, T. M., 1992, Magmatic gas source for the stratospheric SO_2 cloud from the June 15, 1991 eruption of Mount Pinatubo: Geology, v. 20, p. 867–870.

Williams, S. N., Sturchio, N. C., Calvache V., M. L., Mendez F., R., Londono C., A., and Garcia P., N., 1990, Sulfur dioxide from Nevado del Ruiz volcano, Colombia: Total flux and isotopic constraints on its origin: Journal of Volcanology and Geothermal Research, v. 42, p. 53–68.

Williams, S. N., Schaefer, S. J., Calvache, M. L. and Lopez, D., 1992, Global carbon dioxide emission to the atmosphere by volcanoes: Geochimica et Cosmochimica Acta, v. 56, p. 1765–1770.

Wilson, L., Sparks, R. S. J., and Walker, G. P. L., 1980, Explosive volcanic eruptions—IV. The control of magma properties and conduit geometry on eruption column behaviour: Geophysical Journal Royal Astronomical Society, v. 63, p. 117–148.

Wolfe, E. W., and Hoblitt, R. P., and others, this volume, Overview of the eruptions.

Evolution of a Small Caldera Lake at Mount Pinatubo

By Nora R. Campita,[1] Arturo S. Daag,[1] Christopher G. Newhall,[2]
Gary L. Rowe,[2] and Renato U. Solidum[1]

ABSTRACT

Collapse of Mount Pinatubo's edifice during its explosive eruptions on June 15, 1991, created a 2.5-kilometer-wide caldera. By early September 1991, a lake began to form on the caldera floor, mainly by spring discharge from the walls of the caldera augmented by rainfall and surface runoff. The lake became increasingly acidic with time, with pH changing from 6.0 in October 1991 to 1.9 in December 1992. Lake temperature over the same period remained about 38±2°C, except near fumaroles or hot springs. The high initial pH value reflects water of meteoric origin dominating that from a deeper hydrothermal system. pH decreased as the lake absorbed acid magmatic gases, becoming an acid sulfate-chloride brine. Much of the acidification during the period February 1992 to December 1992 may have occurred between July and October, when a tuff cone and then a dome grew within the lake. A slight overall enrichment of Mg and Si suggests posteruption leaching of rock by hot waters in and beneath the lake.

INTRODUCTION

Mount Pinatubo's violent eruption on June 15, 1991, formed a nearly circular caldera about 2.5 km in diameter, and, beginning in September 1991, a lake began to form from spring discharge, supplemented by rainfall and surface runoff. Over time, such lakes can act as large condensers that absorb the heat and volatile emanations from vents on the caldera floor. In addition, such lakes can also serve as recycling centers for condensed volcanic fluids. Mass balance calculations for Poás Volcano, Costa Rica (Brantley and others, 1987, 1992; Rowe and others, 1992), and for El Chichón Volcano, Mexico (Casadevall and others, 1984), indicate that seepage of caldera (or crater) lake brine and subsequent recycling affects the lake chemistry.

For volcano monitoring purposes, increases in lake water SO_4/Cl ratios prior to or during eruptive activity have

been noted in the crater lakes of Zavaritsky Volcano, Russia (Menyailov, 1975), and Kusatsu-Shirane Volcano, Japan (Takano and Watanuki, 1990), as acid, volcanic gases have interacted with meteoric waters. In addition, the Mg/Cl ratio has been used along with pH variations at the crater lake of Mount Ruapehu, New Zealand, to monitor changes in gas flux from fumaroles and to document high-temperature water/rock interaction episodes (Giggenbach, 1974, 1983; Giggenbach and Glover, 1975). The eruption in the crater lake of Soufrière Volcano in Saint Vincent in 1971–72 revealed that the relative ease of leaching components of the lava was Na>Fe>Mg>K>Ca>>Si (Sigurdsson, 1977).

DEVELOPMENT OF THE CALDERA LAKE

Figures 1A–D show development of Pinatubo's caldera lake. In August 1991, ash emission waned enough to permit views into the new caldera, and numerous springs could be seen discharging from the caldera walls. Spring discharge that collected in the vent was being heated rapidly and flashed to steam in frequent phreatic explosions. As a result, no lake was forming.

However, by early September, as the rainy season progressed and as rocks of the vent area cooled, surface runoff and discharge from these springs began to collect and a shallow lake covered much of the caldera floor. Runoff from a catchment area of about 5 km^2 fed into a lake about 1/15 that size, so a meter of rain (average for the last half of August and early September at nearby Cubi Point and Dizon Mines; Rodolfo and others, this volume) could account for more than 15 m of water accumulation on the bowl-shaped caldera floor. Although our data are crude, we judge that the lake required both surface runoff and spring discharge to counter evaporation and to develop during this early period.

The lake rose slightly over the dry season from October 1991 through July 1992 (for general rainfall information, see fig. 3 of Umbal and Rodolfo, this volume)—adding evidence for the importance of spring discharge. Throughout this period, the surface area of the lake remained approximately constant, about 4×10^5 m^2.

[1] Philippine Institute of Volcanology and Seismology.
[2] U.S. Geological Survey.

Figure 1. Progressive development of the Pinatubo caldera lake. *A*, Pinatubo caldera on August 1, 1991, before the caldera lake began to form (photographed from the northwest by T. Casadevall, USGS). *B*, Pinatubo caldera lake as seen on September 10, 1991 (photographed from the northwest by T. Pierson, USGS). *C*, New lava dome that grew through the caldera lake, as seen on July 29, 1992 (photographed from the northeast by R. Arante, PHIVOLCS). *D*, Pinatubo caldera lake on November 22, 1992. Westward growth of deltas partially surrounded the dome and sharply reduced the surface area of the lake (photographed from the northeast by R. Arante, PHIVOLCS).

C

D

Figure 1.—Continued.

In July–August 1992, renewed eruptions formed a small tuff cone and then a dome in the center of the lake (fig. 1C; see also Daag, Dolan, and others, this volume). By late August, heavy rains had washed so much debris off the caldera walls that deltas extending from the eastern wall reached the still-growing dome, and the area of the lake was reduced to about 2.5×10^5 m^2. A rain gauge installed on the north rim of the caldera recorded 280 mm of rainfall in September 1992, and by late September, deltaic sediment surrounded two-thirds of the dome. The dome stopped growing at the end of October.

Logistical difficulties and hazard precluded detailed bathymetry. However, a view in early August 1991 (fig. 1A) shows an inner crater whose floor was at least 100 m below what was soon to become lake level. Erosion of sediment from the caldera walls probably filled most if not all of the deep crater by October 1991 (end of the 1991 rainy season), so the average depth during the period October 1991 to July 1992 was perhaps in the order of 10 to 20 m. Shorelines in October 1991 and July 1992 photographs suggest that the lake rose several meters during the dry season, a period of relatively modest erosion, so the average lake depth probably increased slightly over that period.

When a new dome began to grow through the lake in July and August 1992, occasional updoming of talus on lake sediment surrounding the dome suggested that the lake was only a few meters deep. In November 1992, various lengths of rope were dropped from a helicopter into what appeared to be the deepest part of the lake, immediately west of the dome. These lengths of rope were weighted by a small sandbag at one end and marked by color-coded floats at the other. Floats at the end of 5-, 10-, 15-, and 20-m ropes disappeared from sight, indicating greater depths; unfortunately, the soundings had to be curtailed when rotor wash splashed acidic water on the helicopter. From these admittedly sparse observations, we judge that the average depth of the caldera lake in 1991–92 was in the order of 10 m, shoaling faster by deposition and evaporation than it deepened by influx of water. Its volume increased to a maximum of about 10^7 m^3 in October 1991 and had decreased to half of that volume or less by late 1992. (Editorial note: By 1994, the lake level was several tens of meters higher than in late 1992.)

SAMPLING AND ANALYTICAL METHODS

Lake water samples from October 1991 to December 1992 were all collected and stored in polyethylene bottles. Temperature and pH were measured in situ (except in November 1991) by use of a portable digital pH-temperature meter. Samples were not filtered in the field but were passed through a 0.1-µm millipore filter before analysis. The methods for chemical analyses of lake water

Table 1. Summary of analytical methods used on caldera lake water of Mount Pinatubo.

Specie	Method
Boron	By mannitol method (PHIVOLCS, PNOC).
Chloride	By Mohr method (PHIVOLCS, PNOC).
Fluorine	By potentiometric method (F-electrode) (PHIVOLCS, PNOC).
Cations	All cations are analyzed by atomic absorption
Li	spectrophotometry (PHIVOLCS, PNOC).
Na	
K	
Rb	
Cs	
Ca	
Mg	
Fe	
SiO$_2$	
SO$_4$	By gravimetric and colorimetric method (PHIVOLCS, PNOC).

samples are summarized in table 1. The results of those analyses are shown in table 2A.

Sample PCL–5 was taken from an area of bubbling at the lake shore and is thus a hybrid of hot spring discharge and the lake as a whole. Regrettably, no samples could be obtained of early springs that gushed from the caldera walls; we can only guess that they were some mixture of meteoric water and geothermal fluids described by Delfin and others (this volume).

RESULTS

Lake temperature and pH are shown in table 2A and in figure 2A. Temperature has remained more or less constant; pH has decreased sharply, from 6.0 in October 1991 to 1.9 in December 1992. Li, Na, K, Cl, SO$_4$, Ca, Mg, and total dissolved solids (TDS) dropped from October 1991 to February 1992 but had increased again by December 1992 (table 2A and fig. 3A). F, Fe, and SiO$_2$ generally increased from October 1991 straight through to December 1992. The Mg/Cl ratio (table 2B and fig. 3B) increased between October 1991 and February 1992 and remained more or less the same in December 1992.

APPARENT ABSORPTION OF SO$_2$ EMISSION BY THE CALDERA LAKE

Correlation spectrometer (COSPEC, Barringer Instruments) measurements of SO$_2$ emission before the climactic eruption showed a dramatic increase, brief stoppage, and then an even greater increase, reaching more than 10,000 t/d on June 10 (Daag, Tubianosa, and others, this volume). SO$_2$ emission during July and August (only 3 measurements

Table 2A. Result of chemical analyses of caldera lake water of Mount Pinatubo.

[nd, not determined; LANL, Los Alamos National Laboratory; PNOC/PV, Philippine National Oil Company/PHIVOLCS]

Code	PCL–1	PCL–2	PCL–3	PCL–4	PCL–5	PCL–6	PCL–6
Date collected..............	10/08/91	11/19/91	02/18/92	02/18/92	02/18/92	12/04/92	12/04/92
Temp (°C)....................	40.0	nd	38.5	38.5	46.5	36.7	36.7
Analyzed by	LANL	PNOC/PV	PNOC/PV	PNOC/PV	PNOC/PV	PNOC/PV	USGS
pH-lab..........................	4.79	5.21					
pH-field	6.00	nd	2.74	2.78	2.73	1.9	1.9
Total acidity................	80.00	110.00	230.00	235.00	230.00	320.00	
Mineral-acid acidity	0.00	30.00	100.00	80.00	90.00	80.00	
Concentration (ppm) ...							
As..........................	0.28						
B...........................	31.10	32.00	24.70	22.40	27.80	28.00	31.00
Br	1.65						
Ca..........................	598.00	597.00	400.00	382.00	444.00	419.00	370.00
Cl...........................	1,029.00	742.00	500.00	467.00	567.00	849.00	825.00
CO_3	0.00						
F	0.13	0.36	0.75	0.73	0.61	1.68	
Fe	<0.1	2.39	16.10	18.70	8.90	15.00	35 (as Fe3+)
HCO_3	0.00						
K...........................	80.00	68.00	45.00	43.00	50.00	58.00	54.00
Li...........................	0.72	0.70	0.50	0.40	0.50	0.70	
Mg.........................	95.50	84.10	72.10	73.50	71.20	121.00	110.00
Mn.........................	13.80						
Na..........................	519.00	402.00	285.00	274.00	319.00	395.00	370.00
NH_4	0.02						
NO_3	<0.05						
PO_4	<0.1						
Si	32.60						
SiO_2......................	70.00	95.00	140.00	147.00	119.00	164.00	167.00
SO_4......................	1,727.00	1,689.00	1,288.00	1,253.00	1,389.00	1,364.00	1,600.00
Sr..........................	3.06						1.4
Total dissolved solids ..	4,109.00	3,510.00	3,237.00	3,130.00	3,462.00	4,185.00	
Charge balance (WATEQ4F)–3.1	−0.1	2.0	3.9	1.2	18.1	6.5	

Table 2B. Molar Mg/Cl ratios for caldera lake water of Mount Pinatubo.

Code	PCL–1	PCL–2	PCL–3	PCL–4	PCL–5	PCL–6
Date	10/08/91	11/19/91	02/18/92	02/18/92	02/18/92	12/04/92
Mg/Cl	0.14	0.17	0.21	0.23	0.18	0.21

before the caldera lake formed) were between 1,000 and 5,000 t/d. However, as soon as the lake formed, SO_2 emission (regrettably, only one measurement) dropped to near the detection threshold for the measurements, about 20 t/d.

The next measurements of SO_2 emission were in July-November 1992, during a period of increased seismicity and then growth of the dome through the caldera lake.

Emissions were <300 t/d before the dome rose above lake level and generally were >300 t/d until the dome stopped growing in October.

These observations suggest that the lake absorbed most of the volcano's SO_2 emission while it covered the vent. SO_2 might also have been absorbed into a shallow hydrothermal system beneath the lake in 1992, but we cannot judge this from COSPEC measurements.

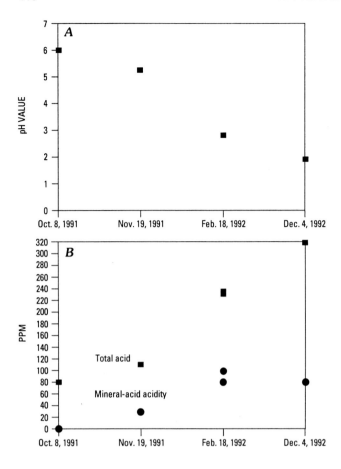

Figure 2. *A*, pH of Pinatubo caldera lake, October 1991–December 1992. *B*, Total acid (T.A.) and mineral-acid acidity (M.A.A.) of the lake during the same period.

DISCUSSION

ACIDITY AND TEMPERATURE

The high initial pH of the lake (6.0) reflects a dominantly meteoric origin of the lake water, with much rainfall in the area from September to October 1991. The subsequent increase in acidity, accompanied by increasing total acidity and mineral-acid acidity (fig. 2B), occurred as the lake and its shallow hydrothermal system absorbed acid magmatic gases.

MAJOR ELEMENTS

The initial enrichment of the caldera lake in Cl, SO_4, and F can be attributed to addition of HCl, HF, and SO_2 from fumaroles during early degassing of the 1991 magma. Solution of anhydrite from rocks within the caldera (mostly pre-1991 rock) could have added SO_4, and the high concentration of B could have been from leaching of volcanic wall rocks (Ellis and Sewell, 1963). Mg, Ca, and K were

probably leached from rock in and around the conduit and carried in residual hydrothermal fluids of Pinatubo.

By February 1992, the volcano had quieted and B, Cl, K, Li, Mg, Na, SO_4, and TDS had decreased correspondingly. The Mg/Cl ratio increased (fig. 3B) as Cl decreased faster than Mg.

The sample of December 1992 reflected dome growth from July to October 1992, during which magmatic gases that were absorbed into the lake and the shallow hydrothermal system raised levels of Cl and F and further decreased pH. Increases in B, K, Mg, Ca, Na, SiO_2, Li, and TDS concentrations (fig. 3A) and the Mg/Cl ratio (fig. 3B) can be explained mainly as interaction of the highly acidic lake water with volcanic materials (Giggenbach, 1974; 1983; Giggenbach and Glover, 1975). At least some of these species could have been derived from interaction of lake water with the high-temperature lava dome that eventually emerged above the caldera lake floor.

Mineralogic controls on lake-water composition were evaluated by use of the equilibrium thermodynamic model WATEQ4F (Ball and Nordstrom, 1991). October–November 1991 waters were approximately saturated with respect to gypsum, but increasing acidity and a concomitant increase in bisulfate concentration caused 1992 lake waters to be undersaturated with respect to gypsum. Lake water from December 1992 (sample PCL–6; USGS analysis, table 2A) was undersaturated with respect to gypsum and severely undersaturated with respect to primary and secondary aluminosilicate phases such as feldspars and clays and also with respect to the aluminum-sulfate-hydroxide mineral natroalunite. Natroalunite is found in sediments of the less acidic crater lakes (pH > 2–3) such as Kusatsu-Shirane, Japan (Takano and Watanuki, 1990), whereas gypsum is characteristic of sediments of the more concentrated crater lakes (pH 0–1) such as Ruapehu (Giggenbach, 1974) and Poás (Brantley and others, 1987). December 1992 lake waters were in equilibrium with amorphous silica.

A lack of speciation data for reduced sulfur species precludes evaluation of the saturation state of native sulfur. However, disproportionation of SO_2 degassed into the lake will result in the production of hydrogen sulfide and sulfuric acid. Subsequent oxidation of the H_2S will cause native sulfur to precipitate in lake sediments and at the lake surface. The calculations suggest that with further degassing, pyroclastic material eroded into the lake will continue to dissolve, thereby increasing the concentrations of all rock-forming elements except silica. Lake sediments will consist mostly of pyroclastic material with small but significant quantities of amorphous silica and elemental sulfur. Future trends in lake water chemistry will be determined by variations in magmatic degassing, the degree of interaction between hot rock and lake water, and the amount of rainfall and spring water-derived recharge received by the lake.

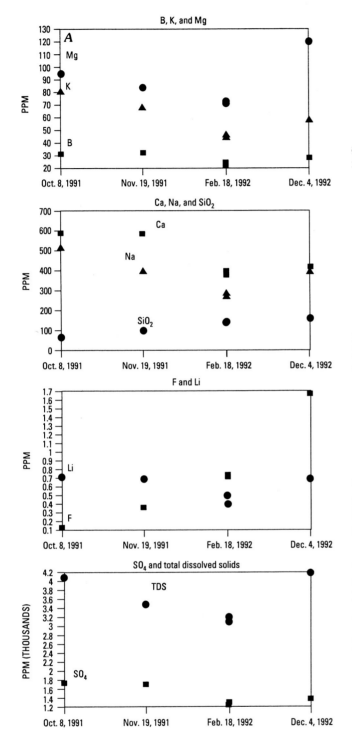

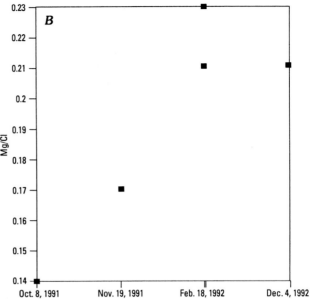

Figure 3. *A*, Concentrations of species in the Pinatubo caldera lake, October 1991–December 1992. *B*, Mg/Cl ratio of Pinatubo caldera lake water for the same period.

CONCLUSIONS

The caldera lake of Mount Pinatubo reflects a combination of magmatic degassing into a hydrothermal system and into the lake itself, water-rock interaction in the hydrothermal system, rainfall and surface runoff, and evaporation.

The composition of the earliest sample suggests a predominance of meteoric water—from surface runoff and springs on the caldera walls, mixed with water from Pinatubo's preexisting hydrothermal system. Magmatic gases acidified the early lake, but, as time passed after the 1991 eruptions, the influence of magmatic gases declined. From the middle through late 1992, magmatic gases were again an important control on lake composition.

Evaporation and dilution by fresh rainwater are probably second-order controls on lake composition that would become increasingly important as volcanic output wanes, but our present data are insufficient to quantify their roles.

ACKNOWLEDGMENTS

We thank the U.S. Navy, the U.S. Marine Corps, and Captain Agustin Consunji of Delta Aviation for help in obtaining the caldera lake water samples. We also thank Fraser Goff (Los Alamos National Laboratory), Aristeo Baltazar and personnel of the Philippine National Oil Corporation (PNOC) Geochemical Laboratory, and Ray van

Hoven (USGS) for help with water analyses. Tom Casadevall (USGS) and David Sussman (UNOCAL) provided valuable reviews.

REFERENCED CITED

Ball, J.W., and Nordstrom, D.K., 1991, User's manual for WATEQ4F, with revised thermodynamic data base and test cases for calculating speciation of major, trace, and redox elements in natural waters: U.S. Geological Survey Open-File Report 91–183, 193 p., one diskette.

Brantley, S.L., Borgia, A., Rowe, G., Fernández, J.F., and Reynolds, J.R., 1987, Poás Volcano acts as a condenser for acid metal-rich brine: Nature, v. 330, p. 470–472.

Brantley, S.L., Rowe, G.L., Konikow, F., and Sanford, W.E., 1992, Toxic waters of Poas Volcano: National Geographic Research and Exploration, v. 8, no. 3, p. 328–337.

Casadevall, T.J., dela Cruz-Reyna, S., Rose, W.I., Jr., Bagley, S., Finnegan, D.L., and Zoller, W.H., 1984, Crater lake and post-eruption hydrothermal activity, El Chichón, Mexico: Journal of Volcanology and Geothermal Research, v. 23, p. 169–191.

Daag, A.S., Dolan, M.T., Laguerta, E.P., Meeker, G.P., Newhall, C.G., Pallister, J.S., and Solidum, R., this volume, Growth of a postclimactic lava dome at Mount Pinatubo, July–October 1992.

Daag, A.S., Tubianosa, B.S., Newhall, C.G., Tuñgol, N.M, Javier, D., Dolan, M.T., Delos Reyes, P.J., Arboleda, R.A., Martinez, M.L., and Regalado, M.T.M., this volume, Monitoring sulfur dioxide emission at Mount Pinatubo.

Ellis, A.J., and Sewell, J.R., 1963, Boron in waters and rocks of New Zealand hydrothermal areas: New Zealand Journal of Science, v. 16, p. 589–606.

Giggenbach, W., 1974, The chemistry of Crater Lake, Mt. Ruapehu (New Zealand) during and after the 1971 active period: New Zealand Journal of Science, v. 17, p. 33–45.

———1983, Chemical surveillance of active volcanoes in New Zealand, in Tazieff, H., and Sabroux, J.C., eds., Forecasting volcanic events: Amsterdam, Elsevier, p. 311–322.

Giggenbach, W., and Glover, R.B., 1975, The use of chemical indicators in the surveillance of volcanic activity affecting the crater lake on Mt. Ruapehu, New Zealand: Bulletin Volcanologique, v. 39, p. 70–81.

Menyailov, I.A., 1975, Prediction of eruptions using changes in composition of volcanic gases: Bulletin of Volcanology, v. 39, p. 112–125.

Rodolfo, K.S., Umbal, J.V., Alonso, R.A., Remotigue, C.T., Paladio-Melosantos, M.L., Salvador, J.H.G., Evangelista, D., and Miller, Y., this volume, Two years of lahars on the western flank of Mount Pinatubo: Initiation, flow processes, deposits, and attendant geomorphic and hydraulic changes.

Rowe, G.L., Jr., Brantley, S.L., Fernandez, M., Fernandez, J.F., Borgia, A., and Barquero, J., 1992, Fluid-volcano interaction in an active stratovolcano: The crater lake system of Poás Volcano, Costa Rica: Journal of Volcanology and Geothermal Research, v. 49, p. 23–51.

Sigurdsson, H., 1977, Chemistry of the crater lake during the 1971-1972 Soufrière eruption: Journal of Volcanology and Geothermal Research, v. 2, no. 2, p. 165–186.

Takano, B., and Watanuki, K., 1990, Monitoring of volcanic eruptions at Yugama crater lake by aqueous sulfur oxyanions: Journal of Volcanology and Geothermal Research, v. 40, p. 71–87.

Umbal, J.V., and Rodolfo, K.S., this volume, The 1991 lahars of southwestern Mount Pinatubo and evolution of the lahar-dammed Mapanuepe Lake.

OBSERVATIONS AND RECONSTRUCTIONS:
THE 1991–92 ERUPTIONS

Several fortuitous circumstances led to a wide variety of observations of the 1991 eruptions. Pinatubo gave observers ample advance warning—eruption precursors and, if those were not enough, the publicized drama of evacuations (Punongbayan, Newhall, Bautista, and others) and a series of preclimactic eruptions (Hoblitt, Wolfe, and others). Observers were ready on the ground both east and west of Mount Pinatubo, and, until the climactic eruption, in helicopters as well (west-side story by Sabit and others; east-side story by Hoblitt, Wolfe, and others; summary by Wolfe and Hoblitt). Others made ground-based and satellite-based weather observations. Oswalt and others describe observations by C-band weather radar at Clark Air Base and Cubi Point Naval Air Station, an infrared telescope at Clark, and rawinsondes and pilot balloons for upper level air temperature, humidity, pressure, and wind speed and direction, until nearly the climax of the eruption. Microbarographs at Cubi and Clark and a temperature recorder and rain gage at Cubi recorded throughout the eruption. Geostationary (GMS) and polar-orbiting (AVHRR) satellites captured most of the eruption (Koyaguchi; Lynch and Stephens; Oswalt and others). Tahira and others describe infrasonic and acoustic gravity waves as recorded in Japan, and Zürn and Widmer report very long-period oscillations recorded by a worldwide network of broadband seismographs.

The general chronology of events is given by Wolfe and Hoblitt in the introductory section. The present section includes an interesting observation by Hoblitt, Wolfe, and others that preclimactic eruptions changed from vertically directed explosions (June 12, 13, early 14) to smaller, more frequent, less energetic eruptions with greater components of pyroclastic surge. The authors present a model in which the rate of pressurization of the tip of the magma column is controlled by competing rates of influx of magma, degassing during repose periods, and eruptions themselves.

Plinian fall deposits from the climactic June 15 eruption are relatively thin and inconspicuous (Paladio-Melosantos and others), partly because Typhoon Yunya, which passed near Pinatubo during the eruption, spread ash in all directions over a broad area. Near-source tephra deposits were also scoured by pyroclastic flows, and tephra along the main lobe was blown into the South China Sea. Regrettably, we had almost no information about offshore ash fall while most papers of this collection were being written. Fortunately, sea-bottom samples that were brought to the surface in 1994 (Wiesner and Wang) provide an important, late-breaking constraint on the volume of tephra fall and of the eruption as a whole.

Pumiceous pyroclastic-flow deposits from the climactic eruption—thick valley fill and thin ridge veneers—formed throughout the period of plinian fall, quite unlike models in which pyroclastic flows follow plinian tephra fall. W.E. Scott and others infer either quasicontinuous low-level collapse of portions of a sustained plinian column or repeated collapse of the entire column. Relatively late-stage, lithic-rich pyroclastic flows probably reflect entrainment of old edifice material as the caldera formed.

Significant, as-yet-unresolved discrepancies exist between estimates of column height from weather radar and those from satellites (the latter, 10 km higher during the 0555 eruption of June 15) (see Hoblitt, Wolfe, and others; Lynch and Stephens). Originally, the discrepancy was explained as a result of temperature disequilibrium and adiabatic cooling within the rising eruption column, but the discrepancy remains even if the temperatures are derived from clouds that were some distance away from the eruption column and drifting with similar vectors to that of the ash cloud (Lynch and Stephens). Eruption columns of June 14 and early June 15, as seen from satellites, were different from those seen from the ground. One explanation is that visible and multi-spectral weather images can show ash in lower concentrations than can weather radar.

Volcanologists routinely correlate deposits with observed eruptions, but most reconstructions of the largest historic and prehistoric eruptions are based on extrapolation of inferences from smaller eruptions. Pinatubo allowed direct correlation of an unusually large and quantified eruption column with features of deposits. Koyaguchi used the rate of expansion of the umbrella cloud, plus grain-size data, to estimate volumetric flow rate of cloud, mass discharge rate, and thence dense-rock-equivalent (DRE) volume of plinian fall deposits. His estimates can be compared to those of Paladio-Melosantos and others as a test of the theoretical approach, or, if one prefers the theoretical approach to sparse field data, his estimates can place upper and lower bounds on the actual volume of tephra. Koyaguchi's estimate for the volume of plinian fall deposit is 2 to 10 km³ DRE, whereas that of Paladio-Melosantos and others is 1.6 to 2.0 km³ DRE. Some of the difference between these two estimates might be explained by W.E. Scott and others' observation that pyroclastic flows (not included in these estimates) were occurring at the same time and could easily have complicated the dynamics of the eruption column.

Also included are two important contributions from far-field, global data. Tahira and others report on-scale, high-resolution infrasonic and acoustic gravity waves in Japan that provide more detail about the intensity of the climactic eruption than was obtained from monitoring near Mount Pinatubo. The infrasonic records show that the strongest phase of the eruption began at 1342, lasted only about 3.5 h, and weakened further after about 10 h. Microbarograph records of acoustic gravity waves recorded in Japan show oscillations at 4.4 mHz, one of two frequencies of Rayleigh wave oscillation noted by the other contributors of global data, Zürn and Widmer. Apparently, the same atmospheric oscillation that was recorded directly by microbarographs as acoustic gravity waves became coupled to the solid earth and was also recorded as long-period seismic waves.

When the dust finally settled, literally and figuratively, a new caldera was 2.5 km in diameter and more than 650 m deep. Low-resolution digital topography for the preeruption and posteruption periods (Jones and Newhall) shows the new caldera and valley-filling pyroclastic-flow deposits relative to the old summit. Major northeast- and southeast-trending conjugate faults that intersect at Pinatubo, and the origins of the Sacobia, Abacan, and Pasig-Potrero Rivers in a single pyroclastic fan, are clearer in this low-resolution digital topography than in previously published 1:50,000-scale topographic maps.

An important observation in the aftermath of the eruption was that still-fluidized or re-fluidized pyroclastic-flow deposits can move farther downslope as secondary pyroclastic flows (Torres and others). Suspected from previous geologic studies but never before observed, these remobilized pyroclastic flows produce deposits that are nearly indistinguishable from those of the primary pyroclastic flows, except that the former are emplaced at slightly lower temperature and are slightly fines-depleted relative to the parent material.

Beginning in July 1992, another intrusion of basaltic magma into the large dacitic reservoir beneath Pinatubo threatened to replicate events of 1991. Magma mixing and ascent of hybrid andesite occurred as in 1991, but this time they did not trigger a large explosive eruption (Daag, Dolan, and others). The residual dacitic magma was probably depleted in volatiles, but an alternate explanation of the different outcomes was that the 1992 intrusion was too slow or too small to serve as a trigger. As of this writing, Mount Pinatubo remains intermittently restless.

Joey Marcial (PHIVOLCS) and Tom Murray (USGS) attempting restoration of seismic signal from Patal Pinto, June 14, 1991. Helicopter support, USAF: photograph by Val Gempis, USAF.

The West-Side Story: Observations of the 1991 Mount Pinatubo Eruptions from the West

By Julio P. Sabit,[1] Ronald C. Pigtain,[1] and Edwin G. de la Cruz[1]

ABSTRACT

On April 2, 1991, during or just after phreatic explosions, vigorous steam vents opened on the upper north slope of the volcano. From April 2 to late May, steam jetted to heights of 300 to 800 meters and sometimes intensified to ash emissions that reached heights of 1,500 to 3,000 meters. Beginning in late May, steaming intensified and included increasing amounts of ash. By June 9, continuous ash emission and occasional small explosions produced ash clouds dense enough to flow slowly down the volcano's western slopes. A series of strong explosive eruptions began at 0851 on June 12, 1991, and, starting at 2218 on June 14, the eruptions slowly became more frequent and culminated in the violent climactic eruption of June 15.

Seismic activity was characterized by swarms of high-frequency volcanic earthquakes in April and May. Daily seismic counts averaged 74 high-frequency volcanic earthquakes in April and 90 high-frequency volcanic earthquakes in May. Early June brought a sudden increase in seismicity, including larger high-frequency volcanic earthquakes with distinct P-phases, low-frequency volcanic earthquakes, and harmonic tremor. On June 8, at a new, more distant seismic station in Poonbato, about 80 percent of the seismicity appeared as low-frequency volcanic earthquakes and low-frequency harmonic tremor. During the explosive eruptions (June 12–15), the Poonbato seismograms were marked by low-frequency volcanic earthquakes and low-frequency harmonic tremor.

INTRODUCTION

Residents of Sitios Tarao and Yamut (fig. 1) felt earthquakes beginning on March 15, 1991, but did not notice any changes in the volume and color of the steaming from a preexisting thermal area on the northwest slope of the volcano. No rock falls or landslides were noted. Then, in the early morning of April 2, 1991, residents at these same sitios felt

more earthquakes than they had experienced during the previous weeks. According to Erro and Palawig, members of the Lubos na Alyansa ng mga Katutubong Ayta ng Sambales (LAKAS) (Negrito People's Alliance of Zambales) (oral commun., 1991), the number of felt earthquakes increased significantly shortly before phreatic explosions began at about 1600 that afternoon.

On April 4, 1991, Sister Emma Fondevilla of LAKAS reported these explosions to the Philippine Institute of Volcanology and Seismology (PHIVOLCS); she also reported felt earthquakes and a strong sulfur odor during and after the explosions. The series of explosions and the strong sulfur odor caused some residents of villages at the northwest, west, and southwest slopes of the volcano to evacuate voluntarily.

PHIVOLCS immediately dispatched a Quick Response Team to conduct an aerial survey and confirmed the reported explosions of Mount Pinatubo on the same day. The aerial survey showed a northeast-trending alignment of nine thermal vents vigorously emitting white to dirty white steam (fig. 2). Our initial interpretation of the activity was that explosions had occurred within the preexisting hydrothermal system.

To judge whether these events presaged more serious eruptions, we began monitoring operations on April 5, 1991. This paper presents the results of that monitoring, on the western side of Mount Pinatubo, before, during, and after the volcano erupted in June 1991.

SEISMIC NETWORK ESTABLISHED

Our first seismic station was established on April 5 at Sitio Yamut, Barangay Villar, Botolan, Zambales, about 7 km west-northwest of Mount Pinatubo (fig. 1, table 1). It was equipped with a portable seismograph (Kinemetrics Ranger, Model PS1A) that had a filter setting fixed at 2 Hz, gain at 48 dB and drum speed of 60 mm/min. A single-sideband radio transmitter was used to transmit reports to the PHIVOLCS central office in Quezon City.

Seismograms from this first station were analyzed immediately after the record was removed from the drum at

---

[1]Philippine Institute of Volcanology and Seismology.

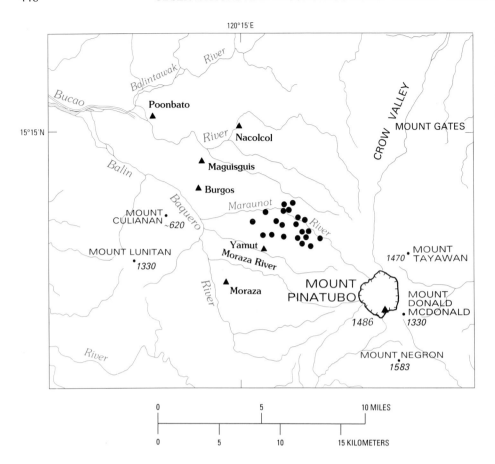

Figure 1. The seismic stations (solid triangles) established at the western side of Mount Pinatubo in April 1991. Earthquake epicenters determined from this network during early April are shown as filled circles.

Figure 2. Steaming vents of Mount Pinatubo. The two most vigorous vents are in two forks of the upper Maraunot River, shown in map view in figure 5. Photograph was taken on April 6, 1991.

about 0600 each day. Seismic observations were correlated with observations of steaming activity. Early seismograms showed numerous high-frequency volcanic earthquakes (HFVQ's), many of which were also felt (table 2), and steaming was strong.

Convinced that Mount Pinatubo was showing signs of unrest, and of the need to locate earthquakes, we established four additional seismic stations between April 8 and 15, at Barangays Maguisguis, Nacolcol, Moraza and Burgos (fig. 1). Each seismic station was equipped with a digital

Table 1. Seismographs installed at the western side of Mount Pinatubo in 1991.

Date installed	Seismic Stations				
	Yamut	Moraza	Burgos	Maguisguis	Nacolcol
April 5	Kinemetrics[1]	—	—	—	—
April 8	Kinemetrics	EDR[2]	EDR	EDR	EDR.
May 2	Kinemetrics	Geotech[3]	Geotech	Geotech	Geotech.
June 6	transferred to Poonbato.		—— removed ——		

[1] Kinemetrics, Ranger seismograph (analog), installed in Yamut from April 5 to June 6; provided the seismic data used in preparing Pinatubo Volcano Updates (PHIVOLCS, 1991).

[2] EDR, Earthquake Data Recorder digital seismographs, donated to PHIVOLCS by the Japan International Cooperation Agency. Data gathered by the seismographs were used to locate earthquakes from April 8–30, 1991.

[3] Geotech, Teledyne Geotech Portacorder RV-320B (analog).

seismograph called an EDR (earthquake data recorder), previously donated by the Japan International Cooperation Agency (JICA) for use at Taal Volcano. Floppy disks with the data were retrieved and taken to Manila, where the data could be played back and reduced.

These were replaced by portacorder seismographs (Teledyne Geotech) during the first week of May (table 1). These portacorders had filter settings of 0.2 Hz for high pass and 12.5 Hz for low pass, gain at 48 dB, and drum speed of 60 mm/min.

Seismic data from these stations, plus visual observations, helped us to assess the daily activity of Mount Pinatubo, and our evaluation of these parameters was disseminated to the media, local officials, and the public through daily volcano updates (PHIVOLCS, 1991).

Table 2. Counts of high-frequency (HF) and low-frequency (LF) volcanic earthquakes as recorded at Yamut, and earthquakes felt on the western side of Pinatubo, April 5–June 6, 1991.

[Note that during this period, the greatest number of felt earthquakes occurred April 5–21, 1991. Int, Rossi-Forel intensity; MD, maximum deflection]

Date	HF	LF	LF+HF	Cumulative total	Remarks	Date	HF	LF	LF+HF	Cumulative total	Remarks
April 5	75	0	75	75	Start 1712; 4 felt at Int II.	May 7	59	0	59	2,166	Three Int I with rumblings.
6	148	0	148	223	0000–1730 only.	8	52	0	52	2,218	No felt event.
7				223	Seismograph out of order but two events felt at Int III.	9	64	0	64	2,282	No felt event.
8	32	1	33	256	1200-0000 only.	10	178	0	178	2,460	Two Int I.
9	83	3	86	342	Three felt at Int I.	11	126	0	126	2,586	One Int II with rumblings.
10	65	2	67	409	One Int I and 1 Int III.	12	139	0	139	2,725	One Int I with rumblings.
11	29	4	33	442	One Int I and 1 Int III.	13	113	0	113	2,838	No felt event.
12	59	2	61	503		14	116	0	116	2,954	No felt event.
13	83	1	84	587	One Int I and 1 Int II.	15	80	0	80	3,034	One Int I with rumblings.
14	85	0	85	672	Four Int I and 1 Int II.	16	55	0	55	3,089	No felt event.
15	135	3	138	810	One Int I.	17	115	0	115	3,204	No felt event.
16	71	2	73	883	One Int I.	18	47	0	47	3,251	No felt event.
17	58	0	58	941	Two Int I with rumblings.	19	75	0	75	3,326	No felt event.
18	85	0	85	1,026	Three Int I with rumblings.	20	62	0	62	3,388	One Int I with rumblings.
19	84	4	88	1,114	Four Int I with rumblings.	21	100	0	100	3,488	No felt event.
20	70	2	72	1,186	One Int I with rumblings.	22	103	0	103	3,591	No felt event.
21	62	0	62	1,248	One Int I with rumblings.	23	52	0	52	3,643	No felt event.
22	47	1	48	1,295	No felt event.	24	31	0	31	3,674	One Int I with rumblings.
23	26	0	26	1,321	No felt event.	25	48	0	48	3,722	No felt event.
24	57	0	57	1,378	No felt event.	26	81	0	81	3,803	No felt event.
25	83	0	83	1,461	Two at MD.	27	56	4	60	3,860	No felt event.
26	76	0	76	1,537	One Int I with rumblings.	28	52	1	53	3,913	One Int I with rumblings.
27	86	0	86	1,623	One int I with rumblings.	29	77	2	79	3,992	One Int I.
28	56	0	56	1,679	No felt event.	30	91	0	91	4,083	One Int II.
29	73	0	73	1,752	No felt event.	31	85	2	87	4,170	No felt event.
30	77	0	77	1,829	No felt event.	June 1	70	2	72	4,242	No felt event.
May 1	74	0	74	1,903	Two Int I with rumblings.	2	90	1	91	4,333	No felt event.
2	48	0	48	1,951	No felt event.	3	59	5	64	4,395	No felt event.
3	43	0	43	1,994	No felt event.	4	50	4	54	4,449	Harmonic tremor.
4	36	0	36	2,030	No felt event.	5	86	4	90	4,539	No felt event.
5	33	0	33	2,063	No felt event.	6	83	3	86	4,625	No felt event.
6	44	0	44	2,107	No felt event.						

7 Seismograph was pulled out from Sitio Yamut and transferred to Poonbato, 22 km northwest of Pinatubo's summit.

DETAILS OF THE PREERUPTIVE SEISMICITY

Mount Pinatubo's seismicity during April, May, and early June was characterized by swarms of HFVQ's. The daily count and amplitudes of earthquakes remained more or less constant from April 5 to May 26. However, on May 27, 1991, low-frequency volcanic earthquakes (LFVQ's) and tremor began and gradually increased in early June until they dominated the seismic data by June 8, 1991. Examples of HFVQ and LFVQ events are shown in figure 3.

HIGH-FREQUENCY VOLCANIC EARTHQUAKES

The first 24 h of seismic recording at the Yamut Station (1712 April 5 to 1730 April 6, 1991), showed 223 HFVQ's (table 2). Twelve of these quakes were recorded at maximum deflection. The rest had trace amplitudes that varied from 5 mm to 20 mm. These had dominant frequencies of 3 Hz to 5 Hz and S-P times (the difference in P-and S-wave arrival times) between 0.88 s and 1.36 s, suggesting that the source was about 4 to 8 km from the Yamut seismic station.

The seismograph malfunctioned, and no records were obtained from 1730 on April 6 to 1200 on April 8, 1991. When the seismograph began operating again on April 8, the Yamut seismic count had decreased from 148 (0000 to 1730, April 6) to 32 earthquakes (1200 April 8 to 0000 April 9) (table 2). Through the balance of April, 26 to 135 HFVQ's occurred each day (average 74), most of which were small and not felt. Around 20 to 25 percent of these earthquakes had trace amplitudes greater than 20 mm and only about 5 percent caused maximum deflection. Some of the larger HFVQ's were felt and heard as rumbling sounds. Measured S-P times were, at Yamut, 1.0–1.4 s; at Moraza, 1.5–1.6 s; at Nacolcol, 1.8–2.0 s; and at Burgos, 1.9–2.0 s. These S-P times suggested that the source was upslope from the Yamut Station (fig. 1).

Earthquake epicenters and· hypocenters were determined graphically, using S-P times. Epicenters from April 8–30, 1991, clustered 5 to 7 km north-northwest of Mount Pinatubo's summit (fig. 1), at depths of 2 to 7 km beneath the surface.

In May, seismic activity increased slightly, to an average of 90 HFVQ's per day (fig. 4). The number of larger HFVQ's (>20 mm trace amplitude) gradually increased from 20 percent of the total count on May 17 to 45 percent by May 31. Felt earthquakes also increased. Residents of Moraza, Villar, and Belbel felt four to five earthquakes per day during this period.

In early June, the number and size of earthquakes continued to increase. About 40 to 45 percent of the recorded earthquakes had trace amplitudes >20 mm and had clear and distinct P-phases.

On June 6, 1991, all of our seismic stations on the western side were pulled out because of the rapidly deteriorating condition of Mount Pinatubo. A temporary seismic station was set up at Barangay Poonbato on the same day, 15 km northwest of Yamut and 22 km northwest of Mount Pinatubo's summit, to continue monitoring the volcano's seismicity. The first records from Poonbato show a few small LFVQ's and HFVQ's. (Note: Because Poonbato is far from Pinatubo, some shallow HFVQ's could have appeared as LFVQ's by the time their signals reached Poonbato, and others could have been missed entirely. Counts from Yamut and Poonbato are not directly comparable.) On June 7, the seismic count remained at a low level. By June 8, around 3 to 5 HFVQ's were recorded, and the number of LFVQ's and periods of tremor were increasing.

LOW-FREQUENCY VOLCANIC EARTHQUAKES AND TREMOR

LFVQ's and tremor were recorded by the Yamut seismic station from April 8–22, 1991 (table 2), during episodes of increased ash emission. The LFVQ's had maximum trace amplitudes of 25 mm, an average duration of 50 to 60 s, and dominant frequencies of 2 to 2.5 Hz. Discontinuous volcanic tremor had trace amplitudes of 0.5 to 1.5 mm, a dominant frequency of 1.0 Hz, and durations that ranged from 50 to 72 s. Curiously, from April 24 to May 26, no LFVQ's or tremor occurred.

On May 27, 1991, the Yamut seismograph recorded four small LFVQ's—with trace amplitudes of 0.5 mm to 1.5 mm and durations from 30 s to 55 s. Beginning June 1, 1991, larger LFVQ's with trace amplitudes ranging from 1.0 mm to 8.0 mm were recorded. By June 3, seismic activity further increased, with five LFVQ's and harmonic tremor lasting about 25 min. On June 4, 5, and 6, the Yamut seismograph continued recording 3 to 4 LFVQ/d, until the Yamut seismograph had to be pulled out and transferred to Poonbato.

At Poonbato, the seismograph continued recording LFVQ's, but, predictably, because of the greater distance from the source, they were fewer and smaller than had been recorded at Yamut. By June 8, LFVQ's and low-frequency harmonic tremor (LFHT) gradually increased and comprised around 80 percent of the recorded seismic events.

The first large eruption began at 0851 on June 12, 1991, preceded by harmonic tremor from 0341 to 0530 of the same day and then several short bursts of volcanic tremor, each lasting about 60 s. The eruption itself was recorded on the seismograph as LFHT. Thereafter, the Poonbato seismograph recorded many LFVQ's and periods of LFHT.

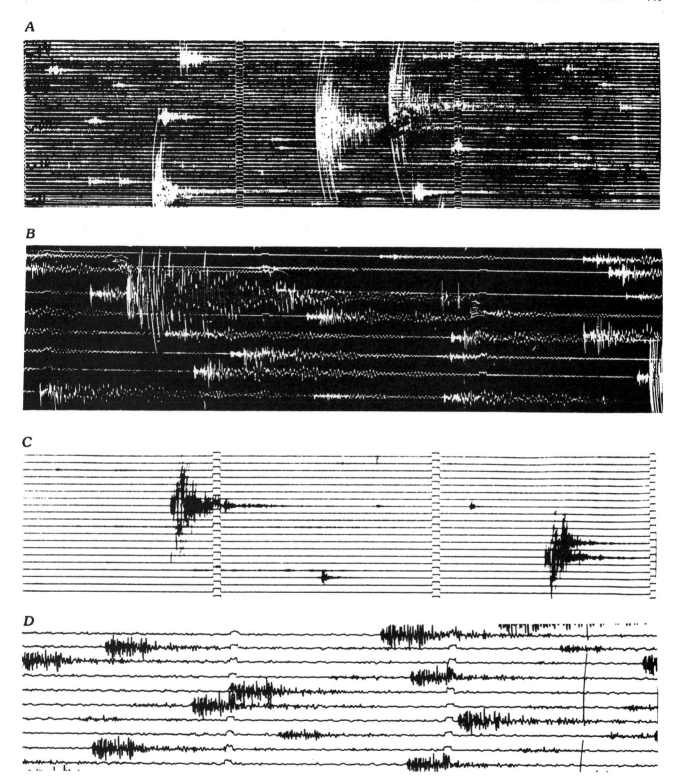

Figure 3. *A*, High-frequency volcanic earthquakes as recorded at Yamut on April 15, 1991. *B*, Low-frequency volcanic earthquakes recorded at the Poonbato station on June 14, 1991. *C*, High-frequency volcanic earthquakes as recorded at station BUGZ on January 30, 1992. *D*, Low-frequency volcanic earthquakes recorded at station BUGZ on September 12, 1992. *A* and *B* were recorded on a smoked drum of a Kinemetrics Ranger seismograph; *C* and *D* were recorded by a Teledyne Geotech Portacorder.

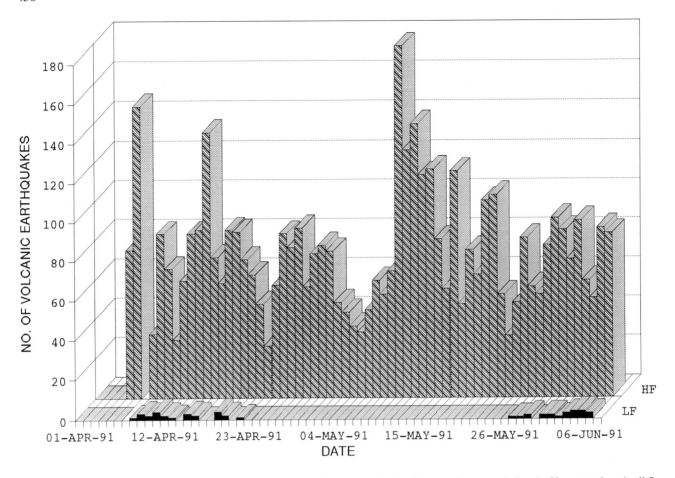

Figure 4.   Numbers of volcanic earthquakes per day (HF=high frequency; LF=low frequency) as recorded at the Yamut station, April 5–
June 6, 1991.

VISUAL OBSERVATIONS

STEAMING ACTIVITY BEFORE APRIL 2, 1991

Before the April 2, 1991, explosions, steaming activity
at Mount Pinatubo was weak and confined within a thermal
area inside a 150-m-diameter depression, 1,180 m above
sea level in the uppermost Maraunot River. The thermal
area had small steaming vents, hot-water seeps, bubbling
pools, and hot ground dotted with small quantities of sulfur
sublimates (Delfin, 1983). According to the Aetas who fre-
quented the area, steam emission was consistently weak
(members of LAKAS, oral commun., 1991).

On August 3, 1990, residents of Yamut and Tarao
reported a ground fracture and steam emission on the upper
slopes of Mount Pinatubo. A PHIVOLCS Quick Response
Team found that a large landslide had occurred on the steep
northwest side of Mount Pinatubo but that the reported
steaming activity corresponded to the preexisting thermal
area (Ramos and Isada, 1990). The team concluded that the
landslide and steaming were not related to any volcanic
activity but, rather, that part of the former thermal area was

exposed when the landslide occurred. No new steaming
vent was noticed.

APRIL 2–MAY 28, 1991: EXPLOSIONS AND INTENSE FUMAROLIC ACTIVITY

At about 1600 on April 2, 1991, phreatic explosions
occurred at and northeast of the thermal area on the upper
northwest slope of Mount Pinatubo. The explosions reportedly lasted until nightfall. Several vigorous new steam vents
developed in a northeast-trending alignment across the
north slope of the volcano (fig. 5). Strong jets of dirty white
steam rose from these vents.

During the rest of April and May, two of these vents
were especially vigorous (fig. 2), both in the upper reaches
of the Maraunot River. Steam routinely rose 300 to 800 m,
and, sometimes, steaming intensified and small quantities
of volcanic ash were ejected with the steam to heights of
1,500 to 3,000 m. Light ash fall occurred on the volcano's
upper northwest, west, and southwest slopes, causing
siltation of the Maraunot-Balin Baquero and the Marella-
Santo Tomas River systems.

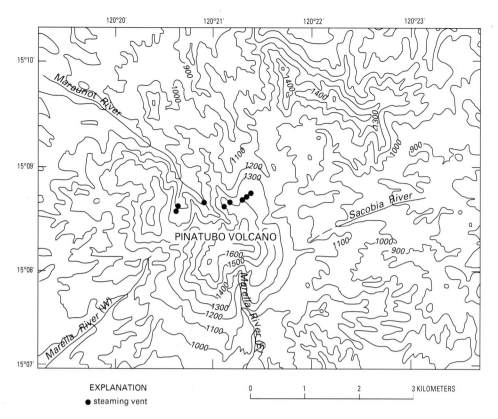

Figure 5. Steaming vents (filled circles) at Mount Pinatubo. These vents formed during the April 2, 1991, explosions. The most vigorous vents are shown by the two larger filled circles in the upper Maraunot River. Contour interval 100 m.

In May, the steaming activity gradually intensified, and ash emissions were occasionally observed. The steam columns changed from white to brown or gray during ash emissions and rose between 1,500 and 5,000 m above the vents. Beginning on May 16, ash emissions were observed frequently. One of the largest of these, on May 28, produced ash clouds that reached a height of 5,000 m and caused light ash fall in Yamut, Villar, Moraza and other areas within 15 km to the northwest, southwest, and west of Mount Pinatubo.

NEAR-ERUPTIVE STEAMING ACTIVITY (LATE MAY TO JUNE 1991)

Steaming progressively intensified and carried increasing amounts of ash after the May 28 ash emission. On June 1, intermittent ash puffs were common, and by June 3, intensified steam and ash emission caused intermittent light ash fall that hampered visual observation of the Pinatubo summit area. On June 5, ash fall reached areas 12 to 15 km west and southwest from the active vents. The activity continued to intensify, and by June 8, ash-laden steam clouds attained a height of 8,000 m. Ash fall from this event reached as far as Poonbato.

OTHER OBSERVATIONS

Steam plumes from the active vents had a strong sulfur odor that could be detected easily within 5 km distance from the volcano's summit. The sulfur-rich gas withered vegetation at and near the active vents.

Ash ejected during the April 2, 1991, explosions and during subsequent ash emissions caused siltation of rivers that flow north and west from the volcano's summit. These rivers are the Maraunot Balin-Baquero River system, Marella-Santo Tomas River system, and the O'Donnell River. The silt destroyed aquatic life in the rivers. Water samples were highly acidic: a water sample collected on April 8 from the Maraunot River, 1.5 km below the preexisting thermal area, had a pH of 2.45, and samples from April 9 and 10 were equally acidic (Campita and Tansinsin, 1992; N.R. Campita written commun., 1993). In comparison, water collected from the Maraunot River in 1983 was neutral to basic with pH values ranging from 7.60 to 8.02 (Clemente, 1984).

JUNE 1991 ERUPTIONS

On June 9, 1991, at 1455 and again at 1504 and 1630, billowing ash clouds flowed down the northwest and west slopes of the volcano, down the Maraunot and Moraza Rivers. These density currents, which looked like pyroclastic flows from our vantage in Poonbato, started from ash columns reaching an altitude of 5,000 m and reached distances of around 4 to 5 km from the active vents. Although subsequent aerial observations found that these density currents had moved slowly and had left only a thin veneer of ash fall in the valleys (Hoblitt, Wolfe, and others, this volume), these currents nonetheless represented a significant intensification of activity.

Continuous emission of ash-laden steam columns followed, but column heights could not be estimated because the ash drifted and dispersed toward the west, nearly reaching our temporary station. At 1330 on June 10, a voluminous ash-laden steam column rose to a height of 4,000 m; other episodes of enhanced emission occurred on June 11 at 0515 and 0540, from which ash rose to 5,000 m and drifted to the southwest. At 1030 on June 11, steam emission gradually decreased, and by 1600, weak, white to dirty white steam clouds rose only 500 to 700 meters above the active vents.

Felt earthquakes at 0341 on June 12 were followed by almost 2 h of harmonic tremor accompanied by continuing ash emission, the column of which rose to 4,000 m and drifted to the northwest. At 0851, an energetic eruption column rose to a height of about 20 km, and pyroclastic surge moved down the northwest, west, and southwest slopes of the volcano, nearly catching news reporters in Sitio Ugik (12 km northwest of the volcano). The eruption column formed an umbrella cloud that slowly extended over the whole western side of the volcano and caused moderate to heavy ash fall in the towns of Cabangan, San Felipe, San Narciso, San Antonio, San Marcelino, Castillejos, and Subic, all in the province of Zambales.

Sometime between 0900 and 1100, 2- to 5-cm-diameter pumice began to fall at Ugik and in Moraza and other barangays at the west and southwest foot of Pinatubo; ash began to fall in San Marcelino, Castillejos, San Antonio, and Cabangan about 0920. By 1100, the fall was nearly finished at Ugik. Another high eruption column observed at 2250 and 2305 was generated by an eruption that began at 2250. Tephra from this column drifted west and southwest from the volcano.

At 0841 on June 13, another large eruption occurred. The eruption cloud rose toward the sky and slowly expanded over the whole western and southern part of the province of Zambales. The eruption produced pyroclastic density currents that cascaded down most river valleys of the west flank, from the Maraunot valley on the north to the Marella valley on the south. Some of these ashy currents reached as far as Sitio Ugik. After the 0841 eruption, Mount Pinatubo's activity dropped back to continuous emission of ash-laden steam columns that rose to heights of 1,500 to 3,000 m.

At 0200 on June 14, steaming activity gradually decreased. From 0500 to 1300, steaming activity was weak, reaching only 300 to 400 meters above the vents. However, an increasing number of shallow earthquakes was recorded by the Poonbato seismograph from 2320 on June 13 to around 0700 on June 14. Harmonic tremor with an average double amplitude of 1 to 2 mm began at 0400. At 1309, Mount Pinatubo erupted anew, this time producing ash

Table 3. Ash emissions and eruptions observed from Poonbato, June 9–15, 1991.

[The eruption times and dates are as observed from Poonbato station through 0831 on June 15, after which the volcano was completely obscured. Estimates of column height through the afternoon of June 14 were based on vertical angles measured at Poonbato; estimates of column height during the evening of June 14 are from weather radar at Clark Air Base. n.d., not determined]

Date	Time (local)	Height (in meters)	Drift
June 9	1455	5,000	West and southwest.
9	1504	5,000	West and southwest.
9	1630	5,000	West and southwest.
10	1330	4,000	West and southwest.
11	0515	5,000	Southwest.
11	0540	5,000	Southwest.
12	0341	4,000	Northwest.
12	0515	5,000	Southwest.
12	0851[1]	20,000	Southwest.
12	2250[1]	20,000	Northwest, west, and southwest.
12	2305	15,000	Northwest, west, and southwest.
13	0841[1]	20,000	Northwest, west, and southwest.
14	1309[1]	20,000	Southwest and southeast.
14	1520[1]	18,000	Southwest.
14	1853[1]	20,000	Southwest and west.
14	1921	20,000	Southwest.
14	2218	20,000	Southwest.
14	2321[1]	n.d.	Southwest.
15	0115[1]	n.d.	Southwest.
15	0140	n.d.	Southwest.
15	0257[1]	n.d.	Southwest.
15	0517	n.d.	Southwest.
15	0549	n.d.	Southwest.
15	0556[1]	n.d.	Southwest.
15	0611	n.d.	Southwest.
15	0730	n.d.	Southwest.
15	0809[1]	n.d.	Southwest.
15	0831	n.d.	Southwest.

[1]Discrete eruption, identified from combined data of seismograms, microbarograph records, weather radar, and visual observations (Hoblitt and others, this volume)

clouds that drifted to the southwest and dropped ash on the towns of Cabangan, San Felipe, San Antonio, San Narciso, San Marcelino, and Castillejos about an hour later. Eruptions producing large pyroclastic surges (R.P. Hoblitt, U.S.G.S., written commun., 1993) were observed at 1520, 1853, and 1921 (latter two might be from a single, continuous event). Starting at 2218, the eruptions became more frequent. Lightning in the mushroom-like eruption cloud helped us to see the drifting eruption cloud. Simultaneously, pyroclastic flows or pyroclastic surges rolled down the northwest, west, and southwest slopes of the volcano and exhibited beautiful fireworks displays like "dancing lights." These pyroclastic materials glowed and caused momentary brightness on the volcano's slopes. Another large explosion was seen at 2321.

Figure 6. A giant pyroclastic surge cascading down the western side of Mount Pinatubo. Photograph taken at Barangay Burgos, Botolan, Zambales, during the early morning of June 15, 1991. (Photograph courtesy of Rene Arante, PHIVOLCS.)

At 0115 on June 15, a vigorous eruption occurred. This time, countless lightning flashes marked the rapidly spreading eruption cloud and covered the sky. At 0140, another mushroom-like eruption column was seen, and the whole volcano area was lit by lightning flashes in pyroclastic density currents that were flowing rapidly down the volcano slopes. Fireworks displays were observed in practically all drainages. Another eruption was observed at 0257. At 0517, we observed large pyroclastic density currents moving down the entire western and southwestern slopes of the volcano; flows in individual drainages could no longer be distinguished.

At 0556, 0611, 0809, and 0831 on June 15, 1991, powerful explosions and giant pyroclastic surges were observed. One large pyroclastic surge down the western slope of Mount Pinatubo, shown in figure 6, invaded Barangay Burgos, 15 km northwest of the volcano. At around 1100, heavy ash fall caused darkness and prevented further visual observation of the June 15 eruptions. Microbarograph records suggest that strong eruption ended about 2230 June 15 (Wolfe and Hoblitt, this volume); seismic records from Poonbato suggest further decline at about 0330 of June 16. The effects of the eruption were aggravated by the passing of typhoon Yunya (local name, Diding), which distributed volcanic ash in all directions and affected places not expecting it, like Metro Manila.

SEISMICITY DURING AND AFTER THE JUNE 12–15 ERUPTIONS

From June 12 to 15, 1991, the seismograms obtained at Poonbato showed tremor and many volcanic earthquakes,

mostly LFVQ's. Most of the harmonic tremor was recorded during eruptions. Only a few HFVQ's were recorded.

Before the 0841 and 1309 eruptions on June 13 and June 14, respectively, the Poonbato seismograph recorded increases in seismic activity. Seismicity increased from 0430 on June 13 until the 0841 eruption of that day (fig. 7). Likewise, seismicity increased from 2320 on June 13 to 0640 on June 14, before the 1309 eruption on June 14. The upsurge in seismicity was characterized by (1) a progressive increase in seismic count, (2) an increase in the duration and trace amplitude (size) of the earthquakes, and (3) increased amplitude and duration of harmonic tremor.

Many felt earthquakes occurred during the June 12 to early June 15 eruptions. These registered at Poonbato at maximum deflection and were felt near the volcano at intensities I–IV (Rossi-Forel scale). Countless additional, and still stronger, earthquakes occurred on June 15–16, 1991, during and just after the climactic eruption, but the analog record obtained at Poonbato was not readable because the seismograms were saturated by large earthquakes and tremor generated by the eruption. Many of the earthquakes that occurred on June 15–16 were felt at intensities I–VII (Rossi-Forel scale) within a distance of 40 km from the volcano, even causing liquefaction at Sasmuan, Pampanga, about 35 km southeast of the volcano.

Seismic activity remained high after the June 12–15, 1991, eruptions. Earthquakes on June 16 remained uncountable. Beginning on June 17, 1991, the Poonbato seismograph recorded an average of 150 HFVQ's per day. Episodes of LFHT were also recorded, particularly during unusually energetic ash emission. On July 4, 1991, there was a significant increase in seismic activity (319 HFVQ's

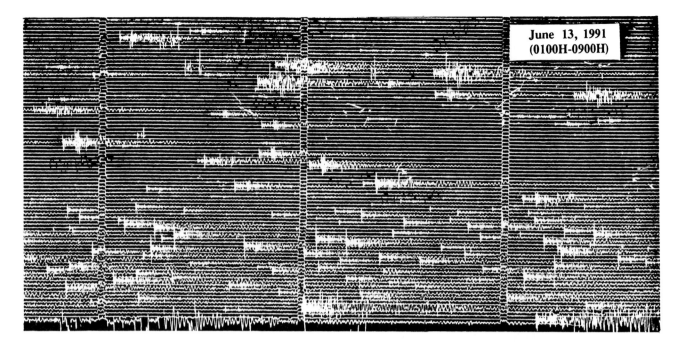

Figure 7.   Increase in seismicity (to 1 to 2 earthquakes per minute) beginning about 0430 on June 13 and continuing until an eruption occurred at 0841 that day. The eruption itself was recorded as harmonic tremor.

recorded at Poonbato). However, on July 5, the seismic count decreased to 120 HFVQ's and by July 13, this further dwindled to 76 HFVQ's (fig. 8).

No LFVQ's were reported by the Poonbato station from June 18 to July 4 because the seismograms were saturated by large-amplitude harmonic tremor and HFVQ's (fig. 8). The overprinting of relatively large HFVQ's and harmonic tremor prevented recognition of smaller LFVQ's on our analog record. By July 5, the size (trace amplitude) of HFVQ's and tremor decreased enough for us to recognize and report LFVQ's.

DISCUSSION

Phreatic explosions of April 2, 1991, occurred when ground water was heated by contact with magma or with hot gases rising from that magma. The hot ground water boiled, vaporized, and blasted through the country rocks. Rock alteration within the preexisting hydrothermal system could have prevented the escape of gases and allowed the buildup of high vapor pressure. Then, a sudden release of pressure may have caused the superheated fluid to expand explosively and produce phreatic explosions. Continued heat transfer to the hydrothermal system caused continued vigorous emission of white to dirty white steam from the newly formed thermal vents.

Stresses from inflating or rising magma caused active fracturing of the country rock and thus swarms of HFVQ's. The onset of LFVQ's on March 27 might indicate that magma had risen to shallow depths by that date.

Beginning June 1, 1991, seismic activity gradually intensified. Although the number of HFVQ's remained more or less constant, between 50 and 90/d, the trace amplitudes (size) of HFVQ's and the daily count of LFVQ's and harmonic tremor increased. Moreover, an increasing number of earthquakes with distinct P-phases was noticed. The fact that the number of HFVQ's recorded at Yamut during this period remained more or less constant, despite generally increasing overall seismicity (Endo and others, this volume; Harlow and others, this volume), might be due to increasing distance of the earthquake source from the Yamut station as the center of seismic activity migrated from the north-northwest cluster of hypocenters to the summit area of Pinatubo (Harlow and others, this volume).

CONCLUSIONS

Significant changes in seismicity and steaming activity at Mount Pinatubo were used to anticipate the onset of the June 1991 eruptions. Principal changes were:

1. An upsurge in seismic activity, including increases in the daily count and trace amplitude (size) of the earthquakes, a shift in the types of volcanic earthquakes, from high frequency to low frequency, and increased occurrence of harmonic tremor.
2. Increased steaming activity, including an increase in the volume and vigor (height) of steam emission, change in the color of steam from white to brown or gray with increasing ash content, and increasingly frequent episodes of enhanced ash emission.

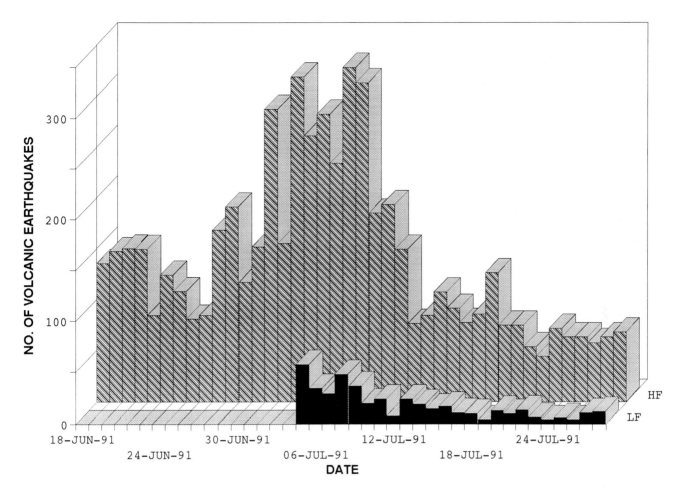

Figure 8. Numbers of volcanic earthquakes per day (HF=high frequency; LF=low frequency) as recorded at Poonbato, June 18 to July 28, 1991. Tremor occurred on all days and was so strong in the last half of June that it obscured many discrete earthquakes; thus, the apparent increase in seismicity from middle to late June is probably an artifact of earthquake counting against strong background tremor.

Even though the observations reported in this paper were relatively simple and used older equipment than observations described in other papers in this volume, they allowed us to keep abreast of the volcano's unstable condition and to issue appropriate warnings.

REFERENCES CITED

Campita, N., and Tansinsin, L., 1992, Water quality effects on Maraunot and other peripheral rivers by the recent volcanic activity of Mount Pinatubo: PHIVOLCS open file report, 24 p.

Clemente, V.C., 1984, A re-evaluation of Mount Pinatubo Geochemistry: Philippine National Oil Company open file memorandum report, 24 p.

Delfin, F.G., 1983, Geology of Mt. Pinatubo Geothermal Prospect: unpublished Philippine National Oil Company report, 45 p.

Endo, E.T., Murray, T.L., and Power, J.A., this volume, A comparison of preeruption real-time seismic amplitude measurements for eruptions at Mount St. Helens, Redoubt Volcano, Mount Spurr, and Mount Pinatubo.

Harlow, D.H., Power, J.A., Laguerta, E.P., Ambubuyog, G., White, R.A., and Hoblitt, R.P., this volume, Precursory seismicity and forecasting of the June 15, 1991, eruption of Mount Pinatubo.

Hoblitt, R.P., Wolfe, E.W., Scott, W.E., Couchman, M.R., Pallister, J.S., and Javier, D., this volume, The preclimactic eruptions of Mount Pinatubo, June 1991.

PHIVOLCS, 1991, Pinatubo Volcano Updates: PHIVOLCS open-file reports, April-December 1991, unpaginated.

Ramos, A., and Isada, M., 1990, Mount Pinatubo not erupting: PHIVOLCS Observer, July–September 1990, v. 6, no. 3, p. 6.

Wolfe, E.W. and Hoblitt, R.P., this volume, Overview of the eruptions.

The Preclimactic Eruptions of Mount Pinatubo, June 1991

By Richard P. Hoblitt,[1] Edward W. Wolfe,[1] William E. Scott,[1]
Marvin R. Couchman,[1] John S. Pallister,[1] and Dindo Javier[2]

ABSTRACT

The preclimactic eruptive activity at Mount Pinatubo in late May and early June of 1991 consisted of dome growth, followed by 4 vertical eruptions, followed by 13 surge-producing eruptions. The dome and pyroclasts from the first vertical eruption are composed almost entirely of andesite. In subsequent eruptions the proportion of andesite pyroclasts declined irregularly as the proportion of dacite pyroclasts increased; pyroclasts in the late surges are composed almost entirely of dacite. That pyroclasts from most preclimactic deposits display a wide density range suggests that the early vanguard magmas were subject to substantial degassing before eruption.

Stratigraphic evidence indicates that eruption magnitude declined successively through the vertical eruptions and generally declined successively through the surge-producing eruptions. This decline in magnitude was accompanied by a general decline in the repose-interval duration, so the successive eruptions were generally smaller and more closely spaced in time. The duration of successive eruptions at first declined but then apparently increased somewhat just before the climactic eruption. This behavior is mimicked by a model wherein the pressurization rate of magma at the top of the magma conduit increases from a value initially inadequate to sustain a continuous eruption and wherein the confining pressure, controlled by degassing at the top of the magma column during repose intervals, decreases with time. The climactic eruption began when the pressurization rate, controlled by the influx of magma from a deep reservoir, exceeded the rate of depressurization, as magma was expelled during eruptions.

INTRODUCTION

One of the greatest challenges in volcano-hazard-mitigation efforts is to forecast eruptive behavior correctly. Because mitigation measures depend on anticipation of eruption types, timing, and the areas likely to be affected, accurate forecasts can minimize the social, political, and economic disruption caused by eruptions. For example, vigorous, convectively driven, vertical eruption plumes disperse tephra over large regions, whereas the effects of pyroclastic density currents are usually far more restricted and far more devastating. Areas subject to pyroclastic density currents need to be evacuated; tephra-prone areas need not be evacuated, but a variety of other mitigative measures should be implemented.

Eruptive behavior remains difficult to forecast with confidence. Forecasts are usually based on data gleaned from deposits produced by past eruptions, from the eruptive behavior of other similar volcanoes, and from historical activity at the volcano under scrutiny. Stratigraphic data are often incomplete, and behavior at different volcanoes or even the same volcano can vary within wide limits. Compare, for example, the Pinatubo eruptive sequence to two other recent sequences: the eruptions of Mount Lamington, Papua New Guinea, in 1951 (Taylor, 1958) and El Chichón volcano, Chiapas, Mexico, in 1982 (Sigurdsson and others, 1984). All three sequences occurred at volcanoes that had been in repose for hundreds of years, all three produced vigorous convectively-driven vertical eruption plumes as well as dilute pyroclastic density currents (surges), and all were driven by sulfur-rich andesite or dacite magmas.

The Pinatubo sequence, to be described in detail below, consisted sequentially of increasing ash emissions, dome growth, 4 vertical eruptions with continued dome growth, 13 pyroclastic surges, and a climactic vertical eruption with associated pyroclastic flows. Postclimactic activity (through 1993) consisted of declining ash emissions and emplacement of a small dome. The El Chichón sequence consisted of an initial major vertical eruption followed 5 days later by two more major eruptions. The second major eruption began with a surge and then progressed into a vertical eruption that also culminated with a surge. The third major eruption began as a vertical eruption and culminated with a minor surge event. Following a week of increasingly vigorous ash emission, Mount Lamington produced a major pyroclastic surge; a large vertical eruption occurred later the same day. Vertical eruptions continued sporadically over the next 1½ months and were succeeded by dome growth.

---

[1]U.S Geological Survey.
[2]Philippine Institute of Volcanology and Seismology.

Despite some similarities, none of the sequences could serve as a reliable predictor of the others. One route to better eruption-pattern forecasts may be through a better understanding of the fundamental processes that control eruptive behavior. To this end, we undertook a reconnaissance study of the deposits produced by Pinatubo's preclimactic eruptions, in the hope that the deposits might provide some insights into the controlling processes. We present here a summary of the preclimactic behavior along with granulometry, component analyses, and densiometry of deposits, as well as some other relevant observations, and discuss these data in terms of processes that may control eruptive behavior.

NOMENCLATURE

In the following discussion, we refer to two basic types of pyroclastic transport processes: pyroclastic fall and pyroclastic density current. We use these terms and their variants in the following way.

Pyroclastic fall.—A rain of pyroclasts that fall to earth after vertical transport within an eruption plume. Particle concentration is low, and interactions between falling particles or particle aggregates are insignificant. The resulting deposit has wide areal extent and is typically well sorted; deposit thickness shows little variation locally, except on slopes that exceed the angle of repose. Topography exerts little or no influence on the areal distribution.

Pyroclastic density current.—A mixture of pyroclasts and gases whose net density is greater than that of the surrounding atmosphere. The mixture flows en masse along the ground like a fluid. We distinguish two types of pyroclastic density current. A pyroclastic flow is a dense, high-concentration gas-pyroclast mixture that typically produces massive, poorly sorted deposits that pond in topographically low areas. A pyroclastic surge is a dilute suspension of pyroclasts in turbulent gas. Particle concentrations are transitional between pyroclastic flow and fall, and deposits are typically relatively thin, show low-angle crossbedding, dune structures, and exhibit topographic control and sorting values intermediate between those of pyroclastic-flow and pyroclastic-fall deposits. We use the modifier "slack" to refer to surges whose densities are just marginally more dense than the surrounding air; these typically produce thin, parallel beds of well-sorted fine ash.

Pyroclastic-surge deposits emplaced on June 14 and 15 within 10 km of Pinatubo's pre-June 15 summit are termed "proximal"; those emplaced at greater distances are termed "distal."

We use the term "pyroclastic fountain" to denote the vertical ejection of a mixture of gas, ash, and larger pyroclasts that falls back to earth, analogous to a vertical water fountain (Hoblitt, 1986, fig. 16*C*). This phenomenon has been termed "column collapse" (Wilson and others, 1980).

Pinatubo's preeruption summit provides a convenient reference point for distance estimates. We refer to this point as the "old summit."

ERUPTION SUMMARY

It is useful to divide the 1991–92 volcanic activity at Mount Pinatubo (fig. 1) into a series of eight behavioral phases (Wolfe and Hoblitt, this volume). Most behavioral changes were abrupt, such as the onset of vertical eruptions on June 12. The change from Phase I to Phase II, however, was gradual, and the boundary differed for different phenomena: ash emissions began to increase in late May, SO_2 emissions reached a peak on May 28 (Daag, Tubianosa, and others, this volume), and seismicity began to escalate on June 1. For convenience, we have chosen June 1 as the boundary between Phase I and Phase II.

Phase	Date	Description
		1991 Eruption Phases
I	March 15– May 31	Felt earthquakes beginning on March 15, phreatic explosions on April 2, continuing emission of steam and minor ash, constant release of seismic energy, and generally increasing SO_2 emission.
II	June 1– June 7	Escalating seismic-energy release, concentration of shallow earthquake hypocenters beneath Pinatubo's summit, diminution in SO_2 emission, inflationary tilt, and increasing ash emission culminating in the birth of a lava dome on June 7.
III	June 7– June 12	Dome growth, increasingly heavy emission of ash, escalating seismic-energy release.
IV	June 12– June 14	Four vertical eruptions with minor pyroclastic flows, continued dome growth, and heavy emission of ash.
V	June 14– June 15	Multiple surge-producing eruptions beginning at about 1516 on June 14 and continuing until the climactic phase.
VI	June 15	The climactic phase of activity, beginning at 1342, consisting of a large vertical eruption with the production of voluminous pumiceous pyroclastic flows. This phase culminated in formation of a collapse caldera.
VII	June 15–early September	Postclimactic phase of activity wherein initially voluminous, continuous ash emissions slowly waned through the middle of late July; intermittent small ash eruptions continued until early September.
		1992 Eruption Phase
VIII	July–Oct.	Growth of small dome of andesite lava.

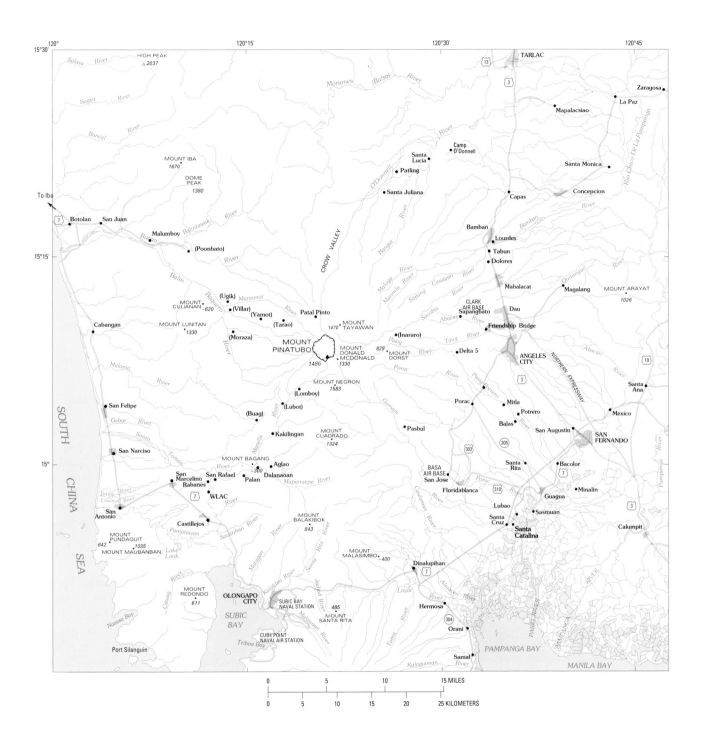

Figure 1. Major landmarks in the vicinity of Mount Pinatubo.

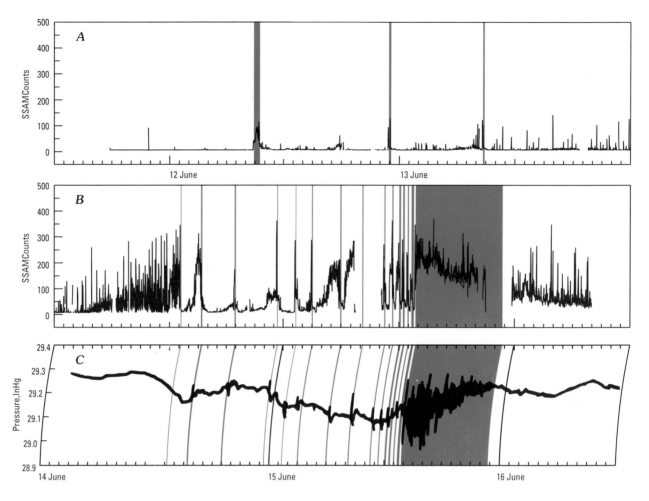

Figure 2. *A,* SSAM record (0.5- to 1.5-Hz band, station Clark) from 1200 on June 11 to 0000 on June 14, 1991; blue bands indicate eruptions. *B,* Same as *A,* from 0000 on June 14 to 1200 on June 16. *C,* Barogram recorded at Clark from about 0200 on June 14 to about 1200 on June 16; blue bands indicate eruptions. Each surge-producing eruption corresponds to a barogram deflection. The deflection corresponding to the 0257 June 15 eruption precedes that eruption by about 30 min; the discrepancy is absent on the Cubi Point barogram (not shown). This suggests that the pen on the Clark barograph malfunctioned temporarily.

We describe briefly Phase I, then focus on Phases II, III, IV, and V; a summary of the narrative is contained in table 1. Eruption durations for Phases IV and V were estimated from seismograms and Seismic Spectral Amplitude Measurement (fig. 2*A,B;* SSAM), which monitors the spectral content of earthquakes (Rogers, 1989; Power and others, this volume). Eruption onset times were determined from seismograms, microbarograph records (fig. 2*C*), and direct visual observations. Plume-height estimates in table 1 are based on C-band radar observations, primarily from Clark Air Base (Clark; fig. 1), 21 km east of Mount Pinatubo. The plume heights for the first three eruptions are minimum estimates. During the first eruption, Clark radar operators were evacuated from their station while the plume was still rising; the plumes of the second and third eruptions exceeded the radar height limit of 24 km. The first eruption almost certainly also exceeded this limit.

PHASE I CHRONOLOGY

On April 2, following some initial seismic activity, small explosions occurred along a 1.5-km-long northeast-trending fracture across Mount Pinatubo's northeast flank. Shortly after the explosions, an aerial inspection revealed nine vigorous fumaroles across the volcano's upper north flank, southwest of the fracture (Sabit and others, this volume). Three of the fumaroles (Wolfe and Hoblitt, this volume) soon became dominant, emitting copious quantities of steam and a minor amount of ash. The estimated SO_2 flux increased from 500 t/d on May 13, the day of the first measurement, to 5,000 t/d on May 28 (Daag, Tubianosa, and others, this volume).

PHASE II CHRONOLOGY

In late May, fumarole emissions, which had been almost entirely steam, began to contain an increasing

Table 1. Chronology of eruptive activity at Mount Pinatubo during June and July 1991. [Explosive-eruption onset times determined from seismograms and, as noted, from direct visual observations. We estimate ±3 min uncertainty in absolute time. Seismic durations determined from seismograms and 1-min-averaged data from Seismic Spectral-Amplitude Measurements (SSAM) system in the 0.5- to 1.5-Hz band (fig. 2A, B; SSAM observations are described by Power and others, this volume). Events prior to 0921 June 12 determined from seismic records of station PIE (9 km northeast of Mount Pinatubo summit; events later than 0921 June 12 determined from seismic records of station CAB (17 km northeast of summit). Plume heights from visual and military weather-radar observations. LP; long-period]

Phase	Date (June 1991)	Onset time	Seismic duration (min)	Event	Plume height (km)	Comment
III	7	~1700		Increased gas and ash emission; dome growth and continuous ash emission begin.	8	Heightened plume (to approximately 8 km) of gas and ash noted at approximately 1700, near culmination of several days of increasing volcano-tectonic earthquakes at depths of 1-5 km beneath the volcano summit and 30 h of gradual outward tilt recorded high on the east flank (Ewert and others, this volume). Tilt leveled off immediately following heightening of plume. Continuous seismicity was recorded at nearby stations from approximately 1640 to 1730; frequent, small earthquakes continued thereafter. Lava dome, within the upper Maraunot River canyon, was first sighted the next morning, but its feeder probably broke the ground surface the preceding evening at ~1700.
	7-12			Continued dome growth and ash emission.	2-3	Dome growth continued through at least June 13. Tephra emission was nearly continuous until the first large vertical eruption on June 12. Tephra plume generally rose to about 2-3 km elevation (approximately 1-2 km above the vent in the upper Maraunot River canyon), and winds distributed tephra generally westward. Seismicity was

Table 1. Chronology of eruptive activity at Mount Pinatubo during June and July 1991—Continued.

Phase	Date (June 1991)	Onset time	Seismic duration (min)	Event	Plume height (km)	Comment
						characterized by frequent small volcano-tectonic earthquakes and LP events interspersed with episodes of escalating tremor that lasted from tens of minutes to as much as 4 h.
	12	0310	31	Increased tremor.		Episode of vigorous tremor, gradually increasing in amplitude. Increased ash emission likely but not observed because of darkness.
IV	12	0341	2	Explosion. Pyroclastic density currents.	?	Explosion indicated by abrupt onset of high-amplitude eruption signature; followed by 1-1.5 h of tremor that waned gradually to low background level. Aerial view in daylight showed that pyroclastic flows had entered the heads of drainages on the northwest, north, and northeast flanks of the volcano and that a hot ash cloud had swept seismic station UBO on the upper east flank.
	12	0851	36	Vertical eruption.	19	Abrupt onset of intense sustained tremor and visual observation of rapidly expanding eruption column. Weather radar indicated tephra-column height ≥ 19 km. Tephra dispersed southwestward; tephra fall reported at Subic Bay and San Miguel. Aerial observation at 0950 showed continuing tephra production from vent and pyroclastic density current in O'Donnell River drainage. Subsequent aerial observation of secondary-explosion plumes suggests hot pyroclastic-flow deposits present in upper 4-5 km of Maraunot drainage.

Table 1. Chronology of eruptive activity at Mount Pinatubo during June and July 1991—Continued.

Phase	Date (June 1991)	Onset time	Seismic duration (min)	Event	Plume height (km)	Comment
	12	Afternoon		Weak tephra emission.	3–4	Intermittent views during afternoon of low tephra plume.
	12	~1725	~50	LP-event swarm.		Increasingly energetic swarm of small LP events; terminated at 1815.
	12	2252	14	Vertical eruption.	≥24	Explosion marked by rapid onset of sustained LP tremor. Weather radar indicated maximum tephra-column height ≥24 km. Tephra dispersed southwestward.
	13	~0600	~120	LP-event swarm.		Increasingly energetic swarm of small LP events peaked at ~0800, then rapidly diminished except for a few relatively large LP events.
	13	0841	5	Vertical eruption.	≥24	Abrupt onset of intense sustained tremor and visual observation of eruption column. Weather radar indicated tephra-column height ≥ 24 km. Tephra dispersed southwestward. Low tephra plume from vent after this eruption reported from Poonbato. Subsequent aerial view (about noon on June 14) showed that pyroclastic-flow deposits from one or more of the vertical eruptions extended 4–5 km down the Maraunot drainage and that the lava dome in the upper Maraunot drainage was intact and larger than when last seen on June 11.
	13	~1200	25 h	LP-event swarm.		About 3 h of relative seismic quiescence (except for occasional relatively large LP events) followed the 0841 eruption. Swarm of low-amplitude LP events,

Table 1. Chronology of eruptive activity at Mount Pinatubo during June and July 1991—Continued.

Phase	Date (June 1991)	Onset time	Seismic duration (min)	Event	Plume height (km)	Comment
						punctuated by occasional larger LP events, began about noon. Intensity of swarm began to increase rapidly at about 0400 June 14, peaking with the vertical eruption at 1309 June 14. No tephra plume visible on June 14 in clear early morning view from east (Clark Air Base) or in noon-time aerial view in upper Maraunot drainage.
	14	1309	3	Vertical eruption.	21	Seismic signal distinct but resembled that of preceding large LP events; large LP events ceased temporarily immediately following this event. Tephra column visible; weather radar recorded slow ascent to maximum height of 21 km, reached about 15 min after onset of explosion. Tephra dispersed southwestward.
V	14	1410	?	Eruptive activity.	15	Aerial view and ground views from the volcano's upper north flank from 1410 through 1435 indicated eruption in progress. Ash clouds appeared to rise from multiple sources, forming broad, relatively low eruption plume unlike the previous high, narrowly based, vertical columns. Weather radar indicated maximum height = 15 km at approximately 1411. Starting time is unknown; tremor, continuous at low level following the 1309 eruption, intensified at approximately 1400, perhaps coincident with onset of this eruptive activity.
	14	1516	7	Surge-producing eruption.	?	Eruption was observed only from west side of volcano. Onset, marked by a strong atmospheric-

Table 1. Chronology of eruptive activity at Mount Pinatubo during June and July 1991—Continued.

Phase	Date (June 1991)	Onset time	Seismic duration (min)	Event	Plume height (km)	Comment
						pressure pulse, followed an intensifying swarm of LP events that began at 1438. No weather-radar observations were reported during the event; subsequently (1530 - 1600) Cubi Point radar reported tephra cloud, with top at 18 km, moving southwestward. Lahar verified in Sacobia River valley adjacent to Clark Air Base; probably generated by pyroclastic flow from this event in Sacobia headwaters.
	14	1700		Low-level eruption.		Weather-radar operator reported that intermittent low tephra emission had been occurring.
	14	1853	5	Surge-producing eruption.	≥24	Explosion signalled by burst of high-amplitude tremor abruptly imposed on much lower-amplitude but gradually intensifying tremor and LP events. Correlative atmospheric-pressure pulse. Weather radar reported tephra-column height ≥ 24 km.
	14	2218	62?	Low-level eruption.	5	Abrupt increase in amplitude of weak tremor and small LP events that followed the 1853 eruption. Weather-radar observers report low (approx. 5 km high) tephra plume between 2218 and 2229; no additional reports until 2320 eruption, but sustained moderate-amplitude tremor and LP events suggest low-level eruption continued.
	14	2320	3	Surge-producing eruption.	21	Abrupt onset of high-amplitude tremor; correlative atmospheric-pressure pulse. High-amplitude tremor for 3 min, sustained moderate-amplitude

Table 1. Chronology of eruptive activity at Mount Pinatubo during June and July 1991—Continued.

Phase	Date (June 1991)	Onset time	Seismic duration (min)	Event	Plume height (km)	Comment
						tremor for an additional 15 min. Weather-radar observers reported tephra plume, low (6-10 km) during period of high-amplitude tremor but gradually ascending to 21 km during moderate-tremor episode. Infrared scanner in Clark Air Base tower detected hot pyroclastic density current descending upper east flank of volcano.
	15	0115	3-23	Surge-producing eruption.	?	Abrupt onset of high-amplitude tremor; correlative atmospheric-pressure pulse. High-amplitude tremor waned after approximately 3 min to moderate tremor, with several brief high-amplitude bursts; this waning tremor persisted for an additional 20 min, then decreased to low-amplitude background-level tremor. Infrared scanner in Clark Air Base tower again detected hot pyroclastic density current descending upper east flank of volcano.
	15	0257	4	Surge-producing eruption.	?	Abrupt onset of high-amplitude tremor; correlative atmospheric-pressure pulse. High-amplitude tremor persisted for approximately 4 min, then rapidly waned in amplitude. No visual or weather-radar observation.
	15	0555	3	Surge-producing eruption.	12	Abrupt onset of high-amplitude tremor; correlative atmospheric-pressure pulse. Explosion followed a progressively intensifying 2-h swarm of LP events. Explosion signal persisted for approximately 3 min, then rapidly waned to low-amplitude tremor.

Table 1. Chronology of eruptive activity at Mount Pinatubo during June and July 1991—Continued.

Phase	Date (June 1991)	Onset time	Seismic duration (min)	Event	Plume height (km)	Comment
						Weather radar showed maximum tephra-plume height of 12 km, and observers to the north, west, and east saw pyroclastic density currents coursing down the volcano flanks. Clark weather-radar observations were discontinued after this event. Ash mixed with rain, transported by winds related to approaching Typhoon Yunya, began falling about 0630 at Clark Air Base; tephra fall continued throughout day, becoming heavier and obscuring daylight about 30–40 min after each surge-producing eruption.
	15	0629	56	Intense tremor.		Episode of sustained high-amplitude tremor from 0629 through 0725, then abrupt decline to extremely low, background-level tremor, which continued until 0731, when signal was lost. No evidence available about whether or not eruptive activity occurred during episode of high-amplitude tremor.
	15	0810	?	Surge-producing eruption.	12?	Gap in seismic data. Explosion documented by atmospheric-pressure pulse, direct observation from Poonbato and Camp O'Donnell, and a video record from Camp O'Donnell. Cubi Point weather-radar observer's notes show tephra-plume height of 12 km at 0824.
	15	1027	4–14	Surge-producing eruption.	15?	Rapid onset of high-amplitude tremor; correlative atmospheric-pressure pulse; direct observation from Poonbato. High-amplitude tremor sustained for

Table 1. Chronology of eruptive activity at Mount Pinatubo during June and July 1991—Continued.

Phase	Date (June 1991)	Onset time	Seismic duration (min)	Event	Plume height (km)	Comment
						about 4 min, then waxed and waned before returning to low background level at about 1041. Cubi Point weather-radar observer's notes show tephra plume to 15 km altitude at 1041.
	15	1117	4-13	Surge-producing eruption.	?	Burst of high-amplitude tremor, with correlative atmospheric-pressure pulse, following approximately 15 min of elevated LP background seismicity. High-amplitude signal sustained for approximately 4 min, followed by 3 min of low-amplitude tremor and 6 min of tremor of variable amplitude that waned to low background level at approximately 1130. Visual observation from north flank (Camp O'Donnell) and Poonbato. Cubi Point weather radar temporarily inoperative.
	15	1158	9	Surge-producing eruption.	8?	Burst of high-amplitude tremor following approximately 15 min of elevated LP background seismicity; correlative atmospheric-pressure pulse. High-amplitude tremor sustained for 2-3 min; then tremor waxed and waned before returning to low background level at approximately 1207. Cubi Point weather-radar observer's notes show tephra plume to 8 km altitude at 1213.
	15	1222	10	Surge-producing eruption.	?	Abrupt onset of high-amplitude tremor; correlative atmospheric-pressure pulse. Cubi Point weather-radar notes ambiguous: at 1225, tephra column, if any, indistinct from tephra already in the atmosphere; at 1250 tephra reported to 8 km

Table 1. Chronology of eruptive activity at Mount Pinatubo during June and July 1991—Continued.

Phase	Date (June 1991)	Onset time	Seismic duration (min)	Event	Plume height (km)	Comment
						altitude.
	15	1252	4	Surge-producing eruption.	?	Rapid onset of high-amplitude tremor; correlative atmospheric-pressure pulse.
	15	1315	13	Surge-producing eruption.	?	Onset of high-amplitude tremor; correlative atmospheric pressure pulse. High-amplitude tremor waned to low to moderate background level at end of event.
VI	15	1342	~9 h	Climactic eruption. Caldera collapse.	34?	Abrupt onset of high-amplitude tremor; correlative atmospheric-pressure pulse. High-amplitude tremor sustained through at least 2015, when seismic system failed. Profound atmospheric-pressure disturbance initiated by this event, diminished to background by approximately 2230. LP (0.5-1.5 Hz) component of seismic signal and local atmospheric-pressure disturbance (fig. 2) both decreased markedly in amplitude about 3 h (approx. 1630) after onset, as did infrasonic signal at distant sations (Tahira and others, this volume). Maximum eruption-column height estimated at approximately 34 km from shadow measurements in satellite image recorded at 1540 (Koyaguchi and Tokuno, 1993). Pumice lapilli noted in falling tephra at Clark Air Base at 1428 and at Cubi Point at 1445 (could have begun earlier in each case). Reports of large earthquakes by 1625. SSAM data (Power and others, this volume) show that high-frequency seismicity (at all recorded frequencies >

Table 1. Chronology of eruptive activity at Mount Pinatubo during June and July 1991—Continued.

Phase	Date (June 1991)	Onset time	Seismic duration (min)	Event	Plume height (km)	Comment
						1.5 Hz) began to increase at approximately 1530 and reached a plateau by 1630 that was sustained at least until the seismic system failed at 2015. These data, combined with the onset of large felt earthquakes, suggest that profound structural adjustment of the volcanic edifice, including foundering of its summit, in response to rapid extraction of magma, began at about 1530, was in full swing by 1630, and was undiminished through at least 2015. Restored seismic recording at approximately midnight showed diminished but still intense seismicity, but the dominant events thereafter were volcano-tectonic earthquakes instead of the LP events and tremor that had dominated since June 12.
VII	15	~2230	~6 wk	Sustained tephra emission; continued but diminishing structural adjustment.		Emission of tephra plume from vents within the newly formed caldera continued until late July with gradually diminishing vigor. Volcano-tectonic earthquakes continued through and beyond this period, with progressively diminishing abundance (Mori, White, and others, this volume).

Figure 3. Steam and ash emission in the uppermost Maraunot valley on May 30, 1991. (View is approximately to the southeast.) Arrow shows approximate site where a lava dome appeared on June 7. (Unless noted otherwise, photographs by R.P. Hoblitt.)

proportion of ash (fig. 3). This change was gradual and variable; periods of increased ash emission alternated with periods of steam emission. Prevailing winds carried ash over the South China Sea, about 33 km to the west. By May 30 we were able to collect ash adhering to vegetation a few kilometers west of the volcano. On May 28, SO_2 emissions reached a peak of 5,000 t/d and then decreased to a minimum of a few hundred tons per day on June 5; on June 7 the rate was about 1,000 t/d (Daag, Tubianosa, and others, this volume).

Seismicity began to change in early June. Previously, most earthquakes were located about 5 km northwest of Pinatubo's summit, at depths of about 3 to 7 km. A lesser number were located just north of the summit, beneath the fumaroles, at a depth of about 1 to 3 km (Harlow and others, this volume). Episodes of low-amplitude volcanic tremor occurred sporadically; RSAM values—used to monitor the rate at which seismic energy is released (Endo and Murray, 1991)—were roughly constant. Beginning on June 1, however, RSAM values began to increase and exhibited greater variation. Tremor episodes became more frequent and more intense, and on June 6 the locus of earthquake activity shifted away from the region to the northwest. At about 0700, seismicity increased rapidly beneath the fumaroles just north of the summit; many of these earthquakes were quite shallow. This period of intense seismicity lasted until 2300 on June 7, when it abruptly diminished (Harlow and others, this volume).

A tiltmeter located on the east flank of Mount Pinatubo detected accelerating inflation on June 6, at about the same time as intense seismicity began beneath the fumaroles (Ewert and others, this volume). This inflationary event ended abruptly at 1700 on June 7, when a sudden gas and

ash emission produced a plume whose top reached about 4 km over the volcano.

PHASE III CHRONOLOGY

The emergence of a lava dome provided the first unequivocal evidence that Pinatubo's activity was driven by rising magma. The inception of dome growth was not witnessed directly. The incipient dome was first sighted at about 0800 on June 8; it was not present when Pinatubo was inspected the previous morning, a time when viewing conditions were good. We conclude that dome growth began late on June 7, probably between 1700, when inflationary tilt ceased, and 2300, when intense seismicity beneath the fumaroles rapidly declined.

Details concerning dome growth are meager because heavy ash emission hindered observation during helicopter inspections. When first sighted (fig. 4), the dome clung to the east wall of the upper Maraunot River canyon. The emergence of the dome at this location (fig. 3) was unexpected because the closest previous activity was a weak fumarole about 100 m to the east. Several vigorous fumaroles were in the general area, the most vigorous of which was about 300 m to the south, in the bed of the uppermost Maraunot River (fig. 3).

Ash emissions, most of which appeared to originate at the east margin of the dome, waxed and waned. In general, however, periods of elevated ash emission were becoming more frequent, and the emission rate during these periods appeared to be increasing. On June 9, ash emission was at times sufficiently heavy to produce ash curtains (fig. 5). During the morning and afternoon of the same day, observers on the west side of Pinatubo (Sabit and others, this

Figure 4. The dome as it was first sighted on June 8, 1991, on the east side of the upper Maraunot valley. (View is approximately to the south.) (Photograph by Valentino M. Gempis, U.S. Air Force.)

Figure 5. Curtains of falling ash on June 9, 1991. (View is approximately to the west.)

volume) reported "pyroclastic flows" moving down the west and northwest flanks of Pinatubo. Subsequent helicopter inspections showed these areas to be dusted with ash but devoid of the destroyed vegetation or deposits typical of pyroclastic flows or surges. Apparently the ash content of the plume occasionally reached concentrations sufficient to produce density currents whose densities marginally exceeded that of the ambient air.

The apparent increase in the rate of ash emission was confirmed by the C-band weather radar at Clark. On June 10, a visually impressive ash cloud was not detected by the radar, but at 1040 on June 11, the ash concentration exceeded the detection threshold.

SO_2 emissions, which began to increase just before the appearance of the dome, continued to increase (Daag, Tubianosa, and others, this volume). The flux rose from about 1,500 t/d on June 8 to >10,000 t/d on June 10, the last SO_2 measurement before the climactic eruption.

Dome growth continued at least until 0700 on June 11—the last inspection during Phase III (fig. 6). Sketch maps of the dome's perimeter on June 8, 9, and 11 (fig. 7A–C) document the growth. Our estimates of the mean dome diameter and height are 200 m and 40±10 m, respectively, on June 11, for a volume of about $1–1.5 \times 10^6$ m^3. We are uncertain whether extrusion was continuous or episodic. However, episodic seismicity during the period of

Figure 6. The dome (indicated by arrow) on June 11, 1991, the last view before the June 12 eruption. The dome's west margin has reached the bottom of Maraunot valley. (View is approximately to the west.)

dome growth (Harlow and others, this volume; Power and others, this volume) suggests that dome growth may have also been episodic.

Once magma reached the surface, seismicity diminished to a low level until the evening of June 8. Seismicity through the balance of Phase III—that is, until June 12—was characterized by swarms of volcano-tectonic earthquakes and gradually increasing tremor, which became continuous on June 10 (Harlow and others, this volume; Power and others, this volume). Overall, the seismic energy release was escalating.

The dome was not sampled directly, because of safety considerations. Samples obtained after the climactic eruption show the dome lava to be andesite, produced by mixing basalt and dacite magmas (Pallister and others, this volume).

PHASE IV CHRONOLOGY

A small explosion at 0341 on June 12 initiated Phase IV. The event was detected seismically at Clark (Harlow and others, this volume) and visually from Poonbato (fig. 1), on the west side of Pinatubo (Sabit and others, this volume). At about 0400, Clark radar detected a cloud over Pinatubo at an altitude of about 11 to 12 km; however, the presence of large thunderheads in the vicinity of Pinatubo at this time makes interpretation of the radar observation ambiguous. At Clark, dawn (~0500) revealed a white plume rising from the volcano. Radar reflections at 0540 were described as the strongest yet seen; they placed the plume top at an altitude of about 3.5 to 4.5 km. The vigor of the plume declined, and by 0630 it was apparent that the plume originated at two closely spaced vents. During a helicopter

inspection about 30 min later, one source was seen to be the east margin of the dome, while the other was at the site of the most vigorous predome fumarole, in the bed of the uppermost Maraunot River (fig. 8). The east and north flanks of the volcano were heavily dusted with ash, at least some of which must have been transported as a weak density current, because the boundary between ash-covered and unaffected terrain was sharp to the east (fig. 9). No unequivocal pyroclastic-flow deposits were recognized in the drainage that originated on the east flank (Sacobia); however, the upper parts of the O'Donnell and Maraunot drainages contained minor pyroclastic-flow deposits that gave way downstream to hot lahar deposits (fig. 10). By 0730, the vigor of Pinatubo's plume had declined further, and the ash content appeared to be low, but ash emission still precluded an inspection of the dome.

The first large vertical eruption began at 0851 on June 12. As seen from Clark, the developing plume was rather broad, with two main protuberances, perhaps a reflection of the two vents noted just before the eruption. Initially, the northern protuberance was dominant, but it soon became subordinate to the southern one (fig. 11). The plume top eventually exceeded the height range of the Cubi Point radar (19 km); Clark radar operators were evacuated from their station when the plume top was at 19 km and still rising. As the eruption continued, a circular collar expanded around the plume at the tropopause (fig. 12; Oswalt and others, this volume).

Our reconnaissance helicopter reached a good vantage point to the north of Pinatubo at about 0945. The course of the upper Maraunot River was marked by diffuse ash clouds that were punctuated in a few places by roiling clouds; apparently, secondary explosions in the river were feeding

Figure 7. Sketches of the dome's perimeter. *A*, June 8. *B*, June 9. *C*, June 11. *D*, June 14. Vertical eruptions of June 12 and 13, from a vent near the dome's southeast margin, removed part of the dome and produced the crescent shape seen on June 14. Contour interval is 20 m.

ash into the adjacent diffuse clouds. At about 0950, a new density current began to move to the north, down the O'Donnell River (fig. 13), and probably down the Maraunot as well. This current dissipated by 1005, after moving down the O'Donnell about 4 km from the vent, and an unknown distance down the Maraunot. We left the area about 1010 to refuel and returned at about 1040. Activity at the vent had now declined to a low level. Secondary explosions in the Maraunot (fig. 14) suggested that pyroclastic flows reached at least 4 km from the vent in that drainage; the explosions and associated ash clouds prevented direct observation of the flow deposits. Explosions were **not** occurring in the O'Donnell. Pyroclastic-flow deposits were absent there, as was any other evidence of their passage. The upper few kilometers of that drainage were, however, heavily dusted with ash, the only apparent product of the density current we witnessed during the previous trip. The density current must have been a slack surge. Upon inspecting other drainages it became apparent that only the Maraunot had received significant pyroclastic-flow deposits during the

eruption. The axis of the tephra deposit crossed the southwest flank of the volcano.

When the 0851 eruption began, high-amplitude eruption signals abruptly ended a period of relative seismic quiescence that began about 3.5 h earlier. The seismically intense part of the eruption lasted 36 min; however, as noted above, visible activity at the vent persisted until at least 1000. In general, the level of seismicity after the eruption was substantially greater than before the eruption (fig. 2A), largely due to the appearance of energetic long-period events (Harlow and others; Power and others, this volume). A swarm of long-period earthquakes began about 1725 and lasted about 50 min.

The second vertical eruption began at 2252, about 14 h after the first. On the basis of seismicity, it lasted about 14 min. Like the first eruption, it began abruptly—ending 4.5 h of relative seismic quiescence (fig. 2A). Clark radar indicated the eruption-column height exceeded 24 km. Tephra was again dispersed to the southwest, this time in the midst of scattered thunderstorms. Direct observations

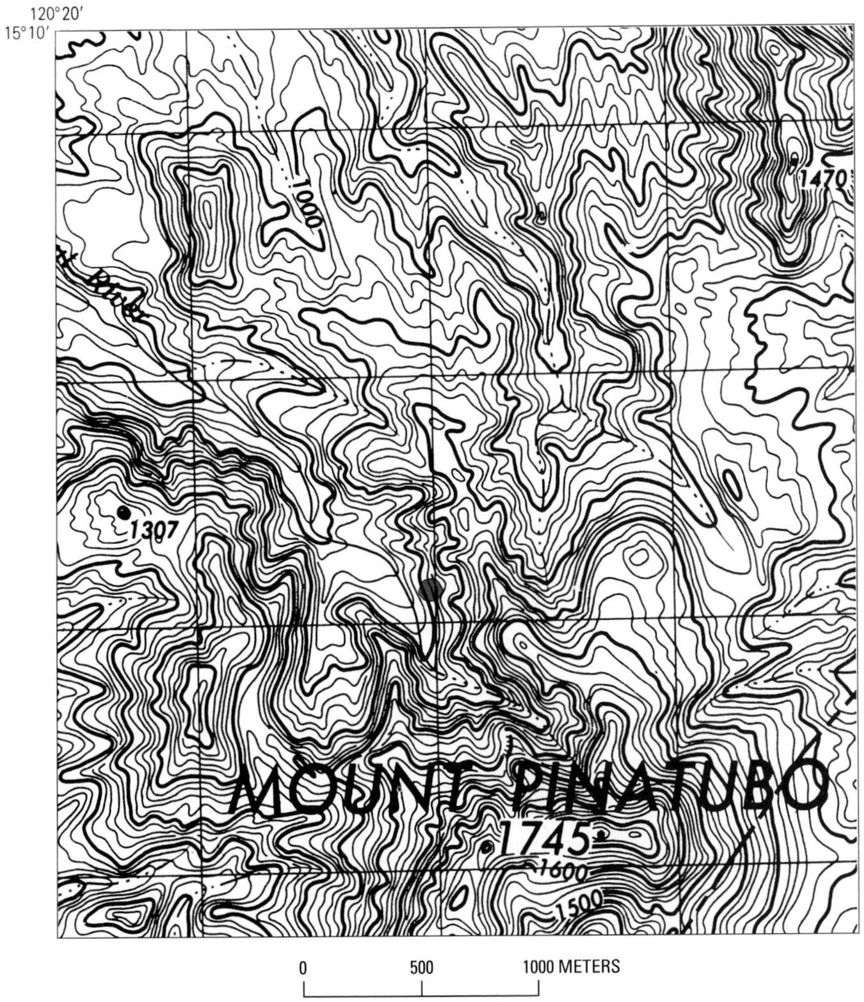

Figure 7.—Continued.

were scanty because of darkness. Abundant lightning was witnessed in the vicinity of Pinatubo, from Clark and from Camp O'Donnell, about 30 km to the north, but at least some this was probably due to the thunderstorms rather than to the eruption. Another swarm of increasingly energetic long-period events began about 0600 the following morning (June 13) and lasted roughly 2 h. We were unable to inspect the volcano before the next eruption began.

A third vertical eruption began in clear weather at 0841 on June 13 (fig. 15) about 10 h after its predecessor. It lasted only about 5 min, on the basis of seismicity. The onset of the eruption was again abrupt, but the period of relative quiescence that preceded it was only about 15 min, much less than those of the first two eruptions (fig. 2A). Clark radar indicated the column height again exceeded 24 km, although the radar reflections were weaker than those of the second eruption. As before, tephra was dispersed to the southwest.

An inspection by helicopter at about 1000 was hindered by a thick haze that hung over Pinatubo. However, we did observe a small pyroclastic-flow deposit (fig. 16) in a ravine in the uppermost Sacobia drainage. It is uncertain whether this flow was the product of the second or third eruption. In any case, it was apparently the first to be emplaced on the east flank of the volcano.

About 3 h of relative seismic quiescence followed the third eruption. Then, at about noon on June 13, a swarm of small long-period events began, occasionally punctuated with larger events. The intensity of the swarm increased gradually until about 0400 on June 14, when the intensity began to increase rapidly. Dawn brought excellent viewing conditions and revealed that, remarkably, steam and ash emissions from Pinatubo were essentially nil (fig. 17). Convinced that an eruption was imminent, we postponed aerial inspection. But the intense seismicity persisted hour after hour, and viewing conditions from Clark slowly deteriorated. Concerned that approaching Typhoon Yunya might prevent observations for an extended period of time, we began to cautiously inspect the volcano at about 1215.

Figure 7.—Continued.

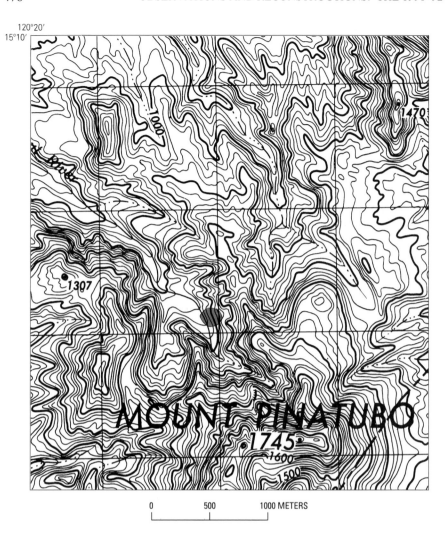

0   500   1000 METERS

The west margin of the dome now nearly spanned the narrow Maraunot River canyon (fig. 18). The vent—dimly visible through ash—had taken a semicircular bite out of the southeast margin of the dome (fig. 7D). The vent diameter was approximately 200 m. New pyroclastic-flow deposits extended down the Maraunot River valley as much as 4.5 km from the vent (figs. 18, 19); the O'Donnell may have received small pyroclastic flows, but, if so, they extended less than 1.5 km north of the vent. The absence of substantial pyroclastic-flow deposits in the O'Donnell drainage is somewhat surprising, because only a low interfluve separated the uppermost O'Donnell drainage from the vent. Clouds prevented us from inspecting the south and west flanks; however, the dearth of pyroclastic-flow deposits in the O'Donnell suggests that the south and west drainages were also little affected by pyroclastic flows.

The swarm of long-period events culminated in the fourth vertical eruption at 1309 (fig. 20). This eruption, in contrast to the previous three, was not preceded by a period of relative seismic quiescence. The eruption lasted only 3

min, then seismicity dropped to low-amplitude tremor. Radar at Clark indicated a column height of 21 km; tephra was again carried to the southwest. The relatively low plume height supports our subjective impression: the 1309 eruption was the least vigorous of the four. This was unexpected, as we mistakenly assumed that the intense precursory seismicity would culminate in a large eruption.

PHASE V CHRONOLOGY

The pace of eruptive activity increased after the 1309 June 14 eruption, just as viewing conditions deteriorated with the approach of Typhoon Yunya. Opportunities for direct observation became progressively less frequent, particularly from Clark, so we became more dependent on instrumental data. We emphasize here the scanty direct observations, and refer the reader to table 1 for some instrumental observations not described in the text.

Following the 1309 eruption, we flew to a site 15.5 km north of Pinatubo in an attempt to repair seismic-

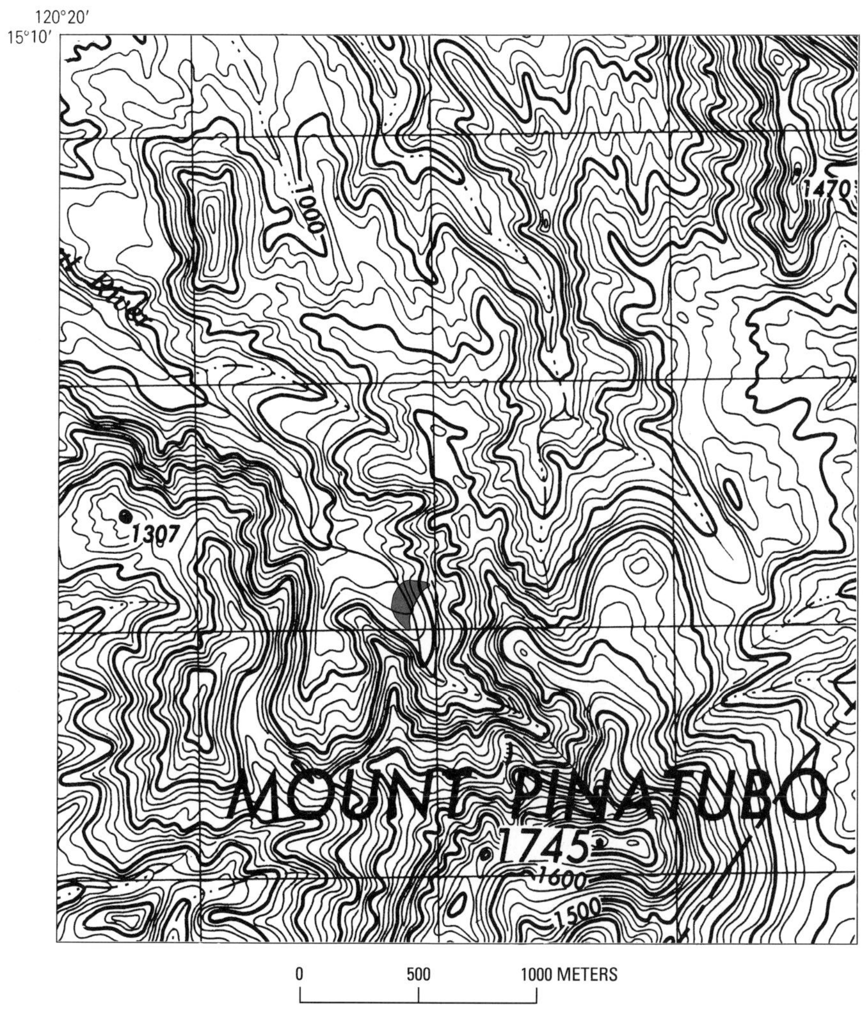

Figure 7.—Continued.

telemetry gear and arrived at about 1410. Viewing conditions were poor; however, it was clear that the volcano was feeding copious quantities of ash into a broad plume, and ash clouds were rising from multiple sources in the vicinity of the Maraunot and O'Donnell drainages. Other drainages were obscured. Apparently, small explosions or low fountaining generated density currents from which the ash clouds were rising. Radar at Clark detected the ash, which reached a peak altitude of about 15 km at 1411. Winds from Typhoon Yunya carried the ash eastward, along with·a thunderstorm that grew amidst the ash. This was to be the first ash to reach Clark, where it fell mixed with rain during the afternoon. Unbeknownst to us, this inspection was to be our last before the climactic eruption. We left the site at about 1435.

This event could be considered either a vestige of the 1309 eruption, a small discrete eruption, or simply a precursor to the larger eruption that occurred a short time later. The start time is uncertain, though it may have been at about 1400, when a modest intensification in the low-amplitude

tremor occurred. Seismicity provides no basis upon which to estimate a duration.

A swarm of long-period events began at 1438 and quickly intensified. At about 1516 a large pyroclastic density current swept out radially from Pinatubo and reached a distance of about 15 km to the west-northwest. Observers to the west had to flee. This behavior differs markedly from that of the Phase IV eruptions, which produced high, relatively narrow eruption columns along with a few channelized density currents of limited extent. The event was not visible from Clark because of the ashy thunderstorm that developed earlier, and, probably for the same reason, Clark radar did not recognize the event as an eruption. Cubi Point radar, however, reported a tephra cloud between 1530 and 1600, moving southwestward, with a top at 18 km. Hot lahars that moved down the Sacobia River past Clark that evening were almost certainly triggered by pyroclastic debris deposited into the Sacobia drainage by this eruption.

The 1516 eruption produced a sizable deflection on the Clark barograph (fig. 2C). The barograph deflections are

Figure 8. Emissions from Mount Pinatubo at about 0700 on June 12. Plume on the left is from the strong, persistent vent to the south of the dome (compare fig. 3); diffuse plume on the right is from the dome. (View is approximately to the south.)

Figure 9. The east flank of Mount Pinatubo, in the upper drainage of the Sacobia River, showing a margin of the area affected by a weak, ash-laden density current, probably generated by the 0341 explosion on June 12. Photograph taken shortly before 0700.

peculiar to the Phase V eruptions; the vertical eruptions of Phase IV did not produce them. At least 13 barograph deflections were recorded during Phase V, and at least five of the eruptions that produced them were witnessed.

The second large Phase V eruption began at 1853 and produced a plume whose height exceeded 24 km. The eruption was not observed directly. The next two eruptions, which began at 2320 on June 14 and 0115 on June 15, were observed from Clark with the aid of an infrared imaging device. The infrared images show ground-hugging density currents moving north, south, and east; density currents presumably also moved westward.

At 0555 on June 15, just after dawn, another eruption produced a radial pyroclastic density current. This eruption was clearly visible from Clark (fig. 21). The density current was driven by a pyroclastic fountain whose projections were briefly visible at the start of the eruption but were soon concealed by ash clouds rising convectively from the density current. This fountain resembled those observed during the postclimactic eruptions of Mount St. Helens in 1980 (Hoblitt, 1986). Ash clouds from throughout the affected area merged into a great cloud that rose over the volcano; briefly, the periphery of this cloud was laced with orange-red lightning. As the cloud rose, surface winds swept up ash from the newly emplaced deposits and carried it back toward the volcano. Radar indicated a maximum plume height of 12 km. A diffuse, light-gray, ground-hugging cloud then formed in the vicinity of Pinatubo and seemingly moved out radially (fig. 22); it certainly moved eastward, because it arrived at Clark about 40 min after the eruption

Figure 10. Hot lahar deposit in the bed of the Maraunot River at about 0700 on June 12. Note steam rising from hot boulders.

Figure 11. Plume from the eruption that began at 0851 on June 12. (View is to the west from Clark Air Base at 0900.) The plume protuberance to the right (north) was initially dominant but was overtaken by the protuberance to the left.

began. Ash mixed with rain began to fall as the cloud arrived. Pinatubo was not seen again from Clark until after the climactic eruption.

For a time, Pinatubo remained visible from the north and west. Another density current was witnessed at about 0810 on June 15 from both Camp O'Donnell and Poonbato (fig. 1); yet another began at 1027 and was witnessed from Poonbato. Poonbato observers regarded the 1027 density current as the largest that they witnessed—marginally larger than the 1516 event of June 14. The last density current to be witnessed began at 1117 on June 15. It was dimly visible from Poonbato and Camp O'Donnell. Subsequent density currents were only detected instrumentally and confirmed at Clark by periods of nearly complete darkness beginning about 35 to 40 min after each eruption; darkness was due to the arrival of wind-drifted ash. Subsequent stratigraphic studies, described below, demonstrate that the Phase V density currents were pyroclastic surges. Accordingly, we refer to the Phase V events as "surge-producing" eruptions.

STRATIGRAPHY OF PHASE IV AND PHASE V DEPOSITS

Some reconnaissance work was conducted in late June and early July of 1991. The new deposits were nearly undissected, so outcrops were available only in the thinner upland or distal deposits. Because of this and safety considerations, most early work was restricted to brief examinations of thin distal deposits.

The majority of fieldwork was conducted in February and March of 1992. By this time, deposits had been subject to one rainy season (July–October) since their emplacement. Access to the proximal deposits was difficult because all access roads were damaged or destroyed by lahars

Figure 12. Plume from the eruption that began at 0851 on June 12. (View is from the north at 0945.) Note the circular collar (indicated by arrow) that developed around the plume where it penetrated the tropopause.

Figure 13. Pyroclastic density current moving to the north of Mount Pinatubo, down the O'Donnell River (center) at about 0950 on June 12. Current probably also moved down the Maraunot River (ash-covered, right). (View is from the north.)

generated by erosion of the fresh deposits during the rainy season. The termini of pyroclastic-flow deposits could be reached by road in the Sacobia, Pasig, and O'Donnell drainages, but further travel on foot was impeded by a remarkably well-developed dendritic erosion pattern. The new deposits had been sculpted into a badlands topography of steep-sided rills and canyons that provided excellent exposures but was exceedingly difficult to traverse on foot. The majority of fieldwork on proximal deposits was conducted during 7 days when helicopter support was available. In an attempt to locate sections containing deposits from as many of the eruptions as possible, we focused on the region to the

southwest of the volcano, where tephras from the June 12–14 eruptions were deposited. Areas of interest were identified on aerial photographs, but the actual study sites were selected from the air, as the helicopter circled the area. Once on the ground, because of restricted mobility and limited time, we could explore only a rather small area. Our results are, therefore, only preliminary.

All proximal sites that we examined lie within the area affected by pyroclastic flows from the climactic eruption. These flows extensively scoured the surfaces over which they passed: consequently, the Phase IV fall deposits and the overlying Phase V surge deposits were found only as

Figure 14. Secondary explosions in the Maraunot River at 1140 on June 12. (View is to the southwest.)

Figure 15. Plume from the eruption that began at 0841 on June 13. (View is to the west from Dau on Clark Air Base at 0848.)

isolated erosional remnants beneath the deposits of the climactic eruption (Phase VI). Such remnants were preserved in topographically protected areas, such as depressions or in the lee of obstacles.

PROCEDURES FOR DEPOSIT ANALYSIS

Andesite, dacite, and accessory component proportions were determined from >4-, >8-, or >16-mm clast fractions by hand sorting them with the aid of a binocular microscope. Andesite and dacite were identified by the presence and absence of clinopyroxene.

Clasts selected for density analysis were washed in water, dried overnight at 110 °C, sprayed with an aerosol containing a trace amount of silicon oil, dried overnight at 110 °C, weighed in air, and weighed in water. Bulk volume was calculated from the weight difference by using Archimedes' principle.

COMPOSITE STRATIGRAPHIC SECTION OF PHASE IV AND PHASE V DEPOSITS

Because deposits were examined and sampled at numerous localities, it is convenient to refer to a composite stratigraphic section (fig. 23). This section also provides the rationale for the deposit sequences used in plots of clast density and component proportions. We include Phase VI fall and flow deposits; these are discussed in detail by W.E. Scott and others (this volume), and Paladio-Melosantos and others (this volume).

Figure 16. The margin of an area affected by a pyroclastic density current produced by the 2252 eruption of June 12 or the 0841 eruption of June 13; photograph taken at 1006 on June 13. (View is to the west at the east flank of Mount Pinatubo, in the uppermost Sacobia drainage.) Smoke rises from vegetation incorporated into a hot deposit in the bottom of the ravine.

Figure 17. Dearth of steam and ash emission from Mount Pinatubo on the morning of June 14, during a period of intense seismicity. (View is from Clark Air Base to the west at 0722.)

PYROCLASTIC-FLOW AND PYROCLASTIC-FALL DEPOSITS OF PHASE IV

The best example of the Phase IV fall deposits was found at site 22 (figs. 23–26), where their total thickness reached 14 cm. Assignment of eruption times to the various beds was aided by the fact that the second eruption of June 12 (beginning at 2251) occurred during a rainstorm. Cohesion of moist particles (rainflushing) produced a fall deposit that is finer grained and more poorly sorted than those above and below it (table 2). The section was sampled for granulometric, component, and density analysis as indicated on figure 26; note that the thickness of the beds decreases upwards and that the 1309 June 14 eruption is

represented only by a thin ash bed, which was not sampled. The sampled fall deposits are dominantly andesite, although the proportion of dacite increases upward (figs. 27 and 28).

Pyroclastic-flow deposits emplaced during the Phase IV vertical eruptions were examined at one site in 1992 (fig. 24, site 24; fig. 23), in the canyon of the Maraunot River. This site received density currents—because the vent was located in the uppermost Maraunot drainage—but stood north of the area that received fall deposits. The lower part of the Phase IV section consists of a thin (10–30 cm) pyroclastic-flow deposit sandwiched between thin ash beds: two ash beds below, three above (fig. 29A). The pyroclastic-flow deposit was almost certainly emplaced sometime

Figure 18. The dome (center) at 1220 on June 14, as seen from the northwest, up the Maraunot River canyon. New pyroclastic-density-current deposits (lower center, light-colored) sit in the canyon bottom and mantle the adjacent terrain.

Figure 19. Newly emplaced pyroclastic-flow deposits partially filling the canyon of the Maraunot River about 5 km from the old summit (site 24, fig. 24). (View is to the southeast at 1221 on June 14.)

during the first (0851) June 12 eruption because, after the eruption, we observed secondary explosions in the Maraunot drainage that extended to the vicinity of site 24 (fig. 14). Component and density data also support correlation of this deposit with the 0851 event (figs. 27 and 28). We interpret the thin ash beds as the products of slack surges of the sort reported by west-side observers on June 9 (Sabit and others, this volume) and the sort we observed during the latter part of the 0851 eruption (fig. 13).

The upper part of the site 24 section is dominated by two thick (3–4 m) pyroclastic-flow deposits (fig. 29B). The

lower one is rich in dense, prismatically jointed blocks of andesite; these are interpreted as fragments of the dome that began to grow on June 7 (Pallister and others, this volume). This must be the pyroclastic-flow deposit photographed in the Maraunot on June 14 (fig. 19); it must have been emplaced during the second (2252) eruption of June 12 or during the June 13 eruption (0841).

The uppermost pyroclastic-flow deposit at site 24 consists of normally graded lithic debris in a pumiceous matrix. Similar deposits are found in all of Pinatubo's major drainages, always at the top of the section produced during the

Figure 20. Plume from the eruption that began at 1309 on June 14; (View is to the west from the southwest end of the airfield at Clark Air Base at 1324.)

climactic eruption (Phase VI). They were probably produced during caldera collapse (W.E. Scott and others, this volume). Numerous preclimactic events were not recorded at this proximal site (fig. 23), probably because of the erosive nature of the Phase VI pyroclastic flows.

PROXIMAL SURGE DEPOSITS OF PHASE V

We found remnants of the Phase V deposits at six (sites 11, 17, 20, 21, 22, 23; fig. 24) of the eight proximal sites that we visited. Phase V deposits (figs. 23, 30A,B) are composed of multiple surge beds, each of which is readily divisible into a light-gray relatively coarse-grained lower layer and a gray, brown, or pink-brown relatively fine-grained upper layer. The fine-grained upper layers resemble the distal Phase V beds that are described below. Individual beds (coarse layer + fine layer) attain a thickness of as much as a few decimeters; the total thickness of Phase V deposits ranged up to about 1 m at our most proximal site (fig. 24, site 11; fig. 30B). In the thicker beds, the coarse-grained

layer consists of a massive, matrix-supported facies that may locally grade upward into a finer grained stratified facies or may grade downward to a friable, fines-deficient, grain-supported facies. All three of the coarse-grained facies were present in a few well-developed examples. Thicker beds exhibit low-angle crossbedding, dune structures, pinch-and-swell structures, or plane-parallel bedding. The coarse-grained layer is in sharp contact with the overlying fine-grained layer, which typically contains accretionary lapilli. Thinner beds consist only of the fine-grained layer or of the stratified facies of the coarse layer and the fine-grained layer. The lower few beds typically contain organic debris, some of which is carbonized. In most sections we examined, the lowermost surge bed is thicker and coarser than those that overlie it. In the most complete sections, the thickness and coarseness of the sequence tend to decrease upwards.

In the exposures we examined, the top of the Phase V sequence almost always consists of one of the fine-grained layers; this is overlain by climactic (Phase VI) pyroclastic-flow deposits. As the contact between the two phases is traced laterally, it typically jumps to a stratigraphically lower or higher fine-grained layer. This jump implies that surge beds were locally scoured away by the climactic pyroclastic flows. Because of this nonconformable contact, we are uncertain whether any of the proximal surge sections we examined are complete. The maximum number of depositional events estimated for any proximal section is 11; 13 surge-producing eruptions were inferred from the barograph record (fig. 2C).

DISTAL SURGE DEPOSITS OF PHASE V

Distal surge deposits typically consist of multiple beds of gray, pinkish-brown, and brown silt-sized ash (fig. 24, site 5; fig. 31). Most bedding is planar in the distal sections, though crossbedding, normal grading, and cut-and-fill structures are locally present in individual beds as much as 14.5 km from the old summit. Individual bed thicknesses range up to a few centimeters. Some beds contain accretionary lapilli, and, locally, some of the lower beds contain uncharred plant debris, dominantly leaf fragments (fig. 31B). Void spaces are locally common in some beds; some of these have irregular margins and constitute the interstices between accretionary lapilli (fig. 31C). Other voids have rounded margins and occur within otherwise structureless ash (fig. 31C). Some contacts are sharp, while others are transitional; thus, it is difficult to establish the exact number of depositional events represented in a given section. Some beds are rather uniform throughout and are delineated by sharp contacts; each of these probably represents a single depositional event. Other beds are graded, either normally or inversely, with respect to grain size or the abundance of accretionary lapilli. At least some of these graded beds are

Figure 21. Photographs of the pyroclastic surge that began at 0555 on June 15, 1991. *A*, 0558. *B*, 0601. (View is to the west from Dau complex on Clark Air Base.)

probably the products of more than one depositional event. Even with the difficulty in counting beds, the number of surge beds clearly varies from outcrop to outcrop but tends to increase with increasing proximity to the volcano. The maximum number of depositional events estimated in any distal section is 10.

To the southwest of Pinatubo, the surge deposits overlie the fall deposits from the June 12–14 vertical eruptions; in other sectors they lie directly on the preexisting surface. The distal surge deposits are everywhere overlain by fall deposits from the June 15 climactic eruption.

ORIGIN OF SPHERICAL VOIDS IN THE FINE-GRAINED SURGE FACIES OF PHASE V DEPOSITS

Spherical voids of the sort present in some of the fine-grained surge beds (fig. 31*B*,*C*) are commonly interpreted as evidence of deposition of hot, wet ash. This interpretation is warranted if there is also evidence of cohesion during deposition—for example, if the deposit is plastered onto vertical surfaces. In our experience, spherical voids can also form by postdepositional wetting of ash deposited at low temperatures in a dry, expanded condition. Vertical

Figure 22. A diffuse, light-gray, ground-hugging cloud approaching Clark at 0622 on June 15. The cloud formed in the vicinity of Mount Pinatubo after the 0555 pyroclastic surge and apparently moved out radially. (View is to the west from Dau on Clark Air Base.)

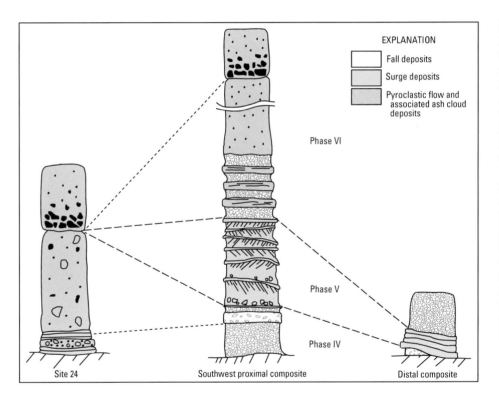

Figure 23. Diagrammatic stratigraphic sections illustrating relations between Phase IV, V and VI units sampled for this study. Site 24: section in Maraunot River canyon (see figs. 24, 29). Southwest proximal composite: Phase IV fall deposits from site 22 (see figs. 24, 25, 26); Phase V surge deposits generalized from sites 11, 17, 20, 21, 22, 23 (see figs. 24, 30); Phase VI fall deposits from site 17 (see W.E. Scott and others, this volume, fig. 19), Phase VI pyroclastic-flow deposits generalized for proximal drainages. Distal composite: generalized distal stratigraphy (see, for example, fig. 31), Phase IV fall deposits lie only to the southwest. Dashed line shows phase boundaries, dotted line shows intraphase correlations.

surfaces, such as tree trunks, were numerous at the distal sites that we visited, but they were not plastered with ash. Thus, voids like those in figure 31C probably formed when dry, powdery ash deposits were moistened after deposition. We are uncertain how the voids form but speculate that the passage of a saturation front through the loose, dry ash realigns particles into a close-packed configuration. The spherical voids reflect the change in interstitial volume as loose, dry ash is converted to compact, wet ash.

DISTRIBUTION OF PHASE V DEPOSITS

From their stratigraphic position, the coarse-grained surge deposits examined in the proximal area must be the products of pyroclastic fountains of the sort we witnessed during the 0555 eruption of June 15. The area swept by these energetic fountain-fed surges is only known approximately. Direct observations from the ground were limited because events occurred at night or in poor weather. For

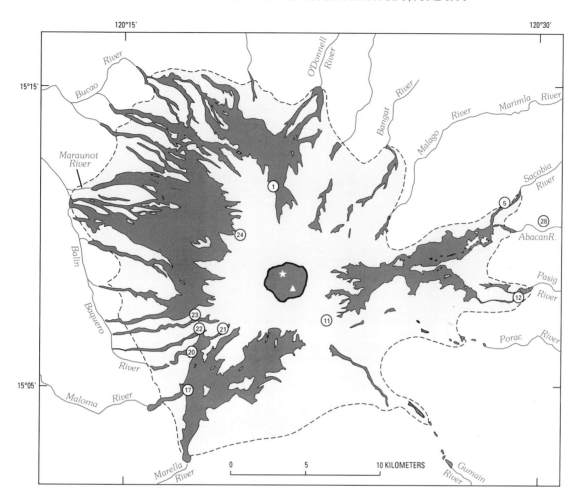

Figure 24. Locations of sites where Phase IV and V deposits were examined. Dashed line shows approximate margin of area affected by pyroclastic flows; dark-gray shading shows thick, ponded pyroclastic-flow deposits (see W.E. Scott and others, this volume). Heavy line outlines caldera; white triangle, old summit; white star, lava dome of June 7–15. Site numbers are the same as those in W.E Scott and others (this volume).

these same reasons, and because winds of Typhoon Yunya deposited ash on Clark Air Base, observations from aircraft were not possible. Were it not for the subsequent pyroclastic flows of the climactic eruption, the margin of devastated vegetation could be used to delineate the area. But in many sectors the pyroclastic flows were as extensive or more extensive than the surges. Photographs and eyewitness accounts provide a few constraints. Photographs taken by the first author at Dau (Clark), when compared with a videotape taken by Maj. Keith McGuire from Camp O'Donnell, indicate that the 0555 surge of June 15 reached a point about 10 km north of Pinatubo's former summit, in the vicinity of the headwaters of the Bucao River. This surge moved about 7 to 8 km to the east, down the Sacobia River valley to the vicinity of Mount Dorst. Its extent to the south was less, probably no more than 2 to 3 km from the old summit. A videotape of this event was taken from the north-

west by one of us (Dindo Javier). Although the travel distance cannot be determined from the tape because of the dearth of topographic markers, eye witnesses estimate that the surge reached within 1 to 2 km of Burgos, about 14 km from the old summit.

Less information is available for the other surges. Witnesses reported that the (first) surge at 1516 on June 14 reached about the same distance to the northwest as the 0555 event. As evidenced by blasted trees, the presence of surge deposits, and the absence of pyroclastic-flow deposits at a site about 1 km north of Mount Cuadrado, the surges collectively reached a maximum distance of about 10 km to the south of the old summit.

The fine-grained facies of the surge deposits, which is found in proximal as well as distal areas, is far more extensive than the coarse-grained facies. Its extent may be appreciated by inspecting an isopach map of tephra layer B

Figure 25. Deposits of June 12–16 at site 22. The Phase IV fall deposits lie directly on soil. The fall beds are overlain successively by surge deposits, fines-poor basal facies of the climactic pyroclastic-flow deposit, and the climactic pyroclastic-flow deposit. Length of ruler is 15 cm.

(Paladio-Melosantos and others, this volume; fig. 6). Layer B is defined as tephra deposited after the first June 12 eruption and before the climactic eruption of June 15; thus it contains contributions from both Phase IV and V. However, the contribution from Phase IV is only significant to the southwest of Pinatubo.

DISCUSSION

In this section we discuss topics relating to the origin of the Phase V surge deposits and to patterns evident in the preclimactic eruptions.

COMPARISON OF PHASE V DEPOSITS TO SIMILAR DEPOSITS AT OTHER VOLCANOES

The Pinatubo surge deposits are similar in many respects to deposits produced by the directed blast at Mount St. Helens, Wash., in 1980 (Hoblitt and others, 1981; Moore and Sisson, 1981; Fisher, 1990) and by the climactic eruption at Mount Lamington, Papua New Guinea, in 1951 (Taylor, 1958). All three deposits are the products of pyroclastic density currents, have low aspect ratios, generally become thinner and finer grained with distance from source, and, at least locally, exhibit features associated with surge deposits. Furthermore, all exhibit similar granulometries (fig. 32A).

There are also differences. The Pinatubo deposits were produced by numerous events separated by tens of minutes to hours—the Mount St. Helens and Mount Lamington examples by single events or, in the case of Mount St. Helens, perhaps by two events separated by tens of seconds to a few minutes. The Mount St. Helens event was laterally directed, the consequence of sector collapse exposing a cryptodome. The Mount Lamington and Pinatubo examples involved vertically directed eruption fountains that collapsed to form radially directed surges. The Mount St. Helens event devastated an azimuthally restricted area of about 600 km², Mount Lamington devastated a radially distributed area of about 230 km², and the Pinatubo events collectively devastated a radially distributed area of about 300 km². The source mechanism for the Pinatubo surges resembles that of Mount Lamington more than that of Mount St. Helens.

GENETIC RELATION BETWEEN DISTAL AND PROXIMAL FINE-GRAINED PHASE V DEPOSITS

Because of the similarity of their grain-size distributions (fig. 32B–D; table 2), the distal Phase V deposits appear to be equivalent to the upper fine-grained facies of the proximal Phase V beds. We know from direct observation that the distal beds were produced dominantly by subvertical fallout from slack surges. Because the distal beds and the fine-grained facies of the proximal beds are granulometrically indistinguishable, and because accretionary lapilli are common in both, we suggest that both were emplaced by the same depositional process. That is, both were emplaced by slack surges whose progenitors were the energetic fountain-fed surges that emplaced the lower coarse-grained facies of the proximal Phase V beds.

INTERPRETATION OF BAROMETRIC DATA

The Phase V surges caused atmospheric-pressure disturbances that were recorded on barographs at Clark Air Base (fig. 2C) and Cubi Point (Oswalt and others, this

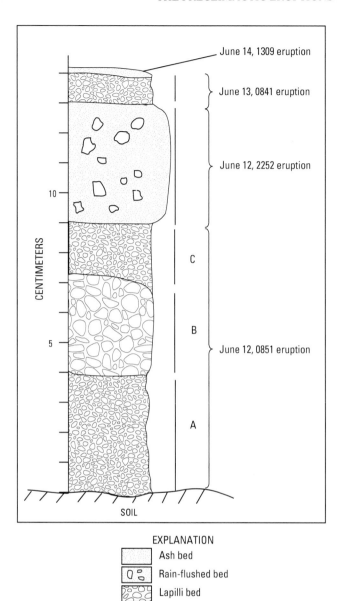

Figure 26. The various preclimactic fall deposits at site 22 (figs. 23, 24). Vertical lines show how units were sampled for component and density analysis; the deposit from the June 14 eruption was not sampled.

volume). Interestingly, the four vertical eruptions of Phase IV did not produce deflections. Most of the surges produced an impulsive compression, followed by a more sustained rarefaction, followed by a second even more sustained compression (fig. 2C). This same behavior was documented for the 1980 lateral blast at Mount St. Helens (Reed, 1980; Banister, 1984). The similarity of the Mount St. Helens and Pinatubo barographs rather strongly suggests that the same mechanisms were operating at both volcanoes. Banister (1984) attributed the initial compression at Mount St. Helens to the lateral explosion and attributed the following sustained rarefaction and compression to the subsequent plinian eruption.

In each of the Pinatubo examples, it seems likely that the initial compression was due to the formation of a fountain-fed pyroclastic surge. Two processes probably contributed to the transient atmospheric compression: (1) the sudden ejection and decompression of eruption products and (2) incorporation and expansion of air as the surge engulfed terrain around the vent. Because the barographs did not record pressure deflections during the Phase IV vertical eruptions, we conclude that the second process—incorporation and expansion of surface air—was more important than ejection and decompression. Expansion of incorporated air and deposition of suspended pyroclasts progressively lowered the density of the surge clouds until they became buoyant and lifted off the ground (Sparks and others, 1986). This great volume of ash-laden gas then rose diapirically. As the diapir rose, surrounding air rushed in to replace it; this probably caused the barometric rarefaction. It seems likely that the second barometric compression was due to the formation of diffuse density currents (slack surges) of the sort that reached Clark about 30 to 40 min after the start of surge-producing eruptions (fig. 22). We can only speculate on the process that produced the slack surges. Perhaps the buoyant mass of ash-laden gas behaves somewhat like a thunderstorm (Barry and Chorley, 1978); that is, it cools adiabatically until condensation occurs. Precipitation in the form of accretionary lapilli would produce a downdraft of relatively cool, ash-laden air that would flow outward as a density current (the slack surge), after it reached the ground. This would account for the abundance of accretionary lapilli in the slack surge deposits.

Whatever the process responsible for the slack surge, it apparently also operated at Mount St. Helens following the lateral blast of May 18, 1980. This is evident from photographs and eyewitness accounts presented by Foxworthy and Hill (1982, p. 56–57). We suggest that this slack surge, rather than the development of a vertical eruption column, was responsible for the second barometric compression observed in the Mount St. Helens example.

Table 2.—Grain-size parameters for deposits produced by the preclimactic eruptions.

[Measures used are those of Inman (1952) and Folk (1980. In ϕ units: Md_ϕ, median; M_Z, graphic mean; σ_ϕ, graphic standard deviation; σ_I, inclusive graphic standard deviation. Dimensionless: Sk_G, graphic skewness; Sk_I, inclusive graphic skewness; K_G, graphic kurtosis]

Site no.	Sample no.	Date emplaced	Deposit type[1]	Md_ϕ	M_Z	σ_ϕ	σ_I	Sk_G	Sk_I	K_G
Phase IV										
22	3-2-92-3-5	6/13, 0841	F	-0.16	-0.27	1.31	1.66	-0.12	0.01	1.66
	3-2-92-3-4	6/12, 2252	F	2.62	2.50	3.38	3.27	-0.05	-0.04	0.86
	3-2-92-3-3	6/12, 0851 C	F	-0.19	-0.18	1.43	1.58	0.01	0.10	1.36
	3-2-92-3-2	6/12, 0851 B	F	-3.02	-2.87	1.69	1.81	0.14	0.19	1.24
	3-2-92-3-1	6/12, 0851 A	F	0.13	0.09	0.85	0.91	-0.07	-0.04	1.06
24	2-28-92-1-1	6/13, 0841 ?	PF	-0.02	-0.45	3.34	3.06	-0.20	-0.16	0.81
	2-28-92-2-6	6/12-6/13	SS	3.83	3.63	1.26	1.57	-0.24	-0.02	1.42
	2-28-92-2-5	"	SS	3.89	4.17	2.30	2.34	0.18	0.19	0.95
	2-28-92-2-4	"	SS	4.46	4.61	2.03	2.24	0.11	0.10	1.22
	2-28-92-2-3	6/12, 0851	PF	0.83	1.05	2.57	2.55	0.13	0.17	0.93
	2-28-92-2-2	6/9-6/12	SS	3.84	4.05	1.82	1.92	0.18	0.22	1.13
	2-28-92-2-1	"	SS	3.88	4.05	2.13	2.21	0.12	0.14	1.04
Phase V										
5	2-21-92-1-6	6/14-6/15	SS	5.37	5.11	1.91	2.26	-0.20	-0.22	1.31
	2-21-92-1-5	"	SS	3.95	4.00	1.64	1.73	0.05	0.09	1.13
	2-21-92-1-4	"	SS	4.51	4.39	2.10	2.20	-0.08	-0.03	1.11
	2-21-92-1-3	"	SS	5.13	4.91	1.77	1.91	-0.18	-0.10	1.13
	2-21-92-1-2	"	SS	4.78	4.56	2.03	2.14	-0.16	-0.08	1.09
	2-21-92-1-1	"	SS	4.19	4.22	1.71	1.86	0.03	0.10	1.09
11	3-18-92-1-10	"	SS?	2.21	2.15	3.17	3.11	-0.03	0.02	1.06
	3-18-92-1-11	"	SS	4.24	4.92	2.26	2.39	0.45	0.30	0.97
	3-18-92-1-9	"	S	-0.36	-0.29	2.22	2.17	0.05	0.10	0.92
	3-18-92-1-12	"	SS	5.01	4.95	2.34	2.48	-0.04	-0.10	1.06
	3-18-92-1-8	"	S	-2.26	-1.69	2.54	2.50	0.33	0.37	0.96
	3-18-92-1-7	"	S	0.43	0.52	1.30	1.32	0.11	0.13	1.01
	3-18-92-1-6	"	S	-0.80	-1.16	2.84	2.58	-0.19	-0.11	0.80
	3-18-92-1-14	"	SS	4.00	4.39	2.59	2.58	0.23	0.18	0.95
	3-18-92-1-5	"	S	0.34	0.43	2.29	2.25	0.06	0.09	0.95
	3-18-92-1-13	"	SS	3.79	4.00	2.93	2.90	0.11	0.08	0.95
	3-18-92-1-4	"	S	-2.50	-1.87	2.23	2.25	0.42	0.44	1.04
	3-18-92-1-3	"	S	-0.06	-0.14	2.22	2.19	-0.05	-0.02	0.96
	3-18-92-1-2	"	S	-0.50	-0.41	2.75	2.61	0.05	0.10	0.85
	3-18-92-1-1	"	S	0.57	0.49	2.22	2.32	-0.05	-0.04	1.17
12	2-25-92-1-11	"	SS	4.12	4.30	2.79	2.74	0.09	0.08	0.95
	2-25-92-1-10	"	SS	5.13	5.16	2.07	2.12	0.02	0.02	0.95
	2-25-92-1-9	"	S	2.51	2.67	1.63	1.68	0.15	0.19	1.05
	2-25-92-1-8	"	SS	3.52	3.46	2.29	2.22	-0.04	0.01	0.88
	2-25-92-1-7	"	SS	5.26	5.24	2.11	2.08	-0.01	0.02	0.89
	2-25-92-1-6	"	SS	3.69	3.70	2.23	2.22	0.01	0.07	0.88
	2-25-92-1-5	"	SS	5.46	5.16	1.76	1.97	-0.25	-0.23	1.32

Table 2.—Grain-size parameters for deposits produced by the preclimactic eruptions—Continued.

Site no.	Sample no.	Date emplaced	Deposit type[1]	Md_ϕ	M_Z	σ_ϕ	σ_I	Sk_G	Sk_I	K_G
	2-25-92-1-4	6/14-6/15	SS	3.90	4.07	1.64	1.80	0.16	0.15	1.23
	2-25-92-1-3	"	SS	4.88	4.79	2.49	2.48	-0.05	-0.04	0.95
	2-25-92-1-2	"	SS	4.20	3.78	3.11	2.80	-0.20	-0.10	0.64
	2-25-92-1-1	"	SS	3.80	4.11	2.09	2.10	0.23	0.23	1.00
17	3-11-92-1-7C	"	SS	3.91	3.63	2.86	2.86	-0.15	-0.12	1.04
	3-11-92-1-7B	"	S-B	3.58	3.81	2.12	2.17	0.17	0.18	1.05
	3-11-92-1-7A	"	S-A	2.24	2.50	2.20	2.24	0.18	0.20	1.18
	3-11-92-1-6B	"	SS	4.65	4.31	2.64	2.70	-0.19	-0.14	1.06
	3-11-92-1-6A	"	S	1.58	1.85	1.85	2.04	0.22	0.29	1.35
	3-11-92-1-5	"	SS	5.21	5.22	1.67	2.84	0.01	0.29	2.42
	3-11-92-1-4	"	SS	5.50	5.48	2.02	2.99	-0.01	0.24	2.00
	3-11-92-1-3	"	SS	4.91	4.99	1.60	1.74	0.08	0.13	1.24
	3-11-92-1-2	"	SS	4.67	4.68	1.77	1.92	0.00	0.05	1.14
20	3-16-92-1-5C	"	SS	3.43	3.67	2.38	2.39	0.15	0.14	1.07
	3-16-92-1-5B	"	S-B	1.66	1.69	1.58	1.65	0.03	0.07	1.10
	3-16-92-1-5A	"	S-A	1.48	1.53	1.76	1.80	0.04	0.07	1.05
	3-16-92-1-4C	"	SS	3.37	3.54	2.70	2.60	0.09	0.09	0.88
	3-16-92-1-4B	"	S-B	2.51	2.84	2.50	2.44	0.20	0.21	1.03
	3-16-92-1-4A	"	S-A	1.64	1.74	1.81	1.91	0.09	0.15	1.12
	3-16-92-1-3B	"	S-B	2.27	2.55	2.34	2.33	0.18	0.20	1.14
	3-16-92-1-3A	"	S-A	1.67	1.83	1.72	1.88	0.14	0.21	1.20
21	2-26-92-2-9B	"	SS	4.57	4.76	2.33	2.27	0.12	0.12	0.88
	2-26-92-2-9A	"	S	1.79	1.71	1.66	1.72	-0.07	-0.02	1.11
	2-26-92-2-8B	"	SS	3.98	4.27	2.36	2.37	0.18	0.15	0.96
	2-26-92-2-8A	"	S	0.74	0.78	1.49	1.47	0.04	0.02	0.94
	2-26-92-2-7B	"	SS	4.49	4.57	2.84	2.73	0.04	0.01	0.85
	2-26-92-2-7A	"	S	1.59	1.51	1.44	1.46	-0.08	-0.07	0.99
	2-26-92-2-6B	"	SS	2.42	2.93	3.07	2.88	0.25	0.25	0.85
	2-26-92-2-6A	"	S	0.51	0.53	1.43	1.48	0.02	-0.02	0.99
	2-26-92-2-5B	"	SS	1.65	1.72	2.43	2.49	0.04	0.13	1.03
	2-26-92-2-5A	"	S	1.09	1.20	1.56	1.51	0.10	0.13	0.95
	2-26-92-2-4B	"	SS	3.71	3.79	2.76	2.74	0.05	0.04	1.01
	2-26-92-2-4A	"	S	0.88	0.99	1.55	1.50	0.11	0.17	0.95
	2-26-92-2-3B	"	SS	2.41	2.66	1.80	1.84	0.21	0.23	1.10
	2-26-92-2-3A	"	S	1.84	1.81	1.77	1.80	-0.02	-0.01	1.05
	2-26-92-2-2B	"	SS	3.10	3.35	2.14	2.10	0.17	0.17	0.86
	2-26-92-2-2A	"	S	1.45	1.58	1.37	1.44	0.14	0.19	1.07
	2-26-92-2-1B	"	SS	3.78	3.90	2.13	2.21	0.08	0.08	1.00
	2-26-92-2-1A	"	S	1.40	1.34	1.96	2.01	-0.05	-0.08	1.07
22	3-2-92-3-7B	"	SS	5.26	5.27	2.15	2.21	0.01	-0.02	0.96
	3-2-92-3-7A	"	S	1.74	1.75	1.58	1.61	0.01	0.04	1.06
	3-2-92-3-6C	"	SS	4.48	4.71	1.87	2.00	0.18	0.15	1.04
	3-2-92-3-6B	"	S-B	1.50	1.40	2.00	2.09	-0.08	-0.11	1.11
	3-2-92-3-6A	"	S-A	-0.09	-0.13	2.12	2.13	-0.03	-0.01	1.01
23	3-17-92-1-9	"	SS	3.71	4.17	2.54	2.47	0.27	0.26	0.89

Table 2.—Grain-size parameters for deposits produced by the preclimactic eruptions—Continued.

Site no.	Sample no.	Date emplaced	Deposit type[1]	Md_ϕ	M_z	σ_ϕ	σ_I	Sk_G	Sk_I	K_G
	3-17-92-1-8B	6/14-6/15	SS	4.00	4.41	2.44	2.43	0.25	0.23	0.94
	3-17-92-1-8A	"	S	2.33	2.36	1.35	1.37	0.03	0.05	0.98
	3-17-92-1-7B	"	SS	3.93	4.33	2.61	2.53	0.23	0.22	0.85
	3-17-92-1-7A	"	S	1.28	1.22	1.68	1.64	-0.06	-0.04	0.92
	3-17-92-1-6C	"	SS	2.98	3.47	2.53	2.46	0.29	0.29	0.96
	3-17-92-1-6B	"	S-B	1.42	1.42	1.69	1.69	0.00	-0.03	0.98
	3-17-92-1-6A	"	S-A	1.02	0.94	1.59	1.65	-0.07	-0.11	1.11
	3-17-92-1-5B	"	S-B	1.53	1.56	1.63	1.61	0.03	0.04	0.96
	3-17-92-1-5A	"	S-A	1.24	1.25	1.40	1.43	0.01	-0.01	1.03
	3-17-92-1-4B	"	S-B	1.30	1.53	1.59	1.59	0.21	0.21	0.98
	3-17-92-1-4A	"	S-A	1.15	1.11	1.61	1.62	-0.04	-0.04	1.02
	3-17-92-1-3	"	S	0.90	0.80	2.06	2.16	-0.08	-0.12	1.11
	3-17-92-1-2C	"	SS	4.53	4.50	2.79	2.70	-0.02	0.00	0.90
	3-17-92-1-2B	"	S-B	1.35	1.42	1.58	1.57	0.06	0.08	0.96
	3-17-92-1-2A	"	S-A	0.62	0.74	1.86	1.83	0.10	0.10	0.94
	3-17-92-1-1C	"	SS	5.08	4.97	2.44	2.61	-0.07	-0.12	1.14
	3-17-92-1-1B	"	S-B	1.92	1.88	1.77	1.74	-0.03	-0.02	0.96
	3-17-92-1-1A	"	S-A	0.84	0.76	2.21	2.15	-0.05	-0.07	0.95
28	2-20-92-2	"	SS	5.40	5.25	1.97	2.34	-0.12	-0.16	1.38

[1] F, fall; S, surge, coarse facies; S-B, surge, upper part of coarse facies; S-A, surge, lower part of coarse facies; SS, surge, fine (slack) facies; PF, pyroclastic flow.

ORIGIN OF PHASE V STRATA

We envision three stages to each Phase V event: (1) development of an energetic fountain-fed surge that swept out radially from the vicinity of the vent until its density dropped below that of ambient air; (2) buoyant rise of a great diapir of ash and hot gas, coupled with a "back draft" as air flowed inward to replace that which had risen; and (3) development of a slack surge beneath the diapir. Each stage is probably recorded in some fashion in the stratigraphic record. The fountain-fed surges must have emplaced at least the lower part of the coarse-grained facies of the proximal Phase V beds. And slack surges most likely emplaced the fine-grained facies in both the proximal and distal areas. Back-draft deposits, if they exist, must lie at the top of the coarse-grained proximal facies, either as some minor bed forms not yet recognized or as the stratified beds observed in some sections.

COMPOSITION AND DENSITY VARIATIONS IN DEPOSITS OF PHASES IV, V, AND VI

The most striking feature of the component data is the overall decrease in the proportion of andesite in the preclimactic eruptions (fig. 27). Tephra erupted during the first June 12 eruption was almost entirely andesite, while the proportion of dacite increased in the next two vertical eruptions. Dacite proportions increased irregularly through the surge events; the uppermost bed analyzed at site 11, the most proximal site, was devoid of andesite. This decrease helps correlate the preclimactic pyroclastic-flow deposits in the Maraunot River valley (site 24) with specific eruptions, as shown in figure 27. As indicated by the absence of dacite pyroclasts, the first (lowest) pyroclastic-flow deposit was almost certainly produced during the first vertical eruption (0851 on June 12). We know from photographs that the second pyroclastic flow was emplaced before the June 14 vertical eruption. The component data therefore indicate that this second flow was emplaced either during the second vertical eruption of June 12 or, more likely, during the June 13 eruption.

The component data show that the vanguard magma—the first magma to reach the surface—was andesite. This was progressively replaced by dacite magma. The dearth of andesite pyroclasts in the deposits of the last few preclimactic eruptions suggests that the climactic eruption began shortly after the vanguard magma was fully replaced by dacite magma.

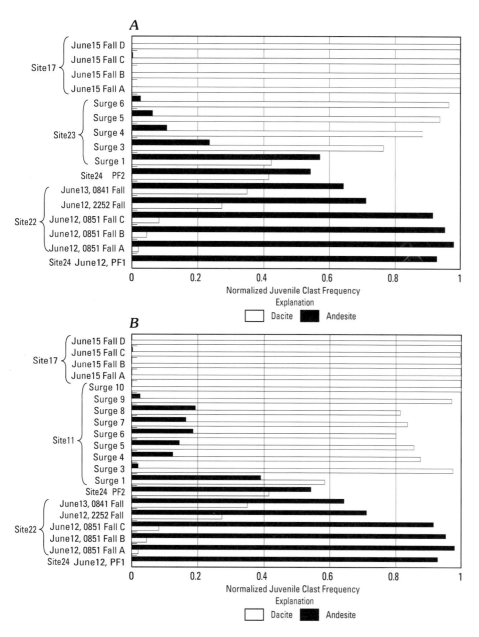

Figure 27. Normalized frequency of juvenile andesite and dacite clasts for two composite stratigraphic sections from the June 1991 eruptions. *A*, Composite section including component data from surge beds at site 23. *B*, Composite section including component data from surge beds at site 11. The proportion of juvenile basalt, a minor component associated with andesite, is not shown.

Only one type of dacite pumice is present in the preclimactic deposits. The climactic (Phase VI) deposits, in contrast, contain two types: white, phenocryst-rich pumice, the dominant type, and tan, phenocryst-poor pumice (Pallister and others, this volume). The preclimactic dacite pumice closely resembles the white, phenocryst-rich type found in the climactic deposits.

The most striking feature of the density distributions (fig. 28) is the high proportion of dense lithic clasts in the preclimactic deposits relative to the climactic deposits. The only apparent difference between the preclimactic dacite pumice and dacite lithic clasts is their vesicularity. The same is true of the andesite scoria and andesite lithics. We strongly suspect that this reflects extensive preeruption

degassing of magma at the top of the magma column (Hoblitt and Harmon, 1993) during the periods of repose between preclimactic eruptions.

PATTERNS OF ERUPTIVE BEHAVIOR

The "vigor" of the preclimactic explosive eruptions apparently decreased from the first vertical eruption of June 12 until the last surge of June 15, shortly before the start of the climactic eruption. This assertion is based on several indirect lines of evidence because we have no single objective measure of vigor. Perhaps the most compelling evidence comes from the strata produced by the eruptions. In

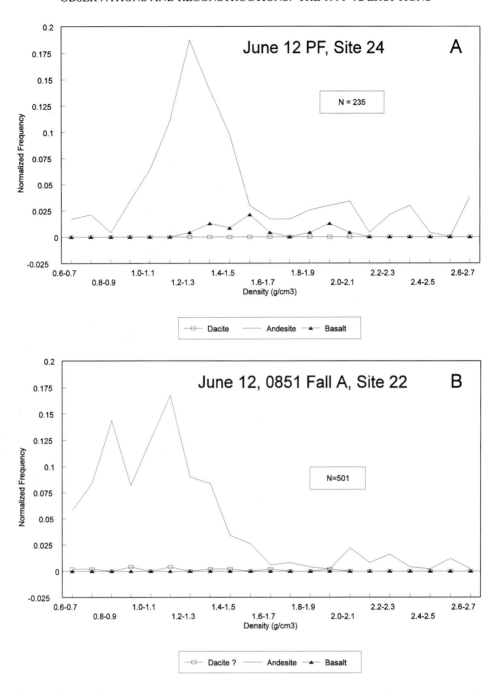

Figure 28. Density distributions of lapilli from various stratigraphic units produced by the June 1991 eruptions. Graphs arranged in the stratigraphic sequence used in figure 27A. *A*, The lower of two preclimactic pyroclastic-flow (PF) deposits at site 24. *B–D*, Sample fractions A, B, and C (site 22, fig. 26) of the June 12, 0851 fall deposit. *E*, Fall deposit from the June 12, 2252 eruption (site 22, fig. 26). *F*, Fall deposit from the June 13, 0841 eruption (site 22, fig. 26). *G*, The upper of two preclimactic pyroclastic-flow deposits at site 24. *H–L*, Five sequential surge beds from site 23. *M–P*, Four sequential sample fractions from the June 15 climactic fall deposit at site 17.

both Phase IV and Phase V deposits—the sequence of fall deposits from the vertical eruptions and the sequence of surge deposits from the eruption fountains—the beds tend to thin and become finer grained upwards. This tendency suggests that, within each of the two phases, the quantity of ejecta declined from one eruption to the next, as did the height to which it was ejected. The progression in eruption-plume heights estimated from weather-radar observations (fig. 33) supports the latter suggestion: heights decrease irregularly from >24 km for the June 12 and 13 eruptions

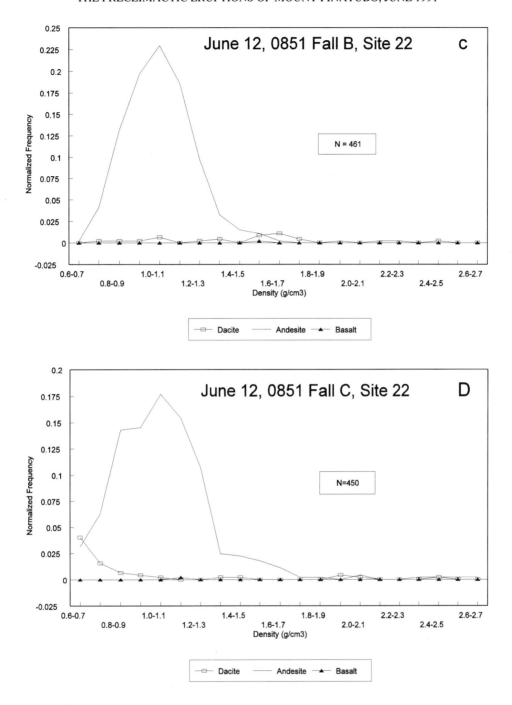

Figure 28.—Continued.

(the radar system had a ceiling of 24 km) to less than 10 km just before the climactic eruption.

Eruption-duration estimates based on SSAM records (fig. 34) also suggest waning eruptive vigor. These estimates assume that periods of high-amplitude, low-frequency seismicity correspond to eruptive activity. On this basis, eruption duration dropped precipitously during the first two vertical eruptions and then remained roughly the same or perhaps increased somewhat just before the climactic eruption.

Another notable feature of the preclimactic eruptive sequence is the general decline in the duration of repose between successive eruptions (fig. 35).

In continuous eruptions, the change from convecting vertical columns to pyroclastic fountains (column collapse) is usually rationalized in terms of either an increase in the vent diameter, a decrease in volatile content, or both (Wilson and others, 1980). We concluded above that the magma involved in the preclimactic eruptions had undergone substantial degassing; this would both lower the

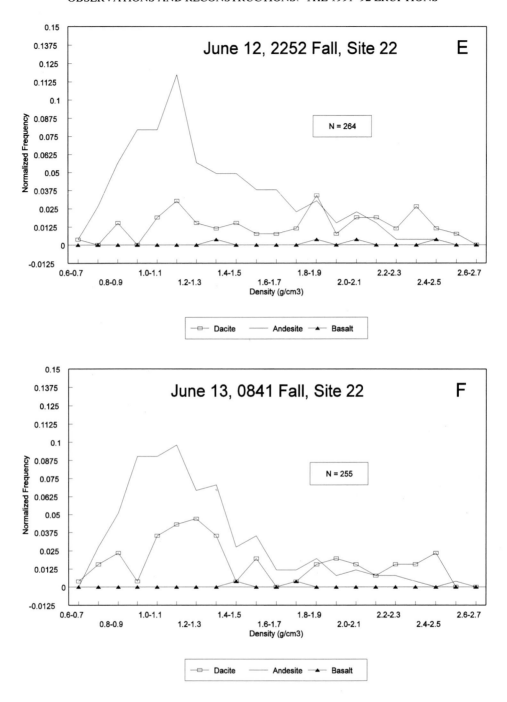

Figure 28.—Continued.

volatile content and increase the mean pyroclast density. One could argue that degassed magma became progressively more abundant from eruption to eruption until the net volatile content was insufficient to produce a convective eruption plume. If this were true, we would expect mean clast density to increase from eruption to eruption. With data currently available, the evidence for a systematic increase in mean clast density is not compelling (fig. 36). Some surge beds do indeed have greater mean densities than the preceding fall deposits, but others do not. The

observed variations could be due to density segregation during transport rather than to initial density distributions; a substantially greater quantity of density data, collected from a broad area, would be necessary to resolve the ambiguity.

The volatile content may have been increasing, rather than decreasing, as the proportion of dacite increased through the preclimactic eruptions. This is because the andesite magma was a mixture composed of dacite and basalt magmas (Pallister and others, 1992; Pallister and others, this volume), and basalt magmas are generally believed

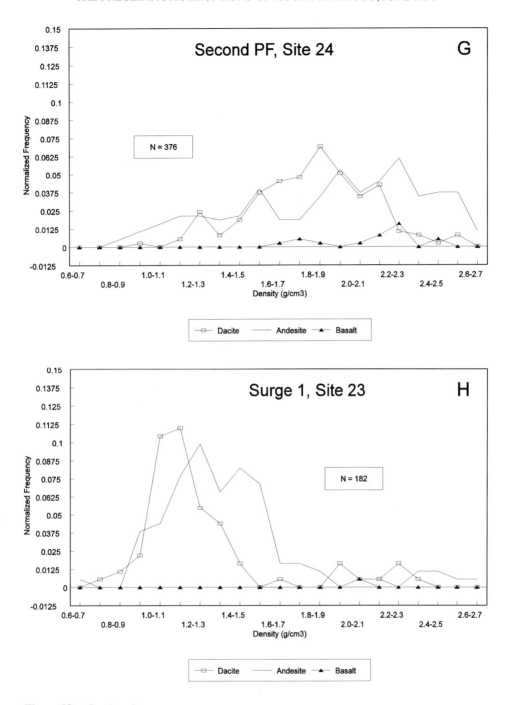

Figure 28.—Continued.

to possess a lower volatile content than dacite magmas. Thus, the undegassed hybrid andesite magma quite likely had a lower volatile content than that of the undegassed dacite magma. Because the proportion of dacite increased through the preclimactic eruptions, the mean volatile content should have been increasing.

An increase in vent diameter seems to be a more viable explanation for the change in eruptive behavior, although this explanation seems inconsistent with a progressive decline in quantity of ejecta.

Perhaps the change in behavior from Phase IV to Phase V was indeed somehow related to an increase in vent diameter or a decrease in the volatile content, although we have here a series of eruptions rather than a single continuous eruption. It is not obvious, however, how a systematic change in either volatile content or vent diameter could

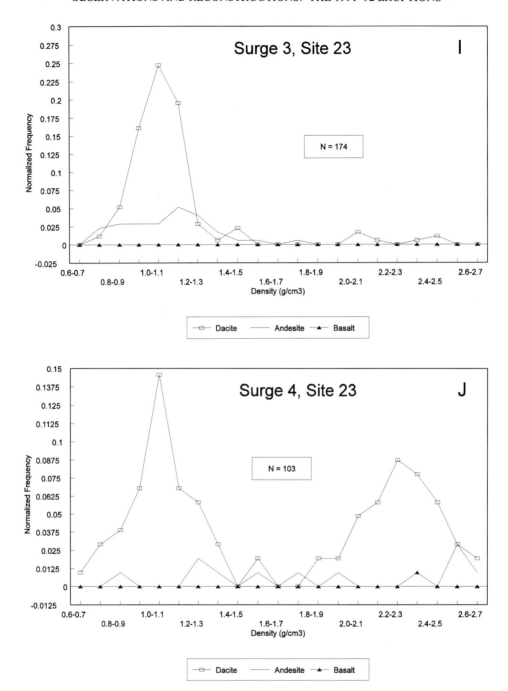

Figure 28.—Continued.

produce all of the patterns exhibited by the preclimactic eruptions.

A PHENOMENOLOGICAL MODEL

Any model of Mount Pinatubo's eruptive behavior must account for: (1) the occurrence of a series of preclimactic eruptions, (2) the progressive decline in vigor of these eruptions, (3) the decline in the duration of successive repose intervals, and (4) the initial decline in eruption duration, perhaps followed by an increase in duration just before the culminating eruption. We present here a phenomenological model that accounts for these gross behavioral patterns.

WHY NUMEROUS PRECLIMACTIC ERUPTIONS?

Why did Pinatubo produce a series of smaller eruptions instead of building directly to the climactic eruption?

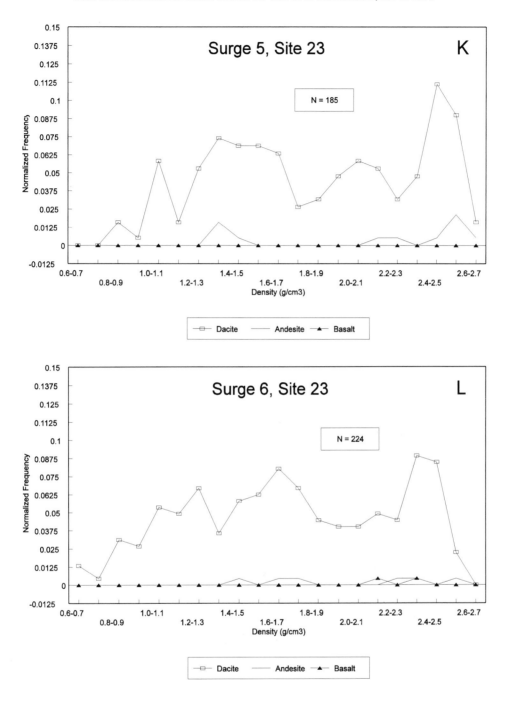

Figure 28.—Continued.

Magma obviously rose to the surface with sufficient pressure to begin and sustain brief eruptions. Eruptions ended because magma pressure dropped below some minimum value necessary to sustain an eruption. But, repeatedly, the pressure increased sufficiently to begin new eruptions. Because the pressure dropped when material was ejected during eruptions, it is reasonable to assume that the pressure increase during repose intervals was caused by new magma flowing into the conduit. This implies that the conduit—the

plumbing system that delivered magma to the surface—dilated elastically in response to the magma influx during repose intervals. During eruptions the conduit contracted, as stored elastic strain energy forced magma to the surface. The volumes of magma expelled by the preclimactic eruptions are small; the largest (the first June 12 eruption) is equivalent to a sphere with a diameter of about 200 m (Paladio-Melosantos and others, this volume).

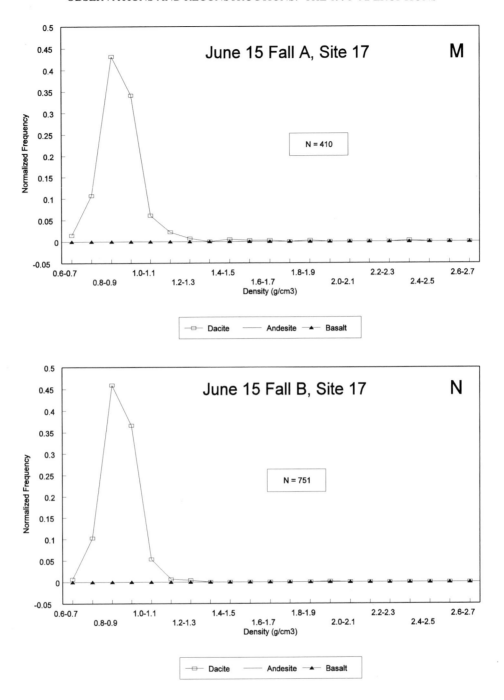

Figure 28.—Continued.

DECLINE IN ERUPTIVE VIGOR OF PRECLIMACTIC ERUPTIONS

Why did the eruptive vigor decline through the preclimactic eruptions? In response to magma influx, the pressure in the conduit rose to the value necessary to overcome the confining pressure at the top of the magma column. Apparently, the confining pressure at the top of the magma column declined from eruption to eruption. So in each eruption, successively less magma was forced to the surface, at successively lower pressure.

A decline in confining pressure after the first eruption is expectable. Before Mount Pinatubo awakened in April of 1991, the bulk of its edifice consisted of a plug dome, a relic of activity about 5 centuries earlier (Newhall and others, this volume). The old dome was presumably connected to the parent magma body by a conduit whose magma had long since cooled and solidified. When new magma began

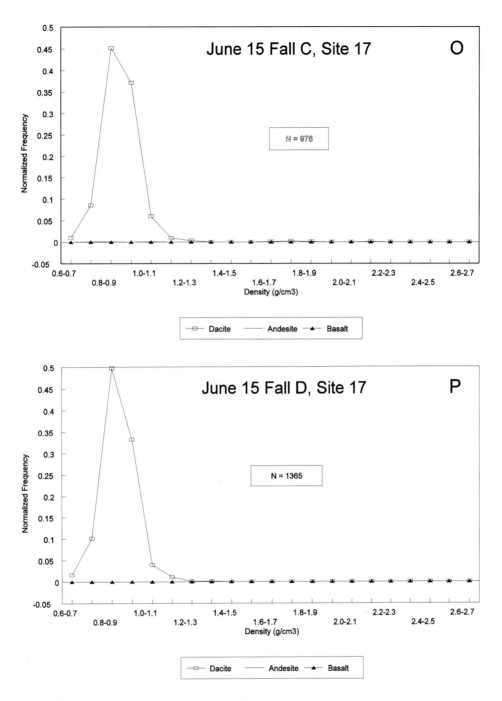

Figure 28.—Continued.

to force its way upward, apparently in response to a magma-mixing event deep within the magma chamber (Pallister and others, this volume), it had to force open a new pathway to the surface. The magma was volatile saturated in the deep magma reservoir (Gerlach and others, this volume), but vesicle growth accelerated as the magma ascended and additional water was exsolved from the melt. When the magma reached the depth at which growing vesicles became interconnected, juvenile gases were able to escape

upward through fractures (Eichelberger and Westrich, 1983; Eichelberger and others, 1986; Hoblitt and Harmon, 1993), even though the magma had not yet reached the surface. These gases were probably responsible for the high SO_2 values measured in late May (Daag, Tubianosa, and others, this volume). At first the vanguard magma degassed freely. But as degassing continued, vesicles in the magma collapsed and microlites began to form; these increased magma viscosity. The degassed magma then became an

Figure 29. June 1991 strata at site 24; see figure 23 for a diagram of stratigraphic relations. *A*, Phase IV strata at the base of the June 1991 section: a thin (10–30 cm) pyroclastic-flow deposit (behind shovel handle) sandwiched between five thin ash beds (three above, two below). These units are underlain by alluvium and overlain by the two pyroclastic-flow deposits shown in *B*. *B*, Two (3- to 4-m thick) pyroclastic-flow deposits that compose the upper part of the June 1991 section. Lower flow (Phase IV) is rich in dense, prismatically jointed blocks of andesite interpreted as fragments of the lava dome that began to grow on June 7; the same flow deposit, undissected, is shown in figure 19. Upper flow consists of normally graded lithic debris in a pumiceous matrix, probably produced during caldera collapse in Phase VI. Minor resistant bed that caps the right side of the section may be a posteruption debris-flow deposit.

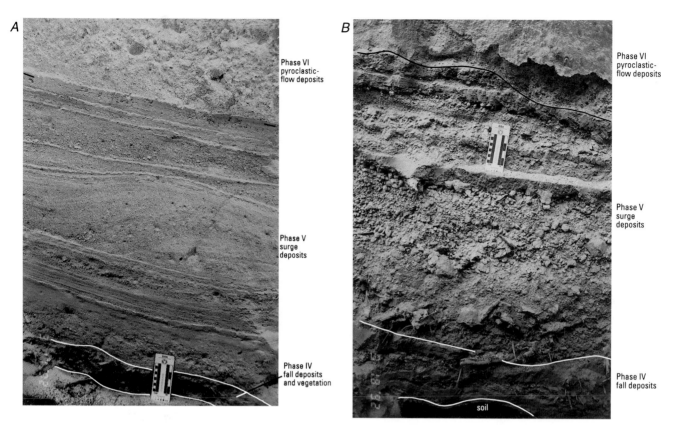

Figure 30. Proximal Phase V surge beds overlain by Phase VI climactic pyroclastic-flow deposits at (*A*) site 23 and (*B*) site 11. Subdivisions on the left side of the scale are centimeters.

impermeable, viscous plug retarding the escape of gases and slowing the magma's upward progress. This explanation is consistent with the decreasing SO_2 flux observed in late May and early June (Daag, Tubianosa, and others, this volume). But because undegassed magma continued to flow from depth and to accumulate behind the degassed plug, the pressure behind the viscous plug increased, even though the rate of gas escape dropped. Eventually, on June 7, the magma pressure increased sufficiently to force the degassed vanguard magma to the surface, and an andesite dome began to grow. The pressure beneath the growing dome continued to increase because the rate of extrusion was less than the rate of intrusion beneath the dome. Finally, at 0851 on June 12, the magma pressure increased sufficiently to overcome the confining pressure, and the first vertical eruption began. About 36 min later (table 1), the pressure dropped to a value less than that required to sustain an eruption. The vent choked with pyroclasts, and the eruption ended.

The system then began to repeat the behavior it followed before the first eruption—that is, degassing of the top of the magma column, continuing influx of new magma into the conduit, and consequent pressure increase. However, the system had changed somewhat. The original narrow, tenacious plug of degassed magma had been at least partly

replaced with a pyroclast-filled vent. The pressure threshold for initiating the second eruption would be lower than that of the first. But what about subsequent eruptions?

Two processes began at the end of each eruption: repressurization due to magma influx, and degassing at the top of the magma column. Degassing increases viscosity and, so, increases the confining pressure. Gas loss is a diffusive process. For simplicity, we assume here that gas loss from the top of the magma conduit approximates the diffusive loss of gas from a semiinfinite medium whose surface gas concentration is maintained at zero (Crank, 1990, p. 32). That is, we assume that the total gas loss and, therefore, the confining pressure, is proportional to the square root of time elapsed since the end of the previous eruption. At first, confining pressure increases more rapidly than magma pressure in the conduit. But because the rate of gas loss decreases with time, the rate of confining-pressure growth also decreases with time. Eventually, the magma pressure exceeds the confining pressure, and another eruption begins. After the first eruption, the time dependence of gas loss from the top of the magma column probably changed little from eruption to eruption. If the magma-pressurization rate remained constant, the confining pressure would remain constant, and the vigor of successive eruptions would all be about the same, as would the

Figure 31. Distal Phase V surge deposits. *A*, Surge sequence (behind ruler; length, 15 cm) at site 5, overlain by fall and ash-cloud deposits from the climactic eruption. *B*, Lower part of the surge sequence at site 5. Note leaf and other organic material in the unit just above the soil. *C*, Closeup view of the section shown in *B*. Note accretionary lapilli in the upper unit and spherical voids in the lower units.

durations of successive eruptions and repose intervals. But if the magma-pressurization rate was increasing with time, progressively less time would be available for gas loss to increase the confining pressure before the magma pressure exceeded the confining pressure. The magma pressure would exceed the confining pressure at successively lower values, eruptions would begin at successively lower magma pressures, and the vigor of successive eruptions would

C

Figure 31.—Continued.

decline. Because the vigor of the preclimactic eruptions successively declined, we assume that the magma-pressurization rate was increasing with time.

DECLINE IN DURATION OF SUCCESSIVE REPOSE INTERVALS

The duration of repose intervals declined from eruption to eruption because the magma-pressurization rate was increasing with time. The higher rate reduced the repose interval in two ways. First, it reduced the time necessary for the magma pressure to rise to the confining pressure. Second, it reduced the time available for the confining pressure to increase, so confining pressure was less for each successive eruption.

INITIAL DECLINE IN ERUPTION DURATION, PERHAPS FOLLOWED BY AN INCREASE JUST BEFORE THE CLIMACTIC ERUPTION

Eruptions persist as long as magma is delivered to the surface at a pressure that exceeds some minimum threshold value. Ejection of magma in an eruption causes pressure to drop, but this drop is offset by the influx of new magma. If the pressurization rate is increasing with time, as we have suggested, successive eruption durations should increase. But eruption duration also depends on the confining pressure threshold, and these thresholds decreased in successive eruptions. The effect of a decreasing confining-pressure threshold dominates eruption duration at first, but the increasing pressurization rate eventually becomes dominant, and eruption durations lengthen. This dominance becomes absolute when the pressurization rate—the rate at which magma pressure rises as a result of the influx of new magma—exceeds the depressurization rate—the rate at which magma pressure drops as a result of the ejection of magma in an eruption. When this occurs, a sustained climactic eruption begins.

THE MODEL

We now have the elements necessary to construct a simple phenomenological model.

Let:

P = Pressure at the top of the magma column;

dP/dt = Total rate of change of P;

dP_{in}/dt = Rate of change of P due to influx of new magma;

dP_{out}/dt = Rate of change of P due to ejection of magma during eruptions;

P_{close} = Pressure value below which eruption will terminate;

P_{open} = Confining pressure threshold; that is, the pressure at which an eruption will begin.

We assume:

dP_{in}/dt is directly proportional to time elapsed since the volcano reawakened;

dP_{out}/dt is constant;

P_{close} is constant;

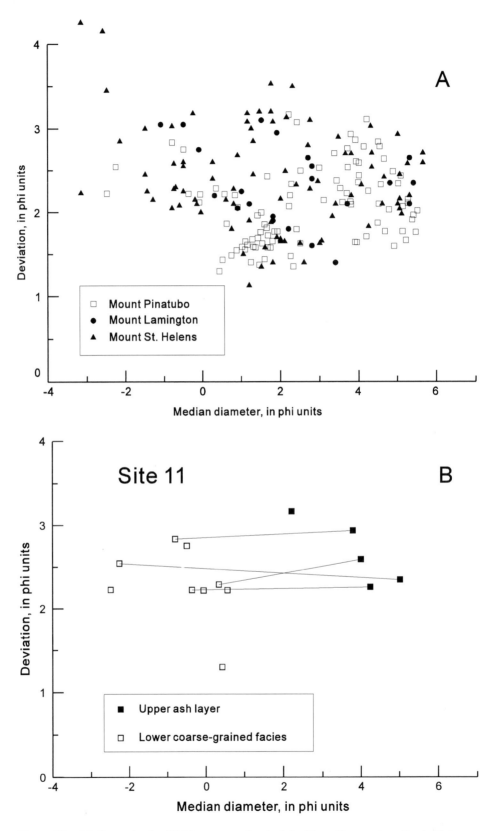

Figure 32. Median grain size (Md$_\phi$) versus sorting (σ_ϕ) for Phase V surge deposits. *A*, Plot comparing Pinatubo data to that of the 1980 Mount St. Helens and 1951 Mount Lamington (Papua New Guinea) surge deposits. *B*, Proximal data, site 11. Lines connect upper ash layer (filled squares) and lower coarse-grained facies (open squares) from individual surge beds. *C*, Proximal data, site 23. *D*, Data from both distal and proximal surge deposits.

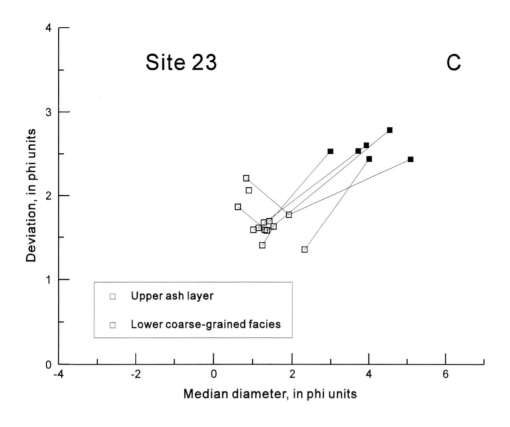

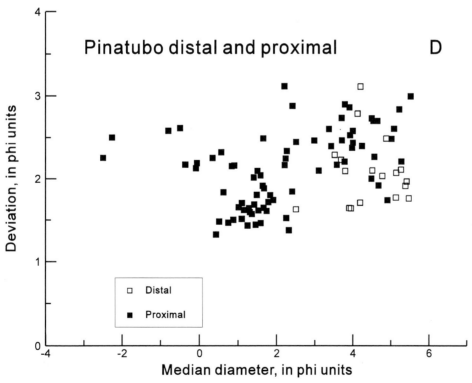

Figure 32.—Continued.

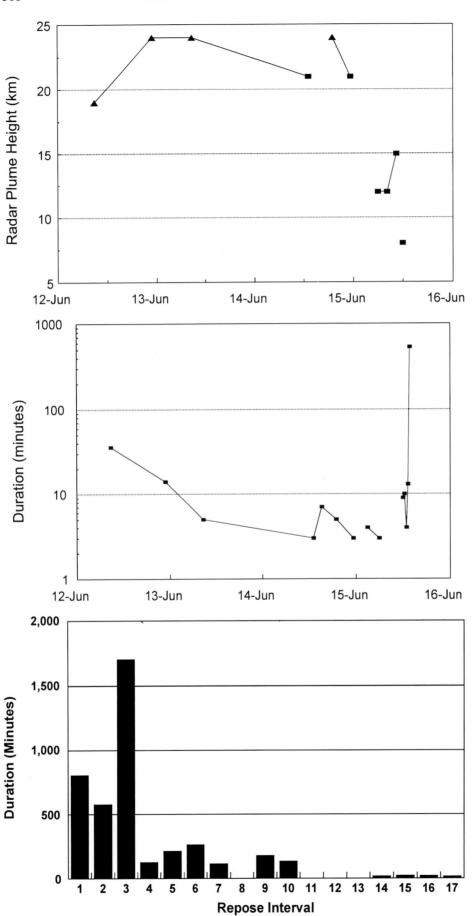

Figure 33. Plume height (in kilometers, as determined by weather radar) versus time for the preclimactic eruptions. Triangles indicate plume height exceeded radar observation limit. Line gaps indicate one or more eruptions for which data are unavailable.

Figure 34. Eruption duration (based on SSAM data) versus time for the preclimactic and climactic eruptions. Line gaps indicate one or more eruptions for which data are unavailable.

Figure 35. The duration, in minutes, of the successive preclimactic repose intervals. Bars absent where duration data unavailable.

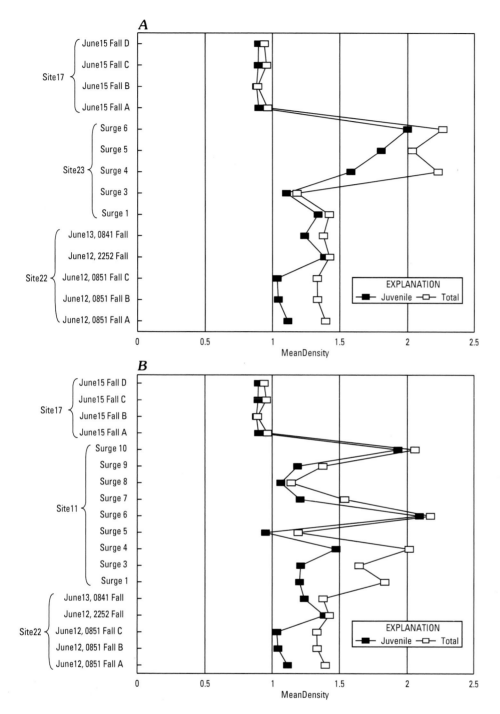

Figure 36. Mean clast density for two composite stratigraphic sections, both of which employ the same Phase IV and Phase VI fall deposits. Phase V surge deposit density data from (*A*) site 23 and (*B*) site 11. Filled squares, mean density of juvenile clasts; open squares, mean density of juvenile and nonjuvenile clasts.

P_{open} is directly proportional to the square root of time elapsed since the end of the previous eruption.

We increase pressure according to $dP/dt = dP_{in}/dt$ until $P = P_{open}$, and then we vary pressure according to $dP/dt = dP_{in}/dt - dP_{out}/dt$ until $P = P_{close}$. Using arbitrary values for the various parameters, we obtain the pressure-variation pattern shown in figure 37. This pattern mimics the gross features of the Pinatubo eruption sequence. The successive repose intervals decrease monotonically, while the succes-

sive eruption durations decrease to a minimum and then increase prior to the climactic eruption. The vigor of the eruptions, proportional to the magma pressure, declines before the climactic eruption. The eruptive style would change from vertical convection to pyroclastic fountaining when the ejecta velocity fell below some threshold value necessary for convective rise.

In this scheme, eruptive behavior is controlled primarily by the magma-pressurization rate at the top of the magma column and the rate of gas loss from the top of this

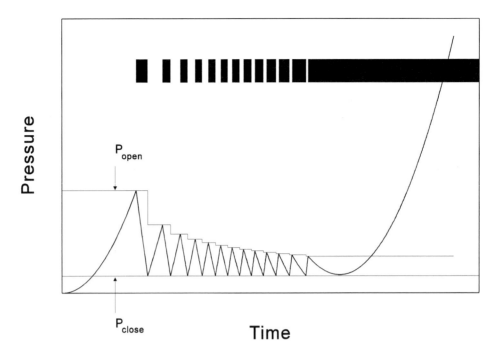

Figure 37. Pressure at the top of the magma column versus time for a phenomenological model designed to mimic Pinatubo's eruptive behavior. The rate of pressurization increases directly with time. The pressure threshold that must be reached before an eruption can start is shown by the line labeled P_{open}. Once an eruption begins, pressurization is offset by a constant-rate depressurization. As long as the depressurization rate exceeds the pressurization rate, the pressure declines until P_{close} is reached. P_{close} is the minimum pressure necessary to sustain an eruption. Climactic eruption begins when the rate of pressurization exceeds depressurization rate. Black bars show eruption duration, the spaces between the black bars are repose intervals.

column. To produce the pattern of preclimactic eruptions observed at Pinatubo, it is necessary for the magma-pressurization rate to increase with time. Such behavior is probably common for a reawakening volcano, because the magma-pressurization rate is directly related to the magma-supply rate, which is apt to increase with time. Rising magma will erode the conduit margins and increase its mean cross-sectional area and thus increase the flow rate. Flow rate is very sensitive to conduit geometry. In a cylindrical conduit, for example, the flow rate is proportional to the fourth power of the radius (Bird and others, 1960). Clearly, the flow rate increased as Pinatubo went from repose to weeks of accelerating unrest, to days of discontinuous eruptions, to the 9-h climactic eruption. We have restricted our discussion to the period during which the flow rate was increasing with time—the preclimactic period. However, the flow rate must have eventually decreased with time.

The laws that govern eruption patterns are certainly more complex than those we present here. Yet, even the simple assumptions we employ yield patterns that qualitatively mimic Pinatubo's patterns, so our phenomenological model may serve as the foundation for a more physically based model of broader applicability.

ACKNOWLEDGMENTS

Most of the observations and data presented in this paper were acquired with helicopter support provided by the U.S. Air Force, U.S. Navy, U.S. Marine Corps, and the Philippine Air Force. Without exaggeration, this study would not have been possible without that support. Logistical support during the eruptions was provided by the U.S. Air Force and after the eruptions by the Philippine Air Force. Bobbie Myers and B. Arlene Compher kindly assisted with illustration preparation. Roger Denlinger and Dan Dzurisin contributed valuable discussions. Kathy Cashman, Bill Rose, and Chris Newhall reviewed the manuscript and provided numerous insightful comments and suggestions.

REFERENCES CITED

Banister, J.R., 1984, Pressure wave generated by the Mount St. Helens eruption: Journal of Geophysical Research, v. 89, no. D3, p. 4895–4904.

Barry, R.G., and Chorley, R.J., 1978, Atmosphere, weather and climate (3d ed.): London, Methuen & Co., Ltd., 432 p.

Bird, R.B., Stewart, W.E., and Lightfoot, E.N., 1960, Transport phenomena: New York, John Wiley & Sons, 780 p.

Crank, John, 1990, The mathematics of diffusion (2d ed.): New York, Oxford University Press, 414 p.

Criswell, C.W., 1987, Chronology and pyroclastic stratigraphy of the May 18, 1980, eruption of Mount St. Helens, Washington: Journal of Geophysical Research, v. 92, no. B10, p. 10237–10266.

Daag, A.S., Tubianosa, B.S., Newhall, C.G., Tuñgol, N.M, Javier, D., Dolan, M.T., Delos Reyes, P.J., Arboleda, R.A., Martinez, M.L., and Regalado, M.T.M., this volume, Monitoring sulfur dioxide emission at Mount Pinatubo.

Eichelberger, J.C., Carrigan, C.R., Westrich, H.R., and Price, R.H., 1986, Non-explosive silicic volcanism: Nature, v. 323, p. 598–602.

Eichelberger, J.C., and Westrich, H.R., 1983, Behavior of water in rhyolitic magma at shallow depth: Eos, Transactions, American Geophysical Union, v. 64, p. 338.

Endo, E.T., and Murray, Tom, 1991, Real-time Seismic Amplitude Measurements (RSAM): A volcano monitoring and prediction tool: Bulletin of Volcanology, v. 53, p. 533–545.

Ewert, J.W., Lockhart, A.B., Marcial, S., and Ambubuyog, G., this volume, Ground deformation prior to the 1991 eruptions of Mount Pinatubo.

Fisher R.V., 1990, Transport and deposition of a pyroclastic surge across an area of high relief: The 18 May 1980 eruption of Mount St. Helens, Washington: Geological Society of America Bulletin, v. 102, p. 1038–1054.

Folk, R.L., 1980, Petrology of sedimentary rocks: Austin, Hemphills, 184 p.

Foxworthy, B.L., and Hill, Mary, 1982, Volcanic eruptions of 1980 at Mount St. Helens—The first 100 days: U.S. Geological Survey Professional Paper 1249, 125 p.

Gerlach, T.M., Westrich, H.R., and Symonds, R.B., this volume, Preeruption vapor in magma of the climactic Mount Pinatubo eruption: Source of the giant stratospheric sulfur dioxide cloud.

Harlow, D.H., Power, J.A., Laguerta, E.P., Ambubuyog, G., White, R.A., and Hoblitt, R.P., this volume, Precursory seismicity and forecasting of the June 15, 1991, eruption of Mount Pinatubo.

Hoblitt, R.P., 1986, Observations of the eruptions of July 22 and August 7, 1980, at Mount St. Helens, Washington: U.S. Geological Survey Professional Paper 1335, 44 p.

Hoblitt, R.P., and Harmon, R.S., 1993, Bimodal density distribution of cryptodome dacite from the 1980 eruption of Mount St. Helens, Washington: Bulletin of Volcanology, v. 55, p. 421–37.

Hoblitt, R.P., Miller, C.D., Vallance, J.W., 1981, Origin and stratigraphy of the deposit produced by the May 18 directed blast, in Lipman, P.W., and Mullineaux, D.R., eds., The 1980 eruptions of Mount St. Helens, Washington: U.S. Geological Survey Professional Paper 1250, p. 401–419.

Inman, D.L., 1952, Measures for describing the size distribution of sediments: Journal of Sedimentary Petrology, v. 22, no. 3, p. 125–145.

Koyaguchi, Takehiro, and Tokuno, Masami, 1993, Origin of the giant eruption cloud of Pinatubo, June 15, 1991: Journal of Volcanology and Geothermal Research, v. 55, p. 85–96.

Moore, J.G., and Sisson, T.W., 1981, Deposits and effects of the May 18 pyroclastic surge, in Lipman, P.W., and Mullineaux,

D.R., eds., The 1980 eruptions of Mount St. Helens, Washington: U.S. Geological Survey Professional Paper 1250, p. 421–438.

Newhall, C.G., Daag, A.S., Delfin, F.G., Jr., Hoblitt, R.P., McGeehin, J., Pallister, J.S., Regalado, M.T.M., Rubin, M., Tamayo, R.A., Jr., Tubianosa, B., and Umbal, J.V., this volume, Eruptive history of Mount Pinatubo.

Oswalt, J.S., Nichols, W., and O'Hara, J.F., this volume, Meteorological observations of the 1991 Mount Pinatubo eruption.

Paladio-Melosantos, M.L., Solidum, R.U., Scott, W.E., Quiambao, R.B., Umbal, J.V., Rodolfo, K.S., Tubianosa, B.S., Delos Reyes, P.J., Alonso, R.A., and Ruelo, H.R., this volume, Tephra falls of the 1991 eruptions of Mount Pinatubo.

Pallister, J.S., Hoblitt, R.P., Meeker, G.P., Knight, R.J., and Siems, D.F., this volume, Magma mixing at Mount Pinatubo: Petrographic and chemical evidence from the 1991 deposits.

Pallister, J.S., Hoblitt, R.P., and Reyes, A.G., 1992, A basalt trigger for the 1991 eruptions of Pinatubo Volcano?: Nature, v. 356, p. 426–428.

Power, J.A., Murray, T.L., Marso, J.N., and Laguerta, E.P., this volume, Preliminary observations of seismicity at Mount Pinatubo by use of the Seismic Spectral Amplitude Measurement (SSAM) system, May 13–June 18, 1991.

Reed, J.W., 1980, Air pressure waves from Mt. St. Helens eruptions: Albuquerque, N. Mex., Sandia National Laboratories, SAND80–1970A, 14 p.

Rogers, J.A., 1989, Frequency-domain detection of seismic signals using a DSP co-processor board, in Lee, W.H.K., ed., IASPEI software volume I: Toolbox for seismic data acquisition, processing, and analysis: International Association of Seismology and Physics of the Earth's Interior, p. 151–172.

Sabit, J.P., Pigtain, R.C., and de la Cruz, E.G., this volume, The west-side story: Observations of the 1991 Mount Pinatubo eruptions from the west.

Sigurdsson, H., Carey, S.N., and Espindola, J.M., 1984, The 1982 eruptions of El Chichón volcano, Mexico: Stratigraphy of pyroclastic deposits: Journal of Volcanology and Geothermal Research, v. 23, p. 11–37.

Scott, W.E., Hoblitt, R.P., Torres, R.C., Self, S, Martinez, M.L., and Nillos, T., Jr., this volume, Pyroclastic flows of the June 15, 1991, climactic eruption of Mount Pinatubo.

Sparks, R.S.J, Moore, J.G., and Rice, C.J., 1986, The initial giant umbrella cloud of the May 18th, 1980, explosive eruption of Mount St. Helens: Journal of Volcanology and Geothermal Research, v. 28, p. 257–274.

Taylor, G.C., 1958, The 1951 eruption of Mount Lamington, Papua: Australian Bureau of Mineral Resources, Geology and Geophysics Bulletin 38, 129 p.

Wilson, Lionel, Sparks, R.S.J., and Walker, G.P.L., 1980, Explosive volcanic eruptions—IV: The control of magma properties and conduit geometry on eruption column behavior: Geophysical Journal of the Royal Astrological Society, v. 63, p. 117–148.

Tephra Falls of the 1991 Eruptions of Mount Pinatubo

By Ma. Lynn O. Paladio-Melosantos,[1] Renato U. Solidum,[1] William E. Scott,[2]
Rowena B. Quiambao,[1] Jesse V. Umbal,[3] Kelvin S. Rodolfo,[3] Bella S. Tubianosa,[1]
Perla J. Delos Reyes,[1] Rosalito A. Alonso,[4] and Hernulfo B. Ruelo[5]

ABSTRACT

Tephra falls of varying character and volume occurred between April 2 and early September 1991, from eruptions of Mount Pinatubo. From April 2 to June 12, first phreatic explosions and later ash emissions related to emplacement of a lava dome produced mostly thin and fine-grained deposits over several hundred square kilometers west and south of the vent. A brief explosive eruption on the morning of June 12 deposited about 0.014 cubic kilometers of andesitic scoria, ash, and accidental lithic fragments southwest of the volcano (layer A). Several similar events over the next 2 days, followed by numerous pyroclastic-surge-producing explosions between the afternoon of June 14 and early afternoon of June 15, emplaced a 0.2-cubic-kilometer, laminated, mostly fine-grained ash-fall deposit (layer B) over broad areas around the volcano. The wide dispersal of layer B was induced by ash clouds convecting upward from pyroclastic surges that moved radially outward about 10 kilometers from the vent and the onset of low-altitude northerly to westerly winds as a tropical storm approached the area. The most voluminous deposit of the 1991 eruption sequence is a dacitic pumice-fall deposit (layer C) that was produced by the climactic eruption during the afternoon of June 15. A densely settled area of about 2,000 square kilometers was blanketed by 10 to 25 centimeters of rain-soaked tephra; 189 people were killed by collapsing buildings, and damage to utilities and agricultural lands was extensive. Most of Luzon and a roughly 4-million-square-kilometer area of the South China Sea and Southeast Asia were affected by tephra fall. The bulk volume of layer C probably lies between 3.4 and 4.4 cubic kilometers, ranking it among the five largest of the 20th century. The climactic eruption also produced voluminous pyroclastic-flow deposits and a 2.5-kilometer-diameter caldera. Slowly diminishing ash emissions continued from several vents in the caldera for about 6 weeks following the climactic eruption and produced a fine-grained laminated tephra deposit (layer D), which has a bulk volume of about 0.2 cubic kilometer.

Grain-size analyses of samples of layer C display well-known features of plinian tephra-fall deposits as distance from the vent increases, including decrease in median grain size, decrease in maximum pumice size, and improvement in sorting. Component analyses show that pumice dominates in grain-size fractions coarser than 1 millimeter, whereas crystals dominate in finer fractions. Lithic fragments make up a few percent or less of each fraction.

Deposits of layer C typically have normal grading, which suggests that eruption intensity peaked early and then decreased until ending prior to cessation of pyroclastic-flow activity. Various lines of evidence imply that peak activity of the climactic eruption was sustained for approximately 3 hours, and a waning level of activity continued for 6 hours more.

INTRODUCTION

All phases of the 1991 eruptions of Mount Pinatubo between April 2 and midsummer produced tephra falls of varying character and volume that climaxed with the great plinian eruption of June 15. The climactic eruption produced 5 to 6 km^3 of pyroclastic-flow deposits that partly buried valleys within 12 to 16 km of the volcano (W.E. Scott and others, this volume), but tephra-fall deposits were dispersed far and wide. Ash was carried westward across the South China Sea, where trace amounts fell in parts of Vietnam, Malaysia, and Borneo (Smithsonian Institution, 1991). The magnitude of this eruption earned it a place among the largest eruptions of this century (Self and others, this volume; W.E. Scott and others, this volume).

Because the repose period of volcanoes characterized by large-magnitude eruptions is usually on the order of

[1] Philippine Institute of Volcanology and Seismology.

[2] U.S. Geological Survey.

[3] Department of Geological Sciences, M/C 186, University of Illinois at Chicago, 801 W. Taylor St., Chicago, IL 60607; and Philippine Institute of Volcanology and Seismology.

[4] National Institute of Geological Sciences, University of the Philippines, Diliman, Quezon City, Philippines.

[5] Renison Goldfields Consolidated Exploration Pty., Limited, Australia.

hundreds or even thousands of years (Simkin and Siebert, 1984), the documentation of the Pinatubo eruptions provided a rare opportunity to observe and study the physical processes attendant to this size of eruption. This paper focuses on the stratigraphy, character, distribution, and volume of tephra-fall deposits formed by the 1991 eruptions. Although tephra fall from an eruption like that of June 15 would be of enormous interest for calibrating models that relate eruption intensity to grain size and distribution of deposits (Walker, 1981; Sparks, 1986; Carey and Sigurdsson, 1986, 1989; Carey and Sparks, 1986), tephra dispersal during the climactic eruption of Mount Pinatubo was complicated by the passage of Typhoon Yunya. Furthermore, much of the tephra fell in the South China Sea, with little or no documentation.

HUMAN IMPACTS

Impacts of tephra fall are the most wide reaching among those directly resulting from explosive eruptions, as is illustrated by the Pinatubo experience with unquestionable clarity. Heavy tephra fall darkened central Luzon for most of the afternoon of June 15 during the climactic eruption. More than 30 cm of tephra-fall deposits accumulated close to the volcano, while a densely settled area of about 2,000 km² received 10 to 25 cm. Fall deposits that were wetted by typhoon rains collapsed buildings and damaged public utilities and agricultural lands. Roof collapse accounted for 189 fatalities, or 61 percent of the total number recorded during the first 3.5 months after the eruption (Magboo and others, 1992). The estimated cost of damage to property is P10.62 billion (U.S. $400 million). The June 15 eruption itself affected about 216,000 families (National Disaster Coordinating Council, 1991). In addition, ash far from the volcano damaged aircraft (Casadevall and others, this volume) and ships.

METHODOLOGIES

Field investigations after the climactic eruption included mapping the extent of the tephra-fall deposits, measuring the thickness of individual beds or total thickness, and describing and sampling stratigraphic sections. Descriptors of grain size follow Fisher and Schmincke (1984): <0.0625 mm, fine ash; 0.0625–2 mm, coarse ash; 2–64 mm, lapilli; >64 mm, bombs. Where more detailed descriptors are necessary, we employ the Wentworth scale, for instance, medium-sand-size ash. Thickness data from more than 250 localities were used to construct isopach maps for the different eruptive events and to estimate areal extent and volume of tephra-fall deposits.

Tephra samples were collected from many localities around the volcano out to 45 km. About 20 percent of these were taken through the section from a known area so that

bulk density could be determined after drying and weighing. Twelve sections of the climactic tephra-fall deposit at distances of 15 to 35 km from the vent were sampled for grain-size and component analysis. We collected 20 to 100 g of layers of fine ash and 500 to 1,000 g of layers of coarser grained tephra. Maximum pumice size was measured from a sampling area of 50×50 cm. In the laboratory, about one-quarter of each sample was manually sieved and separated into 7 to 10 fractions at 1-φ intervals down to 3 (0.125 mm) or 4 φ (0.0625 mm). The remainder was stored for future reference or study. The different fractions were weighed and examined petrographically to determine percentage frequency of pumice, crystals, and lithic fragments.

CHRONOLOGY AND DESCRIPTION OF PRECLIMACTIC TEPHRA FALLS AND DEPOSITS

The detailed chronology of the Pinatubo eruptions is discussed by other authors (Wolfe and Hoblitt; Hoblitt, Wolfe, and others; Sabit and others; all this volume); a summary of eruptive events follows that concentrates on tephra falls and their deposits. We refer to the eruptive phases of Wolfe and Hoblitt (this volume), but do not use the phases as a primary means for subdividing tephra-fall deposits because the events that define the phases (such as changes in seismicity or eruptive behavior) are not always recognizable in the tephra-fall sequence. Rather, we employ a modified version of the informal stratigraphic subdivision of tephra-fall deposits proposed by Koyaguchi and Tokuno (1993). Figure 1 shows a typical section of tephra-fall deposits southwest of the volcano and unit designations for tephra layers as used in this report. Koyaguchi (this volume) subdivides layer C into a lower, coarser grained layer C1 and an upper, finer grained layer C2.

APRIL 2

The first sign of Pinatubo's awakening in 1991 was reported by local residents in early April (early part of phase I of Wolfe and Hoblitt, this volume). Steam and ash emissions from the volcano were ascribed to phreatic explosions (Daag and others, 1991; Pinatubo Volcano Observatory Team, 1991; Punongbayan and others, 1991, this volume; Hoblitt, Wolfe, and others, this volume). Between 4 and 9 vents were active at any time. Ash clouds were observed to rise 100 to 900 m above the vents (Philippine Institute of Volcanology and Seismology, 1992). The explosion on April 2 emplaced poorly sorted debris up to 3 m thick within 100 m of the vents; the debris consisted chiefly of angular blocks to coarse ash of pink and gray hornblende andesite and dacite. Beyond the zone of coarse debris and extending out several hundred meters was a deposit of

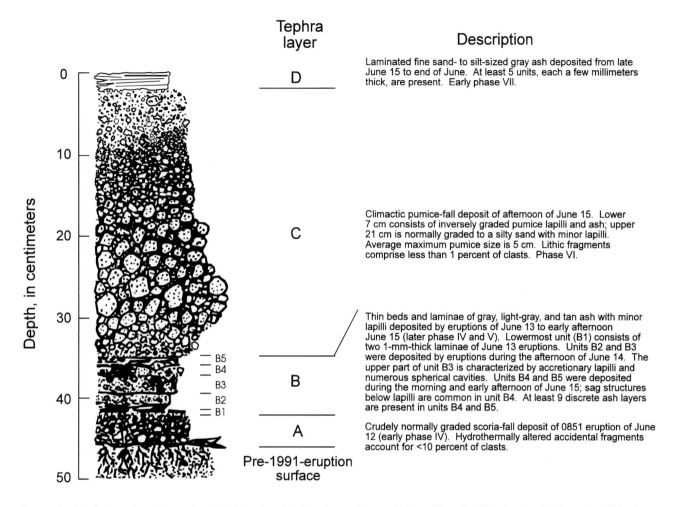

Figure 1. Typical stratigraphic section of 1991 tephra-fall deposits southwest of Mount Pinatubo. Sketch of section located at Sitio (hamlet) Upper Kakilingan, San Marcelino (location KAK, fig. 7), about 13 km south-southwest of the volcano. Eruptive phases are those defined by Wolfe and Hoblitt (this volume).

coarse ash up to a few centimeters thick (C.G. Newhall, U.S. Geological Survey, written commun., 1993). Traces of fine-grained ash reportedly reached Sitio Yamut in Barangay Burgos, Botolan, about 7 km from the vents. Most of the ash of April 2 and of emissions in the weeks following was deposited on the west-southwest flanks of the volcano, thinly covering an area of about 100 km² (fig. 2); however, the actual distribution of tephra fall from these events is poorly known. Petrographic analyses showed the tephra to be vitric-crystal ash composed of glass, plagioclase, hornblende, magnetite, biotite, and clinopyroxene with fragments of hornblende andesite and dacite, soil, and hydrothermally altered material (A.G. Reyes, Philippine National Oil Company, written commun., 1991). Owing to the fresh appearance of the glass, Reyes suggested that the tephra was partly of magmatic origin, but the presence of pyrite also indicated a hydrothermal origin. Other investiga-

tors concluded that there was no firm evidence that the glass was a juvenile component (Wolfe and Hoblitt, this volume).

MAY AND EARLY JUNE

Ash emissions intensified in late May and early June, affecting increasingly larger areas on the west side of the volcano with fine ash fall (later phase I through phase III of Wolfe and Hoblitt, this volume; Sabit and others, this volume). An explosion on June 7 that preceded growth of the June lava dome in the headwaters of the Maraunot River ejected ash to an altitude of 8 km. Succeeding ash emissions that accompanied dome growth between June 7 and the early morning of June 12 produced ash columns 2 to 5 km high. By June 12 more than 500 km² on the west side of the volcano had received a thin coating of fine ash (fig. 2).

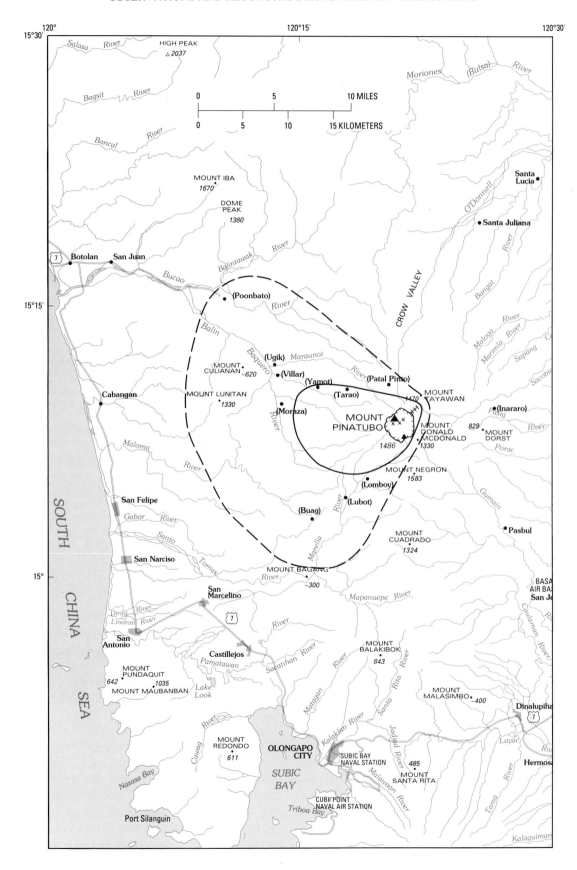

← **Figure 2.** Outline of areas affected by tephra falls between April 2 and early June 12, 1991 (phases I through III of Wolfe and Hoblitt, this volume). Area that received tephra fall from April 2 phreatic eruption is within solid line; dashed line shows area that received light, intermittent tephra fall up until early June 12. Cross-hachured line, fissure of April 2 eruption; ×, fumarole (from Wolfe and Hoblitt, this volume, fig. 6). In this and tephra-isopach maps (figs. 3, 6, and 7), small solid circle marks pre-June 15 summit of Mount Pinatubo, solid triangle marks site of June lava dome and presumed vent for explosive eruptions, and hachured line shows rim of caldera formed during climactic eruption.

MORNING OF JUNE 12

A more explosive phase of activity began on June 12 at 0851 (local time), signaled by a brief subplinian eruption that sent a column to an altitude of at least 19 km (beginning of phase IV of Wolfe and Hoblitt, this volume). Tephra from this eruption, which we call layer A (Koyaguchi and Tokuno, 1993; Koyaguchi, this volume), reached the South China Sea (fig. 3; Oswalt and others, this volume, figs. 2 and 3). A normally graded bed more than 5 cm thick of gray scoriaceous lapilli and ash and lithic fragments accumulated within 10 km southwest of the vent (figs. 1 and 4B). Farther downwind, layer A is brownish-gray, crystal-rich, sand-size ash with scattered coarser clasts. Phenocrysts consist of plagioclase, hornblende, quartz, magnetite, biotite, olivine, and pyroxene (see Pallister and others and Hoblitt, Wolfe, and others, both this volume, for petrologic data). Tephra fall affected nine municipalities in the Province of Zambales with up to 1 cm of ash and covered more than 2,000 km^2 of land. The bulk volume of tephra emplaced during the 0851 eruption on June 12 is about 14 million m^3 (0.014 km^3), as calculated from the isopach map of figure 3 by using the root-area method (table 1; fig. 5; Pyle, 1989; Fierstein and Nathenson, 1992). This bulk volume converts to a dense-rock-equivalent (DRE) or magma volume of about 4 to 6.5 million m^3 (assumes bulk density of tephra-fall deposit of 0.8 to 1.2 g/cm^3 and magma density of 2.6 g/cm^3). For comparison, the June lava dome had a volume less than 2 million m^3 (Hoblitt, Wolfe, and others, this volume).

LATE JUNE 12 TO EARLY AFTERNOON JUNE 15

Tephra falls produced by multiple short-lived explosive events from late on June 12 to early afternoon on June 15 deposited tephra over an ever-widening area as eruptions changed in character and the approach of Typhoon Yunya modified wind directions. Events on late June 12, morning of June 13, and early afternoon of June 14 produced narrow convecting columns that rose to ≥24 km, ≥24 km, and 21 km, respectively (phase IV of Wolfe and Hoblitt, this volume). Deposits of individual events have not been subdivided except in a few localities (Hoblitt, Wolfe, and others,

this volume); they were probably finer grained and thinner than those of layer A but covered essentially the same area southwest of the vent. Beginning midafternoon of June 14 and continuing for almost 24 h, a series of explosions generated pyroclastic surges that moved outward more than 10 km before rising into broad convecting ash clouds (Hoblitt, Wolfe, and others, this volume; phase V of Wolfe and Hoblitt, this volume). Some ash from these clouds began drifting southeastward as low-and mid-level winds shifted to more westerly directions (see later discussion of meterological controls on tephra distributions; Oswalt and others, this volume); much fell wet as rains of Typhoon Yunya reached the area. As the climactic eruption began, the mostly fine-grained tephra-fall deposits related to events of late June 12 to early afternoon June 15, here referred to as layer B, covered more than 4,000 km^2 of Luzon; populated areas had by now received as much as 5 cm (figs. 1, 4, and 6). For the first time, tephra falls were affecting densely populated areas in sectors east of the volcano.

In distal areas, layer B consists of finely to very finely laminated light-gray, grayish-brown, and tan fine ash with minor coarser particles. The lower part of the layer is typically more brown and the upper part more gray. The laminations reflect deposition from numerous ash-fall events. Accretionary lapilli and spherical to elongate vesicles indicate the presence of water during deposition of some units. Sag structures caused by impact of coarser clasts are common in the upper units. The bulk volume of layer B as calculated from the isopach map (fig. 6) is about 0.17 km^3 (table 1, fig. 5), or about 0.1 km^3 DRE. The total volume, including the surge deposits, is uncertain, as the thickness of surge deposits in proximal areas varies greatly in relation to local topography (Hoblitt, Wolfe, and others, this volume). These variations may not be accurately represented by the exponentially changing thickness modeled in the volume calculations.

TEPHRA-FALL DEPOSITS OF THE CLIMACTIC ERUPTION OF JUNE 15

On the basis of seismic, barograph, and other records, the climactic eruption of June 15 began at 1342, its peak was sustained for about 3 h, and a waning level of activity continued for about 6 h more (phase VI of Wolfe and Hoblitt, this volume). The eruption column attained a maximum height of about 35 km and spread out broadly in the stratosphere, eventually reaching 250 km upwind (northeast) of the vent (Koyaguchi and Tokuno, 1993). About 7,500 km^2 of Luzon was covered by more than 1 cm of tephra (fig. 7), and almost the entire 105,000-km^2 island received at least a trace.

The tephra-fall deposit of the climactic eruption is the most extensive, and the second most voluminous, deposit of the 1991 Pinatubo eruptions. Only the pyroclastic-flow

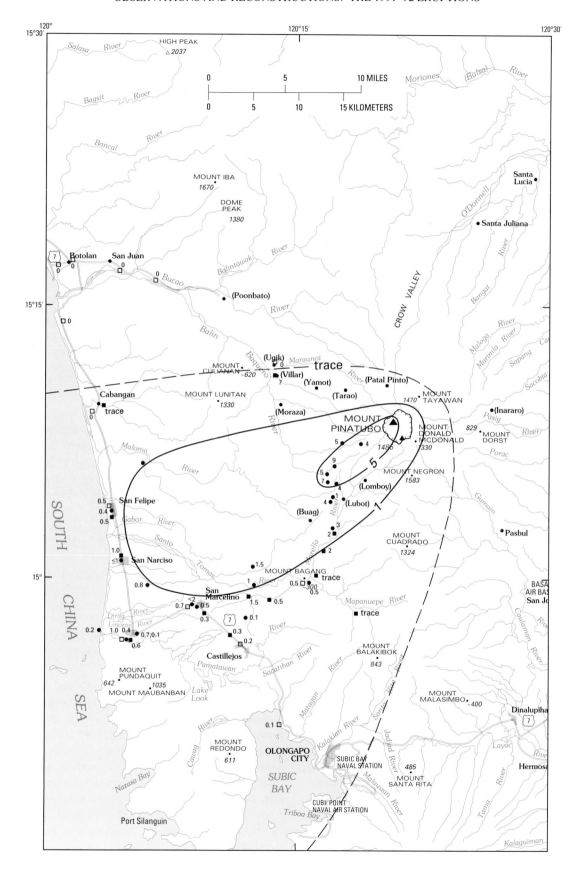

Table 1. Parameters used to calculate bulk volume of tephra-fall deposits from 1991 eruptions of Mount Pinatubo.

[Multiple entries for layer C, the climactic tephra-fall deposit, represent several possible combinations of $A_{ip}{}^{1/2}$ and thickness of the distal isopach of figure 8 as portrayed on figure 9. Included for comparison are data for other large eruptions of the 20th century and the Taupo ultraplinian eruption of 1,820 ^{14}C yr B.P. (Fierstein and Nathenson, 1992); Quizapu volume recalculated to 9.5 km[3] by Hildreth and Drake (1992), who used a three-segment distribution; Hudson data from Scasso and others (1994). T_0, extrapolated maximum thickness at vent; k, slope of line on a plot of logarithm of thickness versus square root of area (for distributions that define two or three straight-line segments, k is for the near-vent segment); k_1 and k_2, slopes of distal segment(s) for distributions that define two or three straight-line segments; $A_{ip}{}^{1/2}$ and $A_{ip}{}^{1/2}{}_1$, square root of area at the points of interception between segments of a two- or three-segment distribution; T_{distal}, thickness at distal isopach used for various estimates of layer C volume; –, not applicable]

Tephra layer	T_0 (cm)	k	k_1	k_2	$A_{ip}{}^{1/2}$ (km)	$A_{ip}{}^{1/2}{}_1$ (km)	Volume (km[3])	T_{distal} (cm)
A	9	0.114	–	–	–	–	0.014	–
B	101	0.108	–	–	–	–	0.17	–
C	39	0.022	–	–	–	–	1.7	–
C	39	0.022	0.0083	–	119	–	3.0	–
C	39	0.022	0.0083	0.0038	119	538	3.4	0.001
C	39	0.022	0.0083	0.0021	119	538	3.9	0.005
C	39	0.022	0.0083	0.0016	119	538	4.4	0.01
D	166	0.130	–	–	–	–	0.20	–
Santa María, Guatemala, 1902	154	0.021	0.0075	–	250	–	7.8	–
Quizapu, Chile, 1932	1300	0.098	0.0067	–	48	–	9.9	–
Novarupta A, Alaska, 1912	336	0.058	0.0140	–	47	–	5.1	–
Novarupta B, Alaska, 1912	137	0.053	0.016	–	46	–	2.7	–
Taupo, New Zealand, 1,820 B.P.	197	0.022	–	–	–	–	7.8	–
Hudson, Chile, 1991	1737	0.180	0.0437	0.0066	19	56	7.6	–

← **Figure 3.** Distribution of layer A, which consists of tephra-fall deposits from the June 12, 1991, eruption that began at 0851 (beginning of phase IV of Wolfe and Hoblitt, this volume). Isopachs are in centimeters; dashed line shows generalized outer limit of identifiable deposits. Measurement sites (with thickness in centimeters) and sources of data: filled circles, PHIVOLCS, USGS, Pinatubo Lahar Hazards Task Force; filled squares, Koyaguchi and Tokuno (1993); open squares, E. Listanco *in* Koyaguchi and Tokuno (1993).

deposits emplaced during the climactic eruption are more voluminous. We refer to the entire climactic tephra-fall deposit as layer C (fig. 1), in contrast to Koyaguchi and Tokuno (1993), who arbitrarily divided the deposit into a lower, coarser grained layer C and an upper, finer grained layer D. Koyaguchi (this volume) now refers to these layers as C1 and C2, respectively. Maximum thickness measured is 33 cm at a site 10.5 km southwest of the vent (fig. 4*B*); the vent is thought to have been at or near the site of the June lava dome. The climactic tephra-fall deposit is typically thin or absent nearer to the volcano, owing to erosion by and (or) incorporation into moving pyroclastic flows that were being generated broadly concurrently with at least parts of the tephra fall (W.E. Scott and others, this volume).

VOLUME

The volume of the climactic tephra-fall deposit was initially poorly constrained because such a large fraction fell into the South China Sea (Scott and others, 1991; Koyaguchi and Tokuno, 1993). The on-land bulk volume

enclosed within the 1-cm isopach is about 0.7 km[3]. The thinnest isopach that we can close with confidence is 15 cm; the thickest is 30 cm (fig. 7). Extrapolation of the trend of thickness against square-root area of the closed isopachs yields a minimum volume estimate of 1.7 km[3] and a maximum thickness at the vent (T_0) of 39 cm (fig. 5). The low slope of this trend compared with those of layers A, B, and D and the modest maximum thickness for a plinian deposit reflect the broad dispersal of the climactic deposit. The very high eruptive column coupled with strong winds of the tropical storm are thought to be two major contributing factors to this great dispersal. Note that the Santa María, Guatemala, eruption of 1902 and the ultraplinian Taupo, New Zealand, eruption of 1820 ^{14}C yr B.P. had similar slopes of tephra distribution (see factor k; table 1).

Tephra thicknesses measured in marine cores and estimated from sediment-trap data (Wiesner and Wang, this volume) define 2- and 1-cm isopachs, constrain the area of the 0.1-cm isopach, and further refine volume estimates. The isopachs define a second segment on a plot of log thickness versus square root of area (k_1; table 1, fig. 8) that increases the total volume to 3.0 km[3]. But the extrapolation of the second segment to low thicknesses appears to underestimate the wide dispersal of fine-grained, thin ash in distal areas.

Unmeasured tephra-fall deposits, described as light, were reported from several areas in Southeast Asia and the southern Philippines (fig. 9). The area of a line enclosing these sites is about 3.8×10^6 km[2]. Most of this area west of Pinatubo lies within regions over which the eruption cloud

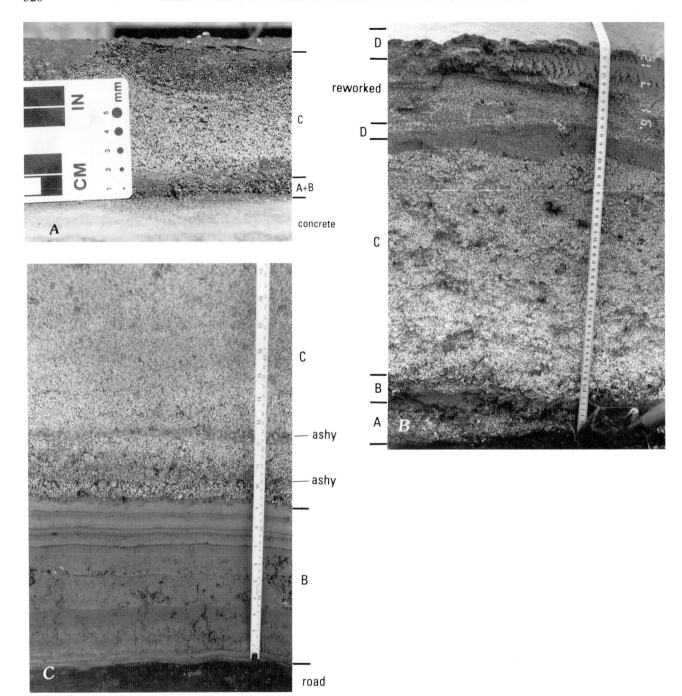

Figure 4. Photographs of tephra-fall deposits of 1991 eruptions. *A*, Section on abutment of bridge across Santo Tomas River north of San Narciso, Zambales; 32 km west-southwest of vent. Layer A is 8 mm of sand-sized ash; layer B is 4 mm of mostly fine ash. Note weak normal grading of layer C and scattered coarse clasts on surface of deposit. (Photograph courtesy of J.J. Major, U.S. Geological Survey, no. R7/21, June 30, 1991.) *B*, Tephra-fall deposits on unimproved road about 10.5 km southwest of vent, west of Marella River. Layer A, about 4 cm thick, consists of coarse ash and fine lapilli; layer B consists of several thin layers of ash totalling 2.5 to 5 cm thick; layer C is 33 cm thick and is the thickest section of the climactic pumice-fall deposit yet found. Note normal grading overall, but 2-cm pumice lapillus in upper left; layer D

consists of two 3- to 4-cm-thick beds of fine ash separated by a bed of water-reworked pumiceous ash. (Photograph by W.E. Scott, no. WES–91–14–3; July 12, 1991.) *C*, Tephra-fall deposits on unimproved road about 9 km southeast of vent, north side of Gumain River. Layer B is 23 cm thick and consists of numerous graded ash beds; layer C is 31 cm thick and has two zones in lower part with minor fine ash coatings. (Photograph by W.E. Scott, no. WES–91–6–9, June 28, 1991.) *D*, Section at mouth of Pasig River canyon about 15 km east of vent. Layer B is 10 cm thick and layer C is about 18 cm thick; note ash-rich zones that stand out owing to increased cohesiveness. (Photograph by W.E. Scott, no. WES–92–19–25, March 14, 1992.)

Figure 4.—Continued.

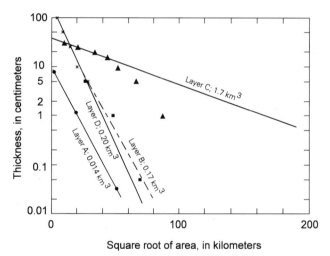

Figure 5. Thickness of tephra-fall deposits (layers A–D) plotted against square root of on-land isopach area. Volumes calculated from data in table 1 by using method of Fierstein and Nathenson (1992); volume shown for layer C is a minimum value calculated for a slope extrapolated from area of isopachs that close on land. Three data points that fall off curve for layer C are for on-land areas within isopachs that extend into the South China Sea (fig. 7).

passed, as imaged by geostationary and polar-orbiting satellites. But many areas in the southern Philippines that reportedly received light ash fall lay south and east of this eruption cloud. If we disregard this eastern area, a 3.1×10^6-km^2 area of tephra fall remains. The thickness of tephra in the distal areas was too thin to cause significant and newsworthy problems. The 4 mm of ash that fell in Metro Manila caused darkness for several hours during the late afternoon, closed the airport for several days, and created a noticeable and persistent haze as wind resuspended the ash. Because nothing approaching these conditions was

reported from Southeast Asia or southern Philippine sites and because Wiesner and Wang's (this volume) data constrain the area of the 1-mm isopach to about 3×10^5 km^2, we infer thicknesses substantially less than 1 mm. Furthermore, as satellite images show, the ash cloud traveled westward as several lobes (fig. 9) rather than as a broad homogeneous mass, so distal isopachs probably aren't simple curves. Therefore, we use a range of estimates of thickness—0.01 to 0.1 mm (10–100 μm)—and area—3.1 and 3.8×10^6 km^2—for a distal isopach in order to include these many uncertainties.

Consistent with the inferred thickness range, the grain size of tephra-fall deposits in Southeast Asia was very fine. Casadevall and others (this volume) infer grain diameters were <30 μm along airline flight paths near Vietnam, on the basis of the lack of abrasion on aircraft windshields. A tephra sample collected in Singapore contained 1- to 10-μm-diameter grains of glass and crystals (C.G. Newhall, written commun., 1994). Deposits of such fine ash a few grains thick can form continuous films on smooth surfaces. We spread fine ash on glass slides at these thicknesses and created very noticeable films.

Extrapolation of distal segments on figure 8 from the 1-mm isopach to various thickness and area estimates define k_2 values (table 1) that yield total bulk volume estimates of 3.4 to 4.4 km^3. The volume added by these distal segments accounts for about 12 to 32 percent of these totals.

We have bulk-density determinations from about 20 percent of the sites at which we measured tephra thickness; we can use these to convert bulk volume to magma volume. Samples include columnar sections of layer C, as well as combined samples of layers A and B, combined samples of layers A, B, and C, and samples of layer D. Most were thoroughly wetted by rains before they were sampled, except for some samples of layer D. Samples were weighed after

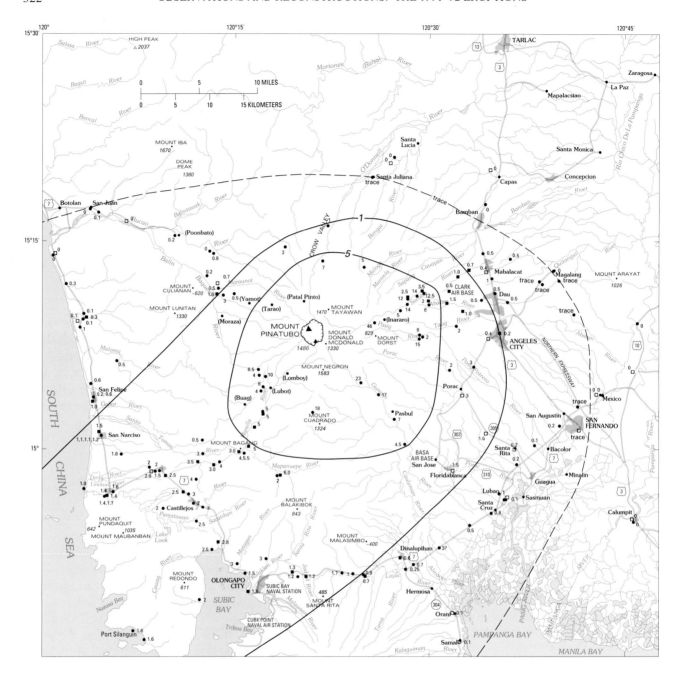

Figure 6. Distribution of tephra-fall deposits of layer B, which were produced by eruptions between late June 12 and early afternoon June 15, 1991 (later part of phase IV and phase V of Wolfe and Hoblitt, this volume). Isopachs are in centimeters; dashed line shows generalized outer limit of identifiable deposits. Sources of data as in figure 3.

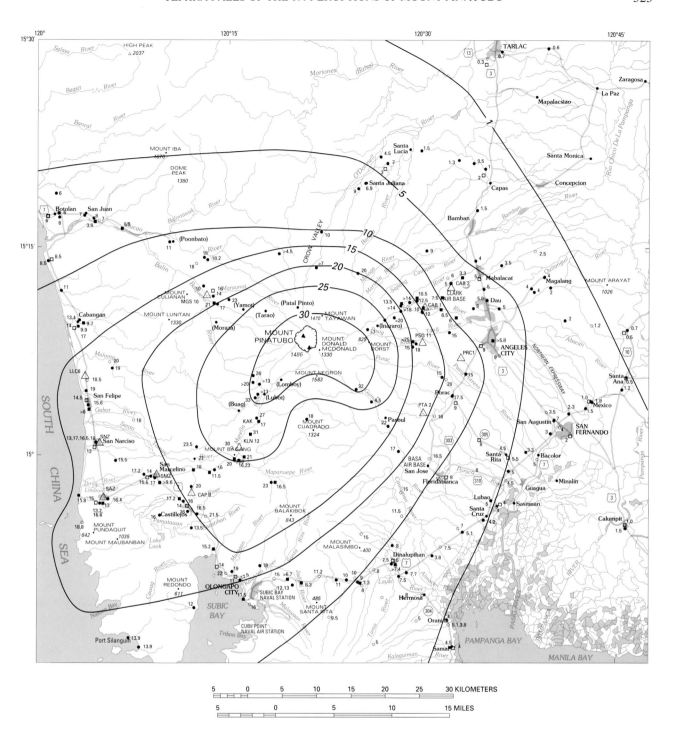

Figure 7. Distribution of tephra-fall deposits of the climatic eruption of June 15 (phase VI of Wolfe and Hoblitt, this volume), layer C, and locations of sections (triangles) sampled for grain-size and component data. KAK is location of section sketched in figure 1. Isopachs are in centimeters; sources of data as in figure 3, but open circles show total thickness of section (in centimeters), which may also include layers A and (or) B.

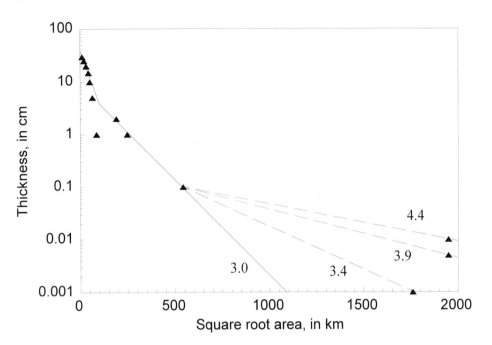

Figure 8. Log of thickness versus square root of area for layer C that includes proximal on-land data from Luzon (fig. 5), marine data from Weisner and Wang (this volume), and three possible distributions for distal tephra-fall deposits (dashed lines) based on two values for area of distal isopach (3.1 and 3.8 million km^2; fig. 9), and three values for thickness of distal isopach (0.01, 0.05, and 0.1 mm). Numerals in boxes are calculated bulk volumes in cubic kilometers.

oven drying or sun drying; their bulk-density values are normally distributed about a mean value of 1.1 g/cm^3 (fig. 10). The wide range in bulk density is the result of differences in (1) grain size that are related to distance from the vent and the varied character of layers A to D and (2) amount of compaction, owing to collection over a period of 9 months following the eruption. Because the samples represent a wide range in grain sizes and because thickness measurements were made over a substantial time period while compaction was occurring, we assume that the mean bulk-density estimate is representative of layer C. Wiesner and Wang (this volume) report an estimated bulk density of 1.1 g/cm^3 for fine-grained tephra that accumulated in a submarine sediment trap. Therefore, we use a mean bulk density of 1.1 g/cm^3 to convert the bulk volume range of 3.4 to 4.4 km^3 for layer C to a magma volume of 1.6 to 2.0 km^3 (assumes magma density of 2.4 g/cm^3). Such a range suggests that the climactic tephra-fall deposit is, at most, about equal to, and may be as little as one-half, the magma volume of the climactic pyroclastic-flow deposits (2.1 to 3.3 km^3; W.E. Scott and others, this volume).

Koyaguchi (this volume) estimates a magma volume of 2 to 10 km^3 (4.4 to 22 km^3 bulk volume if bulk density=1.1 g/cm^3) was injected into the stratosphere during the climactic eruption on the basis of a fluid-dynamic model. The upper part of this range would imply a volume significantly higher than our estimate. In terms of the distribution model (fig. 8), the 2-, 1-, and 0.1-cm isopachs would have to enclose areas about 4 to 5 times larger than those determined by Wiesner and Wang (this volume). Alternatively, as inferred by Koyaguchi (this volume), there maybe a substantial fraction of very fine ash that fell at great distances and is not accounted for in our distribution model. But

another 18 km^3 (bulk volume) of ash would have covered an area equivalent to the entire northern hemisphere with 70 µm of ash—an event that probably could not have gone unnoticed.

GRAIN-SIZE VARIATIONS

The climactic tephra-fall deposit is typically normally graded, although some sections have an inversely graded basal portion overlain by a thicker, normally graded upper portion (figs. 1, 4A–C, and 11; table 2). Coarser grained portions of proximal sections consist of friable coarse ash and lapilli with sparse bombs. Farther away, grain size decreases, and the coarsest portion consists of coarse ash with mostly granule-size lapilli. The upper portions of all sections are ash rich, contain scattered lapilli, and typically have sufficient fine ash to make the deposit slightly cohesive. In areas close to pyroclastic-flow deposits, thin ash-rich zones occur within layer C (fig. 4D). These zones are inferred to reflect addition of fine-grained material from ash clouds of pyroclastic flows (W.E. Scott and others, this volume).

Grain-size distributions of the 22 analyzed samples of layer C, which were collected between 15 and 35 km from the vent, are typically unimodal (fig. 11). A second mode in the fine-ash fraction (finer than 4 ϕ) is evident in some distributions; many would likely be expressed as a gradually fining tail, were the finer fractions analyzed. But in a few samples, the second mode approaches about 10 weight percent and is separated from the main mode by low values in the 4-ϕ fraction, which suggests fine ash additions from ash clouds of pyroclastic flows or rain flushing of fine ash from the eruption cloud. The minor coarse mode in sample

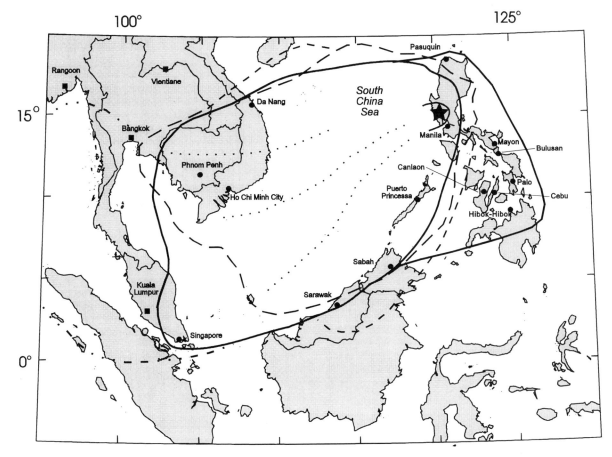

Figure 9. Southeast Asian distribution of distal tephra-fall deposits and edge of ash and aerosol cloud at various times. Solid lines show approximate extents of reported tephra fall. The large area trending west from the Philippines encloses an area of about 3.1 million km², the area covering most of the Philippines represents an additional 0.7 million km² in which minor ash fall was reported, even though much of the area is east of the major eruption cloud as viewed on satellite imagery. Long-dashed line is outline of maximum eruption-cloud extent through 1700 on June 16 from satellite data of Japan Meteorological Agency (JMA; Self and others, this volume); short-dashed line is outline of cloud extent at 0731 on June 16 from NOAA imagery (Lynch and Stephens, this volume) where it extends beyond cloud as shown in JMA imagery; dashed-dotted line shows limits of aerosol cloud on June 18 as depicted in TOMS imagery (Bluth and others, 1992) where it extends beyond limits of previous clouds. Dotted lines are axes of lobes of the ash cloud as shown on JMA imagery (Self and others, this volume). Star, Mount Pinatubo; solid line close to Pinatubo is 1-cm isopach; small filled squares, major cities without reported tephra fall; filled circles, localities that reported tephra fall. Exact sites of tephra-fall observation in states of Sabah and Sarawak, Malaysia, are not known.

MGS–10C1 is attributed to its proximity to the volcano. At such a distance, outsized pumice clasts are very common.

Median grain size ranges from fine-granule-size lapilli to medium-sand-size ash $-1.1-1.8\ \phi$, or 2.2–0.3 mm; figs. 4 and 11; table 2). Where several samples were collected vertically through normally graded sequences (identified as sublayers C1, C2, and so forth, from base upward), the median grain size typically fines upward by 1 to 2 ϕ units. All layer-C samples analyzed are well sorted (σ_ϕ=1-2), whether the entire section (total deposit thickness=10 to 18 cm) or 2 to 4 vertically arrayed subsamples (total deposit thickness=15 to 25 cm) were sampled. Sorting improves slightly (about 0.3 µ unit) upward in most sampled sections as median grain size decreases. Although samples represent a limited range in distance from the volcano (15 to 35 km),

the Pinatubo deposit displays well several features found in many plinian tephra-fall deposits as distance from the vent increases including (1) decrease in median grain size, (2) decrease in maximum pumice size, and (3) improvement in sorting (fig. 12; table 2; Walker, 1971).

COMPONENT ANALYSES

The frequency of pumice, crystals, and lithic fragments in the climactic pumice-fall deposit was determined by point counting each size fraction of 18 of the 22 grain-size samples (fig. 11) in order to characterize the deposit. In most samples, the number of clasts in -3-ϕ (8 mm) and coarser fractions is small (a few to several tens); therefore, the frequency percentages in these fractions have high

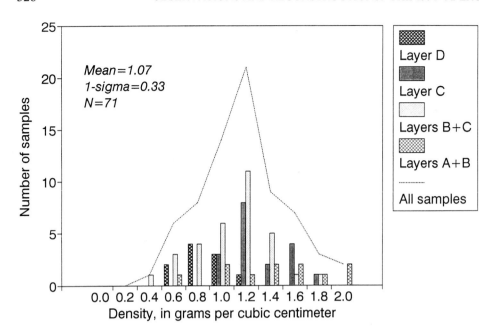

Figure 10. Histogram of dry bulk densities of tephra-fall deposits, which includes 11 samples of layers A+B or B, 18 samples of layer C, 32 samples of layers B+C, and 10 samples of layer D.

Table 2. Statistical measures of grain size (Inman, 1952) of layer C samples, the climactic tephra-fall deposit.

[Sample codes that end in C represent composite samples of the entire deposit; codes that end in a numeral represent samples of sublayers with 1 being a basal sample. Thickness is for total deposit. Maximum pumice size was measured in a 50×50 cm area. Sample locations shown on figure 7. –, no data]

Sample	Medium grain size (md$_\phi$)	Sorting (σ_ϕ)	Maximum pumice (cm)	Total thickness (cm)
CAB 1C	0.19	1.68	2.5	15
CAB 2C	0.20	1.58	3.9	10
SNZ C	0.71	1.24	–	18
SAZ C	1.13	1.11	2.9	16
SMZ C	0.71	1.03	–	17
PRC 1C	0.33	1.33	3.0	15
PTA 2C2	1.14	1.22	–	18
PTA 2C1	0.11	1.46	4.4	18
LLC 6C3	1.76	1.01	–	18
LLC 6C2	1.02	1.21	–	18
LLC 6C1	−0.50	1.36	2.5	18
CAP 8C3	1.41	1.14	–	20
CAP 8C2	0.28	1.35	–	20
CAP 8C1	−0.38	1.40	3.9	20
MGS 10C2	0.42	1.61	–	15
MGS 10C1	−1.11	1.93	9.8	15
PSG 11C2	−0.52	1.65	–	18
PSG 11C1	−0.58	1.55	4.1	18
KLN 12C4	1.29	1.45	–	25
KLN 12C3	1.05	1.31	–	25
KLN 12C2	0.77	1.22	–	25
KLN 12C1	0.11	1.18	–	25

uncertainties and are not considered in the following discussion.

Two varieties of pumice occur in layer C—a white, phenocryst-rich type and a buff to gray, phenocryst-poor type—and were differentiated in the component analyses. They vary only in appearance and are chemically similar (Pallister and others, 1992, this volume). As grain size decreases below 0 ϕ (1 mm), pumice is fragmented into glass and crystals, and discrimination of the two pumice types becomes difficult. Most fine-grained glass is counted here as phenocryst-rich pumice. Phenocryst-rich pumice is the dominant component in fractions coarser than 0 ϕ (fig. 11). The frequency of phenocryst-poor pumice is subordinate to the phenocryst-rich variety, typically <10 percent of the pumice content of a given size fraction. A few sites show a slight decrease in the frequency of phenocryst-poor pumice upwards through the deposit, but such changes are typically not significant at a 2-σ level of uncertainty.

David and others (this volume) also determined the proportion of the two pumice types in layer C east of the volcano. Their results are similar to ours, but they report a greater range in the content of phenocryst-poor pumice, 5 to 25 percent. Their study showed a decreasing amount of phenocryst-poor pumice upward in the deposit, which suggests its content decreased as the eruption progressed. However, as mentioned above, layer C fines upward, and the decreasing-upward amount of phenocryst-poor pumice may also be attributed to nonrecognition in the upper, finer-grained portion.

Crystals occur in the 0-ϕ fraction of all samples and increase in frequency in progressively finer fractions (fig. 11). They comprise as little as 1 percent of the total components in the 0-ϕ fraction to greater than 90 percent in 3-ϕ (0.25 mm) fraction. Plagioclase and hornblende are the

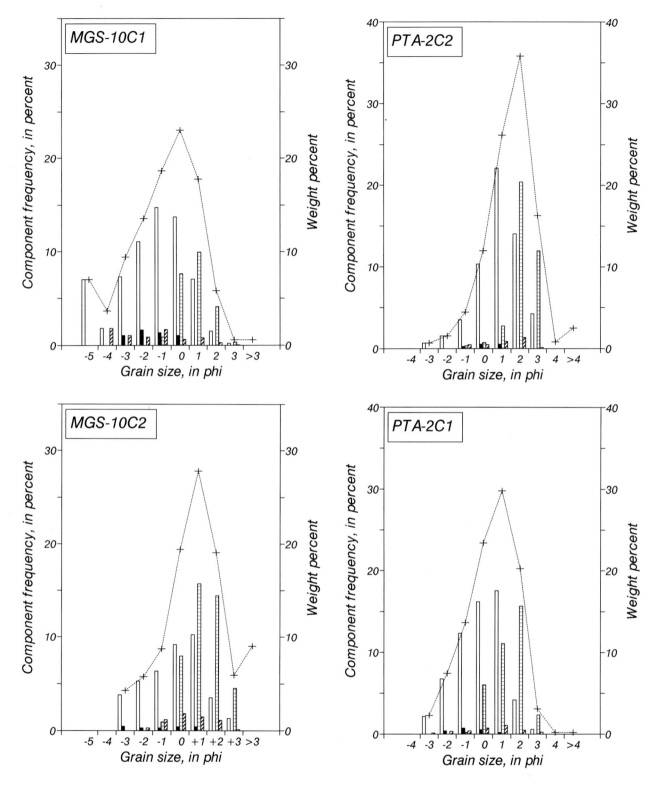

Figure 11. Grain-size distribution (in weight percent; dashed line) of layer C and frequency distribution (bar graph) of its component pumice (phenocryst rich and phenocryst poor), crystals, and lithic fragments. Sample locations shown in figure 7. Sublayers C1, C2, and so forth are numbered from base upward; samples SMZ–C and CAB–1C include the entire layer.

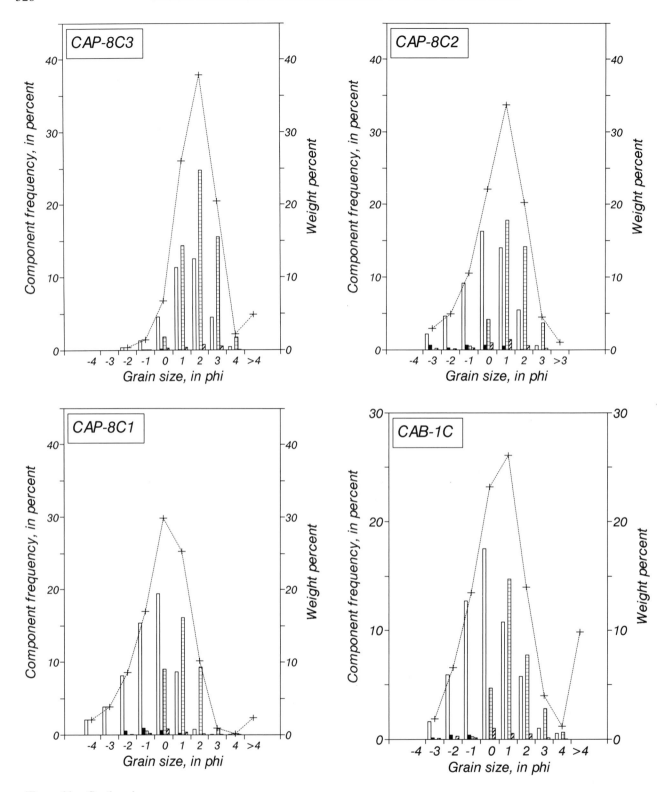

Figure 11.—Continued.

most common phases, while quartz, biotite, magnetite, olivine, and anhydrite are present in minor amounts. The felsic minerals almost wholly comprise the crystal component of the 0-ϕ fraction; an increasing amount of the mafic minerals was observed in finer fractions.

Lithic fragments are found in minor amounts (fig. 11). In size fractions with a statistically significant number of clasts, frequency of lithics ranges from <1 to 13 percent. Average frequency of lithics in a given size fraction for all samples averages 2.5 to 5 percent. No definitive trends are

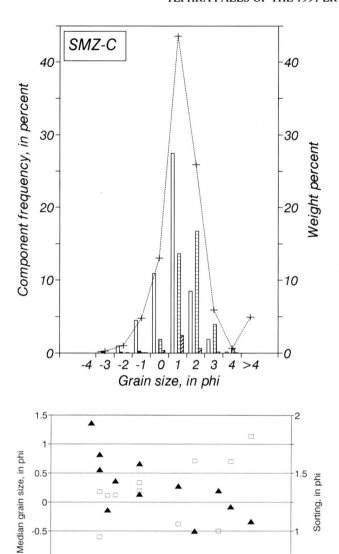

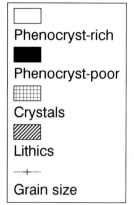

Figure 11.—Continued.

Figure 12. Plot of median grain size (Md$_\phi$) and sorting (σ_ϕ) of layer C against distance from vent. For sections that were sampled in sublayers, only the basal (C1) sample is plotted.

apparent in their distribution with respect to distance from source (15 to 35 km), position in the section, or grain-size fraction. The lithic components consist of red to brown pre-1991 hornblende andesite in various stages of alteration, and dark-gray, angular fragments of pre-1991 and early-June-1991 andesite.

INFERENCES REGARDING DYNAMICS OF JUNE 15 CLIMACTIC ERUPTION

The grain size and grading of the climactic tephra-fall deposit provide insight into eruption processes and

dynamics. The nearly ubiquitous normal grain-size distribution (unimodal) of the samples indicates a uniform mode of transport—fallout from a turbulent cloud—throughout the climactic eruption. Of the 22 samples analyzed, only one proximal sample showed a minor secondary coarse mode.

The normal grading, or mostly normal grading with a thin inversely graded base, that characterizes the deposit in all sectors likely reflects an initially rapid increase of eruption intensity. Coarse particles attained their maximum dispersal early in the eruption; the small amount of fines in basal samples is consistent with an initially high eruption intensity (great column height) and resulting effective winnowing of finer particles by wind. Following this reasoning, the gradual decrease in grain size upward would indicate a decrease in eruption intensity, although coarse particles do occur throughout the deposit, albeit sporadically (figs. 4A and B). Observations of seismicity and atmospheric disturbances during the eruption have been interpreted by most workers as indicating an initial 3-h period of intense activity starting at about 1340 followed by about 6 h of markedly lower and slowly decreasing activity (Power and others, Tahira and others, Wolfe and Hoblitt, W.E. Scott and others, all this volume). Two alternative correlations of this eruptive behavior with variations in the tephra-fall deposit are possible: (1) the coarser, basal part of layer C represents the 3-h peak, and the upper part represents all or part of the 6-h period of declining activity or (2) all of layer C accumulated during the 3-h peak, after which the fine ash of layer D began to accumulate.

Several observations lend support to the second option or a variation thereof:

1. Koyaguchi and Tokuno (1993) show from satellite imagery that the stratospheric cloud expanded upwind (northeast) until 1940, 6 h after the climactic eruption began. They considered this support for 5 h of intense activity, but Koyaguchi (this volume) concludes that this upwind expansion could have continued after its source eruption had ceased, and he thinks a 3-h duration is more likely.

2. Eyewitnesses from the Pinatubo Volcano Observatory (J.W. Ewert and A.B. Lockhart, U.S. Geological Survey, oral commun., 1994) left Clark Air Base about 1500 and drove slowly eastward 15 km to Pampanga Agricultural College, arriving about 1630. They recall that the coarsest grained phase of the tephra fall had largely stopped by the time they left the base, that during their drive the tephra fall consisted chiefly of sand-size material with minor mud, and that on the morning of June 16 their vehicle had little tephra on the windshield, whereas those of vehicles that had been at the college throughout the afternoon of June 15 had several centimeters. These recollections support the interpretation that the part of the eruption that produced layer C was in the initial 3-h period.

3. W.E. Scott and others (this volume) interpret the appearance of ashy zones in the middle and upper parts of layer C near the margins of the pyroclastic-flow deposits as reflecting the onset of substantial production of pyroclastic flows and attendant ash clouds. This effect may also have contributed to the increase in ash content upward in more distal sections of tephra-fall deposits. The production of pyroclastic flows ceased during caldera formation, and only postclimactic layer D overlies the pyroclastic-flow deposits. W.E. Scott and others (this volume) marshall evidence that suggests but is not conclusive that caldera formation occurred 3 to 5 h after the beginning of the climactic eruption. No doubt, most of layer C accumulated during the 3-h peak, but deposition may have continued through all or part of the 6-h period of declining activity.

POSTCLIMACTIC TEPHRA FALL

As the climactic eruption waned during the evening of June 15 and for about 6 weeks following, ash plumes that rose as high as 18 to 20 km billowed from one or more vents on the caldera floor (first half of phase VII of Wolfe and Hoblitt, this volume). Ash also was derived from secondary pyroclastic flows (Torres and others, this volume) and secondary explosions generated by interaction of surface and ground water with the still-hot pyroclastic-flow deposits. The deposits of layer D represented by the isopachs in figure 13 exclude sediments that we infer as having been emplaced by secondary processes. We identify these secondary deposits chiefly on the basis of crossbedding, which is characteristic of surges driven by secondary explosions, large content of clasts derived from pre-1991 deposits that underlay the pyroclastic-flow deposits, and a coarse-grained (very coarse ash to bombs) component derived from the pyroclastic-flow deposits. The secondary emplacement processes also produced distal fine-grained ash that, because it can't be differentiated from ash-fall deposits derived from the vent, is included in the measurements in figure 13. Therefore, the calculated volume of layer D, 0.2 km³ bulk, or about 0.1 km³ DRE (fig. 5; table 1), should be regarded as a maximum value for postclimactic ash erupted from caldera vents.

The tephra of layer D is dominantly gray fine-grained ash (figs. 4B and D; W.E. Scott and others, this volume, fig. 17). Accretionary lapilli formed of fine ash are common throughout the layer, as are zones of vesicular ash formed by rainfall on dry powdery ash. Layer D is laminated to thinly bedded; in proximal areas where it approaches 1 m thick, individual beds rarely exceed 1 cm and are more typically a few millimeters thick. Southwesterly and east-northeasterly winds dominated during this period of ash emission, as is evident from the distribution shown in figure 13. Although most of layer D accumulated in sparsely settled areas and areas already devastated by the climactic eruption, some densely settled areas received up to 10 cm of ash. Fine ash falling continuously for several days at a time and being constantly resuspended by wind and vehicles made life miserable for many residents in northern Pampanga and southern Tarlac Provinces.

METEOROLOGICAL CONTROLS ON DISTRIBUTION OF TEPHRA-FALL DEPOSITS

The Asian monsoon imparts seasonal variation to the wind regime in central Luzon, which is dominated by the northeasterly to easterly trade winds. From June through September, a southwesterly monsoonal flow affects the lower troposphere, while easterly winds continue in the upper troposphere, as well as in the stratosphere, which lies above about 17 km (Pettersen, 1969). These dominant wind directions were used to delimit two main tephra-fall zones in the hazard map constructed in May and June 1991 (Punongbayan and others, this volume). At the time of increasing unrest in May and June, the southwesterly flow was anticipated to begin at any time. Meteorological data for assessing likely tephra-fall patterns during the 1991 eruptions were provided by weather-forecast offices at Cubi Point Naval Air Station and Clark Air Base (fig. 14). There was, however, a gap in the data on June 14–15 when the two stations were unable to take measurements of winds aloft. The only available wind information for this time period is the 24-h wind forecast.

Distribution of tephra-fall deposits from April to early June, and the tephra-fall deposit of June 12, layer A (figs. 2 and 3), conformed well to the hazard zone depicted for easterly winds. Weather radar at Cubi Point and satellite images showed the tephra cloud of June 12 drifting in a generally west-southwest direction (Oswalt and others, this volume, figs. 2–5), consistent with observed wind directions (fig. 14C). Likewise, layer D, the fine-grained ash-fall deposit of the weeks following the climactic eruption, was distributed

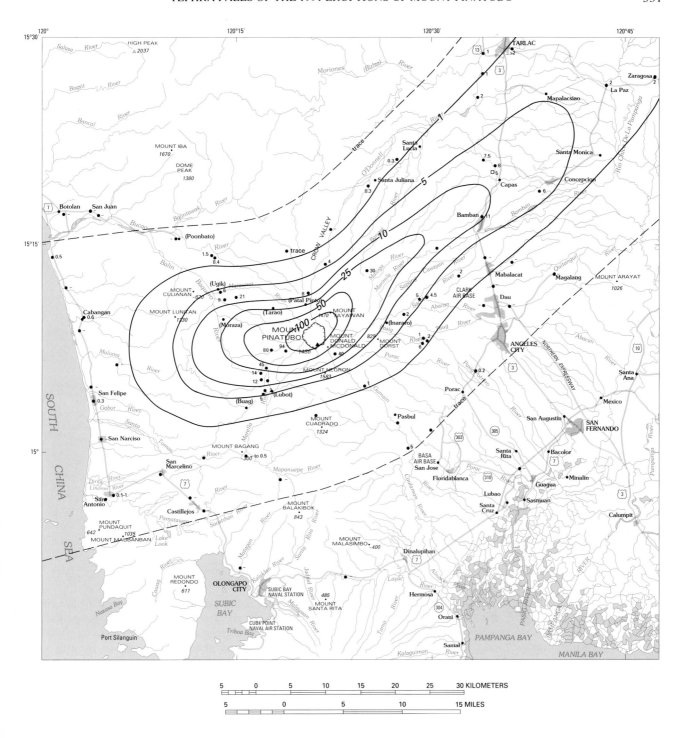

Figure 13. Distribution of tephra layer D, deposits of ash falls that immediately followed the climactic eruption and continued for about 6 weeks (first half of phase VII of Wolfe and Hoblitt, this volume). Isopachs in centimeters; dashed line shows generalized outer limit of identifiable deposits. Sources of data as in figure 3.

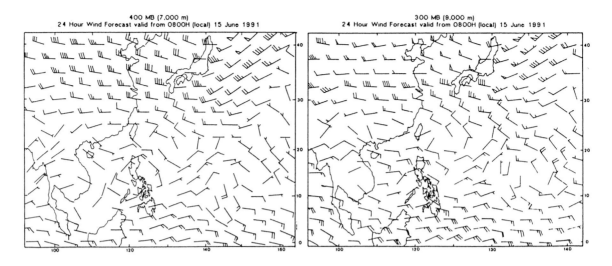

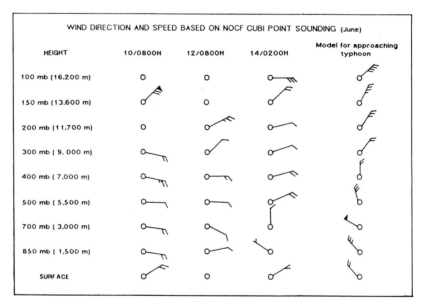

Figure 14. Wind data and wind forecasts for June 10 to 15, 1991 (Weather Office, U.S. Naval Air Station, Cubi Point, written commun., 1992). 24-h wind forecast maps of Southeast Asia for June 15, 1991, at (A) 400 mbar, about 7 km, and (B) 300 mbar, about 9 km in altitude. Bar indicates wind direction; barb is upwind. Full barb represents wind speed of 10 knots (18.5 km/h); half barb, 5 knots (9.3 km/h); triangle, 50 knots (92.6 km/h). C, Summary of observed wind directions at different altitudes above Cubi Point between June 10 and early June 14, 1991, and forecast winds for June 15 based on model for approaching typhoon. Symbols as in A and B.

both west to southwest and northeast of the vent (fig. 13). A southwesterly flow from June 17 to 25 transported much of the ash in the northeast-trending lobe of layer D that affected populated areas in northern Pampanga and southern Tarlac Provinces.

In contrast, the distribution of layer B is very broad and extends much farther to the south (fig. 6), which reflects both changing eruption style and winds. The narrow convecting eruption columns of the first few events that contributed to layer B rose to altitudes of more than 24 km and deposited tephra in much the same fallout zone as layer A. The change to surge-producing explosions on June 14–15 produced lower (typically <20 km), broader (≥10 km radius) eruption clouds (Hoblitt, Wolfe, and others, this volume) that dispersed tephra more widely around the volcano. Concurrently, winds below 5 km were shifting to more northerly and northwesterly directions as the typhoon approached (figs. 14*A* and *C*), which directed much fallout to the south and east.

The magnitude and distribution pattern of the tephra-fall deposit of the climactic eruption, layer C, did not conform to the predictions portrayed on the hazard map for two reasons. First, the magnitude of the tephra fall was underestimated, as field studies had found little evidence of tephra-fall deposits in Pinatubo's record of past eruptions (Punongbayan and others, this volume). Rapid erosion of the climactic 1991 fall deposit, coupled with its essential absence near the volcano (W.E. Scott and others, this volume), probably explains the meager evidence of past tephra-fall events. More importantly, the eye of the typhoon, which had been reduced to a tropical storm shortly after making landfall, was passing just 75 km north of Pinatubo during the climactic eruption (Oswalt and others, this volume, fig. 7). Its passage influenced the tephra-fall distribution by introducing complications to the normal wind pattern as shown in the wind forecast for June 15 (figs. 14*A* and *B*). The forecast shows typical cyclonic winds associated with the storm (counterclockwise rotation) extending over central Luzon from the surface up to about 7,000 m; at 9,000 m and above, the prevailing wind over central Luzon was easterly to northeasterly. Therefore, the portion of the tephra cloud within the effective range of the cyclonic winds was below 9,000 m, between one-third to one-fourth of the 34-km-high eruption column. These cyclonic winds were responsible for deposition of tephra farther south and east than would have occurred under more typical wind conditions.

Another important influence on the distribution of layer C was the great distance ≥250 km) that the mushrooming top of the eruption column flowed radially outward as a density current in the stratosphere (Koyaguchi and Tokuno, 1993; Koyaguchi, this volume). This broad stratospheric plume, in combination with the low-altitude cyclonic circulation, dispersed fine ash fallout over broad areas north, east, and south of the volcano, as well as far downwind (southwest; fig. 9). The major downwind part of the tephra and aerosol cloud displays three axes of dispersal, west-, west-southwest-, and southwest-trending. Also plotted on figure 9 are localities that received fine ash fall, as reported by observers from the Philippine Institute of Volcanology and Seismology, but stand east and south of the cloud's path. Low-level cyclonic winds of the passing typhoon were probably responsible for transporting a small amount of ash to these areas, which extend over much of the Philippine archipelago.

SUMMARY

Tephra falls of changing character and volume occurred between April 2 and midsummer during the 1991 eruptions of Mount Pinatubo. These changes reflect the evolving eruptive behavior that built to a high-intensity plinian eruption and produced voluminous pyroclastic-flow deposits and a small caldera.

1. From April 2 to June 12, first phreatic explosions and later ash emissions related to emplacement of a lava dome produced mostly thin and fine-grained tephra-fall deposits that covered several hundred square kilometers west and south of the vent.

2. A brief explosive eruption on the morning of June 12 deposited about 14 million m^3 of andesitic scoria, ash, and accidental lithic fragments southwest of the volcano (layer A). This event initiated a series of short-lived eruptions that led up to the climactic eruption.

3. Several events similar to that of the morning of June 12 occurred over the next 2 days, but each produced ejecta of smaller volume and finer grain size than layer A. These were followed by numerous pyroclastic-surge-producing eruptions between the afternoon of June 14 and early afternoon of June 15 (Hoblitt, Wolfe, and others, this volume). Together these events emplaced about 0.17 km^3 of laminated, mostly fine-grained ash-fall deposits (layer B) over broad areas around the volcano. The wide dispersal of layer B was induced by ash clouds convecting upward from the pyroclastic surges that moved radially outward ≥10 km from the vent, and by the onset of low-altitude northerly to westerly winds as a tropical storm approached the area.

4. The most voluminous fall deposit of the 1991 eruption sequence is a 3.4- to 4.4-km^3 (bulk) dacitic pumice-fall deposit (layer C) that was produced by the climactic eruption during the afternoon of June 15. This volume probably ranks fifth among 20th century tephra-fall deposits. The climactic eruption also emplaced 5 to 6 km^3 (bulk) of pumiceous pyroclastic-flow deposits and ended with formation of a 2.5-km-wide caldera (W.E. Scott and others, this volume). Most of Luzon and a 3- to 4-million-km^2 area of the South China Sea and Southeast Asia were affected by tephra fall.

5. Grain-size analyses of samples of layer C display well known features of plinian tephra-fall deposits as distance from the vent increases, including decrease in median grain size, decrease in maximum pumice size, and improvement in sorting.

6. Component analyses show that pumice dominates in grain-size fractions coarser than 1 mm, whereas crystals dominate in finer fractions. Lithic fragments make up a few percent or less of each fraction.

7. The thickness and distribution of layer C was not well forecast in the initial hazard assessment because (1) tephra-fall deposits of past eruptions were not well preserved, as evidenced by the rapid erosion of layer C; (2) the eruption column reached a very high altitude (35 km) and mushroomed out widely in the stratosphere, even in the upwind direction (about 250 km from the vent),

which helped to create a broad distribution; and (3) the coincidental passage of Typhoon (later tropical storm) Yunya to the northeast of the volcano caused a shift in low- and middle-tropospheric winds so that tephra was transported farther south than forecast.

8. Deposits of layer C typically have normal grading, which suggests that eruption intensity peaked early and then decreased until ending prior to cessation of pyroclastic-flow activity. Various lines of evidence imply that layer C was deposited in 3 to perhaps as much as 9 h.

9. Slowly diminishing ash emissions continued from several vents in the caldera for about 6 weeks following the climactic eruption and produced a fine-grained laminated tephra-fall deposit (layer D) that has a bulk volume of about 0.2 km^3.

ACKNOWLEDGMENTS

We thank the numerous people who contributed field measurements, descriptions, and samples of tephra-fall deposits, including colleagues at the Philippine Institute of Volcanology and Seismology, Pinatubo Lahar Hazards Taskforce, Mines and Geosciences Bureau, National Institute of Geological Sciences at the University of the Philippines, and U.S. Geological Survey. Personnel at the weather stations at the U.S. Naval Air Station at Cubi Point and Clark Air Base shared weather data and observations; Dr. Emmanuel Anglo and Dr. Jorge de las Alas of the Institute of Meteorology and Oceanography at the University of the Philippines also provided climate information. The manuscript benefited greatly from reviews by Steve Carey, Takehiro Koyaguchi, Chris Newhall, and Ed Wolfe.

REFERENCES CITED

Bluth, G.J.S., Doiron, S.D., Schnetzler, C.C., Krueger, A.J., and Walter, L.S., 1992, Global tracking of the SO$_2$ cloud from the June, 1991 Mount Pinatubo eruptions: Geophysical Research Letters, v. 19, p. 151–154.

Carey, S., and Sigurdsson, H., 1986, The 1982 eruptions of El Chichón volcano, Mexico: Observations and numerical modeling of tephra fall distribution: Bulletin of Volcanology, v. 48, p. 127–141.

———1989, The intensity of plinian eruptions: Bulletin of Volcanology, v. 51, p. 28–40.

Carey, S., and Sparks, R.S.J., 1986, Quantitative models of the fallout and dispersal of tephra from volcanic eruption columns: Bulletin of Volcanology, v. 48, p. 109–125.

Casadevall, T.J., Delos Reyes, P.J., and Schneider, D.J., this volume, The 1991 Pinatubo eruptions and their effects on aircraft operations.

Daag, A.S., Tubianosa, B.S., and Newhall, C.G., 1991, Monitoring sulfur dioxide emissions at Pinatubo volcano, central Luzon, Philippines [abs.]: Eos, Transactions, American Geophysical Union, v. 72, p. 61.

David, C.P.C., Dulce, R.G., Nolasco-Javier, D.D., Zamoras, L.R., Jumawan, F.T., and Newhall, C.G., this volume, Changing proportions of two pumice types from the June 15, 1991, eruption of Mount Pinatubo.

Fierstein, Judy, and Hildreth, Wes, 1992, The plinian eruptions of 1912 at Novarupta, Katmai National Park, Alaska: Bulletin of Volcanology, v. 54, p. 646–684.

Fierstein, Judy, and Nathenson, Manuel, 1992, Another look at the calculation of fallout tephra volumes: Bulletin of Volcanology, v. 54, p. 156–167.

Fisher, R.V., and Schmincke, H.U., 1984, Pyroclastic rocks: Heidelberg, Springer-Verlag, 472 p.

Hildreth, Wes, and Drake, R.E., 1992, Volcán Quizapu, Chilean Andes: Bulletin of Volcanology, v. 54, p. 93–125.

Hoblitt, R.P., Wolfe, E.W., Scott, W.E., Couchman, M.R., Pallister, J.S., and Javier, D., this volume, The preclimactic eruptions of Mount Pinatubo, June 1991.

Inman, D.L., 1952, Measures for describing the size distribution of sediments: Journal of Sedimentary Petrology, v. 22, p. 125–145.

Koyaguchi, T., this volume, Volume estimation of tephra-fall deposits from the June 15, 1991, eruption of Mount Pinatubo by theoretical and geological methods.

Koyaguchi, Takehiro, and Tokuno, Masami, 1993, Origin of the giant eruption cloud of Pinatubo, June 15, 1991: Journal of Volcanology and Geothermal Research, v. 55, p. 85–96.

Lynch, J.S., and Stephens, G., this volume, Mount Pinatubo: A satellite perspective of the June 1991 eruptions.

Magboo, F.P., Abellanossa, I.P., Tayag, E.A., Pascual, M.L., Magpantay, R.L., Viola, G.A., Surmieda, R.S., Roces, M.C.R., Lopez, J.M., Carino, W., White, M.E., and Dayrit, M.M., 1992, Destructive effects of Mount Pinatubo's ashfall on houses and buildings [abs]: International Scientific Conference on Mt. Pinatubo, 27-30 May 1992, Department of Foreign Affairs, Pasay, Metro Manila, Philippines, p. 22.

National Disaster Coordinating Council, 1991, Year-end report: Department of National Defense, Quezon City.

Oswalt, J.S., Nichols, W., and O'Hara, J.F., this volume, Meteorological observations of the 1991 Mount Pinatubo eruption.

Pallister, J.S., Hoblitt, R.P., Meeker, G.P., Knight, R.J., and Siems, D.F., this volume, Magma mixing at Mount Pinatubo: Petrographic and chemical evidence from the 1991 deposits.

Pallister, J.S., Hoblitt, R.P., and Reyes, A.G., 1992, A basalt trigger for the 1991 eruptions of Pinatubo volcano?: Nature, v. 356, p. 426–428.

Pettersen, S., 1969, Introduction to meteorology: New York, McGraw-Hill Book Co., 333 p.

Philippine Institute of Volcanology and Seismology, 1992, Pinatubo awakes from 4 century slumber: Quezon City, PHIVOLCS Press, 36 p.

Pinatubo Volcano Observatory Team, 1991, Lessons from a major eruption: Mt. Pinatubo, Philippines: Eos, Transactions, American Geophysical Union, v. 72, p. 545–555.

Punongbayan, R.S., Ambubuyog, G., Bautista, L., Daag, A., Delos Reyes, P.J., Laguerta, E., Lanuza, A., Marcial, S., Tubianosa, B., Newhall, C.G., Power, J., Harlow, D., Ewert, J., Gerlach, T., Hoblitt, R.P., Lockhart, A., Murray, T., and Wolfe, E., 1991, Eruption precursors or a false alarm? A process of elimination [abs]: Eos, Transactions, American Geophysical union, v. 72, p. 61.

Punongbayan, R.S., Newhall, C.G., Bautista, M.L.P., Garcia, D., Harlow, D.H., Hoblitt, R.P., Sabit, J.P., and Solidum, R.U., this volume, Eruption hazard assessments and warnings.

Pyle, D.M., 1989, The thickness, volume, and grainsize of tephra fall deposits: Bulletin of Volcanology, v. 51, p. 1–15.

Sabit, J.P., Pigtain, R.C., and de la Cruz, E.G., this volume, The west-side story: Observations of the 1991 Mount Pinatubo eruptions from the west.

Self, S., Zhao, J-X., Holasek, R.E., Torres, R.C., and King, A.J., this volume, The atmospheric impact of the 1991 Mount Pinatubo eruption.

Scasso, R.A., Corbella, Hugo, and Tiberi, Pedro, 1994, Sedimentological analysis of the tephra from the 12–15 August 1991 eruption of Hudson volcano: Bulletin of Volcanology, v. 56, p. 121–132.

Scott, W.E., Hoblitt, R.P., Daligdig, J.A., Besana, G., and Tubianosa, B.S., 1991, 15 June 1991 pyroclastic deposits at Mount Pinatubo, Philippines [abs.]: Eos, Transactions, American Geophysical Union, v. 72, p. 61–62.

Scott, W.E., Hoblitt, R.P., Torres, R.C., Self, S, Martinez, M.L., and Nillos, T., Jr., this volume, Pyroclastic flows of the June 15, 1991, climactic eruption of Mount Pinatubo.

Simkin, T., and Siebert, L., 1984, Explosive eruptions in space and time: Duration, intervals, and a comparison of the world's active volcanic belts, in Geophysics Study Committee and others, Explosive volcanism: Inception, evolution, and hazards: Washington, D.C., National Academy Press, p. 110–121.

Smithsonian Institution, 1991, Pinatubo: Bulletin of the Global Volcanism Network, v. 16, no. 5, p. 2–8.

Sparks, R.S.J., 1986, The dimensions and dynamics of volcanic eruption columns: Bulletin of Volcanology, v. 48, p. 3–15.

Torres, R.C., Self, S., and Martinez, M.L., this volume, Secondary pyroclastic flows from the June 15, 1991, ignimbrite of Mount Pinatubo.

Walker, G.P.L., 1971, Grainsize characteristics of pyroclastic deposits: Journal of Geology, v. 79, p. 696–714.

——1981, Generation and dispersal of fine ash and dust by volcanic eruptions: Journal of Volcanology and Geothermal Research, v. 11, p. 81–92.

Williams, S.N., and Self, Stephen, 1983, The October 1902 plinian eruption of Santa María volcano, Guatemala: Journal of Volcanology and Geothermal Research, v. 16, p. 33–56.

Wolfe, E.W. and Hoblitt, R.P., this volume, Overview of the eruptions.

Dispersal of the 1991 Pinatubo Tephra in the South China Sea

By Martin G. Wiesner[1] and Yubo Wang[1]

ABSTRACT

Sea-bottom cores taken across the central South China Sea have revealed that the 1991 Pinatubo tephra blankets an area of at least 37×10^4 square kilometers extending from 10° to 16°N. and 111° to 120°E. The ash is dispersed in a westerly elongated lobe reflecting the prevailing direction of the upper-level winds. Thickness varied from 6.2 centimeters close to Luzon Island to <0.1 centimeter at the continental slope off southeastern Vietnam. The total bulk volume of the airfall ash is calculated at 2.7 cubic kilometers, with the submarine ash volume approximating 0.9 cubic kilometers. The core data underestimate the proximal volume of the tephra deposit because the extrapolated maximum thickness at the source (19.6 cm) is only about 50 percent of the actual value published previously. Bimodal grain-size distributions in the medial part of the tephra layer suggest that both fine- and coarse-mode constituents were transferred to the deep sea in an aggregated form at high sinking speeds. This phenomenon explains the lack of any significant lateral advection of the pyroclasts despite strong ocean surface currents. The bimodality further indicates that secondary thickening may have taken place, but to date no such thickness 'bulge' has been found.

INTRODUCTION

Atmospherically transported volcanic ash is one of the major lithogenic constituents of the unconsolidated sediments in the South China Sea (Niino and Emery, 1961; Chen, 1978; Chen and Zhou, 1992). On the basis of the dispersal patterns of pumice and glass shards in the uppermost sediment sections, Wang and others (1992) have shown that in historic times pyroclasts were mainly sourced from the volcanoes on the Philippine island arc, where the total eruption frequency varies between 4 and 10 per decade (Simkin and Siebert, 1994). Large-magnitude prehistoric eruptions are recorded by numerous discrete ash layers, but their areal extent and volumes are still largely unknown (Wiesner and others, 1994). One of these layers has recently been dated at 33–40 ka B.P. (Chen and Zhou, 1992) and could be traced across the central and southwestern parts of the basin at considerable thicknesses (>4 cm; Wiesner and others, 1994). This tephra may have originated from the largest eruption in the history of Mount Pinatubo, the Inararo episode (>35 ka B.P.; Newhall and others, this volume).

Because most downwind ash plumes derived from the Philippines appear to have been carried across the sea (Kennett, 1981), reconstructions of past eruption intensities require extensions of land-based studies in the marine realm. However, oceanic currents, subaqueous gravity mass flows, or benthic activity may significantly affect the ultimate depositional position and thickness of tephra on the sea floor. This short note reports preliminary data on the thickness and dispersal of volcanic ash in the deep South China Sea injected by the June 1991 cataclysmic eruption of Mount Pinatubo and presents tentative estimates of the total and submarine airfall ash volumes.

PROCEDURES

In May 1994, R/V *Sonne* conducted research operations in the South China Sea to investigate the sedimentary record of short-term Quaternary variabilities of the monsoonal climate in Southeast Asia (Sarnthein and others, 1994). Bottom sediments were sampled by a deep-sea spade box corer assembly to obtain undisturbed sediment surfaces. The core box is $50 \times 50 \times 60$ cm in size and fitted beneath a gimbal-mounted sliding ram over a trapezoidal frame that can be variously weighted to facilitate penetration in sediments of varying character. The corer is also equipped with a pinger system through which lowering, bottom contact, and heaving operations are directly monitored on the ship's 3.5-kHz echo sounder, thereby allowing controlled sampling. Upon penetration, two lids automatically close the top of the box and thus seal the sample against resuspended bottom sediment ('bow wave' effect). After retrieval of the device, the box's front panel is removed to immediately determine the thickness of individual layers and to enable sampling of the sediments in a vertical section.

Of the 65 surface sediments cored, only 10 samples contained the discrete Pinatubo ash layer (a more detailed

[1]Institute of Biogeochemistry and Marine Chemistry, University of Hamburg, Bundesstraße 55, D–20146 Hamburg, Germany.

Table 1. Locations, water depths, and thickness of the Pinatubo ash layer recovered by sea-bottom cores in the South China Sea (SCS–C = sediment trap array).

[Coring locations are shown in figure 1]

Station	Core #	Latitude (N)	Longitude (E)	Water Depth (m)	Thickness (cm)
17921–1	1	14°54.7	119°32.3	2507	6.2
17920–1	2	14°35.1	119°45.1	2507	6.0
17922–1	3	15°25.0	117°27.5	4221	2.0
17923–1	4	15°08.3	117°25.2	1839	2.1
17953–2	5	14°35.8	115°08.6	4309	0.8
17953–3	6	14°33.0	115°08.6	4307	0.8
SCS–C		14°36.1	115°06.4	4270	0.8
17954–1	7	14°45.5	111°31.6	1517	<0.1
17955–1	8	14°07.3	112°10.6	2404	0.1
17956–1	9	13°50.9	112°35.3	3387	0.1
17958–1	10	11°37.1	115°04.9	2581	0.1

sampling of the tephra will be carried out in 1996). The ash was megascopically clearly distinguishable from the underlying and overlying dark olive-green or brownish pelagic sediments by its pale-gray or dark-gray color. Thickness was measured prior to subcoring with a slide caliper and averaged across several points of the core section to account for irregularities in the surface morphology of the hemipelagic sediments (small-scale sediment ridges or microdepressions). Core locations and thickness data are listed in table 1.

First data on the pyroclast properties of the ash bed were obtained for core nos.1, 2, 5, and 6. The samples were subjected to grain-size fractionation carried out by standard dry sieving (<4 phi) and settling (>4 phi) techniques at 0.5-phi intervals down to 2 µm. Component abundances were determined by particle counting in each of the grain-size intervals greater than 6 µm, using a standard petrographic microscope. Electron microprobe analyses were performed on a CAMEBAX 724 at 15 keV with 15 nA sample current (spot size 20 µm on phenocrysts and point beam on glass bubble walls).

ATMOSPHERIC INJECTION AND DOWNWIND PROGRESS OF THE PINATUBO ASH

Magmatic explosions at Mount Pinatubo commenced on June 12, 1991, with a series of vertical and lateral blasts that culminated in a paroxysmal explosion on June 15 at 1342 (Koyaguchi and Tokuno, 1993; Wolfe and Hoblitt, this volume). Ash was ejected to a maximum altitude of 35–40 km, and at about 1420 the cloud spread out laterally into a giant umbrella region between 25 and 30 km altitude. Radial expansion velocity was 125 m/s until 1440, and subsequently the downwind part of the cloud was advected across the South China Sea by high-velocity stratospheric winds (Global Volcanism Network Bulletin, 1991;

Koyaguchi and Tokuno, 1993). About 21 h after the onset of strongest activity, the ash cloud covered the major part of the South China Sea, horizontally extending from 5° to 20°N. and 105° to 120°E. (Global Volcanism Network Bulletin, 1991). By June 18, the plume had already passed the South China Sea and was centered over the Bay of Bengal (Bluth and others, 1992; Stowe and others, 1992).

AREAL EXTENT, THICKNESS, AND TEXTURE OF THE ASH IN THE DEEP SEA

The cores containing the visible Pinatubo ash layer define a tephra blanket that covers an area of at least 37×10^4 km², extending from 10° to 16° N. and 111° to 120°E. (fig. 1). Maximum thickness recovered was 6.2 cm to the east of the Manila Trench, thinning rapidly to 0.8 cm over a distance of about 590 km in a westerly direction (fig. 1). Farther downwind beyond 116°E., ash thicknesses were 0.1 cm except for core no.7 at the northwestern flank of the ash lobe, where the tephra was found to occur in thin scattered patches on the pelagic sediment surface (fig. 1 and table 1).

The progressive decrease in thickness with distance from source is accompanied by significant changes in grain-size and phenocryst abundance. Close to the coast of Luzon (core nos.1 and 2), the deposits consisted of coarse- to medium-sand-sized pyroclasts grading upward into silty ash (fig. 2). The finer grained section was approximately 1 cm in thickness, and the overall textural appearance of both ash beds is quite similar to the June 15 and postclimactic onshore deposits (layers C and D of Paladio-Melosantos and others, this volume). Total phenocryst abundance (plagioclase, hornblende, biotite, quartz, and heavy minerals) averaged 45 percent by volume, and the maximum clast size was 2,000 µm. Toward the medial parts of the ash lobe, normal grading becomes less distinct and the mean grain-size decreases. At core nos. 5 and 6, the bulk of the tephra was silt- to clay-sized, with the phenocrysts comprising about 26 percent of the ash. This is nearly two times lower than noted for the airfall deposits on Luzon Island (Pallister and others, 1992), indicating a loss of crystals during transport. The crystal fraction was dominated by plagioclase (average composition An_{32-35}) and hornblende (Mg# = 67), the latter being occasionally rimmed by clear cummingtonite. Glass shards were rhyolitic (77% SiO_2) in composition with an average refractive index of 1.482. Maximum clast size was 400 µm.

In general, the ash beds were characterized by a sharp basal contact (fig. 2). At thicknesses of less than about 0.1 cm, small-scale burrows and other structures induced by bottom-dwelling organisms were frequent (fig. 2A), but for the massive ash layers (>2 cm) the tephra surface was undisturbed and did not appear to have been significantly reworked by benthic organisms (fig. 2B). The tephra was usually capped by a film of fluffy pelagic material (fig. 2A),

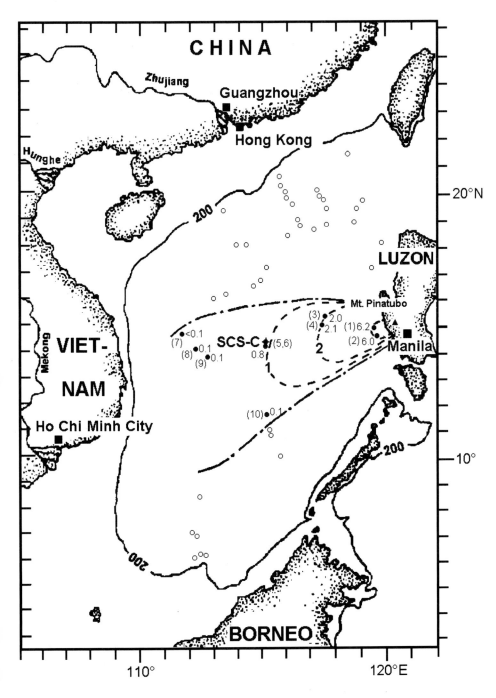

Figure 1. Thickness of Pinatubo fallout tephra (in centimeters) determined from sea-bottom cores (core number in parentheses) and location of the sediment trap system (SCS–C). Dashed-dotted line marks the limit of the discrete ash layer; open circles are cores that lack visible tephra (a detailed description of these cores is given in Sarnthein and others, 1994). Depth contour is 200 m.

an observation that is in agreement with the relatively low fluxes of particulate matter in the central South China Sea, which averaged 82 mg/m²/day for the years 1992–94 (Wiesner and others, in press). Similar flux rates have been reported for the northern South China Sea (Jennerjahn and others, 1992). Assuming a packing density of 2 g/cm³ for particulate matter (Honjo, 1986), the vertical accumulation rate would be 0.015 mm/yr. Because not all of this material reaches the sea floor, due to organic matter mineralization and carbonate dissolution (Wiesner and others, in press), the total thickness of pelagic sediment deposited since the ces-

sation of the eruption should be less than 0.045 mm, which is too thin to measure directly. Higher rates of sediment accumulation during the period following the fallout until May 1994 were obvious for core nos. 1 and 2, where the ash beds were topped by up to about 1 cm of pelagic sediment (fig. 2B). We suggest that this is probably related to lateral advection of suspended riverine material from the nearby Philippines and enhanced in-situ particle production in the surface waters due to upwelling off western Luzon (Pohlmann, 1987).

Figure 2. *A*, Surface view of core no. 9 showing the distal pale-gray Pinatubo ash layer (1 mm in thickness) covered by thin patches of brownish pelagic sediment (length of box edge is 50 cm). Note the presence of scattered burrows and the sharp basal contact of the ash in the lower left region of the core. *B*, Archive box of core no. 2 showing the proximal ash layer in a vertical section. The tephra consists of a normally graded, coarse-grained basal layer 5 cm in thickness topped by 1 cm of fine pale-gray ash. Note the sharp contacts of the tephra layer to the underlying and overlying brownish hemipelagic sediment. Distortion at the upper left side of the section is due to pushing of the archive box into the core.

VOLUME ESTIMATES

On the basis of observed thickness, we have constructed three isopachs (1, 2, and 6 cm; fig. 1) and, by using a planimeter, we calculated the areas contoured by these isopleths at 61,500 km^2 (1-cm segment) and 36,100 km^2 (2-cm segment). The limited number and geographic extent of cored ash beds, however, permit excursions of the 1-cm and 2-cm isopach contours. Considering exponential thinning of the ash layer and interpolation of the thickness between the various core sites along downwind traverses, we calculate the precision to be less than ±5,000 km^2 per segment. Plotting the above data as thickness (log) versus area$^{1/2}$ (fig. 3) and using the equations given by Fierstein and Nathenson (1992), we calculate the slope of the straight line between the two points to be 0.0120. Extrapolating the straight line to area = 0, the maximum thickness (T_0) of airfall ash deposited at the source would be 19.6 cm (fig. 3), and for the total ash volume (V) being defined as $2T_0 \times k^{-2}$ (Fierstein and Nathenson, 1992), we arrive at about 2.7 km^3. However, the value for T_0 is much lower than the actual maximum thickness on land (39 cm; Paladio-Melosantos and others, this volume), suggesting an inflection of the straight line to occur closer to the source. Our data therefore underestimate the proximal ash volume.

Replacing T_0 by the maximum thickness recovered by the cores (6.2 cm) yields a volume of the submarine tephra that approximates 0.9 km^3.

At this stage, it is not possible to give a better estimate for the ash volume deposited in the South China Sea because of insufficient data on the distal and southern parts of the tephra lobe. If it is assumed that core nos. 8, 9, and 10 mark the western boundary of a 0.1-cm thickness contour (fig. 1) and these sites are connected by a straight line, the square root of the area covered by the 0.1-cm segment would be 339 km. Since this value plots below the regression line in figure 2, we conclude that the 0.1-cm isopach must extend much farther to the west.

INTERPRETATION OF DATA

The westerly elongated lobe of the Pinatubo tephra appears to reflect the upper-level wind field that prevailed during the eruption. Upper tropospheric and stratospheric winds were directed to the west and southwest at speeds of 18–24 m/s (H. Houben, NASA, written commun., 1993). Below the 6-km altitude across the South China Sea, winds were constantly blowing to the northeast (H. Houben, NASA, written commun., 1993). The onset of these southwest monsoonal winds took place by the end of May, with

Figure 2.—Continued.

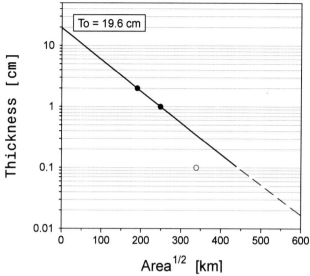

Figure 3. Log thickness versus area$^{1/2}$ plot for the 1- and 2-cm isopachs. Open circle is derived from the assumption that the box core nos. 8, 9, and 10 mark the western limit of the 0.1-cm isopleth. T_0, maximum thickness of airfall ash deposited at the source.

speeds increasing from 4 to 6 m/s in June and reaching a maximum of around 8 m/s in August (Wiesner and others, in press). Long-term records of the oceanic circulation in the South China Sea (see Shaw and Chao, 1994, for summary) have shown that in response to the southwest-monsoon wind vectors, surface currents flow to the east and northeast from June through August at relatively high velocities of 0.4–0.8 m/s; undercurrents flow into the same direction but at greatly reduced speeds (<0.05 m/s). Therefore, ash particles arriving at the sea surface would have been transported back towards the Philippines or into the direction of the Bashi Strait between Taiwan and Luzon Island at 35–70 km/day. The lack of any shift or bending of the axis of the sea-floor ash lobe to northerly directions sug-

gests, however, that if any lateral advection did occur, the quantities of ash transported must have been very low.

Further evidence for the ash deposits being hardly distorted by the current system is provided by data obtained from fully automated collection devices (sediment traps) that were operating during the Pinatubo eruption at 14°36' N., 115°06' E. (fig. 1). The traps were placed along a fixed mooring array at depths of 1,190 and 3,730 m and are designed to collect particulate matter settling through the water column for a designated period in a preprogrammed sequence (Honjo and Doherty, 1988). Within less than 3 days after the release of the major eruption plume, each of these traps simultaneously intercepted a total amount of 9 kg/m^2 of ash (Wiesner and others, in press). On the basis of the mean grain density (2.34 g/cm^3) and average porosity (52.5%) of this material, Wang (1994) calculated a depositional thickness of 0.8 cm. This value is compatible with the data derived from the cores taken close to the trap site (table 1) and indicates that, at least in the medial parts of the lobe, the ash was not redistributed by bottom currents. Furthermore, the ash intercepted by the traps was strongly bimodal in grain size, peaking at 11 and 88 μm (Wang, 1994). Such a distribution is usually indicative of particle aggregation (Sorem, 1982; Cornell and others, 1983; Brazier and others, 1983), and this process may have brought about sufficiently high settling velocities to rapidly carry both fine- and coarse-grained pyroclasts through the strong surficial water currents. The fact that it took only about 58 h from the initial atmospheric injection of ash at Mount Pinatubo to the first registration of the pyroclasts by the traps (Wiesner and others, in press) reveals that subaqueous settling rates at the trap site must have been greater than 1,550 m/day.

The bimodality of the medial ash adds further uncertainty to the volume estimate. The formation of ash clusters may produce secondary thickening of deposits, as has been recognized, for example, in the medial parts of the ash lobe from the 1980 eruption of Mount St. Helens (Carey and Sigurdsson, 1982). This will tend to increase the dispersal of intermediate-thickness isopachs and correspondingly reduce the areal extent of thinner isopachs, leading to steeper slopes in the log of thickness versus area$^{1/2}$ plot. At present, however, the spacing of the sea-bottom cores is not close enough to detect the existence of such a phenomenon in the South China Sea.

The large discrepancy that exists between the horizontal extension of the subaerial ash plume and the deep-sea tephra needs further investigation. We suppose that only small amounts of ash were released from the northern and southern margins of the ash cloud and were easily incorporated into the pelagic sediments by the benthic fauna or diluted by the background sedimentation. Preliminary microscopic inspection of cores adjacent to the ash lobe proved the presence of finely dispersed pumice fragments, glass shards, and phenocrysts largely less than 20 μm in size. On the basis of the data of Ledbetter and Sparks (1979) it roughly takes about 1 year for a 20-μm sized pyroclast with a density of 2.5 g/cm^3 to settle individually through a 4,000-m water column. During this period, pelagic sedimentation in the South China Sea would produce a 0.0015-cm-thick layer (see above); consequently, ash beds of this thickness or less would be hardly recognizable in the cores. In reality, however, vertical settling of fine particles occurs by co-aggregation with larger, faster settling particles, thereby accelerating the sinking speed of fines by many orders of magnitude (Honjo, 1986). This implies that if pelagic dilution has affected the formation of the discrete Pinatubo tephra it should have been effective at much lower thicknesses than indicated by the above value. We believe, therefore, that the northern and southwestern boundaries of the tephra blanket are relatively well defined and that the quantities of ash deposited in the northern and southern South China Sea may not significantly add to the total tephra volume. Our 2.7-km^3 estimate is slightly lower than the 3.4–4.4 km^3 range favored by Paladio-Melosantos and others (this volume).

Within the proximal part of the ash lobe, the massive and 'instantaneous' sedimentation of pyroclasts caused a mass mortality of benthic biota followed by a stepwise recolonization of the ash substrate (Hess and Kuhnt, in press). Even 3 years after the eruption, the benthic community structure was still far from its background levels (Hess and Kuhnt, in press) and, as a consequence, the ash layer was hardly affected by benthic reworking.

ACKNOWLEDGMENTS

We are indebted to the officers and crew of the research vessel R/V *Sonne* for their assistance in coring operations. M. Holmes, C.G. Newhall, K.S. Rodolfo, and W.E. Scott are thanked for their constructive reviews that helped to improve the manuscript. Financial support by the Bundesministerium für Forschung und Technologie (Bonn) is gratefully acknowledged.

REFERENCES CITED

Bluth, G.J.S., Doiron, S.D., Schnetzler, C.C., Krueger, A.J., and Walter, L.S., 1992, Global tracking of the SO$_2$ clouds from the June 1991 Mount Pinatubo eruptions: Geophysical Research Letters, v. 19, p. 151–154.

Brazier, S., Sparks, R.S.J., Carey, S.N., Sigurdsson, H., and Westgate, J.A., 1983, Bimodal grain-size distribution and secondary thickening in air-fall ash layers: Nature, v. 301, p. 115–119.

Carey, S.N., and Sigurdsson H., 1982, Influence of particle aggregation on deposition of distal tephra from the May 18, 1980 eruption of Mount St. Helens volcano: Journal of Geophysical Research, v. 87, p. 7061–7072.

Chen, P.Y., 1978, Minerals in bottom sediments of the South China Sea: Geological Society of America Bulletin, v. 89, p. 211–222.

Chen, W., and Zhou, F, 1992, A study of volcanic glass in the northern South China Sea during the last 100 ka, *in* Jin, X., Kudrass, H.R., and Pautot, G., eds., Marine geology and geophysics of the South China Sea: Beijing, China Ocean Press, p. 174–178.

Cornell, W., Carey, S., and Sigurdsson, H., 1983, Computer modelling of tephra-fallout for the Y–5 Campanian ash layer: Journal of Volcanology and Geothermal Research, v. 17, p. 89–109.

Fierstein, J., and Nathenson, M., 1992, Another look at the calculation of fallout tephra volumes: Bulletin of Volcanology, v. 54, p. 156-167.

Global Volcanism Network Bulletin, 1991, Pinatubo: Washington, D.C., Smithsonian Institution, v. 16, nos. 2–10.

Hess, S., and Kuhnt, W., in press, Deep sea benthic foraminiferal recolonization of the 1991 Pinatubo ash layer in the South China Sea: Marine Micropaleontology.

Honjo, S., 1986, Oceanic particles and pelagic sedimentation in the western North Atlantic Ocean, *in* Vogt, P.R., and Tucholke, B.E., eds., The geology of North America, Volume M, The western North Atlantic Region: Washington, D.C., Geological Society of America, p. 469–478.

Honjo, S., and Doherty, K.W., 1988, Large aperture time series oceanic sediment traps: Design objectives, construction and application: Deep-Sea Research, v. 35, p. 133–149.

Jennerjahn, T.C., Liebezeit, G., Kempe, S., Xu, L., Chen, W., and Wong, H.K., 1992, Particle flux in the northern South China Sea, *in* Jin, X., Kudrass, H.R., and Pautot, G., eds., Marine geology and geophysics of the South China Sea: Beijing, China Ocean Press, p. 228–235.

Kennett, J.P., 1981, Marine tephrochronology, *in* Emiliani, C., ed., The sea: The oceanic lithosphere (Volume 7): Chicago, Wiley, p. 1373–1436.

Koyaguchi, T., and Tokuno, M., 1993, Origin of the giant eruption cloud of Pinatubo, June 15, 1991: Journal of Volcanology and Geothermal Research, v. 55, p. 85–96.

Ledbetter, M.T., and Sparks, R.S.J., 1979, Duration of large-magnitude explosive eruptions deduced from graded bedding in deep-sea ash layers: Geology, v. 7, p. 240–244.

Newhall, C.G., Daag, A.S., Delfin, F.G., Jr., Hoblitt, R.P., McGeehin, J., Pallister, J., Regalado, M.T.M., Rubin, M., Tamayo, R.A., Jr., Tubianosa, B., and Umbal, J.V., this volume, Eruptive history of Mount Pinatubo.

Niino, H., and Emery, K.O., 1961, Sediments of shallow portions of East China Sea and South China Sea: Geological Society of America Bulletin, v. 72, p. 731–762.

Paladio-Melosantos, M.L., Solidum, R.U., Scott, W.E., Quiambao, R.B., Umbal, J.V., Rodolfo, K.S., Tubianosa, B.S., Delos Reyes, P.J., and Ruelo, H.R., this volume, Tephra falls of the 1991 eruptions of Mount Pinatubo.

Pallister, J.S., Hoblitt, R.P., and Reyes, A.G., 1992, A basalt trigger for the 1991 eruptions of Pinatubo volcano?: Nature, v. 356, p. 426–428.

Pohlmann, T., 1987, Three-dimensional circulation model of the South China Sea, *in* Nihoul, J.C.J., and Jamert, B.M., eds., Three-dimensional models of marine and estuarine dynamics: Amsterdam, Elsevier, p. 245–268.

Sarnthein, M., Pflaumann, U., Wang, P.X., and Wong, H.K., 1994, Preliminary report on SONNE-95 cruise 'Monitor Monsoon' to the South China Sea: Berichte-Reports, Geologisch-Paläontologisches Institut Universität Kiel, v. 68, 225 p.

Shaw, P.-T., and Chao, S.-Y., 1994, Surface circulation in the South China Sea: Deep-Sea Research, v. 41, p. 1663–1683.

Simkin, T., and Siebert, L., 1994, Volcanoes of the world: Tucson, Geoscience Press, 349 p.

Sorem, R.K., 1982, Volcanic ash clusters: Tephra rafts and scavengers: Journal of Volcanology and Geothermal Research, v. 13, p. 63–71.

Stowe, L.L., Carey, R.M., and Pellegrino, P.P., 1992, Monitoring the Mt. Pinatubo eruption aerosol layer with NOAA/11 AVHRR data: Geophysical Research Letters, v. 19, p. 159–162.

Wang, H., Zhou, F., and Lin, Z., 1992, Volcanic clasts in the peri-platform carbonate ooze near Zhongsha Islands and their bearing on paleoenvironment, *in* Ye, Z., and Wang, P., eds., Contributions to late Quaternary paleoceanography of the South China Sea: Qingdao, Qingdao Ocean Press, p. 42–55. (in Chinese)

Wang, Y., 1994, Petrographie und Korngrößencharakteristika vulkanischer Aschen des Mount Pinatubo (Luzon, Philippinen) im Südchinesischen Meer: M.Sc. thesis, University of Hamburg, 90 p., unpublished.

Wiesner, M.G., Wang, Y., and Wong, H.K., 1994, Ash layers in the South China Sea: Berichte-Reports, Geologisch-Paläontologisches Institut Universität Kiel, v. 68, p. 195–196.

Wiesner, M.G., Zheng, L. Wong, H.K., Wang, Y., and Chen, W., in press, Fluxes of particulate matter in the South China Sea, *in* Ittekkot, V., and Honjo, S., eds., Particle flux in the ocean: Chichester, Wiley.

Wolfe, E.W., and Hoblitt, R.P., this volume, Overview of the eruptions.

Pyroclastic Flows of the June 15, 1991, Climactic Eruption of Mount Pinatubo

By William E. Scott,[1] Richard P. Hoblitt,[1] Ronnie C. Torres,[2][3]
Stephen Self,[3] Ma. Mylene L. Martinez,[2] and Timoteo Nillos, Jr.[2]

ABSTRACT

About 5.5 cubic kilometers of pyroclastic-flow deposits were emplaced during the climactic eruption of Mount Pinatubo volcano on June 15, 1991, which, combined with plinian pumice-fall deposits, distinguishes the event as one of the five greatest eruptions of the 20th century. Pyroclastic flows traveled as much as 12 to 16 kilometers from the vent in all sectors, impacted directly an area of almost 400 square kilometers, and profoundly altered the landscape. In proximal areas, flows were highly erosive and left little deposit, but, in medial and distal areas, they created broad, thick valley fills and fans of ponded pyroclastic-flow deposits as well as veneers on ridges and uplands.

Pyroclastic-flow deposits comprise three facies. First, by far the most voluminous are massive pumiceous deposits that form the valley fills and fans. Little evidence exists for individual flow or emplacement units except in some upland areas and near distal limits. Second, stratified pumiceous pyroclastic-flow deposits veneer uplands in medial areas and consist of numerous beds several centimeters thick. Stratified deposits grade laterally into, and are therefore cogenetic with, massive pyroclastic-flow deposits. Third, a prominent lithic-rich facies formed of clasts of Pinatubo's former summit dome overlies, or is interbedded in the upper parts of, pumiceous pyroclastic-flow deposits in medial areas of all major drainages.

Stratigraphic evidence indicates that pyroclastic-flow deposits were emplaced during most, and perhaps all, of the plinian pumice fall. We infer that pyroclastic flows were generated by either quasicontinuous, low-level collapse of portions of a sustained plinian column or by repeated brief collapse of the entire column. Either origin is different from classic models of an eruption featuring a period of high plinian column followed by column subsidence into low pyroclastic fountains that fed the bulk of the pyroclastic-flow deposits. The climactic eruption culminated with formation of a caldera 2.5 kilometers in diameter, the onset of which is marked stratigraphically by the abrupt appearance of the lithic-rich facies. We hypothesize that foundering of the caldera (1) produced abundant lithic clasts that were incorporated into the final pyroclastic flows, (2) choked the conduit, and (3) shut down the eruption.

INTRODUCTION

The climactic eruption of Mount Pinatubo on June 15, 1991, is one of the largest volcanic eruptions of the 20th century, as measured by the volume of its products. Relative to past large eruptions, events at Pinatubo were well monitored, and many aspects of eruption dynamics are known or can be inferred with some confidence, even though the climax occurred under poor viewing conditions, owing to the arrival of Typhoon Yunya. This report considers the role of pyroclastic flows in the climactic eruption only; small-volume pyroclastic flows associated with eruptions of June 12–13 (Hoblitt, Wolfe, and others, this volume; Torres and others, this volume), pyroclastic-surge deposits of June 14–15 (Hoblitt, Wolfe, and others, this volume), and secondary pyroclastic flows of 1991 and 1992 (Torres and others, this volume) are discussed elsewhere in this volume. This report includes discussion of (1) origin and behavior of pyroclastic flows, (2) distribution and volume of resulting pyroclastic-flow deposits, (3) types and facies of pyroclastic-flow deposits, (4) timing of pyroclastic flows relative to the major pumice fall of June 15 (Paladio–Melosantos and others, this volume) and other events, and (5) a model for formation of the 2.5-km-diameter caldera.

Owing to logistical and safety considerations, field investigations for this and a companion study (Hoblitt, Wolfe, and others, this volume) were limited, and we consider both to be reconnaissance studies. Fieldwork was restricted to sites around the margins of the area swept by pyroclastic flows that were accessible on foot and by vehicle and to scattered sites closer to the volcano reached by

[1]U.S. Geological Survey.

[2]Philippine Institute of Volcanology and Seismology.

[3]Department of Geology and Geophysics, Hawaii Center for Volcanology, University of Hawaii at Manoa, Honolulu, HI 96822.

helicopter (fig. 1). Aerial photographs and views were useful in mapping the approximate extent of various units, but
readers should keep in mind the lack of ground-based
observations over much of the area.

Pyroclastic flow, as used in this report, is a general
term for hot, gravity-driven density currents of gas and particles that range from dense flows, whose flow regime is
thought to be dominantly laminar, to turbulent flows having
a lower particle concentration. Specific depositional facies
are identified by modifiers that describe dominant clast
lithology or sedimentary structures (for example, massive
and stratified pumiceous pyroclastic-flow deposits; lithic-
rich facies). We avoid the term ignimbrite (see Cas and
Wright, 1987), even though appropriate, because the terms
pyroclastic flow and pyroclastic-flow deposit were used
widely in the Philippines before and after the eruption. Ash
clouds, as used here, are composed dominantly of fine-
grained particles (fine sand and silt) that are elutriated from
and rise above moving pyroclastic flows and can affect
areas beyond the margins of flows (see Cas and Wright,
1987). Ash-cloud deposits may originate by both surge (laterally moving, low-particle-concentration, turbulent currents) and fall (vertical fallout) processes from ash clouds.

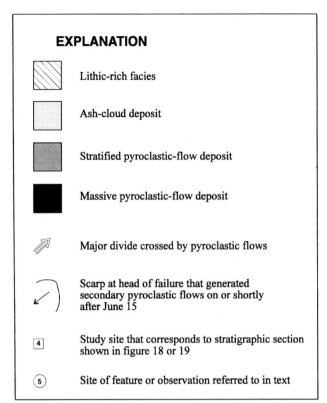

EXPLANATION

Lithic-rich facies

Ash-cloud deposit

Stratified pyroclastic-flow deposit

Massive pyroclastic-flow deposit

Major divide crossed by pyroclastic flows

Scarp at head of failure that generated
secondary pyroclastic flows on or shortly
after June 15

Study site that corresponds to stratigraphic section
shown in figure 18 or 19

Site of feature or observation referred to in text

DISTRIBUTION AND VOLUME OF PYROCLASTIC-FLOW DEPOSITS

Areal distribution, thickness, and volume of pyroclastic-flow deposits emplaced on June 15 provide perspectives
on flow behavior and the major landscape changes they created, as well as key information for various measures of the
size of the eruption. We refer to *proximal*, *medial*, and *distal*
areas that lie at increasing distance from the current caldera
rim. With variations of about 1 km, depending on azimuth,
proximal areas lie within about 2 km, medial areas about 2
to 5 km, and distal areas about 5 to 15 km from the caldera
rim. These modifiers were selected solely to fit the needs of
this discussion; Hoblitt, Wolfe, and others (this volume)
selected a twofold distance classification in their discussion
of preclimactic events. The vent that produced the pyroclastic flows is thought to have been at or near the site of the
June lava dome (Hoblitt, Wolfe, and others, this volume),
which was located near the head of the Maraunot valley
about 1.3 km northwest of the old summit (fig. 1).

AREAL DISTRIBUTION AND THICKNESS

Pyroclastic flows of June 15 and their associated ash
clouds traveled as far as 12 to 16 km from the vent in all
sectors and impacted directly an area of almost 400 km^2
(fig. 1). The great bulk of their deposits is sandy and pumiceous; a minor and areally restricted component forms a
lithic-rich facies, which is composed dominantly of gravel-

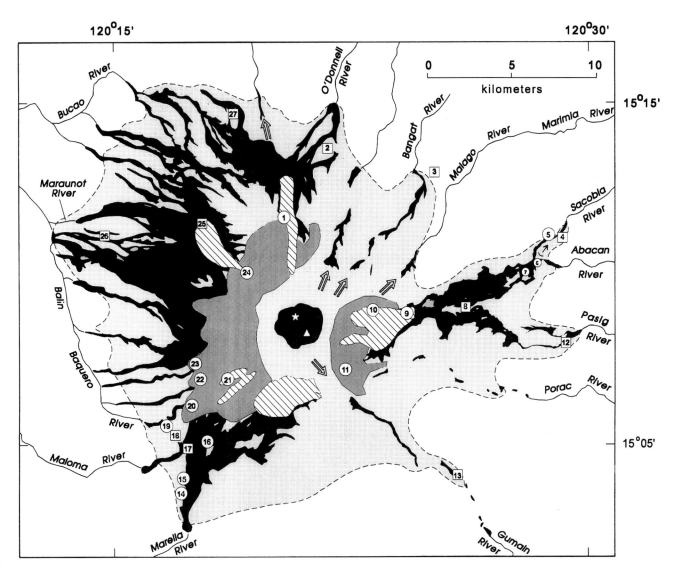

Figure 1. Distribution of pyroclastic-flow and related deposits from the climactic June 15, 1991, eruption of Mount Pinatubo, geographic features referred to in text, and study sites. Triangle marks the former summit; heavy line outlines caldera; star marks approximate location of lava dome of June 7 and presumed location of initial vent. Locations of contacts are based on limited field study only and are inferred largely from aerial observations and photographic evidence. Distal margins of stratified pyroclastic-flow deposits and lithic-rich facies are poorly constrained; inferred presence of lithic-rich facies in eastern part of Marella valley is based on comparisons with other major valleys. Extent of ash-cloud deposits is largely inferred from outer limit of singed or killed vegetation. Patchy distribution of pyroclastic-flow deposits in northeast and southeast sectors (as seen in late June and early July 1991) reflects significant erosion of deposits from narrow canyon floors during heavy typhoon rains that accompanied eruption. Scarp symbol between sites 5 and 6 is drawn off massive pyroclastic-flow deposit for clarity; actual scarp lay in fill of Sacobia valley adjacent to symbol.

Figure 2. Aerial views of Maloma and Marella River valleys comparing preeruption terrain (*A*) and posteruption terrain with fill of pyroclastic-flow deposits (*B*). Orientation of both views is northeast toward Mount Pinatubo, but *B* was taken from a position slightly east of and at higher altitude than that of *A*. Note hill surrounded by pyroclastic-flow deposits near center of *B* and compare with *A*. Main (east fork) channel of Marella River to east (right) of knob has been filled to a depth of about 200 m. (Both photographs by R.P. Hoblitt; *A*, June 3, 1991; *B*, June 23, 1991.)

size, dense lithic clasts. The remainder of this discussion of thickness and volume pertains to pumiceous pyroclastic flows and their deposits; lithic-rich facies will be discussed in a different section.

Distribution of pumiceous pyroclastic flows and their deposits was controlled chiefly by preexisting topography. The asymmetric distribution of deposits shown in figure 1 results from several conditions. Terrain in the west has relatively low relief, and terrain in the east is mountainous. High terrain south and southeast of the volcano may have deflected south- and some southeast-directed flows southwest into the Marella valley. Likewise, north- and northeast-directed flows may have been deflected by high terrain northeast of the volcano northwest into the O'Donnell and Bucao drainages. Furthermore, the vent was probably located in the deeply incised valley of the northwest-flowing Maraunot River. The possibility that strong typhoon

winds enhanced pyroclastic-flow deposition on the west side can be rejected, as forecast wind data for June 15 show that winds below 7 to 9 km in altitude were blowing eastward, and only winds at higher altitudes were blowing westward (Paladio-Melosantos and others, this volume).

The preeruption landscape in the west sector consisted of variably dissected fans and terraces of older pyroclastic and epiclastic deposits (fig. 2*A*). Terrain varied from little-dissected plains to highly dissected badlands. Relief generally decreased outward from several hundred meters in proximal areas to several tens of meters in distal areas. A few prominent isolated hills rose hundreds of meters above this dissected surface. The new June pyroclastic flows were able to spread out across this broad terrain and bury large parts of the preexisting landscape (figs. 1, 2*B*). Medial areas contain typically thin veneers (less than several meters) of pyroclastic-flow deposits on undulating upland surfaces

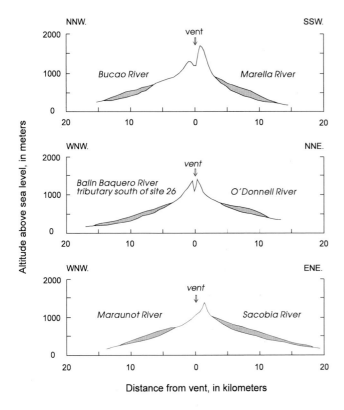

Figure 3. Generalized profiles of former river channels and approximate depth of pyroclastic-flow deposits; approximate azimuth from vent labeled. Thin deposits not shown. Varied profiles near vent reflect rugged topography of preeruption Mount Pinatubo.

with thicker deposits (tens of meters) in some valley bottoms; mean surface gradients of valley fills over areas of several square kilometers are typically 0.08–0.12 (4–7°). Thin, discontinuous deposits occur on steep slopes of canyon walls, ridges, and isolated hills. Distal areas contain broad, thick (up to 200 m) fills of ponded pyroclastic-flow deposits in the southwest sector (Marella River valley), west to northwest sector (unnamed Balin Baquero tributaries and Maraunot valleys), and north sector (Bucao and O'Donnell valleys). Narrow canyon-filling deposits tens of meters thick extend outward from the thick ponds (figs. 1, 3); ash-cloud deposits mantle much of the intervening broad divides and tributary canyons.

In contrast, terrain in much of the east sector is mountainous, with local relief typically 300 to 500 m, steep, rugged ridges, and deep, narrow canyons. The mountainous terrain is bisected by the broad Sacobia and Pasig River valleys, which contained a dissected fill of older pyroclastic and epiclastic deposits. Pyroclastic flows in this area overwhelmed much of the mountainous terrain in medial areas as well as large parts of distal areas, with deposition occurring chiefly in the Sacobia and Pasig valleys but also in several narrow canyons such as the Bangat, Malago, Porac, and Gumain (fig. 1).

Figure 4. Aerial view to northwest showing ponded pyroclastic-flow deposits in upper Bucao River valley leading into narrow outlet channels. Note ridges of preeruption badlands in lower left. (Photograph by W.E. Scott, no. WES–91–12–6, July 5, 1991.)

The thickest pumiceous pyroclastic-flow deposits occur in broad ponds that cover areas of 10 to 30 km^2 and bury former valley floors to depths of 100 to 200 m. Terraces and interfluves between the former valley floors have thinner, discontinuous accumulations. Figure 3 shows generalized profiles of former river channels and estimated surface altitude of pyroclastic-flow deposits. Gradients of most major channels and pond surfaces are about 0.05 (3°); those in the Marella valley are steeper, about 0.07 (4°).

Thinner tongues of pyroclastic-flow deposits lead downstream from the ponds and terminate where valley gradients are typically about 0.01 to 0.02 (0.5–1.0°; figs. 3, 4). In some narrow tributaries of the Balin Baquero and Bucao Rivers, pyroclastic-flow deposits terminate on steeper gradients of 0.02 to 0.04 (1–2°). Pyroclastic-flow deposits in narrow canyons in northeast and southeast sectors also terminate on relatively steep slopes of 0.03 to 0.4 (1.5–2°); their termination probably reflects increased resistance to flow in these narrow, sinuous canyons.

BLOCKED DRAINAGES

Thick valley fills of pyroclastic-flow deposits blocked numerous tributaries to major drainages and created undrained or poorly drained areas, some of which became temporary lakes (fig. 5). Most lakes were shallow and short lived; many drained early in the rainy season of 1991 as they overflowed and outlet channels were rapidly incised through poorly consolidated pyroclastic-flow deposits. Such events produced lahars (Pierson and others, this volume; Umbal and others, this volume). Some lakes filled and drained repeatedly as secondary pyroclastic flows, generated by avalanching of thick pyroclastic-flow deposits (Torres and others, this volume), or lahars rebuilt blockages.

Figure 5. Aerial view to west of blocked tributary along west (left) wall of lower Sacobia River valley (between sites 5 sites 6) showing a new lake, a scarp at head of failure that produced secondary pyroclastic flow shortly after original emplacement, and secondary phreatic explosion craters in right center. Height of failure scarp estimated to be 12 to 20 m. The lake drained during a rainstorm on July 25, 1991, and generated a lahar. (Photograph by W.E. Scott, no. WES–91–10–29, July 2, 1991.)

Figure 6. Aerial view of Holocene(?) diamicts exposed along valley wall on proximal west-southwest flank of Pinatubo. Note etching and fluting of surface by passage of pyroclastic flows, which moved from left to right. Height of slope in foreground is about 50 m. (Photograph by W.E. Scott, no. WES–92–22–20, March 17, 1992.)

Lakes and poorly drained areas in tributaries were also sites of secondary steam explosions driven by heated surface and shallow ground water; such areas were characterized by numerous craters with aprons of surge and fall deposits (fig. 5). Explosions began shortly after the June 15 eruption and continued through at least the next three rainy seasons.

BEHAVIOR OF PUMICEOUS PYROCLASTIC FLOWS

Pumiceous pyroclastic flows were highly mobile in proximal areas and much less so in distal areas. High mobility of pumiceous pyroclastic flows in proximal areas is illustrated by (1) the paucity of deposits in proximal areas, even on relatively flat terrain, and (2) the observation that flows crossed several drainage divides (fig. 1) and descended valleys that do not head directly on the volcano. Pumiceous pyroclastic flows swept over the steep slopes of proximal areas, removed virtually all vegetation and much soil and unconsolidated colluvium, and left only minor deposits in an area of almost 30 km² (fig. 1). Most of the 1991 deposit in this area consists of fine-grained fall deposits from ash plumes that issued from the caldera for several weeks after June 15. The passage of energetic flows etched and fluted canyon walls—some as high as 200 m—composed of pre-1991 volcanic diamicts (fig. 6). Clearly, at least parts of some proximal flows were expanded in order to affect areas high above canyon floors, and all were sufficiently mobile to pass through the area and leave only small patches of pyroclastic-flow deposits in protected areas. We infer that much of the proximal area was in the deflation zone, the zone in which density currents of pyroclasts and

gas were rapidly descending from the eruption column and segregating into density-stratified flows (Walker, 1985). If this inference is correct, the deflation zone extended 3 to 4 km out from the vent.

Pyroclastic flows crossed the 200- to 400-m-high ridge 3 km northeast of the former summit of Pinatubo and flowed down the east tributaries of the O'Donnell River (Bangat and several unnamed streams between the Bangat and main O'Donnell) and the Malago River (western tributary of the Marimla River). The ridge altitude lay close to that of the vent. Flows also crossed lower divides (about 100 m) to enter the Gumain and Porac River canyons in the southeast sector. As previously discussed, the entire volume of flow material directed toward these high ridges did not pass over the ridge because much was transported into west-flowing drainages. Density stratification of flows would favor deflection of relatively dense, lower portions of flows. Blocking of flows (Valentine, 1987) by ridges would accentuate this process by increasing flow density. Although we have no field observations from the obstructing ridge tops, we infer that less dense upper portions of flows rode up and over ridges and then deflated and became concentrated in the narrow valley bottoms.

Yet as reflected by the following two proxies of mobility, pumiceous pyroclastic flows at Pinatubo were not as spectacularly mobile as some other examples. Ratios of vertical fall (measured from the vent altitude) to distance traveled for Pinatubo flows are about 1:11 to 1:14, whereas those of many other ancient and modern flows are as small as 1:30 to 1:75 (Sparks, 1976). Aspect ratios, which are represented by the fraction of average deposit thickness divided by a measure of areal extent (for instance, the diameter of a circle with an area equal to the that of the pyroclastic-flow deposits (Walker, 1983)), further suggest that Pinatubo pyroclastic flows were of comparatively moderate mobility. Aspect ratios for the entire 1991 sequence of Pinatubo flows range from about 1:480 to 1:1,200 (table 1), which fall in the upper range of values obtained for other pyroclastic-flow deposits (Walker, 1983). The larger fraction, or higher aspect ratio, was calculated by using the area of thick valley-fill deposits and thick upland veneer; the lower aspect ratio includes the area of thin veneer. The higher ratio is similar to the aspect ratio calculated by Walker (1983) for the Valley of Ten Thousand Smokes (Alaska) ignimbrite, a so-called high-aspect-ratio ignimbrite.

But both measures used here, vertical-fall-to-runout ratio and aspect ratio, employ assumptions that create uncertainties for comparisons. Actual fall-to-runout ratios for flows probably are larger than those measured, for many flows probably originated at varying heights above the vent from fountains or collapsing portions of eruption columns (Sparks and Wilson, 1976; Hoblitt, 1986). Aspect ratios suffer from comparing sequences containing variable numbers of flows, which obviously influences thickness comparisons. In addition, thin distal or upland veneers of older deposits may have been eroded prior to study, and this erosion would lead to an underestimated deposit area and higher ratios. Redistribution of primary deposits by secondary pyroclastic flows (Torres and others, this volume) beyond the border of primary flows would increase an area estimate and lead to lower ratios. Allowing for these uncertainties, the pyroclastic-flow deposits at Pinatubo probably represent flows of low to intermediate mobility.

VOLUME ESTIMATES OF PYROCLASTIC-FLOW DEPOSITS

Because pyroclastic-flow deposits were emplaced on a highly irregular landscape, measuring their volume is subject to substantial uncertainties. No detailed posteruption topographic maps are yet available that would allow volume determination by differencing preeruption and posteruption topography. We obtained two estimates of volume in the months after the eruption, but recent detailed photogrammetric measurements made by the U.S. Army Corps of Engineers as part of their study of sediment problems are probably the most accurate.

Our estimates used two different maps of valley-filling pyroclastic-flow deposits, one made by Scott and the other made by geologists at the Philippine Institute of Volcanology and Seismology (PHIVOLCS). Both maps used posteruption aerial photographs to map the extent of pyroclastic-flow deposits on the preeruption topographic base. All photographs were uncontrolled, so contacts had to be

transferred by visual inspection. Locating contacts between pyroclastic-flow deposits and the preeruption surface entailed identifying features that had been buried versus those that had not. In areas of well-defined valleys with terraces, spurs, and hills at various heights, the altitude of the contact could locally be determined to within one contour interval (20 m). Contacts were then extended by interpolation between these relatively well-defined points. Uncertainties are greatest in western sectors, where pyroclastic-flow deposits bury broad areas of tens of square kilometers. Altitudes were estimated across these surfaces by constructing broadly convex contours from areas along margins where altitudes could be determined with more certainty. The differences in the two estimates stem largely from different judgments made in placing contacts and in the different methods used to calculate the volume.

Scott used a cross-section method. Figure 7 shows estimated 100-m contours on the surface of 1991 deposits in the Sacobia and Pasig valleys. Mean thickness of 1991 pyroclastic-flow deposits was calculated for cross sections at each 100-m contour interval and multiplied by the area of 1991 deposit between that and adjacent contours to obtain a volume for that reach of the valley. Reach volumes were summed to provide an estimate of the total volume of pyroclastic-flow deposits in the valley. Table 1 lists the estimated volume of pyroclastic-flow deposits for the Sacobia, Pasig, and other drainages.

In addition, veneers of stratified and massive pyroclastic-flow deposits drape broad upland regions between major valleys in medial areas but contribute little to the total volume. If we assume a reasonable mean thickness (5 m) for the 30-km^2 area patterned in figure 1, the volume is appreciably less than 10 percent of the total estimate of about 5 km^3. A much thinner and discontinuous veneer drapes the rugged uplands in east sectors and elsewhere within the margin of ash-cloud deposits shown on figure 1. Mean deposit thickness is probably much less than 1 m in this large area, which accounts for only minor additional volume.

The PHIVOLCS group digitized (at 100-m contours) the preeruption topography and their estimates of posteruption topography, aided by numerous field measurements and estimates of thickness of pyroclastic-flow deposits exposed in gullies and canyons after the substantial erosion that occurred during the rainy season of 1991. From these data, a PC-based contouring program calculated a volume of 4.2 to 4.5 km^3.

The Portland District of the U.S. Army Corps of Engineers (K. Eriksen and M. Pearson, oral commun., 1993) used controlled aerial photographs to map photogrammetrically the boundaries of valley-filling pyroclastic-flow deposits on the preeruption base. They also accounted for much of the veneer of deposits between major valleys. They digitized preeruption topography (20-m contours) and followed a cross-section method similar to, but more detailed

Table 1. Area, thickness, and volume estimates of pyroclastic-flow deposits derived by using valley-cross-section method (see text).

Valley	Area (km²)	Mean thickness (m)[1]	Bulk volume (km³)	USACOE volume (km³)[2]	Aspect ratio[3]
O'Donnell	7.1	39	0.28	0.24	
Bangat, Malago, and others in northeast sector.	3.0	20	.06		
Sacobia, Abacan, and Pasig	21.8	48	1.05		
Sacobia and Abacan				.60	
Pasig				.30	
Gumain and Porac	2.3	15	.03		
Marella and Maloma	21.6	58	1.26	1.40	
Balin Baquero, Maraunot, and Bucao.				3.00	
Upper Balin Baquero tributaries.	8.2	20.7	.17		
Major Balin Baquero tributary south of Maraunot.	23.4	32.3	.76		
Maraunot	10.6	44.1	.47		
Tributaries of Bucao southwest of main fork.	13.2	37.8	.50		
Bucao	9.9	37.9	.38		
Veneer on medial uplands	30	5	.15		
Proximal area of thin, discontinuous deposits.	25	<1	<.02		
Total (± estimated volume uncertainty).	176.1	29.0	5.11 ± 1.0	5.54	1:480[4]
Thin, discontinuous distal deposits.	~150	<1	<0.15		1:1,200[4,5]
Preferred total volume (± uncertainty).			5.5 ± 0.5		

[1] Calculated by dividing volume by area.

[2] U.S. Army Corps of Engineers, Portland District (Karl Eriksen and Monty Pearson, oral commun., 1993).

[3] See text for discussion of aspect ratio.

[4] Based on an average thickness derived from preferred volume divided by area values.

[5] Calculated by adding area of distal deposits to total and adjusting mean thickness to account for increased area.

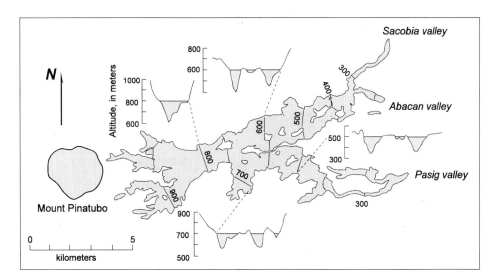

Figure 7. Map and cross sections of pyroclastic-flow deposits in Sacobia and Pasig River valleys used in valley-cross-section method of estimating volume of deposits. Cross sections at 300- and 400-m contours not shown. Heavy line outlines caldera.

than, that described above to measure volumes. Their total volume measurement is about 5.5 km^3 (table 1).

An early volume estimate of 7.1 km^3 made by geologists at PHIVOLCS (J. Daligdig and G. Besana, written commun., *in* Pierson and others, 1992) is considered to be too large. Overestimates of volume in the O'Donnell, Sacobia, Abacan, and Pasig valleys account for most of the discrepancy with the other data sets.

The wide range of volume estimates, about 4 to 7 km^3, is thought to reflect chiefly uncertainties in locating contacts. Significantly, the Corps of Engineers' method, which most reduced this uncertainty through use of controlled photographs and analytical plotters, gave a figure in the middle of the range, 5.5 km^3. But uncertainties in thickness of upland veneer deposits and ash-cloud deposits that cover broad areas still introduce an error of a few tenths of a cubic kilometer, as do uncertainties created by the 20-m contour interval on the preeruption base. Therefore we judge a realistic estimate of the total bulk volume of pyroclastic-flow deposits to be 5.5±0.5 km^3.

We have no measurements of in-situ bulk density from which we can calculate dense-rock or magma volume, because deposits were too friable to sample accurately. We poured representative samples of Pinatubo pyroclastic-flow deposits into 500-mL beakers in the laboratory and got initial bulk density values of 1.1 g/cm^3; with shaking and tamping the density increased to 1.3 to 1.4 g/cm^3. The maximum density we could attain was 1.5 g/cm^3 by tamping each increment as it was added, but the resulting degree of compaction appeared greater than that observed in deposits in the field. Bulk-density values obtained from similar deposits at other volcanoes range from about 0.7 to 1.5 g/cm^3 (Smith, 1960; Riehle, 1973; Riehle and others, 1992; Wilson and Head, 1981). Therefore, we consider that 1.0 to 1.3 g/cm^3 appears a reasonable range of bulk density, which indicates a dense-rock volume of 2.1 to 3.3 km^3. The maximum, probably unreasonably high, dense-rock volume

obtainable by using a density of 1.5 g/cm^3 and bulk volume of 6 km^3 is 3.8 km^3. All estimates use a density of dacitic magma of 2.4 g/cm^3.

FACIES OF PYROCLASTIC-FLOW DEPOSITS

Pyroclastic-flow deposits of June 15 comprise three facies having contrasting clast composition and sedimentary structures—*massive* and *stratified pumiceous facies* and *lithic-rich facies*. The two pumiceous facies, consisting of pumiceous ash, glass shards, crystals, and subordinate pumice lapilli and blocks, constitute the bulk of the pyroclastic-flow deposits. They show no conspicuous compositional grading and are light gray to pinkish light gray when dry, becoming darker gray on wetting. Lithic clasts of older Pinatubo lavas are widely dispersed through the pumiceous units. Most of the pumiceous facies consists of massive deposits with little if any discernible bedding; thickness ranges from tens of centimeters to 200 m. Associated with the massive flow deposits in medial areas are stratified pyroclastic-flow deposits that have conspicuous beds a few to tens of centimeters thick of alternating lapilli and ash. The stratified pyroclastic-flow deposits veneer broad, undulating uplands and are interbedded with and transitional to massive pyroclastic-flow deposits that occupy low areas. Lithic-rich facies are confined to proximal, medial, and near-distal areas of major drainages. They include proximal to medial clast-supported breccias of dense lithic clasts from the former summit dome in a pumiceous matrix, and medial to near-distal pyroclastic-flow deposits that are rich in lithic clasts. In distal areas, *ash-cloud deposits* are interbedded with and veneer areas adjacent to massive pyroclastic-flow deposits. They were deposited by surge and fall from ash clouds elutriated from moving pumiceous pyroclastic flows.

MASSIVE PUMICEOUS PYROCLASTIC-FLOW DEPOSITS

Massive pumiceous pyroclastic-flow deposits constitute the majority of pyroclastic-flow deposits emplaced on June 15. They form large valley ponds in the western sector, broad, thick fills in the Sacobia and Pasig valleys, as well as thinner, narrower fills in canyons in the northeast and southeast sectors (fig. 1). Massive pyroclastic-flow deposits also lie in many lows in uplands of medial areas, where they are associated with stratified pyroclastic-flow deposits. The massive deposits are friable, except where they are moist; fresh exposures of still-hot material easily ravel and collapse. No evidence of welding has been observed. Basal layers (layer 2a of Sparks and others, 1973), from which coarse clasts have been excluded by high shear forces, are displayed locally at the bases of some massive pyroclastic-flow deposits. Ground layers (layer 1 of Sparks and others, 1973), fines-depleted sediments that may underlie the pyroclastic-flow deposit proper, have been observed widely in medial and distal areas. Most are thin (<20 cm) and relatively fine grained (coarse ash, locally containing fine lapilli); the scarcity of coarse-grained ground layers suggests that massive flows were emplaced at comparatively low velocity (Wilson and Walker, 1982).

As of November 1992, exposures revealed only sparse contacts between flow units, although fine ash that typically coats exposures makes observing or tracing flow contacts difficult. Crude bedding is defined by local concentrations of coarse pumice clasts (fig. 8). Sharper contacts, as illustrated in figure 9, are created by a slight enrichment of fines at the top of a flow, which contrasts with the better sorted, relatively fines-depleted base (layer 1) of an overlying flow (sample 23, fig. 10A). Such contact relations suggest sufficient time elapsed between flows for fine ash to settle out (layer 3 of Sparks and others, 1973) and for the surface to firm up enough that the subsequent flow did not disrupt it. Multiple flow units, each unit 1 to 5 m thick, are found mostly in narrow canyon fills downstream from major ponds and in medial-area uplands between ponds, where massive deposits lie in topographic lows. The lack of discernible contacts in deep exposures (up to 80 m) in major ponds (although few ponds have been scrutinized closely) suggests that either single flow units are at least tens of meters thick or the ponded deposits accumulated without sufficiently long time breaks to allow for development of marked contacts. Poorly defined or cryptic contacts in massive ponded deposits would result from the homogeneity of pyroclastic-flow material and the inflated, gas-charged character of recently deposited flow units, which would foster some mixing with successive flows.

Massive pyroclastic-flow deposits are composed dominantly of sand-sized material, which accounts for 70 to 85 wt% of analyzed samples (fig. 10A). The samples are not large enough to represent a statistically valid sample of

Figure 8. Thick (about 20 m), massive pumiceous pyroclastic-flow deposits exposed in canyon cut by west fork of the Marella River (east of site 17, fig. 1); flow from right to left. On figure 2B this section is located between hill surrounded by pyroclastic-flow deposits and point formed by ridge of old terrain left of and slightly below hill. Note pumice-concentration zone (P); otherwise flow is massive without apparent flow-unit breaks. Lahar deposits and moist pyroclastic-flow deposits comprise darker colored material near top of exposure. (Photograph by W.E. Scott, no. WES–92–19–9, March 11, 1992.)

grain sizes coarser than 16 mm (–4 φ), but such excluded coarse material probably constitutes no more than an additional 5 to 20 wt% of sampled volumes. Fine ash (finer than 0.0625 mm, or 4 φ) accounts for only a few percent to a maximum of 18 wt%, which is lower than that of many other pumiceous pyroclastic-flow deposits (table 2), and may account, in part, for the comparatively low mobility of the Pinatubo flows. Possible reasons for this are discussed in another section. The median[4] (Md_ϕ) grain size of most massive pyroclastic-flow deposits analyzed is coarse sand (0 to 1 φ); few are medium to fine sand (1 to 3 φ; table 3, fig. 11). Sorting is typically poor (graphic standard deviation σ_ϕ=1.71–3.07 φ); almost all samples are near symmetrical to fine skewed. Coarse pumice lapilli (64–256 mm) and blocks (>256 mm) are scattered through pyroclastic-flow deposits, with local concentrations (fig. 8), especially near flow margins. Vertical fluid-escape pipes are ubiquitous and conspicuously fines depleted. Compared with certain other pyroclastic-flow deposits as compiled on sorting versus median grain-size plots (Sparks, 1976), massive deposits at Pinatubo are somewhat better sorted (fig. 11).

---

[4]Grain-size parameters used in this discussion are those of Inman (1952) and Folk (1980); modifiers indicating degree of sorting are those of Cas and Wright (1987).

Figure 9. Contact between two pyroclastic-flow units in tributary of Balin Baquero River about 5 km west-southwest of caldera rim (site 23, fig. 1). Locality is in upland area where ridges are veneered with stratified pyroclastic-flow deposits and valley bottoms contain massive pyroclastic-flow deposits. *A*, Outcrop view of two flow units separated by sharp contact. Shovel head is at contact; handle is 43 cm long. Unit above shovel is overlain by darker colored lahar deposits. *B*, Closeup view of contact in *A* showing fine-ash-enriched top of lower pyroclastic-flow deposit (23-mm-diameter coin for scale) contrasting with relatively fine-ash-depleted base of overlying deposit (sample 23, fig. 10*A*; b in fig. 11). (Photograph by W.E. Scott, no. WES–92–22–13, March 17, 1992.)

STRATIFIED PUMICEOUS PYROCLASTIC-FLOW DEPOSITS

Stratified pumiceous pyroclastic-flow deposits, typically less than 5 m thick, veneer much of the dissected uplands in medial areas between extensive ponds of massive pyroclastic-flow deposits (fig. 1). Within the uplands, stratified deposits are locally interbedded with thin, massive pyroclastic-flow deposits and grade laterally into massive pyroclastic-flow deposits as thick as tens of meters that occupy valley floors and other topographic lows. The resulting differential draping reduced and muted former topographic relief in the uplands (fig. 12), but posteruption erosion has largely followed preexisting drainage lines. Stratified deposits are being gullied and stripped from divides, and thick fills along valley bottoms are being intricately incised to near or locally below preeruption levels. The landscape in these areas is now a complex mosaic of 1991 and older deposits.

Stratification is variable in both thickness and clarity (fig. 13*A*). Centimeter-scale stratification is defined by tens of alternating lenses and beds of (1) subround to round pumice lapilli, (2) ash, and (3) slightly cohesive, poorly sorted ash and scattered lapilli that are similar in texture to massive pyroclastic-flow deposits. Some beds of sand-size ash display lamination and cross-lamination on a millimeter scale (fig. 13*B*). Bedding attitudes in stratified pyroclastic-flow deposits range from planar beds that roughly parallel the underlying ground surface (figs. 12, 13*A*) to longitudinal-dune crossbeds with dips of about 5° to 20° that define bedforms tens of centimeters to 1 m or more high and meters to tens of meters long (fig. 13*C*). Contacts are chiefly gradational at scales of millimeters to centimeters; graded contacts at tops of lapilli beds may, in part, represent infilling by overlying ash of originally better sorted lapilli beds. Erosional discontinuities, in which a bed or bed set is locally scoured, occur in many sections. Grain size and sorting of individual beds vary greatly in stratified pyroclastic-

Table 2.   Percent fine ash (silt- plus clay-size fraction; finer than 4 ϕ, or 62.5 μm) of pumiceous pyroclastic-flow deposits from historical and prehistorical eruptions.

Location and deposit	Wt % silt and clay	Reference
Vulsini (Italy) Ignimbrites C and D	10-40	Sparks (1976).
Las Canadas (Tenerife) Granadilla pumice deposit	30	Booth (1973).
Rabaul (Papua New Guinea) Holocene ignimbrite	20-80	Walker (1981); Heming and Carmichael (1973).
Novarupta (Alaska) A.D. 1912 ignimbrite	30-60	Fierstein and Hildreth (1992).
Long Valley (California) Bishop Tuff	10-40	Sheridan (1971).
Mount Mazama (Oregon) Climactic ignimbrite	5-50	Druitt and Bacon (1986).
Mount Pinatubo (Philippines) Pumiceous pyroclastic-flow deposits	2-18	This report.

flow deposits. Lapilli beds are well to poorly sorted (σ_ϕ=1.68 and 2.55 ϕ); sandy beds are well sorted (σ_ϕ=1.31–1.63 ϕ). Representative samples of stratified deposits plotted in figures 10B and 11 are typically better sorted than massive pyroclastic-flow deposits of similar mean grain size and fall near the better sorted edge of the field of pyroclastic-flow deposits of Sparks (1976) on a plot of median grain size versus sorting. Deposits having similar grain size, sorting, and stratification include ignimbrite-veneer deposits at Taupo, New Zealand (Walker and others, 1981; Wilson and Walker, 1982), proximal bedded pyroclastic-flow deposits at Mount St. Helens, Washington (Rowley and others, 1985), high-energy proximal ignimbrite at the Valley of Ten Thousand Smokes (Fierstein and Hildreth, 1992), and stratified lithofacies in the Neapolitan Yellow Tuff, Italy (Cole and Scarpati, 1993).

Stratification changes laterally from sharply to vaguely defined as stratified pyroclastic-flow deposits grade into massive pyroclastic-flow deposits. Such stratigraphic relations, which we have observed most clearly within broad, rolling upland areas (fig. 14), especially on the west side of Pinatubo, demonstrate that both stratified and massive deposits were produced contemporaneously from the same flow; this point supports the inference of Cole and Scarpati (1993) that a pyroclastic flow can entail a continuum of flow processes, from dense, nonturbulent portions that form

massive deposits to turbulent, low-particle-concentration portions that form stratified deposits. In addition, the relation of stratified and massive deposits suggests that the massive pyroclastic-flow deposits accumulated gradually, either from numerous discrete flows or from quasisteady flow, even though there is no evidence for fluctuating depositional conditions in the massive deposits themselves, except locally, as discussed previously. The textural and compositional homogeneity of the deposits would make recognition of contacts difficult. In addition, the inflated gas-charged character of recently deposited flow units would have fostered erosion and mixing by successive flows, further obscuring contacts. Our view of gradual accumulation of the massive fills fits the model of progressive deposition of ignimbrite proposed recently by Branney and Kokelaar (1992).

The widespread presence of stratified pyroclastic-flow deposits at the Valley of Ten Thousand Smokes and Pinatubo, both of which are high-aspect-ratio ignimbrites (see "Behavior of Pumiceous Pyroclastic Flows"), counters the inference of Walker and others (1981) that stratified deposits like their ignimbrite-veneer deposits are well developed only in low-aspect-ratio ignimbrites. Perhaps low-aspect-ratio ignimbrites have stratified deposits developed over a larger proportion of their area, but the presence

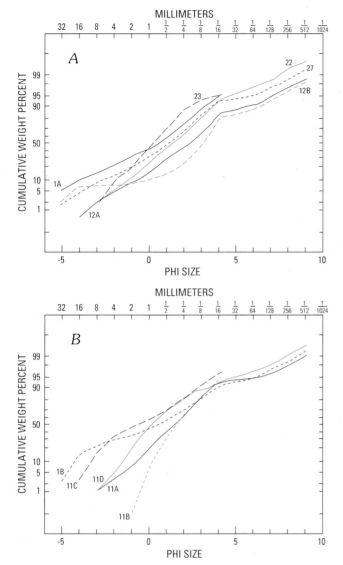

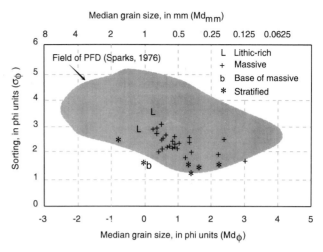

Figure 11. Median grain size (Md_ϕ) plotted against sorting (σ_ϕ) for various facies of pyroclastic-flow deposits defined in text. Stippled area encloses field of pyroclastic-flow deposits compiled by Sparks (1976).

Figure 12. Stratified pyroclastic-flow deposits draping rolling preeruption upland surface between Gumain and Sacobia River basins about 2.5 km southeast of caldera rim (site 11, fig. 1); view to east. (Photograph by R.P. Hoblitt, March 18, 1992.)

Figure 10. Cumulative grain-size plots of pyroclastic-flow deposits. Numbers refer to sites shown on figure 1; letters indicate multiple samples from same site. *A*, Representative massive pyroclastic-flow deposits. *B*, Individual beds of stratified pyroclastic-flow deposits (11A–D) and matrix of lithic-rich facies (1B).

of stratified deposits alone is not firm evidence of a low-aspect-ratio ignimbrite.

LITHIC-RICH FACIES

Lithic-rich facies of pyroclastic-flow deposits range between two endmembers that contrast in clast size and amount of matrix—coarse-grained, clast-supported lithic breccias and matrix-supported, though matrix-poor, lithic-rich pyroclastic-flow deposits. The lithic-rich facies is confined to major drainages that head on the volcano and extends up to 4 to 8 km from the caldera rim (fig. 1). Stratigraphically, lithic-rich facies occur at or near the top of the pumiceous pyroclastic-flow deposits; pumiceous

pyroclastic-flow deposits that overlie lithic-rich facies rarely exceed a few meters.

Lithic breccia, typically less than 5 m thick, displays an interlocked framework of angular to subangular gray to red lithic clasts chiefly decimeters and rarely ≥1 m in length (fig. 15A). The lithic clasts are fragments of Pinatubo's old summit lava dome; the matrix consists of pumiceous ash and lapilli with variable amounts of pulverized summit-dome lava. The two samples of matrix for which we have grain-size distributions are somewhat coarser and more poorly sorted than matrix of massive pumiceous pyroclastic-flow deposits (sample 1B, fig. 10B; fig. 11).

Some of the lithic breccias at Pinatubo resemble near-vent lithic breccias associated with other pumiceous pyroclastic-flow deposits as described by Druitt and Sparks

Table 3. Grain-size parameters of pyroclastic-flow deposits plotted in figures 10 and 11.

[Measures used are those of Inman (1952) and Folk (1980). In ϕ units: Md_ϕ, median; M_Z, graphic mean; σ_ϕ, graphic standard deviation; σ_I, inclusive graphic standard deviation. Dimensionless: Sk_G, graphic skewness; Sk_I, inclusive graphic skewness; K_G, graphic kurtosis]

Sample number[a]	Location[b]	Type[c]	Md_ϕ	M_Z	σ_ϕ	σ_I	Sk_G	Sk_I	K_G
6-27-91 (27*)	D Bucao	M	0.98	0.73	2.36	2.57	-0.16	-0.11	1.24
2-23-92-2 (5)	D Sacobia	M	0.48	0.29	2.13	2.17	-0.13	-0.13	1.12
2-23-92-3 (5)	D Sacobia	M	0.39	0.12	2.04	2.04	-0.20	-0.18	1.03
920225-1 (12)	D Pasig	M	1.35	1.40	2.05	2.36	0.03	0.11	1.31
920226-1 (21)	M Balin Baquero	M	0.80	0.65	2.20	2.48	-0.10	-0.06	1.31
2-27-92-1 (6)	D Abacan	M	0.88	0.77	2.20	2.42	-0.07	0.00	1.21
920302-10 (22*)	M Balin Baquero	M	1.27	1.18	1.86	1.93	-0.07	-0.05	1.07
920302-11 (22)	M Balin Baquero	S	2.23	2.12	1.63	1.85	-0.10	-0.01	1.29
920304-1 (1)	M O'Donnell	M	0.52	0.35	2.54	2.64	-0.10	-0.01	1.08
920304-2 (1B*)	M O'Donnell	L	0.27	-0.24	3.52	3.36	-0.21	-0.09	0.83
3-4-92-1-3 (1A*)	M O'Donnell	M	0.33	0.01	2.72	2.70	-0.18	-0.19	1.02
920311-1 (17)	D Marella	M	0.71	0.67	2.25	2.53	-0.03	0.07	1.27
920311-2 (17)	D Marella	M	0.50	0.31	2.50	2.50	-0.11	-0.10	1.04
920311-3 (17)	D Marella	M	0.65	0.57	2.23	2.42	-0.05	0.00	1.24
3-14-92-1-3 (12A*)	D Pasig	M	2.16	2.01	1.95	2.28	-0.12	-0.03	1.30
3-14-92-1-4 (12)	D Pasig	M	2.34	2.20	2.51	2.80	-0.09	-0.07	1.37
3-14-92-1-5 (12B*)	D Pasig	M	2.98	2.81	1.71	2.61	-0.15	-0.21	2.65
920317-1 (23)	M Balin Baquero	M	0.93	0.88	2.17	2.48	-0.03	0.01	1.30
920317-2 (23*)	M Balin Baquero	B	0.14	0.19	1.52	1.69	0.05	0.09	1.25
920318-1 (11D*)	M Sacobia/Gumain	S	0.00	0.16	1.68	2.02	0.14	0.29	1.44
920318-2 (11B*)	M Sacobia/Gumain	S	1.38	1.49	1.31	1.39	0.12	0.19	1.10
920318-3 (11A*)	M Sacobia/Gumain	S	1.31	1.25	1.67	1.68	-0.05	-0.05	0.99
920318-5 (11C*)	M Sacobia/Gumain	S	-0.78	-0.66	2.55	2.40	0.07	0.12	0.79
92-0228-0416 (24)	M Maraunot	M	1.29	1.25	2.40	2.42	-0.03	0.00	1.00
92-0228-0417 (24)	M Maraunot	M	0.22	-0.08	2.89	2.76	-0.02	-0.13	0.91
92-0228-0418 (24)	M Maraunot	M	0.86	0.95	2.60	2.81	0.05	0.11	1.20
92-0302-XX15 (22)	M Balin Baquero	S	1.66	1.73	1.56	1.71	0.07	0.15	1.18
92-0304-XX08 (1)	M O'Donnell	L	-0.16	-0.09	2.88	3.01	0.03	0.10	1.08
92-1123-0303 (25)	D Maraunot	M	0.86	0.77	2.35		-0.05		
92-1123-0401 (25)	D Maraunot	M	0.33	-0.08	2.94		-0.21		
92-1124-0101 (9)	M Sacobia	M	0.80	0.54	2.45		-0.16		
92-1124-0102 (9)	M Sacobia	M	1.30	1.43	2.58		0.02		
92-1125-0201 (16)	M Marella	M	0.59	0.39	2.67		-0.11		
92-1125-0202 (16)	M Marella	M	0.46	0.06	3.07		-0.20		

Table 3.   Grain-size parameters of pyroclastic-flow deposits plotted in figures 10 and 11—Continued.

[a] Numerals and letters in parentheses correspond with those in figures 1 and 10; * denotes samples plotted in figure 10.
[b] Name of drainage basin: M, medial; and D, distal.
[c] Type of pyroclastic-flow deposit: M, massive pumiceous; B, fines-depleted layer 1 of massive pumiceous; S, stratified pumiceous; L, lithic-rich facies.

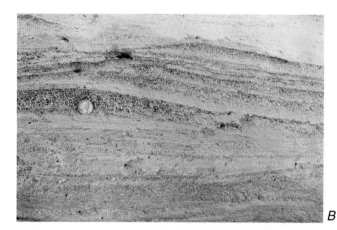

Figure 13.   Stratified pyroclastic-flow deposits exposed in gullies near site 11 (*A* and *C*) and site 23 (*B*). *A*, Planar bedded to weakly crossbedded deposit showing centimeter-scale beds of alternating ash and fine lapilli; flow direction from left to right. Grain-size analyses of samples of some of these beds are plotted in figures 10 and 11. (Photograph by W.E. Scott, no. WES–92–24–30, March 18, 1992.) *B*, Crossbedded stratified pyroclastic-flow deposits showing small-scale dune forms (23-mm-diameter coin for scale); flow direction from left to right. (Photograph by W.E. Scott, no. WES–92–22–9, March 17, 1992.) *C*, Large-scale dune form in stratified pyroclastic-flow deposits showing foreset beds overlain by topset beds that become inclined foresets to right; flow direction from left to right; handle of shovel at base of exposure is 43 cm long. (Photograph by W.E. Scott, no. WES–92–24–2, March 18, 1992.)

(1982) at Santorini (Greece) and Druitt and Bacon (1986) at Mount Mazama (Oregon) but differ in that they occur stratigraphically very near or at the top of pumiceous pyroclastic-flow deposits rather than more widely interbedded through the proximal sequence of pyroclastic-flow deposits (fig. 15*B*). Also, lithic breccias at Pinatubo form a single bed rather than a stratified sequence of units as described elsewhere. Although lithic breccias lie directly

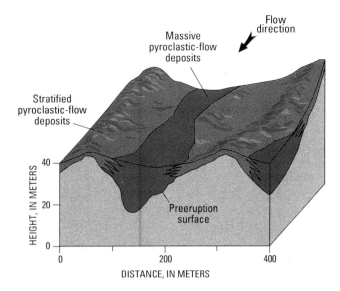

Figure 14. Simplified block diagram showing relationship of stratified and massive pyroclastic-flow deposits in dissected upland terrain of medial area. Veneer (thickness is exaggerated) of stratified pyroclastic-flow deposits (medium gray) on divides intertongues with thick, massive pyroclastic-flow deposits (dark gray) in valley bottom.

on the eroded preJune-15 surface in many proximal areas, downstream tracing of these deposits demonstrates that they lie above all or nearly all of the pumiceous pyroclastic-flow deposits. Earlier pumiceous pyroclastic flows passing these proximal sites left virtually no deposits, whereas lithic breccias represent a lag from later flows; that is a size fraction that could not be transported farther.

The lithic-rich pyroclastic-flow deposits are intimately associated with lithic breccias but are distinguished from the breccias by having sufficient matrix to support coarse clasts. They appear to have moved as single flow units rather than to have accumulated as a lag. Lithic-rich pyroclastic-flow deposits typically overlie or grade upward from lithic breccia in medial areas and extend farther downstream into near-distal areas. Although we have no systematic measurements, clast size appears to decrease upward from lithic breccia to lithic-rich pyroclastic-flow deposits and to decrease downstream in lithic-rich pyroclastic-flow deposits.

ASH-CLOUD DEPOSITS

Distal areas adjacent to pyroclastic-flow deposits contain massive to bedded ash deposits that originated by surge and fall from ash clouds elutriated from moving pumiceous pyroclastic flows. These ash-cloud deposits are typically tens of centimeters thick and are light gray to light pinkish gray; grain size of individual beds ranges from sand with scattered fine lapilli to silt with various mixtures between these endmembers (fig. 16). Stratification is highly

Figure 15. Typical character and stratigraphic position of lithic-rich facies. *A*, Closeup view of lithic-rich facies exposed along O'Donnell River about 5 km north-northeast of caldera rim (site 1, fig. 1) showing clast-supported fabric; much of light-colored area is surface mineral efflorescence. Shovel handle is 43 cm long. Photograph by W.E. Scott, no. WES–92–14–29, March 4, 1992.) *B*, View upstream of lithic-rich facies overlying June 15 massive pyroclastic-flow deposits in 10-m-deep tributary to west fork of Marella River; site is about 4.5 km southwest of caldera rim (site 21, fig. 1). Clasts of lithic-rich facies form lag on eroding slopes, but in situ breccia forms unit about 2- to 3-m thick at top of section (L). (Photograph by W.E. Scott, no. WES–92–9–15, February 26, 1992.)

variable—from faint to well defined, from plane-parallel lamination to low-angle cross lamination. Such deposits mantle valley slopes and uplands adjacent to pyroclastic-flow deposits (fig. 1) and are locally intercalated with massive pyroclastic-flow deposits (see "Relative Timing of Events of the Climactic Eruption of June 15"). As the valleys filled with pyroclastic-flow deposits, ash-cloud deposits of earlier flows were buried by subsequent flows (fig. 16). Ash-cloud deposits are typically restricted to within a few hundred meters of pyroclastic-flow deposits but may extend 1 km or more away on surfaces between channelized deposits that lead downstream from major ponds, such as

Figure 16. Massive to weakly stratified ash-cloud deposits (ac) lying above tephra layers C and B, which lie on former unimproved road surface and below massive pyroclastic-flow deposits (pfl) near mouth of Pasig River canyon (site 12, fig. 1). Total thickness of 1991 deposits is about 1 m. This section is portrayed diagrammatically in figure 18 (section *C*). View is downstream. Photograph by W.E. Scott, no. WES–92–88–5, February 25, 1992.)

between Balin Baquero and Bucao River tributary valleys in the west sector (fig. 1).

STRATA THAT OVERLIE PYROCLASTIC-FLOW DEPOSITS

Pyroclastic-flow deposits of June 15 are mantled by a variable thickness of younger strata that originated by several processes.

1. Ash fell from nearly ubiquitous plumes that rose from the vent for several weeks following the climactic eruption (Wolfe and Hoblitt, this volume). Prevailing winds transported ash chiefly to southwest and northeast sectors, where a mantle up to 1 m thick accumulated in upper tributary valleys of the Balin Baquero and

O'Donnell Rivers; elsewhere, typical thicknesses are 10 to 50 cm. The resulting deposits (layer D of Paladio-Melosantos and others, this volume) contain numerous thin, planar beds and laminae of fine sandy and silty ash as well as lapilli- and coarse-ash-size aggregates of this same material (fig. 17). Some of these aggregates were probably formed as rain fell on recently deposited loose, dry ash, whereas others are nearly spherical and probably formed as accretionary lapilli and ash within ash clouds. The fine ash provides a key stratigraphic marker for separating deposits of June 15, 1991, from younger lahar, phreatic-explosion, and secondary-pyroclastic-flow deposits.

2. Secondary (rootless) phreatic explosions driven by heating of ground and surface waters by pyroclastic-flow deposits have occurred from June 15, 1991, to the time of preparation of this report, chiefly during rainy periods (Pierson and others, this volume). These explosions have created craters and produced surge and fall deposits up to several meters thick that discontinuously mantle pyroclastic-flow deposits (figs. 5, 17).

3. Secondary pyroclastic flows were generated shortly after emplacement of primary flow deposits as well as during the 1991 and subsequent rainy seasons (Torres and others, this volume). The secondary pyroclastic-flow deposits typically overlie erosion surfaces cut into the primary flow deposits. Observations during the first few days after June 15 revealed several large light-colored areas of secondary pyroclastic-flow deposits downstream from steep scarps (figs. 1, 5). These light-colored areas contrasted markedly with the gray surface of surrounding pyroclastic-flow deposits that had been wetted by rains of Typhoon Yunya. The largest of these secondary flows occurred in the large pond of pyroclastic-flow deposits on the west flank of Pinatubo, just south of the preeruption course of the Maraunot River (fig. 1). The head scarp was 10 to 20 m high, and the deposits extended about 7 km downstream to or beyond the limit of primary pyroclastic-flow deposits. The most distal deposits in this drainage may therefore be related to the secondary flows rather than to primary ones.

4. Debris-flow deposits emplaced on June 15 and intermittently up to the present widely mantle pyroclastic-flow deposits. Debris-flow deposits up to several meters thick buried much of the surface of ponded pyroclastic-flow deposits before incision of the drainage network; terminal pyroclastic-flow deposits in many valleys were buried to depths of tens of meters. Lahar deposits resemble the pyroclastic-flow deposits from which they were derived in texture (Pierson and others, this volume), but the lahar deposits are denser and have a firmer consistency in the field.

Figure 17. Postclimactic deposits overlying June 15 massive pyroclastic-flow deposits in Sacobia River valley about 0.5 km south of area shown in figure 5 (site 6, fig. 1). Fine-grained, laminated ash-fall deposits of layer D lie to right of trowel, which is 27 cm long. Coarser grained bedded sediments above layer D were emplaced by fall and base surges generated by secondary (rootless) phreatic explosions in hot pumiceous pyroclastic-flow deposits. (Photograph by W.E. Scott, no. WES–92–19–16, March 13, 1992.)

RELATIVE TIMING OF EVENTS OF THE CLIMACTIC ERUPTION OF JUNE 15

The climactic eruption of June 15 occurred under extremely poor viewing conditions, owing to the arrival of Typhoon Yunya (Oswalt and others, this volume). Satellite observations provide information on behavior of the stratospheric portion of the eruption column (Koyaguchi and Tokuno, 1993; Koyaguchi, this volume; Lynch and Stephens, this volume), but detailed observations of pyroclastic flows, surges, and other events were limited (Hoblitt, Wolfe, and others, this volume; Sabit and others, this volume). Hence, the stratigraphic record provides key evidence for understanding the eruption's course. Deposits of June 12 to early afternoon June 15 are clearly differentiated from deposits of the climax on the basis of lithology and character (Hoblitt, Wolfe, and others, this volume; Pallister and

others, this volume). A variably complete sequence of these deposits mantled much of the landscape within 20 km of the vent by early afternoon of June 15, when the sustained plinian eruption commenced. In distal areas, except along the southwest-trending andesitic scoria-fall deposit of June 12 (layer A of Koyaguchi and Tokuno, 1993; and Paladio-Melosantos and others, this volume), preclimactic deposits are dominantly sand size and finer (layer B of Koyaguchi and Tokuno, 1993; and Paladio-Melosantos and others, this volume); in medial areas they comprise a sequence of numerous graded beds with coarse-grained bases. In contrast, the June 15 pumice-fall deposit (layer C of Paladio-Melosantos and others, this volume) is coarse grained, forming a normally graded bed of dacitic pumice lapilli, ash, and minor lithic debris. The pyroclastic-flow deposits of the climax, as discussed in previous sections, are distinguished by their great thickness and uniform dacitic composition. These climactic June 15 deposits are capped by a variable thickness of one or more types of younger sediments (see "Strata that Overlie Pyroclastic-Flow Deposits").

Stratigraphic relations between layer C and the pyroclastic-flow deposits convey information about the timing and dynamics of the climactic eruption. Below, we summarize data from numerous medial and distal localities that cover all sectors of the volcano. Little information comes from proximal areas, owing to the absence of deposits. In general, stratigraphic sections in distal areas consist of pyroclastic-flow and related deposits overlying or interbedded with layer C, whereas sections in medial areas are composed entirely of pyroclastic-flow deposits with no trace of layer C.

EVIDENCE FROM DISTAL AREAS

Stratigraphic relations at the mouth of the Pasig River canyon (fig. 1, site 12) are typical of those in distal areas (fig. 18). The sections are situated on a vegetated terrace about 100 m south of and 10 to 20 m above the pyroclastic-flow margin (section A), on or near the former valley floor as exposed in what is now a terrace at the edge of the recently downcut channel (section B), and at the margin of pyroclastic-flow deposits on a former road surface along the north valley wall (section C). Each section has a basal layer about 5 to 10 cm thick of fine-grained ash-fall deposits (layer B) related to pyroclastic surges of June 14 and 15 (Hoblitt, Wolfe, and others, this volume). A thin (20 cm) clayey lahar deposit is interbedded in the basal deposits in section B.

The lower, coarse-grained portion of normally graded layer C forms the base of climactic deposits in all three sections. The normal grading of layer C is interrupted by thin beds of silt- and fine-sand-sized ash that may reflect fallout from ash clouds of pyroclastic flows. We infer that

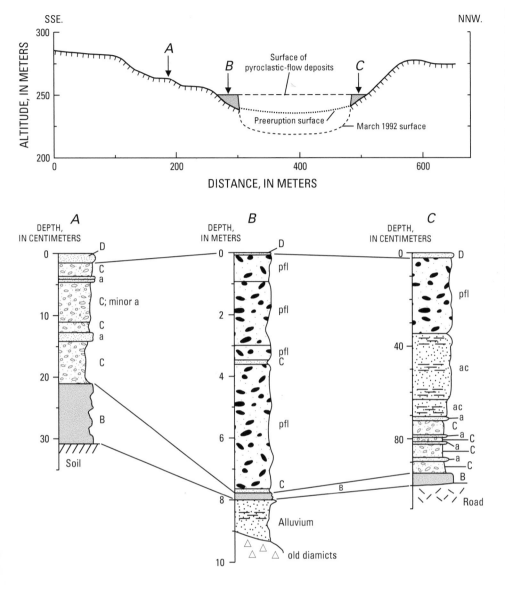

Figure 18. Stratigraphic relations of pumice-fall and pyroclastic-flow deposits at mouth of Pasig River canyon (site 12, fig. 1). Figure 15 is photograph of section *C*. Correlation lines joining columns separate tephra layers D and B from deposits of climactic eruption. Deposits of June 15 climactic eruption: ac, ash-cloud deposits; pfl, pumiceous pyroclastic-flow deposits (solid-symbol pattern); C, tephra layer C (open-symbol pattern); a, ash-rich beds within pumice-fall deposits. Unit a is inferred to be distal equivalent of deposits of pyroclastic flows that terminated upvalley from sections. Note different depth scales for each column.

pyroclastic flows terminated upvalley but their ash clouds continued downvalley to this site. Total thickness of layer C at sections A and C is about 20 to 25 cm, including the finer grained beds. At section C, layer C was followed by ash-cloud deposits and a pyroclastic-flow deposit, whereas section A lay above the reach of pyroclastic flows and received only minor ash-cloud deposits. At section B, in contrast, only about 10 cm of layer C, including an upper finer grained ash bed, accumulated before burial by about 4 m of pyroclastic-flow deposits. Another 5 to 10 cm of layer C accumulated before emplacement of three more pyroclastic-flow deposits. In summary, (1) some of layer C preceded pyroclastic flows reaching this site, (2) ash-rich beds in layer C record pyroclastic flows that terminated farther up valley, (3) the first pyroclastic flow reached this site after about half of layer C had accumulated, (4) layer C continued to accumulate while pyroclastic flows did not reach the

site, and then (5) several more pyroclastic flows reached the site before activity waned, and a thin layer of fine ash (layer D of Paladio-Melosantos and others, this volume) accumulated.

This basic sequence is repeated in other distal areas (fig. 19, sections 18, 26), but at one locality on the Maloma-Marella drainage divide (section 17) there is evidence of pyroclastic flows occurring earlier in the sequence. Here, a total of about 10 to 15 cm of coarse-grained layer C is divided into four parts of subequal thickness by three cohesive beds that we interpret as ash-cloud facies of pyroclastic flows. In nearby areas beyond the pyroclastic-flow margin, layer C is 30 cm or more thick; thus, the pyroclastic flows that produced the ash-cloud deposits must have occurred during the early part of the pumice fall.

In other distal areas, pyroclastic-flow deposits or their ash-cloud facies overlie all or part of layer C (fig. 19,

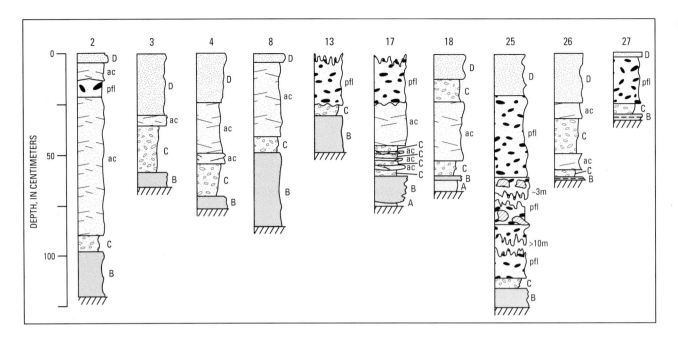

Figure 19. Stratigraphic relations of 1991 pyroclastic-fall and pyroclastic-flow deposits in distal areas. Sections are located on figure 1; symbols same as on figure 18. Unit pfl with lined-symbol pattern (section 25) is lithic-rich facies of climactic pyroclastic-flow deposits. Pyroclastic-flow deposits at top of sections 13 and 17 are several meters thick. Unit B in sections 18 and 26 may include fall deposits of June 12–13 eruptions; unit A is the scoria-fall deposit of June 12.

sections 2, 3, 4, 8, 13, 25, 27). In some of these sections (2, 8, 13, 25, and 27), layer C is thin compared with its thickness in nearby areas that were not swept by pyroclastic flows or their ash clouds. We don't know if this is due to erosion of an initially thicker layer C by subsequent flows, or if flows occurred concurrently with the later part of the pumice fall and the upper part of layer C was incorporated in the flows. Some distal sites, therefore, give evidence that pyroclastic flows were being produced concurrently with the pumice fall, but emplacement of much of the distal pyroclastic-flow deposits may have occurred late in the pumice fall. Such evidence does not preclude pyroclastic flows being generated early in the pumice fall, because we presume that early flows would have encountered the greatest resistance to flow in the narrow, sinuous, preeruption canyons and terminated upvalley. But as filling progressed, resistance to flow would have decreased, and subsequent flows could extend farther downvalley.

EVIDENCE FROM MEDIAL AND PROXIMAL AREAS

The absence of layer C in medial areas, either below thick sequences of flow deposits in valleys or below thin stratified pyroclastic-flow deposits in uplands, is consistent with but not unambiguous evidence for pyroclastic flows coinciding with pumice fall. Nearly continuous or pulsating flows would incorporate falling tephra. If distinct pyroclastic flows occurred, succeeding flows could sweep away any tephra that fell in the likely brief (minutes?) time intervals between flows. Coarse pyroclasts might even become embedded in the inflated, gas-charged tops of recently deposited units. In areas of stratified pyroclastic-flow deposits, turbulent flow conditions would ensure removal of loose pumice-fall deposits.

Whether or not initial pumice-fall deposits mantled the surface of medial areas before the arrival of the first pumiceous pyroclastic flows is difficult to assess. Our only evidence is negative. We have inspected six 1-km² medial sites in which pyroclastic-flow deposits are relatively thin (<15 m) and highly dissected and have found no trace of layer C, even in relatively protected areas behind ridges where the flows killed trees but did not remove them. The bases of pyroclastic-flow deposits lie either on a differentially eroded sequence of preclimactic fall and surge deposits or at some level in the soil or subsoil. Fine-grained surge beds of layer B were relatively resistant to erosion, but once they were removed the underlying coarse-grained layer A was typically entirely stripped away. These scenarios suggest that layer C would have suffered a similar fate. As of this writing, erosion had not yet exposed bases of the thickest sequences of ponded pyroclastic-flow deposits in medial areas, which may have provided sites for preserving early pumice-fall deposits. The presence of at least part of layer C below thick pyroclastic-flow deposits in section 25 (fig. 19) in the near-distal Maraunot area suggests that some pumice-fall deposit will probably be found as erosion exposes more widely the base of the pyroclastic-flow deposits. In summary, we cannot yet determine the relative timing of the beginning of pyroclastic-flow activity with regard to the start of pumice fall, but strong stratigraphic evidence demonstrates that pumiceous pyroclastic flows were generated during much of the pumice fall.

MODEL FOR ERUPTIVE EVENTS OF THE JUNE 15 CLIMAX

Our preliminary investigations of pyroclastic-flow and related deposits of the June 15 climactic eruption of Pinatubo volcano have led us to several key findings that are summarized in the context of the following model (fig. 20).

SIMULTANEOUS PLINIAN PUMICE FALL AND PYROCLASTIC FLOWS

Stratigraphic relations described in the previous section are consistent with broadly synchronous emplacement of pyroclastic-flow and pumice-fall deposits during much of the climactic eruption. Pyroclastic flows must therefore have been spawned either by (1) gravitational collapse of dense portions of a sustained plinian column that was simultaneously producing the pumice fall (fig. 20A) or by (2) alternating collapse and convective rise of the column, which is not depicted in figure 20. Hourly satellite observations during the climactic phase on June 15 (Koyaguchi and Tokuno, 1993) do not show changes in the stratospheric portion of the column directly above the vent that might be interpreted as reflecting extended periods of column collapse. Likewise, the normal grading of layer C (Paladio-Melosantos and others, this volume) suggests that the eruption column was not interrupted by prolonged periods of collapse. If the column had repeatedly collapsed and reformed over long time intervals, layer C should consist of alternating fine- and coarse-grained beds. Thus, if the column was alternating between collapse and convective rise, periods of collapse must have been brief (minutes?). Either origin for pyroclastic flows is different from that inferred initially by Scott and others (1991), who, from preliminary investigations prior to appreciable incision of the deposits, thought that pyroclastic flows largely followed deposition of layer C, as is consistent with classic models of late-stage column collapse (for example, Sparks and Wilson, 1976).

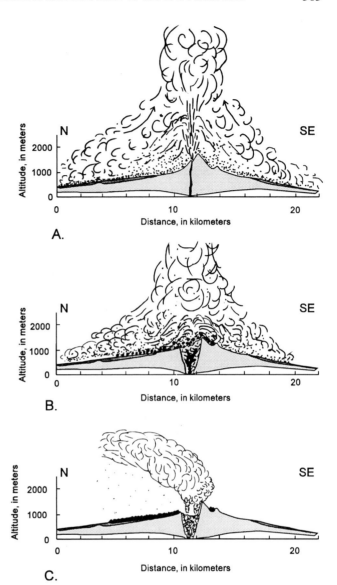

Figure 20. Schematic model for events of June 15 climax as viewed along section from O'Donnell valley (on left) across Mount Pinatubo and into upper Gumain valley (on right); see discussion of model in text. *A*, Conditions that prevailed during most of climactic eruption including plinian column, parts of which were collapsing to form pyroclastic flows. Single vent is shown in position of June 7–15 dome. Long arrows illustrate low-level air circulation into the convectively rising column. *B*, Conditions as caldera is foundering and single vent is destroyed; geometry of caldera is inferred. Large dots (not to scale) represent production of lithic clasts that were left near vent as lithic-rich facies. *C*, Postclimactic conditions as ash plumes rise from several vents in caldera floor. Lithic-rich facies (thickness exaggerated) locally mantle pumiceous pyroclastic-flow deposits (thickness approximately to scale) in proximal and medial areas.

The possibility that pyroclastic flows were generated simultaneously with the pumice fall should provide a challenge for modelers of eruption columns.

LAYER-BY-LAYER ACCUMULATION OF PYROCLASTIC-FLOW DEPOSITS

Broadly synchronous emplacement of fall and flow deposits for several hours and stratigraphic relations between stratified and massive pyroclastic-flow deposits imply that major fills of massive pyroclastic-flow deposits accumulated gradually. Regardless of whether flows were closely spaced discrete events or a quasisteady flow, we infer that multiple, relatively thin (tens of centimeters to less than a few meters?) beds of pyroclastic-flow material came to rest to form a thickening fill (progressive aggradation in the model of Branney and Kokelaar, 1992), even though little stratigraphic evidence for such emplacement is readily visible in the massive deposits themselves. The compositional and textural homogeneity of all flow deposits makes contact recognition difficult. In addition, the inflated gas-charged character of deposits would have encouraged erosion and mixing by subsequent units, further obscuring contacts. Other than pumice-concentration zones and contacts with overlying lithic-rich facies, the only abrupt contacts were formed in areas where substantial time elapsed between flow units to allow fine ash to settle out and the surface to firm up. Apparently only distal areas and some medial uplands met this requirement.

RELATIVELY FINES-DEPLETED PYROCLASTIC-FLOW DEPOSITS

As noted in the discussion of massive pyroclastic-flow deposits, the content of fine ash (silt and clay size) in June 15 pyroclastic-flow deposits, 2 to 18 wt%, is lower than that of certain other comparable pumiceous pyroclastic-flow deposits (table 2). Such a low content of fine ash might partly explain the relatively low mobility of Pinatubo pyroclastic flows. Unless such fines depletion is a sampling artifact, why is fine ash apparently underrepresented in the Pinatubo pyroclastic-flow deposits? Perhaps a typical proportion of fine ash was not produced by fragmentation processes at Pinatubo, but such would be inconsistent with the high volatile content of the magma (Rutherford and Devine, this volume). Rather, this high volatile content ensured great fragmentation, high explosivity, and mixing with air, which together facilitated segregation of fine ash in the eruption column. Likewise, the great inflation, turbulence, and mobility inferred for pyroclastic flows in proximal areas promoted fine-ash segregation (fig. 20A). Such processes also would have produced relatively low emplacement temperatures of pyroclastic-flow deposits, as indicated by the absence of welding. The fine ash from pyroclastic flows could then have been swept by low-level circulation of air into the rapidly convecting column (for example, Fisher, 1979). Once entrained, fine ash would be taken to high altitudes, from which it could be dispersed far downwind.

LATE CALDERA COLLAPSE AND FORMATION OF LITHIC-RICH FACIES

The sudden and apparently short-lived production of abundant lithic clasts late in the emplacement of pyroclastic-flow deposits suggests that vent conditions changed markedly near the end of the eruption. Low lithic contents in the pumice-fall deposit (Paladio-Melosantos and others, this volume) and pumiceous pyroclastic-flow deposits suggest that little vent erosion occurred during most of the eruption. Drawing on the model of Druitt and Sparks (1984), we hypothesize that a large amount of lithic debris was created as the summit dome foundered (fig. 20B). Collapse would have disrupted the vent and forced fragmenting pyroclastic material to escape through growing fractures and faults in the old summit dome. Large amounts of dome rock would have been entrained during this process. Some of the dense lithic material settled through the flows near the vent, as predicted in various models for formation of co-ignimbrite lag deposits (Druitt and Sparks, 1982; Walker, 1985), and some traveled farther to form lithic-rich pyroclastic-flow deposits. In contrast to the radial distribution of pumiceous pyroclastic-flow deposits, lithic debris from the disintegrating summit was preferentially funneled through the deep canyons of major drainages that headed on the former summit dome (fig. 1). Reduced height of pyroclastic-flow formation in the eruption column, or perhaps a transition to low pyroclastic fountains issuing from numerous vents in the foundering caldera, may have localized the lithic-rich facies. In addition, the flows would have been denser than the earlier pumiceous pyroclastic flows and thus strongly channeled by topography.

The stratigraphic position of lithic-rich facies at or near the top of the sequence of pumiceous pyroclastic-flow deposits suggests that once caldera collapse began, pyroclastic-flow activity quickly waned, and the most intense phase of the eruption soon ended. Perhaps foundering rubble created a plug that largely sealed the vent and greatly reduced eruption rate. Williams and Self (1983) inferred that such a mechanism terminated the plinian eruption of Santa María, Guatemala, in 1902. Thus, unlike the Druitt and Sparks model (1984), caldera collapse at Pinatubo did not lead to subsequent emplacement of even greater volumes of pyroclastic-flow deposits. Instead, the eruptive activity waned over a period of weeks as ash issued from several vents in the caldera floor (fig. 20C). These ash emissions deposited layer D over broad areas southwest and northeast of the volcano (Paladio-Melosantos and others, this volume).

We infer that caldera collapse occurred during all or part of the intense, 6-h episode of large, regional, volcano-

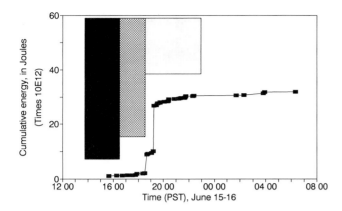

Figure 21. Plot of cumulative seismic energy release for sequence of large earthquakes that occurred between about 1530 on June 15 and 0700 on June 16, 1991. Magnitudes were converted to energy release by using equation in Mori, White, and others (this volume), who estimated that total seismic energy release in the 7-week period beginning at 1530 on June 15 was 6.3×10^{13} J. Dark-gray band, 3-h period of high-amplitude infrasonic record (Tahira and others, this volume) and domination of seismic record by low-frequency earthquakes (Power and others, this volume). Cross-hatched and light-gray bands, 6-h period of decreasing amplitude of barometric (Oswalt and others, this volume) and infrasonic records; boundary between these bands corresponds to marked drop in amplitude of Cubi Point barometric record.

tectonic earthquakes (29 events between m_b 4.5 and 5.7) on June 15 that began about 1530 (Mori, White, and others, this volume; Wolfe and Hoblitt, this volume). These events account for about 50% of the total energy released by earthquakes in the Pinatubo region during the 7-week period that began with this episode (fig. 21). Most of the energy release in this episode is represented by the two largest events, m_b 5.5 (1841) and 5.7 (1915); the location of the second is at or very near the volcano. B.C. Bautista and others (this volume) conclude that the earthquakes of this episode, as well as most of the 7-week period, occurred on regional tectonic faults that were reactivated by the stress change induced by magma withdrawal. The earthquake episode began about 2 h into the climactic eruption (fig. 21); about 1 h later the high-amplitude phases of the infrasonic record ended (Tahira and others, this volume) and high-frequency, volcano-tectonic earthquakes began to dominate the seismic record (Power and others, this volume). Caldera collapse may have occurred at about this time (1630), thus emplacing the lithic-rich facies, ending the 3-h plinian eruption and initiating a 6-h period of diminishing activity as observed by geophysical techniques. The only primary deposit emplaced during this 6-h period would be the lower part of fine-grained, tephra layer D (Paladio-Melosantos and others, this volume).

Alternately, as permitted by stratigraphic relations (fig. 19), the change at 1630 might represent a shift to a low,

persistently collapsing eruption column and continued generation of pyroclastic flows after pumice fall had ceased. Barograph records from Clark Air Base and Cubi Point (Oswalt and others, this volume) indicate a period of declining atmospheric-pressure disturbances after about 1630; the record from Cubi Point shows an additional drop in amplitude at about 1830, just as two large earthquakes dominate a period of conspicuous seismic-energy release (fig. 21). Perhaps caldera collapse occurred at about this time as the crust was responding markedly to the withdrawal of magma. Could these large tectonic earthquakes have even triggered collapse of the caldera? We have few constraints on such speculations, but if collapse occurred at about 1830, the eruption may have progressed through (1) a 3-h-long, high-intensity, plinian phase that emplaced the pumice-fall deposit and some pyroclastic-flow deposits, (2) a phase as long as 2 h characterized by continued generation of pyroclastic flows and ending with caldera collapse and emplacement of the lithic-rich facies of pyroclastic-flow deposits; accompanying tephra fall would be of co-ignimbrite origin and produce fine-grained deposits, and (3) a phase lasting several hours of declining infrasonic and barographic signals accompanied by ash emission from the caldera and continued accumulation of fine-grained tephra layer D. Thus, the emplacement of the 3.7 to 5.3 km^3 (dense-rock equivalent) of pumice-fall deposits and pumiceous pyroclastic-flow deposits (tables 4 and 5) occurred in 3 to 5 h; an average volume eruption rate of 2×10^5 to 5×10^5 m^3/s (mass eruption rate of $5-12 \times 10^8$ kg/s).

We know little about the caldera structure. But in figures 20B and C we infer a funnel-shaped structure (for example, Aramaki, 1984) because of the relatively small diameter of the caldera (2.5 km) compared to estimates of depth of the top of the magma chamber (6 km, Mori, Eberhart-Phillips, and Harlow, this volume; 7 to 8 km, Rutherford and Devine, this volume; 10 km, Pallister and others, this volume). One implication of caldera formation occurring while emplacement of pyroclastic flows was ending, as described above, is that thickly (hundreds of meters) ponded pyroclastic-flow deposits are probably not present in the caldera. Such a condition is consistent with the relatively low crater lake temperatures (Campita and others, this volume) and lack of widespread fumaroles, pumice, and secondary phreatic explosion craters on the caldera floor.

Topographic changes induced by caldera collapse roughly balance the volume of erupted products (table 4). The estimate of the topographic volume lost by caldera collapse (2.5 km^3) is less than the combined volume, in dense-rock equivalent (or DRE), of pyroclastic-flow deposits (2.1–3.3 km^3) and the pumice-fall deposit (1.6–2.0 km^3; Paladio-Melosantos and others, this volume), which totals 3.7–5.3 km^3. But, the rock mass that collapsed into the vent must have increased in volume, owing to dilation. The funnel-shaped collapse as depicted in figure 20C (cone with

Table 4. Volume relations (in rock or magma volume) among climactic eruptive products and topographic change caused by caldera collapse.

	Volume (km^3)	Total volume (km^3)
Topographic volume lost by caldera collapse (assumes caldera floor is 800 m in altitude)	2.5	
Volume change accounted for by 10% dilation of collapsed mass (assumes collapse affects cone of a . radius=1.25 km and a height=10 km)	1.6	
		4.1
Magma volume of pyroclastic-flow deposits	2.1-3.3	
Magma volume of pumice-fall deposits (layer C)[1]	1.6-2.0	
		3.7-5.3

[1] Paladio-Melosantos and others (this volume).

Table 5. Comparison of climactic 1991 Pinatubo eruption and Fierstein and Hildreth's (1992) reconstruction of episode I of 1912 Novarupta eruption, Alaska.

Aspect	Novarupta (Episode 1)	Pinatubo
Duration, in hours	11-16	3-5
Bulk volume of pyroclastic-flow deposits (km^3)	11 ± 4	5.5 ± 0.5
Ratio of vertical fall to distance traveled of pyroclastic flows	[1]1:33	1:11 to 1:14
Aspect ratio of pyroclastic-flow deposits	[2]1:400	1:480 to 1:1,200
Bulk volume of pumice-fall deposit (km^3)	8.8	[3]3.4-4.4
Total bulk volume (km^3)	15.8-23.8	8.4-10.4
Flow deposits as percent of total bulk volume	44-63	53-64
Magma volume (km^3)	8-12	3.7-5.3
Mean mass eruption rate (kg/s)	5x10^8	4.9-11.8x10^8
Column height (km)	23-26	[4]34

[1] For pyroclastic-flow deposit in Valley of Ten Thousand Smokes; calculated from map measurements.
[2] Walker (1983).
[3] Paladio-Melosantos and others (this volume).
[4] Maximum height from Koyaguchi and Tokuno (1993).

radius=1.25 km; height=10 km, to top of magma chamber) has a volume of about 16 km^3. An estimate of 10% dilation (Aramaki, 1984) would account for an additional 1.6 km^3 of volume loss and roughly balance erupted volume.

COMPARISON WITH OTHER 20TH-CENTURY PYROCLASTIC-FLOW ERUPTIONS

Four other explosive eruptions of this century emplaced bulk volumes of silicic pyroclastic deposits similar to or greater than Pinatubo's estimated 8.4 to 10.4 km^3—1902 Santa María, Guatemala, 7.8 km^3 (Fierstein and Nathenson, 1992) or 20.2 km^3 (Williams and Self, 1983); 1912 Novarupta, Alaska, 28±4 km^3 (Fierstein and Hildreth, 1992); 1932 Volcán Quizapu, Chile, 9.5 km^3 (Hildreth and Drake, 1992), and 1991 Cerro Hudson, Chile, 7.6 km^3 (Scasso and others, 1994). Of these large-volume eruptions, the products of Santa María, Quizapu, and Hudson were virtually all plinian pumice-fall deposits, with only minor pyroclastic-flow deposits. The Novarupta eruption was most like the Pinatubo eruption in that both produced appreciable proportions of flow deposits. Moreover, the simultaneous emplacement of fall and flow deposits at Pinatubo is similar to events of episode I (phases A and B) of the Novarupta eruption as reconstructed by Fierstein and Hildreth (1992). Parallels between these events and their deposits are striking (table 5). Significant contrasts between the two events include (1) pronounced compositional changes, greater volume of products, and a more complex conclusion (episodes II and III) of the Novarupta eruption and (2) caldera collapse around the vent and emplacement of lithic-rich facies at Pinatubo late in the pyroclastic-flow sequence.

ACKNOWLEDGMENTS

Much of the fieldwork upon which this study is based would not have been possible without the logistical and helicopter support provided by the U.S. Navy, U.S. Marine Corps, U.S. Air Force, and the Philippine Air Force. Numerous U.S. Geological Survey and Philippine Institute of Volcanology and Seismology colleagues shared important ideas, data, and feedback, of whom only a few can be thanked here: Glenda Besana and Jesse Daligdig worked on the initial volume estimates, John Pallister cooperated in much of the fieldwork in 1992 and contributed many insightful observations and suggestions, and Marvin Couchman performed the density determinations and most of the grain-size analyses. Karl Eriksen and Monty Pearson of the Portland District U.S.Army Corps of Engineers shared their volume estimates. Richard Waitt, Brian Hausback, Chris Newhall, and Ed Wolfe reviewed the manuscript and made many helpful suggestions.

REFERENCES CITED

Aramaki, Shigeo, 1984, Formation of Aira caldera, southern Kyushu, ~22,000 years ago: Journal of Geophysical Research, v. 89, p. 8485–8501.

Bautista, B.C., Bautista, M.L.P., Stein, R.S., Barcelona, E.S., Punongbayan, R.S., Laguerta, E.P., Rasdas, A.R., Ambubuyog, G., and Amin, E.Q., this volume, Relation of regional and local structures to Mount Pinatubo activity.

Booth, B., 1973, The Granadilla pumice deposit of southern Tenerife, Canary Islands: Proceedings of the Geological Association of London, v. 84, p. 353–370.

Branney, M.J., and Kokelaar, Peter, 1992, A reappraisal of ignimbrite emplacement: Progressive aggradation and changes from particulate to non-particulate flow during emplacement of high-grade ignimbrite: Bulletin of Volcanology, v. 54, p. 504–520.

Campita, N.R., Daag, A.S., Newhall, C.G., Rowe, G.L., Solidum, R., this volume, Evolution of a small caldera lake at Mount Pinatubo.

Cas, R.A.F., and Wright, J.V., 1987, Volcanic successions, modern and ancient: London, Allen and Unwin Ltd., 528 p.

Cole, P.D., and Scarpati, C., 1993, A facies interpretation of the eruption and emplacement of the upper part of the Neapolitan Yellow Tuff, Campi Flegrei, southern Italy: Bulletin of Volcanology, v. 55, p. 311–326.

Druitt, T.H., and Bacon, C.R., 1986, Lithic breccia and ignimbrite erupted during the collapse of Crater Lake caldera, Oregon: Journal of Volcanology and Geothermal Research, v. 29, p. 1–32.

Druitt, T.H., and Sparks, R.S.J., 1982, A proximal ignimbrite breccia facies on Santorini, Greece: Journal of Volcanology and Geothermal Research, v. 13, p. 147–171.

———1984, On the formation of calderas during ignimbrite eruptions: Nature, v. 310, p. 679–681.

Fierstein, Judy, and Hildreth, Wes, 1992, The plinian eruptions of 1912 at Novarupta, Katmai National Park, Alaska: Bulletin of Volcanology, v. 54, p. 646–684.

Fierstein, Judy, and Nathenson, Manuel, 1992, Another look at the calculation of fallout tephra volumes: Bulletin of Volcanology, v. 54, p. 156–167.

Fisher, R.V., 1979, Models for pyroclastic surges and pyroclastic flows: Journal of Volcanology and Geothermal Research, v. 6, p. 305–318.

Folk, R.L., 1980, Petrology of sedimentary rocks: Austin, Hemphills, 184 p.

Heming, R.F., and Carmichael, I.S.E., 1973, High-temperature pumice flows from the Rabaul caldera, Papua New Guinea: Contributions of Mineralogy and Petrology, v. 38, p. 1–20.

Hildreth, Wes, and Drake, R.E., 1992, Volcán Quizapu, Chilean Andes: Bulletin of Volcanology, v. 54, p. 93–125.

Hoblitt, R.P., 1986, Observations of the eruptions of July 22 and August 7, 1980, at Mount St. Helens, Washington: U.S. Geological Survey Professional Paper 1335, 44 p.

Hoblitt, R.P., Wolfe, E.W., Scott, W.E., Couchman, M.R., Pallister, J.S., and Javier, D., this volume, The preclimactic eruptions of Mount Pinatubo, June 1991.

Inman, D.L., 1952, Measures for describing the size distribution of sediments: Journal of Sedimentary Petrology, v. 22, p. 125–145.

Koyaguchi, T., this volume, Volume estimation of tephra-fall deposits from the June 15, 1991, eruption of Mount Pinatubo by theoretical and geological methods.

Koyaguchi, Takehiro, and Tokuno, Masami, 1993, Origin of the giant eruption cloud of Pinatubo, June 15, 1991: Journal of Volcanology and Geothermal Research, v. 55, p. 85–96.

Lynch, J.S., and Stephens, G., this volume, Mount Pinatubo: A satellite perspective of the June 1991 eruptions.

Mori, J., Eberhart-Phillips, D., and Harlow, D.H., this volume, Three-dimensional velocity structure at Mount Pinatubo, Philippines: Resolving magma bodies and earthquakes hypocenters.

Mori, Jim, White, R.A., and others, this volume, Volcanic earthquakes following the 1991 climactic eruption of Mount Pinatubo: Strong seismicity during a waning phase.

Oswalt, J.S., Nichols, W., and O'Hara, J.F., this volume, Meteorological observations of the 1991 Mount Pinatubo eruption.

Paladio-Melosantos, M.L., Solidum, R.U., Scott, W.E., Quiambao, R.B., Umbal, J.V., Rodolfo, K.S., Tubianosa, B.S., Delos Reyes, P.J., and Ruelo, H.R., this volume, Tephra falls of the 1991 eruptions of Mount Pinatubo.

Pallister, J.S., Hoblitt, R.P., Meeker, G.P., Knight, R.J., and Siems, D.F., this volume, Magma mixing at Mount Pinatubo: Petrographic and chemical evidence from the 1991 deposits.

Pierson, T.C., Daag, A.S., Delos Reyes, P.J., Regalado, M.T.M., Solidum, R.U., and Tubianosa, B.S., this volume, Flow and deposition of posteruption hot lahars on the east side of Mount Pinatubo, July–October 1991.

Pierson, T.C., Janda, R.J., Umbal, J.V., and Daag, A.S., 1992, Immediate and long-term hazards from lahars and excess sedimentation in rivers draining Mount Pinatubo, Philippines: U.S. Geological Survey Water-Resources Investigations Report 92–4039, 35 p.

Power, J.A., Murray, T.L., Marso, J.N., and Laguerta, E.P., this volume, Preliminary observations of seismicity at Mount Pinatubo by use of the Seismic Spectral Amplitude Measurement (SSAM) system, May 13–June 18, 1991.

Riehle, J.R., 1973, Calculated compaction profiles of rhyolitic ashflow tuffs: Geological Society of America Bulletin, v. 84, p. 2193–2216.

Riehle, J.R., Miller, T.F., McGimsey, R.G., and Keith, T.E.C., 1992, A compaction profile from the 1912 ash-flow sheet, Katmai National Park, Alaska [abs.]: Eos, Transactions, American Geophysical Union, v. 73, no. 43, p. 636.

Rowley, P.D., MacLeod, N.S., Kuntz, M.A., and Kaplan, A.M., 1985, Proximal bedded deposits related to pyroclastic flows of May 18, 1980, at Mount St. Helens, Washington: Geological Society of America Bulletin, v. 96, p. 1373–1383.

Rutherford, M.J., and Devine, J.D., this volume, Preeruption pressure-temperature conditions and volatiles in the 1991 dacitic magma of Mount Pinatubo.

Sabit, J.P., Pigtain, R.C., and de la Cruz, E.G., this volume, The west-side story: Observations of the 1991 Mount Pinatubo eruptions from the west.

Scasso, R.A., Corbella, Hugo, and Tiberi, Pedro, 1994, Sedimentological analysis of the tephra from the 12–15 August 1991

eruption of Hudson volcano: Bulletin of Volcanology, v. 56, p. 121–132.

Scott, W.E., Hoblitt, R.P., Daligdig, J.A., Besana, G., and Tubianosa, B.S., 1991, 15 June 1991 pyroclastic deposits at Mount Pinatubo, Philippines [abs]: Eos, Transactions, American Geophysical Union, v. 72, p. 61–62.

Sheridan, M.F., 1971, Particle-size characteristics of pyroclastic tuffs: Journal of Geophysical Research, v. 76, p. 5627–5634.

Smith, R.L., 1960, Ash flows: Geological Society of America Bulletin, v. 71, p. 795–842.

Sparks, R.S.J., 1976, Grain-size variations in ignimbrites and implications for the transport of pyroclastic flows: Sedimentology, v. 23, p. 147–188.

Sparks, R.S.J., Self, S., and Walker, G.P.L., 1973, Products of ignimbrite eruptions: Geology, v. 1, p. 115–118.

Sparks, R.S.J., and Wilson, Lionel, 1976, A model for the formation of ignimbrite by gravitational column collapse: Journal of the Geological Society of London, v. 132, p. 441–452.

Tahira, M., Nomura, M., Sawada, Y., and Kamo, K., this volume, Infrasonic and acoustic-gravity waves generated by the Mount Pinatubo eruption of June 15, 1991.

Torres, R.C., Self, S., and Martinez, M.L., this volume, Secondary pyroclastic flows from the June 15, 1991, ignimbrite of Mount Pinatubo.

Umbal, J.V., and Rodolfo, K.S., this volume, The 1991 lahars of southwestern Mount Pinatubo and evolution of the lahar-dammed Mapanuepe Lake.

Valentine, G.A., 1987, Stratified flow in pyroclastic surges: Bulletin of Volcanology, v. 49, p. 616–630.

Walker, G.P.L., 1981, Generation and dispersal of fine ash and dust by volcanic eruptions: Journal of Volcanology and Geothermal Research, v. 11, p. 81–92.

––––––1983, Ignimbrite types and ignimbrite problems: Journal of Volcanology and Geothermal Research, v. 17, p. 65–88.

––––––1985, Origin of coarse lithic breccias near ignimbrite source vents: Journal of Volcanology and Geothermal Research, v. 25, p. 157–171.

Walker, G.P.L., Wilson, C.J.N., and Froggatt, P.C., 1981, An ignimbrite veneer deposit: The trail-marker of a pyroclastic flow: Journal of Volcanology and Geothermal Research, v. 9, p. 409–421.

Williams, S.N., and Self, Stephen, 1983, The October 1902 plinian eruption of Santa María volcano, Guatemala: Journal of Volcanology and Geothermal Research, v. 16, p. 33–56.

Wilson, C.J.N., and Walker, G.P.L., 1982, Ignimbrite depositional facies: The anatomy of a pyroclastic flow: Journal of the Geological Society of London, v. 139, p. 581–592.

Wilson, Lionel, and Head, J.W., 1981, Morphology and rheology of pyroclastic flows and their deposits, and guidelines for future observations, in Lipman, P.W., and Mullineaux, D.R., eds., The 1980 eruptions of Mount St. Helens, Washington: U.S. Geological Survey Professional Paper 1250, p. 513–524.

Wolfe, E.W. and Hoblitt, R.P., this volume, Overview of the eruptions.

Preeruption and Posteruption Digital-Terrain Models of Mount Pinatubo

By John W. Jones[1] and Christopher G. Newhall[1]

ABSTRACT

Digital terrain-elevation data for Mount Pinatubo, gathered before and after its June 15, 1991, eruption, show broad features of both the preeruption geology and eruption-related changes. Two major conjugate faults, one trending northeast and the other southeast, seem to cut the ancestral and modern Pinatubo edifices and to bound the north edge of pyroclastic-flow deposits in the Sacobia River valley. The June 1991 eruption created a 2.5-kilometer-wide collapse caldera and filled valleys around Pinatubo with about 5.5±0.5 cubic kilometers of pyroclastic-flow deposits. Products of prehistoric eruptions that were larger than that of 1991 appear as rough-textured inliers (kipukas) surrounded by smooth-textured 1991 pyroclastic-flow deposits. The new summit elevation of Mount Pinatubo is approximately 1,485 meters above sea level, reduced from a preeruption elevation of 1,745 meters; the elevation of the caldera lake is between 820 and 840 meters above sea level, or about 650 meters below the highest point on the new caldera rim.

Attempts to quantify volume loss in the summit region and volume gain in the valleys by subtracting preeruption digital data from posteruption data were unsuccessful. Interferences may have included low resolution of the data (3-arc-second pixels), imprecision in registration of preeruption and posteruption data, loss of forest cover during the eruption, syneruption or posteruption erosion and (or) subsidence, and artifacts from cloud cover or splicing of data sets. Difficulties encountered here are a useful caution against use of similar methods without independent corroboration of volume changes.

INTRODUCTION

Before its June 15, 1991, eruption, Mount Pinatubo consisted of a rounded, steep-sided, domelike mass that rose about 700 m above a broad, gently sloping, deeply dissected apron of pyroclastic and epiclastic materials. Some relics of older volcanic edifices, including an ancestral Mount Pinatubo, lay south, east, and northeast of Mount Pinatubo (Delfin, 1983, 1984; Newhall and others, this volume). In comparison to well-known stratocones such as Mayon or Fuji, Mount Pinatubo was small and inconspicuous, but its extensive pyroclastic apron told of large prehistoric explosive eruptions.

Eruption of about 5 km^3 of magma on June 15, 1991 (W.E. Scott and others, this volume) created a new, 2.5-km-diameter collapse caldera centered slightly northwest of the preeruption summit. The preeruption summit was included in the area of collapse, so the posteruption height of Mount Pinatubo was substantially reduced. Valleys that had existed in the pyroclastic apron were largely filled by eruptive products; valleys that had been carved into older volcanic terrain and partly filled by prehistoric eruptions of Mount Pinatubo were partly filled once again.

Preliminary posteruption digital topography was compared with preeruption topography to correlate preeruption and posteruption features and to estimate the volumes lost from the summit area and added to the surrounding pyroclastic apron. The resolution of our posteruption topography is relatively poor, so much of the discussion that follows is qualitative.

METHODS

DATA PREPROCESSING

The Defense Mapping Agency provided us with tapes containing preeruption and posteruption digital-terrain elevation data (DTED), with 3-arc-second (90-m) lattice resolution. The data span 15 minutes of latitude and 30 minutes of longitude, centered on Mount Pinatubo. Posteruption data are spliced together from several data sets that were collected from November 1991 to November 1992. Data were read from tape into the Earth Resource Data Analysis System (ERDAS), which had an appropriate translator for DTED format files. In ERDAS, we used the nearest-neighbor resampling method (Lillesand and Kiefer, 1987) to

[1]U.S. Geological Survey.

rectify data from latitude/longitude referencing with the 3-arc-second lattice resolution to the UTM coordinate system with a 100-m resolution.

After some initial display, contouring, and examination of the posteruption DTED, some artifacts in the data became readily apparent. For example, phantom linear hills appeared where none actually exist. Therefore, the projected files were transferred to ARC/INFO geographic information system software and edited by use of grid-edit tools in ARC (Environmental Systems Research Institute, Inc.).

Various filtering and visual editing methods to mitigate the effects of the artifacts were tested and found to be inadequate. Finally, artifacts in the posteruption DTED were minimized by inserting terrain information from the preeruption DTED, where necessary, into the posteruption DTED. Because of this procedure, posteruption topography is locally incorrect, but contours of the map as a whole approximate the present terrain better than they would had the artifacts been allowed to remain. The largest patch of inserted, preeruption topography is northeast of the caldera.

Similar substitution of preeruption terrain information into the posteruption data file was done for 10-km-wide strips at the east and west edges of the 15' × 30' area. Thus, no change, real or otherwise, can be described in these margins, which are mostly areas of alluvial fans and likely posteruption sediment deposition.

An additional correction was required for a slight discrepancy in registration of the preeruption and posteruption DTED. We tried shifting the posteruption DTED zero, one, and two pixels north, south, east, west, and diagonally. A shift of posteruption data by one pixel (100 m) to the east minimized residual differences between the two data sets for points which we knew did not change in elevation.

CONTOUR MAP PRODUCTION

Contour maps were produced for two areas: one, 6 km on a side, and the other, encompassing 15 minutes of latitude and about 30 minutes of longitude, both approximately centered on the caldera. Files of these areas were reformatted into a triangulated irregular network (TIN) data structure, in which triangles fitted to the input DTED lattice were subdivided three times in order to produce relatively smooth contour lines in the final maps. A contour interval of 20 m was selected to match the interval used on preeruption published topographic maps.

For both the preeruption and posteruption cases, maps were plotted at 1:250,000 and 1:100,000 scales for the area of 15 minutes × 30 minutes and at an approximate scale of 1:15,000 (actual scale 1:15,030) for the 6 km × 6 km area. The posteruption contour map still required some interactive editing in order to overcome the influence of artifacts in the digital elevation data. This editing was done by comparing of apparent posteruption topography with the broad features of topography known to us from fieldwork in the area.

WATERSHED BOUNDARIES

We delineated watershed boundaries, by hand, on the contour map created from the preeruption digital terrain data. Similar preeruption watershed boundaries had been drawn on published preeruption topographic maps, and we checked our new boundaries against previously drawn boundaries for consistency. The fit was good. The hand-drawn boundaries (fig. 1) were, in turn, digitized as vectors, tagged, and gridded into raster file format using the ARC/INFO software. A resolution of 100 m was used in the gridding process to provide for overlay analysis of the terrain files with the watershed boundaries.

ESTIMATES OF VOLUME CHANGE

Apparent loss of volume from Mount Pinatubo and gains of volume on surrounding terrain were estimated by computer subtraction of the preeruption DTED from the posteruption DTED. Inputs for this procedure were the pixel-shifted, artifact-cleaned DTED. Elevation data were then transferred into PCI software (PCI Inc.), where they were differenced on a cell-by-cell basis. An elevation change was calculated for each 100 m × 100 m cell. Then, the volume of change for each cell was estimated by multiplying each pixel's elevation change by 10^4 m^2, and these results were summed over each watershed area, including the caldera (table 1). Positive and negative changes are listed separately to show any anomalous data.

Independent estimates of volumes of pyroclastic-flow deposits (volume gain) were made by sketching 1991 valley fill from aerial oblique photographs and video onto published preeruption topographic maps, differencing preeruption and posteruption cross-valley profiles, and multiplying by the length of each applicable reach (Daligdig and others, 1991; W.E. Scott and others, 1991; this volume). Estimates by the U.S. Army Corps of Engineers (1994) were made by similar methods but from vertical aerial photographs with better photogrammetric control. Estimates by Daag (1994) for volume gain on the Sacobia-Abacan-Pasig pyroclastic fan used July 1991 aerial oblique stereo photographs, from which levels of fill were marked on preeruption topographic sheets, contoured, and digitized. The preeruption topography was then subtracted digitally from the posteruption topography.

An independent estimate of the volume of caldera collapse (volume lost from the summit area), by Scott and others (this volume) was made by drawing the new rim and floor of the caldera onto a topographic profile of the preeruption volcano, assuming 800 m as the elevation of the

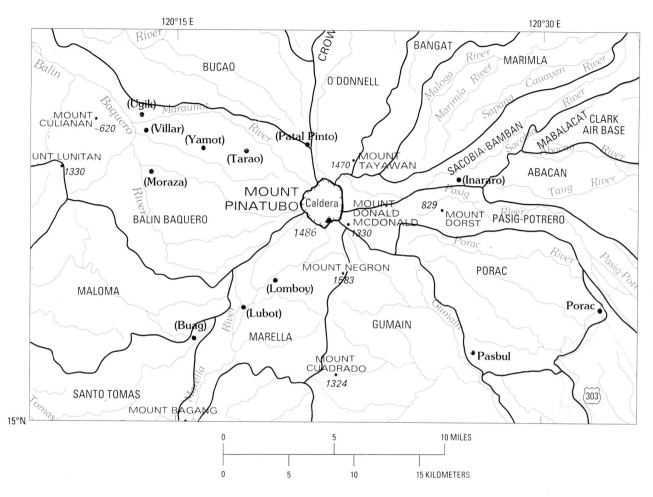

Figure 1.   Pinatubo watersheds and other geographic features noted in text.

caldera floor, and summing the volumes of disk-shaped layers within this simple model.

RESULTS AND DISCUSSION

GENERAL FEATURES OF PREERUPTION AND POSTERUPTION TOPOGRAPHY

Largely because of the coarse resolution of our digital topography (figs. 2A,B), and the opportunity to view this in shaded relief (figs. 3A,B), several large-scale features of interest can be seen. Many of these features were previously known but are particularly well displayed in the digital topography.

- Radial drainage defines the slopes of ancestral Pinatubo northeast and southeast of Mount Pinatubo, on either side of the Sacobia pyroclastic fan. However, that drainage pattern is interrupted southwest of the Gumain River. Disappearance of ancestral Pinatubo drainage patterns southwest of the Gumain River could be explained in two ways: (1) terrain to the southwest of the Gumain River is younger than and buries ancestral Pinatubo or

(2) the dividing line is a fault scarp against which lavas and other deposits of ancestral Pinatubo accumulated but did not overflow. We prefer the latter explanation, partly because (1) limited K-Ar ages (Bruinsma, 1983) suggest that Mount Negron and Mount Cuadrado are, if anything, slightly older than the ancestral Mount Pinatubo and partly because (2) this same southeast-trending line could be an extension of the Iba and Maraunot faults described by de Boer and others (1980), Delfin (1983, 1984), and B.C. Bautista and others (this volume).

- A major northeast-southwest lineament bounds the north side of the Sacobia pyroclastic and alluvial fans, passes through or near the caldera, and may continue southwest to the southwest corner of our 15 minute × 30 minute area (figs. 2, 3). This feature is also seen in SLAR imagery and is called the Sacobia lineament (fig. 4 of Newhall and others, this volume). B.C. Bautista and others (this volume) describe posteruption seismicity in at least two clusters along this lineament, and describe a regional tectonic setting of east-west compression in which such a lineament could be a conjugate strike-slip fault with right-lateral displacement.

Table 1. Comparison of various estimates of volume loss (−) and volume gain (+) resulting from eruption of Mount Pinatubo on June 15, 1991.

[–, not estimated]

Watershed	Estimates of Volume Change (this study)				Independent Estimates of Volume Change			
	(1) Positive volume change[1] (km³)	(2) Negative volume change[1] (km³)	(3) Tree and erosion corrections[2] (km³)	(4) Corrected volume change[3] (km³)	(5) Punongbayan and others[4] (km³)	(6) Scott and others (this volume) (km³)	(7) U.S. Army Corps of Engineers (1994) (km³)	(8) Daag (1994) (km³)
O'Donnell............................	0.76	−1.01	0.11+0.04	0.91	+0.6[5]	+0.28	+0.24[5]	–
Bangat	0.36	−0.72	0.04	0.40	–	+0.06	–	–
Marimla..............................	0.70	−0.88	0.02	0.72	–	–	–	–
Sacobia-Bamban	0.78	−0.49	0.22+0.14	1.14	+0.9	+0.60	+0.60[6]	+0.88
Mabalacat............................	0.02	−0.05	0	0.02	–	–	–	–
Abacan	0.11	−0.10	0	0.11	+0.2	+0.10	–	+0.01
Pasig-Potrero.......................	1.02	−0.33	0.10+.02	1.14	+0.5	+0.30	+0.30	+0.40
Porac..................................	1.35	−0.65	0.02	1.37	–	–	–	–
Gumain..............................	1.81	−1.49	0.02	1.83	+.05[7]	+0.03[7]	–	–
Marella	1.62	−1.53	0.30+0.21	2.13	+1.3[8][9]	+1.26[8][9]	+1.40[8][9]	–
Santo Tomas (excl. Marella) .	0.12	−0.06	0	0.12	–	–	–	–
Maloma	0.12	−0.18	0.03	0.15	–	–	–	–
Balin Baquero	1.42	−1.27	0.63+0.60	2.65	–	+1.40	–	–
Bucao	0.64	−0.72	0.34	0.98	+3.1[10]	+0.88	+3.00[10]	–
Caldera	0.06	−1.64	0.02	−1.62	–	−2.5	–	–
Total volume change	10.77	−11.12	–	+12.05	–	+2.4	–	–

[1] Major apparent volume losses and gains appeared in watersheds where we know that little actual topographic change occurred (Bangat, Marimla, Porac, Gumain). Most of the overestimates of volume gain, we think, resulted from slight misregistration of the preeruption and posteruption data, and are especially pronounced in rugged terrain where even slight misregistration can result in large error. Most underestimates of volume gain, we think, resulted from misregistration of preeruption and posteruption data, flattening of forest by the eruption, and posteruption erosion of pyroclastic deposits.

[2] The first figure for each watershed, tree correction, is the estimated volume of the preeruption forest, as differenced from treetops to actual ground surface. For all watersheds except the caldera, this correction is taken to be 15 m (an average height of the forest canopy in valleys) times the area of each watershed that was buried by valley-filling pyroclastic-flow deposit. Areas of valley-filling pyroclastic-flow deposit are from W.E. Scott and others, this volume. For the caldera, this correction is taken to be 5 m (height of a scrub forest) times the entire area of that watershed. The second figure, where present, is a correction for the volume of 1991-92 erosion, as estimated by the U.S. Army Corps of Engineers (1994). The erosion correction for the Balin Baquero is for the Balin Baquero and Bucao Rivers combined.

[3] Corrected volume change for all areas except the caldera is calculated as column 1 plus column 3. Note that negative volume changes are suppressed except in the caldera. Volume change for the caldera is calculated as column 2 plus column 3. **We do not believe either corrected or uncorrected estimates and show them only to illustrate that, when examined by watershed, some are obviously false.**

[4] PHIVOLCS has reestimated volumes several times. This is the latest. The earliest estimates were by J. Daligdig and G. Besana in 1991; this revision is by N.M. Tungol and R.A. Arboleda.

[5] Includes Bangat watershed.

[6] Includes Abacan watershed.

[7] Includes Porac watershed.

[8] Includes upper Santo Tomas watershed.

[9] Includes Maloma watershed.

[10] Includes Balin Baquero watershed.

- The Sacobia, Abacan, and Pasig Rivers all drain a single pyroclastic and associated alluvial fan (Punongbayan and others, 1994a,b). Frequent captures of one of these streams into another since June 1991 are simply a reflection of the fact that this is a single fan without any fundamental division therein. Those attempting to forecast volumes of lahars still to descend from this fan are now treating this Sacobia pyroclastic fan as a single system with largely stochastic direction of flow into one channel or another.

- Were it not for a low ridge of bedrock southwest of Mount Pinatubo that divides the Balin Baquero and Marella drainages, there would essentially be a single pyroclastic fan around the entire west half of Mount Pinatubo. This broad fan is in notable contrast to the east side of Mount Pinatubo, where the Sacobia pyroclastic fan lies in an erosion- and fault(?)-controlled valley within ancestral Pinatubo. Is this difference between the west and the east sides a result of a progressive offset of modern Pinatubo toward the west from ancestral Pinatubo, a result of differences in underlying, pre-Pinatubo basement terrain, or, perhaps, a result of some prehistoric breaching of the west side of ancestral Pinatubo or of the west caldera wall(s) of modern Pinatubo? We do not know.

- In the posteruption topography, patches of slightly higher preeruption terrain (kipukas) are visible with a rough, dissected texture surrounded by the relatively smooth surface of 1991 pyroclastic-flow deposits. These areas are relics of older eruptions that emplaced larger

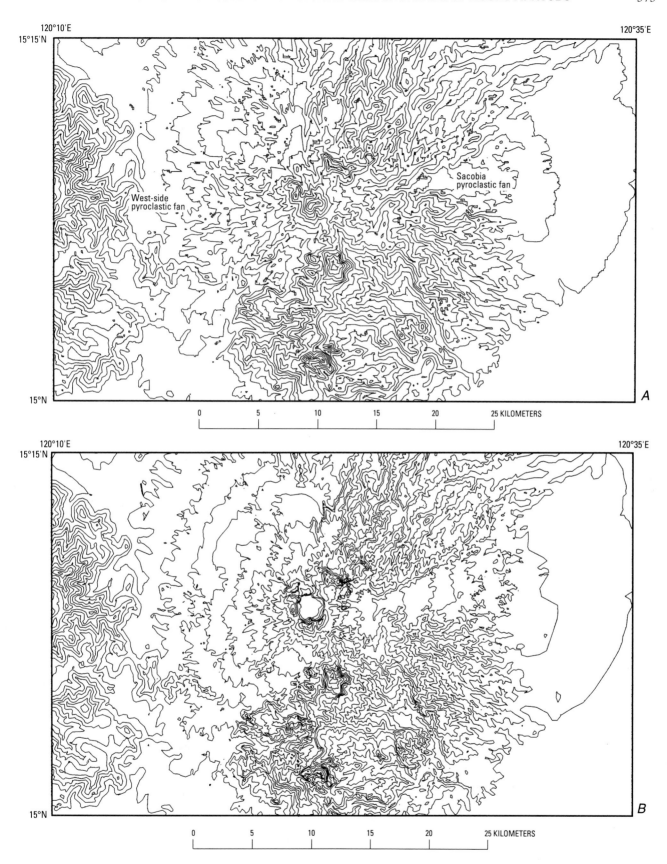

Figure 2. Pinatubo topography, 15 minute × 30 minute area. Contour interval 20 m. A, Preeruption. B, Posteruption, a composite of data gathered from November 1991 to November 1992. Sacobia and west-side pyroclastic fans are shown. Refer to figures 3A,B for identification of additional features.

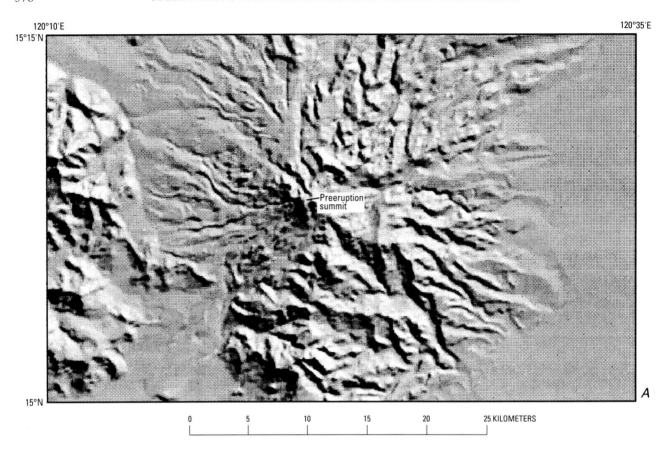

Figure 3. Shaded relief views of Mount Pinatubo. A, Preeruption view showing preeruption summit. B, Posteruption view showing major fault lineaments (Iba-Maraunot and Sacobia), ancestral Pinatubo edifice, modern Pinatubo, and 1991 caldera.

volumes of deposits than that of 1991. We know that prominent dissected terrain on the east, between the Sacobia and Pasig-Potrero Rivers, consists principally of thick pyroclastic-flow units of the >35,000-yr-old Inararo eruptive period (Newhall and others, this volume). We have very few samples from the kipukas on the west and north-northwest, so we cannot say whether these deposits are of the Inararo or the more recent Crow Valley and Maraunot eruptive periods (Newhall and others, this volume).

• The 1991 caldera is approximately 2.5 km in diameter, and its caldera lake surface lies between 820 and 840 m above sea level. The highest point on the caldera rim, and thus the new elevation of Mount Pinatubo, is approximately 1,485 m, or about 260 m below the preeruption height. The depth of the caldera is thus 650 m plus at least 30 m (depth of the caldera lake in late 1992) and an unknown thickness of loose caldera-filling debris.

RELATION OF THE NEW CALDERA AND 1992 DOME VENT TO PREERUPTION TERRAIN AND 1991 DOME VENT

The 1991 caldera is centered about 750 m northwest of the preeruption summit and about 350 m southeast of where a precursory lava dome began to form on June 7, 1991 (figs. 4, 5). The 1991 caldera is more or less concentric within the modern Pinatubo edifice that in turn lies concentrically within the prehistoric Tayawan caldera (fig. 3B) (Delfin, 1983, 1984; Newhall and others, this volume). Evidence for a prehistoric but relatively young caldera only slightly north of the 1991 caldera was discussed in Newhall and others (this volume). A fissure and line of phreatic explosion pits that formed on April 2, 1991 (Sabit, this volume; Wolfe and Hoblitt, this volume), extended northeastward from the northeast margins of both this young prehistoric and the 1991 calderas. A line of vigorously steaming vents that extended southwest of this fissure immediately following the phreatic explosions of April 2 trended nearly through the center of the 1991 caldera.

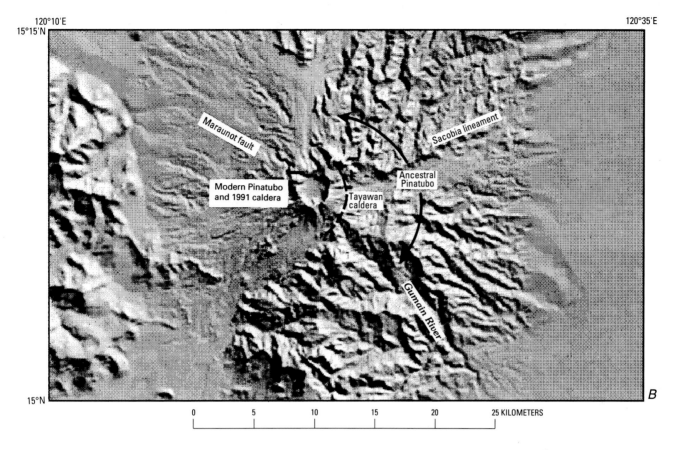

120°10'E
15°15'N

120°35'E

15°N

B

0   5   10   15   20   25 KILOMETERS

Figure 3.—Continued.

The location of the July–October 1992 dome can be seen in posteruption topography, centered within the 1991 caldera. Thus, judging from locations of the June 1991 and July–October 1992 domes, the active vent shifted about 350 m to the southeast between those episodes of dome growth, probably during the caldera-forming eruption of June 15, 1991.

VOLUMES LOST FROM THE SUMMIT AND GAINED ON THE SLOPES

In principle, differencing of preeruption and posteruption DTED should yield reliable volume changes. Qualitatively, the results are reasonable (fig. 6). Those valleys which we know to have received thick pyroclastic fill show significant volume gain, and the new caldera shows significant volume loss. However, quantitative estimates (fig. 5 and table 1) are inconsistent with volumes that others have previously estimated (Daag, 1994; Punongbayan and others, 1994a; U.S. Army Corps of Engineers, 1994; W.E. Scott and others, this volume).

Major volume gains and losses even appeared in watersheds where little actual topographic change occurred (columns 1 and 2, table 1). Why is there such a poor result from our digital reestimation?

Random discrepancies (without tendency to inflate or deflate actual volume change) could have been caused by low resolution of the available DTED, imprecise registration of preeruption and posteruption DTED, or artifacts resulting from splicing of multiple data sets. Because severe errors are apparent in watersheds that have steep terrain and that received only minimal new deposit, we think that imprecise registration is a major problem. The problem remained even after we shifted the posteruption data 1 pixel to the east, as described earlier.

Systematic underestimates of valley filling could have been caused by

- loss of vegetation cover, if preeruption DTED reflected vegetation cover. After the eruption, trees and tall grasses were generally flattened. Several watersheds that had been partly forested and that experienced tree-flattening pyroclastic surges but relatively little deposition "lost" significant volumes from preeruption to posteruption time. We have attempted to correct for this factor.

- posteruption erosion. Posteruption erosion that occurred before the data were gathered would appear as a loss of volume; redeposition of that sediment would have

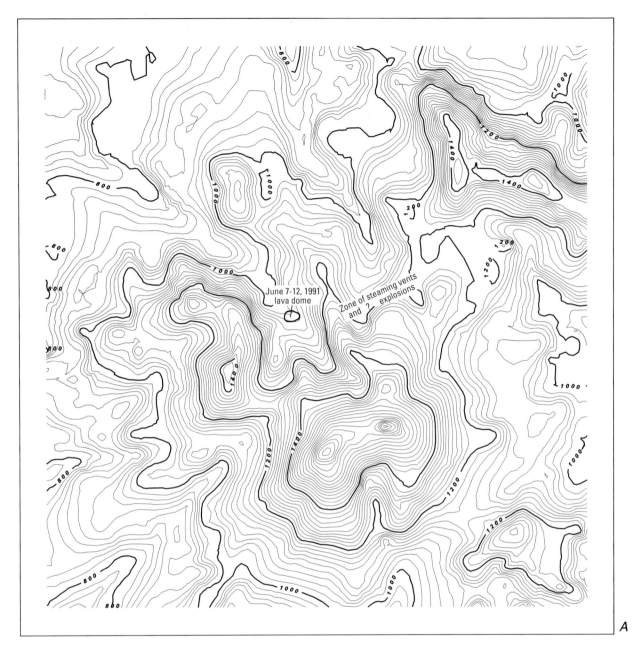

Figure 4. Pinatubo topography, 6 km × 6 km area centered on caldera. A, Preeruption view showing location of the June 1991 dome. B, Posteruption view showing location of the 1992 dome. Note that the 1991 caldera is offset north-northwest from the preeruption summit, and the 1992 dome is offset southeast from the 1991 dome.

appeared within the same watersheds, were it not for the fact that the eastern and western edges of the posteruption data set were patched in from the preeruption data set. We know that about 1×10^9 m^3 of pyroclastic flow material had been eroded from 1991 deposits by late 1992 (U.S. Army Corps of Engineers, 1994), so significant underestimation from erosion was likely.

- syneruption or posteruption subsidence outside the area of the caldera itself. We know that volcanoes experience subsidence following major eruptions, particularly if caldera collapse does not occur. Some broad subsidence

might also occur even when caldera collapse does occur. This would be an interesting topic for future study; we opted not to pursue it here because other uncertainties are still too high.

Systematic overestimates of valley filling (or an underestimate of volume loss, as in the caldera) could have been caused by

- artifacts of weather or ash clouds in posteruption data. Some early problems with clouds required repeated posteruption data collection.

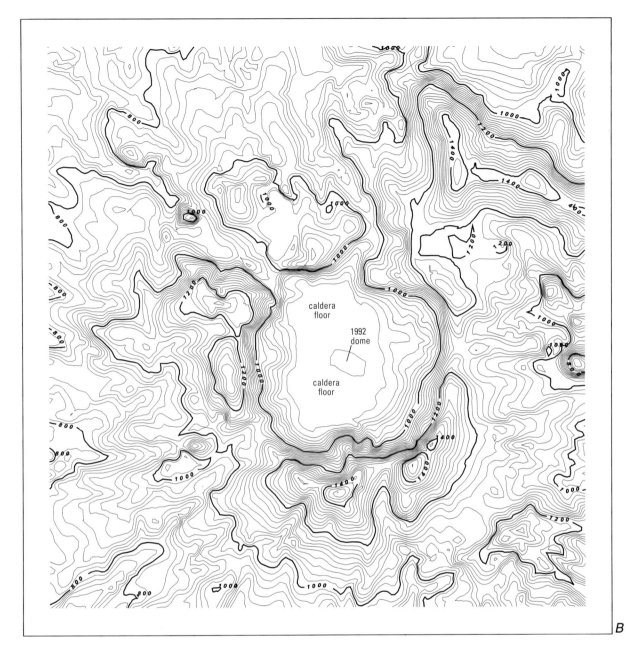

B

Figure 4.—Continued.

• inclusion of tephra fall in estimates, in addition to the pyroclastic-flow deposits that are the sole component of previous estimates.

Serious, nonsystematic error from inaccurate registration of preeruption and posteruption data seems likely, especially in rugged terrain. In addition, because the apparent net volume change for the entire area (sum of columns 1 and 2) is much lower than we expect from field investigations, and, indeed, is negative, whereas the true change is probably positive, we also suspect some serious, systematic source of negative elevation change, certainly including posteruption erosion and possibly also including loss of

forest. Preeruption topography may have been along the treetops; the eruption blew down trees, and the apparent ground surface dropped by the height of the trees. Further evidence for the influence of forest cover is found in the texture of preeruption and posteruption elevation data. Preeruption data lack the texture and erosional features visible in posteruption data (fig. 3A,B).

We cannot correct for imprecise registration of data. To correct for a possible forest artifact, we artificially reset all negative elevation changes except in the caldera region to zero before calculating volume changes. In addition, where pyroclastic-flow deposits filled valleys, our estimate of positive elevation change might be too low, by the height of the

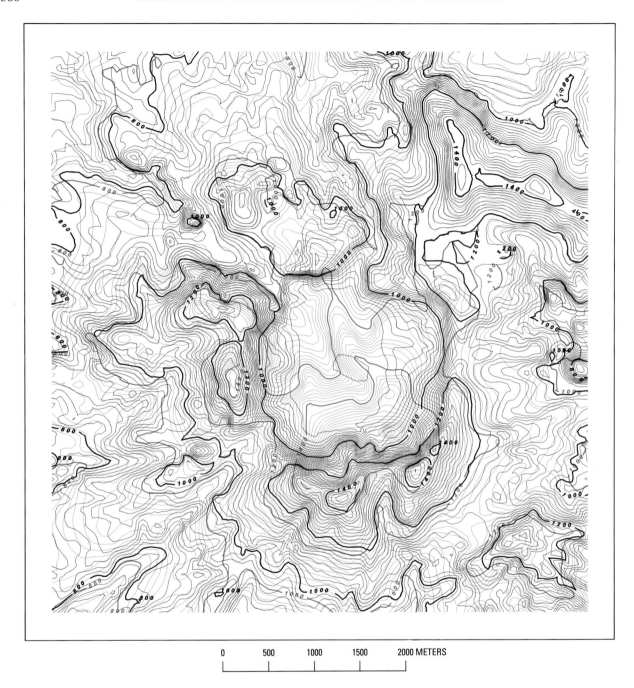

0   500   1000   1500   2000 METERS

Figure 5. Pinatubo topography, 6 km × 6 km area centered on the caldera, with preeruption 20-m elevation contours (blue) overprinted with posteruption 20-m contours (black). Major changes in the summit region are as seen in figure 4, but here are superimposed on each other for ease of checking change at specific points. Closely matching preeruption and posteruption topography northeast of the caldera is in an area that experienced little change during the eruption; the consis-

tency is partly illusory, though, because a patch of preeruption topography was forced onto posteruption topography, as discussed in text, to avoid an even larger error. Several unpatched discrepancies are still present, including apparent postcaldera "hills" at center right and 1.5 km northwest of the caldera rim. Discrepancies are artifacts rather than real change; they can be ignored in map form but have not been successfully removed from estimates of volume change.

120°10'E 120°35'E
15°15'N

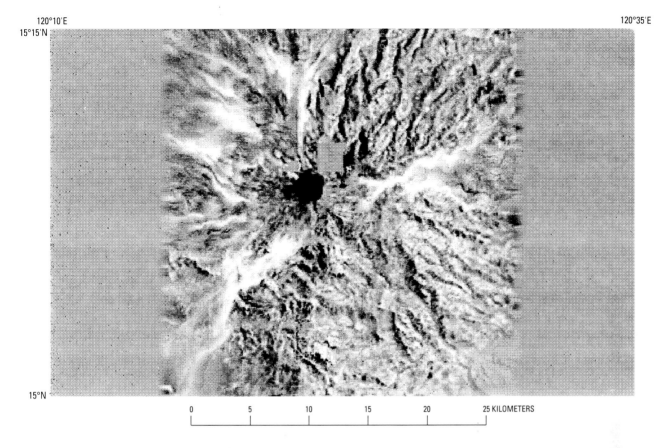

15°N

0 5 10 15 20 25 KILOMETERS

Figure 6. Gray-scale map of volume gain and loss around Mount Pinatubo. An intermediate gray indicates no change, while black shows maximum loss and white shows maximum gain. As in figure 5, most of the change is real, but two important artifacts are easily seen. The first is the patch created by the authors northeast of the caldera; the second consists of 10-km-wide strips along the east and west edges of the figure. In both the patch and the strips, preeruption topography was ported directly to the posteruption data set, so no change appears, and those parts of the image are featureless gray. The strips, placed in the posteruption data prior to any processing by the authors, remove most areas of lahar deposition from calculations, even though much of that lahar deposit was derived from 1991 products whose volume we tried to estimate.

preeruption trees. Column 3 of table 1 includes a volume correction of 15 m (average height of trees) times the area of pyroclastic valley fill in each watershed, to be added to the previous estimate of volume gain. For the area of the former summit (now caldera), a 5-m average thickness of preeruption vegetation cover is multiplied by the entire area of that watershed. Column 3 also includes, for major drainages, a correction for the volume of 1991–92 erosion as estimated by the U.S. Army Corps of Engineers (1994). Column 4 is the sum of columns 1 and 3 except in the case of the caldera, for which column 2 and column 3 are summed.

Previously published, independent estimates are qualitatively consistent with what one can readily observe, and with each other, though estimates of Daligdig and others (1991) are generally higher than the other three estimates. Estimates of volume change from this paper, even after attempted corrections, are not consistent with what one can observe in the field. **Therefore, we believe that our new estimates of volume change are incorrect and should not be used.**

Ironically, we originally thought that it would be easier to quantify volume changes in the uplands, where changes in elevation were greatest, than it would be to quantify volume changes in the lowlands, where deposition from lahars increased elevations by only a few meters. The highlands turned out to be difficult; we do not have enough data for areas of lahar deposition to test whether volume change in those lowlands is, in fact, quantifiable from available DTED.

In short, our attempt at volume estimation from DTED was unsuccessful, and it raises some important cautions for similar applications where there is no independent information by which to corroborate DTED-based estimates.

CONCLUSIONS

Major features of preeruption topography show well in low-resolution (3-arc-second) digital data. Two major conjugate faults, one trending northeast and the other southeast, seem to bound and cut the ancestral Pinatubo edifice and to localize the present Sacobia pyroclastic fan. Products of prehistoric eruptions from the modern Pinatubo, larger than those of 1991, appear as rough-textured inliers (kipukas) on posteruption topography.

Massive topographic changes caused by the 1991 eruption are readily apparent in a comparison of preeruption and posteruption digital topography. The principal features are a new summit caldera and infilling of most valleys surrounding Mount Pinatubo by pyroclastic flows.

Our attempts to quantify volume changes by subtraction of 3-arc-second DTED were unsuccessful, especially in steep terrain. Higher resolution data, including posteruption data for the entire 15 minute by 30 minute area, would allow more precise registration of control points (points that were known to have not changed in elevation), more precise evaluation of the role of forest loss, an automatic correction for posteruption erosion, and thus more precise estimation of actual volume changes.

REFERENCES CITED

Bautista, B.C., Bautista, M.L.P., Stein, R.S., Barcelona, E.S., Punongbayan, R.S., Laguerta, E.P., Rasdas, A.R., Ambubuyog, G., and Amin, E.Q., this volume, Relation of regional and local structures to Mount Pinatubo.

Bruinsma, J.W., 1983, Results of potassium-argon age dating on twenty rock samples from the Pinatubo, southeast Tongonan, Bacon-Manito, Southern Negros, and Tongonan, Leyte areas, Philippines: Proprietary report by Robertson Research Ltd. to the Philippine National Oil Company-Energy Development Corporation, 24 p.

Daag, A.S., 1994, Geomorphic developments and erosion of the Mount Pinatubo 1991 pyroclastic flows in the Sacobia watershed, Philippines: A study using remote sensing and geographic information system (GIS): unpublished M.S. thesis, International Institute for Aerospace Survey and Earth Sciences, Enschede, The Netherlands, 106 p.

Daligdig, J.A., Besana, G.M., and Punongbayan, R.S., 1991, Overview and impacts of the 1991 Mt. Pinatubo eruption [abs.]: Abstracts, GEOCON '91, Geological Society of the Philippines, 4th Annual Geological Convention, p. 8.

de Boer, J., Odom, L.A., Ragland, R.C., Snider, F.G., and Tilford, N.R., 1980, The Bataan orogene: Eastward subduction, tectonic rotations, and volcanism in the western Pacific (Philippines): Tectonophysics, v. 67, p. 251–282.

Delfin, F.G., Jr., 1983, Geology of the Mt. Pinatubo geothermal project: Philippine National Oil Company, unpublished report, 35 p. plus figures.

———1984, Geology and geothermal potential of Mt. Pinatubo: Philippine National Oil Company, unpublished report, 36 p.

Lillesand, T.M., and Kiefer, R.W., 1987, Remote sensing and image interpretation: New York, John Wiley and Sons, 721 p.

Newhall, C.G., Daag, A.S., Delfin, F.G., Jr., Hoblitt, R.P., McGeehin, J., Pallister, J.S., Regalado, M.T.M., Rubin, M., Tamayo, R.A., Jr., Tubianosa, B., and Umbal, J.V., this volume, Eruptive history of Mount Pinatubo.

Punongbayan, R.S., Tuñgol, N.M., Arboleda, R.A., Delos Reyes, P.J., Isada, M., Martinez, M., Melosantos, M.L.P., Puertollano, J., Regalado, T.M., Solidum, R.U., Tubianosa, B.S., Umbal, J.V., Alonso, R.A., and Remotigue, C., 1994a, Impacts of the 1993 lahars, long-term lahars hazards and risks around Pinatubo Volcano, in Philippine Institute of Volcanology and Seismology, Lahar Studies: Quezon City, PHIVOLCS Press, p. 1–40.

Punongbayan, R.S., Arboleda, R.A., Tuñgol, N.M., and Solidum, R.U., 1994b, Secondary explosions and secondary pyroclastic flows as rapid mechanisms for stream piracy: PHIVOLCS, unpublished report, 19 p.

Sabit, J.P., Pigtain, R.C., and de la Cruz, E.G., this volume, The west-side story: Observations of the 1991 Mount Pinatubo eruptions from the west.

Scott, W.E., Hoblitt, R.P., Daligdig, J.A., Besana, G., and Tubianosa, B.S., 1991, 15 June 1991 pyroclastic deposits at Mount Pinatubo, Philippines [abs]: Eos, Transactions, American Geophysical Union, v. 72, p. 61–62.

Scott, W.E., Hoblitt, R.P., Torres, R.C., Self, S, Martinez, M.L., and Nillos, T., Jr., this volume, Pyroclastic flows of the June 15, 1991, climactic eruption of Mount Pinatubo.

U.S. Army Corps of Engineers, 1994, Mount Pinatubo recovery action plan: Long term report: Portland, Oreg., 162 p. + 5 appendices.

Wolfe, E.W. and Hoblitt, R.P., this volume, Overview of the eruptions.

Volume Estimation of Tephra-Fall Deposits from the June 15, 1991, Eruption of Mount Pinatubo by Theoretical and Geological Methods

By Takehiro Koyaguchi[1]

ABSTRACT

The major June 15, 1991, eruption of Mount Pinatubo started with generation of pyroclastic flows in the afternoon, which were subsequently followed by a plinian eruption. This sequence of events is reflected in the stratigraphy of the fall deposits, which consists of a fine ash layer, a lapilli layer commonly including pumice grains of >1 centimeter in diameter, and a lapilli-bearing volcanic sand layer, in ascending order. A giant disc-shaped cloud covering an area of 60,000 square kilometers appeared in the satellite images at 1440 Philippine local time. The cloud expanded radially for 5 hours to an area more than 300,000 square kilometers. According to eyewitness accounts, there was heavy ash fall after 1400, intermittent lapilli fall starting at about 1420, and heavy and continuous lapilli fall starting widely at about 1500. The occurrence of the giant cloud roughly corresponded to the initiation of the intermittent lapilli fall. The satellite and grain-size data indicate that the volumetric flow rate of the umbrella cloud during the climactic plinian phase was 3 to 10×10^{10} cubic meters per second. According to the results of calculations on the dynamics of the eruption cloud, the volumetric flow rate of the eruption cloud is accounted for by a plinian eruption with a magma discharge rate of 4×10^8 to 2×10^9 kilograms per second. The total amount of the ejecta injected into the stratosphere during the climactic plinian eruption was 2 to 10 cubic kilometers dense-rock equivalent. The upper estimate is greater than the upper volume estimate of the tephra-fall deposits including fine ash observed in the South China Sea and Southeast Asia (2 cubic kilometers dense-rock equivalent). The discrepancy may be partly due to fine particles that were extensively dispersed in the atmosphere and could not be counted in the deposit.

INTRODUCTION

This study is an extension of recent work on the dynamics of eruption clouds of the 1991 Pinatubo eruption on the basis of the satellite images (Koyaguchi and Tokuno, 1993). According to the preliminary analyses of Koyaguchi and Tokuno on the basis of a fluid dynamic model of an eruption column (Woods, 1988), the expansion rate of the eruption cloud is accounted for by an injection of pyroclasts into the stratosphere at the rate of 10^9 kg/s. They concluded that the total amount of the pyroclasts is more than 1.8×10^{13} kg (7 km^3 dense-rock equivalent, DRE), derived simply by multiplying the magma discharge rate by the time during which the eruption cloud continued to expand upwind (5 hours). This value is several factors greater than the published total volume of tephra-fall deposits (3 km^3; Pinatubo Volcano Observatory Team, 1991). The discrepancy would be partly due to uncertainties of both the methods for estimating volumes of discharged magma and tephra-fall deposits, but there is another important possibility. The volume of deposits does not necessarily represent that of discharged magma if the ejecta consisted of fine particles and were widely dispersed in the atmosphere. If this possibility played a significant role, the discrepancy between the two quantities gives important information about the environmental effects of the eruption as a result of the dispersal of fine volcanic ash. I examine this problem of the volume estimations by different methods in more detail in this paper. I emphasize at the outset that the purpose of this study is not to judge which method gives the correct value of the volume of tephra deposits but to clarify the difference between implications of volume estimates based on fluid dynamics of eruption clouds and on geological methods. First, I describe tephra-fall deposits and the satellite data to clarify the chronological relation between the growth of the eruption cloud and the deposits. Second, the total volume of ejecta and that of the tephra-fall deposits is estimated on the basis of a dynamical model of the eruption cloud as well as geological methods. The sources of uncertainties of each step of these estimates are evaluated. Finally, the implications of the results of each volume estimate are discussed.

---

[1]Earthquake Research Institute, University of Tokyo, Tokyo 113, Japan.

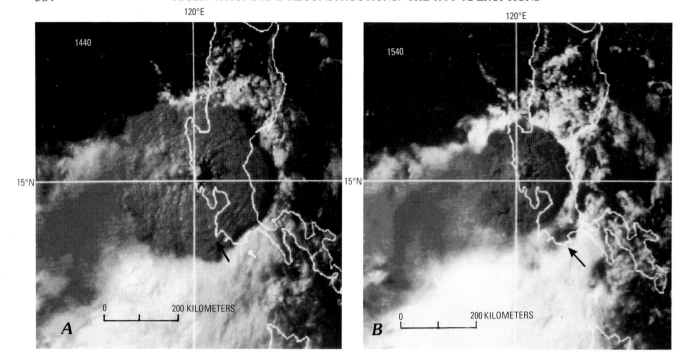

Figure 1.   Visible satellite images at 1440 and 1540 Philippine local time, on June 15, 1991.

OBSERVATIONS

SATELLITE DATA

The major Pinatubo eruptions were recorded hourly in multispectral images by the Japanese Geostationary Meteorological Satellite Himawari 4 (Koyaguchi and Tokuno, 1993). Repetitive explosive eruptions at the volcano culminated in the afternoon of June 15. Eruption clouds repetitively pierced low-level white cloud in the morning of that day. A remarkable disc-shaped cloud was observed at 1440 (local Philippine time, fig. 1). This cloud expanded up to 280 km in diameter (60,000 km^2), the center of which was located about 30 km west of Pinatubo. One hour later (1540) the cloud further expanded up to 400 km in diameter (120,000 km^2). There was a several-kilometer-high dome-shaped swell at the center of the cloud at 1440 and 1540. The widths of shadows of the eruption cloud onto the surrounding white clouds indicate that the altitude of the cloud was approximately 25 km at its eastern edge and 34 km at its center at 1540 (Tanaka and others, 1991). The cloud expanded up to 250 km upwind until 1940, covering an area of 300,000 km^2, and subsequently the east end of the cloud moved westward, at which time the cloud reached a stagnation point upwind but continued to grow downwind and crosswind. It is inferred from the position of the center of the cloud at 1440 and the wind velocity at the altitude of the cloud (typically 10 to 20 m/s southwestward) that the cloud began to expand at about 1420 or before. Average radial expansion velocity of the giant cloud was up to 10^2 m/s before 1440 and approximately 14 m/s between 1440 and 1540.

Infrared images indicate that the surface temperature of the eruption cloud before 1340 was approximately −80°C, which is slightly higher than the temperature at the tropopause (−83°C). The surface temperature of the giant cloud after 1440 was higher than the tropopause temperature by more than 10°C; the east half (upwind) of the cloud was mostly −80 to −70°C, that of the west half was slightly higher than −70°C, and there was a small hot spot (a few tens of kilometers in diameter) with temperatures up to −27°C at the center of the cloud at 1440. At 1540 the area of the surface temperature higher than −60°C with some hot spots of approximately −30°C spread in all directions. The area with surface temperatures higher than −60°C further expanded up to 100 km upwind at 1640 (Tokuno, 1991a,b; Tanaka and others, 1991; Koyaguchi and Tokuno, 1993).

TEPHRA-FALL DEPOSITS

The tephra-fall deposits can be divided into 5 units (fig. 2), designated as layers A, B, C$_1$, C$_2$, and D in ascending order. This lettering scheme is consistent with that used by Paladio-Melosantos and others (this volume); in it, layers C$_1$, C$_2$, and D correspond to layers C, D, and E, respectively, of Koyaguchi and Tokuno (1993). Layer A is a lapilli or volcanic sand layer enriched in grayish pumice grains and lithic fragments. It is the deposit of the eruption on

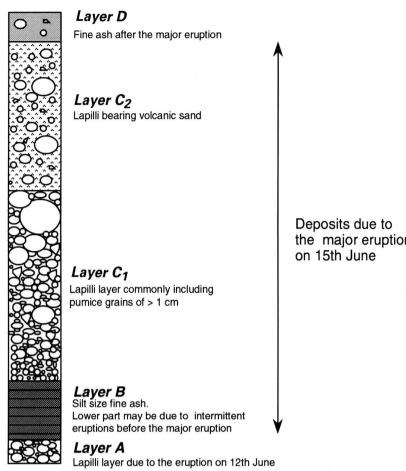

Layer D
Fine ash after the major eruption

Layer C₂
Lapilli bearing volcanic sand

Layer C₁
Lapilli layer commonly including
pumice grains of > 1 cm

Layer B
Silt size fine ash.
Lower part may be due to intermittent
eruptions before the major eruption

Layer A
Lapilli layer due to the eruption on 12th June

10 cm

Deposits due to
the major eruption
on 15th June

Loc. 1101

Figure 2. A columnar section for the tephra-fall deposits of the Pinatubo 1991 eruptions at location 1101, approximately 13 km southwest of the vent.

June 12. Layer B is a silt-sized fine ash layer, the lower part of which was deposited between the eruptions of June 12 and the climactic eruption of June 15. Layer C_1 is a lapilli layer commonly including pumice grains of >1 cm in diameter. Layer C_2 is a lapilli-bearing volcanic sand layer. The uppermost part of layer B and all of layers C_1 and C_2 correlate with the sequence from heavy, fine ash fall to lapilli fall during the major eruption on June 15. Layer D is a fine ash layer, and it originated mainly from fallout of continuous, minor eruptions occurring during the period of June 16 to early July.

Figure 3 shows isopach maps of layers A, B', C_1, and C_2. In contrast to the narrow distribution of layer A, the isopach maps of layers B, C_1, and C_2 show wide distributions in all directions including upwind (northeast) of the vent, although they are slightly elongated in the wind direction. This observation is consistent with the observation in the satellite images that the eruption clouds of the major eruption on June 15 expanded in all directions.

The broad distribution of layer B (fig. 3) suggests that the fine ash in layer B is not distal deposits of a small eruption but originated from a large eruption cloud expanding against wind. Judging from this distribution and the depletion of coarse material, layer B is considered to be derived from an ash cloud that was produced from pyroclastic flows covering several tens of square kilometers (Koyaguchi and Tokuno, 1993). There were also pulses of fine ash fall, mainly downwind from Pinatubo, due to intermittent minor eruptions on June 13, 14, and in the morning of June 15 (Hoblitt, Wolfe, and others, this volume; Paladio-Melosantos and others, this volume), which could not be distinguished in this study from the fine ash deposits from the initial phase of the major eruption.

According to the eyewitness accounts of the major eruption collected at Subic Bay Naval Station, Clark Air Base, Magalang, and Angeles (summarized by Koyaguchi and Tokuno, 1993), fine ash fall with rain started in the morning of June 15 and became heavy after 1400, even in the region upwind of the vent. Intermittent lapilli fall started between 1400 and 1445 and became very heavy and continuous after 1500 even in the region upwind the volcano. Considering the time scale for pumice grains to reach the ground from the bottom of the cloud (approximately 20 minutes for grains of more than 1 cm in diameter; Walker

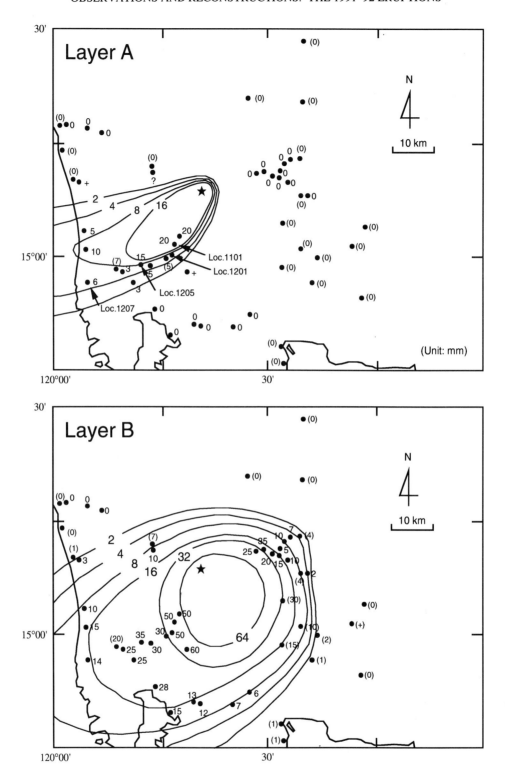

Figure 3. Isopach maps of layers A, B, C$_1$, and C$_2$ (after Koyaguchi and Tokuno, 1993). Star indicates the locality of Mount Pinatubo. Thicknesses are in millimeters. Data in parentheses are unpublished data by E.L. Listanco (Earthquake Research Institute and PHIVOLCS), from June 1991. The other data were collected in December, 1991. At some localities, thicknesses of layer C$_2$ were less in December 1991 than in June 1991, as a result of erosion and compaction. In such cases, data of Listanco from June 1991 were used. +, trace.

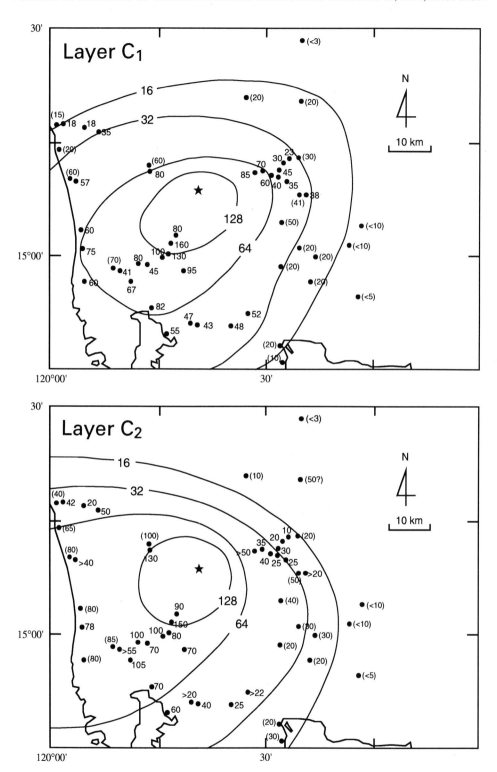

Figure 3.—Continued.

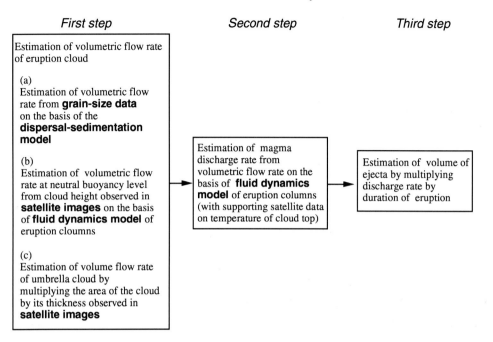

Figure 4. Flowchart of the volume estimation of ejecta and deposits.

VOLUME ESTIMATION OF EJECTA

The volume estimation of ejecta by means of a dynamic model of the eruption cloud is composed of three steps (fig. 4). The first step is the estimation of the volumetric flow rate of the cloud. There are at least three independent ways for estimating the volumetric flow rate of the cloud; one method is an estimation from grain-size data of deposits on the basis of a dispersal-sedimentation model of

and others, 1971; Wilson, 1972), I infer that the pumice grains started falling from the eruption cloud between 1400 and 1425. Thus, the deposition of layer C_1 roughly corresponded to the occurrence of the giant cloud (1420 or before). The dynamics of the expanding umbrella cloud during the major eruption will be discussed on the basis of the granulometric data of layer C_1 and the satellite images from 1440 in later sections.

tephra, and the other two methods are estimations from satellite images on the basis of fluid dynamics models. The second step involves estimation of the magma discharge rate on the basis of the volumetric flow rate of the eruption clouds and fluid dynamics models. Finally, the total amount of the ejecta can be obtained by multiplying the discharge rate by the duration of the eruption in the third step. Temperature data in the satellite images also give important constraints on the dynamics of the eruption cloud. We will describe each step of these estimations below.

ESTIMATION OF THE VOLUMETRIC FLOW RATE OF AN UMBRELLA CLOUD FROM GRAIN SIZE DATA

Theoretical analyses (Sparks, 1986; Woods, 1988) have revealed that an eruption column convectively rises by buoyancy to a height where the column density equals that

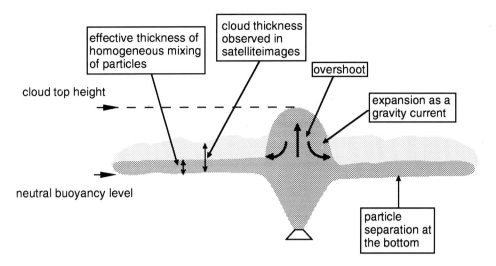

Figure 5. Schematic illustration of an umbrella eruption cloud showing the definitions of cloud thickness.

of the atmosphere and spreads laterally as a gravity current just above the neutral buoyancy level to form an umbrella cloud (fig. 5). The dispersal of tephra-fall deposits is generally controlled by many factors such as dynamics of the eruption cloud, its interaction with wind, and the overall grain-size distribution of the ejecta (Carey and Sparks, 1986; Bursik and others, 1992; Sparks and others, 1992; Koyaguchi, 1994).

Koyaguchi (1994) developed a method for estimating the volumetric flow rate of an umbrella cloud from grain-size data of tephra-fall deposits. The following five assumptions are made:

1. Tephra is homogenized by turbulence in the umbrella cloud.
2. Particles fall out at their terminal velocities from the bottom of the umbrella cloud, where turbulence diminishes and the vertical velocity component is negligible.
3. The rate of change in thickness of the umbrella cloud (or the rate of change in effective thickness where the sediments are homogenized by turbulence) is small compared with the rate of change in the radius of the cloud. Note that this assumption does not necessarily require a constant thickness throughout the cloud.
4. The umbrella cloud expands steadily. This assumption may be relevant if the time scale for the cloud to reach localities near the vent is much shorter than the duration of the eruption; it is not relevant for very distal regions.
5. The expansion velocity of the cloud is so much greater than wind velocity that the cloud expands radially from the center line of the vent.

The model with these assumptions is simplified in order to elucidate basic relations between eruption cloud dynamics and grain-size variation of tephra-fall deposits. These assumptions will be evaluated later.

The principle of the present dispersal-sedimentation model is schematically illustrated in figure 6. As a specific volume of an umbrella cloud is conveyed laterally, particles are removed from the cloud, so that both the total mass of particles, M, and the probability distribution function (abbreviated as *pdf* hereafter) of settling velocity, $f(v)$, in the specific volume will change with distance. The function $f(v)$ is directly related to the grain size distribution if the density and the shape of the particles are uniform. Based on the third and fourth assumptions, the volumetric flow rate of the umbrella cloud, Q, can be approximated as

$$Q \sim 2\pi r \frac{dr}{dt} h \qquad (1)$$

where r is the distance from the center of the umbrella cloud, h is the thickness of the umbrella cloud, and t is the time. When the wind velocity is negligible compared with the expansion velocity (the fifth assumption), r approximates the distance from the center line of the vent. According to Martin and Nokes' theory (1988) and the first and second assumptions, the decay of a mass of particles with a given settling velocity in the specific volume occurs at a rate given by

$$\frac{d(Mf(v))}{dt} = \frac{dr}{dt} \frac{d(Mf(v))}{dr} = -\frac{v}{h}(Mf(v)) \qquad (2)$$

From equations 1 and 2, we obtain the simple solution

$$f(v) = \frac{M_0 f_0(v) \exp\left(-\dfrac{\pi v r^2}{Q}\right)}{M} \qquad (3)$$

where M_0 and $f_0(v)$ are the total mass of particles and the *pdf* in the specific volume of the cloud at $r=0$, respectively. Equation 3 means that the concentration of the coarser particles decreases more rapidly than the finer ones with distance in the specific volume of the cloud, and therefore the *pdf* changes with distance. Because the sedimentation rate is proportional to the terminal velocity of a particle (see fig. 6), the *pdf* in the deposit, $f_{sed}(v)$, should be expressed as

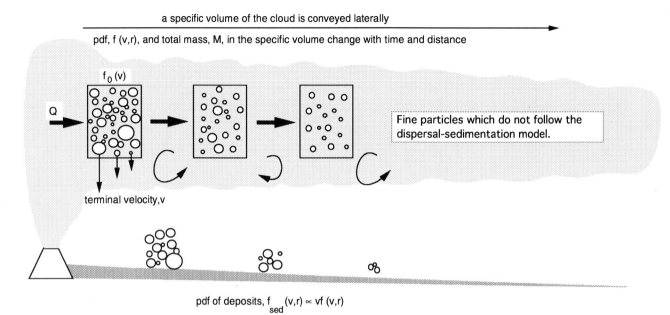

a specific volume of the cloud is conveyed laterally

pdf, f (v,r), and total mass, M, in the specific volume change with time and distance

$f_0(v)$

Q

Fine particles which do not follow the dispersal-sedimentation model.

terminal velocity, v

pdf of deposits, $f_{sed}(v,r) \propto vf(v,r)$

Figure 6. Schematic illustration of the dispersal-sedimentation model from an umbrella cloud. Variables are defined in the text.

$$f_{sed}(v,r) = \frac{vf(v)}{\int_0^\infty vf(v)dv} = \frac{vf_0(v)\exp\left(-\dfrac{\pi vr^2}{Q}\right)}{\int_0^\infty vf_0(v)\exp\left(-\dfrac{\pi vr^2}{Q}\right)dv} \quad (4)$$

The *pdf* in the deposits is a function of the initial *pdf* of the ejecta. From equation 4, we obtain the ratio of the *pdf* in the deposits of two localities at distances from the vent r_1 and r_2, ln R_f, as:

$$\ln R_f = -\frac{\pi(r_1^2 - r_2^2)v}{Q} + \ln A \quad (5a)$$

where

$$A \equiv \frac{\int_0^\infty vf_0(v)\exp\left(-\dfrac{\pi vr_2^2}{Q}\right)dv}{\int_0^\infty vf_0(v)\exp\left(-\dfrac{\pi vr_1^2}{Q}\right)dv} \quad (5b)$$

The parameter A represents the ratio of the total mass of deposits between the two localities, and so it is a constant for a given pair of localities. An important implication of this result is that the first term of the right side of equation 5a is independent of the initial *pdf* and that the logarithm of the ratio of *pdf* has a linear relation with the settling velocity. Because the distance from the vent, r, is a known factor, we can estimate the value of Q from the slope of the line in the diagram of the logarithm of the ratio of *pdf* versus the settling velocity.

The ratio of *pdf* on the basis of the terminal velocity, R_f, is obtained from grain-size data as:

$$R_f = \frac{g(\phi)_{at\ r=r_1}}{g(\phi)_{at\ r=r_2}} \quad (6)$$

where $g(\phi)$ is the *pdf* of grain size based on the phi scale, ϕ. The value of $g(\phi)$ is given by dividing the mass fraction of the sample in the range with a certain interval of phi scale by the interval of the phi scale. On the other hand, the settling velocity of a particle with the size of ϕ at high Reynolds number is calculated from the following formula:

$$v = C_d\left(\frac{\sigma gd}{\rho}\right)^{1/2} = C_d\left(\frac{\sigma g2^{-\phi}}{10\rho}\right)^{1/2} \equiv B2^{-\phi/2} \quad (7)$$

where d is the clast diameter, ρ is the atmospheric density, σ is the clast density, g is the acceleration due to gravity, and C_d is the drag coefficient, which is approximately unity. B is the parameter which relates grain size in phi scale to terminal velocity for each grain. The values of B vary considerably depending on the atmospheric density (2.0×10^{-5} g/cm³ in the stratosphere and 1.3×10^{-3} g/cm³ near the surface). It should also be noted that equation 7 is valid only for particles larger than a few hundreds of microns and that for smaller particles it should be replaced by a different relationship. Because of these uncertainties in estimating the value of B, it would be practical to draw the diagram of ln R_f versus $2^{-\phi/2}$ instead of that of ln R_f versus v. The diagram of ln R_f versus $2^{-\phi/2}$ can be obtained from grain size data alone without any assumptions. In this case, the slope of the line

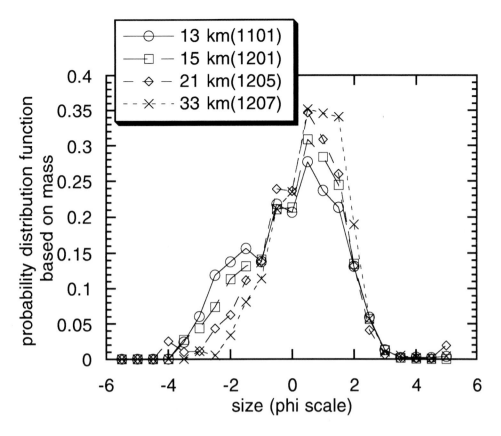

Figure 7. Probability distribution functions of grain size for the tephra-fall deposits of the major eruption (layer C_1) for the localities with variable distances from the vent. The numbers in parentheses are locality numbers from layer A in figure 3.

represents $\pi(r_1^2 - r_2^2)B/Q$, so that we can obtain the value of B/Q for two given localities. Consequently, the value of Q is estimated according to the assumed values of B.

The *pdf* of the grain size of layer C_1 is shown in figure 7. Fine particles ($\phi > 3$) make up only a small fraction of the lapilli fall deposits except for those which fell with rain. The size distribution of layer C_1 collected along the downwind axis systematically changes with distance; the fraction of coarser grains decreases with increasing distance (fig. 7). The value of B/Q is calculated from the diagram of $\ln R_f$ versus $2^{-\phi/2}$ to be 5.7×10^{-14}, 6.2×10^{-14}, and 4.2×10^{-14} cm^{-2} for the pairs of localities of 15 km versus 13 km, 21 km versus 13 km, and 33 km versus 13 km, respectively (fig. 8). The volumetric flow rate of the umbrella cloud is, therefore, estimated to be about 4×10^{16} to 5×10^{16} cm^3/s when the value of B for pumice grains in the stratosphere (2,200 cm/s) is adopted. By use of the obtained value of B/Q, the initial grain size distribution in the umbrella cloud can be calculated from the actual grain-size distribution data (Koyaguchi, 1994) as:

$$g_0(\phi) = \frac{g(\phi)\exp\left(\dfrac{\pi r^2 B 2^{-\phi/2}}{Q}\right)}{2^{-\phi/2}\displaystyle\int_{\infty}^{-\infty}\frac{g(\phi)\exp\left(\dfrac{\pi r^2 B 2^{-\phi/2}}{Q}\right)}{2^{-\phi/2}}\,d\phi} \qquad (8)$$

In order to confirm the above estimation of B/Q, the initial size distributions calculated from the actual grain-size data of different localities are compared. They converge onto a single distribution pattern for $B/Q = 5 \times 10^{-14}$ cm^{-2} (fig. 9).

The present model is based on a number of simplifications that should be evaluated thoroughly. The most important effects would be those of wind. The major eruption occurred when a tropical typhoon, international code name Yunya, was passing 50 km north of Pinatubo. The lower tropospheric winds were strongly affected by the typhoon winds, while mid tropospheric and stratospheric winds (9,000 m and above) were essentially unaffected by the typhoon (Paladio-Melosantos and others, this volume). Because the eruption cloud expanded in the stratosphere, it blew generally southwestward, consistent with the wind direction in the stratosphere. There were some influences of wind at lower altitudes such as those of the typhoon. For example, the center of the isopach map of layer B and C_1 is located about 10 km south of the vent. This may be due to a local influence of the wind within the typhoon.

The effect of the wind in the stratosphere will be evaluated first. The satellite images suggest that the center of the eruption cloud drifted southwestward at the velocity of approximately 10 m/s between 1440 and 1940 (Koyaguchi and Tokuno, 1993), while the cloud was radially expanding. Because of this effect, the distance from the vent does not give the distance from the center of the eruption cloud in the

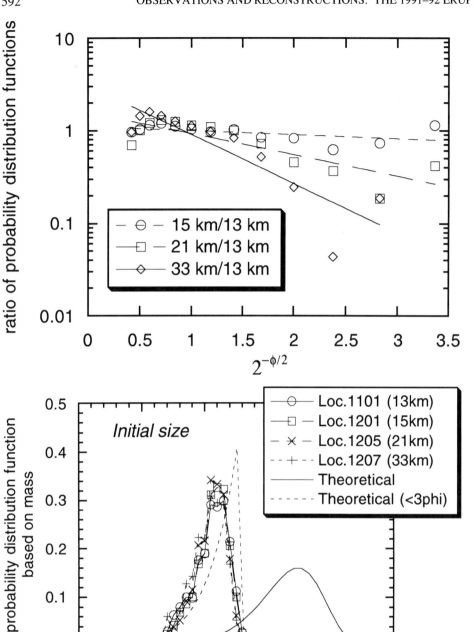

Figure 8. The ratio of the probability distribution functions of grain size from two localities against the settling velocities for determining the volumetric flow rate of an umbrella cloud of the 1991 Pinatubo eruption. Probability distribution functions of grain size from localities that are 15, 21, and 33 km from the vent are normalized to the pdf from a locality 13 km from the vent.

Figure 9. Initial probability distribution functions of grain size for layer C_1, calculated from the actual grain size data. The theoretical probability distribution function which explains exponential thinning behavior from r=0 to infinity for k=2.5×10⁻⁷ cm⁻¹ and B/Q=5×10⁻¹⁴ cm⁻² is shown by a solid curve; part of the theoretical probability distribution function for fine particles (ϕ>3) is shown by a dashed curve. Variables are defined in the text.

region more than a few tens of kilometers away from the vent, where the expansion velocity is comparable to wind velocity. If it is assumed that the thickness of the cloud is constant with time, the downwind distance from the vent, r_{vent}, is approximated by

$$r_{vent} = r\left(1 + \frac{\pi rhV_w}{Q}\right) \qquad (9)$$

where V_w is the wind velocity. The downwind distances of 10 km and 30 km from the vent give approximately 3% and 10% overestimation of the distance from the center of the

cloud, respectively, when Q is 5×10¹⁶ cm³/s, h is 5 km and V_w is 10 m/s. Although the evaluation of this effect on the estimation of Q is not straightforward, because the error of r depends on Q, the overestimation of r can lead to up to a few tens of percentage overestimation of Q when the pair of r_{vent}=30 km and 10 km is used. Consequently, I propose the value of 3 to 4×10¹⁶ cm³/s as the estimate of the volumetric flow rate from grain-size data.

Another important process in the stratosphere is that turbulent fluctuations in the ambient atmosphere can mix the cloud with the surrounding air so that the cloud may lose its coherence. This process would result in change in the effective thickness of the cloud. Furthermore, if the turbulence at the base of the cloud is sufficiently strong, fine particles may be kept in suspension and may not follow the second assumption. This effect would lead to the loss of fine particles in tephra-fall deposits.

The wind at the lower altitude is likely to affect the motion of tephra between the base of the cloud and its arrival on the ground. From equation 7, the settling velocities of particles of a few millimeters in diameter are estimated to be of order 1 to 10 m/s in the lower troposphere, which may be smaller than the wind velocity within the typhoon. This means that these particles can be dispersed laterally over considerable distances while they are falling through a distance of a few tens of kilometers from the umbrella cloud to the ground. These effects of turbulence at the higher and lower altitudes play an important role particularly in the dispersal of fine particles (see fig. 6). The grain-size distribution of coarser particles would give a more reliable estimation of the volumetric flow rate than would the grain-size distribution of fine particles.

ESTIMATION OF THE VOLUMETRIC FLOW RATE OF AN UMBRELLA CLOUD FROM SATELLITE DATA, METHOD 1

Recent theoretical studies of a fluid dynamics model on eruption columns (Wilson and others, 1978; Settle, 1978; Sparks, 1986; Woods, 1988) allow us to relate eruption conditions (for example, discharge rate, temperature, and H_2O content of magma) to dimension and dynamics of eruption clouds (for example, height, depth, temperature, and volumetric flow rate of the cloud). The volumetric flow rate at the neutral buoyancy level can be estimated from satellite data on the basis of the fluid dynamics models. The methods and the values of constants are the same as Woods' model (1988) for steady plinian eruption columns except that the tropical atmospheric conditions are taken into account in this study. Generally speaking, the conditions of plinian eruptions are governed by magma discharge rate, temperature, and H_2O content of magma at the surface. The magma discharge rate is determined by two independent parameters, such as vent radius and initial velocity at the vent, when the temperature and H_2O content of magma are fixed. We evaluate plausible ranges of temperature and water content of magma from viewpoints of petrological features of the Pinatubo pumice first, before the results of numerical calculations for variable magma discharge rates with different pairs of initial velocity and vent radius are shown.

Temperature of magma.—At least two important phenocrystic (or microphenocrystic) phases in the Pinatubo pumice constrain the preeruption temperature of magma: cummingtonite and hemoilmenite. According to the experimental results by Rutherford (1993), the presence of cummingtonite requires an H_2O-rich fluid and pressure in the 150- to 350-MPa range at approximately 1060K (787°C) or less. On the other hand, considering the miscibility gap between ilmenite-hematite solid solution, high hematite content of hemoilmenite indicates that the temperature of magma was higher than 1070K (Imai and others, 1993; this volume). It is, therefore, concluded that the preeruption temperature of the magma was around 1060 to 1070K. The small discrepancy in the crystallization temperature between cummingtonite and hemoilmenite may be due to heterogeneity in the magma chamber or different stages of crystallization between the two minerals. It should be noted that the preeruption temperature does not necessarily represent the temperature of erupting magma at the surface; the temperature of vesiculating magma is significantly cooled by exsolution and expansion of gas phase. If there is good thermal equilibration between the magma and adiabatically expanding gas, the magma with several weight percent H_2O is cooled by a few hundreds of degrees (Williams and McBirney, 1979). The presence of lithic fragments also decreases the effective magma temperature. However, I assumed that this effect is negligible in the present case, because lithic fragments are rarely observed in the tephra-fall deposits (<1%). Consequently, the initial temperature of the eruption cloud at the surface would be approximately 800 to 900K. In the following numerical calculations, a wider range, between 800 and 1200K, is assumed in order to elucidate the effects of magma temperature on the dynamics of the eruption column.

H_2O content of magma.—The presence of hydrous phenocrysts (hornblende, biotite, and cummingtonite) indicates that the magma contained considerable water just prior to the eruption. Ion probe analyses of melt inclusions in plagioclase and quartz yield H_2O contents of 5.5 to 6.5 wt%, which is slightly less than the solubility limit at 200 MPa (Rutherford, 1993). The estimation of H_2O content by means of ion probe analyses of melt inclusions may overestimate H_2O content in the magma because of the effect of overgrowth of the host crystals. In the following calculations, H_2O contents between 1 and 6 wt% are assumed.

The results of the numerical calculations based on the fluid dynamic model (Woods, 1988) suggest that the volumetric flow rate at the neutral buoyancy level is closely related to the total height of the cloud (Bursik and others, 1992). One of the most interesting results is that the relation between the two quantities is not much affected by magmatic temperature, H_2O content, or other conditions of eruption such as vent radius and initial velocity except for some conditions that are close to the critical conditions for column collapse (fig. 10). The total height of the cloud is estimated from the widths of shadows in the satellite images to be approximately 34 km (Tanaka and others, 1991). Although this method is the simplest way among the three methods, it has a disadvantage. The height estimates on the basis of the widths of shadows are largely dependent on the

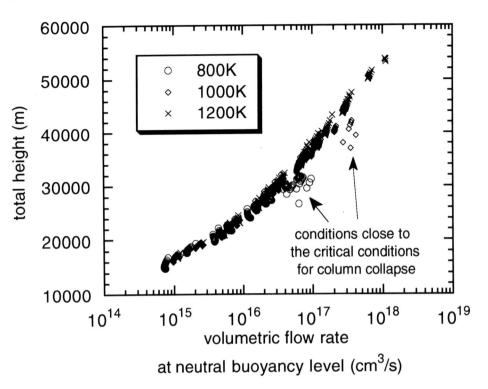

Figure 10. The results of numerical calculations on the fluid dynamics model of eruption clouds showing the relation between the total height of the cloud versus the volumetric flow rate at the neutral buoyancy level. The model by Woods (1988) is applied to the tropical atmospheric conditions. The temperature and H_2O content of magma and the initial velocity are systematically varied in the ranges from 800 to 1200 K, 1 to 6 wt%, and 0 to 350 m/s, respectively.

shape of the edge of clouds and contain considerable errors, which can be as large as the thickness of cloud (up to several kilometers). Taking this uncertainty into account, a volumetric flow rate at the buoyancy level is estimated from the height estimate to be approximately 5×10^{16} to 1×10^{17} cm^3/s. It should be noted that the above value represent the volumetric flow rate at 1540, because the height estimate is based on the satellite image at 1540.

ESTIMATION OF THE VOLUMETRIC FLOW RATE OF AN UMBRELLA CLOUD FROM SATELLITE DATA, METHOD 2

The volumetric flow rate of the umbrella cloud can be estimated from the satellite data in another way. The satellite data indicate that the umbrella cloud expanded from 60,000 km^2 to 120,000 km^2 between 1440 and 1540 and continued to expand until by 1940 it covered an area of 300,000 km^2. This fact suggests that the cloud expanded at the rate of 1.7×10^{11} cm^2/s between 1440 and 1540 and almost steadily at the rate of 1.3×10^{11} cm^2/s between 1540 and 1940. The volumetric flow rate of the umbrella cloud is obtained by multiplying the above rate by the cloud thickness. The largest source of the uncertainty of this method is the cloud thickness, which is obtained only by an indirect method. The maximum thickness of the umbrella cloud would be given by the difference between the altitudes of the cloud top and the neutral buoyancy level (see fig. 5). The difference between the altitudes of the cloud top and the neutral buoyancy level is also related to the volumetric flow rate at the neutral buoyancy level, and this relationship

is not dependent on the magmatic or other conditions of eruption (fig. 11); it is constant (approximately 5 km) when the volumetric flow rate is less than 5×10^{15} cm^3/s, and it increases from 5 km to 10 km as the volumetric flow rate increases from 10^{16} to 10^{17} cm^3/s. According to the satellite images at 1540, there was a small dome-shaped swell (several kilometers high), which suggests that most of the cloud was thinner than the maximum thickness by several kilometers. I infer that the actual thickness of the cloud was 3 to 6 km. The volumetric flow rate between 1440 and 1540 is estimated to be 5×10^{16} to 1×10^{17} cm^3/s by multiplying the spreading rate of the cloud (1.7×10^{11} cm^2/s) by the thickness (3 to 6 km). Recent fluid dynamic considerations suggest that if an umbrella cloud expands as a gravity current, the thickness of the current decreases with time even for a constant volumetric flow rate (A.W. Woods, Institute of Theoretical Geophysics, Cambridge, written commun., 1993). Judging from the expansion rate (1.3×10^{11} cm^2/s), I think it is reasonable to assume that the thickness of the cloud decreased after 1540, although no thickness estimate based on satellite images after 1540 is available. It is suggested that the volumetric flow rate decreased considerably after 1540.

SUMMARY OF ESTIMATION OF THE VOLUMETRIC FLOW RATE OF AN UMBRELLA CLOUD

The two estimations from the satellite image give consistent values of 5×10^{16} to 1×10^{17} cm^3/s, despite the fact that they are based on different kinds of data. On the other

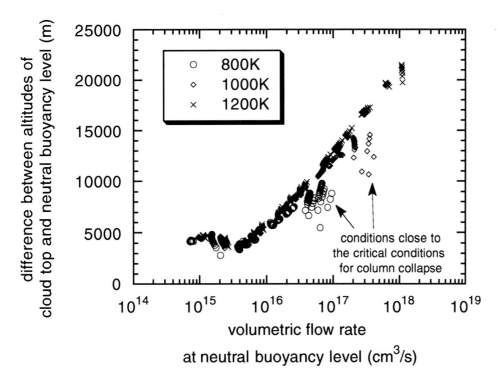

Figure 11. The results of numerical calculations on the fluid dynamics model of eruption clouds showing the difference between the altitudes of the cloud top and the neutral buoyancy level against the volumetric flow rate at the neutral buoyancy level. The ranges of the temperature and H_2O content of magma and the initial velocity are the same as in figure 10.

hand, the volume flow rate on the basis of the grain-size data gives slightly smaller values (3×10^{16} to 4×10^{16} cm^3/s). It should be noted that the meanings of the volumetric flow rates obtained by the above three methods are slightly different. The estimate based on grain size gives the expansion rate of the region where particles are homogenized by turbulence, the estimate based on cloud height represents the volumetric flow rate at the neutral buoyancy level, and the estimate based on cloud thickness and area gives the actual volumetric flow rate of the cloud. Generally, the volumetric flow rate of the eruption cloud increases as its altitude increases because of decompression and entrainment of the ambient air. Because the umbrella cloud is diverted between the altitudes of the neutral buoyancy level and the cloud top, the actual volumetric flow rate would be between those at the neutral buoyancy level and the cloud top. Considering that the umbrella cloud flows as a gravity current, the main part of the current would be diverted just above the neutral buoyancy level (see fig. 5). Consequently, it is considered that the estimate based on grain size and that based on cloud area and thickness roughly represent the volumetric flow rate at the neutral buoyancy level at least near the vent, and they may increase to some extent with distance due to the entrainment of the ambient air during the expansion.

There are at least four possible explanations for the lower estimate of the volumetric flow rate from the grain-size data than estimates from the satellite data. First, the estimates based on satellite data represent the volumetric flow rate at 1540 or the averaged volumetric flow rate between 1440 and 1540, while the estimate based on the grain-size data represents the average volumetric flow rate during the deposition. The lower estimate based on grain

size data may reflect the decrease of volumetric flow rate with time after 1540. Secondly, the effect of the entrainment during the expansion is more significant in the estimation from satellite images of the cloud that expanded more than 10^2 km in diameter than in the estimation from grain-size data near the vent. Third, the average clast density may be greater than that of pumice grains, which can cause the underestimation of the value of B and hence the underestimation of Q. Finally, pyroclastic flows may have been forming at the same time as the plinian fall layers (W.E. Scott and others, this volume), so some of the mass that contributed to rise of the eruption is not represented in the plinian fall deposits.

RELATION BETWEEN THE VOLUMETRIC FLOW RATE OF THE UMBRELLA CLOUD AND THE MAGMA DISCHARGE RATE

The volumetric flow rate at the neutral buoyancy level and that of the cloud top against variable magma discharge rate are shown in figure 12. The relation between the volumetric flow rate of the umbrella cloud and the magma discharge rate depends on the magmatic temperature at the vent but is insensitive to the changes in H_2O content. If the magmatic temperature (at the vent) is 800K, the magma discharge rate is 4 to 8×10^8 kg/s for the volumetric flow rate of 3 to 4×10^{16} cm^3/s. The discharge rate is up to 2×10^9 kg/s if the upper estimate of the volumetric flow rate on the basis of the satellite data (1×10^{17} cm^3/s) is applied. Because the magma discharge rate rather sensitively depends on the volumetric flow rate as well as magmatic temperature at the

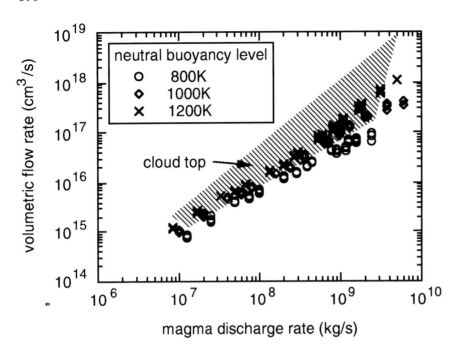

Figure 12. The results of numerical calculations on the fluid dynamics model of eruption clouds showing the relation between the volumetric flow rate at the neutral buoyancy level and the magma discharge rate. The results of the volumetric flow rate at the cloud top are shown by a shaded zone for comparison. The ranges of the temperature and H_2O content of magma and the initial velocity are the same as in figure 10.

vent, the second step of the volume estimation can be a source of serious errors. Furthermore, in contrast to the estimation of volumetric flow rate in the first step, the estimation of the magma discharge rate from the volumetric flow rate is not supported by different kinds of observations. The uncertainties of the results of numerical calculations largely depend on applied fluid dynamics models. Accordingly, the self consistency of the fluid dynamics model will be checked on the basis of the satellite data below.

Constraints from temperature data in the satellite images.—The temperature of an eruption cloud top is a function of many parameters related to the dynamics of the event, such as temperature and H_2O content of magma and the vent radius. Woods and Self (1992) showed that the cloud top temperature deviates from the local temperature at the same altitude and thus could not be used to infer plume elevation. Cloud top temperature increases with decreasing initial velocity and with increasing vent radius if discharge rate and temperature and H_2O content of magma are fixed. One of the most important features of the dynamics of plinian eruptions is that the column is unstable and collapses if the vent radius is greater than a critical value for a constant initial velocity or if initial velocity is less than a critical value for a constant vent radius (Woods, 1988). Because of this constraint, cloud top temperature has an upper limit for a given discharge rate, temperature, and H_2O content of magma (Koyaguchi and Tokuno, 1993). Figure 13 shows the numerical results of cloud top temperature against magma discharge rate for magma temperature of 800, 1000 and 1200K, water content from 1 to 6 wt%, and variable pairs of vent radius and initial velocity on the basis of Woods' model (1988). The maximum cloud temperature for each magma temperature increases as the magma

discharge rate increases. For the cloud top temperature to be as high as that in the satellite image (243K), the magma discharge rate should be greater than 7×10^8, 1×10^9 or 3×10^9 kg/s if the magma temperatures are 800, 1000 or 1200K, respectively. It is notable that the discharge rates estimated from the cloud temperature increase as assumed magmatic temperature increases (fig. 13), whereas those estimated from the volumetric flow rate decrease with the increasing magmatic temperature (fig. 12). The discharge rates estimated from the two different methods agree fairly well when the assumed temperature is 800K (4 to 8×10^8 kg/s vs >7×10^8 kg/s), while they give conflicting results when the assumed temperature is 1200K (2 to 3×10^8 kg/s vs >3×10^9 kg/s). We emphasize that the temperature data in the satellite images should be used only as a supporting constraint at present. Because the fluid dynamics model assumes an steady eruption column which has horizontally uniform temperature and other physical quantities, the numerically obtained temperature cannot be compared directly to the observed cloud top temperatures.

DURATION OF THE MAJOR ERUPTION AND ESTIMATION OF TOTAL VOLUME OF EJECTA

Koyaguchi and Tokuno (1993) assumed that the duration of the eruption was approximately 5 hours on the basis of the duration in which the eruption cloud continued to expand upwind. However, the duration of the cloud expansion does not necessarily indicate the duration of the eruption, because the cloud can continue to expand even after the supply at its source ceases. Tahira and others (this volume) reported that infrasonic waves and acoustic-gravity waves generated by the major eruption of Mount Pinatubo

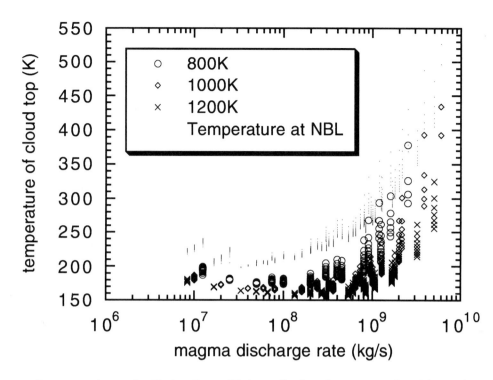

Figure 13. The results of numerical calculations on the fluid dynamics model of eruption clouds showing the relation between the temperature of the cloud top and the magma discharge rate. The temperature of the cloud at the neutral buoyancy level (NBL) is shown by small dots for comparison. The ranges of the temperature and H_2O content of magma and the initial velocity are the same as figure 10.

on June were observed at Kariya, Japan. High amplitudes of these waves lasted about 3.5 hours. Because it is considered that the infrasonic waves reflect the intensity of magma discharge at the source, I adopt the duration of the high amplitudes of the infrasonic waves (3.5 hours) as the duration of the eruption.

The total volume of the ejecta can be estimated by multiplying the magma discharge rate by the duration of the eruption. It should be noted here that the magma discharge rates based on the satellite images represent those at 1540 or those between 1440 and 1540, while those based on grain-size data represent the average magma discharge rate during the deposition. It would be more reasonable to use the average discharge rate for the calculation of the total volume rather than the magma discharge rate at one moment. Accordingly, applying the values based on grain-size data, the total volume of the ejecta is 2 to 4 km³ DRE (5 to 10 km³ bulk volume assuming that the bulk density of the deposit is 1,000 kg/m³). If the values based on satellite images are used, the total volume is estimated to be up to 10 km³ DRE (25 km³ bulk volume).

GEOLOGICAL ESTIMATION OF THE VOLUME OF THE DEPOSITS

It is quite difficult to estimate the whole amount of the ejecta by geological methods, because fine particles are likely to be dispersed in the atmosphere and cannot be counted in the deposits. Pyle (1989) and Fierstein and Nathenson (1992) showed that many deposits display an exponential decrease in thickness with square root of the

area enclosed by an isopach contour. This means that the thickness, T, should be in the form as:

$$T(S) = T_0\exp\left(-kS^{1/2}\right) \qquad (10)$$

where S is the area enclosed by an isopach contour, T_0 is the thickness at $S=0$, and k is the decay constant. They proposed that the volume of the tephra-fall deposits, V, can be estimated by integrating this empirical relationship from $S=0$ to $S=\infty$ as:

$$V = 2T_0/k^2 \qquad (11)$$

The validity of this empirical relation has been tested for the tephra-fall deposits during eruptions where the distal ash falls on land (for example, Mount St. Helens; Fierstein and Nathenson, 1992). However, this method should be applied with caution to eruptions where data from distal areas are sparse. In the case of the Pinatubo eruption, most of tephra-fall deposits fell in the South China Sea, where few thickness data are available. Figure 14 shows the relation between the logarithm of thickness and the square root of the area enclosed by an isopach contour of layers C_1 and C_2 (fig. 3) and their best fit lines. The decay constants of layers C_1 and C_2 are 2.7×10^{-7} and 2.4×10^{-7} (cm⁻¹), respectively. The sum of the volumes of the two layers is calculated from equation 11 to be 1.2 km³ (the volumes of layers C_1 and C_2 are 0.5 and 0.7 km³, respectively). The areas enclosed by isopach contours less than 32 mm thick determined were by smoothly extrapolating the isopach contours obtained on land toward the South China Sea, and so the calculation would contain considerable errors. Similarly, the volume calculated from isopachs of the Pinatubo Volcano

Observatory Team (1991) is 1.5 km³,which agrees well with the present results. An early extrapolation to include distal deposits (3 km³; Pinatubo Volcano Observatory Team, 1991), has more recently been updated by use of the dual-slope exponential thinning relationship (3.4 to 4.4 km³ bulk volume; Paladio-Melosantos and others, this volume; W.E. Scott and others, this volume).

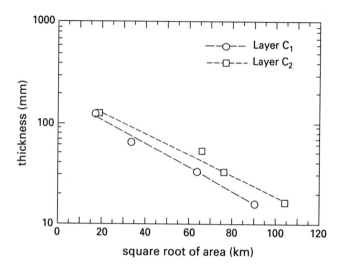

Figure 14. The logarithm of thickness versus the square root of the area enclosed by an isopach contour for layers C_1 and C_2 (fig. 3). Uses terrestrial data only.

Discrepancies between various estimates of volumes are shown schematically in figure 15. By integrating equation 5b we can obtain a volume estimate from on-land grain-size data on the basis of the dispersal-sedimentation model, which yields an anomalously low estimate. Fine particles ($\phi>3$) make up only a small fraction of the Pinatubo tephra-fall deposits on land (fig. 7; see also fig. 9 for estimated initial grain-size distribution). The extreme depletion of the fine particles is not predicted by the present dispersal-sedimentation model alone, unless the ejecta were originally depleted in the fine particles. However, enough fine ash fell with rain or as accretionary lapilli that I do not suggest an initial paucity of fine particles. An alternative is that fine particles did not follow the present model but were instead widely dispersed in the atmosphere (see fig. 6). If the turbulence at the bottom of the cloud and between the cloud and the ground is sufficiently strong, fine particles may be kept in suspension and dispersed laterally over considerable distances.

Generally speaking, the total sedimentation rate and thickness at a given distance are determined by a specific range of grain size; they are determined by those of the larger particles in the proximal region but by those of the finer particles in the distal region. According to Koyaguchi (1994), the thickness-distance relation is basically determined by the sedimentation of coarse particles ($\phi<0$) in the region within 30 km from the vent when $Q>10^{16}$ cm³/s. Therefore, the thickness-distance relation on land around

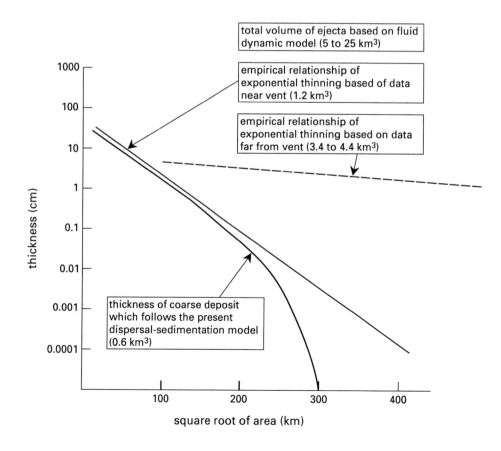

Figure 15. Volume estimation that is consistent with the dispersal-sedimentation model and volume estimates that are based on the assumption of the exponential thinning behavior.

Pinatubo was mainly determined by the observed coarse particles ($\phi<0$), but it is insensitive to the amount of fine particles. Koyaguchi (1994) analytically proved that the dispersal-sedimentation model successfully explained exponential thinning behavior from $r=0$ to $r=\infty$ only for an initial grain-size distribution that is approximately log-normal with $\sigma_\phi = 2.5$; otherwise, the model holds only in a limited range of distance. In figure 9 the theoretical initial grain-size distributions that predict exponential thinning behavior from the dispersal-sedimentation model are calculated for $B/Q = 5\times10^{-14}$ cm^{-2} and k=2.5×10^{-7} cm^{-1}. The initial size distribution estimated from the actual grain-size data shows a substantially coarser distribution than the theoretical initial grain-size distribution. The difference between the theoretical and actual initial grain size is explained by the depletion of fine particles. If particles of size $\phi>3$ are omitted from the theoretical initial grain-size distribution, the *pdf* of the rest shows a better agreement with the distribution deduced from the actual grain size data (fig. 9). Thus, the volume of the particles which follows the dispersal-sedimentation model is understandably smaller than the volume estimated from the data on land on the basis of the exponential thinning relationship.

Volumes estimated from the fluid dynamic model and satellite data are higher than those estimated by Paladio-Melosantos and others (this volume) using the method of Fierstein and Nathenson (1992), even when fine ash observed far from the vent, in Southeast Asia and the South China Sea, is taken into account. Again, I judge that separation of fine ash explains the discrepancy.

CONCLUDING REMARKS

The behavior of fine particles is an important problem not only in volcanology but also in relation to environmental effects of volcanic eruptions or to aviation safety. Walker (1980) developed another method for estimating the amount of lost fine vitric particles on the basis of the degree of the crystal concentration in the deposits. Crystal concentration studies are being carried out for the Pinatubo tephra-fall deposits in order to confirm the present conclusions. In the present study, the numerical calculations on the dynamics of the eruption columns were carried out using a wide ranges of magmatic temperatures, H$_2$O contents, and other conditions of eruption, and the results were cross-checked by different kinds of observations. Nevertheless, the following uncertainties may still modify the present results to some extent:

1. A steady plinian eruption was assumed, while the eruption may have been intermittent or fluctuating.
2. The effects of thermal disequilibrium between pyroclasts and entrained air or moisture of the atmosphere are not taken into account.

ACKNOWLEDGMENTS

Comments by S. Carey (Univ. of Rhode Island), A.W. Woods (Institute of Theoretical Geophysics, Cambridge), W.E. Scott (U.S. Geological Survey), Makoto Tahira (Aichi Univ. of Education), Toshitsugu Fujii (Univ. of Tokyo, Earthquake Research Institute), C.G. Newhall (U.S. Geological Survey), and an anonymous reviewer were helpful. E.L. Listanco (Earthquake Research Institute and PHIVOLCS) kindly allowed me to use his unpublished data. I wish to thank all of these people.

REFERENCES CITED

Bursik, M.I., Sparks, R.S.J., Gilbert, J.S., and Carey, S.N., 1992, Sedimentation of tephra by volcanic plumes. I. Theory and its comparison with a study of the Fogo A plinian deposit, Sao Miguel (Azores): Bulletin of Volcanology, v. 54, p. 324–344.

Carey, S.N., and Sparks, R.S.J., 1986, Quantitative models of the fallout and dispersal of tephra from volcanic eruption columns: Bulletin of Volcanology, v. 48, p. 109–125.

Fierstein, J., and Nathenson, M., 1992, Another look at the calculation of fallout tephra volumes: Bulletin of Volcanology, v. 54, p. 156–167.

Hoblitt, R.P., Wolfe, E.W., Scott, W.E., Couchman, M.R., Pallister, J.S., and Javier, D., this volume, The preclimactic eruptions of Mount Pinatubo, June 1991.

Imai, A., Listanco, E.L. and Fujii, T., 1993, Petrologic and sulfur isotopic significance of highly oxidized and sulfur-rich magma of Mount Pinatubo, Philippines: Geology, v. 21, p. 699–702.

———this volume, Highly oxidized and sulfur-rich dacitic magma of Mount Pinatubo: Implication for metallogenesis of porphyry copper mineralization in the Western Luzon arc.

Koyaguchi, T., 1994, Grain size variation of tephra derived from volcanic umbrella clouds: Bulletin of Volcanology, v. 56, p. 1–9.

Koyaguchi, T., and Tokuno, M., 1993, Origin of the giant eruption cloud of Pinatubo, June 15, 1991: Journal of Volcanology and Geothermal Research, v. 55, p. 85–96.

Martin, D., and Nokes, R., 1988, Crystal settling in a vigorously convecting magma chamber: Nature, v. 332, p. 534–536.

Paladio-Melosantos, M.L., Solidum, R.U., Scott, W.E., Quiambao, R.B., Umbal, J.V., Rodolfo, K.S., Tubianosa, B.S., Delos Reyes, P.J., and Ruelo, H.R., this volume, Tephra falls of the 1991 eruptions of Mount Pinatubo.

Pinatubo Volcano Observatory Team, 1991, Lessons from a major eruption: Mt. Pinatubo, Philippines: Eos, Transactions, American Geophysical Union, v. 72, p. 545–555.

Pyle, D.M., 1989, The thickness, volume and grainsize of tephra fall deposits: Bulletin of Volcanology, v. 51, p. 1–15.

Rutherford, M.J., 1993, Experimental petrology applied to volcanic processes: Eos, Transactions, American Geophysical Union, v. 74, p. 49–55.

Scott, W.E., Hoblitt, R.P., Torres, R.C., Self, S, Martinez, M.L., and Nillos, T., Jr., this volume, Pyroclastic flows of the June 15, 1991, climactic eruption of Mount Pinatubo.

Settle, M., 1978, Volcanic eruption clouds and thermal power output of explosive eruptions: Journal of Volcanology and Geothermal Research, v. 3, p. 309–324.

Sparks, R.S.J., 1986, The dimension and dynamics of volcanic eruption columns: Bulletin of Volcanology, v. 48, p. 3–15.

Sparks, R.S.J., Bursik, M.I., Ablay, G.J., Thomas, R.M.E., and Carey, S.N., 1992, Sedimentation of tephra by volcanic plumes. II. Controls on thickness and grain-size variation of tephra fall deposits: Bulletin of Volcanology, v. 54, p. 685–695.

Tahira, M., Nomura, M., Sawada, Y., and Kamo, K., this volume, Infrasonic and acoustic-gravity waves generated by the Mount Pinatubo eruption of June 15, 1991.

Tanaka, S., Sugimura, T., Harada, T., and Tanaka, M., 1991, Satellite observation of the diffusion of Pinatubo volcanic dust to the stratosphere: Journal of the Remote Sensing Society of Japan, v. 11, p. 91–99. (In Japanese with English abstract.)

Tokuno, M., 1991a, Observation of volcanic eruption clouds from Mt. Pinatubo, Philippines by GMS–4: Meteorological Satellite Center Technical Note, v. 23, p. 1–14.

————1991b, GMS–4 observations of volcanic eruption clouds from Mt. Pinatubo, Philippines: Journal of the Remote Sensing Society of Japan, v. 11, p. 81–89. (In Japanese with English abstract.)

Walker, G.P.L., 1980, The Taupo pumice: Products of the most powerful known (ultraplinian) eruption?: Journal of Volcanology and Geothermal Research, v. 8, p. 69–94.

Walker, G.P.L., Wilson, L., and Boswell, E.L.G., 1971, Explosive volcanic eruptions. I. The rate of fall of pyroclasts: Geophysical Journal of the Royal Astronomical Society, v. 22, p. 377–383.

Williams, H., and McBirney, A.R., 1979, Volcanology: San Francisco, Freeman, Cooper and Company, 397 p.

Wilson, L., 1972, Explosive volcanic eruptions. II. The atmospheric trajectory of pyroclasts: Geophysical Journal of the Royal Astronomical Society, v. 30, p. 381–392.

Wilson, L., Sparks, R.S.J., Huang, T.C., and Watkins, N.D., 1978, The control of volcanic column heights by eruption energetics and dynamics: Journal of Geophysical Research, v. 83, p. 1829–1836.

Woods, A.W., 1988, The fluid dynamics and thermodynamics of eruption columns: Bulletin of Volcanology, v. 50, p. 169–193.

Woods, A.W., and Self, S., 1992, Thermal disequilibrium at the top of volcanic clouds and its effect on estimates of the column height: Nature, v. 355, p. 628–630.

Infrasonic and Acoustic-Gravity Waves Generated by the Mount Pinatubo Eruption of June 15, 1991

By Makoto Tahira,[1] Masahiro Nomura,[1] Yosihiro Sawada,[2] and Kosuke Kamo[3]

ABSTRACT

Remarkable infrasonic waves produced by the Mount Pinatubo eruption of June 15, 1991, were recorded by the infrasonic observation system at Kariya, Japan, 2,770 kilometers to the northeast of the volcano. Waves which arrived 2 hours, 45 to 54 minutes after major eruptions are interpreted as the A1 waves, or the waves that propagated along the shorter great circle path in the atmosphere. A strong infrasonic signal lasting almost 10 hours was recorded from the climactic eruption of Mount Pinatubo on June 15. These data provide new information about the explosive sequence of the eruptions at its climactic stage. In addition, a weak and long-lasting wave train was recorded about 35 hours after the onset of the climactic stage. These waves arrived from the northeast and southwest and are interpreted as the A2 and A3 waves generated by the climactic eruptions.

Acoustic-gravity waves were also recorded at five Japanese stations operated by the Japan Meteorological Agency. The complicated and elongated waveform indicates superposition of several wave trains excited by different explosions. The longest period of oscillation is 13.9 minutes, which is more than twice the longest period observed from the Mount St. Helens eruption in 1980. Tentative energy estimation based on the microbarographic data gave an explosive equivalent of approximately 70 megatonnes of TNT.

INTRODUCTION

The eruption of Mount Pinatubo in 1991 produced various oscillations both in the atmosphere and on the ground. Seismic disturbances lasting approximately 3 h were

detected around the world (Kanamori and Mori, 1992; Widmer and Zürn, 1992; Zürn and Widmer, this volume). These waves have dominant periods at around 228 and 270 s, that have been interpreted by Kanamori and Mori (1992) as Rayleigh waves excited by acoustic coupling in the vicinity of the source. Traveling ionospheric disturbances were recorded in Taiwan (Cheng and Huang, 1992) and Japan (Igarashi and others, 1994). We will present in this paper atmospheric pressure waves detected in Japan.

A large volcanic eruption can be a source of atmospheric pressure waves that propagate around the globe, as reported, for example, from the eruptions of Krakatau in 1883 (Simkin and Fiske, 1983), Bezymianny in 1956 (Passechnik, 1958), Mount St. Helens in 1980 (Donn and Balachandran, 1981; Sawada and others, 1982), and El Chichón in 1982 (Mauk, 1983). These waves are dispersive acoustic-gravity waves, composed of various modes that propagate in the atmosphere under the effects of gravity and compressibility. At the higher end of the spectrum there are infrasound waves, which propagate along acoustic rays determined by the temperature and wind profile in the atmosphere, almost free from the effect of gravitational force. Examples of such waves have been reported from eruptions of Mount Agung in 1963 (Goerke and others, 1965), of Alaskan volcanoes in 1966 through 1968 (Wilson and others, 1966; Wilson and Forbes, 1969), of Izu-oshima in 1986 and 1987 (Tahira and others, 1990), from a large number of summit explosions of Sakurajima volcano (Tahira, 1982, 1988a; Ishihara and others, 1986), and other volcanic eruptions in Japan (Tahira and others, 1988).

Remarkable infrasound wave trains were detected by the observation system at Kariya located 2,770 km from Mount Pinatubo. Preliminary analysis (Kamo and others, 1994) shows an association of these waves with the volcanic activity. In this paper, the data recorded at Kariya will be presented in detail, and used to infer a time history of the volcanic activity, especially after the onset of the climactic stage on June 15, when visual and instrumental observations near the volcano were not possible due to the severe eruptions. Acoustic-gravity waves were recorded at five microbarographic stations in Japan from the climactic eruption of Mount Pinatubo. These waves have been briefly

[1]Department of Earth Sciences, Aichi University of Education, Kariya, Aichi, 448 Japan.

[2]Volcanological Division, Seismological and Volcanological Department, Japan Meteorological Agency, 1-3-4 Ote-machi, Chiyoda-ku, Tokyo, 100 Japan.

[3]Department of General Education, Kumamoto Institute of Technology, 4-22-1, Ikeda, Kumamoto, 860 Japan.

described by Igarashi and others (1992) and Tahira and Sawada (1992). Igarashi and others (1992) analyzed the acoustic-gravity wave data and compared them with the ionospheric events detected over Japan. In the present paper, the data will be analyzed in more detail and compared with the data recorded from the Mount St. Helens eruption in 1980. A tentative estimation of energy will also be attempted. Time is expressed in Philippine local time throughout the paper.

INFRASONIC WAVES

DATA ACQUISITION AND PROCESSING

A tripartite array of sensitive capacitor microphones is installed on the campus of the Aichi University of Education in Kariya, Japan (35.05°N. lat, 137.05°E. long). Kariya is 2,770 km northeast of Mount Pinatubo (fig. 1). The map of the microphone array at Kariya is shown in the inset of figure 1. The frequency range of the microphones is 0.04–1 Hz. Detection of weak infrasonic signals in these frequencies is often made difficult by noise caused by atmospheric turbulence. A device to reduce the wind noise has been developed by Tahira (1981) and adopted in the observations at Kariya. This is a multi-pipe line microphone attached to the infrasonic sensor. The principle of this noise reducer is to take an average of the pressure fluctuations along a line to cancel out uncorrelated wind noise, and is similar to the one proposed by Daniels (1959). The shape, however, is different from the original device given by Daniels. Our line microphone consists of a number of parallel pipes of the same diameter, as schematically shown in figure 2, while Daniels' line microphone is a single pipe whose diameter changes stepwise at the inlet openings.

Another source of noise is microbaroms, or acoustic waves released into the atmosphere by ocean waves in stormy regions (Donn, 1967; Donn and Naini, 1973). A stationary front called *Baiu front* was situated off the Pacific coast of the Japanese islands at the time of the Mount Pinatubo eruption, and low-pressure systems along the front were located to the east and west of Kariya. Microbaroms of considerable strength arrived from these directions and tended to mask the weak infrasonic signals from Mount Pinatubo. This background noise can be removed by using a numerical filter, since the pressure oscillations due to microbaroms are nearly monochromatic with dominant periods of 3 to 5 s, which are shorter than the infrasonic signals of interest.

The output signals from the individual microphones are transmitted to one location by cables and continuously recorded on analog magnetic tape. The data for the time period from 1800 on June 13 through 1500 on June 17 were digitized with a sampling frequency of approximately 8.1 Hz. A numerical low-pass filter was applied to the digitized data to eliminate the noise due to microbaroms.

The effect of the low-pass filter is to remove components of periods shorter than 6 s and retain the periods longer than 12 s, as shown by the dotted line in figure 3. The dashed line in figure 3 shows the factory calibration of the response of the microphone, and the overall frequency response is given by the solid line. The sensitivity of the microphone has a maximum of 295 mV/Pa at a frequency of 0.09 Hz after numerical filtering, and this factor was used in the present analysis to convert voltage output to pressure change. Use of a constant conversion factor will cause underestimation in amplitudes for frequencies lower than 0.09 Hz, but no correction was made because amplification of lower frequency components will enhance not only infrasonic signals but also uncorrelated noise.

After applying the numerical filter, we computed the cross-correlation coefficients between all the pairs of the microphone outputs for 2-min time windows and averaged their maximum values over the three pairs. Traveling signals were identified when this averaged maximum cross-correlation coefficient reached some threshold. We adopted a value of 0.7 for this paper. The time lag that maximizes the correlation coefficient is taken as the optimum time lag of the traveling wave from one microphone site to another, and the direction of arrival was computed from the geometry of the tripartite microphone array.

RECORDED INFRASONIC WAVES

The arrivals of several wave trains were identified as a result of successive analysis of the data for every 2 min during the period mentioned above. We concentrate here on the waves arriving from the southwest and northeast quadrants because our interest is in infrasound generated by eruptions of Mount Pinatubo, which is located to the southwest of Kariya. The circles in figure 4 show the time and direction of arrivals of infrasonic signals from 1700 on June 14 through 1500 on June 17. The data before 1700 on June 14 are not presented here because there was no evidence that coherent wave trains arrived from the direction of interest. Five clear wave trains were recorded on June 15 and 16 from the southwest (see upper panel of figure 4). The average direction of arrival of these waves is 196.5° as measured clockwise from the north. This is 24.5° off the great circle bearings to Mount Pinatubo that is shown by the arrow at the right-hand edge of the figure.

The lower panel of figure 4 indicates that clear infrasonic waves were recorded again on June 17. The directions are mostly in the east-northeast, but there are a considerable number of points in the southwest quadrant as well. These two directions are the antipodal and direct-path direction of Mount Pinatubo as shown by the arrows on the right of the panel. The onset time of these June 17 signals is not clear because of the sparsity of data at the beginning, but we can state that a continuous infrasonic wave train from the east-northeast started at 0110. This wave train lasted

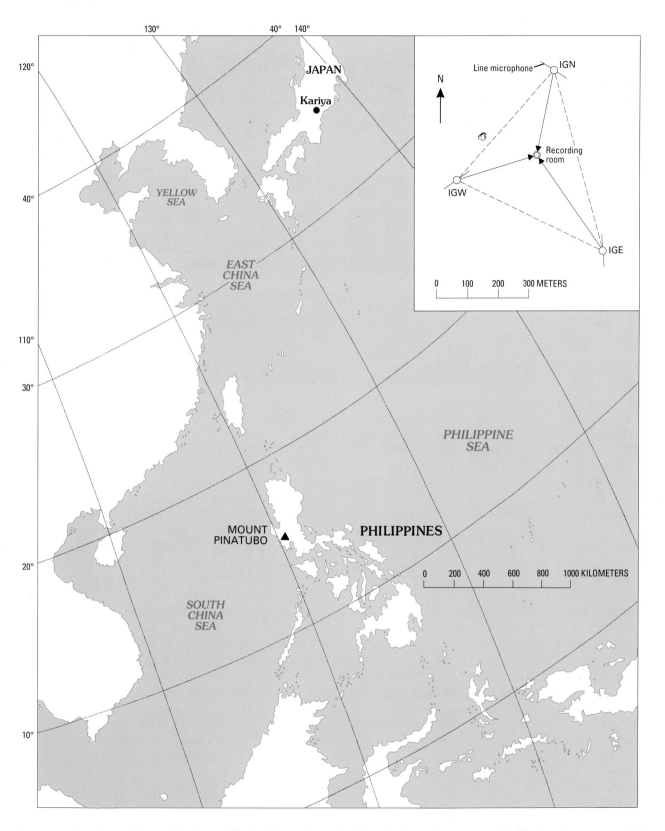

Figure 1.   Locations of Mount Pinatubo and Kariya, Japan, where the infrasonic observations were made. The inset shows a map of the tripartite array of low-frequency microphones at Kariya.

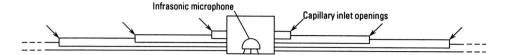

Figure 2. Schematic view of the multipipe line microphone used in the Kariya infrasonic system. Only a few pipes are drawn here for simplicity; the prototype has 13 pipes on each side and takes the spatial average of the pressure field over a length of 100 m.

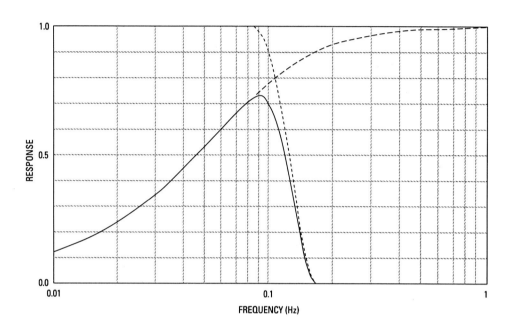

Figure 3. The response function of the infrasonic system. The dashed line shows the factory calibration of the sensitivity of the microphone; the dotted line shows the response of the numerical filter applied to eliminate microbaroms. The overall effect is a band-pass filter as shown by the solid line.

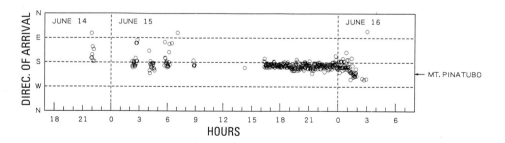

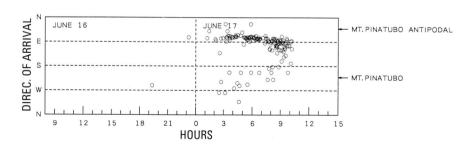

Figure 4. Sequence of arrivals of infrasonic signals at Kariya from 1700 on June 14 through 1500 on June 17. The abscissa of the data points indicates the time when the maximum cross correlation function was 0.7 or over, and the ordinate shows the direction of arrival.

Table 1.   Summary of the infrasonic signals observed at Kariya from the eruption of Mount Pinatubo on June 15, 1991.

[See fig. 5]

Signal	Arrival time of infrasonic signal	Eruption time	Velocity (m/s)	Duration (min)	Direction of arrival	Maximum amplitude (Pa)	Path
A	June 15, 0214	June 14, 2321	267	34	185°	0.17	
B	June 15, 0406	June 15, 0115	270	34	195°	0.48	
C	June 15, 0542	June 15, 0257	272	36	189°	0.43	A1
D	June 15, 0850	June 15, 0556	265	?	191°	0.26	
E	June 15, 1627	Start at June 15, 1342	280	577	194°	1.73	
F	June 17, 0110?	Start at June 15, 1342	291?	about 530	68°	0.13	A2
	June 17, 0300?		319?		230°		A3

approximately for 9 h. The signal from the southwest is recorded intermittently from around 0300 on June 17 and overlaps with the signal from the northeast. The infrasonic events described above are referred to as signals A–F as shown in table 1.

Waveforms of events A–F are reproduced in figure 5. High coherency among the three traces in each panel is clearly observed even by visual inspection, with one exception in signal D, where the coherence decayed suddenly around 0858. The decrease of coherence in signal D is due to the increase of irregular wind noise. The noisy condition at Kariya lasted more than 7 h until 1614 on the same day, when a strong signal was observable again, as illustrated in figure 5 (signal E). Signals A–C reveal quite similar characteristics, with durations of 34 to 36 min and maximum amplitudes 0.25–0.48 Pa (zero-to-peak). On the other hand, signals E and F produced oscillations lasting many hours, suggesting a continuous or repeated excitation at the source. Natures of these signals are summarized in table 1.

The temporal variation of the amplitude of the infrasonic waves is shown in figure 6. One can notice the sudden increase in the signal amplitude at 1627 on June 15 in figure 6. This is the onset of the largest activity of infrasonic disturbances recorded at Kariya. The strong signal frequently exceeding 1 Pa in amplitude continued to arrive for approximately 3 h and was followed by less intense waves until 0151 on June 16. The onset of the largest activity at 1627 on June 15 is also shown in figure 5 (signal E). The last signal (F) is long-lasting but small in amplitude, of the order of 0.1 Pa, as shown in figure 5 (signal F) and figure 6.

ASSOCIATION OF INFRASONIC SIGNALS WITH MOUNT PINATUBO ERUPTIONS

The arrows in figure 6 indicate the eruption times of Mount Pinatubo as reported by PHIVOLCS (R. S. Punongbayan, written commun., 1991). The infrasonic signals A–D can be correlated with major eruptions of Mount Pinatubo,

assuming a time lag of 2 h and 45 to 54 min, as shown by the dashed lines in figure 6. The abrupt increase of the amplitude of the infrasonic waves at 1627 on June 15 occurred 2 h and 45 min after the onset of the largest eruption at Mount Pinatubo, reported at 1342 (Smithsonian Institution, 1991; Punongbayan and others, 1992). The velocities calculated from these time lags and the distance between Mount Pinatubo and Kariya are 265 to 280 m/s (table 1).

Tahira (1982, 1988a,b) has studied the propagation paths in the atmosphere of infrasonic waves generated by summit explosions of Sakurajima volcano, 710 km away from Kariya, and has shown that three types of sound channels exist, depending mainly on the vertical profile of wind components along the direction of propagation. The lowest channel is formed between the ground surface and upper troposphere when strong winds are blowing in the direction of propagation at upper tropospheric heights. The velocity in such cases is nearly the same as the speed of sound at the surface, or 340 to 350 m/s. Infrasonic signals can also be channeled between the upper stratosphere and the ground in the presence of tail winds in the upper stratosphere, and the empirical value of the travel speed for such waves is approximately 310 m/s. The apparent speed is reduced further when acoustic waves are refracted back toward the ground in the thermosphere, say, above 110 km, giving rise to values 260 to 280 m/s (Tahira, 1988a,b). Both the troposphere and the stratosphere in the Northern Hemisphere were in a summer regime at the time of Mount Pinatubo eruption in 1991, with weak westerlies in the troposphere and stable easterlies in the stratosphere at middle latitudes. Therefore, the only acoustic duct to propagate infrasound directly from Mount Pinatubo to Kariya was that formed between the lower thermosphere and the ground. Thus, the computed velocities of the signals A–E (table 1) coincide well with the empirical values for long-range infrasonic waves. We can conclude that the sources of the signals in question are indeed the eruptions of Mount Pinatubo

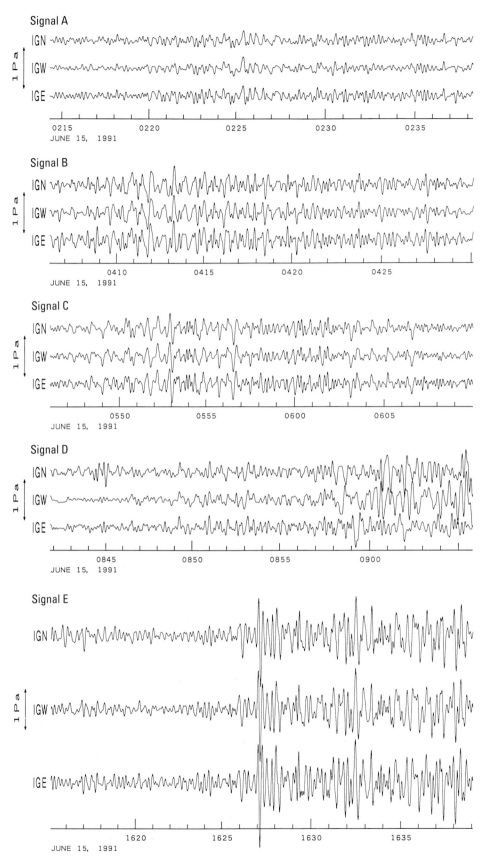

Figure 5.   Infrasonic signals recorded at Kariya from the eruptions of Mount Pinatubo on June 15. The three traces in each panel show the output of the three microphones of the tripartite array, as shown in figure 1.

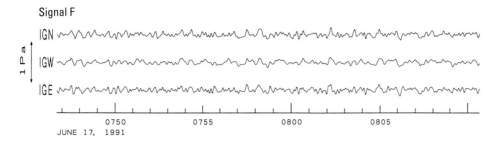

Signal F

JUNE 17, 1991

Figure 5.—Continued.

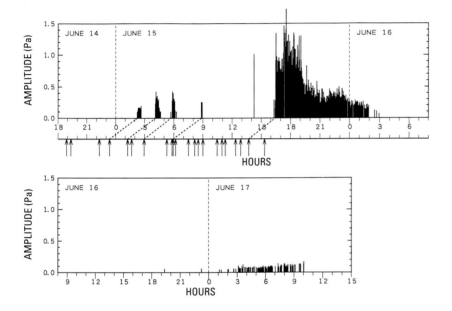

Figure 6. Time sequence of the amplitude of infrasonic waves produced by the eruptions of Mount Pinatubo. The ordinate is the maximum amplitude observed in successive 2-min periods. The arrows indicate the time of major eruptions observed at Mount Pinatubo, and the dashed lines connect the onset of the infrasonic waves with the source eruption.

(dashed lines in fig. 6). These waves are so-called A1 waves, or the waves propagated essentially along the shorter great circle path.

The weak wave train (F) mostly arrived from the direction of northeast, which is close to the antipodal direction of Mount Pinatubo as shown by the arrow in lower panel of figure 4. Therefore, this signal is inferred to be the A2 wave, or the wave propagated via an antipodal great circle path to Kariya, which originated by the climactic eruptions of Mount Pinatubo starting at 1342 on June 15. The deviation of the arrival direction from the antipodal bearing is of the order of a few tens of degrees and almost the same as for the cases of the A1 waves (A–E) for the direct path. The propagation velocity of the A2 wave was 291 m/s, taking 0110 on June 17 as the onset time of the signal (F). This is a little faster than the speed obtained for signals A–E but is still reasonable for long-distance propagation of infrasonic waves and thus supports a Mount Pinatubo source and an antipodal path.

Figure 4 shows another wave train arriving from the southwest and overlapping with the antipodal wave train discussed above. We suggest this is the A3 wave, which is the A1 wave from the climactic eruptions of Mount

Pinatubo that traveled once around the globe. The direction of arrival roughly coincides with the direction of Mount Pinatubo as shown by the arrow in lower panel in figure 4. Since the travel path of the A3 wave is only 15 percent longer than that of the A2 wave, we would expect the amplitude to be roughly the same. The velocity is 314 to 319 m/s, taking the onset time of the signal as 0300 to 0330 on June 17. The velocities of both the A2 and A3 waves are somewhat higher than the empirical values for waves propagating in the thermospheric wave duct.

Figure 7 shows the Mercator projection of the great circle passing through Mount Pinatubo and Kariya. The stratospheric zonal winds for June in mid-latitudes are also shown by the thick arrows in figure 7. It is inferred from this figure that the A2 waves were channeled in the stratospheric acoustic duct while propagating in mid-latitudes of the Northern Hemisphere, where easterly winds prevail in the stratosphere. Likewise, the A3 waves must have traveled through the stratospheric duct in mid-latitudes of the Southern Hemisphere, where the stratospheric prevailing winds are westerlies. The higher velocities of the A2 and A3 waves are thus interpreted by the propagation in the stratospheric duct in some part of the great circle paths.

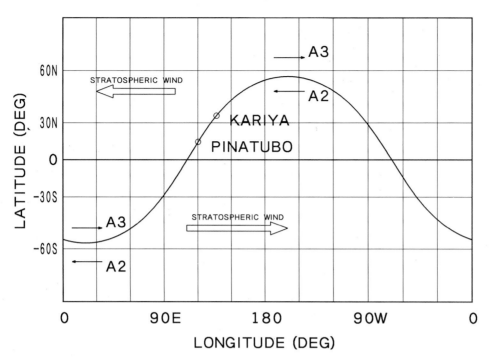

Figure 7. The Mercator projection of the great circle passing through Mount Pinatubo and Kariya. The thick arrows show the stratospheric prevailing winds in June in mid-latitudes.

Furthermore, the winter stratospheric westerlies in the Southern Hemisphere are generally stronger and extend to lower altitudes than the summer easterlies in the Northern Hemisphere (for example, figure 12.3 of Holton, 1992). Thus, the A3 waves could be reflected at lower altitudes, say 20 to 30 km, while the A2 wave must have gone up to 40 to 50 km to be reflected back toward the ground. The higher apparent velocity of the A3 waves is partly explained by the stratospheric wind structure. In conclusion, we can infer that signal F is a composite of A2 and A3 wave trains generated by the eruptions of Mount Pinatubo in its climactic stage.

ACOUSTIC-GRAVITY WAVES

A long train of acoustic-gravity waves was recorded at five microbarographic stations in Japan (Wakkanai, Kushiro, Akita, Tokyo, and Yonago), which are operated by the Japan Meteorological Agency (Igarashi and others, 1994; Tahira and Sawada, 1992). The locations of the microbarographic stations are shown in figure 8, where the arcs indicate the distance from Mount Pinatubo. Since the microbarographic data were provided only as paper records, we digitized the data at a sampling period of 3 s for detailed analysis.

The recorded waveform is reproduced in figure 9 after applying a high-pass filter to eliminate the incoherent noise with periods longer than 34 min and retain periods shorter than 17 min. The indicated time in figure 9 corresponds to the record at Tokyo (in Philippine local time). The time axes of other traces were shifted appropriately to get the best phase match with each other. Coherent waves are noticed

from 1603 through 1950 on June 15. The propagation speed computed from the time shift is 300 m/s, assuming the waves propagated from the direction of Mount Pinatubo. The group velocity of the leading part of the wave is computed as 354 m/s, if we assume that this wave train was produced by the eruptions starting at 1342 on June 15. This is a little faster than the theoretical prediction of acoustic-gravity waves as given, for example, by Harkrider (1964). However, the waves recorded after 1618 have a reasonable group velocity, 320 m/s if associated with the eruptions after 1342. Therefore, the disturbances shown in figure 9 are very likely the acoustic-gravity waves produced by the climactic eruptions of Mount Pinatubo on June 15, 1991.

The waveform is somewhat complicated and looks quite different from the acoustic-gravity waves normally observed from large volcanic explosions, such as the eruption of Mount St. Helens in 1980 (Donn and Balachandran, 1981; Sawada and others, 1982). This is interpreted as being due to the superposition of several acoustic-gravity wave trains excited by different explosions.

Power spectra of the acoustic-gravity waves are shown in figure 10. Common spectral peaks are noticed at the frequencies 1.2, 1.9, 2.4, 2.9, and 4.4 mHz. The corresponding periods are 13.9, 8.8, 6.9, 5.7, and 3.8 min, respectively. The waves from Mount Pinatubo contain oscillations with much longer periods than those recorded from the Mount St. Helens eruption in 1980, for which the reported longest period is 6.5 min (Donn and Balachandran, 1981).

Posey and Pierce (1971) proposed a simple empirical equation to estimate the source energy from the microbarographic records:

$$E = 13P[R\sin(l/R)]^{0.5}H(c\tau)^{1.5} / (4.22 \times 10^{15})$$

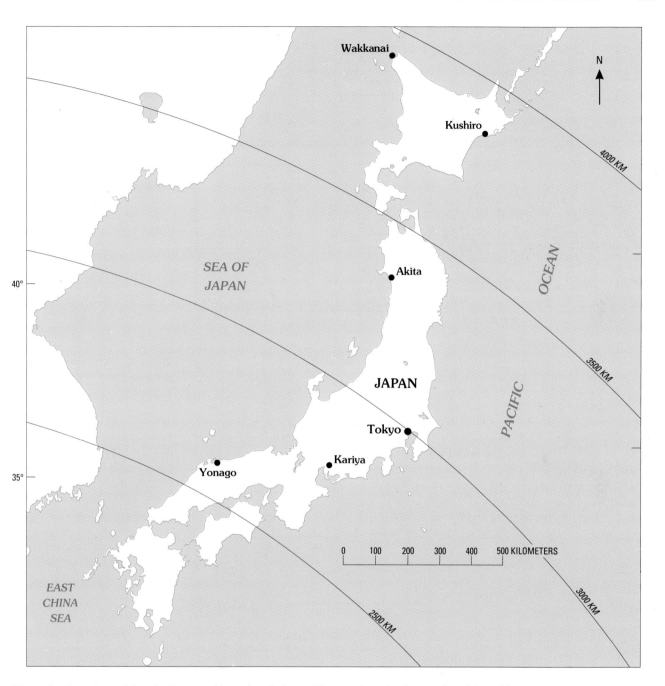

Figure 8. Locations of the microbarographic stations in Japan. The arcs show the distance from Mount Pinatubo.

where E is the energy release in explosive equivalent of TNT (megatonnes), P the first peak-to-peak amplitude (pascals), R the radius of the Earth (meters), l the great circle distance between the source and receiver (meters), H the lower atmospheric scale height (meters), c the representative sound speed (meters per second), and τ the time interval (seconds) between the first and second peaks.

We attempted to estimate the source energy of the Mount Pinatubo eruptions by applying this method to the recorded acoustic-gravity waves. Application of this method requires identification of the leading part of individ-

ual wave trains in the microbarogram. We do not have any objective means at present to separate one wave train from another. One possible way is to apply a numerical low-pass filter to the records, since the acoustic-gravity waves at great distances are highly dispersive, with the lowest frequency component normally arriving first. We expect from the spectral analysis (fig. 10) that the leading portion of the wave trains from significant explosions have periods longer than about 8 min, so that we applied a numerical filter which suppresses periods shorter than 4 min and gradually decreases that suppression until it retains the whole

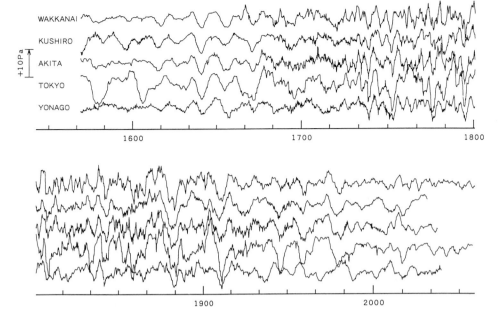

Figure 9. Microbarographic waves observed at five stations in Japan on June 15, 1991. The data were numerically filtered in order to eliminate the uncorrelated low-frequency noise with periods longer than 34 min. The time axis corresponds to the record at Tokyo, and the traces at other stations were shifted to obtain the best phase match with each other.

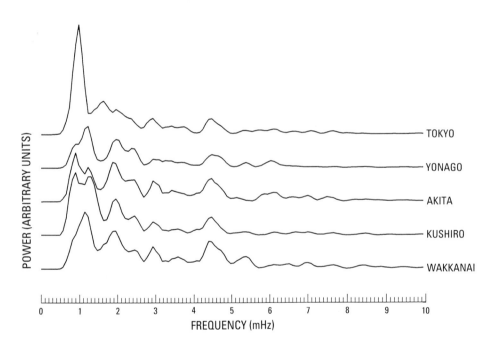

Figure 10. The power spectra of the microbarographic data shown in figure 9.

signal for periods longer than 8 min. The resulting low-pass filtered signals are shown in figure 11.

The filtered waveform suggests onsets of the new oscillations at around 1603, 1715, 1750, 1835, and 1858. In particular, two large waves are noticeable at 1835 to 1855 and 1858 to 1920. We calculated the source energy for these waves, assuming that these portions of the wave packet indicate the arrival of a new wave train, and obtained explosive equivalent of 27.2 and 21.1 MT of TNT. Both of these values are smaller than the energy of the Mount St. Helens eruption in 1980, which was estimated by use of the same technique as 35 MT (Donn and Balachandran, 1981). How-

ever, the total energy release, if one integrates the results for other portions of the microbarogram mentioned above, amounts to about 70 MT, and this is twice the energy of the Mount St. Helens eruption in 1980.

DISCUSSION

It is of interest to compare the infrasonic signals with the acoustic-gravity waves from the climactic eruptions of Mount Pinatubo, since there have been very few occasions that both the infrasonic and acoustic-gravity waves are recorded simultaneously at long distances. High amplitudes

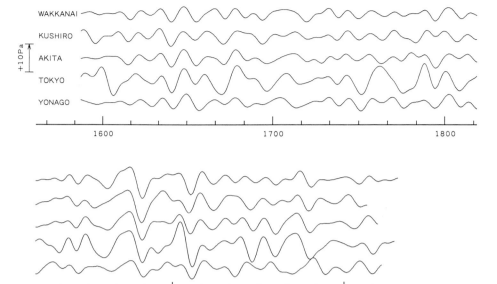

Figure 11. Microbarographic records after applying a low-pass filter that increasingly attenuates wave components with periods of 8 min down to 4 min and wholly eliminates wave components with periods shorter than 4 min. Components with periods longer than 8 minutes are not attenuated.

of infrasonic waves lasted about 3.5 h (fig. 6), and this duration roughly coincides with the duration of acoustic-gravity waves (about 3 h and 50 min, fig. 9). It is also consistent with the barogram obtained at Clark Air Base (Kanamori and Mori, 1992). However, apparent inconsistency is noticed between the fine structure of infrasonic amplitudes and acoustic-gravity waves. The acoustic-gravity waves in figure 11 suggest arrival of the strongest disturbance at around 1835 at Tokyo. The corresponding arrival time of infrasonic waves at Kariya should be 1854, assuming that the infrasonic waves starting at 1627 correspond to the acoustic-gravity waves starting at 1618. The infrasonic amplitude is large indeed at around 1854, but it is not the largest at this time. This fact suggests different source mechanisms for different types of waves.

It should be noted, however, that some ambiguity is included in the infrasonic amplitudes. For example, focusing or defocusing of acoustic rays at the observation site is one of the causes of amplitude variations. Attenuation of infrasonic waves due to viscosity of air becomes more and more serious for waves reflected at higher altitudes in the thermosphere, and the reflection level in turn can be controlled by the atmospheric tidal winds (Donn and Rind, 1972). Since the atmospheric tidal oscillations are highly variable (Tsuda and others, 1983), and no observations were available on the tidal conditions at the time of Mount Pinatubo eruption, some ambiguity in the infrasonic amplitudes is unavoidable. Therefore, discussion based on the fine structure of the amplitude sequence may not be adequate, although the amplitude envelope in figure 6 is a good indication of the explosive sequence of Mount Pinatubo eruptions.

Some other problems regarding the observation of infrasonic waves were left unresolved. It is shown in

figure 4 that the direction of arrival of the infrasonic signals does not strictly coincide with the great circle direction to Mount Pinatubo. One possible explanation is that small errors in time lags at the tripartite array could result from different phase characteristics of the electronics. A small error in time lags can produce a relatively large shift in the determination of the direction of arrival, because the dimension of the tripartite at Kariya is small compared to the wavelength of the infrasonic signals in question. Deflection of the direction of arrival from the great circle bearings can also result from lateral winds in the propagation paths, especially when the source-receiver distance is large. For example, a direction deflection of 40° was observed for infrasonic waves from the eruption of Mount Agung in 1963 at North American stations (Goerke and others, 1965).

A number of eruptions are reported from 0556 through 1342 on June 15, as indicated by the arrows in figure 6. Unfortunately, none of these were detected at Kariya because of the strong wind noise. Wind noise is reduced considerably by the line microphone, but infrasonic signals are still apt to be masked under severe wind conditions.

The eruption energy was estimated by using the acoustic-gravity waves recorded in Japan. Some ambiguity remains, however, in the separation of wave trains; we evaluated here energy for only five apparent wave trains, but it is somewhat difficult to pick all the wave trains from the records. Therefore, the resulting energy estimation should be taken as a minimum of the released energy. A more comprehensive technique will be required for more accurate estimation of the eruption energy, also using the microbarographic records.

It is interesting to note that the power spectra of the acoustic-gravity waves recorded in Japan (fig. 9) clearly show oscillations at a frequency of 4.4 mHz, which coincides with the frequency of the Rayleigh waves widely recorded from Mount Pinatubo eruptions (Kanamori and Mori, 1992; Widmer and Zürn, 1992; Zürn and Widmer, this volume). This fact supports the excitation mechanism of the Rayleigh waves proposed by Kanamori and Mori (1992) and Zürn and Widmer (this volume), stating that the atmospheric oscillations set off by the eruption of Mount Pinatubo are the cause of these peculiar Rayleigh waves. It is not clear at present, however, why the Earth's atmosphere oscillates at this particular frequency; it is somewhat larger than the Brunt-Väisälä frequency (3.3 to 3.5 mHz) in the lower stratosphere. Resonant interaction of the waves due to the nonlinear effect (for example, Yeh and Liu, 1970) could be a possible mechanism, but details are left for future studies.

CONCLUSION

The eruption of Mount Pinatubo on June 15, 1991, produced remarkable infrasonic and acoustic-gravity waves, both of which were recorded by ground-based observation systems in Japan. Several infrasonic wave trains recorded at Kariya, 2,770 km from Mount Pinatubo, were clearly associated with reported eruptions. In particular, infrasonic signals of large amplitudes, often exceeding 1 Pa, were recorded starting at 1627 on June 15 and lasted for approximately 3.5 h. These strong signals were then followed by less intense, but continuous, waves which lasted until 0151 on June 16. These signals provide information describing the explosive sequence of the volcanic activity at its climactic phase, at a time when most visual and instrumental observations near the volcano were interrupted due to the severe eruption.

Detailed analysis of the infrasonic record at Kariya indicated the arrival of weak and long-lasting infrasonic waves from the northeast and southwest quadrants, approximately 35 hours after the onset of the climactic eruption. These waves are inferred to be a composite of A2 and A3 waves.

Acoustic-gravity waves recorded at five meteorological stations in Japan were also associated with the climactic eruption of Mount Pinatubo. The duration of the waves is 3 h and 50 min and is comparable to the duration of strong infrasonic signals at Kariya. The longest period obtained from the spectral analysis is 13.9 min, which is more than twice the longest period observed from the Mount St. Helens eruption. The waveform is complicated because of superposition of several wave trains excited by different explosions. An energy estimation of the eruptions was made after identifying the leading part of individual acoustic-gravity wave trains by applying a low-pass numerical filter. The source energy was estimated to be the explosive equivalent of approximately 70 MT of TNT. This is taken as a minimum of the released energy but is twice as large as that released from the Mount St. Helens eruption in 1980.

ACKNOWLEDGMENT

The authors are grateful to the Japan Meteorological Agency for furnishing the microbarographic data. They also thank J. Mori, J. Kienle, C.G. Newhall, and J.S. Oswalt for reviewing the original draft and giving invaluable comments.

REFERENCES CITED

Cheng, K., and Huang, Y.-I., 1992, Ionospheric disturbances observed during the period of Mount Pinatubo eruptions in June 1991: Journal of Geophysical Research, v. 97, p. 16995–17004.

Daniels, F.B., 1959, Noise-reducing line microphone for frequencies below 1 cps: Journal of the Acoustical Society of America, v. 31, p. 529–531.

Donn, W.L., 1967, Natural infrasound of five seconds period: Nature, v. 215, p. 1469–1470.

Donn, W.L., and Balachandran, N.K., 1981, Mount St. Helens eruption of 18 May 1980: Air waves and explosive yield: Science, v. 213, p. 539–541.

Donn, W.L., and Naini, B., 1973, Sea wave origin of microbaroms and microseisms: Journal of Geophysical Research, v. 78, p. 4482–4488.

Donn, W.L., and Rind, D., 1972, Microbaroms and temperature and wind of the upper atmosphere: Journal of Atmospheric Science, v. 29, p. 156–172.

Goerke, V.H., Young, J.M., and Cook, R.K., 1965, Infrasonic observations of the May 16, 1963, volcanic eruption on the Island of Bali: Journal of Geophysical Research, v. 70, p. 6017–6022.

Harkrider, D.G., 1964, Theoretical and observed acoustic-gravity waves from explosive sources in the atmosphere: Journal of Geophysical Research, v. 69, p. 5295–5321.

Holton, J.R., 1992, An Introduction to Dynamic Meteorology (3d ed.): Academic Press, 507 p.

Igarashi, K., Kainuma, S., Nishimuta, I., Okamoto, S., Kuroiwa, H., Tanaka, T., and Ogawa, T., 1992, Ionospheric disturbances over Japan caused by the eruption of Mount Pinatubo on June 15, 1991: Proceedings, International Symposium on Middle Atmosphere Science, Kyoto, p. 219–220.

———1994, Ionospheric and atmospheric disturbances around Japan caused by the eruption of Mount Pinatubo on June 15, 1991: Journal of Atmospheric and Terrestrial Physics, v. 56, p. 1227–1234.

Ishihara, K., Iguchi, M., and Tahira, M., 1986, Observations of air waves associated with volcanic explosions: Report, 5th Joint Observations, Sakurajima Volcano (October–December, 1982), p. 131–138. (In Japanese.)

Kamo, K., Ishihara K., and Tahira, M., 1994, Infrasonic and seismic detection of explosive eruptions at Sakurajima Volcano, Japan, and the PEGASAS-VE early warning system, *in* Casadevall, Thomas, ed., Proceedings, First International

Symposium on Volcanic Ash and Aviation Safety: U.S. Geological Survey Bulletin 2047, p. 357–365.

Kanamori, H., and Mori, J., 1992, Harmonic excitation of mantle Rayleigh waves by the 1991 eruption of Mount Pinatubo, Philippines: Geophysical Research Letters, v. 19, p. 721-724.

Mauk, F.J., 1983, Utilization of seismically recorded infrasonic-acoustic signals to monitor volcanic explosions: The El Chichón sequence 1982—A case study: Journal of Geophysical Research, v. 88, p. 10385–10401.

Passechnik, I.P., 1958, Seismic and air waves which arose during an eruption of the Volcano Bezymyanny, on March 30, 1956: Izvestia, Geophysics Series, p. 1121–1126. (English translation by A.S. Ryall.)

Posey, J.W., and Pierce, A.D., 1971, Estimation of nuclear explosion energies from microbarographic records: Nature, v. 232, p. 253.

Punongbayan, R.S., Newhall, C.G., and Listanco, E.L., 1992, Brief notes on the 1990–1991 Pinatubo volcano events and corresponding scientific responses: Bulletin of the Volcanological Society of Japan, v. 37, p. 55–59.

Sawada, Y., Wakui, S., and Komiya, M., 1982, Atmospheric pressure waves generated by the May 18, 1980 eruption of Mount St. Helens: Bulletin of the Volcanological Society of Japan, v. 27, p. 195–202. (In Japanese.)

Simkin, T., and Fiske, R.S., 1983, Krakatau 1883, The volcanic eruption and its effects: Washington, Smithsonian Institution Press, p. 367–395.

Smithsonian Institution, 1991, Pinatubo: Bulletin of the Global Volcanism Network, v. 16, no. 5, p. 2–8.

Tahira, M., 1981, A study of the infrasonic wave in the atmosphere. Multi-pipe line microphone for infrasonic observation: Journal of the Meteorological Society of Japan, v. 59, p. 477–486.

————1982, A study of the infrasonic wave in the atmosphere. II, Infrasonic waves generated by the explosions of the volcano Sakura-jima: Journal of the Meteorological Society of Japan, v. 60, p. 896–907.

————1988a, A study of the long range propagation of infrasonic waves in the atmosphere. I, Observation of the volcanic infrasonic waves propagating through the thermospheric duct: Journal of the Meteorological Society of Japan, v. 66, p. 17–26.

————1988b, A study of the long range propagation of infrasonic waves in the atmosphere. II, Numerical study of the waveform deformation along thermospheric ray paths: Journal of the Meteorological Society of Japan, v. 66, p. 27–37.

Tahira, M., Ishihara, K., and Iguchi, M., 1988, Monitoring volcanic eruptions with infrasonic waves: Proceedings of the Kagoshima International Conference on Volcanoes, Tokyo, National Institute for Research Advancement, and Kagoshima, Kagoshima Prefectural Government, p. 530–533.

Tahira, M., Ishihara, K., and Ukai, E., 1990, Infrasonic waves observed at Kariya in association with the 1986 and 1987 eruptions of Izu-Oshima volcano: Bulletin of the Volcanological Society of Japan, v. 35, p. 11–25. (In Japanese.)

Tahira, M., and Sawada, Y., 1992, Pressure waves produced by the Mt. Pinatubo eruptions of 15 June 1991: Proceedings, Annual Meeting of the Volcanological Society of Japan, p. D31–P85.

Tsuda, T., Aso, T., and Kato, S., 1983, Seasonal variation of solar atmospheric tides at meteor heights: Journal of Geomagnetism and Geoelectricity, v. 35, p. 65–86.

Yeh, K.C., and Liu, C.H., 1970, On resonant interactions of acoustic gravity waves: Radio Science, v. 5, p. 39–48.

Wilson, C.R., Nichparenko, S., and Forbes, R.B., 1966, Evidence of two sound channels in the polar atmosphere from infrasonic observations of the eruption of an Alaskan volcano: Nature, v. 211, p. 163–165.

Wilson, C.R., and Forbes, R.B., 1969, Infrasonic waves from Alaskan volcanic eruptions: Journal of Geophysical Research, v. 74, p. 4511–4522.

Widmer, R., and Zürn, W., 1992, Bichromatic excitation of long-period Rayleigh and air waves by the Mount Pinatubo and El Chichon volcanic eruptions: Geophysical Research Letters, v. 19, p. 765–768.

Zürn, W., and Widmer, R., this volume, Worldwide observation of bichromatic long-period Rayleigh waves excited during the June 15, 1991, eruption of Mount Pinatubo.

Worldwide Observation of Bichromatic Long-Period Rayleigh Waves Excited During the June 15, 1991, Eruption of Mount Pinatubo

By W. Zürn[1] and R. Widmer[2]

ABSTRACT

During the climactic phase of the June 15, 1991, Mount Pinatubo eruption, an essentially bichromatic signal with frequencies of 3.68 and 4.44 millihertz was recorded on gravimeters and very long period (VLP) seismometers worldwide. The narrow-band nature of this signal distinguishes these recordings from the usual broadband signal recorded after large earthquakes. Group velocity estimates and particle motions show that the signals propagate as Rayleigh waves. The bichromatic spectra, which had not been recognized during previous plinian eruptions, can be explained only by source models that provide for harmonic forcing of the solid Earth. In this article we summarize the results of the data analysis and focus on the constraints the signals place on models of the source. Two models suggest that the eruption excites atmospheric oscillations that have vanishing horizontal wavenumber and that these oscillations, by exerting a harmonic pressure on the surface of the Earth, are responsible for the bichromatic Rayleigh waves. In order to account for the observed phase coherence of the signal over many hours, one of the models invokes positive feedback between the atmospheric resonances and the plume. Atmospheric oscillations that are associated with severe convective storms have bichromatic spectra very similar to seismic spectra from the Mount Pinatubo eruption, a fact that corroborates both tentative source models.

THE SEISMIC SIGNALS FROM MOUNT PINATUBO

A new powerful source of low-frequency seismic energy has been detected independently by two research teams on the occasion of the climactic eruption of Mount Pinatubo on June 15, 1991 (Kanamori and Mori, 1992; Widmer and Zürn, 1992a,b). The seismic signal from the climactic phase of the eruption that was recorded at teleseismic distances (>20° away) is dominated by a long-lasting (>6 h) narrow-band signal with peaks at 3.68 and 4.44 mHz. These oscillations last from 0700 to 2000 G.m.t. on June 15, 1991 (local time, 1500 on June 15, 1991, to 0400 on June 16). Figure 1 shows the data from the very long period, high-gain channel of the vertical component seismometer (STS–1; Wielandt and Streckeisen, 1982) of station HRV at Cambridge, Mass., from this day. Other records can be found in figures 1, 3, 4, and 7 of Kanamori and Mori (1992) and in figure 1 of Widmer and Zürn (1992a). Essentially all modern vertical long-period seismographs, including superconducting gravimeters at Miami and Strassbourg, observed these oscillations. At 0059 G.m.t. an M_s 6.1 earthquake occurred in the Caucasus followed immediately by an M_s 6.3 earthquake in the South Sandwich Islands at 0113 G.m.t. On the low-pass filtered seismograms, Rayleigh wave trains up to R_4 of the latter event can be identified before the onset of the Pinatubo signal. Between 0113 and 2300 G.m.t. on June 15 (between 0913 local time on June 15 and 0700 on June 16) no earthquake larger than M_s 5.5 occurred worldwide. However, 36 events with the epicenter located on Luzon Island and magnitudes less than M_s 5.5 are listed in the catalog Preliminary Determination of Earthquake Epicenters (USGS, 1991) in this time period, demonstrating the crisis. The higher frequency signals between 1100 and 1200 G.m.t. are body and Rayleigh waves from the largest of these events. Since there is no sharp onset to the low-frequency signal arriving at approximately 0700 G.m.t., the standard techniques for locating earthquakes could not be used. Although we had immediately suspected that the signal was linked to the eruption of Mount Pinatubo, we had to use a cross-correlation method to confirm this source. All available seismograms were cross-correlated with the vertical seismogram from station KMI (Kunming, China), and from the lags determined from these correlograms a group velocity of 3.78 km/s was estimated, which corresponds to the group

[1]Black Forest Observatory Schiltach, Heubach 206, D–77709 Wolfach, Germany.

[2]Geophysical Institute, Karlsruhe University, Hertzstr. 16, D–76187 Karlsruhe, Germany.

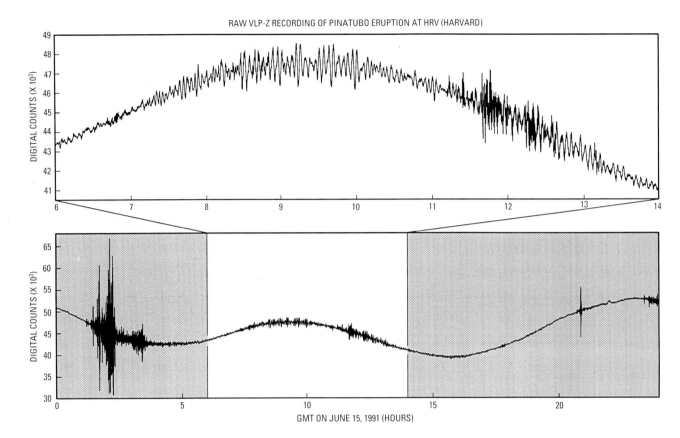

Figure 1. The 24-h record from the IRIS (Incorporated Research Institutions for Seismology) station HRV at Cambridge, Mass., for June 15, 1991. Superimposed on the tides are the first two earthquakes (0059 G.m.t. and 0113 G.m.t.) followed by the signal from Mount Pinatubo (0700 G.m.t. or 1500 Philippine time). Philippine local time is 8 h ahead of G.m.t. In the upper panel the signal from the Pinatubo eruption is shown at higher resolution. The peak amplitude of the signal is approximately 0.5 μgals.

velocity of Rayleigh waves with the periods of the bichromatic signal (see fig. 2 of Widmer and Zürn, 1992a). We also find that the horizontal particle motion at the closest station (KMI) lines up with the direction to Pinatubo and thereby provides a second piece of evidence for the Rayleigh wave hypothesis. KMI was the only station where these signals were clearly visible on the horizontal components. This is presumably because of the generally much higher noise level on horizontal component recordings. Unfortunately no long-period records from the immediate vicinity of the volcano are available to study the ground motion in the source region in more detail.

Figure 2 shows the amplitude spectra of the 42 best vertical component VLP recordings from the global networks. The time period spans from 0600 to 1400 G.m.t. (1400 to 2200 local time) and the time series were Hanning tapered to reduce spectral leakage. The spectra are essentially bichromatic with two sharp peaks at 3.68 and 4.44 mHz (see also fig. 2 of Kanamori and Mori, 1992, and fig. 3 of Widmer and Zürn, 1992a). When Rayleigh waves travel around the Earth more than once, destructive and constructive interference produces spectra with discrete resonance peaks at the periods of the spheroidal free oscillations of the Earth. The amplitude pattern among the emerging peaks depends on the source mechanism and on epicentral distance. In figure 2 we see a strong variation from station to station in the relative amplitudes of the three fundamental spheroidal modes in the 4.44 mHz band. Because we know both the location of the source (Pinatubo) and the coordinates of the stations, we can use this amplitude variation to learn about the characteristics of the source. In order to test the model of an isotropically radiating source, we calculated spherical earth Green's functions for a vertical force couple acting at Pinatubo. A comparison with the observed spectra shows good agreement in the relative amplitudes of the three modes in the 4.44-mHz band and thereby confirms our hypothesis. We do not make use of the phase spectrum, because we lack a model of the time dependence of the source. Using a traveling wave analysis, Kanamori and Mori (1992) have determined the phase arrival times that are also found to be consistent with an isotropically radiating source. Figure 3 shows a world map with the location of the seismic stations that contributed to figure 2 spectra and demonstrates the global nature of the low-frequency

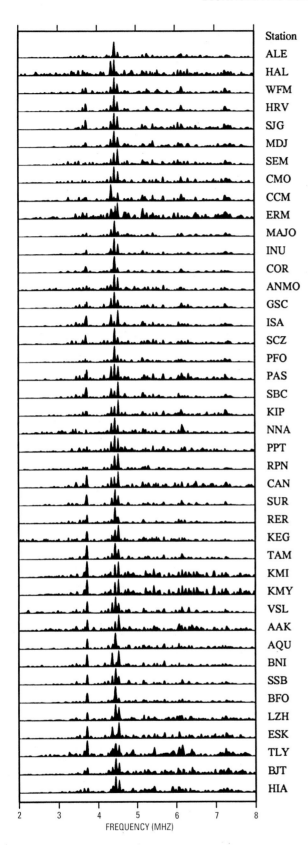

Station

ALE
HAL
WFM
HRV
SJG
MDJ
SEM
CMO
CCM
ERM
MAJO
INU
COR
ANMO
GSC
ISA
SCZ
PFO
PAS
SBC
KIP
NNA
PPT
RPN
CAN
SUR
RER
KEG
TAM
KMI
KMY
VSL
AAK
AQU
BNI
SSB
BFO
LZH
ESK
TLY
BJT
HIA

FREQUENCY (MHZ)

← **Figure 2.** Linear amplitude spectra for the main eruption of Mount Pinatubo (0600 to 1400 G.m.t.). The 42 records are vertical components, taken from regional and global digital seismograph networks. Two resonances at 3.68 and 4.44 mHz are visible on all spectra. Because the time period contains recurring Rayleigh waves up to at least R_5, we can resolve individual fundamental spheroidal modes—that is, the peaks of the resonance around 4.44 mHz are $_0S_{36}-_0S_{38}$. The spectra are normalized to have the same maximum amplitude, and the station codes are given at the right margin. The spectra are arranged so that the azimuth from Pinatubo increases from top to bottom (see also table 1). Note the similarity between stations KMI and KMY. These two stations are separated by only 2.8 km.

observations discussed here. Table 1 lists the station codes, networks, station coordinates, and the azimuths and epicentral distances away from Pinatubo.

The Pinatubo source has excited the elasto-gravitational spheroidal fundamental modes $_0S_{28}$ and the triplet $_0S_{36}$, $_0S_{37}$, and $_0S_{38}$. Thus, the source spectrum encompasses three fundamental spheroidal modes near 4.44 mHz and only one at 3.7 mHz. Considering that the fundamental spheroidal modes between 3 and 10 mHz are spaced at regular 100-μHz (0.1-mHz) intervals, the above observation shows that the source spectrum is narrower at 3.7 mHz than at 4.44 mHz. From the spectral stack in figure 4 it is clear that, additionally, Pinatubo emitted low-frequency seismic energy up to 7.2 mHz; so the bichromatism is not perfect.

We see two reasons that the 36 events on Luzon mentioned above cannot be responsible for the observed long-period signal. First, if they were normal earthquakes, they were not energetic enough. The second reason, which also applies to "slow" and "silent" earthquakes (Beroza and Jordan, 1990), is that a seismic source that is localized in space and time possesses a broad frequency and wavenumber spectrum; hence, a broad spectrum of free oscillations is excited. One remaining possibility for the production of a narrow-band spectrum is a periodic occurrence of point sources with time. The listed earthquakes in the time period of our observations did not show such a periodicity but occurred in a random sequence.

We studied the phase coherence of the signals by using a version of a well-known method called "summation dial" by Bartels (1938) and "phasor walkout" by Rydelek and Sacks (1988). It is also similar to "complex demodulation" described by Bolt and Brillinger (1979). The contributions of successive samples in a time series to its Fourier transform at a given test frequency are added graphically in the complex domain ("phasor walk"). The final value is the complex Fourier coefficient of the whole series. In the idealized case of a time series consisting of a single sine wave the phasor walk consists of a straight line if the test frequency is equal to the signal frequency. If the test frequency

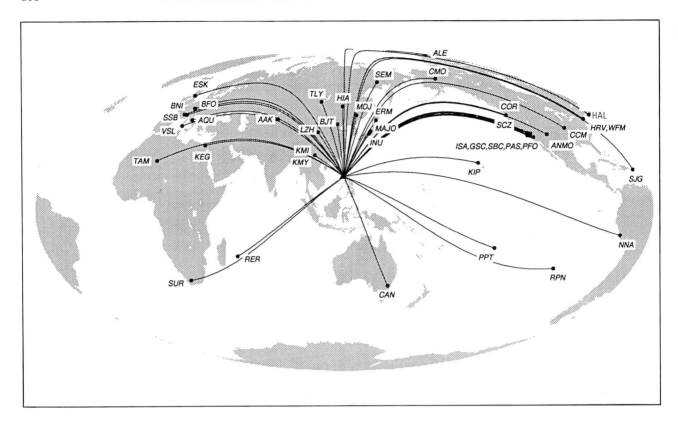

Figure 3. Molweide projection of the world with the locations of the seismic observatories that recorded the bichromatic signal that originated from Mount Pinatubo. The distribution of the stations demonstrates the global nature of the signal. The minor arcs connecting Pinatubo with the different stations are plotted to indicate the azimuthal coverage of the source (see table 1).

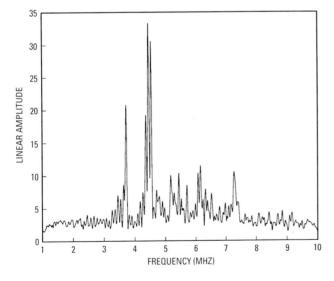

← **Figure 4.** Stack of the 42 spectra shown in figure 2. The spectra were normalized to unit maximum amplitude prior to summation. Several secondary resonances are visible between 4.5 and 7.2 mHz. At 7.4 mHz there seems to be a cutoff frequency beyond which no low-frequency seismic energy was transmitted. We also note the pronounced asymmetry of the resonance peak at 3.7 mHz. This asymmetry is expected if the 3.7 mHz resonance is the high-end cutoff frequency for freely propagating gravity waves as postulated by Kanamori and Mori (1992).

is moved away from the frequency of the signal, the phasor walk shows more and more curvature. If multiple reexcitations of the harmonic signal under investigation occur without phase locking, the phasor walk will consist of several straight segments with sharp angles between them. If there

is no harmonic signal detectable at this frequency, the plot looks like a random walk in two dimensions.

Figures 5A and 5B show these walkouts for the stations ALE (4.440 mHz) and BFO (3.723 mHz), respectively. These are examples in which the amplitude spectra consist of large isolated peaks and no leakage effects occur from neighboring peaks. The steady, linear growth of the phasor demonstrates that the source radiated in phase for at least 8 h. When the peak at 4.44 mHz ($_0S_{37}$) is smaller than its neighbors, $_0S_{36}$ and $_0S_{38}$, this is a sign for destructive interference between the different Rayleigh wave trains. This interference can be observed in the phasor walks, and it makes the picture more complicated. Another complication

Table 1. Seismic stations that recorded the low-frequency signals from the Mount Pinatubo eruption.

[Listed are station code, and geographic coordinates of the stations, as well as the distance and azimuth from Mount Pinatubo. The entries are sorted according to the azimuth from Mount Pinatubo as in fig. 2]

Station code	Network	Station latitude (degrees)	Station longitude (degrees)	Epicentral distance (degrees)	Azimuth from Pinatubo (degrees)
ALE	IDA	82.5	297.6	82.4	0.4
HAL	IDA	44.6	296.4	120.1	3.4
WFM	Geoscope	42.6	288.5	121.2	10.3
HRV	IRIS	42.5	288.4	121.3	10.4
SJG	IDA	18.1	293.9	146.2	11.4
MDJ	CDSN	44.6	129.6	30.5	12.8
SEM	Geoscope	62.9	152.4	52.8	17.6
CMO	IDA	64.9	212.2	77.1	25.8
CCM	IRIS	38.1	268.8	119.1	28.3
ERM	IDA	42.0	143.2	33.2	31.5
MAJO	IRIS	36.5	138.2	26.6	33.1
INU	Geoscope	35.4	137.0	25.1	33.2
COR	IRIS	44.6	236.7	96.9	40.1
ANMO	IRIS	34.9	253.5	113.0	40.6
GSC	TerraScope	35.3	243.2	106.0	45.6
ISA	TerraScope	35.7	241.5	104.6	46.0
SCZ	Geoscope	36.6	238.6	102.1	46.4
PFO	IDA	33.6	243.5	107.1	46.9
PAS	TerraScope	34.1	241.8	105.6	47.2
SBC	TerraScope	34.4	240.3	104.4	47.6
KIP	IDA	21.4	202.0	76.8	71.0
NNA	IDA	−12.0	283.2	162.9	81.7
PPT	Geoscope	−17.6	210.4	94.4	107.0
RPN	IDA	−27.1	250.7	132.3	113.1
CAN	Geoscope	−35.3	149.0	57.2	152.4
SUR	IDA	−32.4	20.8	106.1	240.0
RER	Geoscope	−21.2	55.7	73.1	241.8
KEG	MedNet	29.9	31.8	81.4	298.8
TAM	Geoscope	22.8	5.5	106.0	299.6
KMI	CDSN	25.1	102.7	19.4	303.9
KMY	IDA	25.1	102.7	19.4	303.9
VSL	MedNet	39.5	9.4	95.9	313.6
AAK	IRIS	42.6	74.5	48.0	314.5
AQU	MedNet	42.4	13.4	92.0	315.0
BNI	MedNet	45.0	6.7	95.2	319.5
SSB	Geoscope	45.3	4.5	96.4	320.5
BFO	BFO	48.3	8.3	92.7	321.9
LZH	CDSN	36.1	103.8	25.7	327.7
ESK	IDA	55.3	356.8	95.2	331.6
TLY	IRIS	51.7	103.6	39.0	343.4
BJT	CDSN	40.0	116.2	25.2	352.2
HIA	CDSN	49.3	119.7	34.2	359.1

could arise due to air waves, which might be present in the later parts of the seismograms of the closest stations (see below). From figures 2 and 4 we can also identify resonances at frequencies up to 7.3 mHz, and the phasor walkouts for the frequencies 6.156 and 7.270 mHz at station HRV are shown in figures 5*C* and 5*D*. The steady, nearly linear growth indicates that these secondary resonances remained phase coherent for at least 6 h.

OTHER OBSERVATIONS

BAROGRAMS FOR MOUNT PINATUBO

Kanamori and Mori (1992, fig. 6) and Oswalt and others (this volume) show the microbarogram from Clark Air Base, located about 21 km from the volcano. This record shows pressure pulses from individual explosions as well as continuous oscillations with peak amplitudes of about 300 Pa during the climactic phase of the eruption. Unfortunately, the time resolution is not good enough to determine the periods of these oscillations. Digital air-pressure records with 0.1-Pa resolution from Piñon Flat Observatory (station PFO) in southern California and the Black Forest Observatory (station BFO) in southwestern Germany did not show any unambiguous signals from this eruption.

Tahira and others (this volume), however, report acoustic-gravity waves with a dominant frequency at 1 mHz and a clear, isolated resonance at 4.4 mHz. This observation was made at several observatories in Japan so that the phase speed and direction of propagation could be determined and the source traced back to Pinatubo. Thirty-five hours after the onset of the climactic eruption of Pinatubo, a second disturbance was observed and identified as the airwave, A2, that traveled in the opposite direction from Pinatubo to Japan. The spectra of the direct air wave, A1, show no indication of a resonance at 3.7 mHz, while in the case of El Chichón, the 3.7-mHz signal was well observable in the late-arriving air wave. (See fig. 6 of Widmer and Zürn, 1992a.)

INFRARED SATELLITE IMAGE OF PINATUBO'S ERUPTION CLOUD

In infrared images from the NOAA–10 polar orbiting weather satellite taken on June 15 (Lynch and Stephens, this volume), numerous faint concentric bands can be distinguished in the eruption cloud from Mount Pinatubo. These images provide further evidence for narrow-band, low-frequency atmospheric oscillations.

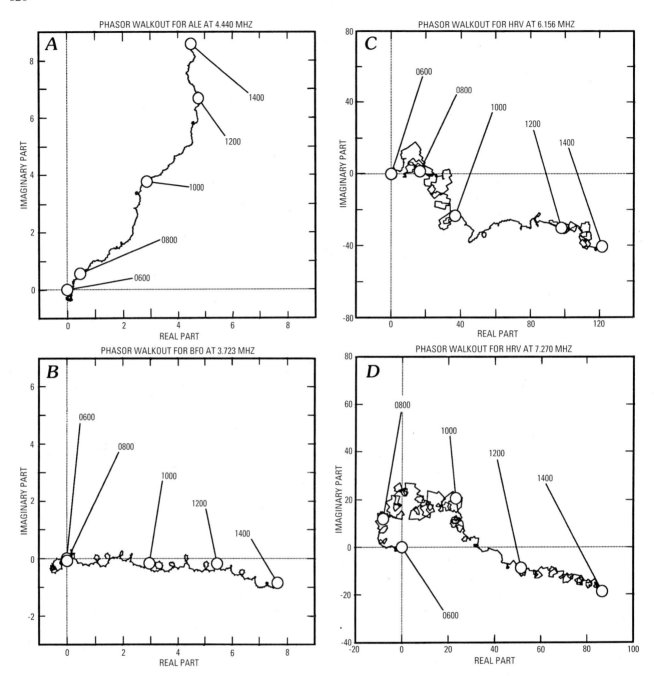

Figure 5. *A*, Phasor walkout for the 4.440 mHz spectral line at the International Deployment of Accelerometers (IDA) station at Alert, Canada (ALE). The steady growth of the Fourier coefficient with time demonstrates the phase coherence of the signal. The time period analyzed spans from 0600 to 1400 G.m.t. (1400 to 2200 local time). *B*, Phasor walkout for the 3.723-mHz spectral line at the Black Forest Observatory (BFO), Germany for the same time window as in figure 5*A*. *C*, Phasor walkout for the 6.156-mHz spectral line at station HRV. The relatively large lateral excursions in this phasor walk are caused by the high-amplitude signal at 3.7 and 4.4 mHz in the analyzed seismogram. However, the overall linear evolution of the phasor walk indicates that at least some of the secondary peaks in figure 4 correspond to phase-coherent resonances (see also fig. 5*D*). *D*, Phasor walkout for the 7.270-mHz spectral line at station HRV. Pinatubo radiated coherently at this frequency from at least 0800 to 1400 G.m.t. (1600 to 2200 local time).

SEARCH FOR PINATUBO-TYPE SEISMIC
SIGNALS FROM OTHER ERUPTIONS

We searched the digital VLP seismic data sets for other Pinatubo-type signals—that is, quasi-harmonic wave trains of long duration. For recent eruptions of Bezymianny (Kamtchatka, February 11, 1979), Mount St. Helens (May 18, 1980), Galunggung (Java, 1982), Colo (Sulawesi, July 28, 1983), Mount Etna (Sicily, September 24, 1986), Redoubt (Alaska, December 15, 1989, January 8, 1990, April 21, 1990), Avachinsky (Kamtchatka, January 13, 1991), and Mount Spurr (Alaska, August 18, 1992), this kind of wave was not observed in the seismic records. However, our search was successful in the case of El Chichón (southern Mexico).

The El Chichón eruption of April 4, 1982, excited Rayleigh and air waves visible on the records from the digital global seismic networks (figs. 4, 5, and 6 of Widmer and Zürn, 1992a). Again, the spectrum turned out to be bichromatic, with frequencies of 3.70 and 5.14 mHz. The lower frequency is nearly the same as for Pinatubo, while the higher one is significantly different. Due to the relatively short duration of the signal in this case (approximately 1 h) we can distinguish two clear phases in the record section (fig. 4 of Widmer and Zürn, 1992a) of the El Chichón eruption: one with a group velocity of 3.7 km/s and the other with 300 m/s. Both have the same frequency contents. On the basis of their group velocity, the two wave groups were identified as Rayleigh and air waves. (In the case of Pinatubo the air-wave arrival, if it exists, is masked by the late-arriving Rayleigh waves.)

KRAKATOA (1883), TUNGUSKA (1908),
MOUNT ST. HELENS (1980), AND
ATMOSPHERIC NUCLEAR EXPLOSIONS

No VLP seismograms are available for the great Krakatoa explosion on August 27, 1883, or the Tunguska meteorite impact on June 30, 1908. However, these events were recorded on microbarographs worldwide and led to several theoretical investigations of the propagation of pulses in the atmosphere. From the published records one can clearly see that these signals were not of the Pinatubo or El Chichón type but consisted of a short, dispersed wave train with a broad spectrum (Pekeris, 1939, 1948). For the case of Krakatoa, however, Kanamori and others (1992) identified atmospheric pressure oscillations with periods of about 300 s (~3.3 mHz) in a barogram from Batavia (Java) about 200 km from Krakatoa. The period cannot be measured accurately because of lacking resolution. Short, dispersive wave trains were also typical for barograms and seismograms from atmospheric nuclear explosions, well documented in the literature (for example, Gossard and

Hooke, 1975). Again, the observations were made in the farfield of the events.

When Mount St. Helens exploded on May 18, 1980, modern digital long-period seismographs recorded seismic body waves and dispersive Rayleigh wave trains very similar to the ones observed after earthquakes. Kanamori and Given (1982) and Kanamori and others (1984) analyzed these records to determine the source mechanism of the event. Pinatubo-like resonant behavior was not observed by these authors then. In addition to the Rayleigh waves dispersive air wave-trains with speeds around 300 m/s, very similar to air waves from atmospheric nuclear explosions, were observed worldwide on barographs, seismographs and gravimeters (Bolt and Tanimoto, 1981; Müller and Zürn, 1983). Recently, Kanamori and others (1992) claimed the observation of an atmospheric oscillation with a 300 s period (~3.3 mHz) on a digital long-period vertical seismograph at station LON, Longmire, Washington State, only 67 km from Mount St. Helens. As in the case of Krakatoa, these oscillations were only observed fairly close to the volcano.

CONVECTIVE STORMS

In the literature of atmospheric acoustic-gravity waves, we find one set of observations that appears very closely related to the bichromatic signal observed during the Pinatubo and El Chichón eruptions. Georges (1973) reports in a study of infrasound emitted from severe convective storms that ionosounders, which use radio waves to map the position of reflective horizons in the ionosphere, find bichromatic waves with frequencies of 3.7 and 4.8 mHz above convective storms. Georges also finds that the line at 3.7 mHz remains unchanged from storm to storm, whereas the higher frequency mode at 4.8 mHz varies by an appreciable (but unspecified) amount from event to event. Analysis of the ionosounder observations also shows that these waves travel in vertical direction and only exist above severe convective storms. Barograms from the vicinity of the storms are not bichromatic but show rather broad spectral features. It appears that the bichromatic oscillations in the ionosphere do not propagate very far away from their source. Considering that the lower frequency reported by Georges (1973) coincides nicely with the lower frequencies observed on seismographs for Pinatubo and El Chichón and that Georges' higher frequency is between the two higher frequencies observed for the two volcanoes, we conjecture that the bichromatism of Pinatubo, El Chichón, and severe convective storms has the same physical explanation.

JUNE 10, 1991

On June 10, 1991, between 1500 and 1800 G.m.t. a clear oscillation with a frequency of 3.7 to 3.8 mHz was observed at the Black Forest Observatory, at the Geoscope station SSB (near Lyon, southern France), and at MedNet station BNI (Italy, close to French border) on vertical seismographs. All our attempts to trace this signal to the then-active volcanoes Mount Pinatubo or Mount Unzen (Japan) failed. The signal was not as large at station BFO as the signal on June 15, 1991, but it was clearly above the noise level. No signal could be seen on the records of other European stations (for example MedNet stations besides BNI: ESK, GRF, KEV, KONO) or any other station of the global networks. Volcanoes being ruled out, we turned to meteorological possibilities, but we have not yet identified the source.

SOURCE MECHANISM

Periodic phenomena associated with volcanoes and geysers are well known. Examples are harmonic tremor (around 1 Hz), geyser repose periods (Old Faithful, Yellowstone), periodic modulations of seismic shock activity (see Martinelli, 1991), and periodic temperature variations in eruption clouds and pulsating or periodic eruptions (mentioned in Williams and McBirney, 1979). The periods corresponding to the above phenomena range from seconds to many hours. Another interesting periodic phenomenon was found by Woods and Caulfield (1992) in laboratory simulations of eruption mechanisms, where, under certain conditions, periodic release of buoyant parcels (called "thermals" by these authors) of material was observed.

However, a comparison of the observed resonance frequencies for Pinatubo and El Chichón with cutoff frequencies found in dispersion calculations for a stratified atmosphere (Harkrider, 1964) led Kanamori and Mori (1992) to suggest that these are free oscillations of the atmosphere excited by the eruption. According to their analysis, the pressure variations observed at Clark Air Base are large enough to explain the observed seismic amplitudes when a source area of radius 37 km is assumed. Underlying this analysis is the assumption that the periods of these pressure variations are the same as in the seismic records. The validity of this assumption may be questioned on the basis of figure 10 of Tahira and others (this volume), who show microbarograph power spectra of acoustic-gravity waves recorded at five different sites in Japan. The energy associated with the 4.4-mHz resonance in these spectra accounts for only approximately 15 percent of the total energy between 0.2 and 10 mHz. Because we do not know how the different spectral components decay with the propagation distance, we can only speculate about the spectrum of the atmospheric pressure variations in the vicinity of Pinatubo. The energy at 3.7 mHz produced at the source was effectively radiated as Rayleigh waves and not as air waves, while the reverse situation is true for the energy around 1 mHz. Most likely the cause for this lies with the energy distribution with height at the source for the different frequency bands.

We, too, favor an explanation where the frequencies are determined by the atmosphere and not by any oscillating phenomena inside the volcanoes. It would be very hard to explain the almost identical frequency of 3.7 mHz for Pinatubo, El Chichón, and the convective storms. However, in order to explain the phase coherence of the source, one has to assume that the oscillator was either only excited once or that reexcitation occurred periodically, perfectly in phase with itself. Single excitation can be ruled out because of the long duration of the climactic phases of the two eruptions. To assume that the reexcitation of the atmospheric oscillations occurred with perfect phase coherence is considered highly unlikely, unless the reexcitation is strongly influenced by the atmospheric oscillations themselves. We therefore propose a source model with positive feedback between the atmospheric free modes and the eruption process or the plume.

Much work has been done on the theory describing the propagation of acoustic-gravity waves in the atmosphere—for example Pekeris (1939, 1948), Lamb (1945), Harkrider (1964), Yeh and Liu (1974), and Francis (1975). From the dispersion relation for the simplistic case of an isothermal atmosphere overlying a rigid halfspace, one finds that three types of waves can exist: (1) the Lamb waves, propagating along the rigid surface, (2) acoustic waves, and (3) gravity waves. The frequency wavenumber domain in this case is split in two regions where freely propagating waves can exist: acoustic waves with frequencies above the acoustic cutoff frequency and gravity waves with frequencies below the Brunt-Väisälä frequency. At the cutoff frequencies, the horizontal wavenumbers (k_H) of the two types of waves vanish, and, thus, the direction of propagation at these frequencies is vertical. This implies that the long-term response in the nearfield of an impulsive source will consist only of a narrow-band signal with all energy concentrated at one or both cutoff frequencies. Energy at other frequencies has a finite horizontal wavenumber and will therefore propagate away from the source region. Numerical values of these cutoff frequencies for acoustic and acoustic-gravity branches for simple atmospheres correspond approximately to the two frequencies observed for the two eruptions, but it is very likely that the atmospheres over violently erupting volcanoes are not this simple. The same certainly applies to severe storms, possibly in a different way.

It is clear that atmospheric modes that strongly excite Rayleigh waves cannot be modes that have vanishing pressure amplitudes at the Earth's surface. From plots of the eigenfunctions given by Francis (1975), it appears that the fundamental gravity mode near its cutoff frequency has very small amplitude at ground level, and all its energy is concentrated around the thermopause at 200 km in altitude. This corroborates the interpretation of Chimonas and Peltier

(1974), who suggest that the bichromatic spectra are due to the fundamental and first higher acoustic modes for $k_H=0$.

A more detailed theoretical investigation of the bichromatic spectrum observed in the ionosphere above severe convective storms was done by Jones and Georges (1976), who compute the transfer function for vertically propagating ($k_H=0$) acoustic waves between the altitudes of 10 and 300 km. They find three effects that shape the transfer function. At the low-frequency end, the transfer function is bounded by the acoustic cutoff frequency; at frequencies higher than 6 mHz, the transfer function rapidly rolls off due to absorption in the thermosphere; between these two frequencies, the spectrum is modulated by acoustic resonances due to the temperature stratification. The frequencies for these resonances are 3.7, 4.6, 5.3, and 6.6 mHz. From the vertical distribution of the displacement amplitudes, Jones and Georges (1976) identify the three higher frequencies as second-, third-, and fourth-order acoustic resonances between the ground and the base of the thermosphere. The lowest frequency resonance at 3.7 mHz occurs as a result of trapping in the temperature maximum at 50 km in altitude. Jones and Georges also find that the frequencies of the three higher resonances are a strong function of the temperature structure in the thermosphere (above 100 km in altitude) but that the frequency of the 3.7-mHz resonance is controlled by the mesospheric temperature profile. While the work of Jones and Georges (1976) is the most detailed theoretical investigation of the bichromatic ionospheric signal that we know of, it is not entirely evident how these results relate to the bichromatic seismic signal observed during the Pinatubo eruption. More work is needed in this field.

If we model the eruption as the superposition of a random sequence of individual explosions, we could obtain the farfield response (assuming linearity) by a convolution of the source time history and the Green's function of the earth-atmosphere system. The Green's function, however, is essentially what was recorded after the atmospheric nuclear explosions conducted around 1960. They have a very broad frequency content similar to the farfield signals from the very short (<4 min) climactic phase of the eruption of Mount St. Helens, the Tunguska event, and the eruption of Krakatoa.

In our source model, the volcanic eruption column and the surrounding atmosphere constitute a self-excited oscillator. This oscillator generates Rayleigh waves by periodically pushing on the surface of the Earth, and it generates Lamb waves (or higher modes) by periodically pushing on the neighboring atmosphere (El Chichón). We propose that the type of atmospheric oscillations that participate in the feedback correspond to waves with vanishing horizontal wavenumber ($k_H=0$), which are traveling vertically, up and down between the ground and the thermopause (at approximately 120 km altitude). These obviously do not propagate away from the source region. The bichromatic spectra now

lead to an interpretation as the cutoff frequencies for the acoustic and gravity waves or for the fundamental and first higher acoustic mode, as suggested by Chimonas and Peltier (1974) and Jones and Georges (1976).

To explain why in the case of Mount St. Helens, Krakatoa, and the severe convective storms, the 3.7-mHz signal can be observed only in the nearfield, we propose that in these cases the vertically propagating waves were directly observed, while in the case of the air waves from El Chichón, the local resonances coupled with the Lamb waves, which then propagated around the globe. If this is correct, we would also expect to see these waves in the case of Pinatubo. As indicated previously, the suspected arrival of the bichromatic air wave is masked by late-arriving Rayleigh waves, and the only observation of the narrowband air waves emitted from Pinatubo that we know of was made with microbarographs in Japan.

A highly speculative physical model that can lead to positive feedback may consist of the following two elements:

1. The rising and expansion of the plume excites a broad spectrum of acoustic and gravity waves. The response of the atmosphere over the volcano, however, is dominated by the $k_H=0$ components of the atmospheric transfer function. This is the forcing of the atmosphere by the plume. (Note that waves for which the horizontal wave number is $k_H>0$ leave the source region laterally and hence cannot interact with the plume or participate in a feedback mechanism.)

2. If we assume that the atmosphere surrounding the plume undergoes harmonic pressure fluctuations, and if we further assume that the plume has a different compressibility than the surrounding atmosphere, then the plume will experience a harmonically varying buoyancy force. This will lead to a harmonic modulation in the rise and expansion of the plume. Such a harmonically forced plume will preferentially excite acoustic and gravity waves at the forcing frequency and can lead to positive feedback. This is the forcing of the plume by the atmosphere.

Of course, the phase of the feedback signal must be right for such a mechanism to work, but possibly this is a self-organized process. Thus, the atmosphere-plume system constitutes a self-excited oscillator, which then radiates seismic and air waves. The energy for this radiation must be provided by the volcano, of course. Presently, the evidence that Kanamori and Mori (1992) and we have collected is circumstantial, and our mechanisms need further verification.

In conclusion, we point out that the bichromatic character of the Pinatubo and El Chichón signals is unique to these two eruptions. The observations we have presented cannot have been caused by a small number of very large explosions. A very strong oscillator was excited by the eruption that radiated the signals observed in the ground and the atmosphere. Explosions could be superimposed on this process but certainly did not dominate the farfield signals.

ACKNOWLEDGMENTS

We thank the operators of the global (IDA, IRIS, Geoscope) and regional (CDSN, MedNet) digital seismic networks for providing excellent low-frequency seismic data, NOAA for making available the infrared satellite image, T.M. Georges for sending us reprints and helping with the literature search, M. Tahira for an early version of his manuscript, D. Seidl for helpful comments, and K.H. Glassmeier for looking at magnetograms and for his interest. J. Mori and C. Newhall are gratefully acknowledged for constructive reviews. The figures were drafted with the programs PLOTXY by Robert Parker, and GMT by Paul Wessel and Walter Smith, and we would like to thank them for sharing these programs. This research was supported by Deutsche Forschungsgemeinschaft within the Sonderforschungsbereich 108: "Stress and Stress Release in the Lithosphere," Project D9. R. Widmer was partially supported through a fellowship from the Swiss Academy of Sciences.

REFERENCES CITED

Bartels, J., 1938, Random fluctuations, persistence and quasipersistence in geophysical and cosmical periodicities: Terrestrial Magnetism and Atmospheric Electricity (forerunner of Journal of Geophysical Research), v. 40, no. 1, p. 1–60.

Beroza, G.C., and Jordan, T.H., 1990, Searching for slow and silent earthquakes using free oscillation data: Journal of Geophysical Research, v. 95, no. B3, p. 2485–2510.

Bolt, B.A., and Brillinger, D.R., 1979, Estimation of uncertainties in eigenspectral estimates from decaying geophysical time series: Geophysical Journal Royal Astronomical Society, v. 59, p. 593–603.

Bolt, B.A., and Tanimoto, T., 1981, Atmospheric oscillations after the May 18, 1980 eruption of Mount St. Helens: Eos, Transactions, American Geophysical Union, v. 62, p. 529–530.

Chimonas, G., and Peltier, W.R., 1974, On severe storm acoustic signals observed at ionospheric heights: Journal of Atmospheric and Terrestrial Physics, v. 36, p. 821–828.

Francis, S.H., 1975, Global propagation of atmospheric gravity waves: A review: Journal of Atmospheric and Terrestrial Physics, v. 37, p. 1011–1054.

Georges, T.M., 1973, Infrasound from convective storms: Examining the evidence: Reviews of Geophysics and Space Physics, v. 11, no. 3, p. 571–594.

Gossard, E.E., and Hooke, W.H., 1975, Waves in the Atmosphere: Amsterdam, Elsevier, 400 p.

Harkrider, D.G., 1964, Theoretical and observed acoustic-gravity waves from explosive sources in the atmosphere: Journal of Geophysical Research, v. 69, no. 24, p. 5295–5321.

Jones, R.M., and Georges, T.M., 1976, Infrasound from convective storms. III. Propagation to the ionosphere: Journal of the Acoustical Society of America. v. 59, p. 765–779.

Kanamori, H., and Given, J.W., 1982, Analysis of long-period seismic waves excited by the May 18, 1980 eruption of Mount St. Helens—A terrestrial monopole?: Journal of Geophysical Research, v. 87, p. 5422–5432.

Kanamori, H., Given, J.W., and Lay, T., 1984, Analysis of seismic body waves excited by the Mount St. Helens eruption of May 18, 1980: Journal of Geophysical Research, v. 89, p. 1856–1866.

Kanamori, H., and Mori, J., 1992, Harmonic excitation of mantle Rayleigh waves by the 1991 eruption of Mount Pinatubo, Philippines: Geophysical Research Letters, v. 19, no. 7, p. 721–724.

Kanamori, H., Mori, J., and Harkrider, D.G., 1992, Excitation of atmospheric oscillations by volcanic eruptions [abs.]: Eos, Transactions, American Geophysical Union, v. 73, no. 43, p. 634.

Lamb, H., 1945, Hydrodynamics (chapters 309–316: Atmospheric waves), New York, Dover, p. 541–561.

Lynch, J.S., and Stephens, G., this volume, Mount Pinatubo: A satellite perspective of the June 1991 eruptions.

Martinelli, B., 1991, Fluidinduzierte Mechanismen für die Entstehung von vulkanischen Tremor-signalen: Dissertation 9376, ETH Zürich, Switzerland, 164 p.

Müller, T. and Zürn, W., 1983, Observation of gravity changes during the passage of cold fronts: Journal of Geophysics, v. 53, p. 155–162.

Oswalt, J.S., Nichols, W., and O'Hara, J.F., this volume, Meteorological observations of the 1991 Mount Pinatubo eruption.

Pekeris, C.L., 1939, The propagation of a pulse in the atmosphere: Proceedings of the Royal Society of London, A, v. 171, p. 434–451.

———1948, The propagation of a pulse in the atmosphere, Part II: The Physical Review, v. 73, no. 2, p. 145–154.

Rydelek, P.A., and Sacks, I.S., 1988, A test for completeness of earthquake catalogs: Nature, v. 337, p. 251–253.

Tahira, M., Nomura, M., Sawada, Y., and Kamo, K., this volume, Infrasonic and acoustic-gravity waves generated by the Mount Pinatubo eruption of June 15, 1991.

U.S. Geological Survey, 1991, Preliminary determination of epicenters, Monthly listing for June 1991: U.S. Geological Survey, 28 p.

Widmer, R., and Zürn, W., 1992a, Bichromatic excitation of long-period Rayleigh and air waves by the Mount Pinatubo and El Chichón volcanic eruptions: Geophysical Research Letters, v. 19, no. 8, p. 765–768.

Widmer, R., and Zürn, W., 1992b, Excitation of atmospheric normal modes by Plinian eruptions [abs.]: Eos, Transactions, American Geophysical Union, v. 73, no. 43, p. 635.

Wielandt, E., and Streckeisen, G., 1982, The leaf-spring seismometer: Design and performance: Bulletin of the Seismological Society of America, v. 72A, no. 6, p. 2349–2368.

Williams, H., and McBirney, A., 1979, Volcanology: San Francisco, Freeman, Cooper & Co., 400 p.

Woods, A.W., and Caulfield, C.-C.P., 1992, A laboratory study of explosive volcanic eruptions: Journal of Geophysical Research, v. 97, no. B5, p. 6699–6712.

Yeh, K. C., and Liu, C. H., 1974, Acoustic-gravity waves in the upper atmosphere: Reviews of Geophysics and Space Physics, v. 12, no. 2, p. 193–216.

Meteorological Observations of the 1991 Mount Pinatubo Eruption

By J. Scott Oswalt,[1] William Nichols,[2] John F. O'Hara[1]

ABSTRACT

Unusually complete meteorological observations were made of eruptions of Mount Pinatubo by two military weather offices that were located within 40 kilometers of the volcano. Surface-weather and radar observations, weather satellite and NASA space shuttle imagery, and atmospheric pressure and temperature measurements recorded preliminary eruptions of June 12–14, 1991, major eruptions of June 15–16, and subsequent lesser eruptions through early September 1991. Observations also helped to record and define the macroscale and microscale atmospheric processes around Mount Pinatubo.

Weather radar was able to track both eruption-column rise rates and horizontal drift of ash clouds from some preparoxysmal and postparoxysmal eruptions. During the second eruption of June 12, 1991, radar indicated an apparent column rise rate in excess of 400 meters per second. Radar observations suggested that higher eruption columns correlated with greater particle size and density within the column.

High-resolution weather satellite imagery showed the existence of gravity waves within spreading ash clouds. Multichannel NOAA imagery distinctly differentiated between fine, drifting ash and high-level clouds. Radar height measurements were typically 10 to 15 percent lower than ash cloud heights inferred from satellite temperature analyses.

From 1530 on June 14 through the paroxysmal stage of June 15, microbarographs proved to be reliable indicators of explosive eruptions. Atmospheric compression waves created by the explosions caused obvious impulses on the barographic recorders. Compression and rarefaction extremes of 8 to 12 millibars were evident on the barograms during the climactic eruptions.

Typhoon Yunya passed within 75 kilometers of Mount Pinatubo at approximately 1100 on June 15. Copious rainfall from the typhoon increased the weight of tephra accumulations and the destruction they caused. Modification of winds aloft by Typhoon Yunya also spread ash over a much wider range of azimuths than would have otherwise received ash.

Postparoxysmal eruptions and residual heat contributed to moisture condensation and atmospheric instability in the vicinity of Mount Pinatubo. This process increased local rainfall, which in turn increased the occurrence of lahars and secondary phreatic explosions.

INTRODUCTION

Due to the coincidental colocation of two U.S. military weather observing sites within 40 km of Mount Pinatubo, the logistics support capability of their host military bases, and preeruption volcano monitoring, the June 1991 Mount Pinatubo eruption may have been the most closely observed major explosive eruption in history.

The purpose of this paper is to detail the capabilities and observations of the two weather sites, which together provided considerable observational data not normally available by other means. We detail the primary observational tools and methodology used. These include weather-observing technicians, weather radars, infrared sensor, satellite imagery, rawinsondes and pilot balloons, microbarographs, and aircraft pilot reports.

We then relate our combined observations through three eruption stages: the preparoxysmal stage of June 12–15, during which time the observational value of the weather sites became readily apparent, the paroxysmal stage of June 15–16, and the postparoxysmal stage of June 16 to November 15, 1991, during which time significant changes to the microscale and mesoscale climate near Mount Pinatubo were observed. We also discuss a significant tephra-fall forecast error and other remarkable ramifications of the eruption interaction with Typhoon Yunya.

Finally, we consider the characteristics of major eruption ash clouds, the utility of weather radar for volcanic ash detection, the viability of atmospheric pressure sensors as explosive eruption detectors, the use of satellite imagery in posteruption monitoring, secondary phreatic explosions, and minor problems created by windblown, suspended ash.

As military meteorologists and oceanographers fortunate enough to experience the 1991 Mount Pinatubo eruption, the authors seek to provide the geophysics community

---

[1] U.S. Navy.
[2] U.S. Air Force.

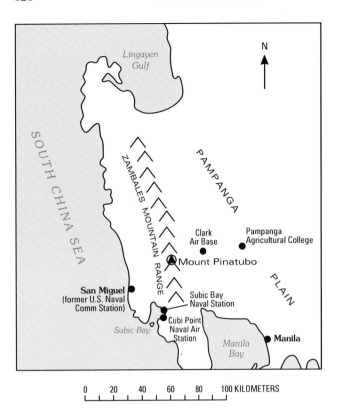

Figure 1. Geography of west-central Luzon Island, showing the three observational sites: Clark Air Base, Pampanga Agricultural College, and Cubi Point Naval Air Station.

with our observations of this major event. Because we lack resources to thoroughly research certain topics we address, we recognize possible omissions and ask readers to contact us with questions or suggestions.

OBSERVATIONAL INSTRUMENTS AND METHODS

Certified military surface weather observers.—Technicians were stationed at Clark Air Base (Clark AB), 15 km east-northeast of Mount Pinatubo, and at Cubi Point Naval Air Station (Cubi Point NAS, part of the U.S. Naval Facilities complex at Subic Bay), 39 km south-southwest of Mount Pinatubo (fig. 1). Both observational sites were manned 24 hours per day, primarily in support of aircraft operations. Hourly and special criteria observations were recorded routinely and disseminated electronically. Recorded weather parameters included time of observation, cloud amount(s) and type(s), horizontal visibility, obstructions to vision, wind direction, average sustained wind speed, wind gust speed, ambient air temperature and dewpoint, atmospheric pressure, and significant weather.

Due to the close, unprotected location of Clark AB at the foot of Mount Pinatubo, evacuation of personnel and instruments was required during some periods of volcanic

Table 1. Military weather radar specifications.

Radar	Clark Air Base AN/FPS–77	Cubi Point Naval Air Station AN/FPS–106
Frequency (megahertz)	5450–5650	5450–5650
Wavelength (centimeters)	5.3–5.5	5.3–5.5
Power Output (kilowatt peak)	174–350	250–350
Pulse Repetition Rate (pps)	186–324	324
Pulse Length (microsecond)	2±0.4	2±0.1
Maximum Range (nautical miles)	200	200
Maximum Height (feet)	80,000	60,000
(meters)	24,400	18,300

activity, notably at the beginning of the paroxysmal stage of June 15 and lasting until June 19. A log was kept of visual observations from an evacuation site 38 km east of the volcano.

Weather radars.—The AN/FPS–77 at Clark AB and the AN/FPS–106 at Cubi Point NAS were 1950's technology, C-band weather radars with very similar performance specifications (table 1). Both had Plan Position Indicator (PPI) display for horizontal return and Range Height Indicator (RHI) display for vertical return, which provided three-dimensional coverage within the performance range limits. Regrettably, neither the Clark nor the Cubi Point radar could produce hard copy of screen displays; our illustrations of radar signals were traced from the radar screen. The Clark radar had superior coverage due to its unobstructed view of Mount Pinatubo's summit, closer location to the mountain, and greater height range 24,000 m (80,000 ft). Both radar antennae were housed in protective domes, which allowed them to operate (rotation and angle inclination) free from ash intrusion into their drive mechanisms.

Infrared sensor.—The Infrared Detecting Set (AN/AAQ–9 in military nomenclature), while not normally used for weather observations, was utilized by Clark personnel to assist in volcano monitoring efforts. Its spectral response of 7.6 to 11.75 μm provided thermal imaging useful for observing volcanic activity at night.

Satellite imagery.—At Clark AB, a Defense Meteorological Satellite Program (DMSP) polar orbiting satellite receiver provided near-realtime high-resolution visual and infrared (IR) imagery. At Clark AB and Cubi Point NAS, geostationary satellite receivers provided near-realtime, low-resolution visual and IR imagery. At the Joint Typhoon Warning Center in Guam, a geostationary satellite receiver, a National Oceanographic and Atmospheric Administration (NOAA) TIROS–N polar orbiting satellite receiver, and a DMSP satellite receiver provided high-resolution, near-realtime visual and IR imagery.

National Aeronautics and Space Administration.—A space shuttle mission underway during the eruption provided realtime visual observation and postevent videography from low Earth orbit.

Rawinsonde and pilot balloons.—The Rawinsonde Set (AN/GMD–5 in military nomenclature) was located at Clark AB, and VAISALA Mini-Rawinsonde (AN/UMQ–12 in military nomenclature) was located at Cubi Point NAS. These systems provided telemetered upper level temperature, humidity, pressure, and wind speed and direction data. Pilot balloons were used to measure vertical visibility and winds aloft by visual tracking.

Microbarograph.—Microbarographs were located at both Clark AB and Cubi Point NAS and provided instantaneous measurement and continuous recording of surface atmospheric pressures. The device is designed for shipboard as well as land use and includes a viscous damping mechanism to prevent motion contamination of the recording.

Pilot reports.—Postevent commercial airline reports detailed ash plume encounters and, in some cases, ash damage. Local military personnel were aware of the ash hazard and associated avoidance areas; therefore, no airborne ash encounters were reported by military aircraft.

OBSERVATIONS AND DISCUSSION

PREPAROXYSMAL STAGE, JUNE 12–15

At approximately 0851 local time (G.m.t. plus 8 h) on June 12, the Clark weather radar RHI display indicated a strong, coherent echo with well-defined edges located directly over the Mount Pinatubo vent. Visual confirmation of an eruption resulted in the immediate evacuation of the remaining U.S. Air Force security and command personnel, including the weather office, from Clark. In the course of a routine hourly weather observation, the Cubi Point weather observer saw the initial eruption as the tephra column rose above the foreground ridge line, which obscures the line of sight from Cubi Point to the volcano summit. The column rose to a level exceeding the height range of the Cubi Point radar (18 km) and was measured at greater than 19 km by the Clark radar just prior to weather office evacuation. Lightning was observed within the tephra column as well as from the column outward to clear air.

When the column reached a neutrally buoyant level (Sparks and others, 1994), it spread horizontally and radially, as shown by the Cubi Point weather radar PPI display depicted in figure 2 and by visual satellite imagery shown in figure 3. It created a broad umbrella or canopy of ash and resembled a "mushroom" cloud or severe thunderstorm. This ash umbrella spread south-southwest past Subic Bay and east over Clark Air Base (figs. 2 and 3), but only a minute trace of ash fell at Cubi Point, where it was monitored on sheets of black and white paper placed outdoors. Significant tephra fall was reported by townships located directly downwind (west-southwestward) of the low to middle tropospheric winds. The former San Miguel Naval Communications Station (fig. 1) reported a sky darkened by

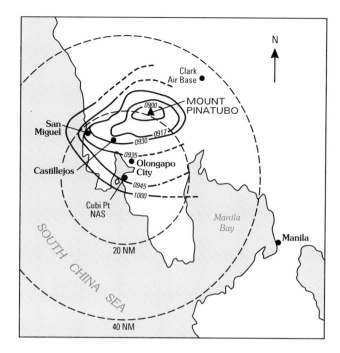

Figure 2. Cubi Point weather radar Plan Position Indicator (PPI) display of the 0851 June 12 eruption ash cloud. Because the radar is used primarily for aviation service, distances are measured in nautical miles (NM). The Plan Position Indicator was set at 60-nm range gate. Edge contours for the ash cloud are in local time.

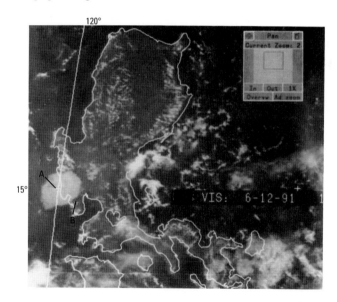

Figure 3. Geostationary weather satellite visual image of the 0851 June 12 eruption. The spreading ash cloud (*A*) covers the entire Subic Bay (*B*). Picture time is 0931.

tephra fall, and accumulations of up to 5 cm. The mixture of ash and small, coarse pumice fragments was easily removed from most surfaces.

Fine ash suspended at higher levels was eventually carried westward over the South China Sea, as reported by commercial aircraft and detected by analysis of satellite

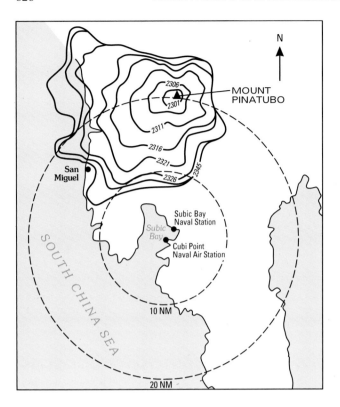

Figure 4. Cubi Point weather radar Plan Position Indicator (PPI) display of the 2252 June 12 eruption ash cloud. In this illustration the PPI was set to a 30 nm range gate. Ash cloud edge contours are in local time.

imagery. Larger and heavier pumiceous tephra concentrations precipitated relatively quickly, while lighter ash particles remained suspended significantly longer and were transported greater distances before settling out of the atmosphere. Observed winds aloft, radar observations of the ash cloud, and local pilot reports enabled military flight weather forecasters to provide accurate recommendations for safe emergency aircraft sortie (evacuation) operations.

Several additional explosive eruptions between June 12 and the paroxysmal stage of June 15 were observed either visually, by radar, by barogram pressure impulse, or by some combination of these three methods. Typically, eruptions reached the tropopause (~17 km) in 3 to 5 minutes. However, faster rates were observed, such as the eruption detected by both radars at 2252 on June 12 (fig. 4), in which the column rose to 9 km, temporarily slowed its ascent, then rose to above 24 km in approximately 30±5 s, indicating upward velocity well in excess of 400 m/s. (Note: During this event, we made the assumption that the radar was, in fact, sensing a rising ash column. Other than the anomalous ascension rate, the radar signature was essentially identical to other eruptive events. Though we are confident of the observation, it is difficult to conceive how such a rise rate might be possible. We believe that, ultimately, it is most important to report this observation, despite uncertainty about its explanation.)

A series of eruptions the afternoon and early evening of June 14 yielded the first major ash fall at Subic Bay Naval Station. As a result of a laterally directed eruption detected at 1530 (Hoblitt, Wolfe, and others, this volume), 2.5 to 5.0 cm of very fine ash was deposited over the base. With local rain showers spawned by approaching tropical cyclone "Yunya" (locally "Diding"), much of the ash fall was wet, had a consistency resembling moist clay, and adhered to exposed surfaces. Having been briefed by U.S. Geological Survey personnel on the effects of volcanic ash fall on aircraft, military authorities directed the closure of the Cubi Point airfield. Closure led to the grounding of a contract commercial cargo plane (DC–10), which subsequently sustained significant damage from prolonged tephra fall on June 15. This ash-fall event also introduced the problem of ash-induced radar attenuation, as the suspended ash caused the leading edges of the ash cloud radar echo to gradually become diffuse and difficult to track. A temporary, simultaneous outage of both the weather radars early that evening, coupled with our uncertainty about the winds aloft (attempts to launch thin latex weather balloons into an ash-filled sky proved futile) resulted in inability to gauge the extent, dimensions, density, and movement of the ash cloud.

Barogram indicators of eruptions on June 12, 13, and early June 14 had been either very subtle or impossible to discern. When the explosions became laterally directed, commencing with the blast at 1530 on June 14, the barograph pressure recorders indicated obvious impulsive pressure fluctuations (compression and rarefaction) in response to atmospheric compression waves radiating outward from the explosions. Thereafter, the barograph was a reliable indicator of explosive events and was used to determine actual times of explosive eruptions.

Weather radar proved useful as verification of eruptive events. Clark AB and Cubi Point NAS noted generally excellent agreement in quantitative and qualitative radar observations. Signal return displays of vertical eruption columns and laterally spreading ash clouds were generally sharp and distinct, being slightly more coherent than a thunderstorm but less than a building or mountain. Both sites also observed that higher eruption columns produced more intense radar returns. This may be attributable to greater particle size and concentration. Use of the iso-echo (contouring of the return-signal strength) display allowed detection of renewed eruptive pulses within the echo display of ash from previous eruptive pulses, and prevented the radar return of new eruptions from being masked by the return of previous events. The largest eruptive events created eruption columns far exceeding the height detection capabilities of the weather radars.

The high-resolution DMSP satellite imagery received at Clark was able to detect moderate to large ash clouds, and enhanced multichannel imagery received at other global sites, was successfully used to differentiate between ash and

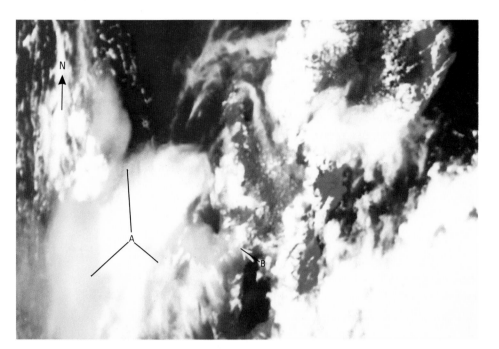

Figure 5. NOAA–11 weather satellite multichannel (channels 1–3) image of the ash cloud (*A*) from the 0851 June 12 eruption drifting over the South China Sea. Mount Pinatubo is at point *B*. Picture time is 1457 June 12.

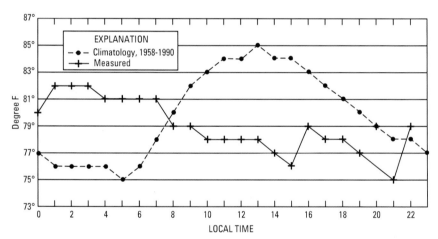

Figure 6. Comparison of observed versus climatological (long-term average) daytime ambient air temperatures at Cubi Point Naval Air Station for June 15. The Fahrenheit temperature scale is used to increase resolution.

water or ice (fig. 5). Low-resolution imagery was unable to detect all but the very largest ash cloud of June 15. Ash-cloud height estimates derived from temperature analyses of ash-cloud tops showed only fair correlation with radar observations (comparisons were only possible with smaller post-paroxysmal eruptions that were within radar height range). Satellite height analyses were typically 10 to 15 percent lower than radar height measurements.

PAROXYSMAL STAGE, JUNE 15–16

An explosive eruption observed by the radars at approximately 1038 on June 15 signaled the onset of vigorous sustained eruptive activity that included the climactic explosions. Reduced light level and suspended ash were observed at Cubi Point shortly after sunrise, and by 1200 the sky was totally obscured and darkened by tephra fall and suspended ash. The evacuation site for the Clark weather office at the Pampanga Agricultural College near Mount Arayat (fig. 1) experienced low light levels as well, as tephra fall began there at approximately 1440. The insolation blockage at Cubi Point was quite dramatic and compounded by a lack of any reflected ground-based light source. The effect was analogous to visibility in a totally darkened room. Without a light source, visibility was reduced to near zero. Insolation blockage also caused a significant decrease in normal ambient air temperature, as measured at Cubi Point (fig. 6). While no measuring equipment was available at the Agricultural College, weather observers there noted that the ambient air temperature felt exceptionally cool that afternoon and evening.

Enveloped in rain, suspended ash, and falling pumice, radar could no longer track the outer edges of the ash cloud. Signal attenuation by tephra fall and suspended ash prohibited all but the strongest returns. Therefore, we were limited to radar detection of activity occurring over the vent, where particle size and concentration were greatest.

Tephra fall continued at Cubi Point throughout the day, varying from completely dry ash through a wet, cement-like mud, to muddy water. Particle size gradually increased, with coarse pumice fragments reaching a maximum of approximately 5 cm in diameter at about 1530. Postevent examination of deposits at Clark yielded pumice fall fragments similar size; locally, fragments 15 to 30 cm in diameter were left by volcanic mudflows (lahars).

Unhangared aircraft at Cubi Point NAS suffered damage due to the weight of accumulated wet tephra. For example, both the commercial DC-10 mentioned previously and a military C–130 transport airplane fell back on the ends of their fuselages as the accumulated ash radically altered their centers of gravity. Civil aviation pilots reported ash-cloud encounters that resulted in engine damage, exterior abrasion, and interior contamination by ash (Casadevall and others, this volume). Tephra fall also created considerable problems for ships in port at Subic Bay. The accumulation of saturated ash and pumice on decks created alarming weight increase and caused potential stability problems by shifting centers of gravity. The sulfur content of the debris corroded exposed sensitive electronic equipment. Exposed rotating machinery, such as radars and wind speed indicators, experienced mechanical failure or was damaged from bending or jamming of the drive mechanisms or rotational surfaces. The ability of ships to safely navigate out of the ash cloud was significantly impacted by low visibility and inoperative navigation radars. Tephra was ingested by some engines (four marine diesel engines required replacement), and pumice material floating on the water limited ships' ability to take in engine cooling water and to distill fresh water. Wet ash accumulations caused power transmission lines, insulators, and transformers to short circuit. The power loss rendered the Cubi Point water treatment plant and pumping stations inoperative. Subsequent electrical power restoration was a lengthy and tedious process of thoroughly cleaning and reenergizing individual segments of the basewide grid.

Throughout this tephra-fall episode, frequent lightning was observed across the celestial dome at both Cubi Point and the Agricultural College. The lightning often appeared red, green, or blue and was not a product of cumulonimbus clouds (thunderstorms) but of the frictional effect of ash fall. Satellite imagery and radar observations suggest that thunderstorm activity did not occur. There were no convective cloud formations evident within the ash cloud, and the dynamic processes required for thunderstorm occurrence were disrupted by the enormous volume of ash suspended throughout the tropospheric layer and by tephra-induced

subsidence. In the absence of this eruption, the passage of a tropical cyclone taking a track such as Typhoon Yunya (fig. 7) would have very likely brought significant thunderstorm activity to central Luzon.

Barograms give a fairly accurate indication of eruptive activity throughout the paroxysmal stage (fig. 8). However, the low temporal resolution of the recordings hampers a detailed analysis of the pressure fluctuation maxima occurring between approximately 1400 and 2200 (0600 and 1400 G.m.t.) on June 15. The maxima shows short-duration pressure extremes of approximately 8 to 12 mbar and corresponded to the onset and duration of considerable sensible seismicity at both observing stations.

Constant, low-frequency rumble, distinctly different from the nearly continuous thunder, emanated from the general direction of the volcano during the period of frequent magnitude 3 to 5.6 earthquakes and vigorous atmospheric pressure fluctuations.

High-resolution imagery also resolved gravity waves propagating outward through the laterally spreading ash umbrella (figs. 9 and 10). Another feature clearly evident on some high-resolution imagery is the eruption column rising above, or overshooting, the ash umbrella (fig. 9). It appears that the rising eruption column is ascending significantly higher than the level of neutral buoyancy, where the ash umbrella is spreading. This is indicated in figure 9 by the shadow the eruption column is casting on top of the ash umbrella as the column rises above it (Tahira and others, this volume; Lynch and Stephens, this volume)

TROPICAL CYCLONE YUNYA

In anticipation of significant tephra fall accompanying a Mount Pinatubo eruption, volcanologists, meteorologists, and various key decisionmakers were interested in defining a tephra-fall hazard area. On the basis of climatology of the monsoon wind regime of central Luzon, tephra-fall hazard areas were generalized as west-southwest (in response to the late northeast monsoon season) and northeast (in response to the southwest monsoon season) of the tephra source. Numerical forecasts based on observed winds aloft were in general agreement with climatology studies and hazard area delineations. Excellent verification of forecast tephra response to northeast monsoonal flow was observed during pre-paroxysmal eruptions on 12 and 13 June, and this appeared to lend support to defined tephra-fall hazard areas.

On June 13, a tropical disturbance in the southern Philippine Sea was upgraded to Tropical Storm Yunya (fig. 7). Yunya tracked to the northwest, skirting the eastern shores of the Visayan Islands and southern Luzon while intensifying to typhoon strength (fig. 11). By 1400 on June 14, Typhoon Yunya was located over the north coast of Catanduanes Island, close enough to central Luzon to begin

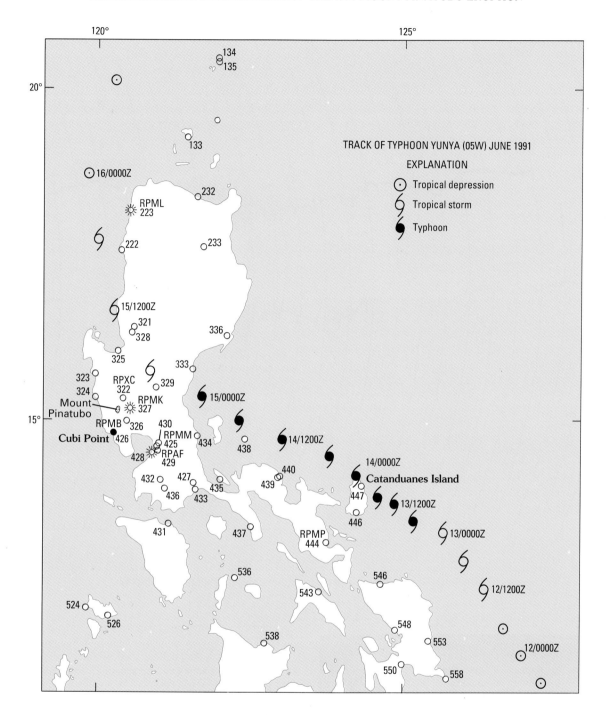

Figure 7. Best track for Typhoon Yunya, as provided by the Joint Typhoon Warning Center, Guam. Positions are Greenwich Mean Time with the day shown first. Local time is G.m.t. + 8 h. Other numbered and lettered symbols are weather stations with various measuring capabilities.

inducing a departure from the climatologically normal monsoon wind regime. The typhoon's proximity and cyclonic (counterclockwise) wind flow induced a southward bias in the normal middle to upper tropospheric winds, which was evidenced by the first significant ash fall over Subic Bay during the June 14 eruptions. Typhoon Yunya made landfall over central Luzon near the Dingalan Bay (120 km east-northeast of Clark) at 0800 on June 15 and decreased to

tropical storm intensity in response to land interaction and shearing. The low-level circulation of Tropical Storm Yunya then proceeded along the favored track of storms entering Luzon over the Dingalan Bay: northwest over the Pampanga plain region (during which time its center passed within 75 km northeast of the erupting volcano). Yunya exited Luzon at the Lingayen Gulf at approximately 1800 on June 15, proceeding north-northwest along the northern

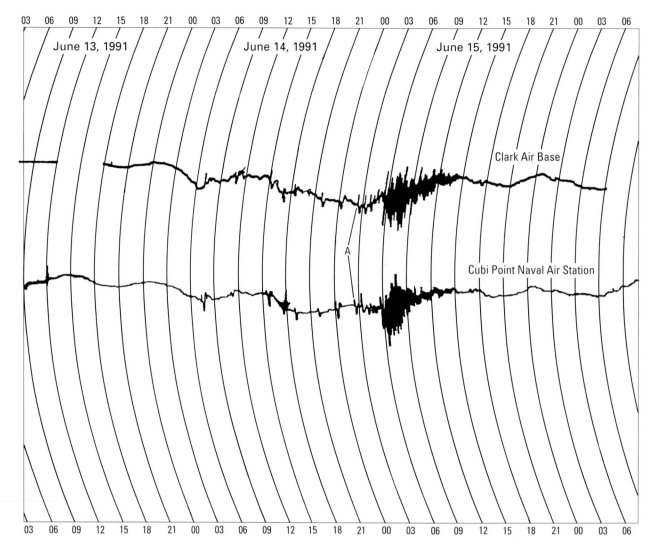

Figure 8.   Surface atmospheric pressure barograms from Clark AB and Cubi Point NAS for the period June 13 to June 16. Barogram time scales are Greenwich Mean Time (G.m.t.). The onset of the paroxysmal stage is indicated at point (*A*).

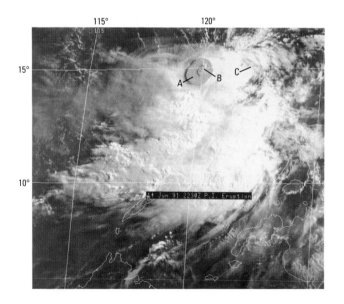

← **Figure 9.**   Geostationary weather satellite visual image of the 0555 June 15 eruption. Picture time is 0630 June 15. Notable features include well-defined concentric gravity waves (*A*) and the eruption column (*B*) "overshooting" the broad ash canopy. The exposed low-level cyclonic circulation of Typhoon Yunya (*C*) is visible to the east of the ash cloud.

Luzon coast (fig. 9). Meanwhile, the upper-level circulation of the system sheared from the lower level and moved westward, and the associated high-level cloudiness intermingled with the ash cloud as both moved out over the South China Sea.

Because of Typhoon Yunya's passage, the entire paroxysmal eruption occurred during a radical departure from the usual monsoon wind regime. This anomaly, coupled with uncertainty about ejecta volume (experts had previously warned of this uncertainty), rendered wind climatology-

Figure 10. Geostationary weather satellite visual image of the climactic eruption series of June 15. Picture time is 1630 June 15. Gravity waves (*A*) are discernible in the massive ash cloud, which is casting a well-defined shadow (*B*) to the east.

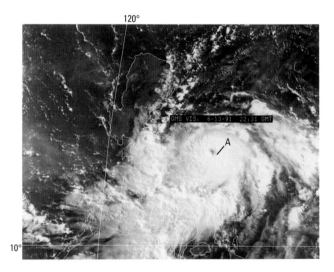

Figure 11. Geostationary weather satellite visual image of Typhoon Yunya (*A*) approaching Luzon. Picture time is 1431 on June 13.

based tephra-fall forecast maps virtually irrelevant. Typhoon Yunya's modification of the seasonal winds caused a southward displacement of light to moderate tephra accumulations, transported by the middle tropospheric to low stratospheric winds, and a northeastward displacement of moderate to heavy accumulations transported by low-tropospheric winds. Due to the highly improbable simultaneous volcanic eruptions and influence of Typhoon Yunya on the wind patterns, plus the enormous amount of volcanic ejecta, Clark, and particularly Subic Bay, received substantially more tephra fall than was generally expected.

A comparison between a climatology-based tephra-fall hazard map and an analysis of actual tephra deposits from the June 12–15 eruptions illustrates the effect of Typhoon Yunya in combination with the huge volume of material ejected from Mount Pinatubo (fig. 12).

The tephra accumulation at Subic Bay was all the more significant in view of previous contingency plans and actions by military authorities. With Clark highly susceptible to tephra fall, and possibly even to pyroclastic flows, an emergency evacuation plan had been developed identifying U.S. Naval Facilities Subic Bay as the prime evacuation site for Clark personnel. Subic Bay had been deemed safe from all but the most improbable catastrophic mountain collapse and was south of the expected tephra-fall hazard area associated with the northeast monsoon. Due to increasing eruptive and seismic activity, Clark had evacuated all military dependents and nonessential military and civilian personnel (14,500) to Subic Bay on June 10. The huge increase in the base population exacerbated the problems created by the tephra fall of June 14–15.

The considerable rainfall produced by Yunya saturated most of the tephra fall, and the cement-like consistency of the accumulated tephra made removal very difficult and tedious. Local analysis showed that the moistening of loose, dry tephra increased its weight by 23 percent, and the added weight created by the rainfall certainly contributed to widespread structural failures at Clark, Subic Bay, and communities throughout central Luzon.

Were the eruption and the passage of tropical cyclone Yunya related? A remarkable but probably coincidental

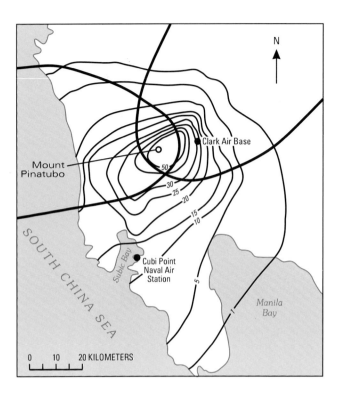

Figure 12. Ash-fall-hazard planning map versus isopachs of ash fall from the June 12–16 eruptions. The unit of ash depth is centimeters. Due to uncertainty about winds on any given day, other ash fall planning maps declined to attempt specific definitions of hazard areas (Punongbayan and others, this volume). Isopachs for individual eruptions are presented in Paladio-Melosantos and others (this volume).

correlation exists between the time of Typhoon Yunya's closest approach to Mount Pinatubo (fig. 7) and the pressure fluctuation maxima. Although the Yunya-induced mean atmospheric pressure fall is unremarkable in terms of tropical cyclone passage (approximately 6.3 mbar), there is, nonetheless, an association between the time of pressure minimum (about 1100 on June 15, on the Clark Air Base barograph), the closest approach of Yunya to Mount Pinatubo (within 75 km at approximately 1400, June 15, as tracked by weather satellites), and strong barometric pressure fluctuations associated with eruptions (strongest between 1038 and 2200 on June 15). Both the eruption and the tropical cyclone were well along in their development before they coincided and presumably could have occurred independently of each other. We know of no evidence for any causal relation between the two, but, clearly, the passage of Yunya greatly increased damage from the eruption.

POSTPAROXYSMAL STAGE

Significant changes to microscale and mesoscale weather patterns near Mount Pinatubo were observed through mid-September. Prominent among these was the formation of "volcanic" thunderstorms. During most of the less vigorous postparoxysmal eruptions, cumulus cloud complexes formed near the top of the buoyant ash plume, where minute ash particles and other material provided abundant hygroscopic nuclei. These cloud complexes frequently developed into cumulonimbus clouds (thunderstorms). The thunderstorms often drifted away from their source region at the top of the plume, producing sometimes significant amounts of localized rainfall, "mudfall," and ash fall along their drift tracks before dissipating. As an example, a deluge at Clark on September 1, which yielded 74 mm of rain in 2 hours, probably was a Pinatubo-induced thunderstorm. New thunderstorms often developed near the top of the ash plume as previous cells drifted away.

As the southwest monsoon (rainy season) took effect in mid-July, visual and radar confirmation of this phenomenon was restricted because of extensive cloudiness and precipitation. Differentiation between buoyant ash plumes and convective cloud formations was then achieved by comparing Clark radar displays with concurrent seismometer data.

Even in the absence of an observable buoyant plume, significantly greater than average occurrences of afternoon thunderstorms and rain showers were observed near the vent and surrounding hot pyroclastic-flow deposits, which together provided the initial buoyancy needed to lift moist, unstable, tropical air to a level where it continued to rise due solely to its own instability (level of free convection). Significant flooding and lahar events were attributable to rain showers and thunderstorms caused by residual heat from Mount Pinatubo.

Rainwater interacting with hot pyroclastic-flow deposits has resulted in steam explosions and jets, ejecting tephra to 15 km and greater, and creating significant localized tephra fall. On some occasions, explosions occurred days after rainfall ended.

CONCLUSIONS AND RECOMMENDATIONS

MODELING OF WIND AND TEPHRA FALL DURING THE JUNE 12–15 ERUPTIONS

Tephra fall distribution from the June 12–15 eruptions was controlled by prevailing winds, which were strongly modified by the approach and passage of Typhoon Yunya. On June 12, Yunya was not yet a factor; on June 14 and 15, it was a major factor. It would be interesting to examine these events with a three-dimensional model of the wind field and attempt to reproduce the observed patterns of tephra fall (fig. 12). For more detailed information about tephra fall, see Paladio-Melosantos and others (this volume) and Wiesner and Wang (this volume).

WEATHER RADAR

Weather radars (even 1950's vintage) proved to be extremely useful in detecting initial and secondary explosions and in tracking drifting airborne ash clouds or plumes, whenever the concentration of airborne ash (threshold unknown) was sufficient. Radar information concerning eruptions, tephra-bearing secondary phreatic explosions, and ash-cloud dimensions and trajectories, was critical for ash warnings. Considering the limitations of the radars used, it seems likely that newer Doppler weather radars would be far more capable in detecting eruptions and tracking airborne ash. A Doppler weather radar should be able to differentiate ash particles from water; this capability would make it useful in areas where clouds and precipitation curtail other means of monitoring. Research could be conducted to determine ash-cloud dispersion versus particle size and fall rate. Doppler also presents obvious potential for study of eruption column dynamics. Newer technology may allow remote radar monitoring and provide recording capability for postevent analysis. We think it should be feasible to configure a mobile van with a relatively small and inexpensive commercial Doppler weather radar, which could then be deployed to various areas of interest.

DEVELOPMENT OF ALGORITHMS OR NOMOGRAMS FOR REAL-TIME FORECASTS OF TEPHRA FALL

Harris and Rose (1983) and Rose and Kostinski (1994) have noted that quantification of volume, particle size, ash-cloud height, and other pertinent parameters either can be obtained by radar observation or can be reasonably assumed, with the resultant data integrated with the winds aloft for the development of a tephra-fall forecast. Research to improve characterization of ash clouds by radar observations is ongoing (Rose and Kostinski, 1994). We recommend that working algorithms or nomograms that incorporate results from this research be developed for use by weather radar operators.

BAROGRAPH

The ability to readily detect explosive eruptions by barograph was shown. Eruptions from June 14 and 15 produced stronger barograph signals than those of June 12–13, perhaps related to stronger laterally directed components (pyroclastic surges and pyroclastic flows) in the June 14–15 eruptions. A relatively inexpensive monitoring system could be constructed by placing atmospheric pressure sensors near potentially explosive volcanoes. While such a system would not provide eruption warning, it would give instantaneous notification of explosive (especially, laterally directed) eruptions and provide a first guess on the eruption strength as based on the magnitude of the pressure impulse. A digital system could provide convenient long-term data storage but would require a sampling rate high enough to resolve discrete explosions.

SATELLITE IMAGERY

Multichannel satellite imagery can distinguish between ash and water droplets or ice crystals (clouds) (fig. 5). Satellite imagery is most useful in posteruption monitoring of ash cloud or plume dimensions and trajectories, in order to warn aircraft that are transiting affected areas. The key to making these warnings effective lies in disseminating information to the global aviation community quickly and updating this information whenever new satellite imagery or other data become available.

MAJOR ASH-CLOUD CHARACTERISTICS

Major eruption columns approximated the physical structure associated with "super" thunderstorms. The magnitudes of several eruptions were great enough to allow penetration of the stratosphere, with the "anvil" or "mushroom" top (ash umbrella) forming around a level of neutral buoyancy. Fluctuation of the ash about this level was detected in high-resolution satellite imagery as concentric gravity waves in the ash umbrella (Lynch and Stephens, this volume; Tahira and others, this volume). Analysis of these waves may be useful in determining relative eruption intensities and magnitude changes.

VOLCANICALLY ENHANCED RAINFALL AND SECONDARY PHREATIC EXPLOSIONS

By creating vertical instability and triggering rain shower and thunderstorm formation, the volcano generated its own microscale and mesoscale weather patterns. After the main eruption, and before the onset of extensive cloudiness from the southwest monsoon, it was quite common to see thunderstorms and rain showers develop exclusively over Mount Pinatubo. With the return to the dry northeast monsoon in mid-October, we continued to observe enhanced cumuliform cloudiness over the vent region. We believe the additional heat emitted from both the vent and the deep, proximal pyroclastic-flow deposits created the vertical instability that caused the cumuliform development.

Enhanced rainfall increased the occurrence of secondary phreatic explosions. These explosions are caused by rain water or lahars coming into contact with still-hot pyro-clastic-flow deposits. The explosions have lifted tephra to over 15 km, and subsequent tephra fall has locally reduced visibility and ambient light level. These secondary explosions pose a continuing hazard to aviation.

SUSPENDED WIND-BLOWN ASH

During the dry northeast monsoon, fine ash was lifted and entrained by low-level winds from local ridge lines and large expanses where vegetation had not stabilized tephra deposits. Sufficient wind speed (>8 m/s) was the primary factor in lifting the ash. Although the concentration of airborne ash was enough to significantly lower near-surface visibility (below visual flight rules), no problems with aircraft engine performance were reported. As vegetation recovered and loose ash was washed away by rainfall or was otherwise removed, the problem of airborne ash diminished relatively rapidly. No recurrences of significant wind-blown ash were observed during the late-1992 northeast monsoon.

ACKNOWLEDGMENTS

We are grateful for the encouragement and assistance provided by C. Newhall, W. Duffield, R. Hoblitt, T. Casadevall, and others from the U.S. Geological Survey and by R. Punongbayan and the Philippine Institute of Volcanology and Seismology. This study has also benefited substantially from work and discussions with K. Rodolfo, J. Umbal, M. Paladio, and others from the 1991 Pinatubo Lahar Hazards Task Force. Illustrations provided by L. Carr, U.S. Navy, S. Runco, National Aeronautics and Space Administration, and R. Hudson, U.S. Air Force, were an invaluable addition to this work. Finally, critical assistance by W. Johnson, U.S. Navy, is greatly appreciated.

REFERENCES CITED

Casadevall, T.J., Delos Reyes, P.J., and Schneider, D.J., this volume, The 1991 Pinatubo eruptions and their effects on aircraft operations.

Harris, D.M., and Rose, W.I., 1983, Estimating particle sizes, concentrations, and total mass of ash in volcanic clouds using weather radar: Journal of Geophysical Research, v. 88, no. C15, p. 10969–10983

Hoblitt, R.P., Wolfe, E.W., Scott, W.E., Couchman, M.R., Pallister, J.S., and Javier, D., this volume, The preclimactic eruptions of Mount Pinatubo, June 1991.

Lynch, J.S., and Stephens, G., this volume, Mount Pinatubo: A satellite perspective of the June 1991 eruptions.

Paladio-Melosantos, M.L., Solidum, R.U., Scott, W.E., Quiambao, R.B., Umbal, J.V., Rodolfo, K.S., Tubianosa, B.S., Delos Reyes, P.J., and Ruelo, H.R., this volume, Tephra falls of the 1991 eruptions of Mount Pinatubo.

Punongbayan, R.S., Newhall, C.G., Bautista, M.L.P., Garcia, D., Harlow, D.H., Hoblitt, R.P., Sabit, J.P., and Solidum, R.U., this volume, Eruption hazard assessments and warnings.

Rose, W.I., and Kostinski, A.B., 1994, Radar remote sensing of volcanic clouds, in, Casadevall, T.J., ed., Volcanic ash and aviation safety: Proceedings from the First International Symposium on Volcanic Ash and Aviation Safety: U.S. Geological Survey Bulletin 2047, p. 391–396.

Sparks, R.S.J., Bursik, M.I., Carey, S.N., Woods, A.W., and Gilbert, J.S., 1994, The controls of eruption column dynamics on the injection and mass loading of ash into the atmosphere, in, Casadevall, T.J., ed., Volcanic ash and aviation safety: Proceedings from the First International Symposium on Volcanic Ash and Aviation Safety: U.S. Geological Survey Bulletin 2047, p. 81–86.

Tahira, M., Nomura, M., Sawada, Y., and Kamo, K., this volume, Infrasonic and acoustic-gravity waves generated by the Mount Pinatubo eruption of June 15, 1991.

Wiesner, M.G., and Wang, Y., this volume, Dispersal of the 1991 Pinatubo tephra in the South China Sea.

Mount Pinatubo: A Satellite Perspective of the June 1991 Eruptions

By James S. Lynch[1] and George Stephens[1]

ABSTRACT

The June 15, 1991, eruption of Mount Pinatubo was the largest explosive volcanic event detected on this planet with satellite imagery. From June 9 to June 17, many of the Mount Pinatubo eruptions were detected and monitored by NOAA and Japanese operational meteorological satellites (NOAA–10, –11, –12, and GMS). This paper summarizes the satellite information and shows selected examples of the enhanced imagery.

INTRODUCTION

In June 1991, Mount Pinatubo (15.1°N. lat, 120.4°E. long) in the Philippines produced a series of violent eruptions culminating in a massive event on June 15. The paroxysmal event was the largest volcanic eruption ever detected on planet Earth with satellite data. On a geological basis this eruption was one of the largest volcanic events of the century—smaller than Novarupta/Katmai (Alaska, USA, 1912), similar to Santa María (Guatemala, 1902) and Cerro Azul/Quizapu (Chile, 1932), and larger than those of El Chichón (Mexico, 1982) and Mount St. Helens (Washington, USA, 1980) (Smithsonian Institution, 1991).

The June eruptions of Mount Pinatubo were monitored by the National Oceanic and Atmospheric Administration's (NOAA's) polar orbiting satellites (NOAA–10, –11, and –12) and the Japanese Geostationary Meteorological Satellite (GMS). Realtime access to GMS imagery provided hourly visible and infrared imagery at approximately 8-km resolution and provided 3-h infrared imagery at about 12-km resolution. Near-realtime and delayed access to imagery from the NOAA spacecraft provided 1.1-km resolution tape recorded coverage over preselected regions and daily coverage at 4.0-km resolution in four or five multispectral bands (two visible and two or three infrared bands; table 1). A priority request for the higher resolution tape-recorded coverage over the Philippines was made by National Environmental Satellite, Data, and Information Service

(NESDIS) investigators days before significant eruptions began.

METHODS FOR ESTIMATION OF MAXIMUM PLUME HEIGHT

Physical scientists around the world have tried a number of approaches to estimate the maximum plume height on the basis of satellite imagery. Four basic techniques are available:

Wind Correlation.—The current methodology used at NESDIS is to correlate ash drift vectors (away from the immediate vicinity of the rising column) with observed atmospheric wind patterns. In general, the ash moves downwind in a direction and at a rate closely matching the prevailing wind. Due to vertical wind shear, the direction and (or) speed of the wind are usually unique at each altitude, so tracking of the plume in successive satellite images allows estimation of plume altitude. This technique is best obtained from nearly continuous geostationary imagery and has proven highly successful since 1990 at NESDIS.

Temperature correlation.—The traditional approach has been to relate plume-top temperatures derived from infrared imagery to environmental temperatures measured by rawinsondes. Simplified parcel theory (that is, assuming 100 percent entrainment of the ambient atmosphere) is not applicable to ascending hot ash plumes. Simplified approaches fail for large eruptions penetrating into the stratosphere because, above the tropopause, ambient temperature increases with height. Additional problems are introduced by increased transmissivity through thin or

Table 1. Wavelengths measured from the Advanced Very High Resolution Radiometer aboard NOAA's polar orbiting satellites

Channels	Wavelengths (μm)	Band
1..........	0.58–0.68	Visible.
2..........	0.725–1.10	Near infrared.
3..........	3.55–3.93	Infrared.
4..........	10.30–11.30	Infrared.
5..........	11.50–12.50	Infrared (not available on NOAA–10).

[1]National Oceanic and Atmospheric Administration, National Environmental Satellite, Data, and Information Service, Washington, D.C.

thinning ash plumes. Woods and Self (1992) described some of the problems associated with using temperatures to estimate column height. In NESDIS operations, temperature correlations are used only as a last resort in the case of weak vertical wind shear.

Stereoscopy.—A technique available when two or more satellites are simultaneously viewing an event is stereoscopy. Geometric and trigonometric calculations can precisely determine column height by using (1) the precise location of the plume and satellites and (2) topographical data. The National Aeronautics and Space Administration (NASA) pioneered this technique in the research environment by monitoring severe thunderstorms with two U.S. geostationary satellites. It is possible that similar approaches can be used on satellites with dual forward-viewing and downward-viewing imagers.

Shadow measurements.—A technique under consideration, but not in operational use, is shadow analysis. Geometric and trigonometric calculations can determine column height precisely by using (1) the precise location of the plume, sun, and satellite; (2) the length of the shadows; and (3) topographical data. Complications arise when shadows fall on a lower cloud layer or on a dark surface such as the ocean. The technique is only useful during daytime hours and with solar zenith angles large enough to produce shadows. NOAA has experimented with this technique for estimating the altitude of thunderstorm canopies.

The heights used in this paper are all derived from wind correlation.

SEQUENCE OF ERUPTIVE EVENTS

Mount Pinatubo began a series of minor ash eruptions on June 9, which became violent on June 12. After a 28-h respite, frequent brief eruptions began on June 14 and culminated with the massive eruptions on June 15. From June 12 to 16, a total of 19 discrete eruptions sent ash and gas into the stratosphere (table 2 and fig. 1).

From June 9 until the paroxysmal event on June 15, each eruption lasted less than 3 h, and extended over relatively small areas (fig. 2). During daylight hours, visible satellite imagery suggested that most of the ash plumes consisted of steam and "light colored" particulates. Beginning with the eruption of 0553 on June 15, however, the ash particulates became very "dark colored," possibly as a result of the injection of larger amounts of particulate matter (fig. 3). (Note: all times are given as Philippine time, G.m.t. + 8 h.)

Satellite imagery suggests that the cataclysmic stage of eruptions began about 1027 on June 15 and lasted for over 21 hours. During this period, a stratospheric ash canopy spread across a 2,700,000-km² area (fig. 4). Surface observations (eyewitness reports and instrument records—seismic, infrasonic, and barographs) suggest that five separate eruptive events occurred between 1031 and approximately 1315 and became continuous at approximately 1342.

The paroxysmal eruption(s) appeared to have three distinct phases in the satellite imagery:

Phase 1.—The first phase, by far the most violent phase from a satellite perspective, appeared continuously in infrared imagery from 1031 until 2231 on June 15 (fig. 5). An extensive ash shield achieved heights of 25 to 30 km. In

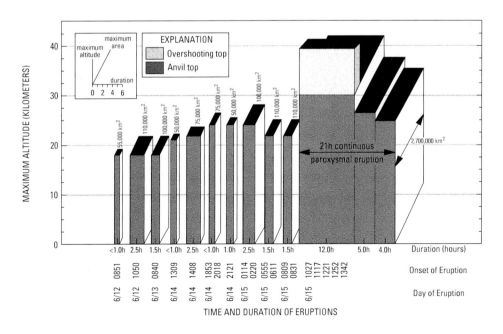

Figure 1. The duration, vertical extent, and areal coverage of the 19 eruptions of Mount Pinatubo from June 9 to 17, 1991, as detected from satellite imagery (see table 2). The term "overshooting top" refers to the central ash column in the immediate vicinity of the volcano that punches high into the stratosphere before sinking back to the level of neutral buoyancy, where it then spreads laterally.

Table 2. Preliminary summary of 1991 eruptions of Mount Pinatubo.

Date	Time of eruption[1]	Time of detection[2]	Duration[3] (h)	Maximum plume height[4] (km)	Direction of movement	Maximum horizontal extent (km[2]) within 48 h of eruption[5]
Minor Explosive Eruptions						
June 9	–	0831	<1.0	~2	NW.	10,000 within 2 h.
June 11	–	1631	<1.0	~3	WSW.	10,000 within 2 h.
Major Explosive Eruptions						
June 12	0850	0931	<1.0	17–19	WSW	55,000 within 8 h.
June 12	2251	1131	2.5	17–19	WSW	110,000 within 8 h.
June 13	0840	0931	1.5	17–19	WSW.	100,000 within 6 h.
June 14	1307	1331	<1.0	20–22	WSW.	50,000 within 4 h.
June 14	1410	1431	2.5	20–22	WSW.	75,000 within 5 h.
June 14	1851	1931	<1.0	23–25	WSW.	75,000 within 6 h.
June 14	2018	–	–	Indistinguishable from 1851 eruption		
June 14	2318	2331	1.0	23–25	WSW.	50,000 within 3 h.
June 15	0113	0131	2.5	23–25	WSW.	100,000 within 4 h.
June 15	0255	–	–	Indistinguishable from 0113 eruption		
June 15	0553	0631	1.5	20–22	WSW.	110,000 within 3 h.
June 15	0611	–	–	Indistinguishable from 0553 eruption		
June 15	0809	0831	1.5	20–22	WSW.	110,000 within 3 h.
June 15	0831	–	–	Indistinguishable from 0809 eruption		
Paroxysmal Eruption(s), Nearly Continuous[6]						
June 15	1027	1031	21.0	25–40	WSW.	1,000,000 within 12 h.
June 15	1117	–	–	Indistinguishable from 1027 eruption		1,500,000 within 18 h.
June 15	1221	–	–	–		2,200,000 within 24 h.
June 15	1252	–	–	–		2,700,000 within 36 h.
June 15	1342	–	–	–		

[1]Times of the onset of eruptions are from Sabit (this volume) and Hoblitt, Wolfe, and others (this volume). Eruptions reported by Sabit (this volume) at 1455–1630 on June 9 and 0515–0540 on June 11 were not seen in satellite images; plumes that were seen in satellite images on those dates were probably from continuous, low level ash emission.
[2]Satellite observations of eruptive episodes were derived from visible and infrared satellite imagery.
[3]Duration of event (to nearest 0.5 h) was derived from visible and infrared satellite imagery.
[4]Plume heights are from PHIVOLCS, USGS, and NOAA/NESDIS (satellite estimates were based on plume motion, correlated with nearby rawinsonde measurements).
[5]Maximum horizontal extent of "opaque" or "semiopaque" plume determined from infrared satellite imagery.
[6]Paroxysmal eruption(s) defined here as the period of a nearly continuous, high ash canopy as seen in GMS and polar orbiting satellite imagery. Surface observations (eyewitness reports and instrument records) indicate discrete eruptions from 1027 through 1315 followed by continuous, strong eruption that began at approximately 1342.

the immediate vicinity of the volcano, the central ash column reached 35 to 40 km; this "overshooting top" was fixed over the volcano for the entire 12-h period. The ash from this phase accounted for over 95 percent of the plume's 2,700,000-km[2] areal coverage.

Phase 2.—The second phase of the eruption was visible as a "ball shaped" plume fixed over the volcano. This phase was seen from 2231 on June 15 until 0331 on June 16. The plume was approximately 26 to 28 km in height.

Phase 3.—The third phase of the eruption, apparent from 0331 through 0731 on June 16, was noted by a "wedge shaped" plume fixed on the volcano (fig. 6). During this phase, ash reached heights of 23 to 25 km.

Large amounts of ash fell across most the island of Luzon and resulted in the closure of Clark Air Base, Cubi

Point Naval Air Station, and Manila International Airport. Significant ashfall was reported across the South China Sea and Indochina. Some ash was reported as far away as Thailand.

Following the June 15 eruptions, the Philippine Institute of Volcanology and Seismology (PHIVOLCS) and the U.S. Geological Survey (USGS) reported nearly continuous minor eruptions (10 to 20 eruptions per day) producing ash to 4 to 5 km in altitude. Most of these events were not detected with infrared satellite imagery, though an occasional plume did appear in the visible imagery. Numerous "secondary explosions" were caused by rain mixing into hot ash and pyroclastic deposits. Occasionally, the eruptions would reach altitudes of 10 to 15 km.

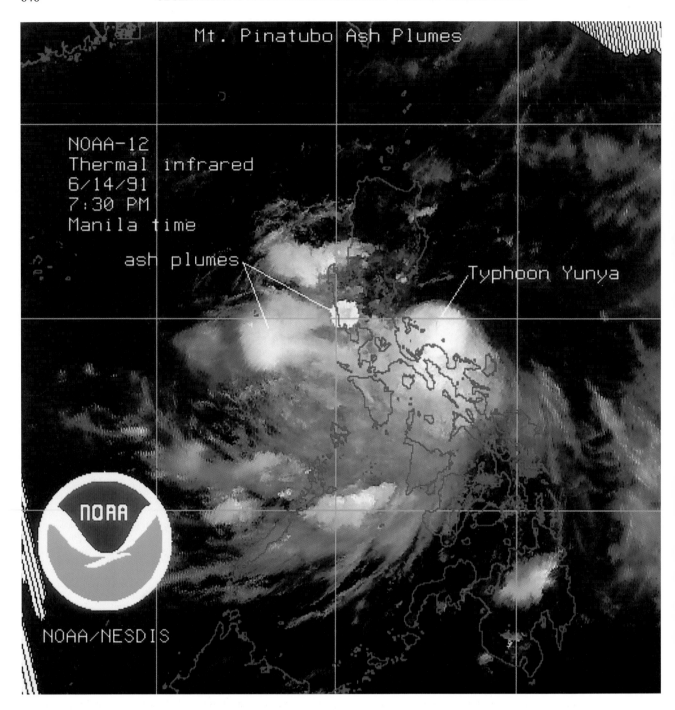

Figure 2. NOAA–12 thermal-infrared image taken at 1930 on June 14, 1991, showing three distinct ash plumes. One (the brightest), directly over the volcano, is associated with the 1851 eruption (table 2). Two plumes over the South China Sea west of Luzon are associated with the 1307 and 1410 eruptions. Typhoon Yunya is shown approaching southeastern Luzon. This image was processed from 4-km resolution data. Temperatures derived from the brightest, or coldest, part of the ash plume are around −80°C, and portions of the ash cloud tracked west-southwestward at 30 m/s, corresponding to an altitude of 20 to 25 km.

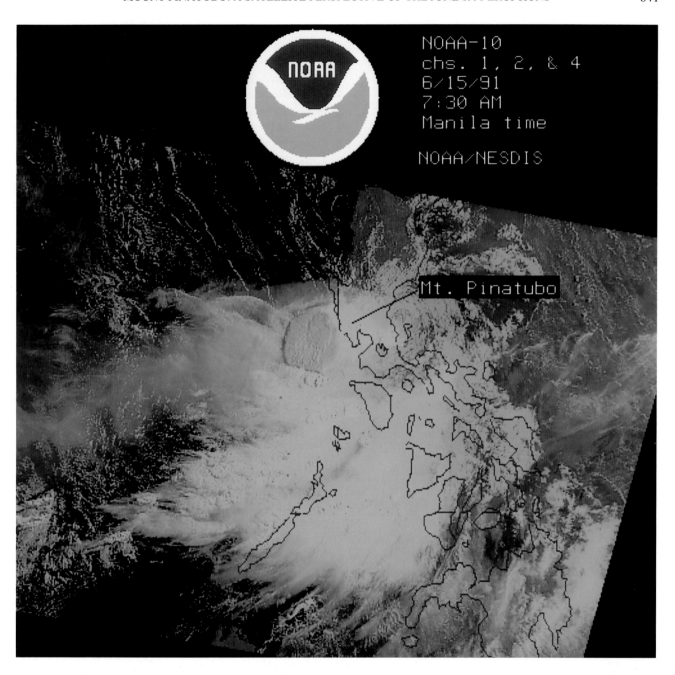

Figure 3. NOAA–10 false-color multispectral (visible and infrared) image, processed from 1.1-km resolution data, taken at 0730 on June 15, 1991. The darker, circular area near the volcano shows darker particulate matter associated with the 0553 and 0611 eruptions in table 2. The more diffuse plume over the South China Sea west of Luzon is ash from earlier eruptions.

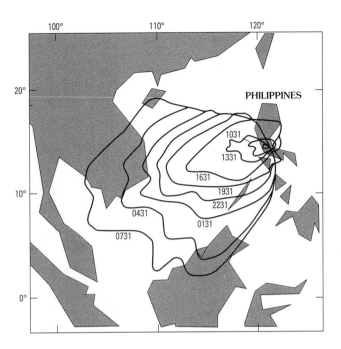

Figure 4. Rate of expansion of the Mount Pinatubo ash plume following the paroxysmal eruption of June 15–16, 1991. The outer edge of the ash cloud was derived in realtime from 12-km resolution GMS infrared imagery available at 3-h intervals. Gray areas are landmasses. A similar figure has been published by Tokuno (1991).

ENCOUNTER WITH THE TYPHOON

Typhoon Yunya approached eastern Luzon on June 14 with sustained winds of 45 m/s (90 knots) and made its closest approach to Mount Pinatubo as a tropical storm around 1400 on June. Yunya's center (or eye) passed within 75 km of Mount Pinatubo during the first phase of the paroxysmal eruption. Many of the rain bands affected the entire island of Luzon, including the volcano, throughout the entire cataclysmic event(s). Upon entering the South China Sea, the typhoon rapidly dissipated. Other than to state that tropical cyclones typically regenerate over the

South China Sea, this paper will not discuss or speculate on the causes for Yunya's demise.

Torrential rains mixed with pyroclastic flows and ash deposits and caused numerous lahars. Buildings and homes collapsed under the terrific weight of rain-soaked ash. Most of the fatalities, injuries, and property damage that occurred on June 15 and 16 throughout Luzon were the result of collapsed buildings (C.B. Bautista and others, this volume; Spence and others, this volume) and lahars.

NOAA RESPONSE

NOAA and the Federal Aviation Administration (FAA) established a cooperative effort in 1988 to detect and monitor explosive volcanic eruptions. The primary focus of this effort has been to warn air traffic managers and en-route commercial aircraft of volcanic hazards in the Flight Information Regions assigned to the United States. However, provisions have been made to relay information for explosive eruptions throughout the world.

The NOAA-FAA Volcanic Hazards Alert Program was activated for Mount Pinatubo at the request of the U.S. Department of Defense and USGS. A total of 47 volcanic hazards alerts, consisting of realtime satellite analyses, trajectory forecasts, and other ancillary information, were issued during the June eruptions of Mount Pinatubo.

Immediately following the paroxysmal eruption on June 15, due to the magnitude of the eruption and expected multiyear effects, NOAA activated teams to track aerosol movement, changes in atmospheric chemistry (particularly ozone depletion), and climate response.

The aerosol cloud was tracked around the world by use of the NOAA polar orbiting satellites and both European and NOAA geostationary satellites (METEOSAT and GOES–7). The plume took 3 weeks to completely encircle the planet between 30°N. and 20°S. lat (fig. 7). Subsequent to this, the aerosol cloud was tracked and monitored by using the operational Aerosol Optical Thickness Product, which is derived from AVHRR data from the NOAA polar orbiting satellites.

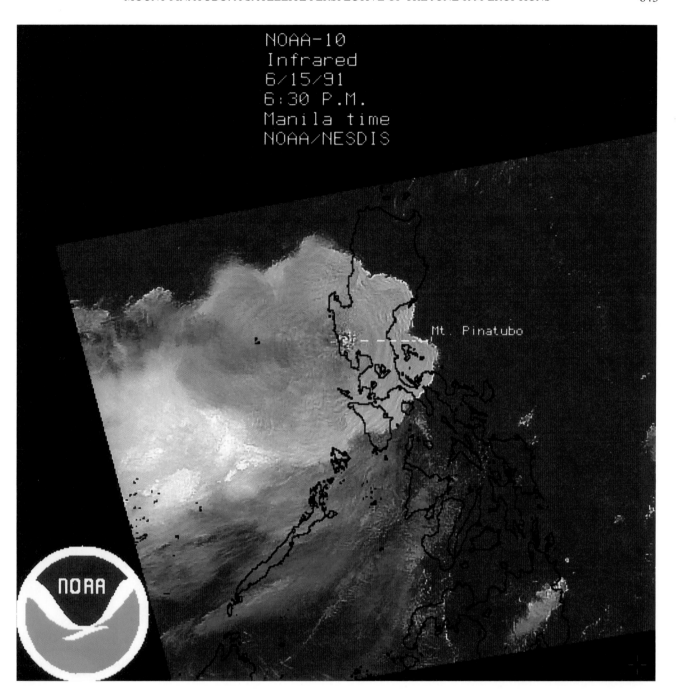

Figure 5. NOAA–10 false-color multispectral (infrared) image processed from 1.1-km data (reduced to 2.2-km) taken at 1830 on June 15. The extensive canopy of ash over Luzon and the South China Sea was produced after nearly 5 h of continuous eruption during the paroxysmal event. The total ash plume continued to grow for a total of 18 h. Here, the "overshooting top" can be seen over the volcano. Even though prevailing winds are from the northeast, the force of the eruption was great enough that some ash was forced upwind. The concentric rings around the central core appear to be gravity waves propagating outward. The majority of the plume spread west-southwestward at nearly 35 to 45 m/s and plume-top temperatures were as low as −88°C; the winds correlated with rawinsonde data to heights of 25 to 30 km, and the temperatures were colder than any environmental observations in either the troposphere or stratosphere. The slightly darker portion of the image extending westward from the volcano tracked westward at approximately 50 to 60 m/s and plume-top temperatures were near −60°C; the winds correlated with rawinsonde data to heights of 35 to 40 km.

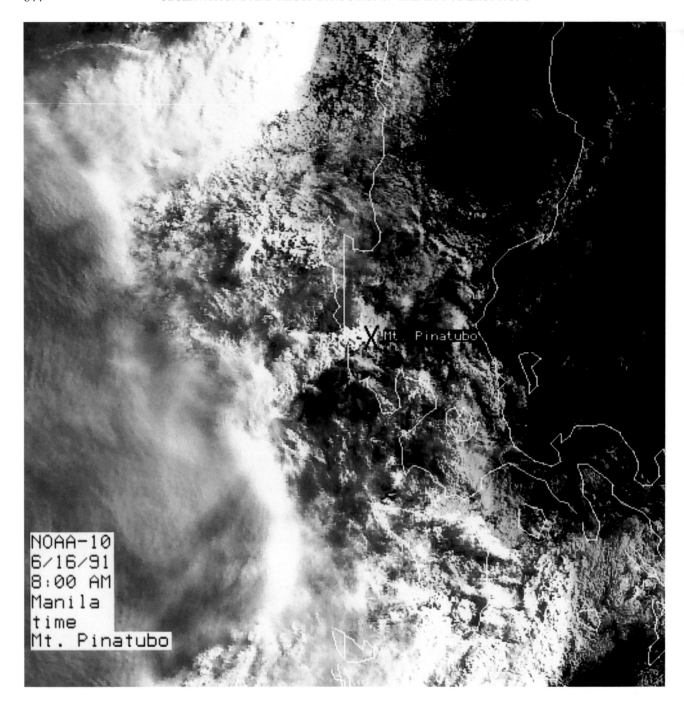

Figure 6. NOAA–10 false-color multispectral (visible and infrared) composite image processed from 1.1-km data at 0800 on June 16. A wedge-shaped ash cloud, locked over the volcano, is depicted during the third phase of the paroxysmal eruption. Material can be seen rising from the volcano and spreading to form the blue-gray plume and ash cloud.

REFERENCES CITED

Bautista, C.B., this volume, The Mount Pinatubo disaster and the people of central Luzon.

Hoblitt, R.P., Wolfe, E.W., Scott, W.E., Couchman, M.R., Pallister, J.S., and Javier, D., this volume, The preclimactic eruptions of Mount Pinatubo, June 1991.

Sabit, J.P., Pigtain, R.C., and de la Cruz, E.G., this volume, The west-side story: Observations of the 1991 Mount Pinatubo eruptions from the west.

Smithsonian Institution , 1991, Pinatubo: Washington D.C., Smithsonian Institution, Bulletin of the Global Volcanism Network, v. 16 (June 30, 1991), p. 2–5.

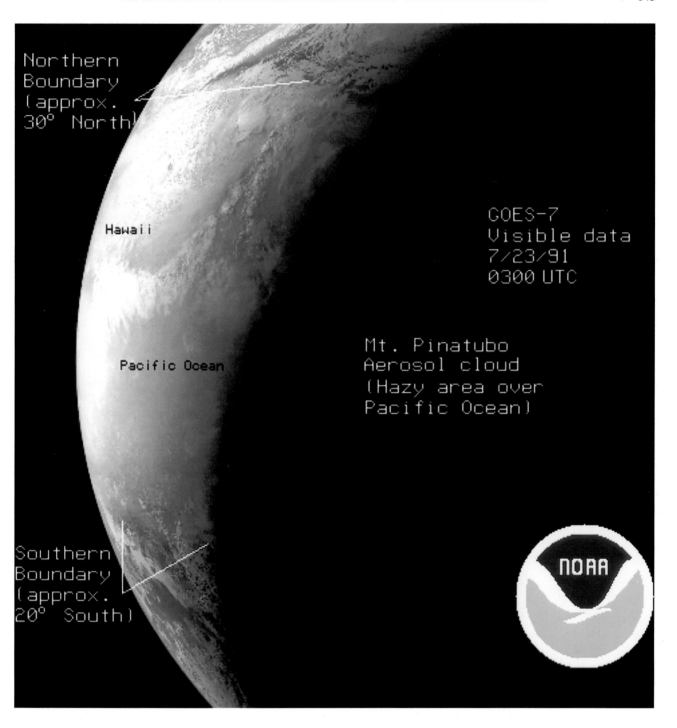

Figure 7.   GOES-7 visible image of the Pacific Ocean on June 23, 1991, at 0300 UTC (1100 Philippine time). The hazy area in a band from 20° S. lat to 30° N. lat is caused by stratospheric aerosols from the Pinatubo eruption(s).

Spence, R.J.S., Pomonis, A., Baxter P.J., Coburn, A.W., White M., and Dayrit, M., this volume, Building damage caused by the Mount Pinatubo eruption of June 14–15, 1991.

Tokuno, M., 1991, GMS–4 Observations of Volcanic Eruption Clouds from Mt. Pinatubo, Philippines: Meteorological

Satellite Center Technical Note Number 23, Japan Meteorological Agency, p. 1–23.

Woods, A.W., and Self, S., 1992, Thermal disequilibrium at the top of volcanic clouds and its effects of column height, Nature, v. 355, p. 628–630.

Growth of a Postclimactic Lava Dome at Mount Pinatubo, July–October 1992

By Arturo S. Daag,[1] Michael T. Dolan,[2] Eduardo P. Laguerta,[1] Gregory P. Meeker,[3] Christopher G. Newhall,[3] John S. Pallister,[3] and Renato U. Solidum[1]

ABSTRACT

An andesitic lava dome grew on the floor of the new caldera of Mount Pinatubo between July and October of 1992. Growth was continuous, with rates of about 8×10^4 cubic meters per day for the first 10 days and about 3×10^4 cubic meters per day thereafter until the end of October. Activity during 1993 was limited to phreatic explosions on and around the dome.

The 1992 dome is similar in composition and texture to another dome that formed in the same area during the week preceding the climactic June 15, 1991, dacitic eruption. Samples of both domes contain disequilibrium mineral assemblages and quenched mafic inclusions. These textural features, as well as chemical compositions of minerals and whole rocks, indicate that both were produced by mixing of basalt and dacite. Least squares modeling of major element compositions and differences in trace element abundances of 1991 and 1992 samples suggest that the 1992 andesite represents a new episode of magma mixing. Breakdown of cummingtonite in a sample of the 1992 dome indicates that more time elapsed between mixing and eruption in 1992 than in 1991.

In 1991, mixing of basalt and dacite and ascent of mixed andesite was followed within a week by major explosive eruptions of dacitic magma. In contrast, ascent and eruption of andesite in 1992 did not trigger explosive eruptions of dacitic magma, possibly because volatiles in the dacitic reservoir were depleted by the 1991 eruptions.

INTRODUCTION

An andesitic lava dome was extruded onto the north flank of Mount Pinatubo between June 7 and June 12, 1991, preceding the large explosive eruptions of andesite and then dacite on June 12–15. This dome was destroyed during the climactic eruption of June 15, 1991. Minor dacitic eruptions continued until early September of 1991, after which the volcano entered a period of temporary repose. Geologic evidence for dome growth late in the latest prehistoric eruptive period (Buag eruptive period; Newhall and others, this volume) suggested that further dome growth was possible.

On July 3–6, 1992, seismicity signaled renewed magma ascent, the onset of a new period of dome growth, and, potentially, more explosive eruptions. Although the topography and presumably the internal structure of the summit area were strongly modified by caldera collapse on June 15, 1991, the 1992 dome grew in the same geographic position as the 1991 dome. In addition to being erupted in the same place, available samples and observations indicate that the 1992 dome is compositionally similar to its 1991 counterpart. These similarities pose an important question: Does the 1992 dome represent a new magma mixing episode, and, if so, why has it not triggered another explosive eruption? This paper describes the growth and composition of the 1992 lava dome and attempts to answer this question.

METHODOLOGY

Difficult access and safety considerations prevented detailed measurements and detailed sampling of the 1992 dome. The following discussion is based on aerial oblique photographs and video (high and low angle) and one brief visit to the dome for sampling. Dimensions of the dome are estimated from photographs; the caldera and caldera lake were used for scale. They are approximate, as are calculated volumes. Judging from repeat trials in volume estimation by different coauthors, we believe the precision on volume estimates is ±30 percent. The samples described in this report were obtained from the dome during a brief visit by helicopter on December 4, 1992.

Seismic records are principally from station FNGZ, 9 km southeast of the dome, and are supplemented after August 9 by records from station CRAZ, newly installed on the north caldera rim, 1 km north of the dome (see Mori,

[1] Philippine Institute of Volcanology and Seismology.

[2] Department of Geology and Geological Engineering, Michigan Technological University, Houghton, MI 49931 USA.

[3] U.S. Geological Survey.

Figure 1. Small tuff cone being formed on July 9, 1992. View is from the northwest.

White, and others, this volume; Ramos and others, this volume). SO$_2$ gas monitoring was by an airborne COSPEC (Correlation Spectrometer; Barringer Instruments) (Daag, Tubianosa, and others, this volume).

CHRONOLOGY OF 1992 DOME GROWTH

May to June 1992: Sporadic low-frequency harmonic tremor (average frequency, 1 Hz).

July 3: The first low-frequency volcanic earthquake during this period of seismic unrest.

July 6: Sudden increase in the number of small, shallow, low-frequency volcanic earthquakes and in low-level, low-frequency tremor, suggesting possible magma ascent. A low-frequency appearance raised immediate concern about the potential for explosive activity as well as dome growth.

July 7: Aerial reconnaissance showed a brown, turbulent plume of sediment in the caldera lake; the plume was elongate in a west-southwest to east-northeast direction. The sediment plume suggested either heating and turbulent convection of lake water or sublacustrine mud explosions.

July 9: Explosions from beneath the shallow caldera lake (possibly beginning on July 8) built a small, roughly 50-m-diameter tuff cone above lake level by 0830 on July 9 (Lieutenant Scheidegger and Major Caldwell, U.S. Marine Corps, oral commun., 1992). By the time we observed the island at 1400 on the same day (fig. 1), it was circular, approximately 70 to 100 m in diameter, and 5 m high. Small phreatic explosions from a 10- to 20-m-wide vent on the north-northwest side of the cone were shooting fountains of steam and mud 50 to 200 m into the air at intervals of sev-

eral minutes. Steaming from fumaroles around the south shore of the caldera lake was notably less than it had been before July. Nighttime observations by Captain C. Owens (U.S. Marine Corps, oral commun., 1992) from 1900 to 2100 showed that explosions were continuing. The products were not hot enough to appear bright when viewed through night-vision goggles.

July 10: The tuff ring had grown to 100 to 150 m in diameter and 5 to 10 m high. The size and frequency of explosions might have decreased; only one of two observation flights saw explosions, which were smaller than those seen on July 9. Video footage from that flight showed interbedded light and dark layers in the tuff cone, possibly representing an alternation of juvenile and accidental material. Steaming along the south shore of the caldera lake that had noticeably diminished on July 9 returned to precrisis (pre-July) levels. Seismicity was limited to a few high-frequency volcanic earthquakes.

July 11–13: Observations were interrupted by passage of Typhoon Eli (local name, Konsing). Occasional low-frequency harmonic tremor.

July 14: A new dome was sighted during a flight at 1000; seismicity indicates that it may have begun to grow late on July 13 (Ramos and others, this volume). A steep-sided jumble of jagged, light-brown, ash-dusted blocks occupied what had been a water-filled vent for phreatic explosions and was steaming profusely. When first sighted, the dome was about 50 m in diameter and 5 to 10 m high and had a volume of between 10^4 and 10^5 m^3 (table 1).

July 15: Jagged spines developed on the dome, and portions of the uplifted tuff cone slumped into the lake (fig. 2).

Figure 2. Early dome growth, July 15, 1992. Dome had been growing for 1 or 2 days and was shouldering aside (and burying) unconsolidated debris of the tuff cone. View is from the north.

Table 1. Growth of the 1992 lava dome.

[Dimensions in meters; volume in cubic meters. Estimation errors are ±30%]

Date	North-South	East-West	Height	Volume
July 14	50	50	5–10	10^4–10^5
July 16	100	100	30	0.2×10^6
July 24	200	200	50	0.8×10^6
August 5	200	200	50	1×10^6
August 26	300	300	50–100	$<2\times10^6$
September 2	300	350	50–100	2×10^6
September 22–26	300	400	70–100	$>2\times10^6$
September 28	300	400	100	3×10^6
October 28–30	350	450	150(?)	4×10^6

July 16: Two days after it was first sighted, the dome had roughly doubled in diameter (now about 100 m), grown to a height of about 30 m, and occupied about 60 percent of the 150-m-diameter island. Steep sides and jagged spines suggested a dacitic composition, though later sampling showed only andesite.

July 23: The tuff cone was now completely covered (or shed) by the new dome, and a talus slope was developing on the north side of the dome.

July 24: The dome had grown to 200 m in diameter and 50 m high and had two 25-m-wide lobes of lava on its west side.

July 28: The dome had become structurally complex, with numerous fractures and fumaroles. The most profuse steaming was along a north-south fracture cutting at least the northern half of the dome (fig. 3). Blocks of talus several meters in diameter lay on the lake floor, partly submerged, immediately east, north, and southwest of the dome.

July 30: The dome appeared to have flattened, due to lateral spreading.

August 5–9: The dome appeared to have a central vent, which was later seen to be a spreading center with a north-south axis. Smooth faces of fresh, viscous lava formed where lava had risen and peeled apart along a north-south fissure and had not yet been broken by endogenous growth. The dome as a whole was also slightly elongated north to south, with an average diameter of about 200 m and a height of about 50 m. On August 5, the amplitude of low-frequency volcanic earthquakes began to increase; earthquake traces were emergent, with indistinct P-wave arrivals at FNGZ. A new seismic station (CRAZ) was installed on the north caldera rim on August 9 to get a closer look at this seismicity.

August 10–11: Uplift of the north side of the dome and adjacent lake bottom, probably a result of sublacustrine endogenous dome growth, had created a 50-m-wide "beach" on which several previously submerged "marker blocks" were recognized (fig. 4). The northern half of the dome was higher than the southern half.

August 12 and subsequent days: Intense rain raised the lake level and submerged the "beach" on the north side of the island, as well as other shorelines of the caldera lake.

August 15: A new family of shallow, low-frequency volcanic earthquakes appeared, with a dominant frequency of 1.25 Hz. P-wave arrivals were emergent but recognizable.

August 24: Another family of low-frequency volcanic earthquakes appeared, with a dominant frequency of 2 Hz. P-wave arrivals were emergent but recognizable.

Figure 3. Expanded dome, July 28, 1992, as seen from the north. Note individual boulders lying partially submerged in shallow water around the dome.

Figure 4. Dome on August 10, 1992, with an expanding northern lobe, including a small beach, and what appears to be a crater-like feature above the original vent. View is from the northeast.

August 26: Growth had shifted to a new, eastern lobe. Uplift on the east side of the dome had created a new, 50-m-wide eastern "beach." The dome continued to grow and by now had 3 peaks (north, southwest, and east) and a small topographic low (crater?) within the southwestern high. North-south and east-west dimensions were both about 300 m; the maximum height was 50 to 100 m.

August 27–September 2: Heavy rains caused the lake level to rise. However, continued uplift of the eastern lobe of the dome kept the eastern "beach" above lake level. On or before August 29, westward aggradation of an apron of

debris that had washed from the eastern wall of the caldera, combined with uplift of the dome itself, joined the island to the eastern lake shore. On August 30, a new family of low-frequency volcanic earthquakes began, and these were of larger amplitude than any earlier in the 1992 unrest. Dominant frequency was 1.25 Hz.

September 2: The eastern lobe continued to grow, increasing the east-west diameter to about 350 m and the volume to about 2×10^6 m^3. The eastern shore of the lake also continued to aggrade, so the eastern lobe of the dome was surrounded by a debris apron, which sloped gently toward the lake.

September 13: The appearance of yet another family of low-frequency volcanic earthquakes of small magnitude was recorded only at the CRAZ station.

September 18: The eastern lobe was areally smaller but steeper and taller than the northern lobe. Most steaming was from vents in the southwestern area.

September 22–26: The dome continued to grow slowly and now measured about 300 m north-south (unchanged) and about 400 m east-west (increased). The highest point on the dome was on the southeastern lobe, about 70 to 100 m above lake level; two high areas on the northern lobe were about 50 m above the lake.

September 28: Horizontal dimensions of the dome were unchanged: the height appeared to have increased slightly.

October 1–10: No significant changes were noted.

October 15: Renewed uplift of the northern lobe caused several huge blocks to spall off and tumble to the north foot of that lobe. Uplift accompanied an increase in small earthquakes, recorded at CRAZ.

October 23: Seismicity increased sharply; low-frequency volcanic earthquakes were occurring at a rate of 1 per minute (see fig. 8).

October 28–30: A fresh spine of dark lava protruded 30 to 50 m above the north-central lobe. The south-central part of the dome also expanded (fig. 5). Total volume of dome was now about 4×10^6 m^3.

October 31: Sharp decrease in seismicity, and the apparent end of 1992 dome growth.

February 1993: Shallow seismicity occurred beneath the caldera, and numerous phreatic explosion pits formed in the alluvial apron north, east, and south of the dome. The greatest concentration of explosion pits was near the southeastern corner of the caldera floor, about 400 m north of a long-standing line of fumaroles that apparently lies along a regional fault. No change was noticed in the dome itself.

March–July 1993: Episodic tremor took place, possibly associated with gas venting or hydrothermal explosions from the dome or the caldera floor. The dome itself appears not to have grown since October 1992 (J. Sincioco, PHIVOLCS, oral commun., 1993).

SIZE, SHAPE, AND VOLUME THROUGH TIME

Changes in the size, shape, and volume of the dome that took place between July and October of 1992 are summarized in figures 6A–F and 7. The most notable feature is that the locus of growth shifted several times, ultimately forming two high lobes (north and east) and a third south-central lobe that rises above the original vent. Uplifting of "beaches," first on the north and later on the east, suggests

that the north and east extrusions were not simply lateral breakouts from the central lobe but, rather, a sequel to intrusions and endogenous growth beneath those northern and eastern lobes.

The dome grew continuously at roughly 0.1×10^6 m^3/d for the first few days, 0.08×10^6 m^3 for the next week, and then $0.03–0.04 \times 10^6$ m^3/d thereafter until growth abruptly ended in late October (fig. 7). Similar rates of dome growth were observed at Santiaguito dome in Guatemala following the major plinian eruption of its parent volcano, Santa María, in 1902 (Rose, 1972, 1973), at Mount St. Helens, U.S.A., following its plinian eruption in 1980 (Swanson and Holcomb, 1989), at Bezymianny, Kamchatka Peninsula, following its major eruption in 1956 (Kirsanov, 1979), and at Unzen Volcano, Japan, in 1991 and early 1992 (Nakada, 1992b). However, all four of these other domes grew for a longer time than the Pinatubo dome, so they are much larger (Santiaguito, 1×10^9 m^3 in 70 years of growth, Rose, 1987; Mount St. Helens, 7.4×10^7 m^3 in 6 years of growth, Swanson and Holcomb, 1989; Bezymianny, 1.6×10^9 m^3 in the first 27 years of growth, Kirsanov, 1979; Unzen, 8.2×10^7 m^3 of lava extrusion in the first 9 months of growth, of which about half later collapsed and produced pyroclastic flows, Nakada, 1992b). Dome growth at Mount Lamington, Papua New Guinea, was even faster during a 1.4-yr period after its major eruption of 1951 (1.0×10^9 m^3 in 1.4 years of growth, Taylor, 1958; Swanson and others, 1987).

SURFACE FEATURES

Much growth of the 1992 dome was exogenous, and surface features reflected extrusion of viscous lava that spread bilaterally from its vent on top of the dome. Known as "spreading centers" (Swanson and others, 1987), "crease structures" (Anderson and Fink, 1992), "cleavage canals" (Omori, 1916), and "petal structures" (Nakada, 1992a,b), such structures have conspicuously smooth, nonscoriaceous surfaces where lava peeled apart as it spread.

Outside the spreading centers, the dome's surface was initially scoriaceous and showed evidence of individual lava-flow lobes. Later endogenous growth created blocky, fracture-crossed surfaces surrounded by talus. Small spines, each roughly 10 m high, formed from time to time along the edges of spreading centers, only to collapse during continued spreading or during the next period of endogenous growth. The tallest and longest lived spine rose about 50 m above the rest of the northern lobe; it was formed partly by extrusion during October 1992 and later accentuated by collapse of surrounding material (fig. 5).

Figure 5. Dome on November 21, 1992, essentially unchanged since late October. The spine on the northern lobe formed partly by extrusion during October 1992; it was later accentuated by collapse of surrounding material. The area southwest of the central vent has been built up slightly, and the eastern lobe is also crumbling, leaving a fragile core. Samples described in this report were collected along the upper right (north and northeast) sides of the dome. View is from the southeast.

DOME GROWTH, SEISMICITY, AND SO₂ EMISSION

During and slightly after the period of fastest dome growth (until middle August 1992), SO$_2$ discharge averaged 580±360 t/d (Daag, Tubianosa, and others, this volume). In contrast, the few measurements for late August-early November, when dome growth was slower, averaged 270±40 t/d (same reference).

Most of the earthquakes before and during dome growth appeared to be low-frequency events and are shown as such on figure 8. However, close inspection of CRAZ records shows that many had a high-frequency onset followed by a lower frequency coda. These may be high-frequency, brittle-fracture events at shallow depth, with direct arrivals followed by lower frequency, ground-filtered surface waves. Alternatively, both the high- and the low-frequency contents may have been source effects of

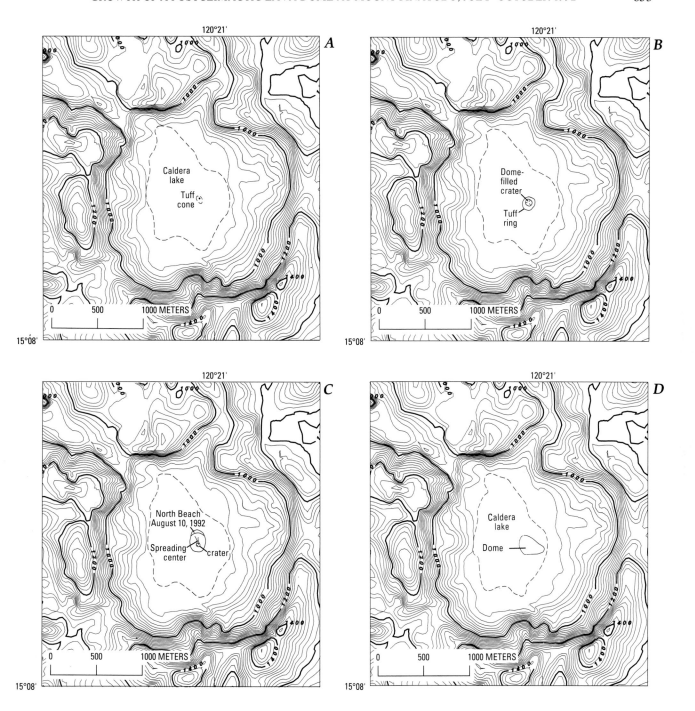

Figure 6. Sketches of the dome growth, July–October 1992. *A*, July 9, initial tuff cone. *B*, July 14, early dome set within tuff cone. *C*, Late July–early August, northward expansion of dome. *D*, Late August, eastward expansion of the dome and notable westward aggradation of the nearby shoreline, after heavy rains. *E*, Late September, minor additional growth of dome, but remarkable aggrada-tion of lakeshore, partly surrounding the dome. *F*, Renewed growth of the northern lobe, with some growth in the southwest portion also possible, followed by crumbling of outer portions, leaving a prominent spine on the northern lobe. Steep topography is the inner caldera wall; dashed line is the shoreline of the caldera lake. Topographic base from Jones and Newhall (this volume).

fracturing in the brittle carapace of a mushy dome. Tremor that occurred shortly before and shortly after the dome began to grow had a dominant frequency of 1 Hz at the FNGZ station but was associated with steam venting and

phreatic explosions. Again, path and source effects are hard to distinguish. Part of the low-frequency appearance may be a real source effect; part may be the result of attenuation of higher frequency components with travel distance, even to

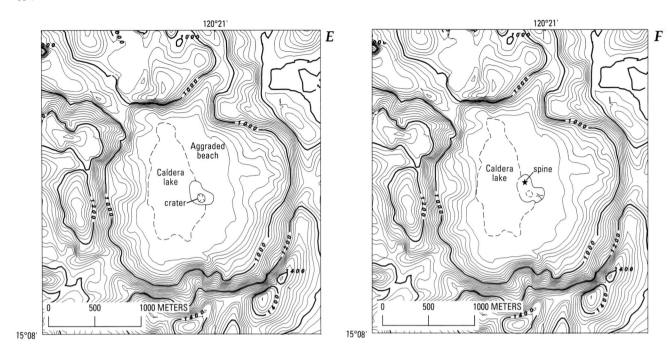

Figure 6.—Continued.

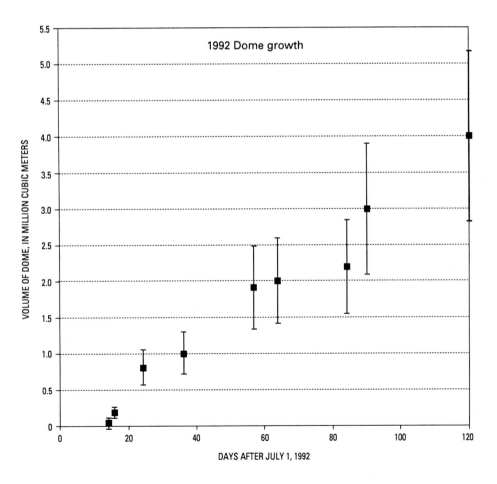

Figure 7. Cumulative dome volume, July–October 1992. An initial period of rapid growth (steep slope) was followed by slower growth until extrusion abruptly stopped in late October.

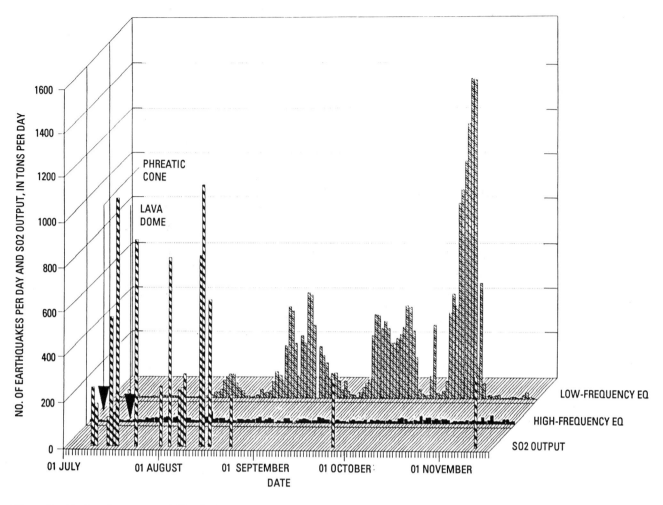

Figure 8.   Earthquake counts and SO$_2$ discharge versus time, through dome growth, July–October 1992.

the closest station, CRAZ, 1 km from the dome. Details of seismicity during dome growth are given by Ramos and others (this volume).

During the period of rapid dome growth and high SO$_2$ discharge (late July through early August), earthquake counts were relatively low; during the period of slow dome growth and lower SO$_2$ discharge, earthquake counts were relatively high (fig. 8). Although the correlation is imperfect in day-to-day detail, we think that the change from high SO$_2$ discharge, low seismicity, and rapid dome growth to low SO$_2$ discharge, high seismicity, and slow dome growth reflects progressive solidification of magma in the shallow part of the conduit and within the dome itself. Solidification may have resulted from early volatile loss, and solidification may also have trapped remaining volatiles. Thus, during the late stage of 1992 dome growth, gas-depleted, increasingly viscous magma had difficulty breaking new paths to the surface.

PETROLOGY AND CHEMISTRY

The 1992 dome consists mainly of andesitic lava that is host to numerous inclusions. The three samples described here represent the host andesite (CN12492–1h), abundant basaltic andesite inclusions (CN12492–1i), and sparse dacite inclusions (CN12492–2). The 1992 dome andesite and the mafic inclusion samples are similar in composition, mineral assemblage, and texture to andesite and mafic inclusions from the lava dome that grew in the head of the Maraunot River valley during the week preceding the climactic June 15, 1991, eruption (Pallister and others, this volume). The dacite inclusion sample is similar in composition to dacite erupted during the climactic eruption of June 15, 1991. However, subtle differences in composition and texture suggest that the 1992 dome represents a new mixing event rather than left over hybrid magma from 1991.

MIXED ANDESITE (CN12492–1h)

The dominant rock type of the 1992 dome is sparsely vesicular (10–20%) hornblende-plagioclase andesite with accessory olivine, augite, biotite, quartz, ilmenite, magnetite, and anhydrite (figs. 9A–D). This is a disequilibrium assemblage of minerals; the olivine and augite are too magnesian to be early liquidus phases, as they have magnesium numbers ($100*Mg/Mg+Fe^{+2}$) as high as 86 for olivine and 87 for augite (table 2). In addition, many of the minerals have reacted with the host melt, as described below. Plagioclase occurs as relatively large (1–3 mm) euhedral to subhedral oscillatory-zoned phenocrysts and microphenocrysts. Most phenocrysts contain thin bands that are rich in glass+vapor inclusions; however, some have highly resorbed core regions and are overgrown by clear oscillatory-zoned margins (fig. 9B). The average composition is about An_{37}, although core compositions and some zonal bands approach An_{60}. Hornblende also occurs in two populations: unreacted, oscillatory-zoned phenocrysts (0.5–2 mm) and microphenocrysts (<0.3 mm) that have relatively high TiO_2 (~2%, table 2; fig. 10) and large (1–3 mm), variably resorbed phenocrysts with lower TiO_2 and Al_2O_3 and with augite-rich rims. In contrast to the occurrence of hornblende crystals with cummingtonite rims in the 1991 andesite, unreacted cummingtonite is not seen in the 1992 samples. Instead, the 1992 andesite contains fibrous aggregates of hypersthene that rim low-TiO_2 hornblende. The occurrence of these rims on only the outer growth surfaces of hornblende phenocrysts and the similarity in thickness to cummingtonite rims in other Pinatubo rocks suggest that they are pseudomorphs after cummingtonite (fig. 9C). Accessory quartz is present as sparse large (~1 mm) grains that are resorbed (rounded or cuspate forms) and contain glass+vapor inclusions. Biotite is present both as cores to hornblende crystals and as isolated subhedral phenocrysts (hexagonal plates with variably resorbed margins). Augite forms oscillatory-zoned microphenocrysts (~0.2 mm). Olivine ($Fo_{84–86}$) is present both as glomerocrystic clots with augite and as isolated subhedral-anhedral phenocrysts. The olivine phenocrysts are bordered by aggregates of small (<0.1×0.5 mm) hornblende laths. Anhydrite is present as sparse but relatively large (~1 mm) and highly resorbed crystals. Apatite is also present both as large euhedral phenocrysts that occur with anhydrite and as elongate microphenocrysts in the groundmass and included in plagioclase. Both ilmenite and magnetite are present as small phenocrysts and microphenocrysts (<10 μm to 0.5 mm). All ilmenite crystals are bordered by magnetite rims (fig. 9D), and many grains display exsolution lamellae.

Like the 1991 andesite (Pallister and others, 1992; this volume), the 1992 andesite was produced by magma mixing. Evidence for mixing includes the wide range in phenocryst species and compositions, presence of minerals that do not normally coexist in equilibrium (for example, quartz

and olivine), and multiple populations of the same minerals (for example, resorbed and pristine types of both hornblende and plagioclase). Development of disequilibrium-reaction textures (hornblende borders on olivine, augite rims on hornblende, hypersthene pseudomorphs after cummingtonite, magnetite rims on ilmenite) is consistent with the suggested mixing.

BASALTIC-ANDESITE INCLUSION (CN12492–1i)

About 5 to 10 percent of the 1992 dome consists of dark, mostly subrounded inclusions of vesicular olivine-augite-hornblende basaltic andesite (fig. 9A). Our sample contains abundant (~40%) irregular and interconnected vesicles in a porous (diktytaxitic) groundmass texture. In contrast to the plagioclase-phyric texture of the host andesite, plagioclase is present only as microlites and as sparse xenocrysts with resorbed cores. The basaltic andesite contains phenocrysts of olivine (<0.5 to 2.0 mm), oscillatory- or sector-zoned augite (<0.5 mm), and oscillatory-zoned hornblende (fig. 9E). Most of the hornblende is similar in composition to the high-TiO_2 type found in the host andesite (fig. 10); the olivine and augite phenocrysts are magnesian and overlap in composition with those in the host andesite (table 2). Hornblende occurs as phenocrysts (to 2 mm) and grades in size to elongate microphenocrysts (<0.3

Figure 9. Photographs of 1992 samples. A, Hand sample of → 1992 andesite (at top, with large plagioclase phenocrysts) showing contact with basaltic andesite inclusion (dark, plagioclase-poor material at bottom of image). B, Photomicrograph (5.5 mm wide) of 1992 andesite showing olivine (OL) with hornblende rim, augite (AG), plagioclase (PL) with and without resorbed cores, and hornblende (HB) phenocrysts. C, Photomicrograph (1.3 mm wide) of hornblende (HB) phenocryst with hypersthene-rich reaction rim (CM) in 1992 andesite; hypersthene is probably pseudomorphous after cummingtonite. D, Photomicrograph (1.3 mm wide) of ilmenite (IL) with magnetite (MT) rim (center of image) in 1992 andesite. Rounded quartz (QZ), irregular vesicles (VES), and plagioclase (PL) also shown. E, Photomicrograph (5.5 mm wide) of 1992 basaltic andesite inclusion showing olivine (OL) with hornblende (HB) rim, augite (AG) phenocrysts and microphenocrysts, and lath to needle-shaped hornblende crystals in a glassy groundmass. F, Photomicrograph (0.34 mm wide) showing "hopper" shapes of plagioclase (PL) microlites in glass (GL) from 1992 basaltic andesite. VES, vesicle. G. Back-scattered electron image (0.875 mm wide) of ilmenite grains with exsolved magnetite and magnetite rims in 1992 basaltic andesite. H, Titanium distribution map (0.14 mm wide), showing magnetite (MT) grain in contact with ilmenite (IL). Grain contact is thin dark line. Lighter areas indicate higher Ti concentration. Note that Ti abundance in magnetite increases toward ilmenite contact. Map produced digitally from wavelength spectrometer output while rastering electron-microprobe stage.

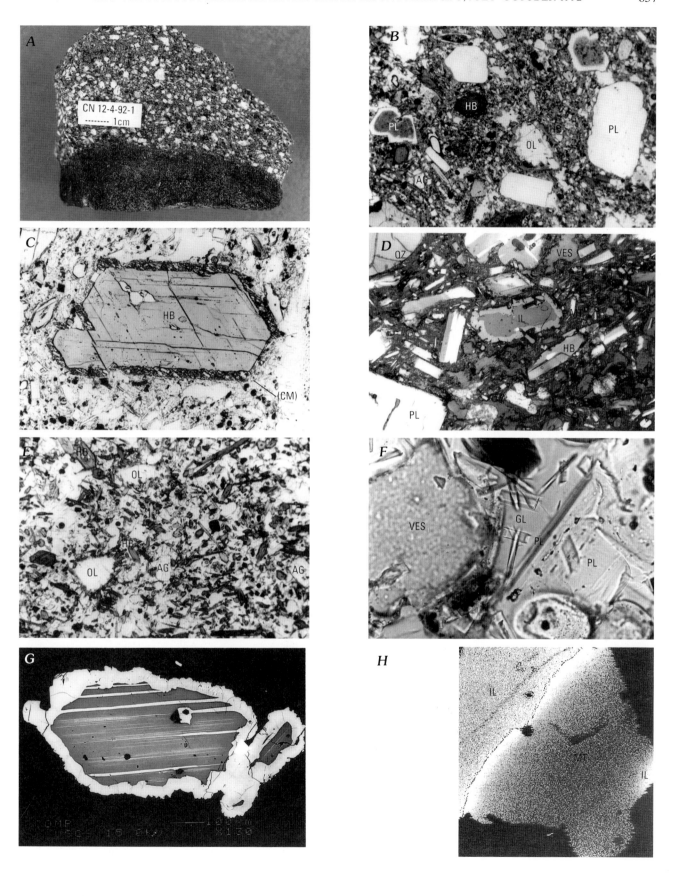

Table 2. Average mineral compositions and representative microprobe analyses from 1992 dome samples.

[%Fo = 100Mg/(Mg+Fe^{+2}+Mn); Mg# = 100Mg/(Mg+Fe^{+2}); %An = 100Ca/(Ca+Na+K); SD = standard deviation, Max. = maximum; Min. = minimum; n = number of analyses; mag. = magnetite; ilm. = ilmenite; ulv. = ulvospinel. Fe$_2$O$_3$, FeO, and mineral partitions for oxides by method of Stormer (1983). Temperatures and oxygen fugacities by method of Andersen and Lindsley (1988). Distance = distance of magnetite analysis point from contact with adjacent ilmenite grain. Dash indicates below detection limit]

Olivine

	Basaltic andesite	Andesite
Avg. %Fo	85.5	85.3
SD	1.3	0.6
Max.	86.5	85.8
Min.	80.2	84.6
n	21	5
Point no.	ol3–core2	hol3–4–3
MgO	46.36	45.44
SiO$_2$	40.54	40.45
CaO	0.16	0.19
MnO	0.18	0.20
FeO	13.78	13.67
NiO	0.10	0.13
Total	101.11	100.07
%Fo	85.7	85.6

Augite

	Basaltic Andesite		Andesite	
Avg. Mg#	78.5		82.3	
SD	4.1		3.0	
Max.	88.3		87.0	
Min.	71.4		75.9	
n	63		16	
Point no.	1i 3–1	1i 3–9	ipx1–4	ipx1–13
Na$_2$O	0.24	0.36	0.29	0.27
MgO	17.00	13.67	15.70	14.50
Al$_2$O$_3$	1.87	2.87	3.17	1.73
SiO$_2$	53.23	51.42	52.04	52.81
CaO	22.24	21.96	23.24	21.79
TiO$_2$	0.19	0.35	0.30	0.23
Cr$_2$O$_3$	0.56	–	0.76	–
MnO	0.11	0.43	0.07	0.34
FeO	4.03	8.58	4.16	8.21
Total	99.46	99.63	99.72	99.88
Mg#	88.30	74.00	87.10	75.90

Hypersthene — Andesite (Fibrous rim on hornblende)

Point no.	1h 7–2
MgO	23.84
Al$_2$O$_3$	0.93
SiO$_2$	53.58
CaO	1.59
TiO$_2$	0.22
Cr$_2$O$_3$	–
MnO	0.94
FeO	19.02
Total	100.12
Mg#	69.1

Hornblende

	Basaltic andesite		Andesite		Dacite
Type	High-Ti	Low-Ti	High-Ti	Low-Ti	Low-Ti
Avg. Mg#	67.4	67.1	65.6	67.0	66.8
SD	4.6	1.7	2.6	0.9	2.4
Max.	72.8	70.1	68.8	68.4	69.5
Min.	58.7	64.6	62.5	65.5	60.9
n	42	13	6	7	16
Point no.	pd11	92icor	1h 7–1	1h 6–1	dchb2–8
F	0.07	0.08	–	–	0.17
Na$_2$O	1.24	2.31	1.21	2.35	1.24
MgO	15.59	14.13	15.49	14.00	15.33
A$_2$O$_3$	6.85	13.30	7.09	13.00	6.98
SiO$_2$	48.81	41.61	48.38	42.05	49.41
K$_2$O	0.22	0.91	0.24	0.75	0.23
CaO	10.65	11.66	10.27	11.62	10.65
TiO$_2$	0.74	2.08	0.77	2.05	0.76
MnO	0.54	0.12	0.50	0.09	0.42
FeO	12.62	11.78	12.74	11.64	12.67
Total	97.33	97.98	96.69	97.55	97.86
TiO$_2$/Al$_2$O$_3$	0.107	0.157	0.109	0.158	0.109
Mg#	68.8	68.1	68.4	68.2	68.3

Plagioclase

	Basaltic andesite (microlites)	Andesite (phenocrysts)	Dacite (phenocrysts)
Avg.%An	56.48	36.84	39.02
SD	12.48	4.73	5.89
Max.	73.85	59.46	57.79
Min.	31.78	32.16	32.31
n	21	43	64
Point no.	pl–tr20	pt20#30	d7–32–5
Na$_2$O	3.97	6.87	6.76
Al$_2$O$_3$	29.12	26.16	26.69
SiO$_2$	52.30	59.29	58.59
K$_2$O	0.26	0.30	0.26
CaO	12.99	7.54	8.08
FeO	0.61	0.16	0.14
Total	99.24	100.31	100.53
% An	63.4	37.1	39.2

Oxides in 1992 Dacite Inclusion

	Magnetite	Ilmenite	Magnetite	Ilmenite
Point no.	G2–6	G2–AV	G3–9	G3–AV
Distance	10 µm		133 µm	
MgO	1.47	1.27	1.39	1.17
Al$_2$O$_3$	1.36	0.29	1.76	0.32
TiO$_2$	6.93	27.46	3.98	27.65
Cr$_2$O$_3$	0.13	0.09	0.23	0.08
FeO^T	85.25	66.09	87.74	66.30
MnO	0.46	0.24	0.48	0.23
Sum	95.61	95.44	95.58	95.75
FeO	35.35	22.18	32.94	22.55
Fe$_2$O$_3$	55.46	48.80	60.90	48.62
Total	101.17	100.33	101.68	100.62
X ulv.	0.19		0.11	
X ilm.		0.52		0.52
T(°C)	917±27		794±25	
log(f$_{O_2}$)	−9.58±0.12		−11.34±0.14	

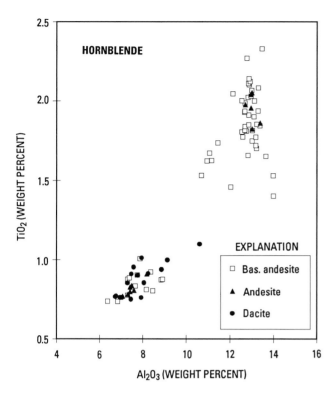

Figure 10. Variation of TiO_2 and Al_2O_3 in hornblende phenocrysts in samples from the 1992 Pinatubo dome. Note the bimodal distribution and lack of high-TiO_2 and high-Al_2O_3 hornblende in the dacite sample.

mm) in the groundmass. Sparse low-TiO_2 hornblende xenocrysts (to 2 mm) are extensively replaced by fine-grained augite. As in the host andesite, olivine occurs both as glomerocrysts with augite and as isolated subhedral phenocrysts bordered by hornblende. Sparse quartz xenocrysts are highly resorbed and rimmed by glass and augite. Apatite needles are common in the groundmass glass and as inclusions in plagioclase microlites. Magnetite microphenocrysts (<0.1 mm) are the only primary oxide; however, a few large (to 0.7 mm) ilmenite xenocrysts are present. These are rimmed by magnetite and contain exsolution lamellae of magnetite. A sulfide mineral is present as small (<10 µm) blebs within the groundmass and as inclusions in some magnetite, augite, and hornblende crystals.

The basaltic andesite inclusions show evidence of rapid microlite growth at a late stage, presumably due to rapid cooling upon entrainment in the cooler andesite host magma. The diktytaxitic groundmass texture consists of randomly oriented microlites (hornblende, plagioclase, augite, oxides) enclosed in clear to brown vesicular glass. Elongate hornblende microlites and abundant plagioclase microlites attest to a late period of rapid crystal nucleation. Small (<50 µm) plagioclase microlites have "hopper" shapes, and "swallow-tail" terminations are observed on some hornblende needles (fig. 9F). These textures indicate

undercooling and rapid crystal growth. Similar textures are seen in basaltic inclusions from the 1991 dome and are described in more detail by Pallister and others (this volume).

DACITE INCLUSION (CN12492–2)

Sparse dacitic inclusions occur within the dome andesite. The sampled inclusion is a dense hornblende-plagioclase dacite with <5% small, irregular vesicles and a microlite-charged groundmass. Although the groundmass appears partly isotropic in polarized light, electron imagery shows it to be microcrystalline and composed mainly of intergrown silica, albite, and potassium feldspar. The microcrystalline groundmass suggests that the dacite had cooled and solidified prior to inclusion in the hotter andesite. The dacite contains about 40% phenocrysts, mainly of oscillatory zoned plagioclase and hornblende, and accessory quartz, biotite, ilmenite, magnetite, resorbed anhydrite, and trace amounts of apatite and zircon. Plagioclase forms large oscillatory zoned phenocrysts with average compositions of An_{35-40}. These crystals typically contain several resorption surfaces, each of which is bounded by more calcic zonal bands that approach An_{60} (table 2).

Individual magnetite grains lack exsolution lamellae, yet they range in TiO_2 from 4 to 7 wt%; (table 2). Maximum TiO_2 abundances occur along grain boundaries with ilmenite and decrease exponentially away from those contacts (Figure 9H). Fe-Ti-oxide thermobarometry yields apparent temperatures of 900 to 1000°C and oxygen fugacities of 10^{-10} to 10^{-9} for these high-TiO_2 zones. Thermobarometric calculations that use the low-TiO_2 core areas of magnetite grains yield apparent temperatures and oxygen fugacities in the range 760 to 800°C and 10^{-12} to 10^{-11} (table 2). Although these results do not represent equilibrium conditions, they suggest that the dacite was heated upon inclusion in the hotter andesite but that there was insufficient time between inclusion and eruption for complete reequilibration. The TiO_2 gradient within magnetite grains may represent diffusion of titanium into magnetite between the times of inclusion and eruption.

Hornblende phenocrysts in the dacite are of the low-TiO_2 type (table 2; fig. 10). They show oscillatory zoning and are mantled either by 5- to 10-µm-thick reaction rims or by 50- to 150-µm-thick reaction rims. The thin rims are present where hornblende is in direct contact with the groundmass but are not present where hornblende is in contact with other minerals. They occur not only along the outer growth surfaces of hornblende phenocrysts but also along broken surfaces that cut through phenocryst cores. The thin rims contain blocky microlites of augite, hypersthene, and glass, and in some areas also contain plagioclase and oxides. They are similar to the depressurization reaction rims described in dacite of Mount St. Helens by

Rutherford and Hill (1993). The thick rims are composed of fibrous aggregates of hypersthene, glass, and minor oxides. As in the 1992 andesite, the similarity in thickness of the hypersthene-rich rims to cummingtonite rims in 1991 and earlier dacites of Mount Pinatubo and the occurrence of hypersthene-rich rims that are only present on the outer growth surfaces of crystals suggest that they represent pseudomorphs after former cummingtonite rims.

The mineral assemblage and phenocryst-rich character of the dacite is similar to the dominant type of dacite erupted on June 15, 1991. However, the microlite-rich groundmass and evidence of late-stage reaction of hornblende, cummingtonite, and anhydrite show that this dacite ascended more slowly and degassed en route to the surface. In fact, the microcrystalline groundmass and Fe-Ti oxide evidence of heating suggest that our sample of the dacite was not magmatic when included.

CHEMISTRY

Major element compositions for the 1992 dome andesite and its dominant inclusion type are given in table 3. The bulk composition of the 1992 dome sample is broadly similar to the average composition of the 1991 andesite, although the 1992 sample has slightly higher SiO_2 and lower MgO, CaO, and total iron. Our analyzed 1992 inclusion is texturally similar to basaltic inclusions of the 1991 dome but has more SiO_2, Al_2O_3 and Na_2O, less MgO, CaO and total iron, and a lower magnesium number (62) than do basaltic inclusions of the 1991 dome. Olivine crystals in the 1992 samples are about Fo_{86} in composition, similar to those in the 1991 samples and indicating that the primary mafic mixing endmember was basaltic, with a magnesium number of about 68 and a temperature of >1200°C (see discussion of olivine-melt equilibria in Pallister and others, this volume). The lower magnesium number of the basaltic andesite and the presence of sparse plagioclase, low-TiO_2 hornblende, and ilmenite xenocrysts suggest that our 1992 inclusion represents basaltic magma that had already mixed with a small amount of dacite prior to entrainment in the andesite.

Least squares modeling of major and trace element data shows that the 1991 andesite composition can be produced by mixing of dacite and basalt in the proportions 64:36 (Pallister and others, this volume). Application of the same 1991 dacite-basalt model to produce the 1992 andesite composition yields proportions of 72:28, with an unacceptably high sum of squared residuals of 0.6 (table 4). On the basis of our experience modeling the 1991 rocks and the analytical uncertainties reported in table 2, we expect the sum of squared residuals for an acceptable model to be less than 0.2. When this criterion is used, none of the two-component mixing models that use a 1991 composition produce an acceptable fit. Although the fit of 1991-based models can

be improved by adding or subtracting minor amounts of the observed mineral phases, an acceptable fit can also be obtained by using only 1992 compositions. For example, the best-fit two-component mixing model is obtained by mixing 62% 1992 dacite with 38% 1992 basaltic andesite.

We reject the 1991-based models for the following reasons: The composition of the 1992 basaltic andesite inclusion is distinct from compositions of 1991 magmas, and it cannot be produced by any combination of 1991 or 1992 endmember magmatic compositions (table 4) or by any reasonable combination of mixing and removal or addition of major phenocryst phases. In addition, trace-element abundances and patterns of each of the 1992 samples are distinct from their 1991 counterparts. In particular, the 1992 samples show enrichments in the light rare earth elements compared to their 1991 counterparts (fig. 11). As in the 1991 samples, the most mafic composition is also the most enriched with respect to incompatible element abundance. This relation rules out major phase fractionation to produce the more felsic magmas. The basaltic andesite is enriched in light rare earth elements and depleted in heavy rare earth elements relative to the 1991 basalt; it is a crossing pattern, which also indicates that the basaltic andesite is not a simple fractionate of the 1991 basalt.

The two-component 1992-based mixing model (62% 1992 dacite with 38% 1992 basaltic andesite) is capable of reproducing 30 of the 36 oxide and elemental abundances in table 3 within two standard deviations of the combined errors. With respect to the analyzed 1992 andesite sample, the model solution is depleted by 3 to 15% in K_2O, Rb, Cs, Sc, Cr, and Zn, and possibly in As and Ni. This indicates that although our 1992 samples of dacite and basaltic andesite are close, they are not completely representative of the true mixing endmembers, which were not sampled. This is not surprising, given the textural evidence that the 1992 dacite was not magmatic when included. However, the textural and chemical evidence that the 1992 andesite is a mixed magma is conclusive. We suspect that it was produced by mixing of a basaltic magma that was parental to the 1992 basaltic andesite inclusion with dacitic magma that is represented by the 1992 dacite inclusion.

DISCUSSION

The 1992 dome is similar to the dome that formed just before the major explosive eruptions of 1991. Both domes are composed mainly of andesite that formed by mixing of basalt and dacite. However, in sharp contrast to events of 1991, extrusion of the 1992 dome was not followed by significant explosive activity.

Is the 1992 andesite left-over mixed magma from the same batch that was first erupted on June 7, 1991, or does it represent a new episode of magma mixing in 1992? If new mixing has occurred, the dacitic magma reservoir is still

Table 3. Major and trace element analyses of 1992 dome rocks compared to average compositions of 1991 eruptive products.

[Major oxides analyzed by David Siems using X-ray fluorescence method and reported in wt% normalized volatile free. Trace element abundances determined by James Budahn and Roy Knight using instrumental neutron activation analysis. Uncertainties shown in parentheses as coefficients of variation (in percent). Elemental data in ppm, except S (wt%) and Au (ppb). LOI = loss on ignition at 925° C. Sulfur determined by J. Curry of USGS on separate sample splits using a combustion-infrared technique. Standard deviations for average values of 1991 samples given in Pallister and others (this volume)]

		1991 basalt	1991 dacite	1991 andesite	1992 basaltic andesite		1992 dacite		1992 andesite		Model[1]	
Sample No.					CN12492–1i		CN12492–2		CN12492–1h		38:68	Percent difference
Lab No.					D–523548		D–523550		D–523547			
SiO_2	(0.2)	51.2	64.6	59.7	53.9		65.2		61.0		60.9	−0.1
Al_2O_3	(0.3)	14.8	16.5	16.0	16.2		16.3		16.1		16.3	1
$FeTO_3$	(0.2)	9.01	4.37	6.13	8.33		4.29		5.80		5.82	0.4
MgO	(1)	8.85	2.39	4.62	6.25		2.30		3.92		3.80	−3
CaO	(0.3)	10.15	5.23	6.94	9.32		5.03		6.59		6.66	1
Na_2O	(1)	3.05	4.49	3.98	3.20		4.57		3.93		4.05	3
K_2O	(0.2)	1.45	1.54	1.60	1.44		1.48		1.62		1.46	−10
TiO_2	(1)	0.90	0.53	0.66	0.87		0.50		0.64		0.64	0.0
P_2O_5	(0.2)	0.35	0.20	0.26	0.36		0.19		0.26		0.26	−3
MnO	(5)	0.16	0.10	0.13	0.15		0.10		0.12		0.12	−2
Sum		100	100	100	100		100		100		100	
LOI		0.15	0.77	0.91	0.29		0.16		0.24		0.21	
[2]FeO		3.76	2.14	3.01	2.71		1.98		2.47		2.26	
[3]Mg#		68.6	54.8	62.5	62.3		54.1		59.8		59.0	−1
H_2O+		0.58	0.68	0.73	0.44		0.17		0.28		0.27	
$H_2O−$		0.07	0.12	0.20	0.10		0.10		0.29		0.10	
CO_2		0.01	<0.01	<0.00	<0.01		<0.01		<0.01			
S (wt%)		<0.05	0.12	0.17	0.08		<0.05		<0.05			
Sc		34.6	8.82	18.5	28.9	(1)	9.3	(1)	17.2	(1)	16.7	−3
Cr		342	37.3	150	176	(1)	37.0	(1)	97	(1)	90	−7
Co		36.3	10.5	20.5	29.0	(1)	10.7	(1)	17.9	(1)	17.7	−1
Ni		105.2	16.4	50.2	69.8	(7)	20.7	(9)	33.7	(8)	39.4	17
Zn		78.3	53.2	67.1	66.9	(2)	54.7	(2)	65.0	(2)	59.3	−9
As		1.3	3.5	2.4	1.9	(11)	2.5	(8)	3.1	(6)	2.2	−26
Rb		46.1	40.1	44.5	50.4	(3)	39.0	(3)	51.6	(3)	43.3	−16
Sr		615	581	571	604	(3)	586	(2)	563	(3)	593	5
Zr		114	107	107	112	(8)	101	(5)	124	(7)	105	−15
Sb		0.17	0.34	0.26	0.17	(14)	0.24	(12)	0.23	(10)	0.22	−5
Cs		3.51	2.83	3.27	4.01	(2)	2.50	(1)	3.61	(1)	3.07	−15
Ba		310	465	393	330	(2)	454	(1)	415	(2)	407	−2
La		22.2	15.6	18.9	20.6	(1)	14.5	(1)	17.6	(1)	16.8	−4
Ce		48.1	32.7	39.7	43.8	(2)	29.2	(2)	36.5	(2)	34.7	−5
Nd		25.2	15.3	19.8	24.4	(3)	14.8	(3)	18.9	(3)	18.4	−2
Sm		5.40	3.15	4.18	5.58	(1)	3.16	(1)	4.26	(1)	4.08	−4
Eu		1.56	0.85	1.13	1.52	(1)	0.85	(1)	1.13	(1)	1.11	−2
Gd		4.71	2.76	3.67	5.13	(5)	2.92	(5)	4.00	(5)	3.76	-6
Tb		0.634	0.379	0.476	0.657	(2)	0.372	(2)	0.489	(2)	0.480	−2
Tm		0.282	0.203	0.244	0.335	(9)	0.215	(8)	0.267	(5)	0.261	−2
Yb		1.71	1.30	1.49	1.92	(2)	1.32	(2)	1.60	(2)	1.55	−3
Lu		0.254	0.195	0.216	0.278	(2)	0.196	(2)	0.233	(2)	0.227	−3
Hf		2.41	2.97	2.70	2.64	(2)	2.96	(1)	2.87	(1)	2.84	−1
Ta		0.179	0.330	0.263	0.204	(3)	0.322	(1)	0.267	(2)	0.277	4
W		<2.2	1.3	–	1.5	(18)	0.8	(16)	0.5	(25)	1.1	
Au (ppb)		<1.2	9.2	7.8	<0.04		77.5	(2)	<1.5			
Th		5.87	4.64	5.33	5.69	(1)	4.51	(1)	5.16	(1)	4.96	−4
U		1.54	1.59	1.57	1.69	(3)	1.59	(3)	1.76	(3)	1.63	−7

[1] Least squares model (38% 1992 basaltic andesite + 62% 1992 dacite), deviation of 1992 andesite from model composition expressed as percent difference.

[2] Titration value.

[3] Calculated assuming $Fe_2O_3/FeO=0.123$.

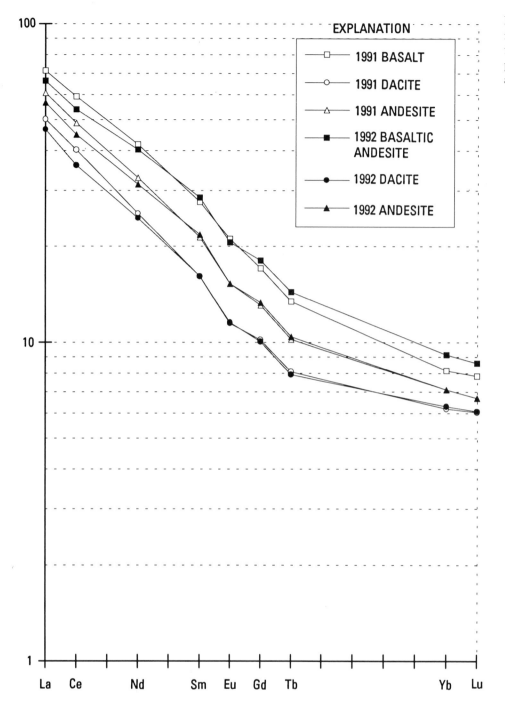

Figure 11. Chondrite-normalized rare earth element patterns for 1992 Pinatubo samples compared to patterns for average abundances in 1991 samples.

being replenished by mafic magma, and a strong possibility of further eruptions remains. On the other hand, if the 1992 andesite is part of the 1991 batch that was stored in the reservoir or conduit system, its eruption would not indicate replenishment and accompanying disturbance of the main dacite reservoir at depth.

The slightly more siliceous composition of the 1992 samples, our inability to produce the 1992 compositions by two-component mixing involving 1991 components, and the distinct trace element patterns of the 1992 samples

indicate that they are not unerupted aliquots of the magma that formed the June 7–12, 1991, andesite dome and the June 12, 1991, andesite scoria. Instead, they represent the products of a new mixing event.

Unreacted cummingtonite rims on hornblende, and unhomogenized magnetite-ilmenite pairs, suggest that less than 4 days elapsed between the final stages of magma mixing and eruption in 1991 (Rutherford and others, 1993; Rutherford and Devine, this volume). Conversely, the breakdown of cummingtonite rims observed in the 1992

Table 4. Results of least squares models to produce the 1992 andesite and 1992 basaltic andesite compositions by simple two-component mixing.

[Preferred model solution shown in bold; values for mixing components are in percent]

	Mixing components				Sum of squared residuals
	1991 dacite	1991 basalt	1992 dacite	1992 basaltic andesite	
1992 andesite			**62**	**38**	**0.09**
1992 andesite	63			37	0.32
1992 andesite		29	71		0.39
1992 andesite	72	28			0.60
1992 basaltic andesite.		24	76		3.94
1992 basaltic andesite.	24	76			3.90

andesite sample indicates that more than 4 days elapsed between mixing and eruption in 1992.

We envision that **some** of the "1992" dacitic magma rose from the main reservoir (>8 km depth) and reached depths where hornblende was not stable (<4 km, Rutherford and Devine, this volume). Ascent took place sometime after June 15, 1991, and before July 1992. The dacite body cooled around its margins and developed a microcrystalline groundmass. Then, in mid-1992, magnesian olivine basalt (Mg#=68) intruded the >8-km-deep dacitic reservoir and mixed first with a small amount of similar "1992" dacitic magma to form basaltic andesite and then with a larger volume of the same "1992" dacitic magma to form andesitic magma. Days or perhaps weeks later, the andesitic magma rose toward the surface, picked up the xenolith that is our only surviving sample of "1992" dacite, and was erupted to form the 1992 lava dome. Preservation of Ti-zonation in magnetite (fig. 9H) indicates that the xenolith was included and heated by the andesitic magma for only a brief period before the eruption.

The lack of another explosive dacite eruption in 1992–93, despite evidence for renewed replenishment and magma mixing, suggests that eruption from the large residual dacite reservoir (Mori, Eberhardt-Phillips, and Harlow, this volume) is no longer easily triggered. The climactic eruption of June 15, 1991, certainly depleted volatiles in the dacitic reservoir, perhaps to such a low level that large eruptions will be unlikely until volatiles can reconcentrate in the upper parts of that reservoir. Alternatively, the mixing episode in 1992 may have been too small to trigger an explosive eruption from the dacitic reservoir. The total volume of andesitic magma erupted in 1991 (June 7–12 dome and June 12 tephra) was between 10^7 and 10^8 m^3; the volume of andesitic magma erupted to form the 1992 dome **and** left in a hypothetical conduit beneath that dome was of the same order of magnitude. Thus, we suspect that volatile depletion in the dacitic reservoir is preventing further large eruptions,

but we cannot rule out that a larger intrusion of basalt than occurred in 1992 could still trigger a large eruption from the dacitic reservoir.

ACKNOWLEDGMENTS

We are grateful to the Philippine Air Force and the U.S. Marine Corps for photographic air support. Various colleagues kindly shared their photos, video, and sketches. We are also grateful to Ms. Tina Monzon-Palma (Channel 5, Manila) and the late Captain Agustin Consunji for helicopter access to the crater to sample the 1992 lava dome. Don Swanson and Jon Fink provided helpful reviews.

REFERENCES CITED

Andersen, D.J., and Lindsley, D.H., 1988, Internally consistent solution models for Fe-Mg-Mn-Ti oxides: Fe-Ti oxides: American Mineralogist, v. 73, p. 714–726.

Anderson, S.W., and Fink, J.H., 1992, Crease structures: Indicators of emplacement rates and surface stress regimes of lava flows: Geological Society of America Bulletin, v. 104, p. 615–625.

Daag, A.S., Tubianosa, B.S., Newhall, C.G., Tuñgol, N.M, Javier, D., Dolan, M.T., Delos Reyes, P.J., Arboleda, R.A., Martinez, M.L., and Regalado, M.T.M., this volume, Monitoring sulfur dioxide emission at Mount Pinatubo.

Jones, J.W., and Newhall, C.G., this volume, Preeruption and posteruption digital-terrain models of Mount Pinatubo.

Kirsanov, I.T., 1979, Extrusive eruptions on Bezymianny volcano in 1965–1974 and their geologic effect: Moscow, Akademia Nauk SSSR, Sibirskoye Otdeleniye, Institut Geologii i Geofiziki, Nauka, p. 50–69. [translated by D.B. Vitaliano.]

Mori, J., Eberhart-Phillips, D., and Harlow, D.H., this volume, Three-dimensional velocity structure at Mount Pinatubo, Philippines: Resolving magma bodies and earthquakes hypocenters.

Mori, J., White, R.A., Harlow, D.H., Okubo, P., Power, J.A., Hoblitt, R.P., Laguerta, E.P., Lanuza, L., and Bautista, B.C., this volume, Volcanic earthquakes following the 1991 climactic eruption of Mount Pinatubo, Philippines: Strong seismicity during a waning eruption.

Nakada, S., 1992a, Photographic records of eruption products at Unzen Volcano during May 1991-May 1992, in Yanagi, T., Okada, H., and Ohta, K., eds., Unzen Volcano: The 1990-1992 Eruption: Fukuoka, Japan, The Nishinippon and Kyushu University Press, p. 12–20.

———1992b, Lava domes and pyroclastic flows of the 1991–1992 eruption at Unzen Volcano, in Yanagi, T., Okada, H., and Ohta, K., eds., Unzen Volcano: The 1990–1992 Eruption, The Nishinippon and Kyushu University Press, Fukuoka, p. 56–66.

Newhall, C.G., Daag, A.S., Delfin, F.G., Jr., Hoblitt, R.P., McGeehin, J., Pallister, J.S., Regalado, M.T.M., Rubin, M., Tamayo, R.A., Jr., Tubianosa, B., and Umbal, J.V., this volume, Eruptive history of Mount Pinatubo.

Omori, F., 1916, The Sakura-jima eruptions and earthquakes: Tokyo, Bulletin of the Imperial Earthquake Investigation Committee, v. 8, p. 181–321.

Pallister, J.S., Hoblitt, R.P., Meeker, G.P., Knight, R.J., and Siems, D.F., this volume, Magma mixing at Mount Pinatubo: Petrographic and chemical evidence from the 1991 deposits.

Pallister, J.S., Hoblitt, R.P., and Reyes, A.G., 1992, A basalt trigger for the 1991 eruptions of Pinatubo volcano?: Nature, v. 356, p. 426–428.

Ramos, E.G., Laguerta, E.P., and Hamburger, M.W., this volume, Seismicity and magmatic resurgence at Mount Pinatubo in 1992.

Rose, W.I., Jr., 1972, Santiaguito volcanic dome, Guatemala: Geological Society of America Bulletin, v. 83, p. 1413–1434.

———1973, Pattern and mechanism of volcanic activity at the Santiaguito volcanic dome: Bulletin Volcanologique, v. 37, p. 73–94.

———1987, Volcanic activity at Santiaguito volcano, 1976-1984, in Fink, J.H., ed., The emplacement of silicic domes and lava flows: Geological Society of America Special Paper 212, p. 17–27.

Rutherford, M.J., Baker, Leslie, and Pallister, J.S., 1993 Petrologic constraints on timing of magmatic processes in the 1991 Pinatubo volcanic system [abs.]: Eos, Transactions, American Geophysical Union, v. 74, p. 671.

Rutherford, M.J., and Devine, J.D., this volume, Preeruption pressure-temperature conditions and volatiles in the 1991 dacitic magma of Mount Pinatubo.

Rutherford, M.J., and Hill, P.M., 1993, Magma ascent rates from amphibole breakdown: An experimental study applied to the 1980–1986 Mount St. Helens eruptions: Journal of Geophysical Research, v. 98, p. 19667–19685.

Stormer, J.C., 1983, The effects of recalculation on estimates of temperature and oxygen fugacity from analyses of multicomponent iron-titanium oxides: American Mineralogist, v. 68, p. 586–594.

Swanson, D.A., Dzurisin, D., Holcomb, R.T., Iwatsubo, E.Y., Chadwick, W.W., Jr., Casadevall, T.J., Ewert, J.W., and Heliker, C.C., 1987, Growth of the lava dome at Mount St. Helens, Washington (USA), 1981–1983, in Fink, J.H., ed., The emplacement of silicic domes and lava flows: Geological Society of America Special Paper 212, p. 1–16.

Swanson, D.A., and Holcomb, R.T., 1989, Regularities in growth of the Mount St. Helens dacite dome, 1980-1986, in Fink, J.H., ed., Lava flows and domes: Emplacement mechanisms and hazard implications: New York, Springer-Verlag, IAVCEI Proceedings in Volcanology No. 2, p. 3–24.

Taylor, G.A.M., 1958, The 1951 eruption of Mount Lamington, Papua: Australian Bureau of Mineral Resources Bulletin, v. 38, 117 p.

Secondary Pyroclastic Flows from the June 15, 1991, Ignimbrite of Mount Pinatubo

By Ronnie C. Torres,[1][2] Stephen Self,[2] and Ma. Mylene L. Martinez[1]

ABSTRACT

Secondary pyroclastic flows, some leaving deposits up to 10 kilometers long, 1 kilometer wide, and 10 meters thick, occurred at Pinatubo for more than 2 years after the June 15, 1991, eruption. Avalanching of the valley-ponded facies of the 1991 ignimbrite generates these secondary pyroclastic flows and leaves an avalanche scarp at their origin. Material is transported in hot, high-concentration laminar flow over a gentle slope, driven by gravity and gas fluidization, and is deposited as massive, valley-filling ignimbrite. Secondary pyroclastic flows are very similar to vent-derived pyroclastic flows and their deposits are virtually indistinguishable from primary ignimbrite in the field. However, in deposits so far studied, secondary ignimbrite tends to be fines depleted and its coarse clasts have nonuniform thermoremanent magnetic polarity. Secondary pyroclastic flows differ from other secondary movements in ignimbrite noted by previous workers, principally in that they occur long after the deposition of primary ignimbrite. Such flows, apparently a common mechanism for redistributing ignimbrite after initial deposition, present poseruption hazards that were previously unrecognized.

INTRODUCTION

Explosive activity of Mount Pinatubo climaxed on June 15, 1991, when thick, nonwelded ignimbrite was emplaced in all major river valleys and on low-lying pyroclastic fans around the volcano. The eruption coincided with the onset of the rainy season, and new drainage systems were rapidly reestablished within ignimbrite fans, preferentially over former river axes. Intense and prolonged rainfall after the climactic eruption generated hot lahars, and, in some instances, avalanches of hot primary ignimbrite. Several of the latter events involved collapse of large masses of primary ignimbrite that, after transport as secondary pyroclastic flow, were deposited as secondary ignimbrites. We use the term "secondary pyroclastic flow" for the moving system and "secondary ignimbrite" for the deposit.

Secondary pyroclastic flows were first observed on August 12–13, 1991, and more have been documented over the succeeding 2 years, up to the time of this writing (August 1993).

This paper describes some of the largest secondary pyroclastic flows, remobilized from thick, nonwelded, vent-derived ignimbrite and transported as dry, gas-fluidized flows. We also describe their deposits, contrasting them with the primary ignimbrite at Mount Pinatubo. Deposits described herein are part of a dynamic system involving large ignimbrite sheets and lahar conveyance channels. The exact units and the outcrops we describe will most likely not remain exposed for long and will be eroded or buried by new events.

In effect, ignimbrite emplacement is a continuum that begins with the vent-derived pyroclastic flow and continues during remobilization as secondary and tertiary pyroclastic flows long after the initial deposition. The second and third orders in this ignimbrite continuum have been overlooked in previous studies of ignimbrite sheets because the deposits are so alike. However, secondary and even tertiary pyroclastic flows are hazardous, and we hope that this paper will alert future workers to this hazard.

PREVIOUS DESCRIPTIONS OF SECONDARY MOVEMENTS AND DEPOSITS ASSOCIATED WITH IGNIMBRITES

Secondary mass flow (Chapin and Lowell, 1979; Ellwood, 1982) and secondary flow structures (Wolff and Wright, 1981) have been recognized in high-temperature rheomorphic ignimbrites and other welded tuffs. Such flow is thought to occur during or shortly after emplacement, while the material is still a coherent viscous fluid (Schmincke and Swanson, 1967; Ellwood, 1982; Branney and Kokelaar, 1992), and permit, for instance, slumping down the oversteepened sides of paleovalleys toward the

---

[1]Philippine Institute of Volcanology and Seismology.

[2]Hawaii Center for Volcanology, Dept. of Geology and Geophysics, School of Ocean and Earth Science and Technology (SOEST), University of Hawaii, 2525 Correa Rd., Honolulu, HI 96822.

valley axis (Chapin and Lowell, 1979). Fierstein and Hildreth (1992) recognized segregation (pseudoflow) units at the Valley of Ten Thousand Smokes that they attribute to internal shear at the margin and termini of the main ignimbrite during the last stages of deposition.

A different kind of secondary flow was inferred at Mount St. Helens, where parts of the main blast flow condensed into secondary pyroclastic flows down South Coldwater Creek, and were emplaced contemporaneously with blast deposits (Hoblitt and others, 1981; Fisher and others, 1987). However, because blast material might not have come to rest before evolving into pyroclastic flows, Walker and others (in press) propose that flows like those in the South Coldwater valley be called drain-down deposits and that the term secondary pyroclastic flows be reserved for phenomena such as those observed at Pinatubo, which occur well after deposition of the primary ignimbrite.

Still another kind of secondary, pumice-rich flow deposit occurs in the Valley of Ten Thousand Smokes, Alaska. It strongly resembles the primary ignimbrite except for a lack of physical evidence for hot emplacement and a ratio of dacite to rhyolite that suggests remobilization of tephra rather than of primary ignimbrite (Hildreth, 1983).

THE 1991 PRIMARY IGNIMBRITE

By far the most voluminous pyroclastic-flow deposits of the 1991 eruptions are pumice-rich ignimbrites emplaced during the climactic phase on June 15. These are mostly massive, valley-confined, and reach 200 m in thickness in places; 40 to 80 m in thickness is more common. The bulk of the volume was deposited within 3 to 9 km from the vent, but some deposits extend 16 km from the vent.

The 1991 primary ignimbrite at Mount Pinatubo infilled and buried the deeply incised slopes and all preexisting major river valleys: Sacobia, Pasig-Potrero, Marella, Balin Baquero, Maraunot, Bucao, and O'Donnell (fig. 1; see also W.E. Scott and others, this volume). Peaks of former highlands were left protruding out of the ignimbrite fans like isolated islands. Small deposits of ignimbrite were also emplaced within the Bangat and Gumain River valleys.

The fresh ignimbrite is typically hot, pumiceous, fines rich, unconsolidated, and extremely unstable along oversteepened sides, where small avalanches commonly occur. Deposits contain >60% pumice clasts in the -2ϕ to -4ϕ size range, and the remainder are dense juvenile and accidental lithic clasts. The pumice component includes coarsely vesiculated, phenocryst-rich and finely vesiculated, phenocryst-poor types. Pumice-rich ignimbrites are moderately sorted to poorly sorted ($\sigma_\phi = 1.5$ to 3.5) and have a grain-size distribution dominated by the finer fraction (fig. 2). Median grain sizes range from 0 to 2.5 ϕ (fig. 3).

Within the pumice-rich ignimbrite, a lithic-rich facies occurs as pods, flow-segregated layers, and distinct flow units (W.E. Scott and others, this volume). The lithic clasts, many of which bear hydrothermal alteration, are dominantly dacitic, derived from pre-1991 lava domes and lava flows of Mount Pinatubo. Although some dense clasts occur in secondary ignimbrites, pumice-rich ignimbrites are the source of virtually all of the remobilized material discussed here.

Numerous fumaroles and phreatic explosion craters formed on the surface of the 1991 ignimbrite. Secondary explosions and hot avalanches persist to this time of writing (August 1993), although they significantly decreased within a year after the eruption. Fumarolic activity and secondary explosions were most intense along axes of former river valleys, where the ignimbrite is also thickest. A year and a half after the eruption, temperatures of fumaroles in primary ignimbrite of the Marella River valley were as high as 390°C at 1 m in depth. Actual emplacement temperature must have been higher. Juvenile pumice clasts exhibit uniform both normal and reversed thermoremanent magnetic (TRM) polarity (fig. 4 and C.G. Newhall, written commun., 1991).

OBSERVATIONS OF SECONDARY PYROCLASTIC FLOWS

Collapses of primary ignimbrite, generating secondary pyroclastic flows, are contemporaneous with lahar-forming events. Avalanches originate in the medial and distal parts of valley-filling primary ignimbrite and travel along active lahar channels. All large secondary pyroclastic flows have an associated headscarp region in which interconnected or en-echelon scarps suggest multiple avalanche failures of the primary ignimbrite. Details of some individual events are given below.

THE AUGUST 12–13, 1991, MARAUNOT EVENT

The first known occurrence of a secondary pyroclastic flow was August 12–13, 1991, following days of heavy rains and lahars events. A 20-m-thick mass, occupying 1.25 to 2.0 km² of the medial part of the ignimbrite fan at the former Maraunot River valley, was remobilized and transported about 10 km downslope (Pinatubo Volcano Observatory Team, 1991; Smithsonian Institution, 1991). The redeposited material could be distinguished from the surrounding primary ignimbrite by its relatively undissected surface and lighter tone, which indicated new deposition of pyroclastic materials. This deposit could be traced upslope until it terminated in a headscarp. The development of a deep headscarp and long scar paved the way for the establishment of a new river course that now feeds the Bucao

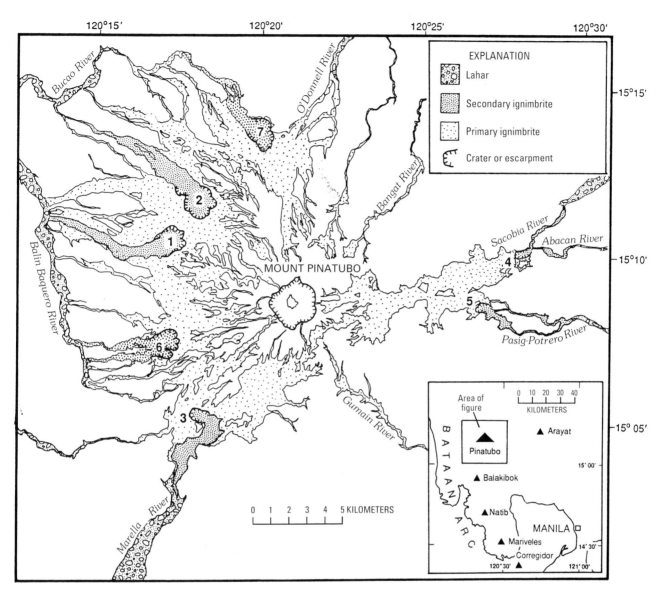

Figure 1. Distribution of primary and secondary ignimbrites after the 1991 eruption of Mount Pinatubo. River valleys affected by major secondary pyroclastic flows are (1) north Balin-Baquero, (2) Maraunot, (3) Marella, (4) Sacobia, (5) Pasig-Potrero, (6) south Balin-Baquero, and (7) upper Bucao. Secondary ignimbrites are shown as densely dotted areas. (Distribution of primary ignimbrite is by R.S. Punongbayan, in PHIVOLCS-NEDA (1992)).

drainage system. The volume of this secondary ignimbrite is estimated to be 0.04 to 0.05 km³ on the basis of the calculated volume loss at the headscarp. However, some of the volume lost at the headscarp may have been transported earlier as lahars.

During August 12–13, a ground-based weather radar at Cubi Point detected five ash clouds that reached 10 to 15 km in altitude (Smithsonian Institution, 1991). Correlative peaks appear in Realtime Seismic Amplitude Measurements (RSAM) at 0106, 0338, 0537, 1410, and 2250 on August 13, 1991 (Philippine Institute of Volcanology and Seismology (PHIVOLCS) Pinatubo Volcano Daily Update, August 14, 1991). These high ash clouds were assumed to have come from the crater, but, in retrospect, it is quite

possible that they were generated by secondary explosions and (or) progressive avalanche failures in ignimbrite of the Maraunot River valley. Three felt earthquakes were recorded on this day and one of these, felt at Poonbato (on the Bucao River) with intensity I on the Rossi-Forel scale, coincided with the ash cloud at 0537.

THE SEPTEMBER 4, 1991, MARELLA EVENT

Another secondary pyroclastic flow was generated on September 4, 1991, on the Marella River ignimbrite fan (fig. 5). The upper 20 to 25 meters of 100- to 200-m thick ignimbrite broke loose and flowed about 5 km (Pinatubo

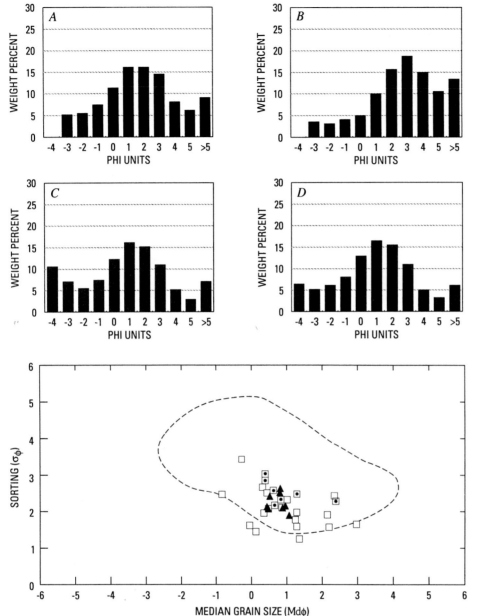

Figure 2. Grain-size distribution of the primary ignimbrite. Samples consist of pumice-rich ignimbrite at (*A*) Sacobia, (*B*) Pasig-Potrero, (*C*) Marella, and (*D*) Maraunot.

Figure 3. Median grain diameter (Md_ϕ) and sorting (σ_ϕ) of the primary ignimbrite (squares with dots) and secondary ignimbrites (solid triangles). Undotted squares are primary ignimbrite of W.E. Scott and others (this volume). Ignimbrite field (broken lines) after Sparks (1976) is delineated for comparison.

Volcano Observatory Team, 1991). The flow event was not actually observed, but the timing was reckoned from the conspicuous buildup of a 15-km-high ash plume at about 1400 (PHIVOLCS Pinatubo Volcano Daily Update, September 5, 1991), which was also documented by the NOAA–11 polar-orbiting weather satellite (Smithsonian Institution, 1991). The headscarp and a fresh deposit of pyroclastic materials were found 2 days after the event. Neither the seismicity nor aerial investigation of the crater gave any indication of a renewed magmatic eruption, so the ash plume was not vent-derived. In fact, the eruption alert level was lowered by PHIVOLCS on September 4 from Level 5 (eruption in progress) to Level 3 with a corresponding reduction of the boundary of the danger zone from 20 to 10 km in radius from the vent (PHIVOLCS Pinatubo Volcano

Daily Update, September 5, 1991). However, the secondary pyroclastic flows traveled beyond the prescribed 10-km danger zone. The ash cloud and already overcast weather in the Pinatubo area caused darkness over Clark Air Base and vicinity for about 3 h. Light to moderate ash fall affected the towns to the east and northeast of the Pinatubo caldera.

THE APRIL 4, 1992, SACOBIA-ABACAN EVENT

A secondary pyroclastic flow was directly observed on April 4, 1992, in the Sacobia River valley. The event was probably initiated at 1514, during which time a secondary explosion and (or) avalanche triggered an ash column at least 1 to 2 km high (PHIVOLCS Pinatubo Volcano Daily

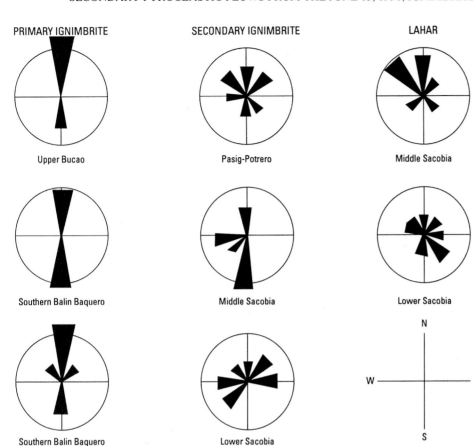

Figure 4. Thermoremanent magnetic (TRM) polarity distribution of clasts of primary and secondary ignimbrites and lahars at Mount Pinatubo. Most clasts used in TRM polarity determination consist of pumice, except a set of samples from the southern Balin-Baquero watershed, which are dense juvenile fragments. Primary ignimbrite exhibits both normal and reversed TRM polarity. Secondary ignimbrite and lahars generally display a non-uniform distribution. A secondary ignimbrite in the middle Sacobia valley shows a distinct reversed mode. Circle radius represents five samples.

Update, April 5, 1992). Eyewitnesses in the Sacobia River valley, upstream from Clark Air Base, described the secondary pyroclastic flow as a valley-confined "boiling ash flow."

This secondary pyroclastic flow originated just above the divergence of the Sacobia and Abacan valleys. Flows traveled down both valleys, 4 km down the Sacobia and about 3 km down the Abacan, and caused about 5 m of aggradation along both rivers (figs. 6 and 7). Several 5- to 7-m high dams that had been built to trap sediment and slow lahars along the Abacan and Sacobia Rivers were completely buried by the secondary ignimbrite. The vegetation on both sides of the channel was singed or smothered by wet fallout. A temperature of 240°C was measured in an exposed section 3 months after emplacement. Fumaroles on the secondary ignimbrite surface steamed for several months.

Associated ash and muddy rain fell heavily at 1545 and reduced the visibility over Clark Air Base from about 1600 to 2030. Wetted ash fall caked on the valley walls along paths of the secondary pyroclastic flow. A 2- to 3-cm-thick laminated and accretionary lapilli-rich fallout deposit draped the topography surrounding the headscarp area at Sacobia and Abacan Rivers; correlative fallout thins to <1 cm near the terminus of the secondary ignimbrite along the Sacobia channel at 3 km from the source.

THE JULY 13, 1992, PASIG-POTRERO EVENT

A secondary ignimbrite was emplaced in the Pasig-Potrero River valley on the late afternoon and evening of July 13, 1992 (fig. 8). An ash column at least 4 km high was observed from Clark Air Base and from Barangay Mancatian (fig. 7B), along the Pasig-Potrero River, and light ash fell over Clark Air Base and adjacent areas from 1830 to 1930 (PHIVOLCS Pinatubo Volcano Daily Update, July 14, 1992).

On July 15, we observed a new secondary ignimbrite that was traceable back to a 0.5-km-wide and 1.0-km-long escarpment in primary ignimbrite of the Pasig-Potrero watershed (fig. 7C). About 0.010 to 0.015 km³ of primary ignimbrite was remobilized and transported 5 to 6 km downriver. The bulk of the flow, rafting abundant pumice, went down the main valley of the Pasig-Potrero River. A fines-rich facies spilled from the rim of the headscarp, into a northern tributary of the Pasig Potrero River, and joined the main flow at a confluence 4 km from the source. The fines-rich facies covered the pumice-rich main-channel facies from their confluence point to their full runout distance. Near the foot of Mount Cutuno (fig. 7A), the main mass of secondary ignimbrite blocked a tributary to the main Pasig-Potrero River, and created a small impoundment of water.

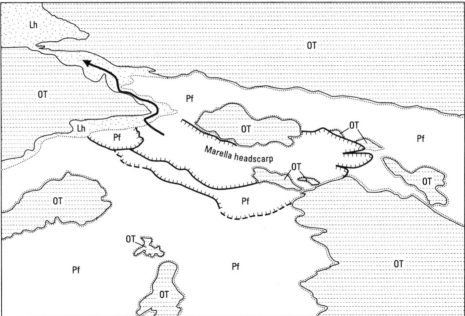

Figure 5. Oblique aerial photograph of the headscarp area of the Marella secondary pyroclastic flow. Sketch made from the above photograph shows the configuration of the headscarp and the distribution of the pre-1991 topography (old topography, OT), 1991 primary ignimbrite (pyroclastic flow, Pf), and lahar (Lh). Arrow indicates downstream direction. Small patches of proximal facies or secodary ignimbrite are not mappable at this scale. Medial and distal ignimbrite were eroded or buried by subsequent lahar events.

Figure 6. Photographs taken before (above) and a week after (below) the April 4, 1992, secondary pyroclastic-flow event in the Sacobia river valley, which caused channel aggradation of 5 to 7 m. The secondary pyroclastic flow buried the earlier lahar deposits. This is just upstream from the terminus of 1991 primary pyroclastic flows in the Sacobia valley. The channel walls consist of deposits of lahars and pyroclastic flows from several prehistoric eruptive episodes.

The secondary ignimbrite contains partly to completely charred wood chunks. Temperature measurements of the secondary ignimbrite taken 5 days after emplacement ranged from 240 to 260°C at 1 m in depth.

INFERRED EVENTS

Secondary pyroclastic flows were unknown hazards at Mount Pinatubo before the eruption. Because there was no specific effort to monitor their occurrence, it is very likely that some events, particularly those of late June and July 1991, which occurred at night or during inclement weather, escaped observation. In many cases, secondary pyroclastic flows were inferred after a new avalanche scar, an undissected pumice plain, or a fresh fallout deposit was identified

by aerial or ground surveys. However, field inspection of ignimbrite fans was not possible on a regular basis, especially during the first 2 months after emplacement of primary ignimbrite. Sporadic ash venting, overcast or stormy weather, frequent phreatic explosions, and widespread fumarolic activity made the ignimbrite fans largely inaccessible even by helicopter, as most pilots refused to operate under these conditions. Consequently, the timing of some secondary pyroclastic-flow events cannot be established.

One large inferred event left an escarpment in the northern portion of the ignimbrite fan of the Balin Baquero watershed, about 1 to 2 km southwest of the Maraunot headscarp, that was noted from posteruption radar imagery flown in November 1991 (see fig. 4B of Newhall and others, this volume). Because the headscarp and valley morphology are very similar to those created by secondary pyroclastic flows in the Maraunot and Marella River watersheds, we believe that the same process operated in the northern Balin Baquero watershed. Although it is difficult to fix a specific date for the secondary pyroclastic flow in the northern Balin Baquero, greater incision of its headscarp and valley floor hints that it occurred earlier than the nearby, August 1991 Maraunot secondary pyroclastic flow.

Two additional secondary ignimbrites—one in the southern part of the Balin Baquero watershed and another in the upper Bucao watershed—were identified during an aerial survey a few days after heavy rainfall on September 20–21, 1992. At 0957 on September 21, an approximately 18-km-high ash plume was observed by the weather radar station at Cubi Point Naval Air Station, south of Pinatubo. Fallout from this ash cloud affected wide areas, including Manila, 90 km southeast of Pinatubo. The southern Balin Baquero and upper Bucao secondary pyroclastic flows each remobilized 0.035 to 0.045 km^3 of primary ignimbrite. It is not certain whether these two secondary pyroclastic flows are coeval or not. No other distinct buildup of ash clouds was noted between July 13, 1992, and the time these deposits were found in late September 1992.

Smaller secondary pyroclastic flows with runout distances of about 1 km were also noted during fieldwork in the middle Sacobia valley in November 1992 and in the Balin Baquero and Marella valleys during fieldwork in July 1993. These smaller secondary pyroclastic flows are surely common, but their deposits are quite ephemeral, as lahars rework or bury them within a short time.

CHARACTERISTICS OF THE SECONDARY IGNIMBRITE

Secondary ignimbrites of Pinatubo are distributed along major river valleys that drain from the 1991 pyroclastic fans, namely the Sacobia, Pasig-Potrero, Marella, Balin Baquero, Maraunot and Bucao Rivers. The largest secondary pyroclastic flows occurred along the western flanks of

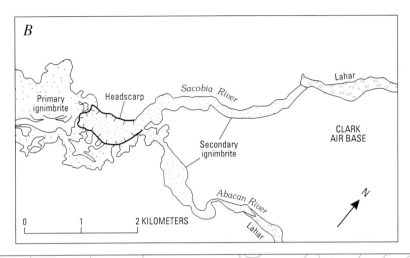

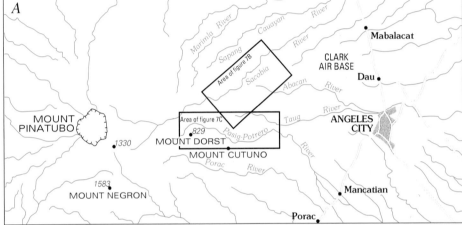

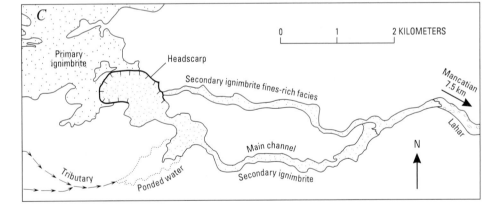

Figure 7. *A,* Location of the Sacobia, Abacan, and Pasig-Potrero River valleys. *B,* Primary and secondary ignimbrites of the Sacobia and Abacan Rivers. A 10- to 15-m-thick block of primary ignimbrite was remobilized as secondary pyroclastic flow into both the Sacobia and Abacan Rivers. The degradation at the headscarp area caused the Abacan River to be cut off from the upper drainage system it shared with the Sacobia River. *C,* Secondary pyroclastic flow that travelled 5 to 6 km from the source headscarp along the main channel of the Pasig-Potrero River. A fines-rich facies spilled from a notch in the escarpment. This secondary pyroclastic flow also dammed a tributary to the main channel, which broke out 6 weeks later and caused devastating lahars in the Mancatian area.

Mount Pinatubo in the Marella, Balin Baquero, and Maraunot River valleys, which also received the most extensive and thickest primary ignimbrite. These large secondary ignimbrites have volumes in the range 0.01 to 0.05 km³; smaller secondary ignimbrites also occur, but, as noted above, are easily lost from the geologic record. Deposit characteristics are similar regardless of size.

Because secondary pyroclastic flows at Pinatubo followed valleys, it is suggested that they were dense flows. In at least one instance, along the Pasig-Potrero River, a dilute, fines-rich facies overflowed banks near the source. The

flow did not erode substrate over which it flowed. Compared to other pyroclastic flows of comparable volume, Pinatubo secondary pyroclastic flows had a low ratio of vertical drop (H) to runout distance (L) (fig. 9, after Hayashi and Self, 1992), which indicates unusual mobility.

The main body of the secondary ignimbrite is dry, massive, loosely consolidated, and poorly sorted (σ_ϕ = 2.0–3.0). Grain-size distribution of the secondary ignimbrite (fig. 10) is very similar to that of the primary ignimbrite (fig. 2; comparison in fig. 12). Calculated grain-size parameters, such as median grain diameter and sorting, plot within

Figure 8. Photographs of the main channel of the Pasig-Potrero River taken a month before (above) and 5 days after (below) the July 14, 1992, secondary pyroclastic-flow event, near the confluence of the main channel and the fines-rich arm of that flow (fig. 7C). The secondary pyroclastic flow caused about 10 to 12 m of aggradation in the main channel. Channel walls consist of old columnarly jointed ignimbrites. Recent lahars form the terraced deposits in the channel.

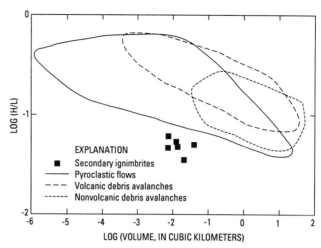

Figure 9. Log of H/L versus log of volume (in cubic kilometers) of secondary pyroclastic flows. H/L is the ratio of the vertical drop (H) to runout distance (L) of the secondary pyroclastic flows. Fields of pyroclastic flows, volcanic debris avalanches, and non-volcanic debris avalanches after Hayashi and Self (1992).

the field of the pumice-rich primary ignimbrite at Mount Pinatubo. A major mode appears in the 1- to 2-ϕ (crystal-dominated) size range. Internally, the secondary pyroclastic-flow deposits strongly resemble a massive primary ignimbrite. A section of a proximal facies of secondary ignimbrite (fig. 11) exhibits inverse grading, gas escape pipes, surface fumaroles, and a dense lithic-rich basal layer. Medial and distal facies are generally massive and have a poorly defined basal layer (layer 2a of Sparks, 1976) and dense lithic segregations. Incorporated accretionary lapilli remained as coherent particles near the base. Secondary deposits lack the characteristic cohesiveness of laharic debris-flow deposits found in the same area.

Clasts in the secondary ignimbrite consist dominantly of pumice, which is strongly concentrated in the upper part

of the flow. Charred wood debris on the surface was oriented perpendicular to the flow direction along the flow axis but subparallel along the margin. The "frozen" fabric of the rafted pumice and floating wood debris indicates that the maximum flow velocity was along the axis of flow and decreased toward the margin. Insignificant overbanking around bends of the channel (as deduced from absence of swash marks and nearly horizontal cross-sectional profiles) suggests a very low flow velocity, approximately a few meters per second.

TRM polarity distribution of the juvenile pumice clasts revealed no consistent orientation (fig. 4). Temperatures of 240–260°C were measured at points of steam emission on the surface of secondary ignimbrite at Pasig-Potrero valley five days after its emplacement. A temperature of 240°C was determined from an exposed 5- to 7-m-thick section of secondary ignimbrite 3 months after it was deposited at Sacobia valley. Together, these observations suggest typical emplacement temperatures of between 250°C and 300°C.

Secondary pyroclastic flows were commonly accompanied by secondary explosions and ash columns. Several of these ash columns were detected by weather radar and ground-based observation at heights comparable to violent vent ejections. Some events were not observed at all because they occurred during the night or cloudy weather. However, ash plumes were inferred from new ash layers on the surface of the ignimbrite fan and traces of ash on downwind vegetated areas. Though widely dispersed, the fallout deposits are preserved only near their source escarpments and are ephemeral elsewhere.

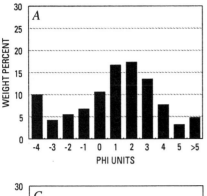

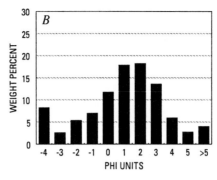

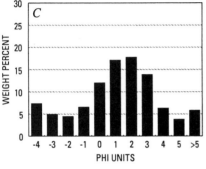

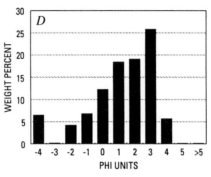

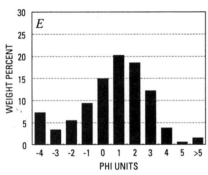

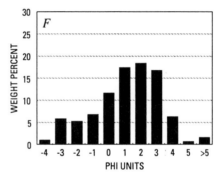

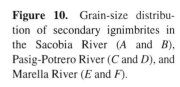

Figure 10. Grain-size distribution of secondary ignimbrites in the Sacobia River (*A* and *B*), Pasig-Potrero River (*C* and *D*), and Marella River (*E* and *F*).

DISCUSSION

POSSIBLE MECHANISMS OF SECONDARY PYROCLASTIC FLOWS

The generation of secondary pyroclastic flows involves a massive remobilization of hot, uncompacted primary ignimbrite. We envisage that two processes govern secondary pyroclastic-flow generation: overall reduction of friction within the primary ignimbrite and subsequent gravitational collapse. A decrease in the yield strength may result from an increase in pore pressure; gravitational collapse may be facilitated by the decrease in yield strength and by reduction of lateral support during channel erosion. The various thermodynamic and hydrologic conditions needed to cause a significant pore pressure increase and slope instability are yet to be resolved. Nevertheless, general aspects of the generation, triggering, and transport of secondary pyroclastic flows are outlined below.

Much of the pore fluid in the ignimbrite consists of water and steam in the originally dry interior of the primary

ignimbrite. The manner by which water is introduced is unclear, but possibilities include upflow of ground water at the base of the ignimbrite induced by static loading on a porous and saturated river bed, percolation of rain water through cracks and open fumarolic pipes, and seepage from lahars, streamflows, and springs. As water recharges the hot primary ignimbrite, it is heated and a significant proportion is vaporized. As a result, the ignimbrite body becomes a pressurized system and remains as such prior to collapse. Poor internal permeability (due to a high proportion of fines), high lithostatic gradient, and a confining layer of fine ashfall prevent a significant release of pore pressure. Pore water vaporization decreases the stability of primary ignimbrite, priming it for collapse.

Heavy rainfall, channel erosion, secondary explosions, and earthquakes are possible triggers for collapse of the primary ignimbrite and, thus, for the generation of secondary pyroclastic flows. The close association of heavy rain and active lahars with secondary pyroclastic flows suggests a dominant role for rain and lahars in triggering secondary pyroclastic flows. Rain and lahars on the surface can load

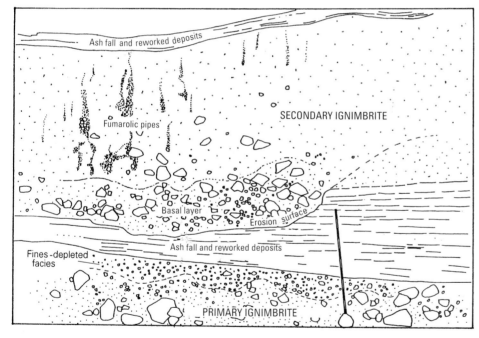

Figure 11. Vertical section of secondary ignimbrite and underlying lithic-rich primary ignimbrite in the Marella River valley. Sketch is made from the above photograph. The section is close to the valley wall and the source escarpment. The secondary ignimbrite is deposited on an erosional surface cut into ash fall and reworked deposits overlying a fines-depleted facies of the primary ignimbrite. The massive section of secondary ignimbrite also shows a basal layer and fumarolic pipes. When the photograph was taken (November 1992), the underlying primary ignimbrite displayed active fumaroles that deposited yellow sublimates on the wall of the section. About 10 to 20 cm of ash fall and reworked deposits cover the section and drape the surrounding topography. Upright scale is 1 m long.

and perhaps destabilize an unstable block of primary ignimbrite. Lahars also cause deep incision on ignimbrite fans, erode channel walls into oversteepened and overhanging sections, and induce localized slope failures along the channel. The failing mass disintegrates into a small-scale cohesionless turbulent flow, accompanied by a convection of fine elutriates. Scarp collapse exposes fresh sections of primary ignimbrite, which sometimes react explosively in contact with lahars. Secondary explosions promote additional slope instability by forming steep escarpments and phreatic explosion craters. The combined effects of channel erosion and secondary explosions critically reduce the lateral

support of potential failure surfaces. We also envisage that some avalanching involves rapid decompression of pore pressure at the collapsing front and triggers a runaway upslope propagation of the failure surface.

Felt earthquakes occurred on the same day as the Maraunot event, but many other earthquakes of similar magnitude were not followed by secondary pyroclastic flows. We do not claim an association, but we can envision, where unstable blocks are primed for failure, that ground shaking could induce compaction of the primary ignimbrite and, therefore, increase the pore pressure. If the pore pressure approximated the lithostatic load, the ignimbrite would lose its yield strength, undergo liquefaction with gaseous fluids as fluidizing agents, and flow downvalley over a very gentle gradient.

Transport of remobilized ignimbrite on a very gentle slope for several kilometers requires a mechanism that sustains fluidity. Characteristically low H/L ratios, the abundance of fumaroles on secondary ignimbrite surfaces, and the common association of high ash columns suggest that high gas pressure fluidizes the secondary pyroclastic flows until they are degassed to the point that they begin to deposit, from the bottom up. The absence of basal erosion, the massive character of deposits, channel-confined flow, and apparent slow flow velocities all suggest a high-concentration laminar flow in the depositional regime. Density and size segregation occur during transport of secondary pyroclastic flows, where dense block-size lithics settle near the headscarp area while large pumices are carried downstream and progressively segregated to the top of the flow.

THE DISTINCTION BETWEEN PRIMARY AND SECONDARY IGNIMBRITE

The recognition of secondary ignimbrite at Pinatubo makes detailed interpretation of older ignimbrites more difficult. Having very similar sedimentary characteristics, primary and secondary ignimbrites are hardly distinguishable in the field unless one has knowledge of the depositional site before and after the events. However, some points of distinction can be made, including slight fines depletion, random TRM polarity of clasts in secondary ignimbrites, and, in well-incised sections, the nature of the underlying unit.

Secondary ignimbrite tends to be fines depleted relative to the primary ignimbrite from which it was derived (fig. 12). Some of the primary fine fraction is lost as elutriate that forms associated ash columns and secondary co-ignimbrite ash. Fines depletion also suggests that the secondary flowage at Pinatubo is gentle and does not grind up a new generation of fines to replace that being lost.

TRM polarity distribution of the clasts provides a more promising, yet not infallible, way to distinguish primary and secondary ignimbrite deposits. The uniform or bimodal TRM polarity of pumice clasts in primary ignimbrite at Pinatubo is in contrast to the generally random distribution in the secondary ignimbrite. Elsewhere, in primary ignimbrite that is dominated by magnetite and lacks self-reversing ferrian ilmenite (Nord and Lawson, 1989), the magnetic distinction between primary and secondary ignimbrite would be even more pronounced.

Because the chronology of events is known at Mount Pinatubo, identification of the underlying unit provides an added basis for recognition of secondary flow deposits. Primary ignimbrite usually overlies either the preeruption surface, the June 12 pyroclastic flow, the blast deposits, or the plinian fallout. By contrast, most secondary ignimbrites were emplaced on lahar or stream deposits that postdate the June 15 eruption.

IMPACTS OF SECONDARY PYROCLASTIC FLOWS

There were no known fatalities during major secondary flow events at the Maraunot, Marella, and Balin Baquero events. These events occurred in very remote locations and at a time when people were still reluctant to venture near designated danger zones. However, the April 4, 1992, event, which was smaller than earlier flows, nearly caught people who were working in the Sacobia and Abacan River channels. The more serious impacts of secondary pyroclastic flows at Mount Pinatubo, such as burial and burning, are limited to the channel and areas proximal to the source. Light fallout of very fine ash from secondary co-ignimbrite and phreatic explosion columns affects distal areas. Ash plumes formed during secondary pyroclastic flows can also pose serious hazards to air traffic above the vicinity of Mount Pinatubo, as they can rise to the cruising altitudes of commercial airplanes (see description of August 1993 encounter, Casadevall and others, this volume).

Changes of morphology in the headscarp region and the emplacement of thick secondary ignimbrites alter the channel conditions for lahars and indirectly trigger other hazards. The April 4, 1992, event allowed the Sacobia to capture all flow from the Abacan River. Thus, lahar hazard was sharply reduced along the Abacan and raised along the Sacobia. No true debris flow has occurred in the Abacan channel since the April 4 secondary pyroclastic flow. Rapid aggradation of the channel can dam tributary streams and create a less efficient lahar-conveyance system. After the July 13, 1992, secondary pyroclastic flow in the Pasig-Potrero watershed, lahars stopped for more than a month, even during periods when the nearby Sacobia River had large lahars. On August 29, 1992, a large-volume, sustained lahar came down the Pasig-Potrero River, due to newly reintegrated drainage and breakout of water that had ponded behind a dammed tributary at the foot of Mount Cutuno (Arboleda and Martinez, this volume). Residents of the Mancatian area were surprised, and much damage occurred.

CONCLUSIONS

Secondary pyroclastic flows at Pinatubo are a documented example of a little known process of ignimbrite remobilization in a relatively dry state. Secondary movements in this sense had been suspected at the Valley of Ten Thousand Smokes, Alaska, and may be common after large ignimbrite-producing eruptions. However, distinguishing between primary, vent-derived ignimbrite and secondary

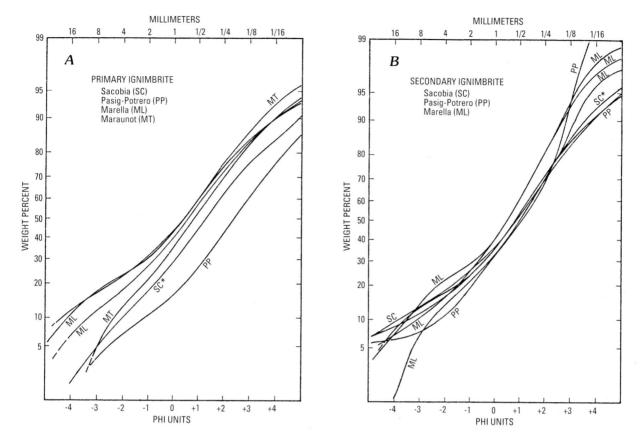

Figure 12. Size frequency curves of the (*A*) primary and (*B*) secondary ignimbrites plotted as cumulative weight percent on normal probability scale versus the particle size in φ units and millimeters. A primary and secondary ignimbrite from the Sacobia valley, shown as SC*, were found to be in depositional contact and are shown individually, for comparison.

ignimbrite is difficult. So far, at Pinatubo, the distinction lies in the recognition of fines depletion and non-uniform TRM polarity of clasts in secondary ignimbrite. Problems may persist when the loss of fines during transport is negligible or when the primary ignimbrite is emplaced at relatively low temperature and (or) contains abundant self-reversing ferrian ilmenite.

Secondary pyroclastic flows at Pinatubo were generated by massive avalanching of the primary ignimbrite. Pore fluid vaporization provides the mechanism for a positive pore pressure within the ignimbrite, while erosion at the toe of the slope of ignimbrite further decreases the stability of the mass. Failure is triggered by heavy rain, lahars, erosion of critical footslopes, secondary explosions, and perhaps also by earthquakes.

Secondary pyroclastic flows differ from secondary movements in ignimbrites noted by previous workers in that they (1) occur sporadically for months or years after deposition of the primary ignimbrite, (2) are generated in all parts of ignimbrite fans, (3) involve both gravity and gas fluidization, and (4) exhibit internal sedimentary characteristics very similar to valley-ponded primary ignimbrite. Unlike rheomorphic ignimbrite that requires a steep slope or temperatures sufficiently high to permit movement of the

agglutinating pumice and glass shards, secondary pyroclastic flows may be generated on gentle slopes and at much lower temperatures.

Primary and secondary ignimbrite deposition represents a continuum of flow events beginning from the initial emplacement and continuing long after deposition. Final distribution of ignimbrite will be the result of a series of events of reworking and remobilization, which persist until the thermal energy of the system, the internal fluid pressure, the available water, or the channel gradient is no longer capable of mobilizing the pyroclastic materials. Many secondary ignimbrites may well have been mapped in the field as primary ignimbrite, and, for most purposes, they are the same. However, at active volcanoes, secondary pyroclastic flows need to be recognized as an important redepositional process and a potentially serious, lingering hazard of explosive volcanism.

ACKNOWLEDGMENTS

We thank Wes Hildreth, Jim Riehle, Ray Punongbayan, George Walker, Tess Regalado, and Chris Newhall for valuable comments; the helicopter pilots, officers, and staff of the Philippine Air Force at Clark Air Base for providing aerial views and access to remote sections; Chris Newhall and Jim Anderson for lending us fluxgate magnetometers; our Aeta field guides, especially Boy Tanglaw; and our colleagues at the Philippine Institute of Volcanology and Seismology (PHIVOLCS), especially the staff of the Pinatubo Volcano Observatory and the Lahar Monitoring Group, for generously sharing information. The study was supported by the U.S. National Science Foundation (NSF) and PHIVOLCS. This is SOEST contribution no. 3556.

Note: An ash column of the type observed during secondary pyroclastic flow generation occurred on 11 July 1995, reaching 9 to 10 km in altitude. This event demonstrates that this phenomenon has persisted and still poses a considerable hazard on the ground and in the air 4 years after the main eruption.

REFERENCES CITED

Branney, M.J., and Kokelaar, P., 1992, A reappraisal of ignimbrite emplacement: Progressive aggradation and changes from particulate to non-particulate flow during emplacement of high-grade ignimbrite, Bulletin of Volcanology, v. 54, p. 504–520.

Casadevall, T.J., Delos Reyes, P.J., and Schneider, D.J., this volume, The 1991 Pinatubo eruptions and their effects on aircraft operations.

Chapin, C.E., and Lowell, G.R., 1979, Primary and secondary flow structures in ash-flow tuffs of the Gribbles Run paleovalley, central Colorado, in Chapin, C.E., Elston, W.E., eds., Ash flow tuffs: Geological Society of America Special Paper 180, p. 137–154.

Ellwood, B.B., 1982, Estimates of flow directions for calc-alkaline welded tuffs and paleomagnetic data reliability from anisotropy of magnetic susceptibility measurements: Central San Juan Mountains, southwest Colorado: Earth and Planetary Science Letters, v. 59, p. 303–314.

Fierstein, J., and Hildreth, W., 1992, The plinian eruptions of the 1912 at Novarupta, Katmai National Park, Alaska: Bulletin of Volcanology, v. 54, p. 646–684.

Fisher, R.V., Glicken, H.X., and Hoblitt, R.P., 1987, May 18, 1980, Mount St. Helens deposits in South Coldwater creek,

Washington: Journal of Geophysical Research, v. 92, no. B10, p. 10267–10283.

Hayashi, J.N., and Self, S., 1992, A comparison of pyroclastic flow and debris avalanche mobility: Journal of Geophysical Research, v. 97, p. 9063–9071.

Hildreth, W., 1983, The compositionally zoned eruption of the 1912 in the Valley of Ten Thousand Smokes, Katmai National Park, Alaska: Journal of Volcanology and Geothermal Research, v. 18, p. 1–56.

Hoblitt, R.P., Miller, C.D., and Vallance, J.W., 1981, Origin and stratigraphy of the deposit produced by the May 18 directed blast, in Lipman, P.W., and Mullineaux, D.R., eds., The 1980 eruptions of Mount St. Helens, Washington: U.S. Geological Survey Professional Paper 1250, p. 401–419.

Newhall, C.G., Daag, A.S., Delfin, F.G., Jr., Hoblitt, R.P., McGeehin, J., Pallister, J.S., Regalado, M.T.M., Rubin, M., Tamayo, R.A., Jr., Tubianosa, B., and Umbal, J.V., this volume, Eruptive history of Mount Pinatubo.

Nord, G.L., Jr., and Lawson, C.A., 1989, Order-disorder transition-induced twin domains and magnetic properties in ilmenite-hematite: American Mineralogist, v. 74, p. 160–176.

PHIVOLCS-NEDA, 1992, Pinatubo Lahar Hazards Map: Manila, Philippine Institute of Volcanology and Seismology and the National Economic Development Authority, 1 sheet at scale 1:200,000; 4 sheets, Quadrants 1–4, at scale 1:100,000.

Pinatubo Volcano Observatory Team, 1991, Lessons from a major eruption: Mt. Pinatubo, Philippines: Eos, Transactions, American Geophysical Union, v. 72, no. 49, p. 545, 552–553, 555.

Schmincke, H-U., and Swanson, D.A., 1967, Laminar viscous flowage structures in ash-flow tuffs from Gran Canaria, Canary Islands: Journal of Geology, v. 75, p. 641–664.

Scott, W.E., Hoblitt, R.P., Torres, R.C., Self, S, Martinez, M.L., and Nillos, T., Jr., this volume, Pyroclastic flows of the June 15, 1991, climactic eruption of Mount Pinatubo.

Smithsonian Institution, 1991, Pinatubo: Bulletin of the Global Volcanism Network, v. 16, no. 8, p. 4-6.

Sparks, R.S.J., 1976, Grain size variations in ignimbrites and implications for the transport of pyroclastic flows: Sedimentology, v. 23, p. 147–188.

Walker, G.P.L., Hayashi, J.N., and Self, S., in press, Travel of pyroclastic flows as transient waves: implications for the energy line concept and particle-concentration assessment: Journal of Volcanology and Geothermal Research.

Wolff, J.A., and Wright, J.V., 1981, Rheomorphism of welded tuffs: Journal of Volcanology and Geothermal Research, v. 10, p. 13–34.

A STORY IN THE ROCKS

The eruption of Pinatubo carried samples of ash far and wide, but a cooperative effort by field geologists and petrologists spread samples of Pinatubo pumice even farther and almost as fast. A number of fascinating discoveries ensued.

The lava dome that preceded explosive eruptions of 1991 consisted of hybrid andesite, the product of mixing of an unusually hydrous olivine-pyroxene basalt (present as quenched inclusions) and phenocryst-rich dacite (mostly assimilated into the mixture) (Pallister and others; Bernard and others; Hattori). Andesitic pumice of the June 12 eruption, compositionally identical to andesite of the dome, was gradually replaced by dacite in eruptions of June 13–15 (Hoblitt, Wolfe, and others; Pallister and others). Compositions intermediate between andesite and dacite were not found; rather, proportions of the two lithologies changed.

The climactic eruption produced two types of dacitic pumice--one was tan-gray and phenocryst poor, the other was white and phenocryst rich. The white was always dominant (80–90%); the proportion of tan-gray pumice decreased from about 20% to 10% from bottom to top in the tephra-fall deposit (David and others). The differences are mainly textural; hence, Pallister and others make an intriguing suggestion that intense explosions shattered phenocrysts of the once-phenocryst-rich magma into tiny crystal fragments of the phenocryst-poor magma. Luhr and Melson suggest that mechanical fragmentation occurred upon violent mixing of melt-rich dacitic magma into crystal-rich dacitic magma.

The dacitic magma was volatile saturated, exceptionally rich in water (Rutherford and Devine), and had a high oxygen fugacity (previously cited papers, plus Gerlach and others; Hattori; Imai and others). Most authors estimated the preeruption temperature of the dacitic magma to be about 780–800°C, on the basis of Fe-Ti oxide thermometry and the presence of cummingtonite as a stable phase. Luhr and Melson give a higher range, 840–890°C, based on the oxide thermometers of Ghiorso and Sacks (1991) and Anderson and others (1993). They attribute the discrepancy to either a late-stage heating event that affects Fe-Ti oxides but not cummingtonite or to the need for a correction for highly oxidizing conditions (as applied by Rutherford and Devine and other authors). Preeruption pressure is estimated to have been about 2.2±0.5 kbar (7–11 km depth) from the Al-in-hornblende geobarometer, confirmed and refined to about 2.0 kbar by experimentally determined phase relations for the cummingtonite-bearing dacite (Rutherford and Devine).

Anhydrite was a relatively abundant, primary phenocryst phase. Sulfur isotopic composition of single grains suggested that anhydrite of the June 12 (vent-clearing) eruption was both primary and hydrothermal in origin, while that of the June 15 eruption was mostly primary (McKibben and others); sulfur of anhydrite in basalt was so light (about 1 per mil) that it is probably not the sole source for the bulk sulfur isotopic composition of June 15 dacite of 5 to 7 per mil. The presence of anhydrite inclusions in hornblende and plagioclase phenocrysts suggests to Fournelle and others that anhydrite was present in magma at depth and that prehistoric sulfur-rich eruptions might be found by a search for such inclusions. Pasteris and others note an absence of anhydrite inclusions in their samples and conclude (on that basis in part) that the dacitic magma reached saturation with respect to a CO_2-SO_2-H_2O fluid *before* anhydrite saturation. The two conclusions are potentially compatible because, as concluded by Gerlach and others, a separate volatile phase probably existed at depth, potentially before anhydrite crystallization and yet still early and deep enough for subsequent anhydrite to be included within plagioclase and hornblende.

The high oxygen fugacity and high sulfur content of Pinatubo magma are closely related; the former makes the latter possible. Indeed, Pinatubo magma may be a prime example of an important class of hydrous, highly oxidizing, sulfur-rich, anhydrite-bearing magmas that produce eruption clouds so rich in SO_2 that they can influence global climate (Gerlach and others; Luhr and Melson; Self and others). The high oxygen fugacity might also inhibit early, deep sulfide fractionation and allow abundant sulfur to remain in the magma until that magma is emplaced at shallow depths, where porphyry copper deposits can then form (Hattori; Imai and others).

The immediate source of sulfur of the eruption cloud is still under debate, as discussed in an earlier section by Gerlach and others and in this section by Rutherford and Devine, Pallister and others, Bernard and others, McKibben and others, Fournelle and others, and Pasteris and others. Evidence for a separate volatile phase at depth is strong; some contribution from anhydrite breakdown during magma ascent also seems possible.

Oxygen isotope data (Fournelle and others) and compositions and textures of olivines, pyroxenes, and spinels (Pallister and others) argue against significant contamination from the underlying Zambales Ophiolite Complex; beryllium isotopes show no evidence for large input from marine sediments (Bernard and others). Strontium-rubidium and neodymium-samarium isotope ratios for dacite and andesite pairs are identical; basalt has slightly more radiogenic strontium and slightly less radiogenic neodymium (Bernard and others). Bernard and others considered lead isotopes in andesite and dacite to be identical and to thus reflect a common origin. In contrast, Castillo and Punongbayan found that dacite was slightly less enriched in incompatible trace elements and in $^{87/86}$Sr a $^{206/204}$Pb and slightly higher in $^{143/144}$Nd than the andesite. Castillo and Punongbayan attributed the difference to variable mixtures of depleted basalts (mantle source) and an enriched component from subducted sediments.

In an unusual confluence of geophysical and petrologic studies, seismic monitoring captured an intrusion of basalt into a silicic magma reservoir (White), thereby providing unusual time constraints on magma mixing and disequilibrium reequilibration processes (Pallister and others; Rutherford and Devine). Estimates from this natural laboratory are broadly consistent with those from controlled laboratory experiments, and efforts to resolve minor differences in time estimates may bring yet more insights into these processes.

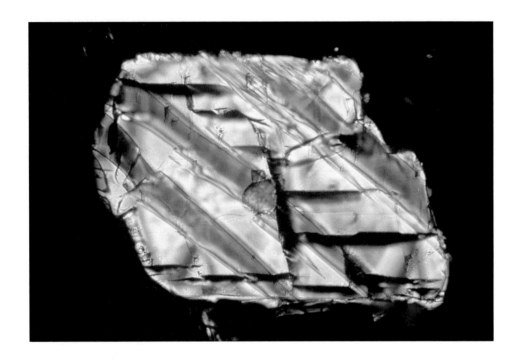

Subhedral anhydrite in phenocryst-rich dacite pumice of the climactic, June 15, 1991, eruption. Crossed nicols; field of view, 0.34 mm maximum dimension. Photograph by John Pallister (USGS).

Changing Proportions of Two Pumice Types from the June 15, 1991, Eruption of Mount Pinatubo

By Carlos Primo C. David,[1] Rosella G. Dulce,[2] Dymphna D. Nolasco-Javier,[1] Lawrence R. Zamoras,[1] Ferdinand T. Jumawan,[1] and Christopher G. Newhall[3]

ABSTRACT

Pumice erupted by Mount Pinatubo on June 15, 1991, was of two types: a white, phenocryst-rich variety and a tan phenocryst-poor variety. The volumetric proportion of phenocryst-poor pumice decreased through the course of the eruption, from about 20 percent at the base of the plinian fall unit to about 10 percent at its top. The progressive change might reflect early tapping of a hotter, deeper, phenocryst-poor part of the magma reservoir or, alternatively, early and unusually explosive tapping of the top of the magma reservoir. Abundant broken crystal fragments in the groundmass of the phenocryst-poor groundmass and preservation of the same phenocryst assemblage in the two pumice types favor the second explanation.

INTRODUCTION

Several workers have noted that June 15 eruption of Mount Pinatubo produced two distinct pumice types, one rich in phenocrysts (fig. 1) and the other poor in phenocrysts (fig. 2) (Bernard and others, this volume; Imai and others, this volume; Luhr and Melson, this volume; Pallister and others, 1992; this volume). These workers agree that the two pumice types are chemically similar dacites (64–65 wt% SiO_2) and that the principal difference between the two types is a textural difference, probably resulting from fragmentation of crystals in the phenocryst-poor type. Yet, there is also permissive evidence of some degree of crystal resorption by disequilibrium melting (Luhr and Melson, this volume; Pallister and others, this volume).

The dominant pumice type, which constitutes about 80–90 vol% of the lapilli and blocks observed in June 15 tephra-fall deposits, is white, contains large interconnected

vesicles, and large (>3-mm-diameter) phenocrysts. The subordinate type is tan, contains small (<2-mm-diameter) spherical vesicles and has few crystals larger than 1 mm. In addition, fluidal banding appears more common in the phenocryst-rich than in the phenocryst-poor pumice. The two pumice types are readily distinguished in the field by color and by texture.

In this short paper we report the abundance of the two pumice types in several stratigraphic sections through the June 15 tephra-fall deposit, and we examine whether and how the proportions of each type changed through the course of the eruption.

SAMPLES AND METHODS

Ten samples each of phenocryst-rich and phenocryst-poor pumice were collected from the June 1991 pyroclastic-flow deposit of Mount Pinatubo, in the valley of the Sacobia River, about 500 m northwest from Gate 14 of Clark Air Base (fig. 3). Thin sections of these samples were prepared and examined to compare with the results of other workers and to verify our ability to distinguish the two pumice types in the field. Both pumice types contain similar crystal assemblages that are dominantly plagioclase and hornblende, with subordinate quartz, cummingtonite (as rims on hornblende crystals), biotite, magnetite, ilmenite, anhydrite, apatite, olivine (rimmed by hornblende), and trace zircon. Modal analyses of our thin sections are not presented here because of plucking of large grains from the phenocryst-rich pumice during section preparation and because of difficulty in optical counting the abundant, but very small, crystal fragments in the phenocryst-poor pumice.

Samples of tephra-fall deposit, for component analysis, were taken at seven sites that are roughly 25 km east and northeast of Mount Pinatubo's crater. Sampling 6 months after the eruption, we encountered some difficulty in finding complete and undisturbed sections of the June 15 coarse tephra deposits. At each site, the coarse tephra layer was divided into three stratigraphic intervals: bottom (L1), middle (L2), and top (L3). In some cases where the June 15

[1] University of the Philippines, National Institute of Geological Sciences, Diliman, Quezon City.

[2] Philippine National Oil Company, Geothermal Division, Fort Bonifacio, Metro Manila.

[3] U.S. Geological Survey.

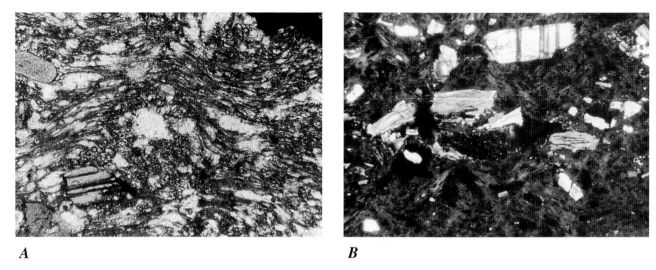

A   *B*

Figure 1. Phenocryst-rich pumice of June 15, 1991. *A,* Prominent stretching of the glassy matric of the phenocryst-rich pumice (sample CR–9, uncrossed polars, 63×). *B,* Phenocryst-rich pumice under crossed polars (sample CR–9, crossed polars, 25×). Vesicles in the phenocryst-rich pumice are larger and interconnected along prominent flow structures of the glassy matrix.

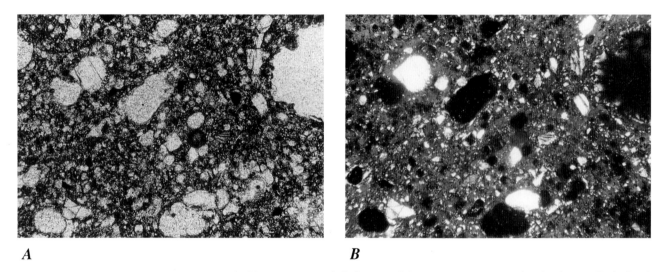

A   *B*

Figure 2. Phenocryst-poor pumice of June 15, 1991. The characteristic features of the phenocryst-poor pumice showing smaller isolated, nonelongated vesicles and a glassy matrix with generally no observable flow structure (sample A–7; *A,* uncrossed polars and *B,* crossed polars, 63×).

tephra layer was less than 3.5 cm thick, the section is only divided into two intervals, L1 and L2 (fig. 4). Approximately 200 pumice fragments >2 mm in diameter were counted per sample layer.

PROPORTIONS OF TWO PUMICE TYPES

The dominant pumice type in the June 15 coarse tephra layer is the phenocryst-rich type (table 1). It comprises 80-90% of the coarse tephra deposits, while the phenocryst-poor type comprises the remaining 10–20%. In sections less than 3.5 cm thick, the phenocryst-poor percentage is greater in the bottom interval L1, than in the top interval L2. In

sections having three sampled intervals (Floridablanca, Porac, and Clark), the phenocryst-poor percentage at the bottom interval (L1) is greater than in the topmost layer (L3), and the middle layer (L2) contains varying proportions of the two types.

We think these results are representative of tephra fall on all sides of the volcano, including down the principal axis (southwest from Pinatubo). Field counts of pumice types by R.P. Hoblitt (USGS, written commun., 1992) and subsequent, similar counts from tephra sampled in San Antonio, Zambales (southwest of Pinatubo), over several time increments during the climactic eruption (K. Rodolfo and C. Arcilla, Univ. of Illinois, Chicago, written commun., 1993) both gave results consistent with ours. However, for

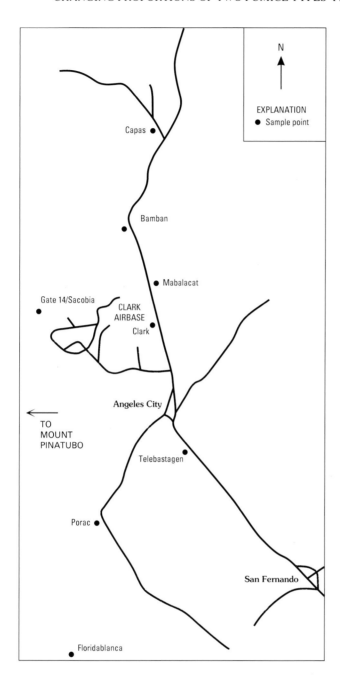

Figure 3. Location map for June 15 coarse tephra samples. Scale is approximately 1:250,000.

completeness, we note an inconsistency with the results of an early, impromptu field count of pumice types by several of us at Dalanaoan (south-southwest of Pinatubo). That initial survey, done without size grading or washing of pumices, suggested the opposite sense of change, in which the phenocryst-poor variety increased upward in the section, from 10% near the base to 21% near the top of the coarse plinian pumice-fall deposit. If subsequent investigations bear out this inconsistency, we would have to conclude that

local wind patterns were a factor and that the tephra southwest of the volcano is not strictly contemporaneous with that east of the volcano.

DISCUSSION

We explore three hypotheses for the origin of the two pumice types. The first two hypothesis call on a magma

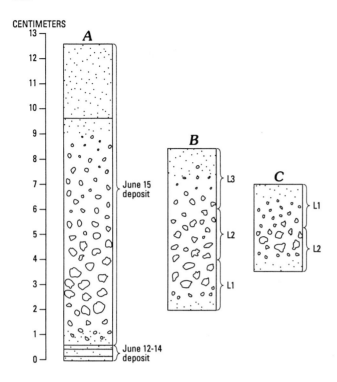

Figure 4. *A*, Generalized tephra section, as seen in Castillejos, Zambales, southwest of Pinatubo (not sampled; shown here for reference). *B*, Generalized tephra section for Floridablanca, Porac, and Clark Air Base. *C*, Generalized tephra section for Bamban, Capas, Mabalacat, and Telebastagen. All materials are pumiceous ash and lapilli, with variations in grain size as shown.

reservoir that was vertically zoned and tapped in such a way as to produce a dominant (and increasing) percentage of the phenocryst-rich type through the course of the eruption. At Pinatubo, we have no evidence of gross compositional zonation, but we can imagine some differences in temperature or volatile content of the magma that could have resulted in different textures of pumice. Higher temperature would inhibit crystallization (and enhance resorption of crystals formed at lower temperature) and thus would result in fewer phenocrysts. However, we can rule out a region of sustained higher temperature, because similar Fe-Ti oxide temperatures have been determined for the two pumice types (Pallister and others, this volume; Rutherford and others, this volume; Bernard and others, this volume; Luhr and Melson, this volume; Imai and others, this volume). This relation argues strongly against thermal zoning.

Alternatively, extreme volatile concentration at the top of the reservoir could have inhibited phenocryst development, and thus the earliest tapped magma (from the top of the reservoir) would have produced a relatively high percentage of phenocryst-poor pumice. Circumstantial evidence of volatile zonation includes the exceptionally high volatile content of erupted magma (Gerlach and others, this volume) and the fact that the erupted magma represents only a relatively small percentage of the whole Pinatubo magma reservoir (Mori, Eberhart-Phillips, and Harlow, this volume; Pallister and others, this volume). However, this hypothesis is not consistent with the presence of the same

Table 1. Relative proportions of phenocryst-rich and phenocryst-poor pumice, June 15, 1991, tephra fall.

Location	Thickness (in cm)	Layer	Avg. size (in mm)	Phenocryst poor (N)	Phenocryst rich (N)	Lithics (N)	Phenocryst poor (%)[1]	Phenocryst rich (%)[1]
Floridablanca....	4.5	L3	4.0	24	141	3	14	86
		L2	6.5	32	186	3	15	85
		L1	5.0	38	180	24	17	83
Porac................	6.5	L3	4.0	32	200	4	14	86
		L2	5.0	48	226	15	18	82
		L1	5.0	46	177	27	21	79
Clark................	4.0	L3	3.0	30	192	3	14	86
		L2	4.0	48	192	17	25	75
		L1	4.5	30	130	21	19	81
Bamban	3.0	L2	3.0	32	184	1	15	85
		L1	4.5	64	180	17	26	74
Capas...............	2.5	L2	2.5	25	183	3	12	88
		L1	3.0	42	180	4	19	81
Mabalacat........	3.5	L2	4.5	8	177	0	4	96
		L1	4.5	14	166	8	8	92
Telebastagen.....	3.5	L2	3.0	27	188	19	13	87
		L1	4.0	26	179	9	13	87

[1] Pumice counts (N) were normalized to 100% before relative proportions (%) were counted. Lithics were excluded from percent calculation.

phenocryst mineral assemblage in both types, including late-grown cummingtonite, which indicates that both pumice types had initially crystallized to the same extent, nor does it explain the fragmental textures in the phenocryst-poor pumice.

Our third and favored explanation is that the magma reservoir could have been more or less homogeneous until processes associated with the eruption itself changed the texture of some of the magma and produced two different pumice types (Pallister and others, this volume; Bernard and others, this volume). Unusually abundant, broken crystal fragments in the phenocryst-poor pumice suggest that unusually strong explosions might have produced that pumice type while more sustained ejection produced the phenocryst-rich pumice. In this scenario, the phenocryst-rich pumice is a vesicular equivalent of the reservoir dacite, and the phenocryst-poor pumice is a severely shocked derivative of the phenocryst-rich magma.

Our petrographic examination cannot resolve any significant difference other than the smaller average grain size and greater abundance of small crystal fragments in the phenocryst-poor variety. Thus, tentatively, we accept fragmentation as the simplest explanation for differences between the two pumice types. However, we cannot rule out the possibility that extreme volatile concentration near the top of the reservoir might also have caused some resorption and thus enhanced the effect of fragmentation on crystal size reduction.

CONCLUSION

The phenocryst-rich type, always dominant, increased in abundance through the course of the eruption. One scenario for this change, consistent with but unproven by our data, is that the uppermost, most volatile rich magma was erupted most explosively and a greater percentage of the phenocrysts were fragmented. Then, as magma with the highest volatile concentrations was depleted, less powerful explosions erupted the remaining magma with correspondingly less fragmentation of phenocrysts.

ACKNOWLEDGMENTS

We thank Richard Hoblitt and John Pallister for constructive reviews and John Pallister for further assistance during the revision process.

REFERENCE CITED

Bernard, A., Knittel, U., Weber, B., Weis, D., Albrecht, A., Hattori, K., Klein, J., and Oles, D., this volume, Petrology and geochemistry of the 1991 eruption products of Mount Pinatubo.

Gerlach, T.M., Westrich, H.R., and Symonds, R.B., this volume, Preeruption vapor in magma of the climactic Mount Pinatubo eruption: Source of the giant stratospheric sulfur dioxide cloud.

Imai, A., Listanco, E.L., and Fujii, T., this volume, Highly oxidized and sulfur-rich dacitic magma of Mount Pinatubo: Implication for metallogenesis of porphyry copper mineralization in the Western Luzon arc.

Luhr, J.F., and Melson, W.G., this volume, Mineral and glass compositions in June 15, 1991, pumices: Evidence for dynamic disequilibrium in the dacite of Mount Pinatubo.

Pallister, J.S., Hoblitt, R.P., Meeker, G.P., Knight, R.J., and Siems, D.F., this volume, Magma mixing at Mount Pinatubo: Petrographic and chemical evidence from the 1991 deposits.

Mori, J., Eberhart-Phillips, D., and Harlow, D.H., this volume, Three-dimensional velocity structure at Mount Pinatubo, Philippines: Resolving magma bodies and earthquakes hypocenters.

Pallister, J.S., Hoblitt, R.P., and Reyes, A.G., 1992, A basalt trigger for the 1991 eruptions of Pinatubo volcano?: Nature, v. 356, p. 426–428.

Rutherford, M.J., and Devine, J.D., this volume, Preeruption pressure-temperature conditions and volatiles in the 1991 dacitic magma of Mount Pinatubo.

Magma Mixing at Mount Pinatubo: Petrographic and Chemical Evidence from the 1991 Deposits

By John S. Pallister,[1] Richard P. Hoblitt,[1] Gregory P. Meeker,[1] Roy J. Knight,[1] and David F. Siems[1]

ABSTRACT

Four principal juvenile magmatic components are present in the 1991 deposits of Mount Pinatubo: hybrid andesite (59–60 weight percent SiO_2), olivine-clinopyroxene basalt (50–52 weight percent SiO_2) with abundant hornblende microphenocrysts, and phenocryst-rich and phenocryst-poor dacite (both 64.5±0.3 weight percent SiO_2). The first eruptions produced hybrid andesite as lava that fed the June 7–12 dome and then as scoria in the pre-paroxysmal vertical columns and surges of June 12–15. Olivine-clinopyroxene basalt occurs only as undercooled and quenched inclusions in the June 7–12 dome. Dacitic pumice first appeared as a minor component of the vertical eruption on June 12 and increased in abundance in the subsequent 1991 eruptions. Phenocryst-rich and phenocryst-poor dacitic pumice are the primary (~85 percent) and secondary (~15 percent) juvenile components of the June 15 climactic eruption. The phenocryst-poor dacite contains abundant small (<5 micrometers) crystal fragments, most of which were produced by mechanical breakage of phenocrysts during magma ascent.

The hybrid andesite contains a wide variety of phenocrysts: plagioclase, hornblende, cummingtonite, biotite, augite, olivine, Fe-Ti oxides, quartz, apatite, and anhydrite. Comparison of mineral compositions and textures indicates that the olivine (Fo_{86-89}), clinopyroxene (Mg# 75–85), and many of the hornblende phenocrysts were derived from basaltic magma similar to the inclusions in the June 7–12 dome, whereas the plagioclase, cummingtonite, and anhydrite phenocrysts are from June-15 type dacitic magma. Ilmenite grains in the andesite were probably derived from the dacitic magma, as indicated by similarity in composition to those in the June 15 dacite and by the presence of magnetite rims. Olivine-melt equilibria indicate the basaltic magma was relatively primitive (Mg# 68) and hot (~1,200°C) prior to mixing. Fe-Ti-oxide thermometry and cummingtonite stability indicate that dacite was relatively cool (~780°C) and highly oxidized (nickel-nickel oxide +3 log units) prior to mixing. The magnetite-rimmed ilmenites in the andesite yield intermediate Fe-Ti oxide temperatures (~950°C) and lower oxygen fugacity (nickel-nickel oxide +2 log units) than the dacite does. Evidence of disequilibrium in the hybrid andesite includes the wide variety and varied compositions of minerals and matrix glass and the presence of magnetite rims on ilmenites, hornblende rims on olivine, and clinopyroxene rims on quartz. Preservation of these disequilibrium features indicates that little time passed between magma mixing and eruption.

Origin of the andesite by magma mixing is also indicated by major and trace element compositions of pumice and scoria samples and by variable matrix glass compositions. Mass balance indicates that the andesite represents a 64:36 mix of dacite and basalt. Phase equilibria and seismic data suggest that magma mixing took place at about 8 to 9 kilometers in depth, within a large (>50 cubic kilometers) magma reservoir beneath the volcano. Buoyant ascent of hybrid andesitic magma in early June established a conduit to the surface and thereby triggered the eruption of about 5 cubic kilometers of the relatively cool, fluid-saturated dacitic magma on June 15. Without addition of basalt, the viscous, crystal-rich dacite would probably not have erupted in 1991.

Basalt may also be the ultimate source for 20 megatonnes of SO_2 that was added to the stratosphere by the June 15, 1991, eruption. We suspect that the high-level dacitic part of the >50 cubic kilometer magma reservoir was enriched in sulfur by multiple episodes of magma mixing in the past. Similar intrinsic conditions for magmas at Pinatubo and El Chichón suggest that volcanoes that give rise to sulfur-rich explosive eruptions are underlain by long-lived magma reservoirs with relatively cool, oxidized, fluid saturated, and crystal-rich upper zones that act as traps for volatiles that are supplied by basaltic magmas from greater depth.

INTRODUCTION

In a preliminary report (Pallister and others, 1992), we described evidence for mixing of basaltic and dacitic magmas shortly before the climactic eruption of Mount

---

[1] U.S. Geological Survey.

Pinatubo on June 15, 1991. Herein, we present more detailed petrographic and geochemical data for products of the 1991 eruptions; these data strongly confirm the magma mixing hypothesis. Not only was the mixed magma erupted, but both endmembers are also present in the 1991 deposits. We previously suggested that subordinate "crystal-poor dacite" of the June 15 eruption was produced by disequilibrium melting of phenocrysts during basalt underplating of crystal-rich dacite. In this report, we present new electron-microbeam images that show the so called crystal-poor dacite of June 15 to contain abundant micrometer-sized crystal fragments. Accordingly, we now refer to this pumice as *phenocryst*-poor instead of *crystal*-poor, and we favor a dominantly mechanical process of crystal size reduction. We follow Wilcox (1954) in placing the distinctions between phenocrysts and microphenocrysts at 0.3 mm (longest dimension) and in placing the distinctions between microphenocrysts and microlites at 0.03 mm. Analytical methods employed in this study are described in the appendix.

ERUPTIVE SEQUENCE

The 1991 eruptive sequence was reviewed by Wolfe (1992) and is presented in more detail in other chapters of this volume. In the discussion that follows, we will review only those events that are most relevant to the petrology of the deposits. The 1991 eruptions ended a 500-year hiatus in eruptive activity at Pinatubo volcano (Newhall and others, this volume). In contrast to the common silicic to mafic eruptive sequence observed at many volcanoes, the 1991 eruptions tapped andesite before dacite. The initial 1991 eruptions formed an andesitic lava dome at the head of the Maraunot River valley (fig. 1; Hoblitt, Wolfe, and others, this volume), beginning on June 7. The andesitic lava dome continued to grow until 0850 on June 12, when the first large explosive eruption destroyed the southern part of the dome and generated a vertical column that rose to more than 19 km in altitude. Ash from the column was transported mainly southwest from the volcano, and small pyroclastic flows descended the upper Maraunot River valley. The northern part of the lava dome survived the first June 12 eruption as well as vertically directed explosive eruptions of andesite beginning at 2251 on June 12, 0840 on June 13, and 1307 on June 14.

A series of at least 13 lateral blasts occurred on June 14 and 15 (Hoblitt, Wolfe, and others, this volume). Proportions of clast types in the resulting blast deposits (fig. 2) indicate that the dominant magmatic component changed from andesite to dacite during this interval. The lateral blasts became more frequent and release of seismic energy increased exponentially during the morning hours of June 15, leading to the climactic eruption, which was under way by early afternoon. Tephra-fall deposits from the June 12–

14 vertical eruptions are considerably thinner and are more areally restricted than the June 15 fall deposit (fig. 2A). Juvenile components in the June 12–14 tephra deposit are dominantly andesitic scoria. Dacite makes up less than 10 percent of juvenile clasts in the 0851 June 12 tephra, 28 percent of the 2252 June 12 tephra, and increases to 35 percent in the June 13 tephra (Hoblitt, Wolfe, and others, this volume). The June 13 deposit is recognized in proximal sections because it overlies an ash bed that was flushed by rain that fell during the night of June 12 (fig. 2B). Taken together, the June 12–14 eruptions involved at least an order-of-magnitude less magma than the June 15 eruption (Paladio-Melosantos and others, this volume). Estimates of 3.7 to 5.3 km^3 (W. E. Scott and others, this volume) or possibly more (Koyaguchi, this volume) have been made for the volume of dacitic magma that was vented on June 15. The June 15 eruption resulted in collapse of a caldera 2.5 km in diameter (Wolfe and others, this volume; W.E. Scott and others, this volume), which decapitated the upper part of the Maraunot drainage and destroyed the June 7–12 lava dome.

SAMPLE LOCALITIES AND GEOLOGIC CONTEXT

Pinatubo volcano lies in the Bataan frontal arc, approximately 120 km inboard of the Manila trench and 100 km above the Wadati-Benioff zone (Defant and others, 1988; Newhall and others, this volume). The volcano was constructed atop the southern part of the broadly exposed (3,500 km^2), east-dipping Zambales ophiolite complex (Evans and others, 1991). The 1991 caldera is nested within a 3×5-km caldera formed earlier in the volcano's history (fig. 1). Exploration drilling indicates that the Zambales ophiolite complex makes up the southeast wall of the older caldera and that, if present, the ophiolite must underlie more than 2 km of dacitic rocks that were drilled within the caldera (Delfin, 1983; Delfin and others, this volume).

Our samples of the June 7–12 dome were taken from prismatically jointed and scoriaceous blocks in a series of pyroclastic-flow deposits in the headwaters of the Maraunot River (figs. 1 and 3). Two pyroclastic-flow deposits are visible at the sample locality (fig. 3), and both contain blocks of dome rock. The lower deposit is correlated with the 0841 eruption of June 13 on the basis of similarity in clast-type distribution to that seen in the tephra-fall deposits (Hoblitt, Wolfe, and others, this volume). An underlying pyroclastic-flow deposit exposed upstream (and not visible in fig. 3) is correlated with the 0851 eruption of June 12; it is overlain and underlain by thin ash beds and contains breadcrust-surfaced blocks of andesitic scoria similar to lapilli in the fall deposit. The upper deposit in figure 3 contains abundant prismatically jointed dome blocks, possibly mined from the June 7–12 lava dome during caldera collapse on June 15.

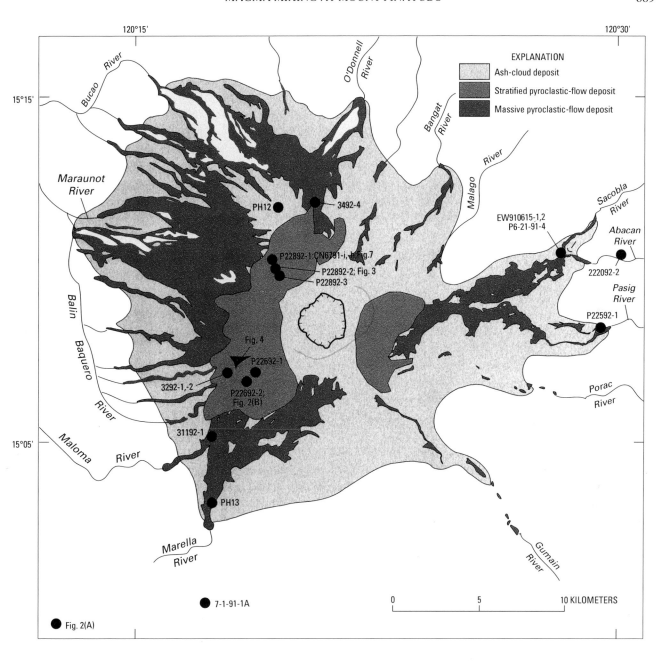

Figure 1. Distribution of 1991 pyroclastic-flow deposits from Mount Pinatubo (from W.E. Scott and others, this volume), outline of pre-1991 buried caldera (in blue, from Delfin, 1983), outline of 1991 caldera (hachured), and sample localities for this study. Sample localities indicated by dots. Field photograph localities indicated by figure numbers. Small arrowhead shows perspective of oblique aerial photograph in figure 4. Areas outside of shaded region were blanketed by tephra fall.

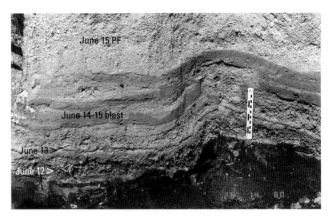

A

B

Figure 2. *A*, Section of 1991 tephra-fall deposits at bridge abutment, Santo Tomas River (fig. 1). Lower, dark layer at base is organic-rich pre-1991 soil, the thin lapilli-rich layer at center is the June 12, 1991, fall deposit, which is overlain by 3-cm-thick laminated ash from the June 14–15 June blasts, which is itself overlain by the white June 15 pumiceous fall deposit. *B*, Dark pre-1991 soil overlain first by June 12 and 13 tephra, both rich in andesitic scoria

lapilli, then by silt-to lapilli-rich blast beds of June 14–15, which contain both andesitic scoria and dacitic pumice. The blast beds are truncated and overlain by a white June 15 pyroclastic-flow deposit. The feature above and to the left of the scale is a lapilli-filled trough, possibly produced by lateral drag of an object across the surface during blast emplacement. Smaller divisions on the scale are 1 cm long.

Figure 3. Pyroclastic-flow deposits in the upper Maraunot River valley. Photograph of the outcrop was taken on February 28, 1992. Prominent bouldery layer is a June 15 pyroclastic-flow deposit, rich in blocks from the June 7–12 lava dome; it is underlain by a pyroclastic-flow deposit that is correlated with the 0841 eruption of June 13. A June 12 pyroclastic-flow deposit underlies both of these deposits upstream (out of view to the right).

Tephra-fall deposits from the June 12–13 vertical eruptions are exposed downwind from the volcano, to the west and southwest. At distances of less than 10 km from the caldera, June 15 pyroclastic flows scoured to bedrock much of the west flank of the volcano (fig. 4). Consequently, proximal June 12–14 fall and blast deposits are preserved only in the lee of bedrock ridges. Our most detailed petrographic and microprobe work on the June 12 tephra was done on scoria lapilli (sample 7–1–91–1A) collected on July 1, 1991, from near the axis of the June 12 fall deposit near Mount Bagang, 16 km southwest of the caldera (fig. 1). Another scoria sample collected from the same area by

A.G. Reyes is petrographically and chemically similar, as are scoria lapilli from several proximal tephra sections that we collected in 1992.

We sampled pumice blocks and lapilli from the June 15 eruption at a number of localities around the volcano. Our most detailed petrographic and microprobe data are for several samples collected in June and July of 1991, shortly after the climactic eruption. Subsequent petrographic examination and bulk rock analysis of a larger suite of samples collected in 1992 indicate that our 1991 samples are representative of the principal eruptive products.

Figure 4. View up southwest flank of Mount Pinatubo showing dissected 1991 pyroclastic-flow deposits. Photograph taken March 2, 1992, from a position 5.5 km from southwest caldera rim. Banana plants grew from root stock remaining on lee sides of ridges (facing camera). Fore sides of ridges were scoured to basement by the pyroclastic flows.

Table 1. Modal data for 1991 Pinatubo samples based on optical point counting of representative thin sections at a magnification of 400 in reflected and transmitted light (in volume percent, vesicle free).

[Groundmass consists mainly of glass in samples EW910615–1, 7–1–91–1A, and CN6791–i, and glass with abundant microlites in samples EW910615–2, PH–13d, and CN6791–h. Tr., trace amount; ( ), highly resorbed xenocrysts; dash indicates none detected; PR, phenocryst rich; PM, phenocryst moderate; PP, phenocryst poor]

Eruption date	6/15/91	6/15/91	6/15/91	6/12/91	6/7/91	6/7/91
Unit	**PR dacite**	**PM dacite**	**PP dacite**	**Andesite**	**Andesite**	**Basalt**
Sample No.	EW910615-1	EW910615-2	PH13d	7-1-91-1A	CN6791-h	CN6791-i
Rock type	pumice	pumice	pumice	scoria	lava	inclusion
Points counted	3,206	1,802	1,618	1,606	2,061	2,280
Plagioclase	31	20	11	17	31	22
Hornblende	12	6	3	12	12	31[1]
Cummingtonite	1	<1	tr.	<1	-	-
Biotite	tr.	tr.	-	<1	1	-
Augite	-	-	-	4	5	12
Olivine	tr.	tr.	-	2	2	3
Bronzite	tr.	-	-	-	-	-
Fe-Ti oxides	1	<1	<1	<1	2	2
Quartz	1	<1	1	1	<1	(1)
Apatite	tr.	tr.	tr.	tr.	tr.	<1
Anhydrite	<1	tr.	<1	1	tr.	-
Sulfides	-	-	-	tr.	tr.	tr.
Phenocrysts	47	26	15	38	52	71
Groundmass	53	74	85	62	48	29
Vesicles	61	50	59	53	4	23

[1]Does not include ~1% hornblende xenocrysts.

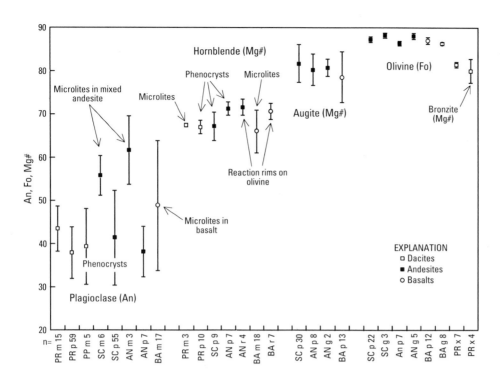

Figure 5. Average composition of principal minerals in samples of 1991 Pinatubo eruptive products, expressed as An content of plagioclase, Fo content of olivine, and Mg# of augite, hornblende, and bronzite. Analytical data for cummingtonite and biotite are summarized in table 2 and oxide data are given in table 5. The average compositions and $\pm 1\sigma$ range given are from table 2. Symbols: PR, phenocryst-rich dacite of June 15; PP, phenocryst-poor dacite of June 15; SC, scoria from air-fall tephra of June 12; AN, dense andesite lava from June 7–12 dome; BA, basalt inclusion from June 7–12 lava dome; m, microphenocryst or microlite; p, phenocryst; g, glomerocryst; r, reaction rim; x, xenocryst.

PETROGRAPHY AND MINERAL CHEMISTRY

Modal data for the 1991 magmatic samples are summarized in table 1, and petrographic descriptions are given herein. Modes were counted optically at a magnification of 400 by use of reflected and transmitted light on doubly polished thin sections. The presence of extremely fine (<1 µm) glass filaments, microlites or very small (<5 µm) crystal fragments, and low contrast between epoxy mounting medium and glass in some samples made quantitative optical analysis of groundmass phases and vesicle content difficult. As a result, the abundances of "groundmass" in table 1 include both glass and microlites (or microlite-sized broken crystal fragments), and the abundance of vesicles is approximate.

Chemical variations of the principal mineral phases are shown graphically in figure 5 in terms of An (anorthite) content of plagioclase, Fo (forsterite) content of olivine, and Mg# for pyroxenes and hornblende.[2] Microprobe analyses for average mineral compositions are listed in table 2.

JUNE 7–14 HYBRID ANDESITE

Samples of the June 7–12 lava dome, taken from the Maraunot River pyroclastic-flow deposits, are remarkable both in the variety of phenocrysts present and in the devel-

opment of disequilibrium textures. These andesite samples contain plagioclase, hornblende, augite, olivine, quartz, anhydrite, magnetite, ilmenite, cummingtonite, and biotite (table 1, fig. 6A). These phenocrysts are set in a pilotaxitic (microlite-charged) groundmass (fig. 6B). Andesitic scoria from tephra-fall deposits of June 12 are identical to the dome andesite samples with respect to phenocryst assemblages, mineral compositions, bulk compositions, and several reaction textures. The most distinct differences between the dome and tephra samples lie in the greater vesicularity of the scoria (table 1) and in the degree of groundmass crystallization. In contrast to the pilotaxitic texture of the dome andesite, the June 12 scoria has a glassy microvesicular groundmass.

Because the dome and tephra samples share common petrographic and compositional features and are believed to represent the same magma, they are described together here. Olivine phenocrysts in the hybrid andesite are rimmed by aggregates of hornblende laths (fig. 6C). Individual hornblende laths in the rims are commonly 0.05 mm across and 0.2 mm in length, comparable in size to hornblende microphenocrysts in the groundmass of the andesite and similar in composition to the more magnesian of the hornblende phenocrysts and microphenocrysts (table 2). Two types of olivine phenocrysts are observed in the dome andesite, and both are also present in the scoria. The first type is prismatic and subhedral to euhedral, consisting of single millimeter-sized crystals with optical continuity throughout, such as seen in figure 6C and D. The second type is glomerocrystic, consisting of aggregates of smaller olivine grains or, less commonly, olivine and clinopyroxene

---

[2] Magnesium numbers (Mg#) calculated as 100 times the molecular ratio of Mg/(Mg+Fe^{+2}+Mn).

Table 2. Average compositions, standard deviations, and maximum and minimum values of mineral endmember components for minerals in 1991 Pinatubo samples.

[Microprobe analytical procedures given in the appendix. Avg., average; n, number of analyses averaged; SD, standard deviation; qz, quartz; hb, hornblende; cm, cummingtonite; PR, phenocryst-rich; PP, phenocryst poor; PM, phenocryst moderate. Dash indicates not determined]

AVERAGE MICROPROBE ANALYSES OF OLIVINES

	June 15 PR dacite EW910615-1 Xenocrysts		June 12 tephra 7-1-91-1A Prismatic phenocrysts		June 12 tephra 7-1-91-1A Glomerocryst		June 7-12 andesitic lava CN6791-d Prismatic phenocrysts		June 7-12 andesitic lava CN6791-d Glomerocrysts		Basaltic inclusion CN6791-i Prismatic phenocrysts		Basaltic inclusion CN6791-i Glomerocrysts	
	Avg.	SD (n=7)	Avg.	SD (n=22)	Avg.	SD (n=3)	Avg.	SD (n=7)	Avg.	SD (n=5)	Avg.	SD (n=12)	Avg.	SD (n=8)
MgO	43.83	0.62	47.51	0.66	48.41	0.71	46.99	0.30	48.05	0.50	47.34	0.74	47.05	0.55
SiO_2	39.32	.40	40.60	.46	40.13	.69	40.14	.18	40.64	.24	40.47	.32	40.51	.52
CaO	.03	.03	.13	.02	.12	.01	.13	.01	.12	.01	.13	.03	.13	.01
MnO	.27	.07	.13	.02	.08	.02	.14	.03	.13	.03	.14	.04	.15	.02
FeO	17.52	.52	12.27	.59	11.46	.47	13.06	.51	11.51	.69	12.46	.65	13.13	.26
NiO	-	-	.09	.06	-	-	-	-	-	-	.05	.08	.00	.00
Total	100.96	.42	100.64	.83	100.21	.40	100.46	.27	100.45	.28	100.57	.59	100.97	.83
Mg	1.645	.016	1.742	.016	1.779	.031	1.735	.010	1.760	.014	1.740	.018	1.727	.010
Si	.990	.008	.999	.004	.989	.012	.994	.003	.999	.004	.998	.003	.998	.005
Ca	.001	.001	.003	.001	.003	.000	.003	.000	.003	.000	.003	.001	.003	.000
Mn	.006	.001	.003	.000	.002	.000	.003	.001	.003	.001	.003	.001	.003	.001
Fe	.369	.013	.252	.013	.236	.010	.271	.011	.237	.015	.257	.015	.271	.007
Ni	-	-	.000	.001	-	-	.000	.000	-	-	.000	.001	-	-
Sum	3.010	.008	3.000	.004	3.010	.012	3.006	.003	3.001	.004	3.002	.003	3.002	.005
O	4.000		4.000		4.000		4.000		4.000		4.000		4.000	
[1]%Fo	81.4	.6	87.2	.7	88.2	.6	86.9	.0	88.0	.7	87.0	.8	86.3	.4
Max. %Fo	82.5		88.3		88.9		86.9		89.2		88.1		86.9	
Min. %Fo	80.7		85.9		87.8		85.9		87.5		85.9		85.8	

[1]%Fo= 100 X Mg/(Mg+Mn+Fe). Dash indicates not determined.

Table 2. Average compositions, standard deviations, and maximum and minimum values of mineral endmember components for minerals in 1991 Pinatubo samples—Continued.

AVERAGE MICROPROBE ANALYSES OF CHROME SPINEL INCLUSIONS IN OLIVINES

	June 15 PR dacite Xenocrysts EW910615-1	June 12 tephra In olivine prisms 7-1-91-1A		June 7 andesitic lava In olivine prisms CN6791-d	Basaltic inclusion In olivine prisms CN6791-i	
	Avg. (n=2)	Avg.	SD (n=5)	Avg. (n=2)	Avg.	SD (n=5)
MgO	3.85	11.17	0.78	12.53	10.72	0.88
Al_2O_3	4.54	16.47	2.19	17.20	15.69	2.13
TiO_2	1.91	.92	.23	.54	.84	.09
Cr_2O_3	11.33	34.81	3.43	41.93	36.59	1.81
MnO	.14	.08	.04	.06	.09	.07
^{2}FeO	27.58	17.18	.77	15.35	17.84	.94
Fe_2O_3	50.07	16.50	2.65	10.76	15.91	2.09
Total	99.41	97.11	.75	98.37	97.68	1.38
Mg	.209	.548	.027	.599	.527	.032
Al	.195	.639	.073	.650	.609	.066
Ti	.052	.023	.006	.013	.021	.002
Cr	.327	.906	.089	1.064	.954	.049
Mn	.004	.002	.001	.001	.002	.002
Fe^{+2}	.841	.473	.030	.412	.492	.032
Fe^{+3}	1.373	.409	.071	.260	.395	.060
Sum	3.000	3.000		2.999	3.000	
O	4.000	4.000		4.000	4.000	
$Mg/(Mg+Fe^{+2})$	0.199	0.537	0.028	0.593	0.517	0.032
$Cr/(Cr+Al)$	0.626	0.586	0.046	0.621	0.610	0.032
$Fe^{+3}/(Fe^{+3}+Cr+Al)$	0.725	0.209	0.038	0.132	0.202	0.031

$^2Fe_2O_3$ and FeO calculated on basis of $R^{+2}R^{+3}_2O_4$ formulae.

Table 2. Average compositions, standard deviations, and maximum and minimum values of mineral endmember components for minerals in 1991 Pinatubo samples—Continued.

AVERAGE MICROPROBE ANALYSES OF PYROXENE

| | June 15 PR dacite EW910615-1 | | June 12 tephra 7-1-91-1A scoria | | June 7-12 andestic lava CN6791-d | | | | Basaltic inclusion CN6791-i | | | |
| | Bronzite Xenocryst | | Phenocrysts[3] | | Phenocrysts | | Glomerocryst | Rim on qz. | Phenocrysts | | Rims on hb and qz | |
	Avg. (n=4)	SD	Avg. (n=30)	SD	Avg. (n=8)	SD	Avg.(n=2)	Avg.(n=2)	Avg. (n=13)	SD	Avg. (n=7)	SD
Na_2O	<0.1		0.14	0.07	0.21	0.05	0.20	0.17	0.22	0.09	0.26	0.07
MgO	29.94	1.37	15.83	1.14	15.23	1.12	14.93	15.67	14.51	1.15	13.74	1.27
Al_2O_3	1.11	.78	2.80	1.27	4.29	1.55	4.46	2.15	3.55	.59	2.67	1.67
SiO_2	55.04	.83	51.55	1.57	49.88	1.63	50.53	51.82	50.65	.44	51.31	.98
CaO	.27	.10	21.48	1.08	21.98	.55	21.44	21.90	22.33	.59	21.66	.89
TiO_2	.00	.00	.37	.25	.76	.43	.72	.35	.45	.13	.45	.34
Cr_2O_3	.06	.11	.12	.15	.16	.24	.01	.00	.20	.25	.00	.00
MnO	.45	.17	.12	.12	.06	.03	.06	.07	.18	.20	.33	.35
FeO	12.82	1.56	6.21	1.46	6.57	1.08	6.27	5.79	6.87	1.74	8.53	1.68
Total	99.69	.66	98.68	.60	99.14	.76	98.64	97.94	98.96	.47	98.96	1.07
Na			.010	.005	.015	.004	.014	.013	.016	.006	.019	.005
Mg	1.592	.059	.883	.054	.847	.057	.831	.877	.810	.060	.771	.069
Al	.047	.033	.123	.057	.189	.069	.196	.096	.157	.026	.118	.073
Si	1.964	.025	1.922	.042	1.862	.053	1.886	1.944	1.897	.018	1.932	.048
Ca	.010	.004	.858	.043	.880	.018	.857	.880	.896	.021	.874	.037
Ti	.000	.000	.010	.007	.021	.012	.020	.010	.013	.004	.013	.010
Cr	.002	.003	.004	.004	.005	.007	.000	.000	.006	.007	.000	.000
Mn	.014	.005	.004	.004	.002	.001	.002	.002	.006	.006	.011	.011
Fe	.383	.050	.194	.047	.205	.035	.196	.182	.215	.056	.269	.054
Sum	4.012	.009	4.009	.013	4.027	.012	4.003	4.004	4.016	.006	4.006	.009
O	6.000		6.000		6.000		6.000	6.000	6.000		6.000	
[4]Mg#	80.0	2.7	81.7	4.4	80.3	3.6	80.8	82.6	78.6	5.9	73.4	5.8
Max. Mg#	83.0		88.2		84.7		82.2	85.7	85.0		79.6	
Min. Mg#	76.6		74.8		75.2		79.4	79.5	68.9		66.0	
[5]Ca#			44.3	2.2	45.5	1.2	45.5	45.4	46.6	1.2	45.7	1.9
Max. Ca#			47.2		46.8		45.6	45.5	48.4		48.0	
Min. Ca#			36.3		43.3		45.4	45.3	44.5		43.5	

[3]Phenocryst category includes microphenocrysts.   [4]Mg#= 100 X Mg/(Mg+Mn+total Fe).   [5]Ca#= 100 X Ca/(Mg+Ca+total Fe).

Table 2. Average compositions, standard deviations, and maximum and minimum values of mineral endmember components for minerals in 1991 Pinatubo samples—Continued.

AVERAGE MICROPROBE ANALYSES OF AMPHIBOLE

| | June 15 PR dacite EW910615-1 | | | | | | June 12 tephra 7-1-91-1A scoria | | | | June 7-12 andesitic lava CN6791-d | | | | | | Basaltic inclusion CN6791-i | | | | | |
| | Microlites (n=3) | | Phenocrysts (n=10) | | Cumm. rims (n=8) | | Phenocrysts (n=9) | | Xenocrysts (n=2) | | Phenocrysts (n=7) | | Rims on olivine (n=4) | | Xenocrysts (n=3) | | Microphenocrysts (n=18) | | Rims on olivine (n=7) | | Xenocrysts (n=3) | |
	Avg.	SD	Avg.	SD	Avg.	SD	Avg.	SD	Avg.	SD	Avg.	SD	Avg.	SD	Avg.	SD	Avg.	SD	Avg.	SD	Avg.	SD
Na_2O	1.21	0.05	1.18	0.09	0.17	0.12	1.91	0.52	1.34	0.14	2.39	0.04	2.32	0.04	2.03	0.61	2.32	0.06	2.26	0.07	2.13	0.35
MgO	15.43	.18	15.35	.52	21.07	.48	14.40	.79	14.71	.02	15.10	.38	15.25	.29	15.23	.18	13.73	1.25	15.26	.77	14.61	.57
Al_2O_3	7.20	.31	7.03	.58	1.69	.50	10.66	2.31	7.45	.29	12.71	.23	12.72	.13	11.02	2.89	13.12	.87	12.44	1.07	11.18	1.95
SiO_2	48.86	.51	48.49	1.00	54.02	1.03	44.05	3.32	46.16	.63	41.66	.28	41.69	.19	44.10	3.66	41.10	1.01	43.00	1.22	43.79	2.53
K_2O	.19	.02	.23	.05	.05	.00	.58	.29	.26	.01	.78	.07	.71	.07	.61	.33	.81	.07	.71	.15	.75	.32
CaO	9.95	.07	10.45	.20	1.93	.58	10.84	.51	10.56	.05	11.49	.28	11.04	.35	11.30	.70	11.70	.14	11.22	.44	11.28	.67
TiO_2	.77	.05	.79	.08	.13	.05	1.51	.62	.88	.19	2.03	.10	1.88	.19	1.66	.74	1.99	.30	1.24	.28	1.88	1.00
MnO	.42	.06	.46	.05	.87	.13	.22	.19	.37	.02	.04	.02	.04	.03	.18	.22	.07	.06	.09	.05	.19	.19
FeO	12.85	.14	12.99	.48	16.74	.36	12.34	1.26	12.72	.20	10.81	.55	10.75	.75	11.17	1.75	12.50	1.55	11.21	.48	11.81	1.98
Total	96.87	.31	96.98	1.07	96.61	.67	96.53	.72	94.45	.29	97.00	.40	96.43	.22	97.30	.33	97.32	.69	97.43	.33	97.63	.53
Na	.343	.016	.334	.027	.048	.033	.550	.153	.391	.043	.685	.011	.669	.011	.578	.177	.669	.015	.643	.022	.608	.100
Mg	3.355	.026	3.346	.116	4.526	.073	3.184	.162	3.307	.008	3.327	.076	3.373	.063	3.329	.047	3.041	.253	3.338	.155	3.197	.107
Al	1.238	.058	1.212	.102	.287	.086	1.868	.421	1.324	.057	2.216	.044	2.224	.022	1.907	.511	2.301	.164	2.153	.192	1.936	.339
Si	7.128	.045	7.091	.081	7.785	.083	6.530	.420	6.963	.069	6.160	.036	6.186	.022	6.464	.487	6.111	.101	6.310	.153	6.430	.359
K	.036	.003	.044	.010	.000	.000	.111	.055	.050	.001	.146	.014	.135	.013	.114	.063	.153	.015	.132	.028	.140	.061
Ca	1.555	.013	1.637	.038	.299	.091	1.723	.096	1.706	.002	1.820	.041	1.755	.056	1.777	.122	1.864	.031	1.765	.074	1.775	.108
Ti	.084	.006	.087	.009	.014	.005	.170	.071	.100	.022	.225	.011	.210	.021	.184	.082	.222	.033	.137	.031	.208	.109
Mn	.052	.007	.057	.006	.106	.015	.027	.023	.047	.003	.005	.002	.005	.003	.022	.027	.009	.007	.011	.006	.024	.023
Fe	1.568	.020	1.590	.059	2.018	.046	1.530	.150	1.605	.019	1.337	.070	1.334	.094	1.369	.204	1.556	.206	1.376	.064	1.451	.244
Sum	15.358	.018	15.399	.050	15.082	.056	15.694	.242	15.493	.037	15.922	.035	15.893	.016	15.744	.269	15.927	.044	15.864	.055	15.768	.171
O	23.000		23.000		23.000		23.000		23.000		23.000		23.000		23.000		23.000		23.000		23.000	
Al(IV)	0.872		0.909		0.215		1.466		1.038		1.84		1.814		1.533		1.888		1.689		1.569	
Al(VI)	0.365		0.302		0.072		0.398		0.286		0.375		0.411		0.372		0.411		0.462		0.366	
$^a Fe_2O_3$	0		0.62		0.96		0.93		1.43		2.5		2.47		1.44		2.87		2.47		0.57	
$^a FeO$	12.85		12.43		15.88		11.51		11.43		8.56		8.53		9.87		9.92		8.99		11.29	
Mg#	67.4	.4	67.0	1.6	68.1	.5	67.2	3.3	66.7	.4	71.3	1.5	71.6	1.8	70.6	3.6	66.0	4.9	70.6	1.9	68.5	4.6
Max. Mg#	67.8		69.9		68.7		72.4		66.9		73.2		73.3		73.2		73.4		73.6		73.3	
Min. Mg#	67.0		64.9		67.3		62.4		66.4		69.2		69.9		66.6		54.1		68.1		64.2	

$^a Fe_2O_3$ and FeO calculated by charge balance on 15 cation basis. P_2O_5 less than 0.1% and Cr_2O_3 not detected.

Table 2. Average compositions, standard deviations, and maximum and minimum values of mineral endmember components for minerals in 1991 Pinatubo samples—Continued.

AVERAGE MICROPROBE ANALYSES OF BIOTITE

| | June 15 PR and PP dacite | | June 12 tephra | |
| | EW910615-1 & PH13D | | 7-1-91-1A scoria | |
	Avg.	SD (n=7)	Avg.	SD (n=6)
Na$_2$O	0.59	0.05	0.58	0.04
MgO	15.95	.29	15.79	.21
Al$_2$O$_3$	14.20	.18	14.64	.37
SiO$_2$	38.30	.58	38.08	.20
K$_2$O	7.46	.31	7.04	.49
TiO$_2$	3.33	.16	3.28	.09
MnO	.04	.02	.06	.02
FeO	13.52	.36	13.78	.38
[7]Total	93.39	1.13	93.25	.80
Na	.188	.015	.183	.013
Mg	3.878	.069	3.840	.043
Al	2.729	.017	2.814	.050
Si	6.246	.026	6.212	.024
K	1.552	.075	1.466	.107
Ti	.408	.016	.402	.010
Mn	.005	.003	.008	.003
Fe	1.844	.040	1.880	.048
Sum	16.850	.064	16.804	.065
O	24.000		24.000	
Mg#	67.7	.8	67.0	.8
	69.4		68.3	
	67.0		66.2	

[7]CaO not detected.

Table 2. Average compositions, standard deviations, and maximum and minimum values of mineral endmember components for minerals in 1991 Pinatubo samples—Continued.

AVERAGE MICROPROBE ANALYSES OF PLAGIOCLASE

| | June 15 PR dacite EW910615-1 | | | | June 15 PP dacite PH13d | | June 12 tephra 7-1-91-1A scoria | | | | June 7-12 andesitic lava CN6791-d | | | | Basaltic inclusion CN6791-i | |
| | Microlites (n=15) | | Phenocrysts (n=59) | | Microlites (fragments) (n=5) | | Microlites (n=6) | | Phenocrysts (n=55) | | Microlites (n=3) | | Phenocrysts (n=7) | | Microlites (n=17) | |
	Avg.	SD	Avg.	SD	Avg.	SD	Avg.	SD	Avg.	SD	Avg.	SD	Avg.	SD	Avg.	SD
Na_2O	6.37	0.60	7.00	0.67	6.77	0.98	4.70	0.40	6.51	1.23	4.14	0.87	6.86	0.66	5.61	1.52
Al_2O_3	26.30	.98	25.35	1.06	26.04	1.56	27.49	1.83	26.24	1.73	29.64	1.67	25.67	1.02	27.26	2.93
SiO_2	57.73	1.45	59.06	1.82	58.93	2.40	54.99	2.29	57.75	2.93	52.15	2.14	58.17	1.51	56.31	4.28
K_2O	.22	.05	.30	.08	.25	.08	.33	.28	.28	.09	.10	.07	.24	.05	.41	.32
CaO	8.84	1.06	7.73	1.23	7.94	1.76	10.79	1.36	8.30	2.18	12.01	1.49	7.66	1.16	9.78	3.14
FeO	.18	.08	.07	.04	.11	.04	.61	.20	.19	.18	.33	.09	.08	.03	.47	.09
Total	99.64	.53	99.52	1.03	100.04	.53	98.95	.80	99.28	.68	98.38	.75	98.69	.79	99.82	1.01
Na	.555	.051	.609	.056	.586	.082	.415	.032	.568	.104	.369	.077	.602	.057	.489	.129
Al	1.394	.056	1.341	.062	1.370	.090	1.477	.100	1.395	.104	1.607	.095	1.369	.057	1.450	.167
Si	2.596	.056	2.650	.062	2.630	.090	2.507	.097	2.602	.109	2.399	.088	2.632	.056	2.538	.167
K	.013	.003	.017	.005	.014	.004	.019	.016	.016	.005	.006	.004	.014	.003	.023	.018
Ca	.426	.052	.372	.061	.380	.086	.527	.068	.402	.109	.592	.075	.371	.057	.474	.156
Fe	.007	.003	.003	.001	.004	.002	.023	.008	.007	.007	.013	.003	.003	.001	.018	.004
Sum	4.991	.008	4.992	.012	4.984	.013	4.971	.037	4.992	.011	4.985	.010	4.991	.008	4.993	.016
O	8.000		8.000		8.000		8.000		8.000		8.000		8.000		8.000	
%An[8]	43.4	5.2	37.9	6.0	39.3	8.8	55.8	4.6	41.4	11.0	61.6	7.9	38.2	5.9	48.8	15.0
Max. %An	52.8		53.0		48.5		60.6		63.6		70.7		51.2		68.5	
Min. %An	37.9		27.0		29.0		49.2		27.9		56.3		33.8		20.4	

[8] %An= 100 X Ca/(Ca+Na+K)

grains (fig. 6*E*). Both types of olivine have the same restricted range in composition (Fo_{86-89}; fig. 5).

Initially, we were concerned that the Pinatubo olivines might be xenocrysts from the Zambales ophiolite complex beneath the volcano, a possibility also considered by Fournelle (1991) and Fournelle and others (this volume). However, spinel inclusions within the olivines are intermediate pleonaste-chromite solid solutions with relatively high Fe_2O_3 and intermediate TiO_2, typical of abundances in arc basalts and unlike the spinels in most ophiolitic peridotites, including the Zambales (compare Haggerty, 1976; Dick and Bullen, 1984; Evans and Hawkins, 1989). The TiO_2 abundances of the spinel inclusions overlap with those in the cumulus (gabbroic) suite of the ophiolite (Evans, 1987), but high Fe_2O_3 distinguishes the volcanic spinels from those in the ophiolite. In addition, low nickel and high calcium abundance, subhedral prismatic form of many crystals, presence of glass inclusions, and presence of oscillatory zoned magnesian augite in the same samples indicate that the olivine is of volcanic origin. Additional proof became available in 1992 when we found basalt inclusions with identical magnesian olivine and augite phenocrysts.

Clinopyroxene is present as small (<0.5 mm), stubby euhedral prisms in the hybrid andesite. Micrometer-scale oscillatory zoning is clearly visible in transmitted light, especially when crystals are viewed down the *c*-axis (fig. 6*B*). Most crystals show normal overall zoning; Mg# ranges from as high as 88 for cores to as low as 75 for crystal edges and averages about 81. Sector-zoned augite microphenocrysts are present in some samples. The augite overlaps in other major and minor element abundances with oscillatory-zoned augites in the basalt inclusions (table 2).

Hornblende occurs as small (<1 mm) phenocrysts and microphenocrysts, larger (1–2 mm) phenocrysts, sparse xenocrysts with augite rims, and (as noted previously) in reaction rims surrounding olivine. Many phenocrysts are oscillatory zoned, and some grains have resorbed inclusion-rich cores that are overgrown by oscillatory-zoned margins with higher Ti, Al, and Mg# (fig. 6*F*). Glass inclusions with vapor bubbles are common in the phenocrysts. The hornblendes in the scoria are olive green to brown; those in the dome andesite are red-brown to tan, indicative of late oxidation. Dehydration rims are lacking on most of the hornblendes, so rapid ascent is suggested (compare Rutherford and Hill, in press). Compositions of hornblende phenocrysts in the andesitic scoria show a considerable range in Mg#, overlapping with those in both the andesitic lava and June 15 dacite (fig. 5). The hornblendes in both types of andesite, however, are distinct from those in the June 15 dacite in other major and minor element abundances. In particular, the analyzed phenocrysts in the andesite have significantly higher Al, Ti, Na, and K, and lower Si and Mn (table 2). In contrast, they overlap in all major and minor element abundances with hornblende phenocrysts and microphenocrysts in the basaltic inclusions. Therefore, we believe that most

were derived from a mafic source similar to the basaltic inclusions.

Cummingtonite is a minor phase in the andesites. It occurs as partial rims on a few broken hornblende phenocrysts. Surprisingly little, if any, reaction of cummingtonite with the host melt is evident. Biotite occurs mainly as cores to hornblende phenocrysts and less commonly as separate small grains in the groundmass of the andesitic scoria.

Plagioclase occurs as large (1–10 mm) oscillatory-zoned, euhedral or broken phenocrysts in the hybrid andesite. Those in the dome lava tend to be larger and less commonly broken than those in the scoria. Small glass inclusions (most <30 μm) with vapor bubbles are common within interior regions of the phenocrysts as well as being concentrated with microlites of apatite and other phases to form dusty resorption zones near the crystal margins (fig. 6*G,H*). Compositions of the plagioclase phenocrysts in the andesite average about An_{40} and range widely (An_{28-64}). Although the average compositions overlap with those in the June 15 dacite, phenocrysts in the andesite extend to more calcic compositions, owing to thin (10–30 μm) rims of calcic plagioclase (An_{58-65}) over more sodic oscillatory-zoned interiors (typically An_{30-40} or An_{35-55}) (fig. 6*H*). Microlites in the andesites are relatively calcic, with average compositions of about An_{60}, similar to the calcic rims on the phenocrysts (fig. 5).

Quartz is a minor component of the hybrid andesite. It forms resorbed anhedral to subhedral grains that are typically 1 to 5 mm in diameter. These grains also contain glass inclusions with vapor bubbles, and some of the inclusions exceed 100 μm in diameter. Apatite is a ubiquitous microlite phase in all of the 1991 Pinatubo rocks; it occurs as inclusions in plagioclase, hornblende, and anhydrite. Apatite also forms sparse small phenocrysts and microphenocrysts, both as isolated euhedra in the groundmass and in association with anhydrite, as described below.

Anhydrite forms euhedral phenocrysts in the June 12 scoria. The anhydrite phenocrysts are up to 1 mm in length; they commonly include apatite microphenocrysts (fig. 6*I*) or are in growth contact with apatite phenocrysts. In contrast, anhydrite in the June 7–12 dome andesite occurs as highly resorbed anhedral masses with dusty reaction rims that contain oxide and sulfide minerals (fig. 6*J*). We attribute the difference to reaction of anhydrite with host melt that had cooled slowly and degassed near the surface.

Both ilmenite and magnetite occur as small (0.01–0.5 mm) grains in the hybrid andesites. Some of the magnetites in the hybrid andesite contain minute sulfide blebs (<20 μm), similar to those in magnetites of the basaltic inclusions. The ilmenites occur only as core regions of composite grains with magnetite borders. As is evident in figure 6*K*, the magnetite borders are overgrowths that nucleated and grew on ilmenite grain margins that were in contact with the hybrid melt. Where the ilmenite grain margins were embedded in silicate minerals no magnetite

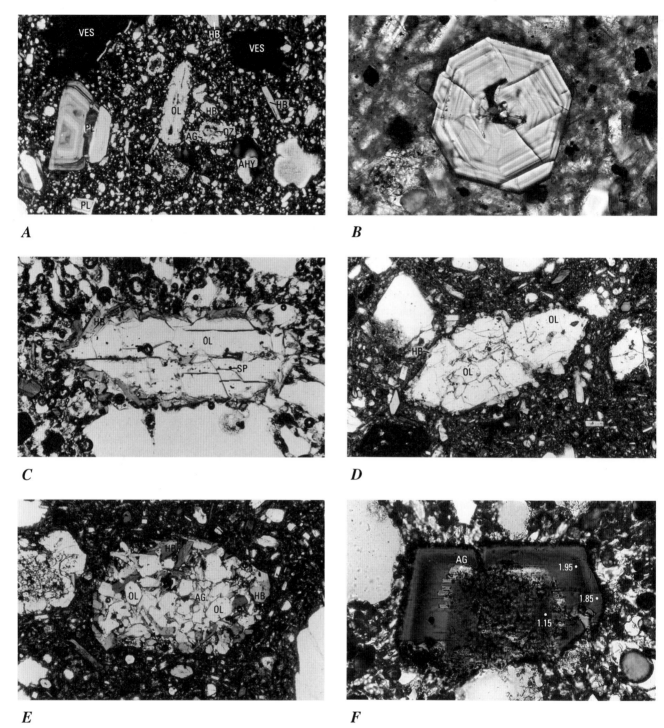

Figure 6. Microscopic images from thin sections of June 7–12 hybrid andesite. Symbols: T, transmitted light; R, reflected light; X, cross polarized; P, plane polarized; *xx* mm, width of frame in millimeters. AG, augite; AHY, anhydrite; AP, apatite; GL, glass; HB, hornblende; OL, olivine; PL, plagioclase; QZ, quartz; RXN, reaction rim; VES, vesicle; MT, magnetite; IL, ilmenite; SP, chrome spinel. *A*, Photomicrograph showing varied mineral assemblage in June 7–12 dome andesite sample CN6791–d (T, X, 5.5 mm). *B*, Oscillatory and normally zoned augite (Mg# 76–86) microphenocryst set in a pilotaxitic, microlite-charged groundmass. Section is perpendicular to the *c*-axis (T, P, 0.34 mm).

C, Fo_{87} olivine, rimmed by hornblende in andesitic scoria sample 7–12–91–1A from the June 12, 1991, tephra-fall deposit (T, P, 1.3 mm). *D*, Fo_{87} olivine showing euhedral prismatic crystal form and thin hornblende rims in June 7–12 lava dome sample CN6791–1d (T, P, 2.7 mm). *E*, Glomerocrystic clot of Fo_{88} olivine and Mg# 81 augite with hornblende rim in June 7–12 lava dome sample CN6791–1d (T, P, 2.7 mm). *F*, Resorbed low-Al-Ti hornblende with oscillatory-zoned high-Al-Ti overgrowth in June 7–12 andesite scoria sample 7–1–91–1A (T, P, 0.68 mm). Small augite inclusions decorate the margin of the resorbed core. Three TiO_2 values are indicated (wt%).

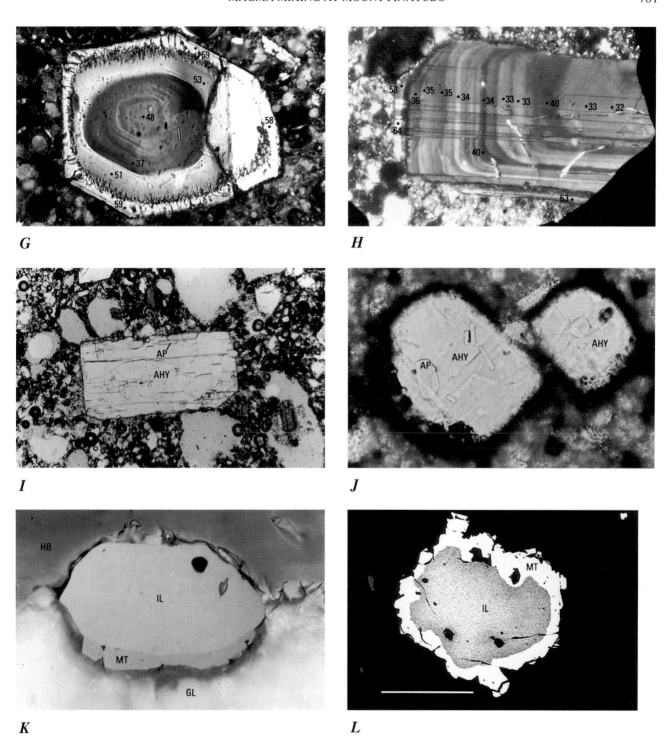

Figure 6 (Continued). *G*, Oscillatory-zoned plagioclase with calcic rim and glass inclusions along resorption boundary, June 12 scoria sample 7–1–91–1A. Numbers give An contents at analyzed spots (T, X, 0.68 mm). *H*, Oscillatory zoned plagioclase with calcic rim in June 12 scoria sample 7–1–91–1A. Numbers give An contents at analyzed spots. Note wide range in composition between zoning bands (T+R, X, 0.68 mm). *I*, Euhedral anhydrite phenocryst with apatite inclusions in June 12 scoria sample 7–1–91–1A (T, P, 1.3 mm). *J*, Resorbed anhydrite microlites with apatite inclusions in June 7–12 lava dome sample CN6791–1d (T, P,

0.34 mm). *K*, Ilmenite grain with magnetite overgrowth on the margin in contact with groundmass glass (but not on the margin embedded in hornblende), June 12 scoria sample P22892–2A4 (T+R, P, 0.13 mm). *L*, Back-scattered electron image of ilmenite grain with magnetite overgrowth in June 12 scoria sample 7–1–91–1A. Bright areas indicate higher atomic number (greater iron content). Note zoning to lower Fe-ilmenite within about 10 μm of contact with magnetite rim. Line is 100 μm long; frame width is about 0.3 mm.

Figure 7. Boulder derived from the June 7–12 lava dome within the June 15 pyroclastic-flow deposit, upper Maraunot River valley near locality P22892–1. Arrowheads indicate a few of the largest of the relatively abundant inclusions of dark basalt. Hammer is 37 cm long.

crystallized. Ilmenite cores of composite grains in scoria from the June 12 tephra preserve weak zonation to higher titanium within 5 to 10 µm of magnetite overgrowths (fig. 6L). Ilmenite cores to composite grains in the June 7–12 dome rocks contain exsolution-oxidation lamella of hematite, consistent with slower cooling of the dome samples. The compositions of the oxides, and implications for temperature and oxygen fugacity, will be discussed in a subsequent section.

BASALTIC INCLUSIONS FROM THE JUNE 7–12 DOME

Inclusions of olivine-augite basalt with abundant hornblende microphenocrysts are common in blocks of the June 7–12 dome andesite from the upper Maraunot River valley (fig. 7). These inclusions range in size and shape from relatively small (millimeter scale), irregular blebs identifiable in thin section to ellipsoidal bodies up to 40 cm across.

Most macroscopic ones are ellipsoidal, have sharp margins, and are 5 to 10 cm in diameter. Basaltic inclusions with irregularly deformed margins (fig. 8A) and banded basalt-andesite scoria blocks were found in the 0851 June 12 pyroclastic-flow deposit from the upper Maraunot valley. Banded scoria lapilli with dark bands rich in olivine and clinopyroxene are also common in the June 12–14 scoria-fall deposits.

We describe here a single 8-cm-diameter ellipsoidal basalt inclusion within dense dome andesite (sample CN6791–i, fig. 8B). This inclusion is representative of many others found in the Maraunot dome blocks. The most obvious distinguishing feature of the inclusion is the absence of the large plagioclase phenocrysts that are common in the andesitic host rock and the presence of a diktytaxitic groundmass texture.

The basalt contains 0.5- to 2-mm olivine phenocrysts and 0.1- to 0.5-mm augite phenocrysts and microphenocrysts. The larger size of the olivine crystals suggests that they were the initial liquidus phase in the basalt. Hornblende occurs as abundant microlites and as small (0.1–0.7 mm) subhedral to euhedral phenocrysts. Xenocrysts of hornblende and quartz are present in some thin sections, and both phases are rimmed by augite. Compositions of all three phenocryst phases and crystal habits of the olivine and augite are identical to those in the hybrid andesite. Olivine (Fo$_{86–88}$; fig. 5, table 2) is rimmed by hornblende microphenocrysts (fig. 9A) and augite (Mg# 69–85) shows micrometer-scaled oscillatory zoning (fig. 9B). As in the andesite, the olivine occurs both as subhedral prisms and as glomerocrysts with augite. Spinel inclusions within the olivines are likewise similar in composition to those in the olivines of the hybrid andesite (table 2). These similarities leave little doubt that the basaltic inclusions represent samples of the source magma for the olivine and augite phenocrysts in the hybrid andesite.

A two-stage crystallization history is evident from the phenocrysts. Olivine and augite phenocrysts formed during an initial stage of anhydrous crystallization. That no hydrous phases were stable is indicated by reaction rims of augite on large hornblende xenocrysts (fig. 9E,F). These reaction rims are composed of two zones of augite microphenocrysts. The outer zone consists of acicular crystals (Mg# 66–68) oriented parallel to the elongation of the relict amphibole. The inner zone is composed of optically continuous and more magnesian augite (Mg# 75–79) that is intergrown with oxide minerals. Major and minor element abundances of the reaction-rim augites overlap with those of the zoned augite phenocrysts. Compositions of the resorbed cores of the amphibole xenocrysts are similar to late-crystallized hornblende microlites (table 2). If they had been included after the magma was hydrated, they would not have been resorbed, nor would augite rims have formed. Consequently, the initial phase of crystallization involved only the anhydrous phases olivine and augite.

A

Figure 8. *A*, Deformed basalt inclusion with irregular margin in andesitic scoria from June 12 pyroclastic-flow deposit. *B*, Oval-shaped undercooled inclusion (sample CN6791–i) in dense andesitic lava from the June 7–12 dome. Both samples are from pyroclastic-flow deposits in the upper Maraunot River valley. Labeled quartz (QZ) and hornblende (HB) are xenocrysts discussed in text.

B

Most of the quartz xenocrysts are relatively large (1–4 mm in diameter) and subrounded. As is the case with quartz in the andesite, these are also rimmed by anhydrous, augite-rich reaction zones, and, like the hornblende xenocrysts, they were probably also included at an early stage. Pyroxene-rich reaction boundaries form on quartz grains in static dissolution experiments when the basalt host is below its liquidus, and they effectively armor the quartz from further dissolution (Donaldson, 1985). Accordingly, it is difficult to estimate how long the quartz grains were included in the basaltic magma.

The second stage of crystallization took place after the fugacity of water built up in the melt. Hornblende crystals nucleated and grew to form borders surrounding olivine phenocrysts. Hornblende, plagioclase, and magnetite microphenocrysts and microlites crystallized to produce the diktytaxitic groundmass texture and drive the matrix melt to

an evolved rhyolitic composition (74 wt% SiO_2). The degree of undercooling increased during the final stage of crystallization. Rapid growth of elongate crystal faces produced glass-inclusion-rich, hopper-shaped plagioclase microlites (fig. 9*G*) and swallow tails on hornblende microphenocrysts (fig. 9*H*). Mottled plagioclase microlites, consisting of irregular zones of variable An content, grew quickly and included irregular melt channels (fig. 9*I*). Skeletal magnetite grains also grew during this period of strong undercooling (fig. 9*J*). A trace amount of magmatic copper sulfide was trapped as small (<15 μm) blebs, typically within or adjacent to larger magnetite grains but also within hornblende, and rarely as free grains in glass.

The diktytaxitic groundmass texture probably formed when the basalt magma was intruded into cooler dacitic to andesitic magma. This texture is characterized by abundant irregular and interconnected void space into which small

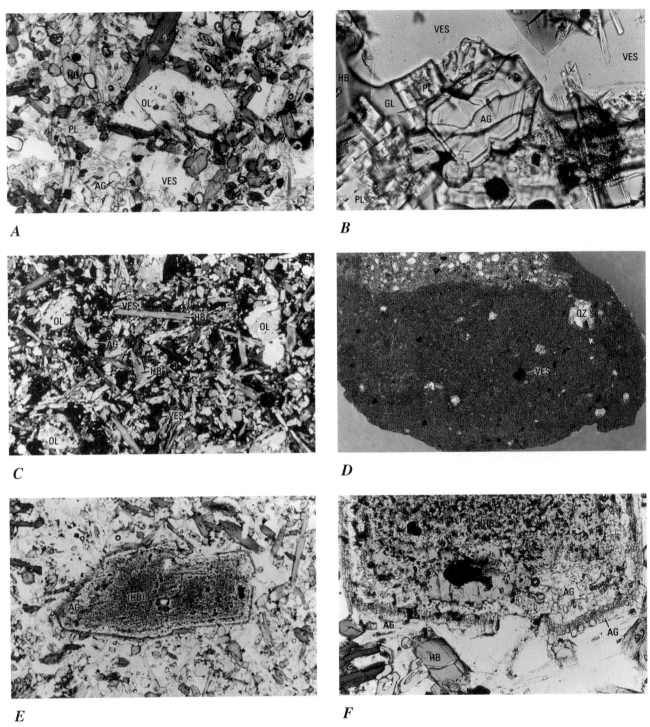

Figure 9. Microscopic and hand-sample images of basaltic inclusions from June 7–12 lava dome. Symbols: T, transmitted light; R, reflected light; X, cross polarized; P, plane polarized; *xx* mm, width of frame in millimeters; AG, augite; HB, hornblende; (HB), resorbed hornblende xenocryst; OL, olivine; PL, plagioclase; QZ, quartz; VES, vesicle; MT, magnetite; GL, glass. *A,* Fo₈₇ olivine phenocryst with hornblende rim in sample CN6791–i (T, P, 2.7 mm). *B,* Oscillatory and normally zoned augite (Mg# 70–85) microphenocryst in sample CN6791–i (T, P, 0.34 mm). *C,* Diktytaxitic textured basaltic inclusion CN6791–i showing acicular hornblende and plagioclase microphenocrysts, augite microphenocrysts and small phenocrysts, olivine phenocrysts with hornblende rims, and irregular vesicle space (T, X, 5.5 mm). *D,* Basaltic inclusion sample P22892–1a showing sparse spherical to irregular vesicles developed in porous diktytaxitic groundmass. Large white grains are quartz and feldspar xenocrysts. *E,* High-Al-Ti hornblende xenocryst with clinopyroxene reaction rim, basaltic inclusion sample CN6791–1i. Relict hornblende composition is similar to unreacted groundmass hornblende (table 2) (T, P, 5.5 mm). *F,* Margin of resorbed hornblende xenocryst from *E,* showing inner (Mg# 80) and outer (Mg# 66) rims of irregular, then fibrous, augite (T, P, 1.3 mm).

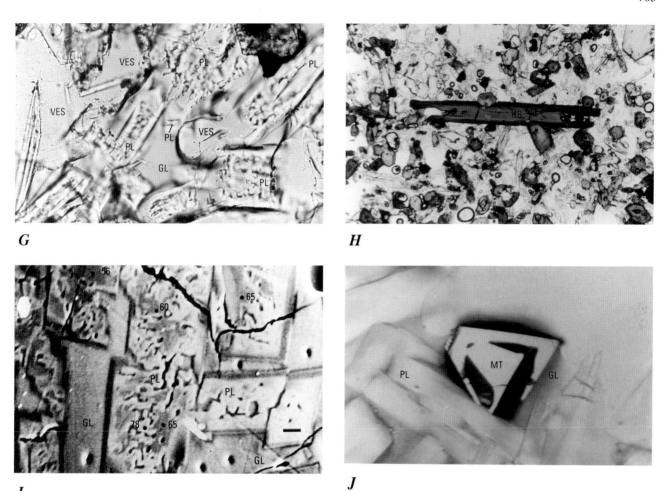

G

H

I

J

Figure 9—Continued. G, Hollow, glass-inclusion-rich plagioclase microlites, interstitial glass, and void space in basaltic inclusion sample CN6791–1i. Note also hopper-shaped plagioclase
microlite at center frame (T, P, 0.34 mm). H, Swallow-tailed hornblende phenocryst in basaltic inclusion CN6791–1i (T, P, 2.7 mm).
I, Back-scattered electron image of mottled plagioclase and 73

wt% SiO_2 matrix glass in basaltic inclusion sample CN6791–1i.
Scale given by 5-μm bar, frame width about 100 μm. Numbers
give An content of plagioclase. Dark pits in matrix glass locate
areas analyzed (areas damaged by the electron beam). J, Skeletal
magnetite grain in basaltic inclusion sample CN6791–1i (R+T, P,
0.25 mm).

crystals of plagioclase and hornblende protrude (fig. 9B,C).
In addition to the irregular network of diktytaxitic voids,
there are sparse 1- to 2-mm spherical vesicles (fig. 9D) that
are most common within cores of the inclusions. This
implies two-stage volatile exsolution. Perhaps, initial exsolution of a supercritical fluid phase created the diktytaxitic
interstices, and then saturation and separation of a vapor
phase formed the spherical vesicles. The diktytaxitic voids
must have been pressurized for the spherical vesicles to
have developed in what would otherwise have been an open
medium.

The inclusions lack well-defined chilled margins, possibly because it was not possible to establish strong thermal
gradients in such relatively small bodies. Alternatively,
magmatic erosion may have removed chill margins. The
presence of groundmass glass, undercooling textures, vesicle concentrations, deformed margins, and banded scoria

provide strong evidence that the basalt was magmatic when
it was included in the andesite. For comparison, similar textural evidence from a variety of mafic inclusions is
reviewed by Bacon (1986) and Stimac and others (1990) as
evidence for mingling of magmas.

JUNE 15 DACITIC PUMICE

Dacitic pumice from the June 15 climactic eruption
can be divided into two principal types: white phenocryst-
rich, and tan phenocryst-poor. There is considerable variation in phenocryst content, ranging from about 50 percent to
as little as 15 percent in optical modes (table 1). This variation principally reflects smaller crystals in the tan pumice,
as many small crystal fragments and microlites are difficult
to discern optically and are included in the groundmass
fraction in optical modes. Back-scattered electron images of
pumice bubble walls reveal abundant small (<10 μm to

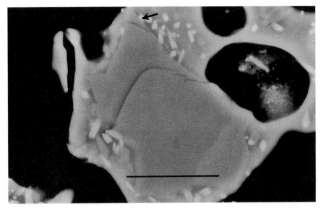

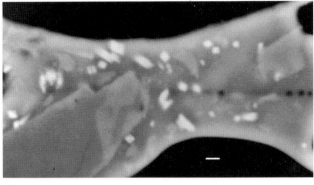

A

B

Figure 10. Back-scattered electron images of crystal-fragment-charged matrix glass filaments in phenocryst-poor dacite pumice sample PH13d from a pyroclastic-flow deposit. Image density inversely proportional to mean atomic number (lighter areas are higher atomic number). Small white grains are amphibole, lighter gray fragments are plagioclase, and darker gray material between grains is glass. *A*, A relatively large (15 µm) plagioclase fragment with arcuate internal fracture and faint composition zonation band (at lower right), surrounded by numerous small (<2 µm) crystal fragments. Fragment at upper edge of feldspar (at arrowhead) can be fit back onto main grain. *B*, Enlargement of bubble wall, which is charged with crystal fragments. Light (higher calcium) compositional zone in large plagioclase grain truncated at grain margin. Scale bar in *A* is 10 µm; bar in *B*, 1 µm.

A ———— 2MM

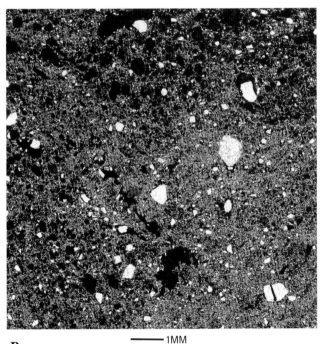

B ———— 1MM

Figure 11. Wavelength-dispersive X-ray intensity maps showing aluminum distribution in 1-cm² thin-section areas of phenocryst-rich (*A*) and phenocryst-poor (*B*) pumice. Maps were generated by driving the microprobe stage over a 1,000×1,000 point grid with 10-µm node spacing and 20-ms dwell time and storing the data in a 1,000×1,000 pixel image. Individual pixels in the image represent activation areas of ~3 µm². White grains are plagioclase, dark gray grains are hornblende, gray network is matrix glass, and black areas represents epoxy-filled vesicles. Abundance of plagioclase is similar in both *A* and *B* on a vesicle-free basis, as determined from pixel counts.

<1 µm) crystal fragments, including plagioclase fragments with compositional zones that are truncated at grain margins (fig. 10*A,B*). Preliminary analysis of microprobe elemental distribution maps of phenocryst-rich and phenocryst-poor pumice sections (fig. 11*A,B*) suggests that the total crystal abundances of the two types approach one another when microscopic crystal fragments in the groundmass are taken into account (fig. 11; Meeker and others, 1993).

Sparse, large (1–2 mm) plagioclase crystals in the phenocryst-poor pumice are rounded, with oscillatory zones

Table 3. Phenocryst pumice densities (D, in grams per cubic centimeter) and abundances of phenocryst-rich and phenocryst-poor dacitic lapilli in pyroclastic flow deposits of June 15, 1991.

[Component proportions were determined from <4-mm clast fractions by hand sorting with the aid of a binocular microscope. Densities determined by the method of Hoblitt, Wolfe, and others (this volume). n, number of lapilli]

Sample	Phenocryst-poor dacite		Phenocryst-rich dacite		Total
	Avg. D	n	Avg. D	n	n
6-27-91	0.864	5	0.982	16	21
2-23-92-2	.943	4	1.005	11	15
2-23-92-3		0	0.996	14	14
2-25-92-1	.806	2	1.002	15	17
2-26-92-1	.722	3	.934	15	18
2-27-92-1		0	.989	18	18
3-02-92-10		0	1.015	4	4
3-02-92-11		0	0.944	20	20
3-04-92-1	.866	2	.940	16	18
3-04-92-1-3	.827	5	.933	9	14
3-04-92-2		0	.992	8	8
3-11-92-1	.934	3	.961	13	16
3-11-92-2		0	.979	18	18
3-11-92-3	1.041	4	.980	11	15
3-14-92-1-3	.747	4	.935	17	21
3-14-92-1-4	.744	4	.923	21	25
3-14-92-1-5	.622	9		0	9
3-17-92-1		0	1.010	23	23
3-17-92-2		0	1.170	4	4
3-18-92-1	.811	4	.925	22	26
3-18-92-5	.722	2	.917	22	24
Total n:		51 (15%)		297 (85%)	348
Avg. D:	.819		.977		
Std. dev.:	.108		.055		
Vol%:	14.6		85.3		

that are truncated at the rounded margins (fig. 12*A*,*B*). Secondary electron images of external surfaces of these crystals are pitted, a feature we initially interpreted as solution pitting (Pallister and others, 1992). However, rounding of these large crystals could also be related to a physical process of abrasion and fragmentation, just as the larger and more equant feldspar fragments in the matrix of figure 10 have been broken into smaller and more angular fragments. But, given the evidence of extreme fragmentation seen in the groundmass, how did the large crystals survive? They could be relicts of "outsized" megacrysts (to 1 cm) that are observed in some of the phenocryst-rich pumice blocks. Alternatively, the large crystals could have been entrained at a late stage of the fragmentation process.

The phenocryst-rich pumice typically has large (>1 mm) stretched vesicles. In contrast, the phenocryst-poor pumice has smaller (<1 mm), spherical vesicles and forms breadcrust-surfaced blocks. Inclusions of the white phenocryst-rich pumice in the tan phenocryst-poor pumice are relatively common, and the two types are mingled together in banded pumice blocks. Densities and relative abundances of phenocryst-rich and phenocryst-poor pumice lapilli in 21 bulk samples from June 15 pyroclastic-flow deposits are summarized in table 3. There is considerable variation in the relative abundance from site to site, possibly due to flow segregation. On average, 85 percent (by volume) of the pumice lapilli in these samples is the phenocryst-rich type and 15 percent is phenocryst-poor, similar to our previous estimate of 80 percent and 20 percent (Pallister and others, 1992). The average bulk density of the phenocryst-rich pumice is 0.98, and that of the phenocryst-poor pumice is 0.82. An upward decrease in the abundance of phenocryst-

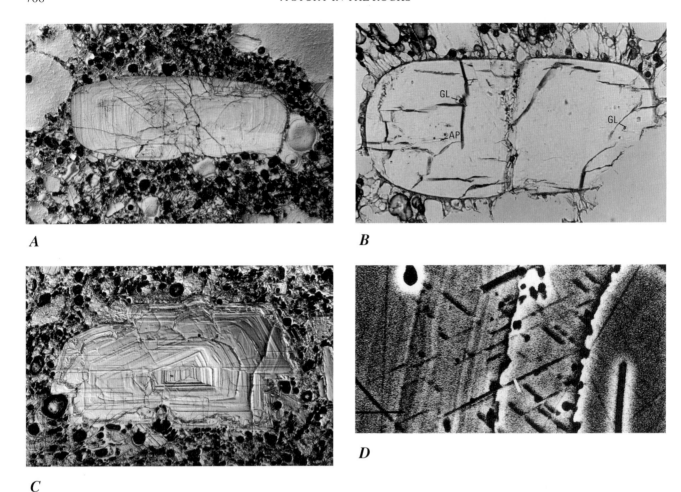

A

B

C

D

Figure 12. Microscopic images of plagioclase phenocrysts in June 15 dacitic pumice. Abbreviations as in figure 6. *A*, Large, rounded plagioclase in phenocryst-poor pumice. Normarski differential contrast illumination, showing oscillatory zoning profiles. Note termination of zoning bands at rounded crystal margins (R, X, 2.7mm). *B*, Rounded plagioclase crystal in phenocryst-poor pumice. Note small apatite (AP) and bubble-bearing glass inclu-

sions (T, P, 1.3 mm). *C*, Euhedral, oscillatory-zoned, plagioclase phenocryst in phenocryst-rich pumice. Normarski differential contrast illumination showing zoning profiles and internal resorption discontinuities (R, X, 2.7 mm). *D*, Back-scattered electron image of oscillatory zoning in plagioclase crystal from phenocryst-rich pumice sample PH13a. Scale given by 10-μm bar; width of field about 80 μm.

poor pumice lapilli has been noted in several sections through the tephra-fall deposits (David and others, this volume; Paladio-Melosantos and others, this volume; R.P. Hoblitt, unpub. data, 1993). This trend would normally indicate a decreasing supply of phenocryst-poor magma during the June 15, 1991, eruption. Although this may be the case, the possibility of sorting within the eruptive column has not been evaluated.

Phenocrysts in both types of June 15 dacitic pumice are principally plagioclase and hornblende. Lesser amounts of cummingtonite, biotite, quartz, ilmenite, magnetite, apatite, and anhydrite are present (table 1). In contrast to the resorbed and broken crystals in the phenocryst-poor pumice, plagioclase in the phenocryst-rich pumice forms large (1–2 mm, rarely to 5 mm) euhedral crystals or euhedral crystals that are broken into only a few pieces. Most of these crystals show only subtle internal evidence of

resorption, such as unconformities in zoning. A subordinate population of crystals shows a distinct dark resorption band that is rich in inclusions and typically located about four fifths of the way to the crystal margins.

The similarity of the average composition (about An_{40}), and compositional range (An_{27-53}) in plagioclase phenocrysts from the dacite to the average and range in the hybrid andesite (fig. 5, table 2) suggests that the dacitic magma was the sole source for plagioclase in the andesite. As in the andesite, the plagioclase displays complex oscillatory zonation (fig. 12*C*). Individual zones are recognized at 1-μm scale in back-scattered electron images (fig. 12*D*) and variations as great as 20 mol% anorthite component (An_{29-50}) and commonly 5 to 10 mol% An are seen between point analyses of adjacent bands or sets of bands (fig. 13). Large changes in An content sometimes, but not always, correlate with unconformities in zoning patterns.

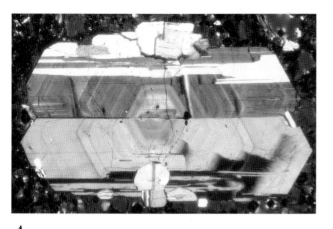

A   *B*

Figure 13. Euhedral, oscillatory zoned plagioclase crystal (*A*), from phenocryst-rich dacitic pumice sample EW910615–1. Area of microprobe traverse *B* is shown in lower right quadrant. Numbers indicate percent An content at indicated spots. Point-beam analyses were used; no migration of sodium was detected.

Edge compositions for two euhedral crystals in the phenocryst-rich pumice range from An_{34} to An_{37}. Three analyzed microphenocrysts (<100 µm) in the phenocryst-rich pumice are normally zoned (An_{53-38}).

Glass inclusions are common within the plagioclase phenocrysts; they are rhyolitic in composition and contain as much as 6.4 wt% H_2O but less than 100 ppm sulfur (Westrich and Gerlach, 1992; Rutherford and Devine, this volume). The inclusions occur both as small (<10 µm) spheres concentrated parallel to zoning bands and as larger (to 100 µm) and typically more polygonal bodies that often crosscut zoning bands. Vapor bubbles are ubiquitous in the larger inclusions and are present in at least some of the smaller inclusions. A few inclusions contain a third (crystalline) phase that we have not identified. As pointed out by Westrich and Gerlach (1992), the large volume of vapor bubbles in plagioclase and other phenocrysts suggests that the dacitic magma was volatile saturated. Experimental work by Rutherford and Devine (this volume) confirms that the magma was either saturated or close to saturation with a water-rich (95 vol% H_2O, 5 vol% SO_2) fluid phase.

Hornblende forms euhedral or broken phenocrysts (typically 0.5–1 mm) in the phenocryst-rich dacitic pumice and forms smaller broken crystal fragments in the phenocryst-poor pumice. Euhedral crystals are rimmed by 10- to 50-µm-thick cummingtonite borders (fig. 14*A*), and some crystals show weak internal zonation, although not as pronounced as that in the higher titanium hornblendes of the andesite and basalt. The hornblende contains moderately abundant glass inclusions that range in size from <1 to 100 µm and that contain high-silica rhyolite glass and vapor bubbles. Hornblende compositions are distinct from most of those in the hybrid andesite and basalt, with lower Mg#, Al, Ti, Na, and K (fig. 5, table 2). As in the andesite, the hornblende phenocrysts lack dehydration reaction rims. Lack of reaction rims on similar composition dacite from Mount St.

Helens indicates ascent to the surface in a few days or less (Rutherford and Hill, 1993).

Biotite is a minor component in the dacite. It commonly occurs as cores to hornblende phenocrysts and in rare glomerocrysts with plagioclase and hornblende. Quartz forms sparse, highly fractured grains with large glass inclusions. Small (<0.3 mm) grains of both ilmenite and magnetite are common within the groundmass glass and included within the silicate phenocrysts. Apatite is ubiquitous as microlites but also forms larger (~0.1 mm) euhedral phenocrysts associated with anhydrite (fig. 14*B*). Anhydrite forms euhedral crystals that contain apatite inclusions and is similar in appearance to the anhydrite crystals found in the andesite scoria. Although apatite is a common inclusion in the silicate phenocrysts, anhydrite rarely occurs as an inclusion phase. Anhydrite typically forms subhedral crystals surrounded by vesicular matrix glass. A single biotite-hornblende-plagioclase-anhydrite glomerocryst was observed in thin sections from 25 samples of June 15 dacite, and a small anhydrite inclusion was found at the contact between a hornblende phenocryst and its cummingtonite rim in a dacite pumice lapilli from a surge bed that was deposited on June 14.

Olivine and bronzite occur as sparse, highly resorbed anhedral grains. The olivine grains are bordered first by bronzite with fine-grained magnetite inclusions, and then by an outer zone composed of aggregates of hornblende and biotite microphenocrysts (fig. 14*C*). These olivine crystals are distinctly less magnesian than those in the scoria and basalt (fig. 5). A spinel inclusion in one of the olivines is exceedingly iron-rich (about 70 wt% total iron, table 2) and Cr- and Al-poor (11.3 wt% Cr_2O_3, 4.5 wt% Al_2O_3) and has likely undergone subsolidus alteration. We suspect that these olivine xenocrysts are the relicts of grains that were entrained during earlier (pre-1991) episodes of magma mixing.

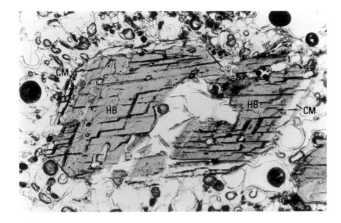

A

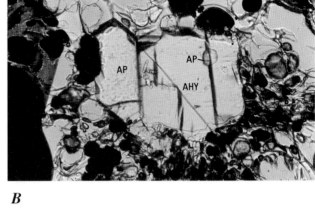

B

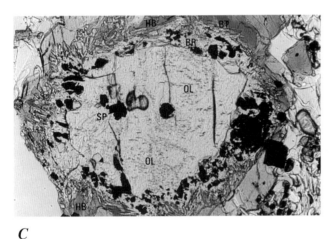

C

Figure 14. Photomicrographs of phenocryst-rich dacitic pumice sample EW910615–1 from a June 15 pyroclastic-flow deposit in the Sacobia River valley. Symbols: T, transmitted light; R, reflected light; X, cross polarized; P, plane polarized; *xx* mm, width of frame in millimeters; AHY, anhydrite; AP, apatite; CM, cummingtonite; BT, biotite; HB, hornblende; OL, olivine; BR, bronzite; MT, magnetite; SP, chrome spinel. *A*, Euhedral (now fractured) hornblende phenocryst with clear cummingtonite rim (T, P, 1.3 mm). *B*, Euhedral anhydrite(AHY)-apatite(AP) phenocryst composite. Note apatite inclusion within anhydrite (T, P, 0.68 mm). *C*, Fo$_{80}$ olivine mantled by bronzite + magnetite and an outer rim of hornblende + biotite (T, P, 0.68 mm).

GLASS COMPOSITIONS

Major element and volatile abundances for numerous glass inclusions and for matrix glass in the June 15 dacite are reported by Westrich and Gerlach (1992) and by Rutherford and Devine (this volume). In this report we focus on variations in the major element composition of *matrix* glass within our samples of (1) June 12 andesitic scoria, (2) a basalt inclusion from the June 7–12 andesitic dome, and (3) phenocryst-rich and phenocryst-poor dacitic pumice from the June 15 eruption. Average compositions and standard deviations for matrix glasses are listed in table 4 and compared graphically in figure 15. All glass analyses utilized a point electron beam; analyses were corrected for migration of alkalis, silica, and alumina by use of an iterative counting and correction procedure similar to that of Nielson and Sigurdsson (1981). Use of a point beam was required by the high vesicularity and thin bubble walls of pumice samples and by the presence of abundant microscopic crystal fragments in the phenocryst-poor pumice and microlites in the andesite. The analytical procedure is described in more detail in the appendix.

Major element compositions of matrix glasses do not lie along simple mixing trends but are complicated by the effects of crystal growth and resorption following mixing. Matrix glass in the basalt is rhyolitic, with relatively high K$_2$O (fig. 15*A*). The matrix melt in the basaltic inclusions must have been low originally in SiO$_2$ but must have undergone extreme fractionation by growth of abundant hornblende, plagioclase, and magnetite microlites upon inclusion in the cooler dacite or andesite. The high K$_2$O content of the glass is consistent with an overall relative enrichment of incompatible elements in the basalt. Dark bands within the andesitic scoria are composed of lower silica glass that is enriched in olivine and clinopyroxene phenocrysts. Consequently, we attribute much of the variation in the matrix glass of the andesite scoria to insufficient time for homogenization following mixing of rhyolitic melt from the dacitic magma and more mafic melt from the basalt.

Matrix glass in the phenocryst-poor dacite shows the widest variation in composition; that of the phenocryst-rich dacite is the most restricted and most evolved of all the 1991 samples. Because both of the dacitic types are identical in major element composition, we initially suggested that the wide range in glass compositions from the phenocryst-poor pumice was produced by disequilibrium melting of phenocrysts, a process that would shift the glass composition back toward the lower silica bulk-rock compositions

Table 4. Average compositions and standard deviations for matrix glass in 1991 Pinatubo samples.

[Analytical procedures given in the appendix. MnO less than 0.1 wt%. PR, phenocryst rich; PP, phenocryst poor; Avg., average; n, number of analyses averaged; SD, standard deviation; pl, plagioclase; hb, hornblende; cm, cummingtonite; ag, augite]

	June 12 tephra 7-1-91-1A scoria Avg. SD (n=12)		June 15 PR dacite EW910615-1 Avg. SD (n=24)		June 15 PP dacite PH13d (near pl) Avg. SD (n=23)		PH13d (near hb, cm, ag) Avg. SD (n=17)		PH13d (misc. matrix) Avg. SD (n=6)		Basalt inclusion CN6791-i Avg. SD (n=30)	
SiO$_2$	70.39	1.18	76.73	0.76	69.37	1.92	75.38	1.50	72.65	1.37	73.20	1.23
Al$_2$O$_3$	14.96	.70	12.55	.23	14.87	1.21	12.68	.45	13.55	1.11	13.97	.37
FeO	2.41	.14	.74	.04	2.56	.53	1.26	.43	1.67	.72	1.89	.13
MgO	.72	.16	.10	.02	1.51	.81	.49	.37	.75	.53	.08	.06
CaO	2.57	.16	1.20	.04	3.60	.71	1.54	.45	2.35	.91	.66	.11
Na$_2$O	2.81	.56	3.66	.62	3.47	.75	3.97	.28	3.74	.75	3.79	.65
K$_2$O	3.18	.17	3.07	.20	2.36	.27	3.01	.16	2.75	.23	5.02	.28
TiO$_2$	.31	.05	.06	.03	.26	.06	.11	.08	.16	.10	.25	.04
Total	97.35	.97	98.11	1.03	98.00	1.32	98.44	1.08	97.62	2.07	98.86	.96

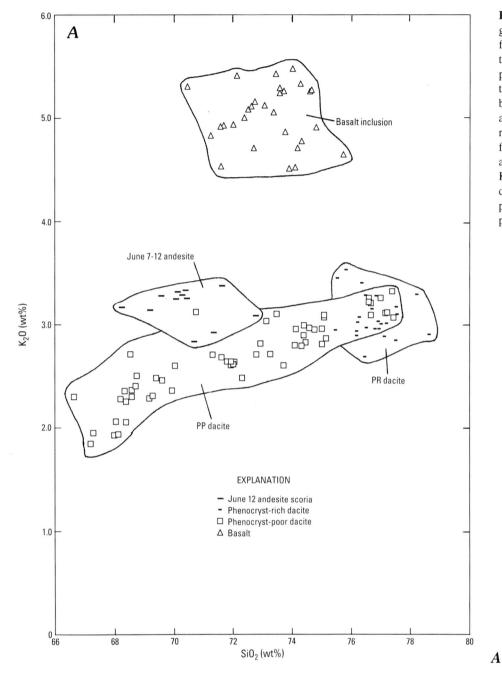

Figure 15. SiO₂ variation diagrams for K₂O (*A*) and FeO (*B*) for matrix glasses in 1991 eruptive products. Compositions of plagioclase and hornblende from the phenocryst-rich dacite, and bulk analyses of the 1991 dacite, andesite, and basalt (with all Fe recalculated as FeO) are shown for comparison in *B*. Point-glass analyses were corrected for Na, K, Al, and Si migration as described in the appendix. PR, phenocryst rich, PP, phenocryst poor.

(Pallister and others, 1992). Support for this hypothesis is evident in figure 15*B*; the less evolved glass compositions trend back toward the bulk composition of the June 15 dacite. Least-squares mass balance of the major element compositions indicates that addition of up to 22 wt% plagioclase +9 wt% hornblende +1 wt% oxides to the high-silica glass would explain the average chemical trend.

However, the compositional effects of dissolution of phenocrysts and contamination of the matrix glass by tiny phenocryst fragments has proven to be difficult to distinguish. Analysis points were located by use of high-magnification back-scattered electron images to avoid the abundant microscopic crystal fragments in the matrix glass; but even

with this precaution, adjacent or underlying crystal fragments were sometimes excited and contributed to the analyses. Dispersion of the compositional trend in figure 15*B* toward plagioclase or hornblende is explained by excitation of adjacent or underlying fragments of these crystals, yet the overall trend back toward the bulk dacite composition would require systematic contamination by a consistent modal ratio of about 2 parts plagioclase to 1 part hornblende (22% to 9%, see above). Either we are seeing modal melting, or many areas that appear to be free of crystal fragments in our back-scattered electron images are actually composed of a well-mixed "slurry" of even smaller crystal particles and glass. Higher resolution microbeam and X-ray

Figure 15.—Continued.

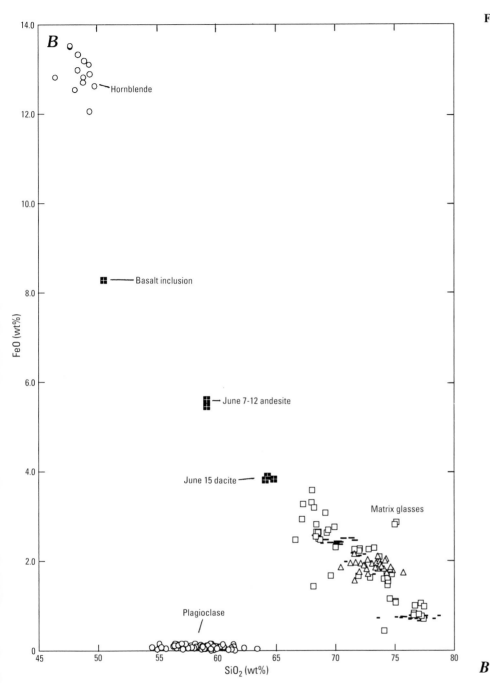

B

diffraction work is under way to evaluate this possibility. At present, although we acknowledge that crystal breakage would facilitate dissolution by increasing surface area and that some melting appears to have occurred, abundant broken crystal fragments in the phenocryst-poor pumice and the lack of higher Fe-Ti-oxide temperatures in the phenocryst-poor pumice indicate that most of the difference between the two pumice types is due to mechanical breakage of crystals (figs. 10, 11; Meeker and others, 1993).

THERMOBAROMETRY

FE-MG EXCHANGE EQUILIBRIA

Fe-Mg exchange between olivine and melt provides a means to estimate the Mg# and temperature of melt that coexisted with the olivines in the June 7–12 Pinatubo rocks. As previously noted, the forsterite content of olivines in the June 7–12 andesite and basalt samples overlap; they average Fo$_{87}$ (table 2). There is little range in composition; the most magnesian core composition for olivine prisms is similar in the andesite and basalt samples, Fo$_{88-89}$. Using the distribution coefficient for (FeO/MgO)^{olivine}/ (FeO/MgO)^{melt} of 0.3 (Roeder and Emslie, 1970; Ulmer, 1989), and assuming that Fe$_2$O$_3$/FeO=0.123, as discussed later, we calculate an average Mg# of 67 for equilibrium melts and a slightly higher value of 69 during initial crystallization of Fo$_{88}$ olivine cores. The whole-rock Mg# of the least evolved basalt inclusion is 69 (see table 7), consistent with the olivine cores being early equilibrium phases and with the basalt being representative of the mafic mixing end-member for the hybrid andesites. Equilibrium fractional crystallization of the liquid line of descent for the basalt at atmospheric pressure, as determined by the CHAOS program of Neilsen (1990), predicts that Fo$_{88}$ olivine would initially crystalize at 1,260°C and would remain as the only liquidus phase until 1,210°C, where An$_{70}$ plagioclase would appear, to be joined by Mg# 85 clinopyroxene at 1,170°C. The predicted mineral compositions are in good agreement with the most magnesian olivine and clinopyroxene and the most calcic plagioclase analyzed in the basalt (table 2). However, the order of crystallization inferred from mineral textures (olivine and clinopyroxene preceding plagioclase) is different, probably as a result of crystallization at either high pressure (Bender and others, 1978; Grove and Kinzler, 1986) or under hydrous conditions (Spulber and Rutherford, 1983; Sisson and Grove, 1993). Higher pressure crystallization would also shift the distribution coefficient for FeO/MgO in olivine to higher values (Ulmer, 1989) but we cannot rigorously evaluate this effect because we do not know the primary oxidation state of iron in the basalt melt. Our selection of 0.123 for Fe$_2$O$_3$/FeO may have been fortuitous; selection of a higher ratio would require a higher distribution coefficient (and consequently higher pressure) to reproduce the Fo$_{87-88}$ olivines. Presence of H$_2$O would lower the liquidus temperature of the basalt and would also favor early crystallization of clinopyroxene (Sisson and Grove, 1993). Rutherford and others (1993) estimate that addition of 2.5 wt% H$_2$O to the basaltic composition from Mount Pinatubo would result in an olivine-augite cotectic temperature of 1,200°C.

FE-TI OXIDES

Average compositions of coexisting ilmenite and magnetite from the 1991 Pinatubo samples, and temperatures based on the Andersen and Lindsley (1988) thermometer, are listed in table 5. Results are given for the dacite and andesite; the basalt samples contain only one oxide (magnetite). Iron distribution and mole fractions of ulvospinel and ilmenite were calculated by the method of Stormer (1983). Similar temperatures and oxygen fugacities were obtained by using the mineral projections and oxide equilibria

routines in the QUILF program of Andersen and others (1993). Only homogeneous ilmenite-magnetite grain pairs that are touching or are within a few micrometers of one another in thin section were used for thermometry. Despite relatively low Mn abundances, Mg/Mn ratios for the Pinatubo oxides overlap the equilibrium envelope of Bacon and Hirschmann (1988), as seen in figure 16 (inset).

The average temperature and oxygen fugacity (in log units) determined for four magnetite-ilmenite pairs in the June 15 phenocryst-rich dacitic pumice is 816±22°C and 11.08±0.12 respectively, indistinguishable from results for the phenocryst-poor pumice. The temperatures have been reduced by 30°C (yielding an average of 786±22°C) prior to plotting in figure 16 to correct for overestimation of Andersen-Lindsley-Stormer temperatures at high oxygen fugacity (~3 log units above Ni-NiO) (Geschwindt and Rutherford, 1993; Rutherford and Devine, this volume).

Apparent temperatures and oxygen fugacities were also calculated for the magnetite-rimmed ilmenites of the June 12 scoria (sample 7-1-91-1A). Only grains without oxidation lamellae were used for thermometry; grains with lamellae from the more slowly cooled dome samples were not used. As described previously, the magnetite rims are overgrowths of new crystals that nucleated and grew on ilmenite grains during magma mixing. The rimmed crystals indicate changing conditions in the magma, a disequilibrium situation. The zoning of core ilmenite to higher titanium within about 10 μm of the magnetite rim (fig. 6L) suggests diffusive exchange with the growing rim. Temperatures and oxygen fugacities for the pairs core versus rim and high-titanium zone versus rim yield temperatures of 961°C at log f_{O_2}=−9.1 and 930°C at log f_{O_2}=−9.7, respectively. These values are intermediate between those determined for the dacite and those inferred from olivine-melt equilibria (~1,200°C) at the reduced conditions implied by the presence of sulfides in the basalt. We can test whether the Fe-Ti-oxide temperatures and oxygen fugacities are reasonable by calculating the oxidation state of iron for the bulk sample according to the method of Sack and others (1980) and comparing the result to that obtained from direct analysis. Of the 6.1 wt% Fe$_2$O$_3$ total in a bulk analysis of sample 7-1-91-1A, we calculate 2.3 wt% as Fe$_2$O$_3$ and 3.3 wt% as FeO at T=950°C and log f_{O_2}=−9.5. This result is similar to the distribution of iron in the bulk sample that was determined analytically (see table 7), circumstantial evidence that the Fe-Ti-oxide temperatures and oxygen fugacities are representative of the mixed magma just before eruption.

CUMMINGTONITE STABILITY

The presence of cummingtonite rims on hornblende, in combination with the composition of plagioclase rims and microlites in the June 15 dacitic pumice samples, provides a

Table 5. Microprobe compositions of Fe-Ti oxides, iron recalculation and formulas by the method of Stormer (1983), and calculated temperatures and oxygen fugacities for mineral pairs according to the method of Andersen and Lindsley (1988).

[Avg., average; *n*, number of analyses; X ulv, mole fraction ulvospinel; X ilm, mole fraction ilmenite]

	June 15 PR dacite (EW910615-1)														
Grain no.	10-MAG		10-ILM		14-MAG		14-ILM		16-MAG		16-ILM		17-MAG		17-ILM	
Index no.	50		50		53		53		56		56		59		59	
	Avg.	SD	Avg.	SD	Avg.	SD	Avg.	SD	Avg.	SD	Avg.	SD	Avg.	SD	Avg.	SD
n	7		6		5		4		5		5		4		5	
MgO	1.11	.04	1.00	.02	1.14	.01	1.03	.01	1.13	.01	1.01	.02	1.12	.02	1.01	.03
Al_2O_3	2.09	.02	.39	.00	2.07	.02	.38	.01	2.05	.02	.38	.01	2.09	.05	.37	.02
TiO_2	4.48	.13	28.57	.14	4.41	.03	28.74	.21	4.49	.07	29.11	.10	4.16	.03	28.06	.21
Cr_2O_3	.18	.02	.08	.01	.19	.01	.08	.01	.19	.01	.08	.02	.16	.01	.08	.02
FeO (total)	86.93	.42	65.78	.57	86.95	.24	65.67	.31	87.30	.36	66.01	.17	83.71	.29	64.40	.52
MnO	.43	.01	.22	.02	.43	.01	.22	.01	.44	.02	.21	.01	.42	.01	.21	.01
Sum	95.22	.75	96.04	.81	95.19	.37	96.12	.62	95.60	.55	96.80	.37	91.66	.54	94.13	.92
FeO	34.60		24.04		34.46		24.20		34.66		24.57		33.68		23.73	
Fe_2O_3	58.15		46.39		58.33		46.08		58.50		46.05		55.60		45.19	
Total	101.04		100.69		101.03		100.73		101.46		101.41		97.23		98.65	
X ulv	0.132				0.129				0.131				0.127			
X ilm			0.540				0.542				0.546				0.540	
T (°C)	819 ±23				815 ±23				817 ±22				812 ±23			
log fO_2	-11.02 ±0.12				-11.09 ±0.12				-11.07 ±0.12				-11.12 ±0.12			

	June 15 PP dacite (PH13D)												June 12 tephra (7-1-91-1A scoria)					
Grain no.	7-MAG		7-ILM		8-MAG		8-ILM		10-MAG		10-ILM		19-MAG (rim)		19-ILM-core		19-ILM-edge	
Index no.	62		62		65		65		68		68		69		71		74	
	Avg.	SD	Avg.	SD	Avg.	SD	Avg.	SD	Avg.	SD	Avg.	SD	Avg.	SD	Avg.	SD	Avg.	SD
n	5		4		5		4		5		5		5		4		4	
MgO	1.10	.04	1.01	.02	1.40	.21	1.06	.06	1.11	.05	1.00	.02	2.87	.05	1.57	.27	2.28	.06
Al_2O_3	1.90	.04	.35	.02	2.03	.05	.35	.01	1.89	.06	.36	.01	2.73	.02	.44	.04	.51	.02
TiO_2	4.31	.11	28.67	.11	4.48	.03	28.92	.19	4.26	.23	28.60	.05	8.27	.43	28.08	.89	30.85	.16
Cr_2O_3	.20	.01	.06	.01	1.15	.04	.05	.02	.16	.01	.06	.03	.09	.01	.09	.01	.06	.02
FeO (total)	87.12	.23	66.00	.12	84.16	.39	65.98	.98	87.74	.16	65.97	.24	80.49	.13	65.33	1.20	62.37	.50
MnO	.45	.02	.23	.01	.46	.02	.23	.01	.42	.03	.23	.03	.57	.01	.21	.02	.27	.01
Sum	95.08	.56	96.32	.39	93.68	.87	96.59	1.37	95.58	.55	96.22	.51	95.02	.76	95.72	2.51	96.34	.89
FeO	34.35		24.16		33.53		24.28		34.48		24.21		35.03		22.65		23.76	
Fe_2O_3	58.64		46.50		56.27		46.34		59.20		46.51		50.52		47.44		42.91	
Total	100.95		100.98		99.32		101.23		101.52		100.97		100.08		100.48		100.64	
X ulv	0.126				0.135				0.123				0.232					
X ilm			0.539				0.542				0.539				0.524		0.567	
T (°C)	811 ±23				824 ±23				808 ±23				967 ±29				936 ±25	
log fO_2	-11.14 ±0.12				-10.96 ±0.12				-11.19 ±0.12				-8.97 ±0.13				-9.55 ±0.12	

means to fix the temperature and P_{H_2O} of the dacite during its final stage of crystallization prior to eruption. As is explained elsewhere (Geschwindt and Rutherford, 1993; Rutherford, 1993; Rutherford and Devine, this volume), the field of cummingtonite+hornblende (coexisting with plagioclase, magnetite, ilmenite, and melt) is limited to temperatures of <800°C and P_{H_2O} of about 100–400 MPa (1–4 kbar) in these dacitic magmas. The Fe-Ti-oxide temperature of about 780°C constrains the magma to the higher temperature range of cummingtonite stability and, with coexisting quartz and An_{34-40} plagioclase rims, to a more limited P_{H_2O} range of about 200–320 MPa (see fig. 2 of Rutherford, 1993, and fig. 4 of Rutherford and Devine, this volume).

BULK-ROCK CHEMISTRY

Sample localities and descriptions for analyzed samples of 1991 eruptive products are given in table 6, and major and trace element analyses are listed in table 7. Most of the analyzed samples were derived from single pumice or scoria blocks or from single basalt inclusions (as noted in table 6). Bulk-rock compositions are typical of calc-alkalic arc rocks, with relatively low TiO_2 and low abundances of high-field-strength (HFS) elements (fig. 17). Pinatubo rocks are enriched in light rare earth elements (LREE) compared to those from nearby volcanoes in the Bataan arc. In this respect, the 1991 Pinatubo magmas are more like basalts from the back-arc Mount Arayat volcano to the east than to

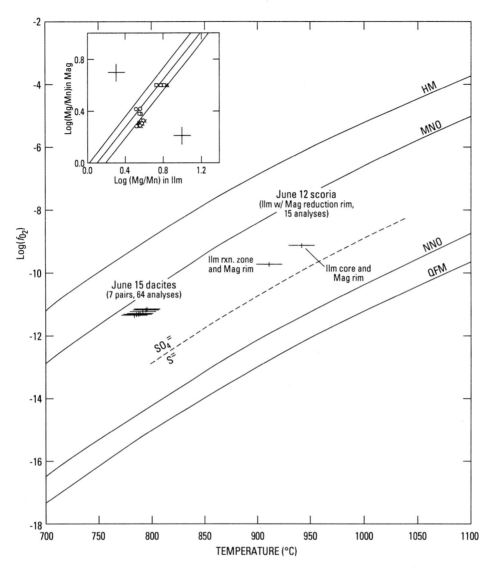

Figure 16. Oxygen fugacity - temperature diagram showing results from Fe-Ti-oxide thermometry on 1991 Pinatubo samples. Buffer curves from Rutherford (1993). Inset shows Mg/Mn distribution in Pinatubo oxides compared to equilibrium field of Bacon and Hirschmann (1988); crosses show typical analytical errors in Mg/Mn for June 15 dacite and June 12 scoria. Mag, magnetite; Ilm, ilmenite; HM, hematite-magnetite; MNO, $MnO-Mn_3O_4$; NNO, Ni-NiO; QFM, quartz-fayalite-magnetite; rxn, reaction.

the andesites and dacites of the adjacent frontal arc Natib and Mariveles stratovolcanoes to the south (compare REE data in fig. 17 and table 7 with data of Defant and others, 1991).

The dacites of Mount Pinatubo are unusually depleted with respect to incompatible element abundances. This is especially evident from the fact that the dacites have lower rare earth element abundances than the basalt inclusions. The depletion of incompatible elements in the dacites relative to the basalts indicates that the basalts cannot be parental to the dacites by major phase crystal fractionation. In addition, the parallelism of the elemental distribution patterns (fig. 17) rules out control by fractionation of trace element enriched minor phases, such as zircon, apatite, or other non-uniformly enriched REE-bearing minerals. In this respect, the dacites of Mount Pinatubo are similar to dacites from Mount St. Helens, where an origin by large-degree partial melting of mafic crustal rocks is favored by Smith and Leeman (1987).

The compositions of the phenocryst-rich and phenocryst-poor dacitic pumice types are identical at the ± 1-σ level for all major and trace elements (table 7). This finding is consistent with a closed-system process of crystal size reduction, such as proposed above.

Three data clusters that correspond to basalt, andesite, and dacite are apparent on a silica variation diagram (fig. 18). The single data points at 62.4 wt% SiO_2 correspond to a rare intermediate-composition lapillus from the tephra fall of June 13 (sample P22692–2A2). The linear trends on this diagram suggest a magma mixing origin for the andesitic bulk compositions. We tested the magma mixing hypothesis by using least-squares modeling (Wright and Doherty, 1970) of the average major element compositions of the basalt, andesite, and dacite. The model solution was then applied to the trace element data. The result: the average andesite major element and trace element composition can be produced from 36.5 wt% basalt + 63.5 wt% dacite, with a squared sum of residuals for the major elements of <0.05

Table 6. Localities and notes for analytical samples.

[Abbreviations: p.f., pyroclastic flow; qz., quartz; xenos., xenocrysts; NF, north fork]

Field No.	Drainage	Latitude (N.)	Longitude (E.)	Date erupted	Rock name	Notes	Material analyzed
Basalt inclusions in June 7-12 andesite lava							
P-22892-1a	Maraunot	15° 10' 5"	120° 19' 20"	6/7-12/91	Basalt inclusion (15 cm)	In 2-m andesite block in 6/15 lithic p.f. deposit. Qz. xenos. removed	Single inclusion
P-22892-2a	Maraunot	15° 10' 0"	120° 19' 21"	6/7-12/91	Basalt inclusion (40 cm)	From contact between 6/12 and 6/15 lithic p.f. deposits	Single inclusion
CN6791-i	Maraunot	15° 10' 5"	120° 19' 20"	6/7-12/91	Basalt inclusion	In prismatically jointed block from p.f. deposit	Single inclusion
June 7-12 hybrid andesite (lava and scoria)							
CN6791-d	Maraunot	15° 10' 5"	120° 19' 20"	6/7-12/91	Andesite lava block	Prismatically jointed block from p.f. deposit	Single lava block
P22892-1a2h	Maraunot	15° 10' 5"	120° 19' 20"	6/7-12/91	Andesite lava block	2-m block in 6/15 lithic p.f. deposit	Single lava block
7-1-91-1A	Mapanuepe	15° 15' 0"	120° 17' 1"	6/12/91	Andesite scoria lapilli	Collected 7/1/91, near Mt. Bagang	Composite of 5 lapilli
P-22892-3	Maraunot	15° 9' 48"	120° 19' 28"	6/12/91	Andesite scoria block	From base of thin 6/12 p.f. deposit	Single scoria block
P-22892-3c	Maraunot	15° 9' 48"	120° 19' 28"	6/12/91	Andesite scoria (breadcrusted)	From base of thin 6/12 p.f. deposit	Single scoria block
P-22692-2a-2	Marella	15° 6' 40"	120° 18' 28"	6/12-13/91	Andesite scoria lapillus (4 cm)	Rare leucocratic component, ~1% of lapilli	Single lapillus
P-22692-2a-3	Marella	15° 6' 40"	120° 18' 28"	6/12-13/91	Andesite scoria lapillus (10 cm)	Unbanded. Similar to matrix of banded scoria	Single lapillus
P-22692-2a-4	Marella	15° 6' 40"	120° 18' 28"	6/12-13/91	Andesite lapilli	Dense component. Few percent of lapilli	Composite of 10 lapilli
June 15 phenocryst-rich dacite							
EW910615-1, 1x	Sacobia	15° 11' 21"	120° 29' 10"	6/15/91	Phenocryst-rich dacite	Pumice blocks from p.f. deposit	Single pumice block
PH12a	Bucao	15° 11' 37"	120° 19' 26"	6/15/91	Phenocryst-rich dacite	Collected 6/27/91 from p.f. deposit	Single pumice block
PH12c	Bucao	15° 11' 37"	120° 19' 26"	6/15/91	Phenocryst-rich dacite	Collected 6/27/91 from p.f. deposit	Single pumice block
PH12d	Bucao	15° 11' 37"	120° 19' 26"	6/15/91	Phenocryst-rich dacite	Collected 6/27/91 from p.f. deposit	Single pumice block
PH13a	Marella	15° 3' 4"	120° 17' 21"	6/15/91	Phenocryst-rich dacite	Collected 7/8/91 at terminus of p.f.	Single pumice block
PH13b	Marella	15° 3' 4"	120° 17' 21"	6/15/91	Phenocryst-rich dacite	Collected 7/8/91 at terminus of p.f.	Single pumice block
June 15 phenocryst-poor dacite							
EW910615-2, 2x	Sacobia	15° 11' 21"	120° 29' 10"	6/15/91	Phenocryst-poor dacite	Pumice blocks from p.f. deposit	Single pumice block
PH12a	Bucao	15° 11' 37"	120° 34' 26"	6/15/91	Phenocryst-poor dacite	Collected 6/27/91 from p.f. deposit	Single pumice block
PH12e	Bucao	15° 11' 37"	120° 34' 26"	6/15/91	Phenocryst-poor dacite	Collected 6/27/91 from p.f. deposit	Single pumice block
PH12f	Bucao	15° 11' 37"	120° 34' 26"	6/15/91	Phenocryst-poor dacite	Collected 6/27/91 from p.f. deposit	Single pumice block
PH12g	Bucao	15° 11' 37"	120° 34' 26"	6/15/91	Phenocryst-poor dacite	Collected 6/27/91 from p.f. deposit	Single pumice block
PH13c	Marella	15° 3' 4"	120° 17' 21"	6/15/91	Phenocryst-poor dacite	Collected 7/8/91 at terminus of p.f. deposit	Single pumice block
PH13d	Marella	15° 3' 4"	120° 17' 21"	6/15/91	Phenocryst-poor dacite	Collected 7/8/91 at terminus of p.f. deposit	Single pumice block
P6-21-91-4	Sacobia	15° 11' 21"	120° 29' 10"	6/15/91	Phenocryst-poor dacite	Breadcrusted bombs	Single pumice block
P-2592-1e	Pasig	15° 7' 58"	120° 29' 0"	6/15/91	Phenocryst-poor dacite	From 6/15/92 p.f. deposit that overlies lithified lahar deposit	Single pumice block

Table 7. Major and trace element analyses of 1991 Pinatubo rocks.

[Oxides analyzed by X-ray fluorescence, in wt% normalized volatile free. Element symbols followed by coefficients of variation (cv, in %) were determined by instrumental neutron activation analysis, others by energy-dispersive X-ray fluorescence. LOI=loss on ignition at 925°C. Sulfur was determined by J. Curry of the U.S. Geological Survey on separate sample splits using a combustion-infrared technique. Elemental data in ppm, except S (wt%) and Au (ppb). SD, standard deviation]

	Lab No.	Date	SiO_2	Al_2O_3	$FeTO_3$[1]	MgO	CaO	Na_2O	K_2O	TiO_2	P_2O_5	MnO	Sum	LOI	Sum[2]	Fe_2O_3	FeO[3]	Mg#[4]	H_2O^+	H_2O^-	CO_2	F	S%	Cl	Sc	cv	Cr	cv	Co	cv	Ni	cv	Cu	cv	Zn
Basalt																																			
P22892-1a	D-516226	6/7-12/91	52.0	14.9	8.72	8.64	9.77	3.11	1.50	0.89	0.32	0.16	100	0.15	99.43	4.47	3.77	68.8	.55	<0.05	.01	-	<0.05	-	33.7	1	327	2	35.2		115	5	74	5	78.4
P22892-2a	D-516227	6/7-12/91	51.3	14.9	8.98	8.69	10.12	3.04	1.52	.91	.35	.16	100	.14	99.94	4.80	3.74	68.3	.61	<.05	<.01	-	<.05	-	34.9	1	331	3	35.7	1	95.5	3	84	3	76.7
CN679i	D-504517	6/7-12/91	50.5	14.7	9.32	9.21	10.55	3.02	1.34	.90	.35	.16	100	.17	99.66	5.07	3.78	68.7	.58	.07	.01	-	<.05	-	35.1	1	367	3	38.1	1	105	3	81	3	79.7
Average (n=3)			51.2	14.8	9.01	8.85	10.15	3.05	1.45	.90	.35	.16	100	.15	99.68	4.78	3.76	68.6	.58	.07	.01	-	<.05	-	34.6		342	3	36.3	2	105.2	4	80	5	78.3
SD (n=3)			.8	.1	.30	.31	.39	.05	.10	.01	.02	.00		.02		.30	.02	.3	.03		.00				.8		22		1.6		9.8		5		1.5
Hybrid Andesite																																			
CN679id	D-504516	6/7-12/91	59.2	16.0	6.31	4.77	7.23	3.88	1.56	.65	.26	.12	100	.07	99.91	3.01	2.96	62.7	.22	.07	<.01	-	<.05	-	18.9	1	153	1	20.7	1	50.8	9	106	9	68.1
P22892-1a2h	D-516225	6/7-12/91	59.6	15.8	6.27	4.76	6.94	3.90	1.62	.69	.26	.13	100	.23	99.94	3.05	2.88	62.8	.36	.08	.01	-	<.05	-	18.7	1	153	1	20.4	1	54.6	4	46	4	66.2
7-1-91-1-1a	D-500918	6/12/91	59.2	16.1	6.13	4.86	7.17	3.91	1.56	.64	.29	.12	100	1.04	98.91	2.42	3.22	63.8	.58	.13	<.01	-	-	-	17.4	1	139	1	19.3	1	43.6	6	64	6	56.1
P22892-3	D-516228	6/12/91	59.2	15.9	6.30	4.76	6.99	4.16	1.66	.66	.26	.14	100	1.56	99.74	2.82	3.03	62.7	.92	.40	<.01	-	.29	-	18.6	1	154	1	21.9	1	59.8	3	56	3	79.4
P22892-3c	D-516229	6/12/91	59.6	16.0	6.19	4.64	6.95	3.95	1.65	.68	.27	.13	100	.80	99.81	2.80	3.00	62.5	.74	.23	<.01	-	.11	-	18.2	1	145	1	19.7	1	42.6	5	53	5	64.6
P22692-2a2	D-516230	6/12-13/91	62.4	16.2	5.30	3.46	5.90	4.20	1.64	.58	.23	.11	100	1.75	99.48	2.30	2.59	59.2	1.17	.39	<.01	-	.22	-	-		-		-		-		87		-
P22692-2a3	D-516231	6/12-13/91	59.2	15.9	6.32	4.87	7.18	3.86	1.56	.68	.26	.13	100	1.05	99.67	2.69	3.19	63.1	.99	.21	<.01	-	.14	-	19.	1	156	1	20.7	1	49.5	2	45	2	67.9
P22692-2a4	D-516232	6/12-13/91	59.0	16.2	6.19	4.82	7.18	3.97	1.56	.69	.26	.13	100	.79	99.59	2.55	3.21	63.4	.83	.10	<.01	-	.10	-	-		-		-		-		38		-
Average (n=8)			59.7	16.0	6.13	4.62	6.94	3.98	1.60	.66	.26	.13	100	.91	99.63	2.71	3.01	62.5	.73	.20	<.01	-	.17	-	18.5	1	150	1	20.5	1	50.2	5	62	5	67.1
SD (n=8)			1.1	.1	.34	.47	.44	.13	.04	.04	.02	.01		.58		.27	.21	1.4	.32	.13			.08		.6		7		.9		6.5		23		7.5
Phenocryst-rich dacite																																			
EW910615-1	D-379505	6/15/91	64.8	16.6	4.30	2.28	5.28	4.47	1.52	.52	.19	.09	100	.97	99.44	1.90	2.10	54.2	.88	.09	<.01	<.01	.14	.06	8.79	1	37.9	1	10.6	1	17.0	26	28		49.9
EW910615-1x	D-502278	6/15/91	64.5	16.4	4.54	2.47	5.20	4.44	1.58	.55	.21	.10	100	1.08	99.34	1.99	2.22	54.8	.88	.14	<.01	.02	.11	.06	9.27	1	41.6	1	11.0	1	18.4	12	-		54.1
PH12B	D-502282	6/15/91	64.6	16.5	4.35	2.43	5.25	4.51	1.54	.52	.20	.10	100	.87	99.39	1.89	2.16	55.3	.78	.12	<.01	.20	-	.06	8.94	1	39.8	1	10.5	1	20.4	6	-		50.7
PH12C	D-502283	6/15/91	64.3	16.8	4.31	2.35	5.33	4.58	1.48	.50	.19	.10	100	.59	99.28	1.93	2.09	54.8	.46	.14	<.01	.02	.13	.05	8.76	1	36.1	1	10.1	1	17.6	9	-		50.2
PH12D	D-502284	6/15/91	65.0	16.0	4.59	2.50	4.97	4.35	1.64	.65	.20	.10	100	.77	99.30	2.04	2.23	54.8	.73	.13	<.01	.03	.10	.05	8.76	1	36.1	1	10.9	1	15.3	14	-		55.5
PH13A	D-502288	6/15/91	64.4	16.9	4.24	2.32	5.30	4.60	1.50	.50	.20	.09	100	.54	99.37	1.90	2.06	54.9	.48	.09	<.01	.02	.10	.05	8.66	1	34.9	1	10.2	1	15.0	14	-		56.5
PH13B	D-502289	6/15/91	64.7	16.6	4.29	2.37	5.27	4.48	1.49	.50	.20	.10	100	.56	99.34	1.91	2.14	55.1	.53	.12	<.01	.07	.12	.07	8.57	1	35.0	1	9.9	1	11.0	13	-		55.7
Average (n=7)			64.6	16.5	4.37	2.39	5.23	4.49	1.54	.53	.20	.10	100	.77	99.35	1.94	2.14	54.8	.68	.12	<.01	.05	.12	.06	8.82	1	37.3	1	10.5	1	16.4	13	28		53.2
SD (n=7)			.2	.3	.13	.08	.12	.08	.06	.06	.01	.00		.21		.06	.07	.4	.18	.02			.02	.01	.23		2.6		.4		3.0		-		2.9
Phenocryst-poor dacite																																			
EW910615-2	D-379506	6/15/91	64.3	16.6	4.37	2.53	5.47	4.42	1.47	.50	.18	.09	100	1.01	99.57	1.95	2.12	56.2	.79	.19	<.01	<.01	.14	.04	9.51	1	49.0	1	11.0	1	17.0	17	50		49.8
EW910615-2x	D-502279	6/15/91	64.4	16.8	4.17	2.37	5.36	4.57	1.49	.49	.20	.09	100	.90	99.60	1.92	1.98	55.8	.78	.16	<.01	.02	.12	.04	9.11	3	36.7	3	10.4	1	19.2	4	-		49.2
PH12A	D-502281	6/15/91	64.2	16.7	4.32	2.55	5.45	4.48	1.50	.50	.20	.09	100	.81	99.56	2.01	2.03	56.8	.76	.08	<.01	.20	.09	.04	9.95	1	49.0	1	11.5	1	20.0	2	-		51.9
PH12E	D-502286	6/15/91	64.0	16.7	4.44	2.66	5.44	4.49	1.51	.52	.21	.10	100	.67	99.15	2.18	1.97	57.1	.54	.14	<.01	.02	.06	.04	9.97	3	47.0	3	11.3	1	20.9	9	-		49.7
PH12F	D-502286	6/15/91	64.4	16.7	4.27	2.47	5.38	4.50	1.49	.50	.20	.10	100	.63	99.41	1.95	2.04	56.2	.50	.11	<.01	.02	.11	.04	9.48	1	41.5	1	10.8	1	14.0	19	-		50.3
PH12G	D-502287	6/15/91	64.7	16.8	4.16	2.28	5.27	4.61	1.45	.49	.19	.09	100	.60	99.07	1.98	1.91	55.0	.40	.11	<.01	.02	.10	.04	8.25	1	33.5	1	9.6	1	12.0	13	-		53.7
PH13C	D-502290	6/15/91	64.3	16.6	4.39	2.55	5.38	4.47	1.52	.51	.20	.10	100	.74	99.53	2.15	1.97	56.4	.55	.12	<.01	.02	.08	.06	9.74	1	45.4	1	11.1	1	21.0	15	-		58.9
PH13D	D-500917	6/15/91	64.1	17.1	4.28	2.38	5.28	4.62	1.47	.48	.19	.10	100	1.32	98.99	2.06	1.91	55.2	.49	.10	<.01	-	-	.04	9.40	1	37.6	1	11.1	1	14.5	15	81		58.3
P6-21-91-4	D-502291	6/15/91	64.6	16.7	4.22	2.36	5.30	4.54	1.48	.49	.19	.10	100	.62	99.42	2.04	1.92	53.9	.58	.09	<.01	.02	.08	.04	9.03	1	37.8	1	10.5	1	10.9	10	32		61.2
P22592-1e	D-516233	6/15/91	64.8	16.7	4.25	2.23	5.15	4.58	1.49	.51	.19	.10	100	1.25	99.42	2.04	1.92	53.9	1.06	.22	.01	.02	.05	.04	9.38	1	41.9	1	10.8	1	16.6	12	54		53.7
Average (n=10)			64.4	16.7	4.29	2.44	5.35	4.53	1.49	.50	.20	.10	100	.86	99.38	2.03	1.98	55.8	.65	.13	<.01	.02	.09	.04	9.38		41.9		10.8		16.6	12	54		53.7
SD (n=10)			.3	.1	.09	.14	.10	.07	.02	.01	.01	.01		.26		.08	.07	1.0	.20	.05			.03	.01	.54		5.8		.6		3.9		25		4.6

[1]Total iron expressed as Fe2O3. [2]Unnormalized sum including LOI. [3]Titration value. [4]Calculated assuming Fe2O3/FeO=0.123.

Table 7. Major and trace element analyses of 1991 Pinatubo rocks—Continued.

	cv	As	cv	Rb	cv	Sr	cv	Zr	cv	Sb	cv	Cs	cv	Ba	cv	La	cv	Ce	cv	Nd	cv	Sm	cv	Eu	cv	Gd	cv	Tb	cv	Tm	cv	Yb	cv	Lu	cv	Hf	cv	Ta	cv	Th	cv	U	cv
P22892-1a	6	1.47	11	46.3	2	590	8	106	4	.196	12	3.52	1	322	2	21.8	1	46.2	3	23.8	5	5.04	1	1.52	1	4.64	5	.601	2	.273	9	1.68	1	.253	2	2.49	1	.176	2	5.69	1	1.55	3
P22892-2a	2	1.17	11	48.0	2	662	3	115	11	.140	18	3.50	2	295	3	22.1	1	48.6	1	25.6	3	5.37	1	1.55	1	4.63	7	.629	1	.277	9	1.69	1	.253	3	2.38	1	.186	1	5.97	1	1.49	2
CN6791i	4	1.32	14	44.0	2	592	4	120	21	.162	13	3.50	1	313	2	22.6	1	49.4	2	26.3	1	5.78	1	1.61	1	4.87	4	.671	2	.297	9	1.76	1	.256	4	2.36	1	.176	2	5.96	1	1.58	4
Average	4	1.32	12	46.1	2	615	5	114	12	.166	14	3.51	1	310	1	22.2	1	48.1	2	25.2	1	5.40	1	1.56	1	4.71	5	.634	2	.282	9	1.71	1	.254	2	2.41	1	.179	2	5.87	1	1.54	3
SD (n=3)		.2		2.0		41		7		.028		.01		14		.4		1.7		1.3		.37		.05		.14		.035		.013		.04		.002		.07		.006		.16		.05	
CN6791d	3	2.19	7	47.2	1	605	2	108	2	.156	8	3.36	1	401	1	19.4	1	39.4	1	19.8	2	4.23	1	1.19	1	3.63	10	.481	1	.248	9	1.52	1	.219	2	2.77	1	.253	1	5.40	1	1.58	1
P22892-1a2h	5	2.56	5	45.4	3	562	1	113	4	.240	18	3.43	1	406	4	19.2	1	40.9	1	18.6	1	4.14	1	1.13	1	3.74	5	.487	1	.242	8	1.50	1	.218	5	2.85	1	.271	1	5.57	1	1.58	7
7-1-91-1-1a	3	2.21	7	39.0	3	515	3	101	5	.486	3	2.88	1	380	1	17.5	1	35.8	1	18.0	2	3.97	1	1.03	1	3.35	8	.459	1	.228	8	1.44	1	.209	4	2.60	1	.267	1	4.84	1	1.50	3
P22892-3	6	2.5	7	46.0	1	617	1	104	5	.272	5	3.35	1	404	2	19.5	1	41.7	1	20.7	3	4.24	1	1.16	1	3.68	5	.477	1	.249	9	1.52	2	.217	5	2.72	1	.274	1	5.44	1	1.56	1
P22892-3c	3	2.77	6	43.7	2	563	2	107	5	.214	9	3.46	1	397	1	18.9	1	41.4	1	20.7	1	4.23	1	1.14	1	3.93	5	.476	2	.244	8	1.48	2	.214	4	2.70	1	.277	1	5.52	1	1.66	3
P22692-2a2	-		-		-		-		-		-		-		-		-		-		-		-		-		-		-		-		-		-		-		-	
P22692-2a3	6	2.16	4	45.4	2	566	1	109	3	.206	7	3.11	1	368	2	18.6	1	39.2	3	20.8	6	4.24	1	1.14	1	3.69	7	.473	1	.251	7	1.50	1	.220	4	2.56	4	.238	1	5.18	1	1.52	4
P22692-2a4	-		-		-		-		-		-		-		-		-		-		-		-		-		-		-		-		-		-		-		-		-	
Average	4	2.4	7	44.5	2	571	2	107	4	.262	8	3.27	1	393	1	18.9	1	39.7	1	19.8	2	4.18	1	1.13	1	3.67	7	.476	1	.244	8	1.49	1	.216	4	2.70	2	.263	1	5.33	1	1.57	3
SD (n=8)		.2		2.9		36		4		.116		.23		15		.7		2.2		1.2		.11		.05		.19		.009		.008		.03		.004		.11		.015		.27		.06	
EW910615-1	4	3.23	3	38.0	3	569	3	108	3	.277	8	2.75	1	455	1	15.9	1	34.4	2	14.3	1	2.93	1	.838	1	2.84	4	.370	1	.185	10	1.20	1	.196	1	3.07	1	.328	1	4.69	1	1.69	3
EW910615-1x	3	3.83	4	42.2	2	567	3	109	8	.655	2	2.92	1	476	2	16.	1	33.1	1	15.6	3	3.22	1	.878	1	2.76	6	.394	1	.216	9	1.34	2	.197	1	3.01	1	.344	1	4.86	1	1.57	1
PH12B	4	3.43	4	41.5	2	591	2	117	7	.327	6	3.00	1	477	1	15.8	1	32.8	1	16.6	2	3.25	1	.862	1	2.84	5	.376	1	.211	8	1.34	2	.198	1	3.07	1	.339	1	4.74	1	1.65	2
PH12C	4	3.44	3	39.8	2	602	2	111	6	.286	6	2.75	1	456	1	15.4	1	31.6	1	16.1	3	3.14	1	.853	1	2.87	6	.387	2	.202	9	1.31	2	.192	12	2.96	1	.321	1	4.49	1	1.50	2
PH12D	2	3.81	3	43.1	1	551	1	107	3	.302	5	3.17	1	505	1	16.	1	33.2	1	14.9	3	3.31	1	.831	1	2.85	5	.387	1	.204	10	1.32	1	.204	1	2.85	1	.356	1	5.01	1	1.83	2
PH13A	5	3.26	4	37.8	2	608	1	100	4	.274	3	2.57	1	431	1	14.9	1	31.9	1	14.7	2	3.04	1	.843	1	2.54	7	.377	1	.202	10	1.27	1	.190	2	2.85	1	.319	1	4.28	1	1.40	2
PH13B	5	3.33	3	38.5	1	580	1	98	5	.273	10	2.64	1	452	1	15.1	1	31.9	1	15.1	2	3.13	1	.821	1	2.64	6	.363	1	.205	7	1.32	1	.190	4	2.78	1	.306	1	4.42	1	1.49	3
Average	4	3.5	3	40.1	2	581	2	107	5	.342	5	2.83	1	465	1	15.6	1	32.7	1	15.3	2	3.15	1	.847	1	2.76	6	.379	1	.203	9	1.30	1	.195	3	2.97	1	.330	1	4.64	1	1.59	2
SD (n=7)		.2		2.1		20		6		.139		.21		24		.5		1.0		.8		.13		.019		.13		.011		.010		.05		.005		.11		.017		.26		.14	
EW910615-2	4	3.34	2	39.9	2	567	2	102	13	.275	4	2.75	1	433	2	15.1	1	31.2	1	15.1	2	3.05	1	.846	1	2.76	9	.392	1	.207	8	1.32	2	.196	3	2.85	1	.304	1	4.39	1	1.51	2
EW910615-2x	4	3.44	2	38.9	2	632	3	109	3	.264	4	2.79	1	459	2	15.5	1	32.6	1	14.0	2	3.25	1	.875	1	2.93	6	.392	2	.211	10	1.28	2	.206	2	2.91	1	.304	1	4.52	1	1.60	2
PH12A	4	3.15	4	38.6	2	620	3	113	3	.277	4	2.70	1	436	1	15.3	1	32.1	1	15.3	2	3.24	1	.894	1	2.67	6	.410	1	.204	5	1.25	2	.205	3	3.07	1	.308	1	4.31	1	1.50	3
PH12E	4	3.24	4	39.5	1	573	3	108	3	.262	4	2.75	1	448	1	15.6	1	32.1	1	15.3	2	3.30	1	.880	1	2.78	9	.396	1	.218	9	1.36	2	.201	1	2.85	1	.300	1	4.50	1	1.55	2
PH12F	3	3.49	3	39.9	1	588	1	121	2	.296	6	2.77	1	463	1	15.5	1	32.1	1	15.2	5	3.31	1	.874	1	2.94	5	.389	1	.217	10	1.32	2	.207	1	3.16	2	.324	1	4.50	1	1.56	1
PH12G	9	3.01	5	37.0	2	570	1	106	6	.234	4	2.65	1	451	1	14.9	1	31.2	1	15.2	2	3.04	1	.809	1	2.67	9	.367	1	.190	11	1.21	2	.186	2	2.85	1	.308	1	4.31	1	1.40	1
PH13C	6	3.68	4	41.3	1	591	3	105	4	.321	3	2.86	1	436	8	16.1	1	33.9	1	15.8	2	3.34	1	.877	1	2.89	9	.389	1	.210	10	1.33	1	.204	2	3.03	2	.321	1	4.77	1	1.57	2
PH13D	2	3.34	5	42.0	2	590	2	104	3	.261	5	2.77	1	454	2	18.8	1	40.2	1	16.7	3	3.61	1	.927	1	3.44	9	.435	1	.236	10	1.49	4	.219	4	3.14	1	.337	1	4.97	1	1.62	3
P6-21-91-4	9	3.36	6	37.6	6	586	1	115	2	.246	4	2.71	1	454	1	15.3	1	32.5	1	15.1	2	3.18	1	.859	1	2.87	6	.391	2	.218	11	1.33	1	.200	2	3.04	1	.324	1	4.45	1	1.50	5
P22592-1e	-		-		-		-		-		-		-		-		-		-		-		-		-		-		-		-		-		-		-		-		-	
Average	5	3.3	4	39.4	2	591	2	109	4	.271	4	2.75	1	448	1	15.8	1	33.1	1	15.3	2	3.26	1	.871	1	2.88	7	.396	1	.212	9	1.32	1	.203	2	2.99	1	.314	1	4.52	1	1.53	3
SD (n=10)		.2		1.6		22		6		.026		.06		11		1.2		2.8		.7		.17		.032		.23		.018		.013		.08		.009		.13		.012		.22		.07	

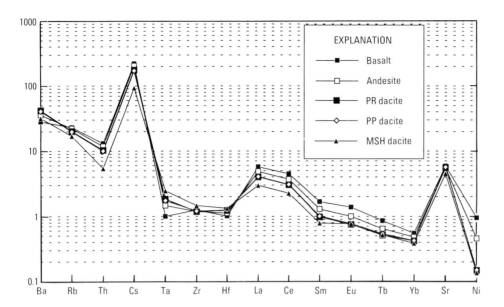

Figure 17. Abundance of trace elements in 1991 Pinatubo samples normalized to midocean ridge basalt (MORB) and compared to average dacite of Mount St. Helens (MSH) of Smith and Leeman (1987). PR, phenocryst-rich; PP, phenocryst poor.

(table 8). All but five of the 27 analyzed trace element abundances are within one standard deviation of the model solution, and all are within two standard deviations. The largest deviation is for K_2O, which has a residual of 0.9 wt%. The model solution for the rare earth elements is graphically illustrated in figure 19.

DISCUSSION AND SUMMARY

Samples from the 1991 eruptions of Mount Pinatubo provide unequivocal petrographic and chemical evidence for magma mixing involving basalt that was initially at ~1200°C and dacite at ~780°C. Eruption of the hybrid andesite magma first (on June 7–14) and preservation of disequilibrium textures indicate that mixing of basalt and dacite took place *shortly* before, and probably triggered, the climactic dacitic eruption on June 15, 1991. Recent experimental calibration of reequilibration rates of minerals in the mixed andesite of Mount Pinatubo provides an estimate of the time interval between mixing and eruption (Rutherford and others, 1993). The preservation of thin cummingtonite rims on hornblende, inhomogeneous glass, and magnetite rims on ilmenite suggest that mixing occurred within four days of eruption, and the lack of dehydration reaction rims on hornblende phenocrysts indicates that ascent to the surface also took place in less than four days. Consistent with these results, deep (>30 km) long-period earthquakes were recorded during two periods (May 26–28 and May 31–June 6) at Pinatubo and are related to ascent of basaltic magma to the magma reservoir within a time span of days to hours before eruption (White, this volume).

THE MAGMA RESERVOIR

Our best evidence for the site of magma mixing comes from evaluation of phase equilibria and seismic data. A high Mg# and lack of plagioclase as a liquidus phase in the basalt indicate that it originated at great depth and possibly under hydrous conditions and that it was little modified by shallow fractionation. Studies of melt and vapor inclusions in phenocrysts and phase equilibria (Westrich and Gerlach, 1992; Rutherford and Devine, this volume) indicate that the June 15 dacitic magma was saturated with a water-rich fluid phase prior to eruption. An apparent equilibration pressure of 220±50 MPa for the dacite is calculated by Rutherford and Devine (this volume) from the Al-in-hornblende geobarometer of Johnson and Rutherford (1989). A similar result (225±50 MPa) may be calculated from the hornblende data in table 2; however, several members of the required equilibrium assemblage are either lacking (sphene, sanidine) or resorbed (quartz). An independent pressure estimate is provided by the coexistence of cummingtonite and $An_{34–40}$ plagioclase as late magmatic phases, which indicates that the June 15 dacite last equilibrated at pressures of 200–320 MPa. Comparison of natural glass compositions to those from experimental samples of June 15 dacite favors equilibration under water-saturated conditions at 780°C and further restricts the equilibration pressure to about 200 MPa (Rutherford and Devine, this volume). On the basis of geothermal exploration and drilling (Delfin, 1983; Delfin and others, this volume), we infer that the volcano is underlain by about 2 km thickness of vesicular dacite (with density of ~2.0 g cm^3) and 2 km of dacitic to andesitic lavas and intrusions (~2.6 g cm^3), which overlie ophiolitic rocks (~2.8 g cm^3). Given this rock column and assuming that $P_{H_2O}=P_{total}$, 200 MPa is equivalent to a depth of about 8 km beneath the pre-1991 summit (~6 km depth, relative to sea level).

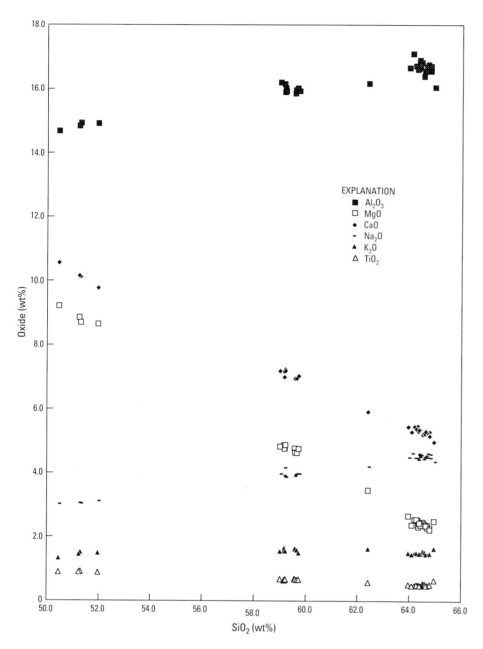

Figure 18. SiO$_2$ variation diagram for 1991 Pinatubo major element analyses. Samples fall into basalt (50–52 wt% SiO$_2$), andesite (59–60 wt% SiO$_2$), and dacite (64–65 wt% SiO$_2$) on an anhydrous basis and display linear trends with respect to SiO$_2$ abundance. The single sample at 62.4 wt% SiO$_2$ is from a nonbanded, light-gray lapillus from the June 13, 1991, fall deposit.

Hypocenter locations for earthquakes recorded from the time the seismic net was established on May 8, 1991, until June 15 were mainly at shallow depths of <4 km. After the June 15 eruption, numerous earthquakes took place at greater depth, eventually outlining a broad envelope of seismicity that extends to depths of 20 km (fig. 20). It is tempting to take the dense dome-shaped distribution of hypocenters seen in the east-west projection of figure 20 as the outline of a large (200 km^3) magma chamber at depth. This would be analogous to the approach used by Scandone and Malone (1985) to infer the shape of the magma reservoir beneath Mount St. Helens. However, in detail the picture appears to be more complex. The distribution of deep hypocenters is not symmetrical about a vertical axis through the volcano but consists of two dense dikelike or pipelike

bodies (prominent in fig. 20) and one less-dense body to the north, each of which extends from about 5 to 20 km in depth. A sparse shroud of hypocenters appears to connect the pipe-like distributions, especially around the northern side of the hypocenter envelope, and outlines a virtually earthquake-free interior zone. Tomographic inversion of the seismic data shows a cylindrical low-velocity zone that is inferred to represent magma, at depths of >6 km (relative to sea level) (Mori, Eberhart-Phillips, and Harlow, this volume). This body is 4 to 5 km in diameter and has a volume of 40 to 90 km^3 for the depth interval 6 to 11 km.

We suggest that the low-velocity zone represents the source for the 1991 eruptions. The region within the envelope of hypocenters is inferred to be an arc pluton that is melt-rich only at depths of more than 8 km beneath the level

Table 8. Least-squares mixing model for hybrid andesite.

[Major element least-squares model proportions were used to calculate trace element abundances for model solution. Percentages in parentheses indicate analytical uncertainty based on average counting statistics. Residuals in bold print are outside 1σ of andesite mean. SD, standard deviation; *n*, number of samples]

Least-squares mixing model for average compositions
36.45% basalt +63.55% dacite= andesite

	Model solution	Average andesite	Residuals	SD (*n*=8) Avg. andesite
SiO_2	59.7	59.7	0.1	1.1
Al_2O_3	15.92	16.02	-.09	.14
[1]$FeTO_3$	6.06	6.13	-.07	.34
MgO	4.74	4.62	.12	.47
CaO	7.02	6.94	.08	.44
Na_2O	3.97	3.98	-.01	.13
K_2O	1.51	1.60	**-.09**	.04
TiO_2	.67	.66	.01	.04
P_2O_5	.25	.26	-.01	.02
MnO	.12	.13	-.01	.01
Squared sum of oxide residuals			.045	
Sc (1%)	18.2	18.5	-.26	.59
Cr (1%)	148	150	-1.73	6.57
Co (1%)	19.9	20.5	-.56	.91
Ni (5%)	48.7	50.2	-1.40	6.54
Cu(23%)	46.8	61.9	-15.0	23.3
Zn (4%)	62.4	67.1	-4.70	7.50
As (6%)	2.7	2.4	**.3**	.2
Rb (2%)	42.3	44.5	-2.14	2.90
Sr (2%)	593	571	22	36
Zr (4%)	110	107	3	4
Sb (8%)	.28	.26	.02	.12
Cs (1%)	3.08	3.27	-.19	.23
Ba (2%)	408	393	**16**	15
La (1%)	18.0	18.9	**-.9**	.7
Ce (1%)	38.3	39.7	-1.4	2.2
Nd (3%)	18.9	19.8	-.8	1.2
Sm (1%)	3.97	4.18	**-.21**	.11
Eu (1%)	1.11	1.13	-.03	.05
Gd (7%)	3.47	3.67	**-.20**	.19
Tb (1%)	.47	.48	.00	.01
Tm (8%)	.23	.24	-.01	.01
Yb (1%)	1.45	1.49	-.04	.03
Lu (4%)	.22	.22	.00	.00
Hf (1.5%)	2.77	2.70	.07	.11
Ta (1%)	.28	.26	.01	.01
Th (1%)	5.09	5.33	-.23	.27
U (3%)	1.57	1.57	.01	.06

[1]Refers to total iron expressed as Fe2O3

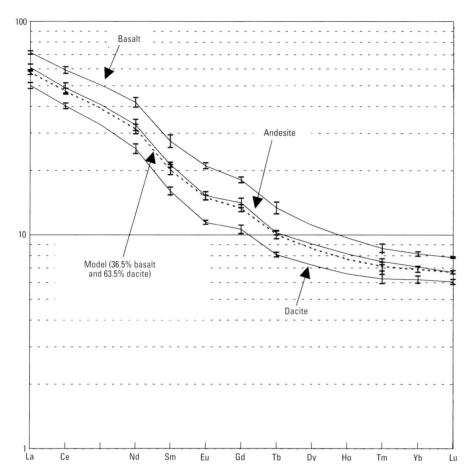

Figure 19. Chondrite-normalized rare earth element diagram showing average patterns for 1991 Pinatubo basalt, andesite, and dacite, and pattern calculated for 36.5 wt% average basalt and 63.5 wt% average dacite (proportions derived from major-element least-squares model in table 8).

of the pre-1991 summit. Much of the magma in the reservoir was "uneruptable" because of high viscosity caused by high-silica matrix melt and crystal contents that exceeded 50 percent (Marsh, 1981). Had not basaltic magma been added to the system, the dacite probably would not have erupted in 1991.

BASALT TRIGGERING OF THE JUNE 15 ERUPTION

Mixing of basalt and dacite could trigger an explosive eruption by increasing magmatic pressure as a result of addition of mass, fluid, or heat, the latter leading to convective overturn and increased volatile exsolution (Sparks and others, 1977; Williams and McBirney, 1979, p. 74–86; Cas and Wright, 1988, p. 40–41). For Pinatubo we must first explain why these two magmas, presumably with vastly different physical properties, would mix at all; then we must explain how the andesitic magma, which would normally be more dense, could reach the surface before the dacitic magma could. The problem of mixing dissimilar magmas is well known (for example, Sparks and others, 1984; Huppert and others, 1984; Sparks and Marshall, 1986). These studies point out that thorough mixing requires turbulent flow, which is favored by similar viscosities and high flow rates.

Consequently, mixing models typically call on cooling of the mafic magma and heating of the felsic magma such that the densities and viscosities of the two approach one another. However, the ability to mix also depends on the relative volumes of the two components. If the volume and thermal mass of the mafic magma component is small, it tends to quench when intruded into cooler felsic magma, but when the volume of the mafic magma is relatively large (about 50%; Sparks and Marshall, 1986), mixing is possible.

We suggest the following magma mixing model for the 1991 eruptions. Basalt began to leak into a crystal-rich dacite reservoir at about 8 km in depth beneath Pinatubo, possibly starting on or before April 2, 1991, the date of the first phreatic explosions. Initial batches of basaltic magma were small and quenched in the dacite to form fluid-saturated diktytaxitic pillows, which were erupted later as inclusions in the June 7–12 dome lava. A small amount of dacite was entrained in the basalt, so hornblende and quartz xenocrysts were produced. This dacite was derived either from wallrocks at greater depth or from the dacitic magma. With continued influx, basalt accumulated at the base of the dacitic reservoir and cooled more slowly. As it cooled the basalt liberated heat to the overlying dacite and crystallized augite, magnetite, and high-titanium hornblende. By virtue of cooling and crystallization, the viscosity of the basalt increased

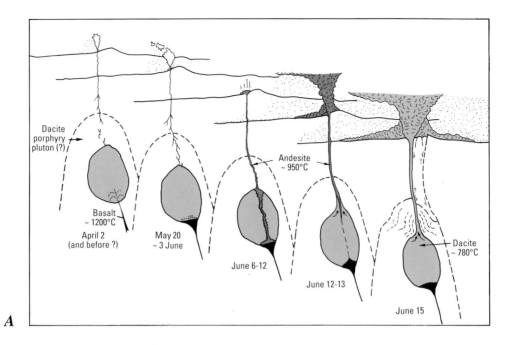

A

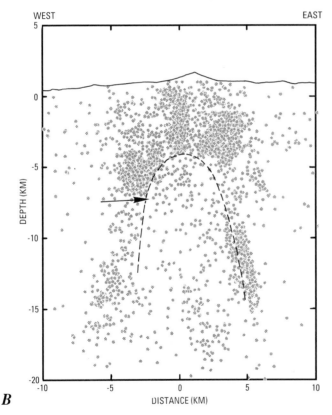

B

Figure 20. *A*, Speculative model for the Pinatubo pluton, as based on depth constraints from phase equilibria and projection of earthquake hypocenters for the period May 8–August 19, 1991, onto an east-west plane (*B*) (compare, Mori, Eberhart-Phillips, and Harlow, this volume).

to the point that mixing became feasible. Although much of the heat evolved to the dacite was probably transported away from the interface by convection, limited heating and disequilibrium melting of phenocrysts may also have lowered the bulk viscosity of the dacite.

We suspect that at this point either extensive fluid exsolution from the residual liquid in the basalt lowered its bulk density below that of the dacite (compare, Eichel-

berger, 1980; Huppert and others, 1982) and caused it to rise in a turbulent plume of basalt foam and mix, or a particularly violent new influx of basalt stirred the two components together. The latter interpretation would be consistent with ascent of a voluminous batch of basaltic magma accompanying the higher rate episode of deep, long-period earthquakes recorded by White (this volume) on May 31–June 6. In either case, mixing was exclusively a mechanical process in which phenocrysts and matrix melts were intimately mingled together to produce a *commingled magma* in the sense of Sparks and Marshall (1986). The lack of reequilibration of glass and minerals in the andesite indicates that ascent to the surface began almost immediately upon mixing. The fluid-saturated dacitic component was heated to ~940°C without chemical equilibration. At 220 MPa and 940°C, water has a density of 0.38 g/cm³ (Burnham and others, 1969). Consequently, only a relatively small increase in the volume percent of fluid bubbles would be required to lower the bulk density of the mixture below that of the dacite.

As the buoyant plume of mixed andesite made its way upward, it would have encountered semirigid (possibly thixotropic?) dacite crystal mush in the upper part of the magma reservoir overlain by solidified dacite porphyry, such that fracture transport would be possible. The andesitic magma continued to propagate upward and established a relatively narrow conduit to the surface, whereas, under the same pressure gradient, the more viscous dacite could not rise en masse until a larger diameter conduit was established. Some dacitic magma was entrained into the rising column of andesite, probably by viscous coupling at the base of the conduit. It was first erupted as a minor fraction of pumice in the moderate-sized June 12 vertical eruption. The conduit was enlarged as more magma was transported to the surface. A larger diameter conduit and slightly reduced load pressure (due to vesiculation at shallow levels) allowed progressively greater amounts of the viscous dacitic magma to enter and rise through the conduit, and contribute increasing volumes to the June 13–14 eruptions, and, finally, to dominate the June 15 eruption.

ORIGIN OF CRYSTAL FRAGMENTS IN THE PHENOCRYST-POOR DACITIC PUMICE

The reduction in grain size seen in the phenocryst-poor pumice of June 15 took place during ascent through the conduit system, below the fragmentation level, because crystal fragments are present in the matrices of intact pumice blocks. The apparent concentration of phenocryst-poor pumice in the lower part of the tephra-fall deposits of June 15, but the lack of pumice with the same type of texture in the blast deposits of June 14–15 (Hoblitt, Wolfe, and others, this volume; David and others, this volume), suggests that the process of crystal size reduction took place mainly during the early climactic phase of the June 15 eruption. Our electron imagery and Fe-Ti-oxide thermometry indicate that most of the reduction in grain size was brought about by mechanical fragmentation of phenocrysts rather than melting, but details of the process are unknown. Working models include (1) shear-fragmentation of phenocrysts in magma adjacent to conduit walls, (2) strain imposed by viscous coupling of matrix melt and crystals during boiling

of the fluid phase (vesiculation), (3) explosion of gas-saturated inclusions in phenocrysts (Gerlach and others, this volume), and (4) shock fragmentation, possibly related to the choked-flow mechanism for long-period volcanic earthquakes of Chouet and others (1994), although probably involving a magmatic fluid phase rather than ground water. Occurrence of both textural types of dacitic pumice in deposits from previous eruptive periods suggests a common process that has occurred repeatedly during the 30,000-year history of the volcano.

PROSPECTS FOR FUTURE ERUPTIONS

The magmatic model developed above has obvious implications for the future. The hot basaltic magma added to the reservoir in 1991 could rejuvenate the magma system and lead to additional eruptions. Volcanic seismicity continued in 1992 (Ramos and others, this volume) and a new lava dome began to grow in the caldera during the summer of 1992. The dome is hybrid andesite, very similar in phenocryst assemblage, texture, and composition to the June 7–12, 1991, lava dome (Daag, Dolan, and others, this volume). Is the eruption of this second batch of mixed magma likely to trigger another large dacitic eruption? Clearly, it has not had that effect in the interval between the appearance of the 1992 dome and proofing of this paper (August 1995). We suspect that the main dacitic reservoir was partially depleted in eruptible magma by the June 15, 1991, eruption and that some unknown amount of time must pass before the volatile content can build up to an eruptible level again.

Will Pinatubo volcano erupt again? The record of past activity indicates that the answer is yes. Pinatubo has been the site of periodic eruptions of dacite and hybrid andesite over a time span of more than 30,000 years. Hydrous cummingtonite-bearing magma, compositionally and texturally similar to the 1991 dacite, has been erupted repeatedly (Newhall and others, this volume; Pallister and others, 1993). Pumice from previous episodes ranges over several wt% in SiO_2, and banded pumice and mafic scoria with olivine and augite xenocrysts are present. These relations indicate that repeated mixing of dacite and basalt has taken place at the volcano. Pinatubo has produced large-volume dacitic flowage deposits and domes of hybrid andesite in the past. There is no reason to think that future eruptions will be appreciably different.

SULFUR-RICH EXPLOSIVE ERUPTIONS— THE ROLE OF BASALT AND MAGMA MIXING

The June 15, 1991, eruption, venting about 20 Mt of SO_2 into the stratosphere (Bluth and others, 1992), resulted in an increase in optical depth of the atmosphere and decrease in the global mean temperature sufficient to suppress the global warming trend through the early and mid-1990's (Hansen and others, 1992). In terms of climatic effect, it is particularly important to understand the processes that lead to high-sulfur explosive eruptions and to determine the frequency of such eruptions in the past. Sources for the sulfur emissions *during* the June 15 eruption of Pinatubo have been attributed either to degassing of an exsolved fluid phase in the dacite (Westrich and Gerlach, 1992) or to breakdown of anhydrite during the eruption (Rutherford and Devine, 1991; this volume). But where was the sulfur *before* the eruption and what process led to the enrichment of the magma in sulfur in the first place?

Either the sulfur source is exotic, as in the evaporite bed hypothesis at El Chichón (Duffield and others, 1984), or it is a primary phase of the magmatic (and related hydrothermal) system. Lacking evidence for sedimentary anhydrite at Pinatubo, we tend to favor a magmatic source. The anhydrite in the pumice of Mount Pinatubo is considered as either a primary magmatic phase (Bernard and others, 1991; this volume), or it represents xenocrysts entrained from the hydrothermal system (A.R. Reyes, written commun., 1991), or both (McKibben and others, 1992; this volume). As previously noted, anhydrite occurs primarily as subhedral to euhedral crystals in the June 12 scoria and June 15 pumice. It typically includes apatite microlites and, as at El Chichón (Luhr and others, 1984), it is found in growth contact with apatite phenocrysts. Anhydrite is only rarely found in growth contact with silicate phenocrysts. These relations are consistent with growth mainly from a separate fluid phase in the magma. The euhedral character of most anhydrite crystals and the association with euhedral apatite favor magmatic crystallization over entrainment of xenocrysts, as does the high sulfur, chlorine, and fluorine content of apatite microphenocrysts (Imai and others, 1993; Pallister and others, 1993). However, the association with apatite does not rule out origin from the hydrothermal system for some crystals. Anhydrite and apatite are common in a broad region where temperatures exceed 220°–300°C at depths of >1.8 km below the pre-1991 summit (Delfin, 1983; Delfin and others, this volume). This broad anhydrite-bearing region attests to the long-term evolution of sulfur-rich fluids from the Pinatubo magmatic system. Sulfur isotopic data for anhydrite crystals from the June 12, 1991, scoria are bimodal, with $\delta^{34}S$ modes at +6 and +10, possibly reflecting both magmatic (+6) and hydrothermal (+10) sources (McKibben and others, 1992; this volume). Anhydrite from the June 15 dacite is unimodal at $\delta^{34}S$ of +7, consistent with a dominantly magmatic source (McKibben and Eldridge, 1993).

We previously suggested that basaltic magma may have contributed sulfur to the dacite (Pallister and others, 1992). Matthews and others (1992) also favor this model, noting that mixing of basalt with cool dacite at shallow levels would lead to separation of a sulfur-rich vapor phase.

Westrich and Gerlach (1992) discounted this possibility for the June 15 eruption because of predicted slow rates of bubble rise (from Stokes Law calculations) and the low oxygen fugacity of most basaltic magmas. We agree that these are problems for basaltic magma as the *immediate* source of sulfur during the June 15 eruption. However, we are more concerned with the long-term source for the sulfur, and we note that Gerlach and others (this volume) now also favor basalt as the long-term source. Because the solubility of sulfur is enhanced by high iron and calcium content of the melt (Matthews and others, 1992), basalt is a more effective transport agent for sulfur than dacite is. Assuming average abundances of 1,000 ppm sulfur in basaltic magma (Devine and others, 1984; Carroll and Rutherford, 1987), we calculate that it would require about 4 km^3 of basalt to account for the 1991 emissions. Experimental data on sulfur solubility indicate that a hydrous basalt that is oxidized above the nickel-nickel oxide buffer curve (NNO, fig. 16) could have carried in excess of 1 wt% SO_3 (Luhr, 1990). Consequently, if the basalt were hydrated and oxidized at depth, the required volume to balance the 1991 emissions would be lowered to less than 1 km^3. In either case, these are relatively small volumes compared to the potential size and consequent longevity of the inferred magma reservoir, especially considering that many batches of basalt may have been added to the magmatic system over the >30,000-year life of the volcano.

We suggest that a large dacitic magma reservoir has been present beneath Pinatubo for an extended period of time and that basalt has repeatedly mixed with dacite in the reservoir (Newhall and others, this volume). Periodic addition and mixing of basaltic magma probably triggered many of the previous eruptions of crystal-rich dacite and hybrid andesite and also maintained a relatively elevated abundance of sulfur in resident dacitic magma within the shallow reservoir. Copper and zinc abundances are elevated in the 1991 dacite (table 7), possibly as a result of fluid transport from previous batches of basaltic magmas. Rare grains of olivine, pyroxene, and sulfide in the 1991 dacite (for example, Bernard and others, this volume; Hattori, this volume) may also record previous mixing events.

We note that the last climatically important eruption, that of El Chichón in 1980, involved sulfur-rich trachyandesite magma that was also cool (810°C ±40°; Rye and others, 1984), crystal rich (58 wt% crystals), and gas saturated and that a hydrous and oxidized basaltic parent magma is suggested to transport sulfur from depth (Luhr, 1990). Perhaps one of the characteristic features of volcanoes that give rise to sulfur-rich explosive eruptions is the presence of a long-lived magma reservoir with a relatively cool, oxidized, and crystal-rich upper zone, which acts as a trap for volatiles that are supplied by mafic magmas from greater depth.

ACKNOWLEDGMENTS

We thank Agnes Reyes of the Philippine National Oil Company for providing splits of her samples of the June 12, 1991, eruption and for her collaboration in our initial report on the 1991 eruptions. We also thank Ray Punongbayan and the staff of the Philippine Institute of Volcanology and Seismology (PHIVOLCS) for support while in the Philippines; we especially enjoyed working with Mylene Martinez, Timoteo Nillos, and Ronnie Torres of PHIVOLCS. The first author expresses his gratitude to the staff and students of the geology department of the University of the Philippines for their hospitality and interest during his visit to Manila. We thank the Philippine Air Force and the U.S. Marines for logistical and helicopter support. Finally, we thank Esperanza Soratos for her able assistance at the Pinatubo Volcano Observatory and for sharing a lively sense of humor with us during the 1992 field season. Discussions with Ken Hon, Chris Newhall, Malcolm Rutherford, and Terry Gerlach were of considerable help in interpreting the data presented here. We also appreciate the constructive and insightful manuscript reviews by Wes Hildreth, Jim Luhr, Chris Newhall, and Malcolm Rutherford.

APPENDIX

ANALYTICAL PROCEDURES

Silicate mineral and glass analyses in tables 2 and 4 were obtained on an ARL-SEMQ microprobe equipped with six scannable wavelength-dispersive crystal spectrometers. An accelerating potential of 15 kV and beam current of 10 nA were used on all samples. Natural and synthetic silicate standards were used, and off-peak background corrections were applied to standards and unknowns. Microprobe automation employed the OPUS program of Meeker and Quick (1991), and data reduction was done online by using the CITZAF routine of Armstrong (1988). On the basis of replicate analyses of secondary standards over the 4-year period 1989–92, since the new automation and data reduction software has been implemented, analytical reproducibility is estimated to be ±1–2 percent of the reported amounts for major elements and equal to or less than the standard deviation that is based on counting statistics for minor elements (typically ±10 percent of reported values in the range 0.2 to 1.0 wt%).

Oxide analyses in table 5 were obtained on a JEOL–8900 microprobe equipped with five scannable wavelength-dispersive crystal spectrometers. An accelerating potential of 15 kV and beam current of 25 nA were used. Natural and synthetic silicate and oxide standards were used, and off-peak background corrections were applied to standards and unknowns. Data reduction was by the ZAF method.

Microprobe analysis of glass filaments in the vesicular samples from Pinatubo required the use of a point electron beam. Beam damage and heating resulted in significant migration of the alkalis and silica (fig. A1). Consequently, corrections for migration of Na, K, and Si had to be applied. We used a correction procedure similar to that described by Neilsen and Sigurdsson (1981) in which count data were collected in 1-s increments and then exponential decay functions were fit to the resulting curves. This procedure was done online, and intercept values for count rate at t=0 for Na, K, and Si were used in the data reduction routine (CITZAF). Curve fits were excellent, typically yielding $r^2 >$ 0.9 for Na and K such that the analytical uncertainty was affected as much by the loss in total counts (typically half the starting amount in a 20-s count window) as by curve fitting. We estimate analytical uncertainties of ±1 percent of the reported SiO_2, and ±5 percent of the reported Na_2O and K_2O for any one analysis, based on replicate analyses of high-silica rhyolite glass standards.

Major element analyses reported in table 7 were determined by X-ray fluorescence methods as described by Taggart and others (1987). Trace elements were determined principally by instrumental neutron activation analysis as described by Baedecker and McKown (1987). On the basis of replicate analyses of standards (Taggart and others, 1987), analytical reproducibility of major element abundances is estimated to be better than ±0.4 percent of the reported values for SiO_2 and Al_2O, and for other elements better than ±2 percent of the values in the range 1–10 wt%, and better than ±6% for abundances less than 1 wt%. Coefficients of variation ($100\sigma/\bar{x}$ in percent) based on counting statistics are given for instrumental neutron activation data in table 7.

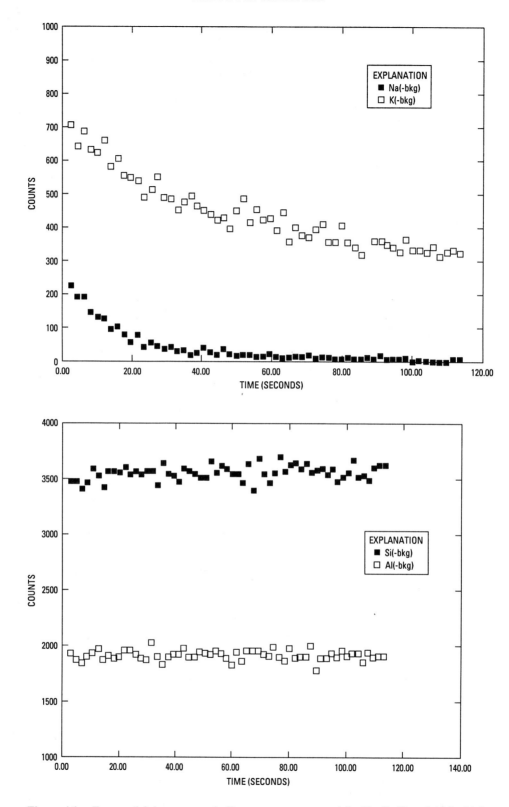

Figure A1. Exponential decay curves in X-ray counts per second for Na, K, Si, and Al for high-silica rhyolite matrix glass in phenocryst-rich dacite sample EW910615–1. Accelerating potential 15 kV, beam current 10 nA. Count data corrected for background counts (-bkg).

REFERENCES CITED

Andersen, D.J., and Lindsley, D.H., 1988, Internally consistent solution models for Fe-Mg-Mn-Ti oxides: Fe-Ti oxides: American Mineralogist, v. 73, p. 714–726.

Andersen, D.J., Lindsley, D.H., and Davidson, P.M., 1993, QUILF: A PASCAL program to assess equilibria among the Fe-Mg-Mn-Ti oxides, pyroxenes, olivine, and quartz: Computers and Geosciences, v. 19, p. 1333–1350.

Armstrong, J.T., 1988, Analysis of silicate and oxide minerals: Comparison of Monte Carlo, ZAF, and $\theta(\rho z)$ procedures: Microbeam Analysis, v. 1988, p. 239–246.

Bacon, C.R., 1986, Magmatic inclusions in silicic and intermediate volcanic rocks: Journal of Geophysical Research, v. 91, p. 6091–6112.

Bacon, C.R., and Hirschmann, M.M., 1988, Mg/Mn partitioning as a test for equilibrium between coexisting Fe-Ti oxides: American Mineralogist, v. 73, p. 57–61.

Baedecker, P.A., and McKown, D.M., 1987, Instrumental neutron activation analysis of geochemical samples, in Baedecker, P.A., ed., Methods for geochemical analysis: U.S. Geological Survey Bulletin 1770, p. H1–H14.

Bender, J.F., Hodges, F.N., and Bence, A.E., 1978, Petrogenesis of basalts from the project FAMOUS area: Experimental study from 0 to 15 kbars: Earth and Planetary Science Letters, v. 41, p. 277–302.

Bernard, A., Demaiffe, D., Mattielli, N., and Punongbayan, R.S., 1991, Anhydrite-bearing pumices from Mount Pinatubo: Further evidence for the existence of sulfur-rich silicic magmas: Nature, v. 354, p. 139–140.

Bernard, A., Knittel, U., Weber, B., Weis, D., Albrecht, A., Hattori, K., Klein, J., and Oles, D., this volume, Petrology and geochemistry of the 1991 eruption products of Mount Pinatubo.

Bluth, G.J.S., Doiron, S.D., Schnetzler, C.C., Krueger, A.J., and Walter, L.S., 1992, Global tracking of the SO_2 clouds from the June, 1991 Mount Pinatubo eruptions: Geophysical Research Letters, v. 19, p. 151–154.

Burnham, C.W., Holloway, J.R., and Davis, N.F., 1969, Thermodynamic properties of water to 1000°C and 10,000 bars: Geological Society of America Special Paper 132, 96 p.

Carroll, M.R., and Rutherford, M.J., 1987, The stability of igneous anhydrite: Experimental results and implications for sulfur behavior in the 1982 El Chichón trachyandesite and other evolved magmas: Journal of Petrology, v. 28, p. 781–801.

Cas, R.A.F., and Wright, J.V., 1988, Volcanic successions, modern and ancient: London, Unwin Hyman, 528 p.

Chouet, B.A., Page, R.A., Stephens, C.D., and Lahr, J.C., 1994, Precursory swarms of long-period events at Redoubt Volcano (1989–1990), Alaska: Their origin and use as a forecasting tool: Journal of Volcanology and Geothermal Research, v. 62, p. 95–135.

Daag, A.S., Dolan, M.T., Laguerta, E.P., Meeker, G.P., Newhall, C.G., Pallister, J.S., and Solidum, R., this volume, Growth of a postclimactic lava dome at Mount Pinatubo, July–October 1992.

David, C.P.C., Dulce, R.G., Nolasco-Javier, D.D., Zamoras, L.R., Jumawan, F.T., and Newhall, C.G., this volume, Changing proportions of two pumice types from the June 15, 1991, eruption of Mount Pinatubo.

Defant, M.J., De Boer, J.Z., and Oles, D., 1988, The western central Luzon arc, the Philippines: Two arcs divided by rifting?: Tectonophysics, v. 145, p. 305–317.

Defant, M.J., Maury, R.C., Ripley, E.M., Feigensen, M.D., and Jaques, D., 1991, An example of island-arc petrogenesis: Geochemistry and petrology of the southern Luzon arc, Philippines: Journal of Petrology, v. 32, p. 455–500.

Delfin, F.G., Jr., 1983, Geology of the Mt. Pinatubo geothermal prospect: unpublished Philippine National Oil Company report, 35 p.

Delfin, F.G., Jr., Villarosa, H.G., Layugan, D.B., Clemente, V.C., Candelaria, M.R., Ruaya, J.R., this volume, Geothermal exploration of the pre-1991 Mount Pinatubo hydrothermal system.

Devine, J.D., Sigurdsson, H., and Davis, A.N., 1984, Estimates of sulfur and chlorine yield to the atmosphere from volcanic eruptions and potential climatic effects: Journal of Geophysical Research, v. 89, p. 6309–6325.

Dick, H.J.B., and Bullen, T., 1984, Chromian spinel as a petrologic indicator in abyssal and alpine-type peridotites and spatially associated lavas: Contributions to Mineralogy and Petrology, v. 86, p. 54–76.

Donaldson, C.H., 1985, The rates of dissolution of olivine, plagioclase, and quartz in a basalt melt: Mineralogical Magazine, v. 49, p. 683–693.

Duffield, W.A., Tilling, R.I., and Canul, R., 1984, Geology of El Chichón volcano, Chiapas, Mexico, Journal of Volcanology and Geothermal Research, v. 20, p. 117–132.

Eichelberger, J.C., 1980, Vesiculation of mafic magma during replenishment of silicic magma reservoirs: Nature, v. 288, p. 446–450.

Evans, C.A., 1987, Oceanic magmas with alkalic characteristics; evidence from basal cumulate rocks in the Zambales ophiolite, Luzon, Philippine Islands: Geological Society of America Special Paper 215, p. 139–150.

Evans, C.A., Casteneda, G., and Franco, H., 1991, Geochemical complexities preserved in the volcanic rocks of the Zambales ophiolite, Philippines: Journal of Geophysical Research, v. 96, p. 16251–16262.

Evans, C.A., and Hawkins, J.W. Jr., 1989, Compositional heterogeneities in the upper mantle peridotites from the Zambales Range ophiolite, Luzon, Philippines: Tectonophysics, v. 168, p. 23–41.

Fournelle, J.H., 1991, Anhydrite and sulfide in pumices from the 15 June 1991 eruption of Mt. Pinatubo: Initial examination: Eos, Transactions, American Geophysical Union, v. 72, p. 68.

Fournelle, J, Carmody, R., and Daag, A.S., this volume, Anhydrite-bearing pumices from the June 15, 1991, eruption of Mount Pinatubo: Geochemistry, mineralogy, and petrology.

Gerlach, T.M., Westrich, H.R., and Symonds, R.B., this volume, Preeruption vapor in magma of the climactic Mount Pinatubo eruption: Source of the giant stratospheric sulfur dioxide cloud.

Geschwindt, C.-H., and Rutherford, M.J., 1993, Cummingtonite and the evolution of the Mount St. Helens (Washington) magma system: An experimental study: Geology, v. 20, p. 1011–1014.

Grove, T.L., and Kinzler, R.J., 1986, Petrogenesis of andesites: Annual Reviews of Earth and Planetary Science, v. 14, p. 417–454.

Haggerty, S.E., 1976, Opaque mineral oxides in terrestrial igneous rocks, *in* Rumble, Douglas, III, ed., Reviews in mineralogy, v. 3, Oxide minerals: Blacksburg, Va., Mineralogical Society of America, p. Hg101–Hg277.

Hansen, J., Lacis, A., Ruedy, R., and Sato, M., 1992, Potential climate impact of Mount Pinatubo eruption: Geophysical Research Letters, v. 19, p. 215–218.

Hattori, K., this volume, Occurrence and origin of sulfide and sulfate in the 1991 Mount Pinatubo eruption products.

Hoblitt, R.P., Wolfe, E.W., Scott, W.E., Couchman, M.R., Pallister, J.S., and Javier, D., this volume, The preclimactic eruptions of Mount Pinatubo, June 1991.

Huppert, H.E., Sparks, R.S.J., and Turner, J.S., 1982, Effects of volatiles on mixing in calc-alkaline magma systems: Nature, v. 297, p. 554–557.

————1984, Some effects of viscosity on the dynamics of replenished magma chambers: Journal of Geophysical Research, v. 89, p. 6857–6877.

Imai, A., Listanco, E.L., and Fujii, T., 1993, Petrologic and sulfur isotopic significance of highly oxidized and sulfur-rich magma of Mt. Pinatubo, Philippines: Geology, v. 21, p. 699–702.

Johnson, M.C., and Rutherford, M.J., 1989, Experimental calibration of the aluminum-in-hornblende geobarometer with application to Long Valley caldera (California) volcanic rocks: Geology, v. 17, p. 837–841.

Koyaguchi, T., this volume, Volume estimation of tephra-fall deposits from the June 15, 1991, eruption of Mount Pinatubo by theoretical and geological methods.

Luhr, J.F., 1990, Experimental phase relations of water- and sulfur-saturated arc magmas and the 1982 eruptions of El Chichón volcano: Journal of Petrology, v. 31, p. 1071–1114.

————1991, Volcanic shade causes cooling: Nature, v. 354, p. 104–105.

Luhr, J.F., Carmichael, I.S.E., and Varekamp, J.C., 1984, The 1982 eruptions of El Chichón volcano, Chipas, Mexico: Mineralogy and petrology of the anhydrite-bearing pumices: Journal of Volcanology and Geothermal Research, v. 23, p. 69–108.

Marsh, B.D., 1981, On the crystallinity, probability of occurrence, and rheology of lava and magma: Contributions to Mineralogy and Petrology, v. 78, p. 85–98.

Matthews, S.J., Jones, A.P., and Bristow, C.S., 1992, A simple magma-mixing model for sulfur behaviour in calc-alkaline volcanic rocks: Mineralogical evidence from Mount Pinatubo 1991 eruption: Journal of the Geological Society, London, v. 149, p. 863–866.

McKibben, M.A., and Eldridge, C.S., 1993, Sulfur isotopic systematics of the June 1991 eruptions of Mount Pinatubo: A SHRIMP ion microprobe study: Eos, Transactions, American Geophysical Union, v. 74, p. 668.

McKibben, M.A., Eldridge, C.S., and Reyes, A.G., 1992, Multiple origins of anhydrite in Mt. Pinatubo pumice: Eos, Transactions, American Geophysical Union, v. 73, p. 633–634.

————this volume, Sulfur isotopic systematics of the June 1991 Mount Pinatubo eruptions: A SHRIMP ion microprobe study.

Meeker, G.P., Pallister, J.S., and Hoblitt, R.P., 1993, Phenocryst-rich and -poor pumices from Mount Pinatubo: Eos, Transactions, American Geophysical Union, v. 74, p. 668.

Meeker, G.P., and Quick, J.E., 1991, OPUS: Old probe updated-software: Microbeam Analysis, v. 1991, p. 353.

Mori, J., Eberhart-Phillips, D., and Harlow, D.H., this volume, Three-dimensional velocity structure at Mount Pinatubo, Philippines: Resolving magma bodies and earthquake hypocenters.

Neilsen, R.L., 1990, Simulation of igneous differentiation processes, *in* Nicholls, J., and Russell, J.K., eds., Reviews in mineralogy, v. 27: Blacksburg, Va., Mineralogical Society of America, p. 65–105.

Newhall, C.G., Daag, A.S., Delfin, F.G., Jr., Hoblitt, R.P., McGeehin, J., Pallister, J.S., Regalado, M.T.M., Rubin, M., Tamayo, R.A., Jr., Tubianosa, B., and Umbal, J.V., this volume, Eruptive history of Mount Pinatubo.

Nielsen, C.H., and Sigurdsson, H., 1981, Quantitative methods for electron microprobe analysis of sodium in natural and synthetic glasses: American Mineralogist, v. 66, p. 547–552.

Paladio-Melosantos, M.L., Solidum, R.U., Scott, W.E., Quiambao, R.B., Umbal, J.V., Rodolfo, K.S., Tubianosa, B.S., Delos Reyes, P.J., and Ruelo, H.R., this volume, Tephra falls of the 1991 eruptions of Mount Pinatubo.

Pallister, J.S., Hoblitt, R.P., and Reyes, A.G., 1992, A basalt trigger for the 1991 eruptions of Pinatubo volcano?: Nature, v. 356, p. 426–428.

Pallister, J.S., Meeker, G.P., Newhall, C.G., and Hoblitt, R.P., 1993, 30,000 years of the "same old stuff" at Pinatubo: Eos, Transactions, American Geophysical Union: v. 74, p. 667–668.

Ramos, E.G., Laguerta, E.P., and Hamburger, M.W., this volume, Seismicity and magmatic resurgence at Mount Pinatubo in 1992.

Roeder, P.L., and Emslie, R.F., 1970, Olivine-liquid equilibrium: Contributions to Mineralogy and Petrology, v. 29, p. 275–289.

Rutherford, M.J., 1993, Experimental petrology applied to volcanic processes: Eos, Transactions, American Geophysical Union: v. 74, p. 49 and p. 55.

Rutherford, M.J., Brown, L., and Pallister, J.S., 1993, petrologic constraints on timing of magmatic processes in the 1991 Pinatubo volcanic system: Eos, Transactions, American Geophysical Union, v. 74, p. 671.

Rutherford, M.J., and Devine, J.D., 1991, Pre-eruption conditions and volatiles in the 1991 Pinatubo magmas: Transactions of the American Geophysical Union (Eos), v. 72, p. 62.

————this volume, Preeruption pressure-temperature conditions and volatiles in the 1991 dacitic magma of Mount Pinatubo.

Rutherford, M.J., and Hill, P.M., 1993, Magma ascent rates and magma mixing from amphibole breakdown: Experiments and the 1980–1986 Mount St. Helens eruptions: Journal of Geophysical Research, v. 98, p. 19667–19685.

Rye, R.O., Luhr, J.F., and Wasserman, M.D., 1984, Sulfur and oxygen isotopic systematics of the 1982 eruptions of El Chichón volcano, Chiapas, Mexico: Journal of Volcanology and Geothermal Research, v. 12, p. 109–123.

Sack, R.O., Carmichael, I.S.E., Rivers, M., and Ghiorso, M.S., 1980, Ferric-ferous equilibria in natural silicate liquids at 1 bar: Contributions to Mineralogy and Petrology, v. 75, p. 369–376.

Scandone, R., and Malone, S.D., 1985, Magma supply, magma discharge, and readjustment of the feeding system of Mount St. Helens during 1980: Journal of Volcanology and Geothermal Research, v. 23, p. 239–262.

Scott, W.E., Hoblitt, R.P., Torres, R.C., Self, S, Martinez, M.L., and Nillos, T., Jr., this volume, Pyroclastic flows of the June 15, 1991, climactic eruption of Mount Pinatubo.

Sigurdsson, H., Devine, J.D., and Davis, A.N., 1985, The petrologic estimate of volcanic degassing: Jökull, v. 35, p. 1–8.

Sisson, T.W., and Grove, T.L., 1993, Experimental investigations of the role of H_2O in calc-alkaline differentiation and subduction zone magmatism: Contributions to Mineralogy and Petrology, v. 113, p. 143–166.

Smith, D.R., and Leeman, W.P., 1987, Petrogenesis of Mount St. Helens dacitic magmas: Journal of Geophysical Research, v. 92B, p. 10313–10334.

Sparks, R.J.S., Huppert, H.E., and Turner, J.S., 1984, The fluid dynamics of evolving magma chambers: Philosophical Transactions of the Royal Society of London, v. A310, p. 511–534.

Sparks, R.J.S., and Marshall, L.A., 1986, Thermal and mechanical constraints on mixing between mafic and silicic magmas: Journal of Volcanology and Geothermal Research: v. 29, p. 99–124.

Sparks, S.R.J., Sigurdsson, H., and Wilson, L., 1977, Magma mixing: A mechanism for triggering explosive eruptions: Nature, v. 267, p. 315–318.

Spulber, S.D., and Rutherford, M.J., 1983, The origin of rhyolite and plagiogranite in oceanic crust: An experimental study: Journal of Petrology, v. 24, p. 1–25.

Stimac, J.A., Pearce, T.H., Donnelly-Nolan, J.M., and Hearn, B.C., 1990, The origin and implications of undercooled andesitic inclusions in rhyolites, Clear Lake Volcanics, California: Journal of Geophysical Research, v. 95B, p. 17729–17746.

Stormer, J.C., 1983, The effects of recalculation on estimates of temperature and oxygen fugacity from analyses of multicomponent iron-titanium oxides: American Mineralogist, v. 68, p. 586–594.

Taggart, J.E., Lindsay, J.R., Scott, B.A., Vivit, D.V., Bartel, A.J., Stewart, K.C., 1987, Analysis of geologic materials by X-ray fluorescence spectrometry, in Baedecker, P.A., ed., Methods for geochemical analysis: U.S. Geological Survey Bulletin 1770, p. E1–E19.

Ulmer, P., 1989, The dependence of the Fe^{2+}-Mg cation partitioning between olivine and basaltic liquid on pressure, temperature and composition, an experimental study to 30 kbars: Contributions to Mineralogy and Petrology, v. 101, p. 261–273.

Westrich, H.R., and Gerlach, T.M., 1992, Magmatic gas source for the stratospheric SO_2 cloud from the June 15, 1991 eruption of Mount Pinatubo: Geology, v. 20, p. 867–870.

White, R.A., this volume, Precursory deep long-period earthquakes at Mount Pinatubo, Philippines: Spatio-temporal link to a basalt trigger.

Wilcox, R.E., 1954, Petrology of Parícutin volcano, Mexico: U.S. Geological Survey Bulletin 965–C, p. 281–353.

Williams, H., and McBirney, A.R., 1979, Volcanology: San Francisco, Freeman, Cooper and Co., 397 p.

Wolfe, E.W., 1992, The 1991 eruptions of Mount Pinatubo, Philippines: Earthquakes and Volcanoes, v. 23, p. 5-37.

Wolfe, E.W. and Hoblitt, R.P., this volume, Overview of the eruptions.

Wright, T.L., and Doherty, P.C., 1970, A linear programming and least squares method for solving petrologic mixing problems: Geological Society of America Bulletin, v. 81, p. 1995–2008.

Mineral and Glass Compositions in June 15, 1991, Pumices: Evidence for Dynamic Disequilibrium in the Dacite of Mount Pinatubo

By James F. Luhr[1] and William G. Melson[1]

ABSTRACT

Whole-rock compositions, textures, and mineral and glass compositions were investigated in two pumices from the June 15 eruption: white, phenocryst-rich, and gray, phenocryst-poor. The gray pumice is slightly poorer in SiO_2 and K_2O and richer in V, Cr, Co, Ni, Cu, and Zn compared to the white pumice, although more extensive sample suites described elsewhere in this volume show the two pumice types to be virtually identical in composition. The mineral assemblage is similar in both cases (plagioclase, hornblende, titanomagnetite, ilmenite, cummingtonite, biotite, quartz, apatite, anhydrite, sulfides, rare zircon, and xenocrystic olivine plus chromite, augite, and orthopyroxene), but the textures and mineral abundances are very different. The white pumice has large, euhedral phenocrysts surrounded by vesiculated, compositionally homogeneous matrix glass. In the gray pumice, however, most phenocrysts show broken margins, and the vesiculated matrix glass is compositionally heterogeneous and contains abundant tiny crystals; some of these appear to be microlites, but others are clearly broken fragments. Cummingtonite rims are common on hornblende in the white pumice but are rare in the gray pumice.

Compositions of coexisting Fe-Ti oxides indicate temperatures of 845–893°C and 836–842°C for the white and gray pumices, respectively, on the basis of two different published algorithms. Calculated oxygen fugacities fall about 2.3 log units above the Ni-NiO solid oxygen buffer and thus reflect highly oxidized conditions that are consistent with the presence of primary anhydrite. The Fe-Ti oxide temperatures are not consistent with the presence of cummingtonite in the pumices of Mount Pinatubo; experimental studies have shown that cummingtonite is not stable above about 800°C. This discrepancy may reflect a late-stage heating event that caused the Fe-Ti oxides to reequilibrate but

was too rapid for the kinetically inhibited reaction of cummingtonite. Alternatively, this discrepancy may reflect inadequate calibration of the Fe-Ti oxide geothermometers for the highly oxidized conditions of the magmas of Mount Pinatubo.

Plagioclase phenocrysts in the white pumice show complex zoning patterns with abrupt outward rises in calcium followed by more gradual declines. Some of these calcium spikes can be correlated from crystal to crystal and may record major magma-mixing events in the earlier history of the magma. Detailed traverses across the outer 100 micrometers of 15 plagioclase growth rims show considerable variability, but the outermost rims are reasonably homogeneous in composition at $An_{40.2\pm1.8}$. Surprisingly, application of plagioclase-melt models by using this rim composition and the average matrix glass of the white pumice shows that the plagioclase rims are too calcic to have been in equilibrium with the matrix glass at any likely temperature. These anomalously calcic rims may have grown under disequilibrium conditions induced by water loss from the magma system during eruptive ventings preceding the June 15 climax.

The gray pumice appears to record quenched disequilibrium that resulted as a dacitic melt-rich magma violently invaded the main, coarsely crystalline dacitic body and began to blend with its high-silica rhyolite melt. The violence of mixing may have caused shattering of phenocrysts from the main dacite. Small microlites were growing from the partially blended melt when the system was frozen upon eruption. As indicated by the plagioclase-melt modeling discussed above, even the dominant coarse-textured dacite from the June 15 eruption may have been out of chemical equilibrium prior to eruption.

The 1991 dacite of Mount Pinatubo and the 1982 trachyandesite of El Chichón are examples of an important class of hydrous, sulfur-rich, anhydrite-bearing magmas that erupt from subduction-zone volcanoes. These eruptions have the potential to inject large quantities of sulfur gases into the stratosphere, where they can play a significant role in modifying the Earth's climate. Recognition of ancient,

[1]Department of Mineral Sciences, NHB–119, Smithsonian Institution, Washington, DC 20560.

high-sulfur eruptions in the geologic record is complicated by the rapid dissolution of anhydrite from vesicular pumices in surface waters. The pumices of Mount Pinatubo and El Chichón are strikingly different in whole-rock composition and mineralogy but share several features that might allow analogous eruptions to be identified in the ancient record: high oxidation states, low preeruption temperatures, high crystal contents, abundant hornblende (plus biotite), sulfate-rich apatite, and most importantly, inclusions of anhydrite that are trapped in other phenocrysts and thereby preserved.

INTRODUCTION

The climactic eruption of Mount Pinatubo on June 15, 1991, was remarkable both for the mass of magma ejected and for the mass of sulfur gases released to the atmosphere. The eruption ejected at least 4 km^3 of dacitic magma, an amount that makes it one of the largest eruptions of the 20th century (Scott and others, 1991); it ranks behind the 1912 Novarupta-Katmai (Alaska) eruption but is of roughly similar magnitude to the 1932 Cerro Azul-Quizapu (Chile) and 1902 Santa María (Guatemala) eruptions, given the uncertainty of tephra volume estimates (Fierstein and Nathenson, 1992). Approximately 20 million tons of SO_2 were lofted into the stratosphere by the June 15 eruption (Bluth and others, 1992), the largest stratospheric sulfur injection since satellite-based measurements began in 1978. The stratospheric SO_2 combined with water vapor to form a cloud of submicrometer-sized sulfuric acid aerosol droplets that were predicted to lower surface air temperatures by an average of about 0.5°C through 1993 (Hansen and others, 1992b), a prediction that has been borne out by globally integrated air temperature measurements to date (Hansen and others, 1992a; Kerr, 1993). Although quantitative estimates are complicated by the heavy rainfall caused by Typhoon Yunya during the June 15 eruption, studies of leachable sulfate from fresh ash-fall deposits (Bernard and others, this volume) indicate that a substantial additional mass of sulfur gases was released by the eruption but quickly precipitated as tiny sulfate minerals that were adsorbed onto ash particles and carried to the ground. Thus, the total release of sulfur by the eruption was significantly greater than the 20 million tons of SO_2 measured in the stratosphere.

Fresh pumices from the 1991 eruption contain euhedral-subhedral microphenocrysts of the water-soluble sulfate mineral anhydrite (Bernard and others, 1991; see also papers in this volume: Bernard and others, Fournelle and others, Hattori, McKibben and others, Pallister and others, Rutherford and Devine). Although anhydrite has long been reported in volcanic rocks, it has usually been interpreted as xenocrystic (Yoshiki, 1933; Kôzu, 1934; Kwano, 1948; Taneda, 1949; Katsui, 1958; Taylor, 1958; Nicholls, 1971;

Yagi and others, 1972; Arculus and others, 1983). Anhydrite was first recognized as an important, stable igneous mineral after the 1982 eruption of the Mexican volcano El Chichón (Luhr and others, 1984). The El Chichón eruption injected about 7 million tons of SO_2 into the stratosphere, which is second only to Pinatubo in the 14-year record of satellite-based SO_2 measurements (Bluth and others, 1992). Leachate studies of fresh El Chichón ashes indicated that a roughly equivalent mass of sulfur gases was rapidly converted to sulfates, adsorbed onto ash particles, and carried to the Earth's surface (Varekamp and others, 1984). El Chichón and Mount Pinatubo represent a newly recognized class of subduction-zone volcanoes whose eruptions of water- and sulfur-rich, oxidized, anhydrite-bearing magmas are capable of playing an important role in short-term modification of the Earth's climate.

The purpose of this study is to contribute toward documentation of the mineralogy, petrology, and geochemistry of the two major types of pumice ejected during the climactic eruption of Mount Pinatubo on June 15, 1991. Our focus is on (1) geochemical and textural comparisons of the white and gray pumices, (2) estimation of preeruptive temperature and oxygen fugacity from analysis of coexisting Fe-Ti oxide compositions and whole-rock ferric/ferrous ratios, (3) compositional zoning profiles in plagioclase crystals, (4) estimation of preeruptive water contents in the melt from plagioclase-glass compositional relations, and (5) glass compositions from the pumice matrices and inclusions in phenocrysts. We emphasize the similarities and differences between the dacite of Mount Pinatubo and the trachyandesite of El Chichón, which we hope will aid in the recognition of sulfur-rich eruptions of this type in ancient deposits from which the primary anhydrite crystals have long since been leached by surface waters.

SAMPLES STUDIED

Our study is based on two pumices that were collected on about July 18, 1991, from the surface of the June 15 pyroclastic-flow deposit just outside the rear gate of Clark Air Base (the area called Mactan in other reports: 15°10.2'N., 120°29.0'E.). The deposit was about 5 m thick; at the time of collection the interior still had a temperature of several hundred degrees centigrade, but the surface was cool and rainwashed. One sample is white and has relatively large and abundant phenocrysts (USNM# 116534–2) and will subsequently be referred to as the "white" pumice. It is equivalent to the "phenocryst-rich" pumices discussed elsewhere in this volume. Prior to our studies, the white-pumice sample was approximately 12×12×6 cm in size and weighed about 300 g. The other specimen (USNM# 116534–1) is light gray with phenocrysts that are both smaller and less abundant than in the white pumice; it will be referred to as the "gray" pumice and is equivalent to the "phenocryst-poor" types discussed elsewhere in the volume. The sample was 10×8×6 cm in size and weighed about 200 g. As discussed by Pallister and others (this volume), the white pumices are the dominant variety and account for about 85 percent of the June 15 lapilli, with the gray types, which typically have breadcrust surfaces, accounting for the remainder. Inclusions of white pumice are commonly found in gray pumices, and the two types can be mingled together to form banded pumices. Investigations of vertical sections through the tephra-fall sequence demonstrate that the two pumice types were ejected together throughout the eruption, with the gray pumices decreasing slightly in abundance with time from about 20 percent of the lapilli near the base of the sequence to about 13 percent near the top (David and others, this volume).

ANALYTICAL TECHNIQUES

Slabs were cut from the two pumices for preparation of polished sections and for whole-rock analysis. The powders were prepared by grinding small pumice chips in a shatter-box with alumina puck and container until the powder passed completely through a 100-mesh nylon seive. The powders were dried at 110°C and then at 1,000°C, with weight losses reported as H_2O^- and LOI (loss on ignition), respectively, in table 1. The dehydrated powder was then combined with Li-tetraborate and fused to a glass disk for X-ray fluorescence (XRF) determination of TiO_2, Al_2O_3, $Fe_2O_3^{total}$, MnO, Na_2O, K_2O, and P_2O_5 abundances using the Smithsonian Institution's Philips PW–1480 spectrometer (table 1). Values for SiO_2, FeO, CaO, and MgO were determined by wet chemical techniques on the original powder. Analyses of S, Cl, and nine other trace elements were performed by XRF on separate disks prepared by pressing rock powder dried at 110°C with cellulose. Twenty additional trace element abundances were determined by instrumental neutron activation (INA) analysis of the original powders at Washington University (Lindstrom and Korotev, 1982).

Electron microprobe analyses were conducted on the Smithsonian's 9-spectrometer ARL-SEMQ instrument, with 15 kV accelerating potential, a specimen current of 15 nA on brass, natural and synthetic standards, on-peak background corrections, and Bence-Albee interelement corrections. The beam was focused and stationary for analysis of Fe-Ti oxides and plagioclase, focused and manually moved for analysis of all glass inclusions and matrix glass in the gray pumice, and defocused and manually moved for the matrix glass in the white pumice.

WHOLE-ROCK COMPOSITIONS

The June 15 pumices from table 1 both are classified as medium-K dacites by the scheme of Gill (1981). The whole-rock sulfur contents of the white and gray pumices (0.17 and 0.03 wt% SO_3, respectively) can be used to estimate the amount of anhydrite in the pumices, assuming that anhydrite (with 59 wt% SO_3) contains virtually all of the sulfur in the samples. This calculation indicates 0.29 and 0.05 wt% anhydrite for the white and gray pumices, respectively. Other whole-rock pumice analyses from Bernard and others (this volume), Fournelle and others (this volume), and Pallister and others (this volume) show 0.25–0.48 wt% SO_3 (0.42–0.81 wt% anhydrite) for white pumices and 0.13–0.35 wt% SO_3 (0.22-0.59 wt% anhydrite) for gray pumices; these values are all higher than those measured in our study. The relatively large ranges for SO_3 may reflect primary variations in the otherwise compositionally homogeneous dacites of Mount Pinatubo. Given the rapid dissolution of anhydrite in surface waters and the fact that Typhoon Yunya was crossing Luzon at the time of the climactic eruption, however, this spread of sulfate contents is perhaps more likely a result of different degrees of anhydrite dissolution, with the samples analyzed in our study having undergone the greatest posteruption leaching.

In comparing the white and gray pumice analyses in table 1, the white pumice is seen to be slightly less mafic: richer in Ba, Ta, and U and poorer in MgO, V, Ni, Cu, Sc, Cr, and Co. When evaluated against the larger data set of Pallister and others (this volume), however, no systematic differences between the white and gray pumices are found.

MINERAL ASSEMBLAGES
AND TEXTURES

Both pumice types contain the same assemblage of stable minerals: plagioclase, hornblende, titanomagnetite, ilmenite, cummingtonite, biotite, quartz, apatite, anhydrite, sulfides, and zircon. Cummingtonite occurs exclusively as rims on hornblende; it is common in the white pumice but is rare in the gray pumice. The typical rounding of quartz crystals indicates that they may have become unstable shortly before the eruption. The fact that major element compositions of glass inclusions within quartz crystals are virtually indistinguishable from compositions of glass inclusions within other phenocrysts and the matrix glass of the white pumice (see table 4; Westrich and Gerlach, 1992; Gerlach and others, this volume) demonstrates that the quartz crystals grew from the dacite of Mount Pinatubo and are not xenocrystic. Hattori (this volume) gives a detailed treatment of sulfides in the 1991 products; those present in the dacites are Cu-Fe sulfides and pyrrhotite. Minor zircon has been identified as inclusions in both anhydrite and Fe-Ti oxides (Bernard and others, 1991; Matthews and others,

Table 1. Whole-rock analyses of June 15, 1991, pumices.

["White" is phenocryst-rich white pumice, USNM# 116534–2. "Gray" is phenocryst-poor gray pumice, USNM# 116534–1]

	White	Gray		White	Gray
Major oxides[1]			INA analysis[2] (ppm)		
SiO_2	64.19	64.09	Sc	9.21	9.53
TiO_2	.50	.50	Cr	37.1	47.8
Al_2O_3	16.69	16.85	Co	10.7	11.2
Fe_2O_3	1.93	2.04	As	3.0	2.9
FeO	2.18	2.07	Br	.6	.5
MnO	.10	.10	Sb	.23	.22
MgO	2.39	2.58	Cs	2.72	2.76
CaO	5.13	5.17	La	15.6	15.4
Na_2O	4.42	4.46	Ce	30.0	30.4
K_2O	1.51	1.50	Nd	14	16
P_2O_5	.19	.19	Sm	2.90	2.94
SO_3	.17	.03	Eu	.84	.86
H_2O^-	.10	.07	Tb	.38	.40
LOI	.60	.38	Yb	1.20	1.27
			Lu	.20	.20
Total wt.%	100.10	100.03	Hf	3.11	3.10
			Ta	.34	.30
XRF spectroscopy (ppm)			Au	6	9
F	155	39	Th	4.30	4.27
Cl	432	234	U	1.58	1.38
V	78	89			
Ni	18	24			
Cu	31	71			
Zn	50	54			
Rb	39	40			
Sr	535	544			
Y	14	14			
Zr	114	116			
Ba	478	454			

[1] Wet chemical techniques were used in the determination of SiO_2, FeO, CaO, and MgO. A specific ion electrode was used to determine F. H_2O^- represents moisture lost at 110°C, whereas LOI (loss on ignition) represents weight loss at 1,000°C. Other values (excepting INA analysis) were measured by X-ray fluorescence (XRF) spectroscopy at the Smithsonian Institution.

[2] Values for instrumental neutron activation (INA) analysis were determined at Washington University (Lindstrom and Korotev, 1982). Typical uncertainties for XRF and INA data can be found in Pier and others (1992).

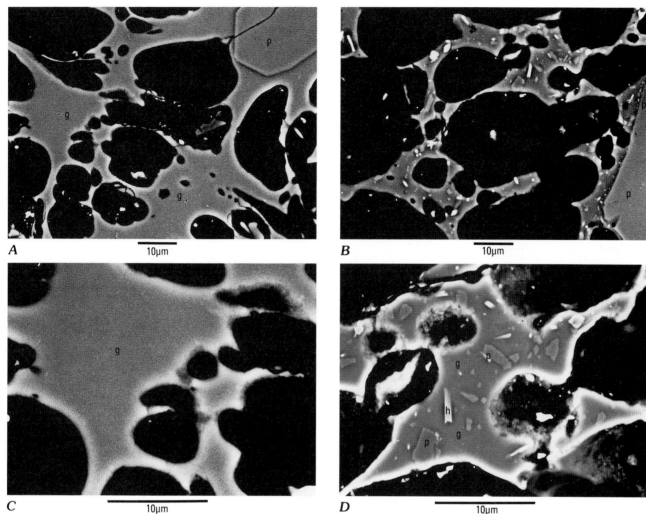

Figure 1. Back-scattered electron images of white (*A* and *C*) and gray pumices (*B* and *D*). Abbreviations: g, glass; p, plagioclase; h, hornblende. The white "p" and black "p" on the plagioclase of B identify darker (calcium-rich) and lighter (sodium-rich) composi- tional zones, respectively, that are separated by a planar interface and truncated by the broken crystal margin. *C* is the largest glass patch from *A* at higher magnification; *D* is one of the largest glassy areas of *B* at higher magnification.

1992). Rare xenocrysts in the dacites include olivine (with chromite inclusions), augite, and orthopyroxene. Pallister and others (this volume) determined that these xenocrysts do not have appropriate compositions to have originated in the andesites and basalts erupted during the days prior to June 15 and suggested that they may represent crystals from the underlying Zambales ophiolite complex, a source also advocated by Fournelle and others (this volume).

Despite their similarities in mineralogy, the white and gray pumices differ greatly in texture. The white pumices have large, relatively unshattered phenocrysts of plagio- clase, hornblende, and quartz surrounded by clear vesicu- lated glass that is largely free of microlites. Some plagioclase and quartz grains exceed 3 mm in diameter. In the gray pumices, most crystals show at least one broken margin, and crystal sizes are generally less than 1 mm. The vesiculated glassy matrix is charged with small crystals. Back-scattered electron photomicrographs of the vesicu-

lated matrix glasses from the white and gray pumices are shown in figure 1 at two different magnifications; similar photos and observations are found in Pallister and others (this volume). Figure 1*A* shows a euhedral plagioclase phe- nocryst surrounded by vesiculated, crystal-free glass. The largest glass patch from this scene is magnified in figure 1*C*. Figure 1*B* shows a broken plagioclase phenocryst with truncated compositional zonation surrounded by vesicu- lated glass containing many small plagioclase and horn- blende crystals. One of the largest glassy areas is magnified in figure 1*D*. The small crystals in the gray-pumice matrix glass appear to include both euhedral microlites, such as the one labeled hornblende with slight swallow-tail termina- tions in figure 1*D*, and anhedral fragments broken from larger crystals. Although hornblende is common among these tiny crystals, we have not identified cummingtonite either as euhedral microlites or as crystal fragments. It

Table 2. Modal abundances by point counting and least-squares calculation.

[Mineral abbreviations: Plag, plagioclase; Hbd, hornblende; Oxid, Fe-Ti oxides (titanomagnetite plus ilmenite); Anhy, anhydrite; Apat, apatite; Qtz, quartz. Textural abbreviations: ph, phenocrysts (>0.3 mm); mp, microphenocrysts (>0.03 mm, <0.3 mm: after Wilcox, 1954); tr, trace. Both samples also contain trace amounts of cummingtonite rims on hornblende, biotite, sulfides, zircon inclusions, and xenocrysts of olivine with chromite inclusions, augite, and orthopyroxene]

		White Pumice			Gray Pumice		
		point counting[1]		lsq model[2]	point counting		lsq model
		vol.%	wt.%	wt.%	vol.%	wt.%	wt.%
Plag	ph	37.4	38.4	--	6.2	6.6	--
	mp	4.0	4.1	--	17.0	18.2	--
	total	41.4	42.5	39.0	23.2	24.8	33.5
Hbd	ph	5.8	7.2	--	.9	1.2	--
	mp	3.2	3.9	--	3.0	3.8	--
	total	9.0	11.1	13.5	3.9	5.0	12.4
Oxid	ph	.3	.6	--	--	--	--
	mp	1.1	2.0	--	1.5	2.9	--
	total	1.4	2.6	2.1	1.5	2.9	1.8
Anhy	mp	.3	.3	.3	tr	tr	.1
Apat	mp	tr	tr	.4	tr	tr	.4
Qtz	ph	.7	.7	--	--	--	--
Groundmass		47.2	42.8	44.7	71.4	67.3	51.8
Vesicles		45.8	--	--	52.6	--	--

[1] Point counting was conducted in combined transmitted and reflected light: 1,229 points for white pumice and 1,452 points for gray pumice. Values in vol.% were converted to wt.% by using the following densities (g/cm^3): plag, 2.67; hbd, 3.2; apat, 3.2; anhy, 2.95; oxid, 4.9; and groundmass, 2.35.

[2] Least-squares (lsq) models were calculated by first determining values for anhydrite and apatite based on the whole-rock SO_3 and P_2O_5 concentrations, respectively. After subtraction of appropriate CaO values, the remaining major elements were fit by a least-squares model using whole-rock, glass, titanomagnetite, and ilmenite compositions from this paper, and average plagioclase and hornblende data from Pallister and others, this volume. Both models gave good fits, with low values of Σr^2: 0.09 and 0.22 for the white and gray models, respectively.

seems to be restricted to overgrowths on the rims of hornblende phenocrysts.

Mineral abundances were determined for the two pumices by both point counting and least-squares modeling, with results given in table 2. A single polished section of

each pumice was point counted under combined transmitted-reflected light. Following the criteria of Wilcox (1954), phenocrysts (>0.3 mm), microphenocrysts (<0.3 mm, >0.03 mm), and groundmass (<0.03 mm) were distinguished and counted separately. The point-counting results (volume percent) were then converted to weight percent for comparison with weight percent modes calculated by least-squares modeling (see table 2 for methods).

The grain-size differences of the two pumice types are clearly reflected in the point-count data. The white pumice contains 44.2 vol% phenocrysts and 47.2% groundmass glass plus crystals, whereas the gray pumice has only 7.1% phenocrysts and 71.4% groundmass material. Although the point-counted and modeled modes are quite similar for the white pumice, the results are quite discordant for the gray pumice. This difference stems from the fact that the groundmass material for the white pumice is largely homogeneous glass, whereas for the gray pumice it consists of heterogeneous glass with abundant crystal fragments and microlites. An attempt was made to count vesicles, although distinguishing thin glass septa from vesicles, even in reflected light, is very difficult. A more reliable estimate of the vesicularity can be obtained from the average densities calculated for vesicle-free pumices from modes and phase densities in table 2 (~2.65 g/cm^3), coupled with actual pumice densities measured by Pallister and others (this volume): white = 0.819 g/cm^3 and gray = 0.977 g/cm^3. These data indicate vesicularities of 31 vol% and 37 vol% for the white and gray pumices, respectively, compared with the point-counted values of 46 vol% and 53 vol% in table 2.

FE-TI OXIDE COMPOSITIONS AND ESTIMATES OF TEMPERATURE AND OXYGEN FUGACITY

Discrete crystals of titanomagnetite and ilmenite are present in both pumices. These Fe-Ti oxides are homogeneous and unzoned within each sample and show no significant differences between the two pumices. Electron microprobe analyses of the oxides in both samples are given as mean and 1σ values in table 3, along with recalculations of Fe_2O_3, FeO, and mineral formulas after Stormer (1983). These mean analyses were used to calculate temperatures and oxygen fugacities based on two different algorithms, Andersen and others (1993) and Ghiorso and Sack (1991), with the results listed in table 3; these methods yield temperatures of 845°C and 893°C, respectively, for the white pumice, and 836°C and 842°C, respectively, for the gray pumice. Similar temperatures are obtained from the same geothermometers when using Fe-Ti oxide data in Pallister and others (this volume) and Rutherford and Devine (this volume). As noted in both of those studies, however, these temperatures are inconsistent with the experimentally based upper temperature limit of about 790°C for cummingtonite stability at pressures of 2–3 kbar (Geschwind and Rutherford, 1992).

This discrepancy might reflect a late-stage heating event that caused the Fe-Ti oxide compositions to reequilibrate but was unable to melt the cummingtonite or convert it to orthopyroxene + quartz as a result of sluggishness of those reactions. This heating might have been related to the intrusion of basalt into the Pinatubo system, which has been invoked as a trigger to the 1991 eruptions (Pallister and others (1992; this volume). Experimental data in Fonarev and Korolkov (1980) show that metastable cummingtonite compositions persisted up to 21 days at temperatures of 760–780°C. A late-stage heating event might also explain the corroded outlines of quartz crystals in the dacites of Mount Pinatubo; the experimental data of Rutherford and Devine (this volume) show the upper thermal stability limit of quartz in water-saturated systems to be 790°C at $P_{H_2O}=3$ kbar and 820°C at $P_{H_2O}=2$ kbar. Late-stage heating might also provide an explanation for the apparent disequilibrium between plagioclase and matrix glass, discussed in a later section.

An alternative explanation for the discrepancy between Fe-Ti oxide temperatures and the thermal stability limit of cummingtonite has been put forward by Geschwind and Rutherford (1992) and Rutherford and Devine (this volume). They noted that the highly oxidized conditions pertinent to the magmas of Mount Pinatubo lie outside the experimental calibration of the Fe-Ti oxide geothermometers and showed that for experimental charges equilibrated at these high-f_{O_2} conditions, the geothermometer of Andersen and others (1993) yields temperatures that are about 30°C too high. They have suggested, therefore, that Fe-Ti oxide temperatures calculated from this geothermometer for highly oxidized systems should be adjusted downward by 30°C. Although the analyses in table 3 would still yield temperatures for the geothermometer of Andersen and others (1993) that are up to 20°C too high for cummingtonite stability, this difference is not significant when considering likely errors for Fe-Ti oxide geothermometry. The formulation of Ghiorso and Sack (1991), however, yields an unacceptably high corrected temperature of 963°C for the white pumice.

The algorithms of Andersen and others (1993) and Ghiorso and Sack (1991) yielded estimates for oxygen fugacity that are listed in table 3 both as -log f_{O_2} values and as log-unit deviations from the Ni-NiO (NNO) solid oxygen buffer. The latter method of comparison is convenient in that it removes temperature-dependent variations in the calculated oxygen fugacities. The two algorithms are very consistent in indicating oxygen fugacities about 2.3 log units above NNO. These values are consistent with the presence of anhydrite in the 1991 pumices, given the experimentally determined lower f_{O_2} limit for anhydrite stability of 1–1.5 log units above NNO (Carroll and Rutherford, 1987).

Sack and others (1980) and Kilinc and others (1983) presented an alternative method for estimating magmatic oxygen fugacity based on the Fe^{3+}/Fe^{2+} value and major element composition of glass quenched at known temperature. We have not determined the Fe^{3+}/Fe^{2+} values for matrix glasses in this study but, rather, use the whole-rock

Table 3. Electron microprobe analyses of titanomagnetite and ilmenite.

[Mineral abbreviations: tmt, titanomagnetite; ilm, ilmenite; ulv, ulvöspinel. n indicates the number of individual spot analyses included in the mean and 1σ values. T and $-\log f_{O_2}$ values were calculated from the mean analyses using programs provided by Andersen and others (1993: QUILF Version 4.1, selected reactions: FeMgIlSp, FeMnIlSp, FeTi, and MH) and Ghiorso and Sack (1991). ΔNNO values are deviations from the Ni-NiO buffer of Huebner and Sato (1970), calculated for P=2,000 bar. Their expression for the NNO buffer is: $\log f_{O_2} = -24930/T + 9.36 + 0.046 \times (P-1)/T$, with T in Kelvin and P in bars]

Mineral	White Pumice		Gray Pumice	
	tmt	ilm	tmt	ilm
n	6	8	6	8
Major oxides	**Mean weight percent**			
SiO_2	.09	.06	.11	.07
TiO_2	4.51	28.07	4.37	28.48
Al_2O_3	1.76	.31	1.65	.27
FeO^{total}	84.94	65.41	85.97	65.56
MnO	.50	.31	.50	.33
MgO	1.18	1.04	1.27	1.09
Total	92.98	95.20	93.87	95.80
	Standard Deviation			
SiO_2	.02	.01	.04	.03
TiO_2	.15	.31	.09	.73
Al_2O_3	.10	.04	.07	.04
FeO^t	.82	.45	.78	.80
MnO	.02	.02	.01	.03
MgO	.04	.04	.21	.07

Iron recalculation and mineral formulas after Stormer (1983)

FeO	32.95	23.15	33.01	23.42
Fe_2O_3	57.78	46.96	58.86	46.83
Total	98.78	99.90	99.77	100.51
Cations				
Si	.0034	.0016	.0043	.0019
Ti	.1293	.5409	.1240	.5454
Al	.0792	.0093	.0733	.0082
Fe^{3+}	1.6555	.9057	1.6701	.8972
Fe^{2+}	1.0493	.4962	1.0410	.4987
Mn	.0161	.0067	.0160	.0071
Mg	.0672	.0396	.0713	.0415
Total	3.0000	2.0000	3.0000	2.0000
X_{ulv}	.1316		.1251	
X_{ilm}		.5336		.5375

Table 3. Electron microprobe analyses of titanomagnetite and ilmenite—Continued.

<u>Andersen and others (1993)</u>

T (°C)	845	836
-log f_{O2}	10.57	10.71
ΔNNO	2.29	2.33

<u>Ghiorso and Sack (1991)</u>

T (°C)	893	842
-log f_{O2}	-9.60	-10.60
ΔNNO	2.34	2.32

compositions from table 1 to make these calculations. The algorithm of Kilinc and others (1983) yields oxygen fugacities 2.5 and 2.7 log units above NNO for the white and gray pumices, respectively, which are indistinguishable from those estimates that are based on Fe-Ti oxide compositions.

On the basis of phase equilibrium experiments, Rutherford and Devine (this volume) have modeled the magma of Mount Pinatubo as a vapor-saturated system at a pressure of about 2 kbar. At that pressure and an oxygen fugacity 2.3 log units above NNO, thermodynamic data of Helgeson and others (1978) can be used to calculate the fugacity ratio f_{SO_2}/f_{H_2S} for the gas phase, which increases from about 5 to 32 over the temperature range from 800°C to 900°C. This calculation shows that SO_2 was the dominant sulfur gas species in the magma, consistent with the release of some 20 million tons of stratospheric SO_2 as detected by the satellite-borne Total Ozone Mapping Spectrometer (TOMS) shortly after the eruptions (Bluth and others, 1992). The trachyandesite erupted by El Chichón volcano in 1982 was slightly less oxidized (NNO + 1 log unit), and although it was anhydrite saturated and also produced a large cloud of SO_2 in the stratosphere, calculations similar to those described above indicate that H_2S was the dominant sulfur species, with the fugacity ratio f_{SO_2}/f_{H_2S} of about 0.06 (Luhr, 1990). Comparison of the Pinatubo and El Chichón systems demonstrates the sensitivity of sulfur speciation to oxygen fugacity and the fact that SO_2 is not necessarily the dominant sulfur species in the gas phase of all anhydrite-saturated magmas. Regardless of the original sulfur speciation, however, the ultimate fate of gaseous sulfur carried to the stratosphere is to become oxidized and hydrated to form aerosol droplets of sulfuric acid (McKeen and others, 1984).

MATRIX GLASS COMPOSITIONS

As shown in figures 1A and C, the matrix of the white pumice contains relatively broad septa of homogeneous, crystal-free glass that could be analyzed easily by microprobe. Analysis 1 in table 4 gives the mean and 1σ values for 11 moving, defocused spot analyses of the matrix glass. As shown on figure 2, a plot of SiO_2 versus K_2O, this mean composition is very similar to white-pumice matrix glass analyses reported elsewhere (Gerlach and others, this volume; Matthews and others, 1992; Pallister and others, this volume; Rutherford and Devine, this volume), and this similarity demonstrates the homogeneous nature of the matrix glass.

The white-pumice matrix glass analysis listed in table 4 has 1.35 wt% corundum in the CIPW norm; normative corundum is characteristic of melt compositions that have undergone considerable hornblende fractionation (Cawthorn and Brown, 1976; Zen, 1986). Some workers have interpreted normative corundum in glass analyses as a reflection of sodium loss during analysis and have corrected the analyses by adding sodium until the normative corundum disappears (Merzbacher and Eggler, 1984). Although this technique may be appropriate for hornblende-free systems, it is clearly inappropriate for hornblende-rich rocks such as the dacites of Mount Pinatubo.

Glass inclusions are a conspicuous feature of many phenocrysts in the white pumice. Analyses 2, 3, and 4 in table 4 are moving, focused spot analyses on large glass inclusions in titanomagnetite, quartz, and plagioclase, respectively. As shown in figure 2, the glass inclusion in quartz and several other glass-inclusion analyses from the literature are virtually identical to the matrix glass, whereas other glass-inclusion analyses from this study and the literature fall to both higher and lower SiO_2 contents (fig. 2: open circles). In general, though, glass inclusions trapped in phenocrysts of the white pumice show compositions very similar to the enclosing matrix glass.

In contrast to the matrix glasses in the white pumice, gray-pumice matrix glasses range widely in composition (analyses 5–7: table 4). Analysis 5 is virtually identical to the white-pumice matrix glasses, whereas analyses 6 and 7 and three other gray-pumice matrix glass analyses from Pallister and others (this volume) range downward over nearly 10 wt% in SiO_2 (fig. 2: filled squares). Two analyses of glass inclusions in plagioclase from the gray pumice (analy-

Table 4. Electron microprobe analyses of glass and plagioclase (in weight percent).

	SiO_2	TiO_2	Al_2O_3	FeO^{total}	MnO	MgO	CaO	Na_2O	K_2O	Total
1	77.91	.15	12.89	.85	.05	.24	1.21	3.70	3.00	100.00
	.65	.02	.31	.04	.03	.04	.07	.17	.08	
2	76.67	.18	13.26	1.67	.03	.23	1.22	3.72	2.75	99.73
3	74.43	.11	12.80	.77	.08	.21	1.10	3.70	2.84	96.04
4	72.47	.15	14.15	.85	.11	.20	1.29	3.67	3.10	95.99
5	77.34	.16	12.23	.86	.01	.23	1.29	3.03	3.13	98.28
6	73.18	.24	14.01	1.43	.04	.70	2.62	3.74	2.79	98.75
7	68.49	.34	14.65	2.35	.06	1.41	4.16	4.73	2.39	98.58
8	74.21	.13	12.78	.96	.10	.12	.72	3.58	3.48	96.08
9	69.38	.17	14.42	.65	.11	.15	1.96	3.68	2.82	93.34
10	57.67	.02	25.97	.20	.00	.02	7.88	6.49	.28	98.53
	.93	.02	.57	.03	.00	.01	.33	.29	.06	

White-pumice glasses:

1. Mean and 1σ values for 11 moving, defocused (5 μm) spot analyses on matrix glass. When normalized anhydrous with $Fe^{3+} = 0.44 \, xFe^{total}$ (appropriate for oxygen fugacity of NNO+2.3 log units: after Kilinc and others, 1983), the analysis shows 1.35 wt.% normative corundum (c).
2. Moving, focused spot analysis of glass inclusion in titanomagnetite (1.95 wt.% c).
3. Moving, focused spot analysis of glass inclusion in quartz (1.71 wt.% c).
4. Moving, focused spot analysis of glass inclusion in plagioclase (2.52 wt.% c).

Gray-pumice glasses

5. Moving, focused spot analysis of matrix glass (1.55 wt.% c).
6. Moving, focused spot analysis of matrix glass (0.09 wt.% c).
7. Moving, focused spot analysis of matrix glass (7.29 wt.% normative diopside: di).
8. Moving, focused spot analysis of glass inclusion in plagioclase (1.90 wt.% c).
9. Moving, focused spot analysis of glass inclusion in plagioclase (1.87 wt.% c).

White-pumice plagioclase rims

10. Mean and 1 standard deviation values for outermost rims of 15 plagioclase crystals from white pumice, those whose profiles are shown in figure 4. Formula gives 40.17 ± 1.79 mol.% An.

ses 8 and 9: table 4) fall in the high-silica part of this range (fig. 2: open squares). A similar relation was noted in some of the pumices erupted from Nevado del Ruiz in 1985 (Melson and others, 1990) and interpreted to indicate remelting and partial assimilation of a highly crystalline, cooler carapace by a hotter andesitic magma, which itself was probably heated by and partially mixed with invading basaltic magma.

The K_2O contents of the whole-rock pumices (1.5 wt%: table 1) and matrix glasses (table 4) can be used to estimate the abundance of glass because most of the K_2O is partitioned into glass during crystallization. Among the

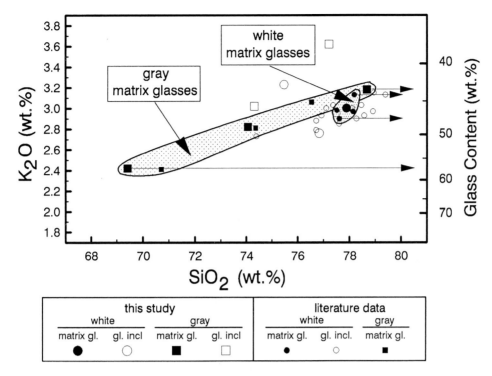

Figure 2. Matrix glass (solid symbols) and glass inclusion (open symbols) analyses from this study and literature sources (Gerlach and others, this volume; Matthews and others, 1992; Pallister and others, this volume; Rutherford and Devine, this volume) shown as SiO_2 versus K_2O, with all data normalized anhydrous using $Fe^{3+}=0.44 \times Fe^{total}$, which is appropriate for log f_{O_2}=NNO+2.3 (Kilinc and others, 1983). Stippled balloons enclose all analyses for matrix glasses from the white and gray pumices. Estimates of glass content on the right-hand y-axis were calculated assuming that both pumices contain 39 wt% plagioclase (with 0.28 wt% K_2O) and 13.5 wt% hornblende (with 0.22 wt% K_2O) and that all remaining K_2O in the whole-rock analyses (1.38 wt% K_2O=1.52 wt% (table 1) minus 0.14 wt% from plagioclase and hornblende) resides in matrix glass. Thus, matrix glass content (wt%)=100×1.38/(K_2O in glass). The horizontal arrows mark the lowest and highest estimated glass contents for the white (44–48 wt%) and gray (43–57 wt%) pumices.

major crystalline phases, only plagioclase and hornblende contain significant K_2O: 0.28 and 0.22 wt%, respectively. The least-squares model for the white pumice shown in table 2 indicates 39 wt% plagioclase and 13.5 wt% hornblende. For simplicity we have assumed the same amounts of plagioclase and hornblende in both the white and gray pumices; the calculated glass abundances are very insensitive to variations in these modal estimates. The right-hand y-axis to figure 2 shows the glass contents estimated for various matrix-glass K_2O values, with horizontal arrows marking the limits for the white (44–48 wt% glass) and gray pumices (43–57 wt% glass). The gray pumices appear to represent mixtures of different magma batches with quite different crystallinities.

COMPOSITIONAL ZONING IN PLAGIOCLASE

Plagioclase compositions were studied in considerable detail in the white pumice. Care was taken to identify

crystals with prominent growth bands outward to a euhedral face in contact with vesiculated glass. A few plagioclase analyses were also made for the gray pumice, but no large phenocrysts are present, and the broken margins of most crystals made the identification of growth rims difficult.

For about a dozen euhedral crystals in the white pumice, automated traverses were made from rim to rim at a spacing of 2 μm. Three examples of relatively symmetrical compositional traverses are shown in figure 3. Traverse 105 was made across a 215-μm euhedral crystal. The core of $An_{40–42}$ is surrounded by a mantle of $An_{45–47}$, from which the composition drops sharply outward to An_{37} before rising to An_{40} at the very rim. Traverse 109 was made across a 450-μm euhedral crystal. It shows most of the same general features found in traverse 105 (as indicated by dashed correlation lines) but also shows a core area that rises up to An_{48}, apparently reflecting plagioclase crystallization prior to nucleation of the crystal shown in traverse 105. A still earlier stage of plagioclase growth is evident in traverse 110, taken across a 1,100-μm euhedral phenocryst. Its core area ranges from An_{31} to An_{39}. An abrupt increase to An_{45}

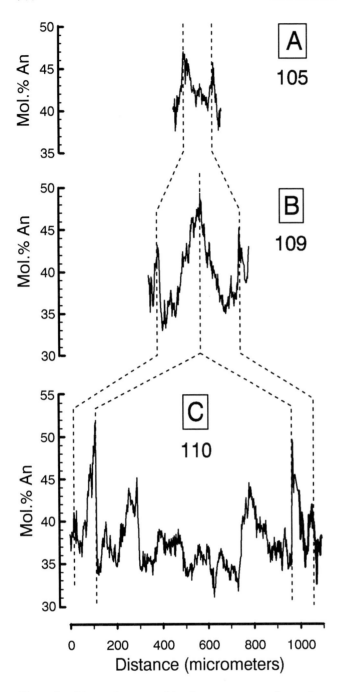

Figure 3. Rim-to-rim compositional traverses across three euhedral plagioclase crystals in the white pumice, with a point spacing of 2 μm. Traverse A (105) is across a 215-μm euhedral crystal; Traverse B (109) is across a 450-μm crystal; Traverse C (110) is across a 1,100-μm crystal. Dashed lines show tentative correlations of calcium-rich zones. Mol% An=100×Ca/(Ca+Na).

mafic magma into a more evolved magma body, such as the scenario advocated by Pallister and others (1992; this volume) for triggering of the 1991 eruption. If so, these recurrent calcium spikes provide evidence for repeated mixing events in the history of the 1991 dacite.

Although plausible correlations could be made from zone to zone among the three crystals shown in figure 3, other crystals could not be correlated with them easily. Our study of this problem suffers from reliance on conventional polished sections. A proper investigation of correlated zoning changes in plagioclase would involve traverses made across crystals that were separated from the rock, oriented crystallographically, and polished down to a plane through the crystal center. Our results offer encouragement to a future study of this type, which might reveal important details of the early magmatic history of the dacite of Mount Pinatubo. Also, the detailed stratigraphy revealed by combined electron microprobe and Normarski- and laser-interference imaging as used by Pearce and others (1987) would be informative.

Coexisting plagioclase and glass compositions can be used to estimate the preeruptive water contents of the melt (Housh and Luhr, 1991) and, accordingly, one of the main purposes of analyzing plagioclase in the 1991 pumices was to determine the composition of plagioclase that was in equilibrium with the matrix glass just prior to eruption. In order to address this question, 15 traverses of 100-μm length and 2-μm spacing were made across euhedral plagioclase rims in the white pumice. These traverses are shown in figure 4. Traverses A and B, C and D, and N and O are for different rims of the same crystal. These pairs show broadly similar patterns of compositional variation, but not all details of the traverses can be correlated. The problem is even greater when comparing traverses from one crystal to another. Although most of the traverses in figure 4 were intentionally placed above and below others of broadly similar form, when taken as a group, it is clear that these crystal rims show quite a diversity of compositional zoning. It appears that zoning in plagioclase may be affected mostly by the local environment, with only major compositional changes (such as the large Ca spikes of fig. 3) correlatable from crystal to crystal. We note, however, that crystals from quite different preeruption crystallization sites could have been mixed during eruption and possibly before eruption by convection and other movements in the magma.

Despite the dissimilar zoning profiles in plagioclase crystal rims, the outermost rim compositions show relatively little variation. The composition of the outermost plagioclase rim for each traverse is labeled on figure 4. These range from $An_{37.2}$ to $An_{43.0}$. The mean An content is 40.2 with 1σ of 1.8. The mean analysis and 1σ values are given in table 4.

is followed outward by a general decline to An_{35} before another abrupt increase to An_{50-52}. This latter zone is tentatively correlated with the core of the crystal in traverse 109. Abrupt outward increases in An content such as those in traverses 109 and 110 were observed in many other plagioclase crystals. They may reflect mixing of calcium-rich

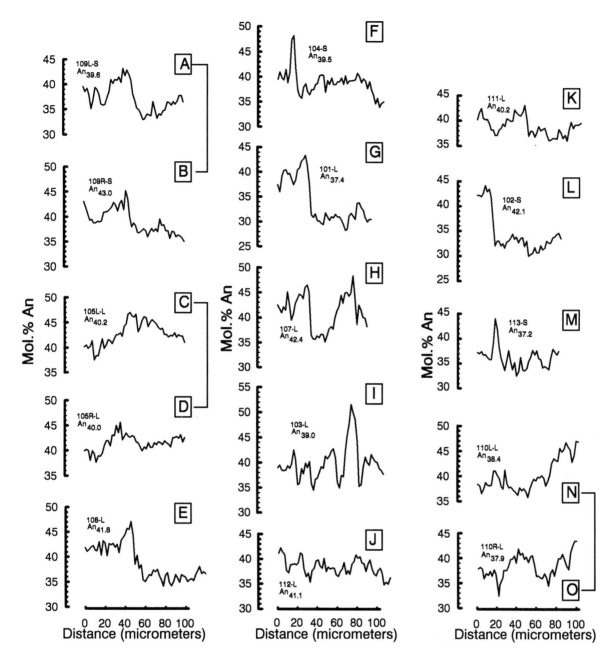

Figure 4. Compositional traverses across the outermost 100 μm of 15 different plagioclase rims in the white pumice, with a point spacing of 2 μm. Lines connecting traverses A, and B, C and D, and N and O indicate that these pairs are for different faces of the same crystal. Small labels give the traverse number and the An content of the outermost rim. Mol.% An=100×Ca/(Ca+Na).

PLAGIOCLASE-MELT EQUILIBRIA AND ESTIMATED WATER CONTENTS IN THE MELT

Housh and Luhr (1991) gave two different algorithms for calculating water contents of melts that are based on solution of albite and anorthite exchange reactions between coexisting plagioclase and melt. The calculated melt water contents are strongly dependent upon the assumed temperature of equilibration, which must be known independently.

Figure 5 shows the results of these calculations for the white-pumice matrix-glass analysis from table 4, with plagioclase composition plotted against calculated melt water content. The solid lines show predicted values for the albite equation, and the dashed lines show solutions for the anorthite exchange reaction. Different temperatures of solution are indicated, and they bracket the temperature range discussed above, which is from 780°C to 900°C. The large dots show values of mole fraction anorthite in plagioclase (X_{An}) and melt H_2O content for which albite and anorthite

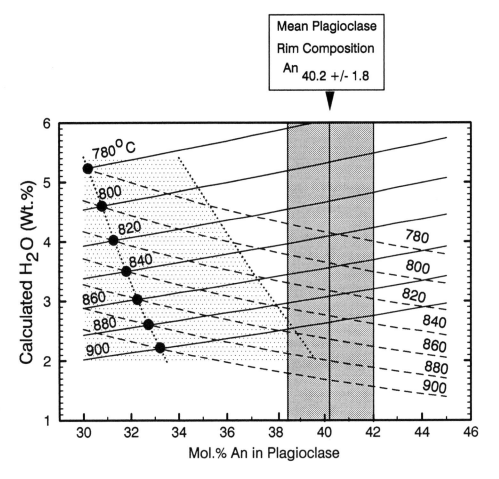

Figure 5. Plagioclase composition (Mol% An=100×Ca/(Ca+Na)) versus melt water contents calculated from the albite (solid lines) and anorthite (dashed lines) exchange reactions between plagioclase and melt from Housh and Luhr (1991) at the indicated temperatures. The matrix glass composition for the white pumice from table 4 was used in all calculations. The large black dots and connecting dotted line indicate the locus of $T-X_{An}^{pl}-H_2O$ points for which the two equations give the same melt water contents. The dotted line bounding the lightly stippled field is the locus of points of maximum An content agreement for the two equations when the errors given by Housh and Luhr (1991) for the albite (±0.54 wt% H_2O) and anorthite (±0.33 wt% H_2O) reactions are considered. Thus, the lightly stippled field represents acceptable agreement between the two equations. The vertical line and bounding heavily stippled field show the mean plagioclase rim composition and 1σ variations from table 4. Note the minimal overlap of the two stippled fields, which indicates apparent lack of equilibrium between the matrix glass and plagioclase rims.

solutions converge. Housh and Luhr (1991) estimated the errors on melt water content to be ±0.54 wt% and ±0.33 wt% for the albite and anorthite equations, respectively, so an exact match of the two curves is an unrealistic expectation. The lightly stippled field bounded by the other dotted line shows the range of permissible An-H_2O solutions, given these error limits. The vertical line and bounding stippled field indicate the mean plagioclase rim composition for the white pumice (table 4) and 1σ variations.

Even considering the errors indicated by the lightly stippled field, the plagioclase rim compositions are much too calcic to have been in equilibrium with the mean glass composition at any likely temperature and melt water

content. The model of Housh and Luhr (1991) was calibrated with experiments that bracket the plagioclase and melt compositions found in the pumices of Mount Pinatubo, and thus is expected to be applicable. It might be argued that the zone of plagioclase that actually equilibrated with the matrix glass composition under preeruptive conditions is not the outermost layer but lies inward from the rim. In this scenario, the outermost layers would have grown after or during eruptive degassing. In that case, however, one would expect strong normal zoning of the rims if there was an approach to equilibrium crystallization, which is not seen in most crystals (figure 4). The lack of consistency in the zoning patterns of plagioclase rims makes it very difficult to evaluate this hypothesis. Clearly, zones of appropriate composition (An_{31-34}) can be found in individual plagioclase crystals, but these cannot be correlated from one crystal to another.

If we accept a temperature of 780°C for the magma prior to eruption, consistent with the stability of cummingtonite, and use the mean glass and plagioclase analyses from table 4, the albite equation indicates about 6.2 wt% H_2O in the melt, whereas the anorthite equation yields just 4.2 wt% H_2O. As discussed below, the albite value is actually quite close to melt water content estimates based on ion-microprobe analysis of trapped glass inclusions. The dilemma posed by these results is the wide discrepancy between the albite and anorthite solutions, a discrepancy that is much greater than estimated errors for the technique. These problems in application of the Housh and Luhr (1991) model to the white pumices of Mount Pinatubo appear to indicate that the plagioclase rims and matrix glass were not in chemical equilibrium at the time of eruptive quenching, a surprising conclusion, given the compositional homogeneity of both phases.

We can propose one viable mechanism to account for these seemingly anomalous calcic rim compositions. During cooling experiments, Lofgren (1980) produced reversely zoned plagioclase margins with disequilibrium compositions by suddenly lowering temperature to a new static value, which induced crystallization. He noted that high degrees of undercooling can also be produced by sudden loss of water from the system, a relevant process for the Pinatubo case. Growth zoning under such conditions is controlled by interface-kinetic and diffusion-rate factors and can lead to compositions far from those expected under equilibrium crystallization. Smith and Brown (1988) reviewed the problem of plagioclase zone compositions at large degrees of undercooling and found that initial plagioclase compositions are consistently more calcic than expected from equilibrium relations. The calcic rims of the plagioclases in the white pumice of Mount Pinatubo may reflect such a process, induced by rapid water loss during the eruptive events preceding June 15, 1991.

Two papers in this volume reported estimates for preeruptive water contents of the melt based on direct ion microprobe measurement of H_2O in trapped glass inclusions: Gerlach and others, and Rutherford and Devine. Many host crystals develop cracks during eruptive cooling, and when these intersect the melt/glass inclusions, leakage of H_2O and other volatiles can result. In addition, minerals with strong cleavages, such as hornblende and plagioclase, can be particularly prone to leakage. Thus, the most reliable data are expected for strong minerals with poor cleavage, such as quartz or olivine. All three of the above studies reported ion microprobe H_2O values for glass inclusions in quartz, and these range from 4.4 to 6.3 wt%; it is expected that the highest values are the most reliable. As seen on figure 5, melt water contents approaching 6 wt% would only

be consistent with both the albite *and* anorthite solution models at 780°C if the plagioclase rim composition was in the range of An_{30-33}. As shown in figures 3 and 4 and discussed above, however, plagioclase of this composition is not common in the outer 100 μm of plagioclase crystals in the white pumice.

DISCUSSION

ORIGIN OF THE GRAY PUMICES: MAGMA MIXING AND DISEQUILIBRIUM

The abundance of broken crystals, the crystal-rich nature of the matrix glass, and the strong compositional heterogeneity in that glass are important clues for understanding the origin of the gray pumices of June 15. Pallister and others (1992) interpreted the gray pumices to represent a thermal boundary layer between normal white-pumice-type dacite and intruding basalt. Inclusions of the latter were found in mixed andesitic scoriae erupted on June 12, 1991, just prior to the climactic eruption. In their view, the gray pumices are reheated white-pumice-type dacites in which many crystals dissolved and drove the melt to a more mafic composition. This early model has been revised in Pallister and others (this volume); it now recognizes the importance of crystal fragmentation in the gray pumices in producing the shattered phenocrysts and crystal-rich matrix glass. They advocated fragmentation of the gray-pumice crystals upon ascent of the dacite in the conduit system during the early phase of the June 15 eruption, with fragmentation caused either by choked flow in the conduit or by violent vesiculation of the magma.

We envision an important role for magma mixing in the origin of the gray pumice, but we believe that the mixing was between the white-pumice dacite and a more melt-rich magma, at either higher temperature and (or) higher water content, with a dacitic melt composition close to that of analysis 7 (table 4). This melt-rich magma invaded and mixed with the dominant, white-pumice variety just prior to eruption. The lack of equilibration in the gray pumices, indicated most dramatically by the wide range of matrix glass compositions, suggests that this mixing event happened just hours or at most a few days before the cataclysmic eruption. It may ultimately be possible to correlate this proposed mixing event with characteristics of the preeruptive seismicity (Harlow and others, this volume; White and others, this volume). The minute quench crystals present in the gray-pumice matrix (fig. 1D) may reflect crystallization brought on by the intimate and possibly highly turbulent mixing of the two magmas. The mixing appears to have been sufficiently violent that most crystals in the gray pumices, except for the smallest quench crystals, were fragmented, even down to the scale of several micrometers (fig. 1D). The very similar compositions of Fe-Ti oxides (table 3) and other minerals between the white and gray pumices may indicate that most of the minerals in the latter grew in white-pumice-type dacite prior to the mixing event and that only the silica-poor matrix glass compositions from the gray pumices provide direct evidence for the nature of the invading magma. Failure of the Housh and Luhr (1991) plagioclase-melt model when applied to the white pumice from June 15 indicates that even the main phenocryst-rich dacite from Pinatubo was not in chemical equilibrium upon eruptive quenching. The anomalously calcic plagioclase rims may have grown under the influence of kinetic factors attendant on the initiation of supercooling during H_2O degassing.

IDENTIFICATION OF ANCIENT HIGH-SULFUR ERUPTIONS

The 1991 dacite of Mount Pinatubo and the 1982 tra-chyandesite of El Chichón are two examples of anhydrite-bearing magmas that released large masses of sulfur gases to the Earth's atmosphere. Given the poor preservation potential of primary igneous anhydrite, it is important to find other parameters that can be used to identify similar eruptions in ancient deposits. The magmas of Mount Pinatubo and El Chichón differ in many other elemental and mineralogical aspects. The magma of Mount Pinatubo is a medium-K dacite that is also relatively poor in other incompatible trace elements (table 1). In sharp contrast, the magma of El Chichón is a high-K trachyandesite that is strongly enriched in other incompatible elements (Luhr and others, 1984). Among the minerals present in June 15 dacites of Mount Pinatubo, quartz, cummingtonite rims on hornblende, ilmenite, zircon, and olivine, chromite, and orthopyroxene xenocrysts are not found in the trachyandesite of El Chichón, and the latter has stable sphene and augite that are not present in the former. Those differences aside, the Pinatubo and El Chichón rocks show other important similarities. Both rock types are very crystal rich (58 wt% for El Chichón, 55 wt% for white pumices of Mount Pinatubo: table 2), have relatively low preeruption temperatures of ~800°C, are dominated by phenocrysts of complexly zoned plagioclase and hornblende, contain minor biotite, and have apatites that are unusually rich in sulfate: 0.34 wt% SO_3 for apatites from the trachyandesite of El Chichón (Luhr and others, 1984) and up to 0.78 wt% SO_3 for apatites from the dacite of Mount Pinatubo (Imai and others, this volume; Pallister and others, 1993). Hornblende is by far the most abundant mafic phenocryst in both cases. Both rocks also have high values of Fe^{3+}/Fe^{total}: 0.44 for the white pumice of Mount Pinatubo (table 1) and 0.41 for the trachyandesite of El Chichón. The abundance of hornblende and the presence of biotite, along with the high crystal contents and low temperatures, demonstrate the water-rich nature of both magmas. The high Fe^{3+}/Fe^{total} values are also

consistent with the abundance of hydrous mafic minerals as argued by Carmichael (1967). Thus, the magmas of Mount Pinatubo and El Chichón were alike only in being strongly oxidized and very water- and sulfur-rich. Ancient analogs can be identified by their high crystal contents, high ratios of hornblende to anhydrous mafic minerals, low calculated preeruption temperatures, and sulfate-rich apatites. Their expected high Fe^{3+}/Fe^{total} values will be quickly overprinted by posteruption oxidation. In pumices from both El Chichón (Luhr and others, 1984) and Pinatubo (Fournelle and others, this volume), anhydrite occurs as occasional inclusions within phenocrystic minerals. These inclusions will survive in the geologic record and provide convincing evidence for ancient high-sulfur eruptions.

ACKNOWLEDGMENTS

We thank Chris Newhall for providing specimens from the June 15 eruption. Gene Jarosewich and Joe Nelen performed the wet chemical and XRF analyses and provided essential assistance in electron microprobe analyses. John Pallister, Malcolm Rutherford, and Chris Newhall reviewed the manuscript and offered many useful comments and suggestions to help resolve earlier conflicts with other studies in this volume.

REFERENCES CITED

Andersen, D.J., Lindsley, D.H., and Davidson, P.M., 1993, QUILF: A PASCAL program to assess equilibria among the Fe-Mg-Mn-Ti oxides, pyroxenes, olivine, and quartz: Computers and Geosciences, v. 19, p. 1333–1350.

Arculus, R.J., Johnson, R.W., Chappell, B.W., McKee, C.O., and Sakai, H., 1983, Ophiolite-contaminated andesites, trachybasalts, and cognate inclusions of Mount Lamington, Papua New Guinea: Anhydrite-amphibole-bearing lavas and the 1951 cumulodome: Journal of Volcanology and Geothermal Research, v. 18, p. 215–247.

Bernard, A., Demaiffe, D., Mattielli, N., and Punongbayan, R.S., 1991, Anhydrite-bearing pumices from Mount Pinatubo: Further evidence for the existence of sulfur-rich silicic magmas: Nature, v. 354, p. 139–140.

Bernard, A., Knittel, U., Weber, B., Weis, D., Albrecht, A., Hattori, K., Klein, J., and Oles, D., this volume, Petrology and geochemistry of the 1991 eruption products of Mount Pinatubo.

Bluth, G.J.S., Doiron, S.D., Schnetzler, C.C., Krueger, A.J., and Walter, L.S., 1992, Global tracking of the SO_2 clouds from the June, 1991 Mount Pinatubo eruptions: Geophysical Research Letters, v. 19, no. 2, p. 151–154.

Carmichael, I.S.E., 1967, The iron-titanium oxides of salic volcanic rocks and their associated ferromagnesian silicates: Contributions to Mineralogy and Petrology, v. 14, p. 36–64.

Carroll, M.R., and Rutherford, M.J., 1987, The stability of igneous anhydrite: Experimental results and implications for sulfur behavior in the 1982 El Chichón trachyandesite and other evolved magmas: Journal of Petrology, v. 28, p. 781–801.

Cawthorn, B.G., and Brown, P.A., 1976, A model for the formation and crystallization of corundum-normative calc-alkaline magmas through amphibole fractionation: Journal of Geology, v. 84, p. 467–476.

David, C.P.C., Dulce, R.G., Nolasco-Javier, D.D., Zamoras, L.R., Jumawan, F.T., and Newhall, C.G., this volume, Changing proportions of two pumice types from the June 15, 1991, eruption of Mount Pinatubo.

Fierstein, J., and Nathenson, M., 1992, Another look at the calculation of fallout tephra volumes: Bulletin of Volcanology, v. 54, p. 156–167.

Fonarev, V.I., and Korolkov, G.J., 1980, The assemblage orthopyroxene + quartz. The low-temperature stability limit: Contributions to Mineralogy and Petrology, v. 73, p. 413–420.

Fournelle, J, Carmody, R., and Daag, A.S., this volume, Anhydrite-bearing pumices from the June 15, 1991, eruption of Mount Pinatubo: Geochemistry, mineralogy, and petrology.

Gerlach, T.M., Westrich, H.R., and Symonds, R.B., this volume, Preeruption vapor in magma of the climactic Mount Pinatubo eruption: Source of the giant stratospheric sulfur dioxide cloud.

Geschwind, C.-H., and Rutherford, M.J., 1992, Cummingtonite and the evolution of the Mount St. Helens (Washington) magma system: An experimental study: Geology, v. 20, p. 1011–1014.

Ghiorso, M.S., and Sack, R.O., 1991, Fe-Ti oxide geothermometry: Thermodynamic formulation and estimation of intensive variables in silicic magmas: Contributions to Mineralogy and Petrology, v. 108, p. 485–510.

Gill, J.B., 1981, Orogenic andesites and plate tectonics: New York, Springer-Verlag, 391 p.

Hansen, J., Lacis, A., Lerner, J., Rind, D., Sato, M., Wilson, H., and McCormick, P., 1992a, Climate impact of Mount Pinatubo eruptionm [abs.]: Eos, Transactions, American Geophysical Union, v. 73, no. 43, p. 632.

Hansen, J., Lacis, A., Ruedy, R., and Sato, M., 1992b, Potential climate impact of Mount Pinatubo eruption: Geophysical Research Letters, v. 19, no. 2, p. 215–218.

Harlow, D.H., Power, J.A., Laguerta, E.P., Ambubuyog, G., White, R.A., and Hoblitt, R.P., this volume, Precursory seismicity and forecasting of the June 15, 1991, eruption of Mount Pinatubo.

Hattori, K., this volume, Occurrence and origin of sulfide and sulfate in the 1991 Mount Pinatubo eruption products.

Helgeson, H.C., Delany, J.M., Nesbitt, H.W., and Bird, D.K., 1978, Summary and critique of the thermodynamic properties of rock-forming minerals: American Journal of Science, v. 278A, p. 1–229.

Housh, T.B., and Luhr, J.F., 1991, Plagioclase-melt equilibria in hydrous systems: American Mineralogist, v. 76, p. 477–492.

Huebner, J.S., and Sato, M., 1970, The oxygen fugacity-temperature relationships of manganese oxide and nickel oxide buffers: American Mineralogist, v. 55, p. 934–952.

Imai, A., Listanco, E.L., and Fujii, T., this volume, Highly oxidized and sulfur-rich dacitic magma of Mount Pinatubo, Philippines: Implication for metallogenesis of porphyry copper mineralization in the western Luzon arc.

Katsui, Y., 1958, Groundmass anhydrite in the olivine basalt from the Rishiri volcano, Hokkaido: Journal of the Japanese Association of Mineralogy, Petrology, and Economic Geology, v. 42, p. 188–191.

Kerr, R.A., 1993, Pinatubo global cooling on target: Science, v. 259, p. 594.

Kilinc, A., Carmichael, I.S.E., Rivers, M.L., and Sack, R.O., 1983, The ferric-ferrous ratio of natural silicate liquids equilibrated in air: Contributions to Mineralogy and Petrology, v. 83, p. 136–140.

Kôzu, S., 1934, The great activity of Komagataké in 1929: Mineralogische und Petrographische Mitteilung, v. 45, p. 133–174.

Kwano, Y., 1948, A new occurrence of phenocrystic anhydrite crystals in the glassy rocks from Himeshima, Oita Prefecture, Japan: Geological Survey of Japan Report, v. 126, p. 1–7.

Lindstrom, D.J., and Korotev, R.L., 1982, TEABAGS: Computer programs for instrumental neutron activation analysis: Journal of Radioanalytical Chemistry, v. 70, p. 439–458.

Lofgren, G., 1980, Experimental studies on dynamic crystallization of silicate melts, in Hargraves, R.B., ed., Physics of magmatic processes: Princeton, N.J., Princeton University Press, p. 487–551.

Luhr, J.F., 1990, Experimental phase relations of water- and sulfur-saturated arc magmas and the 1982 eruptions of El Chichón Volcano: Journal of Petrology, v. 31, no. 5, p. 1071–1114.

Luhr, J.F., Carmichael, I.S.E., Varekamp, J.C., 1984, The 1982 eruptions of El Chichón Volcano, Chiapas, México: Mineralogy and petrology of the anhydrite-bearing pumices: Journal of Volcanology and Geothermal Research, v. 23, p. 69–108.

Matthews, S.J., Jones, A.P., and Bristow, C.S., 1992, A simple magma-mixing model for sulfur behaviour in calc-alkaline volcanic rocks: Mineralogical evidence from Mount Pinatubo 1991 eruption: Journal of the Geological Society of London, v. 149, p. 863–866.

McKeen, S.A., Liu, S.C., and Kiang, C.S., 1984, On the chemistry of stratospheric SO_2 from volcanic eruptions: Journal of Geophysical Research, v. 89, no. D3, p. 4873–4881.

McKibben, M.A., Eldridge, C.S., and Reyes, A.G., this volume, Sulfur isotopic systematics of the June 1991 Mount Pinatubo eruptions: A SHRIMP ion microprobe study.

Melson, W.G., Allan, J.F., Jerez, D.R., Nelen, J., Calvache, M.L., Williams, S.N., Fournelle, J., and Perfit, M., 1990, Water contents, temperatures, and diversity of the magmas of the catastrophic eruption of Nevado del Ruiz, Colombia, November 13, 1985: Journal of Volcanology and Geothermal Research, v. 41, p. 97–126.

Merzbacher, C., and Eggler, D.H., 1984, A magmatic geohygrometer: Application to Mount St. Helens and other dacitic magmas: Geology, v. 12, p. 587–590.

Nicholls, I.A., 1971, Calcareous inclusions in lavas and agglomerates of Santorini volcano: Contributions to Mineralogy and Petrology, v. 30, p. 261–276.

Pallister, J.S., Hoblitt, R.P., and Reyes, A.G., 1992, A basalt trigger for the 1991 eruptions of Pinatubo volcano?: Nature, v. 356, p. 426–428.

Pallister, J.S., Hoblitt, R.P., Meeker, G.P., Knight, R.J., and Siems, D.F., this volume, Magma mixing at Mount Pinatubo: Petrographic and chemical evidence from the 1991 deposits.

Pallister, J.S., Meeker, G.P., Newhall, C.G., Hoblitt, R.P., and Martinez, M., 1993, 30,000 years of the "same old stuff" at

Pinatubo [abs.]: Eos, Transactions, American Geophysical Union, v. 74, no. 43, p. 667–668.

Pearce, T.H., Russel, S.K., and Watson, I., 1987, Laser-interference and Normarski-interference imaging of zoning profiles in plagioclase phenocrysts from the May 18, 1980 eruption of Mount St. Helens, Washington: American Mineralogist, v. 72, p. 1131–1143.

Pier, J.G., Luhr, J.F., Podosek, F.A., and Aranda-Gómez, J.J., 1992, The La Breña–El Jagüey Maar Complex, Durango, Mexico: II. Petrology and geochemistry: Bulletin of Volcanology, v. 54, p. 405–428.

Rutherford, M.J., and Devine, J.D., this volume, Preeruption pressure-temperature conditions and volatiles in the 1991 dacitic magma of Mount Pinatubo.

Sack, R.O., Carmichael, I.S.E., Rivers, M., and Ghiorso, M.S., 1980, Ferric-ferrous equilibria in natural silicate liquids at 1 bar: Contributions to Mineralogy and Petrology, v. 75, p. 369–376.

Scott, W.E., Hoblitt, R.P., Daligdig, J.A., Besana, G., and Tubianosa, B.S., 1991; 15 June 1991 pyroclastic deposits at Mount Pinatubo, Philippines [abs.]: Eos, Transactions, American Geophysical Union, v. 72, no. 44, p. 61–62.

Smith, J.V., and Brown, W.L., 1988, Feldspar minerals, vol. 1: New York, Springer-Verlag, p. 436–445.

Stormer, J.C., Jr., 1983, The effects of recalculation on estimates of temperature and oxygen fugacity from analyses of multicomponent iron-titanium oxides: American Mineralogist, v. 68, p. 586–594.

Taneda, S., 1949, On the anhydrite from Himeshima lava: Journal of the Japanese Association of Mineralogy, Petrology, and Economic Geology, v. 33, p. 69–73.

Taylor, G.A., 1958, The 1951 eruption of Mount Lamington, Papua: Bureau of Mineral Resources Geology and Geophysics Bulletin, v. 38, p. 1–117.

Varekamp, J.C., Luhr, J.F., and Prestegaard, K.L., 1984, The 1982 eruptions of El Chichón volcano (Chiapas, México): Character of the eruptions, ash-fall deposits, and gas phase: Journal of Volcanology and Geothermal Research, v. 23, p. 39–68.

Westrich, H.R., and Gerlach, T.M., 1992, Magmatic gas source for the stratospheric SO_2 cloud from the June 15, 1991, eruption of Mount Pinatubo: Geology, v. 20, p. 867–870.

White, R.A., this volume, Precursory deep long-period earthquakes at Mount Pinatubo, Philippines: Spatio-temporal link to a basalt trigger.

Wilcox, R.E., 1954, Petrology of Parícutin volcano, Mexico: U.S. Geological Survey Bulletin 965–C, p. 281–349.

Yagi, K., Takeshita, H., and Oba, Y., 1972, Petrological study of the 1970 eruption of Akita-Komagatake volcano, Japan: Journal of the Faculty of Science, Hokkaido University, ser. IV, v. 15, p. 109–138.

Yoshiki, B., 1933, Activity of Akita-Komagatake volcano: Journal of the Japanese Association of Mineralogy, Petrology, and Economic Geology, v. 9, p. 153–160.

Zen, E-an, 1986, Aluminum enrichment in silicate melts by fractional crystallization: Some mineralogic and petrographic constraints: Journal of Petrology, v. 27, p. 1095–1117.

Preeruption Pressure-Temperature Conditions and Volatiles in the 1991 Dacitic Magma of Mount Pinatubo

By Malcolm J. Rutherford[1] and Joseph D. Devine[1]

ABSTRACT

The pumice erupted from Mount Pinatubo June 14–15 1991, is composed of approximately 80 to 90 percent white phenocryst-rich dacite and approximately 10 to 20 percent tan, finer grained, fragmental-looking dacite of the same bulk composition. The phenocryst phase assemblage of plagioclase (An_{34-66}), hornblende, cummingtonite, biotite, quartz, magnetite, anhydrite, ilmenite, and apatite occurs in both pumice types. Although there is both normal and reverse chemical zoning in plagioclase, and slight zoning in some hornblende crystals, the compositions of phenocrysts in contact with matrix glass (melt) are relatively uniform. Microprobe analyses of melt inclusions trapped in plagioclase, hornblende, quartz, and cummingtonite crystals indicate that they are all volatile-rich (H_2O=5.1 to 6.4 weight percent by the difference method), high-SiO_2 rhyolite glasses similar to the matrix glass on an anhydrous basis. Ion probe analyses confirm the H_2O content of 5.5 to 6.4 weight percent for these preeruption melts, and infrared spectroscopic analyses indicate that dissolved CO_2 is less than 20 parts per million. The average sulfur content of the melt inclusions ranges from 55 to 77±28 parts per million, which is 19 to 40 parts per million in excess of the matrix glass concentration.

Conditions in the preeruption white dacitic magma, as deduced from iron-titanium geothermometry and Al-in-hornblende geobarometry, are 780±10° Celsius, an oxygen fugacity of NNO+3, and a total pressure of 220±50 megapascals, which is equivalent to a 7 to 11 kilometer depth beneath Mount Pinatubo. This pressure is confirmed by experimentally determined stability limits for cummingtonite and cummingtonite + quartz at 780° Celsius in the Mount Pinatubo dacite composition. The preeruption magma is required to be very H_2O rich in order to stabilize cummingtonite, and to explain approximately 6.4 weight percent volatiles in the melt inclusions. The composition of the melt in equilibrium with the natural phenocrysts is reproduced at approximately 200 MPa under H_2O-saturated conditions and is not reproduced at higher temperature, or H_2O pressure; it is produced in 300-MPa experiments at X_{H_2O} in the fluid equal to 0.7. It is concluded that the 1991 dacitic magma of Mount Pinatubo was essentially volatile saturated with an H_2O-rich fluid just prior to the eruption.

INTRODUCTION

The 1991 eruptions of Mount Pinatubo began with a series of steam and ash emissions in April that were followed by lava dome extrusion in the period of June 7–12. An explosive eruption on June 12 produced a plinian column and a tephra deposit and destroyed parts of the dome (Pallister and others, this volume). These events culminated in a series of plinian and lateral blast eruptions during June 14–15, with a paroxysmal eruption occurring in the early hours of June 15. This was one of the largest explosive eruptions of the past century (Pinatubo Volcano Observatory Team, 1991).

The June 7–11 lava dome and the June 12 plinian eruptions involved magma of overall andesitic composition that was apparently a mixture of basaltic and dacitic components (Pallister and others, 1992; this volume). The June 14–15 eruptions, however, produced mainly dacitic pumice, approximately 80 to 90 percent of which is white and coarsely crystalline (~35 percent crystals with phenocrysts up to 5 mm). The remainder of the erupted pumice (10 to 20 percent) is tan and finer grained but is generally similar in composition and mineralogy to the coarsely crystalline dacite. The tan, fine-grained pumice contains crystals that are both smaller and more angular than those contained in the white pumice and are interpreted to be fragments of larger crystals. Pallister and others (1992) report finding inclusions of the white, phenocryst-rich variety in phenocryst-poor pumice and also find the two types mingled together in banded pumices. Fragments of both pumice types were found in geographically diverse tephra samples supplied to us for study.

The 1991 Pinatubo eruptions are of great interest in the field of petrology and magma dynamics as well as to the

[1]Department of Geological Sciences, Brown University, Providence, RI 02912.

atmospheric chemistry and climate dynamics communities. Petrological and volcanological interest stems from the many unique aspects of the magma system and the eruptions. For example, the dacitic magmas are very sulfur rich and oxidized compared to most calc-alkaline dacites (Bernard and others, 1991; Rutherford and Devine, 1991). The dacites also contain a very distinctive phenocryst phase assemblage including plagioclase, hornblende, cummingtonite, quartz, biotite, anhydrite, magnetite, and ilmenite, an assemblage which should contain sufficient petrogenetic information to reveal the preeruptive history of the magma. The presence of mixed basalt and dacite magmas in some samples and the possibility that the tan, fine-grained pumice is from the basalt-dacite thermal boundary layer (Pallister and others, 1992) suggest a rare opportunity for a pressure-temperature-time study of magma interaction in a well-defined, two-magma system. The origin of the crystal fragmentation in the tan pumice must also be explained. Finally, there are interesting questions concerning the origin of the sulfur in the magma and the origin of the huge mass of SO_2 injected into the stratosphere during the June 14–15 plinian eruptions. Several alternatives have been suggested for the sulfur in the magma, including build-up through crystal fractionation, vapor transfer from underlying basalt, and assimilation of overlying anhydrite and apatite-bearing hydrothermal vein deposits (Matthews and others, 1992; McKibben and others, 1992; Pallister and others, 1992), but convincing evidence does not yet exist for any of these models.

The June 14–15 Plinian-style eruptions injected an estimated 20 Mt of potentially climate-altering SO_2 into the stratosphere (Bluth and others, 1992), giving rise to predictions of its atmospheric and climatological importance. The origin of the atmospheric sulfur remains an interesting and unsolved problem, because it is clear that it was not dissolved in the preeruption melt phase of the erupted magma (Rutherford and Devine, 1991; Westrich and Gerlach, 1992). Westrich and Gerlach (1992) have suggested that the preeruption magma contained an excess sulfur-bearing volatile phase, a phase which could have carried the excess SO_2. A source involving degassing of associated basalt is suggested by Pallister and others (1992) and Matthews and others (1992). Another alternative is proposed by Baker and Rutherford (1992), who present evidence suggesting that the atmospheric sulfur may have formed by breakdown of anhydrite accompanying H_2O degassing of magma during its ascent.

The present study of crystal-melt-fluid equilibria in the phenocryst-rich dacite has been carried out in order to evaluate the various models of magma petrogenesis, magma degassing, and eruption processes proposed for the 1991 Mount Pinatubo eruption. Hydrothermal experiments have been carried out on the natural dacite to simulate phase equilibria in the preeruption magma storage region, and both natural and experimental samples have been analyzed by microbeam and spectroscopic techniques.

ANALYTICAL AND EXPERIMENTAL METHODS

ANALYTICAL METHODS

Pumice fragments were collected at sites 10 to 15 km south, east, and northeast of the volcanic center by T. Casadevall and E. Endo of the U.S. Geological Survey, who kindly supplied samples for this study. The major element compositions of phenocryst phases, glassy melt inclusions trapped in phenocrysts, and matrix glasses from the 1991 dacites (white and tan) were determined by using established electron microprobe techniques (Rutherford and Devine, 1988). Sodium-loss during microprobe analysis of glasses was monitored and accounted for online by using the method of Nielsen and Sigurdsson (1981). Crystals and coexisting melt (glass) in experimental charges were analyzed by using the same analytical routines.

Sulfur and chloride analyses were also obtained with the electron microprobe. The wavelength of sulfur X-rays produced by electron bombardment of geologic glasses depends on the oxidation state of sulfur in the sample (Carroll and Rutherford, 1988). Extended sulfur peak searches of Pinatubo glasses revealed that the dissolved sulfur exists predominantly in the oxidized state (S^{6+}), as would be expected in a highly oxidized melt (see below) in equilibrium with magmatic anhydrite. Sulfur analyses took into account this wavelength shift. The error bars on the glass sulfur analyses are large because the concentrations are close to the 30-ppm (parts per million) detection limit.

The H_2O content of a few plagioclase and quartz melt inclusions was determined by using the ion probe facility at Woods Hole Oceanographic Institution and methods developed by G. Layne. A new set of hydrothermally fused rhyolitic glasses was analyzed by FTIR and used as standards for these ion probe measurements. The small size of Mount Pinatubo melt inclusions generally prevented quantitative H_2O and CO_2 analyses in our micro-FTIR facility, but we were able to determine that the preeruption CO_2 content of the Mount Pinatubo melt was below the detection limit of about 20 ppm.

EXPERIMENTAL METHODS

Hydrothermal experiments have been carried out on a crushed sample of 1991 white pumice at various pressure, temperature, f_{H_2O} (water fugacity), and f_{O_2} (oxygen fugacity) conditions that bracket those suggested for the preeruption magma by geothermometry and geobarometry on the natural phenocryst assemblage. Samples consisting of 50 to 100 mg of powder and a weighed quantity of $H_2O \pm CO_2$ (as

AgC_2O_4) were loaded in Ag or AgPd tubes (one end sealed) and placed in 5-mm sealed Ag tubes along with a second tube containing a solid-phase oxygen buffer. $Re\text{-}ReO_2$ (ReO) and $MnO\text{-}Mn_3O_4$ (MNO) buffers were used. The experiments were carried out in Rene pressure vessels with an H_2O pressure medium. Pressures were recorded with a pressure transducer checked against a factory-calibrated Heise gauge and are considered accurate to ±1 MPa. The temperature in the sample position has been checked by melting experiments on NaCl and Au and is within ±5°C of the recorded temperature.

EXPERIMENTAL AND ANALYTICAL RESULTS

PETROGRAPHY

The June 15 phenocryst-rich white dacite, which represents about 80 to 90 percent of the erupted material, contains about 35 volume percent phenocrysts (<5 mm) in a vesiculated glassy matrix. The main phenocryst phases, plagioclase and hornblende, coexist with smaller amounts (<3 percent) of cummingtonite, biotite, quartz, magnetite, ilmenite, anhydrite, apatite, and zircon. The phenocryst phases are generally subhedral to euhedral in appearance, with the exceptions of quartz and biotite, which are more rounded. Many phenocrysts have been fractured (fig. 1), and the fragments have suffered different degrees of separation.

Plagioclase in the white pumice is concentrically and variably zoned, with cores as rich in anorthite as An_{66} and as low as An_{33}. The composition of plagioclase rims in contact with matrix glass averages $An_{41.1\pm4.2}$ (table 1), but values as low as An_{33} have been measured. Some of the variability on this average is undoubtedly the result of difficulty in analyzing thin rims on crystals with strong chemical zonation. Analyses of plagioclase rims in contact with glass in the tan pumice (An_{39}) show that they are similar in composition to the rims of white pumice plagioclase. In contrast, the average composition of the "cores" of crystal fragments in the tan pumice (An_{40}) are not as anorthite rich on average as the cores of white pumice phenocrysts (~An_{46}). This is considered to be a sampling artifact. Specifically, the "cores" of small, tan pumice plagioclase fragments that have at least one planar face (that is, rim) against glass are themselves near-rim samples of formerly large crystals.

Pleochroic green to tan hornblende in the white dacite is relatively uniform in composition (table 1); early-crystallizing or xenocrystic cores can be recognized by high Al_2O_3 and MgO contents. Cummingtonite typically occurs as euhedral overgrowths on some hornblende crystal faces, and biotite is commonly present as euhedral to anhedral inclusions. Magnetite (<1 mm) typically contains apatite

and rare chalcopyrite inclusions and may be intergrown with ilmenite. All grains analyzed for the purpose of oxygen barometry were pairs in contact. Subhedral anhydrite phenocrysts occur either alone in contact with melt or occasionally in aggregates with apatite. Chalcopyrite occurs exclusively as inclusions in other minerals and was not observed in contact with glass in the groundmass.

Both hornblende and plagioclase contain abundant glassy melt inclusions whose average size is small (<30 μm) compared with other tephra we have studied. Overnight leaching of plagioclase mineral separates in HBF_4 reveals that many plagioclase melt inclusions are intersected by minute cracks (which are etched by the acid). Melt inclusions in hornblende may also be intersected by cracks or cleavage planes that potentially could have resulted in the loss of some of the original volatile content from melt inclusions during eruption. Melt-inclusion analyses must therefore be interpreted in light of this possibility.

MELT INCLUSIONS VERSUS MATRIX GLASS

The major and volatile element compositions of melt inclusions and matrix glasses have been determined for samples of the white June 15 dacite pumice from each of the three sites sampled. No differences in the glass (or mineral) chemistry have been observed between the different sites or from sample to sample beyond that which exists in a given sample. When the analyses are normalized to an anhydrous basis, the average hornblende, plagioclase, and quartz melt-inclusion compositions are essentially identical to the high-SiO_2 rhyolite matrix glass (table 2). Although inclusions in cummingtonite are rare and small, and therefore difficult to analyze, they also have the same high-SiO_2 rhyolite chemical composition.

The sulfur and chlorine contents of the melt inclusions and matrix glass, also determined by electron microprobe, show very small differences between the different melt-inclusion populations and the matrix glass. In fact, the chlorine content of the matrix glass and melt inclusions is equal (1,200±120 ppm) within the limits of the electron microprobe analyses, although the thin strands of matrix glass are difficult to analyze. The average sulfur abundance in hornblende melt inclusions (77±29 ppm) is similar to that determined by Westrich and Gerlach (1992) and about 40 ppm above the average matrix glass composition (36±28 ppm). The average sulfur content of plagioclase melt inclusions is somewhat lower (55±23 ppm). A comparison of melt-inclusion sulfur content with variations in the host hornblende composition indicates that a few anomalous inclusions in terms of sulfur abundance are associated with compositionally anomalous hornblende. The cores of some hornblende phenocrysts have higher MgO (and lower K_2O) than the rims; melt inclusions in these cores contain higher sulfur contents. The higher MgO hornblendes may be xenocrysts;

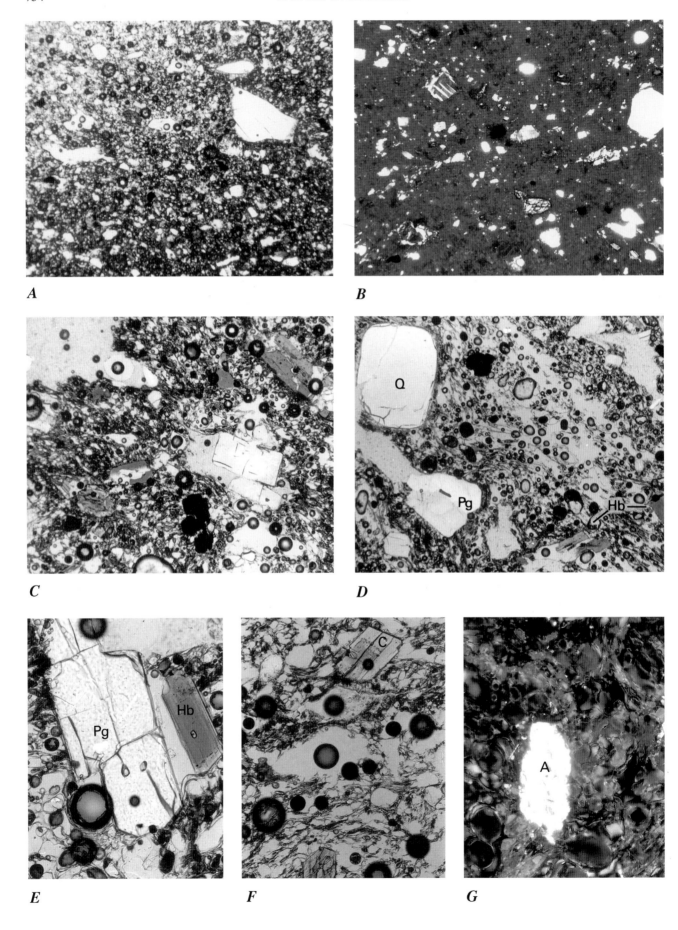

A

B

C

D

E

F

G

they chemically resemble hornblendes in basalt inclusions in the dacite (Pallister and others, 1992; this volume). These MgO-rich hornblendes probably crystallized at >780°C on the basis of the higher melt inclusion sulfur abundance (~80 to 100 ppm) and available data on sulfur solubility (see next section). It should be noted, however, that most melt inclusions contain a high-SiO_2 melt similar in composition to the matrix glass, and the sulfur abundance in these inclusions is only 20 to 40 ppm in excess of that present in the matrix glass.

The H_2O and CO_2 contents of melt inclusions were determined as accurately as possible (given their small size) by using a combination of infrared spectroscopy (FTIR), ion probe, and electron microprobe techniques. No doubly polished sections of plagioclase or quartz melt inclusions sufficiently large for quantitative FTIR analyses of H_2O and CO_2 were obtained. However, infrared spectroscopy of inclusion-bearing quartz and plagioclase showed CO_2 abundances below detection (<20 ppm) and H_2O contents of >5.0 weight percent. Using a new set of FTIR-calibrated standards, ion probe analyses of melt inclusions in plagioclase and quartz yielded H_2O contents of 6.4 to 6.6 and 7.0 weight percent, respectively. The volatile content ($H_2O \pm CO_2$) of the trapped melts estimated by these techniques is consistent with the electron microprobe "volatiles by difference" method within the error limits of microprobe analyses (table 2); they indicate an H_2O abundance of 5.1 to 6.4 weight percent. The lower H_2O abundances may reflect some H_2O loss during the eruption. It is concluded that samples of melt containing ~6.4 weight percent H_2O, ~1,200 ppm chlorine and ~70 ppm sulfur were trapped by growing hornblende, plagioclase, and cummingtonite phenocrysts in the 1991 Mount Pinatubo magma storage region prior to the eruption. The similarity in major and minor element chemistry, particularly K_2O, of the melt inclusions and matrix glasses indicates that no significant crystallization occurred after melt entrapment. The apparent lack of chlorine loss from the interstitial melt (matrix glass) as the vapor phase grew during the eruption may be explained, at least partly, by the large decrease in D_{Cl} (Cl concentration

in melt/Cl concentration in vapor) that occurs with H_2O dilution of the vapor phase. Significantly lower amounts (30 percent) of Li, Be, and B and a slightly higher H_2O in the quartz melt inclusions, as determined by ion microprobe, suggest that these melts may have been trapped in a separate magmatic stage compared to melts trapped by hornblende and plagioclase. On the basis of the rounded quartz crystal outlines and the slightly lower MgO, FeO and CaO in the trapped melts, the quartz growth took place in a more evolved melt that was part of an earlier more crystal-rich magma.

FE-TI OXIDES: TEMPERATURES AND OXYGEN FUGACITIES

The average composition of magnetite and ilmenite obtained from analyses of 12 crystal pairs in contact with melt are given in table 1. All analyses cluster tightly around the average, as indicated by the low standard deviations. The compositions of each oxide pair yielded a temperature of 810°C ±10° when the geothermometer calibration of Andersen and Lindsley (1988) and the solution model of Stormer (1983) were used. The use of other solution models (Andersen and Lindsley, 1988) produces less than 10°C variation in the calculated temperatures. The average $\log f_{O_2}$ indicated by these oxides is −11.1, three log units above the Ni+NiO oxygen buffer (NNO+3.0). This solution is outside the limits of the Andersen and Lindsley calibration, however, so these temperature and f_{O_2} estimates need to be confirmed. In a study of iron-titanium oxide temperatures and cummingtonite stability, Geschwind and Rutherford (1992) found that in this f_{O_2} range (NNO+3), the Andersen and Lindsley (1988) algorithm yields oxide temperatures that are high by 30°C. This is now confirmed by the results of experiments in which the Re+ReO₂ oxygen buffer is used (table 3). This buffer, which is calculated to lie about 2 to 3 log units above NNO (Pownceby and O'Neill, 1991), was used in this study, and the iron-titanium oxides from some experiments have been analyzed. When the Andersen and Lindsley (1988) algorithm is applied, these oxide analyses yield temperatures 30±5°C above the experimental temperature at a $\log f_{O_2}$ of −10 to −11, depending on the temperature (fig. 2). On the basis of these data, it is concluded that the white pumice from Mount Pinatubo was at 780±10°C just prior to eruption. The oxides of the tan, crystal-poor dacitic magma yield identical temperature-f_{O_2} conditions.

An independent estimate of magmatic temperature can be obtained from sulfur concentrations in the melt inclusions and experimental sulfur-solubility data if it is assumed that the silicate melt was sulfate saturated when the melt inclusions were trapped. This appears to be a reasonable assumption, given the compositional similarity of the melt inclusion and matrix glasses. Figure 3, modified from Carroll and Rutherford (1987), shows that sulfur solubility

Table 1. Phenocrysts in white dacite from the 1991 eruption of Mount Pinatubo.[1]

	Plagioclase[1]		Hornblende	
	Core	Rim	Cores	Rims
SiO_2	56.80(258)	58.19(146)	47.37(139)	47.76(103)
TiO_2	n.d.	n.d.	.99(15)	.89(11)
Al_2O_3	27.29(177)	26.48(93)	8.03(115)	7.81(74)
FeO^*	.19(8)	.19(4)	13.31(56)	12.98(75)
MgO	n.d.	n.d.	15.32(92)	15.51(89)
CaO	9.00(195)	8.09(85)	10.75(26)	10.80(33)
Na_2O	5.81(98)	5.81(98)	1.33(19)	1.27(12)
K_2O	.24(8)	.30(10)	.28(11)	.27(7)
MnO	n.d.	n.d.	.53(7)	.46(6)
Cr_2O_3	n.d.	n.d.	n.d.	n.d.
Total	99.23	99.47	97.90	97.76
No. analyses	19	19	10	26

	Cummingtonite	Biotite[2]	Magnetite[3]	Ilmenite[4]
SiO_2	54.03(50)	37.77(35)	n.d.	n.d.
TiO_2	.21(7)	3.52(16)	4.29(12)	28.09(100)
Al_2O_3	2.06(35)	15.40(27)	1.90(3)	.37(6)
FeO^*	17.17(25)	14.03(28)	87.36(34)	65.61(87)
MgO	21.12(26)	16.59(19)	1.08(2)	1.04(10)
CaO	1.99(54)	.09(4)	n.d.	n.d.
Na_2O	.32(10)	.70(9)	n.d.	n.d.
K_2O	.03(2)	8.21(24)	n.d.	n.d.
MnO	.88(9)	.14(4)	0.47(4)	.27(5)
Cr_2O_3	n.d.	n.d.	.22(5)	.09(3)
Total	97.81	96.45	95.32	95.47
No. analyses	4	13	12	12

[1] Numbers in parentheses are estimated standard deviations in terms of least units cited; thus, 56.80(258) indicates a standard deviation of 2.58 weight percent. n.d., not determined.

[2] The average Al_2O_3 content of amphibole host mineral in contact with biotite mineral inclusions is 9.27(105) weight percent (8 analyses).

[3] Fe_2O_3 = 59.72 weight percent and FeO = 33.63 weight percent (using the solution model of Stormer, 1983).

[4] Fe_2O_3 = 47.21 weight percent and FeO = 23.13 weight percent (using the solution model of Stormer, 1983).

increases significantly with temperature and f_{O_2} above 800°C. Experiments at 800°C yield an estimated sulfur-solubility of ~60±20 ppm in the anhydrite-saturated Mount Pinatubo melts over a range of f_{O_2} from NNO+1 to MNO. If the Mount Pinatubo magma melt inclusions had equilibrated with anhydrite at temperatures above 800°C, they should contain measurably higher sulfur contents than those observed. Melt inclusions with large vapor bubbles were purposely avoided because of the presumed secondary leak-

age problem; therefore, no variation in the sulfur content was observed that could be attributed to this leakage.

HORNBLENDE CHEMISTRY AND GEOBAROMETRY

The composition of hornblende in the phenocryst-rich dacite of Mount Pinatubo (hereafter referred to as Mount

Table 2. Melt inclusions and matrix glasses in white dacite from the 1991 eruption of Mount Pinatubo.[1]

	Plagioclase[1] melt inclusions[2]	Hornblende melt inclusions	Cummingtonite melt inclusions	Quartz melt inclusions	Matrix glass
SiO_2	72.46(91)	72.63(69)	73.66(90)	73.60(115)	76.44(22)
TiO_2	.14(4)	.12(4)	.13(2)	.13(2)	.14(4)
Al_2O_3	12.19(32)	12.27(28)	12.01(17)	11.90(14)	12.63(8)
FeO^*	.78(16)	.97(20)	.99(11)	.70(6)	.79(10)
MgO	.21(5)	.20(2)	.19(4)	.18(2)	.23(3)
CaO	1.09(11)	1.35(11)	1.22(4)	1.03(8)	1.28(5)
Na_2O	3.87(33)	3.73(15)	3.92(13)	3.79(20)	4.10(13)
K_2O	2.84(9)	2.83(18)	2.71(15)	2.83(5)	2.94(5)
MnO	.06(3)	.07(5)	.06(2)	.05(4)	.05(3)
Total	93.64	94.17	94.89	94.21	398.60
Vol. by dif., wt%	6.36	5.83	5.11	5.79	trace
No. analyses	15	5	3	7	6
S, ppm	55(23)	77(29)	n.d.	68(21)	36(28)
Cl, ppm	1279(100)	1178(148)			1202(122)
No. analyses	24	25		9	35
SiO_2[4]	77.38	77.13	77.63	78.12	77.50
TiO_2	.15	.13	.14	.14	.14
Al_2O_3	13.02	13.03	12.66	12.63	12.81
FeO^*	.83	1.03	1.04	.74	.80
MgO	.22	.21	.20	.19	.24
CaO	1.16	1.43	1.29	1.09	1.29
Na_2O	4.13	3.96	4.13	4.02	4.16
K_2O	3.03	3.01	2.86	3.00	2.98

[1] Analyses by electron microprobe; FeO^*, total iron as FeO; n.d., not determined. Sodium loss during analysis accounted for online by using the method of Nielsen and Sigurdsson (1981).

[2] Numbers in parentheses are standard deviations (see table 1).

[3] Low analytical total due to the fact that glass shards analyzed were extremely thin; in some cases, the beam penetrated the underlying epoxy mounting medium, and low totals have resulted. The volatile content of matrix glasses should be negligible (a few tenths of a weight percent H_2O).

[4] Glass analyses normalized to 100 weight percent (anhydrous basis).

Table 3. Hydrothermal experiments on 1991 dacite of Mount Pinatubo.

Run No.	Starting[1] Material	T (°C)	P (MPa)	X_{H_2O} in fluid	Time in days	Products[2]
				Re + ReO₂-Buffered Experiments		
502	1	925	220	1.0	1.5	G
498	1	900	220	1.0	2	G,Opx
476	1	860	90	1.0	4	P,Opx,M,I,G,(Hb)
506	1	870	220	1.0	3	Hb
436	1	850	160	1.0	5	Hb,Pg,Opx,M,I,G
474a	1	840	220	1.0	6	Hb,Pg,Opx,M,I,G
477	1	840	220	1.0	4	Hb,Pg,Opx,M,I,G
500	1	840	220	0.5	8	Hb,Pg,Opx,M,I,G
473	2	810	220	1.0	10	Hb,Pg,Opx,Bi,M,I,G
493a	1	810	220	0.75	14	Hb,Pg,Opx,(C),Bi,Q,M,I,G
493b	2	810	220	0.75	14	Hb,Pg,Opx,Bi,Q,M,I,G
492a	2	810	220	0.50	14	Hb,Pg,Opx,Bi,Q,M,I,G
492b	1	810	220	0.50	14	Hb,Pg,Opx,C,Bi,M,I,G
503	2	780	220	1.0	22	Hb,Pg,C,Bi,Q,M,I,G
478	2	780	220	0.75	21	Hb,Pg,Opx,(C),Bi,Q,M,I,G
479	1	780	220	0.5	25	Hb,Pg,Opx,Bi,Q,M,I,G
485	2	780	200	1.0	25	Hb,Pg,(Opx),C,Bi,Q,M,I,G
483a	2	780	300	1.0	20	Pg,C,(Hb),Bi,Q,M,I,G
483b	2	780	300	1.0	20	Pg,C,(Hb),Bi,Q,M,I,G
481a	1	780	390	1.0	6	Hb,Pg,Bi,M,I,G
481b	2	780	390	1.0	6	Hb,Pg,Bi,M,I,G
512	2	780	300	0.84	7	Hb,Pg,C,Bi,M,I,G
516a	1	780	300	0.7	8	Hb,Pg,C,M,I,G
516b	2	780	300	0.7	8	Hb,Pg,Opx,C,M,I,G
505	1	780	350	1.0	6	Hb,Pg,C,Bi,M,I,G
480	1	760	390	1.0	7	Hb,Pg,Bi,M,I,G
				MnO+Mn₃O₄-Buffered Experiments		
466	1	820	220	1.0	10	Pg,Hb,Opx,Bi,(Q),M,I,G
440a	2	800	200	1.0	10	Pg,Hb,Opx,Bi,Q,M,I,G
440b	1	800	200	1.0	10	Pg,Hb,(C),Opx,Bi,Q,M,I,G
442a	2	800	300	1.0	4	Pg,Hb,Opx,Bi,Q,M,I,G
442b	1	800	300	1.0	4	Pg,Hb,C,Opx,Bi,Q,M,I,G
469	1	780	220	1.0	14	Pg,Hb,C,Bi,Q,M,I,G
497a	2	780	220	0.5	16	Pg,Hb,(C),Opx,Bi,Q,M,I,G
497b	1	780	220	0.5	16	Pg,Hb,C,Opx,Bi,Q,M,I,G

[1] Starting material: 1=disaggregated white, crystal-rich pumice from Mount Pinatubo; 2=white pumice annealed under high temperature (860°C) at an H_2O pressure of 100–130 MPa for 4–6 days.

[2] Product abbreviations are: Pg=plagioclase, Hb=hornblende, C=cummingtonite, Bi=biotite, Opx=low-Ca pyroxene, Q=quartz, M=magnetite, I=ilmenite, and G=glass. Parentheses indicate a phase present only in trace amounts.

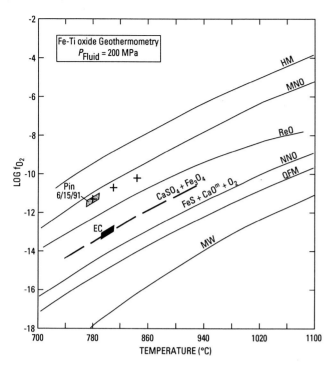

Figure 2. Oxygen fugacity-temperature diagram showing the results from iron-titanium oxide geothermometry (Andersen and Lindsley, 1988, algorithm) for the white pumice (Pin) from the 1991 eruption of Mount Pinatubo. The +'s show the results when the algorithm was used for oxides in the Re+ReO$_2$-buffered experiments at 780, 810, and 840°C. As discussed in the text, a –30°C correction must be applied to algorithm results in order to reproduce experimental temperatures under these oxidizing conditions. ReO is the calculated position of the Re-ReO$_2$ buffer. The 1982 El Chichón (EC) temperature and the FeS-CaSO$_4$ equilibrium are from Carroll and Rutherford (1987). HM, hematite-magnetite; MNO, manganite-hausmanite; NNO, nickel-nickel oxide; QFM, quartz-fayalite-magnetite; WM, wustite-magnetite, (all are standard oxygen buffers).

Pinatubo dacite) is slightly variable, with Mg# (100×Mg/Mg+Fe atomic) varying from 69 to 62 and Al$_2$O$_3$ contents varying from 7.0 to 11.0 weight percent. The hornblende in contact with melt is relatively uniform in composition however, with an average Mg# of 66±1 and an Al$_2$O$_3$ content of 7.9±0.8 weight percent. The variable core compositions are not necessarily representative of earlier high-temperature crystallization; cores have both higher and lower Mg# than the rims. Some of these crystal cores with atypically high Al$_2$O$_3$ contents and irregular zoning may be xenocrystic, as indicated above. In contrast, cummingtonite is present as rims on many hornblendes and as small euhedral crystals in contact with melt, and is quite uniform in composition (table 1).

Using the Johnson and Rutherford (1989) calibration of the Al-in-hornblende geobarometer, the hornblende in equilibrium with melt in the phenocryst-rich 1991 Mount Pinatubo dacite (7.8±0.9 weight percent Al$_2$O$_3$) indicates a pressure of 220±50 MPa. This calibration is used because the experiments were done at the same temperature (760 to 780°C) as that determined for the white Mount Pinatubo dacites; the Schmidt (1992) calibration at 650 to 700°C yields a pressure of ~320 MPa. Alkali feldspar, which is rare in the Pinatubo dacite, is required in the geobarometer phenocryst-melt assemblage, however, so these pressure estimates also need to be confirmed. Unfortunately, the experiments performed as part of this project have not yielded new Al-in-hornblende data because of the hornblende composition variability in the natural starting material and because of cummingtonite overgrowth interference with hornblende-melt reequilibration.

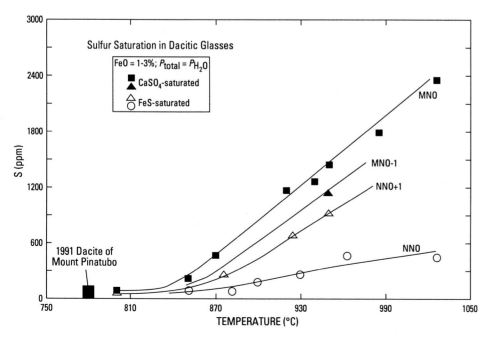

Figure 3. Sulfur solubility versus temperatures for dacitic composition melts modified from Carroll and Rutherford (1987). New experimental data are plotted at 800°C. MNO and NNO oxygen buffer curves as in figure 2; NNO+1, one log unit above the NNO buffer.

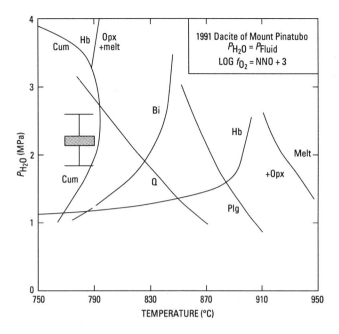

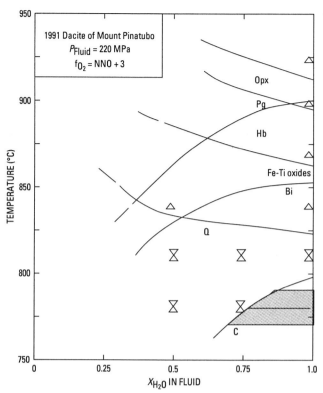

Figure 4. Pressure-temperature diagram for the dacite composition of Mount Pinatubo under H_2O-saturated conditions. Triangular symbols show the location of experiments (table 3) and point in the direction of approach to the final conditions; filled symbols are used where phenocryst phases found in the natural pumice are stable. The curved phase boundaries indicate the appearance of a phase in this dacitic composition. The shaded symbols at 780°C and 220 MPa mark (with error bars) the conditions in the preeruptive 1991 dacitic magma from iron-titanium oxide geothermometry and Al-in-hornblende geobarometry (see text). Hb, hornblende; Opx, orthopyroxene; Pg, plagioclase; Bi, biotite; Q, quartz; C, cummingtonite. NNO+3, 3 log units above the Ni+NiO oxygen buffer.

Figure 5. Temperature-X_{H_2O} (mole fraction H_2O in the fluid) diagram for the 1991 dacite composition of Mount Pinatubo at 220 MPa and a NNO+3 (Re+ReO$_2$ buffer) oxygen fugacity. Symbols as in figure 4. The shaded area indicates where the Pinatubo phenocryst phase assemblage is stable at 780°C.

EXPERIMENTAL PHASE EQUILIBRIA

Phase equilibrium experiments were performed on samples of the white phenocryst-rich pumice of Mount Pinatubo (hereafter referred to as Mount Pinatubo pumice) to determine the water pressure required to stabilize the natural phenocryst assemblage. Most experiments were done by using the Re+ReO$_2$ oxygen buffer, but MNO was also used to explore the effect of small changes in f_{O_2} under the very oxidizing conditions indicated by the natural iron-titanium oxides. Figure 4 shows the pressure-temperature dacite phase equilibria determined for water-saturated conditions. The iron-titanium oxide temperature and total pressure from Al-in-hornblende geobarometry are indicated. Water-saturated conditions could have existed in the Mount Pinatubo preeruption magma since, at 220 MPa, the rhyolitic composition melt would contain 6.4 weight percent H_2O (Fogel, 1989; Silver and others, 1989), equivalent to the H_2O content determined from melt-inclusion analyses. Figure 4 indicates that the natural phenocryst phase assemblage is stable between 150 and 320 MPa at 780°C. The

stability of cummingtonite is particularly sensitive to water pressure (P_{H_2O}) and temperature changes in this general pressure-temperature range; it is replaced by low-CaO pyroxene above 800°C and at pressures above 350 MPa. These are the conditions of maximum cummingtonite stability. Quartz disappears from the Mount Pinatubo dacite phase assemblage at water pressures above 300 MPa at 780°C. Thus, the Mount Pinatubo dacite phenocryst assemblage could not have existed above 800°C under any set of P_{total} or P_{H_2O} conditions in the preeruption magma, and the maximum pressure range possible in a 780°C magma reservoir is from 150 to 300 MPa under water-saturated conditions.

The stability of cummingtonite at any total pressure is further restricted if the magma is not water saturated. Figure 5 shows the results of water-undersaturated experiments at a total pressure of 220 MPa. At this pressure and 780°C, cummingtonite is stable in melts only where the X_{H_2O} (molar) in the coexisting fluid is >0.75. The results of H_2O-undersaturated experiments at 300 MPa (table 3) show that cummingtonite is only stable when the X_{H_2O} in the fluid is >0.70.

RESIDUAL MELT COMPOSITIONS

The preeruption conditions in the white Mount Pinatubo dacitic magma as indicated by the above methods can be tested by comparing experimental glass compositions with the matrix and melt-inclusion glasses. This is a sensitive test because the melt composition responds rapidly to changes in intensive parameters, coming to equilibrium as indicated by the similar compositions in melting and crystallization experiments. Equilibrium melt-phase compositions appear to be achieved by reaction with crystal rims, even though equilibrium among the crystalline phases is not totally achieved, as indicated by some residual chemical zonation, or is difficult to demonstrate. However, because of the highly variable zoning in the natural Mount Pinatubo plagioclase starting material and because of the slow reequilibration rates of these crystals at 780°C, it has not yet been possible to determine the equilibrium experimental plagioclase compositions with sufficient confidence that they can be used to constrain conditions in the preeruption magma.

The glasses in ReO-buffered experiments carried out under different pressure-temperature and P_{H_2O} conditions are compared with Mount Pinatubo melt-inclusion and matrix glasses in figure 6. These plots of Al_2O_3 versus SiO_2 illustrate the fact that an H_2O- saturated experiment at ~200 to 220 MPa reproduces a melt composition equivalent to the tight cluster of natural glasses. Glasses produced in higher P_{H_2O} experiments are not similar to the natural glasses. Similarly, the experimental glass composition moves away from the natural glass compositions with increases in temperature (constant P_{total} and X_{H_2O}) and decreases in X_{H_2O} at 220 MPa. The latter is somewhat surprising in the light of previous studies of X_{H_2O} on crystallization but is tentatively attributed to destabilization of cummingtonite in our water-undersaturated experiments. The glass produced at 300 MPa, 780°C, and X_{H_2O} in the fluid=0.70 is an excellent match with the melt-inclusion glasses, including the dissolved volatile content, although cummingtonite appears incipiently unstable in these experiments. Thus, the glass data suggest that the Mount Pinatubo magma last equilibrated at 220 MPa under H_2O-saturated conditions or at a pressure between 220 and 300 MPa; the X_{H_2O} in the fluid would decrease as the pressure increases, reaching 0.70 at 300 MPa.

DISCUSSION

The phenocryst-melt assemblage in the white dacitic pumice of Mount Pinatubo is interpreted to be an equilibrium assemblage. The compositions of the phenocryst phases, including the iron-titanium oxides, cummingtonite, and the *rims* of plagioclase and hornblende, are uniform within a sample and from sample to sample. In addition, the matrix glass is uniform in composition throughout these phenocryst-rich pumice samples and is identical to the melt-inclusion glasses on an anhydrous basis. As a result, it is possible to recognize and determine the equilibrium conditions in the preeruption phenocryst-rich dacite. The temperature of the magma reservoir was 780°C±10° according to iron-titanium oxide geothermometry (table 1, fig. 2), where the error represents a 1 sigma standard deviation on the oxide analyses. Given that this temperature is also supported by the low sulfur content of melt inclusions compared to experimental sulfur-solubility data (fig. 3), and the thermal limit on cummingtonite stability (Geschwind and Rutherford, 1992; fig. 4), this temperature is considered to be well constrained. The magma was also very oxidized (NNO+3 log units), sufficiently so to stabilize anhydrite.

The total pressure in the 1991 Mount Pinatubo magma storage zone is estimated to have been 220±50 MPa (Al-in-hornblende geobarometry), or approximately 7 to 11 km depth. This pressure is supported by cummingtonite stability data (fig. 5), which show that the total pressure had to be in the 150 to 350 MPa range at 780°C, assuming the magma was water saturated, and is also supported by the comparison between natural glass and the experimental glass compositions at 200 to 220 MPa. If the magma was water undersaturated, ($P_{H_2O}<P_{total}$), then the maximum pressure for cummingtonite would be less than 350 MPa. The residual melt compositions produced experimentally indicate that a range of conditions between 220 MPa, H_2O saturated, and 280 MPa with X_{H_2O}=0.70 are possible; the lower pressure estimate is favored on the basis of the Al-in-hornblende barometry and because there is no detectable CO_2 in the melt trapped during the most recent phenocryst growth event (hornblende + plagioclase + cummingtonite growth). Assuming the preeruption magma was near volatile saturation, a volatile species other than SO_2 is necessary to lower the X_{H_2O} in the fluid if the total pressure in the magma is above 220 MPa; CO_2 is the only reasonable candidate, and apparently it was not present. The petrological estimate of magma reservoir pressure/depth compares well with seismic activity following the June 15 eruption, which appears to outline an aseismic zone (fig. 7) below about 7 km (Wolfe, 1992; Mori, Eberhart-Phillips, and Harlow, this volume). The 7 to 11 km depth estimate for the dacite magma storage region also compares well with a large low-velocity seismic zone identified as a magma reservoir by Mori, Eberhart-Phillips, and Harlow (this volume).

A complete description of conditions in the preeruption magma storage region requires an estimate of the P_{H_2O} necessary to stabilize the observed phenocryst-melt assemblage, and if P_{H_2O} is less than P_{total}, an indication whether other volatile species are sufficiently abundant to make P_{fluid} equal to P_{total} is needed. The ion probe analyses of melt inclusions trapped in phenocrysts of the white pumice indicate an H_2O content of 6.4 weight percent. As discussed in the previous section, the microprobe "volatiles by

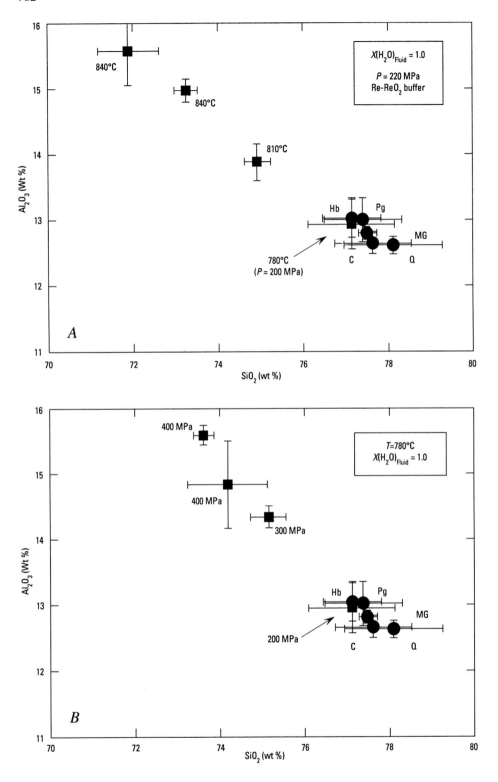

Figure 6. Comparisons of natural and experimental glass compositions on a plot of Al_2O_3 versus SiO_2. Melt inclusion (indicated by host mineral abbreviations as in figure 4) and matrix glass compositions (MG) are from table 2. The experiments, indicated by the experimental temperature in (*A*), the pressure in MPa (*B*), or the X_{H_2O} in the coexisting fluid (*C*), are listed in table 3.

difference" estimates of 5.1 to 6.4 weight percent dissolved H_2O are consistent with the ion probe data. Interestingly, the Mount Pinatubo plagioclase rim and matrix glass compositions used with the Housh and Luhr (1991) plagioclase-melt model yield H_2O estimates of 6.4 and 5.9 weight percent using their AB and AN models, respectively. If the preeruption melt was saturated with H_2O at 220 MPa, it would contain ~6.4 weight percent H_2O according to water solubility experiments (Fogel, 1989; Silver and others, 1989). This is sufficiently close to the measured concentration to suggest that the magma may have been fluid saturated. Certainly if there was an excess fluid (vapor) phase, it

Figure 6.—Continued.

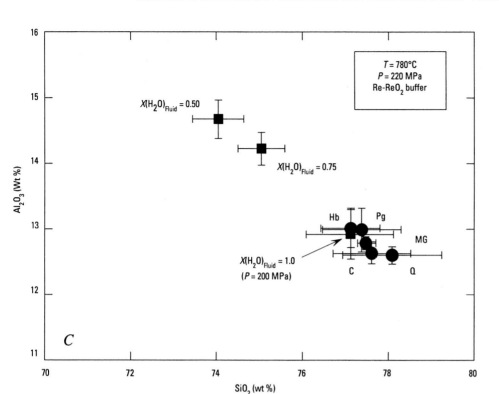

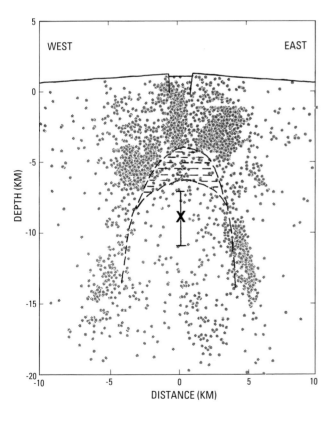

← **Figure 7.** A cross section through Mount Pinatubo (modified after Pallister and others, this volume, and Mori, Eberhart-Phillips, and Harlow, this volume) showing seismicity from May 8 to August 19, 1991, projected on to the plane of the east-west cross section. The X marks the 7 to 11 km source depth for the June 14–15 magma storage region; on the basis of the petrology of the eruption products; on the basis of the geobarometry, the horizontally ruled portion of the aseismic magma storage zone appears not to have erupted on June 14–15.

had to be very H_2O rich. This conclusion, required by the similarity of measured and calculated H_2O-saturation abundances and by the cummingtonite stability at 780°C, is supported by analytical and solubility data for CO_2 and SO_2. Less than detectable CO_2 concentrations in melt inclusions require a P_{CO_2} of <25 MPa (Fogel and Rutherford, 1990). Similarly, experimental data on SO_2 concentrations in H_2O-rich fluids coexisting with anhydrite suggest that the P_{SO_2} would have been ~11 MPa in a 220-MPa pressure fluid (Baker and Rutherford, 1992). Thus, we conclude here that the Mount Pinatubo preeruption dacitic magma was very close to saturation, if not in fact saturated, with an H_2O-rich fluid at a depth equivalent to 220 MPa.

The above conclusion that the Mount Pinatubo dacitic magma was volatile saturated suggests that the huge excess of SO_2 injected into the stratosphere during the June 14–15 eruptions might be explained by the release of an H_2O-rich, sulfur-bearing vapor phase that coexisted with the preeruption magma (Westrich and Gerlach, 1992). The 20 Mt of SO_2 injected into the stratosphere (Bluth and others, 1992) is far in excess of the SO_2 lost by melt (melt-inclusion sulfur minus matrix-glass sulfur is <40 ppm) in the approximately 5 km³ (DRE) of erupted Mount Pinatubo magma (~0.52 Mt of SO_2). While a vapor phase in the preeruption dacitic magma is possible and even likely, it may not be the main source of the excess atmospheric sulfur for several reasons. The excess sulfur injected into the stratosphere is equivalent to ~1,000 ppm sulfur (2,000 ppm SO_2) in the 5 km³ of erupted magma. If a magma density of 2.30 g/cm³ and an SO_2 vapor density of 0.7 g/cm³ (at 220 MPa) are assumed, 2,000 ppm of SO_2 is equivalent to ~0.6 percent of the magma by volume. But the vapor at depth is approximately 99 percent H_2O according to recent analyses of the Baker and Rutherford (1992) experiments, which means that the magma would have contained >20 volume percent vapor (vapor bubbles) at depth. It seems likely that this gas-charged magma should have been erupted during the June 7–13 period when there was both extrusion and explosive eruption of andesite formed by the very recent mixing (Rutherford and others, 1993) of dacitic and basaltic magma beneath Mount Pinatubo. It is also difficult to produce the H_2O for a water-rich fluid phase of this size. The erupted magma would have had to suffer approximately 15 volume percent crystallization after reaching H_2O saturation in order to generate the excess volatiles. Similarly, other magma(s) at depth would only release their H_2O as they crystallized beyond the point of H_2O saturation. An alternate source for the excess sulfur is suggested by the experiments of Baker and Rutherford (1992), which indicate that anhydrite in an H_2O-rich magma would break down during ascent from depth. If degassing in the Mount Pinatubo dacitic magma began more than 3 h prior to the eruption, at least some of the excess sulfur must have come from anhydrite breakdown in the Mount Pinatubo magma during the eruption.

The origin of the sulfur in the Mount Pinatubo magma system is a problem not directly addressed by this experimental study, but it may be related to the highly oxidized nature of the preeruption magma. The iron-titanium oxide phenocrysts indicate that the magma is among the most oxidized (NNO+3 log units) of calc-alkaline magmas that have been analyzed (for example see Carmichael, 1991). In addition, experimental studies have shown that sulfur solubility may be high in an oxidized melt (fig. 3), depending on the temperature. Luhr (1990) developed a model for the sulfur in the El Chichón magma system that evoked differentiation of an equally oxidized high-temperature (>1,000°C) magma that would have been capable of carrying the sulfur in solution in the melt. A similar origin for the high amount of sulfur carried by the Mount Pinatubo magma is proposed by Bernard and others (1991) and Westrich and Gerlach (1992) to produce anhydrite and an excess sulfur-rich vapor. The problem with this model is that highly oxidized basaltic and andesitic magmas would be required in order to carry the observed sulfur abundances in Mount Pinatubo dacites

(fig. 3), and these oxidized magmas have not erupted at Mount Pinatubo (Pallister and others, this volume) or elsewhere (Carmichael, 1991). High-temperature calc-alkaline magmas are generally more reduced than associated low-temperature silicic magmas, in keeping with their mantle origin.

Another possible source of the Mount Pinatubo sulfur is the associated basalt that appeared in the mixed magma of the June 7–11 dome. Pallister and others (1992) and Matthews and others (1992) suggest that the anhydrite and sulfur-rich vapor, respectively, resulted from degassing of basalt that intruded the dacitic magma storage region. Although it is certainly possible that cooling and crystallization of the basalt could have produced a sulfur-rich vapor that was released during the June 14–15 eruption, it is unlikely that this vapor was the source of the anhydrite in the dacite. The ubiquitous presence of anhydrite in the dacite, the presence of anhydrite inclusions in hornblende, and the uniform sulfur abundance (anhydrite saturated) in melt inclusions suggest that the sulfur-rich character of the dacite was developed significantly prior to the eruption. At least some of the anhydrite may have originated from an adjacent hydrothermal zone containing anhydrite and apatite, according to the range of sulfur isotopic compositions found in individual crystals of the June 12 mixed magma (McKibben and others, 1992; this volume).

CONCLUSIONS

The main conclusions reached as a result of this analytical and phase equilibrium study of the 1991 Mount Pinatubo eruption products are the following:

1. Although there is evidence of disequilibrium in the form of zoned and cored crystals in the phenocryst-rich dacite pumice, possibly resulting from magma mixing some significant time before the eruption, most phenocrysts are uniform in composition from sample to sample and appear to be in equilibrium with the matrix glass. The rims of chemically zoned plagioclase and hornblende phenocrysts are uniform in composition. In contrast, clear evidence of disequilibrium and magma mixing is present in the andesitic magmas erupted immediately before the June 15 dacites (Pallister and others, 1992; this volume).

2. Melt inclusions trapped in growing phenocrysts in the dacite are H_2O-rich (~6.4 weight percent), high-SiO_2 rhyolite glasses similar (on an anhydrous basis) to the matrix glass regardless of the trapping phase. This similarity suggests one major melt-entrapment event in the magma, which is surprising, given the complex compositional zoning of some phenocrysts. The sulfur content of the melt inclusions indicates entrapment at ~800°C and a P_{SO_2} of 1 MPa; the H_2O content indicates a P_{H_2O} of ~220 MPa.

3. Phenocryst geothermometry and geobarometry using iron-titanium oxides and the Al-in-hornblende method indicate preeruption equilibration conditions in the dacite of 780°C ±10°, an oxygen fugacity that equals NNO+3 log units, and a total pressure of 220 ±50 MPa. Similarity of the P_{H_2O} and P_{total} estimates suggests that the preeruption magma was volatile saturated.

4. The highly oxidized character of the dacite probably reflects both the H_2O-rich nature of the magma and the episodic long-lived character of the storage region (Pallister and others, 1992).

5. Phase equilibrium studies of the white pumice of Mount Pinatubo confirm the geothermometry and geobarometry estimates of conditions in the preeruption magma. In particular, the pressure must have been in the 150- to 320-MPa range for H_2O saturation at 780°C in order to stabilize cummingtonite plus quartz in this bulk composition, and the 320- MPa upper limit decreases for H_2O-undersaturated conditions.

6. The preeruption June 14–15 Mount Pinatubo magma was H_2O rich and essentially volatile saturated, according to analytical data and phase equilibrium constraints, and could have contained an excess vapor phase. However, a vapor phase of sufficiently large volume to explain the excess SO_2 injected into the stratosphere on June 14–15 is considered unlikely because of the physical problems involved in building up this H_2O-rich volatile phase and in maintaining it during the June 7–13 eruptions.

ACKNOWLEGMENTS

The authors thank John Pallister, Terry Gerlach, and Chris Newhall for thoughtful reviews that significantly improved the manuscript and thank Elliot Endo and Tom Casadevall for collecting the samples. This research was supported by NSF grant EAR–9105110.

REFERENCES CITED

Andersen, D.J., and Lindsley, D.H., 1988, Internally consistent solution model for Fe-Mg-Mn-Ti oxides: Fe-Ti oxides: American Mineralogist, v. 73, p. 714–726.

Baker, L., and Rutherford, M.J., 1992, Anhydrite breakdown as a possible source of excess sulfur in the 1991 Mount Pinatubo eruption: Eos, Transactions, American Geophysical Union, v. 73, no. 43, p. 625.

Bernard, A., Demaiffe, D., Mattielli, N., and Punongbayan, R.S., 1991, Anhydrite-bearing pumices from Mount Pinatubo: Further evidence for the existence of sulfur-rich silicic magmas: Nature, v. 354, p. 139–140.

Bluth, G.J.S., Doiron, S.D., Schnetzler, C.C., Krueger, A.J., and Walter, L.S., 1992, Global tracking of the SO_2 clouds from the June 1991 Mount Pinatubo eruptions: Geophysical Research Letters, v. 19, no. 2, p. 151–154.

Carmichael, I.S.E., 1991, The redox states of basic and silicic magmas: A reflection of their source regions: Contributions to Mineralogy and Petrology, v. 106, p. 129–141.

Carroll, M.R., and Rutherford, M.J., 1987, The stability of igneous anhydrite: Experimental results and implications for sulfur behavior in the 1982 El Chichón trachyandesite and other evolved magmas: Journal of Petrology, v. 28, p. 781–801.

———1988, Sulfur speciation in hydrous experimental glasses of varying oxidation state: Results from measured wave-length shifts of sulfur X-rays: American Mineralogist, v. 73, p. 845–849.

Fogel, R.A., 1989, The role of C-O-H-Si volatiles in planetary igneous and metamorphic processes: Providence, R.I., Brown University, Ph.D. thesis, 200 p.

Fogel, R.A., and Rutherford M.J., 1990, The solubility of carbon dioxide in rhyolitic melts: A quantitative FTIR study: American Mineralogist, v. 75, p. 1311–1326.

Geschwind, C.H.G., and Rutherford, M.J., 1992, Cummingtonite and the evolution of the Mount St. Helens magma system: An experimental study: Geology, v. 20, p. 1011–1014.

Housh, T.B., and Luhr, J.F., 1991, Plagioclase-melt equilibria in hydrous systems: American Mineralogist, v. 76, p. 477–492.

Johnson, M.C., and Rutherford, M.J., 1989, Experimental calibration of the Al-in-hornblende geobarometer with application to Long Valley Caldera, Calif., volcanic rocks: Geology, v. 17, p. 837–841.

Luhr, J.F., 1990, Experimental phase relations of water- and sulfur-saturated arc magmas and the 1982 eruptions of El Chichón volcano: Journal of Petrology, v. 31, p. 1071–1114.

Matthews, S.J., Jones, A.P., and Bristow, C.S., 1992, A simple magma-mixing model for sulfur behaviour in calc-alkaline volcanic rocks: Mineralogical evidence from Mount Pinatubo 1991 eruption: Journal of the Geological Society of London, v. 149, p. 863–866.

McKibben, M.A., Eldridge, C.S., and Reyes, A.G., 1992, Multiple origins of anhydrite in Mt. Pinatubo pumice: Eos, Transactions, American Geophysical Union, v. 73, no. 43, p. 633.

———this volume, Sulfur isotopic systematics of the June 1991 Mount Pinatubo eruptions: A SHRIMP ion microprobe study.

Mori, J., Eberhart-Phillips, D., and Harlow, D.H., this volume, Three-dimensional velocity structure at Mount Pinatubo, Philippines: Resolving magma bodies and earthquakes hypocenters.

Nielson, C.H., and Sigurdsson, H.R., 1981, Quantitative methods of electron microprobe analysis of sodium in natural and synthetic glasses: American Mineralogist v. 66, p. 547–552.

Pallister, J.S., Hoblitt, R.P., Meeker, G.P., Knight, R.J., and Siems, D.F., this volume, Magma mixing at Mount Pinatubo: Petrographic and chemical evidence from the 1991 deposits.

Pallister, J.S., Hoblitt, R.P., and Reyes, A.G., 1992, A basalt trigger for the 1991 eruptions of Pinatubo volcano?: Nature, v. 356, p. 426–428.

Pinatubo Volcano Observatory Team, 1991, Lessons from a major eruption: Mt. Pinatubo, Philippines, Eos, Transactions, American Geophysical Union, v. 72, no. 49, p. 545–555.

Pownceby, M., and O'Neill, H., 1991, The NiPd redox sensor: Calibration from EMF measurements and application to the Re-ReO$_2$ oxygen buffer. Eos, Transactions, American Geophysical Union, v. 72, no. 44, p. 522.

Rutherford, M.J., and Devine, J.D., 1988, The May 1, 1980 eruption of Mount St. Helens: 3, Stability and chemistry of amphibole in the magma chamber: Journal of Geophysical Research, v. 93, no. 11, p. 949–959.

———1991, Pre-eruption conditions and volatiles in the 1991 Pinatubo magma: Eos, Transactions, American Geophysical Union, v. 72, no. 44, p. 62.

Rutherford, M.J., Baker, L., and Pallister, J.S., 1993, Petrologic constraints on timing of magmatic processes in the 1991 Mount Pinatubo volcanic system: Eos, Transactions, American Geophysical Union, v. 74, no. 43, p. 671.

Schmidt, M.W., 1992, Amphibole composition in tonalite as a function of pressure: An experimental calibration of the Al-in-hornblende barometer: Contributions to Mineralogy and Petrology, v. 110, p. 304–310.

Silver, L.A., Ihinger, P.D., and Stolper, E., 1989, The influence of bulk composition on the speciation of water in silicate glasses: Contributions to Mineralogy and Petrology, v. 104, p. 142–162.

Stormer, J.C., 1983, The effects of recalculation on estimates of temperature and oxygen fugacity from analyses of multicomponent iron-titanium oxides: American Mineralogist, v. 68, p. 586-594.

Westrich, H.R., and Gerlach, T.M., 1992, Magmatic gas source for the stratospheric SO_2 cloud from the June 15, 1991 eruption of Mount Pinatubo: Geology, v. 20, p. 867–870.

Wolfe, E.W., 1992, The 1991 eruptions of Mount Pinatubo, Philippines: Earthquakes and Volcanoes, v. 23, no. 1, p. 5–37.

Petrology and Geochemistry of the 1991
Eruption Products of Mount Pinatubo

By Alain Bernard,[1] Ulrich Knittel,[2] Bernd Weber,[2] Dominique Weis,[1] Achim Albrecht,[3]
Keiko Hattori,[4] Jeffrey Klein,[5] and Dietmar Oles[2]

ABSTRACT

The June 1991 activity of Mount Pinatubo erupted about 5 cubic kilometers of magma, mainly during the paroxysmal June 15 eruption. Most of the erupted magma was dacite with an overall constant composition (64.5 weight percent SiO_2) but with variable amounts of phenocrysts. Small volumes of andesite (56 weight percent SiO_2) characterized the June 7–12 dome-building phase and the June 12 eruption. This andesite contains inclusions of a primitive basalt (50.6 weight percent SiO_2). The dacite is extraordinarily enriched in sulfur (1,474 to 2,211 ppm sulfur), most of which is present as phenocrysts and microphenocrysts of anhydrite. The dacite also contains xenocrystic olivine, which, because of abundant chromite inclusions, is considered to be of juvenile nature. Major and trace element compositions of dacite and andesite display the same geochemical features as the other silicic rocks from the Bataan arc volcanoes of the Philippines. Strontium and neodymium isotopic compositions of all 1991 dacite and andesite and their phenocrysts are identical, with $^{87}Sr/^{86}Sr$ = 0.7042 to 0.7043 and $^{143}Nd/^{144}Nd$ = 0.51286 to 0.51298. The basalt contains slightly more radiogenic strontium and slightly less radiogenic neodymium. The isotopic signatures are in agreement with the volcano's location, which is relatively far from the North Palawan continental terrane (NPCT) collision zone as compared with the Macolod corridor, Luzon, or the Mindoro segment. These compositions suggest also that an unusually large contamination by marine sediments is not likely to explain the large sulfur enrichment observed in the Pinatubo magma. Dacite and andesite also have identical lead isotopic compositions characterized by high $^{208}Pb/^{204}Pb$ and $^{207}Pb/^{204}Pb$ for relatively low $^{206}Pb/^{204}Pb$, while the basalt contains slightly more radiogenic lead. The basalt is likely to have been generated by partial melting of peridotite because of its high MgO content and high forsterite content of olivine phenocrysts and the abundance of chromite inclusions. It is close to the primitive basalts found in the Macolod corridor.

INTRODUCTION

The 1991 eruption of Mount Pinatubo in the Philippines ranks among one of the largest volcanic eruptions of this century, with about 5 km^3 of magma erupted (dense rock equivalent, DRE). The extraordinary feature of this eruption lies in the very large amount of SO_2 (20 Mt) injected into the stratosphere, well above the 7 Mt of SO_2 produced by the eruption of El Chichón volcano in 1982 (Bluth and others, 1992). Shortly after the eruption, the volcanic cloud produced by the conversion of this SO_2 into sulfuric acid aerosols spread rapidly in both hemispheres and significantly reduced the amount of sunlight reaching the Earth's surface (McCormick and Veiga, 1992; Stowe and others, 1992). Mean atmospheric and sea surface temperatures measured by satellites have now demonstrated that the cooling effect culminated in 1992 with a maximum temperature decrease of 0.6°C, which temporarily counteracted the effect of the greenhouse warming (Yan and others, 1992; Kerr, 1993). Due to enhanced concentrations of volcanic material at high altitude, a significant depletion in stratospheric ozone above Antarctica also occurred in 1991 (Hofman and others, 1992).

The high sulfur content of the Pinatubo magma is manifested by the presence of anhydrite phenocrysts in the pumices (Bernard and others, 1991; Fournelle, 1991; Knittel and others, 1991). Primary anhydrite is exceptional and has been reported only for very few young volcanic rocks (for example, Mount Lamington, Arculus and others, 1983; El Chichón, Luhr and others, 1984). Commonly, arc volcanics

[1]Department of Geology, CP 160/02, Université Libre de Bruxelles, 50 Ave. F.D. Roosevelt, B–1050 Brussels, Belgium.

[2]Institut für Mineralogie und Lagerstättenlehre, Wüllnerstrasse 2, D–52056 Aachen, Germany. Now at: Institut für Geowissenschaften/Mineralogie Saarstr. 21, D–55099 Mainz, Germany.

[3]Department of Chemistry, Rutgers University, Piscataway, NJ 08855; now at: EAWAG/ETH, Umweltphysik, 8600 Duebendorf, Switzerland.

[4]Department of Geology, University of Ottawa, Ottawa, K1N 6N5 Canada.

[5]Department of Physics, The University of Pennsylvania, Philadelphia, PA 19104.

contain less than 200 ppm sulfur (S) (Gill, 1981) and typically contain less than 40 ppm S (Ueda and Sakai, 1984), while those from Pinatubo contain 1,500 to 2,400 ppm S (Bernard and others, 1991, Pallister and others, 1992).

A number of questions arise from the observation of this unusual eruption. The first question is what is the frequency of these sulfur-rich eruptions in the geological record? Up to now, the answers are only tentative:

1. Other silicic volcanoes may also be enriched in sulfur, but this is not known because most of them have never been studied in detail. Furthermore, anhydrite may have been overlooked or misinterpreted (as a product of hydrothermal alteration), or anhydrite that was originally present may have been leached away because this mineral is not very stable in tropical environments (Arculus and others, 1983; Luhr and others, 1984). Another problem is sample preparation. We have noticed that water-based thin-section preparation can remove anhydrite.

2. Sulfur may have been distributed unevenly in the subducted lithosphere and hence in the magma sources. The enrichment in sulfur may have an accidental origin, such as by contamination from metalliferous sediments. These sediments are usually only of limited extension and may have contributed only to the source region of a specific volcano.

Another question is what is the likely source for the sulfur emitted into the atmosphere? Contrasting with the widespread occurrence of anhydrite, glass inclusions trapped in phenocrysts of dacite at Pinatubo show no evidence of sulfur enrichment. These inclusions contain only 60–90 ppm S and reveal that the preeruptive sulfur content dissolved in the silicate melt was low and close to the solubility limit for sulfur under the conditions of the magma reservoir (Westrich and Gerlach, 1992). The exsolution of the bulk sulfur dissolved in the melt during the eruption clearly cannot account for the 20 Mt of SO_2 produced. The alternative sources proposed for this excess sulfur are (1) breakdown of anhydrite during the eruption as was first suggested by Sigurdsson and others (1985) for the El Chichón eruption (Rutherford and Devine, 1991), or (2) presence of sulfur as a separate discrete phase before the eruption (Luhr, 1990; Westrich and Gerlach, 1992).

Other aspects of interest concern the origin of the huge volume of relatively homogeneous dacite and its relationship to basalt, thought to have triggered the eruption (Pallister and others, 1992; this volume).

We present here the results of a geochemical investigation carried out on the 1991 Pinatubo products. The main purpose of this investigation was to find geochemical evidence to explain the origin of this sulfur-rich magma in order to establish whether it is possible to identify ancient eruptions of anhydrite-bearing magmas on the basis of geochemical or petrological features.

GEOLOGIC SETTING

Mount Pinatubo, located on Luzon Island at long 120° 22' E., lat 15° 08' N., is the northernmost of several large stratovolcanoes forming the volcanic front along the Bataan segment of the Taiwan-Luzon arc (Defant and others, 1989) (figs. 1, 2). Volcanism of this arc is related to the subduction of the oceanic crust of the South China Sea along the Manila trench. Presently the arc is situated approximately above the 100-km depth contour line of the Wadati-Benioff zone (Cardwell and others, 1980; de Boer and others, 1980). West of Mount Pinatubo, the Zambales Ophiolite Complex is traversed by the west-northwest-trending Iba fracture zone, one of two major fracture zones that cut the ophiolite into three blocks (fig. 2). Thus, Mount Pinatubo appears to be located on the intersection of the Iba fracture zone and the north-south Bataan lineament (Wolfe and Self, 1983).

With an altitude of 1,745 m prior to the 1991 eruption, Mount Pinatubo was the highest volcanic edifice of this volcano chain. The volcano consists of an older, andesitic and dacitic stratovolcano and a younger, dominantly dacitic dome complex (Newhall and others, this volume). The last eruptions prior to 1991 produced voluminous ash-flow tuffs extending up to 20 km from the dome complex (Wolfe and Self, 1983). There are no records of historic eruptions prior to 1991 (since the 16th century); however, roaring vents near the summit and deposition of sulfur have been reported repeatedly. Dates obtained by the ^{14}C method indicate that the last volcanic activity prior to the 1991 eruption occurred about 500 years ago (Newhall and others, this volume).

THE 1991 ERUPTION

Increased solfataric activity in August 1990 (PHIVOLCS, 1990), possibly in response to the July 16, 1990, earthquake along the Philippine-Digdig fault, was the first indication for a reawaking of Mount Pinatubo after about 500 years of dormancy. The eruptive activity started on April 2, 1991, with small explosions that created six new vents on the northern slope of the main cone. Starting May 16, small amounts of ash were ejected. This activity intensified on May 28 and reached a first culmination by the extrusion of a dome on June 7. Powerful eruptions producing andesitic ash started in the morning of June 12. The eruption reached its climax on June 15, when, within 12 h, an estimated 5 km^3 of dacitic magma was erupted (W.E. Scott and others, this volume). Moderate to mild ash ejection continued until the end of August. (The account of the eruption

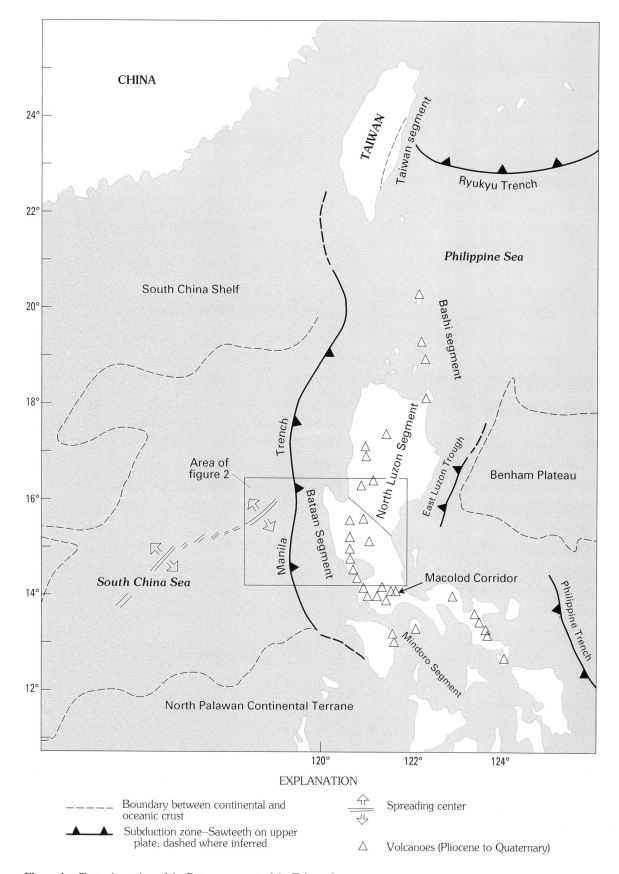

Figure 1. Tectonic setting of the Bataan segment of the Taiwan-Luzon arc.

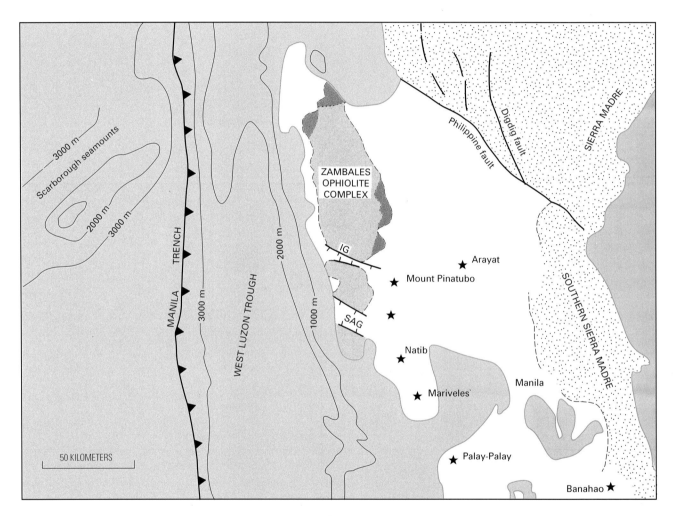

Figure 2. The geology of the Bataan segment of the Taiwan-Luzon arc. The Zambales Ophiolite Complex is after Evans and others (1991). Offshore, the Scarborough seamounts mark the trace of the extinct spreading center of the South China Sea. Bathymetric contour interval 1,000 m. IG, Iba graben; SAG, San Antonio graben.

is summarized from Pinatubo Volcano Observatory Team, 1991, and Sabit, 1992).

SAMPLES

Samples Pt6 through Pt32 were collected by Oles within 2 weeks of the June 15th eruption and samples Pin1 through Pin3 as well as Pin12 and Pin14 were collected during the same period by Bernard, who also collected samples Pin4 through Pin10 in September 1991. Additional samples were donated by Mylene Martinez (PHIVOLCS, samples Pt40, Pt41) and Chris Newhall (Pt51) and collected by Weber (Pt48, Pt52) in June 1992. Hattori collected samples of andesite with basalt inclusions in August 1992 (Pt50, Pt53). As pointed out by Newhall (oral commun., 1993), we cannot be absolutely sure that Pt50 and Pt53 are from the 1991 eruption; however, the chemical and mineralogical similarity of these last two samples with those analyzed by Pallister and others (this volume), together with their freshness, strongly supports this assumption.

ANALYTICAL PROCEDURES

XRF ANALYSES

Major elements were determined at Aachen University (samples with prefix "Pt") on fused discs with sample/flux ratio = 1:10. Trace elements were determined on undiluted pressed powder pellets. Samples with prefix "Pin" were analyzed at Liège University (Belgium) with the same analytical procedure.

STRONTIUM, NEODYMIUM, AND LEAD ISOTOPIC ANALYSIS

Most strontium (Sr) and neodymium (Nd) isotopic determinations and all lead (Pb) isotopic analyses were carried out at the Belgian Centre for Geochronology, Université Libre de Bruxelles, according to the following procedures.

Sample preparation.—About 3 to 4 g of rock powder were leached in hot, <100°C water in ultrasonic baths in order to dissolve the anhydrite. The rock was weighed before and after the H_2O bath in order to estimate how much anhydrite had been extracted. No precipitate was observed in any of the three solutions to which we added HCl 6 N in order to dissolve the anhydrite and to prevent gypsum precipitation. The rock powders were then leached for half an hour in warm 2.5 N HCl to eliminate secondary phases, if present, and to make sure that we would be left only with the primary silicates and glass. An additional leaching procedure (warm HCl 6 N) was applied to sample Pin4 to check the efficiency of the leaching procedure (see table 9). No difference was observed in any of the isotopic compositions analyzed, and samples Pin4 and Pin4a overlapped entirely within error limits. Again the powders were dried on the hotplate and weighed repeatedly until stable weight was obtained. About 100 to 200 mg of sample powder were dissolved in a mixture 6:1:1 HF-HNO_3-$HClO_4$ in a Teflon vessel.

Chemical separation.—Strontium and rare earth elements (REE) were separated with 2.5 N HCl and 4 N HCl, respectively, on a cation exchange column (Dowex 50W × 8). Neodymium was separated with 0.3 N HCl on a column prepared following Richard and others (1976). The procedure is described in detail by Weis and others (1987). Both columns were calibrated with ^{139}Ce. Lead was separated from the same starting sample solution by anion exchange columns with HBr-HCl following a method adapted from Manhès and others (1978). Lead and uranium (U) concentrations were measured by isotope dilution on the same solution (split before loading on columns and spiked with a mixed ^{235}U–^{206}Pb spike). Uranium was separated in an HNO_3 medium. All these operations were done in an overpressurized (>5 mm mercury) ultraclean laboratory and in

laminar-flow cabinets. Blanks for the whole chemical procedure were <1 ng for Pb and <3 ng for Nd and Sr.

Mass spectrometry.- All the isotopic compositions, Sr, Nd, and Pb, were measured on the VG Sector54 7 collectors in dynamic mode for Sr (single tantalum (Ta) filaments) and Nd (triple, rhenium (Re) center and Ta side filaments) and in static mode for Pb (single Re filaments). The 2σ errors on the mean for the isotopic compositions are less than 3×10^{-5} for Nd and Sr isotopic ratios (that is, less than 0.01‰) and less than 0.1‰ for ^{206}Pb/^{204}Pb and ^{207}Pb/^{204}Pb, and less than 0.15‰ for ^{208}Pb/^{204}Pb. The Nd and Sr isotopic ratios were corrected for mass fractionation by using ^{146}Nd/^{144}Nd = 0.7219 and ^{86}Sr/^{88}Sr = 0.1194. Analyses of the NBS987 Sr standard gave ^{87}Sr/^{86}Sr = 0.710230±5 ($2\sigma_m$ on 67 measurements) and for the nNdb solution (Wasserburg and others, 1981), 0.51190±2 ($2\sigma_m$ on 15 measurements). Lead isotope results were corrected for mass discrimination (1.3±0.3‰ per mass unit, Catanzaro and others, 1968) by repeated analyses of NBS 981.

Lead concentrations were measured with a Finnigan MAT–260 single collector mass spectrometer with double Re filaments. Rb and Sr concentrations were measured by X-ray fluorescence analysis to a precision of ±2 percent of values above 15 ppm.

Additional Sr and Nd isotopic data were collected by M. Feigenson (Rutgers University) and K. Hattori (University of Ottawa). At Rutgers, the measured ^{87}Sr/^{86}Sr ratio for NBS SRM 987 was 0.710248. ^{143}Nd/^{144}Nd was normalized to ^{146}Nd/^{144}Nd = 0.7219, and a value of ^{143}Nd/^{144}Nd = 0.511852 was obtained for the La Jolla standard. At Ottawa, isotopic determinations were made using Re double filaments for Sr and Nd isotopes in a Finnigan MAT-261 multicollector solid-source mass spectrometer. The Eimer and Amend standard gave ^{87}Sr/^{86}Sr = 0.70802 normalized to ^{86}Sr/^{88}Sr = 0.1194. ^{143}Nd/^{144}Nd for the La Jolla standard was 0.51186 normalized to ^{146}Nd/^{144}Nd = 0.7219.

BERYLLIUM ISOTOPES

The procedures for beryllium (Be) separation outlined here are based on methods developed for meteorite work. Adjustments were necessary because of the larger sample masses required (5 g) and the high aluminum (Al) concentrations. After addition of 1.5 mg ^{9}Be carrier, open acid digestion using HF, $HClO_4$, HNO_3, and anion exchange to remove iron (Fe), the sample solutions were introduced into a large cation exchange column (allowing primary separation of Be from Al). After a buffered precipitation of Be and Al, still present as hydroxides, a second cation exchange procedure using small columns was performed to purify Be, which again was precipitated as hydroxide. This hydroxide was converted to BeO, which was loaded into the ion source of the Tandem Accelerator of the University of Pennsylvania, where the ^{10}Be/^{9}Be ratio was determined (Klein and

Middleton, 1984). A blank measurement for the outlined procedure gave 6×10^{-15} for the $^{10}Be/^{9}Be$ ratio, a relative error of 7 percent (see table 10).

PETROGRAPHY

The June 7–12 dome-building phase and the June 12 eruptions were characterized by the emission of andesite, which is considered to have been generated by mixing of basalt and dacite (Pallister and others, 1992; this volume). The basalt is represented by basaltic inclusions in the andesite.

The paroxysmal eruptions of June 15 produced two types of pumice: (1) white, phenocryst-rich pumice with large vesicles (>1 mm) and (2) gray, phenocryst-poor pumice with small vesicles (<1 mm). Both pumice types have identical chemical compositions (Pallister and others, 1992; table 1).

Rutherford (1993) and Rutherford and Devine (this volume) estimate that the preeruption pressure of the dacite at Pinatubo was 2.2 kbar with P_{H_2O}=1.78 kbar. The oxygen fugacity is estimated to have been about 3 log units above the Ni-NiO (NNO) buffer (similar values are also calculated by Imai and others, this volume). The presence of cummingtonite rims around amphibole shows that the temperature was less than 810°C (Geschwind and Rutherford, 1992) and is estimated at 780°C.

DACITES

Phenocryst-rich dacitic pumice.— The phenocryst-rich dacitic clasts are strongly porphyritic and vesicular pyroclastic rocks composed of euhedral and fragmented or fractured phenocrysts set in a vesicular, glassy groundmass that contains single glass shards and, to a lesser extent, fragments of the phenocryst assemblage. The pumiceous clasts contain 63 vol% vesicles, 16 vol% phenocrysts and 21 vol% glassy matrix (table 2). These data are compatible with the density of the pumiceous clasts (table 2) and close to the values reported by Pallister and others (this volume).

The phenocryst assemblage is dominated by plagioclase (29 vol%, on a vesicle-free basis), green hornblende (12 vol%, sometimes with cummingtonite rims), iron-titanium (Fe-Ti) oxides (0.8 vol%), and quartz (<0.1 vol%). Almost all samples contain microphenocrysts of anhydrite (0.1 vol%). These frequently contain inclusions of glass (often with a bubble) and apatite or are intergrown with apatite (fig. 3A). Biotite and clinopyroxene are present only in some of the analyzed clasts. Some samples contain glomerocrysts of plagioclase or polymineralic clots of quartz and plagioclase.

An interesting aspect of the dacites that erupted in 1991 is the occurrence of xenocrystic olivine (Reyes, Philippine National Oil Co., unpub. report, 1991; Knittel and others, 1991), the primitive nature of which is indicated by abundant chromite inclusions. These xenocrysts are commonly rimmed by brown amphibole. Occasionally the overgrowths also contain orthopyroxene. A few samples contain xenomorphic olivine grains without amphibole rims (fig. 3B).

The cryptocrystalline, highly vesiculated groundmass is dominated by single glass shards and light-colored matrix glass that encloses microphenocrysts and fragments of the phenocryst assemblage. Some clasts show a distinct lamination of the vesiculated groundmass. Sulfides were never observed in the groundmass. Very few micronic inclusions of pyrrhotite or chalcopyrite were observed in few plagioclase, amphibole, and magnetite phenocrysts (see detailed study of Hattori, this volume).

Phenocryst-poor dacitic pumice.—Although phenocryst-poor and phenocryst-rich dacitic pumices show similar chemical composition, there are various differences concerning mineralogical and textural appearance. The phenocryst content is much lower (10 vol%) and, on a vesicle-free basis, includes plagioclase (19 vol%), hornblende (4 vol%), and Fe-Ti oxides (0.4 vol%) (table 2). Other phenocrysts are rare. Only a few of the clasts studied contain quartz (<0.1 vol%), anhydrite (<0.1 vol%), biotite, and xenocrystic olivine.

Most phenocrysts in phenocryst-poor dacites are fragments of plagioclase and hornblende, much smaller than phenocrysts of the phenocryst-rich dacites. Euhedral crystals are significantly less abundant in phenocryst-poor dacite than they are in the phenocryst-rich dacite, and fragmentation is clearly evident from zoned plagioclase crystals in which the core is located at the grain edge. Therefore, we do not concur with the initial suggestion of Pallister and others (1992), who interpreted subhedral plagioclase to result from resorption. Note that Pallister and others (this volume) now consider fragmentation as the primary factor in generation of phenocryst-poor dacite.

The cryptocrystalline groundmass is dominated by highly vesiculated matrix glass that encloses microphenocrysts and fragments of the phenocryst assemblage. The average size of vesicles in phenocryst-poor pumice is smaller and results in a much denser appearance than that of the phenocryst-rich pumice. In addition, laminated texture of the groundmass is less common in the phenocryst-poor pumice than it is in the phenocryst-rich pumice.

DOME ANDESITE

The andesites are highly porphyritic rocks containing between 39 and 42 vol% phenocrysts, which are embedded in a fine-grained matrix (49–58 vol%). In contrast to the dacites, the andesites have significantly lower vesicle contents (3–7 vol%). This low vesicle content may not reflect the primary gas content of the andesite, as the dome

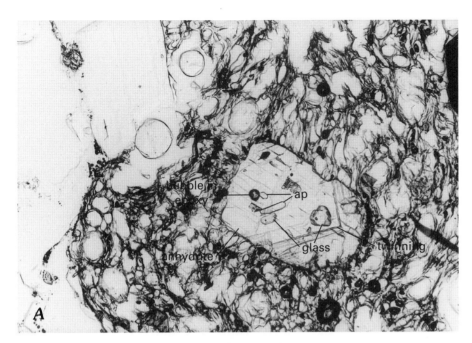

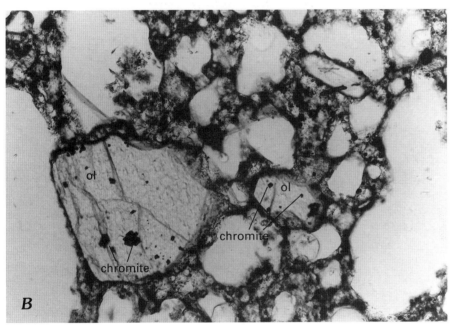

Figure 3. *A*, Anhydrite with glass (with bubble) and apatite inclusion. Anhydrite is about 0.43 mm in diameter. Plane polarized light. *B*, Xenomorphic olivine with chromite inclusions (black) in type 1 dacite. Olivine grain is about 0.35 mm across. Plane polarized light. *C*, Quartz xenocryst with corona of clinopyroxene, in basalt sample Pt53. The quartz grain has a diameter of 1.2 mm. Crossed polars. *D*, Contact between basalt (coarse grained, light colored) and andesite (fine grained, dark matrix). Note: In the basalt, the matrix between the plagioclase crystals (white) and amphiboles (dark gray) is filled by glass (light gray) and vesicles (white). Width of the photograph is 1.33 mm.

andesite may have lost its volatile content by degassing before final emplacement.

The phenocryst assemblage comprises plagioclase, hornblende, clinopyroxene, olivine, and Fe-Ti oxides. Quartz xenocrysts and anhydrite are common but are not present in all samples. As in dacite, plagioclase is the dominant phenocryst (13–24 vol%) followed by hornblende (9–12 vol%), clinopyroxene (7–11 vol%), olivine (1–4 vol%), and Fe-Ti oxides (1–2 vol%). Quartz xenocrysts and anhydrite comprise up to 1.5 and 0.3 vol%, respectively, of some samples.

The fine-grained matrix is composed of clinopyroxene, plagioclase, and Fe-Ti microphenocrysts with abundant gray to light-brown matrix glass. Similar to the dacites, the andesitic groundmass contains fragments of phenocrysts.

BASALT

Basalt has been found as inclusions of variable size (millimeters to several centimeters) in andesite. The mineralogy is dominated by long, prismatic crystals of brown hornblende (30–35 vol%), which is very unusual for basaltic rocks. However, unlike in lamprophyres, the amphibole is not an early phase but postdates olivine and clinopyroxene.

Figure 3.—Continued

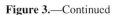

Other phases are clinopyroxene (14–19 vol%), plagioclase (9–11 vol%), olivine (5–6 vol%), and Fe-Ti oxides and chromite (sum, 0.3–0.7 vol%). While clinopyroxene forms well-developed, euhedral crystals, the olivine is always marginally replaced by amphibole. Olivine usually contains tiny chromite inclusions. Pyrrhotite occurs as inclusions in the silicates (Hattori, this volume). The matrix of the mafic minerals is formed by small laths of plagioclase with brown glass and cavities in the interstices.

The basalt also contains several large crystals which show signs of instability and may be either xenocrysts or are derived from a small amount of evolved (dacitic?)

magma mixed into the basalt. One of these phases is mottled, sieve-textured euhedral plagioclase. Some samples also contain quartz crystals rimmed by clinopyroxene coronas (fig. 3C). Commonly, relatively large glass patches are associated with the quartz xenocrysts. Furthermore, there are decomposed mafic minerals. Textural evidence suggests that they are amphiboles that have been replaced by clinopyroxene, and perhaps also clinopyroxenes.

The contacts between basalt and andesite are generally sharp (fig. 3D). Cuspate and interfingering contact relations between the two suggest that both rocks were partially molten when they came into contact.

MINERALOGY

Plagioclase.—The plagioclase crystals in dacite are relatively sodic, with rims often more calcic than cores (cores An_{30-40} and rims up to An_{45}, fig. 4). In detail, the plagioclase phenocrysts show complex zoning (see Pallister and others, this volume). Plagioclase in evolved dacites of Mount St. Helens have similar compositions (Smith and Leeman, 1987), but those in dacites from northern Luzon are less albite rich and show much greater compositional variability within single samples (Knittel and others, unpub. data, 1992).

Amphibole.—The green amphiboles in dacite are fairly homogeneous. With 6.4–6.9 Si (on the basis of 23 O) and around 1.6 Ca, they approach the composition of edenite (fig. 5, table 3). The amphibole in the andesite appears to be even slightly higher in Si (though this could be a sample bias; only one andesite was investigated). The high Si content is typical for amphibole in evolved rocks. The brown amphibole in basalt is distinctly less Si rich (Si~6) but contains more Ca (Ca~1.8).

Clinopyroxene.— The clinopyroxenes show little compositional variability. They are magnesian salites generally low in titanium (Ti) and sodium (Na) (table 4).

Anhydrite.— Anhydrite contains significant SrO (~0.23 wt%, table 5), which is, however, only half the SrO content of El Chichón apatite (0.58 wt% SrO; Luhr and others, 1984).

Biotite.—Biotite is intergrown with amphibole in rims around olivine, in andesite sample Pin6. This biotite is characterized by extremely high but variable Mg/(Mg+Fe) of 0.74–0.89 (table 6). The highest values are probably due to the fact that this biotite replaces Mg-rich olivine without significant addition or loss of Fe and Mg. Mg/(Mg+Fe) of biotite crystals in dacite is about 0.67–0.70 (table 6; see also Imai and others, this volume). These values are still high for such evolved rocks but may reflect the high oxidation state, hence high Fe^{3+} and low Fe^{2+} contents of the melt (see Imai and others, this volume, for a more detailed discussion).

Olivine.— Olivine is a major constituent of the basalts and andesites. Xenocrystic olivines are also found in several samples of dacitic pumice. Petrographic features of the olivines have already been described. Regardless of the host rock, olivine core compositions are uniformly Fo_{86-88} and rims have Fo_{85-87} (table 7). Pallister and others (this volume) and Imai and others (this volume) found relatively Fe-rich olivine in dacite ($Fo_{~81}$), but we found only the Fo-rich olivines in three dacite samples; we have at present no explanation for this discrepancy. Because the olivine grains commonly are replaced by amphibole along the margins and some grains in dacite appear to have experienced some kind of abrasion or resorption as indicated by their subhedral shape, the "original" rims may have been more Fe rich.

Chromite.—The olivine grains contain tiny, commonly euhedral chromite crystals, characterized by Cr/(Cr+Al) = 0.64–0.68, $Mg/(Mg+Fe^{2+})$= 0.33–0.48 (table 8) and Fe^{3+} contents that are high relative to chromites from midocean ridge basalt (MORB) (Dick and Bullen, 1984) and primitive basalts from the Macolod corridor, located to the south (fig. 6). The chromite inclusions in basalt and andesite have relatively constant Fe^{3+} contents, while dacites have slightly more oxidized chromites that may have reequilibrated. This is suggested by the observation that the most highly oxidized chromite (filled stars in fig. 6) occurs within a marginally oxidized olivine grain. Similarly high Fe^{3+} contents have been reported by Nye and Reid (1986) for chromite inclusions in olivine from primitive basalts from Okmok Island, Aleutians (fig. 6). Chromite phenocrysts in basalt that are not included in olivine are strongly zoned, with Al/Mg-rich cores surrounded by magnetite rims.

GEOCHEMISTRY

Clasts of both types of dacite have virtually identical compositions, with 64.5 wt% SiO_2 on average. Al_2O_3 is slightly higher in the phenocryst-poor dacite (16.7 versus 16.0 wt%), a fact that may reflect a sampling or analytical bias because all phenocryst-rich dacites analyzed in Liège fall at the low end of the range in Al_2O_3 observed for the samples analyzed in Aachen. A clear analytical bias exists for barium (Ba), where all Liège analyses show higher values than the Aachen analyses. Ba data given by Pallister and others (this volume) show better agreement with the Liège data. With regard to all other major and trace element abundances, the results obtained in the two laboratories are compatible within analytical uncertainty.

Sulfur content in the dacite ranges from 1,470 to 2,210 ppm. The only andesite analyzed for sulfur (Pin6) is slightly enriched in sulfur relative to the dacites. Two small clasts (5–10 cm in diameter) collected in September 1991 (Pin9 and Pin10, with 421 and 353 ppm S, respectively) clearly show the effect of leaching of anhydrite by rainwater during the first rainy season following the eruption.

Harker variation diagrams of selected major and trace elements of the 1991 Pinatubo rocks are shown in figures 7 and 8. In these figures, the Pinatubo compositions are compared to other volcanics from the Luzon arc reported by Defant (1985) and Defant and others (1991). The 1991 dacites and andesite compare well with the most silicic rocks of the Bataan arc-front volcanoes segment, but it should be noted that volcanics with SiO_2>62 wt% are quite rare in the Bataan segment. The dacites of Pinatubo display the same geochemical characteristics that Defant and others (1991) observed for the Bataan arc-front volcanoes segment: significantly lower K_2O values than the volcanics from the Bataan behind-arc volcanoes, Macolod, and Mindoro segments. In detail, however, the Pinatubo dacites are slightly

Table 1. Major and trace element compositions of the 1991 Mount Pinatubo eruptive rocks.

[Oxide values in weight percent; elemental values in parts per million; n.a., not analyzed; LOI, loss on ignition]

Rock type Sample Location	Pt6 Subic	Pt9a Subic	Pt15 Poonbato	Pt16 Poonbato	Pt26 Clark	Pt31 Dizon	Pin1 Marella	Pin2 Marella	Pin3 Marella	Pin4 Marella	Pin5 Sacobia	Pin7 Marella	Pin8 Maraunot	Pin9 Marella	Pin10 Clark	Average, Phenocryst rich
							Phenocryst-rich pumice									
SiO_2	65.18	64.85	64.66	64.67	64.86	64.50	64.24	64.52	64.41	63.60	64.19	63.83	64.36	64.53	64.97	64.50
TiO_2	.62	.54	.48	.49	.50	.49	.47	.50	.49	.47	.46	.48	.46	.47	.47	.49
Al_2O_3	15.35	15.54	16.94	16.89	17.01	16.66	15.75	15.40	15.63	15.92	16.10	15.64	15.83	15.71	15.58	16.00
FeO	5.00	4.63	4.21	4.19	4.44	4.16	4.15	4.18	4.26	4.07	3.87	4.07	3.80	4.13	3.97	4.21
MnO	.11	.11	.10	.10	.11	.09	.11	.11	.11	.10	.10	.10	.10	.10	.10	.10
MgO	2.55	2.43	2.16	2.23	2.36	2.15	2.60	2.37	2.52	2.54	2.52	2.81	2.44	2.60	2.53	2.45
CaO	4.92	4.83	5.29	5.32	5.54	5.22	5.30	5.16	5.18	5.38	5.24	5.46	5.08	4.68	5.12	5.18
Na_2O	4.30	4.55	4.66	4.79	4.64	4.48	4.67	4.80	4.44	4.64	4.74	4.86	5.08	4.88	4.73	4.68
K_2O	1.71	1.68	1.42	1.46	1.35	1.52	1.57	1.55	1.57	1.48	1.48	1.51	1.52	1.56	1.58	1.53
P_2O_5	.22	.20	.17	.18	.17	.15	.15	.15	.17	.14	.17	.16	.16	.15	.17	.17
LOI	n.a.	n.a.	n.a.	n.a.	n.a.	.44	.73	.57	1.02	.96	1.06	1.11	1.02	.43	.45	—
Total	99.96	99.36	100.09	100.32	100.98	99.52	99.75	99.32	99.80	99.30	99.92	100.02	99.84	99.24	99.68	99.81
Rb	43	45	37	37	34	40	40	42	39	37	38	39	39	39	42	39
Sr	473	485	572	570	570	547	552	557	548	565	573	563	568	538	548	549
Ba	422	411	373	383	364	364	509	508	492	471	478	454	479	468	464	443
Nb	5	5	5	5	5	4	4	4	4	4	5	3	3	4	4	4
Zr	127	123	119	106	125	105	107	112	103	122	108	107	116	114	115	114
Y	14	12	11	11	12	13	12	12	13	14	12	13	12	12	11	12
Cr	21	140	129	19	22	27	40	34	26	27	27	35	24	31	27	42
Pb	10	9	10	7	8	13	11	13	7	10	11	11	8	11	10	10
Cu	4	5	0	3	0	56	37	16	17	27	29	23	29	49	27	21
Ni	15	13	11	12	13	17	15	12	12	15	16	18	15	16	16	14
Cl	n.a.	n.a.	n.a.	n.a.	n.a.	n.a.	473	476	523	438	480	420	467	525	501	—
V	n.a.	n.a.	n.a.	n.a.	n.a.	n.a.	77	78	69	81	73	79	72	86	83	—
S	n.a.	n.a.	n.a.	n.a.	n.a.	n.a.	1,615	1,474	1,911	2,161	1,628	2,211	1,637	421	353	—

Table 1. Major and trace element compositions of the 1991 Mount Pinatubo eruptive rocks—Continued.

Rock type	Phenocryst-poor pumice					Banded pumice	Mingled pumice	Andesite			Basalt	Ash	
Sample	Pt9B	Pt32	Pt48	Pt52	Average, phenocryst-poor	Pt40	Pt41	Pt50	Pt51	Pin 6	Pt53	PIN12	PIN14
Location	Subic	Dizon	Sacobia	Maraunot		Maraunot	Maraunot			Marella			
SiO_2	65.39	65.19	63.88	64.65	64.78	64.67	60.15	60.36	59.76	59.73	50.60	61.13	62.82
TiO_2	.50	.49	.52	.50	.50	.49	.60	.64	.62	.63	.90	.45	.41
Al_2O_3	17.07	16.69	16.56	16.36	16.67	16.43	16.29	15.96	15.89	15.23	13.92	16.23	16.38
FeO	4.28	4.22	4.43	4.13	4.27	4.20	5.51	6.10	5.83	5.88	9.04	3.97	3.45
MnO	.10	.10	.10	.10	.10	.10	.12	.12	.12	.15	.16	.11	.10
MgO	2.26	2.32	2.50	2.15	2.31	2.23	4.04	4.38	4.26	5.07	9.15	2.79	2.75
CaO	5.23	5.25	5.57	5.16	5.30	5.18	6.79	6.98	6.82	6.63	10.10	5.87	5.92
Na_2O	5.09	4.71	4.36	4.45	4.65	4.49	4.01	3.84	3.86	3.99	2.78	5.26	5.10
K_2O	1.56	1.52	1.55	1.43	1.52	1.48	1.56	1.59	1.52	1.33	1.41	1.38	1.44
P_2O_5	.18	.20	.17	.15	.18	.15	.21	.23	.22	.17	.30	.17	.17
LOI	n.a.	n.a.	1.15	.10	—	.59	.95	.18	.12	1.35	.23	1.22	1.10
Total	101.66	100.69	99.64	99.08	100.27	100.01	99.28	100.20	98.94	100.17	98.59	98.58	99.64
Rb	37	39	39	40	39	40	44	47	47	34	48	43	37
Sr	542	543	544	536	541	543	568	558	566	490	531	617	605
Ba	362	381	347	357	362	362	349	360	377	418	317	429	368
Nb	5	5	3	4	4	4	3	4	3	4	3	4	3
Zr	118	111	105	109	111	112	108	109	109	87	104	111	113
Y	11	11	13	13	12	13	15	15	14	17	20	12	13
Cr	130	136	82	41	97	41	136	156	149	131	394	38	34
Pb	9	11	13	11	11	12	9	11	8	6	8	14	9
Cu	42	19	47	34	36	67	32	87	73	29	94	98	73
Ni	16	17	40	n.a.	24	40	17	47	56	56	131	20	19
Cl	n.a.	n.a	n.a.	n.a.	—	n.a.	n.a.	n.a.	n.a.	659	n.a.	1,540	1,022
V	n.a.	n.a.	n.a.	n.a.	—	n.a.	n.a.	n.a.	n.a.	126	n.a.	89	75
S	n.a.	n.a.	n.a.	n.a.	—	n.a.	n.a.	n.a.	n.a.	2,491	n.a.	2,827	2,633

enriched in MgO, Na_2O, and Sr and are depleted in Al_2O_3 and yttrium (Y).

It is interesting to compare the Pinatubo dacite with that of Mount St. Helens, because many data are available for these (for example, Smith and Leeman, 1987). Compared to 1980 Mount St. Helens dacite, Pinatubo dacite is enriched in SiO_2 and lower in Al_2O_3 and FeO. These differences exist also if older Mount St. Helens dacites are

considered, except for a few lavas of the Kalama period (Smith and Leeman, 1987). Large-ion lithophile element (LILE) abundances in Pinatubo dacite are slightly higher (in agreement with slightly higher K_2O), zirconium (Zr) abundances are identical, and niobium (Nb) abundances are lower than in dacite from Mount St. Helens.

ISOTOPES

STRONTIUM AND NEODYMIUM ISOTOPIC COMPOSITIONS

Sr and Nd isotopic compositions were measured on type 1 (phenocryst-rich) and type 2 (phenocryst-poor) dacites of Pinatubo. Both types give identical compositions within the error limits of the analytical method, even with some of the analyses carried out in different laboratories (table 9). These compositions are in close agreement with the data reported by Castillo and Punongbayan (this volume). A sample of the andesite found as an inclusion in a dacitic pumice clast (Pin6, table 1) and similar to the andesite scoria described by Pallister and others (1992) had Sr and Nd isotopic compositions identical to those of the dacites (table 9). In contrast, the basalt contains slightly more radiogenic Sr and slightly less radiogenic Nd than the dacites (table 9).

$^{87}Sr/^{86}Sr$ compositions were also obtained for mineral separates (table 9). Because of the small amounts of anhydrite present in the samples, this mineral was extracted by leaching with distilled water. Since no other water-soluble mineral was observed in the dacite, and because the anhydrite is Sr rich (SrO up to 0.23 wt%), the analyses of the leachates are considered to be representative of the Sr isotopic compositions of this sulfate. In the four leachate samples analyzed, the anhydrite has the same Sr compositions

Table 2. Modal abundances of phenocryst-rich and phenocryst-poor dacites.

[Phenocryst and matrix abundances are in volume percent on a vesicle-free basis. Modal abundances were measured with a JEOL 733 scanning electron microscope and a VOYAGER-NORAN image analysis system on polished pumice samples previously impregnated by a fluid epoxy resin under vacuum. Vesicles' abundances were easily measured from back-scattered electron images by using the high atomic number contrast between the low-atomic-number epoxy resin filling the pores and the silicate matrix. Phenocrysts and microphenocrysts (>5 μm) were measured on the same areas by automated phase analysis with the energy dispersive spectrometer (counting time: 10 s/phase). Glassy matrix abundances were obtained as the difference between the total area of the image and the sum of the fractional areas occupied by vesicles+phenocrysts+microphenocrysts]

	Type 1 dacite (phenocryst rich)	Type 2 dacite (phenocryst poor)
Plagioclase	29	19
Hornblende	12	4
Fe-Ti oxides	.8	.4
Quartz	<.1	<.1
Apatite	.2	.1
Anhdrite	.1	<.1
Biotite	—	—
Clinopyroxene	—	—
Sum of phenocrysts	42	24
Glassy matrix	58	76
Vesicles	63	56
Measured density (g/cm³)	.92	1.07

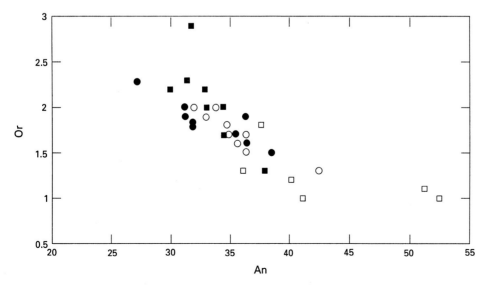

Figure 4. Orthoclase (Or) versus anorthite (An) content (percent) of plagioclase from dacite (filled symbols = core compositions, open symbols = rim compositions). Squares, phenocryst-poor dacites; circles, phenocryst-rich dacites.

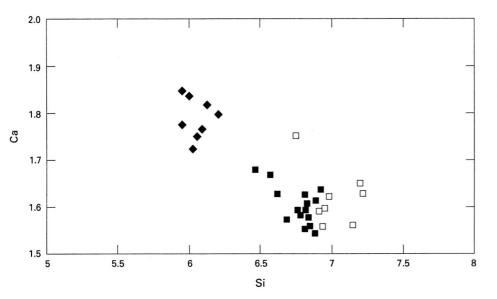

Figure 5. Silicon versus calcium content for amphiboles (number of atoms on the basis of 23 oxygen atoms). Diamonds, basalt; open squares, andesite; filled squares, dacite.

Table 3. Microprobe analytical compositions of amphibole.

[Oxide values are in weight percent; n.a., not analyzed]

Rock type Sample	Dacite Pt16		Andesite Pt43 core	Basalt Pt53		
	Core	Cummingtonite rim		core	rim	Olivine rim
SiO₂............................	45.02	52.59	49.67	40.85	39.88	40.88
TiO₂............................	1.67	.34	.76	1.57	1.59	1.66
Al₂O₃	12.53	2.53	5.87	14.43	13.89	13.87
MgO............................	15.06	20.98	15.63	14.49	14.98	15.52
CaO	10.95	1.66	10.51	11.72	11.59	11.00
MnO............................	.15	.96	.66	.22	.16	.15
FeO............................	10.45	17.69	13.39	11.89	11.80	11.48
Na₂O	2.26	.47	1.16	2.72	2.59	2.62
K₂O	.46	.02	.22	.83	.75	.75
F................................	n.a.	n.a.	n.a.	n.a.	n.a.	n.a.
Cl..............................	n.a.	n.a.	n.a.	n.a.	n.a.	n.a.
H₂O calculated.............	2.09	2.08	2.07	2.05	2.01	2.04
Total	100.64	99.32	99.94	100.77	99.24	99.97
Number of cations on the basis of 23 oxygens						
Si	6.462	7.592	7.209	5.983	5.938	6.010
Al...............................	2.119	.430	1.004	2.491	2.437	2.403
Al^IV	1.538	.408	.791	2.017	2.062	1.990
Al^VI	.581	.023	.213	.473	.376	.412
Ti	.180	.037	.083	.173	.178	.184
Mg..............................	3.223	4.517	3.383	3.165	3.326	3.402
Ca	1.684	.257	1.634	1.839	1.849	1.732
Mn..............................	.018	.117	.081	.027	.020	.019
Fe...............................	1.255	2.138	1.627	1.458	1.471	1.413
Na	.629	.132	.326	.772	.748	.747
K................................	.084	.004	.041	.155	.143	.141
Sum	17.774	15.654	16.393	18.553	18.547	18.141

Table 4. Microprobe analytical compositions of pyroxene.

[Oxide values are in weight percent; n.a., not analyzed; Wo, wollastonite; En, enstatite; Fs, ferrosilite]

Rock type Sample	Clinopyroxene								Orthopyroxene		
	Andesitic pumice Pt41			Andesite Pt46					Andesite Pt46		
	core	rim	core	rim	core	rim	core	rim	#1	#2	#3
SiO$_2$...............	52.57	50.74	51.54	49.53	53.09	52.43	51.92	52.10	55.04	55.64	56.09
TiO$_2$................	.19	.43	.35	.51	.18	.23	.26	.27	.02	.10	.04
Al$_2$O$_3$	1.89	3.60	3.00	4.01	1.90	1.91	2.24	2.24	.60	1.08	.50
Fe$_2$O.................	2.88	3.14	2.44	4.65	2.28	2.57	3.34	3.49	.00	.00	.00
FeO.................	1.98	2.64	3.29	2.19	2.42	2.87	2.22	2.21	15.97	14.82	13.00
MnO	.12	.12	.15	.21	.12	.19	.14	.17	.66	.75	.59
MgO	17.93	15.62	16.71	15.51	17.79	17.42	17.70	17.37	28.72	24.48	30.71
CaO	21.83	22.65	21.62	22.12	22.04	21.73	21.59	21.92	.14	.89	.16
Na$_2$O	.21	.31	.22	.28	.23	.19	.14	.24	.01	.21	.00
K$_2$O.................	.00	.00	.00	.00	.03	.02	.03	.00	.00	.00	.00
Cr$_2$O$_3$	n.a.	n.a.	n.a.	n.a.	n.a.	n.a.	n.a.	n.a.	.00	.01	.01
Total	99.61	99.24	99.31	99.01	100.08	99.57	99.58	100.01	101.16	97.99	101.09
Number of cations on the basis of 6 oxygens											
Si	1.922	1.877	1.899	1.843	1.932	1.924	1.904	1.905	1.965	2.035	1.974
Ti	.005	.012	.010	.014	.005	.006	.007	.007	.001	.003	.001
Al.....................	.081	.157	.130	.176	.081	.083	.097	.097	.025	.047	.021
Fe^{3+}	.079	.087	.068	.130	.063	.071	.092	.096	.000	.000	.000
Fe^{2+}	.060	.082	.101	.068	.074	.088	.068	.068	.477	.453	.383
Mn	.004	.004	.005	.007	.004	.006	.004	.005	.020	.023	.018
Mg	.977	.861	.918	.860	.965	.953	.968	.947	1.528	1.334	1.611
Ca	.855	.898	.854	.882	.859	.854	.848	.859	.006	.035	.006
Na	.015	.022	.016	.020	.016	.014	.010	.017	.001	.015	.000
K.......................	.000	.000	.000	.000	.001	.001	.001	.000	.000	.000	.000
Cr.....................	–	–	–	–	–	–	–	–	.000	.000	.000
Wo	43.3	46.5	43.9	45.3	43.7	43.3	42.8	43.5	.3	1.9	.3
En	49.5	44.6	47.2	44.2	49.1	48.3	48.9	47.9	75.3	72.3	79.7
Fs......................	7.2	9.0	8.9	10.5	7.2	8.4	8.3	8.6	24.5	25.8	19.8
Mg/(Mg+Fe)	87.6	83.4	84.5	81.3	87.6	85.7	85.8	85.2	75.46	73.70	80.07

Table 5. Microprobe analytical compositions of anhydrite.

[Values are in weight percent]

Rock type Sample	Dacite Pt31
CaO.........................	42.68
SrO	.23
BaO.........................	.07
Na$_2$O......................	.01
K$_2$O	.00
FeO	.03
SO$_3$	57.22
Total..................	100.24

as the whole rocks. This similarity supports the assumption of a primary magmatic origin for this sulfate, as suggested in the previous sections.

Data obtained from other phenocrysts, plagioclase, and amphibole show likewise that they have crystallized in isotopic equilibrium with the silicate melt.

^{87}Sr/^{86}Sr and ^{143}Nd/^{144}Nd isotopic compositions of the 1991 Pinatubo products are at the high ^{143}Nd/^{144}Nd and low ^{87}Sr/^{86}Sr ends of the spectra observed for the Bataan segment (fig. 9). Relative to most other volcanoes of the Bataan segment, they show a very slight shift toward high ^{87}Sr/^{86}Sr (fig. 9). Another Bataan segment sample that shows such a shift is from Mount Natib (Defant and others, 1991), a nearby volcano that appears to be heterogeneous with respect to Sr isotopic composition (Knittel and Defant, 1988, reported ^{87}Sr/^{86}Sr ratios of 0.7042–0.7051 from Mount Natib).

Knittel and others (1988), Knittel and Defant (1988) and Defant and others (1990, 1991) have shown that recent volcanism along the Luzon volcanic arc shows systematic regional variation. In northern Luzon (North Luzon segment) the volcanics have Sr-Nd isotopic compositions similar to older (middle Tertiary) plutonic rocks and to recent

Table 6. Microprobe analytical compositions of biotite.

[Oxide values are in weight percent; n.a., not analyzed]

Rock type Sample	Andesite (biotite in rim around olivine)		Dacite (phenocryst)
	PIN6		Pt23
	3	4	
SiO$_2$	40.61	38.55	38.62
TiO$_2$	.52	1.82	3.04
Al$_2$O$_3$	14.22	14.49	13.35
Cr$_2$O$_3$	.53	.20	n.a.
FeO	5.28	11.65	14.44
MnO	.06	.12	.09
MgO	24.31	18.82	17.59
CaO	.00	.00	.00
Na$_2$O	.86	.78	.76
K$_2$O	8.90	8.66	9.06
H$_2$O	3.81	3.68	4.05
Total	99.10	98.77	101.00
Number of cations on the basis of 22 oxygens			
Si	5.764	5.659	5.650
Ti	.056	.201	.330
Al	2.379	2.507	2.330
Cr	.059	.023	n.a.
Fe	.627	1.430	1.770
Mn	.008	.015	.010
Mg	5.143	4.117	3.840
Ca	.000	.001	.000
Na	.236	.221	.220
K	1.612	1.622	1.690
OH	4.000	4.000	3.960
Total	19.884	19.796	19.770

volcanics associated with westward subduction along the Philippine trench (eastern coast of Luzon, fig. 1; Knittel-Weber and Knittel, 1990). Toward the south (Bataan and Mindoro segments), the volcanics have progressively less radiogenic Nd and more radiogenic Sr values. Knittel and others (1988) and Defant and others (1991) suggested that these differences can be attributed to the incorporation of more crustal material into the source region of these mantle-derived magmas. The likely source of the crustal material is subducted fragments from the North Palawan continental terrane (NPCT), which collided with the Philippine archipelago during the late Miocene, or sediments derived from the NPCT (fig. 1).

The 1991 Pinatubo dacites and their phenocrysts have $^{87}Sr/^{86}Sr$ and $^{143}Nd/^{144}Nd$ isotopic signatures in agreement with the volcano's location relatively far from the NPCT, as compared with the Macolod corridor or the Mindoro segment. Therefore, it is unlikely that the sulfur enrichment observed in the Pinatubo magma is resulting from unusually large contamination by marine sediments (evaporites).

LEAD ISOTOPIC COMPOSITION

Lead isotopic compositions were measured on two crystal-rich dacites (Pin4 and Pin7), an andesite (Pin6) and a basalt inclusion (Pt53). They are listed in table 9 and are presented graphically in figure 10. The dacite and the andesite have virtually the same composition, while the basalt contains slightly more radiogenic Pb.

Mukasa and others (1987) have shown that volcanic rocks from the Philippines extruded after the collision episode and close to the NPCT collision zone (that is, southern Luzon, Panay, and Mindoro) are characterized by high $^{208}Pb/^{204}Pb$ and $^{207}Pb/^{204}Pb$ and relatively low $^{206}Pb/^{204}Pb$. Lead isotopic compositions obtained from various older rocks and rocks remote from the collision zone (the West

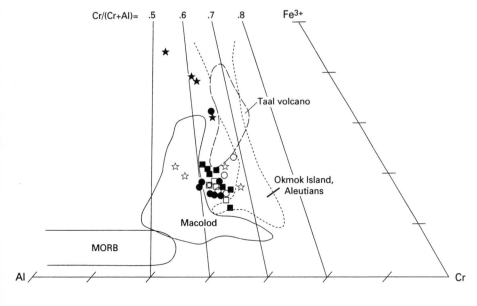

Figure 6. Compositional variation of chromite inclusions in olivine in the different Pinatubo eruption products, plotted in the Cr-Al-Fe^{+3} triangle. Chromite inclusions in olivine from basalt are shown as filled squares (Pt42) and circles (Pt53). Those from andesite are shown by open squares and circles (two different parts of sample Pt42), and those from dacite as filled (Pt21) and open (Pt9) stars. Fields for Cr-spinels from Macolod basalts (Knittel and Oles, 1995), Taal basalt (U. Knittel, unpub. data, 1992), and basalts from Okmok Island (Nye and Reid, 1986) are shown for comparison; spinels from MORB typically plot below and to the left of the Macolod field (Dick and Bullen, 1984).

Table 7. Microprobe analytical compositions of olivine.

[Oxide values are in weight percent]

Rock type Sample	Dacite						Andesite				Basalt	
	Pt26 K1		Pt41 K2		Pt46 K3		Pt42 K5		K2		Pt53 K5	
	core	rim	core	rim	core	rim	core	rim	core	rim	core	rim
SiO₂................	41.06	41.27	40.90	40.46	40.07	40.50	42.05	41.63	41.08	41.10	40.50	41.27
FeO................	11.28	12.94	11.00	13.08	13.31	13.40	10.46	11.02	11.47	13.09	11.02	12.66
MnO...............	.16	.26	.21	.24	.24	.25	.19	.21	.22	.24	.21	.22
MgO...............	46.38	45.85	47.28	46.25	45.98	46.39	45.88	46.33	47.74	46.87	48.34	47.04
CaO................	.19	.25	.16	.18	.18	.20	.16	.15	.17	.21	.19	.22
NiO................	.23	.16	.25	.18	.21	.14	.28	.25	.22	.12	.22	.16
Total.............	99.30	100.73	99.80	100.39	99.99	100.88	99.02	99.59	100.90	101.63	100.48	101.57
Number of cations on the basis of 4 oxygens												
Si....................	1.018	1.017	1.010	1.003	.999	1.001	1.040	1.027	1.005	1.005	.995	1.008
Fe...................	.234	.267	.227	.271	.278	.277	.216	.227	.235	.268	.226	.259
Mn..................	.003	.005	.004	.005	.005	.005	.004	.004	.005	.005	.004	.005
Mg..................	1.715	1.684	1.740	1.709	1.710	1.709	1.691	1.704	1.741	1.709	1.770	1.712
Ca...................	.005	.007	.004	.005	.005	.005	.004	.004	.004	.006	.005	.006
Ni...................	.005	.003	.005	.004	.004	.003	.006	.005	.004	.002	.004	.003
Total.............	2.980	2.983	2.990	2.997	3.001	3.000	2.961	2.971	2.994	2.995	3.004	2.993
Fo...................	88.0	86.3	88.5	86.3	86.0	86.1	88.7	88.2	88.1	86.4	88.7	86.9

Table 8. Microprobe analytical compositions of chromite inclusions in olivine.

[Oxide values are in weight percent]

Rock type Sample	Dacitic pumice Pt40	Mingled pumice		Andesite		Basalt	
		Pt41	Pt41	Pt42	Pt42	Pt53	Pt54
SiO₂.............................	0.12	0.23	0.34	0.10	0.09	0.47	0.16
TiO₂.............................	.97	1.10	1.01	.80	.85	1.17	.57
Al₂O₃............................	16.26	17.17	14.24	15.37	12.09	20.64	17.46
Fe₂O₃............................	15.17	13.99	16.69	11.84	18.13	25.26	9.23
Cr₂O₃............................	36.05	37.61	35.60	41.73	37.48	19.96	43.34
FeO.............................	22.94	19.77	27.73	19.74	23.61	24.03	16.09
MgO............................	8.09	10.62	5.20	10.00	7.03	8.16	12.49
Total........................	99.60	100.49	100.81	99.58	99.28	99.69	99.34
Cations on the basis of 3 cations and 4 oxygens							
Si................................	.004	.007	.011	.003	.003	.015	.005
Ti................................	.024	.026	.025	.020	.022	.028	.014
Al................................	.631	.646	.562	.590	.483	.787	.654
Fe³⁺.............................	.376	.336	.421	.290	.463	.615	.221
Cr................................	.938	.950	.943	1.074	1.005	.511	1.088
Fe²⁺.............................	.631	.528	.777	.537	.669	.650	.427
Mg...............................	.397	.506	.260	.485	.355	.394	.591

Philippines volcanic arc) contain less radiogenic Pb (Mukasa and others, 1987). Despite the relatively unradiogenic Sr and radiogenic Nd, the 1991 Pinatubo rocks fall well into the field of recent volcanic arcs (postcollision) located close to the NPTC collision zone. Similar Pb isotopic ratios were also observed in lavas and mantle nodules from Batan Island, a volcanic island located north of Luzon (fig. 10), and interpreted to indicate the presence of lead derived from subducted sediments in the source region (Vidal and others, 1989).

¹⁰BE

The presence of significant amounts of subducted sediment potentially can be monitored by ¹⁰Be, a short-lived

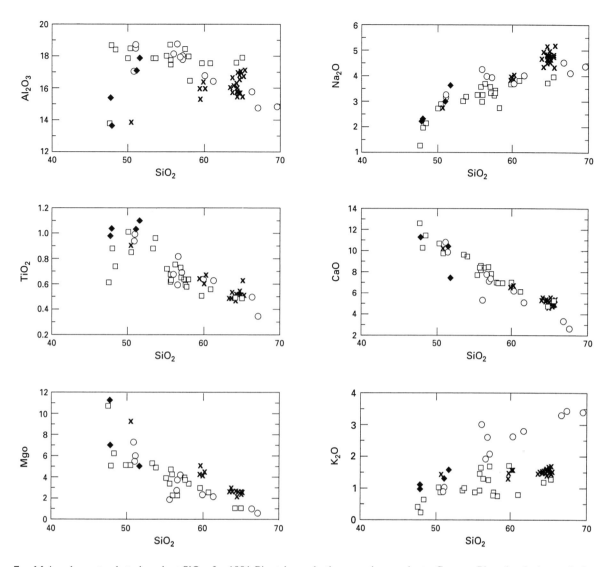

Figure 7. Major elements plotted against SiO_2, for 1991 Pinatubo and other eruption products. Crosses, Pinatubo dacite, andesite, and basalt (together); diamond, Bataan behind-arc volcano lavas; square, Bataan arc-front volcano lavas; circle, Macolod corridor lavas.

cosmogenic nuclide (half-life 1.5 m.y.) that is produced in the atmosphere, from where it is removed by rain or adsorbed by solid particles. ^{10}Be concentrations in marine sediments are particularly high because of low sedimentation rates (on average $5,000\times10^6$ atoms $^{10}Be/g$ sample in young sediment, Brown, 1984). For ^{10}Be to be found in arc magmas, subduction of the uppermost sediment layer and incorporation of that sediment into the magma source in <9 m.y. is required (Morris and others, 1990). Thus, lack of detectable ^{10}Be does not imply the lack of a sediment component in the source region; ^{10}Be could also be undetected because the uppermost sediment layer was scraped off or because subducted components were recycled too slowly (Woodhead and Fraser, 1985).

Two samples of phenocryst-rich pumice, Pt12 and Pt25, which were also analyzed for their Sr and Nd isotopic composition (table 9), were chosen for ^{10}Be analysis. These samples contain 1.96 and 2.77×10^6 atoms $^{10}Be/g$ sample

(table 10), respectively, which is higher than the value of 0.8×10^6 obtained for Mount Mayon (Tera and others, 1986), the only other Philippine volcano for which ^{10}Be data are available. These data are evidence that the Pinatubo dacites contain a sedimentary component that has been recycled through the subduction system.

SULFUR ISOTOPES

$\delta^{34}S$ for whole-rock pumice, water-soluble anhydrite, and insoluble sulfur were obtained on both types of dacites and are listed in table 11. These data show a relatively large spread in $\delta^{34}S$ values, ranging from +6.9 to +12.2‰, and bracket $\delta^{34}S$ values reported elsewhere for the dacite: +7.8 to +9.0‰ (Imai and others, this volume). This range extends even to lower values ($^{34}S/^{32}S = 6.3‰$) for soluble sulfate measured on the andesite (Pin6) and suggests a

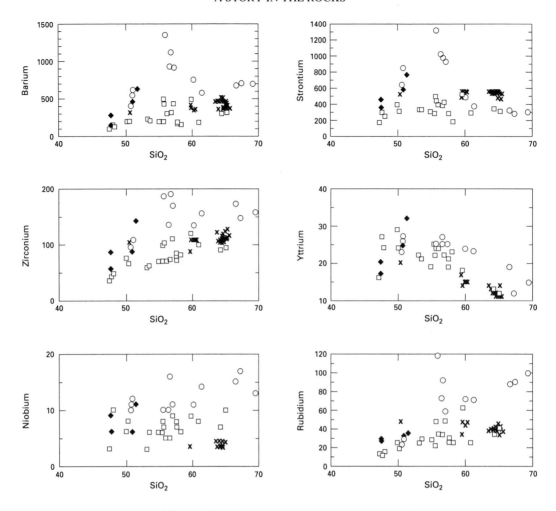

Figure 8.   Trace elements plotted against SiO_2, for 1991 Pinatubo and other eruption products. Symbols as in figure 7.

relatively heterogeneous composition for sulfur. Individual anhydrite phenocrysts of June 12 products show a very broad range of $\delta^{34}S$ values from 3 to 16, but phenocrysts of June 15 products show a narrower range of $\delta^{34}S$ values (McKibben and others, this volume). McKibben and others conclude that contamination of the magma by extraneous sulfur at shallow levels seems necessary to explain the observations on June 12 samples. This conclusion is also made by Hattori (this volume) on the basis of a study of sulfides in the 1991 Pinatubo eruption products.

Two fresh ash samples collected on July 4, 1991 (Pin12 and Pin14), were analyzed also. The sulfur content of these ash samples (table 1) is markedly higher than the sulfur content of the dacite because a significant amount of sulfate derived from the gas phase in the eruptive column is adsorbed on these samples. In consequence, the measured $\delta^{34}S$ is a mixed contribution of about 30 to 35 percent of sulfate from the gas phase and 65 to 70 percent of sulfate derived from the anhydrite microphenocrysts. These values are within the range of data obtained on the dacite.

DISCUSSION

GENESIS OF THE BASALT

The basalt, found as inclusions in the June 7–12, 1991, andesite lava, is likely to have been generated by partial melting of peridotite because of its high Mg content, the high Fo content of olivine phenocrysts, and the abundance of chromite inclusions in olivine. Furthermore, typical olivine core compositions are in equilibrium with a melt having $Mg/(Mg+Fe^{2+}) = 69$ (assuming a partition coefficient K_d $Fe/Mg_{(olivine/liquid)} = 0.3$; Roeder and Emslie, 1970), and this is the observed $Mg/(Mg+Fe^{2+})$ of the basalt, assuming $Fe^{2+}/(Fe^{3+}+Fe^{2+}) = 0.85$ (Nicholls and Whitford, 1976). This equilibrium between core compositions and the melt in turn implies that the basalt was not highly oxidized or was oxidized only at a late stage. This conclusion is supported by the presence of abundant sulfides in the basalt (Hattori, this volume).

The fact that the basalt appears to be in equilibrium with olivine Fo_{88} would suggest that the magma is not far removed from its primary composition by fractional

Table 9. Strontium, neodymium, and lead isotopic compositions of the 1991 Mount Pinatubo rocks and phenocrysts.

[ID, isotopic dilution; XRF, X-ray fluorescence; 2σ, 2 standard deviations]

Sample	Rb (ppm)	Sr (ppm)	$^{87}Sr/^{86}Sr$	2σ	$^{143}Nd/^{144}Nd$	2σ	Pb ID	Pb XRF	$^{206}Pb/^{204}Pb$	2σ	$^{207}Pb/^{204}Pb$	2σ	$^{208}Pb/^{204}Pb$	2σ
Pin4 dacite														
Whole-rock	37	565	0.704206	7	0.512915	22	9	9.9	18.430	13	15.598	14	38.623	39
Anhydrite			.704214	34										
Double leached .			.704202	6	.512910	13	5.3	7.2	18.440	16	15.608	20	38.628	45
Pin6 andesite														
Whole-rock	34	490	.704194	5	.512924	8	7.1	5.5	18.426	14	15.584	15	38.586	42
Anhydrite			.704201	20										
Pin7 dacite														
Whole-rock	38	562	.704225	5	.512914	17	8.8	11	18.435	15	15.594	17	38.612	42
Anhydrite			.704201	15										
Pt53 basalt														
Whole-rock			.704330	7	.512786	8	6.3	8	18.482	15	15.621	14	38.719	39
Pt29 dacite														
Whole-rock			.70422	1	.512863	11								
Amphibole			.70433	3	.512903	5								
Plagioclase			.70420	1	.512864	11								
Leachate			.70421	3										
Pt25 dacite														
Whole-rock			.70425	1	.512891	4								
Amphibole			.70419	3	.51298	1								
Plagioclase			.70422	1	.51290	6								
Pt12 dacite														
Whole-rock			.70422	1	.512901	8								

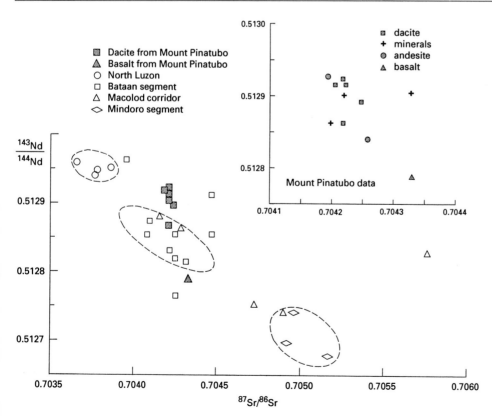

Figure 9. $^{143}Nd/^{144}Nd$ versus $^{87}Sr/^{86}Sr$ for volcanic rocks from the southern part of the Taiwan-Luzon arc (North Luzon, Bataan, and Mindoro segments, and Macolod corridor) compared to Pinatubo samples. Inset shows all Pinatubo data in detail (this paper; Castillo and Punongbayan, this volume). Main fields for data from the North Luzon, Bataan, and Mindoro segments are indicated. Data sources: North Luzon—Defant and others (1990); Bataan and Mindoro segments—Knittel and others (1988), Defant and others (1991), Mukasa and others (1994); Macolod corridor—Knittel and others (1988), Knittel and others (1992).

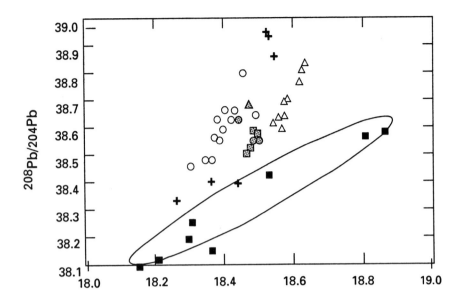

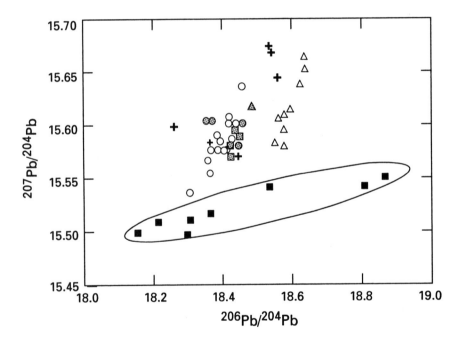

Figure 10. Lead isotopic compositions of 1991 Pinatubo eruption products (this study; Castillo and Punongbayan, this volume) compared to other eruption products from the Taiwan-Luzon arc. Data sources: Batan Island—Vidal and others (1989), McDermott and other (1993); Taal volcano, precollision volcanics and postcollision rocks far from the collision zone—Mukasa and others (1987, 1994); sediments—McDermott and others (1993), Mukasa and others (1987). The observed steep trends indicate mixing between "precollision-type" mantle and subducted sediment.

EXPLANATION

▨ Dacite from
   Mount Pinatubo

◉ Andesite from
   Mount Pinatubo

▲ Basalt from
   Mount Pinatubo

○ Batan Island (Vidal and
   others, 1989; McDermott
   and others, 1993

△ Taal

+ Sediments

■ Precollision (Panay) and
   Postcollision far from
   the collision zone

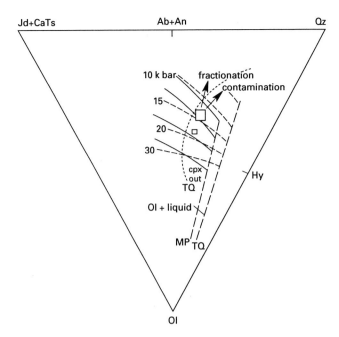

Figure 11. Basalt from Pinatubo (large square) plotted into the (Jd+CaTs)-Qz-Ol (jadeite+Ca tschermakite - quartz - olivine) projection of the basalt tetrahedron of Falloon and others (1988). Shown are melt compositions derived from moderately depleted MORB-pyrolite (MP) and highly depleted Tinaquillo lherzolite (TQ). Subvertical lines are melts in equilibrium with olivine. For melts derived from TQ, compositions to the right of the dotted "cpx-out" line are in equilibrium with olivine+orthopyroxene, and those to the left with olivine+orthopyroxene+clinopyroxene. The small square represents Pinatubo basalt with 10 percent olivine added (see text). The arrows indicate the effect that fractionation and contamination have on compositions plotting in the same area as basalt of Mount Pinatubo. The composition of the basalt suggests earlier equilibration at about 14 kbar.

Table 10. Beryllium isotopic compositions of whole-rock pumice.

[Total procedure blank is 0.06×10^{-13}, error ±7%]

Sample	Sample mass	^9Be carrier mass	$^{10}Be/^9Be$	^{10}Be
Pt12........	4.9952 g	1.7456 mg	0.80×10^{-13}	1.96×10^6
Pt25........	5.0483 g	1.7464 mg	1.20×10^{-13}	2.77×10^6

Table 11. Sulfur isotopic compositions of dacite, andesite, and ash samples from Mount Pinatubo.

[Dacite I = phenocryst rich; dacite II = phenocryst poor; FR = fresh ash fall, July 4, 1991]

	Whole rock	Soluble sulfate	Insoluble
PT9B dacite II	11.67	12.21	8.23
PT14 dacite I	8.53	—	—
PT15 dacite I	7.85	8.72	—
PT32 dacite I	8.50	—	—
Pin3 dacite I.........	—	7.1	—
Pin7 dacite I.........	—	6.9	—
Pin4 dacite I.........	—	7.0	—
Pin6 andesite	—	6.3	—
FR11	—	9.4	—
FR9	—	9.7	—

crystallization. However, the relatively low nickel (Ni) content of the whole rock (131 ppm Ni) and the olivines (0.25% NiO at Fo_{88}) suggests that the melt does not represent a primary composition but is slightly evolved.

Despite this limitation, it is tempting to plot the composition into the (Jd+CaTs)-Qz-Ol projection of the basalt tetrahedron of Falloon and others (1988) to obtain a rough estimate of the pressure at which the melt equilibrated with the residue. The position of the basalt in the plot suggests a value of about 14 kbar (fig. 11) and also suggests that the melt was in equilibrium with olivine+orthopyroxene± spinel, that is, with a harzburgitic assemblage. Since the basalt has fractionated some olivine, the equilibrium pressure must have been slightly higher (addition of 10% olivine would give an equilibrium pressure of about 18 kbar; addition of more olivine does not seem to be justified in view of the high Mg# of the basalt). In particular, the effect of olivine fractionation may have been increased by some assimilation, which is indicated by the presence of rare quartz, amphibole, and large plagioclase crystals in the basalt.

Primitive basalts are very rare in island arcs. Mg/$(Mg+Fe^{2+})$ ratios of typical eruption products of arc-front volcanoes of the Bataan segment are 0.50 to 0.55 (Defant and others, 1991). However, primitive basalts have been erupted by numerous small eruption centers within the Macolod corridor in southwestern Luzon (Oles, 1988; Förster and others, 1990; Knittel and others, 1992). Olivine core compositions in these basalts reach Fo_{89}, and Mg/$(Mg+Fe^{2+})$ ratios of the more primitive basalts are 64 to 73. Chromite inclusions have similar, though usually slightly lower, Cr/(Cr+Al) than those in Pinatubo olivine xenocrysts (fig. 6).

The basalt inclusions of Mount Pinatubo closely resemble the primitive Macolod basalts in their major element composition, except for their lower Al_2O_3 content (fig. 12), a feature they share with basalts erupted from Taal, located in the western part of the Macolod corridor (Miklius and others, 1991). However, Taal basalts are distinctly more CaO rich compared to the Pinatubo basalt.

The scarcity of primitive basalts in arcs suggests that such magma may reach the surface in this environment only under special circumstances. The Macolod corridor (fig. 1) is a graben striking northeast-southwest that is roughly perpendicular to the volcanic arc (Voss, 1971; Oles, 1988; Förster and others, 1990). The presence of primitive basalt in the 1991 Pinatubo eruption products suggests that this volcano also may be located in an extensional setting. Indeed, Mount Pinatubo is situated at the intersection of the volcanic front and the Iba fracture zone, a major east-southeast-trending structure that cuts the Zambales Ophiolite

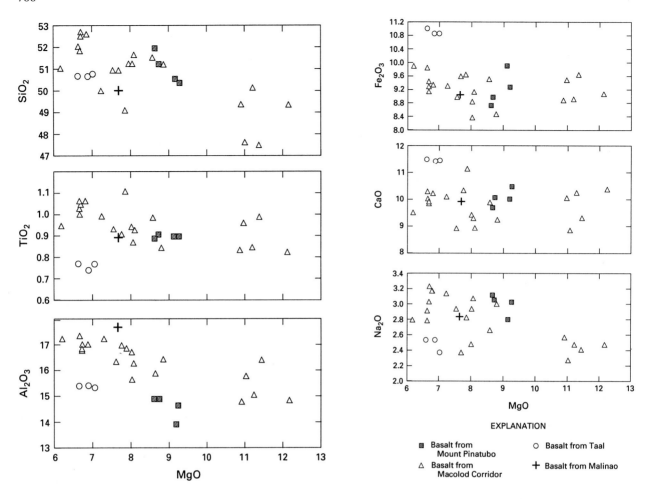

Figure 12. Major element compositions of basalts (in weight percent) from Pinatubo (this paper; Pallister and others, this volume) compared to other basalts from the Philippines (Macolod basalts—Knittel and Oles, in press; basalts from Taal—Miklius and others, 1991; basalt from Malinao (southern Luzon, Bicol arc—Knittel-Weber and Knittel, 1990).

Complex (fig. 2). Accordingly, we suggest that the magmatic activity of Mount Pinatubo is controlled by its position relative to the underlying plate and its location within what may be a tensional regime. The Iba fracture zone may be part of an incipient horst and graben system which, like the Macolod corridor, accommodates the differential movements resulting from eastward-directed subduction along the Manila trench to the north and westward-directed subduction along the Philippine trench to the south.

Additional evidence linking the basalt genesis to the setting of the Iba fracture zone is provided by the REE pattern of the basalt (data in Pallister and others, this volume). Pallister (written commun., 1993) noted that the Pinatubo basalt is highly enriched in the light REE (LREE) (La/Yb as high as 13), while other volcanic rocks along the volcanic front of the Bataan segment are only slightly enriched in LREE (La/Yb mostly about 4; Defant and others, 1991) (fig. 13). Similar high fractionation of LREE and heavy REE (HREE) is observed for some basalts of Mount Arayat, the only major stratovolcano located behind the volcanic

front of the Bataan segment, 45 km east-northeast of Mount Pinatubo. Most Arayat basalts have La/Yb = 5 to 10; only three samples have La/Yb = 14 to 18; Defant and others, 1991; Bau and Knittel, 1993). Mount Arayat may owe its existence to some "weak" zone within the crust, though we note that it does not lie along strike of the Iba fracture zone.

GENESIS OF THE DACITE

There is no generally accepted model for the genesis of dacite at convergent plate boundaries, and possibly there is no single unique mechanism responsible for the generation of such rocks. Commonly, the following mechanisms for genesis of arc dacite are considered (after Reid and Cole, 1983).

1. Fractional crystallization of basic precursors.
2. Mixing of rhyolitic and more basic magmas (Graham and Worthington, 1988); alternatively, a basic magma may have been massively contaminated by silicic rocks.

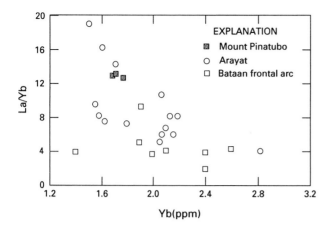

Figure 13. Lanthanum/ytterbium versus ytterbium for basalt of Pinatubo (this paper, Pallister and others, this volume), basalts of Arayat (Bau and Knittel, 1993), and various volcanic rocks from the volcanic front of the Bataan segment (Defant and others, 1991).

3. Generation of dacitic melt by melting of basaltic rocks either in amphibolite facies at crustal levels (Giese and others, 1986; Smith and Leeman, 1987) or in eclogite facies within the subducted slab (Drummond and Defant, 1990; Defant and Drummond, 1993).

The relatively high concentrations of incompatible trace elements in the basalt (K, Rb, Ba, Zr; fig. 8) rule out that the dacites are derived by fractional crystallization from the basalt that is found as inclusions in the andesite. For the same reason, Smith and Leeman (1987) concluded that the Mount St. Helens dacite is not derived from basalts erupted from Mount St. Helens. In addition, the difference in isotopic composition between dacites and basalt rules out a cogenetic relation. Alternatively, there is evidence that some volcanoes produced different types of basalt (for example, at Mount St. Helens; Leeman and others, 1990) and, thus, the dacites of Pinatubo could be derived from a basaltic precursor that is as yet not discovered at Mount Pinatubo. However, REE data for the dacite provided by Pallister and others (this volume) and Castillo and Punong-bayan (this volume) show that the dacite is depleted in HREE, and model calculations carried out by Martin (1987) show that such patterns probably cannot be generated by fractional crystallization of mantle-derived basalt.

In view of the suggestion of Pallister and others (1992) that the eruption was triggered by the intrusion of basaltic magma into a reservoir containing evolved melt, could magma mixing be a process for the generation of the dacite? According to Pallister and others (1992), dacite was one of the endmembers, and mixing produced andesite. Therefore, if the Pinatubo dacite had been produced by mixing, this must have occurred prior to the events that led to the 1991 eruption. In addition, the most evolved rocks known to have been erupted from Mount Pinatubo are dacites with 64 to 66 wt% SiO_2 (Defant, 1985; Pallister and others, this volume).

The Pinatubo dacites have some features pointing to an origin by melting of basaltic precursors, which typically are rich in Al_2O_3 (fig. 14) and are characterized by normative Ab/An >1 and low normative Or content (fig.15; Helz, 1976; Rapp and others, 1991). Surprisingly, the Pinatubo dacites have about the lowest Al_2O_3 contents of all dacites in Luzon (fig. 7 and Knittel and others, unpub. data, 1992), but many Archean trondhjemites, believed to have been generated by melting of basaltic protoliths, likewise have slightly lower Al_2O_3 than predicted by the experiments (Rapp and others, 1991).

Compared to lavas of other volcanoes from the volcanic front of the Bataan segment, the Pinatubo dacites are notable for their high Sr and low Y contents (fig. 16). In recent papers, Drummond and Defant (1990) and Defant and Drummond (1990) identified these features as characteristics of arc magmas that contain significant melt contributions from subducted oceanic crust that has been metamorphosed to eclogite.

In principle, the hypothesis that the Pinatubo dacite was generated by partial melting of eclogite in the subducting slab can be tested on the basis of the REE contents of these rocks, because garnet is the only petrologically significant phase able to fractionate the middle REE (MREE) from the HREE and to cause significant HREE depletion (see models calculated by Martin, 1987). The Pinatubo dacites show significant fractionation between LREE and HREE abundances (La_N/Yb_N =8, where the subscript "N" indicates a value that is normalized to chondritic values) and HREE depletion relative to "common" arc rocks (Yb_N = 6). We note, however, that the LREE/HREE fractionation and the HREE depletion of the Pinatubo dacite is smaller than in any of the rocks considered by Drummond and Defant (1990) to be partial melts derived from eclogite.

To further evaluate the origin of the dacites, we have calculated hypothetical REE patterns for the source of the Pinatubo dacite, assuming simple batch melting and that the residue consists only of garnet and clinopyroxene. A simple melting model seems to be justified, as small-volume melts probably cannot reach crustal levels because of the interaction with peridotitic wall rocks within the mantle.

A major problem of such calculations is choosing the distribution coefficients for REE in garnet and siliceous melts, because the range of values is wide (for example, Irving and Frey, 1978). Previously, low (Kay, 1978) as well as high (Martin, 1987) $K_{d(REE)}$ have been used to calculate models supporting the eclogite melting hypothesis. Though the data given by Irving and Frey (1978) suggest that REE are significantly more compatible in garnets found in siliceous rocks, we have considered two extreme data sets. Calculations assuming low $K_{d(REE)}$ values (those of Kay, 1978) require about 30% of garnet in the residue to give REE pattern with Yb_N around 10 (fig. 17), a value commonly observed for MORB including South China Sea basalts (which are represented by the East Taiwan Ophiolite

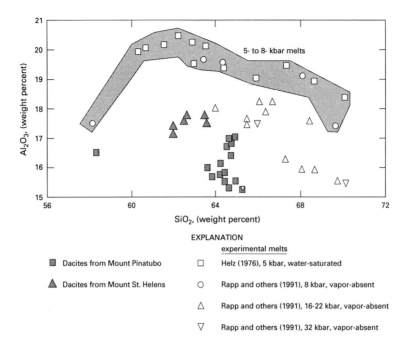

Figure 14. Al_2O_3 versus SiO_2 plot comparing dacites from Pinatubo and experimentally produced melts in the range SiO_2=56–71 wt% (Helz, 1976; Rapp and others, 1991). Dacites from Mount St. Helens (Smith and Leeman, 1987) are plotted for comparison. Note that melts produced at 5 to 8 kbar have significantly higher Al_2O_3 contents than both Mount Pinatubo and Mount St. Helens dacites.

EXPLANATION

experimental melts

■ Dacites from Mount Pinatubo   □ Helz (1976), 5 kbar, water-saturated

▲ Dacites from Mount St. Helens   ○ Rapp and others (1991), 8 kbar, vapor-absent

                                       △ Rapp and others (1991), 16-22 kbar, vapor-absent

                                       ▽ Rapp and others (1991), 32 kbar, vapor-absent

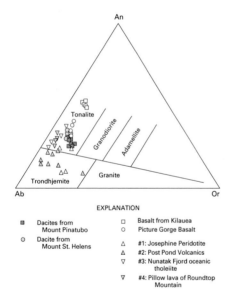

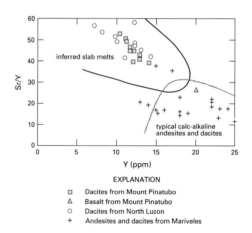

Figure 16. Strontium/yttrium plotted against yttrium for dacites from Mount Pinatubo (this paper), dacites from northern Luzon (Knittel and others, in press), and andesites and dacites from Mariveles (Defant, 1985). Fields for typical calc-alkaline lavas, considered to be derived from basic precursors, and for melts thought to have been generated by melting of MORB in eclogite facies, are after Drummond and Defant (1990) and Defant and Drummond (1990).

Figure 15. Dacites from Pinatubo and Mount St. Helens, plotted in the normative albite-orthoclase-anorthite triangle and compared to compositions produced by melting basalts experimentally. CIPW norms were calculated with FeO assumed to be $0.85FeO_{total}$. Data sources: Kilauea basalt and Picture Gorge basalt from Helz, 1976; basalt numbers 1–4 from Rapp and others (1991).

according to Chung and Sun, 1992; REE data from Jahn, 1986). A higher garnet content in the residue would produce unrealistic REE patterns with $La_N/Sm_N>1$ and $Sm_N/Yb_N<1$ (that is a wavy REE pattern); lower garnet content in the residue produces Yb_N lower than in MORB. Calculations

using high $K_{d(REE)}$ constrain the amount of residual garnet to about 10%; otherwise, the hypothetical source would be highly HREE enriched. MORB-type REE patterns are obtained with approximately 10% garnet and 30% partial melting (fig. 17).

To summarize, the source models calculated on the basis of the REE content of the Pinatubo dacite require either (1) a low garnet content of the residue, which is not

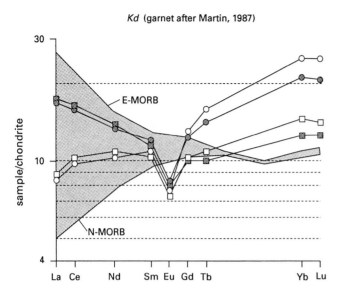

Kd (garnet after Martin, 1987)

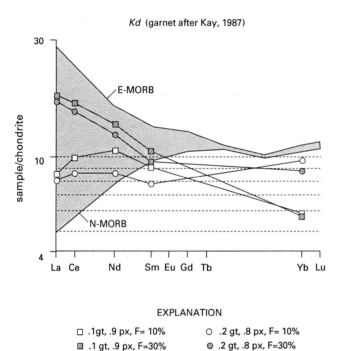

Kd (garnet after Kay, 1987)

EXPLANATION

☐ .1gt, .9 px, F= 10%   ○ .2 gt, .8 px, F= 10%

▨ .1 gt, .9 px, F=30%   ◉ .2 gt, .8 px, F=30%

Figure 17. Calculated REE patterns of hypothetical sources of dacite from Pinatubo, when it is assumed that the melts were generated by melting of eclogite. Shaded fields are patterns for MORB with boundaries based on E- and N-MORB from the South China Sea (Jahn, 1986; Chung and Sun, 1992). Degree of partial melting and residue compositions are given in the explanation (gt=garnet, px=clinopyroxene, F=melt fraction). Chondritic values after Nakamura (1974). K_d, distribution coefficient.

supported by the experimental results of Rapp and others (1991), or (2) that low $K_{d(REE)}$ apply for the genesis of Mount Pinatubo dacite, a condition that is not supported by what is known about the variation of $K_{d(REE)}$ in garnet.

Alternatively the source could have HREE contents significantly different from those of MORB. Thus, while the calculations do not disprove the eclogite melting hypothesis, due to the uncertainties involved, they provide no evidence for such a model.

The low garnet content in the residue, suggested by the calculations, would, however, be compatible with the assumption that the dacites were generated by melting of garnet-bearing amphibolite at crustal levels. Radiogenic isotopes would probably not show the involvement of crustal rocks, because there is no evidence for the presence of crustal rocks old enough to have evolved radiogenic isotopic composition. Sr isotopic compositions of two amphibolites from northern Luzon, which have been carried up by a pluton, have $^{87}Sr/^{86}Sr$ of about 0.7039 to 0.7040 (U. Knittel, unpub. data, 1992).

It is also possible that no single process is responsible for the genesis of Pinatubo dacite but that several processes have worked in combination, such as extensive interaction between mantle-derived magma and crustal materials (for example, the MASH model of Hildreth and Moorbath, 1988). This model requires an efficient transfer of Be from the basalt to the dacite to account for the observed ^{10}Be.

Tentatively we suggest the following petrogenetic model for Mount Pinatubo. Initially Pinatubo was an andesitic and dacitic stratovolcano (Newhall and others, this volume), built by lavas derived from primitive basalt generated by partial melting of the mantle wedge. Differentiation of the basalt is assumed to have taken place at the boundary between mantle and crust. Upon activation of the Iba fracture zone, deep faults provided conduits that allowed the primitive basalt to reach crustal levels. Due to their high density, these basalts did not reach the surface. During storage (and differentiation) of the basalt at crustal levels, the heat transfer to the host rocks caused them to melt partially. Possibly these anatectic melts mixed with differentiated melts derived from the basalt, to result in a homogeneous dacitic magma. Repeated injection of new primitive basalt into this reservoir may have triggered repeated violent eruptions.

The problem of the genesis of two texturally different types of dacite, phenocryst-rich and phenocryst-poor ones, cannot be solved in this model. Probably their genesis is related to the eruption dynamics.

MAGMA MIXING

The presence of olivine xenocrysts in dacite suggest that a small amount of mantle-derived basalt was mixed into the dacite; mixing of more substantial amounts of basalt with dacite resulted in the formation of andesite. (Pallister and others, 1992, calculated that a 40:60 mixture of basalt and dacite closely approximates the composition of the andesite).

Alternative sources for the olivine in dacite are unlikely: olivines from mantle peridotite and plutonic rocks typically have lower CaO content (<0.10 wt% CaO; Simkin and Smith, 1970) than those from Mount Pinatubo (0.19–0.25 wt% CaO; table 1), which are typical for volcanic rocks. Furthermore, olivines in plutonic mafic/ultramafic rocks (including metamorphic mantle peridotite) commonly lack tiny chromite inclusions, because this mineral forms discrete grains in deep-seated rocks, whereas chromite inclusions are common in olivine from primitive volcanic rocks. The ophiolitic basalts also are unlikely sources for the olivine xenocrysts because those basalts are mostly non-porphyritic, and clinopyroxene dominates over olivine among the phenocrysts (Evans and others, 1991). Hence, we would expect to see clinopyroxene xenocrysts also, if ophiolitic basalts were the source of xenocrystic olivine.

Therefore, the primitive basalts that were discovered as inclusions in andesite are the most likely source of the olivine xenocrysts (Knittel and others, 1991; Pallister and others, 1992). This conclusion is supported by the similar composition of the olivines and their chromite inclusions in dacite and basalt.

The presence of slightly contaminated basalt (with a few large quartz, amphibole, and plagioclase xenocrysts) in relatively homogeneous hybrid andesite, and the presence of olivine in dacite, suggest that mixing occurred in several stages. Following Koyaguchi (1986), we suggest that basalt intruding the magma reservoir ponded below dacite. Because of the faster convective overturn of the basalt (due to its lower viscosity), small batches of dacite were successively entrained into the basalt to result ultimately in the formation of the hybrid andesite. This andesite may have started vesiculation (high water content is indicated by the high amphibole content) and thereby reduced its density. This less-dense andesite may have become gravitationally unstable and started to rise. Thus, it disrupted the stratification of the magma chamber and trapped some basalt from the boundary layer between the hybrid andesite and the basalt. At the margins of the rising "diapir" some andesite was mixed into the dacite to produce the banded and mingled dacite. Some of this mixture was subsequently sufficiently diluted so that only the olivine xenocrysts, being isolated by amphibole rims, are still recognizable in dacite.

The evidence for the involvement of a basaltic component and mixing processes led Pallister and others (1992) to suggest that the eruption may have been triggered by injection of basalt into a reservoir containing dacitic melt. The evidence discussed above requires that some time must have elapsed between the injection of basalt into the reservoir and the eruption, in order for the andesite to have formed. It has been suggested that quartz may not survive a substantial period of time in disequilibrium with basalt. However, the clinopyroxene corona may have efficiently isolated the quartz from the basalt. Therefore, in view of the lack of other time constraints, we suggest that the events leading to the 1991 eruption were triggered by the July 16, 1990, earthquake. Influx of basalt into the dacite reservoir may initially have led only to increased solfataric activity in August 1990 (PHIVOLCS, 1990).

SEDIMENT INPUT

^{10}Be contents of Pinatubo samples are high, 1.96×10^6 and 2.77×10^6 atoms of ^{10}Be per gram of sample. Such high contents have been found outside arc settings only in rare cases where secondary, mostly atmospheric, alteration was likely (Tera and others, 1986). For the Pinatubo samples investigated here, such contamination can be ruled out because the samples were collected within a few days after the eruption. ^{10}Be contamination via some kind of hydrothermal exchange with meteoric water may be possible, but is very unlikely in view of the huge volume of the Pinatubo eruption and the very high water to rock ratios required (Tera and others, 1986). Pristine ^{10}Be content of volcanic rocks has been attributed to involvement of subducted sediments in the melting process (Tera and others, 1986; Morris and others, 1990).

Most of the western Pacific arc volcanoes are characterized by relatively low ^{10}Be, as in Japan (mostly less than 0.7×10^6 atoms ^{10}Be/g), the Marianas (0.1–0.5×10^6 atoms ^{10}Be/g), and the Sunda arc (0.1–0.6×10^6 atoms ^{10}Be/g) (fig. 18) (Tera and others, 1986; Woodhead and Fraser, 1985). The high values at Pinatubo suggest either that sediment is particularly rapidly recycled into the source region of Mount Pinatubo or that the magma contains a relatively large sediment component. The slight shift towards elevated ^{87}Sr/^{86}Sr relative to ^{143}Nd/^{144}Nd (see McCulloch and others, 1980), and the radiogenic Pb, suggest involvement of sediments and igneous rock that have experienced seawater alteration.

While the data leave little doubt about the presence of a recycled sedimentary component, the interpretation of these data is not straightforward, because the basalt found as inclusion in the dome andesite is not a potential precursor of the dacite. Hence the dacite may be derived from other sources, such as lower crustal sources, in which case the ^{10}Be must have been transferred from the basalt to the dacite by fluid phases.

On a more regional scale, the presence of ^{10}Be in the Pinatubo magmatic system supports the conclusion of Hayes and Lewis (1984) that sediment is subducted along the Manila trench. Furthermore, the presence of ^{10}Be in the Pinatubo pumice may be considered as evidence that the Manila trench subduction system is still active, despite limited seismic activity.

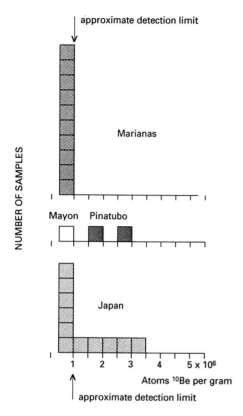

Figure 18. ^{10}Be content in lavas from several arcs of the western Pacific. Data sources: Pinatubo—this study; other data—Tera and others, 1986.

ORIGIN OF THE SULFUR ENRICHMENT

The Pinatubo dacite, with a S content of 1,500 to 2,400 ppm, is extraordinarily enriched in S as compared with arc volcanic rocks, which typically contain less than 40 ppm S (Ueda and Sakai, 1984). The isotopic composition of this S is, however, not unusual for interoceanic arc magmas (fig. 19). All the Pinatubo δ^{34}S values are characterized by much higher ^{34}S/^{32}S ratios than those reported for tholeiitic ocean ridge and ocean island basalt (δ^{34}S = −0.5 to +1.0‰) (Harmon and Hoefs, 1986; Sakai and others, 1982, 1984). They are, however, typical of other interoceanic arc volcanics, which are characterized by a large range of δ^{34}S (Ueda and Sakai, 1984; Woodhead and others, 1987) (fig. 19).

The two potential sources usually proposed for this S enrichment are S-rich crustal sediments such as marine evaporites or hydrothermal sulfide deposits in subducted oceanic crust. If the sulfur enrichment observed in the Pinatubo dacite is a consequence of the assimilation of evaporite beds by the magma on its way to the surface, the δ^{34}S should reflect much higher values, close to those reported for Oligocene evaporites of around 20 to 22‰ (Claypool and others, 1980). Moreover, there is no geologic evidence for the presence of evaporitic beds in the basement of the volcano or on the Luzon arc. It is therefore unlikely that contamination by evaporites is the source of this

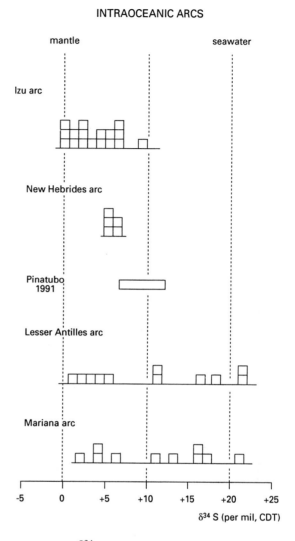

Figure 19. δ^{34}S values of dacites and andesites from Pinatubo compared to other intraoceanic arc magmas (modified from Woodhead and others, 1987).

unusual S enrichment. The relatively low ^{87}Sr/^{86}Sr isotopic ratios also preclude significant contributions from marine evaporites. On the other hand, sulfur of magmatic origin occurs in anhydrite in hydrothermally altered volcanic rocks in the basement of Mount Pinatubo, so crustal contamination as a mechanism of S enrichment cannot be excluded (Hattori, this volume).

Alternatively, hydrothermal sulfide deposits, generated on the oceanic crust, could have been subducted along the Manila trench and could be responsible for the unusual S enrichment. The isotopic compositions of these hydrothermal sulfide range from δ^{34}S = +4.3 to +10.7‰, being generally between +6 to +8‰ (Hallbach and others, 1989; Kusakabe and others, 1990). These relatively heavy S compositions are compatible with the δ^{34}S observed for the Pinatubo dacites.

Mass balance calculations suggest that during the eruption about half of the total S content of the magma was released as a gas phase and the remainder is essentially present as anhydrite phenocrysts (Westrich and Gerlach, 1992). In consequence, determination of the sulfur isotopic composition of the gas phase is essential to evaluate the $\delta^{34}S$ of the bulk magma. If the sulfate adsorbed on two ash samples (Pin12 and Pin14) is representative of the gas phase, the isotopic composition of leachates suggests that the vapor had roughly the same composition as the S left in the magma as anhydrite. Thus, on the basis of these S isotopic compositions, anhydrite breakdown as the source of stratospheric S cannot be confirmed or ruled out.

CONCLUSIONS

The mineralogical, geochemical, and isotopic data presented reveal no unusual features that may be connected with the extreme sulfur enrichment of the 1991 Pinatubo eruption products. While there is evidence for the involvement of recycled crustal material in the generation of the Pinatubo dacites, the extent of this contribution seems to be of similar magnitude as in other volcanoes of the Bataan segment. In particular, there is no evidence for the involvement of marine evaporites. Hence, the process leading to this unusual sulfur enrichment cannot be detected by classical petrological studies. In consequence, it would be extremely difficult to detect a similar sulfur-rich magma from old eruptions if anhydrite is completely removed, because no uncommon features are displayed except for the presence of anhydrite and a relatively high f_{O_2}. Clearly this leaves open the possibility that sulfur-rich eruptions are far more frequent than hitherto expected.

A few volcanoes are found to emit excess amounts of sulfur into the atmosphere. Basaltic magma as a source for this excess sulfur has repeatedly been invoked (see Andres and others, 1991). In the case of the 1991 Pinatubo eruption, there is clear evidence for the involvement of a basaltic component, the volume of which, however, is unknown. Similar basalt may have been added to the magma reservoir during previous mixing episodes, well before the June 12–15 eruption, but its role as a source for sulfur is doubtful for several reasons. First, the high f_{O2} of the dacite is not compatible with the exsolution of volatiles from a reduced basaltic melt, as also suggested by Westrich and Gerlach (1992). Second, the homogeneous distribution of anhydrite throughout a minimum of 5 km^3 of dacite is difficult to understand if the S had been recently introduced as a volatile phase from the bottom of the dacite magma. A homogeneous distribution of the anhydrite is more plausible if it formed following saturation of the melt in sulfate.

ACKNOWLEDGMENTS

We thank Ray Punongbayan and PHIVOLCS scientists for support in the field. Data collection in Aachen and participation in the International Scientific Conference on Mount Pinatubo in Manila was made possible by grants of Deutsche Forschungsgemeinschaft (grant Kn237/5–1 and travel grant). Fieldwork by Oles was largely sponsored by Professor H. Förster. Fieldwork by Bernard was supported by grants from Fonds National de la Recherche Scientifique (Belgium). M.D. Feigenson kindly determined the Sr and Nd isotopic composition of the two samples analyzed for ^{10}Be. We thank J.P. Mennessier (Brussels) for his help processing the anhydrite and lavas for chemical analysis, J.C. Duchesne and G. Bologne (Liège) for X-ray fluorescence analyses, and S. Wauthier (Louvain La Neuve) for the electron microprobe analyses. We also acknowledge preprints of other contributions to the Pinatubo volume that we received from J.S. Pallister, P.R. Castillo, M.A. McKibben, and A. Imai. C. Arcilla commented on several early versions of the manuscript. Finally we wish to thank our reviewers, T.M. Gerlach and M.J. Defant, largely for encouragement, and J.S. Pallister for a very detailed and constructive review; largely due to his review, the number of errors, misinterpretations, and omissions was kept at a minimum.

REFERENCES CITED

Andres, R.J., Rose, W.I., Kyle, P.R., de Silva, S., Francis, P., Gardeweg, M., and Moreno Roa, H., 1991, Excessive sulfur dioxide emissions from Chilean volcanoes: Journal of Volcanology and Geothermal Research, v. 46, p. 323–329.

Arculus, R.J., Johnson, R.W., Chappell, B.W., McKee, C.O., and Sakai, H., 1983, Ophiolite-contaminated andesites, trachybasalts, and cognate inclusions of Mt. Lamington, Papua New Guinea anhydrite-amphibole-bearing lavas and the 1951 cumulodome: Journal of Volcanology and Geothermal Research, v. 18, p. 21–247.

Bau, M., and Knittel, U., 1993, Significance of slab-derived partial melts and aqueous fluids for the genesis of tholeiitic and calc-alkaline island-arc basalts: Evidence from Mt. Arayat, Philippines: Chemical Geology, v. 105, p. 233–251.

Bernard, A., Demaiffe, D., Mattielli, N., and Punongbayan, R.S., 1991, Anhydrite-bearing pumices from Mount Pinatubo: Further evidence for the existence of sulphur-rich silicic magmas: Nature, v. 354, p. 139–140.

Bluth, G.J.S., Doiron, S.D., Schnetzler, C.C., Krueger, A.J., and Walter, L.S., 1992, Global tracking of the SO_2 clouds from the June, 1991 Mount Pinatubo eruptions: Geophysical Research Letters, v. 19, p. 151–154.

Brown, L., 1984, Applications of accelerator mass spectrometry: Annual Review of Earth and Planetary Science, v. 12, p. 39–69.

Cardwell, R.K., Isacks, B.L., and Karig, D.E., 1980, The spatial distribution of earthquakes, focal mechanism solutions, and subducted lithosphere in the Philippine and Northeastern Indonesian Islands, *in* Hayes, D.E., ed., The tectonic and geologic evolution of Southeast Asia: American Geophysical Union Monograph 23, p. 1–35.

Castillo, P.R., and Punongbayan, R.S., this volume, Petrology and Sr, Nd, and Pb isotopic geochemistry of Mount Pinatubo volcanic rocks.

Catanzaro, E.J., Murphy, T.J., Shields, W.R., and Garner, E.L., 1968, Absolute isotopic abundance ratios of common, equalatom, and radiogenic lead isotope standards: Journal of Research of the National Bureau of Standards, v. 72A, p. 261–267.

Chung, S.-l., and Sun, S.-s., 1992, A new genetic model for the East Taiwan Ophiolite and its implications for Dupal domains in the Northern Hemisphere: Tectonophysics, v. 109, p. 133–145.

Claypool, G.E., Holser, W.T., Kaplan, I.R., Sakai, H., and Zak, T., 1980, The age curves of sulfur and oxygen isotopes in marine sulfate and their mutual interpretation: Chemical Geology, v. 28, p. 199–260.

de Boer, J., Odom, L.A., Ragland, P.C., Snider, F.G., and Tilford, N.R., 1980, The Bataan Orogene: Eastward subduction, tectonic rotations, and volcanism in the western Pacifics (Philippines): Tectonophysics, v. 67, p. 251–282.

Defant, M.J., 1985, The potential origin of the potassium depth relationship in the Bataan orogene, the Philippines: Tallahassee, Florida State University, Ph.D. dissertation, 622 p.

Defant, M.J., and Drummond, M.S., 1990, Derivation of some modern arc magmas by melting of young subducted lithosphere: Nature, v. 347, p. 662–664.

————1993, Mount St. Helens: Potential example of the partial melting of the subducted lithosphere in a volcanic arc: Geology, v. 21, p. 547–550.

Defant, M.J., Jacques, D., Maury, R.C., de Boer, J.Z., and Joron, J.L., 1989, Geochemistry of the Luzon arc, Philippines: Geological Society of America Bulletin, v. 101, p. 663–672.

Defant, M.J., Maury, R.C., Joron, J.-L., Feigenson, M.D., Leterrier, J., Bellon, H., Jacques, D., and Richard, M., 1990, The geochemistry and tectonic setting of the northern section of the Luzon arc (the Philippines and Taiwan): Tectonophysics, v. 183, p. 187–205.

Defant, M.J., Maury, R.C., Ripley, E.M., Feigenson, M.D., and Jacques, D., 1991, An example of island-arc petrogenesis: Geochemistry and petrology of the southern Luzon Arc, Philippines: Journal of Petrology, v. 32, p. 455–500.

Dick, H.J.B., and Bullen, T., 1984, Chromian spinel as a petrogenetic indicator in abyssal and alpine-type peridotites and spatially associated lavas: Contributions to Mineralogy and Petrology, v. 86, p. 54–76.

Drummond, M.S., and Defant, M.J., 1990, A model for trondhjemite- tonalite-dacite genesis and crustal growth via slab-melting: Archean to modern comparisons: Journal of Geophysical Research, v. 95, p. 21503–21521.

Evans, C.A., Casteneda, G., and Franco, H., 1991, Geochemical complexities preserved in the volcanic rocks of the Zambales Ophiolite, Philippines: Journal of Geophysical Research, v. 96, p. 16251–16262.

Falloon, T.J., Green, D.H., Hatton, C.J., and Harris, K.L., 1988, Anhydrous partial melting of a fertile and depleted peridotite

from 2 to 30 kb and application to basalt petrogenesis: Journal of Petrology, v. 29, p. 1257–1282.

Förster, H., Oles, D., Knittel, U., Defant, M.C., and Torres, R.C., 1990, The Macolod Corridor: A rift crossing the Philippine island arc: Tectonophysics, v. 183, p. 265–271.

Fournelle, J.H., 1991, Anhydrite and sulfide in pumices from the 15 June 1991 eruption of Mt. Pinatubo; initial examination [abs.]: Eos, Transactions, American Geophysical Union, v. 72, no. 44, p. 68.

Geschwind, C.-H., and Rutherford, M.J., 1992, Cummingtonite and the evolution of the Mount St. Helens (Washington) magma system: An experimental study: Geology, v. 20, p. 1011–1014.

Giese, U., Knittel, U., and Kramm, U., 1986, The Paracale Intrusion: Geologic setting and petrogenesis of a trondhjemite intrusion in the Philippine island arc: Journal of Southeast Asian Earth Sciences, v. 1, p. 235–245.

Gill, J.B., 1981, Orogenic andesites and plate tectonics: Heidelberg, Springer, 390 p.

Graham, I.J., and Worthington, 1988, Petrogenesis of Tauhara dacite (Taupo Volcanic Zone, New Zealand)—evidence for magma mixing between high-alumina andesite and rhyolite: Journal of Volcanology and Geothermal Research, v. 35, p. 279–294.

Hallbach, P., Nakamura, K.-i., Wahsner, M., Lange, J., Sakai,H., Kaselitz, L., Hansen, R.-D., Yamano, M., Post, J., Prause, B., Seifert, R., Michaelis, W., Teichmann, F., Kinoshita, M., Marten, A., Ishibashi, J., Czerwinski, S., and Blum, N., 1989, Probable modern analogue of Kuroko-type massive sulfide deposits in the Okinawa Trough back-arc basin: Nature, v. 338, p. 496–499.

Harmon, R.S., and Hoefs, J., 1986, S-isotope relationships in late Cenozoic destructive plate margin and continental intra-plate volcanic rocks: Terra Cognita, v. 6, p. 182.

Hattori, K., this volume, Occurrence and origin of sulfide and sulfate in the 1991 Mount Pinatubo eruption products.

Hayes, D.E., and Lewis, S.D., 1984, A geophysical study of the Manila Trench, Luzon, Philippines. 1, Crustal structure, gravity, and regional tectonic evolution: Journal of Geophysical Research, v. 89, p. 9171–9195.

Helz, R.T., 1976, Phase relations of basalt in their melting ranges at P_{H_2O} = 5 Kb. Part II, Melt compositions: Journal of Petrology, v. 17, p. 139–193.

Hildreth, W., and Moorbath, S., 1988, Crustal contributions to arc magmatism in the Andes of central Chile: Contributions to Mineralogy and Petrology, v. 98, p. 455–489.

Hofman, D.J., Oltmans, S.J., Harris, J.M., Solomon, S., Deshler, T., and Johnson, B.J., 1992, Observation and possible causes of new ozone depletion in Antartica in 1991: Nature, v. 359, p. 283–287.

Imai, A., Listanco, E.L, and Fujii, T., this volume, Highly oxidized and sulfur-rich dacitic magma of Mount Pinatubo: Implication for metallogenesis of porphyry copper mineralization in the Western Luzon arc.

Irving, A.J., and Frey, F.A., 1978, Distribution of trace elements between garnet megacrysts and host volcanic liquids of kimberlitic to rhyolitic composition: Geochimica and Cosmochimica Acta, v. 42, p. 771–787.

Jahn, B.-m., 1986, Mid-ocean ridge or marginal basin origin of the East Taiwan Ophiolite: Chemical and isotopic evidence: Contributions to Mineralogy and Petrology, v. 92, p. 194–206.

Kay, R.W., 1978, Aleutian magnesian andesites: Melts from subducted Pacific ocean crust: Journal of Volcanology and Geothermal Research, v. 4, p. 117–132.

Kerr, R.A., 1993, Pinatubo global cooling on target: Science, v. 259, p. 594.

Klein, J., and Middleton, R., 1984, Accelerator mass spectrometry at the University of Pennsylvania: Nuclear Instruments and Methods, v. B5, p. 129–133.

Knittel., U., and Defant, M.J., 1988, Sr isotopic and trace element variations in Oligocene to Recent igneous rocks from the Philippine island arc: Evidence for recent enrichment in the sub-Philippine mantle: Earth and Planetary Science Letters, v. 87, p. 87-99.

Knittel, U., Defant, M.J., and Razcek, I., 1988, Recent enrichment in the source region of arc magmas from Luzon island, Philippines: Sr and Nd isotopic evidence: Geology, v. 16, p. 73–76.

Knittel, U., Hegner, E., Satir, M., Oles, D., and Förster, H., 1992, $^{87}Sr/^{86}Sr$ and $^{143}Nd/^{144}Nd$ isotopic ratios in rift basalts from the Taiwan-Luzon arc: implications for the composition of the sub-arc mantle: European Journal of Mineralogy, Supplement, v. 1, p. 150.

Knittel, U., and Oles, D., 1995, Basaltic volcanism associated with extensional tectonics in the Taiwan-Luzon island arc: Evidence for non-depleted sources and subduction zone enrichment, in Smellie, J.L., ed., Volcanism Associated with extension at Consuming Plate Margins: Geological Society Special Publication 81, p. 77–93.

Knittel, U., Oles, D., Förster, H., and Punongbayan, R.S., 1991, Pinatubo pumice contains anhydrite: Eos, Transactions of the American Geophysical Union, v. 72, p. 522.

Knittel, U., Trudu, A., Winter, W., Yang, T.F., and Gray, C., in press, Volcanism above a subducted extinct spreading center: A reconnaissance study of the North Luzon Segment of the Taiwan-Luzon Volcanic Arc (Philippines): Journal of Southeast Asian Earth Sciences.

Knittel-Weber, C., and Knittel, U., 1990, Petrology and genesis of the volcanic rocks on the eastern flank of Mount Malinao, Bicol arc (southern Luzon, Philippines): Journal of Southeast Asian Earth Sciences, v. 4, p. 267–280.

Koyaguchi, T., 1986, Evidence for two-stage mixing in magmatic inclusions and rhyolite lava domes on Niijima Island, Japan: Journal of Volcanology and Geothermal Research, v. 29, p. 71–98.

Kusakabe, M., Mayeda, S., and Nakamura, E., 1990, S, O, and Sr isotope systematics of active vent materials from the Mariana backarc basin spreading axis at 18°N: Earth and Planetary Science Letters, v. 100, p. 275–282.

Leeman, W.P., Smith, D.R., Hildreth, W., Palacz, Z., and Rogers, N., 1990, Compositional diversity of Late Cenozoic basalts in a transect across the southern Washington Cascades: Implications for subduction zone magmatism: Journal of Geophysical Research, v. 95, p. 19561–19582.

Luhr, J.F., 1990, Experimental phase relations of water- and sulfursaturated arc magmas and the 1982 eruptions of El Chichón Volcano: Journal of Petrology, v. 31, p. 1071–1114.

Luhr, J.F., Carmichael, I.S.E., and Varekamp, J.C., 1984, The 1982 eruptions of El Chichón Volcano, Chiapas, Mexico: Mineralogy and petrology of the anhydrite-bearing pumices: Journal of Volcanology and Geothermal Research, v. 23, p. 69–108.

Manhès, G., Minster, J.F., and Allègre, C.J., 1978, Comparative uranium-thorium-lead and rubidium-strontium study of the St. Severin amphoterite: Consequences for early solar system chronology: Earth and Planetary Science Letters, v. 39, p. 14–24.

Martin, H., 1987, Petrogenesis of Archaean trondhjemites, tonalites, and granodiorites from Eastern Finland: Major and trace element geochemistry: Journal of Petrology, v. 28, p. 921–953.

McCormick, M.P., and Veiga, R.E., 1992, SAGE II measurements of early Pinatubo aerosols: Geophysical Research Letters, v. 19, p. 155–158.

McCulloch, M.T., Gregory, R.T., Wasserburg, G.J., and Taylor, H.P., 1980, A neodymium, strontium, and oxygen isotopic study of the Cretaceous Samail ophiolite and implications for the petrogenesis and seawater-hydrothermal alteration of oceanic crust: Earth and Planetary Science Letters, v. 46, p. 201–211.

McDermott, F., Defant, J.J., Hawkesworth, C.J., Maury, R.C., and Joron, J.L., 1993, Isotope and trace element evidence for three component mixing in the genesis of the North Luzon arc lavas (Philippines): Contributions to Mineralogy and Petrology, v. 113, p. 9–23.

McKibben, M.A., Eldridge, C.S., and Reyes, A.G., this volume, Sulfur isotopic systematics of the June 1991 Mount Pinatubo eruptions: A SHRIMP ion microprobe study.

Miklius, A., Flower, M.F.J., Huijsmans, J.P.P., Mukasa, S.B., and Castillo, P., 1991, Geochemistry of lavas from Taal volcano, southwestern Luzon, Philippines: Evidence for multiple magma supply systems and mantle source heterogeneity: Journal of Petrology, v. 32, p. 593–627.

Morris, J.D., Leeman, W.P., and Tera, F., 1990, The subducted component in island arc lavas: Constraints from Be isotopes and B-Be systematics: Nature, v. 344, p. 31–36.

Mukasa, S.B., Flower, M.F.J., Miklius, A., 1994, The Nd-, Sr-, and Pb-isotopic character of lavas from Taal, Laguna de Bay and Arayat Volcanoes, Southwestern Luzon, Philippines: implications for arc petrogenesis: Tectonophysics, v. 235, p. 205–221.

Mukasa, S.B., McCabe, R., and Gill, J.B., 1987, Pb-isotopic compositions of volcanic rocks in the West and East Philippine island arc: Presence of the Dupal isotopic anomaly: Earth and Planetary Science Letters, v. 84, p. 153–164.

Nakamura, H., 1974, Determination of REE, Ba, Mg, Na, and K in carbonaceous and ordinary chondrites: Geochimica et Cosmochimica Acta, v. 38, p. 757–775.

Newhall, C.G., Daag, A.S., Delfin, F.G., Jr., Hoblitt, R.P., McGeehin, J., Pallister, J.S., Regalado, M.T.M., Rubin, M., Tamayo, R.A., Jr., Tubianosa, B., and Umbal, J.V., this volume, Eruptive history of Mount Pinatubo.

Nicholls, I.A., and Whitford, D.J., 1976, Primary magmas associated with Quaternary volcanism in the western Sunda arc, Indonesia, in, Johnson, R.W., ed., Volcanism in Australasia: Amsterdam, Elsevier, p. 77-90.

Nye, C.J., and Reid, M.R., 1986, Geochemistry of primary and least fractionated lavas from Okmok Volcano, central

Aleutians: Implications for arc magma genesis: Journal of Geophysical Research, v. 91, p. 10271-10287.

Oles, D., 1988, Das San Pablo-Vulkangebiet (Philippinen): Graben-vulkanismus im Luzon-Inselbogen (The San Pablo Volcanic Field: volcanism in a horst and graben setting in the Luzon volcanic arc): Aachen, Rheinisch Westfälische Technische Hochschule, Ph.D. dissertation, 156 p.

Pallister, J.S., Hoblitt, R.P., Meeker, G.P., Knight, R.J., and Siems, D.F., this volume, Magma mixing at Mount Pinatubo: Petrographic and chemical evidence from the 1991 deposits.

Pallister, J.S., Hoblitt, R.P., and Reyes, A.G., 1992, A basalt trigger for the 1991 eruptions of Pinatubo volcano? Nature, v. 356, p. 426–428.

PHIVOLCS, 1990, Mount Pinatubo not erupting, PHIVOLCS Observer, July–September 1990, v. 6, no. 3, p. 6.

Pinatubo Volcano Observatory Team, 1991, Lessons from a major eruption: Mount Pinatubo, Philippines: Eos, Transactions of the American Geophysical Union, v. 72, p. 545–555.

Rapp, R.P., Watson, E.B., and Miller, C.F., 1991, Partial melting of amphibolite/eclogite and the origin of Archean trondhjemites and tonalites: Precambrian Research, v. 51, p. 1–25.

Reid, F.W., and Cole, J.W., 1983, Origin of dacites of Taupo Volcanic Zone, New Zealand: Journal of Volcanology and Geothermal Research, v. 18, p. 191–214.

Richard, P., Shimizu, N., and Allègre, C.J., 1976, $^{143}Nd/^{146}Nd$, a natural tracer: an application to oceanic basalts: Earth and Planetary Science Letters, v. 31, p. 269-278.

Roeder, P.L., and Emslie, R.F., 1970, Olivine-liquid equilibrium: Contributions to Mineralogy and Petrology, v. 29, p. 275-289.

Rutherford, M.J., 1993, Experimental petrology applied to volcanic processes: Eos, Transactions of the American Geophysical Union, v. 74, p. 49-52.

Rutherford, M.J., and Devine, J.D., 1991, Pre-eruption conditions and volatiles in the 1991 Pinatubo magma [abs.]: Eos, Transactions, American Geophysical Union, v. 72, no. 44, p. 62.

———this volume, Preeruption pressure-temperature conditions and volatiles in the 1991 dacitic magma of Mount Pinatubo.

Sabit, J.P., 1992, Pinatubo volcano's 1991 eruptions: Unpublished manuscript distributed at the International Scientific Conference on Mt. Pinatubo, Manila, Department of Foreign Affairs, May 1992.

Sabit, J.P., Pigtain, R.C., and de la Cruz, E.G., this volume, The west-side story: Observations of the 1991 Mount Pinatubo eruptions from the west.

Sakai, H., Casadevall, T.J., and Moore, J.G., 1982, Chemistry and isotope ratios of sulphur in basalts and volcanic gases at Kilauea volcano, Hawaii: Geochimica et Cosmochimica Acta, v. 46, p. 729–738.

Sakai, H., Des Marais, D.J., Ueda, A., and Moore, J.G., 1984, Concentration and isotopic ratios of carbon, nitrogen, and sulphur in ocean-floor basalts: Geochimica et Cosmochimica Acta, v. 48, p. 2433–2441.

Scott, W.E., Hoblitt, R.P., Torres, R.C., Self, S, Martinez, M.L., and Nillos, T., Jr., this volume, Pyroclastic flows of the June 15, 1991, climactic eruption of Mount Pinatubo.

Sigurdsson, H., Devine, J.D., and Davis, A.N., 1985, The petrologic estimation of volcanic degassing: Jokull, v. 35, p. 1–8.

Simkin, T., and Smith, J.V., 1970, Minor element distribution in olivine: Journal of Geology, v. 78, p. 304-325.

Smith, D.R., and Leeman, W.P., 1987, Petrogenesis of Mount St. Helens dacitic magmas: Journal of Geophysical Research, v. 92, p. 10313–10334.

Stowe, L.L., Carey, R.M., and Pellegrino, P.P., 1992, Monitoring the Mt. Pinatubo layer with NOAA/11 AVHRR data: Geophysical Research Letters, v. 19, p. 159–162.

Tera, F., Brown, L., Morris, J., Sacks, I.S., Klein, J., and Middleton, R., 1986, Sediment incorporation in island-arc magmas: Inferences from ^{10}Be: Geochimica et Cosmochimica Acta, v. 50, p. 535–550.

Ueda, A., and Sakai, H., 1984, Sulfur isotope study of Quaternary volcanic rocks from the Japanese Island Arc: Geochimica et Cosmochimica Acta, v. 48, p. 1837–1848.

Vidal, Ph., Dupuy, C., Maury, R., and Richard, M., 1989, Mantle metasomatism above subduction zones: Trace-element and radiogenic isotope characteristics of peridotite xenoliths from Bataan Island (Philippines): Geology, v. 17, p. 1115–1118.

Voss, F., 1971, Die geologische und morphologische Struktur des aktivsten Vulkangebietes der Philippinen (The geologic and morphologic structure of the most active volcanic field in the Philippines): Die Erde, v. 102, p. 308–316.

Wasserburg, G.J., Jacobsen, S.B., DePaolo, D.J., McCulloch, M.T., and Wen, T., 1981, Precise determination of Sm/Nd ratios, Sm and Nd isotopic abundances in standard solutions: Geochimica et Cosmochimica Acta, v. 45, p. 2311–2323.

Weis, D., Demaiffe, D., Cauët, S., and Javoy, M., 1987, Sr, Nd, O and H isotopic ratios in Ascension lavas and plutonic inclusions: Cogenetic origin: Earth and Planetary Science Letters, v. 82, p. 255–268.

Westrich, H.R., and Gerlach, T.M., 1992, Magmatic gas source for the stratospheric SO_2 cloud from the June 15, 1991, eruption of Mount Pinatubo: Geology, v. 20, p. 867–870.

Wolfe, J.A., and Self, S., 1983, Structural lineaments and Neogene volcanism in southwestern Luzon, in Hayes, D.E., ed., The tectonic and geologic evolution of Southeast Asian seas and islands, Part 2: American Geophysical Union, Geophysical Monograph 27, p. 157–172.

Woodhead, J.D., and Fraser, D.G., 1985, Pb, Sr and ^{10}Be studies of volcanic rocks from the northern Mariana Islands—Implications for magma genesis and crustal recycling in the Western Pacific: Geochimica et Cosmochimica Acta., v. 49, p. 1925–1930.

Woodhead, J.D., Harmon, R.S., and Fraser, D.G., 1987, O, S, Sr and Pb isotope variations in volcanic rocks from the northern Mariana Islands: Implications for crustal recycling in intraoceanic arcs: Earth and Planetary Science Letters, v. 83, p. 39–52.

Yan, X-H., Ho, C-R, Zheng, Q., and Klemas, V., 1992, Temperature and size variabilities of the Western Pacific warm pool: Science, v. 258, p. 1643–1645.

Petrology and Sr, Nd, and Pb Isotopic Geochemistry of Mount Pinatubo Volcanic Rocks

By Paterno R. Castillo[1] and Raymundo S. Punongbayan[2]

ABSTRACT

Dacitic ash and pumice and andesitic scoria erupted during the June 1991 Mount Pinatubo volcanic activity are geochemically similar to the calc-alkaline volcanic rocks typical of island-arc settings. Their isotopic signature overlaps with that of other volcanic rocks from the Luzon arc, particularly those from the northern segment. In detail, the dacite is slightly less enriched in incompatible trace elements, $^{87}Sr/^{86}Sr$, and $^{206}Pb/^{204}Pb$ and is slightly higher in $^{143}Nd/^{144}Nd$ than the andesite. These data clearly indicate that the dacite and andesite have different magma sources—probably variable mixtures of depleted basalts of the Zambales ophiolite complex (or their mantle source), and an enriched component from subducted sediments. Direct sediment involvement in the generation of the Mount Pinatubo volcanic rocks was minimal; the presence of ^{10}Be, high $\delta^{34}S$, and recent light element enrichment were most probably acquired through metasomatism of the ophiolitic basement basalts, or their mantle source, by fluids derived from the subducted slabs.

INTRODUCTION

Mount Pinatubo is located approximately 90 km northwest of Manila in a chain of mostly inactive volcanoes on the west side of the central part of Luzon Island, Philippines (fig. 1). This volcanic chain is the Bataan segment (Defant and others, 1989) of the Luzon arc, which extends from Taiwan in the north to Mindoro Island in the south. Volcanism along the Luzon arc is the result of the eastward subduction of the South China Sea plate under northern Luzon and the Philippine Sea plate, and it is generally believed that sediments above the subducting slab are involved in the generation of Luzon arc magmas (Knittel and others, 1988; Defant and others, 1988; McDermott and others, 1993; Mukasa and others, 1994; Bernard and others, this volume). The types and actual mechanisms of sediment involvement in the generation of the magmas, however, are still a subject of intense scrutiny (Chen and others, 1990; Maury and others, 1992; McDermott and others, 1993).

After about 500 years of dormancy (Newhall and others, this volume), Mount Pinatubo erupted about 5 km³ of magma during its June 1991 eruptive activity (W.E. Scott, and others, this volume). Most of the rock erupted was dacitic ash and pumice, of which two general types were recognized (Pallister and others, 1992). One group is phenocryst-rich and has large (>1 mm) irregular vesicles, whereas the other group is phenocryst-poor and has smaller (<1 mm) spherical vesicles. Despite their textural differences, how-

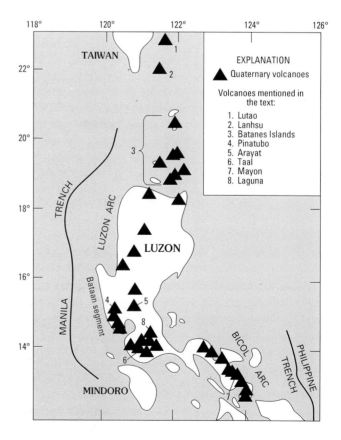

Figure 1. Locations of Mount Pinatubo and other volcanic centers in the northern Philippines, western Pacific.

[1] Scripps Institution of Oceanography, La Jolla, CA 92093.
[2] Philippine Institute of Volcanology and Seismology.

ever, both pumice types are mineralogically and chemically very similar. In addition to the pumice and ash, a lesser amount of andesitic scoria was erupted during the early, less-intense stages of the volcanic activity, from June 12 to June 14 (Hoblitt, Wolfe, and others, this volume). Pallister and others (1992) believe that the andesite is a mixture of crystal-rich dacite and an olivine-hornblende basalt and that the mixing event occurred when a mafic magma intruded the Mount Pinatubo chamber containing silicic magma; the injection of the mafic magma may have also triggered the eruptions.

In this paper, we present the major element, rare earth element (REE), and strontium (Sr), neodymium (Nd), and lead (Pb) isotopic compositions of andesite blocks from the small lava dome built in the summit of the volcano during the period June 7 to 11 (Hoblitt, Wolfe, and others, 1991; Wolfe and Hoblitt, this volume) and of representative pumice, ash, and basement rocks. The main objective of our study is to use these analytical data to constrain the petrogenesis and source composition of the June 1991 Mount Pinatubo eruptive products.

SAMPLES AND METHODS

The analyzed samples consist of ash from the northeastern slope of the volcano, a medium- to coarse-grained (>1 mm) crystal-rich pumice, two medium-grained (<5 mm) crystal-rich pumices, and two fine-grained (<1 mm) crystal-poor pumices from near Clark Air Base, and two samples of the June 1991 andesite dome. Two samples of magnesian basalts from the basement terrane northwest of Pinatubo were also analyzed. Major element contents of these samples were analyzed by the X-ray fluorescence method, REE contents by inductively-coupled plasma mass spectrometry, and Sr, Nd, and Pb isotopic compositions by thermal ionization mass spectrometry. Analytical results and precision of the analyses are presented in table 1.

RESULTS

As has been shown previously (Bernard and others, 1991; Pallister and others, 1992), the different pumice types and ash are all dacites (~64 wt% SiO_2) that have very similar, if not identical, major element compositions. The variation in their REE contents is barely outside analytical error, with the coarse-grained pumice generally showing the highest REE content. The REE concentration pattern of dacite of Mount Pinatubo overlaps with the fairly wide field for other Luzon arc volcanic rocks (fig. 2). The pattern is different from that of Mount Mayon volcanic rocks (P.R. Castillo and C.G. Newhall, unpub. data, 1993), which were generated as a result of subduction of the Philippine Sea plate along the eastern margin of southern Luzon (fig. 1).

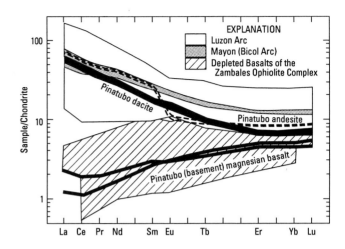

Figure 2. Chondrite-normalized REE concentration patterns of the 1991 Mount Pinatubo eruptive products and basement samples. Fields for volcanic rocks for Mount Mayon in the Bicol arc (P.R. Castillo and C.G. Newhall, unpub. data, 1993), the Luzon arc from the literature (Knittel and others, 1988; Defant and others, 1988, 1989, and 1990; McDermott and others, 1993), and some of the variably depleted basalts from the Zambales ophiolite complex (Evans and others, 1991; F. Florendo and J. Hawkins, unpub. data, 1993) are shown for comparison. See figure 1 for location of samples.

The two samples of the dome are andesites (~60 wt% SiO_2) and have identical major element compositions. They are more enriched in light and heavy REE than the dacites (fig. 2). Bernard and others (this volume) report that the incompatible element contents of the olivine-hornblende basalt inclusions within the andesite dome are also as enriched, if not more enriched than those of the dacite. In general, both the dacite and the andesite are calc-alkaline rocks that are typical of island arcs, and, in particular, are typical of the nearby volcanoes along the Bataan segment (Defant and others, 1988, 1990; Knittel and others, 1988).

The magnesian basalts represent part of the basement of Mount Pinatubo. They have high (>10 wt%) MgO contents and are very depleted in light REE, very much like some of the Eocene volcanic rocks from the Zambales ophiolite complex located north and west of Mount Pinatubo (fig. 2). Basalts of the Zambales ophiolite complex were derived from variably depleted mantle sources including a source like that for mid-ocean ridge basalt (Evans and others, 1991; F. Florendo and J. Hawkins, unpub. data, 1993). On the basis of their REE concentrations and model calculations, the dacites and andesites could not have been derived from these basalts through simple crystal fractionation. This notion is strongly supported by the isotopic data, which clearly suggest that the three rock groups have compositionally distinct mantle sources.

The dacites are very similar, if not identical, in their Sr, Nd, and Pb isotopic compositions, and so are the andesites (fig. 3) (S.B. Mukasa analyzed a split of our ash sample and

Table 1. Chemical and isotopic analyses of representative Mount Pinatubo samples.

[Major elements are measured in weight percent; REE's are measured in parts per million. Analytical errors based on repeated analyses of standards are better than 5% for the major and rare earth elements, ± 0.00002 for $^{87}Sr/^{86}Sr$, ± 0.000016 for $^{143}Nd/^{144}Nd$, ± 0.012 for $^{206}Pb/^{204}Pb$, ± 0.013 for $^{207}Pb/^{204}Pb$, and 0.030 for $^{208}Pb/^{204}Pb$. Data are reported relative to standards values: $^{87}Sr/^{86}Sr$ = 0.71025 for NBS 987 Sr and $^{143}Nd/^{144}Nd$ = 0.51186 for La Jolla Nd; Pb isotopic compositions were fractionation corrected by using the NBS 981 Pb values of Todt and others (1983)]

Sample Lithology Location	Old lava-1, basement basalt (NW. side)	Old lava-2, basement basalt (NW. side)	Pin-Sc1, 6/7–12 dome scoria (W. flank)	Pin-Sc2, 6/7–12 dome scoria (W. flank)	Ash deposit, 6/15 ash (NE. flank)	Phenocryst-rich 6/15 coarse pumice (NW. flank)	Phenocryst-rich1 6/15 fine pumice (near Clark Air Base)	Phenocryst-rich2 6/15 fine pumice (near Clark Air Base)	Phenocryst-poor1 6/15 pumice (near Clark Air Base)	Phenocryst-poor2 6/15 pumice (near Clark Air Base)
Major Element Contents										
SiO$_2$	50.45	51.18	59.50	59.75	62.66	64.50	64.14	63.95	64.18	64.24
TiO$_2$	.23	.25	.68	.67	.50	.49	.53	.50	.47	.49
Al$_2$O$_3$	14.43	15.37	15.15	15.48	17.25	16.82	16.83	17.19	17.14	16.95
Fe$_2$O$_3$	8.69	8.17	6.68	6.53	4.37	4.25	4.31	4.07	4.09	4.29
MnO	.14	.12	.14	.15	.10	.09	.09	.09	.09	.09
MgO	11.31	9.48	4.94	4.68	2.79	2.45	2.45	2.44	2.37	2.67
CaO	13.18	13.35	7.15	6.98	5.57	5.24	5.37	5.34	5.28	5.39
Na$_2$O	1.40	1.94	3.84	3.96	5.15	4.44	4.62	4.78	4.70	4.21
K$_2$O	.07	.04	1.66	1.60	1.40	1.53	1.46	1.47	1.49	1.49
P$_2$O$_5$	.11	.10	.25	.20	.21	.19	.19	.18	.19	.19
Total	100.0	100.0	100.0	100.0	100.0	100.0	100.0	100.0	100.0	100.0
REE Contents										
La	.55	.36	17.9	17.5	13.3	14.9	13.5	14.0	13.6	14.1
Ce	1.19	.82	36.1	35.6	27.0	29.6	26.8	27.7	26.9	28.0
Pr	.19	.14	4.43	4.38	3.33	3.62	3.33	3.38	3.32	3.46
Nd	1.05	.86	19.5	19.3	13.5	14.7	13.5	13.3	13.3	14.1
Sm	.45	.41	4.05	3.94	2.77	2.86	2.69	2.56	2.71	2.82
Eu	.17	.18	.60	.66	.89	.91	.83	.85	.81	.86
Tb	.13	.15	.35	.36	.38	.38	.36	.36	.36	.39
Er	.78	.84	1.49	1.48	1.17	1.20	1.16	1.10	1.11	1.20
Yb	.78	.86	1.47	1.47	1.18	1.21	1.15	1.09	1.10	1.18
Lu	.13	.14	.24	.24	.19	.20	.18	.17	.18	.10
Isotopic Compositions										
$^{87}Sr/^{86}Sr$	.70390	.70392	.70426	.70425	.70419	.70422			.70422	
$^{143}Nd/^{144}Nd$	.513044	.513023	.512837	.512851	.512909	.512918			.512922	
$^{206}Pb/^{204}Pb$	18.243		18.443	18.433		18.417			18.412	
$^{207}Pb/^{204}Pb$	15.517		15.603	15.584		15.583			15.576	
$^{208}Pb/^{204}Pb$	38.118		38.659	38.602		38.575			38.553	

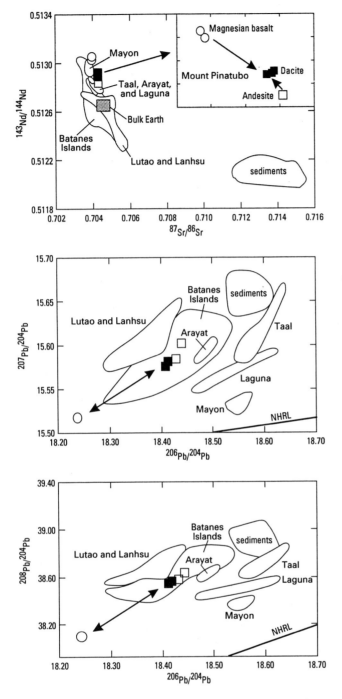

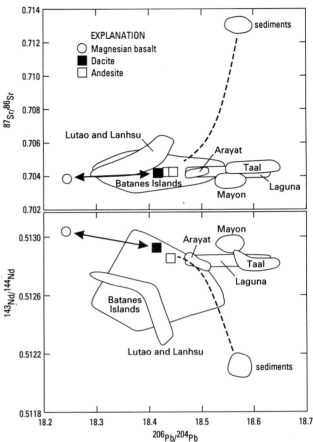

Figure 4. $^{87}Sr/^{86}Sr$ and $^{143}Nd/^{144}Nd$ against $^{206}Pb/^{204}Pb$ diagrams for the samples analyzed. Solid arrows and dashed lines illustrate binary mixing curves between the magnesian basalt and sediments necessary to produce the Mount Pinatubo volcanic rocks. Data suggest that the magnesian basalt and the sediments that were mixed to produce the Mount Pinatubo samples do not have identical Sr/Pb and Nd/Pb concentration ratios. Sources of data as in figures 2 and 3.

Figure 3. $^{143}Nd/^{144}Nd$ against $^{87}Sr/^{86}Sr$ (upper panel) and $^{207}Pb/^{204}Pb$ and $^{208}Pb/^{204}Pb$ against $^{206}Pb/^{204}Pb$ (lower two panels) diagrams for the samples analyzed. Only volcanic rocks from Luzon, Batanes, Lutao, and Lanhsu with Sr, Nd and Pb isotopic ratios measured on the same samples are shown for clarity of presentation. Data are from P.R. Castillo and others (unpub. data, 1993); McDermott and others (1993); Mukasa and others (1994). Additional Pb isotopic data from Sun (1980) for Lanhsu are included. Inset in the top diagram shows isotopic data for Mount Pinatubo samples in greater detail. Arrows indicate linear mixing relationship among the magnesian basalt (o), dacite (■), and andesite (□). NHRL, Northern Hemisphere Regression Line (see text).

his Sr, Nd, and Pb isotopic ratios are very similar to ours—written commun., 1993). That both the $^{87}Sr/^{86}Sr$ and $^{143}Nd/^{144}Nd$ ratios of the dacites and andesites plot above the bulk earth values suggests that their source materials have a long-term depletion of Rb with respect to Sr and Nd with respect to Sm. This is inconsistent with their REE concentration patterns, which show high Nd/Sm and, by inference, Rb/Sr (fig. 2). In detail, the isotopic signatures of the dacites and the andesites are distinct from one another, with the dacites having higher $^{143}Nd/^{144}Nd$ and lower $^{87}Sr/^{86}Sr$ and $^{206}Pb/^{204}Pb$ ratios. Interestingly, the $^{143}Nd/^{144}Nd$ and $^{206}Pb/^{204}Pb$ of the dacites, respectively, are among the highest and lowest values reported for volcanic rocks from the Bataan and southern segments of the Luzon arc (Bernard and others, this volume; Mukasa and others, 1994). Both the dacites and andesites, however, plot within the field for Batanes

Islands and very close to the field for Lutao and Lanhsu Islands in the ^{87}Sr/^{86}Sr against ^{206}Pb/^{204}Pb diagram (fig. 4).

The magnesian basalts also have identical ^{87}Sr/^{86}Sr and ^{143}Nd/^{144}Nd ratios. They have the highest ^{143}Nd/^{144}Nd and lowest ^{87}Sr/^{86}Sr and ^{206}Pb/^{204}Pb ratios among the samples analyzed, and, altogether, the analyses define consistent linear arrays in all permutations of isotopic ratio diagrams (figs. 3 and 4). The magnesian basalt isotopic signature overlaps with those of the volcanic and ophiolitic rocks of the Taiwan Coastal Range (Sun, 1980; Chen and others, 1990; P.R. Castillo and T. Lee, unpub. data, 1993). Their ^{143}Nd/^{144}Nd also overlap with, though are generally lower than, those of the volcanic rocks from the Zambales ophiolite complex (Evans and others, 1991). It is important to note, however, that the ^{143}Nd/^{144}Nd of the ophiolitic volcanic rocks generally decrease toward the south, and the magnesian basalts are located south of the ophiolitic samples. Moreover, the Pb isotopic ratios of a dolerite from the Zambales ophiolite complex are fairly unradiogenic and close to the composition of the magnesian basalts (^{206}Pb/^{204}Pb=18.19—Hamelin and others, 1984). (The few reported ^{87}Sr/^{86}Sr ratios for volcanic rocks of the Zambales ophiolite complex clearly have been affected by alteration—Evans and others, 1991.) The similarity of the Nd and Pb isotopic ratios and REE concentration pattern of the magnesian basalts of Mount Pinatubo and volcanic rocks of the Zambales ophiolite complex, in addition to their geographical proximity, strongly suggest that they are petrogenetically related.

Another interesting isotopic feature shared by the dacite and andesite is that their Pb isotopic ratios plot above the Northern Hemisphere regression line (NHRL of fig. 3), which is a line fitted through Pb isotopic data for mid-ocean ridge basalts from the Pacific and North Atlantic and some oceanic islands (Hart, 1984). This Pb isotopic signature, together with their ^{87}Sr/^{86}Sr and ^{143}Nd/^{144}Nd ratios and calc-alkaline affinity, appears to be a characteristic feature of many Philippine volcanic rocks (McDermott and others, 1993; Mukasa and others, 1987; 1994).

DISCUSSION

Our data show that the 1991 Mount Pinatubo eruptive products are similar to some Luzon arc volcanic rocks. Another important point detailed by the data is that the dacite, andesite, and magnesian basalt are all isotopically collinear, which strongly suggests a mixing relationship among them. If the andesite is already a mixture of the dacite and the olivine-hornblende basalt inclusion in the andesite (Pallister and others, 1992; this volume), then the inclusion should plot at the end of the mixing line, opposite the magnesian basalt. Indeed, Pb and Sr isotopic ratios for the basaltic inclusion (U. Knittel and others, unpub. data, 1993) are the highest among those for the 1991 Mount Pinatubo

eruptive products and plot at the end of the mixing array away from the magnesian basalt.

Available data also clearly show that the dacite magma is not a mantle source end-component because it is not a direct melt from the mantle and its isotopic composition trends toward the more extreme isotopic composition of the magnesian basalt. The dacite is most likely a crystal fractionation product of a more mafic magma whose source is a product of mixing between the mantle source of the geochemically depleted and isotopically nonradiogenic basement magnesian basalt and an isotopically radiogenic and geochemically enriched end-component (figs. 3 and 4). It is also possible that the dacite magma was produced by melting of the basaltic basement due to the influx of isotopically and geochemically enriched melts generated above the subducting slab. Of course, there are other possible scenarios to generate the dacite magma, but it is rather premature to constrain the best possible model(s) for its generation at this stage of the project because the available data are limited. For simplicity, the remainder of the discussion focuses on the isotopically and geochemically enriched end-component of the mixing proposed by the isotopic data (figs. 3 and 4) and on the most probable mechanism of mixing this component with the source (or melt) of the depleted magnesian basalt. A more detailed discussion of the mixing process is beyond the scope of this paper.

Mantle materials that are enriched in incompatible trace elements but have ^{87}Sr/^{86}Sr and ^{143}Nd/^{144}Nd ratios that record a time-integrated history of depletion of these elements are the source of many Luzon Island volcanic rocks; these are most probably generated by recent incorporation by the depleted upper mantle of sediment-derived components coming from the subducted slabs (Knittel and others, 1988; Defant and others, 1990; Mukasa and others, 1994). Such sediment-derived components most probably also comprise the enriched endmember that was mixed with the basement magnesian basalt source to produce the parental magma of the dacite of Mount Pinatubo. This hypothesis is strongly supported by the presence of excess ^{10}Be and δ^{34}S in the dacite (Knittel and others, 1992; Bernard and others, this volume). This hypothesis is also consistent with the Pb isotopic compositions of the South China Sea and other western Pacific sediments (Sun, 1980; McDermott and others, 1993; P.R. Castillo and T. Lee, unpub. data, 1993) because the Pb isotopic ratios of the dacite and andesite of Mount Pinatubo indeed line up between the sediments and the basement magnesian basalt (fig. 3). Diagrams of ^{206}Pb/^{204}Pb against ^{87}Sr/^{86}Sr and ^{143}Nd/^{144}Nd ratios are also consistent with such a mixing relationship (fig. 4).

A direct sediment involvement, however, is most probably minimal. Sediments contain a lot more Pb than the ophiolitic basalts (Sun, 1980), so even a small amount of sediment mixed with the basalt source will produce a large shift in the isotopic ratios of the resultant product toward the Pb isotopic composition of the sediments. Yet data show

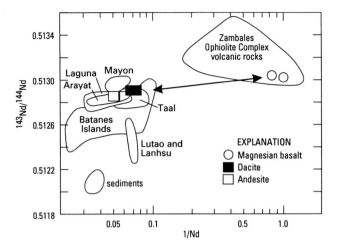

Figure 5. $^{143}Nd/^{144}Nd$ against 1/Nd diagram for the samples analyzed. Note that the sediment field is not collinear with the Pinatubo samples and (or) any of the Philippine volcanic rocks. Sources of data as in figures 2 and 3.

that the Pb isotopic composition of the dacites is only roughly halfway between those of the sediments and the magnesian basalt (fig. 3). The same is true for the $^{143}Nd/^{144}Nd$ isotopes of these samples because the South China Sea sediments contain much more Nd than the ophiolitic basalts (Chen and others, 1990). In fact, the $^{143}Nd/^{144}Nd$ compositions of the dacite, andesite, and magnesian basalt argue against sediment involvement. Figure 5 is a diagram of $^{143}Nd/^{144}Nd$ against 1/Nd in which mixing would be represented by a straight line, as is shown again by the alignment of the magnesian basalt, dacite, and andesite (the olivine-hornblende basalt inclusion is not plotted in the diagram because data are lacking). This Mount Pinatubo mixing line, however, does not pass through the sediment field, so it is suggested that sediments are not directly involved in the mixing process.

One may argue that the dacite and andesite are both differentiated rocks and, thus, their 1/Nd values must not be used in the $^{143}Nd/^{144}Nd$ against 1/Nd diagram. These differentiated rocks have to be corrected for crystal fractionation first, and then the higher 1/Nd values of their more mafic parental magmas must be used instead. In other words, the 1/Nd values of the dacite and andesite parental magmas may be high enough for them to plot between the magnesian basalts and sediments in figure 5. However, this scenario implies that the dacite and andesite, which have ~25 to 30 ppm Nd, must have fractionally crystallized from parental magmas that have only ~2 ppm, and this is very unlikely. Fractional crystallization, though, potentially can generate the high-Nd dacite and andesite from low-Nd parental magmas if accompanied by assimilation of the country rocks. The presence of a third, high $^{143}Nd/^{144}Nd$ end-component with high Nd content may also explain the

nonlinear arrangement of the magnesian basalt, dacite, andesite, and sediments. Unfortunately, these alternative explanations neither can be supported nor ruled out by the few available data; the one thing clearly suggested by figure 5 is that direct sediment involvement in the generation of the dacite and andesite is minimal.

On the basis of these results, we propose the following tentative model for the origin of Mount Pinatubo eruptive products. The source of the dacite magma, or, more appropriately, of the parental magma of the dacite, represents a mixture of the variably depleted mantle source (or melt) of volcanic rocks of the Zambales ophiolite complex and metasomatic fluids coming from the dehydrating slab being subducted. This fluid carries with it the geochemical and isotopic signature of the sediments in proportion to the differential mobility of incompatible elements in hydrous fluids (Tatsumi and others, 1986; Spivack and others, 1992; You and others, 1992). As such, a "sediment component" in the elemental and isotopic compositions is more readily seen in S and Be than in Sr, in Sr more than in Pb, and particularly more than in Nd. Subsequent small degrees of partial melting of the mixed source will further enrich the incompatible trace element concentrations in the resultant melt but will no longer affect the isotopic composition. Further enrichment of the incompatible elements can also be accomplished during percolation of melt through the mantle (Navon and Stolper, 1987) and finally by crystal fractionation. As Pallister and others (1992) proposed earlier, the dacitic magma most probably had been crystallizing in a magma chamber for some period of time (>400 years) when an olivine-hornblende basalt magma intruded. Hybrid andesite was formed and the 1991 eruptions began. The mantle source of the olivine-hornblende basalt is similar to that of the dacite, except that the olivine-hornblende basalt magma apparently has more sediment component than the dacite magma.

SUMMARY AND CONCLUSIONS

Dacitic ash and pumice, as well as andesitic scoria extruded during the 1991 Mount Pinatubo eruptive activity, and ophiolitic magnesian basalt from the volcano basement, have been analyzed for major element, REE and Sr, Nd, and Pb isotopic compositions. Not surprisingly, the samples analyzed are similar to the lavas along the Luzon volcanic arc, to which Mount Pinatubo belongs.

Data indicate that the samples analyzed could have been generated by a simple mixing of the variably depleted mantle source of the ophiolitic basement basalt and an enriched, metasomatic fluid from the subducting slab. The metasomatic fluid carries with it some of the elemental and isotopic signatures of the sediments being subducted. The elemental, though not the isotopic, composition of the melt generated from the mixed mantle source most probably can be further enriched by small degrees of partial melting plus subsequent elemental enrichment processes during ascent of the melt through the mantle and the crust. Mixing of melts inside the shallow magma chamber, as evidently shown by the dacite-andesite-olivine-hornblende basalt melt mixing relationship, can further complicate the composition of the volcanic rocks.

ACKNOWLEDGMENT

We thank PHIVOLCS, USGS, and MGB (Mines and Geosciences Bureau) members of the Pinatubo Volcano Observatory Team for helping us collect samples; P. Janney, F. Florendo, C. MacIsaac, and B. Hanan for assistance during the analyses; and J. Hawkins, C.-F. You, U. Knittel, and M. Defant for reviews of the manuscript. This work is supported by grant NSF INT91–96017 and NSF EAR93–13571 to P.R. Castillo.

REFERENCES CITED

Bernard, A., Demaiffe, D., Mattielli, N., and Punongbayan, R.S., 1991, Anhydrite-bearing pumices from Mount Pinatubo: Further evidence for the existence of sulphur-rich silicic magmas: Nature, v. 354, p. 139–140.

Bernard, A., Knittel, U., Weber, B., Weis, D., Albrecht, A., Hattori, K., Klein, J., and Oles, D., this volume, Petrology and geochemistry of the 1991 eruption products of Mount Pinatubo.

Chen, C.-H., Shieh, U.-N., Lee, T., Chen, C.H., and Mertzman, S.A., 1990, Nd-Sr-O isotopic evidence for source contamination and unusual component under Luzon arc: Geochimica et Cosmochimica Acta, v. 54, p. 2473–2483.

Defant, M.J., de Boer, J.Z., and Oles, D., 1988, The western central Luzon volcanic arc, the Philippines: Two arcs divided by rifting?: Tectonophysics, v. 145, p. 305–317.

Defant, M.J., Jacques, D., Maury, R.C., de Boer, J.Z., Joron, J.L., 1989, Geochemistry of the Luzon arc, Philippines: Geological Society of America Bulletin, v. 101, p. 663–672.

Defant, M.J., Maury, R.C., Joron, J.-L., Feigenson, M., Leterrier, J., Bellon, H., Jacques, D., and Richard, M., 1990, The geochemistry and tectonic setting of the northern section of the Luzon arc (the Philippines and Taiwan): Tectonophysics, v. 183, p. 187–205.

Evans, C.A., Castaneda, G., and Franco, H., 1991, Geochemical complexities preserved in the volcanic rocks of the Zambales ophiolite, Philippines: Journal of Geophysical Research, v. 96, p. 16251–16262.

Hamelin, B., Dupre, B., and Allegre, C.J., 1984, The lead isotope systematics of ophiolite complexes: Earth and Planetary Science Letters, v. 67, p. 351–366.

Hart, S.R., 1984, A large-scale isotope anomaly in the Southern Hemisphere mantle: Nature, v. 309, p. 753–757.

Hoblitt, R.P., Wolfe, E.W., Lockhart, A.B., Ewert, J.E., Murray, T.L., Harlow, D.H., Mori, J., Daag, A.S., and Tubianosa, B.S., 1991, 1991 eruptive behavior of Mount Pinatubo, Philippines [abs.]: Eos, Transactions, American Geophysical Union, v. 72, p. 61.

Hoblitt, R.P., Wolfe, E.W., Scott, W.E., Couchman, M.R., Pallister, J.S., and Javier, D., this volume, The preclimactic eruptions of Mount Pinatubo, June 1991.

Knittel, U., Defant, M.J., and Raczek, I., 1988, Recent enrichment in the source region of arc magmas from Luzon Island, Philippines: Sr and Nd isotopic evidence: Geology, v. 16, p. 73–76.

Knittel, U., Hattori, K., Hoefs, J., and Oles, D., 1992, Isotopic compositions of anhydrite phenocryst-bearing pumices from the 1991 eruption of Mount Pinatubo, Philippines [abs.]: Eos, Transactions, American Geophysical Union, v. 73, p. 342.

Maury, R.C., Defant, M.J., and Joron, J.-L., 1992, Metasomatism of the sub-arc mantle inferred from trace elements in Philippine xenoliths: Nature, v. 360, p. 661–663.

McDermott, F., Defant, M.J., Hawkesworth, C.J., Maury, R.C., and Joron, J.L., 1993, Isotope and trace element evidence for three component mixing in the genesis of the North Luzon arc lavas (Philippines): Contributions to Mineralogy and Petrology, v. 113, p. 9–23.

Mukasa, S.B., Flower, F.J., and Miklius, A., 1994, The Nd-, Sr- and Pb-isotopic character of lavas from Taal, Laguna de Bay and Arayat volcanoes, SW. Luzon, Philippines: Implications for arc magma petrogenesis: Tectonophysics, v. 235, p. 205–221.

Mukasa, S.B., McCabe, R., and Gill, J.B., 1987, Pb-isotopic compositions of volcanic rocks in the west and east Philippine island arcs: Presence of the Dupal isotopic anomaly: Earth and Planetary Science Letters, v. 84, p. 153–164.

Navon, O., and Stolper, E., 1987, Geochemical consequences of melt percolation: The upper mantle as a chromatographic column: Journal of Geology, v. 95, p. 285–307.

Newhall, C.G., Daag, A.S., Delfin, F.G., Jr., Hoblitt, R.P., McGeehin, J., Pallister, J.S., Regalado, M.T.M., Rubin, M., Tamayo, R.A., Jr., Tubianosa, B., and Umbal, J.V., this volume, Eruptive history of Mount Pinatubo.

Pallister, J.S., Hoblitt, R.P., Meeker, G.P., Knight, R.J., and Siems, D.F., this volume, Magma mixing at Mount Pinatubo: Petrographic and chemical evidence from the 1991 deposits.

Pallister, J.S., Hoblitt, R.P., and Reyes, A.G., 1992, A basalt trigger for the 1991 eruptions of Pinatubo volcano?: Nature, v. 356, p. 426–428.

Scott, W.E., Hoblitt, R.P., Torres, R.C., Self, S., Martinez, M.L., and Nillos, T., this volume, Pyroclastic flows of the June 15, 1991, climactic eruption of Mount Pinatubo.

Spivack, A.J., You, C.-Y., Gieskies, J., Rosenbauer, R., and Bischoff, J., 1992, Experimental study of B geochemistry: Implications for Be-B systematics in the subduction zones [abs.]: Eos, Transactions, American Geophysical Union, v. 73, p. 638.

Sun, S.-S., 1980, Lead isotopic study of young volcanic rocks from mid-ocean ridges, ocean islands and island arcs: Philosophical Transactions, Royal Society of London, v. 297, p. 409–445.

Tatsumi, Y., Hamilton, D.L, and Nesbitt, R.W., 1986, Chemical characteristics of fluid phase released from a subducted lithosphere and origin of arc lavas: Evidence from high-pressure experiments and natural rocks: Journal of Volcanology and Geothermal Research, v. 29, p. 293–309.

Todt, W., Dupre, B., and Hofmann, A.W., 1983, Pb isotope measurements using a multi collector: Applications to standards and basalts: Terra Cognita, v. 3, p. 140.

Wolfe, E.W., and Hoblitt, R.P., this volume, Overview of the eruptions.

You, C.-F., Rosenbauer, R., Xu, X., Ku, T.-L., Gieskes, J., and Bischoff, J., 1992, Distribution of Be-10 and Be-9 in the sedimentary column at Site 808, Nankai Trough, Japan [abs.]: Eos, Transactions, American Geophysical Union, v. 73, p. 638.

Occurrence and Origin of Sulfide and Sulfate in the 1991 Mount Pinatubo Eruption Products

By Keiko Hattori[1]

ABSTRACT

Sulfide phases are found in a wide variety of the 1991 eruption products from Mount Pinatubo. Sulfides in early-formed phenocrysts (olivine and augite) in basalt fragments and dome-forming andesite are globular nickel-bearing pyrrhotite, whereas some sulfides in the glass are irregularly shaped and copper rich. Sulfides in dacitic pumice are mostly copper-rich sulfides (chalcopyrite, $CuFeS_2$, with or without an exsolution product of bornite, Cu_5FeS_4). These sulfides in dacitic pumice contain significant Zn (up to 1.3 weight percent), Se, Ag, As, and Cd. Sulfides in the glass of dacitic pumice exhibit desulfidation reaction rims.

Anhydrite is commonly surrounded by the matrix glass in gray, white, and banded pumice. Smooth contact between the anhydrite and glass confirms that the dacitic melt was in equilibrium with anhydrite immediately before eruption.

The occurrence of sulfide globules in the eruption products indicates that an immiscible sulfide liquid formed in silicate melts and that the melts were once reduced, with an oxygen fugacity below the redox boundary of dissolved sulfur. Later formation of anhydrite in dacitic melt requires an addition of sulfur and oxidation of the magma. It is proposed that supercritical fluid released from ascending mafic melt beneath Mount Pinatubo provided volatile elements and sulfur. Sulfur that discharged from the hot dry melt was mostly SO_2. The reduction of sulfur to H_2S in the cool (~800°C), wet dacite caused oxidation of this dacitic magma. H_2S formed in this way was initially precipitated in the dacite as sulfide minerals together with other volatile elements. Continued influx of SO_2 led oxidation of the dacite and an increase in the sulfur solubility of the melt, which caused partial resorption of sulfide minerals and led to excess sulfur, which was precipitated as anhydrite. The proposed model is consistent with compositions of iron-titanium oxides, abundant fluid inclusions in phenocrysts, high contents of volatile elements and hydrophyllic metals in sulfides, strontium isotopic compositions of anhydrite, and sulfur isotopic values of the bulk rocks.

INTRODUCTION

The eruption of Mount Pinatubo in June, 1991, introduced ~20 Mt SO_2 into the stratosphere and had a significant effect on the global climate (Bluth and others, 1992; Grant and others, 1992; Gleason and others, 1993). The sulfur-rich nature of the magma is reflected in the occurrence of anhydrite phenocrysts in dacitic pumices (Bernard and others, 1991). Several models have been proposed for the origin of this sulfur, including incorporation of subducted sulfide-sulfate deposits on the East China plate (Whitney, 1992), assimilation of sulfide deposits from the underlying Zambales Ophiolite Complex (Fournelle, 1991), assimilation of hydrothermal sulfate in preexisting volcanic rocks (McKibben and others, 1992; this volume), and the introduction of sulfur gases from underlying mafic magma (Pallister and others, 1992; Matthews and others, 1992). Discussions on the origin of sulfur are so far based mostly on chemical and isotopic data for bulk rocks. This paper describes the occurrence and composition of sulfur-bearing phases in various types of eruption products with an emphasis on sulfide minerals and proposes a model for the formation of high-sulfur dacitic magma at Pinatubo.

SAMPLE DESCRIPTION

The 1991 eruption products are dacitic pumice, dome-forming andesite, and basalt. The latter two are volumetrically insignificant, but they are important because they provide evidence supporting an injection of mafic melt as a trigger of the eruption of a semi-solidified dacitic magma (Pallister and others, 1992; Matthews and others, 1992). Basalt commonly occurs as angular to rounded inclusions in andesite that range in diameter from several centimeters to meters. Small inclusions of basalt (<5 cm) are rarely found in gray pumice.

[1]Ottawa-Carleton Geoscience Centre, *and* Department of Geology, University of Ottawa, Ottawa, Ontario, K1N 6N5, Canada.

Samples used for this study include six of white pumice; six of gray pumice, three of banded pumice, nine of dome andesite, and four of basalt enclosed in andesites. Three to 10 thin and thick polished sections were made from each sample. The usage of gray and white pumice follows the description by Pallister and others (1992); "gray" pumice actually varies from gray to tan. White pumice is well vesiculated, porphyritic, and contains coarse plagioclase and hornblende phenocrysts. Inclusions are rare, but angular fragments of gray pumice are observed. Gray pumice is poorly vesiculated and has phenocrysts that are much smaller than those of the white pumice (<3 mm). Gray pumices are typically heterogeneous in texture, containing different types of fragments including white pumice, andesite, basalt, and hydrothermally altered volcanic rocks. The altered volcanic fragments do not contain glass, show extensive biotitization and silicification, and appear to have been derived from old volcanic rocks. The banded pumice contains alternating layers of various thickness (1 cm to several centimeters wide) of white pumice and gray pumice; boundaries between bands may be sharp or diffused, and some show intricate mingling textures. Gray layers contain both olivine and hornblende, and they are interpreted to be gray pumice because of the presence of plagioclase phenocrysts and fine grain size of phenocrysts, similar to that of gray pumice. The lack of calcic rims and dusty zones in the plagioclase phenocrysts and the absence of plagioclase microlites argue against an alternative possibility—that the gray layers are andesite—because calcic rims, dusty zones, and microlites are common in the plagioclase population of the dome-forming andesite. Volcanic bombs are usually gray pumice, and one such bomb (~25 cm long) of gray pumice was also examined in this study.

Our samples include two white pumice and one gray pumice specimens collected shortly after the eruption in June 1991 and one specimen of 1991 andesite dome collected along the Maraunot River in the spring of 1992. The rest of the samples were collected in August 1992. Most andesites and basalts were collected along the Maraunot River, ~4 km west of the caldera wall, and one of the basalt samples was from a fragment more than 2 m in diameter, which contains angular andesite fragments. Most pumice samples were collected along the Sacobia River near the upper end of Clark Air Base. There are no apparent differences in mineralogy and textures of samples collected in 1991 and 1992. Anhydrite is well preserved in pumice and andesite samples collected in August 1992. Good preservation of anhydrite in samples collected in 1992 was unexpected, because heavy rainfall in the area could have leached anhydrite from the volcanic rocks, as observed at El Chichón (Luhr and others, 1984). Its preservation at Pinatubo may be attributed to short exposure time of samples at the surface as a result of the daily occurrences of lahars, which continuously brought buried eruption products to the surface.

The 1991 eruption also ejected older volcanic rocks. All samples, however, have been identified with confidence as products of the 1991 eruption because of their highly angular shape and large sizes and the lack of evidence of weathering and devitrification in the glass. Older eruption products are commonly semirounded, and they show various degrees of cloudiness of matrix glass due to devitrification and nucleation of fine dusts of hematite, and Fe-Ti oxide microphenocrysts usually show well-developed exsolution lamellae.

ANALYTICAL PROCEDURES

Major chemical compositions of sulfides were determined by use of a JEOL 6400 digital scanning electron microprobe (SEM), which has a 40° take-off angle for X-rays and is interfaced to a Link X-ray analyzer system (eXL LZ5). Analytical conditions were 20 kV accelerating potential, 39 mm distance between the specimen and the analyzer, 0.8 nA absorbed current on a Faraday cup, and a counting time of 140 to 200 s. Raw X-ray spectra were reduced to elemental concentrations by use of Link ZAF4-FLS analytical software. Analytical standards were natural pyrite for Fe and S, chalcopyrite for Cu, Ni metal for Ni, Co metal for Co, and synthetic ZnS for Zn. Analyses are believed to be accurate to ±2 percent of the amount present; detection limits are ~0.1 percent.

More than 200 grains of sulfides were subjected to probe analysis, but it was difficult to obtain satisfactory analytical results for all grains because of their small size (less than several micrometers in diameter) and interference from the host phases. The compositions of representative sulfides that are large enough to provide quantitative data are given in table 2.

Trace elements of sulfides in thick sections (~100 μm) were determined by a proton-induced X-ray emission microprobe (PIXE). The analytical technique for trace elements in sulfide minerals has been well established (Cabri and others, 1984; Campbell and others, 1989; Czamanske and others, 1992). Operating conditions are similar to those of earlier workers; 3 meV of proton energy, 45° take-off angle, ~8 nA of specimen current, 600 to 900 s counting time, and 4×5 μm beam size. Aluminum absorbers 249 and 352 μm thick were used for pyrrhotite and copper-rich sulfides, respectively. A synthetic pyrrhotite standard (in wt%; Fe=60.93, S=38.87, Se=0.09, and Pd=0.11) was used for calibration. X-rays were detected by a Kevex silicon detector fitted with a beryllium proton shield of 50 μm thick, and data reduction of the raw X-ray spectra were performed by the GUPIX program of Maxwell and others (1989). Detection limits are three times the errors obtained from the background.

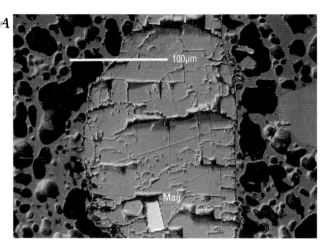

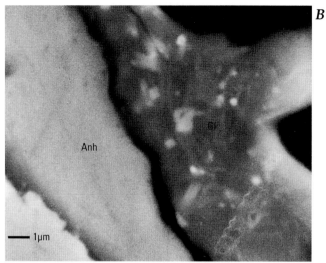

Figure 1. Back-scattered electron images of anhydrite in pumice. *A*, Anhydrite enclosing low-titanium magnetite (Mag) in banded pumice. *B*, Contact between anhydrite (Anh) and glass in white pumice. Note smooth boundary between the two phases. White specks in glass are microphenocrysts.

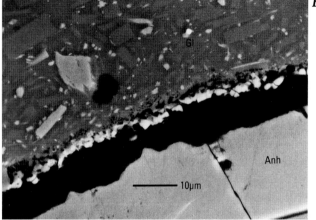

Figure 2. Back-scattered electron images of anhydrite (Anh) in dome andesite. *A*, The reaction rim, only 10 μm wide, is well exposed in the lower right because anhydrite was plucked out during the section preparation (sample CN1B–C). *B*, Boundary between anhydrite (Anh) and glass (Gl) in sample 92814–01B. Minute minerals between glass and anhydrite are apatite, calcite, calcic plagioclase and other calc-silicate minerals. Most of them are plucked out during the section preparation.

OCCURRENCE OF SULFUR PHASES

ANHYDRITE

The occurrence of anhydrite in dacitic pumice has been reported by Knittel and others (1991), Bernard and others (1991), Pallister and others (1991, 1992), and others. This study confirmed its common occurrence in gray and white pumice from Pinatubo. Banded pumice also contains anhydrite in both gray and white pumice layers. Anhydrite crystals are commonly coarse, <0.6 mm, and some grains enclose microphenocrysts of apatite and low-Ti (<1 wt% TiO_2) magnetite (fig. 1*A*). In all pumice samples, the

contact between the grains and glass is smooth and there is no evidence of reactions between the two phases (fig. 1*A*,*B*), confirming that anhydrite was a stable liquidus phase in the dacitic melt immediately before eruption. An anhydrite inclusion in hornblende was reported by Fournelle (1991), but none were observed during this study, so it is suggested that such occurrences are extremely rare.

Anhydrite grains up to 0.4 mm in length also occur in the dome andesite. These grains display reaction rims, several micrometers wide along the contact with glass, of calcite, apatite, calcic plagioclase, and other calc-silicates (fig. 2*A*,*B*). Quantitative determination of these phases was impossible because of their small sizes, <0.5 μm.

SULFIDES

All sulfide grains are fine grained and are found in all sections examined in this study, from a variety of eruption products. Occurrences are summarized in table 1. Grains enclosed in phenocrysts are usually small, rarely exceeding 40 μm, but one globular sulfide of cubanite composition (~120 μm) was found in unexsolved rhombohedral ilmenite-hematite solid solution (ilmenite$_{ss}$; fig. 3). Sulfides in the eruption products display four different shapes: spherical or globular sulfide enclosed in phenocrysts and glass (figs. 3, 4), symplectic minute droplets in phenocrysts (fig. 5), and irregular-shaped sulfides in glass (fig. 6). Globular sulfides and symplectic droplets in basalt and andesite are mostly Ni-pyrrhotite, whereas irregularly shaped sulfides in glass contain high copper and commonly show exsolution of bornite, Cu_5FeS_4. Most sulfides in dacitic pumice, independent of shapes and host phases, are Cu-rich sulfides. Several sulfide grains in phenocrysts occur together with glass inclusions (fig. 7).

Basalt.—Sulfide grains are relatively common in basalt fragments in the dome andesite. Phenocrysts of olivine, clinopyroxene, hornblende, ilmenite$_{ss}$, and ulvöspinel-magnetite solid solution (magnetite$_{ss}$) contain sulfides; inclusions in olivine are extremely rare, but a few inclusions were observed. Most sulfides in phenocrysts are small (<10 μm), globular, Ni-bearing pyrrhotite with low copper (<0.1 wt%) (fig. 5). Spherical nickel-bearing pyrrhotite also occurs in the glass (fig. 8). These occurrences suggest that a sulfide liquid had the composition of Ni-Fe monosulfide solid solution. Copper-bearing pyrrhotite and an exsolution product of cubanite ($CuFe_2S_3$) occur in apparently late phenocryst phases, such as magnetite$_{ss}$ and hornblende, which rims olivine. Copper-rich sulfides (atomic ratio of Cu/Fe >1) are not common in basalts and are only found in glass. Chalcopyrite is also found with a complex mixture of iron-rich orthopyroxene and magnetite that appears to replace Ni-pyrrhotite and olivine (fig. 9).

Andesite.—The dome andesite contains sulfides and anhydrite. Sulfides are similar to those of basalt; these sulfides are usually small (<20 μm), but some are ~0.1 mm (fig. 3). Sulfides occur in olivine, augite, hornblende, and oxides (ilmenite$_{ss}$, magnetite$_{ss}$), K-feldspar (confirmed with electron microprobe), and glass. Sulfides in augite, hornblende, oxides are mostly globular Ni-bearing pyrrhotite. Some in hornblende rimming olivine and oxides are copper bearing. One grain of Ni-pyrrhotite in magnetite$_{ss}$ is surrounded by perovskite, possibly a reaction product (fig. 10).

Several Ni-pyrrhotite grains in phenocrysts are replaced by copper-rich sulfide along cleavages and cracks (fig. 11). Sulfides in glass have angular to irregular shapes and they are all copper rich (chalcopyrite with or without exsolution of bornite) (fig. 12). It is common to find pyrrhotite only in phenocrysts and chalcopyrite only in glass (fig. 13). The occurrences and phases of sulfides demon-

Table 1. Summary of the occurrence of sulfides in the Pinatubo eruption products.

[Chr, chromite; Cpx, augite; Hbl, hornblende; Il, ilmenite-hematite solid solution; K-fs, K-feldspar; Mag, titanomagnetite (ulvöspinel-magnetite solid solution); Ol, olivine; Qtz, quartz. Most sulfides in the gray pumice were small and analyses were qualitative—adequate to identify the species (below) but not good enough for inclusion in table 2]

Host	Abundance of sulfides in the host	Shape of sulfides	Cu/Fe atomic ratio of bulk composition of S-phase[1]
Gray pumice			
Glass	very rare	angular	~1
Hbl	minor	globular, <5 μm	~1
Mag	less common	globular, <4 μm	~1
Pl	rare	tabular, 3 × 8 μm	~1
White pumice			
Glass	minor	irregular, <40 μm with reaction rim.	>1
Hbl, Mag.	minor	globular, <14 μm	~1, 0
Qtz	rare	angular, <20 μm	>1
Pl	minor	angular, <10 μm	>1, ~1
Banded pumice			
Glass	minor	irregular, <30 μm with reaction rim.	0, ~1
Hbl	minor	globular, <15 μm	0
Mag	minor	globular, angular	0, >1
K-fs	minor	irregular	~1
Andesite			
Glass	minor	angular	~1, >1
K-fs	minor	angular	~1
Hbl	common	globular	0–1
Il	minor	globular	0–1
Mag	common	globular	0–1
Cpx	not common	globular	0
Ol	rare	globular	0
Chr	none		
Basalt			
Glass	common	globular	0
		irregular, <40 μm	>1
Hbl	common	globular	0–1
Il	rare	globular	0–1
Mag	rare	globular	~1
Cpx	rare	globular	0
Ol	rare	globular	0

[1] Grains containing bornite exsolution obviously have Cu/Fe atomic ratios higher than 1. Grains containing cubanite exsolution have the ratio close to 1. The ratios of Fe-sulfides with no exsolution phases are shown as 0, although some contain several wt% Cu.

strate that early sulfides are Ni-pyrrhotite and late sulfides are Cu-sulfides.

Dacitic Pumices.—Most sulfide grains in white pumice are rich in copper (Cu/Fe atomic ratio >1) and a high copper phase is formed as an exsolution product (figs. 14,

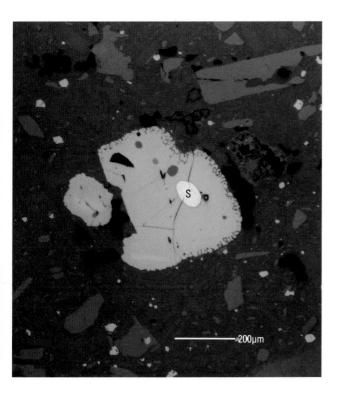

Figure 3. Photomicrograph of a rounded cubanite grain (S; 120 μm long) enclosed in ilmenite$_{ss}$ in dome andesite (sample 920814–10).

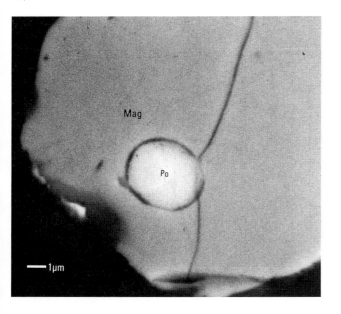

Figure 4. Back-scattered electron image of spherical grain of pyrrhotite (Po) in magnetite$_{ss}$ (Mag) in dome andesite (sample CN1B–C–X).

15). The sulfides were originally formed as a high-temperature Cu-Fe-S intermediate solid solution. Due to their small sizes and fine lamellae width (figs. 14, 15, 16), it was impossible to determine the precise compositions of the Cu-rich phase, but it is likely bornite (Cu$_5$FeS$_4$), on the

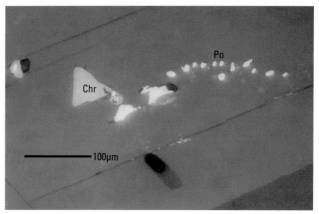

Figure 5. Photomicrograph showing symplectic chains of Ni-bearing pyrrhotite (Po) droplets in hornblende in basalt. Olivine and hornblende commonly enclose chromite (Chr) (sample 920814–11A). Scale bar corresponds to 100 μm.

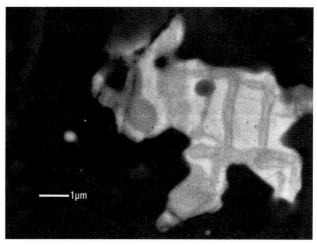

Figure 6. Back-scattered electron image of irregularly shaped sulfide grain containing exsolution product of bornite (bright part) in chalcopyrite (darker matrix) in the glass of dome andesite (sample CN1B–D–C).

basis of the Cu-Fe-S phase relations and the degree of electron back-scattering from the phase. Pyrrhotite is very rare in pumice samples, but rounded grains of pyrrhotite were found within hornblende and magnetite$_{ss}$ phenocrysts in gray pumice.

Sulfides in phenocrysts are usually globular to spherical (fig. 14) and many in plagioclase are angular (fig. 15). Sulfides in matrix glass generally have an angular shape (figs. 16, 17). Some are exceptionally high in copper, forming a bornite matrix with exsolved chalcopyrite inclusions. Several Cu-Fe sulfides in glass are high in zinc, up to 1.3 wt% (table 3). Ni-pyrrhotite grains are extremely rare but are found in gray pumice and in bands of gray pumice within banded pumice. Most sulfide grains enclosed in glass

Table 2. Chemical compositions of representative sulfides in Pinatubo eruption products.

[Cpx, augite; Gl, glass; Hbl, hornblende; Il, ilmenite-hematite solid solution; K-fs, K-feldspar; Mag, ulvöspinel-magnetite solid solution; Ol, olivine; Ox, oxides; Pl, plagioclase]

Sample no.	Host	Weight percent						Atomic percent					Cu/Fe	Remarks
		S	Fe	Ni	Co	Cu	Sum	S	Fe	Ni	Co	Cu		
White Pumice														
PP1-A1	Gl	34.83	33.26	0.00	0.00	31.41	99.50	49.92	27.37	0.00	0.00	22.72	0.83	
PP1A-3	Gl	34.35	33.12	0.00	0.00	30.96	98.43	49.79	27.56	0.00	0.00	22.65	0.82	
PP1-A,x	Gl	34.99	33.53	0.00	0.23	30.73	99.48	50.08	27.55	0.00	0.18	22.19	0.81	
10-PP1A4	Gl	35.17	33.43	0.19	0.00	31.71	100.50	49.91	27.23	0.15	0.00	22.71	0.83	
PP1A4	Hbl	36.30	56.83	0.00	0.25	0.65	94.03	52.32	47.02	0.00	0.20	0.47	0.91	
32PP1-BE1	Gl	35.54	41.74	0.00	0.00	21.69	98.97	50.45	34.02	0.00	0.00	15.54	0.46	1.3% Zn
33PP1-BE2	Gl	34.44	40.56	0.00	0.00	19.87	94.76	50.83	34.37	0.00	0.00	14.80	0.43	1.0% Zn
PP2C-B	Pl	31.67	30.43	0.28	0.00	34.24	96.62	47.58	26.24	0.23	0.00	25.95	0.99	
PP2D,B	Pl	31.49	29.98	0.00	0.00	33.81	95.28	47.89	26.16	0.00	0.00	25.95	0.99	
4PP1-C,A	Gl	26.55	14.83	0.00	0.00	55.84	97.22	41.98	13.46	0.00	0.00	44.55	3.31	Cu-rich part of a grain
PP1-C,A	Gl	31.08	27.38	0.00	0.00	37.13	95.59	47.42	23.99	0.00	0.00	28.59	1.19	matrix of the grain
PP1-C	Hbl	35.09	39.47	0.43	0.00	22.50	97.49	50.55	32.65	0.00	0.00	16.36	0.50	
PP1-AD1	Hbl	32.73	33.82	0.31	0.00	31.33	98.19	48.01	28.48	0.24	0.00	23.19	0.81	
PP1-AD2	Hbl	32.72	33.74	0.00	0.00	30.85	97.31	48.36	28.63	0.00	0.00	23.01	0.80	
PP1-AE	Mag	32.88	33.90	0.00	0.00	32.80	99.58	47.72	28.25	0.00	0.00	24.03	0.85	
29PP1BD1	Hbl	33.93	35.56	0.00	0.00	31.74	101.23	48.22	29.02	0.00	0.00	22.76	0.78	
30PP1BA1	Gl	34.30	29.92	0.00	0.00	33.82	98.04	50.04	25.06	0.00	0.00	24.90	0.99	
31PP1BA2	Gl	31.98	29.26	0.00	0.00	33.34	94.58	48.75	25.61	0.00	0.00	25.65	1.00	
Gray Pumice														
511-812-4B	Gl	31.61	30.98	0.00	0.00	32.84	95.53	47.92	26.96	0.00	0.00	25.12	0.93	
513-812-4C	Ox	31.97	35.14	0.00	0.00	30.76	97.87	47.25	29.81	0.00	0.00	29.81	0.77	
Banded pumice														
528-813-17C1	Gl	39.64	61.12	0.00	0.00	0.00	100.76	53.04	46.95	0.00	0.00	0.00	0.00	
536-813-17A1	K-fs	34.42	30.42	0.00	0.00	34.52	99.36	49.67	25.20	0.00	0.00	25.13	1.00	
549-813-17B1	Mag	32.74	33.17	0.44	0.00	32.05	98.41	48.00	27.93	0.35	0.00	23.72	0.85	
529-813-17	Gl	35.68	59.31	0.00	0.00	0.65	95.64	50.93	48.60	0.00	0.00	0.47	0.01	
530-813-17	Gl	34.10	36.38	0.00	0.00	26.94	97.43	49.72	30.45	0.00	0.00	19.82	0.65	
Andesite														
CN1-BD2	Il	27.13	60.04	0.47	0.00	0.75	88.39	43.60	55.39	0.42	0.00	0.59	0.01	
CN1-BB1	Gl	36.18	51.48	1.49	0.00	6.19	95.34	51.93	42.44	1.17	0.00	4.48	0.11	
CN1-BCX	Ox	35.04	55.25	0.29	0.00	3.79	94.37	50.91y	46.09	0.23	0.00	2.77	0.06	
CN1-BB2	Gl	32.05	29.10	0.00	0.00	32.21	93.36	49.30	25.70	0.00	0.00	25.01	0.97	
CN1B-A1	K-fs	34.45	30.35	0.00	0.00	34.48	99.28	49.73	25.15	0.00	0.00	25.11	0.75	
401-814-1BB1	Ol/Hbl	39.65	54.34	4.34	0.00	0.00	98.33	54.16	42.61	3.23	0.00	0.00	0.00	beside chromite
402-814-1BB	Hbl	34.72	35.20	0.43	0.00	28.42	99.18	49.95	29.07	0.00	0.00	20.64	0.71	
404-814-1BB2	Gl	29.03	25.33	0.00	0.00	39.63	94.00	45.66	22.89	0.00	0.00	31.46	1.38	center
405-814-1BB2	Gl	30.00	16.56	0.00	0.00	53.69	100.25	45.05	14.28	0.00	0.00	40.68	2.85	branch part

Table 2. Chemical compositions of representative sulfides in Pinatubo eruption product—Continued.

Sample no.	Host	Weight percent						Atomic percent						Remarks
		S	Fe	Ni	Co	Cu	Sum	S	Fe	Ni	Co	Cu	Cu/Fe	
Andesite—Continued														
407–814–1BC	Hbl	33.38	27.83	0.00	0.00	37.27	98.47	48.97	23.44	0.00	0.00	27.59	1.18	
409–814–1BA1	Mg	33.42	38.13	0.00	0.00	28.97	100.51	47.79	31.31	0.00	0.00	20.90	0.67	
412–814–10	Ox	35.64	39.87	0.00	0.00	27.40	102.91	49.26	31.64	0.00	0.00	19.11	0.60	in fig. 3.
414–814–10	Mag	32.68	31.76	0.00	0.00	37.95	102.39	46.64	26.05	0.00	0.00	27.33	1.05	
814–10	Mag	34.94	39.09	0.00	0.00	26.86	100.90	48.90	31.40	0.00	0.00	18.97	0.60	
525–814–3D	Gl	34.26	35.69	0.00	0.00	27.95	97.90	49.76	29.75	0.00	0.00	20.49	0.69	branch part.
527–814–3B1	Hbl/Gl	34.27	30.69	0.00	0.00	33.97	98.93	49.65	25.52	0.00	0.00	24.83	0.97	
Basalt fragment														
417–814–11BA2	K-fs	39.44	59.69	0.46	0.00	0.74	100.33	53.06	46.10	0.34	0.00	0.50	0.01	
418–814–11BA3	Gl	39.85	58.57	0.00	0.00	0.94	99.36	53.89	45.47	0.00	0.00	0.64	0.01	
419–814–11CA	Hbl	39.29	60.37	1.20	0.00	0.00	100.86	52.67	46.46	0.88	0.00	0.00	0.00	
420–814–11CB	Hbl	37.54	45.93	1.04	0.00	17.10	101.61	51.35	36.07	0.78	0.00	11.71	0.33	beside chromite.
548–814–11A1	Hbl	37.23	57.08	1.62	0.00	2.52	98.45	51.59	45.42	1.23	0.00	1.76	0.04	
549–814–11AB	Cpx	38.93	58.02	1.18	0.00	0.00	98.13	53.41	45.71	0.88	0.00	0.00	0.00	
437–814–2AA	Il	39.05	58.96	1.33	0.00	0.00	99.34	53.04	45.98	0.99	0.00	0.00	0.00	
432–814–2AE	Mag/Gl	30.44	17.02	0.44	0.00	52.24	100.14	45.56	14.63	0.36	0.00	39.45	2.70	Cu-rich part.
431–814–2AE	Mag/Gl	29.54	33.92	0.00	0.00	33.63	97.09	44.77	29.52	0.00	0.00	25.72	0.87	Cu-poor part.
433–814–2AE	Mag/Gl	35.21	31.62	0.00	0.00	35.03	101.86	49.56	25.56	0.00	0.00	24.88	0.97	another part.
435–814–2AD1	Hbl	37.35	48.13	2.80	0.00	8.29	96.57	52.26	38.66	2.14	0.00	5.85	0.15	
436–814–2AD2	Hbl	34.49	37.86	0.71	0.00	23.42	96.48	50.40	31.77	0.56	0.00	17.27	0.54	
434–814–2AB	Ox	36.67	46.37	0.00	0.00	19.48	102.52	50.35	36.41	0.00	0.00	13.44	0.37	

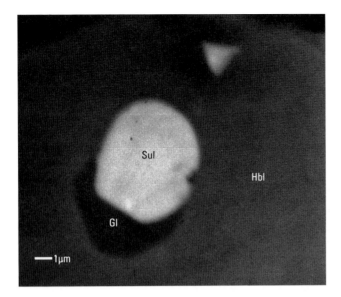

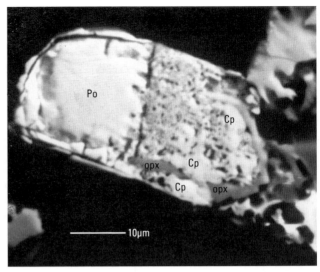

Figure 7. Back-scattered electron image of sulfide inclusions (Sul) in hornblende (Hbl) in white pumice (sample PP1–92526–C). Dark area (Gl) surrounding sulfide contains K, Cl, Al, Si, and Fe, and it is interpreted to be a glass inclusion.

Figure 9. Reflected light photomicrograph of a symplectic mixture of very fine grained magnetite, orthopyroxene (opx) and chalcopyrite (Cp) replacing Ni-bearing pyrrhotite (Po). Quantitative determination of silicate phase (dark part in Po) was difficult because of the small sample size, but the X-ray spectrum suggests it is olivine. Sample (920814–11) is from a large basalt fragment in andesite.

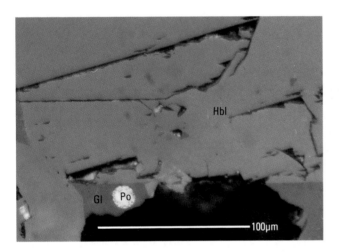

Figure 8. Reflected light photomicrograph of spherical grain of pyrrhotite (Po) in basalt glass (Gl) adjacent to hornblende (Hbl).

show reaction rims that are composed of a fine mixture of oxides and sulfides. The presence of reaction rims reflects desulfidation (figs. 16, 17), which indicates that sulfide was not stable in the dacitic magma immediately before eruption. The narrow width of rims and dispersed occurrences of remnant sulfides prevented their quantitative compositional analysis, but they appear to be fine iron oxides.

Selenium contents of sulfides are high with Se/S ratios in a range of 20×10^{-5} to 100×10^{-5} (table 3). The ratios are slighter higher than the values of meteorites ($\sim33\times10^{-5}$; Wedephol, 1972), sulfides in mantle xenoliths ($\sim30\times10^{-5}$; Hattori and others, 1992), and sulfides in midocean ridge

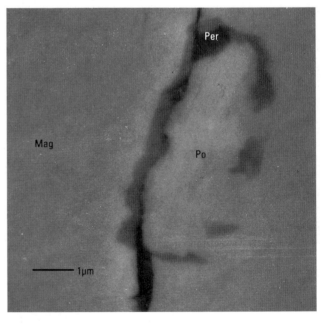

Figure 10. Back-scattered electron image of Ni-bearing pyrrhotite (Po) in magneite$_{ss}$ (Mag). Dark phase between sulfide and oxides is perovskite (Per). Sample (CH1BE–W) is a dome andesite.

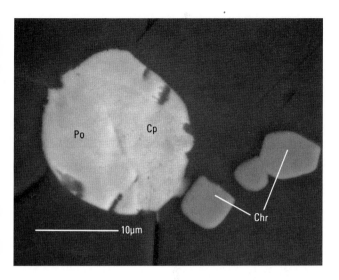

Figure 11. Back-scattered electron image of globular Ni-bearing pyrrhotite (Po), partially replaced by Cu-rich sulfides (Cp) along the cracks of olivine rimmed by hornblende. Chromite (Chr) grains at the right. Note that the rim of pyrrhotite is partially corroded.

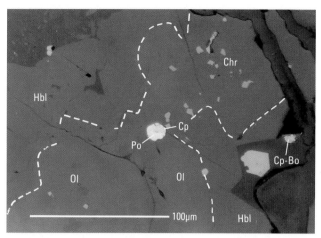

Figure 13. Back-scattered electron image of Ni-bearing pyrrhotite (Po) in olivine (Ol) rimmed by hornblende (Hbl) and Cu-rich sulfides (Cp, Bo) in glass. Minute chromite grains (Chr) are visible in olivines. A closeup view of the Po-Cp grain is shown in figure 11. Scale bar is 200 µm.

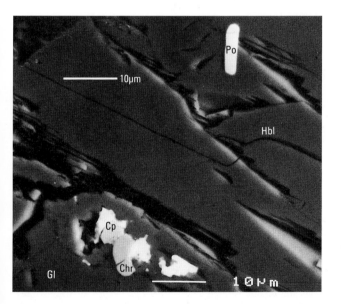

Figure 12. Back-scattered electron image of two kinds of sulfides. Sulfides enclosed in hornblende (Hbl) are rounded Ni-rich pyrrhotite (Po), and sulfides in glass (Gl) are irregularly shaped, Cu-rich sulfide (Cp). Fine grains of chromite (Chr) are common. Sample (CH1BB–A) is dome andesite.

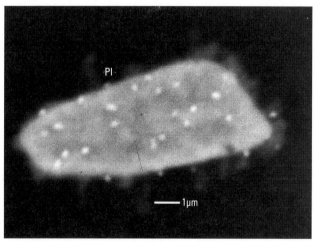

Figure 14. Back-scattered electron image showing typical occurrence of Cu-rich sulfides in white pumice. Sulfides in 2-mm-long plagioclase (Pl) phenocryst. The bright portion is the exsolution product of bornite; the dark portion is chalcopyrite (sample PKN 29–2).

Figure 15. Back-scattered electron image of sulfide in 3-mm-long plagioclase (Pl) phenocryst (sample PP1–C–A).

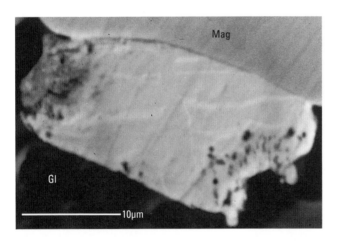

Figure 16. Back-scattered electron image of sulfides showing desulfidization rim around the grain in the glass of white pumice. The sulfide on the contact with magnetite (mag) and glass (Gl) displays exsolution lamellae of bornite (bright) from chalcopyrite (dark) (sample PKN29–2–A).

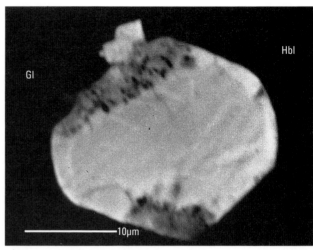

Figure 17. Back-scattered electron image of sulfide in white pumice with oxidized rim (sample PP1–A–A). Gl, glass; Hbl, hornblende.

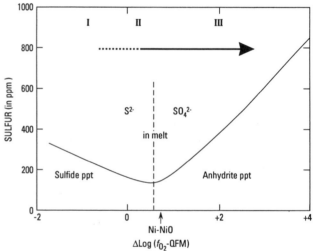

Figure 18. Sulfur solubility in dacitic melt at 900°C (modified after Carroll and Rutherford, 1988). Dotted line is the redox boundary of predominant sulfur species in the melt. Thick arrow shows the change of the Pinatubo dacitic melt. At stage I, an immiscible sulfide liquid forms and scavenges nickel and copper from the silicate melt. At stage II, SO_2 that has been released from mafic melt is reduced to H_2S as it oxidizes the dacitic melt. The added H_2S forms sulfides in the dacitic melt that are high in Cu, Cd, Zn, and Se. At stage III, continuing addition of SO_2 from mafic melt increases f_{O_2} and results in partial desulfidation of previously formed sulfides and precipitation of anhydrite in the dacitic melt. ppm, parts per million; ppt, precipitate.

Table 3. Trace element contents in sulfides in white pumice (in ppm).

Sample	Phase[1]	Host[2]	Zn	As	Se	Mo	Ag	Cd	Sb	Te	S/Se	Se/S × 10^-5	Shape
White pumice													
UK29–3	Cp	Gl	2,233	24	314	<6	36	20	<15	35	1,074	93.1	Angular
PP1BB	Cp	Gl	2,569	<8	299	<6	32	24	<16	<22	1,192	83.9	Angular
PP1BB	Cp	Gl	2,884	<9	266	8	10	14	<15	<23	1,244	80.4	Angular
PP1BA	CpB	Gl	212	<3	170	<4	242	42	<22	<34	2,031	49.2	Angular
PP1C,F[3]	Cp	Hbl	1,247	12	61	<5	21	<10	<17	74	2,741	36.5	Globular
PP1A1A[3]	CpB	Gl	2,209	8	276	44	91	22	<18	39	1,501	66.6	Angular
PPA1A[3]	CpB	Gl	2,184	22	333	149	173	16	36	43	1,146	87.3	Angular
Andesite													
4–92814–10C	Cub	Il	919	18	26	<5	unc[4]	30	<25	<35	11,540	8.7	Globular, fig. 3
Basalt fragment													
92814–11AK	CubP	Gl	998	unc[4]	107	<18	<28	<30	62	<60	2,388	41.9	Globular
92814–11AC		Hbl/Gl	536	20	104	<16	25	<35	<47	<70	2,921	34.2	Angular
92814–2BZ	CuPo	Gl	927	41	76	64	46	<35	62	<65	3,778	26.5	Irregular
92814–2BA	CubP	Gl	684	28	78	<16	46	<29	43	<48	4,125	24.2	Irregular
92814–11AD	Po	Hbl	148	13	57	<19	30	<35	<65	<73	5,897	17.0	Globular
92814–11AD	Po	Gl	453	unc[4]	20	26	<17	<14	<29	<36	17,485	5.7	Globular

[1] Sulfide phase: CpB, a mixture of chalcopyrite and exsolution product of bornite; Cp, chalcopyrite; Cub, cubanite; CubP, cubanite-pyrrhotite; CuPo, Cu-bearing pyrrhotite; Po, pyrrhotite.
[2] Phase containing sulfides: Gl, glass; Hbl, Hornblende; Il, unexsolved ilmenite-hematite solid solution.
[3] The grains were too small and the total of Cu, Fe, and S are in a range of 50 to 90%. Trace element concentrations are accordingly normalized.
[4] unc, the presence of the element is uncertain, and the measured concentration is between the detection limit and three times the measurement error.

basalts (~20×10^{-5}; Hamlyn and Keays, 1986; and 30×10^{-5} to 60×10^{-5}; Hattori and others, 1992).

Sulfides in gray phenocryst-poor pumices are rare, but small grains (<1 μm) are found in phenocrysts of magnetite$_{ss}$, hornblende, and plagioclase. Sulfides are extremely rare in the glass of these pumices. Gray pumices with higher phenocryst contents (>10 vol%) contain more sulfide grains than do pumices of the phenocryst-poor variety.

DISCUSSION

SULFUR IN SILICATE MELTS

The solubility of sulfur in silicate melts is controlled by temperature, pressure, oxygen fugacity, and melt composition. The solubility of sulfur at constant temperature is higher in melts with higher iron; mafic melts have a higher solubility of sulfur than less mafic melts (Buchanan and Nolan, 1979). Sulfur in reduced melt is predominantly S^{2-} with minor HS^-, whereas sulfur in oxidized melt is SO_4^{2-} (fig. 18; Katsura and Nagashima, 1974; Carroll and Rutherford, 1988). The redox boundary of dissolved sulfur species, S^{2-} and SO_4^{2-}, is one log f_{O_2} unit above the Ni-NiO (NNO) buffer (Carroll and Rutherford, 1988). At constant temperature and pressure, the solubility of sulfur decreases steadily as f_{O_2} increases up to the redox boundary, and then solubility increases sharply with a further increase in f_{O_2} (fig. 18).

Sulfide liquid formed in mafic silicate melts is enriched in Ni, Co, and Cu because the partition coefficients for these elements between sulfide liquid and silicate melt are high (>250 for Ni and Cu and 50 for Co at 1200°C; Rajamani and Naldrett, 1978). Sulfide liquid that formed in evolved silicate melts tends to be low in nickel and cobalt and high in copper because of preferential incorporation of nickel and cobalt into early silicate minerals.

The evolution of sulfur in the magma may be evaluated from the occurrences of sulfides and anhydrite, assuming that the basalt samples represent mafic melt. Mafic magma was present beneath the dacitic magma chamber shortly before the 1991 eruption (Pallister and others, 1992; Matthews and others, 1992).

SULFUR IN MAFIC MELT AT PINATUBO

Occurrences of globular sulfide grains in phenocrysts and in the glass of basalt fragments suggest that they formed from sulfide liquid and that the mafic melt was saturated with sulfur. Common occurrence of sulfides in oxides is consistent with the experimental results of the solubility of sulfur. Formation of oxides, loss of iron from the melt, probably prompted the separation of sulfide liquid in the silicate melt.

The formation of liquid sulfide in silicate melt indicates that the predominant dissolved sulfur species in the melt was S^{2-} and that the f_{O_2} of the melt was lower than the redox boundary of sulfur species in the melt (f_{O_2} = NNO +1; fig. 18; Nagashima and Katsura, 1973; Carroll and Rutherford, 1988). The mafic magma was not sufficiently oxidized to contain significant SO_4^{2-}, so it could not form anhydrite. The lack of anhydrite in basalt is consistent with this interpretation.

The composition of globular sulfides, Ni-pyrrhotite, indicates that the sulfide liquid was Ni-Fe monosulfide solid solution, similar to most sulfides in many igneous rocks, including midoceanic ridge basalts (Mathez, 1976), oceanic island basalts (Desborough and others, 1968), back-arc basin basalts (Francis, 1990), island arc basalts and andesites (Heming and Carmichael, 1973; Ueda and Itaya, 1981), and continental arc rocks (Anderson, 1974; Whitney and Stormer, 1983).

The contents of copper in the Pinatubo monosulfide solid solution are low, <1 wt%. The low concentration of copper in iron sulfides in the Pinatubo eruption products suggests that the mafic melt was not originally high in copper, as the Cu/Fe ratios of sulfides reflect the ratios of silicate melt. This implies that high copper in later formed sulfides requires a mechanism for the enrichment of copper other than igneous processes.

The Cu/Ni ratio in melts increases during fractional crystallization. Formation of sulfide liquid containing 30 wt% Cu requires a melt with 0.15 wt% Cu, using the partition coefficient for copper between sulfide liquid and silicate melt (Rajamani and Naldrett, 1978). The concentration of copper would not increase to this high level in silicate melts by fractional crystallization alone. If fractional crystallization was the cause, the primary mafic melt should have had exceptionally high copper. As discussed, the possibility is rejected because early-formed sulfides are low in copper. Copper, therefore, must have been added to the mafic melt from an external source after crystallization of these phenocrysts.

The occurrence of highly irregular shaped Cu-rich sulfides in the glass (fig. 6) is very unusual for sulfides in volcanic rocks. The shape suggests that they were not formed as sulfide liquid in a silicate melt. Instead, they formed after the glass was already semisolidified. This interpretation implies that the formation of Cu-rich sulfides may have taken place after the entrapment of mafic melt into the andesitic or dacitic melts.

Fe-Ti oxide assemblages, however, failed to demonstrate reduced f_{O_2} of the mafic melt. The estimated temperatures and f_{O_2} using the solution model by Frost and Lindsley (1992) are similar to those of dacitic melt, ~800°C and two log units above NNO (table 4). The temperatures recorded in oxides are very low for temperatures of mafic magma. Obviously, oxides in basalt have undergone subsolidus equilibration after the entrapment of the magma in cooler

Table 4. Temperature and f_{O_2} estimates from Fe-Ti oxides.

[And, andesite; Bas, basalt; BP, banded pumice; GP, gray pumice; WP, white pumice; Δ_{QFM}, difference between the f_{O_2} and QFM buffer in log units (QFM = quartz-fayalite-magnetite). Contents of Fe_2O_3 and components of ulvöspinel (Usp) and ilmenite (Ilm) calculated using QUILF program by Andersen and others (1993)]

Sample	Rock	SiO$_2$	Al$_2$O$_3$	TiO$_2$	V$_2$O$_3$	FeO	Fe$_2$O$_3$	MnO	MgO	Usp	Ilm	Temp. (°C)	Δ_{QFM}
UK29AX	WP	0.00	1.73	4.38	0.48	32.59	52.80	0.43	0.91	0.127		825	+3.1
		.15	.45	28.60	.34	24.32	43.51	.00	.79		0.557		
UK29AY	WP	.09	1.86	4.68	.48	32.83	56.65	.43	1.09	.136		845	+3.0
		.38	.40	28.82	.23	23.44	45.18	.37	1.17		.539		
UK29AZ	WP	.22	1.80	4.58	.44	33.06	56.64	.41	.84	.133		826	+3.1
		.15	.29	28.60	.19	23.82	43.82	.19	.96		.549		
92812–3A	GP	.00	1.92	4.02	.56	32.81	58.16	.00	1.08	.121		817	+3.2
		.00	.66	28.41	.00	23.99	45.08	.00	.88		.547		
92813–01B (in plagioclase)	GP	.00	2.95	7.24	.25	33.60	50.97	.41	2.25	.215		931	+2.6
		.00	.75	29.08	.25	22.92	43.82	.28	1.65		.543		
92813–17E	BP	.00	2.09	4.67	.35	33.42	59.91	.38	1.71	.129		821	+3.1
		.00	.38	29.35	.42	24.43	46.32	.00	1.10		.542		
92814–3C	And	.00	2.10	5.41	.26	33.06	56.32	.55	1.63	.153		856	+2.9
		.00	.51	29.43	.00	24.11	44.57	.00	1.32		.550		
92814–11AF	Bas	.22	1.71	4.00	.36	32.70	58.94	.44	1.00	.113		796	+3.2
		.19	.32	31.36	.00	25.82	40.22	.49	1.06		.590		
92814–11BY (in hornblende)	Bas	.00	3.30	11.95	.33	37.63	42.17	.45	2.64	.357		973	+2.0
		.00	.59	35.75	.00	27.11	33.19	.35	2.62		.648		

andesitic and dacitic magmas. Subsolidus equilibration of oxide composition is known in many slowly cooled igneous rocks (Morse, 1980; Frost and Lindsley, 1991) and experiments have documented fast subsolidus equilibration (Hammond and Taylor, 1982). The once-reduced nature of the mafic melt is, however, evident from low abundances of oxides in basalt. Formation of oxides requires high Fe^{3+}/Fe^{2+}. It is also supported by higher Ti and lower Fe^{2+} of oxides enclosed in phenocrysts (table 4). Phenocryst phases surrounding the oxides prevented them from equilibrating with the matrix glass. Reduced nature of the melt at an earlier time is also supported by the occurrence of a symplectic mixture of magnetite and Fe-rich orthopyroxene, which appear to have replaced pyrrhotite and olivine (fig. 9). The assemblage is indicative of progressive subsolidus oxidation of olivine (Johnston and Stout, 1984).

SULFUR IN DACITIC MELT

Like the mafic melt, the dacitic melt also once had a low f_{O_2} promoting the formation of sulfide liquid during phenocryst crystallization, as documented by globular sulfide inclusions (fig. 18). During that time, sulfur in the melt was predominantly S^{2-}. Fe-Ti oxides again fail to reflect the reduced condition because of later reequilibration of their compositions. They only record the condition just prior to eruption, ~800°C and 3 log units above NNO (table 4).

Under this condition, the predominant dissolved sulfur in the melt was SO_4^{2-} (fig. 18) and anhydrite could have formed as a stable phase, as supported by the stable occurrence of anhydrite in the glass (fig. 1A,B).

ORIGIN OF SULFUR AND ANHYDRITE

The solubility of SO_4^{2-} sharply increases with the increase in f_{O_2} (fig. 18). Formation of anhydrite in dacitic melt requires higher concentration of sulfur in the melt than does the formation of sulfide. The relatively reduced nature of the original dacitic melt, indicated by the occurrence of globular sulfides, suggests that the formation of anhydrite in the Pinatubo dacite melt requires two conditions: the oxidation of the residual melt and an addition of sulfur. The addition of sulfur is essential because the silicate melt should have lost much of its sulfur during the transition from the reduced to oxidized state, as the solubility of sulfur in the silicate melt is minimal at the redox boundary (Nagashima and Katsura, 1973). The addition of external sulfur took place during and after crystallization of hornblende, plagioclase, and magnetite, but before the eruption. Most apatite apparently crystallized after oxidation of the residual dacitic melt because apatite in the dacite is high in SO_3, up to 0.8 wt% (Imai and others, 1992; this volume).

Proposed sources of sulfur in high-sulfur magma include (1) subducted seafloor sulfide deposits (Whitney,

1984; 1992), (2) evaporite beds (Rye and others, 1984), (3) sulfide deposits in underlying ophiolite complexes (Fournelle, 1991), (4) sulfate minerals from preexisting volcanic rocks and hydrothermal deposits (McKibben and others, 1992), and (5) SO_2 released from underlying mafic magma (Pallister and others, 1992; Hattori, 1993). High sulfur in the dacitic melt at source (model 1) is not applicable because this study shows that the dacitic magma acquired sulfur from an external source during and after crystallization of phenocrysts. In addition, solubility of sulfur as S^{2-} is lower than the solubility of sulfur as SO_4^{2-} (fig. 18). The content of sulfur in the melt remained low until the melt was oxidized.

Model 2, incorporation of evaporite, proposed for the El Chichón eruption product, is rejected because of the lack of evaporites in the Pinatubo area. There is a thin sequence of Eocene to Pliocene shallow marine sedimentary rocks between the volcanic rocks and Zambales Ophiolite Complex, and the Complex occurs below the sea level beneath the volcano (Delfin, 1984; Delfin and others, this volume). The shallow marine rocks may contain evaporite minerals, but this possibility is discounted for three reasons. First, the assimilation of an evaporite would lower Se/S because of its low Se/S ratios ($<0.1\times10^{-5}$; Measures and others, 1980). Second, evaporites generally contain low amounts of copper. Third, incorporation of young evaporites would raise $^{87}Sr/^{86}Sr$ ratios, given their high $^{87}Sr/^{86}Sr$ (~0.709). The ratios of anhydrite in pumices (0.70421±0.00001; this study) and bulk pumice values (0.70422 to 0.70426; this study) are similar to other volcanic rocks in the Bataan arc (Knittel and others, 1992).

Model 3, incorporation of ophiolite-derived sulfide, is also discounted because it would not provide sulfur with high Se/S and an oxidizing agent for the magma. The two possible sources for additional sulfur that would accompany oxidation of dacitic magma are sulfate minerals from the old volcanic rocks (model 4) and sulfur released from underlying mafic magma (model 5). In either model, sulfur ultimately originated from underlying mafic magma. In the former model (model 4), sulfur was indirectly supplied from the mafic magma to the dacitic melt. This model is appealing, as intense hydrothermal activity had been noted at Pinatubo for a long time. Isotope studies and gas analyses have indicated that the hydrothermal activity was linked to the discharge of volatiles from magma at depth (Ruaya and others, 1992). The preexisting volcanic rocks were extensively altered to form secondary minerals, including anhydrite and sulfides (Delfin, 1984, Delfin and others, this volume; author's examination of Philippine National Oil Company drill chips in 1992). In addition, two porphyry copper deposits that are the product of magmatic-hydrothermal activity (the Dizon mine and Pisumpan deposit; Sillitoe and Gappe, 1984) are located within Mount Pinatubo, and they contain abundant high-temperature hydrothermal anhydrite. Minor porphyry copper deposits

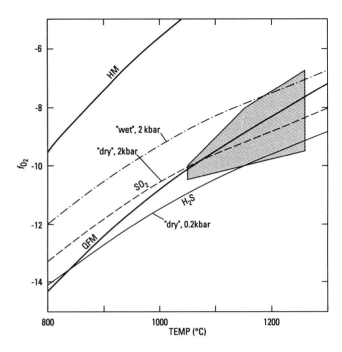

Figure 19. Equal concentrations of SO_2 and H_2S at 200 atm (dashed curve) and 2 kbar (dashed-dotted curve). "Dry" and "wet" fluids have X_{H_2O} 0.01 and 1, respectively, where X is volume ratio. Thick curves are quartz-fayalite-magnetite (QFM) and hematite-magnetite (HM) buffers. Oxidation conditions of mafic melts are shown by the screened area (from Wallace and Carmichael, 1992). Note that fluids released from progressively "drier" and hotter melts have progressively higher SO_2/H_2S ratios.

might have been present directly beneath the caldera. Assimilation of such rocks into the dacitic magma chamber could have enriched sulfur and oxidized the magma.

In the latter model (model 5), sulfur was directly added to the dacitic magma from the underlying mafic melt. The mafic melt at Pinatubo was saturated with sulfur, as evidenced by the occurrences of sulfide globules in phenocrysts. When mafic melt ascends, the vapor phase separates because the solubilities of CO_2 and H_2O decrease with pressure decrease (Holloway, 1976; Burnham and Ohmoto, 1980). Separation of any volatile gases from a melt will cause degassing of other gases, such as H_2S and SO_2, because they have mutual solubilities. For example, degassing of CO_2 or H_2O is accompanied by release of SO_2 from magmas before the saturation of SO_2 itself in the magma because SO_2 will be partitioned to the CO_2 phase from the silicate melt as CO_2 exsolves. Sulfur released from magmas is mostly SO_2 and H_2S. The ratio of SO_2/H_2S of the fluids depends on temperature, f_{O_2}, and f_{H_2O} (fig. 19). That ratio in high-temperature fluids discharged from mafic melts is generally high, as is well documented from volcanic gas data (Gerlach, 1986). H_2S may be predominant in cooler and more hydrated magmas, where SO_2 will be converted to H_2S (fig. 19).

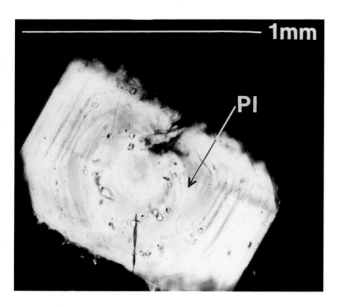

Figure 20. Transmitted light photomicrograph of fluid inclusions in plagioclases (Pl) in dacitic pumice. The inclusions usually occur near the rims of phenocrysts, but some occur in the center of phenocrysts, as shown here.

Underlying mafic magmas at Pinatubo likely released SO_2-rich supercritical fluids during their ascent. The fluids were likely absorbed by the overlying semisolidified dacitic magma. As dacitic magma is cooler and more hydrated than mafic magma, SO_2 would be converted to H_2S (fig. 19). The net effect is oxidation of the dacitic magma. The H_2S produced in the dacitic magma would form sulfides in the dacite melt (stage II of fig. 18) together with elements transported by the fluids, such as Zn, Cd, Cu, and Se. Continual addition of SO_2 would cause further oxidation of the dacitic magma, desulfidation of sulfides, and crystallization of anhydrite.

Sulfur isotopic data of the eruption products support both models. A variation in $\delta^{34}S$ values for individual grains of anhydrite (+3 to +16 ‰; McKibben and others, 1992; this volume) appears to support the incorporation of various types of sulfur from host rocks. Most $\delta^{34}S$ values are in a range between +6 and +8 (Knittel and others, 1992; McKibben and others, 1992; this volume; Bernard and others, this volume). Sulfates formed by hydrolysis of SO_2 have similar $\delta^{34}S$ values (Hattori and Cameron, 1986). The isotopic compositions outside this range could be attributed to a fluctuation in $\delta^{34}S$ of gases discharged from mafic magma, because $\delta^{34}S$ values of sulfur gases released from melt vary, depending upon temperature, f_{O_2}, and f_{H_2O} of the melts (Ueda and Sakai, 1984). Apparent isotopic equilibrium between sulfide and anhydrite in the dacitic magma (McKibben and others, this volume) may be attributed to the isotopic equilibration of sulfur between H_2S and SO_2 in the gas.

At present, there is no evidence to reject the former model, but the latter model is favored because of abundant occurrences of fluid inclusions in the phenocryst rims of hornblende, plagioclase, and quartz in the dacitic pumice (fig. 20), and because of $^{87}Sr/^{86}Sr$ ratios of anhydrite. Phenocrysts contain numerous fluid inclusions, some of which are high in CO_2. They most commonly align with growth bands of phenocrysts near rims. The occurrence supports that fluid percolation from mafic melt into dacitic melt started while phenocrysts were still being formed in the dacitic magma chamber.

As mentioned, $^{87}Sr/^{86}Sr$ ratios of anhydrite are identical to the bulk rocks (this study), whereas alteration products (epidote and anhydrite) have various $^{87}Sr/^{86}Sr$ ratios, up to 0.708 (Hattori and others, 1992).

The percolation of supercritical fluid is also in accord with the occurrence of copper-rich sulfides and their trace elements in the glass. When fluid separation takes place in silicate melts, volatiles such as Se and Cl are enriched in the fluid phase (Greenland and Aruscavage, 1986; Candela and Holland, 1984). Metals that have affinity with these volatiles are also enriched in the fluid. Experimental studies document preferential incorporation of copper into a fluid (Candela and Holland, 1984; Urabe, 1985). Significant copper and zinc are also recorded in volcanic gases emanating from silicate magma (Mizutani, 1970). The enrichment of copper in the upper part of a magma chamber is attributed to vapor transport (Lowenstern and others, 1991). Other elements that have affinity with Cl, such as Cd, Zn, and Ag, are known to be incorporated into a fluid phase (Symonds and others, 1990).

ORIGIN OF SULFUR DISCHARGED TO THE STRATOSPHERE

Various models have been proposed for the origin of sulfur discharged to the stratosphere: (1) breakdown of anhydrite in the magma (Rutherford and Devine, 1991; Baker and Rutherford, 1992), (2) breakdown/absorption of sulfides in melts (Whitney, 1992), and (3) release of gas present in a separate vapor phase (Westrich and Gerlach, 1992). The lack of reaction rims around anhydrite in dacitic pumices argues against the first model. It is also difficult to envisage a conversion of sulfur from a solid phase to SO_2 even at magmatic temperatures in the short period during the height of eruption. In addition, thin reaction rims of sulfides and preservation of most sulfides in the glass phase are not in accord with the second hypothesis. The occurrence of sulfur-bearing phases documented in this paper support the proposal by Westrich and Gerlach (1992) that the SO_2 was most likely derived by release of sulfur present in a vapor phase in the melt.

SUMMARY

The presence of globular and irregularly shaped sulfides in all types of eruption products indicates that an immiscible sulfide liquid formed in the silicate melts during phenocryst crystallization, that sulfur in the melt at that time was predominantly S^{2-}, and that f_{O_2} was below the sulfide/sulfate redox boundary. Later formation of anhydrite in the dacitic melt required oxidation of the melt, as well as addition of sulfur from an external source. High concentrations of Cu, Zn, and Se in late-formed sulfides suggest addition of these elements together with sulfur. It is proposed that the dacitic melt incorporated a supercritical fluid from the underlying mafic magma. The fluid had high SO_2/H_2S and was enriched in volatile and chalcophile metals. Conversion of SO_2 to H_2S in cool, hydrated dacitic magma caused oxidation of the dacitic magma chamber. While f_{O_2} was still below the redox boundary, Cu-rich sulfides formed in the melt. Once f_{O_2} became higher than the redox boundary, incorporation of sulfur into the dacitic magma caused crystallization of anhydrite.

ACKNOWLEDGMENTS

The author thanks R.S. Punongbayan for providing the lodging and transportation for the author's field work, C.G. Newhall and his family for their hospitality at Clark Air Base and their assistance for sampling in the summer of 1992, Ulrich Knittel (Institut für Mineralogie und Lagerstättenlehre, Technische Hochschule Aachen) for providing samples, H.P. Ferrer (Philippine National Oil Company) for permission to examine thin sections and for provision of drill chips in the area, H.G. Aniceto-Villarosa for assisting the examination of drill chip samples, Peter Jones (Ottawa-Carleton Geoscience Centre) for his superb operation of the SEM laboratory, and J.L. Campbell and W.J. Teesdale (Guelph University) for assisting in PIXE analysis. Comments by E.M. Cameron (Geological Survey of Canada), J.A. Donaldson (Ottawa-Carleton Geoscience Centre), J.H. Fournelle (Johns Hopkins University/University of Wisconsin, Madison), Ulrich Knittel, J.F. Luhr (Smithsonian Institution), M.A. McKibben (University of California, Riverside), N. Metrich (Centre National de la Recherche Scientifique, Saclay, France), and C.G. Newhall (USGS) are greatly appreciated. The project was supported by a grant from Natural Science and Engineering Research Council of Canada.

REFERENCES CITED

Andersen, D.J., Lindsley, D.H., and Davidson, P.M., 1993, QUILF: A PASCAL program to assess equilibria among Fe-Mg-Ti oxides, pyroxenes, olivine and quartz: Computers in Geosciences, v. 19, p. 1333–1350.

Anderson, A.T., 1974, Chlorine, sulfur and water in magmas and oceans: Geological Society of America Bulletin, v. 85, p. 1485–1492.

Baker L., and Rutherford, M.J., 1992, Anhydrite breakdown as a possible source of excess Sulfur in the 1991 Mount Pinatubo eruption [abs.]: Eos, Transactions, American Geophysical Union, v. 73, no. 43, p. 625.

Bernard, A., DeMaiffe, D., Mattielli, N., and Punongbayan, R.S., 1991, Anhydrite-bearing pumice from Mount Pinatubo: Further evidence for the existence of sulfur-rich magmas: Nature, v. 354, p. 139–140.

Bernard, A., Knittel, U., Weber, B., Weis, D., Albrecht, A., Hattori, K., Klein, J., and Oles, D., this volume, Petrology and geochemistry of the 1991 eruption products of Mount Pinatubo.

Bluth, G.J.S., Doiron, S.D., Schnetzler, C.C., Krueger, A.J., and Walter, L.S., 1992, Global tracking of the SO_2 clouds from the June 1991 Mount Pinatubo eruptions: Geophysical Research Letters, v. 19, p. 151–154.

Buchanan, D.L., and Nolan, J., 1979, Soluility of sulfur and sulfide immiscibility in synthetic tholeiitic melts and their relevance to Bushveld Complex rocks: Canadian Mineralogist, v. 17, p. 483–494.

Burnham, C.W., and Ohmoto, H., 1980, Late-state processes of felsic magmatism: Society of Mining Geologists of Japan Special Issue, v. 8, p. 1–11.

Cabri, L.J., Blanck, H., El Goresy, A., LaFlamme, J.H.G., Nobiling, R., Sizgoric, M.B., and Traxel, K., 1984, Quantitative trace-element analyses of sulfides from Sudbury and Stillwater by proton microprobe: Canadian Mineralogist, v. 22, p.521–542.

Campbell, J.L., Maxwell, J.A., Teesdale, W.J., Wang, J.X., and Cabri, L.J., 1989, Micro-PIXE as a complement to electron probe microanalysis in mineralogy: Nuclear Instruments and Methods in Physics Research, section B, v. 44, p. 347–356.

Candela, P.A., and Holland, H.D, 1984, The partitioning of copper and molybdenum between melts and aqueous fluids: Geochimica et Cosmochimica Acta, v. 48, p. 373–380.

Carroll, M.R., and Rutherford, M.J., 1988, Sulfur speciation in hydrous experimental glasses of varying oxidation state: Results from measured wavelength shifts of sulfur X-rays: American Mineralogist, v. 73, p. 845–849.

Czamanske, G.K., Kunilov, V.E., Zientek, M.L., Cabri, L.J., Likhachev, A.P., Calk, L.C., and Oscarson, R.L., 1992, A proton-microprobe study of magmatic sulfide ores from the Noril'sk-Talnakh district, Siberia: Canadian Mineralogist, v. 30, p. 249–287.

Delfin, F.G., Jr., 1984, Geology and geothermal potential of Mt. Pinatubo: Philippine National Oil Company, unpublished report, 36 p.

Delfin, F.G., Jr., Villarosa, H.G., Layugan, D.B., Clemente, V.C., Candelaria, M.R., Ruaya, J.R., this volume, Geothermal exploration of the pre-1991 Mount Pinatubo hydrothermal system.

Desborough, G.A., Anderson, A.T., and Wright, T.L., 1968, Mineralogy of sulfides from certain Hawaiian basalts: Economic Geology, v. 63, p. 636–644.

Fournelle, J.H., 1991, Anhydrite and sulfide in pumices from the 15 June 1991 eruption of Mount Pinatubo: Initial examination [abs.]: Eos, Transactions, American Geophysical Union, v. 72, p. 68.

Francis, R.D., 1990, Sulfide globules in mid-ocean ridge basalts (MORB) and the effect of oxygen abundance in Fe-S-O liquids on the ability of those liquids to partition metals from MORB and komatiite magmas: Chemical Geology, v. 85, p. 199–213.

Frost, B.R., and Lindsley, D.H., 1991, Occurrence of iron-titanium oxides in igneous rocks, *in* Lindsley, D.H., ed., Oxide minerals: Petrologic and magnetic significance: Mineralogical Society of America, Reviews in Mineralogy, v. 25, p. 433–468.

———1992, Equilibria among Fe-Ti oxides, pyroxenes olivine, and quartz, pt. II. Application: American Mineralogist, v. 77, p. 1004–1020.

Gerlach, T.M., 1986, Exsolution of H_2O, CO_2, and S during eruptive episodes at Kilauea volcano, Hawaii: Journal of Geophysical Research, v. 91, p. 12177–12185.

Gleason, J.F., Bhartia, P.K., Herman, J.R., McPeters, R., Newman, P., Stolarski, R.S., Flynn, L., Labow, G., Larko, D., Seftor, C., Wellemeyer, C., Komhyr, W.D., Miller, A.J., and Planet, W., 1993, Record low global ozone in 1992: Science, v. 260, p. 523–526.

Grant, W.B., Browell, E.V., Fishman, J., Brackett, V.G., Fenn, M.A., Butler, C.F., Myor, S.D., Nowicki, G.D., Veiga, R.E., Nganga, D., Minga, A., and Cros, B., 1992, Perturbation of tropical stratospheric ozone by Mt. Pinatubo aerosols-Experimental data [abs.]: Eos, Transactions, American Geophysical Union, v. 73, no. 14, p. 632.

Greenland, L.P., and Aruscavage, P., 1986, Volcanic emission of Se, Te and As from Kilauea volcano, Hawaii: Journal of Volcanology and Geothermal Research, v.27, p. 195–201.

Hamlyn, P.R., and Keays, R.R., 1986, Sulfur saturation and second-stage melts: application to the Bushveld platinum metal deposits: Economic Geology, v. 81, p. 1431–1445.

Hammond, P.A., and Taylor, L.A., 1982, The ilmenite/titanomagnetite assemblage: Kinetics of reequilibration: Earth and Planetary Sciences Letters, v. 61, p. 143–150.

Hattori, K., 1993, High-sulfur magma, a product of fluid discharge from underlying mafic magma: Evidence from Mount Pinatubo, Philippines: Geology, v. 21, p. 1083–1086.

Hattori, K., Arai, S., Francis, D.M., and Clarke, D.B., 1992, Variations of siderophile and chalcophile trace elements among mantle-derived sulfides detected by in-situ micro-PIXE analysis [abs.]: Eos, Transactions, American Geophysical Union, v. 73, no. 14, p. 344.

Hattori, K., and Cameron, E.M., 1986, Archaean magmatic sulphate: Nature, v. 319, p. 45–47.

Heming, R.F., and Carmichael, I.S.E., 1973, High-temperature pumice flows from the Rabaul caldera, Papua New Guinea: Contributions to Mineralogy and Petrology, v. 38, p. 1–20.

Holloway, J.R., 1976, Fluids in the evolution of granitic magmas: Consequences of finite CO_2 solubility: Geological Society of America Bulletin, v. 87, p. 1513–1518.

Imai, A., Listanco, E.L., and Fujii, T., 1992, Petrographic and sulfur isotopic significance of highly oxidized and sulfur-rich magma of Mt. Pinatubo, Philippines [abs.]: Abstracts, International Scientific Conference on Mt. Pinatubo, Manila, 27–30 May 1992, p. 6–7.

———this volume, Highly oxidized and sulfur-rich dacitic magma of Mount Pinatubo: Implication for metallogenesis of porphyry copper mineralization in the Western Luzon arc.

Johnston, A.D., and Stout, J.H., 1984, Development of orthopyroxene-Fe/Mg ferrite symplectites by continuous olivine oxidation: Contributions to Mineralogy and Petrology, v. 88, p. 196–202.

Katsura, T., and Nagashima, S., 1974, Solubility of sulfur in some magmas at 1 atm: Geochimica et Cosmochimica Acta, v. 38, p. 517–531.

Knittel, U., Hattori, K., Hoefs, J., and Oles, D., 1992, Isotopic compositions of anhydrite phenocryst-bearing pumices from the 1991 eruption of Mount Pinatubo, Philippines [abs.]: Eos, Transactions, American Geophysical Union, v. 73, no. 14, p. 342.

Knittel, U., Oles, D., Forster, H., and Punongbayan, R.S., 1991, Pinatubo pumice contains anhydrite [abs.]: Eos, Transactions, American Geophysical Union, v. 72, p. 522.

Lowenstern, J.B., Mahood, G.A., Rivers, M.L., and Sutton, S.R., 1991, Evidence for extreme partitioning of copper into a magmatic vapor phase: Science, v. 252, p. 1405–1409.

Luhr, J.F., Carmichael, I.S.E., and Varekamp, J.C., 1984, The 1982 eruptions of El Chichón volcano, Chiapas, Mexico: Mineralogy and petrology of the anhydrite-bearing pumices: Journal of Volcanology and Geothermal Research, v. 23, p. 69–108.

Mathez, E.A., 1976, Sulfur solubility and magmatic sulfides in submarine basalt glass: Journal of Geophysical Research, v. 81, p. 4269–4276.

Matthews, S.J., Jones, A.P., and Bristow, C.S., 1992, A simple magma-mixing model for sulphur behavior in calc-alkaline volcanic rocks: Mineralogical evidence from Mount Pinatubo 1991 eruption: Journal of the Geological Society of London, v. 149, p. 863–866.

Maxwell, J.A., Campbell, J.L., and Teesdale, W.J., 1989, The Guelph PIXE software package: Nuclear Instruments and Methods in Physics Research, section B, v. 43, p. 218–230.

McKibben, M.A., Eldridge, C.S., and Reyes, A.G., 1992, Multiple origins of anhydrite in Mt. Pinatubo pumice [abs.]: Eos, Transactions, American Geophysical Union, v. 73, p. 633–634.

———this volume, Sulfur isotopic systematics of the June 1991 Mount Pinatubo eruptions: A SHRIMP ion microprobe study.

Measures, C.I., McDuff, R.E., and Edmond, J.M., 1980, Selenium redox chemistry at GEOSECS I re-occupation: Earth and Planetary Science Letters, v. 49, p. 102–108.

Mizutani, Y., 1970, Copper and zinc in fumarolic gases of Showashinzan volcano, Hokkaido: Geochemical Journal (Japan), v. 4, p. 87–91.

Morse, S.A., 1980, Kiplapait mineralogy, II: Fe-Ti oxide minerals and the activities of oxygen and silica: Journal of Petrology, v. 21, p. 685–719.

Nagashima, S., and Katsura, T., 1973, The solubility of sulfur in Na_2O-SiO_2 melts under various oxygen partial pressures at 1100°C, 1250°C and 1300°C: Bulletin of Chemical Society of Japan, v. 46, p. 3099–3103.

Pallister, J.S., Hoblitt, R.P., and Reyes, A.G., 1992, A basalt trigger for the 1991 eruptions of Pinatubo volcano?: Nature, v. 356, p. 426–428.

Rajamani, V., and Naldrett, A.J, 1978, Partitioning of Fe, Co, Ni and Cu between sulfide liquid and basaltic melts and the composition of Ni-Cu sulfide deposits: Economic Geology, v. 73, p. 82–93.

Ruaya, J.R., Ramos, M.N., and Gonfiantini, R., 1992, Assessment of magmatic components of the fluids at Mt. Pinatubo volcanic-geothermal system, Philippines from chemical and isotopic data: Report of the Geological Survey of Japan, v. 279, p. 141–151.

Rutherford, M.J., and Devine, J.D., 1991, Pre-eruption conditions and volatiles in the 1991 Pinatubo magma [abs.]: Eos, Transactions, American Geophysical Union, v. 72, p. 62.

Rye, R.O., Luhr, J.F., and Wasserman, M.D., 1984, Sulfur and oxygen isotopic systematics of the 1982 eruptions of El Chichón volcano, Chiapas, Mexico: Journal of Volcanology and Geothermal Research, v. 23, p. 109–123.

Sillitoe, R.H., and Gappe, I.M., Jr., 1984, Philippine porphyry copper deposits: Geologic setting and characteristics: Committee for Coordination of Joint Prospecting for Mineral Resources in Asian Offshore Areas (CCOP) Project Office Technical Paper 14, 89 p.

Symonds, R.B., Rose, W.I., Gerlach, T.M., Briggs, P.H., and Harmon, R.S., 1990, Evaluation of gases, condensates, and SO_2 emissions from Augustine volcano, Alaska: The degassing of a Cl-rich volcanic system: Bulletin of Volcanology, v. 52, p. 355–374.

Ueda, A., and Itaya, T., 1981, Microphenocrystic pyrrhotite from dacite rocks of Satsuma-Iwojima, Southwest Kyushu, Japan and the solubility of sulfur in dacite magma: Contributions to Mineralogy and Petrology, v. 78, p. 21–26.

Ueda, A., and Sakai, H., 1984, Sulfur isotope study of Quaternary volcanic rocks from the Japanese Islands Arc: Geochimica et Cosmochimica Acta, v. 48, p. 1837–1848.

Urabe, T., 1985, Aluminous granite as a source magma of hydrothermal ore deposits: An experimental study: Economic Geology, v. 80, p. 148–157.

Wallace, P., and Carmichael, I.S.E., 1992, Sulfur in basaltic magmas: Geochimica et Cosmochimica Acta, v. 56, p. 1863–1874.

Wedepohl, K.L., ed., Handbook of geochemistry, 1972, ch. 34, Selenium: New York, Springer-Verlag, v. II-3, 34 B–O.

Westrich, H.R., and Gerlach, T.M., 1992, Magmatic gas source for the stratospheric SO_2 cloud from the June 15, 1991, eruption of Mount Pinatubo: Geology, v. 20, p. 867–870.

Whitney, J.A., 1984, Fugacities of sulfurous gases in pyrrhotite-bearing silicic magmas: American Mineralogist, v. 69, p. 69–78.

———1992, Origin and evolution of sulfur in silicic magmas [abs.]: Eos, Transactions, American Geophysical Union, v. 73, no. 14, p. 366–367.

Whitney, J.A., and Stormer, J.C., Jr., 1983, Igneous sulfides in the Fish Canyon Tuff and the role of sulfur in calc-alkaline magmas: Geology, v. 11, p. 99–102.

Sulfur Isotopic Systematics of the June 1991 Mount Pinatubo Eruptions: A SHRIMP Ion Microprobe Study

By Michael A. McKibben,[1] C. Stewart Eldridge,[2] and Agnes G. Reyes[3]

ABSTRACT

The SHRIMP ion microprobe has been used to make in situ analyses of individual anhydrite and sulfide crystals in the June 1991 eruption products of Mount Pinatubo. In airfall pumice from the June 12 eruption, anhydrite crystals exhibit a broad, bimodal distribution of $\delta^{34}S$ values (46 analyses on 23 crystals; range 3 to 16 per mil; modes of 6.5 per mil and 10.5 per mil). Chalcopyrite crystals have $\delta^{34}S$ values with a mode of -1 per mil (5 analyses of 5 crystals, range -2 to 0 per mil). In crystal-rich ash-flow pumice from the main June 15 eruption, anhydrite crystals exhibit a narrow, unimodal distribution of $\delta^{34}S$ values (44 analyses on 22 crystals, range 5 to 11 per mil, mode and average of 7 per mil). A single analysis of a chalcopyrite crystal yielded $\delta^{34}S=0$ per mil.

Under the preeruptive temperature and oxygen fugacity conditions of both stages of the eruption, the chalcopyrites are in isotopic equilibrium with the $\simeq 7$ per mil mode of anhydrites, and both phases are likely primary. The isotopically heavier anhydrites in the June 12 pumice may be xenocrysts acquired at a shallow level during this early vent-clearing eruption; the isotopically heaviest anhydrites have $\delta^{34}S$ values similar to those of hydrothermal vein anhydrites in a drillcore sample recovered from the geothermal system on the flank of Mount Pinatubo (9 analyses of 5 crystals, range 17–22 per mil, average 19 per mil).

The $\delta^{34}S$ value of primary anhydrite places constraints on the sulfur isotopic composition of SO_2 gas resulting from the eruptions. If magma prior to the eruptions coexisted with an exsolved vapor phase containing SO_2, then isotopic systematics require that the SO_2 had a $\delta^{34}S$ value of $\simeq 3.5$ per mil. If most of this SO_2 was erupted and quantitatively oxidized to H_2SO_4 in the stratosphere, then Mount Pinatubo aerosols could have a similar sulfur isotopic composition. Given estimates that a significant amount of the total sulfur in the eruption was present as a gas phase prior to eruption, the bulk sulfur isotopic composition of the eruption (crystals + melt + gas) would have been 5.0 to 6.5 per mil.

If, instead, most of the SO_2 was generated by rapid irreversible breakdown of anhydrite during ascent decompression and eruption, then the sulfur isotopic composition of the total eruption and the erupted SO_2 would have been similar to that of the primary anhydrite, 7 per mil. Quantitative conversion of this SO_2 to H_2SO_4 in the stratosphere could have yielded sulfate aerosols with a similar isotopic composition.

Two analyses of a primary chalcopyrite grain in a quenched basalt inclusion from the June 7–14 hybrid andesite dome yielded $\delta^{34}S$ values of 1 per mil and likely reflect the sulfur isotopic composition of the mafic magma underplating the dacite. Long-term degassing of basalt magma may have been an important ultimate source of reduced sulfur to the dacite. However, Rayleigh effects upon degassing of a mixed SO_2/H_2S vapor from basalt magma would partition bulk sulfur into the dacite with a total $\delta^{34}S$ value no greater than 3 per mil, significantly less than the 5 to 7 per mil bulk sulfur isotopic composition of the eruption. The further ^{34}S enrichment of the dacite is best explained by passive (noneruptive) steady-state degassing of H_2S and SO_2 from the dacite magma, in dynamic isotopic equilibrium with primary anhydrite, over long time periods.

INTRODUCTION

Anhydrite is a characteristic and important mineralogical component of fresh tephra from Mount Pinatubo and other arc volcanoes that erupt oxidized, hydrous sulfur-rich magma (Luhr and others, 1984; Rye and others, 1984; Fournelle, 1990; Bernard and others, 1991). Anhydrite is not well preserved in older pyroclastic rocks because it weathers out of tephra soon after eruption. Experimental studies have demonstrated the relatively high solubility of anhydrite and the stability of dissolved SO_4^{2-} species in oxidized

---

[1] Department of Earth Sciences, University of California, Riverside, CA 92521.

[2] Geology Department and Research School of Earth Sciences, Australian National University, Canberra, A.C.T. 0200, Australia.

[3] Geothermal Division, Philippine National Oil Corporation, Fort Bonifacio, Metro Manila, Philippines; now at Institute of Geological and Nuclear Sciences, POB 30-368, Gracefield Road, Lower Hutt, New Zealand.

melts (Carroll and Rutherford, 1987, 1988; Luhr, 1990). It is therefore commonly assumed that anhydrite in recent pyroclastic materials is a juvenile magmatic phase.

Bernard and others (1991) argued that anhydrites in pumices from the June 15, 1991 eruption of Mount Pinatubo were primary phenocrysts, mainly on the basis of their intergrowth textures with apatite. However, anhydrite and apatite are also common phases in hydrothermal veins within the geothermal systems that occur on the flanks of most Philippine volcanoes (Reyes, 1990). Such vein anhydrite ultimately owes its origin to degassing of magma batches: the magmatic SO_2 (and H_2S) becomes oxidized and deposited as hydrothermal anhydrite in the geothermal systems that commonly develop in the shallow portions of volcanic piles. An explored geothermal field containing abundant hydrothermal anhydrite existed on the northwest flank of Mount Pinatubo (Delfin and others, this volume; Cabel, 1990).

Beyond the question of a juvenile versus xenocrystic origin of anhydrite, there remains debate over its role in the mechanism of SO_2 release from arc magmas. Three mechanisms have been proposed: (1) preeruptive exsolution of an SO_2—bearing magmatic vapor phase from the dacitic melt (Luhr, 1990; Westrich and Gerlach, 1992; Gerlach and others, this volume); (2) rapid breakdown of anhydrite to SO_2 in the magma during ascent decompression and eruption (Rutherford and Devine, 1991, this volume; Baker and Rutherford, 1992); and (3) decomposition of anhydrite to SO_2 in the eruption plume (Devine and others, 1984, Sigurdsson, 1990).

For Mount Pinatubo, Westrich and Gerlach (1992), Matthews and others (1992), and Gerlach and others (this volume) have argued against mechanism 2 based on a reported lack of textural evidence for anhydrite breakdown and a lack of corresponding oxidation rims on iron-bearing phases. They also argue against mechanism 3 on the basis of the high temperatures required for decomposition (>1,300°C) and thermodynamic and kinetic barriers to anhydrite reduction in a cold, oxygenated plume.

Westrich and Gerlach (1992) and Gerlach and others (this volume) argue in favor of mechanism 1 for the June 1991 Mount Pinatubo eruption on the basis of mass balance considerations. Differences between the sulfur content of glass inclusions trapped in phenocrysts (preeruptive melt) and that of interstitial glass in pumice (degassed melt) are negligible and cannot account for the 20 million tonnes of "excess" SO_2 that were estimated from satellite spectral data to have been erupted into the stratosphere (Bluth and others, 1991). (Luhr and others (1984) and Luhr (1990) encountered a similar "excess" sulfur problem in applying the sulfur-content difference method to glasses from the 1982 El Chichón eruption). Furthermore, they note that the amount of SO_2 erupted from Mount Pinatubo is an order of magnitude greater than that which could have been dissolved in the melt at depth. Westrich and Gerlach (1992)

and Gerlach and others (this volume) argue that this "excess" sulfur problem requires that magmatic water, CO_2, and SO_2 must have exsolved at depth to form a free gas phase within the magma, prior to eruption. They cited the presence of large vapor bubbles in glass inclusions in phenocrysts and the nearly sulfur-saturated composition of the inclusion glass as evidence for the presence of this early exsolved gas phase.

Whatever the mechanism for the shallow generation of the erupted SO_2, it has been argued that the ultimate source of sulfur in the 1991 Pinatubo eruptions involved long-term degassing of sulfur from basaltic magma underplating the dacitic magma (Pallister and others, 1991, 1992, this volume; Matthews and others, 1992).

Knowledge of the sulfur isotopic compositions of sulfur-bearing minerals in the 1991 Mount Pinatubo eruption products may contribute to resolving the origin and release of sulfur in dacite from Mount Pinatubo, because different sources of melt sulfur and (or) different mechanisms of generating SO_2 may result in distinctive isotopic signatures being preserved in these minerals. For this reason, we conducted a SHRIMP ion microprobe study of the in situ sulfur isotopic compositions of anhydrite and sulfide crystals in the June 1991 eruption products of Mount Pinatubo. We also analyzed sulfur-bearing hydrothermal vein minerals in drillcores from the Pinatubo geothermal field (Delfin and others, this volume; Cabel, 1990).

SHRIMP SULFUR ISOTOPIC MICROANALYSIS

Microbeam techniques of stable isotopic analysis offer unique insights into geochemical processes, because individual crystals and crystal growth zones often exhibit more isotopic variability than is found with conventional bulk isotopic sampling and analytical techniques. Sulfur isotopic microanalyses are routine on the SHRIMP (Sensitive High Resolution Ion Microprobe) facility at the Australian National University. Standards exist, fractionation is well understood, and correction procedures have been developed for the common metallic sulfides and sulfates (Eldridge and others 1987, 1989; McKibben and Eldridge, in press). Analytical work has been done on fine-grained natural sulfides and sulfates from numerous ore deposits (Eldridge and others, 1988, 1989, 1993; McKibben and Eldridge, 1995; McKibben and others, 1993), from sedimentary and volcanic rocks in the Salton Sea and Valles Caldera geothermal systems (McKibben and Eldridge, 1989, 1990), and from mantle rocks (Eldridge and others, 1991; Rudnick and others, 1993). Although the analytical precision for $\delta^{34}S$ on the SHRIMP is typically 2‰ at the 2σ level (compared to 2σ precisions of ≃0.2‰ for conventional bulk sampling and analysis), the SHRIMP studies have shown that intra- and intercrystalline $\delta^{34}S$ variations in minerals can be more than an order of magnitude greater than this over distances of less than 200 μm. Therefore, the advantages of the spatial resolution of the SHRIMP in resolving such variations far outweigh its limits in precision.

The SHRIMP ion beam excavates a sharp elliptical crater on the target mineral's surface that is typically 20–30 μm wide and 5 μm deep. The ability of the SHRIMP to achieve such a small sampling volume while maintaining the textural integrity of the sample makes it an ideal facility for analyzing individual crystals and crystal growth zones in fine-grained materials such as tephra.

Sulfur isotopic analysis on the SHRIMP requires samples that are polished to levels appropriate for reflected-light microscopy, usually with final polishing grits in the 0.05-μm size range. Anhydrite-bearing pyroclastic samples should be prepared by using a minimum of water because of the high solubility of anhydrite under ambient conditions. However, there is negligible sulfur isotopic fractionation between anhydrite and aqueous SO_4^{2-} at room temperature (Ohmoto and Rye, 1979), so any partial dissolution of anhydrite during sample preparation does not measurably modify the sulfur isotopic composition of the remaining material.

ANALYZED SAMPLES

We conducted SHRIMP $\delta^{34}S$ analyses on sulfide and sulfate minerals in four different Mount Pinatubo samples, as described briefly below and in table 1. More detailed petrographic and geochemical data from similar samples are found elsewhere in this volume (Pallister and others; Gerlach and others; Delfin and others). SHRIMP data for these samples are listed in table 1, and histograms of the $\delta^{34}S$ values are plotted for each of the four samples in figure 1.

PUMICE FROM THE JUNE 12, 1991, ERUPTION

The June 12 phase of the 1991 Mount Pinatubo eruptions was the major vent-clearing eruption that preceded the climactic June 15 main eruption (Wolfe and Hoblitt, this volume). We obtained 46 SHRIMP $\delta^{34}S$ analyses on 23 crystals of anhydrite and 5 SHRIMP $\delta^{34}S$ analyses on 5 crystals of chalcopyrite in the pumice from this eruption. Modally, the pumice contains 64.5% colorless glass, 20% plagioclase, 10% green hornblende, 1.3% K-feldspar, 1% biotite, 1% magnetite, 1% anhydrite, 0.7% augite, 0.4% quartz, and 0.1% olivine. The rare olivine xenocrysts are rimmed by hornblende laths. Pallister and others (1991, 1992, and this volume) have described the products of this stage of the eruption as hybrid mixed dacite-basalt, on the basis of incompatible phenocryst assemblages and disequilibrium textures.

The anhydrite crystals occur as isolated anhedral to subhedral grains from 100 to 500 μm in diameter, with

crystal edges often in direct contact with glass (fig. 2A). Some anhydrite crystals contain inclusions of euhedral apatite. Analyzed chalcopyrite crystals occur as anhedral grains from 50 to 80 μm in diameter, either isolated in glass or associated with pyrrhotite within hornblende and other mafic phases (fig. 2B). Pyrrhotite, though generally more common than chalcopyrite, occurred as crystals that were too small (<30 μm) to be analyzed.

Most anhydrite crystals were large enough to permit analysis of both cores and rims (table 1, fig. 2A). In contrast, chalcopyrite crystals were too small to allow more than a single analytical crater, which nearly spanned the entire crystal (fig. 2B).

The anhydrite $\delta^{34}S$ values exhibit a 13‰ range in isotopic compositions (fig. 1A), more than six times the typical 2σ analytical precision. The mean of the anhydrite $\delta^{34}S$ data is 8.3‰, with a standard deviation of 2.9. A bimodal distribution in the anhydrite $\delta^{34}S$ values is evident, with modes of 6.5‰ and 10.5‰. This apparent bimodality is real and statistically significant: a chi-squared test indicates a likelihood of less than 5/1,000 (e^2=0.0045) for randomly deriving the observed distribution from a single normally distributed population of $\delta^{34}S$ values with the same mean and standard deviation. Though not shown explicitly, the data from this sample yield anhydrite core and rim $\delta^{34}S$ subpopulations that are statistically identical, each showing the same range and bimodality. There is no obvious petrographic difference between the two anhydrite modes.

A positive correlation exists between anhydrite core and rim $\delta^{34}S$ values (fig. 3A). These relations indicate that almost all of the anhydrite crystals are not isotopically zoned at a level detectable by SHRIMP (2‰ precision at 2σ level): only two crystals exhibit a difference of more than 4‰ between core and rim values. Figure 3A also indicates that the observed 13‰ range in values is significant and not due to random analytical uncertainties: there are uniformly ^{34}S-depleted and uniformly ^{34}S-enriched anhydrite crystals.

Due to their smaller grain size and much lower modal abundance, the number of chalcopyrite analyses is few in comparison to anhydrite (table 1). The chalcopyrite $\delta^{34}S$ values cluster tightly, with an average of −1‰ (fig. 1A).

PUMICE FROM THE JUNE 15, 1991, ERUPTION

The main June 15 stage of the 1991 eruption sequence produced mostly phenocryst-rich dacite tephra and lesser phenocryst-poor dacite tephra (Wolfe and Hoblitt, this volume; Pallister and others, this volume). We obtained 44 SHRIMP $\delta^{34}S$ analyses on 22 crystals of anhydrite in a phenocryst-rich dacite pumice from this eruption (table 1). Owing to the small size (<30 μm) of most sulfide grains in this sample, we obtained only a single SHRIMP $\delta^{34}S$ analysis for one chalcopyrite grain.

Table 1.   SHRIMP $\delta^{34}S$ analytical data for 1991 Mount Pinatubo samples ($\delta^{34}S$ of samples are in per mil).

Pumice of the June 12 Eruption[1]

	$^{34}S/^{32}S$ of Sample	Standard error (1σ)	$^{34}S/^{32}S$ of Standard	Standard error (1σ)	$\delta^{34}S$ sample	Error (2σ)
Anhydrite						
core xl[2] 1	0.038654	0.000042	0.038496	0.000017	10	2
rim xl 1	0.038651	0.000038	0.038496	0.000017	10	2
core xl 2	0.038740	0.000035	0.038496	0.000017	13	2
rim xl 2	0.038880	0.000035	0.038496	0.000017	16	2
core xl 3	0.038821	0.000037	0.038496	0.000017	15	2
rim xl 3	0.038620	0.000033	0.038496	0.000017	10	2
core xl 4	0.038629	0.000046	0.038496	0.000017	10	3
rim xl 4	0.038654	0.000044	0.038496	0.000017	10	2
core xl 5	0.038651	0.000042	0.038543	0.000019	9	2
rim xl 5	0.038730	0.000047	0.038543	0.000019	11	3
core xl 6	0.038813	0.000051	0.038543	0.000019	13	3
rim xl 6	0.038584	0.000051	0.038543	0.000019	7	3
core xl 7	0.038706	0.000040	0.038543	0.000019	11	2
rim xl 7	0.038713	0.000040	0.038543	0.000019	11	2
core xl 8	0.038665	0.000040	0.038543	0.000019	10	2
rim xl 8	0.038535	0.000048	0.038543	0.000019	6	3
core xl 9	0.038527	0.000040	0.038543	0.000019	6	2
rim xl 9	0.038532	0.000047	0.038543	0.000019	6	3
core xl 10	0.038684	0.000040	0.038496	0.000019	11	2
rim xl 10	0.038678	0.000036	0.038496	0.000019	11	2
core xl 11	0.038574	0.000039	0.038496	0.000019	8	2
rim xl 11	0.038637	0.000047	0.038496	0.000019	10	3
core xl 12	0.038589	0.000047	0.038496	0.000019	9	3
rim xl 12	0.038662	0.000044	0.038496	0.000019	11	3
core xl 13	0.038639	0.000035	0.038496	0.000019	10	2
rim xl 13	0.038531	0.000039	0.038496	0.000019	7	2
core xl 14	0.038578	0.000034	0.038496	0.000019	8	2
rim xl 14	0.038619	0.000040	0.038496	0.000019	10	2
core xl 15	0.038515	0.000040	0.038596	0.000019	4	2
rim xl 15	0.038602	0.000043	0.038596	0.000019	7	2
core xl 16	0.038567	0.000053	0.038596	0.000019	6	3
rim xl 16	0.038587	0.000045	0.038596	0.000019	6	2
core xl 17	0.038626	0.000034	0.038596	0.000019	7	2
rim xl 17	0.038635	0.000035	0.038596	0.000019	7	2
core xl 18	0.038480	0.000038	0.038596	0.000019	3	2
rim xl 18	0.038487	0.000032	0.038596	0.000019	4	2
core xl 19	0.037807	0.000034	0.037783	0.000020	7	2
rim xl 19	0.037752	0.000034	0.037783	0.000020	6	2
core xl 20	0.037770	0.000037	0.037783	0.000020	6	2
rim xl 20	0.037733	0.000031	0.037783	0.000020	5	2
core xl 21	0.037770	0.000036	0.037783	0.000020	6	2
rim xl 21	0.037810	0.000029	0.037783	0.000020	7	2
core xl 22	0.037758	0.000036	0.037783	0.000020	6	2
rim xl 22	0.037705	0.000025	0.037783	0.000020	4	2
core xl 23	0.037715	0.000028	0.037783	0.000020	5	2
rim xl 23	0.037819	0.000033	0.037783	0.000020	7	2
Chalcopyrite						
xl 1	0.042064	0.000036	0.042400	0.000020	-1	2
xl 2	0.042014	0.000034	0.042400	0.000020	-2	2
xl 3	0.042097	0.000035	0.042400	0.000020	0	2
xl 4	0.042120	0.000045	0.042400	0.000020	0	2
xl 5	0.042043	0.000048	0.042400	0.000020	-2	2

Table 1. SHRIMP δ^{34}S analytical data for 1991 Mount Pinatubo samples (δ^{34}S of samples are in per mil)—Continued.

Pumice of the June 15 Eruption[3]

	^{34}S/^{32}S of Sample	Standard error (1σ)	^{34}S/^{32}S of Standard	Standard error (1σ)	δ^{34}S sample	Error (2σ)
Anhydrite						
core xl 1	0.038600	0.000032	0.038631	0.000022	6	2
rim xl 1	0.038597	0.000032	0.038631	0.000022	5	2
core xl 2	0.038913	0.000032	0.038737	0.000020	11	2
rim xl 2	0.038888	0.000033	0.038737	0.000020	10	2
mid xl 2	0.038763	0.000031	0.038660	0.000011	9	2
mid xl 2	0.038835	0.000041	0.038660	0.000011	11	2
core xl 3	0.038553	0.000044	0.038475	0.000020	8	2
rim xl 3	0.038607	0.000044	0.038475	0.000020	9	3
core xl 4	0.038493	0.000038	0.038475	0.000020	6	3
rim xl 4	0.038603	0.000039	0.038475	0.000020	9	2
core xl 5	0.038687	0.000032	0.038631	0.000022	8	2
core xl 6	0.038822	0.000032	0.038811	0.000016	7	2
rim xl 6	0.038901	0.000032	0.038811	0.000016	9	2
core xl 7	0.038864	0.000035	0.038811	0.000016	8	2
rim xl 7	0.038847	0.000031	0.038811	0.000016	7	2
core xl 8	0.038855	0.000035	0.038811	0.000016	7	2
rim xl 8	0.038854	0.000034	0.038811	0.000016	7	2
core xl 9	0.038941	0.000032	0.038811	0.000016	10	2
rim xl 9	0.038844	0.000031	0.038811	0.000016	7	2
core xl 10	0.038705	0.000033	0.038660	0.000011	8	2
rim xl 10	0.038596	0.000033	0.038660	0.000011	5	2
core xl 11	0.038636	0.000039	0.038631	0.000022	6	2
core xl 12	0.038688	0.000032	0.038631	0.000022	8	2
mid xl 12	0.038669	0.000034	0.038631	0.000022	7	2
rim xl 12	0.038640	0.000032	0.038631	0.000022	7	2
core xl 13	0.038668	0.000034	0.038631	0.000022	7	2
core xl 14	0.038733	0.000030	0.038631	0.000022	9	2
rim xl 14	0.038698	0.000030	0.038631	0.000022	8	2
core xl 15	0.038684	0.000034	0.038631	0.000022	8	2
core xl 16	0.038508	0.000034	0.038475	0.000020	7	2
rim xl 16	0.038499	0.000044	0.038475	0.000020	6	3
core xl 17	0.038753	0.000031	0.038811	0.000016	5	2
core xl 17	0.038861	0.000029	0.038811	0.000016	8	2
rim xl 17	0.038776	0.000034	0.038811	0.000016	5	2
core xl 18	0.038821	0.000034	0.038811	0.000016	7	2
core xl 19	0.038485	0.000039	0.038475	0.000020	6	2
rim xl 19	0.038553	0.000044	0.038475	0.000020	7	2
core xl 20	0.038658	0.000030	0.038660	0.000011	6	2
mid xl 20	0.038692	0.000029	0.038660	0.000011	7	2
rim xl 20	0.038613	0.000034	0.038660	0.000011	5	2
rim xl 20	0.038642	0.000029	0.038660	0.000011	6	2
core xl 21	0.038654	0.000030	0.038660	0.000011	6	2
rim xl 21	0.038605	0.000031	0.038660	0.000011	5	2
core xl 22	0.038688	0.000031	0.038660	0.000011	7	2
Chalcopyrite						
rim xl 1	0.043263	0.000040	0.043552	0.000023	0	2

Table 1.   SHRIMP δ^{34}S analytical data for 1991 Mount Pinatubo samples (δ^{34}S of samples are in per mil)—Continued.

Basalt Inclusion[4]

	^{34}S/^{32}S of Sample	Standard error (1σ)	^{34}S/^{32}S of Standard	Standard error (1σ)	δ^{34}S sample	Error (2σ)
Chalcopyrite						
core xl 1	0.043337	0.000036	0.043603	0.000020	1	2
core xl 1	0.043360	0.000035	0.043603	0.000020	1	2

Drillcore PIN 2D[5]

	^{34}S/^{32}S of Sample	Standard error (1σ)	^{34}S/^{32}S of Standard	Standard error (1σ)	δ^{34}S sample	Error (2σ)
Anhydrite						
core xl 1	0.039240	0.000039	0.038803	0.000019	18	2
rim xl 1	0.039209	0.000035	0.038803	0.000019	17	2
core xl 2	0.039339	0.000038	0.038803	0.000019	20	2
rim xl 2	0.039315	0.000039	0.038803	0.000019	20	2
core xl 3	0.039258	0.000040	0.038803	0.000019	18	2
rim xl 3	0.039213	0.000043	0.038803	0.000019	17	2
core xl 4	0.039322	0.000041	0.038803	0.000019	20	2
rim xl 4	0.039343	0.000045	0.038803	0.000019	20	3
xl 5	0.039421	0.000048	0.038803	0.000019	22	3
Pyrite						
xl 1	0.042715	0.000042	0.043365	0.000019	-3	2
xl 2	0.042800	0.000042	0.043365	0.000019	-1	2
xl 3	0.042654	0.000042	0.043365	0.000019	-5	2
xl 4	0.042757	0.000032	0.043365	0.000019	-2	2
xl 5	0.042772	0.000032	0.043365	0.000019	-2	2

Footnotes:

[1] Tephra fall pumice fragment collected at approximately 0930 on Mount Bagang by Agnes G. Reyes (then of Geothermal Division of Philippine National Oil Corporation).

[2] xl = crystal.

[3] Fragment of crystal rich dacite pumice from the main eruption, collected from the Sacobia River valley pyroclastic flow near Mactan Gate on 6/21/91 by Ed Wolfe of the USGS; provided by Terry Gerlach (USGS) and Hank Westrich (Sandia National Laboratory).

[4] Inclusion of basalt derived from the 6/7/91 through 6/14/91 hybrid andesite dome, collected from the Maraunot valley pyroclastic flow by John Pallister of the USGS (his sample number P22892 1a)

[5] Hydrothermal vein material from a depth of 1437 m in the geothermal exploration well, collected by Agnes G. Reyes and provided by Tom Casadevall (USGS).

Representative photomicrographs of anhydrite and chalcopyrite from this pumice sample are shown in figure 4. In reflected light many of the anhydrite crystals are *distinctly* rounded, more so than in the June 12 pumice. This rounding occurs in direct contact with glass and thus is not due to partial dissolution of the anhydrite during sample preparation. Either the anhydrite partially dissolved in the melt or the silicic, viscous nature of the June 15 dacite melt promoted abrasion of the relatively soft anhydrite phenocrysts. This rounding differs from textures described by Bernard and others (1991), who observed sharp, euhedral contacts between June 15 anhydrite crystals and glass in transmitted light.

The analyzed chalcopyrite grain exhibits exsolution lamellae of bornite (fig. 4B), which likely formed by breakdown of magmatic Cu-Fe-S intermediate solid solution to these two phases plus pyrrhotite upon cooling to $< 350°C$ (see Sugaki and others, 1975). Because equilibrium $\delta^{34}S$ differences between chalcopyrite and bornite are less than 1‰ at this temperature, and because the SHRIMP craters were placed so as to avoid the visible bornite lamellae, the SHRIMP $\delta^{34}S$ value of this chalcopyrite should be representative of the $\delta^{34}S$ value of the original magmatic sulfide (intermediate solid solution).

A single population of anhydrite $\delta^{34}S$ values was observed in this sample, with a mode of 7‰, a mean of 7.3‰ and a standard deviation of 1.6‰ (fig. 1B). This mode is virtually identical to the mode of the isotopically lighter anhydrites in the June 12 pumice samples (fig. 1A). No significant isotopic zoning is evident in any analyzed anhydrite crystal. A plot of core versus rim $\delta^{34}S$ values in June 15 anhydrite shows no statistically significant correlation (fig. 3B), exhibiting a tighter clustering than anhydrites in the June 12 pumice (fig. 3A).

Using conventional bulk sampling techniques, Imai and others (1993) and Bernard and others (this volume) obtained $\delta^{34}S$ values for the June 15 crystal-rich dacite that averaged 8.2‰ (Imai and others, acid leachates) and 7.4‰ (Bernard and others, water-soluble sulfate), in excellent agreement with our SHRIMP values (fig. 1B). The $\delta^{34}S$ value for chalcopyrite in the pumice from the June 15 eruption (fig. 1B) is within the range of $\delta^{34}S$ values for chalcopyrite from pumice of the June 12 eruption (fig. 1A).

BASALT INCLUSION

Pallister and others (1991, 1992, this volume) argue that the basalt inclusions found in some of the Pinatubo pyroclastic flows represent samples of an underplating mafic magma that may have slowly degassed sulfur into the overlying dacite and whose sudden intrusion to shallower levels in 1991 may have triggered the eruption of the dacite. Presumably, the sulfur isotopic composition of primary sulfide in the basalt inclusions may represent the isotopic composition of reduced sulfur present in the deep mafic melt.

Pyrrhotite and chalcopyrite grains are common in the basalt inclusion, but only one 25-μm grain of chalcopyrite (fig. 5) was large enough to be analyzed during an analytical session with the new SHRIMP II, which has a spatial resolution of 10–15 μm. Two SHRIMP craters on this grain yielded identical $\delta^{34}S$ values of 1‰ (fig. 1C), statistically indistinguishable from the $\delta^{34}S$ values for chalcopyrite in pumice samples from the June 12 and 15 eruptions (figs. 1A and 1B).

VEIN IN DRILLCORE

Geothermal exploration wells drilled into the flank of Mount Pinatubo in 1988 encountered excess enthalpy fluids with highly variable compositions, gas contents, and generally acidic pH values (Buenviaje, 1991; Michels and others, 1991; Delfin and others, this volume). Drillcores from these wells preserve evidence of past hydrothermal mineral deposition from liquid-dominated fluids. Veins and massive alteration zones are generally dominated by calcite at $<300°C$ and anhydrite at $>300°C$.

Within well PIN–2D, drillcore from a depth of 1,437 m contains a thick vein of intergrown coarsely crystalline anhydrite and lesser pyrite in bleached wall rock. Fluid inclusions in the anhydrite yield homogenization temperatures of 278–308°C and freezing point depressions of 1.1–4.8°C.

Pyrite crystals from the vein exhibit a narrow isotopic range with a mean $\delta^{34}S$ value of $-2‰$ (fig. 1D), statistically indistinguishable from $\delta^{34}S$ values of the chalcopyrites in the pumices and basalt inclusion. Coexisting vein anhydrite $\delta^{34}S$ values are slightly more variable, with an average of 18.7‰ (fig. 1D). They are significantly heavier than $\delta^{34}S$ values for anhydrites in pumice of the June 15 eruption, but the isotopically lightest values of the drillcore vein anhydrites are statistically indistinguishable from the isotopically heaviest values of anhydrites in pumice of the June 12 eruption.

DISCUSSION

PRIMARY ANHYDRITE AND SULFIDE

The unimodal anhydrite $\delta^{34}S$ population in pumice of the June 15 eruption (fig. 1B) implies that an isotopically uniform and well-mixed primary SO_4^{2-} component was present in magma erupted during the main eruption. In contrast, the broader bimodal distribution of anhydrite $\delta^{34}S$ values in pumice of the June 12 eruption (fig. 1A) implies that isotopically distinct sources of SO_4^{2-} contributed to this earlier, vent-clearing eruption.

A primary, juvenile origin for the $\simeq 7\permil$ mode anhydrite population is supported by the apparent sulfur isotopic equilibrium between these anhydrites and coexisting sulfide in both pumice samples. In both the June 12 and June 15 magmas, oxygen fugacity was relatively high and melt SO_4^{2-} was much greater than melt HS^- (Rutherford and Devine, this volume), so the sulfur isotopic composition of primary melt sulfide should have been fixed (buffered) by the much larger masses of phenocryst $CaSO_4$ and primary melt SO_4^{2-}. Spectroscopic studies (Carroll and Rutherford, 1988) suggest that SO_4^{2-} in melt and SO_4^{2-} in anhydrite have the same chemical structure, so it can be assumed that sulfur isotopic fractionation between dissolved and crystalline SO_4^{2-} should be negligible at magmatic temperatures.

In pumice of the June 15 eruption, the mean observed sulfur isotopic difference between anhydrite and chalcopyrite is $7–8\permil$ (fig. 1B). This is in excellent agreement with the expected equilibrium fractionation of $7–8\permil$ at $780°C$ (fig. 6), the estimated magma temperature prior to eruption as calculated from Fe-Ti oxide geothermometry (Rutherford and Devine, this volume).

In pumice of the June 12 eruption, the mean observed isotopic difference between chalcopyrite and the isotopically lighter mode of anhydrite is $7–8\permil$, in reasonably good agreement with the expected equilibrium fractionation of $6–7\permil$ at $950°C$ (fig. 6), the estimated hybrid magma temperature prior to eruption (Rutherford and Devine, this volume). The mean observed difference of $11–12\permil$ between chalcopyrite and the isotopically heavier mode of anhydrite corresponds to a sulfur isotope equilibration temperature of $<600°C$ (fig. 6), well below the estimated preeruption temperatures of Mount Pinatubo magmas.

ORIGIN OF ISOTOPICALLY HEAVY ANHYDRITE

Because the anhydrite in pumice of the June 12 eruption is not isotopically homogeneous, and because the isotopically heavier anhydrites are not in isotopic equilibrium with sulfide at the estimated preeruptive magma temperature, the heavier anhydrites must have been derived from extraneous sources. They must have been acquired relatively late (shallow) by the magma, such that complete mixing and isotopic reequilibration with primary SO_4^{2-} and HS^- could not occur prior to eruption.

One shallow source that could have provided late-stage extraneous SO_4^{2-} is subvolcanic igneous rock derived from crystallization of earlier SO_4^{2-}—bearing dacite magma batches. However, primary anhydrite in such dacitic subvolcanic rocks would not necessarily be isotopically distinct from the primary anhydrite in the 1991 dacite magma. Also, the isotopically heavy anhydrites show no evidence of intergrowths with older dacitic material.

Another possible source of shallow extraneous SO_4^{2-} may be local concentrations of hydrothermal anhydrite that occur within the flanks of Mount Pinatubo (Delfin and others, this volume; Cabel, 1990). Such anhydrite ultimately owes its origins to degassing of SO_2 (and H_2S) from earlier magma batches into overlying hydrothermal/ground-water systems, which commonly occur within the edifices of arc volcanoes in tropical latitudes (Reyes, 1990; Giggenbach, 1992).

This hydrothermal xenocryst hypothesis poses some textural difficulties, however. Hydrothermal anhydrite in such deposits often occurs intergrown with sulfide, apatite, calcite, and other authigenic phases, as well as acid-altered volcanic rock, but only apatite has been commonly seen intergrown with anhydrite crystals in Mount Pinatubo pumices. It may be possible that the dacite melt was saturated in anhydrite and apatite but strongly undersaturated in other hydrothermal phases, a situation that would have promoted the generation of partially resorbed hydrothermal xenoliths containing only anhydrite and apatite. Nonetheless, firm textural evidence for the xenolithic nature of the isotopically heavy June 12 anhydrites is lacking.

The isotopic compositions of vein anhydrite and pyrite in the drillcore sample are consistent with an ultimate origin by degassing of magmatic SO_2 from earlier magma batches into hydrothermal fluids. Authigenic anhydrite forming in magma-hydrothermal systems is commonly enriched in ^{34}S compared with the magmatic SO_2 from which it was derived (Rye, 1993). In the presence of water at temperatures below $350°C$, SO_2 hydrolyzes according to the reaction

$$4SO_2+4H_2O=3H^++3HSO_4^-+H_2S$$

(Ohmoto and Rye, 1979). Because of the strong relative partitioning of sulfur isotopes among SO_2, H_2S, and SO_4^{2-} at lower temperatures (fig. 6), aqueous SO_4^{2-} generated by the breakdown reaction tends to be enriched in ^{34}S relative to the original SO_2 in magmatic vapor. Likewise, any aqueous H_2S generated by the breakdown reaction tends to be depleted in ^{34}S relative to the original SO_2. The magnitudes of the relative enrichments and depletions of these decomposition products depend on the initial magmatic vapor

Figure 1A. SHRIMP $\delta^{34}S$ data (table 1) for chalcopyrite and → anhydrite from pumice of the June 12, 1991 eruption of Mount Pinatubo (sample collected on Mount Bagang). Chalcopyrite data shown as solid bars; anhydrite data as cross-hatched bars. The chalcopyrite data have a mode of $-1\permil$; anhydrite data are bimodal at 6.5 and $10.5\permil$. B, SHRIMP $\delta^{34}S$ data (table 1) for chalcopyrite and anhydrite from pumice of the June 15, 1991 eruption of Mount Pinatubo (sample collected in Sacobia River Valley). Chalcopyrite data shown as solid bars; anhydrite data as cross-hatched bars. The anhydrite data are unimodal at $7\permil$.

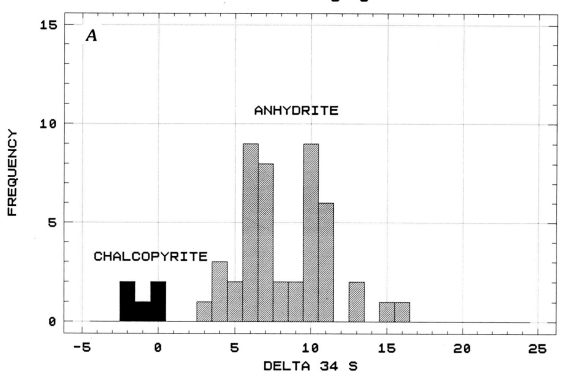

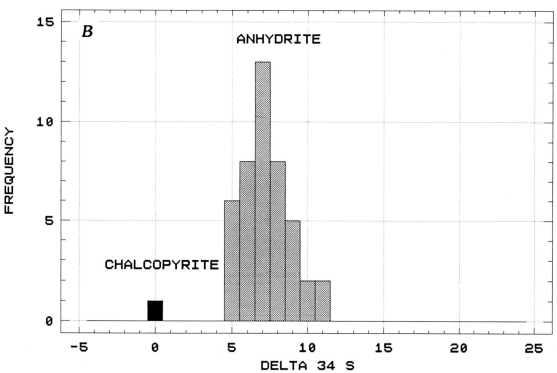

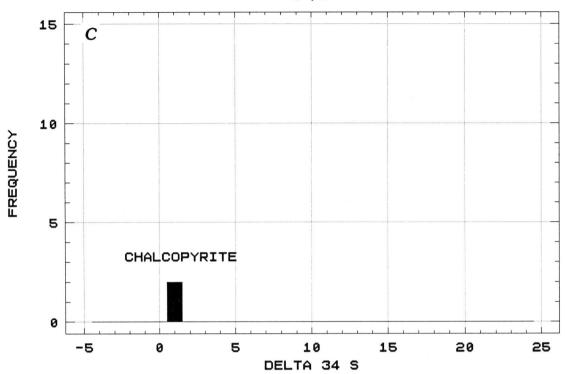

SULFIDE IN BASALT INCLUSION
Maraunot Valley pyroclastic flow

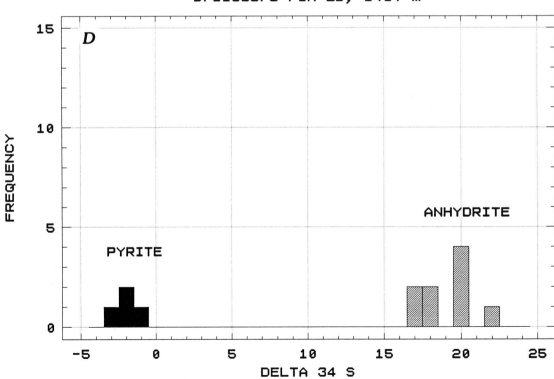

HYDROTHERMAL MINERALS IN VEIN
Drillcore PIN-2D, 1437 m

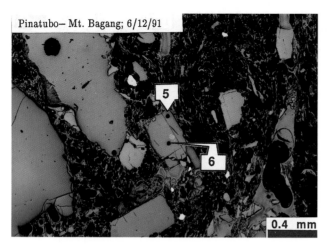

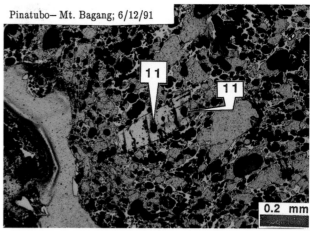

Figure 2A. Representative photomicrographs (polished thin sections in reflected light) of SHRIMP analytical craters in anhydrite, in pumice of the June 12, 1991 eruption of Mount Pinatubo (sample collected from Mount Bagang). Numbers in boxes are $\delta^{34}S$ values (in ‰ relative to Canyon Diablo troilite (CDT) meteorite standard) for the craters identified by the arrows; craters are 20–30

µm in diameter and about 5 µm deep. Anhydrite crystals are anhedral to subhedral, with some crystal boundaries in direct contact with glass. Cleavage is often visible in the anhydrite crystals, as are hexagonal cross sections of apatite crystals. Dark areas are voids; dusty, less-reflective areas are voids filled with epoxy that was used to impregnate the samples prior to polishing.

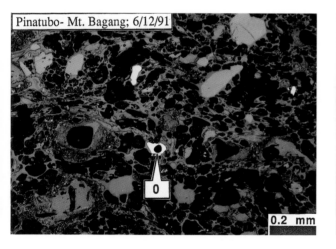

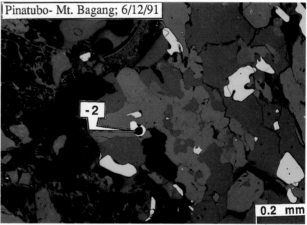

Figure 2B. Representative photomicrographs (polished thin sections in reflected light) of SHRIMP analytical craters in chalcopyrite, in pumice of the June 12, 1991 eruption of Mount Pinatubo (sample collected from Mount Bagang). Numbers in boxes are $\delta^{34}S$ values (in ‰ relative to CDT standard) for the craters identified by the arrows; craters are 20–30 µm in diameter and about

5 µm deep. Chalcopyrite crystals are anhedral and occur as isolated crystals in glass or in mafic phases associated with subhedral embayed magnetite and sometimes pyrrhotite. Dark areas are voids; dusty, less-reflective areas are voids filled with epoxy that was used to impregnate the samples prior to polishing.

← **Figure 1—Continued.** *C*, SHRIMP $\delta^{34}S$ data (table 1) for chalcopyrite in basalt inclusion from June 7–14 hybrid dome. *D*, SHRIMP $\delta^{34}S$ data (table 1) for hydrothermal vein pyrite and anhydrite from drillcore sample at 1,437-m depth in well PIN–2d. Pyrite data shown as solid bars; anhydrite data as cross-hatched bars. The anhydrite data have an average of 19‰.

PINATUBO ANHYDRITES (6/12/91)
Mount Bagang

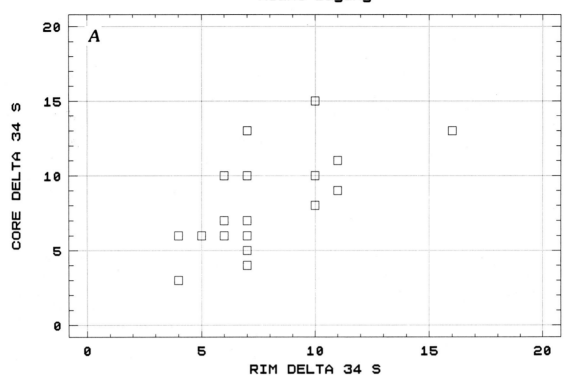

PINATUBO ANHYDRITES (6/15/91)
Sacobia River Valley

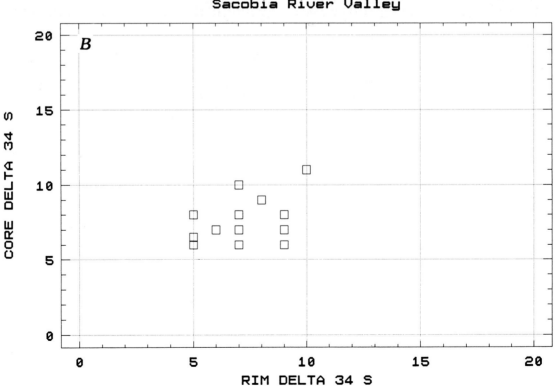

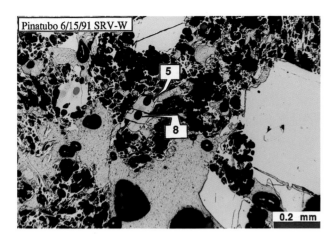

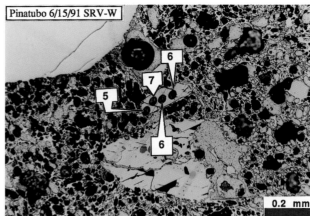

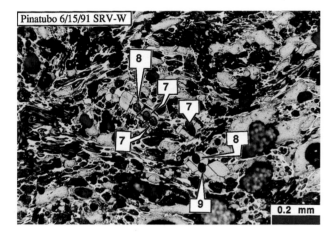

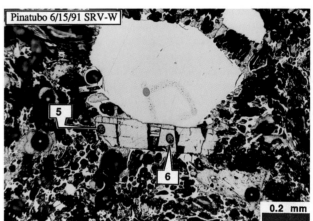

Figure 4A. Representative photomicrographs (polished thin sections in reflected light) of SHRIMP analytical craters in anhydrite, from pumice of June 15, 1991 eruption of Mount Pinatubo (sample collected from Sacobia River valley). Numbers in boxes are $\delta^{34}S$ values (in ‰ relative to CDT standard) for the craters identified by the arrows; craters are 20–30 μm in diameter and about 5 μm deep.

Anhydrite crystals are anhedral to subhedral, with some rounded crystal boundaries in direct contact with glass. Cleavage is visible in some anhydrite crystals, as are hexagonal cross sections of apatite crystals. Dark areas are voids; dusty, less-reflective areas are voids filled with epoxy that was used to impregnate the samples prior to polishing.

← **Figure 3A.** Plot of core versus rim SHRIMP $\delta^{34}S$ data (in ‰, table 1) for anhydrite crystals from pumice of the June 12, 1991 eruption of Mount Pinatubo (sample collected from Mount Bagang). There are distinctly ^{34}S-depleted crystals and distinctly ^{34}S-enriched crystals. Only two crystals show a difference >4‰ between core and rim $\delta^{34}S$ values, indicative of intracrystalline isotopic zoning at a level detectable by SHRIMP. *B*, Plot of core versus rim SHRIMP $\delta^{34}S$ data (in ‰, table 1) for anhydrite crystals in pumice from June 15, 1991 eruption of Mount Pinatubo (sample collected from Sacobia River valley). The data cluster much more tightly than those from June 12 anhydrites. No crystals show a difference >4‰ between core and rim $\delta^{34}S$ values.

SO_2/H_2S ratio and the temperature-oxygen fugacity path that the fluid follows as it cools (Rye, 1993). As temperatures cool to 400–200°C, the isotopically enriched SO_4^{2-} and isotopically depleted H_2S are often coprecipitated as hydrothermal anhydrite and sulfides.

The relatively enriched vein anhydrite $\delta^{34}S$ values and relatively depleted vein pyrite $\delta^{34}S$ values (fig. 1*D*) are consistent with such an origin for sulfur in the Mount Pinatubo drillcore vein minerals. The observed relative proportions of anhydrite being greater than pyrite in the vein are also consistent with the stoichiometry of the SO_2 decomposition reaction. The observed sulfur isotopic fractionation of 21‰ between the vein anhydrite and pyrite corresponds to a sulfur isotopic equilibration temperature of 285°C (Ohmoto and Rye, 1979), in excellent agreement with the fluid inclusion temperatures.

If such isotopically enriched hydrothermal vein anhydrite was assimilated during the early June 1991 Mount

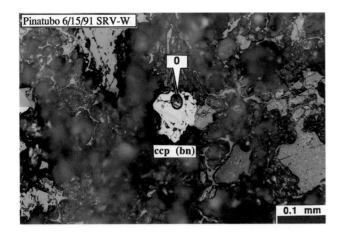

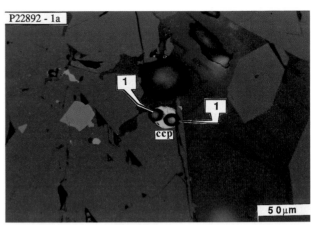

Figure 4B. Photomicrograph (polished thin section in reflected light) of SHRIMP analytical crater in chalcopyrite, from pumice of June 15, 1991 eruption of Mount Pinatubo (sample collected from Sacobia River valley). Number in box is $\delta^{34}S$ value (in ‰ relative to CDT standard) for the crater identified by the arrow; crater is 20–30 μm in diameter and about 5 μm deep. Chalcopyrite crystal (ccp) is anhedral, with darker bornite (bn) exsolution lamellae, and occurs as isolated crystal in glass. Dark areas are voids; dusty, less-reflective areas are voids filled with epoxy that was used to impregnate the samples prior to polishing.

Figure 5. Photomicrograph (polished thin section in reflected light) of SHRIMP II analytical craters in chalcopyrite (ccp) (with minor bornite lamellae not clearly visible), in basalt inclusion from June 7–14 hybrid dome (sample collected from pyroclastic flow in Maraunot valley). Numbers in boxes are $\delta^{34}S$ values (in ‰ relative to CDT standard) for the craters identified by the arrows; each crater is 10–15 μm in diameter and about 5 μm deep. Chalcopyrite crystal is anhedral and occurs as isolated crystal in feldspar. Dark areas are voids; dusty, less-reflective areas are voids filled with epoxy that was used to impregnate the samples prior to polishing.

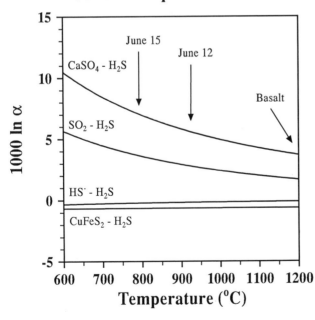

Figure 6. Curves showing equilibrium sulfur isotopic fractionation (in ‰) between relevant S-bearing phases and H_2S as a function of temperature (Ohmoto and Rye, 1979) over the range applicable to June 1991 Mount Pinatubo eruption products. Preeruptive temperatures for Mount Pinatubo magmas taken from Pallister and others (this volume) and Rutherford and Devine (this volume).

Pinatubo eruptions, then this could explain the isotopically heavy mode of anhydrites in pumice of the June 12 eruption. Assimilation presumably would have sampled a variety of hydrothermal anhydrites that had been deposited over a range of SO_2 decomposition conditions with a corresponding variety of ^{34}S enrichments. For this reason the $\delta^{34}S$ range and mode of any anhydrite xenocrysts in pumice of the June 12 eruption should not necessarily coincide precisely with those of anhydrites in the single drillcore sample.

Because isotopically heavy anhydrites constitute nearly half of those analyzed in pumice of the June 12 eruption, a significant component of extraneous SO_4^{2-} must have existed in this vent-clearing eruption. These heavy anhydrites are clearly out of isotopic equilibrium with the other anhydrites and sulfides. Their heaviest $\delta^{34}S$ values are statistically indistinguishable from the lightest values of hydrothermal vein anhydrites from one drillcore sample. Although a hydrothermal xenocryst origin for heavy anhydrites in the June 12 pumice is unsupported by firm textural evidence, no alternative hypothesis can meet all of the sulfur isotopic constraints.

In considering the feasibility of nonprimary sources of "excess" sulfur for the 1991 Mount Pinatubo eruptions, Gerlach and others (this volume) argue convincingly that any SO_4^{2-} that was dissolved in shallow hydrothermal *fluids* is of insufficient mass to have been a significant source of sulfur to the main June 15 eruption. However, this does not mean that hydrothermal vein anhydrite, whose equivalent SO_4^{2-} mass could have vastly exceeded the mass of SO_4^{2-} dissolved in hydrothermal fluids, could not have contributed anhydrite xenocrysts to the vent-clearing June 12 and later stages of the eruption.

CONSTRAINTS ON ISOTOPIC COMPOSITION OF SO_2 GAS AND H_2SO_4 AEROSOLS

The sulfur isotopic composition of the erupted SO_2 and resultant H_2SO_4 aerosols can be estimated from the primary anhydrite $\delta^{34}S$ values and preeruption temperatures. The estimated values depend on assumptions regarding the mechanism of SO_2 generation.

Gerlach and others (this volume) and Westrich and Gerlach (1992) present a vapor saturation and accumulation model in which a hydrous CO_2- and SO_2-bearing vapor phase coexisted with the dacite magma *prior* to its ascent and eruption. If sulfur isotopic equilibrium between melt anhydrite and vapor SO_2 was attained prior to eruption, and if irreversible anhydrite breakdown during eruption (Rutherford and Devine, this volume) did not contribute significantly to the SO_2 content of the eruption, then the sulfur isotopic composition of the SO_2 gas in the eruption cloud and the total sulfur isotopic composition of the bulk eruption (solids plus gas) can be estimated.

For a preeruption temperature of 780°C for the main June 15 eruption, the equilibrium isotopic fractionation between anhydrite and SO_2 is 3.5‰ (fig. 6). If the $\delta^{34}S$ value of primary anhydrite in melt was 7‰ (fig. 1B), then any coexisting SO_2 gas must have had a $\delta^{34}S$ value of 3.5‰.

Comparison of the total sulfur content of the erupted dacite with the amount of sulfur erupted as SO_2 gas (from data in table 1 of Gerlach and others, this volume) implies that 16 to 57 mass percent of the total erupted sulfur was contained within the vapor phase prior to eruption, with the remainder occurring as anhydrite crystals in the melt. Therefore, the $\delta^{34}S$ value of total sulfur in preeruptive magma plus vapor would have been 5.0 to 6.5‰.

If the erupted SO_2 gas was quantitatively oxidized to H_2SO_4 in the stratosphere, then sulfate aerosols from the June 15 eruption should have a $\delta^{34}S$ value identical to that of the SO_2, 3.5‰. This estimated value is in excellent agreement with a 3.3‰ $\delta^{34}S$ value for sulfate deposits collected from the engines of aircraft that experienced performance problems at altitudes of 10 to 12.5 km in June 1992 (Casadevall and Rye, 1994). However, there is no absolute certainty that these engine sulfate deposits were derived solely from quantitative oxidation of June 15 Mount Pinatubo SO_2.

Rutherford and Devine (this volume) argue that decomposition of anhydrite during ascent decompression and eruption is another potential source of SO_2 during the June 15 eruption. If such decomposition occurred rapidly and irreversibly, such as via the reaction

$$3FeO_{melt} + CaSO_{4_{crystal}} \rightarrow SO_{2_{gas}} + CaO_{melt} + Fe_3O_{4_{melt}}$$

then the $\delta^{34}S$ value of the SO_2 gas product would have been identical to that of the primary anhydrite, 7‰.

Bernard and others (this volume) analyzed two "fresh" Mount Pinatubo ash samples (collected on July 4, 1991) for soluble sulfate by using conventional bulk sampling techniques. They obtained $\delta^{34}S$ values of 9.4 and 9.7‰, significantly heavier than values for anhydrite in pumice of the June 15 eruption. They noted that the ash samples had total sulfur contents that were markedly higher than dacite pumice of the June 15 eruption and therefore concluded that the ash had adsorbed significant amounts of SO_4^{2-} derived from SO_2 in the eruption cloud. If so, this implies that either the SO_2 was isotopically heavy to begin with ($\delta^{34}S > 9‰$) and was quantitatively converted to SO_4^{2-} or that oxidation of SO_2 to SO_4^{2-} in the eruption plume favors the heavier sulfur isotope and produces a significant sulfur isotopic enrichment in the SO_4^{2-} product that is subsequently adsorbed on ash. Given the constraints on the $\delta^{34}S$ value of erupted SO_2 as discussed above, only the latter hypothesis (preferential oxidation of $^{34}SO_2$ over $^{32}SO_2$) appears feasible as an explanation for the isotopically heavy SO_4^{2-} adsorbed on ash of the June 15 eruption. This implies that quantitative conversion of SO_2 to SO_4^{2-} aerosols in the atmosphere may not occur, in which case the original sulfur isotopic signature of volcanic SO_2 may not necessarily be preserved in aerosols.

Direct analysis of the $\delta^{34}S$ values of aircraft-collected stratospheric Pinatubo sulfate aerosols might help resolve the uncertainties and apparent conflicts over the isotopic composition of the erupted SO_2 and the mechanism of SO_2 generation. Such analyses might also indicate the extent of any isotopic fractionation during conversion of SO_2 to adsorbed sulfate on ash in the eruption plume or to H_2SO_4 aerosols over longer residence times in the stratosphere. Once such processes are better known, it may prove possible to use the $\delta^{34}S$ composition of sulfate aerosols as a precise "fingerprint" to distinguish among sources such as volcanic eruptions, combustion of fossil fuels, terrestrial biogenic sources, and seasalt (see Calhoun and others, 1991; Nriagu and others, 1991).

SULFUR IN BASALT AND IMPLICATIONS FOR SULFUR IN DACITE

The single analyzed grain of sulfide in the basalt inclusion ($\delta^{34}S = 1‰$, fig. 1C) is primary in appearance, occurring within plagioclase (fig. 5). Given that such basalt is thought to be quenched inclusions of an eruption-triggering (fresh) basaltic magma batch (Pallister and others, this volume), the analyzed sulfide is likely representative of the sulfur isotopic composition of sulfide that was dissolved in the underplating basalt melt. The sulfur isotopic composition of sulfides in gabbros from the nearby and possibly underlying Zambales Ophiolite Complex is also 1‰ (J. Fournelle, Univ. of Wisconsin, and R. Carmody, USGS, written commun., 1994). Mafic to ultramafic igneous rocks typically contain primary sulfide minerals with $\delta^{34}S$ values between −1 and 2‰ (Ohmoto and Rye, 1979).

Because sulfur isotopic fractionation between melt HS−, chalcopyrite, pyrrhotite, and H_2S vapor is negligible at >1200°C (fig. 6; Ohmoto and Rye, 1979), crystal fractionation of solid sulfides or loss of H_2S gas during basalt magma crystallization should produce negligible Rayleigh effects. If the basalt became oxidized at some point (e.g., via interaction with the base of the more hydrous dacite "mush") such that SO_2 could form in addition to H_2S in an exsolving basaltic gas phase, then preferential partitioning of ^{34}S-enriched SO_2 into the gas could have occurred upon relatively low extents of basalt degassing, even at the high temperatures required (fig. 6). However, even under such oxidizing conditions, isotopic mass balance requires that the basalt could not have degassed sulfur (as a supercritical fluid with relatively high SO_2/H_2S) that was more than 2‰ heavier than the initial sulfide in the basalt melt at 1200°C (Ohmoto and Rye, 1979, eq. 10.6). Rayleigh effects with progressive degassing would make this constraint even less than 2‰. This means that any vapor degassed into the dacite by the basalt was constrained to have had a bulk $\delta^{34}S$ value no greater than 3‰, given a 1‰ value for sulfide in underplating basalt.

These constraints make the source of the isotopically enriched primary SO_4^{2-} in the dacite problematical. If the vapor saturation and accumulation model (Westrich and Gerlach, 1992; Gerlach and others, this volume) is correct, then the bulk sulfur isotopic composition of the June 1991 dacite eruption (gas plus solids) was 5.0 to 6.5‰, as was calculated above. If the anhydrite decomposition model of Rutherford and Devine (this volume) is correct, then the bulk sulfur isotopic composition of the preeruptive dacite magma must have approached that of the primary anhydrites, 7‰. Either model yields a dacite bulk sulfur isotopic composition that is significantly heavier than the maximum possible (3‰) from degassing of the underlying basalt melt. Other processes in addition to degassing of basalt are thus required to explain the bulk sulfur isotopic enrichment of the dacite.

Given the high anhydrite/pyrite ratio and ^{34}S-enriched nature of anhydrite in hydrothermal deposits occurring in the flanks of Mount Pinatubo, it is possible that repeated cycles of heating and *noneruptive* intrusion of the dacite into the volcanic pile could have allowed assimilation of such anhydrite by the magma. This could have provided sufficient time for mixing and isotopic homogenization of the resultant anhydrite "phenocrysts." However, if this

process occurred many times and was a major means of sulfur isotopic enrichment of the dacite, then xenoliths of acid-altered volcanic rocks with other hydrothermal mineral phases should have been abundant; these are not clearly manifested in the dacite.

Crystal fractionation and gravity settling of ^{34}S-depleted sulfide crystals from the dacite could result in limited ^{34}S enrichment of the remaining magma at temperatures appropriate for dacite of the June 15 eruption (fig. 6). However, given that inclusions of quenched underplating basalt magma were entrained in the dacite eruption products, one could expect that inclusions of any gravity-settled basal cumulate sulfides should also be seen in the basalt inclusions; none have been reported.

If assimilation or crystal fractionation mechanisms are thus infeasible or unlikely, then the only other means of generating a preeruptive dacite with bulk primary $\delta^{34}S$ >3‰ is via passive (noneruptive) steady-state loss of SO_2 or H_2S vapor from the dacite, in dynamic equilibrium with melt SO_4^{2-} and anhydrite phenocrysts.

A STEADY-STATE LONG-TERM DEGASSING MODEL

Over long time periods, H_2S and SO_2 from underplating basalt (total $\delta^{34}S$ between 1 and 3‰) could have been degassed passively into the more hydrous dacite; such degassing could cause generation of SO_2 and $CaSO_4$ in the dacite melt via reactions such as

$$H_2S_{gas} + 2H_2O_{melt} \rightarrow SO_{2_{melt}} + 3H_{2_{melt}}$$
$$SO_{2_{gas}} + H_2O_{melt} + CaO_{melt} \rightarrow CaSO_{4_{crystal}} + H_{2_{melt}}$$

As shown in figure 6, if these redox reactions attain isotopic equilibrium under passive conditions, they result in the preferential partitioning of ^{34}S from basaltic H_2S and SO_2 into SO_4^{2-}; the SO_4^{2-} then dissolves in the dacite melt and ultimately precipitates as anhydrite phenocrysts. The remaining relatively ^{34}S depleted H_2S and SO_2 will also dissolve in the dacite melt initially, but at some point a slow buildup of these volatiles could cause passive vapor saturation in the dacite. The gases would then be able to diffuse slowly through and out of the dacite magma over long time periods. The dacite melt could maintain a high oxidation state because the hydrogen gas produced by the reactions above should also diffuse slowly out of the melt, once vapor saturation is obtained.

If basaltic heat and H_2S-SO_2 gas are thus added slowly to the base of the dacite, it might reach and maintain vapor saturation passively, allowing steady-state *noneruptive* loss of the ^{34}S-depleted gases and hydrogen from the dacite over long time periods. This steady-state flux of ^{34}S-depleted gas diffusing up through the dacite would be accompanied by a buildup of ^{34}S-enriched SO_4^{2-} in the melt as anhydrite phenocrysts; the result would be a dynamic isotopic

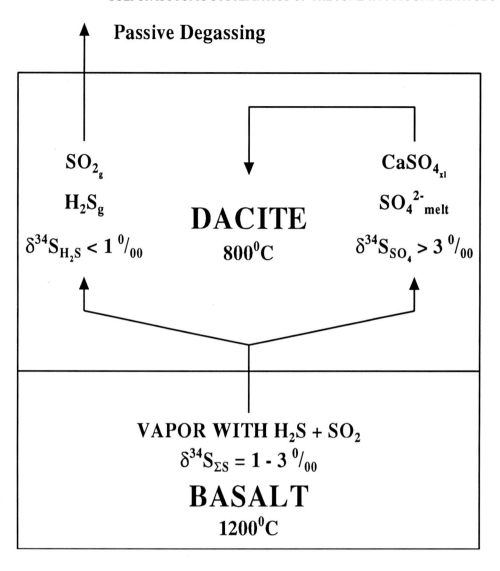

Figure 7. Schematic model for degassing of basaltic H_2S-SO_2 vapor into dacite and its subsequent partitioning into relatively ^{34}S-depleted SO_2-H_2S vapor and relatively ^{34}S-enriched melt SO_4^{2-} and $CaSO_4$ phenocrysts. Long-term steady-state passive (noneruptive) degassing of dacite, with vapor in dynamic isotopic equilibrium with SO_4^{2-} in anhydrite, results in a dacite melt that is more enriched in ^{34}S than the basaltic vapor. See text for detailed discussion.

equilibrium. Eventually, the sulfur isotopic compositions of H_2S and SO_2 in the dacite would be buffered by the large accumulated reservoir of ^{34}S-enriched anhydrite phenocrysts.

A schematic illustration of this model is shown in figure 7. This steady-state passive vapor loss model is consistent with the model for preeruptive vapor saturation and accumulation proposed by Westrich and Gerlach (1992) and Gerlach and others (this volume) and provides an explanation for the otherwise problematic ^{34}S-enriched nature of the dacite magma relative to the underplating basalt magma. Under conditions where a sudden large increase in basaltic heat and gas input causes the dacite to undergo catastrophic vapor oversaturation, a combination of SO_2-bearing vapor and high $\delta^{34}S$ SO_4^{2-}—rich magma would erupt.

CONCLUSIONS

Application of ion microprobe analytical techniques to the June 1991 eruption products of Mount Pinatubo has allowed the sulfur isotopic systematics of the volcanic system to be worked out in a manner that is not possible by use of conventional bulk techniques. The pumices contain a primary population of anhydrite phenocrysts with $\delta^{34}S \simeq 7‰$ that are in sulfur isotopic equilibrium with primary sulfides having $\delta^{34}S \simeq -1‰$. Pumice from the early vent-clearing June 12 eruption contains a significant component of isotopically heavy anhydrite ($\delta^{34}S \simeq 10‰$) that may represent xenocrysts derived from hydrothermal anhydrite deposits occurring within the flanks of the volcano.

Isotopic mass balance implies that the bulk sulfur isotopic composition of the preeruptive magma (solids + melt

±gas) was between 5 and 7‰ and that the SO_2 cloud from the June 15 eruption had a $\delta^{34}S$ value between 3.5 and 7‰, depending on whether the SO_2 existed in equilibrium with the magma in a preeruptive vapor phase or, instead, was generated rapidly by irreversible anhydrite decomposition during magma ascent and eruption. H_2SO_4 aerosols produced in the stratosphere from the erupted SO_2 may have a similar isotopic composition, although subsequent ^{34}S enrichment during atmospheric oxidation of SO_2 cannot be ruled out.

Basalt inclusions in the June 7–14 hybrid dome contain sulfide with $\delta^{34}S \simeq 1‰$, implying that any H_2S/SO_2 vapor degassed into the dacite by the basalt must have had a bulk $\delta^{34}S$ value between 1 and 3‰. Because the eruption was significantly more enriched in ^{34}S than this, other sulfur isotopic partitioning processes must have affected the dacite magma over relatively long time periods prior to the June 15 eruption. In light of textural and sulfur isotopic constraints, assimilation of ^{34}S-enriched hydrothermal anhydrite and (or) crystal fractionation of ^{34}S-depleted sulfide seem to be unlikely mechanisms for generating ^{34}S-enriched dacite. Instead, a mechanism involving long-term passive degassing of basaltic sulfur into the dacite and the attainment of steady-state dynamic isotopic equilibrium between a diffusing ^{34}S-depleted vapor phase and ^{34}S-enriched anhydrite phenocrysts can explain the sulfur isotope systematics.

ACKNOWLEDGMENTS

Research was sponsored by National Science Foundation grant EAR 93–04434 to M.A. McKibben. C.S. Eldridge acknowledges the support of the Australian Research Council and a Visiting Fellowship from the Research School of Earth Sciences. A.G. Reyes acknowledges past support from the Geothermal Division of the Philippine National Oil Corporation. This research project owes its origins to Werner F. Giggenbach, who first suggested that we apply the SHRIMP to the "sulfur problem" in volcanic eruptions. T.J. Casadevall, T.M. Gerlach, J.S. Pallister, and H.R. Westrich graciously provided many of the samples for this study. Comments on earlier versions of the manuscript by T.J. Casadevall, D.E. Crowe, A.Imai, J.F. Luhr, C.G. Newhall, and W.C. Shanks, and discussions with L. Baker, T.J. Casadevall, J.Fournelle, T.M. Gerlach, J.S. Pallister, R.O. Rye, M.J. Rutherford, R.I. Tilling, and H.R. Westrich, all resulted in a much improved manuscript. In particular, Chris Newhall prompted a critical reevaluation and improvement of many of the senior author's early ideas and biases.

REFERENCES CITED

Baker, L., and Rutherford, M.J., 1992, Anhydrite breakdown as a possible source of excess sulfur in the 1991 Mount Pinatubo eruption: Eos, Transactions, American Geophysical Union, v. 73 (Suppl.), p. 625.

Bernard, A., Demaiffe, D., Mattielli, N., and Punongbayan, R.S., 1991, Anhydrite-bearing pumices from Mount Pinatubo: Further evidence for the existence of sulphur-rich silicic magmas: Nature, v. 354, p. 139–140.

Bernard, A., Knittel, U., Weber, B., Weis, D., Albrecht, A., Hattori, K., Klein, J., and Oles, D., this volume, Petrology and geochemistry of the 1991 eruption products of Mount Pinatubo.

Bluth, G.J.S., Doiron, S.D., Schnetzler, C.C., Kreuger, A.J., and Walter, L.S., 1991, Global tracking of the SO_2 clouds from the June 1991 Mount Pinatubo eruptions: Geophysical Research Letters, v. 19, p. 151–154.

Buenviaje, M.M., 1991, Geochemical characteristics of acid fluids in Mt. Pinatubo, Philippines: Stanford University, Proceedings, Sixteenth Workshop on Geothermal Reservoir Engineering, Report SGP–TR–134, p. 267–273.

Cabel, A.C., Jr., 1990, Petrology of the permeable section of drillhole PIN–3D, Mt. Pinatubo geothermal prospect, Luzon, Philippines: University of Auckland, New Zealand, Geothermal Institute, Report No. 90.04, 30 p.

Calhoun, J.A., Bates, T.S., and Charlson, R.J., 1991, Sulfur isotopic measurements of submicrometer sulfate aerosol particles over the Pacific Ocean: Geophysical Research Letters, v. 18, no. 10, p. 1877–1880.

Carroll, M.R., and Rutherford, M.J., 1987, The stability of igneous anhydrite: Experimental results and implications for sulfur behavior in the 1982 El Chichón trachyandesite and other evolved magmas: Journal of Petrology, v. 28, p. 781–801.

———1988, Sulfur speciation in hydrous experimental glasses of varying oxidation state: Results from measured wavelength shifts of sulfur X-rays: American Mineralogist, v. 73, p. 845-849.

Casadevall, T.J., and Rye, R.O., 1994, Sulfur in the atmosphere: Sources and effects on jet-powered aircraft [abs.]: Eos, Transactions, American Geophysical Union, v. 75 (Suppl.), p. 78.

Delfin, F.G., Jr., Villarosa, H.G., Layugan, D.B., Clemente, V.C., Candelaria, M.R., Ruaya, J.R., this volume, Geothermal exploration of the pre-1991 Mount Pinatubo hydrothermal system.

Devine, J.D., Sigurdsson, H., Davis, A.N., and Self, S., 1984, Estimates of sulfur and chlorine yield to the atmosphere from volcanic eruptions and potential climatic effects: Journal of Geophysical Research, v. 89, p. 6309–6325.

Eldridge, C.S., Compston, W., Williams, I.S., Both, R.A., Walshe, J.L., and Ohmoto, H., 1988, Sulfur isotope variability in sediment-hosted massive sulfide deposits as determined using the ion microprobe SHRIMP: I. An example from the Rammelsberg orebody: Economic Geology, v. 83, p. 205–211.

Eldridge, C.S., Compston, W., Williams, I.S., Harris, J.W., and Bristow, J.W., 1991, Isotope evidence for the involvement of recycled sediments in diamond formation: Nature, v. 353, p. 649–653.

Eldridge, C.S., Compston, W., Williams, I.S., and Walshe, J.L., 1989, Sulfur isotope analyses on the SHRIMP ion microprobe, *in* Shanks, W.C., III, and Criss, R.E., eds., New frontiers in stable isotope analysis, U.S. Geological Survey Bulletin 1890, p. 163–174.

Eldridge, C.S., Compston, W., Williams, I.S., Walshe, J.L., and Both, R.A., 1987, In situ microanalysis for $^{34}S/^{32}S$ ratios using the ion microprobe SHRIMP: International Journal of Mass Spectrometry and Ion Processes, v. 76, p. 65–83.

Eldridge, C.S., Williams, N., and Walshe, J.L., 1993, Sulfur isotopic variability in sediment-hosted massive sulfide deposits as determined using the ion microprobe SHRIMP: II. A study of the H.Y.C. deposit at McArthur River, Northern Territory, Australia: Economic Geology, v. 88, p. 1–26.

Fournelle, J., 1990, Anhydrite in Nevado del Ruiz November pumice: Relevance to the sulfur problem: Journal of Volcanology and Geothermal Research, v. 42, p. 189–201.

Gerlach, T.M., Westrich, H.R., and Symonds, R.B., this volume, Preeruption vapor in magma of the climactic Mount Pinatubo eruption: Source of the giant stratospheric sulfur dioxide cloud.

Giggenbach, W.F., 1992, Magma degassing and mineral deposition in hydrothermal systems along convergent plate boundaries: Economic Geology, v. 87, p. 1927–1944.

Imai, A., Listanco, E.L., and Fujii, T, 1993, Petrologic and sulfur isotopic significance of highly oxidized and sulfur-rich magma of Mt. Pinatubo, Philippines: Geology, v. 21, p. 699–702.

Luhr, J.F., 1990, Experimental phase relations of water- and sulfur-saturated arc magmas and the 1982 eruptions of El Chichón volcano: Journal of Petrology, v. 31, p. 1071–1114.

Luhr, J.F., Carmichael, I.S.E., and Varenkamp, J.C., 1984, The 1982 eruptions of El Chichón volcano, Chiapas, Mexico: Mineralogy and petrology of the anhydrite-bearing pumices: Journal of Volcanology and Geothermal Research, v. 23, p. 69–108.

Matthews, S.J., Jones, A.P., and Bristow, C.S., 1992, A simple magma-mixing model for sulfur behavior in calc-alkaline volcanic rocks: Mineralogical evidence from Mount Pinatubo 1991 eruption: Journal of the Geological Society of London, v. 149, p. 863–866.

McKibben, M.A., and Eldridge, C.S., 1989, Sulfur isotopic variations among minerals and fluids in the Salton Sea geothermal system: A SHRIMP ion microprobe and conventional study of active ore genesis in a sediment-hosted environment: American Journal of Science, v. 289, p. 661–707.

———1990, Radical sulfur isotopic zonation of pyrite accompanying boiling and epithermal gold deposition: A SHRIMP study of the Valles Caldera: Economic Geology, v. 85, p. 1917–1925.

———1995, Microscopic sulfur isotopic variations in ore minerals from the Viburnum Trend, southeastern Missouri, U.S.A.: A SHRIMP study: Economic Geology, vol. 90, p. 228–245.

McKibben, M.A., Eldridge, C.S., Bethke, P.M., and Rye, R.O., 1993, SHRIMP study of extreme bacteriogenic sulfur isotope enrichments in the moat sediments of the Creede Caldera

[abs.]: Geological Society of America, Abstracts with Program, v. 25, no. 6, p. A–316.

Michels, D.E., Clemente, V.C., and Ramos, M.N., 1991, Hydrochemical features of a geothermal test well in a volcanic caldera, Mt. Pinatubo, Philippines: Stanford University, Proceedings, Sixteenth Workshop on Geothermal Reservoir Engineering, Report SGP–TR–134, p. 261–266.

Nriagu, J.O., Coker, R.D., and Barrie, L.A., 1991, Origin of sulfur in Canadian Arctic haze from isotope measurements: Nature, v. 349, p. 142–145.

Ohmoto, H., and Rye, R.O., 1979, Isotopes of sulfur and carbon, *in* Barnes, H.L., ed., Geochemistry of hydrothermal ore deposits (2d. ed.): Wiley, p. 509–567.

Pallister, J.S., Hoblitt, R.P., Meeker, G.P., Knight, R.J., and Siems, D.F., this volume, Magma mixing at Mount Pinatubo: Petrographic and chemical evidence from the 1991 deposits.

Pallister, J.S., Hoblitt, R.P., Newhall, C.G., and Scott, W.E., 1992, A basalt trigger for the 1991 eruptions of Pinatubo volcano? Yes. [abs.]: Eos, Transactions, American Geophysical Union, v. 73 (Suppl.), p. 347.

Pallister, J.S., Hoblitt, R.P., and Reyes, A.G., 1991, A basalt trigger for the 1991 eruptions of Pinatubo volcano?: Nature, v. 356, p. 426–428.

Reyes, A.G., 1990, Petrology of Philippine geothermal systems and the application of alteration mineralogy to their assessment: Journal of Volcanology and Geothermal Research, v. 43, p. 279–309.

Rudnick, R.L., Eldridge, C.S., and Bulanova, G.P., 1993, Diamond growth history from in situ measurement of Pb and S isotopic compositions of sulfide inclusions: Geology, v. 21, p. 13–16.

Rutherford, M.J., and Devine, J.D., this volume, Preeruption pressure-temperature conditions and volatiles in the 1991 dacitic magma of Mount Pinatubo.

Rye, R.O., 1993, The evolution of magmatic fluids in the epithermal environment: the stable isotope perspective: Economic Geology, v. 88, p. 733–753.

Rye, R.O., Luhr, J.F., Wasserman, M.D., 1984, Sulfur and oxygen isotope systematics of the 1982 eruptions of El Chichón volcano, Chiapas, Mexico: Journal of Volcanology and Geothermal Research, v. 23, p. 109–123.

Sigurdsson, H., 1990, Assessment of the atmospheric impact of volcanic eruptions, *in* Sharpton, V.L., and Ward, P.D., eds., Global catastrophes in Earth history: Geological Society of America Special Paper 247, p. 99–110.

Sugaki, A., Shima, H., Kitakaze, A., and Harada, H., 1975, Isothermal phase relations in the system Cu-Fe-S under hydrothermal conditions at 350° and 300°C: Economic Geology, v. 70, p.806–823.

Westrich, H.R., and Gerlach, T.M., 1992, Magmatic gas source for the stratospheric SO_2 cloud from the June 15, 1991, eruption of Mount Pinatubo: Geology, v. 20, p. 867–870.

Wolfe, E.W. and Hoblitt, R.P., this volume, Overview of the eruptions.

Anhydrite-Bearing Pumices from the June 15, 1991, Eruption of Mount Pinatubo: Geochemistry, Mineralogy, and Petrology

By John Fournelle,[1] Rebecca Carmody,[2] and Arturo S. Daag[3]

ABSTRACT

A distinctive feature of the dacitic pumices from the June 15, 1991, eruption of Mount Pinatubo is the presence of anhydrite, both as microphenocrysts and as inclusions. Over 97 percent of the sulfur present in the dacitic pumice is sulfate. Sulfides are present in minor amounts, mainly as inclusions. Mineral indicators yield the following preeruptive conditions in the dacitic magma: a temperature of 805 to 830° Celsius at a pressure 200–250 megapascals and an oxygen fugacity of 2.5 log units above the Ni-NiO buffer. The presence of cummingtonite rims and groundmass crystals indicates that parts of the magma had cooled to less than 800° Celsius prior to eruption.

Oxygen isotope data indicate that there was no significant contamination of dacitic magma by the underlying Zambales Ophiolite Complex.

Anhydrite is present as inclusions within hornblende and plagioclase crystals; these inclusions confirm the existence of anhydrite in the magma to 9 kilometers in depth. Careful scrutiny of ancient volcanic deposits for anhydrite inclusions may provide evidence for sulfur-rich eruptions in the past even though the anhydrite phenocrysts in these rocks have vanished.

INTRODUCTION

The eruption of Mount Pinatubo on June 15, 1991, produced the largest stratospheric SO_2 cloud in recorded history. Approximately 20 Mt were measured by the Total Ozone Mapping Spectrometer aboard the Nimbus-7 satellite (Bluth and others, 1992), nearly three times that of the 1982 eruption of El Chichón. Increasing attention is being paid to such SO_2-rich volcanic eruptions, particularly for their influence on global climate.

A major question is the source of the sulfur and its relation to convergent margin volcanism. Experimental studies have shown that significant amounts (thousands of parts per million) of dissolved sulfate can be present in a wet, oxidized silicate melt saturated with anhydrite ($CaSO_4$) (Carroll and Rutherford, 1987; Luhr, 1990).

It generally has been accepted that the sulfur ultimately comes from subducted sulfide-bearing oceanic crust, although this belief does not preclude some possible involvement of crustal sulfates or sulfides. The presence of mafic xenocrysts in pumice at Pinatubo led Fournelle (1991) to suggest possible involvement of underlying Zambales ophiolitic rocks, a suggestion similar to that by Arculus and others (1983) for ophiolite contamination of Mount Lamington (Papua New Guinea) anhydrite-bearing trachybasalts and andesites.

REGIONAL SETTING: THE ZAMBALES RANGE

Mount Pinatubo is located south of the Masinloc massif of the Zambales Range, and east of the Cabangan massif. The Zambales Range has been recognized to consist of ophiolitic fragments, dipping to the east; several areas have been examined in detail (Geary and Kay, 1983; Abrajano and others, 1989; Evans and Hawkins, 1989). Abrajano and Pasteris (1989) described two distinct sulfide associations present in the Acoje massif farther to the north.

A geothermal well slant-drilled from the north flank of Mount Pinatubo encountered mafic material at 1 km in depth, as did another well on its east flank (Delfin and others, this volume). Rocks from the ophiolite complex include microdiorite and diabase dikes, completely altered rocks, hornfels, basalts and gabbros, and monzodiorites extending from ~1.1 to ~2.7 km below the surface (Santos and Diomampo, National Institute of Geological Sciences, Univ. of the Philippines, unpub. report, 1992).

[1] Department of Earth and Planetary Sciences, Johns Hopkins University, Baltimore, MD 21218. Now at: Department of Geology and Geophysics, University of Wisconsin, 1215 W. Dayton St., Madison, WI 53706.
[2] U.S. Geological Survey.
[3] Philippine Institute of Volcanology and Seismology.

SAMPLE DESCRIPTIONS

Three pumices from the eruption have been examined. Two samples, P1 and P2, were collected from the upper surface of pyroclastic-flow deposits in the Sacobia River valley near Clark Air Base 1 month after the June 15 eruption. These have been studied by several researchers and labeled, respectively, type 2 and type 1 pumices by Imai and others (1993). They also have been cataloged at the Smithsonian Institution: USNM# 116534–1 and 116534–2 (Luhr and Melson, this volume).

Sample P1 is a dark-gray, crystal-poor pumice with cataclastic features, whereas P2 is a white pumice with more (~20 vol%) crystals. A third sample (P4), collected from a pumice-fall layer atop a roof in Angeles City at 1930 on June 15 (Rosalinda M. Temprosa, ICLARM, written commun., 1993), has ~20 vol% of both light- and dark-colored crystals.

Samples of the hybrid andesite (CN6791d) and a basalt inclusion (CN6791i) were examined only for their oxygen isotope compositions. Their major and trace element compositions are reported elsewhere (Pallister and others, this volume).

Several other samples have been examined also. P3 is a Holocene sample from Porac, Pampanga. In hand sample it resembles P2, with ~20 vol% crystals (mostly plagioclase and amphibole).

ZAM1 is a sample of the nearby Zambales Range country rock, from a location 19 km northwest of the Mount Pinatubo crater, near the Balin Baquero River. This area is mapped as "peridotite" by Abrajano and others (1989). This rock is dark, coarse grained, and cut by thin light-green veins.

WHOLE-ROCK CHEMICAL COMPOSITIONS: MAJOR AND TRACE ELEMENTS

Major and minor element whole-rock compositions of these samples were determined by X-ray fluorescence (XRF) analysis at the Department of Geosciences, Franklin and Marshall College, Lancaster, Pa. Major element contents were determined by using the standard $LiBO_4$ fused disk method. Trace element (including sulfur) compositions were determined by the pressed pellet method. Ferrous iron was determined by standard wet chemical method, and loss on ignition (LOI) by heating to 900°C.

Monosulfides (chalcopyrite, sphalerite, pyrrhotite) and disulfides (pyrite) were extracted with hot 6 N HCl and $HCl-CrCl_2$ solution, respectively. The H_2S produced was bubbled through a $AgNO_3$ trap and the sulfide collected as Ag_2S.

The major, minor, and trace element compositions are given in table 1. Pumices P1, P2, P3, and P4 are dacitic (64–65 wt% SiO_2) with 4.4–4.7 wt% Na_2O and 1.5–1.6 wt% K_2O. They are medium-K calc-alkaline in Miyashiro's (1974) classification.

For the large-ion lithophile elements (LILE's), P1-P4 have 501–556 ppm Ba, 519–561 ppm Sr, and 45–52 ppm Rb; for compatible trace elements, Cr is ~60 ppm, Ni ~20 ppm, V ~90 ppm; and for the incompatibles, Y is ~12 ppm and Zr 84–104 ppm.

Whole-rock sulfur contents are highest in the two crystal-rich pumices, 1,200 ppm (P2) and 910 ppm (P4), and low in crystal-poor P1, at 480 ppm. All but one of these samples were collected a month after the eruption. During this time they were exposed to abundant rainwater, and some anhydrite originally present may have dissolved. Anhydrite also may have broken down during deuteric reactions with magmatic water. These values therefore represent minimal sulfur compositions. They are lower than those reported by others, such as 1,500–2,400 ppm (Bernard and others, 1991; Pallister and others, this volume).

One sample, P4, was collected immediately after the eruption; its sulfur content (910 ppm) should not reflect anhydrite dissolution by meteoric water. The differences of sulfur content between the various Pinatubo samples reported in this volume may also reflect some differences in the distribution of anhydrite within the magma.

The Holocene P3 sample contained only 70 ppm S, as might be expected, given the dissolution of anhydrite by meteoric water.

The Zambales country rock is a gabbro with an Mg# (molar Mg•100/Mg+Fe) of 70. It is moderately oxidized, with 30% of the iron being ferric, and contains 236 ppm monosulfide and 479 ppm disulfide. It is depleted in alkali elements (K, Ba, Rb), and Sr, Y, and Zr are low, whereas Cr and Ni are relatively high.

STABLE ISOTOPES

Whole-rock powders of four Pinatubo dacitic pumices (P1–P4), dome hybrid andesite (CN6791d) and its basalt inclusion (CN6791i), and Zambales Ophiolite Complex sample (ZAM1) were analyzed for their oxygen isotope compositions (fig. 1; table 2). Analyses were performed at the U.S. Geological Survey in Reston, Va. The technique for extraction of oxygen from whole-rock powders follows that of Clayton and Mayeda (1963) with some modifications, including use of ClF_3 reagent. Oxygen isotope ratios were measured on a Finnigan MAT 251 mass spectrometer.

Ag_2S precipitates from ZAM1 monosulfide and disulfide were combined with Cu_2O and combusted at 1,050°C. Sulfate was leached from sample P2 with HCl and precipitated as $BaSO_4$. The $BaSO_4$ was combined with Cu_2O and silica glass powder and combusted at 1,150°C. SO_2 produced from both Ag_2S and $BaSO_4$ was purified of CO_2, H_2O, and O_2 by vacuum distillation. Sulfur isotope ratios

Table 1. Pinatubo whole-rock chemical analyses.

[ICLARM, International Center for Living Aquatic Resources Management, Manila; LOI, loss on ignition. Major and minor trace element values are in weight percent; trace element values are in parts per million. NA, not applicable]

Source	Pinatubo	Pinatubo	Pinatubo	Pinatubo	Pinatubo	Zambales
Sample	P1	P2	P4	P4	P3	ZAM1
Type	pumice	pumice	pumice	glass separate	pumice	gabbro
Eruption date	6/15/91	6/15/91	6/15/91	6/15/91	Holocene	NA
Location	Sacobia River	Sacobia River	Angeles	Angeles	Porac	Balin Baquero
Date collected	mid-July	mid-July	6/15/91	6/15/91	early May	unknown
Collected by	C. Newhall	C. Newhall	ICLARM	ICLARM	C. Newhall	A. Daag
SiO_2	64.72	63.76	65.24	75.94	64.74	49.47
TiO_2	.46	.52	.52	.15	.52	.37
Al_2O_3	16.43	16.11	15.64	13.99	16.05	14.97
Fe_2O_3	1.97	2.15	2.28	.33	2.06	2.37
FeO	2.03	2.18	2.25	.54	2.24	5.76
FeO (total)	3.80	4.11	4.31	.83	4.09	7.89
MnO	.10	.10	.10	.04	.10	.16
MgO	2.25	2.39	2.31	.29	2.35	10.39
CaO	4.98	5.03	4.70	1.89	4.68	13.89
Na_2O	4.69	4.52	4.49	4.14	4.42	1.17
K_2O	1.52	1.53	1.64	2.79	1.61	.02
P_2O_5	.17	.16	.18	.08	.18	.00
LOI	.94	1.44	1.07	.76	1.98	1.91
Total	100.25	99.88	100.42	100.94	100.91	100.48
S (ppm)	480	1,200*	910	170	70	715**
SO3	.120	.300	.228	.042	.018	.179
Ba	501	527	566	749	533	10
Cr	66	60	53	19	68	76
Ni	23	17	19	4	21	95
Rb	45	47	52	86	50	6
Sr	561	550	519	345	530	94
V	89	91	90	11	96	251
Y	11	13	12	<1	12	19
Zr	95	84	104	52	100	47

*P2 Includes 4 ppm monosulfide and 22 ppm disulfide.
**ZAM1: Includes 236 ppm monosulfide and 479 ppm disulfide.

Table 2. Pinatubo and Zambales stable isotope compositions.

[Values are in per mil]

Sample no.	P1	P2	P4	CN6791i	CN6791d	P3	ZAM1
Type	dacite pumice	dacite pumice	dacite pumice	basalt inclusion	hybrid andesite	dacite pumice	ophiolite
Eruption	6/15/91	6/15/91	6/15/91	6/7–12/91	6/7-12/91	Holocene	
Ave. $\delta^{18}O$	7.1	7.4	7.4	6.9	7.3	7.0	4.6
Ave. $\delta^{34}S$	n.a.	8.2 (sulfate)	n.a.	n.a.	n.a.	n.a.	1.0 (monosulfide)
							0.9 (disulfide)

were measured on a 6-in., 60°-sector Nuclide Corporation mass spectrometer.

Isotope compositions are reported in the usual δ notation in per mil (‰) relative to the standard mean ocean water (SMOW) standard for oxygen isotope ratios and relative to the Canyon Diablo troilite (CDT) standard for sulfur isotope ratios. Oxygen and sulfur isotope compositions were measured on two (10–15 mg) aliquots of each sample.

Estimated uncertainty on these measurements is ±0.25‰ and ±0.20‰ for oxygen and sulfur isotope values respectively.

The oxygen isotope compositions of the seven analyzed samples are included in table 2 and shown in figure 1. The most obvious contrast is between the Pinatubo samples, which cluster in the range of 6.9 to 7.4‰, and the Zambales Ophiolite Complex sample at 4.6‰. This lower value falls

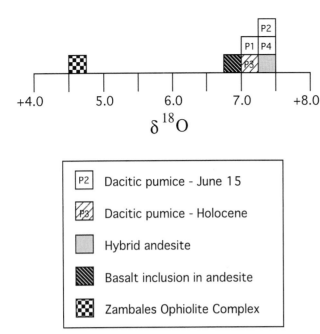

Figure 1. Histogram of oxygen isotope compositions for whole-rock powders of Pinatubo and Zambales samples.

within the range of analyses (2.3 to 5.9‰) of Zambales samples reported by Sturchio and others (1989). The Pinatubo pumices are enriched in ^{18}O relative to the average midocean ridge basalt (MORB) $\delta^{18}O$ of 5.6‰, whereas the Zambales sample is depleted in ^{18}O relative to MORB, probably due to oxygen isotope exchange with meteoric water (Sturchio and others, 1989). This contrast between the Pinatubo pumices and the Zambales Ophiolite Complex suggests that the Pinatubo magma passed through the Zambales Ophiolite Complex without exchanging or assimilating oxygen from it. The Pinatubo dacitic pumice $\delta^{18}O$ values overlap the anhydrite-bearing 1982 El Chichón trachyandesite $\delta^{18}O$ of 7.2 to 8.1‰ (Rye and others, 1984).

Among the Pinatubo samples, the basaltic inclusion (CN6791i) is the least enriched in ^{18}O, whereas dacitic pumices P2 and P4 are the most enriched. The data in table 2 suggest a subtle contrast in $\delta^{18}O$ between the phenocryst-rich type 1 pumice samples, P2 and P4, and the cataclastic, phenocryst-poor type 2 pumice, P1, with the type 1 pumices being ~0.3‰ heavier than the type 2 pumice. We believe that this contrast is real, although it should be confirmed by additional oxygen isotope analyses on other type 1 and type 2 samples (the 0.3‰ contrast observed is only slightly larger than the analytical uncertainty). This difference could have resulted from an additional boiling event suffered by the type 2 (phenocryst-poor) magma that was not experienced by the type 1 (phenocryst-rich) magma. Pallister and others (this volume) indicate that such a boiling event could account for the abundance of broken crystals in the phenocryst-poor magma. Removal of additional water from the type 2 magma relative to the type 1 magma would cause a

slight depletion of ^{18}O in the type 2 magma because a water-rich fluid phase concentrates ^{18}O relative to feldspars and presumably silicic melt (O'Neil and Taylor, 1967) at the preeruptive temperature of 805–830°C (see below).

The oxygen isotope data also support the hypothesis that the hybrid andesite ($\delta^{18}O$=7.3‰) represents a mixture of basalt similar to the inclusion CN6791i ($\delta^{18}O$=6.9‰) and dacitic magma; we would further add that the dacite may have been the heavier ($\delta^{18}O$=7.4‰) type 1 magma.

SULFUR ISOTOPES

Sulfur isotope composition ($\delta^{34}S$) of monosulfides and disulfide from the Zambales Ophiolite Complex sample ZAM1 are 1.0‰ and 0.9‰, respectively. These values are comparable, although slightly lighter than the range of 2 to 3‰ found by Abrajano and Pasteris (1989) in their study of the Acoje critical zone of the Zambales. The sulfur isotope composition of sulfide from the Zambales Ophiolite Complex is slightly heavier than that of Pinatubo chalcopyrite analyzed by McKibben and others (1992, this volume) and McKibben and Eldridge (1993), which gave sulfur isotope compositions in the range of −2 to 9‰. As expected, sulfate in the Pinatubo pumices is more enriched in ^{34}S than the sulfides with $\delta^{34}S$ values of ~7 to 9‰ reported by McKibben and Eldridge (1993) and Imai and others (1993) and is in agreement with an analysis of $\delta^{34}S$=8.2‰ for sulfate from pumice sample P2.

PETROGRAPHY—MINERAL AND GLASS CHEMISTRY

Mineral and glass compositions of the 1992 pumices were determined by using two electron microprobes: a JEOL 8600 in the Department of Earth and Planetary Sciences of the Johns Hopkins University, Baltimore, Md., and a Cameca SX50 in the Department of Geology and Geophysics at the University of Wisconsin, Madison. Operating conditions for both were a 15-keV accelerating voltage with a 20-nA beam current, using a rastered beam for glass analyses and a fixed beam for minerals. Data was reduced with the CITZAF matrix correction program (Armstrong, 1988) for the 8600 and the PAP $\phi(\rho z)$ routine for the SX50. Representative compositions of silicate and oxide minerals and glasses are given in table 3 and compositions of sulfides in table 4.

These pumices contain phenocrysts or microphenocrysts (in estimated decreasing abundance) of plagioclase, hornblende, ilmenite, magnetite, biotite, cummingtonite, anhydrite, apatite, orthopyroxene, quartz, and zircon.

Phenocryst-poor (type 2) pumice sample P1 consists mainly of glass and crystal shards, with few phenocrysts of plagioclase, hornblende (some with anhydrite inclusions),

Table 3. Pinatubo mineral and glass compositions.

[Plag, plagioclase; Hb, hornblende; Bi, biotite; Cum, cummingtonite; Opx, orthopyroxene; Cpx, clinopyroxene; Ol, olivine; Ap, apatite; Mt, magnetite; Ilm, ilmenite; xl, crystal; mega, megacryst; gmass, groundmass; pheno, phenocryst; n.a., not applicable; n.d., not determined. Analytical conditions described in text. Analysis of fluorine done without pulse height discrimination, so values in biotite and apatite are higher than actually present]

Sample	P2	P2	P2	P2	P2	P4	P4	P1	P1	P2	P2	P4	P2	P4	P4	P1
Mineral description	Matrix glass	Melt inclusion	Plag with CaSO$_4$	Plag with CaSO$_4$	Plag gmass	Hb pheno with melt inclusions	Hb pheno with melt inclusions	Hb gmass	Cum gmass	Cum edge of clot	Opx euhedral xls in clot	Ap gmass on Mt	Ap with CaSO$_4$ (fig. 6C)	Mt with t17	Ilm with t16	Ilm gmass
ID	IX.9-42	IX.26-16	IX.3-26	IX.3-28	IX.3-41	X.21-t3	X.21-t35	IX.3-t18	IX.3-51	IX.3-25	IX.3-10	X.21-14	IX.9-37	X.21-t16	X.21-t17	IX.3-t16
SiO$_2$	78.55	74.84	59.76	54.70	60.60	45.44	48.20	47.43	55.95	54.57	55.86	0.00	0.13	0.11	0.09	0.00
TiO$_2$	.00	.00	.02	.06	.02	1.86	1.02	.83	.17	.25	.10	.00	.00	3.89	28.52	31.18
Al$_2$O$_3$	12.99	13.06	25.95	28.90	25.63	12.26	8.67	8.02	1.23	1.44	1.79	.00	.02	1.99	.41	.33
Cr$_2$O$_3$	.07	.00	.07	.00	.01	.08	.07	.03	.07	.00	.00	.00	.00	.19	.08	.00
FeO	.64	1.27	.18	.22	.27	8.64	12.74	13.28	16.99	17.84	10.49	.83	.10	86.14	62.75	62.12
MnO	.02	.15	.02	.01	.00	.26	.56	.57	1.12	.87	.10	.00	.14	.59	.24	.38
MgO	.20	.22	.00	.03	.00	15.86	14.82	14.43	21.98	21.39	31.80	.00	.09	1.2	1.27	1.31
CaO	1.08	1.35	7.66	11.00	7.22	11.68	10.76	10.67	1.63	1.27	.72	56.03	53.71	.04	.08	.04
Na$_2$O	3.57	3.80	6.90	5.23	6.42	2.05	1.45	1.39	.16	.22	.04	.00	.09	.07	.11	.03
K$_2$O	2.60	2.81	.31	.16	.32	.52	.10	.14	.18	.00	.00	.00	.01	.00	.00	.00
P$_2$O$_5$	.00	.07	n.d.	n.d.	n.d.	.04	.14	n.d.	n.d.	n.d.	n.d.	42.17	42.74	n.d.	n.d.	n.d.
SO$_3$	.00	.01	n.d.	n.d.	n.d.	.27	.06	n.d.	n.d.	n.d.	n.d.	n.d.	.04	n.d.	n.d.	n.d.
Cl	.11	.13	n.d.	n.d.	n.d.	.00	.01	n.d.	n.d.	n.d.	n.d.	.25	1.47	n.d.	n.d.	n.d.
F	.00	.00	n.d.	n.d.	n.d.	n.d.	n.d.	n.d.	n.d.	n.d.	n.d.	2.25	1.94	n.d.	n.d.	n.d.
Total	99.81	97.68	100.87	100.31	100.49	98.85	98.57	96.79	99.48	97.85	100.90	101.53	99.29	94.53	93.66	95.39
Mg#	36	24	n.a.	n.a.	n.a.	77	67	66	70	68	86	n.a.	n.a.	n.a.	n.a.	n.a.
An#	n.a.	n.a.	38	54	38	n.a.	n.a.	n.a.	n.a.	n.a.	n.a.	n.a.	n.a.	n.a.	n.a.	n.a.

Table 4. Pinatubo sulfide compositions.

[iss, Cu-Fe-S intermediate solid solution; mss, solid solution between $Fe_{1-x}S$ and $Ni_{1-x}S$; Mt, magnetite]

Sample	P4	P4	P2	P2	P2	P2
Sulfide (phase)	CuFeS (iss)	CuFeS (iss)	(Fe,Ni)S (mss)	(Fe,Ni)S (mss)	CuFeS (cubanite)	(Fe,Ni)S (mss)
Host	Mt	Mt	Bronzite	Bronzite	Bronzite	Bronzite
Size, association	15 μm	15 μm	with Mt	with Mt	with Mt	with Mt
Figure 10 label			P2a 1	P2a 2	P2a 3	P2a 4
S	32.97	32.74	37.28	34.91	35.67	39.12
Fe	35.80	34.38	47.07	54.04	37.38	53.40
Ni	.09	.10	13.91	8.14	1.22	8.21
Cu	30.66	32.54	.07	.26	24.78	.28
Total	99.52	99.76	98.33	97.35	99.05	101.01

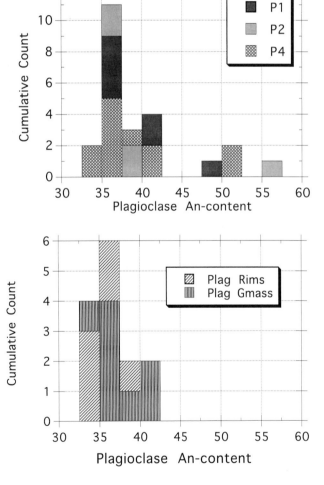

Figure 2. Histogram of plagioclase anorthite content. Top, all plagioclase compositions (core, rims, groundmass) shown. Bottom, rims and groundmass compositions plotted; the mean is An$_{36}$.

Fe-Ti oxides, cummingtonite, quartz, and biotite. No anhydrite microphenocrysts were found in the two thin sections examined.

Phenocryst-rich (type 1) pumice P2 contains plagioclase and hornblende phenocrysts; anhydrite is present both as microphenocrysts (~250 μm) and as inclusions in plagioclase and hornblende. Biotite, cummingtonite, Fe-Ti oxides, apatite, quartz, zircon, and bronzite are also present. Sample P4 contains plagioclase, hornblende, biotite, Fe-Ti oxides, anhydrite, apatite, and quartz.

Plagioclase is the dominant phenocryst in the dacitic pumices and ranges in composition from An$_{32}$ to An$_{55}$, with most in the range An$_{35-40}$ (table 3; fig. 2). Hornblende, cummingtonite, biotite, and pyroxenes have Mg numbers ranging from 64 to 77, with the mean ~70 (table 3; fig. 3). Bronzite of Mg# 86 is also present.

Groundmass crystals of plagioclase (An$_{40-48}$), hornblende (Mg# 68), magnetite, and biotite (Mg# 68) are abundant in sample P1. In P2 the groundmass crystals are mainly plagioclase (An$_{34-41}$), hornblende (Mg# 63–70), and cummingtonite (Mg# 70); initial examination suggests that they are less common than in sample P1.

Some sulfides also were found. Round blebs (<10 μm diameter) of chalcopyrite (Cu-Fe-S intermediate solid solution, iss) are present as inclusions in magnetite and ilmenite in P4 (fig. 4). A variety of sulfides have been found in P2, including FeNiS mss (solid solution between $Fe_{1-x}S$ and $Ni_{1-x}S$), and cubanite in a complex mafic assemblage. Energy dispersive spectroscopy (EDS) analysis also revealed a 10-μm grain of a Cu-Zn-S phase elsewhere in the matrix glass. Bernard and others (1991) and McKibben and others (this volume) found small amounts of pyrrhotite in Pinatubo pumice.

Matrix and melt inclusion glasses from P2 were analyzed by electron microprobe. The matrix glass is rhyolitic (75–80 wt% SiO$_2$, average = 76.94 wt% for 11 analyses) with low sulfur content (below the detection limit) and ~1,200 ppm Cl. Average Na$_2$O is 3.47 wt%, and K$_2$O is 2.94 wt%. The totals are ~98 to 100 wt%. (See table 3.)

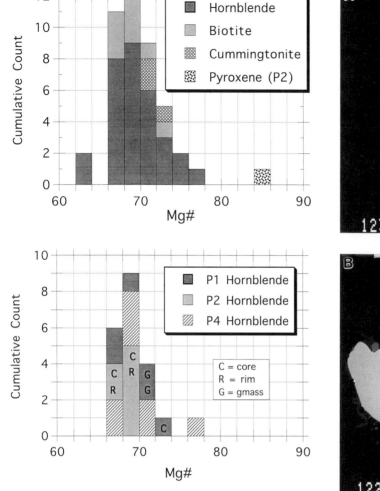

Figure 3. Histograms of Mg number of Fe-Mg phases. Top, hornblende, biotite, cummingtonite, and pyroxenes from all samples are plotted. Bottom, hornblendes are plotted by sample number (some cores, rims, and groundmass grains are differentiated).

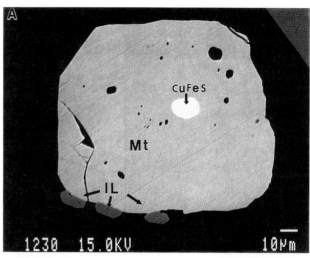

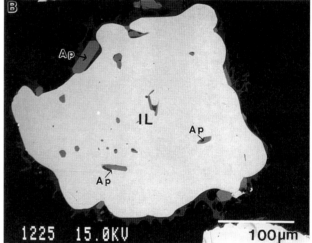

Figure 4. Backscattered electron images of P4 sulfide (Cu-Fe-S intermediate solid solution, iss) inclusions in magnetite (*A*) and ilmenite (*B*). In A, magnetite is X_{usp}=0.13, and ilmenite is X_{il}=0.60. Both oxides are mantled with matrix glass. IL, ilmenite; Mt, magnetite; Ap, apatite; CuFeS, Cu-Fe-S iss.

Matrix glass was separated from P4 (with heavy liquids and a Frantz magnetic separator) and analyzed by XRF (table 1). It is similar in composition to the rhyolitic matrix glass analyzed by electron microprobe. Its ferric/ferrous content was determined for calculation of melt f_{O_2} state. Trace element contents differ significantly from whole-rock values: Ba (749 ppm) and Rb (86 ppm) are higher than the whole-rock content, whereas lower values are noted for Cr (19 ppm), Ni (4 ppm), Sr (345 ppm), V (11 ppm), Y (<1 ppm), and Zr (52 ppm).

Melt inclusions examined by electron microprobe are slightly lower in SiO_2 than matrix glass is (69–77 wt%, average = 73.76 for five analyses), with other oxides as well

as sulfur at similar levels to those of the matrix glass (average Na_2O = 3.35 wt%, K_2O = 2.61 wt%), and having 96 to 97 wt% totals.

Holocene sample P3 contains large (to 2 mm) phenocrysts of plagioclase with healed fractures; hornblende is also present.

Anhydrite and apatite are intimately related frequently (also noted by Bernard and others, 1991). Blocks (200 μm) of anhydrite and apatite lie side by side in the matrix glass. Apatite inclusions are also present in anhydrite. This anhydrite-apatite association was noted in El Chichón pumices (Luhr and others, 1984).

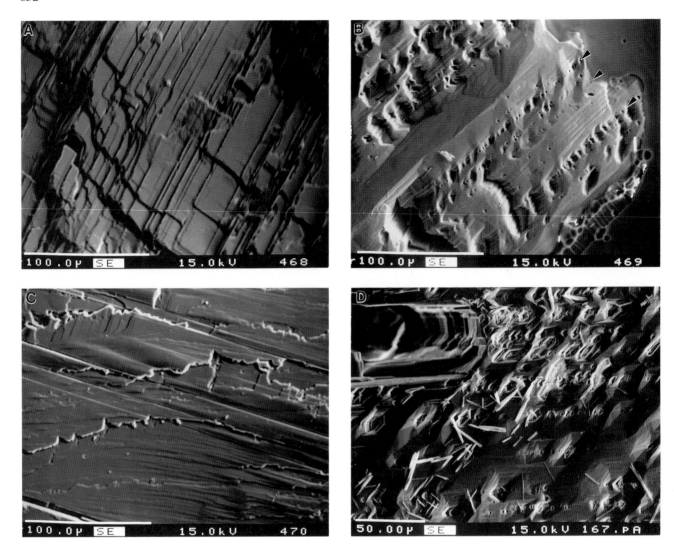

Figure 5.   Secondary electron images of anhydrite dissolution results after 2 months in distilled water. *A* and *C*, Freshly fractured and unreacted surfaces shown. *B* and *D*, Dissolution pits are evident in the reacted grains, with tracks of pits marking possible dislocation trails in *B* (arrowheads). Surface of anhydrite in *D* contains acicular crystals believed to be gypsum.

ANHYDRITE: WATER DISSOLUTION EXPERIMENT

A simple experiment was performed to determine what type of surface features might be produced in anhydrite by dissolution with water, as possibly might be observed in some of the Pinatubo pumices. Anhydrite (UW collection #8912A, from Bancroft, Ontario) was examined by X-ray diffraction (XRD) to verify its identity; it was then crushed and sieved; 20 to 30 grains of the 1000- to 700-µm fraction were put in a beaker with ~125 mL distilled water. The beaker was stationary, unsealed, with no movement of the contents (preventing mechanical breakage). Every 2 to 3 weeks, the water was decanted (some had evaporated), with fresh distilled water added. At the end of 2 months, the water was decanted, and the grains were air dried and mounted for examination. A control group was briefly cleaned ultrasonically in alcohol to remove fragments from surfaces and was then air dried and mounted.

The results of anhydrite sitting in water for 2 months are shown in figure 5*B,D* with controls in figure 5*A,C*. The original fractured surfaces (steplike) have remained, with some reduction of their roughness. Notable is the development of distinct, euhedral dissolution pits along the exposed surfaces (fig. 5*B*). These features resemble both natural and experimental dissolution of feldspars, where selective dissolution occurred in etch pits along apparent dislocations (Berner and Holdren, 1977). On some exposed surfaces (fig. 5*D*), small acicular crystals are present, possibly gypsum (Hardie, 1967).

We suspect that magmatic dissolution of anhydrite might have similar features present at high surface-energy locations, such as dislocations.

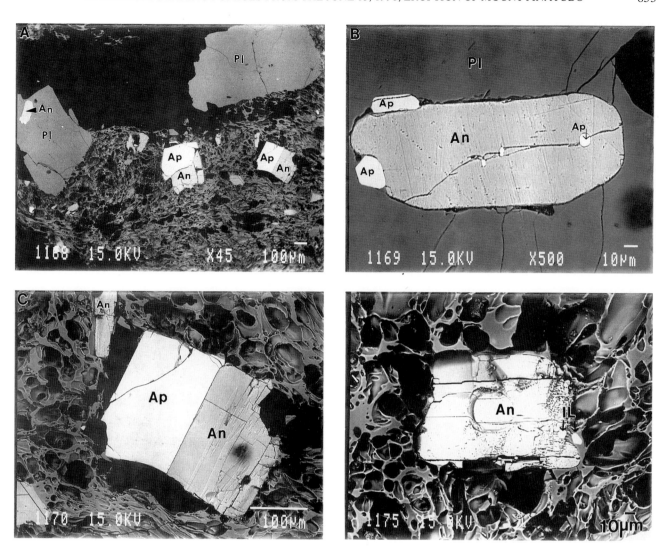

Figure 6. Backscattered electron images showing P2 pumice textures. Anhydrite is present in three forms: as inclusions in plagioclase (*A, C*), in the matrix glass (*B, D*), and in intimate relation to apatite (*A, B*). The anhydrite in *D* appears to have a round feature in its middle (a core) and to be dissolving along its edges.

ANHYDRITE IN THE 1991 DACITIC PUMICE

ANHYDRITE INCLUSIONS

Virtually all of the anhydrite trapped within plagioclase and amphibole in sample P2 is rectangular with rounded corners (figs. 6*A,B* and 7*B,D*). Similar features were seen in anhydrite in pumice of Nevado del Ruiz (within plagioclase and orthopyroxene; Fournelle, 1990; unpub. data, 1990).

ANHYDRITE PHENOCRYSTS

The pumices with the highest whole-rock sulfur content also have the greatest abundance of anhydrite microphenocrysts. A mass balance calculation for sample

P2 indicates ~0.5 wt% anhydrite in the pumice. The total sulfur content of P2 is 1,200 ppm, with only 27 ppm present as sulfide (4.3 ppm monosulfide, 22.4 disulfide).

Anhydrite in sample P2 has been examined in detail (figs. 6 and 7). Many of the anhydrite phenocrysts in this pumice are euhedral (fig. 6*A,C*) and have no evidence of dissolution.

Some the anhydrite crystals have irregular edges (fig. 7*A*), rounded corners (fig. 7*C*), or features somewhere in between (fig. 6*D*). Could these anhedral features be indicative of dissolution of the anhydrite?

First, consider the anhydrite in figure 7*A*, which has one irregular face. This face also has contiguous matrix glass, so deuteric, meteoric, or lab dissolution seems unlikely. The irregular surface, however, resembles the gross features (steplike outline) of the experimentally fractured and then weathered (simulated) anhydrite, although

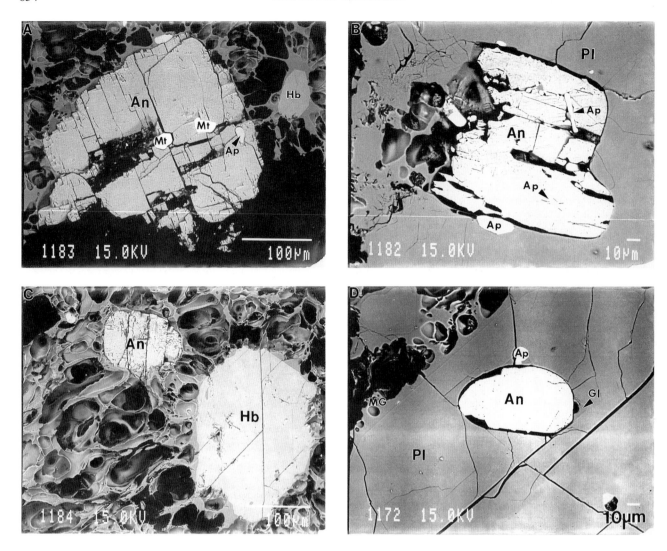

Figure 7. Backscattered electron images of anhydrite in P2 pumice. Anhydrite in matrix glass (*A, C*) has rounded edges, and *A* in particular is anhedral. Anhydrite in *B* and *D* is rounded inclusions in plagioclase. Apatite inclusions in the anhydrite are common. An, anhydrite; Ap, apatite; Hb, hornblende; Gl, matrix glass; Mt, magnetite; Pl, plagioclase.

caution must be taken in extrapolating the three-dimensional data (dissolution pits along a surface) to the two dimensions of a crystal surface intersecting the plane of the thin section. This anhydrite from Pinatubo apparently was fractured (along cleavage planes) at some point in its magmatic history; some dissolution is possible, although not certain.

Second, several of the anhydrite crystals have rounded corners, particularly the anhydrite inclusions. We believe these features are not indicative of dissolution, on the basis of the textures of anhydrite in sulfate-water-saturated melt experiments. Luhr (1990) studied sulfate-saturated basalt and trachyandesite at 800 to 1,000°C, 200 to 400 MPa. Anhydrite was present as an equilibrium phase. Backscattered electron (BSE) photomicrographs (his figs. 5 and 6) show that anhydrite at 200 MPa is present in irregular clus-

ters of spherical grains 5 to 10 μm in size; some oval-shaped grains were also present. At 400 MPa, euhedral prisms and euhedral blades were present as well as spherical grains.

Two anhydrite inclusions are present within hornblende, which is itself within plagioclase (fig. 8*A,B*). This situation indicates that the anhydrite had been trapped in the dacitic magma at a depth where the hornblende was stable, that is, with a temperature less than ~900°C and minimal melt water content of at least ~2 wt% (Merzbacher and Eggler, 1984; Johnson and Rutherford, 1989a). Plagioclase-bearing anhydrite inclusions in sample P2 range in core composition from An_{38} to An_{55}, with rims of An_{38}. The occurrence of the plagioclase also sets a maximum temperature range for the hornblende + anhydrite trapping event, given the limited range of plagioclase growth in water-rich melts.

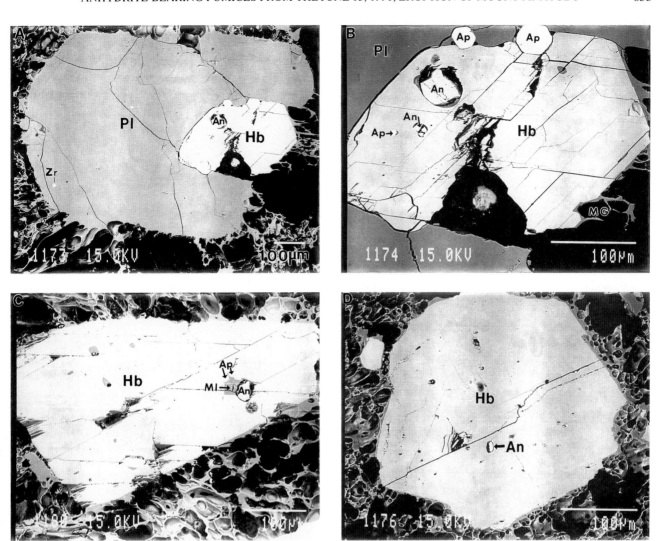

Figure 8.   *A, B.* Backscattered electron images of P2 plagioclase that contains hornblende with an anhydrite inclusion. The aluminum content of this hornblende suggests a pressure of 1.8 to 2.2 kbar (rim-core). Note the ubiquitous apatite in *B*. *C*, Hornblende in P2 plagioclase with anhydrite inclusion. *D*, Hornblende in P1 plagioclase with anhydrite inclusion. An, anhydrite; Ap, apatite; Hb, hornblende; Pl, plagioclase; MI, melt inclusion; Zr, zircon.

SCANNING ELECTRON MICROSCOPE EXAMINATION OF ANHYDRITE

Pumice fragments were glued to a slide, carbon coated, and examined by scanning electron microscope (SEM); no water was involved in the sample preparation. A concentration of a Ca-S phase, presumably $CaSO_4$, was found by EDS on the rim of plagioclase phenocryst (fig. 9). These are apparently growing into a vesicle; no matrix glass is attached to this face of the plagioclase. The hexagonal morphology of these crystals is striking. Anhydrite is orthorhombic, although also characterized as pseudohexagonal. The possible $CaSO_4$ hexagonal phases are gamma-$CaSO_4$ or hemihydrate (bassanite), $CaSO_4 \cdot \frac{1}{2}H_2O$ (Deer and others, 1962).

There are several possible implications: primary (magmatic) anhydrite may have recrystallized to the hemihydrate polymorph after cooling, with just enough H_2O present for the transformation but not enough to dissolve the grain, or these grains may have formed directly from a separate sulfur-rich gas phase. R.B. Symonds (USGS, oral commun., 1992) has described sulfuric acid droplets precipitating on the walls of silica sampling tubes in fumaroles at Mount St. Helens. T.M. Gerlach (USGS, oral commun., 1992) has suggested the possibility that sulfuric acid droplets precipitated on plagioclase may have reacted to form the $CaSO_4$.

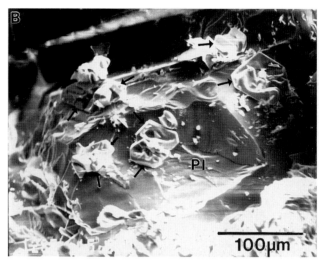

Figure 9. Scanning electron microscope photographs of hexagonal Ca-S crystals (indicated by arrows) residing on P2 plagioclase in a coarse pumice fragment. *A*, Backscattered electron image. *B*, Secondary electron image. Pl, plagioclase.

XENOCRYSTS

Most of the minerals present in the dacitic pumice are consistent with the equilibrium phase assemblages in rhyolitic melt as established by several experimental studies (for example, Johnson and Rutherford, 1989a). Luhr and Melson (this volume), however, present evidence for plagioclase-melt disequilibrium.

Although relatively sparse, the presence of xenocrysts raises the question whether the dacitic magmatic system may have been open to input from other sources.

Hornblende, biotite and cummingtonite in dacite P2 range from Mg# 64 to 73, with most 68 to 70. Present within the thin section are a mat of euhedral (~20×50 μm) bronzite (En_{84-86}) grains surrounded by a ~100-μm-wide corona of hornblende and cummingtonite (Mg# 70–73) and magnetite (fig. 10A). Within this bronzite are two ~20-μm-wide sulfide grains, Fe-Ni-S and Cu-Fe-S phases, in composite relation to magnetite (fig. 10B,C). Bronzite and Fe-Ni sulfides are uncommon in dacites.

The bronzite and Fe-Ni sulfides suggest interaction of the dacitic magma with either mafic magma or mafic rocks. Two identified candidates are the basaltic magma suggested by Pallister and others (1992) to have triggered the June 1991 eruptions, or the Zambales ophiolite.

GABBRO OF THE ZAMBALES OPHIOLITE COMPLEX

The Zambales sample (ZAM1) consists of calcic plagioclase (An_{75}), clinopyroxene (Mg# 77) and orthopyroxene (Mg# 71), with some ilmenite and magnetite (see table 5 for compositions). Sulfides are also present (table 6): pyrite, pyrrhotite, chalcopyrite, pendlandite, and sphalerite. They exist as complex intergrowths (fig. 11A,B) and as individual grains. Lamellar pyrrhotite is present in clinopyroxene and fills sites of presumably original exsolution lamellae (fig. 11C,D). Secondary silicate minerals are chlorite, edenite-pargasite, subcalcic ferroaugite, and sphene.

Pallister and others (1992, this volume) examined basaltic inclusions present in the dacite at Pinatubo and found several mafic phases: Mg-rich olivine (Fo_{86-89}) with Cr-spinels, clinopyroxene (Mg# 69–85), and plagioclase (An_{20-69}).

The bronzite found in dacite pumice P2 is more magnesian (Mg# 86) than the orthopyroxene (Mg# 71) found in the Zambales sample examined here. The bronzite's source remains to be identified.

PREERUPTIVE INTENSIVE PARAMETERS

Minerals present in the June 15 Pinatubo dacitic pumices have been evaluated for evidence of preeruptive magmatic conditions. Adjacent grains of ilmenite (X_{Ilm} = .55–.60) and magnetite (X_{Ulsp} = .12-.13) from P4 yield temperatures of 805 to 830°C and log f_{O_2} of −11.0 to −11.3 (2.3 to 2.5 log units above the NNO buffer when calculated by using the geothermometer code of Ghiorso and Sack, 1991). These grains pass the Mg/Mn equilibrium Fe-Ti oxide partitioning test (Bacon and Hirschmann, 1988). The f_{O_2} is corroborated by the ferric/ferrous content of separated matrix

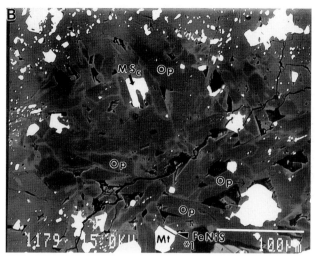

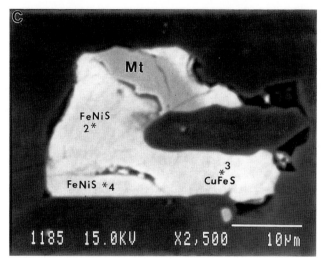

Figure 10. Backscattered electron images of (*A*) P2 bronzite surrounded by corona of hornblende and iron oxides, with cummingtonite on rim, (*B*) a closeup of the bronzite showing magnetite-sulfide grains, and (*C*) a closeup of one magnetite-sulfide intergrowth (MS$_c$). Locations of microprobe analyses of sulfides indicated by nos. 1–4. Cm, cummingtonite; Hb, hornblende; MS, magnetite-sulfide intergrowths; Mt, magnetite, Op, orthopyroxene.

Table 5. Zambales Ophiolite Complex mineral compositions.

[Plag, plagioclase; Opx, orthopyroxene; Amphib, amphibole; Cpx, clinopyroxene; Mt, magnetite; Ilm, ilmenite; n.a., not applicable]

Mineral	Plag	Cpx	Chlorite	Opx	Amphib	Sphene	Cpx	Ilmenite	Magnetite
ID	Ill.10–4	Ill.10–5	Ill.10–12	Ill.10–18	Ill.11–8	Ill.11–12	Ill.11–13	Ill.11–20	Ill.11–17
Description	with sulfides		in plag	opx	edenite lamellae	in cpx	subcalcic ferroaugite	with Mt	with Ilm
SiO$_2$	49.76	51.71	27.42	53.98	44.98	31.78	49.48	0.03	0.14
TiO$_2$	.07	.74	.03	.24	2.79	32.95	.09	49.67	.41
Al$_2$O$_3$	32.00	2.90	19.31	1.46	10.28	3.04	2.95	.10	.30
Cr$_2$O$_3$	.00	.04	.00	.03	.01	.00	.00	.00	.13
FeO*	.22	7.63	19.74	17.89	12.58	1.56	24.60	46.66	91.67
MnO.................	.00	.20	.22	.36	.15	.01	1.87	2.44	.00
MgO.................	.00	14.07	18.13	24.64	13.70	.25	8.03	.06	.09
CaO.................	14.69	22.06	.16	.85	11.54	27.66	9.14	.00	.05
Na$_2$O	3.01	.25	.00	.02	2.15	.00	.00	.07	.09
K$_2$O.................	.03	.00	.00	.00	.00	.00	.00	.00	.00
Total	99.78	99.60	85.01	99.47	98.18	97.25	96.16	99.03	92.88
Mg#.................	n.a.	77	62	71	66	22	37	n.a.	n.a.
An#	75	n.a.	n.a.	n.a.	n.a.	n.a.	n.a.	n.a.	n.a.

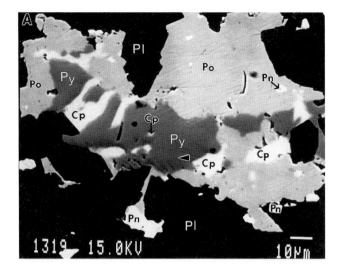

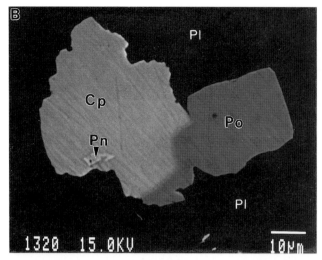

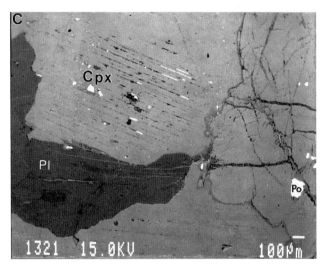

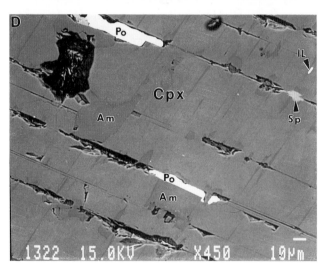

Figure 11. Backscattered electron images of sulfides and silicates from Zambales gabbro sample ZAM1. *A, B.* Sulfides present in An_{75} plagioclase (Pl) host; complex intergrowth of pyrrhotite (po), pyrite (py), pentlandite (pn), and chalcopyrite (cp).

C, D. Clinopyroxene (cpx) with exsolution lamellae replaced with sulfide (po); also replacements of clinopyroxene with edenite-pargasite, subcalcic ferroaugite, and sphene (sp). IL, ilmenite; Am, amphibole.

Table 6. Zambales Ophiolite Complex sulfide compositions.

[n.d., not determined]

Mineral	Pyrite	Chalcopyrite	Pentlandite	Pyrrhotite	Sphalerite
ID	Ill.10–6	Ill.10–10	Ill.10–12	Ill.10–14	Ill.11–25
Description			5 μm		
Fe	48.34	32.74	31.43	61.54	12.78
S	54.60	34.90	33.32	39.62	33.02
Cu..............	n.d.	33.14	n.d.	.20	n.d.
Ni	n.d.	n.d.	32.44	.23	n.d.
Zn..............	n.d.	n.d.	n.d.	n.d.	53.81
Total	102.94	100.78	97.19	101.59	99.61

glass from pumice P4 (table 1), which yields an f_{O_2} of 2 to 3 log units above the NNO buffer when calculated by the model of Kilinic and others (1983).

Rutherford (1991) found cummingtonite rims on Pinatubo hornblende and used them to constrain maximum temperature to 800°C and pressure between 200 and 400 MPa (Geschwind and Rutherford, 1992).

The Al-in-hornblende geobarometer of Johnson and Rutherford (1989b) has been applied to the dacitic pumices, although it is not strictly applicable, given the lack of sphene and alkali feldspar (the K_2O level in the Pinatubo dacite is significantly lower than that of the experimental compositions used for their calibration).

P1, P2, and P4 hornblende compositions have been evaluated, and 20 give values ranging from 160 to 360 MPa (fig. 12). All of the hornblendes from dacitic pumice sam-

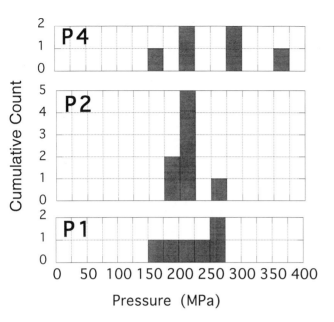

Figure 12. Histogram of pressures from aluminum contents of Pinatubo hornblendes that is based upon the experimental calibration of Johnson and Rutherford (1989a). Hornblende in dacite pumices P1 and P2 crystallized at pressures between 160 and 260 MPa, with most in range 200 to 250 MPa. There is evidence in P4 hornblendes for somewhat higher pressures, with the maximum near (greater than) the seismically determined base of the magma reservoir.

ples P1 and P2 fall in the range from 160 to 260 MPa, with a concentration at 200 to 250 MPa. The Mg# of these hornblendes ranges from 66 to 71.

The P4 pumice contains hornblendes of the similar composition and inferred pressure range, as well as three others giving pressures of 280, 290, and 360 MPa. These contain no gross evidence suggestive of disequilibrium (that is, Mg# <68, no Cr-spinels or oxides).

Three other analyzed hornblende crystals have aluminum contents that would suggest high pressures (310, 450, and 530 MPa—not plotted in fig. 12); however these have Mg# of 68 to 77 and Cr-spinels or other oxides. These latter hornblendes may have formed by reaction of silicic melt with magnesium-rich olivine or bronzite, and equilibrium may not have been achieved with the required silicic mineral assemblage.

Anhydrite exists as inclusions within hornblende (partially enclosed by An_{39} plagioclase in P2); the hornblende aluminum contents give a core pressure of 220 MPa, with 180 MPa on the rim. Another hornblende with an anhydrite inclusion gives a pressure of 240 to 280 MPa.

The presence of relatively large (to 50 μm) oxide inclusions in some hornblende crystals suggests changing conditions in the magma reservoir; hornblende destabilizes with increasing temperature or decreasing X_{H_2O} in the melt (Rutherford and Devine, 1988).

Cummingtonite, stable only below 800°C (Geschwind and Rutherford, 1992), either rims hornblende or is in the groundmass in the samples examined. The lack of larger cummingtonite crystals suggests that parts of the dacitic magma cooled below 800°C only just prior to eruption.

SULFUR-MAGMA PHASE EQUILIBRIA

With estimates of temperature, pressure, and oxygen fugacity, the Pinatubo dacite may be evaluated in light the experimental phase equilibria studies of Carroll and Rutherford (1987) and Luhr (1990). Their experiments showed that sulfur solubility increased with increasing temperature, pressure, and oxygen fugacity in fluid-saturated anhydrite-bearing andesitic to dacitic magmas. Their data have been plotted in figure 13A, with sulfur solubility versus pressure, for temperatures between 800 and 900°C and for f_{O_2} 3 to 5 log units above the NNO buffer.

The Pinatubo melt (glass) compositions are rhyolitic, compared to the andesitic to rhyolitic glass compositions of the experiments (most are dacitic to rhyodacitic). The results are assumed to be similar. Sulfur present in the matrix glass (60 ppm: Westrich and Gerlach, 1992) is consistent with a equilibration pressure of ~150 MPa at 800°C (fig. 13A). Rutherford (1993), Imai and others (1993), and Gerlach and others (this volume) indicate that the Pinatubo melt was at or near water saturation before eruption.

In light of these experimental results, we want to evaluate the question raised by Westrich and Gerlach (1992): is it possible to dissolve in the Pinatubo rhyolitic melt (at the determined T, P, P_{H_2O} and f_{O_2}) the amount of sulfur inferred from the SO_2 released into the atmosphere?

There is generally good agreement on the preeruptive intensive parameters deduced from mineral compositions in the Pinatubo dacites (table 7), although some differences exist, depending upon exactly which timeframe is being considered. One timeframe is that of cummingtonite rim growth immediately preceding eruption, whereas another is that of the host hornblende crystallization some time prior.

According to Westrich and Gerlach (1992), in order to account for the amount of SO_2 released into the atmosphere by direct loss from the Pinatubo melt, it would have had to contain ~1,000 ppm S prior to the sulfur's release. Could this much sulfur be dissolved in a rhyolitic melt under the conditions summarized in table 7?

The experimental data from Carroll and Rutherford (1987) and Luhr (1990) plotted in figure 13 suggest that in order to dissolve this much sulfur in the melt, higher temperatures and (or) pressures are required—for example, 850°C at 500 MPa or 900°C at 250 MPa. An additional constraint is the magma reservoir's depth of 6 to 11 km

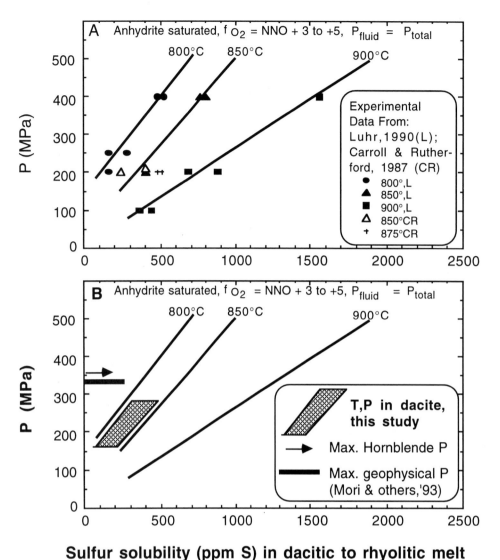

Figure 13. *A,* Sulfur solubility in dacitic to rhyolitic melt: concentration in parts per million sulfur versus pressure, for 800-900°C, fluid-saturated (mostly water), and $f_{O_2} \geq 3$ log units above the NNO buffer. Isotherms based upon the experimental data of Carroll and Rutherford (1987) and Luhr (1990). *B,* Inferred temperature and pressure field of dacitic magma before cooling below 800°C and cummingtonite crystallization, as derived from Fe-Ti oxides and Al-hornblende geobarometry.

Sulfur solubility (ppm S) in dacitic to rhyolitic melt

Table 7. Summary of temperature, pressure, and oxygen fugacity calculations for the June 15 dacite.

[n.d., not determined]

	Temperature (°C)	Pressure (MPa)	$\log f_{O_2}$
Rutherford (1991).	800±20	225±50	-11.0
Imai and others (1993).	800–850	n.d.	-11.0 to –11.5
This study	805–830	160–280* 170–360**	-11.0 to –11.3
Pallister and others, this volume.	776±22	200±50	–11.3

*P1 and P2 hornblendes.
**P4 hornblendes.

(seismically determined by Mori and others, 1993), and thus a maximum pressure of ~330 MPa.

Considering the pressure and temperature conditions summarized in table 7, a maximum temperature of 850°C at 330 MPa would permit ~600 ppm S to be dissolved in the melt. For the average pressure (200–250 MPa) and temperature (805–830°C) estimates, the water-rich rhyolitic melt could have held ~400 ppm S (fig. 13*B*). This is close to the maximum melt-inclusion sulfur content (330 ppm) in hornblende reported by Westrich and Gerlach (1991).

Only, at most, half of the SO$_2$ erupted could then have been theoretically dissolved as sulfate in the melt at the maximum depth and temperature. The discrepancy is larger, for more sulfur must be present in the preeruptive system to account for the anhydrite present in the pumices after eruption, as the anhydrite does not appear to be xenocrystic (McKibben and Eldridge, 1993). Pallister and others (this volume) report an average of 1,200 ppm S from seven analyses of the phenocryst-rich dacite, similar to the results found in this study (P2 and P4). Bernard and others (1991) reported a maximum of 2,210 ppm S in one dacite. This represents mainly anhydrite phenocrysts and inclusions, and these values are minimum because of a possible anhydrite loss during a month of rain on the pumices. It is unlikely that the dacitic magma could have held all this sulfur as dissolved sulfate in the melt. The suggestion of Westrich and Gerlach (1992) and Gerlach and others (this volume) that a volatile phase containing most of the SO_2 separated from the melt at depth, prior to melt-inclusion formation, is an attractive hypothesis.

Baker and Rutherford (1992) suggested anhydrite breakdown as possibly being involved in the SO_2-rich eruption. The textures of anhydrite in this study support the suggestion by Westrich and Gerlach (1992) and Gerlach and others (this volume) that the SO_2 vented to the atmosphere was not significantly augmented by breakdown of anhydrite immediately prior or during the eruption.

CONCLUSION

The June 1991 eruption of Mount Pinatubo presents an opportunity to study sulfur-rich magmas in a convergent-margin volcano. Such eruptions are of increasing concern for their climatic effects. Experimental phase equilibria have shown that oxidized andesitic-dacitic magmas have significant sulfur solubility and that anhydrite is a stable phase in these magmas.

Anhydrite is present in the Pinatubo pumices as microphenocrysts and as inclusions within silicate phenocrysts (hornblende and plagioclase). Assuming the anhydrite is not xenocrystic, its presence within hornblende indicates that the melt was sulfate saturated up to 280 MPa.

Many of the anhydrite microphenocrysts are euhedral. Most of the inclusions, as well as some microphenocrysts, have rounded edges, which we suggest are equilibrium features and not caused by dissolution. Dissolution of a minor amount of anhydrite microphenocrysts, however, cannot be ruled out.

An additional occurrence of anhydrite or a hydrated $CaSO_4$ phase was found in the pumice, possibly formed by crystallization from a gas phase or by reaction of condensed sulfuric acid with plagioclase.

Mineral geothermometers and barometers in the Pinatubo June 15 pumice suggest preeruptive dacitic magma temperatures of 805 to 830°C with f_{O_2} ~2.5 log units above NNO. Assuming the Al-hornblende geobarometer to be valid for these low K_2O samples, hornblende crystallized over a range of pressures from 160 to 360 MPa, with most in the range 200 to 250 MPa. Rims and groundmass grains generally give lower values, 160 to 190 MPa. Cummingtonite rims and groundmass grains suggest that the magma had cooled to below 800°C just prior to eruption.

Oxygen isotope data indicate that assimilation of silicate material from the underlying Zambales Ophiolite Complex is insignificant. Assimilation of significant amounts of Zambales sulfide is also unlikely, in light of the sulfur isotope compositions of the Zambales sulfides compared with the Pinatubo dacite sulfate $\delta^{34}S$.

Additionally, oxygen isotopic analyses of the two types of June 15 dacite suggest slight differences possibly related to an additional boiling event in the phenocryst-poor dacitic magma. The earlier hybrid andesite could have resulted from a mixture of basalt- and phenocryst-rich dacite.

Phase equilibria suggest that a quantity of sulfur equal to that vented into the stratosphere could not have been dissolved in the melt at 805 to 830°C and 200 to 250 MPa.

Finally, the presence of anhydrite inclusions within other phases is significant and suggests that it may be possible to study these more enduring trapped crystals from old volcanic deposits for evidence of SO_2-rich eruptions long after anhydrite phenocrysts have dissolved.

ACKNOWLEDGMENTS

The authors thank Chris Newhall for samples and discussions, Lou Walter (NASA) and Steve Self (University of Hawaii) for encouragement, Tom Moritz (California Academy of Sciences) for sample P4, Steve Sylvester (Franklin and Marshall College) and Maya Wheelock (Johns Hopkins University) for XRF assistance, Amy Gribb (University of Wisconsin–Madison) for XRD work, and the American Geophysical Union for funds to attend the Chapman Conference on Climate, Volcanism, and Global Change. Microprobe analyses were supported by National Science Foundation grants EAR–8916850 and DPP–9117576 to Bruce Marsh of the Johns Hopkins University. Thanks also go to Chris Newhall, Terry Gerlach, James Luhr, and Akira Imai for their reviews, which helped improve the manuscript.

REFERENCES CITED

Abrajano, T.A., Jr., and Pasteris, J.D., 1989, Zambales ophiolite, Philippines, II. Sulfide petrology of the critical zone of the Acoje massif: Contributions to Mineralogy and Petrology, v. 103, p. 64–77.

Abrajano, T.A., Pasteris, J.D., and Bacuta, G.C., 1989, Zambales ophiolite, Philippines, I. Geology and petrology of the critical zone of the Acoje massif: Tectonophysics, v. 168, p. 65–100.

Arculus, R.J., Johnson, R.W., Chappell, B.W., McKee, C.O., and Sakai, H., 1983, Ophiolite-contaminated andesites, trachybasalts, and cognate inclusions of Mount Lamington, Papua New Guinea: Anhydrite-amphibole-bearing lavas and the

1951 cumulodome: Journal of Volcanology and Geothermal Research, v. 18, p. 215–247.

Armstrong, J.T., 1988, Quantitative analysis of silicate and oxide materials: Comparison of Monte Carlo, ZAF, and $\phi(\rho z)$ procedures, *in* Newbury, D.E., ed., Microbeam analysis, Proceedings of the 23rd Annual Conference of the Microbeam Analysis Society, August 8–12, 1988: San Francisco, Calif., San Francisco Press, Inc., p. 239–246.

Bacon, C.R., and Hirschmann, M.M., 1988, Mg/Mn partitioning as a test for equilibrium between coexisting Fe-Ti oxides: American Mineralogist, v. 73, p. 57–61.

Baker, L., and Rutherford, M.J., 1992, Anhydrite breakdown as a possible source of excess sulfur in the 1991 Mount Pinatubo eruption [abs.]: Eos, Transactions, American Geophysical Union, v. 73, p. 62 5.

Bernard, A., Demaiffe, D., Mattielli, N., and Punongbayan, R.S., 1991, Anhydrite-bearing pumices from Mount Pinatubo: Further evidence for the existence of sulphur-rich silicic magmas: Nature, v. 354, p. 139–140.

Berner, R.A., and Holdren, G.R., Jr., 1977, Mechanism of feldspar weathering: Some observational evidence: Geology, v. 5, p. 369–372.

Bluth, G.J.S., Doiron, S.D., Schnetzler, C.C., Krueger, A.J., and Walter, L.S., 1992, Global tracking of the SO_2 clouds from the June, 1991 Mount Pinatubo eruptions: Geophysical Research Letters, v. 19, no. 2, p. 151–154.

Carroll, M.R., and Rutherford, M.J., 1987, The stability of igneous anhydrite: Experimental results and implications for sulfur behavior in the 1982 El Chichón trachyandesite and other evolved magmas: Journal of Petrology, v. 28, no. 5, p. 781–801.

Clayton, R.N., and Mayeda, T.K., 1963, The use of bromine pentafluoride in the extraction of oxygen from oxides and silicates for isotopic analysis: Geochimica et Cosmochimica Acta, v. 27, p. 43–52.

Deer, W.A., Howie, R.A., and Zussman, J., 1962, Rock-forming minerals: Non-silicates, v. 5: New York, John Wiley, p. 202–225.

Delfin, F.G., Jr., Villarosa, H.G., Layugan, D.B., Clemente, V.C., Candelaria, M.R., Ruaya, J.R., this volume, Geothermal exploration of the pre-1991 Mount Pinatubo hydrothermal system.

Evans, C., and Hawkins, J.W., Jr., 1989, Compositional heterogeneities in upper mantle peridotites from the Zambales Range Ophiolite, Luzon, Philippines: Tectonophysics, v. 168, p. 23–41.

Fournelle, J., 1990, Anhydrite in Nevado del Ruiz November 1985 pumice: Relevance to the sulfur problem: Journal of Volcanology and Geothermal Research, v. 42, p. 189–201.

———1991, Anhydrite and sulfide in pumices from the 15 June 1991 eruption of Mount Pinatubo: initial examination [abs.]: Eos, Transactions, Americal Geophysical Union, v. 72, p. 68.

Geary, E.E., and Kay, R.W., 1983, Petrological and geochemical documentation of ocean floor metamorphism in the Zambales ophiolite, Philippines, *in* Hayes, D., ed., The tectonic and geologic evolution of Southeast Asian seas and islands, pt. 2: American Geophysical Union, p. 139–156.

Gerlach, T.M., Westrich, H.R., and Symonds, R.B., this volume, Preeruption vapor in magma of the climactic Mount Pinatubo

eruption: Source of the giant stratospheric sulfur dioxide cloud.

Geschwind, C.-H., and Rutherford, M.J., 1992, Cummingtonite and the evolution of the Mount St. Helens (Washington) magma system: An experimental study: Geology, v. 20, p. 1011–1014.

Ghiorso, M.S., and Sack, R.O., 1991, Fe-Ti oxide geothermometry: thermodynamic formulation and the estimation of intensive variables in silicic magmas: Contributions to Mineralogy and Petrology, v. 108, p. 485–510.

Hardie, L.A., 1967, The gypsum-anhydrite equilibrium at one atmosphere pressure: American Mineralogist, v. 52, p. 171–200.

Imai, A., Listanco, E.L., and Fujii, T., 1993, Petrologic and sulfur isotopic significance of highly oxidized and sulfur-rich magma of Mt. Pinatubo, Philippines: Geology, v. 21, p. 699–702.

———this volume, Highly oxidized and sulfur-rich dacitic magma of Mount Pinatubo: Implication for metallogenesis of porphyry copper mineralization in the Western Luzon arc.

Johnson, M.C., and Rutherford, M.J., 1989a, Experimentally determined conditions in the Fish Canyon Tuff, Colorado, magma chamber: Journal of Petrology, v. 30, p. 711–737.

———1989b, Experimental calibration of the aluminum-in-hornblende geobarometer with application to Long Valley caldera (California) volcanic rocks: Geology, v. 17, p. 837–841.

Kilinic, A., Carmichael, I.S.E., Rivers, M.L., and Sack, R.O., 1983, Ferric-ferrous ratio of natural silicate liquids equilibrated in air: Contributions to Mineralogy and Petrology, v. 83, p. 136–140.

Luhr, J.F., 1990, Experimental phase relations of water- and sulfur-saturated arc magmas and the 1982 eruptions of El Chichón volcano: Journal of Petrology, v. 31, no. 5, p. 1071–1114.

Luhr, J.F., Carmichael, I.S.E., and Varekamp, J.C., 1984, The 1982 eruptions of El Chichón volcano, Chiapas, Mexico: Mineralogy and petrology of the anhydrite-bearing pumices: Journal of Volcanology and Geothermal Research, v. 23, p. 69–108.

Luhr, J.F., and Melson, W.G., this volume, Mineral and glass compositions in June 15, 1991, pumices: Evidence for dynamic disequilibrium in the dacite of Mount Pinatubo.

McKibben, M.A. and Eldridge, C.S., 1993, Sulfur isotopic systematics of the June 1991 eruption of Mount Pinatubo: a SHRIMP ion microprobe study: Eos, Transactions, American Geophysical Union, v. 74, p. 668.

McKibben, M.A., Eldridge, C.S., and Reyes, A.G., 1992, Multiple origins of anhydrite in Mount Pinatubo pumice: Eos, Transactions, American Geophysical Union, v. 73, p. 633–634.

———this volume, Sulfur isotopic systematics of the June 1991 Mount Pinatubo eruptions: A SHRIMP ion microprobe study.

Merzbacher, C., and Eggler, D.H., 1984, A magmatic geohygrometer: application to Mount St. Helens and other dacitic magmas: Geology, v. 12, p. 587–590.

Miyashiro, A., 1974, Volcanic rock series in island arcs and active continental margins: American Journal of Science, v. 274, p. 321–355.

Mori, J., Eberhart-Phillips, D., and Harlow, D., 1993, 3-Dimensional velocity structure at Mount Pinatubo, Philippines [abs.]: Resolution of magma bodies and earthquake hypocenters: Eos, Transactions, American Geophysical Union, v. 74, p. 667.

O'Neil, J.R., and Taylor, H.P., Jr., 1967, The oxygen isotope and cation exchange of feldspars: American Mineralogist, v. 52, p. 1414–1437.

Pallister, J.S., Hoblitt, R.R., and Reyes, A.G., 1992, A basalt trigger for the 1991 eruptions of Pinatubo volcano?: Nature, v. 356, p. 426–428.

Pallister, J.S., Hoblitt, R.P., Meeker, G.P., Knight, R.J., and Siems, D.F., this volume, Magma mixing at Mount Pinatubo: Petrographic and chemical evidence from the 1991 deposits.

Rutherford, M.J., 1991, Pre-eruption conditions and volatiles in the 1991 Pinatubo magma [abs.]: Eos, Transactions, American Geophysical Union, v. 72, p. 62.

————1993, Experimental petrology applied to volcanic processes [abs.]: Eos, Transactions, American Geophysical Union, v. 74, p. 49–52.

Rutherford, M.J., and Devine, J.D., 1988, The May 18, 1980 eruption of Mount St. Helens, 3. Stability and chemistry of amphibole in the magma chamber: Journal of Geophysical Research, v. 93, p. 11949–11959.

Rye, R.O., Luhr, J.F., and Wasserman, M.D., 1984, Sulfur and oxygen isotope systematics of the 1982 eruptions of El Chichón volcano, Chiapas, Mexico: Journal of Volcanology and Geothermal Research, v. 23, p. 109–123.

Sturchio, N.C., Abrajano, T.A., Jr., Murowchick, J.B., and Muehlenbachs, K., 1989, Serpentinization of the Acoje massif, Zambales ophiolite, Philippines: Hydrogen and oxygen isotope geochemistry: Tectonophysics, v. 168, p. 101–107.

Westrich, H.R., and Gerlach, T.M., 1991, Concentrations of sulfur in Mount Pinatubo glass inclusions [abs.]: Eos, Transactions, American Geophysical Union, v. 72, p. 62.

————1992, Magmatic gas source for the stratospheric SO_2 cloud from the June 15, 1991 eruption of Mount Pinatubo: Geology, v. 20, p. 867–870.

Highly Oxidized and Sulfur-Rich Dacitic Magma of Mount Pinatubo: Implication for Metallogenesis of Porphyry Copper Mineralization in the Western Luzon Arc

By Akira Imai,[1] Eddie L. Listanco,[2][3] and Toshitsugu Fujii[2]

ABSTRACT

Dacitic pumices from pyroclastic-flow deposits and tephra fall of the June 15, 1991, eruption of Mount Pinatubo, Philippines, are rich in sulfur, and the presence of microphenocrystic anhydrite suggests that sulfur existed dominantly as oxidized species in the magma. A high sulfur content in the magma is corroborated by unusually high sulfur contents (up to 0.78 weight percent as SO_3) in apatite microphenocrysts and apatite inclusions in other phenocrystic minerals; the oxidized state of that sulfur is consistent with highly oxidized magma, which, by extrapolation from the two-oxide method, was close to the manganosite-hausmanite buffer. The highly oxidized state of the magma may have caused the extraordinarily high sulfur content of that magma by prohibiting sulfide fractionation and by increasing solubility of sulfur as oxygen fugacity increased. Hornblende geobarometry indicates pressure of about 2 kilobars for phenocryst formation. The mineralogical similarity of pumice from Mount Pinatubo to intrusives that are genetically related to porphyry copper deposits suggests magmatic water saturation and high oxygen fugacity at the time of emplacement at shallow crustal levels.

INTRODUCTION

Calc-alkaline dacitic pumices from Mount Pinatubo have unusually high sulfur contents (Bernard and others, 1991) compared with ordinary calc-alkaline magmas in subduction-related arcs; other unusually high sulfur contents have been noted in subalkaline rocks such as the trachyandesite at El Chichón volcano, Mexico (Luhr and others, 1984) and the trachybasalt at Mount Lamington, Papua New Guinea (Arculus and others, 1983). Primary magmatic anhydrite microphenocrysts are a notable feature

of pumice from Mount Pinatubo (hereafter referred to as Pinatubo pumice, for simplicity). We report mineral compositions and sulfur isotopic determinations and their implications for the probable cause of unusually high sulfur contents in Mount Pinatubo's dacitic magma. Samples were collected from the equivalent of tephra layer C (Koyaguchi and Tokuno, 1993) (tephra layer C_1 of Koyaguchi, this volume) at Mabalacat, Pampanga (fig. 1). The strong similarity of the dacite of Mount Pinatubo (hereafter referred to as Pinatubo dacite) to intrusives genetically related to porphyry copper deposits is noted. Characterization of the Pinatubo pumice will, we hope, contribute to an understanding of the mechanism by which magmas become enriched in sulfur, one of the most important elements in magmatic-hydrothermal metallic deposits such as porphyry copper deposits.

GEOLOGIC BACKGROUND

Mount Pinatubo is one of the active volcanoes in the western Luzon volcanic chain, which is associated with eastward subduction of the Eurasian plate at the Manila trench (PHIVOLCS, 1991) (fig. 1). Porphyry copper-gold deposits of Miocene to Pliocene age are genetically associated with hydrous intermediate to silicic magmatism along this same arc. This metallogenic province extends from the Lepanto-FSE deposit in the northern Luzon, past the Dizon deposit in west-central Luzon, only 19 km south of Mount Pinatubo, to the Taysan deposit in southern Luzon. The basement underlying Mount Pinatubo includes the Zambales ophiolite complex.

PETROGRAPHY

The June 1991 eruption of Mount Pinatubo produced voluminous dacitic pyroclastic-flow and tephra deposits. Essential ejecta can be classified into two types, namely type 1 (white and phenocryst rich) and type 2 (yellowish and phenocryst poor) (Listanco, 1991; Pallister and others,

[1] Geological Institute, University of Tokyo.
[2] Earthquake Research Institute, University of Tokyo.
[3] Present address: Philippine Institute of Volcanology and Seismology.

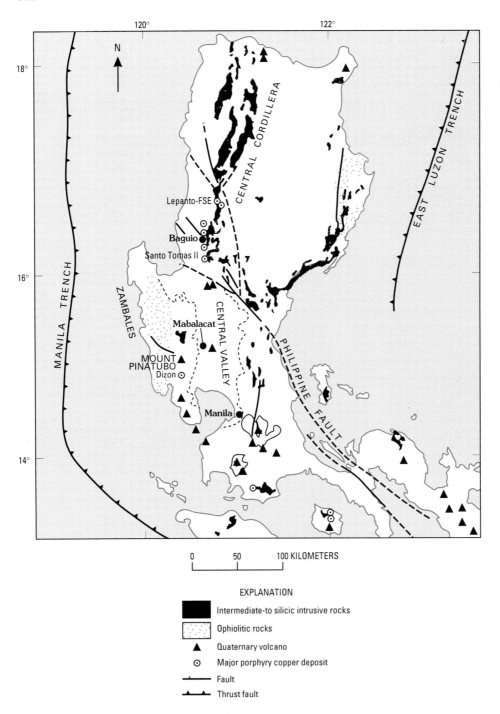

Figure 1. Distribution of Quaternary volcanoes, major porphyry copper deposits, Tertiary intermediate to silicic intrusives, and ophiolitic rocks, Luzon and vicinity, Philippines. Samples for the present study were collected in Mabalacat, Pampanga.

EXPLANATION

◼ Intermediate-to silicic intrusive rocks

▨ Ophiolitic rocks

▲ Quaternary volcano

⊙ Major porphyry copper deposit

— Fault

—▲— Thrust fault

1992). Both have broadly similar mineralogic constituents. Phenocrysts in type 2 pumice are mostly shattered. Major phenocrystic phases in both types include plagioclase and hornblende with subordinate quartz, biotite, titanomagnetite, ferrian ilmenite, and anhydrite. Cummingtonite-rimmed hornblende phenocrysts commonly occur in type 1 pumice (fig. 2A); in type 2 pumice, rare, discrete, shattered hypersthene occurs instead of cummingtonite. Olivine occurs in type 2 pumice (Pallister and others, 1992); in

addition, hornblende-rimmed olivine is also found in type 1 pumice. In both instances, olivine is inferred to be a xenocryst.

Apatite, titanomagnetite, and ferrian ilmenite occur as discrete microphenocrysts and as inclusions in various phenocrysts (fig. 2B, C). Trace amounts of biotite also occur as discrete microphenocrysts and as inclusions in and (or) attached to other phenocrysts (fig. 2D).

Figure 2. Photomicrographs of pumice from the June 15, 1991, eruption of Mount Pinatubo. *A*, Cummingtonite (cm) -rimmed hornblende (hb) phenocryst in type 1 pumice with silica-poor compositional domain (dom). *B*, Apatite (ap) inclusion (with the highest SO₃ contents, table 1, ap *h) and titanomagnetite (mt) inclusion in hornblende (hb) phenocryst in type 1 pumice. *C*, Coexisting titanomagnetite (mt) and ferrian ilmenite (il) microphenocrysts in type 1 pumice. *D*, Biotite (bt) microphenocryst attached to hornblende (hb) phenocryst in type 1 pumice.

TEMPERATURE, OXYGEN FUGACITY, AND SULFUR SPECIES OF MOUNT PINATUBO DACITIC MAGMAS

The presence of primary, magmatic anhydrite indicates that sulfur existed dominantly as oxidized species under highly oxidized conditions at least since the phenocrystic minerals crystallized. This is corroborated by unusually high SO_3 contents (up to 0.78 weight percent) of apatite microphenocrysts and apatite inclusions in phenocrystic minerals such as hornblende (table 1; fig. 2*B*). In type 1 pumice, the sulfur content in apatite is generally higher in the inclusions in other phenocrysts than in the discrete microphenocrysts (table 1). No apatite inclusion was analyzed in type 2 pumice due to scarcity of phenocrysts. Apparently, the activity of sulfate in the magma decreased during phenocryst formation prior to eruption, probably due to degassing of SO_2.

Considering the miscibility gap for the ilmenite-hematite solid-solution series, unusually high hematite contents of ferrian ilmenite (about 43 mole percent; table 1) indicate a primary magmatic origin at a temperature higher than about 800°C (Lindsley, 1976). Extrapolation of the two-oxide method (Buddington and Lindsley, 1964; Spencer and Lindsley, 1981; Andersen and Lindsley, 1988) by utilizing ferrian ilmenite and titanium-poor titanomagnetite compositions gives a temperature of about 800°C and an oxygen fugacity of about 10^{-11} bar (Andersen and Lindsley, 1988), close to the manganosite-hausmanite buffer. The fact that discrete microphenocrystic iron-titanium oxides are identical in composition to inclusions in hornblende and plagioclase (fig. 2*C*, table 1) indicates that the high oxygen fugacity existed even before crystallization of those phenocrystic minerals.

Assuming a temperature of about 800°C extrapolated from the two-oxide method, the atomic ratio of Mg/ (Mg+Fe) (abbreviated as X_{Mg}) of microphenocrystic biotite

Table 1. Microprobe analyses of phenocrystic minerals of Pinatubo pumice of June 15, 1991.

[All compositions are mean values, except for apatite inclusions in phenocrystic hornblende and plagioclase in type 1 pumice (indicated by *h and *p, respectively) which have high SO_3 contents. Numbers in parentheses after mean values are standard deviations; *dom* indicates a compositional domain that was found within a phenocryst. Dashes indicate that the oxide was below or around detection limit. Abbreviations: ol, olivine; pl, plagioclase; hb, hornblende; cm, cummingtonite; opx, orthopyroxene (hypersthene); bt, biotite; ap, apatite; mt, titanomagnetite; il, ferrian ilmenite. X_{Mg} denotes the atomic Mg/ (Mg + Fe) of mafic silicates. The Fe_2O_3* contents of magnetite and ferrian ilmenite were calculated assuming stoichiometry. The mole fractions of ulvospinel of titanomagnetite ($X_{usp}(mt)$) in type 1 pumice and type 2 pumice are 0.121±0.002 and 0.120±0.002, respectively, while the mole fractions of hematite in ferrian ilmenite ($X_{hm}(il)$) in type 1 and type 2 pumice are 0.434±0.015 and 0.428±0.012, respectively. Microprobe analyses were done with the JEOL 733 MKII of the Geological Institute, University of Tokyo, using 15keV, 12nA, and a beam diameter of 5 μm]

	Type 1 (phenocryst-rich pumices)						
No. of points analyzed	ol	pl	pl dom	hb	hb dom	hb tremolitic rim	cm
	9	135	13	104	6	2	11
SiO_2	39.80(0.09)	59.68(1.04)	55.26(1.01)	48.55(1.07)	44.51(.28)	52.04(.60)	55.04(.60)
TiO_2	–	–	–	.89(.18)	1.67(.24)	.48(.12)	.20(.06)
Al_2O_3	–	25.12(.64)	28.17(.65)	7.46(.81)	11.01(.49)	4.79(.56)	1.64(.43)
Cr_2O_3	–	–	–	–	–	–	–
V_2O_3	–	–	–	–	–	–	–
FeO	17.04(.54)	.17(.04)	.18(.04)	12.21(.65)	12.12(1.53)	9.29(.17)	15.17(.94)
MnO	.40(.08)	–	–	.51(.06)	.37(.13)	.41(.01)	.94(.08)
MgO	43.22(.16)	–	–	15.19(.72)	13.82(1.06)	18.15(.21)	21.17(.44)
NiO	.14(.03)	–	–	–	–	–	–
CaO	.09(.03)	7.33(.70)	10.76(.80)	10.41(.32)	10.73(.28)	10.77(.45)	1.66(.47)
Na_2O	–	6.61(.41)	4.93(.41)	1.22(.14)	1.78(.11)	.78(.11)	.25(.07)
K_2O	–	.28(.04)	.15(.02)	.25(.04)	.40(.07)	.15(.02)	.02(.01)
P_2O_5							
Fe_2O_3*							
SO_3	–	–	–	–	–	–	–
Cl	–	–	–	.04(.01)	.05(.02)	.05(.01)	.02(.01)
F	–	–	–	.08(.09)	.13(.08)	.08(.08)	.08(.04)
Total	100.69	99.19	99.45	96.80	96.56	97.57	96.16
Oxygen atoms	4	8	8	23	23	23	23
Si	1.002(.010)	2.674(.038)	2.491(.041)	7.087(.114)	6.573(.064)	7.467(.103)	7.895(.055)
Ti	–	–	–	.098(.021)	.186(.028)	.051(.014)	.021(.006)
Al	–	1.327(.038)	1.501(.039)	1.283(.146)	1.914(.072)	.801(.093)	.277(.074)
Cr	–	–	–	–	–	–	–
V	–	–	–	–	–	–	–
Fe^{2+}	.359(.010)	.006(.001)	.007(.002)	1.491(.084)	1.499(.198)	1.103(.022)	1.818(.099)
Mn	.009(.002)	–	–	.063(.008)	.046(.016)	.049(.001)	.114(.009)
Mg	1.622(.006)	–	–	3.303(.134)	3.040(.209)	3.838(.051)	4.527(.066)
Ni	.003(.001)	–	–				
Ca	.002(.001)	.352(.034)	.521(.042)	1.628(.051)	1.700(.038)	1.637(.065)	.255(.075)
Na	–	.574(.034)	.434(.035)	.346(.042)	.510(.031)	.211(.030)	.070(.019)
K	–	.016(.002)	.009(.001)	.046(.009)	.075(.012)	.027(.004)	.004(.002)
P	–	–	–	–	–	–	–
Fe^{3+}							
S	–	–	–				–
Cl	–	–	–	.011(.003)	.012(.006)	.011(.002)	.004(.002)
F	–	–	–	.036(.044)	.058(.039)	.038(.038)	.023(.019)
X_{Mg}	.819(.004)			.689(.014)	.670(.044)	.777(.001)	.714(.010)
X_{usp} (mt)							
X_{mt} (mt)							
X_{il} (il)							
X_{hm} (il)							

Table 1. Microprobe analyses of phenocrystic minerals of Pinatubo pumice of June 15, 1991—Continued.

	Type 1 (phenocryst-rich pumices)						
No. of points analyzed	bt 9	ap inclusion 9	ap discrete 17	ap *h	ap *p	mt 23	il 36
SiO$_2$	38.54(.27)	0.23(.09)	0.13(.04)	0.37	0.23	0.13(.10)	.07(.08)
TiO$_2$	3.42(.06)	–	–	–	–	4.27(.17)	28.80(.88)
Al$_2$O$_3$	14.48(.10)	–	–	–	–	1.88(.11)	.35(.02)
Cr$_2$O$_3$	–	–	–	–	–	.20(.04)	.06(.03)
V$_2$O$_3$	–	–	–	–	–	.50(.04)	.45(.04)
FeO	12.61(.23)	.48(.11)	.39(.18)	.48	.40	33.16(.36)	23.81(.63)
MnO	.11(.04)	.20(.04)	.19(.03)	.14	.23	.45(.08)	.24(.04)
MgO	16.26(.13)	.16(.07)	.14(.01)	.14	.13	1.17(.05)	1.09(.08)
NiO	–	–	–	–	–	–	–
CaO	.12(.08)	54.51(.67)	54.85(.59)	53.33	54.65	–	–
Na$_2$O	.75(.05)	.15(.06)	.08(.01)	.29	.19	–	–
K$_2$O	7.48(.15)	–	–	–	–	–	–
P$_2$O$_5$	–	40.29(.41)	40.78(.31)	39.65	40.12		
Fe$_2$O$_3$*						57.86(.55)	45.20(1.70)
SO$_3$	–	.39(.17)	.13(.03)	.78	.43	–	–
Cl	.10(.01)	1.24(.08)	1.16(.14)	1.19	1.38	–	–
F	.22(.12)	1.68(.50)	1.79(.23)	1.74	1.92	–	–
Total	94.01	99.33	99.64	98.47	99.75	99.62	100.06
Oxygen atoms	22	12	12	12	12	4	3
Si	5.721(.032)	.019(.007)	.011(.003)	.031	.019	.005(.004)	.001(.002)
Ti	.382(.007)	–	–	–	–	.121(.004)	.553(.016)
Al	2.534(.020)	–	–	–	–	.084(.005)	.010(.000)
Cr	–	–	–	–	–	.006(.001)	.000(.000)
V	–	–	–	–	–	.015(.001)	.010(.000)
Fe^{2+}	1.565(.027)	.034(.008)	.027(.013)	.046	.028	1.046(.006)	.509(.013)
Mn	.014(.006)	.014(.003)	.014(.002)	.020	.016	.014(.003)	.005(.000)
Mg	3.597(.026)	.021(.009)	.018(.002)	.014	.016	.065(.003)	.041(.003)
Ni	–	–	–	–	–	–	–
Ca	.019(.013)	4.923(.044)	4.945(.018)	4.861	4.961	–	–
Na	.215(.015)	.025(.011)	.013(.002)	.047	.032	–	–
K	1.416(.029)	–	–	–	–	–	–
P	–	2.873(.033)	2.903(.014)	2.854	2.876	–	–
Fe^{3+}						1.643(.012)	.869(.031)
S	–	.025(.011)	.008(.002)	.049	.028	–	–
Cl	.024(.002)	.177(.012)	.166(.022)	.172	.198	–	–
F	.101(.058)	.449(.136)	.477(.065)	.471	.514	–	–
X$_{Mg}$	.697(.003)						
X$_{usp}$ (mt)						.121(.003)	
X$_{mt}$ (mt)						.791(.003)	
X$_{il}$ (il)							.508(.015)
X$_{hm}$ (il)							.435(.015)

Table 1. Microprobe analyses of phenocrystic minerals of Pinatubo pumice of June 15, 1991—Continued.

Type 2 (phenocryst-poor pumices)								
No. of points analyzed	pl	pl dom	hb	hb dom	opx	bt	ap	mt	il
	47	4	50	6	4	4	9	22	22
SiO₂	59.47(.94)	54.49(1.71)	48.30(1.07)	42.74(.31)	56.04(.36)	38.04(.11)	0.09(.01)	0.04(.02)	0.01(.02)
TiO₂	–	–	.90(.17)	2.06(.12)	.24(.03)	3.38(.00)	–	4.27(.10)	29.14(.63)
Al₂O₃	25.39(.68)	28.75(1.13)	7.65(.87)	12.70(.43)	2.19(.25)	14.25(.09)	–	1.84(.08)	.33(.03)
Cr₂O₃	–	–	–	–	–	–	–	.20(.03)	.07(.04)
V₂O₃	–	–	–	–	–	–	–	.48(.06)	.44(.04)
FeO	.18(.03)	.17(.04)	12.41(.58)	9.88(.83)	16.07(.08)	12.64(.55)	.28(.02)	33.09(.53)	24.02(.43)
MnO	–	–	.54(.07)	.15(.03)	.98(.08)	.13(.02)	.17(.01)	.49(.05)	.26(.05)
MgO	–	–	15.16(.59)	15.25(.48)	21.15(.05)	15.78(.03)	.14(.00)	1.31(.03)	1.12(.08)
NiO	–	–	–	–	–	–	–	–	–
CaO	7.41(.80)	11.12(1.44)	10.52(.28)	11.45(.21)	2.09(.01)	.07(.02)	54.60(.21)	–	–
Na₂O	6.55(.40)	4.68(.78)	1.24(.19)	2.38(.06)	.26(.01)	.66(.02)	.07(.14)	–	–
K₂O	.28(.05)	.14(.02)	.26(.05)	.42(.02)	.23(.01)	7.73(.04)	–	–	–
P₂O₅	–	–	–	–	–	–	40.43(.14)	–	–
Fe₂O₃*								58.92(.69)	44.44(1.46)
SO₃	–	–	–	–	–	–	.11(.01)	–	–
Cl	–	–	.04(.01)	.01(.01)	–	.09(.01)	1.32(.03)	–	–
F	–	–	.06(.07)	.04(.04)	–	.20(.08)	1.77(.19)	–	–
Total	99.27	99.38	97.09	97.10	99.25	93.01	97.75	100.64	99.89
Oxygen atoms	8	8	23	23	6	22	12	4	3
Si	2.664(.039)	2.464(.067)	7.040(.132)	6.251(.058)	2.041(.014)	5.731(.011)	.008(.001)	.001(.001)	.000(.000)
Ti	–	–	.098(.019)	.226(.013)	.007(.001)	.383(.000)	–	.120(.003)	.562(.012)
Al	1.341(.037)	1.533(.066)	1.315(.154)	2.189(.068)	.094(.011)	2.530(.015)	–	.081(.003)	.010(.000)
Cr	–	–	–	–	–	–	–	.006(.001)	.001(.001)
V	–	–	–	–	–	–	–	.015(.001)	.010(.001)
Fe²⁺	.007(.001)	.007(.001)	1.513(.072)	1.208(.104)	.490(.002)	1.592(.072)	.020(.003)	1.033(.018)	.514(.009)
Mn	–	–	.067(.009)	.018(.004)	.030(.002)	.017(.003)	.012(.002)	.015(.001)	.006(.001)
Mg	–	–	3.294(.113)	3.324(.093)	1.148(.003)	3.543(.005)	.017(.002)	.073(.016)	.043(.003)
Ni	–	–	–	–	–	–	–	–	–
Ca	.356(.039)	.539(.072)	1.643(.047)	1.795(.029)	.081(.007)	.011(.004)	4.970(.041)	–	–
Na	.569(.034)	.410(.067)	.351(.055)	.674(.015)	.018(.000)	.194(.007)	.012(.004)	–	–
K	.016(.003)	.008(.001)	.048(.010)	.077(.003)	.011(.000)	1.486(.008)	–	–	–
P	–	–	–	–	–	–	2.906(.012)	–	–
Fe³⁺								1.656(.009)	.856(.025)
S	–	–	–	–	–	–	.008(.002)	–	–
Cl	–	–	.011(.003)	.003(.003)	–	.022(.004)	.190(.009)	–	–
F	–	–	.031(.035)	.021(.022)	–	.097(.041)	.476(.091)	–	–
X_Mg			.685(.065)	.734(.014)	.701(.002)	.690(.010)			
X_usp (mt)								.120(.002)	
X_mt (mt)								.787(.002)	
X_il (il)									.514(.012)
X_hm (il)									.428(.012)

(fig. 2D) (about 0.7), high for a dacite, suggests a high oxygen fugacity (Wones and Eugster, 1965). However, an assumption of the presence of K-feldspar can result in an overestimate of the oxygen fugacity (Imai and others, 1993). The X_{Mg} values of calcium-poor mafic silicates (biotite, cummingtonite, and hypersthene) are almost identical (around 0.7), except for that of olivine ($X_{Mg} = 0.82$) (table 1). The high X_{Mg} of olivine and existence of rimming hornblende suggest that olivine was not in equilibrium with dacitic melt and thus was probably xenocrystic.

Geschwind and Rutherford (1992) experimentally demonstrated that cummingtonite is stable only below 790–800°C. Because cummingtonite occurs only as a rim on hornblende phenocrysts in Pinatubo pumice, the temperature at which cummingtonite formed might be lower than those temperatures at which the other phenocrysts formed.

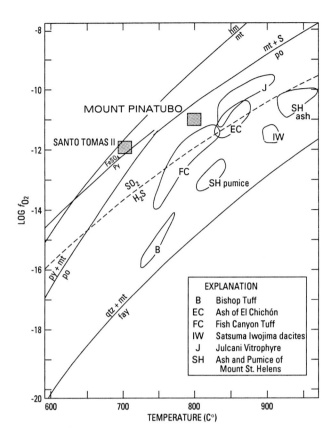

Figure 3. Estimated temperature and oxygen fugacity (f_{O_2}) for Pinatubo pumices of June 15, 1991, and some other volcanoes in the world. After Whitney (1984) and Ueda and Itaya (1981). Abbreviations: fay, fayalite; hm, hematite; mt, magnetite; po, pyrrhotite; py, pyrite; qtz, quartz.

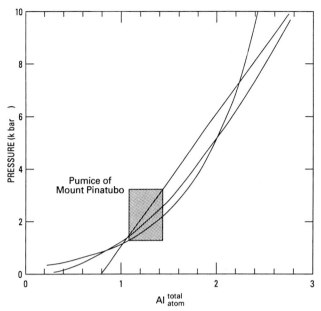

Figure 4. Hornblende geobarometry (Hammarstrom and Zen, 1986) applied to Mount Pinatubo pumice. Regression lines suggested by Hammarstrom and Zen (1986) are indicated for reference. The total atomic aluminum contents per half unit cell of amphibole formula of phenocrystic hornblende in type 1 and type 2 pumice are 1.28 ± 0.15 and 1.31 ± 0.15, respectively.

Estimated temperature and oxygen fugacity for dacitic magma erupted on June 15 is plotted along with other previously published estimates for andesites and rhyolites (Whitney, 1984) (fig. 3). Dacites and andesites from the other areas contain microphenocrystic pyrrhotite and plot near or below the $SO_2 = H_2S$ boundary curve, whereas Mount Pinatubo dacite is more highly oxidized, with sulfur existing dominantly as oxidized species such as SO_2. This high oxygen fugacity, because it has prohibited sulfide fractionation (Ueda and Sakai, 1984) and because sulfate solubility is high in oxidized silicate melt (Carroll and Rutherford, 1987, 1988; Luhr, 1990), is the possible cause of extraordinarily high sulfur content.

DEPTH AND WATER FUGACITY OF MOUNT PINATUBO DACITIC MAGMAS

Phenocrystic minerals were formed at pressures of around 2 kbar, on the basis of igneous hornblende geobarometry (Hammarstrom and Zen, 1986) (fig. 4). These depth estimates seem consistent with an earthquake-poor, low-velocity zone (a magma chamber?) between 6 and 11 km in depth (Mori, Eberhardt-Phillips, and Harlow, this volume). Silica-poor hornblende is found as compositional domains in ordinary hornblende phenocrysts in both pumice types (table 1) (fig. 2A), and anorthite-rich plagioclase domains are found in plagioclase phenocrysts (table 1). These compositional domains might be remnants of the early history of magmatic differentiation, suggestive of the probable existence of less differentiated magma at depth.

Considering the solubility of water in dacitic magma with respect to pressure and the igneous-hornblende stability field with respect to magmatic water content, the Mount Pinatubo dacitic magma was apparently close to or at water saturation prior to eruption (Naney and Swanson, 1980). In addition to the occurrence of cummingtonite, which indicates high magmatic water fugacity, near or at water saturation (Wood and Carmichael, 1973; Geschwind and Rutherford, 1992), studies on glass inclusions and matrix glasses of the Mount Pinatubo pumice (Westrich and Gerlach, 1991; Gerlach and others, this volume) suggest magmatic vapor saturation prior to eruption.

IMPLICATION FOR POSSIBLE METALLOGENY OF PORPHYRY COPPER MINERALIZATION

A chain of late Miocene to Pliocene porphyry copper-gold deposits, including the Lepanto-FSE, the Santo Tomas II, and the Dizon deposits, delineates a metallogenic province extending from northern Luzon to west-central Luzon (fig. 1). The Dizon deposit of 2.7 Ma (Malihan, 1987) is located only 19 km south of Mount Pinatubo, and several prospects of porphyry-type deposits are known in the vicinity. Shallow-depth intermediate to slicic intrusives that are genetically related to porphyry copper deposits in the western Luzon arc are usually corroded quartz-phenocryst-bearing hornblende-quartz-diorite porphyry to hornblende-andesite porphyry (Balce, 1979; Imai, unpub. data, 1991). Hornblende-andesite porphyry at the Santo Tomas II deposit in northern Luzon (Imai, unpub. data, 1993) is characterized by tremolite-rimmed hornblende phenocrysts which, at the low pressure (< 1 kbar) that is implied by fluid-inclusion compositions and amphibole geobarometry, suggests water saturation in the magma at the time of its emplacement. The porphyry of Santo Tomas II also has coexisting magnetite and titanohematite (X_{ilm} = 42 mole percent) in high temperature sulfide-rare quartz stockworks (T>700°C on the basis of Lindsley, 1976), and high X_{Mg} of tremolite and biotite. Again, high oxygen fugacity, at the hematite-magnetite buffer, is suggested. Lastly, anhydrite is a common gangue and alteration constituent.

These characteristics of the porphyry copper-related hydrous magmatic system of northern Luzon are also common in other porphyry copper systems, including several in Papua New Guinea and vicinity (Mason, 1978; Chivas, 1981). Petrologic similarity between Mount Pinatubo dacitic magma and such intrusives is notable.

Oblique subduction is thought to be favorable to porphyry copper metallogeny in the Chilean Andean cordillera (see Davidson and Mpodozis, 1991). The Philippine archipelago is regarded as the product of the late Cenozoic oblique convergence between the Philippine Sea plate and the Eurasian margin (for example, Rangin, 1991). With a tectonic stress field that produces shallow reservoirs of hydrous magma, the Western Luzon arc of both the present and ancient (late Miocene to Pliocene) times has been favorable for porphyry copper mineralization.

Because sulfur in the dacitic magma of June 15 existed dominantly as oxidized species (see above), the sulfur isotopic composition of bulk hydrochloric-acid leachate from pumice (+7.8 to +9 per mil) (table 2) can be regarded as the sulfur isotopic value of the magma within ±2 per mil (see McKibben and others, this volume, for detailed sulfur isotopic data). That sulfur isotopic ratios of Pinatubo dacite are similar to those of the aforementioned porphyry copper-gold deposits of northern Luzon (+9 to +10 per mil,

estimated for copper-iron sulfide-bearing anhydrite veins at the Santo Tomas II and Lepanto-FSE deposit, Imai, unpub. data, 1991) indicates the same origin of sulfur. The sulfur isotopic compositions of these two deposits are almost identical; copper-iron sulfides with sulfur isotopes of +3 per mil (Lepanto-FSE) to +4 per mil (Santo Tomas II) coexist with gangue anhydrite with +13.5 per mil at both deposits. In addition, the sulfide minerals from the Dizon deposit also yield similar sulfur isotopic compositions (+3 to +5 per mil), but no hypogene anhydrite has been analyzed. These sulfur isotopes, heavier than those of midoceanic ridge basalts that derived directly from the mantle (Moore and Fabbi, 1971), are likewise a common feature of Quaternary volcanic rocks and high-temperature volcanic gases from subduction-related arc settings (Ueda and Sakai, 1984; Rye and others, 1984). Furthermore, magnetite-series granitoids, formed in an ancient island-arc and (or) continental-margin setting in the present Japanese islands, are also enriched in heavy sulfur (Sasaki and Ishihara, 1979). Enrichment of heavy sulfur in hydrous magmas in an arc setting is thus a common feature of both plutonic and volcanic rocks and suggests a heavy sulfur source (1) beneath the arc's mantle wedge, such as sea-water-derived sulfur in the subducted slab (Sasaki and Ishihara, 1979), and (or) (2) possible assimilation from hydrothermally altered sulfate-bearing country rocks.

CONCLUSIONS

The dacitic magma erupted by Mount Pinatubo on June 15, 1991, was highly oxidized, close to the manganosite-hausmanite buffer. This condition was attained prior to formation of the phenocrystic minerals. Sulfur has existed dominantly as oxidized species in the magma since the crystallization of the phenocrysts.

The bulk sulfur isotopic composition of the Mount Pinatubo magma is almost identical with the isotopic composition of the late Cenozoic porphyry copper-gold deposits in northern Luzon. Cummingtonite rims on phenocrystic hornblende and high X_{Mg} of biotite and hornblende are also common to the Pinatubo magma and to the intrusive rocks associated with northern Luzon porphyry copper deposits. We infer that in both cases, magma was saturated with water at the time of emplacement at shallow crustal levels.

Table 2. Whole rock sulfur isotopic compositions of Mount Pinatubo pumices of June 15, 1991.

[Sulfur isotopic ratios were determined for hydrochloric acid leachate from pumice]

	δS (per mil)	average ±1 σ
Type 1	+7.8, +7.9, +8.2, +8.7	+8.2 ±0.4
Type 2	+8.7, +9.0	+8.9 ±0.2

ACKNOWLEDGMENTS

We thank R.S. Punongbayan, E.G. Domingo, G.P. Yumul, Jr., J.W. Hedenquist, S. Aramaki, T. Koyaguchi, H. Shimazaki, I. Kushiro, N. Shikazono, and M. Toriumi for their encouragement and discussions. S. Yui helped in ZAF correction of Fe-Ti oxide microprobe analyses as part of a Grant-in-Aid for Scientific Research (No. 02302032) to him. Constructive review comments by T.M. Gerlach, J. Luhr, and C.G. Newhall are appreciated.

REFERENCES CITED

Andersen, D.J., and Lindsley, D.H., 1988, Internally consistent solution models for Fe-Mg-Mn-Ti oxides: Fe-Ti oxides: American Mineralogist, v. 73, p. 714–726.

Arculus, R.J., Johnson, R.W., Chappell, B.W., McKee, C.O., and Sakai, H., 1983, Ophiolite-contaminated andesites, trachybasalts, and cognate inclusions of Mount Lamington, Papua New Guinea: Anhydrite-amphibole-bearing lavas and the 1951 cumulodome: Journal of Volcanology and Geothermal Research, v. 18, p. 215–247.

Balce, G.R., 1979, Geology and ore genesis of the porphyry copper deposits in Baguio district, Luzon island, Philippines: Journal of the Geological Society of the Philippines, v. 23, p. 1–43.

Bernard, A., Demaiffe, D., Mattielli, N., and Punongbayan, R.S., 1991, Anhydrite-bearing pumices from Mount Pinatubo: Further evidence for the existence of sulphur-rich silicic magmas: Nature, v. 354, p. 139–140.

Buddington, A.F., and Lindsley, D.H., 1964, Iron titanium oxide minerals and synthetic equivalents: Journal of Petrology, v. 5, p. 310–357.

Carroll, M.R., and Rutherford, M.J., 1987, The stability of igneous anhydrite: Experimental results and implications for sulfur behavior in the 1982 El Chichón trachyandesite and other evolved magmas: Journal of Petrology, v. 28, p. 781–801.

———1988, Sulfur speciation in hydrous experimental glasses of varying oxidation state: Results from measured wavelength shifts of sulfur X-rays: American Mineralogist, v. 73, p. 845–849.

Chivas, A.R., 1981, Geochemical evidence for magmatic fluids in porphyry copper mineralization: I. Mafic silicates from the Koloula igneous complex: Contributions to Mineralogy and Petrology, v. 78, p. 389–403.

Davidson, J., and Mpodozis, C., 1991, Regional geologic setting of epithermal gold deposits, Chile: Economic Geology, v. 86, p. 1174–1186.

Gerlach, T.M., Westrich, H.R., and Symonds, R.B., this volume, Preeruption vapor in magma of the climactic Mount Pinatubo eruption: Source of the giant stratospheric sulfur dioxide cloud.

Geschwind, C-H., and Rutherford, M.J., 1992, Cummingtonite and the evolution of the Mount St. Helens (Washington) magma system: An experimental study: Geology, v. 20, p. 1011–1014.

Hammarstrom, J.M., and Zen, E., 1986, Aluminum in hornblende: An empirical igneous geobarometer: American Mineralogist, v. 71, p. 1297–1313.

Imai, A., Listanco, E.L., and Fujii, T., 1993, Petrologic and sulfur isotopic significance of highly oxidized and sulfur-rich magma of Mt. Pinatubo, Philippines: Geology, v. 21, p. 699–702.

Koyaguchi, T., this volume, Volume estimation of tephra-fall deposits from the June 15, 1991, eruption of Mount Pinatubo by theoretical and geological methods.

Koyaguchi, T., and Tokuno, M., 1993, Origin of the giant eruption cloud of Pinatubo, June 15, 1991: Journal of Volcanology and Geothermal Research, v. 55, p. 85–96.

Lindsley, D.H., 1976, Experimental studies of oxide minerals, in Rumble, D., III, ed., Oxide minerals: Washington, D.C., Mineralogical Society of America, p. L61–L88.

Listanco, E.L., 1991, Some characteristics of tephra fall from the 1991 eruption of Pinatubo volcano, Philippines [abs.]: Abstracts, GEOCON '91, Annual Meeting of the Geological Society of the Philippines, p. 10.

Luhr, J.F., 1990, Experimental phase relations of water- and sulfur-saturated arc magmas and the 1982 eruptions of El Chichón volcano: Journal of Petrology, v. 31, p. 1071–1114.

Luhr, J.F., Carmichael, I.S.E., and Varekamp, J.C., 1984, The 1982 eruptions of El Chichón volcano, Chiapas, Mexico: Mineralogy and petrology of the anhydrite-bearing pumices: Journal of Volcanology and Geothermal Research, v. 23, p. 69–108.

Malihan, T.D., 1987, The gold-rich Dizon porphyry copper mine in the western central Luzon Island, Philippines: Its geology and tectonic setting: Proceedings of the Pacific Rim Congress '87, Australian Institutes of Mining and Metallurgy, Parkville, Victoria, Australia, p. 303–307.

Mason, D.R., 1978, Compositional variation in ferromagnesian minerals from porphyry copper-generating and barren intrusions of the Western Highlands, Papua New Guinea: Economic Geology, v. 73, p. 878–890.

McKibben, M.A., Eldridge, C.S., and Reyes, A.G., this volume, Sulfur isotopic systematics of the June 1991 Mount Pinatubo eruptions: A SHRIMP ion microprobe study.

Moore, J.G., and Fabbi, B.P., 1971, An estimate of the juvenile sulfur content of basalt: Contributions to Mineralogy and Petrology, v. 33, p. 118–127.

Mori, J., Eberhart-Phillips, D., and Harlow, D.H., this volume, Three-dimensional velocity structure at Mount Pinatubo, Philippines: Resolving magma bodies and earthquakes hypocenters.

Naney, M.T., and Swanson, S.E., 1980, The effects of Fe and Mg on crystallization in granitic systems: American Mineralogist, v. 65, p. 639-653.

Pallister, J.S., Hoblitt, R.P., and Reyes, A.G., 1992, A basalt trigger for the 1991 eruptions of Pinatubo volcano?: Nature, v. 356, p. 426–428.

Philippine Institute of Volcanology and Seismology, 1991, Volcanoes of the Philippines: Quezon City, Philippines, PHIVOLCS Press, 41 p.

Rangin, C., 1991, The Philippine Mobile Belt: A complex plate boundary: Journal of Southeast Asian Earth Sciences, v. 6, p. 209–220.

Rye, R.O., Luhr, J.F., and Wasserman, M.D., 1984, Sulfur and oxygen isotope systematics of the 1982 eruptions of El Chichón volcano, Chiapas, Mexico: Journal of Volcanology and Geothermal Research, v. 23, p. 109–123.

Sasaki, A., and Ishihara, S., 1979, Sulfur isotopic composition of magnetite-series and ilmenite-series granitoids in Japan: Contributions to Mineralogy and Petrology, v. 68, p. 107–115.

Spencer, K.J., and Lindsley, D.H., 1981, A solution model for coexisting iron-titanium oxides: American Mineralogist, v. 66, p. 1189–1201.

Ueda, A., and Itaya, T., 1981, Microphenocrystic pyrrhotite from dacitic rocks of Satsuma-Iwojima, southwest Kyushu, Japan, and the solubility of sulfur in dacitic magma: Contributions to Mineralogy and Petrology, v. 78, p. 21–26.

Ueda, A., and Sakai, H., 1984, Sulfur isotope study of Quaternary volcanic rocks from the Japanese Island Arc: Geochimica et Cosmochimica Acta, v. 48, p. 1837–1848.

Westrich, H.R., and Gerlach, T.M., 1992, Magmatic gas source for the stratospheric SO_2 cloud from the June 15, 1991, eruption of Mount Pinatubo: Geology, v. 20, p. 867–870.

Whitney, J.A., 1984, Volatiles in magmatic systems, *in* Henley, R.W., and others, eds., Fluid-mineral equilibria in hydrothermal systems: Reviews in Economic Geology, v. 1, p. 155–175.

Wones, D.R., and Eugster, H.P., 1965, Stability of biotite: Experiment, theory, and application: American Mineralogist, v. 50, p. 1228–1272.

Wood, B.J., and Carmichael, I.S.E., 1973, P_{total}, P_{H_2O}, and the occurrence of cummingtonite in volcanic rocks: Contributions to Mineralogy and Petrology, v. 40, p. 149–158.

Relative Timing of Fluid and Anhydrite Saturation: Another Consideration in the Sulfur Budget of the Mount Pinatubo Eruption

Jill Dill Pasteris,[1] Brigitte Wopenka,[1] Alian Wang,[1] and Teresa N. Harris[1]

ABSTRACT

Raman microsampling spectroscopy and petrographic analysis were done on four major types of inclusions in quartz and plagioclase phenocrysts in phenocryst-rich dacite from the June 15, 1991, eruption of Mount Pinatubo, Philippines: (1) glass inclusions, typically containing one or more bubbles, (2) vapor-dominated inclusions, (3) aqueous inclusions consisting of liquid and vapor ± solids, and (4) solid inclusions. In quartz-hosted glass inclusions with bubbles, Raman analysis revealed only CO_2 in some enclosed bubbles, but not SO_2 or H_2S (below detection limit of 1 bar partial pressure); dissolved H_2O, but not CO_2 (less than the detection limit of 900 parts per million) was detected in the glass phase. The observed bubble:glass ratios in such inclusions, the Raman-inferred CO_2 pressures in some bubbles, the documentation by Raman spectroscopy of gas leakage from bubbles over time, and the petrographic documentation of physical breaching of many glass inclusions suggest that these inclusions may represent the two-phase entrapment at depth of coexisting melt and supercritical fluid but that the fluid was not retained. No volatile species were detected in the vapor-dominated inclusions, so very low gas pressures are indicated. Raman spectroscopic analysis revealed anhydrite among the solid phases in fracture-lining aqueous inclusions that are found only in quartz, implying an elevated sulfate content in late-stage aqueous fluids. Solid inclusions of apatite, zircon, and amphibole were identified spectroscopically. Of note is the lack of anhydrite inclusions in our samples. This observation, coupled with reports of high SO_3 in apatite inclusions, lack of sulfur isotopic zonation in anhydrite phenocrysts, and relatively low bulk-sulfur contents in Mount Pinatubo dacites (only 25% of those in El Chichón eruptive rocks), suggests that the Mount Pinatubo dacitic magma may have reached saturation with respect to a CO_2-SO_2-H_2O fluid before anhydrite saturation occurred. Since sulfur strongly partitions into a fluid over a melt phase, early saturation of the magma with a supercritical aqueous fluid could have effectively stripped sulfur from the melt. In this case, neither the bulk-sulfur content nor the abundance of anhydrite crystals in eruptive rocks could be used as an indicator of the original sulfur content of the system. If fluid saturation indeed occurred early in Mount Pinatubo's magmatic history, then the $(H_2O+CO_2)/SO_2$ ratio in the melt may have controlled the timing of the removal of sulfur from the melt via fluid exsolution and the proportion of the available sulfur that ultimately reached the atmosphere.

INTRODUCTION

Mount Pinatubo is one of the active members of the Luzon volcanic chain, which is part of the Bataan frontal arc. It lies about 120 km east of the Manila trench, above the eastward subducting Eurasian plate. Mount Pinatubo is also within the southern part of the areally extensive Zambales ophiolite. The volcano's eruptions in June 1991 ended a 500-year hiatus in activity. The climactic June 15th eruption produced an estimated 4 to 5 km³ dense-rock equivalent of dominantly dacitic pumice and tephra (W.E. Scott and others, this volume) and released about 20 Mt SO_2 into the atmosphere (Bluth and others, 1992). One of the working hypotheses for explaining the large amount of sulfur released into the atmosphere is that a separate, sulfur-bearing volatile phase coexisted with the dacitic magma at depth prior to eruption (see Westrich and Gerlach, 1992; Gerlach and others, this volume). Various sources of the sulfur have been considered, including an underlying basalt reservoir and a hydrothermal system (Gerlach and others, this volume; Pallister and others, this volume; Rutherford and Devine, this volume).

The goals of our study of fluid and solid inclusions in quartz and plagioclase phenocrysts are to provide more data by which to evaluate the hypothesis of a coexisting fluid phase during eruption and to determine what kinds of information can be obtained on the available samples by Raman microsampling spectroscopy. Raman analysis of volatile-bearing and solid inclusions can help to track the movement of sulfur within the volcanic system and provide

[1] Department of Earth and Planetary Sciences, Washington University, Campus Box 1169, St. Louis, MO 63130–4899.

information on the state of gas saturation of the magma during the time of phenocryst precipitation. Combined Raman, petrographic, and petrologic data are used to infer the mechanism by which gas saturation was reached.

ANALYTICAL TECHNIQUE

Raman spectroscopy is an optical technique that monitors the inelastic scattering of monochromatic visible light as it interacts with covalently bonded molecules in solids, liquids, or gases. The position of the peaks and the number of peaks in the Raman spectrum are controlled by the energy and symmetry, respectively, of the molecular vibrations. In Raman microsampling spectroscopy, the sample is both viewed and excited in a research-grade optical microscope. The same high-magnification, high-numerical-aperture objective is used to optically image the sample (via a video camera and a TV monitor), focus the laser (to about 1 μm spot diameter), and transmit the scattered radiation to a monochromator. Nondestructive analysis is performed at ambient atmosphere and temperature and can be done on materials as small as 1 μm in diameter on or below the optically transparent surface of a sample. Geological samples require minimal or no sample preparation in that (1) unmounted polished wafers, polished thin sections, loose grains on a glass slide, and irregular masses of material all can be analyzed and (2) no carbon coating or other sample preparation is required; unopened fluid inclusions can be analyzed nondestructively. The inelastically scattered Raman radiation is monitored by a photon detector and recorded in terms of intensity (number of photons per second) as a function of Stokes Raman shift (relative wavenumbers, Δcm^{-1}). The Raman shift is the difference in frequency between the exciting laser radiation and the Raman-scattered radiation.

Raman microsampling spectroscopy was done with either a single-channel Jobin-Yvon RAMANOR U–1000 or a multichannel Jobin-Yvon S–3000 laser Raman microprobe, both marketed by Instruments SA. In both of our instruments, the 514.5-nm green line of an argon-ion laser is focused into the sample by a modified research-grade microscope. The beam can be focused into an inclusion or portion of an inclusion, typically below the surface of the polished sample. The configuration of both of our instruments is such that 180° backscattered radiation is collected. The higher the numerical aperture (N.A.) of the objective, the better is the optical throughput. Given the fact that the Raman effect is an extremely weak phenomenon, it is desirable to use the objective with the highest numerical aperture possible. However, the size of an inclusion and its depth below the surface (necessitating a specific free-working distance) also dictate the choice of an objective. We used the following objectives from several manufacturers: 80×

Nachet (N.A. 0.90), 80× ultra-long-working-distance Olympus (N.A. 0.75), 100× Olympus (N.A. 0.95), and 160× Leitz (N.A. 0.95). The laser power at the surface of the sample usually was fixed at 15 mW, but lower power was used when the inclusion showed signs of laser heating (see below).

The U–1000 consists of a 1-m double monochromator with two plane holographic gratings (1,800 grooves/mm). The detector is a thermoelectrically cooled RCA C–31034–2 photomultiplier tube. Spectra are obtained in the dispersive scanning mode with variable step size and counting time. Typical analytical conditions for analysis of gases were 0.5 cm^{-1} stepping interval and 10 s dwell time per point. The volatiles that typically were analyzed for are CO_2, CO, N_2, H_2S, SO_2, CH_4, and H_2O. For analysis of solids, the stepping interval was increased to 1 cm^{-1}, and dwell times varied between 1 and 10 s, depending on the strength of the signal.

The S–3000 is a triple-monochromator system that consists of a 320-mm double-monochromator with two 600 grooves/mm gratings (acting as a two-stage foremonochromator in subtractive mode) and a 1-m third-stage monochromator with interchangeable plane holographic gratings. For the analyses performed in this study, we used a grating with 600 grooves/mm in the third-stage monochromator. The detector is a 1-in, 1,024-element, proximity-focused, intensified optical diode array that is cooled by a Peltier-effect thermoelectric photocathode cooler. To cover the total spectral range of interest (from ~100 Δcm^{-1} to ~4,000 Δcm^{-1}), data were acquired in four different spectral windows centered at 830, 1,600, 2,600, and 3,500 Δcm^{-1}. The acquisition time was chosen such that a satisfactory signal-to-noise ratio was achieved but such that the detector was not oversaturated in the wavenumber region of interest. We typically took 20 acquisitions at 10 s for each window.

The choice of which of these instruments to use is governed by the type of spectral information required and the amount of time available. As a scanning monochromator with gratings of high groove density, the U–1000 is the instrument of choice when higher spectral resolution and exact knowledge of band position are required. It is particularly useful when the Raman bands for gases are used to infer gas pressure and when exact peak positions for solids (such as zircons in this study) are required. Analyses on the U–1000 are time consuming, however, requiring about 10 min for each gas to be analyzed. The S–3000 is the instrument of choice for reconnaissance analysis of a wide range of gas species and rapid (1–2 min) identification of solid phases. A total of only about 25 min, including the time necessary for background measurements, is required for a full-spectrum analysis under the conditions listed in the preceding paragraph.

SAMPLES

The pumices that we studied (obtained from H.R. Westrich, Sandia National Laboratory, and T.M. Gerlach, U.S. Geological Survey) were collected from the surface of the pyroclastic-flow deposits in the Sacobia River valley, northeast of Mount Pinatubo. The pumices are of white, phenocryst-rich dacite, which comprises about 80 vol% of the June 15th tephra. Grain separates of quartz phenocrysts (provided by H.R. Westrich and T.M. Gerlach) were made into doubly polished thick wafers, which were cemented onto a glass plate with UV-curing epoxy, whereas the porous pumice samples were vacuum impregnated with blue epoxy before they were made into doubly polished thick rock wafers. The rock wafers were cemented onto glass plates by use of acetone-soluble superglue, which permitted them to be released as free-standing wafers. (Despite their epoxy impregnation, the rock wafers were still too friable to be analyzed by microthermometry on a gas-flow stage.) All samples were appropriate for both petrography and in-situ Raman microprobe analysis.

The typical phenocryst phases in these rocks, as reported in petrologic studies documenting a large number of samples, are plagioclase (An$_{34-66}$), hornblende, cummingtonite, biotite, quartz, magnetite, anhydrite, ilmenite, and apatite (Rutherford and Devine, this volume; Pallister and others, this volume). Petrography of our samples indicates that plagioclase phenocrysts are about 50 times more abundant than quartz and that the quartz grains typically show resorption features. We did not find any anhydrite phenocrysts in our samples, which could be due to the fact that water was used in some of the sample preparation.

Our Raman study concentrated on fluid and solid inclusions in quartz and plagioclase phenocrysts. These two mineral hosts were chosen for their optical clarity, for the abundance of plagioclase phenocrysts in pumice samples, and for the availability of grain separates of quartz phenocrysts. Inclusions were inventoried and analyzed in four thin sections and doubly polished wafers and in 35 quartz grains in grain mounts (table 1). Approximately 30 solid and 30 fluid inclusions were analyzed by Raman spectroscopy.

THE INCLUSIONS

Four major types of inclusions were recognized optically in the quartz, plagioclase, and hornblende phenocrysts: (1) glass inclusions, usually containing one or more bubbles, (2) vapor-dominated inclusions, (3) aqueous inclusions, typically consisting of a large bubble, a small liquid phase, and multiple solid phases, and (4) solid inclusions. Only in quartz were all four types found. Raman microprobe analysis was done on examples of each of these types of inclusions in quartz and plagioclase phenocrysts.

GLASS INCLUSIONS

The glass inclusions are by far the dominant type in both the quartz and plagioclase phenocrysts, as well as in the unanalyzed hornblende grains. In quartz, glass inclusions tend to be square with rounded corners, reflecting the symmetry of the ß-quartz phase in which they were trapped initially. In hornblende, the glass inclusions are irregularly to rectangularly rounded. In plagioclase, they are ovoid to rectangular. Inclusion diameters typically range from 5 to 60 μm, but a few inclusions exceed 100 μm. Most of the glass inclusions are colorless and contain at least one bubble. Much less common are glass inclusions that are light brown (± small bubbles) and inclusions whose glass phase appears to be flecked with minute dark particles (glass undergoing devitrification?). The number and size of the bubbles vary greatly among glass inclusions, as does the total volume percent of bubbles, even among inclusions within the same phenocryst grain (fig. 1A). Even in the smallest glass inclusions (<20 μm), there is a variable bubble:glass ratio.

There are three major differences in the occurrence of glass inclusions hosted by quartz and by plagioclase phenocrysts. First, only in plagioclase do arrays of glass inclusions outline growth zones, which arrangement clearly identifies them as primary inclusions (fig. 1B). Such inclusions usually are smaller than the average for plagioclase. Furthermore, the cores of some plagioclase phenocrysts are densely populated with what appear to be primary glass inclusions (fig. 1C). The rest of the glass inclusions in plagioclase, as well as those in quartz, appear to be isolated individuals (fig. 1D), randomly arranged in clusters, or aligned along fractures (assumed or readily visible). Second, single bubbles are common in glass inclusions in hornblende and plagioclase (fig. 1C,D), whereas multiple bubbles (typically 4–8) are common in glass inclusions in quartz (fig. 1A,E,F). In the latter, bubbles can comprise 50+ vol% of the inclusion, in many cases showing a bimodal size distribution; the larger bubbles are on the order of 15 μm. Third, the association of glass inclusions with fractures is particularly strong in quartz. Many glass inclusions reside within brownish ribbonlike features inclined to the quartz sample's surface, which are interpreted as incompletely healed fractures (fig. 1G). Short, dark fractures in many cases crosscut the glass inclusions. There is no apparent relation between the volume percent of bubbles in glass and the presence of a fracture. Only in quartz phenocrysts are "halos" seen around glass inclusions (fig. 1G). These halos are recognized by an optical interface exactly concentric with the square inclusion and located 5 to 10 μm outward from the edge of the glass. The halo zone is perfectly transparent and optically continuous with the rest of the host quartz. It is recognized as a separate entity because it has even less color than the immediately surrounding quartz and because its outer interface in many cases marks the

Table 1. Description of inclusions within quartz and plagioclase phenocrysts from the June 15, 1991, crystal-rich dacite.

	Glass Inclusions		Vapor-dominated Inclusions		Aqueous Inclusions		Solid Inclusions
	in quartz	in plagioclase	in quartz	in plagioclase	in quartz	in plagioclase	
Abundance	Dominant inclusion type in both phases		Rare	Rare	Rare (in 1 out of 7 phenocrysts).	Not observed	Solid inclusions common; **apatite** more abundant in plagioclase; **zircon** abundant in quartz, rare in plagioclase; **amphiboles** less common.
Occurrence	Isolated; in clusters; along fractures.	Isolated; in clusters; arrays along growth zones.	Common association with aqueous inclusions; associated with glass inclusions; may dominate grain.	Isolated or in arrays of two or three.	Secondary, along fractures.	Not observed	**Apatites:** mostly isolated; some aggregates of prisms. **Zircons:** mostly isolated; some in dispersed groups.
Shape	Square with rounded corners.	Ovoid to rectangular.	Subhedral to euhedral.	Approximately square.	Ovoid to euhedral	Not observed	**Apatites:** stubby grains and blades; needles. **Zircons:** stubby ovoids to needles.
Size	Mostly 5–60 μm; some > 100 μm		5–15 μm	3–12 μm	Most <30 μm	Not observed	**Apatites:** 5–50 μm long; aspect 1:3 to 1:15. **Zircons:** a few to 80 μm.
Color	Mostly colorless, some light brown, or "flecked" (slightly devitrified?).		Dark (high relief)	Dark (high relief).		Not observed	**Apatites:** light yellow-green. **Zircon:** light tan. **Amphiboles:** green, pleochroic.
Other comments	Many bubbles; bimodal size; ≤50 vol% bubbles.	Single bubbles common.	May be vapor member of 2-phase vapor-liquid aqueous fluid.	May be gas lost from glass inclusion.	Bubble ≤90 vol%; birefringent and isotropic daughter crystals.	Not observed	**No anhydrite** inclusions detected optically or spectroscopically.
Raman results	P_{CO_2} <20 bar; P_{SO_2}, P_{H_2S} <1 bar; in some, no gas detected.	Low-level plagioclase fluorescence precludes gas detection.	CO_2, SO_2, H_2S below 1 bar partial pressure; H_2O not detected (see text).		No detection in bubbles; H_2O in liquid phase; anhydrite daughter crystal.	Not observed	Spectra of **zircons** in quartz show an up-shift of bands; spectral differences for zircons in same quartz; normal spectra for zircons in plagioclase.
Photograph	Figure 1A,E,F,G,H	Figure 1B,C,D	Figure 1I,J	Figure1D	Figure 1I	Not observed	Figure 1H, K, L, M,N.
Spectrum	Figures 2, 3A		Figures 3B, 4				Figure 4A.

transition into a fluid-filled decrepitation zone. The Raman spectra of the halos around those inclusions are indistinguishable from the Raman spectrum of the quartz host. It is tempting to label these fracture-associated glass inclusions in quartz as secondary, that is, trapped after the phenocryst already had formed and subsequently had undergone brittle fracture. However, it is possible that the difference in the thermal coefficient of expansion between quartz and the glass/melt inclusions induced fracturing during cooling of the phenocryst.

Both the obvious sharp, dark fractures that intersect many glass inclusions and the halos around inclusions are evidence of breaching of the glass and release of fluids in quartz phenocrysts. At low magnification, the halo-rimmed inclusions are seen to reside within brownish zones indicating major fractures. We infer that each halo, with its enhanced optical clarity (fig. 1G), represents a zone of localized healing within the fracture. The healing was induced by fluid released from the glass inclusion. Such healing could have occurred only at elevated temperature, not during sample preparation. The small dark fractures emanating from many glass inclusions are common in quartz phenocrysts from volcanic rocks and may represent the approximately 1% decrease in volume that occurs in the transition of ß to α quartz (see Roedder, 1984, p. 69). In our rock wafers that were vacuum impregnated with blue epoxy, there is also evidence of breaching of glass inclusions in plagioclase: Veils of blue within phenocrysts indicate throughgoing fractures; spherical blue blebs commonly occur within glass inclusions, usually with no optical indication of any fracture connection to the sample's surface.

Most of the above-listed breaches are believed to have occurred at elevated temperatures, as previously explained. However, some fracturing and release of fluid apparently occurred in response to sample preparation: Raman analyses for CO_2 on the same bubbles in several glass inclusions over a 3-month period showed a noticeable upshift in the Raman peak position and a marked decrease in the peak's intensity, both indicating a drop in CO_2 pressure (fig. 2).

Most of the Raman microprobe analyses of bubbles in glass inclusions in quartz revealed no gases. It should be noted that the Raman spectrum of water vapor in the bubbles cannot be distinguished from the spectrum of water dissolved in the enclosing glass matrix of the bubbles (fig. 3A). Thus, we cannot comment on the presence of water within the bubbles in glass. Some bubbles, however, show the Raman spectrum of CO_2 (fig. 2). Moreover, in some of the gas-bearing bubbles, the specific peak position for CO_2 (based on calibrations of pure CO_2 done in our laboratory by J.C. Seitz, 1993) indicates internal gas pressures on the order of a couple of tens of bars partial pressure of CO_2 (assuming that the CO_2 peak positions are minimally affected by the presence of low-density water vapor). No SO_2 (dominant sulfur gas species under preeruption conditions) or H_2S (dominant sulfur gas species at room

temperature, in presence of water) was detected in any of the bubbles. If the latter species are present in the bubbles, they are below our detection limits of approximately 1 bar partial pressure. Because no CO_2 was detected in the glass, its concentration is below our detection limit, which is estimated to be on the order of 900 ppm. The natural low-level fluorescence of plagioclase increased the spectroscopic background and thus precluded the Raman detection of any gas species in glass inclusions hosted by plagioclase.

In rare cases, birefringent phases occur in the glass-dominated inclusions. The individual tablet-shaped solids enclosed in some glass inclusions in quartz (fig. 1H) have been identified as plagioclase by Raman spectroscopy. In several plagioclase phenocrysts, there are glass inclusions that contain single bubbles that are defined by a low-birefringent thin shell that is not quite round in cross section. Raman analysis of the shell produces no bands other than those of the plagioclase host. Thus, the shell is believed to represent plagioclase that has nucleated and grown from the glass.

VAPOR-DOMINATED INCLUSIONS

One-phase inclusions that appear to be totally vapor occur in low abundance in both plagioclase and quartz phenocrysts. In plagioclase, approximately square, dark (high-relief) inclusions of 3 to 12 μm in diameter occur as isolated

---

Figure. 1. (Following pages.) Photomicrographs of inclusions within phenocrysts of Mount Pinatubo dacite. *A,* Glass inclusions in quartz showing different numbers and sizes of bubbles and different bubble:glass ratios. *B,* Growth zones in plagioclase lined by glass inclusions; two apatite inclusions on left. *C,* Core of plagioclase phenocryst densely populated by single-bubble glass inclusions with constant bubble:glass ratio. *D,* Examples of isolated inclusions in plagioclase: glass inclusion with a single gas bubble in the middle, intersected by narrow fracture (arrow); three small one-phase, approximately square, high-relief vapor inclusions to the left. *E,F,* Multiple bubbles in glass inclusions in quartz; note bimodal size distribution in F and high bubble:glass ratio in both inclusions. *G,* Glass inclusions in quartz: both inclusions show light halos that are interpreted as partial healing of a decrepitation zone; left inclusion is crosscut by a long fracture. *H,* Glass-crystal inclusion in quartz; tablet-shaped birefringent daughter mineral is plagioclase. *I,* One-phase euhedral vapor inclusion (bottom left) coexisting with multiphase aqueous inclusion (top; smaller, similar inclusion on right) in quartz (see text and Raman spectra of larger aqueous inclusion in figures 3B and 4A). *J,* One-phase rounded to elliptical vapor inclusions in quartz; cloudiness induced by density of inclusion population. *K,* Zircon inclusions (needles) and apatite inclusions (blades) in plagioclase phenocryst. *L,* Large apatite needle isolated in plagioclase phenocryst; two tubular structures in core give no Raman signature and are inferred to be voids created during rapid growth. *M,* Amphibole inclusion (blade) in quartz phenocryst. *N,* Zircon inclusion in quartz.

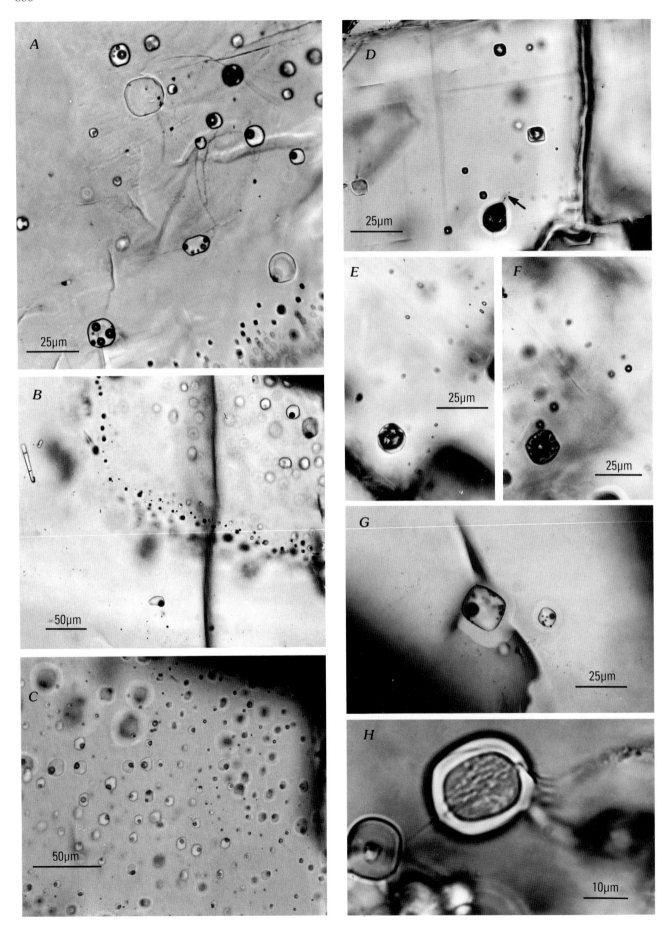

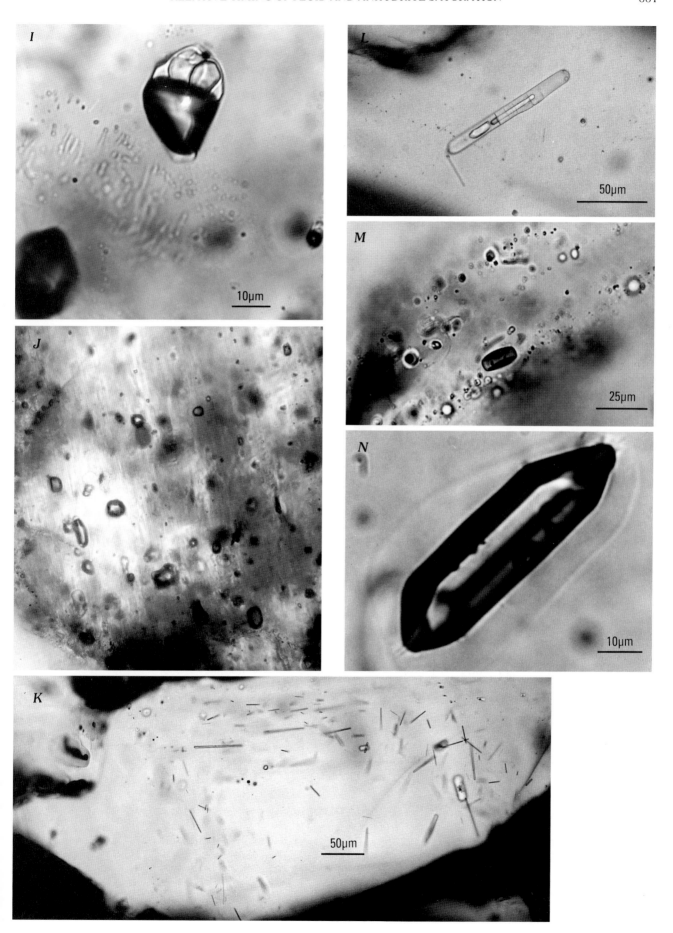

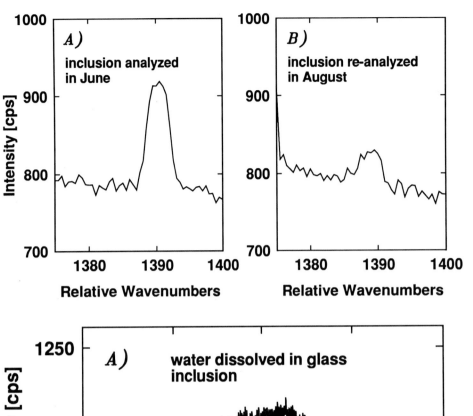

Figure 2. Raman spectra of CO_2 in same bubble of glass inclusion in quartz, providing evidence for leakage over time. *A*, Original analysis. *B*, Repeated analysis 3 months later, in which same analytical settings were used. cps, counts per second.

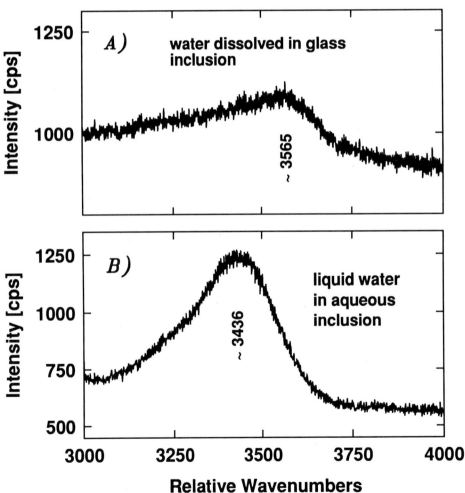

Figure 3. Raman spectral sensitivity to different forms of H_2O in inclusions in Mount Pinatubo phenocrysts. Raman spectra of (*A*) water dissolved in glass inclusion in quartz and (*B*) liquid water in multiphase aqueous inclusion in quartz (shown in fig. 1I). cps, counts per second.

individuals or in arrays of two or three inclusions (fig. 1*D*). In quartz, one-phase inclusions occasionally occur in small numbers associated exclusively with glass inclusions. More typically in quartz, one-phase inclusions occur closely associated with vapor-dominated (>90 vol%) and other less vapor-rich aqueous inclusions that also may have daughter minerals (see below). In such cases, the 5- to15-µm one-phase vapor inclusions are subhedral to euhedral, approaching the negative crystal shape of ß-quartz dipyramids (fig. 1*I*). In these cases, the aqueous and vapor inclusions appear to line secondary fractures. No CO_2, H_2O, SO_2, or H_2S was detected in the one-phase vapor inclusions. If present, these gases occur at levels below the detection limits of about 1 bar partial pressure for each species listed, except H_2O, whose detection limit is higher but unquantified. These inclusions show no physical evidence of leakage, and we infer that they represent the trapping of very low density vapor.

Several quartz grains are almost homogeneously translucent, due to their incorporation of a myriad of one-phase, rounded to elliptical high-relief inclusions of a wide range of sizes (fig. 1*J*). In several grains, not all the inclusions have one phase; some inclusions have a few volume percent liquid (probably water) at one or both extremities (similar to the aqueous inclusions described below), and some of these show visible fractures projecting from them. It is difficult to characterize such inclusions as primary or secondary. The optical properties of such grains are so poor as to preclude Raman analysis of the inclusions. Our interpretation is that these quartz grains precipitated in the presence of a vapor-saturated fluid.

AQUEOUS INCLUSIONS

A small proportion of inclusions is recognized by the presence of a colorless liquid phase (identified as water by Raman analysis) along the walls or in the extremities of the inclusion. Such aqueous inclusions typically are less than 30 μm in diameter and are ovoid to euhedral (doubly terminated prismatic; some dipyramids) in shape. They were observed only in quartz phenocrysts (in grain separates and in situ in rock wafers) and only in about one out of every seven crystals. The quartz grains containing aqueous inclusions also have a small number of square glass inclusions elsewhere in the grain. Where aqueous inclusions occur, they usually dominate the inclusion population of that phenocryst grain or particular volume within the phenocryst by densely populating what appear to be secondary fractures. The aqueous inclusions display various amounts of liquid water or brine, typically less than 30 vol%. Most aqueous inclusions are dominated by a vapor bubble, which can comprise more than 90 vol% in liquid-vapor inclusions and up to 70 vol% in liquid-vapor-solid inclusions (fig. 1*I*). In the bubble phase of such inclusions, all of the gases analyzed for, including H_2O, were below our Raman detection limits. The small amounts of visible liquid gave a strong Raman band for the O-H stretching vibration of liquid water in the spectral region 3,000 to 3,600 Δcm^{-1} (fig. 3*B*). However, there was no obvious Raman band in the 950 to 1,150 Δcm^{-1} region for dissolved sulfate.

Solid, apparent daughter phases were investigated in about 10 different aqueous inclusions, which can be subdivided into two groups on the basis of whether any of the solids showed birefringence. The inclusions with birefringent solids tend to have the negative-crystal shape of doubly terminated prisms and to be smaller (<15 μm) than those with only isotropic solids. In some cases, only a birefringent solid is present, whereas in other cases, it appears that there are isotropic solids at one end and birefringent ones at the other end of the inclusion. It is difficult to optically distinguish a liquid phase in such inclusions, but they have a bubble whose size ranges from 10 to 80 vol%. There are population clusters whose proportions of solid phases all appear to be very similar. Raman spectroscopy was done on five of the small birefringence phases, but no bands other than those for quartz were detected. Thus, the daughter phases either are ionic crystals (poor Raman scatterers), are weak Raman-scatterers, and (or) have Raman bands in spectral positions masked by those for quartz. A simple ionic phase, such as NaCl, is unlikely because the crystals are birefringent.

In one quartz phenocryst, Raman analyses were done on five different aqueous inclusions showing only isotropic phases. These particular inclusions were chosen for their size and good optics and appear representative of that subgroup of aqueous inclusions: euhedral but not prismatic, vapor dominated, and exhibiting multiple solids and a liquid phase. The most diverse suite of solids was analyzed in the inclusion shown in figure 1*I*. The largest solid (top of inclusion, immediately left of center) is isotropic, colorless, square, and of low relief. It yields no Raman bands except those of the host quartz and is inferred to be halite. The small, dark daughter crystal (near upper apex of inclusion) appears to heat under the laser beam and to move around in the inclusion. Laser power was reduced to minimize this effect, but no additional Raman bands were detected; its identity remains unknown. The other colorless, isotropic crystal (top, immediately right of center) that has higher relief than the inferred halite has Raman bands at about 170 and 240 Δcm^{-1}, in addition to any bands that might be masked by those of the quartz host. Its identity also has not yet been established. Of greatest interest in the present study are the two smallest colorless daughter crystals (top apex). These gave Raman bands at 1,018 and 1,130 Δcm^{-1}, which are diagnostic of anhydrite (fig. 4). Their identification as anhydrite was unexpected because the crystals show no birefringence. However, the crystals do appear to demonstrate the retrograde solubility of anhydrite, because additional precipitation seems to occur when the laser beam irradiates the area occupied by these crystals and presumably causes localized heating. The lack of detection of the Raman band for dissolved sulfate in the associated liquid is explained by the fact that anhydrite reaches saturation at very low levels of sulfate concentration. The other four aqueous inclusions also showed the Raman bands of anhydrite but no birefringence. It is inferred that the anhydrite

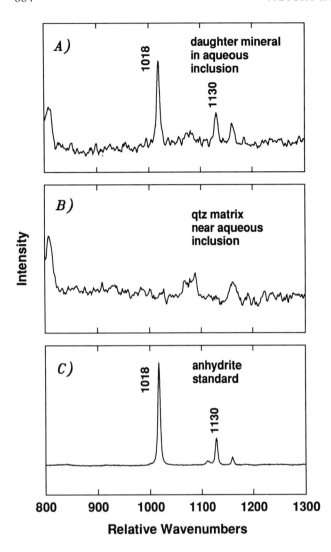

Figure 4. Raman spectra of (*A*) colorless daughter mineral (anhydrite) in multiphase aqueous inclusion in quartz (shown in fig. 1I), (*B*) quartz matrix near the same inclusion, and (*C*) anhydrite standard.

exists as very fine-grained material that does not exhibit strong birefringence.

SOLID INCLUSIONS

Inclusions of several types of transparent and opaque phases are common within plagioclase, quartz, and hornblende phenocrysts. Raman analysis was done on several types of transparent solids. Apatite is a common inclusion phase; it is markedly more abundant in plagioclase than in quartz. Apatite inclusions occur as slightly yellow-green, low-birefringent, stubby prisms and needles on the order of 5 to 50 μm long, with an aspect ratio of about 1:3 to 1:15, and they typically do not display crystallographic orientation within their host. Isolated apatite grains are the more common occurrence, but low-density aggregates of apatite

prisms also occur (fig. 1*K*). In some cases, apatite prisms are intergrown with opaque phases. Some of the wider apatite prisms (several micrometers in diameter) have tubular core structures (fig. 1*L*). Raman analyses of the tubules show no bands other than those for the host apatite, so it is suggested that the tubules are structures that developed during rapid growth (see Wyllie and others, 1962; Gardner, 1972) and are not elongated mineral or fluid inclusions. Of considerably lesser abundance are inclusions identified by Raman spectroscopy as amphibole (fig. 1*M*). Despite intensive petrographic and spectroscopic analysis, we found no inclusions of anhydrite in either plagioclase or quartz phenocrysts.

The solid inclusions that proved most interesting petrologically are zircons. In contrast to other reports of no or only rare zircon inclusions (Matthews and others, 1992), in our samples, zircon is abundant as inclusions within quartz phenocrysts; some quartz phenocrysts contain 10 or more zircon grains of various shapes. The grains range from stubby ovoids of a few micrometers length to elongated prisms (10^+ μm long) that, in many cases, show good terminations (fig. 1*N*). The prisms show parallel extinction but only moderate birefringence, due to their thinness. In contrast, zircon inclusions are rare in plagioclase. Raman analysis indicates that where needlelike and bladelike solid inclusions occur together in a phenocryst (fig. 1*K*), the needles typically are zircon and the blades apatite.

Three spectral features of the zircon inclusions investigated in quartz and plagioclase are of note: (1) the Raman peak positions of several bands characteristic of zircon (for example, Knittle and Williams, 1993) are upshifted, by as much as 10 cm^{-1}, in essentially all zircon inclusions in quartz, (2) in several cases, the Raman peak positions are different from one zircon inclusion to the next within the same quartz phenocryst host, and (3) only the zircon inclusions (rare) in plagioclase have a "normal" Raman spectrum. We believe that the spectral upshifting is due to chemical substitution in the zircons. It was difficult to obtain electron microprobe analyses of the zircon inclusions because it is difficult to grind and polish the sample so as to intersect (but not eliminate!) such small individual grains. The electron microprobe analysis (by K. Bartels, Washington University) of a zircon inclusion whose main Raman band lies at 1,114 Δcm^{-1} revealed about 1.5 wt% HfO_2, which is not compositionally unusual. We infer that the spectroscopically important substitution may be by a light element, and are doing further studies to document the chemistry of the zircons.

DISCUSSION

The goal of this paper is to integrate our data on fluid and solid inclusions with other available petrologic information in order to better understand the evolution of the magmatic system at Mount Pinatubo. The andesitic to dacitic eruptive sequence in mid-June was reversed from the more usual volcanic compositional progression from more felsic to more mafic. Data from coexisting iron-titanium-oxide phases indicate that, at eruption, the dacitic magma had a temperature of 780 ±10°C and an oxygen fugacity (f_{O_2}) of 3 log units above that of the nickel-nickel oxide buffer (f_{O_2} = NNO + 3; Pallister and others, this volume; Rutherford and Devine, this volume); these factors indicate that the dacites from Mount Pinatubo are more oxidized than the trachyandesites from El Chichón. The preeruption pressure at Mount Pinatubo is estimated to be about 2,200 bar (Rutherford and Devine, this volume). Gerlach and others (this volume) calculate that the fluid phase that evolved from the magma was approximately 85 mol% H_2O, 10 mol% CO_2, and 5 mol% SO_2. These data provide important constraints on the timing of volatile saturation and the sulfur budget at Mount Pinatubo, which can be evaluated further on the basis of information from solid and fluid inclusions.

EVIDENCE FROM THE SOLID PHASES

It is difficult to determine the relative timing of phenocryst precipitation at Mount Pinatubo on the basis of textural relations alone. However, it seems significant that plagioclase is so much more abundant than quartz and that quartz grains typically show resorption. A difference in the timing of the initial precipitation of plagioclase and quartz also is evidenced by the differences in the types of their inclusions—for instance, the higher bubble:glass ratio in glass inclusions in quartz and the occurrence of late-stage aqueous inclusions exclusively in quartz. These observations suggest that quartz phenocrysts postdate most of the plagioclase phenocrysts. However, plagioclase (and hornblende) almost certainly continued to precipitate after quartz reached saturation.

Spectral differences between zircon inclusions in plagioclase and quartz and spectral variations among zircon inclusions within the same quartz grain are evidence that zircons from different petrologic environments were encapsulated during the growth of these phenocrysts. In the context of what is known about the petrology and geophysics of the volcanic system at Mount Pinatubo, at least the following possibilities should be considered to explain the different spectral signatures of the zircon inclusions: (1) zircons reached saturation early in two different melts that presumably mixed before the major phenocryst phases crystallized in the dacite (there is no additional petrologic evidence to substantiate such mixing; J. Pallister, USGS, oral commun., 1993), (2) zircons reached saturation at different levels in a compositionally zoned magma chamber and thereby produced different minor- and trace-element signatures in grains that subsequently were encapsulated in growing

phenocrysts (no additional petrologic evidence for this), (3) some of the zircons crystallized directly from a silicate melt, whereas others precipitated from a supercritical aqueous fluid that coexisted with the melt, (4) the zircons and their complex chemistry reflect only local equilibria at the interface between siliceous melt and phenocrysts precipitating from it (Michael, 1988; Bacon, 1989), and (5) some of the zircons are inherited from wallrocks that were mechanically and (or) chemically incorporated into the magma. The latter may be true for the equidimensional smallest zircons but not for the abundant needlelike zircons with excellent terminations (fig. 1*N*).

It appears significant that, although anhydrite is recognized as a phenocryst phase that precipitated from the magma (Bernard and others, 1991; Pallister and others, 1992; Imai and others, 1993; Rutherford and Devine, this volume; Pallister and others, this volume; Fournelle and others, this volume), it does not occur as inclusions within plagioclase or quartz phenocrysts in the samples we studied and indeed is relatively rare as inclusions in other Mount Pinatubo rocks (Pallister and others, this volume; Rutherford and Devine, this volume; Fournelle and others, this volume; J. Fournelle, University of Wisconsin, written commun., 1994). It thus seems possible that, in *most* parts of the magmatic system, anhydrite reached saturation only after quartz phenocrysts had begun to precipitate, meaning that anhydrite was a late phenocryst phase. From our Raman data, we know that anhydrite definitely had reached saturation by the time of the development of late-stage aqueous fluids. Our hypothesis of late saturation with anhydrite, which addresses the overall petrologic character of the dacite, does not preclude the possibility of early, local saturation of the magma with small amounts of anhydrite (see point 4 above concerning zircon). Nor is our hypothesis in conflict with the late entrapment of anhydrite phenocrysts in early-saturating phases such as hornblende and plagioclase (Fournelle and others, this volume), which continued to precipitate throughout much or all of the crystallization history of the dacite.

STATE OF FLUID SATURATION

Several petrologic studies (Gerlach and others, this volume; Pallister and others, this volume; Rutherford and Devine, this volume) have provided chemical evidence that the Mount Pinatubo dacitic magma became saturated with an aqueous fluid during its evolution, but the timing of the saturation still remains unconstrained. On the basis of mass balance, Gerlach and others (this volume) argue convincingly that the preeruptive melt coexisted with a separate fluid phase. The best candidate inclusions from which to evaluate the state of fluid saturation of the dacitic magma are the glass inclusions with bubbles, which are enclosed within quartz and plagioclase phenocrysts. The large bubble-to-glass ratio (bubbles representing up to 50+ vol% of the inclusion) clearly indicates that the bubbles cannot simply be due to contraction during cooling of the glass. The spherical shape of the bubbles reflects the molten state of the host when the bubbles were formed or trapped. The common occurrence of multiple bubbles probably reflects the high viscosity of the rhyolitic melt, which hindered coalescence of the gas packets. (In Lowenstern's (1994) heating experiments on hydrous peralkaline rhyolitic melt inclusions, nucleation of multiple bubbles commonly occurred during cooling.) The common bimodal distribution of bubble sizes in Mount Pinatubo glass inclusions further suggests that there were two events of bubble formation or entrainment. If the bubbles in glass were formed in a closed system, the two most reasonable mechanisms by which they formed are (1) by the degassing of volatiles from the melt inclusions as a result of the decreases in pressure (and temperature) that occurred during the phenocrysts' rapid rise to the surface (exsolution model) and (or) (2) by the trapping of a magmatic volatile phase that coexisted with the melt at depth (magmatic entrapment model).

The exsolution model can be evaluated from data on the solubility of CO_2 in rhyolitic melts under the pressure-temperature conditions pertinent to Mount Pinatubo (Fogel and Rutherford, 1990) and infrared data on the concentration of CO_2 dissolved in the glass inclusions as they exist now (Wallace, 1993; Rutherford and Devine, this volume). An upper bound on the pressure of exsolved gas in the glass inclusions can be calculated by overestimating the amount of CO_2 expelled during decompression and underestimating the relative volume of bubbles. The rhyolitic melt that was trapped in the phenocrysts has an estimated density of 2.4 g/cm^3 (Westrich and Gerlach, 1992). If the inclusions represent CO_2-saturated melt entrained at 800°C and 2.2 kbar pressure, then the initial concentration of dissolved CO_2 was about 1,800 ppm by weight (Fogel and Rutherford, 1990). This value is appropriate for a very low-H_2O rhyolite, rather than an H_2O-saturated melt like that at Mount Pinatubo. From Fourier-transform infrared (FTIR) analyses, Wallace (1993) reports a range of 280 to 415 ppm CO_2 in Mount Pinatubo glass inclusions in quartz that were chosen because they show no visible cracks and have either no bubbles or only very small bubbles (glass presumably not degassed). In contrast, Rutherford and Devine's (this volume) FTIR analyses of Mount Pinatubo glass inclusions in quartz and plagioclase indicated dissolved CO_2 concentrations of less than 20 ppm (glass presumably degassed). For the purpose of modeling, we assume an upper-bound loss of 900 ppm by weight of CO_2 from the time of trapping, and the retention of that CO_2 in bubbles in the glass. If the bubble(s) represent 20 vol% of the glass inclusion, then the partial pressure of CO_2 in each bubble should be about 6 bar—if no leakage has occurred. This pressure is significantly lower than that indicated by Raman spectroscopy of some bubbles in glass inclusions within quartz, which appear to

contain up to a couple of tens of bars partial pressure of CO_2.

On the other hand, if the bubbles represent a magmatic fluid coexisting with the melt at 2 kbar pressure and 800°C, then isochoric cooling would produce an inclusion that now should consist of a vapor phase that is enriched in CO_2 compared to the surrounding phase of liquid water, which should comprise several tens of vol% of the total "bubble" (Roedder, 1984, p. 226). In fact, no liquid is visible in the bubbles, nor did the Raman spectra that were obtained for H_2O in several inclusions show a band structure indicative of liquid water (see Wopenka and others, 1990). Therefore, the Raman data indicate much lower pressures and densities than those expected for the magmatic entrapment of fluids.

Thus, because neither of the endmember models presented above explains the spectroscopic observations, a more complex mechanism or open-system behavior is suggested. One possibility is that the fluid species diffusively reequilibrated with melt surrounding individual phenocrysts (Qin and others, 1992). However, the rapidity of the cooling process and the short time since eruption do not support such a model. On the other hand, there is overwhelming evidence of physical leakage from the bubble-bearing glass inclusions: abundant fractures transecting glass inclusions, partially healed decrepitation halos around glass inclusions (fig. 1G), decreases in CO_2 pressure in individual bubbles over time as documented by repeated Raman analyses (monitoring shifts in band position and decreases in intensity; fig. 2), and observations of the infiltration of colored epoxy into glass inclusions induced by vacuum impregnation of the rock before it was sectioned. Thus, the gas pressures currently monitored in the bubbles in glass inclusions cannot be used to constrain the pressure at which trapping occurred. Leakage of volatiles from glass inclusions is a common phenomenon (Anderson, 1991; Lowenstern and others, 1991; Tait, 1992). Anderson (1991) and Tait (1992) present detailed discussions of leakage mechanisms in phenocrysts, and Anderson and Brown (1993) discuss the possible significance of coexisting bubble-free and bubble-bearing glass inclusions.

In summary, there is petrographic and Raman spectroscopic evidence that a separate fluid phase coexisted with the dacitic melt early in its crystallization history. In addition, there is good petrographic evidence for fluid saturation during quartz crystallization, and the Raman data support this interpretation. In quartz, we infer that the bimodal distribution in the size of bubbles in glass inclusions reflects the entrapment of magmatic fluid together with melt, followed by the exsolution of dissolved volatiles from the melt during rise and depressurization (large and small bubbles, respectively). For plagioclase, the petrographic evidence for fluid saturation is not so compelling as for quartz, but the water contents of the glass inclusions (Gerlach and others, this volume; Pallister and others, this volume; Rutherford and Devine, this volume) indicate that the magma already was saturated with an aqueous fluid during the precipitation of plagioclase.

Obviously, the loss of fluid from the bubbles in glass makes it difficult to prove that magmatic volatiles were trapped together with melt. However, the experimental results and photographic documentation of Lowenstern (1994) on hydrous peralkaline melts support our interpretation of coentrapment of volatiles and melt. In his review paper on evidence from fluid inclusions of immiscibility in magmatic systems, Roedder (1992) cites several studies in which liquid water was documented in the "shrinkage bubbles" within glass inclusions; such high-density fluid indicates entrapment at elevated pressure and depth.

FLUID EVOLUTION

One of the most obvious, and yet critical, observations about the dacitic magmatic system at Mount Pinatubo is that fluid saturation was reached with respect to an *aqueous* fluid. Despite the attention that has been focused on the number of million tons of SO_2 that were released into the atmosphere, volatile saturation may have been initiated by CO_2 and ultimately became dominated by water (Gerlach and others, this volume). Furthermore, Gerlach and others (this volume) estimate a vapor:melt distribution coefficient of almost 800 for sulfur. The strong preferential partitioning of sulfur into a supercritical aqueous fluid phase and the vast quantities of magmatic water released into the atmosphere imply that the timing of water saturation and the evolution of this aqueous fluid had a large impact on the sulfur budget. We infer that it was this dense, sulfur-and CO_2-bearing aqueous fluid that was so poorly retained in the bubbles within glass inclusions in the early phenocrysts such as plagioclase and quartz.

Those fluid inclusions that are preserved in the phenocrysts, however, provide a means of tracking the evolution of this aqueous fluid during the course of crystallization and rise of the magma. For instance, the attainment of two-phase equilibria in the water-rich fluid of the Mount Pinatubo magmatic system is recorded by the liquid-bearing aqueous inclusions in quartz. Regardless of the mechanism that induced phase separation into a liquid-like and a vaporlike fluid, chloride, sulfate, and other dissolved nonvolatile components would have partitioned selectively into the liquid phase, whereas the gases (CO_2, SO_2) would have partitioned selectively into the vapor phase. This partitioning is reflected in the large volumetric proportion of solids in some of the liquid-bearing inclusions (fig. 1*I*).

One means of generating coexisting liquid-rich and vapor-rich inclusions is to decrease the pressure and temperature of an originally supercritical aqueous fluid, which could still be present even below the dacite solidus. Such a decrease, particularly in pressure, would bring the initially one-phase supercritical fluid into the two-phase field of liquid + vapor. The localization of liquid- and vapor-rich inclusions along fractures in some quartz grains indicates that those host crystals *already had crystallized* before two-phase separation occurred in the aqueous fluid coexisting with the magma. The low density of the aqueous inclusions, as indicated by their large volumetric proportion of vapor, does not necessarily mean that the quartz hosts were near the surface during trapping or that they have leaked (no physical evidence of leakage in most cases). The low pressure required for the separation of an initial supercritical fluid into vapor and liquid could have been induced by decompression associated with the ascent and eruption of the magma. In contrast, those quartz phenocrysts that are totally clouded by one-phase vapor or very vapor-rich inclusions (fig. 1*J*) probably *crystallized* during the stage of vapor effervescence, when they effectively trapped the low-density vapor phase. Some of these inclusions have obvious fractures, as mentioned above.

Another explanation for the high-salinity aqueous inclusions is that they represent hydrous saline melt, documenting high-temperature immiscibility that arose in the course of differentiation. Such immiscibility is common in porphyry copper systems (Roedder, 1992). Among the necessary criteria for saline melt inclusions are very high temperatures of homogenization (greater than 500°C) and the coexistence of silicate melt inclusions (Roedder, 1992; Lowenstern, 1994). The inclusions that we studied have not been subjected to microthermometric analysis. Although there are melt inclusions in the same quartz grains that contain abundant aqueous inclusions, the spatial and genetic relations between the two types of inclusions cannot be ascertained. In some quartz grains, the rare round, low-relief, isotropic inclusions (no Raman signature) and other optically complex inclusions containing bleblike phases may represent the entrapment of a saline melt. We cannot rule out the possibility that some of the quartz grains that were investigated represent xenocrysts derived from the wallrocks. However, such contamination is believed to be minimal, given that the volcano already had "cleared its throat" during earlier eruptions in mid-June.

One-phase vapor inclusions in plagioclase and quartz appear to have monitored two distinct processes. As discussed above, some appear to represent the vapor member of an aqueous brine that has segregated into liquid + vapor (fig. 1*I*). Others probably represent the capture of volatiles that leaked from glass inclusions and were subsequently trapped along fractures (fig. 1*D*). One-phase vapor inclusions in plagioclase phenocrysts are restricted to the latter type, whereas one-phase inclusions in quartz are mostly of the former type.

AQUEOUS FLUID VERSUS ANHYDRITE SATURATION: A DIFFERENT MODEL FOR THE BEHAVIOR OF SULFUR IN MELTS

In the experiments of Carroll and Rutherford (1987, 1988) and Luhr (1990) on the solubility of sulfur in oxidized calc-alkaline melts, the silicate melts are saturated with respect to both anhydrite and an aqueous fluid phase. The typical application of these experiments to anhydrite-bearing calc-alkaline systems involves the assumption that the magmas first reached saturation with respect to anhydrite and subsequently with respect to water (Carroll and Rutherford, 1987; Luhr, 1990). In the El Chichón eruptive rocks, that saturation sequence is evidenced by the abundance of anhydrite phenocrysts and, especially, anhydrite inclusions in silicate phenocrysts (Luhr and others, 1984). However, an unexpected lack of anhydrite has been noted in evolved volcanic rocks from various localities whose magmas were sufficiently oxidized to have stabilized a sulfate phase. Carroll and Rutherford (1987) have suggested that the latter magmas may have had bulk-sulfur concentrations that were too low to support much precipitation of anhydrite.

The petrographic and Raman evidence in the Mount Pinatubo samples studied by our group and others suggests the consideration of another model, namely, that H_2O-CO_2-SO_2 fluid saturation was reached before anhydrite saturation. This model would explain several otherwise anomalous observations that have been made on the Mount Pinatubo dacites, some of them already highlighted in the literature.

1. Late saturation with anhydrite would explain the dearth of anhydrite inclusions in early phenocryst phases such as plagioclase.
2. Late precipitation of anhydrite also would explain the high concentrations of sulfur detected in many apatite grains from Mount Pinatubo. Pallister and others (1993) report SO_3 concentrations of up to 1.1 wt% in unspecified apatites from the dacite. It seems particularly significant that early apatite grains, trapped within other phases, consistently have higher SO_3 contents (≈ 0.39 wt%) than do discrete apatites (≈ 0.13 wt%; Imai and others, this volume). As soon as anhydrite had reached fluid saturation, sulfur would have partitioned preferentially into that phase over apatite.
3. The Mount Pinatubo magmas are estimated to have equilibrated at several log f_{O_2} units higher than those at El Chichón (Carroll and Rutherford, 1987; Luhr, 1990; Westrich and Gerlach, 1992; Imai and others, 1993; Gerlach and others, this volume; Pallister and others, this volume; Rutherford and Devine, this volume; Imai and others, this volume), and sulfur solubility in silicate melt has been shown to increase with f_{O_2}, although not greatly at the pertinent temperature of 800°C (Carroll and

Rutherford, 1987, 1988; Luhr, 1990). This high oxidation state makes it reasonable to assume that the Mount Pinatubo magma could have accommodated at least as much, and perhaps somewhat more, dissolved sulfur as the El Chichón magma. It therefore is possible that the Mount Pinatubo magma could have reached anhydrite saturation at a later fractionation stage than the El Chichón magma. There is no conclusive evidence, however, about the actual original sulfur contents of the two magmatic systems.

4. McKibben and others (this volume) have interpreted the lack of sulfur isotopic zoning in anhydrite crystals from Mount Pinatubo (June 12th and 15th eruptions: McKibben and others, this volume; McKibben and Eldridge, 1993) as evidence that extensive SO_2 degassing, if it occurred, must have occurred before much anhydrite had precipitated.
5. The reported 0.14 wt% bulk-sulfur content in phenocryst-rich dacite from Mount Pinatubo (Pallister and others, this volume) compared to the 0.6 wt% sulfur in the El Chichón eruptive rocks (Luhr and others, 1984; Carroll and Rutherford, 1987; Bernard and others, 1991) also supports a model in which much of the initial sulfur content of the Mount Pinatubo magmatic system was extracted early.
6. The relatively constant sulfur concentration among glass inclusions within a variety of phenocrysts is interpreted by Rutherford and Devine (this volume) as evidence of anhydrite saturation of the magma. However, the sulfur partitioning imposed on the system by the coexistence of a fluid phase also could have buffered the sulfur concentration in the melt.

Although the activities of both calcium and sulfate control anhydrite saturation, it is reasonable to consider the possibility that early development of a fluid phase could delay the precipitation of anhydrite through effects on the fugacity of SO_2 (f_{SO_2}). Such effects are most easily envisioned if the preeruptive magmatic system initially was undersaturated with respect to both a fluid and anhydrite. Changes in the activities of the dissolved species in the melt due to a decrease in temperature, fractionation of early phases, and the introduction of sulfur and (or) other volatiles from outside the magma chamber could bring the melt incrementally closer to saturation with respect to each of those phases. If fluid saturation were reached before the saturation activity for anhydrite were reached, then the development of a fluid phase, into which SO_2 would partition preferentially over the melt, could buffer the f_{SO_2} in the magma at a level too low for anhydrite precipitation. As the magma cooled and depressurized, however, the cosaturation conditions for anhydrite and a fluid could be fulfilled, and anhydrite could be allowed to precipitate after the onset of fluid saturation. In addition, the f_{SO_2} of the melt also would be affected once eruption occurred and permitted open-system behavior with respect to the fluid phase, at which time SO_2 could be removed effectively from the system.

The above hypothesis of early saturation of the magma with respect to a water-dominated fluid would have important petrologic consequences for water-rich, highly oxidized, calc-alkaline magmas: Magmatic systems that had high concentrations of volatiles early in their evolution, due to their initial melt chemistry or to external introduction of volatiles, could segregate a separate fluid phase before much crystallization had occurred. The specific composition of the fluid phase would reflect both the relative concentrations and solubilities of volatiles in the magma; in most magmas, the fluid phase rapidly would become dominated by water. As soon as a separate fluid developed, sulfur species would be among those that were selectively extracted into it from the magma. One possible consequence of the temporal segregation of sulfur into an early fluid phase (subsequently released into the atmosphere) and later anhydrite and matrix glass phases (retained in the rock) is that the abundance, size, and timing of precipitation of anhydrite crystals in recent or historic hydrous calc-alkaline volcanics are not necessarily good indicators of the sulfur or sulfate concentration in the original melt.

A corollary to the above reasoning is that the ratio of other volatiles (for example, H_2O and CO_2) to SO_2 in the early melt of a given magmatic system could control the timing of sulfur removal from the magma via fluid exsolution, as well as the proportion of the available sulfur that reaches the atmosphere (compared to the proportion retained in the rocks as quenched glass and anhydrite). Due to their abundance and solubility, certain volatile species would induce fluid segregation from a magma. Magmas that reached saturation early with respect to water or a more chemically complex fluid would be stripped of a greater proportion of their total sulfur and thereby retain a lower bulk-rock sulfur content in their extrusive products than would magmas that reached saturation with respect to anhydrite earlier than to a fluid.

In this regard, it is important to consider the role of CO_2 in the initiation of fluid immiscibility in the Mount Pinatubo magmatic system. Gerlach and others (this volume) estimate that up to 10% of the bulk volatile fluid was CO_2, and CO_2 is much less soluble in siliceous melts than is H_2O. Thus, the *initiation* of fluid saturation could have been induced by CO_2. In bubbles within glass inclusions hosted by quartz, the Raman detection of CO_2, together with the lack of Raman detection or optical observation of H_2O, suggests that CO_2 was a major component of the early magmatic fluid phase at Mount Pinatubo. Finally, if crustally assimilated sulfur was an important component in this magma, then the timing of sulfur assimilation was important because of its effect on the $(H_2O+CO_2)/SO_2$ ratio. Our model of early saturation with respect to a fluid phase, specifically before anhydrite saturation, is consistent with the calculations of Gerlach and others (this volume) indicating

that the dacitic magma reached fluid saturation before eruption.

Although our model explains many features of the Mount Pinatubo phenocryst-rich dacites, it is not applicable to all oxidized, calc-alkaline systems. For instance, this model seems inappropriate for the El Chichón eruptive rocks. A difference in the relative timing of saturation with respect to water and anhydrite might account for two petrographic differences between the samples we studied from those two eruptions. Our Mount Pinatubo thin sections had less anhydrite and a higher volumetric proportion of bubbles in the glass inclusions (possibly an effect of cooling history) than did the two samples from El Chichón (obtained from J.F. Luhr, Smithsonian Institution). Our conclusion is that the relative, as well as absolute, concentrations in a melt of such volatile species as H_2O, CO_2, and SO_2 will control the timing of saturation with respect to a fluid phase, anhydrite, and other volatile-containing phases. Further experimental studies of oxidized, calc-alkaline systems may reveal how small differences in initial composition can strongly control their eventual fractionation paths and their products of crystallization.

SUMMARY AND CONCLUSIONS

Detailed petrographic and Raman spectroscopic analyses of inclusions suggest that much of the plagioclase crystallized before quartz, that anhydrite was a very late phenocryst phase, and that the dacitic magma reached aqueous fluid saturation relatively early in its crystallization. In glass inclusions in quartz, the high bubble-to-glass ratio and, particularly, the Raman-determined gas contents and pressures in some bubbles suggest that these inclusions represent the simultaneous trapping of a melt and a supercritical fluid phase at depth. This trapping was followed rapidly by exsolution of volatiles during the rise and depressurization of the magma; thus, some glass inclusions in quartz show a bimodal distribution of bubble sizes. It is unfortunate, but typical, that quartz and its glass inclusions were unreliable containers for fluid, as evidenced by petrographic observation of inclusion rupture and by Raman documentation of low to below-detectable gas contents in the bubbles in glass. It therefore is not possible to use the measured gas contents and pressures to infer the pressure-temperature conditions of initial fluid trapping. For inclusions in plagioclase, the physical evidence of fluid trapping is not so compelling as for quartz, but the high water contents of the glass inclusions (Gerlach and others, this volume; Pallister and others, this volume; Rutherford and Devine, this volume) are in accord with water saturation of the magma at the time of plagioclase precipitation.

In our model, the evolving supercritical, CO_2- and sulfur-bearing aqueous fluid that coexisted with the melt subsequently underwent two-phase separation into liquid and vapor due to an increase in fluid salinity resulting from fractionation (not probable), the rise and depressurization of the magma (more probable), or both. This episode in the development of the fluid is recorded in the late-stage coexistence of all-vapor and liquid-vapor fluid inclusions in quartz. The continued high concentration of chlorine and sulfur in the fluid phase is documented by the abundance of chloride and sulfate daughter minerals in the liquid-rich aqueous inclusions.

Our Raman spectroscopic identifications and petrographic observations, coupled with the observations of other petrologists, indicate a very low abundance of anhydrite inclusions within phenocrysts; however, the latter contain abundant inclusions of hornblende, plagioclase, and especially apatite and zircon. These observations led us to suggest that anhydrite might have been among the last phenocryst phases to reach saturation. Raman evidence of the entrapment of CO_2-rich fluid in phenocrysts, especially quartz, suggests consideration of the possibility that saturation with a CO_2-bearing aqueous fluid was reached early in the evolution of the dacitic magma at Mount Pinatubo (before the eruption began) and that partitioning of sulfur into this fluid delayed the precipitation of anhydrite. Other mineral-chemical and sulfur isotopic data lend support to this hypothesis. We further postulate that the early release of an H_2O-CO_2-SO_2 fluid and its effective extraction of SO_2 from the coexisting melt in part might account for the large volume of SO_2 released into the atmosphere during the June 15th eruption. The importance of the timing of volatile saturation, as a means of extracting specific elements, has been recognized in porphyry copper systems (for example, Cline and Bodnar, 1991; Lowenstern, 1993). Furthermore, as Imai and others (1993) discuss, there are several important links between the Mount Pinatubo volcano and nearby porphyry copper deposits in the Philippines.

The observations and interpretations presented above could be explained better and tested if the results of several additional types of studies were available: (1) experimental monitoring of oxidized, sulfur-rich melts that are saturated with respect to an aqueous fluid but not anhydrite could provide important information on sulfur partitioning, especially if the volatile species in equilibrium with the melts were analyzed; (2) as suggested by the work of Gerlach and others (this volume), the partitioning of sulfur gas species should be investigated in systems consisting of a silicate melt and an aqueous fluid, in which the CO_2/H_2O ratio in the fluid is varied; and (3) the Raman spectral sensitivity of inclusion phases to minor compositional differences should be investigated further.

ACKNOWLEDGMENTS

This research was supported by National Science Foundation grant GER–9023520 to J.D. Pasteris. The authors thank Terrence Gerlach, Henry Westrich, James Luhr, Akira Imai, and Paul Pohwat for providing samples for this study and Karen Bartels for electron microprobe analysis. The authors also benefited from discussions with Alfred Anderson, Charles Bacon, Harvey Belkin, John Fournelle, Terrence Gerlach, Michael McKibben, James Luhr, John Pallister, Paul Wallace, and Henry Westrich. Reviews on an earlier version of this manuscript by Terrence Gerlach, Jacob Lowenstern, James Luhr, Gwendolyn Miner, Christopher Newhall, and Malcolm Rutherford are gratefully acknowledged. The authors retain full responsibility, however, for all data and interpretations.

REFERENCES CITED

Anderson, A.T., 1991, Hourglass inclusions: Theory and application to the Bishop rhyolitic tuff: American Mineralogist, v. 76, p. 530–547.

Anderson, A.T., and Brown, G.G., 1993, CO_2 contents and formation pressures of some Kilauean melt inclusions: American Mineralogist, v. 78, p. 794–803.

Bacon, C.R., 1989, Crystallization of accessory phases in magmas by local saturation adjacent to phenocrysts: Geochimica et Cosmochimica Acta, v. 53, p. 1055–1066.

Bernard, A., Demaiffe, D., Mattielli, and Punongbayan, R.S., 1991, Anhydrite-bearing pumices from Mount Pinatubo: Further evidence for the existence of sulfur-rich silicic magmas: Nature, v. 354, p. 139–140.

Bluth, G.J.S., Doiron, S.D., Schnetzler, C.C., Krueger, A.J., and Walter, L.S., 1992, Global tracking of the SO_2 clouds from the June, 1991 Mount Pinatubo eruptions: Geophysical Research Letters, v. 19, no. 2, p. 151–154.

Carroll, M.R., and Rutherford, M.J., 1987, The stability of igneous anhydrite: Experimental results and implications for sulfur behavior in the 1982 El Chichon trachyandesite and other evolved magmas: Journal of Petrology, v. 28, pt. 5, p. 781–801.

————1988, Sulfur speciation in hydrous experimental glasses of varying oxidation state: Results from measured wavelength shifts of sulfur X-rays: American Mineralogist, v. 73, p. 845–849.

Cline, J.S., and Bodnar, R.J., 1991, Can economic porphyry copper mineralization be generated by a typical calc-alkaline melt?: Journal of Geophysical Research, v. 96, p. 8113–8126.

Fogel, R.A., and Rutherford, M.J., 1990, The solubility of carbon dioxide in rhyolitic melts: A quantitative FTIR study: American Mineralogist, v. 75, p. 1311–1326.

Fournelle, J, Carmody, R., and Daag, A.S., this volume, Anhydrite-bearing pumices from the June 15, 1991, eruption of Mount Pinatubo: Geochemistry, mineralogy, and petrology.

Gardner, P.M., 1972, Hollow apatites in a layered basic intrusion, Norway: Geological Magazine, v. 109, p. 385–391.

Gerlach, T.M., Westrich, H.R., and Symonds, R.B., this volume, Preeruption vapor in magma of the climactic Mount Pinatubo eruption: Source of the giant stratospheric sulfur dioxide cloud.

Imai, A., Listanco, E.L., and Fujii, T., 1993, Petrologic and sulfur isotopic significance of highly oxidized and sulfur-rich magma of Mt. Pinatubo, Philippines: Geology, v. 21, p. 699–702.

———this volume, Highly oxidized and sulfur-rich dacitic magma of Mount Pinatubo: Implication for metallogenesis of porphyry copper mineralization in the Western Luzon arc.

Knittle, E., and Williams, Q., 1993, High-pressure Raman spectroscopy of $ZrSiO_4$: Observation of the zircon to scheelite transition at 300 K: American Mineralogist, v. 78, p. 245–252.

Lowenstern, J.B., 1993, Evidence for a copper-bearing fluid in magma erupted at the Valley of Ten Thousand Smokes, Alaska: Contributions to Mineralogy and Petrology, v. 114, p. 409–421.

———1994, Chlorine, fluid immiscibility and degassing in peralkaline magmas from Pantelleria, Italy: American Mineralogist, v. 79, p. 353–369.

Lowenstern, J.B., Mahood, G.A., Rivers, M.L., and Sutton, S.R., 1991, Evidence for extreme partitioning of copper into a magmatic vapor phase: Science, v. 252, p. 1405–1409.

Luhr, J.F., 1990, Experimental phase relations of water- and sulfur-saturated arc magmas and the 1982 eruptions of El Chichon volcano: Journal of Petrology, v. 31, pt. 5, p. 1071–1114.

Luhr, J.F., Carmichael, I.S.E., and Varekamp, J.C., 1984, The 1982 eruptions of El Chichon volcano, Chiapas, Mexico: Mineralogy and petrology of the anhydrite-bearing pumices: Journal of Volcanology and Geothermal Research, v. 23, p. 69–108.

Matthews, S.J., Jones, A.P., and Bristow, C.S., 1992, A simple magma-mixing model for sulfur behaviour in calc-alkaline volcanic rocks: Mineralogical evidence from Mount Pinatubo 1991 eruption: Journal of the Geological Society, v. 149, p. 863–866.

McKibben, M.A., and Eldridge, C.S., 1993, Sulfur isotopic systematics of the June 1991 eruptions of Mount Pinatubo: A SHRIMP ion microprobe study [abs.]: Eos, Transactions, American Geophysical Union, v. 74, no. 43, p. 668.

McKibben, M.A., Eldridge, C.S., and Reyes, A.G., this volume, Sulfur isotopic systematics of the June 1991 Mount Pinatubo eruptions: A SHRIMP ion microprobe study.

Michael, P.J., 1988, Partition coefficients for rare earth elements in mafic minerals of high silica rhyolites: The importance of accessory mineral inclusions: Geochimica et Cosmochimica Acta, v. 52, p. 275–282.

Pallister, J.S., Hoblitt, R.P., Meeker, G.P., Knight, R.J., and Siems, D.F., this volume, Magma mixing at Mount Pinatubo: Petrographic and chemical evidence from the 1991 deposits.

Pallister, J.S., Hoblitt, R.P., and Reyes, A.G., 1992, A basalt trigger for the 1991 eruptions of Pinatubo volcano?: Nature, v. 356, p. 426–428.

Pallister, J.S., Meeker, G.P., Newhall, C.G., Hoblitt, R.P., and Martinez, M., 1993, 30,000 years of the "same old stuff" at Pinatubo [abs.]: Eos, Transactions, American Geophysical Union, v. 74, no. 43, p. 667.

Qin, Z., Lu, F., Anderson, A.T., Jr., 1992, Diffusive reequilibration of melt and fluid inclusions: American Mineralogist, v. 77, p. 565–576.

Roedder, Edwin, 1984, Fluid Inclusions, in Ribbe, P.H., ed., Reviews in mineralogy, v. 12: Washington, D.C., Mineralogical Society of America, 644 p.

———1992, Fluid inclusion evidence for immiscibility in magmatic differentiation: Geochimica et Cosmochimica Acta, v. 56, p. 5–20.

Rutherford, M.J., and Devine, J.D., this volume, Preeruption pressure-temperature conditions and volatiles in the 1991 dacitic magma of Mount Pinatubo.

Scott, W.E., Hoblitt, R.P., Torres, R.C., Self, S, Martinez, M.L., and Nillos, T., Jr., this volume, Pyroclastic flows of the June 15, 1991, climactic eruption of Mount Pinatubo.

Tait, S., 1992, Selective preservation of melt inclusions in igneous phenocrysts: American Mineralogist, v. 77, p. 146–155.

Wallace, P., 1993, Pre-eruptive gas saturation in the June 15, 1991, Mount Pinatubo dacite: New evidence from CO_2 contents of melt inclusions [abs.]: Eos, Transactions, American Geophysical Union, v. 74, no. 43, p. 668.

Westrich, H.R., and Gerlach, T.M., 1992, Magmatic gas source for the stratospheric SO_2 cloud from the June 15, 1991, eruption of Mount Pinatubo: Geology, v. 20, p. 876–870.

Wopenka, B., Pasteris, J.D., and Freeman, J.J., 1990, Analysis of individual fluid inclusions by Fourier transform and Raman microspectroscopy: Geochimica et Cosmochimica Acta, v. 54, p. 519–533.

Wyllie, P.J., Cox, K.G., and Biggar, G.M., 1962, The habit of apatite in synthetic systems and igneous rocks: Journal of Petrology, v. 3, p. 238–243.

LAHARS, LAHARS, AND MORE LAHARS

Geologists have long known that fresh pyroclastic debris on the slopes of volcanoes will be eroded and transported onto adjoining alluvial fans. In April and May 1991, both the geologic record at Pinatubo and experience at other tropical volcanoes practically shouted that rain-induced lahars (hyperconcentrated streamflows and debris flows) would be major hazards should Pinatubo erupt. But the scales and rates of lahar processes at Pinatubo have surprised even veteran observers.

Pyroclastic flows of June 15, 1991, filled canyons to depths of up to 200 m (typically, 50–100 m) and left nearly flat, featureless landscapes save for a few islands (kipukas) of older terrain whose tops remained unburied. About 5 to 6 km^3 of fresh pyroclastic-flow deposit and about 0.5 km^3 of tephra-fall deposit rested briefly on Pinatubo's slopes, awaiting heavy typhoon and monsoonal rains. Burial and stripping of vegetation, plus a cover of fine ash, increased runoff dramatically. Rains from Typhoon Yunya during the climactic eruption itself followed in short order by more typhoons and the southwest monsoon, quickly began their work. Drainages were reintegrated within the first 2 months, and more than 0.8 km^3 of debris was moved onto the fans within just 3 months. By the end of the second year, lahar deposits exceeded 1.4 km^3, and by the end of the third year, about 1.9 km^3 had reached the fans. Rates of sediment yield during 1991, as high as 4×10^6 m^3 of sediment per square kilometer of watershed, and as high as 1,600 m^3 of sediment per square kilometer per millimeter of rainfall (Janda and others), were an order of magnitude higher than previous records from Mount St. Helens and Sakurajima Volcano; sediment yields in 1992 and 1993, though declining, began to include significant components of pre-1991 debris.

Papers in this volume address only the first 2 to 3 years of posteruption lahars, and thus only the first chapters of the ultimate lahar story. Papers are arranged chronologically, geographically, and topically. Syneruption lahars associated with Typhoon Yunya (Major and others) were followed by other 1991 lahars of the east side (Pierson and others; K.M. Scott and others) and west side (Rodolfo and others; Umbal and Rodolfo). Umbal and Rodolfo describe the evolution and periodic breakout of a lahar-dammed lake on the southwest side of the volcano; Scott and others present channel cross sections that graphically show filling of lowland channels. Differences from one watershed to the next are addressed by several authors.

Papers on 1992 lahars on the east side present results of more systematic monitoring than was possible in 1991. Instrumental monitoring (Marcial and others) included rainfall monitoring and use of acoustic flow monitors (designed by R.G. LaHusen) to quantify lahars. Even relatively light rainfall (0.3–0.4 mm/min for 30 min) triggered lahars in the Sacobia River during 1991 and 1992, and there was an essentially linear relation between rainfall and instrumentally recorded lahar runoff. In addition, both instantaneous and time-integrated records of acoustic flow sensors correlated well with manually observed discharge at nearby observation posts (Tuñgol and Regalado; Martinez and others; Arboleda and Martinez). Cumulative records from the acoustic flow monitors are in remarkably good agreement with postseason measurements of the volume of lahar deposits.

For the west side of Pinatubo, Rodolfo and others make interesting comparisons between lahars of 1991 and 1992—comparisons that are possible because those authors began systematic observations shortly after the eruption and continued them doggedly until lahars threatened their watchpoints late in 1993 and spread over such a broad area that observations from a single streambank became meaningless.

Filling of downstream channels, followed by overbank flow into surrounding fields and villages, caused massive damage. Continued deposition, at progressively higher points on the alluvial fans, is increasing the threat in some drainages that lahars will move into adjoining drainages and threaten towns that to date have still escaped major damage. More than 50,000 persons have been displaced, about 400 km^2 of rich agricultural land have been covered, and an even larger area is now flooded during heavy rains or left dry by choking of irrigation canals. Transportation has been disrupted repeatedly, and the specter of further lahars has slowed economic recovery of the area. Indeed, direct and indirect damage from lahars has probably exceeded that from the eruption several times over.

Intricately dissected pyroclastic-flow deposits, Sacobia River.

Two hot lahars, marked by trains of steam plumes, flowing down forks of the Marella River, September 19, 1991.

Watershed Disturbance and Lahars on the East Side of Mount Pinatubo During the mid-June 1991 Eruptions

By Jon J. Major,[1] Richard J. Janda,[1][3] and Arturo S. Daag[2]

ABSTRACT

Densely populated alluvial fans on the east side of Mount Pinatubo were affected by widespread lahars triggered during the June 15, 1991, eruptions. The lahars were triggered by heavy rainfall. However, the lahars were not generated because of uncommonly heavy precipitation, but rather because of radical alteration of watershed hydrology by volcanic deposits in conjunction with heavy rainfall. Fine-grained fall and surge deposits related to eruptions that *preceded* the climactic-phase activity damaged vegetation, reduced the infiltration capacity of hillslope surfaces, and smoothed the natural-scale hillslope roughness. These effects led to enhanced overland flow that instigated hillslope and channel erosion, triggered minor slope failures, and initiated the peak-discharge lahars. Sediment mobilized by rilling and shallow landsliding of the mantle of pumice tephra deposited by the climactic-phase eruption contributed to lahars interbedded with pyroclastic valley fill and to pumice-bearing recessional flow that followed peak-discharge.

The peak-discharge lahars that flooded fan channels in Tarlac and Pampanga Provinces varied in rheology, magnitude, and timing. Below fanheads, lahars were dominantly hyperconcentrated streamflow or flow transitional to hyperconcentrated streamflow; above fanheads, lahars were dominantly debris flow. Along the mountain front from the Gumain River to the Sacobia River, peak discharge preceded deposition of broadly distributed plinian pumice fall; between the Sacobia River and the O'Donnell River, peak discharge largely followed pumice fall. Variations in rheology, magnitude, and timing of peak flows reflect the interplay of variations in rainfall intensity and timing, variations in the degree of watershed disturbance, and characteristic response times of the watersheds. Interplay of these factors affects the amount of water that moves from hillslopes to channels to fanheads, how rapidly it moves, and how easily sediment is entrained and transported.

Damage to habitation and infrastructure on the alluvial fans resulted primarily from lateral bank erosion and from aggradation of mainstem channels that induced backflooding of tributary channels. Overbank flooding by lahars was generally localized and of secondary importance except along distal alluvial plains, where primarily agricultural land was inundated.

INTRODUCTION

Heavy rainfall during the explosive eruptions of Mount Pinatubo in mid-June 1991 produced lahars that damaged habitation, infrastructure, and arable land on densely populated alluvial fans that flank the eastern sector of the volcano and the surrounding terrain. These lahars had sediment and fluid-dynamic characteristics very similar to lahars that have formed as a result of explosive eruptions at snow-clad volcanoes, for example during the 1989-90 eruptions of Redoubt Volcano, Alaska (Alaska Volcano Observatory, 1990; Brantley, 1990; Dorava and Meyer, 1994), the 1980 and 1982 eruptions of Mount St. Helens (Janda and others, 1981; Waitt and others, 1983; Pierson and Scott, 1985; Major and Voight, 1986; K.M. Scott, 1988), and during the 1971 eruption of Hudson Volcano, Chile (Best, 1992), despite having markedly different initiation mechanisms. However, unlike lahars generated during eruptions of snow-clad volcanoes, where interaction among volcanic products and snow and ice (for example, Major and Newhall, 1989; Trabant and others, 1994) can greatly increase the volume of water typically delivered to watershed channels, the lahars at Mount Pinatubo were generated by commonplace, albeit heavy, tropical rainfall. Although a typhoon crossed central Luzon island coincident with the June 15 eruption, that typhoon was not especially intense (Oswalt and others, this volume). The lahars concurrent with the June 15 eruptions of Mount Pinatubo were not the result of an uncommon meteorological event coincident with the eruption but were, as we will argue, the result of a fundamental alteration

---

[1]U.S. Geological Survey.

[2]Philippine Institute of Volcanology and Seismology.

[3]Deceased.

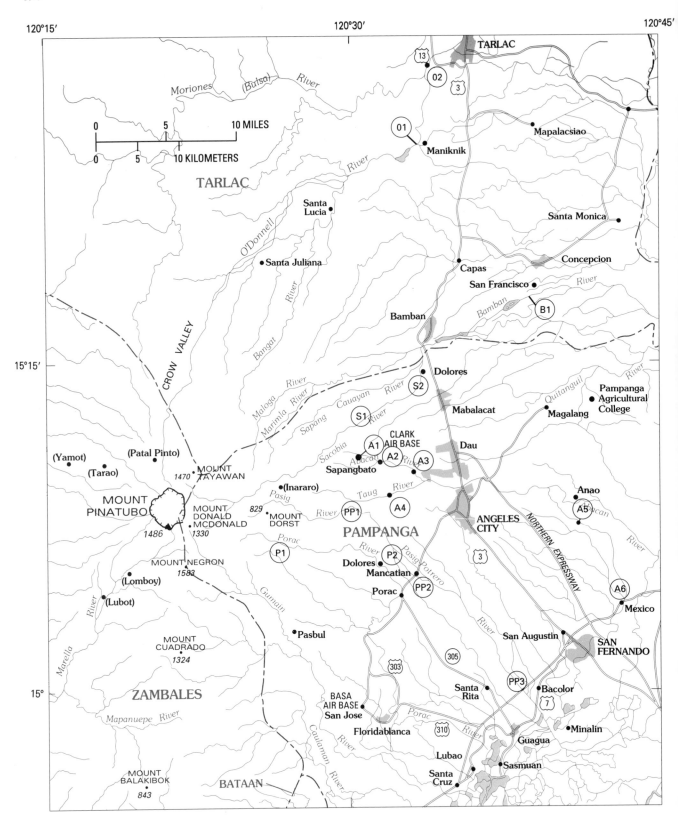

Figure 1. Drainage network around Mount Pinatubo. Cited outcrops are indicated. Villages in parentheses were destroyed during the eruption.

of watershed hydrology that acted in conjunction with heavy rainfall.

The purposes of this paper are to document and discuss the effects of, and the deposits of, lahars that occurred on the east side of Mount Pinatubo, in Tarlac and Pampanga Provinces (fig. 1), before, during, and shortly after the climactic June 15, 1991, eruption, and to explore the central cause of these lahars. In particular, we discuss the sedimentology and behavior of lahars generated in several watersheds, the relation between the lahars and tephras deposited by the mid-June eruptions, and the geomorphological consequences of those lahars.

Annual monsoon rains have mobilized vast amounts of sediment and generated numerous lahars that have greatly modified the east-side alluvial fans since the June 1991 eruptions (Pierson and others, this volume; K.M. Scott and others, this volume; Dolan, 1993). This paper discusses lahar deposits and channel conditions as they existed in late June to mid-July 1991. Many deposits documented in this paper have been eroded or have been buried by as much as several tens of meters of sediment transported by subsequent lahars. Geographically, this paper will cover an area bounded on the north by the O'Donnell-Tarlac River and on the south by the Porac River (fig. 1). We do not discuss the Gumain River system. Papers by Rodolfo and others (this volume) and Umbal and Rodolfo (this volume) discuss lahars affecting basins west of Mount Pinatubo.

PHYSIOGRAPHIC SETTING

Mount Pinatubo is located in the Philippines approximately 100 km northwest of Manila in central Luzon (fig. 1). The volcano stood 1,745 m above sea level prior to the June 15 eruption; the highest point on its crater rim now stands about 1,485 m above sea level (Jones and Newhall, this volume). The volcano is surrounded by steeply dissected mountains to the west, south, and northeast. The provinces of Tarlac and Pampanga are drained by the O'Donnell-Tarlac, Sacobia-Bamban, Abacan, Pasig-Potrero, Porac, and Gumain watersheds (fig. 1). Prior to the June 1991 eruptions, the Abacan and Pasig-Potrero Rivers did not head on the flanks of the volcano; their watersheds were developed in steeply dissected terrain that fronted Mount Pinatubo. Above 1,000 m altitude, watersheds had hillslope angles that generally ranged from nearly 20° to more than 45°; channel gradients commonly were about 6° to 18°. Between altitudes of about 1,000 m and 200 m, channel gradients flattened to about 1°. Below 200 m, broad, coalescing alluvial fans having low gradients (fig. 2) that range from about 0.02 m/m at fan heads to less than 0.0002 m/m along distal alluvial plains dominate the topography (Pierson and others, 1992; this volume). Stream channels generally were well incised above fanheads; on the alluvial fans, channels were only mildly confined, except

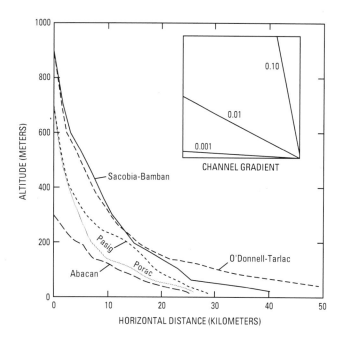

Figure 2. Preeruption longitudinal profiles of major river channels east of Mount Pinatubo that conveyed lahars during the June 15, 1991, eruption. Horizontal distance is measured from the channel head.

where artificially manipulated, by river banks less than a few meters high. Drainage basin characteristics above fanheads are provided in table 1.

TERMINOLOGY

The term *lahar* can mean many things to many people; the literature is replete with confusing, and sometimes conflicting, usage. Participants of a recent Geological Society of America Penrose Conference proposed the following usage (Smith and Fritz, 1989): "...lahar: a general term for a rapidly flowing mixture of rock debris and water (other than normal streamflow) from a volcano. A lahar is an event; it can refer to one or more discrete processes but does not refer to a deposit." This usage allows a single event to encompass a variety of rheological behaviors. In the following discussion we shall emphasize the effects, behavior, and consequences of two types of highly concentrated suspensions of sediment and water: *debris flows*, defined as flowing mixtures of sediment and water that commonly consist of a broad distribution of grain sizes and have sediment concentrations that usually exceed about 60 percent by volume, and *hyperconcentrated flows*, defined as flowing mixtures of sediment and water commonly composed of a narrower distribution of solid particles, dominantly sand sized, which commonly have sediment concentrations that range between 20 and 60 percent by volume (for example, Beverage and Culbertson, 1964; Pierson and Scott, 1985;

Table 1. Characteristics of drainage basins above fanheads east of Mount Pinatubo.

Watershed		Area (km²)	Total lengths of channels (km)	Drainage density (km/km²)	Area covered by pyroclastic flow (km²)	Percent of watershed covered by pyroclastic flow	Estimated equilibrium response time (hrs)[1]
O'Donnell River above 120 m		136	140.6	1.03	~7.7	6	3-3.5
Marimla River above Bamban River		56.4	68.7	1.22	~0.5	<1	1.5-2
Sacobia River:	above 200 m	42.5	53.6	1.26	~12.3	29	2.5-3
	above Abacan	33.1	43.2	1.31	~11.5	35	
Abacan River above Taug River (includes Taug River basin)		38.5	63.5	1.65	~3.1	8	1-1.5
Pasig-Potrero River above 240 m		22.7	26.1	1.15	~7.5	33	1.5-2
Porac River above 120 m		30.8	31.2	1.01	~0.3	<1	1.5-2

[1] Computation of response time assumes full watershed contribution above the chosen outlet point and is divided into two contributing processes: turbulent overland flow off hillslopes and turbulent channel flow. The overland flow model assumes roughness factors ranging from 0.1 to 10 (Dunne and Dietrich, 1980) and a precipitation intensity >> infiltration rate. A precipitation intensity of 150 mm/day, approximately double the mean precipitation intensity measured at Clark Air Base from June 15-17, 1991, was assumed for the calculation. The channel contribution assumes a mean flow velocity of 3 m/s. Chosen outlet points include: Sacobia River above Bamban River; Abacan River above Angeles; Pasig-Potrero River above Mancatian; and Porac River above Porac (fig. 1). The outlet points for the O'Donnell River and the Marimla River are as shown in table.

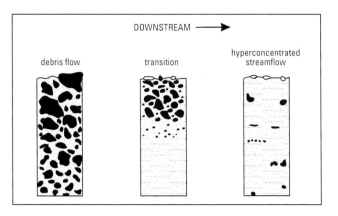

Figure 3. Deposit facies associated with lahars (from Pierson and Scott, 1985). Debris-flow deposits commonly are poorly sorted and have broad distributions of grain size; the mean grain size of the nongravel fraction commonly ranges from about 1 mm to more than 4 mm (very coarse sand to pebbles). Hyperconcentrated streamflow deposits commonly are massive to slightly laminated and have a narrower distribution of grain sizes; the mean grain size commonly ranges from about 0.25 mm to 0.5 mm (medium to coarse sand).

Smith and Lowe, 1991; Pierson and others, this volume). Sedimentologic characteristics of deposits inferred to have resulted from these processes are illustrated in figure 3.

Wentworth sediment-size terms (silt, sand, gravel) are used to describe the textures of deposits. Division of the Wentworth size terms into finer designations is well established (Folk, 1974). Pyroclastic size terms (ash, lapilli, bombs), on the other hand, represent broad categories of grain size that are insufficiently divided into finer designations (Fisher and Schminke, 1984). The use of Wentworth sediment-size terms may irritate some readers, especially when the terms are applied to pyroclastic deposits. However, the Wentworth terminology accurately describes the texture of *sediments*, is unambiguous, and holds no genetic connotations.

THE EXPLOSIVE ERUPTIONS AND THEIR DEPOSITS

CHRONOLOGY OF EVENTS AND CHARACTERISTICS OF DEPOSITS

The first in a series of powerful explosions occurred on June 12, 1991 (Philippine Volcano Observatory Team, 1991; Hoblitt, Wolfe, and others, this volume). Explosive pulses during the early part of this phase of activity produced pyroclastic flows that traveled up to 6 km from the old summit on the north, northwest, and west flanks of the volcano. Small-volume pyroclastic flows also descended the south and east flanks of the volcano. Sandy tephra fell mainly north to southwest of the volcano; little if any tephra

fell to the east during these explosions (Hoblitt, Wolfe, and others, this volume). Eruptions on the afternoon of June 14 marked the end of a phase of strong vertical eruptions and the beginning of activity dominated by laterally directed blasts (PVO Team, 1991; Hoblitt and others, 1991; Hoblitt, Wolfe, and others, this volume; Wolfe and Hoblitt, this volume). Several of these blasts produced various pyroclastic density currents that traveled as much as 10 km from the vent, and heavy fall that blanketed many sectors of the surrounding terrain (Hoblitt, Wolfe, and others, this volume; W.E. Scott and others, 1991; this volume). Explosive activity culminated about 1400 (all times listed are local time) on June 15 with a sustained pumice-rich, climactic eruption that persisted into the evening (Hoblitt and others, 1991; Hoblitt, Wolfe, and others, this volume). Following the climactic phase of activity, voluminous ash emission continued, and slowly decreased, over a period of several weeks.

Explosive activity on June 15 produced voluminous deposits of tephra fall and of various pyroclastic density currents (W.E. Scott and others, 1991; this volume; Paladio-Melosantos and others, this volume). Pyroclastic currents swept all sectors to a maximum distance of 16 km from the vent. Pyroclastic-flow deposits accumulated chiefly between 5 and 15 km of the vent and broadly buried upland channels to depths ranging from a few centimeters to more than 200 m; large areas are buried to an average thickness of 30 to 50 m (W.E. Scott and others, 1991; this volume). On the east flank of Mount Pinatubo, mainstem valleys in the headwaters of the O'Donnell, Sacobia, and Pasig River watersheds are deeply filled (table 1; W.E. Scott and others, this volume, figs. 1 and 2). Valleys northeast and southeast of the volcano contain relatively minor pyroclastic-flow deposits (W.E. Scott and others, 1991).

Tephra-fall and surge deposits associated with preclimactic eruptive activity are chiefly fine grained (fig. 4). East of the volcano these deposits consist primarily of normally graded fine sand to silt (Hoblitt, Wolfe, and others, this volume). The climactic phase of the eruption, on the other hand, produced a dominantly coarse, granular pumice fall (W.E. Scott and others, 1991; Paladio-Melosantos and others, this volume). East of the volcano this fall deposit consists chiefly of a normally graded bed of pumiceous and minor lithic small pebbles, granules, and sand, depending on proximity to the cone (fig. 4). Tephra-fall thickness east of the volcano ranges from about 1 to 5 cm distributed broadly across the distal alluvial fans to about 50 cm on slopes in upland areas (see figs. 3, 4 of Paladio-Melosantos and others, this volume). Pumiceous pyroclastic flows generated by the climactic eruption are synchronous with, or lie stratigraphically above, the pumice fall (W.E. Scott and others, this volume).

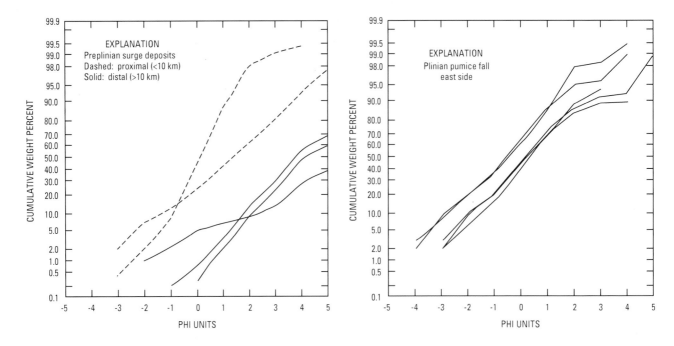

Figure 4. Cumulative grain-size distributions of the fine-grained preplinian and plinian tephra-fall deposits east of Mount Pinatubo. Data from R.P. Hoblitt (written commun., 1993) and Paladio-Melosantos and others (this volume).

WATERSHED DISTURBANCE

Assorted volcanic processes disturbed watersheds in the vicinity of Mount Pinatubo to varying degrees. Pyroclastic-flow deposits are more extensive in watersheds west of Mount Pinatubo than in watersheds to the east (W.E. Scott and others, this volume). Minor volumes of pyroclastic debris were deposited in the upper O'Donnell and Sacobia watersheds on June 11 (Hoblitt, Wolfe, and others, this volume). Activity on June 12 and 13 did not affect east-side watersheds. Small explosions and low fountaining of ejecta on the afternoon of June 14 dispersed pyroclastic density currents into the O'Donnell basin. Many subsequent explosions leading to the climactic activity dispersed pyroclastic density currents radially and deposited sediment from pyroclastic surges in nearly all watersheds.

Thicknesses of surge deposits in catchments are not well known. Proximal surge deposits, within 10 km of the old summit, are as much as a meter thick; individual beds of distal surge deposits typically range from nearly 0 to a few centimeters thick (Hoblitt, Wolfe, and others, this volume).

Tephra fall was relatively evenly distributed across all sectors. The first broadly distributed, fine-grained fallout to affect the east side of Mount Pinatubo was associated with an eruption at about 1500 on June 14 (Hoblitt, Wolfe, and others, this volume). Fine-grained tephra associated with preclimactic eruptions on the morning of June 15 was also broadly distributed to the east and obscured views of the volcano from early morning until after the climactic eruption.

Pyroclastic flows associated with the climactic-phase eruption extended down several mainstem channels on the east side, deeply filling the upper valleys (W.E. Scott and others, this volume). Plinian pumice fall on the east side is commonly 10 to 50 cm thick in the upper basins (Paladio-Melosantos and others, this volume). We will show, however, that peak discharge of several of the lahars that inundated channels on the east side of Mount Pinatubo occurred *before* deposition of the plinian pumice fall and the extensive pyroclastic valley fill. The pyroclastic-surge deposits that accumulated in the 24-h period from the afternoon of June 14 until the beginning of the climactic eruption on the afternoon of June 15 were responsible for altering watershed hydrology and establishing conditions conducive to the generation of lahars once rainfall commenced.

RAINFALL

The climate of central Luzon is characterized by distinct wet and dry seasons. The climate generally is dry when dominated by easterly flow of the North Pacific trade winds. Between May and October, however, the moist, tropical southwest monsoon dominates (Pierson and others, this volume), and many typhoons commonly pass across, or near, the Philippines and draw in even more moist air (Pierson and others, this volume). More detailed discussions of the central Luzon climate are found in Pierson and others (1992, this volume), Rodolfo (1991), Rodolfo and others (this volume), and Umbal and Rodolfo (this volume).

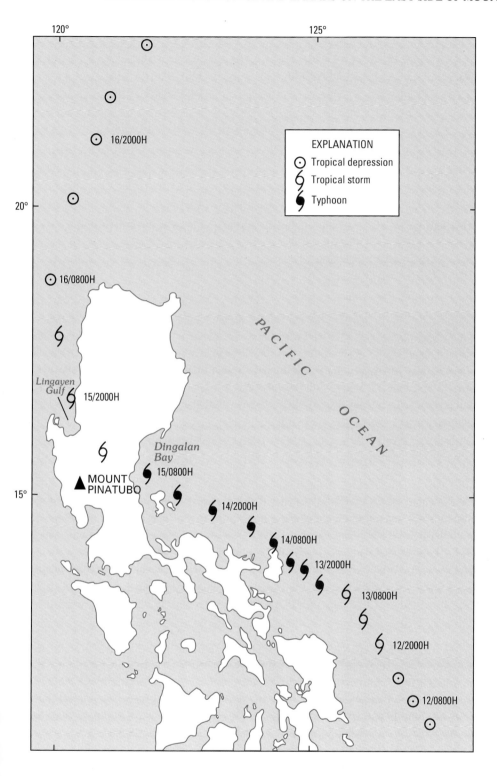

Figure 5. Path of Typhoon Yunya (after Oswalt and others, this volume). Circles with dots indicate tropical depression; unfilled circles indicate tropical storm; and filled circles indicate typhoon. Date and local time are given as 15/0800H, meaning 0800 on June 15, 1991.

On June 11 and 12, 1991, a small tropical depression over the Pacific Ocean east of the Philippines developed into Typhoon Yunya (Oswalt and others, this volume). By 0800 on June 14, the typhoon, still at sea southeast of central Luzon, packed winds that peaked at 195 km/hr; nevertheless, Yunya was considered only a modest-sized typhoon (Oswalt and others, this volume). Twenty-four hours later,

the typhoon went ashore about 90 km east of Clark Air Base (fig. 5). The typhoon tracked northwesterly and rapidly lost organization as it passed over land and began interacting with the eruptive column from Mount Pinatubo (Oswalt and others, this volume). By 1200 on June 15 it passed within 60 km northeast of Clark, and by about 1700 the typhoon, then downgraded to a tropical storm, passed Baguio, 150

Table 2. Daily precipitation (in millimeters) measured at stations in the vicinity of Mount Pinatubo, Pampanga Province.

Station	Coordinates (lat/long)	Elevation (m)	Day in June 1991					
			12	13	14	15	16	17
Magalang	15°13'N./120°42'E.	38	0	6.6	26.4	19.8	12.9	17.5
San Fernando	15°02'N./120°42'E.	1	0	1.4	42.0	8.2	19.4	8.6

km northeast of Mount Pinatubo. Yunya exited the Philippines approximately 185 km northwest of Clark at about 2000.

Rainfall from Typhoon Yunya triggered several lahars at Mount Pinatubo; however, the amounts and intensities of rainfall produced by the typhoon in upland areas are unknown. Measurements at Clark indicated that 150 mm of rain fell between the mornings of June 15 and June 17, but the onset and intensities of the rainfall are not known (Maj. W. Nichols, U.S. Air Force, oral commun., 1991). Eyewitnesses report that rain did not begin to fall at Clark until late-morning on June 15 and that heavy rainfall did not begin until around 1400 (R.P Hoblitt, U.S. Geological Survey, oral commun., 1993). Distributing this rainfall evenly over 2 days yields a minimum rainfall intensity of 75 mm per day. Daily rainfall of this magnitude is not uncommon during the southwest monsoon. Rainfall data collected at Clark from June 20 to September 9, 1991, show that daily rainfall exceeded 70 mm once, and on 6 days exceeded 60 mm (U.S. Air Force, unpub. data, 1991). Despite ambiguities of onset and intensity, we infer that the rainfall associated with the typhoon was not an exceptional meteorological event.

Rainfall in central Luzon is characterized by high spatial and temporal variability. Records from Magalang and San Fernando (fig. 1) illustrate the variability of rainfall across distal reaches of alluvial fans east of Mount Pinatubo from June 12 to 17, 1991 (table 2). Between June 14 and 16, precipitation at these stations ranged from 8.2 to 42.0 mm per day. Two-day moving averages of the rainfall collected at the gages between June 14 and 17 range from 13.8 to 25.1 mm per day, far less than the estimated 75 mm per day at Clark during the same period. Precipitation on lowland alluvial fans has little bearing, however, on precipitation responsible for triggering lahars in upland catchments, owing to orographic and rain-shadow effects. This has been illustrated clearly elsewhere, particularly in tropical climates in which the geomorphic setting is characterized by rugged topography (for example, Giambelluca and others,

1984; MacArthur and others, 1992). A study of rainfall-triggered lahars at Mayon volcano, southern Luzon, showed that orographic effects enhanced the rainfall accumulation on the slopes of the volcano commonly more than two, and sometimes as much as five, times the accumulation recorded on lowland alluvial fans only 12 km away (Rodolfo and Arguden, 1991). That study also showed that rainfall intensity, and, particularly, the short-term intensity of rainfall that accumulates during the most intense 10 min, were important factors in triggering lahars at Mayon.

Intensities, durations, and amounts of precipitation in the upper basins of the major river systems that drain Mount Pinatubo probably exceeded those recorded on the distal fans. Intense convective storm cells commonly formed in close spatial and temporal association with localized volcanic heat sources prior to the June 15 eruptions (Pierson and others, 1992; Oswalt and others, this volume), and it is likely that similar locally developed storm cells were induced by pyroclastic debris in catchment headwaters and by the large convecting plume of ash discharging from the volcano on June 15. As an example of the intensity of rain that can fall in the upper catchments, we cite a storm that occurred on July 9, 1991. A rain gage installed at approximately 1,060 m in altitude near Mount Cuadrado in the Gumain River watershed (fig. 1) recorded 25.4 mm of rain 13 min after the onset of a local storm and 50.8 mm of rainfall after 60 min (R.J. Janda, U.S. Geological Survey, unpub. data, 1991). Four rain gages within or close to the east-side watersheds, installed subsequent to the June 1991 eruptions, show large variations in measured daily rainfall during the rainy season (Pierson and others, this volume). In fact, high-intensity rainfall was sufficiently localized in July, August, and September 1991 that lahars commonly occurred when little or no rainfall was recorded at one or more of the rain gages (Pierson and others, this volume). Although we have no information on rainfall in the upper basins during the June 15 eruptions, we infer that the rainfall intensity was least 75 mm/day and probably substantially greater.

THE LAHARS: CHARACTERISTICS, TIMING, AND GEOMORPHIC CONSEQUENCES

Owing to variations in the volcanic processes that affected watersheds, to variations in the degree of disturbance by those processes, and to variations in rainfall intensity and timing, the lahars that inundated channels on the eastern flank of Mount Pinatubo varied in rheology, magnitude, and timing. The downstream geomorphic consequences of the lahars are directly related to characteristics and timing of the lahars, which in turn are directly linked to initiation mechanisms, sediment characteristics, and watershed hydrology.

O'DONNELL-TARLAC RIVER SYSTEM

Deposits in the lower O'Donnell-Tarlac system record passage of several lahars having characteristics of hyperconcentrated flow or of flow undergoing transition from debris flow to hyperconcentrated flow (fig. 6; table 3). At an irrigation dam near Maniknik (location O1, fig. 1), three thin, sharply defined deposits of pebbly to pebbly-muddy sand support pumice and (or) lithic pebbles. The basal deposit contains almost exclusively pumice pebbles, whereas the upper two deposits are dominantly lithic rich. The middle unit ranges widely in thickness, is discontinuous, and exhibits soft sediment deformation, indicative that it was not sufficiently consolidated before the overlying unit was deposited. The sequence of lahar deposits overlies a discontinuous deposit of June 15 pumice fall, which in turn lies directly on preeruption alluvium.

By the time the lahars reached Tarlac, 8 km farther downstream (location O2, fig. 1), they had dropped nearly all of their gravel in the channel and were evolving toward hyperconcentrated and (or) sediment-laden streamflow (fig. 6; table 3). Here, flood-plain deposits of slightly pebbly to slightly pebbly-muddy sand support few clasts larger than 3 cm diameter, although scattered pumice cobbles as large as 20 cm were present on the surface. These deposits exhibit a greater variety of sedimentary textures than those upstream, ranging from massive and poorly sorted to well laminated and trough crossbedded. Fall deposits were not observed within this section.

Precisely when lahars flooded the lower O'Donnell River is ambiguous. Residents near Maniknik report that the first notable flow arrived between the evening of June 15 and early on June 16. Residents 15 km upstream report arrival of the first flow at 1900 on June 15 and subsequent flows on June 18. All eyewitness reports indicate several rises in stage having durations of 1 to 2 h, with intervening recessions of nearly 1 h. All accounts consistently report that multiple waves of flow swept the channel within a span of several hours and that the first flow did not reach the lower O'Donnell River until well after the onset of climactic eruptive activity.

SACOBIA-BAMBAN RIVER SYSTEM

Exposures in the Sacobia-Bamban valley reveal passage of several lahars prior to, during, and well after onset of the climactic phase of eruptive activity. The deposits suggest that flows progressively diluted downvalley and that the peak-discharge lahar prior to the onset of climactic activity either did not extend far onto the alluvial fan, deposited its sediment load entirely within the active channel, or diluted to sediment-laden streamflow that passed easily through the distal fan environment.

Incision in valley fill adjacent to Clark Air Base (location S1, fig. 1) reveals multiple debris-flow deposits beneath climactic-phase pyroclastic-flow deposits. Exposures also show that debris flows occurred before and after deposition of the pumice fall (fig. 6). The basal debris-flow deposit, of which 40 cm was exposed in early July 1991, consisted of a sandy gravel that contained dispersed lithic pebbles as large as 4 cm (sample S1D, fig. 6, table 3). By late July, channel incision exposed more than 4 m of that basal deposit. Deeper in the section, it consisted of dark, lithic boulders dispersed in a lithic sand matrix (K.M. Scott, U.S. Geological Survey, written commun., 1993).

The basal debris-flow deposit is separated from an overlying debris-flow deposit by more than a centimeter of gray laminated silt that represents fall and surge deposits associated with eruptions that occurred during the 24 h preceding the climactic eruption (Hoblitt, Wolfe, and others, this volume; W.E. Scott and others, this volume). Until mid-morning of June 15, little widespread tephra had fallen to the east; however, density currents easily transported fine-grained sediments along confined channels to distances in excess of 10 km (Hoblitt, Wolfe, and others, this volume). Judging by the thickness of silt that overlies the basal debris-flow deposit, we conclude that the debris flow occurred before mid-morning of June 15. R.P. Hoblitt (U.S. Geological Survey, oral commun., 1991) observed a hot lahar flowing along the Sacobia River near Clark Air Base on the evening of June 14, and we believe the basal deposit in this section records that lahar.

A debris flow that occurred between the evening of June 14 and early afternoon on June 15 deposited sediment that lies between silt layers below the pumice fall. Because this deposit lies between a relatively thick silt at its base and only a couple of millimeters of silt at its top, we speculate that it records flow along the valley during the late morning to early afternoon on June 15. This sandy gravel deposit, which is slightly finer grained than the upper part of the underlying debris-flow deposit and which locally thickens and truncates along channels eroded into the underlying deposit, has sedimentary characteristics that suggest that it

O'DONNELL-TARLAC

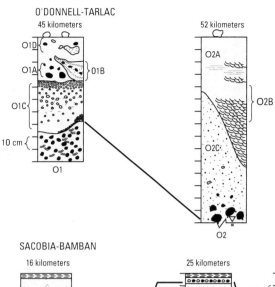

SACOBIA-BAMBAN

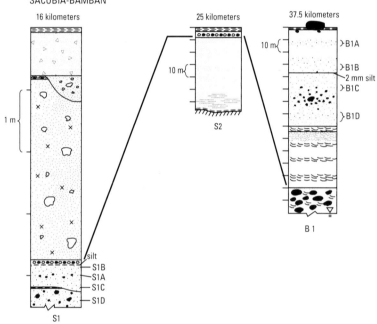

ABACAN

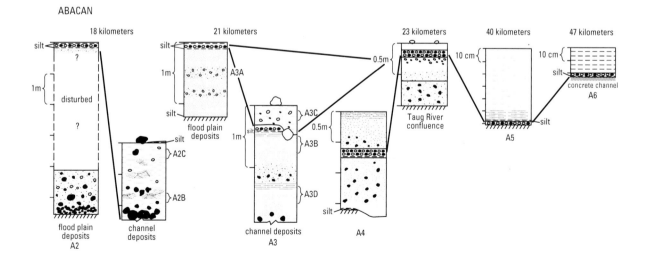

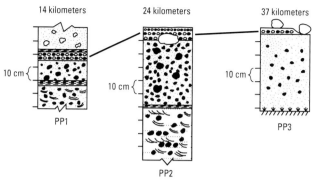

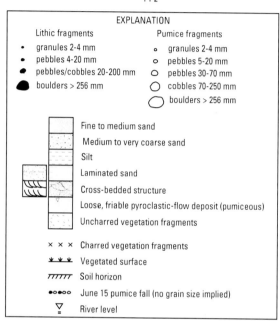

Figure 6. Stratigraphic sections of the mid-June lahar and related deposits along river channels east of Mount Pinatubo. Values (45 km) indicate the distance from the former summit of the volcano; code names (PP1) indicate cited outcrops (see fig. 1). Sample numbers refer to data in table 3. The line between sections correlates the position of the plinian pumice fall.

resulted from a debris flow that may have been evolving toward hyperconcentrated streamflow (figs. 3, 6; samples S1A,B,C, table 3).

A several-meter-thick sequence of pumiceous pyroclastic-flow deposits overlies the pumice fall in the lower Sacobia valley. Within that sequence of typically loose, silty sediment is a deposit of well-indurated, poorly sorted, ungraded, pumiceous silty sand that contains dispersed inflated pumice pebbles around which are localized zones of fines-depleted sand. This deposit fills shallow channels incised into underlying debris and truncates locally over minor topographic rises. These deposits are found near the distal end of the pyroclastic fill in the Sacobia valley. We

find it unlikely that the pyroclastic flows at this distance would be particularly erosive, especially on such low channel gradients. Indeed, a 3-m-thick pyroclastic-flow deposit overlies easily erodible pumice fall at this site. We conclude that water eroded the channels observed within the sequence and that the well-indurated deposit that fills those channels resulted from shallow slurries of reworked pyroclastic debris. Clearly, significant quantities of water were moving off hillslopes and through valleys during the emplacement of the pyroclastic valley fill.

Peak discharge inundated the flood plain on the lower Sacobia River near Dolores (Mabalacat Municipality) (location S2, fig. 1) during early morning to early afternoon on June 15. The flood-plain deposit from that flow consists of massive slightly pebbly sand that is moderately laminated at its base (fig. 6). It underlies pumice fall and overlies several millimeters of gray silt. Stratigraphic position of the peak-flow deposit between fall deposits of June 15 eruptions constrains the timing of the event. Deposit textures suggest the peak flow wave gradually diluted downvalley; here it was composed of a relatively dilute leading edge followed by a hyperconcentrated wave.

Lahars from the Sacobia River valley passed from a naturally confined channel along the mountain front and across the alluvial fan into the artificially contained Bamban River on the alluvial plain. Although lahars generally remained within the containment structures along the upper reaches of the plain, they broadly inundated arable land on the distal plains. Deposits of two lahars that inundated the flood plain are exposed near San Francisco (location B1, fig. 1), nearly 40 km from the volcano's former summit. Here, deposits consist of generally thin beds (<50 cm thick) of lithic pebbly to slightly pebbly sand (fig. 6; table 3) inferred to be deposited by hyperconcentrated flows or by flows transitional to hyperconcentrated flow. These deposits overlie pumice-rich alluvium that mantles preeruption lithic alluvium.

Conspicuously absent from the Bamban River exposure are the pumice fall and prepumice-fall lahar deposits correlative with the lithic-rich peak-flow deposits in the Sacobia valley. Beyond the containment dikes, an 8-mm-thick pumice-fall deposit is preserved. Eyewitnesses report that the pumice-rich alluvium at the base of the section was deposited on the evening of June 15 and that superposed sediments were deposited later that evening, about 2300 h. Streamflow that deposited the pumice alluvium possibly eroded and reworked the pumice fall within the confined reach. Absence of prepumice-fall deposits on the flood plain may indicate that the earlier peak-discharge lahar, recorded in the Sacobia valley, did not get this far down valley (unlikely), that it deposited its sediment load broadly within the active channel, or that it transformed completely to sediment-laden streamflow that passed easily through the distal fan environment and onto the alluvial plains, leaving little recognizable sediment.

Table 3. Characteristics of mid-June 1991 lahar deposits on the eastern sector of Mount Pinatubo.

[See figure 6 for positions of samples in deposits. Sediment statistic data exclusive of clasts >32 mm. M_z and σ_I are the mean grain diameter and sorting coefficient defined by Folk (1974); D_{50} is the median grain diameter. Mud refers to particles <63 μm]

River system	Location (fig. 1)	Sample	Distance from summit (km)	Deposit thickness (cm)	Largest clast within deposit (cm) P¹	L²	Modal clast (cm) P	L	Sediment content (wt %) gravel	sand	mud	M_z mm	Phi units	D_{50} mm	Phi units	Sorting σ_I
O'Donnell-Tarlac	O1	D³	45	4-34	2	6.5	0.3	-	0	92.8	7.2	0.26	1.95	0.30	1.75	1.06
		A³			-	-	-	-	38.0	61.6	.4	1.67	-0.74	1.34	-.42	2.12
		B		<5-16	-	1.5	-	.3	1.0	83.1	15.9	.22	2.21	.27	1.88	1.50
		C		42	7	-	.5-2	-	17.1	62.1	20.8	.33	1.59	.32	1.65	2.40
	O2	A	52	25-43	0.3	-	-	-	.1	88.8	11.1	.26	1.95	.30	1.75	1.29
		B		0-80	3	-	-	-	.5	91.1	8.4	.29	1.81	.30	1.72	1.31
		C		<20-100	1.5	>4	-	-	.7	83.9	15.4	.26	1.94	.30	1.76	1.75
Sacobia-Bamban	S1	B³	16	35	-	3	-	.5	18.4	78.6	3.0	.82	.28	.90	.15	1.59
		A³			-	-	-	-	64.2	31.8	4.0	3.29	-1.72	5.28	-2.40	2.70
		C³			-	-	-	-	45.1	45.9	9.0	1.23	-.30	1.72	-.78	2.29
		D		>40	-	4	-	.7	37.5	58.1	4.4	2.51	-1.33	1.21	-.27	2.19
	B1	A³	37.5	35	-	1.2	-	.3	4.8	87.2	8.0	.34	1.56	.34	1.57	1.66
		B³			-	100 on surface	-	-	.3	99.6	.1	.35	1.52	.35	1.52	0.97
		C⁴		42	-	5	-	.5-1.5	13.1	86.9	.0	.66	.61	.55	.85	1.40
		D⁴			-	-	-	-	.3	91.3	8.4	.27	1.88	.27	1.88	1.46
Abacan	A1	A	13	33	4	-	-	-	20.0	63.7	16.3	.45	1.14	.49	1.03	2.63
		B		<5-20	5	-	-	-	21.2	58.8	20.0	.46	1.13	.43	1.21	2.87
	A2	A⁵	18	260	-	6	-	1.5-3	3.2	93.8	3.0	.56	.84	.72	.47	1.41
		C^{3,5}		250	-	>50	-	-	20.9	79.0	.1	.88	.19	.89	.17	1.74
		B^{3,5}			-	-	-	-	58.9	40.2	1.0	3.43	-1.78	3.63	-1.86	2.31
	A3	A	21	108	.8	-	0.3	-	1.1	95.4	3.5	.54	.89	.57	.80	1.15
		C		28	38	0.5	<1	.3	2.0	94.0	4.0	.44	1.17	.47	1.09	1.24
		B		75	14	1.5	-	.3-6	15.0	85.0	.0	.84	.25	.90	.16	1.30

Table 3. Characteristics of mid-June 1991 lahar deposits on the eastern sector of Mount Pinatubo—Continued.

		>65		27	<1	D							
A5	40	57		—	—	16.7	80.1	3.2	.94	.09	1.09	-.12	1.37
A6	47	20		.3	.3	.0	95.6	4.4	.23	2.12	.23	.13	.91
				—	—	.0	38.5	61.5	.06	4.01	.06	4.17	.86
Pasig-Potrero PP2	24	50	30	5	1-3	41.9	55.3	2.8	1.51	-.59	1.45	-.54	2.32
PP3	37	60		2	1-2	36.2	62.9	.9	1.08	-.11	1.07	-.10	2.20

[1] Pumice.
[2] Lithic.
[3] Samples from same deposit.
[4] Samples from same deposit.
[5] Fluvial deposit.

Flow velocity of at least one lahar was estimated from the geometry of a mudline preserved on the concrete piers of the San Francisco bridge. The mudline extended nearly 3 m above pier footings on the upstream side of the bridge and nearly 2 m above footings on the downstream side. Eyewitnesses report that a pumice-rich flood scoured the channel slightly, exposing the bridge footings, and that a later flow left the mudline. From the mudline geometry, we infer that at least one of the later lahars had an instantaneous minimum wave velocity of about 4 to 4.5 m/s (see Major and Iverson, 1993). Mean flow velocity and mean discharge are not known, however.

ABACAN RIVER SYSTEM

Stratigraphic relations among deposits of pyroclastic flows, ash-cloud surges, lahars, and tephra fall in the upper Abacan watershed record sustained rainfall runoff and active formation of lahars concurrent with emplacement of the pyroclastic valley fill. Despite stratigraphic complexities, deposits near a gap in the divide that separates the Sacobia River valley from the Abacan watershed (location A1, figs. 1, 7) indicate that pyroclastic activity associated with the June 15 eruption was sustained for many hours. Near-surface pyroclastic deposits, which represent the latter phases of that sustained activity as well as secondary pyroclastic flows (W.E. Scott and others, this volume), lie stratigraphically above the pumice fall but below the silty fall of waning-phase eruptive activity. Debris-flow deposits, composed of well-indurated, poorly sorted, pumiceous, muddy sandy gravels that have vesiculated matrices (table 3), are found interbedded with pyroclastic deposits, filling depressions and thinning over topographic rises. One extensive debris-flow deposit mantled the deposits of all but the last pyroclastic flow to spill through the gap. Morphology, stratigraphy, and sedimentology of deposits in the upper Abacan watershed show that several lahars were concurrent with pyroclastic activity. However, none of these lahar deposits record the peak flow that passed down the Abacan River channel, which occurred prior to the deposition of the pyroclastic valley fill.

Near Sapangbato (fig. 1), adjacent to Clark Air Base, two tributaries join to form the mainstem Abacan River. Stratigraphic relations indicate that lahar deposition along the northern tributary occurred primarily after the pumice fall; however, isolated areas from which channel alluvium was stripped to bedrock and subsequently replaced by pumice fall attest to an important phase of prepumice-fall activity. These stratigraphic relations contrast sharply with those observed in the southern tributary channel. In that channel the peak-discharge lahar deposit contains almost no pumice and is overlain by pumice fall. In both channels, pumice-bearing alluvium was deposited after the pumice fall. Below the confluence (location A2, fig. 1) channel- and peak-flow-

Figure 7. Aerial view of the upper and lower gap in the watershed divide between the Sacobia and Abacan River valleys. The view is looking from the Abacan watershed toward the Sacobia valley. Note the thick pyroclastic fill that spilled through the gap. Photograph by R.P. Hoblitt, March 18, 1992.

facies deposits reflect variable character and timing of flows (fig. 6). Channel-facies deposits consist primarily of laminated to crossbedded, lithic, sandy- to sandy-gravel alluvium that contains frothy pumice fragments (fig. 6; table 3). Gray silt mantles some of these channel deposits that record dominantly fluvial deposition during a period of sustained flow that followed the pumice fall. Pumice-rich alluvium mantled only the valley floor. The peak-flow deposit, on the other hand, was preserved on the valley side more than 6 m above the channel bed (figs. 6, 8). That deposit consisted of massive lithic sand underneath pumice fall and waning-phase silty fall. Near the channel bed, however, a lithic debris-flow deposit that contains minor angular fragments of pumice overlies preeruption deposits. This deposit does not overlie pumice fall nor was pumice fall found on its surface. There was no obvious contact in the intervening zone between the deposit at the limit of peak flow and the deposit near the channel bed. However, the valley margin had been disturbed by later flow. Sustained recessional flow possibly removed whatever pumice fall may have mantled the lower deposit.

Although houses upstream were inundated locally by lahars, severe channel erosion occurred along the Abacan River in Sapangbato (fig. 9). Sections of the channel were scoured to bedrock, suggesting that at least a meter or more of alluvium had been removed, and several buildings were undercut, indicative of several meters of lateral bank erosion (fig. 10). Neither exhumed bedrock nor freshly exposed bank faces in Sapangbato were covered by pumice fall. These observations, combined with the absence of

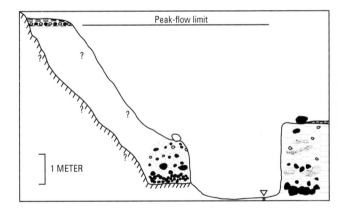

Figure 8. Stratigraphic sequence of deposits exposed in Abacan River valley at Sapangbato (location A2), Pampanga Province. View is looking downflow and represents a half cross section of the valley; deposits on the right of the figure are situated in the middle of the valley. See figure 6 and text for descriptive details. Horizontal dimensions not to scale.

pumice fall on channel-facies deposits, suggest that at least some of the erosion occurred during the period of sustained recessional flow that followed the peak-flow lahar. Above Sapangbato, particularly in the northern tributary of the Abacan River, pumice fall lies directly on some bedrock from which alluvium has been stripped. That stratigraphic relationship and the absence of prepumice-fall deposits in that channel suggest that water passed along the northern tributary before deposition of the pumice fall and scavenged sediment from the channel. This flow may have combined

Figure 9. Homes along the northern tributary of the Abacan River above Sapangbato inundated by lahars after deposition of the plinian pumice fall. View is downstream. Photograph taken July 8, 1991.

Figure 10. Lateral bank erosion caused by recessional phase of the June 15 lahar on Abacan River in Sapangbato, Pampanga Province. Photograph taken July 6, 1991.

with flow coming down the southern tributary and formed the peak-flow lahar whose deposits were preserved at the distal end of Sapangbato. In Sapangbato, thick channel fill and an absence of pumice fall mantling eroded channel banks suggest that channel aggradation during recessional flow contributed to extensive lateral bank erosion. Increased sediment supply and rapid aggradation lead to smoothing of the channel bed (Dietrich and others, 1989), which in turn alters bedforms that can promote flow diversion and lead to bank erosion. From the observed stratigraphic relations and from general hydraulic responses of bedload-dominated rivers to increased sediment supply (for example, Richards, 1982), we speculate that bed sediment was scoured and mobilized mainly by the peak flow and that lateral bank erosion was accomplished primarily by the recessional flow.

Flood-plain sediments along the Abacan River 3 km below Sapangbato (location A3, fig. 1) record peak-flow deposition by a hyperconcentrated flow (figs. 6, 11; table 3). A massive to mildly laminated, slightly pebbly sand that contains lithic and pumice pebbles, deposited approximately 4 m above the channel bed, lies between gray silt deposits and is overlain by pumice fall. These stratigraphic relations show that peak flow passed this reach between the evening of June 14 and early afternoon on June 15. Personnel at Clark report hearing thunderous noises along the Abacan valley around 1400 on June 15 (R.P. Hoblitt, U.S. Geological Survey, oral commun., 1993). Peak-flow depth along this reach was about 2 m above the channel bed (at the time of investigation in late June 1991 the channel contained an estimated 3 m of fill), and the flood-plain deposit lies about 2 m above that limit. The position of the deposit, which lies at a sharp bend in the channel, suggests a minimum peak-flow velocity of about 6 m/s if the implied

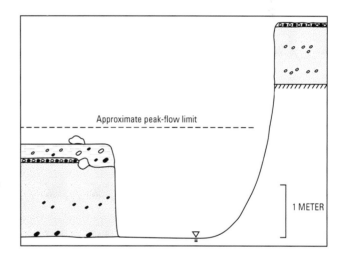

Figure 11. Stratigraphic sequence of deposits exposed in the Abacan River channel adjacent to Clark Air Base (location A3). The section view is oblique to the downflow direction. See figure 6 and text for descriptive details. Horizontal dimensions not to scale.

potential energy resulted from the complete conversion of the kinetic energy of the flow.

Channel deposits along this reach of the Abacan River consisted of primarily massive, pebbly to distinctly laminated, slightly pebbly lithic sands (fig. 6; table 3) that record hyperconcentrated flow and subsequent sediment-laden streamflow. Two hyperconcentrated-flow deposits underlie the pumice fall; streamflow deposits overlie the pumice fall.

Stratigraphic relations among deposits at the confluence of the Taug and Abacan Rivers, in Angeles City, record the passage of at least three waves of flow prior to deposition of the pumice fall (fig. 6). On the flood plain, poorly sorted, lithic sandy gravel overlies preeruption

Figure 12. View of Abacan River at Angeles City. Note erosion of left channel bank. Photograph taken June 25, 1991.

volcaniclastic deposits and is sharply overlain by massive sand and inversely graded pebbly sand. This sequence of deposits lies beneath the pumice fall. The basal unit is distinctly coarser than both the flood-plain and channel deposits observed a few kilometers upstream on the Abacan River. Those upstream deposits exhibit greater similarity to the massive and inversely graded sands at this section. The basal deposit here records a debris flow along the Taug River that preceded lahars on the Abacan River. Stratigraphic relations observed on Taug River 3 km upstream from the confluence (location A4, fig. 1) reveal a generally massive, poorly sorted lithic pebbly sand having matrix-supported clasts that lies between gray silt deposits and underlies pumice fall (fig. 6). This deposit nonconformably overlies the substrate (W.E. Scott, U.S. Geological Survey, written commun., 1992), indicating that the responsible flow was erosive along the channel bottom. Postpumice-fall flows on the Taug River generally were shallow and fluvial.

Several meters to tens of meters of lateral bank erosion occurred along the Abacan River in Angeles City (fig. 12). Buildings erected on terraces several meters above the channel collapsed after their foundations were undercut. Similar to the erosion documented in Sapangbato, the extensive lateral erosion in Angeles City occurred primarily during recessional flow after deposition of the pumice fall. Eroded banks and associated talus are not draped with pumice fall. Scour around piers and impacts from boulders and other debris caused structural failure of nearly all bridges across the river in Angeles City. The Northern Expressway bridge downstream from Angeles City escaped collapse; however, channel aggradation nearly buried that bridge.

The most areally extensive deposition by lahars in the Abacan watershed occurred in distributary channels across the alluvial plain beyond Angeles City (fig. 13). By the time peak flow reached the bridge between Anao and Mexico

(location A5, fig. 1), some 40 km river distance from the former summit of Mount Pinatubo, it had been significantly diluted by loss of sediment and it succeeded the pumice fall. There, pumice fall lies on flood-plain soil and beneath gray silt (fig. 6). Laminated sand overlies the silt (fig. 6; table 3). The top of that sand deposit is more than 2 m above the channel bed and about 1.2 m below the peak-flow limit preserved on fence posts.

In the village of Mexico, several kilometers farther downstream, peak flow was about 1 to 2 m deep according to eyewitnesses, and it deposited a laminated very fine sand and silt that overlies gray silt and pumice fall (location A6, fig. 6; table 3). This deposit, as well as that near Anao, shows clearly that along the alluvial plains the peak-discharge lahar associated with the June 15 eruption attenuated, slowed, and diluted.

PASIG-POTRERO RIVER SYSTEM

At the confluence of the primary tributaries that join to form the mainstem Pasig River (location PP1, fig. 1), a lahar deposit is interbedded with tephra fall and underlies thick pyroclastic fill (fig. 6; W.E. Scott, U.S. Geological Survey, written commun., 1992). At this location a muddy silt rests directly on an old road surface and on the former channel bed. In the former channel, this muddy silt is overlain by a generally thin, poorly sorted, lithic pebbly sand that lies directly below several centimeters of pumice fall. The pumice fall is overlain by several meters of pyroclastic debris as well as laharic debris from late-season monsoon-triggered flows. That the pebbly sand deposited below the pumiceous tephra, interpreted to be the deposit of a debris flow, is not found on terraces above the former channel indicates that the primary lahar was confined to the canyon at this location.

Deposits near Mancatian (location PP2, fig. 1) confirm that the primary lahar on the Pasig-Potrero River was a debris flow (fig. 6; table 3). Here, a massive, poorly sorted sandy pebble gravel was deposited on the flood plain more than 4 m above the channel bed. Pebbles are composed of both fresh and weathered lithics and of rare pumice. This deposit overlies silt that in turn rests on preeruption flood plain alluvium. The debris-flow deposit is overlain by pumice fall. These stratigraphic relations as well as an absence of silt on top of the debris-flow deposit suggest that peak discharge in the middle reach of the Pasig-Potrero River occurred in the early afternoon on June 15. Residents of Bacolor, about 13 km downstream, reported a sudden rise of stage at 1500. Within the main channel, and partly on the flood plain, the draping of preeruption volcaniclastic deposits and the pumice fall with a veneer of pumice sand, pebbles, and cobbles transported by subsequent water floods indicates that multiple waves of elevated discharge passed

Figure 13. Downstream aerial view of Abacan River below Angeles City. Beyond Angeles City, the lahar branched into many distributary channels, spilled overbank, and inundated parts of the alluvial plain, as seen on the right side of photograph. Photograph taken June 25, 1991.

through the channel subsequent to the onset of the June 15 eruptions (fig. 14).

The debris flow and early recessional-phase flow scoured nearly 1 m of alluvium from the channel and flood plain in the Mancatian reach, exhuming the underlying, relatively indurated, volcaniclastic deposits prior to deposition of the pumice fall (fig. 14). The minimum depth of the peak flow was about 4.5 m if we assume that the channel bed was not significantly lowered. The peak flow was sufficiently powerful to transport meter-sized boulders, to collapse the concrete bridge that crossed the river, and to transport slabs of that bridge, more than several meters long, nearly 2 km downvalley.

An aerial reconnaissance on July 3, 1991, revealed that the Pasig-Potrero channel was eroded along a several-kilometer reach. Several nickpoints were developed on the main stem of the river, and tributary valleys above Mancatian were freshly incised. The aerial investigation confirmed the passage of multiple flows. Consistent with the stratigraphic evidence, the peak discharge occurred before the pumice fall, and subsequent elevated discharges passed through the system during the late stages of, and after, pumice-fall deposition.

At the Santa Barbara bridge near Bacolor (location PP3, fig. 1), peak discharge is recorded by an ungraded, poorly sorted, lithic sandy pebble-gravel debris-flow deposit that lies directly on the formerly vegetated flood plain. This deposit, which is finer grained than that at Mancatian (table 3), is overlain by pumice fall. Pumice boulders are scattered on top of the fall deposit and locally on top of the debris-flow deposit where the fall has been eroded (fig. 6). This stratigraphic succession lies approximately 1.75 m above the former channel bed and records passage of at

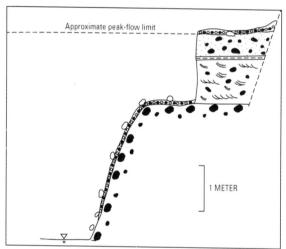

Figure 14. Stratigraphic sequence of deposits exposed on the right-bank flood plain of the Pasig-Potrero River at Mancatian (location PP2). See figure 6 and text for descriptive details. Horizontal dimensions not to scale.

least two flows. At least one, and perhaps several, wave(s) of streamflow having a lower stage than the primary lahar reworked deposits within and near the channel and deposited the pumice clasts on top of the pumice fall.

Flow lines on the bridge piers attest to the differences in character and stage between the peak and recessional flows in this channel. An upper mudline reached nearly 3 m above the channel bed on the upstream face of the pier and about 2.4 m on the downstream face. A lower trimline, nearly 1.5 m above the channel bed on both the upstream and downstream faces of the pier, reflected the limit to which dilute flow(s) removed the muddy veneer.

Differences in mudline elevation suggest that the instantaneous velocity of the peak discharge was about 3 to 3.5 m/s.

The peak flow along the Pasig-Potrero River spilled over a containment dike on the northeast side of the channel inundating arable land about 0.5 km upstream from Bacolor. Below Bacolor, aggradation of the mainstem channel and flow upstream along low-gradient distributary channels blocked drainage canals leading from rice fields to the river. As a result, parts of Bacolor and the surrounding area were inundated by backflooding of tributary canals. Although this lahar did not directly inundate large sections of inhabited or arable lands, it wreaked considerable havoc indirectly by inducing backflooding of channels along the alluvial plains.

PORAC RIVER SYSTEM

Residents of Porac report that silty tephra fall on June 15 was followed by a lahar at 1330, which was followed shortly thereafter by pumice fall (W.E. Scott, U.S. Geological Survey, written commun., 1992). Deposits in Porac and those a few kilometers upstream near Dolores (Porac Municipality) are consistent with those reports and further show that both debris flow and hyperconcentrated flow passed along the Porac River. Near Dolores (location P2, fig. 1), older flood-plain deposits are mantled by a few centimeters of silty tephra fall, which is overlain by a thin, massive to slightly inversely graded sand and a pebbly muddy sand (W.E. Scott, U.S. Geological Survey, written commun., 1992). These deposits are covered by pumice fall. The pumice fall is overlain by poorly sorted coarse sand containing conspicuous lithic pebbles and massive to laminated sand that contains scattered lithic pebbles. The post-pumice-fall deposits represent a debris flow followed by more dilute recessional flow that passed through the channel on June 30 (W.E. Scott, U.S. Geological Survey, written commun., 1992).

Complex stratigraphy in the upper Porac River basin (location P1, fig. 1) indicates contemporaneous emplacement of pyroclastic flows and at least one lahar and also shows that some lahars were generated well after tephra-mantled hillslopes were incised by overland flow. Deposits related to the peak-discharge lahar, which occurred prior to the deposition of the pumice fall, were not observed, however. Several meters of multiple postpumice-fall pyroclastic-flow deposits fill the upper valley. At least one debris-flow deposit is interbedded with this fill, another mantles the fill, and channel fill inset against the pyroclastic deposits has the characteristics of a debris-flow deposit. An extensive network of rills eroded into the tephra mantle scores the surrounding hillsides (fig. 15). These rills, some as deep as a few tens of centimeters, extend downslope to the valley fill where they are partly filled by debris-flow deposits. Only a few major rills cut across the surface of the

valley fill; minor rills are truncated. These relations show that debris was mobilized along the channel after the formation of the rill network on the hillslopes and that overland flow had diminished before deposition of the valley fill was complete.

Aerial reconnaissance on July 3 revealed that most of the low-order channels in the Porac watershed were scoured, reminiscent of erosion observed along lahar-impacted steepland channels following the 1985 eruption of Nevado del Ruiz Volcano, Colombia (Pierson and others, 1990). Although the sediment mobilized by the rilling of the hillslope tephra mantle contributed to downstream channel aggradation, it was not the source of sediment for the peak-discharge lahar. We suspect that the lithic-rich peak flow, which preceded deposition of the pumice fall, resulted from erosion and mobilization of channel sediments along the low-order channels.

DISCUSSION

The geomorphological responses of watersheds on the east side of Mount Pinatubo and the resulting timing and magnitudes of lahars reflect volcanic as well as climatic effects. Widespread lahars and floods of a magnitude and scale similar to, or slightly larger than, those of June 1991 last occurred during an eruptive period about 500 years ago (Newhall and others, this volume). Although isolated large, sediment-routing floods and debris flows undoubtedly occurred during the five-century repose, the watersheds generally have routed precipitation from hillslopes to channels and out onto alluvial fans without generating extensive laharic flows. Precipitation in June 1991 was apparently not uncommon, yet widespread debris flows and floods were generated, many prior to the plinian pumice fall. Clearly there is a link with renewed eruptive activity. Unlike lahars generated during eruptions of snow-clad volcanoes, where interaction among volcanic products and snow and ice can greatly increase the amount of water typically delivered to watershed channels, at Mount Pinatubo we cannot invoke an unusual increase in the volume of water delivered to the watersheds as the principle basis for generating these lahars. The lahars were triggered not solely by heavy precipitation but by substantive changes in watershed hydrology, which altered the manner and the rate that water was delivered from the hillslopes to the channels, in concert with coincidental heavy rains. Both conditions were necessary; neither was sufficient.

Volcanic events can profoundly alter the hydrologic regime of watersheds in two primary ways: (1) they can damage, strip, or remove vegetation and (2) they can deposit fine-grained, low-permeability sediment that commonly mantles the topography (for example, Segerstrom, 1950; Waldron, 1967; Kadomura and others, 1983; Collins and Dunne, 1986). Damage and removal of vegetation

Figure 15. Network of rills established on tephra-mantled hillslopes by rain that fell during the June 15 eruption. Photograph taken June 25, 1991.

reduces interception of precipitation by the canopy, so more water falls unimpeded onto the landscape. Deposition of low-permeability sediment buries ground litter, which reduces infiltration capacity of the substrate and also reduces the scale of hillslope roughness by smoothing the landscape. Infiltration rates can be reduced by more than an order of magnitude following tephra deposition (Leavesley and others, 1989). Each of these types of disturbance promotes increased overland flow, which delivers greater amounts of water from hillslopes to channels at a rate faster than would normally occur under preeruption conditions, and, through commonly associated rilling (for example, Collins and Dunne, 1986) and related shallow landsliding, enhances sediment delivery. Volcanically disturbed watersheds are therefore subject to larger peak flows and to greater volumes of runoff than would occur under preeruption conditions, and opportunities for sediment erosion and transport are substantially increased (Leavesley and others, 1989). Enhanced overland flow resulting from accumulated tephra fall in watersheds around Parícutin volcano, Mexico (Segerstrom, 1950), Irazú volcano, Costa Rica (Ulate and Corrales, 1966; Waldron, 1967), and Usu volcano, Japan (Kadomura and others, 1983), triggered several debris flows, even during relatively light rainfall.

Although the climactic-phase activity produced the most dramatic landscape disturbance, the hydrology of watersheds at Mount Pinatubo was severely altered, and substantial overland flow was generated, *before* the onset of the climactic eruption. Stratigraphic relations show clearly that many of the peak-discharge lahars were generated prior to deposition of the widely distributed pumice fall related to the climactic eruption. Catchment headwaters were impacted by several pyroclastic surges, associated tephra

fall, and minor pyroclastic flows in the days, and particularly in the 24 h, preceding the climactic eruption. These volcanic events damaged, removed, or buried vegetation and deposited generally thin, but variable, layers of chiefly fine-grained sediments, dominated by fine sand and silt, across the landscape. The occurrence of widespread, preclimactic lahars illustrates the delicate sensitivity of watershed hydrology to disturbance by even relatively thin, fine-grained volcanic deposits.

Significant rainfall runoff persisted into the phase of climactic eruptive activity, then rapidly diminished. An extensive rill network and numerous shallow landslides were observed in the mantle of pumice tephra on the hillslopes (figs. 15, 16), yet valley fill truncates many rills, and by late July 1991, the thick pyroclastic debris filling several valleys still lacked an integrated drainage system (K.M. Scott and others, this volume). The fact that only major rills cut across valley-bottom deposits indicates that most of the observed rill network was developed in the short time between deposition of the pumice tephra during the climactic eruption and deposition of the thick valley fill. Despite the fact that deposits from the climactic eruption completely destroyed vegetation in catchment headwaters and severely disrupted watershed hydrology, the magnitude of overland flow had diminished before final deposition of the valley fill; otherwise, a more extensive drainage network would have developed on that fill. An extensive, integrated drainage network rapidly developed on the thick valley fill once the seasonal rains began in earnest (K.M. Scott and others, this volume). The reduction of overland flow likely reflects migration of Typhoon Yunya and a decrease in precipitation intensity. The sediment mobilized from these rills, and from the shallow slides,

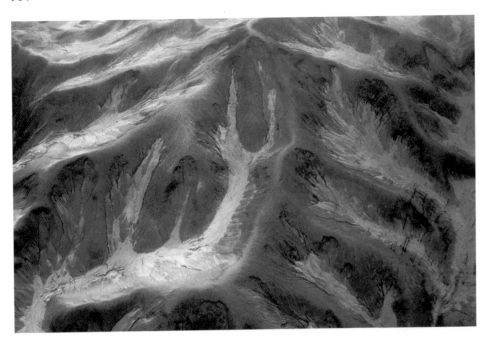

Figure 16. Shallow landslides in pumice-tephra mantle. Photograph taken June 25, 1991.

accounts for the numerous debris-flow deposits observed interbedded with the pyroclastic valley fill, but these were not the sources of sediment for the peak-discharge lahars.

Along the mountain front from the Porac River to the Sacobia River, peak discharge preceded deposition of the broadly distributed pumice fall (table 4). Between the Sacobia River and the O'Donnell River, peak discharge largely followed pumice fall; however, we were unable to examine proximal areas. The timing of peak discharge in each watershed probably reflects localization of variable-intensity eruption-induced rainfall, northward tracking of Typhoon Yunya, and perhaps variations in the intensity to which catchment headwaters were disturbed prior to the climactic activity. The characteristic response time of each watershed, the time required for the whole watershed to contribute to steady-state discharge at some outlet point, is probably a second-order effect influencing the variation in the timing of peak discharge. Transit times for overland flow off hillslopes and for channel flow to the chosen outlet point (see Dunne and Leopold, 1978; Dunne and Dietrich, 1980) reveal gross similarity of estimated response times (table 1). Thus, under equivalent conditions, a parcel of fluid from the farthest point in the basin should reach a comparable outlet point in a roughly equivalent amount of time for each watershed, except for the larger O'Donnell watershed, where the response time is slightly longer.

Peak flow in the Abacan River valley may have been affected by the emplacement of fill in the Sacobia River valley. We have established that peak discharge along the Abacan River and its tributaries occurred before deposition of the pumice fall and that channel erosion probably was associated with peak flow, at least in the upper basin. The position of peak-flow deposits along valley walls in Sapangbato (more than 6 m above the channel floor), the magnitude of

channel erosion by peak flow, and the thickness of subsequent channel deposits attest to large, and sustained, discharges of water and sediment moving through this valley. Yet, the Abacan River does not head on the flanks of Mount Pinatubo, and its watershed above Sapangbato has an area of only about 14 km^2. The magnitude and physical effects of the peak flow suggest a larger source of water than is probably deliverable by runoff within that part of the watershed. To estimate a possible peak discharge by rainfall runoff, we used a unit hydrograph developed for the upper Abacan watershed by the U.S. Army Corps of Engineers (S.L. Stockton, USCOE, written commun., 1993) and an assumed design storm of 150 mm of rainfall over 6 hours. In the calculation of runoff from that storm (Dunne and Leopold, 1978, p. 385–387) we allow a 1-h hydrograph lag but do not allow runoff removal by infiltration. We estimate that the watershed above Sapangbato was capable of delivering a peak water discharge of about 20 m^3/s from such a storm. Even if we allow peak flow to have a sediment concentration of about 66 volume percent, the peak discharge deliverable from such a storm in the upper Abacan watershed would be only about 60 m^3/s, much less than the peak discharge that is suggested by channel conditions. Thus, the upper Abacan watershed was either subject to a more intense storm or there was contribution of water from another source. By late July 1991, flow from the Sacobia River valley, whose watershed encompasses some 30 km^2 above Clark Air Base, was diverted completely into the Abacan watershed (K.M. Scott and others, this volume). We hypothesize that the peak discharge on the upper Abacan River, which preceded the pumice fall, may have resulted from temporary diversion of flow from the Sacobia River valley to the Abacan River valley. If so, temporary diversion of flow must have occurred before, or during, the

Table 4. Summary of characteristics and timing of peak-discharge lahars on the east side of Mount Pinatubo.

Watershed	Location (fig. 1)	Distance from summit (km)	Arrival of peak[1]	Type of flow[2]	Dominant lithology[3]	Depth (m)	Velocity (m/s)
O'Donnell	Maniknik	45	apf	hcf/ff	l,p	-	-
	Tarlac	52	apf	df/hcf/ff	l,p	>1.5	-
Sacobia-Bamban	Clark	16	bpf	df/hcf	l	>3	-
	San Francisco	38	apf	hcf/ff	l,p	~3	~4
Abacan	Sapangbato	18	bpf	df	l	~6	-
	Angeles	24	bpf	df/hcf	l	~5	~6
	Anao	40	apf	ff	l	>3	<4
	Mexico	47	apf	ff	l	~1	<4
Pasig-Potrero	Mancatian	24	bpf	df	l	>4.5	-
	Bacolor	37	bpf	df	l	~2.5	~3
Porac	headwaters	10	bpf	df	-	-	-
	Porac	22	bpf	df/hcf	l	~2	-

[1] apf = after pumice fall; bpf = before pumice fall.

[2] df = debris flow; hcf = hyperconcentrated flow; ff = fluvial flow.

[3] l = lithic; p = pumice.

earliest phases of the onset of climactic eruptive activity. W.E. Scott and others (this volume) have shown that within 5 km of the volcano, deposition of the thick pyroclastic valley fill and the pumice fall were synchronous; beyond about 5 km, pumice fall preceded emplacement of the pyroclastic fill. If our hypothesis regarding the origin of peak discharge on the upper Abacan River is correct, it indicates that some 40 to 60 m of fill must have rapidly choked the Sacobia River valley and allowed water to spill across the drainage divide before pumice fall began along the mountain front. Thick fluvial deposits (more than 2 m) in the Abacan River channel that do not underlie pumice fall indicate sustained elevated discharge, which suggests that flow from the Sacobia watershed may have been actively diverted for several hours. Pyroclastic-flow deposits that cap the fill in the Sacobia River valley indicate that pyroclastic activity also was sustained for several hours and outlasted any diversion of water, because no well-defined channel was incised on that fill or through the watershed divide by July 8, 1991. The primary drawback to our hypothesis on the cause of peak flow in the Abacan River valley is that peak-flow deposits contain little pumice. It is difficult to reconcile substantial water sweeping across extensive pyroclastic valley fill without entraining much pumice, unless a part of that valley fill was not pumiceous. W.E. Scott and others (this volume) have noted zones of lithic debris in at least the upper sections of the valley fill, but they attribute this debris to collapse of the caldera, which occurred well after the onset of climactic activity. Observations of recently exposed stratigraphy in the Sacobia River valley fill near the Abacan watershed gap (T.C. Pierson, U.S. Geological Survey, oral commun., 1993) reveal, however, that several meters of bouldery, lithic debris lie directly on the preeruption valley floor and that the overlying pyroclastic deposits are interbedded with several thin debris-flow deposits. Apparently there is no thick lithic fill beneath the pumiceous valley fill near the gap in the Abacan River-Sacobia River divide. Although these observations do not preclude a thicker lithic fill farther up the Sacobia valley, and a possible diversion of flow farther upstream, the origin of the peak discharge on the Abacan River remains enigmatic.

Despite their minuscule magnitude in comparison with subsequent lahars, the mid-June lahars seriously disturbed channels downstream. Overall, little direct damage resulted from lahars spilling overbank and flooding populous areas, although localized overbank flooding along fan channels inundated homes and on distal alluvial plains inundated arable land, particularly along the Bamban and Abacan Rivers. Instead, lahars generally aggraded channels, and this aggradation triggered lateral bank erosion. Along the alluvial plains, aggradation of mainstem channels led to backflooding of tributary channels.

Hydrograph peaks attenuated and slowed, and flows diluted as they traveled downvalley (table 4). The best example is illustrated by flow along the Abacan River. The hydrograph peak at Sapangbato was nearly 6 m above the present channel bed, and it passed before the pumice fall was completely deposited. Three kilometers downstream, peak-flow depth was no more than 2 m above the present channel bed and probably no more than 5 m above the original channel bed. By the time the peak traveled nearly 20 km farther downvalley, it had been significantly diluted by loss of sediment, its depth was little more than 3 m, and it succeeded the pumice fall. Along the distal plains, peak flow was less than 2 m deep and transported only very fine sand and silt. Above Angeles City, where peak flow preceded pumice fall, we have estimated peak-flow velocity to be about 6 m/s. Eyewitnesses report that peak flow reached the town of Mexico at approximately 1600 on June 15. If we assume that widespread pumice fall began in the Pampanga Province around 1400, the arrival time of peak flow in Mexico yields a mean velocity of less than 4 m/s below Angeles City. Shunting of flow into several distributary channels and dilution of flow through loss of sediment probably contributed to the attenuation and slowing of peak flow below Angeles City.

CONCLUSIONS

Lahars were generated in all major watersheds east of Mount Pinatubo as a result of the June 1991 eruptions. Although these lahars were triggered by heavy rainfall associated with Typhoon Yunya, and perhaps with localized convective storm cells, the rainfall that triggered these lahars was not uncommon for this tropical climate. Heavy rainfall alone was not responsible for generating the lahars. Another necessary condition for the formation of these lahars was alteration of watershed hydrology by volcanic deposits. Deposits associated with pyroclastic density currents and chiefly fine-grained tephra fall that disturbed catchment headwaters prior to the climactic phase of eruptive activity on June 15 played a key role in the formation of the peak-discharge lahars that impacted channels along the mountain front from the Porac River to the Sacobia River. Those volcanic deposits destroyed or damaged vegetation, reduced infiltration capacity of the substrate, and smoothed the natural-scale roughness of the hillslopes. As a result, overland flow increased, and a higher percentage of the rain that fell on the landscape was delivered from the hillslopes to the channels at a rate faster than usual. The rapid delivery of large volumes of water from hillslopes to channels triggered hillslope and bank erosion, channel scour, and possibly minor slope failures, which led to lithic-rich debris flows and more dilute hyperconcentrated flows. Unconsolidated channel sediment probably was the dominant source of sediment for these lahars because many channel banks were composed of pumiceous debris (Newhall and others, this volume). Pyroclastic flows, ash-cloud surges, and pumice fall deposited during the climactic eruptive activity completely destroyed vegetation and produced the most dramatic landscape disturbance in catchment headwaters. Sustained rainfall during this activity led to extensive, shallow slope failures and to extraordinary overland flow that developed an extensive rill network on the pumice-tephra mantle. Sediment mobilized from these rills and shallow landslides generated several pumice-rich lahars that are found interbedded with the pyroclastic valley fill and contributed to the postpumice-fall lahars on the O'Donnell River and to the pumice-bearing recessional-flow deposits observed in several channels. Truncation of rills by valley fill and the lack of an integrated drainage on that valley fill show that excessive overland flow had diminished before the caldera formed and pyroclastic flows ended. Although the climactic eruptive activity produced the most dramatic and widespread landscape disturbance and severely disrupted watershed hydrology, stratigraphic constraints on the timing of peak discharge in several eastern watersheds illustrate the delicate sensitivity of watershed hydrology to disturbance by even relatively thin, fine-grained volcanic deposits.

The timing of peak discharge with respect to the onset of the climactic phase of activity on June 15 varied in a northerly direction across watersheds and downstream within watersheds. From the Porac River to the Sacobia River, peak discharge along the mountain front occurred before the plinian pumice fall; between the Sacobia River and O'Donnell River, peak discharge largely followed the pumice fall. On the alluvial plains, peak flows generally followed the pumice fall. The timing of these lahars probably reflects localization of variable-intensity, eruption-induced rainfall, northward movement of Typhoon Yunya, and perhaps variations in the degree of disturbance of catchment headwaters by preclimactic eruptive activity. Characteristic response times of watersheds do not appear to differ significantly, and their influence on the variation in lahar timing is probably more subtle than is the influence of variations in rainfall. The interplay of these various factors affects the amount of water that moves from hillslopes to channels to fanheads, how rapidly it moves, and its ability to entrain and transport sediment.

Peak flows commonly attenuated, slowed, and diluted as they moved across the alluvial fans and onto the alluvial plains. Along the mountain front, peak flows were as deep as 6 m or more and had instantaneous velocities of about 6 m/s. In distal reaches, peak flows commonly were less than 1 to 3 m deep, and mean peak-flow velocities were less than 4 m/s. The sedimentology of downstream deposits suggests that many lahars were rapidly transformed to hyperconcentrated and normal streamflows beyond the mountain front; however, some of the lahars remained as debris flows across the alluvial fans.

The mid-June lahars damaged inhabited areas on the densely populated alluvial fans of Tarlac and Pampanga Provinces primarily by triggering lateral bank erosion that undermined buildings and by aggrading mainstem channels that led to backflooding of tributary channels. Although some areas were directly buried by lahars, burial was generally of secondary importance to damage triggered by bank erosion and by induced backflooding, except along distal alluvial plains where primarily agricultural land was inundated. As devastating as these lahars were, they had far less social and economic impact than subsequent lahars triggered by the seasonal monsoon rains.

ACKNOWLEDGMENTS

Reconnaissance fieldwork upon which this report is based was accomplished between late June and mid-July 1991, when J.J. Major and R.J. Janda visited the Philippines under the auspices of the USGS-USAID Volcano Disaster Assistance Program. The emphasis of the fieldwork was to establish the nature and magnitude of watershed disturbance and the character of downstream deposits in order to provide a first-approximation forecast of subsequent hydrologic hazards. Under very trying circumstances, Dick Janda was a master at honing in upon, and elucidating, the critical essence of outcrop and landscape relations regardless of stratigraphic and morphologic complexity. His enthusiasm, critical eye, and firm belief that colleagues should have the opportunity to experience such trying situations led to the development of this paper. He will be sorely missed. Logistical support for the study was provided by the U.S. Air Force, U.S. Navy, and the Pinatubo Lahar Task Force. We thank W.E. Scott, K.M. Scott, and R.P. Hoblitt for several enlightening discussions of volcanic and hydrologic events at Mount Pinatubo, and Tom Dunne for discussions concerning watershed hydrology. We thank K.M. Scott, C.G. Newhall, G.A. Smith, Kelin Whipple, and K.S. Rodolfo for critical reviews of an early draft.

REFERENCES CITED

Alaska Volcano Observatory, 1990, The 1989-1990 eruption of Redoubt Volcano: Eos, Transactions, American Geophysical Union, v. 71, no. 7, p. 265–275.

Best, J.L., 1992, Sedimentology and event timing of a catastrophic volcaniclastic mass flow, Volcan Hudson, Southern Chile: Bulletin of Volcanology, v. 54, p. 299–318.

Beverage, J.P., and Culberston, J.K., 1964, Hyperconcentrations of suspended sediment: American Society of Civil Engineers, Journal of Hydraulics Division, v. 90, HY6, p. 117–128.

Brantley, S.R., ed., 1990, The eruption of Redoubt Volcano, Alaska, December 14, 1989–August 31, 1990: U.S. Geological Survey Circular 1061, 33 p.

Collins, B.D., and Dunne, Thomas, 1986, Erosion of tephra from the 1980 eruption of Mount St. Helens: Geological Society of America Bulletin, v. 97, p. 896–905.

Dietrich, W.E., Kirchner, J.W., Ikeda, Hiroshi, and Iseya, Fujiko, 1989, Sediment supply and the development of the coarse surface layer in gravel-bedded rivers: Nature, v. 340, p. 215–217.

Dolan, M.T., 1993, Buried in mud: A look at the sedimentological and morphological characteristics of lahars on Pinatubo Volcano's eastern slopes: Eos, Transactions, American Geophysical Union, v. 74, no. 43, p. 671.

Dorava, J.M., and Meyer, D.F., 1994, Hydrologic hazards in the lower Drift River basin associated with the 1989–90 eruptions of Redoubt Volcano, Alaska: Journal of Volcanology and Geothermal Research, v. 62, p. 387–407.

Dunne, Thomas, and Dietrich, W.E., 1980, Experimental investigation of Horton overland flow on tropical hillslopes. 2. Hydraulic characteristics and hillslope hydrographs: Zeitschrift für Geomorphologie, Supplement-Band 35, p. 60–80.

Dunne, Thomas, and Leopold, L.B., 1978, Water in environmental planning: W.H. Freeman, 818 p.

Fisher, R.V., and Schminke, H.U., 1984, Pyroclastic rocks: Springer-Verlag, 472 p.

Folk, R.L., 1974, Petrology of sedimentary rocks: Austin, Tex., Hemphill Publishing Co., 182 p.

Giambelluca, T.W., Lau, L.S., Fok, Y.S., and Schroeder, T.A., 1984, Rainfall frequency study for Oahu: U.S. Army Corps of Engineers, Honolulu district, report R–73, 34 p.

Hoblitt, R.P., Wolfe, E.W., Lockhart, A.B., Ewert, J.E., Murray, T.L., Harlow, D.H., Mori, J., Daag, A.S., and Tubianosa, B.S., 1991, 1991 eruptive behavior of Mount Pinatubo: Eos, Transactions, American Geophysical Union, v. 72, no. 4, p. 61.

Hoblitt, R.P., Wolfe, E.W., Scott, W.E., Couchman, M.R., Pallister, J.S., and Javier, Dindo, this volume, The preclimactic eruptions of Mount Pinatubo, June 1991.

Janda, R.J., Scott, K.M., Nolan, K.M., and Martinson, H.A., 1981, Lahar movement, effects, and deposits, in Lipman, P.W., and Mullineaux, eds., The 1980 eruptions of Mount St. Helens, Washington: U.S. Geological Survey Professional Paper 1250, p. 461–478.

Jones, J.W., and Newhall C.G., this volume, Preeruption and posteruption digital terrain models of Mount Pinatudo.

Kadomura, Hiroshi, Imagawa, Tashiaki, and Yamamoto, Hiroshi, 1983, Eruption-induced rapid erosion and mass movements on Usu volcano, Hokkaido: Zeitschrift für Geomorphologie, Supplement Band 46, p. 123–142.

Leavesley, G.H., Lusby, G.C., and Lichty, R.W., 1989, Infiltration and erosion characteristics of selected tephra deposits from the 1980 eruption of Mount St. Helens, Washington, USA: Hydrological Sciences Journal, v. 34, p. 339–353.

MacArthur, R.C., Hamilton, D.L., Harvey, M.D., and Kekaula, H.W., 1992, Analyses of special hazards and flooding problems in tropical island environments: Proceedings of American Society of Civil Engineers Hydraulic Engineering Water Forum '92, Baltimore, Maryland, August 2–6, p. 1061–1066.

Major, J.J., and Iverson, R.M., 1993, Is the dynamic behavior of a debris flow recorded by its deposit?: Eos, Transactions, American Geophysical Union, v. 74, no. 43, p. 315.

Major, J.J., and Newhall, C.G., 1989, Snow and ice perturbation during historical volcanic eruptions and the formation of lahars and floods—a global review: Bulletin of Volcanology, v. 52, p. 1–27.

Major, J.J., and Voight, Barry, 1986, Sedimentology and clast orientations of the May 18, 1980 southwest flank lahars, Mount St. Helens, Washington: Journal of Sedimentary Petrology, v. 56, p. 691–705.

Newhall, C.G., Daag, A.S., Delfin, F.G., Hoblitt, R.P., McGeehin, J., Pallister, J.S., Regalado, M.T.M., Rubin, M., Tubianosa, B.S., Tamayo, R.A., and Umbal., J.V., this volume, Eruptive history of Mount Pinatubo.

Oswalt, J.S., O'Hara, J.F., and Nichols, William, this volume, Meteorological observations of the 1991 Mount Pinatubo eruption.

Paladio-Melosantos, M.L., Solidum, R.U., Scott, W.E., Quiambao, R.B., Umbal, J.V., Rodolfo, K.S., Tubianosa, B.S., and Delos Reyes, P.J., this volume, Tephra falls of the 1991 eruptions of Mount Pinatubo.

Pierson T.C., Daag, A.S., Delos Reyes, P.J., Regalado, M.T.M., Solidum, R.U., and Tubianosa, B.S., this volume, Flow and deposition of post-eruption hot lahars on the east side of Mount Pinatubo, July–October, 1991.

Pierson, T.C., Janda, R.J., Thouret, J.C., and Borrero, C.A., 1990, Perturbation and melting of snow and ice by the 13 November 1985 eruption of Nevado del Ruiz, Colombia, and consequent mobilization, flow and deposition of lahars: Journal of Volcanology and Geothermal Research, v. 41, p. 17–66.

Pierson, T.C., Janda, R.J., Umbal, J.V., and Daag, A.S., 1992, Immediate and long-term hazards from lahars and excess sedimentation in rivers draining Mount Pinatubo, Philippines: U.S. Geological Survey Water-Resources Investigation Report 92–4039, 35 p.

Pierson, T.C., and Scott, K.M., 1985, Downstream dilution of a lahar: Transition from debris flow to hyperconcentrated streamflow: Water Resources Research, v. 21, no. 10, p. 1511–1524.

Pinatubo Volcano Observatory Team, 1991, Lessons from a major eruption: Mount Pinatubo, Philippines: Eos, Transactions, American Geophysical Union, v. 72, no. 49, p. 545–555.

Richards, Keith, 1982, Rivers: Form and process in alluvial channels: Methuen, 361 p.

Rodolfo, K.S., 1991, Climatic, volcaniclastic, and geomorphic controls on the differential timing of lahars on the east and west sides of Mount Pinatubo during and after its June 1991 eruption: Eos, Transactions, American Geophysical Union, v. 72, no. 44, p. 62.

Rodolfo, K.S., and Arguden, A.T., 1991, Rain-lahar generation and sediment-delivery systems at Mayon Volcano, Philippines, in Fisher, R.V., and Smith, G.A., eds., Sedimentation in volcanic settings: SEPM Special Publication No. 45, p. 71–87.

Rodolfo, K.S., Umbal, J.V., Alonso, R.A., Remotigue, M.C., Paladio, M.L., Salvador, J.H.G., Evangelista, D., and Miller, Y., this volume, Two years of lahars on the western flank of Mount Pinatubo: Initiation, flow processes, deposits, and attendant geomorphic and hydraulic changes.

Scott, K.M., 1988, Origins, behavior, and sedimentology of lahars and lahar-runout flows in the Toutle-Cowlitz River system: U.S. Geological Survey Professional Paper 1447–A, 74 p.

Scott, K.M., Janda, R.J., de la Cruz, E., Gabinete, E., Eto, I., Isada, M., Sexon, M., and Hadley, K.H., this volume, Channel and sedimentation responses to large volumes of 1991 volcanic deposits on the east flank of Mount Pinatubo.

Scott, W.E., Hoblitt, R.P., Daligdig, J.A., Besana, G., and Tubianosa, B.S., 1991, 15 June 1991 pyroclastic deposits at

Mount Pinatubo, Philippines: Eos, Transactions, American Geophysical Union, v. 72, no. 44, p. 61–62.

Scott, W.E., Hoblitt, R.P., Torres, R.C., Self, S., Martinez, M.L., and Nillos, T., this volume, Pyroclastic flow of the June 15, 1991, paroxysmal eruption, Mount Pinatubo volcano, Philippines.

Segerstrom, Kenneth, 1950, Erosion studies at Parícutin, State of Michoacan, Mexico: U.S. Geological Survey Bulletin 965–A, 164 p.

Smith, G.A. and Fritz, W.J., 1989, Volcanic influences on terrestrial sedimentation: Geology, v. 17, p. 375–376.

Smith, G.A., and Lowe, D.R., 1991, Lahars: Volcano-hydrologic events and deposition in the debris flow-hyperconcentrated flow continuum, in Fisher, R.V., and Smith, G.A., eds., Sedimentation in volcanic settings: SEPM Special Publication No. 45, p. 59–70.

Trabant, D.C., Waitt, R.B., and Major, J.J., 1994, Disruption of Drift glacier and origin of floods during the 1989-90 eruptions of Redoubt Volcano, Alaska: Journal of Volcanology and Geothermal Research, v. 62, p. 369–385.

Ulate, C.A., and Corrales, M.F., 1966, Mud floods related to the Irazú volcano eruptions: American Society of Civil Engineers Journal of Hydraulics Division, v. 92, no. HY6, p. 117–129.

Umbal, J.V., and Rodolfo, K.S., this volume, The 1991 lahars of southwestern Mount Pinatubo and evolution of the lahar-dammed Mapanuepe Lake.

Waitt, R.B., Pierson, T.C., MacLeod, N.S., Janda, R.J., Voight, B., and Holcomb, R.T., 1983, Eruption-triggered avalanche, flood, and lahar at Mount St. Helens—effects of winter snowpack: Science, v. 221, p. 1394–1396.

Waldron, H.H., 1967, Debris flow and erosion control problems caused by the ash eruptions of Irazú volcano, Costa Rica: U.S. Geological Survey Bulletin 1241–I, 37 p.

Wolfe, E.W., and Hoblitt, R.P., this volume, Overview of the eruptions.

Flow and Deposition of Posteruption Hot Lahars on the East Side of Mount Pinatubo, July–October 1991

By Thomas C. Pierson,[1] Arturo S. Daag,[2] Perla J. Delos Reyes,[2] Ma. Theresa M. Regalado,[2] Renato U. Solidum,[2] and Bella S. Tubianosa[2]

ABSTRACT

Monsoon and typhoon rains in the first rainy season following the June 15, 1991, eruption of Mount Pinatubo generated more than 200 hot (~50 °C) lahars in drainage basins on the east side of the volcano. Channelized lahars having peak discharges on the order of 100 to 1,000 cubic meters per second typically were noncohesive pumiceous debris flows, some of which transformed to hyperconcentrated flows prior to final deposition. Nearly all of the sediment in the lahar deposits (>90 percent) during this first rainy season was eroded from the thick pyroclastic-flow deposits filling valleys in the upper reaches of the watersheds. Lahar deposition occurred primarily on low-gradient, coalescing alluvial fans 15 to 50 kilometers downstream from the caldera at the base of the volcano, where deposit thicknesses generally ranged from 0.5 to 5 meters (mean thicknesses about 1.5 to 2 meters). Total depositional volume on the east-side alluvial fans in 1991 was about 0.38 cubic kilometers, which is almost one-third of the potential contributing volume from the source pyroclastic sediments. Sediment yields in 1991 were on the order of 1 million cubic meters per square kilometer per year for three of the five east-side drainages, nearly an order of magnitude greater than the maximum sediment yield computed following the May 18, 1980, eruption at Mount St. Helens, Washington, U.S.A.

INTRODUCTION

PHYSIOGRAPHIC SETTING

Mount Pinatubo volcano, situated at approximately 15° N. lat and about 100 km northwest of Manila in central Luzon, Philippines, stood 1,745 m above sea level prior to June 15, 1991. Its deeply dissected, forested flanks were drained by eight major river systems, five of which flowed eastward: the O'Donnell-Tarlac, Sacobia-Bamban, Abacan, Pasig-Potrero, and Porac-Gumain systems (fig. 1). The steep upper slopes of the volcano above about 1,000 m in altitude were deeply dissected, having slopes as steep as 60° to 70° and channel slopes ranging from about 7° to 20°. On the lower flanks of the mountain, from about 1,000 m down to 200 to 300 m in altitude, channel slopes flattened out to about 1° (gradient 0.01–0.02 m/m), and down to this point, streams had been deeply to moderately incised in older pyroclastic deposits. Heavily populated and intensively farmed alluvial fans and alluvial plains with gradients from about 0.02 to less than 0.0002 m/m extended beyond the volcano's lower flanks. On this alluvial apron, all of the rivers had been (at least in part) artificially straightened and constrained by earthen dikes.

CLIMATE

Mount Pinatubo's climate is controlled by the easterly flow of the North Pacific trade winds and the Northeast Monsoon for most of the year, but from May through October a southwesterly flow of moist air, the Southwest Monsoon, dominates. These months generally are the rainiest of the year; August is usually the wettest month. Although major tropical storms (typhoons) may also strike during this season, it is more common for typhoons to pass to the north of Luzon. These typhoons exert a strong pull on the southwesterly flow and cause heavier than normal rainfall in the Mount Pinatubo area (Maj. W. Nichols, U.S. Air Force, oral commun., 1993).

Monsoonal rainfall is highly variable in space and time, and mountains about 20 km to the west and about 40 km to the southwest of Mount Pinatubo act as partial orographic barriers to complicate rainfall patterns further. Clark Air Base weather radar indicated that storm cells vary in size but are typically on the order of 10 km in diameter. The monsoon rains on the east side of Mount Pinatubo usually begin with heavy showers and thunderstorms for about 15 to 45 min, followed by light to moderate more continuous

[1] U.S. Geological Survey.

[2] Philippine Institute of Volcanology and Seismology.

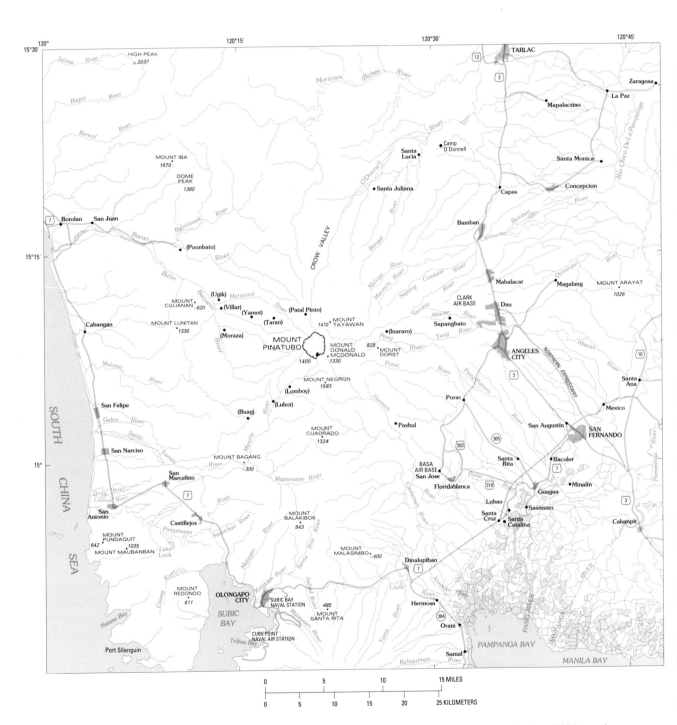

Figure 1.   Preeruption drainage map of Mount Pinatubo and vicinity, showing main rivers affected by the June 1991 eruptions.

rainfall for up to 10 h (Maj. W. Nichols, U.S. Air Force, oral commun., 1993).

LAHARS FOLLOWING THE JUNE 15 ERUPTION

Numerous debris flows and hyperconcentrated flows, collectively termed *lahars* (Smith and Lowe, 1991), were triggered during and following the climactic explosive eruption of Mount Pinatubo on June 15, which erupted a total bulk volume of 8.4 to 10.4 km^3. The eruption removed part of the mountain's summit and left a caldera about 2.5 km in diameter. It deposited abundant, loose volcaniclastic debris that would later be the primary source sediment for lahars—5 to 6 km^3 of pumiceous pyroclastic-flow deposits in the heads of valleys draining the volcano and about 0.2 km^3 of tephra *on the volcano's flanks* (out of the 0.7 km^3 that fell on land) (W.E. Scott and others, 1991; this volume; Pierson and others, 1992; Paladio-Melosantos and others, this volume). About one quarter of this erupted material was deposited on the east flank of the volcano (fig. 2).

During the 1991 monsoon season, lahars eroded and transported large volumes of 1991 eruption deposits within all the major watersheds draining the volcano. These lahars were triggered by (1) monsoonal rainstorms, sometimes enhanced by greater than normal southwesterly air flow during the passage of typhoons farther to the north, (2) volcanically induced convective rainstorms over localized heat sources (main vent and thick pyroclastic-flow deposits), and (3) breakouts from debris-dammed lakes (Rodolfo, 1991; Punongbayan and others, 1991; Umbal and others, 1991; Pierson and others, 1992; Oswalt and others, this volume; Rodolfo and others, this volume; K.M. Scott and others, this volume). By the end of October 1991 lahars had buried hundreds of square kilometers of agricultural land, caused extensive damage to homes, roads, and bridges in lowland areas surrounding the volcano, and displaced tens of thousands of people (Pierson and others, 1992; Rodolfo and others, this volume).

TYPES OF LAHARS

Lahars are discrete flow events involving highly concentrated mixtures of volcaniclastic sediment and water (Smith and Lowe, 1991) that can be triggered during or after volcanic eruptions by a number of mechanisms (Neall, 1976; Major and Newhall, 1989). Two rheologically distinct types of sediment-water flow, *debris flow* and *hyperconcentrated flow* (see Pierson and Scott, 1985; Smith, 1986; Pierson and Costa, 1987), occur as lahars at Mount Pinatubo. Differentiation of lahar types was based on field observations and evidence of relative yield strength and viscosity (see Pierson and Costa, 1987).

Debris flows are liquefied slurries of poorly sorted sediment (particles ranging from clay to usually the largest

clasts available, including boulders) that (1) have the consistency of wet concrete, (2) possess high viscosity and a finite yield strength (a non-Newtonian "plastic" fluid), and (3) usually exhibit laminar flow, often with higher flow velocities than for similarly scaled water flows (Costa, 1984; Johnson, 1984; Pierson and Costa, 1987). Sediment concentrations are so high that sediment and water move en masse, and particles have great difficulty segregating by size or settling out of suspension. Fluid bulk densities of debris-flow slurries typically range from about 1.8 to 2.3 g/cm^3 for grains with typical "lithic" particle densities of 2.4 to 2.7 g/cm^3, giving volumetric sediment concentrations in the range of 50 to 75 percent, depending on grain-size distribution. The 1991 Mount Pinatubo deposits involved high proportions of lower density pumice, but volumetric sediment concentrations were similar. Debris-flow deposits characteristically exhibit massive internal structure (lack of internal stratification), very poor to extremely poor sorting, dense packing (often highly consolidated, even in fresh deposits), inverse to normal grading, and support of coarse clasts within a finer grained matrix.

Hyperconcentrated flows are dense suspensions of sediment in water, but concentrations are low enough for coarser sediment particles to be able to settle out of suspension when flow velocities decrease. These flows appear more viscous than normal-concentration streamflow, somewhat resembling dirty motor oil; flow is characteristically turbulent, but some turbulence is dampened by the higher fluid viscosity (Beverage and Culbertson, 1964; Pierson and Scott, 1985). Sediment concentrations typically range from about 20 to 60 percent by volume (depending on grain-size distribution), overlapping in range with debris flows (Pierson and Costa, 1987). Hyperconcentrated sediment-water mixtures possess a low yield strength (Hampton, 1972; Kang and Zhang, 1980), but normal-density gravel is not carried in suspension as it is in debris flows. Hyperconcentrated-flow deposits are characterized by (1) coarse sand to fine gravel textures, (2) poor sorting, (3) horizontal bedding (often very faint and thicker than typical fluvial laminae), (4) absence of crossbedding, and (5) intrastratal occurrence of small gravel lenses or outsized gravel clasts. All of these features suggest rapid aggradational deposition from suspension or traction (Pierson and Scott, 1985; Smith, 1986).

All of the 1991 lahars had raised temperatures from inclusion of hot pyroclastic-flow deposits. The maximum temperature measured in moving flows was about 60°C, and some fresh lahar deposits were as hot as 80°C. Recent hot lahars have also been described at Mayon volcano (Philippines) by Arguden and Rodolfo (1990), where effects of the raised temperatures were noted in flow behavior and deposit characteristics. Without "cold" lahars of similar composition at Mount Pinatubo for comparison, there was no way to determine whether slurry temperature significantly altered debris-flow rheology or sedimentology.

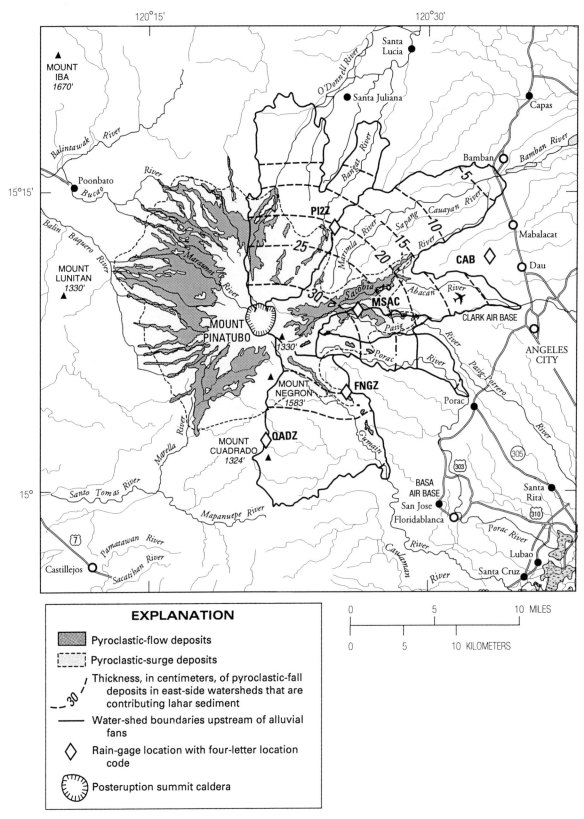

Figure 2. Distribution of 1991 volcanic deposits (thick, pumiceous pyroclastic-flow deposits; thin, pyroclastic-surge deposits; and June 15 coarse tephra-fall deposits) acting as primary lahar source sediments in east-side watersheds at Mount Pinatubo (after W.E. Scott and others, this volume; Paladio-Melosantos and others, this volume). Tephra that fell on slopes close to the caldera was largely swept away and mixed with subsequent flow and surge deposits. Distribution of thin, fine-grained ash layers that fell before or after the main June 15 eruption is not shown (see W.E. Scott and others, this volume). Locations of rain gages and watershed boundaries are also indicated.

PURPOSE AND SCOPE

The hydrologic response of volcanic landscapes following explosive eruptions in tropical or semitropical regions (areas of potentially heavy rainfall) has been studied and described for several cases in Southeast Asia and Central America (Schmidt, 1934; Segerstrom, 1950, 1960, 1966; Ulate and Corrales, 1966; Waldron, 1967; Ollier and Brown, 1971; Kuenzi and others, 1979; Smart, 1981; Hamidi, 1989; Rodolfo, 1989; Rodolfo and others, 1989; Arguden and Rodolfo, 1990; Rodolfo and Arguden, 1991). Except for a general description of long-term sedimentation following the 1902 Santa María (Guatemala) eruption (Kuenzi and others, 1979), the hydrologic response to large eruptions (≥ 10 km^3 of ejecta) had not been documented prior to this volume. This study, together with companion papers from this volume (Arboleda and Martinez; Major and others; Marcial and others; Martinez and others; Rodolfo and others; K.M. Scott and others; Tuñgol and Regalado; Umbal and Rodolfo), examines part of the hydrologic response of Mount Pinatubo following the eruption of June 15, 1991. Better understanding of the type, distribution, magnitude, and frequency of hydrologic processes at Mount Pinatubo is required for effective long-term hazard prediction (see Pierson and others, 1992).

This paper focuses on lahars triggered in eastward- and southeastward-draining watersheds after the main sequence of eruptions in June and covers the period from mid-July (when recording instruments were installed) to October 1991. The gap from June 15 to July 18 is not covered, but relatively few lahars were noted during this period. Data were collected on lahar occurrence in the Sacobia, Abacan, and Gumain Rivers, where radiotelemetered acoustic flow sensors (Hadley and LaHusen, 1991, 1995; Marcial and others, this volume) were deployed. Flow data were received and stored on computers at the Pinatubo Volcano Observatory (PVO), located in the base operations building at Clark Air Base, 25 km east of the volcano. The Pasig-Potrero River was an active contributor of lahars (Arboleda and Martinez, this volume), but continuous monitoring was not carried out there. Little information was available on lahar occurrence in the O'Donnell-Tarlac basin during the 1991 monsoon season, but lahar activity there appeared to be relatively minor.

SEDIMENT SOURCE AREAS

Lahar sediment at Mount Pinatubo came from five distinct sources in 1991: (1) coarse tephra dropped from the eruption columns of the mid-June eruptions, (2) pyroclastic-flow deposits, (3) fine-grained tephra from ash-cloud deposits, phreatic explosions, and eruptive events postdating June 15, (4) 1991 lahar deposits, and (5) unconsolidated volcaniclastic deposits predating June 15. The spatial

relations between these source materials and processes responsible for sediment mobilization are schematically depicted in figure 3 and discussed below. Grain-size distributions of representative samples of the different units are shown in figure 4.

TEPHRA DEPOSITS

Coarse tephra (predominantly coarse ash to lapilli) fell on all surfaces on the east flank of Mount Pinatubo during the June 15 eruption, achieving accumulations of about 10 to 50 cm (fig. 2). On the broad, thick valley fills it is mostly interbedded with the pumiceous pyroclastic-flow deposits (W.E. Scott and others, this volume). On valley sides and upland surfaces not swept by the pyroclastic flows, it accumulated as a single, relatively porous layer (Paladio-Melosantos and others, this volume).

The coarse tephra was capped on the east side of the volcano by thinly bedded fine tephra (fine sandy to silty ash) that originated from frequent ventings from the caldera, phreatic explosions, and secondary pyroclastic-flow ash clouds. It was carried predominantly northeastward by prevailing winds (W.E. Scott and others, this volume). The deposits contained abundant accretionary lapilli and were as thick as 1 m. Owing to its grain-size distribution (fig. 4) and relatively loose packing, this layer of fine deposits absorbed water during the rainy season up to nearly the point of saturation and then could liquefy when disturbed—by being walked on or presumably by earthquakes. Overflights by helicopters in the upper Sacobia watershed revealed relatively flat areas where lateral-spread mass movements had mobilized this layer (W.E. Scott, U.S. Geological Survey, oral commun., 1993).

Two typical hydrologic effects of such tephra mantles are (1) the formation of incipient crusts on tephra deposits, which decrease the infiltration capacity of the ground surface, and (2) the destruction or burial of vegetation. The hydrologic response of ash-covered upland slopes is altered so that proportionately more rainfall runs off as surface flow and that it runs off more quickly (Segerstrom, 1950; Waldron, 1967; Collins and Dunne, 1986; Leavesley and others, 1989). The prodigious volumes of loose, erodible sediment widely distributed over the landscape acted both as a sediment source for lahars and as a runoff-enhancing infiltration barrier that also helped to generate lahars.

Erosion from steep tephra-covered hillslopes in the watersheds occurred as rill and gully erosion (fig. 5), with some minor shallow landsliding. Integrated rill and gully networks were well established by the middle of the monsoon season. By October, runoff from monsoonal rainfall had extensively dissected the tephra mantle with rills tens of centimeters wide, many of which had cut down to the preeruption ground surface. These rills extended fully from ridge crests to bases of slopes. Roughly 20 to 50 percent of

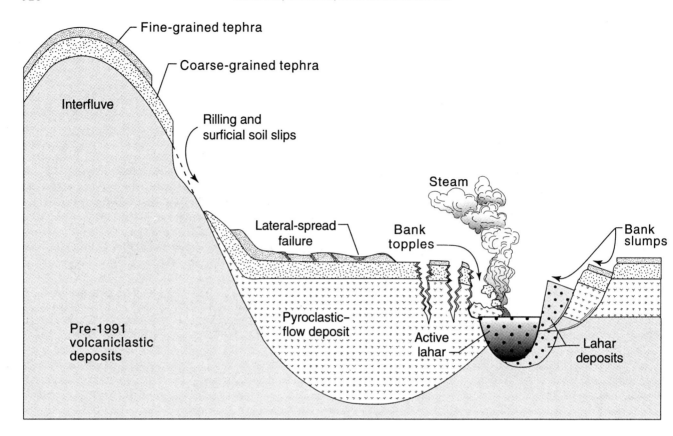

Figure 3. Schematic representation of spatial relations between different types of volcanic deposits acting as lahar source sediments in upper watersheds on the east flank of Mount Pinatubo.

these deposits had been eroded by the end of the first monsoon season, on the basis of visual estimates.

PYROCLASTIC-FLOW DEPOSITS

VOLUME, CHARACTERISTICS, AND PHREATIC ACTIVITY

Pyroclastic-flow deposits extended as far as 16 km from the vent in east-side drainages and caused profound changes to the landscape by filling entire valleys and changing steep, deeply dissected terrain to broad, gently sloping (up to about 3°) plains. Total accumulated volume in eastward-draining watersheds was about 1.4 km^3 (1.4×10^9 m^3), with most of that accumulating in the heads of the Sacobia and Pasig River valleys. Total area covered was about 34 km^2 (W.E. Scott and others, this volume). Mean total thickness of the deposits, by drainage basin, ranged from 15 to 48 m, with maximum localized thickness in the upper Sacobia valley in excess of 200 m (W.E. Scott and others, this volume). The great thickness of these deposits has caused diversions of surface drainage and the blockage of tributary valleys, resulting in the formation of unstable lakes, including several in the Sacobia basin.

Massive, very poorly sorted pumice layers tens of meters thick are the predominant facies composing most of

the thick valley fills. Thinner layers of pumice-fall and ash-cloud deposits are interbedded within the main valley fills, and lithic breccias overlie the pumiceous deposits near the vent (W.E. Scott and others, this volume). The pumiceous deposits are predominantly sand (fig. 4) and have little gravel coarser than cobble size. Fine ash (finer than 0.0625 mm) makes up from a few percent up to a maximum of 18 percent of the deposit by weight (W.E. Scott and others, this volume).

The pyroclastic flows were very hot when emplaced, and heat retained in the deposits throughout 1991 profoundly affected subsequent geomorphic processes. The preeruption magma temperature at Mount Pinatubo was estimated to be 800±20°C (Rutherford and Devine, 1991), and the emplacement temperature of the pyroclastic flows was probably not much below this. Temperatures around 300°C were measured 30 cm below the deposit surface shortly after deposition (R. Hoblitt, U.S. Geological Survey, oral commun., 1991). During the 1991 monsoon season and continuing through the 1992 and 1993 seasons, numerous secondary phreatic explosions occurred almost daily, particularly during or shortly after rainstorms, due to rapid vaporization of water trapped beneath the hot pyroclastic valley fills (see Rowley and others, 1981; Moyer and Swanson, 1987). Explosions commonly sent ash eruption columns

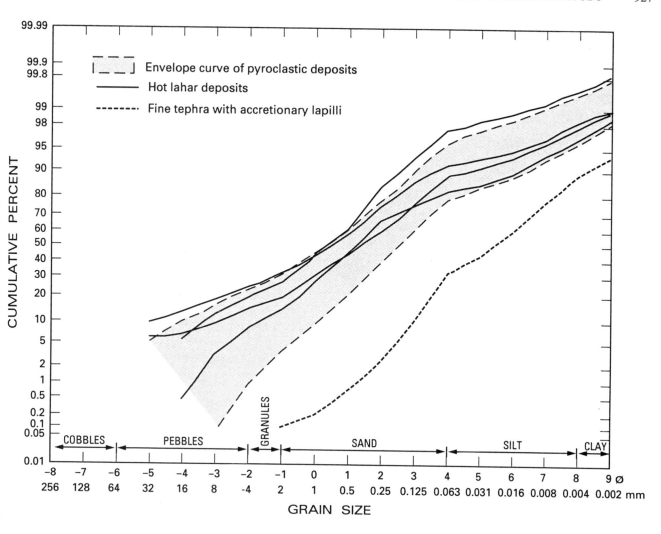

Figure 4.   Grain-size distribution of representative samples of lahar source sediments from the east side of Mount Pinatubo. Pyroclastic-flow deposits sampled and analyzed by W.E. Scott and others (this volume). Hot lahar and fine tephra samples collected along Sacobia River at west end of Clark Air Base.

3 to 4 km high, sometimes higher, and left explosion craters up to hundreds of meters in diameter. This activity regularly destabilized the valley-fill deposits during the rainy season and at times remobilized them as secondary pyroclastic flows (Torres and others, this volume). Erosion of the valley fills was facilitated by this phreatic activity through dilation of pyroclastic deposits (making them more vulnerable to erosion), disruption of surface crusting, and transport of loose debris directly into active channels.

EROSION

Erosion of pyroclastic-flow valley-fill deposits has been by both bank erosion and vertical incision. In the narrow, vertically walled gullies and channels developed on the fill surfaces (fig. 6), sections of banks (typically 1 to 10 m³) topple or slide into the channels when undercut by lahars.

Videotapes and direct observations of various types of flows in the lower part of the Sacobia valley fill showed that moderate to high discharges of hyperconcentrated flow or dilute debris flow—flows with fully developed turbulence—were most effective in erosively undercutting banks. On channel bends where lahars continuously impinged on vertical banks, lateral channel shifts of tens of meters per day were common. Bank erosion supplies large volumes of sediment directly to the lahars and results in progressive increases in lahar discharge with distance downstream. Collapse of hot pyroclastic-flow deposits into a passing flow causes especially vigorous steaming (fig. 7). Rapid (hourly to daily) fluctuations of channel bed levels (up to about 10 m) due to alternating bed incision and aggradation were observed during the 1991 rainy season in the lower valley fill in the Sacobia River valley. Bank erosion was particularly active during channel aggradation.

Figure 5. Rill erosion in coarse tephra at ridge crest 9 km southeast of summit caldera. Slope here is about 20°; headcut in center of photograph is about 20 cm wide. Tephra is about 40 cm thick at this site.

LAHAR DEPOSITS

Repeated cutting and filling of channels by lahars on the surface of the pyroclastic-flow deposits resulted in multiple periods of lahar deposition during the aggradational phases. This resulted in a gradual replacement of pyroclastic-flow deposits with lahar deposits, particularly in the downstream portions of the valley fills. By October 1991, a large proportion (estimated 30 to 50 percent) of the distal Sacobia valley-fill pyroclastic-flow deposits (those adjacent to the west end of Clark Air Base) had been replaced by lahar deposits. Subsequent erosion of these deposits resulted in remobilization of deposits previously eroded from farther upstream. Bank collapse in the cooler lahar deposits was accomplished by sapping (erosion by outward-seeping ground water) at the foot of banks, as well as by undercutting from passing lahars. Lahar deposits in the valley-fill channels had the same general appearance as the primary pyroclastic-flow deposits, but minor differences

included (1) lower temperatures (50–80°C at the distal end of the Sacobia valley fill); (2) slightly higher content of silt- and clay-sized particles (fig. 4) and consequently more cohesion; (3) slightly darker, browner coloration; and, if the deposits were fresh, (4) the presence of interstitial water.

CHARACTERISTICS OF LAHAR-TRIGGERING RAINFALL

Four automatic tipping-bucket rain gages (designated PI2Z, MSAC, FNGZ, and QADZ) were situated on divides of eastward-draining watersheds on Mount Pinatubo, no more than 9 km from the nearest other gage, during at least part of the 1991 monsoon season; a fifth, manual gage (CAB) was located farther east at Clark Air Base (fig. 2; table 1). The four automatic gages telemetered realtime rainfall data to PVO, while the CAB gage was read at 6-h intervals by U.S. Air Force meteorologists. These five rain gages covered a combined catchment area of about 300 km^2 in terrain that is steep and rugged, and where precipitation is significantly influenced by orographic effects.

Measured daily rainfall amounts in 1991 at CAB were about average for the monsoon season at Mount Pinatubo (Maj. W. Nichols, U.S. Air Force, oral commun., 1991; U.S. Air Force, Environmental Technical Applications Center, Scott Air Base, unpub. report, 1988). August was the wettest month (fig. 8). A total of 60 rainstorms were recorded by the automatic gages during the period from July 18 to October 31, 1991, not all of which triggered lahars; a storm is here defined as a more or less continuous period of rainfall bounded by dry intervals of at least 6 h. Storms were both localized convective storms and more regional storms, usually related to the passage of a typhoon weather system nearby. Most storms in 1991 seemed to be the localized type, seldom delivering more than 100 mm of rain or having durations in excess of 10 to 15 h. Larger, apparently more regional storms typically delivered more than 200 mm of rain and had durations well in excess of 24 h (table 2).

Spatial distribution of total storm rainfall was highly variable at the five east-side rain gages in 1991, with consistent differences of at least several hundred percent between some of the gages (fig. 9). Similar disparities occur in storm durations and intensities (table 2). Heaviest rainfall was usually recorded at the QADZ gage (69 percent of storms), probably because of its higher altitude and the southwesterly tracks of most of the storms. Heaviest rainfall totals were recorded 21 percent of the time at PI2Z and 10 percent of the time at FNGZ. Of the storms analyzed, the MSAC rain gage never had the highest rainfall totals of the four automatic gages, which is likely a reflection of its low altitude and the rain-shadow effect of the ridge on which the QADZ gage is located.

Fifty-two of the 60 recorded storms triggered lahars in east-side watersheds in 1991 (table 2). Eighty-five percent

Figure 6. Erosion of thick, pumiceous pyroclastic-flow deposits filling the Sacobia River valley in early September 1991. *A*, Oblique aerial view of channel development approximately 7 km downstream from caldera rim. View looking south. White circular area in upper left center is hot, dry pumice ejecta from a phreatic explosion. *B*, Steep unstable bank of channel about 9 km downstream from the caldera rim. Bank collapse has exposed light-colored, dry, still-hot deposits (lower right); bank is probably 4 to 5 m high. Note the relative thinness (tens of centimeters) of the wetted zone on top of the pyroclastic-flow deposits. Dark hill (left center) is crest of ridge from preexisting topography.

A

B

of the lahar-triggering storms had durations less than 20 h (most less than 10 h) and are considered to be localized convective storms. This localization is evident in the spatial distribution of rainfall as reflected in differences in storm totals shown in table 2. In 23 percent of the cases, lahars were triggered when no rain at all was recorded at one of the 4 automatic gages, and no rain was recorded at two of the four gages for 12 percent of the lahar-triggering storms (table 2). Given the typically small cell diameter for con-

vective storms (about 10 km), which is close to the average distance separating adjacent gages (9.3 km), it is unlikely that many of the rainstorm characteristics measured at a single rain gage would be representative of the rainfall that actually triggered lahars. Therefore, rainfall totals and intensities recorded at east-side rain gages must be considered to be only minimum values for any given storm.

In order to test whether rainfall-intensity thresholds for triggering lahars could be defined by using data from all

Figure 7. Steam generated by collapse of very hot bank of pyroclastic-flow deposit into lahar in Sacobia River at west end of Clark Air Base, August 24, 1991. View looking downstream. Bank is 5 to 6 m high.

Table 1. Site characteristics of rain gages on the east flank of Mount Pinatubo during the period July 18–October 31, 1991.

[Gage locations are shown in figure 2]

Rain gage	Distance from caldera rim (km)	Direction from caldera rim	Approximate altitude (m)	Watershed position	Beginning/end of record
QADZ	9.5	South	1,060	West boundary, Gumain R. basin	July 18, 1991/ ongoing
FNGZ	9.2	Southeast	740	East boundary, Gumain R. basin	July 31, 1991/ ongoing
MSAC	8.7	East	560	South boundary, Sacobia R. basin	Sept. 11, 1991/ ongoing
PI2Z	10.2	Northeast	710	North boundary, Marimla R. basin	July 18, 1991/ ongoing
CAB	21	East-northeast	150	Between Sacobia and Abacan R. channels, Clark Air Base	prior to June 15, 1991/ Sept. 30, 1991

four automatic gages, the highest rainfall intensities (both average storm intensity and maximum rainfall-burst intensity) from among the four rain gages were used to construct rainfall intensity-duration plots (fig. 10) for the 1991 rainstorms (see Caine, 1980; Rodolfo and Arguden, 1991; Rodolfo and others, this volume; Tuñgol and Regalado, this volume). Storms were differentiated as those causing no lahars, those causing at least one minor lahar, and those causing at least one major lahar (see table 2 for definitions of major and minor lahars). Although the big storms clearly

triggered major lahars, major lahars were also triggered by seemingly small, low-intensity storms. These anomalies can be explained if it is assumed that the "small" storms are really just the edges of bigger but localized, more intense convective storms that have slipped in between the rain gages. The upper limit of storm intensities <u>not</u> triggering lahars would be a more valid basis for definition of a lahar-trigger threshold, but not enough nonlahar storms were recorded in 1991 to do that. Either a denser rain-gage network or a longer record is required to characterize lahar-

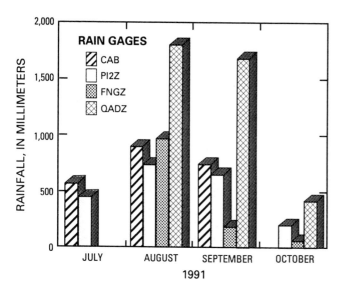

Figure 8. Monthly rainfall totals for the 1991 monsoon season on the east side of Mount Pinatubo. Asterisked totals are incomplete for the month because of gage or telemetry malfunction. The PI2Z total for the last part of July was proportionally extrapolated to a full month on the basis of the partial-to-complete ratio at the CAB gage.

triggering rainfall accurately in these upper basins. It also follows that rainfall data from any one telemetered rain gage would be unreliable for forecasting lahars for the purposes of warning downstream communities; the use of all four gages together improves reliability somewhat.

"Volcanic thunderstorms" were a phenomenon that occurred throughout the 1991 monsoon season, locally augmenting the normally occurring rainfall in the vicinity of Mount Pinatubo and triggering lahars. U.S. Navy and Air Force meteorologists repeatedly observed (on weather radar) a spatial and temporal association between the formation of intense convective storm cells and localized volcanic heat sources, such as pyroclastic- flow deposits and the eruptive vent (Pierson and others, 1992; Oswalt and others, this volume). Thus, some lahar-triggering rainstorms can be generated directly by volcanic activity if atmospheric conditions are favorable.

LAHAR OCCURRENCE

Most of the lahars occurring between July and October 1991 on the east side of Mount Pinatubo appear to have been directly triggered by rainfall, although the connection between measured rainfall and lahar magnitude or frequency is not clear (figs. 10, 11). Outbreaks of lakes temporarily dammed by pyroclastic-flow or lahar deposits occurred during or following periods of heavy rain and apparently augmented rainfall runoff in triggering lahars on at least three occasions: July 25 on the Sacobia River (W.E.

Scott and others, this volume), August 21 on the Sacobia River (Pierson and others, 1992), and September 7 on the Pasig-Potrero River (Pierson and others, 1992; K.M. Scott and others, this volume). In the last case, the lahar caused fatalities. Major lahars occurred on several other occasions when little or no rainfall was recorded (table 2), but, as previously noted, these may have been due to undetected localized storms.

Acoustic flow sensors (Bautista and others, 1991; Hadley and LaHusen, 1991, 1995; Marcial and others, this volume) were installed along river channels to record ground vibrations produced by passing lahars. Locations of two flow sensors along the Sacobia River (USAC and LSAC, situated 8.7 km apart), one along the Abacan River (ABAC) at the point of stream capture from the upper Sacobia basin, and one along the Gumain River (GUMA) are shown in figure 2. Acoustic signals were transmitted back to PVO via a ground-based radiotelemetry system. When plotted against time, signal intensity plots resemble hydrographs (fig. 12). Between July 18 and October 28, a total of 219 lahars were recorded. Eyewitnesses noted the occurrence of additional lahars between June 15 and July 18, before the flow sensors began operating, and some small lahars probably occurred in November.

LAHAR RHEOLOGY AND COMPOSITION

Direct sampling of lahars during the monsoon season was limited, but videotapes and observations of numerous flows in the Sacobia and Abacan Rivers indicate that lahars ranged from turbulent, erosive hyperconcentrated flows (fig. 13) to viscous, usually laminar debris flows (fig. 14). A sample of muddy streamflow from the Sacobia River collected between lahars has finer sediment and is much less concentrated (table 3).

Dip samples were collected directly from the surface of an apparently laminarly flowing debris flow in the Sacobia River on August 20 and from the surface of a very turbulent lahar, which had the appearance of a dilute debris flow, in the Abacan River on September 4. The samples are representative of the sampled flows, because very little material moving in the channels was coarser than fine gravel. Both flows were intensely steaming hot lahars with temperatures of about 50–60°C, but no unusual aspects of flow behavior were observed that might be attributable to temperature. The Sacobia slurry samples were composed predominantly of sand-size particles with gravel-fraction particle densities ranging from 0.96 to 1.47 g/cm^3, owing to the high pumice content (table 3). Sample bulk densities of 1.66 to 2.02 g/cm^3 correspond to sediment concentration fractions of 0.73 to 0.85 by weight (0.55 to 0.70 by volume). Although the Abacan River lahar was more turbulent than the Sacobia flow (for approximately equivalent distances downstream from head of basin), sediment

Table 2. Characteristics of rainstorms that triggered lahars on the east side of Mount Pinatubo (rain gages PI2Z, MSAC, FNGZ, and QADZ) for period July 18 –October 31, 1991.

["Storms" defined as periods of rainfall separated by more than 6 h with no rain. "Major" lahars defined as having durations greater than 2 h with moderate to high acoustic amplitude. "Minor" lahars have either durations less than 2 h or low acoustic amplitudes. Acoustic flow sensors: ABAC, Abacan; GUMA, Gumain; LSAC, lower Sacobia. –, no data or activity.]

Rain gage	Date	Storm duration (h)	Total rainfall (mm)	Average intensity (mm/h)	Maximum intensity (mm/h)		Lahar activity
PI2Z	7/18/91	4.7	39.1	8.4	29.6	(1 h)	Major – LSAC
QADZ		–	0	–	–		
PI2Z	7/19	12.8	26.0	2.1	9.3	(1 h)	Minor – LSAC
QADZ		–	0	–	–		
PI2Z	7/22	5.7	121.0	21.4	57.3	(1.5 h)	Major – LSAC
PI2Z	7/23	19.6	53.1	2.7	6.6	(4.7 h)	Minor – LSAC
PI2Z	7/24	–	0	–	–		Major – ABAC
PI2Z	7/25 (Storm 1)	1.4	7.1	5.0	5.0	(1.4 h)	Minor – LSAC / Major – ABAC
PI2Z	7/25 (Storm 2)	6.0	12.0	2.1	4.9	(0.8 h)	Major – ABAC / Minor – LSAC
PI2Z	7/26	16.0	45.0	≥2.8	–		Major – LSAC / Minor – ABAC
PI2Z	7/31	–	0	–	–		Major – LSAC
FNGZ		–	0	–	–		Major – ABAC
QADZ		6.7	26.9	4.1	76.1	(0.3 h)	
PI2Z	8/1	4.0	23.9	6.0	14.8	(0.3 h)	Major – LSAC
FNGZ		2.8	19.1	6.8	12.0	(0.5 h)	Minor – ABAC
QADZ		5.4	29.0	5.4	90.1	(0.1 h)	
PI2Z	8/5	4.5	56.1	12.4	12.4	(4.5 h)	Major – LSAC / Major – ABAC
PI2Z	8/8	.4	7.1	16.7	16.7	(0.4 h)	Minor – ABAC
QADZ		2.2	3.0	1.3	1.3	(2.2 h)	Minor – GUMA
FNGZ		–	0	–	–		
PI2Z	8/10	3.0	12.0	3.9	3.9	(3.0 h)	Minor – ABAC
FNGZ		–	0	–	–		Minor – GUMA
QADZ		1.9	12.0	6.5	6.5	(1.9 h)	
PI2Z	8/11	2.2	25.0	11.7	33.4	(0.3 h)	Major – LSAC
FNGZ		2.2	83.0	37.8	82.5	(0.1 h)	Minor – ABAC
QADZ		1.9	32.0	17.3	125.1	(0.1 h)	Minor – GUMA
P12Z	8/13	4.0	10.1	2.5	–		Major – LSAC
FNGZ		–	0	–	–		Major – ABAC
QADZ		4.8	38.0	7.9	9.1	(0.9 h)	Minor – GUMA
PI2Z	8/14 (Storm 1)	4.0	19.1	4.7	4.7	(4.0 h)	Major – LSAC
FNGZ		–	0	–	–		Major – ABAC
QADZ		4.8	51.0	10.6	24.1	(0.8 h)	Minor – GUMA

Table 2.   Characteristics of rainstorms that triggered lahars on the east side of Mount Pinatubo (rain gages PI2Z, MSAC, FNGZ, and QADZ) for period July 18 –October 31, 1991—Continued.

Rain gage	Date	Storm duration (h)	Total rainfall (mm)	Average intensity (mm/h)	Maximum intensity (mm/h)		Lahar activity
PI2Z	8/14	4.0	25.0	6.3	58.9	(0.2 h)	Major – LSAC
FNGZ	(Storm 2)	2.8	7.1	2.5	2.5	(2.8 h)	Major – ABAC
QADZ		3.0	32.0	10.9	36.4	(0.2 h)	Minor – GUMA
PI2Z	8/15–16	13.8	34.8	2.5	–		Major – LSAC
FNGZ		13.0	29.0	2.2	6.0	(1.0 h)	Minor – ABAC
QADZ		17.8	68.0	3.8	18.3	(0.2 h	Minor – GUMA
PI2Z	8/16	14.0	39.2	2.8	–		Major – LSAC
FNGZ	(Storm 2)	13.5	76.1	5.7	18.0	(0.5 h)	Minor – ABAC
QADZ		4.8	36.1	7.4	10.2	(0.1 h)	Minor – GUMA
PI2Z	8/17	11.9	12.0	.9	–		Major – LSAC
FNGZ		10.5	12.0	1.1	–		
QADZ		14.5	34.0	2.4	3.8	(0.1 h)	
PI2Z	8/18	21.4	19.1	.9	–		Major – LSAC
FNGZ		23.0	39.1	1.7	6.0	(0.5 h)	Minor – ABAC
QADZ		25.9	116.1	4.4	7.2	(1.4 h)	
PI2Z	8/19–21	40.5	142.1	3.5	21.4	(0.3 h)	Major – LSAC
FNGZ		45.6	551.1	12.1	119.7	(0.2 h)	Major – ABAC
QADZ		59.9	485.1	8.0	60.0	(0.5 h)	Major – GUMA
PI2Z	8/21	–	0	–	–		Major – LSAC
FNGZ	(Storm 2)	2.0	12.0	6.0	8.3	(0.5 h)	Minor – ABAC
QADZ		1.8	18.0	10.2	21.4	(0.4 h)	
PI2Z	8/22	–	0	–	–		Major – LSAC
FNGZ		–	0	–	–		Minor – ABAC
QADZ		4.6	23.9	4.9	60.0	(1.0 h)	Major – GUMA
PI2Z	8/23	–	0	–	–		Major – LSAC
FNGZ		–	0	–	–		Minor – ABAC
QADZ		9.6	47.1	4.9	21.3	(05 h)	Minor – GUMA
P12Z	8/24	10.0	58.0	5.8	20.0	(0.1 h)	Major – LSAC
FNGZ		7.0	32.0	4.6	23.6	(0.4 h)	Major – ABAC
QADZ		4.4	50.1	11.3	77.0	(0.1 h)	
P12Z	8/25	5.0	72.0	14.5	58.3	(0.1 h)	Major – LSAC
FNGZ		–	0	–	–		Major – ABAC
QADZ		9.1	199.1	21.7	58.9	(0.2 h)	Major – GUMA
PI2Z	8/26	1.0	2.0	2.0	2.0	(1.0 h)	Minor – LSAC
FNGZ		–	0	–	–		Major – ABAC
QADZ		6.3	18.0	2.8	2.8	(6.3 h)	
PI2Z	8/27	–	0	–	–		Major – LSAC
FNGZ		–	0	–	–		Major – ABAC
QADZ		.7	17.0	25.4	25.4	(0.7 h)	

Table 2. Characteristics of rainstorms that triggered lahars on the east side of Mount Pinatubo (rain gages PI2Z, MSAC, FNGZ, and QADZ) for period July 18 –October 31, 1991—Continued.

Rain gage	Date	Storm duration (h)	Total rainfall (mm)	Average intensity (mm/h)	Maximum intensity (mm/h)		Lahar activity
PI2Z	8/28	1.0	3.9	3.9	3.9	(1.0 h)	Minor – LSAC
FNGZ		–	0	–	–		Major – ABAC
QADZ		–	0	–	–		
PI2Z	8/29	2.9	6.0	2.0	7.4	(0.3 h)	Major – LSAC
FNGZ		–	0	–	–		Major – ABAC
QADZ		3.3	33.1	10.1	13.9	(0.6 h)	
PI2Z	8/30	7.0	28.0	3.9	14.8	(0.3 h)	Minor – LSAC
FNGZ		–	0	–	–		Major – ABAC
QADZ		6.5	40.0	6.1	17.3	(0.6 h)	Minor – GUMA
PI2Z	8/31	1.0	10.1	10.1	10.1	(1.0 h)	Major – LSAC
FNGZ		–	0	–	–		
QADZ		4.0	20.0	5.0	12.4	(0.1 h)	
PI2Z	9/1	3.4	57.2	17.0	55.3	(0.3 h)	Major – LSAC
QADZ		9.8	117.7	12.0	141.8	(0.1 h)	Major – ABAC
							Major – GUMA
PI2Z	9/2	–	0	–	–		Minor – LSAC
FNGZ		–	0	–	–		Minor – ABAC
QADZ		4.8	27.1	5.7	15.0	(0.6 h)	Minor – GUMA
PI2Z	9/3	3.0	17.0	5.7	14.0	(0.5 h)	Major – LSAC
FNGZ		1.5	6.0	3.9	3.9	(1.5 h)	Major – ABAC
QADZ		3.8	40.0	10.6	38.9	(0.2 h)	
PI2Z	9/4	5.3	26.0	4.9	26.1	(0.2 h)	Major – LSAC
FNGZ		5.0	13.1	2.7	3.0	(3.0 h)	Major – ABAC
QADZ		4.4	71.0	16.1	16.1	(4.4 h)	Major – GUMA
PI2Z	9/6–8	22.8	94.3	4.1	21.6	(0.6 h)	Major – LSAC
FNGZ		11.4	21.3	1.9	6.5	(1.0 h)	Major – ABAC
QADZ		60.7	183.6	3.0	26.8	(0.8 h)	Minor – GUMA
P12Z	9/10	6.0	75.6	12.6	29.6	(1.4 h)	Minor – LSAC
FNGZ		2.0	5.7	2.8	6.0	(0.9 h)	Major – ABAC
QADZ		7.5	43.9	5.8	50.1	(0.4 h)	Minor – GUMA
PI2Z	9/12	–	0	–	–		Minor – ABAC
FNGZ		1.0	8.2	8.2	8.2	(1.0 h)	
QADZ		2.8	3.2	1.1	–		
PI2Z	9/13	–	0	–	–		Major – LSAC
FNGZ		3.5	9.6	2.7	11.7	(0.4 h)	Major – ABAC
QADZ		29.0	37.6	1.3	35.1	(0.4 h)	

Table 2. Characteristics of rainstorms that triggered lahars on the east side of Mount Pinatubo (rain gages PI2Z, MSAC, FNGZ, and QADZ) for period July 18 –October 31, 1991—Continued.

Rain gage	Date	Storm duration (h)	Total rainfall (mm)	Average intensity (mm/h)	Maximum intensity (mm/h)		Lahar activity
PI2Z	9/14	6.9	27.2	3.9	–		Minor – LSAC
FNGZ		2.0	8.8	4.4	–		Minor – ABAC
QADZ		9.2	60.0	6.6	50.2	(0.8 h)	
MSAC		3.8	15.3	4.1	–		
PI2Z	9/15–20	75.3	138.0	1.9	30.4	(0.8 h)	Minor – LSAC
FNGZ		102.5	63.9	.6	–		Major – ABAC
QADZ		117.6	483.2	4.1	20.8	(4.4 h)	Minor – GUMA
MSAC		66.3	190.1	2.8	32.6	(0.6 h)	
PI2Z	9/22–23	17.6	23.9	1.4	25.5	(0.4 h)	Minor – ABAC
FNGZ		3.6	3.9	1.1	–		
QADZ		20.4	18.4	0.9	5.5	(0.8 h)	
PI2Z	9/27–	26.7	37.6	1.4	–		Minor – ABAC
FNGZ	10/1	12.8	10.1	.8	–		
QADZ		103.8	314.1	3.0	25.2	(1.2 h)	
PI2Z	10/3	1.9	58.7	30.9	157.2	(0.1 h)	Minor – ABAC
FNGZ		1.6	6.0	3.8	–		
QADZ		6.2	18.7	3.0	30.6	(0.5 h)	
PI2Z	10/4–5	.7	28.8	40.6	40.6	(0.7 h)	Minor – LSAC
FNGZ		36.2	17.2	.5	6.3	(1.7 h)	Major – ABAC
QADZ		6.5	15.3	2.4	9.5	(0.5 h)	
PI2Z	10/13	–	–	–	–		Major – ABAC
FNGZ		4.3	10.9	2.5	–		
QADZ		6.3	53.9	8.5	50.4	(0.2 h)	
MSAC		3.6	56.4	15.9	29.5	(1.1 h)	
PI2Z	10/14	.6	5.2	8.2	–		Minor – LSAC
FNGZ		–	–	–	–		Minor – ABAC
QADZ		9.7	38.0	3.9	17.8	(0.5 h)	
MSAC		.8	21.3	25.5	48.0	(0.3 h)	
PI2Z	10/15	1.6	10.9	6.8	12.4	(0.6 h)	Minor – ABAC
FNGZ		–	–	–	–		
QADZ		–	–	–	–		
MSAC		–	0	–	–		
PI2Z	10/16	–	–	–	–		Major – ABAC
FNGZ		–	–	–	–		
QADZ		9.5	8.7	.9	.9	(9.5 h)	
MSAC		7.1	55.6	7.9	37.3	(0.8 h)	
PI2Z	10/28	30.3	78.8	2.7	–		Minor – LSAC
FNGZ		14.0	13.5	.9	–		Major – ABAC
QADZ		36.5	221.6	6.1	21.4	(1.2 h)	
MSAC		12.4	66.5	5.4	27.4	(0.5 h)	

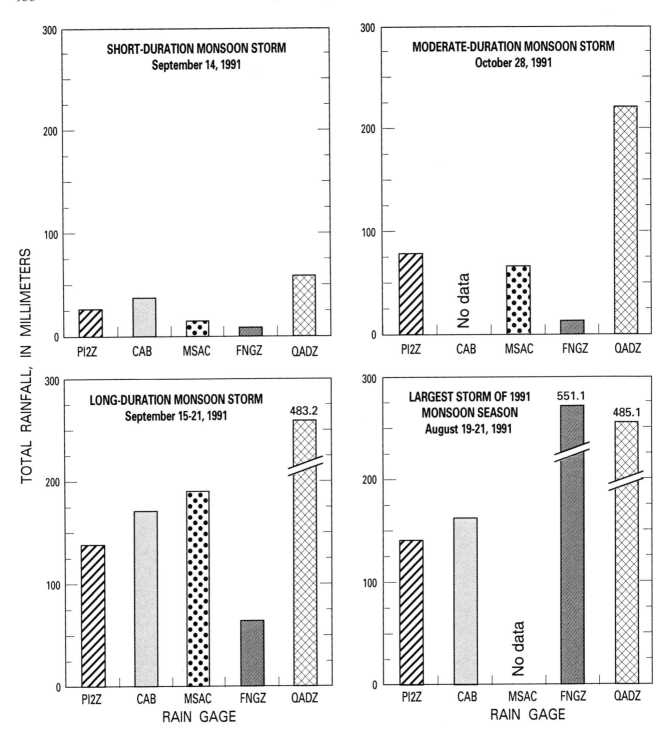

Figure 9.   Comparison of total storm rainfall recorded at the four automatic rain gages and CAB for representative storms of different durations. Each of these storms triggered lahars.

concentrations were not much less. The two Abacan samples from the west end of Clark Air Base both had sediment concentration fractions of 0.72 by weight; slightly different average particle densities changed the volume fractions to

0.51 and 0.55 (table 3). A sample from the same flow collected about 9 km downstream 30 min later had a sediment concentration fraction of only 0.44 by weight. In general, the lahars became more dilute with distance from source,

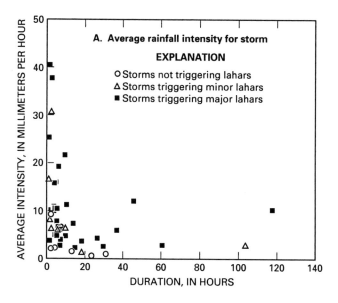
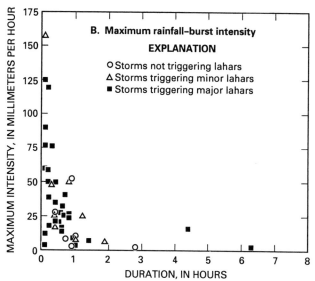

Figure 10. Intensity-duration plots for 1991 rainstorms; data points for each storm are the highest intensity values from the PI2Z, MSAC, FNGZ, or QADZ rain gage. Note that the different storm types do not readily fall into separate fields. *A*, Average storm intensity versus duration. *B*, maximum burst intensity versus duration (bursts typically 0.1- to 1-h duration).

and transformations to hyperconcentrated flow commonly occurred as lahars reached or flowed out onto the alluvial fans.

Extremely pumice-rich debris-flow depositional units within 1991 lahar deposits in the east-side drainages confirm evidence from surface dip samples that the valley-fill pumiceous pyroclastic-flow deposits were the dominant sediment source for the 1991 lahars. Lithic-rich deposits (having particles of normal density) were rare and were limited to relatively thin sandy layers deposited by hyperconcentrated flows or normal streamflow (fig. 15). This is in contrast to field observations of 1992 and 1993 lahar deposits, which more typically showed a marked differentiation between lithic-rich and pumice-rich debris-flow facies within single lahar depositional units; this differentiation suggests density stratification or segregation within flows and a mixed source area for later lahars (1991 pumiceous deposits *and* older lithic-rich volcaniclastic deposits).

Grain-size distributions of the dip samples were quite similar to the distributions of the pyroclastic-flow source deposits upstream (fig. 16). Distribution similarities were especially close for grain sizes finer than medium sand; gravel fractions fluctuated more. All the lahar samples were very poorly sorted, as is typical for debris-flow samples (table 3). Most of the samples are positively skewed (excess fine material) and leptokurtic (tails of distribution more poorly sorted than central portion).

LAHAR FREQUENCY

Three acoustic flow sensors, the LSAC, ABAC, and GUMA stations (fig. 2) detected passage of 219 lahars

down the Sacobia, Abacan, and Gumain Rivers through the 1991 monsoon season. They were in operation for 105, 92, and 55 days, respectively, between July 18 and October 31, and they recorded 93, 95, and 31 lahars each. A fourth flow sensor (USAC) was located in the upper part of the Sacobia basin. These totals break down to average daily rates of 0.9, 1.0, and 0.6 lahars per day, respectively. Lahars also occurred in the O'Donnell-Tarlac and Pasig-Potrero Rivers but could not be documented on a continuous basis. During August, the occurrence of three to five lahars per day in a channel were common (fig. 12); the maximum recorded in a single day was 7. The lower rate in the Gumain River could have been due to either instrument insensitivity (the flow sensor was located farther from the channel than at either of the other two sites) or the relatively smaller supply of source sediment in that basin (Pierson and others, 1992).

LAHAR MAGNITUDE

Most of the posteruption lahars occurring after the typhoon that accompanied the June 15, 1991, eruption (Major and others, this volume) were triggered by localized monsoonal rainstorms, although some were triggered by more regional storms, and several were augmented by lake breakouts. The 1991 lahars from east-side drainages (contributing catchment areas of approximately 50 to 250 km²) generally were of moderate magnitude compared with other lahars worldwide, particularly those generated by rapid snowmelt (Pierson, 1995). Typical posteruption lahars were estimated to be 2 to 3 m deep, 20 to 50 m wide, and moving at 4 to 8 m/s (surface velocity, in channels up to 10 km upstream of fan heads) with peak discharges in the

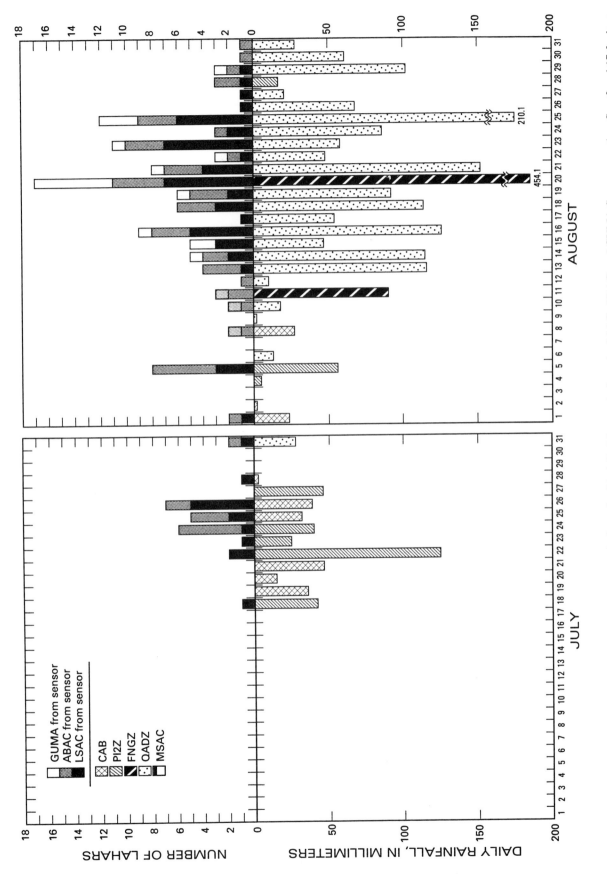

Figure 11. Comparison of maximum daily rainfall (rain gage indicated) with daily lahar totals from the LSAC, ABAC, and GUMA flow sensors (see figs. 2 and 17 for locations) for most of the 1991 monsoon season. Flows counted as separate lahars if peaks separated by at least 1 h and acoustic intensities (at 10–100 Hz) greater than 1,200 acoustic units (for LSAC and ABAC) or 800 units (for GUMA).

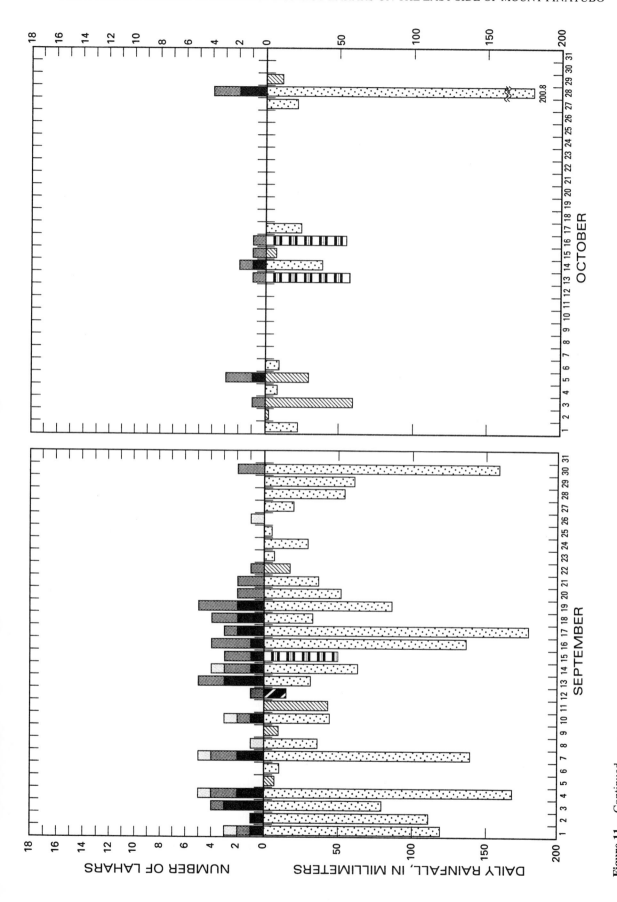

Figure 11.—Continued.

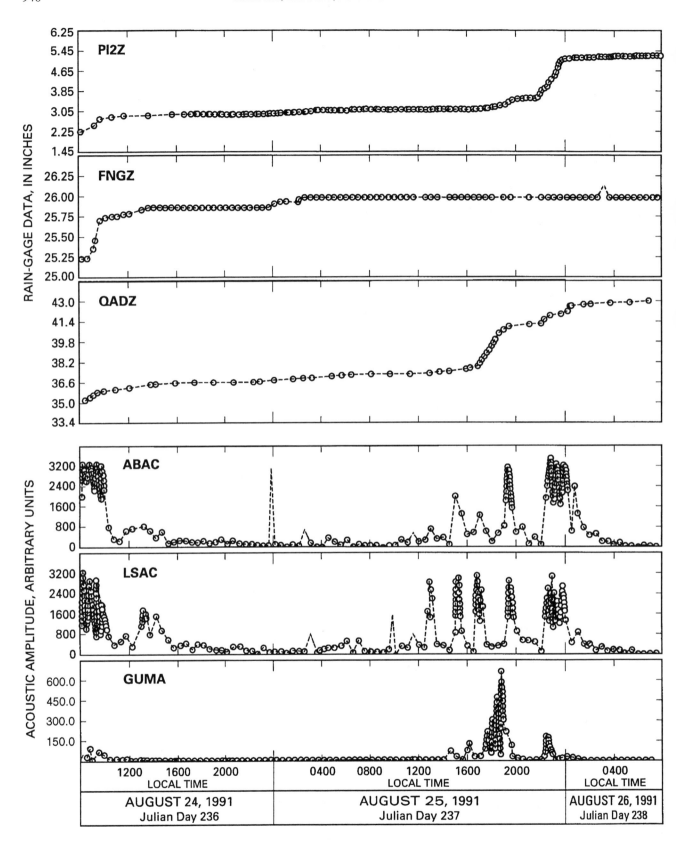

Figure 12. Rain-gage and acoustic flow-sensor data on same time base for a part of August 1991, showing the spatially varied nature of the rainfall and the uncertain relation between lahars and measured rainfall (some lahars apparently beginning <u>before</u> rainfall begins).

Figure 13. Example of *hyperconcentrated flow* occurring in the Sacobia River about 20 km downstream from the caldera on September 26, 1993; flow is from left to right. Flow depth is 45 to 50 cm; width is about 30 m. Surface velocity is varying between 2.5 and 3.1 m/s, and fluid temperature is ambient. Sediment concentration of the flowing mixture is 39 percent sediment by weight (23 percent by volume). This flow was actively eroding laterally as the picture was taken; sections of the bank in the foreground were toppling into the flow every few minutes.

range of about 200 to 1,200 m³/s. A few lahars in late July were as deep as 5 m and as fast as 11 m/s, with peak discharges possibly as large as 5,000 m³/s. Average celerity (measured peak to peak) of four sharply peaked lahars between the USAC and LSAC stations (8.7 km apart) varied between 3.6 and 12.1 m/s. The acoustic intensity of the lahars on the flow sensor records had not been calibrated to actual discharge in 1991.

The 1991 lahars commonly were multipeaked and had flow durations of 2 to 4 h. If a right triangle is taken as a simple model for hydrograph shape, the range of common discharges above would result in an approximate average lahar volume per event (water and sediment) of between 0.7×10^6 and 8.6×10^6 m³. Because debris-flow hydrograph recessional limbs typically show an exponential rather than a linear decrease in discharge (Weir, 1982), the above values probably overestimate lahar volumes somewhat; 0.5×10^6 to 5×10^6 m³ might be a more realistic range.

LAHAR DEPOSITION

During the 1991 monsoon season, phenomenally large volumes of sediment were transported out of the upper east-side watersheds at Mount Pinatubo, primarily by lahars, and deposited (1) in and adjacent to valley-confined stream channels downstream from major pyroclastic valley fills, (2) on the low-gradient coalescing alluvial fans at the foot of the volcano, and (3) slightly beyond the fans on very flat alluvial plains and in coastal marshes (fig. 17). Most of the deposition occurred on the distal ends of the fans, with thinner fine-grained deposits extending out in channels onto the

alluvial plain. Individual depositional units of debris flows, found predominantly in the channels upstream of fans and on the proximal parts of the fans, were generally 1 to 2 m thick. No crusting or discoloration was observed that might be attributed to slurry temperature, as has been reported elsewhere (Arguden and Rodolfo, 1990). Aggradation by deposition in confined valleys, which was as much as 25 m, and accompanying geomorphic changes have been reported by Punongbayan and others (1991) and by K.M. Scott and others (this volume).

GROWTH OF ALLUVIAL FANS

Alluvial fan surfaces, primarily from about 25 to 45 km away from the caldera, were the east-side areas most seriously affected by lahar deposition during the 1991 monsoon season (figs. 17, 18). Sediment deposition on the east-side fans generally ranged from 0.5 to 5 m in thickness and appeared to average about 2 m. This aggradation caused rivers to avulse out of existing channels (including engineered channels) and to spread the deposition in broad, braided channels over large areas of countryside (fig. 18). In general, deposition occurred primarily on the distal ends of fans during the early part of the monsoon season. Later in the season, areas progressively farther upstream became inundated, although some of the down-fan depositional areas expanded as well.

Deposits of the 1991 lahars, including those of the syneruption events (Major and others, this volume) covered more than 200 km², most of it prime agricultural land (table 4). Many more tens of square kilometers were inundated by backflooding that was caused by the blockage of

A

B

Figure 14. Example of *debris flow* occurring in the Sacobia River at two adjacent sites about 16 km downstream from the caldera on August 20, 1991. Flow is estimated to be about 2 m deep and 15 m wide; surface velocity varied between 4.8 and 5.9 m/s (at site in photograph *A*), and fluid temperature was approximately 50 to 60°C. On the basis of these flow parameters, flow should be supercritical. Sediment concentration of the fluid mixture was 85 percent sediment by weight (70 percent by volume). The flow in photograph *A* (immediately upstream of photograph *B*) appeared to be flowing laminarly (toward the camera) and appeared to have a "rigid plug" in the center part of the flow. The flow in photograph *B* (moving left to right) erupted in violently turbulent standing waves (about 1 to 2 m high) where the channel is slightly more constricted and slightly steeper than in *A*. Such turbulence was very seldom observed in debris flows having such high sediment concentrations, although nonbreaking standing waves were not uncommon.

preeruption stream channels and canals by the aggrading lahar deposits (fig. 19).

Lahar deposition on farmland and in towns has forced tens of thousands of residents into refugee evacuation centers. Lahars buried or otherwise adversely affected 111 barangays (villages) and 3 separate military installations on the east side of the volcano. In addition, bank erosion from lateral migration of the Abacan River channel destroyed

hundreds of homes and businesses around Angeles City (fig. 17). Most major highway bridges on all sides of the mountain had been destroyed by the end of 1991. The Sacobia-Bamban channel aggraded about 20 m in just over 2 months (E. de la Cruz, Philippine Institute of Volcanology and Seismology, oral commun., 1991), and the Highway 3 (MacArthur Highway) bridge crossing the channel was destroyed by a lahar on August 21 (fig. 20).

Figure 15. Still-warm, unstratified debris-flow deposit from the August 20–21 lahars at the San Francisco bridge, which is about 41 km downstream from the caldera on the Bamban River. Photograph taken on August 23. The debris-flow deposit is extremely pumice rich and is sandwiched between two thin, darker, lithic-rich sand units deposited during less concentrated flow. The shovel grip is 14 cm wide.

DEPOSITIONAL VOLUMES AND SEDIMENT YIELDS

The total volume of volcaniclastic sediment deposited in 1991 in the east-side basins is about 0.38 km^3 (3.8×10^8 m^3), which is equivalent to about 30 percent of the total volume of the 1991 pyroclastic source deposits in the upper basins. Approximate depositional volumes and sediment yields for the eastward-draining basins are shown for basins

individually in table 4. Volumes are derived from contour maps of deposit thicknesses based on aerial photography, preeruption topography (maps with 10- and 20-m contour intervals), and field measurements or estimates of thickness. Given the imprecise nature of preeruption map information and the local variability of deposit thickness, error bars of at least ±20 percent should be attached to these volumes. The computed volumes of sediment eroded from east-side watersheds in 1991 are at the upper limit or exceed the range of volumes of eroded sediment that were predicted by Pierson and others (1992) (table 4).

Annual total sediment yields, computed on the basis of 1 year's data, are on the order of 10^6 m^3/km^2·y for three of the basins (table 4). These values are extraordinarily high—about an order of magnitude higher than those computed for Mount St. Helens following the May 18, 1980, eruption (Janda and others, 1984). Such large amounts of sediment being transported have had significant effects on channel morphology (K.M. Scott and others, this volume).

CONCLUSIONS

In the first rainy season following the June 15, 1991, eruption of Mount Pinatubo, seasonally typical monsoonal rainfall on the east side of the volcano generated hundreds of hot lahars. The primary sediment sources for the lahars in each watershed were the thick pyroclastic-flow deposits that filled the heads of the valleys tens to hundreds of meters deep. Lahar deposition occurred mainly on the apron of coalescing alluvial fans surrounding the volcano. Deposition began early in the season at the downstream ends of the fans and progressed up-fan during the later part of the 1991 monsoon season.

Extraordinarily large volumes of volcaniclastic sediment have been transported from the flanks of the volcano to the surrounding alluvial fans and plains. After only one rainy season, the lahar deposits covered more than 200 km^2 with several meters of coarse volcaniclastic sediment. Total lahar-deposit volume on the east-side fans at the end of 1991 was more than one-third of a cubic kilometer, which is equivalent to about 30 percent of the pyroclastic-flow valley-fill deposits (primary source sediments) in the heads of the drainage basins. Sediment yields for three of the watersheds, on the order of 10^6 m^3/km^2•y, are about an order of magnitude higher than yields computed for the Toutle River at Mount St. Helens following the 1980 eruption there, which had been among the highest ever computed for moderately sized drainage basins.

Table 3. Physical properties of lahar dip samples from the August 20 Sacobia River lahar and the September 4 Abacan River lahar compared to a low-flow sample (September 8, Bamban River).

[-, no data]

Sample	Low Flow, Bamban River	Lahar 1A (debris flow), Sacobia River	Lahar 1B (debris flow), Sacobia River	Lahar 2 (debris flow), Sacobia River	Lahar 3A (transitional debris flow), Abacan River	Lahar 3B (transitional debris flow), Abacan River	Lahar 4B (hyperconc. flow), Abacan River
Distance downstream from caldera rim (km)	37	17	17	28	18	18	27
Wet sample bulk density (g/cm³)	1.08	2.02	-	1.66	1.75	1.59	-
Particle density: whole sample mean (g/cm³)	1.93	2.48	-	2.19	2.45	2.08	-
Particle density: 8 mm, 16 mm grains only (g/cm³)	-	1.47	1.00	0.96	0.97	1.04	0.92
Sediment concentration (volume fraction)	0.09	0.70	-	0.55	0.51	0.55	-
Sediment concentration (weight fraction)	0.16	0.85	0.79	0.73	0.72	0.72	0.44
Percent gravel	0	20.3	17.9	42.0	23.4	10.4	21.2
Percent sand	43.8	63.5	65.8	45.1	62.2	74.0	59.7
Percent silt	47.5	14.3	14.2	11.4	12.9	13.6	17.1
Percent clay	8.7	1.9	2.1	1.5	1.5	2.0	2.0
Slurry temperature (°C)	ambient	~50	~50	~50	48	48	-
Mean grain size, M_z (ϕ)	5.68	1.18	1.38	-0.26	0.67	1.71	1.51
Sorting, σ_I (ϕ)	3.06	3.32	3.06	3.96	3.33	2.57	3.30
Skewness, Sk_I	0.29	-0.07	0.01	-0.05	-0.04	0.12	-0.11
Kurtosis, K_G	1.24	1.40	1.30	0.76	1.33	1.42	1.08

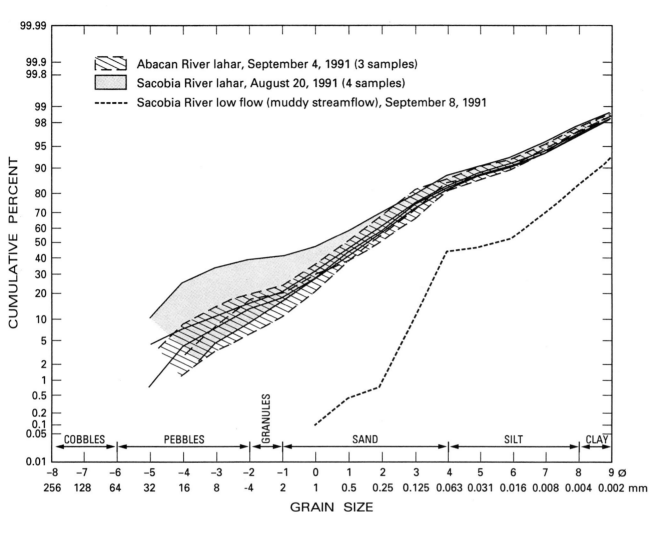

Figure 16. Cumulative grain-size distribution curves for samples of flowing lahars on August 20 (Sacobia River) and September 4 (Abacan River), and one sample of low-discharge "normal" flow on September 8 (Bamban River at San Francisco bridge).

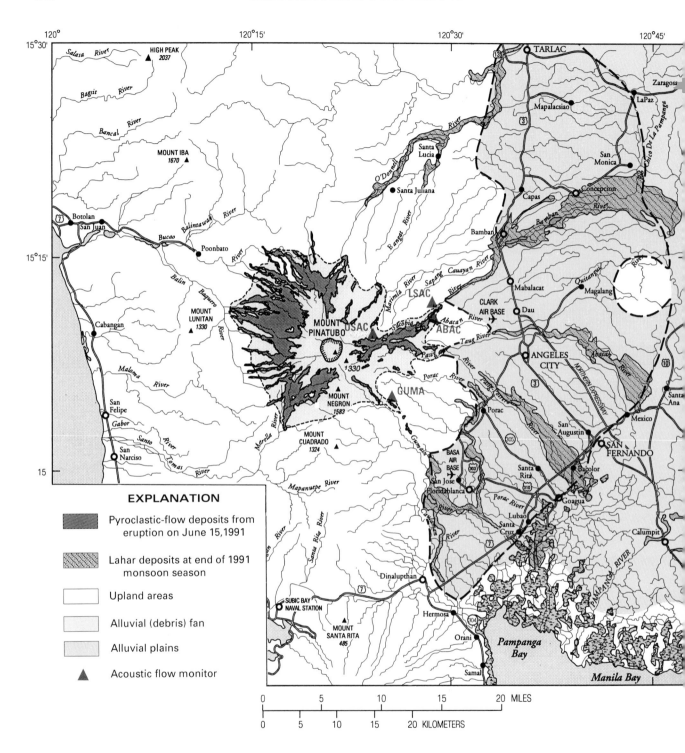

Figure 17. Pattern of lahar deposition primarily on alluvial fans on the east side of Mount Pinatubo following the 1991 monsoon season, relative to the distribution of pyroclastic source sediment (thick, pumiceous pyroclastic-flow deposits and thin pyroclastic surge deposits). Average thickness of deposits on fans is approximately 2 m. Alluvial fans and alluvial plains are defined on the basis of topography and drainage patterns.

Figure 18. Lahar deposition on alluvial fans on the east side of Mount Pinatubo. *A*, Oblique aerial view (looking west, upstream) of lahar deposition along Bamban River about 35 km northeast of caldera rim on August 15, 1991. Left-bank channel levee was breached on June 15. *B*, Lahar deposition on either side of levee-confined Pasig-Potrero River, about 34 km southeast of the caldera rim on August 23, 1991. View is looking southwest, across the fan.

A

B

Figure 19. Backflooding and transient lake caused by the damming of the Marimla River by lahar deposits along the main Sacobia-Bamban channel, August 23, 1991. That main channel visible in distance. View is looking east. Breakout of a former lake here augmented the lahar on August 21 that destroyed the Bamban bridge.

Figure 20. Site of former Highway 3 (MacArthur Highway) bridge at Bamban town (concrete bridge piers visible in channel) and remaining spans of railway bridge across Sacobia-Bamban River on August 23, 1991. View is upstream, looking southwest. The highway bridge was destroyed by a lahar on August 21; the railway lost one bridge span (far right of photograph) to lahars on June 15 and long segments of earthen embankment (upper right-center and upper left of photograph) on August 21.

REFERENCES CITED

Arboleda, R.A., and Martinez, M.L., this volume, 1992 lahars in the Pasig-Potrero River system.

Arguden, A.T., and Rodolfo, K.S., 1990, Sedimentologic and dynamic differences between hot and cold laharic debris flows of Mayon Volcano, Philippines: Geological Society of America Bulletin, v. 102, p. 865–876.

Bautista, B.C., Bautista, L.P., Marcial, S.S., Melosantos, A.A., and Hadley, K.C., 1991, Instrumental monitoring of Mount Pinatubo lahars, Philippines [abs.]: Eos, Transactions, American Geophysical Union, v. 72, no. 44, p. 63.

Beverage, J.P., and Culbertson, J.K., 1964, Hyperconcentrations of suspended sediment: American Society of Civil Engineers Proceedings, Journal of the Hydraulics Division, v. 90, no. HY6, p. 117–128.

Caine, Nel, 1980, The rainfall intensity-duration control of shallow landslides and debris flows: Geografiska Analer, v. 62, p. 23–27.

Collins, B.D., and Dunne, Thomas, 1986, Erosion of tephra from the 1980 eruption of Mount St. Helens: Geological Society of America Bulletin, v. 97, p. 896–905.

Costa, J.E., 1984, Physical geomorphology of debris flows, in Costa, J.E., and Fleisher, P.J., eds., Developments and applications of geomorphology: Berlin, Springer, p. 268–317.

Table 4. Sediment volumes delivered to depositional areas (primarily alluvial fans) and sediment yields for east-side drainages at Mount Pinatubo, June–October 1991, based on areas of deposits and measured or estimated deposit thicknesses contoured over the depositional areas.

[Source volume data from Pierson and others (1992), Swiss Disaster Relief (1993), U.S. Army Corps of Engineers (unpublished data), and W.E. Scott and others (this volume); -, no data]

Watershed	Approx. drainage area contrib. to lahars (km²)	Deposition area (km²)	Average deposit thickness (m)	Approx. deposit volume (km³)	Approx. source volume (km³)	Ratio of deposit volume/ source volume	Sediment yield (m³/km²/y)	Range of deposit volumes predicted by Pierson and others (1992, tables 3 and 4) (km³)
O'Donnell-Tarlac	281	-	-	0.06	0.2	0.30	0.2 x 10⁶	0.03 - 0.06
Sacobia-Bamban (incl. Marimla)	152	81	1.9	.15	.6	.25	1.0 x 10⁶	.06 - .13
Abacan	68	34	1.8	.06	.1	.60	.8 x 10⁶	.008 - .02
Pasig-Potrero	20	38	1.3	.05	.3	.17	2.5 x 10⁶	.03 - .06
Porac-Gumain	219	38	1.6	.06	.05	1.2	.3 x 10⁶	.004 - .01

Hadley, K.C., and LaHusen, R.G., 1991, Deployment of an acoustic flow-monitor system and examples of its application at Mount Pinatubo, Philippines [abs.]: Eos, Transactions, American Geophysical Union, v. 72, no. 44, p. 67.

Hadley, K.C., and LaHusen, R.G., 1995, Technical manual for the experimental acoustic flow monitor: U.S. Geological Survey Open-File Report 95–114, 24 p.

Hamidi, Sumarna, 1989, Lahar of Galunggung Volcano from 1982 through 1986, in Proceedings of International Symposium on Erosion and Volcanic Debris Flow Technology, July 31–August 3, 1989, Yogyakarta, Indonesia: Jakarta, Ministry of Public Works, p. VP1-1–VP1-23.

Hampton, M.A., 1972, The role of subaqueous debris flow in generating turbidity currents: Journal of Sedimentary Petrology, v. 42, p. 775–793.

Janda, R.J., Meyer, D.F., and Childers, Dallas, 1984, Sedimentation and geomorphic changes during and following the 1980-1983 eruptions of Mount St. Helens, Washington: Shinsabo, v. 37, no. 2, p. 10–21.

Johnson, A.M., 1984, Debris flow, in Brunsden, Denys, and Prior, D.B., eds., Slope instability: Chichester, John Wiley and Sons, p. 257–361.

Kang, Z., and Zhang, S., 1980, A preliminary analysis of the characteristics of debris flow, in Proceedings of the International Symposium on River Sedimentation: Beijing, Chinese Society for Hydraulic Engineering, p. 225–226.

Kuenzi, W.D., Horst, O.H., and McGehee, R.V., 1979, Effect of volcanic activity on fluvial-deltaic sedimentation in a modern arc-trench gap, southwestern Guatemala: Geological Society of America Bulletin, v. 90, p. 827–838.

Leavesley, G.H., Lusby, G.C., and Lichty, R.W., 1989, Infiltration and erosion characteristics of selected tephra deposits from the 1980 eruption of Mount St. Helens, Washington, USA: Hydrological Sciences Journal, v. 34, p. 339–353.

Major, J.J., Janda, R.J., and Daag, A.S., this volume, Watershed disturbance and lahars on the east side of Mount Pinatubo during the mid-June 1991 eruptions.

Major, J.J., and Newhall, C.G., 1989, Snow and ice perturbations during historical volcanic eruptions and the formation of lahars and floods: Bulletin of Volcanology, v. 52, p. 1–27.

Marcial, S.S., Melosantos, A.A., Hadley, K.C., LaHusen, R.G., and Marso, J.N., this volume, Instrumental lahar monitoring at Mount Pinatubo.

Martinez, M.L., Arboleda, R.A., Delos Reyes, P.J., Gabinete, E., and Dolan, M.T., this volume, Observations of 1992 lahars along the Sacobia-Bamban River system.

Moyer, T.C., and Swanson, D.A., 1987, Secondary hydroeruptions in pyroclastic-flow deposits: Examples from Mount St. Helens: Journal of Volcanology and Geothermal Research, v. 32, p. 299–319.

Neall, V.E., 1976, Lahars—Global occurrence and annotation bibliography: Wellington, New Zealand, Victoria University, Publication no. 5, 18 p.

Ollier, C.D., and Brown, M.J.F., 1971, Erosion of a young volcano in New Guinea: Zeitschrift für Geomorphologie, v. 15, p. 12–28.

Oswalt, J.S., Nichols, W., and O'Hara, J.F., this volume, Meteorological observations of the 1991 Mount Pinatubo eruption.

Paladio-Melosantos, M.L., Solidum, R.U., Scott, W.E., Quiambao, R.B., Umbal, J.V., Rodolfo, K.S., Tubianosa, B.S., Delos

Reyes, P.J., and Ruelo, H.R., this volume, Tephra falls of the 1991 eruptions of Mount Pinatubo.

Pierson, T.C., 1995, Flow characteristics of large eruption-triggered debris flows at snow-clad volcanoes: constraints for debris-flow models: Journal of Volcanology and Geothermal Research, v. 66, nos. 1–4, p. 283–294.

Pierson, T.C., and Costa, J.E., 1987, A rheologic classification of subaerial sediment-water flows: Geological Society of America Reviews in Engineering Geology, v. 7, p. 1–12.

Pierson, T.C., Janda, R.J., Umbal, J.V., and Daag, A.S., 1992, Immediate and long-term hazards from lahars and excess sedimentation in rivers draining Mount Pinatubo, Philippines: U.S. Geological Survey Water-Resources Investigations Report 92–4039, 35 p.

Pierson, T.C., and Scott, K.M., 1985, Downstream dilution of a lahar: Transition from debris flow to hyperconcentrated streamflow: Water Resources Research, v. 21, p. 1511–1524.

Pinatubo Volcano Observatory Team, 1991, Lessons from a major eruption: Mt. Pinatubo, Philippines: Eos, Transactions, American Geophysical Union, v. 72, no. 49, p. 545, 552–553, 555.

Punongbayan, R.S., Delacruz, Edwin, Gabinete, Elmer, Scott, K.M., Janda, R.J., and Pierson, T.C., 1991, Initial stream channel responses to large 1991 volumes of pyroclastic-flow and tephra-fall deposits at Mount Pinatubo, Philippines [abs.]: Eos, Transactions, American Geophysical Union, v. 72, no. 44, p. 63.

Rodolfo, K.S., 1989, Origin and early evolution of lahar channel at Mabinit, Mayon Volcano, Philippines: Geological Society of America Bulletin, v. 101, p. 414–426.

———1991, Climatic, volcaniclastic, and geomorphic controls on the differential timing of lahars on the east and west sides of Mount Pinatubo during and after its June 1991 eruptions [abs.]: Eos, Transactions, American Geophysical Union, v. 72, no. 44, p. 62.

Rodolfo, K.S., and Arguden, A.T., 1991, Rain-lahar generation and sediment-delivery systems at Mayon Volcano, Philippines, in Fisher, R.V., and Smith, G.A., eds., Sedimentation in volcanic settings: SEPM (Society for Sedimentary Geology) Special Publication No. 45, p. 71–87.

Rodolfo, K.S., Arguden, A.T., Solidum, R.U., and Umbal, J.V., 1989, Anatomy and behavior of a post-eruptive rain lahar triggered by a typhoon on Mayon Volcano, Philippines: International Association of Engineering Geologists Bulletin, v. 40, p. 79–90.

Rodolfo, K.S., Umbal, J.V., Alonso, R.A., Remotigue, C.T., Paladio-Melosantos, M.L., Salvador, J.H.G., Evangelista, D., and Miller, Y., this volume, Two years of lahars on the western flank of Mount Pinatubo: Initiation, flow processes, deposits, and attendant geomorphic and hydraulic changes.

Rowley, P.D., Kuntz, M.A., and MacLeod, N.S., 1981, Pyroclastic-flow deposits, in Lipman, P.W., and Mullineaux, D.R., eds., The 1980 eruptions of Mount St. Helens, Washington: U.S. Geological Survey, Professional Paper 1250, p. 489–512.

Rutherford, M.J., and Devine, J.D., 1991, Preeruption conditions and volatiles in the 1991 Pinatubo magma [abs.]: Eos, Transactions, American Geophysical Union, v. 72, no. 44, p. 62.

Schmidt, K.G., 1934, Die Schuttströme am Merapi auf Java nach dem Ausbruch von 1930: De Ingenieur in Nederlandsch-Indië, v. 1, no. 7–9, p. 96–171.

Scott, K.M., Janda, R.J., de la Cruz, E., Gabinete, E., Eto, I., Isada, M., Sexon, M., and Hadley, K.C., this volume, Channel and sedimentation responses to large volumes of 1991 volcanic deposits on the east flank of Mount Pinatubo.

Scott, W.E., Hoblitt, R.P., Daligdig, J.A., Besana, G., and Tubianosa, B.S., 1991, 15 June1991 pyroclastic deposits at Mount Pinatubo, Philippines [abs.]: Eos, Transactions, American Geophysical Union, v. 72, no. 44, p. 61.

Scott, W.E., Hoblitt, R.P., Torres, R.C., Self, S, Martinez, M.L., and Nillos, T., Jr., this volume, Pyroclastic flows of the June 15, 1991, climactic eruption of Mount Pinatubo.

Segerstrom, Kenneth, 1950, Erosion studies at Parícutin, State of Michoacán, Mexico: U.S. Geological Survey Bulletin 965–A, p. 1–64.

———1960, Erosion and related phenomena at Parícutin in 1957: U.S. Geological Survey Bulletin 1104–A, p. 1–18.

———1966, Parícutin, 1965—Aftermath of eruption: U.S. Geological Survey Professional Paper 550–C, p. C93–C101.

Smart, G.M., 1981, Volcanic debris control, Gunung Kelud, East Java, in Erosion and sediment transport in Pacific Rim steeplands: International Association of Hydrologic Sciences, Publication No. 132, p. 604–623.

Smith, G.A., 1986, Coarse-grained nonmarine volcaniclastic sediment: Terminology and depositional process: Geological Society of America Bulletin, v. 97, p. 1–10.

Smith, G.A., and Lowe, D.R., 1991, Lahars: volcano-hydrologic events and deposition in the debris flow—hyperconcentrated flow continuum, in Fisher, R.V., and Smith, G.A., eds., Sedimentation in volcanic settings: SEPM (Society for Sedimentary Geology), Special Publication No. 45, p. 59-70.

Swiss Disaster Relief, 1993, Lahars in the O'Donnell River system, Mt. Pinatubo, Philippines—Hazard assessment and engineering measures: Government of Switzerland, Swiss Disaster Relief report, 68 p.

Torres, R.C., Self, S., and Martinez, M.L., this volume, Secondary pyroclastic flows from the June 15, 1991, ignimbrite of Mount Pinatubo.

Tuñgol, N.M., and Regalado, M.T.M., this volume, Rainfall, acoustic flow monitor records, and observed lahars of the Sacobia River in 1992.

Ulate, C.A., and Corrales, M.F., 1966, Mud floods related to the Irazú Volcano eruptions: American Society of Civil Engineers Proceedings, Journal of the Hydraulics Division, v. 92, no. HY–6, p. 117–129.

Umbal, J.V., and Rodolfo, K.S., this volume, The 1991 lahars of southwestern Mount Pinatubo and evolution of the lahar-dammed Mapanuepe Lake.

Umbal, J.V., Rodolfo, K.S., Alonso, R.A., Solidum, R.U., Paladio, M.L., Tamayo, R., Angeles, M.B., Tan, R., Remotique, C., and Jalique, V., 1991, Lahars remobilized by breaching of a lahar-dammed non-volcanic tributary, Mount Pinatubo, Philippines [abs.]: Eos, Transactions, American Geophysical Union, v. 72, no. 44, p. 63.

Waldron, H.H., 1967, Debris flow and erosion control problems caused by the ash eruptions of Irazú Volcano, Costa Rica: U.S. Geological Survey Bulletin 1241–I, 37 p.

Weir, G.J., 1982, Kinematic wave theory of Ruapehu lahars: New Zealand Journal of Science, v. 25, p. 197–203.

The 1991 Lahars of Southwestern Mount Pinatubo and Evolution of the Lahar-Dammed Mapanuepe Lake

By Jesse V. Umbal[1][2] and Kelvin S. Rodolfo[1]

ABSTRACT

The 1991 eruption of Mount Pinatubo deposited more than 6 cubic kilometers of pyroclastic debris on the volcano's slope, of which 1.3 cubic kilometers was emplaced as the Marella pyroclastic fan covering 25 square kilometers of the southwest sector. The fan is drained by the Marella River, along which numerous hot lahars, generated by intense monsoon rains from July to November 1991, transported about 185 million cubic meters (14 percent) of the new pyroclastic-fan deposits downstream and thus affected an area of 46 square kilometers.

The Mapanuepe River, which drains 85 square kilometers of mountainous watershed south of Mount Pinatubo, joins the Marella River to form the Santo Tomas River, the master stream that, en route to the South China Sea, follows the margin of a broad, 235-square-kilometer alluvial plain. Geomorphic and sedimentologic evidence indicates that the Mapanuepe valley was the site of previous eruption-related lahar-dammed lakes prior to the 1991 eruption.

Enhanced surface runoff from the tephra-fall- and pyroclastic-flow-impacted watershed, coupled with the steeper stream gradient of the Marella channel, enabled hot laharic debris flows and hyperconcentrated streamflows to block outflow from the Mapanuepe River. A bedrock constriction just below the confluence of these two rivers created a backwater effect that promoted deposition from hyperconcentrated streamflows. Later, the constriction was repeatedly plugged by hot debris flows, which caused subsequent lahars to flow upstream and deposit in the Mapanuepe valley. As each episode of lahar activity in the Marella waned, rising lake waters eventually overtopped and breached the debris dam and released floodwaters that bulked to hyperconcentrated streamflows by incorporating sediments from previous lahar deposits. At its maximum extent in 1991 the lake flooded an area of 6.7 square kilo-

meters and impounded an estimated 75 million cubic meters of water.

Approximately 15 to 25 percent of the materials remaining in the new Marella pyroclastic fan will need to be remobilized before lahars aggrade the Marella-Santo Tomas channel to the level of the highest prehistoric river terraces. Typically, sediment delivery rates are highest during the first few years after the eruption and are expected to decline exponentially over the next 5 to 10 years. Typhoons delivering intense rain over the volcano, however, may modify this predicted schedule by generating lahars of catastrophic proportions.

During this period of prolonged lahar activity at Pinatubo, Mapanuepe Lake will continue to expand in pace with the aggradation at the confluence. Like all other previous lahar-dammed lakes that developed in the Mapanuepe valley, Mapanuepe Lake will eventually cease to exist as it is gradually filled with sediment. Prior to this, however, the lake may experience episodes of lake breakouts and will continue to be a hazard to downstream communities long after frequent lahar activity has ceased in the Santo Tomas River.

INTRODUCTION

The June 15, 1991, eruption of Pinatubo volcano emplaced more than 6 km^3 of tephra (W.E. Scott and others, this volume) on the Abacan-Sacobia, Bucao-Balin-Baquero, O'Donnell, and Marella watersheds (fig. 1). Pyroclastic deposits, in areas more than 100 m thick, virtually buried most of the preeruption drainages on the eastern, southwestern, and northwestern slopes of the volcano. These materials, metastably perched on the flanks of the volcano, were remobilized by heavy rains into hot lahars that affected all major drainages of Mount Pinatubo. Numerous hot lahars, principally of the hyperconcentrated-streamflow type, were generated during the eruption and the following southwest monsoon and typhoon season, from late June to November (Janda and others, 1991; Rodolfo, 1991). High sediment influx caused rapid aggradation of lahar-affected channels, which caused encroachment and burial of adjoining

[1]Department of Geological Sciences M/C 186, University of Illinois at Chicago, 801 W. Taylor St., Chicago, IL 60607–7059.

[2]Philippine Institute of Volcanology and Seismology.

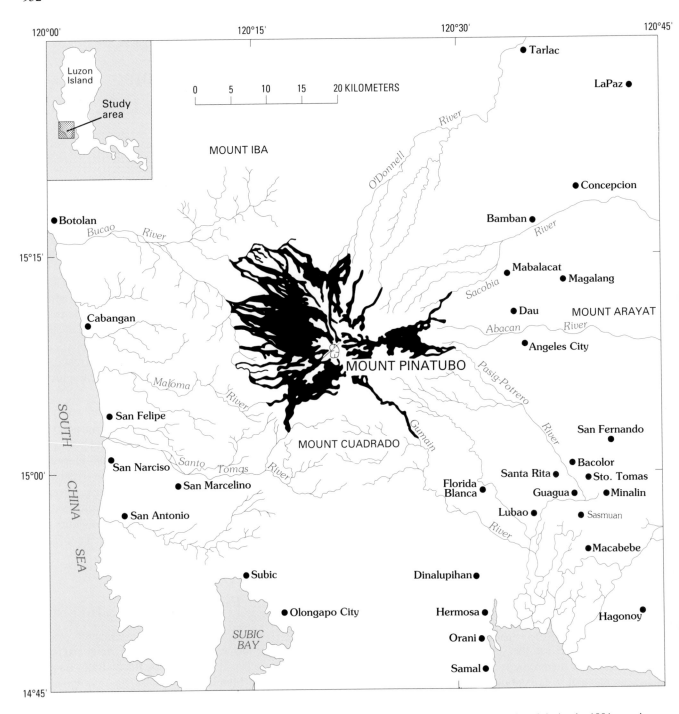

Figure 1. Mount Pinatubo and vicinity and the distribution of the pyroclastic-flow deposits (black) emplaced during its 1991 eruption.

lowlands at the foot of the volcano (Punongbayan and others, 1991; Pierson and others, 1992; Rodolfo and Umbal, 1992; Major and others, this volume; Rodolfo and others, this volume; K.M. Scott and others, this volume). Owing to the extraordinary thickness of the accumulated pyroclastic debris, destructive lahars are expected to occur for at least 5 more years (Pierson and others, 1992; Rodolfo and Umbal, 1992; Rodolfo and others, this volume). This scenario provides us the opportunity to study the generation and behav-

ior of hot and cold lahars and to determine how rivers respond to prolonged periods of high sediment influx.

High sediment discharge and rapid aggradation along lahar-affected channels blocked some tributaries whose origin was not Pinatubo and formed new lakes (Umbal and others, 1991; Pierson and others, 1992; Rodolfo and Umbal, 1992; Rodolfo and others, this volume; K.M. Scott and others, this volume). When accumulating water overtops and rapidly erodes the lahar dams, escaping floodwaters bulk up

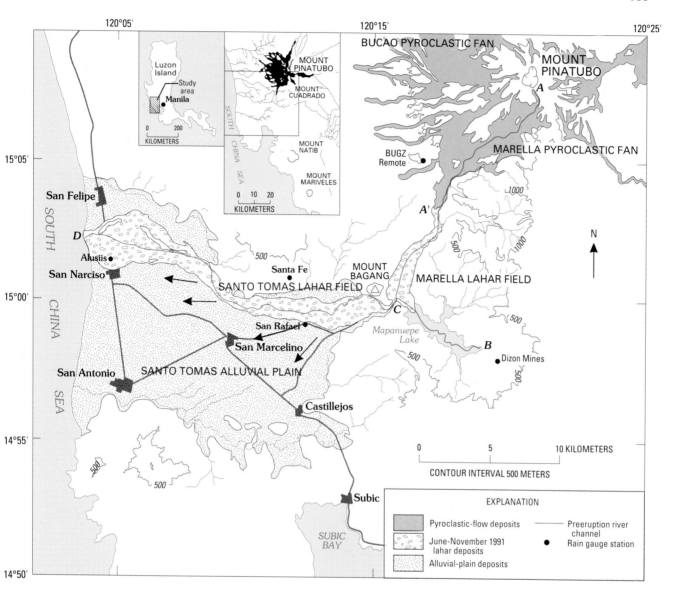

Figure 2. Areas affected by pyroclastic flows, lahar deposition, and lake flooding in the southwest sector of Mount Pinatubo. Arrows indicate possible avulsion sites for future lahars along the Santo Tomas River system as delineated by the Pinatubo Lahar Hazard Taskforce (as of August 23, 1991). Lettered sections shown in figure 6.

to hyperconcentrated streamflows by incorporating sediments from previous lahar deposits. The largest of these lahar-dammed lakes is Mapanuepe Lake, on the southwest flank of the volcano at the confluence of the active Marella lahar channel and the Mapanuepe River (fig. 2). Several lake breakout episodes during 1991 caused loss of life and damage to property. Here we describe the lahar activity along the Marella and Santo Tomas Rivers during the 1991 rainy season and the subsequent growth and development of Mapanuepe Lake, with emphasis on the lahar-damming and lake breakout processes. Such processes and the hazards they pose may well persist even after eruptive and lahar activity have waned or ceased.

GEOLOGIC AND TOPOGRAPHIC SETTING OF MOUNT PINATUBO

Mount Pinatubo, a hornblende andesite-dacite stratovolcano and dome complex, is situated at the northern end of a 50-km long, south-trending line of calcalkaline volcanic centers. Radiometrically dated rocks along the entire line range in age from early Pliocene to Holocene (De Boer and others, 1980). The oldest dated Pinatubo rocks are Pleistocene andesitic lava flows and intrusives from an ancestral Pinatubo stratovolcano (Newhall and others, this volume). Recent prehistoric volcanism was silicic and violent, as testified by extensive ash- and pumice-flow sheets on all flanks of the volcano, in places up to 300 m thick (Delfin, 1983). Six eruptive episodes between >35,000 and 500 years ago are recognized from radiocarbon dating of charred wood in these deposits (Newhall and others, this volume). At the southwestern sector of the volcano, most of the pyroclastic-flow and lahar deposits exposed along major channels are products from the last two major eruptive episodes of Pinatubo: the Maraunot eruptive period, which occurred 3,900 to 2,300 years ago, and the Buag eruptive period, which occurred about 500 years ago (Newhall and others, this volume).

Mount Pinatubo, formerly the highest peak along the Pinatubo-Mariveles volcanic trend, is inferred to rest on an older 3- to 4-km-wide caldera (Delfin, 1983; Wolfe and Self, 1983) and stood 1,745 m in altitude prior to the 1991 eruption. Slopes near the summit areas steep as 65°. The volcano is irregular in plan, being partially nestled in other topographic elements of the Zambales Mountains along the west coast of Luzon. Its maximum radius is about 25 km in the northeastern quadrant, where the regional slopes decrease from about 12° near the crater rim to about 2° where the volcaniclastic apron meets the swampy flats of Luzon's central valley. Where the northern and western flanks abut mafic-ultramafic blocks of the Eocene Zambales Ophiolite Complex, edifice radii are typically 10 to 15 km, and regional slopes are somewhat gentler, ranging from 10° near the summit to about 2° near the foot. Prior to the eruption, Mount Pinatubo was densely and deeply dissected by a system of eight major, radial drainages ending in broad, gently sloping aprons of laterally coalesced alluvial fans that extended beyond the flanks of the volcano. Voluminous pyroclastic flows that funneled through these drainages during the 1991 eruption partially or completely filled the upper stretches of the channels, and subsequent lahars aggraded most of the lower reaches.

REGIONAL CLIMATIC SETTING

The climate of west-central Luzon is governed by seasonal changes in positions of dominant air masses and attendant wind shifts, with the annual rainfall averaging 3,800 mm. During the dry months of October to May, west-flowing tropical maritime air is orographically uplifted by the Sierra Madre Mountains along the east coast of Luzon, the result of which is heavy rainfall there and pronounced dehumidification of the air before it reaches central Luzon. The west coast of Luzon typically receives less than 20 percent of its rains from October through May (Huke, 1963).

During the wet months of June to September, southwesterly winds of the South Indian Ocean Monsoon cross the Equator and blow maritime equatorial air across the South China Sea to the Philippines. As the precipitation records for central Luzon show, the initial Southwest Monsoon rains can arrive as early as April and begin in earnest

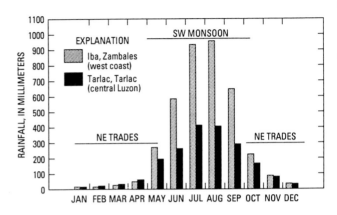

Figure 3. Comparative long-term average monthly rainfall in central Luzon for both the west coast (Iba, Zambales) and central valley (Tarlac, Tarlac) sectors. Note the exceedingly high amount of precipitation during the southwest monsoon.

in June. In addition, on average, 17 typhoons enter Philippine space every year, mostly during the period of June to October, and about 20 percent of these strike central Luzon (Huke, 1963). Together, the Southwest Monsoon and typhoons deliver 82 percent of the annual precipitation at Iba, on the west coast of Zambales, and 70 percent of the annual precipitation at Tarlac, in the central valley, in the 4 months from June to September (fig. 3). The most intense precipitation is from typhoons or from strong southwesterly airflow that is drawn in by typhoons.

THE 1991 ERUPTIVE PRODUCTS OF PINATUBO

A chronology of the 1991 eruption of Mount Pinatubo is discussed by the Pinatubo Volcano Observatory Team (1991), Hoblitt and others (1991), Pierson and others (1992), Wolfe (1992), and Wolfe and Hoblitt (this volume). Of the large volume of pyroclastic materials erupted, about 1.3 km^3 were emplaced on the southwest slopes of the volcano (figs. 2 and 4) to cover an area of about 22 km^2 (W.E. Scott and others, this volume), referred to here as the Marella pyroclastic fan. The thickness of the deposit at this sector averages about 60 m but may be up to 150 m within river valleys. The materials are mostly lapilli- and ash-sized, porphyritic biotite-hornblende quartz dacite pumice, with very few bomb-sized fragments. Temperature of the deposits, taken from distal pyroclastic flows in the Marella watershed 2 weeks after the eruption, were as high as 260°C at depths of a few decimeters (A. Bernard, Université Libre de Bruxelles, oral commun., 1991) and would doubtless be higher at greater depths. The materials are very unstable, and large mass failures, triggered by strong steam explosions, were remobilized as secondary pyroclastic flows that traveled distances of more than 4 km (Pierson and others,

Figure 4. Aerial view of the Marella pyroclastic field, with Mount Pinatubo erupting in the background, taken 12 days after the June 15 eruption. Only a few ridges jut out from the central portion of the pyroclastic field; most drainages have been buried (photograph courtesy of W.E. Scott).

1992; Torres and others, this volume). In view of the extreme thickness, it will probably take several years for the pyroclastic deposits to cool down completely; thus, secondary pyroclastic flows will continue to pose serious hazards, particularly on the upper slopes of the volcano.

STREAM MORPHOLOGY BEFORE THE 1991 ERUPTION

THE SANTO TOMAS RIVER AND ALLUVIAL PLAIN

The southwest flanks of Mount Pinatubo, and the older terranes south of the volcano comprising a 232 km[2] area above 50 m in altitude, is drained by the Santo Tomas River, the master stream of the Mapanuepe and Marella Rivers (figs. 2 and 5). The river, with an average stream gradient of 0.2°, traversed the northern margin of a broad, 235-km[2]

alluvial plain (referred to here as the Santo Tomas plain) for 24 km before emptying into the South China Sea. Braided along most of its length, the river is relatively shallow, generally less than 0.5 m at its deepest portion, with channels 1 to 2 km wide flanked by banks less than 2 m high. Five towns with an aggregate population of 140,000 people lie on the alluvial plain.

Outcrops along the river banks show typical braided stream deposits, ranging in thickness from 1 to 3 m, overlying beds of pumiceous sandy debris-flow and hyperconcentrated-streamflow deposits. In a quarry near the town of Castillejos (fig. 2), ^{14}C-dated charcoal samples collected from the uppermost interbedded pumiceous sandy debris-flow and hyperconcentrated-streamflow deposits indicate an age of 2,880±70 ^{14}C years (Newhall and others, this volume); this age suggests that the last eruption-related aggradation by lahars of the southern sector of the alluvial plain occurred during the Maraunot eruptive period. Similar pumiceous laharic deposits comprising the terrace at sitio (hamlet) Palan (figs. 5 and 6C, section H-H') were dated to be 2,950±50 ^{14}C years.

We have not directly dated the lower terraces nestled within the embankment of the Santo Tomas River (fig. 6C, terraces buried by 1991 laharic deposits from sections G–G' down to O–O'). However, geomorphically correlative terraces found in the Marella channel at Aglao (fig. 5) (Javellosa, 1984) were dated to be 760±60 ^{14}C years, during the Buag eruptive period.

THE MARELLA RIVER

The Marella River, the main lahar feeder for the Santo Tomas River system, drains a 31-km[2] area of the southwestern slopes of Mount Pinatubo. Topography of the upper Marella watershed was characterized by smooth and less dissected slopes, due to recent burial by pyroclastic flows during the Buag eruptive period. Before the 1991 eruption, tributaries of the Marella River that drained Pinatubo slopes showed steps in their longitudinal profiles (fig. 6B) at the fronts of underlying lava flows, and the main tributary extended up to the landslide-scarred southern face of the summit dome of the volcano. Four major breaks in slope, at altitudes 980, 580, 310, and 280 m, became the loci for pyroclastic-flow deposition during the 1991 eruption. From the summit down to 800 m, where the underlying rocks are predominantly andesitic to dacitic lava flows, channels were steep with deep, narrow, V-shaped valleys. Below this point, as the river eroded into thick (>100 m), unwelded dacitic pumice pyroclastic-flow deposits, the channel valley was U-shaped and bordered by 40- to 90-m-high channel banks. Distal pyroclastic-flow units interbedded with laharic deposits, exposed along the channel side walls below 200 m in altitude, were dated to have been emplaced during the Maraunot eruptive period (Newhall and others, this vol-

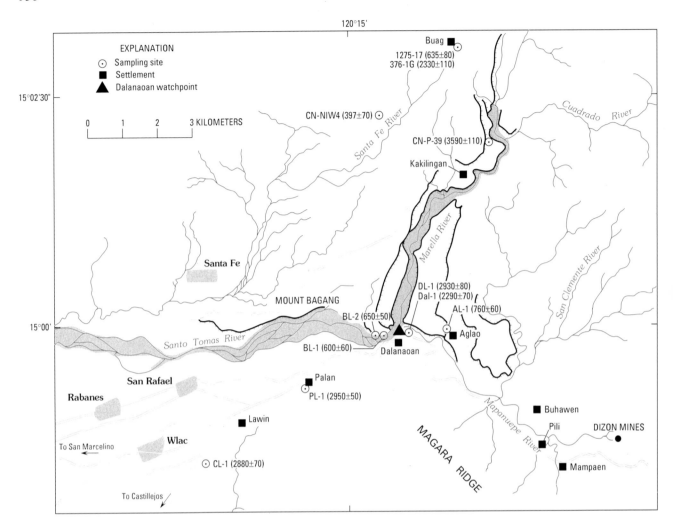

Figure 5. The confluence site of the Santo Tomas, Marella, and Mapanuepe Rivers, sampling sites, and localities mentioned in the text. The shaded areas correspond to the braided channel complex of the Santo Tomas and Marella Rivers, while hachured lines correspond to channels with vertical channel walls or embankments.

For sampling sites, the values enclosed by parenthesis correspond to the radiocarbon age of the deposits. Samples were from lahar and fluvial deposits, except for CN–P–39 and Dal–1, which were from pyroclastic-flow deposits.

ume). Downstream from the 200-m altitude, the channel widens from an initial width of 700 m to almost 2 km at its confluence with the Mapanuepe River (fig. 6B, sections A–A' to F–F').

Nestled within the broad valley floor of the Marella were lahar and fluvial terraces, each representing different paleoflood/aggradation levels that appear to coincide with periods of explosive eruptions of Mount Pinatubo. Five pre-1991 eruption terrace levels were identified, the highest of which rose 20 to 35 m above the channel floor (fig. 6B) and had several upland communities built upon it. We have not directly dated the uppermost terrace level, but the age of deposits exposed along the channel sidewalls of the Marella and what appears to be its downstream continuity with the highest terraces along the Santo Tomas River suggests an age not older than the latter part of the Maraunot eruptive period and, possibly, even younger. The terrace where

barangay (village) Aglao formerly stood (fig. 5), which is several meters lower than the highest terrace, was formed during the Buag eruptive period.

MAPANUEPE RIVER

An 88-km^2 area of ultramafic and older volcanic terrane south of Pinatubo was (and is still) drained by the Mapanuepe River, constituting the main tributary of the Santo Tomas River not originating on Mount Pinatubo. Upper tributary channels of the Mapanuepe River have V-shaped valleys that progressively become U-shaped toward their junctions with the main channel. Flows from the eastern watershed basin of Mapanuepe are temporarily impounded in an artificial lake constructed by Dizon Mines, which utilizes the water in the operation of its open-pit copper-gold mine. Downstream from the Dizon water-retaining dam, the 30- to 45-m-wide Mapanuepe River followed a west-northwesterly, slightly sinuous course along a 500- to 700-m-wide flood plain for 7 km before joining the Marella River to form the Santo Tomas River. The average Mapanuepe stream gradient along this stretch was about 0.06°, an order of magnitude lower than the 0.6° stream gradient of the Marella along an equivalent reach above the junction (fig. 6A).

Available subsurface data from exploratory drill holes and ground-water surveys conducted by Dizon Mines from 1978 to 1985 showed the thickness of the sediment fill within Mapanuepe valley to range from 20 m to more than 60 m, with a trend of gradually increasing thickness downstream towards its junction with the Marella (T.D. Malihan, 1984, and A.C. Olarte, 1985; unpublished Benguet Corporation memorandum reports). Subsurface materials were predominantly coarse alluvium, consisting of a heterogeneous mixture of pebbles, cobbles, and boulders in a loose matrix of sand, silt, and clay. However, several horizons, consisting mostly of fine sand, silt, and clay materials with very little coarse fragments, interbedded with the coarse sediments, were intercepted during the drilling operations. In some areas these sand-silt-clay horizons were more than 11 m thick (A.C. Olarte, 1985; unpublished Benguet Corporation memorandum report). Similar fine-grained deposits were encountered at depths of 5 to 15 m from the surface during recent ground water well explorations by Dizon Mines in Pili and Mampaen, at sites 4 and 5 km upstream from Mapanuepe's junction with the Marella channel, after the June 15, 1991, eruption (V. Tombokon, Benguet Corporation, oral commun., 1992).

Outcrop sections at the southern bank of Mapanuepe (fig. 5), near the vicinity of the junction, showed two thick (10–17 cm) clay-silt layers superimposed between alternating beds of pumice-bearing, lithic-rich debris-flow deposits and pumiceous sandy debris-flow, hyperconcentrated-streamflow, and fluvial deposits. Incipient soil formation was commonly observed on the upper portion of the clay-silt layers and above the pumice-bearing, lithic-rich debris-flow deposits. The basal portion of the section, overlying ophiolitic bedrock materials, consists of a yellowish-brown pumiceous pyroclastic-flow deposit that was dated to be $2,990\pm70$ ^{14}C yr B.P. (fig. 5; Newhall and others, this volume). Charcoal samples from the uppermost pumiceous lahars (about 6 m above the pyroclastic-flow deposit) had similar age, about $2,930\pm80$ ^{14}C years. Similar clay-silt and sandy silt layers, showing soft sediment deformation features, interbedded with pumiceous sandy debris-flow, hyperconcentrated-streamflow, and fluvial deposits, were logged at the other side of the river in a terrace near Aglao but were younger, about 760 ± 60 ^{14}C years old.

The thick sediment fill, consisting of several horizons of fine-grained sediments interbedded with coarse sediments, indicates that the Mapanuepe valley was the site of a previous lake. Stratigraphic and sedimentologic evidence

indicates that the lake developed as a consequence of repeated damming of Mapanuepe River that was due to rapid aggradation along the Marella. Aggradation of the Marella, in turn, appears to be a response to increased sediment influx following explosive eruption of Mount Pinatubo. We infer, from available age data, that a lake existed in the area prior to the 1991 eruption on at least two occasions: during the Maraunot and Buag eruptive periods. The Maraunot eruptive period, because of its long duration (perhaps, more than a thousand years; Newhall and others, this volume), may have included several episodes of lake formation and filling.

Additional evidence that the Mapanuepe valley was the site of a previous lake is provided by the stream gradient of the Mapanuepe River, which is anomalously low for a river in a tectonically active terrain of relatively high regional slope (Javellosa, 1984). Apparently, the stream is still in the process of adjusting its gradient in response to the modification imposed by the recent lake-sediment infill.

Recent archeological digs in the Mapanuepe valley uncovered an old boat hull with associated Chinese pottery (V. Tombokon, Benguet Corporation, oral commun., 1992). Boat transport in the area would not have been possible or necessary along the narrow and shallow pre-1991 Mapanuepe channel. However, boat transport would have been both possible and necessary if a large lake had hindered movement of people in the area, as Mapanuepe Lake does today. If the pottery discovered in the site turns out to be similar to that found in a 13th to 15th century age archeological site at sitio (hamlet) Buag in Barangay Kakilingan, San Marcelino (Newhall and others, this volume), then this inferred lake would have been produced, Mapanuepe fashion, during the Buag eruptive period. The absence of any reference to the existence of a lake in the historical accounts of the region indicates that the lake was already filled, probably within the last 100 to 200 years after the Buag eruptive period, prior to Spanish exploration of the region during the 18th century.

THE CONFLUENCE SITE

Bedrock constricted the channel width from more than a kilometer to about 60 to 75 m along the initial 160-m-long segment of the Santo Tomas River downstream from the junction of the Marella and Mapanuepe Rivers (fig. 6B, section F–F' and fig. 6C, section G–G'). Steep channel walls, >35 m high, bordered this narrow reach of the Santo Tomas River. The stream gradient along this stretch increased to about 1°, significantly higher than the stream gradient of both the Marella and Mapanuepe Rivers. Javellosa (1984) attributed this steepening of the stream gradient to recent uplift along an active fault that transects northwestward across the confluence site, but it could also have been due to increased flow discharge along this stretch of the channel as

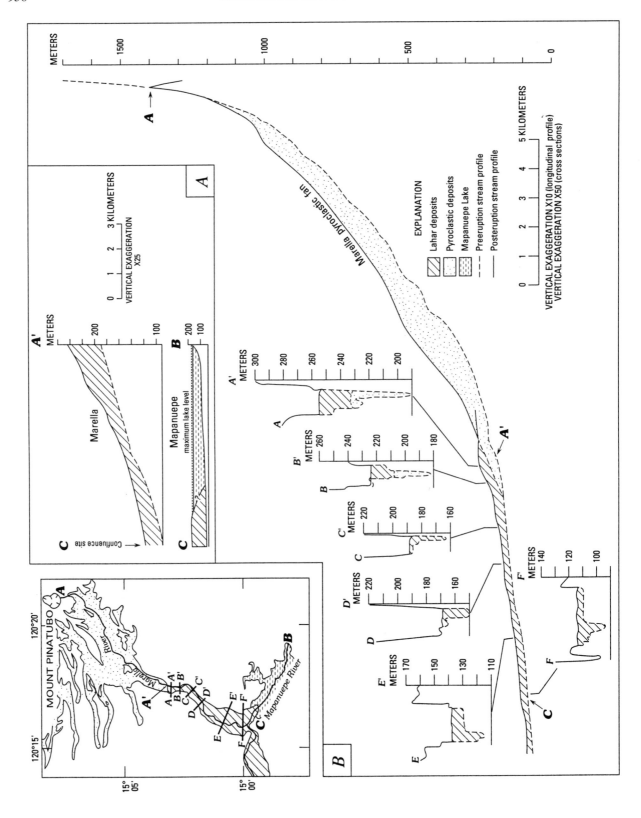

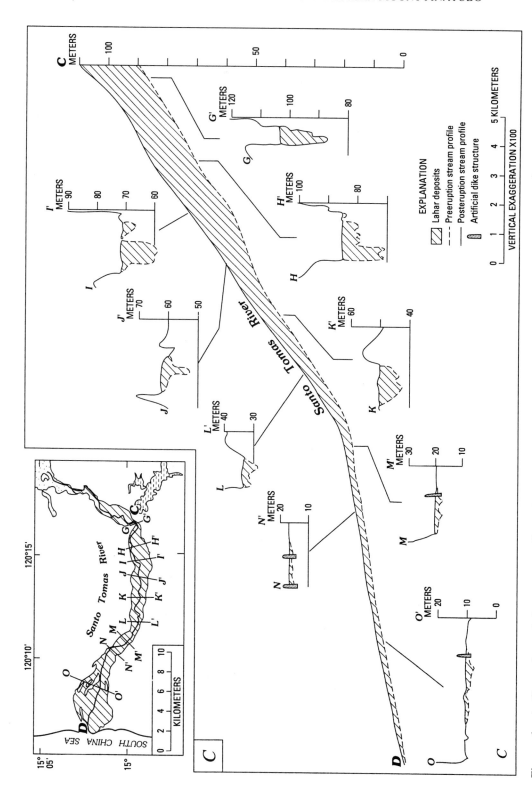

Figure 6. *A,* Comparison of the stream gradient of the Marella and Mapanuepe Rivers along an equivalent reach from the channel junction. The steeper gradient of the Marella caused flows to have momentum sufficient to block Mapanuepe outflow, and in some instances, to enable incursion of Marella flow upstream into the Mapanuepe River. *B,* Longitudinal profile and cross sections of the Marella River showing preeruption channel conditions and the levels of 1991 pyroclastic-flow and lahar deposition. *C,* Longitudinal profile and cross sections of the Santo Tomas River showing preeruption channel conditions and the levels of 1991 lahar deposition. Areas of pyroclastic-flow and lahar deposition are strongly controlled by the presence of major breaks in slope. See figure 2.

Figure 7. One of the small lakes formed by pyroclastic-flow deposits damming a non-Pinatubo tributary at its junction with major Pinatubo drainage. This picture was taken in the Bucao watershed.

flows moved from an initially wide conduit to one that is narrow.

A 5- to 10-m-thick sequence of fluvial, lahar, and pyroclastic-flow deposits, underlain by resistant, andesitic dike-intruded, Eocene gabbroic ophiolites, comprises the uppermost part of exposed sections at the right sidewall of the channel constriction. Two charcoal samples were taken from separate pumiceous laharic deposits immediately below fluvial deposits at the right bank of the channel (fig. 5); one is 600±50 and the other is 650±50 ^{14}C years old (Newhall and others, this volume). These dates suggest an eastward shift of the channel confluence from Mount Bagang to its pre-1991 eruption position in the last 600 years. Entrenched channels at the Santo Tomas side of the hooklike channel embankment, east of Mount Bagang, represent the relics of these abandoned channels.

GEOMORPHIC CHANGES SINCE THE JUNE 15, 1991, ERUPTION

Most of the southwest drainages that existed before the 1991 eruption, from the summit down to 500 m in altitude, were buried during the climactic phase of the eruption (fig. 4). From 500 m down to the 180-m altitude, distal pyroclastic flows were confined within the remnant tributary channels of the Marella, and the local stream gradient steepened by about 0.5° (fig. 6B). Surface runoff on the pyroclastic fan was funneled through these remnant channels, which became the loci for the subsequent headward erosion of the Marella into the new pyroclastic-fan system.

Tributaries of the Marella that drained Mounts Negron and Cuadrado were blocked by pyroclastic-flow deposits, and small lakes about 300 m across and 10 m deep formed

as a result (fig. 7). Within the pyroclastic fan, steam venting pockmarked the surface of the fan with numerous explosion pits that accumulated water during the rainy season. Failure and the release of impounded water from these ponds facilitated the initial posteruption channelization on the Marella pyroclastic fan.

Drainage on the pyroclastic fan was not fully integrated until late July. As a result, lahars did not occur along the Marella during the first month following the eruption, despite several episodes of intense rainfall. In contrast, at the eastern side of Mount Pinatubo, where most preeruption channels were only partially filled by pyroclastic flows, lahars were already occurring in earnest during the first month. By late July, aerial surveys showed the surface of the pyroclastic fan to be incised by dense, pinnate rill networks connected to several incipient, headward-eroding channels roughly following the preeruption tributary routes of the Marella (fig. 8). Aerial surveys in late November showed significant widening and deepening of these youthful channels, although most had not yet eroded to the preeruption base level.

LAHAR ACTIVITY ALONG THE SANTO TOMAS RIVER SYSTEM

The first lahars along the southwestern drainages of the volcano occurred on the early morning of June 15, contemporaneous with the late pre-paroxysmal eruptions (Hoblitt, Wolfe, and others, this volume). These lahars were triggered by rains from Typhoon Yunya and from convective storm cells generated by the eruptions. At least three debris-flow events were documented from deposits, now buried, that were progressively transformed downstream

Figure 8. July 15, 1991, photograph of the Marella pyroclastic field showing initial incision of the deposits.

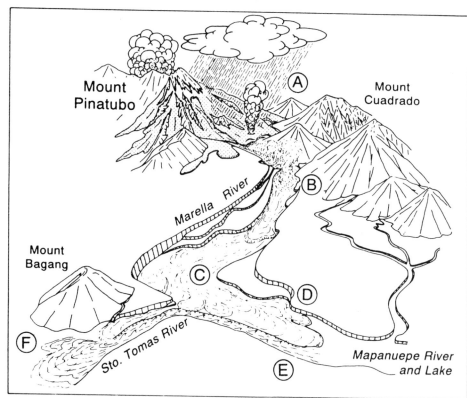

Figure 9. Idealized sketch of lahar activity, Mapanuepe damming, and lahar mobilization by lake breakouts, southwest sector of Mount Pinatubo. A, Heavy rain on the Marella pyroclastic field generates lahars and rootless fumaroles. B, Hot, steaming lahars cascade down the Marella River. C, Marella lahars spread and deposit along the wide channel reach and choke the constriction at the Mapanuepe junction. D, At the confluence, Marella lahars intrude upstream along the Mapanuepe River, damming it. E, After Marella lahars cease, rising Mapanuepe Lake waters breach the dam. F, Breakout waters remobilize deposits of previous lahars along the Santo Tomas channel and erode its bank as they descend toward the South China Sea. Note terraces along the Marella River from past eruptions. Stippled areas are pyroclastic deposits of the 1991 eruption.

into hyperconcentrated streamflows and normal streamflows. The components were mainly dense, lithic fragments, and hence, were cold. These lahars destroyed the bridge connecting Barangay (village) Santa Fe to the town of San Marcelino (fig. 2).

Most lahars after the June 15, 1991, eruption were also triggered by intense rainfall (fig. 9). Preliminary studies suggest that average rainfall intensities of about 0.3 to 0.4 mm/min over periods of 30 to 40 min were sufficient to trigger lahars in the Marella watershed (Pinatubo Lahar Hazard Taskforce, unpublished report dated August 28, 1991; Regalado and Tan, 1992; Tuñgol and Regalado, this volume). These values are considerably lower than those observed at Mayon volcano (Rodolfo and Arguden, 1991),

Figure 10. Hot lahar occurrences in the Marella were always accompanied by vigorous hydrothermal venting on the pyroclastic field. (Photograph taken August 14, 1991, by R.U. Solidum.)

probably due in part to differences in the characteristics of source materials for the two volcanoes. Pinatubo pyroclastic-flow deposits are fine grained and pumiceous and thus were more easily remobilized than Mayon pyroclastic-flow deposits, which consist of coarse, relatively dense basaltic andesite (typical density range from 2.4 to 2.8 g/cm^3).

Pinatubo's eruption also produced a volumetrically larger proportion of tephra-fall deposits, which facilitated lahar initiation by reducing rainwater infiltration and increasing surface runoff. In addition, eroded June 15, 1991, tephra-fall deposits mantling the Marella pyroclastic field were being continuously replenished by ash winnowed from frequent steam explosions in the pyroclastic field. In contrast, the rainfall data for Mayon lahars were collected when most of the tephra-fall deposit mantling the pyroclastic deposits was already depleted by the first monsoon rains after the September 1984 eruption. Daido (1985) observed similar dependence on the availability of fine-grained tephra-fall deposits for the occurrence of laharic debris flow at Sakurajima volcano, Japan.

Lahars may undergo flow transformations from debris flows to hyperconcentrated streamflows to normal streamflows (Pierson and Scott, 1985; Rodolfo and Arguden, 1991; Smith and Lowe, 1991), so hyperconcentrated flows observed at the lower reaches of a channel may actually represent the diluted portion of a debris flow initiated upstream. Flow transformation of Pinatubo lahars is possible, indeed, likely, considering the distance of the observation point to the area of initiation (more than 15 km) and the presence of tributaries joining the Marella that drain nonvolcanic areas and contribute flows of low-sediment concentration. Some laharic hyperconcentrated-streamflow deposits below our Dalanaoan watchpoint (fig. 5) were traced and observed to interfinger with laharic debris-flow

deposits upstream of the Marella channel. Inclusion of rainfall data that coincided with observed hyperconcentrated streamflows at our observation post would therefore be a reasonable approximation of rainfall threshold values for initiation of both hyperconcentrated streamflows and debris flows in the Marella watershed.

There were numerous occasions, however, when lahars occurred without any significant rainfall recorded by the rain gauges. This could be related either to the spatial variability of rainfall intensity within the volcano's watershed, which could not be monitored in detail by the existing rain-gauge network at Pinatubo (Pierson and others, this volume; Tuñgol and Regalado, this volume), or to lahar-triggering mechanisms other than rainfall.

Vigorous steam explosions in the pyroclastic fan area almost always accompany lahar occurrences (fig. 10). Most of these explosions occur as lahars erode their banks and come in contact with still-hot pyroclastic flow material. However, some lahars might have been triggered by steam explosions, where ground vibration from the explosions was sufficient to trigger mass failure of water-saturated pyroclastic materials, or where explosions directly breached debris-dammed lakes (Umbal and others, 1991). Lahars whose deposits are characterized by a hummocky surface near their origin may have been triggered by steam explosions.

Lahars after late July were hot and steaming, with measured temperatures in the range of 34–60°C. Observed flow behavior and the character of the deposit indicate that these events were predominantly hyperconcentrated flows, though bigger flows had distinct debris-flow phases (Rodolfo and others, this volume). In contrast, lahars triggered by lake breakouts were cold and, at most, bulked to hyperconcentrated flows. Measured velocities typically

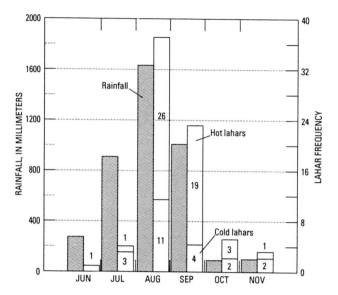

Figure 11. Monthly precipitation and lahar frequency for 1991. Note that the greatest number of lahar events coincided with the period of maximal rainfall.

ranged from 2 to 6 m/s at slopes less than 1°. Peak discharges averaged 150 m³/s, but events with discharge exceeding 1,000 m³/s were not uncommon.

Deposits consist mostly of sand-sized juvenile pumice, with very minor boulder-sized clasts and lithic fragments. Thin hard-crusted surfaces, which are resistant to rain drop impact, mantle the surface of these hot-lahar deposits. Thick deposits remained warm several days after emplacement, and in a 5-m-thick laharic debris-flow deposit, the measured temperature at a depth of 1 m was 40°C 7 days after it was emplaced.

During the 1991 rainy season, about 50 hot lahars were visually observed (fig. 11), not including nocturnal events recorded by USGS-PHIVOLCS remote flow sensors. Most of these events occurred in August and September, during the seasonal peak in rainfall. We estimated that more than 185×10^6 m³ of debris was remobilized during this period and that it affected an area of 46 km² and aggraded the Santo Tomas and lower Marella Rivers by 1 to 24 m (fig. 6*B*,*C*). Over the same period, 18 cold-lahar events were documented, of which all but one were related to breaching of Mapanuepe Lake. The only exception was a similar lake breakout originating from a dammed tributary near Mount Cuadrado.

LAKE DEVELOPMENT

DAMMING PROCESSES

Rapid aggradation by Marella lahars blocked outflow from Mapanuepe River and caused a lahar-dammed lake to form (fig. 12). Several factors contributed to the damming

process: the strong discharge and high sediment load of Marella flows, the steep gradient of the Marella channel, and the presence of a bedrock constriction near the confluence of the two rivers. The steeper gradient of the Marella caused Marella lahars to have a greater momentum as they impinged upon and thereby constricted the width of outflow from the Mapanuepe River. A bedrock constriction near the downstream side of the confluence of the Marella and Mapanuepe Rivers created a backwater effect that facilitated rapid deposition at the confluence site. On several occasions, the 45-m-wide channel was repeatedly plugged during the passage of large laharic debris flows, which at times reached lobe dimensions of more than 5 m thick and 100 m wide; at other times, the channel was filled by gradual sediment deposition at the junction during hyperconcentrated-streamflow events.

The strong color contrast between the dark-gray, sediment-rich Marella discharge and the mostly clear-water Mapanuepe discharge facilitated observation of flow interactions and analysis of the damming processes at the channel confluence during hyperconcentrated flow events. Little mixing was involved as flows converged; plumes from each channel were still identifiable almost a kilometer downstream from the confluence. Within the upstream region adjacent to the junction corner, flows experienced a decrease in velocity resulting from the mutual deflection of flows away from the junction corner. This decrease in flow velocity induced deposition of a portion of the total sediment load of Marella flow near the upstream corner of the channel junction. For channel confluences where one tributary forms a linear extension with the channel used below the confluence, this region of flow-velocity reduction is referred to as the "region of flow stagnation," the geometry of which depends mainly on the discharge ratio between the confluent channels and the junction angles (Best, 1987). Increased discharge from the Marella gradually constricted the width of the Mapanuepe outflow and at the same time extended the region of flow stagnation to eventually encompass the entire width of the Mapanuepe channel. Due to its high sediment concentration, Marella flows debouched into Mapanuepe waters as underflows and formed a steep-fronted delta that prograded into the Mapanuepe channel. When deposition outpaced the rise in water level in the Mapanuepe channel, Mapanuepe flow was dammed. At this stage of development, the dam grew continuously by successive spillovers from the Marella that, with sufficient duration of sustained high Marella discharge, remained stable even after Marella flow waned. Otherwise, the underdeveloped blockage was swamped and destroyed by the onrush of ponded Mapanuepe waters as Marella flow diminished.

During the earliest damming episodes, potential sites for deposition from hyperconcentrated streamflows and the configuration of the blockage were partly determined by the geometry of the stream junction. This geometry changed

Figure 12. Oblique aerial photographs of the Mapanuepe-Marella-Santo Tomas confluence site. *A,* June 23, 1991, prior to major lahar activity. View is from the east, looking down the Mapanuepe and the Santo Tomas Rivers. *B,* September 10, 1991, shortly before the lake attained its maximum extent. View is from the east-southeast, looking across the Mapanuepe impoundment; most sediment flow is from the right (Marella River) to the upper left (Santo Tomas River).

during and after each major lahar event. Further complexities were imposed by divergence of Marella flow upstream of the confluence, particularly during the later part of the 1991 monsoon season when, in some instances, more than one active channel of the Marella joined the Mapanuepe and thereby increased the number of potential dam sites.

Variable rainfall intensities on the Marella and Mapanuepe watersheds may have influenced discharge rates of the channels and thus affected damming at the channel confluence of the two rivers. A comparative plot of daily rainfall during the month of August indicates that damming occurred whenever rainfall over the Marella watershed

exceeded that of the Mapanuepe (fig. 13). Whether the high amount of rainfall over the Marella watershed translated to a higher discharge in the Marella channel compared to that of Mapanuepe is not yet clear, given the much larger watershed of the latter. The timing by which rainfall contribution is delivered into the main channels, however, may be significant in determining the variability of discharge between the two rivers. The response of a stream hydrograph to rainfall and storm events depends on many factors, but in general becomes increasingly delayed as watershed area increases, primarily as a result of increased channel transmission loss as channel routes becomes longer (Pilgrim and others,

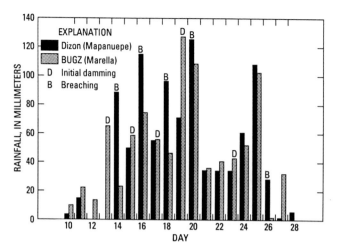

Figure 13. Comparative daily amounts of precipitation for the Mapanuepe and Marella watersheds during the month of August and the relation between damming (D) and breaching (B) of the lake.

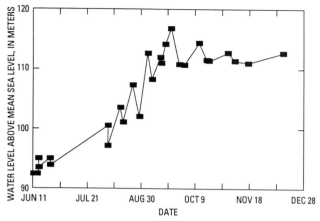

Figure 14. Water-level changes of Mapanuepe Lake during 1991. Sharp drops in water level coincide with initial breaching of the dam.

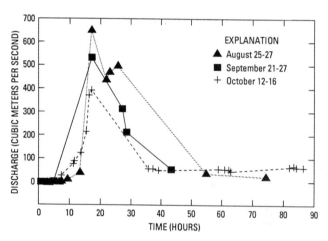

Figure 15. Hydrographs for the three lake breakout events in 1991. Discharge measurements for the September 21–27 event were made for 42 h, until discharge returned to near its prebreakout level. Measurements during August 25–27 were made for 74 h, also to prebreakout level, and measurements during October 12–16 were stopped after 88 h, even though discharge was still slightly elevated.

1982). The Marella watershed, because it is smaller and has shorter tributaries, brought runoff to the main channel and thence to the confluence of the Marella and the Mapanuepe Rivers before runoff arrived from the Mapanuepe watershed. The early arrival of peak flows from the Marella at the channel confluence favored conditions for the damming of the Mapanuepe. In addition, reduced infiltration rates on the Marella pyroclastic fan favored surface runoff into the Marella channel, where bulking with new pyroclastic materials further enhanced the volume of surface runoff.

LAKE BREAKOUT PROCESSES

As Marella flow waned after each blockage, continued drainage from the Mapanuepe watershed raised water in the lake until it overtopped and eroded the debris dam (fig. 14). Overtopping and breaching almost always occurred at the southern margin of the blockage, where the dam is at its lowest height. The breach started initially as a small rivulet that gradually increased in size by lateral erosion of the channel bank and headward erosion from local steepening on the lahar-dam surface. During the first 10 h after the initial overtopping, discharge from the outlet increased gradually followed by a rapid climb to peak flow 6 h later (fig. 15). This exponential increase in discharge may have resulted from mass failure of the debris dam and accelerated erosion rates. The water-saturated condition of the blockage makes it prone to mass failure, as shown by frequent slumping along the sides of the channel breach. Ground water issued from the slip faces of these slumps. Lateral erosion rates were high, partly due to the loose, sandy, noncohesive character of the blockage material, and increased with increasing discharge. In the three lake breakouts that we documented, measured peak discharges ranged from 400 to

650 m^3/s, and lateral erosion rates were as high as 6 m/min during a 15-min peak flow period. In the absence of significant rainfall, lake discharge gradually declined and returned to normal levels of 8 to 20 m^3/s about 35 to 50 h after the initial overtopping. The volume of water released during these breaching episodes ranged from 0.1 to 30×10^6 m^3.

Total draining of the lake, however, was never attained, because successive hot lahars repeatedly plugged the breached section after each lake breakout episode. As a result, lake area expanded intermittently during most of the rainy season. At its largest extent, on September 21, 1991, the lake flooded an area of 6.7 km^2 and impounded an estimated 75×10^6 m^3 of water (fig. 16), but declined thereafter as the monsoon season came to a close. The last breakout

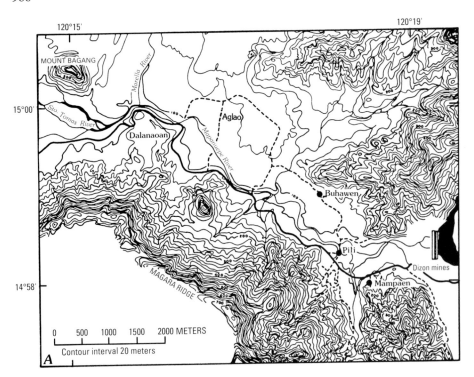

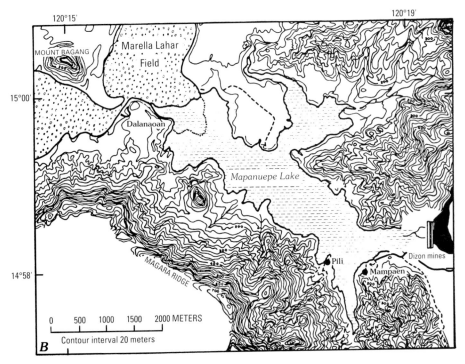

Figure 16. Mapanuepe River and the confluence site (*A*) before the 1991 eruption and (*B*) after the 1991 lahar season. At its largest extent the lake covered an area of 6.7 km² and had a stored water volume of 75×10^6 m³.

occurred on December 3, and then the lake outlet remained open until the end of 1991. During late December discharge from the lake averaged 6 to 10 m³/s, and lake level stabilized at about 111 m in altitude.

The time lag between damming and breaching was relatively short during the initial development of the lake, occurring only a few hours after Marella flow waned. In contrast, subsequent breaching episodes lasted days and even weeks. This may in part be due to the initial restriction of the lake within the narrow confines of the preeruption Mapanuepe channel and thus to relatively rapid rates of water-level rise. As the lake expanded to cover a wider area, the rates of lake-level rise decreased, even though the lake probably received the same amount of water from tributaries as in previous damming and lake breakout episodes. Thick tephra-fall deposits in Mapanuepe's catchment area, by reducing the infiltration rate, may have initially increased surface runoff entering the lake and contributed to the rapid rise in lake level during the initial stage of lake development. In addition, the predominantly hyperconcentrated flow events during the first month after the eruption formed smaller blockages that could immediately be overtopped and eroded by the rising lake water level. As lahar frequency and the occurrence of debris flows increased with the advent of the monsoon rains and improved drainage integration in the Marella watershed, aggradation at the confluence progressed faster than the rise in lake level, so a considerably longer period between damming and breaching episodes resulted.

The volume of water released during lake breakouts and the accompanying peak discharges varied for each breaching episodes. The highest measured peak discharge occurred on September 21, when the lake area and water volume were at maximum. To some extent, peak discharge may have been influenced by the volume of water stored in the lake. However, the relatively small peak discharge (<100 m^3/s) and the long time interval between initial overtopping and peak flow (>10 h) from lake breakouts in October, despite the large volume of water ($>40\times10^6$ m^3) remaining in the lake, suggest the interplay of other factors. Decline in rainfall as the end of the monsoon season approached may have contributed in two ways: (1) it enabled excess water to drain from the porous blockage materials and thereby increased its stability; and (2) it reduced tributary input into the lake during lake breakouts.

SOCIOECONOMIC IMPACTS

Lake water rise drowned the upstream communities of Lower Aglao (figs. 12B and 16), Buhawen (fig. 16), and Pili (figs. 16 and 17), necessitating the evacuation of more than 500 people. Mining activities at Dizon Mines were disrupted, particularly the transport of ores and supplies, when more than 7 km of roads and several bridges linking the mine and upland settlements to the nearest town were submerged. Anticipating future rise in the lake, mine management constructed a new road network along the ridge crest south of the lake. Future encroachment of the lake into the mining complex, however, may necessitate the transfer of some mining infrastructure to higher ground.

At least 15 houses at Barangay San Rafael (fig. 2) in the municipality of San Marcelino were destroyed during lake breakouts. Additionally, lateral erosion and deposition by lahars damaged 35 km^2 of agricultural land. Other areas that were not directly affected by lahars, such as barangays Santa Fe and Aglao, were isolated for most of the rainy season. Low-lying areas near the sea coast, particularly Barangay Alusiis (fig. 2) in the town of San Narciso, were repeatedly inundated by floodwaters, and houses were partly buried under more than a meter of sediment.

ANTICIPATED LAHAR ACTIVITY

By the end of 1991, about 185×10^6 m^3 of pyroclastic materials, or 14 percent of the estimated volume of the Marella pyroclastic fan, was transported as lahars, aggrading the Santo Tomas and Marella Rivers by 1 m to as much as 24 m at the Marella-Mapanuepe junction. Typical lahar sedimentation is greatest during the first few years after an eruption and decreases exponentially thereafter (Pierson and others, 1992), and we expect that aggradation along the Santo Tomas River will follow the same general pattern. In the future, lahar activity might increase should renewed explosive volcanic activity introduce new materials into the pyroclastic fan systems, or should a strong typhoon bring unusually intense rains to Pinatubo's slopes. In view of post-1991 activity of Mount Pinatubo (Daag, Dolan, and others, this volume), a resurgence of explosive activity appears to be remote. In contrast, the possibility of a typhoon delivering intense rain over Mount Pinatubo is quite high. On average, three to four typhoons pass the vicinity of the volcano in any given year. The amount of rainfall that can be delivered during a typhoon can be estimated by considering the range of extreme daily precipitation in the area. Based on a 28-year record of maxima in daily rainfall, the probabilities that at least one rain event on the volcano will exceed a daily total of 200 mm, 300 mm, and 400 mm at least once in a given year are 82, 36, and 18 percent, respectively (U.S. Navy, unpub. data, 1992). As of this writing, the highest known 24-h rainfall in the area is 744 mm, recorded on July 25–26, 1994, on Mount Cuadrado (Janda and others, this volume), during strong southwesterly air flow. Considering the limited span of the observation period, and the expected duration of lahar activity on the volcano, rainfall from future typhoons could equal or even exceed this value. In the 375-year recorded history of Mayon Volcano on southern Luzon, the most lethal event was a typhoon-triggered rain lahar that killed 1,500 people in November 1875, 3 years after a major eruption (Ramos-Villarta and others, 1985).

The estimated magnitude and level of lahar inundation in an area depend primarily on the availability of materials for remobilization as lahars. Compared to previous eruptions of Mount Pinatubo, the 1991 eruption is one of the smallest that can be identified in the geologic record (Newhall and others, this volume). The 1991 eruption may be comparable to the eruptions of the Buag eruptive period, so the extent and level of posteruption lahar deposition may be similar to that after the Buag eruption(s).

For the Santo Tomas River system, our worst-case scenario for lahar deposition estimates that lahar aggradation would reach the level of the highest prehistoric river terraces, known to have formed during the Maraunot eruptive period. At least 15 to 25 percent of the remaining pyroclastic materials in the Marella pyroclastic fan would be required for aggradation to reach the level of these terraces.

Figure 17. Upstream view of the lake showing flooded areas of Dizon Mines. The water retention dam of the mine can be seen at the right-hand side of the photograph.

Several meters of fluvial deposits in these terraces, in erosional contact with the underlying laharic deposits, suggest that an unknown thickness of lahar deposits was eroded away before fluvial sediments were deposited. If the high aggradation rate that we have documented during this first year after the eruption were to continue, then lahars may actually fill the channels of the Marella and Santo Tomas Rivers by 1994 or 1995. This would not bode well for the inhabitants in the Santo Tomas plain, as lahars in the near future may overtop and escape from the present lahar field by following any of the lahar routes previously delineated for the Santo Tomas River system (fig. 2).

If left to evolve naturally, the lahar-dammed Mapanuepe Lake would continue to expand in pace with the aggradation at the confluence site. Natural diversion of the Marella River into the lake would cause the Marella-side lake margin to prograde upstream, at the same time gradually filling the lake. In time, the lake would shrink in size as it gradually became filled with sediment—within the next 100 to 200 years if sediment delivery proceeds as predicted. Interspersed within this period of lake development, the lake might experience episodes of lake breakouts and continue to be a long-term hazard to downstream communities on the Santo Tomas plain. However, as of this time of writing, an artificial channel through bedrock near the Mapanuepe dam has, at least temporarily, controlled the rise of lake level. Sediments will continue to fill the lake but the risk of lake breakouts and downstream flooding has been minimized.

ACKNOWLEDGMENTS

We thank all our fellow members of the Pinatubo Lahar Hazards Taskforce, especially R.A. Alonso; M.B. Angeles, R.A. Arboleda, M. Bumanlag, V. Jalique, M.L. Paladio, C. Remotigue, G. Salvador, R.U. Solidum, R. Tamayo, and R. Tan. Director J.B. Muyco of the Philippine Mines and Geosciences Bureau, Director B. Austria of the Philippine National Institute of Geological Sciences, Director R.S. Punongbayan of the Philippine Institute of Volcanology and Seismology, Mr. F. Salonga of the Philippine Shipyard and Engineering Corporation, Undersecretary M. Pablo of the Philippine Department of Public Works and Highways, Congresswoman K. Gordon of the First District of Zambales, Mayor R.J. Gordon and the City of Olongapo, the U.S. Navy—Cubi Point Weather Station, and the U.S. Marine Air and Ground Task Force provided invaluable logistical support. PHIVOLCS and U.S. Geological Survey personnel of the Pinatubo Volcano Observatory were generous in sharing information. We thank the following reviewers: K.M. Scott, G.A. Smith, R.U. Solidum, K. Whipple, and C.G. Newhall for their critical comments, which greatly improved the paper. We would also like to extend our appreciation to M. Rubin for the ^{14}C dates, which proved valuable and critical to the outcome of this study. The research was supported by National Science Foundation SGER Grant EAR–9116724 and EAR Grant 9205132.

Editors' note: As of August 1995, the artificial channel continues to stabilize the level of Mapanuepe Lake. Massive lahar deposition downstream from the confluence of the Marella and Mapanuepe Rivers has completely filled and overflowed both natural stream banks and man-made levees.

REFERENCES CITED

Best, J.L., 1987, Flow dynamics of river channel confluences: Implications for sediment transport and bed morphology, *in* Ethridge, F.G., Flores, R.M., and Harvey, M.D., eds., Recent developments in fluvial sedimentology: Society of Economic Paleontologists and Mineralogists Special Publication No. 39, p. 27–36.

Daag, A.S., Dolan, M.T., Laguerta, E.P., Meeker, G.P., Newhall, C.G., Pallister, J.S., and Solidum, R., this volume, Growth of a postclimactic lava dome at Mount Pinatubo, July–October 1992.

Daido, A., 1985, Effect of volcanic ash on occurrence of mud-debris flow: Proceedings of the International Symposium on Erosion, Debris Flow and Disaster Prevention, Tsukuba, Japan, September 3–5, 1985.

De Boer, J., Odom, L.A., Ragland, P.C., Snider, F.G., and Tilford, N.R., 1980, The Bataan orogene: Eastward subduction, tectonic rotations, and volcanism in the western Pacific (Philippines): Tectonophysics, v. 67, no. 3–4, p. 251–282.

Delfin, F.G., Jr., 1983, Geology of the Mount Pinatubo geothermal prospect: unpublished report, Philippine National Oil Company, 35 p.

Hoblitt, R.P., Wolfe, E.W., Lockhart, A.B., Ewert, J.E., Murray, T.L., Harlow, D.H., Mori, J., Daag, A.S., Tubianosa, B.S., 1991, 1991 eruptive behavior of Mount Pinatubo, Philippines [abs.]: Eos, Transactions, American Geophysical Union, v. 72, no. 44, p. 61.

Hoblitt, R.P., Wolfe, E.W., Scott, W.E., Couchman, M.R., Pallister, J.S., and Javier, D., this volume, The preclimactic eruptions of Mount Pinatubo, June 1991.

Huke, R.E., 1963, Shadows on the land: An economic geography of the Philippines: Manila, Bookmark, 428 p.

Janda, R.J., Daag, A.S., Delos Reyes, P.J., Newhall, C.G., Pierson, T.C., Punongbayan, R.S., Rodolfo, K.S., Solidum, R.U., and Umbal, J.V., this volume, Assessment and response to lahar hazard around Mount Pinatubo.

Janda, R.J., Major, J., Scott, K.M., Besana, G., Daligdig, J.A., and Daag, A.S., 1991, Lahars accompanying the mid-June 1991 eruptions of Mount Pinatubo, Tarlac, Pampanga provinces, the Philippines [abs.]: Eos, Transactions, American Geophysical Union, v. 72, no. 44, p. 62–63.

Javellosa, R.S., 1984, Morphogenesis and neotectonism of the Santo Tomas Plain, southwestern Zambales, Luzon, Philippines: unpublished M.Sc. thesis, International Institute for Aerospace Survey and Earth Sciences (ITC), Enschede, The Netherlands, 103 p.

Major, J.J., Janda, R.J., and Daag, A.S., this volume, Watershed disturbance and lahars on the east side of Mount Pinatubo during the mid-June 1991 eruptions.

Newhall, C.G., Daag, A.S., Delfin, F.G., Jr., Hoblitt, R.P., McGeehin, J., Pallister, J.S., Regalado, M.T.M., Rubin, M., Tamayo, R.A., Jr., Tubianosa, B., and Umbal, J.V., this volume, Eruptive history of Mount Pinatubo.

Pierson, T.C., Daag, A.S., Delos Reyes, P.J., Regalado, M.T.M., Solidum, R.U., and Tubianosa, B.S., this volume, Flow and deposition of posteruption hot lahars on the east side of Mount Pinatubo, July–October, 1991.

Pierson, T.C., Janda, R.J., Umbal, J.V. and Daag, A.S., 1992, Immediate and long-term hazards from lahars and excess sedimentation in rivers draining Mount Pinatubo, Philippines: U.S. Geological Survey Water-Resources Investigations Report 92–4039, 32 p.

Pierson, T.C. and Scott, K.M., 1985, Downstream dilution of a lahar: Transition from debris-flow to hyperconcentrated-streamflow: Geological Society of America Bulletin, v. 21, p. 1511–1524.

Pilgrim, D.H., Cordery, I., and Baron, B.C., 1982, Effects of catchment size on runoff relationship: Journal of Hydrology, v. 58, p. 205–221.

Pinatubo Volcano Observatory Team, 1991, Lessons from a major eruption: Mount Pinatubo, Philippines: Eos, Transactions, American Geophysical Union, v. 72, no.49, p. 545, 552–553, 555.

Punongbayan, R.S., de la Cruz, E., Gabinete, E., Scott, K.M., Janda, R.J., and Pierson, T.C., 1991, Initial channel response to large 1991 volumes of pyroclastic flow and tephra fall deposits at Mount Pinatubo, Philippines [abs.]: Eos, Transactions, American Geophysical Union, v. 72, no. 44, p. 63.

Ramos-Villarta, S.C., Corpuz, E.G., and Newhall, C.G., 1985, Eruptive history of Mayon Volcano, Philippines: Philippine Journal of Volcanology, v. 2, no. 1–2, p. 1–35.

Regalado, M.T., and Tan, R., 1992, Rain generation at Marella River, Mount Pinatubo: unpublished report, University of the Philippines, 15 p.

Rodolfo, K.S., 1991, Climatic, volcaniclastic, and geomorphic controls on the differential timing of lahars on the east and west sides of Mount Pinatubo during and after its June 1991 eruptions [abs.]: Eos, Transactions, American Geophysical Union, v. 72, no. 44, p. 62.

Rodolfo, K.S. and Arguden, A.T., 1991, Rain-lahar generation and sediment-delivery systems at Mayon Volcano, Philippines, *in* Fisher, R.V., and Smith, G.A., eds., Sedimentation in volcanic settings: Society of Economic Paleontologists and Mineralogists Special Publication No. 45, p. 71–88.

Rodolfo, K.S., and Umbal, J.V., 1992, Catastrophic lahars on the western flanks of Mount Pinatubo, Philippines: Proceedings of the Workshop on the Effects of Global Climate Change on Hydrology and Water Resources at the Catchment Scale, February 3-6, 1992, Tsukuba Science City, Japan, Japan-U.S. Committee on Hydrology, Water Resources, and Global Change (JUCHWR), p. 493–510.

Rodolfo, K.S., Umbal, J.V., Alonso, R.A., Remotigue, C.T., Paladio-Melosantos, M.L., Salvador, J.H.G., Evangelista, D., and Miller, Y., this volume, Two years of lahars on the western flank of Mount Pinatubo: Initiation, flow processes, deposits, and attendant geomorphic and hydraulic changes.

Scott, K.M., Janda, R.J., de la Cruz, E., Gabinete, E., Eto, I., Isada, M., Sexon, M., and Hadley, K.C., this volume, Channel and sedimentation responses to large volumes of 1991 volcanic deposits on the east flank of Mount Pinatubo.

Scott, W.E., Hoblitt, R.P., Torres, R.C., Self, S, Martinez, M.L., and Nillos, T., Jr., this volume, Pyroclastic flows of the June 15, 1991, climactic eruption of Mount Pinatubo.

Smith, G.A., and Lowe, D.R., 1991, Lahars: Volcano-hydrologic events and deposition in the debris flow-hyperconcentrated flow continuum, in Fisher, R.V., and Smith, G.A., eds., Sedimentation in volcanic settings: Society of Economic Paleontologists and Mineralogists Special Publication No. 45, p. 59–70.

Torres, R., Self, S., and Martinez, M., this volume, Secondary pyroclastic flows from the June 15, 1991, ignimbrite of Mount Pinatubo.

Tuñgol, N.M., and Regalado, M.T.M., this volume, Rainfall, acoustic flow monitor records, and observed lahars of the Sacobia River in 1992.

Umbal, J.V., Rodolfo, K.S., Alonso, R.A., Paladio, M.L., Tamayo, R., Angeles, M.B., Tan, R., and Jalique, V., 1991, Lahars remobilized by breaching of a lahar-dammed non-volcanic tributary, Mount Pinatubo, Philippines [abs.]: Eos, Transactions, American Geophysical Union, v. 72, no. 44, p. 63.

Wolfe, E.W., 1992, The 1991 eruptions of Mount Pinatubo, Philippines: Earthquakes and Volcanoes, v. 23, no. 1, p. 5–37.

Wolfe, E.W. and Hoblitt, R.P., this volume, Overview of the eruptions.

Wolfe, J.A., and Self, S., 1983, Subduction, arc volcanism, and hydrothermal mineralization: The Manila trench sector, Philippines: Philippine Journal of Volcanology, v. 1, no. 1, p. 11–40.

Channel and Sedimentation Responses to Large Volumes of 1991 Volcanic Deposits on the East Flank of Mount Pinatubo

By Kevin M. Scott,[1] Richard J. Janda,[1] Edwin G. de la Cruz,[2] Elmer Gabinete,[2] Ismael Eto,[2] Manuel Isada,[2] Manuel Sexon,[2] and Kevin C. Hadley[1]

ABSTRACT

With the beginning of the 1991 rainy season in late July, devastatingly high sediment yields began from the east flank of Mount Pinatubo. Sediment was derived from the deposits of pyroclastic flows, which filled valleys locally to depths of over 100 m during the mid-June eruption, and from airfall deposits that mantled upland slopes. This paper focuses on early sedimentation responses and on the characteristics of the flows and deposits that are important for immediate control measures. Long-term responses were forecast for individual drainages on the basis of the type and thickness of eruptive products and on bedrock characteristics.

A critical factor in the immediate planning of mitigation measures is a trend toward posteruption fill in fanhead areas, indicated by measured channel cross sections that record high rates of upstream deposition. With the consequent potential for channel avulsions near the fanhead, the entire arc of active fans is at risk from future flows. This trend is validated by the pattern of sedimentation after previous eruptions.

At Mount Pinatubo, the flows causing widespread damage on populated fan surfaces are noncohesive (low in clay-size sediment). The flows are dominantly in the range of hyperconcentrated flow (containing 20 to 60 percent sediment by volume), and sand-size sediment is the dominant constituent. Some large flows leave the mountain front as debris flows that, because of their noncohesive character, are rapidly diluted to erosive hyperconcentrated flow. Consequently, mitigation plans must assess both the loss of channel capacity and a high potential for levee erosion and breaching.

A related consideration is the texture of fan deposits available for levee construction. Noncohesive deposits of identical hyperconcentrated flows associated with previous eruptions dominate the sediment composing the fans; they

are the most accessible levee-construction materials but are, however, extremely erodible. Coarser, less-erodible materials for levee facing or wing dikes are less readily available.

INTRODUCTION

Between 5 and 6 km^3 of eruptive products were emplaced in watersheds surrounding Mount Pinatubo during the eruption climaxing June 15, 1991 (see W.E. Scott and others, this volume). Most of this volume was emplaced as thick valley fills of pyroclastic-flow deposits. Part of a smaller volume of airfall deposits on Luzon, estimated as approximately 1 km^3 (Paladio-Melosantos and others, this volume), mantled upland areas. The potential for destructive redistribution of this material by sediment-laden, water-mobilized flows into habitated areas was immediately clear; how, where, and over what timeframe it would occur was not.

The greatest concern was for the downstream, highly populated areas on the east side of the volcano (fig. 1). The risks, however, would probably vary with factors that would become clear only after onset of the rainy season. Widely disparate thicknesses of pyroclastic-flow and airfall deposits occurred on watersheds that would differ greatly in their erosional and rainfall-runoff responses.

Various types of sediment-laden flow transport much of the sediment from volcaniclastic terranes and have done so from Mount Pinatubo. In debris flows, sediment moves the interstitial water, rather than the reverse, as in normal streamflow, and the flows have the appearance and impact forces of flowing concrete. At Mount Pinatubo, debris flows evolve downstream into more dilute types of flow, especially hyperconcentrated flow, as described in the section on flow processes. Both debris flows and hyperconcentrated flows are commonly known by the Indonesian term "lahar." This general term is used here only when there is no evidence for a specific flow type such as debris flow or hyperconcentrated flow. All types of sediment-laden flow can produce harmful effects, and these include, in addition to loss of human life and habitation, the loss of all types of

---

[1]U.S. Geological Survey.
[2]Philippine Institute of Volcanology and Seismology.

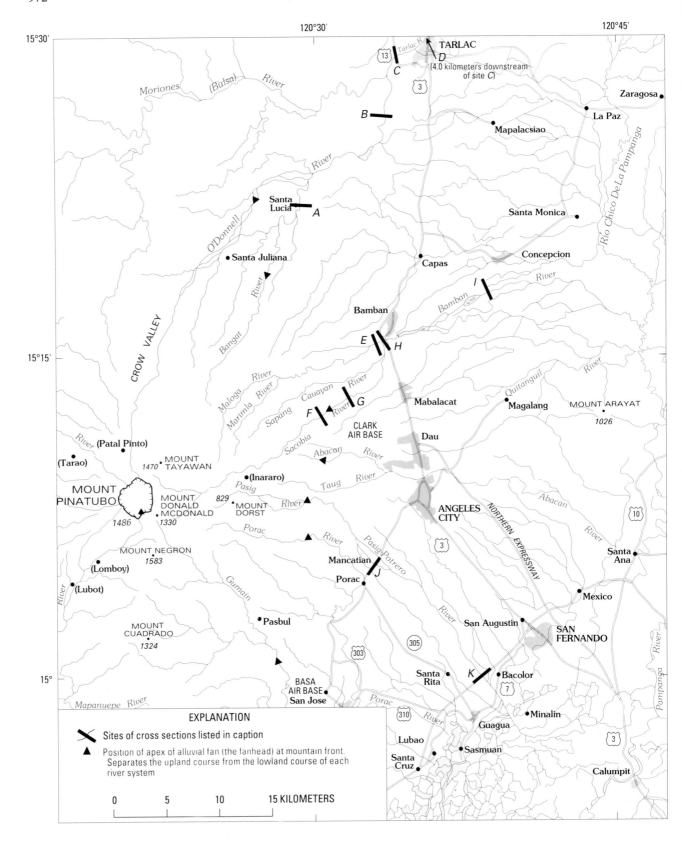

EXPLANATION

✕ Sites of cross sections listed in caption

▲ Position of apex of alluvial fan (the fanhead) at mountain front.
    Separates the upland course from the lowland course of each
    river system

```
0        5        10      15 KILOMETERS
```

infrastructure, the loss of channel and irrigation-canal capacity, burial of agricultural lands, and siltation of aquaculture ponds.

Scour and fill are short-term erosion and deposition, respectively, commonly in channels. Degradation and aggradation are long-term erosion and deposition, respectively, in channels or on any part of the landscape. Much of the fill discussed here will ultimately prove to be aggradation.

The purpose of this contribution is to summarize the important effects of sediment transport and deposition, as they concern mitigation measures, and as they both appeared to us initially and again with the subsequent perspective of the 1991 and 1992 rainy seasons. The object is to provide planners with information critical to the engineering mitigation now being planned; that is, to define the type, behavior, and the depositional pattern of the sediment-conveying flows. Other long-term evaluations of the situation at Pinatubo have illustrated potential inundation areas in the absence of any engineering mitigation (Punongbayan and others, 1991; Pierson and others, l992). These studies define probable long-term trends, both areally and over time, whereas this paper summarizes other aspects of the sediment-distribution system that are important in planning present mitigation measures. Rodolfo (1991) has contrasted early responses on the east and west sides of the volcano.

DRAINAGE SYSTEM ON THE EAST SIDE OF MOUNT PINATUBO

The major drainages are shown on figure l. From north to south, the systems of greatest interest are the O'Donnell-Tarlac River system (downstream portion known as the Tarlac River), Sacobia-Bamban River system (downstream portion known as the Bamban River), the Abacan River system, and the Pasig-Potrero River system (downstream portion most commonly known as the Pasig-Potrero River). The compound systems may be referred to in part or entirety by their compound names, but upstream sites may be referenced by the names of the upstream drainages: the

---

← **Figure 1.** Drainage system of the eastern side of Mount Pinatubo. Cross-section sites are as follows: *A*, Bangat River at Santa Lucia (Bangat Bridge); Bangat River is a tributary of O'Donnell-Tarlac River. *B*, O'Donnell-Tarlac River upstream of Tarlac. *C*, O'Donnell-Tarlac River at Tarlac (Agana Bridge). *D*, O'Donnell-Tarlac River at Tarlac (downstream of site C). *E*, Marimla River above confluence with Sacobia-Bamban River. *F*, Sacobia River above Clark Air Base. *G*, Sacobia River near Clark Air Base. *H*, Sacobia-Bamban River at Bamban. *I*, Sacobia-Bamban River at San Francisco Bridge. *J*, Pasig-Potrero River at Mancatian. *K*, Pasig-Potrero River at Santa Barbara Bridge. Villages in parentheses were destroyed by the eruption.

O'Donnell, Sacobia, or Pasig River, for example. The two major drainages farther south are the Porac and Gumain Rivers, which are now joined in a leveed drain leading to the axial, north-south drainage of central Luzon.

POSTERUPTION CHARACTERISTICS OF SLOPES AND CHANNELS

The rainfall-runoff characteristics of the landscape on the east side of the volcano were radically changed by the slope-mantling airfall deposits. Rilling in the deposits began almost immediately, and rills and small gulleys rapidly cut through the airfall and into the preexisting ground surface. Shallow, lahar-conveying channels gradually migrated upstream on the valley-fill deposits of the pyroclastic flows.

Primary pyroclastic flow deposits filled east-flank valleys to depths of as much as 140 m (R.S. Punongbayan, oral commun., 1992). Before rainy-season onset, little drainage integration had occurred on the fill surfaces, and many channels flowed into phreatic explosion pits. Explosion pits were first caused by infiltrating surface drainage, but they subsequently guided drainage development over the short term. A form of beaded drainage (fig. 2) was observed in the first overflights and aerial photographs. This drainage was largely independent of the preexisting pattern, other than in general direction. Some drainage followed primary flow lineations preserved on deposit surfaces of the pyroclastic flows.

Significant flooding and some lahars had extended to eastern populated areas in response to the storm shortly before the eruption, coincident with it, and with later storms that predated the rainy season. The timing of these events relative to the tephra fall and pyroclastic flows was complex (see Janda and others, 1991; Major and others, this volume). These events, however, were not the precursors of major trends.

The most obvious early channel response at bridge crossings nearest the mountain front was scour. Scour was observed as a response to the syneruptive lahars throughout the upper lowland course of the Pasig-Potrero River (J.J. Major, written commun., 1992). Scour also affected the fanhead area of the Sacobia-Bamban River. Scour of 2 to 5 m occurred between July 15 and 22 opposite the site of Clark Air Base, before most of the headwater drainage was recaptured on July 25 (see section on channel diversions). Another episode of 6 to 7 m of scour occurred there in late August and early September (Pierson and others, 1992). Bridge-pier footings were visibly exposed to a depth of about 2.5 m during mid-July at the site on the Sacobia-Bamban River labeled I on figure 1.

Scour proved to be a misleading response in light of the devastating volumes of fill that began in upland and many lowland areas with the onset of the 1991 rainy season in mid-July. Some remaining bridge piers were buttressed

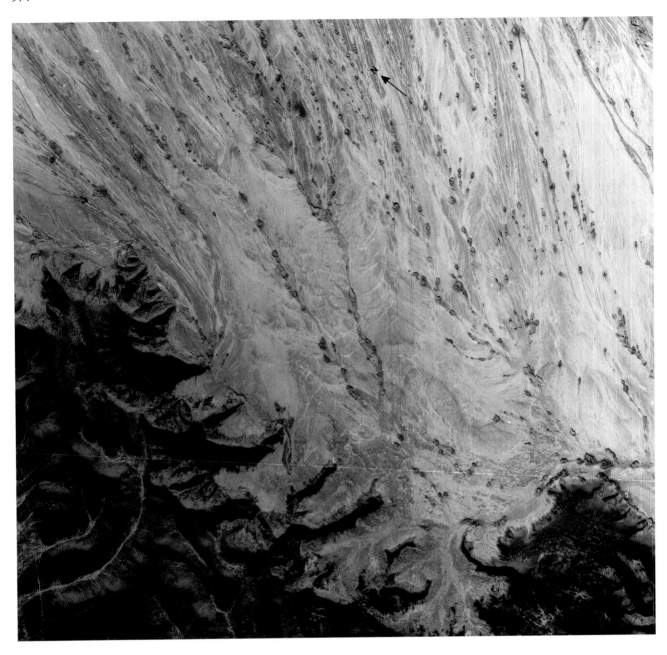

Figure 2. Drainage pattern on surface of deposits of pyroclastic flows. Note beaded drainage pattern connecting phreatic explosion pits. The latter are here aligned along primary flow lines on the surface of the deposits. Drainage is from upper left to lower right. Because cloud cover obscured similar features on the east flank of Mount Pinatubo before they were destroyed by erosion, this photograph is from the Marella River drainage, southwest of Mount Pinatubo. Photograph is approximately 0.5 km in width and was taken July 5, 1991, by the U.S. Air Force.

against erosion but would be mainly subject to unremitting burial. The longer term trends, which appeared as the rainy season progressed, are discussed below.

GEOMORPHIC SETTING

Although the vegetative cover is normally dense, the eastern side of Mount Pinatubo (preeruption altitude of 1,745 m) is in many ways similar to the dissected uplands and alluvial fans of semiarid regions (see Ely and Baker, 1990). The principles of flood hydrology on semiarid alluvial fans are closely applicable. The factor in common to both settings is rare periods of intense sediment yield. Terrace levels and alluvial fills record previous eruptive periods at Mount Pinatubo, just as the same features record the alluviation by rare, catastrophic flows in semiarid areas.

Incised channels on alluvial fans east of Pinatubo reveal sequences of flow deposits from previous eruptions. This is a critical observation in assessing the distribution of future flows and possible structural measures for their control.

Upland terrane, whether consisting of bedrock or unconsolidated deposits, will be referred to here as "upland," and areas below the mountain front (bedrock-alluvium contact) will be distinguished as "lowland." The lowland is a piedmont formed of coalesced alluvial fans. Well-defined fanhead areas separate the upland portions of drainage systems from their lowland extensions (fig. 1). Implications of the alluvial fan environment include the fanwide distribution of flood risk, beginning at the apex of the arc in the fanhead area. Fanhead areas are characterized by incised channels, known as fanhead trenches, that record low-flow channel degradation between major flows. Rapid fill at the fanhead can direct large flows to any part of an active fan. During a sedimentation episode as profound as that beginning at Mount Pinatubo, repeated cycles of trench filling, avulsion, degradation, and repeated filling will expectedly distribute sediment across much of a fan surface.

CLIMATE

The tropical climate is dominated by several air masses during the year, the most important of which at Mount Pinatubo is the southwest monsoon that typically begins in late May or early June and lasts until September or October (see Umbal and Rodolfo, this volume). The 1991 rainy season began in July, but the 1992 rainy season arrived in late June. Mean annual precipitation at Clark Air Base, altitude 146 m, is 1,950 mm. On the basis of long-term trends, 57 percent of the precipitation occurs in July, August, and September. Precipitation is substantially greater at higher altitudes. The Clark site represents a typical middle altitude location, between the Sacobia and Abacan Rivers, near the mountain front on the east side of the volcano.

Typhoons are rare early in the year, but their frequency increases gradually through April and May. They are frequent from June through December and are an especially characteristic threat from September to November, a period that is late in the rainy season and during the "transition season" following the main rainy season. Detailed long-term weather records are available from the location of Clark Air Base (U.S. Air Force, 1991).

FLOW TYPE AND DEPOSIT SEDIMENTOLOGY

Granular debris flows are the initial process by which volcanic detritus is redistributed from uplands to lowlands, primarily during rainy seasons. The flows are dominated by sand-size sediment (0.0625 to 2.0 mm); they are uniformly low in clay-size sediment (finer than 0.004 mm) and commonly contain less than 1 or 2 percent of that size fraction. The flows are distinct in both origin and behavior from debris flows with more than 3 to 5 percent of clay, and the two types can be distinguished as noncohesive and cohesive, respectively (Scott and others, 1992).

Both cohesive and noncohesive flows commonly recur at many volcanoes, but noncohesive flows dominate at Pinatubo. A significant behavioral distinction is that noncohesive flows transform from debris flows to hyperconcentrated flow (20 to 60 percent sediment by volume) to normal streamflow (less than 20 percent sediment) as they move downstream. Figure 3 illustrates three textural fields, based on data from Mount St. Helens, through which cumulative curves undergo downward inflection as these changes in the flow occur. Textures of three hyperconcentrated flows from the Pasig-Potrero River plot within the field for hyperconcentrated flows at Mount St. Helens.

A distinctive "transition facies" records the downstream dilution from noncohesive debris flow to hyperconcentrated flow (Scott, 1988, fig. 37); many examples of this facies were seen in the 1991–92 deposits of the Abacan and Sacobia-Bamban Rivers. Although nearly all hyperconcentrated flows in the Cascade Range evolved from upstream debris flows, that common origin there and the evidence from the transition facies do not mean that origin was the rule, or was even common, for hyperconcentrated flows during the 1991 and 1992 rainy seasons at Pinatubo. Observations of 1992 deposit sequences reveal many deposits of hyperconcentrated flows that may not have entrained enough sediment to reach the level of sediment concentration for rheological debris flow (fig. 4).

Regardless of their origin as primary hyperconcentrated flows or as the runouts of debris flows, flows with hyperconcentrations of sediment volumetrically dominate the flow system and depositional record downstream. Noncohesive debris flows and hyperconcentrated flows at Mount Pinatubo are highly erosive, on the basis of their observed ability to scour older deposits and to erode levees. Likewise, the sandy deposits of noncohesive flows are significantly more erodible than those either coarser or with more clay. Gravel-size fractions (coarser than 2 mm), which would increase erosion resistance of levee materials, are mainly deposited in upland channels as the noncohesive flows lose competence and transform to more dilute flow types. Thus, mainly erodible, sand-dominated deposits are locally available for levee construction at downstream sites, where the confinement of flow is most needed.

The deposits of hyperconcentrated flow are distinctive: a dominant size mode in the sand range, massive or poorly stratified beds typically 0.5 to 2.0 m thick, and sorting (a measure of size dispersion) intermediate between that of debris flow deposits and the sorting of deposits of normal streamflow. The loss of normal-density, gravel-size clasts

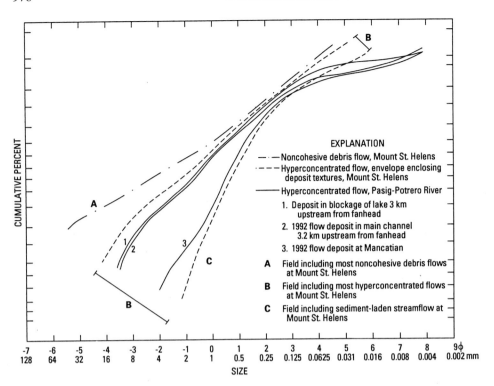

Figure 3. Cumulative curves illustrating textures (particle sizes and their distribution) of three deposits of hyperconcentrated flow in the Pasig-Potrero River (curves 1, 2, and 3), compared with a typical deposit of noncohesive debris flow at Mount St. Helens. Also shown are fields containing cumulative curves of deposits of (A) noncohesive debris flow, (B) hyperconcentrated flow, and (C) normal, sediment-laden streamflow, all based on data from Mount St. Helens. ϕ, a measure of particle size (see Folk, 1980).

shows that the flow has either lost, or never possessed, the strength necessary to suspend such clasts.

The textures of typical Pinatubo hyperconcentrated flows are shown in figure 3. Note that the downward inflection of the cumulative curves, and thus the transformation in flow type, is reflected primarily by loss of coarser fractions. A measure of sorting, graphic deviation (σ_G, measured in ϕ units; Folk, 1980), typically varies from 1 ϕ to about 2 ϕ in hyperconcentrated flows. An exact range cannot be specified because of the strong influence of fine-sediment content on shear strength.

Hyperconcentrated flow deposits in the eastern drainages of Mount Pinatubo are dominated by sand-size phenocrysts from the pyroclastic flows with an admixture of mineral grains from older deposits. Pumice clasts are present but are volumetrically minor; most are preserved in coarse deposits near the source. Cumulative curves of deposits show that Pinatubo flows contain slightly more fine sediment that their St. Helens analogs (fig. 3). Most fine sediment at Mount Pinatubo, including devitrified glass and comminuted mineral and vitric material, is flushed through the fan environment to be deposited in axial lowland drainages and deltaic mudflats. This mineralogic fractionation is a partial analog of an example in the geologic record (Cather and Folk, 1991).

FIRST APPROXIMATION OF WATERSHED RESPONSES

Between the time of the climactic eruption, June 15, 1991, and the then-unknown time of rainy-season onset, it was necessary to characterize and prioritize how each drainage basin would respond to increased runoff. Resources were few, access problems were large because of destroyed bridges, and the rationale for moving additional groups into evacuation camps was weak unless the threat was immediate. The east-side watersheds were grouped by one of us (Janda) into three categories of response (table 1). This grouping, slightly modified from the initial version, was intended to be a first approximation, and, as such, it was successful. The categories were:

A. Watersheds in which the sedimentation response would be immediate and persistent. Each was characterized by thick, fine-grained airfall distributed relatively uniformly on an erodible terrane. Pyroclastic flows entered some headwaters, but sediment yields from erosion of their deposits would not dominate the systems.

B. Watersheds in which the flow response could be delayed but would also be large and persistent. Each drainage was characterized by large deposits of pyroclastic flows, abundant airfall deposits, and an erodible upland terrane.

Figure 4. Deposits of 1992 hyperconcentrated flows in the Sacobia River inset against deposits from an older eruptive episode. Arrows mark the surface of the 1992 deposits; about 15 m of 1992 deposits occur in vertical section above the river. Location is 1 km upstream from the fanhead of the Sacobia River, November 15, 1992.

Table 1. Sedimentation response of watersheds on east side of Mount Pinatubo, as based on thickness and type of eruptive products and erodibility of terrane.

[See text for discussion of A–C]

Drainage	Upland drainage area (km²)	Slope (m/m)
A. Immediate, persistent response[1]		
Bangat (or Bangut) River (major tributary of O'Donnell-Tarlac River).	51.6	0.04
Marimla River (tributary of Sacobia-Bamban River).	65.1	.04
B. Delayed, massive, persistent response.		
O'Donnell-Tarlac River	111.2	.08
Sacobia-Bamban River	[2]43.5	.06
Abacan River (before stream piracy of July 25, 1991).	[2,3]47.5	.04
Pasig-Potrero River	23.9	.04
C. Immediate but less-severe response		
Abacan River (after stream piracy of July 25, 1991).	[2,3]14.2	.04
Porac River	29.9	.06
North Fork Gumain River	47.0	.07
Middle Fork Gumain River		
West Fork Gumain River	117.1 (both forks)	.07

[1]Category A is similar to category C, but it was assumed that response would be greater in category A because of larger tephra volumes.

[2]Most of 33.3 km² diverted from Abacan River back to Sacobia River July 25, 1991.

[3]Excludes Taug River (Sapang Bayo Creek).

Thick deposits of pyroclastic flows created the potential for large sediment yields and massive downstream aggradation once drainage was integrated across the deposit surfaces.

C. Watersheds in which the response would be immediate but less severe and of shorter duration than in the preceding categories. This response would be a function of the relative paucity of pyroclastic flows entering the drainages as well as the resistant underlying bedrock.

DEVELOPMENT OF DRAINAGE PATTERN AND FLOW-MARGIN LAKES

With the onset of the 1991 rainy season in July, overflights showed an almost immediate transition from sheet wash and shallow, lahar-conveying channels on the valley-fill deposits to channels rapidly working their way upstream, almost exclusively by headcutting. Headcuts in the pyroclastic-flow deposits in the Sacobia River valley were estimated at greater than 10 m in height.

The longitudinal rates of headcutting, when seen during brief intervals of clear weather, were well in excess of 10 m per hour in some cases. Drainage of adjacent uplands was already integrated with the surfaces of the flow deposits, so that the process accelerated as the contributing drainage area increased exponentially.

The effects of secondary pyroclastic flows weeks, months, and more than a year later were an unanticipated circumstance (see Torres and others, this volume). The largest was reported in mid-August 1991 (U.S. Department of

Figure 5. Lake in tributary of Sacobia River dammed by pyroclastic flow deposits in Sacobia River (light area trending southwest-northeast at lower right). Note slumping on both sides of the blockage. Length of lake is approximately 0.4 km. Photograph taken July 7, 1991, by the U.S. Air Force.

Defense, unclassified data, 1991) on the west side of Mount Pinatubo. That flow averaged almost a kilometer in width and extended at least 9 km. The largest 1991 flow on the east side occurred in the upper Pasig-Potrero River in early September and was probably more than 5 km in length. Another of similar size occurred July 13, 1992, in the same drainage. The thicknesses of these secondary flows were as much as several tens of meters. Triggering mechanisms are discussed by Torres and others (this volume).

Because the secondary pyroclastic flows occurred early in both rainy seasons, their occurrence triggered numerous lahars. An effect of the deposits of both flow types was the damming of flow from lateral tributaries. Several small lakes formed as a result of runoff gradually accumulating behind flow deposits emplaced in mid-June 1991. A typical example is shown in figure 5. This lake had discharged by July 21, 1991, and the blockage is seen in figure 5 (taken July 7) to be eroding both from upstream headcutting from a large explosion pit in deposits in the Sacobia River and by wave erosion or saturation-induced slumping of the blockage face.

The most significant blockage on the eastern side of Mount Pinatubo was that on the south side of the Pasig-Potrero River just 3 km upstream from the fanhead. A shallow lake was first observed there on July 24, 1991, and its clear water indicated recent impoundment. When next seen on August 1 the water was highly discolored by algal growth. The original blockage was a primary pyroclastic flow, lahars, or both. When the lake discharged in September 1991, the large secondary pyroclastic flow noted above had raised the blockage considerably, and the lake volume was greatly increased. The lake reformed, due to blockage by the 1992 secondary pyroclastic flow and associated lahars, and that lake discharged during a period of intense rainfall in late August 1992 (see Arboleda and Martinez, this volume).

CHANNEL DIVERSIONS

So thick were the fills of pyroclastic flows in some upland areas that, in many cases on the west side of the volcano (Rodolfo, 1991) and several cases on the east side, watersheds were filled and drainage divides were crossed. Drainage could potentially shift from the Sacobia-Bamban River to the Pasig-Potrero River, from the Sacobia-Bamban to the Abacan River, and from the Pasig-Potrero to the Abacan and Porac Rivers. Reversals of any diversions were clearly also possible but would be less likely once an integrated channel system was established. The distribution of even more extensive deposits from previous eruptive periods shows that similar diversions, of even larger scope, occurred during those periods.

The best documented channel shift of this type was drainage from the Sacobia-Bamban River system entering the Abacan River about July 15, 1991, and then mainly reverting to its preeruption course on July 25. As first observed on July 6 by J.J. Major and R.J. Janda (Major, written commun., 1992), the potential for diversion was clear but it had not yet occurred. A small channel that was headcutting from the Abacan River through the drainage divide had the potential to intersect the yet-embryonic drainage of the Sacobia-Bamban River. However, study of deposits in the Abacan drainage led Major and others (this volume) to suggest that a significant volume of flow may have entered the Abacan watershed from the Sacobia-Bambam River prior to the main pyroclastic flows of June 15.

Most of the upper Sacobia-Bamban River drainage (33.3 km[2]; table 1) was diverted to the Abacan River after July 6. Long periods, hours to days, of lahar surges were generated by relatively gentle rains (up to 6 to 8 cm per day on the eastern, lee side of the volcano). These flows traveled through Angeles City, reaching as far as the town of Mexico, 43 km downstream. As recorded by acoustic flow monitors (fig. 6), the rapid headcutting of the pyroclastic-flow deposits in the Sacobia-Bamban River then triggered recapture of most of the upper part of the river's drainage.

The dramatic shift in acoustic signal at 2100 hours on July 25 records the exact time of the capture (fig. 6). Almost continuous flow of lahars then began in the Sacobia-Bamban River. The importance and at least short-term irreversibility of the capture were not confirmed until ground observations early on July 26. Figure 6 illustrates the suddenness of the diversion. For a detailed discussion of flow sensing with acoustic flow monitors, see Marcial and others (this volume).

A subordinate, southern part of the Sacobia-Bamban River continued drainage to the Abacan, but this pattern ceased in response to a secondary pyroclastic flow in the Sacobia-Bamban system in April 1992. All flow from the upper Sacobia-Bamban drainage was subsequently directed to the downstream part of that system.

CHANNEL CROSS SECTIONS

Repeated surveys of channel cross sections at bridge locations were critical in evaluating the locations and magnitudes of early channel changes. These data record the trends in erosion and deposition and thus yield important data for the planning of countermeasures. Selected examples of these surveys are shown in figures 7, 8, and 9. The significance of the changes is discussed and interpreted in the following sections by major drainage, from north to south. Survey sites and drainages are shown on figure 1. Burial of the section reference point at one site (Sacobia-Bamban River at Bamban) required a change from the local datum to mean sea level (MSL).

O'DONNELL-TARLAC RIVER SYSTEM (INCLUDING TRIBUTARY BANGAT RIVER)

Bangat River at Santa Lucia (site A).—Although the Bangat River does not head near the crater, airfall and subordinate pyroclastic-flow deposits, emplaced on erodible bedrock (table 1), were rapidly flushed into downstream reaches. Nearly all channel capacity at the bridge site was lost with the 3 m of fill that occurred by August 23, 1991, one month into the first rainy season (site A, fig. 7). The magnitude of this response was expected (table 1).

O'Donnell-Tarlac River upstream of Tarlac (site B).— This site was surveyed only in 1992 (site B, fig. 7). Little change occurred early in the rainy season, but a uniform 1.5 m of fill occurred between August 13 and 20. The immediately following scour may have been natural, in response to dilute recession flows. Because scour this uniform across a channel is so atypical, it was more probably due to channel dredging. The channel level on October 14, near the end of the rainy season, was similar to that near the start of the rainy season.

O'Donnell-Tarlac River at Tarlac (where it is known as Tarlac River) (site C).—Gradual net scour occurred here, probably due to nearby channel dredging, during the 1991 rainy season (site C–1, fig. 7). By the early part of the 1992 season (site C–2, fig. 7), a uniform meter of fill had occurred, followed by almost 3 m of scour probably in response to dredging. In spite of the disturbance of the natural pattern by engineering work, the general lack of extensive fill both here and at site B shows that early channel response in the downstream part of this drainage was relatively benign compared to more southerly east-side drainages. However, the extensive fill observed upstream in the O'Donnell-Tarlac River can be expected to extend to these downstream reaches.

O'Donnell-Tarlac River at Tarlac (downstream of site C, also where drainage is known as Tarlac River) (site D).—The events at site C are confirmed by the same pattern of scour and fill at this location (site D; D–1 and D–2,

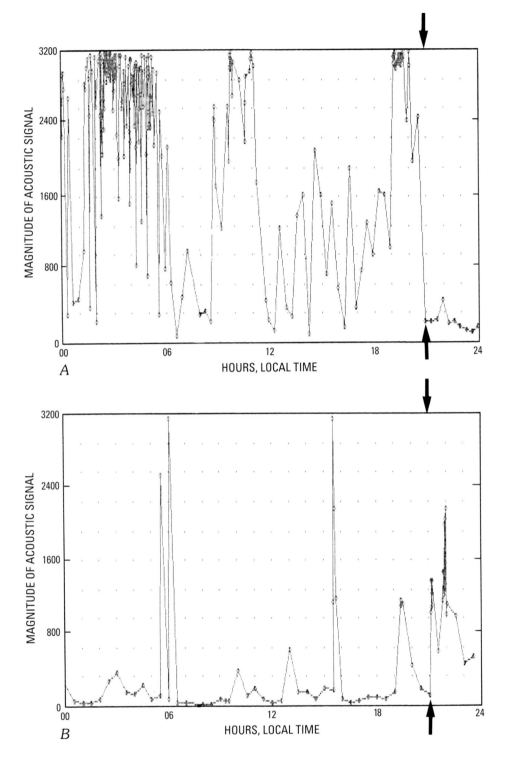

Figure 6. Avulsion of flow in upper Sacobia River from flow routed down Abacan River to flow routed normally to downstream Sacobia River, as recorded by acoustic flow monitors. *A*, Record of acoustic flow monitor recording flow in Abacan River on July 25, 1991. *B*, Record of acoustic flow monitor recording flow in Sacobia River on July 25, 1991. Arrows record time of completed channel diversion, 2100 hours, July 25, 1991.

fig. 7). Timing and causes of the channel changes are also similar, but changes are less pronounced, probably indicating a diminution of effects in a downstream direction. As at site C, fill will occur in the future, the amount dependent on engineering works and the effects of tributaries to increase sediment conveyance through these reaches.

SACOBIA-BAMBAN RIVER SYSTEM (INCLUDING TRIBUTARY MARIMLA RIVER)

Marimla River above confluence with Sacobia-Bamban River (site E).—Fill occurred rapidly at the start of the 1991 rainy season, and most channel capacity was lost by August 17 (site E, fig. 8). The rapidity of the fill reflects the higher base level and backwater effects caused by the extreme aggradation in the Sacobia-Bamban River. The confluence of the two streams is immediately downstream.

Sacobia River above Clark Air Base (site F).—Although not evident in the 1992 cross sections (site F, fig. 8), extensive fill at this site is documented by estimated levels of the fill surface late in 1992 compared with levels observed during the 1991 rainy season. At least 10 m, and locally as much as 15 to 20 m, of fill has occurred during this period at and near this location. The 2 to 5 m of scour seen in the cross sections as occurring during late July 1992 is a perturbation in the longer trend, evidenced in the subsequent fill at the following downstream site.

Sacobia River near Clark Air Base (site G).—The devastating volume of fill beginning in the upper part of the river system is illustrated by change at this site. Slightly over 10 m of 1992 fill is recorded at one side of the channel (site G, fig. 8). Approximately the same amount was observed late in 1991. Times of the 1992 surveys indicate that much of the fill occurred during major storm periods in late August and early September 1992. Similarity of the level of the low-water channel in late 1992 to the channel level prior to those runoff periods is mainly due to cutting by recession flows.

Sacobia-Bamban River at Bamban (site H).—Fill at this site of a major highway bridge began with the eruption, with approximately 10 m occurring before the first survey on July 23, 1991 (H–1, fig. 8); the bridge shown in the 1991 survey was originally at a height of 20 m; it failed from inundation by sediment-laden flows in late August 1991.

A new bridge was constructed above the remaining supports of the old bridge, and the new, higher bridge failed in the same way during the late August 1992 storm period. Almost 5 m of fill occurred across the channel during that 3-day period (H–2, fig. 8). Unfortunately, benchmarks for the two destroyed bridges cannot be compared, and the 1991 and 1992 surveys cannot be tied to the same datum. At a minimum, 25 m of fill has occurred locally at this site since the eruption. By the end of the 1992 rainy season, scour related to downstream channel maintenance had exposed the tops of the supports of the second bridge.

Sacobia-Bamban River at San Francisco Bridge (site I).—Scour at this site, 9.2 km downstream from site H, is in marked contrast to the upstream fill. The late 1991 scour is due to removal of sediment from the channel. Fill early in the 1991 (I–1, fig. 8) and 1992 (I–2, fig. 8) rainy seasons is the dominant natural process but is clearly of less magnitude than that upstream. This decrease is explained mainly by upstream losses of flow diverted from the leveed channel. Even so, the total cross-sectional area of fill at a given point in both the main channel and the diversions generally decreases downstream.

The channel downstream from Clark Air Base will be especially prone to diversion and levee breaks. This risk is due to the tendency toward radial drainage on the fan surface and to the trend toward fill in the engineered channels.

PASIG-POTRERO RIVER SYSTEM

Pasig-Potrero River at Mancatian (site J).—The 15 m of scour recorded here during August 1991 (J–1, fig. 9) is due to natural processes and channel maintenance. Scour began with flows associated with the eruption (see Major and others, this volume), and the Mancatian bridge was destroyed shortly thereafter. The early flows were almost certainly more dilute and thus more erosive than those later in the June-September period, reflecting the gradual integration of the drainage system on the vast pyroclastic flow deposits upstream and the consequent increase in sediment conveyance. With this general increase in sediment concentration of the flows, unremitting fill occurred between September 4 and November 21, 1991. This behavior coincided generally with the expected watershed response (table 1).

The 1991 filling continued unabated throughout most of the 1992 rainy season (J–2, fig. 9). Mining of large volumes of sand from the channel immediately downstream of Mancatian occurred during 1992 and slowed the rate of fill there. Channel dredging downstream also acted to slow the rate of 1992 filling, especially late in the rainy season as recorded in the surveys on September 1 and November 4, 1992. Runoff was also relatively low following the early September storm period noted in the section on flow-margin lakes.

The channel reaches from above Mancatian to the vicinity of Bacolor (fig. 1) are prone to channel diversions and levee breaks for the same reasons cited above in the case of the Sacobia-Bamban River below Clark Air Base.

Pasig-Potrero River at Santa Barbara Bridge (site K).—This site, located far from the mountain front and near the axial drainage of central Luzon, illustrates the fate of large leveed channels at the distal segments of river systems. This wide channel filled very rapidly early in the 1991 rainy season (K–1, fig. 9), and the bridge would have been destroyed but for loss of flow from the channel upstream.

This situation continued in 1992 (K–2, fig. 9). Large flows left the leveed channel system a short distance downstream of Mancatian and caused substantial damage in several small lowland communities. The channel at the Santa Barbara Bridge was maintained with sufficient capacity to retain the bridge, but most flows during major storm periods broke out of the leveed channel upstream. The channel at the bridge and for a considerable distance upstream is perched on fill at a level several meters above the surrounding countryside, so that losses from the channel are continuous and cause local flooding of areas adjacent to the channel.

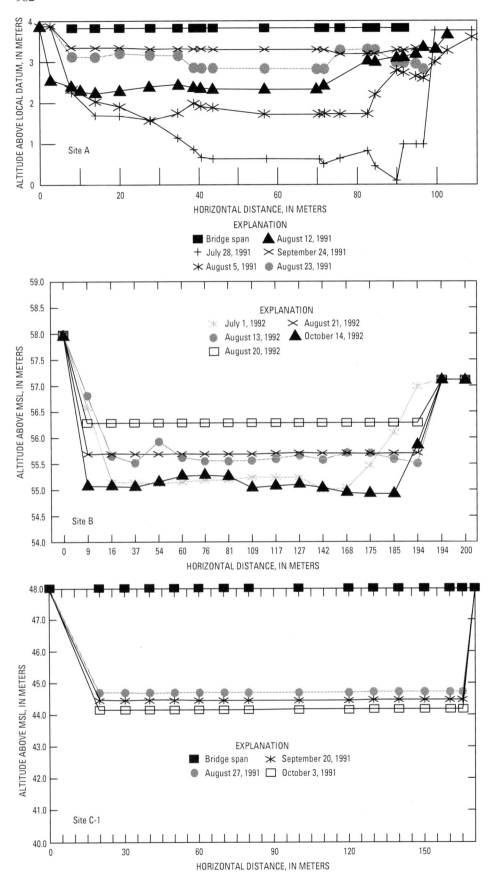

Figure 7. Cross-section changes in the O'Donnell-Tarlac River system, including the tributary Bangat River. Sites are as follows (locations on fig. 1): A, Bangat River at Santa Lucia, 1991; B, O'Donnell-Tarlac River upstream of Tarlac, 1992; C–1, O'Donnell-Tarlac River at Tarlac, 1991; C–2, same as site C–1, 1992; D–1, O'Donnell-Tarlac River downstream of site C, 1991; D–2, same as site D–1, 1991–92. Note variation in horizontel scale, sites B, D–1, and D–2.

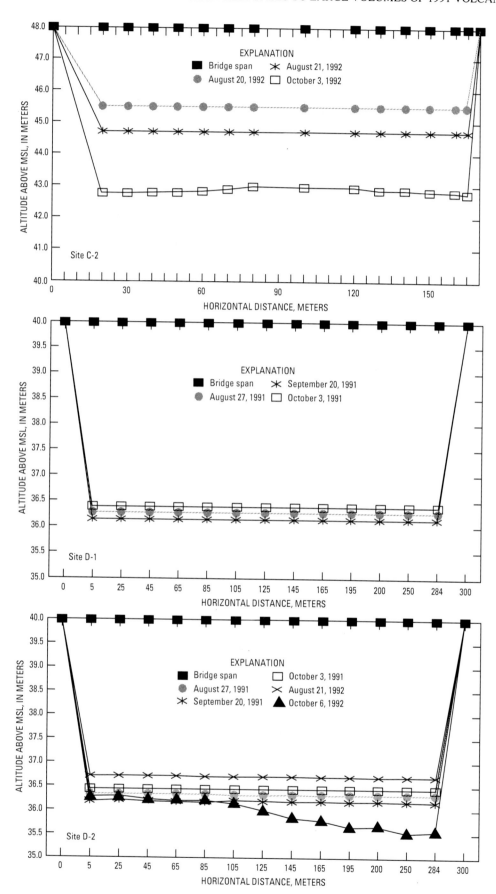

Figure 7.—Continued.

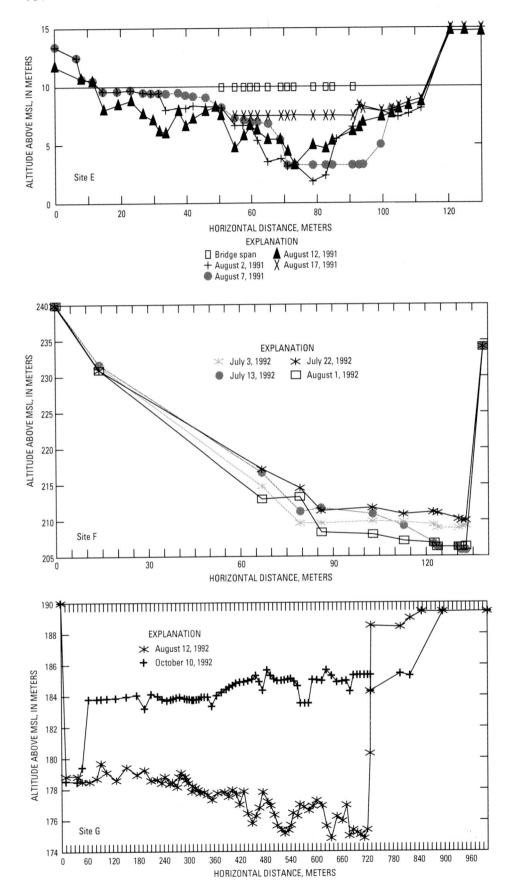

Figure 8. Cross-section changes in the Sacobia-Bamban River system, including the tributary Marimla River. Sites are as follows (locations on fig. 1): E, Marimla River above confluence with Sacobia-Bamban River, 1991; F, Sacobia River above Clark Air Base, 1992; G, Sacobia River near Clark Air Base, 1992; H–1, Sacobia-Bamban River at Bamban, 1991; H–2, same as site H–1, 1992; I–1, Sacobia-Bamban River at San Francisco Bridge, 1991; I–2, same as site I–1, 1992.

Figure 8.—Continued.

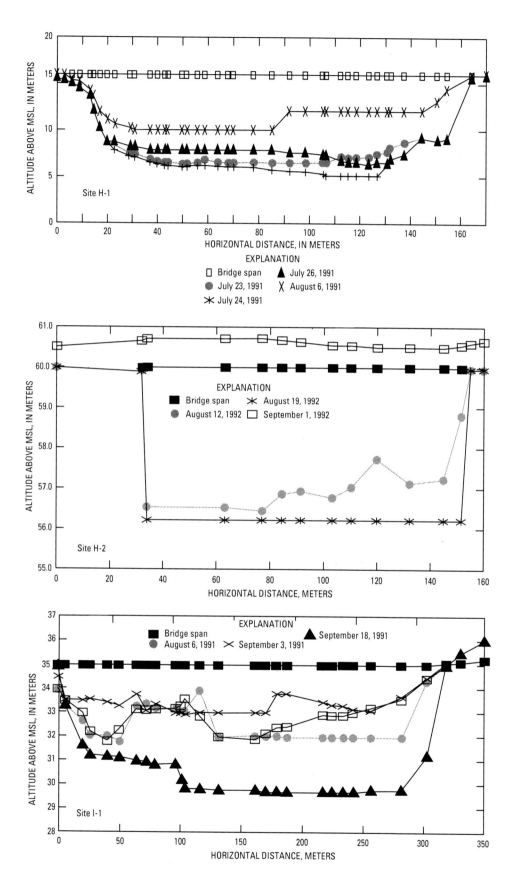

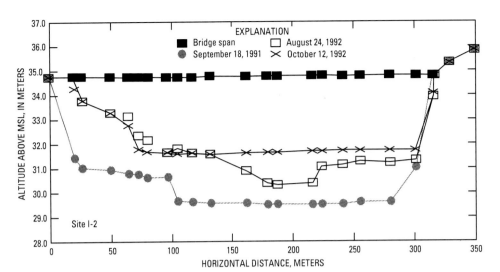

Figure 8.—Continued.

LATERAL EROSION AT CROSS-SECTION SITES

Lateral or bank erosion is not addressed by the cross-section data because the survey sites are located predominantly at bridge locations, which are stable, bedrock-side-sloped sites or are massively protected against lateral erosion. Generally, however, more bridges fail from lateral erosion than from any other cause, but this commonly occurs because of lateral erosion of a pier footing. At nearly all cross-section sites at bridges or bridge sites, aggradation precludes either scour or lateral erosion of pier footings. Lateral change seen in the successive cross sections is relatively insignificant and represents the vagaries of intrachannel flow movement. However, rapidly aggrading channels are generally associated with high rates of lateral erosion, and observations away from the cross-section sites confirm this association. The hyperconcentrated flows conveying most of the volcaniclastic sediment from the uplands clearly can erode unprotected levees very rapidly.

CONCLUSIONS

Fluvial redistribution of the extensive eruptive products of the 1991 eruption will continue to pose widespread risk to lowland areas surrounding Mount Pinatubo. The trends in channel filling, depositional site, and the types of flows are important factors in assessing the engineering alternatives for sediment-control measures. The most critical observations for immediate mitigation plans can be summarized in the following two categories:

1. Factors related to the alluvial-fan environment. In distributing the flood risk on active alluvial fans in semiarid areas, modern hydrologic practice assumes that risk extends across the entire surface of the fan. In all probability, this assumption extends to the tropical analogs of alluvial fans that surround Mount Pinatubo. The populated lowlands around the volcano are largely underlain by fluvially redistributed volcaniclastic sediment from eruptions prior to 1991.

Trends in channel cross sections show, after local 1991 syneruptive scour, large and continuing amounts of fill concentrated (and probably increasingly concentrated) in fanhead and upper fan areas. At least 25 m of fill is documented in a major river channel (Sacobia-Bamban River at Bamban). This trend will probably continue, and fanhead channels will fill and channel avulsion will occur. Repetitions of periods of aggradation and avulsion at the fanhead can distribute sediment across any part of an active fan. Thus, a levee system conveying flow from the mountain front is likely to be subject to burial or bypass.

These observations do not imply that existing fan channels with significant capacity will not continue to convey flows, over an intermediate time span of several years, to midfan locations where temporary levee systems can then route flow to sediment-storage sites. They do imply, however, that these channels are likely to be filled during the present aggradational cycle.

Any analysis of the situation must be viewed as a first approximation, with successive refinements to follow; thus, any mitigation plan will view effective sediment control as a "moving target." Control must be a running battle because the only permanent solution—huge sediment-retention structures tied to bedrock at the mountain front, with channels conveying the remaining flow downstream—is unlikely to be economically or topographically feasible at the scale required.

2. Factors related to flow type and deposit texture. The flows that convey volcaniclastic sediment to the populated lowlands around Mount Pinatubo are predominantly hyperconcentrated streamflows, and their deposits are dominated by sand-size sediment. The flows to date have been highly erosive, as observed in their ability to

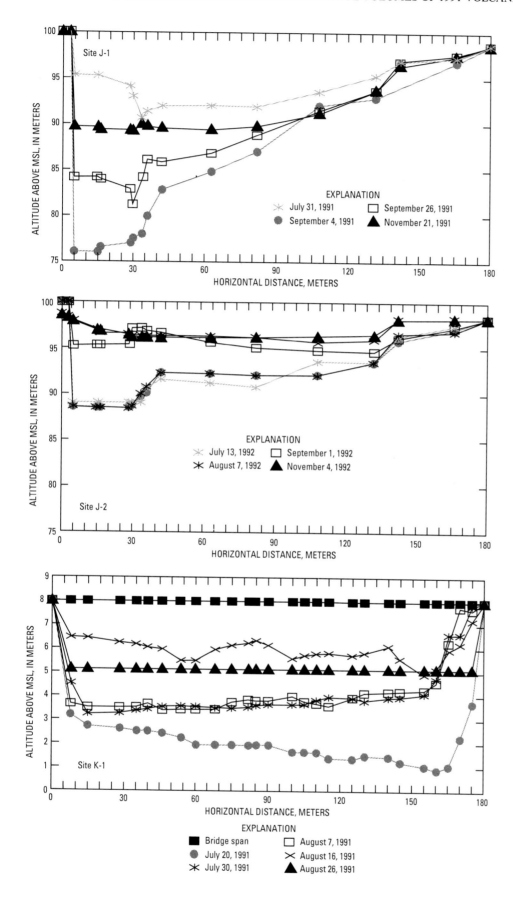

Figure 9. Cross-section changes in the Pasig-Potrero River system. Sites are as follows (locations on fig. 1): J–1, Pasig-Potrero River at Mancatian, 1991; J–2, same as site J–1, 1992; K–1, Pasig-Potrero River at Santa Barbara Bridge, 1991; K–2, same as site K–1, 1992.

Figure 9.—Continued.

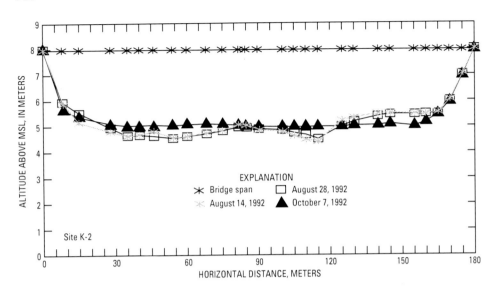

Site K-2

EXPLANATION
— Bridge span
— August 14, 1992
— August 28, 1992
— October 7, 1992

cut levees and by evidence seen in deposits of their ability to scour. The deposits of the flows are highly erodible because of the relative paucity of either cohesive fine material or material coarser than sand. Consequently, because the populated lowlands are widely underlain by the deposits of similar flows from previous eruptions, the availability of effectively erosion-resistant materials for levee construction is severely limited. This limitation is crucial to the economics of mitigation plans.

REFERENCES CITED

Arboleda, R.A., and Martinez, M.L., this volume, 1992 lahars in the Pasig-Potrero River system.

Cather, S.M., and Folk, R.L., 1991, Pre-diagenetic sedimentary fractionation of andesitic detritus in a semi-arid climate: an example from Eocene Datil Group, New Mexico, *in* Fisher, R.V., and Smith, G.A., eds., Sedimentation in volcanic settings: SEPM (Society for Sedimentary Geology) Special Publication 45, p. 211–226.

Ely, L.L., and Baker, V.R., 1990, Large floods and climate change in the southwestern United States: Hydraulics/hydrology of arid lands, Proceedings of the International Symposium, San Diego, Calif., July 30–August 2, 1990: Hydraulics Division, American Society of Civil Engineers, p. 651–656.

Folk, R.L., 1980, Petrology of sedimentary rocks: Austin, Tex., Hemphill Publishing Company, 182 p.

Janda, R.J., Major, J.J., Scott, K.M., Besana, G.M., Daligdig, J.A., and Daag, A.S., 1991, Lahars accompanying the mid-June 1991 eruptions of Mount Pinatubo, Tarlac and Pampanga Provinces, The Philippines [abs.]: Eos, Transactions, American Geophysical Union, v. 72, no. 44, p. 62.

Major, J.J., Janda, R.J., and Daag, A.S., this volume, Watershed disturbance and lahars on the east side of Mount Pinatubo during the mid-June 1991 eruptions.

Marcial, S.S., Melosantos, A.A., Hadley, K.C., LaHusen, R.G., and Marso, J.N., this volume, Instrumental lahar monitoring at Mount Pinatubo.

Paladio-Melosantos, M.L., Solidum, R.U., Scott, W.E., Quiambao, R.B., Umbal, J.V., Rodolfo, K.S., Tubianosa, B.S., Delos Reyes, P.J., and Ruelo, H.R., this volume, Tephra falls of the 1991 eruptions of Mount Pinatubo.

Pierson, T.C., Janda, R.J., Umbal, J.E., and Daag, A.S., 1992, Immediate and long-term hazards from lahars and excess sedimentation in rivers draining Mt. Pinatubo, Philippines: U.S. Geological Survey Water-Resources Investigations Report 92–4039, 35 p.

Punongbayan, R.S., Besana, G.M., Daligdig, J.A., Daag, A.S., and Rimando, R.E., 1991, Mudflow hazard map: Philippine Institute of Volcanology and Seismology, 1 sheet.

Rodolfo, K.S., 1991, Climatic, volcaniclastic, and geomorphic controls on the differential timing of lahars on the east and west sides of Mount Pinatubo during and after its June 1991 eruptions [abs.]: Eos, Transactions, American Geophysical Union, v. 72, no. 44, p. 62.

Scott, K.M., 1988, Origins, behavior, and sedimentology of lahars and lahar-runout flows in the Toutle-Cowlitz River system, Mount St. Helens, Washington: U.S. Geological Survey Professional Paper 1447–A, 74 p.

Scott, K.M., Pringle, P.T., and Vallance, J.W., 1992, Sedimentology, behavior, and hazards of debris flows at Mount Rainier, Washington: U.S. Geological Survey Open-File Report 90–385, 106 p.

Scott, W.E., Hoblitt, R.P., Torres, R.C., Self, S, Martinez, M.L., and Nillos, T., Jr., this volume, Pyroclastic flows of the June 15, 1991, climactic eruption of Mount Pinatubo.

Torres, R.C., Self, S., and Martinez, M.L., this volume, Secondary pyroclastic flows from the June 15, 1991, ignimbrite of Mount Pinatubo.

Umbal, J.V., and Rodolfo, K.S., this volume, The 1991 lahars of southwestern Mount Pinatubo and evolution of the lahar-dammed Mapanuepe Lake.

U.S. Air Force, 1991, Terminal forecast reference notebook: Detachment 5, 20th Weather Squadron, Clark Air Base, Republic of the Philippines, individually paginated sections.

Two Years of Lahars on the Western Flank of Mount Pinatubo: Initiation, Flow Processes, Deposits, and Attendant Geomorphic and Hydraulic Changes

By Kelvin S. Rodolfo,[1] Jesse V. Umbal,[1][2] Rosalito A. Alonso,[1] Cristina T. Remotigue,[3] Ma. Lynn Paladio-Melosantos,[2] Jerry H. G. Salvador,[4] Digna Evangelista,[4] and Yvonne Miller[5]

ABSTRACT

Two years after the eruption of Mount Pinatubo, lahars continued to occur along major drainages of the volcano. In Zambales Province on the western flank, pyroclastic debris is funneled as lahars through the Bucao and Santo Tomas River systems, and to a limited extent, along the Maloma River. Lahars at Pinatubo are initiated in a variety of ways. Primary, hot flows are triggered by rain with threshold values of 0.3 to 0.4 millimeters per minute sustained over periods >30 minutes. Others result from mass failures, induced by steam explosions, of partially water-saturated pyroclastic materials blocking upper tributary channels in the pyroclastic fans. Hot lahars display transitional flow types: precursory and waning stage hyperconcentrated streamflow, and debris-flow phases that coincide with peak discharges ranging from 200 to >1,000 cubic meters per second. Interactions of primary lahar channels with tributaries from outside Pinatubo, and variable discharge from contributing tributaries, result in multiple peak discharges and complex flow behavior. Primary lahars aggrade with sufficient rapidity (about 7 to 20 meters per year) to block tributaries that joined the lahar channels from watersheds outside the volcano, forming lahar-dammed lakes. Breakouts from these lakes generate cold, hyperconcentrated lahars.

All laharic deposits consist of poorly sorted pumiceous sand with very little coarse and fine material. The total sediment transported by Zambales drainages during 1991–92 was about 8.8×10^8 cubic meters, several orders of magnitude greater than observed at other volcanoes. The lahars have filled channels almost to capacity, so future flows are expected to avulse out of the present lahar field along new routes.

INTRODUCTION

The 1991 eruption of Mount Pinatubo deposited an estimated 5 to 6 km^3 of pyroclastic materials on the slopes of the volcano (W.E. Scott and others, this volume). Roughly two-thirds was emplaced on the western slopes that are situated entirely in Zambales Province, where it buried the preexisting topography except for a few isolated peaks and ridges (fig. 1). In 1991 and 1992 some of this metastably perched pyroclastic material was reworked as landslides and secondary pyroclastic flows that were induced in part by steam explosions on the still-hot pyroclastic fans. Much more debris, however, was remobilized into lahars by intense rains during the 1991 and 1992 monsoon seasons from June to October, and the major drainages were aggraded rapidly (Janda and others, 1991; Punongbayan and others, 1991; Rodolfo, 1991; Rodolfo and Umbal, 1992; Pierson and others, this volume; K.M. Scott and others, this volume). The new blanket of fine ash and the destruction of vegetative cover reduced soil infiltration capacity and evaporation and thus greatly enhanced sediment movement (Pierson and others, 1992; Major and others, this volume). Sediment yield during these two rainy season was high, about 1.3×10^9 m^3, several orders of magnitude greater than those previously observed at other volcanoes (Vessel and Davies, 1981; Swanson and others, 1982; Lehre and others, 1983).

Sediments eroded from the western pyroclastic fans were funneled principally into the Bucao and Santo Tomas River systems (Rodolfo, 1991; Pierson and others, 1992; Rodolfo and Umbal, 1992). Lahars of the 1991 and 1992 monsoon season rapidly filled these channels, and some flows overtopped channel banks and encroached adjacent

---

[1] Department of Geological Sciences M/C 186, University of Illinois at Chicago, 801 W. Taylor St., Chicago IL 60607–7059.

[2] Philippine Institute of Volcanology and Seismology.

[3] National Institute of Geological Sciences, University of the Philippines, Diliman, Quezon City, Philippines.

[4] Mines and Geosciences Bureau, Department of Environment and Natural Resources, North Avenue, Diliman, Quezon City, Philippines.

[5] Department of Mineralogy, University of Geneva, 13 Rue de Maraichers, 1211 Geneva 4, Switzerland.

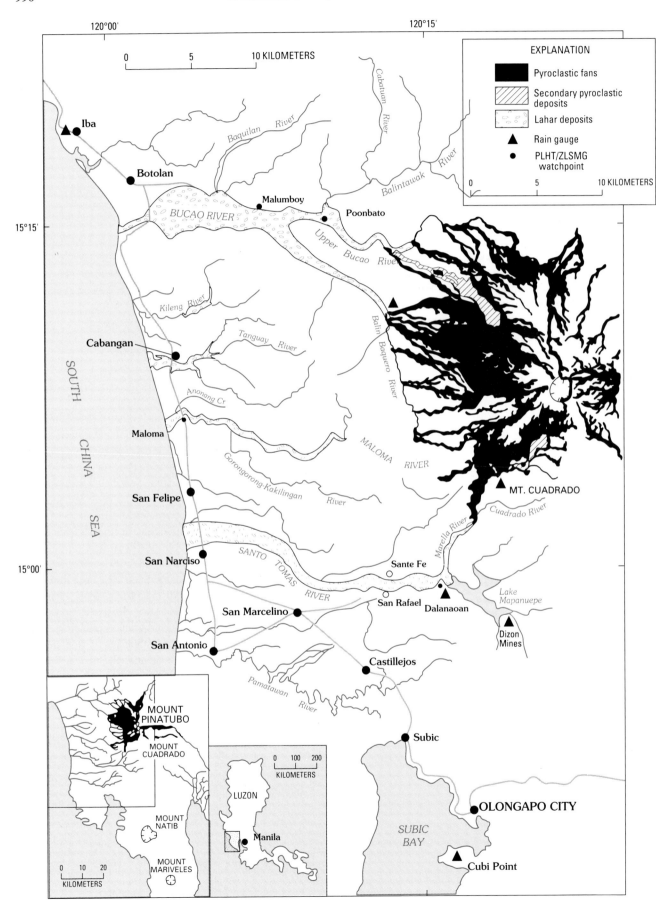

inhabited areas and agricultural fields. Considering the vast quantity of fresh pyroclastic debris, lahars and excess sedimentation may continue for 5 to 10 years or longer (Pierson and others, 1992; Rodolfo and Umbal, 1992).

Similar rapid filling in western drainages outside the volcano watersheds, mainly from remobilized tephra, caused channel shifting and flooding of adjacent areas, especially in 1991; however, erodible pyroclastic materials on the slopes of these non-Pinatubo drainages is limited, so these adverse effects are expected to be a diminishing and short-term problem. With adequate maintenance, these channels should return to their preeruption hydraulic conditions within a few years.

In this report we describe the processes of initiation and transport of lahars on the western sector of Mount Pinatubo and the response of the rivers, both within and outside the western Pinatubo watershed, to the excess sedimentation resulting from the 1991 Mount Pinatubo eruptions. Data for this study were gathered during the 1991 and 1992 monsoon seasons.

TERMINOLOGY

The *lahar* is one of the most destructive phenomena associated with composite volcanoes. In this report, following the lead of a recent gathering of volcaniclastic sedimentologists (Smith and Fritz, 1989, p. 375), *lahar* means "…a rapidly flowing mixture of rock debris and water (other than normal stream flow) from a volcano. A lahar is an event; but it can refer to one or more discrete processes [such as debris flow and hyperconcentrated streamflow], but does not refer to a deposit." Lahars exhibit complex flow behavior, changing in character from debris flow to hyperconcentrated streamflow, and vice versa (Pierson and Scott, 1985; Rodolfo and Arguden, 1991; Smith and Lowe, 1991). Debris flows are non-Newtonian fluids having water contents generally less than 25 percent by weight, and they move as fairly coherent masses in generally laminar fashion (Pierson, 1986; Pierson and Costa, 1987). Hyperconcentrated streamflows, as originally defined by Beverage and Culbertson (1964), are stream flows having sediment concentrations between 40 and 80 percent by weight. They move in fluid fashion (Pierson and Costa, 1987), even though Kang and Zhang (1980) have reported that they possess measurable yield strengths (generally less than 400 dyn/cm^2). It is not unusual for a single lahar to involve more than one debris flow phase, with transitional as well as precursor and waning-stage hyperconcentrated-streamflow phases. Given appropriate circumstances, either flow type can erode or deposit along any reach of its channel, and so the morphologic and sedimentologic effects of lahars can be very complex.

GEOLOGIC AND TOPOGRAPHIC SETTING

Mount Pinatubo, an andesite-dacite dome complex and stratovolcano, lies at the northern end of a 50-km-long Pliocene to Holocene calcalkaline volcanic belt related to eastward subduction of the South China Sea plate at the Manila Trench along western Luzon Island (de Boer and others, 1980; Wolfe and Self, 1983). The oldest Pinatubo rocks, isolated ridges of andesitic lava flows and pyroclastic deposits exposed on the middle and lower slopes of the volcano, are remnants of an "ancestral" Pleistocene stratovolcano (Delfin, 1983; Newhall and others, this volume). Silicic "modern" eruptions emplaced extensive ash- and pumice-flow deposits in all sectors of the volcano during at least six episodes since >35 ka (Newhall and others, this volume). On the western side of the volcano, pyroclastic flows from pre-1991 eruptions extended more than 17 km from the volcano, and major pyroclastic aprons, distributed within radial distances of 10 to 15 km, impinged upon ridges of the Zambales Ophiolite complex that had been uplifted in late Miocene time.

Prior to the eruption, the summit of the volcano stood at 1,745 m in altitude. The topography of the upper western flank of the volcano was characterized by steep to moderate slopes, deeply incised by a dense, radial network of drainages (table 1). Valleys at the upper slopes were deep, narrow, and U-shaped and were steep and steplike in longitudinal profiles (fig. 2). Most interfluve slopes ranged from 10° near the summit to about 2° downslope, but were steepest in the southern sector, where slopes were 28° within a kilometer from the summit and decreased to 8° near the base of the edifice. Along their lower reaches the valleys presented gently sloping, concave-upward profiles with broad, generally shallow banks flanked by multileveled river terraces, each approximately 5 to 10 m high. Most of the upper drainages were buried by pyroclastic flows during the climactic June 15, 1991, eruption, and subsequent lahars have aggraded the lower reaches almost to the level of the highest terraces.

PINATUBO DRAINAGES

In Zambales, the principal channels affected by lahars were those that tap the new pyroclastic-flow deposits. These are the Bucao River system that drains the west and northwest sectors of the volcano, the Santo Tomas River system, including the Marella River, that drains the southwest

---

← **Figure 1.** Distribution of pyroclastic fans and major drainages on the western flank of Mount Pinatubo.

Table 1. Watershed areas and estimated volumes of pyroclastic deposits of major drainages on the western flank of Mount Pinatubo and the contiguous Zambales Range.

Drainage system	Watershed area (km²)	Pyroclastic-flow deposits (km³)		
		(Besana and Daligdig, 1991, unpub. PHIVOLCS report)	(Scott and others, 1991)	(Torres and others, 1992, unpub. PHIVOLCS report)
1. Bucao	659	3.1	2.5	
Pinatubo Drainages				
Balin Baquero	64	.4		1.6
Villar	36			
Maraunot	42	.5		.7
Upper Bucao	128	.4		.5
Miscellaneous tributaries.		1.2		
Non-Pinatubo Drainages.				
Balintawak-Cabatuan	175			
Baquilan	60			
Malumboy	16			
Unnamed tributaries	138			
2. Santo Tomas	232			
Pinatubo Drainages				
Marella	31	1.3	1.0	.6
Santo Tomas	40			
Non-Pinatubo Drainages.				
Mapanuepe	88			
Cuadrado	22			
Negron	8			
Unnamed tributaries	43			
3. Maloma	153			<<.1
Maloma	111			
Gorongorong-Kakilingan.	42			
4. Tanguay	77			
5. Kileng	41			

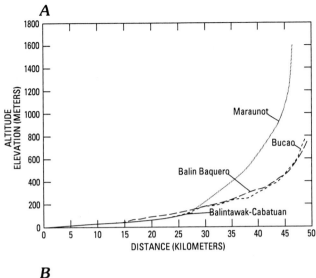

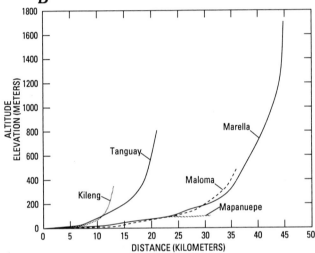

Figure 2. Longitudinal profiles of major drainages on the western flank of Mount Pinatubo. *A*, Major tributaries of the Bucao River system. *B*. Major tributaries of the Santo Tomas River system, the Maloma River, and non-Pinatubo drainages.

THE BUCAO RIVER SYSTEM

The Bucao River system, a broad watershed with an area of 659 km², drains the west and northwest sector of Mount Pinatubo and surrounding ultramafic terrain. Its Pinatubo portion, with an area of 270 km², is by far the largest catchment basin on the volcano. The southern part of the Bucao watershed was drained by nine unnamed Pinatubo tributaries of the upper Balin Baquero River, many of which were buried in pyroclastic flows during the eruption. Immediately downstream and to the north of these tributaries, the Balin Baquero River is joined by its largest Pinatubo tributary, the Maraunot River, which alone drained an area of 42 km² prior to the eruption. The upper Bucao River, which has a catchment area of 128 km² situated to the north, joins

sector, and, to a much lesser extent, the Maloma River. The Bucao and Santo Tomas Rivers are expected to continue experiencing destructive lahars for several years. In contrast, due to its limited access to new pyroclastic-flow deposits, the Maloma River did not have lahars as frequently as the other two but may become more active in the future if it beheads and captures part of the watershed area of the Marella River (Pierson and others, 1992; Rodolfo and Umbal, 1992).

the Balin Baquero River 26 km west-northwest of the former summit.

Several large rivers and many small creeks draining the surrounding ophiolitic Zambales Mountains join the main lahar avenues. The largest of these non-Pinatubo tributaries, the Balintawak-Cabatuan River, with a watershed area of 175 km^2 northwest of Pinatubo, joins the Bucao River 3 km above the Balin Baquero junction, in the vicinity of the former *Barangay* (village) Poonbato. Normal streamflow from this river dilutes the lahars from Pinatubo drainages and is a major agent in redistributing channelized sediments along the lower reaches of the Bucao River during the dry seasons.

The Bucao Valley is narrowest (only a kilometer wide) immediately below the Balin Baquero junction, at the site of the former flood-plain *sitio* (hamlet) of Malumboy (now buried in lahar deposits). It widens to about 3 km, including lahar and fluvial terraces that flank both sides of the valley, downstream to the Baquilan River junction 9 km from the South China Sea. Along the Malumboy-Baquilan reach, the Bucao valley is confined between the steep flanks of the ophiolitic Zambales Mountains. The Baquilan River, with a 60-km^2 watershed north of the Bucao River, is its second largest tributary from outside the Pinatubo drainage system. It joins the Bucao River about 5 km downstream of the Balin Baquero confluence, where it further dilutes the flows. Downstream from the Baquilan confluence, the alluvial plain of the Bucao River widens to the north and is shared by other coastal rivers, but the Bucao River braids over the 1- to 2-km-wide southern portion of the plain and reaches the coast south of the Municipality of Botolan. This town has an aggregate population of 35,752, including the inhabitants of 22 barangays that survived the pyroclastic flows of 1991 and the subsequent lahars. In all, eight other barangays of Botolan no longer exist. Poonbato, Malumboy, and parts of two other upland barangays were destroyed by lahars, the rest by pyroclastic flows.

THE MALOMA RIVER SYSTEM

The watershed of the Maloma River and its tributaries, 153 km^2 in area, drains almost exclusively ultramafic terrain. About 100 m wide along its middle reaches, the flood plain broadens downstream to about 500 m at its junction, 8 km from the coast, with the Gorongorong-Kakilingan River. This largest tributary has a 42-km^2 watershed south of the trunk stream. The Gorongorong-Kakilingan River and other streams in the vicinity that are incised into ultramafic rocks are characterized by narrow, V-shaped valleys and sharp channel bends imposed by the structural fabric of the bedrock, mainly northwest-trending normal faults and fractures related to the Iba fracture zone (de Boer and others, 1980). In 1991, these bends became temporary impoundment sites for remobilized tephra-fall deposits

eroded from its slopes. The broad, flat Maloma valley, old (pre-1991) pumiceous fluvial and debris-flow deposits exposed in its channel walls, and the continuing headward reestablishment of its tributaries up to the lower slopes of the volcano indicate that the river once drained larger portions of the southwest flank of the volcano, until a major, pre-1991 explosive eruption modified the terrain and diverted most flow away from the Maloma River.

THE SANTO TOMAS RIVER SYSTEM

The primary lahar avenue of the Santo Tomas River System, which has an aggregate watershed area of 232 km^2 above 50 m in altitude, is the Marella River, which drains a 31-km^2 area of the southwest slopes of Mount Pinatubo (fig. 1; table 1). About 19 km from the preeruption summit, the Marella River was joined by the Mapanuepe River, with a watershed of 88 km^2, forming the Santo Tomas River. Partially contained by a narrow bedrock constriction at the junction, the lower Marella River acted as a natural debris basin for the 1991 and 1992 lahars, which aggraded rapidly and blocked the Mapanuepe River, forming Mapanuepe Lake.

Below the junction, the Santo Tomas River flows along the northern margin of a broad, 235-km^2 alluvial plain that is populated by some 140,000 inhabitants in five municipalities, each with numerous barangays and sitios. The southern edge of the plain is marked by the irregular course of the Pamatawan River as it follows the northern bases of the ophiolitic southern Zambales Mountains. Along this margin, several small, isolated hills trending west from Castillejos have peak altitudes of 80 to 143 m; otherwise, local relief is under 5 m, due mainly to a deranged system of numerous shallow, small creeks that drain the greater, southern portion of the plain, which is underlain by the deposits of pre-1991 lahars and normal streamfloods. The preeruption channel morphology of the Santo Tomas River System and the interplay between Marella lahars and Mapanuepe River discharge are discussed in detail elsewhere in this volume (Umbal and Rodolfo).

NON-PINATUBO DRAINAGES

The coastal slopes of the Zambales Range between the Bucao and Maloma River systems are drained by the Tanguay and Kileng Rivers (table 1), neither of which extends eastward into the new pyroclastic-flow deposits on Pinatubo. Along these rivers, especially along the Tanguay River, mobilization of fresh tephra-fall deposits caused heavy siltation and flooding in 1991 and, to a much lesser extent, in 1992. These problems should lessen because the watersheds have lost most of their tephra-fall

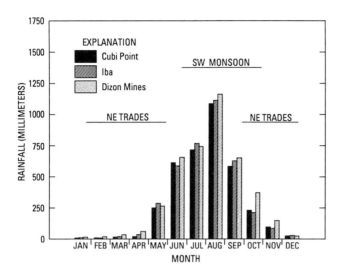

Figure 3. Monthly average rainfall in Zambales based on data from the Philippine Atmospheric, Geophysical, and Astronomical Services Administration at Iba (1956–85), the U.S. Navy Meteorological Station at Cubi Point (1955–87), and Dizon Mines (1977–90).

accumulations; however, the accumulated silt should continue to cause serious flooding.

CLIMATIC SETTING

The annual climate of central Luzon is characterized by distinct dry and wet seasons. On the western side of Mount Pinatubo, the dry season lasts from November to April, and the wet season occurs from May or June to October. About 70 to 80 percent of the 3,800 to 3,900 mm of annual rainfall in Zambales is delivered by the Southwest Monsoon, which normally begins in June and lasts until September (fig. 3; Huke, 1963; Pierson and others, 1992). During this season, average daily rainfall is about 24 mm but may exceed 150 mm. August is the wettest month, when daily precipitation averages 36 mm but may exceed 180 mm. However, the greatest 24-h rainfall recorded in the region was 442 mm, on May 19, 1966, delivered by a typhoon prior to the onset of the monsoon season (unpub. data, U.S. Navy). Much of the annual rain is brought by three or four typhoons, which may occur in any given month but typically occur from July to November and are most frequent in August. Orographic uplift causes more monsoonal rain to fall on the volcano than on the lowlands (fig. 3), and more rain falls on the western than on the eastern slopes (Rodolfo, 1991; Rodolfo and Umbal, 1992). Short, intense rainstorms are common, particularly during late afternoons when the diurnal wind shift brings warm, moist, marine air inland.

DISTRIBUTION AND VOLUME OF PYROCLASTIC MATERIALS ON THE WESTERN FLANK

About two-thirds of the estimated 5 to 6 km^3 of pyroclastic materials produced during the 1991 eruption was emplaced on the western slopes of Mount Pinatubo (table 1). The largest fan of pyroclastic-flow deposits, in the northwest sector, is drained by the Bucao and Maloma River systems, and a smaller fan in the southwest sector is drained by the Marella River (fig. 1). These deposits, in some areas more than 150 m thick, buried most preeruption drainages. At their farthest extents, the June 15, 1991, pyroclastic flows on the west flank of the volcano traveled more than 17 km from the summit.

Of the roughly 1 km^3 of tephra fall deposited on land, about 75 percent was emplaced along a west-southwest-trending axis by prevailing winds (Nichols and others, 1991; W.E. Scott and others, 1991; Paladio-Melosantos and others, this volume; Oswalt and others, this volume). In Zambales, these deposits ranged in thickness from 5 to 35 cm. After two rainy seasons, an estimated 75 to 80 percent of the tephra emplaced on moderate to steep mountain slopes had been eroded.

CHRONOLOGY AND FREQUENCY OF SYNERUPTIVE AND POSTERUPTIVE LAHARS

The first lahar on the western sector occurred along the Maraunot River on the afternoon of June 12, 1991, triggered by a local downpour on tephra-fall deposits that had been emplaced starting on May 16 and pyroclastic-flow deposits that had been emplaced on the morning of June 12. Intense rains, from the passage of Typhoon Yunya and from eruption-induced convective storm cells, triggered lahars of varying magnitude that affected all major rivers on the volcano during the initial paroxysmal explosions on the early morning of June 15 (Janda and others, 1991; Pierson and others, 1992; Major and others, this volume). Along the Marella-Santo Tomas River, stratigraphic sections showed three successive layers of debris-flow deposits of predominantly dense old lithic fragments, including boulders more than a meter in diameter, between the June 12–14 and the June 15 tephra falls. Each debris-flow unit was separated by a thin layer of silt and clay and graded laterally downstream into hyperconcentrated streamflow and normal streamflow deposits. In most sections, the uppermost debris-flow unit was capped by a centimeter-thick laminated layer of silt and fine sand. We interpret the June 15 lahar event along the Marella-Santo Tomas River as comprising three pulses or surges of debris flows that became diluted downstream to hyperconcentrated streamflows and normal streamflows, as has occurred on other volcanoes (Pierson and Scott, 1985; Rodolfo and Arguden, 1991; Smith and Lowe, 1991). The laminated silt and sand layer on top of the uppermost debris-flow unit may correspond to overbank deposits from a normal flood or hyperconcentrated-streamflow event that may have resulted from the sudden release of waters from a tributary, most likely the Mapanuepe River, that was temporarily dammed by lahars from the Marella River. These series of flow events destroyed the bridge between Barangays Santa Fe and San Rafael, east of San Marcelino municipality (fig. 1).

In the month following the climactic eruption, rainfall from orographic uplift of the prevailing easterly tradewinds was restricted mainly to the eastern side of the volcano. On the west side during this period, not as much rain fell, and the new pyroclastic fans still had no organized systems of valleys and tributaries. Consequently, runoff was not quickly delivered and concentrated, and only a few cold, hyperconcentrated flows occurred along the Bucao and Santo Tomas River systems (Rodolfo, 1991; Rodolfo and Umbal, 1992; Umbal and Rodolfo, this volume). Small hot lahars did not occur in Zambales until July 26.

On the afternoon of August 5, 1991, the first major hot lahars aggraded the Santo Tomas channel just below the Marella-Mapanuepe junction with more than 5 m of debris and overtopped channel banks at Nagpare, a sitio of Botolan municipality along the lower reaches of the Bucao River, leaving pumiceous debris-flow deposits more than a meter thick. As the southwest monsoon season intensified, lahars became more frequent in both rivers.

A total of 73 lahars were documented along the Santo Tomas River in 1991, not including nocturnal events recorded by the BUGZ flow sensor but not verified on the ground (Umbal and Rodolfo, this volume). These consisted of 50 hot lahars, from the Marella River, and 23 cold lahars, of which 18 resulted from lake breakouts from Mapanuepe Lake and 1 from a dammed tributary from Mount Cuadrado (fig. 4). We documented 15 lahars along the Bucao River until August 17, when floodwaters eroded sections of the National Road at the southern approach to Tanguay bridge, and lahars buried the barangay road at Nagpare, Botolan, cutting off access to our Malumboy and Poonbato stations for the duration of the 1991 rainy season. Seven of these early lahars were cold and eight were hot. If individual lahars along each of the numerous Bucao tributaries could have been tabulated, this number doubtless would have been significantly higher than that of the Marella River. We completed a permanent station at Malumboy, immediately downstream of the Balin Baquero confluence, prior to the onset of the 1992 monsoons, and continuously monitored the Bucao River lahars in 1992; however, it was not possible to document the details of lahar activity along the inaccessible upper Bucao and Balin Baquero Rivers.

In 1992, fewer lahars occurred on both the Santo Tomas and the Bucao Rivers; however, most of the flows

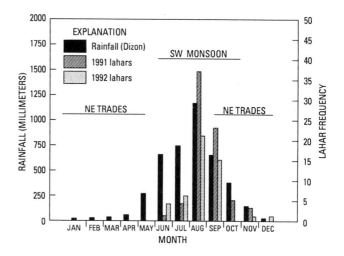

Figure 4. Monthly lahar frequencies in the Santo Tomas River in 1991 and 1992. Monthly rainfall is the long-term average at Dizon Mines (1977–1990), the rainfall monitoring station closest to the volcano.

were considerably larger and were mostly debris flows, whereas the 1991 events were mostly hyperconcentrated flows. At the Santo Tomas River we documented a total of 47 lahars, of which 39 were hot and 8 were cold hyperconcentrated flows resulting from lake breakouts. Two of these lake breakouts originated from a dammed tributary at Mount Cuadrado, whereas the rest originated from Mapanuepe Lake. Two of the events were induced by excavation of an artificial channel through the debris dam.

At the Bucao River in 1992, we observed 41 lahars originating either from the upper Bucao or Balin Baquero tributaries. Barangay Poonbato was buried in 17 m of sediment by several major debris flows, including two documented events on August 20 and September 10 that respectively left deposits 3 m and 4.5 m thick. For most of the season, lahar flow below the Balintawak junction was generally restricted to a principal channel 10 to 20 m south of the north bank that varied in width from 20 to 100 m before dissipating into a broad, braided surface a kilometer downstream of our Malumboy watchpoint. Cool Balintawak-Cabatuan discharge dominated the northern portion of the channel, and lahars from the upper Bucao and Balin Baquero Rivers generally were restricted to the southern portion; consequently, flow temperatures rarely exceeded 37°C where we could measure them at Malumboy on the north bank.

In the area of the Bucao–Balin Baquero confluence, and extending downstream to Malumboy, lahars from the latest pre-1991 eruption had formerly left a broad, vegetated alluvial fan (fig. 5A), into which was incised a central Bucao channel bounded along the south bank by two major and two lesser terraces, each 5 to 10 m high (fig. 6A). The channel was completely filled in by 1991 lahar deposits, which restored and smoothed the fan surface above the base

Figure 5. Eastward views of the Bucao–Balin Baquero lahar fan. *A*, June 1991, before major lahars. Note the lahar terraces and the vegetation cover. Mount Pinatubo is in the background. The Balin Baquero River discharges onto the fan in the right portion of the frame, to the right of the prominent, vegetated terrace in the distance. *B*, January 1992. *C*, September 1992, after the lahar season.

of the uppermost terrace (figs. 5*B*, 6*B*). Thus, the 1992 lahars from the Balin Baquero River generally fanned out widely over the smooth, restored fan surface (figs. 5*C*, 6*C*) into shallow (<1 m), hyperconcentrated sheet flows. These merged laterally northward with the Bucao-Balintawak flows along 3 km of fan margin and generally remained restricted to the southern portion of the principal flow channel. On September 5, however, a major debris flow from the Balin Baquero River raised the middle of the fan and the central portions of the lahar avenue at Malumboy by over-topping both margins of its principal channel. Most subse-

quent Balin Baquero lahars followed the new channel, where they remained segregated from Bucao flows before coalescing with them more than a kilometer downstream of Malumboy.

We have little data regarding the 1991 and 1992 lahars of the Maloma River. Eyewitness accounts suggest that most 1991 lahars were cold, hyperconcentrated flows resulting from remobilized tephra-fall deposits. Possible exceptions were the September 2 and 14 lahar events, which were described as steaming and were debris flows, judging from their deposits along the middle reach of the channel. In

Figure 5.—Continued.

Figure 6. Southward view from Malumboy watchpoint of the central portion of the Bucao–Balin Baquero lahar fan. *A*, June 1991, before major lahars. Note the 5-m-high terraces along the south bank. The light color of the mountains is due to fresh ash fall. *B*, Same view, January 1992. *C*, Same view, September 1992, after the lahar season.

contrast, most 1992 lahars were hot, and abundant charred wood and charcoal in their deposits indicate that the upper tributaries of the Maloma had eroded into the new pyroclastic fan.

LAHAR TRIGGERING MECHANISMS

Pinatubo lahars were triggered mainly by intense rainfall brought about by monsoon rains and passing typhoons. If the telemetered data from the BUGZ station of the U.S. Geological Survey-Philippine Institute of Volcanology and Seismology (USGS-PHIVOLCS) rain-gauge network are representative of rainfall over the Marella pyroclastic fan, intensity threshold values for initiation of Marella lahars were 0.3 to 0.4 mm/min, sustained over periods lasting longer than 30 min (Pinatubo Lahar Hazard Taskforce, 1991, memorandum dated August 28, 1991; Regalado and Tan, 1992). Tuñgol and Regalado (this volume) cite a similar threshold of 0.3 mm/min over periods lasting longer than 30 minutes for lahars along the Sacobia River on the east side of the volcano. These rainfall intensities are below normal maxima for monsoonal rains and typhoons in this

Figure 6.—Continued.

B

C

region. They are significantly lower than the threshold defined by Caine (1980) from a worldwide compilation of both volcanic and nonvolcanic debris flows, and by Rodolfo and Arguden (1991) for laharic debris flows of Mayon Volcano. This discrepancy is due in part to the fact that the Pinatubo lahars include not only debris flows but also hyperconcentrated flows, which are triggered by rainfall of lower intensities and durations. In any case, the abundant, erodible volcaniclastic sediment was easily and frequently mobilized by the typhoons and monsoonal rains of 1991 and 1992, in a remarkably rapid response to the drastic

geomorphologic changes wrought by the Pinatubo eruption on the watersheds and on the hydraulic characteristics of the river channels.

Lahars were also triggered by lake breakouts from non-Pinatubo tributaries blocked by pyroclastic-flow deposits and aggrading lahar channels, especially at the Marella-Mapanuepe junction of the Santo Tomas River system, as described in detail by Umbal and Rodolfo (this volume; also see Umbal and others, 1991; Rodolfo and Umbal, 1992; Pierson and others, 1992; K.M. Scott and others, this volume, Tuñgol and Regalado, this volume). Lake

Figure 7. Hummocky topography of the proximal portion of the June 19, 1992, lahar along the middle reach of the Marella channel, approximately 13 km away from the volcano. Most likely, a steam explosion caused a landslide, which was transformed downstream into a lahar. The size and frequency of hummocks decrease downstream. For scale, note the meterstick in the center of the photograph.

breakouts were initiated by erosion of the blockage by overtopping lake waters that generated normal streamfloods, sometimes with hyperconcentrated-streamflow stages of short duration. At times, rapid aggradation by heavy lahar flow from the upper Bucao blocked the Balintawak River, but the blockage could not last for more than a few days before it was overcome by large volumes of accumulated Balintawak waters that eventually broke out and remobilized the channel deposits. Occasionally, heavy lahar activity along the Balin Baquero also dammed the combined Bucao-Balintawak discharge for short periods and generated streamfloods and cold lahars upon breakout. The longest lasting episode commenced with damming on August 29. The dam was partially breached 2 days later, but limited blockage persisted until September 4.

In some instances, lahars may have been triggered by steam explosions that generated strong ground vibrations, which induced failure of partially water-saturated pyroclastic materials, particularly near the vicinity of pyroclastic-flow- and lahar-dammed tributaries. One such event was the June 19, 1992, Marella lahar, as evidenced by (1) the lack of triggering rainfall, (2) a strong steam explosion minutes prior to the lahar, and (3) flow-sensor seismic records showing a large initial amplitude followed by a gradual decline in high-frequency signals. This event, consisting of a precursory hyperconcentrated flow followed immediately by a debris flow with a lobate front several centimeters high consisting mainly of pebble- to cobble-sized fragments, left a proximal hummocky deposit (fig. 7). Alternate explanations are that a lake breakout may have triggered the explosion, or, conversely, that the explosion triggered the lake breakout.

LAHAR FLOW CHARACTERISTICS

The majority of 1991 lahars observed at our watchpoints on the Bucao and Santo Tomas Rivers were hyperconcentrated flows, although larger events had distinct debris-flow phases. During 1992, however, most lahars, particularly those along the Marella channel, were dominated by debris-flow phases.

The pattern of flow successions of most Marella hot lahars was from initial muddy streamflow, to hyperconcentrated flow, to debris flow, and finally to waning-stage hyperconcentrated flow. Transitions between phases were characterized by changes in flow behavior, flow temperature, and flow depth. Sediment concentrations in sampled lahars increased correlatively during the transitions from hyperconcentrated to debris flow. Corresponding to these transitions, hydrographs showed measured discharge increasing to single peaks, followed by gradual declines back to normal streamflow conditions (fig. 8A).

Variable contributions by non-Pinatubo tributaries resulted in multiple peak discharges (fig. 8B), and complex flow behavior. In some instances, multiple peaks may have resulted from (1) variable or nonsynchronous rainfall on the catchment basins of individual tributaries and (2) repeated damming at channel bends and constrictions by lahars, and breaching of these dams by subsequent lahar or lake-breakout flow, as frequently occurred at the bedrock constriction of the Marella-Mapanuepe junction. Minor pulses in discharge may also have resulted from internally driven flow instabilities (Davies and others, 1991).

Discharge and lahar-generation patterns were especially complicated in the Bucao River system. It is the only

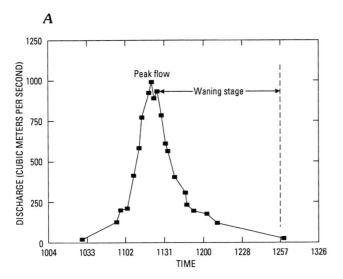

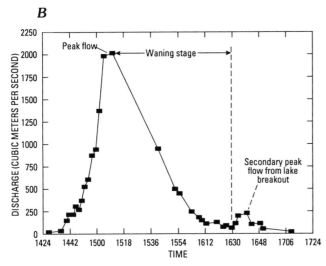

Figure 8. *A*, Hydrograph of the June 28, 1992, lahar along the Santo Tomas River showing the typical sharp increase in discharge followed by a gradual decline. Peak discharge in the hydrograph coincided with the transition from hyperconcentrated streamflow to debris flow. *B*, Hydrograph of the July 29, 1992, lahar at the Santo Tomas River. The second peak at the tapering end of the graph coincides with the arrival of discharge from the Mapanuepe River, which had been temporarily blocked by Marella lahars.

system on Pinatubo in which two major lahar avenues merge, the Balin Baquero River, which drains the largest pyroclastic fan on the volcano, and the upper Bucao river, which has a Pinatubo watershed area comparable in size to that of the Marella River. Furthermore, the Balintawak-Cabatuan watershed contributes enormous volumes of normal streamflow to the Bucao River from outside the volcano's drainage network, and its hydrographs are complicated because this large area does not receive its rainfall uniformly over time. Mixing of Balintawak-Cabatuan muddy streamflow with lahars from the upper Bucao reaches was minimal because discharge from both rivers

was strong. This lack of mixing resulted in two contrasting flow types, muddy, reddish, lateritic streamflow from the Balintawak-Cabatuan River and gray, mainly hyperconcentrated lahars from the upper Bucao River (or from the combined laharic discharge of the Bucao and Balin Baquero rivers), flowing in the same channel and remaining distinct and immiscible for several kilometers downstream from the confluences.

When strong rain was localized on the upper Bucao watershed, the resulting lahars could temporarily block the Balintawak-Cabatuan River for up to 2 days, and subsequent rupture of the blockage by accumulated water behind it caused cold lahars. Similarly, rainfall localized on the Balin Baquero watershed at times generated lahars that blocked the combined Bucao-Balintawak discharge and caused hot lahars to reach our Malumboy watchpoint, and subsequent rupture of blockage delivered peak discharges at times when no rain was falling on the volcano.

Along the Zambales lahar avenues, pulses or surges of muddy streamflows were precursors of incoming lahars (fig. 9). These surges had minimal sediment concentrations, generally less than 15 percent by volume, and flow temperatures a few degrees higher than ambient temperature, which normally was 24°C. The typical 2.5- to 3.5-m/s velocities of these surges enabled them to overtake and override the surface of normal streamflows, which had velocities below 1 m/s.

A gradual increase in flow depth and increased surge frequency signaled the onset of hyperconcentrated flow, which we designated as such primarily from its visual aspect and erosive behavior: increased turbulence, continuous migration across the width of the channel accompanied by strong lateral erosion at typical rates of 0.5 to 3 m/min, and the formation of standing waves as high as 2 to 3 m and 15 to 20 m long (fig. 10), in phase with antidune bedforms of lesser height. The wave dimensions are useful tools for indirectly measuring the depths and velocities of these dangerous flows. Empirically, we observed that flow depths were about 1.5 to 2 times the wave height. Flow velocities U, which theoretically equate to wavelength L according to the equation $U^2 = gL/2\pi$ that governs gravity waves (Kennedy, 1969), including atmospheric and deep-water ocean waves, agreed well with velocities measured by natural and artificially introduced floats.

Bulk densities of hyperconcentrated-streamflow phases of three Santo Tomas lahars systematically sampled in June and July 1992 ranged between 1.20 and 1.35 g/cm³. Sediment concentrations were low in comparison to the arbitrary 40 to 80 weight-percent values of Beverage and Culbertson (1964), ranging between 32 and 49 weight percent and averaging about 40 percent. Corresponding percentages by volume were 18 to 31 percent with an average of 24 percent. These low values may reflect the fact that virtually all of these materials were pumice fragments with dry densities of 0.9 to 1.4 g/cm³.

Figure 9. Precursory muddy streamflow surge, 25 cm high, overtaking and overriding the preceding flow. Photograph was taken along the Santo Tomas River near the PLHT-ZLSMG watchpoint at Dalanaoan, San Marcelino. The stage marker is 8 m high.

As a hyperconcentrated lahar progressed in intensity and discharge, its standing waves started to break, roaring continuously like ocean surf, and the flow became very turbulent and erosive. The flow no longer cut sideways so strongly; instead, it cut downward into the channel bed. The material eroded from the bed became part of the lahar, making it more dense and energetic, so it could flow faster. We regard this as a positive-feedback phenomenon: because the flow contained more and more solid volcanic material, it pressed more and more heavily on the channel bed, and the increased weight and speed of the lahar allowed it to incise the channel more deeply. Velocities at this stage ranged from 3 to 6 m/s, and flow samples showed increased flow densities, on the order of 1.10 to 1.25 g/cm^3. At the Marella-Santo Tomas confluence, measured flow temperatures also showed systematic increases during this stage, from 22°C to 33°C.

The end of the transition from the hyperconcentrated-flow phase to the debris-flow phase was manifested by a gradual dampening of turbulence to a smooth slurrylike consistency of the flow surface. Discharge increased, commonly accompanied by a rise in flow level; however, during several episodes along the Bucao River, the flow stage at the north bank outside the principal channel instead dropped by 1 or 2 dm, and the channel narrowed because its deepening increased its capacity so it could accommodate all of the flow. The onset of a debris-flow phase, with the exception of the June 19, 1992, lahar event, lacked bouldery fronts typical of debris flows elsewhere (Johnson, 1970; Okuda and others, 1980; Pierson, 1981; Johnson and Rodine, 1984; Pierson, 1986; Umbal, 1986; Ledda and Rodolfo, 1990; Ledda, 1991), probably because boulders are not abundant in the source area.

The reduction of turbulence in the debris flows was probably due to increased sediment content, which constituted as much as 85 percent by volume. Measured flow densities, however, were typically between 1.4 and 1.5 g/cm^3 and never greater than 1.70 g/cm^3—less than the typical 2.0 g/cm^3 observed in debris flows elsewhere (Costa, 1984). This is because the pumiceous solid phases floating in the sampled flow surfaces had typical dry densities of 1.1 to 1.5 g/cm^3, some values being even less than 1 g/cm^3. By weight, sediment contents of 10 debris flows systematically sampled in June and July 1992 were 57 to 91 percent and averaged 69 percent. Corresponding volume-percent values were 39 to 86 percent and averaged 57 percent.

Average debris-flow temperatures were slightly above 45°C, but some were as high as 70°C, and earlier flows in 1991 may have been even hotter. Large, fragile masses of lahar deposits scoured from the channels, some as big as 5 m across, were commonly observed bobbing in the debris flows. Many meter-sized pieces were transported intact for distances of more than 3 km despite flow velocities in excess of 6 m/s at slopes of less than 1°, a feat attesting to the laminarity of the flows. Debris-flow phases, corresponding to peak-flow discharges, lasted on the average between 15 to 30 minutes, but major events lasted up to an hour.

Increased turbulence and gradual drops in flow temperature and stage characterized waning stages of flow. Strong vertical and lateral scouring of the channel accompanied this hyperconcentrated waning flow (the "lahar runout" of Scott, 1988). Nickpoints and other irregularities on the channel floor served as loci for headward erosion, and they migrated upstream at rates of several decimeters per minute (fig. 11). Remnants of these waning-stage, headward-eroding nickpoints are near-vertical faces with more than a meter of drop in channel floor levels.

Figure 10. Standing waves in hyperconcentrated streamflow in the Santo Tomas River near the **PLHT-ZLSMG** watchpoint at Dalanaoan, San Marcelino, on August 14, 1991. *A*, Amplitudes of 1.5 m and wavelengths of 6 m before the waves. *B*, Waves breaking during transition to debris flow.

Figure 11. Headward-eroding nickpoints of a waning-stage hyperconcentrated streamflow along the lower reach of the Marella channel. The stage marks on the tree are each 0.1 m; the vertical drop between the upstream and downstream channel floor at this stage is about 1.5 m. The nickpoints migrated upstream at rates of 0.33 to 0.50 m/min for 45 minutes.

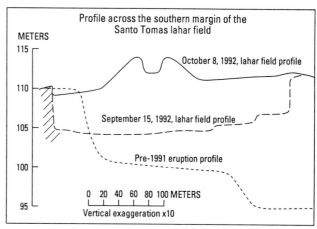

Figure 12. Cross-sectional profile across a debris-flow lobe.

MORPHOLOGY AND SEDIMENTOLOGY OF THE DEPOSITS

Debris-flow deposits at Pinatubo are commonly lobate. Each lobe has a convex-upward cross-sectional profile with a narrow U-shaped channel incision at the center of the lobe (fig. 12). Slump features are common along the inner banks of these channel incisions. Individual lobes vary in width from less than a meter to as much as 200 to 400 m, and thicknesses range from 5 to 8 m. Some of the small lobes exhibit steep, narrow levees along their margins, where most of the coarser particles congregated. Near the source area, where debris-flow deposits predominate, the surface topography is generally irregular and gently undulating. At the distal portion of the lahar field, where debris flows had been progressively diluted to hyperconcentrated flows and normal streamflows, surface topography is generally smooth and flat. Thin, indurated crusts, littered with pebble- to cobble-sized pumice fragments, armored and non-armored earthballs incorporated from eroded older deposits, and charred and uncharred plant debris, mantle the surfaces of most debris-flow deposits. Successive fillings and undercuttings of the active channel commonly formed several lahar-terrace levels.

Deposits of laharic debris flows and hyperconcentrated flows at Pinatubo are poorly sorted and sand rich with minor coarse and clay-sized components (fig. 13). The low densities of pumice fragments buoyed them toward the upper portion of debris flows, so inversely graded beds resulted. Hyperconcentrated-flow deposits generally lack flow structures except for crude stratification with lenses of coarse-grained, clast-supported pumice. Water-escape structures are common in debris-flow deposits, and in hyperconcentrated-flow beds as well.

In contrast, most beds resulting from lake breakouts are better sorted, fines-depleted and stratified streamflood deposits (fig. 14), commonly with crossbedding, although poorly stratified hyperconcentrated-flow beds are also present. Well-sorted, clast-supported accumulations of pumice fragments are common along portions of the channel where flows overtopped channel banks. Fine-grained deposits accumulated locally in stagnant areas such as abandoned channels and interlobe margins but are thickest within the boundaries of aggrading lahar channels and lahar-dammed lakes. Sediment components of the breakout streamflow deposits are mostly juvenile pumice fragments, with minor lithic fragments, mostly andesite, dacite, and gabbro, that were incorporated, together with old pumice fragments, from lateral and vertical channel scouring.

INITIAL DRAINAGE DEVELOPMENT ON THE PYROCLASTIC WATERSHEDS

A ubiquitous layer of fine tephra fall initially mantled the surfaces of the pyroclastic fans, where it reduced water infiltration and increased the surface runoff, which first flowed as sheetwash and later developed into dense, pinnate rill networks (fig. 15). These incipient, deranged, intricate rill networks followed local topographic lows and left temporary accumulations in local depressions such as steam explosion craters and interlobe contacts between the deposits of individual pyroclastic flows. The covering of flat areas on the pyroclastic fans by coalescing small debris-flow tongues and lobes emanating from these incipient rill networks indicates that rill-gully flows were already sediment-enriched even at this stage. The early period of drainage development on the pyroclastic fan was one of constant micropiracy between various rills and channels, and the

Figure 13. Representative outcrops of recent laharic deposits showing interlayered debris flow (unstratified) and hyperconcentrated streamflow (stratified) deposits. In *A*, handle of shovel is 0.5 m long. In *B*, pencil is 15 cm long; note the thinly laminated deposits of hyperconcentrated, waning-stage hot lahar sandwiched between two debris-flow deposits at the top of the section.

abandonment of the less efficient ones, generally those with low gradients. Small lakes and ponds had been formed where pyroclastic-flow deposits blocked tributaries from adjacent older terrain. The sudden release of water from these ponds facilitated channel incision on the pyroclastic fan and may have been a factor in determining which among the incipient channels would develop into main tributaries.

Two months after the eruption, the networks of rills incised into the surfaces of the pyroclastic fans were feeding semipermanent, youthful, headward-eroding stream channels. Most of these incipient dominant streams developed along the margin of the pyroclastic fan where it abuts pre-1991 eruption topography; the streams then converged into the remnant trunk channels of previous major tributaries below altitudes of about 300 m. By the end of the 1991 rainy season, deep, narrow tributaries, about 15 to 70 m wide and less than 20 m deep, constituted the primary conduits for lahars funneled toward the Bucao and Santo Tomas Rivers. For a majority of these tributaries, the depth of ver-

tical dissection was still well above those of pyroclastic-flow aggradation, although a few tributaries, particularly those within the Maraunot and upper Bucao watershed area, had eroded below the pre-1991 surface level by the end of the 1991 monsoon season (C.G. Newhall, oral commun., 1991). By the end of the 1992 rainy season, pyroclastic fans were densely dissected by deep, narrow channels (fig. 16), but some new upper tributary channels had widened enough to develop meanders. The channels that had formed along the contacts of pyroclastic fans with the older terrain adjacent to Pinatubo tended to erode laterally into the less resistant pyroclastic materials, thus shifting into the pyroclastic fields.

Fine ash winnowed from pyroclastic deposits by steam explosions replenished the primary ash deposits that had been eroded from the surface of the pyroclastic field. This mantle of fine materials minimized infiltration of rainwater into the pyroclastic materials and maintained relatively high rates of surface runoff on the pyroclastic fan.

Figure 14. Typical deposits from Mapanuepe Lake breakouts. Note meter scale in *A*. Shovel in *B* is 0.7 m long.

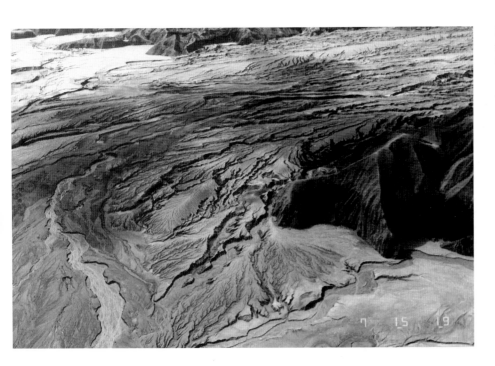

Figure 15. Incipient drainages on the Marella pyroclastic fan a month after the eruption.

Figure 16. Dense, narrow rills and gully systems incised into the loose pyroclastic materials of the Marella pyroclastic fan after two monsoon seasons.

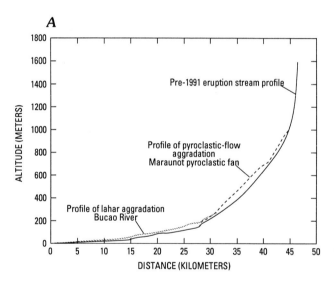

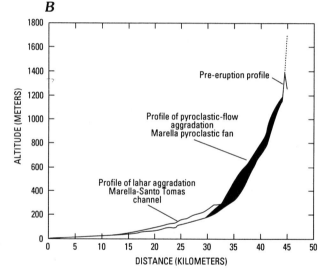

Figure 17. *A*, Longitudinal profile of the Maraunot-Bucao River showing the level of pyroclastic-flow and lahar aggradation along the channel stretch after two monsoon seasons following the June 1991 eruption. *B*, Longitudinal profile of the Marella-Santo Tomas River showing the level of pyroclastic-flow and lahar aggradation along the channel stretch two monsoon seasons after the June 1991 eruption. Note very limited changes in stream level at the lower reach of both rivers.

CHANNEL RESPONSES TO LAHARS

PINATUBO DRAINAGES

The overall trend for channels draining the western pyroclastic fans has been dominated by erosion along the upper reaches, aggradation and erosion by laterally migrating channels along the middle reaches, and continuous aggradation along the lower reaches (fig. 17). Deposition and erosion, however, can occur locally along any reach, in response to changes in channel constrictions and bends, and in response to local changes in slope, as at migrating nickpoints. The boundaries of these zones change with each major flow event and are also sporadically altered by mass failures on the pyroclastic fans. For the Marella channel and the upper main tributaries of the Bucao River, the middle, transitional zones lie between 180 and 300 m in altitude.

Figure 18 documents about 20 m/yr of aggradation at the constricted portion of the Marella-Mapanuepe River junction during the last two monsoon season. The overall trend has been of continuous channel fill temporarily interrupted by episodes of scour during lake breakouts, mainly

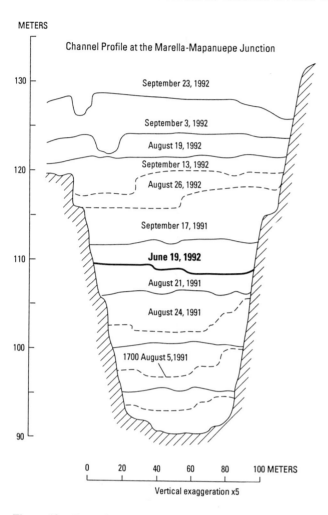

METERS

Channel Profile at the Marella-Mapanuepe Junction

September 23, 1992

September 3, 1992

August 19, 1992

September 13, 1992

August 26, 1992

September 17, 1991

June 19, 1992

August 21, 1991

August 24, 1991

1700 August 5,1991

0   20   40   60   80   100 METERS

Vertical exaggeration x5

Figure 18. Rate of aggradation at the PLHT/ZLSMG watchpoint at Dalanaoan, San Marcelino, after two monsoon seasons. The pre-1991 lahar profile is hachured, and the pre-1992 lahar profile (bold date) is shown by a heavy line. Floodwaters escaping from Mapanuepe Lake periodically scoured the channel (dashed profiles). The channel was finally abandoned after the lahar on September 23, 1992, and a new channel was incised to the west, near Mount Bagang.

from the lahar-dammed Mapanuepe Lake. Repeated plugging led to the abandonment of the channel and the formation of a new channel midway between the old channel and Mount Bagang, an isolated ultramafic block, during the early part of August 1992.

The large width of the Bucao River and the soft, quick condition of the lahar deposits prevented us from conducting a similar detailed, repeated channel cross-sectional survey after each flow event. Qualitatively, however, the overall trend has been one of continued aggradation on the depositional fan between the Bucao-Balintawak junction and Malumboy, except along the north bank, where scour by normal Balintawak discharge maintains the principal channel and minimizes aggradation. Since the eruption, the 1.5-km-wide reach at Malumboy has aggraded at average

rates of 7 m/yr. At the National Highway bridge across the Bucao River, 2.5 km from the coast, aggradation rates were 1.75 m/yr.

Downstream along the lower channel reaches, aggradation systematically decreased in response to changes in stream gradient and channel morphology. Where gradients are fairly uniform, thicknesses of lahar deposits do not vary much, changing dramatically only at major breaks in the stream profiles. Aggradation rates were relatively high upstream of channel constrictions, where repeated channel plugging caused flows to deposit. Downstream, flows spread in the wider channel, where their thickness, basal shear stress, and competence all decreased, so deposition began. Debris flows were deposited en masse, in contrast to more fluid hyperconcentrated flows and normal streamflows, which spread out and traveled farther downstream, carrying decreasing, finer-grained loads because the coarser sediments were selectively deposited en route. In consequence, the rivers experienced relatively rapid aggradation within the region of active debris-flow deposition and progressively less aggradation downstream beyond this point.

A combination of the foregoing factors may account for the disparate accumulation of volcanic debris along the Santo Tomas River below the Marella-Mapanuepe junction (fig. 17B). Gradually decreasing gradient and channel widening promoted debris-flow deposition at rates of more than 7 m/yr above 30 m in altitude. Channel infilling dropped dramatically below this altitude, tapering to thicknesses of 2 to 3 m/yr at 20 m in altitude. The low density and fine grain size of channel deposits along this lower segment results in continued sediment movement even during low flow conditions. The inflection in the stream profile at 50 m in altitude was smoothed by aggradation in 1991 and 1992, but the break in slope at 20 m altitude still persisted at the end of that period. Part of the channel condition along the lower reach may have resulted from human intervention, primarily dredging in order to maintain channel capacity.

At the Maloma River, a concrete dike extending 400 m upstream from the National Highway bridge was breached by floodwaters that reoccupied an old meander route that had been abandoned when the dike was constructed. By the end of the 1992 monsoon season, aggradation at the bridge was about 3 m above the preeruption stream level. A sharp curvature of the channel a kilometer upstream from the Maloma bridge may direct future flows to cut a route joining it to Anonang Creek. Aggradation of the Maloma River forced its Gorongorong-Kakilingan tributary to shift its channel course and flood an estimated 0.6-km² area south of the Maloma channel margin.

Only minor lahars, consisting principally of remobilized tephra fall, flowed along the Maloma River in 1991, because only about 5 km² of new pyroclastic deposits on the Pinatubo slopes are drained by the head of the northeasternmost Maloma tributary. However, continued headward

Figure 19. A view of Mapanuepe Lake, southwest sector of Mount Pinatubo. The lake was formed when the Mapanuepe River was blocked by rapidly aggrading Marella lahar deposits (light-colored deltaic margin in middle ground). The rising lake waters had already submerged three barangays at the time of this photograph.

erosion of this tributary into the pyroclastic field caused lahars to increase in frequency and extend farther downstream in 1992, where they delivered charcoal that local inhabitants began to harvest at the National Highway bridge in September.

LAHAR-DAMMED TRIBUTARIES

The rapid aggradation along the Bucao and Santo Tomas Rivers (about 20 m/yr at our Dalanaoan watchpoint and 7 m/yr at our Malumboy watchpoint) has blocked the mouths of non-Pinatubo tributaries and has caused ponds and lakes to form. The largest of these lahar-dammed lakes is Mapanuepe Lake in the southwest sector of the volcano, at the Marella-Mapanuepe confluence (fig.19). By the end of the 1992 monsoon season, the lake, with an average depth of 25 m, covered an area of about 8 km². The evolution of this lake is discussed in detail in a separate study by Umbal and Rodolfo (this volume). In 1992 another lahar-dammed lake of moderate size was situated upstream of the Marella-Mapanuepe junction at the confluence of the Mount Cuadrado tributary to the Marella River. Along the Bucao River system, the largest lahar-dammed lake was formed at the confluence of the Balintawak-Cabatuan River with the upper Bucao River. At its maximum extent, the area of this lake was about 4 km². With an average depth of only about 15 m, it was shallower than Mapanuepe Lake.

Waters accumulating behind these natural dams eventually rose to levels sufficient to overtop and erode the lahar blockages. Intervals between damming and initial incision of a blockage depended on the height and stability of the

blockage, and the rate of lake-level rise, which was dependent on the size of the river watershed and lake area. Lake breakouts from Mapanuepe Lake were able to remobilize more than a million cubic meters of the deposits of previous lahars in less than 2 days, after which discharge from the lake would slow down or stop altogether. Continued aggradation along the Bucao and Santo Tomas Rivers has prevented complete draining of the lakes, which have steadily increased in size, keeping pace with the rising channel-floor levels at the river junctions. These lakes along the margins of the lahar channels will survive as long as lahars of significant size continue to occur, possibly for more than 5 years.

NON-PINATUBO DRAINAGES

Even before the eruption, areas adjacent to the Maloma, Tanguay, and Kileng Rivers were lined with concrete and earthen flood-prevention structures to contain the perennial monsoonal flooding. After the eruption, flooding was exacerbated by rapid siltation of the channels from tephra-fall deposits eroded from the river watersheds. Siltation was so severe along the Tanguay and Kileng Rivers that the channel shifted toward the southern margin of the flood plain. Aggradation of the Tanguay River at the National Highway exceeded 3 m by the end of August 1991, reaching the level of the bridge. Overtopping floodwaters repeatedly eroded the southern approach of the bridge and swept away several hastily constructed bridges. This same section of the National Highway was again rendered impassable by floodwaters for much of the 1992 monsoon season.

Table 2. Area affected and amounts of sediment remobilized by lahars along major drainages on the western flank of Mount Pinatubo during the 1991 and 1992 monsoon seasons.

Drainage system		Area (km^2)	Volume (10^6 m^3)
Bucao	(1991)	53.3	250
	(1992)	55.7	229
Santo Tomas	(1991)	46.0	185
	(1992)	37.5	194
Maloma		10.1	20

ESTIMATED VOLUME OF REMOBILIZED SEDIMENTS

The amount of sediment transported during a single lahar is extremely variable, depending primarily on the intensity and duration of rainfall on the watershed area and upon the resulting relative dominance of debris flow and hyperconcentrated flow. Average discharge by individual lahars, reconstructed from field assessments, was on the order of 200 to 300 m^3/s, but values exceeding 1,000 m^3/s were not uncommon. Small- to moderate-sized lahars remobilized between 2×10^5 to 10^6 m^3 of sediment; larger events lasting up to 4 h may have transported about ten times this amount.

Approximately 8.8×10^8 m^3 of sediments were remobilized by lahars along the Bucao, Santo Tomas, and Maloma Rivers during the 1991 and 1992 monsoon seasons (table 2), not including the sediments that reached the ocean. The aggregate area affected by these lahars is about 112 km^2, much of which was prime agricultural land. The sediment contributions by the major rivers are compared to the extrapolated amount of materials remaining to be mobilized in table 3. The total annual sediment yield by Pinatubo drainages for 1991 and 1992 is within the range of values predicted by Pierson and others (1992). An estimated 2.5×10^7 m^3 of sediments, mainly tephra-fall deposits, were eroded from river watersheds outside the Pinatubo drainage basin and deposited downstream, where they affected an area of roughly 6.5 km^2 (table 4). This amount, after correction was made for the estimated proportion of older materials, comprised 75 to 80 percent of the 1991 tephra fall emplaced within the river catchment basins (table 4). Most of the remaining tephra-fall deposits are situated on stable, flat or gently sloping terrain. Those remaining on steep to moderate slopes are already sheltered by rejuvenated vegetation that intercepts rains and minimizes raindrop impact. Thus, potential erodible deposits in the watersheds outside Mount Pinatubo already have been largely depleted.

DISCUSSION AND CONCLUSIONS: PROSPECTS FOR THE FUTURE

For almost all lahar-affected channels, including the Bucao River, sediment yields were lower in 1992 than in 1991. The only exception is the Santo Tomas River, which had a slight increase in sediment yield in 1992. Whether this apparent overall decrease indicates that sediment delivery rates by lahars are starting to taper is still an open question (Janda and others, this volume). Several lines of evidence suggest that this may not be the case.

Fewer lahars occurred in 1992 than in 1991; however, most were debris flows that on average transported more sediments per event than lahars in 1991. This increase in sediment transport may have been due to rainfall delivered by fewer but heavier storms in 1992. Another factor may be the continued enhancement of drainage integration that delivered materials more efficiently to the main channels.

In 1991 and 1992, the Pacific-wide weather condition termed the El Nino–Southern Oscillation (ENSO) caused the southwest monsoon rains to arrive late and to be unusually mild. The 1992 monsoon season was relatively short, and its monthly rainfalls were less than both the 1991 and the normal long-term values for the region. This ENSO episode was continuing in 1993, so the outlook for the timing and severity of the succeeding monsoon seasons was uncertain.

The worst-case scenario for lahars still had not happened by 1992. The largest, most destructive lahars would be triggered if a major typhoon were to approach Pinatubo, as last happened June 15, 1991. If no exceptionally heavy and protracted rainfall were to occur for several more years, it is possible, but not likely, that the reestablishment of vegetation cover would prevent such a worst case from happening at all.

In 1991 and 1992, lahars were confined mainly to the pre-1991-eruption river channels. By the end of the 1992 lahar season, however, these channels largely had been filled, so future lahars were expected to seek paths through previously unaffected areas. Prediction of the future avenues on the coastal plain of the Bucao River is relatively straightforward, owing to its well-defined topographic containments. The low river gradient west of the Baquilan confluence would probably not permit extensive debris flows near the coast, so the southern part of the plain was expected to continue aggrading slowly with the deposits of hyperconcentrated lahars and normal streamflows. Without engineering intervention, this portion should eventually aggrade sufficiently to cause flows to encroach on the southern outskirts of Botolan. A strong dike, anchored to the bedrock ridge west of Baquilan and properly footed to withstand the lateral erosion characteristic of hyperconcentrated lahars, could be extended to the coast south of Botolan to save that municipality from slow burial. The success of such a structure would be enhanced by periodic dredging

Table 3. Predicted and actual quantities of sediment remobilized during the 1991 and 1992 lahar seasons.

[Pyroclastic-flow volumes are the low estimates of Pierson and others (1992). Lahar volumes are from the PHIVOLCS Lahar Monitoring Team (1992) and the Zambales Lahar Scientific Monitoring Group (1992). Volumes are in millions of cubic meters]

Watershed	Volume of pyroclastic-flow deposits	Erosion intensity factor	Pyroclastic-flow volume expected to be eroded	Preeruption volume that could be eroded*	Total volume of sediments expected to be eroded	Volume of 1991 lahar deposits	Percent of expected volume to be remobilized	Volume of 1992 lahar deposits	Percent of expected volume to be remobilized	Total volume of lahar deposits 1991–92	Percent of expected volume to be remobilized
Marella-Santo Tomas	1,000	0.5	500	50	550	185	34	194	35	379	69
Bucao-Balin Baquero	2,500	.5	1,250	125	1,375	250	18	229	17	479	35
O'Donnell/Bangut-Tarlac	300	.4	120	12	132	100	76	unknown		100	76
Sacobia-Bamban	600	.4	240	24	264	100	38	70	27	170	64
Abacan	100	.4	40	4	44	60	136	negligible		60	136**
Pasig-Potrero	300	.4	120	12	132	50	38	38	29	88	67
Porac-Gumain	30	.7	21	2	23	60	260	negligible		60	260**
Total	4,830	–	2,291	229	2,520	805	32	531	21	1,336	53

* Estimated to be 10 percent of the volume of pyroclastic flow that could be eroded.
** The volume of remobilized sediments exceeded forecasts.

Table 4. Amounts and distributions of 1991 tephra-fall deposits and amounts subsequently remobilized by 1991 and 1992 monsoon rains in the Kileng and Tanguay watersheds.

Drainage system	Watershed (km²)	Volume of air fall (10⁶ m³)	Area affected by siltation (km²)	Volume of sediments (10⁶ m³)	Remobilized air fall (percent)
1. Kineng	41	4.2	1.8	3.3	75
2. Tanguay	77	14.1	4.7	11.4	80

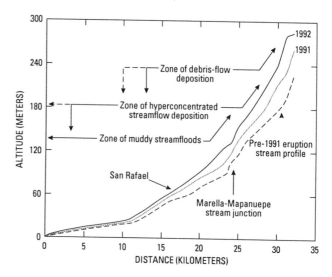

Figure 20. Longitudinal profiles of the Santo Tomas River below 250 m in altitude, showing the extent of aggradation by the various lahar types. The dashed arrows denote possible downstream extensions of each flow type in the future.

of the river from its mouth to the National Highway to facilitate disposal in the ocean of sediment that would otherwise contribute to aggradation in the channel and on the plain. A prominent submarine canyon funnels Bucao sediment discharge westward into deep water. A recent hydrographic survey by the Mines and Geosciences Bureau (A. Bravo and E. Domingo, oral commun., 1993) verified that the near-shore gradient of the canyon axis is about 110 m/km and that the canyon head extends shoreward to the vicinity of the southern part of the river mouth. Its presence should facilitate the disposal of dredging spoil.

The Maloma headwaters, which barely reached the new pyroclastic fan after the eruption, have already extended headward to a significant degree and may become a more active lahar conduit in the future, with serious consequences for the more than 1,000 inhabitants of Barangay Maloma. Increased lahar activity would enhance flooding and affect several sitios on the Maloma River flood plain. Detailed surveillance of the continuing headward extension of the Maloma headwaters, and the degree to which it might capture runoff from the Marella watershed, is also necessary in view of the consequences for Barangay Maloma and for the towns and villages of the Santo Tomas flood plain as well.

Along the Santo Tomas River, the record of aggradation since 1991 is clear (fig. 20). In 1992, the zone of maximum deposition, dominated by debris flows, had extended 3 km farther downstream than in 1991, to the vicinity of Barangay San Rafael. Greater aggradation rates in proximal compared to distal reaches cause downstream increases in gradient over time and possibly contribute to the downstream increases in debris-flow runout, a trend that may continue.

In late 1992, a dike along the southern bank of the Santo Tomas River was being planned to protect the municipalities of Castillejos, San Marcelino, and San Narciso with a dike, anchored to the east on the edge of the ophiolitic mountains west of Barangay San Rafael and extending to the coast (F. Soriquez, Department of Public Works and Highways, written commun., 1993). Such a dike would have provided an excellent test of the following hypothesis:

The eastern end of this structure will be most severely tested by laharic debris flows from the Marella River, and it must be built higher than the sum of (1) debris flows themselves (as thick as 9 m); (2) expectable run-up (when flows encounter a barrier, their inertia causes them to rise against gravity g by a vertical distance h that depends on flow velocity U according to Chow's 1959 equation $U^3 = 2gh$). A debris flow with the velocity of 7 m/s, which is typical for this area, will run up about 2.5 m vertically against a dike); and (3) continued aggradation of up to 15 m/yr. The structure must also be built to withstand impact forces of 10 to 1,000 t/m² (Pierson and others, 1992), the buoyant lift due to the high flow densities of lahars, and lateral and vertical erosion by waning-stage hyperconcentrated flows that can undermine dike footings to undetermined depths.

Unfortunately, the dike was not built to such specifications and was breached and overtopped at both its eastern and western ends in 1993.

Future lahar paths through the Santo Tomas River sector are much more difficult to predict than they are for the Bucao alluvial plain. The numerous pre-1991 eruption creeks incised into the southern portion of the Santo Tomas alluvial plain are all potential avenues for avulsing lahars, especially creeks from the southern margin of the 1992 lahar field in the vicinities of Barangay San Rafael, and due north of San Marcelino.

ACKNOWLEDGMENTS

We thank all our fellow members of the Pinatubo Lahar Hazards Taskforce and Zambales Lahar Scientific Monitoring Group, especially M.B. Angeles, R.A. Arboleda, M. Bumanlag, R. Giron, V. Jalique, N. Lacadin, R.U. Solidum, R. Tamayo, R. Tan, J. Tor, and R. Viran, and field drivers F. Bunuan, S. Fragante, E. Gutierrez, P. Mabasa, R. Manuel, B. Rizaldo, R. Unating, and, especially, R. Rivera, who also performed as videocamera operator, electronics and electrical specialist, general handyman, cook, and morale booster at Malumboy. Director J.B. Muyco and Chief Geologist Edwin Domingo of the Philippine Mines and Geosciences Bureau, Director B. Austria and Professor E. Tamesis of the Philippine National Institute of Geological Sciences, Director R.S. Punongbayan of PHIVOLCS, Mr. F. Salonga of the Philippine Shipyard and Engineering Corporation, Undersecretary M.H. Pablo of the Philippine Department of Public Works and Highways, Mayor R.J. Gordon and the City of Olongapo, Botolan Mayor S. Deloso, Congresswoman K.H. Gordon, Gen. P. Dumlao, Dr. E.M. Sacris of Dizon Mines, the U.S. Navy's Cubi Point weather station, and the U.S. Marine Air and Ground Task Force provided invaluable logistic support. PHIVOLCS and USGS personnel of the Pinatubo Volcano Observatory were generous in sharing information. We also would like to express our appreciation to J.J. Major, G.V. Middleton, C.G. Newhall, and R.S. Punongbayan for their critical comments and suggestion for improving the manuscript. The research was supported by National Science Foundation SGER Grant EAR–9116724 and EAR Grant 9205132.

REFERENCES CITED

Beverage, J.P., and Culbertson, J.K., 1964, Hyperconcentrations of suspended sediment: American Society of Civil Engineers, Journal of the Hydrology Division, v. 90, no. HY6, p. 117–126.

Caine, N., 1980, The rainfall-intensity duration control of shallow landslides and debris flows: Geografiska Annaler, v. 62, p. 23–27.

Chow, V.T., 1959, Open-channel hydraulics: New York, McGraw-Hill, 680 p.

Costa, J.E., 1984, Physical geomorphology of debris flows, in Costa, J.E., and Fleischer, P.J., eds., Developments and applications of geomorphology: Berlin, Springer-Verlag, p. 268–317.

Davies, T.R., Phillips, C.J., Pearce, A.J., and Zhang, X.B., 1991, New aspects of debris flow behaviour: Proceedings of the Japan-U.S. workshop on snow avalanche, landslide, debris flow prediction and control: Tsukuba, Japan, Organizing committee of the Japan-U.S. workshop on snow avalanche, landslide, debris flow prediction and control, p. 443–452.

de Boer, J., Odom, L.A., Ragland, P.C., Snider, F.G., and Tilford, N.R., 1980, The Bataan orogene: Eastward subduction, tectonic rotations, and volcanism in the western Pacific (Philippines): Tectonophysics, v. 67, no. 3–4, p. 251–282.

Delfin, F.G., Jr., 1983, Geology of the Mount Pinatubo geothermal prospect: Unpublished report, Philippine National Oil Company, 35 p.

Huke, R.E., 1963, Shadows on the land: An economic geography of the Philippines: Manila, Bookmark, 428 p.

Janda, R.J., Daag, A.S., Delos Reyes, P.J., Newhall, C.G., Pierson, T.C., Punongbayan, R.S., Rodolfo, K.S., Solidum, R.U., and Umbal, J.V., this volume, Assessment and response to lahar hazard around Mount Pinatubo.

Janda, R.J., Major, J.J., Scott, K.M., Besana, G.M., Daligdig, J.A., and Daag, A.S., 1991, Lahars accompanying the mid-June 1991 eruptions of Mount Pinatubo, Tarlac, Pampanga provinces, the Philippines [abs.]: Eos, Transactions, American Geophysical Union, v. 72, no. 44, p. 62–63.

Johnson, A.M., 1970, Physical processes in geology: San Francisco, Freeman, Cooper and Company, 577 p.

Johnson, A.M., and Rodine, J.R., 1984, Debris flow, in Brunsden, D., and Prior, D.B., eds., Slope instability: New York, John Wiley and Sons, p. 257–362.

Kang, Z., and Zhang, S., 1980, Preliminary analysis of the characteristics of debris flows, in Proceedings of the International Symposium on River Sedimentation: Beijing, Chinese Society for Hydraulic Engineering, p. 225–226.

Kennedy, J.F., 1969, The formation of sediment ripples, dunes and antidunes: Annual Review of Fluid Mechanics, v. 1, p.147–168.

Ledda, P.G.P., 1991, Flow properties and deposits of lahars at Mayon volcano, Philippines: Chicago, University of Illinois at Chicago, unpublished M.S. thesis, 108 p.

Ledda, G.P., and Rodolfo, K.S., 1990, Flow processes and deposits of lahars along Mabinit Channel, Philippines [abs.]: Eos, Transactions, American Geophysical Union, v. 71, no. 7, p. 513.

Lehre, A., Collins, B.D., and Dunne, T., 1983, Post-eruption sediment budget for the North Fork Toutle River drainage, June 1980–June 1981: Zeitschrift Geomorphologie Neues, v. 46, p. 143–163.

Major, J.J., Janda, R.J., and Daag, A.S., this volume, Watershed disturbance and lahars on the east side of Mount Pinatubo during the mid-June 1991 eruptions.

Newhall, C.G., Daag, A.S., Delfin, F.G., Jr., Hoblitt, R.P., McGeehin, J., Pallister, J.S., Regalado, M.T.M., Rubin, M., Tamayo, R.A., Jr., Tubianosa, B.S., and Umbal, J.V., this volume, Eruptive history of Mount Pinatubo.

Nichols, W.D., O'Hara, J.F., and Oswalt, J.S., 1991, Atmospheric impact of the Mount Pinatubo eruption (abs.): Eos, Transactions, American Geophysical Union, v. 72, no. 44, p. 65.

Okuda, S., Suwa, H., Okunishi, K., Yokoyama, K., Nakano, M., 1980, Observations on the motion of a debris flow and its geomorphological effects: Zeitschrift für Geomorphologie Supplement Band, v. 35, p. 142–163.

Oswalt, J.S., Nichols, W., and O'Hara, J.F., this volume, Meteorological observations of the 1991 Mount Pinatubo eruption.

Paladio-Melosantos, M.L., Solidum, R.U., Scott, W.E., Quiambao, R.B., Umbal, J.V., Rodolfo, K.S., Tubianosa, B.S., Delos Reyes, P.J., and Ruelo, H.R., this volume, Tephra falls of the 1991 eruptions of Mount Pinatubo.

Pierson, T.C., 1981, Dominant particle support mechanisms in debris flows at Mount Thomas, New Zealand, and implications for flow mobility: Sedimentology, v. 28, p. 49–60.

———1986, Flow behavior of channelized debris flows, Mount St. Helens, Washington, in Abrahams, A.D., ed., Hillslope processes: Boston, Allen and Unwin, p. 269–296.

Pierson, T.C., and Costa, J.E., 1987, A rheologic classification of subaerial sediment-water flows, in Costa, J.E., and Wieczorek, G.F., eds., Debris flows/avalanches: Processes, recognition and mitigation: Geological Society of America Reviews in Engineering Geology, v. 7, p. 93–105.

Pierson, T.C., Daag, A.S., Delos Reyes, P.J., Regalado, M.T.M., Solidum, R.U., and Tubianosa, B.S., this volume, Flow and deposition of posteruption hot lahars on the east side of Mount Pinatubo, July–October 1991.

Pierson, T.C., and Scott, K.M., 1985, Downstream dilution of a lahar: Transition from debris flow to hyperconcentrated streamflow: Water Resources Research, v. 21, p. 1511–1524.

Pierson, T.C., Janda, R.J., Umbal, J.V., and Daag, A.S., 1992, Immediate and long-term hazards from lahars and excess sedimentation in rivers draining Mount Pinatubo, Philippines: U.S. Geological Survey Water-Resources Investigations Report 92–4039, 41 p.

Punongbayan, R.S., de la Cruz, E., Gabinete, E., Scott, K.M., Janda, R.J., and Pierson, T.C., 1991, Initial channel response to large 1991 volumes of pyroclastic flow and tephra fall deposits at Mount Pinatubo, Philippines [abs.]: Eos, Transactions, American Geophysical Union, v. 72, no. 44, p. 63.

Regalado, M.T.M., and Tan, R., 1992, Rain generation at Marella River, Mount Pinatubo: unpublished report, University of the Philippines, 15 p.

Rodolfo, K.S., 1991, Climatic, volcaniclastic, and geomorphic controls on the differential timing of lahars on the east and west sides of Mount Pinatubo during and after its June 1991 eruptions [abs.]: Eos, Transactions, American Geophysical Union, v. 72, no. 44, p. 62.

Rodolfo, K.S., and Arguden, A.T., 1991, Rain-lahar generation and sediment-delivery systems at Mayon Volcano, Philippines, in Fisher, R.V., and Smith, G.A., eds., Sedimentation in volcanic settings: Society of Economic Paleontologists and Mineralogists Special Publication 45, p. 71–88.

Rodolfo, K.S., and Umbal, J.V., 1992, Catastrophic lahars on the western flanks of Mount Pinatubo, Philippines: Proceedings of the Workshop on the Effects of Global Climate Change on Hydrology and Water Resources at the Catchment Scale, February 3-6, 1992, Tsukuba Science City, Japan, Japan-U.S. Committee on Hydrology, Water Resources, and Global Change (JUCHWR), p. 493–510.

Scott, K.M., 1988, Origins, behavior, and sedimentology of lahars and lahar-runout flows in the Toutle-Cowlitz River system: United States Geological Survey Professional Paper 1447–A, 74 p.

Scott, K.M., Janda, R.J., de la Cruz, E., Gabinete, E., Eto, I., Isada, M., Sexon, M., and Hadley, K.C., this volume, Channel and sedimentation responses to large volumes of 1991 volcanic deposits on the east flank of Mount Pinatubo.

Scott, W.E., Hoblitt, R.P., Daligdig, J.A., Besana, G.M., and Tubianosa, B.S., 1991, 15 June 1991 pyroclastic deposits at Mount Pinatubo, Philippines [abs.]: Eos, Transactions, American Geophysical Union, v. 72, no. 44, p. 61.

Scott, W.E., Hoblitt, R.P., Torres, R.C., Self, S., Martinez, M.L., and Nillos, T., Jr., this volume, Pyroclastic flows of the June 15, 1991, climactic eruption of Mount Pinatubo.

Smith, G.A., and Fritz, W.J., 1989, Penrose Conference Report: Volcanic influence to terrestrial sedimentation: Geology, v. 66, p. 433–434.

Smith, G.A., and Lowe, D.R., 1991, Lahars: Volcano-hydrologic events and deposition in the debris flow-hyperconcentrated flow continuum, in Fisher, R.V., and Smith, G.A., eds., Sedimentation in volcanic settings: Society of Economic Paleontologists and Mineralogists Special Publication 45, p. 59–70.

Swanson, F.J., Collins, B., Dunne, T., and Wiecherski, B.P., 1982, Erosion of tephra from hillslopes near Mount St. Helens and other volcanoes, in Proceedings of the Symposium on Erosion Control in Volcanic Areas: Tsukuba, Public Works Research Institute, Technical Memorandum 1908, p. 183–221.

Tuñgol, N.M., and Regalado, M.T.M., this volume, Rainfall, acoustic flow monitor records, and observed lahars of the Sacobia River in 1992.

Umbal, J.V., 1986, Mayon lahars during and after the 1984 eruption: Philippine Journal of Volcanology, v. 3, no. 2, p. 38–59.

Umbal, J.V., and Rodolfo, K.S., this volume, The 1991 lahars of southwestern Mount Pinatubo and evolution of the lahar-dammed Mapanuepe Lake.

Umbal, J.V., Rodolfo, K.S., Alonso, R.A., Paladio, M.L., Tamayo, R., Angeles, M.B., Tan, R., and Jalique, V., 1991, Lahars remobilized by breaching of a lahar-dammed non-volcanic tributary, Mount Pinatubo, Philippines [abs.]: Eos, Transactions, American Geophysical Union, v. 72, no. 44, p. 63.

Vessel, R.K., and Davies, D.K., 1981, Non-marine sedimentation in an active forearc basin, in Ethridge, F.G., and Flores, R.M., eds., Recent and ancient nonmarine depositional environments: Models for exploration: Society of Economic Paleontologists and Mineralogists Special Publication 31, p. 31–45.

Wolfe, J.A., and Self, S., 1983, Subduction, arc volcanism, and steam mineralization: The Manila trench sector, Philippines: Philippine Journal of Volcanology, v. 1, no. 1, p. 11–40.

Instrumental Lahar Monitoring at Mount Pinatubo

By Sergio Marcial,[1] Arnaldo A. Melosantos,[1] Kevin C. Hadley,[2]
Richard G. LaHusen,[2] and Jeffrey N. Marso[2]

ABSTRACT

Rain gauges and experimental acoustic flow monitors (AFM's) were installed near drainages of Mount Pinatubo after the eruption of June 15, 1991. The AFM's use exploration-class geophones to detect ground vibration caused by passing flows, principally debris flows and hyperconcentrated streamflows. Data from the rain gauges and AFM's are telemetered to the Pinatubo Volcano Observatory, where they are used by the Philippine Institute of Volcanology and Seismology for realtime detection and warning of lahars. Although some lahars were missed during the period of initial installation in 1991 and during periods of sensor or telemetry malfunction, every lahar that passed an operating sensor was detected, and civil defense was warned. Civil defense, in turn, used these alerts to query manned watchpoints and to raise public lahar-alert levels.

INTRODUCTION

The explosive eruption of Mount Pinatubo on June 15, 1991, deposited 5–6 km^3 of loose, unconsolidated, pyroclastic-flow and tephra-fall debris on the slopes of the volcano (W.E. Scott and others, this volume). This debris, when eroded by rainwater, flows downslope as lahars, which erode river banks on the upper slopes and bury residential and agricultural lands in the adjoining lowlands. By the end of the 1992 rainy season, lahar deposits had covered an area of at least 360 km^2 (Mercado and others, this volume) and affected densely populated municipalities around Mount Pinatubo.

Because many people are reluctant to leave threatened, lowland communities until lahars are imminent, the Philippine Institute of Volcanology and Seismology (PHIVOLCS) and others set up warning systems to provide as much as several hours of time for them to move to safer ground. Those warning systems have included rain gauges, conventional 1-Hz seismometers that are also used for monitoring volcanic and tectonic seismicity, conventional 10-to 300-Hz exploration seismometers and associated acoustic flow monitor (AFM) circuitry, trip wires, and manned watchpoints.

This paper describes a network of high-frequency AFM's and automated rain gauges that were brought to Pinatubo by the Volcano Crisis Assistance Team (VCAT) of the U.S. Geological Survey and that are now operated and maintained by PHIVOLCS. Data are received at PHIVOLCS' Pinatubo Volcano Observatory, which passes lahar warnings to the Watch Point Center (WPC) of the Regional Disaster Coordinating Council (RDCC), at Camp Olivas in San Fernando, Pampanga.

PREVIOUS INSTRUMENTAL MONITORING OF LAHARS

In the Philippines, previous instrumental monitoring of lahars relied on conventional 1-Hz seismometers, also used for volcano monitoring (Bautista and others, 1986). Relatively high-frequency tremor at Mayon Volcano was correlated visually with lahars in progress. Signals were strongest when lahars passed along gullies near the seismometers, and peak signals corresponded to peak discharge. Successful warnings were given to concerned authorities and to the public.

However, because 1-Hz seismometers record signals between 1 and 10 Hz that can travel through a whole volcano, it was difficult to pinpoint the exact channel along which a lahar was flowing. It was also difficult, at times, to distinguish lahar signals from other sources of tremor (for example, pyroclastic flows and jetting of steam from summit fumaroles).

Trip wires, though not previously used in the Philippines, have been used elsewhere with mixed success. In pairs or series, they can detect the approximate magnitude of a flow (according to the level of the wire that has been broken) and track flows past upstream and then downstream sets of wires. Such trip wires are also subject to vandalism or accidental breakage, cannot provide information about the flow after they have been broken, and must be restrung each time they are broken.

[1]Philippine Institute of Volcanology and Seismology.
[2]U.S. Geological Survey.

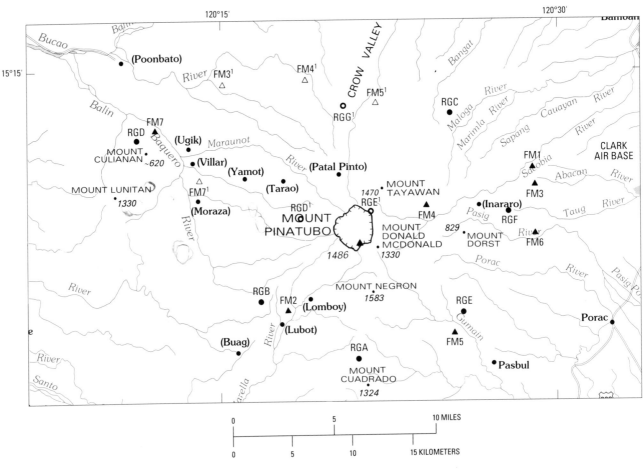

Figure 1. Location of rain gauges and flow monitors in 1991–92 (solid symbols) and 1993 (open symbols). For 1991–92, RG (rain gauge) A, Mount Cuadrado; RG B, Marella, at the BUGZ seismic station; RG C, at PI2 seismic station; RG D, Mount Culianan; RG E, Gumain; RG F, Sacobia. FM (flow monitor) 1, Lower Sacobia; FM 2, Marella-Santo Tomas; FM 3, Abacan Gap; FM 4, Upper Sacobia; FM 5, Gumain; FM 6, Pasig-Potrero; FM 7, Balin Baquero. New stations for 1993 were RG D[1], Kamanggi; RG E[1], Summit rim, east-southeast of CRAZ seismic station; RG G[1], O'Donnell-Upper Bucao; FM 3[1], Lopoton Creek; FM 4[1], Upper Bucao; FM 5[1], O'Donnell; and FM 7[1], Maraunot.

1991 INSTRUMENTAL MONITORING NETWORK

Three weeks after the paroxysmal June 15, 1991, eruption, the Volcano Crisis Assistance Team (VCAT) of the U.S. Geological Survey (USGS) brought a newly developed set of AFM's capable of recording acoustic vibrations in nearby rocks and soil as lahars flow along channels (Hadley and LaHusen, 1991; in press). The flow sensors were supplemented by automated rain gauges, so early warnings could be made of possible flows and rainfall and lahars could be correlated (Bautista and others, 1991; Tuñgol and Regalado, this volume). All components were designed to be relatively inexpensive, quick to install, and durable. Each field station is powered by a 12-V battery charged by a solar panel. Data from both the AFM's and rain gauges are telemetered to the Pinatubo Volcano Observatory (PVO, on Clark Air Base, and also in PHIVOLCS' Main Office, Quezon City), where they are received by a computer and made available in realtime in both graphical and numerical formats.

The combined network of flow sensors and rain gauges serves three roles: to provide immediate warnings of lahar hazards, to collect data for studies of the hydrologic aftermath of the eruption, and to test technical aspects of the system itself.

The rain gauges were installed high in the lahar source areas, free of nearby obstructions. In 1991, six rain gauges and seven flow monitors were deployed at strategic places around the volcano (fig. 1; table 1). Sites for rain gauges were selected on relatively high ground in the lahar source regions, with helicopter access and line-of-sight telemetry to PVO or to a repeater site.

Flow monitor sites were selected along the middle reaches of major lahar-bearing rivers, below most of the lahar source region yet far enough above population sites as

Table 1. Rain gauge and acoustic flow sensor stations at Mount Pinatubo.

Rain gauge		Acoustic flow monitor	
Channel	Location	Channel[1]	Location
1991–92			
A	Mount Cuadrado (south)	010,011,012	Lower Sacobia (east-northeast)
B	BUGZ (southwest	020,021,022	Marella (southwest)
C	PI2 (northeast)	030,031,032	Abacan Gap (east-northeast)
D	Mount Culianan (northwest)	040,041,042	Upper Sacobia (east-northeast)
E	Gumain (southeast)	050,051,052	Gumain (southeast)
F	Middle Sacobia (east-northeast)	060,061,062	Pasig-Potrero (east-southeast)
		070,071,072	Balin Baquero (northwest)
1993			
A	Mount Cuadrado (south)	010,011,012	Lower Sacobia (east-northeast)
B	BUGZ (southwest)	020,021,022	Marella (southwest)
C	PI2 (northeast)	030,031,032	Lopoton Creek (northwest)
D	Kamanggi (northwest)	040,041,042	Upper Bucao (north-northwest)
E	Summit Rim (north-northeast)	050,051,052	O'Donnell (north-northeast)
F	Middle Sacobia (east-northeast)	060,061,062	Pasig-Potrero (east-southeast)
G	O'Donnell-Upper Bucao (north)	070,071,072	Maraunot (northwest)

[1]Each acoustic flow sensor (1–7, middle digit here) can be queried on three channels, for example, 010, 011, and 012, representing the full band, low band, and high band of data.

to maximize lead time in warnings. Sites are on canyon walls or hilltops, safely above flows but close enough to record high-frequency ground vibration (typically, <100 m and, sometimes, <20 m from the stream). Telemetry is generally repeated through rain gauge sites. Flow sensors were installed along the Marella River, on the Balin Baquero River just downstream from its confluence with the Maraunot River, in the upper and lower reaches of the Sacobia River, along the Abacan River just downstream from its divergence from the Sacobia River, and along the Pasig-Potrero and Gumain Rivers (fig. 1). In 1992, sensors were in the same places but, owing to difficulties of access and maintenance, some malfunctioned, especially late in the rainy season. Details are given under "Installation and Maintenance of Field Sites." In 1993, several sensors were moved to higher and safer locations (table 1).

MONITORING INSTRUMENTS

RAIN GAUGES

The rain gauge counts tips of a magnetic-switch tipping bucket (fig. 2A); field calibration showed that each tip of gauges that were in the field in 1991–92 represented rainfall of 1 ± 0.1 mm. Because ash from mild eruptions and secondary explosions affects operation of the buckets and leads to erroneous readings, one of us (R.G. LaHusen) designed an Ash-Resistant Rain Gauge (ARRG) in which rain falls through a 12-cm-diameter orifice and collects in a vertical PVC pipe, closed at the bottom. A spillover outlet about halfway up the column (fig. 2A) leads to the tipping

bucket mechanism in a separate, ash-protected enclosure. The PVC pipe is filled with water to spilling level, and subsequent rain falls into the water column to cause an equivalent overflow to the tipping bucket. Any ash entering the ARRG settles to the bottom of the PVC tube and safely away from the tipping mechanism. Because the volume of ash is negligible in comparison to the volume of rainwater, it does not seriously inflate the rain measurements.

The rain gauges use a microprocessor-controlled digital telemetry platform. Each time the bucket tips, the magnetic switch senses the tip and sends a signal to a digital counter. Data are transmitted every 30 min or every 10 tips, whichever comes first. In practice, data are transmitted every 30 min during periods of light or no rain and every 10 tips during heavy rainfall. Occasionally, when heavy rains persist for several days, frequent transmission, clouds, and ash on the solar panel prevent recharging and drain a battery completely.

Rainfall is plotted as a cumulative value, in inches (fig. 3A), and is displayed for any period of time within a given year. Typically, rainfall over the previous few minutes, hours, and days has been of the most utility for forecasting lahars.

ACOUSTIC FLOW MONITOR

The AFM (fig. 2B) uses inexpensive, ruggedized, L10AR digital exploration geophones to monitor ground vibration caused by lahars, streamflow, and other sources of similar frequency (Hadley and LaHusen, in press). The geophones have a -3-dB flat response between 10 Hz and

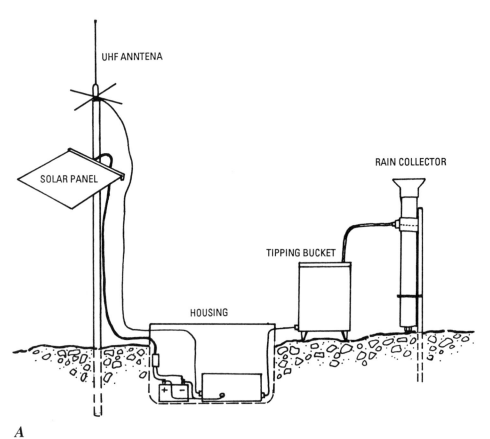

Figure 2. Setup of (*A*) a rain gauge station and (*B*) an acoustic flow monitor station.

UHF ANNTENA

SOLAR PANEL

RAIN COLLECTOR

TIPPING BUCKET

HOUSING

A

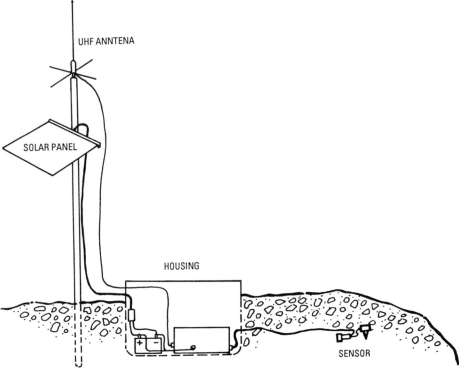

UHF ANNTENA

SOLAR PANEL

HOUSING

SENSOR

B

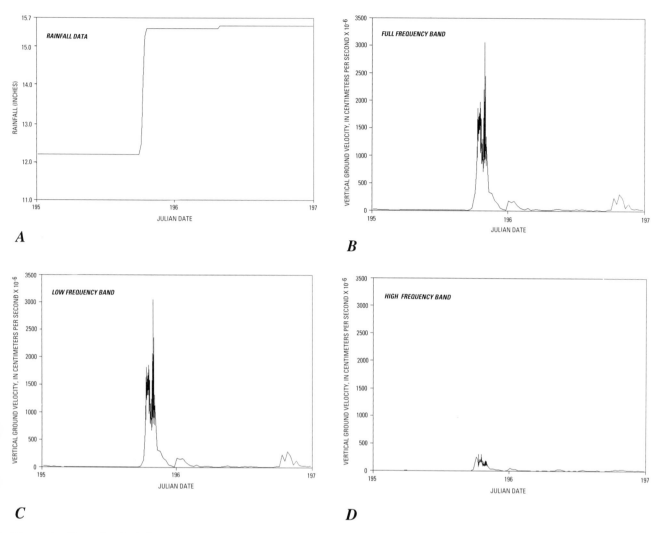

Figure 3. Example of rainfall and flow data, July 14, 1992. *A*, Cumulative rainfall. *B*, Full-frequency band (10–300 Hz). *C*, Low-frequency band (10–100 Hz). *D*, High-frequency band (100–300 Hz). Data were from RG F (Sacobia) and FM 1 (Lower Sacobia); flow observed 10 min later, about 2 km downstream from FM 1, was transitional between a debris flow and a hyperconcentrated streamflow.

300 Hz and a maximum dynamic range of about 3.8×10^{-3} cm/s vertical ground velocity (Fig. 3*B*). Ground vibration is monitored continuously in three frequency bands: the full band (10–300 Hz), a low band (10–100 Hz), and a high band (100–300 Hz).

A microprocessor-controlled digital telemetry platform measures the amplitude of the signal every second and then sends the data to the base station, at the Pinatubo Volcano Observatory on Clark Air Base, once every 30 min in normal mode. If the signal in a preselected frequency band exceeds a preprogrammed threshold value (adjusted for each site) for longer than 30 s (in 1993, >45 s), the AFM sends immediate reports that are flagged as "alerts." It continues to send flagged data at 1-min intervals for as long as the signal level in that frequency band remains above the threshold level. When the signal drops below the threshold, the AFM resumes normal operation, transmitting every 30 min. The triggering duration was set to 30 (later, 45) s to avoid recording volcanic and tectonic earthquake spikes, and the geophone was buried at least 1 m deep to avoid unnecessary noise from wind and rain.

When plotted against time, AFM data define a pseudohydrograph (fig. 3*B*) in which signal strength is a proxy for relative discharge. Potentially, with calibration to actual measured discharge at each AFM site, absolute discharge can be estimated from AFM data (Tuñgol and Regalado, this volume).

The acoustic flow monitor overcomes several difficulties that are inherent in the use of trip wires. AFM's provide enough data before and after passage of a lahar that the signal requires no corroboration, and information about the lahar is provided even after a trip wire in the same place would have broken. Furthermore, there is no need to visit each site after each lahar.

ROUTINE DATA RETRIEVAL

At the base station, a laptop computer displays raw data, time of transmission, and a channel identity that identifies the field station from which the data came and the type of data (for example, the low-frequency band of vibration, or rain gauge bucket tip count). Data are plotted onscreen and on an attached printer; they are written from the hard disk to a floppy and transferred to another computer for graphing or analysis with another software package, BOB (Murray, 1990). The BOB program is used primarily to display data from various rain gauges and flow sensors on single screens and pages so that the times and magnitudes of events can be compared from one part of the volcano to another.

REMOTE CONTROL OF FIELD SITE PARAMETERS

The base station computer is also used to issue commands to the field stations, to reset event-detection parameters or transmit intervals, or to interrogate a unit for up-to-the minute data. Communication between the rain gauges, flow monitors, and the base computer is through two-way UHF radio. The transmitter for one site also serves as the repeater for other sites. Each rain gauge is identified by a unique letter of the alphabet, and each acoustic flow monitor by a unique number (table 1).

Instrument control menus are available for setting and checking of operating parameters such as clock time of field instrument, repeater paths, routine transmission interval, next transmission time, duration for event detection, and threshold levels. When threshold levels of activity are exceeded, the instrument begins frequent "event mode" transmissions. The threshold for rainfall is fixed at 10 tips within less than the routine 30 min, and the thresholds for AFM amplitude are set between 150 and 1,000 AFM units (cm/s $\times 10^{-6}$) for 30 (later, 45) s or longer. AFM parameters had to be varied from site to site and empirically determined. The threshold settings during 1991 and 1992 were for Upper Sacobia, 300; Lower Sacobia, 700 in 1991 and 900 in 1992; Abacan, 1,000; Marella, 500; Pasig-Potrero, 300; Gumain, 200; and Maraunot, 150 AFM units.

There are also special commands for checking the condition of field instruments. Only one station and one authorized operator control the network; one additional, receive-only station receives and graphs data but cannot change the parameters of the field instruments.

INSTALLATION AND MAINTENANCE OF FIELD SITES

The lahar monitoring system was designed to operate under conditions of heavy rainfall, lightning, wind, and volcanic ash. Care was taken in site selection, installation, and maintenance to keep the system operating despite the hostile environment. Areas prone to erosion or lightning strikes were avoided where possible.

Precautions notwithstanding, we have had trouble with some sites. The biggest problems have been lightning, theft, and ash fall from secondary explosions. Even though each station uses one or two 100 A-h, 12-V batteries, connected in parallel and charged by an 18-W solar panel, stations close to the pyroclastic deposits often need to have ash cleaned from their solar panels. Solar panels are inclined 15 degrees to the south (Philippine standard) for optimum charging and also for protection against strong winds; steeper inclinations have been required in areas of unusually heavy ash fall from secondary explosions.

A partial list of problems encountered to date includes, in 1992, the following: the Culianan rain gauge and the Pasig-Potrero flow sensor were struck by lightning, batteries drained on the middle Sacobia rain gauge when ash covered the solar panel, and the lower Sacobia flow sensor was stolen. Wet ash also got into the BUGZ rain gauge and shorted out the battery. During the 1993 rainy season, the magnetic switch on the BUGZ rain gauge malfunctioned, the Summit rain gauge was damaged by lightning or buildup of static electricity, and the battery of the lower Sacobia flow sensor was stolen. Before a new battery could be brought to the site, the rest of the unit was stolen. Midway through the 1994 lahar season, telemetry line-of-sight problems are blocking data from (mostly new) sites on the west side, and ash from secondary explosions clogged the mechanism of the middle Sacobia rain gauge and covered the solar panels of the Pasig-Potrero flow sensor and (briefly) the upper Sacobia rain gauge. An alternate flow sensor that was installed at a police-manned lahar watchpoint was offline briefly when it was disconnected from our solar panel to charge a police battery.

PRELIMINARY RESULTS

In monitoring lahars of Mount Pinatubo from 1991 through 1993, the acoustic flow monitor system proved to be effective in lahar detection and warning. Some lahars were missed before instruments were installed and during instances of battery or other electronic failure; however, to our knowledge, no lahar passed by an operating flow sensor without detection. Empirically, flow sensors begin reporting increased vibration several minutes (as much as 1 km) before a lahar reaches the AFM station and continue above background until the tail of a lahar has passed. Data from the lower Sacobia and Pasig-Potrero flow sensors during the 1992 rainy season are given in Martinez and others (this volume) and Arboleda and Martinez (this volume), respectively; data from the 1993 lahar season are given in PHIVOLCS (1994).

Personnel who monitor the data at PVO can detect lahars 0.5 to 1 h earlier than watchers at manned, civil defense watchpoints, which are (necessarily) several kilometers downstream of our unattended instruments. PVO provides early warning to watchpoints manned by police and army personnel and receives from them, in return, a modest amount of information about the flow height, speed, and other flow characteristics. Utilization of PHIVOLCS' warnings by civil defense leaders is generally good but not perfect (Janda and others, this volume). Visual observations by PHIVOLCS staff were used to calibrate the Lower Sacobia flow sensor in 1992–93 (Martinez and others, this volume; Tuñgol and Regalado, this volume).

Data from several lahars during 1991 and 1992 suggest that energy from debris flows and hyperconcentrated streamflows is concentrated in the low band of the AFM's (10–100 Hz), while that from normal streamflow is concentrated in the high band (100–300 Hz). For example, on July 14, 1992, a flow past the Lower Sacobia flow monitor from 1800 to 2230 was seen, just downstream at the Mactan watchpoint, to be transitional between debris flow and hyperconcentrated flow. Its energy was strongly concentrated in the low band (fig. 3C,D).

Because the number of rain gauges per watershed (1 or 2) is small and rainfall is often localized, some lahars occur and are detected by AFM's without a corresponding rainfall signal. In addition, some lahars are caused by breakouts of small lakes, with or without concurrent rain. These "lahars without recorded rainfall" occur often enough that PVO reports them to the Office of Civil Defense without waiting for a rainfall signal. For example, at 1500 on October 12, 1991, a lahar was recorded by the Pasig-Potrero flow sensor without recorded or observed rains and was reported promptly to RDCC. No damage or casualties were reported.

Conversely, it is rare that heavy recorded rainfall does not result in a lahar, especially if the ground surface has been saturated by rain in preceding days. We now use such records to indicate that a lahar is likely. In one example, at 1800 on September 7, 1991, a general warning of lahars was issued on the basis of heavy rain at the Sacobia rain gauge. Less than an hour after the warning was issued, a lahar from the Pasig-Potrero River damaged San Antonio, Bacolor, and killed several people. Unfortunately, at this time the Pasig-Potrero flow sensor had not yet been installed. Field investigation conducted after the event revealed that the lahar event was brought about by the breaching of a newly formed lake in the upper reaches of the Pasig-Potrero River, presumably by overtopping and (or) erosion related to the heavy rainfall. Because the network of rain gauges is sparse, rainfall data cannot always tell us the magnitude of lahars or even the river channel in which they are flowing, but the AFM's provide this missing information.

Potential improvements in the network include for some AFM's to be placed at manned watchpoints and at variable distances back from the river channel. Such placement is needed to calibrate AFM data to actual discharge. Further correlation of rainfall and AFM signals and a denser network of rain gauges are needed for prediction of lahar discharge based upon rainfall. Finally, all stations need annual maintenance and special protection against moisture and ash in the electronics and against ash on the solar panels.

CONCLUSIONS

Judging from results from 1991 through 1993, a network of telemetered rain gauges and acoustic flow monitors is yielding important data for understanding and warning of Pinatubo lahars. Key factors in its success are (1) the small size and low power requirements of units, allowing them to be installed high on the volcano close to where lahars form, and (2) the choice of acoustic sensors rather than trip wires for flow monitoring.

At its best, the system provides civil defense and lowland residents with up to several hours of warning of impending and actual lahars. Typically, the system provides 0.5 to 1 h of extra lead time before flows are observed by manned watchpoints. At its worst, near the end of the 1992 lahar season, many stations were inoperative because of battery drain, site erosion, lightning strikes, and theft. New and repaired stations were installed in time for the 1993 and 1994 rainy seasons.

ACKNOWLEDGMENTS

The authors thank the Pinatubo Volcano Observatory staff for their valuable help and the U.S. Navy, U.S. Marines and the Philippine Air Force for their helicopter support for installations, repair, and maintenance of rain gauges and flow monitors. We also thank USAID Philippines and USAID Office of Foreign Disaster Assistance for financial support. Lastly, we thank M.L.P. Bautista, M.L.P. Melosantos, B.S. Tubianosa, R.U. Solidum, C.G. Newhall, R.S. Punongbayan, A.B. Lockhart, and M. Reid for constructive comments on our paper.

REFERENCES CITED

Arboleda, R.A., and Martinez, M.L., this volume, 1992 lahars in the Pasig-Potrero River system.

Bautista, B.C., Bautista, M.L.P., and Garcia, D.C., 1986, Seismic monitoring: A useful tool for mudflow detection at Mayon volcano, Albay, Philippines: Philippine Journal of Volcanology, v. 3, no. 2, p. 90–108.

Bautista, B.C., Bautista, M.L.P., Marcial, S.S., Melosantos, A.A., and Hadley, K.C., 1991, Instrumental monitoring of Mount Pinatubo lahars, Philippines [abs.]: Eos, Transactions, American Geophysical Union, v. 72, no. 44, p. 63.

Hadley, K.C., and LaHusen, R.G., 1991, Deployment of an acoustic flow-monitor system and examples of its application at Mount Pinatubo, Philippines [abs.]: Eos, Transactions, American Geophysical Union, v. 72, no. 44, p. 67.

———in press, Technical manual for an experimental acoustic flow monitor: U.S. Geological Survey Open-File Report.

Janda, R.J., Daag, A.S., Delos Reyes, P.J., Newhall, C.G., Pierson, T.C., Punongbayan, R.S., Rodolfo, K.S., Solidum, R.U., and Umbal, J.V., this volume, Assessment and response to lahar hazard around Mount Pinatubo.

Martinez, M.L., Arboleda, R.A., Delos Reyes, P.J., Gabinete, E., and Dolan, M.T., this volume, Observations of 1992 lahars along the Sacobia-Bamban River system.

Mercado, R.A., Lacsamana, J.B.T., and Pineda, G.L., this volume, Socioeconomic impacts of the Mount Pinatubo eruption.

Murray, T.L., 1990, A user's guide to the PC-based time-series data-management and plotting program BOB: U.S. Geological Survey Open-File Report 90–56, 53 p.

PHIVOLCS, 1994, Lahar studies: Quezon City, PHIVOLCS, 80 p. Special report of UNESCO-sponsored studies.

Scott, W.E., Hoblitt, R.P., Torres, R.C., Self, S, Martinez, M.L., and Nillos, T., Jr., this volume, Pyroclastic flows of the June 15, 1991, climactic eruption of Mount Pinatubo.

Tuñgol, N.M., and Regalado, M.T.M., this volume, Rainfall, acoustic flow monitor records, and observed lahars of the Sacobia River in 1992.

Rainfall, Acoustic Flow Monitor Records, and Observed Lahars of the Sacobia River in 1992

By Norman M. Tuñgol[1] and Ma. Theresa M. Regalado[1]

ABSTRACT

Instrumental records and field observations of 1992 lahars show a strong correlation between acoustic flux (a measure of ground shaking near the lahar channel) and discharge rates as observed at the Mactan watchpoint. Integration of this correlation over time suggests that the cumulative volume of lahars during the 1992 rainy season was about 1.0×10^8 cubic meters, a figure in reasonable congruence with the mapped 7×10^7 cubic meters of 1992 lahar deposits.

Correlation of acoustic flow data and rainfall suggests that increase in lahar magnitude is roughly proportional to the amount of triggering and sustaining rainfall. Furthermore, lahars along the Sacobia River seem to be triggered by rainfall with intensities greater than $I=46D^{-1.5}$, where I is the rainfall intensity in millimeters per minute and D is the duration of rainfall expressed in minutes. This remarkably low trigger threshold is due at least in part to low permeability and infiltration capacity of the fine ash-rich Pinatubo pyroclastic-flow deposits.

INTRODUCTION

Virtually all of the lahars that flowed down the Sacobia-Bamban River were derived from the Sacobia pyroclastic fan, which is drained by the Sacobia, Abacan and Pasig-Potrero Rivers (fig. 1). Pyroclastic deposits in the Sacobia and the Abacan watersheds have an estimated volume of approximately 7×10^8 m³, of which about 40 percent, or about 3×10^8 m³, was expected to be eroded within a decade (Pierson and others, 1992). During 1991 alone, at least 1.6×10^8 m³ of lahar deposit filled channels and buried nearby fields (Martinez and others, this volume). After the April 4, 1992, secondary pyroclastic flow, which cut off the Abacan River from the Sacobia pyroclastic fan, the Sacobia River became the sole channel draining the northeast lobe of that fan and carried about 0.7×10^8 m³ of additional debris in 1992.

Lahars along the Sacobia River are generally induced and triggered by rainfall. Large-scale lahar damming and lake breakouts similar to those of the Mapanuepe Lake, southwest of Mount Pinatubo (Umbal and Rodolfo, this volume), are not a significant lahar-triggering mechanism in the Sacobia watershed. Rain-induced debris flows have been the subject of many scientific papers, but only a few scientists have studied the detailed relationship of rainfall to debris flows in a tropical volcanic setting. Caine (1980) synthesized published records of 73 rainfall events associated with shallow landsliding and debris-flow activity from all over the world and came up with a limiting threshold for the initiation of debris flows, expressed in terms of the intensity and duration of the triggering rainfall. This data base included undifferentiated volcanic and nonvolcanic events. A similar study of the volcanic debris flows of Mayon volcano, Philippines, by Rodolfo and Arguden (1991) identified a higher threshold than that of Caine. For the lahars of Pinatubo, Umbal and Rodolfo (this volume) found that an average rainfall intensity of 0.3 to 0.4 mm/min over a period of 30 to 40 min was sufficient to trigger lahars on the Marella watershed in August 1991.

OBJECTIVES AND SCOPE OF THIS PAPER

This study adopted the approach of Regalado and Tan (1992), who correlated rainfall with lahar generation along the Marella River in 1991 by using rain-gauge and acoustic flow monitor (AFM) data. Here, we use instrumental and field data acquired along the Sacobia River during 1992 to (1) correlate the Sacobia AFM records with estimated flow discharge, (2) estimate the cumulative volume of 1992 lahars along the Sacobia River by using calibrated flow monitor data, (3) estimate the minimum intensity and duration of rainfall that triggered lahars along the Sacobia River, and (4) correlate the total amount of rainfall with the magnitudes of the associated lahars (rainfall-runoff relation).

[1]Philippine Institute of Volcanology and Seismology.

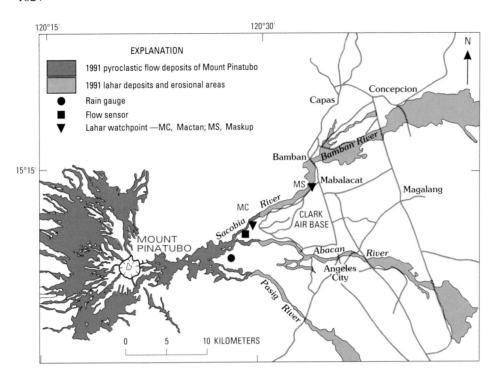

Figure 1. The 1992 lahar monitoring network along the Sacobia-Bamban River, northeast of Mount Pinatubo. The flow sensor shown in this figure is FM 1, Lower Sacobia, and the rain gauge is RG F, Sacobia, of Marcial and others (this volume).

SACOBIA LAHAR MONITORING IN 1992

For the 1992 rainy season, the system of lahar monitoring in the Sacobia watershed, as shown in figure 1, consisted of:

1. a rain gauge installed at midslope on the volcano to measure the amount of rain falling on the pyroclastic flow deposits;
2. an acoustic flow monitor (Hadley and LaHusen, 1991) installed near the lower portion of the pyroclastic fan, just below the lahar source area but well upstream from population centers, and
3. manned watchpoints for visual observations, measurements, and sampling of active lahars.

THE SACOBIA RAIN GAUGE

The Sacobia rain gauge is located about 10 km east of the crater, within the Sacobia pyroclastic fan, which provides the source materials for the lahars along the Sacobia River. The instrument consists of a standard tipping-bucket recorder that has been modified to accommodate ash fall (Hadley and LaHusen, 1991; Marcial and others, this volume). The data are transmitted to a computer at the Pinatubo Volcano Observatory at Clark Air Base (PVO-CAB) and recorded as the cumulative number of tips through time (1 tip = 1 mm of rainfall; Marcial and others, this volume). The data may be viewed in graph form on the computer screen. The Sacobia rain gauge, together with weather forecasts from the Philippine Atmospheric, Geo-

physical and Astronomical Services Administration (PAGASA) and the weather station of the U.S. Naval Air Station at Cubi Point, served as a lahar early-warning system during the 1992 lahar season.

THE SACOBIA ACOUSTIC FLOW MONITOR

The Sacobia acoustic flow monitor is located about 4 km downstream from the rain gauge, near the lower end of the Sacobia pyroclastic fan (fig. 1) (Marcial and others, this volume). The flow monitor records vertical ground velocities (in cm/sec×10^{-6}) generated by the passage of a flowing mass along the river channel (Hadley and LaHusen, 1991, 1995; R. LaHusen, written commun., 1993). The flow monitor data, like those of the rain gauge, are radiotelemetered to PVO-CAB. Normally, the instrument sends data every 30 min. When it detects a sustained high-amplitude flow, however, an alarm is set off, and it starts transmitting every minute until the flow vibrations drop below the alarm threshold. Like those of the rain gauge, realtime data are available in graphical as well as numerical formats.

The flow monitor has been programmed to record acoustic amplitudes in different frequency bands: the low-frequency range (10–100 Hz), the high-frequency range (100–300 Hz), and the broad-range band (10–300 Hz). A lahar, being much more debris laden than normal streamflow or surface runoff, generates a low rumbling noise that correlates best in the acoustic records with the signals in the low-frequency range (fig. 2; Suwa and Okuda, 1985; Zhang, 1990; Hadley and LaHusen, 1991).

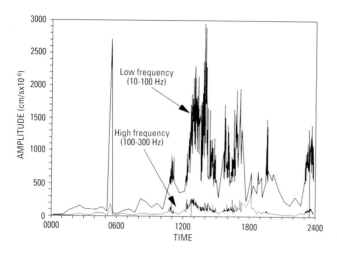

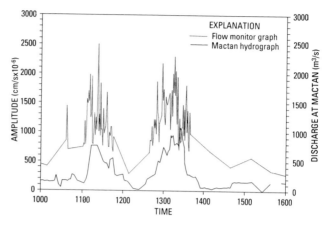

Figure 2. Acoustic-flow-monitor record of the August 28, 1992, flow events. The three events between 1000 and 1800 were debris flows dominated by low-frequency vibration, so the low-frequency wave band (10–100 Hz) recorded these lahars better than did the higher frequency band (100–300 Hz).

Figure 3. Low-frequency flow sensor record and the corresponding hydrograph at the Mactan watchpoint for the lahars on August 29, 1992.

LAHAR WATCHPOINTS

During the 1992 lahar season, the Philippine Institute of Volcanology and Seismology (PHIVOLCS) maintained two observation posts along the Sacobia River: one near the Mactan Gate of Clark Air Base, about 15 km from the crater, and another at Sitio Maskup, Barangay Dolores, Mabalacat, Pampanga, about 8 km farther downstream. During rainy weather and lahars, flow discharges were estimated from active channel widths, estimated flow depths, and surface velocities (measured by timing floating objects as they traveled a known distance). Whenever possible, samples of the flows were collected and their temperatures measured, but this was not practicable during big lahar events.

Observations at the Mactan watchpoint were used in correlating field and instrumental data, as this watchpoint is closest to the instruments. Some change of flow magnitude might occur along the 2- to 2.5-km reach from the flow monitor site to the Mactan watchpoint, but proportionality between flow magnitude past the AFM site and flow magnitude at Mactan watchpoint was probably consistent enough to make correlation of AFM records and observations at Mactan meaningful.

CALIBRATION OF AFM RECORDS AND CORRELATION WITH RAINFALL

CALIBRATING THE ACOUSTIC FLOW MONITOR

To correlate flow monitor records and observed discharge and thereby "calibrate" the Sacobia AFM, we had first to determine flow traveltimes between the AFM and our watchpoint at Mactan. Figure 3 shows a striking correspondence between the graph of the low-frequency flow monitor record and the hydrograph at the Mactan watchpoint for lahars of August 29, 1992. Distinct flow pulses can be seen and correlated, from which we see that flows took between 5 and 15 min to travel from the flow monitor site to the watchpoint. However, not all flow signatures were as easily distinguishable as those of the August 29 events. Small-amplitude flow signals, particularly those below the alarm threshold amplitude and thus recorded at 30-min intervals, were correlated with lahars observed at Mactan within 40 min.

For quantitative comparisons, we integrated the data over defined flow events or segments of flow events, rather than use instantaneous seismic amplitudes and discharge measurements. We define "acoustic flux" as the summed product of each amplitude reading and its duration and estimated this flux for 31 flow segments from 16 flow events in July and August (table 1 and fig. 4). Discharge and integrated volume measurements of the same flows were made at Mactan.

The slope of the best fit between discharge and acoustic flux (fig. 4A) will vary greatly with distance of the sensor from the river channel, channel roughness, and type of flow. Therefore, the "calibration" shown here cannot be transferred to other sites even at Pinatubo, much less at other volcanoes. However, provided that the Sacobia AFM instrument is left unchanged, future AFM data can be a reasonable proxy for discharge, especially when darkness, severe weather, steam, and hazard prevent direct measurement of that discharge.

Table 1. Flow measurements made at the Mactan watchpoint correlated with the low-frequency (10–100 Hz) flow sensor data for observed events from July 1 to August 29, 1992.

[Values are for the specified time periods of actual observations at Mactan watchpoint. Amplitude and discharge are averages for those periods; acoustic flux is a product of amplitude and duration. Volumes were calculated by integrating instantaneous discharge through the hydrograph and thus differ slightly from the product of average discharge and duration. HF, hyperconcentrated streamflow; DF, debris flow; NSF, normal streamflow]

Date	Flow event	Flow sensor					Mactan watchpoint				
		Time		Acoustic flux $\times 10^3$	Duration (min)	Average amplitude (cm/s $\times 10^{-6}$)	Volume ($m^3 \times 10^3$)	Duration (min)	Average discharge (m^3/s)	Flow type	Flow temperature (°C)
		From	To								
July 1	1	1312	1442	16	90	186	152	90	28	HF-DF	50
		1442	1542	57	60	951	406	70	97	DF	
		1542	1642	10	60	166	57	34	28	HF-DF	
July 13	2	1908	2042	112	94	1,196	619	94	110	DF	72
July 20	3	1211	1653	80	282	284	236	278	14	DF	55
		1653	1827	109	94	1,156	584	91	107	DF-HF	62
July 26		1541	1711	2	90	27	7	88	1	HF-DF	60–70
	4	1711	1811	20	60	339	80	46	29	HF-DF	
July 29	5	1541	1811	36	150	237	187	131	24	DF	
July 30	6	1403	1441	27	38	700	252	46	91	DF	
		1441	1541	12	60	193	163	60	45	DF	
August 1		1611	1711	1	60	22	28	58	8	HF-NSF	
August 3	7	1511	1741	39	150	261	234	146	27	DF-HF	
August 7	8	1411	1641	17	150	112	66	142	8	DF	
August 15	9	0640	0910	39	150	257	71	140	8	DF	86
August 16	10	2341	0041	10	60	162	90	59	25	DF	59
August 20	11	1622	1643	11	21	506	52	17	51	DF	
		1643	1810	14	87	161	37	70	9	DF	
August 26	12	1134	1440	16	186	88	39	184	4	HF	50–60
August 27	13	0917	1240	51	203	250	1,010	202	83		
		1610	1810	26	120	216	163	109	25		
August 28	15	1040	1216	46	96	482	305	108	47	DF	
		1216	1538	206	202	1,018	3,460	199	290	DF	
		1538	1714	91	96	949	965	106	152	DF	
August 29	16	0939	1036	26	57	464	521	56	155	HF	
		1036	1039	4	3	1,246	125	8	260	HF	
		1039	1101	15	22	698	141	15	156	HF	
		1101	1209	76	68	1,115	1,750	67	435	DF	
		1209	1240	9	31	289	194	33	98	DF-HF	
		1240	1409	106	89	1,185	2,250	67	559	DF	
		1409	1739	84	210	401	1,260	215	97	DF	

CORRELATING FLOW MONITOR AND RAIN GAUGE DATA

Any marked deviation from the background amplitude in the low-frequency flow monitor record may correspond to a lahar along the Sacobia. In 1992, however, some flows with AFM amplitudes of less than 100 units were muddy streamflows (nonlahars) rather than hyperconcentrated streamflows or debris flows. To avoid consideration of muddy streamflow, we considered only events with peak amplitude of at least of 100 units.

Rainfall events that were recorded by the rain gauge from June 1 to September 3, 1992, were classified into lahar-related and nonlahar events. Slightly modifying the definitions of Rodolfo and Arguden (1991), we define *lahar-triggering* rainfall as a rainfall event that includes no pauses longer than 30 min and results in a flow that registers an amplitude of at least 100 units in the low-frequency flow monitor record; *sustaining rainfall* is the additional rain that falls during the flow. The *total rainfall* is the sum of triggering and sustaining rainfall. Other rainfall events are considered *nonlahar* rain.

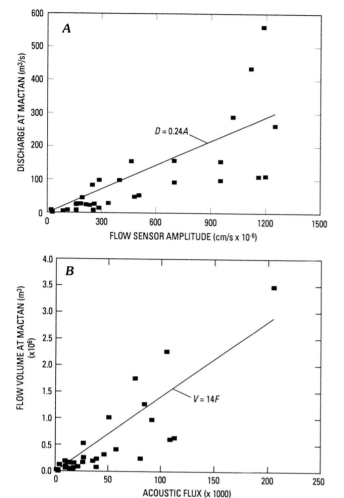

Figure 4. *A*, Observed discharge (*D*) at Mactan versus low-frequency flow sensor amplitude (*A*) for observed flows in July and August 1992. *B*, The measured flow volume (*V*) versus the "acoustic flux" (*F*) for observed flows in July and August 1992. All values are averaged over the duration of each event. Acoustic flux is a product of amplitude and duration.

DISCUSSION

ESTIMATING LAHAR MAGNITUDE FROM AFM DATA

Figure 4*A* shows a general increase in average discharge with the average amplitude recorded in the low-frequency band by the flow monitor for 31 defined flow segments, representing 17 flow events, in July and August 1992. The relationship may be approximated by the equation

$$Q=0.24A \qquad \text{(correlation coefficient } (r) =0.76),$$

where Q is the discharge at Mactan in cubic meters per second, and A is the average amplitude for each event. An

exponential correlation actually yields a higher r value of 0.81, but the best-fit equation,

$$Q=10.7e^{0.029A},$$

indicated improbably high discharges for amplitudes of 2,000 and above.

Figure 4*B* shows that the flow volume at Mactan (discharge integrated over time) increased in rough proportion to the acoustic flux (AFM amplitude integrated over the same period). The correlation is best represented by a linear fit:

$$V=14F \qquad (r=0.61),$$

where V is the estimated volume of each flow event in cubic meters, and F is the instrumental acoustic flux. This equation gives a total volume of about 1.0×10^8 m^3 for all 58 lahars detected by the flow sensor from June to early September (table 2). An exponential best-fit curve,

$$V=6891e^{2.46 \times 10F},$$

has a lower r (.56) and gives unrealistically high flow volumes for modest acoustic flux values.

The 1992 lahar deposits were estimated to be about 7×10^7 m^3 on the basis of aerial photographs and rough thickness estimates in the field. We judge this to be in remarkably good congruence with the 1.0×10^8 m^3 estimated from the linear equation, especially in light of the fact that the volume of a lahar shrinks as water drains from it.

TRIGGERING RAINFALL

Figure 5 presents data from 58 lahar-triggering and 94 nonlahar rainfall events as recorded by the Sacobia rain gauge from June 1 to September 3, 1992. A power curve,

$$I = 46D^{-1.5}$$

(where I is the intensity in millimeters per minute, and D is the duration in minutes of the rain event), plots above all but 5 of 94 nonlahar rainfall events. This suggests that rain falling at a rate of at least 0.3 mm/min for 30 min or 0.1 mm/min for 1 hour is enough to trigger lahars along the Sacobia River.

Because we had only one rain gauge in the entire 50 km^2 Sacobia watershed, we need to ask whether the **recorded** rainfall is representative of the **actual** rains that triggered and sustained the lahars. The average size of convective storm cells at Pinatubo was about 10 km in diameter (Pierson and others, this volume), and thus roughly the size of the Sacobia watershed, but some rain was much more localized and was either heavier or lighter than that for the watershed as a whole (Hadley and LaHusen, 1991; Pierson and others, this volume). This is apparent in figure 5, which

Table 2A. Lahars detected by the Sacobia flow monitor from June 4 to September 3, 1993, and the calculated flow volume for each event.

[This table includes many events that occurred during times when we were not present at the Mactan watchpoint. Accordingly, the acoustic fluxes and volumes in table 2A are generally higher than those in table 1]

Date	Flow monitor record					Calculated volume (m³×10⁶)	Triggering rainfall			Sustaining rainfall			Total rainfall		
	Time From	Time To	Average amplitude	Duration (min)	Acoustic flux ×10³		Amount (mm)	Duration (min)	Intensity (mm/min)	Amount (mm)	Duration (min)	Intensity (mm/min)	Amount (mm)	Duration (min)	Intensity (mm/min)
June 4	1313	1413	245	60	15	0.2	4	38	0.1	8	82	0.1	12	120	0.1
June 19	1743	2043	269	180	48	0.7	8	48	.2				8	48	.2
June 27	1712	2343	501	391	196	2.7	15	15	1.0	22	266	.08	37	281	.1
June 28	0907	1342	458	275	126	1.8	11	13	.8	6	367	.02	17	380	.05
July 1	1242	1642	481	240	115	1.6	2	8	.2	24	112	.2	26	120	.2
July 11	1342	2342	730	600	438	6.1	6	48	.1	77	626	.1	82	674	.1
July 13	1812	2242	719	270	194	2.7	6	18	.4	77	63	1.2	83	81	1.0
July 16	1342	1612	286	150	43	0.6	1	20	.05	9	160	.06	10	180	.06
July 20	0641	0711	272	30	11	0.2	2	27	.07				2	27	.07
July 20	1011	1041	162	30	5	0.07	3	98	.03	2	19	.1	5	117	.04
July 20	1142	0011	582	749	436	6.1				30	551	.05	30	551	.05
July 21	0311	0642	239	211	50	0.7	14	178	.08	15	102	.1	29	280	.1
July 21	1211	1511	153	180	28	0.4				12	40	.3	12	40	.3
July 23	1811	2011	240	120	29	0.4				4	20	.2	4	20	.2
July 26	1711	1841	323	90	29	0.4	9	17	.5	1	2	.5	10	19	.5
July 27	0041	0311	544	150	82	1.1	6	10	.6	28	92	.3	34	102	.3
July 28	1511	1911	588	240	141	2.0	1	3	.3	33	150	.2	34	153	.2
July 29	1511	1841	368	210	77	1.1	21	57	.4						
July 30	1311	1511	523	120	63	0.9	12	38	.3	9	43	.2	30	100	.3
August 3	1411	1741	474	210	100	1.4	11	38	.3	0.2	2	.1	11	40	.3
August 4	1251	1541	826	170	140	2.0	17	78	.2	11	82	.1	28	160	.2
August 6	1311	1341	117	30	4	0.05	17	57	.3	9	43	.2	26	100	.3
August 7	1311	1541	471	150	71	1.0									
August 10	0111	0241	272	90	24	0.3	12	38	.3				12	38	.3
August 15	0540	0840	389	180	70	1.0	22	67	.3	7	173	.04	29	240	.1
August 15	1240	1440	343	120	41	0.6	1	7	.2	3	13	.2	4	20	.2
August 16	2302	0011	504	69	35	0.5				3	17	.2	3	17	.2
August 17	1241	1311	109	30	3	0.05	1	7	.2	2	13	.2	3	20	.2
August 17	1900	2011	704	71	50	0.7	1	7	.2	3	13	.2	4	20	.2
August 17	2141	2241	266	60	16	0.2	4	68	.06				4	68	.06
August 18	0811	1241	251	270	68	0.9	3	20	.1	9	180	.05	12	200	.06
August 18	1341	1741	262	240	63	0.9	4	7	.6	14	233	.06	18	240	.08

Table 2A. Lahars detected by the Sacobia flow monitor from June 4 to September 3, 1993, and the calculated flow volume for each event—Continued.

Date	Time From	Time To	Average amplitude	Duration (min)	Acoustic flux $\times 10^3$	Calculated volume $(m^3 \times 10^6)$	Triggering Amount (mm)	Triggering Duration (min)	Triggering Intensity (mm/min)	Sustaining Amount (mm)	Sustaining Duration (min)	Sustaining Intensity (mm/min)	Total Amount (mm)	Total Duration (min)	Total Intensity (mm/min)
August 20	0211	0341	407	90	37	0.5	8	18	.4	.3	2	.1	8	20	.4
August 20	0511	0640	477	89	43	0.6	7	18	.4	4	42	.09	11	60	.2
August 20	0910	1010	274	60	16	0.2				4	5	.8	4	5	.8
August 20	1219	1240	264	21	6	0.08	4	26	.2				4	26	.2
August 20	1541	1710	523	89	46	0.7	7	28	.2	2	12	.2	9	40	.2
August 20	1810	2010	544	120	65	0.9	3	17	.2	15	83	.2	18	100	.2
August 20	2140	2210	137	30	4	0.06	4	47	.08				4	47	.08
August 21	0612	0810	145	118	17	0.2	10	59	.2	.2	2	.1	10	61	.2
August 21	1240	1640	301	240	72	1.0	11	67	.2	9	233	.04	20	300	.07
August 25	1110	1140	337	30	10	0.1				3	20	.2	3	20	.2
August 26	1038	1240	262	122	32	0.4	10	45	.2				10	45	.2
August 27	0830	1240	408	250	102	1.4	12	17	.7	7	43	.2	19	60	.3
August 27	1510	1910	285	240	68	1.0	5	37	.1	3	20	.2	8	57	.1
August 28	0210	0510	121	180	22	0.3	10	77	.1	.3	3	.1	10	80	.1
August 28	0710	0004	586	1,014	594	8.3	7	57	.1	139	900	.2	146	957	.2
August 29	0004	2339	583	1,436	837	11.7				232	1,439	.2	232	1,439	.2
August 30	0009	0009	388	1,440	559	7.8				59	1,440	.04	59	1,440	.04
August 31	0009	0008	728	1,439	1,048	14.7				62	1,439	.04	62	1,439	.04
September 1	0008	2310	382	1,382	528	7.4				47	1,362	.04	47	1,362	.04
September 2	0040	0410	173	210	36	0.5	3	87	.04	7	193	.03	10	280	.04
September 2	0610	0840	139	150	21	0.3	4	117	.03	4	123	.03	8	240	.03
September 2	1040	1140	159	60	10	0.1	4	127	.03	1	20	.05	5	147	.03
September 2	1240	0110	348	750	261	3.7	3	67	.04	19	733	.03	22	800	.03
September 3	0310	0410	128	60	8	0.1	1	17	.05	1	43	.03	2	60	.03
September 3	0610	1140	131	330	43	0.6	3	108	.03	incomplete data.					
September 3	1240	2110	431	510	220	3.1									
				Total		105.2									

Table 2B. Rainfall events that did not result in lahars.

[These events define the triggering threshold for lahars (fig. 5)]

Date	Amount (mm)	Duration (min)	Intensity (mm/min)	Date	Amount (mm)	Duration (min)	Intensity (mm/min)	Date	Amount (mm)	Duration (min)	Intensity (mm/min)
June 3	1	20	0.05	July 18	12	40	.3	August 18	4	20	.2
June 7	7	40	.2	July 19	3	20	.2	August 18	3	20	.2
June 15	3	40	.08	July 21	3	40	.08	August 18	5	19	.3
June 17	4	40	.1	July 21	3	20	.2	August 18	3	20	.2
June 17	1	20	.05	July 22	1	20	.05	August 18	3	18	.2
June 20	2	20	.1	July 24	1	20	.05	August 19	8	40	.2
June 20	3	20	.2	July 24	1	20	.05	August 19	3	20	.2
June 23	9	20	.4	July 27	3	21	.1	August 19	8	40	.2
June 23	4	20	.2	July 29	1	20	.05	August 19	5	24	.2
June 24	2	20	.1	July 30	3	20	.2	August 19	4	20	.2
June 25	1	20	.05	July 30	4	20	.2	August 20	3	20	.2
June 25	1	20	.05	July 31	1	20	.05	August 20	4	20	.2
June 26	11	60	.2	August 1	4	20	.2	August 21	4	20	.2
June 26	4	20	.2	August 4	2	20	.1	August 21	6	40	.2
June 26	1	20	.05	August 5	1	20	.05	August 21	1	20	.05
June 27	5	20	.2	August 7	3	40	.08	August 21	4	80	.05
June 27	3	20	.2	August 8	1	20	.05	August 21	1	20	.05
June 28	3	14	.2	August 10	5	20	.2	August 22	4	20	.2
June 28	2	20	.1	August 11	1	20	.05	August 24	5	102	.05
June 29	3	20	.2	August 11	1	20	.05	August 24	1	20	.05
June 30	1	20	.05	August 13	1	6	.2	August 24	11	60	.2
June 30	1	20	.05	August 14	7	40	.2	August 24	1	14	.07
June 30	1	20	.05	August 14	4	20	.2	August 25	1	20	.05
July 2	2	40	.05	August 15	3	23	.1	August 25	1	20	.05
July 3	3	20	.2	August 15	3	20	.2	August 25	5	20	.2
July 12	1	20	.05	August 16	4	20	.2	August 26	3	20	.2
July 12	1	18	.06	August 16	3	20	.2	August 27	4	3.	1.3
July 14	2	20	.1	August 16	4	20	.2	August 27	4	20	.2
July 15	3	20	.2	August 16	3	20	.2	August 27	4	20	.2
July 16	4	20	.2	August 16	5	20	.2	August 27	2	20	.1
July 18	12	8	1.5	August 17	4	20	.2	September 3	2	60	.03
July 18	2	20	.1								

shows that many lahars occurred even when the **recorded** rainfall was comparable to that which at other times did not cause lahars; in fact, a few lahars have occurred after practically no **recorded** rain at all! In some and probably most of these instances, localized lahar-triggering rain occurred elsewhere in the watershed but not at the site of the rain gauge. Given this problem, we have estimated the threshold for lahar triggering (fig. 5) to be the maximum rainfall that **does not** trigger lahars, rather than the minimum (apparent) rainfall that **does** trigger lahars. With a large enough data set, the effects of localized rainfall become apparent, and we are reasonably confident that rainfall greater than the threshold curve in figure 5 usually will trigger a lahar.

The threshold defined in figure 5 for Mount Pinatubo is much lower than that of Rodolfo and Arguden (1991) for debris flows at Mayon:

$$I=2.16D^{-0.38}$$

(converted from their equation $I=27.3D^{-0.38}$, where their I was in millimeters per hour and their D was in hours). The threshold for Mayon was determined exclusively for debris flows, while the present study includes both debris flows and hyperconcentrated streamflows, and the threshold for hyperconcentrated streamflow is surely lower than that for debris flows. It is also likely, however, that the pyroclastic materials of Pinatubo are also mobilized more easily

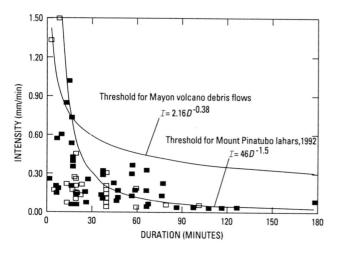

Figure 5. Intensity (*I*) versus duration (*D*) of 58 lahar-triggering (filled squares) and 97 nonlahar (empty squares) rainfall events, derived from the flow monitor and rain-gauge data, from June 1 to September 3, 1992. The minimum threshold curve, described in the text, is the curve of rainfall above which lahars will nearly always result.

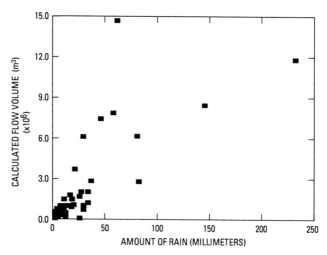

Figure 6. Total rainfall and the volume of the resulting lahars, the latter being calculated from the relationship between acoustic flux and observed discharge (fig. 4*A*), from June 1 to September 3, 1992.

because they contain more fine ash, which causes lower permeability and infiltration capacity. Low permeability and low infiltration rates enhance runoff that eventually bulks up through headward and lateral erosion to form lahars.

RAINFALL AND LAHAR VOLUME

Figure 6 shows that the volume of each lahar event, as estimated from flow monitor data, is directly proportional to the total amount of rainfall that triggered and sustained the flow (*r*=0.77). Regalado and Tan (1992) defined a similar linear relationship between rainfall and the magnitude of lahars, as observed and as recorded by the flow monitor, for the Marella watershed in 1991. Their graph is reproduced here (fig. 7) for comparison. For 1992 events in the Sacobia watershed, instrumentally recorded rainfall also correlated well with the frequency of lahar occurrences (Martinez and others, this volume). Close correlation between rainfall and lahars was already known from field observations; the apparently linear relation between rainfall and (inferred) lahar volume is a new and potentially useful finding. Although the volume of lahars generated by a specified amount of rainfall will certainly decrease over time, as the watershed "heals," the relationship presented here might help to predict the volume of lahars that would result from various assumptions about future rainfall.

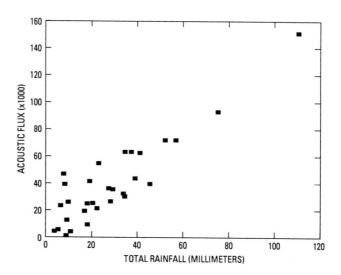

Figure 7. Rainfall versus acoustic flux for August-September 1991 events in the Marella watershed (from Regalado and Tan, 1992). For these events, acoustic flux could not be converted to volume of lahars because considerable deposition and erosion of sediment, and addition of tributary flow, took place between the flow monitor and the nearest observation point, Dalanaoan, San Marcelino.

CONCLUSIONS

1. Lahars along the Sacobia River are triggered by rainfall exceeding the threshold curve

$$I=46D^{-1.5},$$

or above about 0.3 mm/min for 30 min.

2. The total volume of a flow increases linearly with total rainfall (that is, the sum of triggering and sustaining rainfall).

3. Sacobia flow monitor records and field estimates of discharge during 1992 were sufficiently well correlated that we could use flow monitor records to estimate discharge during periods for which we had no direct observations.

4. With this continuous estimator of discharge, we estimated the cumulative flow volume of lahars along the Sacobia channel at Mactan to be 1×10^8 m^3, a figure that correlates well with 7×10^7 m^3 of observed lahar deposits.

5. Future use of rain gauges to forecast lahar discharge should increase the density of rain gauges in each watershed. Future use of acoustic flow sensors to estimate lahar discharge should colocate the flow sensor at the observation post [editor's note: this was done at Mactan in 1993], make corrections for flow densities, and establish new calibrations at each flow sensor site.

ACKNOWLEDGMENTS

Rick LaHusen developed the AFM and modified the rain gauge used for instrumental data collection, and Kevin Hadley, Jeff Marso, and the PHIVOLCS Instrumentation Group (Joey Marcial, Arnold Melosantos, Ronald Pigtain, Arnold Chu, and Ramses Valerio) installed and repaired the instruments under difficult and hazardous conditions. Field observations of lahars were by members of the 1992 Pinatubo Lahar Monitoring Group: Onie Arboleda, Art Daag, Edwin de la Cruz, Peejay Delos Reyes, Atoy Garduque, Rey Macaspac, Mylene Martinez, Jed Paladio-Melosantos, Alex Pataray, Weng Quiambao, Bella Tubianosa, Mike Dolan, Rick Dinicola, Yvonne Miller, and the Newhall family. Staff of the Pinatubo Volcano Observatory—Jimmy Sincioco, Mike Eto, Manny Isada and Alex Ramos—watched the instruments for lahar alerts and watched out for our team of field observers. Onie Arboleda, Mylene Martinez, Ishmael Narag, Weng Quiambao, and Obet Tan offered critical help in data interpretation, as did Rick LaHusen, Chris Newhall, Tom Pierson, R.S. Punongbayan, Kelvin Rodolfo, and Rene Solidum in reviews.

REFERENCES CITED

Caine, N., 1980, The rainfall intensity-duration control of shallow landslides and debris flows: Geografiska Annaler, v. 62, p. 23–27.

Hadley, K.C., and LaHusen, R.G., 1991, Deployment of an acoustic flow-monitor system and examples of its application at Mount Pinatubo, Philippines [abs.]: Eos, Transactions, American Geophysical Union, v. 72, p. 67.

Hadley, K.C., and LaHusen, R.G., 1995, Technical manual for an experimental acoustic flow monitor: U.S. Geological Survey Open-File Report 95–114, 24 p.

Marcial, S.S., Melosantos, A.A., Hadley, K.C., LaHusen, R.G., and Marso, J.N., this volume, Instrumental lahar monitoring at Mount Pinatubo.

Martinez, M.L., Arboleda, R.A., Delos Reyes, P.J., Gabinete, E., and Dolan, M.T., this volume, Observations of 1992 lahars along the Sacobia-Bamban River system.

Pierson, T.C., Daag, A.S., Delos Reyes, P.J., Regalado, M.T.M., Solidum, R.U., and Tubianosa, B.S., this volume, Flow and deposition of posteruption hot lahars on the east side of Mount Pinatubo, July–October 1991.

Pierson, T.C., Janda, R.J., Umbal, J.V., and Daag, A.S., 1992, Immediate and long-term hazards from lahars and excess sedimentation in rivers draining Mount Pinatubo, Philippines: U.S. Geological Survey Water Resources Investigation Report 92–4039, 37 p.

Regalado, M.T., and Tan, R., 1992, Rain-lahar generation at Marella River, Mount Pinatubo: unpublished report, Quezon City, National Institute of Geological Sciences, University of the Philippines, 17 p.

Rodolfo, K.S., and Arguden, A.T., 1991, Rain-lahar generation and sediment-delivery systems at Mayon Volcano, Philippines: Sedimentation in Volcanic Settings, SEPM Special Publication no. 45, p. 71–87.

Suwa, H., and Okuda, S., 1985, Measurement of debris flows in Japan: Proceedings of the IVth International Conference and Field Workshop on Landslides, Tokyo, p. 391–400.

Umbal, J.V., and Rodolfo, K.S., this volume, The 1991 lahars of southwestern Mount Pinatubo and evolution of the lahar-dammed Mapanuepe Lake.

Zhang, S., 1990, Geosound characteristics and measurement of debris flow at Jiangjia Gully, in Wu, J., Kang, Z., Tian, L., and Zhang, S., eds., Debris flow observations and research in Jiangjia Gully, Yunnan: Beijing, Science Press, p. 141–164.

Observations of 1992 Lahars along the Sacobia-Bamban River System

By Ma. Mylene L. Martinez,[1] Ronaldo A. Arboleda,[1] Perla J. Delos Reyes,[1] Elmer Gabinete,[1] and Michael T. Dolan[2]

ABSTRACT

Hot lahars occurred in the Sacobia-Bamban River system from June to September 1992. Intensified monsoon rains during the third week of August through the first week of September 1992 initiated the largest debris flows. Approximately 0.7×10^8 cubic meters of 1992 lahar debris, added to 1.0×10^8 cubic meters of debris from 1991 lahars, covered 90 square kilometers by the end of 1992.

Field observations of active lahars showed severalfold attenuation between Mactan (16 kilometers from the summit) and Maskup (24 kilometers from the summit), especially in those flows which had peak discharge of <200 cubic meters per second at Mactan. These smaller lahars also traveled about half the speed of lahars with discharge rates >1,000 cubic meters per second. Some flows transformed from debris flows to hyperconcentrated streamflows and muddy streamflows between the two observation posts.

Channel aggradation, most pronounced in the middle and upper parts of the Sacobia-Bamban fan, caused numerous avulsions and thus severe damage to homes and agricultural lands in the towns of Bamban and Mabalacat; the town of Concepcion, badly hit in 1991, was largely spared by upstream avulsions in 1992.

INTRODUCTION

The June 1991 eruptions of Mount Pinatubo deposited an estimated 5 to 7×10^9 m^3 of pyroclastic materials in the major valleys radiating from the volcano's vent (Daligdig and others, 1992; W.E. Scott and others, this volume). These unconsolidated, loosely compacted pyroclastic-flow deposits are highly erodible and are the primary source of Pinatubo lahars.

About 6×10^8 m^3 of these pyroclastic-flow deposits were emplaced on the eastern slopes of Mount Pinatubo, forming the Sacobia pyroclastic fan (fig. 1) (W.E. Scott and

others, this volume). The deposit separates into two main lobes downslope, a southeast lobe drained by the Pasig-Potrero River and a northeast lobe that, initially, was drained by both the Abacan and Sacobia Rivers. A secondary pyroclastic flow on April 4, 1992 (Torres and others, this volume) eroded a narrow divide between the Sacobia and Abacan rivers, allowing the Sacobia to capture all flow from the northeast lobe of the Sacobia pyroclastic fan.

From December to May of each year, the climate of Mount Pinatubo is dominated by the northeast monsoon and is relatively dry. Much greater precipitation occurs from June to September with the onset of the southwest monsoon. The heaviest rains of all are brought by typhoons that pass over or northeast of the Pinatubo area during these months and October. Typhoons passing northeast of Pinatubo increase rainfall by enhancing prevailing southwest monsoonal air flow.

Heavy precipitation and runoff erode material from the upper slopes, especially by downcutting and widening of gullies on the pyroclastic valley fill. Mixtures of water and sediment, with temperature ranging from 38° to 59° C and 40 to 90 percent sediment (by weight) flowed as far as 40 to 60 km from their source, burying agricultural lands and communities. This paper summarizes observations of active lahars, lahar channels, and lahar damage along the Sacobia-Bamban River during the 1992 rainy season.

PREERUPTION CHANNEL MORPHOLOGY

From the volcano's summit down to 1,000 m in elevation, the preeruption channel of the Sacobia River occupied a deep, narrow, V-shaped canyon, and the channel gradient decreased from 30° to 10°. From 1,000 m down to about 200 m in elevation, the channel slope decreased to between 6° and 1°. Along this reach it cuts into 600- to 3,000-year-old nonwelded pyroclastic-flow deposits (Newhall and others, this volume; Umbal and Rodolfo, this volume). From 200 m (opposite the Mactan gate of Clark Air Base) to 100 m in elevation (8 km downstream), the average slope

[1] Philippine Institute of Volcanology and Seismology.
[2] Michigan Technological University, Houghton, MI 49931.

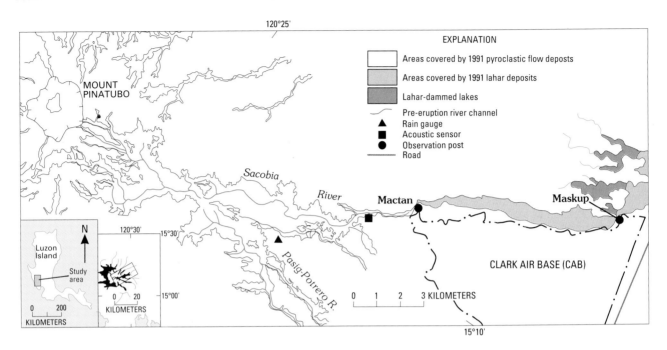

Figure 1. Distribution of the 1991 pyroclastic flow and lahar deposits on the east flank of Mount Pinatubo and locations of lahar observation posts and remote instruments.

decreases to 0.62° (see fig. 2), and the channel becomes box shaped. In this reach the channel is at least 500 m wide and erodes into old lahar terraces and well-indurated fluvial deposits. At 100 m in elevation, it reaches a 30-m-wide constriction (Maskup), below which the Sacobia channel is joined by two tributaries, Sapang Kawayan and the Marimla River, and the resulting river is known as the Bamban River. The pre-1992 channel depth at Maskup was 7 m.

1991 LAHARS OF THE SACOBIA-BAMBAN RIVER

The first lahar along Sacobia-Bamban channel occurred on June 14, 1991, and larger lahars occurred during the climactic eruption of June 15; these covered the broad floor of the Sacobia channel upstream from the town of Bamban (Major and others, this volume). More devastating lahars occurred during periods of intense monsoonal rainfall a few months after the climactic eruption (Pierson and others, this volume).

On the basis of data from acoustic flow sensors (Hadley and Lahusen, 1991), lahars in the Sacobia and Abacan occurred (on average) once every 1.2 days, and 3 to 5 lahar events per day were common during heavy rains in August (Pierson and others, 1992).

Initial channel response near the fanhead was aggradational, with deposition of 3 m at Mactan (figs. 1, 2). However, this initial phase of deposition was followed by rapid downcutting from late August to early September. Down-

stream, in the 8-km-long, 500- to 800-m-wide channel from Mactan to the constriction at Maskup (figs. 1, 2), deposition was at least 5 to 8 m and perhaps as much as 15 m (C.G. Newhall, written commun., 1992). Still farther downstream, at the Bamban Bridge, initial channel response in July was scouring, but, by late August, 20 m of aggradation had occurred at the Bamban Bridge (Pierson and others, this volume; K.M. Scott and others, this volume). The 1991 lahar deposits covered a total area of about 80 km^2 of agricultural lands and villages (Pierson and others, 1992) (fig. 3A).

Sapang Kawayan, the Marimla River, and a smaller tributary of the Sacobia just downstream from Maskup (fig. 3A), were dammed by aggradation in the Sacobia. Lakes formed behind these natural dams. Sediment that had dammed the Marimla River was breached on August 21, 1991, and the outbreak flood contributed to the lahars that destroyed the Bamban River Bridge along the MacArthur Highway (fig. 3A) (Pierson and others, 1992).

1992 LAHAR MONITORING NETWORK FOR THE SACOBIA CHANNEL

During the 1992 rainy season, two observation posts were established along the Sacobia channel to monitor discharge and flow characteristics. The first post (Mactan) is about 0.5 km downstream from the Mactan Gate of Clark Air Base and about 16 km from Pinatubo's summit. Located at the south bank of the channel at about 240 m in elevation,

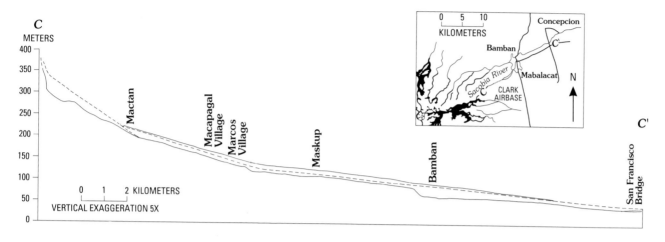

Figure 2. Longitudinal profile of the Sacobia-Bamban River, starting at 350 m in elevation, 12 km from the summit of Mount Pinatubo, and extending down to less than 50 m in elevation. Preeruption profile is shown by the lower solid line; 1991 lahar deposit by the dashed line; 1992 lahar deposit by the upper solid line. Note that vertical exaggeration is 5× and that only the lower, gently-sloping reaches of the Sacobia are shown.

the post overlooks a 60-m-wide constriction in the channel. The second observation post (Maskup), is located at another constriction at 100 m in elevation, about 8 km downstream from Mactan and 4 km upstream from the town of Bamban. The Maskup post is also located on the south bank of the channel and overlooks a 25- to 30-m-wide constriction.

Observations were by the PHIVOLCS Lahar Monitoring Team, often under trying conditions of rainfall, steam and secondary explosions, ashfall, and darkness. Estimated flow velocities are surface velocities; flow depths were estimated by measuring visible channel depth before and during each event and then subtracting the latter from the former. Such estimates of flow depth are necessarily uncertain; actual depths could have been greater (if the flow was scouring) or less (if the flow was depositing). Channel widths were usually taken to be the width of the freely flowing surface or, if that could not be seen through the steam, the width of the channel as measured before each event.

Information presented in this paper about 1992 rainfall came from a lone rain gauge installed near the divide between the northeast and southeast lobes of the Sacobia pyroclastic fan, and instrumental records of 1992 lahars came from an acoustic flow sensor that was installed in 1991 about 2 km upstream from Mactan (fig. 1) (Hadley and LaHusen, 1991; Marcial and others, this volume). An alarm automatically sounded at our Pinatubo Volcano Observatory on Clark Air Base when preset thresholds of rainfall or flow were exceeded.

SUMMARY OF 1992 LAHAR EVENTS

Lahars along the Sacobia-Bamban channel were triggered by three different types of rainfall: (1) ordinary, short-duration, localized rain showers on the upper slopes of the volcano; (2) heavy rains brought about by a passing typhoon; and (3) heavy precipitation lasting for days, brought by the southwest monsoon.

LAHARS ASSOCIATED WITH LIGHT RAIN SHOWERS: JUNE TO EARLY AUGUST

During June, lahars were infrequent (on average, one every 6 days). Rain showers of at least 6 to 8 mm in 40 min generated hot lahars, with peak discharges of up to 75 m^3/s and temperatures of 38° to 59° C. Rainfall and lahars began to increase in July (fig. 4A,B), but lahars were still relatively small, with peak discharges of about 150 m^3/s. All lahars were channel confined, and none resulted in any serious damage. However, on several occasions, they resulted in 1 to 2 m of aggradation or scouring of the channel, depending on the rheological character of the flow.

Small flows with discharges of 100 m^3/s or less at the Mactan observation point generally did not reach the downstream observation post at Maskup. Often, observers looking 2 to 3 km upstream from Maskup could see steam rising from hot lahars that approached, but never reached, Maskup. Channel gradient decreases and channel width increases between Mactan and Maskup, so many small lahars deposited their sediment before reaching Maskup.

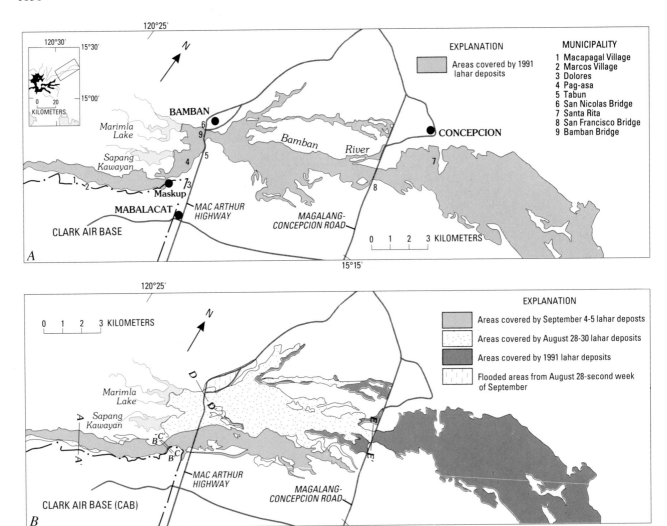

Figure 3. *A*, Lahar-affected areas along the Sacobia-Bamban River in 1991. Major settlements and significant points along the river are numbered. The two main roads connecting Manila with northern Luzon, the MacArthur Highway and Magalang-Concepcion Road, were cut and threatened, respectively, by major lahar avulsions. *B*, Distribution of 1992 lahar deposits superimposed on 1991 lahar deposits. Early 1992 lahar deposition was on and north of 1991 deposits; deposition in late August and September was on and south of 1991 deposits. Cross-sectional profiles are illustrated on figure 8.

Small lahars that did reach Maskup showed a marked decrease in flow magnitude and sediment concentration relative to Mactan. A sample of hot debris flow taken at Mactan at 2350 on August 16 contained 77.5 percent (by weight) sediment content. By the time this event reached Maskup, it was only muddy streamflow, 0.2 m deep and with less than 10 percent by weight sediment. As with flows that failed to reach Maskup, most sediment was deposited upstream from Maskup. Small lahars of this type also occurred on June 28 and August 3.

TYPHOON-RELATED LAHARS

Typhoons passed near Pinatubo on June 27, July 11, and July 20. Resulting lahars had peak discharges of 60 to 250 m³/s at Mactan. The debris flows observed at Mactan were sustained and reached the Maskup observation post, though their peak discharge had decreased by at least 50 percent and surface velocities had decreased from 4 m/s to 2.5 m/s.

At 1838 on July 20, a 4-m-deep debris flow with 4 m/s velocity and a peak discharge of 250 m³/s was observed at Mactan. Its temperature ranged from 55° to 62° C. The same flow at Maskup (at 1915) was 1 m deep, moving at 2.5 m/s, and had a peak discharge of about 75 m³/s. Cumulative rainfall during this event was small: only 24 mm in 11 hours.

In general, typhoon-generated lahars during the early part of the 1992 season were relatively small and channel confined. Most flow events were short, ranging from 3 to 7 hours, on the basis of flow sensor records.

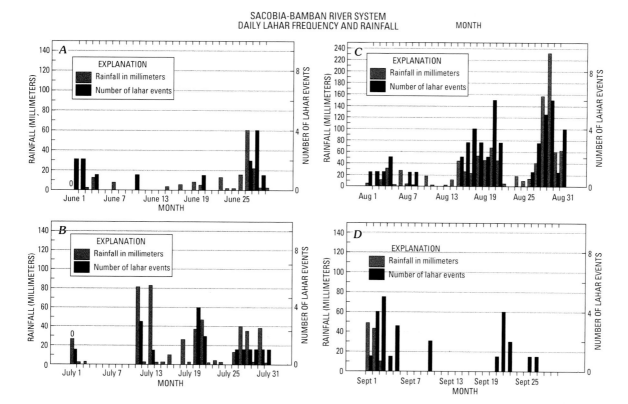

Figure 4. Daily rainfall and number of lahar events for *A*, June; *B*, July; *C*, August; and *D*, September based on the Sacobia flow sensor and rain gauge. The peak of monsoon rains occurred from the end of the second week through the last week of August. Lahars became more frequent starting on the last week of July and progressively increased and peaked in the third and fourth weeks of August and the first week of September. The August 28–30 and September 4–5 lahar events were the most devastating for 1992. High rainfall and lahars on June 27, July 11, and July 20 reflect the passage of typhoons.

INTENSE MONSOONAL RAINS: THIRD WEEK OF AUGUST TO EARLY SEPTEMBER

Toward the end of the second week of August, lahars began to occur at a rate of one to two per day. Most were still small, and those which were debris flows at Mactan had usually transformed into hyperconcentrated streamflows or muddy streamflows by the time they reached the Maskup observation post. From the third week of August until the first week of September, as typhoons passed over the north end of Luzon, monsoonal rains intensified and triggered debris flows with discharges at Mactan ranging from 600 to 1,100 m³/s. These debris flows avulsed near Mabalacat and Bamban, resulting in casualties and damage to properties in several barangays that had been only minimally affected by the 1991 lahars. Some areas that were not affected in 1991 were buried by 3 to 4 m of lahar deposits in 1992 (fig. 3*B*).

On August 20, five lahar events were recorded by the acoustic flow sensor. The initial flow started as early as 0515 and lasted almost 2 hours. Four more distinct events followed until 2300, each lasting an average of 2 hours. The biggest debris flow, which occurred between 1845 and 2100, avulsed on the south bank of the channel near Maskup. Farmland along the channel was partially buried by a thin layer of debris-flow deposits. This was the first lahar event for 1992 that overtopped the channel bank. Within the channel at Maskup, 2- to 3-m-thick lahar deposits were emplaced, leaving only 1 to 2 m of freeboard.

On August 26, a typhoon that passed over the northern portion of Luzon induced continuous heavy monsoonal rains in the Pinatubo area. Soon after, small to moderate-sized hot lahars (still less than 300 m³/s), with temperatures of 50° to 60° C, started cascading down the major drainages of Mount Pinatubo.

From 0940 to 1815 on August 27, at least six debris-flow and hyperconcentrated streamflow events were observed at the Mactan post. The first event, which occurred at 0940 with peak discharge of about 240 m³/s and a surface velocity of 5.5 m/s, was the largest flow for the day. The rest of the peak discharges ranged from 100 to 200 m³/s. The lahars were still mainly channel confined and no significant damage was reported.

Three big multipulse debris flows were generated by continuous rains on August 28. One pulse, passing Mactan at 1313 with a discharge of 760 m³/s, passed Maskup at 1350 with a discharge of 390 m³/s. Hot debris flows almost overtopped the south bank terrace of the constriction, and freeboard virtually vanished.

The first debris flow to hit the town of Bamban started at about 2300 on August 28 and continued until 0330 on August 29. The Bamban channel had been deepened during the early part of 1992, prior to the onset of the rainy season, when the Department of Public Works and Highways (DPWH) built 4-m-high protective earth levees on both sides of the channel. During the early morning lahar event of August 29, compounded by breaching of the lahar-dammed Marimla Lake, the north levee (dike) of the Bamban River was breached, and parts of Bamban town were buried beneath 2 to 3 m of sediment (fig. 3B). Beyond the area of direct deposition, flooding reached depths of 0.5 to 1 m.

The Bamban channel was completely filled, so when another debris flow arrived at around 1000, it overtopped dikes on the southern side of the channel as well. Lahars avulsed into Barangay Pag-asa (fig. 3B) and spread toward the east-southeast. By 1030, most houses along this path were either partially to totally buried by lahars (fig. 3B).

About the same time, another debris flow passed the Mactan post with an estimated peak discharge of 700 to 850 m³/s and surface velocity of 8.5 m/s. By the time it reached the Maskup observation post at 1113, the flow had a maximum discharge of 250 m³/s and surface velocity of 4.6 m/s. It spread directly toward Barangay Pag-asa and buried houses up to their roof levels (2 to 3 m deep). A 3-km stretch of the MacArthur Highway, which connects the towns of Mabalacat and Bamban, was buried with 3 m of lahar deposit. Throughout the duration of these flows, there was a noticeable absence of lahars at San Francisco Bridge, only 9 km downstream from Maskup (fig. 3B), because breaches and avulsions were already routing flow outside the engineered channel. Portions of the Magalang-Concepcion Road were also flooded by water overflowing from heavily silted irrigation canals.

From 1300 to 1400 on August 29, hot debris flows continued to bury some barangays of Bamban and Mabalacat. By this time, the field monitoring group had already abandoned its watchpoint at Maskup, because lahars had inundated portions of the access road to the watchpoint. At 1325, even larger debris flows (peak discharge of 1,100 m³/s) passed Mactan and, within an hour, buried Barangay Tabun (fig. 3B) beneath 2 to 3 m of deposit.

Lahars consisting mostly of hyperconcentrated streamflow continued to pass Maskup along the Sacobia channel on August 30–31. The events of August 31 appeared on flow sensor records as intermittent peaks against a high background, but, by this time, field observation and discharge measurements at Maskup had been temporarily

discontinued because of the high risk of working in the area. Also, the access road to the post had been closed due to the avulsion and deposition of lahars in Barangays Pag-asa and Dolores.

Starting on September 1, rains poured down anew on the slopes of Mount Pinatubo. On September 4, a series of multipulse lahars was recorded by the Sacobia flow sensor. The debris flow that reached the Maskup observation post at 1030 overtopped both the north and south river banks and deposited about 2 to 3 m of hot debris. Our newly built observation post was buried to that depth, as were several houses along the northeast perimeter wall of Clark Air Base (fig. 3B). At about 1100, part of this wall collapsed from the lateral pressure of the lahar debris. Houses in Dolores, a barangay of Mabalacat immediately northeast of the Clark Air Base wall, were buried in 2- to 4-m-thick lahar deposits (fig. 3B). Across and 2 km beyond the MacArthur Highway, the advancing lahar spread thinly over additional houses and agricultural lands. By September 5, flows waned and stopped. Thereafter, small, mostly hyperconcentrated streamflows continued intermittently through the end of September. The latter flows were relatively insignificant and had no damaging effect on downstream communities.

MAJOR IMPACTS OF THE 1992 LAHARS

1991 lahars left about 1.0×10^8 m³ of deposits on approximately 80 km² of alluvial fan and plain of the Sacobia-Bamban River, much in distal areas near Concepcion, Tarlac. Lahars during 1992 buried many areas near Bamban anew, and lateral spillover raised the area of severe damage to about 90 km². Along the 8-km stretch of channel between Mactan and Maskup, aggradation averaged 6 m, and the volume of 1992 lahar deposits was at least 7×10^7 m³ (PHIVOLCS, unpub. data, 1992).

The Bamban Bridge on the MacArthur Highway (fig. 3A) was a vital link in the major transportation route between south and north Luzon and connected the towns of Mabalacat, Pampanga, and Bamban, Tarlac. After the original bridge was washed away during a major lahar event on August 21, 1991, DPWH dredged this portion of the Bamban channel and maintained two temporary bridges to keep the MacArthur Highway open. But, by August 29, 1992, lahar deposition and scouring of bridge approaches made the new bridges impassable, and lahars of September 4 buried the new bridges beneath 3 to 5 m of fresh deposit.

The three Sacobia tributaries that were blocked in 1991 were impounded again by 1992 lahar deposits. Then, when the lahar dam of Marimla Lake was breached on the early morning of August 28, 1992, floodwaters flowed into San Nicolas Creek (fig. 3A) and thence into lowlying portions of Bamban. The town was submerged in at least half a meter of floodwater for several days. Continuous outflow from Marimla lake undercut and destroyed a 3-km stretch

Table 1. Summary of discharge measurements taken from active lahar flows from two observation posts along the Sacobia River.

Date	Mactan		Maskup		Percent attenuation of discharge $(M-m)/M \times 100$	Lag time at peak flow (min)	Total rain (mm)	Front velocity (m/s)
	Peak discharge (M) (m³/s)	Surface velocity (m/s)	Peak discharge (m) (m³/s)	Surface velocity (m/s)				
July 20	260	4	75	2.5	71	38	20.9	3.5
August 3	80	5	2	1.5	98	45	10.8	3.0
August 16	150	3	2	1.5?	99	50	8.0	2.7
August 27	220	5.5	110	6	49	44	18.8	3.0
August 28	760	8.5	390	11	49	37	32.9	3.6
August 29	760	8.5	260	5	66	32	42.4	4.2
	950	8	370	5	61	21	29.1	6.4

of concrete pavement of the MacArthur Highway in Bamban.

DISCUSSION

LAHAR FREQUENCY AND RAINFALL

Figures 4A–D and 5 compare lahar occurrence to daily and monthly rainfall from May to September of 1992, as recorded by the Sacobia flow sensor and rain gauge. The frequency of lahar events increased from June to the early part of August and peaked during the latter part of August when rains were heaviest. From visual observations of dark clouds, we think that all of the 1992 lahars were triggered by rainfall, even though four events in early June had no significant recorded rainfall (fig. 4A). Because only one rain gauge was present, in the middle Sacobia watershed, some localized rains farther upslope or far from the rain gauge may not have been recorded.

More detailed correlation of rainfall and lahar generation (Tuñgol and Regalado, this volume) suggests that a rainfall rate of about 9 mm in 30 min is the minimum amount required to generate lahars in the 53-m² catchment of the upper Sacobia River.

FLOW VELOCITIES

We caught the flow front of only one flow at both observation posts, on June 28. The flow was small (peak discharge at Mactan was <200 m³/s), and its front took 55 min to travel 8 km from Mactan to Maskup. Peaks of flows with peak discharge <200 m³/s took as much as 50 min to travel that distance, while peaks of larger flows (200 to 1,200 m³/s) traveled that distance in 20-30 min.

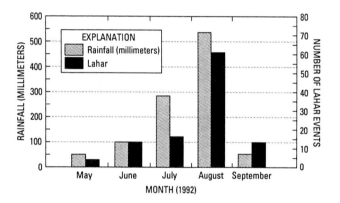

Figure 5. Monthly rainfall and number of lahar events for the period May to September 1992. September rainfall is incomplete due to failure of the instrument on September 3.

PEAK DISCHARGES, AND DOWNSTREAM CHANGES IN FLOW MAGNITUDE

The largest discharge we were able to estimate was 1,100 m³/s, at Mactan on August 29. Judging from flow sensor records (Tuñgol and Regalado, this volume), discharge during flows of August 31 might have been greater.

Table 1 and figure 6 show discharge measurements that can be compared from Mactan to Maskup. Peak discharges at Mactan are followed by peak discharge at Maskup 20 to 50 min later. The change, or percent attenuation, in peak discharge at these two points is $[(M-m)/M] \times 100$, where M is peak discharge at Mactan and m is peak discharge at Maskup. Most lahars attenuated over this reach by 30 to 50 percent, by upstream sedimentation and probably also by longitudinal stretching of peak flow.

Figure 7 shows apparent relationships of rainfall and peak discharge to percent attenuation and the traveltime ("lag time") of peak flow between Mactan and Maskup.

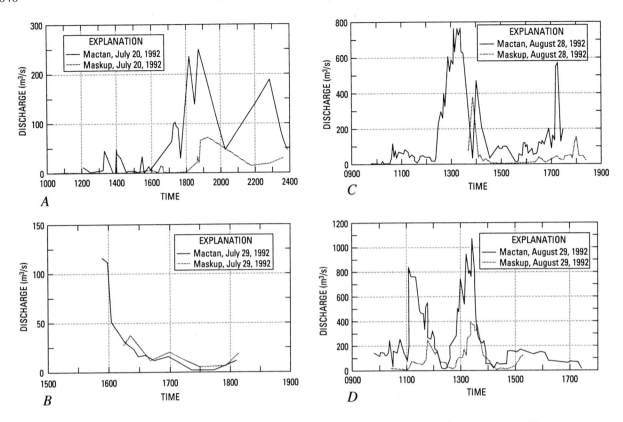

Figure 6. Pairs of hydrographs of selected lahars, as estimated at the Mactan and Maskup observation posts. Note downstream attenuation and traveltimes of 20 to 50 min from Mactan to Maskup.

Although our data are sparse, it appears that percent attenuation is less when there was more rain in the lahar source area and a greater peak discharge (fig. 7A, C). It also appears that flows associated with more rain, and larger flows, traveled faster than smaller ones (fig. 7B, D).

CHANNEL RESPONSE: EROSION AND DEPOSITION BY LAHARS

Channel cross sections were established in early July at four locations along Sacobia-Bamban River and resurveyed after each major lahar event to monitor deposition and changes in channel morphology (fig. 8). Reconnaissance field inspections were also conducted along the river after each significant event to estimate changes in channel bed elevations. At Macapagal Village, 5 to 6 km upstream from Maskup (fig. 1), up to 10 m of deposition occurred between August 12 and October 10, consistent with the observation that lahars attenuated markedly between Mactan and Maskup. At Maskup, alternating aggradation and scouring took place between July 2 and August 26, with net aggradation of about 3 m (added to 2 m from June) (fig. 8, section B-B'). Minor lahar avulsion on August 20 almost buried the reference benchmark. Because of this, the

Maskup line was transferred about 10 m downstream from its original location after the August 26 survey. Another meter of aggradation occurred from August 26–29 (fig. 8, section C–C'). Because benchmarks on both the north and south banks were buried during the September 4–5 lahars, future surveys cannot be tied to those of 1992. Total 1992 aggradation at Maskup was >10 m.

Surveys of a cross section adjacent to the new Bamban Bridge (fig. 8, section D–D') showed about 3.5 m of deposition from August 19 to September 1. During the same period, no significant aggradation took place at the San Francisco Bridge (fig. 8, section E–E') because major deposition and avulsions occurred 6 km short of the latter bridge. Continued dredging of the channel near that bridge from August 13 until the last survey on September 18 appears as 1.5 m of apparent scouring.

At both Mactan and Maskup, channels also experienced repeated episodes of incision, channel widening, and backfilling in a matter of minutes—much more than is apparent from figure 8. Net fill of 3 to 4 m at Maskup occurred by tens of episodes of alternating fill and scour. Hyperconcentrated flows were especially effective agents of lateral erosion, while debris flows tended to deepen channels before they slowed and deposited their sediment (Pierson and Scott, 1985; Punongbayan and others, 1992).

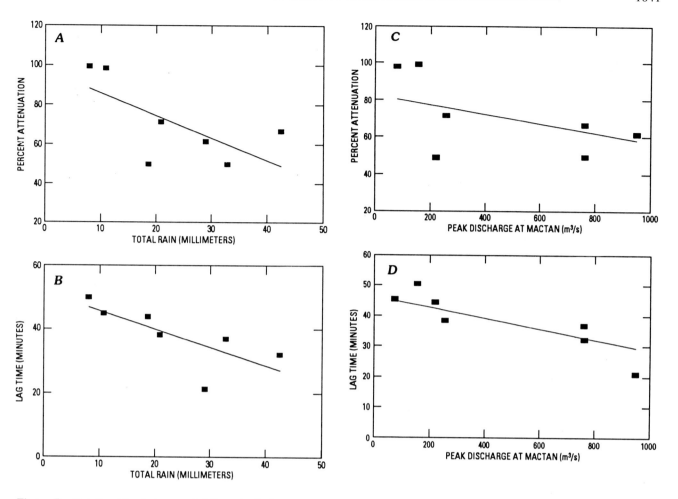

Figure 7. Relationships between rainfall, peak discharge, percent attenuation, and the traveltime of peak flow between Mactan and Maskup. *A*, Percentage attenuation of peak discharge versus rainfall during the flow. *B*, Traveltime ("lag time") of peak flow versus rainfall. *C*, Peak discharge at Mactan versus percentage attenuation. *D*, Peak discharge at Mactan versus "lag time."

AVULSIONS

Most lahars from early June until the second week of August were channel confined. Their damage was limited to minor lateral erosion of the protective earth dikes built by the DPWH along Bamban channel. However, these early, small-magnitude lahars gradually aggraded and decreased the carrying capacity of downstream channels. Then, when relatively large lahars (debris flows) of 1,000 to 1,200 m³/s started flowing down the Sacobia River on August 29, discharges far exceeded channel capacity. Major avulsions (channel overflow) occurred at constrictions in the Sacobia channel, for example, at Maskup and in channel bends at Barangays Pag-asa and Dolores (fig. 3*A,B*). These avulsions, farther upstream than those of 1991, focused lahar deposition between Mactan, Maskup, and Bamban rather than downstream. Relatively little deposition occurred in the town of Concepcion, as feared after serious 1991 impacts, but occurred instead near the towns of Bamban and Mabalacat.

CONCLUSIONS

Intense monsoon rains and rain from typhoons were the primary causes of lahars of the Sacobia-Bamban River in 1992. Approximately 7×10^7 m³ of lahar sediment was deposited during 1992 in addition to 1×10^8 m³ in 1991. By the end of 1992, 90 km² of Sacobia alluvial plain and fan was covered with lahar deposits. The focus of lahar deposition in 1992 was on the upper part of the Sacobia alluvial fan, 20 to 30 km from Mount Pinatubo's summit.

Flows attenuated from the head of the fan (Mactan) to another observation point 8 km downstream (Maskup). A number of flows also transformed from debris flows to hyperconcentrated streamflows and muddy streamflows over that same reach. Complex alternation of deposition and scouring ended with net deposition along most of the fan, locally as great as 10 m.

Because of this continued channel filling, the communities along Sacobia-Bamban River that were identified in the lahar hazards map of PHIVOLCS (1992) are still vulnerable to lahar encroachment and heavy siltation.

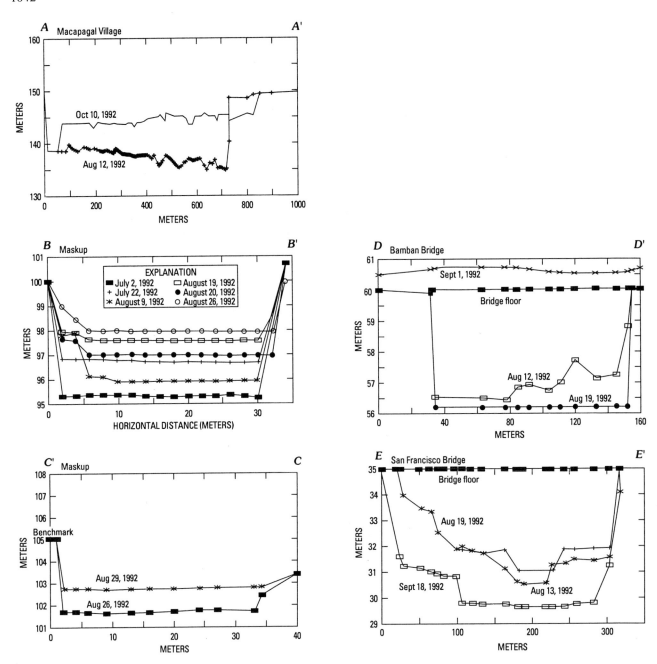

Figure 8. Cross-sectional profiles across the Sacobia-Bamban River showing changes in channel bed elevations due to aggradation or scouring. Dates of the 1992 channel surveys are indicated. Locations of the sections are shown in figure 3B. Cross section C–C' was established 10 m downstream from B–B' after a major lahar avulsion on August 20. Figure 3B shows location of cross section lines.

ACKNOWLEDGMENTS

We thank our colleagues for dedicated participation in the 1992 lahar monitoring activity: N.M. Tuñgol, T.M. Regalado, B.S. Tubianosa, A.S. Daag, R.C. Torres, M.L.O. Paladio, R.B. Quiambao, E.G. dela Cruz, A.L. Pataray, R.G. Garduque, R.M. Macaspac, and C.G. Newhall and family. We also thank the PVO-CAB staff, headed by Mr. J. Sincioco with I. Eto, M. Isada, and A. Ramos, for the assistance that they have extended for the group. Flow sensor and rain gauge data were provided by PHIVOLCS' Volcano Monitoring and Eruption Prediction Division. Air support was provided by the Philippine Air Force. Special thanks go to Director R.S. Punongbayan, R.U. Solidum, J.A. Daligdig, and R. Dinicola for their valuable suggestions.

REFERENCES CITED

Daligdig, J.A., Besana, G.M., and Punongbayan, R.S., 1992, Overview and impacts of the 1991 Mt. Pinatubo eruptions: Proceedings of the 4th Annual Geological Convention, 1991: Geological Society of the Philippines, p. 29–55.

Hadley, K.C., and Lahusen, R., 1991, Deployment of an acoustic flow sensor monitor system and examples of its application at Mt. Pinatubo, Philippines [abs.], Eos, Transactions, American Geophysical Union, v. 72, no. 44, p. 67.

Major, J.J., Janda, R.J., and Daag, A.S., this volume, Watershed disturbance and lahars on the east side of Mount Pinatubo during the mid-June 1991 eruptions.

Newhall, C.G., Daag, A.S., Delfin, F.G., Jr., Hoblitt, R.P., McGeehin, J., Pallister, J.S., Regalado, M.T.M., Rubin, M., Tamayo, R.A., Jr., Tubianosa, B., and Umbal, J.V., this volume, Eruptive history of Mount Pinatubo.

PHIVOLCS, 1992, Pinatubo lahar hazards map: Manila, National Economic Development Authority, Manila, 1 sheet at 1:200,000; 4 sheets, quadrants 1–4, at 1:100,000.

Pierson, T.C., and Scott, K.S., 1985, Downstream dilution of a lahar: Transition from debris flows to hyperconcentrated streamflow: Water Resources Research, v. 21, no. 10, p. 1151–1524.

Pierson, T.C., Janda, R.J., Umbal, J.V., and Daag, A.S., 1992, Immediate and long-term hazards from lahars and excess sedimentation in rivers draining Mount Pinatubo, Philippines: U.S. Geological Survey Water-Resources Investigations Report 92–4039, 37 p.

Punongbayan, R.S., Umbal, J., Torres, R., Daag, A.S., Solidum, R., Delos Reyes, P., Rodolfo, K.S., and Newhall, C.G., 1992, A technical primer on Pinatubo lahars: Quezon City, PHIVOLCS, 20 p.

Scott, K.M., Janda, R.J., de la Cruz, E., Gabinete, E., Eto, I., Isada, M., Sexon, M., and Hadley, K.C., this volume, Channel and sedimentation responses to large volumes of 1991 volcanic deposits on the east flank of Mount Pinatubo.

Scott, W.E., Hoblitt, R.P., Torres, R.C., Self, S, Martinez, M.L., and Nillos, T., Jr., this volume, Pyroclastic flows of the June 15, 1991, climactic eruption of Mount Pinatubo.

Torres, R.C., Self, S., and Martinez, M.L., this volume, Secondary pyroclastic flows from the June 15, 1991, ignimbrite of Mount Pinatubo.

Tuñgol, N.M., and Regalado, M.T.M., this volume, Rainfall, acoustic flow monitor records, and observed lahars of the Sacobia River in 1992.

Umbal, J.V., and Rodolfo, K.S., this volume, The 1991 lahars of southwestern Mount Pinatubo and evolution of the lahar-dammed Mapanuepe Lake.

1992 Lahars in the Pasig-Potrero River System

By Ronaldo A. Arboleda[1] and Ma. Mylene L. Martinez[1]

ABSTRACT

Localized rainshowers, monsoonal rains, rains from tropical cyclones, and release of impounded water generated 62 lahars in the Pasig-Potrero River System during the 1992 rainy season. Most were triggered by rainfall exceeding the power function (of intensity and duration) $I=0.7D^{-0.37}$. Following a secondary pyroclastic flow on July 13, 1992, lahars from the upper part of the Pasig-Potrero watershed were temporarily diverted into an impounded lake, and lahars reaching downstream areas were small and generally erosive. When heavy rainfall and erosion on August 29, 1992, breached the impoundment and reestablished direct drainage of the upper part of the watershed, major lahars (peak discharge, 1,400 cubic meters per second) filled downstream channels, avulsed, and buried nearby villages and farmland. About 38×10^6 cubic meters of sediment was delivered downstream during the 1992 lahar season.

INTRODUCTION

The 1991 eruption of Mount Pinatubo deposited about 1×10^9 m^3 of pyroclastic-flow materials on its eastern slopes, of which about 3×10^8 m^3 is in the Pasig-Potrero watershed (W.E. Scott and others, this volume; Pierson and others, 1992). About 40 percent of this, or roughly 1.2×10^8 m^3, is expected to be washed down as lahars over the next few years (Pierson and others, 1992).

In 1991, lahars in the Pasig-Potrero River started to occur during and shortly after the climactic eruption on June 15 (Major and others, this volume). The most destructive lahar occurred on September 7, 1991; it buried parts of the town of Bacolor in Pampanga with 1 to 3 m of volcanic debris. This lahar was associated with the breakout of a transient debris-dammed lake in the headwaters of the river. By the end of the 1991 lahar (rainy) season, an estimated 50×10^6 m^3 of volcanic materials had been deposited on the alluvial fan of the Pasig-Potrero River.

The Pasig-Potrero River continued to be an active lahar channel in 1992, conveying sediments to its lower and middle reaches and damaging previously unaffected areas. This paper describes events that occurred along the Pasig-Potrero River during the 1992 lahar season.

METHODS

Visual observation and monitoring at the Mancatian watchpoint (fig. 1) by the Philippine Institute of Volcanology and Seismology's (PHIVOLCS') eastside lahar monitoring team started on July 1, 1992. Unpublished reports that were written by the team after each major lahar event were a primary reference during preparation of this paper. Information about events prior to this date, and other events that escaped monitoring, is from interviews, post-event field surveys, and data supplied by the Philippine National Police (PNP) Watchpoint Center (WPC). WPC data were obtained through the two-way radio linking the Pinatubo Volcano Observatory at Clark Air Base with the PNP.

CHRONOLOGY OF 1992 LAHARS IN THE PASIG-POTRERO RIVER SYSTEM

The 1992 lahars in the Pasig-Potrero River were triggered by (1) localized rainshowers before the rainy season, (2) monsoonal rains and thunderstorms during the rainy season, (3) extended monsoonal rains ("siyam-siyam") heightened by a tropical cyclone, and (4) release of impounded water.

A localized afternoon rainshower on April 11 produced the first 1992 lahar of the Pasig-Potrero River. This and all subsequent lahars up to the onset of the rainy season in early June were triggered by localized rainshowers, most of which occurred in the late afternoon or early evening. Almost all of the flows were small, distinct, single-pulse, muddy to hyperconcentrated streamflow events lasting for a few minutes and resulting in channel scour along the middle and lower reaches of the channel. Discharge ranged from 2 to 20 m^3/s. The May 10 event was a small, hyperconcentrated streamflow that was sustained for 1½ h by 41 mm of rainfall during a period of 2 h.

[1]Philippine Institute of Volcanology and Seismology.

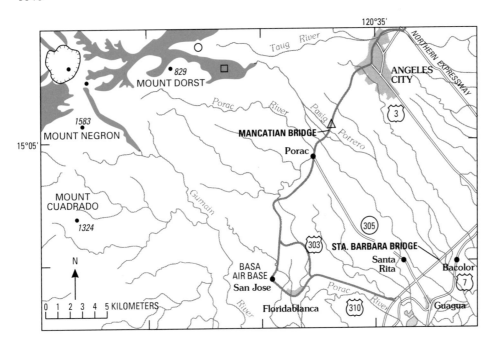

Figure 1. Lahar monitoring network of the Pasig-Potrero River system. The open circle marks the location of the rain gauge (RG), and the open square marks the location of the acoustic flow sensor (FS). The open triangle near Mancatian Bridge was the PHIVOLCS watchpoint (now covered). Red areas are the 1991 pyroclastic-flow deposits.

During the early part of the 1992 rainy season (early to late June), before the southwest monsoon intensified, short-duration monsoonal rains triggered lahars from 0.2 to 0.5 m deep and 5 to 20 m wide with peak discharges ranging from 2 to 50 m³/s. These were mostly hot, hyperconcentrated streamflows and were predominantly erosional at the Mancatian watchpoint, located about 10 km downstream from the toe of the nearest 1991 pyroclastic-flow deposit (fig. 1).

The first cyclone to hit the country in 1992, tropical depression Asyang, did not intensify the southwest monsoon. It produced 60 mm of rainfall that generated only a small, hyperconcentrated streamflow. On July 11, Typhoon Konsing spawned 80 mm of rain that triggered a pumice-rich, multipulsed, hot, hyperconcentrated streamflow and debris flow. The flow reeroded and carried 4-m-long chunks of freshly deposited lahar terrace. The peak flow, with a depth of at least 2.5 m, was laminar with an average velocity of 10 m/s and a discharge of at least 700 m³/s. Samples taken during the waning stage of the peak flow yielded con-·centrations of 42 to 52 percent sediment by weight.

Intensified southwest monsoon winds, bringing 83 mm of rain on July 13, triggered a moderate-sized (50 to 150 m³/s), multipulsed, hot, hyperconcentrated streamflow and also caused secondary explosions and a secondary pyroclastic flow from still-hot 1991 pyroclastic-flow deposits in the headwaters of the Pasig-Potrero River (Torres and others, this volume). The secondary pyroclastic flow partially filled the Pasig-Potrero channel from the foot of Mount Dorst down to 250 m in altitude and blocked a tributary at the foot of Mount Cutuno (fig. 2). Damming of this tributary led to formation of a small transient lake at that location.

In spite of the heavy rains spawned by tropical depression Ditang on July 20, no lahar was observed at the Mancatian watchpoint, and nothing was registered on the Pasig-Potrero acoustic flow monitor (AFM). These rains might have triggered a lahar in the uppermost headwaters of the Pasig-Potrero, but any such flow failed to reach Mancatian, because it would have been blocked by or dissipated on deposits of the July 13 secondary pyroclastic flow. Continuous rain eventually created new rills and gulleys on the fresh deposits of the secondary pyroclastic flow, reintegrating drainage, so that monsoonal rains induced by the tail end of tropical depression Ditang the next day brought a small, hot debris flow past the Mancatian watchpoint. Flow width ranged from 3 to 18 m, while flow depth ranged from 0.5 to 1.7 m; peak discharge was 58 m³/s. The sediment concentration in one sample was 68 percent by weight. Estimated temperature was 40–50°C.

The southwest monsoon was strongest from the last week of July to the last week of August, but lahars generated during this period remained abnormally minor. Lahars produced from July 27 to August 28 were small, with flow width ranging from 2 to 25 m, flow depth from 0.2 to 1.5 m, and velocities from 2 to 5 m/s. Discharge ranged from 4 to 50 m³/s. These lahars were erosional in the middle reaches, past Mancatian. Flows were not steaming and were only lukewarm to the touch (30–40°C). Apparently, lahars that reached the Mancatian watchpoint were formed only from the distal portion of the July 13 secondary pyroclastic flow deposit; flows from higher in the watershed were temporarily diverted into the small lake near Mount Cutuno during late July or early August.

A slightly larger lahar occurred on August 4, triggered by a sudden heavy thunderstorm over the alluvial fan itself, downslope from the rain gauge. Its peak flow at Mancatian was 3 m deep, 15 m wide, and had a velocity of 4 m/s. Rain

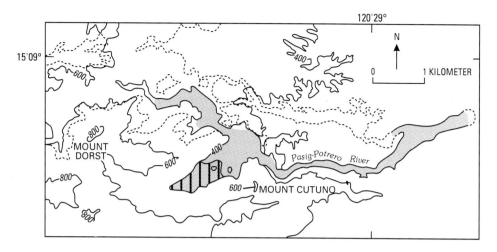

Figure 2. July 13, 1992 secondary pyroclastic-flow deposit (stippled). Areas enclosed by the dotted lines are the 1991 primary pyroclastic-flow deposits. The striped area is where the 1992 transient lake was formed.

that began with tropical depression Gloring on August 17 and lasted up to August 22 produced generally small, hyperconcentrated streamflows.

Continuous heavy rain in the Pasig-Potrero watershed, starting on August 27, initially triggered only small muddy to hyperconcentrated streamflows. However, at about noon of August 29, a major debris flow reached Mancatian. The debris flow was hot, laminar, pumice-rich, and carried chunks of collapsed new lahar terraces as much as 5 m long. Flow width ranged from 30 to 45 m, flow depth from 2 to 4 m, and velocity from 4 to 10 m/s. Peak discharge was estimated to be about 1,400 m³/s. The flow was uniform, maintaining its laminar character for about 9 h. This lahar filled much of the Pasig-Potrero channel with sediment, first in the leveed reach below Mancatian and then upstream even above Mancatian.

The August 29 lahar incised the deposit of the July 13 secondary pyroclastic flow and thereby cleared the channel to convey lahars from the headwaters of the watershed. Subsequent lahars triggered by monsoonal rains, sometimes strengthened by passing weak tropical cyclones, were small- to moderate-sized hot, hyperconcentrated streamflow or debris flow. Typically, these lahars after August 29 were 0.3 to 1.5 m deep, 5 to 15 m wide, and had velocities of 2 to 4 m/s.

The last significant lahar event in 1992 occurred on October 26, during the passage of Typhoon Paring. This was a relatively small, hot, hyperconcentrated streamflow that resulted in minor scouring of the channel bed along the alluvial fan.

LAHAR FREQUENCY

In 1992, 62 lahars were observed or detected by the Pasig-Potrero acoustic flow monitor. From April until early July, lahars occurred at the rate of 1 every 5 days. From July 18 to August 28, when the southwest monsoon was at its strongest, there was a lahar, on average, every 1.6 days.

Table 1. List of lahar events from May 10 to August 28, 1992. Lahars from July 21 to August 27 are believed to have originated in the lower part of the Pasig-Potrero watershed, below most of the July 13 secondary pyroclastic-flow deposit.

Date	Rainfall intensity (mm/min)	Rainfall duration (min)
May 10	0.82	36
May 20	.68	31
June 19	.39	11
June 27	.11	222
July 11	.16	168
July 13	1.28	47
July 21	.25	47
July 27	.25	42
July 28	.76	40
August 4	.22	22
August 18	.13	60
August 19	.25	13
August 26	.25	40
August 27	.68	13
August 28	.17	474

Some days during this period had 2 distinct lahars. Probably, more lahars would have been observed during this period were it not for the channel blockage by the July 13 secondary pyroclastic-flow deposit and later diversion of flow into the temporary lake. After a channel was reestablished through the secondary pyroclastic-flow deposit on August 29, and when the southwest monsoon had considerably weakened, the occurrence rate was one every 3 days.

THRESHOLD RAINFALL

Table 1 is a list of rainfall-triggered lahars at the Pasig-Potrero River from May 10, 1992, to August 28, 1992. Ideally, we would have determined the amount of rainfall needed to trigger lahars on the basis of events

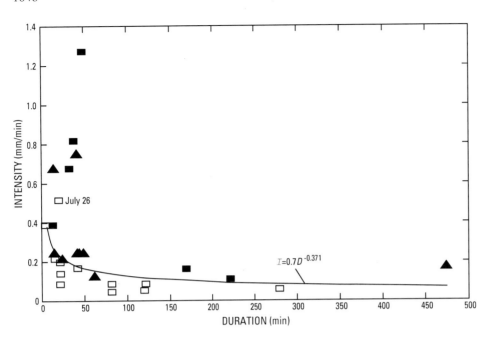

Figure 3. Threshold rainfall curve for 1992 Pasig-Potrero lahars. Rainfall indicated by shaded rectangles generated lahar events before July 13; rainfall indicated by shaded triangles produced lahar events from July 21 to August 28, 1992. Open rectangles indicate rainfall that (apparently) did not trigger any lahar. The curve represents the maximum rainfall that **does not** trigger lahars, rather than the minimum (apparent) rainfall that **does** trigger lahars (Tuñgol and Regalado, this volume).

occurring before the secondary pyroclastic flow of July 13. However, because the number of data points preceding July 13 is small, we have considered events both before and after the secondary pyroclastic flow. The results is a higher overall threshold than probably would have been estimated had the secondary pyroclastic flow not occurred.

The method of data treatment by Rodolfo and Arguden (1991) was adopted. Rainfall events selected were those preceded by a pause of at least 1 h and immediately followed by a flow as recorded by the flow sensor. Only those flows with AFM amplitudes of 100 units were considered (see related work by Tuñgol and Regalado, this volume).

A power curve fitting routine (fig. 3) gives the resulting equation for the threshold rainfall for the 1992 Pasig-Potrero lahars:

$$I = 0.7D^{-0.37}$$

where I = rainfall intensity in millimeters per minute, and

D = duration in minutes.

Lahars postdating August 29 were not considered because the Pasig-Potrero acoustic flow monitor failed during the morning of August 30.

Rainfall on July 26 is above our threshold curve yet did not result in a lahar (fig. 3). Either that rainfall was localized within a small area near the rain gauge or, as we suspect, any lahar that had been generated was trapped by temporary diversion of drainage into the small lake near Mount Cutuno. The first streams that were reestablished through the July 13 secondary-pyroclastic-flow deposit drained into the lake (C. Newhall, written commun., 1993). Interestingly, the threshold for triggering lahars in the Pasig-Potrero was similar to that in the neighboring Sacobia drainage (Tuñgol and Regalado, this volume), despite

temporary disruption of drainage in the Pasig-Potrero by the secondary pyroclastic flow of July 13.

CHANNEL RESPONSE AND IMPACTS

Muddy to hyperconcentrated streamflows from April to June 1992 were erosional, so scour resulted along the upper and middle reaches of the channel. Erosion of the channel floor was noticeable at the Mancatian watchpoint and at other observation points along the middle and lower reaches of the channel. Significant aggradation occurred during lahars of Typhoon Konsing of July 11, but the deposition was confined within manmade levees between Mancatian and the more distant Santa Barbara bridge. Lahars after the July 13 secondary pyroclastic flow were predominantly erosional throughout this reach, although there were instances of thalweg filling by as much as 0.5 m in some portions of the channel near the Santa Barbara bridge (fig. 1). On August 29, a debris flow generated during the early stage of the event began aggrading the channel about a kilometer downstream of the Santa Barbara bridge. Then, aggradation migrated progressively upstream within the channel and completely filled the levee-bounded channel up to the location of the Mancatian bridge. Because the channel was full, lahars eventually avulsed on both sides of the channel, inundating sugarcane fields and burying parts of Barangay Mitla under 1 to 2 m of debris (fig. 4). Succeeding lahars flowed along the southwest side of the Pasig-Potrero River, following the Sapang Matua and Quiratac Creeks and further aggrading the Mitla area by at least another meter. The September 4 lahar flowed further downstream toward the barangays of Balas and San Isidro (fig. 4). By September 8, lahars that passed through the

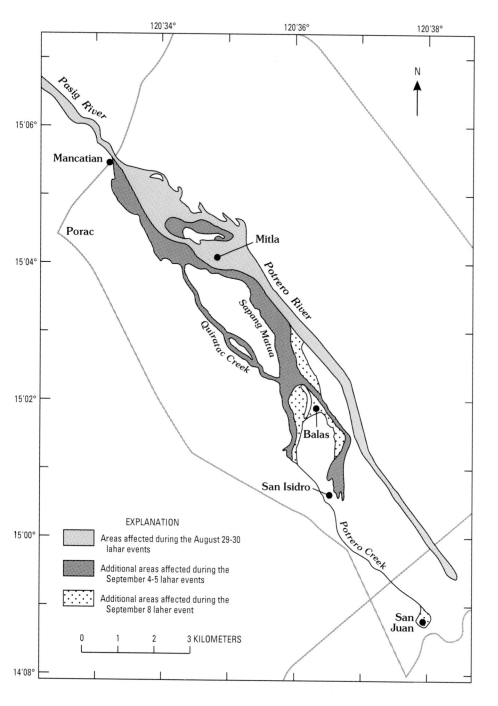

Figure 4. The 1992 lahar deposits of the Pasig-Potrero River system. This map shows the sequence of lahar deposition in the alluvial-fan section of the Pasig-Potrero River system from August 29 to September 8, 1992. Note that the complete areas covered by the September 4–5 and September 8 events are not shown; only those areas which were covered for the first time during those events are shown. Thus, for example, the September 8 event covered areas with coarse stippling *and* a substantial part of the area that had been covered by earlier lahars.

Patutero Creek overflowed at Barangay San Juan, Guagua, covering some rice fields with a thin veneer of lahar deposits (fig. 4). By September 9, the 1992 lahars in the Pasig-Potrero River had affected approximately 9 km^2 of prime agricultural and residential land.

Debris flows of September 21 and 25 aggraded the channel bed near Mancatian to the level of the surrounding areas (A.S. Daag, oral commun., 1992). Near Mitla, lahars breached a portion of the diked channel from the outside, so that by the waning stage of the September 25 lahar, a portion of the flow reentered the original channel. The October 21 lahar flowed through the main channel, and by October 26 downcutting and channel scouring had resumed.

ADDITIONAL OBSERVATIONS AND THE AUGUST 29 LAHAR IN PASIG-POTRERO RIVER

Muddy and hyperconcentrated streamflows from August 13 to 28 were small, lasted no more than 52 min,

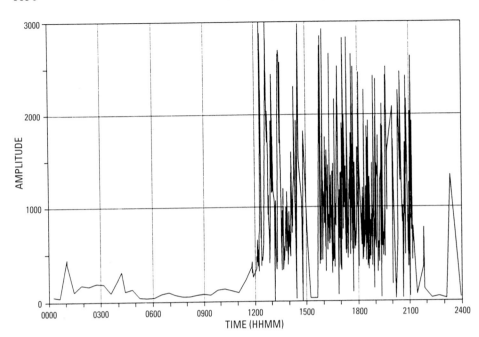

Figure 5. Lahar events in the Pasig-Potrero River on August 29, 1992, as recorded by the acoustic flow monitor. Flows associated with the lake discharge were detected at 1219.

and correlated well with rain that fell on the Pasig-Potrero watershed during this period. On August 29, a major lahar in Pasig-Potrero River began suddenly. Heavy rain had been falling for 2 days before the event, but only sediment-laden streamflow had been reaching Mancatian. About 1200 on August 29, without any significant increase in rainfall, a major debris flow began. After an initial surge of at least 1,400 m³/s, the discharge fluctuated (without noticeable pulses) between 500 to 700 m³/s. This continuous, rapid, uniform flow lasted for about 9 h (fig. 5) as laminar flow, except for short-lived (up to 3 min), localized episodes of turbulence marked by antidunes.

Even when the rainfall intensity changed, there was no corresponding change in the discharge or the character of the flow as observed from the Mancatian watchpoint. From 1552 to 1753, the rainfall intensity increased to 5 mm/min from a previous average of 0.08 mm/min. However, discharge fluctuated only between 580 and 700 m³/s, as noted above. Apparently, the effect of rainfall was being buffered or swamped by some other factor.

During an aerial survey on August 13, the lake at the foot of Mount Cutuno occupied two-thirds of the tributary valley and covered an area of approximately 3×10^5 m² (fig. 6A). When next seen on September 7, the area of the lake had been reduced to about 7×10^4 m² (fig. 6B). During the latter survey, the lake was draining out through a narrow bedrock notch (fig. 6C), which probably controlled discharge from the lake. The amount of water that was released from the lake (on the order of 10^6 m³ of water) is roughly comparable to the volume of water that would have been needed to mobilize discharge of 600 m³/s of lahar for 9 h (between 10^6 and 10^7 m³ of water), especially if one adds in water available in the channel from rains of that day.

The above observations suggest that the lake started discharging on August 29. The releasing water quickly bulked up to form a major debris flow. Because flow was controlled by a bedrock lip, it was sustained for 9 h; without the bedrock lip, flow might have been much more catastrophic than it was.

CONCLUSIONS AND HAZARD ASSESSMENT

By the end of the 1992 lahar season, about 38×10^6 m³ of old and new pyroclastic materials had been delivered to the lowlands along the Pasig-Potrero River. The main bulk of this was carried during the August 29 lahar, the biggest lahar in the Pasig-Potrero River for this year. The volume of 1991 and 1992 lahar deposits, combined, is about 88×10^6 m³.

Despite these voluminous lahars, there is still a great volume of potential lahar materials left at the headwaters of Pasig-Potrero River. With some lahars triggered by only 5 mm of rain in 20 min, we expect that the Pasig-Potrero River will have lahar problems for years to come. Judging from events in 1992, small and highly diluted lahars will be generated during the early stages of the next rainy season, scouring and laterally eroding the channel. The initial channel response will be to convey lahars all the way down to distal reaches of the alluvial fan. Then, subsequent big and sustained lahars will aggrade and reduce channel capacity, with the effect starting in the lower reaches and moving headward. Recessive flows from lahars and small diluted flows will resume scouring and might restore part of the channel conveyance capacity for the lahars in the next rainy season, though not necessarily within the original,

Figure 6. The transient lake at the foot of Mount Cutuno before and after the August 29 lahar. *A,* View (from the north) of the incipiently dissected secondary pyroclastic flow, August 6, 1992, showing diversion of drainage from the Pasig-Potrero watershed diverted around the tip of a ridge and into the transient lake. *B,* The transient lake on September 7, 1992, as seen from the northeast. By that date, much of the lake had drained through the gully visible along the left side of this photograph. *C,* The transient lake on September 7, showing the small notch between a bedrock knob and the canyon wall, through which lake water was draining. View is from the southwest.

A

B

engineered channels. This pattern will probably be repeated in the next several years, possibly accompanied by events such as large secondary explosions or secondary pyroclastic flows, and lake drainage, such as occurred in 1992.

With heavy deposition from 1992 lahars on the upper parts of the Pasig-Potrero alluvial fan, the potential for channel avulsion near and above Mancatian increases. If the lahars escape the channel to the northeast, the barangays of Pasig and Manibaug are threatened. If the lahars avulse to the southwest, then the lahars will inundate the floodplains southwest of the channel before impacting the sand dike

that separates the floodplains of Pasig-Potrero River from those of the Porac River. Breaching or overtopping of the sand dike will depend on the volume of source materials remaining and the channel morphology at that time; in the event of such a breach, the Pasig-Potrero River could be diverted into the Porac River, threatening Porac town proper as well as areas downstream. Hazard will be highest during periods of heavy rain, but lahar watchers must also be on guard for sudden release of any impounded water, even on rainless days.

Figure 6.—Continued.

C

Editor's note added in proof (August 1995): In October 1993, the Pasig-Potrero River captured the uppermost watershed of the Sacobia River. Lahar activity on the Pasig-Potrero in late 1993, all of 1994, and to date in 1995 has been similar to that of 1992, and substantially higher than it would have been had the channel capture not occurred. In 1994, lahar deposits dammed the same tributary near Mount Cutuno that was dammed in 1991 and 1992 (fig. 2), and breakout of that lake in September 1994 generated lahars that buried several barangays north and west of the original channel, including San Antonio, Parulog, and San Vicente, Bacolor.

ACKNOWLEDGMENTS

The authors thank the entire Eastside Lahar Monitoring Team for their support, especially in the data gathering phase. Geodetic surveys by Elmer Gabinete and the Ground Deformation Team, data processing and curve fitting by Ishmael Narag, data retrieval and computer work by Theresa Regalado, material and manpower support from the PVO-CAB staff, photographic documentation by D.V. Javier and A.S. Daag, information from the PNP-WPC police watchpoints, logistical support by the Clark Air Base Command, and reviews of this paper by C.G. Newhall, K.M. Scott, and M.T. Dolan are highly appreciated.

REFERENCES CITED

Major, J.J., Janda, R.J., and Daag, A.S., this volume, Watershed disturbance and lahars on the east side of Mount Pinatubo during the mid-June 1991 eruptions.

Pierson, T.C., 1992, Rainfall-triggered lahars at Mt. Pinatubo, Philippines, following the June 1991 eruption: Landslide News, no. 6, p. 6–9.

Pierson, T.C., Janda, R.J., Umbal, J.V., and Daag, A.S., 1992, Immediate and long-term hazards from lahars and excess sedimentation in the rivers draining Mount Pinatubo, Philippines: U.S. Geological Survey Water-Resources Investigations Report 92–4039. 35 p.

Rodolfo, K.S., and Arguden, A.T., 1991, Rain lahar generation and sediment-delivery systems at Mayon Volcano, Philippines, *in* Fisher, R.V., and Smith, G.A., eds., Sedimentation in volcanic settings: SEPM Special Publication no. 45, p. 71-88.

Scott, W.E., Hoblitt, R.P., Torres, R.C., Self, S, Martinez, M.L., and Nillos, T., Jr., this volume, Pyroclastic flows of the June 15, 1991, climactic eruption of Mount Pinatubo.

Torres, R.C., Self, S., and Martinez, M.L., this volume, Secondary pyroclastic flows from the June 15, 1991, ignimbrite of Mount Pinatubo.

Tuñgol, N.M., and Regalado, M.T.M., this volume, Rainfall, acoustic flow monitor records, and observed lahars of the Sacobia River in 1992.

SELECTED IMPACTS

When Pinatubo was threatening to erupt, reports of effects and recovery strategies from eruptions elsewhere in the world were a useful starting point for contingency planning; after the eruption, the demand for such reports rose sharply. A full account of the effects of the Pinatubo eruption is beyond the scope of this volume but would be valuable for future workers.

For now, an overview of impacts, with special emphasis on people, is told compellingly by C.B. Bautista in the introductory section. Starting in 1991, the Pinatubo story has been a saga of massive evacuations (at times, reaching more than 200,000 persons), massive multiyear dislocations (as of 1993, more than 50,000 persons), and extensive damage to towns, infrastructure, and farms. Initially, the indigenous Aeta population was hit hardest; later, they were joined by an even greater number of lowlanders, routed from their homes and land by lahars. Substantial social changes are underway—some temporary, some irreversible and permanent.

Impacts on buildings, described by Spence and others, were unusually serious because the ash was wet (from typhoon rains) and therefore heavier than it would have otherwise been. Impacts of ash on industries, health, and agriculture are topics for future discussion by others.

Economic impacts of the eruption and lahars are described by Mercado and others, who see a serious but temporary interruption in the economy. Immediate impacts such as loss of income from agriculture and from the U.S. bases are expected to diminish as crops recover and new industry fills the former military bases. Investor confidence and regional infrastructure are also expected to recover.

One unusually wide-reaching impact was that to jet aircraft (Casadevall and others). Commercial and military aircraft in the Philippines were generally alerted to the potential for adverse effects from ash and were accordingly moved out of the way or diverted in flight before the main eruption. However, a number of commercial jet aircraft flying across Southeast Asia either failed to receive or ignored warnings of the approaching ash cloud and suffered substantial damage. Fortunately, no crash resulted. These impacts came virtually on the eve of a scheduled symposium in Seattle, led in part by the USGS, to detail and publicize the potential hazards of ash to aircraft and to stimulate constructive protective measures. Registration for the symposium was good before Pinatubo but jumped even higher as soon as the aviation community learned of encounters with Pinatubo ash.

The last paper in the volume (Self and others) reviews even farther reaching impacts on the global atmosphere. Atmospheric scientists have mobilized in force to quantify and track the unusual "tracer" signal from Pinatubo (including sulfate aerosol, of interest for questions of climate change). Self and others' paper offers a bridge from this geological collection to the extensive atmospheric sciences literature on Pinatubo impacts.

"Hot cars" on Clark Air Base, blanketed by tephra and washed by torrents of water and debris that overflowed storm drains.

Two generations of schools at Santa Barbara, Bacolor. The older is buried in 1991 lahar deposits; the younger was built in 1992.

School is out: kids atop lahar-buried school in Bamban.

Building Damage Caused by the Mount Pinatubo Eruption of June 15, 1991

By Robin J.S. Spence,[1] Antonios Pomonis,[2] Peter J. Baxter,[3] Andrew W. Coburn,[2] Mark White,[4] Manuel Dayrit,[4] and Field Epidemiology Training Program Team[4]

ABSTRACT

This paper presents the results of a survey of the building damage caused by the ash fall from the cataclysmic eruption of Mount Pinatubo. The survey is a damage assessment based on a scale of damage derived from earthquake-damage survey techniques; the damage was related to aspects of the building form and construction technology. Fifty-one houses in the town of Castillejos were surveyed. Of these, one-third either collapsed or were seriously damaged. It was concluded that the roofs that failed did so because the ash load was greater than their vertical load-carrying capacity and that the nature of the roof supporting structure was the principal factor influencing the level of damage sustained; other significant factors included the overall construction type and the roof pitch. The paper also discusses implications for future research on building damage caused by volcanic eruptions. It is suggested that in situ methods for measuring ash loads need to be developed and that a building classification system based on ash-fall vulnerability should be devised for use by disaster planners.

INTRODUCTION

Ash fall is one of the most destructive phenomena associated with massive plinian events like the cataclysmic eruption of Mount Pinatubo on June 15, 1991. Huge quantities of tephra may be transported downwind for tens, or even hundreds, of kilometers and fall in deposits deep enough to cause severe disruption, including damage to, and collapse of, buildings. Although major loss of life from pyroclastic flows was avoided by a successful and timely evacuation of about 60,000 people living within 30 km of the volcano, as of July 17, 1991, the official numbers of casualties were 320 dead and 279 injured, and most of these casualties were believed to have been due to the collapse of roofs under the weight of wet ash on June 15 (Pinatubo Volcano Observatory Team, 1991). Of all the major eruptions in recent times the Mount Pinatubo event provided the first opportunity to study the destructiveness of tephra fallout in a densely populated area, though the impact would have been considerably less without the passage of Typhoon Yunya. We present our findings of a photographic survey of buildings in a representative area of heavy ash fall to draw the attention of emergency planners to the importance of this volcanic hazard and to the need for specific preventive measures to be devised in all areas of active explosive volcanism.

THE SURVEY

Soon after the June 15 eruption, the Field Epidemiology Training Program Team of the Philippine Department of Health carried out photographic surveys of some of the damage. Figure 1 shows typical examples from Olongapo City and the town of Subic. The most detailed survey was in Castillejos, a town with a population of less than 50,000, situated 27 km southwest of Mount Pinatubo, north of the Subic Bay area in one of the worst affected areas. The thickness of the ash fall at this point was about 20 cm (fig. 2). The field survey consisted of choosing a sector of the damaged settlement and recording the state of each building in it photographically, whether damaged or not. Between one and four photographs were taken of each building. Other details were recorded on a specially prepared record sheet. The survey was carried out on June 29th, about 2 weeks after the main eruption; at that stage little work to remove fallen roofs or repair the damage had taken place. Altogether, the Castillejos survey comprised 51 buildings.

Subsequently, the photographs and other information were sent for analysis to Cambridge, where both building typologies and damage levels for all buildings were

---

[1] Department of Architecture, University of Cambridge.

[2] Cambridge Architectural Research Ltd., The Eden Centre, 47 City Road, Cambridge CB1 1DP, U.K.

[3] Department of Community Medicine, University of Cambridge, Institute of Public Health, Forvie Site, Robinson Way, Cambridge CB2 2SR, U.K.

[4] Department of Health, Manila, Philippines.

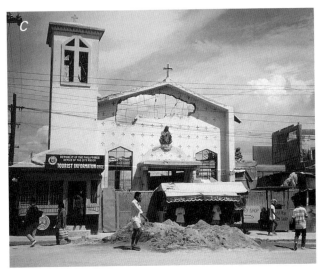

Figure 1.　Examples of ash-fall damage in Olongapo City and the town of Subic at the end of June 1991. *A*, Homes and commercial buildings, Olongapo. *B*, Roadside stores, Olongapo. *C*, Church with complete roof collapse, Olongapo. *D*, Collapsed canopy of gasoline station, Olongapo. *E*, Collapsed roof and upper story of a mixed building with a reinforced concrete lower story and a wooden upper story, plus concrete firewall. *F*, Damage to roof and eaves of school. *G*, Collapsed roof of Olongapo health center.

Figure 1.—Continued.

assessed. The building type analysis included identification of:

- principal constructional materials used
- number of stories
- roof structure, shape, and pitch
- building usage (residential or nonresidential)

For damage assessment, a scale of damage categorization appropriate to ash-fall damage was needed. The MSK earthquake intensity scale (Karnik and others, 1984), which defines six levels of damage, has proven easy and reliable for postearthquake damage assessment, and it was decided to adapt this scale for use in the present survey. The six degrees of damage were defined as:

D0: *No damage.*

D1: *Light roof damage.*—Gutter damage; few tiles dislocated.

D2: *Moderate roof damage.*—Bending or excessive deflection of roof sheeting or purlins; no damage to principal roofing supports.

D3: *Severe roof damage and some damage to vertical structure.*—Severe damage or partial collapse of roof overhangs or verandahs; severe deformation of main roof sheeting; some damage to roof supporting structure, columns, trusses.

D4: *Partial roof collapse and moderate damage to rest of building.*—Collapse of sheeting but not truss; partial collapse of sheeting and some truss failure; failure of supporting structure; moderate damage to other parts of building resulting from roof collapse.

D5: *Complete roof collapse and severe damage to the rest of the building.*—Collapse of roof and supporting structure over more than 50 percent of roof area; partition walls destroyed; external walls destabilized.

Table 1. Distribution of buildings by construction type.

Construction types	Number	Percent
Reinforced concrete	21	41
Timber frame	17	33
Confined masonry	9	18
Steel frame	1	2
Not known	3	6

Table 2. Distribution of buildings by roof support.

Roof support	Number	Percent
Long span	12	24
Short span	37	72
Not known	2	4

CONSTRUCTION TYPOLOGIES AND USAGE

Table 1 shows the distribution of the overall construction types identified. The confined masonry buildings had concrete block walls with reinforced concrete columns in corners or near openings, but there were no confining beams. Some buildings had ground floors of masonry or reinforced concrete and an upper floor of timber frame. In the analysis of damage, these were classified as timber buildings, since it was the upper floor that was exposed to the ash fall.

Table 2 shows the analysis of roof types. In virtually all cases the roof covering material was corrugated galvanized steel sheet, of indeterminate gauge, but the supporting structure varied. In most cases roofs were of domestic scale, with short spans: roof sheets rested on purlins that were supported on closely spaced rafters running down the slope of the roof and resting on external walls or ridge beams. In a

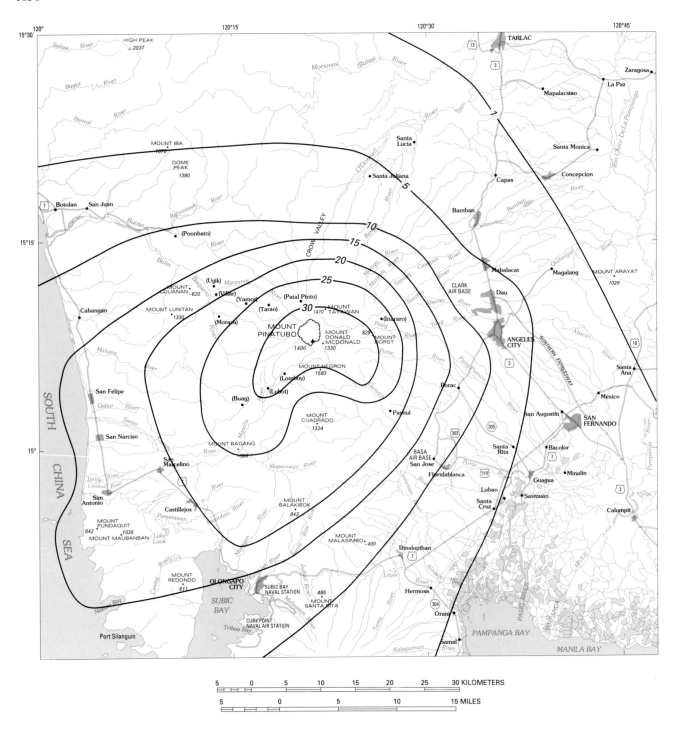

Figure 2. Isopach map of layer C, plinian tephra fall from June 15, 1991. Most of the tephra that caused roof collapse was in this event and layer. An additional tenth (roughly) of these amounts fell on June 12–14, in layers A and B. Thicknesses are in centimeters. Modified from Paladio-Melosantos and others (this volume, fig. 7).

Table 3. Distribution of buildings by roof pitch.

Roof pitch	Number	Percent
≥30°	4	8
20–29°	14	27
10–19°	32	63
Not known	1	2

Table 4. Distribution of the buildings by usage.

Building usage	Number	Percent
Residential	33	65
Commercial	7	14
School	4	8
Church	4	8
Public (offices, etc.)	3	6

smaller number of cases, roofs had much larger spans; roof sheets were supported on more substantial purlins that were in turn supported on more widely spaced trusses spanning between walls or columns a distance of 5 m or more apart. Table 2 shows the observed distribution of the buildings between the different principal roof types. The pitch of the roof also varied; in most cases it was below 20°, but in a smaller number of cases it was between 20° and 30° (table 3).

The distribution of buildings by usage category is shown in table 4. Over half were residential buildings, and the rest a mixture.

DAMAGE DISTRIBUTION

The damage distribution over the whole sample is shown in table 5. The variation in the level of damage is very considerable: about one-third of the sample suffered no damage, while another one-third suffered either total or partial collapse.

It is important to try to identify the significance of the various attributes of the construction classification in contributing to damage potential. Because numbers are small, in subsequent analysis, damage has been divided into just two classes: buildings without significant damage (D0+D1) and buildings with partial or complete roof collapse (D4+D5).

Table 6 shows the distribution of damage with aspects of the building construction and usage. It shows that timber frame buildings suffered significantly more damage than reinforced concrete frame buildings, that buildings with steeper roof pitches suffered worse than those with shallower pitches, and that nonresidential buildings were much more vulnerable than residential ones. The most significant indicator of damage, however, was the roof supporting

Table 5. Damage distribution of the whole sample.

[D0, no damage; D1, light roof damage; D2, moderate roof damage; D3, severe roof damage and some damage to vertical structure; D4, partial roof collapse and moderate damage to rest of building; D5, complete roof collapse and severe damage to the rest of the building]

Damage degree	Number	Percent	Cumulative percent
D0	15	29	100
D1	3	6	71 (≥ D1)
D2	8	16	65 (≥ D2)
D3	8	16	49 (≥ D3)
D4	9	18	33 (≥ D4)
D5	8	16	16 (D5)

Table 6. Damage distribution by characteristics of construction and usage.

Construction or usage	D0+D1 (%)	D4+D5 (%)
Construction type:		
Reinforced concrete	57	14
Timber frame	24	24
Roof pitch:		
≥ 20°	22	56
10–19°	44	19
Roof support:		
Long span	17	75
Short span	43	16
Building usage:		
Residential	39	15
Nonresidential	28	67

structure. Only 16 percent of the short-span roofs suffered major damage, with 43 percent having no significant damage; while of the long-span roofs, 75 percent were severely damaged, and only 17 percent were without significant damage. Long-span roofs were, therefore, nearly 5 times more likely to suffer major damage than short-span roofs.

SUMMARY OF DAMAGE

1. Although many roofs had been cleared by the time of the survey, there was evidence from the uncollapsed roofs that the wet ash was able to accumulate to depths of at least 15 cm on metal sheet roofs of pitch up to 25°, without slippage.

2. Out of the total sample of 51 buildings in Castillejos, 17 suffered partial or complete roof damage, while 18 suffered no damage or only light damage.

3. Buildings tended to suffer worse damage if they were (a) constructed with long-span roofs (greater than 5 m clear span), rather than with short-span domestic scale construction, (b) of timber frame rather than reinforced

concrete frame construction, (c) of higher rather than lower roof pitch, or (d) non-residential rather than residential.

4. Some other factors that seem to have contributed to damage, though statistical evidence is inadequate to demonstrate their significance are (a) unbraced supporting walls or columns, and (b) large unsupported roof overhangs.

The principal cause of the damage was that the load of ash on the roof exceeded the strength either of the roof sheets or of the roof supporting structure, or both. This is to be expected, because the load of 15–20 cm of water-saturated ash on the roof would probably have exceeded 2.0 kN/m^2 (approximately 200 kg/m^2), whereas typical design loads for pitched roofs in tropical cyclone areas would rarely approach this level even for engineered structures, and they would be designed principally to resist wind suction forces. Some buildings, by virtue of their form of construction, had a better inherent resistance than others, and this explains in part the large variation in performance, although some roofs may also have been cleared during the first 2 days of the eruption. In particular, it appears that the spacing and span of the principal roof-supporting members was critical. Domestic-scale roofs supported on closely spaced rafters seemed to have a much greater reserve of strength than the longer span roofs. This is perhaps because the longer span roofs would have been designed more precisely for the expected wind and maintenance load. It is not clear why roofs with steeper pitch were more severely damaged, but it is worth noting that none had a slope great enough to overcome the frictional resistance of the roof sheeting material.

The principal design recommendation arising from this study is that, to protect lives, roofs of buildings exposed to possible ash fall should be designed for a superimposed load related to the probable level of ash fall, in a manner analogous to design for snow loading in cold climates. In the case of Castillejos, a design load of 2.0 kN/m^2 would almost certainly have been sufficient to avoid collapse during the 1991 eruption.

To our knowledge, this study is the first published attempt to investigate the impact of heavy ash fall on buildings, and its deficiencies highlight the need for more comprehensive scientific evaluations to be undertaken in future eruptions to enable disaster reduction measures to be developed as one of the goals of the International Decade for Natural Disaster Reduction (IDNDR). The results of surveys of building damage need to be linked to epidemiological surveys designed to define the size of the population at risk in the buildings at the time of the ash fall, the numbers of fatalities, and the numbers and types of injuries encountered. Autopsy studies will be needed to clarify the causes of death, such as severe trauma, asphyxia from inhaling ash, or a combination of the two. Unfortunately, these data were not collected at the time of our survey. Conventional advice to protect property against ash falls is to shovel or sweep by hand as a way of removing ash from roofs, as was done in the eruption of Heimaey, Iceland in 1973 (UNDRO, 1985). Such advice is difficult and risky to follow during a Pinatubo-type eruption. With the heavy fallout of tephra combined with rain and lightning, most people were too afraid to venture onto their roofs, and those who did might have found the ash difficult to shovel and impossible to sweep. Also, experience at Cerro Negro, Nicaragua, and elsewhere has suggested that the risks of falls, death, and injury from working on ash-laden roofs might exceed the risk from spontaneous collapse, so the conventional wisdom about sweeping must be tempered by judgment in any specific circumstances. Some residents reported to one of us how, while sheltering in their houses during the night of June 15, they listened, terrified, to the sound around them of roofs breaking and the accumulated ash sliding inside. Recognition that most of the lightning was intracloud lightning might have eased fears, but probably not by much.

IMPLICATIONS FOR FUTURE RESEARCH

One of the weaknesses of the present survey is that no measurement of the loading actually imposed by the ash on the roofs was possible. It is known that the specific weight of dry ash can vary from 400 to 700 kg/m^3, and that rainwater can increase this by 50–100 percent if the ash becomes saturated (Blong, 1981). Although there was heavy rainfall during the eruption of June 15, the extent to which the ash was saturated, and whether this caused compaction, which would reduce the thickness of the layer, is not recorded. Ash layers of at least 15 cm are certainly observable on some of the roofs in Castillejos that survived the ash fall. In future building surveys, it will be essential to measure not just the thickness, but the insitu specific weight of the ash. A small portable device for this purpose needs to be designed.

Design recommendations such as those given above may help to avoid future life loss in buildings designed in accordance with them. But some assessment of the probable losses of existing settlements at risk, in the Philippines and elsewhere, is also needed to alert civil protection authorities to the scale of evacuation and other protection measures that may be needed. For this purpose, it will be necessary to develop some understanding of the relation between ash-fall loading and the extent of damage, in a manner analogous to the vulnerability functions that have been developed for assessing earthquake damage (Spence and others, 1992). Ideally, such relationships would be built on the basis of extensive field damage data, covering a range of different levels of ash-fall loading and a large range of building types. There is, however, no existing body of such data, and the study reported here provides just one data point for the relation between ash fall and damage level for the two principal building types used locally, so building such relations on the basis of damage data alone will be a lengthy task. However, where enough is known about the structural form and materials of a building at risk, an assessment of its response to any level of loading can be made by calculation of its resistance. It may be possible to extend this approach beyond that of a single building to estimate the damage to a whole population of buildings, if enough is known about the conventional construction technologies in use. But such a predictive approach will depend on the ability to predict future ash-fall loading intensity, not just thickness.

This paper has concentrated on the damage caused by ash fall because it is clear that roof overload was the principal cause of the failures observed in Castillejos. Some earthquake activity also took place during the period up to June 29th, and this activity could in some cases have increased the ash-fall damage, but the ground shaking recorded was not sufficient, in itself, to have caused the damage seen. However, if a predictive approach to damage assessment of buildings is to be developed, it will be important not to ignore the potential building damage from all the hazards associated with possible future eruptions, including earthquakes, tephra bombs, pyroclastic flows, and lahars. as well as ash fall. The distributions of these hazards will differ from that of ash fall, as will the characteristics of buildings that are vulnerable to each hazard. Thus, composite hazard and risk maps will need to be developed. There is a need, however, to develop simple and rapid building and settlement survey techniques to identify buildings at risk and to classify them according to their vulnerability to each of the separate hazards.

Note from the editors: Densities of Pinatubo tephra that were collected at Clark Air Base and weighed later in a laboratory ranged from 1.2 to 1.6 g/cm^3 (dry) and 1.5 to 2.0 g/cm^3 (wet) (G. Heiken and D. Riker, written commun., 1994). If the density of wet tephra in Castillejos, Olongapo, and Subic was approximately 1.5 g/cm^3, a 20-cm thickness of that tephra would have exerted a force of 3 kN/m^2.

REFERENCES CITED

Blong, R.J., 1981, Some effects of tephra falls on buildings, *in* Self, S., and Sparks, R.S.J., eds., Tephra studies: Dordrecht, D. Reidel Publishing Co., p. 405–420.

Karnik, V., Schenkova, Z., and Schenk, V., 1984, Vulnerability and the MSK scale: Engineering Geology, v. 20, p. 161–168.

Pinatubo Volcano Observatory Team, 1991, Lessons from a major eruption: Mount Pinatubo, Philippines: Eos, Transactions, American Geophysical Union, v. 72, p. 545, 552–553, 555.

Office of the United Nations Disaster Relief Coordinator (UNDRO), 1985, Volcanic Emergency Management: New York, United Nations, 86 p.

Paladio-Melosantos, M.L., Solidum, R.U., Scott, W.E., Quiambao, R.B., Umbal, J.V., Rodolfo, K.S., Tubianosa, B.S., Delos Reyes, P.J., Alonso, R.A., and Ruelo, H.R., this volume, Tephra falls of the 1991 eruptions of Mount Pinatubo.

Spence, R.J.S., Coburn, A.W., Pomonis, A., and Sakai, S., 1992, Correlation of building damage with strong ground motion, *in* World Conference of Earthquake Engineering, 10th, Madrid, Spain, Proceedings, v. 1: p. 551–557.

Socioeconomic Impacts of the Mount Pinatubo Eruption

By Remigio A. Mercado,[1] Jay Bertram T. Lacsamana,[1] and Greg L. Pineda[1]

ABSTRACT

The Mount Pinatubo eruptions and their aftereffects, particularly lahars during rainy seasons, not only have taken the lives of many but also have wrought havoc to the infrastructure and to economic activities of Central Luzon. Damage to crops, infrastructure, and personal property totaled at least 10.1 billion pesos ($US 374 million) in 1991, and an additional 1.9 billion pesos ($US 69 million) in 1992. In addition, an estimated 454 million pesos ($US 17 million) of business was foregone in 1991, as was an additional 37 million pesos ($US 1.4 million) of business in 1992. Lahars continue to threaten lives and property in many towns in the provinces of Tarlac, Pampanga, and Zambales.

The actual destruction, coupled with the continuing threat of lahars and ash fall, has disrupted the otherwise flourishing economy of Central Luzon, slowing the region's growth momentum and altering key development activities and priorities. Major resources have been diverted to relief, recovery, and prevention of further damage.

The costs of caring for evacuees (including construction of evacuation camps and relocation centers) was at least 2.5 billion pesos ($US 93 million) in 1991–92, and an additional 4.2 billion pesos ($US 154 million) was spent during the same period on dikes and dams to control lahars.

The longevity and impact of the calamity is so great that the public and private response must go beyond traditional relief and recovery. Return to preeruption conditions is impossible. Instead, responses must create an attractive climate for new investments, provide new livelihood and employment alternatives, promote growth in areas that are safe from future lahars and flooding, and provide an infrastructure that is tough enough to survive future natural disasters.

INTRODUCTION

The eruptions of Mount Pinatubo and their continuing aftereffects have disrupted the flourishing socioeconomic

environment in Central Luzon. Economic growth, which had been spreading to the region from Manila and which was slated for an extra boost from conversion of former U.S. military bases, has been weakened by the eruption. The auspicious development picture for the region has been replaced by uncertainty and delays. For the short and medium term, rehabilitation and reconstruction dominate socioeconomic planning; for the long term, planning needs to take advantage of new opportunities that are presented by massive rebuilding of the socioeconomic infrastructure.

This paper presents actual and projected damage arising from the Mount Pinatubo eruptions, their implications for Central Luzon's development, and broad directions that could be taken to respond to the calamity. This report is neither comprehensive nor complete, as the calamity is still ongoing.

NATURE OF THE DISASTER

During the June 15, 1991, eruption, heavy damage was caused by ash fall, which buried large tracts of land and collapsed roofs of buildings near the volcano. Although ash fell in varying amounts across the whole of Luzon, the most heavily affected provinces were those adjacent to Mount Pinatubo—Pampanga, Tarlac, and Zambales.

Continuing effects are now brought by lahars—rain-induced torrents of loose volcanic debris that flow down the major river systems around the volcano and out into densely populated, adjoining lowlands. Lahars destroy and bury everything along their path: people and animals, farm and forest lands, public infrastructure, natural waterways, houses, and other facilities. Infilling of stream channels has caused overbank flows, drowning of areas behind natural impoundments, and other forms of flooding in low-lying areas.

Secondary explosions also continue—explosions that occur when heavy rain and runoff come in contact with still-hot pyroclastic deposits on the volcano's slopes. These explosions produce fine, powdery ash fall that continues to impact, among other things, the former Clark Air Base.

---

[1]National Economic and Development Authority, Region III, San Fernando, Pampanga, Philippines.

REPORT ON DAMAGE

In the 2-year period since June 1991, the damage from eruptions and their aftereffects has been staggering and debilitating. Worse, it is expected to continue for at least several years more, until lahars no longer occur.

Damage is reported in different ways for different sectors. That for public infrastructure, natural resources, and military facilities is the estimated cost to repair or replace damaged assets. Estimates of damage to trade and industry include the cost to repair or replace facilities and projected income from foregone sales and service. Estimated damage in agriculture is the expected value of yield multiplied by the area damaged.

Except for the basic data on monthly foregone income per industry type (see table 8), which was provided by the National Statistics Office (NSO), all reports on damage provided below came from government agencies and departments involved in rescue, relief, reconstruction, and rehabilitation and were consolidated through the National Disaster Coordinating Council (NDCC). The NDCC used this information to recommend to the President of the Philippines which areas should be declared to be under a state of calamity.

The other legal body that facilitated information gathering on damage was the Presidential Task Force on the Rehabilitation of Areas Affected by the Eruption of Mt. Pinatubo and its Effects, popularly known as Task Force Mt. Pinatubo. Created by President Corazon C. Aquino on June 26, 1991, by Memorandum Order No. 369, Task Force Mt. Pinatubo was mandated to guide all rehabilitation efforts of the government and to coordinate these with the private sector and, whenever necessary, with the international community. In December 1992, the work of Task Force Mt. Pinatubo was taken over by the Mount Pinatubo Assistance, Resettlement and Development Commission (MPC) created under Republic Act No. 7637.

AREAS AND POPULATION AFFECTED

From June 1991 to November 1992, means of livelihood, houses, or both were partially or wholly lost in 364 barangays (villages) (table 1). About 329,000 families (2.1 million people), about one-third of the region's population, lived in these 364 barangays at the time of the 1990 census. In 1991, 4,979 houses were totally destroyed and 70,257 houses were partially damaged. The number decreased in 1992, when 3,281 houses were wholly destroyed and 3,137 units were partially damaged (table 2).

Of the 329,000 families (2.1 million persons) affected, 7,840 families (35,120 persons) were of the Aeta cultural minority (Office for Northern Cultural Communities, unpub. data, August 14, 1991). Although constituting less

Table 1. Total number of barangays affected as of November 17, 1992 (National Disaster Coordinating Council, 1992).

["Affected" refers to a situation where means of livelihood, houses, or both are lost or partially or completely destroyed]

Province	Affected barangays	Number of families
Zambales	96	30,115
Pampanga	173	239,131
Tarlac	88	44,367
Angeles City	5	14,197
Nueva Ecija	2	1,331
Total	364	329,141

Table 2. Total number of houses damaged (National Disaster Coordinating Council, 1992; Presidential Task Force on Mount Pinatubo, 1992; Department of Social Welfare and Development, unpub. data, 1992).

[Partial damage refers to any degree of physical destruction attributed to the disaster. Total destruction is the condition when the house is no longer livable]

Extent of damage	1991	1992	Total
Totally destroyed	4,979	3,281	8,260
Partially damaged	70,257	3,137	73,394
Total	75,236	6,418	81,654

than 2 percent of the total affected population, these cultural minorities have received significant attention.

DAMAGE BY SECTOR

PUBLIC INFRASTRUCTURE

In its damage assessment report as of August 23, 1991, the Department of Public Works and Highways (DPWH) Regional Office III estimated damage to public infrastructure amounting to 3.8 billion pesos (table 3). The gravest destruction was on irrigation and flood control systems, roads and bridges, and school buildings. Additional damage of at least 1 billion pesos was done to roads and bridges by lahars of 1992 (National Disaster Coordinating Council, 1992).

NATURAL RESOURCES

The Mount Pinatubo eruptions buried some 18,000 ha of forest lands in ash fall of about 25 cm (Paladio-Melosantos and others, this volume). The heaviest concentration of ash fall was in the mountains of Botolan and San Marcelino in Zambales, in Porac and Floridablanca in Pampanga, and in Bamban and Capas, Tarlac.

Table 3. Total cost of damage to infrastructure as of August 23, 1991 (National Disaster Coordinating Council, 1992; Presidential Task Force on Mount Pinatubo, 1992; Department of Public Works and Highways, Region III, unpub. data, 1991).

[The prevailing foreign exchange rate during this period was $1 = 27.07 pesos]

Infrastructure subsector/Facility	Damage Cost (in thousand pesos)
Transportation	1,149,908
Communication	13,215
Power and electrification	54,918
Water resources	1,568,642
Social infrastructure	1,045,708
Total	3,832,391

Table 4. Damage to Contract Reforestation and Integrated Social Forestry projects, 1991 (National Disaster Coordinating Council, 1992; Presidential Task Force on Mount Pinatubo, 1992; Department of Environment and Natural Resources, Region III, unpub. data, 1991).

[This damage was caused by ash fall and ash flow. Contract Reforestation—Regular DENR project involving a 3-year plantation and maintenance of forest trees through family-based, community-based, and corporate-based modes of contracting. Integrated Social Forestry (ISF)—community-based/family-based planting of 20% forest trees and 80% agricultural-based crops. Contractors are given a 25-year security of tenure through the certificate of stewardship contract (CSC)]

Province	Contract Reforestation		Integrated Social Foresty	
	Area (hectares)	Value/Amount lost (pesos)	Area (hectares)	Value/Amount lost (pesos)
Zambales	2,108.0	33,576,690.8	799.7	6,136,419.5
Pampanga	2,116.1	19,137,420.8	1,789.8	1,749,500.0
Tarlac	4,842.0	54,440,562.8	1,719.1	760,563.8
Bataan	529.0	8,133,902.2	236.5	1,507,500.0
Total	9,595.1	115,288,576.6	4,545.1	10,153,983.3

Reforestation activities have been seriously set back. Approximately 14,140 ha of newly established plantations were destroyed and some 125 million pesos worth of seedlings were lost (table 4). About 43,800 ha of natural forest cover and old plantations were damaged.

Heavy rains that came after the eruptions caused ash deposits from the mountain slopes to wash down to low-lying areas in the form of lahars. At least eight major river systems have been clogged by lahars (table 5; see also Arboleda and Martinez, this volume; Martinez and others, this volume; Pierson and others, this volume; Rodolfo and others, this volume; K.M. Scott and others, this volume; Umbal and Rodolfo, this volume).

Table 5. Major river systems affected (National Disaster Coordinating Council, 1992; Presidential Task Force on Mount Pinatubo, 1992; PHIVOLCS/NEDA, 1992).

[Total lahar hazard areas include those prone to lahar deposition, siltation and flooding, and bank erosion]

River system	Areas actually affected as of August 1992 (hectares)	Total lahar hazard area (hectares)
Abacan	2,930	4,060
Bucao-Balin Baquero	5,380	8,600
Maloma	1,820	1,700
O'Donnell-Bangut	3,350	11,540
Pasig-Potrero	4,370	10,000
Porac-Gumain	3,140	3,370
Sacobia-Bamban	10,310	25,090
Santo Tomas	4,640	12,590
Total	35,940	76,950

Table 6. Actual damage to agricultural area by commodity as of July 1991 (Department of Agriculture, Region III, unpub. data, 1991; National Disaster Coordinating Council, 1992; Presidential Task Force on Mount Pinatubo, 1992).

Commodity	Area or number damaged	Value (Philippine pesos)
Rice (hectares)	81,895	350,855,594
Vegetables (hectares)	2,486	163,548,456
Rootcrops (hectares)	2,070	182,791,365
Assorted fruit trees (number)	2,646	290,061,075
Fisheries (hectares)	7,129	284,098,228
Livestock and poultry (heads)	778,714	203,191,200
Total		1,474,545,918

AGRICULTURE

About 96,200 ha of agricultural land was seriously affected by ash fall. Damage to crops, livestock, and fisheries was about 1.4 billion pesos (table 6).

Damage from lahars, flooding, and siltation, as of November 17, 1992, was reported to be 778 million pesos (table 7). Of this, crops suffered the biggest damage (547 million pesos), followed by fisheries (165 million pesos), sugarcane (57 million pesos), and livestock (10 million pesos). The estimate of 778 million was later raised to 1,422 million (see table 10).

TRADE AND INDUSTRY

The manufacturing subsector, and consequently the exporting subsector, was heavily damaged. Lost assets for 559 firms totaled 851 million pesos. Foregone production losses for 1991 were reported to be about 45 percent of the potential sales for the year 1991, or 454 million pesos, and 424 million pesos of capital investment was destroyed at 306 surveyed firms. The furniture industry was hardest hit, with damage of 156.5 million in 108 firms. The processed food sector suffered 97 million pesos of loss in 18 firms, and the gifts, toys, and housewares sector lost 60 million pesos in 92 firms.

In 1992, foregone income in the manufacturing subsector was 1.5 million pesos per month, followed by the wholesale and retail subsector with foregone income of 846,000 pesos per month. Foregone income for the financial, real estate, and business services subsector was about 635,000 pesos per month, and that of the transportation, storage, and communication subsector was estimated to be 65 million pesos per month.

Total foregone income during 1992, in all sectors, was 3.1 million pesos. By province, industries in Pampanga and

Tarlac had the greatest share of foregone sales, of 1.7 million and 0.6 million pesos per month, respectively (table 8).

SOCIAL SERVICES SECTOR

Health.—An increase in morbidity and mortality rates occurred mainly in evacuation centers. The leading diseases were acute respiratory infections (ARI), diarrhea, and measles (Department of Health, unpub. data, 1991). The death rate (Aetas and lowlanders combined) was 7 per 10,000 per week during 1991; that for Aetas in 1991 reached as high as 26 per 10,000 per week, and averaged 16 per 10,000 per week (Department of Health, 1992), and was especially high among Aeta children.

Social welfare.—The continuing threat of lahars has required that relief—food, clothing, shelter, and other help—be provided far beyond the period that is normal for typhoons and other calamities. As of October 28, 1993, approximately 1,309,000 people were being served outside evacuation centers. As of the same date, 159 evacuation centers were being maintained by the Department of Social Welfare and Development (DSWD) throughout Region III, housing some 11,455 families or 54,880 persons and providing them with food-for-work or cash-for-work assistance.

Education.—Destruction of about 700 school buildings with 4,700 classrooms displaced an estimated 236,700 pupils and 7,009 teachers. Damage to school buildings was estimated to be 747 million pesos as of August 1991 (table 9), an amount that is growing with continuing lahar activity. (Note: This value is also included within the category of social infrastructure in table 3.) Disruption of schooling is compounded by use of undamaged school buildings as evacuation centers, which forces delays in the opening of classes and causes other disruptions of the school calendar. Initial damage to instructional materials,

Table 7. Existing damage to agricultural commodities (in million pesos; Department of Agriculture, Region III, unpub. data, 1991; National Disaster Coordinating Council, 1992).

[Damage cost = total area damaged × expected yield per hectare. Expected yield is computed by referring to precalamity yield. Postcalamity yield is derived by referring to precalamity yield and subjecting the damaged crops to recovery chances/percentages. The value of the crops with negative chances/percentages is derived by multiplying them by the prevailing market prices of the crops. This value then becomes the damage cost.]

Commodity	1991	1992	Total
Crops (hectares)	987.2	546.8	1,534.0
Livestock (heads)	203.2	9.8	213.0
Fisheries (hectares)	284.1	164.9	449.0
Sugarcane (hectares)		56.9	56.9
Total	1,474.5	778.4	2,252.9

Table 8. Monthly foregone gross income per industry type per affected province for 1992 (National Statistics Office, Region III, unpub. data, 1992)

[The estimated loss to industry was based on the proportion of the affected households and their average expenditure on each type of industry. Does not include construction (52,689 pesos). No provincial breakdown available]

Province	Manufacturing	Wholesale/Retail	Transport/Storage/ Communication	Financing institution/ Real estate/ Business services	Total for province
Bataan	45,000	23,000	25,000	180,000	273,000
Bulacan[1]	23,000	5,100	590	620	29,310
Nueva Ecija	23,000	12,000	2,100	1,700	38,800
Pampanga[2]	1,100,000	540,000	22,000	83,000	1,745,000
Tarlac	230,000	190,000	8,900	190,000	618,900
Zambales	103,000	76,000	6,100	180,000	365,100
Total	1,524,000	846,100	64,690	635,320	3,070,110

[1] Municipality of Calumpit only.
[2] Includes Angeles City.

Table 9. Estimated cost of damage to school buildings by province or city as of August 12, 1991 (National Disaster Coordinating Council, 1992; Presidential Task Force on Mount Pinatubo, 1992; Department of Education, Culture, and Sports, Region III, unpub. data, 1991).

[Ash fall is the major cause for this type of damage]

Province/City	Cost (in thousand pesos)
Zambales	410,000
Bataan	34,000
Olongapo City	140,000
Pampanga	130,000
Tarlac	13,000
Angeles City	12,000
Bulacan	5,050
Nueva Ecija	3,200
Total	747,250

Table 10. Existing sectoral damage and production losses, 1991–92 (in millions of pesos) (National Disaster Coordinating Council, 1992; Presidential Task Force on Mount Pinatubo, 1992; National Economic Development Authority, unpub. data, 1991, 1992).

Sector	1991	1992	Total 1991–92
Public infrastructure	3,830	454	4,284
Agriculture	1,474	1,422	2896
Military facilities	3,842	0	3,842
Trade and industry	851	0	851
Natural resources	125	0	125
Foregone income (trade and industry)	454	37	491
Total	10,576	1,913	12,489

furniture, equipment, and other school supplies was estimated at 93 million pesos (Department of Education, Culture, and Sports, unpub. data, 1991).

MILITARY FACILITIES

Damage to military facilities was considerable, but estimates of that damage are difficult to obtain or make. For the purposes of this report, we use an estimate of 3.8 billion pesos of damage in 1991 and no additional damage in 1992 (table 10). This estimate does not include heavy damage to former U.S. military facilities.

ALL SECTORS

In sum, damage and production losses resulting from the eruption and subsequent lahars were about 10.5 billion pesos in 1991 and 1.9 billion pesos in 1992 (table 10). These values include only damage and losses that are readily quantifiable. Additional losses, not included in these estimates, include human life, social fabric of communities, children's schooling, and a host of other, mostly social, items that are discussed in C.B. Bautista (this volume).

COSTS OF EVACUATIONS AND OTHER RISK MITIGATION

It is beyond the scope of this paper to discuss evacuations and other risk mitigation measures in detail. However, for comparison to estimates of damage, at least 2.5 billion pesos was spent in construction and operation of evacuation sites (Department of Budget and Management, Region III,

unpub. data). About 4.2 billion pesos was spent in 1991–92 for dredging of river channels and for construction of dikes and dams to control lahars (Department of Public Works and Highways, 1992).

IMPACT ON THE REGIONAL ECONOMY

The 1991 Gross Regional Domestic Product (GRDP) of Region III accounted for about 9.4 percent of the Gross Domestic Product (GDP) (Economic and Social Statistics Office, National Statistical Coordination Board, ESSO-NSCB, unpub. data, July 1993). (The GDP is Gross National Product (GNP) less net factor income from the rest of the world.) The average growth of the region's GRDP from 1987 to 1991 was 5 percent per year (NEDA, Agricultural Staff, 1993). The largest contributor from 1987 to 1991 was industry (42 percent), followed by services (35 percent) and agriculture (23 percent) (table 11).

Because of the eruption, the GRDP in 1991 amounted to only 67.2 billion pesos, compared to the 1990 GRDP of 68.8 billion pesos (table 11). This represents a 1.6 billion pesos (2.3 percent) reduction in output. All sectors of the economy were affected by the eruption. Hardest hit were manufacturing, mining and quarrying, agriculture, and private services.

In 1992, GRDP amounted to 72.2 billion pesos, a 7 percent increase from 1991. Industry and services exhibited positive growth rates (Economic and Social Statistics Office, National Statistical Coordination Board, ESSO–NSCB, unpub. data, July 1993). However, agricultural productivity was still below the 1991 level because lahars took additional agricultural lands out of production in 1992.

Table 11. Gross Regional Domestic Product by industrial origin from 1987 to 1992 at constant prices, Region III, Central Luzon (in thousand pesos; Economic and Social Statistics Office, National Statistical Coordination Board, unpub. data, July 1993).

Industrial origin	1987	1988	1989	1990	1991	1992
Gross Regional Domestic Product	57,456,387	61,712,579	64,419,389	68,814,787	67,184,484	72,227,785
Agriculture and forestry	12,943,820	13,241,781	14,462,739	15,849,415	16,043,616	16,038,629
Agriculture..........................	12,928,545	13,230,282	14,450,556	15,833,694	16,033,651	16,032,689
Forestry................................	15,275	11,499	12,183	15,721	9,965	5,940
Industry.....................................	23,567,988	26,618,118	26,751,658	29,187,703	27,745,807	32,505,094
Mining and quarrying...........	1,324,296	1,435,041	1,519,655	1,297,769	1,165,203	1,170,126
Manufacturing	17,237,722	19,960,049	19,802,819	22,691,941	21,018,947	21,731,866
Construction	3,264,967	3,296,771	3,368,449	3,248,637	3,890,864	7,631,570
Electricity, gas, and water.....	1,741,003	1,926,257	2,060,835	1,949,356	1,670,793	1,971,532
Services	20,944,579	21,862,680	23,204,992	23,777,669	23,395,061	23,684,062
Transportation......................	3,444,086	3,600,625	3,766,868	3,781,629	3,727,350	3,769,834
Trade....................................	8,766,074	9,034,766	9,592,306	9,772,620	9,644,546	9,768,813
Finance and housing.............	769,154	826,842	911,405	978,366	964,006	970,116
Real estate............................	3,389,119	3,586,285	3,848,135	3,962,822	3,912,763	3,931,460
Private services.....................	3,188,222	3,326,732	3,534,305	3,643,687	3,471,446	3,562,519
Government services............	1,387,924	1,477,430	1,551,973	1,638,545	1,674,950	1,693,320

RECOMMENDATIONS

The overall impact of the Mount Pinatubo eruptions is the slowing down of the region's growth momentum and alteration of key development activities and priorities. The calamity can, however, be taken as an opportunity, in which rehabilitation and reconstruction can aid in regional development. Specifically, rehabilitation and reconstruction should:

1. Mitigate further destruction, mainly from lahars and flash floods.
2. Normalize and accelerate economic recovery including the creation of an attractive investment climate.
3. Provide adequate livelihood and employment alternatives, especially for displaced farmers and workers (including those from Clark Air Base and the former Subic Bay Naval Station).
4. Promote growth and development in resettlement and new settlement areas that can serve as alternatives to heavily devastated or high risk areas.
5. Ensure the continuous flow of goods and services, especially during relief operations following future calamities.
6. Strengthen public awareness and institutional mechanisms for disaster preparedness.
7. Reduce the infrastructure's susceptibility to damage from lahars and other natural disasters.
8. Prevent future degradation of the environment and rehabilitate damaged ecosystems.

The complexity of these challenges and the expectation of more lahars to come demand no less than a well-coordinated, integrated response from the government sector, non-governmental organizations, and the victims themselves. With unity, selflessness, and honesty of those who serve and are being served, economic growth in the disaster-stricken areas of Central Luzon will become a reality.

REFERENCES CITED

Arboleda, R.A., and Martinez, M.L., this volume, 1992 lahars in the Pasig-Potrero River system.

Bautista, C.B., this volume, The Mount Pinatubo disaster and the people of Central Luzon.

Department of Health, 1992, Annual report for 1992: San Fernando, Department of Health, Region III, unpaginated.

Department of Public Works and Highways, 1992, Highlights of Mount Pinatubo infrastructure accomplishment: unpublished report, September 29, 1992.

Martinez, M.L., Arboleda, R.A., Delos Reyes, P.J., Gabinete, E., and Dolan, M.T., this volume, Observations of 1992 lahars along the Sacobia-Bamban River system.

National Disaster Coordinating Council, 1992, Damage from 1992 lahars of Mount Pinatubo, Update report re: Mt. Pinatubo related activities: unpub. report, Quezon City, National Disaster Coordinating Council, September 21, 1992, unpaginated.

National Economic and Development Authority (NEDA), Agricultural Staff, 1993, Philippine Development Report, 1987–1992: Quezon City, NEDA, September 1993, p. 2.

Paladio-Melosantos, M.L., Solidum, R.U., Scott, W.E., Quiambao, R.B., Umbal, J.V., Rodolfo, K.S., Tubianosa, B.S., Delos Reyes, P.J., and Ruelo, H.R., this volume, Tephra falls of the 1991 eruptions of Mount Pinatubo.

Pierson, T.C., Daag, A.S., Delos Reyes, P.J., Regalado, M.T.M., Solidum, R.U., and Tubianosa, B.S., this volume, Flow and deposition of posteruption hot lahars on the east side of Mount Pinatubo, July–October 1991.

Philippine Institute of Volcanology and Seismology and the National Economic Development Authority (PHIVOLCS/ NEDA), 1992, Pinatubo Lahar Hazards Map: NEDA, Manila, 1 sheet at scale 1:200,000; 4 sheets, Quadrants 1–4, at scale 1:100,000.

Presidential Task Force on Mt. Pinatubo, 1992, Terminal Report on Mt. Pinatubo Rehabilitation and Reconstruction Program. Presidential Task Force on Mt. Pinatubo, Quezon City, November 1992, p. 3, 5–7, 9–13, and 23–26.

Rodolfo, K.S., Umbal, J.V., Alonso, R.A., Remotigue, C.T., Paladio-Melosantos, M.L., Salvador, J.H.G., Evangelista, D., and Miller, Y., this volume, Two years of lahars on the western flank of Mount Pinatubo: Initiation, flow processes, deposits, and attendant geomorphic and hydraulic changes.

Scott, K.M., Janda, R.J., de la Cruz, E., Gabinete, E., Eto, I., Isada, M., Sexon, M., and Hadley, K.C., this volume, Channel and sedimentation responses to large volumes of 1991 volcanic deposits on the east flank of Mount Pinatubo.

Umbal, J.V., and Rodolfo, K.S., this volume, The 1991 lahars of southwestern Mount Pinatubo and evolution of the lahar-dammed Mapanuepe Lake.

The 1991 Pinatubo Eruptions and Their Effects on Aircraft Operations

By Thomas J. Casadevall,[1] Perla J. Delos Reyes,[2] and David J. Schneider[3]

ABSTRACT

The explosive eruptions of Mount Pinatubo in June 1991 injected enormous clouds of volcanic ash and acid gases into the stratosphere to altitudes in excess of 100,000 feet. The largest ash cloud, from the June 15 eruption, was carried by upper level winds to the west and circled the globe in 22 days. The June 15 cloud spread laterally to cover a broad equatorial band from about 10° S. to 20° N. latitude and contaminated some of the world's busiest air traffic corridors. Sixteen damaging encounters were reported between jet aircraft and the drifting ash clouds from the June 12 and 15, 1991, eruptions. Three encounters occurred within 200 kilometers from the volcano with ash clouds less than 3 hours old. Twelve encounters occurred over Southeast Asia at distances of 720 to 1,740 kilometers west from the volcano when the ash cloud was between 12 and 24 hours old. Encounters with the Pinatubo ash cloud caused in-flight loss of power to one engine on each of two different aircraft. A total of 10 engines were damaged and replaced, including all four engines on a single jumbo jet. Following the 1991 eruptions, longer term damage to aircraft and engines related to volcanogenic SO_2 gas has been documented including crazing of acrylic airplane windows, premature fading of polyurethane paint on jetliners, and accumulation of sulfate deposits in engines.

Ash fall in the Philippines damaged aircraft on the ground and caused seven airports to close. Restoration of airport operations presented unique challenges, which were successfully met by officials at Manila International Airport and at Cubi Point Naval Air Station, Subic Bay. Lessons learned in these clean-up operations have broad applicability worldwide.

Between April 12 and June 9, 1991, Philippine aviation authorities issued at least eight aeronautical information notices about the preeruption restless state of Mount Pinatubo. The large number of aircraft affected by the Pinatubo ash clouds indicates that this information either did not reach appropriate officials or that the pilots, air traffic controllers, and flight dispatchers who received this information were not sufficiently educated about the volcanic ash hazard to know what to do with the information.

INTRODUCTION

Jet aircraft are damaged when they fly through clouds containing finely fragmented rock debris and acid gases produced by explosive volcanic eruptions (Casadevall, 1992). Clouds of volcanic ash and corrosive gases cannot be detected by weather radar currently carried aboard airplanes, and such clouds are difficult to distinguish visually from meteorological clouds. In the past 15 years, there have been more than 80 in-flight encounters between volcanic ash clouds and commercial jet aircraft.

The explosive eruptions of Mount Pinatubo volcano in the Philippines in June 1991 injected enormous clouds of volcanic ash and gases into the stratosphere to altitudes in excess of 100,000 ft. Within several days of the June eruptions, at least 16 commercial jet airplanes had been damaged by in-flight encounters with the drifting ash clouds from Pinatubo. Closer to the volcano, ash fall in the Philippines damaged about two dozen aircraft on the ground and affected seven airports. This report describes the effects of the 1991 Pinatubo eruptions on aircraft and airports, seeks to understand why so many encounters occurred, and reviews the solutions to the ash-cloud hazard reached by Philippine authorities.

THE 1991 ERUPTIONS

From June 12 to 16, 1991, Mount Pinatubo erupted at least 16 times to produce eruption columns that penetrated the tropopause at an altitude of 53,000 ft and entered the stratosphere (Wolfe and Hoblitt, this volume). These

[1] U.S. Geological Survey.

[2] Philippine Institute of Volcanology and Seismology, Quezon City, Philippines.

[3] Michigan Technological University, Department of Geological Engineering, Houghton, MI 49931.

eruptions were detected by visual observations, seismic recordings, and barograph records and weather radar observations at Clark Air Base and Cubi Point Naval Air Station. Upper level winds generally carried ash from the volcano to the west and southwest (Oswalt and others, this volume), contaminating airspace in the Hong Kong, Bangkok, Ho Chi Minh, and Singapore Flight Information Regions.

Through July 1991, eruptions from the crater were less frequent and of decreasing vigor. However, beginning late in July 1991 and continuing at least throughout 1994, smaller, secondary explosions began to occur from sites on the slopes of the volcano. Heating of rainwater through contact with the still-hot deposits of pyroclastic material from the June 1991 eruption caused steam explosions that produced ash clouds to 60,000 ft in altitude. Unlike the magmatic eruptions in June 1991 for which pre-eruption warnings could be given (Harlow and others, this volume), secondary explosions took place with no precursory seismic activity.

The first large ash-producing eruption of Mount Pinatubo took place at 0851 local time (0051 G.m.t.) on June 12, 1991. Analysis of satellite images indicates that the ash cloud from the June 12 eruption was carried by upper level winds at speeds of approximately 15 to 20 m/s along a heading of 215° from the volcano (Potts, 1993), into the airspace west of Manila. At least three aircraft flew into this ash cloud (table 1). The first aircraft encounter was on June 12, 1991, at 1220 local (0420 G.m.t.) at a site approximately 170 km southwest of the volcano (fig. 1). The second encounter took place at 1630 local (0830 G.m.t.) and occurred at a site approximately 1,000 km west of the volcano (fig. 1). The position of the third encounter is unknown. While these encounters were reported in the national and international news media, the gravity of the volcanic threat to aviation safety was not fully appreciated until the days after the June 15 eruption.

The most powerful eruptions and the largest ash clouds from Pinatubo were on June 15 (Koyaguchi and Tokuno, 1993; Wolfe and Hoblitt, this volume). The patterns of ash dispersion of these clouds were complicated by the passage of Typhoon Yunya (Oswalt and others, this volume). Under normal weather conditions for the season, ash would have moved to the west-southwest. However, counterclockwise winds associated with the typhoon forced the proximal ash cloud to the south and over the Olongapo-Subic Bay area (Paladio-Melosantos and others, this volume). The heavy tropical rainfall associated with the typhoon saturated the ash as it fell, and loading of airport hangars and facilities with water-saturated ash caused extensive damage to facilities at Cubi Point, Clark, and Basa airports (fig. 2) (see fig. 4 in Casadevall, 1992).

A large mass of the June 15 ash was injected to altitudes above the influence of Typhoon Yunya. The June 15 cloud was carried by upper level winds to the west and circled the globe in approximately 22 days. This cloud spread longitudinally to cover a broad equatorial band from about 10° S. to 20° N. lat, contaminating some of the world's busiest air traffic corridors in the Southeast Asia region. At least 13 aircraft flew into the ash cloud from the June 15 eruption (table 1). Movement of the June ash cloud was detected by the Japanese geostationary meteorological satellite (GMS) (Tanaka and others, 1991; Tokuno, 1991; Potts, 1993; Koyaguchi, this volume); by the total ozone mapping spectrometer (TOMS) aboard the Nimbus-7 polar-orbiting satellite (Bluth and others, 1992); and by the advanced very high resolution radiometer (AVHRR) aboard the NOAA-10 and -11 polar-orbiting satellite (Schneider and Rose, 1992; Potts, 1993; Lynch and Stephens, this volume). Observations of the eruptions from these satellites were important for detecting and tracking the ash clouds. The relevance of these observations to aviation operations is reviewed here.

GEOSTATIONARY METEOROLOGICAL SATELLITE (GMS) OBSERVATIONS

Visible and infrared imagery from the GMS satellite allowed scientists to track the development of the June 15 ash cloud in images made hourly (Tokuno, 1991). Figure 3 shows the development of the ash cloud from the cataclysmic June 15 eruption and the position of the leading edge of the cloud at 3-h intervals. The cloud advanced to the west and south-southwest, in two principal lobes with headings of 250° and 210°. The advance rates for both lobes was approximately 26 m/s averaged over the 24 h after the cataclysmic eruption.

An important feature of the GMS imagery was its immediate availability and ease of interpretation. This permitted prompt mapping of the outline and westward advance of the cloud in near real-time. Using this GMS imagery on June 15–16, 1991, meteorologists of NOAA's Synoptic Analysis Branch (SAB) issued frequent bulletins about the location and movement of the Pinatubo ash clouds to the Federal Aviation Administration (FAA) and to the United States Geological Survey (USGS) (Lynch and Stephens, this volume). This information was used on June 15 and 16 in discussions with flight dispatchers to track the position of the cloud and to verify pilot suspicions about encounters between jetliners and the ash cloud over the South China Sea and Vietnam.

TOTAL OZONE MAPPING SPECTROMETER (TOMS) OBSERVATIONS

The total ozone mapping spectrometer detected the SO_2 gas contained in the Pinatubo cloud, and data from TOMS were used to map the movement of the cloud (Bluth and others, 1992). However, unlike the GMS imagery, which is available hourly, TOMS data are available only

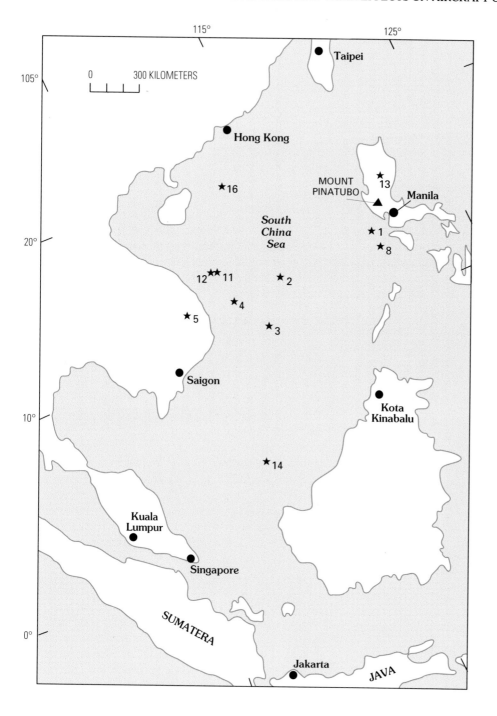

Figure 1. Location of positions of encounters between jet airplanes and the ash clouds from the June 1991 eruptions of Mount Pinatubo. Numbers indicate incidents listed in table 1.

once daily and require extensive computational processing before interpretation is possible.

The leading edge of the SO_2-rich cloud from the June 15–16 eruption, moved in a westerly direction (265°) at approximately 35 m/s, while the axis of the highest density portion of the SO_2 cloud was more southwesterly (250°) and moved at approximately 26 m/s (Bluth and others, 1992). The main mass of the SO_2 cloud was approximately coincident with the southwest (250°) lobe of the June 15 cloud as detected by GMS observations. The absence of detectable SO_2 in the area of the south-

west (210°) lobe seen on the GMS imagery may reflect masking of the this portion of the cloud by the ash or by higher water vapor content in the troposphere.

The SO_2 content of the June 15 cloud was approximately 20 Mt, the largest mass of SO_2 erupted by a volcano since the start of TOMS observations in 1978 and more than 3 times larger than mass of SO_2 from the eruption of El Chichón Volcano, Mexico, in 1982 (Bluth and others, 1992). Most of the SO_2 from Pinatubo was injected into the stratosphere, where some portion is still detectable through June 1994.

Table 1. Encounters between aircraft and volcanic ash clouds from June 1991 eruptions of Pinatubo.

[uk, unknown; na, not applicable; EGT, exhaust gas temperature; all latitudes are north and all longitudes are east]

Incident number	Date	Time (G.m.t.)[1]	Location	Latitude	Longitude	Altitude (feet)	Aircraft type	Comments
91–01	6/12/91	0420	170 km from volcano; 60 nautical miles from LUBANG along air route B460.	14°00'	119°30'	37,000	747–300	During a 3-min encounter with volcanic ash, crew experienced thin haze inside aircraft that smelled like a burning electrical wire. Aircraft landed safely at Manila Airport. Aircraft and engines were inspected and serviced at Manila in accordance with recommended procedures. When aircraft attempted to depart, its four engines had a strong vibration, and aircraft was grounded at Manila for detailed maintenance and replacement of all four engines.
91–02	6/12/91	uk	720 km west of volcano on route from Singapore to Tokyo.	13°50'	113°50'	37,000	747–400	No significant damage to aircraft when inspected on ground in Tokyo.
91–03	6/12/91	1630	Approx. 1,000 km from volcano; between way points ADPIM and LAVEN.	11°10'	112°10'	33,000	DC–10 series 40	Flight from Kuala Lumpur to Tokyo; observed a discharge phenomena on windshield for 20 min. Ground inspection at Narita revealed no damage. Encounters 3 and 11 involved same aircraft.
91–04	6/15/91	1740	Approx. 1,150 km from volcano; between way points SUKAR and CAVOI.	13°10'	110°50'	29,000	747–400	Aircraft encountered ash cloud at 29,000 ft at approximately 600 nm west of volcano. Crew observed St. Elmo's fire on the windshield and a scent similar to an electrical fire in the cockpit for 6 to 8 min as they went through the ash. There was no abnormal indication in the cockpit. The crew observed a green echo, which seemed to be ash on weather radar, but it disappeared when they were clear of the ash. Flight attendants reported thin (whitish) fog in the cabin, most dense in the upper deck compartment, followed by the forward cabin. The flight was continued to Tokyo, where engine inspection revealed that all four engines were damaged and were replaced. First-stage nozzle guide vane cooling air holes were 70–80% blocked. Other damage occurred to the cockpit windows, cabin windows, Pitot static probes, landing light covers, navigation lights, and all leading edge areas.
91–05	6/15/91	1547	Over Vietnam on route from Hong Kong to Singapore; in Bangkok FIR.	13°00'	108°00'	uk	747–SP	Ash and sulfur odor, electrostatic discharge, blue-green light over Vietnam. Ground inspection revealed no significant damage, and aircraft continued in service.
91–06	6/15/91	uk	uk	uk	uk	uk	747–200 freighter	Aircraft flew through "heavy volcanic ash." Cockpit and cabin areas were contaminated with volcanic ash. No additional information available.
91–07	6/15/91	uk	Route between Tokyo and Singapore.	uk	uk	35,000	747–251	Flight from Narita to Singapore was rerouted to Manila due to weather in Singapore area. En route to Manila, encountered volcanic ash cloud at 35,000 ft for approximately 12 min and was then diverted to Taipei. Engines set at cruise. Sparks were noted coming from windows and Crew reported hearing ash hit the aircraft. EGT for all four engines rose 40–50 °C and started to fluctuate. One hour later all EGTs were back to normal. Ground inspection in Taipei revealed no significant damage to exterior or to engines. Aircraft continued in service.
91–08	6/15/91	uk	<200 km from volcano; on approach to Manila from south.	uk	uk	uk	DC–10 series 30	Flight from Sydney to Manila encountered ash on approach to Manila from south. Engines set at low power but found to contain "lots of ash" when inspected after landing. Exterior abrasion visible, including engine cowls.
91–09	6/15/91	uk	Route between Singapore and Osaka.	uk	uk	uk	747–300	Aircraft was in ash cloud for 29 min while en route from Singapore to Osaka. Date of encounter uncertain, probably 6/15; one report indicates 6/19. Inspection of aircraft exterior showed no significant damage. Engines #1 and #4 were replaced; "90% of the first-stage turbine blades have bullseyes on the airfoil's mid-span pressure side and some first-stage vane leading edge ash buildup at 3 o'clock position."

Table 1. Encounters between aircraft and volcanic ash clouds from June 1991 eruptions of Pinatubo—Continued.

Incident number	Date	Time (G.m.t.)[1]	Location	Latitude	Longitude	Altitude (feet)	Aircraft type	Comments
91–10	6/15/91	1545	Route between Hong Kong and Mauritius.	uk	uk	31,000	747–200B	Incident 91–10 involved same aircraft as incident 91–16.
91–11	6/15/91	1730	Approx. 1,050 km from volcano; between way points SUKAR and CAVOI, 120 nautical miles from CAVOI.	15°15'	110°30'	29,000	DC–10 series 40	Flight from Kuala Lumpur to Tokyo; observed a discharge phenomena on windshield for 25 min. Ground inspection at Narita revealed no damage. Encounters 3 and 11 involved same aircraft.
91–12	6/15/91	1910	Approx. 1,050 km from volcano; between way points SUKAR and CAVOI, 120 nautical miles from CAVOI.	15°15'	110°30'	29,000	DC–10 series 40	Flight from Singapore to Osaka; crew observed a discharge phenomena on windshield for 30 min. Ground inspection at Narita revealed no damage.
91–13	6/15/91	0910	Approx. 100 km from volcano; flight from Manila to Hong Kong.	uk	uk	uk	747–428	After takeoff from Manila, airplane skirted a volcanic ash cloud. On the ground in Hong Kong, black marks were noted on the exterior of the left wing. Engines were borescoped and no discrepancies were found. Airplane continued to Delhi. Preparing to leave Delhi, unable to start engine #1. Fuel pump was replaced and additional inspections of airplane revealed no damage. Airplane continued to Paris.
91–14	6/16/91	uk	Route between Kuala Lumpur and Kota Kinabalu.	uk	uk	uk	737–200 freighter	Indications that aircraft flew through volcanic ash cloud were apparent only after aircraft underwent ground inspection in Kuala Lumpur, which revealed abrasion of plexiglass landing light covers and navigation lights, which were totally opaque. Cowling intakes were abraded and rough to the touch, while compressor blades were remarkably clean. Landing gear bays were covered in ash with ash sticking to oily surfaces. No apparent damage to windshields.
91–15	6/17/91 (?)	uk	Flight likely on Tokyo to Singapore route.	uk	uk	uk	DC–10	Airplane reportedly encountered ash from Pinatubo on June 17. #3 engine was reported to have been shut down in flight; ash encounter may have caused in-flight shutdown. Inspection of engines revealed "heavy deposits" of what was presumed to be volcanic ash. No information about flight route, encounter duration, and such.
91–16	6/17/91	0412	930 km from volcano; 50 nautical miles east of way point IDOSI on route A901.	19°30'	112°40'	37,000	747–200B	Flight from Johannesburg to Taipei via Mauritius. Encounter occurred at 37,000 ft 50 nm east of way point IDOSI on route A901; entered a cloud at 0412 G.m.t.; temperature increased from –48°C to –37°C in 2 min; aircraft descended to 29,000 ft and landed at 0540 G.m.t.; engine #1 surged and was shut down; engine #4 lost power; descended to 29,000 ft to restart #1. Aircraft landed safely at Taipei. Service terminated. Engine #1 replaced and aircraft returned to South Africa on 6/21 for further inspection.
91–17	6/15/91	na	Aircraft on ground at Manila International Airport.	14°30'	121°00'	On ground	L–1011	Maintenance crew attempted to remove volcanic ash from window by using wiper blades. Resulted in abrasion of windows, which required replacement.
91–18	6/15/91	na	Aircraft on ground at Cubi Point Naval Air Station.	14°47'	120°16'	On ground	DC–10 series 30	Aircraft landed at Cubi Point Naval Air Station on June 14, 1991, just prior to start of major eruption. Up to 6 in of ash accumulated on aircraft, including wings and horizontal stabilizer, and caused it to tilt back on the tail. Weight of ash was approximately 32 lb/ft². Aircraft suffered some damage to exterior of tail fuselage, rear pressure bulkhead, and APU compartment. See figure 6.

[1] Add 8 h to G.m.t. to attain local time in the Philippines.

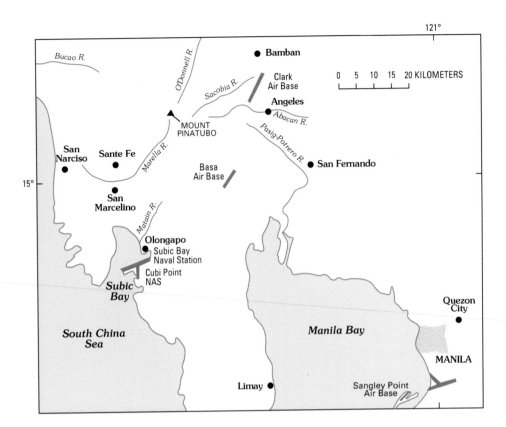

Figure 2. Airports in the Manila area affected by ash from the June 15, 1991, eruptions of Mount Pinatubo.

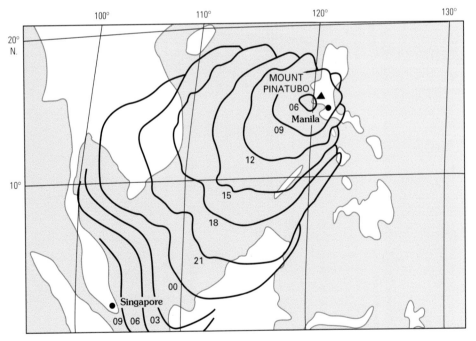

Figure 3. Shape and position of leading edge at 3-h intervals of Pinatubo ash cloud from June 15, 1991, eruptions. Data from GMS–4 visible band (modified from Tokuno, 1991).

ADVANCED VERY HIGH RESOLUTION RADIOMETER (AVHRR) OBSERVATIONS

Multispectral digital imagery from the advanced very high resolution radiometer provides imagery of the Southeast Asia region about every 4 h. AVHRR data have been used previously to discriminate volcanic clouds from meteorological clouds (Prata, 1989; Holasek and Rose, 1991; Wen and Rose; 1994; Schneider and Rose, 1994; Schneider and others, in press). Volcanic clouds can be distinguished from meteorological clouds by calculating an apparent brightness temperature difference (ΔT) between corrected temperature values derived from band 4 (10.5–

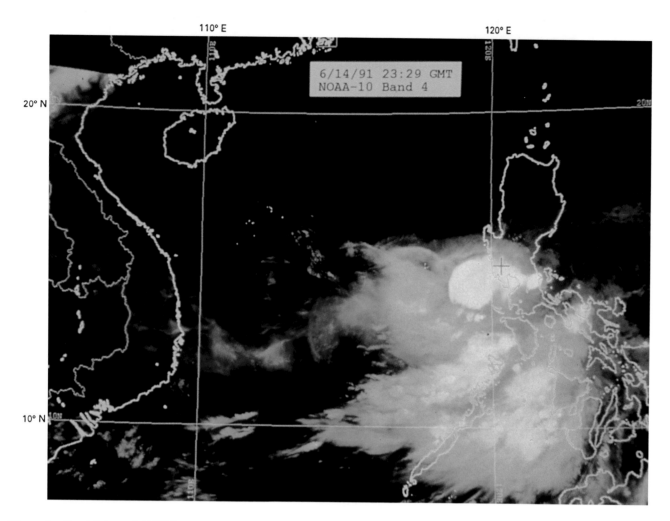

Figure 4. Band 4 thermal AVHRR image collected by NOAA–10 on June 14, 1991 at 2329 G.m.t. (0729 local), more than 18 h before aircraft encounter 91–04 (table 1). Image shows a circular eruption cloud overlying meteorological clouds associated with Typhoon Yunya and diffuse clouds extending from the circular cloud westward toward the coastline of Vietnam.

11.3 μm) and band 5 (11.5–12.5 μm) of the AVHRR. Volcanic clouds and dust clouds have negative band 4 minus band 5 ΔT values (Prata, 1989), while meteorological clouds have positive ΔT values (Yamanouchi and others, 1987). Young volcanic clouds that contain large amounts of water and ice are spectrally similar to meteorological clouds (positive ΔT). As the ash cloud ages and disperses, it develops a fringe of negative ΔT values that extend to the entire volcanic cloud as it continues to dry out (Schneider and others, in press). Two examples demonstrate the method as applied to Pinatubo clouds.

Figure 4 is an AVHRR band 4 thermal image collected by NOAA–10 on June 14, 1991, at 0729 local (2329 G.m.t.), about 18 h before aircraft encounter 91–04. The image shows the top of a circular ash cloud overlying meteorological clouds associated with Typhoon Yunya and also shows diffuse clouds extending from the circular cloud westward toward the coastline of Vietnam. Comparison between the

location of the ash cloud, as determined by AVHRR, and the SO$_2$ cloud measured by TOMS at 1145 local 0345 (G.m.t.) on June 15, 1991 (Bluth and others, 1992), shows a strong correlation (Schneider and Rose, 1992) and suggests that these diffuse clouds are of volcanic origin.

An image collected on June 16, 1991, at 1451 local (0651 G.m.t.), following the climactic eruption (fig. 5A) shows a large cloud mass extending from the Philippines, southwest to the island of Borneo, and west to the Malay Peninsula. By this time, at least eight additional aircraft had flow into the ash cloud (table 1). The apparent ΔT image (fig. 5B) clearly defines a fringe of negative ΔT values that we interpret to be the edge of the volcanic cloud. The interior of this cloud shows positive ΔT and is spectrally similar to meteorological clouds, owing to both the large droplet size and the large amount of water incorporated into the cloud during eruption. These results demonstrate that a simple technique can be used to define the limits of a very large

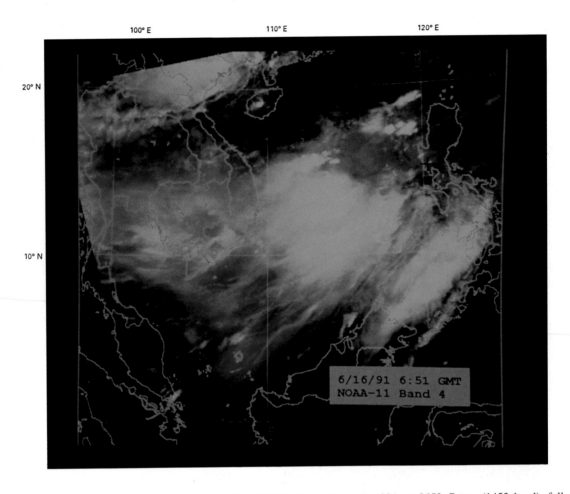

Figure 5A. Band 4 thermal AVHRR image collected by NOAA–11 on June 16, 1991, at 0652 G.m.t. (1452 local), following the climactic eruption, shows a large cloud mass extending from the Philippines southwest to the island of Borneo and west to the Malay Peninsula.

volcanic cloud. In fact, the technique is apparently more sensitive for dispersed regions of the cloud such as the edges than for the dense interior.

As with the TOMS data, AVHRR imagery must be processed prior to interpretation. Current research aims to reduce the processing time in order to make AVHRR imagery useful for operational applications such as defining the extent and movement of volcanic ash clouds (Potts, 1993; Schneider and Rose, in press) and determining the range of particle sizes and concentrations in the cloud (Wen and Rose, 1994).

AIRCRAFT ENCOUNTERS WITH ASH

At least 20 commercial jet aircraft were involved in incidents related to volcanic ash from the June 1991 eruptions. Sixteen in-flight encounters occurred between June 12 and 18; at least two encounters involved loss of

engine power. In addition to the numerous in-flight encounters, about two dozen airplanes on the ground in the Philippines were also damaged by Pinatubo ash (fig. 6).

The costs associated with these incidents are difficult to determine and include direct costs from damage to aircraft in the air and on the ground, delays, cancellations, and rerouting of flights, and closure and clean-up efforts at airports. A figure widely discussed in Manila in 1991 is that costs to aviation, including repair of aircraft damaged in flight, exceeded $100 million (P. Pacete, Manila Airport Operators' Council, oral commun., 1991). This figure does not include the costs of airport repair and cleanup for Manila International Airport and for Clark, Cubi Point, Basa, and Sangley Point military airports. Nor does this figure include costs associated with damages related to the gas cloud such as crazing of acrylic windows, fading of exterior paint, and accumulation of sulfate deposits in engines.

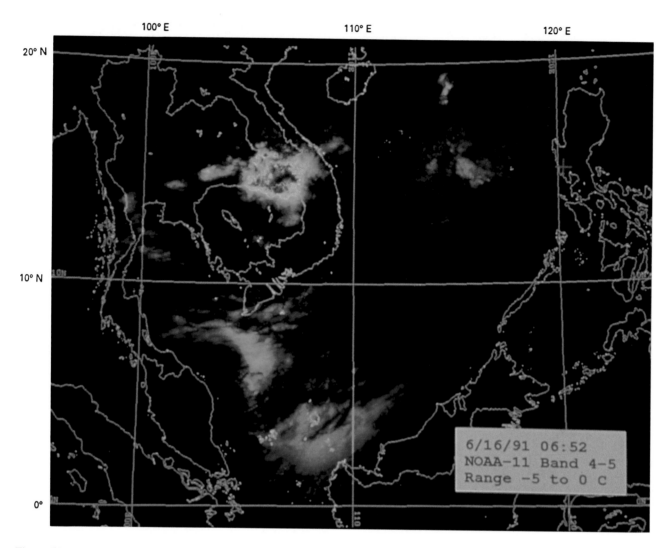

Figure 5B. Apparent brightness temperature difference image clearly defines a fringe of negative temperature difference values, which we interpret to be the edge of the volcanic cloud. The interior of this cloud shows positive temperature difference and is spectrally similar to meteorological clouds, owing to the large amount of water incorporated into the cloud during eruption.

Table 1 summarizes information available to us about the timing and location of each encounter, as well as information about the nature of the encounter and the damage to the airplane. These data are important in efforts to correlate information about the nature and timing of encounters with information about cloud movement and position determined from analysis of satellite images and from cloud trajectory forecasts. Information for table 1 is from many sources including airline companies, engine and airframe manufacturers, and reports from pilots. The detail of information was variable, especially concerning the locations of encounters and damage. In some cases, carriers were reluctant to discuss encounters, owing to concerns over possible future liability. In other cases, pilots may have been unaware that their aircraft had flown through an ash cloud, and damage to

the aircraft might not have been noticed until the aircraft was later inspected on the ground. This partly explains why there are position data for only 11 of the incidents (fig. 1; table 1).

Three encounters occurred within the Manila Flight Information Region (FIR) within 200 km from the volcano (fig. 1). Seven of the eleven encounters occurred within the Hong Kong FIR to the west of Manila at distances of 720 to 1,150 km from the volcano (fig. 1). At least four additional encounters were along Tokyo-Singapore and Hong Kong-Singapore routes and likely were within the Hong Kong FIR. One encounter occurred over Vietnam in the airspace controlled by the Bangkok FIR. One encounter occurred between Kuala Lumpur, Malaysia, and Kota Kinabalu, Sabah, Malaysia.

Figure 6. DC–10 jumbo jet on the ground at Cubi Point Naval Air Station. (U.S. Navy photograph by R. L. Rieger.)

DAMAGE

When a jetliner flying in excess of 400 knots (740 km/h) enters a cloud of finely fragmented rock particles, the principal damage will be abrasion of the exterior, forward-facing surfaces and accumulation of ash into surface openings (Casadevall, 1992). An example of the exterior damage to one jumbo jet after an encounter with a Pinatubo ash cloud is shown schematically in figure 7. Ingestion of ash into the engines will cause abrasion damage, especially to compressor fan blades. Because jet engines operate at temperatures in excess of 700°C, melting of ash and accumulation of this ash in the turbine section is an important problem as well (Przedpelski and Casadevall, 1994). Remelted ash may block the passage of air through the engines and cause the engine to stop. In an least one airplane (incident 91–04 in table 1), first-stage nozzle guide vane cooling holes were 70 to 80 percent blocked.

The majority of the Pinatubo encounters occurred at distances of up to 2,000 km from the volcano with an ash cloud that was at least 12 h old. The aging of the ash cloud allowed the coarser ash to settle from the cloud and prevented some of the more severe damage such as that which occurred to jumbo-jet aircraft from earlier encounters with volcanic ash (Smith, 1983; Tootell, 1985; and Casadevall, 1994). In the Pinatubo case, there were few reports of abrasion of forward-facing cabin windows, so it is suggested that particles larger than about 30 μm in diameter had already settled from the cloud. Particles smaller than this diameter are efficiently swept over the window surface by the slipstream and do not impact the window surface (Pieri and Oeding, 1991).

Longer term damage related primarily to the SO_2 gas and sulfuric acid aerosols produced by the eruption (Self and others, this volume) did not become apparent until months after the eruption. Some Asian-based carriers noted that jet engines on their airplanes have accumulated deposits of sulfate minerals such as anhydrite and gypsum in the turbine. This material blocked cooling holes in the first-stage nozzle guide vane at the inlet to the turbine section of the engine and thereby interfered with the cooling of the turbine. As a result, engines overheated. The sulfate deposits found in the turbine section appear to be related to ingestion and oxidation of SO_2 and sulfuric acid aerosols that originated in the Pinatubo eruption clouds of June 15 (Casadevall and Rye, 1994).

Additional problems related to the acidic aerosols include the increased incidence of crazing of acrylic windows (Berner, 1993) and fading of polyurethane paint on jetliners (T.M. Murray, Boeing, written commun., 1993). Unlike the circumstances involving in-flight encounters with the ash clouds, which were largely restricted to the region west of the volcano, the gas cloud from Pinatubo has been widely dispersed throughout the Northern Hemisphere and has thereby affected aircraft that fly in this airspace. A similar increase in the incidence of window crazing was observed for several years following the eruptions of El Chichón Volcano in 1982 (Rogers, 1984; 1985; Bernard and Rose, 1990). Pinatubo erupted nearly 3 times more SO_2 than did El Chichón (Bluth and others, 1992). Thus, the types of problems related to volcanogenic sulfur gas and sulfuric acid aerosols may be expected to persist longer following the Pinatubo activity than after El Chichón.

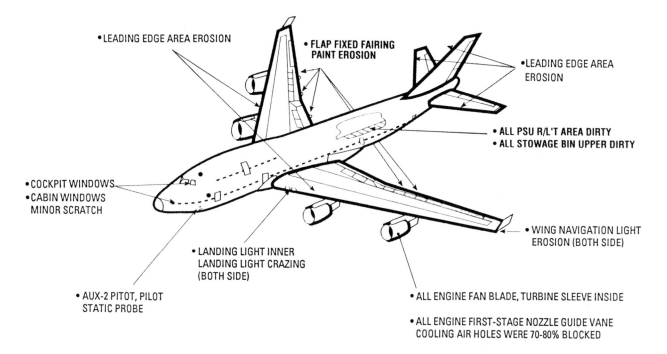

Figure 7. Damage to exterior surfaces of a 747–400 jumbo jet following an encounter with the June 15, 1991, ash cloud from Mount Pinatubo.

In addition to the airplanes that encountered the ash cloud while in flight, about two dozen airplanes were caught unprotected on the ground by the June 15 ash fall. Advance warning of the eruption enabled U.S. Air Force officials to evacuate jet aircraft from Clark Air Base (fig. 2) before the first explosive eruption on June 12. However, a squadron of 11 obsolete F–5 jets of the Philippine Air Force parked at Basa Air Base was covered by ash from the June 15 eruption.

At Cubi Point Naval Air Station (fig. 2) one C–130 transport, one C–141 transport, and one DC–10 cargo jet (fig. 6) were covered by 15 to 20 cm of ash from the June 15–16 eruptions. The accumulation of water-saturated ash on the wings, horizontal stabilizers, and fuselages of these aircraft caused two of the aircraft to rotate back on their tails. The water-saturated ash on the DC–10 weighed approximately 32 lb/ft^2 (R. Rieger, U.S. Navy, oral commun., 1991). The DC–10 suffered minor damage to the tail fuselage, the rear pressure bulkhead, and the auxiliary power unit compartment.

In Manila, at least eight jumbo-jet passenger airplanes were on the ground during the June 15–16 ash fall. Because of ash on the runways, these aircraft could not depart until June 19. Fortunately, ash fall in Manila was relatively light (< 0.5 cm). Since hangar space was limited, most aircraft were adequately protected by plastic sheeting and duct tape placed over windows and openings in the aircraft surfaces and engines. Windshields on several aircraft were abraded when window wipers were used to remove ash.

AIRPORTS

Ash from the June 15–16, 1991, eruption caused the closing of civilian airports at Manila, Puerto Princesa, and Legaspi and military airfields at Clark Air Base, Basa Air Base, Sangley Point Air Base, and Cubi Point Naval Air Station (fig. 2). At Manila's Ninoy Aquino International Airport (NAIA) and Sangley Point Air Base, between 0.5 and 1 cm of fine sand to powder-size ash fell in a mostly dry condition on June 15–16, 1991. This ash caused these airports to close for 4 days until ash could be removed from runways, taxiways, and apron surfaces (Casadevall, 1993). Normal operations at NAIA resumed on July 4, 1991.

Volcanic ash on airport surfaces caused reduced visibility and affected aircraft maneuvering, especially when the ash was wet. Ash on airport surfaces was ingested into engines during taxiing, takeoffs, and landings and also contaminated landing gear assemblies and brakes. Ash fall from a minor eruption on July 17, 1991, deposited less than 1 mm of ash over the Manila area, and NAIA was closed for ash removal from 1700 on July 17 through 1900 on July 18. While landing on July 19, one jumbo jet skidded off the runway because of reduced braking action on wet ash.

The principal damage to airport surfaces and buildings at Clark, Cubi Point, and Basa was caused by the accumulation of from 15 to 20 cm of water-saturated sand to fine gravel-size ash. The weight of the wet ash, combined with the ground shaking from the earthquakes during June 15, led to the collapse of a large number of airport buildings including hangars at Clark, Cubi Point, and Basa (see fig. 4

in Casadevall, 1992). Buildings that remained standing were usually those from which ash was removed as it fell (B. Wood, U.S. Navy, oral commun., 1991). Acidic gases adsorbed to the wet ash formed H_2SO_4, which corroded metal-roofed buildings, airport electrical systems, ground service equipment, and aircraft that were not properly cleaned immediately after the ash fall (C.M. Navarro, Philippine Airlines, oral commun., 1992). The dried ash hardened to a concretelike consistency that made subsequent removal more difficult and costly.

AIRPORT CLEANUP

Airport officials in the Philippines tried a variety of techniques to remove and dispose of volcanic ash. The principal problem was that winds continually resuspended the ash and recontaminated previously cleaned surfaces. The first attempts at NAIA on June 16 to remove the ash utilized a vacuum sweeper and pressurized water. This method left some ash that was resuspended by wind after it dried. Manual sweeping of the runway and taxiway surfaces to accumulate the ash into furrows was finally settled on as the most effective method of ash removal. The furrows of ash accumulated at the edges of runways and taxiways were either collected into bags for removal and disposal or were stabilized by covering with emulsified asphalt, which acted as a binder for the ash. The emulsified asphalt was also sprayed on ash-covered infield areas adjacent to runways, taxiways, and aprons and proved to be the most effective method for restoring airport surfaces to safe operation. Ash removal proceeded in a stepwise manner that focused first on the main runway and taxiways. Once these were cleared, aircraft were towed to and from the runway in order to prevent resuspension of ash by airplane engines.

The greater thickness of ash (15 to 20 cm) in the Olongapo-Subic Bay area required a more intensive effort for ash removal at Cubi Point NAS. Officials at Cubi Point took an aggressive approach to ash removal and utilized a contingent of 1,800 personnel and a full complement of earth-moving equipment of a U.S. Navy Construction Battalion (Seabees) that was in Subic Bay in transit to the U.S.A. from Operation Desert Storm. After experimentation with sweepers and washing, officials settled on a procedure similar to that used at NAIA, where road graders scraped ash into furrows. The accumulated ash was loaded into dump trucks and dumped at the southwest edge of the runway in a designated landfill. The scraping-loading-dumping operation still left a residue of fine ash on the runway surface which was swept and washed onto the grass infield and covered with emulsified asphalt. Partial operations were restored at Cubi Point NAS by June 26, 1991, but ash removal activities continued through 1991. Clark and Basa reopened for limited operations in 1992. However, at Clark, ash falls from secondary explosions have repeatedly closed down even limited operations. Ash from the secondary explosions has rarely affected Cubi Point.

COMMUNICATIONS

Prior to the 1991 eruptions of Pinatubo, there had been no encounters between airplanes and ash clouds in Philippine airspace. During the eruptions of the Philippine volcano Mayon in 1984, the Philippine Bureau of Air Transportation (now the Air Transportation Office) developed Air Traffic Section procedure order 3–84 for dealing with the hazard of ash clouds. Procedure 3–84 specified that communication of information about volcanic activity was to be coordinated with the Philippine Institute of Volcanology and Seismology (PHIVOLCS) and that aeronautical information to pilots and contingency flight routing to bypass areas affected by volcanic ash would be coordinated through the Manila Area Control Center (ACC). These procedures formed the basis for procedures implemented during the Pinatubo emergency. The Manila Flight Information Region (FIR) encompasses Luzon Island and extends for approximately 750 km to the west and borders the Hong Kong, Bangkok, and Ho Chi Minh Flight Information Regions. Control of air traffic and communications with aircraft within the Manila FIR is handled by the Manila ACC.

By international agreement, information about conditions that pilots might encounter while flying is supplied by use of aeronautical information notices known as NOTAMs and meteorological information notices known as SIGMETs. These are issued by authorities in the ACC and are passed by teletype to air traffic controllers and airline flight dispatchers for relay to pilots either during preflight briefings or by radio to flights already en route. Information used in NOTAMs and SIGMETs may come from a variety of sources including reports from pilots, air traffic services, meteorological services, and in the case of volcanic activity, from ground-based volcano observers.

In the Philippines, the first NOTAM alerting pilots about Mount Pinatubo was issued by the Flight Observation Briefing Service (FOBS) at the Air Transportation Office (ATO), Manila International Airport (NAIA) on April 12, 1991 (appendix 1), 10 days after unrest of the volcano was first detected. From April 12 through June 9, at least eight additional NOTAMs were released by ATO. Discussions between the authors and members of the Manila Airport Operators' Council (AOC) and Board of Airline Representatives in July and August 1991 indicated that few airline dispatchers could recall that NOTAMs about the volcano had been issued prior to June 9. Several said that they had paid little attention to the notices because the activity of Mount Pinatubo at that time had consisted only of low-altitude emission of white steam. Since the volcano had not yet revealed an explosive nature, there was a perception that it posed little hazard to aircraft safety. In addition, many airline authorities and aviation officials told us that, at the time, it was not conceivable to them that an ash cloud would move beyond Luzon Island.

The large number of encounters with the Pinatubo ash cloud (table 1), including several encounters that involved airplanes from the same company, reflects a major breakdown in the way that information about the ash cloud hazard was communicated and the ways that users responded to the information. This breakdown occurred at several levels, including between adjacent FIRs as well as within individual airline companies.

To meet the unprecedented demands on the air traffic control system by the Pinatubo eruption, Philippine authorities acted quickly to streamline the collection and issuance of information about the activity of the volcano. By late June, several new communications links were established between aviation authorities, airline companies, meteorological agencies, and volcanologists. Important additions to the existing communication network were the airline companies as represented by the Manila AOC and the Board of Airline Representatives (BAR). To the existing network, the AOC and the BAR were able to add company pilot reports, weather observations, and perhaps most importantly, direct communications with the pilots in the cockpit.

A series of meetings in July and August 1991 between PHIVOLCS scientists and officials of ATO, the Philippine Atmospheric, Geophysical, and Astronomical Services Administration (PAGASA), the Manila Airport Authority, and airline management resulted in the streamlining of information flow. Under the revised notification plan, PHIVOLCS directly relayed information of eruptions to the FOBS office of the ATO. The FOBS/ATO office issues NOTAMs about Pinatubo's activity after receiving information from PHIVOLCS and following consultation with PAGASA to verify the forecast of prevailing wind drift. This interaction is facilitated at Manila International Airport, where the PAGASA and FOBS offices are adjacent to one another.

ALERT CODE

To simplify communications with public officials, a numerical volcano alert code was introduced by PHIVOLCS in May 1991 (Tayag and others, this volume). Aviation officials adopted the alert code and designated corresponding flight restrictions and contingency routings for air traffic under each alert level. Under the original code, the appearance at the surface of new lava indicated an eruption was in progress and required that the original Alert Level 5 be maintained, even though little or no ash might be produced. Revision of the code became necessary after July 1992, when seismicity and nonexplosive growth of a lava dome (Tayag and others, this volume) required that PHIVOLCS maintain the highest alert at Alert Level 5 even

though the eruption produced little or no ash. With the revised code, adopted on November 27, 1992, Alert Level 5 specifies a *hazardous explosive eruption in progress with pyroclastic flows and (or) eruption column rising at least 6 km or 20,000 ft above sea level*. Because dome growth after July 1992 was nonexplosive, PHIVOLCS lowered the alert level during nonexplosive dome growth from Alert Level 5 to 2 (*moderate level of seismic, other unrest, with positive evidence for involvement of magma*) on December 8, 1992, and normal air traffic routing near the volcano was resumed. This alleviated the increased operating costs associated with flying the longer distance contingency routings as required when operating under the original Alert Level 5 (C.M. Navarro, Philippine Airlines, written commun., 1992).

LESSONS LEARNED

The large number of damaging encounters between airplanes and the ash clouds from the 1991 eruptions of Pinatubo prompted volcanologists, meteorologists, and aviation authorities to reevaluate the hazard that these clouds present to aviation safety and how available technology and operational procedures can be applied or modified to mitigate the ash hazard.

DETECTION AND TRACKING OF THE ASH CLOUDS BY REMOTE SENSING METHODS

Satellite-based remote sensing methods provided information for the detection and tracking of the Pinatubo ash clouds. To be of maximum benefit to aviation, these data should be collected at a central station, quickly and succinctly interpreted, and widely broadcast and disseminated in a form that is understandable to users including airline dispatchers and pilots. Delays of minutes to hours reduce the value and utility of the information. On June 15, tracking of ash clouds by meteorologists in the United States and Japan by use of the hourly GMS images gave timely information about ash cloud position. However, this information did not reach Philippine authorities in time to be incorporated into operational applications. Indeed, most countries do not have immediate access to satellite-based sensors to monitor ash cloud movement. It is important for countries with satellite capabilities to determine the extent and movement of ash clouds to pass on such information quickly to countries and agencies at risk from drifting ash clouds. Following the Pinatubo emergency, the World Meteorological Organization and the International Civil Aviation Organization have requested the assistance of the governments of Australia, Japan, and the United States to develop satellite techniques that would provide warning of ash cloud movement that threatens aviation safety in areas where satellite data are not available.

COMMUNICATIONS

Warnings about volcanic eruptions commonly are given too late to prevent in-flight encounters. However, in the Pinatubo case, information about the restless state of the volcano was available in aeronautical notices issued by Philippine authorities up to 2 months before the first eruption in June. Our analysis of the large number of aircraft affected by the Pinatubo ash clouds indicates that information and warnings about the hazard of volcanic ash either did not reach appropriate officials in time to prevent these encounters or that those pilots, dispatchers, and air traffic controllers who received this information were not sufficiently educated about the volcanic ash hazard to know what steps to take to avoid ash clouds.

The key to communicating information about volcanic eruptions in a timely and readily understandable form is to involve all interested groups (geologists, meteorologists, pilots, and air traffic controllers) in the development of information and to streamline the distribution of this information between essential parties. During the Pinatubo crisis, Philippine authorities established practical and straightforward procedures for addressing the volcanic threat. These included regular meetings between all agencies involved with addressing the volcanic threat to aviation safety. Beyond the Philippines, communications must include realtime communications between Flight Information Regions. An important element of any communications plan is frequent exercising of the plan to insure that information users are not caught off-guard by the sudden appearance of information about a restless or erupting volcano.

Applying these lessons, the International Civil Aviation Organization (ICAO) has requested that coordination efforts similar to those used in the Philippines be established for other countries in the Asia-Pacific region (Casadevall and Oliveira, 1993) and in South America. Accuracy and timeliness of SIGMETs are additional issues that have been addressed by ICAO (ICAO, 1992). In particular, ICAO regulations now require that an outlook advisory be included with SIGMETs describing volcanic activity. The outlook is valid for a period of 4 to 12 h and is developed by using information from satellite tracking of the cloud and from trajectory forecast models such as the volcanic ash transport and diffusion model of Heffter and Stunder (1993).

EDUCATION

In our study, we found that often, warnings about volcanic eruptions are passed to decisionmakers including pilots, flight dispatchers, and air traffic controllers who are not adequately informed about the nature of volcanic clouds to know how to use such information to safeguard the airplane and its passengers. Regular and repeated training of

pilots and aviation officials about volcanic hazards must be a component in flight safety training (Boeing, 1992).

We also found that volcanologists were seldom aware of the hazards related to far-drifting clouds of volcanic ash. With their traditional focus on the flowage hazards (ash flows, lava flows, mud flows) that affect the slopes and lower flanks of the volcano itself, volcano scientists have rarely treated volcanic ash clouds in their assessments of volcanic hazards. The hazards posed by drifting ash clouds, not only near an erupting volcano but at considerable distances downwind, must be described and assessed by scientists who evaluate hazards, especially for volcanoes that erupt explosively.

LONG-TERM DAMAGE

In addition to the aircraft damage that was immediately evident in the days following the June 15 eruption, damage related primarily to SO_2 gas has been reported by some airline companies and manufacturers. One year after the eruption, in June 1992, there was an incident involving loss of engine power on a jumbo jet owing to accumulation of sulfate deposits in jet engines. Isotopic studies of these deposits suggest that the sulfate is derived from the ingestion and oxidation of SO_2 and sulfuric acid aerosols that originated in the Pinatubo eruption cloud of June 15 (Casadevall and Rye, 1994). Related problems recognized in 1992 such as the increased incidence of crazing of acrylic windows (Berner, 1993) and fading of polyurethane paint on jetliners are also due to volcanogenic sulfuric acid droplets in the atmosphere. Frequent inspections of aircraft should reveal any corrosion problems due to volcanogenic sulfur gases.

SUMMARY

The 1991 eruptions of Mount Pinatubo produced enormous clouds of volcanic ash and corrosive gases. In the days after the eruptions, these clouds had an immediate impact on air traffic routes and aircraft flying in the Southeast Asia region. In addition, the corrosive gases from Pinatubo continue to affect aircraft through 1993.

Unlike earlier volcanic eruptions such as those from Redoubt Volcano in 1989–90, where jet aircraft flew into ash clouds owing largely to lack of adequate information about the position and nature of the ash cloud (Casadevall, 1994), the clouds from the Pinatubo eruptions were well known and described in the aeronautical information available to pilots and dispatchers. However, for various reasons, this information was not widely incorporated by the airline companies into their operational planning.

Within days after the cataclysmic eruption in mid-June, Philippine authorities acted quickly to establish an interagency plan to streamline the collection and flow of information between field observers and pilots. This system relied heavily on the involvement of the Airline Operators' Council and Board of Airline Representatives of Manila International Airport, in cooperation with scientists of PHI-VOLCS, PAGASA, and the authorities of the Manila Airport Authority. This plan continues in effect more than 4 years after the eruption and has been selected and promoted by ICAO as an example of a particularly effective operational model for other countries of the world facing the volcanic threat to aviation safety (Casadevall and Oliveira, 1993).

ACKNOWLEDGMENTS

Continuing interaction with the aviation community focused our attention on what we hope are the more important and timely issues related to volcanic hazards and aviation safety. In particular, we are grateful for the support and interest in this problem from Captains Ed Miller (Air Line Pilots Association) and Peter Foreman (International Federation of Air Line Pilots Associations), Thomas M. Murray (Boeing), Zygmunt Przedpelski (General Electric, retired), Saburo Onodera (Japan Air Lines), Michael Dunn (CALSPAN), and James Wood (McDonnell-Douglas).

A number of colleagues assisted us with information about effects of Pinatubo activity on airport operations. Eduardo Carrascoso, then General Manager of NAIA, Ruben Gaddi, Manager, Airport Ground Operations Division, NAIA, and C.M. Navarro, Philippine Airlines, provided valuable information about the impacts to Manila International Airport. Captain Bruce Wood and Commander Daniel Harrigan (U.S. Navy) provided information about the effects of ash and about cleanup procedures used at Cubi Point. Information about communications between PHIV-OLCS and Philippine government agencies was retrieved from files consolidated by Nancy Largo, Jason Villegas, Leyo Bautista, and Ting Diao of PHIVOLCS. Information, including a summary of NOTAMs, was kindly provided by Flight Observation Briefing Service with assistance from Pepe Pacete and Chris Senarosa of Flight Operation, Japan Airlines.

Information about aircraft encounters came from many sources. In particular, we acknowledge Al Weaver (Pratt and Whitney), Captain Ernie Campbell (Boeing), and Tom Fox (ICAO-Montreal).

Reviews of this paper by Ray Punongbayan (PHIVOLCS), Chris Newhall, Jim Riehle, and Dennis Krohn substantially improved our presentation and are gratefully acknowledged.

Appendix 1. Chronology of NOTAMs (Notices to Airmen) Issued by Manila Flight Information Region To Warn of Pinatubo Activity (ICAO, Montreal, and JAL, Tokyo, unpub. data, 1991).

[The following chronology indicates the NOTAM number and time of issuance. The activity of Mount Pinatubo began on April 2, 1991, with a series of phreatic explosions from a fissure on the north side of the volcano. The first NOTAM (B513) was issued April 12]

April 2:	First eruption of Pinatubo.
April 12:	First NOTAM (B513) issued: "Pinatubo volcano…steaming activity was moderate with considerable amount of white steam plume with a height from 250 to 400 m from vent and drifting SW direction and in abnormal condition. All pilots are advised to exercise extreme caution and avoid flying over the area."
April 24:	Manila Area Control Center (ACC) replies to carrier inquiry that "…all concerned will be notified within 10 minutes after receiving the actual eruption [notification] from PHIVOLCS."
May 3:	NOTAM B513 is revised as NOTAM B634 with revision of plume height.
June 5:	PHIVOLCS upgraded Pinatubo warning from Alert Level 2 to Alert Level 3.
June 7:	PHIVOLCS upgraded Pinatubo warning from Alert Level 3 to Alert Level 4.
June 8:	JAL notifies AOC carriers at NAIA that PHIVOLCS upgraded Mount Pinatubo warning level to Alert Level 4, meaning that an eruption may occur within 24 hours.
June 9:	Pinatubo erupts: NOTAM B634 is revised and reissued as NOTAM B797 to notify that Pinatubo had erupted at 1425 local time (0625 Z). PHIVOLCS upgrades Pinatubo warning to Alert Level 5, the highest level.
June 10:	NOTAM B808 is revised and reissued as NOTAM 818.
June 12:	Philippine Independence Day.
	NOTAM B818 is revised and reissued as NOTAM B834 and B835 to notify of eruption at 0855 local time (0055 Z).
June 13:	NOTAM 844 issued at 0830 local time (0030 Z) following June 12 eruption; NOTAM 848 issued at 1145 local time (0345 Z), closing airways in aftermath of encounter 91–01.
June 14:	NOTAM B865 issued 10:43 local time (0243 Z) to warn of possible presence of volcanic ash in a designated area and up to 50,000 feet altitude.
	Several NOTAMs are issued between 1309 local time on June 14 and 0556 local time on June 15 (0509 and 2156 Z). Data for additional NOTAMs were typically from pilot reports, Clark Air Base and Cubi Point NAS, and public radio reports. Data from PHIVOLCS were minimal.
	Advisory sent at 1452 local time (0652 Z) indicated "…PRESENCE OF TROPICAL DEPRESSION YUNYA…WILL INVITE VOLCANIC ASH TO MOVE TO SOUTHEAST TO MANILA FROM MT. PINATUBO."
June 15:	NOTAM B882 was issued 0930 local time (0130 Z), closing additional airways.
	NOTAM B884 was issued 1027 local time (0227 Z) notifying of two eruptions of Pinatubo.
	Several NOTAM were issued during the day. Manila Airport began to receive ash fall at 1535 local time (0735 Z) with a continuous light to moderate rain.
	NOTAM B891 issued (1315 Z) to say Manila Airport operations were suspended due to ash fall.
	At 0144 Z, airports at Legaspi and Puerto Princesa report presence of volcanic ash.
	From 1100 to 2100 Z, Manila area experienced about 1 cm of ash fall.
June 18:	NOTAM B912 issued at 1850 local time (1050 Z) to say that Manila Airport was open to propeller aircraft.
	NOTAM B913 issued at 1905 local time (1105 Z) to say airport is open to departing jet aircraft, which are towed to runway from ramp.
June 19:	Five jet aircraft depart between 0800 and 1300 local time (0000–0500 Z).
	At 1800 local time (1000 Z), two aircraft land (one 747, one DC–10).
June 26:	Airport reopened with restrictions.
July 4:	Airport reopened completely for normal operations.
July 17:	Airport closed for 12 hours starting at 1700 local time (0900 Z) because of light ash fall.
July 19–present:	Airport opened for normal operations.

Appendix 2. Acronym glossary.

ACC	Area Control Center
AOC	Airport Operators' Council
APU	auxiliary power unit
ATO	Air Transportation Office
AVHRR	Advanced Very High Resolution Radiometer
BAR	Board of Airline Representatives
EGT	exhaust gas temperature
FIR	Flight Information Region
FOBS	Flight Observation Briefing Service
GMS	Geostationary Meteorological Satellite (Japanese Meteorological Agency)
ICAO	International Civil Aviation Organization
NAIA	Ninoy Aquino International Airport (Manila International Airport; also Villamor Air Base of the Philippine Air Force)
NAS	U.S. Naval Air Station (Cubi Point NAS, for example)
NDCC	National Disaster Control Center
NOAA	National Oceanic and Atmospheric Administration
NOTAM	Notice To Airmen
PAGASA	Philippine Atmospheric, Geophysical, and Astronomical Services Administration
PHIVOLCS	Philippine Institute of Volcanology and Seismology
PVO	Pinatubo Volcano Observatory
RDCC	Regional Disaster Control Center
SAB	Synoptic Analysis Branch
SIGMET	Significant Meteorological Event Notification
TOMS	Total Ozone Mapping Spectrometer
USGS	United States Geological Survey

REFERENCES CITED

Bernard, A., and Rose, Jr., W.I., 1990, The injection of sulfuric acid aerosols in the stratosphere by El Chichón volcano and its related hazards to the international air traffic: Natural Hazards, v. 3, p. 59–67.

Berner, P., 1993, Operators confront mounting window damage: Aviation Equipment Maintenance, November 1993, p. 34–37.

Bluth, G.J.S., Doiron, S.D., Schnetzler, C.C., Krueger, A.J., and Walter, L.S., 1992, Global tracking of the SO_2 clouds from the June, 1991 Mount Pinatubo eruption: Geophysical Research Letter, v. 19, p. 151–154.

Boeing, 1992, Volcanic ash avoidance: Flight crew briefing: Video presentation (33 minutes), Customer Training and Flight Operations Support, Seattle, Wash.

Casadevall, T.J., 1992, Volcanic hazards and aviation safety: Federal Aviation Administration Aviation Safety Journal, v. 2, no. 3, p. 9–17.

———1993, Volcanic ash and airports: U.S. Geological Survey Open-File Report 93–518, 53 p.

———1994, The 1989–1990 eruption of Redoubt Volcano, Alaska: Impacts on aircraft operations, in Miller, T.P., and Chouet, B.A., eds., The 1989–90 eruptions of Redoubt Volcano, Alaska, Journal of Volcanology and Geothermal Research, v. 62, p. 301–316.

Casadevall, T.J., and Oliveira, F.A.L., 1993, Special project in the Asia/Pacific region boosts awareness of danger posed by volcanic ash: International Civil Aviation Organization Journal, v. 48, no. 8, p. 16–18.

Casadevall, T.J., and Rye, R.O., 1994, Sulfur in the atmosphere: Sources and effects on jet-powered aircraft [abs.]: Eos, Transactions, American Geophysical Union, v. 75, no. 16 (Supplement), p. 75.

Harlow, D.H., Power, J.A., Laguerta, E.P., Ambubuyog, G., White, R.A., and Hoblitt, R.P., this volume, Precursory seismicity and forecasting of the June 15, 1991, eruption of Mount Pinatubo.

Heffter, J.L., and Stunder, B.J.B., 1993, Volcanic ash forecast transport and dispersion (VAFTAD) model: Weather and Forecasting, v. 8, p. 533–541.

Holasek, R.E., and Rose, W.I., 1991, Anatomy of 1986 Augustine Volcano eruptions recorded by multispectral image processing of digital AVHRR weather satellite data: Bulletin of Volcanology v. 53, p. 420–435.

International Civil Aviation Organization, 1992, Meteorological service for international air navigation: Annex 3 to the Convention on International Civil Aviation, 11th edition, Montreal, 82 p.

Koyaguchi, T., and Tokuno, M., 1993, Origin of the giant eruption cloud of Pinatubo, June 15, 1991: Journal of Volcanology and Geothermal Research, v. 55, p. 85–96.

Lynch, J.S., and Stephens, G., this volume, Mount Pinatubo: A satellite perspective of the June 1991 eruptions.

Oswalt, J.S., Nichols, W., and O'Hara, J.F., this volume, Meteorological observations of the 1991 Mount Pinatubo eruption.

Paladio-Melosantos, M.L., Solidum, R.U., Scott, W.E., Quiambao, R.B., Umbal, J.V., Rodolfo, K.S., Tubianosa, B.S., Delos Reyes, P.J., and Ruelo, H.R., this volume, Tephra falls of the 1991 eruptions of Mount Pinatubo.

Pieri, D., and Oeding, R., 1991, Grain impacts on an aircraft windscreen: The Redoubt 747 encounter [abs.], in Casadevall, T.J., ed., First International Symposium on Volcanic Ash and Aviation Safety: Program and Abstracts, U.S. Geological Survey Circular 1065, p. 35–36.

Pinatubo Volcano Observatory Team, 1991, Lessons from a major eruption: Mt. Pinatubo, Philippines: Eos, Transactions, American Geophysical Union, v. 72, p. 545, 552–553, 555.

Potts, R.J., 1993, Satellite observations of Mt. Pinatubo ash clouds: Australian Meteorological Magazine, v. 42, p. 59–68.

Prata, A.J., 1989, Infrared radiative transfer calculations for volcanic ash clouds: Geophysical Research Letters, v. 16, p. 1293–1296.

Przedpelski, Z.J., and Casadevall, T.J., 1994, Impact from volcanic ash from 15 December 1989 Redoubt Volcano eruption of GE CF6–80C2 turbofan engines, in Casadevall, T.J., ed., Proceedings of the First International Symposium on Volcanic Ash and Aviation Safety: U.S. Geological Survey Bulletin 2047, p. 129–135.

Rogers, J.T., 1984, Results of El Chichón: Premature acrylic window crazing: Boeing Airliner, April–June, p. 19–25.

———1985, Results of El Chichón—part II: Premature acrylic window crazing status report: Boeing Airliner, April–June, p. 1–5.

Schneider, D.J., and Rose, W.I., 1992, Comparison of AVHRR and TOMS imagery of volcanic clouds from Pinatubo Volcano [abs.]: Eos, Transactions, American Geophysical Union, v. 73, p. 624.

———1994, Observations of the 1989-90 Redoubt Volcano eruption clouds using AVHRR satellite imagery, *in* Casadevall, T.J., ed., Proceedings of the First International Symposium on Volcanic Ash and Aviation Safety: U.S. Geological Survey Bulletin 2047, p. 405–418.

Schneider, D.J., Rose, W.I., and Kelley, L., in press, Tracking of 1992 Crater Peak/Spurr eruption clouds using AVHRR, *in* Keith, T.E.C., ed., The 1992 eruptions of Mt. Spurr, Alaska: U.S. Geological Survey Bulletin.

Self, S., Zhao, J-X., Holasek, R.E., Torres, R.C., and King, A.J., this volume, The atmospheric impact of the 1991 Mount Pinatubo eruption.

Smith, W.S., 1983, High-altitude conk out: Natural History, v. 92, no. 11, p. 26–34.

Tanaka, S., Sugimura, T., Harada, T., and Tanaka, M., 1991, Satellite observation of the diffusion of Pinatubo volcanic dust to the stratosphere: Journal of Remote Sensing Society, Japan, v. 11, p. 91–99 (in Japanese).

Tayag, J., Insauriga, S., Ringor, A., and Belo, M., this volume, People's response to eruption warning: The Pinatubo experience, 1991–92.

Tokuno, M., 1991, GMS–4 observations of volcanic eruption clouds from Mt. Pinatubo, Philippines: Meteorological Satellite Center Technical Note no. 23, p. 1–14.

Tootell, E., 1985, All 4 engines have failed; the true and triumphant story of flight BA 009 and the Jakarta incident: Hutchinson Group Ltd., Auckland, 178 p.

Wen, Shiming, and Rose, W.I., 1994, Retrieval of sizes and total masses of particles in volcanic clouds using AVHRR bands 4 and 5: Journal of Geophysical Research, v. 99, p. 5421–5431.

Wolfe, E.W. and Hoblitt, R.P., this volume, Overview of the eruptions.

Yamanouchi, T., Suzuki, K., and Kawaguchi, S., 1987, Detection of clouds in Antarctica from infrared multispectral data of AVHRR: Journal of the Meteorological Society of Japan, v. 65, p. 949–961.

The Atmospheric Impact of the 1991 Mount Pinatub

Stephen Self,[1] Jing-Xia Zhao,[2] Rick E. Holasek,[1][3] Ronnie C. Torres,[1][4] and Alan J. King[1]

ABSTRACT

The 1991 eruption of Pinatubo produced about 5 cubic kilometers of dacitic magma and may be the second largest volcanic eruption of the century. Eruption columns reached 40 kilometers in altitude and emplaced a giant umbrella cloud in the middle to lower stratosphere that injected about 17 megatons of SO_2, slightly more than twice the amount yielded by the 1982 eruption of El Chichón, Mexico. The SO_2 formed sulfate aerosols that produced the largest perturbation to the stratospheric aerosol layer since the eruption of Krakatau in 1883. The aerosol cloud spread rapidly around the Earth in about 3 weeks and attained global coverage by about 1 year after the eruption. Peak local midvisible optical depths of up to 0.4 were measured in late 1992, and globally averaged values were about 0.1 to 0.15 for 2 years. The large aerosol cloud caused dramatic decreases in the amount of net radiation reaching the Earth's surface, producing a climate forcing that was two times stronger than the aerosols of El Chichón. Effects on climate were an observed surface cooling in the Northern Hemisphere of up to 0.5 to 0.6°C, equivalent to a hemispheric-wide reduction in net radiation of 4 watts per square meter and a cooling of perhaps as large as −0.4°C over large parts of the Earth in 1992–93. Climate models appear to have predicted the cooling with a reasonable degree of accuracy. The Pinatubo climate forcing was stronger than the opposite, warming effects of either the El Niño event or anthropogenic greenhouse gases in the period 1991–93. As a result of the presence of the aerosol particles, midlatitude ozone concentrations reached their lowest levels on record during 1992–93, the Southern Hemisphere "ozone hole" increased in 1992 to an unprecedented size, and ozone depletion rates were observed to be faster than ever before recorded. The atmospheric impact of the Pinatubo eruption has been pro-found, and it has sparked a lively interest in the role that volcanic aerosols play in climate change. This event has shown that a powerful eruption providing a 15 to 20 megaton release of SO_2 into the stratosphere can produce sufficient aerosols to offset the present global warming trends and severely impact the ozone budget.

INTRODUCTION

After 10 weeks of precursory activity, Mount Pinatubo (15°08' N. lat, 120°21' E. long) erupted on June 12–16, 1991, producing one of this century's greatest volcanic eruptions, the largest stratospheric SO_2 cloud ever observed by modern instruments, and the major stratospheric aerosol event since Krakatau exploded in 1883. By far the largest volume of ejecta (perhaps >90% of the total), the highest eruption columns, and the longest duration of stratospheric injection occurred during the 9 h of more-or-less continuous high-output activity from about 1340 to about 2230 on June 15 (Hoblitt, Wolfe, and others, this volume). (All times are local time unless otherwise stated.)

Stratospheric sulfate aerosols generated by the Pinatubo eruption cloud have had a far-reaching impact on the radiation budget, atmospheric and surface temperatures, regional weather patterns, global climatic changes, and atmospheric chemistry, including environmentally important atmospheric effects such as global ozone depletion. In this paper we review the widespread atmospheric impact of the Pinatubo eruption by considering the stratospheric injection and mass of the aerosol-generating sulfur gases (primarily SO_2), the transport of the eruption cloud and conversion of SO_2 to stratospheric sulfate aerosols, and the effects of this aerosol layer on radiation, weather, and climate. Local weather phenomena caused by the eruption are discussed elsewhere (Oswalt and others, this volume). We close with a short retrospective comparing the atmospheric effects of Pinatubo with those of other eruptions of the past century.

For the past 4 years, the Pinatubo stratospheric aerosol cloud has provided an exceptional natural laboratory for atmospheric scientists. The presence of the volcanic aerosol veil with a peak global midvisible optical depth (τ) of at least 0.1 (Sato and others, 1993), initial radiation losses of up to 5% for the first 10 months (Dutton and Christy, 1992),

[1]Hawaii Center for Volcanology and Department of Geology and Geophysics, University of Hawaii at Manoa, Honolulu, HI 96822, USA.

[2]Department of Meteorology, School of Ocean and Earth Science and Technology, University of Hawaii at Manoa, Honolulu, HI 96822, USA.

[3]Now at SETS Technology, Inc., 30 Kalehu Ave. #10, Miliani, HI 96789.

[4]Also at Philippine Institute of Volcanology and Seismology, Quezon City, Philippines.

and the concomitant, measurable climate anomalies such as global surface cooling of perhaps in excess of 0.5°C in 1992 (Dutton and Christy, 1992; Hansen and others, 1993) have produced tremendous excitement in the atmospheric science community. Measurements of Pinatubo aerosols and their effects on the Earth's climate system will enable validation of the new generation of global circulation models, improve our understanding of global aerosol dispersal and decay (and thus stratospheric circulation), and permit testing of models of aerosol formation. They will also provide better knowledge of controls on the global ozone budget, and will throw light on problems such as why surface cooling is clearly documented after some eruptions (for example, Gunung Agung, Bali, in 1963; Hansen and others, 1978) but not others—for example, El Chichón, Mexico, in 1982 (Angell, 1988; Ramanathan, 1988). Intense interest in the atmospheric aftermath of Pinatubo has been expressed by the tremendous proliferation of papers published since 1991 on the aerosol cloud and its effects. A representative portion of this considerable body of information has been canvassed for this study, but the authors are aware that many studies, especially some of those published after Fall 1993, when this paper was written, are not quoted herein; to the authors of those works we extend our apologies. Obviously, the optimum time for complete evaluation of the atmospheric effects of this exceptional eruption will be in a few years, when the results of many studies in progress are available.

ERUPTION CHARACTERISTICS AND VOLATILE RELEASE

It is important both to overall considerations of the size of the Pinatubo eruption and to the volume of magma that accompanied the atmospheric injection of the SO_2 cloud that we briefly consider the magma volume and how the material was ejected into the atmosphere. Moreover, some studies published in the nonvolcanological literature have made various erroneous claims regarding the size of the Pinatubo eruption, suggesting, for instance, that it was the century's largest eruption. We show here that the volume of erupted magma is not necessarily related to the size of the aerosol cloud generated, particularly in the case of Pinatubo (see also Gerlach and others, this volume), and we put the size of the Pinatubo eruption in perspective. As in most great eruptions (Rampino and Self, 1982), plinian eruption columns from a "point source" vent and co-ignimbrite eruption clouds (Woods and Wohletz, 1991; Koyaguchi and Tokuno, 1993; Koyaguchi, this volume) derived from the pyroclastic flows at Pinatubo contributed to the eruption column and cloud, and thus to the mass of SO_2 injected into the atmosphere.

VOLUME OF EJECTA

The present best estimate of 8.4 to 10.4 km³ total bulk volume for Pinatubo 1991 ejecta is the sum of the ~5–6 km³ of bulk volume of ignimbrite (W.E. Scott and others, this volume) and the 3.4 to 4.4 km³ bulk volume of fallout deposits (Paladio-Melosantos and others, this volume). A dense rock equivalent (DRE) maximum of 3.7 to 5.3 km³ (W.E. Scott and others, this volume) is probably as good a figure as will be obtained. This value is in reasonable agreement with an independent volume estimate of ~5.5 km³ based on eruption column heights and durations obtained from satellite images of the eruptive system and models of eruption column dynamics (Holasek, 1995). Somewhat larger values (18 km³ bulk volume, which converts to <9 km³ DRE) are supported by modeling of the giant Pinatubo eruption cloud (Koyaguchi and Tokuno, 1993; Koyaguchi, this volume).

STRATOSPHERIC INJECTION BY THE ERUPTION COLUMN

The maximum and average heights reached by the eruption columns on June 12–15, 1991, are important in assessing the original altitude of injection of the ash and gas cloud. Table 1 shows all times during which emissions reached stratospheric elevations (see also Hoblitt, Wolfe, and others, this volume). The tropopause was at about 17 km over the Philippines during the eruption.

Of most importance to the generation of the aerosol cloud is the period of 21 h from about 0555 on June 15 to 0300 on 16 June that had the highest eruption columns. During this period, a series of explosions beginning at 1027 on June 15 produced a fluctuating eruption column and umbrella clouds above the volcano (see Tokuno, 1991a) and gradually escalated into a 9-h-long climactic phase beginning before 1340. Examination of images from the Japanese Geostationary Meteorological Satellite (GMS-4) by Tokuno (1991A,B), Tanaka and others (1991), Koyaguchi and Tokuno (1993), and this study (fig. 1), provides some details of this part of the eruption, which was ill observed from the ground. It is noted elsewhere (Oswalt and others, this volume) that the closest pass of tropical typhoon Yunya to Pinatubo occurred just before the start of the climactic phase.

The first GMS image of the climactic phase collected at about 1340 (figs. 1A,E) shows the center of the eruption column located nearly over Pinatubo's vent, but it is difficult to pinpoint exactly where the center of the rising column is relative to the volcano in subsequent images as the column rises and grows. This is due to an offset introduced by the geometry of the spacecraft's look angle and position with respect to any non-nadir point on the Earth's surface; the offset increases with altitude of imaged objects above

Table 1. Chronology of ash columns with significant stratospheric injection of ash and volatiles during the June 1991 eruptions of Mount Pinatubo, Philippines.

[Local time = G.m.t. + 8. Dispersal of ash column is west-southwest in all cases. After Hoblitt, Wolfe, and others, this volume; Pinatubo Volcano Observatory Team, 1991; and Smithsonian Institution, 1991]

Date, June 1991	Time (local time)	Seismic duration (min)	Column height (km)	Dispersal and plume size, × hours after	Description of events
12	0852	42	>19	5.5×10^4 km² in 8 h	Eruption with subsequent pyroclastic flows reaching 4–5 km along the NW. flank. Pyroclastic-flow deposits enriched with dense andesitic dome fragments; fallout intense southwestward.
12	2251	14	24–25	1.1×10^5 km² in 8 h	Fallout dispersed W., NW., and SW. with pumice lapilli and coarse ash falling 15–20 km away from the vent; pyroclastic flows on the N., NW., and W. flanks.
13	0841	5	24	1×10^5 km² in 6 h	Formation of 200- to 300-m-diameter crater; partial destruction of andesitic dome; pyroclastic flow traveled 4–5 km NW.
14	1307	2	21	5×10^4 km² in 4 h	Eruption column reached the height of 21 km in about 15 min after onset of explosion; fallout toward SW.
14	1410	?	15–20	6×10^4 km² in 5h	Eruption appeared to issue from multiple vents; pyroclastic flows in the NW. flank reached 15 km downvalley.
14	1853	5	≥24	7.5×10^4 km² in 6 h	Eruption with subsequent pyroclastic flows on the NW. flank.
14	2330	3	≥21	5×10^4 km² in 3 h	Fallout at 25–30 km SW. to WSW. from the vent; pyroclastic flow was detected on the upper E. flank.
15	0114	3–23	23–25	1.5×10^5 km² in 4 h	Eruption with pyroclastic flows on the SW. and upper E. flanks.
15	0555	3	12–20	1.1×10^5 km² in 3 h	Generation of outwardly moving pyroclastic flows initially referred to as blast; ash column was momentarily visible, immediately obscured by elutriating ash.
15	0810	?	12–20	1.1×10^5 km² in 3 h	Gap in the seismic record; pyroclastic flows were observed on N. and NW. flanks.
15	1027	4–14	>20	1.1×10^6 km² in 12 h	Eruption followed by successive explosions, ash cloud buildup, and dispersal.
15	1342	~9 h	≥35	?	Beginning of paroxysmal eruption, consisting of several indistinguishable column formations and accompanied by massive deposition of pyroclastic flows all around the vent, pumice lapilli fallout, and large earthquakes. Formation of the summit caldera.
15	2231	~4 h	26–28	?	Ball-shaped column emanating from the newly formed caldera.
16	0331	~4 h	23–25	?	Wedge-shaped plume; last observed eruption-related ash column with height above 17 km.

the surface (Holasek and Self, 1995). Tokuno (1991b) notes that in some of the satellite images presented in his work, the locus of the growing volcanic umbrella cloud moves off to 10 to 20 km west of Pinatubo, but the significance of this observation is difficult to evaluate without knowing details of the positioning (mapping) system used in creating these images. In our images of the same data produced using the MCIDAS software package (figs. 1B–D and F–H), the rising center of the cloud varies slightly in position relative to Pinatubo's vent, suggesting that the vent-derived plinian eruption column and giant umbrella cloud were augmented

later in the event by co-ignimbrite ash columns lifting off the pyroclastic flows to the west and south of the volcano. In fact, plinian and pyroclastic flow activity were probably concomitant during much of the climactic phase (W.E. Scott and others, this volume). Model calculations show that the width of point-source plinian eruption columns near the top is about twice the height for columns >30 km high (Wilson and Walker, 1987), so that a mean column height of 35 km would be predicted to have an umbrella region about 70 km wide. Because the Pinatubo umbrella cloud reached >500 km in diameter by 1640, a larger surface area for the plume

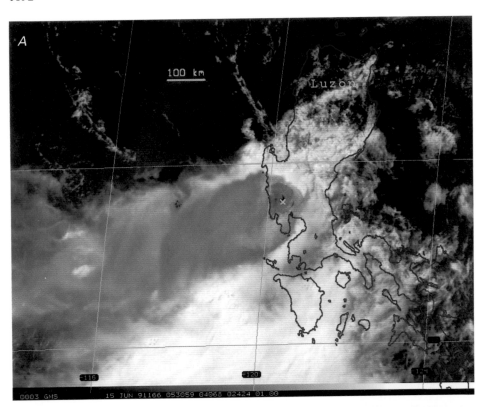

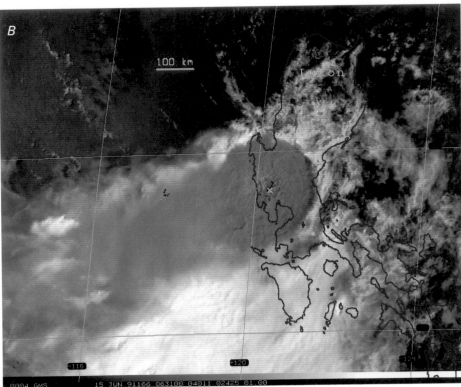

Figure 1. Japanese GMS visible- and thermal-IR wavelength satellite images of top of giant umbrella cloud developing above eruption column of Mount Pinatubo on June 15, 1991. *A–D*, Visible wavelength images spanning 3 h from 1340 to 1640 local time; *E–H*, Thermal-IR images at same times. Scale bar shows size of developing umbrella cloud; x in visible wavelength images marks center of Mount Pinatubo volcano. Color bar in bottom right of thermal-IR images gives instrument-perceived temperatures in degrees Celsius.

source is indicated such as would be provided by the central vent and surrounding pyroclastic flows (Woods and Wohletz, 1991).

Although it is recognized that perhaps 90% of the erupted volume was produced in the 9-h climax, infrasonic records from Japan (fig. 6 in Tahira and others, this volume) and barometric records collected near Pinatubo identify a 3-h period from about 1340 to 1640 during which the output and, by inference, the mass eruption rate were the highest. This period is covered by figure 1 and shows the

Figure 1.—Continued.

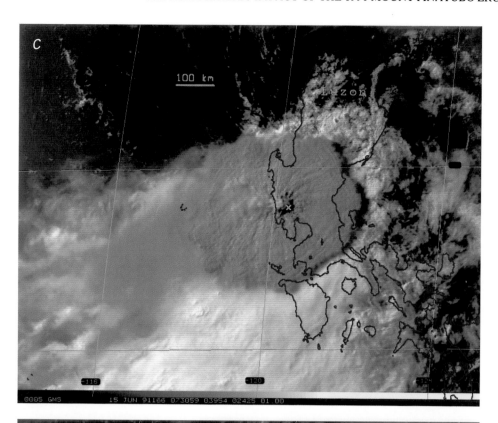

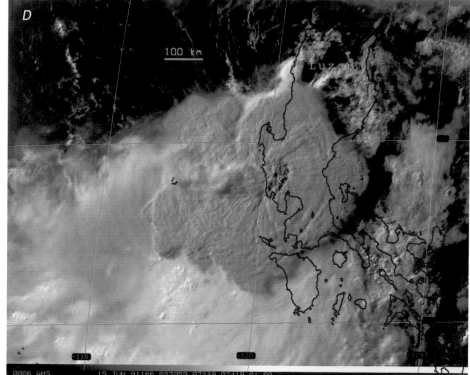

development of the remarkable umbrella cloud that reached 550 km in diameter east to west by 1640 and pushed up to 250 km upwind at an average expansion rate of 20 m/s with a maximum average expansion of the plume between 1420 and 1440 of 125 m/s (Koyaguchi and Tokuno, 1993). The giant umbrella cloud covered an area of 300,000 km² at 1940 and was sustained and growing during the whole 9-h climax, eventually reaching >1,100 km in diameter.

Several studies of available satellite images estimated the top of the umbrella cloud to be at least 35 km during the

Figure 1.—Continued.

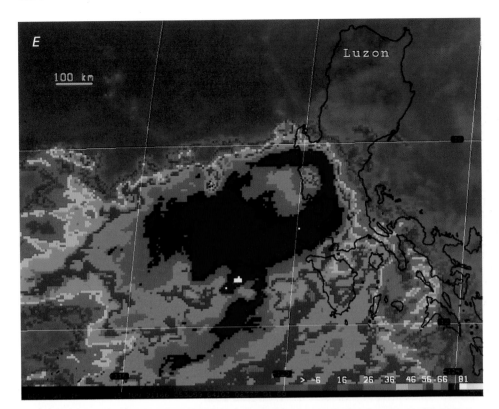

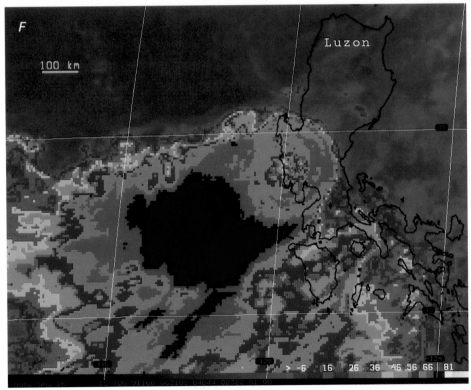

climax (Koyaguchi and Tokuno, 1993; Lynch and Stephens, this volume). This value is corroborated here by measurement of cloud-top temperature on both NOAA Advanced Very High Resolution Radiometer (AVHRR) and GMS thermal infrared (IR) images, taking into account that the central area of the umbrella cloud around the rising column can potentially suffer brief but dramatic undercooling (Woods and Self, 1992). That area had wildly varying temperatures (fig. 1E–H), and, therefore, temperatures were determined on a high-level portion of the umbrella cloud

Figure 1.—Continued.

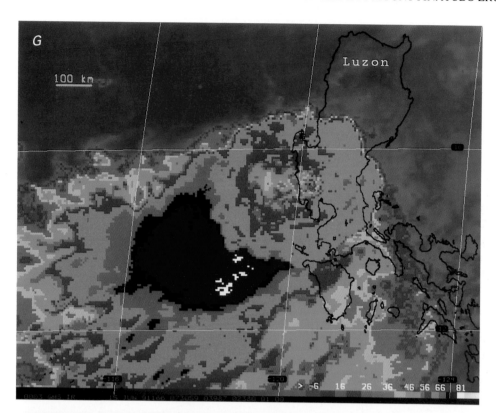

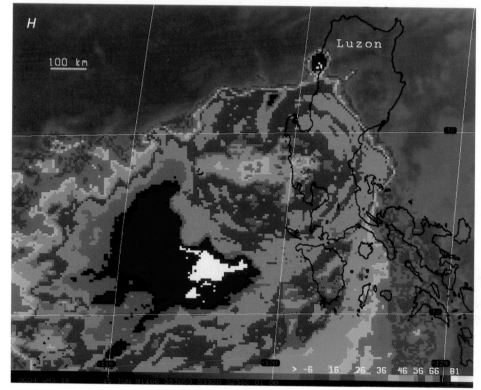

away from this central zone. The instrument-perceived plume-top temperatures were compared with an atmospheric temperature profile for the same period (fig. 2). Values indicate a middle-stratospheric height for the average cloud top (34–37 km), if it is assumed that the cloud is in the stratosphere. The assumption is based on the fact that the umbrella region casts shadows in the visible images on the tropospheric clouds associated with Typhoon Yunya. Holasek (1995) compared several thermal-IR image altitudes with shadow-determined plume heights in visible-IR

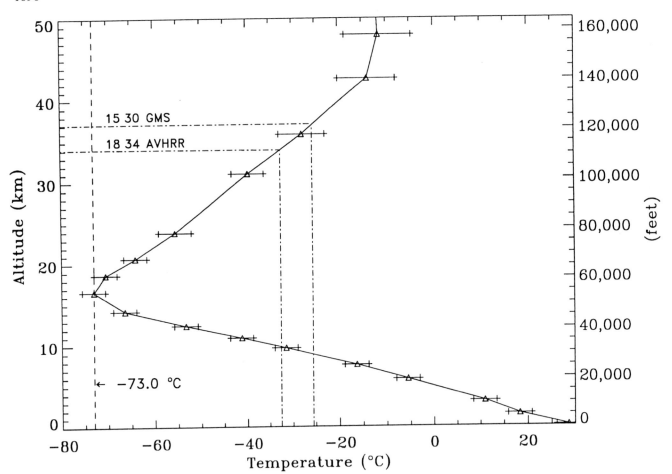

Figure 2. Temperatures retrieved from AVHRR thermal-IR and GMS weather satellite images of the relatively warm ash plume blowing westward off the overshooting top of the Pinatubo umbrella cloud indicate approximate altitude by comparison with National Meteorological Center grid point data for temperatures over the Philippines. GMS data collected at 1541 (1530 image; see fig. 1G) indicate 37 km in altitude and AVHRR data at 1834 indicate 34 km in altitude.

images indicates that the overshooting top of the eruption column was higher than this, perhaps exceeding 40 km. Height determinations by the shadow method on the main umbrella cloud also agree with these values, giving an altitude of 25 km at the eastern edge and 34 km at its center for the 1540 image (Tanaka and others, 1991). Thus, the umbrella cloud probably occupied a 10–15-km-thick section of the atmosphere from the tropopause to 35 km for a period in excess of 12 h. As the cloud subsided and entered higher wind fields at lower altitudes, GMS satellite images of the entrained plume shows that it covered an area of 2.7×10^6 km^2 36 h after the eruption, a size unprecedented in recent times.

DISPERSAL OF THE ASH AND GAS CLOUD

The main ash cloud was transported by the prevailing winds from the east-northeast, probably at levels in the middle stratosphere down to the upper troposphere, typical of tropical circulation at that time of year in the easterly phase of the quasibiennial oscillation, and much of the ash fell at sea where no data were recovered (fig. 3A). The ash cloud could be clearly tracked on GMS weather satellite images until June 17, after which sufficient ash must have fallen out to make the plume hard to define. Movement of the volcanic cloud after this time, as tracked by the SO_2 cloud, could then be followed on Total Ozone Mapping Spectrometer (TOMS) satellite images (fig. 3B) and by aerosol determinations from the AVHRR (Lynch and Stephens, this volume) and Stratospheric Aerosol and Gas Experiment (SAGE) II (McCormick and Veiga, 1992). The circumglobal transport of the developing aerosol cloud is discussed in a following section.

After the climactic phase, eruption columns from Pinatubo explosions reached upper tropospheric to lower stratospheric altitudes on several occasions. Some ash columns associated with secondary pyroclastic flows (Torres and others, this volume) even occurred in 1993. However, these were all of insignificant size compared to the main

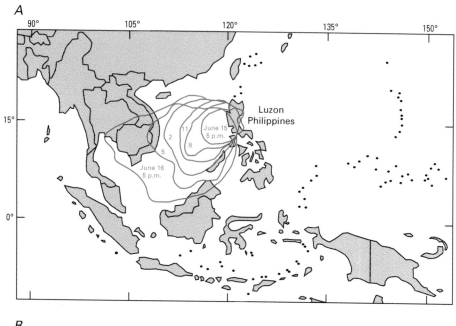

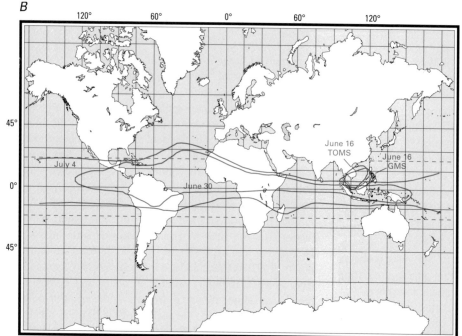

Figure 3. *A,* The spreading of Pinatubo eruption cloud as derived from Japanese GMS satellite images at the given times (Philippine local time). *B,* The transition from ash-laden eruption cloud to SO_2-dominated stratospheric cloud mapped by TOMS satellite. (Data courtesy of Gregg Bluth and Arlin Krueger, NASA Goddard Space Flight Center.)

eruption and did not approach the atmospheric impact of the main eruption, except, perhaps, to pose a threat to aircraft.

VOLATILE EMISSIONS

During the climactic phase of the Pinatubo eruption on June 15, large amounts of volcanic SO_2 and other gases that were released into the atmosphere rapidly produced a large increase in the sulfate aerosol loading of the stratosphere. The TOMS satellite measured the largest SO_2 cloud ever detected during the instrument's 13 years of operation, 20 ($\pm$6) megatons (1 Mt = 10^9 kg) of SO_2, almost all from the

9-h climax (Bluth and others, 1992). The estimate is based on a TOMS measurement of 18.5 Mt about 36 h after the eruption ended, combined with an observed average decrease in SO_2 in the cloud of 1 to 1.5 Mt per day. We note that TOMS measurements may be subject to errors of about 30% (A. Krueger, personal commun., 1993). The stratospheric cloud was observed by TOMS to encircle the Earth in about 22 days (fig. 3*B*).

Other estimates of the amount of SO_2 released from remote sensing data were made from the Microwave Limb Sounder (MLS) on the Upper Atmosphere Research Satellite (UARS), which made its first measurements on September 12, 1991 (Read and others, 1993), and from the spectral

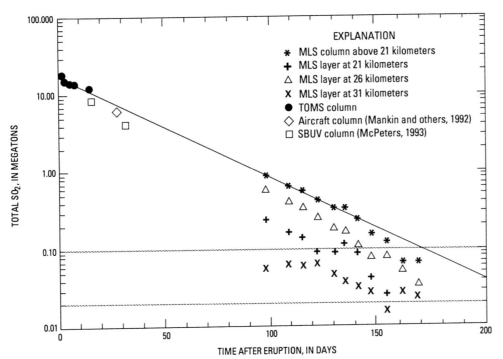

Figure 4. Measurements of the SO_2 produced by the June 15, 1991, Pinatubo eruption from MLS, TOMS, and SBUV determination between June 1991 and March 1993. Upper and lower dotted lines are uncertainty limits of MLS measurements at 21 and 26, and 31 km, respectively. (After Read and others, 1993. See text for discussion.)

scan data of the SBUV–2 instrument on the NOAA–11 satellite (McPeters, 1993) (fig. 4). The initial masses indicated are 17 and 13.5 ±1.5 Mt, respectively, giving an average when combined with the TOMS figure of 17 ±~2 Mt (see also Gerlach and others, this volume). The initial mass of the injection may have been greater than the estimates obtained by extrapolation of the remotely sensed data, because some unknown amount of sulfur was probably scrubbed out of the atmosphere by the falling silicate ash particles. However, the lower estimates of SO_2 above are in better agreement with the measured maximum aerosol masses, as will be shown subsequently in the text.

The origin of this large amount of SO_2 was probably from a water-rich, sulfur-bearing fluid phase coexisting with the dacitic Pinatubo magma (Gerlach and others, this volume; see also several other papers in this volume for discussion of the origin of the erupted sulfur). The Gerlach study shows convincingly that the dissolved sulfur contained in the >5-km^3 magma body before eruption, as estimated by the petrologic method (Devine and others, 1984), was insufficient to account for the SO_2 cloud. Thus, in the Pinatubo case (Hattori, 1993; this volume), and in other eruptions of evolved magmas developed under highly oxidizing conditions, such as with El Chichón, Mexico, in 1982 (Luhr and others, 1984), the erupted magma volume bears little, if any, relation to the size of the volatile release, as discussed in more detail by Gerlach and others (this volume).

As well as sulfur, CO_2, H_2O, and a small amount of chlorine were released to the stratosphere (Gerlach and others, this volume), and some of the water may have been important in early formation of aerosols. Electron microprobe analyses of glass inclusions and matrix glass in Pinatubo dacitic pumice suggests that up to 4.4 wt% H_2O was exsolved from the magma to a gas phase, implying the release of about 250 to 500 Mt of H_2O during the June 15 eruption (Westrich and Gerlach, 1992; Gerlach and others, this volume). Although 3 Mt of chlorine was erupted (Gerlach and others, this volume) and was potentially available for subsequent participation in ozone-destroying reactions (Turco, 1991), observations by airborne infrared Fourier transform spectrometry of the stratospheric cloud 3 weeks after the June 15 eruption showed little increase in HCl above stratospheric background levels (Mankin and others, 1992; Wallace and Livingston, 1992). Erupted chlorine, as HCl, is highly soluble in water and is very efficiently scavenged by water droplets in the eruption column and rapidly returned to the surface of the Earth as precipitation (Tabazadeh and Turco, 1993). Much of the chlorine may have thus been removed from the atmosphere during or shortly after eruption.

STRATOSPHERIC AEROSOLS AND THEIR EFFECTS

GAS-TO-PARTICLE CONVERSION

The large Pinatubo stratospheric SO_2 cloud began to spread rapidly and oxidize to form stratospheric sulfuric acid aerosols. Approximately half of the SO_2 had been converted to sulfuric acid aerosols by 21 to 28 days after the eruption (Winker and Osborn, 1992a). The MLS on the

UARS satellite detected minor amounts of unoxidized SO_2 for up to 170 days after the eruption. Accordingly, the average conversion rate of SO_2 is about 33 days (Read and others, 1993). Multiwavelength stratospheric aerosol extinction measurements from the SAGE II and airborne and ground-based photometers revealed greatly increased extinction with an initial wavelength dependence indicating the presence of very small aerosols created by gas-to-particle conversion (Thomason, 1992; Valero and Pilewskie, 1992; Russell and others, 1993b; Dutton and others, 1994). These newly formed aerosols grew to larger sizes by condensation of sulfuric acid and water vapor and by the coagulation process (Russell and others, 1993a,b; Dutton and others, 1994), leading to optical depth spectra that peaked at midvisible ($\tau \approx 0.5$ μm) or longer wavelength, starting about 2 months after the eruption. Over 90% of the particles collected from the volcanic clouds were composed of H_2SO_4/H_2O solution (Deshler and others, 1992b). The typical size of volcanic aerosols observed after several months of the eruption was in the range of 0.3 to 0.5 μm (Deshler and others, 1992a, 1993; Pueschel and others, 1992; Asano, 1993; Asano and others, 1993). Numerical modeling has reproduced the observed microstructure and optical properties of the Pinatubo aerosols during the period of the formation and growth in the stratosphere (Zhao and others, unpub. data, 1994).

About 20 to 30 Mt of new aerosol produced by the Pinatubo eruption was estimated by use of SAGE II data (McCormick and Veiga, 1992). The mean mass, about 25 Mt of sulfate aerosol, requires that only 13 Mt of SO_2 is available to form it, if it is assumed that the aerosols are 75 wt% H_2SO_4 and 25 wt% H_2O (Hamill and others, 1977). This estimate is somewhat smaller than those of SO_2 release given in figure 4. Infrared absorption by the Pinatubo aerosol also suggests that the composition is 59 to 77% H_2SO_4, the remainder being water (Grainger and others, 1993).

DISPERSAL OF AEROSOL CLOUD

Optical depth from SAGE II satellite measurements and NOAA/NESDIS aerosol optical thicknesses (AOT) derived from reflected solar radiation measurements of the AVHRR instrument on board the NOAA–11 polar orbiting satellite revealed that the Pinatubo aerosol layer circled the Earth in 21 days and had spread to 30° N. lat and about 10° S. lat in the same period (McCormick and Veiga, 1992; Stowe and others, 1992). It had covered 42% of the Earth's surface by mid-August 1991 (fig. 5), with the aerosol cloud at that time having a maximum mean midvisible optical thickness of 0.3 (Stowe and others, 1992). Optical depth or thickness (τ) is defined as the natural logarithm of the ratio of incident to transmitted direct beam radiation of wavelength λ, assuming vertical incidence.

Local aerosol optical depths exceeding $\tau = 0.4$ were measured in July 1991 by an aircraft-borne radiometer over the Caribbean region (Valero and Pilewskie, 1992). Average monthly dispersion rates of the aerosol cloud in the Northern Hemisphere were measured at 5° of latitude per month (Nardi and others, 1993). Pittock (1992) reported the first arrival of the aerosol cloud over Melbourne (37°45' S. lat) on July 19, 1991.

The slow poleward dispersal of the aerosol in the first 10 months can be considered as a series of detrainment events from a tropical reservoir (Trepte and others, 1993). Rosen and others (1992) noted localized fast meridional spreading of the lower part of the aerosol cloud into the northern part of the Northern Hemisphere at 20 km in altitude by October 1991. Trepte and Hitchman (1992) suggest that one reason why the bulk of the aerosol cloud was slow to penetrate to northern latitudes, thereby increasing the lifetime of the dense aerosol over tropical latitudes, was because it was high in the middle stratosphere above the zone affected by the quasibiennial oscillatory easterly shear that transports tropical aerosols polewards. By 1 year after the eruption, the aerosols had covered almost the entire globe, and the concentration has been decreasing exponentially since then. The background aerosol concentration in the stratosphere has not been reached in the more than 3 years since the eruption. SAGE II satellite data provide an overview of the profound changes to the atmospheric aerosol loadings following the eruption (fig. 6).

VERTICAL DISTRIBUTION

Between June and August 1991, SAGE II and AVHRR-derived measurements indicated an aerosol layer located primarily between 20 and 25 km in altitude and between 10° S. to 20° N. lat (McCormick and Veiga, 1992; Long and Stowe, 1994), increasing the stratospheric aerosol loading by two orders of magnitude over preeruption Pinatubo values (fig. 6). Enhanced aerosol concentration to altitudes above 35 km are consistent with estimated eruption column heights. Dustsonde measurements over the mid-North American continent (Deshler and others, 1992b) show that the early volcanic cloud was between 20 and 30 km in altitude, with a double layer structure during the early period (fig. 7) that merged into a single layer in August 1991. The two layers of aerosol can be seen in figure 7, a photograph of the atmosphere over South America taken by space shuttle astronauts in early August 1991.

Lidar studies showed that, in July, aerosols were at 17 to 26 km in altitude near Barbados (Winker and Osborn, 1992b), 21 to 23 km over Hawaii (DeFoor and others, 1992), 15 to 20 km over Germany (Jäger, 1992), and 15 to 16 km in the upper troposphere to lower stratosphere over Colorado (Post and others, 1992). In late August another aerosol layer was seen at 25 km over Colorado and at 20 to

NOAA/AVHRR AEROSOL OPTICAL THICKNESS DIFFERENCE FROM 2 YEAR AVERAGE
4-   JULY   -1991 TO 30-DEC-1993

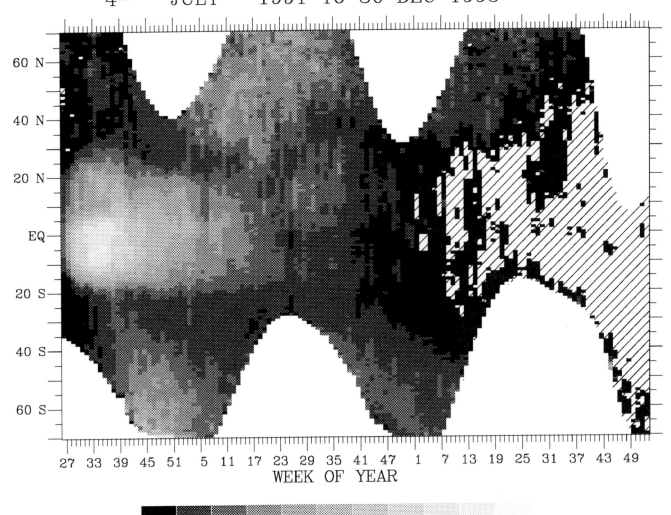

WEEK OF YEAR

.05 .10 .15 .20 .25 .30 .35 .40 .45 .50

Figure 5. Dispersal of Pinatubo aerosols between July 4, 1991, and December 30, 1993, shown by time series of one degree zonally averaged optical thickness (depth) departures following the Mount Pinatubo eruption. The departures are computed from the 2-year weekly mean values observed before the eruption. Derived from AVHRR NOAA–11 satellite data, after Long and Stowe (1994). The plot is made from intensity measurements of reflected solar radiation by the AVHRR. Note Mount Hudson (Chile) aerosol cloud from 40° S. to 60° S. lat beginning week 39.

25 km over Germany. This two-tiered aerosol layer has been monitored after other tropical eruptions (Trepte and others, 1993) and corresponds to typical transport patterns at mid-latitudes in summer, with westerlies below 20 km and easterlies above.

TEMPORAL VARIATION

The Pinatubo aerosol cloud persisted for 3 years at concentration levels well above the preeruption background in the Northern Hemisphere (Dutton and others, 1992) as a rèsult of its original high density. The increase in decay rates of the aerosol cloud depends strongly on location. An e-folding time (time to decay to $1/e$ of the original optical depth) of over 13 months was estimated near the Arctic (Stone and others, 1993), which is slightly longer than for most volcanic aerosol events. SAGE II measurements and analysis (McCormick and others, in press) yield peak surface areas of >50 $\mu m^2/cm^3$ and peak mass mixing ratios of 300 ppbm (parts per billion by mass). The aerosol mass and

91-April-10 to 91-May-13   91-June-15 to 91-July-25

91-August-23 to 91-September-30   93-December-5 to 94-January-16

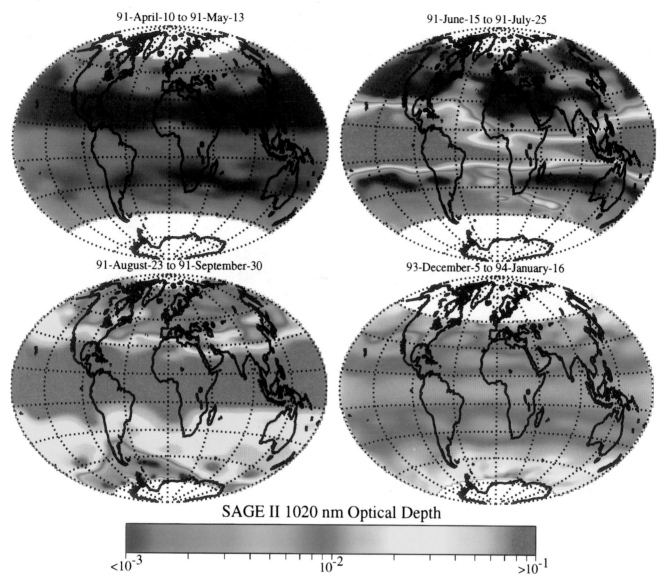

SAGE II 1020 nm Optical Depth

$<10^{-3}$ 10^{-2} $>10^{-1}$

Figure 6. Integrated SAGE II stratospheric optical depth at wavelength 1,020 nm for four periods just preceding and following the eruption of Mount Pinatubo. The stratospheric opacity during the April 10 through May 13, 1991, period was at near-background levels. The data from June 15 through July 25, 1991, show the tropical confinement of the Mount Pinatubo aerosol and the approximately two orders of magnitude increase in the 1-μm optical depth. Some indications of the initial transport to middle and high latitudes are also evident at this time. Later that year, SAGE II measurements showed that significant increases were occurring in middle and high latitudes and, by early 1992, the stratospheric optical depth was at or exceeded 0.1 at all latitudes. The figure for December 5, 1993, through January 16, 1994, shows the result of the gradual removal of aerosol from the stratosphere. While the optical depth has decreased by approximately an order of magnitude by this time, it is also evident that the stratospheric optical depth was still dominated by volcanically derived aerosol.

surface area observed at Laramie were greatly increased after the arrival of the volcanic cloud (fig. 8; Deshler and others, 1993). The sulfuric acid aerosol surface area and mass had maximum values of 40 μm^2/cm^3 and 160 ppbm, respectively, approximately 180 days after the eruption and still remained an order of magnitude higher than ambient levels for at least 2 years after the eruption. Such a great enhancement in aerosol mass and surface area due to the eruption produced significant variations of atmospheric optical properties and ozone abundance.

The time series of optical depths (at 0.5 μm) from AVHRR measurements show the time lags of maximum values at different latitudes, depending on the lateral spreading rates of volcanic clouds (Long and Stowe, 1994). In the tropical region between 20° S. and 30° N. lat, the average optical depth reached its peak value of 0.3 to 0.4

Figure 7. Space Shuttle (Mission STS 43) photograph of the Earth over South America taken on August 8, 1991, showing double layer of Pinatubo aerosol cloud (dark streaks) above high cumulonimbus tops.

about 3 months after the eruption and then gradually decayed afterward with a slight seasonal variation. Because the volcanic cloud spread to the Southern Hemisphere faster than to the Northern Hemisphere, the maximum optical depths in northern middle latitudes appears 6 months later than in the Southern Hemisphere.

OPTICAL PROPERTIES

The widespread dispersal of the aerosol into both hemispheres led to many optical effects such as unusual colored sunrises and sunsets, crepuscular rays, and a hazy, whitish appearance of the sun. These were experienced in Hawaii for much of late 1991, through most of 1992, and, after a lull in the fall of 1992, returned in the early months of 1993, finally dying away in about August.

In the months following the Pinatubo eruption optical depth increases of the stratosphere were the highest ever measured by modern techniques, in the order of 0.3 to 0.4 (Stowe and others, 1992; Valero and Pilewskie, 1992). Optical depths remained high, above 0.1, in 1992 (fig. 6), and the decrease in incident radiation, as measured by atmospheric transmission at Mauna Loa Observatory, gradually grew smaller toward 1993 (see fig. 10) (Dutton and others, 1992).

Optical depths (at $\lambda = 0.55$ μm) of 0.2 were measured at 6° S. lat 5 months after the eruption (Saunders, 1993) and as high as 0.22 in high northern latitudes 6 months after the eruption but were only 0.08 at 18° to 20° S. lat at about the same time (Russell and others, 1993a; Stone and others, 1993). One year after the eruption, τ was measured at 0.1 at 53° N. lat (Ansmann and others, 1993). The global average optical depth probably peaked at about 0.15 in early 1992, consistent with a total global aerosol at that time of about 20 Mt (Rosen and others, 1994).

Optical depths have remained at higher values above background for a longer period following Pinatubo than following El Chichón and are expected to take more than 4 years to reach background levels (Dutton and Christy, 1992). A chronology of radiation changes at Mauna Loa Observatory after both the Pinatubo and El Chichón events (fig. 9, inset) shows that the 1982 perturbation was slightly greater at that site but that the effect decayed faster than the 1991–92 event. The reason that more sudden and large increases in τ were seen from the overall smaller aerosol loading from El Chichón is that the young El Chichón cloud passed directly over the observatory, whereas Pinatubo aerosols had dispersed and settled out of stratosphere somewhat before the cloud passed 22° N. lat. The pattern of the trends through the posteruption years at Mauna Loa Observatory is quite similar, reflecting similar decay histories of the two clouds at that latitude.

In July 1993, τ (0.5 μm) was still 0.02 at Mauna Loa (background = 0.003), about the same as the peak value recorded there after the passage of the Northern Hemisphere portion of the aerosol cloud from the Agung eruption of 1963 (E.G. Dutton, NOAA, CMDL, oral commun.,

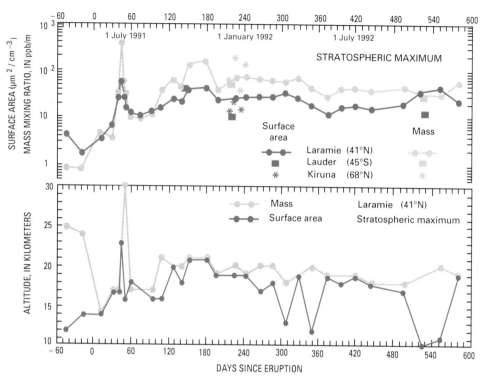

Figure 8. History and altitude of the maximum surface area and mass of sulfuric acid of aerosol from Pinatubo as based on soundings made at Laramie (Wyoming), Kiruna (Sweden), and Lauder (New Zealand). (After Deshler and others, 1993.)

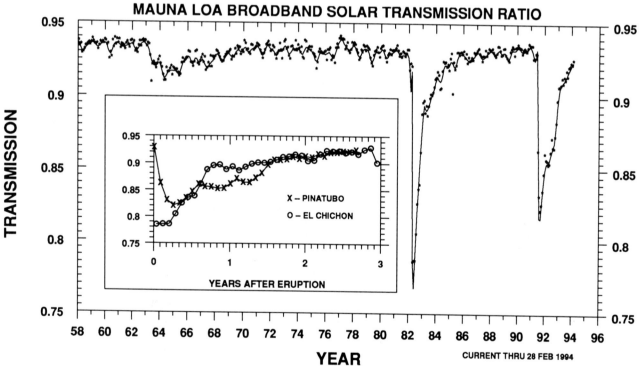

Figure 9. Atmospheric transmission of direct solar radiation at Mauna Loa Observatory, Hawaii. Plotted values are monthly averages; inset shows detail of the post-El Chichón and post-Pinatubo years at an expanded time scale. (Diagram and data courtesy of E. Dutton, NOAA-CMDL.)

1993). At most locations, maximum local values of τ after Pinatubo equaled or exceeded those after El Chichón. The much bigger aerosol mass loading from Pinatubo caused a radiative perturbation to the whole climate system about 1.7 times larger than that of El Chichón (Dutton and Christy, 1992).

It is worth noting that up to the time of writing no studies had been published of radiative properties of the fine

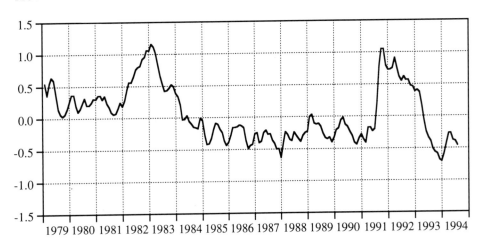

Figure 10. Lower stratospheric temperature anomalies from the microwave sounding unit (MSU) channel 4 for the globe 85° S. lat to 85° N. lat. Anomalies are computed from the 1982 to 1991 base period. (After Christy and Drouilhet, 1994).

silicate dust in the atmosphere due to this eruption or of its role in aerosol-forming processes. Even though it had a shorter residence time than the sulfate aerosols, it may have had a significant effect. Depolarization of airborne lidar data shows that uneven-shaped particles, perhaps ash, were present in the lower part of the aerosol cloud up to 7 to 9 months after the eruption (L.R. Poole, NASA, LARC, oral commun., 1993). Aircraft impactor data also show the coated silicate particles just above the tropopause through March 1992 (Pueschel and others, 1992).

TEMPERATURE, WEATHER, AND CLIMATIC EFFECTS

Radiative forcing of the climate system by stratospheric aerosols depends on the geographic distribution, altitude, size distribution, and optical depth of the aerosols, but tropospheric temperatures are most strongly dependent on the total optical depth (Lacis and others, 1992). The optically dense Pinatubo aerosol cloud caused marked changes in the amount of radiation reaching the Earth's surface; in turn, these changes affected weather and climate over the past 3 years following the eruption.

OBSERVATIONAL DATA

As observed after several eruptions, including Agung in 1963 and El Chichón in 1982, stratospheric warming and lower tropospheric and surface cooling have been documented after the Pinatubo eruption. Labitzke and McCormick (1992) show that warming in the lower stratosphere (16 to 24 km or 30 to 100 mbar) of up to 2 to 3°C occurred within 4 to 5 months of the eruption between the equator and 20° N. lat., and it was also later noticed in middle northern latitudes (Angell, 1993). The warming distribution closely mirrored the dispersal pattern of the aerosol cloud; this mirroring strongly suggests that the warming was due to absorption of radiation by the aerosols. The warming was more intense in southern temperate-polar

latitudes, perhaps due to the presence of aerosols from the Mount Hudson eruption. Such temperature changes can influence stratospheric dynamics (Pitari, 1992). Since the peak of stratospheric warming in late 1991, temperatures in the 18- to 24-km region have cooled considerably, passing the average in early 1993 (fig. 10); temperatures in 1993 were the coldest ever recorded (Christy and Drouilhet, 1994; Monastersky, 1994) and may be related to ozone destruction in the lower stratosphere. Stratospheric temperatures also plummeted and stayed cooler than average for 7 years after the El Chichón eruption.

Several experiments have measured the radiative climate forcing of the Pinatubo aerosols. The NASA Earth Radiation Budget Experiment (ERBE) recently provided the first unambiguous direct measurement of the climate forcing on a large scale in both hemispheres (Minnis and others, 1993), an average radiative cooling of 2.7 W/m^2 by August of 1991. Direct solar beam reductions of 25 to 30% were measured at widely distributed stations by Dutton and Christy (1992), while Stowe and others (1992) showed from AVHRR-derived optical depth measurements that the globally averaged net radiation at the top of the atmosphere may have decreased by about 2.5 W/m^2 in late 1991. These values translate into a global cooling of at least 0.5 to 0.7°C, as seen in the global and Northern Hemisphere temperature records by September 1992 (Dutton and Christy, 1992). A net cooling effect of approximately 0.3°C was estimated as a result of the El Chichón aerosol (Angell and Korshover, 1983; Handler, 1989), but the overall potential cooling caused by the El Chichón cloud was moderated by warming associated with El Niño-Southern Oscillation (Angell, 1988, 1990). Pinatubo had a much larger radiative influence than El Chichón in the Southern Hemisphere (Dutton and Christy, 1992). Pinatubo's cloud caused about 1.7 times the global radiative forcing of El Chichón, making the estimated cooling of 0.5°C a more robust figure.

One possible opposite effect, leading to surface warming, may have been caused by stratospheric to tropospheric transport of aerosols, due to aerosol-induced changes in atmospheric dynamics, and in a theoretical study Jensen and

Toon (1992) suggest that this process may cause higher than usual amounts of cirrus clouds in the upper troposphere. Warmer than average winters and cooler than average summers over continental Northern Hemisphere areas have been documented and modeled after several eruptions, including Pinatubo (fig. 11; Robock and Mao, 1992; Graf and others, 1993), and this appears to be part of the normal Northern Hemisphere response after volcanic aerosol events (Groisman, 1992; Robock and Liu, 1994).

MODELING

Analysis of surface temperatures or weather patterns postdating Pinatubo's eruption are still underway because of the short time elapsed since the presence of the Pinatubo aerosol, but a great deal of attention has been focused on modeling of the potential climatic impact of the eruption. The interest is in Pinatubo as a natural experiment and climatic perturbation—what changes in surface temperature and circulation a large volcanic aerosol event can bring about—and also as a validation for climate models. Can the models predict realistic climate changes, and could eruption-induced cooling offset greenhouse-induced global warming? With a three-dimensional global circulation model, Hansen and others (1992) were able to predict the global cooling in 1991–1993 (fig. 12A) and then check their results against real surface-temperature trends. Using a forcing in the model equivalent to a global mean τ of about 0.15 based on conditions appropriate for the Pinatubo aerosol cloud yielded a model radiative forcing at the tropopause of -4 W/m^2. Hansen and others (1993) show observed maximum average coolings of up to 0.6°C by late 1992 over high-latitude land masses, and less elsewhere (fig. 12B), in agreement with the modeled coolings. It should be noted that temperature data plotted in figure 12 are dominantly from stations on land.

Other attempts to evaluate the climatic response to the Pinatubo aerosol include comparison with the global temperature record of the University of East Anglia, which includes surface air temperature over land and sea (Robock and Mao, 1994), and on Southern Hemisphere sea surface temperatures (Walsh and Pittock, 1992). These studies shows that the temperature anomaly after Pinatubo is about 0.4 to 0.6 K cooler than average over a large part of the Earth (fig. 13).

The superimposition of Pinatubo's aerosol-induced climatic effects on long- and short-term variable trends, such as El Niño-Southern Oscillation and "greenhouse" warming, has led to much discussion as to the cooling effects on the current warming trend. Global temperature trends show a gradual, unsteady rise from the mid-1970's to the mid-1980's (Hansen and Lebedeff, 1987), perhaps due to forcing by greenhouse gases, and continuation of this rise had led to claims of record high temperatures in the early 1990's (see Bassett and Lin, 1993; Hansen and Wilson, 1993). However, the cooling since Pinatubo's eruption has offset the warming trend considerably, such that cooler than normal conditions dominate the Northern Hemisphere (Mo and Wang, 1994).

The Pinatubo aerosol cloud was, like with El Chichón, coincident with sudden warming due to an El Niño event, but this warming event was not nearly as strong as in the El Chichón case. El Niño may have provided a warming of about 0.2°C, partially offsetting the Pinatubo-induced cooling in the tropics, but modeling suggests a coincidence rather than a cause and effect relation between eruptions and El Niño (Robock and Liu, 1994). The Pinatubo climate forcing is stronger than the opposite, warming effect of the El Niño event or anthropogenic greenhouse gases.

The predicted and observed Pinatubo climatic cooling resulted in noticeable changes in the local climate and weather. For example, in 1992, the United States had its third coldest and third wettest summer in 77 years. Floods along the Mississippi River in the summer of 1993 and drought in the Sahel area of Africa may be attributable to climatic shifts caused by the Pinatubo aerosols and aerosol-induced temperature changes (Mo and Wang, 1994; Robock and Liu, 1994). Moreover, the cooling is not spatially uniform, as underlined by the several recent models mentioned above, and many areas have suffered above-average warm conditions such as the 1991 and 1992 winters in Eurasia. Model results (Hansen and others, 1993) show that global circulation model runs do, for various plausible scenarios, predict actual temperature changes to date (fig. 12A). Exactly for how long the Pinatubo-induced cooling will manifest itself as changing surface or regional climate and weather patterns remains to be seen and will be documented in future years.

IMPACT ON STRATOSPHERIC CHEMISTRY AND OZONE

Sulfate aerosols in the stratosphere can catalyze heterogeneous reactions that affect global ozone abundance (Farman and others, 1985; Hofmann and Solomon, 1989; Wolff and Mulvaney, 1991; Prather, 1992). These heterogeneous processes occurring on the surface of sulfate particles can convert stable chlorine reservoirs (such as HCl and ClONO$_2$) into photochemically active chlorine species (Cl$_2$, ClNO$_2$, HOCl) that are active in ozone destruction (Hofmann and Solomon, 1989; Solomon and others, 1993). Increase in aerosol surface area due to the Pinatubo volcanic eruption has had a considerable effect on global ozone (Bhartia and others, 1993; Deshler and others, 1992b; Gleason and others, 1993; Grant and others, 1992; Hofmann and Oltmans, 1993; Hofmann and others, 1994a,b; Schoeberl and others, 1993; Weaver and others, 1993). For example, reduced ozone concentrations with peak decreases as large

A
Winter (DJF) 1991–92
Average Lower Troposphere Temperature Anomalies

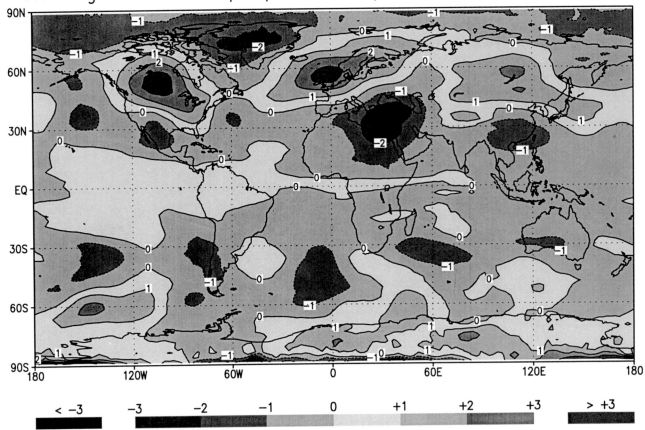

Figure 11. Winter (December, January, February) 1991–92 (*A*) and summer (June, July, August) 1992 (*B*) temperature anomalies (departures from long-term means) in degrees Celsius, demonstrating the Pinatubo-induced winter warming and summer cooling over Northern Hemisphere continental areas. Data are a combination of satellite atmospheric temperature determinations by J. Christy, University of Alabama, Huntsville, and surface temperature data from P.R. Jones, University of East Anglia, U.K. Plots courtesy of A. Robock, University of Maryland.

as 20% at 16 to 25 km in altitude were found in the tropical stratosphere 3 to 6 months after the Pinatubo eruption (Grant and others, 1992). Up to 6% reduction of equatorial total ozone was observed by TOMS measurements following the eruption of Pinatubo (Schoeberl and others, 1993). At the time of maximum aerosol development, up to 20% depletion in ozone was measured over Colorado and Hawaii (Hofmann and others, 1993; 1994a), and mid-latitude ozone abundance reached its lowest level on record during 1992–93. The total ozone amount was 2 to 3% lower than in any earlier year, with the largest decreases in the regions from 10° S. to 20° S. lat and 10° N. to 60° N. lat (Gleason and others, 1993).

Startling decreases in ozone abundance and in rates of ozone destruction were also observed over Antarctica in 1991 and 1992. This ozone decrease may be due in part to the presence of Pinatubo aerosols but also to the extra aerosol injection from the Mount Hudson eruption in Chile during August 1991 (Doiron and others, 1991; Barton and others, 1992). A sharp decrease in ozone at 9 to 11 km in altitude (approximately at the tropopause) in the austral spring of 1991 was noted at the time of arrival of the Pinatubo and Mount Hudson aerosols (Deshler and others, 1992a). The Southern Hemisphere "ozone hole" increased in 1992 to an unprecedented 27×10^6 km^2 in size, and depletion rates were observed to be faster than ever before recorded (Brasseur, 1992; Hofmann and others, 1992; 1994b). In late 1992, weather patterns caused a shift in the polar vortex, and warm ozone-rich tropical air entered the Antarctic atmosphere to partially halt the ozone depletion.

Ozone depletion causes an enhancement in the amount of biologically destructive ultraviolet radiation that reaches the Earth's surface (Smith and others, 1992; Vogelmann and others, 1992). Although the Pinatubo eruption was probably

B

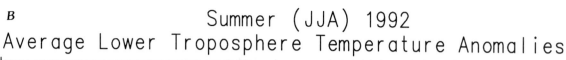

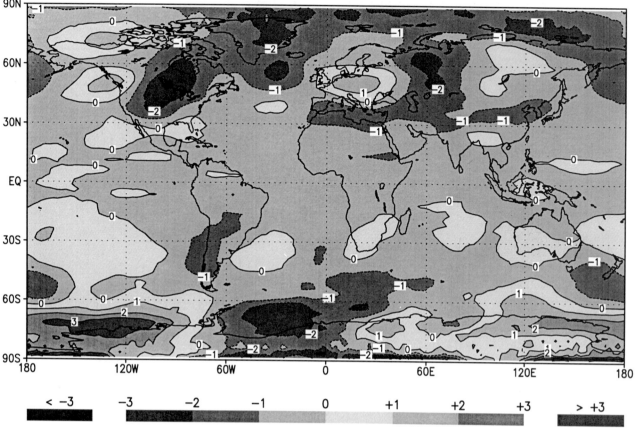

Figure 11.—Continued.

not of sufficient magnitude to cause large decreases in ozone over wide regions (2 to 4% decrease is average), much larger past eruptions (for example, Tambora, Indonesia, in 1815), with an estimated aerosol mass 5 to 10 times that of Pinatubo, may have caused drastic increases in the amount of harmful ultraviolet radiation at the Earth's surface.

PINATUBO IN PERSPECTIVE

The 1991 Pinatubo eruption, about 5 km^3 DRE, could well be the second largest this century, after Katmai-Novarupta, Alaska, in 1912. It is slightly bigger than either the plinian eruption of Santa María, Guatemala, in 1902 or the 4 km^3 DRE plinian eruption of Cerro Azul, Chile, in 1932 (Hildreth and Drake, 1992). For Santa María, Fierstein and Nathenson (1992) suggest a smaller volume, about 4 to 5 km^3 DRE, than the 9 km^3 proposed by Williams and Self (1983).

Aerosols derived from Katmai-Novarupta, which produced about 13 km^3 DRE (Fierstein and Hildreth, 1992), also caused diminution of solar transmission by about 20% at locations north of about 40° N. lat (Kimball, 1918, 1924), but the aerosols did not have widespread climatic influence, because they were contained in northern latitudes (Arctowski, 1915). Although it is not known with certainty, the 1912 eruption may have yielded a smaller amount of SO$_2$ and therefore caused a lower aerosol loading. Estimated optical depths after 1912 for the 30° to 90° latitude zone are only 0.1 or less (Sato and others, 1993).

Pinatubo is also much larger in terms of volume erupted than the other notable eruptions of this century that have caused atmospheric perturbations—for example, Agung in 1963 and El Chichón in 1982, both of which erupted about 0.5 km^3 of magma (Self and King, 1993). However, as the Pinatubo event has shown, magma volume erupted and amount of SO$_2$ released are not always proportional to each other. The small El Chichón eruption yielded just less than half of the amount of SO$_2$ released by Pinatubo (7–8 Mt; Varekamp, 1984) and generated just less than half the amount of aerosol (McCormick and others, in press), but the magma volume erupted was an order of magnitude smaller than Pinatubo's. The relative size and duration of the Pinatubo aerosol perturbation compared to that after El Chichón can be seen on the integrated aerosol backscatter record from Langley, Research Center, Hampton, Va., USA (37.1° N.) (fig. 14); these two events dominate stratospheric loading and chemistry over the past two decades. The peak values after Pinatubo are less than expected because of the delayed spread of the aerosol to these latitudes, while the El Chichón cloud was still near its peak density when it passed over this area.

Certainly, in terms of widespread impact, due to its equatorial location, the early summer date of eruption, and the resulting global spread of the aerosol cloud, the Pinatubo aerosol cloud that enveloped the Earth from the end of June 1991 to late 1993 is the largest since that caused by the approximately 10-km³ DRE Krakatau eruption in late August 1883, which also produced an aerosol veil of global extent. In fact, the maximum 20- to 30-Mt Pinatubo stratospheric aerosol loading may not be that much smaller than Krakatau's, variously estimated at between 30 and 50 Mt (Rampino and Self, 1982, 1984). This conclusion is consistent with estimated global aerosol midvisible optical depths for Krakatau, which were 0.14 in late 1884 to early 1885 (Sato and others, 1993; fig. 15). This value is for the Krakatau aerosol layer after more than 1 year's dispersal, and, presumably, much sedimentation of particles, and peak optical depths may have been considerably larger. The global optical depth is equivalent to a global aerosol loading of about 25 Mt, on the basis of the relationship of Stothers (1984a), similar to that of the Pinatubo peak loading. Both Pinatubo and Krakatau, however, are dwarfed by the eruption of Tambora in 1815, both in volume erupted (50 km³) and the peak optical depth attained by the widespread aerosol cloud, estimated to be >1.0 in northern latitudes 6 months after the eruption (Stothers, 1984b).

CONCLUSIONS

The 1991 eruption of Pinatubo, culminating in a 9-h climactic plinian and pyroclastic-flow-producing phase on June 15, produced about 5 km³ of dacitic magma and is the second largest volcanic eruption of the century in terms of magnitude (volume of magma produced). Eruption columns rising above the vent and off the pyroclastic flows reached in excess of 35 km in altitude and emplaced a giant umbrella cloud in the middle to lower stratosphere that attained a maximum dimension of over 1,100 km in diameter. This cloud injected about 17 Mt of SO_2 into the stratosphere (twice the amount produced by the 1982 El Chichón eruption), and this SO_2 immediately began to convert into H_2SO_4 aerosols, forming the largest perturbation to the

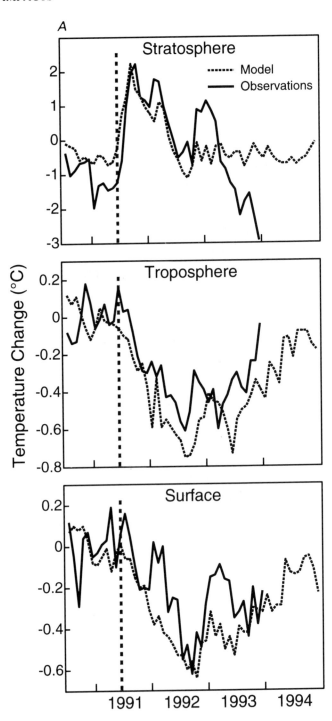

Figure 12A. Observed and modeled monthly temperature change of stratosphere, troposphere, and surface after the Mount Pinatubo eruption. Stratospheric observations are 30-mbar zonal mean temperature at 10° S. lat; model results are 10- to 70-mbar layer at 8° to 16° S. lat. Other results are essentially global, with observed surface temperature derived from a network of meteorological stations. Base period for tropospheric temperatures is 1978–92, while troposphere and surface are referenced to the 12 months preceding the Pinatubo eruption, the latter marked by a vertical dashed line.

OBSERVED TEMPERATURE CHANGES

B

Jun-Jul-Aug 1991   Global Ave=0.53 °C

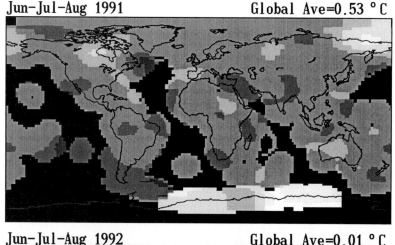

Jun-Jul-Aug 1992   Global Ave=0.01 °C

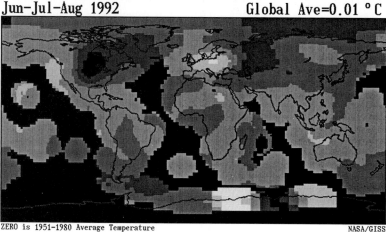

ZERO is 1951-1980 Average Temperature   NASA/GISS

Figure 12B. Model cooling predicted for 1991–92 (Hansen and others, 1993) shown by observed model air temperature anomalies relative to the 1951–80 mean for Northern Hemisphere summers of 1991 and 1992. Figures courtesy of James Hansen and Helene Wilson, NASA Goddard Institute for Space Studies, after Hansen and others (1993).

stratospheric aerosol layer since the aerosol cloud of Krakatau in 1883.

The aerosol cloud spread rapidly around the globe in about 3 weeks and attained global coverage 1 year after the eruption. The SO_2 release was sufficient to generate over 25 Mt of sulfate aerosol, and peak local and regional mid-visible optical depths of up to 0.4 were recorded. Global values after widespread dispersal and sedimentation of aerosol were about 0.1 to 0.15, with a residence time of over 2 years. This large aerosol cloud caused dramatic decreases in the amount of net radiation reaching the Earth's surface. This was certainly the largest atmospheric perturbation by an aerosol cloud in this century, producing a climate forcing two times stronger than the aerosols of El Chichón. The lower stratosphere also warmed immediately after the eruption and has cooled to the lowest temperatures recorded since then, causing changes in atmospheric circulation.

Effects on climate were an observed surface cooling in the Northern Hemisphere of up to 0.5 to 0.6°C, equivalent to a hemispheric reduction in net radiation of 4 W/m², and an overall cooling of perhaps as large as −0.4°C over large

parts of the Earth in 1992–93. Climate models appear to have predicted the cooling currently occurring with a reasonable degree of accuracy. The Pinatubo climate forcing was stronger than the opposite, warming effects of either the El Niño event or anthropogenic greenhouse gases in the period 1991–93.

Atmospheric composition also underwent some remarkable changes that were due to the Pinatubo aerosols, most notably that mid-latitude ozone abundance reached its lowest level on record during 1992–93. The total ozone amount was 2 to 3% lower than in any earlier year, with the largest decreases in the regions from 10° to 20°S. lat and 10° to 60° N. lat. The Southern Hemisphere "ozone hole" increased in 1992 to an unprecedented 27×10^6 km² in size, and depletion rates were faster than ever before recorded.

The atmospheric impact of the Pinatubo eruption has been profound, and it has sparked a lively interest in the role that volcanic aerosols have played in climate change. It has been an extremely important and timely event to the atmospheric sciences, permitting climate models to be tested and tuned, and showing that a powerful eruption

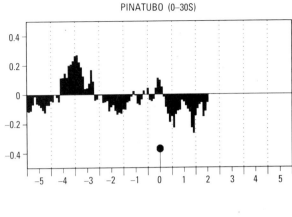

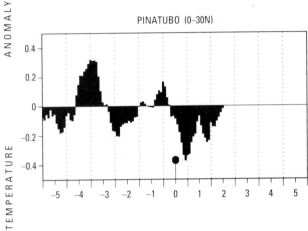

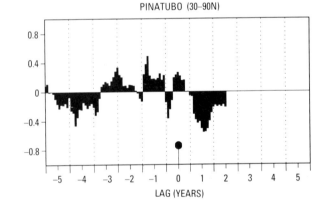

providing a 15 to 20 Mt SO_2 release into the stratosphere can produce sufficient aerosols to offset global warming trends and severely impact the ozone budget.

ACKNOWLEDGMENTS

We thank E.G. Dutton (NOAA Climate Monitoring and Diagnostic Laboratory (CMDL)), J.E. Hansen and H. Wilson (NASA Goddard Institute for Space Studies (GISS)), A. Robock (University of Maryland), and G. Stephens (NOAA-NESDIS) for generously providing data, and M.P. McCormick (NASA Langley Research Center), E.A. Dutton, P.B. Russell (NASA Ames Research Center), C.G. Newhall (USGS), A. Robock, L.S. Walter (NASA Goddard Space Flight Center), and A. Tabazadeh (University of California, Los Angeles) for reviews of earlier versions of the manuscript. This work was supported by NASA grants NAG 5–1839 and NAG W–3721. This is SOEST contribution No. 3563.

Figure 13.   Latitude band anomalies in temperature (Kelvin) plotted with respect to the 5-year mean before the Pinatubo eruption. Data are surface temperature for land-based stations combined with sea surface temperatures, from P.R. Jones (University of East Anglia, U.K.). Year 0 is the eruption year; precise time of eruption indicated at bottom of each graph. Data and diagram courtesy of A. Robock, University of Maryland, from Robock and Mao (1995).

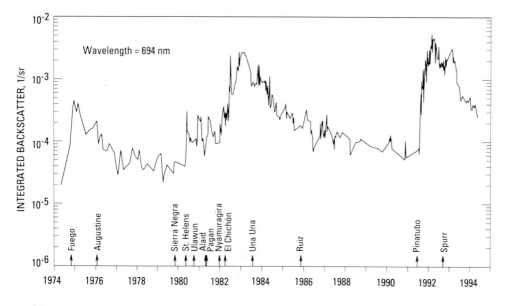

Figure 14. Integrated aerosol backscatter measurements from the tropopause to 30 km obtained by lidar at Hampton, Va. (37° N. lat, 76° W. long), at a wavelength of 694 nm since 1974. The major volcanic eruptions that increased Northern Hemispheric mid-latitude aerosol loading are noted by arrows on the time axis. Data courtesy of M. Osborn and M. P. McCormick, NASA Langley Research Center.

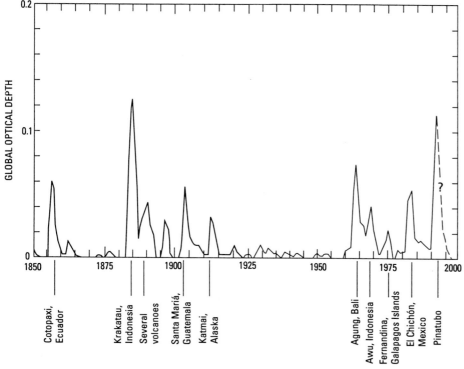

Figure 15. Estimated global stratospheric optical depth at $\lambda = 0.55$ µm for the period 1850 to 1993, after Sato and others (1993). Most peaks are the result of instantaneous volcanic injections of SO_2 into the stratosphere and subsequent rapid formation and monotonic decline of sulfate aerosols.

REFERENCES CITED

Angell, J.K., 1988, Impact of El Niño on the delineation of tropospheric cooling due to volcanic eruptions: Journal of Geophysical Research, v. 93, p. 3697–3704.

——1990, Variation in global tropospheric temperature after adjustment for the El Niño influences, 1958–1989: Geophysical Research Letters, v. 17, p. 1093–1096.

——1993, Comparison of stratospheric warming following Agung, El Chichón and Pinatubo volcanic eruptions: Geophysical Research Letters, v. 20, p. 715–718.

Angell, J.K., and Korshover, J., 1983, Comparison of stratospheric warming following Agung and El Chichón: Monthly Weather Review, v. 111, p. 2129–2135.

Ansmann, A., Wandinger, U., and Weitkamp, C., 1993, One-year observations of Mount-Pinatubo aerosol with an advanced raman lidar over Germany at 53.5°N: Geophysical Research Letters, v. 20, p. 711–714.

Arctowski, H., 1915, Volcanic dust veils and climatic variations: Annals of the New York Academy of Sciences, v. 26, p. 149–174.

Asano, S., 1993, Estimation of the size distribution of Pinatubo volcanic dust from Bishop's ring simulations: Geophysical Research Letters, v. 20, p. 447–450.

Asano, S., Uchiyama, A., and Shiobara, M., 1993, Spectral optical thickness and size distribution of the Pinatubo volcanic aerosols as estimated by ground-based sun photometry: Meteorological Society of Japan Bulletin, v. 71, p. 165–173.

Barton, I.J., Prata, A.J., Watterson, I.G., and Young, S.A., 1992, Identification of the Mount Hudson volcanic cloud over SE Australia: Geophysical Research Letters, v. 19, p. 1211–1214.

Bassett, G.W., and Lin, Z., 1993, Breaking global temperature records after Mt. Pinatubo: Climatic Change, v. 25, p. 179–184.

Bhartia, P.K., Herman J., and McPeters, R.D., 1993, Effect of Mount Pinatubo aerosols on total ozone measurements from backscatter ultraviolet (BUV)experiments: Journal of Geophysical Research v. 98, p. 18547–18554.

Bluth, G.J.S., Doiron, S.D., Schnetzler, C.C., Krueger, A.J., and Walter, L.S., 1992, Global tracking of the SO_2 clouds from the June 1991 Mount Pinatubo eruptions: Geophysical Research Letters, v. 19, p. 151–154.

Brasseur, G., 1992, Ozone depletion: Volcanic aerosols implicated: Nature, v. 359, p. 275–276.

Christy, J.R., and Drouilhet, S.J., 1994, Variability in daily, zonal-mean lower-stratospheric temperatures: Journal of Climate, v. 7, p.106–120.

Defoor, T.E., Robinson, E., and Ryan, S., 1992, Early LIDAR observations of the June 1991 Pinatubo eruption plume at Mauna Loa Observatory, Hawaii: Geophysical Research Letters, v. 19, p. 187–190.

Deshler, T., Adriani, A., Gobbi, G.P., Hofmann, D.J., DiDonfrancesco, G., and Johnson, B.J, 1992a, Volcanic aerosol and ozone depletion within the Antarctic polar vortex during the austral spring of 1991: Geophysical Research Letters, v. 19, p. 1819–1822.

Deshler, T., Hofmann, D.J., Johnson, B.J., and Rozier, W.R., 1992b, Balloonborne measurements of Mt. Pinatubo aerosol size distribution and volatility at Laramie, Wyoming, during the summer of 1991: Geophysical Research Letters, v. 19, p. 199–202.

Deshler, T., Johnson, B.J., and Rozier, W.R., 1993, Balloonborne measurements of Pinatubo aerosol during 1991 and 1992 at 41°N: Vertical profiles, size distribution and volatility: Geophysical Research Letters, v. 20, p. 1435–1438.

Devine, J.D., Sigurdsson, H., Davis, A.N., and Self, S., 1984, Estimates of sulfur and chlorine yield to the atmosphere from volcanic eruptions and potential climatic effects: Journal of Geophysical Research, v. 89, p. 6309–6325.

Doiron, S.D., Bluth, G.J.S., Schnetzler, C.C., Krueger, A.J., and Walter, L.S., 1991, Transport of Cerro Hudson SO_2 clouds: Eos, Transactions, American Geophysical Union, v. 72, no. 45, p. 489, 498.

Dutton, E.G., and Christy, J.R., 1992, Solar radiative forcing at selected locations and evidence for global lower tropospheric cooling following the eruptions of El Chichón and Pinatubo: Geophysical Research Letters, v. 19, p. 2313-2316.

Dutton, E.G., Reddy, R., Ryan, S., and DeLuisi, J., 1994, Features and effects of aerosol optical depth observed at Mauna Loa, Hawaii, 1982–1992: Journal of Geophysical Research, v. 99, p.8295–8306.

Farman, J.C., Gardiner, B.G., and Shanklin, J.D., 1985, Large losses of total O_3 in atmosphere reveal seasonal ClO_x/NO_x interaction: Science, v. 211, p. 832–834.

Fierstein, J., and Hildreth, W., 1992, The plinian eruptions of 1912 at Novarupta, Katmai National Park, Alaska: Bulletin of Volcanology, v. 54, p. 646–684.

Gerlach, T.M., Westrich, H.R., and Symonds, R.B., this volume, Preeruption vapor in magma of the climactic Mount Pinatubo eruption: Source of the giant stratospheric sulfur dioxide cloud.

Gleason, J.F., Bhartia, P.K., Herman, J.R. and others, 1993, Record low global ozone in 1992: Science, v. 260, p. 523–526.

Graf, H-F., Kirchner, I., Robock, A., and Schult, I., 1993, Pinatubo eruption winter climate effects: Model versus observations: Climate Dynamics, v. 9, p. 81–93.

Grainger, R.G., Lambert, A., Taylor, F.W., Remedios, J.J., Rodgers, C.D., and Corney, M., 1993, Infrared absorption by volcanic stratospheric aerosols observed by ISAMS: Geophysical Research Letters, v. 20, p. 1293–1286.

Grant, J., Fishman, J., Browell, E.V., and others, 1992, Observations of reduced ozone concentrations in the tropical stratosphere after the eruption of Mt. Pinatubo: Geophysical Research Letters, v. 19, p. 1109–1112.

Groisman, P.Ya., 1992, Possible regional climate consequences of the Pinatubo eruption: An empirical approach: Geophysical Research Letters, v. 19, p. 1603–1606.

Hamill, P., Kiang, C.S., and Cadle, R.D., 1977, The nucleation of H_2SO_4-H_2O solution aerosol particles in the stratosphere: Journal of the Atmospheric Sciences, v. 34, p. 150–162.

Handler, P., 1989, The effect of volcanic aerosols on global climate: Journal of Volcanology and Geothermal Research, v. 37, p. 233–249.

Hansen, J., Lacis, A., Ruedy, R., and Sato, M., 1992, Potential climate impact of the Mount Pinatubo eruption: Geophysical Research Letters, v. 19, p. 215–218.

Hansen, J., Lacis, A., Ruedy, R., Sato, M., and Wilson, H., 1993, How sensitive is the world's climate?: National Geographic Research and Exploration, v. 9, p. 143–158.

Hansen, J., and Lebedeff, S., 1987, Global trends of measured surface air temperature: Journal of Geophysical Research, v. 92, p. 13345–13372.

Hansen, J.E., Wang, W.-C., and Lacis, A., 1978, Mount Agung eruption provides test of a global climatic perturbation: Science, v. 199, p. 1065–1068.

Hansen, J., and Wilson, H., 1993, Commentary on the significance of global temperature records: Climate Change, v. 25, p. 185–191.

Hattori, K., 1993, High-sulfur magma, a product of fluid discharge from underlying mafic magma: Evidence from Mount Pinatubo, Philippines: Geology, v. 21, p. 1083–1086.

Hildreth, W., and Drake, R.E., 1992, Volcan Quizapu, Chilean Andes: Bulletin of Volcanology, v. 54, p. 93–125.

Hoblitt, R.P., Wolfe, E.W., Scott, W.E., Couchman, M.R., Pallister, J.S., and Javier, D., this volume, The preclimactic eruptions of Mount Pinatubo, June 1991.

Hofmann, D.J., and Oltmans, S.J, 1993, Anomalous Antarctic ozone during 1992: Evidence for Pinatubo volcanic aerosol effects, Journal of Geophysical Research, v. 98, p. 18555–18561.

Hofmann, D.J., Oltmans, S.J., Harris, S., Komhyr, W.D., Lathrop, J.A., Defoor, T., and Kuniyuki, D., 1993, Ozonesonde measurements at Hilo, Hawaii, following the eruption of Pinatubo: Geophysical Research Letters, v. 20, p.1555–1558.

Hofmann, D.J., Oltmans, S.J., Harris, S., Solomon, S., Deshler, T., and Johnson, B.J., 1992, Observation and possible causes of new ozone depletion in Antarctica in 1991: Nature, v. 359, p. 283–287.

Hofmann, D.J., Oltmans, S.J., Komhyr, W.D., Harris, J.M., Lathrop, A., Langford, A.O., Deshler, T., Johnson, B.J., Torres, A., and Matthews, W.A., 1994a, Ozone loss in the lower stratosphere over the United States in 1992–1993: Evidence for heterogeneous chemistry of the Pinatubo aerosol: Geophysical Research Letters, v. 21, p. 65–68.

Hofmann, D.J., Oltmans, S.J., Lathrop, J.A., Harris, J.M., and Vömel, H., 1994b, Record low ozone at the South Pole in the Spring of 1993: Evidence for heterogeneous chemistry of the Pinatubo aerosol: Geophysical Research Letters, v. 21, p. 421-424.

Hofmann, D.J., and Solomon, S., 1989, Ozone destruction through heterogeneous chemistry following the eruption of El Chichón: Journal of Geophysical Research, v. 94, p. 5029–5041.

Holasek, R.E., 1995, Volcanic eruption plumes: Satellite remote sensing observations and laboratory experiments: unpub. Ph.D. thesis, University of Hawaii at Manoa, 252 p.

Holasek, R.E., and Self, S., 1995, GOES weather satellite observations and measurements of the May 18, 1980, Mount St. Helens eruption: Journal of Geophysical Research, v. 100, p. 8469–8487.

Jäger, H., 1992, The Pinatubo eruption cloud observed by LIDAR at Garmisch-Partenkirchen: Geophysical Research Letters, v. 19, p. 191–194.

Jensen, E.J., and Toon, O.B., 1992, The potential effects of volcanic aerosols on cirrus cloud microphysics: Geophysical Research Letters, v. 19, p. 1759–1762.

Kimball, H.H., 1918, Variation in solar radiation intensities: Monthly Weather Review, v. 46, p. 355–356.

———1924, Variation in solar radiation intensities measured at the surface of the Earth: Monthly Weather Review, v. 52, p. 527–529.

Koyaguchi, T., this volume, Volume estimation of tephra-fall deposits from the June 15, 1991, eruption of Mount Pinatubo by theoretical and geological methods.

Koyaguchi, T., and Tokuno, M., 1993, Origin of the giant eruption cloud of Pinatubo, June 15, 1991: Journal of Volcanology and Geothermal Research, v. 55, p. 85–96.

Labitzke, K., and McCormick, M.P., 1992, Stratospheric temperature increases due to Pinatubo aerosols: Geophysical Research Letters, v. 19, p. 207–210.

Lacis, A., Hansen, J., and Sato, M., 1992, Climate forcing by stratospheric aerosols: Geophysical Research Letters, v. 19, p. 1607–1610.

Long, C.S., and Stowe L.L., 1994, Using the NOAA/AVHRR to study stratspheric aerosol optical thicknesses following the Mt. Pinatubo eruption: Geophysics Research Letters, v. 21, p. 2215–2218.

Luhr, J.F., Carmichael, I.S.E., and Varekamp, J.C., 1984, The 1982 eruptions of El Chichón volcano, Chiapas, Mexico: mineral-ogy and petrology of the anhydrite-bearing pumices: Journal of Volcanology and Geothermal Research, v. 23, p. 69–108.

Lynch, J.S., and Stephens, G., this volume, Mount Pinatubo: A satellite perspective of the June 1991 eruptions.

Mankin, W.G., Coffey, M.T., and Goldman, A., 1992, Airborne observations of SO_2, HCl, and O_3 in the stratospheric plume of the Pinatubo volcano in July 1991: Geophysical Research Letters, v. 19, p. 179–182.

McCormick, M.P., Thomason, L.W., and Trepte, C.R., in press, Atmospheric effects of the Mt. Pinatubo Eruption: Nature.

McCormick, M.P., and Veiga, R.E., 1992, SAGE II measurements of early Pinatubo aerosols: Geophysical Research Letters, v. 19, p. 155–158.

McPeters, R.D., 1993, The atmospheric SO_2 budget for Pinatubo derived from NOAA–11 SBUV/2 spectral data: Geophysical Research Letters, v. 20, p. 1971–1974.

Minnis, P., Harrison, E.F., Stowe, L.L., Gibson, G.G., Denn, F.M., Doelling, D.R., and Smith, W.L., Jr., 1993, Radiative climate forcing by the Mount Pinatubo eruption: Science, v. 259, p. 1411–1415.

Mo, K.C., and Wang, X., 1994, The global climate of June-August 1992: Warm ENSO episode decays and colder than normal conditions dominate the Northern Hemisphere: Journal of Climate, v. 7, p. 335–357.

Monastersky, R., 1994, Climate still reeling after Pinatubo blast: Science News, v. 145, p. 70.

Nardi, B., Chanin, M.-L., Hauchecorne, A., Avdyushin, S.I., Tulinov, G.F., Ivanov, M.S., Kuzmenko, B.N., and Mezhuev, I.R., 1993, Morphology and dynamics of the Pinatubo aerosol layer in the Northern Hemisphere as detected from a shipborne Lidar: Geophysical Research Letters, v. 20, p. 1967–1970.

Oswalt, J.S., Nichols, W., and O'Hara, J.F., this volume, Meteorological observations of the 1991 Mount Pinatubo eruption.

Paladio-Melosantos, M.L., Solidum, R.U., Scott, W.E., Quiambao, R.B., Umbal, J.V., Rodolfo, K.S., Tubianosa, B.S., Delos Reyes, P.J., and Ruelo, H.R., this volume, Tephra falls of the 1991 eruptions of Mount Pinatubo.

Pinatubo Volcano Observatory Team, 1991, Lessons form a major eruption: Mt. Pinatubo, Philippines: Eos, Transactions, American Geophysical Union, v. 72, p. 545, 552–553, 555.

Pitari, G., 1992, On the possible perturbation of stratospheric dynamics due to Pinatubo aerosols: Il Nuovo Cimento; Note Brevi, v. 15C, p. 485-489.

Pittock, A.B., 1992, Eruption of Mount Pinatubo: CSIRO Climate Change Research and Progress, Annual Report 1990–1991, p. 17.

Post, M.J., Grund, C.J., Langforn, A.O., and Proffitt, M.H., 1992, Observations of Pinatubo ejecta over Boulder, Colorado by LIDARs of three different wavelengths: Geophysical Research Letters, v. 19, p. 195–198.

Prather, M., 1992, Catastrophic loss of stratospheric ozone in dense volcanic clouds: Journal of Geophysical Research, v. 97, p. 10187–10191.

Pueschel, R.F., Snetsinger, K.G., Russel, P.B., Kinne, S.A., and Livington, J.M., 1992, The effects of the 1991 Pinatubo volcanic eruption on the optical and physical properties of stratospheric aerosols, in Keevalik, S., ed., Proceedings, IRS92: Current problems in atmospheric radiation: A.Deepak Publishing Co., p. 183–186.

Read, W.G., Froidevaux, L., and Waters, J.W., 1993, Microwave limb sounder measurement of SO_2 from Mt. Pinatubo volcano: Geophysical Research Letters, v. 20, p. 1299–1302.

Ramanathan, V., 1988, The greenhouse theory of climate change: A test by inadvertent global experiment: Science, v. 240, p. 293–299.

Rampino, M.R., and Self, S., 1982, Historic eruptions of Tambora (1815), Krakatau (1883), and Agung (1963), their stratospheric aerosols and climatic impact: Quaternary Research, v. 18, p. 127–143.

———1984, Sulphur-rich volcanic eruptions and stratospheric aerosols: Nature, v. 310, p. 677–679.

Robock, A., and Liu, Y., 1994, The volcanic signal in Goddard Institute for Space Studies three-dimensional model simulations: Journal of Climate, v. 7, p. 44–55.

Robock, A., and Mao, J., 1992, Winter warming from large volcanic eruptions: Geophysical Research Letters, v. 19, p. 2405–2408.

———1995, The volcanic signal in surface temperature records: Journal of Climate, v. 8, p. 1086–1103.

Rosen, J., Kjome, N.T., and Fast, H., 1992, Penetration of Mt. Pinatubo aerosols into the north polar vortex: Geophysical Research Letters, v. 19, p. 1751–1754.

Rosen, J., Kjome, N.T., Fast, H., and Larsen, N., 1994, Volcanic aerosols and polar stratospheric clouds in the winter 1992/1993 in the north polar vortex: Geophysical Research Letters, v. 20, p. 61–64.

Russell, P.B., Livingston, J.M., Dutton, E.G., Pueschel, R.F., and others, 1993b, Pinatubo and pre-Pinatubo optical depth spectra: Mauna Loa measurements, comparison, inferred particle size distribution; radiative effects, and relationship to Lidar data: Journal of Geophysical Research, v. 98, p. 22969–22985.

Russell, P.B., Livingston, J.M., Pueschel, R.F., Reagan, J.A., and others, 1993a, Post-Pinatubo optical depth spectra vs. latitude and vortex structure: Airborne tracking sunphotometer measurements in AASEII: Geophysical Research Letters, v. 20, p. 2571–2574.

Sato, M., Hansen, J.E., McCormick, M.P., and Pollack, J.B., 1993, Stratospheric aerosol optical depths, 1850–1990: Journal of Geophysical Research, v. 98, p. 22987–22994.

Saunders, R., 1993, Radiative properties of Mount Pinatubo volcanic aerosols over the tropical Atlantic: Geophysical Research Letters, v. 20, p. 137–140.

Schoeberl, M.R., Bhartia, P.K., and Hilsenrath, E., 1993, Tropical ozone loss following the eruption of Mt. Pinatubo: Geophysical Research Letters, v. 20, p. 29–32.

Scott, W.E., Hoblitt, R.P., Torres, R.C., Self, S, Martinez, M.L., and Nillos, T., Jr., this volume, Pyroclastic flows of the June 15, 1991, climactic eruption of Mount Pinatubo.

Self, S., and King, A.J., 1993, The 1963 eruption of Gunung Agung, Bali, and its atmospheric impact [abs.]: Eos, Transactions, American Geophysical Union, v. 74, no. 43, p. 105.

Smith, R.C., Prezelin, B.B., Baker, K.S., Bidigare, R.R., and others, 1992, Ozone depletion: Ultraviolet radiation and phytoplankton biology in Antarctic waters: Science, v. 255, p. 952–959.

Smithsonian Institution, 1991, Pinatubo: Bulletin of the Global Volcanism Network, Smithsonian Institution, v. 16, no. 5, p. 2–8.

Solomon, S., Sanders, R.W., Garcia, R.R., and Keys, J.G., 1993, Increased chlorine dioxide over Antarctica caused by volcanic aerosols from Mount Pinatubo: Nature, v. 363, p. 245–248.

Stone, R.S., Keys, J., and Dutton, E.G., 1993, Properties and decay of stratospheric aerosols in the Arctic following the 1991 eruptions of Mount Pinatubo: Geophysical Research Letters, v. 20, p. 2539–2362.

Stothers, R.S., 1984a, Mystery cloud of ad 536: Nature, v. 307, p. 344–345.

———1984b, The great Tambora eruption and its aftermath: Science, v. 224, p. 1191–1198.

Stowe, L.L., Carey, R.M., and Pellegrino, P.P., 1992, Monitoring the Mt. Pinatubo aerosol layer with NOAA–11 AVHRR data: Geophysical Research Letters, v. 19, p. 159–162.

Strong, A.E., and Stowe, L.L., 1993, Comparing stratospheric aerosols from El Chichón and Mount Pinatubo using AVHRR data: Geophysical Research Letters, v. 20, p. 1183–1186.

Tabazadeh, A., and Turco, R.P., 1993, Stratospheric chlorine injection by volcanic eruptions: Hydrogen chloride scavenging and implications for ozone: Science, v. 20, p. 1082–1086.

Tahira, M., Nomura, M., Sawada, Y., and Kamo, K., this volume, Infrasonic and acoustic-gravity waves generated by the Mount Pinatubo eruption of June 15, 1991.

Tanaka, S., Sugimura, T., Harada, T., and Tanaka, M., 1991, Satellite observations of the diffusion of Pinatubo volcanic dust to the stratosphere: Journal of the Remote Sensing Society of Japan, v. 11, p. 91–99 [in Japanese with English abstract].

Thomason, L.W., 1992, Observations of a new SAGE II aerosol extinction mode following the eruption of Mt. Pinatubo: Geophysical Research Letters, v. 19, p. 2179–2182.

Tokuno, M., 1991a, GMS–4 observations of volcanic eruption clouds from Mt. Pinatubo, Philippines: Journal of Remote Sensing Society of Japan, v. 11, p. 81–89 [in Japanese with English abstract].

Tokuno, M., 1991b, GMS–4 observations of volcanic eruption clouds from Mt. Pinatubo, Philippines: Meteorological Satellite Center, Tokyo, Japan, Technical Note No. 23, p. 1–14.

Trepte, C.R., and Hitchman, M.H., 1992, Tropical stratospheric deduced from satellite aerosol data: Nature, v. 355, p. 626–628.

Trepte, C.R., Viega, R.E., and McCormick, M.P., 1993, The poleward dispersal of Mount Pinatubo aerosol: Journal of Geophysical Research, v. 98, p. 18563–18573.

Turco, R., 1991, Volcanic aerosols: Chemistry, microphysics, evolution and effects, in Walter, L.S., and de Silva, S., eds., Volcanism-climate interactions: NASA Conference Publication 10062, p. D1–D30.

Valero, F.P.J., and Pilewskie, P., 1992, Latitudinal survey of spectral optical depths of the Pinatubo volcanic cloud—derived particle sizes, columnar mass loadings, and effects on planetary albedo: Geophysical Research Letters, v. 19, p. 163–166.

Varekamp, J.C., Luhr, J.F., and Prestegaard, K.L., 1984, The 1982 eruptions of El Chichón volcano, Chiapas, Mexico: Character of the eruptions, ash-fall deposits, and gas phase: Journal of Volcanology and Geothermal Research, v. 23, p. 39–68.

Vogelmann, A.M., Ackerman, T.P., and Turco, R.P., 1992, Enhancements in biologically effective ultraviolet radiation following volcanic eruptions: Nature, v. 359, p. 47–49.

Wallace, L., and Livingston, W., 1992, The effect of the Pinatubo cloud on hydrogen chloride and hydrogen fluoride: Geophysical Research Letters, v. 19, p. 1209–1211.

Walsh, K., and Pittock, A.B., 1992, Modeling the effects of the Mt. Pinatubo eruption on sea surface temperatures in the southern hemisphere: Australian Meteorological and Oceanographic Society Bulletin, v. 5, p. 31–35.

Weaver, A., Loewenstein, M., Podolske, J.R., and others, 1993, Effects of Pinatubo aerosol on stratospheric ozone at midlatitudes: Geophysical Research Letters, v. 20, p. 2515–2518.

Westrich, H.R., and Gerlach, T.M., 1992, Magmatic gas source for the stratospheric SO_2 cloud from the June 15, 1991, eruption of Mount Pinatubo: Geology, v. 20, p. 867–870.

Williams, S.N., and Self, S., 1983, The October 1902 plinian eruption of Santa María volcano, Guatemala: Journal of Volcanology and Geothermal Research, v. 16, p. 33–56.

Wilson, L., and Walker, G.P.L., 1987, Explosive volcanic eruptions—VI. Ejecta dispersal in Plinian eruptions: The control of eruption conditions and atmospheric properties: Geophysical Journal of the Royal Astronomical Society, v. 89, p. 657–679.

Winker, D.M., and Osborn, M.T., 1992a, Preliminary analysis of observations of the Pinatubo volcanic plume with a polarization-sensitive LIDAR: Geophysical Research Letters, v. 19, p. 155–158.

———1992b, Airborne LIDAR observations of the Pinatubo volcanic plume: Geophysical Research Letters, v. 19, p. 167–170.

Wolff, E.W., and Mulvaney, R., 1991, Reactions on sulphuric acid aerosols and on polar stratospheric clouds in the Antarctic stratosphere: Geophysical Research Letters, v. 18, p. 1007–1010.

Woods, A.W., and Self, S., 1992, Thermal disequilibrium at the top of volcanic clouds and its effect on estimates of the column heights: Nature, v. 355, p. 628–630.

Woods, A.W., and Wohletz, K., 1991, Dimensions and dynamics of co-ignimbrite eruption columns: Nature, v. 350, p. 225–227.

AUTHORS AND THEIR AFFILIATIONS

Albrecht, Achim   Department of Chemistry, Rutgers University, Piscataway, NJ 08855.
                           Current address: EAWAG/ETH, Umweltphysik, 8600
                           Duebendorf, Switzerland

Alonso, Rosalito A.   Department of Geological Sciences M/C 186, University of Illinois
                           at Chicago, 801 W. Taylor St., Chicago IL 60607–7059 USA; now
                           at: National Institute of Geological Sciences, University of the
                           Philippines, Diliman, Quezon City

Ambubuyog, Gemme   PHIVOLCS, C.P. Garcia Avenue, U.P. Diliman, Quezon City,
                           Philippines

Amin, Erlinda Q.   PHIVOLCS, C.P. Garcia Avenue, U.P. Diliman, Quezon City,
                           Philippines

Arboleda, Ronaldo A.   PHIVOLCS, C.P. Garcia Avenue, U.P. Diliman, Quezon City,
                           Philippines   Email: onie@x5.phivolcs.dost.gov.ph

Barcelona, Edito S.   PHIVOLCS, C.P. Garcia Avenue, U.P. Diliman, Quezon City,
                           Philippines; now at: Philippine Bureau of Energy Development

Bautista, Bartolome C.   PHIVOLCS, C.P. Garcia Avenue, U.P. Diliman, Quezon City,
                           Philippines

Bautista, Ma. Leonila P.   PHIVOLCS, C.P. Garcia Avenue, U.P. Diliman, Quezon City,
                           Philippines; temporarily at: Dept. of Earth and Planetary Sciences,
                           Kyoto University, Sakyoku, Kyoto 606-01, Japan   Email:
                           leyo@style.kugi.kyoto-u.ac.jp

Bautista, Cynthia Banzon   Center for Integrative and Development Studies, University of the
                           Philippines, Diliman, Quezon City, Philippines

Baxter, Peter J.   Department of Community Medicine, University of Cambridge,
                           Institute of Public Health, Forvie Site, Robinson Way, Cambridge
                           CB2 2SR, U.K.; also, Department of Community Medicine,
                           University of Cambridge, Fenner's, Gresham Road, Cambridge
                           CB1 2ES, UK

Belo, Mel   PHIVOLCS, C.P. Garcia Avenue, U.P. Diliman, Quezon City,
                           Philippines

Bernard, Alain   Department of Geology, CP 160/02, Université Libre de Bruxelles,
                           50 Ave. F.D. Roosevelt, B–1050 Brussels, Belgium   Email:
                           abernard@ulb.ac.be

Campita, Nora R.   PHIVOLCS, C.P. Garcia Avenue, U.P. Diliman, Quezon City,
                           Philippines

Candelaria, M.R.   Geothermal Division, PNOC-Energy Development Corporation, Ft.
                           Bonifacio, Metro Manila, Philippines 1201

Carmody, Rebecca   USGS, MS 954, National Center, Reston, VA 20192 USA

Casadevall, Thomas J.   USGS, MS 150, 345 Middlefield Rd., Menlo Park, CA 94025
                                 USA   Email: tcasadev@usgs.gov

Castillo, Paterno R.   Scripps Institution of Oceanography A–008, La Jolla, CA 92093–
                                 0220 USA   Email: pcastillo@ucsd.edu

Clemente, V.C.   Geothermal Division, PNOC-Energy Development Corporation, Ft.
                                 Bonifacio, Metro Manila, Philippines 1201

Coburn, Andrew W.   Cambridge Architectural Research Ltd., The Eden Centre, 47 City
                                 Road, Cambridge CB1 1DP, UK

Cola, Raoul M.   College of Public Administration, University of the Philippines,
                                 Diliman, Quezon City, Philippines

Cornelius, Reinold R.   The Pennsylvania State University, Department of Geosciences,
                                 University Park, PA 16802; now at: Cornell University, The
                                 Global Basins Research Network, Snee Hall, Ithaca, NY 14853–
                                 1504   Email: corneliu@maestro. geo. utexas.edu

Couchman, Marvin R.   USGS, Cascades Volcano Observatory, 5400 MacArthur Blvd.,
                                 Vancouver, WA 98661 USA   Email: couchman@
                                 pwavan.wr.usgs.gov

Daag, Arturo S.   PHIVOLCS, C.P. Garcia Avenue, U.P. Diliman, Quezon City,
                                 Philippines; temporarily at: International Training Center for
                                 Aerospace Survey and Earth Sciences, Enschede, The
                                 Netherlands   Email:asdaag@itc.nl

David, Carlos Primo C.   University of the Philippines, National Institute of Geological
                                 Sciences

Davidson, Gail   Alaska Volcano Observatory, Alaska Division of Geological and
                                 Geophysical Surveys, 794 University Ave., Suite 2001, Fairbanks,
                                 AK 99709

Dayrit, Manuel   Field Epidemiology Training Program Team and Assistant Secretary,
                                 Department of Health, Manila, Philippines

de la Cruz, Edwin G.   PHIVOLCS, C.P. Garcia Avenue, U.P. Diliman, Quezon City,
                                 Philippines

Delfin, F.G., Jr.   Geothermal Division, PNOC-Energy Development Corporation,
                                 Ft. Bonifacio, Metro Manila, Philippines 1201   Email: delfin@
                                 edc.energy.com.ph

Delos Reyes, Perla J.   PHIVOLCS, C.P. Garcia Avenue, U.P. Diliman, Quezon City,
                                 Philippines   Email: peejay@x5.phivolcs.dost.gov.ph

Devine, Joseph D.   Department of Geological Sciences, Brown University, Providence,
                                 RI 02912

Dolan, Michael T.   Michigan Technological University, Houghton, MI 49931   Email:
                                 mtdolan@mtu.edu

Dulce, Rosella G.   University of the Philippines, National Institute of Geological
                                 Sciences, Diliman, Quezon City, Philippines; now at: Philippine
                                 National Oil Company-Energy Development Corporation, Ft.
                                 Bonifacio, Metro Manila, Philippines 1201

Eberhart-Phillips, Donna   USGS-Cal Tech, 525 S. Wilson St., Pasadena, CA 91106 USA

Eldridge, C. Stewart   Geology Department and Research School of Earth Sciences,
                                 Australian National University, Canberra, A.C.T. 0200, Australia

Endo, Elliot T.   USGS, Cascades Volcano Observatory, 5400 MacArthur Blvd.,
                                 Vancouver, WA 98661 USA   Email: etendo@
                                 pwavan.wr.usgs.gov

Eto, Ismael — PHIVOLCS, C.P. Garcia Avenue, U.P. Diliman, Quezon City, Philippines

Evangelista, Digna — Mines and Geosciences Bureau, Department of Environment and Natural Resources, North Avenue, Diliman, Quezon City, Philippines

Ewert, J.W. — USGS, Cascades Volcano Observatory, 5400 MacArthur Blvd., Vancouver, WA 98661 USA   Email: jwewert@ pwavan.wr.usgs.gov

Fournelle, John — Department of Earth and Planetary Sciences, Johns Hopkins University, Baltimore, MD 21218; now at: Department of Geology and Geophysics, University of Wisconsin, 1215 W. Dayton St., Madison, WI 53706 USA   Email: johnf@ geology.wisc.edu

Fujii, Toshitsugu — Earthquake Research Institute, University of Tokyo, Bunkyo-ku, Tokyo 113, JAPAN

Gabinete, Elmer — PHIVOLCS, C.P. Garcia Avenue, U.P. Diliman, Quezon City, Philippines

Garcia, Delfin — PHIVOLCS, C.P. Garcia Avenue, U.P. Diliman, Quezon City, Philippines

Gerlach, Terrence M. — USGS, Cascades Volcano Observatory, 5400 MacArthur Blvd., Vancouver, WA 98661 USA   Email: tgerlach@ pwavan.wr.usgs.gov

Hadley, Kevin C. — USGS, Cascades Volcano Observatory, 5400 MacArthur Blvd., Vancouver, WA 98661 USA   Email: kchadley@ pwavan.wr.usgs.gov

Hamburger, Michael W. — Department of Geological Sciences, Indiana University, Bloomington, IN 47405 USA   Email: hamburg@ terra.geology.indiana.edu

Harlow, David H. — USGS, MS 977, 345 Middlefield Rd., Menlo Park, CA 94025

Harris, Teresa N. — Department of Earth and Planetary Sciences, Washington University, Campus Box 1169, St. Louis, MO 63130–4899 USA

Hattori, Keiko — Ottawa-Carleton Geoscience Centre, and Department of Geology, University of Ottawa, Ottawa, Ontario, K1N 6N5, Canada Email: khattori@acadVM1.UOttawa.CA

Hoblitt, Richard P. — USGS, Cascades Volcano Observatory, 5400 MacArthur Blvd., Vancouver, WA 98661 USA   Email: rhoblitt@ pwavan.wr.usgs.gov

Holasek, Rick E. — Hawaii Center for Volcanology and Department of Geology and Geophysics, School of Ocean and Earth Science and Technology, University of Hawaii at Manoa, Honolulu, HI 96822 USA; now at: SETS Technology Inc., 300 Kalehu Ave. #10, Miliani, HI 96789.   Email: rick@baby.pgd. hawaii.edu

Imai, Akira, — Geological Institute, University of Tokyo, 7–3–1 Hongo, Bunkyo-ku, Tokyo 113, JAPAN   Email: akira@tsunami. geol.s.u-tokyo.ac.jp

Insauriga, Sheila — PHIVOLCS, C.P. Garcia Avenue, U.P. Diliman, Quezon City, Philippines

Isada, Manuel — PHIVOLCS, C.P. Garcia Avenue, U.P. Diliman, Quezon City, Philippines

Janda, Richard J. — USGS, Cascades Volcano Observatory, Vancouver, WA 98661 USA (deceased)

Javier, Dindo   PHIVOLCS, C.P. Garcia Avenue, U.P. Diliman, Quezon City, Philippines

Jones, John W.   USGS, 521 National Center, Reston, VA 20192 USA   Email: jwjones@sunbeta.er.usgs.gov

Jumawan, Ferdinand T.   University of the Philippines, National Institute of Geological Sciences, Diliman, Quezon City, Philippines

Kamo, Kosuke   Sakurajima Volcanological Observatory, Disaster Prevention Research Institute, Kyoto University, Sakurajima, Kagoshima, 891–14 Japan; now at: Department of General Education, Kumamoto Institute of Technology, 4–22–1, Ikeda, Kumamoto, 860 Japan

King, Alan J.   Hawaii Center for Volcanology and Department of Geology and Geophysics, School of Ocean and Earth Science and Technology, University of Hawaii at Manoa, Honolulu, HI 96822 USA

Klein, Jeffrey   Department of Physics, The University of Pennsylvania, Philadelphia, PA 19104 USA

Knight, Roy J.   USGS, MS 474, Box 25046 DFC, Denver, CO 80225 USA

Knittel, Ulrich   Institut für Mineralogie und Lagerstättenlehre, Wüllnerstr. 2, D–D–52056, Aachen, Germany   Email: knittel@rwth-aachen.de; also at: Institut für Geowissenschaften/Mineralogie, Saarstr. 21, D–55099, Mainz, Germany

Koyaguchi, Takehiro   Earthquake Research Institute, University of Tokyo, Tokyo 113, Japan   Email: tak@eri.u-tokyo.ac.jp

Lacsamana, Jay Bertram T.   National Economic and Development Authority, Region III, San Fernando, Pampanga, Philippines

Laguerta, Eduardo P.   PHIVOLCS, Lignon Hill Volcano Observatory, Legazpi City, Philippines

LaHusen, Richard G.   USGS, Cascades Volcano Observatory, 5400 MacArthur Blvd., Vancouver, WA 98661 USA   Email: rlahusen@pwavan.wr.usgs.gov

Lanuza, Angelito   PHIVOLCS, C.P. Garcia Avenue, U.P. Diliman, Quezon City, Philippines

Layugan, D.B.   Geothermal Division, PNOC-Energy Development Corporation, Ft. Bonifacio, Metro Manila, Philippines 1201

Listanco, Eddie L.   Earthquake Research Institute, University of Tokyo, Bunkyo-ku, Tokyo 113, Japan; now at: PHIVOLCS, C.P. Garcia Avenue, U.P. Diliman, Quezon City, Philippines Email: edlist@x5.phivolcs.dost.gov.ph

Lockhart, Andrew B.   USGS, Cascades Volcano Observatory, 5400 MacArthur Blvd., Vancouver, WA 98661 USA   Email: ablock@pwavan.wr.usgs.gov

Luhr, James F.   Department of Mineral Sciences, NHB-119, Smithsonian Institution, Washington, DC 20560 USA Email: mnhms033@sivm.si.edu

Lynch, James S.   National Oceanic and Atmospheric Administration, National
                                    Environmental Satellite, Data, and Information Service, World
                                    Weather Building, Washington, DC 20230 USA

Major, Jon J.   USGS, Cascades Volcano Observatory, 5400 MacArthur Blvd.,
                                    Vancouver, WA 98661 USA   Email: jjmajor@
                                    pwavan.wr.usgs.gov

Marcial, Sergio   PHIVOLCS, C.P. Garcia Avenue, U.P. Diliman, Quezon City,
                                    Philippines; now at: King Abdulaziz City for Science and
                                    Technology, Institute of Astronomical and Geophysical Research,
                                    Office Villa 2–059, Prince Abdullah Bin Abdul Aziz Road,
                                    Riyadh 11442, Kingdom of Saudi Arabia   Email:
                                    earthq1%sakacs00.bitnet@vm1.nodak.edu

Marso, Jeffrey N.   USGS, Cascades Volcano Observatory, 5400 MacArthur Blvd.,
                                    Vancouver, WA 98661 USA   Email: jnmarso@
                                    pwavan.wr.usgs.gov

Martinez, Ma. Mylene L.   PHIVOLCS, C.P. Garcia Avenue, U.P. Diliman, Quezon City,
                                    Philippines; temporarily at: Dept. of Geology, Arizona State
                                    University, Tempe, AZ 85287–1404   Email: martinez@asu.edu

McGeehin, John   USGS, MS 971, National Center, Reston, VA 20192 USA

McKibben, Michael A.   Department of Earth Sciences, University of California, Riverside,
                                    CA 92521–0423 USA   Email: mckibben@ucrac1.ucr.edu

Meeker, Gregory P.   USGS, MS 903, Box 25046 DFC, Denver, CO 80225 USA
                                    Email: gmeeker@usgsprobe.cr.usgs.gov

Melosantos, Arnaldo A.   PHIVOLCS, C.P. Garcia Avenue, U.P. Diliman, Quezon City,
                                    Philippines   Email: amiel@x5.phivolcs.dost.gov.ph

Melson, William G.   Department of Mineral Sciences, NHB-119, Smithsonian Institution,
                                    Washington, DC 20560 USA

Mercado, Remigio A.   National Economic and Development Authority, Region III,
                                    San Fernando, Pampanga, Philippines

Miller, Yvonne   Department of Mineralogy, University of Geneva, 13 Rue de
                                    Maraichers, 1211 Geneva 4, Switzerland

Mori, Jim   USGS-Cal Tech, 525 S. Wilson St., Pasadena, CA 91106 USA
                                    Email: mori@bombay.gps.caltech.edu

Murray, Thomas L.   USGS, Cascades Volcano Observatory, 5400 MacArthur Blvd.,
                                    Vancouver, WA 98661 USA   Email: tlmurray@
                                    pwavan.wr.usgs.gov

Newhall, Christopher G.   USGS, Volcano Systems Center/ Geological Sciences, University of
                                    Washington, Box 351310, Seattle, WA 98195 USA
                                    Email: cnewhall@geophys.washington.edu

Nichols, William   U.S. Air Force; now at: National Weather Service, 104 Airport Road,
                                    Dodge City, KS 67801 USA

Nillos, Timoteo, Jr.   PHIVOLCS, C.P. Garcia Avenue, U.P. Diliman, Quezon City,
                                    Philippines

Nolasco-Javier, Dymphna D.   University of the Philippines, National Institute of Geological
                                    Sciences, Diliman, Quezon City, Philippines

Nomura, Masahiro   Department of Earth Sciences, Aichi University of Education,
                                    Kariya, Aichi, 448 Japan

O'Hara, John F., CDR — U.S. Navy, Commander, Naval Meteorology and Oceanography Command, 1020 Balch Blvd., Stennis Space Center, MS 39529–5000 USA

Okubo, P. — USGS, Hawaiian Volcano Observatory, Hawaii National Park, HI 96718 USA   Email: pokubo@liko.wr.usgs.gov

Oles, Dietmar — Institut für Mineralogie und Lagerstättenlehre, Wüllnerstr. 2, D–5100, Aachen, Germany

Oswalt, J. Scott, LT — U.S. Navy, PSC 1008 Box 84, FPO AA 34051–0084 USA; (after Dec 95 at Naval Oceanographic Office, 1002 Balch Blvd., Stennis Space Center, MS 39522–5001 USA)

Paladio-Melosantos, Ma. Lynn O. — PHIVOLCS, C.P. Garcia Avenue, U.P. Diliman, Quezon City, Philippines   Email: lynn@x5.phivolcs.dost.gov.ph

Pallister, John S. — USGS, MS 908, National Center, Reston, VA 20192 USA Email: jpallist@usgs.gov

Pasteris, Jill Dill — Department of Earth and Planetary Sciences, Washington University, Campus Box 1169, St. Louis, MO 63130–4899 USA Email: pasteris@realrock1.wustl.edu

Pierson, Thomas C. — USGS, Cascades Volcano Observatory, 5400 MacArthur Blvd., Vancouver, WA 98116 USA   Email: tpierson@pwavan.wr.usgs.gov

Pigtain, Ronald C. — PHIVOLCS, C.P. Garcia Avenue, U.P. Diliman, Quezon City, Philippines

Pineda, Greg L. — National Economic and Development Authority, Region III, San Fernando, Pampanga, Philippines

Pomonis, Antonios — Cambridge Architectural Research Ltd., The Eden Centre, 47 City Road, Cambridge CB1 1DP, U.K.

Power, John A. — USGS, Alaska Volcano Observatory, 4200 University Drive, Anchorage, AK 99508–4667 USA   Email: jpower@usgs.gov

Punongbayan, Raymundo S. — PHIVOLCS, C.P. Garcia Avenue, U.P. Diliman, Quezon City, Philippines   Email: rsp@x5.phivolcs.dost.gov.ph

Quiambao, Rowena B. — PHIVOLCS, C.P. Garcia Avenue, U.P. Diliman, Quezon City, Philippines; temporarily at: ITC, Postbus 6, Blvd. 1945, 7500AA Enschede, Holland   Email: rowena@itc.nl

Ramos, Emmanuel G. — Department of Geological Sciences, Indiana University, Bloomington, IN 47405 USA; now at: PHIVOLCS, C.P. Garcia Avenue, U.P. Diliman, Quezon City, Philippines Email: eramos@x5.phivolcs.dost.gov.ph

Rasdas, Ariel R. — PHIVOLCS, C.P. Garcia Avenue, U.P. Diliman, Quezon City, Philippines

Regalado, Ma. Theresa M. — PHIVOLCS, C.P. Garcia Avenue, U.P. Diliman, Quezon City, Philippines

Remotigue, Cristina T. — National Institute of Geological Sciences, University of the Philippines, Diliman, Quezon City, Philippines

Reyes, Agnes G. — Geothermal Division, Philippine National Oil Corporation, Ft. Bonifacio, Metro Manila, Philippines; now at: Institute of Geological and Nuclear Sciences, POB 30–368, Gracefield Road, Lower Hutt, New Zealand

Ringor, Anne — PHIVOLCS, C.P. Garcia Avenue, U.P. Diliman, Quezon City, Philippines

Rodolfo, Kelvin S.   Department of Geological Sciences M/C 186, University of Illinois at Chicago, 801 W. Taylor St., Chicago IL 60607–7059 USA

Rowe, Gary L.   USGS, WRD, 975 W 3rd Ave, Columbus, OH 43212 USA
Email: glrowe@qvarsa.er.usgs.gov

Ruaya, J.R.   Geothermal Division, PNOC-Energy Development Corporation, Ft. Bonifacio, Metro Manila, Philippines 1201

Rubin, Meyer   USGS, MS 971, National Center, Reston, VA 20192 USA

Ruelo, Hernulfo B.   Renison Goldfields Consolidated Exploration Party, Limited, Australia

Rutherford, Malcolm J.   Department of Geological Sciences, Brown University, Providence, RI 02912 USA   Email: malcolm_rutherford @brown.edu

Sabit, Julio P.   PHIVOLCS, C.P. Garcia Avenue, U.P. Diliman, Quezon City, Philippines

Salvador, Jerry H.G.   Mines and Geosciences Bureau, Department of Environment and Natural Resources, North Avenue, Diliman, Quezon City, Philippines

Sawada, Yosihiro   Volcanological Division, Seismological and Volcanological Department, Japan Meteorological Agency, 1–3–4 Ote-machi, Chiyoda-ku, Tokyo, 100 Japan

Schneider, David J.   Michigan Technological University, Department of Geological Engineering, Houghton, MI 49931 USA

Scott, Kevin M.   USGS, Cascades Volcano Observatory, 5400 MacArthur Blvd., Vancouver, WA 98661 USA

Scott, William E.   USGS, Cascades Volcano Observatory, 5400 MacArthur Blvd., Vancouver, WA 98661 USA   Email: wescott@ pwavan.wr.usgs.gov

Self, Stephen   Department of Geology and Geophysics, Hawaii Center for Volcanology, University of Hawaii at Manoa, Honolulu, HI 96822 USA   Email: self@soest.hawaii.edu

Sexon, Manuel   PHIVOLCS, C.P. Garcia Avenue, U.P. Diliman, Quezon City, Philippines

Siems, David F.   USGS, MS 973, Box 25046 DFC, Denver, CO 80225 USA

Solidum, Renato U.   PHIVOLCS, C.P. Garcia Avenue, U.P. Diliman, Quezon City, Philippines; temporarily at: Scripps Institution of Oceanography A-008, La Jolla, CA 92093–0220 USA   Email: rsolidum@ ucsd.edu

Spence, Robin J.S.   Department of Architecture, University of Cambridge, Cambridge, UK

Stein, Ross S.   USGS, MS 977, 345 Middlefield Rd., Menlo Park, CA 94025 USA
Email: rstein@isdmnl.wr.usgs.gov

Stephens, George   National Oceanic and Atmospheric Administration, National Environmental Satellite, Data, and Information Service, World Weather Building, Washington, DC 20230 USA

Symonds, Robert B.   USGS, Cascades Volcano Observatory, 5400 MacArthur Blvd., Vancouver, WA 98661   Email: bsymonds@ pwavan.wr.usgs.gov

Tahira, Makoto   Department of Earth Sciences, Aichi University of Education, Kariya, Aichi, 448 Japan   Email: mtahira@ auecc.aichi-edu.ac.jp

Tamayo, Rodolfo A., Jr.   University of the Philippines, National Institute of Geological
                               Sciences, Diliman, Quezon City, Philippines

Tayag, Jean   PHIVOLCS, C.P. Garcia Avenue, U.P. Diliman, Quezon City,
                               Philippines

Torres, Ronnie C.   PHIVOLCS, C.P. Garcia Avenue, U.P. Diliman, Quezon City,
                               Philippines,   Email: torres@x5.phivolcs.dost.gov.ph;
                               temporarily at: Hawaii Center for Volcanology, Department of
                               Geology and Geophysics, School of Ocean and Earth Science and
                               Technology, University of Hawaii, 2525 Correa Rd., Honolulu, HI
                               96822 USA   Email: torres@soest. hawaii.edu

Tubianosa, Bella S.   PHIVOLCS, C.P. Garcia Avenue, U.P. Diliman, Quezon City,
                               Philippines

Tuñgol, Norman M.   PHIVOLCS, C.P. Garcia Avenue, U.P. Diliman, Quezon City,
                               Philippines   Email: norman@x5.phivolcs.dost.gov ph

Umbal, Jesse V.   Department of Geological Sciences M/C 186, University of Illinois
                               at Chicago, 801 W. Taylor St., Chicago IL 60607–7059 USA; now
                               at: PHIVOLCS, C.P. Garcia Avenue, U.P. Diliman, Quezon City,
                               Philippines   Email: jvu@x5. phivolcs.dost.gov.ph

Villarosa, H.G.   Geothermal Division, PNOC-Energy Development Corporation,
                               Ft. Bonifacio, Metro Manila, Philippines 1201

Voight, Barry   The Pennsylvania State University, Department of Geosciences,
                               University Park, PA 16802 USA; also at: U.S. Geological
                               Survey   Email: voight@ems. psu.edu

Wang, Alian   Department of Earth and Planetary Sciences, Washington University,
                               Campus Box 1169, St. Louis, MO 63130-4899 USA

Wang, Yubo   Universität Hamburg, Institute für Biogeochemie und Meereschemie,
                               Bundestrasse 55, D–20146 Hamburg, Germany
                               Email: fg4a111@rrz.cip-1.rrz.uni-hamburg.de

Weber, Bernd   Institut für Mineralogie und Lagerstättenlehre, Wüllnerstr. 2,
                               D-5100, Aachen, Germany

Weis, Dominique   Department of Geology, CP 160/02, Université Libre de Bruxelles,
                               50 Ave. F.D. Roosevelt, B–1050 Brussels, Belgium

Westrich, Henry R.   Sandia National Laboratory, Geochemistry Dept. 6118, Albuquerque,
                               NM 87185 USA

White, Randall A.   USGS, MS 977, 345 Middlefield Rd., Menlo Park, CA 94025 USA
                               Email: white@andreas.wr.usgs.gov

White, Mark   Field Epidemiology Training Program Team, Department of Health,
                               Manila, Philippines, and Centers for Disease Control, Atlanta,
                               Georgia USA

Widmer, R.   Geophysical Institute, Karlsruhe University, Hertzstr. 16, D–76187
                               Karlsruhe, Germany   Email: widmer@
                               gpina.physik.uni-karlsruhe.de

Wiesner, Martin G.   Universität Hamburg, Institute für Biogeochemie und Meereschemie,
                               Bundestrasse 55, D–20146 Hamburg, Germany

Wolfe, Edward W.   USGS, Cascades Volcano Observatory, 5400 MacArthur Blvd.,
                               Vancouver, WA 98661 USA   Email: ewwolfe@
                               pwavan.wr.usgs.gov

Wopenka, Brigitte   Department of Earth and Planetary Sciences, Washington University,
                               Campus Box 1169, St. Louis, MO 63130–4899 USA

Zamoras, Lawrence R.   University of the Philippines, National Institute of Geological
                              Sciences, Diliman, Quezon City, Philippines

Zhao, Jing-Xia   Department of Meteorology, School of Ocean and Earth Science and
                              Technology, University of Hawaii at Manoa, Honolulu, HI 96822
                              USA

Zürn, W.   Black Forest Observatory Schiltach, Heubach 206, D–77709
                              Wolfach, Germany

Bright spirits in spite of lahars: lowland schoolchildren in Santa
Lucia, Capas. Photograph by Tom Pierson.

A long, homeward journey: Aeta couple returning to their land, Botolan. Photograph by R.S. Punongbayan.